Chemistry
of the
Elements

Related Butterworth-Heinemann titles of interest

CHOPPIN, RYDBERG & LILJENZIN
Radiochemistry and Nuclear Chemistry

PERRIN & ARMAREGO
Purification of Laboratory Chemicals, 3rd Edition

RODGER & RODGER
Molecular Geometry

SHAW
Introduction to Colloid and Surface Chemistry, 4th Edition

SLATER
Principles of Ion Exchange Technology

The International System of Units (SI)

SI base units

Physical quantity	Name of SI unit	Symbol for SI unit	Definition of SI unit
length	metre	m	(1)
mass*	kilogramme	kg	(2)
time	second	s	(3)
electric current	ampere	A	(4)
thermodynamic temperature	kelvin	K	(5)
amount of substance	mole	mol	(6)
luminous intensity	candela	cd	(7)

(1) The **metre** is the length equal to 1 650 763.73 wavelengths in vacuum of the radiation corresponding to the transition between the levels $2p_{10}$ and $5d_5$ of the krypton-86 atom.
(2) The **kilogramme** is the unit of mass; it is equal to the mass of the international prototype of the kilogramme.
(3) The **second** is the duration of 9 192 631 770 periods of the radiation corresponding to the transition between the two hyperfine levels of the ground state of the caesium-133 atom.
(4) The **ampere** is that constant current which, if maintained in two straight parallel conductors of infinite length, of negligible circular cross-section, and placed 1 metre apart in a vacuum, would produce between these conductors a force equal to 2×10^{-7} newton per metre of length.
(5) The **kelvin**, unit of thermodynamic temperature, is the fraction 1/273.16 of the thermodynamic temperature of the triple point of water.
(6) The **mole** is the amount of substance of a system which contains as many elementary entities as there are atoms in 0.012 kilogramme of carbon-12. When the mole is used, the elementary entities must be specified and may be atoms, molecules, ions, electrons, other particles, or specified groups of such particles.
(7) The **candela** is the luminous intensity, in the perpendicular direction, of a surface of 1/600 000 square metre of a black body at the temperature of freezing platinum under a pressure of 101 325 newtons per square metre.

SI supplementary units

Physical quantity	Name of SI unit	Symbol for SI unit	Definition of SI unit
plane angle	radian	rad	The radian is the plane angle between two radii of a circle which cut off on the circumference an arc of length equal to the radius.
solid angle	steradian	sr	The steradian is the solid angle which, having its vertex at the centre of a sphere, cuts off an area of the surface of the sphere equal to that of a square with sides of length equal to the radius of the sphere.

SI derived units (A list of some non-SI units and some useful conversion factors is given in Appendix 3, p. 1495.)

Physical quantity	Name of SI unit	Symbol for SI unit	Definition of SI unit
frequency	hertz	Hz[1]	s^{-1}
energy	joule	J	$kg\ m^2\ s^{-2}$
force	newton	N	$kg\ m\ s^{-2} = J\ m^{-1}$
power	watt	W	$kg\ m^2\ s^{-3} = J\ s^{-1}$
pressure	pascal	Pa	$kg\ m^{-1}\ s^{-2} = N\ m^{-2} = J\ m^{-3}$
electric charge	coulomb	C	$A\ s$
electric potential difference	volt	V	$kg\ m^2\ s^{-3}\ A^{-1} = J\ A^{-1}\ s^{-1}$
electric resistance	ohm	Ω	$kg\ m^2\ s^{-3}\ A^{-2} = V\ A^{-1} = S^{-1}$
electric conductance	siemens	S	$kg^{-1}\ m^{-2}\ s^3\ A^2 = \Omega^{-1}$
electric capacitance	farad	F	$A^2\ s^4\ kg^{-1}\ m^{-2} = A\ s\ V^{-1}$
magnetic flux	weber	Wb	$kg\ m^2\ s^{-2}\ A^{-1} = V\ s$
inductance	henry	H	$kg\ m^2\ s^{-2}\ A^{-2} = V\ A^{-1}s$
magnetic flux density (magnetic induction)	tesla	T	$kg\ s^{-2}\ A^{-1} = V\ s\ m^{-2} = Wb\ m^{-2}$

[1] Neither "c.p.s." nor "c/s", but either s^{-1} or Hz.

Origin of SI Prefixes

10^{-1}	d	**deci**	Latin *decimus* tenth	10	da	**deca**	Greek δεκα ten
10^{-2}	c	**centi**	Latin *centum* hundred	10^2	h	**hecto**	Greek ἑκατόν hundred
10^{-3}	m	**milli**	Latin *milli* thousand	10^3	k	**kilo**	Greek κίλιοι thousand
10^{-6}	μ	**micro**	Greek μῖκρός small	10^6	M	**mega**	Greek μέγας great
10^{-9}	n	**nano**	Greek νάνος dwarf	10^9	G	**giga**	Greek γίγᾱς giant
10^{-12}	p	**pico**	Italian *piccolo* small	10^{12}	T	**tera**	Greek τέρας monster
10^{-15}	f	**femto**	Danish *femten* fifteen	10^{15}	P	**peta**	See footnote†
10^{-18}	a	**atto**	Danish *atten* eighteen	10^{18}	E	**exa**	See footnote†

† The almost unbelievably grotesque derivation of these two prefixes is as follows: suppose, without any justification whatever, that tera (meaning $10^{12} = 10^{3 \times 4}$) had been derived from tetra (meaning 4) by omission of a consonant and not from the Greek word for monster. Then the appropriate SI prefix for $10^{15} = 10^{3 \times 5}$ is {(penta minus a consonant) = peta} and for $10^{18} = 10^{3 \times 6}$ is {(hexa minus a consonant) = exa}.

Chemistry
of the
Elements

N. N. GREENWOOD and A. EARNSHAW

Department of Inorganic and Structural Chemistry
University of Leeds, U.K.

Butterworth-Heinemann Ltd
Linacre House, Jordan Hill, Oxford OX2 8DP

℞ A member of the Reed Elsevier plc group

OXFORD LONDON BOSTON
MUNICH NEW DELHI SINGAPORE SYDNEY
TOKYO TORONTO WELLINGTON

First published by Pergamon Press plc 1984
Reprinted with corrections 1985, 1986
Reprinted 1989, 1990, 1993, 1994, 1995

© Butterworth-Heinemann Ltd 1984

British Library Cataloguing in Publication Data
Greenwood, N. N.
  Chemistry of the elements.
  1. Chemical elements
  I. Title II. Earnshaw, A.
  540   QD458

ISBN 0 7506 2832 4

Library of Congress Cataloguing in Publication Data
Greenwood, N. N. (Norman Neill)
  Chemistry of the elements.
  Includes index.
  1. Chemical elements. I. Earnshaw, A. (Alan)
  II. Title.
  QD466.G74 1984 546 83-13346

Printed in Great Britain at the University Press, Cambridge

Foreword

THE publication of a new comprehensive treatment of the chemistry of the elements is an event of major importance for both teachers and students. The majority of the more recent textbooks of inorganic chemistry place considerable emphasis on the theories of inorganic chemistry and do not give a systematic comprehensive treatment of the properties and reactions of the elements and their compounds. The *Chemistry of the Elements* therefore fills a real need for an up-to-date, critical, comprehensive account of the chemistry of the elements. The facts concerning the properties and reactions of substances are the very essence of chemistry. Facts undergo little if any change in contrast to constantly changing theories. Moreover, it is important that students appreciate that a chemist needs a solid background of facts in order to appreciate the need for theories and to be able to judge their usefulness and the limits of their applicability. Anyone who just dips into the book at random will quickly realize that there are numerous facts that remain intriguing and unexplained and which provide a stimulus for the development of new or modified theories. The writing of this book was clearly a prodigious task and the authors are to be congratulated on having presented such a comprehensive in-depth account of the elements in such a readable fashion. It deserves to become the standard reference in inorganic chemistry for both teachers and students for many years to come and I wish it every success.

R. J. GILLESPIE
McMaster University
Hamilton, Ontario, Canada

Preface

IN this book we have tried to give a balanced, coherent, and comprehensive account of the chemistry of the elements for both undergraduate and postgraduate students. This crucial central area of chemistry is full of ingenious experiments, intriguing compounds, and exciting new discoveries. We have specifically avoided the term *inorganic chemistry* since this emphasizes an outmoded view of chemistry which is no longer appropriate in the closing decades of the 20th century. Accordingly, we deal not only with inorganic chemistry but also with those aspects which might be called analytical, theoretical, industrial, organometallic, bio-inorganic, or any other of the numerous branches of the subject currently in vogue.

We make no apology for giving pride of place to the phenomena of chemistry and to the factual basis of the subject. Of course the chemistry of the elements is discussed within the context of an underlying theoretical framework that gives cohesion and structure to the text, but at all times it is the chemical chemistry that is emphasized. There are several reasons for this. Firstly, theories change whereas facts do so less often—a greater permanency and value therefore attaches to a treatment based on a knowledge and understanding of the factual basis of the subject. We recognize, of course, that though the facts may not change dramatically, their significance frequently does. It is therefore important to learn how to assess observations and to analyse information reliably. Numerous examples are provided throughout the text. Moreover, it is scientifically unsound to present a theory and then describe experiments which purport to prove it. It is essential to distinguish between facts and theories and to recognize that, by their nature, theories are ephemeral and continually changing. Science advances by removing error, not by establishing truth, and no amount of experimentation can "prove" a theory, only that the theory is consistent with the facts as known *so far*. (At a more subtle level we also recognize that all facts are theory-laden.)

It is also important to realize that chemistry is not a static body of knowledge as defined by the contents of a textbook. Chemistry came from somewhere and is at present heading in various specific directions. It is a living self-stimulating discipline, and we have tried to transmit this sense of growth and excitement by reference to the historical development of the subject when appropriate. The chemistry of the elements is presented in a logical and academically consistent way but is interspersed with additional material which illuminates, exemplifies, extends, or otherwise enhances the chemistry being discussed.

Chemistry is a human activity and its results have a substantial impact on our daily lives. However, we have not allowed ourselves to become obsessed by "relevance". Today's relevance is tomorrow's obsolescence. On the other hand, it would be obtuse in the modern world not to recognize that chemistry, in addition to being academically stimulating and aesthetically satisfying, is frequently also useful. This gives added point to much of the chemistry of the elements and indeed a great deal of that chemistry has been specifically developed because of society's needs. To many this is one of the most attractive aspects of the subject—its potential usefulness. We therefore wrote to over 500 chemically based firms throughout the world asking for information about the chemicals they manufactured or used, in what quantities, and for what purposes. This produced an immense wealth of technical information which has proved to be an invaluable resource in discussing the chemistry of the elements. Our own experience as teachers had already alerted us to the difficulty of acquiring such topical information and we have incorporated much of this material where appropriate throughout the text. We believe it is important to know whether a given compound was made perhaps once in milligram amounts, or is produced annually in tonne quantities, and for what purpose.

In a textbook devoted to the chemistry of the elements it seemed logical to begin with such questions as: where do the elements come from, how were they made, why do they have their observed terrestrial abundances, what determines their atomic weights, and so on. Such questions, though usually ignored in textbooks and certainly difficult to answer, are ones which are currently being actively pursued, and some tentative answers and suggestions are given in the opening chapter. This is followed by a brief description of chemical periodicity and the periodic table before the chemistry of the individual elements and their group relationships are discussed on a systematic basis.

We have been much encouraged by the careful assessment and comments on individual chapters by numerous colleagues not only throughout the U.K. but also in Australia, Canada, Denmark, the Federal Republic of Germany, Japan, the U.S.A and several other countries. We believe that this new approach will be widely welcomed as a basis for discussing the very diverse behaviour of the chemical elements and their compounds.

It is a pleasure to record our gratitude to the staff of the Edward Boyle Library in the University of Leeds for their unfailing help over many years during the writing of this book. We should also like to express our deep appreciation to Mrs Jean Thomas for her perseverance and outstanding skill in preparing the manuscript for the publishers. Without her generous help and the understanding of our families this work could not have been completed.

N. N. GREENWOOD
A. EARNSHAW

Contents

Chapter 1   Origin of the Elements. Isotopes and Atomic Weights   1

   1.1   Introduction   1
   1.2   Origin of the Universe   1
   1.3   Stellar Evolution and the Spectral Classes of Stars   4
   1.4   Synthesis of the Elements   10
      1.4.1   Hydrogen burning   11
      1.4.2   Helium burning and carbon burning   12
      1.4.3   The α-process   13
      1.4.4   The e-process (equilibrium process)   14
      1.4.5   The s- and r-processes (slow and rapid neutron absorption)   15
      1.4.6   The p-process (proton capture)   16
      1.4.7   The x-process   16
   1.5   Atomic Weights   18
      1.5.1   Uncertainty in atomic weights   19
      1.5.2   The problem of radioactive elements   21
   1.6   Points to Ponder   22

Chapter 2   Chemical Periodicity and the Periodic Table   24

   2.1   Introduction   24
   2.2   The Electronic Structure of Atoms   25
   2.3   Periodic Trends in Properties   27
      2.3.1   Trends in atomic and physical properties   27
      2.3.2   Trends in chemical properties   31
   2.4   Prediction of New Elements and Compounds   32
   2.5   Questions on the Periodic Table   36

Chapter 3   Hydrogen   38

   3.1   Introduction   38
   3.2   Atomic and Physical Properties of Hydrogen   40
      3.2.1   Isotopes of hydrogen   40
      3.2.2   *Ortho-* and *para-*hydrogen   41
      3.2.3   Ionized forms of hydrogen   42
   3.3   Preparation, Production, and Uses   43
      3.3.1   Hydrogen   43
      3.3.2   Deuterium   47
      3.3.3   Tritium   48
   3.4   Chemical Properties and Trends   50
      3.4.1   Protonic acids and bases   51
   3.5   The Hydrogen Bond   57
      3.5.1   Influence on properties   57
      3.5.2   Influence on structure   63
      3.5.3   Strength of hydrogen bonds and theoretical description   66
   3.6   Hydrides of the Elements   69

**Chapter 4   Lithium, Sodium, Potassium, Rubidium, Caesium, and   75
               Francium**

       4.1   Introduction   75
       4.2   The Elements   75
              4.2.1   Discovery and isolation   75
              4.2.2   Terrestrial abundance and distribution   76
              4.2.3   Production and uses of the metals   82
              4.2.4   Properties of the alkali metals   85
              4.2.5   Chemical reactivity and trends   87
              4.2.6   Solutions in liquid ammonia and other solvents   88
       4.3   Compounds   91
              4.3.1   Introduction: the ionic-bond model   91
              4.3.2   Halides and hydrides   94
              4.3.3   Oxides, peroxides, superoxides, and suboxides   97
              4.3.4   Hydroxides   100
              4.3.5   Oxoacid salts and other compounds   101
              4.3.6   Complexes, crowns, and crypts   105
              4.3.7   Organometallic compounds   110

**Chapter 5   Beryllium, Magnesium, Calcium, Strontium, Barium, and   117
               Radium**

       5.1   Introduction   117
       5.2   The Elements   118
              5.2.1   Terrestrial abundance and distribution   118
              5.2.2   Production and uses of the metals   120
              5.2.3   Properties of the elements   121
              5.2.4   Chemical reactivity and trends   123
       5.3   Compounds   123
              5.3.1   Introduction   123
              5.3.2   Hydrides and halides   125
              5.3.3   Oxides and hydroxides   131
              5.3.4   Oxoacid salts and coordination complexes   133
              5.3.5   Organometallic compounds   141
                      Beryllium   141
                      Magnesium   146
                      Calcium, strontium, and barium   153

Chapter 6   Boron   155

       6.1   Introduction   155
       6.2   Boron   156
              6.2.1   Isolation and purification of the element   156
              6.2.2   Structure of crystalline boron   157
              6.2.3   Atomic and physical properties of boron   160
              6.2.4   Chemical properties   161
       6.3   Borides   162
              6.3.1   Introduction   162
              6.3.2   Preparation and stoichiometry   163
              6.3.3   Structures of borides   165
       6.4   Boranes (Boron Hydrides)   171
              6.4.1   Introduction   171
              6.4.2   Structure, bonding, and topology   179
              6.4.3   Properties of boranes   185
              6.4.4   Chemistry of diborane, B_2H_6 186
              6.4.5   Chemistry of *nido*-pentaborane, B_5H_9 193
              6.4.6   Chemistry of *nido*-decaborane, $B_{10}H_{14}$ 198
              6.4.7   Chemistry of *closo*-$B_nH_n^{2-}$ 201
       6.5   Carboranes   202

6.6	Metallocarboranes		209
6.7	Boron Halides		220
	6.7.1	Boron trihalides	220
	6.7.2	Lower halides of boron	225
6.8	Boron–Oxygen Compounds		228
	6.8.1	Boron oxides and oxoacids	229
	6.8.2	Borates	231
	6.8.3	Organic compounds containing boron–oxygen bonds	234
6.9	Boron–Nitrogen Compounds		234
6.10	Other Compounds of Boron		240

Chapter 7   Aluminium, Gallium, Indium, and Thallium   243

7.1	Introduction		243
7.2	The Elements		244
	7.2.1	Terrestrial abundance and distribution	244
	7.2.2	Preparation and uses of the metals	246
	7.2.3	Properties of the elements	250
	7.2.4	Chemical reactivity and trends	252
7.3	Compounds		256
	7.3.1	Hydrides and related complexes	256
	7.3.2	Halides and halide complexes	261
		Aluminium trihalides	262
		Trihalides of gallium, indium, and thallium	266
		Lower halides of gallium, indium, and thallium	270
	7.3.3	Oxides and hydroxides	273
	7.3.4	Ternary and more complex oxide phases	278
		Spinels and related compounds	279
		Sodium-β-alumina and related phases	281
		Tricalcium aluminate, $Ca_3Al_2O_6$	282
	7.3.5	Other inorganic compounds	285
	7.3.6	Organometallic compounds	289

Chapter 8   Carbon   296

8.1	Introduction		296
8.2	Carbon		297
	8.2.1	Terrestrial abundance and distribution	297
	8.2.2	Allotropic forms	303
	8.2.3	Atomic and physical properties	306
	8.2.4	Chemical properties	308
8.3	Graphite Intercalation Compounds		313
8.4	Carbides		318
8.5	Hydrides, Halides, and Oxohalides		322
8.6	Oxides and Carbonates		325
8.7	Chalcogenides and Related Compounds		333
8.8	Cyanides and Other Carbon–Nitrogen Compounds		336
8.9	Organometallic Compounds		345
	8.9.1	Monohapto ligands	347
	8.9.2	Dihapto ligands	357
	8.9.3	Trihapto ligands	362
	8.9.4	Tetrahapto ligands	366
	8.9.5	Pentahapto ligands	367
	8.9.6	Hexahapto ligands	373
	8.9.7	Heptahapto and octahapto ligands	375

Chapter 9   Silicon   379

9.1	Introduction	379

	9.2	Silicon	380
		9.2.1 Occurrence and distribution	380
		9.2.2 Isolation, production, and industrial uses	381
		9.2.3 Atomic and physical properties	382
		9.2.4 Chemical properties	385
	9.3	Compounds	386
		9.3.1 Silicides	386
		9.3.2 Silicon hydrides (silanes)	388
		9.3.3 Silicon halides and related complexes	391
		9.3.4 Silica and silicic acids	393
		9.3.5 Silicate minerals	399
		Silicates with discrete units	400
		Silicates with chain or ribbon structures	403
		Silicates with layer structures	406
		Silicates with framework structures	414
		9.3.6 Other inorganic compounds of silicon	417
		9.3.7 Organosilicon compounds and silicones	419

Chapter 10   Germanium, Tin, and Lead   **427**

	10.1	Introduction	427
	10.2	The Elements	428
		10.2.1 Terrestrial abundance and distribution	428
		10.2.2 Production and uses of the elements	429
		10.2.3 Properties of the elements	431
		10.2.4 Chemical reactivity and group trends	434
	10.3	Compounds	436
		10.3.1 Hydrides and hydrohalides	436
		10.3.2 Halides and related complexes	437
		10.3.3 Oxides and hydroxides	445
		10.3.4 Derivatives of oxoacids	451
		10.3.5 Other inorganic compounds	453
		10.3.6 Metal–metal bonds and clusters	454
		10.3.7 Organometallic compounds	459
		Germanium	459
		Tin	459
		Lead	463

Chapter 11   Nitrogen   **466**

	11.1	Introduction	466
	11.2	The Element	469
		11.2.1 Abundance and distribution	469
		11.2.2 Production and uses of nitrogen	471
		11.2.3 Atomic and physical properties	472
		11.2.4 Chemical reactivity	473
	11.3	Compounds	479
		11.3.1 Nitrides, azides, and nitrido complexes	479
		11.3.2 Ammonia and ammonium salts	481
		Liquid ammonia as a solvent	486
		11.3.3 Other hydrides of nitrogen	489
		Hydrazine	489
		Hydroxylamine	494
		Hydrogen azide	496
		11.3.4 Thermodynamic relations between N-containing species	497
		11.3.5 Nitrogen halides and related compounds	503
		11.3.6 Oxides of nitrogen	508
		Nitrous oxide, N_2O	510
		Nitric oxide, NO	511
		Dinitrogen trioxide, N_2O_3	521
		Nitrogen dioxide, NO_2, and dinitrogen tetroxide, N_2O_4	522

		Dinitrogen pentoxide, N_2O_5, and nitrogen trioxide, NO_3	526
	11.3.7	Oxoacids, oxoanions, and oxoacid salts of nitrogen	527
		Hyponitrous acid and hyponitrites	527
		Nitrous acid and nitrites	531
		Nitric acid and nitrates	536
		Orthonitrates, $M_3^I NO_4$	545

Chapter 12 Phosphorus   **546**

12.1	Introduction		546
12.2	The Element		548
	12.2.1	Abundance and distribution	548
	12.2.2	Production and uses of elemental phosphorus	553
	12.2.3	Allotropes of phosphorus	553
	12.2.4	Atomic and physical properties	557
	12.2.5	Chemical reactivity	559
12.3	Compounds		562
	12.3.1	Phosphides	562
	12.3.2	Phosphine and related compounds	564
	12.3.3	Phosphorus halides	567
		Phosphorus trihalides	568
		Diphosphorus tetrahalides	571
		Phosphorus pentahalides	571
		Pseudohalides of phosphorus(III)	574
	12.3.4	Oxohalides and thiohalides of phosphorus	575
	12.3.5	Phosphorus oxides, sulfides, and oxosulfides	577
		Oxides	577
		Sulfides	581
		Oxosulfides	584
	12.3.6	Oxoacids of phosphorus and their salts	586
		Hypophosphorous acid and hypophosphites [$H_2PO(OH)$ and $H_2PO_2^-$]	591
		Phosphorous acid and phosphites [$HPO(OH)_2$ and HPO_3^{2-}]	591
		Hypophosphoric acid ($H_4P_2O_6$) and hypophosphates	592
		Other lower oxoacids of phosphorus	594
		The phosphoric acids	595
		Orthophosphates	603
		Chain polyphosphates	608
		Cyclo-polyphosphoric acids and *cyclo*-polyphosphates	617
	12.3.7	Phosphorus–nitrogen compounds	619
		Cyclophosphazenes	621
		Phosphazenes	622
		Polyphosphazenes	624
	12.3.8	Organophosphorus compounds	633

Chapter 13 Arsenic, Antimony, and Bismuth   **637**

13.1	Introduction		637
13.2	The Elements		638
	13.2.1	Abundance, distribution, and extraction	638
	13.2.2	Atomic and physical properties	641
	13.2.3	Chemical reactivity and group trends	644
13.3	Compounds of Arsenic, Antimony, and Bismuth		646
	13.3.1	Intermetallic compounds and alloys	646
	13.3.2	Hydrides of arsenic, antimony, and bismuth	650
	13.3.3	Halides and related complexes	651
		Trihalides, MX_3	651
		Pentahalides, MX_5	655
		Mixed halides and lower halides	656
		Halide complexes of M^{III} and M^V	659

Contents

Oxide halides | 665
13.3.4 Oxides and oxo compounds | 668
Oxo compounds of M^{III} | 668
Mixed-valence oxides | 672
Oxo compounds of M^V | 672
13.3.5 Sulfides and related con.pounds | 674
13.3.6 Metal–metal bonds and clusters | 678
13.3.7 Other inorganic compounds | 689
13.3.8 Organometallic compounds | 690
Organoarsenic(III) compounds | 690
Organoarsenic(V) compounds | 693
Physiological activity of arsenicals | 694
Organoantimony and organobismuth compounds | 694

Chapter 14 Oxygen
698

14.1 The Element | 698
14.1.1 Introduction | 698
14.1.2 Occurrence | 700
14.1.3 Preparation | 701
14.1.4 Atomic and physical properties | 704
14.1.5 Other forms of oxygen | 707
Ozone | 707
Atomic oxygen | 712
14.1.6 Chemical properties of dioxygen, O_2 | 713
14.2 Compounds of Oxygen | 718
14.2.1 Coordination chemistry: dioxygen as a ligand | 718
14.2.2 Water | 726
Introduction | 726
Distribution and availability | 727
Physical properties and structure | 729
Water of crystallization, aquo complexes, and solid hydrates | 733
Chemical properties | 735
Polywater | 741
14.2.3 Hydrogen peroxide | 742
Physical properties | 742
Chemical properties | 745
14.2.4 Oxygen fluorides | 748
14.2.5 Oxides | 751
Various methods of classification | 751
Nonstoichiometry | 753

Chapter 15 Sulfur
757

15.1 The Element | 757
15.1.1 Introduction | 757
15.1.2 Abundance and distribution | 759
15.1.3 Production and uses of elemental sulfur | 761
15.1.4 Allotropes of sulfur | 769
15.1.5 Atomic and physical properties | 781
15.1.6 Chemical reactivity | 782
Polyatomic sulfur cations | 785
Sulfur as a ligand | 786
Other ligands containing sulfur as donor atom | 794
15.2 Compounds of Sulfur | 798
15.2.1 Sulfides of the metallic elements | 798
General considerations | 798
Structural chemistry of metal sulfides | 803
Anionic polysulfides | 805
15.2.2 Hydrides of sulfur (sulfanes) | 806

15.2.3   Halides of sulfur   808
       Sulfur fluorides   808
       Chlorides, bromides, and iodides of sulfur   815
15.2.4   Oxohalides of sulfur   819
15.2.5   Oxides of sulfur   821
       Lower oxides   821
       Sulfur dioxide, SO_2 824
       Sulfur oxide ligands   828
       Sulfur trioxide   832
       Higher oxides   833
15.2.6   Oxoacids of sulfur   834
       Sulfuric acid, H_2SO_4 837
       Peroxosulfuric acids, H_2SO_5 and $H_2S_2O_8$ 846
       Thiosulfuric acid, $H_2S_2O_3$ 846
       Dithionic acid, $H_2S_2O_6$ 848
       Polythionic acids, $H_2S_nO_6$ 849
       Sulfurous acid, H_2SO_3 850
       Disulfurous acid, $H_2S_2O_5$ 853
       Dithionous acid, $H_2S_2O_4$ 853
15.2.7   Sulfur–nitrogen compounds   854
       Binary sulfur nitrides   855
       Sulfur–nitrogen cations and anions   864
       Sulfur imides, $S_{8-n}(NH)_n$ 868
       Other cyclic sulfur–nitrogen compounds   869
       Sulfur–nitrogen–halogen compounds   869
       Sulfur–nitrogen–oxygen compounds   875

Chapter 16   Selenium, Tellurium, and Polonium   882

16.1   The Elements   882
     16.1.1   Introduction: history, abundance, distribution   882
     16.1.2   Production and uses of the elements   883
     16.1.3   Allotropy   886
     16.1.4   Atomic and physical properties   889
     16.1.5   Chemical reactivity and trends   890
     16.1.6   Polyatomic cations, M_x^{n+} 896
16.2   Compounds of Selenium, Tellurium, and Polonium   898
     16.2.1   Selenides, tellurides, and polonides   898
     16.2.2   Hydrides   899
     16.2.3   Halides   901
       Lower halides   901
       Tetrahalides   904
       Hexahalides   907
       Halide complexes   908
     16.2.4   Oxohalides and pseudohalides   909
     16.2.5   Oxides   911
     16.2.6   Hydroxides and oxoacids   913
     16.2.7   Other inorganic compounds   916
     16.2.8   Organo-compounds   917

Chapter 17   The Halogens: Fluorine, Chlorine, Bromine, Iodine, and Astatine   920

17.1   The Elements   920
     17.1.1   Introduction   920
       Fluorine   920
       Chlorine   923
       Bromine   924
       Iodine   925
       Astatine   926

Contents

	17.1.2	Abundance and distribution	927
	17.1.3	Production and uses of the elements	928
	17.1.4	Atomic and physical properties	934
	17.1.5	Chemical reactivity and trends	938
		General reactivity and stereochemistry	938
		Solutions and charge-transfer complexes	940
17.2	Compounds of Fluorine, Chlorine, Bromine, and Iodine		943
	17.2.1	Hydrogen halides, HX	943
		Preparation and uses	944
		Physical properties of the hydrogen halides	949
		Chemical reactivity of the hydrogen halides	950
		The hydrogen halides as nonaqueous solvents	954
	17.2.2	Halides of the elements	957
		Fluorides	958
		Chlorides, bromides, and iodides	961
	17.2.3	Interhalogen compounds	964
		Diatomic interhalogens, XY	964
		Tetra-atomic interhalogens, XY_3	968
		Hexa-atomic and octa-atomic interhalogens, XF_5 and IF_7	974
	17.2.4	Polyhalide anions	978
	17.2.5	Polyhalonium cations, XY_{2n}^+	983
	17.2.6	Halogen cations	986
	17.2.7	Oxides of chlorine, bromine, and iodine	988
		Oxides of chlorine	989
		Oxides of bromine	996
		Oxides of iodine	997
	17.2.8	Oxoacids and oxoacid salts	999
		General considerations	999
		Hypohalous acids, HOX, and hypohalites, XO^-	1003
		Halous acids, HOXO, and halites, XO_2^-	1007
		Halic acids, $HOXO_2$, and halates, XO_3^-	1009
		Perhalic acid and perhalates	1013
		Perchloric acid and perchlorates	1013
		Perbromic acid and perbromates	1020
		Periodic acids and periodates	1022
	17.2.9	Halogen oxide fluorides and related compounds	1026
		Chlorine oxide fluorides	1026
		Bromine oxide fluorides	1032
		Iodine oxide fluorides	1034
	17.2.10	Halogen derivatives of oxoacids	1037
17.3	The Chemistry of Astatine		1039

Chapter 18 The Noble Gases: Helium, Neon, Argon, Krypton, Xenon, and Radon   1042

18.1	Introduction		1042
18.2	The Elements		1043
	18.2.1	Distribution, production, and uses	1043
	18.2.2	Atomic and physical properties of the elements	1044
18.3	Chemistry of the Noble Gases		1047
	18.3.1	Clathrates	1047
	18.3.2	Compounds of xenon	1048
	18.3.3	Compounds of other noble gases	1059

Chapter 19 Coordination Compounds   1060

19.1	Introduction	1060
19.2	Types of Ligand	1061
19.3	Stability of Coordination Compounds	1064
19.4	The Various Coordination Numbers	1068

19.5	Isomerism	1077
	Conformational isomerism	1078
	Geometrical isomerism	1078
	Optical isomerism	1079
	Ionization isomerism	1080
	Linkage isomerism	1081
	Coordination isomerism	1081
	Polymerization isomerism	1081
	Ligand isomerism	1082
19.6	The Coordinate Bond	1082
	Valence bond theory	1083
19.7	Crystal Field Theory	1084
19.8	Colours of Complexes	1089
19.9	Thermodynamic Effects of Crystal Field Splitting	1096
19.10	Magnetic Properties	1098
19.11	Ligand Field Theory	1099
19.12	Molecular Orbital Theory	1099

Chapter 20 Scandium, Yttrium, Lanthanum, and Actinium 1102

20.1	Introduction	1102
20.2	The Elements	1103
	20.2.1 Terrestrial abundance and distribution	1103
	20.2.2 Preparation and uses of the metals	1103
	20.2.3 Properties of the elements	1104
	20.2.4 Chemical reactivity and trends	1105
20.3	Compounds of Scandium, Yttrium, Lanthanum, and Actinium	1107
	20.3.1 Simple compounds	1107
	20.3.2 Complexes	1108
	20.3.3 Organometallic compounds	1110

Chapter 21 Titanium, Zirconium, and Hafnium 1111

21.1	Introduction	1111
21.2	The Elements	1112
	21.2.1 Terrestrial abundance and distribution	1112
	21.2.2 Preparation and uses of the metals	1112
	21.2.3 Properties of the elements	1114
	21.2.4 Chemical reactivity and trends	1116
21.3	Compounds of Titanium, Zirconium, and Hafnium	1117
	21.3.1 Oxides and sulfides	1117
	21.3.2 Mixed, or complex, oxides	1121
	21.3.3 Halides	1123
	21.3.4 Compounds with oxoanions	1125
	21.3.5 Complexes	1126
	Oxidation state IV (d^0)	1126
	Oxidation state III (d^1)	1130
	Lower oxidation states	1133
	21.3.6 Organometallic compounds	1133

Chapter 22 Vanadium, Niobium, and Tantalum 1138

22.1	Introduction	1138
22.2	The Elements	1139
	22.2.1 Terrestrial abundance and distribution	1139
	22.2.2 Preparation and uses of the metals	1139
	22.2.3 Atomic and physical properties of the elements	1141
	22.2.4 Chemical reactivity and trends	1142
22.3	Compounds of Vanadium, Niobium, and Tantalum	1143

22.3.1	Oxides		1144
22.3.2	Isopolymetallates		1146
22.3.3	Sulfides, selenides, and tellurides		1150
22.3.4	Halides and oxohalides		1152
22.3.5	Compounds with oxoanions		1157
22.3.6	Complexes		1157
	Oxidation state V (d^0)		1157
	Oxidation state IV (d^1)		1158
	Oxidation state III (d^2)		1160
	Oxidation state II (d^3)		1163
22.3.7	Organometallic compounds		1163

Chapter 23   Chromium, Molybdenum, and Tungsten 1167

23.1	Introduction		1167
23.2	The Elements		1167
	23.2.1	Terrestrial abundance and distribution	1167
	23.2.2	Preparation and uses of the metals	1168
	23.2.3	Properties of the elements	1169
	23.2.4	Chemical reactivity and trends	1170
23.3	Compounds of Chromium, Molybdenum, and Tungsten		1171
	23.3.1	Oxides of chromium, molybdenum, and tungsten	1171
	23.3.2	Isopolymetallates	1175
	23.3.3	Heteropolymetallates	1184
	23.3.4	Tungsten and molybdenum bronzes	1185
	23.3.5	Sulfides, selenides, and tellurides	1186
	23.3.6	Halides and oxohalides	1187
	23.3.7	Complexes of chromium, molybdenum, and tungsten	1191
		Oxidation state VI (d^0)	1191
		Oxidation state V (d^1)	1193
		Oxidation state IV (d^2)	1194
		Oxidation state III (d^3)	1196
		Oxidation state II (d^4)	1200
	23.3.8	Biological activity and nitrogen fixation	1205
	23.3.9	Organometallic compounds	1207

Chapter 24   Manganese, Technetium, and Rhenium 1211

24.1	Introduction		1211
24.2	The Elements		1212
	24.2.1	Terrestrial abundance and distribution	1212
	24.2.2	Preparation and uses of the metals	1212
	24.2.3	Properties of the elements	1214
	24.2.4	Chemical reactivity and trends	1215
24.3	Compounds of Manganese, Technetium, and Rhenium		1217
	24.3.1	Oxides and chalcogenides	1218
	24.3.2	Oxoanions	1221
	24.3.3	Halides and oxohalides	1222
	24.3.4	Complexes of manganese, technetium, and rhenium	1226
		Oxidation state VII (d^0)	1226
		Oxidation state VI (d^1)	1226
		Oxidation state V (d^2)	1227
		Oxidation state IV (d^3)	1228
		Oxidation state III (d^4)	1228
		Oxidation state II (d^5)	1231
		Lower oxidation states	1234
	24.3.5	Organometallic compounds	1234

Chapter 25 Iron, Ruthenium, and Osmium   1242

  25.1 Introduction   1242
  25.2 The Elements   1243
         25.2.1 Terrestrial abundance and distribution   1243
         25.2.2 Preparation and uses of the elements   1243
         25.2.3 Properties of the elements   1247
         25.2.4 Chemical reactivity and trends   1250
  25.3 Compounds of Iron, Ruthenium, and Osmium   1252
         25.3.1 Oxides and other chalcogenides   1254
         25.3.2 Mixed metal oxides and oxoanions   1256
         25.3.3 Halides and oxohalides   1257
         25.3.4 Complexes   1260
                 Oxidation state VIII (d^0)   1260
                 Oxidation state VII (d^1)   1260
                 Oxidation state VI (d^2)   1260
                 Oxidation state V (d^3)   1261
                 Oxidation state IV (d^4)   1262
                 Oxidation state III (d^5)   1263
                 Oxidation state II (d^6)   1268
                 Lower oxidation states   1274
         25.3.5 The biochemistry of iron   1275
                 Haemoglobin and myoglobin   1276
                 Cytochromes   1279
                 Iron–sulfur proteins   1280
         25.3.6 Organometallic compounds   1282
                 Carbonyls   1282
                 Carbonyl hydrides and carbonylate anions   1284
                 Carbonyl halides and other substituted carbonyls   1286
                 Ferrocene and other cyclopentadienyls   1286

Chapter 26 Cobalt, Rhodium, and Iridium   1290

  26.1 Introduction   1290
  26.2 The Elements   1290
         26.2.1 Terrestrial abundance and distribution   1290
         26.2.2 Preparation and uses of the elements   1291
         26.2.3 Properties of the elements   1292
         26.2.4 Chemical reactivity and trends   1294
  26.3 Compounds of Cobalt, Rhodium, and Iridium   1296
         26.3.1 Oxides and sulfides   1296
         26.3.2 Halides   1297
         26.3.3 Complexes   1300
                 Oxidation state IV (d^5)   1300
                 Oxidation state III (d^6)   1301
                 Oxidation state II (d^7)   1311
                 Oxidation state I (d^8)   1316
                 Lower oxidation states   1320
         26.3.4 The biochemistry of cobalt   1321
         26.3.5 Organometallic compounds   1322
                 Carbonyls   1323
                 Cyclopentadienyls   1326

Chapter 27 Nickel, Palladium, and Platinum   1328

  27.1 Introduction   1328
  27.2 The Elements   1329
         27.2.1 Terrestrial abundance and distribution   1329
         27.2.2 Preparation and uses of the elements   1329
         27.2.3 Properties of the elements   1331
         27.2.4 Chemical reactivity and trends   1333

27.3   Compounds of Nickel, Palladium, and Platinum   1335
     27.3.1   The Pd/H$_2$ system   1335
     27.3.2   Oxides and chalcogenides   1336
     27.3.3   Halides   1337
     27.3.4   Complexes   1339
         Oxidation state IV (d^6)   1339
         Oxidation state III (d^7)   1340
         Oxidation state II (d^8)   1342
         Oxidation state I (d^9)   1355
         Oxidation state 0 (d^{10})   1355
     27.3.5   Organometallic compounds   1356
         σ-Bonded compounds   1356
         Carbonyls   1357
         Cyclopentadienyls   1358
         Alkene and alkyne complexes   1360
         π-Allylic complexes   1362

Chapter 28   Copper, Silver, and Gold   1364

28.1   Introduction   1364
28.2   The Elements   1365
     28.2.1   Terrestrial abundance and distribution   1365
     28.2.2   Preparation and uses of the elements   1365
     28.2.3   Atomic and physical properties of the elements   1367
     28.2.4   Chemical reactivity and trends   1368
28.3   Compounds of Copper, Silver, and Gold   1372
     28.3.1   Oxides and sulfides   1373
     28.3.2   Halides   1374
     28.3.3   Photography   1376
     28.3.4   Complexes   1379
         Oxidation state III (d^8)   1379
         Oxidation state II (d^9)   1380
         Electronic spectra and magnetic properties of copper(II)   1386
         Oxidation state I (d^{10})   1386
         Gold cluster compounds   1390
     28.3.5   Biochemistry of copper   1392
     28.3.6   Organometallic compounds   1392

Chapter 29   Zinc, Cadmium, and Mercury   1395

29.1   Introduction   1395
29.2   The Elements   1396
     29.2.1   Terrestrial abundance and distribution   1396
     29.2.2   Preparation and uses of the elements   1396
     29.2.3   Properties of the elements   1399
     29.2.4   Chemical reactivity and trends   1400
29.3   Compounds of Zinc, Cadmium, and Mercury   1403
     29.3.1   Oxides and chalcogenides   1403
     29.3.2   Halides   1406
     29.3.3   Mercury(I)   1408
         Polycations of mercury   1410
     29.3.4   Zinc(II) and cadmium(II)   1411
     29.3.5   Mercury(II)   1413
         Hg^{II}-N compounds   1414
     29.3.6   Organometallic compounds   1416
     29.3.7   Biological and environmental importance   1420

Chapter 30   The Lanthanide Elements ($Z = 58$–71)   1423

    30.1   Introduction   1423
    30.2   The Elements   1425
         30.2.1   Terrestrial abundance and distribution   1425
         30.2.2   Preparation and uses of the elements   1426
         30.2.3   Properties of the elements   1429
         30.2.4   Chemical reactivity and trends   1433
    30.3   Compounds of the Lanthanides   1437
         30.3.1   Oxides and chalcogenides   1437
         30.3.2   Halides   1439
         30.3.3   Magnetic and spectroscopic properties   1441
         30.3.4   Complexes   1444
             Oxidation state IV   1444
             Oxidation state III   1445
             Oxidation state II   1447
         30.3.5   Organometallic compounds   1448
             Cyclopentadienides and related compounds   1448
             Alkyls and aryls   1449

Chapter 31   The Actinide Elements ($Z = 90$–103)   1450

    31.1   Introduction   1450
             Superheavy elements   1453
    31.2   The Elements   1454
         31.2.1   Terrestrial abundance and distribution   1454
         31.2.2   Preparation and uses of the elements   1455
             Nuclear reactors and atomic energy   1456
             Nuclear fuel reprocessing   1461
         31.2.3   Properties of the elements   1463
         31.2.4   Chemical reactivity and trends   1464
    31.3   Compounds of the Actinides   1469
         31.3.1   Oxides and chalcogenides   1470
         31.3.2   Mixed metal oxides   1473
         31.3.3   Halides   1473
         31.3.4   Magnetic and spectroscopic properties   1476
         31.3.5   Complexes   1477
             Oxidation state VII   1478
             Oxidation state VI   1478
             Oxidation state V   1479
             Oxidation state IV   1481
             Oxidation state III   1483
             Oxidation state II   1483
         31.3.6   Organometallic compounds   1484

Appendix 1   Atomic Orbitals   1487

Appendix 2   Symmetry Elements, Symmetry Operations, and Point Groups   1492

Appendix 3   Some Non-SI Units and Conversion Factors   1495

Appendix 4   Abundance of Elements in Crustal Rocks   1496

Appendix 5   Effective Ionic Radii   1497

Appendix 6   Nobel Prize for Chemistry   1498

Appendix 7   Nobel Prize for Physics   1502

Index   1507

Contents

Chapter 30. The Lanthanide Elements (Z = 58–71)   1227

30.1 Introduction
30.2 The elements
30.2.1 Terrestrial abundance and distribution
30.2.2 Preparation and uses of the elements
30.2.3 Properties of the elements
30.2.4 Chemical reactivity and trends
30.3 Compounds of the lanthanides
30.3.1 Oxides and chalcogenides
30.3.2 ......
30.3.3 Oxoacids, and organic compounds
30.3.4 Hydrides
30.3.5 Oxidation state +IV
30.3.6 Oxidation state +II
30.3.7 Coordination chemistry
30.3.8 Organometallic compounds of the lanthanides
30.3.9 ......

Chapter 31. The Actinide Elements (Z = 90–103)   1250

31.1 Introduction
31.2 The elements
31.2.1 Discovery
31.2.2 Terrestrial abundance and distribution
31.2.3 Preparation and uses of the elements
31.2.4 Nuclear properties of actinium
31.2.5 Properties of the elements
31.2.6 Chemical reactivity and trends
31.3 Compounds of the actinides
31.3.1 Oxides and chalcogenides
31.3.2 ......
31.3.3 Halides
31.3.4 Magnetic and spectroscopic properties
31.3.5 Complexes
Oxidation state VII
Oxidation state VI
Oxidation state V
Oxidation state IV
Oxidation state III
Oxidation state II
31.3.6 Organometallic compounds

Appendix 1. Atomic Orbitals   1497

Appendix 2. Symmetry Elements, Symmetry Operations and Point Groups   1499

Appendix 3. Some Non-SI Units and Conversion Factors   1503

Appendix 4. Abundance of Elements in Crustal Rocks   1504

Appendix 5. Effective Ionic Radii   1505

Appendix 6. Nobel Prize for Chemistry   1506

Appendix 7. Nobel Prize for Physics   1507

Index

1

Origin of the Elements. Isotopes and Atomic Weights

1.1 Introduction

This book presents a unified treatment of the chemistry of the elements. At present 107 elements are known, though not all occur in nature: of the 92 elements from hydrogen to uranium all except technetium and promethium are found on earth and technetium has been detected in some stars. To these elements a further 15 have been added by artificial nuclear syntheses in the laboratory. Why are there only 90 elements in nature? Why do they have their observed abundances and why do their individual isotopes occur with the particular relative abundances observed? Indeed, we must also ask to what extent these isotopic abundances commonly vary in nature, thus causing variability in atomic weights and possibly jeopardizing the classical means of determining chemical composition and structure by chemical analysis.

Theories abound, and it is important at all times to distinguish carefully between what has been experimentally established, what is a useful model for suggesting further experiments, and what is a currently acceptable theory which interprets the known facts. The tentative nature of our knowledge is perhaps nowhere more evident than in the first few sections of this chapter dealing with the origin of the chemical elements and their present isotopic composition. This is not surprising, for it is only in the last few decades that progress in this enormous enterprise has been made possible by discoveries in nuclear physics, relativity, and quantum theory.

1.2 Origin of the Universe

One currently popular theory for the origin and evolution of the universe to its present form starts with the "hot big bang".[1] It is supposed that all the matter in the universe was once contained in a primeval nucleus of immense density and temperature which, for some reason, exploded and distributed matter uniformly throughout space. In the beginning,

---

[1] J. SILK, *The Big Bang: The Creation and Evolution of the Universe*, W. H. Freeman, San Francisco, 1980, 394 pp. J. D. BARROW and J. SILK, *The Left Hand of Creation: The Origin and Evolution of the Expanding Universe*, Heinemann, London, 1984, 256 pp.

according to one model, the density of the universe was some 10^{96} g cm^{-3} and the temperature 10^{32} K. After 1 s the temperature is believed to have fallen to ca. 10^{10} K and the expanding universe was populated by the "elementary particles" familiar to chemists—neutrons and protons in approximately equal numbers, and electrons. Soon conditions were right for protons and neutrons to combine to form the nuclei of deuterium and helium: the process of element building had begun. During the small niche in cosmic history from about 10–500 s after the big bang, the entire universe is thought to have behaved as an enormous nuclear fusion reaction and, within this short time, some 25% of the mass of the universe was converted to ^{4}He and a minute 10^{-3}% was converted to ^{2}H (deuterium). It was from this matter that the earliest stars condensed and the subsequent synthesis of the chemical elements ensued.

From the observed present rate of expansion of the universe (about 18 km s^{-1} per 10^6 light years), and assuming that the rate has remained constant, it can readily be calculated that the big bang occurred 1.8×10^{10} y ago. [1 light year $= 9.46 \times 10^{12}$ km; 1 year $= 3.15 \times 10^7$ s; hence $t_0 = 9.46 \times 10^{18}/(18 \times 3.15 \times 10^7)$ y.] One attractive aspect of this theory is that it readily explains the universal presence of isotropic thermal black body radiation at a temperature of ~ 2.7 K. This radiation, discovered in 1965 by A. A. Penzias and R. W. Wilson of Bell Laboratories, USA, and for which they were awarded the 1978 Nobel prize for physics, is seen as the dying remnants of the big bang and is of about the magnitude to be expected of this theory.[2] No other current cosmological theory can interpret this radiation temperature satisfactorily, and for this reason, among others, steady-state theories such as the continuous creation of matter, and other theories such as the cold big bang or polyneutron theory, are less favoured. Fortunately for our purpose an explanation of the present concentration of the chemical elements does not depend on the acceptance of any particular cosmological theory: i.e. we distinguish between the origin of matter (cosmology) and the origin of the chemical elements (nucleogenesis). Likewise, the distribution of isotopes within the universe, the solar system, and on earth is independent of cosmology and can be derived on the basis of observation and the application of accepted physical principles.

Objects for which the abundances of at least some of the elements can be obtained include (i) the sun and the stars, (ii) gaseous nebulae, including some in other galaxies, (iii) the interstellar medium, (iv) cosmic-ray particles, (v) the earth, moon, and meteorites, and (vi) other planets, asteroids, and comets in the solar system. Information on the first three groups depends on inferences from spectroscopic data whereas direct analysis of samples is possible for cosmic rays, meteorites, and at least the surface regions of the earth and the moon. The results indicate extensive differentiation in the solar system and in some stars, but the overall picture is one of astonishing uniformity of composition. Hydrogen is by far the most abundant element in the universe, accounting for some 88.6% of all atoms (or nuclei). Helium is about eightfold less abundant (11.3%), but these two elements together account for over 99.9% of the atoms and nearly 99% of the mass of the universe. Clearly nucleosynthesis of the heavier elements from hydrogen and helium has not yet proceeded very far.

[2] A. A. PENZIAS and R. W. WILSON, A measurement of excess antenna temperature at 4080 Mc/s, *Astrophys. J.* **142**, 419–21 (1965). R. H. DICKE, P. J. E. PEEBLES, P. G. ROLL, and D. T. WILKINSON, Cosmic black-body radiation, *Astrophys. J.* **142**, 414–19 (1965). R. W. WILSON, The cosmic microwave background radiation, pp. 113–33 in *Les Prix Nobel 1978*, Almqvist & Wiksell International, Stockholm 1979. A. A. PENZIAS, The origin of the elements, pp. 93–106 in *Les Prix Nobel 1978* (also in *Science* **105**, 549–54 (1979)).

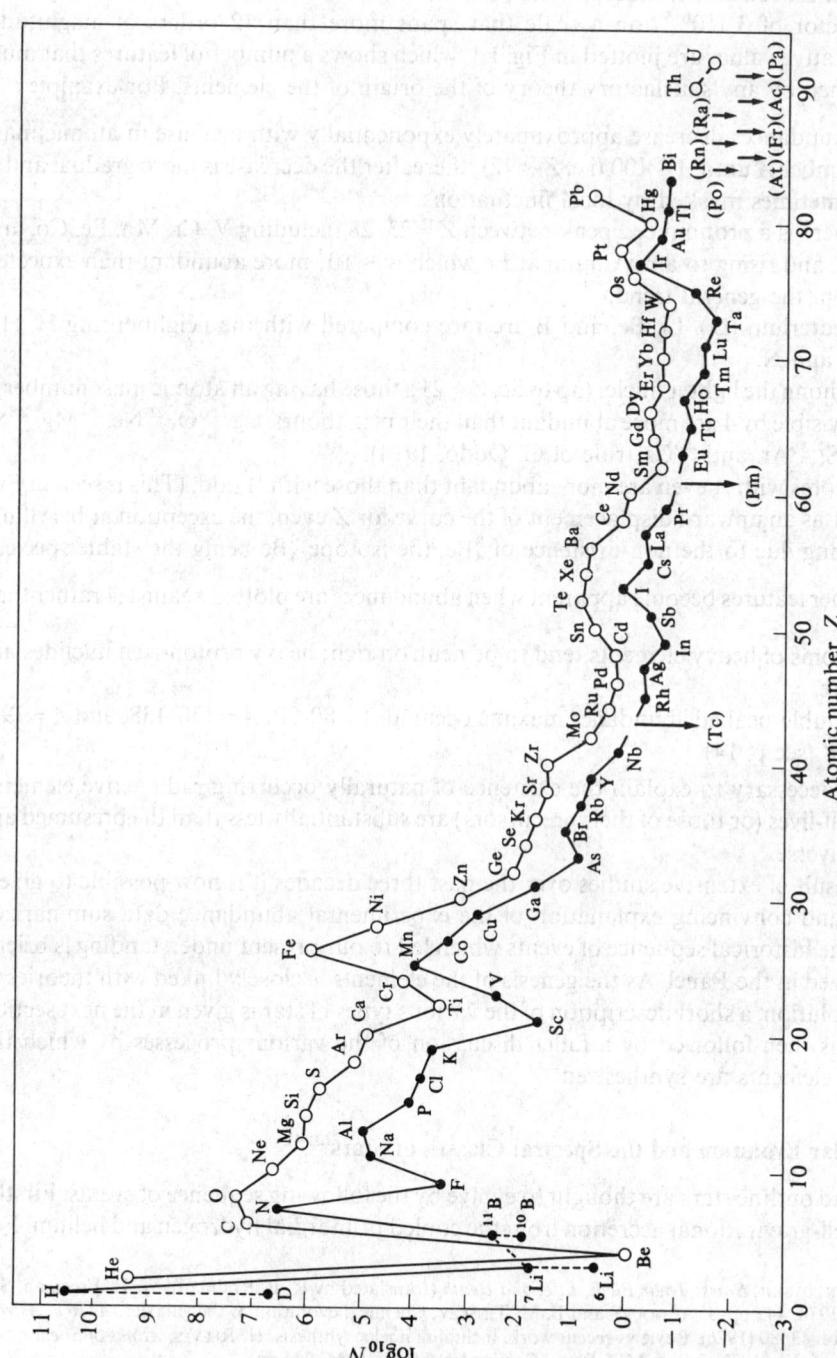

FIG. 1.1 Cosmic abundances of the elements as a function of atomic number Z. Abundances are expressed as numbers of atoms per 10^6 atoms of Si and are plotted on a logarithmic scale. (From A. G. W. Cameron, *Space Sci. Rev.* **15**, 121–46 (1973), as updated by Brian Mason, private communication.)

Various estimates of the universal abundances of the elements have been made and, although these sometimes differ in detail for particular elements, they rarely do so by more than a factor of 3 ($10^{0.5}$) on a scale that spans more than 12 orders of magnitude. Representative values are plotted in Fig. 1.1, which shows a number of features that must be explained by any satisfactory theory of the origin of the elements. For example:

(i) Abundances decrease approximately exponentially with increase in atomic mass number A until $A \sim 100$ (i.e. $Z \sim 42$); thereafter the decrease is more gradual and is sometimes masked by local fluctuations.

(ii) There is a pronounced peak between $Z = 23$–28 including V, Cr, Mn, Fe, Co, and Ni, and rising to a maximum at Fe which is $\sim 10^3$ more abundant than expected from the general trend.

(iii) Deuterium (D), Li, Be, and B are rare compared with the neighbouring H, He, C, and N.

(iv) Among the lighter nuclei (up to Sc, $Z = 21$), those having an atomic mass number A divisible by 4 are more abundant than their neighbours, e.g. ^{16}O, ^{20}Ne, ^{24}Mg, ^{28}Si, ^{32}S, ^{36}Ar, and ^{40}Ca (rule of G. Oddo, 1914).

(v) Atoms with A even are more abundant than those with A odd. (This is seen in Fig. 1.1 as an upward displacement of the curve for Z even, the exception at beryllium being due to the non-existence of $^{8}_{4}Be$, the isotope $^{9}_{4}Be$ being the stable species.)

Two further features become apparent when abundances are plotted against A rather than Z:

(vi) Atoms of heavy elements tend to be neutron rich; heavy proton-rich nuclides are rare.

(vii) Double-peaked abundance maxima occur at $A = 80, 90$; $A = 130, 138$; and $A = 196, 208$ (see p. 14).

It is also necessary to explain the existence of naturally occurring radioactive elements whose half-lives (or those of their precursors) are substantially less than the presumed age of the universe.

As a result of extensive studies over the past three decades it is now possible to give a detailed and convincing explanation of the experimental abundance data summarized above. The historical sequence of events which led to our present understanding is briefly summarized in the Panel. As the genesis of the elements is closely linked with theories of stellar evolution, a short description of the various types of star is given in the next section and this is then followed by a fuller discussion of the various processes by which the chemical elements are synthesized.

1.3 Stellar Evolution and the Spectral Classes of Stars[3, 4]

In broad outline stars are thought to evolve by the following sequence of events. Firstly, there is self-gravitational accretion from the cooled primordial hydrogen and helium. For

---

[3] I. S. SHKLOVSKII, *Stars: Their Birth, Life, and Death* (translated by R. B. Rodman), W. H. Freeman, San Francisco, 1978, 442 pp. J. AUDOUZE and B. M. TINSLEY, Chemical evolution of the galaxies, *A. Rev. Astron. Astrophys.* **14**, 43–80 (1976). Reviews recent work, including nucleosynthesis. H. REEVES, *Atoms of Silence: An Exploration of Cosmic Evolution*, MIT Press, Cambridge, Mass., 1984, 244 pp.

[4] D. H. CLARK and F. R. STEPHENSON, *The Historical Supernovae*, Pergamon Press, Oxford, 1977, 233 pp. A fascinating account of observations of novae and supernovae throughout the centuries, coupled with a valuable description of the associated astrophysics.

Genesis of the Elements—Historical Landmarks

1890s	First systematic studies on the terrestrial abundances of the elements	F. W. Clarke; H. S. Washington, and others
1905	Special relativity theory: $E = mc^2$	A. Einstein
1911	Nuclear model of the atom	E. Rutherford
1913	First observation of isotopes in a stable element (Ne)	J. J. Thompson
1919	First artificial transmutation of an element $^{14}_7N(\alpha, p)^{17}_8O$	E. Rutherford
1925–8	First abundance data on stars (spectroscopy)	C. H. Payne Gaposchkin; H. N. Russell
1929	First proposal of stellar nucleosynthesis by proton fusion to helium and heavier nuclides	R. D'E. Atkinson and F. G. Houtermans
1937	The "missing element" $Z = 43$ (technetium) synthesized by $^{99}_{42}Mo(d, n)^{99}_{43}Tc$	C. Perrier and E. G. Segré
1938	Catalytic CNO process independently proposed to assist nuclear synthesis in stars	H. A. Bethe; C. F. von Weizsäcker
1938	Uranium fission discovered experimentally	O. Hahn and F. Strassmann
1940	First transuranium element $^{239}_{93}Np$ synthesized	E. M. McMillan and P. Abelson
1947	The last "missing element" $Z = 61$ (promethium) discovered among uranium fission products	J. A. Marinsky, L. E. Glendenin, and C. D. Coryell
1948	Hot big-bang theory of expanding universe includes (incorrect) theory of genesis of the elements	R. A. Alpher, H. A. Bethe, and G. Gamow
1952–4	Helium burning as additional process for nucleogenesis	E. E. Salpeter; F. Hoyle
1954	Slow neutron absorption added to stellar reactions	A. G. W. Cameron
1955–7	Comprehensive theory of stellar synthesis of all elements in observed cosmic abundances	E. M. Burbridge, G. R. Burbridge, W. A. Fowler, and F. Hoyle
1965	2.7 K radiation detected	A. P. Penzias and R. W. Wilson

a star the size and mean density of the sun (mass $= 1.991 \times 10^{30}$ kg $= 1$ M$_\odot$) this might take ~ 20 y. This gravitational contraction releases heat energy, some of which is lost by radiation; however, the continued contraction results in a steady rise in temperature until at $\sim 10^7$ K the core can sustain nuclear reactions. These release enough additional energy to compensate for radiational losses and a temporary equilibrium or steady state is established.

When 10% of the hydrogen in the core has been consumed gravitational contraction again occurs until at a temperature of $\sim 2 \times 10^8$ K helium burning (fusion) can occur. This is followed by a similar depletion, contraction, and temperature rise until nuclear reactions involving still heavier nuclei ($Z = 8$–22) can occur at $\sim 10^9$ K. The time scale of these processes depends sensitively on the mass of the star, taking perhaps 10^{12} y for a star of mass 0.2 M$_\odot$, 10^{10} y for a star of 1 solar mass, 10^7 y for mass 10 M$_\odot$, and only 8×10^4 y for a star of 50 M$_\odot$; i.e. the more massive the star, the more rapidly it consumes its nuclear fuel. Further catastrophic changes may then occur which result in much of the stellar material being ejected into space, where it becomes incorporated together with further hydrogen and helium in the next generation of stars. It should be noted, however, that, as iron is at the maximum of the nuclear binding energy curve, only those elements up to iron ($Z = 26$) can be produced by exothermic processes of the type just considered, which occur automatically if the temperature rises sufficiently. Beyond iron, an input of energy is required to promote further element building.

The evidence on which this theory of stellar evolution is based comes not only from known nuclear reactions and the relativistic equivalence of mass and energy, but also from the spectroscopic analysis of the light reaching us from the stars. This leads to the spectral classification of stars, which is the cornerstone of modern experimental astrophysics. The spectroscopic analysis of starlight reveals much information about the chemical composition of stars—the identity of the elements present and their relative concentrations. In addition, the "red shift" or Doppler effect can be used to gauge the relative motions of the stars and their distance from the earth. More subtly, the surface temperature of stars can be determined from the spectral characteristics of their "blackbody" radiation, the higher the temperature the shorter the wavelength of maximum emission. Thus cooler stars appear red, and successively hotter stars appear progressively yellow, white, and blue. Differences in colour are also associated with differences in chemical composition as indicated in Table 1.1.

TABLE 1.1   *Spectral classes of stars*

Class[a]	Colour	Surface (T/K)	Spectral characterization	Examples
O	Blue	> 25 000	Lines of ionized He and other elements; H lines weak	10 Lacertae
B	Blue-white	11 000–25 000	H and He prominent	Rigel, Spica
A	White	7500–11 000	H lines very strong	Sirius, Vega
F	Yellow-white	6000–7000	H weaker; lines of ionized metals becoming prominent	Canopus, Procyon
G	Yellow	5000–6000	Lines of ionized and neutral metals prominent (especially Ca)	Sun, Capella
K	Orange	3500–5000	Lines of neutral metals and band spectra of simple radicals (e.g. CN, OH, CH)	Arcturus, Aldebaran
M	Red	2000–3500	Band spectra of many simple compounds prominent (e.g. TiO)	Betelgeuse, Antares

[a] Further division of each class into 10 subclasses is possible, e.g. .... F8, F9, G0, G1, G2.... The sun is G2 with a surface temperature of 5780 K. This curious alphabetical sequence of classes arose historically and can perhaps best be remembered by the mnemonic "Oh Be A Fine Girl (Guy), Kiss Me".

If the spectral classes (or temperatures) of stars are plotted against their absolute magnitudes (or luminosities) the resulting diagram shows several preferred regions into which most of the stars fall. Such diagrams were first made, independently, by Hertzsprung and Russell about 1913 and are now called HR diagrams (Fig. 1.2). More than 90% of all stars fall on a broad band called the main sequence, which covers the full range of spectral classes and magnitudes from the large, hot, massive O stars at the top to the small, dense, reddish M stars at the bottom. This relationship is illustrated in Fig. 1.3,

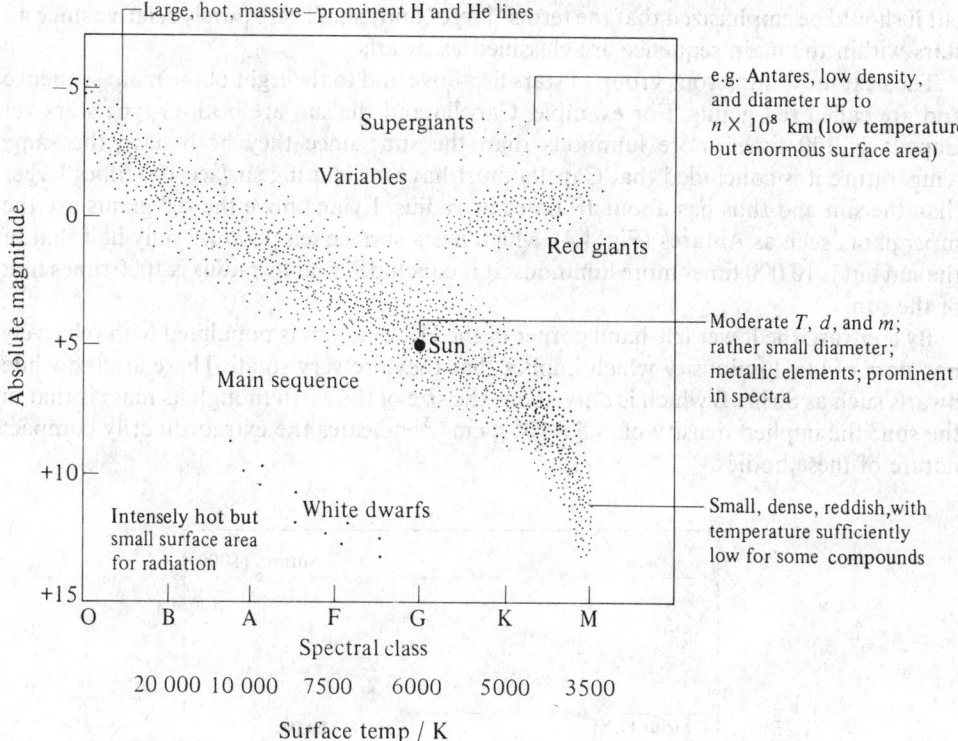

Large, hot, massive—prominent H and He lines

e.g. Antares, low density,
and diameter up to
$n \times 10^8$ km (low temperature
but enormous surface area)

Supergiants

Variables

Red giants

Moderate T, d, and m;
rather small diameter;
metallic elements; prominent
in spectra

●Sun

Main sequence

Intensely hot but
small surface area
for radiation

White dwarfs

Small, dense, reddish,with
temperature sufficiently
low for some compounds

Absolute magnitude

−5

0

+5

+10

+15

Spectral class

O   B   A   F   G   K   M

20 000 10 000   7500   6000   5000   3500

Surface temp / K

FIG. 1.2   The Hertzsprung–Russell diagram for stars with known luminosities and spectra.

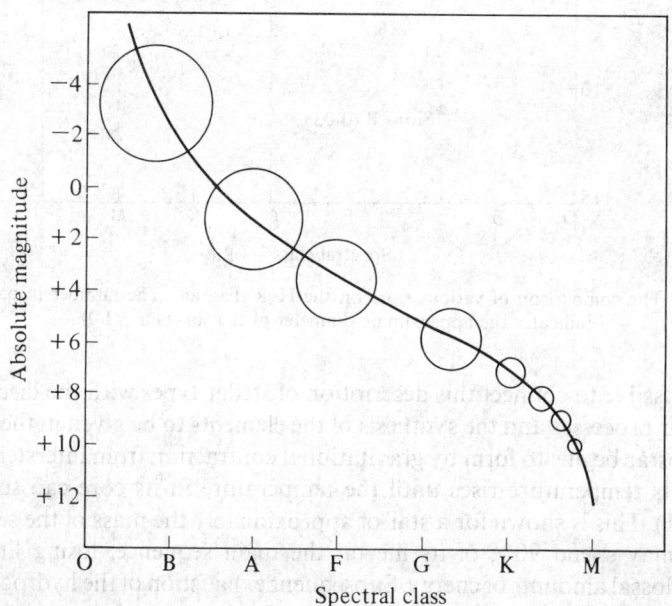

Absolute magnitude

−4
−2
0
+2
+4
+6
+8
+10
+12

Spectral class

O   B   A   F   G   K   M

FIG. 1.3   The main sequence of the H–R diagram showing the relationship between position on
the main sequence and size of star. Note that these are all "dwarfs" compared with the giants and
supergiants.

but it should be emphasized that the terms "large" and "small" are purely relative since all stars within the main sequence are classified as dwarfs.

The next most numerous group of stars lie above and to the right of the main sequence and are called red giants. For example, Capella and the sun are both G-type stars yet Capella is 100 times more luminous than the sun; since they both have the same temperature it is concluded that Capella must have a radiating surface 100 times larger than the sun and thus has about 10 times its radius. Lying above the red giants are the supergiants such as Antares (Fig. 1.4), which has a surface temperature only half that of the sun but is 10 000 times more luminous: it is concluded that its radius is 1000 times that of the sun.

By contrast, the lower left-hand corner of the HR diagram is populated with relatively hot stars of low luminosity which implies that they are very small. These are the white dwarfs such as Sirius B which is only about the size of the earth though its mass is that of the sun: the implied density of $\sim 5 \times 10^4$ g cm^{-3} indicates the extraordinarily compact nature of these bodies.

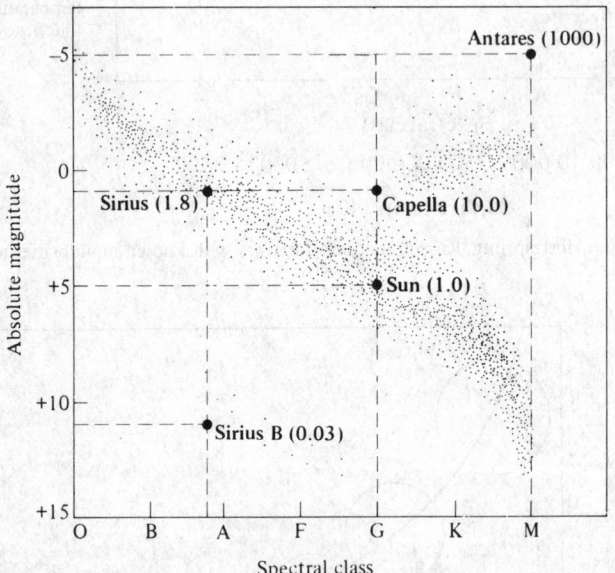

Fig. 1.4   The comparison of various stars on the H–R diagram. The number in parentheses indicates the approximate diameter of the star (sun = 1.0).

It is now possible to connect this description of stellar types with the discussion of the thermonuclear processes and the synthesis of the elements to be given in the next section. When a protostar begins to form by gravitational contraction from interstellar hydrogen and helium, its temperature rises until the temperature in its core can sustain proton burning (p. 11). This is shown for a star of approximately the mass of the sun in Fig. 1.5. Such a star may spend 90% of its life on the main sequence, losing little mass but generating colossal amounts of energy. Subsequent exhaustion of the hydrogen in the core (but not in the outer layers of the star) leads to further contraction to form a helium-burning core which forces much of the remaining hydrogen into a vast tenuous outer

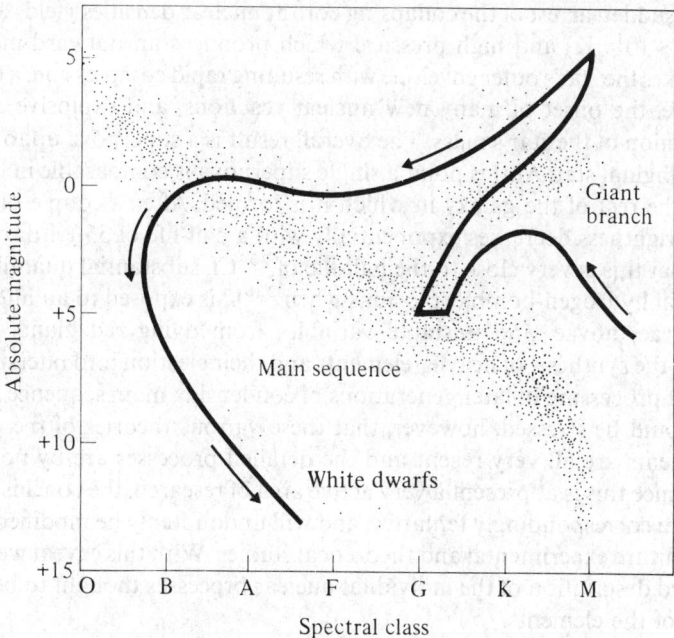

FIG. 1.5   The possible evolutionary track of a star with a mass approximately that of the solar mass. The schematic diagram shows the general progress from pre-main sequence to white-dwarf stage but does not attempt to indicate the possible back-tracking or other erratic routing that may occur.

envelope—the star has become a red giant since its enormous radiating surface area can no longer be maintained at such a high temperature as previously despite the higher core temperature. Typical red giants have surface temperatures in the range 3500–5500 K; their luminosities are about 10^2–10^4 times that of the sun and diameters about 10–100 times that of the sun. Carbon burning (p. 12) can follow in older red giants followed by the α-process (p. 13) during its final demise to white dwarf status.

Many stars are in fact partners in a binary system of two stars revolving around each other. If, as frequently occurs, the two stars have different masses, the more massive one will evolve faster and reach the white-dwarf stage before its partner. Then, as the second star expands to become a red giant its extended atmosphere encompasses the neighbouring white dwarf and induces instabilities which result in an outburst of energy and transfer of matter to the more massive partner. During this process the luminosity of the white dwarf increases perhaps ten-thousandfold and the event is witnessed as a nova (since the preceding binary was previously invisible to the naked eye).

As we shall see in the description of the e-process and the γ-process (p. 14), even more spectacular instabilities can develop in larger main sequence stars. If the initial mass is greater than about 3.5 solar masses, current theories suggest that gravitational collapse may be so catastrophic that the system implodes beyond nuclear densities to become a black hole. For main sequence stars in the mass range 1.4–3.5 $M_\odot$, implosion probably halts at nuclear densities to give a rapidly rotating neutron star (density $\sim 10^{14}$ g cm^{-3}) which may be observable as a pulsar emitting electromagnetic radiation over a wide range of frequencies in pulses at intervals of a fraction of a second. During this process of star implosion the sudden arrest of the collapsing core at nuclear densities yields an enormous temperature ($\sim 10^{12}$ K) and high pressure which produces an outward-moving shock wave. This strikes the star's outer envelope with resulting rapid compression, a dramatic rise in temperature, the onset of many new nuclear reactions, and explosive ejection of a significant fraction of the star's mass. The overall result is a supernova up to 10^8 times as bright as the original star. At this point a single supernova is comparable in brightness to the whole of the rest of the galaxy in which it is formed. After a couple of months the supernova's brightness decreases exponentially with a half-life of 55 ± 1 day. It has been pointed out that this is very close to the half-life of ^{254}Cf, substantial quantities of which are produced in hydrogen-bomb explosions when ^{238}U is exposed to an intense neutron flux. Supernovae, novae, and unstable variables from dying red giants are thus all candidates for the synthesis of heavier elements and their ejection into interstellar regions for subsequent processing in later generations of condensing main sequence stars such as the sun. It should be stressed, however, that these various theories of the origin of the chemical elements are all very recent and the detailed processes are by no means fully understood. Since this is at present a very active area of research, the conclusions given in this chapter are correspondingly tentative, and will undoubtedly be modified and refined in the light of future experimental and theoretical studies. With this caveat we now turn to a more detailed description of the individual nuclear processes thought to be involved in the synthesis of the elements.

1.4   Synthesis of the Elements[5-9]

The following types of nuclear reactions have been proposed to account for the various types of stars and the observed abundances of the elements:

(i) Exothermic processes in stellar interiors: these include (successively) hydrogen burning, helium burning, carbon burning, the α-process, and the equilibrium or e-process.

(ii) Neutron capture processes: these include the s-process (slow neutron capture) and the r-process (rapid neutron capture).

(iii) Miscellaneous processes: these include the p-process (proton capture) and spallation within the stars, and the x-process which involves spallation (p. 17) by galactic cosmic rays in interstellar regions.

[5] D. N. Schramm and R. Wagoner, Element production in the early universe, *A. Rev. Nucl. Sci.* **27**, 37–74 (1977).

[6] E. M. Burbidge, G. R. Burbidge, W. A. Fowler, and F. Hoyle, Synthesis of the elements in stars, *Rev. Mod. Phys.* **29**, 547–650 (1957). This is the definitive review on which all later work has been based.

[7] L. H. Aller, *The Abundance of the Elements*, Interscience, New York, 1961, 283 pp. A good general account which discusses nucleogenesis as well as abundance data.

[7a] L. H. Ahrens (ed.), *Origin and Distribution of the Elements*, Pergamon Press, Oxford, 1979, 920 pp. Proceedings of the Second UNESCO Symposium, Paris, 1977.

[8] R. J. Taylor, Origin of the elements, *Rept. Prog. Phys.* **29**, 489–538 (1966). A thought-provoking account of more recent work which also stresses problems of interpretation and assessment.

[9] R. J. Taylor, *The Origin of Chemical Elements*, Wykeham Publications, London, 1972, 169 pp. An excellent introductory account designed for beginning students in physics.

1.4.1. *Hydrogen burning*

When the temperature of a contracting mass of hydrogen and helium atoms reaches about 10^7 K, a sequence of thermonuclear reactions is possible of which the most important are as shown in Table 1.2.

TABLE 1.2   *Thermonuclear consumption of protons*

Reaction	Energy evolved, Q	Reaction time[a]
$^1H + {}^1H \rightarrow {}^2H + e^+ + \nu_e$	1.44 MeV	1.4×10^{10} y
$^2H + {}^1H \rightarrow {}^3He + \gamma$	5.49 MeV	0.6 s
$^3He + {}^3He \rightarrow {}^4He + 2{}^1H$	12.86 MeV	10^6 y

[a] The reaction time quoted is the time required for half the constitutents involved to undergo reaction—this is sensitively dependent on both temperature and density; the figures given are appropriate for the centre of the sun, i.e. 1.3×10^7 K and 200 g cm^{-3}.

The overall reaction thus converts 4 protons into 1 helium nucleus plus 2 positrons and 2 neutrinos:

$$4{}^1H \longrightarrow {}^4He + 2e^+ + 2\nu_e; \quad Q = 26.72 \text{ MeV}$$

Making allowance for the energy carried away by the 2 neutrinos (2×0.25 MeV) this leaves a total of 26.22 MeV for radiation, i.e. 4.20 pJ per atom of helium or 2.53×10^9 kJ mol^{-1}. This vast release of energy arises mainly from the difference between the rest mass of the helium-4 nucleus and the 4 protons from which it was formed (0.028 atomic mass units). There are several other peripheral reactions between the protons, deuterons, and ^{3}He nuclei, but these need not detain us. It should be noted, however, that only 0.7% of the mass is lost during this transformation, so that the star remains approximately constant in mass. For example, in the sun during each second, some 600×10^6 tonnes (600×10^9 kg) of hydrogen are processed into 595.5×10^6 tonnes of helium, the remaining 4.5×10^6 tonnes of matter being transformed into energy. This energy is released deep in the sun's interior as high-energy γ-rays which interact with stellar material and are gradually transformed into photons with longer wavelengths; these work their way to the surface taking perhaps 10^6 y to emerge.

In fact, it seems unlikely that the sun is a first-generation main-sequence star since spectroscopic evidence shows the presence of many heavier elements thought to be formed in other types of stars and subsequently distributed throughout the galaxy for eventual accretion into later generations of main-sequence stars. In the presence of heavier elements, particularly carbon and nitrogen, a catalytic sequence of nuclear reactions aids the fusion of protons to helium (H. A. Bethe and C. F. von Weizsäcker, 1938) (Table 1.3).

The overall reaction is precisely as before with the evolution of 26.72 MeV, but the 2 neutrinos now carry away 0.7 and 1.0 MeV respectively, leaving 25.0 MeV (4.01 pJ) per cycle for radiation. The coulombic energy barriers in the CN cycle are some 6–7 times greater than for the direct proton–proton reaction and hence the catalytic cycle does not predominate until about 1.6×10^7 K. In the sun, for example, it is estimated that about 10% of the energy comes from this process and most of the rest comes from the straightforward proton–proton reaction.

When approximately 10% of the hydrogen in a main-sequence star like the sun has been

TABLE 1.3.   *C–N–O thermonuclear catalysis*

Nuclear reaction	Q/MeV	Reaction time, $t_{\frac{1}{2}}$ [(a)]
$^{12}C + {}^1H \rightarrow {}^{13}N + \gamma$	1.95	1.3×10^7 y
$^{13}N \rightarrow {}^{13}C + e^+ + \nu$	2.22	7 min
$^{13}C + {}^1H \rightarrow {}^{14}N + \gamma$	7.54	3×10^6 y
$^{14}N + {}^1H \rightarrow {}^{15}O + \gamma$	7.35	3×10^5 y
$^{15}O \rightarrow {}^{15}N + e^+ + \nu$	2.70	82 s
$^{15}N + {}^1H \rightarrow {}^{12}C + {}^4He$	4.96	10^5 y

(a) In the central region of a star like the sun at 1.3×10^7 K and a hydrogen density of 100 g cm^{-3}.

consumed in making helium, the outward thermal pressure of radiation is insufficient to counteract the gravitational attraction and a further stage of contraction ensues. During this process the helium concentrates in a dense central core ($\rho \sim 10^5$ g cm^{-3}) and the temperature rises to perhaps 2×10^8 K. This is sufficient to overcome the coulombic potential energy barriers surrounding the helium nuclei, and helium burning (fusion) can occur. The hydrogen forms a vast tenuous envelope around this core with the result that the star evolves rapidly from the main sequence to become a red giant (p. 8).

1.4.2 *Helium burning and carbon burning*

The main nuclear reactions occurring in helium burning are:

$$^4He + {}^4He \rightleftharpoons {}^8Be \quad \text{and} \quad {}^8Be + {}^4He \rightleftharpoons {}^{12}C^* \longrightarrow {}^{12}C + \gamma$$

The nucleus ^8Be is unstable to α-particle emission ($t_{\frac{1}{2}} \sim 2 \times 10^{-16}$ s) and is only 0.094 MeV more stable than its constituent helium nuclei; under the conditions obtaining in the core of a red giant the calculated equilibrium ratio of ^8Be to ^4He is $\sim 10^{-9}$. Though small, this enables the otherwise improbable 3-body collision to occur. It is noteworthy that from consideration of stellar nucleogenesis F. Hoyle predicted in 1954 that the radioactive excited state of $^{12}C^*$ would be 7.70 MeV above the ground state of ^{12}C some 3 y before it was observed experimentally at 7.653 MeV. Experiments also indicate that the energy difference $Q(^{12}C^* - 3{}^4He)$ is 0.373 MeV, thus leading to the overall reaction energy

$$3{}^4He \longrightarrow {}^{12}C + \gamma; \quad Q = 7.281 \text{ MeV}$$

Further helium-burning reactions can now follow during which even heavier nuclei are synthesized:

$$^{12}C + {}^4He \longrightarrow {}^{16}O + \gamma; \quad Q = 7.148 \text{ MeV}$$
$$^{16}O + {}^4He \longrightarrow {}^{20}Ne + \gamma; \quad Q = 4.75 \text{ MeV}$$
$$^{20}Ne + {}^4He \longrightarrow {}^{24}Mg + \gamma; \quad Q = 9.31 \text{ MeV}$$

These reactions result in the exhaustion of helium previously produced in the hydrogen-burning process and an inner core of carbon, oxygen, and neon develops which eventually undergoes gravitational contraction and heating as before. At a temperature of $\sim 5 \times 10^8$ K carbon burning becomes possible in addition to other processes which must be considered. Thus, ageing red giant stars are now thought to be capable of generating a carbon-rich nuclear reactor core at densities of the order of 10^4 g cm^{-3}. Typical initial reactions would be:

$$^{12}C + {}^{12}C \longrightarrow {}^{24}Mg + \gamma; \qquad Q = 13.85 \text{ MeV}$$
$$\longrightarrow {}^{23}Na + {}^{1}H; \qquad Q = 2.23 \text{ MeV}$$
$$\longrightarrow {}^{20}Ne + {}^{4}He; \qquad Q = 4.62 \text{ MeV}$$

The time scale of such reactions is calculated to be $\sim 10^5$ y at 6×10^8 K and ~ 1 y at 8.5×10^8 K. It will be noticed that hydrogen and helium nuclei are regenerated in these processes and numerous subsequent reactions become possible, generating numerous nuclides in this mass range.

1.4.3 *The α-process*

The evolution of a star after it leaves the red-giant phase depends to some extent on its mass. If it is not more than about 1.4 $M_\odot$ it may contract appreciably again and then enter an oscillatory phase of its life before becoming a white dwarf (p. 8). When core contraction following helium and carbon depletion raises the temperature above $\sim 10^9$ K the γ-rays in the stellar assembly become sufficiently energetic to promote the (endothermic) reaction $^{20}Ne(\gamma, \alpha)^{16}O$. The α-particle released can penetrate the coulomb barrier of other neon nuclei to form ^{24}Mg in a strongly exothermic reaction:

$$^{20}Ne + \gamma \longrightarrow {}^{16}O + {}^{4}He; \qquad Q = -4.75 \text{ MeV}$$
$$^{20}Ne + {}^{4}He \longrightarrow {}^{24}Mg + \gamma; \qquad Q = +9.31 \text{ MeV}$$
i.e. $\qquad 2 {}^{20}Ne \longrightarrow {}^{16}O + {}^{24}Mg + \gamma; \qquad Q = +4.56 \text{ MeV}$

Some of the released α-particles can also scour out ^{12}C to give more ^{16}O and the ^{24}Mg formed can react further by $^{24}Mg(\alpha, \gamma)^{28}Si$. Likewise for ^{32}S, ^{36}Ar, and ^{40}Ca. It is this process that is considered to be responsible for building up the decreasing proportion of these so-called α-particle nuclei (Figs. 1.1 and 1.6). The relevant numerical data (including for comparison those for ^{20}Ne which is produced in helium and carbon burning) are as follows:

Nuclide	(^{20}Ne)	^{24}Mg	^{28}Si	^{32}S	^{36}Ar	^{40}Ca	^{44}Ca	^{48}Ti
Q_α/MeV	(9.31)	10.00	6.94	6.66	7.04	5.28	9.40	9.32
Relative abundance (as observed)	(8.4)	0.78	1.00	0.39	0.14	0.052	0.0011	0.0019

In a sense the α-process resembles helium burning but is distinguished from it by the quite different source of the α-particles consumed. The straightforward α-process stops at ^{40}Ca since $^{44}Ti^*$ is unstable to electron-capture decay. Hence (and including atomic numbers Z as subscripts for clarity):

$$^{40}_{20}Ca + {}^{4}_{2}He \longrightarrow {}^{44}_{22}Ti^* + \gamma$$
$$^{44}_{22}Ti^* + e^- \longrightarrow {}^{44}_{21}Sc^* + \nu_+; \qquad t_{\frac{1}{2}} \sim 47 \text{ y}$$
$$^{44}_{21}Sc^* \longrightarrow {}^{44}_{20}Ca + \beta^+ + \nu_+; \qquad t_{\frac{1}{2}} \sim 3.9 \text{ h}$$
Then $\qquad ^{44}_{20}Ca + {}^{4}_{2}He \longrightarrow {}^{48}_{22}Ti + \gamma$

The total time spent by a star in this α-phase may be $\sim 10^2$–10^4 y (Fig. 1.7).

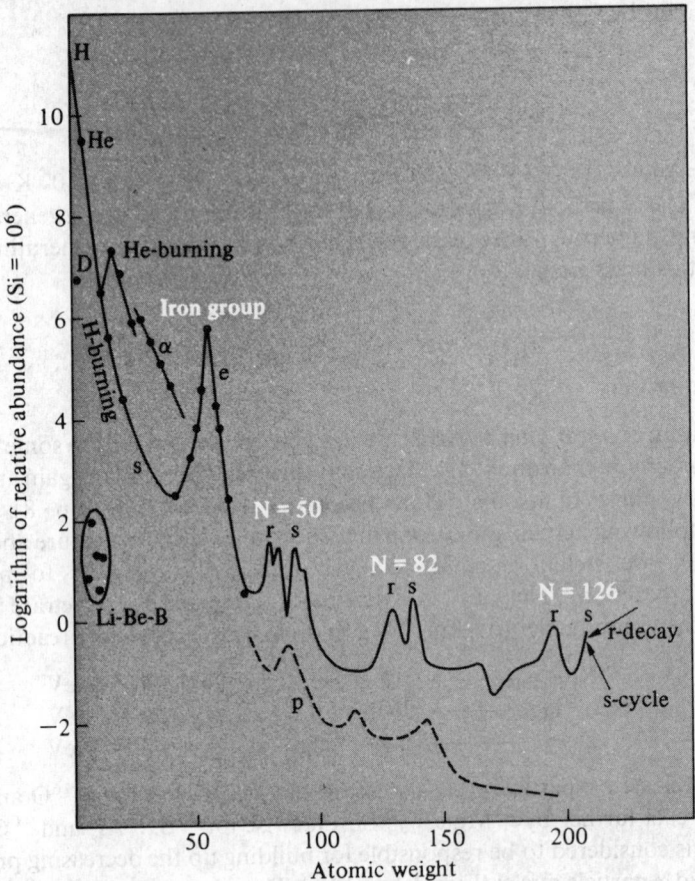

FIG. 1.6 Schematic representation of the main features of the curve of cosmic abundances shown in Fig. 1.1, labelled according to the various stellar reactions considered to be responsible for the synthesis of the elements. (After E. M. Burbidge *et al.*[6].)

1.4.4 *The e-process (equilibrium process)*

More massive stars in the upper part of the main-sequence diagram (i.e. stars with masses in the range 1.4–3.5 $M_\odot$) have a somewhat different history to that considered in the preceding sections. We have seen (p. 5) that such stars consume their hydrogen much more rapidly than do smaller stars and hence spend less time in the main sequence. Helium reactions begin in their interiors long before the hydrogen is exhausted, and in the middle part of their life they may expand only slightly. Eventually they become unstable and explode violently, emitting enormous amounts of material into interstellar space. Such explosions are seen on earth as supernovae, perhaps 10 000 times more luminous than ordinary novae. In the seconds (or minutes) preceding this catastrophic outburst, at temperatures above $\sim 3 \times 10^9$ K, many types of nuclear reactions can occur in great profusion, e.g. (γ, α), (γ, p), (γ, n), (α, n), (p, γ), (n, γ), (p, n), etc. (Fig. 1.7). This enables numerous interconversions to occur with the rapid establishment of a statistical equilibrium between the various nuclei and the free protons and neutrons. This is believed

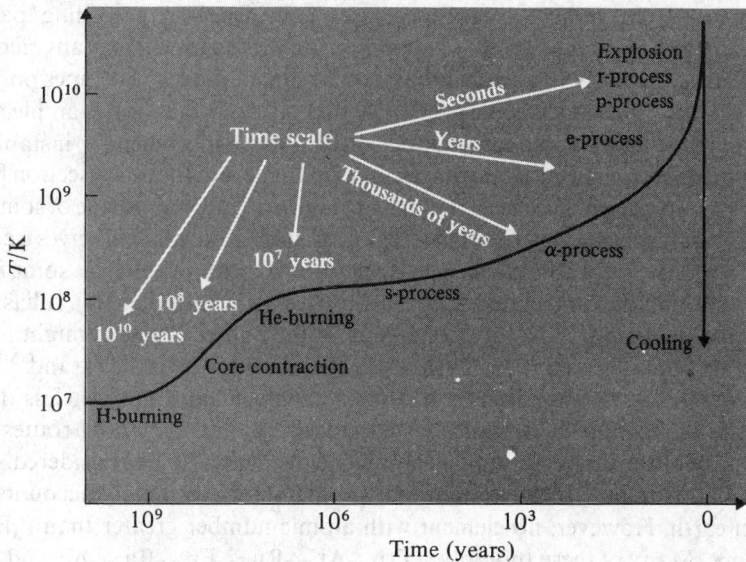

FIG. 1.7   The time-scales of the various processes of element synthesis in stars. The curve gives the central temperature as a function of time for a star of about one solar mass. The curve is schematic.[6]

to explain the cosmic abundances of elements from $_{22}$Ti to $_{29}$Cu. Specifically, since $_{26}^{56}$Fe is at the peak of the nuclear binding-energy curve, this element is considerably more abundant than those further removed from the most stable state.

1.4.5   *The s- and r-processes (slow and rapid neutron absorption)*

Slow neutron capture with emission of γ-rays is thought to be responsible for synthesizing most of the isotopes in the mass range $A = 63$–209 and also the majority of non-α-process nuclei in the range $A = 23$–46. These processes probably occur in red giants over a time span of $\sim 10^7$ y, and production loops for individual isotopes are typically in the range 10^2–10^5 y. Several stellar neutron sources have been proposed, but the most likely candidates are the exothermic reactions ^{13}C$(\alpha, n)^{16}$O and ^{21}Ne$(\alpha, n)^{24}$Mg. In both cases the target nuclei $(A = 4n + 1)$ would be produced by a (p, γ) reaction on the more stable $4n$ nucleus followed by positron emission.

Because of the long time scale involved in the s-process, unstable nuclides formed by (n, γ) reactions have time to decay subsequently by β^- decay (electron emission). The crucial factor in determining the relative abundance of elements formed by this process is thus the neutron capture cross-section of the precursor nuclide. In this way the process provides an ingenious explanation of the local peaks in abundance that occur near $A = 90$, 138, and 208, since these occur near unusually stable nuclei (neutron "magic numbers" 50, 82, and 126) which have very low capture cross-sections (Fig. 1.6). Their concentration therefore builds up by resisting further reaction. In this way the relatively high abundances of specific isotopes such as $_{39}^{89}$Y and $_{40}^{90}$Zr, $_{56}^{138}$Ba and $_{58}^{140}$Ce, $_{82}^{208}$Pb and $_{83}^{209}$Bi can be understood.

In contrast to the more leisured processes considered in preceding paragraphs, conditions can rise (e.g. at $\sim 10^9$ K in supernovae outbursts) where many neutrons are rapidly added successively to a nucleus before subsequent β-decay becomes possible. The time scale for the r-process is envisaged as $\sim 0.01–10$ s, so that, for example, some 200 neutrons might be added to an iron nucleus in 10–100 s. Only when β^- instability of the excessively neutron-rich product nuclei becomes extreme and the cross-section for further neutron absorption diminishes near the "magic numbers", does a cascade of some 8–10 β^- emissions bring the product back into the region of stable isotopes. This gives a convincing interpretation of the local abundance peaks near $A = 80$, 130, and 194, i.e. some 8–10 mass units below the nuclides associated with the s-process maxima (Fig. 1.6). It has also been suggested that neutron-rich isotopes of several of the lighter elements might also be the products of an r-process, e.g. ^{36}S, ^{46}Ca, ^{48}Ca, and perhaps ^{47}Ti, ^{49}Ti, and ^{50}Ti. These isotopes, though not as abundant as others of these elements, nevertheless do exist as stable species and cannot be so readily synthesized by other potential routes.

The problem of the existence of the heavy elements must also be considered. The short half-lives of all isotopes of technetium and promethium adequately accounts for their absence on earth. However, no element with atomic number greater than $_{83}Bi$ has any stable isotope. Many of these (notably $_{84}Po$, $_{85}At$, $_{86}Rn$, $_{87}Fr$, $_{88}Ra$, $_{89}Ac$, and $_{91}Pa$) can be understood on the basis of secular equilibria with radioactive precursors, and their relative concentrations are determined by the various half-lives of the isotopes in the radioactive series which produce them. The problem then devolves on explaining the cosmic presence of thorium and uranium, the longest lived of whose isotopes are ^{232}Th ($t_{\frac{1}{2}}$ 1.4×10^{10} y), ^{238}U ($t_{\frac{1}{2}}$ 4.5×10^9 y), and ^{235}U ($t_{\frac{1}{2}}$ 7.0×10^8 y). The half-life of thorium is commensurate with the age of the universe ($\sim 1.8 \times 10^{10}$ y) and so causes no difficulty. If all the present terrestrial uranium was produced by an r-process in a single supernova event then this occurred 6.6×10^9 y ago (p. 1457). If, as seems more probable, many supernovae contributed to this process, then such events, distributed uniformly in time, must have started $\sim 10^{10}$ y ago. In either case the uranium appears to have been formed long before the formation of the solar system $(4.6–5.0) \times 10^9$ y ago. More recent considerations of the formation and decay of ^{232}Th, ^{235}U, and ^{238}U suggest that our own galaxy is $(1.2–2.0) \times 10^{10}$ y old.

1.4.6 *The p-process (proton capture)*

Proton capture processes by heavy nuclei have already been briefly mentioned in several of the preceding sections. The (p, γ) reaction can also be invoked to explain the presence of a number of proton-rich isotopes of lower abundance than those of nearby normal and neutron-rich isotopes (Fig. 1.6). Such isotopes would also result from expulsion of a neutron by a γ-ray, i.e. (γ, n). Such processes may again be associated with supernovae activity on a very short time scale. With the exceptions of ^{113}In and ^{115}Sn, all of the 36 isotopes thought to be produced in this way have even atomic mass numbers; the lightest is $^{74}_{34}Se$ and the heaviest $^{196}_{80}Hg$.

1.4.7 *The x-process*

One of the most obvious features of Figs. 1.1 and 1.6 is the very low cosmic abundance of the stable isotopes of lithium, beryllium, and boron.[10] Paradoxically, the problem is not to explain why these abundances are so low but why these elements exist at all since their isotopes are bypassed by the normal chain of thermonuclear reactions described on the preceding pages. Again, deuterium and ^{3}He, though part of the hydrogen-burning process, are also virtually completely consumed by it, so that their existence in the universe, even at relatively low abundances, is very surprising. Moreover, even if these various isotopes were produced in stars, they would not survive the intense internal heat since their bonding energies imply that deuterium would be destroyed above 0.5×10^6 K, Li above 2×10^6 K, Be above 3.5×10^6, and B above 5×10^6. Deuterium and ^{3}He are absent from the spectra of almost all stars and are now generally thought to have been formed by nucleosynthesis during the last few seconds of the original big bang; their main agent of destruction is stellar processing.

It now seems likely that the 5 stable isotopes ^{6}Li, ^{7}Li, ^{9}Be, ^{10}B, and ^{11}B are formed predominantly by spallation reactions (i.e. fragmentation) effected by galactic cosmic-ray bombardment (the x-process). Cosmic rays consist of a wide variety of atomic particles moving through the galaxy at relativistic velocities. Nuclei ranging from hydrogen to uranium have been detected in cosmic rays though ^{1}H and ^{4}He are by far the most abundant components [^{1}H: 500; ^{4}He: 40; all particles with atomic numbers from 3 to 9: 5; all particles with $Z \geq 10$: ~ 1]. However, there is a striking deviation from stellar abundances since Li, Be, and B are vastly over abundant as are Sc, Ti, V, and Cr (immediately preceding the abundance peak near iron). The simplest interpretation of these facts is that the (heavier) particles comprising cosmic rays, travelling as they do great distances in the galaxy, occasionally collide with atoms of the interstellar gas (predominantly ^{1}H and ^{4}He) and thereby fragment. This fragmentation or spallation, as it is called, produces lighter nuclei from heavier ones. Conversely, high-speed ^{4}He particles may occasionally collide with interstellar iron-group elements and other heavy nuclei, thus inducing spallation and forming Li, Be, and B (and possibly even some ^{2}H and ^{3}He), on the one hand, and elements in the range Sc–Cr, on the other. As we have seen, the lighter transition elements are also formed in various stellar processes, but the presence of elements in the mass range 6–12 can at present only be satisfactorily accounted for by a low-temperature low-density extra-stellar process. In addition to spallation, interstellar (p, α) reactions in the wake of supernova shock waves may contribute to the synthesis of boron isotopes:

$$^{13}C(p, \alpha)^{10}B \quad \text{and} \quad ^{14}N(p, \alpha)^{11}C \xrightarrow{\beta^-} {}^{11}B.$$

In summary, using a variety of nuclear syntheses it is now possible to account for the presence of the 273 known stable isotopes of the elements up to $^{209}_{83}$Bi and to understand, at least in broad outline, their relative concentrations in the universe. The tremendous number of hypothetically possible internuclear conversions and reactions makes detailed computation extremely difficult. Energy changes are readily calculated from the known relative atomic masses of the various nuclides, but the cross-sections (probabilities) of many of the reactions are unknown and this prevents precise calculation of reaction rates and equilibrium concentrations in the extreme conditions occurring even in stable stars. Conditions and reactions occurring during supernova outbursts are even more difficult to

---

[10] H. REEVES, Origin of the light elements, *A. Rev. Astron. Astrophys.* **12**, 437–69 (1974). An account of the problems concerning the abundance of deuterium, helium-3, lithium, beryllium, and boron.

define precisely. However, it is clear that substantial progress has been made in the last two decades to interpret the bewildering variety of isotopic abundances which comprise the elements used by chemists. The approximate constancy of the isotopic composition of the individual elements is a fortunate result of the quasi-steady-state conditions obtaining in the universe during the time required to form the solar system. It is tempting to speculate whether chemistry could ever have emerged as a quantitative science if the elements had had widely varying isotopic composition, since gravimetric analysis would then have been impossible and the great developments of the nineteenth century could hardly have occurred. Equally, it should no longer cause surprise that the atomic weights of the elements are not necessarily always "constants of nature", and variations are to be expected, particularly among the lighter elements, which can have appreciable effects on physicochemical measurements and quantitative analysis.

1.5   Atomic Weights[11]

The concept of "atomic weight" or "mean relative atomic mass" is fundamental to the development of chemistry. Dalton originally supposed that all atoms of a given element had the same unalterable weight but, after the discovery of isotopes earlier this century, this property was transferred to them. Today the possibility of variable isotopic composition of an element (whether natural or artificially induced) precludes the possibility of defining *the* atomic weight of most elements, and the tendency nowadays is to define *an* atomic weight of an element as "the ratio of the average mass per atom of an element to one-twelfth of the mass of an atom of ^{12}C". It is important to stress that atomic weights (mean relative atomic masses) of the elements are dimensionless numbers and therefore have no units.

Because of their central importance in chemistry, atomic weights have been continually refined and improved since the first tabulations by Dalton (1803–5). By 1808 Dalton had included 20 elements in his list and these results were substantially extended and improved by Berzelius during the following decades. An illustration of dramatic and continuing improvement in accuracy and precision during the past 100 y is given in Table 1 4. In 1874 no atomic weight was quoted to better than one part in 200, but by 1903 33 elements had values quoted to one part in 10^3 and 2 of these (silver and iodine) were quoted to 1 in 10^4. Today the majority of values are known to 1 in 10^4 and 23 elements have a precision approaching or exceeding 1 in 10^6. This improvement was first due to improved chemical methods, particularly between 1900 and 1935 when increasing use of fused silica ware and electric furnaces reduced the possibility of contamination. More recently the use of mass spectrometry has effected a further improvement in precision. Mass spectrometric data were first used in a confirmatory role in the 1935 table of atomic weights, and by 1938 mass spectrometric values were preferred to chemical determinations for hydrogen and osmium and to gas-density values for helium. In 1959 the atomic weight values of over 50 elements were still based on classical chemical methods, but by 1973 this number had dwindled to 9 (Ti, Ge, Se, Mo, Sn, Sb, Te, Hg, and Tl) or to 10 if the coulometric determination for Zn is

[11] N. N. GREENWOOD, Atomic weights, Ch. 8 in Part I, Vol. 1, Section C, of Kolthoff and Elving's *Treatise on Analytical Chemistry*, pp. 453–78, Interscience, New York, 1978. This gives a fuller account of the history and techniques of atomic weight determinations and their significance, and incorporates a full bibliographical list of Reports on Atomic Weights.

TABLE 1.4   *Evolution of atomic weight values for selected elements*[a]; *(the dates selected were chosen for the reasons given below)*

Element	1873–5	1903	1925	1959	1961	1981
H	1	1.008	1.008	1.0080	1.007 97	1.007 94($\pm$7) gmr
C	12	12.00	12.000	12.011 15	12.011 15	12.011 r
O	16	16.00	16.000	**16**	15.9994	15.9994($\pm$3) gr
P	31	31.0	31.027	30.975	30.9738	30.973 76
Ti	50	48.1	48.1	47.90	47.90	47.88($\pm$3)
Zn	65	65.4	65.38	65.38	65.37	65.38
Se	79	79.2	79.2	78.96	78.96	78.96($\pm$3)
Ag	108	107.93	107.880	107.880	107.870	107.8682($\pm$3) g
I	127	126.85	126.932	126.91	126.9044	126.9045
Ce	92	140.0	140.25	140.13	140.12	140.12 g
Pr	—	140.5	140.92	140.92	140.907	140.9077
Re	—	—	188.7[b]	186.22	186.22	186.207
Hg	200	200.0	200.61	200.61	200.59	200.59($\pm$3)

[a] The annotations g, m, and r appended to some values in the final column have the same meanings as those in the definitive table (inside back cover).
[b] The value for rhenium was first listed in 1929.
Note on dates:
  1874   Foundation of the American Chemical Society (64 elements listed).
  1903   First international table of atomic weights (78 elements listed).
  1925   Major review of table (83 elements listed).
  1959   Last table to be based on oxygen = 16 (83 elements listed).
  1961   Complete reassessment of data and revision to $^{12}C = 12$ (83 elements).
  1981   Latest available IUPAC values (87 + 19 elements listed).

counted as chemical. The values for a further 8 elements were based on a judicious blend of chemical and mass-spectrometric data, but the values quoted for all other elements were based entirely on mass-spectrometric data.

Precise atomic weight values do not automatically follow from precise measurements of relative atomic masses, however, since the relative abundance of the various isotopes must also be determined. That this can be a limiting factor is readily seen from Table 1.4: the value for praseodymium (which has only 1 stable naturally occurring isotope) is 100 times more precise than the value for the neighbouring element cerium which has 4 such isotopes.

1.5.1   *Uncertainty in atomic weights*

Numerical values for the atomic weights of the elements are now reviewed every 2 y by the Commission on Atomic Weights and Isotopic Abundances of IUPAC (the International Union of Pure and Applied Chemistry). Their most recent recommendations[12] are tabulated on the inside back cover. From this it is clear that there is still a wide variation in the reliability of the data. Figure 1.8 summarizes the relative uncertainties in the tabulated values. It can be seen that all values are reliable to better than 1 part per 1000 and the majority are reliable to better than 2 parts in 10^4. Boron is

[12] IUPAC Inorganic Chemistry Division, Atomic Weights of the Elements 1981, *Pure Appl. Chem.* **55**, 1101–18 (1983). This is the latest report and, in addition to atomic weights, lists the isotopic composition of each of the elements and discusses current problems.

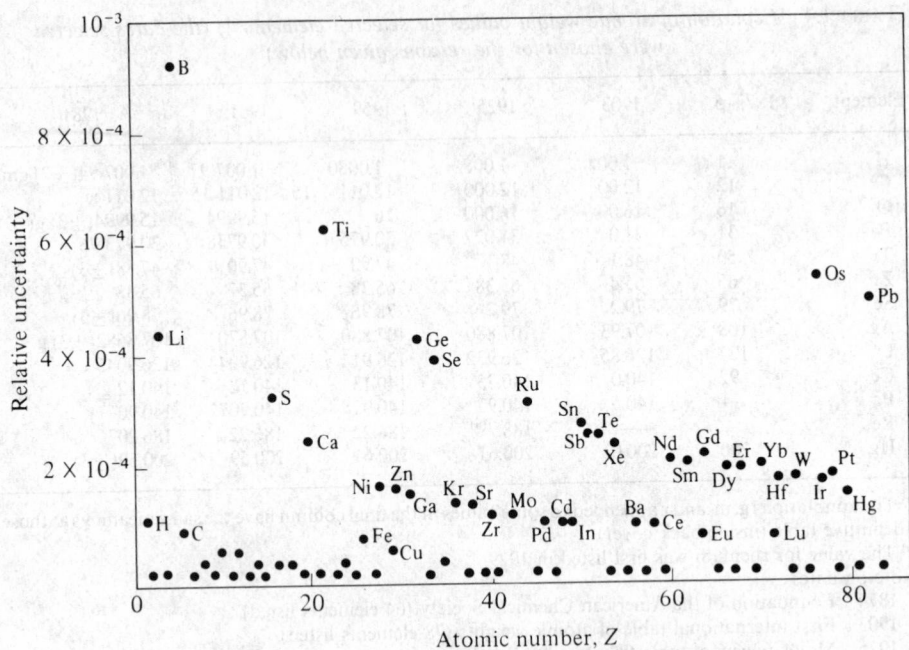

FIG. 1.8 Relative uncertainty of tabulated atomic weights (1979).

prominent on this diagram not because of experimental error (since the use of calibrated mass-spectrometric measurements has yielded results of very high precision) but because the natural variation in the relative abundance of the 2 isotopes ^{10}B and ^{11}B results in a range of values of at least ± 0.003 about the quoted value to 10.81. By contrast, there is no known variation in isotopic abundances for elements such as titanium and osmium, but calibrated mass-spectrometric data are not available, and the existence of 5 and 7 stable isotopes respectively for these elements makes high precision difficult to obtain: they are thus prime candidates for improvement.

Atomic weights are known most precisely for elements which have only 1 stable isotope; the relative atomic mass of this isotope can be determined to at least 1 ppm and there is no possibility of variability in nature. There are 20 such elements: Be, F, Na, Al, P, Sc, Mn, Co, As, Y, Nb, Rh, I, Cs, Pr, Tb, Ho, Tm, Au, and Bi. (Note that all of these elements except beryllium have odd atomic numbers—why?)

Elements with 1 predominant isotope can also, potentially, permit very precise atomic weight determinations since variations in isotopic composition or errors in its determination have a correspondingly small effect on the mass-spectrometrically determined value of the atomic weight. Nine elements have 1 isotope that is more than 99% abundant (H, He, N, O, Ar, V, La, Ta, and U) and carbon also approaches this category (^{13}C 1.11% abundant).

Known variations in the isotopic composition of normal terrestrial material prevent a more precise atomic weight being given for 9 elements and these carry the footnote r in the Table of Atomic Weights. For each of these elements (H, Li, B, C, O, S, Ar, Cu, and Pb) the accuracy attainable in an atomic weight determination on a given sample is greater than that implied by the recommended value since this must be applicable to any sample and so must embrace all known variations in isotopic composition from commercial terrestrial sources. For example, for hydrogen the present attainable accuracy of calibrated mass-spectrometric atomic weight determinations is about ± 1 in the sixth significant figure, but the recommended value of 1.007 94(± 7) is so given because of the natural terrestrial variation in the deuterium content. The most likely value relevant to laboratory chemicals (e.g. H_2O) is 1.007 97, but it should be noted that hydrogen gas used in laboratories is often inadvertently depleted during its preparation by electrolysis, and for such samples the atomic weight is close to 1.007 90. By contrast, intentional fractionation to yield heavy water (thousands of tonnes annually) or deuterated chemicals implies an atomic weight approaching 2.014, and great care should be taken to avoid contamination of "normal" samples when working with or disposing of such enriched materials.

Fascinating stories of natural variability could be told for each of the other 8 elements and, indeed, determinations of such variations in isotopic composition are now an essential tool in unravelling the geochemical history of various ore bodies. For example, the atomic weight of sulfur obtained from virgin Texas sulfur is detectably different from that obtained from sulfate ores, and an overall range of ± 0.01 is found for terrestrial samples; this limits the value quoted to 32.06 though the accuracy of atomic weight determinations on individual samples is $\pm 0.000\ 15$. Boron is even more adversely affected, as previously noted, and the actual atomic weight can vary from 10.807 to 10.819 depending on whether the mineral source is Turkey or the USA.

Even more disconcerting are the substantial deviations in atomic weight that can occur in commercially available material because of inadvertent or undisclosed changes in isotopic composition (footnote m in the Table of Atomic Weights). This situation at present obtains for 7 elements (H, Li, B, Ne, Kr, Xe, and U) and may well also soon affect others (such as He, C, N, and O). The separated or partially enriched isotopes of Li, B, and U are now extensively used in nuclear reactor technology and weaponry, and the unwanted residues, depleted in the desired isotopes, are sometimes dumped on the market and sold as "normal" material. Thus lithium salts may unsuspectingly be purchased which have been severely depleted in ^6Li (natural abundance 7.5%), and a major commercial supplier has marketed lithium containing as little as 3.75% of this isotope, thereby inducing an atomic weight change of 0.53%. For this reason practically all lithium compounds now obtainable in the USA are suspect and quantitative data obtained on them are potentially unreliable. Again, the practice of "milking" fission-product rare gases from reactor fuels and marketing these materials, produces samples with anomalous isotopic compositions. The effect, particularly on physicochemical computations, can be serious and, whilst not wishing to strike an alarmist note, the possibility of such deviations must continually be borne in mind for elements carrying the footnote m in the Table of Atomic Weights.

The related problem arising from radioactive elements is considered in the next section.

1.5.2   *The problem of radioactive elements*

Elements with radioactive nuclides amongst their naturally occurring isotopes have a built-in time variation of the relative concentration of their isotopes and hence a continually varying atomic weight. Whether this variation is chemically significant depends on the half-life of the transition and the relative abundance of the various isotopes. Similarly, the actual concentration of stable isotopes of several elements (e.g. Ar,

Ca, and Pb) may be influenced by association of those elements with radioactive precursors (i.e. ^{40}K, ^{238}U, etc.) which generate potentially variable amounts of the stable isotopes concerned. Again, some elements (such as technetium, promethium, and the transuranium elements) are synthesized by nuclear reactions which produce a single isotope of the element. The "atomic weight" therefore depends on which particular isotope is being synthesized, and the concept of a "normal" atomic weight is irrelevant. For example, cyclotron production of technetium yields ^{97}Tc ($t_{\frac{1}{2}}$ 2.6 × 10^6 y) with an atomic weight of 96.906, whereas fission product technetium is ^{99}Tc ($t_{\frac{1}{2}}$ 2.13 × 10^5 y), atomic weight 98.906, and the isotope of longest half-life is ^{98}Tc ($t_{\frac{1}{2}}$ 4.2 × 10^6 y), atomic weight 97.907.

At least 19 elements not usually considered to be radioactive do in fact have naturally occurring unstable isotopes. The minute traces of naturally occurring ^{3}H ($t_{\frac{1}{2}}$ 12.33 y) and ^{14}C ($t_{\frac{1}{2}}$ 5.73 × 10^3 y) have no influence on the atomic weights of these elements though, of course, they are of crucial importance in other areas of study. The radioactivity of ^{40}K ($t_{\frac{1}{2}}$ 1.28 × 10^9 y) influences the atomic weights of its daughter elements argon (by electron capture) and calcium (by β^- emission) but fortunately does not significantly affect the atomic weight of potassium itself because of the low absolute abundance of this particular isotope (0.0117%). The half-lives of the radioactive isotopes of the 16 other "stable" elements are all greater than 10^{10} y and so normally have little influence on the atomic weight of these elements even when, as in the case of ^{115}In ($t_{\frac{1}{2}}$ 5 × 10^{14} y, 95.7% abundant) and ^{187}Re ($t_{\frac{1}{2}}$ 5 × 10^{10} y, 62.6% abundant), they are the most abundant isotopes. Note, however, that on a geological time scale it has been possible to build up significant concentrations of ^{187}Os in rhenium-containing ores (by β^- decay of ^{187}Re), thereby generating samples of osmium with an anomalous atomic weight nearer to 187 than to the published value of 190.2. Lead was the first element known to be subject to such isotopic disturbances and, indeed, the discovery and interpretation of the significance of isotopes was itself hastened by the reluctant conclusion of T. W. Richards at the turn of the century that a group of lead samples of differing geological origins were identical chemically but differed in atomic weight—the possible variation is now known to span almost the complete range from 204 to 208. Such elements, for which geological specimens are known in which the element has an anomalous isotopic composition, are given the footnote g in the Table of Atomic Weights. In addition to Ar, Ca, Os, and Pb just discussed, such variability affects at least 26 other elements, including Sr (resulting from the β^- decay of ^{87}Rb), Ra, Th, and U. A spectacular example, which affects virtually every element in the central third of the periodic table, has recently come to light with the discovery of prehistoric natural nuclear reactors at Oklo in Africa (see p. 1457). Fortunately this mine is a source of uranium ore only and so will not affect commercially available samples of the other elements involved.

In summary, as a consequence of the factors considered in this and the preceding section, the atomic weights of only the 20 mononuclidic elements can be regarded as "constants of nature". For all other elements variability in atomic weight is potentially possible and in several instances is known to occur to an extent which affects the reliability of quantitative results of even modest precision.

1.6   Points to Ponder

* Gravitational forces are 10^{37} times weaker than electrostatic forces and 10^{39} times

weaker than nuclear forces yet they provide the initial mechanism for heating protostars to 10^7 K. (*Hint*: there are 10^{57} atoms in the sun.)

* Successively higher temperatures are needed for proton–proton reactions, helium-burning nuclear reactions, carbon-burning nuclear reactions, and the e-process.

* Hydrogen, which is unstable with respect to all other nuclides in nature, is the most abundant element in the universe.

* The most stable nuclei occur amongst the isotopes of iron and nickel yet these two elements account for less than 1% of the mass of the universe.

* Deuterium, lithium, beryllium, and boron are virtually non-existent in stars but are a substantial component of cosmic rays.

* Spectroscopic data are used to estimate the concentration of elements in stars though only the photosphere is directly visible.

* At very high stellar temperatures hydrogen is ionized and therefore has no electronic spectrum. At lower temperatures, helium is in its ground state but its electronic emission spectrum then occurs in the far ultraviolet region and so cannot be observed on earth because of absorption in the atmosphere.

2

Chemical Periodicity and the Periodic Table

2.1 Introduction

The concept of chemical periodicity is central to the study of inorganic chemistry. No other generalization rivals the periodic table of the elements in its ability to systematize and rationalize known chemical facts or to predict new ones and suggest fruitful areas for further study. Chemical periodicity and the periodic table now find their natural interpretation in the detailed electronic structure of the atom; indeed, they played a major role at the turn of the century in elucidating the mysterious phenomena of radioactivity and quantum effects which led ultimately to Bohr's theory of the hydrogen atom. Because of this central position it is perhaps not surprising that innumerable articles and books have been written on the subject since the seminal papers by Mendeleev in 1869, and some 700 forms of the periodic table (classified into 146 different types or subtypes) have been proposed.[1-3] A brief historical survey of these developments is summarized in the Panel.

There is no single *best* form of the periodic table since the choice depends on the purpose for which the table is used. Some forms emphasize chemical relations and valence, whereas others stress the electronic configuration of the elements or the dependence of the periods on the shells and subshells of the atomic structure. The most convenient form for our purpose is the so-called "long form" with separate panels for the lanthanide and actinide elements (see inside front cover). The following sections of this chapter summarize:

(a) the interpretation of the periodic law in terms of the electronic structure of atoms;
(b) the use of the periodic table and graphs to systematize trends in physical and chemical properties and to detect possible errors, anomalies, and inconsistencies;
(c) the use of the periodic table to predict new elements and compounds, and to suggest new areas of research.

[1] F. P. VENABLE, *The Development of the Periodic Law*, Chemical Publishing Co., Easton, Pa., 1896. This is the first general review of periodic tables and has an almost complete collection of those published to that time.

[2] E. G. MAZURS, *Graphic Representation of the Periodic System during One Hundred Years*, University of Alabama Press, Alabama, 1974. An exhaustive topological classification of over 700 forms of the periodic table, emphasizing their various characteristics. Good historical sections and extensive bibliography.

[3] J. W. VAN SPRONSEN, *The Periodic System of the Chemical Elements*, Elsevier, Amsterdam, 1969, 368 pp. An excellent modern account of the historical developments leading up to Mendeleev's table, including an assessment of priorities and contributions by others.

Mendeleev's Periodic Table

Precursors and Successors

1772	L. B. G. deMorveau made the first table of "chemically simple" substances. A. L. Lavoisier used this in his *Traité Elémentaire de Chemie* published in 1789.
1817–29	J. W. Döbereiner discovered many triads of elements and compounds, the combining weight of the central component being the average of its partners (e.g. CaO, SrO, BaO, and NiO, CuO, ZnO).
1843	L. Gmelin included a V-shaped arrangement of 16 triads in the 4th edition of his *Handbuch der Chemie*.
1857	J.-B. Dumas published a rudimentary table of 32 elements in 8 columns indicating their relationships.
1862	A. E. B. de Chancourtois first arranged the elements in order of increasing atomic weight; he located similar elements in this way and published a helical form in 1863.
1864	L. Meyer published a table of valences for 49 elements.
1864	W. Odling drew up an almost correct table with 17 vertical columns and including 57 elements.
1865	J. A. R. Newlands propounded his law of octaves after several partial classifications during the preceding 2 y; he also correctly predicted the atomic weight of the undiscovered element germanium.
1868–9	L. Meyer drew up an atomic volume curve and a periodic table, but this latter was not published until 1895.
1869	D. I. Mendeleev enunciated his periodic law that "the properties of the elements are a periodic function of their atomic weights". He published several forms of periodic table, one containing 63 elements.
1871	D. I. Mendeleev modified and improved his tables and predicted the discovery of 10 elements (now known as Sc, Ga, Ge, Tc, Re, Po, Fr, Ra, Ac, and Pa). He fully described with amazing prescience the properties of 4 of these (Sc, Ga, Ge, Po). Note, however, that it was not possible to predict the existence of the noble gases or the number of lanthanide elements.
1894–8	Lord Rayleigh, W. Ramsay, and M. W. Travers detected and then isolated the noble gases (He), Ne, Ar, Kr, Xe.
1913	N. Bohr explained the form of the periodic table on the basis of his theory of atomic structure and showed that there could be only 14 lanthanide elements.
1913	H. G. J. Moseley observed regularities in the characteristic X-ray spectra of the elements; he thereby discovered atomic numbers Z and provided justification for the ordinal sequence of the elements.
1940	E. McMillan and P. Abelson synthesized the first transuranium element $_{93}$Np. Others were synthesized by G. T. Seaborg and his colleagues during the next 15 y.
1944	G. T. Seaborg proposed the actinide hypothesis and predicted 14 elements (up to $Z = 103$) in this group.

2.2 The Electronic Structure of Atoms[4]

The ubiquitous electron was discovered by J. J. Thompson in 1896 some 25 y after the original work on chemical periodicity by D. I. Mendeleev and Lothar Meyer; however, a further 20 y were to pass before G. N. Lewis and then I. Langmuir connected the electron with valency and chemical bonding. Refinements continued via wave mechanics and molecular orbital theory, and the symbiotic relation between experiment and theory still continues today. It should always be remembered, however, that it is incorrect to "deduce" known chemical phenomena from theoretical models; the proper relationship is that the currently accepted theoretical models interpret the facts and suggest new

[4] N. N. GREENWOOD, *Principles of Atomic Orbitals*, revised SI edition, Monograph for Teachers, No. 8, Chemical Society, London, 1980, 48 pp.

experiments—they will be modified (or discarded and replaced) when new results demand it. Theories can never be proved by experiment—only refuted, the best that can be said of a theory is that it is consistent with a wide range of information which it interprets logically and that it is a fruitful source of predictions and new experiments.

Our present views on the electronic structure of atoms are based on a variety of experimental results and theoretical models which are fully discussed in many elementary texts. In summary, an atom comprises a central, massive, positively charged nucleus surrounded by a more tenuous envelope of negative electrons. The nucleus is composed of neutrons (^1_0n) and protons (^1_1p, i.e. $^1_1H^+$) of approximately equal mass tightly bound by the force field of mesons. The number of protons (Z) is called the atomic number and this, together with the number of neutrons (N), gives the atomic mass number of the nuclide ($A = N + Z$). An element consists of atoms all of which have the same number of protons (Z) and this number determines the position of the element in the periodic table (H. G. J. Moseley, 1913). Isotopes of an element all have the same value of Z but differ in the number of neutrons in their nuclei. The charge on the electron (e^-) is equal in size but opposite in sign to that of the proton and the ratio of their masses is $1/1836$.

The arrangement of electrons in an atom is described by means of four quantum numbers which determine the spatial distribution, energy, and other properties, see Appendix 1 (p. 1487). The principal quantum number n defines the general energy level or "shell" to which the electron belongs. Electrons with $n = 1, 2, 3, 4, \ldots$, are sometimes referred to as K, L, M, N, . . ., electrons. The orbital quantum number l defines both the shape of the electron charge distribution and its orbital angular momentum. The number of possible values for l for a given electron depends on its principal quantum number n; it can have n values running from 0 to $n-1$, and electrons with $l = 0, 1, 2, 3, \ldots$, are designated s, p, d, f, . . ., electrons. Whereas n is the prime determinant of an electron's energy this also depends to some extent on l (for atoms or ions containing more than one electron). It is found that the sequence of increasing electron energy levels in an atom follows the sequence of values $n + l$; if 2 electrons have the same value of $n + l$ then the one with smaller n is more tightly bound.

The third quantum number m is called the magnetic quantum number for it is only in an applied magnetic field that it is possible to define a direction within the atom with respect to which the orbital can be directed. In general, the magnetic quantum number can take up $2l + 1$ values (i.e. $0, \pm 1, \ldots, \pm l$); thus an s electron (which is spherically symmetrical and has zero orbital angular momentum) can have only one orientation, but a p electron can have three (frequently chosen to be the x, y, and z directions in Cartesian coordinates). Likewise there are five possibilities for d orbitals and seven for f orbitals.

The fourth quantum number m_s is called the spin angular momentum quantum number for historical reasons. In relativistic (four-dimensional) quantum mechanics this quantum number is associated with the property of symmetry of the wave function and it can take on one of two values designated as $+\frac{1}{2}$ and $-\frac{1}{2}$, or simply α and β. All electrons in atoms can be described by means of these four quantum numbers and, as first enumerated by W. Pauli in his *Exclusion Principle* (1926), each electron in an atom must have a unique set of the four quantum numbers.

It can now be seen that there is a direct and simple correspondence between this description of electronic structure and the form of the periodic table. Hydrogen, with 1 proton and 1 electron, is the first element, and, in the ground state (i.e. the state of lowest energy) it has the electronic configuration $1s^1$ with zero orbital angular momentum.

Helium, $Z = 2$, has the configuration $1s^2$, and this completes the first period since no other unique combination of $n = 1$, $l = m = 0$, $m_s = \pm\frac{1}{2}$ exists. The second period begins with lithium ($Z = 3$), the least tightly bound electron having the configuration $2s^1$. The same situation obtains for each of the other periods in the table, the number of the period being the principal quantum number of the least tightly bound electron of the first element in the period. It will also be seen that there is a direct relation between the various blocks of elements in the periodic table and the electronic configuration of the atoms it contains: the s block is 2 elements wide, the p block 6 elements wide, the d block 10, and the f block 14, i.e. $2(2l + 1)$, the factor 2 appearing because of the spins.

In so far as the chemical (and physical) properties of an element derive from its electronic configuration, and especially the configuration of its least tightly bound electrons, it follows that chemical periodicity and the form of the periodic table can be elegantly interpreted in terms of electronic structure.

2.3 Periodic Trends in Properties[5, 6]

General similarities and trends in the *chemical* properties of the elements had been noticed increasingly since the end of the eighteenth century and predated the observation of periodic variations in *physical* properties which were not noted until about 1868. However, it is more convenient to invert this order and to look at trends in atomic and physical properties first.

2.3.1 *Trends in atomic and physical properties*

Figure 2.1 shows a modern version of Lothar Meyer's atomic volume curve: the alkali metals appear at the peaks and elements near the centre of each period (B, C; Al, Si; Mn, Ru, and Os) appear in the troughs. This finds a ready interpretation on the electronic theory since the alkali metals have only 1 electron per atom to contribute to the binding of the 8 nearest-neighbour atoms, whereas elements near the centre of each period have the maximum number of electrons available for bonding. Elements in other groups fall on corresponding sections of the curve in each period, and in several groups there is a steady trend to higher volumes with the increasing atomic number. Closer inspection reveals that a much more detailed interpretation would be required to encompass all the features of the curve which includes data on solids held by very diverse types of bonding. Note also that the position of helium is anomalous (why?), and that there are local anomalies at europium and ytterbium in the lanthanide elements (see Chapter 30). Similar plots are obtained for the atomic and ionic radii of the elements and an inverted diagram is obtained, as expected, for the densities of the elements in the solid state (Fig. 2.2).

Of more fundamental importance is the plot of first-stage ionization energies of the elements, i.e. the energy I_M required to remove the least tightly bound electron from the neutral atom in the gas phase:

$$M(g) \longrightarrow M^+(g) + e^-; \quad \Delta H^\circ = I_M$$

[5] R. Rich, *Periodic Correlations*, W. A. Benjamin, New York, 1965, 159 pp. A good source of tabular and graphical data illustrating trends in the properties of elements and compounds.

[6] R. T. Sanderson, *Inorganic Chemistry*, Reinhold Publishing Corp., New York, 1967, 430 pp. A textbook stressing periodicity in properties and containing over 70 sets of data displayed on periodic charts.

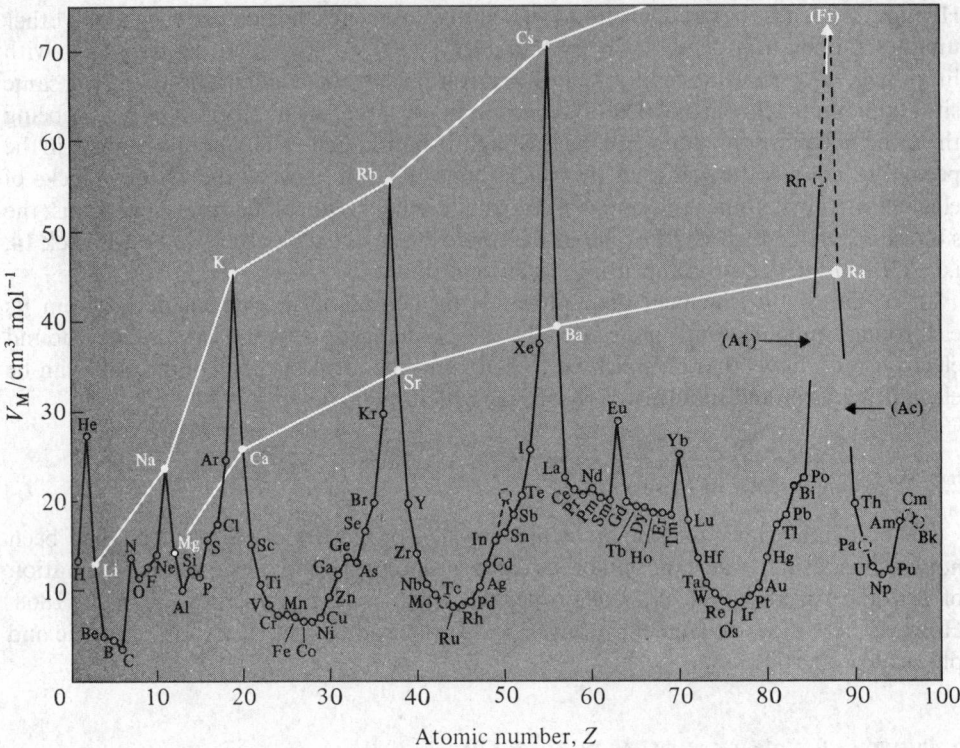

These are shown in Fig. 2.3 and illustrate most convincingly the various quantum shells and subshells described in the preceding section. The energy required to remove the 1 electron from an atom of hydrogen is 13.606 eV (i.e. 1311 kJ per mole of H atoms). This rises to 2370 kJ mol^{-1} for He (1s^2) since the positive charge on the helium nucleus is twice that of the proton and the additional charge is not completely shielded by the second electron. There is a large drop in ionization energy between helium and lithium (1s^{2}2s^1) because the principal quantum number n increases from 1 to 2, after which the ionization energy rises somewhat for beryllium (1s^{2}2s^2), though not to a value which is so high that beryllium would be expected to be an inert gas. The interpretation that is placed on the other values in Fig. 2.3 is as follows. The slight decrease at boron (1s^{2}2s^{2}2p^1) is due to the increase in orbital quantum number l from 0 to 1 and the similar decrease between nitrogen and oxygen is due to increased interelectronic coulomb repulsion as the fourth p electron is added to the 3 already occupying 2p$_x$, 2p$_y$, and 2p$_z$. The ionization energy then continues to increase with increasing Z until the second quantum shell is filled at neon (2s^{2}2p^6). The process is precisely repeated from sodium (3s^1) to argon (3s^{2}3p^6) which again occurs at a peak in the curve, although at this point the third quantum shell is not yet completed (3d). This is because the next added electron for the next element potassium ($Z = 19$) enters the 4s shell ($n + l = 4$) rather than the 3d ($n + l = 5$). After calcium ($Z = 20$) the 3d shell fills and then the 4p ($n + l = 5$, but n higher than for 3d). The implications of this and of the subsequent filling of later s, p, d, and f levels will be elaborated in considerable detail in later chapters. Suffice it to note for the moment that the chemical inertness of the

Trends in Atomic and Physical Properties

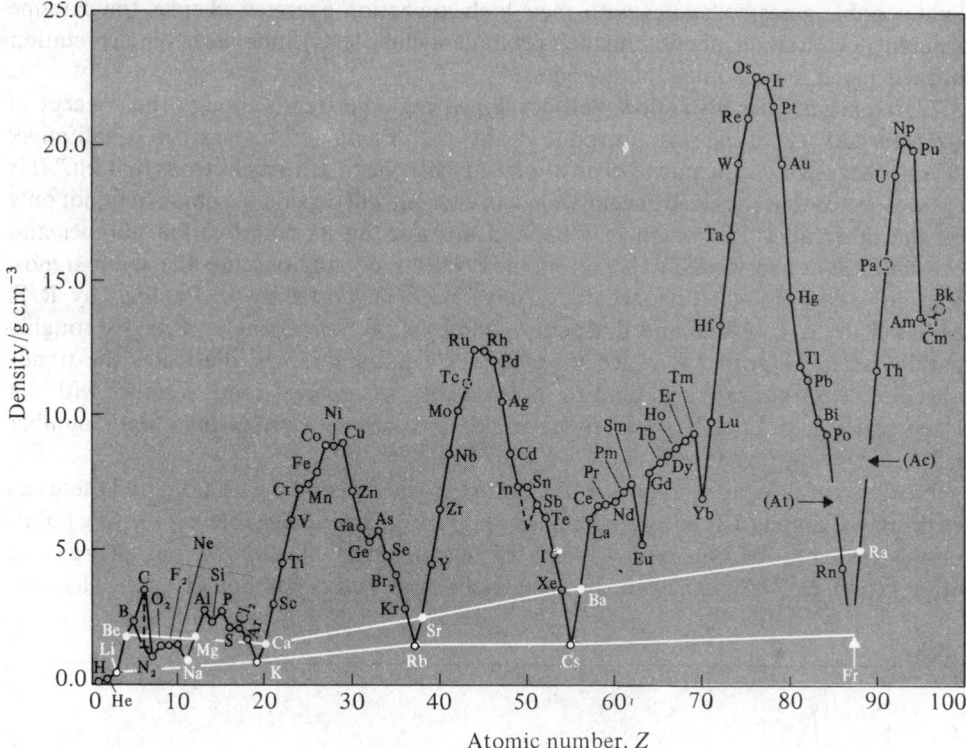

FIG. 2.2   Densities of the elements in the solid state.

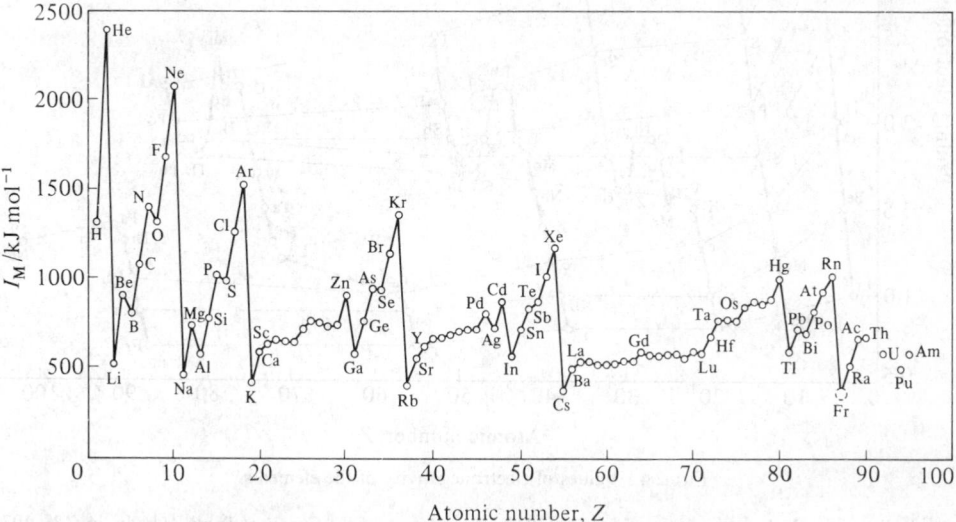

FIG. 2.3   First-stage ionization energies of the elements.

lighter noble gases correlates with their high ionization energies whereas the extreme reactivity of the alkali metals (and their prominent flame tests) finds a ready interpretation in their much lower ionization energies.

Electronegativities also show well-developed periodic trends though the concept of electronegativity itself, as introduced by L. Pauling,[7] is rather qualitative: "Electronegativity is the power of an atom *in a molecule* to attract electrons to itself." It is to be expected that the electronegativity of an element will depend to some extent not only on the other atoms to which it is bonded but also on its coordination number and oxidation state; fortunately, however, these effects do not obscure the main trends. Various measures of electronegativity have been proposed by L. Pauling, by R. S. Mulliken, by A. L. Allred and E. Rochow, and by R. T. Sanderson, and all give roughly parallel scales. Figure 2.4, which incorporates Pauling's values, illustrates the trends observed; electronegativities tend to increase with increasing atomic number within a given period (e.g. Li to F, or K to Br) and to decrease with increasing atomic number within a given group (e.g. F to At, or O to Po).

Numerous other properties have been found to show periodic variations and these can be displayed graphically or by circles of varying size on a periodic table, e.g. melting points of the elements, boiling points, heats of fusion, heats of vaporization, energies of atomization, etc.[6] Similarly, the properties of simple binary compounds of the elements

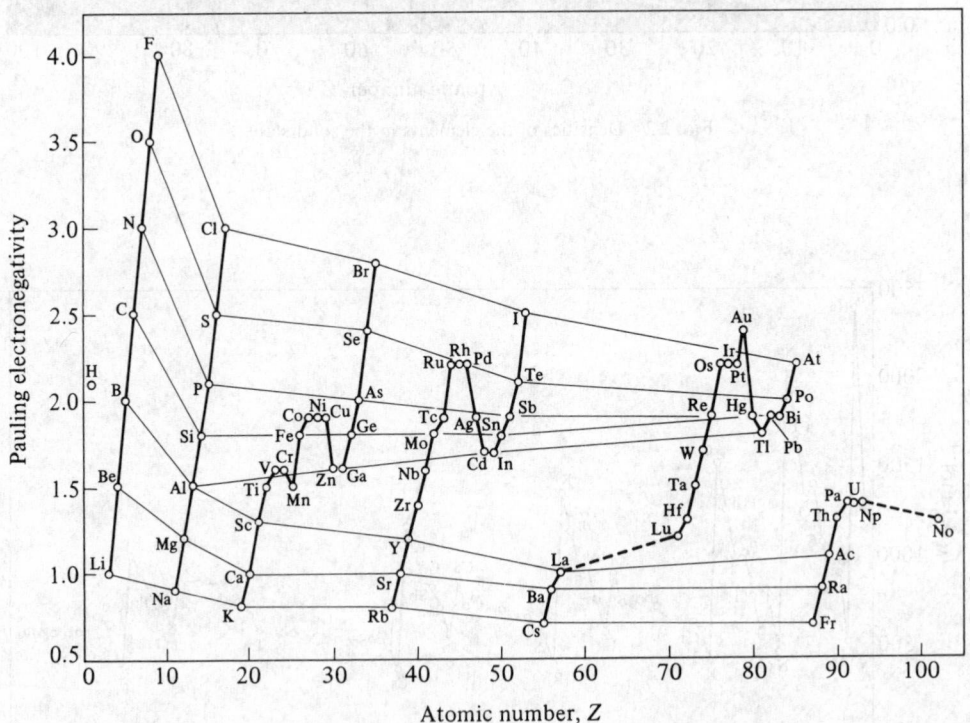

FIG. 2.4   Values of electronegativity of the elements.

[7] L. PAULING, *J. Am. Chem. Soc.* **54**, 3570 (1932); *The Nature of the Chemical Bond*, 3rd edn., pp. 88–107. Cornell University Press, Ithaca, NY, 1960.

can be plotted, e.g. heats of formation, melting points and boiling points of hydrides, fluorides, chlorides, bromides, iodides, oxides, sulfides, etc.[6] Trends immediately become apparent, and the selection of compounds with specific values for particular properties is facilitated. Such trends also permit interpolation to give estimates of undetermined values of properties for a given compound though such a procedure can be misleading and should only be used as a first rough guide. Extrapolation has also frequently been used, and to good effect, though it too can be hazardous and unreliable particularly when new or unsuspected effects are involved. Perhaps the classic example concerns the dissociation energy of the fluorine molecule which is difficult to measure experimentally: for many years this was taken to be ~ 265 kJ mol^{-1} by extrapolation of the values for iodine, bromine, and chlorine (151, 193, and 243 kJ mol^{-1}), whereas the most recent experimental values are close to 154 kJ mol^{-1} (see Chapter 17). The detection of such anomalous data from periodic plots thus serves to identify either inaccurate experimental observations or inadequate theories (or both).

2.3.2 *Trends in chemical properties*

   These, though more difficult to describe quantitatively, also become apparent when the elements are compared in each group and along each period. Such trends will be discussed in detail in later chapters and it is only necessary here to enumerate briefly the various types of behaviour that frequently recur.

   The most characteristic chemical property of an element is its valence. There are numerous measures of valency each with its own area of usefulness and applicability. Simple definitions refer to the number of hydrogen atoms that can combine with an element in a binary hydride or to twice the number of oxygen atoms combining with an element in its oxide(s). It was noticed from the beginning that there was a close relation between the position of an element in the periodic table and the stoichiometry of its simple compounds. Hydrides of main group elements have the formula MH_n where n was related to the group number N by the equations $n = N$ ($N \leqslant IV$), and $n = VIII - N$ for $N > IV$. By contrast, oxygen elicited an increasing valence in the highest normal oxide of each element and this was directly related to the group number, i.e. $M_2O, MO, M_2O_3, \ldots, M_2O_7$. These periodic regularities find a ready explanation in terms of the electronic configuration of the elements and simple theories of chemical bonding. In more complicated chemical formulae involving more than 2 elements, it is convenient to define the "oxidation state" of an element as the formal charge remaining on the element when all other atoms have been removed as their normal ions. For example, nitrogen has an oxidation state of -3 in ammonium chloride [$NH_4Cl - (4H^+ + Cl^-) = N^{3-}$] and manganese has an oxidation state of $+7$ in potassium permanganate {tetraoxomanganate(1−)} [$KMnO_4 - (K^+ + 4O^{2-}) = Mn^{7+}$]. For a compound such as Fe_3O_4 iron has an average oxidation state of $+2.67$ [i.e. $(4 \times 2)/3$] which may be thought of as comprising $1Fe^{2+}$ and $2Fe^{3+}$. It should be emphasized that these charges are formal, not actual, and that the concept of oxidation state is not particularly helpful when considering predominantly covalent compounds (such as organic compounds) or highly catenated inorganic compounds such as S_7NH.

   The periodicity in the oxidation state or valence shown by the elements was forcefully illustrated by Mendeleev in one of his early forms of the periodic system and this is shown in an extended form in Fig. 2.5 which incorporates more recent information. The predictive and interpolative powers of such a plot are obvious and have been a fruitful source of chemical experimentation for over a century.

Other periodic trends which occur in the chemical properties of the elements and which are discussed in more detail throughout later chapters are:

(i) The "anomalous" properties of elements in the first short period (from lithium to fluorine)—see Chapters 4, 5, 6, 8, 11, 14, and 17.
(ii) The "anomalies" in the post-transition element series (from gallium to bromine) related to the d-block contraction—see Chapters 7, 10, 13, 16, and 17.
(iii) The effects of the lanthanide contraction—see Chapters 21–30.
(iv) Diagonal relationships between lithium and magnesium, beryllium and aluminium, boron and silicon.
(v) The so-called inert pair effect (see Chapters 7, 10, and 13) and the variation of oxidation state in the main group elements in steps of 2 (e.g. IF, IF_3, IF_5, IF_7).
(vi) Variability in the oxidation state of transition elements in steps of 1.
(vii) Trends in the basicity and electropositivity of elements—both vertical trends within groups and horizontal trends along periods.
(viii) Trends in bond type with position of the elements in the table and with oxidation state for a given element.
(ix) Trends in stability of compounds and regularities in the methods used to extract the elements from their compounds.
(x) Trends in the stability of coordination complexes and the electron–donor power of various series of ligands.

2.4 Prediction of New Elements and Compounds

Newlands (1864) was the first to predict correctly the existence of a "missing element" when he calculated an atomic weight of 73 for an element between silicon and tin, close to the present value of 72.59 for germanium (discovered by C. A. Winkler in 1886). However, his method of detecting potential triads was unreliable and he predicted (non-existent) elements between rhodium and iridium, and between palladium and platinum. Mendeleev's predictions 1869–71 were much more extensive and reliable, as indicated in the historical panel on p. 25. The depth of his insight and the power of his method remain impressive even today, but in the state of development of the subject in 1869 they were monumental. A comparison of the properties of eka-silicon predicted by Mendeleev and those determined experimentally for germanium is shown in Table 2.1. Similarly accurate predictions were made for eka-aluminium and gallium and for eka-boron and scandium.

Of the remaining 26 undiscovered elements between hydrogen and uranium, 11 were lanthanoids which Mendeleev's system was unable to characterize because of their great chemical similarity and the new numerological feature dictated by the filling of the 4f orbitals. Only cerium, terbium, and erbium were established with certainty in 1871, and the others (except promethium, 1945) were separated and identified in the period 1879–1907. The isolation of the (unpredicted) noble gases also occurred at this time (1894–8).

The isolation and identification of 4 radioactive elements in minute amounts took place

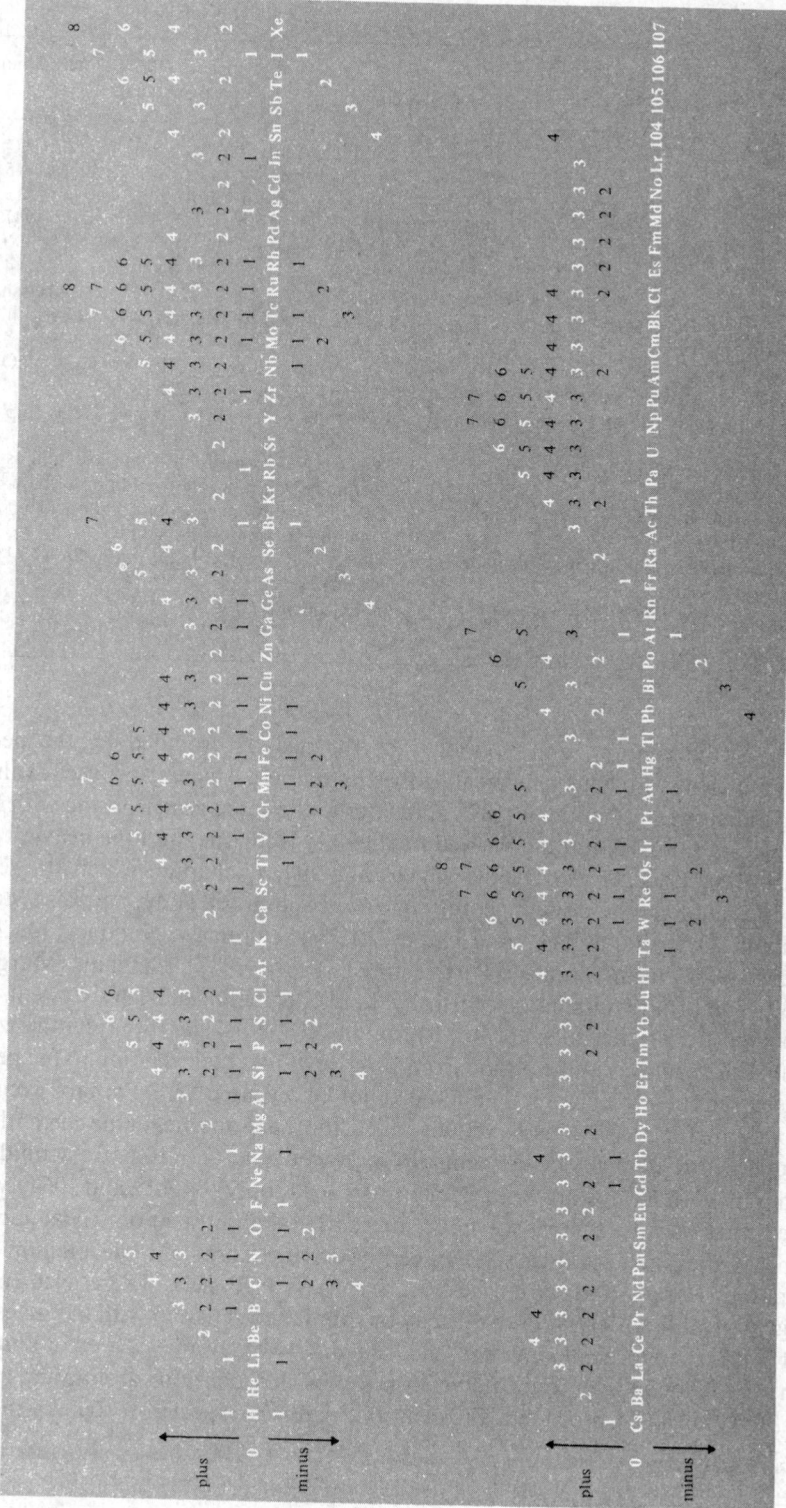

Fig. 2.5   Formal oxidation states of the elements displayed in a format originally devised by Mendeleev in 1889. The more common oxidation states (including zero) are shown in white. Nonintegral values, as in B_5H_9, C_3H_8, HN_3, S_8^{2+}, etc., are not included.

TABLE 2.1

Mendeleev's predictions (1871) for eka-silicon, M		Observed properties (1977) of germanium (discovered 1886)	
Atomic weight	72	Atomic weight	72.59
Density/g cm^{-3}	5.5	Density/g cm^{-3}	5.35
Molar volume/cm^3 mol^{-1}	13.1	Molar volume/cm^3 mol^{-1}	13.57
MP/°C	high	MP/°C	947
Specific heat/J g^{-1} K^{-1}	0.305	Specific heat/J g^{-1} K^{-1}	0.309
Valence	4	Valence	4
Colour	dark grey	Colour	greyish-white
M will be obtained from MO_2 or K_2MF_6 by reaction with Na		Ge is obtained by reaction of K_2GeF_6 with Na	
M will be slightly attacked by acids such as HCl and will resist alkalis such as NaOH		Ge is not dissolved by HCl or dilute NaOH but reacts with hot conc HNO_3	
M, on being heated, will form MO_2 with high mp, and d 4.7 g cm^{-3}		Ge reacts with oxygen to give GeO_2, mp 1086°, d 4.228 g cm^{-3}	
M will give a hydrated MO_2 soluble in acid and easily reprecipitated		"$Ge(OH)_4$" dissolves in conc acid and is reprecipitated on dilution or addition of base	
MS_2 will be insoluble in water but soluble in ammonium sulfide		GeS_2 is insoluble in water and dilute acid but readily soluble in ammonium sulfide	
MCl_4 will be a volatile liquid with bp a little under 100°C and d 1.9 g cm^{-3}		$GeCl_4$ is a volatile liquid with bp 84°C and d 1.8443 g cm^{-3}	
M will form MEt_4 bp 160°C		$GeEt_4$ bp 185°C	

at the turn of the century, and in each case the insight provided by the periodic classification into the predicted chemical properties of these elements proved invaluable. Marie Curie identified polonium in 1898 and, later in the same year working with Pierre Curie, isolated radium. Actinium followed in 1899 (A. Debierne) and the heaviest noble gas, radon, in 1900 (F. E. Dorn). Details will be found in later chapters which also recount the discoveries made in the present century of protactinium (O. Hahn and Lise Meitner, 1917), hafnium (D. Coster and G. von Hevesey, 1923), rhenium (W. Noddack, Ida Tacke, and O. Berg, 1925), technetium (C. Perrier and E. Segré, 1937), francium (Marguerite Perey, 1939), and promethium (J. A. Marinsky, L. E. Glendenin, and C. D. Coryell, 1945).

A further group of elements, the transuranium elements, has been synthesized by artificial nuclear reactions in the period from 1940 onwards; their relation to the periodic table is discussed fully in Chapter 31 and need not be repeated here. Perhaps even more striking today are the predictions, as yet unverified, for the properties of the currently non-existent superheavy elements.[8] Elements up to lawrencium ($Z = 103$) are actinides (5f) and the 6d transition series starts with element 104. So far only 104, 105, and 106 have been synthesized, and, because there is as yet no international agreement on trivial names for these elements, they are here given their agreed systematic names: un-nil-quadium (Unq), un-nil-pentium (Unp), and un-nil-hexium (Unh) (see Panel). These elements are increasingly unstable with respect to α-decay or spontaneous fission with half-lives of the order of 1 s. It is therefore unlikely that much chemistry will ever be carried out on the 6d transition series though their ionization energies, mps, bps, densities, atomic and metallic radii, etc., have all been predicted. Element 112, Uub, is expected to be eka-mercury,

[8] B. FRICKE, Superheavy elements, *Structure and Bonding* **21**, 89 (1975). A full account of the predicted stabilities and chemical properties of elements with atomic numbers in the range $Z = 104$–184.

Nomenclature of Elements having $Z > 100$

Approved by IUPAC *1977*

The principles applied in deriving the names were as follows:

(i) The names should be short, systematic, and obviously related to the atomic numbers of the elements.
(ii) The names should end *ium* whether the element is expected to be metal or otherwise.
(iii) The symbols should consist of three letters to avoid duplicating some of the symbols already used for elements.
(iv) The symbols should be derived directly from the atomic number and be visually related to the names as far as possible.

The existence of this systematic nomenclature for unknown elements does not deny the right of "discoverers" of new elements to suggest other names to IUPAC after their discovery has been established beyond doubt in the scientific community. Such trivial names have already been accepted for elements 101 (mendelevium), 102 (nobelium), and 103 (lawrencium).

The nomenclature rules are:

1. The name is derived directly from the atomic number of the element using the following numerical roots:

0	1	2	3	4	5	6	7	8	9
nil	un	bi	tri	quad	pent	hex	sept	oct	enn

2. The roots are put together in the order of the digits which make up the atomic number and terminated by *ium* to spell out the name. The final *n* of *enn* is elided when it occurs before *nil* and the final *i* of *bi* and *tri* when it occurs before *ium*.
3. The symbol of the element is composed of the initial letters of the numerical roots which make up the name. (The mixture of Greek and Latin roots is necessary to avoid ambiguities at this point.)

In the following examples, hyphens are inserted to assist comprehension and pronunciation but are not part of the names.

Atomic number	Name	Symbol	Atomic number	Name	Symbol
101	un-nil-unium	Unu	120	un-bi-nilium	Ubn
102	un-nil-bium	Unb	121	un-bi-unium	Ubu
103	un-nil-trium	Unt	130	un-tri-nilium	Utn
104	un-nil-quadium	Unq	140	un-quad-nilium	Uqn
105	un-nil-pentium	Unp	150	un-pent-nilium	Upn
106	un-nil-hexium	Unh	200	bi-nil-nilium	Bnn
107	un-nil-septium	Uns	201	bi-nil-unium	Bnu
108	un-nil-octium	Uno	202	bi-nil-bium	Bnb
109	un-nil-ennium	Une	300	tri-nil-nilium	Tnn
110	un-un-nilium	Uun	400	quad-nil-nilium	Qnn
111	un-un-unium	Uuu	500	pent-nil-nilium	Pnn
112	un-un-bium	Uub	900	en-nil-nilium	Enn

followed by the 7p and 8s configurations $Z = 113$–120. On the basis of present theories of nuclear structure an "island of stability" is expected near $^{298}_{114}$Uuq with half-lives in the region of years. Much effort is being concentrated on attempts to make these elements, and oxidation states are expected to follow the main group trends (e.g. Uut:eka-thallium mainly $+ 1$). Other physical properties have been predicted by extrapolation of known periodic trends. Still heavier elements have been postulated, though it is unlikely (on present theories) that their chemistry will ever be studied because of their very short predicted half-lives. Calculated energy levels for the range $Z = 121$–154 lead to the expectation of an unprecedented 5g series of 18 elements followed by 14 6f elements.

In addition to the prediction of new elements and their probable properties, the periodic table has proved invaluable in suggesting fruitful lines of research in the preparation of new compounds. Indeed, this mode of thinking is now so ingrained in the minds of chemists that they rarely pause to reflect how extraordinarily difficult their task would be if periodic trends were unknown. It is the ability to anticipate the effect of changing an element or a group in a compound which enables work to be planned effectively, though the prudent chemist is always alert to the possibility of new effects or unsuspected factors which surprisingly intervene.

Typical examples taken from the developments of the past two or three decades include:

(i) the organometallic chemistry of lithium and thallium (Chapters 4 and 7);
(ii) the use of boron hydrides as ligands (Chapter 6);
(iii) solvent systems and preparative chemistry based on the interhalogens (Chapter 17);
(iv) the development of the chemistry of xenon (Chapter 18);
(v) ferrocene—leading to ruthenocene and dibenzene chromium, etc. (Chapters 8, 25 and 23 respectively);
(vi) the development of solid-state chemistry.

Indeed, the influence of Mendeleev's fruitful generalization pervades the whole modern approach to the chemistry of the elements.

2.5 Questions on the Periodic Table

1. Why did Mendeleev use the sequence of increasing atomic weights to order the elements? Was he justified in inverting the positions of Co/Ni and Te/I; why do the atomic weights of these pairs of elements (and of Ar/K) not follow the sequence of atomic numbers?
2. Why did Mendeleev list thallium with the alkali metals, platinum with the alkaline earth metals, and uranium in Group III?
3. Why was Mendeleev unable to predict the noble gases though he correctly predicted several other undiscovered elements? Likewise for the lanthanides.
4. From the known physical and chemical properties of S, Se, and Te, formulate the properties to be expected for polonium; check with the actual properties given in Chapter 16.
5. From the known formulae and properties of the fluorides and oxides of Sb, Te, and I, predict possible formulae and properties for the fluorides and oxo-compounds of

xenon. Why was the discovery of such compounds delayed until the 1960s despite predictions of their stability by L. Pauling 30 y earlier?

6. Comment on the statement that there could be no satisfactory theory of chemical bonding before relativistic quantum mechanics because the four quantum numbers imply a four-dimensional space–time continuum.

3

Hydrogen

3.1 Introduction

Hydrogen is the most abundant element in the universe and is also common on earth, being the third most abundant element (after oxygen and silicon) on the surface of the globe. Hydrogen in combined form accounts for about 15.4% of the atoms in the earth's crust and oceans and is the ninth element in order of abundance by weight (0.9%). In the crustal rocks alone it is tenth in order of abundance (0.15 wt%). The gradual recognition of hydrogen as an element during the sixteenth and seventeenth centuries forms part of the obscure and tangled web of experiments that were carried out as chemistry emerged from alchemy to become a modern science.[1] Until almost the end of the eighteenth century the element was inextricably entwined with the concept of phlogiston and H. Cavendish, who is generally regarded as having finally isolated and identified the gas in 1766, and who established conclusively that water was a compound of oxygen and hydrogen, actually communicated his findings to the Royal Society in January 1784 in the following words: "There seems to be the utmost reason to think that dephlogisticated air is only water deprived of its phlogiston" and that "water consists of dephlogisticated air united with phlogiston".

The continued importance of hydrogen in the development of experimental and theoretical chemistry is further illustrated by some of the dates listed in the Panel.

Hydrogen was recognized as the essential element in acids by H. Davy after his work on the hydrohalic acids, and theories of acids and bases have played an important role ever since. The electrolytic dissociation theory of S. A. Arrhenius and W. Ostwald in the 1880s, the introduction of the pH scale for hydrogen-ion concentrations by S. P. L. Sørensen in 1909, the theory of acid–base titrations and indicators, and J. N. Brønsted's fruitful concept of acids and conjugate bases as proton donors and acceptors (1923) are other land marks (see p. 51). The discovery of *ortho-* and *para*-hydrogen in 1924, closely followed by the discovery of heavy hydrogen (deuterium) and tritium in the 1930s, added a further range of phenomena that could be studied by means of this element (pp. 40–50). In more recent times, the technique of nmr spectroscopy, which was first demonstrated in 1946 using the hydrogen nucleus, has revolutionized the study of structural chemistry and permitted previously unsuspected phenomena such as fluxionality to be studied. Simultaneously, the discovery of complex metal hydrides such as $LiAlH_4$ has had a major

---

[1] J. W. MELLOR, *A Comprehensive Treatise on Inorganic and Theoretical Chemistry*, Vol. 1, Chap. 3, Longmans, Green & Co., London, 1922.

Hydrogen—Some Significant Dates

1671	R. Boyle showed that dilute sulfuric acid acting on iron gave a flammable gas; several other seventeenth-century scientists made similar observations.
1766	H. Cavendish established the true properties of hydrogen by reacting several acids with iron, zinc, and tin; he showed that it was much lighter than air.
1781	H. Cavendish showed quantitatively that water was formed when hydrogen was exploded with oxygen, and that water was therefore not an element as had previously been supposed.
1783	A. L. Lavoisier proposed the name "hydrogen" (Greek ὕδωρ γείνομαι, water former).
1800	W. Nicholson and A. Carlisle decomposed water electrolytically into hydrogen and oxygen which were then recombined by explosion to resynthesize water.
1810–15	Hydrogen recognized as the essential element in acids by H. Davy (contrary to Lavoisier who originally considered oxygen to be essential—hence Greek ὀξύς γείνομαι, acid former).
1866	The remarkable solubility of hydrogen in palladium discovered by T. Graham following the observation of hydrogen diffusion through red-hot platinum and iron by H. St. C. Deville and L. Troost, 1863.
1878	Hydrogen detected spectroscopically in the sun's chromosphere (J. N. Lockyer).
1895	Hydrogen first liquefied in sufficient quantity to show a meniscus (J. Dewar) following earlier observations of mists and droplets by others, 1877–85.
1909	The pH scale for hydrogen-ion concentration introduced by S. P. L. Sørensen.
1920	The concept of hydrogen bonding introduced by W. M. Latimer and W. H. Rodebush (and by M. L. Huggins, 1921).
1923	J. N. Brønsted defined an acid as a species that tended to lose a proton: $A \rightleftharpoons B + H^+$.
1924	*Ortho-* and *para-*hydrogen discovered spectroscopically by R. Mecke and interpreted quantum-mechanically by W. Heisenberg, 1927.
1929–30	Concept of quantum-mechanical tunnelling in proton-transfer reactions introduced (without experimental evidence) by several authors.
1931	First hydrido complex of a transition metal prepared by W. Hieber and F. Leutert.
1932	Deuterium discovered spectroscopically and enriched by gaseous diffusion of hydrogen and electrolysis of water (H. C. Urey, F. G. Brickwedde, and G. M. Murphy).
1934	Tritium first made by deuteron bombardment of D_3PO_4 and $(ND_4)_2SO_4$ (i.e. $^2D + ^2D = ^3T + ^1H$); M. L. E. Oliphant, P. Harteck, and E. Rutherford.
1939	Tritium found to be radioactive by L. W. Alvarez and R. Cornog after a prediction by T. W. Bonner in 1938.
1946	Proton nmr first detected in bulk matter by E. M. Purcell, H. C. Torrey, and R. V. Pound; and by F. Bloch, W. W. Hansen, and M. E. Packard.
1947	$LiAlH_4$ first prepared and subsequently shown to be a versatile reducing agent; A. E. Finholt, A. C. Bond, and H. I. Schlesinger.
1950	Tritium first detected in atmospheric hydrogen (V. Faltings and P. Harteck) and later shown to be present in rain water (W. F. Libby *et al.*, 1951).

impact on synthetic chemistry and enabled new classes of compound to be readily prepared in high yield (p. 259). It is now realized that hydrogen forms more chemical compounds than any other element, including carbon, and a survey of its chemistry therefore encompasses virtually the whole periodic table. However, before embarking on such a review in Sections 3.4–3.6 it is convenient to summarize the atomic and physical properties of the various forms of hydrogen (Section 3.2), to enumerate the various methods used for its preparation and industrial production, and to indicate some of its many applications and uses (Section 3.3).

3.2 Atomic and Physical Properties of Hydrogen[2]

Despite its very simple electronic configuration ($1s^1$) hydrogen can, paradoxically, exist in over 40 different forms most of which have been well characterized. This multiplicity of forms arises firstly from the existence of atomic, molecular, and ionized species in the gas phase: H, H_2, H^+, H^-, H_2^+, and H_3^+; secondly, from the existence of three isotopes, $_1^1H$, $_1^2H(D)$, and $_1^3H(T)$, and correspondingly of D, D_2, HD, DT, etc.; and, finally, from the existence of nuclear spin isomers for the homonuclear diatomic species, i.e. *ortho-* and *para*-dihydrogen, -dideuterium, and -ditritium.†

3.2.1 *Isotopes of hydrogen*

Hydrogen as it occurs in nature is predominantly composed of atoms in which the nucleus is a single proton. In addition, terrestrial hydrogen contains about 0.0156% of deuterium atoms in which the nucleus also contains a neutron, and this is the reason for its variable atomic weight (p. 20). Addition of a second neutron induces instability and tritium is radioactive, emitting low-energy β^- particles with a half-life of 12.35 y. Some characteristic properties of these 3 atoms are given in Table 3.1, and their implications for stable isotope studies, radioactive tracer studies, and nmr spectroscopy are obvious.

TABLE 3.1   *Atomic properties of hydrogen, deuterium, and tritium*

Property	H	D	T
Relative atomic mass	1.007 825	2.014 102	3.016 049
Nuclear spin quantum number	$\frac{1}{2}$	1	$\frac{1}{2}$
Nuclear magnetic moment/(nuclear magnetons)[a]	2.792 70	0.857 38	2.978 8
NMR frequency (at 2.35 tesla)/MHz	100.56	15.360	104.68
NMR relative sensitivity (constant field)	1.000	0.009 64	1.21
Nuclear quadrupole moment/(10^{-28} m^2)	0	2.766×10^{-3}	0
Radioactive stability	Stable	Stable	β^- $t_{\frac{1}{2}}$ 12.35 y[b]

[a] Nuclear magneton $\mu_N = eh/2m_p = 5.0508 \times 10^{-27}$ J T^{-1}.
[b] E_{max} 18.6 keV; E_{mean} 5.7 keV.

In the molecular form, dihydrogen is a stable, colourless, odourless, tasteless gas with a very low mp and bp. Data are in Table 3.2 from which it is clear that the values for deuterium and tritium are substantially higher. For example, the mp of T_2 is above the bp of H_2. Other forms such as HD and DT tend to have properties intermediate between those of their components. Thus HD has mp 16.60 K, bp 22.13 K, ΔH_{fus} 0.159 kJ mol^{-1}, ΔH_{vap} 1.075 kJ mol^{-1}, T_c 35.91 K, P_c 14.64 atm, and ΔH_{dissoc} 439.3 kJ mol^{-1}. The critical temperature T_c is the temperature above which a gas cannot be liquefied simply by application of pressure, and the critical pressure P_c is the pressure required for liquefaction at this point.

Table 3.2 also indicates that the heat of dissociation of the hydrogen molecule is extremely high, the H–H bond energy being larger than for almost all other single bonds.

† The term dihydrogen (like dinitrogen, dioxygen, etc.) is used when it is necessary to refer unambiguously to the molecule H_2 (or N_2, O_2, etc.) rather than to the element as a substance or to an atom of the element.
2 K. M. MACKAY, The element hydrogen, *Comprehensive Inorganic Chemistry*, Vol. 1, Chap. 1. K. M. MACKAY and M. F. A. DOVE, Deuterium and tritium, *ibid.*, Vol. 1, Chap. 3, Pergamon Press, Oxford, 1973.

TABLE 3.2   *Physical properties of hydrogen, deuterium, and tritium*

Property[a]	H_2	D_2	T_2
MP/K	13.957	18.73	20.62
BP/K	20.39	23.67	25.04
Heat of fusion/kJ mol^{-1}	0.117	0.197	0.250
Heat of vaporization/kJ mol^{-1}	0.904	1.226	1.393
Critical temperature/K	33.19	38.35	40.6 (calc)
Critical pressure/atm[b]	12.98	16.43	18.1 (calc)
Heat of dissociation/kJ mol^{-1} (at 298.2 K)	435.88	443.35	446.9
Zero point energy/kJ mol^{-1}	25.9	18.5	15.1
Internuclear distance/pm	74.14	74.14	(74.14)

[a] Data refer to H_2 of normal isotopic composition (i.e. containing 0.0156 atom % of deuterium, predominantly as HD). All data refer to the mixture of *ortho-* and *para*-forms that are in equilibrium at room temperature.
[b] 1 atm = 101.325 kN m^{-2} = 101.325 kPa.

This contributes to the relative unreactivity of hydrogen at room temperature. Significant thermal decomposition into hydrogen atoms occurs only above 2000 K: the percentage of atomic H is 0.081 at this temperature, and this rises to 7.85% at 3000 K and 95.5% at 5000 K. Atomic hydrogen can, however, be conveniently prepared in low-pressure glow discharges, and the study of its reactions forms an important branch of chemical gas kinetics. The high heat of recombination of hydrogen atoms finds application in the atomic hydrogen torch—dihydrogen is dissociated in an arc and the atoms then recombine on the surface of a metal, generating temperatures in the region of 4000 K which can be used to weld very high melting metals such as tantalum and tungsten.

3.2.2 Ortho- *and* para-*hydrogen*

All homonuclear diatomic molecules having nuclides with non-zero spin are expected to show nuclear spin isomers. The effect was first detected in dihydrogen where it is particularly noticeable, and it has also been established for D_2, T_2, $^{14}N_2$, $^{15}N_2$, $^{17}O_2$, etc. When the two nuclear spins are parallel (*ortho*-hydrogen) the resultant nuclear spin quantum number is 1 (i.e. $\frac{1}{2} + \frac{1}{2}$) and the state is threefold degenerate (2S + 1). When the two proton spins are antiparallel, however, the resultant nuclear spin is zero and the state is non-degenerate. Conversion between the two states involves a forbidden triplet–singlet transition and is normally slow unless catalysed by interaction with solids or paramagnetic species which either break the H–H bond, weaken it, or allow magnetic perturbations. Typical catalysts are Pd, Pt, active Fe_2O_3, and NO. *Para*-hydrogen (spins antiparallel) has the lower energy and this state is favoured at low temperatures. Above 0 K (100% *para*) the equilibrium concentration of *ortho*-hydrogen gradually increases until, above room temperature, the statistically weighted proportion of 3 *ortho*:1 *para* is obtained, i.e. 25% *para*. Typical equilibrium concentrations of *para*-hydrogen are 99.8% at 20 K, 65.4% at 60 K, 38.5% at 100 K, 25.7% at 210 K, and 25.1% at 273 K (Fig. 3.1). It follows that, whereas essentially pure *para*-hydrogen can be obtained, it is never possible to obtain a sample containing more than 75% of *ortho*-hydrogen. Experimentally, the presence of both o-H_2 and p-H_2 is seen as an alternation in the intensities of successive

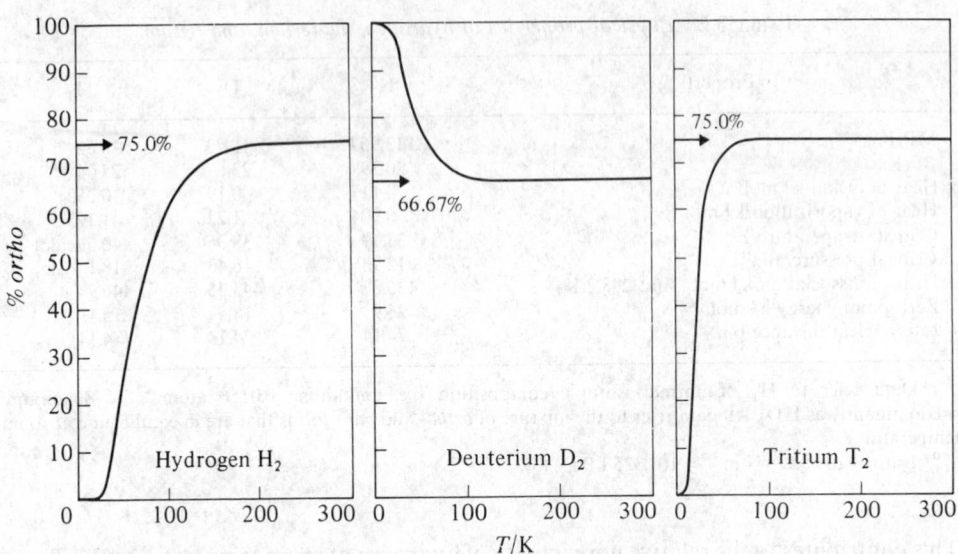

Fig. 3.1   *Ortho–para* equilibria for H_2, D_2, and T_2.

rotational lines in the fine structure of the electronic band spectrum of H_2. It also explains the curious temperature dependence of the heat capacity of hydrogen gas.

Similar principles apply to *ortho*- and *para*-deuterium except that, as the nuclear spin quantum number of the deuteron is 1 rather than $\frac{1}{2}$ as for the proton, the system is described by Bose–Einstein statistics rather than the more familiar Fermi–Dirac statistics. For this reason, the stable low-temperature form is *ortho*-deuterium and at high temperatures the statistical weights are 6 *ortho*:3 *para* leading to an upper equilibrium concentration of 33.3% *para*-deuterium above about 190 K as shown in Fig. 3.1. Tritium (spin $\frac{1}{2}$) resembles H_2 rather than D_2.

Most physical properties are but little affected by nuclear-spin isomerism though the thermal conductivity of *p*-H_2 is more than 50% greater than that of *o*-H_2, and this forms a ready means of analysing mixtures. The mp of *p*-H_2 (containing only 0.21% *o*-H_2) is 0.15 K below that of "normal" hydrogen (containing 75% *o*-H_2), and by extrapolation the mp of (unobtainable) pure *o*-H_2 is calculated to be 0.24 K above that of *p*-H_2. Similar differences are found for the bps which occur at the following temperatures: normal-H_2 20.39 K, *o*-H_2 20.45 K. For deuterium the converse relation holds, *o*-D_2 melting some 0.03 K below "normal"-D_2 (66.7% *ortho*) and boiling some 0.04 K below. The effects for other elements are even smaller.

3.2.3   *Ionized forms of hydrogen*

This section briefly considers the proton H^+, the hydride ion H^-, the hydrogen molecule ion H_2^+, and the triatomic 2-electron species H_3^+.

The hydrogen atom has a high ionization energy (1311 kJ mol^{-1}) and in this it resembles the halogens rather than the alkali metals. Removal of the 1s electron leaves a bare proton which, having a radius of only about 1.5×10^{-3} pm, is not a stable chemical entity in the condensed phase. However, when bonded to other species it is well known in solution and in solids, e.g. H_3O^+, NH_4^+, etc. The proton affinity of water and the enthalpy of solution of H^+ in water have been estimated by several authors and typical values that are currently accepted are:[3]

$$H^+(g) + H_2O(g) \longrightarrow H_3O^+(g); \quad -\Delta H \sim 720 \text{ kJ mol}^{-1}$$
$$H^+(g) \longrightarrow H_3O^+(aq); \quad -\Delta H \sim 1090 \text{ kJ mol}^{-1}$$

It follows that the heat of solution of the oxonium ion in water is ~ 370 kJ mol^{-1}, intermediate between the values calculated for Na^+ (405 kJ mol^{-1}) and K^+ (325 kJ mol^{-1}). Reactions involving proton transfer will be considered in more detail in Section 3.4.1.

The hydrogen atom, like the alkali metals (ns^1) and halogens (ns^2np^5), has an affinity for the electron and heat is evolved in the following process:

$$H(g) + e^- \longrightarrow H^-(g); \quad -\Delta H_{calc} = 72 \text{ kJ mol}^{-1}$$

This is larger than the corresponding value for Li (57 kJ mol^{-1}) but substantially smaller than the value for F (333 kJ mol^{-1}). The hydride ion H^- has the same electron configuration as helium but is much less stable because the single positive charge on the proton must now control the 2 electrons. The hydride ion is thus readily deformable and this constitutes a characteristic feature of its structural chemistry (see p. 72).

The species H_2^+ and H_3^+ need not detain us long for little chemistry is involved. They are, however, important as model systems for chemical bonding theory. The hydrogen molecule ion H_2^+ comprises 2 protons and 1 electron and is extremely unstable even in a low-pressure gas discharge system. There is some uncertainty about the energy of dissociation and the internuclear distance but recent values (with the corresponding values for H_2 in parentheses) are:

$$\Delta H_{dissoc} \ 255(436) \text{ kJ mol}^{-1}; \quad r(H–H) \ 106(74.2) \text{ pm}$$

The triatomic hydrogen molecule ion H_3^+ was first detected by J. J. Thompson in gas discharges and later fully characterized by mass spectrometry; its relative atomic mass, 3.027, clearly distinguishes it from tritium (3.016). The "observed" triangular 3-centre, 2-electron structure is more stable than the hypothetical linear structure, and the comparative stability of the species is shown by the following gas-phase enthalpies:

$$H + H + H^+ = H_3^+; \quad -\Delta H \ 770 \text{ kJ mol}^{-1}$$
$$H_2 + H^+ = H_3^+; \quad -\Delta H \ 337 \text{ kJ mol}^{-1}$$
$$H + H_2^+ = H_3^+; \quad -\Delta H \ 515 \text{ kJ mol}^{-1}$$

3.3 Preparation, Production, and Uses[4, 5]

3.3.1 *Hydrogen*

Hydrogen can be prepared by the reaction of water or dilute acids on electropositive metals such as the alkali metals, alkaline earth metals, the metals of Groups IIIA, IVA, and

[3] R. P. BELL, *The Proton in Chemistry*, 2nd edn., Chapman & Hall, London, 1973, 223 pp.
[4] W. J. GRANT and S. L. REDFEARN, Industrial gases, in R. THOMPSON (ed.), *The Modern Inorganic Chemicals Industry*, pp. 273–301, Chemical Society Special Publication No. 31, London 1977.
[5] R. B. SCOTT, W. H. DENTON, and C. M. NICHOLLS (eds.), *Technology and Uses of Liquid Hydrogen*, Pergamon Press, Oxford, 1964, 415 pp.

the lanthanoids. The reaction can be explosively violent. Convenient laboratory methods employ sodium amalgam or calcium with water, or zinc with hydrochloric acid. The reaction of aluminium or ferrosilicon with aqueous sodium hydroxide has also been used. For small-scale preparations the hydrolysis of metal hydrides is convenient, and this generates twice the amount of hydrogen as contained in the hydride, e.g.:

$$CaH_2 + 2H_2O \longrightarrow Ca(OH)_2 + 2H_2$$

Electrolysis of acidified water using platinum electrodes is a convenient source of hydrogen (and oxygen) and, on a larger scale, very pure hydrogen ($> 99.95\%$) can be obtained from the electrolysis of warm aqueous solutions of barium hydroxide between nickel electrodes. The method is expensive but becomes economical on an industrial scale when integrated with the chloralkali industry (p. 931). Other bulk processes involve the (endothermic) reaction of steam on hydrocarbons or coke:

$$CH_4 + H_2O \xrightarrow{1100°C} CO + 3H_2$$

$$C + H_2O \xrightarrow{1000°C} CO + H_2 \quad \text{(water gas)}$$

In both processes the CO can be converted to CO_2 and more hydrogen generated by passing the gases and steam over an iron oxide or cobalt oxide catalyst at 400°C:

$$CO + H_2O \xrightarrow[\text{catalyst}]{400°C} CO_2 + H_2$$

This is the so-called water–gas shift reaction ($-\Delta G^{\circ}_{298}$ 19.9 kJ mol^{-1}) and very recently it has also been effected by low-temperature homogeneous catalysts in aqueous acid solutions.[6] The extent of subsequent purification of the hydrogen depends on the use to which it will be put.

The industrial production of hydrogen is considered in more detail in the Panel. The largest single use of hydrogen is in ammonia synthesis (p. 482) but other major applications are in the catalytic hydrogenation of unsaturated liquid vegetable oils to solid, edible fats (margarine), and in the manufacture of bulk organic chemicals, particularly methanol (by the "oxo" or hydroformylation process)

$$CO + 2H_2 \xrightarrow[\text{catalyst}]{\text{cobalt}} MeOH$$

Direct reaction of hydrogen with chlorine is a major source of hydrogen chloride (p. 946), and a smaller, though still substantial use is in the manufacture of metal hydrides and complex metal hydrides (p. 69). Hydrogen is used in metallurgy to reduce oxides to metals (e.g. Mo, W) and to produce a reducing atmosphere. Direct reduction of iron ores in steelmaking is also now becoming technically and economically feasible. Another medium-scale use is in oxyhydrogen torches and atomic hydrogen torches for welding and cutting. Liquid hydrogen is used in bubble chambers for studying high-energy particles and as a rocket fuel (with oxygen) in the space programme. Hydrogen gas is potentially a large-scale fuel for use in internal combustion engines and fuel cells if the notional "hydrogen economy" (see Panel on p. 46) is ever developed.

⁶ C.-H. CHENG and R. EISENBERG, Homogeneous catalysis of the water–gas shift reaction using a platinum chloride–tin chloride system, *J. Am. Chem. Soc.* **100**, 5969–70 (1978).

Industrial Production of Hydrogen

Many reactions are available for the preparation of hydrogen and the one chosen depends on the amount needed, the purity required, and the availability of raw materials. Most ($\sim 95\%$) of the hydrogen produced in industry is consumed in integrated plants on site (e.g. ammonia synthesis, petrochemical works, etc.). Even so, vast amounts of the gas are produced for the general market, e.g. $\sim 3 \times 10^9$ m^3 or 0.25 million tonnes yearly in the USA alone. Small generators may have a capacity of 100–4000 m^3 h^{-1}, medium-sized plants 4000–10 000 m^3 h^{-1}, and large plants can produce 10^4–10^5 m^3 h^{-1}. The dominant large-scale process in integrated plants is the catalytic steam–hydrocarbon reforming process using natural gas or oil-refinery feedstock. After desulfurization (to protect catalysts) the feedstock is mixed with process steam and passed over a nickel-based catalyst at 700–1000°C to convert it irreversibly to CO and H$_2$, e.g.

$$C_3H_8 + 3H_2O \xrightarrow[\text{catalyst}]{900\ C} 3CO + 7H_2$$

Two reversible reactions also occur to give an equilibrium mixture of H$_2$, CO, CO$_2$, and H$_2$O:

$$CO + H_2O \rightleftharpoons CO_2 + H_2$$

$$CO + 3H_2 \rightleftharpoons CH_4 + H_2O$$

The mixture is cooled to ~ 350°C before entering a high-temperature shift convertor where the major portion of the CO is catalytically and exothermically converted to CO$_2$ and hydrogen by reaction with H$_2$O. The issuing gas is further cooled to 200° before entering the low-temperature shift convertor which reduces the CO content to 0.2 vol%. The product is further cooled and CO$_2$ absorbed in a liquid contacter. Further removal of residual CO and CO$_2$ can be effected by methanation at 350°C to a maximum of 10 ppm. Provided that the feedstock contains no nitrogen the product purity is about 98%. Alternatively the low-temperature shift process and methanation stage can be replaced by a single pressure–swing absorption (PSA) system in which the hydrogen is purified by molecular sieves. The sieves are regenerated by adiabatic depressurization at ambient temperature (hence the name) and the product has a purity of $\geqslant 99.9\%$.

Hydrogen is also produced as a byproduct of the brine electrolysis process for the manufacture of chlorine and sodium hydroxide (p. 931). The ratio of H$_2$:Cl$_2$:NaOH is, of course, fixed by stoichiometry and this is an economic determinant since bulk transport of the byproduct hydrogen is expensive. To illustrate the scale of the problem: the total UK chlorine production capacity is about 10^6 tonnes per year which corresponds to 28 000 tonnes of hydrogen (340×10^6 m^3). Plants designed specifically for the electrolytic manufacture of hydrogen as the main product, use steel cells and aqueous potassium hydroxide as electrolyte. The cells may be operated at atmospheric pressure (Knowles cells) or at 30 atm (Lonza cells). Theoretically, 2 faradays of electric charge (1 F = 96 485 C mol^{-1}) should produce 1 mole H$_2$ and $\frac{1}{2}$ mole O$_2$ (i.e. 2.35 MA h per 10^3 m^3 H$_2$ at STP). At the electrode current-densities normally used the cell voltage is ~ 2 V which, together with transformer and rectifier losses, gives a practical power consumption of about 5 MWh per 10^3 m^3 of hydrogen.

The various techniques outlined above for producing pure hydrogen are not always practicable when relatively small quantities of the gas are required at remote locations such as weather stations. A hydrogen generator (1.25 m high, 2.16 m long, 1.07 m wide; Pl. 3.1) has recently been developed for the inflation of weather balloons and for other purposes; it is mounted on a 1-tonne general-service trailer and produces hydrogen at a rate of 4.2 m^3 h^{-1} (3 mol min^{-1}). Other units are available producing 1–17 m^3 h^{-1}. During production a 1:1 molar mixture of methanol and water is vaporized and passed over a "base–metal chromite" type catalyst at 400°C where it is cracked into hydrogen and carbon monoxide; subsequently steam reacts with the carbon monoxide to produce the dioxide and more hydrogen:

$$MeOH \xrightarrow[\text{catalyst}]{400\ C} CO + 2H_2$$

$$CO + H_2O \longrightarrow CO_2 + H_2$$

All the gases are then passed through a diffuser separator comprising a large number of small-diameter thin-walled tubes of palladium–silver alloy tightly packed in a stainless steel case. The solubility of hydrogen in palladium is well known (p. 1335) and the alloy with silver is used to prolong the life of the diffuser by avoiding troublesome changes in dimensions during the passage of hydrogen. The hydrogen which emerges is cool, pure, dry, and ready for use via a metering device.

The Hydrogen Economy[7-11]

The growing recognition during the past decade that world reserves of coal and oil are finite and that nuclear power cannot supply all our energy requirements, particularly for small mobile units such as cars, has prompted an active search for alternatives. One solution which has many attractive features is the "hydrogen economy" whereby energy is transported and stored in the form of liquid or gaseous hydrogen. Enthusiasts point out that such a major change in the source of energy, though apparently dramatic, is not unprecedented and has in fact occurred twice during the past 100 y. In 1880 wood was overtaken by coal as the main world supplier of energy and now it accounts for only about 2% of the total. Likewise in 1960 coal was itself overtaken by oil and now accounts for only 17% of the total. (Note, however, that this does not imply a decrease in the total amount of coal used: in 1930 this was 14.5×10^6 barrels per day of oil equivalent and was 75% of the then total energy supply whereas in 1975 coal had increased in absolute terms by 11% to 16.2 b/d oe, but this was only 18% of the total energy supply which had itself increased 4.6-fold in the interim.) Another change may well be in the offing since nuclear power, which was effectively non-existent as an industrial source of energy in 1950, now accounts for 3% of the world supply (10% in the UK) and is expected to overtake coal by 1985 and oil perhaps by the turn of the century. The aim of the "hydrogen economy" is to transmit this energy, not as electric power but in the form of hydrogen; this overcomes the great problem of electricity—that it cannot be stored—and also reduces the costs of power transmission.

The technology already exists for producing hydrogen electrically and storing it in bulk. For example huge quantities of liquid hydrogen are routinely stored in vacuum insulated cryogenic tanks for the US space programme, one such tank alone holding over 3400 m^3 (900 000 US gallons). Liquid hydrogen can be transported by road or by rail tankers of 75.7 m^3 capacity (20 000 US gallons). Underground storage of the type currently used for hydrogen—natural gas mixtures and transmission through large pipes is also feasible, and pipelines carrying hydrogen up to 80 km in the USA and South Africa and 200 km in Europe have been in operation for many years. Smaller storage units based on metal alloy systems have also been suggested, e.g. $LaNi_5$ can absorb up to 7 moles of H atoms per mole of $LaNi_5$ at room temperature and 2.5 atm, the density of contained hydrogen being twice that in the liquid element itself. The danger of leaks and explosions is recognized, as it is for petroleum and natural gas, but no accidents have occurred so far, despite extensive use of liquid hydrogen for Saturn V and other rockets in the US space programme.

The advantages claimed for hydrogen as an automobile fuel are the greater energy release per unit weight of fuel and the absence of polluting emissions such as CO, CO_2, NO_x, SO_2, hydrocarbons, aldehydes, and lead compounds. The product of combustion is water with only traces of nitrogen oxides. Several conventional internal-combustion petrol engines have already been simply and effectively modified to run on hydrogen. Fuel cells for the regeneration of electric power have also been successfully operated commercially with a conversion efficiency of 70%, and test cells at higher pressures have achieved 85% efficiency.

Non-electrolytic sources of hydrogen have also been studied. The chemical problem is how to transfer the correct amount of free energy to a water molecule in order to decompose it. In the last few years about 10 000 such thermochemical water-splitting cycles have been identified, most of

Panel continued on p. 47

[7] D. P. GREGORY, The hydrogen economy, Chap. 23 in *Chemistry in the Environment*, Readings from *Scientific American*, 1973, pp. 219–27.
[8] C. A. McAULIFFE, The hydrogen economy, *Chem. in Br.* **9**, 559–63 (1973).
[9] C. MARCHETTI, The hydrogen economy and the chemist, *Chem. in Br.* **13**, 219–22 (1977).
[10] L. B. McGOWN and J. O'M. BOCKRIS, *How to Obtain Abundant Clean Energy*, Plenum, New York, 1980, 275 pp.
[10a] L. O. WILLIAMS, *Hydrogen Power*, Pergamon Press, Oxford, 1980, 158 pp.
[11] ANON., Hydrogen advocates focus on practical goals, *Chem. Eng. News* **56** (33), 28–31 (1978).

Plate 3.1 Hydrogen generator.

them with the help of computers, though it is significant that the most promising ones were discovered first by the intuition of chemists. Two potentially useful sequences are:

Sequence 1

1. $CaBr_2 + H_2O \xrightarrow{750°C} CaO + 2HBr$

2. $Hg + 2HBr \xrightarrow{100°C} HgBr_2 + H_2$

3. $HgBr_2 + CaO \xrightarrow{25°C} HgO + CaBr_2$

4. $HgO \xrightarrow{500°C} Hg + \frac{1}{2}O_2$

Sequence 2

1. $3FeCl_2 + 4H_2O \xrightarrow{500°C} Fe_3O_4 + 6HCl + H_2$

2. $Fe_3O_4 + \frac{3}{2}Cl_2 + 6HCl \xrightarrow{100°C} 3FeCl_3 + 3H_2O + \frac{1}{2}O_2$

3. $3FeCl_3 \xrightarrow{300°C} 3FeCl_2 + \frac{3}{2}Cl_2$

The stage is thus set, and further work to establish safe and economically viable sources of hydrogen for general energy usage seems destined to flourish as an active area of research for some while.

3.3.2 *Deuterium*

Deuterium is invariably prepared from heavy water, D_2O, which is itself now manufactured on the multitonne scale by the electrolytic enrichment of normal water.[12, 13] The enrichment is expressed as a separation factor between the gaseous and liquid phases:

$$s = (H/D)_g / (H/D)_l$$

The equilibrium constant for the exchange reaction

$$H_2O + HD \rightleftharpoons HDO + H_2$$

is about 3 at room temperature and this would lead to a value of $s = 3$ if this were the only effect. However, the choice of the metal used for the electrodes can also affect the various electrode processes, and this increases the separation still further. Using alkaline solutions s values in the range 5–7.6 are obtained for many metals, rising to 13.9 for platinum cathodes and even higher for gold. By operating a large number of cells in cascade, and burning the evolved H_2/D_2 mixture to replenish the electrolyte of earlier cells in the sequence, any desired degree of enrichment can ultimately be attained. Thus, starting with normal water (0.0156% of hydrogen as deuterium) and a separation factor of 5, the deuterium content rises to 10% after the original volume has been reduced by a factor of 2400. Reduction by 66 000 is required for 90% deuterium and by 130 000 for 99% deuterium. If, however, the separation factor is 10, then 99% deuterium can be obtained by

[12] G. VASARU, D. URSU, A. MIHĂILĂ, and P. SZENTGYÖRGYI, *Deuterium and Heavy Water*, Elsevier, Amsterdam, 1975, 404 pp.
[13] H. K. RAE (ed.), *Separation of Hydrogen Isotopes*, ACS Symposium Series No. 68, 1978, 184 pp.

a volume reduction on electrolysis of 22 000. Prior enrichment of the electrolyte to 15% deuterium can be achieved by a chemical exchange between H_2S and H_2O after which a fortyfold volume reduction produces heavy water with 99% deuterium content. Other enrichment processes are now rarely used but include fractional distillation of water (which also enriches ^{18}O), thermal diffusion of gaseous hydrogen, and diffusion of H_2/D_2 through palladium metal.

Many methods have been used to determine the deuterium content of hydrogen gas or water. For H_2/D_2 mixtures mass spectroscopy and thermal conductivity can be used together with gas chromatography (alumina activated with manganese chloride at 77 K). For heavy water the deuterium content can be determined by density measurements, refractive index change, or infrared spectroscopy.

The main uses of deuterium are in tracer studies to follow reaction paths and in kinetic studies to determine isotope effects.[14] A good discussion with appropriate references is in *Comprehensive Inorganic Chemistry*, Vol. 1, pp. 99–116. The use of deuterated solvents is widespread in proton nmr studies to avoid interference from solvent hydrogen atoms, and deuteriated compounds are also valuable in structural studies involving neutron diffraction techniques.

3.3.3   *Tritium*[15]

Tritium differs from the other 2 isotopes of hydrogen in being radioactive and this immediately indicates its potential uses and its method of detection. Tritium occurs naturally to the extent of about 1 atom per 10^{18} hydrogen atoms as a result of nuclear reactions induced by cosmic rays in the upper atmosphere:

$$^{14}_{7}N + {}^{1}_{0}n = {}^{3}_{1}H + {}^{12}_{6}C$$

$$^{14}_{7}N + {}^{1}_{1}H = {}^{3}_{1}H + \text{fragments}$$

$$^{2}_{1}H + {}^{2}_{1}H = {}^{3}_{1}H + {}^{1}_{1}H$$

The concentration of tritium increased by over a hundredfold when thermonuclear weapon testing began in March 1954 but is now subsiding again as a result of the ban on atmospheric weapon testing and the natural radioactivity of the isotope ($t_{\frac{1}{2}}$ 12.35 y).

Numerous reactions are available for the artificial production of tritium and it is now made on a large scale by neutron irradiation of enriched ^6Li in a nuclear reactor:

$$^{6}_{3}Li + {}^{1}_{0}n = {}^{4}_{2}He + {}^{3}_{1}H$$

The lithium is in the form of an alloy with magnesium or aluminium which retains much of the tritium until it is released by treatment with acid. Alternatively the tritium can be produced by neutron irradiation of enriched LiF at 450° in a vacuum and then recovered from the gaseous products by diffusion through a palladium barrier. As a result of the massive production of tritium for thermonuclear devices and research into energy production by fusion reactions, tritium is available cheaply on the megacurie scale for

[14] L. MELANDER and W. H. SAUNDERS, *Reaction Rates of Isotopic Molecules*, Wiley, New York, 1980, 331 pp.

[15] E. A. EVANS, *Tritium and its Compounds*, 2nd edn., Butterworths, London, 1974, 840 pp. The definitive book on tritium with over 4000 references; mostly concerned with the preparation, safe handling, and uses of tritium-labelled organic compounds.

peaceful purposes.† The most convenient way of storing the gas is to react it with finely divided uranium to give UT_3 from which it can be released by heating above 400°C.

Besides being one of the least expensive radioisotopes, tritium has certain unique advantages as a tracer. Like ^{14}C it is a pure low-energy β^- emitter with no associated γ-rays. The radiation is stopped by ~ 6 mm of air or ~ 6 μm of material of density 1 g cm^{-3} (e.g. water). As the range is inversely proportional to the density, this is reduced to only ~ 1 μm in photographic emulsion ($\rho \sim 3.5$ g cm^{-3}) thus making tritium ideal for high-resolution autoradiography. Moreover, tritium has a high specific activity. The weight of tritium equal to an activity of 1 Ci is 0.103 mg and 1 mmol T_2 has an activity of 58.25 Ci. Tritium is one of the least toxic of radiosiotopes and shielding is unnecessary; however, precautions must be taken against ingestion, and no work should be carried out without appropriate statutory authorization and adequate radiochemical facilities.

Tritium has been used extensively in hydrological studies to follow the movement of ground waters and to determine the age of various bodies of water. It has also been used to study the adsorption of hydrogen and the hydrogenation of ethylene on a nickel catalyst and to study the absorption of hydrogen in metals. Autoradiography has been used extensively to study the distribution of tritium in multiphase alloys, though care must be taken to correct for the photographic darkening caused by emanated tritium gas. Increasing use is also being made of tritium as a tracer for hydrogen in the study of reaction mechanisms and kinetics and in work on homogeneous catalysis.

The production of tritium-labelled organic compounds was enormously facilitated by K. E. Wilzbach's discovery in 1956 that tritium could be introduced merely by storing a compound under tritium gas for a few days or weeks: the β^- radiation induces exchange reactions between the hydrogen atoms in the compound and the tritium gas. The excess of gas is recovered for further use and the tritiated compound is purified chromatographically. Another widely used method of general applicability is catalytic exchange in solution using either a tritiated solution or tritium gas. This is valuable for the routine production of tritium compounds in high radiochemical yield and at high specific activity (> 50 mCi mmol^{-1}). For example, although ammonium ions exchange relatively slowly with D_2O, tritium exchange equilibria are established virtually instantaneously: tritiated ammonium salts can therefore be readily prepared by dissolving the salt in tritiated water and then removing the water by evaporation:

$$(NH_4)_2SO_4 + HTO \rightleftharpoons (NH_3T)_2SO_4 + H_2O, \text{ etc.}$$

For exchange of non-labile organic hydrogen atoms, acid–base catalysis, or some other catalytic hydrogen-transfer agent such as palladium or platinum, is required. The method routinely gives tritiated products having a specific activity almost 1000 times that obtained by the Wilzbach method; shorter times are required (2–12 h) and subsequent purification is easier.

When specifically labelled compounds are required, direct chemical synthesis may be necessary. The standard techniques of preparative chemistry are used, suitably modified for small-scale work with radioactive materials. The starting material is tritium gas which can be obtained at greater than 98% isotopic abundance. Tritiated water can be made

† See also p. 21 for the influence on the atomic weight of commercially available lithium in some countries.

either by catalytic oxidation over palladium or by reduction of a metal oxide:

$$2T_2 + O_2 \xrightarrow{\text{Pd}} 2T_2O$$

$$T_2 + CuO \longrightarrow T_2O + Cu$$

Note, however, that pure tritiated water is virtually never used since 1 ml would contain 2650 Ci; it is self-luminescent, irradiates itself at the rate of 6×10^{17} eV ml^{-1} s^{-1} ($\sim 10^9$ rad day^{-1}), undergoes rapid self-radiolysis, and also causes considerable radiation damage to dissolved species. In chemical syntheses or exchange reactions tritiated water of 1% tritium abundance (580 mCi mmol^{-1}) is usually sufficient to produce compounds having a specific activity of at least 100 mCi mmol^{-1}. Other useful synthetic reagents are NaT, LiAlH$_3$T, NaBH$_3$T, NaBT$_4$, B$_2$T$_6$, and tritiated Grignard reagents. Typical preparations are as follows:

$$LiH + T_2 \xrightarrow{350°C} LiT + HT$$

$$LiBH_4 + T_2 \xrightarrow{200°C} LiBH_3T + HT$$

$$4LiT + AlCl_3 \xrightarrow{\text{ether}} LiAlT_4 + 3LiCl$$

$$B_2H_6 + T_2 \xrightarrow{55°C} B_2H_5T + HT$$

$$3NaBT_4 + 4BF_3 \cdot OEt_2 \longrightarrow 2B_2T_6 + 3NaBF_4$$

$$3T_2O(g) + P(CN)_3 \longrightarrow 3TCN + T_3PO_3$$

$$2AgCl + T_2 \xrightarrow{700°C} 2TCl + 2Ag$$

$$Br_2 + T_2 \xrightarrow{\text{uv}} 2TBr$$

$$P(\text{red}) + I_2 + HTO \longrightarrow TI + HI + \dots$$

$$NH_3 + T_2 \longrightarrow NH_2T + TH$$

$$Mg_3N_2 + 6T_2O \xrightarrow{100°C} 2NT_3 + 3Mg(OT)_2$$

$$M + \frac{x}{2}T_2 \xrightarrow{\text{heat}} MT_x$$

3.4 Chemical Properties and Trends

Hydrogen is a colourless, tasteless, odourless gas which has only low solubility in liquid solvents. It is comparatively unreactive at room temperature though it combines with fluorine even in the dark and readily reduces aqueous solutions of palladium(II) chloride:

$$PdCl_2(aq) + H_2 \longrightarrow Pd(s) + 2HCl(aq)$$

This reaction can be used as a sensitive test for the presence of hydrogen. At higher temperatures hydrogen reacts vigorously, even explosively, with many metals and non-metals to give the corresponding hydrides. Activation can also be induced photolytically, by heterogeneous catalysts (Raney nickel, Pd, Pt, etc.), or by means of homogeneous hydrogenation catalysts. Industrially important processes include the hydrogenation of many organic compounds and the use of cobalt compounds as catalysts in the hydroformylation of olefins to aldehydes and alcohols at high temperatures and pressures (p. 1324):

$$RCH{=}CH_2 + H_2 + CO \longrightarrow RCH_2CH_2CHO \xrightarrow{\text{H}_2} RCH_2CH_2CH_2OH$$

An even more effective homogeneous hydrogenation catalyst is the complex $[RhCl(PPh_3)_3]$ which permits rapid reduction of alkenes, alkynes, and other unsaturated compounds in benzene solution at 25°C and 1 atm pressure (p. 1318). The Haber process, which uses iron metal catalysts for the direct synthesis of ammonia from nitrogen and hydrogen at high temperatures and pressures, is a further example (p. 482).

The hydrogen atom has a unique electronic configuration $1s^1$: accordingly it can gain an electron to give H^- with the helium configuration $1s^2$ or it can lose an electron to give the proton H^+ (p. 42). There are thus superficial resemblances both to the halogens which can gain an electron to give an inert-gas configuration ns^2np^6, and to the alkali metals which can lose an electron to give M^+ (ns^2np^6). However, because hydrogen has no other electrons in its structure there are sufficient differences from each of these two groups to justify placing hydrogen outside either. For example, the proton is so small ($r \sim 1.5 \times 10^{-3}$ pm compared with normal atomic and ionic sizes of ~ 50–220 pm) that it cannot exist in condensed systems unless associated with other atoms or molecules. The transfer of protons between chemical species constitutes the basis of acid–base phenomena (see Section 3.4.1). The hydrogen atom is also frequently found in close association with 2 other atoms in linear array; this particularly important type of interaction is called hydrogen bonding (see Section 3.5). Again, the ability to penetrate metals to form nonstoichiometric metallic hydrides, though not unique to hydrogen, is one of its more characteristic properties as is its ability to form nonlinear hydrogen bridge bonds in many of its compounds. These properties will emerge during the general classification of the hydrides of the elements in section 3.6.

3.4.1 *Protonic acids and bases*[3]

Many compounds that contain hydrogen can donate protons to a solvent such as water and so behave as acids. Water itself undergoes ionic dissociation to a small extent by means of autoprotolysis; the process is usually represented formally by the equilibrium

$$H_2O + H_2O \rightleftharpoons H_3O^+ + OH^-$$

though it should be remembered that both ions are further solvated and that the time a proton spends in close association with any one water molecule is probably only about 10^{-13} s. Depending on what aspect of the process is being emphasized, the species H_3O^+(aq) can be called an oxonium ion, a hydrogen ion, or simply a solvated (hydrated) proton. The equilibrium constant for autoprotolysis is

$$K_1 = [H_3O^+][OH^-]/[H_2O]^2$$

and, since the concentration of water is essentially constant, the ionic product of water can be written as

$$K_w = [H_3O^+] [OH^-] \text{ mol}^2 \, l^{-2}$$

The value of K_w depends on the temperature, being $0.69 \times 10^{-14} \text{ mol}^2 \, l^{-2}$ at $0°C$, 1.00×10^{-14} at $25°C$, and 47.6×10^{-14} at $100°C$. It follows that the hydrogen-ion concentration in pure water at $25°C$ is $10^{-7} \text{ mol } l^{-1}$. Acids increase this concentration by means of the reaction

$$HA + H_2O \rightleftharpoons H_3O^+ + A^-; \quad K = \frac{[H_3O^+] [A^-]}{[HA] [H_2O]}$$

The symbol HA implies only that the species can act as a proton donor: it can be a neutral species (e.g. H_2S), an anion (e.g. $H_2PO_4^-$) or a cation such as $[Fe(H_2O)_6]^{3+}$. The hydrogen-ion concentration is usually expressed as pH (see Panel). In dilute solution the concentration of water molecules is constant at $25°C$ ($55.345 \text{ mol } l^{-1}$), and the dissociation of the acid is often rewritten as

$$HA \rightleftharpoons H^+ + A^-; \quad K_a = [H^+] [A^-]/[HA] \text{ mol } l^{-1}$$

The acid constant K_a can also be expressed by the relation

$$pK_a = -\log K_a$$

Hence, as $K_a = 55.345 \, K$, $pK_a = pK - 1.734$.

Further, as the free energy of dissociation is given by

$$\Delta G° = -RT\ln K = -2.3026RT\log K,$$

the standard free energy of dissociation is

$$\Delta G°_{298.15} = 5.708pK = 5.708(pK_a + 1.734) \text{ kJ mol}^{-1}$$

Textbooks of analytical chemistry should be consulted for further details concerning the ionization of weak acids and bases and the theory of indicators, buffer solutions, and acid–alkali titrations.[16–18]

Various trends have long been noted in the acid strengths of many binary hydrides and oxoacids.[3] Values for some simple hydrides are given in Table 3.3 from which it is clear

TABLE 3.3   *Approximate values of pK_a for simple hydrides*

CH_4	46	NH_3	35	OH_2	16	FH	3
		PH_3	27	SH_2	7	ClH	−7
				SeH_2	4	BrH	−9
				TeH_2	3	IH	−10

[16] A. I. VOGEL, *Quantitative Inorganic Analysis*, 3rd edn., Sections 1.27–1.38, pp. 51–72. Longmans, London, 1964.

[17] R. A. DAY and A. L. UNDERWOOD, *Quantitative Analysis*, 2nd edn., Chap. 4, pp. 74–124, Prentice-Hall, Englewood Cliffs, 1967.

[18] D. ROSENTHAL and P. ZUMAN, Acid–base equilibria, buffers, and titrations in water, Chap. 18 in I. M. KOLTHOFF and P. J. ELVING (eds.), *Treatise on Analytical Chemistry*, 2nd edn., Vol. 2, Part 1, 1979, pp. 157–236. Succeeding chapters (pp. 237–440) deal with acid–base equilibria and titrations in non-aqueous solvents.

The Concept of pH

The now universally used measure of the hydrogen-ion concentration was introduced in 1909 by the Danish biochemist S. P. L. Sørensen during his work at the Carlsberg Breweries (*Biochem. Z.* **21**, 131, 1909):

$$pH = -\log[H^+]$$

The symbol pH derives from the French *puissance d'hydrogène*, referring to the exponent or "power of ten" used to express the concentration. Thus a hydrogen-ion concentration of 10^{-7} mol l^{-1} is designated pH 7, whilst acid solutions with higher hydrogen-ion concentrations have a lower pH. For example, a strong acid of concentration 1 mmol l^{-1} has pH 3, whereas a strong alkali of the same concentration has pH 11 since $[H_3O^+] = 10^{-14}/[OH^-] = 10^{-11}$.

Unfortunately, it is far simpler to define pH than to measure it, despite the commercial availability of instruments that purport to do this. Most instruments use an electrochemical cell such as

$$\text{glass electrode}|\text{test solution}|3.5\ \text{M KCl(aq)}|Hg_2Cl_2|Hg$$

Assuming that the glass electrode shows an ideal hydrogen electrode response, the emf of the cell still depends on the magnitude of the liquid junction potential E_j and the activity coefficients γ of the ionic species:

$$E = E^\circ - \frac{RT}{F}\ln \gamma_{Cl}[Cl^-] + E_j - \frac{RT}{F}\ln \gamma_H[H^+]$$

For this reason, the pH as measured by any of the existing national standards is an operational quantity which has no simple fundamental significance. It is defined by the equation

$$pH(X) = pH(S) + \frac{(E_X - E_S)F}{RT \ln 10}$$

where pH(S) is the *assigned* pH of a standard buffer solution such as those supplied with pH meters.

Only in the case of dilute aqueous solutions (<0.1 mol l^{-1}) which are neither strongly acid or alkaline ($2 < pH < 12$) is pH(X) such that

$$pH(X) = -\log[H^+]\gamma_\pm \pm 0.02$$

where $\gamma_\pm$, the mean ionic activity coefficient of a typical uni-univalent electrolyte, is given by

$$-\log \gamma_\pm = AI^{\frac{1}{2}}(1+I)^{-\frac{1}{2}}$$

In this expression I is the ionic strength of the solution and A is a temperature-dependent constant ($0.51\ l^{\frac{1}{2}}$ mol$^{-\frac{1}{2}}$ at 25°C; $0.50\ l^{\frac{1}{2}}$ mol$^{-\frac{1}{2}}$ at 15°C). It is clearly unwise to associate a pH meter reading too closely with pH unless under very controlled conditions, and still less sensible to relate the reading to the actual hydrogen-ion concentration in solution. For further discussion of pH measurements, see ref. 19.

that acid strength increases with atomic number both in any one horizontal period and in any vertical group. Several attempts have been made to interpret these trends, at least qualitatively, but the situation is complex. The trend to increasing acidity from left to right in the periodic table could be ascribed to the increasing electronegativity of the elements which would favour release of the proton, but this is clearly not the dominant effect within any one group since the trend there is in precisely the opposite direction. Within a group it is the diminution in bond strength with increasing atomic number that prevails, and

[19] C. C. WESTCOTT, *pH Measurements*, Academic Press, New York, 1978, 172 pp. IUPAC Subcommittee on Calibration and Test Materials: Potentiometric ion activities (pH scales and materials), *Pure Appl. Chem.* **50**, 1485–1517 (1978). [More recent discussion on the question of a single primary standard is given in *Chemistry International* (IUPAC) 1980 (No. 6), pp. 23–25.]

Hydrogen

entropies of solvation are also important. It should, perhaps, be emphasized that thermodynamic computations do not "explain" the observed acid strengths; they merely allocate the overall values of ΔG, ΔH, and ΔS to various notional subprocesses such as bond dissociation energies, ionization energies, electron affinities, heats and entropies of hydration, etc., which themselves have empirically observed values that are difficult to compute *ab initio*.

Regularities in the observed strengths of oxoacids have been formulated in terms of two rules by L. Pauling and others:

    (i) for polybasic mononuclear oxoacids, successive acid dissociation constants diminish approximately in the ratios $1:10^{-5}:10^{-10}:\dots$;

    (ii) the value of the first ionization constant for acids of formula $XO_m(OH)_n$ depends sensitively on m but is approximately independent of n and X for constant m, being $\leqslant 10^{-8}$ for $m=0$, $\sim 10^{-2}$ for $m=1$, $\sim 10^3$ for $m=2$, and $>10^8$ for $m=3$.

Thus to illustrate the first rule:

$$H_3PO_4 \rightleftharpoons H^+ + H_2PO_4^-; \quad K_1 = \frac{[H^+][H_2PO_4^-]}{[H_3PO_4]} = 7.11 \times 10^{-3} \text{ mol l}^{-1}; \ pK_1 = 2.15$$

$$H_2PO_4^- \rightleftharpoons H^+ + HPO_4^{2-}; \quad K_2 = \frac{[H^+][HPO_4^{2-}]}{[H_2PO_4^-]} = 6.31 \times 10^{-8} \text{ mol l}^{-1}; \ pK_2 = 7.20$$

$$HPO_4^{2-} \rightleftharpoons H^+ + PO_4^{3-}; \quad K_3 = \frac{[H^+][PO_4^{3-}]}{[HPO_4^{2-}]} = 4.22 \times 10^{-13} \text{ mol l}^{-1}; \ pK_3 = 12.37$$

Qualitatively, a reduction in pK_a for each successive stage of ionization is to be expected since the proton must separate from an anion of increasingly negative charge, though the approximately constant reduction factor of 10^5 is more difficult to rationalize quantitatively.

Acids which illustrate the second rule are summarized in Table 3.4. The qualitative

TABLE 3.4   *Values of pK_a for some mononuclear oxoacids $XO_m(OH)_n$*
$(pK_a \approx 8{-}5m)$

$X(OH)_n$ (very weak)		$XO(OH)_n$ (weak)		$XO_2(OH)_n$ (strong)		$XO_3(OH)_n$ (very strong)	
$Cl(OH)$	7.2	$NO(OH)$	3.3	$NO_2(OH)$	-1.4	$ClO_3(OH)$	(-10)
$Br(OH)$	8.7	$ClO(OH)$	2.0	$ClO_2(OH)$	-1	$MnO_3(OH)$	—
$I(OH)$	10.0	$CO(OH)_2$	$3.9^{(a)}$	$IO_2(OH)$	0.8		
$B(OH)_3$	9.2	$SO(OH)_2$	1.9	$SO_2(OH)_2$	<0		
$As(OH)_3$	9.2	$SeO(OH)_2$	2.6	$SeO_2(OH)_2$	<0		
$Sb(OH)_3$	11.0	$TeO(OH)_2$	2.7				
$Si(OH)_4$	10.0	$PO(OH)_3$	2.1				
$Ge(OH)_4$	8.6	$AsO(OH)_3$	2.3				
$Te(OH)_6$	8.8	$IO(OH)_5$	1.6				
		$HPO(OH)_2$	$1.8^{(b)}$				
		$H_2PO(OH)$	$2.0^{(b)}$				

(a) Corrected for the fact that only 0.4% of dissolved CO_2 is in the form of H_2CO_3; the conventional value is pK_a 6.5.

(b) Note that the value of pK_a for hypophosphorous acid H_3PO_3 is consistent with its (correct) formulation as $HPO(OH)_2$ rather than as $P(OH)_3$, which would be expected to have $pK_a > 8$. Similarly for H_3PO_2, which is $H_2PO(OH)$ rather than $HP(OH)_2$.

explanation for this regularity is that, with increasing numbers of oxygen atoms the single negative charge on the anion can be spread more widely, thereby reducing the electrostatic energy attracting the proton and facilitating the ionization. On this basis one might expect an even more dramatic effect if the anion were monoatomic (e.g. S^{2-}, Se^{2-}, Te^{2-}) since the attraction of these dianions for protons will be very strong and the acid dissociation constant of SH^-, SeH^-, and TeH^- correspondingly small; this is indeed observed and the ratio of the first and second dissociation constants is $\sim 10^8$ rather than 10^5 (Table 3.5).

TABLE 3.5   *First and second ionization constants for H_2S,*
H_2Se, and H_2Te

	pK_1	pK_2	ΔpK
H_2S	7	14	7
H_2Se	4	12	8
H_2Te	3	11	8

The results for dinuclear and polynuclear oxoacids are also consistent with this interpretation. Thus for phosphoric acid, $H_4P_2O_7$, the successive pK_a values are 1.5, 2.4, 6.6, and 9.2; the ~ 10-fold decrease between pK_1 and pK_2 (instead of a decrease of 10^5) is related to the fact that ionization occurs from two different PO_4 units. The third stage ionization, however, is $\sim 10^5$ less than the first stage and the difference between the mean of the first two and the last two ionization constants is $\sim 5 \times 10^5$.

Another phenomenon that is closely associated with acid–base equilibria is the so-called hydrolysis of metal cations in aqueous solution, which is probably better considered as the protolysis of hydrated cations, e.g.:

"hydrolysis": $Fe^{3+} + H_2O \rightleftharpoons [Fe(OH)]^{2+} + H^+$

protolysis: $[Fe(H_2O)_6]^{3+} + H_2O \rightleftharpoons [Fe(H_2O)_5(OH)]^{2+} + H_3O^+$; pK_a 3.05

$[Fe(H_2O)_5(OH)]^{2+} + H_2O \rightleftharpoons [Fe(H_2O)_4(OH)_2]^+ + H_3O^+$; pK_a 3.26

It is these reactions that impart the characteristic yellow to reddish-brown coloration of the hydroxoaquo species to aqueous solutions of iron(III) salts, whereas the undissociated ion $[Fe(H_2O)_6]^{3+}$ is pale mauve, as seen in crystals of iron(III) alum $\{[Fe(H_2O)_6][K(H_2O)_6](SO_4)_2\}$ and iron(III) nitrate $\{[Fe(H_2O)_6](NO_3)_3 \cdot 3H_2O\}$. Such reactions may proceed to the stage where the diminished charge on the hydrated cation permits the formation of oxobridged, or hydroxobridged polynuclear species that eventually precipitate as hydrous oxides (see discussion of the chemistry of many elements in later chapters). A useful summary is in Fig. 3.2. By contrast, extensive studies of the pK_a values of hydrated metal ions in solution has generated a wealth of numerical data but no generalizations such as those just discussed for the hydrides and oxoacids of the non-metals.[20] Typical pK_a values fall in the range 3–14 and, as expected, there is a general tendency for protolysis to be greater (pK_a values to be lower) the higher the cationic charge. For example, aqueous solutions of iron(III) salts are more acidic than solutions of the corresponding iron(II) salts. However, it is difficult to discern any regularities in pK_a

[20] L. G. SILLÉN, *Q. Rev.* (*London*) **13**, 146–68 (1969); *Pure Appl. Chem.* **17**, 55–78 (1968).

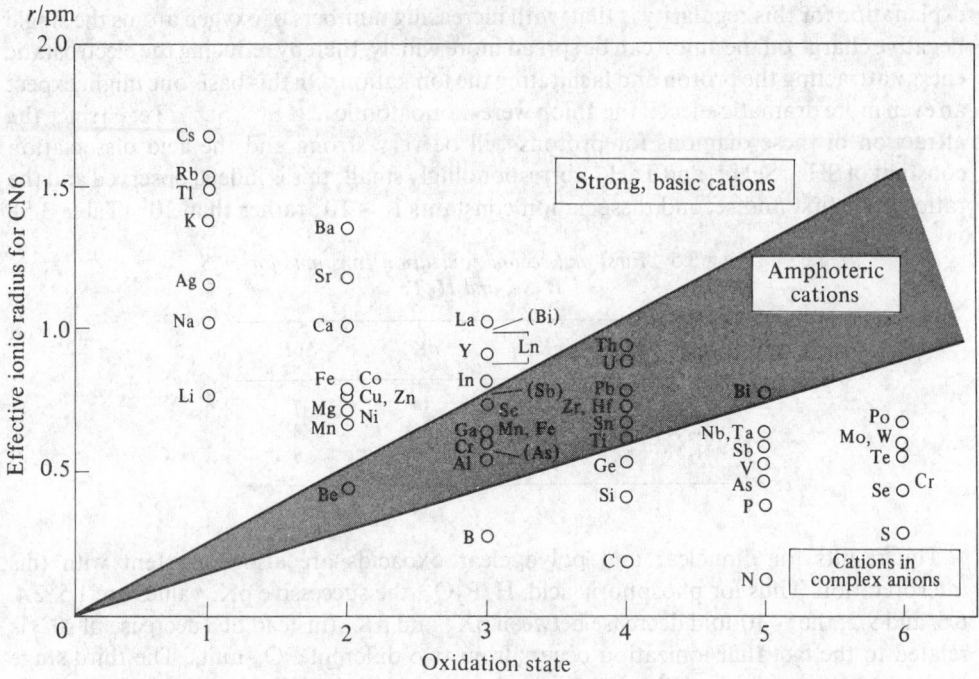

FIG. 3.2 Plot of effective ionic radii *versus* oxidation state for various elements.

for series of cations of the same ionic charge, and it is clear that specific "chemical" effects must also be considered.

Brønsted acidity is not confined to dilute aqueous solutions and the ideas developed in the preceding pages can be extended to proton donors in nonaqueous solutions.[21] In organic solvents and anhydrous protonic liquids the concepts of hydrogen-ion concentration and pH, if not actually meaningless, are certainly operationally in-applicable and acidity must be defined on some other scale. The one most frequently used is the Hammett acidity function H_0 which enables various acids to be compared in a given solvent and a given acid to be compared in various solvents. For the equilibrium between a base and its conjugate acid (frequently a coloured indicator)

$$B + H^+ \rightleftharpoons BH^+$$

the acidity function is defined as

$$H_0 = pK_{BH^+} - \log\{[BH^+]/[B]\}$$

In very dilute solutions

$$K_{BH^+} = [B][H^+]/[BH^+]$$

so that in water H_0 becomes the same as pH. Some values for typical anhydrous acids are in Table 3.6 and these are discussed in more detail in appropriate sections of later chapters.

[21] C. H. ROCHESTER, *Acidity Functions*, Academic Press, London, 1970, 300 pp.

TABLE 3.6   *Hammett acidity functions for some anhydrous acids*

Acid	H_0	Acid	H_0
$HSO_3F + SbF_5 + SO_3$	> 16	HF	10.2
$HF + SbF_5(3M)$	15.2	H_3PO_4	5.0
HSO_3F	12.6	$H_2SO_4(63\%$ in $H_2O)$	4.9
H_2SO_4	11.0	HCO_2H	2.2

It will be noted that addition of SbF_5 to HF considerably enhances its acidity and the same effect can be achieved by other fluoride acceptors such as BF_3 and TaF_5:

$$2HF + MF_n \rightleftharpoons H_2F^+ + MF_{n+1}^-$$

The enhancement of the acidity of HSO_3F by the addition of SbF_5 is more complex and the equilibria involved are discussed on p. 664.

3.5   The Hydrogen Bond[22, 23, 23a]

The properties of many substances suggest that, in addition to the "normal" chemical bonding between the atoms and ions, there exists some further interaction involving a hydrogen atom placed between two or more other groups of atoms. Such interaction is called hydrogen bonding and, though normally weak (10–60 kJ per mol of H bonds), it frequently has a decisive influence on the structure and properties of the substance. A hydrogen bond can be said to exist between 2 atoms A and B when these atoms approach more closely than would otherwise be expected in the absence of the hydrogen atom and when, as a result, the system has a lower total energy. The bond is represented as A–H$\cdots$B and usually occurs when A is sufficiently electronegative to enhance the acidic nature of H (proton donor) and where the acceptor B has a region of high electron density (such as a lone pair of electrons) which can interact strongly with the acidic hydrogen. It will be convenient first to indicate the range of phenomena which are influenced by H bonding and then to discuss more specifically the nature of the bond itself according to current theories. The experimental evidence suggests that strong H bonds can be formed when A is F, O, or N; weaker H bonds are sometimes formed when A is C or a second row element, P, S, Cl, or even Br, I. Strong H bonds are favoured when the atom B is F, O, or N (but never C); the other halogens Cl, Br, I are less effective unless negatively charged and the atoms S and P can also sometimes act as B in weak H bonds.

3.5.1   *Influence on properties*

It is well known that the mps and bps of NH_3, H_2O, and HF are anomalously high when compared with the mps and bps of the hydrides of other elements in Groups V, VI, and

[22] G. C. PIMENTEL and A. L. MCCLELLAN, *The Hydrogen Bond*, W. H. Freeman, San Francisco, 1960, 475 pp. A definitive review with a comprehensive bibliography of 2242 references.
[23] W. C. HAMILTON and J. A. IBERS, *Hydrogen Bonding in Solids*, W. A. Benjamin, New York, 1968, 284 pp.
[23a] J. EMSLEY, Very strong hydrogen bonding, *Chem. Soc. Revs.* **9**, 91–124 (1980).

VII, and the same effect is noted for the heats of vaporization, as shown in Fig. 3.3. The explanation normally given is that there is some residual interaction (H bonding) between the molecules of NH_3, H_2O, and HF which is absent for methane, and either absent or much weaker for heavier hydrides. This argument is probably correct in outline but is deceptively oversimplified since it depends on the assumption that only some of the H bonds in solid HF (for example) are broken during the melting process and that others are broken on vaporization, though not all, since HF is known to be substantially polymerized even in the gas phase. The mp is the temperature at which there is zero free-energy change on passing from the solid to the liquid phase:

$$\Delta G_m = \Delta H_m - T_m \Delta S_m = 0; \quad \text{hence} \quad T_m = \Delta H_m / \Delta S_m$$

It can be seen that a high mp implies either a high enthalpy of melting, or a low entropy of melting, or both. Similar arguments apply to vaporization and the bp, and indicate the difficulties in quantifying the discussion.

Other properties that are influenced by H bonding are solubility and miscibility, heats

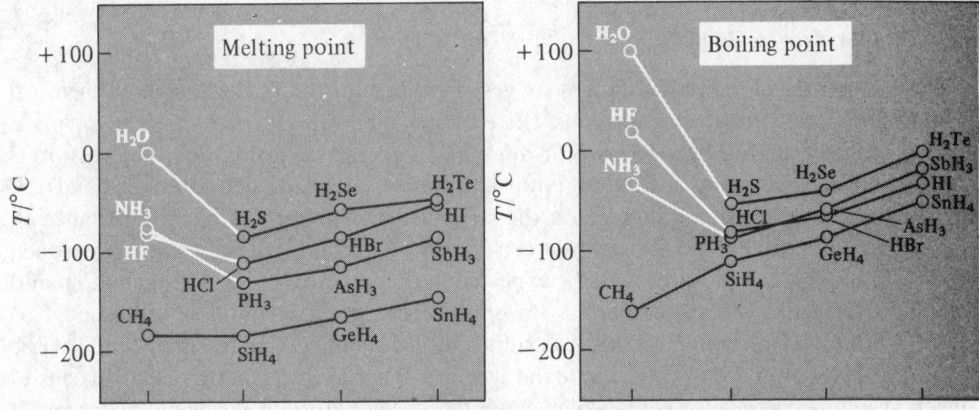

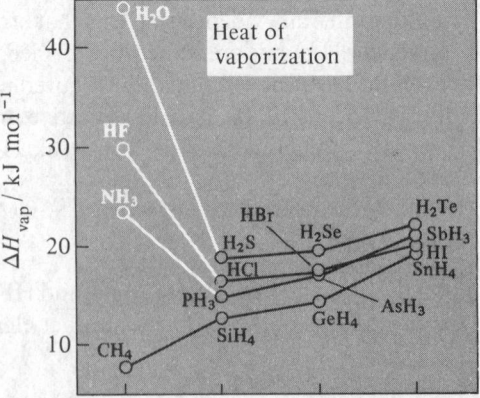

FIG. 3.3   Plots showing the high values of mp, bp, and heat of vaporization of NH_3, H_2O, and HF when compared with other hydrides. Note also that the mp of CH_4 ($-182.5°C$) is slightly higher than that of SiH_4 ($-185°C$).

of mixing, phase-partitioning properties, the existence of azeotropes, and the sensitivity of chromatographic separation. Liquid crystals (or mesophases) which can be regarded as "partly melted" solids also frequently involve molecules that have H-bonded groups (e.g. cholesterols, polypeptides, etc.). Again, H bonding frequently results in liquids having a higher density and lower molar volume than would otherwise have been expected, and viscosity is also affected (e.g. glycerol, anhydrous H_2SO_4, H_3PO_4, etc.).

Electrical properties of liquids and solids are sometimes crucially influenced by H bonding. The ionic mobility and conductance of H_3O^+ and OH^- in aqueous solutions are substantially greater than those of other univalent ions due to a proton-switch mechanism in the H-bonded associated solvent, water. For example, at 25°C the conductance of H_3O^+ and OH^- are 350 and 192 ohm^{-1} cm^2 mol^{-1}, whereas for other (viscosity-controlled) ions the values fall mainly in the range 50–75 ohm^{-1} cm^2 mol^{-1}. (To convert to mobility, v cm^2 s^{-1} V^{-1}, divide by 96 485 C mol^{-1}.) It is also notable that the dielectric constant is not linearly related to molecular dipole moments for H-bonded liquids being much higher due to the orientating effect of the H bonds: large domains are able to align in an applied electric field so that the molecular dipoles reinforce one another rather than cancelling each other due to random thermal motion. Some examples are given in Fig. 3.4, which also illustrates the substantial influence of temperature on the dielectric constant of

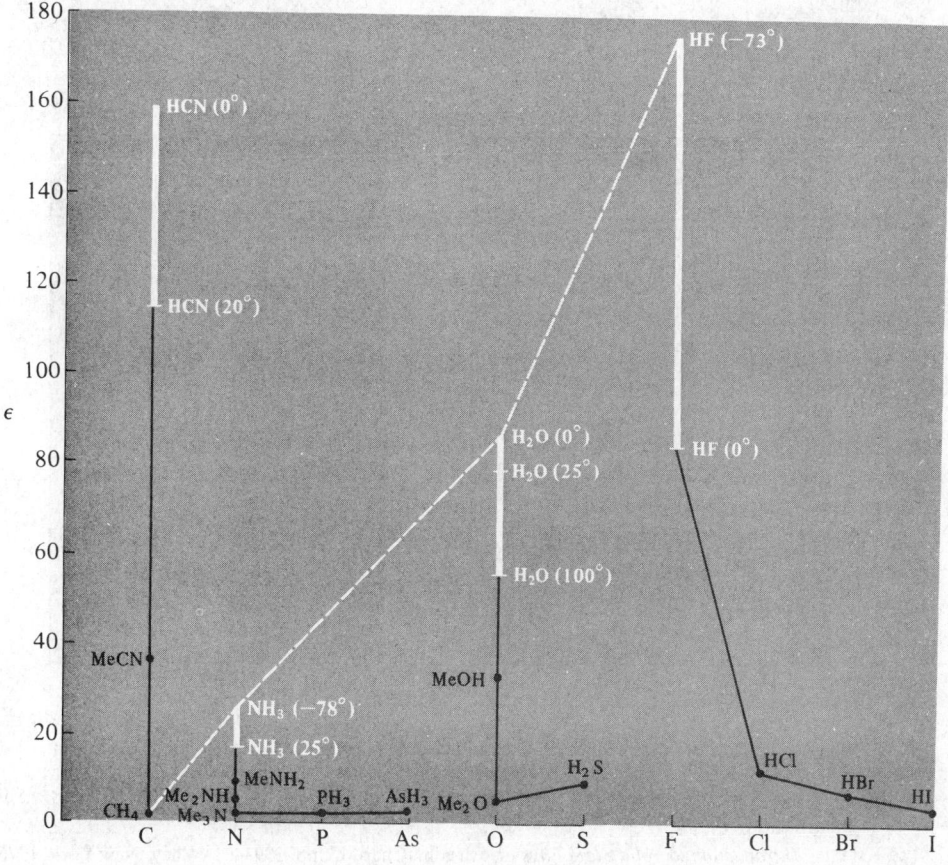

Fig. 3.4 Dielectric constant of selected liquids.

H-bonded liquids presumably due to the progressive thermal dissociation of the H bonds. Even more dramatic are the properties of ferroelectric crystals where there is a stable permanent electric polarization. Hydrogen bonding is one of the important ordering mechanisms responsible for this phenomenon as discussed in more detail in the Panel.

Ferroelectric Crystals[24, 25]

A ferroelectric crystal is one that has an electric dipole moment even in the absence of an external electric field. This arises because the centre of positive charge in the crystal does not coincide with the centre of negative charge. The phenomenon was discovered in 1920 by J. Valasek in Rochelle salt, which is the H-bonded hydrated d-tartrate $NaKC_4H_4O_6.4H_2O$. In such compounds the dipole moment can rise to enormous values of 10^3 or more due to presence of a stable permanent electric polarization (see figure). Before considering the effect further, it will be helpful to recall various definitions and SI units:

electric polarization $P = D - \varepsilon_0 E$ (C m^{-2})
where D is the electric displacement (C m^{-2})
$\quad$ E is the electric field strength (V m^{-1})
$\quad$ ε_0 is the permittivity of vacuum (F m^{-1} = A s V^{-1} m^{-1})

dielectric constant $\varepsilon = \dfrac{\varepsilon_0 E + P}{\varepsilon_0 E} = 1 + \chi$ (dimensionless)

where $\chi = \varepsilon - 1 = P/\varepsilon_0 E$ is the dielectric susceptibility.

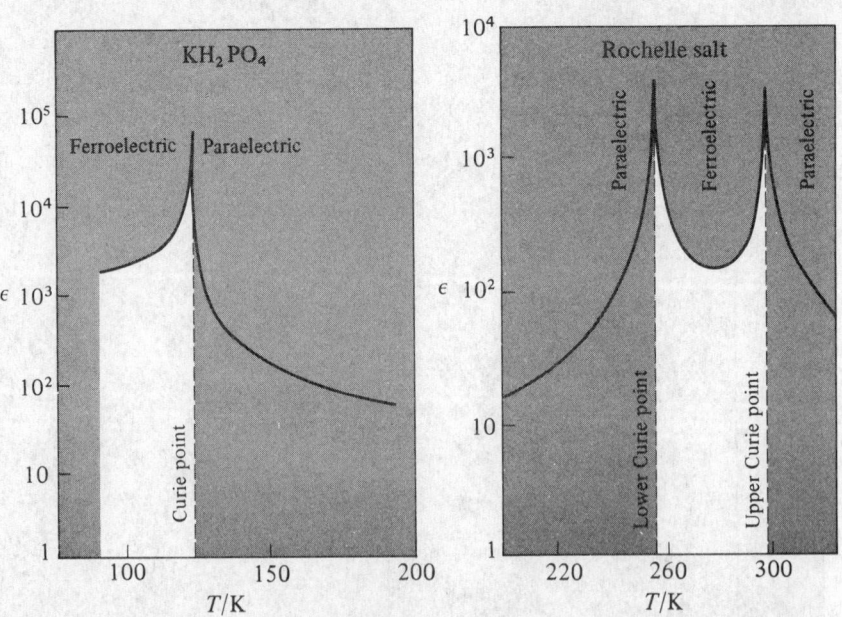

Anomalous temperature dependence of relative dielectric constant of ferroelectrics at transition temperature.

Panel continues

[24] C. KITTELL, *Introduction to Solid State Physics*, 5th edn., Chap. 13, pp. 399–431. Wiley, New York, 1976.
[25] J. C. BURFOOT, *Ferroelectrics*, D, van Nostrand, London, 1967, 261 pp.

There are two main types of ferroelectric crystal:

(a) those in which the polarization arises from an ordering process, typically by H bonding;
(b) those in which the polarization arises by a displacement of one sublattice with respect to another, as in perovskite-type structures like barium titanate (p. 1122).

The ferroelectricity usually disappears above a certain transition temperature (often called a Curie temperature) above which the crystal is said to be paraelectric; this is because thermal motion has destroyed the ferroelectric order. Occasionally the crystal melts or decomposes before the paraelectric state is reached. There are thus some analogies to ferromagnetic and paramagnetic compounds though it should be noted that there is no iron in ferroelectric compounds. Some typical examples, together with their transition temperatures and spontaneous permanent electric polarization P_s, are given in the Table.

TABLE *Properties of some ferroelectric compounds*

Compound	T_c/K	$P_s/\mu C\ cm^{-2}$ [a]	(at T/K)
KH_2PO_4	123	5.3	(96)
KD_2PO_4	213	4.5	—
KH_2AsO_4	96	3.0	(80)
KD_2AsO_4	162		
RbH_2PO_4	147	5.6	(90)
$(NH_2CH_2CO_2H)_3 . H_2SO_4$ [b]	322	2.8	(293)
$(NH_2CH_2CO_2H)_3 . H_2SeO_4$ [b]	295	3.2	(273)
$BaTiO_3$	393	26.0	(296)
$KNbO_3$	712	30.0	(523)
$PbTiO_3$	763	> 50.0	(300)
$LiTaO_3$	890	23.0	(720)
$LiNbO_3$	1470	300.0	—

[a] To convert to the basic SI unit of $C\ m^{-2}$ divide the tabulated values of P_s by 10^2; to convert to the CGS unit of esu cm^{-2} multiply by 3×10^3. For a full compilation see E. C. Subbarao, *Ferroelectrics* **5**, 267 (1973).

[b] Triglycinesulfate and selenate.

In KH_2PO_4 and related compounds each tetrahedral $[PO_2(OH)_2]^-$ group is joined by H bonds to neighbouring $[PO_2(OH)_2]^-$ groups; below the transition temperature all the short O–H bonds are ordered on the same side of the PO_4 units, and by appropriate application of an electric field, the polarization of the H bonds can be reversed, as indicated schematically below:

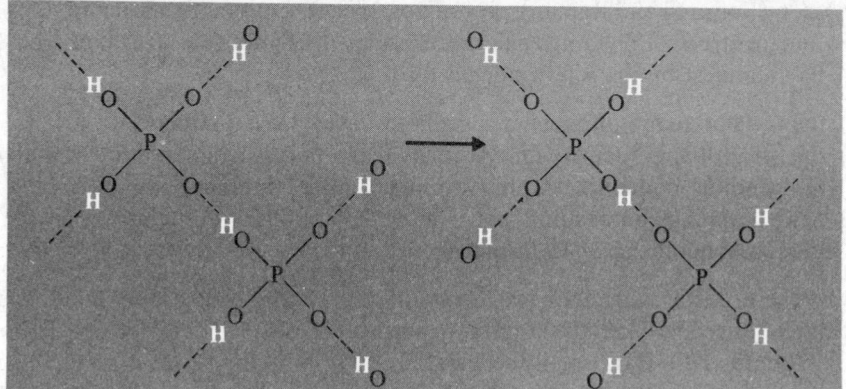

Panel continues

The dramatic effect of deuterium substitution in raising the transition temperature of such compounds can be seen from the Table; this has been ascribed to a quantum-mechanical effect involving the mass dependence of the de Broglie wavelength of hydrogen. Other examples of H-bonded ferroelectrics are $(NH_4)H_2PO_4$, $(NH_4)H_2AsO_4$, $Ag_2H_3IO_6$, $(NH_4)Al(SO_4)_2 \cdot 6H_2O$, and $(NH_4)_2SO_4$. Rochelle salt is unusual in having both an upper and a lower critical temperature between which the compound is ferroelectric.

The closely related phenomenon of antiferroelectric behaviour is also known, in which there is an ordered, self-cancelling arrangement of permanent electric dipole moments below a certain transition temperature; H bonding is again implicated in the ordering mechanism for several ammonium salts of this type, e.g. $(NH_4)H_2PO_4$ 148 K, $(NH_4)D_2PO_4$ 242 K, $(NH_4)H_2AsO_4$ 216 K, $(NH_4)D_2AsO_4$ 304 K, and $(NH_4)_2H_3IO_6$ 254 K. As with ferroelectrics, antiferroelectrics can also arise by a displacive mechanism in perovskite-type structures, and typical examples, with their transition temperatures, are:

$PbZrO_3$ 506 K, $PbHfO_3$ 488 K, $NaNbO_3$ 793, 911 K, and WO_3 1010 K.

Ferroelectrics have many practical applications: they can be used as miniature ceramic capacitors because of their large capacitance, and their electro-optical characteristics enable them to modulate and deflect laser beams. The temperature dependence of spontaneous polarization induces a strong pyroelectric effect which can be exploited in thermal and infrared detection. Many applications depend on the fact that all ferroelectrics are also piezoelectrics. Piezoelectricity is the property of acquiring (or altering) an electric polarization P under external mechanical stress, or conversely, the property of changing size (or shape) when subjected to an external electric field E. Thus ferroelectrics have been used as transducers to convert mechanical pulses into electrical ones and vice versa, and find extensive application in ultrasonic generators, microphones, and gramophone pickups; they can also be used as frequency controllers, electric filters, modulating devices, frequency multipliers, and as switches, counters, and other bistable elements in computer circuits. A further ingeneous application is in delay lines by means of which an electric signal is transformed piezoelectrically into an acoustic signal which passes down the piezoelectric rod at the velocity of sound until, at the other end, it is reconverted into a (delayed) electric signal.

It should be noted that, whereas ferroelectrics are necessarily piezoelectrics, the converse need not apply. The necessary condition for a crystal to be piezoelectric is that it must lack a centre of inversion symmetry. Of the 32 point groups, 20 qualify for piezoelectricity on this criterion, but for ferroelectric behaviour a further criteria is required (the possession of a single non-equivalent direction) and only 10 space groups meet this additional requirement. An example of a crystal that is piezoelectric but not ferroelectric is quartz, and indeed this is a particularly important example since the use of quartz for oscillator stabilization has permitted the development of extremely accurate clocks (1 in 10^8) and has also made possible the whole of modern radio and television broadcasting including mobile radio communications with aircraft and ground vehicles.

Intimate information about the nature of the H bond has come from vibrational spectroscopy (infrared and Raman), proton nmr spectroscopy, and diffraction techniques (X-ray and neutron). In vibrational spectroscopy the presence of a hydrogen bond $A–H \cdots B$ is manifest by the following effects:

  (i) the A–H stretching frequency v shifts to lower wave numbers;
 (ii) the breadth and intensity of $v(A–H)$ increase markedly, often more than tenfold;
(iii) the bending mode $\delta(A–H)$ shifts to higher wave numbers;
(iv) new stretching and bending modes of the H bond itself sometimes appear at very low wave numbers (20–200 cm^{-1}).

Most of these effects correlate roughly with the strength of the H bond and are particularly noticeable when the bond is strong. For example, for isolated non-H-bonded hydrogen groups, $v(O–H)$ normally occurs near 3500–3600 cm^{-1} and is less than 10 cm^{-1} broad; in the presence of O–H$\cdots$O bonding $v_{antisym}$ drops to $\sim$1700–2000 cm^{-1}, is several hundred cm^{-1} broad, and much more intense. A similar effect of

$\Delta v \sim 1500\text{--}2000$ cm^{-1} is noted on F–H$\cdots$F formation and smaller shifts have been found for N–H$\cdots$F ($\Delta v \leqslant 1000$ cm^{-1}), N–H$\cdots$O ($\Delta v \leqslant 400$ cm^{-1}), O–H$\cdots$N ($\Delta v \leqslant 100$ cm^{-1}), etc. A full discussion of these effects, including the influence of solvent, concentration, temperature, and pressure, is given in ref. 22. Suffice it to note that the magnitude of the effect is much greater than expected on a simple electrostatic theory of hydrogen bonding, and this implies appreciable electron delocalization (covalency) particularly for the stronger H bonds.

Proton nmr spectroscopy has also proved valuable in studying H-bonded systems. As might be expected, substantial chemical shifts are observed and information can be obtained concerning H-bond dissociation, proton exchange times, and other relaxation processes. The chemical shift always occurs to low field and some typical values are tabulated below for the shifts which occur between the gas and liquid phases or on dilution in an inert solvent:

Compound	CH_4	C_2H_6	$CHCl_3$	HCN	NH_3	PH_3
δ ppm	0	0	0.30	1.65	1.05	0.78
Compound	H_2O	H_2S	HF	HCl	HBr	HI
δ ppm	4.58	1.50	6.65	2.05	1.78	2.55

The low-field shift is generally interpreted, at least qualitatively, in terms of a decrease in diamagnetic shielding of the proton:[26] the formation of A–H$\cdots$B tends to draw H towards B and to repel the bonding electrons in A–H towards A thus reducing the electron density about H and reducing the shielding. The strong electric field due to B also inhibits the diamagnetic circulation within the H atom and this further reduces the shielding. In addition, there is a magnetic anisotropy effect due to B; this will be positive (upfield shift) if the principal symmetry axis of B is towards the H bond, but the effect is presumably small since the overall shift is always downfield.

Ultraviolet and visible spectra are also influenced by H bonding, but the effects are more difficult to quantify and have been rather less used than ir and nmr. It has been found that the $n \rightarrow \pi^*$ transition of the base B always moves to high frequency (blue shift) on H-bond formation, the magnitude of Δv being $\sim 300\text{--}4000$ cm^{-1} for bands in the region $15\,000\text{--}35\,000$ cm^{-1}. By contrast $\pi \rightarrow \pi^*$ transitions on the base B usually move to lower frequencies (red shift) and shifts are in the range -500 to -2300 cm^{-1} for bands in the region $30\,000\text{--}47\,000$ cm^{-1}. Detailed interpretations of these data are somewhat complex and obscure, but it will be noted that the shifts are approximately of the same magnitude as the enthalpy of formation of many H bonds (83.54 cm^{-1} per atom $\equiv 1$ kJ mol^{-1}).

3.5.2 *Influence on structure*[27, 28]

The crystal structure of many compounds is dominated by the effect of H bonds, and numerous examples will emerge in ensuing chapters. Ice (p. 731) is perhaps the classic

[26] J. A. POPLE, W. G. SCHNEIDER, and H. J. BERNSTEIN, *High Resolution Nuclear Magnetic Resonance*, Chap. 15, McGraw-Hill, New York, 1959.
[27] L. PAULING, *The Nature of the Chemical Bond*, 3rd edn., Chap. 12, Cornell University Press, Ithaca, 1960.
[28] A. F. WELLS, *Structural Inorganic Chemistry*, 4th edn., Oxford University Press, Oxford, 1975.

example, but the layer lattice structure of $B(OH)_3$ (p. 229) and the striking difference between the α- and β-forms of oxalic and other dicarboxylic acids is notable (Fig. 3.5). The more subtle distortions that lead to ferroelectric phenomena in KH_2PO_4 and other crystals have already been noted (p. 61). Hydrogen bonds between fluorine atoms result in the formation of infinite zigzag chains in crystalline hydrogen fluoride with $F–H\cdots F$ distance 249 pm and the angle HFH 120.1°. Likewise, the crystal structure of NH_4HF_2 is completely determined by H bonds, each nitrogen atom being surrounded by 8 fluorines, 4 in tetrahedral array at 280 pm due to the formation of $N–H\cdots F$ bonds, and 4 further away at about 310 pm; the two sets of fluorine atoms are themselves bonded pairwise at 232 pm by $F–H–F$ interactions. Ammonium azide NH_4N_3 has the same structure as NH_4HF_2, with $N–H\cdots N$ 298 pm. Hydrogen bonding also leads NH_4F to crystallize with a structure different from that of the other ammonium (and alkali) halides: NH_4Cl, NH_4Br, and NH_4I each have a low-temperature CsCl-type structure and a high-temperature NaCl-type structure, but NH_4F adopts the wurtzite (ZnS) structure in which each NH_4^+ group is surrounded tetrahedrally by 4 F to which it is bonded by 4 $N–H\cdots F$ bonds at 271 pm. This is very similar to the structure of ordinary ice. Typical values of $A–H\cdots B$ distances found in crystals are given in Table 3.7.

The precise position of the H atom in crystalline compounds containing H bonds has excited considerable experimental and theoretical interest. In situations where a symmetric H bond is possible in principle, it is frequently difficult to decide whether the proton is vibrating with a large amplitude about a single potential minimum or whether it is vibrating with a smaller amplitude but is also statistically disordered between two close sites, the potential energy barrier between the two sites being small.[23, 23a] It now seems well established that the $F–H–F$ bond is symmetrical in $NaHF_2$ and KHF_2, and that the

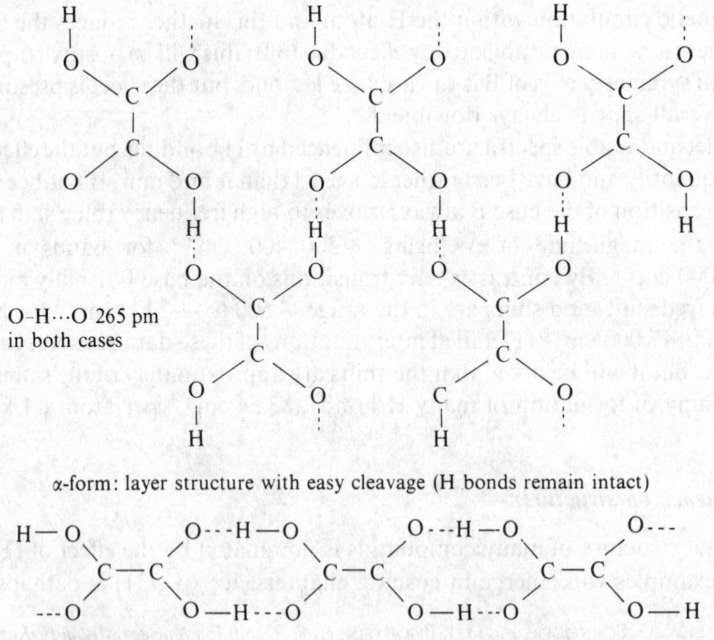

O–H$\cdots$O 265 pm
in both cases

α-form: layer structure with easy cleavage (H bonds remain intact)

β-form: long chains giving crystals that cleave into lathes parallel to the chains

FIG. 3.5 Schematic representation of the two forms of oxalic acid, $(–CO_2H)_2$.

TABLE 3.7 *Length of typical H bonds*[23, 28]

Bond	Length/pm	Σ/pm[a]	Examples
F–H–F	226	(270)	$NaHF_2$, KHF_2
F–H$\cdots$F	245–249	(270)	KH_4F_5. HF
O–H$\cdots$F	265–272	(275)	$CuF_2.2H_2O$, $FeSiF_6.6H_2O$
O–H$\cdots$Cl	292–318	(320)	$HCl.H_2O$, $(NH_3OH)Cl$, $CuCl_2.2H_2O$
O–H$\cdots$Br	304	(335)	$Cs_3Br_3(H_3O)(HBr_2)$
O–H–O	240–250	(280)	Ni dimethylglyoxime, KH maleate, $HCrO_2$, $Na_3H(CO_3)_2.2H_2O$
O–H$\cdots$O	248–290	(280)	KH_2PO_4, $NH_4H_2PO_4$, KH_2AsO_4, AlOOH, αHIO_3, numerous hydrated metal sulfates and nitrates
O–H$\cdots$S	320+	(325)	$MgS_2O_3.6H_2O$
O–H$\cdots$N	268–279	(290)	$N_2H_4.4MeOH$, $N_2H_4.H_2O$
N–H$\cdots$F	262–282	(285)	NH_4F, $N_2H_6F_2$
N–H$\cdots$Cl	300–320	(330)	Me_3NHCl, Me_2NH_2Cl, $(NH_3OH)Cl$
N–H$\cdots$I	346	(365)	Me_3NHI
N–H$\cdots$O	281–304	(290)	HSO_3NH_2, $(NH_4)_2SO_4$, NH_4OOCH, $CO(NH_2)_2$
N–H$\cdots$N	294–315	(300)	NH_4N_3, $NCNC(NH_2)_2$ (i.e. dicyandiamide)
P–H$\cdots$I	424	(405)	PH_4I

[a] Σ = sum of van der Waals' radii (in pm) of A and B (ignoring H which has a value of $\sim$ 120 pm) and using the values F 135, Cl 180, Br 195, I 215; O 140, S 185; N 150, P 190.

O–H–O bond is symmetrical in $HCrO_2$. Other examples are the intra-molecular H bonds in potassium hydrogen maleate (KH *cis*-butenedioate) and its monochloro derivative:

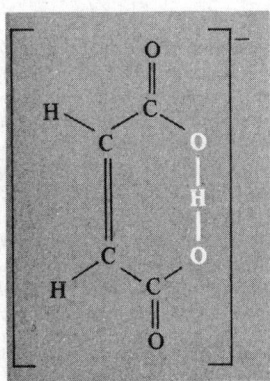 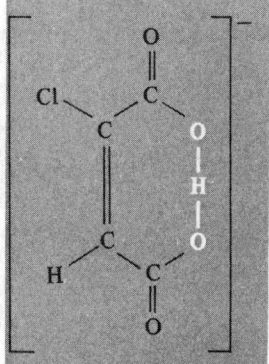

Numerous other examples of H bonding will be found in later chapters.

In summary, we can see that H bonding influences crystal structure by linking atoms or groups into larger structural units. These may be:

finite groups: HF_2^-; $[O_2CO–H\cdots OCO_2]^{3-}$ in $Na_3H(CO_3)_2.2H_2O$;
    dimers of carboxylic acids, etc.;
infinite chains: HF, HCN, HCO_3^-, HSO_4^-, etc.;
infinite layers: $N_2H_6F_2$, $B(OH)_3$, $B_3O_3(OH)_3$, H_2SO_4, etc.;
three-dimensional nets: NH_4F, H_2O, H_2O_2, $Te(OH)_6$, $H_2PO_4^-$ in KH_2PO_4, etc.

H bonding also vitally influences the conformation and detailed structure of the polypeptide chains of protein molecules and the complementary intertwined poly-nucleotide chains which form the double helix in nucleic acids (see Panel).

H Bonds in Proteins and Nucleic Acids (See pp. 498–504 of ref. 27.)

Proteins are built up from polypeptide chains of the type

These chains are coiled in a precise way which is determined to a large extent by N—H···O hydrogen bonds of length 279 ± 12 pm depending on the amino-acid residue involved. Each amide group is attached by such a hydrogen bond to the third amide group from it in both directions along the chain, resulting in an α-helix of pitch (total rise of helix per turn) of about 538 pm, corresponding to 3.60 amino-acid residues per turn. These helical chains can, in turn, become stretched and form hydrogen bonds with neighbouring chains to generate either parallel-chain pleated sheets (repeat distances 650 pm) or antiparallel-chain pleated sheets (700 pm).

Nucleic acids, which control the synthesis of proteins in the cells of living organisms and which transfer heredity information via genes, are also dominated by H bonding. Their structure involves two polynucleotide chains intertwined to form a double helix. The complimentariness in the structure of the two chains is ascribed to the formation of H bonds between the pyrimidine residue (thymine or cytosine) in one chain and the purine residue (adenine or guanine) in the other as illustrated on the page opposite. Whilst there is still some uncertainty as to the precise configuration of the N—H···O and N—H···N hydrogen bonds in particular cases, the extraordinary fruitfulness of these basic ideas has led to a profusion of developments of fundamental importance in biochemistry.

3.5.3 *Strength of hydrogen bonds and theoretical description*

Measurement of the properties of H-bonded systems over a range of temperatures leads to experimental values of ΔG, ΔH, and ΔS for H-bond formation, and these data have been supplemented in recent years by increasingly reliable *ab initio* quantum-mechanical

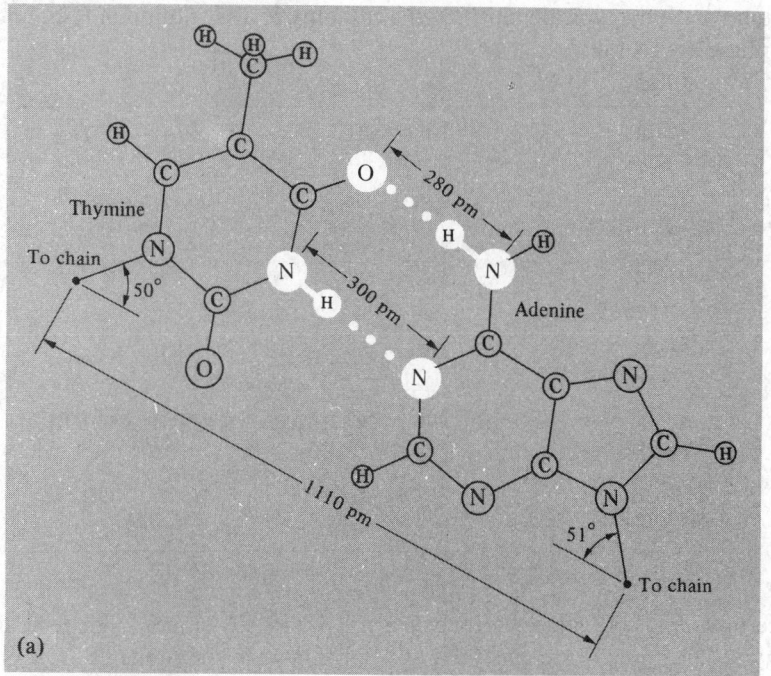

(a)

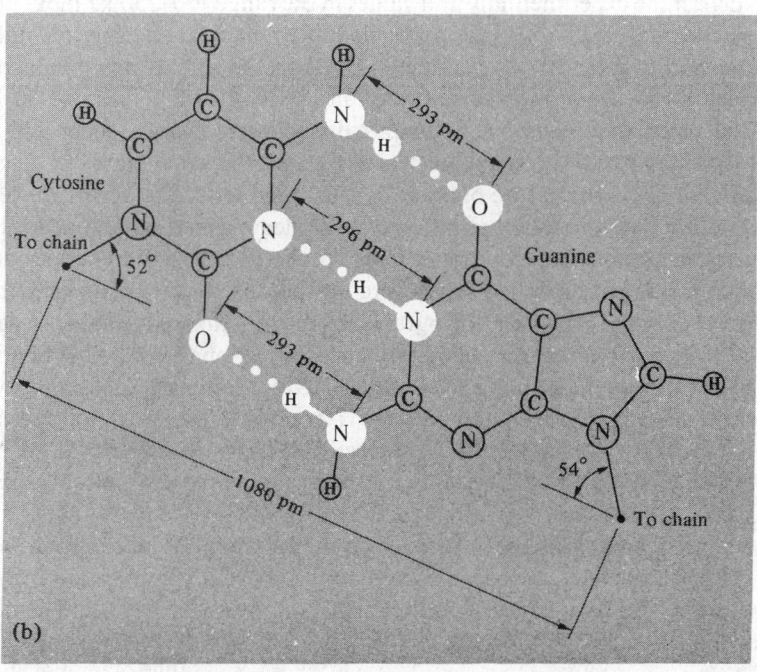

(b)

calculations.[29] Some typical values for the enthalpy of dissociation of H-bonded pairs in the gas phase are as follows:

A H···B $\Delta H_{298}/\text{kJ mol}^{-1}$	$HSH \cdots SH_2$ 7	$NCH \cdots NCH$ 16
A H···B $\Delta H_{298}/\text{kJ mol}^{-1}$	$H_2NH \cdots NH_3$ 17	$MeOH \cdots OHMe$ 19
A H···B $\Delta H_{298}/\text{kJ mol}^{-1}$	$HOH \cdots OH_2$ 22	$FH \cdots FH$ 29
A H···B $\Delta H_{298}/\text{kJ mol}^{-1}$	$ClH \cdots OMe_2$ 30	$HOH \cdots Cl^-$ 55
A H···B $\Delta H_{298}/\text{kJ mol}^{-1}$	$HCONH_2 \cdots OCHNH_2$ 59	$HCOOH \cdots OCHOH$ 59
A H···B $\Delta H_{298}/\text{kJ mol}^{-1}$	$HOH \cdots F^-$ 98	$H_2OH^+ \cdots OH_2$ 151

The uncertainty in these values varies between ± 1 and $\pm 6\,\text{kJ mol}^{-1}$. In general, H bonds of energy $< 25\,\text{kJ mol}^{-1}$ are classified as weak; those in the range 25–35 kJ mol^{-1} are medium; and those having $\Delta H > 35\,\text{kJ mol}^{-1}$ are strong. Until recently, it was thought that the strongest H bond was that in the hydrogendifluoride ion $[F-H-F]^-$; this is difficult to determine experimentally and values in the range 150–250 kJ mol^{-1} have been reported. Theoretical calculations are probably more reliable than experimental values in this case and lead to estimates in the range 217–234 kJ mol^{-1}. It now seems that the H bond between formic acid and the fluoride ion $[HCO_2H \cdots F]^-$ is some 30 kJ mol^{-1} stronger, the calculated value of the bond dissociation energy being 250 kJ mol^{-1} compared with 220 kJ mol^{-1} calculated for HF_2^- on the same basis.[30]

Early discussions on the nature of the hydrogen bond tended to adopt an electrostatic approach in order to avoid the implication of a covalency greater than one for hydrogen. Indeed, such calculations can reproduce the experimental H-bond energies and dipole moments, but this is not a particularly severe test because of the parametric freedom in positioning the charges. However, the purely electrostatic theory is unable to account for the substantial increase in intensity of the stretching vibration $v(A-H)$ on H bonding or for the lowered intensity of the bending mode $\delta(A-H)$. More seriously, such a theory does not account for the absence of correlation between H-bond strength and dipole moment of the base, and it leaves the frequency shifts in the electronic transitions unexplained. Nonlinear $A-H \cdots B$ bonds would also be unexpected, though numerous examples of angles in the range 150–180° are known.[23]

Valence-bond descriptions envisage up to five contributions to the total bond wave function,[22] but these are now considered to be merely computational devices for approximating to the true wave function. Perturbation theory has also been employed and apportions the resultant bond energy between (1) the electrostatic energy of

[29] P. A. KOLLMAN, Hydrogen bonding and donor–acceptor electronic interactions, Chap. 3 in H. F. SCHAEFFER (ed.), *Applications of Electronic Structure Theory*, Plenum Press, New York, 1977.

[30] J. EMSLEY, O. P. A. HOYTE, and R. E. OVERILL, The strongest H bond, *JCS Chem. Comm.* 1977, 225.

interaction between the fixed nuclei and the electron distribution of the component molecules, (2) Pauli exchange repulsion energy between electrons of like spin, (3) polarization energy resulting from the attraction between the polarizable charge cloud of one molecule and the permanent multipoles of the other molecule, (4) quantum-mechanical charge-transfer energy, and (5) dispersion energy, resulting from second-order induced dipole-induced dipole attraction. The results suggest that electrostatic effects predominate, particularly for weak bonds, but that covalency effects increase in importance as the strength of the bond increases. It is also possible to apportion the energy obtained from the most recent *ab initio* SCF-MO calculations in this way.[31] For example, in one particular calculation for the water dimer $HOH \cdots OH_2$, the five energy terms enumerated above were calculated to be: $E_{elec\,stat} - 26.5$, $E_{Pauli} + 18$, $E_{polar} - 2$, $E_{ch\,tr} - 7.5$, E_{disp} 0 kJ mol^{-1}. There was also a coupling interaction $E_{mix} - 0.5$, making in all a total attractive force $\Delta E_0 = E_{dimer} - E_{monomers} = -18.5$ kJ mol^{-1}. To calculate the enthalpy change ΔH_{298} as listed on p. 68, it is also necessary to consider the work of expansion and the various spectroscopic degrees of freedom:

$$\Delta H_{298} = E_0 + \Delta(PV) + \Delta E_{trans} + \Delta E_{vib} + \Delta E_{rot}$$

Such calculations can also give an indication of the influence of H-bond formation on the detailed electron distribution within the interacting components. There is general agreement that in the system $X-A-H \cdots B-Y$ as compared with the isolated species XAH and BY, there is a net gain of electron density by X, A, and B and a net loss of electrons by H and Y. There is also a small transfer of electronic charge (~ 0.05 electrons) from BY to XAH in moderately strong H bonds (20–40 kJ mol^{-1}). In virtually all neutral dimers, the increase in the A–H bond length on H-bond formation is quite small (< 5 pm), the one exception so far studied theoretically being $ClH \cdots NH_3$, where the proton position in the H bond is half-way between completely transferred to NH_3 and completely fixed on HCl.

It follows from the preceding discussion that the H bond can be regarded as a 3-centre 4-electron bond $A-H \cdots B$ in which the 2 pairs of electrons involved are the bond pair in A–H and the lone pair on B. The degree of charge separation on bond formation will depend on the nature of the proton-donor group AH and the Lewis base B. The relation between this 3-centre bond formalism and the 3-centre bond descriptions frequently used for boranes, polyhalides, and compounds of xenon is particularly instructive and is elaborated in Fig. 3.6. Numerous examples are also known in which hydrogen acts as a bridge between metallic elements in binary and more complex hydrides, and some of these will be mentioned in the following section which considers the general question of the hydrides of the elements.

3.6 Hydrides of the Elements[32-34]

Hydrogen combines with many elements to form binary hydrides MH_x (or M_mH_n). All

[31] H. UMEYAMA and K. MOROKUMA, The origin of hydrogen bonding: an energy decomposition study, *J. Am. Chem. Soc.* **99**, 1316–32 (1977).

[32] K. M. MACKAY, *Hydrogen Compounds of the Metallic Elements*, E. and F. N. Spon, London, 1966, 168 pp.; Hydrides, *Comprehensive Inorganic Chemistry*, Vol. 1, Chap. 2, Pergamon Press, Oxford, 1973.

[33] E. WIBERG and E. AMBERGER, *Hydrides of the Elements of Main Groups I–IV*, Elsevier, Amsterdam, 1971, 785 pp.

[34] W. M. MUELLER, J. P. BLACKLEDGE, and G. G. LIBOWITZ (eds.), *Metal Hydrides*, Academic Press, New York, 1968, 791 pp.

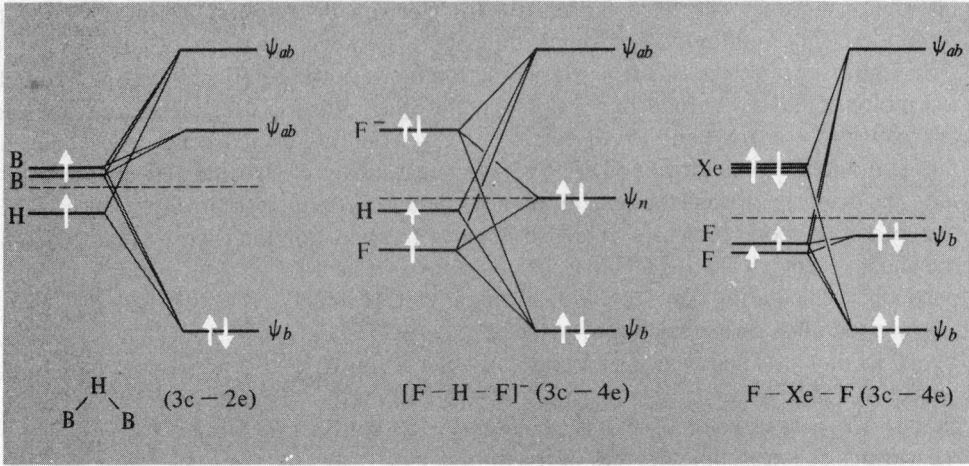

Fig. 3.6   Schematic representation of the energy levels in various types of 3-centre bond. The B–H–B ("electron deficient") bond is non-linear, the ("electron excess") F–Xe–F bond is linear, and the A—H···B hydrogen bond can be either linear or non-linear depending on the compound.

the main-group elements except the noble gases and perhaps indium and thallium form hydrides, as do all the lanthanoids and actinoids that have been studied. Hydrides are also formed by the more electropositive transition elements, notably Sc, Y, La, Ac; Ti, Zr, Hf; and to a lesser extent V, Nb, Ta; Cr; Cu; and Zn. Hydrides of other transition elements are either non-existent or poorly characterized, with the spectacular exception of palladium which has been more studied than any other metal hydride system.[35] The situation is summarized in Fig. 3.7; this indicates the idealized formulae of the known hydrides though many of the d-block and f-block elements form phases of variable compositions.

It has been customary to group the binary hydrides of the elements into various classes according to the presumed nature of their bonding: ionic, metallic, covalent, polymeric, and "intermediate" or "borderline". However, this is unsatisfactory because the nature of the bonding is but poorly understood in many cases and the classification obscures the important point that there is an almost continuous gradation in properties—and bond types(?)—between members of the various classes. It is also somewhat misleading in implying that the various bond types are mutually exclusive whereas it seems likely that more than one type of bonding is present in many cases. The situation is not unique to hydrides but is also well known for binary halides, oxides, sulfides, etc.: this serves to remind us that the various bond models represent grossly oversimplified limiting cases and that in most actual systems the position is more complex. For example, oxides might be classed as ionic (MgO), metallic (TiO, ReO$_3$), covalent (CO$_2$), polymeric (SiO$_2$), or as intermediate between these various classes, though any adequate bonding theory would recognize the arbitrary nature of these distinctions which merely emphasize particular features of the overall assembly of molecular orbitals and electron populations.

The metals in Groups I and II of the periodic table react directly with hydrogen to form white, crystalline, stoichiometric hydrides of formula MX and MX$_2$ respectively. The salt-like character of these compounds was recognized by G. N. Lewis in 1916 and he

[35] F. A. LEWIS, *The Palladium–Hydrogen System*, Academic Press, London, 1967, 178 pp.

H_2																	He
LiH	BeH_2											~25 boranes B_mH_n	Innumerable hydrocarbons C_mH_n	NH_3 N_2H_4 N_3H	OH_2 O_2H_2	FH	Ne
NaH	MgH_2											AlH_3	Si_mH_{2m+2} $m=1-8(+)$	PH_3 P_2H_4	S_nH_2 $n=1-8(+)$	ClH	Ar
KH	CaH_2	ScH_2	TiH_2	VH VH_2	CrH	Mn	Fe	Co	(NiH)	CuH	ZnH_2	GaH_3	Ge_mH_{2m+2} $m=1-5(+)$	AsH_3	SeH_2	BrH	Kr
RbH	SrH_2	YH_2 YH_3	ZrH_2	NbH NbH_2	Mo	Tc	Ru	Rh	$PdH_{<1}$	Ag	(CdH_2) ?	(InH_x) ?	SnH_4 Sn_2H_6	SbH_3	TeH_2	IH	Xe
CsH	BaH_2	LaH_2 LaH_3	HfH_2	TaH	W	Re	Os	Ir	Pt	Au	(HgH_2) ?	(TlH_x) ?	PbH_4	BiH_3	(PoH_2) ?	AtH	Rn
Fr	Ra	AcH_2															

CeH_2 CeH_3	PrH_2 PrH_3	NdH_2 NdH_3	Pm	SmH_2 SmH_3	EuH_2	GdH_2 GdH_3	TbH_2 TbH_3	DyH_2 DyH_3	HoH_2 HoH_3	ErH_2 ErH_3	TmH_2 TmH_3	YbH_2 $YbH_{2.5}$	LuH_2 LuH_3
ThH_2 Th_4H_{15}	PaH_3	UH_3	NpH_2 NpH_3	PuH_2 PuH_3	AmH_2 AmH_3	CmH_2	Bk	Cf	Es	Fm	Md	No	Lr

FIG. 3.7 The hydrides of the elements.

suggested that they contained the hydride ion H^-. Shortly thereafter (1920) K. Moers showed that electrolysis of molten LiH (mp 692°C) gave the appropriate amount of hydrogen at the anode; the other hydrides tended to decompose before they could be melted. As expected, the ionic-bond model is most satisfactory for the later (larger) members of each group, and the tendency towards covalency becomes more marked for the smaller elements LiH, MgH_2, and particularly BeH_2, which is best described in terms of polymeric covalent bridge bonds. X-ray and neutron diffraction studies show that the alkali metal hydrides adopt the cubic NaCl structure (p. 273) whereas MgH_2 has the tetragonal TiO_2 (rutile type) structure (p. 1120) and CaH_2, SrH_2, and BaH_2 adopt the orthorhombic $PbCl_2$-type structure (p. 444). The implied radius of the hydride ion H^- ($1s^2$) varies considerably with the nature of the metal because of the ready deformability of the pair of electrons surrounding a single proton. Typical values are given below and these can be compared with $r(F^-) \sim 133$ pm and $r(Cl^-) \sim 184$ pm.

Compound	MgH_2	LiH	NaH	KH	RbH	CsH	Free H^- (calculated)
$r(H^-)$/pm	130	137	146	152	154	152	208

The closest M–M approach in these compounds is often less than for the metal itself: this should occasion no surprise since this is a common feature of many compounds in which there is substantial separation of charge. For example, the shortest Ca–Ca interatomic distance is 393 pm in calcium metal, 360 pm in CaH_2, 380 pm in CaF_2, and only 340 pm in CaO (why?).

The thermal stability of the alkali metal hydrides decreases from lithium to caesium, the temperature at which the reversible dissociation pressure of hydrogen reaches 10 mmHg being ~ 550°C for LiH, ~ 210°C for NaH and KH, and ~ 170°C for RbH and CsH. The corresponding figures for the alkaline earth metal hydrides are CaH_2 885°C, SrH_2 585°C, and BaH_2 230°C, though for MgH_2 it is only 85°C. Chemical reactivity depends markedly on both the purity and the state of subdivision but increases from lithium to caesium and from calcium to barium with CaH_2 being rather less reactive than LiH. The reaction of water with these latter two compounds forms a convenient portable source of hydrogen, but with NaH the reaction is more violent than with sodium itself. RbH and CsH actually ignite spontaneously in dry air.

Turning next to Group IIIA, Fig. 3.7 indicates that hydrides of limiting stoichiometry MH_2 are also formed by Sc, Y, La, Ac, and by most of the lanthanoids and actinoids. In the special case of EuH_2 (Eu^{II} $4f^7$) and YbH_2 (Yb^{II} $4f^{14}$) the hydrides are isostructural with CaH_2 and the ionic bonding model gives a reasonable description of the observed properties; however, YbH_2 can absorb more hydrogen up to about $YbH_{2.5}$. The other hydrides adopt the fluorite (CaF_2) crystal structure (p. 129) and the supernumerary valence electron is delocalized, thereby conferring considerable metallic conductivity. For example, LaH_2 is a dark-coloured, brittle compound with a conductivity of about 10 ohm^{-1} cm^{-1} ($\sim 1\%$ of that of La metal). Further uptake of hydrogen progressively diminishes this conductivity to $< 10^{-1}$ ohm^{-1} cm^{-1} for the cubic phase LaH_3 (cf. $\sim 3 \times 10^{-5}$ ohm^{-1} cm^{-1} for YbH_2). The other Group IIIA elements and the lanthanoids and actinoids are similar.

There is a lively controversy concerning the interpretation of these and other properties, and cogent arguments have been advanced both for the presence of hydride ions H^- and for the presence of protons H^+ in the d-block and f-block hydride phases.[32, 34] These difficulties emphasize again the problems attending any classification based on presumed bond type, and a phenomenological approach which describes the observed properties is a sounder initial basis for discussion. Thus the predominantly ionic nature of a phase cannot safely be inferred either from crystal structure or from calculated lattice energies since many metallic alloys adopt the NaCl-type or CsCl-type structures (e.g. LaBi, β-brass) and enthalpy calculations are notoriously insensitive to bond type.

The hydrides of limiting composition MH_3 have complex structures and there is evidence that the third hydrogen is sometimes less strongly bound in the crystal. For the earlier (larger) lanthanoids La, Ce, Pr, and Nd, hydrogen enters octahedral sites and LnH_3 has the cubic Li_3Bi structure.[28] For Y and the smaller lanthanoids Sm, Gd, Tb, Dy, Ho, Er, Tm, and Lu, as well as for the actinoids Np, Pu, and Am, the hexagonal HoH_3 structure is adopted. This is a rather complex structure based on an extended unit cell containing 6 Ho and 18 H atoms.[36] The idealized structure has hcp Ho atoms with 12 tetrahedrally coordinated H atoms and 6 octahedrally coordinated H atoms; however, to make room for the bulky Ho atoms, close pairs of tetrahedral H atoms are slightly displaced and there is a more substantial movement of the "octahedral" H atoms towards the planes of the Ho atoms so that 2 of the H atoms are actually in the Ho planes and are trigonal 3-coordinate. The hydrogen atoms are thus of three types having respectively 14, 11, and 9 H neighbours for the distorted trigonal, octahedral, and tetrahedral sites. Each H atom has 3 Ho neighbours at either 210 or 217 or 224–299 pm respectively, and each Ho has 11 hydrogen neighbours, 9 at 210–229 pm and 2 somewhat further away at 248 pm. The 3-coordinate hydrogen is most unusual, the only other hydride in which it occurs being the complex cubic phase Th_4H_{15}.

Uranium forms two hydrides of stoichiometric composition UH_3. The normal β-form has a complex cubic structure and is the only one formed when the preparation is carried out above 200°C. Below this temperature increasing amounts of the slightly denser cubic α-form occur and this can be transformed to the β-phase by warming to 250°C. Both phases have ferromagnetic and metallic properties. Uranium hydride is commonly used as a starting material for the preparation of uranium compounds as it is finely powdered and extremely reactive. It is also used for purifying and regenerating hydrogen (or deuterium) gas.

The hydrides of Ti, Zr, and Hf are characterized by considerable variability in composition and structure. When pure, the limiting phases MH_2 form massive, metallic crystals of fluorite structure (TiH_2) or body-centred tetragonal structure (ZrH_2, HfH_2, ThH_2), but there are also several hydrogen-deficient phases of variable composition and complex structure in which several M–H distances occur.[28, 32, 34] These phases (and others based on Y, Ce, and Nb) have been extensively investigated in recent years because of their potential applications as moderators, reflectors, or shield components for high-temperature, mobile nuclear reactors.

Other hydrides with interstitial or metallic properties are formed by V, Nb, and Ta; they are, however, very much less stable than the compounds we have been considering and have extensive ranges of composition. Chromium also forms a hydride, CrH, though this

[36] M. MANSMANN and W. E. WALLACE, The structure of HoD_3, *J. de Physique*, **25**, 454–9 (1964).

must be prepared electrolytically rather than by direct reaction of the metal with hydrogen. It has the anti-NiAs structure (p. 649). Most other elements in this area of the periodic table have little or no affinity for hydrogen and this has given rise to the phrase "hydrogen gap". The notable exception is the palladium–hydrogen system which is discussed on p. 1335.

The hydrides of the later main-group elements present few problems of classification and are best discussed during the detailed treatment of the individual elements. Many of these hydrides are covalent, molecular species, though association via H bonding sometimes occurs, as already noted (p. 57). Catenation flourishes in Group IV and the complexities of the boron hydrides merit special attention (p. 171). The hydrides of aluminium, gallium, zinc, (and beryllium) tend to be more extensively associated via M–H–M bonds, but their characterization and detailed structural elucidation has proved extremely difficult.

Two further important groups of hydride compounds should be mentioned and will receive detailed attention in later chapters. One is the group of complex metal hydrides of which notable examples are $LiBH_4$, $NaBH_4$, $LiAlH_4$, $Al(BH_4)_3$, etc.[37] The other is the growing number of compounds in which the hydrogen atom is a monodentate or bidentate ligand to a transition element:[38-41] these date from the early 1930s when W. Hieber discovered $[Fe(CO)_4H_2]$ and $[Co(CO)_4H]$ and now cover an astonishing variety of structural types. The modest steric requirements of the H atom enable complexes such as $[ReH_9]^{2-}$ to be synthesized, and bridged complexes such as the linear $[Cr_2(CO)_{10}H]^-$ and bent $[W_2(CO)_9H(NO)]$, are known. The role of hydrido complexes in homogeneous catalysis is also currently exciting considerable attention.

[37] A. HAJOS, *Complex Hydrides*, Elsevier, Amsterdam, 1979, 398 pp.

[38] J. C. GREEN and M. L. H. GREEN, Transition metal hydride compounds, *Comprehensive Inorganic Chemistry*, Vol. 4, Chap. 48, Pergamon Press, Oxford, 1973.

[39] H. D. KAESZ and R. B. SAILLANT, Hydride complexes of transition metals, *Chem. Rev.* **72**, 231–81 (1972).

[40] A. P. HUMPHRIES and H. D. KAESZ, The hydrido-transition metal cluster complexes, *Progr. Inorg. Chem.* **25**, 145–222 (1979).

[41] G. L. GEOFFROY, Photochemistry of transition metal hydride complexes, *Progr. Inorg. Chem.* **27**, 123–51 (1980).

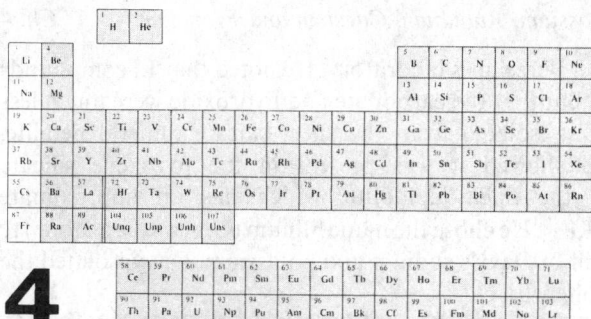

4

Lithium, Sodium, Potassium, Rubidium, Caesium, and Francium

4.1 Introduction

The alkali metals form a homogeneous group of extremely reactive elements which illustrate well the similarities and trends to be expected from the periodic classification, as discussed in Chapter 2. Their physical and chemical properties are readily interpreted in terms of their simple electronic configuration, ns^1, and for this reason they have been extensively studied by the full range of experimental and theoretical techniques. Compounds of sodium and potassium have been known from ancient times and both elements are essential for animal life. They are also major items of trade, commerce, and chemical industry. Lithium was first recognized as a separate element at the beginning of last century but did not assume major industrial importance until about 30 y ago. Rubidium and caesium are of considerable academic interest but so far have few industrial applications. Francium, the elusive element 87, has only fleeting existence in nature due to its very short radioactive half-life, and this delayed its discovery until 1939.

4.2 The Elements

4.2.1 *Discovery and isolation*

The spectacular success (in 1807) of Humphry Davy, then aged 29 y, in isolating metallic potassium by electrolysis of molten caustic potash (KOH) is too well known to need repeating in detail.[1] Globules of molten sodium were similarly prepared by him a few days later from molten caustic soda. Earlier experiments with aqueous solutions had been unsuccessful because of the great reactivity of these new elements. The names chosen by Davy reflect the sources of the elements.

Lithium was recognized as a new alkali metal by J. A. Arfvedson in 1817 whilst he was

---

[1] M. E. WEEKS, *Discovery of the Elements*, Journal of Chemical Education, Easton, 6th edn., 1956, 910 pp.

working as a young assistant in J. J. Berzelius's laboratory. He noted that Li compounds were similar to those of Na and K but that the carbonate and hydroxide were much less soluble in water. Lithium was first isolated from the sheet silicate mineral petalite, $LiAlSi_4O_{10}$, and Arfvedson also showed it was present in the pyroxene silicate spodumene, $LiAlSi_2O_6$, and in the mica lepidolite, which has an approximate composition $K_2Li_3Al_4Si_7O_{21}(OH,F)_3$. He chose the name lithium (Greek λιθος, stone) to contrast it with the vegetable origin of Davy's sodium and potassium. Davy isolated the metal in 1818 by electrolysing molten Li_2O.

Rubidium was discovered as a minor constituent of lepidolite by R. W. Bunsen and G. R. Kirchhoff in 1861 only a few months after their discovery of caesium (1860) in mineral spa waters. These two elements were the first to be discovered by means of the spectroscope, which Bunsen and Kirchhoff had invented the previous year (1859); accordingly their names refer to the colour of the most prominent lines in their spectra (Latin *rubidus*, deepest red; *caesius*, sky blue).

Francium was first identified in 1939 by the elegant radiochemical work of Margueritte Perey who named the element in honour of her native country. It occurs in minute traces in nature as a result of the rare (1%) branching decay of ^{227}Ac in the ^{235}U series:

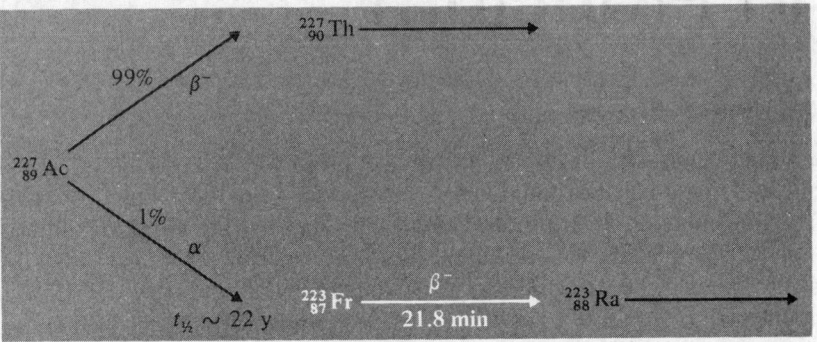

Its terrestrial abundance has been estimated as 2×10^{-18} ppm, which corresponds to a total of only 15 g in the top 1 km of the earth's crust. Other isotopes have since been produced by nuclear reactions but all have shorter half-lives than ^{227}Fr, which decays by energetic β^- emission, $t_{\frac{1}{2}}$ 21.8 min. Because of this intense radioactivity it is only possible to work with tracer amounts of the element.

4.2.2 *Terrestrial abundance and distribution*

Despite their chemical similarity, Li, Na, and K are not closely associated in their occurrence, mainly because of differences in size (see Table on p. 86). Lithium tends to occur in ferromagnesian minerals where it partly replaces magnesium; it occurs to the extent of about 18 ppm by weight in crustal rocks, and this reflects its relatively low abundance in the cosmos (Chapter 1). It is about as abundant as gallium (19 ppm) and niobium (20 ppm). The most important mineral commercially is spodumene, $LiAlSi_2O_6$, and large deposits occur in the USA, Canada, Brazil, Argentina, the USSR, Spain, and the Congo. An indication of the industrial uses of lithium and its compounds is given in the Panel.

Lithium and its Compounds

The dramatic transformation of Li from a small-scale specialist commodity to a multikilotonne industry during the past two decades is due to the many valuable properties of its compounds. About 34 compounds of Li are currently available in bulk, and a similar number again can be obtained in developmental or research quantities. The major industrial use of Li is in the form of lithium stearate which is used as a thickener and gelling agent to transform oils into lubricating greases. These "all-purpose" greases combine high water resistance with good low-temperature properties ($-20°C$) and excellent high-temperature stability ($>150°C$); they are readily prepared from $LiOH \cdot H_2O$ and tallow or other natural fats and have captured nearly half the total market for automotive greases in the USA.

Lithium carbonate is used as a flux in porcelain enamel formulations and in the production of special toughened glasses (by replacement of the larger Na ions): Li can either be incorporated within the glass itself or the preformed Na-glass can be dipped in a molten-salt bath containing Li ions to effect a surface cation exchange. In another application, the use of Li_2CO_3 by primary aluminium producers has risen sharply in recent years since it increases production capacity by 7–10% by lowering the mp of the cell content and permitting larger current flow; in addition, troublesome fluorine emissions are reduced by 25–50% and production costs are appreciably lowered. Estimated world production of Li_2CO_3 in 1977 was over 50 000 tonnes, mainly in the USA and the USSR.

The first commercial use of Li (in the 1920s) was as an alloying agent with lead to give toughened bearings; currently it is used to produce high-strength, low-density aluminium alloys for aircraft construction. With magnesium it forms an extremely tough low-density alloy which is used for armour plate and for aerospace components (e.g. LA 141, d 1.35 g cm^{-3}, contains 14% Li, 1% Al, 85% Mg). Other metallurgical applications employ LiCl as an invaluable brazing flux for Al automobile parts.

LiOH is used for CO_2 absorption in closed environments such as space capsules (light weight) and submarines. LiH is used to generate hydrogen in military, meteorological, and other applications, and the use of LiD in thermonuclear weaponry and research has been mentioned (p. 21). Likewise, the important applications of $LiAlH_4$, Li/NH_3, and organolithium reagents in synthetic organic chemistry are well known, though these account for only a small percentage of the lithium produced. Other specialist uses include the growing market for ferroelectrics such as $LiTaO_3$ to modulate laser beams (p. 61), and increasing use of thermoluminescent LiF in X-ray dosimetry.

Perhaps one of the most exciting new applications stems from the discovery in 1949 that small daily doses (1–2 g) of Li_2CO_3 taken orally provide an effective treatment for manic-depressive psychoses. The mode of action is not well understood but there appear to be no undesirable side effects. The dosage maintains the level of Li in the blood at about 1 mmol l^{-1} and its action may be related to the influence of Li on the Na/K balance and (or) the Mg/Ca balance since Li is related chemically to both pairs of elements.

Looking to the future, Li/FeS_x battery systems are emerging as a potentially viable energy storage system for off-peak electricity and as a non-polluting silent source of power for electric cars. The battery resembles the conventional lead–acid battery in having solid electrodes (Li/Si alloy, negative; FeS_x positive) and a liquid electrolyte (molten LiCl/KCl at 400°C). Other battery systems which have reached the prototype stage include the Li/S and Na/S cells (see p. 801).

Sodium, 22 700 ppm (2.27%) is the seventh most abundant element in crustal rocks and the fifth most abundant metal, after Al, Fe, Ca, and Mg. Potassium (18 400 ppm) is the next most abundant element after sodium. Vast deposits of both Na and K salts occur in relatively pure form as a result of evaporation of ancient seas, a process which still continues today in the Great Salt Lake (Utah), the Dead Sea, and elsewhere (Fig. 4.1). Sodium occurs as rock-salt (NaCl) and as the carbonate (trona), nitrate (saltpetre), sulfate (mirabilite), borate (borax, kernite), etc. There are also unlimited supplies of NaCl in

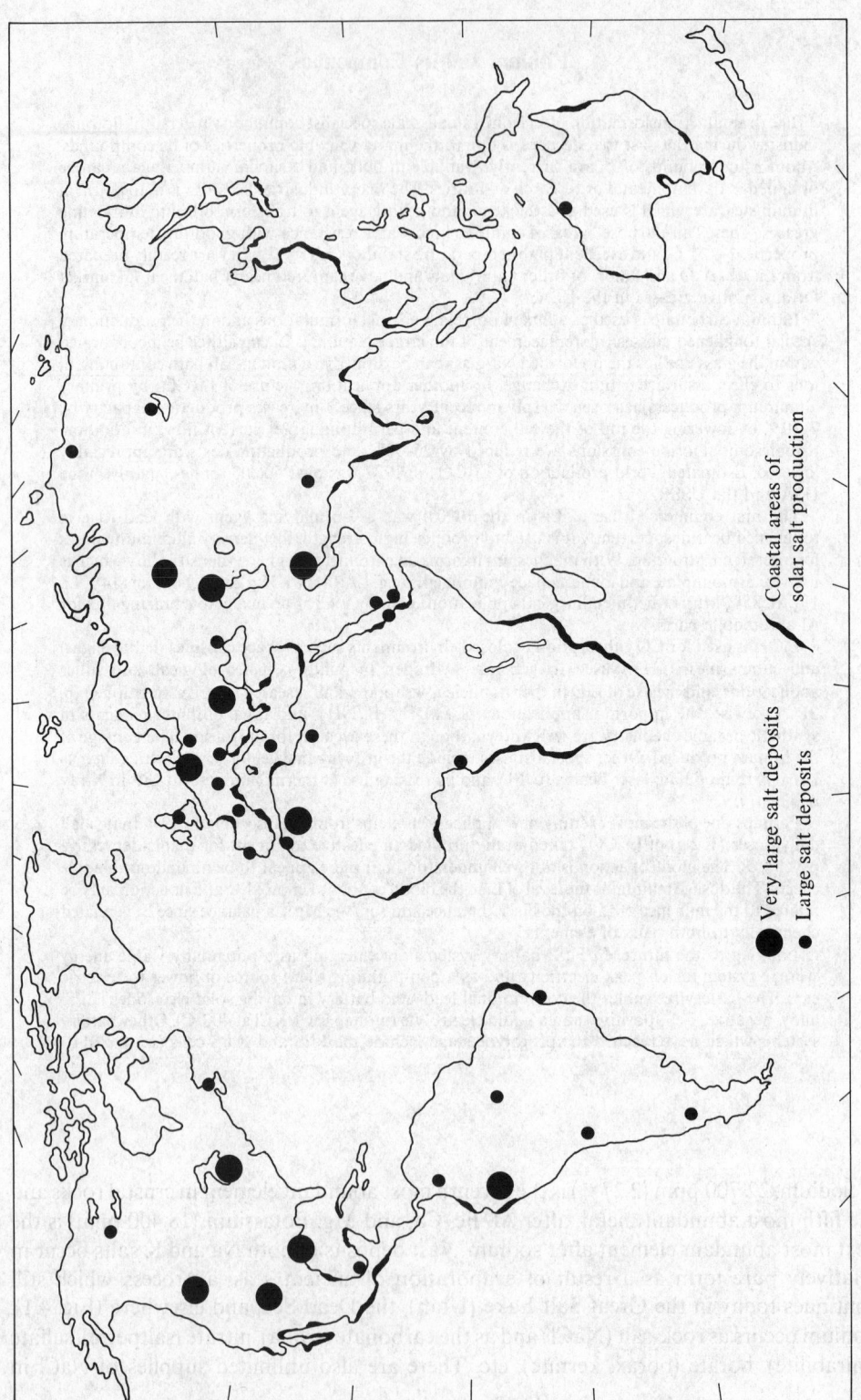

Fig. 4.1 Worldwide distribution of major salt deposits.

natural brines and oceanic waters (~ 30 kg m^{-3}).† The vital importance of NaCl in the heavy chemical industry is indicated in the Panel below.

Potassium occurs principally as the simple chloride (sylvite), as the double chloride $KCl.MgCl_2.6H_2O$ (carnallite), and the anhydrous sulfate $K_2Mg_2(SO_4)_3$ (langbeinite). Some production data are given in the Panel on p. 83.

Rubidium (78 ppm, similar to Ni, Cu, Zn) and caesium (2.6 ppm, similar to Br, Hf, U) are much less abundant than Na and K and have only recently become available in quantity. No purely Rb-containing mineral is known and much of the commercially available material is obtained as a byproduct of lepidolite processing for Li. Caesium occurs as the hydrated aluminosilicate pollucite, $Cs_4Al_4Si_9O_{26}.H_2O$, but the world's only commercial source is at Bernic Lake, Manitoba, and Cs (like Rb) is mainly obtained as a byproduct of the Li industry. The intense interest in Li for thermonuclear purposes since about 1958, coupled with its extensive use in automotive greases (p. 77), has consequently made Rb and Cs compounds much more available than formerly: annual production is in the region of 5 tonnes for each.

Production and Uses of Salt[2-6]

More NaCl is used for inorganic chemical manufacture than is any other material. It is approached only by phosphate rock, and world consumption of each exceeds 150 million tonnes annually. For NaCl, production is dominated by Europe (39%), North America (34%), and Asia (20%), whilst South America and Oceania have only 3% each and Africa 1%. Distribution by countries is shown in Fig. A.

Rock-salt occurs as vast subterranean deposits often hundreds of metres thick and containing $>90\%$ NaCl. The Cheshire salt field (which is the principal UK source of NaCl) is typical, occupying an area of 60 km × 24 km and being some 400 m thick: this field alone corresponds to reserves of $>10^{11}$ tonnes. Similar deposits occur near Carlsbad New Mexico, in Saskachewan Canada, and in many other places (see Fig. 4.1). Production methods vary with locality and with the use to be made of the salt. For example, in the UK 82% is extracted as brine for direct use in the chemical industry and 18% is mined as rock-salt, mainly for use on roads; less than 1% is obtained by solar evaporation. Corresponding data for the USA are shown in Fig. B.

Major sections of the inorganic heavy chemicals industry are based on salt and, indeed, this

Continued on p. 82

† It has been calculated that rock-salt equivalent to the NaCl in the oceans of the world would occupy 19 million cubic km (i.e. 50% more than the total volume of the North American continent above sea-level). Alternatively stated, a one-km square prism would stretch from the earth to the moon 47 times. Note also that, although Na and K are almost equally abundant in the crustal rocks of the earth, Na is some 30 times as abundant as K in the oceans. This is partly because K salts with the larger anions tend to be less soluble than the Na salts and, likewise, K is more strongly bound to the complex silicates and alumino silicates in the soils (ion exchange in clays). Again, K leached from rocks is preferentially absorbed and used by plants whereas Na can proceed to the sea. Potassium is an essential element for plant life and the growth of wild plants is often limited by the supply of K available to them.

² L. F. HABER, *The Chemical Industry during the Nineteenth Century*, Oxford University Press, Oxford, 1958, 292 pp.

³ T. K. DERRY and T. I. WILLIAMS, *A Short History of Chemical Technology*, Oxford University Press, Oxford, 1960, 782 pp.

⁴ A. J. G. NOTHOLT and D. E. HIGHLEY, *Salt*, Mineral Dossier No. 7 of the Mineral Resources Consultative Committee, HMSO, London 1973, 36 pp.

⁵ US Department of the Interior, Bureau of Mines, *Minerals Yearbook 1977*, Vol. 1, *Salt*, pp. 785–96 (1980).

⁶ G. A. COOK, *Survey of Modern Industrial Chemistry*, Ann Arbor, London, 1975, 297 pp.

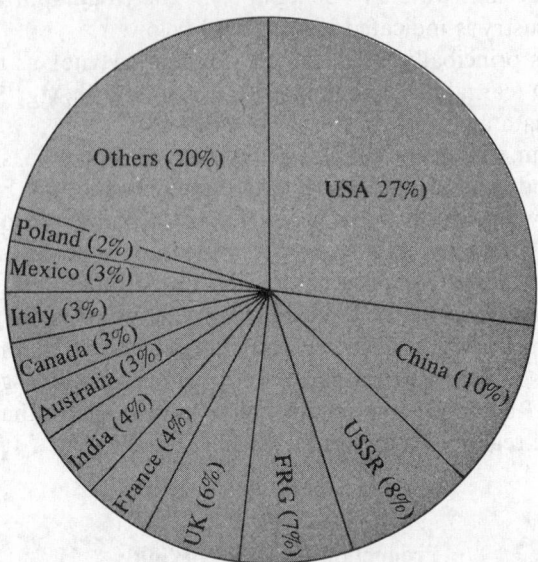

Fig. A   World production of salt by country.

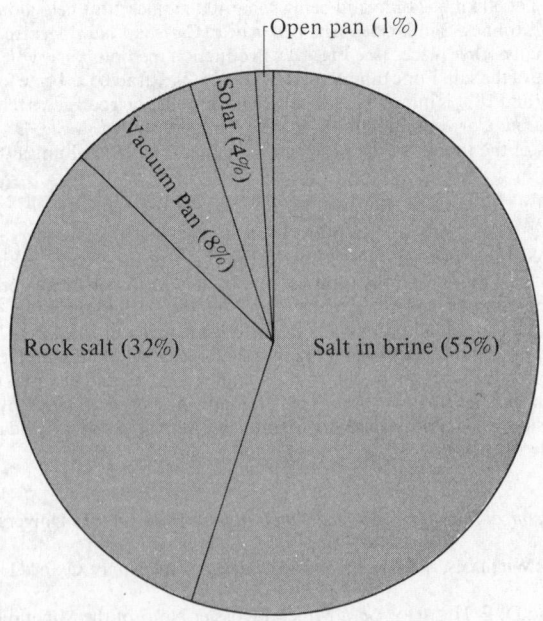

Fig. B   Types of salt production (USA).

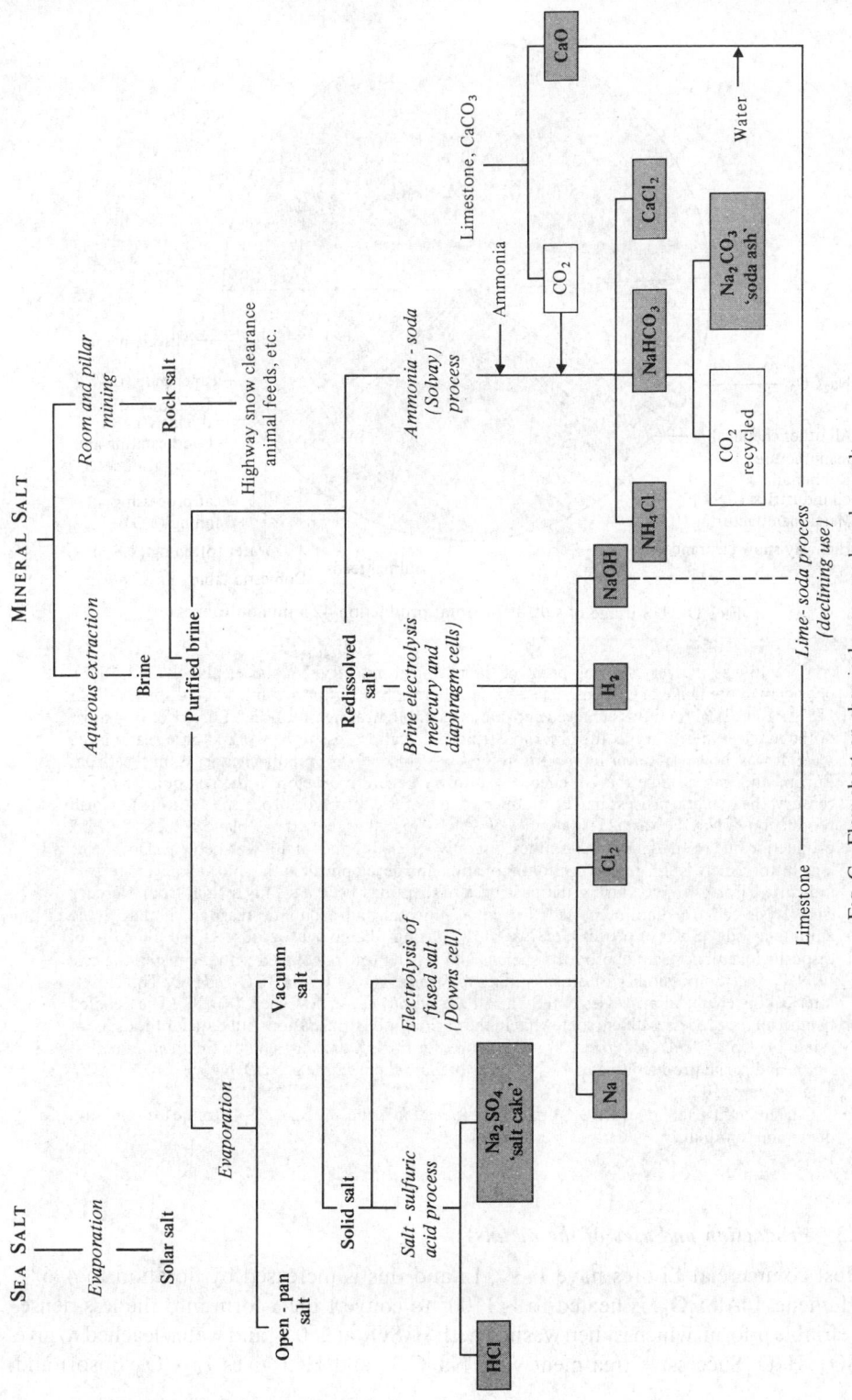

Fig. C   Flow sheet on chemical processes based on salt.

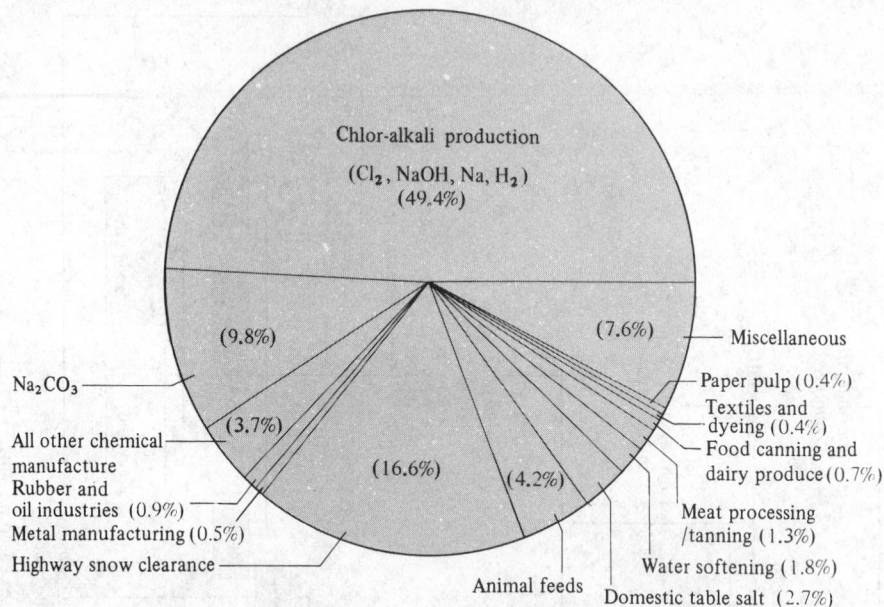

Chlor-alkali production

(Cl_2, NaOH, Na, H_2)

(49.4%)

(9.8%)

(7.6%) ——— Miscellaneous

Na_2CO_3 ———

Paper pulp (0.4%)

Textiles and
dyeing (0.4%)

All other chemical
manufacture

(3.7%)

Food canning and
dairy produce (0.7%)

(16.6%)

(4.2%)

Rubber and
oil industries (0.9%)

Metal manufacturing (0.5%)

Meat processing
/tanning (1.3%)

Highway snow clearance ———

Water softening (1.8%)

Animal feeds   Domestic table salt (2.7%)

FIG. D   US usage of salt, 1974 (total production 42.5 million tonnes).

compound was the very starting point of the chemical industry. Nicolas Leblanc (1742–1806), physician to the Duke of Orleans, devised a satisfactory process for making NaOH from NaCl in 1787 (Patent 1791) and this achieved enormous technological significance in Europe during most of the nineteenth century as the first industrial chemical process to be worked on a really large scale. It was, however, never important in the USA since it was initially cheaper to import from Europe and, by the time the US chemical industry began to develop in the last quarter of the century, the Leblanc process had been superseded by the electrolytic process. Thus in 1874 world production of NaOH was 525 000 tonnes of which 495 000 were by the Leblanc process; by 1902 production had risen to 1 800 000 tonnes, but only 150 000 tonnes of this was Leblanc. Despite its long history, there is still great scope for innovation and development in the chlor-alkali and related industries. For example, during the past decade there has been a steady switch from mercury electrolysis cells to diaphragm cells for environmental and economic reasons. Similarly, the ammonia-soda (Solvay) process for Na_2CO_3 is being phased out because of the difficulty of disposing of embarrasing byproducts such as NH_4Cl and $CaCl_2$, coupled with the increasing cost of NH_3 and the possibility of direct mining for trona, $Na_2CO_3 \cdot NaHCO_3 \cdot 2H_2O$. The closely interlocking chemical processes based on salt are set out in the flow sheet (Fig. C). The detailed balance of the processes differs somewhat in the various industrial nations but data for the usage of salt in the USA (1974) are typical (Fig. D); figures for the UK are very similar. Further discussion on the industrial production and uses of many of these chemicals (e.g. NaOH, Na_2CO_3, Na_2SO_4) is given on p. 102.

Current industrial prices are ~$4 per tonne for salt in brine and $20–30 per tonne for solid salt, depending on quality.

4.2.3   *Production and uses of the metals*

Most commercial Li ores have 1–3% Li and this is increased by flotation to 4–6%. Spodumene, $LiAlSi_2O_6$, is heated to ~1100° to convert the α-form into the less-dense, more friable β-form, which is then washed with H_2SO_4 at 250°C and water-leached to give $Li_2SO_4 \cdot H_2O$. Successive treatment with Na_2CO_3 and HCl gives Li_2CO_3 (insol) and

Production of Potassium Salts[7-9]

Sylvite (KCl) and sylvinite (mixed NaCl, KCl) are the most important K minerals for chemical industry; carnallite is also mined. Ocean waters contain only about 0.06% KCl, though this can rise to as high as 1.5% in some inland marshes and seas such as Searle's Lake, the Great Salt Lake, or the Dead Sea, thereby making recovery economically feasible. Soluble minerals of K are generally referred to (incorrectly) as potash, and production figures are always expressed as the weight of K_2O equivalent. Massive evaporite beds of soluble K salts were first discovered at Stassfurt, Germany, in 1856 and were worked there for potash and rock-salt from 1861 until 1972. World production was 17.8 million tonnes K_2O equivalent in 1970 and 23.7 million tonnes in 1974, of which half was produced in the USSR and Canada (Fig. A).

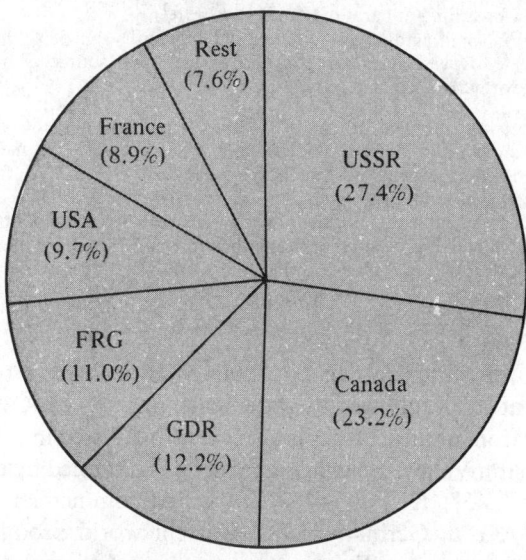

Rest (7.6%)
France (8.9%)
USSR (27.4%)
USA (9.7%)
FRG (11.0%)
Canada (23.2%)
GDR (12.2%)

FIG. A.

In the UK workable potash deposits are confined to the Cleveland–North Yorkshire bed which is ~11 m thick and has reserves of >500 million tonnes. Plates 4.1 and 4.2 illustrate the size of machinery used in the room and pillar mining of evaporite beds. Massive recovery is also possible from brines; e.g. Jordan is at present building a huge plant capable of recovering more than 1.2 million tonnes pa from the Dead Sea. This requires more than 100 km² of evaporating ponds and, when completed, will place Jordan alongside the United States and France in productive capacity.

Potassium is a major essential element for plant growth and potassic fertilizers account for the overwhelming proportion of production (>95%). Again KCl is dominant, accounting for more than 90% of the K used in fertilizers; K_2SO_4 is also used. KNO_3, though an excellent fertilizer, is now of only minor importance because of production costs. [Typical costs of agricultural grade

[7] *Kirk–Othmer Encyclopedia of Chemical Technology*, 2nd edn., 1968, Vol. 16, pp. 361–400.
[8] A. J. G. NOTHOLT, *Potash*, Mineral Dossier No. 16 of the Mineral Resources Consultative Committee, HMSO, London, 1976, 33 pp.
[9] US Department of the Interior, Bureau of Mines, *Minerals Yearbook 1977*, Vol. 1, *Potash*, pp. 739–61 (1980).

material (93–95% pure) per tonne at 1974 prices in the UK were: KCl £40, K_2SO_4 £80, and KNO_3 £117. Prices for technical grade compounds (97–99% pure) were: KCl £97, K_2CO_3 £95, KOH £105, and KNO_3 £130 per tonne.] In addition to its dominant use in fertilizers, KCl is used mainly to manufacture KOH and K. The other industrially important compounds and their uses are as follows:

KOH, used mainly in the preparation of potassium phosphates for liquid detergents, for the production of other chemicals, and the processing of rubber.

K_2CO_3 (from KOH and CO_2), used chiefly in high-quality decorative glassware, in optical lenses, colour TV tubes, and fluorescent lamps; it is also used in china ware, textile dyes, and pigments.

KNO_3, a powerful oxidizing agent now used mainly in gunpowders and pyrotechnics.

$KMnO_4$, an oxidizer, decolorizer, bleacher, and purification agent; its major application is in the manufacture of saccharin.

KO_2, used in breathing apparatus (p. 85).

$KClO_3$, used in small amounts in matches and explosives (pp. 585, 1009).

KBr, used extensively in photography and as the usual source of bromine in organic syntheses; formerly used as a sedative.

It is interesting to note the effect of varying the alkali metal cation on the properties of various compounds and industrial materials. For example, a soap is an alkali metal salt formed by neutralizing a long-chain organic acid such as stearic acid, $CH_3(CH_2)_{16}CO_2H$, with MOH. Potassium soaps are soft and low melting, and are therefore used in liquid detergents. Sodium soaps have higher mps and are the basis for the familiar domestic "hard soaps" or bar soaps. Lithium soaps have still higher mps and are therefore used as thickening agents for high-temperature lubricating oils and greases—their job is to hold the oil in contact with the metal under conditions when the oil by itself would run off.

LiCl. Alternatively, the chloride can be obtained by calcining the washed ore with limestone ($CaCO_3$) at 1000° followed by water leaching to give LiOH and then treatment with HCl. Recovery from natural brines is also extensively used in the USA (Searles Lake, California, and Clayton Valley, Nevada). The metal is obtained by electrolysis of a fused mixture of 55% LiCl, 45% KCl at ~450°C, the first commercial production being by Metallgesellschaft AG, in Germany, 1923. Current world production of Li metal is probably about 1000 tonnes pa. Far greater tonnages of Li compounds are, of course, produced and their major commercial applications have already been noted (p. 77).

Sodium metal is produced commercially on the kilotonne scale by the electrolysis of a fused eutectic mixture of 40% NaCl, 60% $CaCl_2$ at ~580°C in a Downs cell (introduced by du Pont, Niagara Falls, 1921). Metallic Na and Ca are liberated at the cylindrical steel cathode and rise through a cooled collecting pipe which allows the calcium to solidify and fall back into the melt. Chlorine liberated at the central graphite anode is collected in a nickel dome and subsequently purified. Potassium cannot be produced in this way because it is too soluble in the molten chloride to float on top of the cell for collection and because it vaporizes readily at the operating temperatures, creating hazardous conditions. Superoxide formation is an added difficulty since this reacts explosively with K metal. Consequently, commercial production of K relies on reduction of molten KCl with metallic Na at 850°C.† A similar process using Ca metal at 750°C under reduced pressure is used to produce metallic Rb and Cs.

---

† This reduction of KCl by Na appears to be contrary to the normal order of reactivity (K > Na). However, at 850–880° an equilibrium is set up: $Na(g) + K^+(l) \rightleftharpoons Na^+(l) + K(g)$. Since K is the more volatile (p. 86), it distils off more readily, thus displacing the equilibrium and allowing the reaction to proceed. By fractional distillation through a packed tower, K of 99.5% purity can be obtained but usually an Na/K mixture is drawn off because alloys with 15–55% Na are liquid at room temperature and therefore easier to transport.

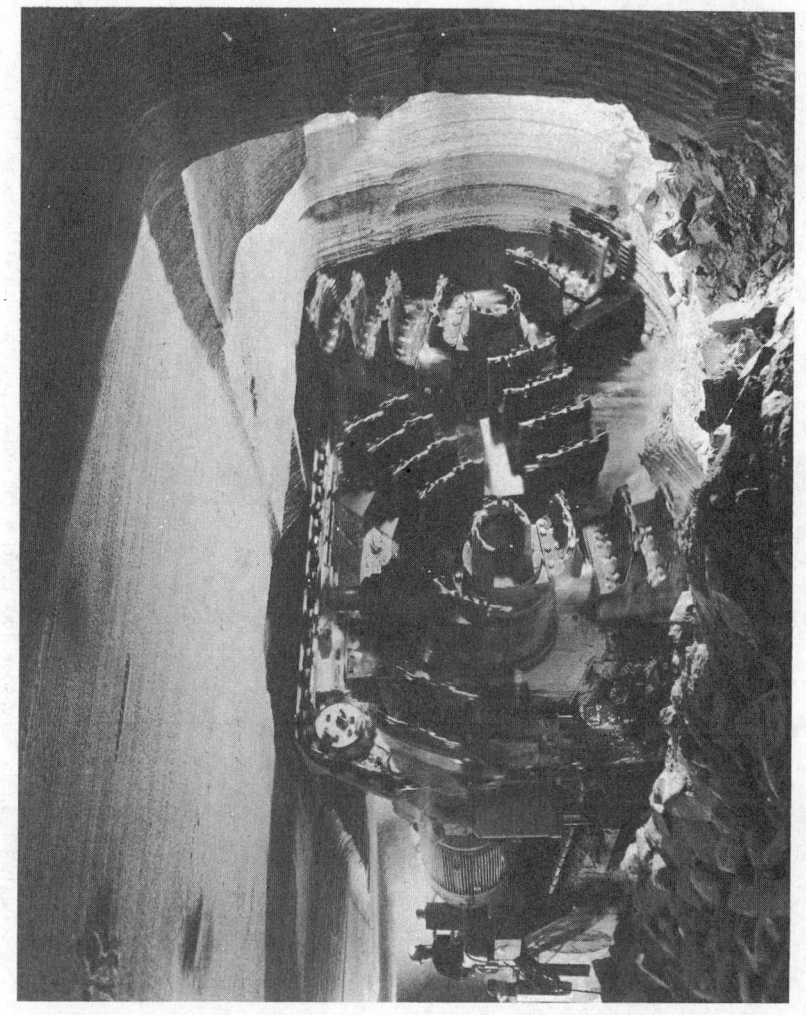

PLATE 4.1 Continuous miners cut through potash ore at the rate of 600 tonnes per hour at Central Canada Potash's mine near Colonsay, Saskatchewan. (Cuts a 3-m high face at 0.3 m per min.)

PLATE 4.2 In the mine, some 1000 m below the surface, patterns such as these are left by the mining machines cutting through the ore. The mill treats 3.63 million tonnes (Mt) of ore pa→1.5 Mt KCl (4.3 kt daily). (Courtesy of Central Canada Potash Co. Limited.)

Industrial uses of Na metal reflect its strong reducing properties. About 60% of world production is used to make $PbEt_4$ (or $PbMe_4$) for gasoline antiknocks via the high-pressure reaction of alkyl chlorides with Na/Pb alloy, though this use is declining for environmental reasons. A further 20% is used to produce Ti, Zr, and other metals by reduction of their chlorides, and about 10% is used to make compounds such as NaH, NaOR, and Na_2O_2. Sodium dispersions are also a valuable catalyst for the production of some artificial rubbers and elastomers. A growing use is as a heat-exchange liquid in fast breeder nuclear reactors where sodium's low mp, low viscosity, and low neutron absorption cross-section combine with its exceptionally high heat capacity and thermal conductivity to make it (and its alloys with K) the most-favoured material. Total annual production of metallic Na in the USA now approaches 200 000 tonnes. Potassium metal, being more difficult and expensive to produce, is manufactured on a much smaller scale. One of its main uses is to make the superoxide KO_2 by direct combustion; this compound is used in breathing masks as an auxiliary supply of O_2 in mines, submarines, and space vehicles:

$$4KO_2 + 2CO_2 \longrightarrow 2K_2CO_3 + 3O_2$$
$$4KO_2 + 4CO_2 + 2H_2O \longrightarrow 4KHCO_3 + 3O_2$$

An indication of the relative cost of the alkali metals in bulk at 1977 prices (USA) is:

Metal	Li	Na	K	Rb	Cs
Price/$ kg^{-1}	17	0.4	44	660	400
Relative cost (per kg)	40	1	10	1650	1000
Relative cost (per mol)	13	1	187	6130	5780

4.2.4 *Properties of the alkali metals*

The Group IA elements are soft, low-melting, silvery-white metals which crystallize with bcc lattices. Lithium is harder than sodium but softer than lead. Atomic properties are summarized in Table 4.1 and general physical properties are in Table 4.2.

Lithium has a variable atomic weight (p. 21) whereas sodium and caesium, being mononuclidic, have very precisely known and invariant atomic weights. Potassium and rubidium are both radioactive but the half-lives of their radioisotopes are so long that the atomic weight does not vary significantly from this cause. The large size and low ionization energies of the alkali metals compared with all other elements have already been noted (pp. 27–29) and this confers on the elements their characteristic properties. The group usually shows smooth trends in properties, and the weak bonding of the single valence electron leads to low mp, bp, and density, and low heats of sublimation, vaporization, and dissociation. Conversely, the elements have large atomic and ionic radii and extremely high thermal and electrical conductivity. Lithium is the smallest element in the group and has the highest ionization energy, mp, and heat of atomization; it also has the lowest density of any solid at room temperature.

All the alkali metals have characteristic flame colorations due to the ready excitation of

TABLE 4.1   *Atomic properties of the alkali metals*

Property	Li	Na	K	Rb	Cs	Fr
Atomic number	3	11	19	37	55	87
Number of naturally occurring isotopes	2	1	$2+1^{(a)}$	$1+1^{(a)}$	1	$1^{(a)}$
Atomic weight	6.941(±3)	22.989 77	39.0983	85.4678(±3)	132.9054	(223)
Electronic configuration	$[He]2s^1$	$[Ne]3s^1$	$[Ar]4s^1$	$[Kr]5s^1$	$[Xe]6s^1$	$[Rn]7s^1$
Ionization energy/kJ mol^{-1}	520.1	495.7	418.6	402.9	375.6	~375
ΔH_{dissoc}/kJ mol^{-1} (M_2)	107.8	73.3	49.9	47.3	43.6	—
Metal radius/pm	152	186	227	248	265	—
Ionic radius (6-coordinate)/pm	76	102	138	152	167	(180)
$E°$/V for $M^+(aq)+e^- \rightarrow M(s)$	−3.03	−2.713	−2.925	−2.93	−2.92	—

(a) Radioactive: ^{40}K $t_{\frac{1}{2}}$ 1.27 × 10⁹ y; ^{87}Rb $t_{\frac{1}{2}}$ 5.7 × 10¹⁰ y; ^{223}Fr $t_{\frac{1}{2}}$ 21.8 min.

TABLE 4.2   *Physical properties of the alkali metals*

Property	Li	Na	K	Rb	Cs
MP/°C	180.5	97.8	63.2	39.0	28.5
BP/°C	1347	881.4	765.5	688	705
Density (20°C)/g cm^{-3}	0.534	0.968	0.856	1.532	1.90
ΔH_{fus}/kJ mol^{-1}	2.93	2.64	2.39	2.20	2.09
ΔH_{vap}/kJ mol^{-1}	148	99	79	76	67
ΔH_f (monatomic gas)/kJ mol^{-1}	162	110	90	88	79

the outermost electron, and this is the basis of their analytical determination by flame photometry or atomic absorption spectroscopy. The colours and principal emission (or absorption) wavelengths are given below but it should be noted that these lines do not all refer to the same transition; for example, the Na D-line doublet at 589.0, 589.6 nm arises from the $3s^1$–$3p^1$ transition in Na atoms formed by reduction of Na^+ in the flame, whereas the red line for lithium is associated with the short-lived species LiOH.

Element	Li	Na	K	Rb	Cs
Colour	Crimson	Yellow	Violet	Red-violet	Blue
λ/nm	670.8	589.2	766.5	780.0	455.5

The reduction potential for lithium appears at first sight to be anomalous and is one of the few properties that does not show a smooth trend with increasing atomic number in the group. This arises from the small size and very large hydration energy of the free gaseous lithium ion. The standard reduction potential $E°$ refers to the reaction $Li^+(aq)+e^- \rightarrow Li(s)$ and is related to the free-energy change: $\Delta G° = -nFE°$. The ionization energy I_M, which is the enthalpy change of the gas-phase reaction $Li(g) \rightarrow Li^+(g)+e^-$, is only one component of this, as can be seen from the following cycle:

$$E^\circ = -\Delta G^\circ / n\mathrm{F}$$

Li⁺(aq) + e⁻ ⟶ Li (s)

$$\Delta G^\circ = \Delta H^\circ - T\Delta S$$

ΔH_{hydr} ΔH_{subl}

$$\Delta H = I_M$$

Li⁺ (g) + e⁻ ⟵ Li (g)

Estimates of the heat of hydration of Li^+ (g) give values near 520 kJ mol⁻¹ compared with 405 kJ mol⁻¹ for Na^+(g) and only 265 kJ mol⁻¹ for Cs^+(g). This factor, together with the much larger entropy change for the lithium electrode reaction (due to the more severe disruption of the water structure by the lithium ion) is sufficient to reverse the position of lithium and make it the most electropositive of the alkali metals (as measured by electrode potential) despite the fact that it is the most difficult element of the group to ionize in the gas phase.

4.2.5 *Chemical reactivity and trends*

The ease of involving the outermost ns^1 electron in bonding, coupled with the very high second-stage ionization energy of the alkali metals, immediately explains both the great chemical reactivity of these elements and the fact that their oxidation state in compounds never exceeds $+1$. The metals have a high lustre when freshly cut but tarnish rapidly in air due to reaction with O_2 and moisture. Reaction with the halogens is vigorous; even explosive in some cases. All the alkali metals react with hydrogen (p. 70) and with proton donors such as alcohols, gaseous ammonia, and even alkynes. They also act as powerful reducing agents towards many oxides and halides and so can be used to prepare many metallic elements or their alloys.

The small size of lithium frequently confers special properties on its compounds and for this reason the element is sometimes termed "anomalous". For example, it is miscible with Na only above 380° and is immiscible with molten K, Rb, and Cs, whereas all other pairs of alkali metals are miscible with each other in all proportions. (The ternary alloy containing 12% Na, 47% K, and 41% Cs has the lowest known mp, $-78°C$, of any metallic system.) Li shows many similarities to Mg. This so-called "diagonal relationship" stems from the similarity in ionic size of the two elements: $r(Li^+)$ 76 pm, $r(Mg^{2+})$ 72 pm, compared with $r(Na^+)$ 102 pm. Thus, as first noted by Arfvedson in establishing lithium as a new element, LiOH and Li_2CO_3 are much less soluble than the corresponding Na and K compounds and the carbonate (like $MgCO_3$) decomposes more readily on being heated. Similarly, LiF (like MgF_2) is much less soluble in water than are the other alkali metal fluorides because of the large lattice energy associated with the small size of both the cation and the anion. By contrast, lithium salts of large, non-polarizable anions such as ClO_4^- are much more soluble than those of the other alkali metals, presumably because of the high energy of solvation of Li^+. For the same reason many simple lithium salts are normally hydrated (p. 101) and the anhydrous salts are extremely hygroscopic: this great affinity for water forms the basis of the widespread use of LiCl and LiBr brines in dehumidifying and air-conditioning units. More subtly there is also a close structural relation between the hydrogen-bonded structures of $LiClO_4 \cdot 3H_2O$ and $Mg(ClO_4)_2 \cdot 6H_2O$ in which the face-shared octahedral groups of $[Li(H_2O)_6]^+$ are replaced alternately by half the number of discrete $[Mg(H_2O)_6]^{2+}$ groups.[10] Lithium sulfate, unlike the other alkali metal sulfates, does not form alums $[M(H_2O)_6]^+[Al(H_2O)_6]^{3+}[SO_4]_2^{2-}$ because the hydrated lithium cation is too small to fill the appropriate site in the alum structure.

Lithium is unusual in reacting directly with N_2 to form the nitride Li_3N; no other alkali metal has this property, which lithium shares with magnesium (which readily forms Mg_3N_2). On the basis of size, it would be expected that Li would be tetrahedrally coordinated by N but, as pointed out by A. F. Wells,[10] this would require 12 tetrahedra to meet at a point which is a geometrical impossibility, 8 being the maximum number theoretically possible; accordingly Li_3N has a unique structure in which one-third of the Li have 2 N atoms as nearest neighbours (at 194 pm) and the remainder have 3 N atoms as neighbours (at 211 pm); each N is surrounded by 2 Li at 194 pm and 6 more at 211 pm.

4.2.6 *Solutions in liquid ammonia and other solvents*[11]

One of the most remarkable features of the alkali metals is their ready solubility in liquid ammonia to give bright blue, metastable solutions with unusual properties. Such solutions have been extensively studied since they were first observed by T. Weyl in 1863,[†] and it is now known that similar solutions are formed by the heavier alkaline earth metals (Ca, Sr, and Ba) and the divalent lanthanoids europium and ytterbium in liquid ammonia. Many amines share with ammonia this ability though to a much lesser extent. It is clear that solubility is favoured by low metal lattice energy, low ionization energies, and high cation solvation energy. The most striking physical properties of the solutions are their colour, electrical conductivity, and magnetic susceptibility. The solutions all have the same blue colour when dilute, suggesting the presence of a common coloured species, and they become bronze-coloured and metallic at higher concentrations. The conductivity of the dilute solutions is an order of magnitude higher than that of completely ionized salts in water; as the solutions becomes more concentrated the conductivity at first diminishes to a minimum value at about 0.04 M and then increases dramatically to approach values typical of liquid metals. Dilute solutions are paramagnetic with a susceptibility appropriate to the presence of 1 free electron per metal atom; this susceptibility diminishes with increase in concentration, the solutions becoming diamagnetic in the region of the conductivity minimum and then weakly paramagnetic again at still higher concentrations.

The interpretation of these remarkable properties has excited considerable interest: whilst there is still some uncertainty as to detail, it is now generally agreed that in dilute solution the alkali metals ionize to give a cation M^+ and a quasi-free electron which is

---

[†] Actually, the first observation was probably made by Sir Humphry Davy some 55 y earlier: an unpublished observation in his Notebook for November 1807 has recently been discovered by Peter Edwards of Cambridge University. It reads "When 8 grains of potassium were heated in ammoniacal gas it assumed a beautiful metallic appearance and gradually became of a pure blue colour".

[10] A. F. WELLS, *Structural Inorganic Chemistry*, 4th edn., Oxford University Press, Oxford, 1975, 1095 pp.
[11] W. L. JOLLY and C. J. HALLADA, Liquid ammonia, Chap. 1 in T. C. WADDINGTON (ed.), *Non-aqueous Solvent Systems*, pp. 1–45, Academic Press, London, 1965. J. C. THOMPSON, The physical properties of metal solutions in non-aqueous solvents, Chap. 6 in J. LAGOWSKI (ed.), *The Chemistry of Non-aqueous Solvents*, Vol. 2, pp. 265–317, Academic Press, New York, 1967.

distributed over a cavity in the solvent of radius 300–340 pm formed by displacement of 2–3 NH_3 molecules. This species has a broad absorption band extending into the infrared with a maximum of ~ 1500 nm and it is the short wavelength tail of this band which gives rise to the deep-blue colour of the solutions. The cavity model also interprets the fact that dissolution occurs with considerable expansion of volume so that the solutions have densities that are appreciably lower than that of liquid ammonia itself. The variation of properties with concentration can best be explained in terms of three equilibria between five solute species M, M_2, M^+, M^-, and e^-:

$$M_{am} \rightleftharpoons M_{am}^+ + e_{am}^-; \quad K \sim 10^{-2}$$
$$M_{am}^- \rightleftharpoons M_{am} + e_{am}^-; \quad K \sim 10^{-3}$$
$$(M_2)_{am} \rightleftharpoons 2M_{am}; \quad K \sim 2 \times 10^{-4}$$

The subscript am indicates that the species are dissolved in liquid ammonia and may be solvated. At very low concentrations the first equilibrium predominates and the high ionic conductivity stems from the high mobility of the electron which is some 280 times that of the cation. The species M_{am} can be thought of as an ion pair in which M_{am}^+ and e_{am}^- are held together by coulombic forces. As the concentration is raised the second equilibrium begins to remove mobile electrons e_{am}^- as the complex M_{am}^- and the conductivity drops. Concurrently M_{am} begins to dimerize to give $(M_2)_{am}$ in which the interaction between the 2 electrons is sufficiently strong to lead to spin-pairing and diamagnetism. At still higher concentrations the system behaves as a molten metal in which the metal cations are ammoniated. Saturated solutions are indeed extremely concentrated as indicated by the following table.

Solute	Li	Na	K	Rb	Cs
$T/°C$	$-33.2°$	$-33.5°$	$-33.2°$	—	$-50°$
$g(M)/kg(NH_3)$	108.7	251.4	463.7	—	3335
$mol(NH_3)/mol(M)$	3.75	5.37	4.95	—	2.34

The lower solubility of Li on a wt/wt basis reflects its lower atomic weight and, when compared on a molar basis, it is nearly 50% more soluble than Na (15.66 mol/kg NH_3 compared to 10.93 mol/kg NH_3). Note that it requires only 2.34 mol NH_3 (39.8 g) to dissolve 1 mol Cs (132.9 g).

Solutions of alkali metals in liquid ammonia are valuable as powerful and selective reducing agents. The solutions are themselves unstable with respect to amide formation:

$$M + NH_3 \longrightarrow MNH_2 + \tfrac{1}{2}H_2$$

However, under anhydrous conditions and in the absence of catalytic impurities such as transition metal ions, solutions can be stored for several days with only a few per cent decomposition. Some reductions occur without bond cleavage as in the formation of alkali metal superoxides and peroxide (p. 98).

$$O_2 \xrightarrow{\ e_{am}^-\ } O_2^- \xrightarrow{\ e_{am}^-\ } O_2^{2-}$$

Transition metal complexes can be reduced to unusually low oxidation states either with or without bond cleavage, e.g.:

$$K_2[Ni(CN)_4] + 2K \xrightarrow{NH_3/-33°} K_4[Ni(CN)_4]; \qquad \text{i.e. Ni(0)}$$

$$[Pt(NH_3)_4]Br_2 + 2K \xrightarrow{NH_3/-33°} [Pt(NH_3)_4] + 2KBr; \qquad \text{i.e. Pt(0)}$$

$$Mn_2(CO)_{10} + 2K \xrightarrow{NH_3/-33°} 2KMn(CO)_5]; \qquad \text{i.e. Mn}(-1)$$

$$Fe(CO)_5 + 2Na \xrightarrow{NH_3/-33°} Na_2[Fe(CO)_4] + CO; \qquad \text{i.e. Fe}(-2)$$

Salts of several heavy main-group elements can be reduced to form polyanions such as $Na_4[Sn_9]$, $Na_3[Sb_3]$, and $Na_3[Sb_7]$ (p. 686).

Many protonic species react with liberation of hydrogen:

$$RC{\equiv}CH + e_{am}^- \longrightarrow RC{\equiv}C^- + \tfrac{1}{2}H_2$$
$$GeH_4 + e_{am}^- \longrightarrow GeH_3^- + \tfrac{1}{2}H_2$$
$$NH_4^+ + e_{am}^- \longrightarrow NH_3 + \tfrac{1}{2}H_2$$
$$AsH_3 + e_{am}^- \longrightarrow AsH_2^- + \tfrac{1}{2}H_2$$
$$EtOH + e_{am}^- \longrightarrow EtO^- + \tfrac{1}{2}H_2$$

These and similar reactions have considerable synthetic utility. Other reactions which result in bond cleavage by the addition of one electron are:

$$R_2S + e_{am}^- \longrightarrow RS^- + \tfrac{1}{2}R_2$$
$$Et_3SnBr + e_{am}^- \longrightarrow Et_3Sn^{\bullet} + Br^-$$

When a bond is broken by addition of 2 electrons, either 2 anions or a dianion is formed:

$$Ge_2H_6 + 2e_{am}^- \longrightarrow 2GeH_3^-$$
$$PhNHNH_2 + 2e_{am}^- \longrightarrow PhNH^{\cdots} + NH_2^-$$
$$PhN{=}O + 2e_{am}^- \longrightarrow PhN^- {-} O^-$$
$$S_8 + 2e_{am}^- \longrightarrow S_8^{2-}$$

Subsequent ammonolysis may also occur:

$$RCH{=}CH_2 + 2e_{am}^- \longrightarrow \{RCH{-}CH_2^{2-}\} \xrightarrow{2NH_3} RCH_2CH_3 + 2NH_2^-$$

$$N_2O + 2e_{am}^- \longrightarrow \{N_2 + O^{2-}\} \xrightarrow{NH_3} N_2 + OH^- + NH_2^-$$

$$NCO^- + 2e_{am}^- \longrightarrow \{CN^- + O^{2-}\} \xrightarrow{NH_3} CN^- + OH^- + NH_2^-$$

$$EtBr + 2e_{am}^- \longrightarrow \{Br^- + Et^-\} \xrightarrow{NH_3} Br^- + C_2H_6 + NH_2^-$$

Solutions of alkali metals in liquid ammonia have been developed as versatile reducing agents which effect reactions with organic compounds that are otherwise difficult or impossible.[12] Aromatic systems are reduced smoothly to cyclic mono- or di-olefins and

---

[12] A. J. BIRCH, Reduction of organic compounds by metal-ammonia solutions, *Qt. Rev.* **4**, 69–93 (1950); A. J. BIRCH and H. SMITH, Reduction by metal-amine solutions, *Qt. Rev.* **12**, 17–33 (1958).

alkynes are reduced stereospecifically to *trans*-alkenes (in contrast to Pd/H_2 which gives *cis*-alkenes).

The alkali metals are also soluble in aliphatic amines and hexamethylphosphoramide, $P(NMe_2)_3$ to give coloured solutions which are strong reducing agents. These solutions appear to be similar in many respects to the dilute solutions in liquid ammonia though they are less stable with respect to decomposition into amide and H_2. Likewise, fairly stable solutions of the larger alkali metals K, Rb, and Cs have been obtained in tetrahydrofuran, ethylene glycol dimethyl ether, and other polyethers. These and similar solutions have been successfully used as strong reducing agents in situations where protonic solvents would have caused solvolysis. For example, naphthalene reacts with Na in tetrahydrofuran to form deep-green solutions of the paramagnetic sodium naphthenide, $NaC_{10}H_8$, which can be used directly in the presence of a bis(tertiary phosphine) ligand to reduce the anhydrous chlorides, VCl_3, $CrCl_3$, $MoCl_5$, and WCl_6 to the zero-valent octahedral complexes $[M(Me_2PCH_2CH_2PMe_2)_3]$, where M = V, Cr, Mo, W. Similarly the planar complex $[Fe(Me_2PCH_2CH_2PMe_2)_2]$ was obtained from *trans*-$[Fe(Me_2PCH_2Ch_2PMe_2)_2Cl_2]$, and the corresponding tetrahedral Co(0) compound from $CoCl_2$.[13]

4.3 Compounds[14, 15]

4.3.1 *Introduction: the ionic-bond model*[16]

The alkali metals form a complete range of compounds with all the common anions and have long been used to illustrate group similarities and trends. It has been customary to discuss the simple binary compounds in terms of the ionic bond model and there is little doubt that there is substantial separation of charge between the cationic and anionic components of the crystal lattice. On this model the ions are considered as hard, undeformable spheres carrying charges which are integral multiples of the electronic charge z_1e^+. Corrections can be incorporated for zero-point energies, London dispersion energies, ligand-field stabilization energies, and non-spherical ions (such as NO_3^-, etc.). The attractive simplicity of this model, and its considerable success during the past 60 y in interpreting many of the properties of simple salts, should not, however, be allowed to obscure the growing realization of its inadequacy.[16, 17] In particular, as already noted, success in calculating lattice energies and hence enthalpy of formation via the Born–Haber cycle, does not establish the correctness of the model but merely indicates that it is consistent with these particular observations. For example, the ionic model is quite successful in reproducing the enthalpy of formation of BF_3, SiF_4, PF_5, and even SF_6

[13] J. Chatt and H. R. Watson, Complexes of zerovalent transition metals with the ditertiary phosphine, $Me_2PCH_2CH_2PMe_2$, *J. Chem. Soc.* 1962, 2545–9.

[14] H. Remy, *Treatise on Inorganic Chemistry* (translated by J. S. Anderson), Vol. 1, Chap. 6, The alkali metals, pp. 153–202, Elsevier, Amsterdam, 1956.

[15] W. A. Hart and O. F. Beumel, Lithium and its compounds, *Comprehensive Inorganic Chemistry*, Vol. 1, Chap. 7, Pergamon Press, Oxford, 1973. T. P. Whaley, Sodium, potassium, rubidium, caesium, and francium, ibid., Chap. 8.

[16] N. N. Greenwood, *Ionic Crystals, Lattice Defects, and Nonstoichiometry*, Butterworths, London, 1968, 194 pp.

[17] D. M. Adams, *Inorganic Solids: An Introduction to Concepts in Solid-State Structural Chemistry*, Wiley, London, 1974, 336 pp.

on the assumption that they are assemblies of point charges at the known interatomic distance, i.e. $B^{3+}(F^-)_3$, etc.,[18] but this is not a sound reason for considering these molecular compounds as ionic. Likewise, the known lattice energy of lithium metal can be reproduced quite well by assuming that the observed bcc arrangement of atoms is made up from alternating ions Li^+Li^- in the CsCl structure;[19] the discrepancy is no worse than that obtained using the same model for AgCl (which has the NaCl structure). It appears that the ionic-bond model is self compensating and that the decrease in the hypothetical binding energy which accompanies the diminution of formal charges on the atoms is accompanied by an equivalent increase in binding energy which could be described as "covalent" (BF_3) or "metallic" (Li metal).

A more satisfactory procedure, at least conceptually, is to describe crystalline salts and other solid compounds in terms of molecular orbitals. Quantitative calculations are difficult to carry out but the model allows flexibility in placing "partial ionic charges" on atoms by modifying orbital coefficients and populations, and it can also incorporate metallic behaviour by modifying the extent to which partly filled individual molecular orbitals are either separated by energy gaps or overlap.

The compounds which most nearly fit the classicial conception of ionic bonding are the alkali metal halides. However, even here, one must ask to what extent it is reasonable to maintain that positively charged cations M^+ with favourably directly vacant p orbitals remain uncoordinated by the surrounding anionic ligands X^- to form extended (bridged) complexes. Such interaction would be expected to increase from Cs^+ to Li^+ and from F^- to I^- (why?) and would place some electron density between the cation and anion. Some evidence on this comes from very precise electron density plots obtained by X-ray diffraction experiments on LiF, NaCl, KCl, MgO, and CaF_2.[20] Data for LiF are shown in Fig. 4.2a from which it is clear that the Li^+ ion is no longer spherical and that the electron density, while it falls to a low value between the ions, does not become zero. Even more significantly, as shown in Fig. 4.2b, the minimum does not occur at the position to be expected from the conventional ionic radii: whatever set of tabulated values is used the cation is always larger than expected and the anion smaller. This is consistent with a transfer of some electronic density from anion to cation since the smaller resultant positive charge on the cation exerts smaller coulombic attraction for the electrons and the ion expands. The opposite holds for the anion. These results also call into question the use of radius-ratio rules to calculate the coordination number of cations and leave undecided the numerical value of the ionic radii to be used (see also p. 72, hydrides). In fact, the radius-ratio rules are particularly unhelpful for the alkali halides, since they predict (incorrectly) that LiCl, LiBr, and LiI should have tetrahedral coordination and that NaF, KF, KCl, RbF, RbCl, RbBr, and CsF should all have the CsCl structure. It may be significant that adoption of the NaCl structure by all these compounds maximizes the p–p orbital overlap along the orthogonal x-, y-, and z-directions, and so favours molecular orbital

[18] F. J. GARRICK, Studies in coordination. Part 4. Some fluorides and chlorides and their complexes, *Phil. Mag.* **14**, 914–37 (1932). It is instructive to repeat some of these calculations with more recent values for the constants and properties used.

[19] C. S. G. PHILLIPS and R. J. P. WILLIAMS, *Inorganic Chemistry*, Vol. 1, Chap. 5, "The ionic model", pp. 142–87, Oxford University Press, Oxford, 1965.

[20] H. WITTE and E. WÖLFEL, X-ray determination of the electron distribution of crystals, *Z. phys. Chem.* **3**, 296–329 (1955). J. KRUG, H. WITTE, and E. WÖLFEL, ibid. **4**, 36–64 (1955). H. WITTE and E. WÖLFEL, Electron distributions in NaCl, LiF, CaF_2, and Al, *Rev. Mod. Phys.* **30**, 51–5 (1958).

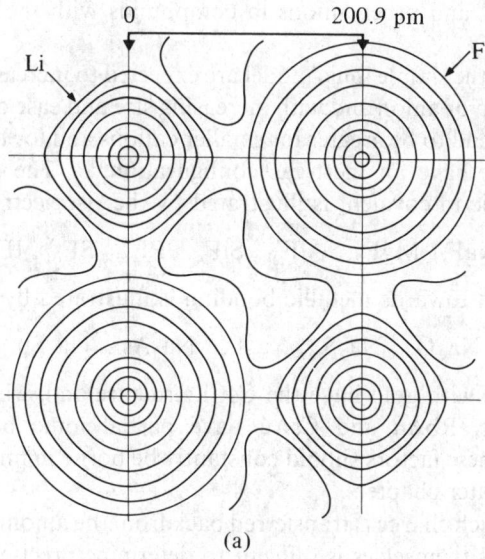

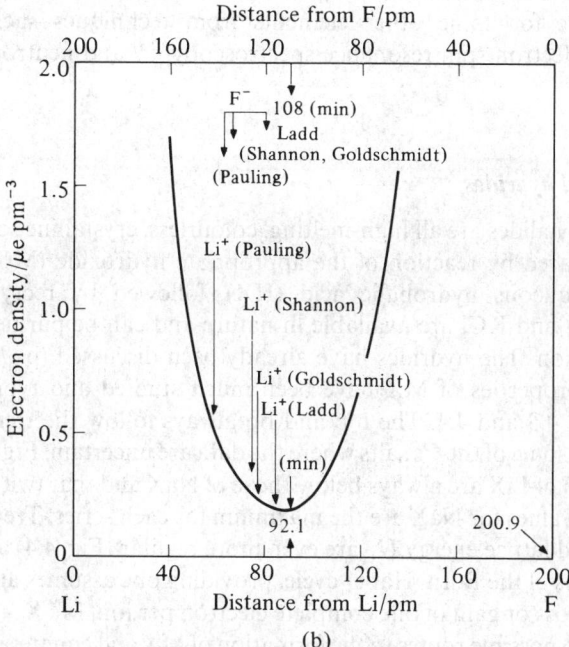

FIG. 4.2   (a) Distribution of electron density (μe/pm^3) in the *xyo* plane of LiF, and (b) variation of electron density along the Li–F direction near the minimum. The electron density rises to 17.99 μe pm^{-3} at Li and to 115.63 μe pm^{-3} at F. (The unit μe pm^{-3} is numerically identical to eÅ^{-3}.)

formation in these directions. Further information on the variation in apparent radius of the hydride, halide, and other anions in compounds with the alkali metals and other cations is in ref. 21.

Deviations from the simple ionic model are expected to increase with increasing formal charge on the cation or anion and with increasing size and ease of distortion of the anion. Again, deviations tend to be greater for smaller cations and for those (such as Cu^+, Ag^+, etc.) which do not have an inert-gas configuration.[16] The gradual transition from predominantly ionic to covalent is illustrated by the "isoelectronic" series:

$$NaF, \quad MgF_2, \quad AlF_3, \quad SiF_4, \quad PF_5, \quad SF_6, \quad IF_7, \quad F_2$$

A similar transition towards metallic bonding is illustrated by the series:

$$NaCl, \quad Na_2O, \quad Na_2S, \quad Na_3P, \quad Na_3As, \quad Na_3Sb, \quad Na_3Bi, \quad Na$$

Alkali metal alloys with gold have the CsCl structure and, whilst NaAu and KAu are essentially metallic, RbAu and CsAu have partial ionic bonding and are n-type semiconductors. These factors should constantly be borne in mind during the discussion of compounds in later chapters.

The extent to which charge is transferred back from the anion towards the cation in the alkali metal halides themselves is difficult to determine precisely. Calculations indicate that it is probably only a few percent for some salts such as NaCl, whereas for others (e.g. LiI) it may amount to more than 0.33 e^- per atom. Direct experimental evidence on these matters is avilable for some other elements from techniques such as Mössbauer spectroscopy,[22] electron spin resonance spectroscopy,[23] and neutron scattering form factors.[24]

4.3.2 *Halides and hydrides*

The alkali metal halides are all high-melting, colourless, crystalline solids which can be conveniently prepared by reaction of the appropriate hydroxide (MOH) or carbonate (M_2CO_3) with aqueous hydrohalic acid (HX), followed by recrystallization. Vast quantities of NaCl and KCl are available in nature and can be purified if necessary by simple crystallization. The hydrides have already been discussed (p. 70).

Trends in the properties of MX have been much studied and typical examples are illustrated in Figs. 4.3 and 4.4. The mp and bp always follow the trend $F > Cl > Br > I$ except perhaps for some of the Cs salts where the data are uncertain. Figure 4.3 also shows that the mp and bp of LiX are always below those of NaX and that (with the exception of the mp of KI) the values for NaX are the maximum for each series. Trends in enthalpy of formation ΔH_f° and lattice energy U_L are even more regular (Fig. 4.4) and can readily be interpreted in terms of the Born–Haber cycle, providing one assumes an invariant charge corresponding to loss or gain of one complete electron per ion, M^+X^-. The Born–Haber cycle considers two possible routes to the formation of MX and equates the corresponding enthalpy changes by applying Hess's law:[16]

[21] O. JOHNSON, Ionic radii for spherical potential ions, *Inorg. Chem.* **12**, 780–85 (1973).

[22] N. N. GREENWOOD and T. C. GIBB, *Mössbauer Spectroscopy*, Chapman & Hall, London, 1971, 659 pp.

[23] P. B. AYSCOUGH, *Electron Spin Resonance in Chemistry*, pp. 300–01, Methuen, London, 1967. P. W. ATKINS and M. C. R. SYMONS, *The Structure of Inorganic Radicals*, pp. 51–73, Elsevier, Amsterdam, 1967.

[24] G. E. BACON, *Neutron Diffraction*, 3rd edn., Oxford University Press, Oxford, 1975, 636 pp.

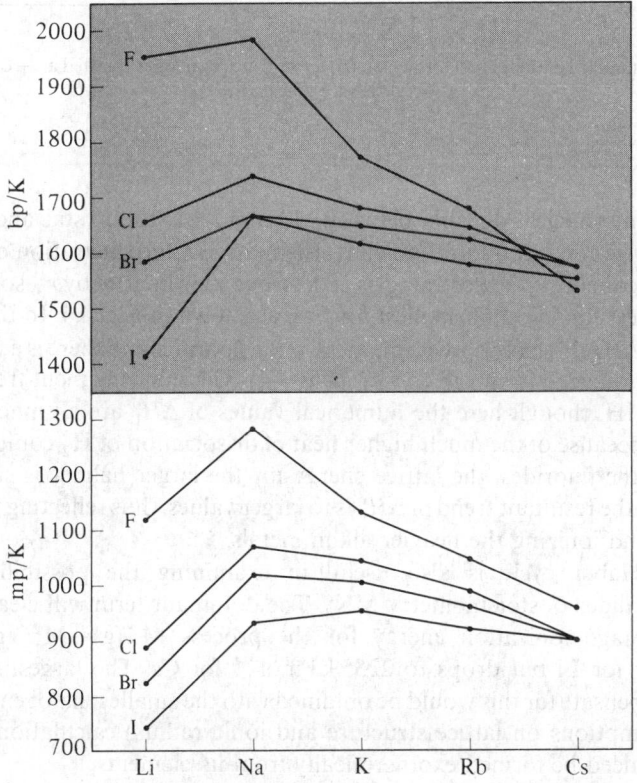

Hence

$$\Delta H_f^\circ(MX) = S_M + \tfrac{1}{2}D_{X_2} + I_M - E_X - U_L$$

where S_M is the heat of sublimation of $M(c)$ to a monatomic gas (Table 4.2), D_{X_2} is the dissociation energy of $X_2(g)$ (Table 4.2), I_M is the ionization energy of $M(g)$ (Table 4.2), and E_X the electron affinity of $X(g)$ (Table 17.4, p. 934). The lattice energy U_L is given approximately by the expression

$$U_L = \frac{N_0 A e^2}{4\pi\varepsilon_0 r_0}\left(1 - \frac{\rho}{r_0}\right)$$

where N_0 is the Avogadro constant, A is a geometrical factor, the Madelung constant (which has the value of 1.7627 for the CsCl structure and 1.7476 for the NaCl structure), r_0

FIG. 4.3 Melting point and boiling point of alkali metal halides.

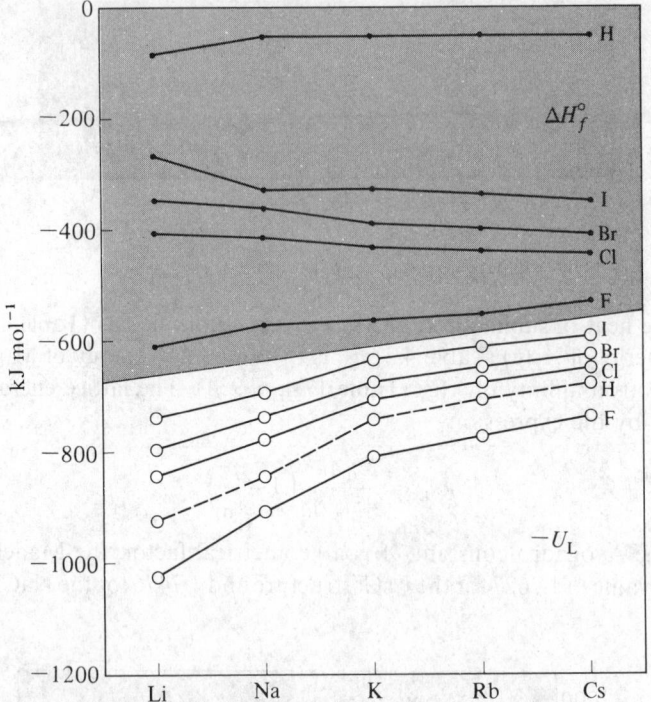

FIG. 4.4   Standard enthalpies of formation (ΔH_f°) and lattice energies (plotted as $-U_L$) for alkali metal halides and hydrides.

is the shortest internuclear distance between M^+ and X^- in the crystal, and ρ is a measure of the close-range repulsion force which resists mutual interpenetration of the ions. It is clear that the sequence of lattice energies is determined primarily by r_0, so that the lattice energy is greatest for LiF and smallest for CsI, as shown in Fig. 4.4. In the Born–Haber expression for ΔH_f° this factor predominates for the fluorides and there is a trend to smaller enthalpies of formation from LiF to CsF (Fig. 4.4). The same incipient trend is noted for the hydrides MH, though here the numerical values of ΔH_f° are all much smaller than those for MX because of the much higher heat of dissociation of H_2 compared to X_2. By contrast with the fluorides, the lattice energy for the larger halides is smaller and less dominant, and the resultant trend of ΔH_f° is to larger values, thus reflecting the greater ease of subliming and ionizing the heavier alkali metals.

The Born–Haber cycle is also useful in examining the possibility of forming alkali–metal halides of stoichiometry MX_2. The dominant term will clearly be the very large second-stage ionization energy for the process $M^+(g) \rightarrow M^{2+}(g) + e^-$; this is 7297 kJ mol^{-1} for Li but drops to 2255 kJ mol^{-1} for Cs. The largest possible lattice energy to compensate for this would be obtained with the smallest halogen F and (making plausible assumptions on lattice structure and ionic radius) calculations indicate that CsF_2 could indeed be formed exothermically from its elements:

$$Cs(s) + F_2(g) = CsF_2(s); \quad \Delta H_f^\circ \simeq -125 \text{ kJ mol}^{-1}$$

However, the compound cannot be prepared because of the much greater enthalpy of formation of CsF which makes CsF_2 unstable with respect to disproportionation:

$$Cs(s) + \tfrac{1}{2}F_2(g) = CsF(s); \quad \Delta H_f^\circ = -530 \text{ kJ mol}^{-1}$$

whence

$$CsF_2(s) = CsF(s) + \tfrac{1}{2}F_2; \quad \Delta H_{disprop} \simeq -405 \text{ kJ mol}^{-1}$$

The alkali metal halides, particularly NaCl and KCl find extensive application in industry (pp. 79 and 83). The hydrides are frequently used as reducing agents, the product being a hydride or complex metal hydride depending on the conditions used, or the free element if the hydride is unstable. Illustrative examples using NaH are:

$$2BF_3 + 6NaH \xrightarrow{200^\circ} B_2H_6 + 6NaF$$

$$BF_3 + 4NaH \xrightarrow{Et_2O/125^\circ} NaBH_4 + 3NaF$$

$$B(OMe)_3 + NaH \xrightarrow{reflux} Na[BH(OMe)_3]$$

$$B(OMe)_3 + 4NaH \xrightarrow{225-275^\circ} NaBH_4 + 3NaOMe$$

$$AlBr_3 + 4NaH \xrightarrow{Me_2O} NaAlH_4 + 3NaBr$$

$$TiCl_4 + 4NaH \xrightarrow{400^\circ} Ti + 4NaCl + 2H_2$$

Sulfur dioxide is uniquely reduced to dithionite (a process useful in bleaching paper pulp, p. 854). CO_2 gives the formate:

$$2SO_2(l) + 2NaH \longrightarrow Na_2S_2O_4 + H_2$$
$$CO_2(g) + NaH \longrightarrow HCO_2Na$$

4.3.3 *Oxides, peroxides, superoxides, and suboxides*

The alkali metals form a fascinating variety of binary compounds with oxygen, the most versatile being Cs which forms 9 compounds with stoichiometries ranging from Cs_7O to CsO_3. When the metals are burned in a free supply of air the predominant product depends on the metal: Li forms the oxide Li_2O (plus some Li_2O_2), Na forms the peroxide Na_2O_2 (plus some Na_2O), whilst K, Rb, and Cs form the superoxide MO_2. Under the appropriate conditions pure compounds M_2O, M_2O_2 and MO_2 can be prepared for all five metals.

The "normal" oxides M_2O (Li, Na, K, Rb) have the antifluorite structure as do many of the corresponding sulfides, selenides, and tellurides. This structure is related to the CaF_2 structure (p. 1407) but with the sites occupied by the cations and anions interchanged so that M replaces F and O replaces Ca in the structure. Cs_2O has the anti-$CdCl_2$ layer structure (p. 1407). There is a trend to increasing coloration with increasing atomic number, Li_2O and Na_2O being pure white, K_2O yellowish white, Rb_2O bright yellow, and

Cs_2O orange. The compounds are fairly stable towards heat, and thermal decomposition is not extensive below about 500°. Pure Li_2O is best prepared by thermal decomposition of Li_2O_2 (see below) at 450°C. Na_2O is obtained by reaction of Na_2O_2, $NaOH$, or preferably $NaNO_2$ with the Na metal:

$$Na_2O_2 + 2Na \longrightarrow 2Na_2O$$
$$NaOH + Na \longrightarrow Na_2O + \tfrac{1}{2}H_2$$
$$NaNO_2 + 3Na \longrightarrow 2Na_2O + \tfrac{1}{2}N_2$$

In this last reaction Na can be replaced by the azide NaN_3 to give the same products. The normal oxides of the other alkali metals can be prepared similarly.

The peroxides M_2O_2 contain the peroxide ion O_2^{2-} which is isoelectronic with F_2. Li_2O_2 is prepared industrially by the reaction of $LiOH.H_2O$ with hydrogen peroxide, followed by dehydration of the hydroperoxide by gentle heating under reduced pressure:

$$LiOH.H_2O + H_2O_2 \longrightarrow LiOOH.H_2O + H_2O$$

$$2LiOOH.H_2O \xrightarrow{\text{heat}} Li_2O_2 + H_2O_2 + 2H_2O$$

It is a thermodynamically stable, white, crystalline solid which decomposes to Li_2O on being heated above 195°C.

Na_2O_2, is prepared as pale-yellow powder by first oxidizing Na to Na_2O in a limited supply of dry oxygen (air) and then reacting this further to give Na_2O_2:

$$2Na + \tfrac{1}{2}O_2 \xrightarrow{\text{heat}} Na_2O \xrightarrow{\frac{1}{2}O_2} Na_2O_2$$

Preparation of pure K_2O_2, Rb_2O_2, and Cs_2O_2 by this route is difficult because of the ease with which they oxidize further to the superoxides MO_2. Oxidation of the metals with NO has been used but the best method is the quantitative oxidation of the metals in liquid ammonia solution (p. 89). The peroxides can be regarded as salts of the dibasic acid H_2O_2. Thus reaction with acids or water quantitatively liberates H_2O_2:

$$M_2O_2 + H_2SO_4 \longrightarrow M_2SO_4 + H_2O_2; \quad M_2O_2 + H_2O \longrightarrow 2MOH + H_2O_2$$

Sodium peroxide finds widespread use industrially as a bleaching agent for fabrics, paper pulp, wood, etc., and as a powerful oxidant; it explodes with powdered aluminium or charcoal, reacts with sulfur with incandescence, and ignites many organic liquids. Carbon monoxide forms the carbonate, and CO_2 liberates oxygen (an important application in breathing apparatus for divers, firemen, and in submarines—space capsules use the lighter Li_2O_2):

$$Na_2O_2 + CO \longrightarrow Na_2CO_3$$
$$Na_2O_2 + CO_2 \longrightarrow Na_2CO_3 + \tfrac{1}{2}O_2$$

In the absence of oxygen or oxidizable material, the peroxides (except Li_2O_2) are stable towards thermal decomposition up to quite high temperatures, e.g. $Na_2O_2 \sim 675°C$, $Cs_2O_2 \sim 590°C$.

The superoxides MO_2 contain the paramagnetic ion O_2^- which is stable only in the presence of large cations such as K, Rb, Cs (and Sr, Ba, etc.). LiO_2 has only been prepared by matrix isolation experiments at 15 K and positive evidence for NaO_2 was first obtained by reaction of O_2 with Na dissolved in liquid NH_3; it can be obtained pure by reacting Na with O_2 at 450°C and 150 atm pressure. By contrast, the normal products of combustion of the heavier alkali metals in air are KO_2 (orange), mp 380°C, RbO_2 (dark brown), mp 412°C, and CsO_2 (orange), mp 432°C. NaO_2 is trimorphic, having the marcasite structure (p. 804) at low temperatures, the pyrite structure (p. 804) between $-77°$ and $-50°$C and a pseudo-NaCl structure above this, due to disordering of the O_2^- ions by rotation. The heavier congeners adopt the tetragonal CaC_2 structure (p. 320) at room temperature and the pseudo-NaCl structure at high temperature.

Sesquoxides "M_2O_3" have been prepared as dark-coloured paramagnetic powders by careful thermal decomposition of MO_2 (K, Rb, Cs). They can also be obtained by oxidation of liquid ammonia solutions of the metals or by controlled oxidation of the peroxides, and are considered to be peroxide disuperoxides $[(M^+)_4(O_2^{2-})(O_2^-)_2]$; however, the fact that they apparently crystallize in the cubic system implies the equivalence of the three O_2^{n-} groups, and the structures merit further investigation.

Ozonides MO_3 have been prepared for Na, K, Rb, and Cs by the reaction of O_3 on powdered anhydrous MOH at low temperature and extraction of the red MO_3 by liquid NH_3:

$$3MOH(c) + 2O_3(g) \longrightarrow 2MO_3(c) + MOH \cdot H_2O(c) + \tfrac{1}{2}O_2(g)$$

Under similar conditions Li gave $[Li(NH_3)_4]O_3$ which decomposed on attempted removal of the coordinated NH_3, again emphasizing the important role of cation size in stabilizing catenated oxygen anions. The ozonides, on standing, slowly decompose to oxygen and the superoxide MO_2, but on hydrolysis they appear to go directly to the hydroxide:

$$MO_3 \longrightarrow MO_2 + \tfrac{1}{2}O_2$$
$$4MO_3 + 2H_2O \longrightarrow 4MOH + 5O_2$$

In addition to the above oxides M_2O, M_2O_2, M_4O_6, MO_2, and MO_3 in which the alkali metal has the constant oxidation state $+1$, rubidium and caesium also form suboxides in which the formal oxidation state of the metal is considerably lower. Some of these intriguing compounds have been known since the turn of the century but only recently have their structures been elucidated by single crystal X-ray analysis.[25] Partial oxidation of Rb at low temperatures gives Rb_6O which decomposes above $-7.3°$C to give copper-coloured metallic crystals of Rb_9O_2:

$$2Rb_6O \xrightarrow{\ -7.3°\ } Rb_9O_2 + 3Rb$$

Rb_9O_2 inflames with H_2O and melts incongruently at 40.2° to give $2Rb_2O + 5Rb$. The structure of Rb_9O_2 comprises two ORb_6 octahedra sharing a common face (Fig. 4.5). It thus has the anti-$[Tl_2Cl_9]^{3-}$ structure. The Rb–Rb distance within this unit is only 352 pm (compared with 485 pm in Rb metal) and the nearest Rb–Rb distance between groups is 511 pm. The Rb–O distance is ~ 249 pm, much less than the sum of the conventional ionic radii (289 pm) and the metallic character of the oxide comes from the excess of at least 5 electrons above that required for simple bookkeeping. Crystalline

[25] A. SIMON, The crystal structure of Cs_7O, Rb_9O_2, *Naturwiss.* **58**, 622–23 (1971); *Z. anorg. Chem.* **395**, 301 (1973). See also A. SIMON, Structure and bonding in alkali metal suboxides, *Struct. Bonding* **36**, 81–127 (1979).

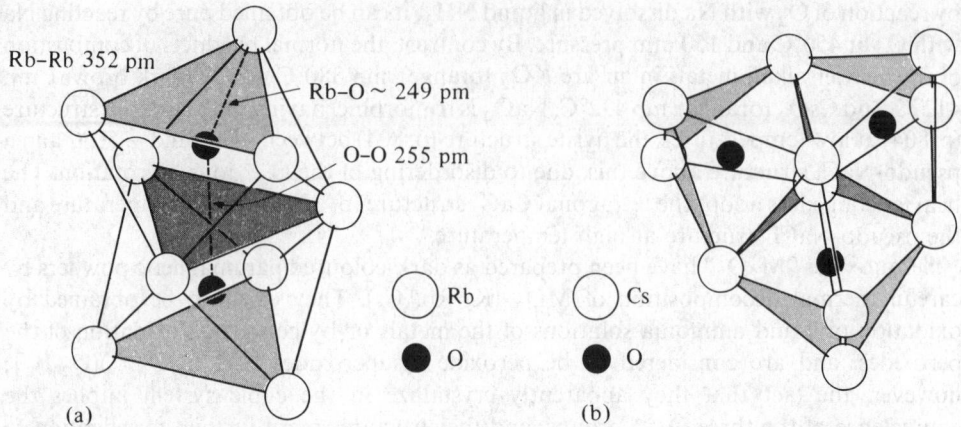

Fig. 4.5   (a) The confacial bioctahedral Rb_9O_2 group in Rb_9O_2 and Rb_6O, and (b) the confacial trioctahedral $Cs_{11}O_3$ group in Cs_7O.

Rb_6O has a unit cell containing 4 formula units, i.e. $Rb_{24}O_4$, and the structure consists of alternating layers of Rb_9O_2 and close-packed metal atoms parallel to (001) to give the structural formula $[(Rb_9O_2)Rb_3]$.

Caesium forms an even more extensive series of suboxides: Cs_7O, bronze-coloured, mp $+4.3°C$; Cs_4O, red-violet, decomposes $> 10.5°$; $Cs_{11}O_3$, violet crystals, mp (incongruent) $52.5°C$; and $Cs_{3+x}O$ a nonstoichiometric phase up to Cs_4O, which decomposes at $166°C$. Cs_7O reacts vigorously with O_2 and H_2O and the unit cell is found to be $Cs_{21}O_3$, i.e. $[(Cs_{11}O_3)Cs_{10}]$. The unit $Cs_{11}O_3$ comprises 3 octahedral OCs_6 groups each sharing 2 adjacent faces to form the trigonal group shown in Fig. 4.5b. These groups form chains along (001) and are also surrounded by the other Cs atoms. The Cs–Cs distance within the $Cs_{11}O_3$ group is only 376 pm, whereas between groups it is 527 pm; this latter distance is also the shortest distance between Cs in a group and the other 10 Cs atoms, and is similar to the interatomic distance in Cs metal. The structures of the other 3 suboxides are more complex but it is salutory to realize that Cs forms at least 9 crystalline oxides whose structures can be rationalized in terms of general bonding systematics.

4.3.4 *Hydroxides*

Evaporation of aqueous solutions of LiOH under normal conditions produces the monohydrate, and this can be readily dehydrated by heating in an inert atmosphere or under reduced pressure. $LiOH.H_2O$ has a crystal structure built up of double chains in which both Li and H_2O have 4 nearest neighbours (Fig. 4.6a); Li is tetrahedrally coordinated by 2OH and $2H_2O$, and each tetrahedron shares an edge (2OH) and two corners ($2H_2O$) to produce double chains which are held laterally by H bonds. Each H_2O molecule is tetrahedrally coordinated by 2Li from the same chain and 2OH from other chains. Anhydrous LiOH has a layer lattice of edge-shared $Li(OH)_4$ tetrahedra (Fig. 4.6b) in which each Li in a plane is surrounded tetrahedrally by 4OH, and each OH has 4Li neighbours all lying on one side; neutron diffraction shows that the OH bonds are normal to the layer plane and there is no H bonding between layers.

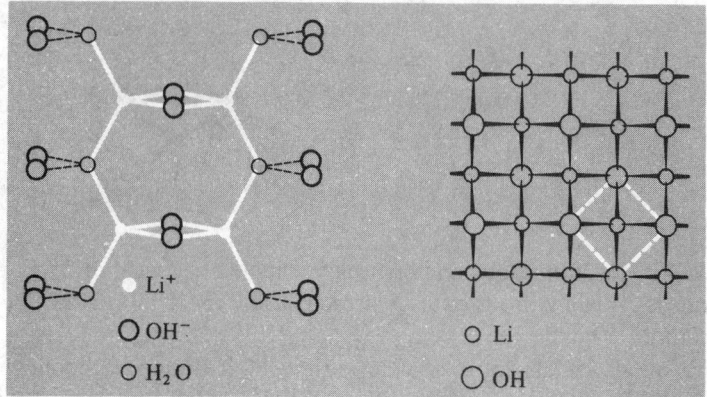

FIG. 4.6   (a) The double-chain structure of LiOH . H_2O, and (b) the layer structure of anhydrous LiOH (see text).

Numerous hydrates have been prepared from aqueous solutions of the heavier alkali metal hydroxides (e.g. NaOH . nH_2O, where $n = 1, 2, 3.5, 4, 5,$ and 7) but little detailed structural information is available. The anhydrous compounds all show the influence of oriented OH groups on the structure,[10] and there is evidence of weak O—H···O bonding for KOH and RbOH. Melting points are substantially lower than those of the halides, decreasing from 471°C for LiOH to 272° for CsOH.

The alkali metal hydroxides are the most basic of all hydroxides. They react with acids to form salts and with alcohols to form alkoxides. They readily absorb CO_2 and H_2S to form carbonates (or hydrogencarbonates) and sulfides (or hydrogensulfides), and are extensively used to remove mercaptans from petroleum products. Amphoteric oxides such as those of Al, Zn, Sn, and Pb react with MOH to form aluminates, zincates, stannates, and plumbates, and even SiO_2 (and silicate glasses) are attacked.

Production and uses of LiOH have already been discussed (p. 77). Huge tonnages of NaOH and KOH are produced by electrolysis of brine (pp. 82, 84) and the enormous industrial importance of these chemicals has already been alluded to.

4.3.5 *Oxoacid salts and other compounds*

Many binary and pseudo-binary compounds of the alkali metals are more conveniently treated within the context of the chemistry of the other element and for this reason discussion is deferred to later chapters, e.g. borides (p. 162), graphite intercalation compounds (p. 313), carbides, cyanides, cyanates, etc. (pp. 318, 336), silicides (p. 386), germanides (p. 455), nitrides, azides, and amides (p. 479), phosphides (p. 562), arsenides (p. 646), sulfides (p. 798), selenides and tellurides (p. 898), polyhalides (p. 978), etc. Likewise, the alkali metals form stable salts with virtually all oxoacids and these are also discussed in later chapters.

Lithium salts show a great propensity to crystallize as hydrates, the trihydrates being particularly common, e.g. LiX.$3H_2O$, X = Cl, Br, I, ClO_3, ClO_4, MnO_4, NO_3, BF_4, etc. In most of these Li is coordinated by $6H_2O$ to form chains of face-sharing octahedra:

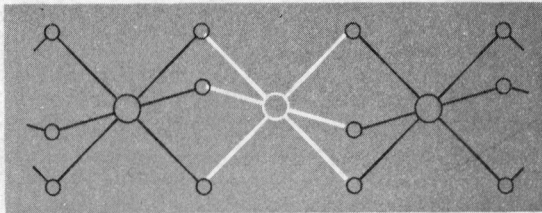

By contrast Li_2CO_3 is anhydrous and sparingly soluble (1.28 wt% at 25°C, i.e. 0.17 mol l^{-1}). The nitrate is also anhydrous but is hygroscopic and much more soluble (45.8 wt% at 25°C, i.e. 6.64 mol l^{-1}).

The heavier alkali metals form a wide variety of hydrated carbonates, hydrogen-carbonates, sesquicarbonates, and mixed-metal combinations of these, e.g. $Na_2CO_3.H_2O$, $Na_2CO_3.7H_2O$, $Na_2CO_3.10H_2O$, $NaHCO_3$, $Na_2CO_3.NaHCO_3.2H_2O$, $Na_2CO_3.3NaHCO_3$, $NaKCO_3.nH_2O$, $K_2CO_3.NaHCO_3.2H_2O$, etc. These systems have been studied in great detail because of their industrial and geochemical significance (see Panel). Some solubility data are in Fig. 4.7, which indicates the considerable solubility of Rb_2CO_3 and Cs_2CO_3 and the lower solubility of the hydrogen carbonates. The complicated phase relationships in the $Na_2CO_3/NaHCO_3/H_2O$ system are shown in more detail in Fig. 4.8. The various stoichiometries reflect differing ways of achieving charge balance, preferred coordination polyhedra, and H bonding. Thus $Na_2CO_3.H_2O$ has two types of 6-coordinate Na, half being surrounded by $1H_2O$ plus 5 oxygen atoms from CO_3 groups and half by $2H_2O$ plus 4 oxygen atoms from CO_3 groups. The decahydrate has octahedral $Na(H_2O)_6$ groups associated in pairs by edge sharing to give $[Na_2(H_2O)_{10}]$. The hydrogencarbonate $NaHCO_3$ has infinite one-dimensional chains of HCO_3 formed by unsymmetrical $O-H\cdots O$ bonds (261 pm) which are held laterally by Na ions. The sesquicarbonates $Na_3H(CO_3)_2.2H_2O$ have short, symmetrical $O-H-O$ bonds (253 pm) which link the carbonate ions in pairs, and longer $O-H\cdots O$ bonds (275 pm) which link these pairs to water molecules. Similar phases are known for the other alkali metals.

Industrial Production and Uses of Sodium Carbonate, Hydroxide, and Sulfate

Na_2CO_3 (soda ash) is interchangeable with NaOH in many of its applications (e.g. paper pulping, soap, detergents) and this gives a valuable flexibility to the chlor–alkali industry. About half the Na_2CO_3 produced is used in the glass industry. One developing application is in the reduction of sulfur pollution resulting from stack gases of power plants and other large furnaces: powdered Na_2CO_3 is injected with the fuel and reacts with SO_2 to give solids such as Na_2SO_3 which can be removed by filtration or precipitation. World production of Na_2CO_3 was 22 million tonnes in 1974: the five leading countries were the USA, the USSR, France, the Federal Republic of Germany, and Japan, and they accounted for over 70% of production. Most of this material was synthetic (Solvay), but the increasing use of natural carbonate (trona) is notable, particularly in the USA where it now accounts for more than half of the Na_2CO_3 produced: reserves in the Green River, Wyoming, deposit alone exceed 10^{10} tonnes and occur in beds up to 3 m thick over an area of 2300 km^2.

Continued

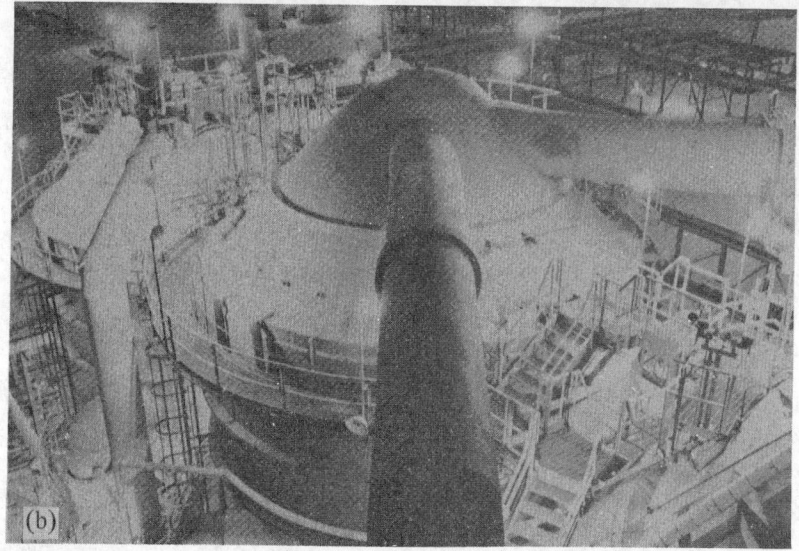

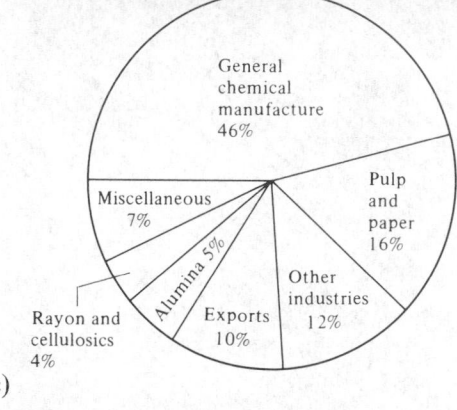

PLATE 4.3   (a) Rows of electrolytic cells in Diamond Shamrock's chlorine–caustic soda plant at La Porte, Texas. (b) Giant tandem evaporators at PPG Industries' Lake Charles plant. (c) Distribution of uses of caustic soda.

Formerly Na_2CO_3 found extensive use as "washing soda" but this market has now disappeared due to the domestic use of detergents. The related compound $NaHCO_3$ is, however, still used, particularly because of its ready decomposition in the temperature range 50–100°C:

$$2NaHCO_3 \longrightarrow Na_2CO_3 + H_2O + CO_2$$

Production in the USA is $\sim 200\,000$ tonnes annually of which 40% is used in baking-powder formulations, 15% in chemicals manufacture, 12% in pharmaceuticals, and 10% in fire extinguishers.

Caustic soda (NaOH) is industry's most important alkali. It is manufactured on a huge scale by the electrolysis of brine (p. 81) and annual production in the USA alone is over 11 million tonnes. The scale of the process is indicated by Pl. 4.3a, which shows the rows of electrolytic cells at a plant in La Porte, Texas. Electrolysis is followed by concentration of the alkali in huge tandem evaporators such as those at PPG Industries' Lake Charles plant shown in Pl. 4.3b. The evaporators, which are perhaps the world's largest, are 41 m high and 12 m in diameter. Almost half the caustic produced is used directly in chemical production (Pl. 4.3c). Principal applications are in acid neutralization, the manufacture of phenol, resorcinol, β-naphthol, etc., and the production of sodium hypochlorite, phosphate, sulfide, aluminates, etc.

Salt cake (Na_2SO_4) is a byproduct of HCl manufacture using H_2SO_4 and is also the end-product of hundreds of industrial operations in which H_2SO_4 used for processing is neutralized by NaOH. For long it had few uses, but now it is the mainstay of the paper industry, being a key chemical in the kraft process for making brown wrapping paper and corrugated boxes: digestion of wood chips or saw-mill waste in very hot alkaline solutions of Na_2SO_4 dissolves the lignin (the brown resinous component of wood which cements the fibres together) and liberates the cellulose fibres as pulp which then goes to the paper-making screens. The remaining solution is evaporated until it can be burned, thereby producing steam for the plant and heat for the evaporation: the fused Na_2SO_4 and NaOH survive the flames and can be reused. Total world production of Na_2SO_4 (1974) was ~ 4.2 million tonnes (45% natural, 55% synthetic). Most of this ($\sim 70\%$) is used in the paper industry and smaller amounts are used in glass manufacture and detergents ($\sim 10\%$ each). The hydrated form, $Na_2SO_4 \cdot 10H_2O$, Glauber's salt, is now less used than formerly. Further information on the industrial production and uses of Na_2CO_3, NaOH, and Na_2SO_4 are given in *Kirk–Othmer Encyclopedia of Chemical Technology*, 2nd edn., Vol. 1, 1963, pp. 707–58; Vol. 18, 1969, pp. 432–515.

Alkali metal nitrates can be prepared by direct reaction of aqueous nitric acid on the appropriate hydroxide or carbonate. $LiNO_3$ is used for scarlet flares and pyrotechnic displays. Large deposits of $NaNO_3$ (saltpetre) are found in Chile and were probably formed by bacterial decay of small marine organisms: the NH_3 initially produced presumably oxidized to nitrous acid and nitric acid which would then react with dissolved NaCl. KNO_3 was formerly prepared by metathesis of $NaNO_3$ and KCl but is now obtained directly as part of the synthetic ammonia/nitric acid industry (p. 482).

Alkali metal nitrates are low-melting salts that decompose with evolution of oxygen above about 500°C, e.g.

$$2NaNO_3 \underset{\sim 500}{\rightleftharpoons} 2NaNO_2 + O_2; \qquad 2NaNO_3 \underset{\sim 800}{\rightleftharpoons} Na_2O + N_2 + \tfrac{5}{2}O_2$$

Thermal stability increases with increasing atomic weight, as expected. Nitrates have been widely used as molten salt baths and heat transfer media, e.g. the 1:1 mixture $LiNO_3 : KNO_3$ melts at 125°C and the ternary mixture of 40% $NaNO_2$, 7% $NaNO_3$, and 53% KNO_3 can be used from its mp 142° up to about 600°C.

The corresponding nitrites, MNO_2, can be prepared by thermal decomposition of MNO_3 as indicated above or by reaction of NO with the hydroxide:

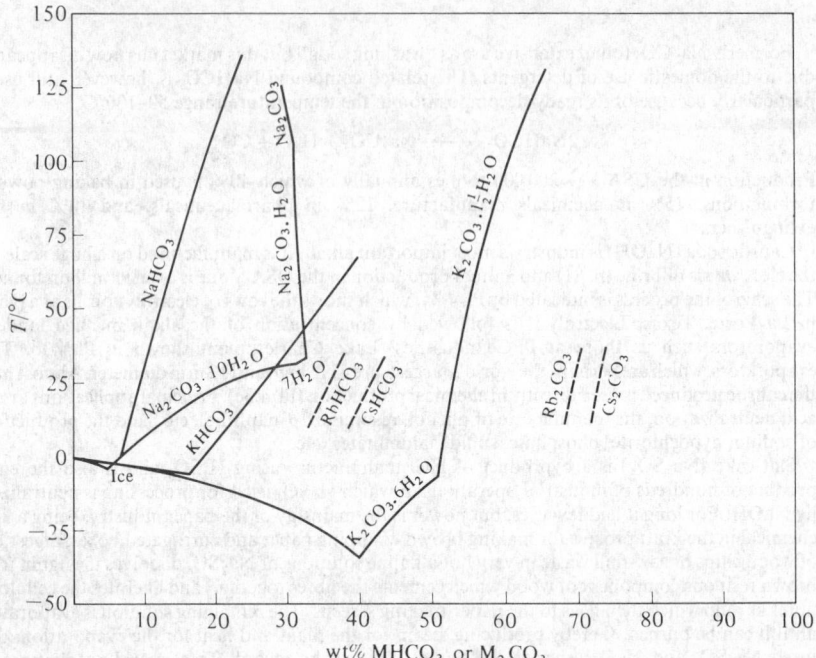

FIG. 4.7   Solubilities of alkali carbonates and bicarbonates. (H. Stephen and T. Stephen, *Solubilities of Inorganic and Organic Compounds*, Vol. 1, Part 1, Macmillan, New York.)

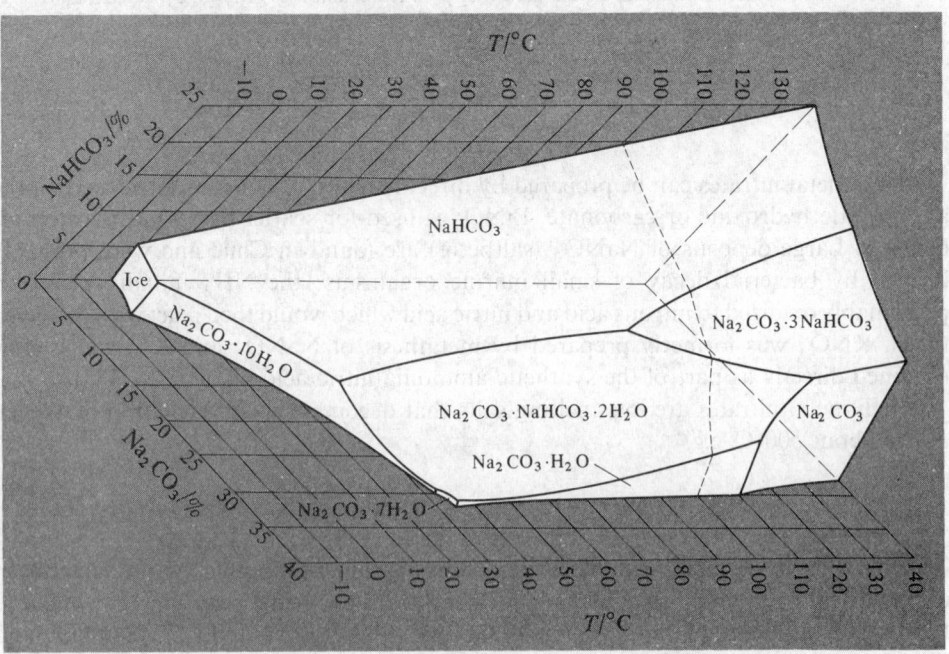

FIG. 4.8   Three-dimensional phase diagram $Na_2CO_3 \cdot NaHCO_3 \cdot H_2O$. (Based on a diagram in *Kirk-Othmer Encyclopedia of Chemical Technology*, Vol. 18, Interscience, New York, 1969, p. 466.)

$$4NO + 2MOH \longrightarrow 2MNO_2 + N_2O + H_2O$$
$$6NO + 4MOH \longrightarrow 4MNO_2 + N_2 + 2H_2O$$

Chemical reduction of nitrates has also been employed:

$$KNO_3 + Pb \longrightarrow KNO_2 + PbO$$
$$2RbNO_3 + C \longrightarrow 2RbNO_2 + CO_2$$

The commercial production of $NaNO_2$ is achieved by absorbing oxides of nitrogen in aqueous Na_2CO_3 solution:

$$Na_2CO_3 + NO + NO_2 \longrightarrow 2NaNO_2 + CO_2$$

Nitrites are white, crystalline hygroscopic salts that are very soluble in water. When heated in the absence of air they disproportionate:

$$5NaNO_2 \longrightarrow 3NaNO_3 + Na_2O + N_2$$

$NaNO_2$, in addition to its use with nitrates in heat-transfer molten-salt baths, is much used in the production of azo dyes and other organo-nitrogen compounds, as a corrosion inhibitor, and in curing meats.

Other oxoacid salts of the alkali metals are discussed in later chapters, e.g. borates (p. 231), silicates (p. 400), phosphites and phosphates (p. 586), sulfites, hydrogen sulfates, thiosulfates, etc. (p. 834) selenites, selenates, tellurites, and tellurates (p. 913), hypohalites, halites, halates, and perhalates (p. 999), etc.

4.3.6 *Complexes, crowns, and crypts*[26-28]

Exciting developments have occurred in the coordination chemistry of the alkali metals during the last few years that have completely rejuvenated what appeared to be a largely predictable and worked-out area of chemistry. Conventional beliefs had reinforced the predominant impression of very weak coordinating ability, and had rationalized this in terms of the relatively large size and low charge of the cations M^+. On this view, stability of coordination complexes should diminish in the sequence $Li > Na > K > Rb > Cs$, and this is frequently observed, though the reverse sequence is also known for the formation constants of, for example, the weak complexes with sulfate, peroxosulfate, thiosulfate, and the hexacyanoferrates in aqueous solutions.[27] It was also known that the alkali metals formed numerous hydrates, or aquo-complexes, as discussed in the preceding section, and there is a definite, though smaller tendency to form ammine complexes such as $[Li(NH_3)_4]I$. Other well-defined complexes include the extremely stable adducts $LiX.5Ph_3PO$, $LiX.4Ph_3PO$, and $NaX.5Ph_3PO$, where X is a large anion such as I, NO_3, ClO_4, BPh_4, SbF_6, $AuCl_4$, etc.; these compounds melt in the range 200–315° and are stable to air and water (in which they are insoluble). They probably all contain the tetrahedral ion $[Li(OPPh_3)_4]^+$ which was established by X-ray crystallography for the compound $LiI.5Ph_3PO$; the fifth molecule of Ph_3PO is uncoordinated. In

[26] P. N. KAPOOR and R. C. MEHROTRA, Coordination compounds of alkali and alkaline earth metals with covalent characteristics, *Coord. Chem. Rev.* **14**, 1–27 (1974).

[27] D. MIDGLEY, Alkali metal complexes in aqueous solution, *Chem. Soc. Revs.* **4**, 549–68 (1975).

[28] N. S. POONIA and A. V. BAJAJ, Coordination chemistry of alkali and alkaline earth cations, *Chem. Revs.* **79**, 389–445 (1979).

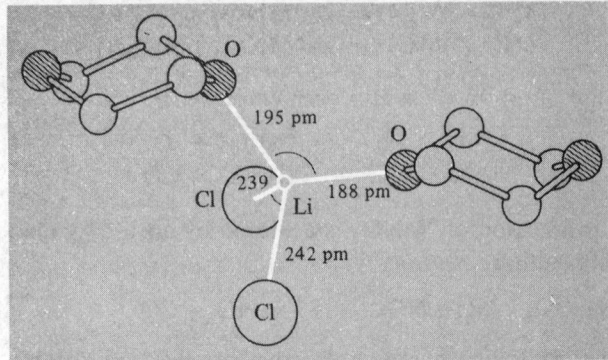

FIG. 4.9   Coordination of 2 Cl atoms and 2 dioxan molecules about Li in LiCl.diox. (After F. Durant, Y. Gobillon, P. Piret, and M. van Meerssche, *Bull. Soc. Chim. Belg.* **75,** 52 (1966).) The angles are OLiO 115.6°, ClLiCl 113.4°, and the 4 ClLiO 105.8°–108.6°.

$LiCl.C_4H_8O_2$, Li is tetrahedrally coordinated by 2Cl and by 2 oxygen atoms from different dioxan molecules (Fig. 4.9); to achieve the correct stoichiometry each Cl and each dioxan is coordinated by 2Li. The pyridine adduct $LiCl.2C_5H_5N.H_2O$ again features tetrahedral Li (Fig. 4.10), the molecules of the complex $[LiCl(OH_2)py_2]$ being held together chiefly by O—H···Cl bonds (313, 315 pm). In LiCl.2en and LiBr.2en ($en = NH_2CH_2CH_2NH_2$), each Li is tetrahedrally surrounded by 4N from 3en—one *gauche* and chelating, the other two *trans* and forming with Li ions infinite chains which are held together by N—H···X bonds. A more complicated structure is found in $NaBr.2MeCONH_2$, each Na being octahedrally coordinated by 4 oxygen atoms in a plane and 2 *trans* Br; the octahedra form infinite chains by sharing common O_2Br faces and there are N—H···Br bonds between the chains: Na–Br 299, 312 pm, Na–O (mean) 235 pm (as in hydrates), and N—H···Br 340–354 pm.

Complexes with chelating organic reagents such as salicylaldehyde and β-diketonates were first prepared by N. Sidgwick and his students in 1925, and many more have since been characterized. Stability, as measured by equilibrium formation constants, is rather

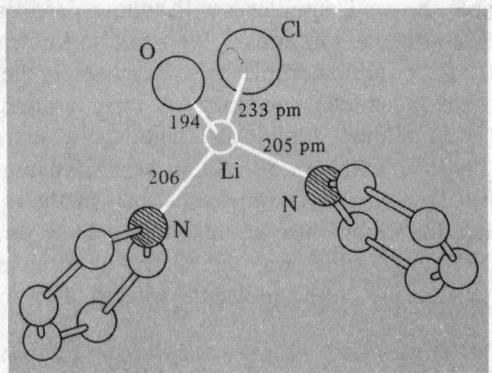

FIG. 4.10   Structure of the tetrahedral complex $[LiCl(NC_5H_5)_2(OH_2)]$. (After F. Durant, P. Piret, and M. van Meerssche, *Acta Cryst.* **22,** 52 (1967).) The angles are ClLiN 107.6°, NLiN 107.5°, NLiO 105.7°, OLiCl 117.9°.

low and almost invariably decreases in the sequence Li > Na > K. The situation changed dramatically in 1967 when C. J. Pedersen announced the synthesis of several macrocyclic polyethers which were shown to form stable complexes with alkali metal and other cations.[29] The stability of the complexes was found to depend on the number and geometrical disposition of the ether oxygen atoms and in particular on the size and shape of potential coordination polyhedra relative to the size of the cation. For this reason stability could peak at any particular cation and, for M^I, this was often K and sometimes Na or Rb rather than Li. Typical examples of such "crown" ethers are given in Fig. 4.11, the numerical prefix indicating the number of atoms in the heterocycle and the suffix the number of ether oxygens. The aromatic rings can be substituted, replaced by naphthalene residues, or reduced to cyclohexyl derivatives. The "hole size" for coordination depends on the number of atoms in the ring and is compared with conventional ionic diameters in Table 4.3.

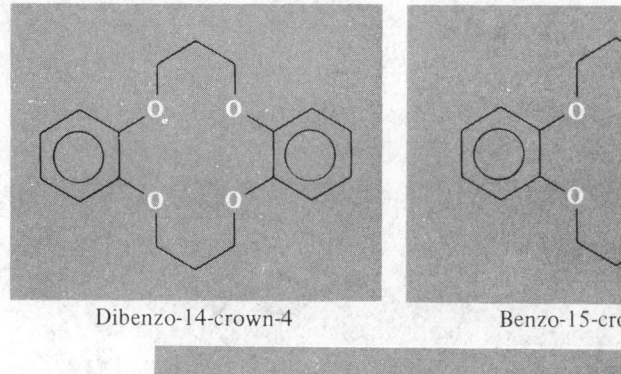

Dibenzo-14-crown-4   Benzo-15-crown-5

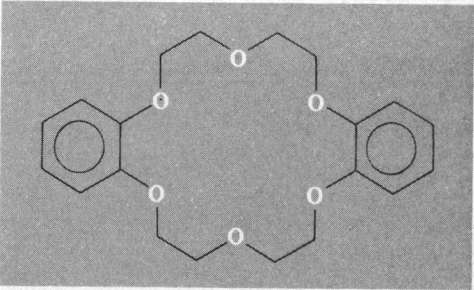

Dibenzo-18-crown-6

FIG. 4.11 Schematic representation of the (non-planar) structure of some typical crown ethers.

TABLE 4.3 *Comparison of ionic diameters and crown ether "hole sizes"*

Cation	Ionic diameter/pm	Polyether ring	"Hole size"/pm
Li^I	152	14-crown-4	120–150
Na^I	204	15-crown-5	170–220
K^I	276	18-crown-6	260–320
Rb^I	304	21-crown-7	340–430
Cs^I	334	—	—

[29] C. J. PEDERSEN, Cyclic polyethers and their complexes with metal salts, *J. Am. Chem. Soc.* **89**, 2495, 7017–36 (1967). See also C. J. PEDERSEN and H. K. FRENSDORF, *Angew. Chem.*, Int. Chem. (Engl.), **11**, 16–25 (1972).

The X-ray crystal structures of many of these complexes have now been determined: representative examples are shown in Fig. 4.12 from which it is clear that, at least for the larger cations, coordinative saturation and bond directionality are far less significant factors than in many transition element complexes.[30, 31] Further interest in these ligands stems from their use in biochemical modelling since they sometimes mimic the behaviour of naturally occurring, neutral, macrocyclic antibiotics such as valinomycin, monactin, nonactin, nigericin, and enneatin.[32] They may also shed some light on the perplexing and remarkably efficient selectivity between Na and K in biological systems.[32, 33]

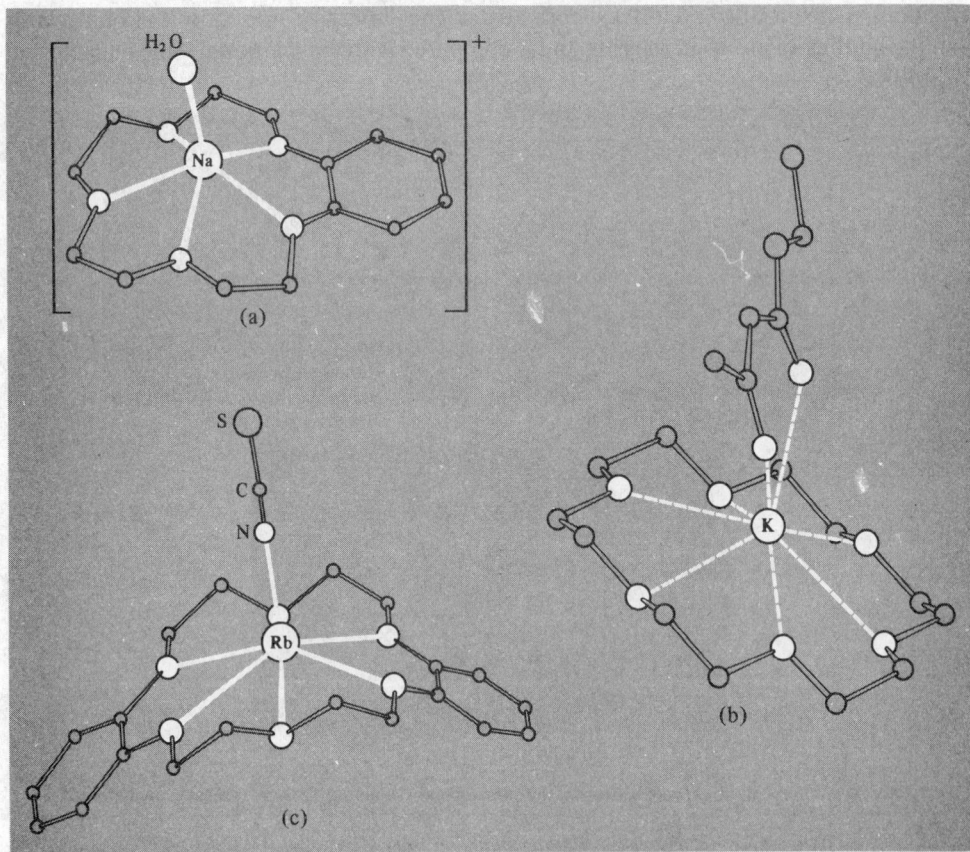

Fig. 4.12   Molecular structures of typical crown-ether complexes with alkali metal cations: (a) sodium-water-benzo-15-crown-5 showing pentagonal-pyramidal coordination of Na by 6 oxygen atoms; (b) 18-crown-6-potassium-ethyl acetoacetate enolate showing unsymmetrical coordination of K by 8 oxygen atoms; and (c) the RbNCS ion pair coordinated by dibenzo-18-crown-6 to give seven-fold coordination about Rb.

[30] J.-M. Lehn, Design of organic complexing agents. Strategies towards properties, *Struct. Bonding* **16**, 1–69 (1973).

[31] M. R. Truter, Structures of organic complexes with alkali metal ions, *Struct. Bonding* **16**, 71–111 (1973).

[32] W. Simon, W. E. Morf, and P. Ch. Meier, Specificity for alkali and alkaline earth cations of synthetic and natural organic complexing agents in membranes, *Struct. Bonding* **16**, 113–60 (1973).

[33] R. M. Izatt, D. J. Eatough, and J. J. Christensen, Thermodynamics of cation-macrocyclic compound interaction, *Struct. Bonding* **16**, 161–89 (1973).

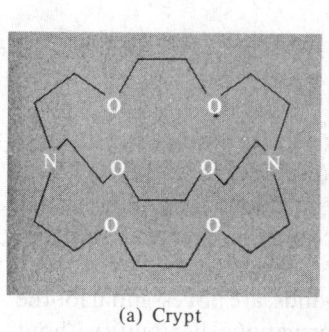

(a) Crypt

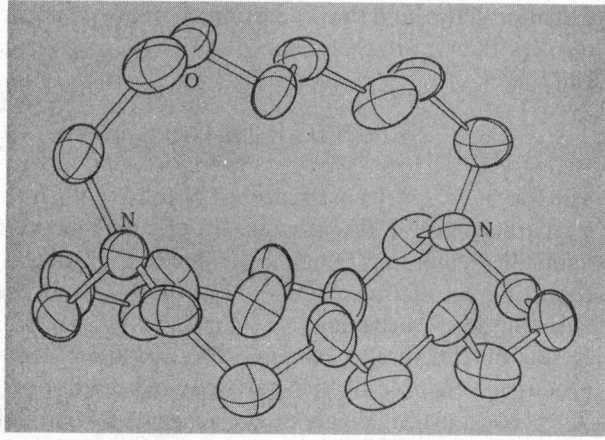

(b) Molecular structure and conformation of
free macrobicyclic crypt ligand

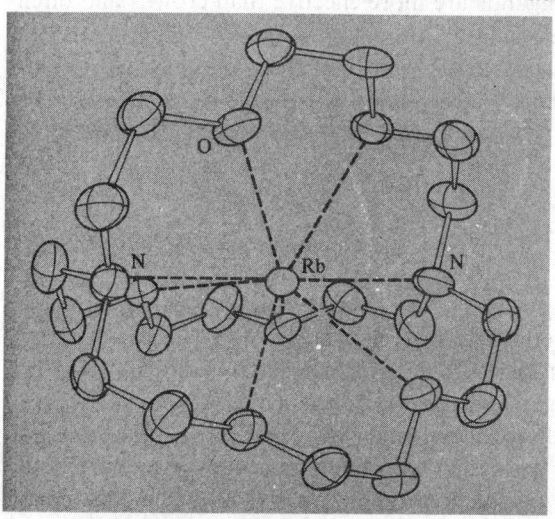

(c) Molecular structure of complex cation
of RbSCN with crypt.

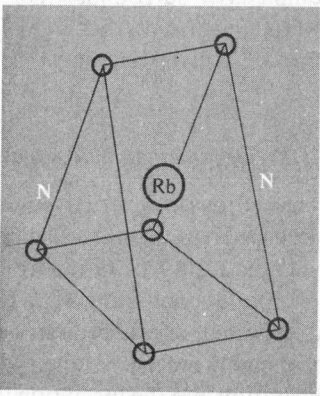

(d) Schematic representation of
bicapped trigonal prismatic
coordination about Rb⁺.

FIG. 4.13   A typical crypt and its complex.

Another group of very effective ligands that have recently been employed to coordinate alkali metal cations are the macrobicyclic crypts of which $N\{(CH_2CH_2O)_2CH_2CH_2\}_3N$ is representative (Fig. 4.13a). This forms a complex $[Rb(crypt)]CNS \cdot H_2O$ in which the ligand encapsulates the cation with a bicapped trigonal prismatic coordination polyhedron (Fig. 4.13c, d). Such complexes are finding increasing use in solvent extraction, phase-transfer catalysis,[34] the stabilization of uncommon or reactive

[34] W. P. WEBER and G. W. GOKEL, *Phase Transfer Catalysis in Organic Synthesis*, Vol. 4 of *Reactivity and Structure*, Springer-Verlag, 1977, 250 pp. R. C. F. JONES, Phase transfer catalysis, Chap. 10 in *General Synthetic Methods*, Vol. 1, pp. 402–23, 1978, Chem. Soc. Specialist Periodical Report. C. M. STARKS and C. LIOTTA, *Phase Transfer Catalysis*, Academic Press, New York, 1978, 365 pp. F. MONTANARI, D. LANDINI, and F. ROLLA, Phase-transfer catalysed reactions, *Topics in Current Chemistry* **101**, 149–201 (1982).

oxidation states, and the promotion of otherwise improbable reactions. For example, Na reacts with crypt in the presence of $EtNH_2$ to give the first example of a sodide salt of Na^-.[35]

$$2Na + N\{(C_2H_4O)_2C_2H_4\}_3N \xrightarrow{EtNH_2} [Na(crypt)]^+Na^-$$

The Na^- is 555 pm from the nearest N and 516 pm from the nearest O, indicating that it is a separate entity in the structure. In organic reactions encapsulation of the cation by a lipophilic ligand may enhance the nucleophilic activity of the anion, assist bimolecular elimination, enhance oxidation by MnO_4^-, and even permit the chemical study of virtually "free" carbanions.

Macrocycles, though extremely effective as polydentate ligands, are not essential for the production of stable alkali metal complexes; additional conformational flexibility without loss of coordinating power can be achieved by synthesizing benzene derivatives with 2–6 pendant mercapto-polyether groups $C_6H_{6-n}R_n$, where R is $-SC_2H_4OC_2H_4OMe$, $-S(C_2H_4O)_3Bu$, etc. Such "octopus" ligands are more effective than crowns and often equally as effective as crypts in sequestering alkali metal cations.[36] Indeed, it is not even essential to invoke organic ligands at all since an inorganic cryptate which completely surrounds Na has been identified in the heteropolytungstate $(NH_4)_{17}Na[NaW_{21}Sb_9O_{86}] \cdot 14H_2O$; the compound was also found to have pronounced antiviral activity.[37]

4.3.7 *Organometallic compounds*[38]

Deviations from strictly ionic bonding, which have been noted in preceding sections, are most apparent with Li and culminate in the diagonal relation with Mg. The existence and synthetic utility of Grignard reagents therefore suggests that analogous alkali–metal compounds containing M–C bonds might be preparable, and such is found to be the case. As expected, organolithium compounds are the most covalent, the most stable, and the least highly reactive of the group, and there is an increasing trend towards ionic charge separation and the formation of reactive carbanions with increasing atomic number from Li to Cs.

Organolithium compounds can readily be prepared from metallic Li and this is one of the major uses of the metal. Because of the great reactivity both of the reactants and the products, air and moisture must be rigorously excluded by use of an inert atmosphere. Lithium can be reacted directly with alkyl halides in light petroleum, cyclohexane, benzene, or ether, the chlorides generally being preferred:

$$2Li + RX \xrightarrow{solvent} LiR + LiX$$

[35] J. L. DYE, J. M. CERASE, M. T. LOK, B. L. BARNETT, and F. J. TEHAN, A crystalline salt of the sodium anion (Na^-), *J. Am. Chem. Soc.* **96**, 608–9, 7203–8 (1974). J. L. DYE, Compounds of alkali metal anions, *Angew. Chem.*, Int. Edn. (Engl.) **18**, 587–98 (1979).

[36] F. VÖGTLE and E. WEBER, Octopus molecules, *Angew. Chem.*, Int. Edn. (Engl.) **13**, 814–15 (1974).

[37] J. FISCHER, L. RICHARD, and R. WEISS, The structure of the heteropolytungstate $(NH_4)_{17}Na[NaW_{21}Sb_9O_{86}] \cdot 14H_2O$: an inorganic cryptate. *J. Am. Chem. Soc.* **98**, 3050–2 (1976).

[38] G. E. COATES, M. L. H. GREEN, and K. WADE, *Organometallic Compounds*, Vol. 1, *The Main Group Elements*, 3rd edn., Chap. 1, The alkali metals, pp. 1–70, Methuen, London, 1967.

Reactivity and yields are greatly enhanced by the presence of 0.5–1% Na in the Li. The reaction is also generally available for the preparation of metal alkyls of the heavier Group IA metals. Lithium aryls are best prepared by metal–halogen exchange using LiBu^n and an aryl iodide, and transmetalation is the most convenient route to vinyl, allyl, and other unsaturated derivatives:

$$LiBu^n + ArI \xrightarrow{\text{ether}} LiAr + Bu^nI$$

$$4LiPh + Sn(CH{=}CH_2)_4 \xrightarrow{\text{ether}} 4LiCH{=}CH_2 + SnPh_4$$

The reaction between an excess of Li and an organomercury compound is a useful alternative when isolation of the product is required, rather than its direct use in further synthetic work:

$$2Li + HgR_2 \text{ (or } HgAr_2) \xrightarrow[\text{benzene}]{\text{petrol or}} 2LiR \text{ (or } LiAr) + Hg$$

Similar reactions are available for the other alkali metals. Metalation (metal–hydrogen exchange) and metal addition to alkenes provide further routes, e.g.

$$2Na + 3C_5H_6 \longrightarrow 2NaC_5H_5 + C_5H_8$$

$$2Cs + CH_2{=}CH_2 \longrightarrow CsCH_2CH_2Cs$$

In the presence of certain ethers such as Me_2O, $MeOCH_2CH_2OMe$, or tetrahydrofuran, Na forms deep-green highly reactive paramagnetic adducts with polynuclear aromatic hydrocarbons such as naphthalene, phenanthrene, anthracene, etc.:

These compounds are in many ways analogous to the solutions of alkali metals in liquid ammonia (p. 88).

The most ionic of the organometallic derivatives of Group IA elements are the acetylides and dicarbides formed by the deprotonation of alkynes in liquid ammonia solutions:

$$Li + HC \equiv CH \xrightarrow{\text{liq NH}_3} LiC \equiv CH + \tfrac{1}{2}H_2$$

$$2Li + HC \equiv CH \xrightarrow{\text{liq NH}_3} Li_2C_2 + H_2$$

The largest industrial use of LiC_2H is in the production of vitamin A, where it effects ethynylation of methyl vinyl ketone to produce a key tertiary carbinol intermediate. The acetylides and dicarbides of the other alkali metals are prepared similarly. It is not always necessary to prepare this type of compound in liquid ammonia and, indeed, further substitution to give the bright red perlithiopropyne Li_4C_3 can be effected in hexane under reflux:[39]

$$4LiBu^n + CH_3C \equiv CH \xrightarrow{\text{hexane}} Li_3CC \equiv CLi + 4C_4H_{10}$$

Organolithium compounds tend to be thermally unstable and most of them decompose to LiH and an alkene on standing at room temperature or above. Among the more stable compounds are the colourless, crystalline solids LiMe (decomp above 200°C), $LiBu^n$ and $LiBu^t$ (which shows little decomposition over a period of days at 100°C). Lithium alkyls are covalent compounds with unusual tetrameric or hexameric structures. For example, LiMe, which is a non-conductor of electricity when fused, contains interconnected tetrameric units $(LiMe)_4$ as shown in Fig. 4.14. The Li_4C_4 cluster consists of a tetrahedron of 4Li with a triply-bridging C above the centre of each face to complete a distorted cube. The clusters are interconnected along cube diagonals via the bridging Me groups and the intercluster Li–C distance (236 pm) is very similar to the Li–C distance within each cluster (231 pm). The carbon atoms are thus essentially 7-coordinate being bonded directly to 3H and 4Li. The Li–Li distance within the cluster is 268 pm, which is virtually identical with the value of 267.3 pm for the gaseous Li_2 molecule and substantially smaller than the value of 304 pm in Li metal (where each Li has 8 nearest neighbours). This strong metal–metal bonding is further revealed by the persistence of the Li_4 group in the fragmentation of lithium methyl under electron impact in the mass spectrometer, the Me groups being lost far more readily than the Li atoms.

Several bonding schemes have been proposed to interpret the structure of Li_4Me_4. One of the simplest envisages localized 4-centre 2-electron Li_3C bonds over each face of the Li_4 tetrahedra formed by overlap of a centrally directed sp^3-type orbital on carbon and one sp^2-type orbital from each of the 3 Li atoms in the face (Fig. 4.15). There is thus one strongly bonding 4-centre orbital above each of the 4 faces, requiring in all 4 pairs of electrons as supplied by the 4 Li atoms and 4 Me groups. The bonding orbitals are thus completely filled and the antibonding orbitals empty. A fuller description, which also takes into account the strong intercluster bonding, can be developed in terms of molecular orbitals embracing the whole molecule.

Higher alkyls of lithium adopt similar structures in which polyhedral clusters of metal atoms are bridged by alkyl groups located over the triangular faces of these clusters. For example, crystalline lithium t-butyl is tetrameric and the structural units $(LiBu^t)_4$ persist in solution; by contrast lithium ethyl, which is tetrameric in the solid state, dissolves in

[39] R. WEST, P. A. CARNEY, and I. C. MINEO, The tetralithium derivative of propyne and its use in the synthesis of polysilicon compounds, *J. Am. Chem. Soc.* **87**, 3788–9 (1965).

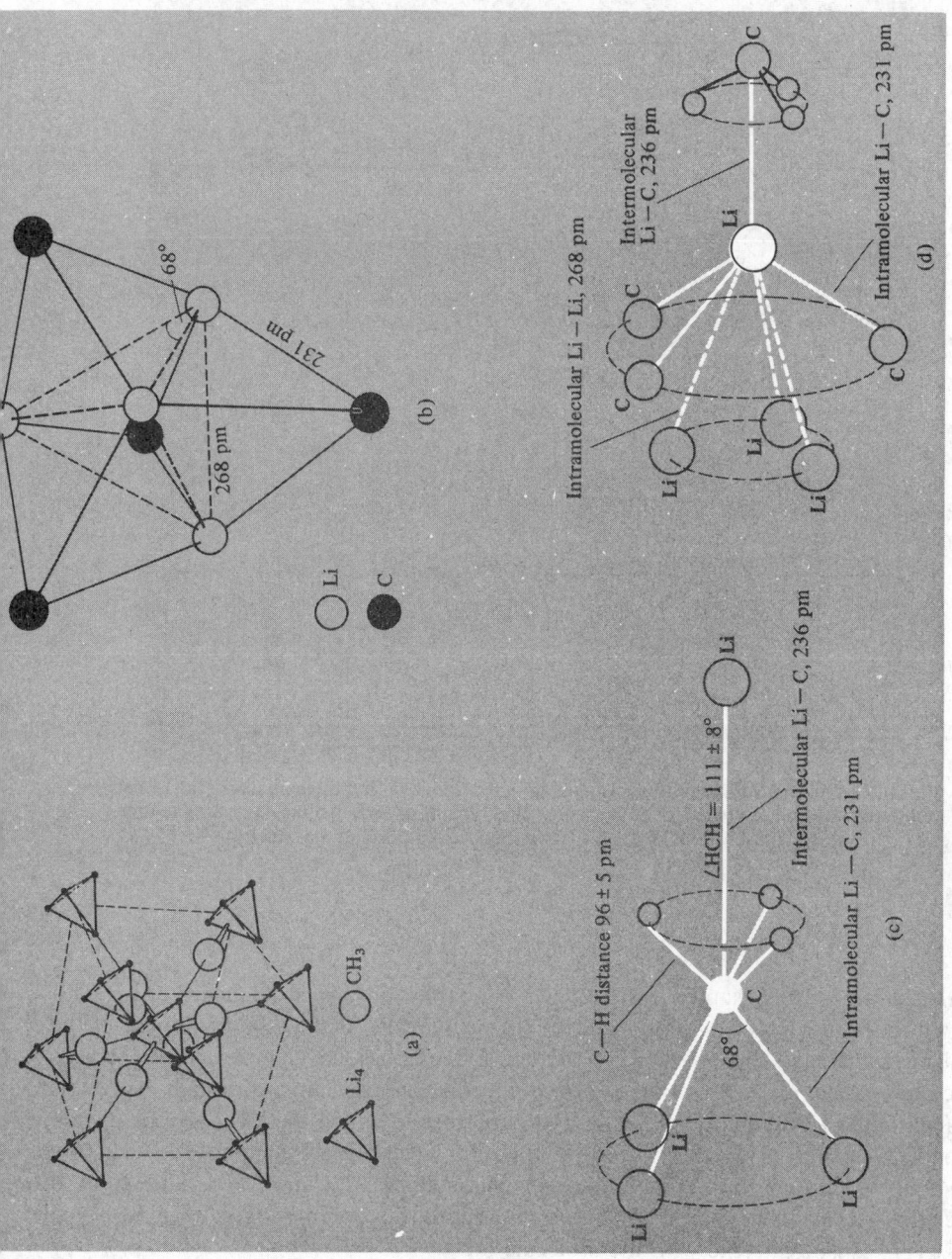

FIG. 4.14 Crystal and molecular structure of (LiMe)$_4$ showing (a) the unit cell of lithium methyl, (b) the Li$_4$C$_4$ skeleton of the tetramer viewed approximately along one of the threefold axes, (c) the 7-coordinate environment of each C atom, and (d) the 7-coordinate environment of each Li atom.[40]

[40] K. WADE, *Electron Deficient Compounds*, Nelson, London, 1971, pp. 203.

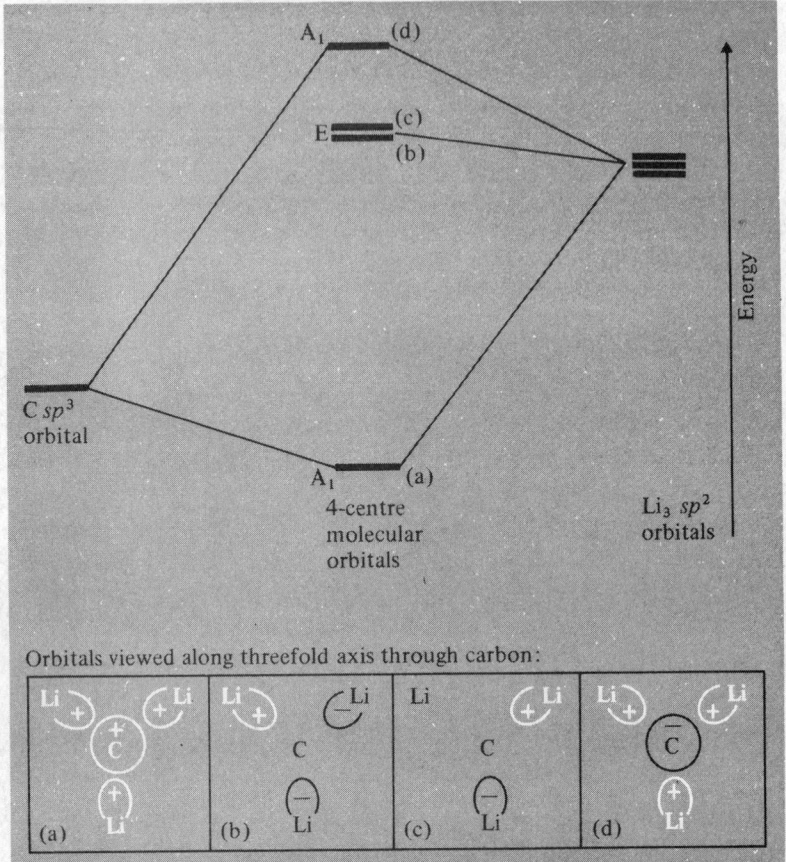

FIG. 4.15　The relative energies and symmetries of the localized 4-centre molecular orbitals that result from interactions between the carbon sp^3 and lithium sp^2 hybrid atomic orbitals pointing towards the centre of an Li_3C trigonal pyramid (of local C_{3v} symmetry).[40]

hydrocarbons as the hexamer $(LiEt)_6$ which probably consists of octahedra of Li_6 with triply bridging $-CH_2CH_3$ groups above 6 of the 8 faces. As the atomic number of the alkali metal increases there is a gradual trend away from these covalent structures towards structures which are typical of partly ionic compounds. Thus, although NaMe is tetrameric like LiMe, NaEt adopts a layer structure in which the CH_2 groups have a trigonal pyramidal array of Na neighbours, and KMe adopts a NiAs-type structure (p. 649) in which each Me is surrounded by a trigonal prismatic array of K. The extent to which this is considered to be $K^+CH_3^-$ is a matter for discussion though it will be noted that CH_3^- is isoelectronic with the molecule NH_3.

The structure of the organometallic complex lithium tetramethylborate $LiBMe_4$ is discussed on p. 142 alongside that of polymeric $BeMe_2$ with which it is isoelectronic.

Organometallic compounds of the alkali metals (particularly LiMe and $LiBu^n$) are valuable synthetic reagents and have been increasingly used in industrial and laboratory-scale organic syntheses during the past 15 y.[41, 42] The annual production of $LiBu^n$ alone has leapt from a few kilograms to almost 1000 tonnes. Large scale applications are as a polymerization catalyst, alkylating agent, and precursor to metalated organic reagents. Many of the synthetic reactions parallel those of Grignard reagents over which they sometimes have distinct advantages in terms of speed of reaction, freedom from complicating side reactions, or convenience of handling. Reactions are those to be expected for carbanions, though free-radical mechanisms occasionally occur.

Halogens regenerate the parent alkyl (or aryl) halide and proton donors give the corresponding hydrocarbon:

$$LiR + X_2 \longrightarrow LiX + RX$$
$$LiR + H^+ \longrightarrow Li^+ + RH$$
$$LiR + R'I \longrightarrow LiI + RR'$$

C–C bonds can be formed by reaction with alkyl iodides or more usefully by reaction with metal carbonyls to give aldehydes and ketones: e.g. $Ni(CO)_4$ reacts with LiR to form an unstable acyl nickel carbonyl complex which can be attacked by electrophiles such as H^+ or $R'Br$ to give aldehydes or ketones by solvent-induced reductive elimination:

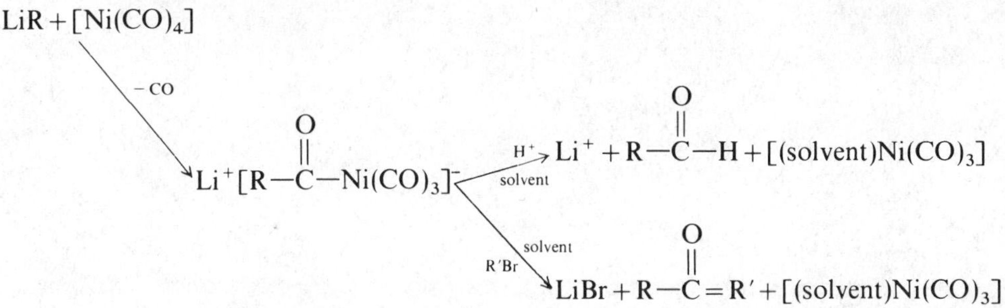

$$LiR + [Ni(CO)_4]$$

$$Li^+[R-\overset{\overset{\textstyle O}{\|}}{C}-Ni(CO)_3]^- \quad \xrightarrow[\text{solvent}]{H^+} \quad Li^+ + R-\overset{\overset{\textstyle O}{\|}}{C}-H + [(\text{solvent})Ni(CO)_3]$$

$$\xrightarrow[R'Br]{\text{solvent}} LiBr + R-\overset{\overset{\textstyle O}{\|}}{C}=R' + [(\text{solvent})Ni(CO)_3]$$

$[Fe(CO)_5]$ reacts similarly. Aldehydes and ketones can also be obtained from N,N-disubstituted amides, and symmetrical ketones are formed by reaction with CO:

$$LiR + HCONMe_2 \longrightarrow LiNMe_2 + RCHO$$
$$LiR + R'CONMe_2 \longrightarrow LiNMe_2 + R'COR$$
$$2LiR + 3CO \longrightarrow 2LiCO + R_2CO$$

Thermal decomposition of LiR eliminates a β-hydrogen atom to give an olefin and LiH, a process of industrial importance for long-chain terminal alkenes. Alkenes can also be produced by treatment of ethers, the organometallic reacting here as a very strong base (proton acceptor):

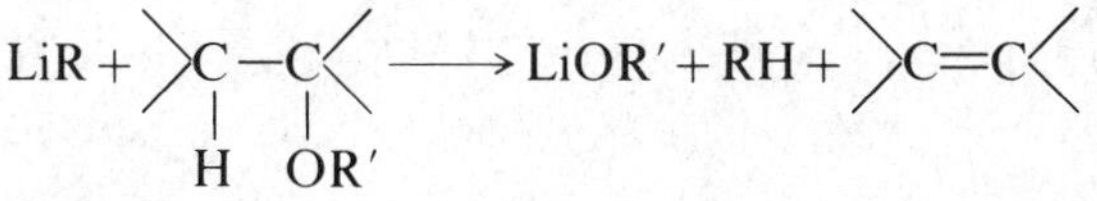

$$LiR + \overset{\quad}{\underset{\underset{H}{|}}{>}C}-\overset{\quad}{\underset{\underset{OR'}{|}}{C}}< \longrightarrow LiOR' + RH + >C=C<$$

Lithium aryls react as typical carbanions in non-polar solvents giving carboxylic acids with CO_2 and tertiary carbinols with aryl ketones:

[41] B. J. Wakefield, *The Chemistry of Organolithium Compounds*, Pergamon Press, Oxford, 1976, 337 pp.
[42] K. Smith. Lithiation and organic synthesis, *Chem. in Br.* **18(1)**, 29–32 (1982).

$$LiAr + CO_2 \longrightarrow ArCO_2Li \xrightarrow{H_2O} LiOH + ArCO_2H$$

$$LiAr + Ar_2'CO \longrightarrow [Ar_2'C(Ar)OLi] \xrightarrow{H_2O} LiOH + Ar_2'C(Ar)OH$$

Organolithium reagents are also valuable in the synthesis of other organometallic compounds via metal–halogen exchange:

$$3LiR + BCl_3 \longrightarrow 3LiCl + BR_3$$
$$(4-x)LiR + SnCl_4 \longrightarrow (4-x)LiCl + SnCl_xR_{4-x} \quad (1 \leq x \leq 4)$$
$$3LiAr + P(OEt)_3 \longrightarrow 3LiOEt + PAr_3$$

Similar reactions have been used to produce organo derivatives of As, Sb, Bi; Si, Ge, and many other elements.

																1 H	2 He
3 Li	4 Be											5 B	6 C	7 N	8 O	9 F	10 Ne
11 Na	12 Mg											13 Al	14 Si	15 P	16 S	17 Cl	18 Ar
19 K	20 Ca	21 Sc	22 Ti	23 V	24 Cr	25 Mn	26 Fe	27 Co	28 Ni	29 Cu	30 Zn	31 Ga	32 Ge	33 As	34 Se	35 Br	36 Kr
37 Rb	38 Sr	39 Y	40 Zr	41 Nb	42 Mo	43 Tc	44 Ru	45 Rh	46 Pd	47 Ag	48 Cd	49 In	50 Sn	51 Sb	52 Te	53 I	54 Xe
55 Cs	56 Ba	57 La	72 Hf	73 Ta	74 W	75 Re	76 Os	77 Ir	78 Pt	79 Au	80 Hg	81 Tl	82 Pb	83 Bi	84 Po	85 At	86 Rn
87 Fr	88 Ra	89 Ac	104 Unq	105 Unp	106 Unh	107 Uns											

58 Ce	59 Pr	60 Nd	61 Pm	62 Sm	63 Eu	64 Gd	65 Tb	66 Dy	67 Ho	68 Er	69 Tm	70 Yb	71 Lu
90 Th	91 Pa	92 U	93 Np	94 Pu	95 Am	96 Cm	97 Bk	98 Cf	99 Es	100 Fm	101 Md	102 No	103 Lr

5

Beryllium, Magnesium, Calcium, Strontium, Barium, and Radium

5.1 Introduction

The Group IIA or alkaline earth metals exemplify and continue the trends in properties noted for the alkali metals. No new principles are involved, but the ideas developed in the preceding chapter gain emphasis and clarity by their further application and extension. Indeed, there is an impressively close parallelism between the two groups as will become increasingly clear throughout the chapter.

The discovery of beryllium in 1798 followed an unusual train of events.[1] The mineralogist R.-J. Haüy had observed the remarkable similarity in external crystalline structure, hardness, and density of a beryl from Limoges and an emerald from Peru, and suggested to L.-N. Vauquelin that he should analyse them to see if they were chemically identical.† As a result, Vauquelin showed that both minerals contained not only alumina and silica as had previously been known, but also a new earth, beryllia, which closely resembled alumina but gave no alums, did not dissolve in an excess of KOH, and had a sweet rather than an astringent taste. (*Caution*: beryllium compounds are now known to be extremely toxic, especially as dusts or smokes;[2] it seems likely that this toxicity results from the ability of Be^{II} to displace Mg^{II} from Mg-activated enzymes due to its stronger coordinating ability.)

Both beryl and emerald were found to be essentially $Be_3Al_2Si_6O_{18}$, the only difference between them being that emerald also contains $\sim 2\%$ Cr, the source of its green colour. The combining weight of Be was ~ 4.7 but the similarity (diagonal relation) between Be and Al led to considerable confusion concerning the valency and atomic weight of Be

---

† A similar observation had been made (with less dramatic consequences) nearly 2000 y earlier by Pliny the Elder when he wrote: "Beryls, it is thought, are of the same nature as the smaragdus (emerald), or at least closely analogous" (*Historia Naturalis*, Book 37).

[1] M. E. WEEKS, *Discovery of the Elements*, 6th edn., Journal of Chemical Education, Easton, Pa, 1956, 910 pp.
[2] J. SCHUBERT, Beryllium and berylliosis, Chap. 34 (1958), in *Chemistry in the Environment*, pp. 321–7, Readings from *Scientific American*, W. H. Freeman, San Francisco, 1973.

(2×4.7 or 3×4.7); this was not resolved until Mendeleev 70 y later stated that there was no room for a tervalent element of atomic weight 14 near nitrogen in his periodic table, but that a divalent element of atomic weight 9 would fit snugly between Li and B. Beryllium metal was first prepared by F. Wöhler in 1828 (the year he carried out his celebrated synthesis of urea from NH_4CNO); he suggested the name by allusion to the mineral (Latin *beryllus* from Greek βηρυλλος). The metal was independently isolated in the same year by A.-B. Bussy using the same method—reduction of $BeCl_2$ with metallic K. The first electrolytic preparation was by P. Lebeau in 1898 and the first commercial process (electrolysis of a fused mixture of BeF_2 and BaF_2) was devised by A. Stock and H. Goldschmidt in 1932. The close parallel with the development of Li technology (pp. 75–77) is notable.

Compounds of Mg and Ca, like those of their Group IA neighbours Na and K, have been known from ancient times though nothing was known of their chemical nature until the seventeenth century. Magnesian stone (Greek Μαγνησία λιθος) was the name given to the soft white mineral steatite (otherwise called soapstone or talc) which was found in the Magnesia district of Thessally, whereas calcium derives from the Latin *calx, calcis*—lime. The Romans used a mortar prepared from sand and lime (obtained by heating limestone, $CaCO_3$) because these lime mortars withstood the moist climate of Italy better than the Egyptian mortars based on partly dehydrated gypsum ($CaSO_4.2H_2O$); these had been used, for example, in the Great Pyramid of Gizeh, and all the plaster in Tutankhamun's tomb was based on gypsum. The names of the elements themselves were coined by H. Davy in 1808 when he isolated Mg and Ca, along with Sr, and Ba by an electrolytic method following work by J. J. Berzelius and M. M. Pontin: the moist earth (oxide) was mixed with one-third its weight of HgO on a Pt plate which served as anode; the cathode was a Pt wire dipping into a pool of Hg and electrolysis gave an amalgam from which the desired metal could be isolated by distilling off the Hg.

A mineral found in a lead mine near Strontian, Scotland, in 1787 was shown to be a compound of a new element by A. Crawford in 1790. This was confirmed by T. C. Hope the following year and he clearly distinguished the compounds of Ba, Sr, and Ca, using amongst other things their characteristic flame colorations: Ba yellow-green, Sr bright red, Ca orange-red. Barium-containing minerals had been known since the seventeenth century but the complex process of unravelling the relation between them was not accomplished until the independent work of C. W. Scheele and J. G. Gahn between 1774 and 1779: heavy spar was found to be $BaSO_4$ and called barite or barytes (Greek βαρύς, heavy), whence Scheele's new base baryta (BaO) from which Davy isolated barium in 1808.

Radium, the last element in the group, was isolated in trace amounts as the chloride by P. and M. Curie in 1898 after their historic processing of tonnes of pitchblende. It was named by Mme Curie in allusion to its radioactivity, a word also coined by her (Latin *radius*, a ray), the element itself was isolated electrolytically via an amalgam by M. Curie and A. Debierne in 1910 and its compounds give a carmine-red flame test.

5.2   The Elements

5.2.1   *Terrestrial abundance and distribution*

Beryllium, like its neighbours Li and B, is relatively unabundant in the earth's crust; it occurs to the extent of about 2 ppm and is thus similar to Sn (2.1 ppm), Eu (2.1 ppm), and As (1.8 ppm). However, its occurrence as surface deposits of beryl in pegmatite rocks (which are the last portions of granite domes to crystallize) makes it readily accessible. Crystals as large as 1 m on edge and weighing up to 60 tonnes have been reported. The most important commercial deposits are in South America and Southern Africa, but total world reserves, even including ores with as little as 0.1% beryl, are only 4×10^6 tonnes. By contrast, world supplies of magnesium are virtually limitless: it occurs to the extent of 0.13% in sea water, and electrolytic extraction at the present annual rate of 100 million tonnes, if continued for a million years, would only reduce this to 0.12%.

Magnesium, like its heavier congeners Ca, Sr, and Ba, occurs in crustal rocks mainly as the insoluble carbonates and sulfates, and (less accessibly) as silicates. Estimates of its total abundance depend sensitively on the geochemical model used, particularly on the relative weightings given to the various igneous and sedimentary rock types, and values ranging from 20 000 to 133 000 ppm are current.[3] Perhaps the most acceptable value is 27 640 ppm (2.76%), which places Mg sixth in order of abundance by weight immediately following Ca (4.66%) and preceding Na (2.27%) and K (1.84%). Large land masses such as the Dolomites in Italy consist predominantly of the magnesian limestone mineral dolomite [$MgCa(CO_3)_2$], and there are substantial deposits of magnesite ($MgCO_3$), epsomite ($MgSO_4 . 7H_2O$), and other evaporites such as carnallite ($K_2MgCl_4 . 6H_2O$) and langbeinite [$K_2Mg_2(SO_4)_3$]. Silicates are represented by the common basaltic mineral olivine [$(Mg,Fe)_2SiO_4$] and by soapstone (talc) [$Mg_3Si_4O_{10}(OH)_2$], asbestos (chrysotile) [$Mg_3Si_2O_5(OH)_4$], and micas. Spinel ($MgAl_2O_4$) is a metamorphic mineral and gemstone. It should also be remembered that the green leaves of plants, though not a commercial source of Mg, contain chlorophylls which are the Mg–porphine complexes primarily involved in photosynthesis.

Calcium, as noted above, is the fifth most abundant element in the earth's crust and hence the third most abundant metal after Al and Fe. Vast sedimentary deposits of $CaCO_3$, which represent the fossilized remains of earlier marine life, occur over large parts of the earth's surface. The deposits are of two main types—rhombohedral calcite, which is the more common, and orthorhombic aragonite, which sometimes forms in more temperate seas. Representative minerals of the first type are limestone itself, dolomite, marble, chalk, and iceland spar. Extensive beds of the aragonite form of $CaCO_3$ make up the Bahamas, the Florida Keys, and the Red Sea basin. Corals, sea shells, and pearls are also mainly $CaCO_3$. Other important minerals are gypsum ($CaSO_4 . 2H_2O$), anhydrite ($CaSO_4$), fluorite (CaF_2: also blue john and fluorspar), and apatite [$Ca_5(PO_4)_3F$].

Strontium (384 ppm) and barium (390 ppm) are respectively the fifteenth and fourteenth elements in order of abundance, being bracketed by S (340 ppm) and F (544 ppm). The most important mineral of Sr is celestite ($SrSO_4$), and strontianite ($SrCO_3$) is also mined. The largest producers are Canada, Mexico, Spain, and the UK, and the world production of these two minerals in 1974 was 10^5 tonnes. It is a critical raw material for the USA which is totally dependent on imports for supplies. The sulfate (barite) is also the most important mineral of Ba: it is mined commercially in over 40 countries throughout the world. Production in 1977 was 5.35 million tonnes, of which 44% was mined in the USA. The major use of $BaSO_4$ (92%) is as a heavy mud slurry in

---

[3] K. K. Turekian, Elements, geochemical distribution of, *McGraw Hill Encyclopedia of Science and Technology*, Vol. 4, pp. 627–30, 1977.

well drilling; production of Ba chemicals accounts for only 3%. A subsidiary mineral is witherite ($BaCO_3$), which was mined principally in the UK until the unique vein in the northern Pennines became exhausted in 1969.

Radium occurs only in association with uranium (Chapter 31); the observed ratio Ra/U is ~1 mg per 3 kg, leading to a terrestrial abundance for Ra of ~10^{-6} ppm. As uranium ores normally contain only a few hundred ppm of U, it follows that about 10 tonnes of ore must be processed for 1 mg Ra. The total amount of Ra available worldwide is of the order of a few kilograms, but its use in cancer therapy is being rapidly superseded by the use of other isotopes, and the annual production of separated Ra compounds is probably now only about 100 g. Chief suppliers are Belgium, Canada, Czechoslovakia, the UK, and the USSR.

5.2.2 *Production and uses of the metals*

Beryllium is extracted from beryl by roasting the mineral with Na_2SiF_6 at 700–750°C, leaching the soluble fluoride with water and then precipitating $Be(OH)_2$ at about pH 12. The metal is usually prepared by reduction of BeF_2 (p. 128) with Mg at about 1300°C or by electrolysis of fused mixtures of $BeCl_2$ and alkali metal chlorides. It is one of the lightest metals known and has one of the highest mps of the light metals. Its modulus of elasticity is one-third greater than that of steel. The largest use of Be is in high-strength alloys of Cu and Ni (see Panel).

Uses of Beryllium Metal and Alloys

The ability of Be to age-harden Cu was discovered by M. G. Corson in 1926 and it is now known that ~2% of Be increases the strength of Cu sixfold. In addition, the alloys (which also usually contain 0.25% Co) have good electrical conductivity, high strength, unusual wear resistance, and resistance to anelastic behaviour (hysteresis, damping, etc.): they are non-magnetic and corrosion resistant, and find numerous applications in critical moving parts of aero-engines, key components in precision instruments, control relays, and electronics. They are also non-sparking and are thus of great use for hand tools in the petroleum industry. A nickel alloy containing 2% Be is used for high-temperature springs, clips, bellows, and electrical connections. Another major use for Be is in nuclear reactors since it is one of the most effective neutron moderators and reflectors known. A small, but important, use of Be is as a window material in X-ray tubes: it transmits X-rays 17 times better than Al and 8 times better than Lindemann glass. A mixture of compounds of radium and beryllium has long been used as a convenient laboratory source of neutrons and, indeed, led to the discovery of the neutron by J. Chadwick in 1932: $^9Be(\alpha,n)^{12}C$.

Magnesium is produced on a large scale (300 000 tonnes in 1979) either by electrolysis or silicothermal reduction. The major producers are the USA (48%), the USSR (24%), and Norway (15%). The electrolytic process uses either fused anhydrous $MgCl_2$ at 750°C or partly hydrated $MgCl_2$ from sea water at a slightly lower temperature.[4] The silicothermal process uses calcined dolomite and ferrosilicon alloy under reduced pressure at 1150°C:

$$2(MgO \cdot CaO) + FeSi \longrightarrow 2Mg + Ca_2SiO_4 + Fe$$

[4] R. D. GOODENOUGH and V. A. STENGER, Magnesium, calcium, strontium, barium, and radium, *Comprehensive Inorganic Chemistry*, Vol. 1, pp. 591–664, Pergamon Press, Oxford, 1973.

Magnesium is industry's lightest constructional metal, having a density less than two-thirds that of Al (see Panel).

Magnesium Metal and Alloys

The principal advantage of Mg as a structural metal is its low density (1.74 g cm^{-3} compared with 2.70 for Al and 7.80 for steel). For equal strength, the best Mg alloy weighs only a quarter as much as steel, and the best Al alloy weighs about one-third as much as steel. In addition, Mg has excellent machinability and it can be cast or fabricated by any of the standard metallurgical methods (rolling, extruding, drawing, forging, welding, brazing, or riveting). Its major use therefore is as a light-weight construction metal, not only in aircraft but also in luggage, photographic, and optical equipment, etc. It is also used for cathodic protection of other metals from corrosion, as an oxygen scavenger, and as a reducing agent in the production of Be, Ti, Zr, Hf, and U. World production exceeds 300 000 tonnes pa.

Magnesium alloys typically contain $>90\%$ Mg together with 2–9% Al, 1–3% Zn, and 0.2–1% Mn. Greatly improved retention of strength at high temperature (up to 450°C) is achieved by alloying with rare-earth metals (e.g. Pr/Nd) or Th. These alloys can be used for automobile engine casings and for aeroplane fuselages and landing wheels. Thus each Volkswagen Beetle used 20 kg of Mg alloy in its die-cast engine block and each Titan ICBM uses nearly 1 tonne. Other uses are in light-weight tread-plates, dock-boards, loading platforms, gravity conveyors, and shovels.

Up to 5% Mg is added to most commercial Al to improve its mechanical properties, weldability, and resistance to corrosion.

For further details see *Kirk–Othmer Encyclopedia of Chemical Technology*, 3rd edn., 1981, Vol. 14, pp. 570–615.

The other alkaline earth metals Ca, Sr, and Ba are produced on a much smaller scale than Mg. Calcium is produced by electrolysis of fused $CaCl_2$ (obtained either as a byproduct of the Solvay process (p. 82) or by the action of HCl or $CaCO_3$). It is less reactive than Sr or Ba, forming a protective oxide-nitride coating in air which enables it to be machined in a lathe or handled by other standard metallurgical techniques. Calcium metal is used mainly as an alloying agent to strengthen Al bearings, to control graphitic C in cast-iron, and to remove Bi from Pb. Chemically it is used as a scavenger in the steel industry (O, S, and P), as a getter for oxygen and nitrogen, to remove N_2 from argon, and as a reducing agent in the production of other metals such as Cr, Zr, Th, and U. Calcium also reacts directly with H_2 to give CaH_2, which is a useful source of H_2. World production of the metal is probably of the order of 1000 tonnes pa.

Metallic Sr and Ba are best prepared by high-temperature reduction of their oxides with Al in an evacuated retort or by small-scale electrolysis of fused chloride baths. They have limited use as getters, and a Ni–Ba alloy is used for spark-plug wire because of its high emissivity.

5.2.3 *Properties of the elements*

Table 5.1 lists some of the atomic properties of the Group IIA elements. Comparison with the data for Group IA elements (p. 86) shows the substantial increase in the ionization energies; this is related to their smaller size and higher nuclear charge, and is particularly notable for Be. Indeed, the "ionic radius" of Be is purely a notional figure since

TABLE 5.1   *Atomic properties of the alkaline earth metals*

Property	Be	Mg	Ca	Sr	Ba	Ra
Atomic number	4	12	20	38	56	88
Number of naturally occurring isotopes	1	3	6	4	7	4[a]
Atomic weight	9.012 18	24.305	40.08	87.62	137.33	226.0254
Electronic configuration	$[He]2s^2$	$[Ne]3s^2$	$[Ar]4s^2$	$[Kr]5s^2$	$[Xe]6s^2$	$[Rn]7s^2$
Ionization energies/ $kJ\,mol^{-1}$	899.2	737.5	589.6	549.2	502.7	509.1
	1757	1450	1145	1064	965	975
Metal radius/pm	112	160	197	215	222	—
Ionic radius (6 coord)/pm	$(27)^b$	72	100	118	135	148
E°/V for $M^{2+}(aq)+2e^- \rightarrow M(s)$	−1.85	−2.37	−2.87	−2.89	−2.91	−2.92

[a] All radioactive: longest $t_{\frac{1}{2}}$ 1600 y.
[b] Four-coordinate.

no compounds are known in which uncoordinated Be has a 2+ charge. In aqueous solutions the reduction potential of Be is much less than that of its congeners, again indicating its lower electropositivity. By contrast, Ca, Sr, Ba, and Ra have reduction potentials which are almost identical with those of the heavier alkali metals; Mg occupies an intermediate position.

The alkaline earth metals are silvery white, lustrous, and relatively soft. Their physical properties (Table 5.2), when compared with those of Group IA metals, show that they have a substantially higher mp, bp, density, and enthalpies of fusion and vaporization. This can be understood in terms of the size factor mentioned in the preceding paragraph and the fact that 2 valency electrons per atom are now available for bonding. Again, Be is notable in melting more than 1100° above Li and being nearly 3.5 times as dense; its enthalpy of fusion is more than 5 times that of Li. Beryllium resembles Al in being stable in moist air due to the formation of a protective oxide layer, and highly polished specimens retain their shine indefinitely. Magnesium also resists oxidation but the heavier metals tarnish readily. Beryllium, like Mg and the high-temperature form of Ca (>450°C), crystallizes in the hcp arrangement, and this confers a marked anisotropy on its properties; Sr is fcc, Ba and Ra are bcc like the alkali metals.

TABLE 5.2   *Physical properties of the alkaline earth metals*

Property	Be	Mg	Ca	Sr	Ba	Ra
MP/°C	1287	649	839	768	727	(700)
BP/°C	~2500	1105	1494	1381	(1850)	(1700)
Density (20°C)/g cm^{-3}	1.848	1.738	1.55	2.63	3.62	5.5
ΔH_{fus}/kJ mol^{-1}	15	8.9	8.6	8.2	7.8	(8.5)
ΔH_{vap}/kJ mol^{-1}	309	127.4	155	158	136	(113)
Electrical resistivity (20°C)/ μohm cm	4.46	4.46	3.5	23	50	(100)

5.2.4 *Chemical reactivity and trends*

Beryllium metal is relatively unreactive at room temperature, particularly in its massive form. It does not react with water or steam even at red heat and does not oxidize in air below 600°, though powdered Be burns brilliantly on ignition to give BeO and Be_3N_2. The halogens (X_2) react above about 600°C to give BeX_2 but the chalcogens (S, Se, Te) require even higher temperatures to form BeS, etc. Ammonia reacts above 1200°C to give Be_3N_2 and carbon forms Be_2C at 1700°. In contrast with the other Group IIA metals, Be does not react directly with hydrogen, and BeH_2 must be prepared indirectly (p. 125). Cold, concentrated HNO_3 passivates Be but the metal dissolves readily in dilute aqueous acids (HCl, H_2SO_4, HNO_3) with evolution of hydrogen. Beryllium is sharply distinguished from the other alkaline earth metals in reacting with aqueous alkalis (NaOH, KOH) with evolution of hydrogen. It also dissolves rapidly in aqueous NH_4HF_2 (as does $Be(OH)_2$), a reaction of some technological importance in the preparation of anhydrous BeF_2 and purified Be:

$$2NH_4HF_2(aq) + Be(s) \longrightarrow (NH_4)_2BeF_4 + H_2(g)$$

$$(NH_4)_2BeF_4(s) \xrightarrow{280} BeF_2(s) + 2NH_4F(subl)$$

Magnesium is more electropositive than the amphoteric Be and reacts more readily with most of the non-metals. It ignites with the halogens, particularly when they are moist, to give MgX_2, and burns with dazzling brilliance in air to give MgO and Mg_3N_2. It also reacts directly with the other elements in Group V and VI (and Group IV) when heated and even forms MgH_2 with hydrogen at 570° and 200 atm. Steam produces MgO (or $Mg(OH_2)$) plus H_2, and ammonia reacts at elevated temperature to give Mg_3N_2. Methanol reacts at 200°C to give $Mg(OMe)_2$ and ethanol (when activated by a trace of iodine) reacts similarly at room temperature. Alkyl and aryl halides react with Mg to give Grignard reagents RMgX (pp. 147–51).

The heavier alkaline earth metals Ca, Sr, Ba (and Ra) react even more readily with non-metals, and again the direct formation of nitrides M_3N_2 is notable. Other products are similar though the hydrides are more stable (p. 70) and the carbides less stable than for Be and Mg. There is also a tendency, previously noted for the alkali metals (p. 97), to form peroxides MO_2 of increasing stability in addition to the normal oxides MO. Calcium, Sr, and Ba dissolve in liquid NH_3 to give deep blue-black solutions from which lustrous, coppery, ammoniates $M(NH_3)_6$ can be recovered on evaporation; these ammoniates gradually decompose to the corresponding amides, especially in the presence of catalysts:

$$[M(NH_3)_6](s) \longrightarrow M(NH_2)_2(s) + 4NH_3(g) + H_2(g)$$

In these properties, as in many others, the heavier alkaline earth metals resemble the alkali metals rather than Mg (which has many similarities to Zn) or Be (which is analogous to Al).

5.3 Compounds

5.3.1 *Introduction*

The predominant divalence of the Group IIA metals can be interpreted in terms of their electronic configuration, ionization energies, and size (see Table 5.1). Further ionization to give simple salts of stoichiometry MX_3 is precluded by the magnitude of the energies involved, the third stage ionization being 14 847 kJ mol^{-1} for Be, 7731 kJ mol^{-1} for Mg, and 4910 kJ mol^{-1} for Ca; even for Ra the estimated value of 3281 kJ mol^{-1} involves far more energy than could be recovered by additional bonding even if this were predominantly covalent. Reasons for the absence of *univalent* compounds MX are less obvious. The first-stage ionization energies for Ca, Sr, Ba, and Ra are similar to that of Li (p. 86) though the larger size of the hypothetical univalent Group IIA ions, when compared to Li, would reduce the lattice energy somewhat (p. 95). By making plausible assumptions about the ionic radius and structure we can estimate the approximate enthalpy of formation of such compounds and they are predicted to be stable with respect to the constituent elements; their non-existence is related to the much higher enthalpy of formation of the conventional compounds MX_2, which leads to rapid and complete disproportionation. For example, the standard enthalpy of formation of hypothetical crystalline MgCl, assuming the NaCl structure, is ~ -125 kJ mol^{-1}, which is substantially greater than for many known stable compounds and essentially the same as the experimentally observed value for AgCl: $\Delta H_f^\circ = -127$ kJ mol^{-1}. However, the corresponding (experimental) value for $\Delta H_f^\circ(MgCl_2)$ is -642 kJ mol^{-1}, whence an enthalpy of disproportionation of -196 kJ mol^{-1}:

$$Mg(c) + Cl_2(g) = MgCl_2(c); \qquad \Delta H_f^\circ = -642 \text{ kJ/mol of } MgCl_2$$
$$2Mg(c) + Cl_2(g) = 2MgCl(c); \qquad \Delta H_f^\circ = -250 \text{ kJ/2 mol of } MgCl$$

$$2MgCl(c) = Mg(c) + MgCl_2(c); \qquad \Delta H_{disprop}^\circ = -392 \text{ kJ/2 mol of } MgCl$$

It is clear that if synthetic routes could be devised which would mechanistically hinder disproportionation, such compounds might be preparable. Although univalent compounds of the Group IIA metals have not yet been isolated, there is some evidence for the formation of Mg^I species during electrolysis with Mg electrodes. Thus H_2 is evolved at the anode when an aqueous solution of NaCl is electrolysed and the amount of Mg lost from the anode corresponds to an oxidation state of 1.3. Similarly, when aqueous Na_2SO_4 is electrolysed, the amount of H_2 evolved corresponds to the oxidation by water of Mg ions having an average oxidation state of 1.4:

$$Mg^{1.4+}(aq) + 0.6H_2O \longrightarrow Mg^{2+}(aq) + 0.6OH^-(aq) + 0.3H_2(g)$$

On the basis of the discussion on p. 92 the elements in Group IIA would be expected to deviate even further from simple ionic bonding than do the alkali metals. The charge on M^{2+} is higher and the radius for corresponding ions is smaller, thereby inducing more distortion of the surrounding anions. This is reflected in the decreased thermal stability of oxoacid salts such as nitrates, carbonates, and sulfates. For example, the temperature at which the carbonate reaches a dissociation pressure of 1 atm CO_2 is: $BeCO_3$ 250°, $MgCO_3$ 540°, $CaCO_3$ 900°, $SrCO_3$ 1289°, $BaCO_3$ 1360°. The tendency towards covalency is greater with Be, and this element forms no compounds in which the bonding is predominantly ionic. Magnesium also has considerable covalent character as reflected in its diagonal relation with Li. For similar reasons Be (and to a lesser extent Mg) forms numerous stable coordination compounds; organometallic compounds are also well characterized, and these frequently involve multicentre (electron deficient) bonding similar to that found in analogous compounds of Li and B.

Many compounds of the Group IIA elements are much less soluble in water than their Group IA counterparts. This is particularly true for the fluorides, carbonates, and sulfates of the heavier members, and is related to their higher lattice energies. These solubility relations have had a profound influence on the mineralization of these elements as noted on p. 119. The ready solubility of BeF_2 ($\sim 20\,000$ times that of CaF_2) is presumably related to the very high solvation enthalpy of Be to give $[Be(H_2O)_4]^{2+}$.

Beryllium, because of its small size, almost invariably has a coordination number of 4. This is important in analytical chemistry since it ensures that EDTA, which coordinates strongly to Mg, Ca (and Al), does not chelate Be appreciably. BeO has the wurtzite (ZnS, p. 1405) structure whilst the other Be chalcogenides adopt the zinc blende modification. BeF_2 has the cristobalite (SiO_2, p. 394) structure and has only a very low electrical conductivity when fused. Be_2C and Be_2B have extended lattices of the antifluorite type with 4-coordinate Be and 8-coordinate C or B. Be_2SiO_4 has the phenacite structure (p. 400) in which both Be and Si are tetrahedrally coordinated, and Li_2BeF_4 has the same structure. $[Be(H_2O)_4]SO_4$ features a tetrahedral aquo-ion which is H bonded to the surrounding sulfate groups in such a way that Be–O is 161 pm and the O–H $\cdots$ O are 262 and 268 pm. Further examples of tetrahedral coordination to Be are to be found in later

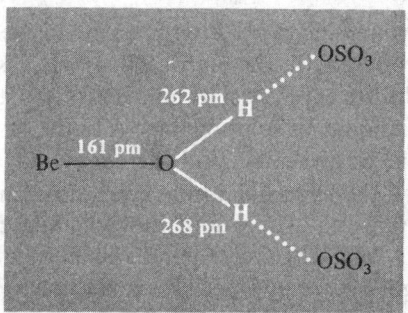

sections; other configurations, involving linear (two-fold) coordination (e.g. $BeBu^t_2$) or trigonal coordination [e.g. cyclic $(MeBeNMe_2)_2$] are rare and most compounds which might appear to have such coordination (e.g. $BeMe_2$, $CsBeF_3$, etc.) achieve 4-coordination by polymerization.

The stereochemistry of Mg and the heavier alkaline earth metals is more flexible than that of Be and, in addition to occasional compounds which feature low coordination numbers (2, 3, and 4), there are many examples of 6, 8, and 12 coordination and even some with 7, 9, or 10.[5]

5.3.2 *Hydrides and halides*

Many features of the structure, bonding and stability of the Group IIA hydrides have already been discussed (p. 70) and it is only necessary to add some comments on BeH_2, which is the most difficult of these compounds to prepare and the least stable. BeH_2

[5] A. F. WELLS, *Structural Inorganic Chemistry*, 4th edn., Oxford University Press, Oxford, 1975, 1095 pp.

(contaminated with variable amounts of ether) was first prepared in 1951 by reduction of $BeCl_2$ with LiH and by the reaction of $BeMe_2$ with $LiAlH_4$. A purer sample can be made by pyrolysis of $BeBu_2^t$ at 210°C and the best product is obtained by displacing BH_3 from BeB_2H_8 using PPh_3 in a sealed tube reaction at 180°C:

$$BeB_2H_8 + 2PPh_3 \longrightarrow 2Ph_3PBH_3 + BeH_2$$

BeH_2 is an amorphous white solid (d 0.65 g cm^{-3}) which begins to evolve hydrogen when heated above 250°C; it is moderately stable in air or water but is rapidly hydrolysed by acids, liberating H_2. Very recently a hexagonal crystalline form (d 0.78 g cm^{-3}) has been prepared by compaction fusion at 6.2 kbar and 130°C in the presence of $\sim 1\%$ Li as catalyst.[6] In all forms BeH_2 appears to be highly polymerized by means of BeHBe 3-centre bonds and its structure is probably similar to that of $BeCl_2$ and $BeMe_2$ (see below). A related compound is the volatile mixed hydride BeB_2H_8, which is readily prepared (in the absence of solvent) by the reaction of $BeCl_2$ with $LiBH_4$ in a sealed tube:

$$BeCl_2 + 2\ LiBH_4 \xrightarrow{120\ C} BeB_2H_8 + 2LiCl$$

BeB_2H_8 inflames in air, reacts almost explosively with water, and reacts with dry HCl even at low temperatures:

$$BeB_2H_8 + 2HCl \longrightarrow BeCl_2 + B_2H_6 + 2H_2$$

The structure of this compound has proved particularly elusive and at least nine different structures have been proposed; it therefore affords an instructive example of the difficulties which attend the use of physical techniques for the structural determination of compounds in the gaseous, liquid, or solution phases. In the gas phase it now seems likely that more than one species is present[7a] and the compound certainly shows fluxional behaviour which makes all the hydrogen atoms equivalent on the nmr time scale.[7b] A linear structure such as (a), with possible admixture of singly bridged B–H–B and triply bridged BeH_3B variants is now favoured, after a period in which triangular structures such as (b) had been vigorously canvassed. Even structure (c), which features planar 3-coordinate Be, had been advocated because it was thought to fit best much of the infrared and electron diffraction data and also accounted for the ready formation of adducts (d) with typical ligands such as Et_2O, thf, R_3N, R_3P, etc. In the solid state the structure has recently been established with some certainty by single-crystal X-ray analysis.[8] BeB_2H_8 consists of helical polymers of BH_4Be units linked by an equal number of bridging BH_4 units (Fig. 5.1). Of the 8 H atoms only 2 are not involved in bonding to Be; the Be is thus 6-coordinate (distorted trigonal prism) though the H atoms are much closer to B (~ 110 pm) than to Be (2 at ~ 153 pm and 4 at ~ 162 pm). The Be $\cdots$ B distance within the helical chain is 201 pm and in the branch is 192 pm. The relationship of this structure to those of $Al(BH_4)_3$ and AlH_3 itself (p. 256) is noteworthy.

[6] G. J. BRENDEL, E. M. MARLETT, and L. M. NIEBYLSKI, Crystalline beryllium hydride, *Inorg. Chem.* **17**, 3589–92 (1978).

[7a] K. BRENDHAUGEN, A. HAARLAND, and D. P. NOVAK, The gas phase electron diffraction pattern of beryllium borohydride, *Acta Chem. Scand.* **A29**, 801–2 (1975).

[7b] D. F. GAINES, J. L. WALSH, and D. F. HILLENBRAND, Gas-phase nuclear magnetic resonance spectroscopic study of the molecular structure of beryllium borohydride Be$(BH_4)_2$, *JCS Chem. Comm.* 1977, 224–5.

[8] D. S. MARYNICK and W. N. LIPSCOMB, Crystal structure of beryllium borohydride, *Inorg. Chem.* **11**, 820–3 (1972). D. S. MARYNICK, Model studies of the electronic structure of solid-state beryllium borohydride, *J. Am. Chem. Soc.* **101**, 6876–80 (1979).

(a) Probable "linear" structure
    (fluxional)

(b) Hypothetical triangular
     structure

(c) Alternative triangular
     structure

(d) Proposed structure for
     LBeB₂H₈

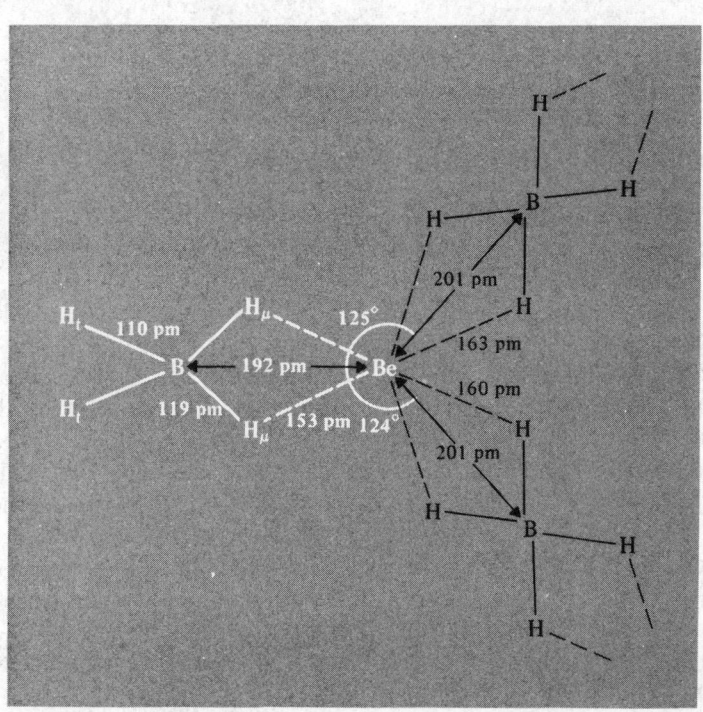

Fig. 5.1 Polymeric structure of crystalline Be(BH₄)₂ showing a section of the ···(H₂BH₂)Be(H₂BH₂)··· helix and one "terminal" or non-bridging {(Hₜ)₂B(H_μ)₂}-group.

Anhydrous beryllium halides cannot be obtained from reactions in aqueous solutions because of the formation of hydrates such as $[Be(H_2O)_4]F_2$ and the subsequent hydrolysis which attends attempted dehydration. Thermal decomposition of $(NH_4)_2BeF_4$ is the best route for BeF_2, and $BeCl_2$ is conveniently made from the oxide

$$BeO + C + Cl_2 \overset{600-800°}{\rightleftharpoons} BeCl_2 + CO$$

$BeCl_2$ can also be prepared by direct high-temperature chlorination of metallic Be or Be_2C, and these reactions are also used for the bromide and iodide. BeF_2 is a glassy material that is difficult to crystallize; it consists of a random network of 4-coordinate F-bridged Be atoms similar to the structure of vitreous silica, SiO_2. Above 270°, BeF_2 spontaneously crystallizes to give the quartz modification (p. 394) and, like quartz, it exists in a low-temperature α-form which transforms to the β-form at 227°; crystobalite and tridymite forms (p. 394) have also been prepared. The structural similarities between BeF_2 and SiO_2 extend to fluoroberyllates and silicates, and numerous parallels have been drawn: e.g. the phase diagram, compounds, and structures in the system $NaF-BeF_2$ resemble those for $CaO-SiO_2$; the system CaF_2-BeF_2 resembles ZrO_2-SiO_2; the compound $KZnBe_3F_9$ is isostructural with benitoite, $BaTiSi_3O_9$, etc.

$BeCl_2$ has an unusual chain structure (a) which can be cleaved by weak ligands such as Et_2O to give 4-coordinate molecular complexes L_2BeCl_2 (b); stronger donors such as H_2O or NH_3 lead to ionic complexes $[BeL_4]^{2+}[Cl]^-_2$ (c).

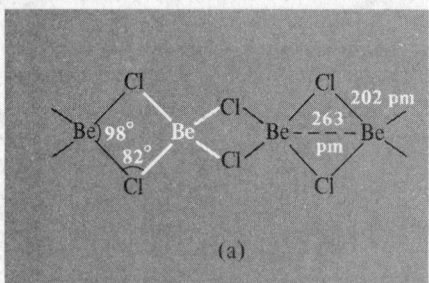 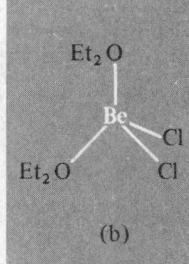 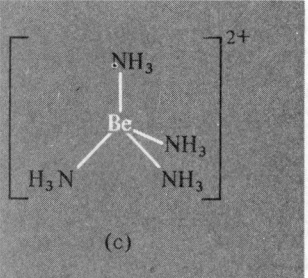

| (a) | (b) | (c) |

In all these forms Be can be considered to use the s, p_x, p_y, and p_z orbitals for bonding; the ClBeCl angle is substantially less than the tetrahedral angle of 109° probably because this lessens the repulsive interaction between neighbouring Be atoms in the chain by keeping them further apart and also enables a wider angle than 71° to be accommodated at each Cl atom, consistent with its predominant use of two p orbitals. The detailed interatomic distances and angles therefore differ significantly from those in the analogous chain structure $BeMe_2$ (p. 142), which is best described in terms of 3-centred "electron-deficient" bonding at the Me groups, leading to a BeCBe angle of 66° and a much closer approach of neighbouring Be atoms (209 pm). In the vapour phase $BeCl_2$ tends to form a bridged sp^2 dimer (d) and dissociation to the linear (sp) monomer (e) is not complete below about 900°; in contrast, BeF_2 is monomeric and shows little tendency to dimerize in the gas phase.

The crystal structures of the halides of the heavier Group IIA elements show some interesting trends (Table 5.3).

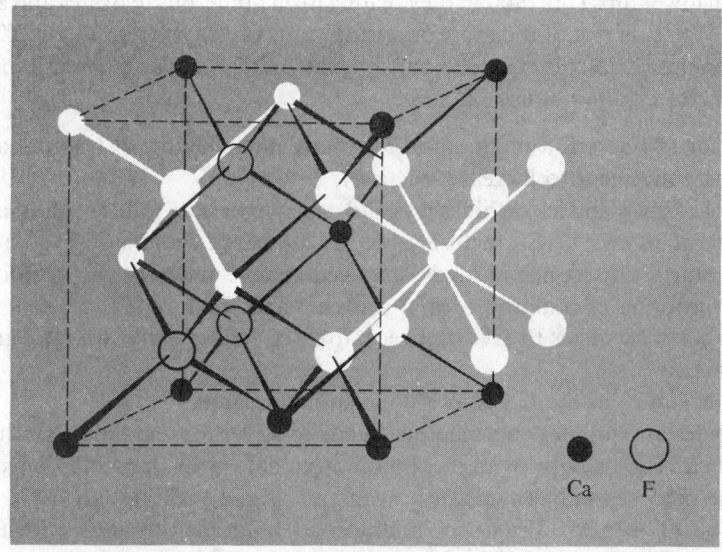

TABLE 5.3 *Crystal structures of alkaline earth halides*[a]

	Be	Mg	Ca	Sr	Ba
F	Quartz	Rutile(TiO_2)	Fluorite	Fluorite	Fluorite
Cl	Chain	$CdCl_2$	Deformed TiO_2	Deformed TiO_2	$PbCl_2$
Br	Chain	CdI_2	Deformed TiO_2	Deformed $PbCl_2$	$PbCl_2$
I	—	CdI_2	CdI_2	SrI_2	$PbCl_2$

[a] For description of these structures see: quartz (p. 394), rutile (p. 1120), $CdCl_2$ (p. 1407), CdI_2 (p. 1407), $PbCl_2$ (p. 444); the fluorite, $BeCl_2$-chain, and SrI_2 structures are described in this subsection.

For the fluorides, increasing size of the metal enables its coordination number to increase from 4 (Be) to 6 (Mg) and 8 (Ca, Sr, Ba). CaF_2 (fluorite) is a standard crystal structure type and its cubic unit cell is illustrated in Fig. 5.2. The other halides (Cl, Br, I) show an increasing trend away from three-dimentional structures, the Be halides forming chains (as discussed above) and the others tending towards layer-lattice structures such as

Fig. 5.2 Unit cell of CaF_2 showing eightfold (cubic) coordination of Ca by 8F and fourfold (tetrahedral) coordination of F by 4Ca. The structure can be thought of as an fcc array of Ca in which all the tetrahedral interstices are occupied by F.

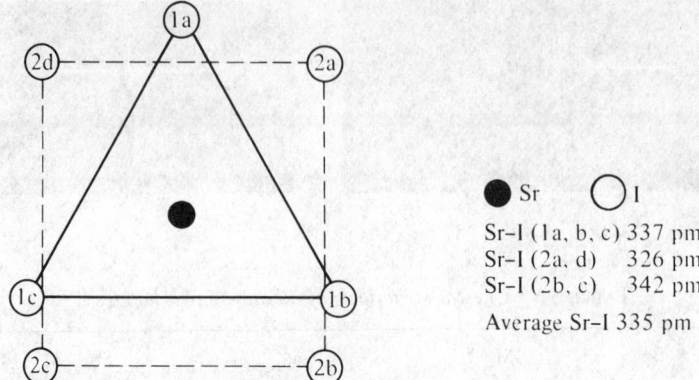

Sr ●   I ○
Sr–I (1a, b, c) 337 pm
Sr–I (2a, d)   326 pm
Sr–I (2b, c)   342 pm
Average Sr–I 335 pm

FIG. 5.3   Structure of SrI_2 showing sevenfold coordination of Sr by I. The planes 1 and 2 are almost parallel (4.5°) and the planes 1a2a2d and 1b2b2c1c are at an angle of 12° to each other.[9]

$CdCl_2$, CdI_2, and PbI_2. SrI_2 is unique in this group in having sevenfold coordination (Fig. 5.3); a similar coordination polyhedron is found in EuI_2, but the way they are interconnected differs in the two compounds.[9]

The most important fluoride of the alkaline earth metals is CaF_2 since this mineral (fluorspar) is the only large-scale source of fluorine (p. 920). Annual world production now exceeds 5 million tonnes the principal suppliers being Mexico (24%), the USSR (10%), Spain and China (8% each), France and Mongolia (7% each), Italy, the UK, and South Africa (5% each). The largest consumer is the USA, though 85% of its needs must be imported. CaF_2 is a white, high-melting (1418°C) solid whose low solubility in water permits quantitative analytical precipitation. The other fluorides (except BeF_2) are also high-melting and rather insoluble. By contrast, the chlorides tend to be deliquescent and to have much lower mps (715–960°); they readily form numerous hydrates and are soluble in alcohols. $MgCl_2$ is one of the most important salts of Mg industrially (p. 120) and its concentration in sea water is exceeded only by NaCl. $CaCl_2$ is also of great importance as noted earlier. Its traditional uses include:

(a) brine for refrigeration plants (and for filling inflated tires of tractors and earth-moving equipment to increase traction);
(b) control of snow and ice on highways and pavements (side walks)—the $CaCl_2$–H_2O eutectic at 30 wt% $CaCl_2$ melts at −55°C (compared with NaCl–H_2O at −18°C);
(c) dust control on secondary roads, unpaved streets, and highway shoulders;
(d) freeze-proofing of coal and ores in shipping and stock piling;
(e) use in concrete mixes to give quicker initial set, higher early strength, and greater ultimate strength.

Production of $CaCl_2$ in the US is in the megatonne region.

The bromides and iodides continue the trends to lower mps and higher solubilities in water and their ready solubility in alcohols, ethers, etc., is also notable; indeed, $MgBr_2$ forms numerous crystalline solvates such as $MgBr_2 . 6ROH$ (R = Me, Et, Pr), $MgBr_2 . 6Me_2CO$, $MgBr_2 . 3Et_2O$, in addition to numerous ammines $MgBr_2 . nNH_3$

[9] E. T. RIETSCHEL and H. BÄRNIGHAUSEN, The crystal structure of SrI_2, *Z. anorg. allgem. Chem.* **368**, 62–72 (1969).

($n = 2$–6). The ability of Group IIA cations to form coordination complexes is clearly much greater than that of Group IA cations (p. 105).

Alkaline earth salts MHX, where M = Ca, Sr, or Ba, and X = Cl, Br, or I can be prepared by fusing the hydride MH_2 with the appropriate halide MX_2 or by heating $M + MX_2$ in an atmosphere of H_2 at 900°. These hydride halides appear to have the PbClF layer lattice structure though the H atoms were not directly located. The analogous compounds of Mg have proved more elusive and the preceding preparative routes merely yield physical mixtures. However, MgClH and MgBrH can be prepared as solvated dimers by the reaction of specially activated MgH_2 with MgX_2 in thf:

$$MgR_2 + LiAlH_4 \xrightarrow{Et_2O} MgH_2 + LiAlH_2R_2$$

$$MgH_2 + MgX_2 \xrightarrow{thf} [HMgX(thf)]_2$$

The chloride can be crystallized but the bromide disproportionates. On the basis of mol wt and infrared spectroscopic evidence the proposed structure is:

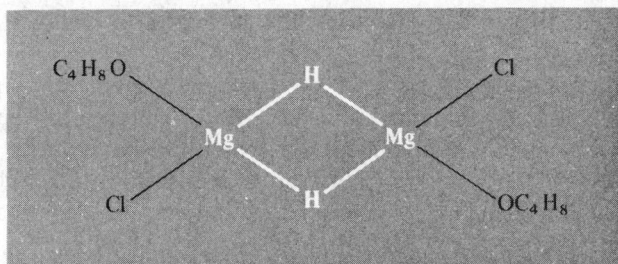

5.3.3 *Oxides and hydroxides*[4, 10]

The oxides MO are best obtained by calcining the carbonates (pp. 124, 133); dehydration of the hydroxides at red heat offers an alternative route. BeO (like the other Be chalcogenides) has the wurtzite structure (p. 1405) and is an excellent refractory, combining a high mp ($2530 \pm 50°C$) with negligible vapour pressure below this temperature; it has good chemical stability and a very high thermal conductivity which is greater than that of any other non-metal and even exceeds that of some metals. The other oxides in the group all have the NaCl structure and this structure is also adopted by the chalcogenides (except MgTe which has the wurtzite structure). Lattice energies and mps are again very high: MgO mp $2826 \pm 30°$, CaO $2613 \pm 25°$, SrO $2430°$, BaO $1923°C$. The compounds are comparatively unreactive in bulk but their reactivity increases markedly with decrease in particle size and increase in atomic weight. Notable reactions (which reverse those used to prepare the oxides) are with CO_2 and with H_2O. MgO is extensively used as a refractory: like BeO it is unusual in being both an excellent thermal conductor and a good electrical insulator, thus finding widepsread use as the insulating radiator in domestic heating ranges and similar appliances. CaO (lime) is produced on an enormous scale in many countries and, indeed, is second only to H_2SO_4 as the largest volume industrial chemical

[10] D. A. EVEREST, *Beryllium*, *CIC*, Vol. 1, pp. 531–90 (1973).

to be manufactured (see Panel on p. 133). Production in 1979 exceeded 18 million tonnes in the USA alone. Its major end uses are as a flux in steel manufacture (45%), in the production of Ca chemicals (10%), in water treatment (10%), sewage treatment and air and water pollution control (5%), pulp and paper manufacture (5%), and the manufacture of non-ferrous metals (5%). Price for bulk quantities is $\sim$\$45 per tonne (1981).

In addition to the oxides MO, peroxides MO_2 are known for the heavier alkaline earth metals and there is some evidence for yellow superoxides $M(O_2)_2$ of Ca, Sr, and Ba; impure ozonides $Ca(O_3)_2$ and $Ba(O_3)_2$ have also been reported.[11] As with the alkali metals, stability increases with electropositive character and size: no peroxide of Be is known; anhydrous MgO_2 can only be made in liquid NH_3 solution, aqueous reactions leading to various peroxide hydrates; CaO_2 can be obtained by dehydrating $CaO_2 . 8H_2O$ but not by direct oxidation, whereas SrO_2 can be synthesized directly at high oxygen pressures and BaO_2 forms readily in air at 500°. Reactions with aqueous reagents are as expected, and the compounds can be used as oxidizing agents and bleaches:

$$CaO_2(s) + H_2SO_4(aq) \longrightarrow CaSO_4(s) + H_2O_2(aq)$$
$$Ca(O_2)_2 + H_2SO_4(aq) \longrightarrow CaSO_4(s) + H_2O_2(aq) + O_2(g)$$

MgO_2 has the pyrite structure (p. 804) and the Ca, Sr, and Ba analogues have the CaC_2 structure (p. 320).

The hydroxides of Group IIA elements show a smooth gradation in properties, with steadily increasing basicity, solubility, and heats of formation from the corresponding oxide. $Be(OH)_2$ is amphoteric and $Mg(OH)_2$ is a mild base which, as an aqueous suspension (milk of magnesia), is widely used as a digestive antacid.† $Ca(OH)_2$ and $Sr(OH)_2$ are moderately strong to strong bases and $Ba(OH)_2$ approaches the alkali hydroxides in strength.

Beryllium salts rapidly hydrolyse in water to give a series of hydroxo complexes of undetermined structure; the equilibria depend sensitively on initial concentration, pH, temperature, etc., and precipitation begins when the ratio $OH^-/Be^{2+}(aq) > 1$. Addition of further alkali redissolves the precipitate and the properties of the resultant solution are consistent (at least qualitatively) with the presence of isopolyanions of the type

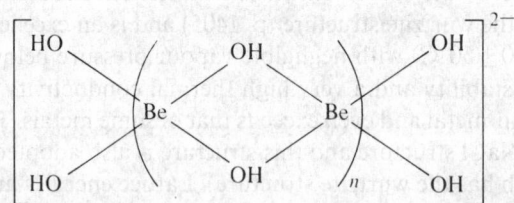

Further addition of alkali progressively depolymerizes this chain anion by hydroxyl addition until ultimately the mononuclear beryllate anion $[Be(OH)_4]^{2-}$ is formed. The analogy with $Zn(OH)_2$ and $Al(OH)_3$ is clear.

---

† Note that, though mild, $Mg(OH)_2$ will neutralize 1.37 times as much acid as NaOH, weight for weight, and 2.85 times as much as $NaHCO_3$.

[11] N.-G. VANNERBERG, Peroxides, superoxides, and ozonides of the metals of Groups IA, IIA, and IIB, *Prog. Inorg. Chem.* **4**, 125–97 (1962).

The solubility of $Be(OH)_2$ in water is only $\sim 3 \times 10^{-4}$ g l^{-1} at room temperature, compared with $\sim 3 \times 10^{-2}$ g l^{-1} for $Mg(OH)_2$ and ~ 1.3 g l^{-1} for $Ca(OH)_2$. Strontium and barium hydroxides have even greater solubilities (8 and 38 g l^{-1} respectively at 20°).

The crystal structures of $M(OH)_2$ also follow group trends.[5] $Be(OH)_2$ crystallizes with 4-coordinate Be in the $Zn(OH)_2$ structure which can be considered as a diamond or cristobalite (SiO_2) lattice distorted by H bonding. $Mg(OH)_2$ (brucite) and $Ca(OH)_2$ have 6-coordinate cations in a CdI_2 layer lattice structure with OH bonds perpendicular to the layers and strong $O-H\cdots O$ bonding between them. Strontium is too large for the CdI_2 structure and $Sr(OH)_2$ features 7-coordinate Sr $(3+4)$, the structure being built up of edge-sharing monocapped trigonal prisms with no H bonds. (The monohydrate has bicapped trigonal prismatic coordination about Sr.) The structure of $Ba(OH)_2$ is complex and not yet full determined.

5.3.4 *Oxoacid salts and coordination complexes*

The chemical trends and geochemical significance of the oxoacid salts of the alkaline earth metals have already been mentioned (p. 119) and the immense industrial importance of the carbonates and sulfates in particular can hardly be over emphasized (see Panel on limestone and lime). Calcium sulfate usually occurs as the dihydrate (gypsum) though anhydrite ($CaSO_4$) is also mined. Alabaster is a compact, massive, fine-grained form of $CaSO_4 . 2H_2O$ resembling marble. When gypsum is calcined at 150–165°C it looses approximately three-quarters of its water of crystallization to give the hemihydrate $CaSO_4 . \frac{1}{2}H_2O$, also known as plaster of Paris because it was originally obtained from gypsum quarried at Montmartre. Heating at higher temperatures yields various anhydrous forms:

$$CaSO_4 . 2H_2O \xrightarrow{\sim 150°} CaSO_4 . \tfrac{1}{2}H_2O \xrightarrow{\sim 200°} \gamma\text{-}CaSO_4 \xrightarrow{\sim 600°} \beta\text{-}CaSO_4 \xrightarrow{\sim 1100°}$$

$$CaO + SO_3$$

Gypsum, though not mined on the same scale as limestone, is nevertheless still a major industrial mineral. World production in 1977 was 65.5 million tonnes, the major producing countries being the USA (18.5%), Canada (10.8%), Iran (10.3%), France (8.9%), the USSR (7.9%), Spain (6.6%), Italy (6.4%), and the UK (5.0%). The price (USA, 1977) was $\sim$ $6 per tonne for crude gypsum and $\sim$ $25 per tonne for calcined gypsum. In the USA about 28% of the gypsum used is uncalcined and most of this is for Portland cement (p. 283) or agricultural purposes. Of calcined gypsum, virtually all (95%) is used for prefabricated products, mainly wall board, and the rest is for industrial and building plasters.

Other oxoacid salts and binary compounds are more conveniently discussed under the chemistry of the appropriate non-metals in later chapters.

Industrial Uses of Limestone and Lime

Limestone rock is the commonest form of calcium carbonate, which also occurs as chalk, marble, corals, calcite, argonite, etc., and (with Mg) as dolomite. Limestone and dolomite are widely used as building materials and road aggregate and both are quarried on a vast scale worldwide. $CaCO_3$ is also a major industrial chemical and is indispensable as the precursor of quick lime (CaO) and slaked lime, $Ca(OH)_2$. These chemicals are crucial to large sections of the chemical, metallurgical, and construction industries, as noted below, and are produced on a scale exceeded by very few other materials.[11a] Thus, world production of lime exceeded 110 million tonnes in 1977, and even this was dwarfed by Portland cement (777 million tonnes in 1977) which is made by roasting limestone and sand with clay (p. 283).

Large quantities of lime are consumed in the steel industry where it is used as a flux to remove P, S, Si, and to a lesser extent Mn. The basic oxygen steel process typically uses 75 kg lime per tonne of steel, or a rather larger quantity (100–300 kg) of dolomitic quick lime, which markedly extends the life of the refractory furnace linings. Lime is also used as a lubricant in steel wire drawing and in neutralizing waste sulfuric-acid-based pickling liquors. Another metallurgical application is in the production of Mg (p. 120): the ferro-silicon (Pidgeon) process uses dolomitic lime and both of the Dow electrolytic methods also require lime.

Pidgeon (ferrosilicon)

$$2(CaO \cdot MgO) + Si/Fe \longrightarrow 2Mg + Ca_2SiO_4/Fe$$

Dow (natural brine):

$$CaO \cdot MgO + CaCl_2 \cdot MgCl_2 + CO_2 \longrightarrow 2CaCO_3 + 2MgCl_2 \text{ (then electrolysis)}$$

Dow (seawater):

$$Ca(OH)_2 + MgCl_2 \longrightarrow Mg(OH)_2 + CaCl_2$$
$$Mg(OH)_2 + 2HCl \longrightarrow 2H_2O + MgCl_2 \text{ (then electrolysis)}$$

Lime is the largest tonnage chemical used in the treatment of potable and industrial water supplies. In conjunction with alum or iron salts it is used to coagulate suspended solids and remove turbidity. It is also used in water softening to remove temporary (bicarbonate) hardness. Typical reactions are:

$$Ca(HCO_3)_2 + Ca(OH)_2 \longrightarrow 2CaCO_3\downarrow + 2H_2O$$
$$Mg(HCO_3)_2 + Ca(OH)_2 \longrightarrow MgCO_3(\downarrow) + CaCO_3\downarrow + 2H_2O$$
$$MgCO_3 + Ca(OH)_2 \longrightarrow Mg(OH)_2\downarrow + CaCO_3\downarrow \text{ etc.}$$

The neutralization of acid waters (and industrial wastes) and the maintenance of optimum pH for the biological oxidation of sewage are further applications. Another major use of lime is in scrubbers to remove SO_2/H_2S from stack gases of fossil-fuel-powered generating stations and metallurgical smelters.

The chemical industry uses lime in the manufacture of calcium carbide (for acetylene, p. 319), cyanamide (p. 341), and numerous other chemicals. Glass manufacturing is also a major consumer, most common glasses having $\sim 12\%$ CaO in their formulation. The insecticide calcium arsenate, obtained by neutralizing arsenic acid with lime, is much used for controlling the cotton boll weevil, codling moth, tobacco worm, and Colorado potato beetle. Lime-sulfur sprays and Bordeaux mixtures $[(CuSO_4/Ca(OH)_2]$ are important fungicides.

The paper and pulp industries consume large quantities of $Ca(OH)_2$ and precipitated (as distinct from naturally occurring) $CaCO_3$. The largest application of lime in pulp manufacture is as a causticizing agent in sulfate (kraft) plants (p. 103). Here the waste Na_2CO_3 solution is reacted with lime to regenerate the caustic soda used in the process:

$$Na_2CO_3 + CaO + H_2O \longrightarrow CaCO_3 + 2NaOH$$

Continued

[11a] R. S. BOYNTON, *Chemistry and Technology of Lime and Limestone*, 2nd edn., Wiley, Chichester, 1980, 575 pp.

About 95% of the $CaCO_3$ mud is dried and recalcined in rotary kilns to recover the CaO. Calcium hypochlorite bleaching liquor (p. 1006) for paper pulp is obtained by reacting lime and Cl_2.

The manufacture of high quality paper involves the extensive use of specially precipitated $CaCO_3$. This is formed by calcining limestone and collecting the CO_2 and CaO separately; the latter is then hydrated and recarbonated to give the desired product:

$$CaCO_3 \xrightarrow{\text{heat}} CaO + CO_2$$

$$CaO + H_2O \longrightarrow Ca(OH)_2$$

$$Ca(OH)_2 + CO_2 \longrightarrow CaCO_3 + H_2O$$

The type of crystals obtained, as well as their size and habit, depend on the temperature, pH, rate of mixing, concentration, and presence of additives. The fine crystals are often subsequently coated with fatty acids, resins, and wetting agents to improve their flow properties. US prices (1978) range from 8–25 cents per kg depending on grade. $CaCO_3$ adds brightness, opacity, ink receptivity, and smoothness to paper and, in higher concentration, counteracts the high gloss produced by kaolin additives and produces a matte or dull finish which is particularly popular for textbooks. Such papers may contain 5–50% by weight of precipitated $CaCO_3$. The compound is also used as a filler in rubbers, latex, wallpaints, and enamels, and in plastics ($\sim 10\%$ by weight) to improve their heat resistance, dimensional stability, stiffness, hardness, and processability.

Domestic and pharmaceutical uses of precipitated $CaCO_3$ include its direct use as an antacid, a mild abrasive in toothpastes, a source of Ca enrichment in diets, a constituent of chewing gum and a filler in cosmetics.

In the dairy industry lime finds many uses. Lime water is often added to cream when separated from whole milk, in order to reduce its acidity prior to pasteurization and conversion to butter. The skimmed milk is then acidified to separate casein which is mixed with lime to produce calcium caseinate glue. Fermentation of the remaining skimmed milk (whey) followed by addition of lime yields calcium lactate which is used as a medicinal or to produce lactic acid on reacidification. Likewise the sugar industry relies heavily on lime: the crude sugar juice is reacted with lime to precipitate calcium sucrate which permits purification from phosphatic and organic impurities. Subsequent treatment with CO_2 produces insoluble $CaCO_3$ and purified soluble sucrose. The cycle is usually repeated several times; cane sugar normally requires ~ 3–5 kg lime per tonne but beet sugar requires 100 times this amount i.e. $\sim \frac{1}{4}$ tonne lime per tonne of sugar.

Beryllium is unique in forming a series of stable, volatile, molecular oxide-carboxylates of general formula $[OBe_4(RCO_2)_6]$, where R = H, Me, Et, Pr, Ph, etc. These white crystalline compounds, of which "basic beryllium acetate" (R = Me) is typical, are readily soluble in organic solvents, including alkanes, but are insoluble in water or the lower alcohols. They are best prepared simply by refluxing the hydroxide or oxide with the carboxylic acid; mixed oxide carboxylates can be prepared by reacting a given compound with another organic acid or acid chloride. The structure (Fig. 5.4) features a central O atom tetrahedrally surrounded by 4 Be. The 6 edges of the tetrahedron so formed are bridged by the 6 acetate groups in such a way that each Be is also tetrahedrally coordinated by 4 oxygens. $[OBe_4(MeCO_2)_6]$ melts at 285° and boils at 330°; it is stable to heat and oxidation except under drastic conditions, is only slowly hydrolysed by hot water, but is decomposed rapidly by mineral acids to give an aqueous solution of the corresponding beryllium salt and free carboxylic acid. The basic nitrate $[OBe_4(NO_3)_6]$ appears to have a similar structure with bridging nitrate groups. The compound is formed by first dissolving $BeCl_2$ in N_2O_4/ethylacetate to give the crystalline solvate $[Be(NO_3)_2 . 2N_2O_4]$; when heated to 50° this gives $Be(NO_3)_2$ which decomposes suddenly at 125°C into N_2O_4 and $[OBe_4(NO_3)_6]$.

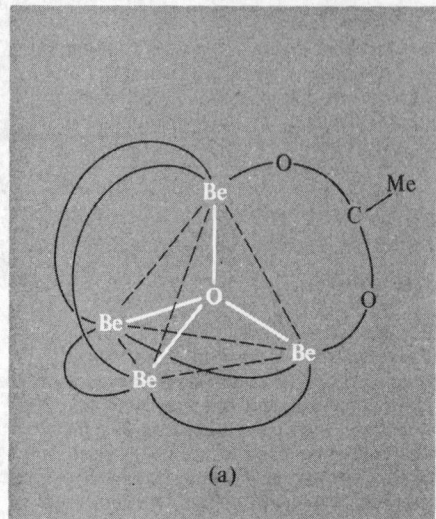 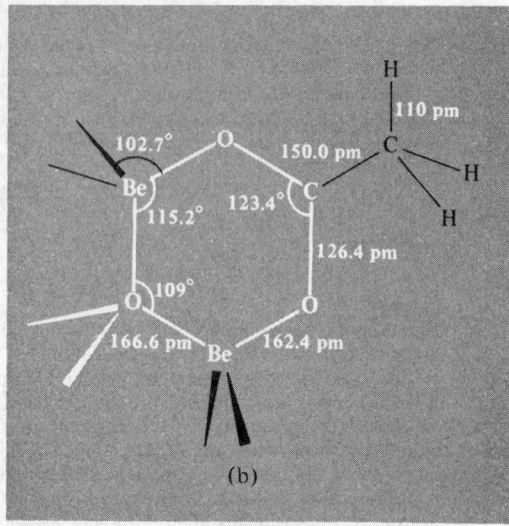

FIG. 5.4   The molecular structure of "basic beryllium acetate" showing (a) the regular tetrahedral arrangement of 4Be about the central oxygen and the octahedral arrangement of the 6 bridging acetate groups, and (b) the detailed dimensions of one of the six non-planar 6-membered heterocycles. (The Be atoms are 24 pm above and below the plane of the acetate group.) The 2 oxygen atoms in each acetate group are equivalent. The central Be–O distances (166.6 pm) are very close to that in BeO itself (165 pm).

In addition to the oxide carboxylates, beryllium forms numerous chelating and bridged complexes with ligands such as the oxalate ion $C_2O_4^{2-}$, alkoxides, β-diketonates, and 1,3-diketonates.[10] These almost invariably feature 4-coordinate Be though severe steric crowding can reduce the coordination number to 3 or even 2; for example, the very volatile dimeric alkoxide (a) was prepared in 1975 and the unique monomeric bis(2,6-di-t-butylphenoxy) beryllium (b) has been known since 1972.

Halide complexes are also well known but complexes with nitrogen-containing ligands are rare. An exception is the blue phthalocyanine complex formed by reaction of Be metal with phthalonitrile, $1,2\text{-}C_6H_4(CN)_2$, and this affords an unusual example of planar

4-coordinate Be (Fig. 5.5). The complex readily picks up 2 molecules of H_2O to form an extremely stable dihydrate, perhaps by dislodging 2 adjacent Be–N bonds and forming 2 Be–O bonds at the preferred tetrahedral angle above and below the plane of the macrocycle.

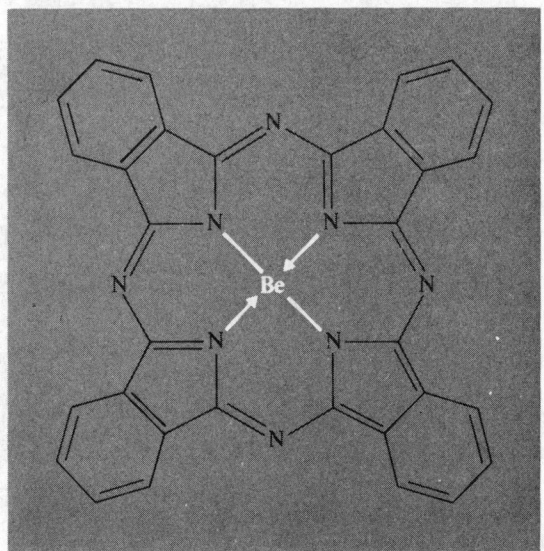

FIG. 5.5   The beryllium phthalocyanine complex.

Magnesium forms few halide complexes of the type MX_4^{2-}, though $[NEt_4]_2[MgCl_4]$ has been reported; examples for the heavier alkaline earths are lacking, though hydrates and other solvates are well known. Oxygen chelates such as those of edta and polyphosphates are of importance in analytical chemistry and in removing Ca ions from hard water. There is no unique sequence of stabilities since these depend sensitively on a variety of factors: where geometrical considerations are not important the smaller ions tend to form the stronger complexes but in polydentate macrocycles steric factors can be crucial. Thus dicyclohexyl-18-crown-6 (p. 107) forms much stronger complexes with Sr and Ba than with Ca (or the alkali metals) as shown in Fig. 5.6.[12] Structural data are also available and an example of a solvated 8-coordinate Ca complex [(benzo-15-crown-5)-Ca(NCS)$_2$.MeOH] is shown in Fig. 5.7 (p. 139). The coordination polyhedron is not regular: Ca lies above the mean plane of the 5 ether oxygens (mean Ca–O 253 pm) and is coordinated on the other side by a methanol molecule (Ca–O 239 pm) and 2 non-equivalent isothiocyanate groups (Ca–N 244 pm) which make angles Ca–N–CS of 153° and 172° respectively.[13] Cryptates (p. 109) are also known and usually follow the stability sequence Mg < Ca < Sr < Ba.[12]

[12] See refs. 26 and 30 of Chapter 4.

[13] J. D. OWEN and J. N. WINGFIELD, A new conformation of the cyclic polyether benzo-15-crown-5 in its solvated complex with calcium isothiocyanate: X-ray crystal structure analysis, *JCS Chem. Comm.* 1976, 318–19.

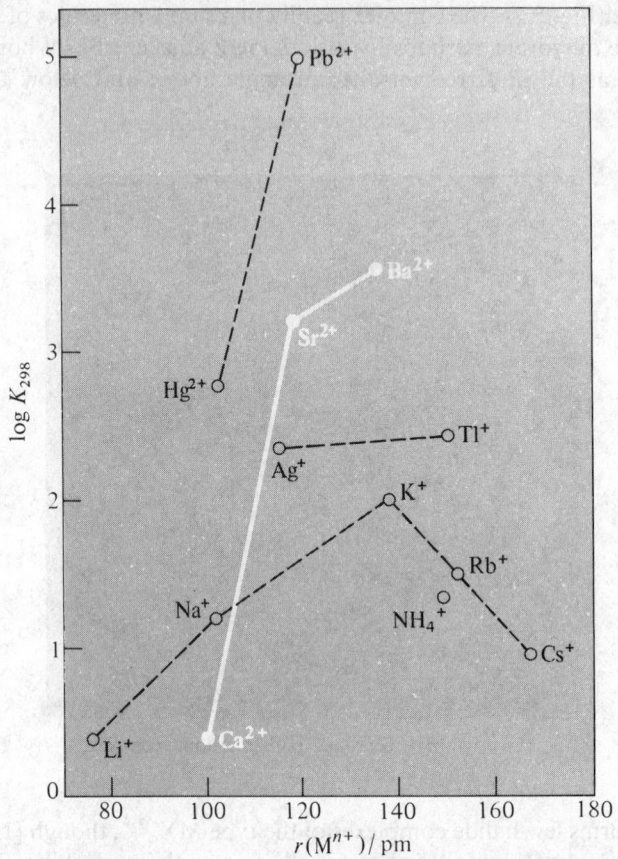

Fig. 5.6   Formation constants K for complexes of dicyclohexyl-18-crown-6 ether with various cations. Note that, although the radii of Ca^{2+}, Na^+, and Hg^{2+} are very similar, the ratio of the formation constants is 1:6.3:225. Again, K^+ and Ba^{2+} have similar radii but the ratio of K is 1:3.5 in the reverse direction.

Preeminent in importance among the macrocyclic complexes of Group IIA elements are the chlorophyls, which are modified porphyrin complexes of Mg. These compounds are vital to the process of photosynthesis in green plants (see Panel). Magnesium and Ca are also intimately involved in biochemical processes in animals: Mg is required to trigger phosphate transfer enzymes, for nerve impulse transmissions, muscle contraction, and carbohydrate metabolism, whereas Ca is required in the formation of bones and teeth, in maintaining the rhythm of the heart, and in blood clotting.[13a–f]

[13a] W. E. C. WACKER, *Magnesium and Man*, Harvard University Press, London, 1980.

[13b] M. N. HUGHES, *The Inorganic Chemistry of Biological Processes*, Wiley, London, 1972, Chap. 8, The alkali metal and alkaline earth metal cations in biology, pp. 256–82.

[13c] G. L. EICHHORN (ed.), *Inorganic Biochemistry*, Elsevier, Amsterdam, 1973, 2 Vols., 1263 pp.

[13d] B. S. COOPERMAN, The role of divalent metal ions in phosphoryl and nucleotidyl transfer, Chap. 2 in H. SIGAL (ed.), *Metal Ions in Biological Systems*, Vol. 5, Dekker, New York, 1976, pp. 80–125.

[13e] K. S. RAJAN, R. W. COLBURN, and J. M. DAVIS, Metal chelates in the storage and transport of neurotransmitters, Chap. 5 in H. SIGAL (ed.), *Metal Ions in Biological Systems*, Vol. 6, Dekker, New York, 1976, pp. 292–321.

[13f] F. N. BRIGGS and R. J. SOLARO, The role of divalent metal in the contraction of muscle fibres, Chap. 6 in H. SIGAL (ed.), *Metal Ions in Biological Systems*, Vol. 6, Dekker, New York, 1976, pp. 324–98.

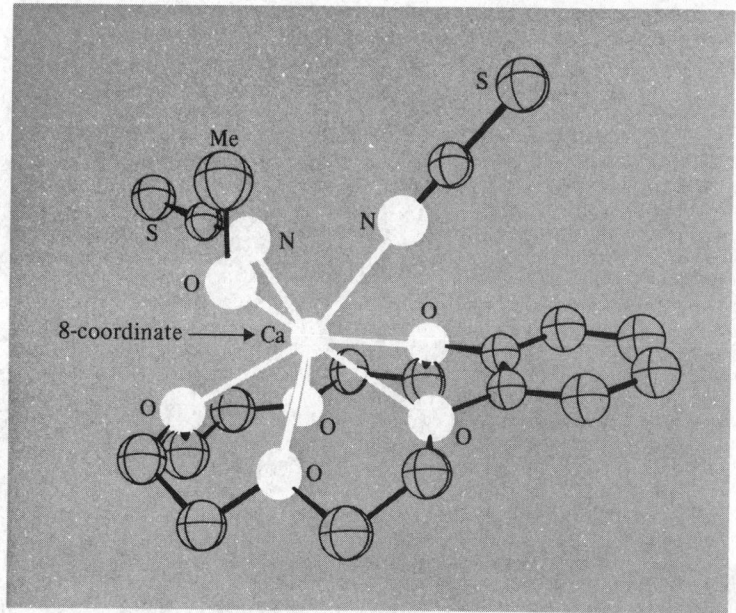

FIG. 5.7   Molecular structure of benzo-15-crown-5-Ca(NCS)$_2$-MeOH.

Chlorophylls and Photosynthesis

Photosynthesis is the process by which green plants convert atmospheric CO_2 into carbohydrates such as glucose. The overall chemical change can be expressed as

$$6CO_2 + 6H_2O \xrightarrow{h\nu} C_6H_{12}O_6 + 6O_2$$

though this is a gross and somewhat misleading over-simplification. The process is initiated in the photoreceptors of the green magnesium-containing pigments which have the generic name chlorophyll (Greek: χλωρός, *chloros* green; φύλλον, *phyllon* leaf), but many of the subsequent steps can proceed in the dark. The overall process is endothermic ($\Delta H° \sim 469$ kJ per mole of CO_2) and involves more than one type of chlorophyll. It also involves a manganese complex of unknown composition, various iron-containing cytochromes and ferredoxin (p. 1280), and a copper containing plastocyanin.

Photosynthesis is essentially the conversion of radiant electromagnetic energy (light) into chemical energy in the form of adenosine triphosphate (ATP) and reduced nicotinamide adenine dinucleotide phosphate (NADP)(p. 614). This energy eventually permits the fixation of CO_2 into carbohydrates, with the liberation of O_2. As such, the process is the basis for the nutrition of all living things and also provides materials for fuel (wood, coal, petroleum), fibres (cellulose) and innumerable useful chemical compounds. About 90–95% of the dry weight of crops is derived from the CO_2/H_2O fixed from the air during photosynthesis—only about 5–10% comes from minerals and nitrogen taken from the soil. The detailed sequence of events is still not fully understood but tremendous advances were made from 1948 onwards by use of the then newly available radioactive $^{14}CO_2$ and paper chromatography. With these tools and classical organic chemistry M. Calvin and his group were able to probe the biosynthetic pathways and thus laid the basis for our present understanding of the complex series of reactions. Calvin was awarded the 1961 Nobel Prize in Chemistry "for his research on the carbon dioxide assimilation in plants".[13g]

[13g] M. CALVIN, The path of carbon in photosynthesis, *Nobel Lectures in Chemistry 1942–62*, Elsevier, Amsterdam, 1964, 618–44.

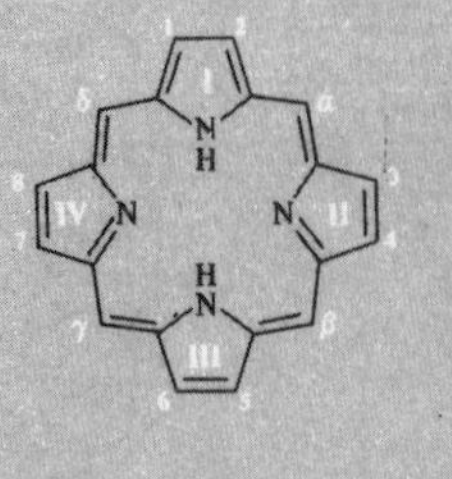

(1) Porphin

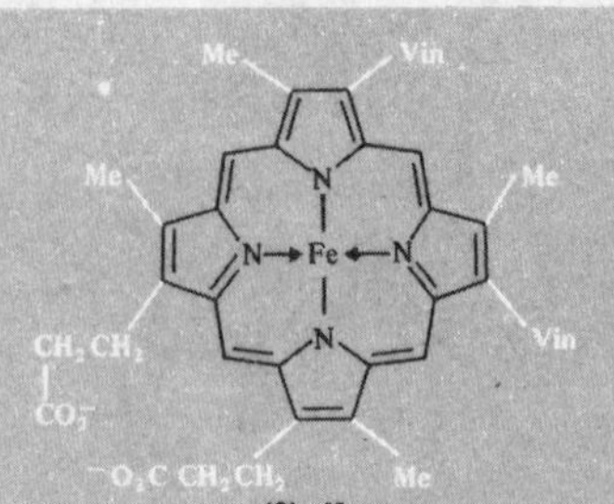

(2) Haem

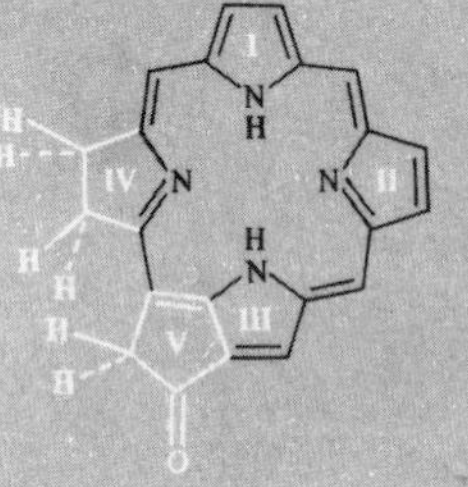

(3) Chlorin

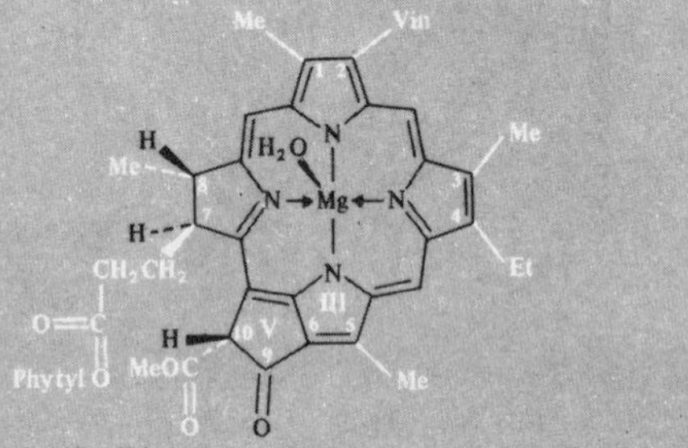

(4) Chlorophyll *a* (the phytyl isoprenoid group is
$-CH_2CH=C(CH_2)_3CH(CH_2)_3CH(CH_2)_3CHMe$)
                Me   Me   Me   Me

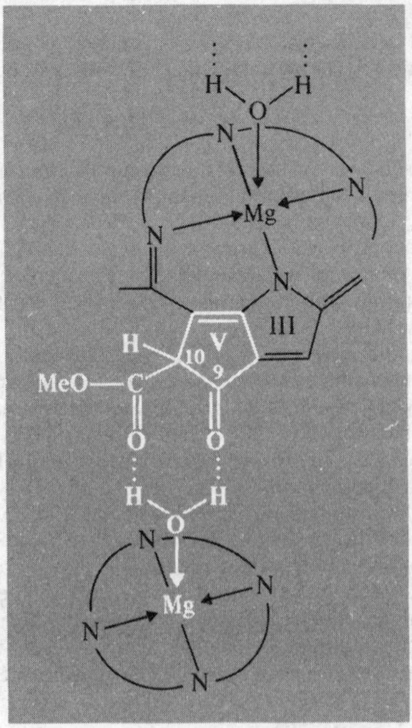

(5)

Chlorophylls are complexes of Mg with macrocyclic ligands derived from the parent tetrapyrrole molecule porphin (structure 1). They are thus related to the porphyrin (substituted porphin) complexes which occur in haem proteins such as haemoglobin, myoglobin and the cytochromes (p. 1279). [Porphyrins derive their name from the characteristic purple-red coloration which these alkaloids give when acidified (Greek πόρφυρ-ος, *porphyros* purple).] The haem group is illustrated in structure 2. When the C=C double bond in the pyrrole-ring IV of porphin is *trans* hydrogenated and when a cyclopentanone ring is formed between ring III and the adjacent (γ) methine bridge then the chlorin macrocycle (structure 3) is produced, and this is the basis for the various chlorophylls. Chlorophyll *a* (Chl *a*) is shown in structure 4; this is the most common of the chlorophylls and is found in all O_2-evolving organisms. It was synthesized with complete chiral integrity by R. B. Woodward and his group in 1960—an achievement of remarkable virtuosity. Variants of chlorophyll are:

Chlorophyll *b*, in which the 3-Me group is replaced by —CHO; this occurs in higher plants and green algae, the ratio Chl *b*:Chl *a* being ~1:3.

Chlorophyll *c*, in which position 7 is substituted by acrylic acid, —CH=CHCO₂H; it occurs in diatoms and brown algae.

Chlorophyll *d*, in which 2-vinyl is replaced by —CHO.

It is important to note that the chlorin macrocycle is "ruffled" rather than completely planar and the Mg atom is ~30–50 pm above the plane of the 4N atoms. In fact the Mg is not 4-coordinate but carries one (or sometimes two) other ligands, notably water molecules, which play a crucial role in interconnecting the basic chlorophyll units into stacks by H bonding to the cyclopenta-none ring V of an adjacent chlorophyll molecule (see structure 5).

The function of the chlorophyll in the chloroplast is to absorb photons in the red part of the visible spectrum (near 680–700 nm) and to pass this energy of excitation on to other chemical intermediates in the complex reaction scheme. At least two photosystems are involved: the initiating photosystem II (P680) which absorbs at 680 nm and the subsequent photosystem I (P700). The detailed redox processes occurring, and the enzyme-catalysed synthetic pathways (dark reactions) in-so-far as they have yet been elucidated, are described in biochemical texts and fall outside our present scope. The Mg ion apparently serves several purposes: (a) it keeps the macrocycles fairly rigid so that energy is not so readily dissipated by thermal vibrations; (b) it coordinates the H_2O molecules which mediate in the H bonding between adjacent molecules in the stack; and (c) it thereby enhances the rate at which the short-lived singlet excited state formed initially by absorption of a photon by the macrocycle is transformed to the corresponding longer-lived triplet state which is involved in the redox chain (since this involves the H bonded system between several individual chlorophyll units over a distance of some 1500–2000 pm). However, it is by no means clear why, of all metals, Mg is uniquely suited for this purpose.

5.3.5　*Organometallic compounds*[14]

Compounds containing M–C bonds are well established for Be and Mg in keeping with their tendency to form covalent bonds but, as with the alkali metals, reactivity within the group increases with increasing electropositivity, and few organometallic compounds of Ca, Sr, or Ba have been isolated.

Beryllium

Beryllium dialkyls (BeR_2, R = Me, Et, Pr^n, Pr^i, Bu^i, etc.) can be made by reacting lithium alkyls or Grignard reagents with $BeCl_2$ in ethereal solution, but the products are difficult

[14] G. E. COATES, M. L. H. GREEN, and K. WADE, *Organometallic Compounds*, Vol. 1, *The Main Group Elements*, 3rd edn., Chap. II, Group II, pp. 71–121, Methuen, London, 1967.

to free from ether and, when pure compounds rather than solutions are required, a better route is by heating Be metal with the appropriate mercury dialkyl:

$$BeCl_2 + 2LiMe \xrightarrow{Et_2O} BeMe_2 . nEt_2O + 2LiCl$$

$$BeCl_2 + 2MeMgCl \xrightarrow{Et_2O} BeMe_2 . nEt_2O + 2MgCl_2$$

$$Be + HgMe_2 \xrightarrow{110°} BeMe_2 + Hg$$

BePh$_2$ (mp 245°) can be prepared similarly, using LiPh or HgPh$_2$; an excess of the former reagent yields Li[BePh$_3$]. Beryllium dialkyls are colourless solids or viscous liquids which are spontaneously flammable in air and explosively hydrolysed by water. BeMe$_2$ (like MgMe$_2$, p. 146) has been shown by X-ray analysis to have a chain structure analogous to that found in BeCl$_2$ (p. 128) though the bonding is probably best described in terms of 2-electron 3–centre bridge bonds involving •CH$_3$ groups rather than that adopted by bridging Cl atoms which each form two 2-electron 2-centre bonds involving a total of 4 electrons per Be–Cl–Be bridge (Fig. 5.8). Each C atom has a coordination number of 5 (cf. bonding in boranes, carbaboranes, etc., p. 179). Higher alkyls are progressively less highly polymerized and the sterically crowded BeBu$^t_2$ is monomeric. As with polymeric BeCl$_2$, addition of strong ligands results in depolymerization and the eventual formation of monomeric adducts, e.g. [BeMe$_2$(PMe$_3$)$_2$], [BeMe$_2$(Me$_2$NCH$_2$CH$_2$NMe$_2$)], etc. Pyrolysis eliminates alkenes and leads to mixed hydrido species of variable composition (see also p. 126).

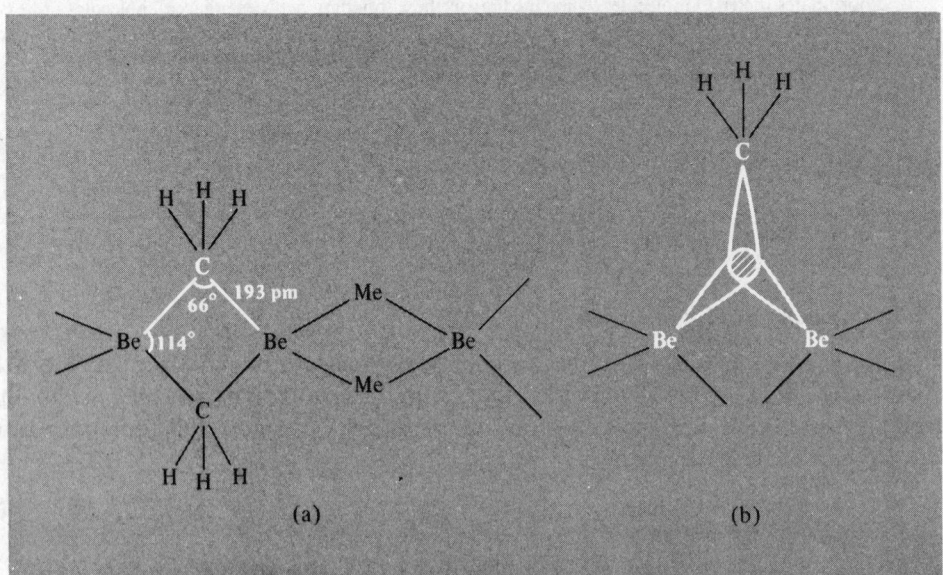

FIG. 5.8   (a) Chain structure of BeMe$_2$ showing the acute angle at the bridging methyl group; the Be···Be distance is 209 pm and the distance between the 2 C atoms across the bridge is 315 pm, and (b) pictorial representation of the 3 approximately sp^3 orbitals used to form one 3-centre bridge bond; this description of the bonding is consistent with the acute bridging angle at C and the close approach to adjacent Be atoms.

Alkylberyllium hydrides of more precise stoichiometry can be prepared by reducing $BeBr_2$ with LiH in the presence of BeR_2, e.g.:

$$BeMe_2 + BeBr_2 + 2LiH \xrightarrow{Et_2O} \underset{\substack{Et_2O \quad H \quad Me}}{\overset{\substack{Me \quad H \quad OEt_2}}{Be \quad Be}} + 2LiBr$$

The coordinated ether molecules can be replaced by tertiary amines. Use of NaH in the absence of halide produces the related compound $Na_2[Me_2BeH_2BeMe_2]$; the corresponding ethyl derivative crystallizes with 1 mole of Et_2O per Na but this can readily be removed under reduced pressure. The crystal structure of the etherate is shown in Fig. 5.9[15] and is important in illustrating once more how misleading it can be to differentiate too sharply between different kinds of bonding in solids, for example: ionic $[Na(OEt_2)]_2^+[Et_2BeH_2BeEt_2]^{2-}$ or polymeric $[Et_2ONaHBeEt_2]_n$. Thus in the structure each Be is surrounded tetrahedrally by 2 Et and 2 bridging H to form a subunit

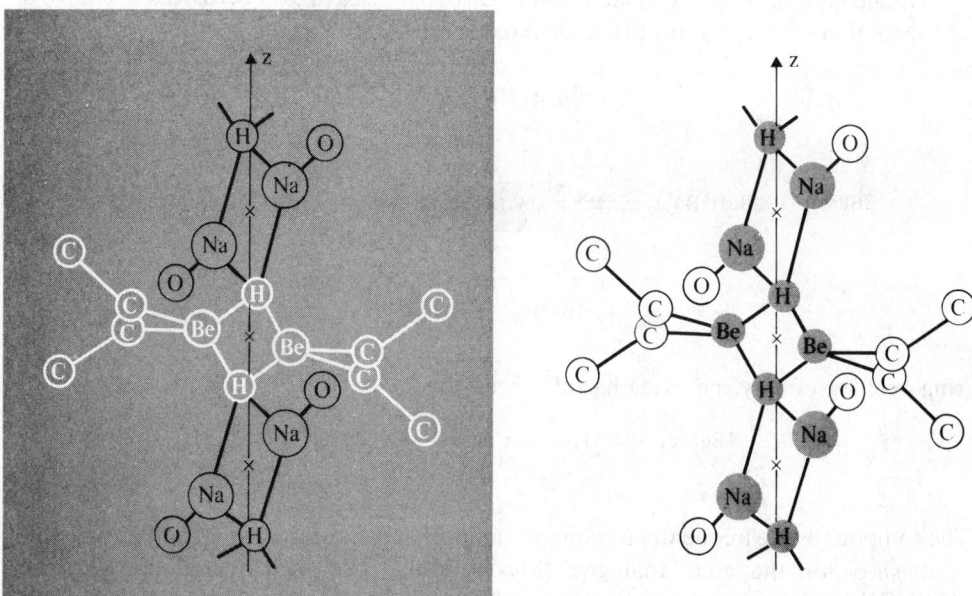

Fig. 5.9 Crystal structure of the etherate of polymeric sodium hydridodiethylberyllate $(Et_2ONaHBeEt_2)_n$ emphasizing two features of the structure (see text).

[15] G. W. Adamson and H. M. M. Shearer, The crystal structure of the etherate of sodium hydridodiethylberyllate, *JCS Chem. Comm.* 1965, 240.

In addition, each H is coordinated tetrahedrally by 2 Be and 2 Na, and each Na is directly bonded to 1 Et_2O. Be–C is 180 pm and Be–H is 140 pm, close to expected values; Na–H is 240 pm, equal to that in NaH. The distance Na$\cdots$Na is 362 pm which is less than in Na metal (372 pm) but greater than in NaH (345 pm), where each Na is surrounded by 6H; Be$\cdots$Be is 220 pm as in Be metal. It is therefore misleading to consider the structure as being built up from the isolated ions $[Na(OEt_2)]^+$ and $[Et_2BeH_2BeEt_2]^{2-}$ and it is perhaps better to regard it as a chain polymer $[Et_2ONaHBeEt_2]_n$ which in plane projection can be written as:

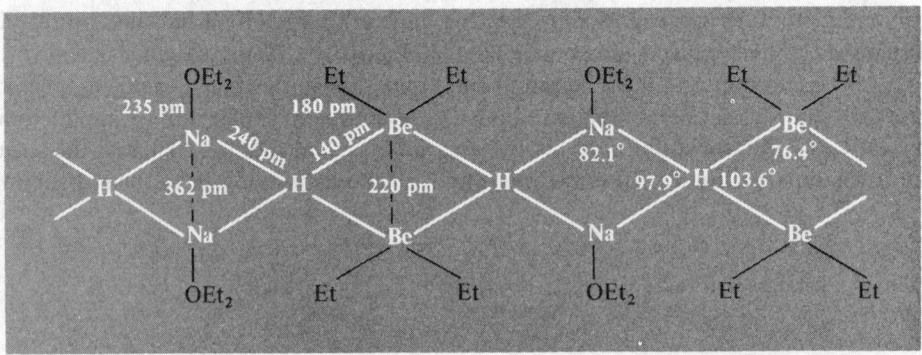

Alkylberyllium alkoxides (RBeOR′) can be prepared from BeR_2 by a variety of routes such as alcoholysis with R′OH, addition to carbonyls, cleavage of peroxides R′OOR′, or redistribution with the appropriate dialkoxide $Be(OR′)_2$, e.g.:

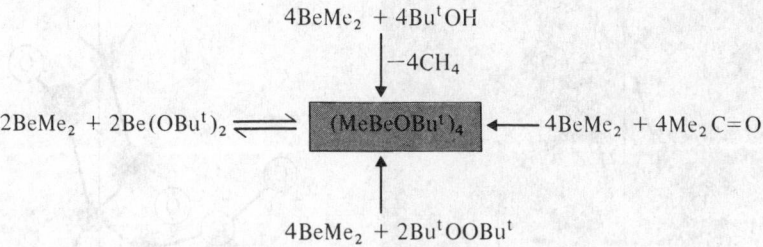

Ring opening of ethylene oxide has also been used:

$$4BeMe_2 + 4CH_2\!\!-\!\!CH_2 \longrightarrow (MeBeOPr^n)_4$$
$$\diagdown O \diagup$$

The compounds are frequently tetrameric and probably have the "cubane-like" structure established for the zinc analogue $(MeZnOMe)_4$. The methylberyllium alkoxides $(MeBeOR′)_4$ are reactive, low-melting solids (mp for R′ = Me 25°, Et 30°, Pr^n 40°, Pr^i 136°, Bu^t 93°). Bulky substituents may reduce the degree of oligomerization, e.g. trimeric $(EtBeOCEt_3)_3$, and reaction with coordinating solvents or strong ligands can also lead to depolymerization, e.g. dimeric $(MeBeOBu^t.py)_2$ and monomeric $PhBeOMe.2Et_2O$:

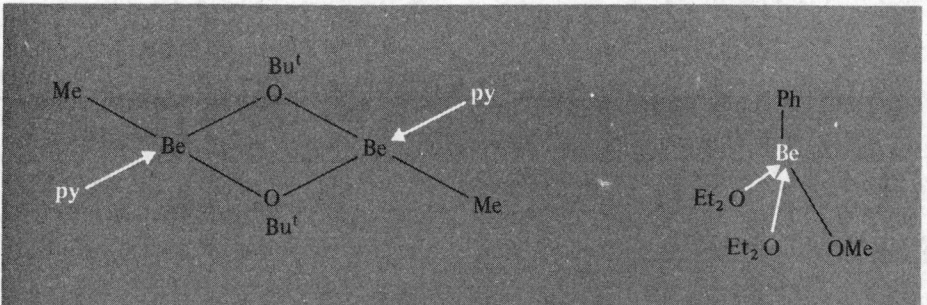

Reaction of beryllium dialkyls with an excess of alcohol yields the alkoxides $Be(OR)_2$. The methoxide and ethoxide are insoluble and probably polymeric, whereas the t-butoxide (mp 112°) is readily soluble as a trimer in benzene or hexane; the proposed structure:

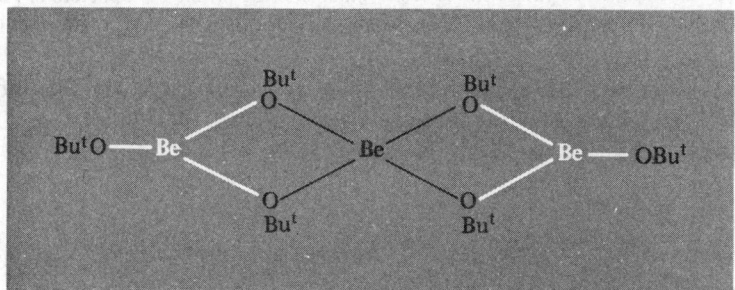

involves both 3- and 4-coordinate Be but is consistent with the observation of 2 proton nmr signals at τ 8.60 and 8.75 with intensities in the ratio 2:1. (A precisely analogous structure has been established by X-ray diffraction analysis for the "isoelectronic" linear trimer $[Be(NMe_2)_2]_3$.)[16]

Beryllium forms a series of cyclopentadienyl complexes $[Be(\eta^5\text{-}C_5H_5)Y]$ with $Y = H$, Cl, Br, $-C\equiv CH$, and BH_4, all of which show the expected C_{5v} symmetry (Fig. 5.10a). If the *pentahapto*-cyclopentadienyl group (p. 369) contributes 5 electrons to the bonding, then these are all 8-electron Be complexes consistent with the octet rule for elements of the first short period.[17] The bis(cyclopentadienyl) compound (mp 59°C), first prepared by E. O. Fischer and H. P. Hofmann in 1959, is also known but does not adopt the ferrocene-type structure (p. 372) presumably because this would require 12 electrons in the valence shell of Be. Instead, the complex has C_s symmetry and is, in fact, $[Be(\eta^1\text{-}C_5H_5)(\eta^5\text{-}C_5H_5)]$, as shown in Fig. 5.10b.[18] The σ-bonded Be–C distance is significantly shorter than the five other Be–C distances and there is some alternation of C–C distances in the σ-bonded cyclopentadienyl group. All H atoms are coplanar with the rings except for the one adjacent to the Be–C_σ bond. For free molecules in the gas phase it seems

[16] J. L. ATWOOD and G. D. STUCKY, Dative metal-nitrogen π-bonding in bis(dimethylamino)beryllium, *Chem. Comm.* 1967, 1169–70.

[17] E. D. JEMMIS, S. ALEXANDRATOS, P. v. R. SCHLEYER, A. STREITWIESER, and H. F. SCHAEFFER, Ab initio SCF-MO study of cyclopentadienylberyllium hydride and of beryllocene, *J. Am. Chem. Soc.* **100**, 5695–700 (1978).

[18] C.-H. WONG, T.-Y. LEE, K.-J. CHAO, and S. LEE, Crystal structure of bis(cyclopentadienyl)beryllium at −120°, *Acta Cryst.* **B28**, 1662–5 (1972).

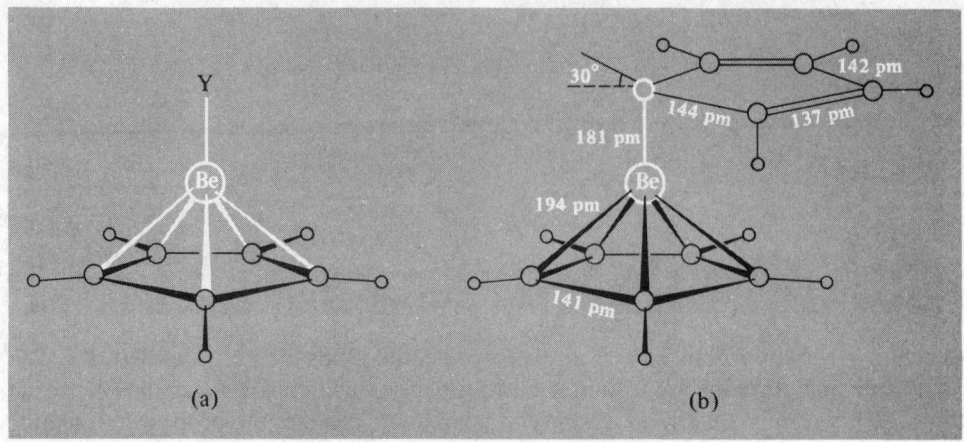

FIG. 5.10 Cyclopentadienyl derivatives of beryllium showing (a) the C_{5r} structure of $[Be(\eta^5\text{-}C_5H_5)Y]$, and (b) the structure of crystalline $[Be(\eta^1\text{-}C_5H_5)(\eta^5\text{-}C_5H_5)]$ at $-120°$ (see text).

unlikely that the two cyclopentadienyl rings are coplanar, and the most recent calculations[19] suggest a dihedral angle between the rings of $117°$ with $Be\text{-}C_\sigma$ 172 pm, $Be\text{-}C_\pi$ 187 pm, and the angle $Be\text{-}C_\sigma\text{-}H$ $108°$.

Magnesium

Magnesium dialkyls and diaryls, though well established, have been relatively little studied by comparison with the vast amount of work which has been published on the Grignard reagents RMgX. The dialkyls (and diaryls) can be conveniently made by the reaction of LiR (LiAr) on Grignard reagents, or by the reaction of HgR_2 ($HgAr_2$) on Mg metal (sometimes in the presence of ether). On an industrial scale, alkenes can be reacted at $100°$ under pressure with MgH_2 or with Mg in the presence of H_2:

$$LiR + RMgX \xrightarrow{\text{Et}_2\text{O}} MgR_2 + LiX$$

$$HgR_2 + Mg \xrightarrow{\text{Et}_2\text{O}} MgR_2 + Hg$$

$$2C_2H_4 + H_2 + Mg \xrightarrow[100°]{\text{pressure}} MgEt_2$$

A suitable laboratory method is to shift the Schlenk equilibrium in a Grignard solution (p. 147) by adding dioxan to precipitate the complex MgX_2. diox; this enables MgR_2 to be isolated by careful removal of solvent under reduced pressure:

$$2RMgX \Longrightarrow MgR_2 + MgX_2 \xrightarrow{C_4H_8O_2} MgR_2 + MgX_2 . C_4H_8O_2$$

$MgMe_2$ is a white involatile polymeric solid which is insoluble in hydrocarbon and only

[19] D. S. MARYNICK, Studies of the molecular and electronic structure of dicyclopentadienylberyllium, *J. Am. Chem. Soc.* **99**, 1436–41 (1977). See also J. B. COLLINS and P. v. R. SCHLEYER, Sandwich-type molecules of first-row atoms. Instability of bis(η^3-cyclopropenyl)beryllium, *Inorg. Chem.* **16**, 152–5 (1977).

slightly soluble in ether. Its structure is very similar to that of $BeMe_2$ (p. 142) the corresponding dimensions for $MgMe_2$ being: Mg–C 224 pm, Mg–C–Mg 75°, C–Mg–C 105°, Mg···Mg 272 pm, and C···C (across the bridge) 357 pm. Precisely analogous bridging Me groups are found in dimeric Al_2Me_6 (p. 289) and in the monomeric compound $Mg(AlMe_4)_2$ which can be formed by direct reaction of $MgMe_2$ and Al_2Me_6:

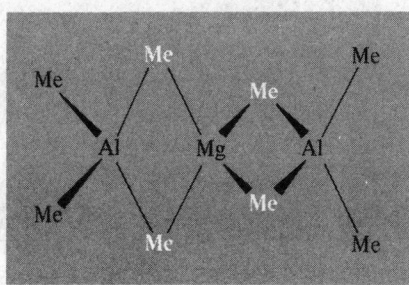

$MgEt_2$ and higher homologues are very similar to $MgMe_2$ except that they decompose at a lower temperature (175–200° instead of $\sim 250°C$) to give the corresponding alkene and MgH_2 in a reaction which reverses their preparation. $MgPh_2$ is similar: it is insoluble in benzene dissolves in ether to give the monomeric complex $MgPh_2 \cdot 2Et_2O$ and pyrolyses at 280° to give Ph_2 and Mg metal. Like $BePh_2$ it reacts with an excess of LiPh to give the colourless complex $Li[MgPh_3]$.

The first organosilylmagnesium compound $[Mg(SiMe_3)_2] \cdot (-CH_2OMe)_2$, was isolated in 1977;[20] it was obtained as colourless, spontaneously flammable crystals by reaction of bis(trimethylsilyl)mercury with Mg powder in 1,2-dimethoxyethane.

Grignard reagents are the most important organometallic compounds of Mg and are probably the most extensively used of all organometallic reagents because of their easy preparation and synthetic versatility. Despite this, their constitution in solution has been a source of considerable uncertainty until recent times.[21] It now seems well established that solutions of Grignard reagents can contain a variety of chemical species interlinked by mobile equilibria whose position depends critically on at least five factors: (i) the steric and electronic nature of the alkyl (or aryl) group R, (ii) the nature of the halogen X (size, electron-donor power, etc.), (iii) the nature of the solvent (Et_2O, thf, benzene, etc.), (iv) the concentration, and (v) the temperature. The species present may also depend on the presence of trace impurities such as H_2O or O_2. Neglecting solvation in the first instance, the general scheme of equilibria can be set out as shown overleaf. Thus "monomeric" (solvated) RMgX can disproportionate to MgR_2 and MgX_2 by the Schlenk equilibrium or can dimerize to $RMgX_2MgR$. Both the monomer and the dimer can ionize, and reassociation can give the alternative dimer R_2MgX_2Mg. Note that only halogen atoms X are involved in the bridging of these species.

Evidence for these species and the associated equilibria comes from a variety of techniques such as vibration spectroscopy, nmr spectroscopy, molecular-weight determinations, radioisotopic exchange using ^{28}Mg, electrical conductivity, etc. In some cases

[20] L. RÖSCH, Bis(trimethylsilyl)magnesium—the first organosilylmagnesium compound isolated, *Angew. Chem.*, Int. Ed. (Engl.) **16**, 247–8 (1977).
[21] E. C. ASHBY, Grignard reagents: compositions and mechanisms of reactions, *Qt. Rev.* **21**, 259–85 (1967).

$$2RMg^+ + 2X^-$$

associate ‖ ionize

R—Mg   Mg—R $\xrightleftharpoons[\text{dissociate}]{\text{dimerize}}$ $\boxed{2RMgX}$ $\xrightleftharpoons[\text{(Schlenk)}]{\text{disproportionate}}$ $MgR_2 + MgX_2$

ionize / associate

dissociate / associate

$$RMg^+ + RMgX_2^- \xrightleftharpoons[\text{ionize}]{\text{associate}}$$

Schematic representation of the principal equilibria in Grignard solutions; solvation of the various species has been omitted for clarity.

equilibria can be displaced by crystallization or by the addition of complexing agents such as dioxan (p. 146) or NEt_3. The crystal structures of several pertinent adducts have recently been determined (Fig. 5.11). None call for special comment except the curious solvated dimer $[EtMg_2Cl_3(OC_4H_8)_3]_2$ which features both 5-coordinate trigonal bipyramidal and 6-coordinate octahedral Mg groups; note also that, whilst 4 of the Cl atoms each bridge 2 Mg atoms, the remaining 2 Cl atoms are triply bridging.

Grignard reagents are normally prepared by the slow addition of the organic halide to a stirred suspension of magnesium turnings in the appropriate solvent and with rigorous exclusion of air and moisture. The reaction, which usually begins slowly after an induction period, can be initiated by addition of a small crystal of iodine; this penetrates the protective layer of oxide (hydroxide) on the surface of the metal. The order of reactivity of RX is $I > Br > Cl$ and alkyl > aryl. The mechanism has been much studied but is not fully understood.[22] The fluorides RMgF (R = Me, Et, Bu, Ph) can be prepared by reacting MgR_2 with mild fluorinating agents such as $BF_3.OEt_2$, Bu_3SnF, or SiF_4.[23] The scope of Grignard reagents in syntheses has been greatly extended by a recently developed method for preparing very reactive Mg (by reduction of MgX_2 with K in the presence of KI).[24] Grignard reagents have a wide range of application in the synthesis of alcohols, aldehydes, ketones, carboxylic acids, esters, and amides, and are probably the most versatile reagents for constructing C–C bonds by carbanion (or occasionally free-radical) mechanisms. Standard Grignard methods are also available for constructing C–N, C–O, C–S (Se, Te), and C–X bonds (see Panel on pp. 150–51).

A related class of compounds are the alkylmagnesium alkoxides: these can be prepared by reaction of MgR_2 with an alcohol or ketone or by reaction of Mg metal with the

[22] H. R. ROGERS, C. L. HILL, Y. FUJIWARA, R. J. RODERS, H. L. MITCHELL, and G. M. WHITESIDES, Mechanism of formation of Grignard reagents. Kinetics of reaction of alkyl halides in diethyl ether with magnesium, *J. Am. Chem. Soc.* **102**, 217–26 (1980), and the three following papers, pp. 226–43.

[23] E. C. ASHBY and J. NACKASHI, The preparation of organomagnesium fluorides by organometallic exchange reactions, *J. Organometall. Chem.* **72**, 203–11 (1974).

[24] R. D. RIEKE and S. E. BALES, Preparation and reactions of highly reactive magnesium metal, *J. Am. Chem. Soc.* **96**, 1775–81 (1974).

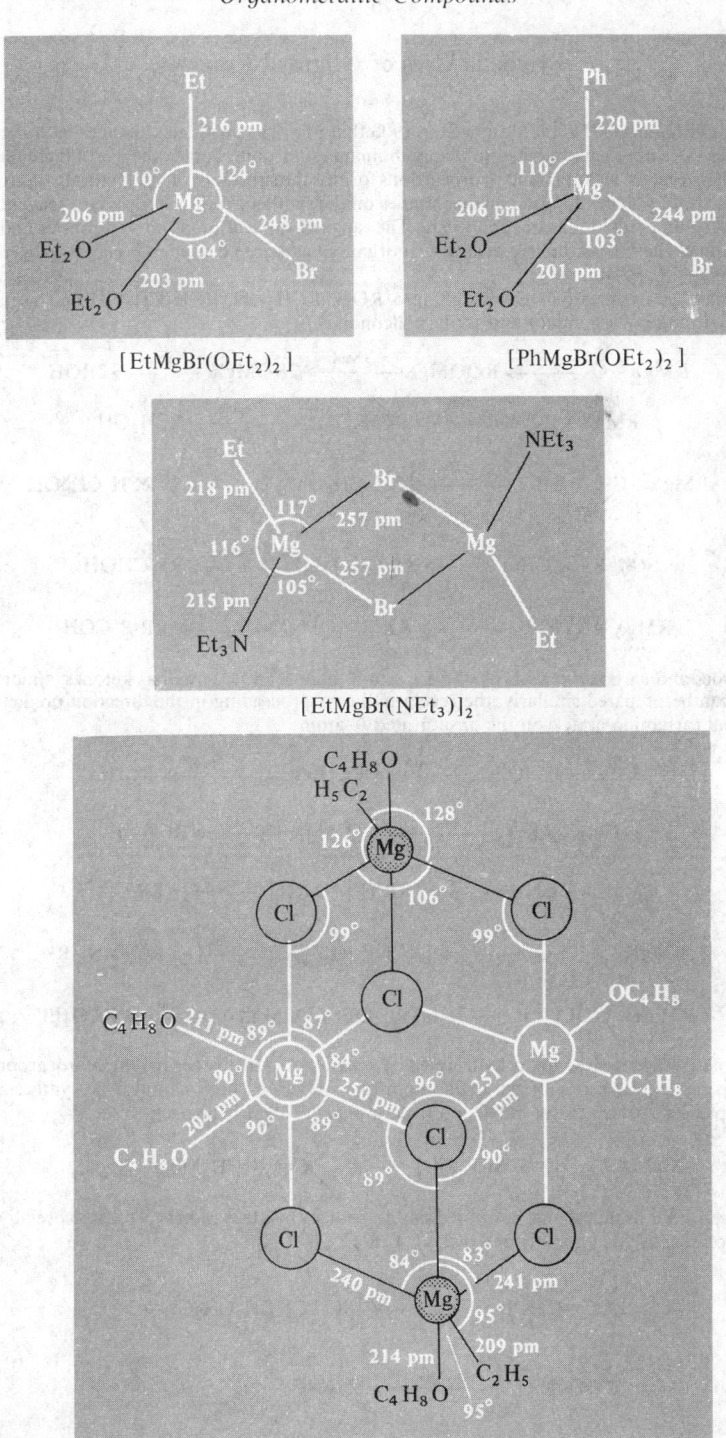

Fig. 5.11 Crystal structures of adducts of Grignard reagents.

Synthetic Uses of Grignard Reagents

Victor Grignard (1871–1935) showed in 1900 that Mg reacts with alkyl halides in dry ether at room temperature to give ether-soluble organomagnesium compounds; the use of these reagents to synthesize acids, alcohols, and hydrocarbons formed the substance of his doctorate thesis at the University of Lyon in 1901, and further studies on the synthetic utility of Grignard reagents won him the Nobel Prize for Chemistry in 1912. The range of applications is now enormous and some indication of the extraordinary versatility of organomagnesium compounds can be gauged from the following brief summary.

Standard procedures convert RMgX into ROH, RCH_2OH, RCH_2CH_2OH and an almost unlimited range of secondary and tertiary alcohols:

$$RMgX + O_2 \longrightarrow ROOMgX \xrightarrow{RMgX} 2ROMgX \xrightarrow{acid} 2ROH$$

$$RMgX + HCHO \longrightarrow RCH_2OMgX \xrightarrow{acid} RCH_2OH$$

$$RMgX + CH_2\!\!-\!\!CH_2 \longrightarrow RCH_2CH_2OMgX \xrightarrow{acid} RCH_2CH_2OH$$
$$\diagdown O \diagup$$

$$RMgX + R'CHO \longrightarrow RR'CHOMgX \xrightarrow{acid} RR'CHOH$$

$$RMgX + R'COR'' \longrightarrow RR'R''COMgX \xrightarrow{acid} RR'R''COH$$

Aldehydes and carboxylic acids having 1 C atom more than R, as well as ketones, amides, and esters can be prepared similarly, the reaction always proceeding in the direction predicted for potential carbanion attack on the unsaturated C atom:

$$RMgX + HC(OEt)_3 \xrightarrow{acid} RCH(OEt)_2 \xrightarrow{acid} RCHO$$

$$RMgX + CO_2 \longrightarrow RCO_2MgX \xrightarrow{acid} RCO_2H$$

$$RMgX + R'CN \longrightarrow [RR'C{=}NMgX] \xrightarrow{acid} RR'C{=}O$$

$$RMgX + R'NCO \longrightarrow [RCNR'(OMgX)] \xrightarrow{acid} RC(O)NHR'$$

$$RMgX + EtOCOCl \longrightarrow [RC(OEt)Cl(OMgX)] \xrightarrow{acid} RCO_2Et$$

Grignard reagents are rapidly hydrolysed by water or acid to give the parent hydrocarbon, RH, but this reaction is rarely of synthetic importance. Hydrocarbons can also be synthesized by nucleophilic displacement of halide ion from a reactive alkyl halide, e.g.

$$MeMgCl + {>}C{=}CHCH_2Cl \longrightarrow {>}C{=}CHCH_2Me + + MgCl_2$$

However, other products may be formed simultaneously by a free-radical process, especially in the presence of catalytic amounts of $CoCl_2$ or CuCl:

$$2\, {>}C{=}CHCH_2Cl \xrightarrow[CoCl_2]{MeMgCl} {>}C{=}CHCH_2CH_2CH{=}C{<}$$

$$2\, {>}C{=}C{<}_{MgX} \xrightarrow{CuCl/thf} {>}C{=}C{<}_{C{=}C{<}}$$

Similarly, aromatic Grignard reagents undergo free-radical self-coupling reactions when treated with MCl_2 (M = Cr, Mn, Fe, Co, Ni), e.g.:

$2PhMgBr + CrCl_2 \longrightarrow 2\text{"MgBrCl"} +$

$Ph_2 + Cr \xleftarrow{\text{heat}}$

Alkenes can be synthesized from aldehydes or ketones using the Grignard reagent derived from CH_2Br_2:

$$CH_2(MgBr)_2 + RR'C{=}O \longrightarrow \left[\begin{array}{c} R \quad\;\; OMgBr \\ \diagdown\;\;C\;\;\diagup \\ \diagup\quad\;\diagdown \\ R' \quad CH_2MgBr \end{array} \right] \longrightarrow RR'C{=}CH_2 + MgO + MgBr_2$$

The formation of C–N bonds can be achieved by using chloramine or *O*-methylhydroxylamine to yield primary amines; aryl diazonium salts yield azo-compounds:

$$RMgX + ClNH_2 \longrightarrow RNH_2 + \text{"MgClX"}$$
$$RMgX + MeONH_2 \longrightarrow RNH_2 + MeOMgX$$
$$RMgX + [ArN_2]X \longrightarrow Rn{=}NAr + MgX_2$$

Carbon–oxygen bonds can be made using the synthetically uninteresting conversion of RMgX into ROH (shown as the first reaction listed above); direct acid hydrolysis of the peroxo compound ROOMgX yields the hydroperoxide ROOH. Carbon–sulfur bonds can be constructed using S_8 to make thiols or thioethers, and similar reactions are known for Se and Te:

$$RMgX + S_8 \longrightarrow RS_xMgX \xrightarrow{RMgX} RSMgX$$

$$RSMgX \begin{cases} \xrightarrow{\text{acid}} & RSH + MgX_2 \\ \xrightarrow{RMgX} & R_2S + MgX_2 + Mg \\ \xrightarrow{R'I} & RSR' + \text{"MgIX"} \end{cases}$$

Formation of C–X bonds is not normally a problem but the Grignard route can occasionally be useful when normal halogen exchange fails. Thus iodination of Me_3CCH_2Cl cannot be achieved by reaction with NaI or similar reagents but direct iodination of the corresponding Grignard effects a smooth conversion:

$$Me_3CCH_2MgCl + I_2 \longrightarrow Me_3CCH_2I + \text{"MgClI"}$$

Further examples of the ingenious use of Grignard reagents will be found in many books on synthetic organic chemistry and much recent work in this area was reviewed in a special edition of *Bull. Soc. Chim. France*, 1972, 2127–86, which commemorated the centenary of Victor Grignard's birth.

appropriate alcohol and alkyl chloride in methylcyclohexane solvent, e.g.:

$$4MgEt_2 + 4Bu^tOH \longrightarrow (EtMgOBu^t)_4 + 4C_2H_6$$

$$2MgMe_2 + 2Ph_2CO \xrightarrow{Et_2O} (MeMgOCMePh_2 . Et_2O)_2$$

$$6Mg + 6Bu^nCl + 3Pr^iOH \longrightarrow (Bu^nMgOPr^i)_3 + 3MgCl_2 + 3C_4H_{10}$$

As with the Grignard reagents, the structure and degree of association of the product depend on the bulk of the organic groups, the coordinating ability of the solvent, etc. This is well illustrated by MeMgOR (R = Pr^n, Pr^i, Bu^t, CMePh$_2$) in thf, Et$_2$O, and benzene:[25] the strongly coordinating solvent thf favours solvated dimers (A) but prevents the formation both of oligomers (B) involving the relatively weak Me bridges and of cubane structures (C) involving the relatively weak triply bonding oxygen bridges.

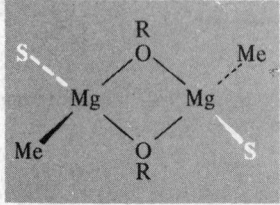

A, solvated dimer.   B, linear oligomer (various isomers are possible, e.g. involving

$$Mg\underset{Me}{\overset{OR}{<}>}Mg \text{ bridges, etc.).}$$

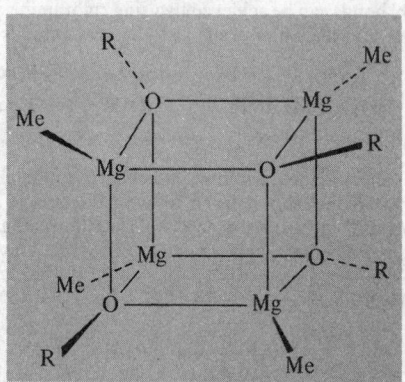

C, cubane tetramer (unsolvated).

By contrast, in the more weakly coordinating solvent Et$_2$O, Me bridges and μ_3-OR bridges can form, leading to linear oligomers and cubanes, provided OR is not too bulky. Thus when R = CMePh$_2$, oligomerization and cubane formation are blocked and MeMgOCMePh$_2$ exists only as a solvated dimer even in Et$_2$O. In benzene, R = Bu^t and Pr^i form cubane tetramers but Pr^n can form an oligomer of 7–9 monomer units. The sensitive dependence of the structure of a compound on solvation energy, lattice energy, and the relative coordinating abilities of its component atoms and groups will be a recurring theme in many subsequent chapters.

Dicyclopentadienylmagnesium [Mg(η^5-C$_5$H$_5$)$_2$], mp 176°, can be made in good yield by direct reaction of Mg and cyclopentadiene at 500–600°; it is very reactive towards air, moisture, CO$_2$, and CS$_2$, and reacts with transition-metal halides to give transition-element cyclopentadienyls. It has the staggered (D_{5d}) "sandwich" structure (cf. ferro-

[25] E. C. ASHBY, J. NACKASHI, and G. E. PARRIS, NMR, infrared, and molecular association studies of some methylmagnesium alkoxides in diethyl ether, tetrahydrofuran, and benzene, *J. Am. Chem. Soc.* **97**, 3162–71 (1975).

cene p. 372) with Mg–C 230 pm and C–C 139 pm;[26] the bonding is thought to be intermediate between ionic and covalent but the actual extent of the charge separation between the central atom and the rings is still being discussed.

Calcium, Strontium, and Barium

Organometallic compounds of Ca, Sr, and Ba are far more reactive than those of Mg and have been much less studied until very recently.[27] The compounds MR_2 (M = Ca, Sr, Ba; R = Me, Et, allyl, Ph, $PhCH_2$, etc.) can be prepared using HgR_2 under appropriate conditions, often at low temperature. Compounds of the type RCaI (R = Bu, Ph, tolyl) have also been known for some time and have recently been isolated as crystals. Calcium (and Sr) dicyclopentadienyl can be made by direct reaction of the metal with either $[Hg(C_5H_5)_2]$ or with cyclo-C_5H_6 itself; cyclopentadiene also reacts with CaC_2 in liquid NH_3 to form $[Ca(C_5H_5)_2]$ and HC≡CH. The barium analogue $[Ba(C_5H_5)_2]$ is best made (though still in small yield) by treating cyclo-C_5H_6 with BaH_2.

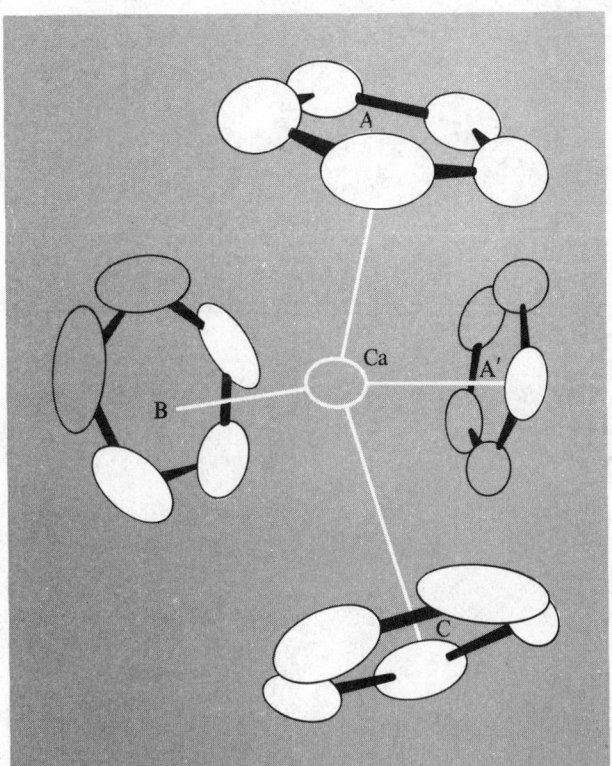

Fig. 5.12   Coordination geometry about Ca in polymeric $[Ca(C_5H_5)_2]$ showing η^5-, η^3-, and η^1-bonding (see p. 154).

[26] W. BÜNDER and E. WEISS, Refinement of the crystal structure of dicyclopentadienylmagnesium, $[Mg(\eta^5\text{-}C_5H_5)_2]$, *J. Organometall. Chem.* **92**, 1–6 (1975).

[27] B. G. GOWENLOCK and W. E. LINDSELL, The organometallic chemistry of the alkaline earth metals, *Organometallic Chemistry Reviews* (*J. Organometall. Chem.*, Library No. 3), 1–73 (1977).

The structure of $[Ca(C_5H_5)_2]$ is unique.[28] Each Ca is surrounded by 4 planar cyclopentadienyl rings and the overall structure involves a complex sharing of rings which bridge the various Ca atoms. The coordination geometry about a given Ca atom is shown in Fig. 5.12: two of the rings (A, C) are η^5, with all Ca–C distances 275 pm. A third ring (B) is η^3 with one Ca–C distance 270, two at 279, and two longer distances at 295 pm. These three polyhapto rings (A, B, C) are arranged so that their centroids are disposed approximately trigonally about the Ca atom. The fourth ring (A') is η^1, with only 1 Ca–C within bonding distance (310 pm) and this bond is approximately perpendicular to the plane formed by the centroids of the other 3 rings. The structure is the first example in which η^5-, η^3-, and η^1-C_5H_5 groups are all present. Indeed, the structure is even more complex than this implies because of the ring-bridging between adjacent Ca atoms; for example, ring A (and A') is simultaneously bonded η^5 to 1 Ca (248 pm from the ring centre) and η^1 to another on the opposite side of the ring, whereas ring C is equally associated in *pentahapto* mode with 2 Ca atoms each 260 pm from the plane of the ring.

[28] R. Zerger and G. Stucky, Unsaturated organometallic compounds of the main group elements. Dicyclopentadienylcalcium, *J. Organometall. Chem.* **80**, 7–17 (1974).

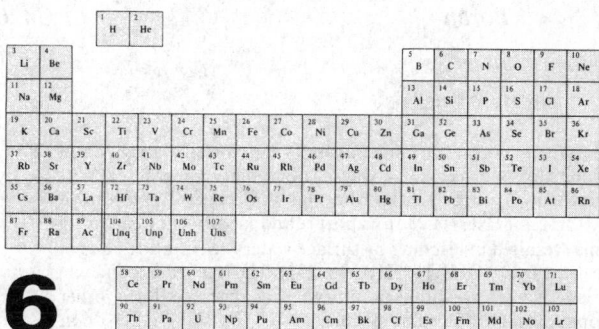

6

Boron

6.1 Introduction

Boron is a unique and exciting element. Over the years it has proved a constant challenge and stimulus not only to preparative chemists and theoreticians, but also to industrial chemists and technologists. It is the only non-metal in Group III of the periodic table and shows many similarities to its neighbour, carbon, and its diagonal relative, silicon. Thus, like C and Si, it shows a marked propensity to form covalent, molecular compounds, but it differs sharply from them in having one less valence electron than the number of valence orbitals, a situation sometimes referred to as "electron deficiency". This has a dominant effect on its chemistry.

Borax was known in the ancient world where it was used to prepare glazes and hard (borosilicate) glasses. Sporadic investigations during the eighteenth century led ultimately to the isolation of very impure boron by H. Davy and by J. L. Gay Lussac and L. J. Thénard in 1808, but it was not until 1892 that H. Moissan obtained samples of 95–98% purity by reducing B_2O_3 with Mg. High-purity boron ($>99\%$) is a product of this century, and the various crystalline forms have been obtained only during the last few decades mainly because of the highly refractory nature of the element and its rapid reaction at high temperatures with nitrogen, oxygen, and most metals. The name *boron* was proposed by Davy to indicate the source of the element and its similarity to carbon, i.e. *bor*(ax + carb)*on*.

Boron is comparatively unabundant in the universe (p. 17); it occurs to the extent of about 9 ppm in crustal rocks and is therefore rather less abundant than lithium (18 ppm) or lead (13 ppm) but is similar to praseodymium (9.1 ppm) and thorium (8.1 ppm). It occurs almost invariably as borate minerals or as borosilicates. Commercially valuable deposits are rare, but where they do occur, as in California or Turkey, they can be vast (see Panel). Isolated deposits are also worked in the USSR, Tibet, and Argentina.

The structural complexity of borate minerals (p. 231) is surpassed only by that of silicate minerals (p. 399). Even more complex are the structures of the metal borides and the various allotropic modifications of boron itself. These factors, together with the unique structural and bonding problems of the boron hydrides, dictate that boron should be treated in a separate chapter. The general group trends, and a comparison with the chemistry of the metallic elements of Group III (Al, Ga, In, and Tl), will be deferred until the next chapter.

Borate Minerals

The world's major deposits of borate minerals occur in areas of former volcanic activity and appear to be associated with the waters from former hot springs. The primary mineral that first crystallized was normally ulexite, $NaCa[B_5O_6(OH)_6].5H_2O$, but this was frequently mixed with lesser amounts of borax, $Na_2[B_4O_5(OH)_4].8H_2O$ (p. 233). Exposure and subsequent weathering (e.g. in the Mojave Desert, California) resulted in leaching by surface waters, leaving a residue of the less-soluble mineral colemanite, $Ca_2[B_3O_4(OH)_3]_2.2H_2O$ (p. 233). The leached (secondary) borax sometimes reaccumulated and sometimes underwent other changes to form other secondary minerals such as the commercially important kernite, $Na_2[B_4O_5(OH)_4].2H_2O$, at Boron, California (Pl. 6.1): This is the world's largest single source of borates and comprises a deposit 6.5 km long, 1.5 km wide, and 25–50 m thick containing material that averages 75% of hydrated sodium tetraborates (borax and kernite). The photograph was taken in mid-1970 when the pit measured 150 m deep, 1.1 km long, and 0.76 km wide. Total annual production in the USA has been fairly constant for several years and amounts to almost 2 million tonnes (equivalent to about 600 000 tonnes of B_2O_3). Production in Turkey has expanded dramatically during the past decade and is now about half that of the USA. Production in the USSR is probably about one-quarter that of the USA but their total reserves are similar. Small-scale production of borates ($\sim$50 000 tonnes pa) is also reported from Argentina.

The main chemical products produced from these minerals are (a) boron oxides, boric acid, and borates, (b) esters of boric acid, (c) refractory boron compounds (borides, etc.), (d) boron halides, (e) boranes and carbaboranes, and (f) organoboranes. The main industrial and domestic uses of boron compounds are:

Heat resistant glasses (e.g. Pyrex), glass wool, fibre glass	30–35%
Detergents, soaps, cleaners, and cosmetics	15–20%
Porcelain enamels	15%
Synthetic herbicides and fertilizers	10%
Miscellaneous (nuclear shielding, metallurgy, corrosion control, leather tanning, flame-proofing, catalysts)	30%

The uses in the glass and ceramics industries reflects the diagonal relation between boron and silicon and the similarity of vitreous borate and silicate networks (pp. 229, 233, and 399). In the UK and continental Europe (but not in the USA) sodium perborate (p. 232) is a major constituent of washing powders since it hydrolyses to H_2O_2 and acts as a bleaching agent in very hot water ($\sim$90°C); in the USA domestic washing machines rarely operate above 70°, at which temperature perborates are ineffective as bleaches.

Details of other uses of boron compounds are noted at appropriate places in the text.

6.2 Boron[1]

6.2.1 *Isolation and purification of the element*

There are four main methods of isolating boron from its compounds:

(i) Reduction by metals at high temperature, e.g. the strongly exothermic reaction

$$B_2O_3 + 3Mg \longrightarrow 2B + 3MgO \quad \text{(Moissan boron, 95–98\% pure)}$$

Other electropositive elements have been used (e.g. Li, Na, K, Be, Ca, Al, Fe), but the product is generally amorphous and contaminated with refractory impurities such as metal borides. Massive crystalline boron (96%) has been prepared by reacting BCl_3 with zinc in a flow system at 900°C.

[1] N. N. GREENWOOD, *Boron*, Pergamon Press, Oxford, 1975, 327 pp.; also as Chap. 11 in *CIC*, Vol. 1, Pergamon Press, Oxford, 1973.

PLATE 6.1   The open pit borate mine and adjacent processing plant at Boron, California; the ore is moved out of the pit by 400-m conveyor belt seen in the centre foreground. (Photograph by courtesy of United States Borax and Chemical Corporation.)

(ii) Electrolytic reduction of fused borates or tetrafluoroborates, e.g. KBF_4 in molten KCl/KF at 800°. The process is comparatively cheap but yields only powdered boron of 95% purity.

(iii) Reduction of volatile boron compounds by H_2, e.g. the reaction of $BBr_3 + H_2$ on a heated tantalum metal filament. This method, which was introduced in 1922 and can now be operated on the kilogram scale, is undoubtedly the most effective general preparation for high purity boron (>99.9%). Crystallinity improves with increasing temperature, amorphous products being obtained below 1000°C, α- and β-rhombohedral modifications between 1000–1200°, and tetragonal crystals above this. BCl_3 can be substituted for BBr_3 but BI_3 is unsatisfactory because it is expensive and too difficult to purify sufficiently. Free energy calculations indicate that BF_3 would require impracticably high temperatures (>2000°).

(iv) Thermal decomposition of boron hydrides and halides. Boranes decompose to amorphous boron when heated at temperatures up to 900° and crystalline products can be obtained by thermal decomposition of BI_3. Indeed, the first recognized sample of α-rhombohedral B was prepared (in 1960) by decomposition of BI_3 on Ta at 800–1000°, and this is still an excellent exclusive preparation of this allotrope.

6.2.2 *Structure of crystalline boron*[1, 2, 2a]

Boron is unique among the elements in the structural complexity of its allotropic modifications; this reflects the variety of ways in which boron seeks to solve the problem of having fewer electrons than atomic orbitals available for bonding. Elements in this situation usually adopt metallic bonding, but the small size and high ionization energies of B (p. 161) result in covalent rather than metallic bonding. The structural unit which dominates the various allotropes of B is the B_{12} icosahedron (Fig. 6.1), and this also occurs in several metal boride structures and in certain boron hydride derivatives. Because of the fivefold rotation symmetry at the individual B atoms, the B_{12} icosahedra pack rather inefficiently and there are regularly spaced voids which are large enough to accommodate additional boron (or metal) atoms. Even in the densest form of boron, the α-rhombohedral modification, the percentage of space occupied by atoms is only 37% (compared with 74% for closest packing of spheres).

The α-rhombohedral form of boron is the simplest allotropic modification and consists of nearly regular B_{12} icosahedra in slightly deformed cubic close packing. The rhombohedral unit cell (Fig. 6.2) has a_0 505.7 pm, α 58.06° (60° for regular ccp) and contains 12 B atoms. It is important to remember that in Fig. 6.2, as in most other structural diagrams in this chapter, the lines merely define the geometry of the clusters of boron atoms; they do not usually represent 2-centre 2-electron bonds between pairs of atoms. In terms of the MO theory to be discussed on p. 179, the 36 valence electrons of each B_{12} unit are distributed as follows: 26 electrons just fill the 13 available bonding MOs

[2] V. I. MATKOVICH (ed.), *Boron and Refractory Borides*, Springer-Verlag, Berlin, 1977, 656 pp.
[2a] GMELIN, *Handbook of Inorganic Chemistry, Boron, Supplement Vol. 2: Elemental Boron. Boron Carbides*, 1981, pp. 242.

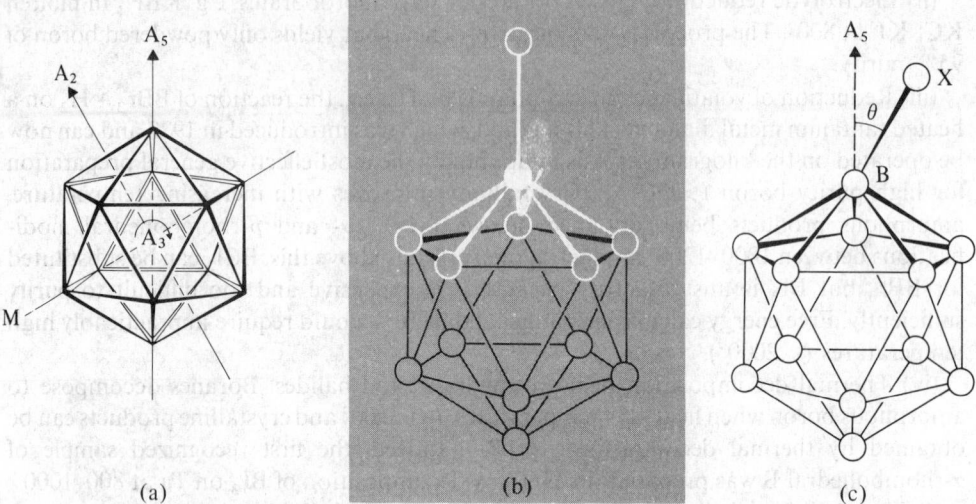

(a)   (b)   (c)

FIG. 6.1   The icosahedron and some of its symmetry elements. (a) An icosahedron has 12 vertices and 20 triangular faces defined by 30 edges. There are 6 fivefold rotation axes defined by the lines linking opposite pairs of the 12 vertices; 10 threefold rotation axes defined by the lines linking the centres of opposite pairs of the 20 triangular faces; and 15 twofold axes defined by the lines connecting opposite pairs of the 30 edges through the centre of symmetry. There are also 15 mirror planes each passing through pairs of opposite edges. (b) The preferred pentagonal pyramidal coordination polyhedron for 6-coordinate boron in icosahedral structures; as it is not possible to generate an infinite three-dimensional lattice on the basis of fivefold symmetry, various distortions, translations, and voids occur in the actual crystal structures. (c) The distortion angle θ, which varies from 0° to 25°, for various boron atoms in crystalline boron and metal borides.

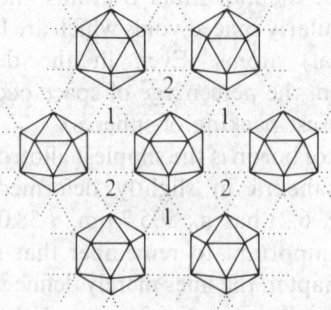

FIG. 6.2   Basal plane of α-rhombohedral boron showing close-packed arrangement of B_{12} icosahedra. The B–B distances within each icosahedron vary regularly between 173–179 pm. Dotted lines show the 3-centre bonds between the 6 equatorial boron atoms in each icosahedron to 6 other icosahedra in the same sheet at 202.5 pm. The sheets are stacked so that each icosahedron is bonded by six 2-centre B–B bonds at 171 pm (directed rhombohedrally, 3 above and 3 below the icosahedron). B_{12} units in the layer above are centred over 1 and those in the layer below are centred under 2.

within the icosahedron and 6 electrons share with 6 other electrons from 6 neighbouring icosahedra in adjacent planes to form the 6 rhombohedrally directed normal 2-centre 2-electron bonds; this leaves 4 electrons which is just the number required for contribution to the 6 equatorial 3-centre 2-electron bonds ($6 \times \frac{2}{3} = 4$).

The thermodynamically most stable polymorph of boron is the β-rhombohedral modification which has a much more complex structure with 105 B atoms in the unit cell (a_0 1014.5 pm, α 65.28°). The basic unit can be thought of as a central B_{12} icosahedron surrounded by an icosahedron of icosahedra; this can be visualized as 12 of the B_7 units in Fig. 6.1b arranged so that the apex atoms form the central B_{12} surrounded by 12 radially disposed pentagonal dishes to give the B_{84} unit shown in Fig. 6.3a. The 12 half-icosahedra are then completed by means of 2 complicated B_{10} subunits per unit cell, each comprising a central 9-coordinate B atom surrounded by 9 B atoms in the form of 4 fused pentagonal rings (Fig. 6.3b). Each of the 9 B atoms is 8-coordinate (6 of them are shared between icosahedra and 3 are unshared). This arrangement corresponds to 104 B ($84 + 10 + 10$) and there is, finally, a 6-coordinate B atom at the centre of symmetry between 2 adjacent B_{10} condensed units, bringing the total to 105 B atoms in the unit cell.

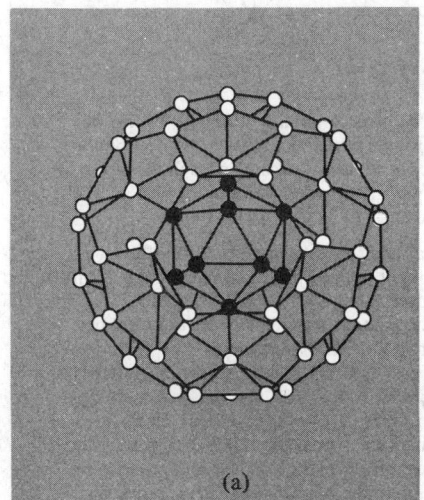
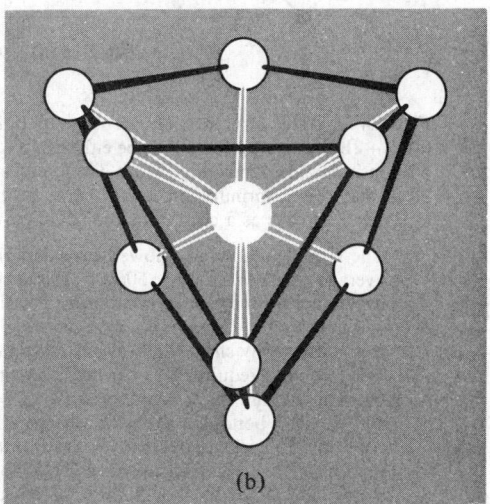

(a)   (b)

FIG. 6.3   (a) The B_{84} unit in β-rhombohedral boron comprising a central B_{12} icosahedron and 12 outwardly directed pentagonal pyramids of boron atoms. The 12 outer icosahedra are completed by linking with the B_{10} subunits shown in (b) as described in the text. The central icosahedron is almost exactly regular with B–B 176.7 pm. The shortest B–B distances (162–172 pm) are between the central icosahedron and the 12 surrounding pentagonal pyramids. The B–B distances within the $12B_6$ pentagonal pyramids (half-icosahedra) are somewhat longer (185 pm) and the longest B–B distances (188–192 pm) occur within the hexagonal rings surrounding the 3-fold symmetry axes of the B_{84} polyhedron. The variations are thought to arise both from electronic bonding factors and from detailed steric accommodations which lead to distortion of the component subunits in the unit cell.

The first crystalline polymorph of B to be prepared (1943) was termed α-tetragonal boron and was found to have 50 B atoms in the unit cell ($4B_{12} + 2B$) (Fig. 6.4). Paradoxically, however, more recent work (1974) suggests that this phase never forms in the absence of carbon or nitrogen as impurity and that it is, in reality, $B_{50}C_2$ or $B_{50}N_2$ depending on the preparative conditions; yields are increased considerably when the

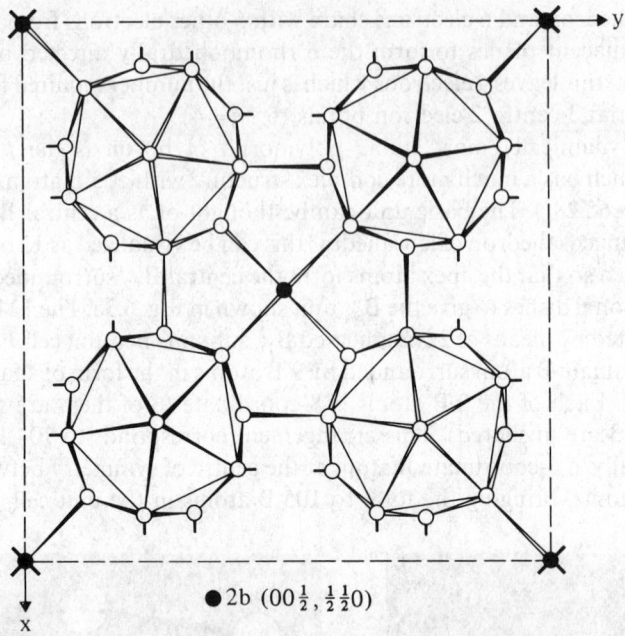

$$\bullet 2b \ (00\tfrac{1}{2}, \ \tfrac{1}{2}\tfrac{1}{2}0)$$

FIG. 6.4 Crystal structure of α-tetragonal boron. This was originally thought to be B_{50} ($4B_{12} + 2B$) but is now known to be either $B_{50}C_2$ or $B_{50}N_2$ in which the 2C (or 2N) occupy the 2(b) positions; the remaining 2B are distributed statistically at other "vacant" sites in the lattice. Note that this reformulation solves three problems which attended the description of the α-tetragonal phase as a crystalline modification of pure B:

1. The lattice parameters showed considerable variation from one crystal to another with average values a 875 pm, c 506 pm; this is now thought to arise from variable composition depending on the precise preparative conditions used.
2. The interatomic distances involving the single 4-coordinate atoms at 2(b) were only 160 pm; this is unusually short for B–B but reasonable for B–C or B–N distances.
3. The structure requires 160 valence electrons per unit cell computed as follows: internal bonding within the 4 icosahedra ($4 \times 26 = 104$); external bonds for the 4 icosahedra ($4 \times 12 = 48$); bonds shared by the atoms in 2(b) positions ($2 \times 4 = 8$). However, 50 B atoms have only 150 valence electrons and even with the maximum possible excess of boron in the unit cell (0.75 B) this rises to only 152 electrons. The required extra 8 or 10 electrons are now supplied by 2C or 2N though the detailed description of the bonding is more intricate than this simple numerology implies.

BBr_3/H_2 mixture is purposely doped with a few per cent of CH_4, $CHBr_3$, or N_2. The work illustrates the great difficulties attending preparative and structural studies in this area. The crystal structures of other boron polymorphs, particularly the β-tetragonal phase with 192 B atoms in the unit cell (a 1012, c 1414 pm), are even more complex and have so far defied elucidation despite extensive work by many investigators.[2a]

6.2.3 *Atomic and physical properties of boron*

Boron has 2 stable naturally occurring isotopes and the variability of their concentration (particularly the difference between borates from California (low in [10]B) and Turkey (high in [10]B)) prevents the atomic weight of boron being quoted more

TABLE 6.1   *Nuclear properties of boron isotopes*

Property	^{10}B	^{11}B
Relative mass ($^{12}C = 12$)	10.012 939	11.009 305
Natural abundance/(%)	19.10–20.31	80.90–79.69
Nuclear spin (parity)	3(+)	$\frac{3}{2}(-)$
Magnetic moment/(nuclear magnetons)[a]	+1.800 63	+2.688 57
Quadrupole moment/barns[b]	+0.074	+0.036
Cross-section for (n, α)/barns[b]	3835(±10)	0.005

[a] 1 nuclear magneton $= 5.0505 \times 10^{-27}$ A m² in SI.

[b] 1 barn $= 10^{-28}$ m² in SI; the cross-section for natural boron ($\sim 20\%$ ^{10}B) is ~ 767 barns.

precisely than 10.81 (p. 20). Each isotope has a nuclear spin (Table 6.1) and this has proved particularly valuable in nmr spectroscopy, especially for ^{11}B. The great difference in neutron absorption cross-section of the 2 isotopes is also notable, and this has led to the development of viable separation processes on an industrial scale. The commercial availability of the separated isotopes has greatly assisted the solution of structural and mechanistic problems in boron chemistry.

Boron is the fifth element in the periodic table and its ground-state electronic configuration is $[He]2s^2 2p^1$. The first 3 ionization energies are 800.5, 2426.5, and 3658.7 kJ mol^{-1}, all substantially larger than for the other elements in Group IIIB. (The values for this and other properties of B are compared with those for Al, Ga, In, and Tl on p. 250). The electronegativity (p. 30) of B is 2.0, which is close to the values for Si (1.8) and Ge (1.8) but somewhat less than the values for C (2.5) and H (2.1). The implied reversal of the polarity of B–H and C–H bonds is an important factor in discussing hydroboration (p. 192) and other reactions.

The determination of precise physical properties for elemental boron is bedevilled by the twin difficulties of complex polymorphism and contamination by irremovable impurities. Boron is an extremely hard refractory solid of high mp, low density, and very low electrical conductivity. Crystalline forms are dark red in transmitted light and powdered forms are black. The most stable (β-rhombohedral) modification has mp 2180°C (exceeded only by C among the non-metals), bp ~ 3650°C, d 2.35 g cm^{-3} (α-rhombohedral form 2.45 g cm^{-3}), $\Delta H_{sublimation}$ 570 kJ per mol of B, electrical conductivity at room temperature 1.5×10^{-6} ohm^{-1} cm^{-1}.

6.2.4 *Chemical properties*

It has been argued[1] that the inorganic chemistry of boron is more diverse and complex than that of any other element in the periodic table. Indeed, it is only during the last two decades that the enormous range of structural types has begun to be elucidated and the subtle types of bonding appreciated. The chemical nature of boron is influenced primarily by its small size and high ionization energy, and these factors, coupled with the similarity in electronegativity of B, C, and H, lead to an extensive and unusual type of covalent (molecular) chemistry. The electronic configuration $2s^2 2p^1$ is reflected in a predominant tervalence and bond energies involving B are such that there is no tendency to form univalent compounds of the type which increasingly occur in the chemistry of Al, Ga, In, and Tl. However, the availability of only 3 electrons to contribute to covalent bonding involving the 4 orbitals s, p_x, p_y, and p_z confers a further range of properties on B leading to electron-pair acceptor behaviour (Lewis acidity) and multicentre bonding. The high affinity for oxygen is another dominant characteristic which forms the basis of the extensive chemistry of borates and related oxo complexes. Finally, the small size of B enables many interstitial alloy-type metal borides to be prepared, and the range of these is considerably extended by the propensity of B to form branched and unbranched chains, planar networks, and three-dimensional arrays of great intrinsic stability which act as host frameworks to house metal atoms in various stoichiometric proportions.

It is thus possible to distinguish five types of boron compound, each having its own chemical systematics which can be rationalized in terms of the type of bonding involved, and each resulting in highly individualistic structures and chemical reactions:

(i) metal borides ranging from M_5B to MB_{66} (or even $MB_{>100}$) (see below);
(ii) boron hydrides and their derivatives including carbaboranes and polyhedral borane–metal complexes (p. 171);
(iii) boron trihalides and their adducts and derivatives (p. 220);
(iv) oxo compounds including polyborates, borosilicates, peroxoborates, etc. (p. 228);
(v) organoboron compounds and B–N compounds (B–N being isoelectronic with C–C) (p. 234).

The chemical reactivity of boron itself obviously depends markedly on the purity, crystallinity, state of subdivision, and temperature. Boron reacts with F_2 at room temperature and is superficially attacked by O_2 but is otherwise inert. At higher temperatures boron reacts directly with all the non-metals except H, Ge, Te, and the noble gases. It also reacts readily and directly with almost all metals at elevated temperatures, the few exceptions being the heavier members of the B subgroups (Ag, Au; Cd, Hg; Ga, In, Tl; Sn, Pb; Sb, Bi).

The general chemical inertness of boron at lower temperatures can be gauged by the fact that it resists attack by boiling concentrated aqueous NaOH or by fused NaOH up to 500°, though it is dissolved by fused $Na_2CO_3/NaNO_3$ mixtures at 900°C. A 2:1 mixture of hot concentrated H_2SO_4/HNO_3 is also effective for dissolving elemental boron for analysis but non-oxidizing acids do not react.

6.3 Borides[1, 2, 2a]

6.3.1 *Introduction*

The borides comprise a group of over 200 binary compounds which show an amazing diversity of stoichiometries and structural types; e.g. M_5B, M_4B, M_3B, M_5B_2, M_7B_3, M_2B, M_5B_3, M_3B_2, $M_{11}B_8$, MB, $M_{10}B_{11}$, M_3B_4, M_2B_3, M_3B_5, MB_2, M_2B_5, MB_3, MB_4, MB_6, M_2B_{13}, MB_{10}, MB_{12}, MB_{15}, MB_{18}, and MB_{66}. There are also numerous nonstoichiometric phases of variable composition and many ternary and more complex phases in which more than one metal combines with boron. The rapid advance in our understanding of these compounds during the past two to three decades has been based mainly on X-ray diffraction analysis and the work has been stimulated not only by the inherent academic challenge implied by the existence of these unusual compounds but also by the extensive industrial interest generated by their unique combination of desirable physical and chemical properties (see Panel).

Properties and Uses of Borides

Metal-rich borides are extremely hard, chemically inert, involatile, refractory materials with mps and electrical conductivities which often exceed those of the parent metals. Thus the highly conducting diborides of Zr, Hf, Nb, and Ta all have mps $> 3000°C$ and TiB_2 (mp $2980°C$) has a conductivity 5 times greater than that of Ti metal. Borides are normally prepared as powders but can be fabricated into the desired form by standard techniques of powder metallurgy and ceramic technology. TiB_2, ZrB_2, and CrB_2 find application as turbine blades, combustion chamber liners, rocket nozzles, and ablation shields. Ability to withstand attack by molten metals, slags, and salts have commended borides or boride-coated metals as high-temperature reactor vessels, vaporizing boats, crucibles, pump impellers, and thermocouple sheaths. Inertness to chemical attack at high temperatures, coupled with excellent electrical conductivity, suggest application as electrodes in industrial processes.

Nuclear applications turn on the very high absorption cross-section of ^{10}B for thermal neutrons (p. 161) and the fact that this property is retained for high-energy neutrons (10^4–10^6 eV) more effectively than for any other nuclide. Another advantage of ^{10}B is that the products of the (n, α) reaction are the stable, non-radioactive elements Li and He. Accordingly, metal borides and boron carbide have been used extensively as neutron shields and control rods since the beginning of the nuclear power industry.

The principal non-nuclear industrial use of boron carbide is as an abrasive grit or powder for polishing or grinding; it is also used on brake and clutch linings. In addition, there is much current interest in its use as light-weight protective armour, and tests have indicated that boron carbide and beryllium borides offer the best choice; applications are in bullet-proof protective clothing and in protective armour for aircraft. More elegantly, boron carbide can now be produced in fibre form by reacting BCl_3/H_2 with carbon yarn at 1600–1900°C:

$$4BCl_3 + 6H_2 + C(fibres) \longrightarrow B_4C(fibres) + 12HCl$$

Fibre curling can be eliminated by heat treatment under tension near the mp, and the resulting fibres have a tensile strength of 3.5×10^5 psi (1 psi $= 6895$ N m^{-2}) and an elastic modulus of 50×10^6 psi at a density of 2.35 g cm^{-3}; the form was 1 ply, 720 filament yarn with a filament diameter of 11–12 μm.[3] The fibres are inert to hot acid and alkali, resistant to Cl_2 up to 700° and air up to 800°C.

Boron itself has been used for over a decade in filament form in various composites; BCl_3/H_2 is reacted at 1300° on the surface of a continuously moving tungsten fibre 12 μm in diameter. US production reached 15 000 kg pa in 1976 and is expected to reach 50 tonnes pa in 1980. The primary use so far has been in military aircraft and space shuttles, but boron fibre composites are being studied as potential reinforcement materials for commercial aircraft such as the DC10, Boeing 737, and European A300 air bus. At the domestic level they are finding increasing application in golf shafts, tennis rackets, and bicycle frames.[4]

6.3.2 *Preparation and stoichiometry*

Eight general methods are available for the synthesis of borides, the first four being appropriate for small-scale laboratory preparations and the remaining four for commercial production on a scale ranging from kilogram amounts to tonne quantities. Because high temperatures are involved and the products are involatile, borides are not easy to prepare pure and subsequent purification is often difficult; precise stoichiometry is

---

[3] W. D. SMITH, Boron carbide fibres from carbon fibres, pp. 541–51 in ref. 2.
[4] V. KRUKONIS, Chemical vapour deposition of boron filament, pp. 517–40 in ref. 2.

also sometimes hard to achieve because of differential volatility or high activation energies. The methods are:

(i) Direct combination of the elements: this is probably the most widely used technique, e.g.

$$Cr + nB \xrightarrow{1150°} CrB_n$$

(ii) Reduction of metal oxide with B (rather wasteful of expensive elemental B), e.g.

$$Sc_2O_3 + 7B \xrightarrow{1800°} 2ScB_2 + 3BO$$

(iii) Co-reduction of volatile mixed halides with H_2 using a metal filament, hot tube, or plasma torch, e.g.

$$2TiCl_4 + 4BCl_3 + 10H_2 \xrightarrow{1300°} 2TiB_2 + 20HCl$$

(iv) Reduction of BCl_3 (or BX_3) with a metal (sometimes assisted by H_2), e.g.

$$nBX_3 + (x+1)M \longrightarrow MB_n + xMX_{3n/x}$$

$$BCl_3 + W \xrightarrow{H_2/1200°} WB + Cl_2 + HCl$$

(v) Electrolytic deposition from fused salts: this is particularly effective for MB_6 (M = alkaline earth or rare earth metal) and for the borides of Mo, W, Fe, Co, and Ni. The metal oxide and B_2O_3 or borax are dissolved in a suitable, molten salt bath and electrolysed at 700–1000° using a graphite anode; the boride is deposited on the cathode which can be graphite or steel.

(vi) Co-reduction of oxides with carbon at temperatures up to 2000°, e.g.

$$V_2O_5 + B_2O_3 + 8C \xrightarrow{1500°} 2VB + 8CO$$

(vii) Reduction of metal oxide (or $M + B_2O_3$) with boron carbide, e.g.

$$Eu_2O_3 + 3B_4C \xrightarrow{1600°} 2EuB_6 + 3CO$$

$$7Ti + B_2O_3 + 3B_4C \xrightarrow{2000°} 7TiB_2 + 3CO$$

Boron carbide (p. 167) is a most useful and economic source of B and will react with most metals or their oxides. It is produced in tonnage quantities by direct reduction of B_2O_3 with C at 1600°: a C resistor is embedded in a mixture of B_2O_3 and C, and a heavy electric current passed.

(viii) Co-reduction of mixed oxides with metals (Mg or Al) in a thermite-type reaction—this usually gives contaminated products including ternary borides, e.g. $Mo_7Al_6B_7$. Alternatively, alkali metals or Ca can be used as reductants, e.g.

$$TiO_2 + B_2O_3 \xrightarrow{\text{molten Na}} TiB_2$$

The various stoichiometries are not equally common, as can be seen from Fig. 6.5; the

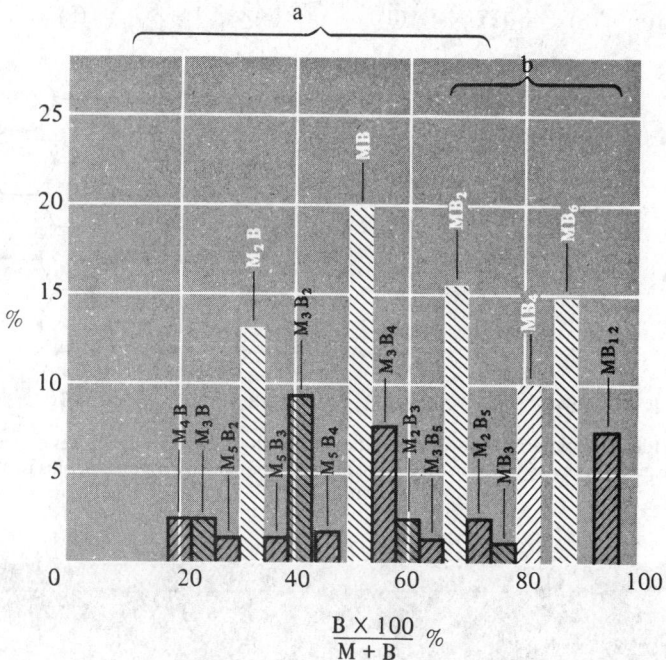

FIG. 6.5   Frequency of various structure occurrences among boride phases: (a) field of borides of
d elements, and (b) field of borides of s, p, and f elements.

most frequently occurring are M_2B, MB, MB_2, MB_4, and MB_6, and these five classes
account for 75% of the compounds. At the other extreme $Ru_{11}B_8$ is the only known
example of this stoichiometry. Metal-rich borides tend to be formed by the transition
elements whereas the boron-rich borides are characteristic of the more electropositive
elements in Groups IA–IIIA, the lanthanides, and the actinides. Only the diborides MB_2
are common to both classes.

6.3.3   *Structures of borides*

The structures of metal-rich borides can be systematized by the schematic arrangements
shown in Fig. 6.6, which illustrates the increasing tendency of B atoms to catenate as their
concentration in the boride phase increases; the B atoms are often at the centres of
trigonal prisms of metal atoms (Fig. 6.7) and the various stoichiometries are
accommodated as follows:

(a) isolated B atoms: Mn_4B; M_3B (Tc, Re, Co, Ni, Pd); Pd_5B_2;
                                     M_7B_3(Tc, Re, Ru, Rh);
                                     M_2B (Ta, Mo, W, Mn, Fe, Co, Ni);
(b) isolated pairs B_2: Cr_5B_3; M_3B_2 (V, Nb, Ta);
(c) zigzag chains of B atoms: M_3B_4 (Ti; V, Nb, Ta; Cr, Mn, Ni);
                                     MB (Ti, Hf; V, Nb, Ta; Cr, Mo, W; Mn, Fe, Co, Ni);
(d) branched chains of B
     atoms: $Ru_{11}B_8$;

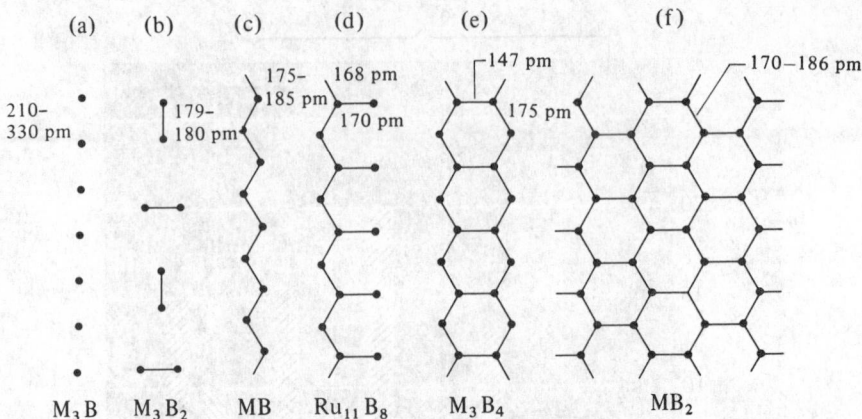

FIG. 6.6   Idealized patterns of boron catenation in metal-rich borides. Examples of the structures (a)–(f) are given in the text. Boron atoms are often surrounded by trigonal prisms of M atoms as shown in Fig. 6.7.

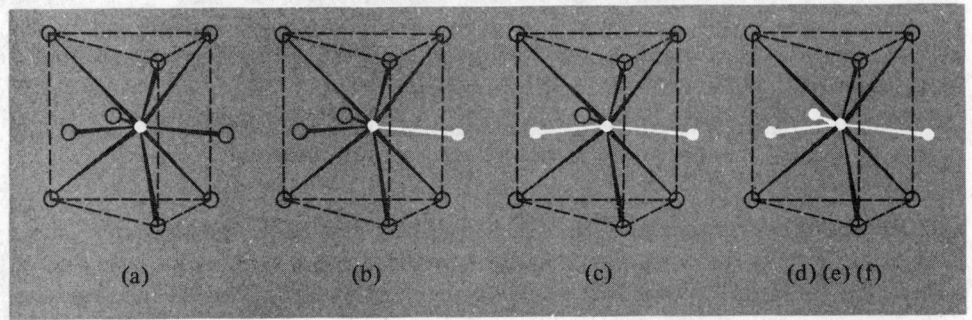

FIG. 6.7   Idealized boron environment in metal-rich borides (see text): (a) isolated B atoms in M_3B and M_7B_3; (b) pairs of B atoms in Cr_5B_3 and M_3B_2; (c) zigzag chains of B atoms in Ni_3B_4 and MB; (d) branched chains in $Ru_{11}B_8$; and (e), (f) double chains and plane nets in M_3B_4, MB_2, and M_2B_5.

(e) double chains of B atoms: M_3B_4 (V, Nb, Ta; Cr, Mn);

(f) plane (or puckered) nets: MB_2 (Mg, Al; Sc, Y; Ti, Zr, Hf; V, Nb, Ta; Cr, Mo, W; Mn, Tc, Re; Ru, Os; U, Pu); M_2B_5 (Ti; Mo, W).

It will be noted from Fig. 6.6 that structures with isolated B atoms can have widely differing interatomic B–B distances, but all other classes involve appreciable bonding between B atoms, and the B–B distances remain almost invariant despite the extensive variation in the size of the metal atoms.

The structures of boron-rich borides (e.g. MB_4, MB_6, MB_{10}, MB_{12}, MB_{66}) are even more effectively dominated by inter-B bonding, and the structures comprise three-dimensional networks of B atoms and clusters in which the metal atoms occupy specific voids or otherwise vacant sites. The structures are often exceedingly complicated (for the reasons given in Section 6.2.2): for example, the cubic unit cell of YB_{66} has a_0 2344 pm and contains 1584 B and 24 Y atoms; the basic structural unit is the 13-icosahedron unit of 156 B atoms found in β-rhombohedral B (p. 159); there are 8 such units (1248 B) in the unit cell and the remaining 336 B atoms are statistically distributed in channels formed by the packing of the 13-icosahedron units.

Another compound which is even more closely related to β-rhombohedral boron is boron carbide, "B_4C"; this is now more correctly written as $B_{13}C_2$,[5] but the phase can vary over wide composition ranges which approach the stoichiometry $B_{12}C_3$. The structure is best thought of in terms of B_{84} polyhedra (p. 159) but these are now interconnected simply by linear C–B–C units instead of the larger B_{10}–B–B_{10} units in β-rhombohedral B. The result is a more compact packing of the 13-icosahedron units so generated and this is reflected in the unit cell dimensions (a 517.5 pm, α 65.74°). A notable feature of the structure (Fig. 6.8) is the presence of regular hexagonal planar rings B_4C_2 (shaded); these can be seen more plainly in Fig. 6.9. Six equatorial B atoms in each icosahedron bond to carbon atoms at the end of the CBC chains with B–C 161 pm and a deviation θ from the normal to the pentagonal pyramid (p. 158) of only 0.21°; the other 6 B atoms form direct B–B bonds to adjacent icosahedra at 172.3 pm with $\theta = 4.8$°. Stringent tests had to be applied to distinguish confidently between B and C atoms in this

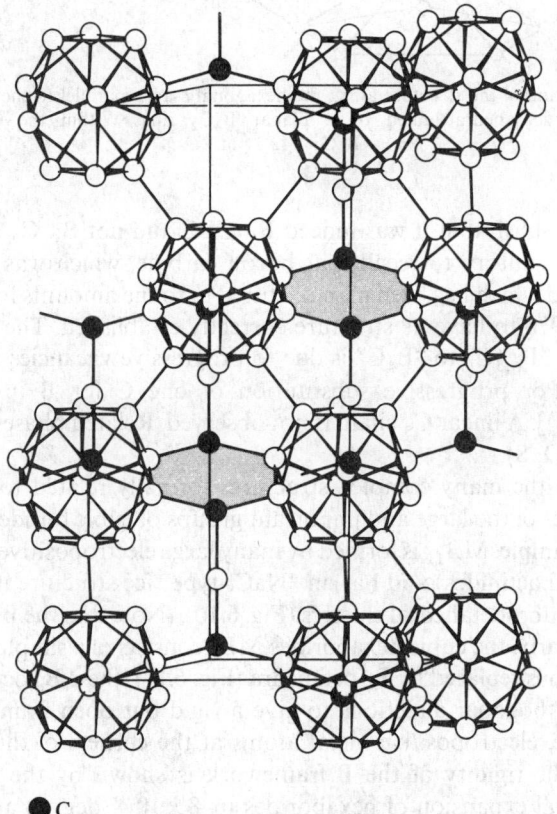

● C

Fig. 6.8   Crystal structure of $B_{13}C_2$ showing the planar hexagonal rings connecting the B_{12} icosahedra. These rings are perpendicular to the C–B–C chains.

[5] G. WILL and K. H. KOSSOBUTZKI, An X-ray diffraction analysis of boron carbide, *J. Less-Common Metals* **47**, 43–48 (1976).

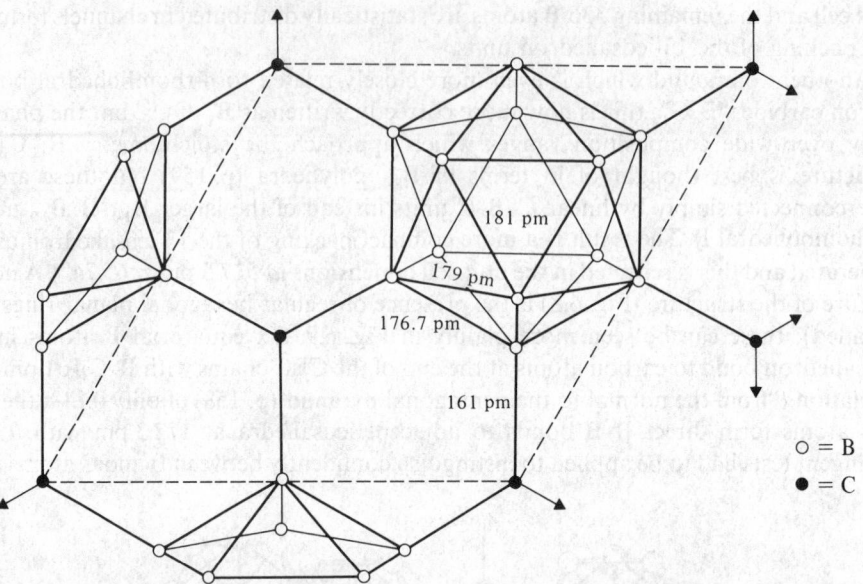

FIG. 6.9   Projection of the structure along the hexagonal c-axis (parallel to the C–B–C chain) showing 2 B_{12} icosahedra linked by a planar B_4C_2 ring. Within the C–B–C chains B–C = 142.3 pm.

structure and to establish that it was indeed $B_{12}CBC$ and not $B_{12}C_3$ as had previously been thought. It is salutory to recall that boron carbide, which was first made by H. Moissan in 1899 and which has been manufactured in tonne amounts for several decades, had to wait until 1976 to have its structure correctly established. The wide variation in stoichiometry from "$B_{6.5}C$" to "B_4C" is due to progressive vacancies in the CBC chain ($B_{12}C_2 \equiv B_6C$) and/or progressive substitution of one C for B in the icosahedron [$(B_{11}C)CBC \equiv B_4C$)]. A linear C_3 chain is not observed. Related phases are $B_{12}PBP$ and $B_{12}X_2$ (X = P, As, O, S).

By contrast with the many complex structures formally related to β-rhombohedral boron, the structures of the large and important groups of cubic borides MB_{12} and MB_6 are comparatively simple. MB_{12} is formed by many large electropositive metals (e.g. Sc, Y, Zr, lanthanides, and actinides) and has an "NaCl-type" fcc structure in which M atoms alternate with B_{12} cubo-octahedral clusters (Fig. 6.10). (Note that the B_{12} cluster is not an icosahedron.) Similarly, the cubic hexaborides MB_6 consists of a simple CsCl-type lattice in which the halogen is replaced by B_6 octahedra (Fig. 6.11); these B_6 octahedra are linked together in all 6 orthogonal directions to give a rigid but open framework which can accommodate large, electropositive metal atoms at the corners of the interpenetrating cubic sublattice. The rigidity of the B framework is shown by the very small linear coefficient of thermal expansion of hexaborides ($6–8 \times 10^{-6}$ deg^{-1}) and by the narrow range of lattice constants of these phases which vary by only 4% (410–427 pm), whereas the diameters of the constituent metal atoms vary by 25% (355–445 pm). Bonding theory for isolated groups such as $B_6H_6^{2-}$ (p. 182) requires the transfer of 2 electrons to the borane cluster to fill all the bonding MOs; however, complete transfer of 2e per B_6 unit is not required in a three-dimensional crystal lattice and calculations for MB_6 (Ca, Sr, Ba)

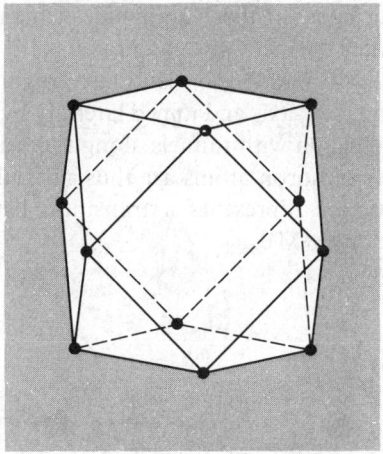

FIG. 6.10 B_{12} Cubo-octahedron cluster as found in MB_{12}. This B_{12} cluster alternates with M
atoms on an fcc lattice as in NaCl, the B_{12} replacing Cl.

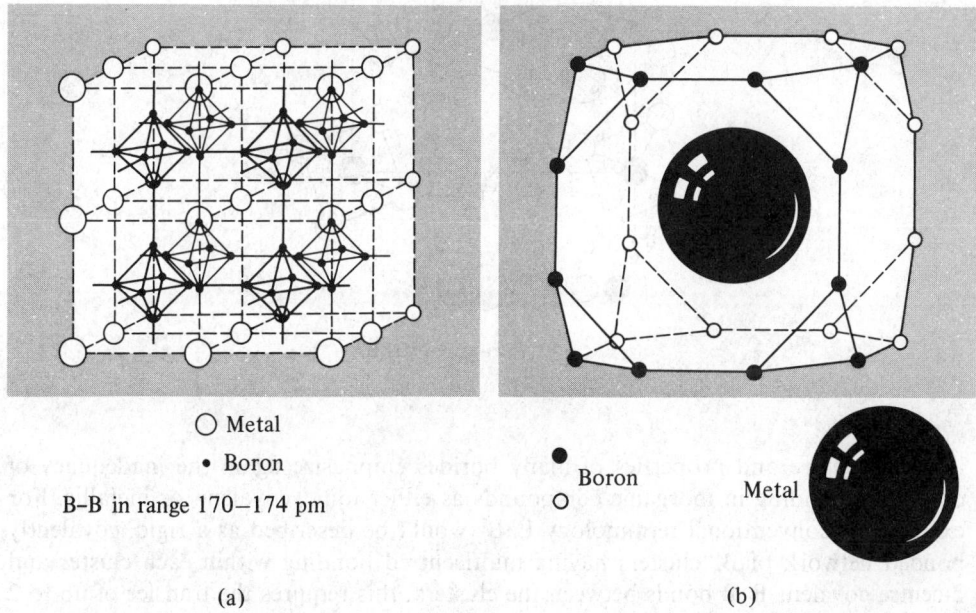

○ Metal

• Boron

B–B in range 170–174 pm

● Boron

○

Metal

(a)   (b)

FIG. 6.11   Cubic MB_6 showing (a) boron octahedra, and (b) 24-atom coordination polyhedron
around each metal atom.

indicate the transfer of only 0.9–1.0e.[6] This also explains why metal-deficit phases
$M_{1-x}B_6$ remain stable and why the alkali metals (Na, K) can form hexaborides. The
$M^{II}B_6$ hexaborides (Ca, Sr, Ba, Eu, Yb) are semiconductors but $M^{III}B_6$ and $M^{IV}B_6$
($M^{III} = $ Y, La, lanthanides; $M^{IV} = $ Th) have a high metallic conductivity at room
temperature (10^4–10^5 ohm^{-1} cm^{-1}).

The "radius" of the 24-coordinate metal site in MB_6 is too large (215–225 pm) to be

[6] P. G. PERKINS, The electronic structures of the hexaborides and the diborides, pp. 31–51 in ref. 2.

comfortably occupied by the later (smaller) lanthanide elements Ho, Er, Tm, and Lu, and these form MB_4 instead, where the metal site has a radius of 185–200 pm. The structure of MB_4 (also formed by Ca, Y, Mo, and W) consists of a tetragonal lattice formed by chains of B_6 octahedra linked along the *c*-axis and joined laterally by pairs of B_2 atoms in the *xy* plane so as to form a 3D skeleton with tunnels along the *c*-axis that are filled by metal atoms (Fig. 6.12). The pairs of boron atoms are thus surrounded by trigonal prisms of metal atoms and the structure represents a transition between the puckered layer structures of MB_2 and the cubic MB_6.

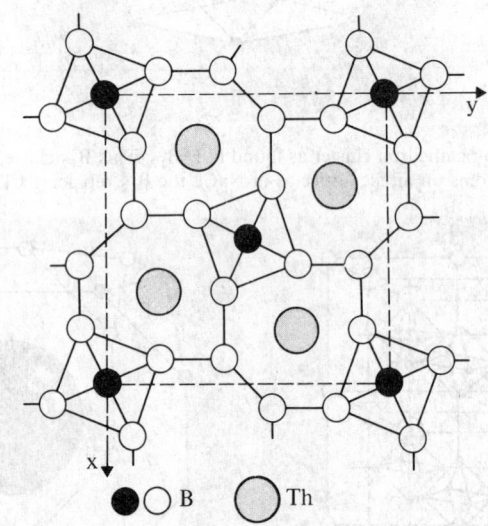

FIG. 6.12   Structure of ThB_4.

The structure and properties of many borides emphasize again the inadequacy of describing bonding in inorganic compounds as either ionic, covalent, or metallic. For example, in conventional terminology LaB_6 would be described as a rigid, covalently bonded network of B_6 clusters having multicentred bonding within each cluster and 2-centre covalent B–B bonds between the clusters; this requires the transfer of up to 2 electrons from the metal to the boron sublattice and so could be said also to involve ionic bonding ($La^{2+}B_6^{2-}$) in addition to the covalent inter-boron bonding. Finally, the third valency electron on La is delocalized in a conduction band of the crystal (mainly metal based) and the electrical conductivity of the boride is actually greater than that of La metal itself so that this aspect of the bonding could be called metallic. The resulting description of the bonding is an *ad hoc* mixture of four oversimplified limiting models and should more logically be replaced by a generalized MO approach.[6] It will also be clear from the preceding paragraphs that a classification of borides according to the periodic table does not result in the usual change in stoichiometry from one group to the next; instead, a classification in terms of the type of boron network and the size and electropositivity of the other atoms is frequently more helpful and revealing of periodic trends.

6.4   Boranes (Boron Hydrides)[1, 7]

6.4.1   *Introduction*

Borane chemistry began in 1912 with A. Stock's classic investigations,[8] and the numerous compounds prepared by his group during the following 20 y proved to be the forerunners of an amazingly diverse and complex new area of chemistry. During the past 25 y the chemistry of boranes and the related carbaboranes (p. 202) has been one of the major growth areas in inorganic chemistry, and interest continues unabated. The importance of boranes stems from three factors: firstly, the completely unsuspected structural principles involved; secondly, the growing need to extend covalent MO bond theory considerably to cope with the unusual stoichiometries; and, finally, the emergence of a versatile and extremely extensive reaction chemistry which parallels but is quite distinct from that of organic and organometallic chemistry. This efflorescence of activity culminated (in the centenary year of Stock's birth) in the award of the 1976 Nobel Prize for Chemistry to W. N. Lipscomb (Harvard) "for his studies of boranes which have illuminated problems of chemical bonding".

Over 25 neutral boranes, B_nH_m, and an even larger number of borane anions $B_nH_m{}^{x-}$ have been characterized; these can be classified according to structure and stoichiometry into 5 series though examples of neutral or unsubstituted boranes themselves are not known for all 5 classes:

closo-boranes (from corruption of Greek κλωβός, *clovo*, a cage) have complete, closed polyhedral clusters of *n* boron atoms;

nido-boranes (from Latin *nidus* a nest) have non-closed structures in which the B_n cluster occupies *n* corners of an (*n*+1)-cornered polyhedron;

arachno-boranes (from Greek ἀράχνη, *arachne*, a spider's web) have even more open clusters in which the B atoms occupy *n* contiguous corners of an (*n*+2)-cornered polyhedron;

hypho-boranes (from Greek ὑφή, *hyphe*, a net) have the most open clusters in which the B atoms occupy *n* corners of an (*n*+3)-cornered polyhedron;

conjuncto-boranes (from Latin *conjuncto*, I join together) have structures formed by linking two (or more) of the preceding types of cluster together.

Examples of these various series are listed below and illustrated in the accompanying structural diagrams. Their interrelations are further discussed in connection with carborane structures 51–74.

Closo-boranes:

$B_nH_n{}^{2-}$ (*n*=6–12) see structures 1–7. The neutral B_nH_{n+2} are not known.

Nido-boranes:

B_nH_{n+4}, e.g. B_2H_6 (8), B_5H_9 (9), B_6H_{10} (10), $B_{10}H_{14}$ (11); B_8H_{12} also has this formula but has a rather more open structure (12) which can be visualized as being formed from $B_{10}H_{14}$ by removal of B(9) and B(10).

$B_nH_{n+3}{}^-$ formed by removal of 1 bridge proton from B_nH_{n+4}, e.g. $B_5H_8{}^-$, $B_{10}H_{13}{}^-$; other anions in this series such as $B_4H_7{}^-$ and $B_9H_{12}{}^-$ are known though the

[7] E. L. MUETTERTIES (ed.), *Boron Hydride Chemistry*, Academic Press, New York, 1975, 532 pp.
[8] A. STOCK, *Hydrides of Boron and Silicon*, Cornell University Press, Ithaca, New York, 1933, 250 pp.

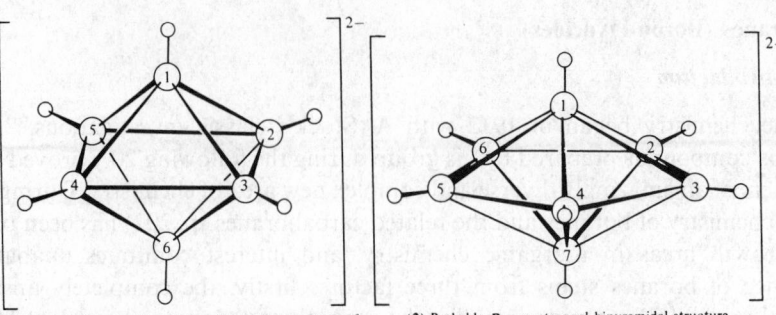

(1) The $B_6H_6^{2-}$ anion. The relationship to the structure of the B_6 network in CaB_6 and the boron cluster in B_5H_9 should be noted

(2) Probable D_{5h} pentagonal bipyramidal structure of the anion $B_7H_7^{2-}$ in solution

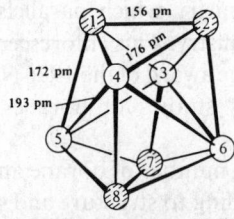

(3) The D_{2d} configuration of the boron atoms in $B_8H_8^{2-}$ showing the two structurally non-equivalent sets of 4 boron atoms

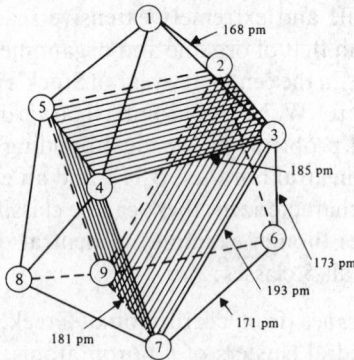

(4) Structure of the boron cluster in $B_9H_9^{2-}$ (interatomic distances ± 1.5 pm) The four unique B-H distances are 107, 110, 127 and 144 ± 15 pm

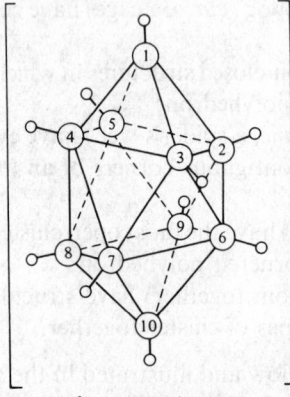

(5) $B_{10}H_{10}^{2-}$: decahydro-*closo*-decaborate (?−)

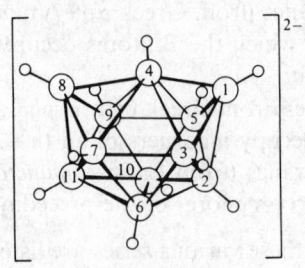

(6) $B_{11}H_{11}^{2-}$

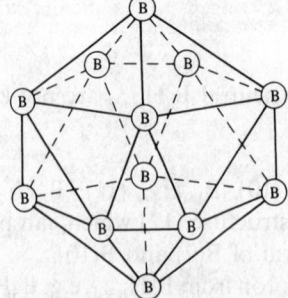

(7) Position of boron atoms and numbering system in the icosahedral borane anion $B_{12}H_{12}^{2-}$ The hydrogen atoms, which are attached radially to each boron atom, are omitted for clarity. There are six B-B distances of 175.5 pm and 24 of 178 pm

parent boranes have proved too fugitive to isolate; BH_4^- can be thought of as formed by addition of H^- to BH_3;

$B_nH_{n+2}^{2-}$, e.g. $B_{10}H_{12}^{2-}$, $B_{11}H_{13}^{2-}$.

Arachno-boranes:

B_nH_{n+6}, e.g. B_4H_{10} (13), B_5H_{11} (14), B_6H_{12} (15), B_8H_{14} (16), $n\text{-}B_9H_{15}$ (17), $i\text{-}B_9H_{15}$;
$B_nH_{n+5}^-$, e.g. $B_2H_7^-$ (18), $B_3H_8^-$ (19), $B_5H_{10}^-$, $B_9H_{14}^-$ (20), $B_{10}H_{15}^-$;
$B_nH_{n+4}^{2-}$, e.g. $B_{10}H_{14}^{2-}$ (21);

Hypho-boranes:

B_nH_{n+8}. No neutral borane has yet been definitely established in this series but the known compounds B_8H_{16} and $B_{10}H_{18}$ may prove to be *hypho*-boranes and several adducts are known to have *hypho*-structures (p. 194).

Conjuncto-boranes:

B_nH_m. At least five different structure types of interconnected borane clusters have been identified; they have the following features:
  (a) fusion by sharing a single common B atom, e.g. $B_{15}H_{23}$ (22);
  (b) formation of a direct 2-centre B–B σ bond between 2 clusters, e.g. B_8H_{18}, i.e. $(B_4H_9)_2$ (23), $B_{10}H_{16}$, i.e. $(B_5H_8)_2$ (3 isomers) (24), $B_{20}H_{26}$, i.e. $(B_{10}H_{13})_2$ (11 possible isomers of which most have been prepared and separated), (e.g. 25a, b, c); anions in this subgroup are represented by the 3 isomers of $B_{20}H_{18}^{4-}$, i.e. $(B_{10}H_9^{2-})_2$ (26);
  (c) fusion of 2 clusters via 2 B atoms at a common edge, e.g. $B_{13}H_{19}$ (27), $B_{14}H_{18}$ (28), $B_{14}H_{20}$ (29), $B_{16}H_{20}$ (30), $n\text{-}B_{18}H_{22}$ (31), $i\text{-}B_{18}H_{22}$ (32);
  (d) fusion of two clusters via 3 B atoms at a common face: no neutral borane or borane anion is yet known with this conformation but the solvated complex $(MeCN)_2B_{20}H_{16} \cdot MeCN$ has this structure (33);
  (e) more extensive fusion involving 4 B atoms in various configurations, e.g. $B_{20}H_{16}$ (34), $B_{20}H_{18}^{2-}$ (35).

Boranes are usually named by indicating the number of B atoms with a latin prefix and the number of H atoms by an arabic number in parentheses, e.g. B_5H_9, pentaborane(9); B_5H_{11}, pentaborane(11). Names for anions end in "ate" rather than "ane" and specify both the number of H and B atoms and the charge, e.g. $B_5H_8^-$ octahydropentaborate-(1 −). Further information can be provided by the optional inclusion of the italicized descriptors *closo*-, *nido*-, *arachno*-, *hypho*-, and *conjuncto*-, e.g.:

$B_{10}H_{10}^{2-}$: decahydro-*closo*-decaborate(2 −) [structure (5)]
$B_{10}H_{14}$: *nido*-decaborane(14) [structure (11)]
$B_{10}H_{14}^{2-}$: tetradecahydro-*arachno*-decaborate(2 −) [structure (21)]
$B_{10}H_{16}$: 1,1′-*conjuncto*-decaborane(16) [structure (24a)]
        [i.e. 1,1′-bi(*nido*-pentaboranyl)]

The detailed numbering schemes are necessarily somewhat complicated but, in all other respects, standard nomenclature practices are followed.[9]

⁹ R. M. ADAMS, Nomenclature of inorganic boron compounds, *Pure Appl. Chem.* **30**, 683–710 (1972).

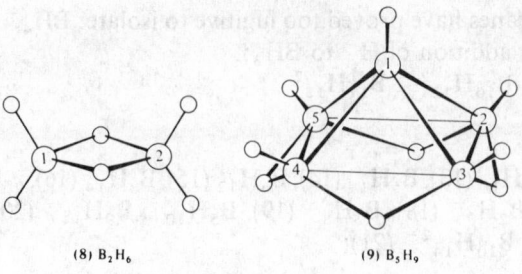

(8) B_2H_6 (9) B_5H_9

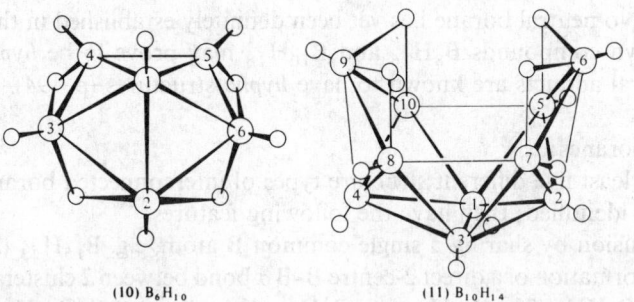

(10) B_6H_{10} (11) $B_{10}H_{14}$

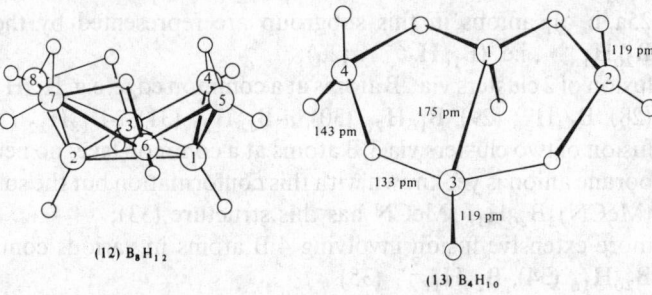

(12) B_8H_{12}

(13) B_4H_{10}

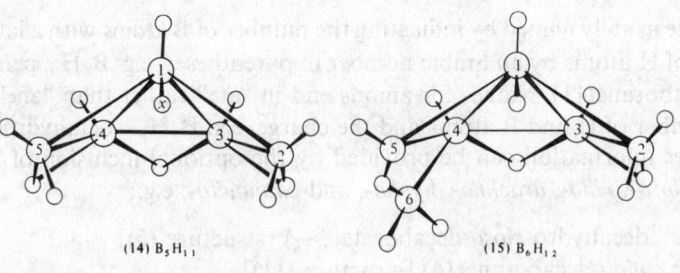

(14) B_5H_{11} (15) B_6H_{12}

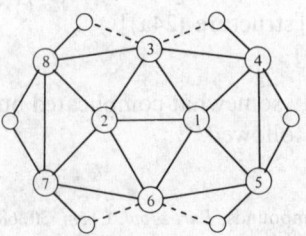

(16) Proposed structure for B_8H_{14}
(terminal H atoms omitted)

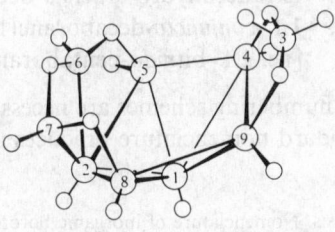

(17) n - B_9H_{15}

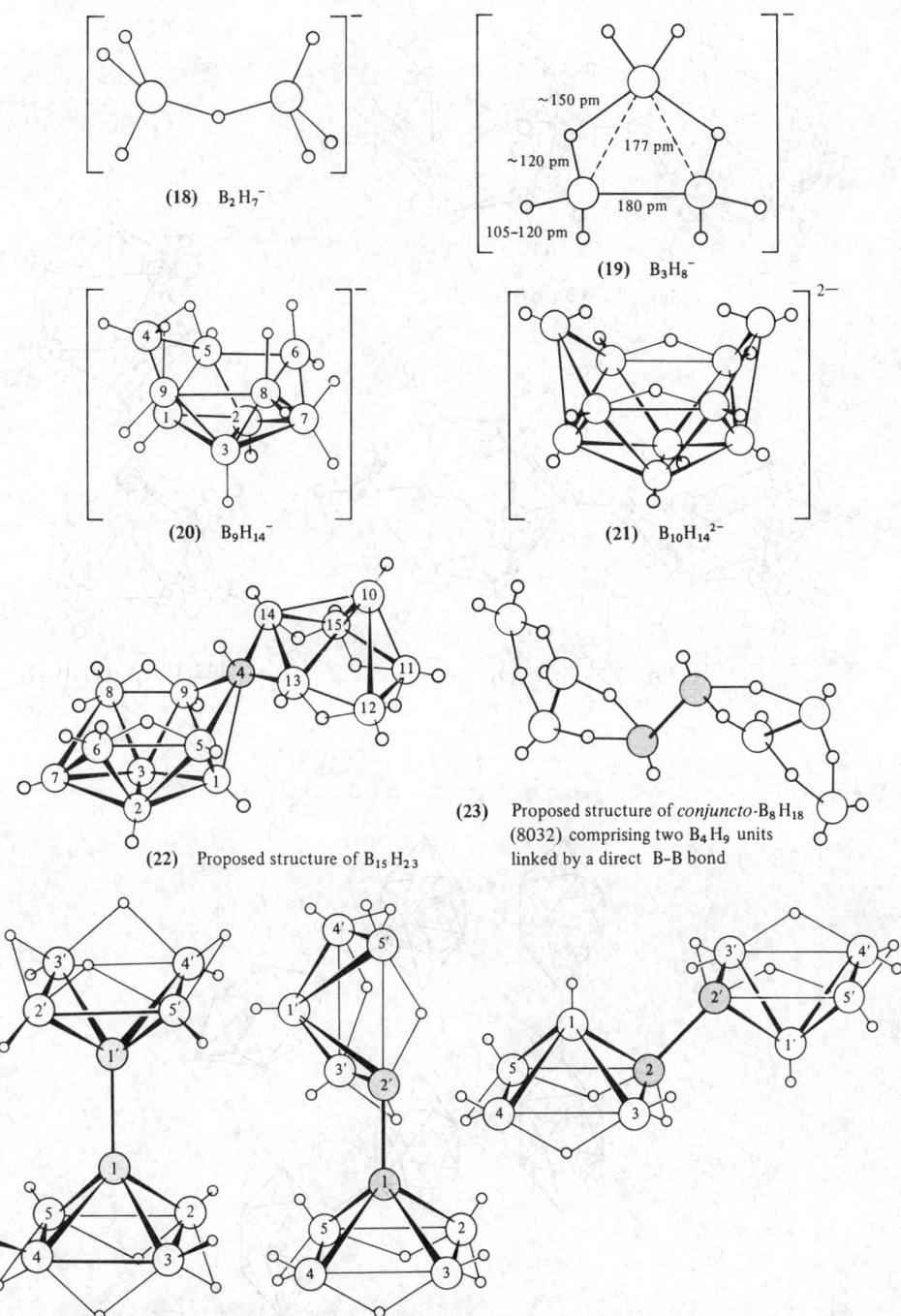

(18) $B_2H_7^-$

(19) $B_3H_8^-$

~150 pm
~120 pm
177 pm
180 pm
105–120 pm

(20) $B_9H_{14}^-$

(21) $B_{10}H_{14}^{2-}$

(22)   Proposed structure of $B_{15}H_{23}$

(23)   Proposed structure of *conjuncto*-B_8H_{18} (8032) comprising two B_4H_9 units linked by a direct B–B bond

(24)   Structures of the three isomers of $B_{10}H_{16}$. The 1,1′ isomer comprises two pentaborane(9) groups linked in eclipsed configuration via the apex boron atoms to give overall D_{4h} symmetry; the B–B bond distances are 174 pm for the linking bond, 176 pm for the slant edge of the pyramids, and 171 pm for the basal boron atoms

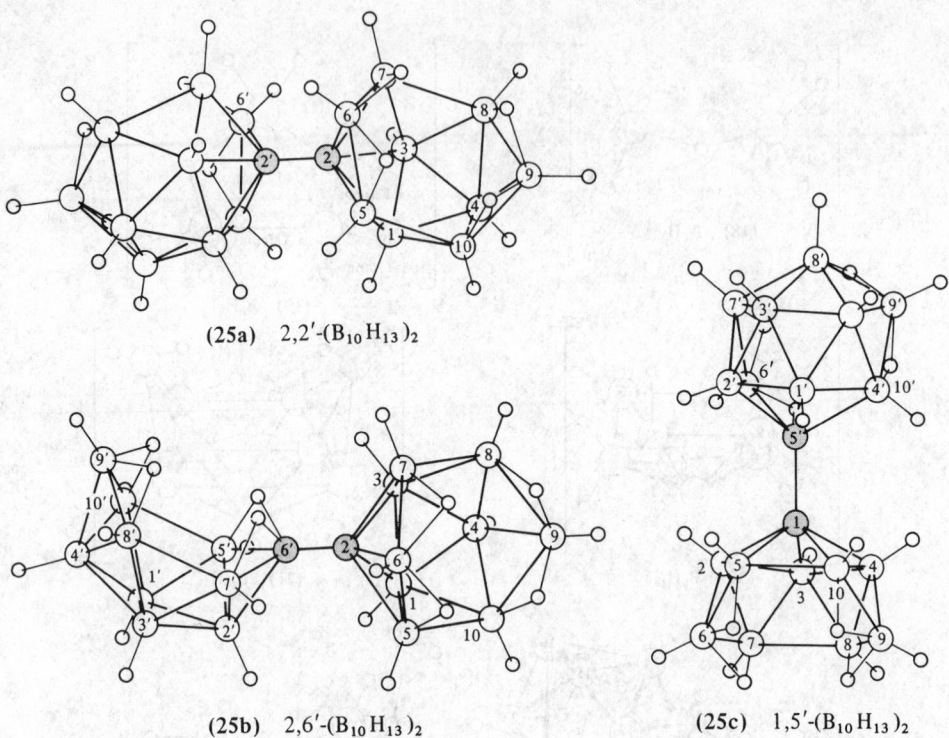

(25a)   2,2'-(B$_{10}$H$_{13}$)$_2$

(25b)   2,6'-(B$_{10}$H$_{13}$)$_2$

(25c)   1,5'-(B$_{10}$H$_{13}$)$_2$

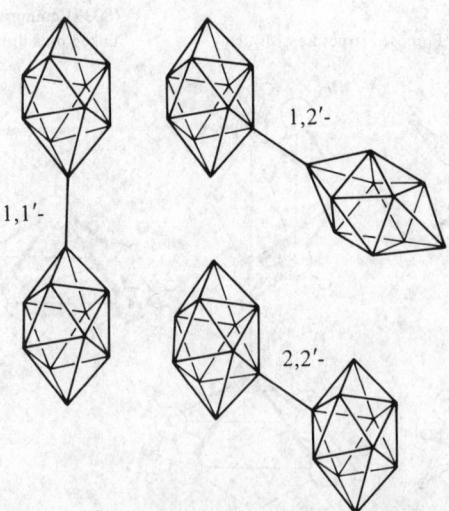

(26)   Proposed structures for the three isomers of [B$_{20}$H$_{18}$]$^{4-}$; terminal hydrogen
atoms omitted for clarity. (See also p. 203.)

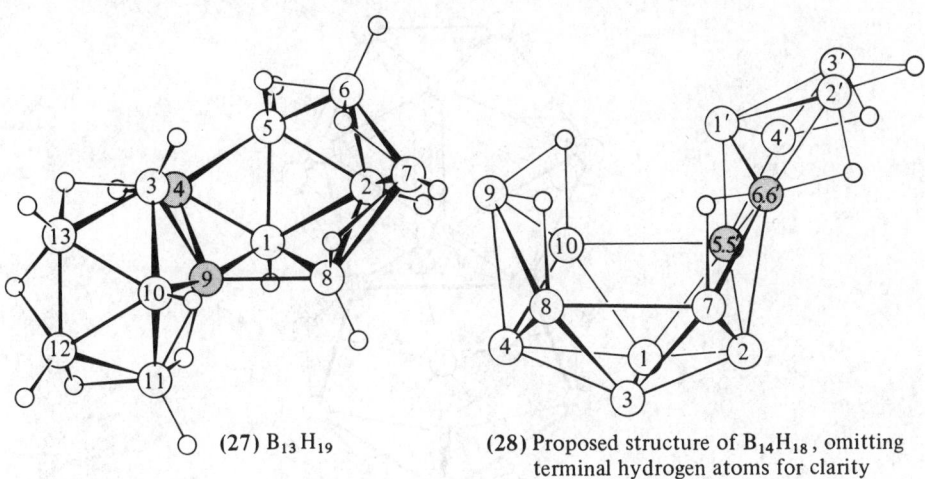

(27) $B_{13}H_{19}$

(28) Proposed structure of $B_{14}H_{18}$, omitting terminal hydrogen atoms for clarity

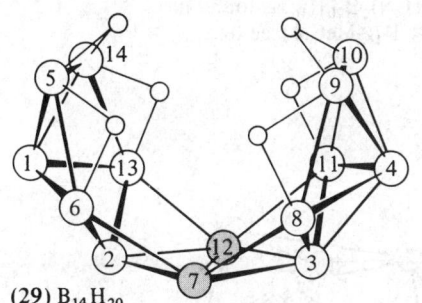

(29) $B_{14}H_{20}$
Terminal hydrogen atoms have been omitted for clarity

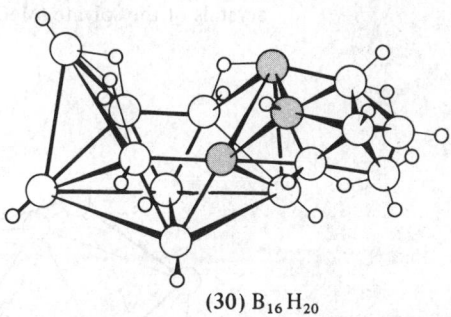

(30) $B_{16}H_{20}$

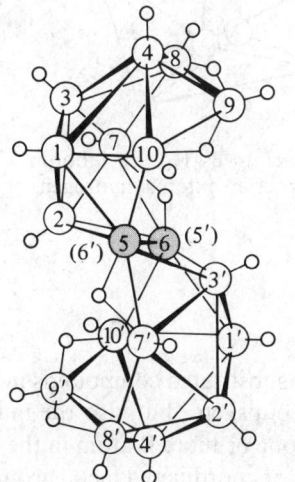

(31) n-$B_{18}H_{22}$ (centrosymmetric)

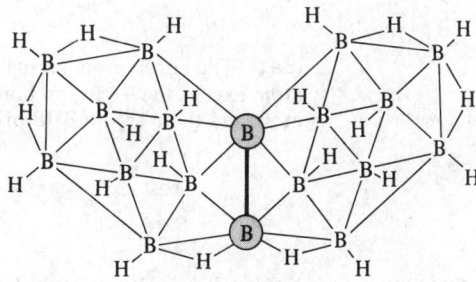

(32) Plane projection of the structure of i-$B_{18}H_{22}$. The two decaborane units are fused at the 5(7′) and 6(6′) positions to give a non-centrosymmetric structure with C_2 symmetry

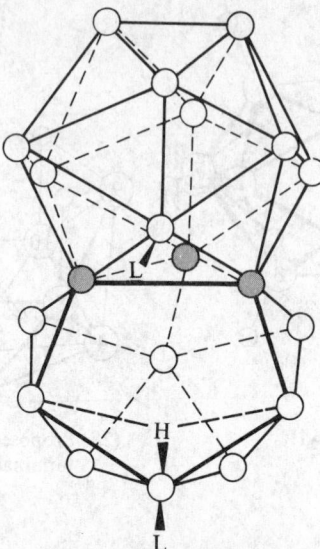

(33) Molecular structure of $(MeCN)_2B_{20}H_{16}$ as found in crystals of the solvate $(MeCN)_2B_{20}H_{16} \cdot MeCN$ (see text)

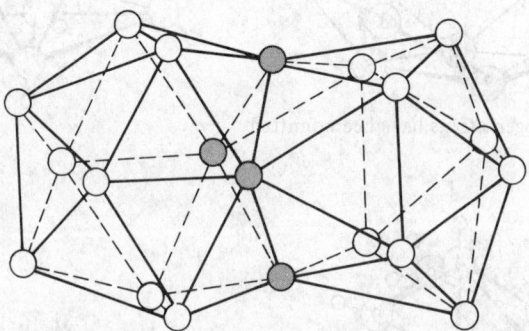

(34) The boron atom arrangement in *closo*-$B_{20}H_{16}$. Each boron atom except the 4 "fusion borons" carries an external hydrogen atom and there are no BHB bridges

Derivatives of the boranes include not only simple substituted compounds in which H has been replaced by halogen, OH, alkyl, or aryl groups, etc., but also the much more diverse and numerous class of compounds in which one or more B atom in the cluster is subrogated by C, P, S, or a wide range of metal atoms or coordinated metal groups. These will be considered in later sections.

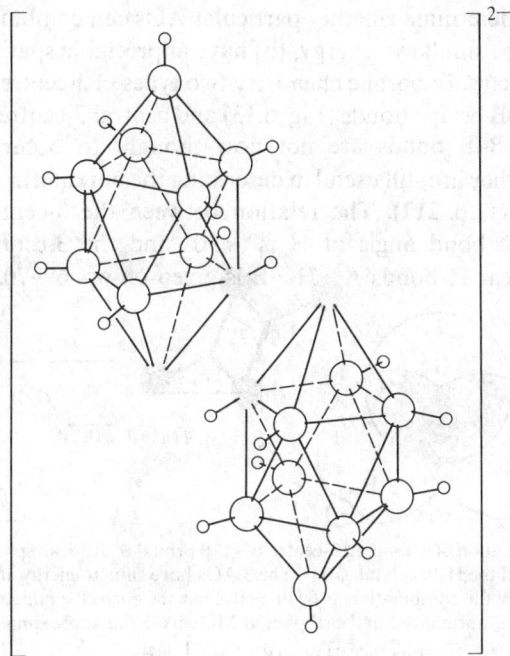

(35)   Structure of the $B_{20}H_{18}^{2-}$ ion. The two 3-centre BBB bonds joining the 2 $B_{10}H_9^-$ units are shown by broad shaded lines

6.4.2   *Structure, bonding, and topology*

The definitive structural chemistry of the boranes began in 1948 with the X-ray crystallographic determination of the structure of decaborane(14); this showed the presence of 4 bridging H atoms and an icosahedral fragment of 10 B atoms. This was rapidly followed in 1951 by the unequivocal demonstration of the H-bridged structure of diborane(6) and by the determination of the structure of pentaborane(9). Satisfactory theories of bonding in boranes date from the introduction of the concept of the 3-centre 2-electron B–H–B bond by H. C. Longuet-Higgins in 1949; he also extended the principle of 3-centre bonding and multicentre bonding to the higher boranes. These ideas have been extensively developed and refined by W. N. Lipscomb and his group during the past 25 y.[10, 11]

In simple covalent bonding theory molecular orbitals (MOs) are formed by the linear combination of atomic orbitals (LCAO); for example, 2 AOs can combine to give 1 bonding and 1 antibonding MO and the orbital of lower energy will be occupied by a pair of electrons. This is a special case of a more general situation in which a number of AOs are combined together by the LCAO method to construct an equal number of MOs of differing energies, some of which will be bonding, some possibly nonbonding and some antibonding. In this way 2-centre, 3-centre, and multicentre orbitals can be envisaged. The

[10] W. N. Lipscomb, Advances in theoretical studies of boron hydrides and carboranes, Chap. 2 in ref. 7, pp. 30–78.

[11] W. N. Lipscomb, *Boron Hydrides*, Benjamin, New York, 1963, 275 pp.

three criteria that determine whether particular AOs can combine to form MOs are that the AOs must (a) be similar in energy, (b) have appreciable spatial overlap, and (c) have appropriate symmetry. In borane chemistry two types of 3-centre bond find considerable application: B–H–B bridge bonds (Fig. 6.13) and central 3-centre BBB bonds (Fig. 6.14). Open 3-centre B–B–B bonds are not now thought to occur in boranes and their anions[10] though they are still useful in describing the bonding in carbaboranes and other heteroatom clusters (p. 217). The relation between the 3-centre bond formation for B–H–B, where the bond angle at H is $\sim 90°$ and the 3-centre bond formation for approximately linear H bonds A—H···B is given on pp. 69–70.

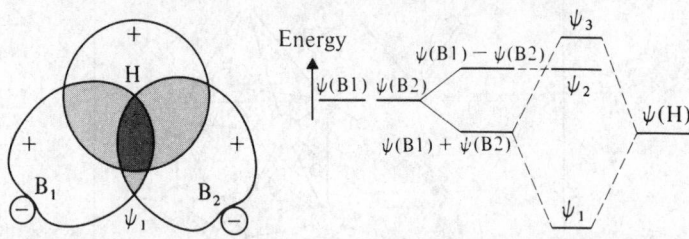

FIG. 6.13   Formation of a bonding 3-centre B–H–B orbital ψ_1 from an sp^x hybrid orbital on each of B(1), B(2) and the H 1s orbital, $\psi(H)$. The 3 AOs have similar energy and appreciable spacial overlap, but only the combination $\psi(B1)+\psi(B2)$ has the correct symmetry to combine linearly with $\psi(H)$. The 3 normalized and orthogonal MOs have the approximate form:

bonding $\psi_1 \approx \tfrac{1}{2}[\psi(B1)+\psi(B2)] + \dfrac{1}{\sqrt{2}}\psi(H)$

nonbonding
(or antibonding) $\psi_2 \approx \dfrac{1}{\sqrt{2}}[\psi(B1)-\psi(B2)]$

antibonding $\psi_3 \approx \tfrac{1}{2}[\psi(B1)+\psi(B2)] - \dfrac{1}{\sqrt{2}}\psi(H)$

Localized 3-centre bond formalism can readily be used to rationalize the structure and bonding in most of the non-*closo*-boranes. This is illustrated for some typical *nido*- and *arachno*-boranes in the following plane-projection diagrams which use an obvious symbolism for normal 2-centre bonds: B–B O—O, B–H$_t$ O—●, (t = terminal), central 3-centre bonds ⟨symbol⟩ , and B–H$_\mu$–B bridge bonds ⟨symbol⟩ . It is particularly

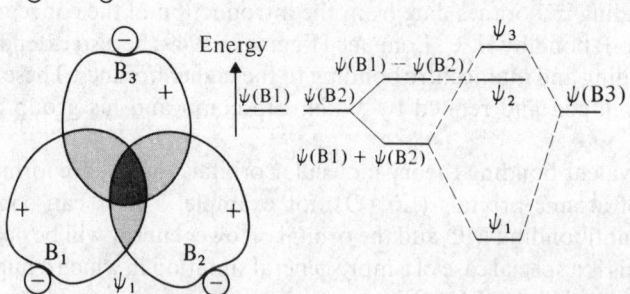

FIG. 6.14   Formation of a bonding, central 3-centre bond ψ_1 and schematic representation of the relative energies of the 3 molecular orbitals $\psi_1, \psi_2,$ and ψ_3. The approximate analytic form of these MOs is:

bonding $\psi_1 \approx [\psi(B1)+\psi(B2)+\psi(B3)]/\sqrt{3}$
antibonding $\psi_2 \approx [\psi(B1)-\psi(B2)]/\sqrt{2}$
antibonding $\psi_3 \approx [\psi(B1)+\psi(B2)-2\psi(B3)]/\sqrt{6}$

important to realize that the latter two symbols each represent a single (3-centre) bond involving one pair of electrons. As each B atom has 3 valence electrons, and each $B-H_t$ bond requires 1 electron from B and one from H, it follows that each $B-H_t$ groups can contribute the remaining 2 electrons on B towards the bonding of the cluster (including B–H–B bonds), and likewise each BH_2 group can contribute 1 electron for cluster bonding. The overall bonding is sometimes codified in a 4-digit number, the so-called *styx* number, where *s* is the number of B–H–B bonds, *t* is the number of 3-centre BBB bonds, *y* the number of 2-centre BB bonds, and *x* the number of BH_2 groups.[11] Examples:

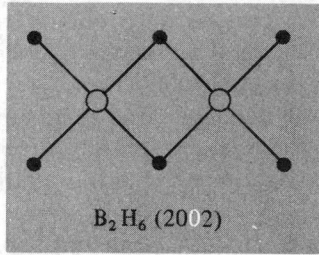

B_2 H_6 (2002)

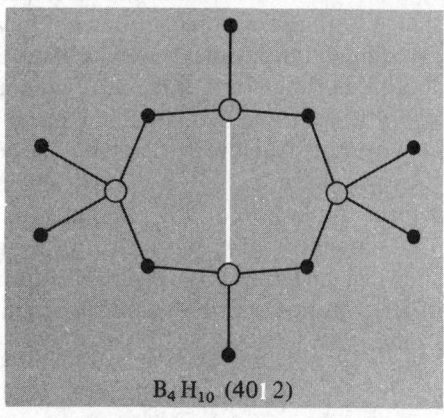

B_4 H_{10} (4012)

Each terminal BH_2 group and each (bridging) H_μ contributes 1 electron to the bridging; these 4 electrons just fill the two B–H–B bonds.

Each of the 4 B and $4H_\mu$ contribute 1 electron to the B–H–B bonds, i.e. 4 pairs of electrons for the 4 (3-centre) bonds. The 2 "hinge" BH_t groups each have 1 remaining electron and 1 orbital which interact to give the 2-centre B–B bond.

In B_5H_9 the bonding can be thought of as involving the structure shown and 3 other equivalent structures in which successive pairs of basal B atoms are combined with the apex B in a 3-centre bond.

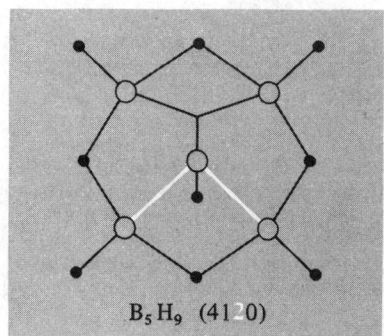

B_5 H_9 (4120)

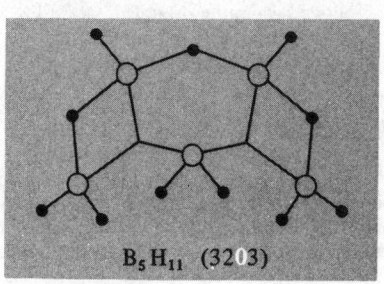

B_5 H_{11} (3203)

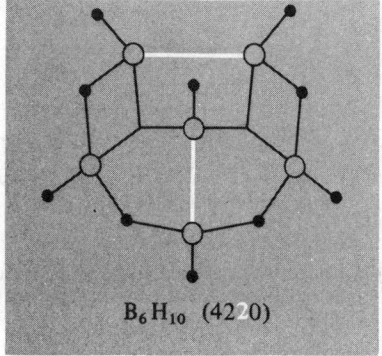

B_6 H_{10} (4220)

Electron counting and orbital bookkeeping can easily be checked in these diagrams: as each B has 4 valency orbitals (s + 3p) there should be 4 lines emanating from each open circle; likewise, as each B atom contributes 3 electrons in all and each H atom contributes 1 electron, the total number of valence electrons for a borane of formula B_nH_m is $(3n + m)$ and the number of bonds shown in the structure should be just half this. It follows, too, that the number of electron-pair bonds in the molecule is n plus the sum of the individual *styx* numbers (e.g. 13 for B_5H_{11}, 14 for B_6H_{10}) and this constitutes a further check.†
An appropriate number of additional electrons should be added for anionic species.

For *closo*-boranes and for the larger open-cluster boranes it becomes increasingly difficult to write a simple satisfactory localized orbital structure, and a full MO treatment is required. Intermediate cases, such as B_5H_9, require several "resonance hybrids" in the localized orbital formation and, by the time $B_{10}H_{14}$ is considered there are 24 resonance hybrids, even assuming that no open 3-centre B–B–B bonds occur. The best single compromise structure in this case is the (4620) arrangement shown at the top of p. 184, but the open 3-centre B–B–B bonds can be avoided if "fractional" central 3-centre bonds replace the B–B and B–B–B bonds in pairs:[10]
A simplified MO approach to the bonding in *closo*-$B_6H_6{}^{2-}$ (structure 1, p. 172) is shown in the Panel. It is a general feature of *closo*-$B_nH_n{}^{2-}$ anions that there are no B–H–B or BH_2 groups and the $4n$ boron atomic orbitals are always distributed as follows:

> n in the $n(B–H_t)$ bonding orbitals
> $(n + 1)$ in framework bonding MOs
> $(2n - 1)$ in nonbonding and antibonding framework MOs

MO Description of Bonding in *closo*-$B_6H_6{}^{2-}$

Closo $B_6H_6{}^{2-}$ (structure 1) has a regular octahedral cluster of 6 B atoms surrounded by a larger octahedron of 6 radially disposed H atoms. Framework MOs for the B_6 cluster are constructed (LCAO) using the 2s, $2p_x$, $2p_y$, and $2p_z$ boron AOs. The symmetry of the octahedron suggests the use of sp hybrids directed radially outwards and inwards from each B along the cartesian axes (see figure on facing page) and 2 pure p orbitals at right angles to these (i.e. oriented tangentially to the B octahedron). These sets of AOs are combined, with due regard to symmetry, to give the MOs indicated. In all, the 24 AOs on the 6 B combine to give 24 MOs of which 7 (i.e. $n + 1$) are bonding framework MOs, 6 are used to form B–H_t bonds, and the remaining 11 are antibonding. The relative energies of these orbitals are shown schematically below:

Energy ↑

t'_{1u} — — — —
e_g — — — —
t_{1g} — — — —
t_{2u} — — — ⎫ 11 antibonding orbitals

e'_g — — — ⎫ 6 outward-pointing orbitals
t''_{1u} — — — ⎬ of correct symmetry to combine
a'_{1g} — ⎭ with 6 H_t to form 6 B–H_t bonds

t_{2g} — — — ⎫ 7 framework (B_6) bonding MOs
t_{1u} — — — ⎬
a_{1g} — ⎭

Continued

---

† Further checks, which can readily be verified from the equations of balance, are (a) the number of atoms in a neutral borane molecule $= 2(s + t + y + x)$, and (b) there are as many framework electrons as there are atoms in a neutral borane B_nH_m since each BH group supplies 2 electrons and each of the $(m - n)$ "extra" H atoms supplies 1 electron, making $n + m$ in all.

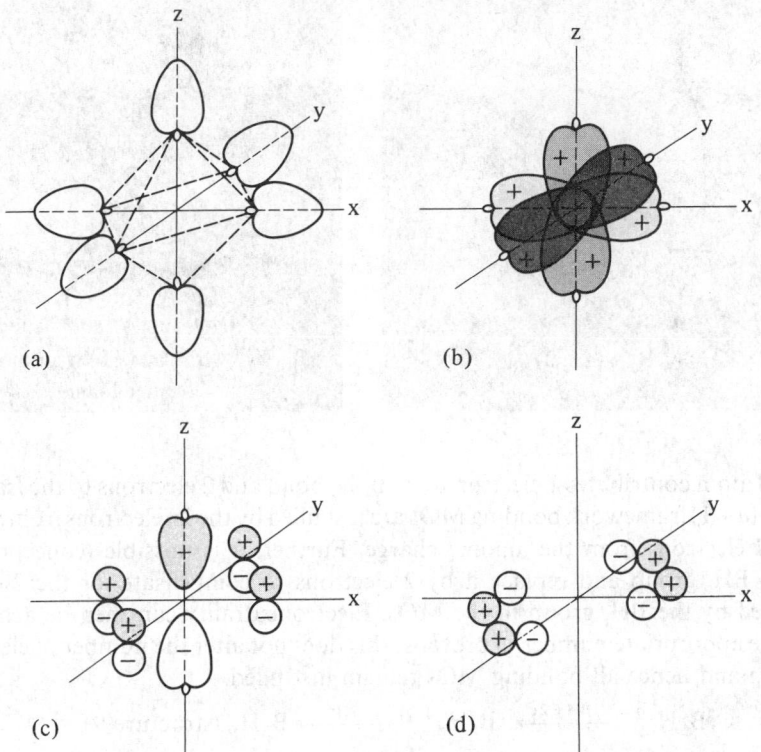

Symmetry of orbitals on the B_6 octahedron. (a) Six outward-pointing (sp) orbitals used for σ bonding to $6H_t$. (b) Six inward-pointing (sp) orbitals used to form the a_{1g} framework bonding molecular orbital. (c) Components for one of the t_{1u} framework bonding molecular orbitals—the other two molecular orbitals are in the yz and zx planes. (d) Components for one of the t_{2g} framework bonding molecular orbitals—the other two molecular orbitals are in the yz and zx planes.

The diagrams also indicate why neutral *closo*-boranes B_nH_{n+2} are unknown since the 2 anionic charges are effectively located in the low-lying inwardly directed a_{1g} orbital which has no overlap with protons outside the cluster (e.g. above the edges or faces of the B_6 octahedron). Replacement of the $6H_t$ by 6 further B_6 builds up the basic three-dimensional network of hexaborides MB_6 (p. 169) just as replacement of the $4H_t$ in CH_4 begins to build up the diamond lattice.

The diagrams also serve, with minor modification, to describe the bonding in isoelectronic species such as *closo*-$CB_5H_6^-$, 1,2-*closo*-$C_2B_4H_6$, 1,6-*closo*-$C_2B_4H_6$, etc. (p. 203). Similar though more complex, diagrams can be derived for all *closo*-$B_nH_n^{2-}$ ($n = 6-12$); these have the common feature of a low lying a_{1g} orbital and n other framework bonding MOs; in each case, therefore $(n+1)$ pairs of electrons are required to fill these orbitals as indicated in Wade's rules (p. 185). It is a triumph for MO theory that the existence of $B_6H_6^{2-}$ and $B_{12}H_{12}^{2-}$ were predicted by H. C. Longuet-Higgins in 1954-5,[12] a decade before $B_6H_6^{2-}$ was first synthesized and some 5 y before the (accidental) preparation of $B_{10}H_{10}^{2-}$ and $B_{12}H_{12}^{2-}$ were reported.[13, 14]

[12] H. C. Longuet-Higgins and M. de V. Roberts, The electronic structure of an icosahedron of boron atoms, *Proc. R. Soc.* A, **230**, 110–19 (1955); see also idem ibid. A, **224**, 336–47 (1954).

[13] J. L. Boone, Isolation of the hexahydroclovohexaborate(2−) anion, $B_6H_6^{2-}$, *J. Am. Chem. Soc.* **86**, 5036 (1964).

[14] M. F. Hawthorne and A. R. Pittochelli, The reactions of bis(acetonitrile)-decaborane with amines, *J. Am. Chem. Soc.* **81**, 5519 (and also 5833–4) (1959); and The isolation of the icosahedral $B_{12}H_{12}^{2-}$ ion, *J. Am. Chem. Soc.* **82**, 3228–9 (1960).

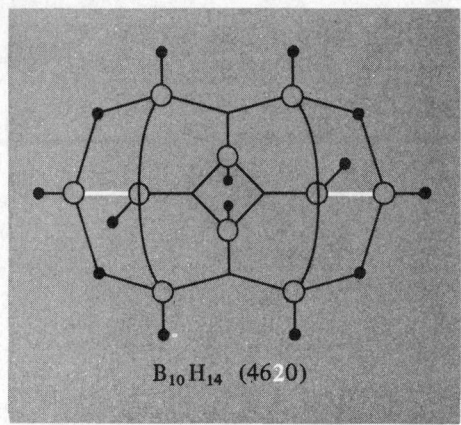

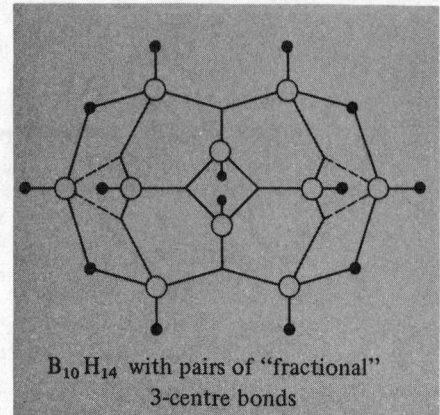

$B_{10}H_{14}$ (4620)

$B_{10}H_{14}$ with pairs of "fractional" 3-centre bonds

As each B atom contributes 1 electron to its $B–H_t$ bond and 2 electrons to the framework MOs, the $(n+1)$ framework bonding MOs are just filled by the $2n$ electrons from nB atoms and the 2 electrons from the anionic charge. Further, it is possible (conceptually) to remove a BH_t group and replace it by 2 electrons to compensate for the 2 electrons contributed by the BH_t group to the MOs. Electroneutrality can then be achieved by adding the appropriate number of protons; this does not alter the number of electrons in the system and hence all bonding MOs remain just filled.

$$B_6H_6^{2-} \xrightarrow{-BH+2e^-} \{B_5H_5^{4-}\} \xrightarrow{4H^+} B_5H_9 \text{ (structure 9)}$$

(structure 1)
$$\Big\downarrow \begin{matrix} (-BH+2e^-) \\ +2H^+ \end{matrix}$$

$$B_6H_6^{2-} \xrightarrow{-2BH+4e^-} \{B_4H_4^{6-}\} \xrightarrow{6H^+} B_4H_{10} \text{ (structure 13)}$$

The structural interrelationship of all the various *closo*-, *nido*-, and *arachno*-boranes thus becomes evident; a further example is shown below:

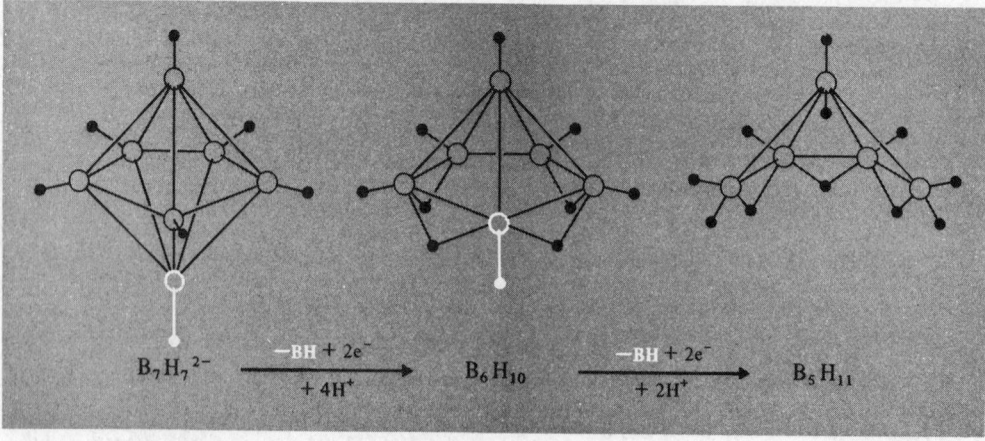

$B_7H_7^{2-} \xrightarrow[+4H^+]{-BH+2e^-} B_6H_{10} \xrightarrow[+2H^+]{-BH+2e^-} B_5H_{11}$

These relationships were codified in 1971 by K. Wade in a set of rules which have been extremely helpful not only in rationalizing known structures, but also in suggesting the probable structures of new species.[15] Wade's rules can be stated in extended form as follows:

closo-borane anions have the formula $B_nH_n{}^{2-}$; the B atoms occupy all n corners of an n-cornered triangulated polyhedron, and the structures require $(n+1)$ pairs of framework bonding electrons;

nido-boranes have the formula B_nH_{n+4} with B atoms at n corners of an $(n+1)$ cornered polyhedron; they require $(n+2)$ pairs of framework-bonding electrons;

arachno-boranes: B_nH_{n+6}, n corners of an $(n+2)$ cornered polyhedron, requiring $(n+3)$ pairs of framework-bonding electrons;

hypho-boranes: B_nH_{n+8}: n corners of an $(n+3)$ cornered polyhedron, requiring $(n+4)$ pairs of framework-bonding electrons.

The rules can readily be extended to isoelectronic anions and carbaboranes $(BH = B^- = C)$ and also to metalloboranes (p. 198), metallocarbaboranes (p. 217) and even to metal clusters themselves, though they become less reliable the further one moves away from boron in atomic size, ionization energy, electronegativity, etc.

More sophisticated and refined calculations lead to orbital populations and electron charge distributions within the borane molecules and to predictions concerning the sites of electrophilic and nucleophilic attack. In general, the highest electron charge density (and the preferred site of electrophilic attack) occurs at apical B atoms which are furthest removed from open faces; conversely the lowest electron charge density (and the preferred site of nucleophilic attack) occurs on B atoms involved in B–H–B bonding. The consistency of this correlation implies that the electron distribution in the activated complex formed during reaction must follow a similar sequence to that in the ground state. Bridge H atoms tend to be more acidic than terminal H atoms and are the ones first lost during the formation of anions in acid–base reactions.

6.4.3 *Properties of boranes*

Boranes are colourless, diamagnetic, molecular compounds of moderate to low thermal stability. The lower members are gases at room temperature but with increasing molecular weight they become volatile liquids or solids (Table 6.2); bps are approximately the same as those of hydrocarbons of similar molecular weight. The boranes are all endothermic and their free energy of formation ΔG_f° is also positive; their thermodynamic instability results from the exceptionally strong interatomic bonds in both elemental B and H_2 rather than the inherent weakness of the B–H bond. (In this the boranes resemble the hydrocarbons.) Thus it has been estimated that typical bond energies in boranes are B–H_t 380, B–H–B 440, B–B 330, and B–B–B 380 kJ mol^{-1}, compared with a bond energy of 436 kJ mol^{-1} for H_2 and a heat of atomization of crystalline boron of 555 kJ per mol of B atoms (i.e. 1110 kJ per mole of 2 B atoms). An alternative set of self-consistent bond enthalpies is also available.[15a]

Boranes are extremely reactive and several are spontaneously flammable in air. *Arachno*-boranes tend to be more reactive (and less stable to thermal decomposition) than

---

[15] K. WADE, Structural and bonding patterns in cluster chemistry, *Adv. Inorg. Chem. Radiochem.* **18**, 1–66 (1976).

[15a] C. E. HOUSECROFT and K. WADE, Bond length-based enthalpies for *nido* and *arachno* boranes B_nH_{n+4} and B_nH_{n+6}, *Inorg. Nucl. Chem. Letters* **15**, 339–42 (1979).

TABLE 6.2   *Properties of some boranes*

Nido-boranes				Arachno-boranes			
Compound	mp	bp	$\Delta H_f^\circ/\text{kJ mol}^{-1}$	Compound	mp	bp	$\Delta H_f^\circ/\text{kJ mol}^{-1}$
B_2H_6	$-164.9°$	$-92.6°$	36	B_4H_{10}	$-120°$	$18°$	58
B_5H_9	$-46.8°$	$60.0°$	54	B_5H_{11}	$-122°$	$65°$	67 (or 93)
B_6H_{10}	$-62.3°$	$108°$	71	B_6H_{12}	$-82.3°$	$\sim85°$ (extrap)	111
B_8H_{12}	Decomp	above $-35°$	–	B_8H_{14}	Decomp	above $-30°$	—
$B_{10}H_{14}$	$99.5°$	$213°$	32	$n\text{-}B_9H_{15}$	$2.6°$	$28°/$ 0.8 mmHg	—

nido-boranes and reactivity also diminishes with increasing mol wt. *Closo*-borane anions are exceptionally stable and their general chemical behaviour has suggested the term "three-dimensional aromaticity".

Boron hydrides are extremely versatile chemical reagents but the very diversity of their reactions makes a general classification unduly cumbersome. Instead, the range of behaviour will be illustrated by typical examples taken from the chemistry of the three most studied boranes: B_2H_6, B_5H_9, and $B_{10}H_{14}$. Nearly all boranes are highly toxic when inhaled or absorbed through the skin though they can be safely and conveniently handled with relatively minor precautions.

6.4.4   *Chemistry of diborane,* B_2H_6

Diborane occupies a special place because all the other boranes are prepared from it (directly or indirectly); it is also one of the most studied and synthetically useful reagents in the whole of chemistry.[1, 16] B_2H_6 gas can most conveniently be prepared in small quantities by the reaction of I_2 on $NaBH_4$ in diglyme [$(MeOCH_2CH_2)_2O$], or by the reaction of a solid tetrahydroborate with an anhydrous acid:

$$2NaBH_4 + I_2 \xrightarrow[(98\% \text{ yield})]{\text{diglyme}} B_2H_6 + 2NaI + H_2$$

$$2NaBH_4(c) + 2H_3PO_4(l) \xrightarrow{(70\% \text{ yield})} B_2H_6(g) + 2NaH_2PO_4(c) + 2H_2(g)$$

When B_2H_6 is to be used as a reaction intermediate without the need for isolation or purification, the best procedure is to add Et_2OBF_3 to $NaBH_4$ in a polyether such as diglyme:

$$3NaBH_4 + 4Et_2OBF_3 \xrightarrow[25°]{\text{diglyme}} 2B_2H_6(g) + 3NaBF_4 + 4Et_2O$$

On an industrial scale gaseous BF_3 can be reduced directly with NaH at $180°$ and the product trapped out as it is formed to prevent subsequent pyrolysis:

$$2BF_3(g) + 6NaH(c) \xrightarrow{180°} B_2H_6(g) + 6NaF(c)$$

---

[16] L. H. LONG, Chap. 22 in *Mellor's Comprehensive Treatise on Inorganic and Theoretical Chemistry*, Vol. 5, Supplement 2, Part 2, pp. 52–162, Longmans, London, 1981.

Care should be taken in these reactions because B_2H_6 is spontaneously flammable; it has a higher heat of combustion per unit weight of fuel than any other substance except H_2, BeH_2, and $Be(BH_4)_2$:

$$B_2H_6 + 3O_2 = B_2O_3 + 3H_2O; \quad \Delta H° = -2165 \text{ kJ mol}^{-1}$$

The pyrolysis of B_2H_6 in sealed vessels at temperatures above 100° is exceedingly complex and a variety of products is formed depending on conditions. The initiating step is the unimolecular dissociation equilibrium:

$$B_2H_6 \xrightleftharpoons{\text{fast}} 2\{BH_3\}$$

(In this and subsequent reactions unstable intermediates that have but transitory existence are placed in brackets {}.) The $\{BH_3\}$ reacts rapidly with B_2H_6 to give $\{B_3H_9\}$ and this species then decomposes more slowly in a rate-controlling step thus explaining the observed order of the reaction which is 1.5 with respect to the concentration of B_2H_6.[17]

$$\{BH_3\} + B_2H_6 \xrightleftharpoons{\text{fast}} \{B_3H_9\}$$

$$\{B_3H_9\} \xrightarrow{\text{rate controlling}} \{B_3H_7\} + H_2$$

The first stable intermediate, B_4H_{10}, is then formed followed by B_5H_{11}:

$$\{BH_3\} + \{B_3H_7\} \rightleftharpoons B_4H_{10}$$

$$B_2H_6 + \{B_3H_7\} \longrightarrow \{BH_3\} + B_4H_{10} \rightleftharpoons B_5H_{11} + H_2$$

A complex series of further steps gives B_5H_9, B_6H_{10}, B_6H_{12}, and higher boranes, culminating in $B_{10}H_{14}$ as the most stable end product, together with polymeric materials BH_x and a trace of icosaborane $B_{20}H_{26}$.

Careful control of temperature, pressure, and reaction time enable the yield of the various intermediate boranes to be optimized. For example, B_4H_{10} is best prepared by storing B_2H_6 under pressure at 25° for 10 days; this gives a 15% yield and quantitative conversion according to the overall reaction:

$$2B_2H_6 \longrightarrow B_4H_{10} + H_2$$

B_5H_{11} can be prepared in 70% yield by the reaction of B_2H_6 and B_4H_{10} in a carefully dimensioned hot/cold reactor at $+120°/-30°$:

$$2B_4H_{10} + B_2H_6 \rightleftharpoons 2B_5H_{11} + 2H_2$$

(Alternative high-yield syntheses of these various boranes via hydride-ion abstraction from borane anions by BBr_3 and other Lewis acids have recently been devised.[17a])

B_5H_9 can be prepared by passing a 1:5 mixture of B_2H_6 and H_2 at subatmospheric

[17] T. P. FEHLNER, Gas phase reactions of borane, BH_3, Chap. 4 in ref. 7, pp. 175–96.

[17a] M. A. TOFT, J. B. LEACH, F. L. HIMPSL, and S. G. SHORE, New, systematic syntheses of boron hydrides via hydride ion abstraction reactions: Preparation of B_2H_6, B_4H_{10}, B_5H_{11}, and $B_{10}H_{14}$, *Inorg. Chem.* **21**, 1952–57 (1982).

pressure through a furnace at $250°$ with a 3-s residence time (or at $225°$ with a 15-s residence time); there is a 70% yield and 30% conversion. Alternatively B_2H_6 can be pyrolysed for 2.5 days in a hot/cold reactor at $180°/-80°$.

The bridge bonds in B_2H_6 are readily cleaved, even by weak ligands, to give either symmetrical or unsymmetrical cleavage products:

Symmetrical (homolytic)

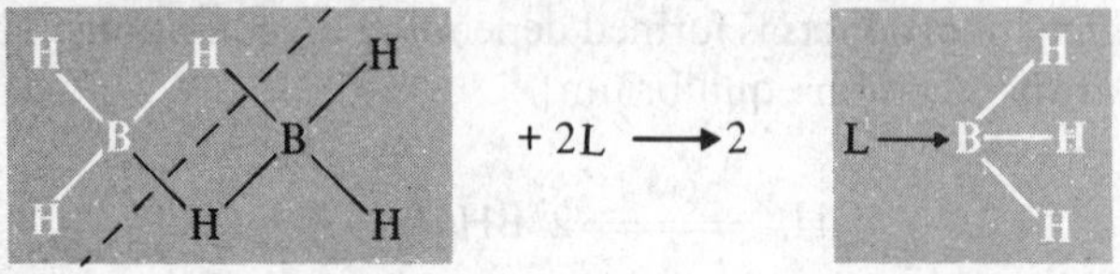

Unsymmetrical (heterolytic)

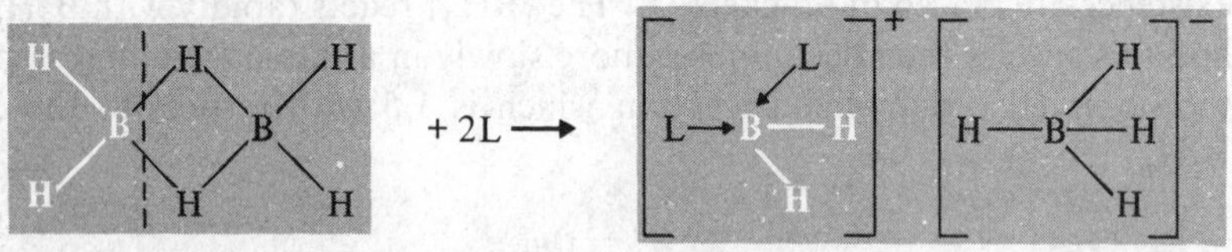

The factors governing the course of these reactions are not fully understood but steric effects play some role;[18] e.g. NH_3, $MeNH_2$, and Me_2NH give unsymmetrical cleavage products whereas Me_3N gives the symmetric cleavage product Me_3NBH_3. Symmetrical cleavage is the more common mode and thermochemical and spectroscopic data lead to the following sequence of adduct stability for LBH_3:

$$PF_3 < CO < Et_2O < Me_2O < C_4H_8O < C_4H_8S < Et_2S < Me_2S < py < Me_3N < H^-$$

The relative stability of the sulfide adducts is notable and many other complexes with N, P, O, S, etc., donor atoms are known. The ligand H^- is a special case since it gives the symmetrical tetrahedral ion BH_4^- isoelectronic with CH_4. The BH_4^- ion itself provides a rare example of a ligand that can be unidentate, bidentate, or tridentate as illustrated by its complexes in Fig. 6.15.

In solution there is normally a rapid interchange between H_t and H_μ in these structures and all H atoms appear the same on an nmr time scale; indeed, this property of fluxionality, which has been increasingly recognized to occur in many inorganic and organometallic systems, was first observed (1955) on the tris-bidentate complex $[Al(\eta-BH_4)_3]$[18a] The properties and synthetic utility of tetrahydroborates are summarized in the Panel.

In addition to pyrolysis and cleavage reactions, B_2H_6 undergoes a wide variety of substitution, redistribution, and solvolytic reactions of which the following are representative:

$$B_2H_6 + HCl \longrightarrow B_2H_5Cl + H_2$$

Continued on p. 191

[18] S. G. SHORE, *Nido-* and *arachno*-boron hydrides, Chap. 3 in ref. 7, pp. 79–174.
[18a] R. A. OGG and J. D. RAY, Nuclear magnetic resonance spectrum and molecular structure of aluminium borohydride, *Disc. Faraday Soc.* **19**, 239–46 (1955).

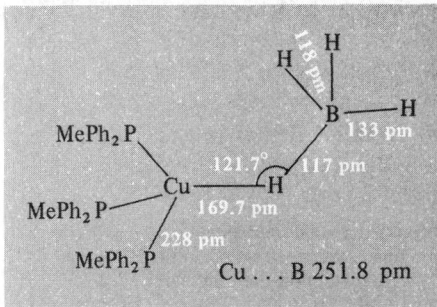

(a) $[Cu^I(\eta^1\text{-}BH_4)(PMePh_2)_3]$

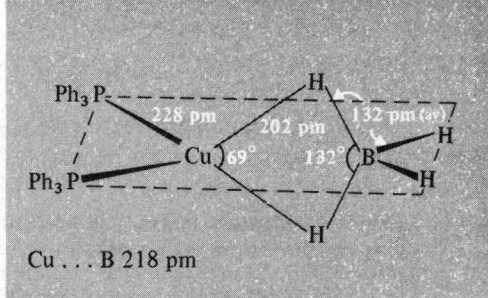

(b) $[Cu^I(\eta^2\text{-}BH_4)(PPh_3)_2]$

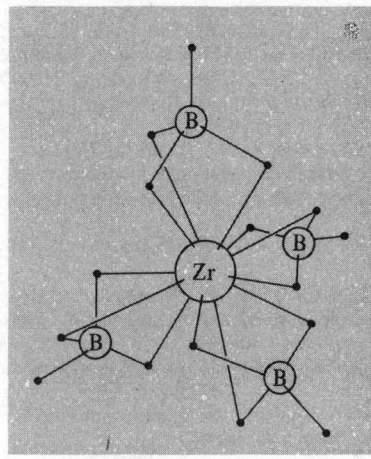

(c) $[Zr^{IV}(\eta^3\text{-}BH_4)_4]$

Fig. 6.15 Examples of BH_4^- as a monodentate, bidentate, and tridentate ligand. Note that, in solution, the BH_4 group in complex (a) becomes bidentate and 1 phosphine ligand is eliminated. Other examples of tridentate BH_4^- are the monomeric 12-coordinate complexes $M(BH_4)_4$ (M = Hf, Np, Pu), whereas the polymeric complexes $M(BH_4)_4$ (M = Th, Pa, U) feature 14-coordinate metal atoms in which M is surrounded by 2 η^3-BH_4^- and 4 bridging η^2-BH_4^- groups.[19]

Tetrahydroborates $M(BH_4)_x$[1, 20]

Tetrahydroborates were first reported in 1940 (M = Li, Be, Al) and since then have been widely exploited as versatile nucleophilic reducing agents which attack centres of low electron density (cf. electrophiles such as B_2H_6 and LBH_3 which attack electron-rich centres). The most stable are the alkali derivatives MBH_4: $LiBH_4$ decomposes above $\sim 380°$ but the others (Na–Cs) are stable up to $\sim 600°$. MBH_4 are readily soluble in water and many other coordinating solvents such as liquid ammonia, amines, ethers ($LiBH_4$), and polyethers ($NaBH_4$). They can be prepared by direct reaction of MH with either B_2H_6 or BX_3 at room temperature though the choice of solvent is often crucial, e.g.:

[19] R. H. BANKS, N. M. EDELSTEIN, B. SPENCER, D. H. TEMPLETON, and A. ZALKIN, Volatility and molecular structure of neptunium(IV) borohydride, *J. Am. Chem. Soc.* **102**, 620–3 (1980), and references therein.
[20] B. D. JAMES and M. G. H. WALLBRIDGE, Metal tetrahydroborates, *Prog. Inorg. Chem.* **11**, 97–231 (1970). A review with 616 references.

$$2LiH + B_2H_6 \xrightarrow{Et_2O} 2LiBH_4$$

$$2NaH + B_2H_6 \xrightarrow{diglyme} 2NaBH_4$$

$$4LiH + Et_2OBF_3 \longrightarrow LiBH_4 + 3LiF + Et_2O$$

$$4NaH + BCl_3 \xrightarrow{Al_2Et_6} NaBH_4 + 3NaCl$$

Reaction of MBH_4 with electronegative elements is also often crucially dependent on the solvent and on the temperature and stoichiometry of reagents. Thus $LiBH_4$ reacts with S at $-50°$ in the presence of Et_2O to give $Li[BH_3SH]$, whereas at room temperature the main products are Li_2S, $Li[B_3S_2H_6]$, and H_2; at $200°$ in the absence of solvent $LiBH_4$ reacts with S to give $LiBS_2$ and either H_2 or H_2S depending on whether S is in excess. Similarly, MBH_4 react with I_2 in cyclohexane at room temperature to give BI_3, HI, and MI, whereas in diglyme B_2H_6 is formed quantitatively (p. 186).

The product of reaction of BH_4^- with element halides depends on the electropositivity of the element. Halides of the electropositive elements tend to form the corresponding $M(BH_4)_x$, e.g. M = Be, Mg, Ca, Sr, Ba; Zn, Cd; Al, Ga, Tl^I; lanthanides; Ti, Zr, Hf, and U^{IV}. Halides of the less electropositive elements tend to give the hydride or a hydrido-complex since the BH_4 derivative is either unstable or non-existent: thus $SiCl_4$ gives SiH_4; PCl_3 and PCl_5 give PH_3; Ph_2AsCl gives Ph_2AsH; $[Fe(\eta^5\text{-}C_5H_5)(CO)_2Cl]$ gives $[Fe(\eta^5\text{-}C_5H_5)(CO)_2H]$, etc.

A particularly interesting reaction (and one of considerable commercial value in the BOROL process for the *in situ* bleaching of wood pulp) is the production of dithionite, $S_2O_4^{2-}$, from SO_2:

$$NaBH_4 + 8NaOH + 9SO_2 \xrightarrow{90\% \text{ yield}} 4Na_2S_2O_4 + NaBO_2 + 6H_2O$$

In reactions with organic compounds, $LiBH_4$ is a stronger (less selective) reducing agent than $NaBH_4$ and can be used, for example, to reduce esters to alcohols. $NaBH_4$ reduces ketones, acid chlorides, and aldehydes under mild conditions but leaves other functions (such as $-CN$, $-NO_2$, esters) untouched; it can be used as a solution in alcohols, ethers, dimethylsulfoxide, or even aqueous alkali (pH > 10). Perhaps the classic example of its selectivity is shown below where an aldehyde group is hydrogenated in high yield without any attack on the nitrogroup, the bromine atom, the olefinic bond, or the thiophene ring:

Industrial interest in $LiBH_4$, and particularly $NaBH_4$,[21] stems not only from their use as versatile reducing agents for organic functional groups and their use in the bleaching of wood pulp, but also for their application in the electroless (chemical) plating of metals. Traditionally, either sodium hypophosphite, NaH_2PO_2, or formaldehyde have been used (as in the silvering of glass), but $NaBH_4$ was introduced on an industrial scale in the early 1960s, notably for the deposition of Ni on metal or non-metallic substrates; this gives corrosion-resistant, hard, protective coatings, and is also useful for metallizing plastics prior to further electroplating or for depositing contacts in electronics. Chemical plating also achieves a uniform thickness of deposit independent of the geometric shape, however complicated. The standard reduction potential under the conditions employed (pH 14) is $E° = -1.24$ V:

$$BH_4^- + 8OH^- \longrightarrow H_2BO_3^- + 5H_2O + 8e^-$$

This is intermediate between -1.57 V for $H_2PO_2^-/HPO_3^{2-}$ and -1.11 V for $HCHO/HCO_2^-$; cf. $Ni(OH)_2/Ni$ -0.72 V and Ni^{2+}/Ni -0.25 V. In practice it is found that $\sim 5\%$ by weight of the deposit is boron and the overall equation can be written as:

$$10NiCl_2 + 8NaBH_4 + 17NaOH + 3H_2O \longrightarrow (3Ni_3B + Ni) + 5NaB(OH)_4 + 20NaCl + 17.5H_2$$

[21] R. WADE, Sodium borohydride and its derivatives, in R. THOMPSON (ed.), *Speciality Inorganic Chemistry*, Royal Soc. Chem., London, 1981, pp. 25–58; see also 74-page trade catalogue on "Sodium Borohydride", Thiokol/Ventron Division, Danvers, Ma, 1979.

It follows that 1 kg $NaBH_4$ precipitates ~ 2.2 kg "$Ni_{10}B$"; this is sufficient for a thickness of 25 μm over an area of 10.7 m^2.

Worldwide production of $NaBH_4$ now approaches 2000 tonnes pa, but it became available on such a scale only recently. The compound was discovered at the University of Chicago in 1942 by H. I. Schlesinger and H. C. Brown during their search for volatile uranium compounds, e.g. $U(BH_4)_4$, but slow declassification of this wartime work delayed publication until 1953. Commercial production of $NaBH_4$ reached its first peak in the mid 1950s during the US programme for high-energy fuels; the present resurgence of scale is a product of the last 15 y. For large-scale production the laboratory syntheses in the opening paragraph above are clearly inadequate. The preferred industrial process, introduced by Bayer in the early 1960s, is the borosilicate route which uses borax (or ulexite), quartz, Na, and H_2 under moderate pressure at 450–500°:

$$(Na_2B_4O_7 + 7SiO_2) + 16\,Na + 8H_2 \longrightarrow 4NaBH_4 + 7Na_2SiO_3$$

The resulting mixture is extracted under pressure with liquid NH_3 and the product obtained as a 98% pure powder (or pellets) by evaporation. An alternative route is:

$$B(OMe)_3 + 4NaH \xrightarrow{250-270} NaBH_4 + 3NaOMe$$

The resulting mixture is hydrolysed with water and the aqueous phase extracted with Pr^iNH_2.

$$B_2H_6 + 3Cl_2 \xrightarrow{\text{low temp}} 2BCl_3 + 6HCl \text{ (also with } F_2)$$

$$B_2H_6 + BCl_3 \longrightarrow B_2H_5Cl + BHCl_2 \text{ (also with } BBr_3)$$

$$B_2H_6 + BMe_3 \longrightarrow Me_nB_2H_{6-n} \ (n = 1\text{-}4)$$

$$B_2H_6 + PbMe_4 \longrightarrow B_2H_5Me + Me_3PbH$$

$$B_2H_6 + 6H_2O \longrightarrow 2B(OH)_3 + 6H_2$$

$$B_2H_6 + 6MeOH \longrightarrow 2B(OMe)_3 + 6H_2$$

$$B_2H_6 + 4MeOH \longrightarrow 2BH(OMe)_2 + 4H_2$$

$$B_2H_6 + NH_3 \longrightarrow H_2B \underset{\underset{H_2}{N}}{\overset{H}{\diamondsuit}} BH_2 + H_2$$

$$1.5B_2H_6 + 3NH_3 \xrightarrow{90°} cyclo\text{-}(HNBH)_3 + 6H_2$$

$$B_2H_6 + 3NH_3 \xrightarrow{\text{heat}} \tfrac{1}{n}[B_2(NH)_3]_n + 6H_2$$

$$2B_2H_6 + 4Me_2PH \longrightarrow cyclo\text{-}(Me_2PBH_2)_4 + 4H_2 \text{ (also trimer)}$$

$$B_2H_6 + Sb_2Me_4 \longrightarrow 2Me_2SbBH_2 + H_2$$

Diborane reacts slowly over a period of days with metals such as Na, K, Ca, or their amalgams and more rapidly in the presence of ether:

$$2B_2H_6 + 2Na \longrightarrow NaBH_4 + NaB_3H_8$$

$B_3H_8{}^-$ prepared in this way was the first polyborane anion (1955); it is now more conveniently made by the reaction

$$B_2H_6 + NaBH_4 \xrightarrow[100°]{\text{diglyme}} NaB_3H_8 + 2H_2$$

The remarkably facile addition of B_2H_6 to alkenes and alkynes in ether solvents at room temperatures was discovered by H. C. Brown and B. C. Subba Rao in 1956:

$$3RCH{=}CH_2 + \tfrac{1}{2}B_2H_6 \longrightarrow B(CH_2CH_2R)_3$$

This reaction, now termed hydroboration, has opened up the quantitative preparation of organoboranes and these, in turn, have proved to be of outstanding synthetic utility.[22, 23] It was for his development of this field that H. C. Brown (Purdue) was awarded the 1979 Nobel Prize in Chemistry. Hydroboration is regiospecific, the boron showing preferential attachment to the least substituted C atom (anti-Markovnikov). This finds ready interpretation in terms of electronic factors and relative bond polarities (p. 161); steric factors also work in the same direction. The addition is stereospecific *cis* (*syn*).

Protonolysis of the resulting organoborane by refluxing it with an anhydrous carboxylic acid yields the alkane corresponding to the initial alkene, and oxidative hydrolysis with alkaline hydrogen peroxide yields the corresponding primary alcohol:

$$B(CH_2CH_2R)_3 \xrightarrow{\text{EtCO}_2\text{H}} 3RCH_2CH_3$$

$$B(CH_2CH_2R)_3 \xrightarrow{\text{NaOH/H}_2\text{O}_2} 3RCH_2CH_2OH$$

Internal alkenes can be thermally isomerized to terminal organoboranes and hence to terminal alkenes (by displacement) or to primary alcohols:

$$RCH{=}CHCH_2CH_3 \xrightarrow{\frac{1}{2}B_2H_6} RCH_2\overset{\overset{\displaystyle BH_2}{|}}{C}HCH_2CH_3 \xrightarrow{150^\circ} R(CH_2)_4BH_2$$

$$RCH_2CH_2CH{=}CH_2 \xleftarrow{\text{R'CH}=CH_2}$$

$$R(CH_2)_4BH_2 \xrightarrow{\text{NaOH/H}_2\text{O}_2} R(CH_2)_4OH$$

Tertiary alcohols can be prepared by transfer of alkyls to CO:

$$BR_3 + CO \xrightarrow[110^\circ]{\text{diglyme}} R_3CBO \xrightarrow{\text{NaOH/H}_2\text{O}_2} R_3COH$$

If the reaction is carried out in the presence of water rather than diglyme a ketone is obtained of chain length $(2n+1)$ C atoms and other variants yield primary alcohols or aldehydes with $(n+1)$ C atoms:

$$BR_3 + CO \xrightarrow[100^\circ]{\text{H}_2\text{O}} RB(OH)CR_2OH \xrightarrow{\text{NaOH/H}_2\text{O}} R_2CO$$

$$BR_3 + CO \xrightarrow[45^\circ]{\text{NaBH}_4/\text{diglyme}} \{\} \xrightarrow{\text{KOH}} RCH_2OH$$

$$BR_3 + CO \xrightarrow[25^\circ]{\text{Li[AlH(OMe)}_3]/\text{thf}} \{\} \xrightarrow{\text{H}_2\text{O}_2} 2ROH + RCHO$$

Numerous other functional groups can be incorporated, e.g. carboxylic acids RCO_2H by

[22] H. C. Brown, *Organic Syntheses via Boranes*, Wiley, New York, 1975, 283 pp., *Boranes in Organic Chemistry*, Cornell University Press, Ithaca, New York, 1972, 462 pp.
[23] D. J. Pasto, Solution reactions of borane and substituted boranes, Chap. 5 in ref. 7, pp. 197–222.

chromic acid oxidation of $B(CH_2R)_3$; paraffins R–R by reaction of BR_3 with alkaline $AgNO_3$; alkyl halides RCl, RBr, and RI by reaction of BR_3 with Et_2NCl, Br_2/OH^-, and I_2/OH^- respectively; primary amines RNH_2 using NH_2Cl or sulfamic acid, NH_2OSO_3H; secondary amines RNHR' using $R'N_3$, etc.

Diborane is an electrophilic reducing agent which preferentially attacks a molecule at a position of high electron density (cf. BH_4^-, which is nucleophilic). Reductions can occur either with or without bond rupture, or by removal of oxygen. Examples of the first category are the reduction of alkenes and alkynes to alkanes mentioned in the preceding paragraph. In the case of heteropolar double and triple bonds the boryl group BH_2 normally adds to the more electron-rich atom, i.e. to the O atom in carbonyls and the N atom in $C \equiv N$ and $C = N$. Thus, after protonolysis, aldehydes yield primary alcohols and ketones yield secondary alcohols, though in the presence of BF_3 complete reduction of $\rangle C = O$ to $\rangle CH_2$ may occur. Likewise, nitrites are reduced to amines, oximes to *N*-alkylhydroxylamines, and Schiff's bases to secondary amines:

$$RC \equiv N \longrightarrow RCH_2NH_2$$

$$RR'C = NOH \longrightarrow RR'CHNHOH$$

$$RC_6H_4CH = NC_6H_4R' \longrightarrow RC_6H_4CH_2NHC_6H_4R'$$

Reductive cleavage of strained rings such as those in cyclopropanes and epoxides occurs readily and acetals (or ketals) are also reductively cleaved to yield an ether and an alcohol:

$$RR'C(OR'')_2 \longrightarrow RR'CHOR'' + R''OH$$

Removal of O atoms can occur either with or without addition of H atoms to the molecule. Thus, phosphine oxides give phosphines and pyridine-*N*-oxide gives pyridine without addition of H atoms, whereas aromatic nitroso compounds are reduced to amines and cyclic diones can be successively reduced by replacement of $\rangle C = O$ by $\rangle CH_2$, e.g.

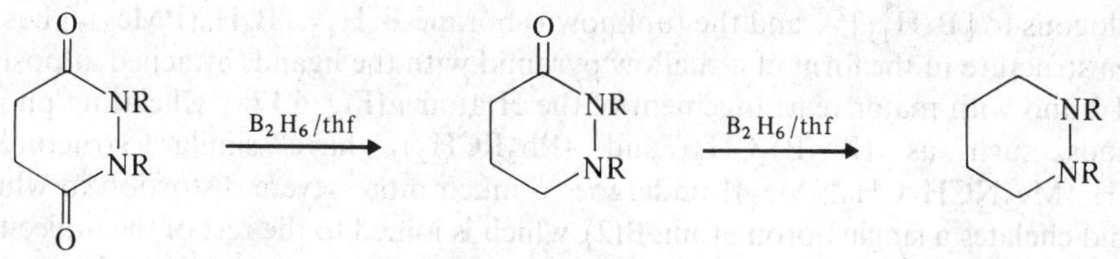

It is fortunate that unstrained cyclic ethers such as thf are not reductively cleaved except under very forcing conditions.

6.4.5   *Chemistry of* nido-*pentaborane*, B_5H_9

Pentaborane(9) is a colourless, volatile liquid, bp 60.0°; it is thermally stable but chemically very reactive and spontaneously flammable in air. Its structure is essentially a square-based pyramid of B atoms each of which carries a terminal H atom and there are 4 bridging H atoms around the base (Fig. 6.16a). Calculations (p. 185) suggest that B(1) has a slightly higher electron density than the basal borons and that H_μ is slightly more positive than H_t. Apex-substituted derivatives $1\text{-}XB_5H_8$ can readily be prepared by

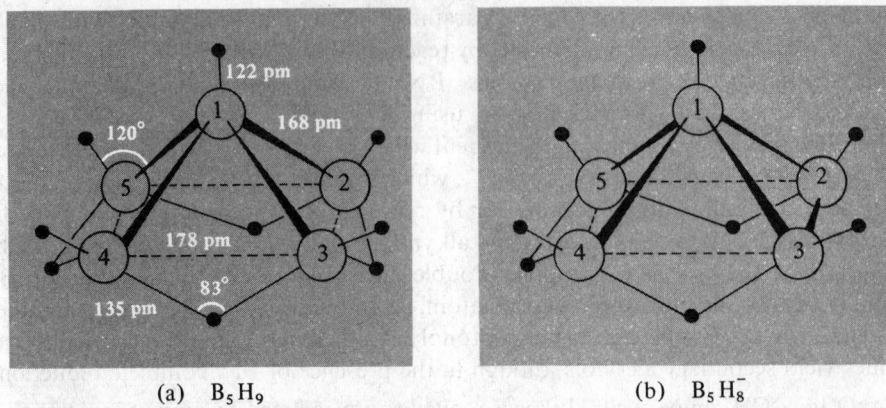

(a) B_5H_9 (b) $B_5H_8^-$

FIG. 6.16   (a) Molecular structure, numbering scheme, and dimensions of B_5H_9; (b) structure of $B_5H_8^-$; this anion is fluxional, the 4 basal borons (and the 3 bridging hydrogens) being equivalent on the nmr time scale.

electrophilic substitution (e.g. halogenation or Friedel–Crafts alkylation with RX or alkenes), whereas base-substituted derivatives $2\text{-}XB_5H_8$ result when nucleophilic reaction is induced by amines or ethers, or when $1\text{-}XB_5H_8$ is isomerized in the presence of a Lewis base such as hexamethylenetetramine or an ether:

$$B_5H_9 \xrightarrow{\text{I}_2/\text{AlX}_3} 1\text{-}IB_5H_8 \xrightarrow[\text{(CH}_2)_6\text{N}_4]{\text{base}} 2\text{-}IB_5H_8 \xrightarrow{\text{SbF}_3} 2\text{-}FB_5H_8$$

Further derivatives can be obtained by metathesis, e.g.

$$2\text{-}ClB_5H_8 + NaMn(CO)_5 \longrightarrow 2\text{-}\{(CO)_5Mn\}B_5H_8 \text{ (also Re)}$$

B_5H_9 reacts with Lewis bases (electron-pair donors) to form adducts, some of which have now been recognized as belonging to the new series of *hypho*-borane derivatives B_nH_{n+8} (p. 173). Thus PMe_3 gives the adduct $[B_5H_9(PMe_3)_2]$ which is formally analogous to $[B_5H_{11}]^{2-}$ and the (unknown) borane B_5H_{13}. $[B_5H_9(PMe_3)_2]$ has a very open structure in the form of a shallow pyramid with the ligands attached at positions 1 and 2 and with major rearrangement of the H atoms (Fig. 6.17a). Chelating phosphine ligands such as $(Ph_2P)_2CH_2$ and $(Ph_2PCH_2)_2$ have similar structures but $[B_5H_9(Me_2NCH_2CH_2NMe_2)]$ undergoes a much more severe distortion in which the ligand chelates a single boron atom, B(2), which is joined to the rest of the molecule by a single bond to the apex B(1) (Fig. 6.17b).[24] With NH_3 as ligand (at $-78°$) complete excision of 1 B atom occurs by "unsymmetrical cleavage" to give $[(NH_3)_2BH_2]^+[B_4H_7^-]$.

B_5H_9 also acts as a weak Brønsted acid and, from proton competition reactions with other boranes and borane anions, it has been established that acidity increases with increasing size of the borane cluster and that *arachno*-boranes are more acidic than *nido*-boranes:

$$\textit{nido}: B_5H_9 < B_6H_{10} < B_{10}H_{14} < B_{16}H_{20} < B_{18}H_{22}$$

$$\textit{arachno}: B_4H_{10} < B_5H_{11} < B_6H_{12} \quad \text{and} \quad B_4H_{10} > B_6H_{10}$$

[24] N. W. ALCOCK, H. M. COLQUHOUN, G. HARAN, J. F. SAWYER, and M. G. H. WALLBRIDGE, Isomeric *hypho*-borane structures $B_5H_9.L$ [L=$(Ph_2P)_2CH_2$, $(Ph_2PCH_2)_2$, and $(Me_2NCH_2)_2$], *JCS Chem. Comm.* 1977, 368–70; *JCS Dalton* 1982, 2243–55.

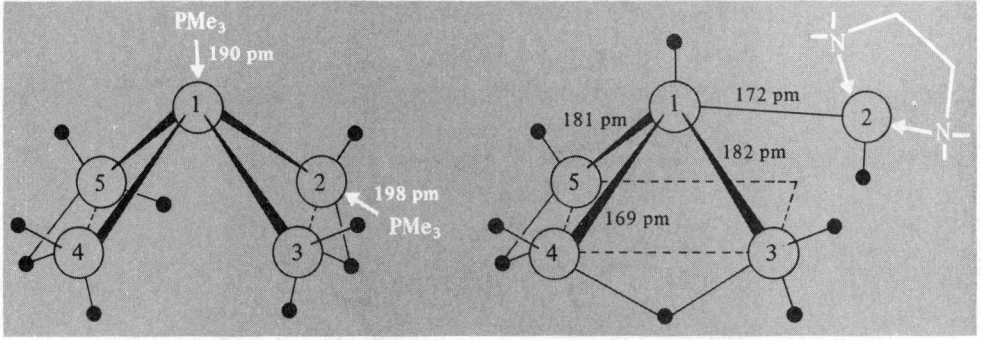

(a) [B₅H₉(PMe₃)₂]   (b) [B₅H₉(Me₂NCH₂CH₂NMe₂)]

FIG. 6.17   Structure of *hypho*-borane derivatives: (a) [B₅H₉(PMe₃)₂]—the distances B(1)–B(2) and B(2)–B(3) are as in B₅H₉ (p. 194) but B(3)···B(4) is 295 pm (cf, B···B 297 pm in B₅H₁₁, p. 181), and (b) [B₅H₉(Me₂NCH₂CH₂NMe₂)]—the distances B(2)···B(3) and B(2)···B(5) are 273 and 272 pm respectively.

Accordingly, B₅H₉ can be deprotonated at low temperatures by loss of H_μ to give $B_5H_8^-$ (Fig. 6.16b) providing a sufficiently strong base such as a lithium alkyl or alkali metal hydride is used:

$$B_5H_9 + MH \longrightarrow MB_5H_8 + H_2$$

Bridge-substituted derivatives of B₅H₉ can then be obtained by reacting MB₅H₈ with chloro compounds such as R₂PCl, Me₃SiCl, Me₃GeCl, or even Me₂BCl to give compounds in which the 3-centre B–H_μ–B bond has been replaced by a 3-centre bond between the 2 B atoms and P, Si, Ge, or B respectively. Many metal–halide coordination complexes react similarly, and the products can be considered as adducts in which the $B_5H_8^-$ anion is acting formally as a 2-electron ligand via a 3-centre B–M–B bond.[25, 26] Thus [Cuᴵ(B₅H₈)(PPh₃)₂] (Fig. 6.18) is readily formed by the low-temperature reaction of KB₅H₈ with [CuCl(PPh₃)₃] and analogous 16-electron complexes have been prepared for many of the later transition elements, e.g. [Cd(B₅H₈)Cl(PPh₃)], [Ag(B₅H₈)(PPh₃)₂], and [Mᴵᴵ(B₅H₈)XL₂], where Mᴵᴵ = Ni, Pd, Pt; X = Cl, Br, I; L₂ = a diphosphine or related ligand. By contrast, [Irᴵ(CO)Cl(PPh₃)₂] reacts by oxidative insertion of Ir and consequent cluster expansion to give [(IrB₅H₈)(CO)(PPh₃)₂] which, though superficially of similar formula, has the structure of an *irida-nido*-hexaborane (Fig. 6.19).[27] In this, the {Ir(CO)(PPh₃)₂} moiety replaces a basal BH_tH_μ unit in B₆H₁₀ (structure 10, p. 174).

Cluster-expansion and cluster-degradation reactions are a feature of many polyhedral borane species. Examples of cluster-expansion are:

$$LiB_5H_8 + \tfrac{1}{2}B_2H_6 \xrightarrow[\text{(fast)}]{Et_2O/-78°} LiB_6H_{11} \xrightarrow{HCl} B_6H_{12}$$

[25] N. N. GREENWOOD and I. M. WARD, Metalloboranes and metal–boron bonding, *Chem. Soc. Revs.* **3**, 231–71 (1974).

[26] N. N. GREENWOOD, The synthesis, structure, and chemical reactions of metalloboranes, *Pure Appl. Chem.* **49**, 791–802 (1977).

[27] N. N. GREENWOOD, J. D. KENNEDY, W. S. McDONALD, D. REED, and J. STAVES, Cage expansion in metallopentaborane chemistry: the preparation and structure of [(IrB₅H₈)(CO)(PPh₃)₂], *JCS Dalton* 1979, 117–23.

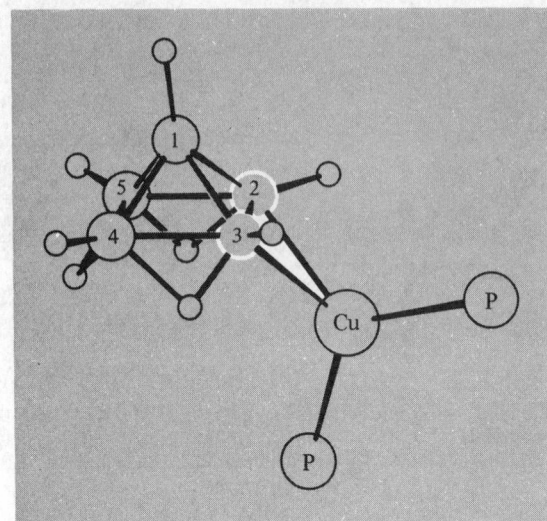

Fig. 6.18   [Cu(B$_5$H$_8$)(PPh$_3$)$_2$].

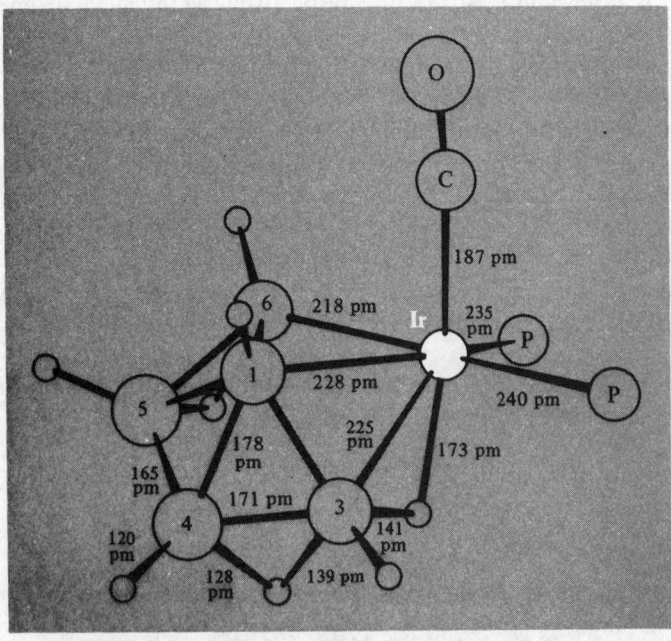

Fig. 6.19   Structure of [(IrB$_5$H$_8$)(CO)(PPh$_3$)$_2$] with the phenyl groups omitted in order to show the structure about the iridium atom and of the metallaborane cluster.

$$LiB_5H_8 + B_2H_6 \xrightarrow[\text{(slow)}]{Me_2O/-78°} \{LiB_7H_{14}\} \longrightarrow B_6H_{10} + LiBH_4$$

Cluster degradation has already been mentioned in connection with the unsymmetrical cleavage reaction (p. 188) and other examples are:

$$[NMe_4]B_5H_8 + 6Pr^iOH \longrightarrow [NMe_4]B_3H_8 + 2B(OPr^i)_3 + 3H_2$$

$$\text{and } B_5H_9 \xrightarrow{TMED} B_6H_9(TMED) \xrightarrow{MeOH} B_4H_8(TMED) + B(OMe)_3 + 2H_2$$

(where Pr^i is CH_3CHCH_3 and TMED is $Me_2NCH_2CH_2NMe_2$).

Subrogation of a {BH} unit in B_5H_9 by an "isoelectronic" organometallic group such as {Fe(CO)$_3$} or {Co(η^5-C$_5$H$_5$)} can also occur, and this illustrates the close interrelation between metalloboranes, metal–metal cluster compounds, and organometallic complexes in general (see Panel).

Metalloboranes, Metal Clusters, and Organometallic Complexes

Copyrolysis of B_5H_9 and [Fe(CO)$_5$] in a hot/cold reactor at 220°/20° for 3 days gives an orange liquid (mp 5°C) of formula [1-{Fe(CO)$_3$}B$_4$H$_8$] having the structure shown in (a).[27a] The isoelectronic complex [1-{Co(η^5-C$_5$H$_5$)}B$_4$H$_8$] (structure b) can be obtained as yellow crystals by pyrolysis at 200° of the corresponding basal derivative [2-{Co(η^5-C$_5$H$_5$)}B$_4$H$_8$] (structure (c)) which is obtained as red crystals from the reaction of NaB$_5$H$_8$ and CoCl$_2$ with NaC$_5$H$_5$ in thf at −20°.[27b] The course of both preparative reactions is obscure and other products are also obtained.

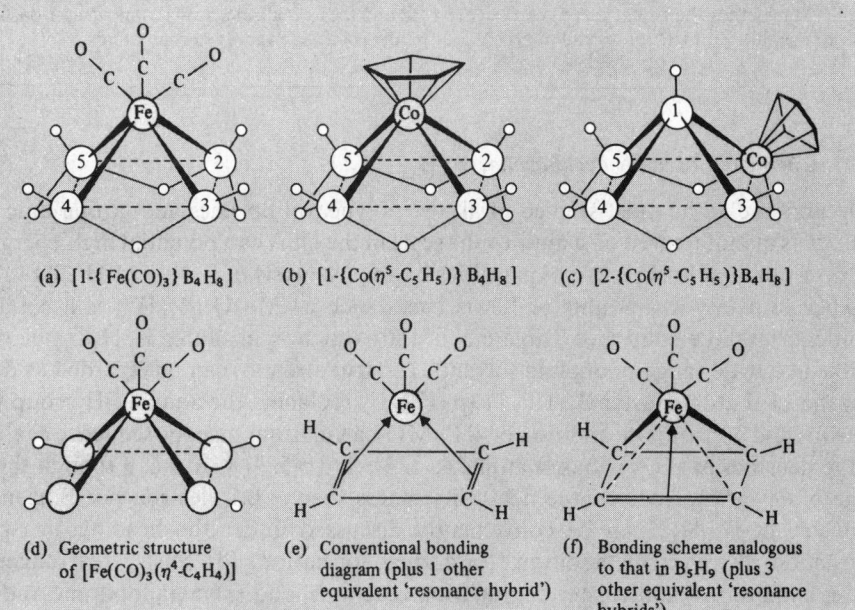

(a) [1-{ Fe(CO)$_3$} B$_4$H$_8$]   (b) [1-{Co(η^5-C$_5$H$_5$)} B$_4$H$_8$]   (c) [2-{Co(η^5-C$_5$H$_5$)}B$_4$H$_8$]

(d) Geometric structure of [Fe(CO)$_3$(η^4-C$_4$H$_4$)]

(e) Conventional bonding diagram (plus 1 other equivalent 'resonance hybrid')

(f) Bonding scheme analogous to that in B_5H_9 (plus 3 other equivalent 'resonance hybrids')

[27a] N. N. GREENWOOD, C. G. SAVORY, R. N. GRIMES, L. G. SNEDDON, A. DAVISON, and S. S. WREFORD, Preparation of a stable small ferraborane B$_4$H$_8$Fe(CO)$_3$, *JCS Chem. Comm.* 1974, 718.
[27b] V. R. MILLER and R. N. GRIMES, Preparation of stable *closo-* and *nido-*cobaltaboranes from Na$^+$B$_5$H$_8{}^-$. Complexes of the formal B$_4$H$_8{}^{2-}$ and B$_4$H$_6{}^{4-}$ ligands, *J. Am. Chem. Soc.* **95**, 5078–80 (1973).

(g) Topology of $[1-\{Fe(CO)_3\}B_4H_8]$
     stressing relation to B_5H_9 (4120)

(h) Topology of $[1-\{Fe(CO)_3\}B_4H_8]$
     stressing relation to $[Fe(CO)_5]$
     and $[Fe(CO)_3(\eta^4\text{-}C_4H_4)]$

As $\{BH_2\}$ is isoelectronic with $\{CH\}$, these metalloborane clusters are isoelectronic with the cyclobutadiene adduct $[Fe(\eta^4\text{-}C_4H_4)(CO)_3]$ (see structure (d) and bonding diagrams (e) and (f)). Likewise the $\{Fe(CO)_3\}$ group or the isoelectronic $\{Co(\eta^5\text{-}C_5H_5)\}$ group can replace a $\{BH\}$ group in B_5H_9 and two descriptions of the bonding are given in the topological diagrams (g) and (h). Diagram (g) emphasizes the relation between $[1-\{Fe(CO)_3\}B_4H_8]$ and B_5H_9 (4120): the Fe atom supplies 2 electrons and 3 atomic orbitals to the cluster (as does BH), thereby enabling it to form 2 Fe–B σ bonds and to accept a pair of electrons from adjacent B atoms to form a 3-centre BMB bond. (Only 1 of 4 equivalent hybrids is shown in (g).) In this description of the bonding the Fe atom is formally octahedral Fe^{II} (d^6). Alternatively, the topology in diagram (h) emphasizes the relation between $[1-\{Fe(CO)_3\}B_4H_8]$ and $[Fe(CO)_3(\eta^4\text{-}C_4H_4)]$ or $[Fe(CO)_5]$: the Fe atom accepts 2 pairs of electrons to form two 3-centre BMB bonds and is formally Fe^0 with a trigonal bipyramidal arrangement of bonds.

In principle it should be possible to replace more than one $\{BH\}$ group in B_5H_9 by a metal centre, and it is notable that the iron carbonyl cluster compound $[Fe_5(CO)_{15}C]$ (p. 355) features the same square-pyramidal cluster in which 5 $\{Fe(CO)_3\}$ groups have replaced the five $\{BH\}$ groups in B_5H_9, and the C atom (in the centre of the base) replaces the 4 bridging H atoms by supplying the 4 electrons required to complete the bonding. An intermediate degree of subrogation is found in the dimetalla species $[1,2-\{Fe(CO)_3\}_2B_3H_7]$.[27c]

Many other equivalent groups can be envisaged and the formalism permits a unified approach to possible synthetic routes and to probable structures of a wide variety of compounds.[15, 25, 28]

6.4.6 *Chemistry of* nido-*decaborane,* $B_{10}H_{14}$

Decaborane is the most studied of all the polyhedral boranes and at one time (mid-1950s) was manufactured on a multitonne scale in the USA as a potential high-energy fuel. It is now obtainable in research quantities by the pyrolysis of B_2H_6 at 100–200°C in the presence of catalytic amounts of Lewis bases such as Me_2O. $B_{10}H_{14}$ is a colourless, volatile, crystalline solid (see Table 6.2, p. 186) which is insoluble in H_2O but readily soluble in a wide range of organic solvents. Its structure (36) can be regarded as derived from the 11 B atom cluster $B_{11}H_{11}{}^{2-}$ (p. 172) by replacing the unique BH group with 2 electrons and appropriate addition of $4H_\mu$. MO-calculations give the sequence of electron charge densities at the various B atoms as 2, 4 > 1, 3 > 5, 7, 8, 10 > 6, 9 though the total range of deviation from charge neutrality is less than ± 0.1 electron per B atom. The chemistry of $B_{10}H_{14}$ can be conveniently discussed under the headings (a) proton abstraction, (b) electron addition, (c) adduct formation, (d) cluster rearrangements, cluster expansions, and cluster degradation reactions, and (e) metalloborane and other heteroborane compounds.

[27c] K. J. HALLER, E. L. ANDERSEN, and T. P. FEHLNER, Crystal and molecular structure of $[(CO)_6Fe_2B_3H_7]$, a diiron analogue of pentaborane(9), *Inorg. Chem.* **20**, 309–13 (1981).

[28] N. N. GREENWOOD and J. D. KENNEDY, Transition-metal derivatives of *nido*-boranes and some related species, Chap. 2 in R. N. GRIMES (ed.), *Metal Interactions with Boron Clusters*, Plenum, New York, 1982.

$B_{10}H_{14}$ can be titrated in aqueous/alcoholic media as a monobasic acid, pK_a 2.70:

$$B_{10}H_{14} + OH^- \rightleftharpoons B_{10}H_{13}^- + H_2O$$

Proton abstraction can also be effected by other strong bases such as H^-, OMe^-, NH_2^-, etc. X-ray studies on $[Et_3NH]^+[B_{10}H_{13}^-]$ establish that the ion is formed by loss of a bridge proton, as expected, and this results in a considerable shortening of the B(5)–B(6) distance from 179 pm in $B_{10}H_{14}$ to 165 pm in $B_{10}H_{13}^-$ (structures 36, 37). Under more forcing conditions with NaH a second H_μ can be removed to give $Na_2B_{10}H_{12}$; the probable structure of $B_{10}H_{12}^{2-}$ is (38) and the anion acts as a formal bidentate (tetrahapto) ligand to many metals (p. 201).

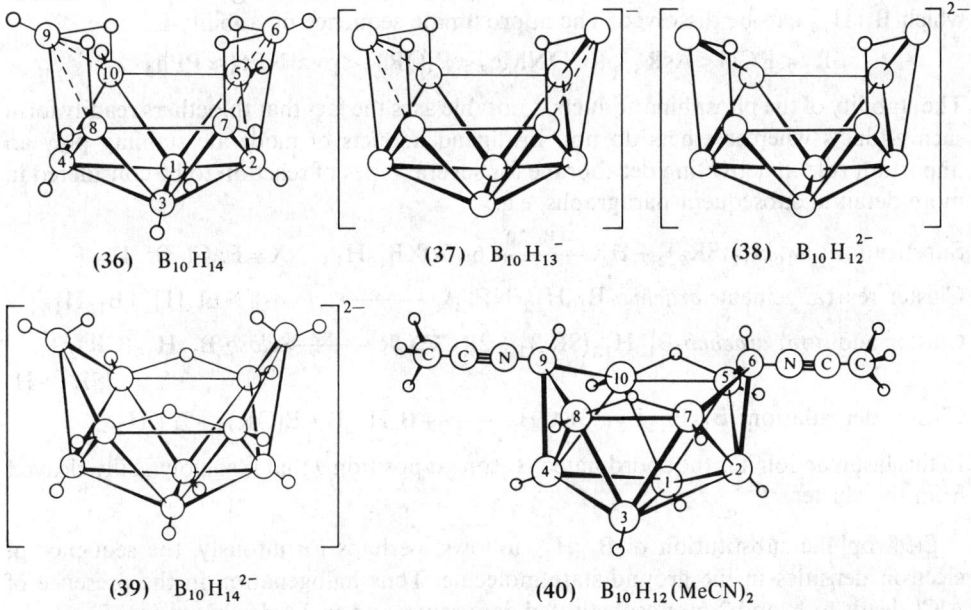

(36) $B_{10}H_{14}$ (37) $B_{10}H_{13}^-$ (38) $B_{10}H_{12}^{2-}$

(39) $B_{10}H_{14}^{2-}$ (40) $B_{10}H_{12}(MeCN)_2$

Electron addition to $B_{10}H_{14}$ can be achieved by direct reaction with alkali metals in ethers, benzene, or liquid NH_3:

$$B_{10}H_{14} + 2Na \longrightarrow Na_2B_{10}H_{14}$$

A more convenient preparation of the $B_{10}H_{14}^{2-}$ anion uses the reaction of aqueous BH_4^- in alkaline solution:

$$B_{10}H_{14} \xrightarrow{+BH_4^- - \{BH_3\}} B_{10}H_{15}^- \underset{+H^+}{\overset{-H^+}{\rightleftharpoons}} B_{10}H_{14}^{2-}$$

Structure (39) conforms to the predicted (2632) topology (p. 181) and shows that the 2 added electrons have relieved the electron deficiency to the extent that the 2 $B-H_\mu-B$ groups have been converted to $B-H_t$ with the consequent appearance of $2BH_2$ groups in the structure. Calculations show that this conversion of a *nido-* to an *arachno-*cluster reverses the sequence of electron charge density at the 2, 4 and 6, 9 positions so that for $B_{10}H_{14}^{2-}$ the sequence is 6, 9 > 1, 3 > 5, 7, 8, 10 > 2, 4; this is paralleled by changes in the chemistry. $B_{10}H_{14}^{2-}$ can formally be regarded as $B_{10}H_{12}L_2$ for the special case of

$L = H^-$. Compounds of intermediate stoichiometry $B_{10}H_{13}L^-$ are formed when $B_{10}H_{14}$ is deprotonated in the presence of the ligand L:

$$B_{10}H_{14} \xrightarrow{Et_2O/NaH} B_{10}H_{13}^- \xrightarrow{Et_2O/L} [B_{10}H_{13}L]^- \xrightarrow{NMe_4^+} [NMe_4][B_{10}H_{13}L]$$

The adducts $B_{10}H_{12}L_2$ (structure 40) can be prepared by direct reaction of $B_{10}H_{14}$ with L or by ligand replacement reactions:

$$B_{10}H_{14} + 2MeCN \longrightarrow B_{10}H_{12}(MeCN)_2 + H_2$$
$$B_{10}H_{12}L_2 + 2L \longrightarrow B_{10}H_{12}L'_2 + 2L$$

Ligands L, L' can be drawn from virtually the full range of inorganic and organic neutral and anionic ligands and, indeed, the reaction severely limits the range of donor solvents in which $B_{10}H_{14}$ can be dissolved. The approximate sequence of stability is:

$$SR_2 < RCN < AsR_3 < RCONMe_2 < P(OR)_3 < py \approx NEt_3 \approx PPh_3$$

The stability of the phosphine adducts is notable as is the fact that thioethers readily form such adducts whereas ethers do not. Bis-ligand adducts of moderate stability play an important role in activating decaborane for several types of reaction to be considered in more detail in subsequent paragraphs, e.g.:

Substitution: $B_{10}H_{12}(SR_2)_2 + HX \xrightarrow{C_6H_6/20°} 6\text{-}(5\text{-})XB_{10}H_{13}$ (X = F, Cl, Br, I)

Cluster rearrangement: *arachno*-$B_{10}H_{12}(NEt_3)_2 \longrightarrow$ (*closo*-)$[NEt_3H]_2^+[B_{10}H_{10}]^{2-}$

Cluster addition: *arachno*-$B_{10}H_{12}(SR_2)_2 + 2RC\equiv CR \longrightarrow$ *closo*-$B_{10}H_{10}(CR)_2 +$
$$2SR_2 + H_2$$

Cluster degradation: $B_{10}H_{12}L_2 + 3ROH \longrightarrow B_9H_{13}L + B(OR)_3 + L + H_2$

In this last reaction it is the coordinated B atom at position 9 that is solvolytically cleaved from the cluster.

Electrophilic substitution of $B_{10}H_{14}$ follows, perhaps fortuitously, the sequence of electron densities in the ground-state molecule. Thus halogenation in the presence of $AlCl_3$ leads to 1- and 2-monosubstituted derivatives and to 2,4-disubstitution. Similarly, Friedel–Crafts alkylations with $RX/AlCl_3$ (or $FeCl_3$) yield mixtures such as 2-$MeB_{10}H_{13}$, 2,4- and 1,2-$Me_2B_{10}H_{12}$, 1,2,3- and 1,2,4-$Me_3B_{10}H_{11}$, and 1,2,3,4-$Me_4B_{10}H_{10}$. By contrast, nucleophilic substitution (like the adduct formation with Lewis bases) occurs preferentially at the 6 (9) position; e.g., LiMe produces 6-$MeB_{10}H_{13}$ as the main product with smaller amounts of 5-$MeB_{10}H_{13}$, 6,5(8)-$Me_2B_{10}H_{12}$, and 6,9-$Me_2B_{10}H_{12}$.

$B_{10}H_{14}$ undergoes numerous cluster-addition reactions in which B or other atoms become incorporated in an expanded cluster. Thus in a reaction which differs from that on p. 199 BH_4^- adds to $B_{10}H_{14}$ with elimination of H_2 to form initially the *nido*-$B_{11}H_{14}^-$ anion (structure 41) and then the *closo*-$B_{12}H_{12}^{2-}$:

$$B_{10}H_{14} + LiBH_4 \xrightarrow{monoglyme/90°} LiB_{11}H_{14} + 2H_2$$
$$B_{10}H_{14} + 2LiBH_4 \longrightarrow Li_2B_{12}H_{12} + 5H_2$$

A more convenient high-yield synthesis of $B_{12}H_{12}^{2-}$ is by the direct reaction of amine-boranes with $B_{10}H_{14}$ in the absence of solvents:

$$B_{10}H_{14} + 2Et_3NBH_3 \xrightarrow{90-100°} [NEt_3H]_2^+[B_{12}H_{12}]^{2-} + 3H_2$$

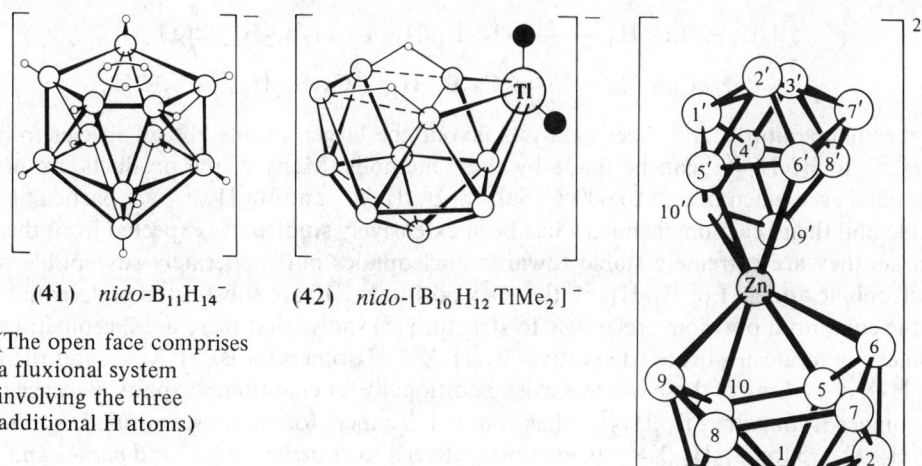

(41)   *nido*-$B_{11}H_{14}^-$ **(42)**   *nido*-$[B_{10}H_{12}TlMe_2]^-$

(The open face comprises
a fluxional system
involving the three
additional H atoms)

(43) $[Zn(B_{10}H_{12})_2]^{2-}$

Heteroatom cluster addition reactions are exemplified by the following:

$$B_{10}H_{14} + Me_3NAlH_3 \xrightarrow{Et_2O} [NMe_3H]^+[AlB_{10}H_{14}]^- + H_2$$

$$B_{10}H_{14} + 2TlMe_3 \xrightarrow{Et_2O} [TlMe_2]^+[B_{10}H_{12}TlMe_2]^-$$

$$B_{10}H_{14} + 2ZnR_2 \xrightarrow{Et_2O} [B_{10}H_{12}Zn(Et_2O)_2] \xrightarrow{H_2O/Me_4NCl} [NMe_4]_2^+[Zn(B_{10}H_{12})_2]^{2-}$$

The structure of the highly reactive anion $[AlB_{10}H_{14}]^-$ is thought to be similar to *nido*-$B_{11}H_{14}^-$ with one facial B atom replaced by Al. The metal alkyls react somewhat differently to give extremely stable metalloborane anions which can be thought of as complexes of the bidentate ligand $B_{10}H_{12}^{2-}$ (structures 42, 43).[25, 28] Many other complexes $[M(B_{10}H_{12})_2]^{2-}$ and $[L_2M(B_{10}H_{12})]$ are known with similar structures except that, where M = Ni, Pd, Pt, the coordination about the metal is essentially square-planar rather than pseudo-tetrahedral as for Zn, Cd, and Hg.

6.4.7 *Chemistry of* closo-$B_nH_n^{2-}$[1, 29, 30]

The structures of these anions have been indicated on p. 172. Preparative reactions are often mechanistically obscure but thermolysis under controlled conditions is the dominant procedure, e.g.

$$5B_2H_6 + 2NaBH_4 \xrightarrow{NEt_3/180} Na_2B_{12}H_{12} + 13H_2$$

[29] E. L. MEUTTERTIES and W. H. KNOTH, *Polyhedral Boranes*, Marcel Dekker, New York, 1968, 197 pp.
[30] R. L. MIDDAUGH, *Closo*-boron hydrides, Chap. 8 in ref. 7, pp. 273–300.

$$5B_2H_6 + 2LiAlH_4 \xrightarrow{\sim 160} Li_2B_{10}H_{10} \ (+ Li_2B_{12}H_{12}, \text{ etc.})$$

$$nCsB_3H_8 \xrightarrow{\text{heat}} Cs_2B_{10}H_{10} + Cs_2B_{12}H_{12} + Cs_2B_9H_9$$

Higher temperatures and ether catalysts favour the larger anions but all species from $B_6H_6{}^{2-}$ to $B_{12}H_{12}{}^{2-}$ can be made by these methods. Many of the products are not degraded even when heated to 600°C. Salts of $B_{12}H_{12}{}^{2-}$ and $B_{10}H_{10}{}^{2-}$ are particularly stable and their reaction chemistry has been extensively studied. As expected from their charge, they are extremely stable towards nucleophiles but moderately susceptible to electrophilic attack. For $B_{10}H_{10}{}^{2-}$ the apex positions 1,10 are substituted preferentially to the equatorial positions; reference to structure (5) shows that there are 2 geometrical isomers for monosubstituted derivatives $B_{10}H_9X^{2-}$, 7 isomers for $B_{10}H_8X_2{}^{2-}$, and 16 for $B_{10}H_7X_3{}^{2-}$. Many of these isomers exist, additionally, as enantiomeric pairs. Because of its higher symmetry $B_{12}H_{12}{}^{2-}$ has only 1 isomer for monosubstituted species $B_{12}H_{11}X^{2-}$, 3 for $B_{12}H_{10}X_2{}^{2-}$ (sometimes referred to as *ortho-*, *meta-*, and *para-*), and 5 for $B_{12}H_9X_3{}^{2-}$.

Oxidation of *closo*-$B_{10}H_{10}{}^{2-}$ with aqueous solutions of Fe^{III} or Ce^{IV} (or electrochemically) yields *conjuncto*-$B_{20}H_{18}{}^{2-}$ (44) which can be photoisomerized to *neo*-$B_{20}H_{18}{}^{2-}$ (45). If the oxidation is carried out at 0° with Ce^{IV}, or in a two-phase system with Fe^{III} using very concentrated solutions of $B_{10}H_{10}{}^{2-}$, the intermediate H-bridged species $B_{20}H_{19}{}^{3-}$ (46) can be isolated. Reduction of *conjuncto*-$B_{20}H_{18}{}^{2-}$ with Na/NH_3 yields the equatorial–equatorial (ee) isomer of *conjuncto*-$B_{20}H_{18}{}^{4-}$ (47), and this can be successively converted by acid catalyst to the ae isomer (49) and, finally, to the aa isomer (48). An extensive derivative chemistry of these various species has been developed. Another important (though mechanistically obscure) reaction of *conjuncto*-$B_{20}H_{18}{}^{2-}$ is its degradation in high yield to n-$B_{18}H_{22}$ by passage of an enthanolic solution through an acidic ion exchange resin; i-$B_{18}H_{22}$ is also formed as a minor product. The relation of these 2 edge-fused decaborane clusters to the B_{20} species is illustrated in structures (31) and (32) (p. 177).

6.5 Carboranes[1, 15, 31-34]

Carboranes burst onto the chemical scene in 1962–3 when classified work that had been done in the late 1950s was cleared for publication. The succeeding 25 y has seen a tremendous burgeoning of activity, and few other areas of chemistry have undergone such spectacular development during this period. Much of the work has been carried out in the USA and the USSR but significant contributions have also come from Czechoslovakia and the UK. Carboranes and their related derivatives the metallocarboranes (p. 209) are now seen to occupy a strategic position in the chemistry of the elements since they overlap and give coherence to several other large areas including the chemistry of polyhedral boranes, transition–metal complexes, metal–cluster compounds, and organometallic chemistry. The field has become so vast that it is only possible to give a few illustrative

[31] R. N. GRIMES, *Carboranes*, Academic Press, New York, 1970, 272 pp.

[32] H. BEALL, Icosahedral carboranes, Chap. 9 in ref. 7, pp. 302–47.

[33] T. ONAK, Carboranes, Chap. 10 in ref. 7, pp. 349–82.

[34] R. E. WILLIAMS, Coordination number-pattern recognition theory of carborane structures, *Adv. Inorg. Chem. Radiochem.* **18**, 67–142 (1976).

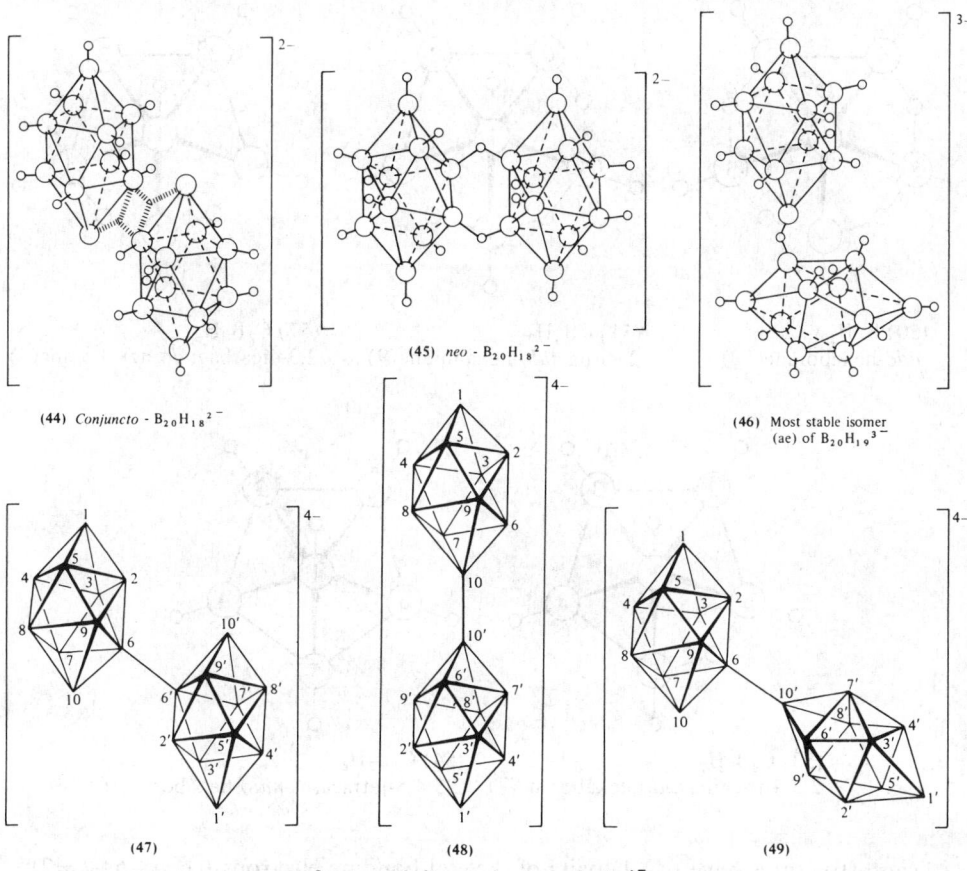

(44) *Conjuncto* - $B_{20}H_{18}^{2-}$

(45) *neo* - $B_{20}H_{18}^{2-}$

(46) Most stable isomer (ae) of $B_{20}H_{19}^{3-}$

(47)

(48)

(49)

Structures of the 3 isomers of *conjuncto* - $B_{20}H_{18}^{4-}$

examples of the many thousands of known compounds, and to indicate the general structural unit a number of C and B atoms arranged on the vertices of a triangulated ordinary substances.

Carboranes (or more correctly carbaboranes) are compounds having as the basic structural unit a number of C and B atoms arranged on the vertices of a triangulated polyhedron. Their structures are closely related to those of the isoelectronic boranes (p. 185) [$BH \equiv B^- \equiv C$; $BH_2 \equiv BH^- \equiv B.L \equiv CH$]. For example, *nido*-$B_6H_{10}$ (structures 10, 50) provides the basic cluster structure for the 4 carboranes CB_5H_9 (51), $C_2B_4H_8$ (52), $C_3B_3H_7$ (53), and $C_4B_2H_6$ (54), each successive replacement of a basal B atom by C being compensated by the removal of one H_μ. Carboranes have the general formula $[(CH)_a(BH)_mH_b]^{c-}$ with a CH units and m BH units at the polyhedral vertices, plus b "extra" H atoms which are either bridging (H_μ) or *endo* (i.e. tangential to the surface of the polyhedron as distinct from the axial H_t atoms specified in the CH and BH groups; H_{endo} occur in BH_2 groups which are thus more precisely specified as BH_tH_{endo}). It follows that the number of electrons available for skeletal bonding is 3e from each CH unit, 2e from each BH unit, 1e from each H_μ or H_{endo}, and ce from the anionic charge. Hence:

total number of skeletal bonding electron pairs $= \frac{1}{2}(3a + 2m + b + c) = n + \frac{1}{2}(a + b + c)$, where $n(= a + m)$ is the number of occupied vertices of the polyhedron.

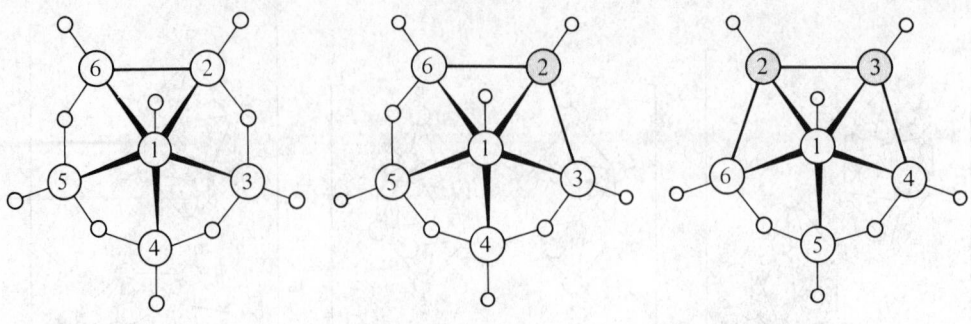

(50) B_6H_{10}
nido-hexaborane(10)

(51) CB_5H_9
2-carba-*nido*-hexaborane(9)

(52) $C_2B_4H_8$
2,3-dicarba-*nido*-hexaborane(8)

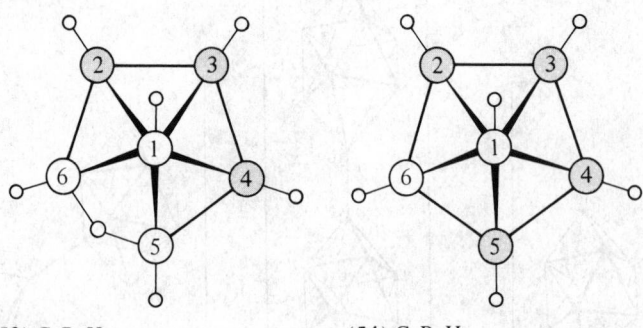

(53) $C_3B_3H_7$
2,3,4-tricarba-*nido*-hexaborane(7)

(54) $C_4B_2H_6$
2,3,4,5-tetracarba-*nido*-hexaborane(6)

closo-structures have $(n+1)$ pairs of skeletal bonding electrons (i.e. $a+b+c=2$).
nido-structures have $(n+2)$ pairs of skeletal bonding electrons (i.e. $a+b+c=4$).
arachno-structures have $(n+3)$ pairs of skeletal bonding electrons (i.e. $a+b+c=6$).

If $a=0$ the compound is a borane or borane anion rather than a carborane. If $b=0$ there are no H_μ or H_{endo}; this is the case for all *closo*-carboranes except for the unique octahedral monocarbaborane, $1\text{-}CB_5H_7$, which has a triply bridging H_μ over one B_3 face of the octahedron. If $c=0$ the compound is a neutral carborane molecule rather than an anion.

Nomenclature follows the well-established oxa–aza convention of organic chemistry. Numbering begins with the apex atom of lowest coordination and successive rings or belts of polyhedral vertex atoms are numbered in a clockwise direction with C atoms being given the lowest possible numbers within these rules.†

Closo-carboranes are the most numerous and the most stable of the carboranes. They are colourless volatile liquids or solids (depending on mol wt.) and are readily prepared from an alkyne and a borane by pyrolysis, or by reaction in a silent electric discharge. This route, which generally gives mixtures, is particularly useful for small *closo*-carboranes ($n=5$–7) and for some intermediate *closo*-carboranes ($n=8$–11), e.g.

---

† As frequently happens in a rapidly developing field, nomenclature and numbering for the carboranes gradually evolved to cope with increasing complexity. Consequently, many systems have been used, often by the same author in successive years, and the only safe procedure is to draw a labelled diagram and convert to the preferred numbering system.

$$nido\text{-}B_5H_9 \xrightarrow[500\ -600]{C_2H_2} closo\text{-}1,5\text{-}C_2B_3H_5 + closo\text{-}1,6\text{-}C_2B_4H_6 + closo\text{-}2,4\text{-}C_2B_5H_7$$
$$(55) \qquad\qquad (56) \qquad\qquad\qquad (57) \qquad\qquad\qquad (58)$$

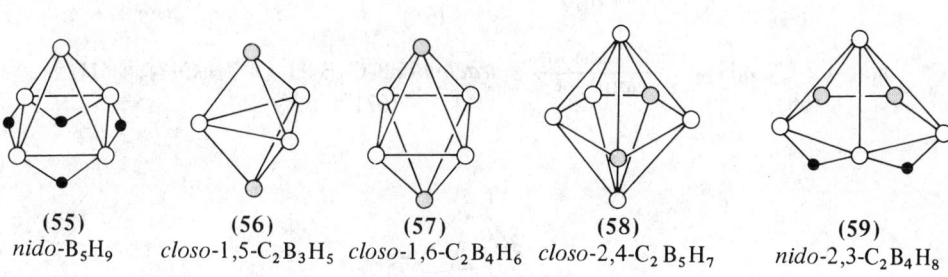

(55)	(56)	(57)	(58)	(59)
$nido\text{-}B_5H_9$	$closo\text{-}1,5\text{-}C_2B_3H_5$	$closo\text{-}1,6\text{-}C_2B_4H_6$	$closo\text{-}2,4\text{-}C_2B_5H_7$	$nido\text{-}2,3\text{-}C_2B_4H_8$

Milder conditions provide a route to *nido*-carboranes, e.g.:

$$nido\text{-}B_5H_9 \xrightarrow[200°]{C_2H_2} nido\text{-}2,3\text{-}C_2B_4H_8 \quad (59)$$

Pyrolysis of *nido*- or *arachno*-carboranes or their reaction in a silent electric discharge also leads to *closo*-species either by loss of H_2 or disproportionation:

$$C_2B_nH_{n+4} \longrightarrow C_2B_nH_{n+2} + H_2$$
$$2C_2B_nH_{n+4} \longrightarrow C_2B_{n-1}H_{n+1} + C_2B_{n+1}H_{n+3} + 2H_2$$

For example, pyrolysis of the previously mentioned *nido*-2,3-$C_2B_4H_8$ gives the 3 *closo*-species shown above, whereas under the milder conditions of photolytic closure the less-stable isomer *closo*-1,2-$C_2B_4H_6$ is obtained. Pyrolysis of alkyl boranes at 500–600° is a related route which is particularly useful to monocarbaboranes though the yields are often low, e.g.:

$$1,2\text{-Me}_2\text{-}nido\text{-}B_5H_7 \longrightarrow$$
$$(60)$$
$$closo\text{-}1,5\text{-}C_2B_3H_5 + closo\text{-}1\text{-}CB_5H_7 + nido\text{-}2\text{-}CB_5H_9 + 3\text{-Me}\text{-}nido\text{-}2\text{-}CB_5H_8$$
$$(56) \qquad\qquad (61) \qquad\qquad (62) \qquad\qquad (63)$$

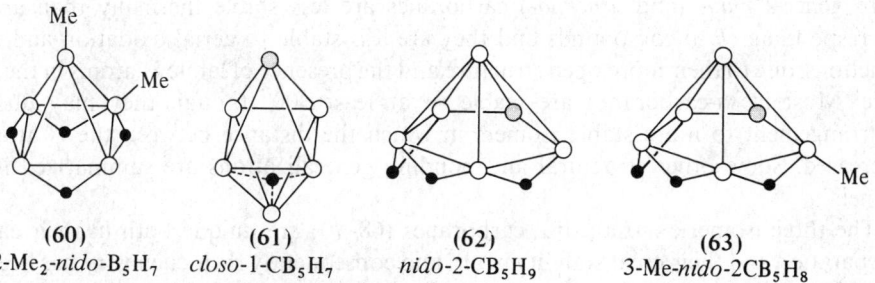

(60)	(61)	(62)	(63)
$1,2\text{-Me}_2\text{-}nido\text{-}B_5H_7$	$closo\text{-}1\text{-}CB_5H_7$	$nido\text{-}2\text{-}CB_5H_9$	$3\text{-Me}\text{-}nido\text{-}2CB_5H_8$

Cluster expansion reactions with diborane provide an alternative route to intermediate *closo*-boranes, e.g.:

$$closo\text{-}1,7\text{-}C_2B_6H_8 + \tfrac{1}{2}B_2H_6 \longrightarrow closo\text{-}1,6\text{-}C_2B_7H_9 + H_2$$
$$closo\text{-}1,6\text{-}C_2B_7H_9 + \tfrac{1}{2}B_2H_6 \longrightarrow closo\text{-}1,6\text{-}C_2B_8H_{10} + H_2$$

Finally, cluster degradation reactions lead to more open structures, e.g.:

$$closo\text{-}1,6\text{-}C_2B_8H_{10} \xrightarrow[\text{OH}^- + 2H_2O]{\text{base hydrol}} arachno\text{-}1,3\text{-}C_2B_7H_{12}^- + B(OH)_3$$
$$\quad\quad (64) \qquad\qquad\qquad\qquad\qquad\qquad\qquad (65)$$

$$nido\text{-}1,7\text{-}C_2B_9H_{12}^- \xrightarrow[\text{+}6H_2O - 6e^-]{\text{chromic acid oxidn.}} arachno\text{-}1,3\text{-}C_2B_7H_{13} + 2B(OH)_3 + 5H^+$$
$$\quad\quad (66) \qquad\qquad\qquad\qquad\qquad\qquad\qquad (67)$$

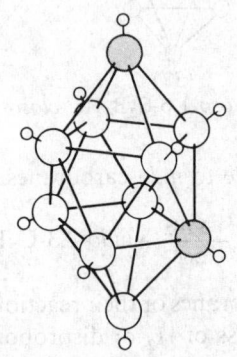

(64)   *closo*-1,6-C$_2$B$_8$H$_{10}$

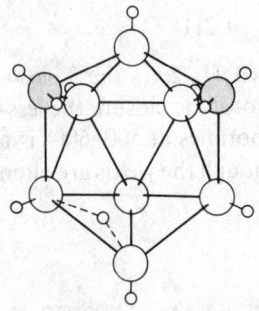 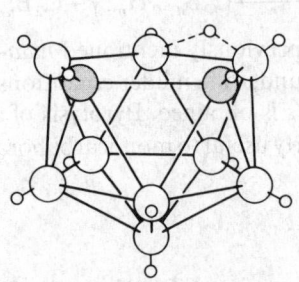 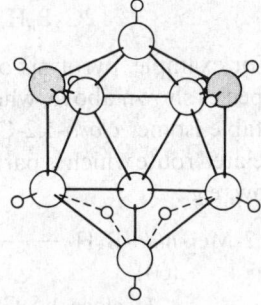

(65) *arachno*-1,3-C$_2$B$_7$H$_{12}^-$ (66) *nido*-1,7-C$_2$B$_9$H$_{12}^-$ (67) *arachno*-1,3-C$_2$B$_7$H$_{13}$

In general *nido*- (and *arachno*-) carboranes are less stable thermally than are the corresponding *closo*-compounds and they are less stable to aerial oxidation and other reactions, due to their more open structure and the presence of labile H atoms in the open face. Most *closo*-carboranes are stable to at least 400° though they may undergo rearrangement to more stable isomers in which the distance between the C atoms is increased. Some other structural and bonding generalizations are summarized in the Panel.

The three isomeric icosahedral carboranes (68–70) are unique both in their ease of preparation and their great stability in air, and consequently their chemistry has been the most fully studied. The 1,2-isomer in particular has been available for over a decade on the multikilogram scale. It is best prepared in bulk by the direct reaction of ethyne (acetylene) with decaborane in the presence of a Lewis base, preferably Et$_2$S:

$$nido\text{-}B_{10}H_{14} + 2SEt_2 \longrightarrow B_{10}H_{12}(SEt_2)_2 + H_2$$

$$B_{10}H_{12}(SEt_2)_2 + C_2H_2 \longrightarrow closo\text{-}1,2\text{-}C_2B_{10}H_{12} + 2SEt_2 + H_2$$

Some Further Generalizations Concerning Carboranes

1. Carbon tends to adopt the position of lowest coordination number on the polyhedron and to keep as far from other C atoms as possible (i.e. the most stable isomer has the greatest number of B–C connections).
2. Boron–boron distances in the cluster increase with increasing coordination number (as expected). Average B–B distances are: 5-coordinate B 170 pm, 6-coordinate B 177 pm, 7-coordinate B 186 pm.
3. Carbon is somewhat smaller than B and interatomic distances involving C are correspondingly shorter (see table).

Distance/pm	5	6	7
B–B	170	177	186
B–C	165	172	—
C–C	145	165	—

Increase (vertical, B–B/B–C/C–C)   *Increase* (horizontal)

4. Negative electronic charge on B is computed to decrease in the sequence: B (not bonded to C) > B (bonded to 1C) > B (bonded to 2C)
Within each group the B with lower coordination number has a greater negative charge than those with higher coordination.
5. CH groups tend to be more positive than BH groups with the same coordination number (despite the higher electronegativity of C). This presumably arises because each C contributes 3e for bonding within the cluster whereas each B contributes only 2e.
6. In *nido*- and *arachno*-carboranes H_μ is more acidic than H_t and is the one removed on deprotonation with NaH.

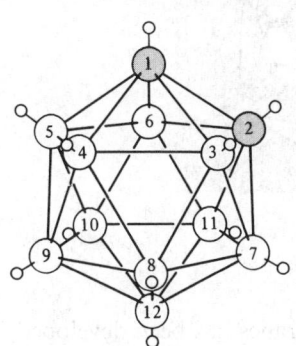 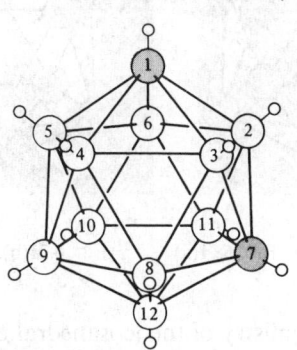 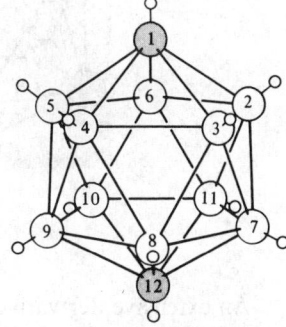

(68) *ortho*-carborane, 1,2-$C_2B_{10}H_{12}$ (mp 320°C)

(69) *meta*-carborane, 1,7-$C_2B_{10}H_{12}$ (mp 265°C)

(70) *para*-carborane, 1,12-$C_2B_{10}H_{12}$ (mp 261°C)

The 1,7-isomer is obtained in 90% yield by heating the 1,2-isomer in the gas phase at 470°C for several hours (or in quantitative yield by flash pyrolysis at 600° for 30 s). The 1,12-isomer is most efficiently prepared (20% yield) by heating the 1,7-isomer for a few seconds at 700°C. The mechanism of these isomerizations has been the subject of considerable speculation but definitive experiments are hard to devise. The "diamond–square–diamond" mechanism has been proposed (Fig. 6.20) for the 1,2⇌1,7

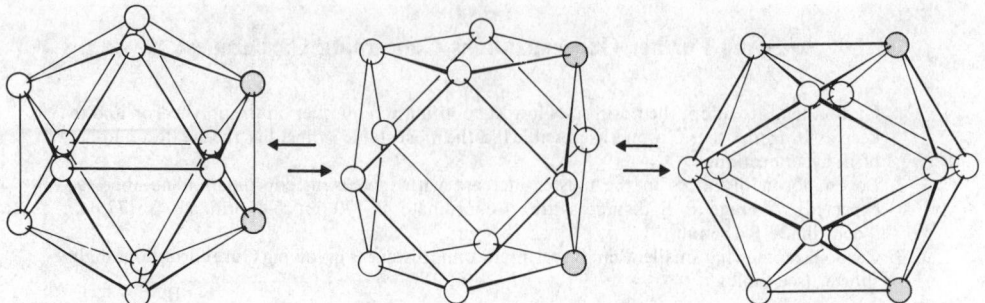

FIG. 6.20 The interconversion of 1,2- and 1,7-disubstituted icosahedral species *via* a cubooctahedral intermediate.

isomerization, but the 1,12 isomer cannot be generated by this mechanism. Moreover, the activation energy required to pass through the cubo–octahedral transition state is likely to be rather high. An alternative proposal, which can lead to both the 1,7 and the 1,12 isomers, is the successive concerted rotation of the 3 atoms on a triangular face as shown in Fig. 6.21a. Yet a third possible mechanism that has been envisaged involves the concerted basal twisting of two parallel pentagonal pyramids comprising the icosahedron (Fig. 6.21b). It is conceivable that the various mechanisms operate in different temperature ranges or that two (or all three) mechanisms are active simultaneously.

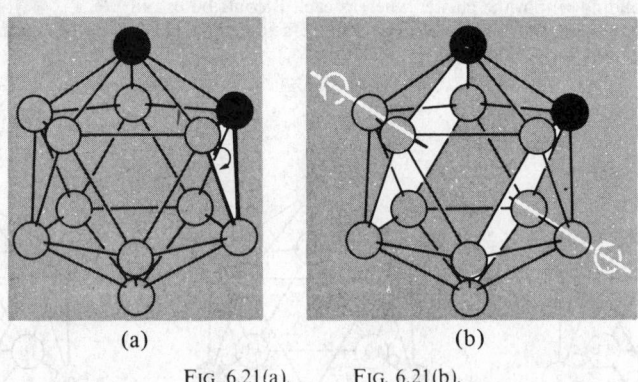

(a)   (b)

FIG. 6.21(a).   FIG. 6.21(b).

An extensive derivative chemistry of the icosahedral carboranes has been developed, especially for $1,2\text{-}C_2B_{10}H_{12}$. Terminal H atoms attached to B undergo facile electrophilic substitution and the sequence of reactivity follows the sequence of negative charge density on the BH_t group:[35]

$$closo\text{-}1,2\text{-}C_2B_{10}H_{12}: \quad (8, 10 \approx 9, 12) > 4, 5, 7, 11 > 3, 6$$

$$closo\text{-}1,7\text{-}C_2B_{10}H_{12}: \quad 9, 10 > 4, 6, 8, 11 > 5, 12 > 2, 3$$

[35] D. A. Dixon, D. A. Kleir, T. A. Halgren, J. H. Hall, and W. N. Lipscomb, Localized orbitals for polyatomic molecules. 5. The *closo*-boron hydrides $B_nH_n{}^{2-}$ and carboranes $C_2B_{n-2}H_n$, *J. Am. Chem. Soc.* **99**, 6226–37 (1977).

Similar reactions occur for other *closo*-carboranes, e.g.:

$$1,6\text{-}C_2B_7H_9 \xrightarrow{Br_2/AlCl_3} 8\text{-}Br\text{-}1,6\text{-}C_2B_7H_8 + HBr$$

$$1,10\text{-}C_2B_8H_{10} \xrightarrow{8Cl_2/CCl_4} 1,10(CH)_2B_8Cl_8 + 8HCl$$

It is noteworthy that, despite the greater electronegativity of C, the CH group tends to be more positive than the BH groups and does not normally react under these conditions.

The weakly acidic CH_t group can be deprotonated by strong nucleophiles such as LiBu or RMgX; the resulting metallated carboranes $LiCCHB_{10}H_{10}$ and $(LiC)_2B_{10}H_{10}$ can then be used to prepare a full range of *C*-substituted derivatives —R, —X, —SiMe$_3$, —COOH, —COCl, —CONHR, etc. The possibility of synthesizing extensive covalent C—C or siloxane networks with pendant carborane clusters is obvious and the excellent thermal stability of such polymers has already been exploited in several industrial applications.

Although *closo*-carboranes are stable to high temperatures and to most common reagents, M. F. Hawthorne showed (1964) that they can be specifically degraded to *nido*-carborane anions by the reaction of strong bases in the presence of protonic solvents, e.g.:

$$1,2\text{-}C_2B_{10}H_{12} + EtO^- + 2EtOH \xrightarrow{85\ C} 7,8\text{-}C_2B_9H_{12}^- + B(OEt)_3 + H_2$$

$$1,7\text{-}C_2B_{10}H_{12} + EtO^- + 2EtOH \longrightarrow 7,9\text{-}C_2B_9H_{12}^- + B(OEt)_3 + H_2$$

Figure 6.22 indicates that, in both cases, the BH vertex removed is the one adjacent to the 2 CH vertices: since the C atoms tend to remove electronic charge preferentially from contiguous B atoms, the reaction can be described as a nucleophilic attack by EtO^- on the most positive (most electron deficient) B atom in the cluster. Deprotonation of the monoanions by NaH removes the bridge proton to give the *nido*-dianions $7,8\text{-}C_2B_9H_{11}^{2-}$ (73) and $7,9\text{-}C_2B_9H_{11}^{2-}$ (74). It was the perceptive recognition that the open pentagonal faces of these dianions were structurally and electronically equivalent to be the pentahapto,cyclopentadienide anion $(\eta^5\text{-}C_5H_5)^-$ (Fig. 6.23) that led to the discovery of the metallocarboranes and the development of some of the most intriguing and far-reaching reactions of the carboranes. These are considered in the next section.

6.6 Metallocarboranes[1, 15, 28, 36-41]

There are six major synthetic routes to metallocarboranes: (i) coordination using *nido*-carborane anions as ligands, (ii) polyhedral expansion, (iii) polyhedral contraction, (iv) polyhedral subrogation, (v) thermal metal transfer, and (vi) direct oxidative insertion of

[36] R. N. GRIMES, Recent studies on metalloboron cage compounds derived from the small carboranes, Second International Meeting on Boron Chemistry, Leeds, 1974, *Pure Appl. Chem.* **39**, 455–74 (1974).

[37] K. P. CALLAHAN and M. F. HAWTHORNE, New chemistry of metallocarboranes and metalloboranes, Second International Meeting on Boron Chemistry, Leeds, 1974, *Pure Appl. Chem.* **39**, 475–95 (1974).

[38] G. B. DUNKS and M. F. HAWTHORNE, *Closo*-heteroboranes exclusive of carboranes, Chap. 11 in ref. 7, pp. 383–430.

[39] K. P. CALLAHAN and M. F. HAWTHORNE, Ten years of metallocarboranes, *Adv. Organometallic Chem.* **14**, 145–86 (1976).

[40] R. N. GRIMES, Reactions of metallocarboranes, Chap. 2 in E. BECHER and M. TSUTSUI (eds.), *Organometallic Reactions and Syntheses* **6**, 63–221 (1977).

[41] F. G. A. STONE, Synthetic applications of d^{10} metal complexes, *J. Organometallic Chem.* **100**, 257–71 (1975).

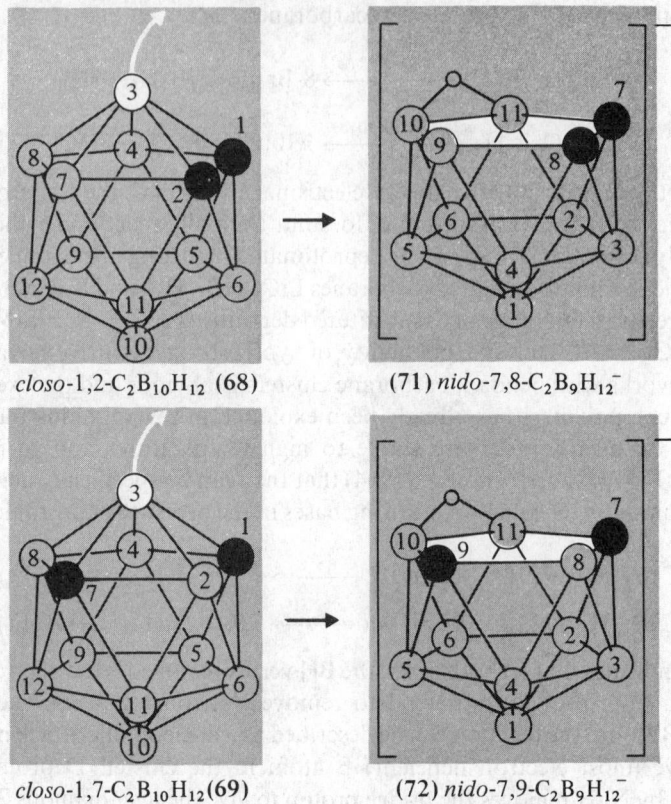

closo-1,2-C$_2$B$_{10}$H$_{12}$ **(68)**   **(71)** *nido*-7,8-C$_2$B$_9$H$_{12}^-$

closo-1,7-C$_2$B$_{10}$H$_{12}$ **(69)**   **(72)** *nido*-7,9-C$_2$B$_9$H$_{12}^-$

FIG. 6.22 Degradation of *closo*-carboranes to the corresponding debor-*nido*-carborane anions.

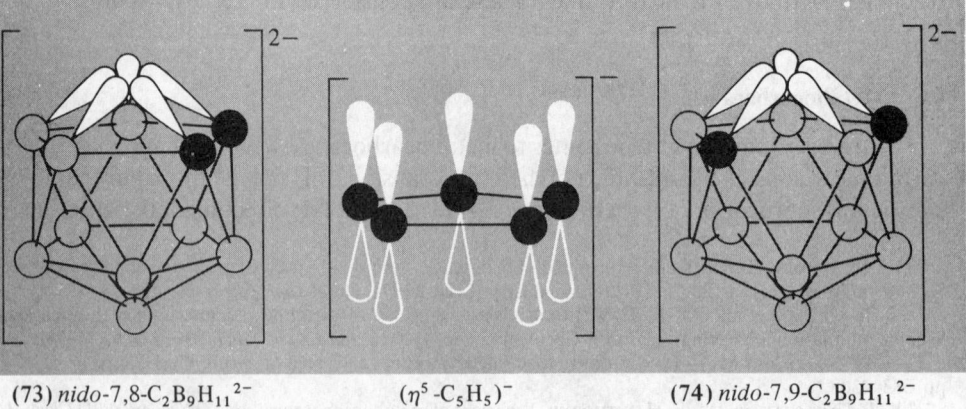

(73) *nido*-7,8-C$_2$B$_9$H$_{11}^{2-}$ $(\eta^5\text{-C}_5\text{H}_5)^-$ **(74)** *nido*-7,9-C$_2$B$_9$H$_{11}^{2-}$

FIG. 6.23 Relation between C$_2$B$_9$H$_{11}^{2-}$ and C$_5$H$_5^-$. In this formalism the *closo*-carboranes C$_2$B$_{10}$H$_{12}$ are considered as a coordination complex between the pentahapto 6-electron donor C$_2$B$_9$H$_{11}^{2-}$ and the acceptor BH^{2+} (which has 3 vacant orbitals). The *closo*-structure can be regained by capping the open pentagonal face with an equivalent metal acceptor that has 3 vacant orbitals.

the metal centre. All except the last were devised by M. F. Hawthorne and his group in the period 1965–74 and have since been extensively exploited by several groups.

(*i*) *Coordination using* nido-*carborane anions as ligands (1965).* Reaction of $C_2B_9H_{11}{}^{2-}$ with $FeCl_2$ in tetrahydrofuran (thf) with rigorous exclusion of moisture and air gives the pink, diamagnetic bis-sandwich-type complex of Fe(II) (structure 75) which can be reversibly oxidized to the corresponding red Fe(III) complex:

$$2C_2B_9H_{11}{}^{2-} + Fe^{2+} \xrightarrow{\text{thf}} [Fe^{II}(\eta^5\text{-}C_2B_9H_{11})_2]^{2-} \underset{+e^-}{\overset{\text{air}}{\rightleftharpoons}} [Fe^{III}(\eta^5\text{-}C_2B_9H_{11})_2]^-$$
$$(75)$$

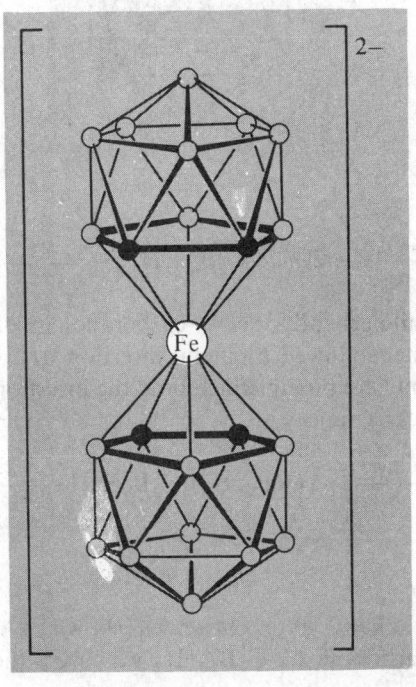

(75)

When the reaction is carried out in the presence of NaC_5H_5 the purple mixed sandwich complex (76) is obtained:

$$C_2B_9H_{11}{}^{2-} + C_5H_5{}^- + Fe^{2+} \xrightarrow[-e^-]{\text{thf}} [Fe^{III}(\eta^5\text{-}C_5H_5)(\eta^5\text{-}C_2B_9H_{11})]$$
$$(76)$$

The reaction is general and has been applied to many transition metals. Variants use metal carbonyls and other complexes to supply the capping unit, e.g.

$$C_2B_9H_{11}{}^{2-} + Mo(CO)_6 \xrightarrow{h\nu} [Mo(CO)_3(\eta^5\text{-}C_2B_9H_{11})]^{2-} + 3CO$$

(*ii*) *Polyhedral expansion (1970).* This entails the 2-electron reduction of a *closo*-carborane with a strong reducing agent such as sodium naphthalide in thf followed by reaction with a transition-metal reagent:

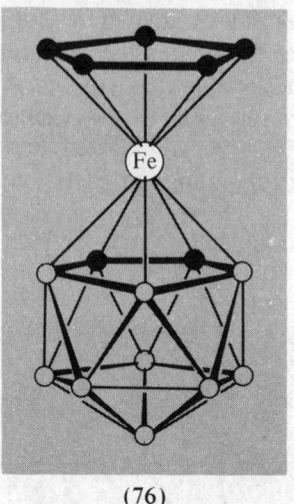

(76)

$$2[closo\text{-}C_2B_{n-2}H_n] \xrightarrow[\text{thf}]{4\text{Na}/\text{C}_{10}\text{H}_8} 2[nido\text{-}C_2B_{n-2}H_n]^{2-} \xrightarrow{M^{m+}} [M(C_2B_{n-2}H_n)_2]^{(m-4)+}$$

The reaction, which is quite general for *closo*-carboranes, involves the reductive opening of an *n*-vertex *closo*-cluster followed by metal insertion to give an $(n+1)$-vertex *closo*-cluster. Numerous variants are possible including the insertion of a second metal centre into an existing metallocarborane, e.g.:

$$closo\text{-}1,7\text{-}C_2B_6H_8 \xrightarrow[\text{CoCl}_2 + \text{NaC}_5\text{H}_5]{2e^-/\text{thf}} [Co(C_5H_5)(C_2B_6H_8)] \xrightarrow[\text{CoCl}_2 + \text{NaC}_5\text{H}_5]{2e^-/\text{thf}}$$

$$[\{Co(C_5H_5)\}_2(C_2B_6H_8)]$$

The structure of the bimetallic 10-vertex cluster was shown by X-ray diffraction to be (77). When the icosahedral carborane $1,2\text{-}C_2B_{10}H_{12}$ was used, the reaction led to the first

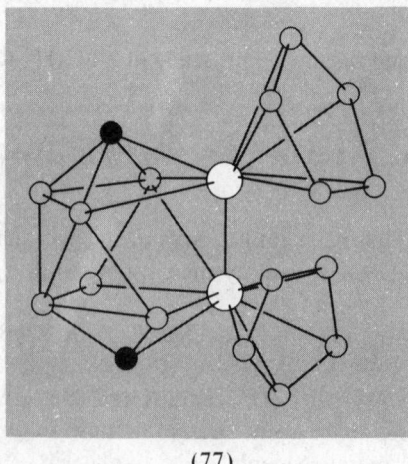

(77)

supraicosahedral metallocarboranes with 13- and 14-vertex polyhedral structures (79)–(82). Facile isomerism of the 13-vertex monometallodicarbaboranes was observed as indicated in the subjoined scheme:

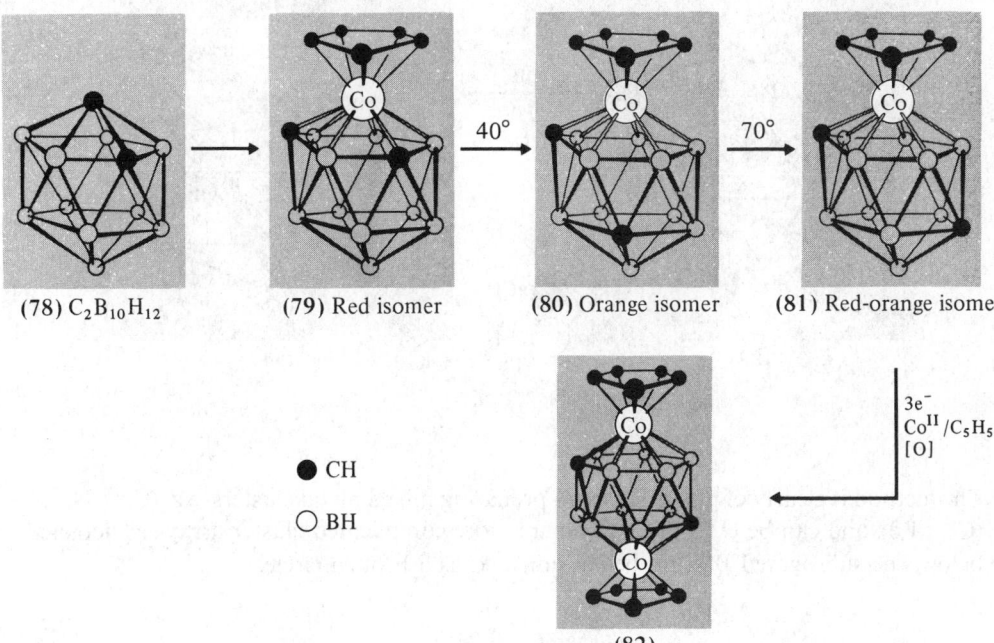

(78) $C_2B_{10}H_{12}$ (79) Red isomer   (80) Orange isomer   (81) Red-orange isomer

● CH

○ BH

$$3e^-$$
$$Co^{II}/C_5H_5$$
$$[O]$$

(82)

(*iii*) *Polyhedral contraction* (*1972*). This involves the clean removal of one BH group from a *closo*-metallocarborane by nucleophilic base degradation, followed by oxidative closure of the resulting *nido*-metallocarborane complex to a *closo*-species with one vertex less than the original, e.g.:

$$[3\text{-}\{Co(\eta^5\text{-}C_5H_5)\}(1,2\text{-}C_2B_9H_{11})] \xrightarrow[\text{(2) }H_2O_2]{\text{(1) }OEt^-} [1\text{-}\{Co(\eta^5\text{-}C_5H_5)\}(2,4\text{-}C_2B_8H_{10})]$$
$$(83) \hspace{6cm} (84)$$

Polyhedral contraction is not so general a method of preparing metallocarboranes as is polyhedral expansion since some metallocarboranes degrade completely under these conditions.

(*iv*) *Polyhedral subrogation* (*1973*). Replacement of a BH vertex by a metal vertex without changing the number of vertices in the cluster is termed polyhedral subrogation. It is an offshoot of the polyhedral contraction route in that degradative removal of the BH unit is followed by reaction with a transition metal ion rather than with an oxidizing agent, e.g.:

$$[Co(\eta^5\text{-}C_5H_5)(C_2B_{10}H_{12})] \xrightarrow[\text{(2) }Co^{II},\ C_5H_5^-]{\text{(1) }OH^-} [\{Co(\eta^5\text{-}C_5H_5)\}_2(C_2B_9H_{11})]$$

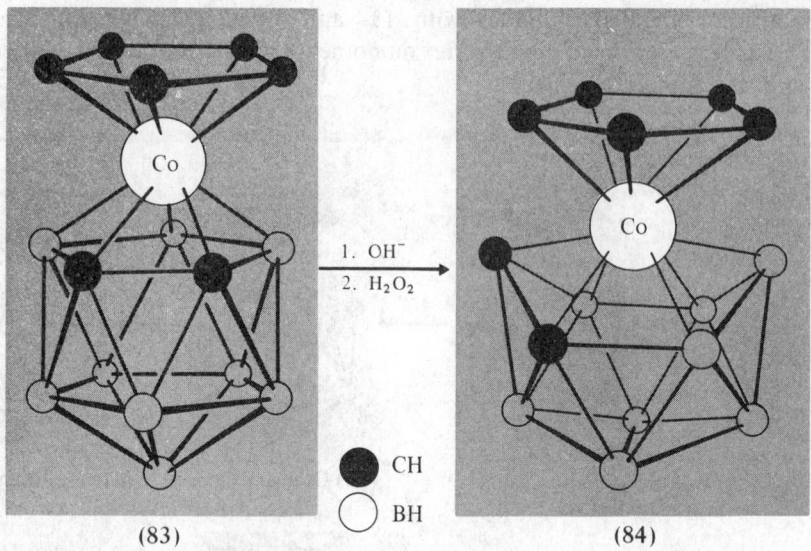

(83)   1. OH⁻
                             2. H₂O₂   (84)

● CH
○ BH

The method is clearly of potential use in preparing mixed metal clusters, e.g. (Co + Ni) or (Co + Fe), and can be extended to prepare more complicated cluster arrays as depicted below, the subrogated B atom being indicated as a broken circle.

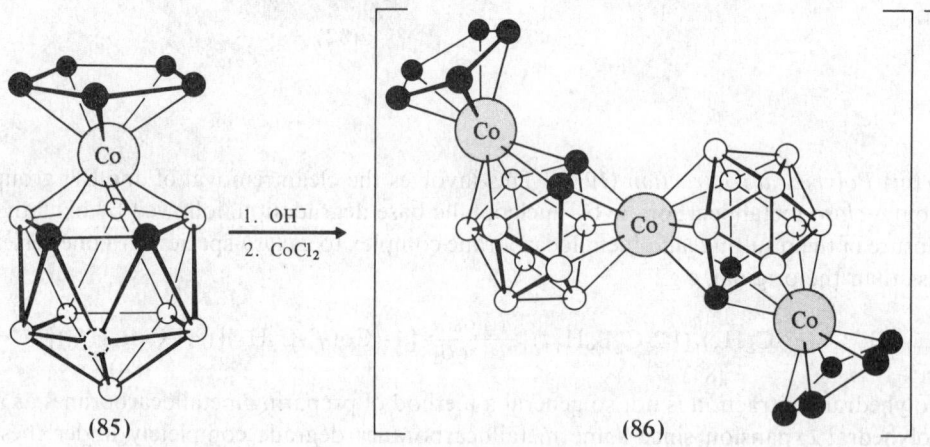

(85)   1. OH⁻
              2. CoCl₂   (86)

(*v*) *Thermal metal transfer (1974).* This method is less general and often less specific than the coordination of *nido*-anions or polyhedral expansion; it involves the pyrolysis of pre-existing metallocarboranes and consequent cluster expansion or disproportionation similar to that of the *closo*-carboranes themselves (p. 205). Mixtures of products are usually obtained, e.g.:

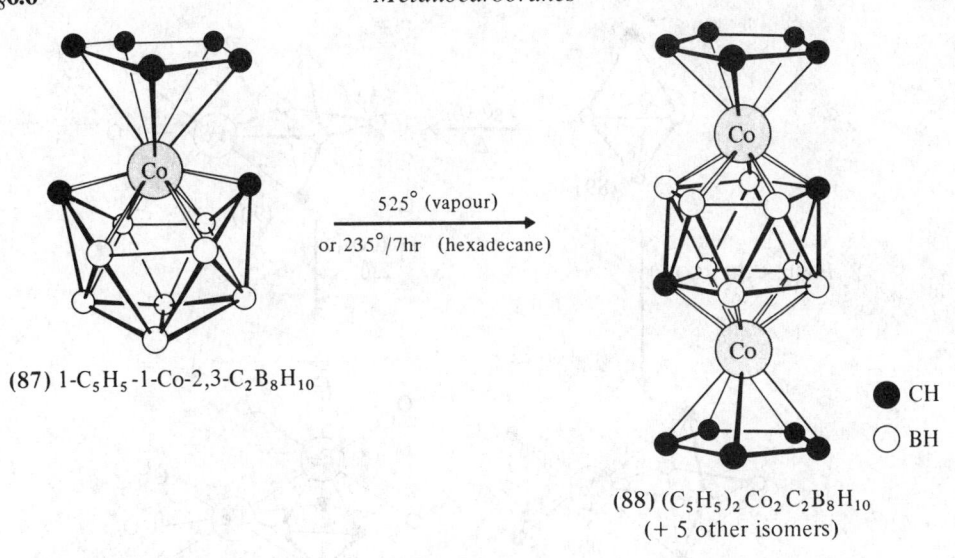

(87) 1-C$_5$H$_5$-1-Co-2,3-C$_2$B$_8$H$_{10}$

$\xrightarrow[\text{or } 235°/7\text{hr (hexadecane)}]{525° \text{ (vapour)}}$

● CH
○ BH

(88) (C$_5$H$_5$)$_2$Co$_2$C$_2$B$_8$H$_{10}$
(+ 5 other isomers)

[(C$_5$H$_5$)$_2$Co]$^+$[(2,3-C$_2$B$_8$H$_{10}$)$_2$Co]$^-$ $\xrightarrow[\text{or } 270° \text{ hexadecane}]{525° \text{ (hot tube)}}$ (C$_5$H$_5$)$_2$Co$_2$C$_2$B$_8$H$_{10}$

A related technique (R. N. Grimes, 1973) is direct metal insertion by gas-phase reactions at elevated temperatures; typical reactions are shown in the scheme (p. 216). The reaction with [Co(η^5-C$_5$H$_5$)(CO)$_2$] also gave the 7-vertex *closo*-bimetallocarborane (94) which can be considered as a rare example of a triple-decker sandwich compound; another isomer (95) can be made by base degradation of [{Co(η^5-C$_5$H$_5$)}(C$_2$B$_4$H$_6$)] followed by deprotonation and subrogation with a second {Co(η^5-C$_5$H$_5$)} unit.[36] It will be noted that the central planar formal C$_2$B$_3$H$_5$$^{4-}$ unit is isoelectronic with C$_5$H$_5$$^-$.

   (*vi*) *Direct oxidative insertion of a metal centre* (F. G. A. Stone et al., *1972*).[41] Nucleophilic zero-valent derivatives of Ni, Pd, and Pt insert directly into *closo*-carborane clusters in a concerted process which involves a net transfer of electrons from the metal to the cage:

$$M^0L_{x+y} + C_2B_nH_{n+2} \longrightarrow [M^{II}L_x(C_2B_nH_{n+2})] + yL,$$

where L = PR$_3$, C$_8$H$_{12}$, RNC, etc. A typical reaction is

$$Pt(PEt_3)_3 + 2,3\text{-Me}_2\text{-}2,3\text{-C}_2B_9H_9 \xrightarrow[\text{petrol}]{-30°} [1\text{-}\{Pt(PEt_3)_2\}\text{-}2,4\text{-}(MeC)_2B_9H_9] + PEt_3$$

Many novel cluster compounds have now been prepared in this way, including mixed metal clusters, and the structures sometimes have unexpectedly more open configurations than simple electron-counting rules would predict.[41]

   The electron-counting rules outlined for boranes (p. 185) and carboranes (pp. 203–4) can readily be extended to the metallocarboranes (see Panel). For bis-complexes of 1,2-C$_2$B$_{10}$H$_{11}$$^{2-}$ which can be regarded as 6-electron penta-hapto ligand, it has been found that "electron-sufficient" (18-electron) systems such as those involving d^6 metal centres (e.g. Fe^{II}, Co^{III}, or Ni^{IV}) have symmetrical structures with the metal atom equidistant from

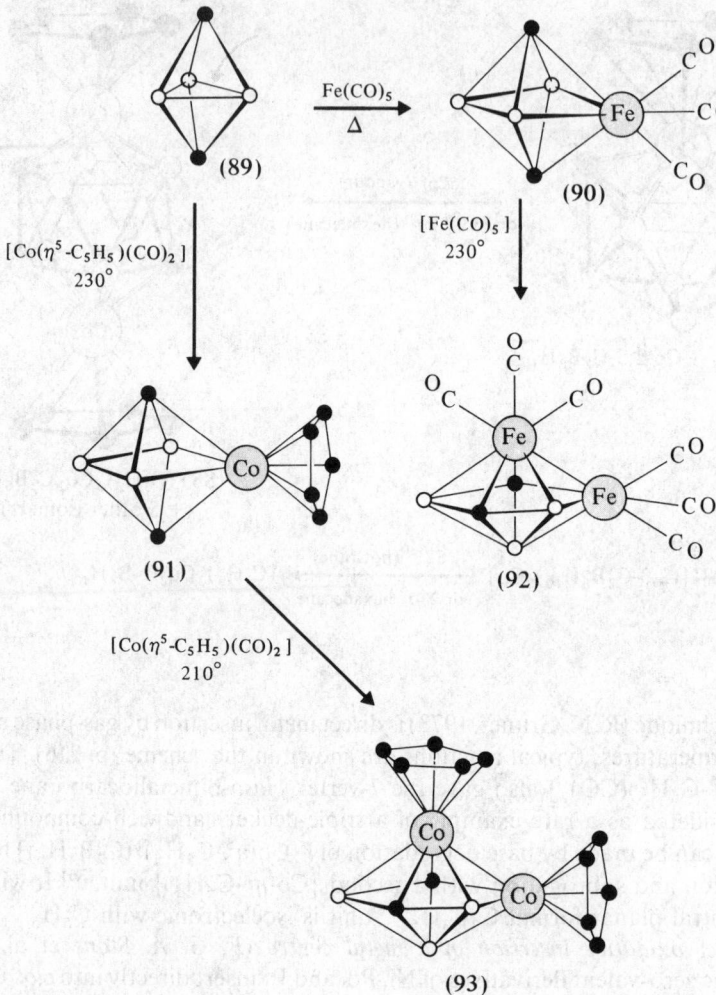

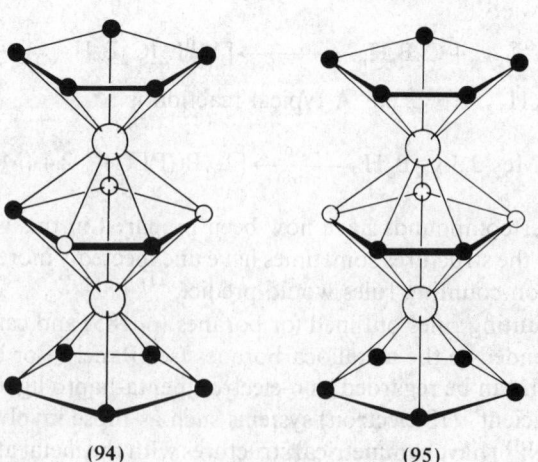

Electron-counting Rules for Metallocarboranes and Other Heteroboranes

As indicated on pp. 182 and 198 each framework atom (except H) uses 3 atomic orbitals (AOs) in cluster bonding. For B, C, and other first-row elements this leaves one remaining AO to bond exopolyhedrally to $-H_t$, $-X$, $-R$, etc. In contrast, transition elements have a total of 9 valence AOs (five d, one s, three p). Hence, after contributing 3 AOs to the cluster, they have 6 remaining AOs which can be used for bonding to external ligands and for storage of nonbonding electrons. In *closo*-clusters the $(n+1)$ MOs require $(2n+2)$ electrons from the B, C, and M vertex atoms. In its simplest form the electron counting scheme invokes only the total number of framework MOs and electrons, and requires no assumptions as to orbital hybridization or formal oxidation state. For example, the neutral moiety $\{Fe(CO)_3\}$ has 8 Fe electrons and 9 Fe AOs: since 3 AOs are involved in bonding to 3 CO and 3 AOs are used in cluster bonding, there remain 3 AOs which can accommodate 6 (nonbonding) Fe electrons, leaving 2 Fe electrons to be used in cluster bonding. Neutral $\{Fe(CO)_3\}$ is thus precisely equivalent to $\{BH\}$, as distinct from $\{CH\}$, which provides 3 electrons for the cluster. Other groups such as $\{Co(\eta^5-C_5H_5)\}$ and $\{Ni(CO)_2\}$ are clearly equivalent to $\{Fe(CO)_3\}$.

An alternative scheme that is qualitatively equivalent is to assign formal oxidation states to the metal moiety and to consider the bonding as coordination from a carborane ligand, e.g. $\{Fe(CO)_3\}^{2+}$ η^5-bonded to a cyclocarborane ring as in $\{C_2B_9H_{11}\}^{2-}$. This is acceptable when the anionic ligand is well characterized as an independent entity, as in the case just cited, but for most metallocarboranes the "ligands" are not known as free species and the presumed anionic charge and metal oxidation state become somewhat arbitrary. It is therefore recommended that the metalloborane cluster be treated as a single covalently bonded structure with no artificial separation between the metal and the rest of the cluster; electron counting can then be done unambiguously on the basis of neutral atoms and attached groups.

To the structural generalizations on carboranes (p. 207) can be added the rule that, in metallocarboranes, the M atom tends to adopt a vertex with high coordination number; M occupancy of a low CN vertex is not precluded, particularly in kinetically controlled syntheses, but isomerization to more stable configurations usually results in the migration of M to high CN vertices.

Other main-group atoms besides C can occur in heteroborane clusters and the electron-counting rules can readily be extended to them.[15] Thus, whereas each $\{BH\}$ contributes 2 e and $\{CH\}$ contributes 3 e to the cluster, so $\{NH\}$ or $\{PH\}$ contributes 4 e, $\{SH\}$ contributes 5 e, $\{S\}$ contributes 4 e, etc. For example, the following 10-vertex thiaboranes (and their isoelectronic equivalents) are known: *closo*-1-SB$_9$H$_9$ ($B_{10}H_{10}^{2-}$), *nido*-6-SB$_9$H$_{11}$ ($B_{10}H_{13}^{-}$), and *arachno*-6-SB$_9$H$_{12}^{-}$ ($B_{10}H_{14}^{2-}$). Similarly, the structures of 12-, 11-, and 9-vertex thiaboranes parallel those of boranes and carboranes with the same skeletal electron count, the S atom in each case contributing 4 electrons to the framework plus an *exo*-polyhedral lone-pair.

the 2 C and 3 B atoms in the pentagonal face. The same is true for "electron-deficient" systems such as those involving d^2 Ti^{II} (14-electron), d^3 Cr^{III} (15-electron), etc., though here the metal-cluster bonds are somewhat longer. With "electron excess" complexes such as $[Ni^{II}(C_2B_{10}H_{11})_2]^{2-}$ and the corresponding complexes of Pd^{II}, Cu^{III}, and Au^{III} (20 electrons), so-called "slipped-sandwich" structures (96) are observed in which the metal atom is significantly closer to the 3 B atoms than to the 2 C atoms. This has been thought by some to indicate π-allylic bonding to the 3 B but is more likely to arise from an occupation of orbitals that are antibonding with respect to both the metal and the cluster thereby leading to an opening of the 12-vertex *closo*-structure to a pseudo-*nido* structure in which the 12 atoms of the cluster occupy 12 vertices of a 13-vertex polyhedron.[41a] A similar type of distortion accompanies the use of metal centres with increasing numbers of

[41a] D. M. P. MINGOS, M. I. FORSYTH, and A. J. WELCH, X-ray crystallographic and theoretical studies on "slipped" metallocarboranes, *JCS Chem. Comm.* 1977, 605–7. See also G. K. BARKER, M. GREEN, F. G. A. STONE, and A. J. WELCH, *JCS Dalton* 1980, 1186–99; D. M. P. MINGOS and A. J. WELCH, ibid. 1674–81.

electron-pairs on the metal and it seems that these electrons may also, at least in part, contribute to the framework electron count with consequent cluster opening.[42] Thus, progressive opening of the cluster is noted for complexes of $1,2\text{-}C_2B_9H_{11}{}^{2-}$ with Re^I (d^6), Au^{III} (d^8), Hg^{II} (d^{10}), and Tl^I ($d^{10}s^2$) as shown in structures (97)–(100). Thus the Re^I (d^6) (97) is a symmetrically bonded 12-vertex cluster with Re–B 234 pm and Re–C 231 pm. The Au^{III} (d^8) complex (98) has the metal appreciably closer to the 3 B atoms (221 pm) than to the 2 C atoms (278 pm). With the Hg^{II} (d^{10}) complex (99) this distortion is even more pronounced and the metal is pseudo-σ-bonded to 1 B atom at 220 pm; there is some additional though weak interaction with the other 2 B (252 pm) but the two Hg$\cdots$C distances (290 pm) are essentially nonbonding. Finally, the Tl^I ($d^{10}s^2$) complex (100),

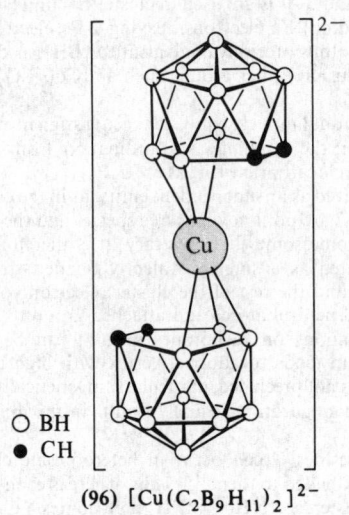

O BH
● CH

(96) $[Cu(C_2B_9H_{11})_2]^{2-}$

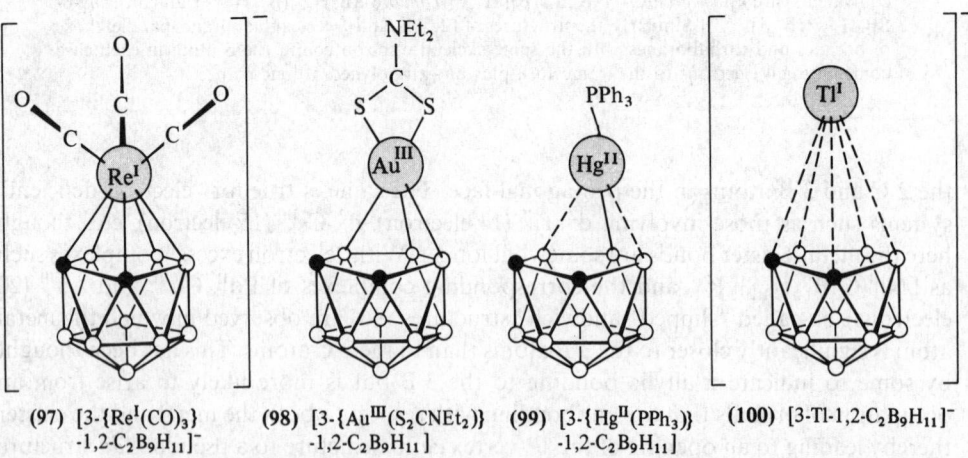

(97) $[3\text{-}\{Re^I(CO)_3\}$ $\text{-}1,2\text{-}C_2B_9H_{11}]^-$

(98) $[3\text{-}\{Au^{III}(S_2CNEt_2)\}$ $\text{-}1,2\text{-}C_2B_9H_{11}]$

(99) $[3\text{-}\{Hg^{II}(PPh_3)\}$ $\text{-}1,2\text{-}C_2B_9H_{11}]$

(100) $[3\text{-}Tl\text{-}1,2\text{-}C_2B_9H_{11}]^-$

[42] H. M. COLQUHOUN, T. J. GREENHOUGH, and M. G. H. WALLBRIDGE, Progressive cage-opening in d^6, d^8, and $d^{10}s^2$ metallocarbaboranes: crystal and molecular structures of $[3\text{-}\{Hg(PPh_3)\}\text{-}1,2\text{-}C_2B_9H_{11}]$ and $[PMePh_3]^+[3\text{-}Tl\text{-}1,2\text{-}C_2B_9H_{11}]$, *JCS Chem. Comm.* **1977**, 737–8; see also H. M. COLQUHOUN, T. J. GREENHOUGH, and M. G. H. WALLBRIDGE, *JCS Chem. Comm.* **1976**, 1019–20; **1977**, 737–8; *JCS Dalton* **1979**, 619–28; *JCS Chem. Comm.* **1980**, 192–4; G. K. BARKER, M. GREEN, F. G. A. STONE, A. J. WELCH, and W. C. WOLSEY, *JCS Chem. Comm.* **1980**, 627–9.

whilst having the Tl atom more symmetrically located above the open face, has Tl-cluster distances that exceed considerably the expected covalent Tl^I-B distance of ~ 236 pm; the shortest Tl B distance is 266 pm and there are two other Tl–B at 274 pm and two Tl–C at 292 pm: the species can thus be regarded formally as being closer to an ion pair $[Tl^+(C_2B_9H_{11})^{2-}]$.

In general, metallocarboranes are much less reactive (more stable) than the corresponding metallocenes and they tend to stabilize higher oxidation states of the later transition metals, e.g. $[Cu^{II}(1,2\text{-}C_2B_9H_{11})_2]^{2-}$ and $[Cu^{III}(1,2\text{-}C_2B_9H_{11})_2]^-$ are known whereas cuprocene $[Cu^{II}(\eta^5\text{-}C_5H_5)_2]$ is not. Likewise, Fe^{III} and Ni^{IV} carborane derivatives are extremely stable. Conversely, metallocarboranes tend to stabilize lower oxidation states of early transition elements and complexes are well established for Ti^{II}, Zr^{II}, Hf^{II}, V^{II}, Cr^{II}, and Mn^{II}: these do not react with H_2, N_2, CO, or PPh_3 as do cyclopentadienyl derivatives of these elements.

Ferrocene can be protonated with strong acids to yield the cationic species $[Fe(\eta^5\text{-}C_5H_5)_2H]^+$, in which the proton is associated with the Fe atom; the carborane analogue $[Fe^{II}(\eta^5\text{-}1,2\text{-}C_2B_9H_{11})_2H]^-$ can be prepared similarly, using HCl or $HClO_4$ and the protonated species is unique in promoting substitution at polyhedral B atoms in good yields, e.g. with weak Lewis bases such as dialkyl sulfides:

$$[Fe^{II}(\eta^5\text{-}1,2\text{-}C_2B_9H_{11})_2H]^- + SR_2 \longrightarrow$$
$$[Fe^{II}(\eta^5\text{-}1,2\text{-}C_2B_9H_{11})(\eta^5\text{-}1,2\text{-}C_2B_9H_{10}SR_2)]^- + H_2$$

Bromination of the corresponding Co^{III} complex in glacial acetic acid gives the hexabromo derivative $[Co^{III}(\eta^5\text{-}8,9,12\text{-}Br_3\text{-}1,2\text{-}C_2B_9H_8)_2]^-$ (structure 101) in which electrophilic substitution has occurred on the triangular face furthest removed from the two C-atoms, i.e. at the three B atoms that are expected to have the greatest electron density. Again, acid catalysed reaction of $K[Co(\eta^5\text{-}1,2\text{-}C_2B_9H_{11})_2]$ with CS_2 in the presence of $HCl/AlCl_3$ gives the novel neutral compound $[Co(\eta^5\text{-}1,2\text{-}C_2B_9H_{10})_2(8,8'\text{-}S_2CH)]$ (structure 102) in which the $8,8'$-B atoms on the two icosahedra have been

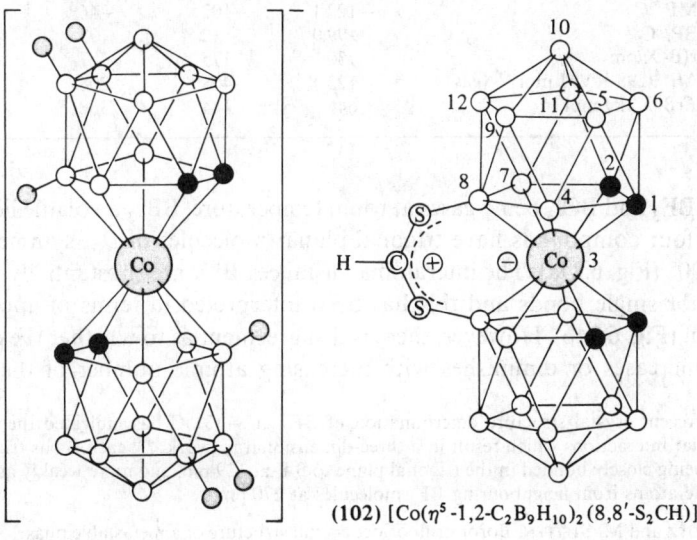

(101) $[Co(\eta^5\text{-}8,9,12\text{-}Br_3\text{-}1,2\text{-}C_2B_9H_8)_2]^-$

(102) $[Co(\eta^5\text{-}1,2\text{-}C_2B_9H_{10})_2(8,8'\text{-}S_2CH)]$

substituted by S and the polyhedra are linked by an S_2CH^+ bridge in a zwitterionic structure.

The chemistry of metallocarboranes of all cluster sizes is still rapidly developing and further unusual reactions are to be expected.

6.7 Boron Halides

Boron forms numerous binary halides of which the monomeric trihalides BX_3 are the most stable and most extensively studied. They can be regarded as the first members of a homologous series B_nH_{n+2}. The second members B_2X_4 are also known for all 4 halogens but only F forms more highly catenated species containing BX_2 groups: B_3F_5, $B_4F_6 \cdot L$, B_8F_{12} (p. 227). Chlorine forms a series of neutral *closo*-polyhedral compounds B_nCl_n ($n = 4, 8-12$) and several similar compounds are known for Br ($n = 7-10$) and I (e.g. B_9I_9). There are also numerous involatile subhalides, particularly of Br and I, but these are of uncertain stoichiometry and undetermined structure.

6.7.1 *Boron trihalides*

The boron trihalides are volatile, highly reactive, monomeric molecular compounds which show no detectable tendency to dimerize (except perhaps in Kr matrix-isolation experiments at 20K).† In this they resemble organoboranes, BR_3, but differ sharply from diborane, B_2H_6, and the aluminium halides, Al_2X_6, and alkyls, Al_2R_6 (p. 289). Some physical properties are listed in Table 6.3; mps and volatilities parallel those of the parent

TABLE 6.3 *Some physical properties of boron trihalides*

Property	BF_3	BCl_3	BBr_3	BI_3
MP/°C	-127.1	-107	-46	49.9
BP/°C	-99.9	12.5	91.3	210
r(B–X/pm	130	175	187	210
ΔH_f° (298 K)/kJ mol^{-1} (gas)	-1123	-408	-208	$+$
E(B–X)/kJ mol^{-1}	646	444	368	267

halogens, BF_3 and BCl_3 being gases at room temperature, BBr_3 a volatile liquid, and BI_3 a solid. All four compounds have trigonal planar molecules of D_{3h} symmetry with angle X–B–X 120° (Fig. 6.24a). The interatomic distances B–X are substantially less than those expected for single bonds and this has been interpreted in terms of appreciable p_π–p_π interaction (Fig. 6.24b). However, there is disagreement as to whether the extent of this π bonding increases or diminishes with increasing atomic number of the halogen; this

---

† A very recent crystal structure determination of BF_3 at $-131°C$ has indicated the presence of weak intermolecular interactions which result in a three-dimensional network:[43] each B has trigonal bipyramidal symmetry, being closely bonded in the trigonal plane to 3 F at 129 pm, and more weakly associated with two axial fluorine atoms from neighbouring BF_3 molecules at 270 pm.

[43] D. MOOTZ and M. STEFFEN, Boron trifluoride: crystal structure of a metastable phase, *Angew. Chem.* Int. Edn. (Engl.), **19**, 483–4 (1980).

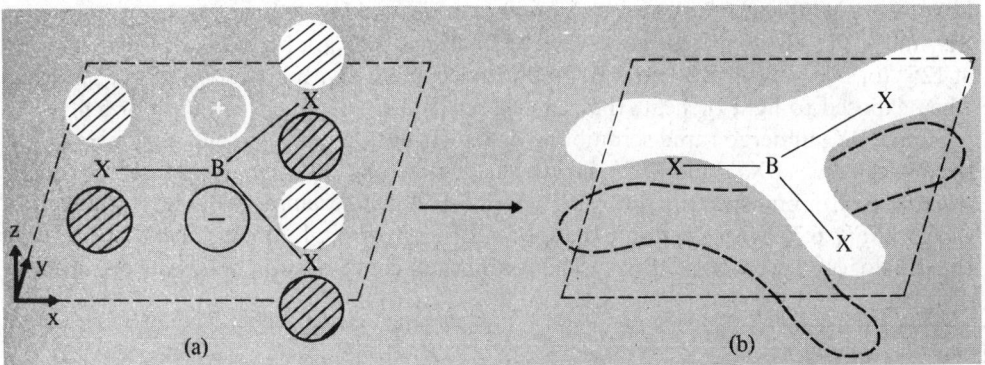

FIG. 6.24   Schematic indication of the p_π–p_π interaction between the "vacant" p_z orbital on B and the 3 filled p_z orbitals on the 3 X atoms leading to a bonding MO of π symmetry about the plane of the molecule.

probably reflects the differing criteria used (extent of orbital overlap, percentage π-bond character, amount of π-charge transfer from X to B, π-bond energy, or reorganization energy in going from planar BX_3 to tetrahedral LBX_3, etc.).[44] For example, it is quite possible for the extent of π-charge transfer from X to B to increase in the sequence $F < Cl < Br < I$ but for the actual magnitude of the π-bond energy to be in the reverse sequence $BF_3 > BCl_3 > BBr_3 > BI_3$ because of the much greater bond energy of the lighter homologues. Indeed, the mean B–F bond energy in BF_3 is 646 kJ mol^{-1}, which makes it the strongest known "single" bond; if $x\%$ of this were due to π bonding, then even if $2.4x\%$ of the B–I bond energy were due to π bonding, the π-bond energy in BI_3 would be less than that in BF_3 in absolute magnitude. The point is one of some importance since the chemistry of the trihalides is dominated by interactions involving this orbital.

BF$_3$ is used extensively as a catalyst in various industrial processes (p. 224) and can be prepared on a large scale by the fluorination of boric oxide or borates with fluorspar and concentrated H_2SO_4:

$$6CaF_2 + Na_2B_4O_7 + 8H_2SO_4 \longrightarrow 2NaHSO_4 + 6CaSO_4 + 7H_2O + 4BF_3$$

Better yields are obtained in the more modern two-stage process:

$$Na_2B_4O_7 + 12HF \xrightarrow{-6H_2O} [Na_2O(BF_3)_4] \xrightarrow{+2H_2SO_4} 2NaHSO_4 + H_2O + 4BF_3$$

On the laboratory scale, pure BF_3 is best made by thermal decomposition of a diazonium tetrafluoroborate (e.g. $PhN_2BF_4 \rightarrow PhF + N_2 + BF_3$). BCl_3 and BBr_3 are prepared on an industrial scale by direct halogenation of the oxide in the presence of C, e.g.:

$$B_2O_3 + 3C + 3Cl_2 \xrightarrow{500°C} 6CO + 2BCl_3$$

^{44} Some key references will be found in D. R. ARMSTRONG and P. G. PERKINS, The ground-state properties of the Group III trihalides, *J. Chem. Soc.* (A), 1967, 1218–22; and in M. F. LAPPERT, M. R. LITZOW, J. B. PEDLEY, P. N. K. RILEY, and A. TWEEDALE, Ionization potentials of boron halides and mixed halides by electron impact and by molecular orbital calculations, *J. Chem. Soc.* (A), 1968, 3105–10.

Laboratory samples of the pure compounds can be made by halogen exchange between BF_3 and Al_2X_6. BI_3 is made in good yield by treating $LiBH_4$ (or $NaBH_4$) with elemental I_2 at 125° (or 200°). Both BBr_3 and BI_3 tend to decompose with liberation of free halogen when exposed to light or heat; they can be purified by treatment with Hg or Zn/Hg.

Simple BX_3 undergo rapid scrambling or redistribution reactions on being mixed and the mixed halides BX_2Y and BXY_2 have been identified by vibrational spectroscopy, mass spectrometry, or nmr spectroscopy using ^{11}B or ^{19}F. A good example of this last technique is shown in Fig. 6.25, where not only the species $BF_{3-n}X_n$ ($n = 0, 1, 2$) were observed but also the trihalogeno species BFClBr.[45] The equilibrium concentration of the various species

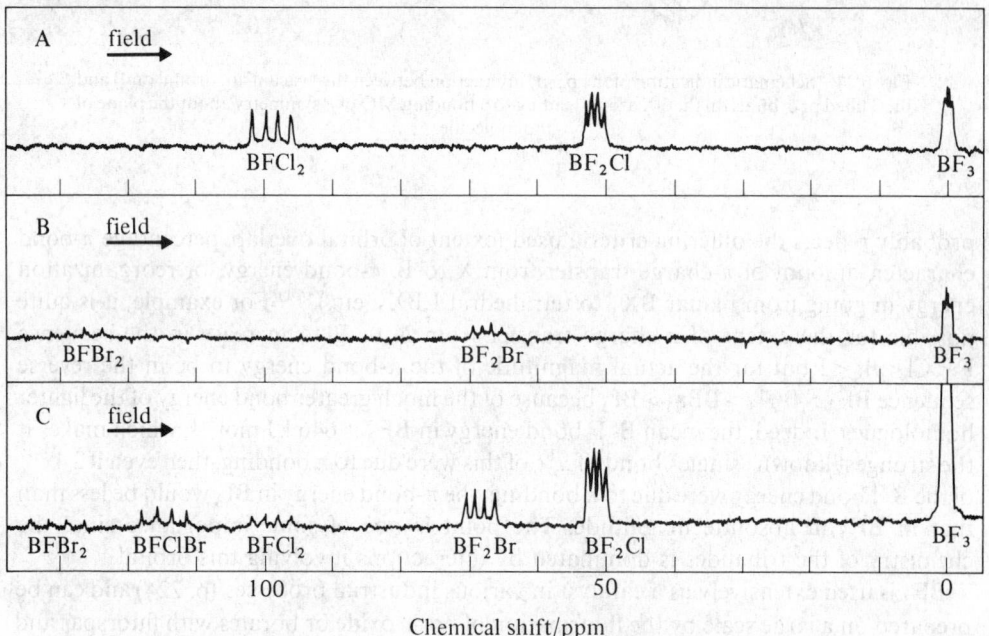

Fig. 6.25   Fluorine-19 nmr spectra of mixtures of boron halides showing the presence of mixed fiuorohalogenoboranes.

are always approximately random (equilibrium constants between 0.5 and 2.0) but it is not possible to isolate individual mixed halides because the equilibrium is too rapidly attained from either direction (< 1 s). The related systems $RBX_2/R'BY_2$ (and $ArBX_2/Ar'BY_2$) also exchange X and Y but not R (or Ar). The scrambling mechanism probably involves a 4-centre transition state (Fig. 6.26). Consistent with this, complexes such as Me_2OBX_3 or Me_3NBX_3 do not scramble at room temperature, or even above, in the absence of free BX_3[46] (cf. the stability of $CFCl_3$, CF_2Cl_2, etc.). Again, species that are expected to form stronger π bonds (such as R_2NBX_2) exchange much more slowly (days or weeks).

The boron trihalides form a great many molecular addition compounds with molecules

[45] T. D. COYLE and F. G. A. STONE, NMR studies on mixed boron halides: detection of the new halide BBrClF, *J. Chem. Phys.* **32**, 1892–3 (1960).

[46] J. S. HARTMAN and J. M. MILLER, Adducts of the mixed trihalides of boron, *Adv. Inorg. Chem. Radiochem.* **21**, 147–77 (1978).

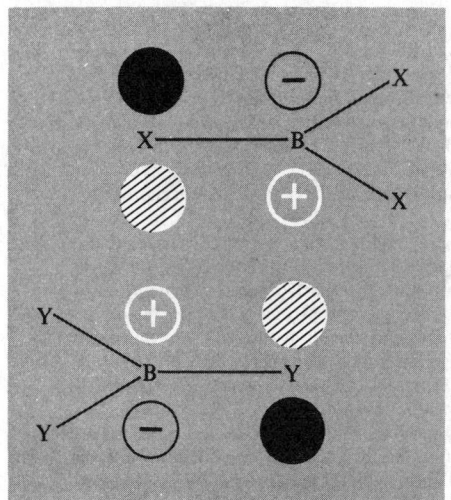
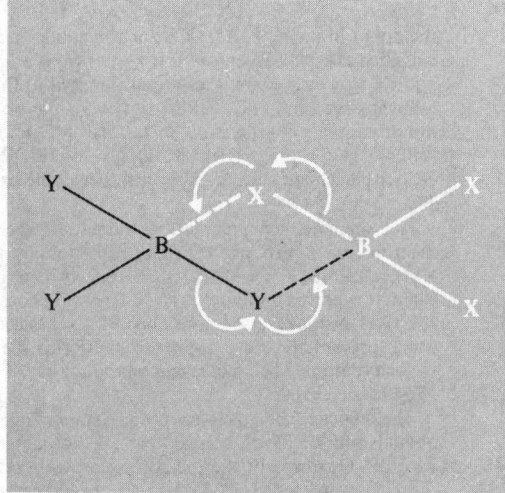

FIG. 6.26   Schematic representation of the favoured 4-centre mechanism for scrambling reactions of boron trihalides.

(ligands) possessing a lone-pair of electrons (Lewis base). Such adducts have assumed considerable importance since it is possible to investigate in detail the process of making and breaking one bond, and to study the effect this has on the rest of the molecule (see Panel). The tetrahalogeno borates BX_4^- are a special case in which the ligand is X^-; they are isoelectronic with BH_4^- (p. 188) and with CH_4 and CX_4. Salts of BF_4^- are readily formed by adding a suitable metal fluoride to BF_3 either in the absence of solvent or in such nonaqueous solvents as HF, BrF_3, AsF_3, or SO_2. The alkali metal salts MBF_4 are stable to hydrolysis in aqueous solutions. Some molecular fluorides such as NO_2F and RCOF react similarly. There is a significant lengthening of the B–F bond from 130 pm in BF_3 to 145 pm in BF_4^-. The other tetrahalogenoborates are less stable but may be prepared using large counter cations, e.g. Rb, Cs, pyridinium, tetraalkylammonium, tropenium, triphenylcarbonium, etc.

Factors Affecting the Stability of Donor Acceptor Complexes

For a given ligand, stability of the adduct LBX_3 usually increases in the sequence $BF_3 < BCl_3 < BBr_3 < BI_3$, probably because the loss of π bonding on reorganization from planar to tetrahedral geometry (p. 221) is not fully compensated for by the expected electronegativity effect. However, if the ligand has an H atom directly bonded to the donor atom, the resulting complex is susceptible to protonolysis of the B–X bond, e.g.:

$$ROH + BX_3 \longrightarrow [ROH . BX_3] \longrightarrow ROBX_2 + HX$$

In such cases the great strength of the B–F bond ensures that the BF_3 complex is more stable than the others. For example, BF_3 forms stable complexes with H_2O, MeOH, Me_2NH, etc., whereas BCl_3 reacts rapidly to give $B(OH)_3$, $B(OMe)_3$, and $B(NMe_2)_3$: with BBr_3 and BI_3 such protolytic reactions are sometimes of explosive violence. Even ethers may be cleaved by BCl_3 to give RCl and $ROBCl_2$, etc.

For a given BX_3, the stability of the complex depends on (a) the chemical nature of the donor atom, (b) the presence of polar substituents on the ligand, (c) steric effects, (d) the stoichiometric ratio of ligand to acceptor, and (e) the state of aggregation. Thus the majority of adducts have as the donor atom N, P, As; O, S; or the halide and hydride ions X^-. BX_3 (but not BH_3) can be classified as type-a acceptors, forming stronger complexes with N, O, and F ligands than with P, S, and Cl. However, complexes are not limited to these traditional main-group donor atoms, and, following the work of D. F. Shriver (1963), many complexes have been characterized in which the donor atom is a transition metal, e.g. $[(C_5H_5)_2H_2W^{IV} \rightarrow BF_3]$, $[(Ph_3P)_2(CO)ClRh^I \rightarrow BBr_3]$, $[(Ph_2PCH_2CH_2PPh_2)_2Rh^I(BCl_3)_2]^+$, $[(Ph_3P)_2(CO)ClIr^I(BF_3)_2]$, $[(Ph_3P)_2Pt^0(BCl_3)_2]$, etc. Displacement studies on several such complexes indicate that BF_3 is a weaker acceptor than BCl_3.

The influence of polar substituents on the ligand follows the expected sequence for electronegative groups, e.g. electron donor properties decrease in the order $NMe_3 > NME_2Cl > NMeCl_2 \gg NCl_3$. Steric effects can also limit the electron-donor strength. For example, whereas pyridine, C_5H_5N, is a weaker base (proton acceptor) than 2-MeC_5H_4N and $2,6\text{-}Me_2C_5H_3N$, the reverse is true when BF_3 is the acceptor due to steric crowding of the α-Me groups which prevent the close approach of BF_3 to the donor atom. Steric effects also predominate in determining the decreasing stability of BF_3 etherates in the sequence $C_4H_8O(\text{thf}) > Me_2O > Et_2O > Pr^i_2O$.

The influences of stoichiometry and state of aggregation are more subtle. At first sight it is not obvious why BF_3, with 1 vacant orbital should form not only $BF_3.H_2O$ but also the more stable $BF_3.2H_2O$; similarly, the 1:2 complexes with ROH and RCOOH are always more stable than the 1:1 complexes. The second mole of ligand is held by hydrogen bonding in the solid, e.g. $BF_3.OH_2\cdots OH_2$; however, above the mp 6.2°C the compound melts and the act of coordinate-bond formation causes sufficient change in the electron distribution within the ligand that ionization ensues and the compound is virtually completely ionized as a molten salt:[47]

The greater stability of the 1:2 complex is thus seen to be related to the formation of H_3O^+, ROH_2^+, etc., and the lower stability of the 1:1 complexes HBF_3OH, HBF_3OR, is paralleled by the instability of some other anhydrous oxo acids, e.g. H_2CO_3. The mp of the hydrate is essentially the transition temperature between an H-bonded molecular solid and an ionically dissociated liquid. A transition in the opposite sense occurs when crystalline $[PCl_4]^+[PCl_6]^-$ melts to give molecular PCl_5 (p. 571) and several other examples are known. The fact that coordination can substantially modify the type of bonding should occasion no surprise: the classic example (first observed by J. Priestley in 1774) was the reaction $NH_3(g) + HCl(g) \rightarrow NH_4Cl(c)$.

The importance of the trihalides as industrial chemicals stems partly from their use in preparing crystalline boron (p. 157) but mainly from their ability to catalyse a wide variety of organic reactions.[48] BF_3 is the most widely used but BCl_3 is employed in special cases. Many of the reactions are of the Friedel–Crafts type and are perhaps not strictly catalytic since BF_3 is required in essentially equimolar quantities with the reactant. The mechanism is not always fully understood but it is generally agreed that in most cases ionic intermediates are produced by or promoted by the formation of a BX_3 complex;

[47] N. N. GREENWOOD and R. L. MARTIN, Boron trifluoride coordination compounds, *Qt. Revs.* **8**, 1–39 (1954).

[48] G. OLAH (ed.), *Friedel–Crafts and Related Reactions*, Interscience, New York, 1963 (4 vols).

electrophilic attack of the substrate by the cation so produced completes the process. For example, in the Friedel–Crafts-type alkylation of aromatic hydrocarbons:

$$RX + BF_3 \rightleftharpoons \{R^+\}\{BF_3X^-\}$$
$$\{R^+\} + PhH \rightleftharpoons PhR + \{H^+\}$$
$$\{H^+\} + \{BF_3X^-\} \rightleftharpoons BF_3 + HX$$

Similarly, ketones are prepared via acyl carbonium ions:

$$RCOOMe + BF_3 \rightleftharpoons \{RCO^+\}\{BF_3(OMe)^-\}$$
$$\{RCO^+\} + PhH \rightleftharpoons PhCOR + \{H^+\}$$
$$\{H^+\} + \{BF_3(OMe)^-\} \rightleftharpoons MeOH . BF_3$$

Evidence for many of these ions has been extensively documented.[47]

$$ROH + BF_3 \rightleftharpoons \{H^+\} + \{BF_3(OR)^-\}$$

$$\{H^+\} + ROH \rightleftharpoons \{ROH_2^+\} \xrightarrow{BF_3} \{R^+\} + H_2OBF_3$$

$$\{R^+\} + ROH \longrightarrow R_2O + \{H^+\}$$

A similar mechanism has been proposed for the esterification of carboxylic acids:

$$\{H^+\} + RCOOH \rightleftharpoons \{RCOOH_2^+\} \xrightarrow{BF_3} \{RCO^+\} + H_2OBF_3$$

$$\{RCO^+\} + R'OH \longrightarrow RCOOR' + \{H^+\}$$

Nitration and sulfonation of aromatic compounds probably occur via the formation of the nitryl and sulfonyl cations:

$$HONO_2 + BF_3 \rightleftharpoons \{NO_2^+\} + \{BF_3(OH)^-\}$$
$$HOSO_3H + BF_3 \rightleftharpoons \{SO_3H^+\} + \{BF_3(OH)^-\}$$

Polymerization of alkenes and the isomerization of alkanes and alkenes occur in the presence of a cocatalyst such as H_2O, whereas the cracking of hydrocarbons is best performed with HF as cocatalyst. These latter reactions are of major commercial importance in the petrochemicals industry.

6.7.2 *Lower halides of boron*

B_2F_4 (mp $-56°$, bp $-34°C$) has a planar (D_{2h}) structure with a rather long B–B bond; in this it resembles the oxalate ion $C_2O_4{}^{2-}$ and N_2O_4 with which it is precisely isoelectronic.

Crystalline B_2Cl_4 (mp $-92.6°C$) has the same structure, but in the gas phase (bp $65.5°$) it adopts the staggered D_{2d} configuration with hindered rotation about the B–B bond (ΔE_r 7.7 kJ mol^{-1}). The structure of gaseous B_2Br_4 is also D_{2d} with B–B 169 pm and ΔE_r 12.8 kJ mol^{-1}. B_2I_4 is presumably similar.

B_2Cl_4 was the first compound in this series to be prepared and is the most studied; it is

F
F
F
F
120°
B
B
172 pm
132 pm
Cl
Cl
Cl
Cl
120°
B
B
175 pm
173 pm
solid (D_{2h})
Cl
Cl
Cl
Cl
120°
B
B
170 pm
175 pm
gas (D_{2d})

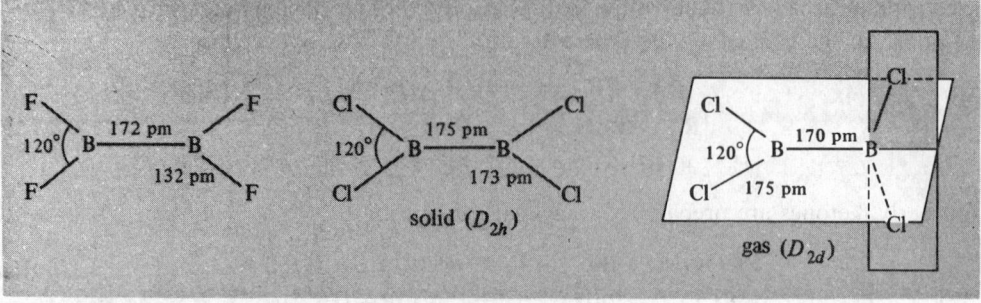

best made by subjecting BCl_3 vapour to an electrical discharge between mercury or copper electrodes:

$$2BCl_3 + 2Hg \longrightarrow B_2Cl_4 + Hg_2Cl_2$$

The reaction probably proceeds by formation of a {BCl} intermediate which then inserts into a B–Cl bond of BCl_3 to give the product directly. Another route is via the more stable $B_2(NMe_2)_4$ (see reaction scheme). Thermal stabilities of these compounds parallel the expected sequence of p_π–p_π bonding between the substituent and B:

$$B_2(NMe_2)_4 > B_2(OMe)_4 > B_2(OH)_4 > B_2F_4 > B_2Cl_4 > B_2Br_4$$

The halides are much less stable than the corresponding BX_3, the most stable member

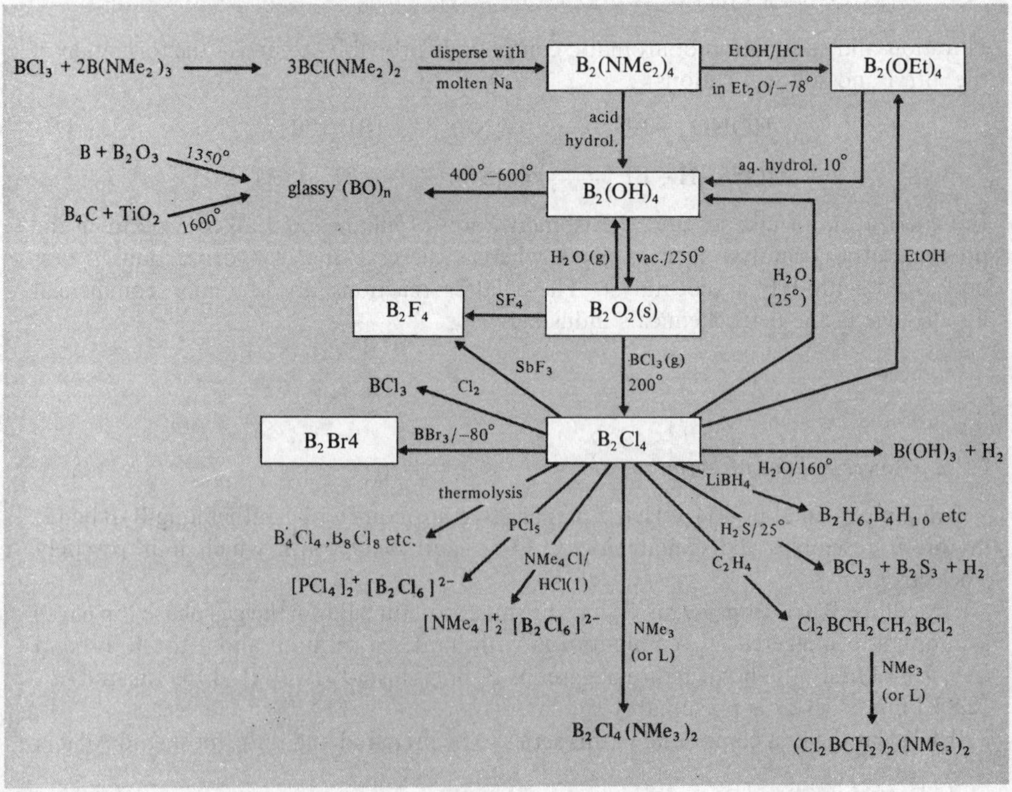

BCl3 + 2B(NMe2)3
3BCl(NMe2)2
disperse with
molten Na
B2(NMe2)4
EtOH/HCl
in Et2O/−78°
B2(OEt)4
acid
hydrol.
B + B2O3
1350°
B4C + TiO2
1600°
glassy (BO)n
400°–600°
B2(OH)4
aq. hydrol. 10°
H2O (g)
vac./250°
H2O
(25°)
EtOH
B2F4
SF4
B2O2(s)
SbF3
BCl3(g)
200°
BCl3
Cl2
B2Br4
BBr3/−80°
B2Cl4
B(OH)3 + H2
H2O/160°
LiBH4
B2H6, B4H10 etc
thermolysis
B4Cl4, B8Cl8 etc.
PCl5
[PCl4]2+ [B2Cl6]2−
NMe4Cl/
HCl(l)
[NMe4]+2 [B2Cl6]2−
NMe3
(or L)
B2Cl4(NMe3)2
H2S/25°
BCl3 + B2S3 + H2
C2H4
Cl2BCH2CH2BCl2
NMe3
(or L)
(Cl2BCH2)2(NMe3)2

B_2F_4 decomposing at the rate of about 8% per day at room temperature. B_2Br_4 disproportionates so rapidly at room temperature that it is difficult to purify:

$$nB_2X_4 \longrightarrow nBX_3 + (BX)_n$$

The compounds B_2X_4 are spontaneously flammable in air and react with H_2 to give BHX_2, B_2H_6, and related hydrohalides; they form adducts with Lewis bases ($B_2Cl_4L_2$ more stable than $B_2F_4L_2$) and add across C–C multiple bonds, e.g.

$$C_2H_2 + B_2Cl_4 \xrightarrow{25°}
\begin{array}{c}
\text{H} \qquad\quad \text{H} \\
\diagdown \qquad\quad \diagup \\
\text{C} = \text{C} \\
\diagup \qquad\quad \diagdown \\
\text{Cl}_2\text{B} \qquad\quad \text{BCl}_2
\end{array}
\xrightarrow{50°}
\begin{array}{c}
\text{Cl}_2\text{B} \qquad\quad \text{BCl}_2 \\
\diagdown \qquad\quad \diagup \\
\text{H}-\text{C}-\text{C}-\text{H} \\
\diagup \qquad\quad \diagdown \\
\text{Cl}_2\text{B} \qquad\quad \text{BCl}_2
\end{array}$$

Other reactions of B_2Cl_4 are shown in the scheme and many of these also occur with B_2F_4.

When BF_3 is passed over crystalline B at 1850°C and pressures of less than 1 mmHg, the reactive gas BF is obtained in high yield and can be condensed out at $-196°$. Cocondensation with BF_3 yields B_2F_4 then B_3F_5 (i.e. F_2B–$B(F)$–BF_2). However, this latter compound is unstable and it disproportionates above $-50°$ according to

$$4(BF_2)_2BF \longrightarrow 2B_2F_4 + B_8F_{12} \quad (103)$$

(103)

The yellow compound B_8F_{12} appears to have a diborane-like structure (103) and this readily undergoes symmetrical cleavage with a variety of ligands such as CO, PF_3, PCl_3, PH_3, AsH_3, and SMe_2 to give adducts $L.B(BF_2)_3$ which are stable at room temperature in the absence of air or moisture.

Thermolysis of B_2Cl_4 and B_2Br_4 at moderate temperatures give a series of *closo*-halogenoboranes B_nX_n where $n = 4$, 8–12 for Cl, and $n = 7$–10 for Br. B_4Cl_4, a pale-yellow-green solid, has a regular *closo*-tetrahedral structure (Fig. 6.27a); it is hyperelectron deficient when compared with the *closo*-boranes $B_nH_n^{2-}$ and the bonding has been discussed in terms of localized 3-centre bonds above the 4 tetrahedral faces supplemented by p_π interaction with p orbitals of suitable symmetry on the 4 Cl atoms: the 8 electrons available for framework bonding from the 4{BCl} groups fill 4 bonding MOs of class A_1 and T_2 and there are 2 additional bonding MOs of class E which have correct symmetry to mix with the Cl p_π orbitals. B_8Cl_8 (dark red or purple crystals) has an irregular dodecahedral (bisphenoid) arrangement of the *closo*-B_8 cluster (Fig. 6.27b) with 14 B–B distances in the range 168–184 pm and 4 substantially longer B–B distances at 193–205 pm. B_9Br_9 is a particularly stable compound; it forms as dark-red crystals together with other subbromides ($n = 7$–10) when gaseous BBr_3 is subjected to a silent electric discharge in the presence of Cu wool, and can be purified by sublimation under

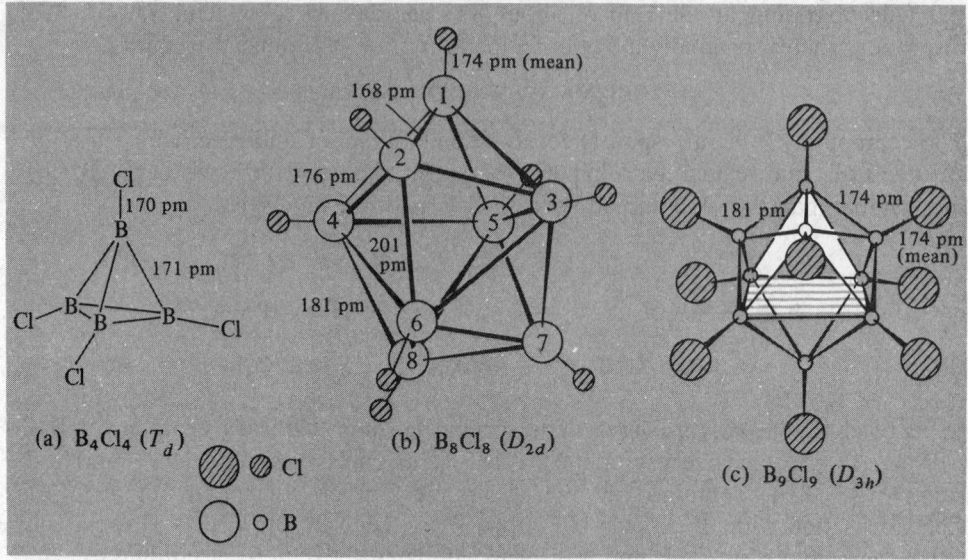

FIG. 6.27 Molecular structures of (a) tetrahedral B_4Cl_4, (b) dodecahedral B_8Cl_8, and (c) tricapped trigonal pyramidal B_9Cl_9 and B_9Br_9. In B_8Cl_8 note that the shortest B–B distances are between two 5-coordinate B atoms, e.g. B(1)–B(2); the longest are between two 6-coordinate B atoms, e.g. B(4)–B(6) and intermediate distances are between one 5- and one 6-coordinate B atom. A similar trend occurs in B_9Cl_9.

conditions (200°C) which rapidly decompose the other products. B_9Br_9 is isostructural with B_9Cl_9 (yellow-orange) (Fig. 6.27c). Many mixed halides $B_nBr_{n-x}Cl_x$ ($n = 9$, 10, 11) have been identified by mass spectrometry and other techniques, but their separation as pure compounds has so far not been achieved. Chemical reactions of B_nX_n resemble those of B_2X_4 except that alkenes do not cleave the B–B bonds in the *closo*-species.[49, 50]

6.8 Boron–Oxygen Compounds

Boron (like silicon) invariably occurs in nature as oxo compounds and is never found as the element or even directly bonded to any other element than oxygen.† The structural chemistry of B–O compounds is characterized by an extraordinary complexity and diversity which rivals those of the borides (p. 162) and boranes (p. 171). In addition, vast numbers of predominantly organic compounds containing B–O are known.

† Trivial exceptions to this sweeping generalization are $NaBF_4$ (ferrucite) and $(K,Cs)BF_4$ (avogadrite) which have been reported from Mt. Vesuvius, Italy.

[49] A. G. MASSEY, Boron subhalides, *Chem. in Br.* **16**, 588–98 (1980). See also A. J. MARKWELL, A. G. MASSEY, and P. J. PORTAL, New routes to halogenated B_8 and B_9 boron cages, *Polyhedron* **1**, 134–35 (1982).
[50] E. H. WONG and R. M. KABBANI, Boron halide clusters and radicals: synthesis and interconversions of the three oxidation states of a nine-boron polyhedron, $B_9X_9{}^{n-}$ (X = Cl, Br, I; $n = 0$, 1, 2), *Inorg. Chem.* **19**, 451–5 (1980). See also E. H. WONG, Nonaiodononaborane(9), B_9I_9. A stable boron iodide cluster, *Inorg. Chem.* **20**, 1300–02 (1981).

6.8.1 *Boron oxides and oxoacids*[51]

The principal oxide of boron is boric oxide, B_2O_3 (mp 450°, bp (extrap) 2250°C). It is one of the most difficult substances to crystallize and, indeed, was known only in the vitreous state until 1937. It is generally prepared by careful dehydration of boric acid $B(OH)_3$. The normal crystalline form (d 2.56 g cm^{-3}) consists of a 3D network of trigonal BO_3 groups joined through their O atoms, but there is also a dense form (d 3.11 g cm^{-3}) formed under a pressure of 35 kbar at 525°C and built up from irregular interconnected BO_4 tetrahedra. In the vitreous state ($d \simeq 1.83$ g cm^{-3}) B_2O_3 probably consists of a network of partially ordered trigonal BO_3 units in which the 6-membered $(BO)_3$ ring predominates; at higher temperatures the structure becomes increasingly disordered and above 450°C polar $-B{\equiv}O$ groups are formed. Fused B_2O_3 readily dissolves many metal oxides to give characteristically coloured borate glasses. Its major application is in the glass industry where borosilicate glasses (e.g. Pyrex) are extensively used because of their small coefficient of thermal expansion and their easy workability. US production of B_2O_3 is ~25 000 tonnes pa.

Orthoboric acid, $B(OH)_3$, is the normal end product of hydrolysis of most boron compounds and is usually made ($\simeq$ 165 000 tonnes pa) by acidification of aqueous solutions of borax. It forms flaky, white, transparent crystals in which a planar array of BO_3 units is joined by unsymmetrical H bonds as shown in Fig. 6.28. In contrast to the

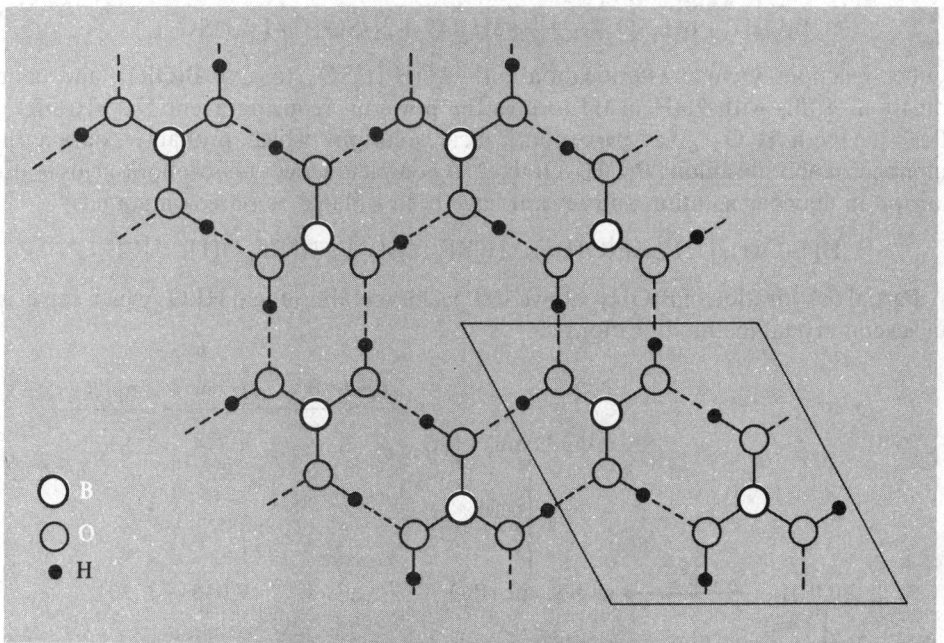

Fig. 6.28 Layer structure of $B(OH)_3$. Interatomic distances are B–O 136 pm, O–H 97 pm, O—H···O 272 pm. Angles at B are 120° and at O 126° and 114°. The H bond is almost linear. The distance between consecutive layers in the crystal is 318 pm.

[51] *Mellor's Comprehensive Treatise on Inorganic and Theoretical Chemistry: Supplement*, Vol. V, *Boron. Part A. Boron–Oxygen Compounds*, Longmans, London, 1980, 825 pp. A series of comprehensive reviews completed at various dates between 1969 and 1976.

short O—H···O distance of 272 pm within the plane, the distance between consecutive layers in the crystal is 318 pm, thus accounting for the pronounced basal cleavage of the waxy, plate-like crystals, and their low density (1.48 g cm^{-3}). $B(OH)_3$ is a very weak monobasic acid and acts exclusively by hydroxyl-ion acceptance rather than proton donation:

$$B(OH)_3 + 2H_2O \rightleftharpoons H_3O^+ + B(OH)_4^-; \quad pK = 9.25$$

Its acidity is considerably enhanced by chelation with polyhydric alcohols (e.g. glycerol, mannitol) and this forms the basis of its use in analytical chemistry; e.g. with mannitol pK drops to 5.15, indicating an increase in the acid equilibrium constant by a factor of more than 10^4:

$B(OH)_3$ also acts as a strong acid in anhydrous H_2SO_4:

$$B(OH)_3 + 6H_2SO_4 \longrightarrow 3H_3O^+ + 2HSO_4^- + [B(HSO_4)_4]^-$$

Other reactions include esterification with ROH/H_2SO_4 to give $B(OR)_3$, and coordination of this with NaH in thf to give the powerful reducing agent $Na[BH(OR)_3]$. Reaction with H_2O_2 gives peroxoboric acid solutions which probably contain the monoperoxoborate anion $[B(OH)_3OOH]^-$. A complete series of fluoroboric acids is also known in aqueous solution and several have been isolated as pure compounds:

$$H[B(OH)_4] \quad H[BF(OH)_3] \quad H[BF_2(OH)_2] \quad H[BF_3OH] \quad HBF_4$$

Partial dehydration of $B(OH)_3$ above 100° yields metaboric acid HBO_2 which can exist in several crystalline modifications:

	CN of B	d/g cm^{-3}	mp/°C
orthorhombic HBO_2	3	1.784	176°
monoclinic HBO_2	3 and 4	2.045	201°
cubic HBO_2	4	2.487	236°

Orthorhombic HBO_2 consists of trimeric units $B_3O_3(OH)_3$ which are linked into layers by H bonding (Fig. 6.29); all the B atoms are 3-coordinate. Monoclinic HBO_2 is built of chains of composition $[B_3O_4(OH)(H_2O)]$ in which some of the B atoms are now 4-coordinate, whereas cubic HBO_2 has a framework structure of tetrahedral BO_4 groups

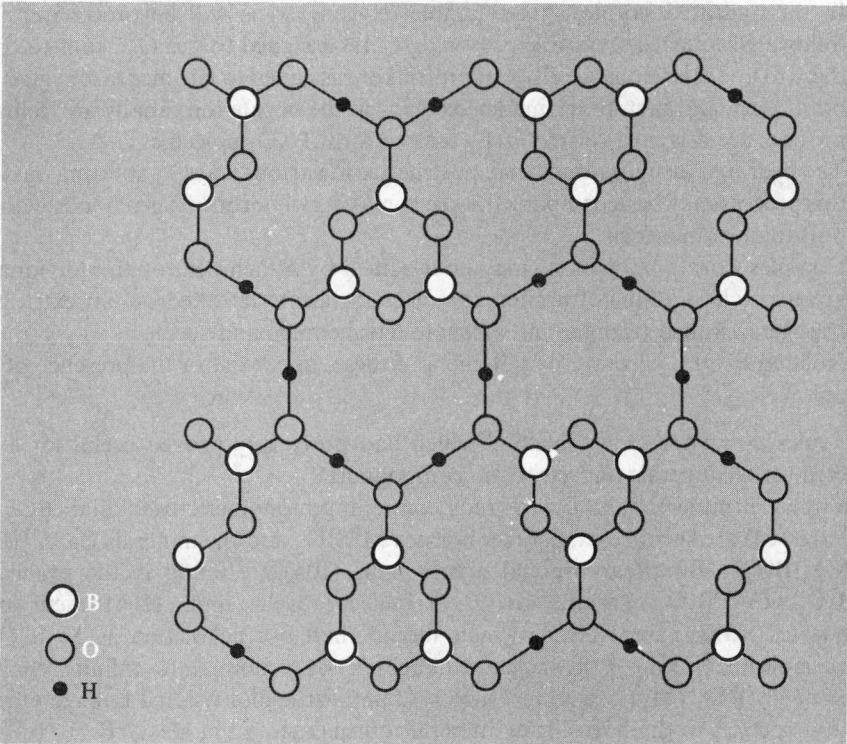

FIG. 6.29   Layer structure of orthorhombic metaboric acid HBO_2(III), comprising units of formula $B_3O_3(OH)_3$ linked by $O\cdots H\cdots O$ bonds.

some of which are H bonded. The increase in CN of B is paralleled by an increase in density and mp.

Boron suboxide $(BO)_n$ and subboric acid $B_2(OH)_4$ were mentioned on p. 226.

6.8.2   *Borates*[51, 52]

The phase relations, stoichiometry, and structural chemistry of the metal borates have been extensively studied because of their geochemical implications and technological importance. Borates are known in which the structural unit is mononuclear (1 B atom), bi-, tri-, tetra-, or penta-nuclear, or in which there are polydimensional networks including glasses. The main structural principles underlying the bonding in crystalline metal borates are as follows:[52]

1. Boron can link either three oxygens to form a triangle or four oxygens to form a tetrahedron.
2. Polynuclear anions are formed by corner-sharing only of boron–oxygen triangles and tetrahedra in such a manner that a compact insular group results.

[52] C. L. CHRIST and J. R. CLARK, A crystal-chemical classification of borate structures with emphasis on hydrated borates, *Phys. Chem. Minerals* **2**, 59–87 (1977). See also J. B. FARMER, Metal borates, *Adv. Inorg. Chem. Radiochem.* **25**, 187–237 (1982).

3. In the hydrated borates, protonatable oxygen atoms will be protonated in the following sequence: available protons are first assigned to free O^{2-} ions to convert these to free OH^- ions; additional protons are assigned to tetrahedral oxygens in the borate ion, and then to triangular oxygens in the borate ion; finally any remaining protons are assigned to free OH^- ions to form H_2O molecules.

4. The hydrated insular groups may polymerize in various ways by splitting out water; this process may be accompanied by the breaking of boron-oxygen bonds within the polyanion framework.

5. Complex borate polyanions may be modified by attachment of an individual side group, such as (but not limited to) an extra borate tetrahedron, an extra borate triangle, 2 linked triangles, an arsenate tetrahedron, and so on.

6. Isolated $B(OH)_3$ groups, or polymers of these, may exist in the presence of other anions.

These rules now supersede others[53] which had previously proved useful for a more restricted set of minerals and synthetic compounds.

Examples of minerals and compounds containing monomeric triangular, BO_3 units (structure 104) are the rare-earth orthoborates $M^{III}BO_3$ and the minerals $CaSn^{IV}(BO_3)_2$ and $Mg_3(BO_3)_2$. Binuclear trigonal planar units (105) are found in the pyroborates $Mg_2B_2O_5$, $Co^{II}{}_2B_2O_5$, and $Fe^{II}{}_2B_2O_5$. Trinuclear cyclic units (106) occur in the metaborates $NaBO_2$ and KBO_2, which should therefore be written as $M_3B_3O_6$ (cf. metaboric acid, p. 231). Polynuclear linkage of BO_3 units into infinite chains of stoichiometry BO_2 (107) occurs in $Ca(BO_2)_2$, and three-dimensional linkage of planar BO_3 units occurs in the borosilicate mineral tourmaline and in glassy B_2O_3 (p. 229).

(104) $[BO_3]^{3-}$ (105) $[B_2O_5]^{4-}$ (106) $[B_3O_6]^{3-}$ (107) $[(BO_2)^-]_n$

Units containing B in planar BO_3 coordination only

Monomeric tetrahedral BO_4 units (108) are found in the zircon-type compound Ta^VBO_4 and in the minerals $(Ta,Nb)BO_4$ and $Ca_2H_4BAs^VO_8$. The related tetrahedral unit $[B(OH)_4]^-$ (109) occurs in $Na_2[B(OH)_4]Cl$ and $Cu^{II}[B(OH)_4]Cl$. Binuclear tetrahedral units (110) have been found in $Mg[B_2O(OH)_6]$ and a cyclic binuclear tetrahedral structure (111) characterizes the peroxoanion $[B_2(O_2)_2(OH)_4]^{2-}$ in "sodium perborate" $NaBO_3.4H_2O$, i.e. $Na_2[B_2(O_2)_2(OH)_4].6H_2O$. A more complex polynuclear structure comprising sheets of tetrahedrally coordinated $BO_3(OH)$ units occurs in the borosilicate mineral $CaB(OH)SiO_4$ and the fully three-dimensional polynuclear structure is found in BPO_4 (cf. the isoelectronic SiO_2), $BAsO_4$, and the minerals $NaBSi_3O_8$ and $Zn_4B_6O_{13}$.

The final degree of structural complexity occurs when the polynuclear assemblages contain both planar BO_3 and tetrahedral BO_4 units joined by sharing common O atoms.

[53] J. O. EDWARDS and V. F. ROSS, The structural chemistry of the borates, Chap. 3 in E. L. MUETTERTIES (ed.), *The Chemistry of Boron and its Compounds*, pp. 155–207, Wiley, New York, 1967.

(108) $[BO_4]^{5-}$ (109) $[B(OH)_4]^-$ (110) $[B_2O(OH)_6]^{2-}$ (111) $[B_2(O_2)_2(OH)_4]^{2-}$

Units containing B in tetrahedral BO_4 coordination only

(112) $[B_5O_6(OH)_4]^-$ (113) $[B_3O_3(OH)_5]^{2-}$ (114) $[B_4O_5(OH)_4]^{2-}$

Units containing B in both BO_3 and BO_4 coordination

The structure of monoclinic HBO_2 affords an example (p. 230). A structure in which the ring has but one BO_4 unit is the spiroanion $[B_5O_6(OH)_4]^-$ (structure 112) which occurs in hydrated potassium pentaborate $KB_5O_8.4H_2O$, i.e. $K[B_5O_6(OH)_4].2H_2O$. The anhydrous pentaborate KB_5O_8 has the same structural unit but dehydration of the OH groups link the spiroanions of structure (112) sideways into ribbon-like helical chains. The mineral $CaB_3O_3(OH)_5.H_2O$ has 2 BO_4 units in the 6-membered heterocycle (113) and related chain elements $[B_3O_4(OH)_3{}^{2-}]_n$ linked by a common oxygen atom are found in the important mineral colemanite $Ca_2B_6O_{11}.5H_2O$, i.e. $[CaB_3O_4(OH)_3].H_2O$. It is clear from these examples that, without structural data, the stoichiometry of these borate minerals gives little indication of their constitution. A further illustration is afforded by borax which is normally formulated $Na_2B_4O_7.10H_2O$, but which contains tetranuclear units $[B_4O_5(OH)_4]^{2-}$ formed by fusing 2 B_3O_3 rings which each contain 2 BO_4 (shared) and 1 BO_3 unit (114); borax should therefore be written as $Na_2[B_4O_5(OH)_4].8H_2O$.

There is wide variation of B–O distances in these various structures the values increasing, as expected, with increase in coordination:

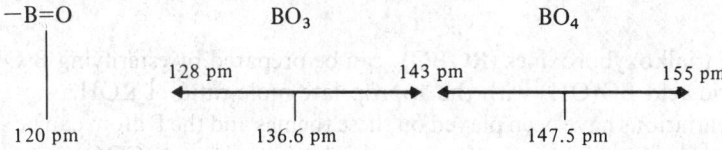

The extent to which B_3O_3 rings catenate into more complex structures or hydrolyse into smaller units such as $[B(OH)_4]^-$ clearly depends sensitively on the activity (concentration) of water in the system, on the stoichiometric ratio of metal ions to boron, and on the temperature ($T\Delta S$).

Many metal borates find important industrial applications (p. 156). Pre-eminent is borax, which is made on a scale approaching 2 million tonnes pa in the USA alone. Main uses are in glass-fibre and cellular insulation, the manufacture of borosilicate glasses and enamels, and as fire retardants. Sodium perborate (for detergents) is manufactured on a 550 000 tonne pa scale.

6.8.3 *Organic compounds containing boron–oxygen bonds*

Only a brief classification of this very large and important class of compounds will be given; most contain trigonal planar B though many 4-coordinate complexes have also been characterized. The orthoborates $B(OR)_3$ can readily be prepared by direct reaction of BCl_3 or $B(OH)_3$ with ROH, and transesterification with R′OH affords a route to unsymmetrical products $B(OR)_2(OR')$, etc. The compounds range from colourless volatile liquids to involatile white solids depending on mol wt. R can be a primary, secondary, tertiary, substituted, or unsaturated alkyl group or an aryl group, and orthoborates of polyhydric alcohols and phenols are also numerous.

Boronic acids $RB(OH)_2$ were first made over a century ago by the unlikely route of slow partial oxidation of the spontaneously flammable trialkyl boranes followed by hydrolysis of the ester so formed (E. Frankland, 1862):

$$BEt_3 + O_2 \longrightarrow EtB(OEt)_2 \xrightarrow{2H_2O} EtB(OH)_2$$

Many other routes are now available but the most used involve the reaction of Grignard reagents or lithium alkyls on orthoborates or boron trihalides:

$$B(OR)_3 + ArMgX \xrightarrow{-50} [ArB(OR)_3]MgX \xrightarrow{H_3O^+} ArB(OH)_2$$

Boronic acids readily dehydrate at moderate temperatures (or over P_4O_{10} at room temperature) to give trimeric cyclic anhydrides known as trialkyl(aryl)boroxines:

$$3RB(OH)_2 \xrightarrow{warm} \quad + 3H_2O$$

The related trialkoxyboroxines $(ROBO)_3$ can be prepared by esterifying $B(OH)_3$, B_2O_3, or metaboric acid $BO(OH)$ with the appropriate mole ratio of ROH.

Endless variations have been played on these themes and the B atom can be surrounded by innumerable combinations of groups such as acyloxy (RCOO), peroxo (ROO), halogeno (X), hydrido, etc., in either open or cyclic arrays. However, no new chemical principles emerge.

6.9 Boron–Nitrogen Compounds

Two factors have contributed to the special interest that attaches to B–N compounds.

Firstly the B–N unit is isoelectronic with C–C and, secondly, the size and electronegativity of the 3 atoms are similar, C being the mean of B and N:

	B	C	N
Number of valence electrons	3	4	5
Covalent single-bond radius/pm	88	77	70
Electronegativity	2.0	2.5	3.0

The repetition of much organic chemistry by replacing pairs of C atoms with the B–N grouping has led to many new classes of compound but these need not detain us. By contrast, key points emerge from several other areas of B–N chemistry and, accordingly, this section deals with the structure, properties, and reaction chemistry of boron nitride, amine-borane adducts, aminoboranes, and the cyclic borazines.

The synthesis of boron nitride, BN, involves considerable technical difficulty; a laboratory preparation yielding relatively pure samples involves the fusion of borax with ammonium chloride, whereas technical-scale production relies on the fusion of urea with $B(OH)_3$ in an atmosphere of NH_3 at 500–950°C. Only a brave (or foolhardy) chemist would attempt to write a balanced equation for either reaction. An alternative synthesis (>99% purity) treats BCl_3 with an excess of NH_3 (see below) and pyrolyses the resulting mixture in an atmosphere of NH_3 at 750°C. The hexagonal modification of BN has a simple layer structure (Fig. 6.30) similar to graphite but with the significant difference that the layers are packed directly on top of each other so that the B atom in one layer is located over an N atom in the next layer at a distance of 333 pm. Cell dimensions and other data for BN and graphite are compared in Table 6.4. Within each layer the B–N distance is only

TABLE 6.4   *Comparison of hexagonal BN and graphite*

	a/pm	c/pm	c/a	Inter-layer spacing/pm	Intra-layer spacing/pm	$d/g\,cm^{-3}$
BN (hexagonal)	250.4	666.1	2.66	333	144.6	2.29
Graphite	245.6	669.6	2.73	335	142	2.255

145 pm; this is similar to the distance of 144 pm in borazine (p. 238) but much less than the sum of single-bond covalent radii (158 pm) and this has been taken to indicate substantial additional π bonding within the layer. However, unlike graphite, BN is colourless and a good insulator; it also resists attack by most reagents though fluorine converts it quantitatively to BF_3 and N_2, and HF gives NH_4BF_4 quantitatively. Hexagonal BN can be converted into a cubic form (zinc-blende type structure) at 1800°C and 85 000 atm pressure in the presence of an alkali or alkaline-earth metal catalyst. The lattice constant of cubic BN is 361.5 pm (cf. diamond 356.7 pm). A wurtzite-type modification (p. 1405) can be obtained at lower temperatures.

Amine-borane adducts have the general formula R_3NBX_3 where R = alkyl, H, etc., and

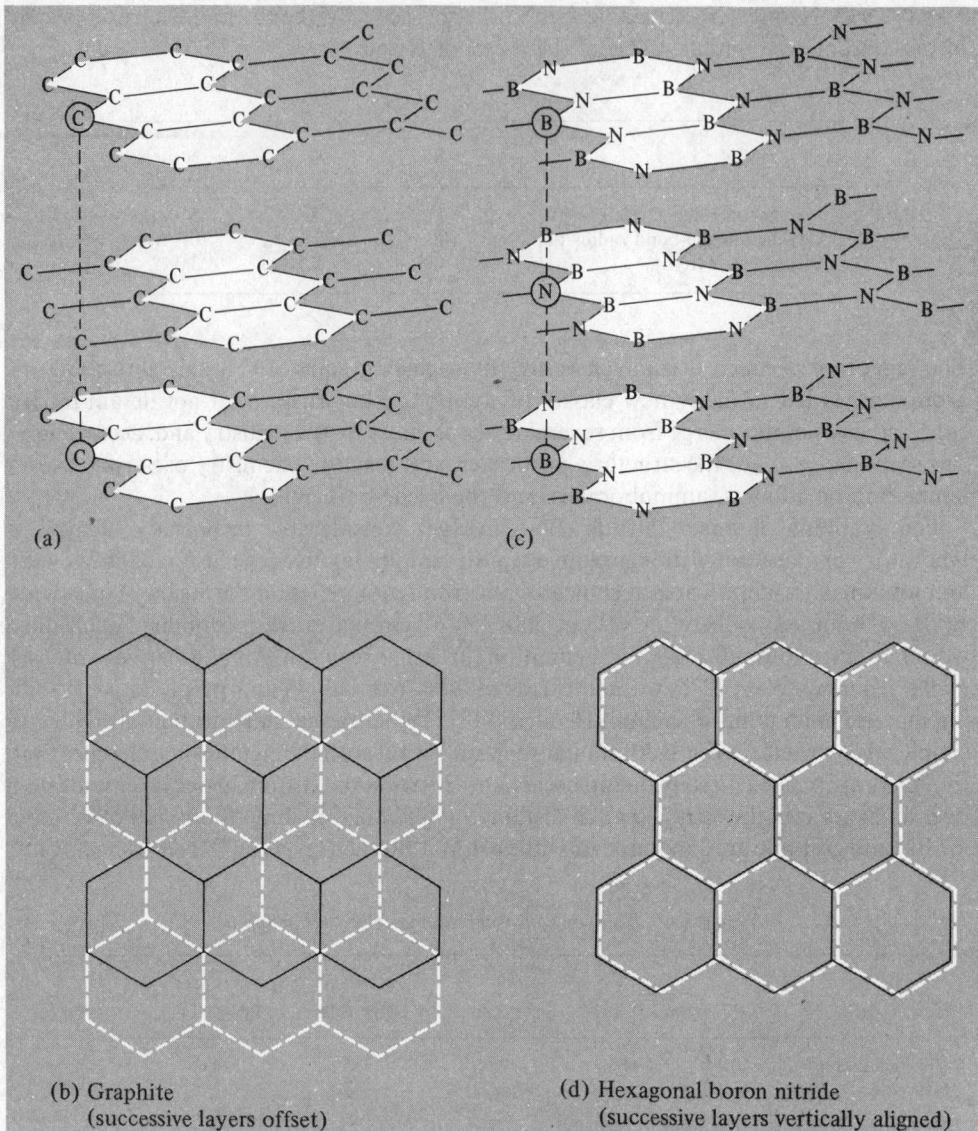

(a)

(c)

(b) Graphite
(successive layers offset)

(d) Hexagonal boron nitride
(successive layers vertically aligned)

FIG. 6.30   Comparison of the hexagonal layer structures of BN and graphite. In BN the atoms of
one layer are located directly above the atoms of adjacent layers with B···N contacts; in graphite
the C atoms in one layer are located above interstices in the adjacent layer and are directly above
atoms in alternate layers only.

X = alkyl, H, halogen, etc. They are usually colourless, crystalline compounds with mp in
the range 0–100° for X = H and 50–200° for X = halogen. Synthetic routes, and factors
affecting the stability of the adducts have already been discussed (p. 188 and p. 223). In
cases where diborane undergoes unsymmetrical cleavage (e.g. with NH_3) alternative
routes must be devised:

$$B_2H_6 + 2NH_3 \longrightarrow [BH_2(NH_3)_2]^+ BH_4^-$$

$$NH_4Cl + LiBH_4 \longrightarrow NH_3BH_3 + LiCl + H_2$$

The nature of the bonding in amine-boranes and related adducts has been the subject of considerable theoretical discussion and has also been the source of some confusion. Conventional representations of the donor–acceptor (or coordinate) bond use symbols such as $R_3N \rightarrow BX_3$ or $R_3\overset{+}{N}-\overset{-}{B}X_3$ to indicate the origin of the bonding electrons and the direction (not the magnitude) of charge transfer. It is important to realize that these symbols refer to the relative change in electron density with respect to the individual separate donor and acceptor molecules. Thus, $R_3\overset{+}{N}$ in the adduct has less electron density on N than has free R_3N, and $\overset{-}{B}X_3$ has more electron density on B in the adduct than has free BX_3; this does not necessarily mean that N is positive with respect to B in the adduct. Indeed, several MO calculations indicate that the change in electron density on coordination merely reduces but is insufficient to reverse the initial positive charge on the B atom. Consistent with this, experiments show that electrophilic reagents always attack N in amine-borane adducts, and nucleophilic reagents attack B.

A similar situation obtains in the aminoboranes where one or more of the substituents on B is an R_2N group (R = alkyl, aryl, H), e.g. Me_2N-BMe_2. Reference to Fig. 6.24 indicates the possibility of some p_π interaction between the lone pair on N and the "vacant" orbital on trigonal B. This is frequently indicated as

$$>N \Rightarrow B<　or　>\overset{+}{N}=\overset{-}{B}<$$

However, as with the amine-borane adducts just considered, this does not normally indicate the actual sign of the net charges on N and B because the greater electronegativity of N causes the σ bond to be polarized in the opposite sense. Thus, N–B bond moments in aminoboranes have been found to be negligible and MO calculations again suggest that the N atom bears a larger net negative charge than the B atom. The partial double-bond formulation of these compounds, however, is useful in implying an analogy to the isoelectronic alkenes. Coordinative saturation in aminoboranes can be achieved not only through partial double bond formation but also by association (usually dimerization) of the monomeric units to form $(B-N)_n$ rings. For example, in the gas phase, aminodimethyl-borane exists as both monomer and dimer in reversible equilibrium:

The presence of bulky groups on either B or N hinders dimer formation and favours monomers, e.g. $(Me_2NBF_2)_2$ is dimeric whereas the larger halides form monomers at least in the liquid phase. Association to form trimers (6-membered heterocycles) is less common, presumably because of even greater crowding of substituents, though triborazane $(H_2NBH_2)_3$ and its *N*-methyl derivatives, $(MeHNBH_2)_3$ and $(Me_2NBH_2)_3$, are known in which the B_3N_3 ring adopts the cyclohexane chair conformation.

Preparative routes to these compounds are straightforward, e.g.:

$$R_2NH_2Cl + MBH_4 \longrightarrow R_2NBH_2 + MCl + H_2 \quad (R = H, \text{ alkyl, aryl})$$
$$R_2NH + R_2'BX + NEt_3 \longrightarrow R_2NBR_2' + Et_3NHX \quad (R' = \text{alkyl, aryl, halide})$$
$$B(NR_2)_3 + 2BR_3 \longrightarrow 3R_2NBR_2, \text{ etc.}$$

In general monomeric products are readily hydrolysed but associated species (containing 4-coordinate B) are much more stable: e.g. $(Me_2NBH_2)_2$ does not react with H_2O at 50° but is rapidly hydrolysed by dilute HCl at 110° because at this temperature there is a significant concentration of monomer present.

The cyclic borazine $(-BH-NH-)_3$ and its derivatives form one of the largest classes of B–N compounds. The parent compound, also known as "inorganic benzene", was first isolated as a colourless liquid from the mixture of products obtained by reacting B_2H_6 and NH_3 (A. Stock and E. Pohland, 1926):

$$3B_2H_6 + 6NH_3 \xrightarrow{180} 2B_3N_3H_6 + 12H_2$$

It is now best prepared by reduction of the *B*-trichloro derivative:

$$3BCl_3 + 3NH_4Cl \xrightarrow[-9HCl]{heat} (BClNH)_3 \xrightarrow{3NaBH_4} B_3N_3H_6 + 3NaCl + \tfrac{3}{2}B_2H_6$$

Borazine has a regular plane hexagonal ring structure and its physical properties closely resemble those of the isoelectronic compound benzene (Table 6.5). Although it is possible to write Kekulé-type structures with $N{\Rightarrow}B$ π bonding superimposed on the σ bonding, the

TABLE 6.5 *Comparison of borazine and benzene*

Property	$B_3N_3H_6$	C_6H_6
Molecular weight	80.5	78.1
MP/°C	−57	6
BP/°C	55	80
Critical temperature	252	288
Density (l at mp)/g cm^{-3}	0.81	0.81
Density (s)/g cm^{-3}	1.00	1.01
Surface tension (mp)/dyne cm$^{-1\,(a)}$	31.1	31.0
Interatomic distances/pm	B–N 144	C–C 142
	B–H 120	C–H 108
	N–H 102	

(a) 1 dyne $= 10^{-5}$ newton.

weight of chemical evidence suggests that borazine has but little aromatic character. It reacts readily with H_2O, MeOH, and HX to yield 1:3 adducts which eliminate $3H_2$ on being heated to 100°, e.g.:

$$B_3N_3H_6 + 3H_2O \xrightarrow{0°} [BH(OH)NH_2]_3 \xrightarrow{100°} [B(OH)NH]_3 + 3H_2$$

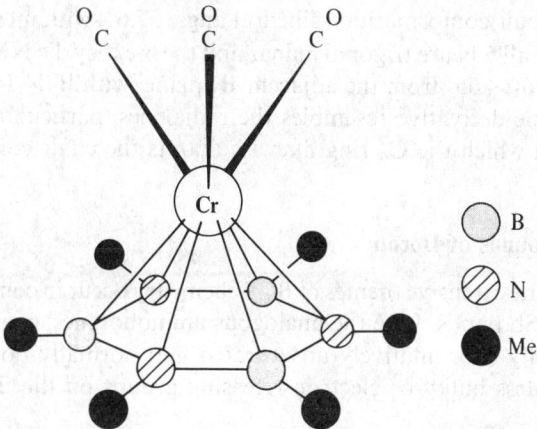

FIG. 6.31　Molecular structure of [Cr(η^6-B$_3$N$_3$Me$_6$)(CO)$_3$].

Numerous other reactions have been documented, most of which are initiated by nucleophilic attack on B. There is no evidence that electrophilic substitution of the borazine ring occurs and conditions required for such reactions in benzenoid systems disrupt the borazine ring by oxidation or solvolysis. However, it is known that the less-reactive hexamethyl derivative B$_3$N$_3$Me$_6$ (which can be heated to 460° for 3 h without significant decomposition) reacts with [Cr(CO)$_3$(MeCN)$_3$] to give the complex [Cr(η^6-B$_3$N$_3$Me$_6$)(CO)$_3$] (Fig. 6.31) which closely resembles the corresponding hexamethyl-benzene complex [Cr(η^6-C$_6$Me$_6$)(CO)$_3$].

　　N-substituted and *B*-substituted borazines are readily prepared by suitable choice of amine and borane starting materials or by subsequent reaction of other borazines with Grignard reagents, etc. Thermolysis of monocyclic borazines leads to polymeric materials and to polyborazine analogues of naphthalene, biphenyl, etc.:

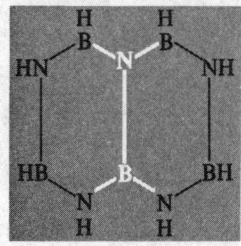
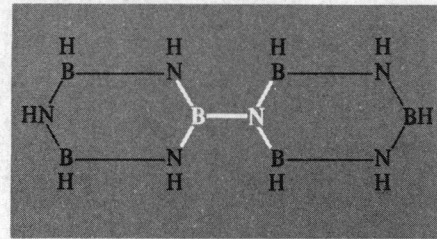

　　We conclude this section on BN compounds by reference to the curious cyclic hexamer [(BNMe$_2$)$_6$]. This has very recently been obtained as orange-red crystals by distilling the initial product formed by dehalogenation of (Me$_2$N)$_2$BCl with Na/K alloy:[54]

$$2(Me_2N)_2BCl \xrightarrow{\text{Na/K}} [B_2(NMe_2)_4] \xrightarrow{\text{thermolysis}} [(BNMe_2)_6]$$

[54] H. Nöth and H. Pommerening, Hexakis(dimethylamino)cyclohexaborane, a boron(I) compound without electron deficiency, *Angew. Chem.*, Int. Edn. (Engl.) **19**, 482–3 (1980).

The B_6 ring has a chair conformation (dihedral angle 57.6°) with mean B–B distances of 172 pm. All 6 B and all 6 N are trigonal planar and the 6-exocyclic NMe_2 groups are each twisted at an angle of ~65° from the adjacent B_3 plane, with B–N 140 pm. Structurally, this cyclohexaborane derivative resembles the radialenes, particularly the isoelectronic $[C_6(=CHMe)_6]$ in which the C_6 ring likewise adopts the chair conformation.

6.10 Other Compounds of Boron

Minor echoes of the extensive themes of B–N chemistry occur in compounds containing B–P, B–As, and B–Sb bonds. Like the analogous aminoboranes, compounds of the type R_2PBR_2' are formally coordinatively unsaturated and normally polymerize to ring or chain structures unless bulky or electron releasing groups on the B atom stabilize the monomer, e.g.:

$$3Et_2POCl + 3LiBH_4 \longrightarrow (Et_2PBH_2)_3 + 3LiCl + 3H_2O$$

$$2Et_2PLi + 2BCl_3 \longrightarrow (Et_2PBCl_2)_2 + 2LiCl$$

$$Et_2PLi + Ph_2BCl \longrightarrow Et_2PBPh_2 + LiCl$$

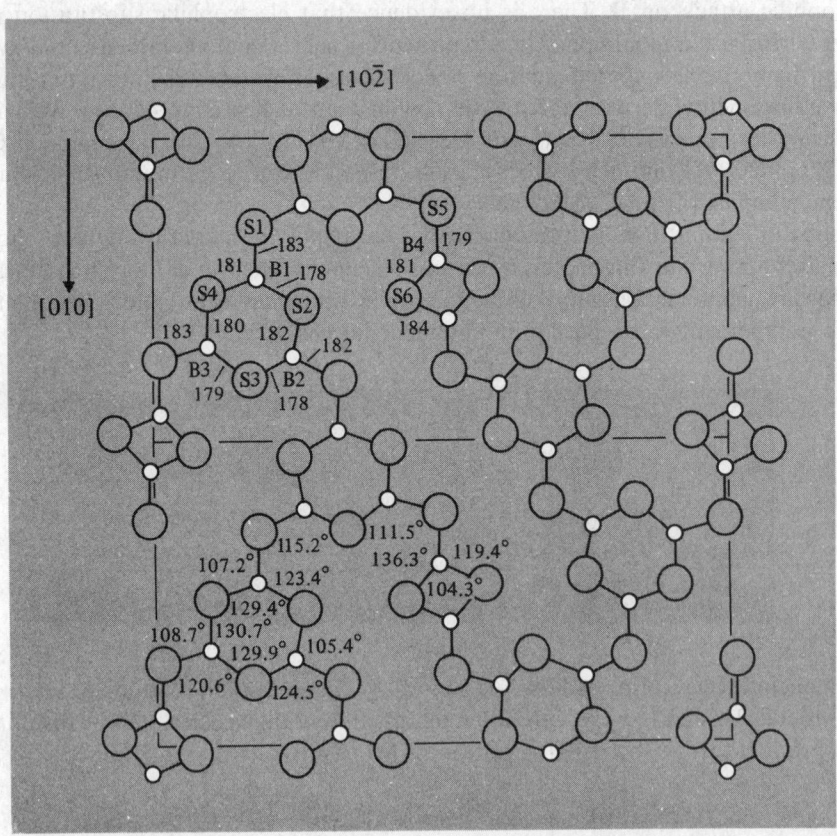

Fig. 6.32   Projection of two unit cells of the layer structure of B_2S_3 perpendicular to the plane of the layer. (Numbers in the upper unit cell indicate interatomic distances in pm; numbers in the lower unit cell indicate angles, in degrees.)

Similarly the vast array of B–O compounds finds no parallel in B–S or B–Se chemistry though thioborates of the type $B(SR)_3$, $R'B(SR)_2$ and $R'_2B(SR)$ are well documented. B_2S_3 itself has been known for many years as a pale-yellow solid which tends to form a glassy phase (cf. B_2O_3 and also B_2Se_3). This absence of a suitable crystalline sample prevented the structural characterization of this compound until as late as 1977. It has now been found that B_2S_3 has a fascinating layer structure which bears no resemblance to the three-dimensionally linked B_2O_3 crystal structure but is slightly reminiscent of BN. The structure (Fig. 6.32) is made up of planar B_3S_3 6-membered rings and B_2S_2 4-membered rings linked by S bridges into almost planar two-dimensional layers.[55] All the boron atoms are trigonal planar with B–S distances averaging 181 pm and the perpendicular interlayer distance is almost twice this at 355 pm.

A second boron sulfide has recently been made by heating B_2S_3 and sulfur to 300° under very carefully defined conditions;[56] the colourless, moisture-sensitive product analyses as BS_2 but X-ray crystal structure analysis reveals that the compound has a porphine-like structure B_8S_{16} as shown in Fig. 6.33.

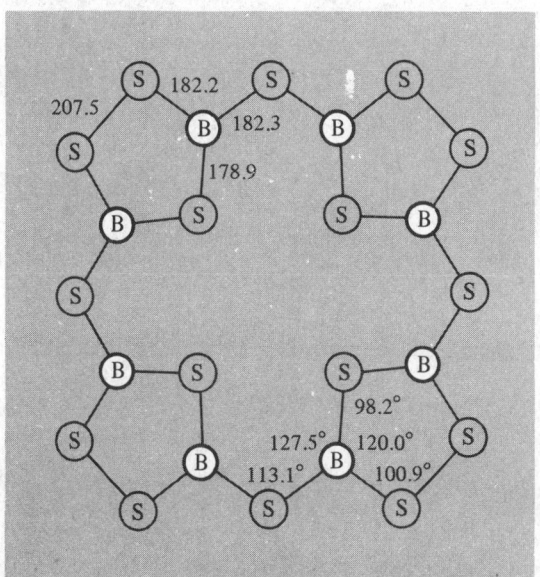

Fig. 6.33 Structure of the porphine-like molecule B_8S_{16}; interatomic distances in pm ($\sigma = 0.3$ pm), angles ($\sigma = 0.1°$).

An alternative, route to B_8S_{16} involves the reaction of dibromotrithiadiborolane with trithiocarbonic acid in an H_2S generator in dilute CS_2 solution:

$$4 \quad \text{[dibromotrithiadiborolane]} \quad + 4(HS)_2CS \longrightarrow B_8S_{16} + 4CS_2 + 8HBr$$

[55] H. DIERCKS and B. KREBS, Crystal structure of B_2S_3: four-membered B_2S_2 rings and six-membered B_3S_3 rings, *Angew. Chem.*, Int. Edn. (Engl.) **16**, 313 (1977).

[56] B. KREBS and H.-U. HÜRTER, B_8S_{16}—An "Inorganic porphine", *Angew. Chem.*, Int. Edn. (Engl.) **19**, 481–2 (1980).

There appears to be no end to the structural ingenuity of boron: whilst it is true that many regularities can now be discerned in its stereochemistry, much more work is still needed to unravel the reaction pathways by which the compounds are formed and to elucidate the mechanisms by which they isomerize and interconvert.

1 H	2 He																
3 Li	4 Be											5 B	6 C	7 N	8 O	9 F	10 Ne
11 Na	12 Mg											13 Al	14 Si	15 P	16 S	17 Cl	18 Ar
19 K	20 Ca	21 Sc	22 Ti	23 V	24 Cr	25 Mn	26 Fe	27 Co	28 Ni	29 Cu	30 Zn	31 Ga	32 Ge	33 As	34 Se	35 Br	36 Kr
37 Rb	38 Sr	39 Y	40 Zr	41 Nb	42 Mo	43 Tc	44 Ru	45 Rh	46 Pd	47 Ag	48 Cd	49 In	50 Sn	51 Sb	52 Te	53 I	54 Xe
55 Cs	56 Ba	57 La	72 Hf	73 Ta	74 W	75 Re	76 Os	77 Ir	78 Pt	79 Au	80 Hg	81 Tl	82 Pb	83 Bi	84 Po	85 At	86 Rn
87 Fr	88 Ra	89 Ac	104 Unq	105 Unp	106 Unh	107 Uns											

58 Ce	59 Pr	60 Nd	61 Pm	62 Sm	63 Eu	64 Gd	65 Tb	66 Dy	67 Ho	68 Er	69 Tm	70 Yb	71 Lu
90 Th	91 Pa	92 U	93 Np	94 Pu	95 Am	96 Cm	97 Bk	98 Cf	99 Es	100 Fm	101 Md	102 No	103 Lr

7

Aluminium, Gallium, Indium, and Thallium

7.1 Introduction

Aluminium derives its name from alum, the double sulfate $KAl(SO_4)_2.12H_2O$, which was used medicinally as an astringent in ancient Greece and Rome (Latin *alumen*, bitter salt). Humphry Davy was unable to isolate the metal but proposed the name "alumium" and then "aluminum"; this was soon modified to aluminium and this form is used throughout the world except in North America where the ACS decided in 1925 to adopt "aluminum" in its publications. The impure metal was first isolated by the Danish scientist H. C. Oersted using the reaction of dilute potassium amalgam on $AlCl_3$. This method was improved in 1827 by H. Wöhler who used metallic potassium, but the first commercially successful process was devised by H. St.-C. Deville in 1854 using sodium. In the same year both he and R. W. Bunsen independently obtained metallic aluminium by electrolysis of fused $NaAlCl_4$. So precious was the metal at this time that it was exhibited next to the crown jewels at the Paris Exposition of 1855 and the Emperor Louis Napoleon III used Al cutlery on state occasions. The dramatic thousand-fold drop in price which occurred before the end of the century (Table 7.1) was due, firstly, to the advent of cheap electric power following the development of the dynamo by W. von Siemens in the 1870s, and, secondly, to the independent development in 1886 of the electrolysis of alumina dissolved in cryolite (Na_3AlF_6) by P. L. T. Héroult in France and C. M. Hall in the USA; both men

TABLE 7.1 *Price of aluminium metal ($ per kg)*

1852	1854	1855	1856	1857	1858	1886	
1200	600	250	75	60	25	17	
		→ Introduction of St. C. Deville's $Na/AlCl_3$ process					
1888	1890	1895	1900	1950	1964	1974	1982
11.5	5.0	1.15	0.73	0.40	0.53	0.75	0.93
→ Introduction of Héroult–Hall electrolysis				↑ minimum			

were 22 years old at the time. World production rose quickly and in 1893 exceeded 1000 tonnes pa for the first time.

Gallium was predicted as eka-aluminium by D. I. Mendeleev in 1870 and was discovered by P. E. Lecoq de Boisbaudran in 1875 by means of the spectroscope; de Boisbaudran was, in fact, guided at the time by an independent theory of his own and had been searching for the missing element for some years. The first indications came with the observation of two new violet lines in the spark spectrum of a sample deposited on zinc, and within a month he had isolated 1 g of the metal starting from several hundred kilograms of crude zinc blende ore. The element was named in honour of France (Latin *Gallia*) and the striking similarity of its physical and chemical properties to those predicted by Mendeleev (Table 7.2) did much to establish the general acceptance of the Periodic Law (p. 24); indeed, when de Boisbaudran first stated that the density of Ga was 4.7 g cm^{-3} rather than the predicted 5.9 g cm^{-3}, Mendeleev wrote to him suggesting that he redetermine the figure (the correct value is 5.904 g cm^{-3}).

TABLE 7.2 *Comparison of predicted and observed properties of gallium*

Mendeleev's predictions (1871) for eka-aluminium, M	Observed properties (1977) of gallium (discovered 1875)
Atomic weight $\sim$68	Atomic weight 69.72
Density/g cm^{-3} 5.9	Density/g cm^{-3} 5.904
MP low	MP/°C 29.78
Non-volatile	Vapour pressure, 10^{-3} mmHg at 1000°C
Valence 3	Valence 3
M will probably be discovered by spectroscopic analysis	Ga was discovered by means of the spectroscope
M will have an oxide of formula M$_2$O$_3$, d 5.5 g cm^{-3}, soluble in acids to give MX$_3$	Ga has an oxide Ga$_2$O$_3$, d 5.88 g cm^{-3}, soluble in acids to give salts of the type GaX$_3$
M should dissolve slowly in acids and alkalis and be stable in air	Ga metal dissolves slowly in acids and alkalis and is stable in air
M(OH)$_3$ should dissolve in both acids and alkalis	Ga(OH)$_3$ dissolves in both acids and alkalis
M salts will tend to form basic salts; the sulfate should form alums; M$_2$S$_3$ should be precipitated by H$_2$S or (NH$_4$)$_2$S; anhydrous MCl$_3$ should be more volatile than ZnCl$_2$	Ga salts readily hydrolyse and form basic salts; alums are known; Ga$_2$S$_3$ can be precipitated under special conditions by H$_2$S or (NH$_4$)$_2$S; anhydrous GaCl$_3$ is more volatile than ZnCl$_2$

Indium and thallium were also discovered by means of the spectroscope as their names indicate. Indium was first identified in 1863 by F. Reich and H. T. Richter and named from the brilliant indigo blue line in its flame spectrum (Latin *indicum*). Thallium was discovered independently by W. Crookes and by C. A. Lamy in the preceding year 1861/2 and named after the characteristic bright green line in its flame spectrum (Greek θαλλός, *thallos*, a budding shoot or twig).

7.2 The Elements

7.2.1 *Terrestrial abundance and distribution*

Aluminium is the most abundant metal in the earth's crust (8.3% by weight); it is exceeded in abundance only by O (45.5%) and Si (25.7%), and is approached only by Fe

(6.2%) and Ca (4.6%). Aluminium is a major constituent of many common igneous minerals including feldspars and micas. These, in turn, weather in temperate climates to give clay minerals such as kaolinite [$Al_2(OH)_4Si_2O_5$], montmorillonite, and vermiculite (p. 406). It also occurs in many well-known though rarer minerals such as cryolite (Na_3AlF_6), spinel ($MgAl_2O_4$), garnet [$Ca_3Al_2(SiO_4)_3$], beryl ($Be_3Al_2Si_6O_{18}$), and turquoise [$Al_2(OH)_3PO_4H_2O/Cu$]. Corundum (Al_2O_3) is one of the hardest substances known and is therefore used as an abrasive; many gemstones are impure forms of Al_2O_3, e.g. ruby (Cr), sapphire (Co), oriental emerald, etc. Commercially, the most important mineral is bauxite $AlO_x(OH)_{3-2x}$ ($0 < x < 1$); this occurs in a wide belt in tropical and subtropical regions as a result of leaching out both silica and other metals from aluminosilicates (see Panel).

Bauxite

The mixed aluminium oxide hydroxide mineral bauxite was discovered by P. Berthier in 1821 near Les Baux in Provence. In temperate countries (such as Mediterranean Europe) it occurs mainly as the "monohydrate" AlOOH (boehmite and diaspore) whereas in the tropics it is generally closer to the "trihydrate" $Al(OH)_3$ (gibbsite and hydrargillite). Since AlOOH is less soluble in aqueous NaOH than is $Al(OH)_3$, this has a major bearing on the extraction process for Al manufacture (p. 246). Typical compositions for industrially used bauxites are Al_2O_3 40–60%, combined H_2O 12–30%, SiO_2 free and combined 1–15%, Fe_2O_3 7–30%, TiO_2 3–4%, F, P_2O_5, V_2O_5, etc., 0.05–0.2%.

World production in 1975 was over 80 million tonne and this is still increasing. Reserves are immense, being of the order of 5×10^9 tonnes in Northern and Western Australia, over 10^9 tonnes in Brazil, Guinea, and Jamaica, and well over 10^8 tonnes in many other African and Central American countries. Australia is also currently the largest producer of alumina (19.3% of world production) followed by the USA 17.8%, the USSR 12.8%, Jamaica 8.4%, Japan 5.9%, the Fed. Rep. of Germany 4.7%, Surinam 4.3%, Canada 4.3%, and France 4.1%. Bauxite is easy to mine by open-cast methods since it occurs typically in broad layers 3–10 m thick with very little topsoil or other overburden. Apart from its preponderant use in Al extraction, bauxite is used to manufacture refractories and high-alumina cements, and smaller amounts are used as drying agents and catalysts in the petrochemicals industry.

Gallium, In, and Tl are very much less abundant than Al and tend to occur at low concentrations in sulfide minerals rather than as oxides, though Ga is also found associated with Al in bauxite. Ga (19 ppm) is about as abundant as N, Nb, Li, and Pb; it is twice as abundant as B (9 ppm) but is more difficult to extract because of the absence of major Ga-containing ores. The highest concentrations (0.1–1%) are in the rare mineral germanite (a complex sulfide of Zn, Cu, Ge, and As); concentrations in sphalerite (ZnS), bauxite, or coal, are a hundredfold less. Gallium always occurs in association either with Zn or Ge, its neighbours in the periodic table, or with Al in the same group. It was formerly recovered from flue dusts emitted during sulfide roasting or coal burning (up to 1.5% Ga) but is now obtained as a byproduct of the vast Al industry. Since bauxites contain 0.003–0.01% Ga, complete recovery would yield some 500–1000 tonnes pa. However, present consumption, though growing rapidly, is only 1% of this and production is of the order of 10 tonnes pa. This can be compared with the estimate of 5 tonnes for the total of Ga metal in the 90 y following its discovery (1875–1965). Its price in 1928 was $50 per g; in 1965 it was $1 per g, similar to the then price of gold ($1.1 per g), and in 1974 it was $0.75 per g.

Indium (0.21 ppm) is similar in abundance to Sb and Cd, whereas Tl (0.7 ppm) is close to Tm and somewhat less abundant than Mo, W, and Tb (1.2 ppm). Both elements are chalcophiles (p. 760), indium tending to associate with the similarly sized Zn in its sulfide minerals whilst the larger Tl tends to replace Pb in galena, PbS. Thallium(I) has a similar radius to Rb^I and so also concentrates with this element in the late magmatic potassium minerals such as feldspars and micas.

Indium is now commercially recovered from the flue dusts emitted during the roasting of Zn/Pb sulfide ores and can also be recovered during the roasting of Fe and Cu sulfide ores. Before 1925 only 1 g of the element was available in the world but production now exceeds 50 000 000 g (i.e. 50 tonnes) each year. Current price (1976) is $50–150 per kg depending on purity.

Thallium is likewise recovered from flue dusts emitted during sulfide roasting for H_2SO_4 manufacture, and from the smelting of Zn/Pb ores. Extraction procedures are complicated because of the need to recover Cd at the same time. As there are no major commercial uses for Tl metal, world production is less than 5 tonnes pa; the price ranges from $20 to $120 per kg depending on purity and amount purchased.

7.2.2 *Preparation and uses of the metals*[1]

The huge difference in scale between the production of Al metal, on the one hand, and the other elements in the group is clear from the preceding section. The tremendous growth of the Al industry compared with all other non-ferrous metals is indicated in Table 7.3 and Al production is now exceeded only by that of iron and steel (p. 1244).

TABLE 7.3   *World production of some non-ferrous metals/million tonnes pa*

Metal	1900	1960	1973
Al	0.0057	4.67	13.6
Cu	0.50	4.40	8.79
Zn	0.48	3.07	6.28
Pb	0.88	2.63	4.42

Production of Al metal involves two stages: (a) the extraction, purification, and dehydration of bauxite, and (b) the electrolysis of Al_2O_3 dissolved in molten cryolite Na_3AlF_6. Bauxite is now almost universally treated by the Bayer process; this involves dissolution in aqueous NaOH, separation from insoluble impurities (red muds), partial precipitation of the trihydrate, and calcining at 1200°. Bauxites approximating to the "monohydrate" AlOOH require higher concentrations of NaOH (200–300 g l^{-1}) and higher temperatures and pressures (200–250°C, 35 atm) than do bauxites approximating to Al(OH)$_3$ (100–150 g l^{-1} NaOH, 120–140°C). Electrolysis is carried out at 940–980°C in a carbon-lined steel cell (cathode) with carbon anodes. Originally Al_2O_3 was dissolved in molten cryolite (Héroult–Hall process) but cryolite is a rather rare mineral and production from the mines in Greenland provide only about 30 000 tonnes pa, quite

---

[1] *Kirk–Othmer Encyclopedia of Chemical Technology*, 3rd edn., Vol. 2, Aluminium and aluminium alloys, pp. 129–88; Aluminium compounds, pp. 188–251. Interscience, New York, 1978.

insufficient for world needs. Synthetic cryolite is therefore manufactured in lead-clad vessels by the reaction

$$6HF + Al(OH)_3 + 3NaOH \longrightarrow Na_3AlF_6 + 6H_2O$$

Typical electrolyte composition ranges are Na_3AlF_6 (80–85%), CaF_2 (5–7%), AlF_3 (5–7%), Al_2O_3 (2–8%—intermittently recharged). See also p. 77 for the beneficial use of Li_2CO_3. The detailed electrolysis mechanism is still imperfectly understood but typical operating conditions require up to 10^5 A at 4.5 V and a current density of 0.7 A cm^{-2}. One tonne Al metal requires 1.89 tonnes Al_2O_3 $\sim$0.45 tonnes C anode material; 0.07 tonnes Na_3AlF_6 and about 15 000 kWh of electrical energy. It follows that cheap electric power is the overriding commercial consideration. World production (1974) approached 14 million tonnes pa, the leading producers being the USA (34%), the USSR (11%), Japan (9%), Canada (8%), the Fed. Rep. of Germany (5%), and Norway (5%). In addition to this primary production, recycling of used alloys probably adds a further 3–4 million tonnes pa to the total Al metal consumed. Some uses of Al and its alloys are noted in the Panel.

Gallium metal is now obtained as a byproduct of the Al industry. The Bayer process for obtaining alumina from bauxite gradually enriches the alkaline solutions from an initial weight ratio Ga/Al of about 1/5000 to about 1/300; electrolysis of these extracts with an Hg electrode gives further concentration, and the solution of sodium gallate is then electrolysed with a stainless steel cathode to give Ga metal. Ultra high-purity Ga for semiconductor uses is obtained by further chemical treatment with acids and O_2 at high temperatures followed by crystallization and zone refining. Gallium has a beautiful silvery blue appearance; it wets glass, porcelain, and most other surfaces (except quartz, graphite, and teflon) and forms a brilliant mirror when painted on to glass. Its main use is in semiconductor technology (p. 290). For example, GaAs (isoelectronic with Ge) can convert electricity directly into coherent light (laser diodes) and is employed in electroluminescent light-emitting diodes (LEDs); it is also used for doping other semiconductors and in solid-state devices such as transistors. The compound $MgGa_2O_4$, when activated by divalent impurities such as Mn^{2+}, is used in ultraviolet-activated powders as a brilliant green phosphor—familiar to users of Xerox copying machines. Another very important application is to improve the sensitivity of various bands used in the spectroscopic analysis of uranium. Minor uses are as high-temperature liquid seals, manometric fluids and heat-transfer media, and for low-temperature solders.

Indium, like Ga, is normally recovered by electrolysis after prior concentration in processes leading primarily to other elements (Pb/Zn). It is a soft, silvery metal with a brilliant lustre and (like Sn) it gives out a high-pitched "cry" when bent. Formerly it was much used to protect bearings against wear and corrosion but the pattern of use has been changing in recent years and now its most important applications are in low-melting alloys and in electronic devices. Thus meltable safety devices, heat regulators, and sprinklers use alloys of In with Bi, Cd, Pb, and Sn (mp 50–100°C) and In-rich solders are valuable in sealing metal–nonmetal joints in high vacuum apparatus. Indium is of particular importance in the manufacture of p–n–p transistor junctions in Ge (p. 429) and to solder semiconductor leads at low temperature; the softness of the metal also minimizes stress in the Ge during subsequent cooling. So-called III–V semiconductors like InAs and InSb are used in low-temperature transistors, thermistors, and optical devices (photoconductors), and InP is used for high-temperature transistors. A further minor use, which

Some Uses of Aluminium Metal and Alloys

Pure aluminium is a silvery-white metal with many desirable properties: it is light, non-toxic, of pleasing appearance, and capable of taking a high polish. It has a high thermal and electrical conductivity, excellent corrosion resistance, is non-magnetic, non-sparking, and stands second only to gold for malleability and sixth for ductility. Many of its alloys have high mechanical and tensile strength. Aluminium and its alloys can be cast, rolled, extruded, forged, drawn, or machined, and they are readily obtained as pipes, tubes, rods, bars, wires, plates, sheets, or foils.

Aluminium resists corrosion not because of its position in the electrochemical series but because of the rapid formation of a coherent, inert, oxide layer. Contact with graphite, Fe, Ni, Cu, Ag, or Pb is disastrous for corrosion resistance; the effect of contact with steel, Zn, and Cd depends on pH and exposure conditions. Protection is enhanced by anodizing the metal; this involves immersing it in 15–20% H_2SO_4 and connecting it to the positive terminal so that it becomes coated with alumina:

$$2Al + 3O^{2-} - 6e^- \longrightarrow Al_2O_3$$

A layer 10–20 μm thick gives excellent protection between pH 4.5–8.7 and is also adequate for external architectural use; thicker layers (50–100 μm) also impart abrasion resistance. The layer can be coloured by incorporating suitable organic or inorganic compounds in the bath and incorporation of photosensitive material enables photographic images to be developed. Decorative engraving using solutions of nitrate or NH_4HF_2 gives the metal a fine silky texture.

TABLE A   *Some aluminium alloys*

1000 Series:	Commercially pure Al (<1% of other elements); good properties except for limited mechanical strength. Used in chemical equipment, reflectors, heat exchangers, buildings, and decorative trim.
2000 Series:	Cu alloys (~5%); excellent strength and machinability, limited corrosion resistance. Used for components requiring high strength/weight ratio, e.g. truck trailer panels, aircraft structure parts.
3000 Series:	Mn alloys (~1.2%); moderate strength, high workability. Used for cooking utensils, heat exchangers, storage tanks, awnings, furniture, highway signs, roofing, side panels, etc.
4000 Series:	Si alloys (≤12%); low mp and low coefficient of expansion. Used for castings and as filler material for brazing and welding; readily anodized to attractive grey colours.
5000 Series:	Mg alloys (0.3–5%); good strength and weldability coupled with excellent corrosion resistance in marine atmospheres. Used for ornamental and decorative trim, street light standards, ships, boats, cryogenic vessels, gun mounts, and crane parts.
6000 Series:	Mg/Si alloys; good formability and high corrosion resistance. Used in buildings, transportation equipment, bridges, railings, and welded construction.
7000 Series:	Zn alloys (3–8%) plus Mg; when heat treated and aged have very high strength. Used principally for aircraft structures, mobile equipment, and equipment requiring high strength/weight ratio.

Many of the mechanical properties of pure Al are greatly improved by alloying it with Cu, Mn, Si, Mg, or Zn (Table A). The example of Cu is particularly important because of the insight which it gives into the subtle solid-state diffusion processes that occur during heat treatment. At room temperature Al dissolves only about 0.1% Cu and this has little effect on its properties. The solubility rises to a maximum of 5.65% Cu at 548°C and this remains in metastable solid solution to give a soft workable alloy when the alloy is rapidly quenched to temperatures below 65°. Subsequent ageing of the shaped material at 100–150° for a few minutes hardens the alloy due to the formation of Guinier–Preston zones: these zones, independently discovered in 1938 by A. Guinier (France) and G. D. Preston (England), are minute discs of material higher in Cu content than the matrix—they are about 4 atoms thick and up to 100 atoms across; they mesh coherently with the host lattice in two directions, the (100) planes, but not in the third. The coherency strains which thereby develop in the lattice are the basis for the hardening of the alloy. Besides its immense technological importance, this phenomenon is particularly significant in being one of the first recognized examples of a single phase which nevertheless varies regularly in composition throughout its extent.

The properties and uses of some typical alloys are summarized in Table A and the distribution of these various uses for the particular case of the USA is shown in the accompanying diagram:

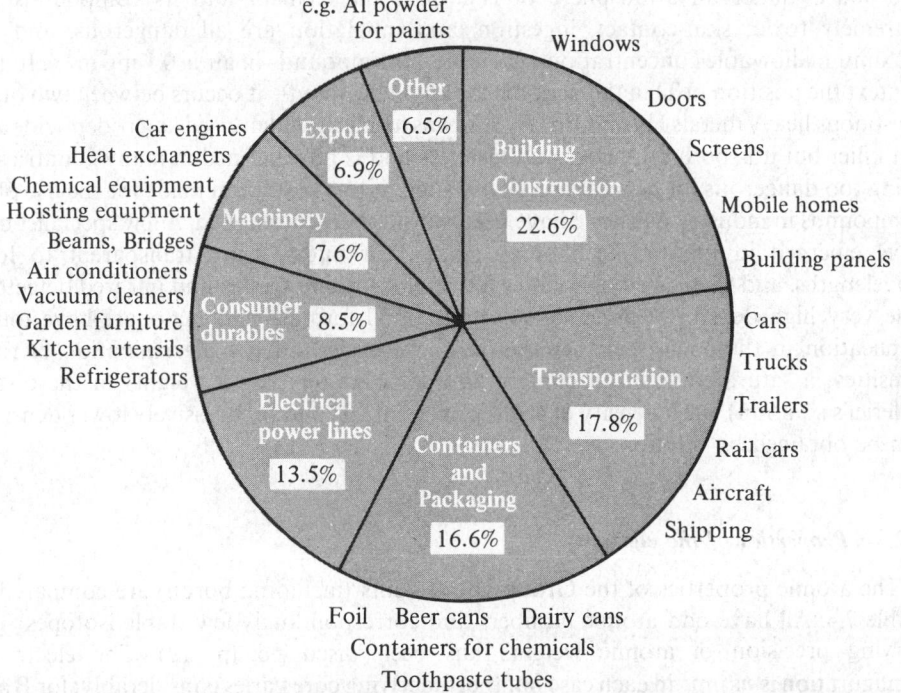

Many of these uses are a matter of everyday observation. In addition we may note that the electrical conductivity of pure Al is 63.5% of the conductivity of an equal *volume* of pure Cu; when the lower density of Al is considered its conductivity is 2.1 times that of Cu on a wt. for wt. basis. This, coupled with its corrosion resistance and ready workability makes it an ideal metal for power lines and, indeed, more than 90% of all overhead electrical transmission lines in the USA are Al alloy.

Aluminium is now extensively used in the construction industry. The first Al-clad skyscraper was completed in 1952 and the metal is now a familiar material for building panels, screens, doors, and decorative trim. Applications in the construction of aircraft are also obvious. Each Boeing 707 contains some 50 tonnes of Al alloy and the air frame of the supersonic Concorde is made of the special alloy Hiduminium RR58 of approximate composition (wt. %) Al 93.5, Cu 2.2, Mg 1.5, Fe 1.2, Ni 1.1, Si 0.2, and others (Mn, Zn, Pb, Sn) 0.3%.

In the aerospace industry Al is extensively used in vehicle structure and skins, in fuel and oxidizer tanks, and in castings for electronic gear. Aluminium powder is also frequently used as an ingredient in solid fuels for extra thrust at blast-off (p. 1015). One of the largest satellites ever sent aloft (Echo II, a sphere 41 m in diameter which can easily be seen from the ground with the naked eye) is made of Al foil 0.46 mm thick laminated to plastic. It is perhaps appropriate to recall that Jules Verne in his prophetic adventure story *From the Earth to the Moon* (1865) selected Al as the perfect material for his projectile. As one of his characters remarked "It is easily wrought, is very widely distributed, forming the basis of most of the rocks, is three times lighter than iron, and seems to have been created for the express purpose of furnishing us with the material for our projectile."

exploits the high neutron capture cross-section of In, is as a component in control rods for certain nuclear reactors.

Technical grade Tl is purified from other flue-dust elements (Ni; Zn, Cd; In; Ge, Pb; As; Se, Te) by dissolving it in warm dilute acid, then precipitating the insoluble $PbSO_4$ and adding HCl to precipitate TlCl. Further purification is effected by electrolysing Tl_2SO_4 in dilute H_2SO_4 with short Pt wire electrodes, followed by fusion of the deposited Tl metal at $350-400°C$ under an atmosphere of H_2. Both the element and its compounds are extremely toxic; skin-contact, ingestion, and inhalation are all dangerous, and the maximum allowable concentration of soluble Tl compounds in air is 0.1 mg m^{-3}. In this context the position of Tl in the periodic table will be noted—it occurs between two other poisonous heavy metals Hg and Pb. Tl_2SO_4 was formerly widely used as a rodenticide and ant killer but it is both odourless and tasteless and is now banned in many countries as being too dangerous for general use. Many suggestions have been made for the use of Tl compounds in industry but none have been substantially developed. A few specialist uses have emerged in infrared technology since TlBr and TlI are transparent to long wavelengths, and there are possibilities for photosensitive diodes and infrared detectors. The very high density of aqueous solutions of Tl formate and malonate have found application in the small-scale separation of minerals and the determination of their densities; a saturated solution containing approximately equal weights of these salts (Clerici's solution) has a density of 4.324 g cm^{-3} at $20°$ and progressively lower densities can be obtained by dilution.

7.2.3 *Properties of the elements*

The atomic properties of the Group III elements (including boron) are compared in Table 7.4. All have odd atomic numbers and correspondingly few stable isotopes. The varying precision of atomic weights has been discussed (p. 19). The electronic configuration is ns^2np^1 in each case but the underlying core varies considerably: for B and Al it is the preceding noble gas core, for Ga and In it is noble gas plus d^{10}, and for Tl noble gas plus $4f^{14}5d^{10}$. This variation has a substantial influence on the trends in chemical properties of the group and is also reflected in the ionization energies of the elements. Thus, as shown in Fig. 7.1, the expected decrease from B to Al is not followed by a further

TABLE 7.4   *Atomic properties of Group III elements*

Property	B	Al	Ga	In	Tl
Atomic number	5	13	31	49	81
No. of nat. occurring isotopes	2	1	2	2	2
Atomic weight	10.81	26.98154	69.72	114.82	204.383
Electronic configuration	[He]2s^{2}2p^1	[Ne]3s^{2}3p^1	[Ar]3d^{10}4s^{2}4p^1	[Kr]4d^{10}5s^{2}5p^1	[Xe]4f^{14}5d^{10}6s^1
Ionization energy/kJ mol^{-1} I	800.5	577.4	578.6	558.2	589.1
II	2426.5	1816.1	1978.8	1820.2	1970.5
III	3658.7	2744.1	2962.3	2704.0	2877.4
Metal radius/pm	(80–90)	143	135 (see text)	167	170
Ionic radius/pm (6-coord.) III	27$^{(a)}$	53.5	62.0	80.0	88.5
I	—	—	120	140	150

$^{(a)}$ Nominal "ionic" radius for B^{III}.

I_M/kJ /mol^{-1}

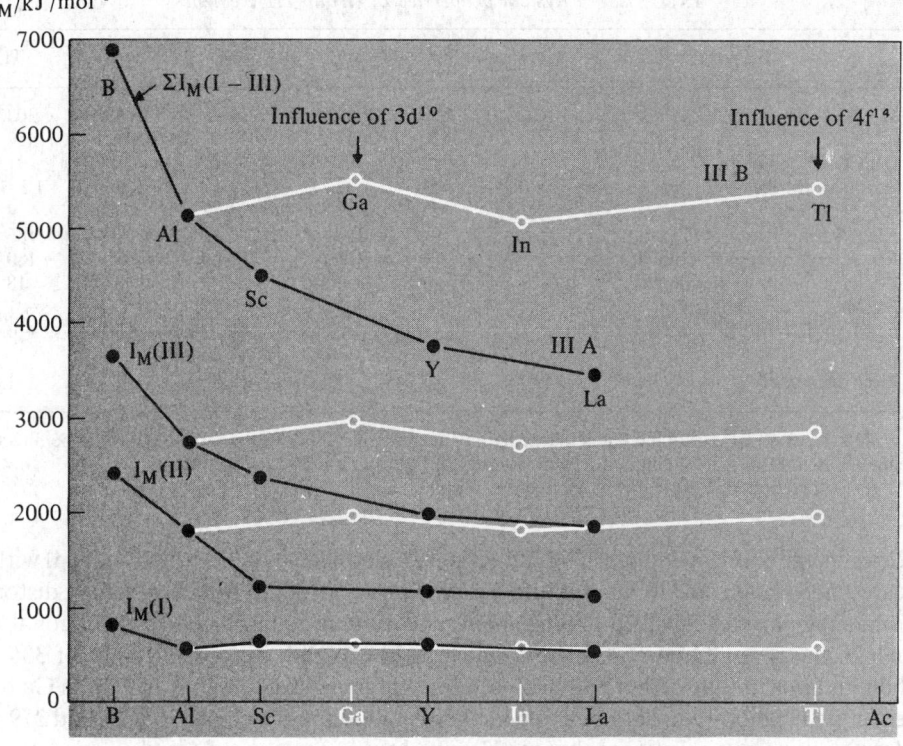

FIG. 7.1   Trends in successive ionization energies $I_M(I)$, $I_M(II)$, and $I_M(III)$, and their sum Σ for elements in Groups IIIA and IIIB.

decrease to Ga because of the "d-block contraction" in atomic size and the higher effective nuclear charge for this element which stems from the fact that the 10 added d electrons do not completely shield the extra 10 positive charges on the nucleus. Similarly, the decrease between Ga and In is reversed for Tl as a result of the further influence of the f block or lanthanide contraction. It is notable that these irregularities for the IIIB elements do not occur for the IIIA elements Sc, Y, and La, which show a steady decrease in ionization energy from B and Al, all 5 elements having the same type of underlying core (noble gas). This has a decisive influence on the comparative chemistry of the two subgroups.

Boron is a covalently bonded, refractory, non-metallic insulator of great hardness and is thus not directly comparable in its physical properties with Al, Ga, In, and Tl, which are all low-melting, rather soft metals having a very low electrical resistivity (Table 7.5). The heats of fusion and vaporization of the metals are also much lower than those of boron and tend to decrease with increasing atomic number. In all these properties the metals resemble the neighbouring metals Zn, Cd, Hg; Sn, Pb, etc., and it is probable that in each case the properties are related to the rather small number of electrons available for metallic bonding. Some have seen this as a manifestation of the "inert-pair effect" (see p. 255). The interatomic distances in these elements are also somewhat longer than expected from general trends.

The crystal structure of Al is fcc, typical of many metals, each Al being surrounded by 12

TABLE 7.5 *Physical properties of Group III elements*

Property	B	Al	Ga	In	Tl
MP/°C	2180	660.37	29.78	156.61	303.5
BP/°C	~3650	2467	2403	2080	1457
Density (20°C)/g cm^{-3}	2.35	2.699	5.904	7.31	11.85
Hardness (Mohs)	11	2.75	1.5	1.2	1.2–1.3
ΔH_{fus}/kJ mol^{-1}	23.60	10.50	5.59	3.26	4.31
ΔH_{vap}/kJ mol^{-1}	504.5	290.8	270.3	231.8	166.1
ΔH_f (monatomic gas)/kJ mol^{-1}	571.1	321.7	286.2	243.1	180.7
Electrical resistivity/μohm cm	6.7×10^{11}	2.655	~27[a]	8.37	18
E° (M^{3+} + 3e$^-$ = M(s))/V	-0.87[b]	-1.66	-0.56	-0.34	$+1.26$
E°(M$^+$ + e$^-$ = M(s))/V	—	0.55	-0.79 (acid) -0.18		-0.34
			-1.39(alkali)		
Electronegativity χ	2.0	1.5	1.6	1.7	1.8

[a] The resistivity of crystalline Ga is markedly anisotropic, the values in the three orthorhombic directions being a 17.5, b 8.20, c 55.3 μohm cm. The resistivity of liquid Ga at 30° is 25.8 μohm cm.
[b] E° for reaction $H_3BO_3 + 3H^+ + 3e^- = B(s) + 3H_2O$.

nearest neighbours at 286 pm. Thallium also has a typical metallic structure (hcp) with 12 nearest neighbours at 340 pm. Indium has an unusual structure which is slightly distorted from a regular close-packed arrangement: the structure is face-centred tetragonal and each In has 4 neighbours at 324 pm and 8 at the slightly greater distance of 336 pm. Gallium has a unique orthorhombic (pseudotetragonal) structure in which each Ga has 1 very close neighbour at 244 pm and 6 further neighbours, 2 each at 270, 273, and 279 pm. The structure is very similar to that of iodine and the appearance of pseudo-molecules Ga$_2$ may result from partial pair-wise interaction on neighbouring atoms of the single p electron outside the [Ar]3d^{10}4s^2 core which immediately follows the first transition series. As such it can be compared with Hg which also has a very low mp and completes the [Xe]4f^{14}5d^{10}6s^2 "pseudo-noble-gas" configuration following the lanthanide elements. Note that all interatomic contacts in metallic Ga are less than those in Al, again emphasizing the presence of a "d-block contraction". Gallium is also unusual in contracting on melting, the volume of the liquid phase being 3.4% less than that of the solid; the same phenomenon occurs with the next element in the periodic table Ge, and also with Sb and Bi, in addition to the well-known example of H$_2$O. In each case, a structural feature in the solid is broken down to permit more efficient packing of atoms in the liquid state.

The standard electrode potentials of the Group IIIB elements reflect the decreasing stability of the +3 oxidation state in aqueous solution and the tendency, particularly of Tl, to form compounds in the +1 oxidation state (p. 255). The trend to increasing electropositivity of the group oxidation state which was noted for Groups IA and IIA does not occur with Group IIIB but is found, as expected, in Group IIIA (Fig. 7.2). Similarly, the steady decrease in electronegativity in the series B > Al > Sc > Y > La > Ac is reversed in Group IIIB and there is a steady *increase* in electronegativity from Al to Tl.

7.2.4 *Chemical reactivity and trends*

The Group III metals differ sharply from the non-metallic element boron both in their

$E°$ (M^{3+}/M)/V

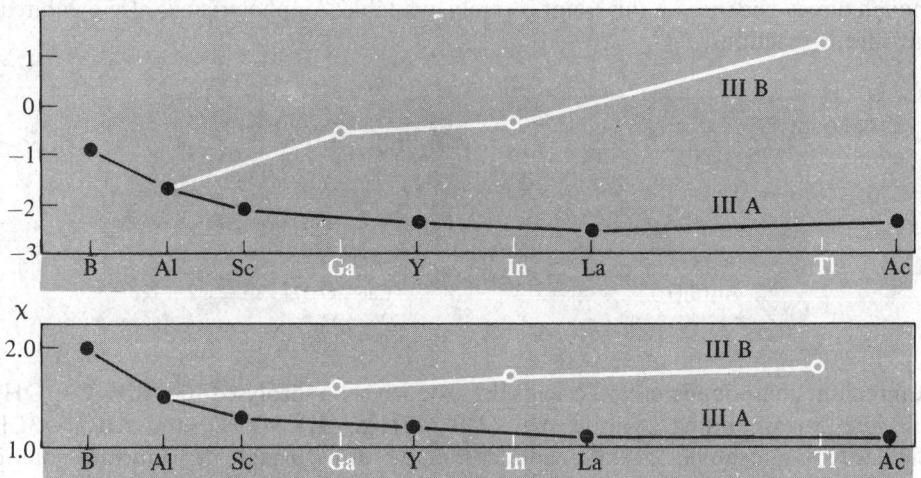

χ

FIG. 7.2   Trends in standard electrode potential $E°$ and electronegativity χ for elements in
Groups IIIA and IIIB.

greater chemical reactivity at moderate temperatures and in their well-defined cationic
chemistry for aqueous solutions. The absence of a range of volatile hydrides and other
cluster compounds analogous to the boranes and carboranes is also notable. Aluminium
combines with most non-metallic elements when heated to give compounds such as AlN,
Al_2S_3, AlX_3, etc. It also forms intermetallic compounds with elements from all groups of
the periodic table that contain metals. Because of its great affinity for oxygen it is used as a
reducing agent to obtain Cr, Mn, V, etc., by means of the thermite process of K.
Goldschmidt. Finely, powdered Al metal explodes on contact with liquid O_2, but for
normal samples of the metal a coherent protective oxide film prevents appreciable
reaction with oxygen, water, or dilute acids; amalgamation with Hg or contact with
solutions of salts of certain electropositive metals destroys the film and permits further
reaction. Aluminium is also readily soluble in hot concentrated hydrochloric acid and in
aqueous NaOH or KOH at room temperature with liberation of H_2. This latter reaction is
sometimes written as

$$Al + NaOH + H_2O \longrightarrow NaAlO_2 + \tfrac{3}{2}H_2$$

though it is likely that the species in solution is the hydrated tetrahydroxoaluminate anion
$[Al(OH)_4]^-$ (aq) or $[Al(H_2O)_2(OH)_4]^-$.

$Al(OH)_3$ is amphoteric, forming both salts and aluminates (Greek ἀμφοτέρως,
amphoteros, in both ways). Thus the freshly precipitated hydroxide is readily soluble in
both acid and alkali:

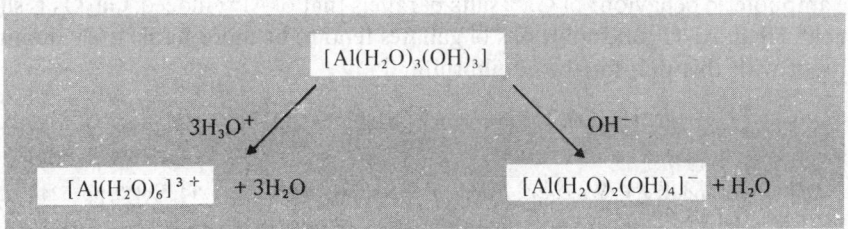

In these reactions the coordination number of Al has been assumed to be 6 throughout though direct evidence on this point is rarely available. Amphoterism is also exhibited in anhydrous reactions, e.g.:

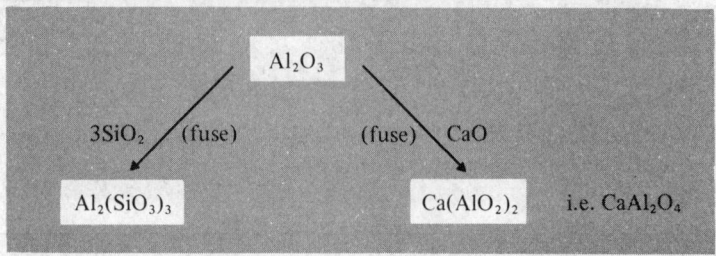

Aluminium compounds of weak acids are extensively hydrolysed to $[Al(H_2O)_3(OH)_3]$ and the corresponding hydride, e.g. $Al_2S_3 \rightarrow 3H_2S$, $AlN \rightarrow NH_3$, and $Al_4C_3 \rightarrow 3CH_4$. Similarly, the cyanide, acetate, and carbonate are unstable in aqueous solution. Hydrolysis of the halides and other salts such as the nitrate and sulfate is incomplete but aqueous solutions are acidic due to the ability of the hydrated cation $[Al(H_2O)_6]^{3+}$ to act as proton donor giving $[Al(H_2O)_5(OH)]^{2+}$, $[Al(H_2O)_4(OH)_2]^+$, etc. If the pH is gradually increased this deprotonation of the mononuclear species is accompanied by aggregation via OH bridges to give species such as

$$[(H_2O)_4Al\diagdown\begin{smallmatrix}OH\\OH\end{smallmatrix}\diagup Al(H_2O)_4]^{4+}$$

and then to precipitation of the hydrous oxide. This is of particular use in water clarification since the precipitating hydroxide nucleates on fine suspended particles which are thereby thrown out of suspension. Still further increase in pH leads to redissolution as an aluminate (Fig. 7.3). Similar behaviour is shown by Be^{II}, Zn^{II}, Ga^{III}, Sn^{II}, Pb^{II}, etc. A detailed quantitative theory of amphoterism is difficult to construct but it is known that amphoteric behaviour occurs when (a) the cation is weakly basic, (b) its hydroxide is moderately insoluble, and (c) the hydrated species can also act as proton donors.[2]

Anhydrous Al salts cannot be prepared by heating the corresponding hydrate for reasons closely related to the amphoterism and hydrolysis of such compounds. For example, $AlCl_3 . 6H_2O$ is, in reality, $[Al(H_2O)_6]Cl_3$ and the strength of the Al–O interaction precludes the formation of Al–Cl bonds:

$$2[Al(H_2O)_6]Cl_3 \xrightarrow{\text{heat}} Al_2O_3 + 6HCl + 9H_2O$$

The amphoteric behaviour of Ga^{III} salts parallels that of Al^{III}; indeed, Ga_2O_3 is slightly *more* acidic than Al_2O_3 and solutions of gallates tend to be more stable than aluminates. Consistent with this pK_a for the equilibrium

$$[M(H_2O)_6]^{3+} \rightleftharpoons [M(H_2O)_5OH]^{2+} + H^+$$

² C. S. G. PHILLIPS and R. J. P. WILLIAMS, *Inorganic Chemistry*, Vol. 1, Chap. 14; Vol. 2, pp. 524–5, Oxford University Press, Oxford, 1966.

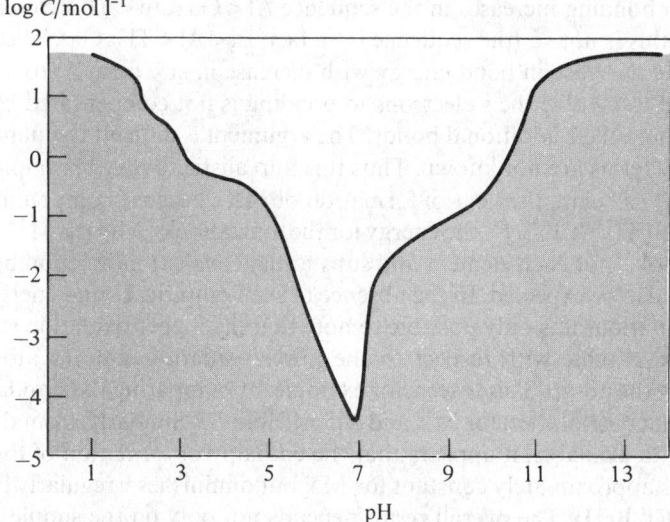

FIG. 7.3   Schematic representation of the variation of concentration of an Al salt as a function of pH (see text).

is 4.95 for Al and 2.60 for Ga. Indium is more basic than Ga and is only weakly amphoteric. The metal does not dissolve in aqueous alkali whereas Ga does. This alternation in the sequence of basicity can be related to the electronic and size factors mentioned on p. 250. Thallium behaves as a moderately strong base but is not strictly comparable with other members of the group because it normally exists as Tl^I in aqueous solution. Thus, Tl metal tarnishes readily and reacts with steam or moist air to give TlOH. The electrode potential data in Table 7.5 show that Tl^I is much more stable than Tl^{III} in aqueous solution and indicate that Tl^{III} compounds can act as strong oxidizing agents.

Compounds of Tl^I have many similarities to those of the alkali metals: TlOH is very soluble and is a strong base; Tl_2CO_3 is also soluble and resembles the corresponding Na and K compounds; Tl^I forms colourless, well-crystallized salts of many oxoacids, and these tend to be anhydrous like those of the similarly sized Rb and Cs; Tl^I salts of weak acids have a basic reaction in aqueous solution as a result of hydrolysis; Tl^I forms polysulfides (e.g. Tl_2S_5) and polyiodides, etc. In other respects Tl^I resembles the more highly polarizing ion Ag^+, e.g. in the colour and insolubility of its chromate, sulfide, arsenate, and halides (except F), though it does not form ammine complexes in aqueous solution and its azide is not explosive.

The stability of the +1 oxidation state in Group III increases in the sequence Al < Ga < In < Tl, and numerous examples of M^I compounds will be found in the following sections. The occurrence of an oxidation state which is 2 less than the group valency is sometimes referred to as the "inert-pair effect" but it is important to recognize that this is a description not an explanation. The phenomenon is quite general among the heavier elements of the p block (i.e. the post-transition elements or B subgroups). For example, Sn and Pb commonly occur in both the +2 and +4 oxidation states; P, As, Sb, and Bi in the +3 and +5, S, Se, Te, Po in the +2, +4, and +6 states, etc. The term "inert-pair effect" is somewhat misleading since it implies that the energy required to involve the

ns^2 electrons in bonding increases in the sequence Al < Ga < In < Tl. Reference to Table 7.4 shows that this is not so (the sequence is, in fact, In < Al < Tl < Ga). The explanation lies rather in the decrease in bond energy with increase in size from Al to Tl so that the energy required to involve the s electrons in bonding is not compensated by the energy released in forming the 2 additional bonds. The argument is difficult to quantify since the requisite energy terms are not known. Thus it is unrealistic to use the simple ionic bond model (p. 91) to calculate the heat of formation of MX_3 because compounds like $TlCl_3$ are not ionic, i.e. $[Tl^{3+}(Cl^-)_3]$—the energy for the ionization of M(g) to M^{3+}(g) is greater than 5000 kJ mol^{-1} for each element and substantial covalent interaction between M^{3+} and X^- would also be expected. In the absence of semi-empirical bond energy data or *ab initio* MO calculations it is only possible to note that the higher oxidation state becomes progressively less stable with respect to the lower oxidation state as atomic number increases within the group. This is seen, for example, by comparing the standard electrode potentials in aqueous solution for M^{III} and M^I in Table 7.5. Similarly, from the somewhat fragmentary data available, it appears that the enthalpy of formation of the anhydrous halides remains approximately constant for MX but diminishes irregularly from Al to Tl for MX_3 (X = Cl, Br, I). The overall result depends not only on the simple Born–Haber terms (p. 94) but also on a combination of several other factors including changes in structure and bond type, covalency effects, enthalpies of hydration, entropy effects, etc., and a quantitative rationalization of all the data has not yet been achieved.

Group IIIB metals furnish a good example of the general rule that an element is more electropositive in its lower than in its higher oxidation state: the lower oxide and hydroxide are more basic and the higher oxide and hydroxide more acidic. The reasons for this behaviour are similar to those already discussed when comparing Group IIA with Group IA (p. 121) and turn on the relative magnitude of ionization energies, cationic size, hydration enthalpy and entropy, etc. Again, the higher the charge on an aquo cation $[M(H_2O)]^{n+}$ the more readily will it act as a proton donor (p. 55).

Other group trends will emerge in subsequent sections.

7.3 Compounds

7.3.1 *Hydrides and related complexes*[3, 4]

The extensive covalent chemistry of the boron hydrides finds no parallel with the heavier elements of Group III. AlH_3 is a colourless, involatile solid which is extensively polymerized via Al–H–Al bonds; it is thermally unstable above 150–200°, is a strong reducing agent, and reacts violently with water and other protic reagents to liberate H_2. Several crystalline and amorphous modifications have been described and the structure of α-AlH_3 has been determined by X-ray and neutron diffraction:[5] each Al is octahedrally surrounded by 6 H atoms at 172 pm and the Al–H–Al angle is 141°. The participation of each Al in 6 bridges, and the equivalence of all Al–H distances suggests that 3-centre 2-

[3] E. WIBERG and E. AMBERGER, *Hydrides of the Elements of Main Groups I–IV*, Chaps. 5 and 6, pp. 381–461, Elsevier, Amsterdam, 1971.

[4] N. N. GREENWOOD, Gallium hydride and its derivatives, Chap. 3 in E. A. V. EBSWORTH, A. G. MADDOCK, and A. G. SHARPE (eds.), *New Pathways in Inorganic Chemistry*, pp. 37–64, Cambridge University Press, Cambridge, 1968.

[5] J. W. TURLEY and H. W. RINN, The crystal structure of aluminium hydride, *Inorg. Chem.* **8**, 18–22 (1969).

electron bonding occurs as in the boranes (p. 179). The closest Al $\cdots$ Al distance is 324 pm, which is appreciably shorter than in metallic Al (340 pm), but there is no direct metal–metal bonding and the density of AlH_3 (1.477 g cm^{-3}) is markedly less than that for Al (2.699 g cm^{-3}); this is because in Al metal all 12 nearest neighbours are at 340 pm whereas in AlH_3 there are 6 Al at 324 and 6 at 445 pm.

AlH_3 is best prepared by the reaction of ethereal solutions of $LiAlH_4$ and $AlCl_3$ under very carefully controlled conditions:[6]

$$3LiAlH_4 + AlCl_3 \xrightarrow{\ Et_2O\ } 4[AlH_3(Et_2O)_n] + 3LiCl$$

The LiCl is removed and the filtrate, if left at this stage, soon deposits an intractable etherate of variable composition. To avoid this, the solution is worked up with an excess of $LiAlH_4$ and some added $LiBH_4$ in the presence of a large excess of benzene under reflux at 76–79°C. Crystals of α-AlH_3 soon form. Slight variations in the conditions lead to other crystalline modifications of unsolvated AlH_3, 6 of which have been identified.

AlH_3 readily forms adducts with strong Lewis bases (L) but these are more conveniently prepared by reactions of the type

$$LiAlH_4 + NMe_3HCl \xrightarrow{\ Et_2O\ } [AlH_3(NMe_3)] + LiCl + H_2$$

$[AlH_3(NMe_3)]$ has a tetrahedral structure and can take up a further mole of ligand to give $[AlH_3(NMe_3)_2]$; this was the first compound in which Al was shown to adopt a 5-coordinate trigonal bipyramidal structure[7]

$$
\begin{array}{c}
H \\
| \\
Me_3N-Al-NMe_3 \\
\diagup \quad \diagdown \\
H \qquad H
\end{array}
$$

Gallium hydride is a viscous liquid, mp $-15°$ which decomposes quantitatively to Ga and H_2 at room temperature. It can be prepared by a low-temperature displacement reaction in the absence of solvent.

$$[GaH_3(NMe_3)](c) + BF_3(g) \longrightarrow GaH_3(l) + [BF_3(NMe_3)](c)$$

The adduct $[GaH_3(NMe_3)]$ is a colourless, crystalline compound mp 70.5° which is formed quantitatively by the reaction of ethereal solutions of $LiGaH_4$ and NMe_3HCl; like the Al analogue it takes up a further mole of ligand to give the trigonal bipyramidal $[GaH_3(NMe_3)_2]$. Numerous complexes have been prepared and the stabilities of the 1:1 adducts decrease in the following sequences:[4]

$$Me_2NH > Me_3N > C_5H_5N > Et_3N > PhNMe_2 \gg Ph_3N$$
$$Me_3N \approx Me_3P > Me_2PH$$
$$Ph_3P > Ph_3N > Ph_3As$$
$$R_3N \text{ or } R_3P > R_2O \text{ or } R_2S$$

[6] F. M. BROWER, N. E. MATZEK, P. F. REIGLER, H. W. RINN, C. B. ROBERTS, D. L. SCHMIDT, J. A. SHOVER, and K. TERADA, Preparation and properties of aluminium hydride, *J. Am. Chem. Soc.* **98**, 2450–3 (1976).

[7] G. W. FRASER, N. N. GREENWOOD, and B. P. STRAUGHAN, Aluminium hydride adducts of trimethylamine: vibrational spectra and structure, *J. Chem. Soc.* (1963) 3742–9. C. W. HEITSCH, C. E. NORDMAN, and R. W. PARRY, The crystal structure and dipole moment in solution of the compound $AlH_3 \cdot 2NMe_3$, *Inorg. Chem.* **2**, 508–12 (1963).

Complexes of the type $[GaH_2X(NMe_3)]$ and $[GaHX_2(NMe_3)]$ are readily prepared by reaction of HCl or HBr on the GaH_3 complex at low temperatures or by reactions of the type

$$2[GaH_3(NMe_3)] + [GaX_3(NMe_3)] \xrightarrow{C_6H_6} 3[GaH_2X(NMe_3)] \quad X = Cl, Br, I$$

The relative stabilities of these various complexes can be rationalized in terms of the factors discussed on p. 223.

InH$_3$ and TlH$_3$ appear to be too unstable to exist in the uncoordinated state though they may have transitory existence in ethereal solutions at low temperatures. A similar decrease in thermal stability is noted for the tetrahydro complexes; e.g. the temperature at which the Li salts decompose rapidly, follows the sequence

$$LiBH_4(380°) > LiAlH_4(100°) > LiGaH_4(50°) > LiInH_4(0°) \approx LiTlH_4(0°)$$

LiAlH$_4$ is a white crystalline solid, stable in dry air but highly reactive towards moisture, protic solvents, and many organic functional groups. It is readily soluble in ether (~ 29 g per 100 g at room temperature) and is normally used in this solvent. LiAlH$_4$ has proved to be an outstandingly versatile reducing agent since its discovery 35 y ago[8] (see Panel). It can be prepared on the laboratory (and industrial) scale by the reaction

$$4LiH + AlCl_3 \xrightarrow{Et_2O} LiAlH_4 + 3LiCl$$

On the industrial (multitonne) scale it can also be prepared by direct high-pressure reaction of the elements or preferably via the intermediate formation of the Na analogue.

$$Na + Al + 2H_2 \xrightarrow[350\,atm]{thf/140°/3\,h} NaAlH_4 \ (99\% \text{ yield})$$

The Li salt can then be obtained by metathesis with LiCl in Et$_2$O. The X-ray crystal structure of LiAlH$_4$ shows the presence of tetrahedral AlH$_4$ groups (Al–H 155 pm) bridged by Li in such a way that each Li is surrounded by 4H at 188–200 pm (cf. 204 pm in LiH) and a fifth H at 216 pm. The bonding therefore deviates considerably from the simple ionic formulation Li$^+$AlH$_4^-$ and there appears to be substantial covalent bonding as found in other complex hydrides (p. 74).

Other complex hydrides of Al are known including Li$_3$AlH$_6$, M^{I}AlH$_4$ (M^I = Li, Na, K, Cs), M^{II}(AlH$_4$)$_2$ (M^{II} = Be, Mg, Ca), Ga(AlH$_4$)$_3$, M^I(AlH$_3$R), M^I(AlH$_2$R$_2$), M^I[AlH(OEt)$_3$], etc. (see Panel). The important complex Al(BH$_4$)$_3$ has already been mentioned (p. 188); it is a colourless liquid, mp $-64.5°$, bp $+44.5°$. It is best prepared in the absence of solvent by the reaction

$$3NaBH_4 + AlCl_3 \longrightarrow Al(BH_4)_3 + 3NaCl$$

Al(BH$_4$)$_3$ was the first fluxional compound to be recognized as such (1955) and its thermal

---

[8] A. E. FINHOLD, A. C. BOND, and H. J. SCHLESINGER, Lithium aluminium hydride, aluminium hydride, and lithium gallium hydride, and some of their applications in organic and inorganic chemistry, *J. Am. Chem. Soc.* **69**, 1199–1203 (1947); see also R. T. SANDERSON, More on complex hydrides, *Chem & Eng. News* 29 Nov. 1976, p. 3.

Synthetic Reactions of LiAlH₄[3, 9]

$LiAlH_4$ is a versatile reducing and hydrogenating reagent for both inorganic and organic compounds. With inorganic halides the product obtained depends on the relative stabilities of the corresponding tetrahydroaluminate, hydride, and element. For example, $BeCl_2$ gives $Be(AlH_4)_2$, whereas BCl_3 gives B_2H_6, and HgI_2 gives Hg metal. The halides of Cu, Ag, Au, Zn, Cd, and Hg give some evidence of unstable hydrido species at low temperatures but all are reduced to the metal at room temperature. The halides of main groups IV and V yield the corresponding hydrides since the AlH_4 derivatives are unstable or non-existent. Thus $SiCl_4$, $GeCl_4$, and $SnCl_4$ yield MH_4 and substituted halides such as R_nSiX_{4-n} give R_nSiH_{4-n}. Similarly, PCl_3 (and PCl_5), $AsCl_3$, and $SbCl_3$ afford MH_3 but $BiCl_3$ is reduced to the metal. $PhAsCl_2$ gives $PhAsH_2$ and Ph_2SbCl gives Ph_2SbH, etc. Less work has been done on oxides but $COCl_2$ yields MeOH, NO yields hyponitrous acid $HON=NOH$ (which can be isolated as the Ag salt), and CO_2 gives $LiAl(OMe)_4$ or $LiAl(OCH_2O)_2$ depending on conditions.

The real importance of $LiAlH_4$ stems from the applications in organic syntheses. Its commercial introduction dates from 1948. By 1951 the number of functional groups that were known to react was 23 and this rose to more than 60 by the 1970s. Despite this, the heyday of $LiAlH_4$ seems to have been reached in the late 1960s and it is now being replaced in many systems by the more selective borohydrides (p. 189) or by organometallic hydrides (see below). Reaction is usually carried out in ether solution, followed by hydrolysis of the intermediate so formed when appropriate. Typical examples are given in Table A.

TABLE A　*Products of reduction by LiAlH₄*

Compound	Product	Compound	Product
Reactive $>C=C<$	$>CH-CH<$	SCR_2-CR_2	$R_2C(SH)CHR_2$
$RCH=CH_2$	$[Al(CH_2CH_2R)_4]^-$	RSSR	RSH
C_2H_2	$[AlH(CH=CH_2)_3]^-$	RCOSR	RCH_2OH
$RC\equiv CH$	$RCH=CH_2$	$RCSNH_2$	RCH_2NH_2
RX	RH (not aryl)	RSCN	RSH
ROH	$[Al(OR)_4]^-$ or	R_2SO	R_2S
	$[AlH(OR)_3]^-$	R_2SO_2	R_2S
RCHO·	RCH_2OH	RSO_2X	RSH
R_2CO	R_2CHOH	$ROSO_2R'$	RH
		$ArOSO_2R'$	ArOH
Quinone	Hydroquinone	RSO_2H	RSSR + RSH
RCO_2H, or		RNC, or	
$(RCO)_2O$,	RCH_2OH	RNCO, or	RNHMe
or RCOX		RNCS	
RCO_2R'	$RCH_2OH + R'OH$	RCN	RCH_2NH_2
Lactones, i.e.	Diols, i.e.		or RCHO
$O(CH_2)_n-C=O$	$HO(CH_2)_{n+1}OH$	$R_2C=NOH$	R_2CHNH_2
$RCONH_2$	RCH_2NH_2	R_3NO	R_3N
	(also 2ry, 3ry)	R_2NNO	R_2NNH_2
Epoxides		RNO_2, or	
OCR_2-CHR	$R_2C(OH)CH_2R$	RNHOH, or	RNH_2
		RN_3	
		$ArNO_2$	$ArN=NAr$

During the last decade $LiAlH_4$ has been eclipsed as an organic reducing agent by the emergence of several cheaper organoaluminium hydrides which are also safer and easier to handle than $LiAlH_4$. Pre-eminent among these are Bu_2^iAlH and $Na[AlEt_2H_2]$ which were introduced

9 N. G. GAYLORD, *Reduction with Complex Metal Hydrides*, Interscience, New York, 1956, 1046 pp.

commercially in the early 1970s, and $Na[Al(OCH_2CH_2OMe)_2H_2]$ which became available in bulk during 1979. All three reagents can be prepared directly:

$$2Me_2CH{=}CH_2 + Al + H_2 \longrightarrow Bu^i_2AlH$$

$$2CH_2{=}CH_2 + Al + 2H_2 \longrightarrow \tfrac{1}{2}(Et_2AlH)_2 \xrightarrow{\text{NaH}} Na[AlEt_2H_2]$$

$$2MeOCH_2CH_2OH + Al + Na \xrightarrow{\text{H}_2/\text{PhMe}} Na[Al(OCH_2CH_2OMe)_2H_2]$$

All three are substantially cheaper than $LiAlH_4$ and are now produced on a far larger scale as indicated in Table B. Data refer to US industrial use in 1980 and even larger markets are available outside the chemical industry (e.g. in polymerization catalysis).

TABLE B (Source: *C. & E. News*, 3 Nov. 1980, pp. 18–20.)

Compound	Production/ kg	Price per kg/$	Price per kg of active H/$
$LiAlH_4$	6000	88.00	835.00
Bu^i_2AlH	195 000	5.10	715.00
$Na[AlEt_2H_2]$	91 000	11.00	605.00
$Na[Al(OCH_2CH_2OMe)_2H_2]$	123 000	6.25	638.00

decomposition led to a new compound which was the first to be discovered and structurally characterized by means of nmr:

$$2Al(BH_4)_3 \xrightarrow{70^\circ} Al_2B_4H_{18} + B_2H_6$$

The binuclear complex is also fluxional and has the structure

$Al_2B_4H_{18}$

At room temperature $Al(BH_4)_3$ reacts quantitatively in the gas phase with Al_2Me_6 (p. 289) to give $[Al(BH_4)_2Me]$ (mp -76°) in which one of the BH_4 groups of the parent compound has been replaced by a Me group:

$$4Al(BH_4)_3 + Al_2Me_6 \xrightarrow{\text{rt}/2\text{ h}} 6[Al(BH_4)_2Me]$$

Electron diffraction studies in the gas phase reveal an unusual structure in which the 5-coordinate Al atom has square-pyramidal geometry (Fig. 7.4a).[9a] The heavy atoms CAlB$_2$ are coplanar and the symmetry is close to C_{2v}. A similar structure has been found by X-ray diffraction for [Ga(BH$_4$)$_2$H], which was prepared by the dry reaction of LiBH$_4$ and GaCl$_3$ at $-45°$C.[9b]

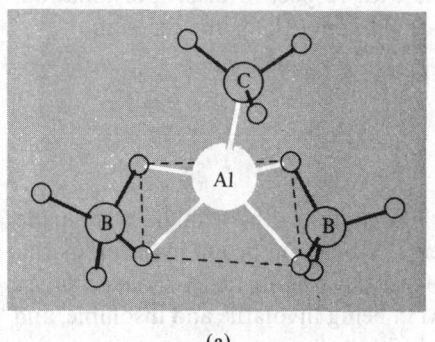
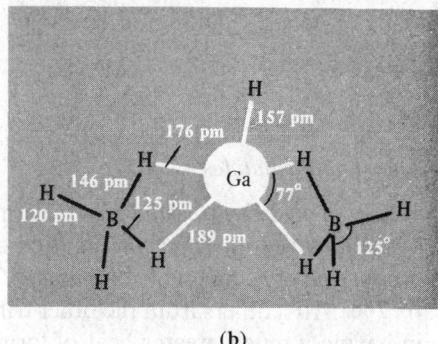

(a)   (b)

FIG. 7.4   (a) Structure of [MeAl(BH$_4$)$_2$] as revealed by electron diffraction, and (b) structure and key dimensions of [HGa(BH$_4$)$_2$] as determined by low temperature X-ray diffractometry.

7.3.2 *Halides and halide complexes*

Several important points emerge in considering the wide range of Group III metal halides and their complexes. Monohalides are known for all 4 metals with each halogen though for Al they occur only as short-lived diatomic species in the gas phase. This may seem paradoxical, since the bond dissociation energies for Al–X are substantially greater than for the corresponding monohalides of the other elements and fall in the range 655 kJ mol^{-1} (AlF) to 365 kJ mol^{-1} (AlI). The corresponding values for the gaseous TlX decrease from 460 to 270 kJ mol^{-1} yet it is these latter compounds that form stable crystalline solids. In fact, the instability of AlX in the condensed phase at normal temperatures is due not to the weakness of the Al–X bond but to the ready disproportionation of these compounds into the even more stable AlX$_3$:

$$\text{AlX(s)} \longrightarrow \tfrac{2}{3}\text{Al(s)} + \tfrac{1}{3}\text{AlX}_3\text{(s)}; \quad \Delta H_{\text{disprop}} \text{ (see Table)}$$

The reverse reaction to give the gaseous species AlX(g) at high temperature accounts for the enhanced volatility of AlF$_3$ when heated in the presence of Al metal, and the ready volatilization of Al metal in the presence of AlCl$_3$. Using calculations of the type outlined on p. 95 the standard heats of formation of the crystalline monohalides AlX and their heats of disproportionation have been estimated as:

[9a] M. T. BARLOW, C. J. DAIN, A. J. DOWNS, P. D. P. THOMAS, and D. W. H. RANKINE, Group III tetrahydroborates. Part 3. The molecular structure of methylaluminium bis(tetrahydroborate) in the gas phase as revealed by electron diffraction, *JCS Dalton* 1980, 1374–8.

[9b] M. T. BARLOW, C. J. DAIN, A. J. DOWNS, G. S. LAURENSON, and D. W. H. RANKIN, Group 3 Tetrahydroborates. Part 4. The molecular structure of hydridogallium bis(tetrahydroborate) in the gas phase as determined by electron diffraction, *JCS Dalton*, 1982, 597–602.

Compound (s)	AlF	AlCl	AlBr	AlI
$\Delta H_f^\circ/\text{kJ mol}^{-1}$	-393	-188	-126	-46
$\Delta H_{\text{disprop}}^\circ/\text{kJ mol}^{-1}$	-105	-46	-50	-59

The crystalline dihalides AlX_2 are even less stable with respect to disproportionation, value of $\Delta H_{\text{disprop}}$ falling in the range -200 to -230 kJ mol^{-1} for the reaction

$$AlX_2(s) \longrightarrow \tfrac{1}{3}Al(s) + \tfrac{2}{3}AlX_3(s)$$

Aluminium trihalides

AlF_3 is made by treating Al_2O_3 with HF gas at $700°$ and the other trihalides are made by the direct exothermic combination of the elements. AlF_3 is important in the industrial production of Al metal (p. 246) and $AlCl_3$ finds extensive use as a Friedel–Crafts catalyst (p. 266). AlF_3 differs from the other trihalides of Al in being involatile and insoluble, and in having a much greater heat of formation (Table 7.6).

TABLE 7.6   *Properties of crystalline AlX_3*

Property	AlF_3	$AlCl_3$	$AlBr_3$	AlI_3
MP/°C	1290	192.4	97.8	189.4
Sublimation pt (1 atm)/°C	1272	180	256	382
$\Delta H_f^\circ/\text{kJ mol}^{-1}$	1498	707	527	310

These differences probably stem from differences in coordination number (6 for AlF_3; change from 6 to 4 at mp for $AlCl_3$; 4 for $AlBr_3$ and AlI_3) and from the subtle interplay of a variety of other factors mentioned below rather than from a discontinuous change in bond type between the fluoride and the other halides. Similar differences dictated by change in coordination number are noted for many other metal halides, e.g. SnF_4 and SnX_4 (p. 443), BiF_3 and BiX_3 (p. 653), etc., and even more dramatically for some oxides such as CO_2 and SiO_2. In AlF_3 each Al is surrounded by a distorted octahedron of 6 F atoms and the 1:3 stoichiometry is achieved by the corner sharing of each F between 2 octahedra. The structure is thus related to the ReO_3 structure (p. 1219) but is somewhat distorted from ideal symmetry for reasons which are not understood. Maybe the detailed crystal structure data are wrong.[10] The relatively "open" lattice of AlF_3 provides sites for water molecules and permits the formation of a range of nonstoichiometric hydrates. In addition, well-defined hydrates $AlF_3 \cdot nH_2O$ ($n = 1$, 3, 9) are known but, curiously, no hexahydrate corresponding to the familiar $[Al(H_2O)_6]Cl_3$.

The complex fluorides of Al^{III} (and Fe^{III}) provide a good example of a family of structures with differing stoichiometries all derived by the sharing of vertices between octahedral $\{AlF_6\}$ units;[10] edge sharing and face sharing are not observed, presumably because of the destabilizing influence of the close (repulsive) approach of 2 Al atoms each of which carries a net partial positive charge. Discrete $\{AlF_6\}$ units occur in cryolite,

---

[10] A. F. WELLS, *Structural Inorganic Chemistry*, 4th edn., Oxford University Press, Oxford, 1975, 1095 pp.

Na_3AlF_6, and in the garnet structure $Li_3Na_3Al_2F_{12}$ (i.e. $[Al_2Na_3(LiF_4)_3]$, see p. 1256) but it is misleading to think in terms of $[AlF_6]^{3-}$ ions since the Al–F bonds are not appreciably different from the other M–F bonds in the structure. Thus the Na_3AlF_6 structure is closely related to perovskite ABO_3 (p. 1122) in which one-third of the Na and all the Al atoms occupy octahedral $\{MF_6\}$ sites and the remaining two-thirds of the Na occupy the 12-coordinate sites. When two opposite vertices of $\{AlF_6\}$ are shared the stoichiometry becomes $\{AlF_5\}$ as in $Tl_2^IAlF_5$ (and $Tl_2^IGaF_5$). The sharing of 4 equatorial vertices of $\{AlF_6\}$ leads to the stoichiometry $\{AlF_4\}$ in Tl^IAlF_4. The same structural motif is found in each of the "isoelectronic" 6-coordinate layer lattices of $K_2Mg^{II}F_4$, $KAl^{III}F_4$, and $Sn^{IV}F_4$, none of which contain tetrahedral $\{MF_4\}$ units.

More complex patterns of sharing give intermediate stoichiometries as in $Na_5Al_3F_{14}$ which features layers of $\{Al_3F_{14}\}$ built up by one-third of the $\{AlF_6\}$ octahedra sharing 4 equatorial vertices and the remainder sharing 2 opposite vertices. Again, Na_2MgAlF_7 comprises linked $\{AlF_6\}$ and $\{MgF_6\}$ octahedra in which 4 vertices of $\{AlF_6\}$ and all vertices of $\{MgF_6\}$ are shared. In all of these structures the degree of charge separation, though considerable, is unlikely to approach the formal group charges: thus AlF_3 should not be regarded as a network of alternating ions Al^{3+} and F^- nor, at the other extreme, of an alternating set of Al^{3+} and AlF_6^{3-}, and lattice energies calculated on the basis of such formal charges placed at the observed interatomic distances are bound to be of limited reliability. Equally, the structure is not well described as a covalently bonded network of Al atoms and F atoms, and detailed MO calculations would be required to assess the actual extent of charge separation, on the one hand, and interatomic covalent bonding, on the other. Similar information might, in principle, be obtainable from very precise electron density plots obtained by X-ray diffraction (p. 92).

The structure of $AlCl_3$ is similarly revealing. The crystalline solid has a layer lattice with 6-coordinate Al but at the mp 192.4° the structure changes to a 4-coordinate molecular dimer Al_2Cl_6; as a result there is a dramatic increase in volume (by 85%) and an even more dramatic drop in electrical conductivity almost to zero. The mp therefore represents a substantial change in the nature of the bonding. The covalently bonded molecular dimers are also the main species in the gas phase at low temperatures (~ 150–$200°$) but at higher temperatures there is an increasing tendency to dissociate into trigonal planar $AlCl_3$ molecules isostructural with BX_3 (p. 220).

By contrast, Al_2Br_6 and Al_2I_6 form dimeric molecular units in the crystalline phase as well as in the liquid and gaseous states and fusion is not attended by such extensive changes in properties. In the gas phase $\Delta H_{dissoc} = 59$ kJ mol^{-1} for $AlBr_3$ and 50 kJ mol^{-1} for AlI_3. In all these dimeric species, as in the analogous dimers Ga_2Cl_6, Ga_2Br_6, Ga_2I_6, and In_2I_6, the M–X_t distance is 10–20 pm shorter than the M–X_μ distance; the external angle X_tMX_t is in the range 110–125° and the internal angle $X_\mu MX_\mu$ is in the range 79–102°.

The trihalides of Al form a large number of addition compounds or complexes and these have been extensively studied because of their importance in understanding the nature of Friedel–Crafts catalysis.[11, 12] The adducts vary enormously in stability from weak interactions to very stable complexes, and they also vary widely in their mode of bonding,

[11] N. N. GREENWOOD and K. WADE, Coordination compounds of aluminium and gallium halides, Chap. 7 in G. A. OLAH (ed.), *Friedel Crafts and Related Reactions*, Vol. 1, pp. 569–622, Interscience, New York, 1963.
[12] K. WADE and A. J. BANISTER, Aluminium, gallium, indium, and thallium, Chap. 12 in *Comprehensive Inorganic Chemistry*, Vol. 1, pp. 993–1172, Pergamon Press, Oxford, 1973.

structure, and properties. Aromatic hydrocarbons and olefins interact weakly though in some cases crystalline adducts can be isolated, e.g. the clathrate-like complex $Al_2Br_6 \cdot C_6H_6$, mp 37° (decomp). With mesitylene, $C_6H_3Me_3$, and the xylenes $C_6H_4Me_2$, the interaction is slightly stronger, leading to dissociation of the dimer and the formation of weak monomeric complexes $AlBr_3L$ both in solution and in the solid state. At the other end of the stability scale NMe_3 forms two crystalline complexes: $[AlCl_3(NMe_3)]$ mp 156.9° which features molecular units with 4-coordinate tetrahedral Al, and $[AlCl_3(NMe_3)_2]$ which has 5-coordinate Al with trigonal bipyramidal geometry and *trans* axial ligands. Alkyl halides interact rather weakly and vibrational spectroscopy suggests bonding of the type $R-X\cdots AlX_3$. However, for readily ionizable halides such as Ph_3CCl the degree of charge separation is much more extensive and the complex can be formulated as $Ph_3C^+AlCl_4^-$. Acyl halides RCOX may interact either through the carbonyl oxygen, $PhC(Cl)=O\rightarrow AlCl_3$, or through the halogen, $RCOX\cdots AlX_3$ or $RCO^+AlX_4^-$. Again, vibrational spectroscopy is a sensitive, though not always reliable, diagnostic for the mode of bonding. X-ray crystal structures of several complexes have been obtained but these do not necessarily establish the predominant species in nonaqueous solvents because of the delicate balance between the various factors which determine the structure (p. 223). Even in the crystalline state the act of coordination may lead to substantial charge separation. For example, X-ray analysis has established that $AlCl_3ICl_3$ comprises chains of alternating units which are best described as ICl_2^+ and $AlCl_4^-$ with rather weaker interactions between the ions (Fig. 7.5). Another instructive example is the ligand $POCl_3$ which forms 3 crystalline complexes $AlCl_3POCl_3$ mp 186.5°, $AlCl_3(POCl_3)_2$ mp 164°(d), and $AlCl_3(POCl_3)_6$ mp 41°(d). Although the crystal structures of these adducts have not been established it is known that $POCl_3$ normally coordinates through oxygen rather than chlorine. Consistent with this there is no exchange of radioactive ^{36}Cl when $AlCl_3$ containing ^{36}Cl is dissolved in inactive $POCl_3$.

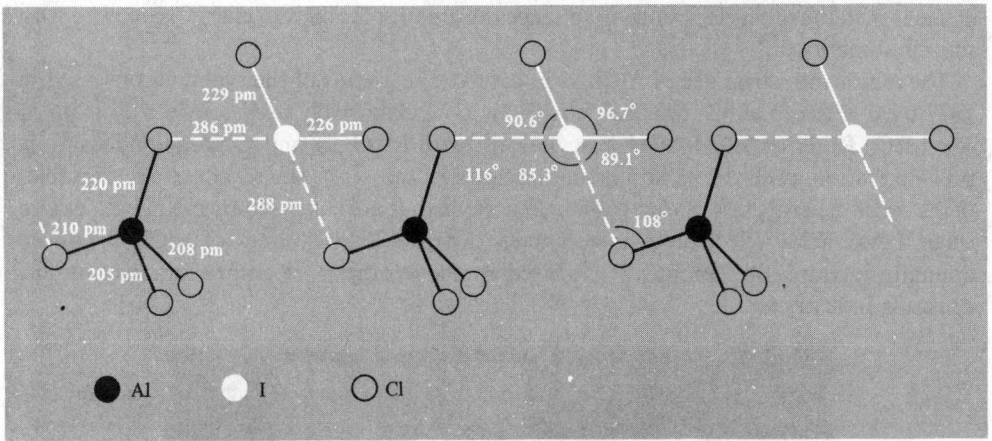

FIG. 7.5   Schematic representation of the structure of $AlCl_3ICl_3$ showing ICl_2^+ and $AlCl_4^-$ units linked into a chain by weaker interactions. The 6 ClAlCl angles at Al are in the range 104.5–113.4°; the AlClI angles at the bridging Cl atoms are also close to tetrahedral and the geometry at I is distorted square planar, with two normal and two long I-Cl distances. Note also that the two Al–Cl$_t$ distances are less than the two Al–Cl$_\mu$ distances, as in the bridge dimeric Al_2Cl_6 itself.

However, such solutions are good electrical conductors and spectroscopy reveals $AlCl_4^-$ as a predominant solute species. The resolution of this apparent paradox was provided by means of ^{27}Al nmr spectroscopy[13] which showed that ionization occurred according to the reaction

$$4AlCl_3 + 6POCl_3 \longrightarrow [Al(OPCl_3)_6]^{3+} + 3[AlCl_4]^-$$

It can be seen that all the Cl atoms in $[AlCl_4]^-$ come from the $AlCl_3$. It was further shown that the same 2 species predominated when Al_2I_6 was dissolved in an excess of $POCl_3$:

$$2Al_2I_6 + 12POCl_3 \rightleftharpoons 4\{AlCl_3\} + 12POCl_2I$$
$$4\{AlCl_3\} + 6POCl_3 \longrightarrow [Al(OPCl_3)_6]^{3+} + 3[AlCl_4]^-$$

No mixed Al species were found by ^{27}Al nmr in this case.

AlCl$_3$ is a convenient starting material for the synthesis of a wide range of other Al compounds, e.g.:

$$AlCl_3 + 3LiY \longrightarrow 3LiCl + AlY_3 \qquad (Y = R, NR_2, N = CR_2)$$
$$AlCl_3 + 4LiY \longrightarrow 3LiCl + LiAlY_4 \qquad (Y = R, NR_2, N = CR_2, H)$$

Similarly, NaOR reacts to give $Al(OR)_3$ and $NaAl(OR)_4$. AlCl$_3$ also converts non-metal fluorides into the corresponding chloride, e.g.

$$BF_3 + AlCl_3 \longrightarrow AlF_3 + BCl_3$$

This type of transhalogenation reaction, which is common amongst the halides of main group elements, always proceeds in the direction which pairs the most electropositive

[13] R. G. KIDD and D. R. TRUAX, Halogen exchange between aluminium halides and phosphoryl chloride, *JCS Chem Comm.* (1969), 160–1.

element with the most electronegative, since the greatest amount of energy is evolved with this combination.[14]

The major industrial use of $AlCl_3$ is in catalytic reactions of the type first observed in 1877 by C. Friedel and J. M. Crafts. $AlCl_3$ is now extensively used to effect alkylations (with RCl, ROH, or $RCH=CH_2$), acylations (with RCOCl), and various condensation, polymerization, cyclization, and isomerization reactions.[15] The reactions are examples of the more general class of electrophilic reactions that are catalysed by metal halides and other Lewis acids (electron pair acceptors). Of the 25 000 tonnes of $AlCl_3$ produced annually in the USA, about 15% is used in the synthesis of anthraquinones for the dyestuffs industry:

$$2C_6H_6 + 2COCl_2 \xrightarrow{AlCl_3} \text{[anthraquinone structure]} + 2HCl$$

A further 15% of the $AlCl_3$ is used in the production of ethyl benzene for styrene manufacture, and 13% in making EtCl or EtBr (for $PbEt_4$):

$$CH_2=CH_2 + PhH \xrightarrow{AlCl_3} PhEt$$

$$CH_2=CH_2 + HX \xrightarrow{AlCl_3} EtX$$

The isomerization of hydrocarbons in the petroleum industry and the production of dodecyl benzene for detergents accounts for a further 10% each of the $AlCl_3$ used.

Trihalides of gallium, indium, and thallium

These compounds have been mentioned several times in the preceding sections. As with AlX_3 (p. 262), the trifluorides are involatile and have much higher mps and heats of formation than the other trihalides;[12] e.g. GaF_3 melts above $1000°$, sublimes at $\sim 950°$ and has the 6-coordinate FeF_3-type structure, whereas $GaCl_3$ melts at $77.8°$, boils at $201.2°$, and has the 4-coordinate molecular structure Ga_2Cl_6. GaF_3 and InF_3 are best prepared by thermal decomposition of $(NH_4)_3MF_6$, e.g.:

$$(NH_4)_3InF_6 \xrightarrow{120-170°} NH_4InF_4 \xrightarrow{300°} InF_3$$

Preparations using aqueous HF on $M(OH)_3$, M_2O_3, or M metal give the trihydrate. TlF_3 is best prepared by the direct fluorination of Tl_2O_3 with F_2, BrF_3, or SF_4 at $300°$. Trends

[14] F. SEEL, *Atomic Structure and Chemical Bonding*, 4th edn. translated and revised by N. N. GREENWOOD and H. P. STADLER, Methuen, London, 1963, pp. 83–4.

[15] G. A. OLAH (ed.), *Friedel–Crafts and Related Reactions*, Vols. 1–4, Interscience, New York, 1963. See especially Chap. 1, Historical, by G. A. OLAH and R. E. A. DEAR, and Chap. 2, Definition and scope by G. A. OLAH.

in the heats of formation of the Group IIIB trihalides show the same divergence from BX_3, AlX_3, and the Group IIIA trihalides as was found for trends in other properties such as I_M, $E°$, and χ (pp. 251–53) and for the same reasons. For example, the data for $\Delta H_f°$ for the trifluorides and tribromides are compared in Fig. 7.6 from which it is clear that the trend noted for the sequence B, Al, Sc, Y, La, Ac is not followed for the Group IIIB trihalides which become progressively less stable from Al to Tl.

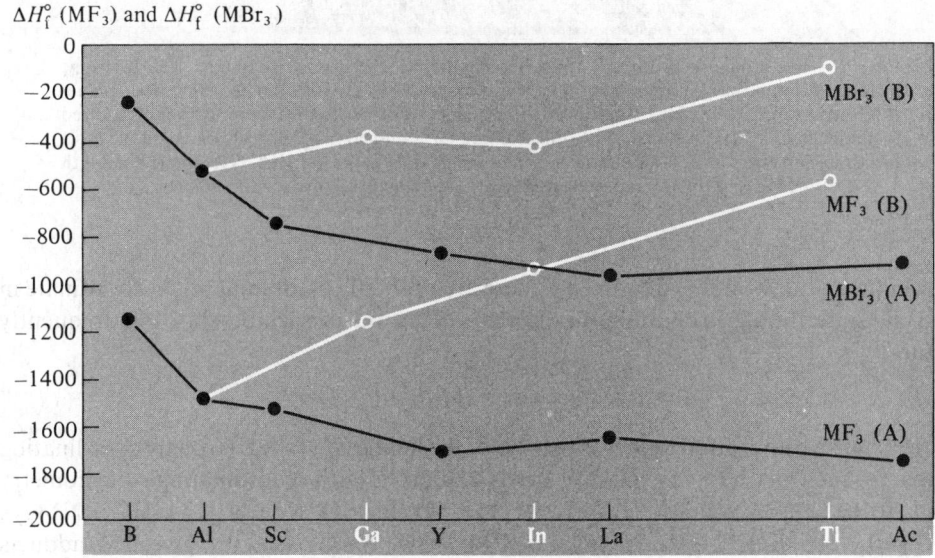

$\Delta H_f°$ (MF_3) and $\Delta H_f°$ (MBr_3)

FIG. 7.6   Trends in the standard enthalpies of formation $\Delta H_f°$ for Groups IIIA and IIIB trihalides as illustrated by data for MF_3 and MBr_3.

The volatile trihalides MX_3 form several ranges of addition compounds MX_3L, MX_3L_2, MX_3L_3, and these have been extensively studied because of the insight they provide on the relative influence of the underlying d^{10} electron configuration on the structure and stability of the complexes. With halide ions X^- as ligands the stoichiometry depends sensitively on crystal lattice effects or on the nature of the solvent and the relative concentration of the species in solution. Thus X-ray studies have established the tetrahedral ions $[GaX_4]^-$, $[InCl_4]^-$, etc., and these persist in ethereal solution, though in aqueous solution $[InCl_4]^-$ looses its T_d symmetry due to coordination of further molecules of H_2O. $[NEt_4]_2[InCl_5]$ is remarkable in featuring a square-pyramidal ion of C_{4v} symmetry (Fig. 7.7) and was one of the first recorded examples of this geometry in nontransition element chemistry (1969), cf $SbPh_5$ on p. 697. The structure is apparently favoured by electrostatic packing considerations though it also persists in nonaqueous solution, possibly due to the formation of a pseudo-octahedral solvate $[InCl_5S]^{2-}$. It will be noted that $[InCl_5]^{2-}$ is not isostructural with the isoelectronic species $SnCl_5^-$ and $SbCl_5$ which have the more common D_{3h} symmetry.

With neutral ligands, L, GaX_3 tend to resemble AlX_3 in forming predominantly MX_3L and some MX_3L_2, whereas InX_3 are more varied.[16] InX_3L_3 is the commonest

[16] A. J. CARTY and D. J. TUCK, Coordination chemistry of indium, *Prog. Inorg. Chem.* **19**, 243–337 (1975).

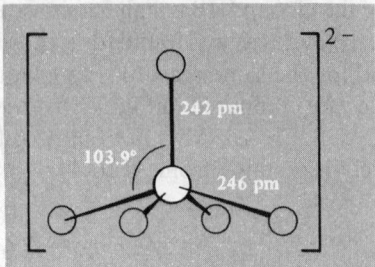

FIG. 7.7 The structure of $InCl_5^{2-}$ showing square-pyramidal (C_{4v}) geometry. The In–Cl_{apex} distance is significantly shorter than the In–Cl_{base} distances and In is 59 pm above the basal plane; this leads to a Cl_{apex}–In–Cl_{base} angle of 103.9° which is very close to the theoretical value required to minimize Cl···Cl repulsions whilst still retaining C_{4v} symmetry (103.6°) calculated on the basis of a simple inverse square law for repulsion between ligands. $[NEt_4]_2[TlCl_5]$ is isomorphous with $[NEt_4]_2[InCl_5]$ and presumably has a similar structure for the anion.

stoichiometry for N and O donors and these are probably predominantly 6-coordinate in the solid state, though in coordinating solvents (S) partial dissociation into ions frequently occurs:

$$InX_3L_3 + S \longrightarrow [InX_2SL_3]^+ + X^-$$

More extensive ionization occurs if, instead of the halides X^-, a less strongly coordinating anion Y^- such as ClO_4^- or NO_3^- is used; in such cases the coordinating stoichiometry tends to be 1:6, e.g. $[InL_6]^{3+}(Y^-)_3$, L = Me_2SO, Ph_2SO, $(Me_2N)_2CO$, $HCO(NMe_2)$, $P(OMe)_3$, etc. Bulky ligands such as PPh_3 and $AsPh_3$ tend to give 1:4 adducts $[InL_4]^{3+}(Y^-)_3$. The same effect of ionic dissociation can be achieved in 1:3 complexes of the trihalides themselves by use of bidentate chelating ligands (B) such as en, bipy, or phen, e.g. $[InB_3]^{3+}(X^-)_3$ (X = Cl, Br, I, NCO, NCS, NCSe). InX_3 complexes having 1:2 stoichiometry also have a variety of structures. Trigonal bipyramidal geometry with axial ligands is found for InX_3L_2, where L = NMe_3, PMe_3, PPh_3, Et_2O, etc. By contrast, the crystal structure of the 1:2 complex of InI_3 with Me_2SO shows that it is fully ionized as $[cis-InI_2(OSMe_2)_4]^+[InI_4]^-$, and fivefold coordination is avoided by a disproportionation into 6- and 4-coordinate species. Complexes having 1:1 stoichiometry are rare for InX_3; $InCl_3$ forms $[InCl_3(OPCl_3)]$, $[InCl_3(OCMe_2)]$, and $[InCl_3(OCPh_2)]$ and py forms a 1:1 (and a 1:3) adduct with InI_3. Frequently, of course, a given donor-acceptor pair combines in more than one stoichiometric ratio.

The thermochemistry of the Group III trihalide complexes has been extensively studied during the past two decades[11, 12, 17] and several stability sequences have been established which can be interpreted in terms of the factors listed on p. 223. In addition, Ga and In differ from B and Al in having an underlying d^{10} configuration which can, in principle, take part in d_π–d_π back bonding with donors such as S (but not N or O); alternatively (or additionally), some of the trends can be interpreted in terms of the differing polarizabilities of B and Al, as compared to Ga and In, the former pair behaving as class-a or "hard" acceptors whereas Ga and In frequently behave as class-b or "soft"

---

[17] N. N. GREENWOOD *et al.*, *Pure Appl. Chem.* **2**, 55–59 (1961); *J. Chem. Soc. A*, 1966, 267–70, 270–3, 703–6; *J. Chem. Soc. A*, 1968, 753–6; 1969, 249–53, 2876–8; *Inorg. Chem.* **9**, 86–90 (1970), and references therein. R. C. GEARHART, J. D. BECK, and R. H. WOOD, *Inorg. Chem.* **14**, 2413–16 (1975).

acceptors. Again, it should be emphasized that these categories tend to provide descriptions rather than explanations. Towards amines and ethers the acceptor strengths as measured by gas-phase enthalpies of formation decrease in the sequence $MCl_3 > MBr_3 > MI_3$ for M = Al, Ga, or In. Likewise, towards phosphines the acceptor strength decreases as $GaCl_3 > GaBr_3 > GaI_3$. However, towards the "softer" sulfur donors Me_2S, Et_2S, and C_4H_8S, whilst AlX_3 retains the same sequence, the order for GaX_3 and InX_3 is reversed to read $MI_3 > MBr_3 > MCl_3$. A similar reversal is noted when the acceptor strengths of individual AlX_3 are compared with those of the corresponding GaX_3: towards N and O donors the sequence is invariably $AlX_3 > GaX_3$ but for S donors the relative acceptor strength is $GaX_3 > AlX_3$. These trends emphasize the variety of factors that contribute towards the strength of chemical bonds and indicate that there are no unique series of donor or acceptor strengths when the acceptor atom is varied, e.g.:

$$\text{towards MeCO}_2\text{Et:} \quad BCl_3 > AlCl_3 > GaCl_3 > InCl_3$$
$$\text{towards py:} \quad AlPh_3 > GaPh_3 > BPh_3 \approx InPh_3$$
$$\text{towards py:} \quad AlX_3 > BX_3 > GaX_3 \quad (X = Cl, Br)$$
$$\text{towards Me}_2\text{S:} \quad GaX_3 > AlX_3 > BX_3 \quad (X = Cl, Br)$$

Regularities are more apparent when the acceptor atom remains constant and the attached groups are varied; e.g., for all ligands so far studied the acceptor strength diminishes in the sequence

$$MX_3 > MPh_3 > MMe_3 \quad (M = B, Al, Ga, In)$$

It has also been found that halide-ion donors (such as X^- in $AlX_4{}^-$ and $GaX_4{}^-$) are more than twice as strong as any neutral donor such as X in M_2X_6, or N, P, O, and S donors in MX_3L.[17]

The trihalides of Tl are much less stable than those of the lighter Group IIIB metals and are chemically quite distinct from them. TlF_3, mp 550° (decomp), is a white crystalline solid isomorphous with β-BiF_3 (p. 653); it does not form hydrates but hydrolyses rapidly to $Tl(OH)_3$ and HF. Nor does it give $TlF_4{}^-$ in aqueous solution, and the compounds $LiTlF_4$ and $NaTlF_4$ have structures related to fluorite, CaF_2 (p. 129): in $NaTlF_4$ the cations have very similar 8-coordinate radii (Na 116 pm, Tl 100 pm) and are disordered on the Ca sites (Ca 112 pm); in $LiTlF_4$, the smaller size of Li ($\sim$83 pm for eightfold coordination) favours a superlattice structure in which Li and Tl are ordered on the Ca sites. Na_3TlF_6 has the cryolite structure (p. 262). $TlCl_3$ and $TlBr_3$ are obtained from aqueous solution as the stable tetrahydrates and $TlCl_3 \cdot 4H_2O$ can be dehydrated with $SOCl_2$ to give anhydrous $TlCl_3$, mp 155°; it has the YCl_3-type structure which can be described as NaCl-type with two-thirds of the cations missing in an ordered manner.

TlI_3 is an intriguing compound which is isomorphous with NH_4I_3 and CsI_3 (p. 979); it therefore contains the linear $I_3{}^-$ ion† and is a compound of Tl^I rather than Tl^{III}. It is obtained as black crystals by evaporating an equimolar solution of TlI and I_2 in concentrated aqueous HI. The formulation $Tl^I(I_3{}^-)$ rather than $Tl^{III}(I^-)_3$ is consistent with the standard reduction potentials $E°(Tl^{III}/Tl^I)$ +1.25 V and $E°(\frac{1}{2}I_2/I^-)$ +0.54 V, which shows that uncomplexed Tl^{III} is susceptible to rapid and complete reduction to Tl^I by I^- in acid solution. The same conclusion follows from a consideration of the $I_3{}^-/3I^-$

---

† This X-ray evidence by itself, does not rule out the possibility that the compound is $[I-Tl^{III}-I]^+I^-$.

couple for which $E° = +0.55$ V. Curiously, however, in the presence of an excess of I^-, the Tl^{III} state is stabilized by complex formation

$$Tl^I I_3(s) + I^- \rightleftharpoons [Tl^{III}I_4]^-$$

Moreover, solutions of TlI_3 in MeOH do not show the visible absorption spectrum of I_3^- and, when shaken with aqueous Na_2CO_3, give a precipitate of Tl_2O_3, i.e.:

$$2TlI_3 + 6OH^- \rightleftharpoons Tl_2O_3 + 6I^- + 3H_2O$$

This is due partly to the great insolubility of Tl_2O_3 (2.5×10^{-10} g l^{-1} at $25°$) and partly to the enhanced oxidizing power of iodine in alkaline solution as a result of the formation of hypoiodate:

$$I_3^- + 2OH^- \rightleftharpoons OI^- + 2I^- + H_2O$$

Consistent with this, even KI_3 is rapidly decolorized in alkaline solution. The example is a salutory reminder of the influence of pH, solubility, and complex formation on the standard reduction potentials of many elements.

Numerous tetrahedral halogeno complexes $[Tl^{III}X_4]^-$ ($X = Cl$, Br, I) have recently been prepared by reaction of quaternary ammonium or arsonium halides on TlX_3 in nonaqueous solution, and octahedral complexes $[Tl^{III}X_6]^{3-}$ ($X = Cl$, Br) are also well established. The binuclear complex $Cs_3[Tl_2^{III}Cl_9]$ is an important structural type which features two $TlCl_6$ octahedra sharing a common face of 3 bridging Cl atoms (Fig. 7.8); the same binuclear complex structure is retained when Tl^{III} is replaced by Ti^{III}, V^{III}, Cr^{III}, and Fe^{III} and also in $K_3W_2Cl_9$ and $Cs_3Bi_2I_9$, etc.

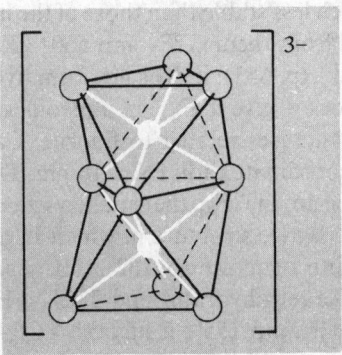

FIG. 7.8 The structure of the ion $[Tl_2Cl_9]^{3-}$ showing two octahedral $TlCl_6$ units sharing a common face: Tl Cl$_t$ 254 pm, Tl Cl$_\mu$ 266 pm. The Tl$\cdots$Tl distance is nonbonding (281 pm, cf. $2 \times Tl^{III} = 177$ pm).

Lower halides of gallium, indium, and thallium

Like AlX (p. 261), GaF and InF are known as unstable gaseous species. The other monohalides are more stable. GaX can be obtained as reactive sublimates by treating GaX_3 with 2Ga: stability increases with increasing size of the anion and GaI melts at $271°$; stability is still further enhanced by coordination of the anion with, for example, AlX_3 to give $Ga^I[Al^{III}X_4]$. Likewise, the very stable "dihalides" $Ga^I[Ga^{III}Cl_4]$, $Ga[GaBr_4]$, and

Ga[GaI$_4$] can be prepared by heating equimolar amounts of GaX$_3$ and Ga, or more conveniently by halogenation of Ga with the stoichiometric amount of Hg$_2$X$_2$ or HgX$_2$. They form complexes of the type [Ga^{I}L$_4$]$^+$[Ga^{III}X$_4$]$^-$ with a wide range of N, As, O, S, and Se donors. Note, however, that the complexes with dioxan, [Ga$_2$X$_4$(C$_4$H$_8$O$_2$)$_2$], do in fact contain Ga^{II} and a Ga–Ga bond, e.g. the chloro complex is a discrete molecule with Ga–Ga 240.6 pm (cf. 239.0 pm in Ga$_2$Cl$_6$$^{2-}$).[17a] As shown in Fig. 7.9, the coordination about each Ga atom is essentially tetrahedral and the compound adopts the unexpected eclipsed structure (Fig. 7.9b) rather than the staggered structure of Ga$_2$Cl$_6$$^{2-}$.

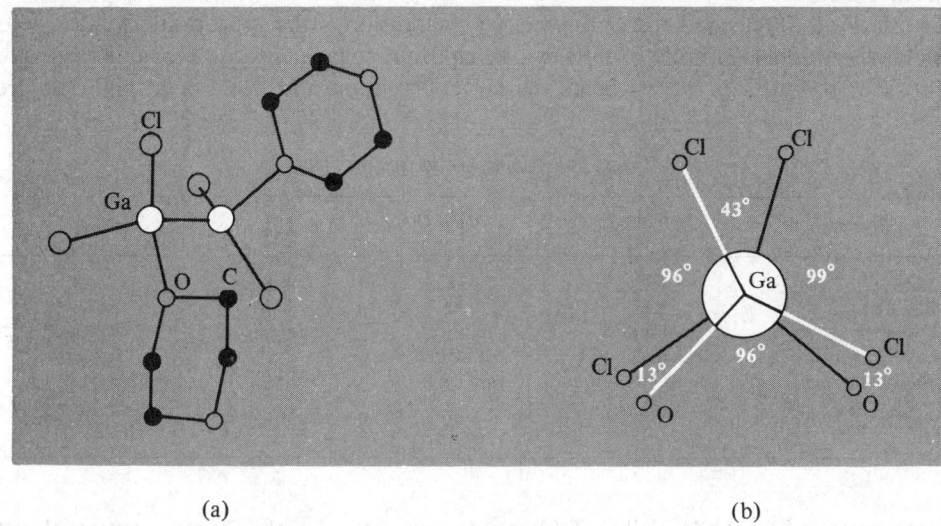

(a)   (b)

FIG. 7.9   Structure of the molecular complex [Ga$_2$Cl$_4$(C$_4$H$_8$O$_2$)$_2$] showing (a) the Ga–Ga bond and tetrahedral coordination about each Ga atom, and (b) the torsion angles in the essentially eclipsed structure.

Indium monohalides, InX, are prepared directly from the elements or by heating In metal with HgX$_2$ at 320–350°. InCl has a deformed NaCl-type structure in which there is a tendency to In–In bonding. Again, InI is the most stable, and mixed halides of the type In^I[Al^{III}Cl$_4$], In^I[Ga^{III}Cl$_4$], and Tl^I[In^{III}Cl$_4$] are known. Numerous intermediate halides have also been reported and structural assignments of varying degrees of reliability have been suggested, e.g. In^I[In^{III}X$_4$] for InX$_2$ (Cl, Br, I); and In$_3^I$[In^{III}Cl$_6$] for In$_2$Cl$_3$. (The previously reported In$_2$Br$_3$ probably does not exist.[17b]) The compounds In$_4$X$_7$ and In$_5$X$_7$ (Cl, Br) and In$_7$Br$_9$ are also known. In all of these halides it is thought that the observed stoichiometry is achieved by varying the ratio of In^I to In^{III}, e.g. [(In$^+$)$_5$(InBr$_4$$^-$)$_2$(InBr$_6$$^{3-}$)], [(In$^+$)$_3$(In$_2Br_6$$^{2-}$)Br$^-$], and [(In$^+$)$_6$(InBr$_6$$^{3-}$)(Br$^-$)$_3$]. Compounds containing In^{II} were unknown until 1976 when the [In$_2$X$_6$]$^{2-}$ dianions having an ethane-like structure were prepared:[18]

[17a] J. C. BEAMISH, R. W. H. SMALL, and I. J. WORRALL, Neutral complexes of gallium(II) containing gallium–gallium bonds, *Inorg. Chem.* **18**, 220–3 (1979).
[17b] J. E. DAVIES, L. G. WATERWORTH, and I. J. WORRALL, Studies on the constitution of indium bromides, *J. Inorg. Nucl. Chem.* **36**, 805–7 (1974).
[18] B. H. FREELAND, J. L. HENCHER, D. G. TUCK, and J. G. CONTRERAS, Preparation and properties of hexahalogenatodiindate(II) anions, *Inorg. Chem.* **15**, 2144–6 (1976).

$$2Bu_4NX + In_2X_4 \xrightarrow{\text{xylene}} [NBu_4]^+{}_2[X_3In\!-\!InX_3]^{2-} \quad (X = Cl, Br, I)$$

The analogous Ga compounds, e.g. $[NEt_4]_2[Cl_3Ga\!-\!GaCl_3]$, have been known for rather longer (1965). Oxidation of $In_2X_6{}^{2-}$ with halogens Y_2 yields the mononuclear mixed halide complexes InX_3Y^- and $InX_2Y_2{}^-$ $(X \neq Y = Cl, Br, I)$.[18a]

Thallium(I) is the stable oxidation state for the halides of this element (p. 255) and some physical properties are in Table 7.7. TlF is readily obtained by the action of aqueous HF on Tl_2CO_3; it is very soluble in water (in contrast to the other TlX) and has a distorted NaCl structure in which there are 3 pairs of Tl–F distances at 259, 275, and 304 pm. TlCl, TlBr, and TlI are all prepared by addition of the appropriate halide ion to acidified solutions of soluble Tl^I salts (e.g. perchlorate, sulfate, nitrate). TlCl and TlBr have the CsCl structure (p. 92) as befits the large Tl^I cation and both salts (and TlI) are

TABLE 7.7   *Some properties of TlX*

Property	TlF	TlCl	TlBr	TlI
MP/°C	322	431	460	442
BP/°C	826	720	815	823
Colour	White	White	Pale yellow	Yellow
Crystal structure	Distorted NaCl	CsCl	CsCl	See text
Solubility/g per 100 g H_2O (°C)	80 (15°)	0.33 (20°)	0.058 (25°)	0.006 (20°)
ΔH_f°/kJ mol^{-1}	-326	-204	-173	-124

photosensitive (like AgX). Yellow TlI has a curious orthorhombic layer structure related to NaCl (Fig. 7.10), and this transforms at 175° or at 4.7 kbar to a metastable red cubic form with 8-iodine neighbours at 364 pm (CsCl type). This transformation is accompanied by 3% reduction in volume. Further application of pressure steadily reduces the volume and at pressures above about 160 kbar, when the volume has decreased by about 35%, the compound becomes a metallic conductor with a resistivity of the order of 10^{-4} ohm cm at room temperature and a positive temperature coefficient. TlCl and TlBr behave similarly. All three compounds are excellent insulators at normal pressures with negligible conductivity and an energy gap between the valence band and conduction band of about 3 eV (~ 300 kJ mol^{-1}), and the onset of metallic conduction is presumably due to the spreading and eventual overlap of the two bands as the atoms are forced closer together.[19]

Several other lower halides of Tl are known: $TlCl_2$ and $TlBr_2$ are $Tl^I[Tl^{III}X_4]$, Tl_2Cl_3 and Tl_2Br_3 are $Tl^I_3[Tl^{III}X_6]$. In addition there is Tl_3I_4, which is formed as an intermediate in the preparation of $Tl^I I_3$ from TlI and I_2 (p. 269).

[18a] J. E. DRAKE, J. L. HENCHER, L. N. KHASROU, D. G. TUCK, and L. VICTORIANO, Coordination compounds of indium. Part 34. Preparative and spectroscopic studies of InX_3Y^- and $InX_2Y_2{}^-$ anions $(X = Y = Cl, Br, I)$, *Inorg. Chem.* **19**, 34–38 (1980).
[19] G. A. SAMARA and H. G. DRICKAMER, Effect of pressure on the resistance of three thallous halides, *J. Chem. Phys.* **37**, 408–10 (1962); see also E. A. PEREZ-ALBUERNE and H. G. DRICKAMER, *J. Chem. Phys.* **43**, 1381–7 (1965).

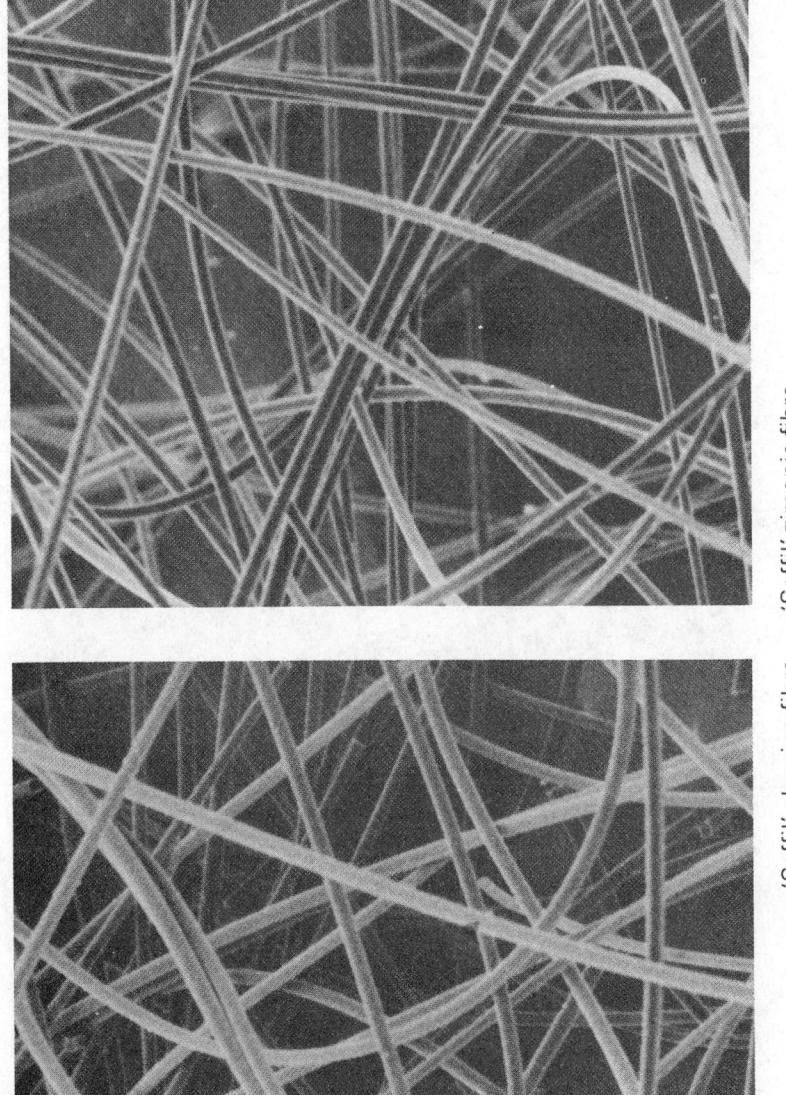

'Saffil' alumina fibre.   'Saffil' zirconia fibre.

PLATE 7.1   Photomicrographs (magnification ×600) of "Saffil" alumina and "Saffil" zirconia fibres. These demonstrate the uniform diameter of "Saffil" fibres and their freedom from shot.

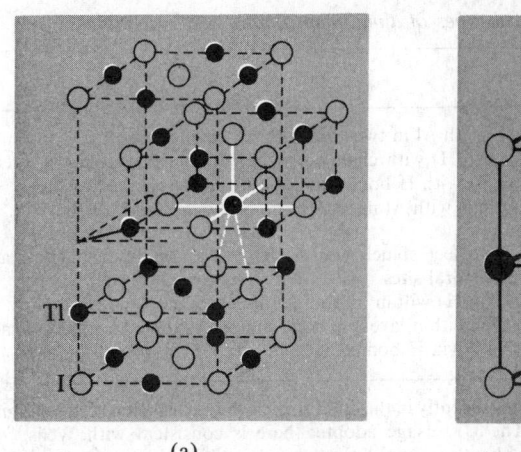 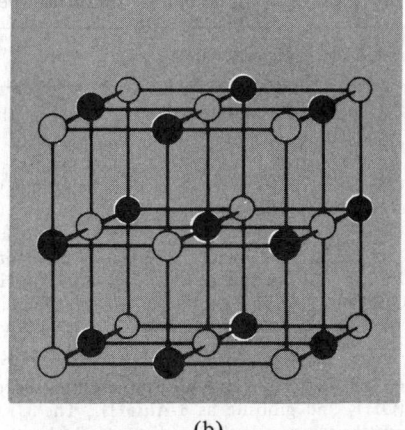

(a)   (b)

FIG. 7.10   Structure of yellow TlI (a) showing its relation to NaCl (b). Tl has 5 nearest-neighbour I atoms at 5 of the vertices of an octahedron and then 2I + 2Tl as next-nearest neighbours; there is 1 I at 336 pm, 4 at 349 pm, and 2 at 387 pm, and the 2 close Tl–Tl approaches one at 383 pm. InBr and InI have similar structures.

7.3.3 Oxides and hydroxides

The structural relations between the many crystalline forms of aluminium oxide and hydroxide are exceedingly complex but they are of exceptional scientific interest and immense technological importance. The principal structural types are listed in Table 7.8 and many intermediate and related structures are also known. Al_2O_3 occurs as the mineral corundum (α-Al_2O_3, d 4.0 g cm^{-3}) and as emery, a granular form of corundum contaminated with iron oxide and silica. Because of its great hardness (Mohs 9),[†] high mp (2045°C), involatility (10^{-6} atm at 1950°), chemical inertness, and good electrical insulating properties, it finds many applications in abrasives (including toothpaste), refractories, and ceramics, in addition to its major use in the electrolytic production of Al metal (p. 246). Larger crystals, when coloured with metal-ion impurities, are prized as gemstones, eg. ruby (Cr^{III} red), sapphire ($Fe^{II/III}$, Ti^{IV} blue), oriental emerald (?Cr^{III}/V^{III} green), oriental amethyst (Cr^{III}/Ti^{IV} violet), and oriental topaz (Fe^{III}, yellow). Many of these gems are also made industrially on a large scale by the fusion process first developed at the turn of the century by A. Verneuil. Pure α-Al_2O_3 is made industrially by igniting $Al(OH)_3$ or $AlO(OH)$ at high temperatures ($\sim 1200°$); it is also formed by the combustion of Al and by calcination of various Al salts. It has a rhombohedral crystal structure comprising a hcp array of oxide ions with Al ordered on two-thirds of the octahedral interstices as shown in Fig. 7.11.[(10)] The same α-M_2O_3 structure is adopted by several other elements with small M^{III} (r 62–67 pm), e.g. Ga, Ti, V, Cr, Fe, and Rh.[‡]

The second modification of alumina is the less compact cubic γ-Al_2O_3 (d 3.4 g cm^{-3}); it

---

† On the Mohs scale diamond is 10 and quartz 7. An alternative measure is the Knoop hardness (kg mm^{-2}) as measured with a 100-g load: typical values on this scale are diamond 7000, boron carbide 2750, corundum 2100, topaz 1340, quartz 820, hardened tool steel 740.

‡ For somewhat larger cations (r 70–96 pm) the C-type rare-earth M_2O_3 structure (p. 1437) is adopted, e.g. for In, Tl, Sc, Y, Sm, and the subsequent lanthanoids, and perhaps surprisingly for Mn^{III} (r 65 pm); the largest lanthanoids La, Ce, Pr, and Nd (r 106–100 pm) adopt the A-type rare-earth M_2O_3 structure (p. 1437).

TABLE 7.8   *The main structural types of aluminium oxides and hydroxides*[a]

Formula	Mineral name	Idealized structure
α-Al_2O_3	Corundum	hcp O with Al in two-thirds of the octahedral sites
α-AlO(OH)	Diaspore	hcp O (OH) with chains of octahedra stacked in layers interconnected with H bonds, and Al in certain octahedral sites
α-Al(OH)$_3$	Bayerite	hcp (OH) with Al in two-thirds of the octahedral sites
γ-Al_2O_3	—	ccp O defect spinel with Al in $21\frac{1}{3}$ of the 16 octahedral and 8 tetrahedral sites
γ-AlO(OH)	Boehmite	ccp O (OH) within layers; details uncertain
γ-Al(OH)$_3$	Gibbsite (Hydrargillite)	ccp OH within layers of edge-shared Al(OH)$_6$; octahedra stacked vertically via H bonds

[a] The Greek prefixes α- and γ- are not used consistently in the literature, e.g. bayerite is sometimes designated as β-Al(OH)$_3$ and gibbsite as α-Al(OH)$_3$. The UK usage adopted here is consistent with Wells[10] and emphasizes the structural relations between the hcp α-series and the ccp γ-series. Numerous other intermediate crystalline phases have been characterized during partial dehydration and designated as γ', δ, ζ, η, θ, κ, κ', ρ, χ, etc.

is formed by the low-temperature dehydration ($<450°$) of gibbsite, γ-Al(OH)$_3$, or boehmite, γ-AlO(OH). It has a defect spinel-type structure (p. 279) comprising a face-centred cubic (fcc) arrangement of 32 oxide ions and a random occupation of $21\frac{1}{3}$ of the 24 available cation sites (16 octahedral, 8 tetrahedral). This structure forms the basis of the so-called "activated aluminas" and progressive dehydration in the γ-series leads to open-structured materials of great value as catalysts, catalyst-supports, ion exchangers, and chromatographic media. Calcination of γ-Al_2O_3 above $1000°$ converts it irreversibly to the more stable and compact α-form (ΔH_{trans} -20 kJ mol^{-1}). Yet another form of Al_2O_3 occurs as the protective surface layer on the metal: it has a defect NaCl-type structure with Al occupying two-thirds of the octahedral (Na) interstices in the fcc oxide lattice. Perhaps the most ingenious and sophisticated development in aluminium technology has been the recent production of Al_2O_3 fibres which can be fabricated into a variety of textile forms, blankets, papers, and boards. Some idea of the many possibilities of such high-temperature inert fabrics is indicated in the Panel opposite.

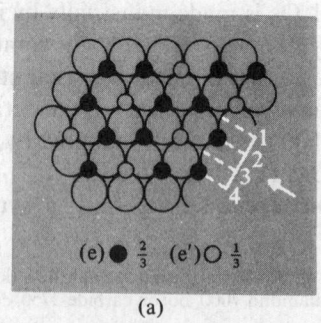

(e) ● $\frac{2}{3}$ (e') ○ $\frac{1}{3}$

(a)

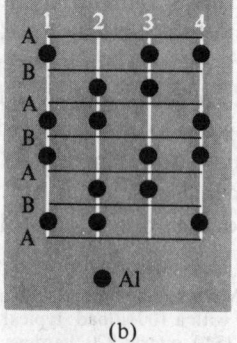

● Al

(b)

FIG. 7.11   Schematic representation of the structure of α-Al_2O_3. (a) pattern of occupancy by Al (●) of the octahedral sites between hcp layers of oxide ion (○), and (b) stacking sequence of successive planes of Al atoms viewed in the direction of the arrow in (a).

Fibrous Alumina and Zirconia

A new family of lightweight inorganic fibres made its commercial debut in 1974 when ICI announced the production of "Saffil", fibrous Al_2O_3 and ZrO_2, made by an undisclosed process on an initial scale of 100 tonnes pa. The fibres, which have no demonstrable toxic effects (cf. asbestos), have a diameter of $\sim 3\ \mu m$ (cf. human hair $\sim 70\ \mu m$), and each fibre is extremely uniform along its length (2–5 cm) as shown in Pl. 7.1. The fibres are microcrystalline (5–50 pm diam) and are both flexible and resilient with a high tensile strength. They have a soft, silky feel and can be made into rope, yarn, cloth, blankets, fibre matts, paper of various thickness, semi-rigid and rigid boards, and vacuum-formed objects of any required shape. The surface area of Saffil alumina is 100–150 $m^2\ g^{-1}$ due to the presence of small pores 2–10 pm diameter between the microcrystals, and this enhances its properties as an insulator, filtration medium, and catalyst support. The fibres withstand extended heating to 1400° (Al_2O_3) or 1600°C (ZrO_2) and are impervious to attack by hot concentrated alkalis and most hot acids except conc H_2SO_4, conc H_3PO_4, and aq HF. This unique combination of properties provides the basis for their use in high-temperature insulation, heat shields, thermal barriers, and expansion joints and seals. Fibrous alumina and zirconia are also valuable in thermocouple protection, electric-cable sheathing, and heating-element supports in addition to their use in the high-temperature filtration of corrosive liquids. Both oxides are stabilized by incorporation of small amounts of other inorganic oxides which inhibit disruptive transformation to other crystalline forms.

The most recent development (Du Pont, 1980) is the use of alumina fibres to strengthen metals. Molten metals (e.g. Al, Mg, Pb) or their alloys are forced into moulds containing up to 70% by volume of α-Al_2O_3 fibre. For example, fibre-reinforced Al containing 55% fibre by volume is 4–6 times stiffer than unreinforced Al even up to 370°C and has 2–4 times the fatigue strength. Potential applications for which high structural stiffness, heat resistance, and low weight are required include helicopter housings, automotive and jet engines, aerospace structures, and lead–acid batteries. For example, fibre reinforced composites of Al or Mg could eventually replace much of the steel used in car bodies without decreasing safety, since the composite has the stiffness of steel but only one-third its density.

Diaspore, α-AlO(OH) occurs in some types of clay and bauxite; it is stable in the range 280–450° and can be made by hydrothermal treatment of boehmite, γ-AlO(OH), in 0.4% aqueous NaOH at 380° and 500 atm. Crystalline boehmite is readily prepared by warming the amorphous, gelatinous white precipitate which first forms when aqueous NH_3 is added to cold solutions of Al salts. In α-AlO(OH) the O atoms are arranged in hcp as shown in Fig. 7.12; continuous chains of edge-shared octahedra are stacked in layers and are further interconnected by H bonds as indicated by the double lines. The underlying hcp structure ensures that diaspore dehydrates directly to α-Al_2O_3 (corundum) which has the same basic hcp arrangement of O atoms. This contrasts with the structure of boehmite, γ-AlO(OH), shown in Fig. 7.13; the structure as a whole is not close-packed but, within each layer the O atoms are arranged in cubic close packing. Dehydration at temperatures up to 450° proceeds via a succession of phases to the cubic γ-Al_2O_3 and the α (hexagonal) structure cannot be attained without much more reconstruction of the lattice at 1100–1200° as noted above.

Bayerite, α-$Al(OH)_3$, does not occur in nature but can be made by rapid precipitation from alkaline solutions in the cold:

$$2[Al(OH)_4]^- aq + CO_2(g) \longrightarrow 2Al(OH)_3 + CO_3^{2-}(aq) + H_2O$$

Gibbsite (or hydrargillite), γ-$Al(OH)_3$, is a more stable form and can be prepared by slow precipitation from warm alkaline solutions or by digesting the α-form in aqueous sodium aluminate solution at 80°. In both bayerite (α) and gibbsite (γ) there are layers of

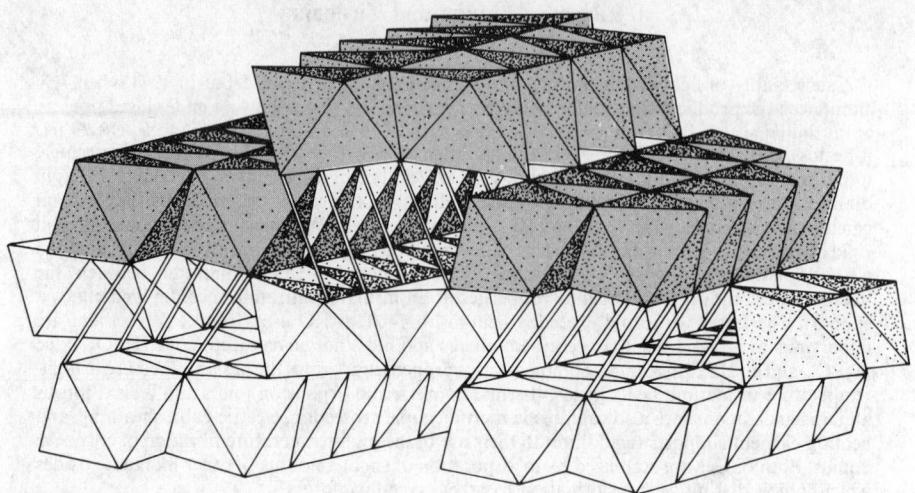

FIG. 7.12   Structure of diaspore, α-AlO(OH), based on hcp O^{2-} and OH^- as described in the text. The H bonds are shown by double lines. The same structure is adopted by α-GaO(OH), α-VO(OH), α-MnO(OH), and α-FeO(OH) (goethite).

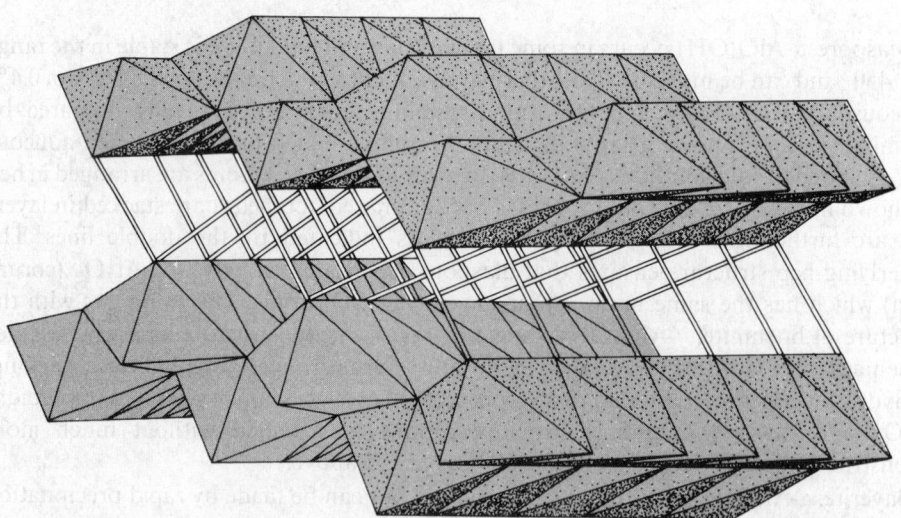

FIG. 7.13   Structure of boehmite, γ-AlO(OH) in which Al^{III} are octahedrally surrounded by $5 O^{2-}$ and one OH^-; the arrangement of O atoms within each layer is ccp and the layers are linked by H bonds (double lines). The same structure is adopted by γ-ScO(OH) and γ-FeO(OH) (lepidocrocite).

composition $Al(OH)_3$ built up by the edge sharing of $Al(OH)_6$ octahedra to give a pair of approximately close-packed OH layers with Al atoms in two-thirds of the octahedral interstices (Fig. 7.14a). The two crystalline modifications differ in the way this layer is stacked; it is approximately hcp in α-$Al(OH)_3$ but in the γ-form the OH groups on the under side of one layer rest directly above the OH groups of the layer below as shown in Fig. 7.14b. A third form of $Al(OH)_3$, nordstrandite, is obtained from the gelatinous hydroxide by ageing it in the presence of a chelating agent such as ethylenediamine, ethylene glycol, or EDTA; this aligns the OH to give a stacking arrangement which is intermediate between those of the α- and γ-forms.

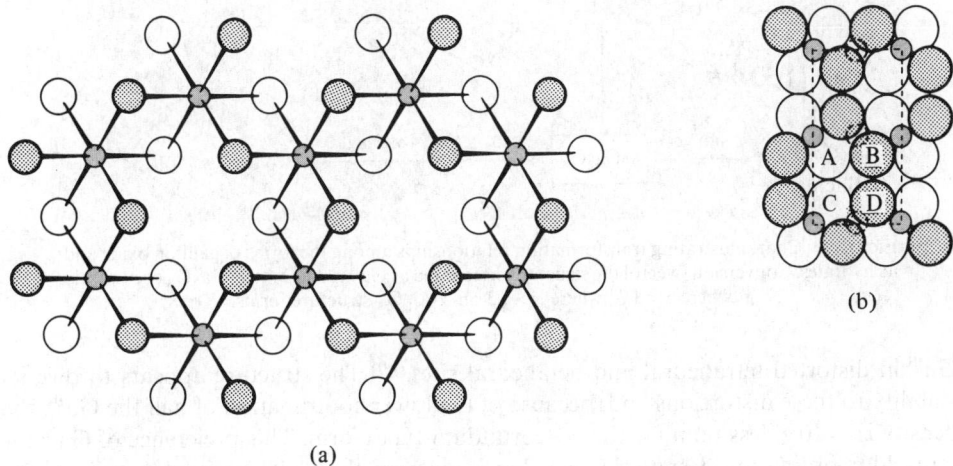

(a)

(b)

FIG. 7.14 (a) Part of a layer of $Al(OH)_3$ (idealized); the heavy and light open circles represent OH groups above and below the plane of the Al atoms. In α-$Al(OH)_3$ the layers are stacked to give approximately hcp. (b) Structure of γ-$Al(OH)_3$ viewed in a direction parallel to the layers; the OH groups labelled C and D are stacked directly beneath A and B. The six OH groups A, B, C, D, and B', D' (behind B and D), form a distorted H-bonded trigonal prism.

As expected from the foregoing structural discussion, gibbsite can be dehydrated to boehmite at $100°$ and to anhydrous γ-Al_2O_3 at $150°$, but ignition above $800°$ is required to form α-Al_2O_3. Numerous recipes have been devised for preparing catalysts of differing reactivity and absorptive power, based on the partial dehydration and progressive reconstitution of the Al/O/OH system. In addition to pore size, surface area, and general reactivity, the basic character of the surface diminishes (and its acidic character increases) in the following series as indicated by the pH of the isoelectric point:

(amorphous Al oxide hydrate) > γ-$AlO(OH)$ > α-$Al(OH)_3$ > γ-$Al(OH)_3$ > γ-Al_2O_3

	boehmite	bayerite	gibbsite	
pH of isoelectric point:	9.45	9.45–9.40	9.20	—	8.00

The binary oxides and hydroxides of Ga, In, and Tl have been much less extensively studied. The Ga system is somewhat similar to the Al system and a diagram summarizing the transformations in the systems is in Fig. 7.15. In general the α- and γ-series have the same structure as their Al counterparts. β-Ga_2O_3 is the most stable crystalline modification (mp $1740°$); it has a unique crystal structure with the oxide ions in distorted ccp and

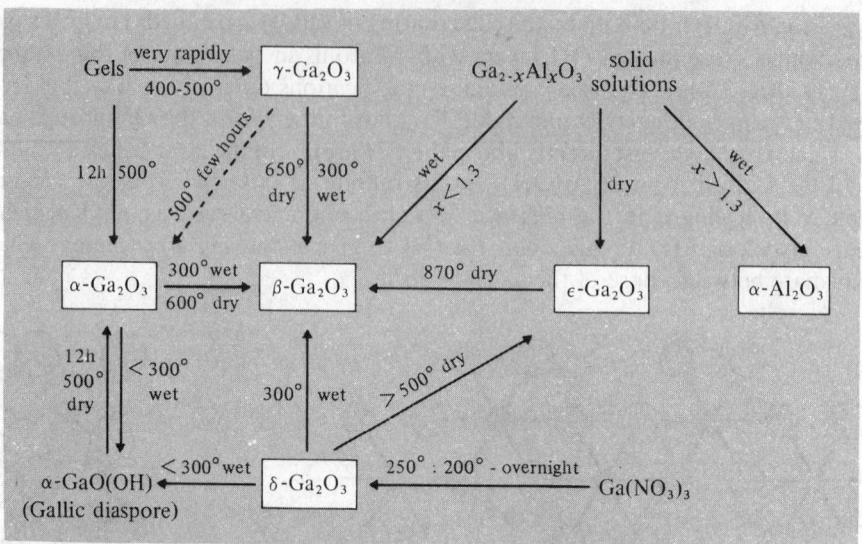

FIG. 7.15 Chart illustrating transformation relationships among the forms of gallium oxide and its hydrates. Conversion (wet) of the structure designated as $Ga_{2-x}Al_xO_3$ to $\beta\text{-}Ga_2O_3$ occurs only where $x < 1.3$; where $x > 1.3$ an $\alpha\text{-}Al_2O_3$ structure forms.

Ga^{III} in distorted tetrahedral and octahedral sites.[20] The structure appears to owe its stability to these distortions and, because of the lower coordination of half the Ga^{III}, the density is $\sim 10\%$ less than for the α- (corundum-type) form. This preference of Ga^{III} for fourfold coordination despite the fact that it is larger than Al^{III} may again indicate the polarizing influence of the d^{10} core; a similar tetrahedral site preference is observed for Fe^{III}.

In_2O_3 has the C-type M_2O_3 structure (p. 1437) and $InO(OH)$ (prepared hydrothermally from $In(OH)_3$ at 250–400°C and 100–1500 atm) has a deformed rutile structure (p. 1120) rather than the layer lattice structure of $AlO(OH)$ and $GaO(OH)$. Crystalline $In(OH)_3$ is best prepared by addition of NH_3 to aqueous $InCl_3$ at 100° and ageing the precipitate for a few hours at this temperature; it has the simple ReO_3-type structure distorted somewhat by multiple H bonds.[10]

Thallium is notably different. Tl_2^IO forms as black platelets when Tl_2CO_3 is heated in N_2 at 700° (mp 596°, d 10.36 g cm^{-3}); it is hygroscopic and gives $TlOH$ with water. $Tl_2^{III}O_3$ is brown-black (mp 716°, d 10.04 g cm^{-3}) and can be made by oxidation of aqueous $TlNO_3$ with Cl_2 or Br_2 followed by precipitation of the hydrated oxide $Tl_2O_3 . 1\frac{1}{2}H_2O$ and desiccation; single crystals have a very low electrical resistivity (e.g. 7×10^{-5} ohm cm at room temperature). A mixed oxide Tl_4O_3 (black) is known and also a violet peroxide Tl^IO_2 made by electrolysis of an aqueous solution of Tl_2SO_4 and oxalic acid between Pt electrodes. $TlOH$ has been mentioned previously (p. 255).

7.3.4 *Ternary and more complex oxide phases*

This section considers a number of extremely important structure types in which Al

[20] S. GELLER, Crystal structure of $\beta\text{-}Ga_2O_3$, *J. Chem. Phys.* **33**, 676–84 (1960).

combines with one or more other metals to form a mixed oxide phase. The most significant of these from both a theoretical and an industrial viewpoint are spinel ($MgAl_2O_4$) and related compounds, Na-β-alumina ($NaAl_{11}O_{17}$) and related phases, and tricalcium aluminate ($Ca_3Al_2O_6$) which is a major constituent of Portland cement. Each of these compounds raises points of fundamental importance in solid-state chemistry and each possesses properties of crucial significance to modern technology.

Spinels and related compounds[21]

Spinels form a large class of compounds whose crystal structure is related to that of the mineral spinel itself, $MgAl_2O_4$. The general formula is AB_2X_4 and the unit cell contains 32 oxygen atoms in almost perfect ccp array, i.e. $A_8B_{16}O_{32}$. In the normal spinel structure (Fig. 7.16) 8 metal atoms (A) occupy tetrahedral sites and 16 metal atoms (B) occupy octahedral sites, and the structure can be regarded as being built up of alternating cubelets of ZnS-type and NaCl-type structures. The two factors that determine which combinations of atoms can form a spinel-type structure are (a) the total formal cation charge, and (b) the relative sizes of the 2 cations with respect both to each other and to the anion. For oxides of formula AB_2O_4 charge balance can be achieved by three combinations of cation oxidation state: $A^{II}B^{III}_2O_4$, $A^{IV}B^{II}_2O_4$, and $A^{VI}B^{I}_2O_4$. The first combination is the most numerous and examples are known with

A^{II} = Mg, (Ca); Cr, Mn, Fe, Co, Ni, Cu; Zn, Cd, (Hg); Sn
B^{III} = **Al, Ga, In**; Ti, V, Cr, Mn, Fe, Co, Ni; Rh

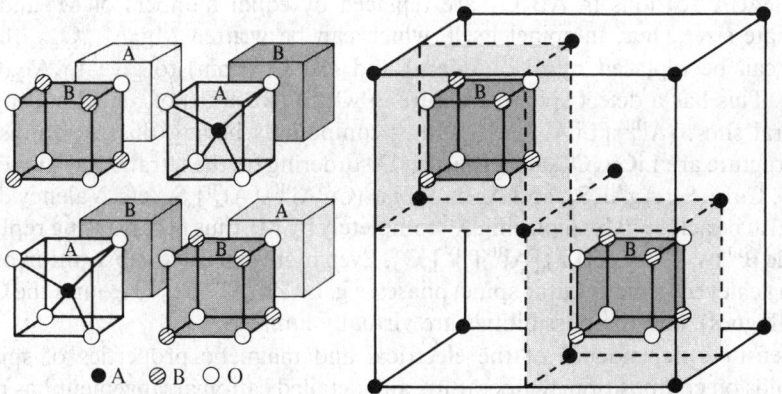

Fig. 7.16   Spinel structure AB_2O_4. The structure can be thought of as 8 octants of alternating AO_4 tetrahedra and B_4O_4 cubes as shown in the left-hand diagram; the 4O have the same orientation in all 8 octants and so build up into a fcc lattice of 32 ions which coordinate A tetrahedrally and B octahedrally. The 4 A octants contain 4 A ions and the 4 B octants contain 16 B ions. The unit cell is completed by an encompassing fcc of A ions (●) as shown in the right-hand diagram; this is shared with adjacent unit cells and comprises the remaining 4 A ions in the complete unit cell $A_8B_{16}O_{32}$. The location of two of the B_4O_4 cubes is shown for orientation.

[21] N. N. GREENWOOD, *Ionic Crystals, Lattice Defects, and Nonstoichiometry*, Butterworths, London, 1968, 194 pp. See also J. K. BURDETT, G. D. PRICE, and S. L. PRICE, *J. Am. Chem. Soc.* **104**, 92–95 (1982).

The anion can be O, S, Se, or Te. Most of the A^{II} cations have radii (6-coordinate) in the range 65–95 pm and larger cations such as Ca^{II} (100 pm) and Hg^{II} (102 pm) do not form oxide spinels. The radii of B^{III} fall predominantly in the range 60–70 pm though Al^{III} (53 pm) is smaller, and In^{III} (80 pm) normally forms sulfide spinels only.†

Many of the spinel-type compounds mentioned above do not have the normal structure in which A are in tetrahedral sites (t) and B are in octahedral site (o); instead they adopt the inverse spinel structure in which half the B cations occupy the tetrahedral sites whilst the other half of the B cations and all the A cations are distributed on the octahedral sites, i.e. $(B)_t[AB]_oO_4$. The occupancy of the octahedral sites may be random or ordered. Several factors influence whether a given spinel will adopt the normal or inverse structure, including (a) the relative sizes of A and B, (b) the Madelung constants for the normal and inverse structures, (c) ligand-field stabilization energies (p. 1097) of cations on tetrahedral and octahedral sites, and (d) polarization or covalency effects.[21]

Thus if size alone were important it might be expected that the smaller cation would occupy the site of lower coordination number, i.e. $Al_t[MgAl]_oO_4$; however, in spinel itself this is outweighed by the greater lattice energy achieved by having the cation of higher charge, (Al^{III}) on the site of higher coordination and the normal structure is adopted: $(Mg)_t[Al_2]_oO_4$. An additional factor must be considered in a spinel such as $NiAl_2O_4$ since the crystal field stabilization energy of Ni^{II} is greater in octahedral than tetrahedral coordination; this redresses the balance, making the normal and inverse structures almost equal in energy and there is almost complete randomization of all the cations on all the available sites: $(Al_{0.75}Ni_{0.25})_t[Ni_{0.75}Al_{1.25}]_oO_4$.

Inverse and disordered spinels are said to have a defect structure because all crystallographically identical sites within the unit cell are not occupied by the same cation. A related type of defect structure occurs in valency disordered spinels where, for example, the divalent A^{II} cations in AB_2O_4 are replaced by equal numbers of M^I and M^{III} of appropriate size. Thus, in spinel itself, which can be written $Mg_8Al_{16}O_{32}$, the $8Mg^{II}$ (72 pm) can be replaced by $4Li^I$ (76 pm) and $4Al^{III}$ (53 pm) to give $Li_4Al_{20}O_{32}$, i.e. $LiAl_5O_8$. This has a defect spinel structure in which two-fifths of the Al occupy all the tetrahedral sites: $(Al_2^{III})_t[Li^IAl_3^{III}]_oO_8$. Other compounds having this cation-disordered spinel structure are $LiGa_5O_8$ and $LiFe_5O_8$. Disordering on the tetrahedral sites occurs in $CuAl_5S_8$, $CuIn_5S_8$, $AgAl_5S_8$, and $AgIn_5S_8$, i.e. $(Cu^IAl^{III})_t[Al_4^{III}]_oS_8$, etc. Valency disordering can also be achieved by replacing A^{II} completely by M^I, thus necessitating replacement of half the B^{III} by M^{IV}, e.g. $(Li^I)_t[Al^{III}Ti^{IV}]_oO_4$. Even more extensive substitution of cations has been achieved in many cubic spinel phases, e.g. $Li_5^IZn_8^{II}Al_5^{III}Ge_9^{IV}O_{36}$ (and the Ga^{III} and Fe^{III} analogues), and the possibilities are virtually limitless.

The sensitive dependence of the electrical and magnetic properties of spinel-type compounds on composition, temperature, and detailed cation arrangement has proved a powerful incentive for the extensive study of these compounds in connection with the solid-state electronics industry. Perhaps the best-known examples are the ferrites, including the extraordinary compound magnetite Fe_3O_4 (p. 1254) which has an inverse spinel structure $(Fe^{III})_t[Fe^{II}Fe^{III}]_oO_4$. It will also be recalled that γ-Al_2O_3 (p. 274) has a

---

† Examples of spinels with other combinations of oxidation state are:

$A^{IV}B_2^{II}X_4^{-II}$: $TiMg_2O_4$, $PbFe_2O_4$, $SnCu_2S_4$
$A^{VI}B_2^IX_4^{-II}$: $MoAg_2O_4$, $MoNa_2O_4$, WNa_2O_4
$A^{II}B_2^IX_4^{-I}$: $NiLi_2F_4$, $ZnK_2(CN)_4$, $CdK_2(CN)_4$

defect spinel structure in which not all of the cation sites are occupied, i.e. $Al^{III}_{21\frac{1}{3}}\square_{2\frac{2}{3}}O_{32}$: the relation to spinel ($Mg^{II}_8Al^{III}_{16}O_{32}$) is obvious, the $8Mg^{II}$ having been replaced by the isoelectronically equivalent $5\frac{1}{3}Al^{III}$. This explains why $MgAl_2O_4$ can form a complete range of solid solutions with $\gamma\text{-}Al_2O_3$: the oxygen builds on to the complete fcc oxide ion lattice and the Al^{III} gradually replaces Mg^{II}, electrical neutrality being achieved simply by leaving 1 cation site vacant for each $3Mg^{II}$ replaced by $2Al^{III}$.

Sodium-β-alumina and related phases[22]

Sodium-β-alumina has assumed tremendous importance as a solid-state electrolyte since its very high electrical conductivity was discovered at the Ford Motor Company by J. T. Kummer and N. Weber in 1967. The compound, which has the idealized formula $NaAl_{11}O_{17}$ ($Na_2O.11Al_2O_3$) was originally thought to be a form of Al_2O_3 and hence called β-alumina (1916); the presence of Na, which was at first either undetected or ignored, is now known to be essential for stability. X-ray analysis shows that the structure is closely related to that of spinel, no fewer than 50 of the 58 atoms in the unit cell being arranged exactly as in spinel. The large Na atoms are situated exclusively in loosely packed planes together with an equal number of O atoms as shown in Fig. 7.17; these planes are 1127 pm apart, being separated by the "spinel blocks". The close-packed oxygen layers above and below the Na planes are mirror images of each other 476 pm apart and they are bound together not only by the Na atoms but by an equal number of Al–O–Al bonds. There are several other sites in the mirror plane which can physically accommodate Na and this permits rapid two-dimensional diffusion of Na within the basal plane; it also explains the very low resistivity of the order of 30 ohm cm. The structure can also accommodate supernumerary Na ions, and the compound, even in the form of single

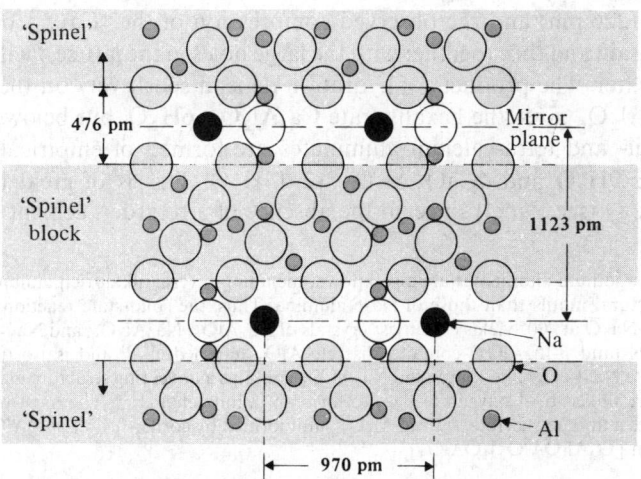

FIG. 7.17   Crystal structure of Na-β-alumina (see text). This section, which is a plane parallel to the *c*-axis, does not show the closest Na–Na distance.

[22] J. T. KUMMER, β-alumina electrolytes, *Prog. Solid State Chem.* **7**, 141–75 (1972). J. H. KENNEDY, The β-aluminas, *Topics in Applied Physics* **21**, 105–41 (1977).

crystals, is massively defective, having typically 20–30% more Na than indicated by the idealized formula; this is probably compensated by additional Al vacancies in the "spinel blocks" adjacent to the mirror planes, e.g. $Na_{2.58}Al_{21.8}O_{34}$.

Sodium-β-alumina can be prepared by heating Na_2CO_3 (or $NaNO_3$ or $NaOH$) with any modification of Al_2O_3 or its hydrates to $\sim 1500°$ in a Pt vessel suitably sealed to avoid loss of Na_2O (as $Na + O_2$).† In the presence of NaF or AlF_3 a temperature of $1000°$ suffices. Na-β-alumina melts at $\sim 2000°$ (probably incongruently) and has d 3.25 g cm^{-3}. The Na can be replaced by Li, K, Rb, Cu^I, Ag^I, Ga^I, In^I, or Tl^I by heating with a suitable molten salt, and Ag^I can be replaced by NO^+ by treatment with molten $NOCl/AlCl_3$. The ammonium compound is also known and H_3O^+-β-alumina can be prepared by reduction of the Ag compound. Similarly, Al^{III} can be replaced by Ga^{III} or Fe^{III} in the preparation, leading to compounds of (idealized) formulae $Na_2O.11Ga_2O_3$, $Na_2O.11Fe_2O_3$, $K_2O.11Fe_2O_3$, etc.

Apart from the intriguing structural implications of these fast-ion solid-state conductors, Na-β-alumina and related phases have been extensively used during the past few years as permeable membranes in the Na/S battery system (p. 801): this requires an air-stable membrane that is readily permeable to Na ions but not to Na atoms or S, that is non-reactive with molten Na and S, and that is not an electronic conductor. Not surprisingly, few compounds have been found to compete with Na-β-alumina in this field.

Tricalcium aluminate, $Ca_3Al_2O_6$

Tricalcium aluminate is an important component of Portland cement yet, despite numerous attempts dating back over the past 50 y, its structure remained unsolved until 1975.[23] The basic unit is now known to be a 12-membered ring of 6 fused $\{AlO_4\}$ tetrahedra $[Al_6O_{18}]^{18-}$ as shown in Fig. 7.18; there are 8 such rings per unit cell surrounding holes of radius 147 pm, and the rings are held together by Ca^{II} ions in distorted sixfold coordination to give the structural formula $Ca_9Al_6O_{18}$. The rather short Ca–O distance (226 pm) and the observed compression of the $\{CaO_6\}$ octahedra may indicate some strain and this, together with the large holes in the lattice, facilitate the rapid reaction with water. The products of hydration depend sensitively on the temperature. Above $21°$ $Ca_3Al_2O_6$ gives the hexahydrate $Ca_3Al_2O_6.6H_2O$, but below this temperature hydrated di- and tetra-calcium aluminates are formed of empirical composition $2CaO.Al_2O_3.5–9H_2O$ and $4CaO.Al_2O_3.12–14H_2O$. This is of great importance in cement technology (see Panel) since, in the absence of a retarder, cement reacts rapidly

---

† Numerous other sodium aluminates are also known, including several alkali-rich aluminates which have quite different structural motifs than those of Na-β-alumina. Thus, the solid-state reaction of α-Al_2O_3 with varying amounts of Na_2O at $700°$ yields colourless crystals of Na_5AlO_4, $Na_7Al_3O_8$, and $Na_{17}Al_5O_{16}$. The first of these compounds (and β-Li_5AlO_4) contains discrete AlO_4 tetrahedra[22a] and is isostructural with the corresponding ferrate Na_5FeO_4. The compound $Na_7Al_3O_8$ contains a novel ring structure made up of 6$\{AlO_4\}$ tetrahedra which are linked by 4 oxygen bridges to form an infinite chain.[22a] The most recent compound, $Na_{17}Al_5O_{16}$ features a unique, discrete, zig-zag Al_5O_{16} unit formed by corner-fusion of 5$\{AlO_4\}$ tetrahedra to give the chain anion $[O_3Al(OAlO_2)_3OAlO_3]^{17-}$.[22b]

[22a] M. G. BARKER, P. G. GADD, and M. J. BEGLEY, Preparation and crystal structures of the first alkali-rich sodium aluminates, $Na_7Al_3O_8$ and Na_5AlO_4, *JCS Chem. Comm.* 1981, 379–81.

[22b] M. G. BARKER, P. G. GADD, and S. C. WALLWORK, A new sodium aluminate, $Na_{17}Al_5O_{16}$, *JCS Chem. Comm.* 1982, 516–517.

[23] P. MONDAL and J. W. JEFFREY, The crystal structure of tricalcium aluminate, $Ca_3Al_2O_6$, *Acta Cryst.* **B31**, 689–97 (1975).

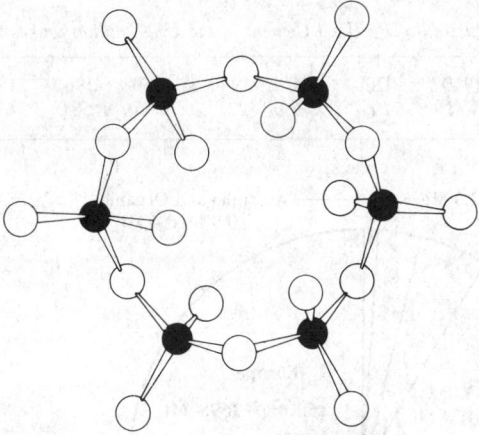

FIG. 7.18   Structure of the $[Al_6O_{18}]^{18+}$ unit in $Ca_3Al_2O_6$ (i.e. $Ca_9Al_6O_{18}$). The Al–O distances are all in the range 175 ± 2 pm.

Portland Cement

The name "Portland cement" was first used by J. Aspdin in a patent (1824) because, when mixed with water and sand the powder hardened into a block that resembled the natural limestone quarried in the Isle of Portland, England. The two crucial discoveries which led to the production of strong, durable, hydraulic cement that did not disintegrate in water, were made in the eighteenth and nineteenth centuries. In 1756 John Smeaton, carrying out experiments in connection with building the Eddystone Lighthouse, recognized the importance of using limes which contained admixed clays or shales (i.e. aluminosilicates), and by the early 1800s it was realized that firing must be carried out at sintering temperatures in order to produce a clinker now known to contain calcium silicates and aluminates. The first major engineering work to use Portland cement was in the tunnel constructed beneath the Thames in 1828. The first truly high-temperature cement (1450–1600°C) was made in 1854, and the technology was revolutionized in 1899 by the introduction of rotary kilns.

The important compounds in Portland cement are dicalcium silicate (Ca_2SiO_4) 26%, tricalcium silicate (Ca_3SiO_5) 51%, tricalcium aluminate $(Ca_3Al_2O_6)$ 11%, and the tetracalcium species $Ca_4Al_2Fe_2^{III}O_{10}$ (1%). The principal constituent of moistened cement paste is a tobermorite gel which can be represented schematically by the following idealized equations:

$$2Ca_2SiO_4 + 4H_2O \longrightarrow 3CaO.2SiO_2.3H_2O + Ca(OH)_2$$
$$2Ca_3SiO_5 + 6H_2O \longrightarrow 3CaO.2SiO_2.3H_2O + 3Ca(OH)_2$$

The adhesion of the tobermorite particles to each other and to the embedded aggregates is responsible for the strength of the cement which is due, ultimately, to the formation of –Si–O–Si–O bonds.

Portland cement is made by heating a mixture of limestone (or chalk, shells, etc.) with aluminosilicates (derived from sand, shales, and clays) in carefully controlled amounts so as to give the approximate composition $CaO \sim 70\%$, $SiO_2 \sim 20\%$, $Al_2O_3 \sim 5\%$, $Fe_2O_3 \sim 3\%$. The presence of Na_2O, K_2O, MgO, and P_2O_5 are detrimental and must be limited. The raw materials are ground to pass 200-mesh sieves and then heated in a rotary kiln to $\sim 1500°$ to give a sintered clinker; this is reground to 325-mesh and mixed with 2–5% of gypsum. An average-sized kiln can produce 1000–2000 tonnes of cement per day and the world's largest kiln (1975) produces 6750 tonnes per day. The vast scale of the industry can be gauged from the US production figures in the table below and world production for 1974 is summarized in the diagram. Price (1977) was $\sim \$40$ per tonne for bulk supplies.

Continued

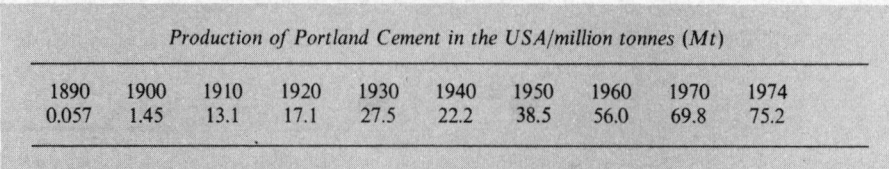

Production of Portland Cement in the USA/million tonnes (Mt)									
1890	1900	1910	1920	1930	1940	1950	1960	1970	1974
0.057	1.45	13.1	17.1	27.5	22.2	38.5	56.0	69.8	75.2

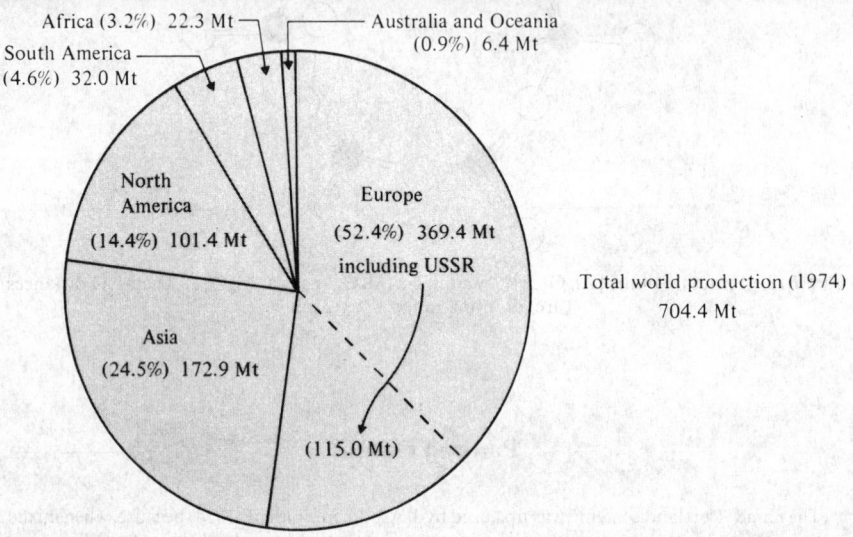

Africa (3.2%) 22.3 Mt

South America (4.6%) 32.0 Mt

Australia and Oceania (0.9%) 6.4 Mt

North America (14.4%) 101.4 Mt

Europe (52.4%) 369.4 Mt including USSR

Asia (24.5%) 172.9 Mt

(115.0 Mt)

Total world production (1974) 704.4 Mt

with water giving a sharp rise in temperature and a "flash set" during which the various calcium aluminate hydrates precipitate and congeal into an unmanageable mass. This can be avoided by grinding in 2–5% of gypsum ($CaSO_4 . 2H_2O$) with the cement clinker; this reacts rapidly with dissolved aluminates in the presence of $Ca(OH)_2$ to give the calcium sulfatoaluminate, $3CaO . Al_2O_3 . 3CaSO_4 . 31H_2O$, which is much less soluble than the hydrated calcium aluminates and therefore preferentially precipitates and prevents the premature congealing.

Another important calcium aluminate system occurs in high-alumina cement (*ciment fondu*). This is not a Portland cement but is made by *fusing* limestone and bauxite with small amounts of SiO_2 and TiO_2 in an open-hearth furnace at 1425–1500°; rotary kilns with tap-holes for the molten cement can also be used. Typical analytical compositions for a high-alumina cement are $\sim 40\%$ each of Al_2O_3 and CaO and about 10% each of Fe_2O_3 and SiO_2; the most important compounds in the cement are $CaAl_2O_4$, $Ca_2Al_2SiO_7$, and $Ca_6Al_8FeSiO_{21}$. Setting and hardening of high-alumina cement are probably due to the formation of calcium aluminate gels such as $CaO . Al_2O_3 . 10H_2O$, and the more basic $2CaO . Al_2O_3 . 8H_2O$, $3CaO . Al_2O_3 . 6H_2O$, and $4CaO . Al_2O_3 . 13H_2O$, though these empirical formulae give no indication of the structural units involved. The most notable property of high-alumina cement is that it develops very high strength at a very early stage (within 1 day). Long exposure to warm, moist conditions may lead to failure but resistance to corrosion by sea water and sulfate brines, or by weak mineral acids, is outstanding. It has also been much used as a refractory cement to withstand temperatures up to 1500°.

7.3.5 *Other inorganic compounds*

At normal temperatures the only stable chalcogenides of Al are Al_2S_3 (white), Al_2Se_3 (grey), and Al_2Te_3 (dark grey). They can be prepared by direct reaction of the elements at $\sim 1000°$ and all hydrolyse rapidly and completely in aqueous solution to give $Al(OH)_3$ and H_2X ($X = S$, Se, Te). The small size of Al relative to the chalcogens dictates tetrahedral coordination and the various polymorphs are related to wurtzite (hexagonal ZnS, p. 1405), two-thirds of the available metal sites being occupied in either an ordered (α) or a random (β) fashion. Al_2S_3 also has a γ-form related to γ-Al_2O_3 (p. 274).

The chalcogenides of Ga, In, and Tl are much more numerous and at least a dozen different structure types have been established by X-ray crystallography.[24] The compounds have been extensively studied not only because of their intriguing stoichiometries, but also because many of them are semiconductors, semi-metals, photo-conductors, or light emitters, and Tl_5Te_3 has been found to be a superconductor at low temperatures. The compounds, as expected from their position in the periodic table, are far from ionic, but formal oxidation states remain a useful device for electron counting and for checking the overall charge balance. Well-established compounds are summarized in Table 7.9. The following points are noteworthy. The hexagonal α- and β-forms of Ga_2S_3 are isostructural with the Al analogues and an additional form, γ-Ga_2S_3, adopts the related defect sphalerite structure derived from cubic ZnS (zinc blende, p. 1405). The same structure is found for Ga_2Se_3 and Ga_2Te_3 but for the larger In^{III} atom octahedral coordination also becomes possible. The corresponding Tl^{III} sesquichalcogenides Tl_2X_3 are either non-existent or of dubious authenticity, perhaps because of the ready reduction to Tl^I (see TlI_3, p. 269).

GaS (yellow, mp 970°) has a hexagonal layer structure with Ga–Ga bonds (248 pm); each Ga is coordinated by 3S and 1Ga, and the sequence of layers along the c-axis is $\cdots$SGaGaS,SGaGaS$\cdots$; the compound can therefore be considered as an example of Ga^{II}. The structures of GaSe, GaTe, red InS, and InSe are similar. By contrast, InTe, TlS (black), and TlSe (black, metallic) have a structure which can be formalized as $M^I[M^{III}X_2]$; each Tl^{III} is tetrahedrally coordinated by 4 Se at 268 pm and the tetrahedra are linked into infinite chains by edge sharing along the c-axis (see structure), whereas each Tl^I lies between these chains and is surrounded by a distorted cube of 8 Se at 342 pm. This explains the marked anisotropy of properties especially the metallic conductivity in the (001) plane and the semiconductivity along the c-axis. Similar edge-linked $\{GeSe_4\}$

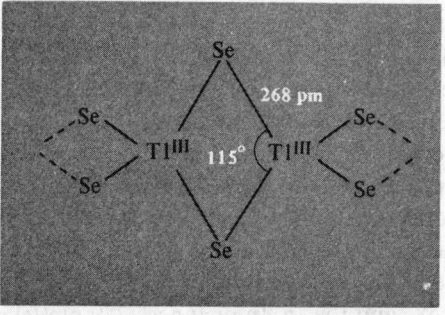

[24] L. I. MAN, R. M. IMANOV, and S. A. SEMILETOV, Types of crystal structure in the chalcogenides of gallium, indium, and thallium, *Sov. Phys. Crystallogr.* **21**, 255–63 (1976).

TABLE 7.9 *Stoichiometries and structures of the crystalline chalcogenides of Group IIIB elements*

Ga_2S	Ga_2Se	
GaS (yellow) layer structure with Ga–Ga bonds	GaSe like GaS	GaTe like GaS
Ga_4S_5		(Ga_3Te_2)
α-Ga_2S_3 (yellow) ordered defect wurtzite (hexagonal ZnS)		
β-Ga_2S_3 defect wurtzite		
γ-Ga_2S_3 defect sphalerite (cubic ZnS)	Ga_2Se_3 defect sphalerite	Ga_2Te_3 defect sphalerite
		Ga_2Te_5 chains of linked $\{GaTe_4\}$ plus single Te atoms
	In_4Se_3 contains $[(In^{III})_3]^V$ groups: $In^I[In_3^{III}]Se_3$	In_4Te_3 like In_4Se_3
InS (red) like GaS	InSe distorted NaCl, somewhat like GaS	InTe like TlSe (cubes and tetrahedra)
In_6S_7 see text	In_6Se_7 like In_6S_7	In_3Te_4
α-In_2S_3 (yellow) cubic γ'-Al_2O_3	α-In_2Se_3 defect wurtzite, but $\frac{1}{16}$ of In octahedral	α-In_2Te_3 defect sphalerite (cubic ZnS)
β-In_2S_3 (red) defect spinel, γ-Al_2O_3	β-In_2Se_3 ordered defect wurtzite (hexagonal ZnS)	β-In_2Te_3
		In_3Te_5
		In_2Te_5
Tl_2S (black) distorted CdI_2 layer lattice (Tl^I in threefold coordination)		
Tl_4S_3 chains of linked $\{Tl^{III}S_4\}$ tetrahedra $(Tl^I)_3[Tl^{III}S_3]$	Tl_5Se_3 complex Cr_5B_3-type structure	Tl_5Te_3 Cr_5B_3 layer structure, CN of Tl varies up to 9 and Te up to 10
TlS (black) like TlSe, $Tl^I[Tl^{III}S_2]$	TlSe (black) chains of edgeshared $\{Tl^{III}Se_4\}$ tetrahedra $Tl^I[Tl^{III}Se_2]$	TlTe variant of W_5Si_3 (complex)
[No Tl_2S_3 known]	Tl_2Se_3	(Tl_2Te_3)
TlS_2 Tl^I polysulfide		
Tl_2S_5 (red and black forms) Tl^I polysulfide		
Tl_2S_9 Tl^I polysulfide		

tetrahedra are found in $Cs_{10}Ga_6Se_{14}$ which was recently obtained as transparent pale-yellow crystals by heating an equimolar mixture of GaSe and Cs in a carefully controlled temperature programme; the compound features the unprecedented finite complex anion $[Se_2Ga(\mu\text{-}Se_2Ga)_5Se_2]^{10-}$ which is 1900 pm long.[24a]

In_6S_7 (and the isostructural In_6Se_7) have a curious structure comprising two separate blocks of almost ccp S which are rotated about the *b*-axis by 61° with respect to each other; the In is in octahedral coordination. The crystal structures of In_4Se_3 and In_4Te_3 show that they can be regarded to a first approximation as $In^I[In_3]^V(X^{-II})_3$ but the compound does not really comprise discrete ions. The triatomic unit $[In^{III}—In^{III}—In^{III}]$ is bent, the angle at the central atom being 158° and the In–In distances 279 pm (cf. 324–326 pm in metallic

[24a] H. J. DEISEROTH and HAN FU-SON, $[Ga_6Se_{14}]^{10-}$: A 1900 pm long, hexameric anion, *Angew. Chem.* Int. Ed. (Engl.) **20**, 962–3 (1981).

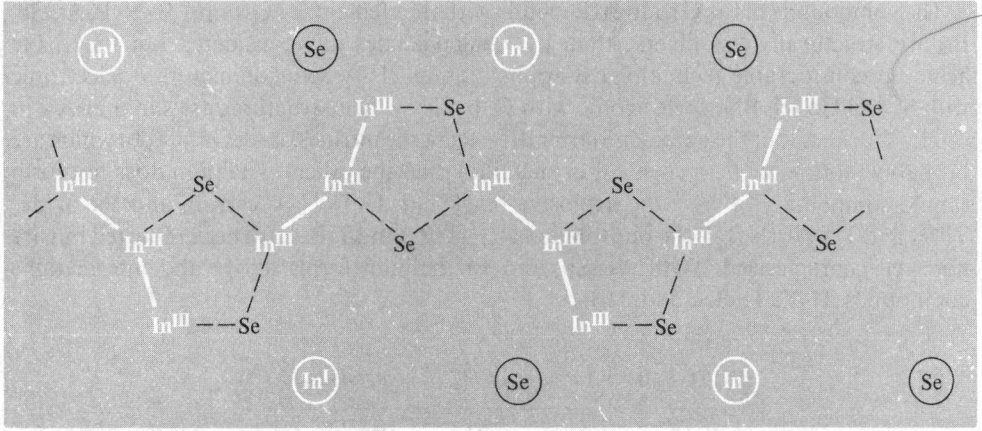

FIG. 7.19   Schematic structure of In_4Se_3.

In). However, it is also possible to discern non-planar 5-membered heterocycles in the structure formed by joining 2 In from 1 $\{In_3\}$ to the terminal In of an adjacent $\{In_3\}$ via 2 bridging Se (or Te) atoms so that the structure can be represented schematically as in Fig. 7.19. The In^{III}–Se distances average 269 pm compared with the closest In^{I}–Se contact of 297 pm. The $[In_3^{III}]^V$ unit can be compared with the isoelectronic species $[Hg_3^{II}]^{II}$. The compound Tl_4S_3, which has the same stoichiometry as In_4X_3, has a different structure (shown schematically below) in which chains of corner-shared $\{Tl^{III}S_4\}$ tetrahedra of over-all stoichiometry $[TlS_3]$ are bound together by Tl^I; within the chains the Tl^{III}–S distance is 254 pm whereas the Tl^I–S distances vary between 290–336 pm. A comparison of the formal designation of the two structures $In^I[(In_3^{III})]^V(Se^{-II})_3$ and $(Tl^I)_3[Tl^{III}S_3]^{-III}$ again illustrates the increasing preference of the heavier metal for the $+1$ oxidation state. The trend continues with the polysulfides $Tl^I S_2$, $Tl_2^I S_5$, and $Tl_2^I S_9$ already alluded to on p. 286.

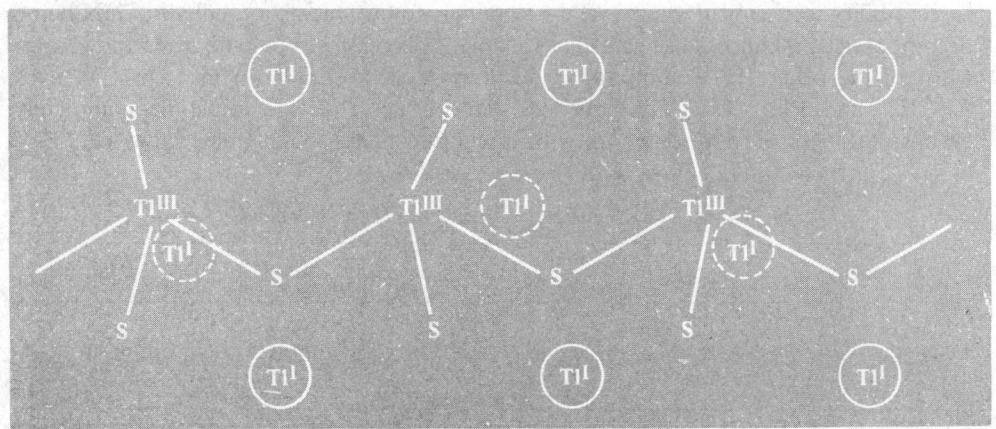

Schematic structure of Tl_4S_3.

The compounds of the Group IIIB metals with the elements of Group VB (N, P, As, Sb, Bi) are structurally less diverse than the chalcogenides just considered but they have achieved considerable technological application as III–V semiconductors isoelectronic with Si and Ge (cf. BN isoelectronic with C, p. 235). Their structures are summarized in Table 7.10: all adopt the cubic ZnS structure except the nitrides of Al, Ga, and In which are probably more ionic (less covalent or metallic) than the others. Thallium does not form simple compounds $M^{III}X^{V}$: the explosive black nitride Tl_3^IN is known, and the azides Tl^IN_3 and $Tl^I[Tl^{III}(N_3)_4]$; the phosphides Tl_3P, TlP_3, and TlP_5 have been reported but are not well characterized. With As, Sb, and Bi thallium forms alloys and intermetallic compounds Tl_3X, Tl_7Bi_2, and $TlBi_2$.

TABLE 7.10   *Structures of III–V compounds MX*[(a)]

X↓M→	B	Al	Ga	In
N	L, S	W	W	W
P	S	S	S	S
As	S	S	S	S
Sb	—	S	S	S

[(a)] L = BN layer lattice (p. 236).
     S = sphalerite (zinc blende), cubic ZnS structure (p. 1405).
     W = wurtzite, hexagonal ZnS structure (p. 1405).

The III–V semiconductors can all be made by direct reaction of the elements at high temperature and under high pressure when necessary. Some properties of the Al compounds are in Table 7.11 from which it is clear that there are trends to lower mp and energy band-gap E_g with increasing atomic number.

Analogous compounds of Ga and In are grey or semi-metallic in appearance and show similar trends (Table 7.12). These data should be compared with those for Si, Ge, Sn, and Pb on p. 434 and for the isoelectronic II–VI semiconductors of Zn, Cd, and Hg with S, Se, and Te (p. 1406). In addition, GaN is obtained by reacting Ga and NH_3 at 1050° and InN by reducing and nitriding In_2O_3 with NH_3 at 630°. The nitrides show increasing susceptibility to chemical attack, AlN being inert to both acids and alkalis, GaN being decomposed by alkali, but not acid, and InN being decomposed by both acids and alkalis. Most of the other III–V compounds decompose slowly in moist air, e.g. AlP gives $Al(OH)_3$ and PH_3. As a consequence, semiconductor devices must be completely encapsulated to prevent reaction with the atmosphere. The great value of III–V

TABLE 7.11   *Some properties of Al III–V compounds*

Property	AlN	AlP	AlAs	AlSb
Colour	Pale yellow	Yellow	Orange	—
MP/°C	>2200 decomp	2000	1740	1060
E_g/kJ mol^{-1}[(a)]	411	236	208	145

[(a)] Energy gap between top of (filled) valence band and bottom of (empty) conduction band (p. 383). To convert from kJ mol^{-1} to eV atom^{-1} divide by 96.485.

TABLE 7.12   *Comparison of some III–V semiconductors*

Property	GaP	GaAs	GaSb	InP	InAs	InSb
MP/°C	1465	1238	712	1070	942	525
E_g/kJ mol$^{-1(a)}$	218	138	69	130	34	17

[a] See note to Table 7.11.

semiconductors is that they extend the range of properties of Si and Ge, and by judicious mixing in ternary phases they permit a continuous interpolation of energy band gaps, current-carrier mobilities, and other characteristic properties. Some of their uses are summarized in the Panel.

Other compounds containing Al–N or Ga–N bonds, including heterocyclic and cluster compounds, are considered in the next section.

7.3.6 *Organometallic compounds*

Many organoaluminium compounds are known which contain 1, 2, 3, or 4 Al–C bonds per Al atom and, as these have an extensive reaction chemistry of considerable industrial importance, they will be considered before the organometallic compounds of Ga, In, and Tl are discussed.

Aluminium trialkyls and triaryls are highly reactive, colourless, volatile liquids or low-melting solids which ignite spontaneously in air and react violently with water; they should therefore be handled circumspectly and with suitable precautions. Unlike the boron trialkyls and triaryls they are often dimeric, though with branched-chain alkyls such as Pr^i, Bu^i, and Me_3CCH_2 this tendency is less marked. Al_2Me_6 (mp 15°, bp 126°) has the methyl-bridged structure shown and the same dimeric structure is found for Al_2Ph_6 (mp 225°).

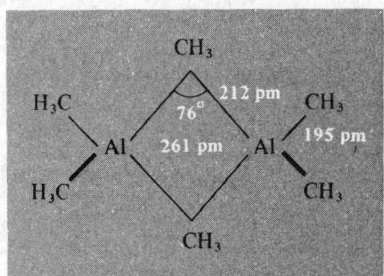
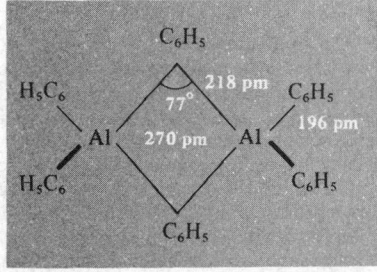

In each case Al–C$_\mu$ is about 10% longer than Al–C$_t$ (cf. Al_2X_6, p. 263; B_2H_6, p. 181). The enthalpy of dissociation of Al_2Me_6 into monomers is 84 kJ mol^{-1}. Al_2Et_6 (mp −53°) and $Al_2Pr_6^n$ (mp −107°) are also dimeric at room temperature.

As with $Al(BH_4)_3$ and related compounds (p. 260), solutions of Al_2Me_6 show only one proton nmr signal at room temperature due to the rapid interchange of bridging and terminal Me groups; at −75° this process is sufficiently slow for separate resonances to be observed.

CTE–K

Applications of III–V Semiconductors

The 9 compounds that Al, Ga, and In form with P, As, and Sb have been extensively studied because of their many applications in the electronics industry, particularly those centred on the interconversion of electrical and optical (light) energy. For example, they are produced commercially as light-emitting diodes (LEDs) familiar in pocket calculators, wrist watches, and the alpha-numeric output displays of many instruments; they are also used in infrared-emitting diodes, injection lasers, infrared detectors, photocathodes, and photomultiplier tubes. An extremely elegant chemical solid-state technology has evolved in which crystals of the required properties are deposited, etched, and modified to form the appropriate electrical circuits. The ternary system $GaAs_{1-x}P_x$ now dominates the LED market for α-numeric and graphic displays following the first report of this activity in 1961. $GaAs_{1-x}P_x$ is grown epitaxially on a single-crystal substrate of GaAs or GaP by chemical vapour deposition and crystal wafers as large as 20 cm² have been produced commercially. The colour of the emitted radiation is determined by the energy band gap E_g; for GaAs itself E_g is 138 kJ mol⁻¹ corresponding to an infrared emission (λ 870 nm), but this increases to 184 kJ mol⁻¹ for $x \sim 0.4$ corresponding to red emission (λ 650 nm). For $x > 0.4$ E_g continues to increase until it is 218 kJ mol⁻¹ for GaP (green, λ 550 nm). Commercial yellow and green LEDs contain the added isoelectronic impurity N to improve the conversion efficiency. A schematic cross-section of a typical $GaAs_{1-x}P_x$ epitaxial wafer doped with Te and N is shown in the diagram: Te (which has one more valence electron per atom than As or P) is the most widely used dopant to give *n*-type impurities in this system at concentration of 10^{16}–10^{18} atoms cm⁻³ (0.5–50 ppm). The p–n junction is then formed by diffusing Zn (1 less electron than Ga) into the crystal to a similar concentration.

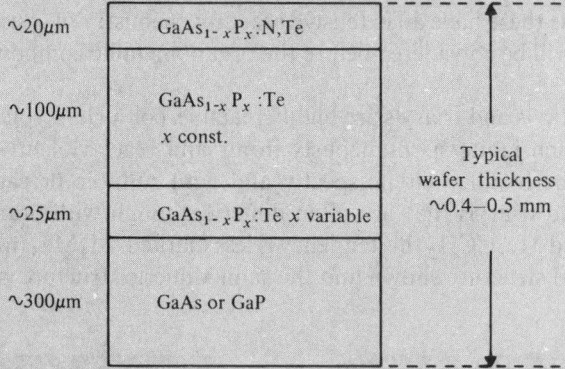

An even more recent application is the construction of semiconductor lasers. In normal optical lasers light is absorbed by an electronic transition to a broad band which lies above the upper laser level and the electron then drops into this level by a non-radiative transition. By contrast the radiation in a semiconductor laser originates in the region of a p–n junction and is due to the transitions of injected electrons and holes between the low-lying levels of the conduction band and the uppermost levels of the valence band. (Impurity levels may also be involved.) The efficiency of these semiconductor injection lasers is very much higher than those of optically pumped lasers and the devices are much smaller; they are also easily adaptable to modulation. As implied by the band gaps on p. 289, emission wavelengths are in the visible and near infrared. A heterostructure laser based on the system $GaAs$–$Al_xGa_{1-x}As$ recently became the first junction laser to run continuously at 3000 K and above (1970).

In the two types of device just considered, namely light emitting diodes and injection lasers, electrical energy is converted into optical energy. The reverse process of converting optical energy into electrical energy (photoconductivity and photovoltaic effects) has also been successfully achieved by III–V semiconductor systems. For example, the small band-gap compound InSb is valuable as a photoconductive infrared detector, and several compounds are being actively studied for use in solar cells to convert sunlight into useful sources of electrical power. The maximum photon flux in sunlight occurs at 75–95 kJ mol⁻¹ and GaAs shows promise, though other factors make Cu_2S–CdS cells more attractive commercially at the present time.

Al_2Me_6 can be prepared on a laboratory scale by the reaction of $HgMe_2$ on Al at $\sim 90°C$. Al_2Ph_6 can be prepared similarly using $HgPh_2$ in boiling toluene or by the reaction of LiPh on Al_2Cl_6. On the industrial (kilotonne) scale Al is alkylated by means of RX or by alkenes plus H_2. In the first method the sesquichloride $R_3Al_2Cl_3$ is formed in equilibrium with its disproportionation products:[†]

$$2Al + 3RCl \xrightarrow[\text{AlCl}_3 \text{ or AlR}_3]{\text{trace of I}_2} R_3Al_2Cl_3 \rightleftharpoons \tfrac{1}{2}R_4Al_2Cl_2 + \tfrac{1}{2}R_2Al_2Cl_4$$

Addition of NaCl removes $R_2Al_2Cl_4$ as the complex $(2NaAlCl_3R)$ and enables $R_4Al_2Cl_2$ to be distilled from the mixture. Reaction with Na yields the trialkyl, e.g.:

$$3Me_4Al_2Cl_2 + 6Na \longrightarrow 2Al_2Me_6 + 2Al + 6NaCl$$

Higher trialkyls are more readily prepared on an industrial scale by the alkene route (K. Ziegler *et al.*, 1960) in which H_2 adds to Al in the presence of preformed AlR_3 to give a dialkyl-aluminium hydride which then readily adds to the alkene:

$$2Al + 3H_2 + 2Al_2Et_6 \xrightarrow{150°} \{6Et_2AlH\} \xrightarrow[70°]{6CH_2CH_2} 3Al_2Et_6$$

Similarly, Al, H_2, and $Me_2C{=}CH_2$ react at $100°$ and 200 atm to give $AlBu^i_3$ in a single-stage process, provided a small amount of this compound is present at the start; this is required because Al does not react directly with H_2 to form AlH_3 prior to alkylation under these conditions. Alkene exchange reactions can be used to transform $AlBu^i_3$ into numerous other trialkyls. $AlBu^i_3$ can also be reduced by potassium metal in hexane at room temperature to give the novel brown compound $K_2Al_2Bu^i_6$ (mp $40°$) which is notable in providing a rare example of an Al–Al bond in the diamagnetic anion $[Bu^i_3AlAlBu^i_3]^{2-}$.[25]

Al_2R_6 (or AlR_3) react readily with ligands to form adducts, $LAlR_3$, and with protonic reagents to liberate alkanes:

$$Al_2R_6 + 6HX \longrightarrow 6RH + 2AlX_3 \quad (X = OH, OR, Cl, Br)$$

Reaction with halides or alkoxides of elements less electropositive than Al affords a useful route to other organometallics:

$$MX_n \xrightarrow{\text{excess AlR}_3} MR_n + \tfrac{n}{3}AlX_3 \quad (M = B, Ga, Si, Ge, Sn, etc.)$$

The main importance of organoaluminium compounds stems from the crucial discovery of alkene insertion reactions by K. Ziegler,[26] and an industry of immense proportions based on these reactions has developed during the past 20–25 y. Two main processes must be distinguished: (a) "growth reactions" to synthesize unbranched long-chain primary alcohols and alkenes (K. Ziegler *et al.*, 1955), and (b) low-pressure polymerization of ethene and propene in the presence of organometallic mixed catalysts

---

[†] It is interesting to note that the reaction of EtI with Al metal to give the sesqui-iodide "$Et_3Al_2I_3$" was the first recorded preparation of an organoaluminium compound (W. Hallwachs and A. Schafarik, 1859).

[25] H. Hoberg and S. Krause, Dipotassium hexaisobutyldialuminate, a complex containing an Al–Al bond, *Angew. Chem.*, Int. Ed. (Engl.) **17**, 949–50 (1979).

[26] K. Ziegler, A forty years' stroll through the realms of organometallic chemistry, *Adv. Organometallic Chem.* **6**, 1–17 (1968).

(1955) for which K. Ziegler (Germany) and G. Natta (Italy) were jointly awarded the Nobel Prize for Chemistry in 1963.

In the first process alkenes insert into the Al–C bonds of monomeric AlR_3 at $\sim 150°$ and 100 atm to give long-chain derivatives whose composition can be closely controlled by the temperature, pressure, and contact time:

$$AlEt_3 \xrightarrow{C_2H_4} Et_2AlCH_2Et \xrightarrow{nC_2H_4} Al \begin{array}{l} (C_2H_4)_x Et \\ (C_2H_4)_y Et \\ (C_2H_4)_z Et \end{array}$$

The reaction is thought to occur by repeated η^2-coordination of the alkene molecules to Al followed by migration of an alkyl group from Al to the alkene carbon atom:

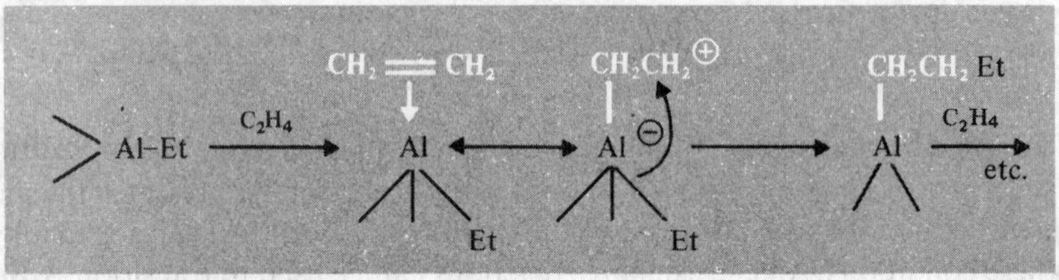

Typically, chain lengths of 14–20 C atoms can be synthesized industrially in this way and then converted to unbranched aliphatic alcohols for use in the synthesis of biodegradable detergents:

$$Al(CH_2CH_2R)_3 \xrightarrow{(i) O_2, (ii) H_3O^+} 3RCH_2CH_2OH$$

Alternatively, thermolysis yields the terminal alkene $RCH=CH_2$.

Even more important is the stereoregular catalytic polymerization of ethene and other alkenes to give high-density polyethene ("polythene") and other plastics. A typical Ziegler–Natta catalyst can be made by mixing $TiCl_4$ and Al_2Et_6 in heptane: partial reduction to Ti^{III} and alkyl transfer occur, and a brown suspension forms which rapidly absorbs and polymerizes ethene even at room temperature and atmospheric pressure. Typical industrial conditions are 50–150°C and 10 atm. Polyethene produced at the surface of such a catalyst is 85–95% crystalline and has a density of 0.95–0.98 g cm^{-3} (compared with low-density polymer 0.92 g cm^{-3}); the product is stiffer, stronger, has a higher resistance to penetration by gases and liquids, and has a higher softening temperature (140–150°). Polyethene is produced in megatonne quantities and used mainly in the form of thin film for packaging or as molded articles, containers, and bottles; electrical insulation is another major application. Stereoregular (isotactic) polypropene and many copolymers of ethene are also manufactured. Much work has been done in an attempt to elucidate the chemical nature of the catalysts and the mechanism of their action; the active site may differ in detail from system to system but there is now general agreement that polymerization is initiated by η^2 coordination of ethene to the partly alkylated lower-valent transition-metal atom (e.g. Ti^{III}) followed by migration of the attached alkyl group from transition-metal to carbon.(An alternative suggestion involves metal–carbene species generated by α-hydrogen transfer.[27])

Coordination of the ethene or propene to Ti^{III} polarizes the C–C bond and allows ready migration of the alkyl group with its bonding electron-pair. This occurs as a concerted process, and transforms the η^2-alkene into a σ-bonded alkyl group. As much as 1 tonne of polypropylene can be obtained from as little as 5 g Ti in the catalyst.

Organometallic compounds of Ga, In, and Tl have been less studied than their Al analogues. The trialkyls do not dimerize and there is a general tendency to diminishing thermal stability with increasing atomic weight of M. There is also a general decrease of chemical reactivity of the M–C bond in the sequence Al > Ga ≈ In > Tl, and this is particularly noticeable for compounds of the type R_2MX; indeed, Tl gives air-stable non-hydrolysing ionic derivatives of the type $[TlR_2]X$, where X = halogen, CN, NO_3, $\frac{1}{2}SO_4$, etc. For example, the ion $[TlMe_2]^+$ is stable in aqueous solution, and is linear like the isoelectronic $HgMe_2$ and $[PbMe_2]^{2+}$.

GaR_3 can be prepared by alkylating Ga with HgR_2 or by the action of RMgBr or AlR_3 on $GaCl_3$. They are low-melting, mobile, flammable liquids. The corresponding In and Tl compounds are similar but tend to have higher mps and bps; e.g.

Compound	GaMe$_3$	InMe$_3$	TlMe$_3$	GaEt$_3$	InEt$_3$	TlEt$_3$
MP	−16°	88.4°	38.5°	−82°	—	−63°
BP	56°	136°	147° (extrap)	143°	84°/12 mmHg	192° (extrap)

The triphenyl analogues are also monomeric in solution but tend to associate into chain structures in the crystalline state as a result of weak intermolecular M⋯C interactions: GaPh$_3$ mp 166°, InPh$_3$ mp 208°, TlPh$_3$ mp 170°. For Ga and In compounds the primary M–C bonds can be cleaved by HX, X_2, or MX_3 to give reactive halogen-bridged dimers $(R_2MX)_2$. This contrasts with the unreactive ionic compounds of Tl mentioned above, which can be prepared by suitable Grignard reactions:

$$TlX_3 + 2RMgX \longrightarrow [TlR_2]X + 2MgX_2$$

[27] M. L. H. GREEN, Studies on the synthesis, mechanism, and reactivity of some organo-molybdenum and -tungsten compounds, *Pure Appl. Chem.* **50**, 27–35 (1978).

A few organometallic compounds of Tl^I are also known. Thus TlC_5H_5 precipitates as air-stable yellow crystals when aqueous TlOH is shaken with cyclopentadiene. In the gas phase $Tl(\eta^5\text{-}C_5H_5)$ has C_{5v} symmetry (Fig. 7.20), whereas in the crystalline phase there are zigzag chains of equispaced alternating C_5H_5 rings and Tl atoms. $In(\eta^5\text{-}C_5H_5)$ is less stable but has the same structure as the Tl compound in each phase. It is best prepared by metathesis of LiC_5H_5 with a slurry of InCl in Et_2O.[27a] The bonding can be described in MO terms (p. 369).

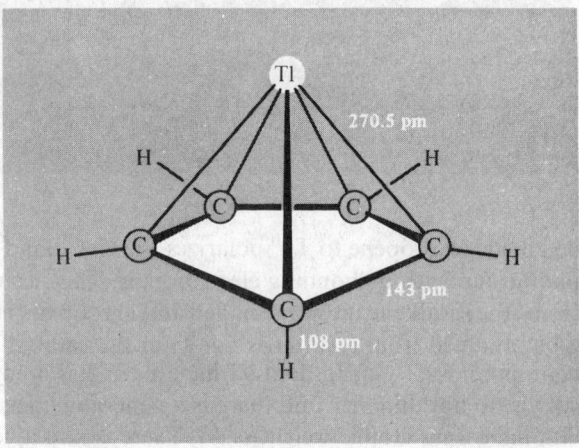

FIG. 7.20 The C_{5v} structure of $Tl(\eta^5\text{-}C_5H_5)$ in the gas phase as revealed by microwave spectroscopy; the Tl atom is 241 pm above the plane of the ring. $In(\eta^5\text{-}C_5H_5)$ is similar with In–C 262 pm (gas-phase electron diffraction); the In atom is 231 pm above the plane of the ring and this distance increases considerably to 319 pm in the crystal.

Finally, attention should be drawn to a remarkable range of heterocyclic and cluster organoaluminium compounds containing various sequences of Al–N bonds[28] (cf. B–N compounds, p. 234). Thus the adduct $[AlMe_3(NH_2Me)]$ decomposes at $70°C$ with loss of methane to give the cyclic amido trimers *cis*- and *trans*-$[Me_2AlNHMe]_3$ (structures 2 and 3) and at $215°$ to give the oligomeric imido cluster compounds $(MeAlNMe)_7$ (structure 6) and $(MeAlNMe)_8$ (structure 7), e.g.:

$$21AlMe_3 + 21NH_2Me \xrightarrow[-21CH_4]{70} 7(Me_2AlNHMe)_3 \xrightarrow[-21CH_4]{215} 3(MeAlNMe)_7$$

Similar reactions lead to other oligomers depending on the size of the R groups and the conditions of the reaction, e.g. *cyclo*-$(Me_2AlNMe_2)_2$ (structure 1) and the imido-clusters $(PhAlNPh)_4$, $(HAlNPr^i)_{4\text{ or }6}$, $(HAlNPr^n)_{6\text{ or }8}$, $(HAlNBu^t)_4$, and $(MeAlNPr^i)_{4\text{ or }6}$ (see structures 4, 5, 7). Intermediate amido-imido compounds have also been isolated from the reaction, e.g. $[Me_2AlNHMe)_2(MeAlNMe)_6]$ (structure 8). All the structures are built up from varying numbers of fused 4-membered and 6-membered AlN heterocycles.

[27a] C. PEPPE, D. G. TUCK and L. VICTORIANO, A simple synthesis of cyclopentadienylindium(I), *JCS Dalton* 1981, 2592.

[28] S. AMIRKHALILI, P. B. HITCHCOCK, and J. D. SMITH, Complexes of organoaluminium compounds. Part 10, *JCS Dalton*, 1979, 1206–12; and references 1–9 therein.

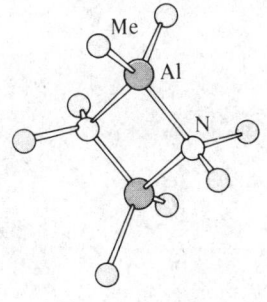

(1) (Me$_2$AlNMe$_2$)$_2$

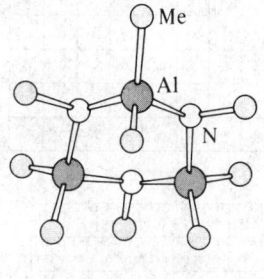

(2) *cis*-(Me$_2$AlNHMe)$_3$

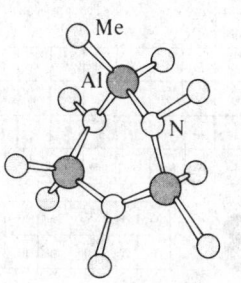

(3) *trans*-(Me$_2$AlNHMe)$_3$

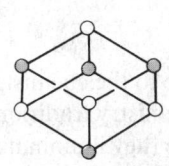

(4) (MeAlNPr^i)$_4$

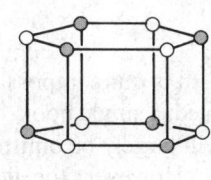

(5) (HAlNPr^i)$_6$

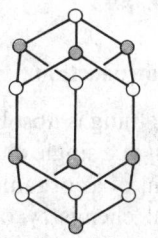

(6) (MeAlNMe)$_7$

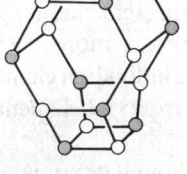

(7) (MeAlNMe)$_8$

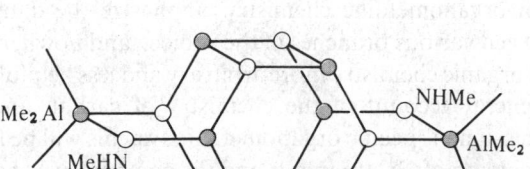

(8) [(Me$_2$AlNHMe)$_2$(MeAlNMe)$_6$]

Several analogous gallium compounds are also known, e.g. [(Me$_2$GaNHMe)$_2$-(MeGaNMe)$_6$] (structure 8).[28]

8

Carbon

8.1 Introduction

One thing is absolutely certain—it is quite impossible to do justice to the chemistry of carbon in a single chapter; or, indeed, a single book. The areas of chemistry traditionally thought of as organic chemistry will largely be omitted except where they illuminate the general chemistry of the element. However, the field of organometallic chemistry is included since this has been one of the most rapidly developing areas of the subject during the past 25 y and has led to major advances in our understanding of the structure, bonding, and reactivity of molecular compounds. In fact, the unifying concepts emerging from organometallic chemistry emphasize the dangers of erecting too rigid a barrier between various branches of the subject, and nowhere is the boundary between inorganic and organic chemistry more arbitrary and less helpful than here. The present chapter gives a general account of the chemistry of carbon and its compounds; a more detailed discussion of specific organometallic systems will be found under the individual elements. Discussion of Group trends and the comparative chemistry of the Group IV elements C, Si, Ge, Sn, and Pb is deferred until Chapter 10.

Carbon was known as a substance in prehistory (charcoal, soot) though its recognition as an element came much later, being the culmination of several experiments in the eighteenth century.[1] Diamond and graphite were known to be different forms of the element by the close of the eighteenth century, and the relationship between carbon, carbonates, carbon dioxide, photosynthesis in plants, and respiration in animals was also clearly delineated by this time (see Panel). The great upsurge in synthetic organic chemistry began in the 1830s and various structural theories developed following the introduction of the concept of valency in the 1850s. Outstanding achievements in this area were F. A. Kekulé's use of structural formulae for organic compounds and his concept of the benzene ring, L. Pasteur's work on optical activity, and the concept of tetrahedral carbon (J. H. van't Hoff and J. A. Le Bel). The first metal carbonyl compounds $Ni(CO)_4$ and $Fe(CO)_5$ were prepared and characterized by L. Mond and his group in 1889–91 and this work has burgeoned into the huge field of metal carbonyl cluster compounds which is still producing results of fundamental importance. Even more extensive is the field of organometallic chemistry which developed rapidly after the seminal papers on the "sandwich" structure of ferrocene (G. Wilkinson, M. Rosenblum, M. C. Whiting, and

---

[1] M. E. WEEKS, *Discovery of the Elements*, Chaps. 1 and 2, pp. 58–89. J. Chem. Educ. Publ., 1956.

Early History of Carbon and Carbon Dioxide

—	Carbon known as a substance in prehistory (charcoal, soot) but not recognized as an element until the second half of the eighteenth century.
BC	"Indian inks" made from soot used in the oldest Egyptian hieroglyphs on papyrus.
AD1273	Ordinance prohibiting use of coal in London as prejudicial to health—the earliest known attempt to reduce smoke pollution in Britain.
~1564	Lead pencils first manufactured commercially during Queen Elizabeth's reign, using Cumberland graphite.
1752/4	CO_2 ("fixed air"), prepared by Joseph Black (aged 24–26), was the first gas other than air to be characterized: (i) chalk when heated lost weight and evolved CO_2 (genesis of quantitative gravimetric analysis), and (ii) action of acids on carbonates liberates CO_2.
1757	J. Black showed that CO_2 was produced by fermentation of vegetables, by burning charcoal, and by animals (humans) when breathing; turns lime water turbid.
1771	J. Priestley established that green plants use CO_2 and "purify air" when growing. He later showed that the "purification" was due to the new gas O_2 (1774).
1779	Elements of photosynthesis elucidated by J. Ingenhousz: green plants in daylight use CO_2 and evolve O_2; in the dark they liberate CO_2.
1789	The word "carbon" (Fr. *carbone*) coined by A. L. Lavoisier from the Latin *carbo*, charcoal. The name "graphite" was proposed by A. G. Werner and D. L. G. Harsten in the same year: Greek γραφίεν (*graphein*), to write. The name "diamond" is probably a blend of Greek διαφανής (*diaphanes*), transparent, and αδαμας (*adamas*), indomitable or invincible, in reference to its extreme hardness.
1796	Diamond shown to be a form of carbon by S. Tennant who burned it and weighed the CO_2 produced; graphite had earlier been shown to be carbon by C. W. Scheele (1779); carbon recognized as essential for converting iron to steel (R.-A.-F de Réaumur and others in the late eighteenth century).
1805	Humphry Davy showed carbon particles are the source of luminosity in flames (lamp black).

R. B. Woodward, 1952) and the "π bonding" of ethylene complexes (M. J. S. Dewar 1951, J. Chatt, and L. A. Duncanson, 1953). The constricting influence of classical covalent-bond theory was finally overcome when it was realized that carbon in many of its compounds can be 5-coordinate (Al_2Me_6, p. 289), 6-coordinate ($C_2B_{10}H_{12}$, p. 207), or even 7-coordinate (Li_4Me_4, p. 113). A compound featuring an 8-coordinate carbon atom is shown on p. 358. In parallel with these developments in synthetic chemistry and bonding theory have been technical and instrumental advances of great significance; foremost amongst these have been the development of ^{14}C radioactive dating techniques (W. F. Libby, 1949), the commercial availability of ^{13}C nmr instruments in the early 1970s, and the industrial production of artificial diamonds (General Electric Company, 1955).

8.2 Carbon

8.2.1 *Terrestrial abundance and distribution*

Carbon occurs both as the free element (graphite, diamond) and in combined form (mainly as the carbonates of Ca, Mg, and other electropositive elements). It also occurs as CO_2, a minor but crucially important constituent of the atmosphere. Estimates of the overall abundance of carbon in crustal rocks vary considerably, but a value of 180 ppm can be taken as typical; this places the element seventeenth in order of abundance after Ba, Sr, and S but before Zr, V, Cl, and Cr.

Some Notable Dates in Carbon Chemistry

1807   J. J. Berzelius classified compounds as "organic" or "inorganic" according to their origin in living matter or inanimate material.

1825–7 W. C. Zeise prepared $K[Pt(C_2H_4)Cl_3]$ and related compounds; though of unknown structure at the time they later proved to be the first organometallic compounds.

1828   The vitalist theory of Berzelius challenged by F. Wöhler (aged 28) who synthesized urea, $(NH_2)_2CO$, from $NH_4(OCN)$.

1830+   Rise of synthetic organic chemistry.

1848   L. Pasteur (aged 26) began work on optically active sodium ammonium tartrate.

1849   First metal alkyls, e.g. $ZnEt_2$, made by E. Frankland (aged 24); he also first propounded the theory of valency (1852).

1858   F. A. Kekulé's structural formulae for organic compounds; ring structure of benzene 1865.

1874   Tetrahedral, 4-coordinate carbon proposed independently by J. H. van't Hoff (aged 22) and J. A. LeBel.

1890   First publication on metal carbonyls $[Ni(CO)_4]$ by L. Mond *et al.*

1891   Carborundum, SiC, made by E. G. Acheson.

1900   First paper by V. Grignard on RMgX syntheses. Nobel Prize 1912.

1924   Solid CO_2 introduced commercially as a refrigerant.

1926 C_8K prepared—the first alkali metal-graphite intercalation compound.

1929   Isotopes of C (^{12}C and ^{13}C) discovered by A. S. King and R. T. Birge in the band spectrum of C_2, CO, and CN (previously undetected by mass spectrometry).

1932   First metal halide-graphite intercalation compound made with $FeCl_3$.

1936   Radiocarbon $^{14}_6C^*$ established as the product of an (n,p) reaction on $^{14}_7N$ by W. E. Burcham and M. Goldhaber.

1940   Chemically significant amounts of ^{14}C synthesized by S. Ruben and M. D. Kamen.

1947–9 Concept and feasibility of ^{14}C dating established by W. F. Libby (awarded Nobel Prize in 1960).

1952   Structure of ferrocene elucidated; organometallic chemistry burgeons: Nobel Prize awarded jointly to E. O. Fischer and G. Wilkinson 1973.

1953   First authentic production of artificial diamonds by ASEA, Sweden; commercial production achieved by General Electric (USA) in 1955.

1955   Stereoregular polymerization of ethene and propene by catalysts developed by K. Ziegler and by G. Natta (Nobel Prize 1963).

1956   Cyclobutadiene–transition metal complexes predicted by H. C. Longuet-Higgins and L. E. Orgel 3 y before they were first synthesized.

1960 π-allylic metal complexes first recognized.

1961 $^{12}C = 12$ adopted as unified atomic weight standard by both chemists and physicists.

1964   6-coordinate carbon established in various carboranes by W. N. Lipscomb and others. (Nobel Prize 1976.)

1966 CS_2 complexes such as $[Pt(CS_2)(PPh_3)_2]$ first prepared in G. Wilkinson's laboratory.

1971 ^{13}C fourier-transform nmr commercially available following first observation of ^{13}C nmr signal by P. C. Lauterbur and by C. H. Holm in 1957.

1976   8-coordinate carbon established in $[Co_8C(CO)_{18}]^{2-}$ by V. G. Albano, P. Chini *et al.* (Cubic coordination of C in antifluorite Be_2C known since 1948.)

Graphite is widely distributed throughout the world though much of it is of little economic importance.[2] Large crystals or "flake" occur in metamorphosed sedimentary silicate rocks such as quartz, mica schists, and gneisses; crystal size varies from fractions of a millimetre up to about 6 mm (average ~ 4 mm) and the deposits form lenses up to 30 m thick stretching several kilometres across country. Average carbon content is 25% but can rise as high as 60% (Malagasy). Beneficiation is by flotation followed by treatment with

---

[2] C. L. MANTELL, *Carbon and Graphite Handbook*, Interscience, New York, 1968, 538 pp.

HF and HCl, and then by heating to 1500°C *in vacuo*. Microcrystalline graphite (sometimes referred to as "amorphous") occurs in carbon-rich metamorphosed sediments and some deposits in Mexico contain up to 95% C. World production has remained fairly constant for the past 15 y and was 440 ktonnes in 1975 (see Panel).

Diamonds are found in ancient volcanic pipes embedded in a relatively soft, dark-coloured basic rock called "blue ground" or "kimberlite", from the South African town of Kimberly where such pipes were first discovered in 1870.[3] Diamonds are also found in alluvial gravels and marine terraces to which they have been transported over geological ages by the weathering and erosion of pipes. The original mode of formation of the diamond crystals is still a subject of active investigation. The diamond content of a typical kimberlite pipe is extremely low, of the order of 1 part in 15 million, and the mineral must be isolated mechanically by crushing, sluicing, and passing the material over greased belts to which the diamonds stick. This, in part, accounts for the very high price of gem-quality

---

Production and Uses of Graphite[4]

There is a world shortage of natural graphite which is particularly marked in North America and Europe. As a result, prices have risen steeply; they vary widely in the range $200–1600 per tonne (1973) depending on crystalline quality; "amorphous" graphite is ~$30 per tonne. The annual world production of 440 ktonnes was distributed as follows in 1975:

	USSR	N. Korea	Mexico	S. Korea	China	Austria
ktonnes	90	77	61	47	50	30.6

	India	Malagasy	FRG	Norway	Sri Lanka	Other
ktonnes	18.9	17.8	16.3	10.0	10.4	11.0

The USA used 86 ktonnes of natural graphite of which 75 ktonnes was imported; in addition, some 293 ktonnes of graphite was manufactured. Natural graphite is used in steelmaking (33%), foundries (18%), refractories (17%), crucibles, retorts, nozzles, etc. (8%), lubricants (5.8%), brake linings (2.8%), carbon products (1.8%), pencils (1.6%), and miscellaneous (12%).

Artificial graphite was first manufactured on a large scale by A. G. Acheson in 1896. In this process coke is heated with silica at ~2500°C for 25–35 h:

$$SiO_2 + 2C \xrightarrow{-2CO} \{SiC\} \xrightarrow{2500°} Si(g) + C(graphite)$$

In the USA artificial graphite is now made on a scale of 325 ktonnes pa (1977) and is used mainly for electrodes (65%), crucibles and vessels (6%), and various unmachined shapes (6%); specialist uses include motor brushes and contacts, fibres and cloth (Pl. 8.1), and refractories of various sorts. Ultra-high-purity graphite is made on a substantial scale for use as a neutron moderator in nuclear reactors. Carbon whiskers grown from highly purified graphite are finding increasing use in high-strength composites; the whiskers are manufactured by striking a carbon arc at 3600°C under 90 atm Ar—the maximum length is ~50 mm and the average diameter 5 μm.

---

[3] E. BRUTON, *Diamonds*, NAG Press, London, 1970, 372 pp.

[4] *Kirk–Othmer Encyclopedia of Chemical Technology*, 3rd edn., Interscience, New York, 1978, Vol. 4: Carbon and artificial graphite, pp. 556–631; Carbon black, pp. 631–66; Diamond, natural and synthetic, pp. 666–89; Natural graphite, pp. 689–709.

diamonds which was, in 1974, ~$850 per g for rough or uncut specimens, i.e. about 1 million times the price of flake graphite. The pattern of world production has changed dramatically over the past few years as indicated in the Panel.

Three other forms of carbon are manufactured on a vast scale and used extensively in industry: coke, carbon black, and activated carbon. The production and uses of these impure forms of carbon are briefly discussed in the Panel on p. 301.

Production and Uses of Natural Diamond[4]

Gemstone diamonds have been greatly prized in eastern countries for over 2000 y though their introduction and recognition in Europe is more recent. The only sources were from India and Borneo until they were also found in Brazil in 1729. In South Africa diamonds were discovered in alluvial deposits in 1867 and the first kimberlite pipe was identified in 1870 with dramatic consequences. A typical pipe is shown in Pl. 8.2. Other finds of economic importance were made in Africa during the first half of this century: Congo 1907; Namibia 1908; Central African Republic 1913; Tanzania 1913 (where large-scale production began in 1940 following the discovery of the enormous Williamson pipe—still the largest in the world and covering an area of 1.4 km^2); Angola 1916; Ghana 1920; and Guinea, Sierra Leone, Liberia, and the Ivory Coast in the 1930s. During the 1950s 99% of the world output of diamonds was from Africa but then the USSR began to emerge as a major producer. This followed the discovery of alluvial diamonds in Siberia in 1948 and the first kimberlite-type pipe at Yakutia later the same year.† Within a decade more than 20 pipes had been located in the great basin of the Vilyui River 4000 km east of the Urals, and Siberia was established as a major producer of both gem-quality and industrial diamonds. However, year-round production in Siberian conditions posed severe developmental problems, and production is now supplemented by newer finds in the Urals near Sverdlovsk. Impressive finds of kimberlite pipes have also been made in North-western Austrlia since 1978. The world's largest producers of diamonds in 1974 were as the Table shows:

1974 diamond production/megacarats[a]

	World	Zaire	USSR	S. Afr.	Botswana	Ghana	Angola	S. Leone	Namibia
Gem quality	12.52	1.10	1.95	3.44	0.41	0.26	1.60	0.67	1.49
Industrial	31.57	11.90	7.85	4.07	2.31	2.32	0.50	1.00	0.08
Total	44.09	13.00	9.80	7.51	2.72	2.58	2.10	1.67	1.57

[a] The weight of diamonds is usually quoted in carats; 1 carat = 0.200 g, hence, to convert megacarats to tonnes divide by 5. Note that this unit differs from the carat used to describe the quality of gold (p. 1367).

It will be seen that whereas South Africa is still the largest producer of gem-quality diamonds it has been surpassed by Zaire and the USSR in total production. Diamond is the hardest and least perishable of all minerals, and these qualities, coupled with its brilliant sparkle, which derives from its transparency and high refractive index, make it the most prized gemstone. However, most naturally occurring diamonds (and all synthetic, high-pressure diamonds) are of "industrial" rather than gem-stone quality. Diamond is used mainly in the form of grit and powder for cutting, drilling, grinding, and polishing.

By far the largest natural diamond ever found (25 January 1905) was the Cullinan; it weighed 3106 carats (621.2 g) and measured ~10 × 6.5 × 5 cm^3. Other famous stones are in the range 100–800 carats though specimens larger than 50 carats are only rarely encountered. The largest synthetic diamonds are about 1 carat. The cost of producing small, synthetic, industrial-quality diamonds is now competitive with the cost of mining them, but gem-quality diamonds are still cheaper to mine than to make.

† Smaller deposits had been known for over a century following von Humboldt's prediction that diamonds should be found in Russia.

CARBON AND GRAPHITE YARNS

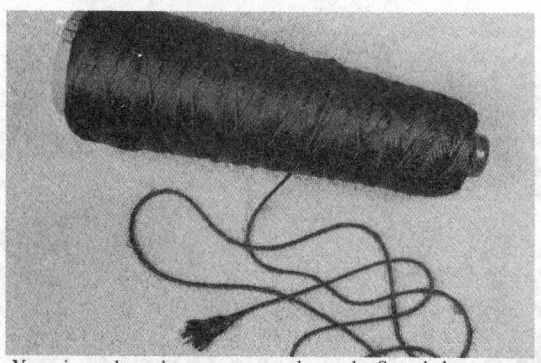

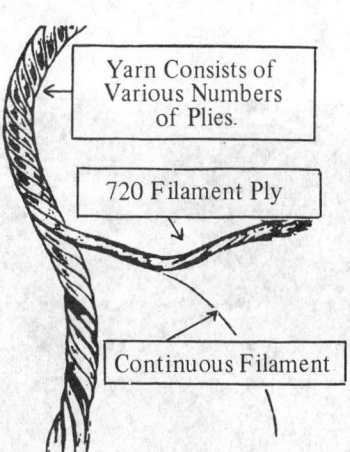

Yarn Consists of Various Numbers of Plies.

720 Filament Ply

Continuous Filament

Yarn is packaged on one pound spools. Spool shown contains 1350 feet of grade GY 2-20 yarn of 99.9% purity, featuring a breaking strength of 40 pounds

CARBON AND GRAPHITE CLOTH

PLATE 8.1. Carbon and graphite products. (From *Graphite Products*, courtesy of Carborundum Co.)

PLATE 8.2. The excavated diamond pipe at Kimberley, South Africa. (Courtesy of Anglo-American Corporation of South Africa Ltd.)

Production and Uses of Coke, Carbon Black, and Activated Carbon[4]

The high-temperature carbonization of coal yields metallurgical coke, a poorly graphitized form of carbon; most of this (92%) is used in blast furnaces for steel manufacture (p. 1244). World production of coke in 1974 exceeded 370 million tonnes and was dominated, as expected, by the large industrial nations. Carbon black (soot) is made by the incomplete combustion of liquid hydrocarbons or natural gas; the scale of operations is enormous, being more than an order of magnitude larger than for coke: world production in 1974 (excluding the USSR and China) was 3590 million tonnes (i.e. 10 million tons daily!) The particle size of carbon black is exceedingly small (0.02–0.30 μm) and its principal application (93%) is in the rubber industry where it is used to strengthen and reinforce the rubber in a way that is not completely understood. For example, each car tyre uses 3 kg carbon black and each truck tyre $\sim$9 kg. Its other main use is as a pigment in inks (2.6%), paints (0.6%), paper (0.1%), and plastics. Price in 1982 was $500–700 per tonne depending on quality.

Activated carbons, being highly specialized products, are produced on a correspondingly smaller scale (e.g. 90 000 tonnes in the USA, 1979). They are distinguished by their enormous surface area which is typically in the range 300–2000 $m^2 g^{-1}$. Activated carbon can be made either by chemical or by gas activation. In chemical activation the carbonaceous material (sawdust, peat, etc.) is mixed or impregnated with materials which oxidize and dehydrate the organic substrate when heated to 500–900°, e.g. alkali metal hydroxides, carbonates, or sulfates, alkaline earth metal chlorides, carbonates, or sulfates, $ZnCl_2$, H_2SO_4, or H_3PO_4. In gas activation, the carbonaceous matter is heated with air at low temperature or with steam, CO_2, or flue gas at high temperature (800–1000°).

Activated carbon is used extensively in the sugar industry as a decolorizing agent and this accounts for some 35% of the output; related applications are in the purification of chemicals and gases including air pollution (25%), in water and waste water treatment (30%), and as a catalyst (10%). Notable catalytic uses are the aerial oxidation in aqueous solution of Fe^{II}, $[Fe^{II}(CN)_6]^{4-}$, $[As^{III}O_3]^{3-}$, and $[N^{III}O_2]^{-}$, the manufacture of $COCl_2$ from CO and Cl_2, and the production of SO_2Cl_2 from SO_2 and Cl_2. The cost of activated carbon (USA, 1978) was $0.5–1.5 per kg depending on the grade.

In addition to its natural occurrence as the free element, carbon is widely distributed in the form of coal and petroleum, and as carbonates of the more electropositive elements (e.g. Group IA, p. 102, Group IIA, p. 119). The great bulk of carbon is immobilized in the form of coal, limestone, chalk, dolomite, and other deposits, but there is also a dynamic equilibrium as a result of the numerous natural processes which constitute the so-called carbon cycle. The various reservoirs of carbon and the flow between them are illustrated in Fig. 8.1 from which it is clear that there are two distinct cycles—one on land and one in the sea, dynamically inter-connected by the atmosphere. CO_2 in the atmosphere ($\sim 6.7 \times 10^{11}$ tonnes) accounts for only 0.003% of carbon in the earth's crust ($\sim 2 \times 10^{16}$ tonnes). It is in rapid circulation with the biosphere being removed by plant photosynthesis and added to by plant and animal respiration, and the decomposition of dead organic matter; it is also produced by the activities of man, notably the combustion of fossil fuels for energy and the calcination of limestone for cement. These last two activities have increased dramatically in recent years as shown in Fig. 8.2, and give some cause for concern. Interchange on a similar scale occurs between the atmosphere and ocean waters, and the total residence time of CO_2 in the atmosphere is $\sim$10–15 y (as measured by ^{14}C experiments).

An increase in the concentration of atmospheric CO_2 has been thought by some to expose the planet to the dangers of a "greenhouse effect" whereby the temperature is raised due to the trapping of the earth's thermal radiation by infrared absorption in the CO_2 molecules. Reliable estimates are extremely difficult to make, particularly as the resultant

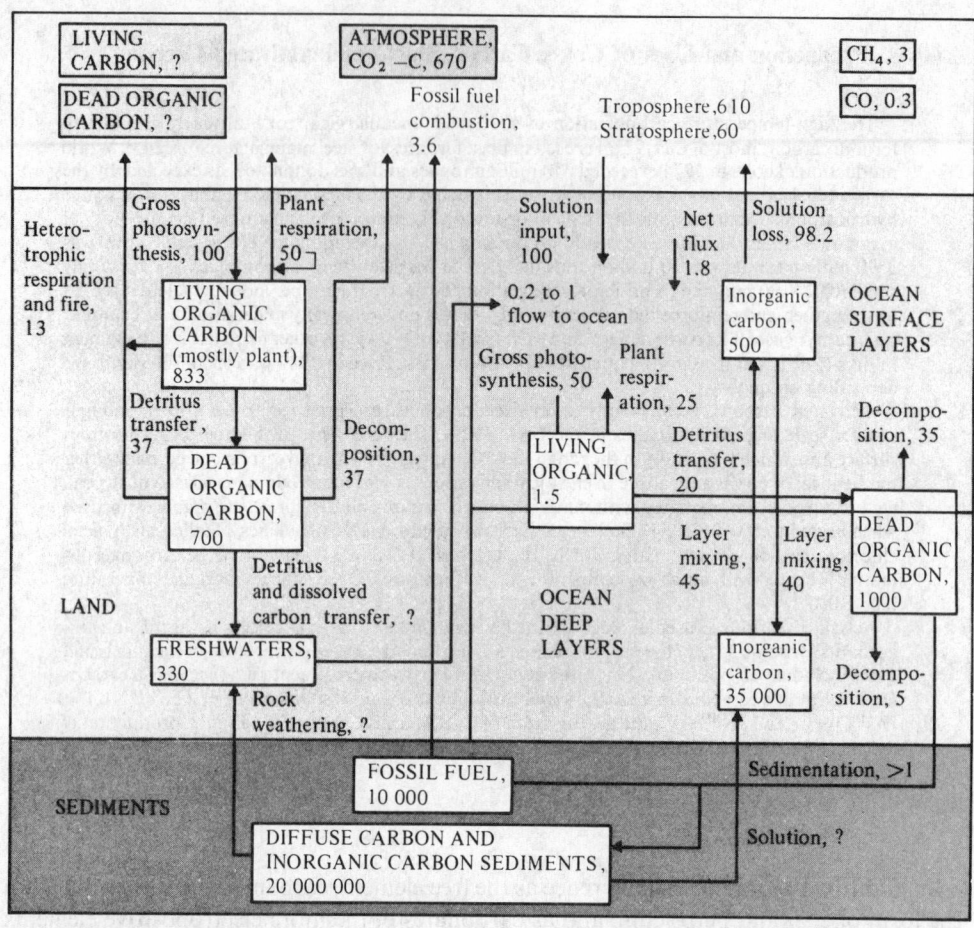

FIG. 8.1 Diagrammatic model of the global carbon cycle. Questions marks indicate that no estimates are available. Figures are in units of 10^9 tonnes of contained carbon but estimates from various sources sometimes differ by factors of 3 or more. The diagram is based on one by B. Bolin[5] modified to include more recent data.[6]

effect is small (much less than the seasonal variation due to the growth of vegetation) and is, in any case, ameliorated by other factors tending to decrease the average temperature. The net effect is unlikely to exceed 0.1° by the turn of the century, but even this may have a significant influence on the extent of the polar ice caps and the general level of oceanic waters. There has also been concern that the increased concentration of CO_2 will significantly lower the pH of surface ocean waters thereby modifying the solution properties of $CaCO_3$ with potentially disastrous consequences to marine life. Informed opinion now discounts such global catastrophes but there has undoubtedly been a measurable perturbation of the carbon cycle in the last 2–3 decades, and the prudent

[5] B. BOLIN, The carbon cycle, *Scientific American*, September 1970, reprinted in *Chemistry in the Environment*, pp. 53–61, W. H. Freeman, San Francisco, 1973.

[6] *SCOPE Report 10 on Environmental Issues*, Carbon, pp. 55–58, Wiley, New York, 1977. SCOPE is the Scientific Committee on Problems of the Environment; it reports to ICSU, the International Council of Scientific Unions.

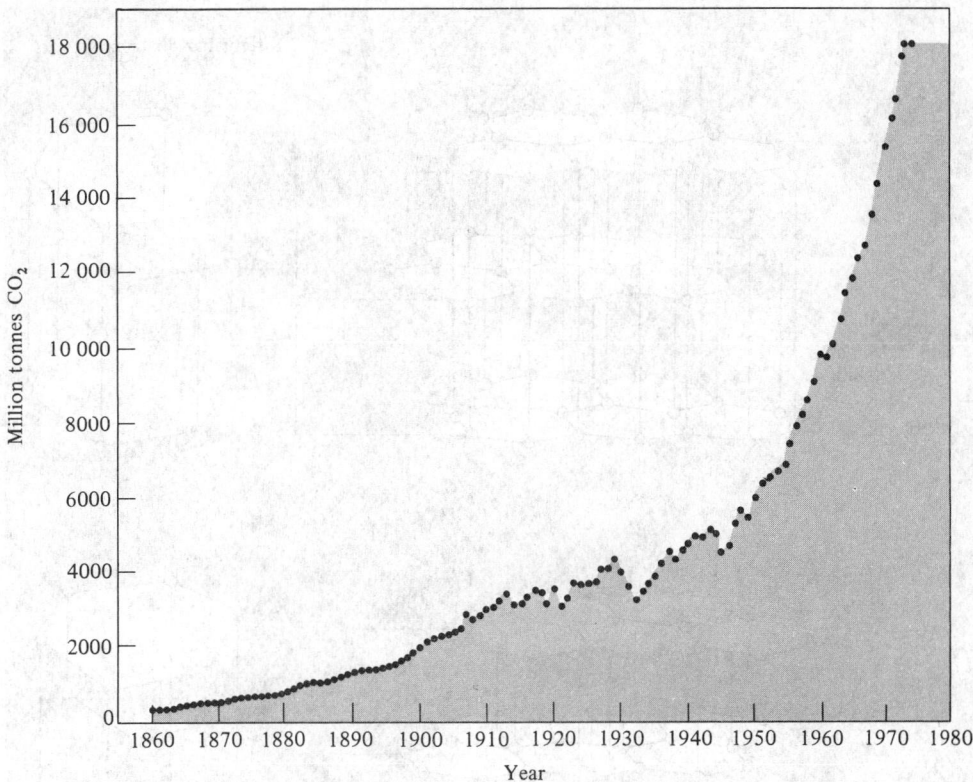

FIG. 8.2 Annual production of CO_2 from combustion of fossil fuels and calcination of limestone for cement.[6]

course is to conserve resources, minimize wasteful practices, and improve efficiency, whilst simultaneously collecting reliable data on the magnitude of the various carbon-containing reservoirs and the rates of transfer between them.[6a]

8.2.2 *Allotropic forms*

Carbon can exist in at least 6 crystalline forms: α- and β-graphite, diamond, Lonsdaleite (hexagonal diamond), chaoite, and carbon(VI). Of these, α- (or hexagonal) graphite is thermodynamically the most stable form at normal temperatures and pressures. The various modifications differ either in the coordination environment of the carbon atoms or in the sequence of stacking of layers in the crystal. These differences have a profound effect on both the physical and the chemical properties of the element.

Graphite is composed of planar hexagonal nets of carbon atoms as shown in Fig. 8.3. In normal α- (or hexagonal) graphite the layers are arranged in the sequence $\cdots ABAB \cdots$ with carbon atoms in alternate layers vertically above each other, whereas in β- (or

[6a] T. M. SUGDEN (Chairman), *Pollution in the Atmosphere*, Final Report of a Royal Society Study Group, June 1978, The Royal Society, London, 25 pp. See also B. BOLIN, E. T. DEGENS, S. KEMPE, and P. KETNER (eds.), The global carbon cycle, *SCOPE* **13**, Wiley, Chichester 1979, 491 pp.

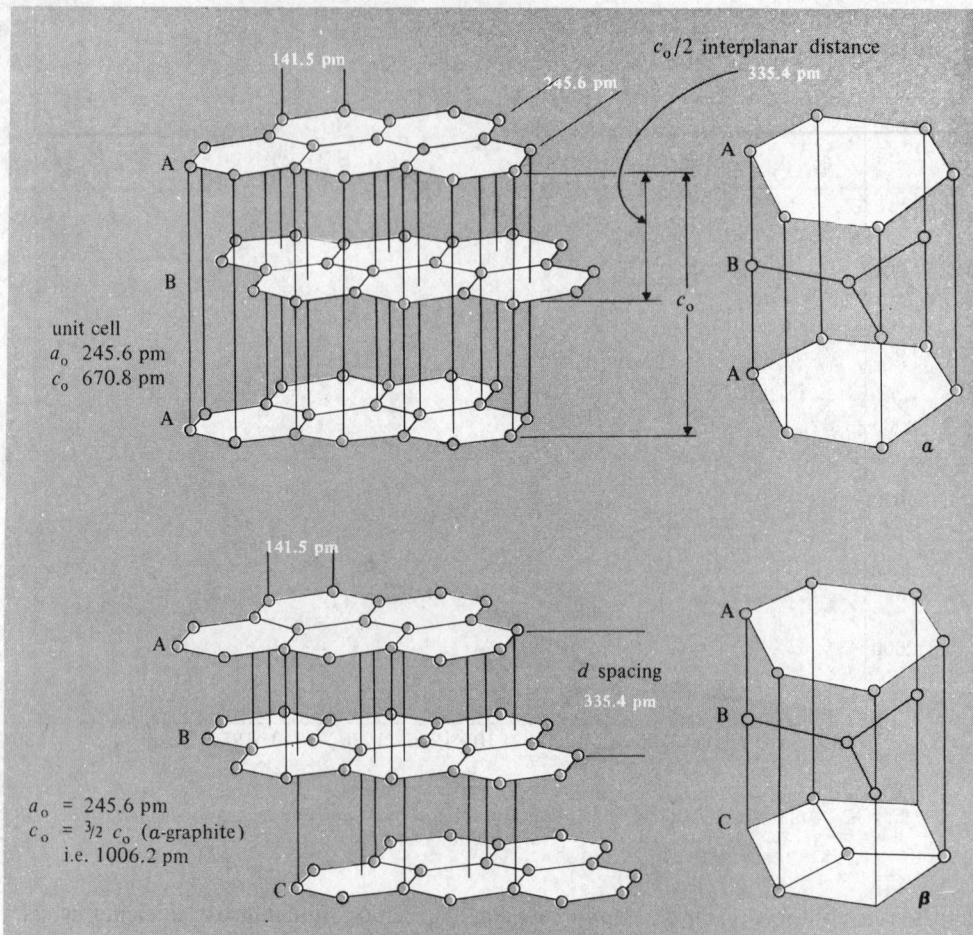

FIG. 8.3　Structure of the α (hexagonal) and β (rhombohedral) forms of graphite.

rhombohedral) graphite the stacking sequence is $\cdots$ABCABC$\cdots$. In both forms the C–C distance within the layer is 141.5 pm and the interlayer spacing is much greater, 335.4 pm. The two forms are interconvertible by grinding ($\alpha \rightarrow \beta$) or heating above 1025°C ($\beta \rightarrow \alpha$), and partial conversion leads to an increase in the average spacing between layers; this reaches a maximum of 344 pm for turbostratic graphite in which the stacking sequence of the parallel layers is completely random. The enthalpy difference between α- and β-graphite is only 0.59 ± 0.17 kJ mol^{-1}.

In diamond, each C atom is tetrahedrally surrounded by 4 equidistant neighbours at 154.45 pm, and the tetrahedra are arranged to give a cubic unit cell with a_0 356.68 pm as in Fig. 8.4. Note that, although the diamond structure itself is not close-packed, it is built up of 2 interpenetrating fcc lattices which are off-set along the body diagonal of the unit cell by one-quarter of its length. Nearly all naturally occurring diamonds are of this type but contain, in addition, a small amount of nitrogen atoms (0.05–0.25%) in platelets of approximate composition C_3N (type Ia) or, very occasionally, dispersed throughout the

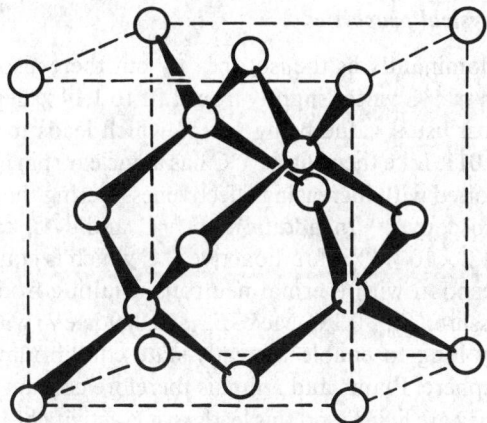

FIG 8.4   Structure of diamond showing the tetrahedral coordination of C; the dashed lines indicate the cubic unit cell containing 8 C atoms.

crystal (type Ib). A small minority of natural diamonds contain no significant amount of N (type IIa) and a very small percentage of these (including the highly valued blue diamonds, type IIb), contain Al. The exceedingly rare hexagonal modification of diamond, Lonsdaleite, was first found in the Canyon Diablo Meteorite, Arizona, in 1967: each C atom is tetrahedrally coordinated but the tetrahedra are stacked so as to give a hexagonal wurtzite-like lattice (p. 1405) rather than the cubic sphalerite-type lattice (p. 1405) of normal diamond. Lonsdaleite can be prepared at room temperature by static pressure along the c-axis of a single crystal of α-graphite, though it must be heated above 1000° under pressure to stabilize it (a_0 252 pm, c_0 412 pm, d_{obs} 3.3 g cm^{-3}, d_{calc} 3.51 g cm^{-3}).

Two other crystalline forms of carbon have been discovered during the past decade. Chaoite, a new white allotrope, was first found in shock-fused graphitic gneiss from the Ries Crater, Bavaria, in 1968; it can be synthesized artificially as white dendrites of hexagonal symmetry by the sublimation etching of pyrolytic graphite under free vaporization conditions above $\sim 2000°$C and at low pressure ($\sim 10^{-4}$ mmHg). The crystals were only 0.5 μm thick and 5–10 μm long and had a_0 894.5 pm, c_0 1407.1 pm and d_{calc} 3.43 g cm^{-3}. Finally, a new hexagonal allotrope, carbon(VI), was obtained in 1972 together with chaoite when graphitic carbons were heated resistively or radiatively at $\sim 2300°$C under any pressure of argon in the range 10^{-4} mmHg to 1 atm; laser heating was even more effective (a_0 533 pm, c_0 1224 pm, $d > 2.9$ g cm^{-3}). The detailed crystal structures of chaoite and carbon(VI) have not yet been determined but they appear to be based on a carbyne-type motif $-C\equiv C-C\equiv C$;[6b] both are much more resistant to oxidation and reduction than graphite is and their properties are closer to those of diamond. Indeed, it now seems likely that there is a sequence of at least 6 stable carbyne allotropes in the region between stable graphite (2300°C) and the mp of carbon ($\sim 3500°$C at 2×10^4 Pa)—(see p. 308) [10^5 Pa = 1 bar = 0.987 atm].

The structural differences between graphite and diamond are reflected in their differing physical and chemical properties, as outlined in the two following sections.

[6b] A. G. WHITTAKER, Carbon: a new view of its high-temperature behaviour, *Science* **200**, 763–4 (1978). See also Anon, Natural carbynes (are) less rare than (previously) thought, *Chem. Eng. News*, 29 Sept., p. 12 (1980).

8.2.3 *Atomic and physical properties*

Carbon occurs predominantly as the isotope ^{12}C but there also is a small amount of ^{13}C; the concentration of ^{13}C varies slightly from 1.01 to 1.14% depending on the source of the element, the most usual value being 1.11% which leads to an atomic weight for "normal" carbon of 12.011. Like the proton, ^{13}C has a nuclear spin quantum number $I = \frac{1}{2}$, and this has been exploited with increasing effectiveness during the past decade in fourier transform nmr spectroscopy.[6c] In addition to ^{12}C and ^{13}C, carbon dioxide in the atmosphere contains $1.2 \times 10^{-10}\%$ of radioactive ^{14}C which is continually being formed by the $^{14}_{7}N(n,p)^{14}_{6}C$ reaction with thermal neutrons resulting from cosmic ray activity. ^{14}C decays by β^- emission (E_{max} 0.156 MeV, E_{mean} 0.049 MeV) with a half-life of 5730 y, and this is sufficiently long to enable a steady-state equilibrium concentration to be established in the biosphere. Plants and animals therefore contain $1.2 \times 10^{-10}\%$ of their carbon as ^{14}C whilst they are living, and this leads to a β-activity of 15.3 counts per min per gram of contained C. However, after death the dynamic interchange with the environment ceases and the ^{14}C concentration decreases exponentially. This is the basis of W. F. Libby's elegant radio-carbon dating technique for which he was awarded the Nobel Prize for Chemistry in 1960. It is particularly valuable for archeological dating.[6d] The practical limit of the technique is about 50 000 y since by this time the ^{14}C activity has fallen to about 0.2% of its original value and becomes submerged in the background counts. ^{14}C is also extremely valuable as a radioactive tracer for mechanistic studies using labelled compounds, and many such compounds, particularly organic ones, are commercially available (p. 329).

Carbon is the sixth element in the periodic table and its ground-state electronic configuration is $[He]2s^2 2p^2$. The first 4 ionization energies of C are 1086.1, 2351.9, 4618.8, and 6221.0 kJ mol^{-1}, all much higher than those for the other Group IV elements Si, Ge, Sn, and Pb (p. 431). Excitation energies from the ground-state to various low-lying electron configurations of importance in valence theory are also well established:

Configuration	$2s^2 2p^2$	$2s^2 2p^2$	$2s^2 2p^2$	$2s^1 2p^3$	$2s^1 2p^3$
Term symbol	^3P	^1D	^1S	$^5S°$	$^5S_{valence\ state}$
Energy/kJ mol^{-1}	0.000	121.5	258.2	402.3	~632

Of these, all are experimentally observable except the $^5S_{valence\ state}$ level which is a calculated value for a carbon atom with 4 unpaired and uncorrelated electron spins; this is a hypothetical state, not amenable to experimental observation, but is helpful in some discussions of bond energies and covalent bonding theory.

The electronegativity of C is 2.5, which is fairly close to the values for other members of the group (1.8–1.9) and for several other elements: B, As (2.0); H, P (2.1); Se (2.4); S, I (2.5); many of the second- and third-row transition metals also have electronegativities in the range 1.9–2.4.

The "single-bond covalent radius" of C can be taken as half the interatomic distance in diamond, i.e. $r(C) = 77.2$ pm. The corresponding values for "doubly-bonded" and "triply-

[6c] G. C. LEVY, R. L. LICHTER, and G. L. NELSON, *Carbon-13 Nuclear Magnetic Resonance Spectroscopy*, Wiley, New York, 1980, 338 pp.

[6d] J .M. MICHELS, *Dating Methods in Archeology*, Seminar Press, New York, 1973, 230 pp., S. FLEMING, *Dating in Archeology: A Guide to Scientific Techniques*, Dent, London, 1976, 272 pp.

bonded" carbon atoms are usually taken to be 66.7 and 60.3 pm respectively though variations occur, depending on details of the bonding and the nature of the attached atom (see also p. 311). Despite these smaller perturbations the underlying trend is clear: the covalent radius of the carbon atom becomes smaller the lower the coordination number and the higher the formal bond order.

Some properties of α-graphite and diamond are compared in Table 8.1. As expected from its structure, graphite is less dense than diamond and many of its properties are markedly anisotropic. It shows ready cleavage parallel to the basal plane, and this accounts for its flaky appearance, softness, and use as a lubricant. By contrast, diamond can be cleaved in many directions, thus enabling many facets to be cut in gem-stones, but it is extremely hard and involatile because of the strong C–C bonding throughout the crystal. Interestingly, diamond has the highest thermal conductivity of any known substance (more than 5 times that of Cu) and for this reason the points of diamond cutting tools do not become overheated. Diamond also has one of the lowest known coefficients of thermal expansion: 1.06×10^{-6} at room temperature.

TABLE 8.1 *Some properties of α-graphite and diamond*

Property	α-Graphite	Diamond
Density/g cm^{-3}	2.266 (ideal) varies from 2.23 (petroleum coke) to 1.48 (activated C)	3.514
Hardness/Mohs	<1	10
MP/K (see text)	4100±100 (at 9 kbar)	4100±200 (at 125 kbar)
ΔH_{subl}/kJ mol^{-1}	715[a]	~710[a]
Refractive index, n (at 546 nm)	2.15 (basal) 1.81 (c-axis)	2.41
Band gap E_g/kJ mol^{-1}	—	~580
ρ/ohm cm	$(0.4-5.0) \times 10^{-4}$ (basal) 0.2–1.0 (c-axis)	$10^{14}-10^{16}$
$\Delta H_{combustion}$/kJ mol^{-1}	393.51	395.41
ΔH_f°/kJ mol^{-1}	0.00 (standard state)	1.90

[a] Sublimation to monatomic C(g).

The optical and electrical properties of the two forms of carbon likewise reflect their differing structures. Graphite is a black, highly reflecting semi-metal with a resistivity $\rho \sim 10^{-4}$ ohm cm within the basal plane though this increases by a factor of ~5000 along the c-axis. Diamond, on the other hand, is transparent and has a high refractive index; there is a band energy gap of ~580 kJ mol^{-1} so that diamond has a negligible electrical conductivity, the specific resistivity being of the order $10^{14}-10^{16}$ ohm cm. (For other properties and industrial applications of diamond, see ref. 6e.)

As may be seen from the heats of combustion, α-graphite is more stable than diamond at room temperature, the heat of transformation being about 1.9 kJ mol^{-1}. However, as the molar volume of diamond (3.418 cm^3) is much smaller than that of graphite (5.301 cm^3), it follows that diamond can be made from graphite by application of a suitably high pressure, provided that the temperature is also sufficiently high to permit movement of the

[6e] J. E. FIELD (ed.), *The Properties of Diamond*, Academic Press, London, 1979, 660 pp.

atoms. Such transformations were first successfully achieved in 1953–5, using pressures up to 100 kbar and temperatures in the range 1200–2800 K; the presence of molten-metal catalysts such as Cr, Fe, or Ni was also found to be necessary, suggesting that the transformation may proceed via the formation of unstable metal carbide intermediates. Two versions of the high-pressure phase-diagram of carbon are shown in Fig. 8.5. The one generally accepted until very recently shows a low-pressure triple point (α-graphite/liquid C/gaseous C) at ~4100 K and 9 kbar; with increasing pressure the mp of graphite increases to a maximum of ~4600 K at 70 kbar and then decreases to the second triple point (α-graphite/diamond/liquid C) at ~4100 K and 125 kbar. However, the recent discovery of a series of carbyne-like allotropes suggests that graphite is, in fact unstable above 2600 K at any pressure and so can not be melted. Instead it transforms into a series of carbyne-like allotropes before finally melting at 3800 K and 0.2 bar (Fig. 8.5b).[6b]

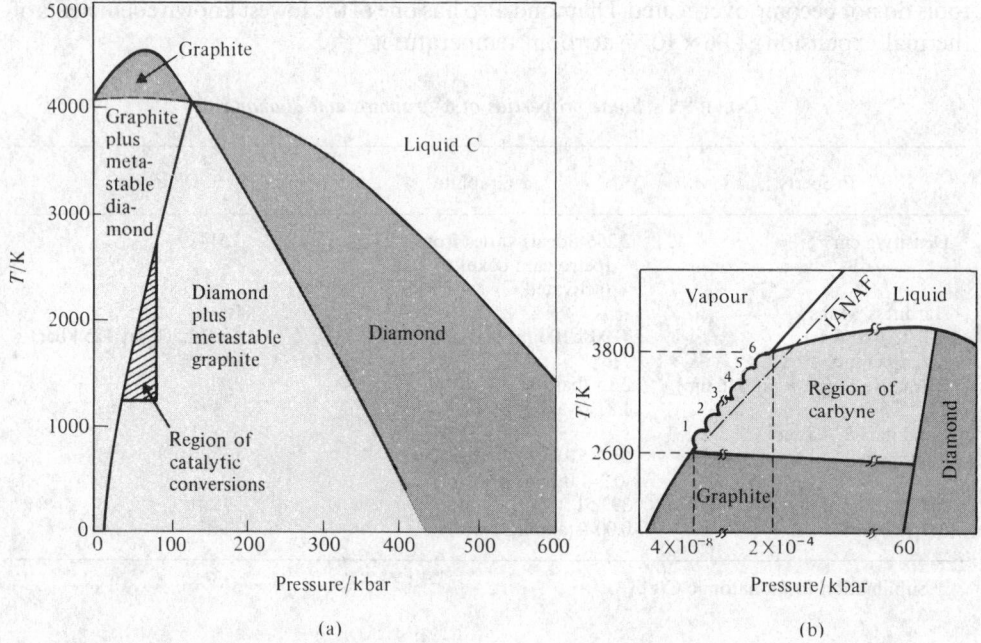

FIG. 8.5(a)   High-pressure phase diagram of carbon. (Based on data published by F. P. Bundy, *J. Chem. Phys.* **38**, 618–30, 631–43 (1963).) (b) A. G. Whittaker's proposed new phase diagram for carbon above 2000 K incorporating (schematically) up to 6 carbyne-type allotropes. Form 1 (transition temperature 2600 K) is probably chaoite; this transforms to form 2 at 2800 and to form 3 at 3050 K.

8.2.4   *Chemical properties*

Carbon in the form of diamond is extremely unreactive at room temperature. Graphite, although thermodynamically more stable than diamond under normal conditions, tends to react more readily due to its more vulnerable layer structure. For example, it is oxidized by hot concentrated HNO_3 to mellitic acid, $C_6(CO_2H)_6$, in which planar-hexagonal C_{12} units are preserved. Graphite reacts with a suspension of $KClO_4$ in a 1:2 mixture (by volume) of conc HNO_3/H_2SO_4 to give "graphite oxide" an unstable, pale lemon-coloured product of variable stoichiometry and structure. Similar products can be prepared by anodic oxidation of graphite or by reaction with $NaNO_3/KMnO_4/conc\ H_2SO_4$. Graphite oxide decomposes slowly at 70°C, and at 200° it undergoes a spectacular deflagration with the formation of CO, CO_2, H_2O, and soot. Infrared and X-ray evidence suggest that the structure-motif is a puckered hexagonal network of C_6 rings predominantly in the "chair" conformation but with a few remaining C=C bonds; in addition there are terminal and bridging O atoms and pendant OH groups; keto-enol tautomerism is implied and the empirical formula can be represented as $C_6O_x(OH)_y$, with $x \sim 1.0$–1.7 and $y \sim 2.25$–1.7.

Graphite reacts with an atmosphere of F_2 at temperatures between 400–500°C to give "graphite monofluoride" CF_x ($x \sim 0.68$–0.99). The reaction is catalysed by HF and can then occur at much lower temperatures (leading, on occasion, to the destruction of graphite electrodes during the preparation of F_2 by the electrolysis of KF/HF melts, p. 929). At $\sim 600°$ the reaction proceeds with explosive violence to give a mixture of CF_4, C_2F_6, and C_5F_{12}. The colour of CF_x depends on the fluorine content becoming progressively lighter from black ($x \sim 0.7$) to grey ($x \sim 0.8$), silver ($x \sim 0.9$) and transparent white ($x > 0.98$).† The structure has not been definitely established but the idealized layer lattice shown in Fig. 8.6 accounts for the observed interplanar spacings, infrared data, colour, and lack of electrical conductivity ($\rho > 3000$ ohm cm). CF is very unreactive, but when heated slowly between 600–1000° it gradually liberates fluorocarbons, C_nF_{2n+2}.

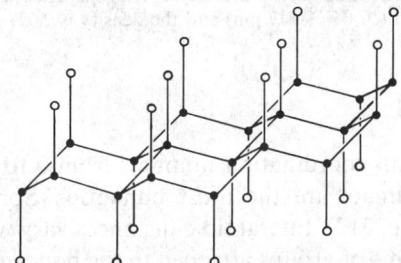

Fig. 8.6 Idealized structure of CF showing puckered layer lattice of fused C_6 rings in "chair" conformation and axial F atoms. The spacing between successive C layers is ~ 817 pm (cf. graphite 335.4 pm) and the density 2.43 g cm^{-3}.

When gaseous mixtures of F_2/HF are allowed to react with finely powdered graphite at room temperature an inert bluish-black compound with a velvety appearance is formed with a composition which varies in the range C_4F to $C_{3.57}F$. The in-plane C–C distance remains as in graphite but the interlayer spacing increases to 534–550 pm depending on the F content. The infrared and X-ray data are best interpreted in terms of the structure shown in Fig. 8.7. The electrical conductivity, though less than that of graphite, is still appreciable, the resistivity being ~ 2–4 ohm cm.

At high temperatures, C reacts with many elements including H (in the presence of a finely divided Ni catalyst), F (but not the other halogens), O, S, Si (p. 386), B (p. 167), and many metals (p. 318). It is an active reducing agent and reacts readily with many oxides to liberate the element or form a carbide. These reactions, which reflect the high enthalpy of formation of CO and CO_2, are of great industrial importance (p. 326).

---

† Very recent work[7] suggests that the colour depends on the reaction temperature rather than the composition. More precise evidence on the structure of $(CF)_n$ was obtained and the new graphite fluoride $(C_2F)_n$ was also characterized and its layer structure determined.

7 Y. KITA, N. WATANABE, and Y. FUJII, Chemical composition and crystal structure of graphite fluoride, *J. Am. Chem. Soc.* **101**, 3832–41 (1979).

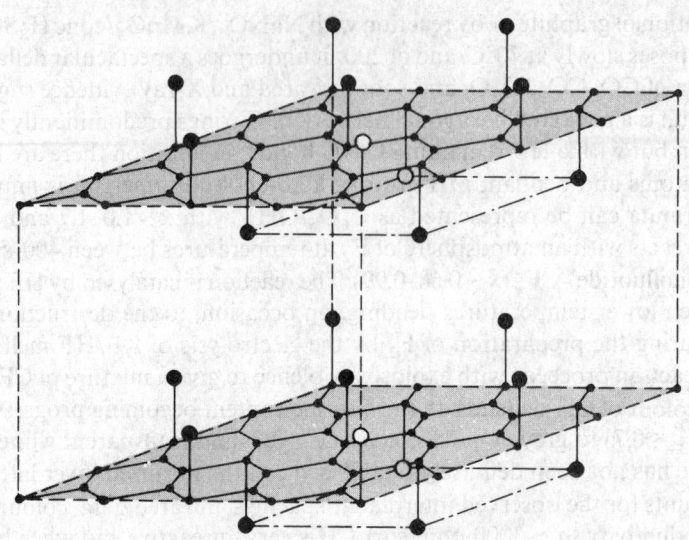

Fig. 8.7   Proposed structure for C_4F showing retention of the planar graphite sheets but with regularly spaced F atoms above and below each layer. The spacing between successive C layers is ~534 pm (cf. CF ~817 pm) and the density is 2.077 g cm^{-3}.

Carbon is known with all coordination numbers from 0 to 8 though compounds in which it is 3- or 4-coordinate are the most numerous. Some typical examples are summarized in the Panel (p. 312). Interatomic distances vary with the type of bond and the nature of the other atoms or groups attached to the bonded atoms. For example, the formally single-bonded C–C distance varies from 146 pm in Me–CN to 163.8 pm in $Bu_2^nPhC–CPhBu_2^n$.[7a] Some typical examples are in Fig. 8.8. Note that because of the breadth of some of these ranges the interatomic distance between quite different pairs of atoms can be identical. For example, the value 133 pm includes C–F, C–O, C–N, and C–C; likewise the value of 185 pm includes C–Br, C–S, C–Se, C–P, and C–Si. The conventional classification into single, double, and triple bonds is adopted for simplicity, but bonding is frequently more subtle and more extended than these localized descriptions imply. Bond energies are listed on p. 435, where they are compared with those for other elements of Group IV. It should perhaps be emphasized that interatomic distances are experimentally observable, whereas bond orders depend on theoretical models and the estimation of bond energies in polyatomic molecules also depends on various assumptions as to how the total energy is apportioned. Nevertheless, taken together, the data indicate that an increase in the order of a bond between 2 atoms is accompanied both by a decrease in bond length and by an increase in bond energy. Similarly, for a given bond order between C and a series of other elements (e.g. C–X), the bond energy increases as the bond length decreases.

[7a] W. LITTKE and U. DRÜCK, Discovery of an exceptionally long unbridged-carbon bond, *Angew. Chem.*, Int. Edn. (Engl.) **18**, 406–7 (1979).

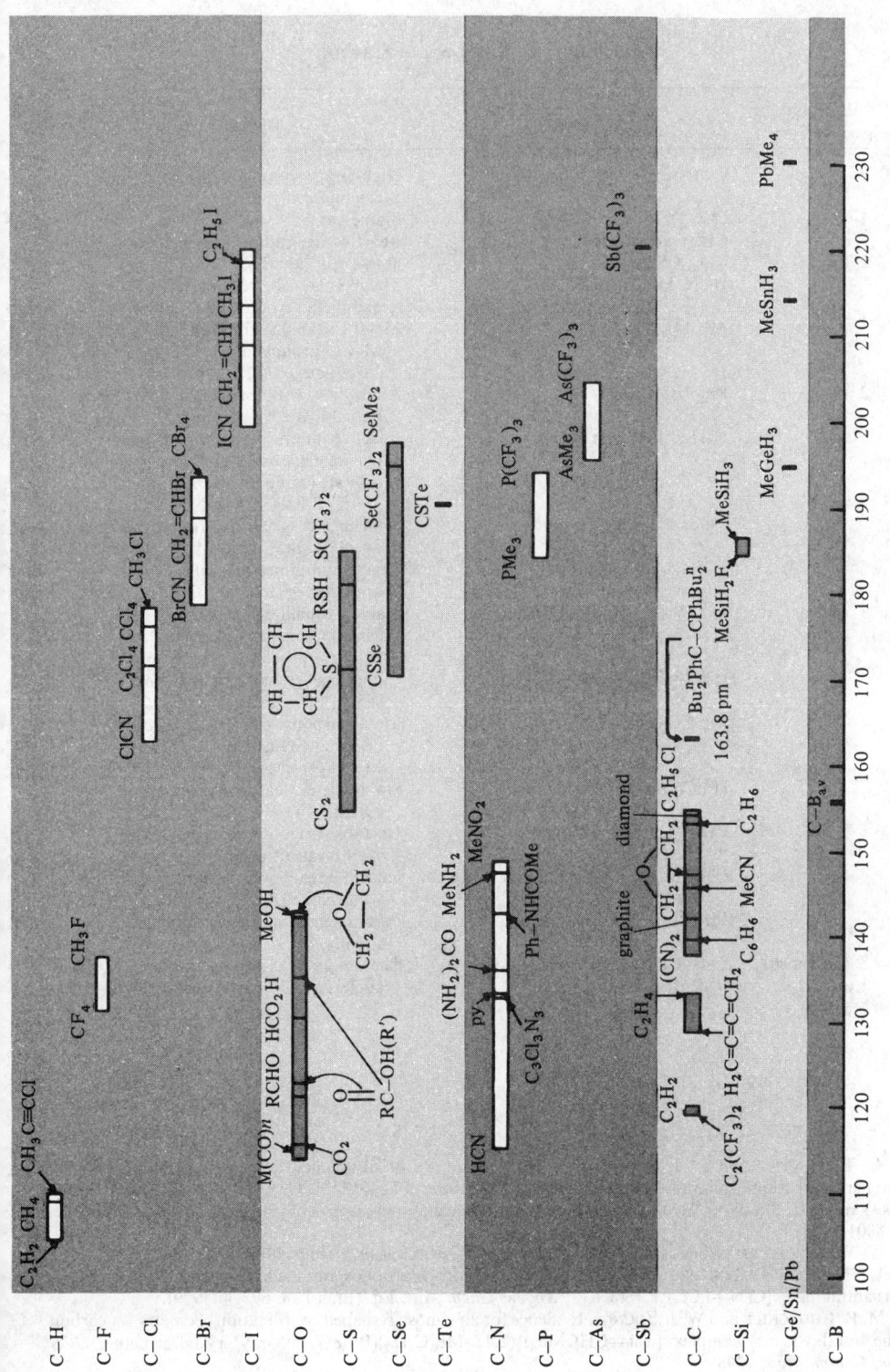

FIG. 8.8   Some interatomic distances involving carbon.

Coordination Numbers of Carbon

CN	Examples	Comments
0	C atoms	High-temperature, low-pressure, gas phase
1	CO	Stable gas
	$CH^{\bullet}$ (carbynes)	Reactive free-radical intermediates
2 (linear)	CO_2, CS_2	Stable gas, liquid
	HCN, $HC\equiv CH$, NCO^-, NCS^-	The ions are isoelectronic with CO_2 and COS respectively
	$M(CO)_n$	Metal carbonyls with terminal M–CO groups (angle MCO sometimes $<180°$)
2 (bent)	$Ph_3P:C:PPh_3$	A bis(ylide) with angle PCP 130.1° (and 143.8)[7b]
	$:CH_2$, $:CX_2$ (carbenes)	Reactive intermediates with 1 lone-pair of electrons and 1 vacant orbital; (carbenes are bent for X = H, F, OH, OMe, NH_2, but linear if X is less electronegative, e.g. BH_2, BeH, or Li)[7c]
	$-CPh_2-$ (methylenes)	Reactive intermediates with 2 unpaired electrons
3 (planar)	COXY (X = H, hal, OH, O^- R, Ar)	Stable oxohalides, carbonates, carboxylic acids, aldehydes, ketones, etc.
	$[C(N=PCl_3)_3]^+[SbCl_6]^-$	Colourless crystals prepared from $[C(N_3)_3]^+ + PCl_3$[7d]
	$M_m(CO)_n$	Metal carbonyl clusters with doubly bridging CO groups M–C(O)–M
	$[PhC(OMe)M(CO)_5]$	Stable metal—carbene complexes, e.g. M = Cr, W
	CH_3^+ (carbonium ions)	Unstable reaction intermediates with 1 vacant orbital
3 (pyramidal)	CH_3^-, CPh_3^- (carbanions), RMgX	Unstable reaction intermediates with 1 lone pair of electrons
	$Ph_3C^{\bullet}$, $R_3C^{\bullet}$ (free radicals)	Paramagnetic species of varying stability
3 (T-shaped)	$Ta(=CHCMe_3)_2(2,4,6,-Me_3C_6H_2)-$ $(PMe_3)_2]$	The unique H is equatorial and angle $Ta=C-CMe_3$ is 169°[7e]

Continued

[7b] A. T. VINCENT and P. J. WHEATLEY, Crystal structure of bis(triphenylphosphoranylidene)methane [hexaphenylcarbodiphosphorane], $Ph_3P:C:PPh_3$, *JCS Dalton*, 1972, 617–22. G. E. HARDY, J. I. ZINK, W. C. KASKA, and J. C. BALDWIN, Structure and triboluminescence of polymorphs of $(Ph_3P)_2C$, *J. Am. Chem. Soc.* **100**, 8001–2 (1978).

[7c] W. W. SCHOELLER, When is a singlet carbene linear?, *JCS Chem. Comm.* 1980, 124–5.

[7d] U. MÜLLER, I. LORENZ, and F. SCHMOCK, Tris(trichlorophosphoranediylamino)carbenium hexachloroantimonate, $[C(N=PCl_3)_3]^+[SbCl_6]^-$, *Angew. Chem.*, Int. Ed. (Engl.) **18**, 693–4 (1979).

[7e] M. R. CHURCHILL and W. J. YOUNGS, Evidence for an almost T-shaped coordination geometry for carbon in the bis(alkylidene) complex $[Ta(=CHCMe_3)_2(2,4,6-Me_3C_6H_2)(PMe_3)_2]$: X-ray crystal structure, *JCS Chem. Comm.* 1978, 1048–9.

4 (tetrahedral)	CX_4, etc.	4-coordinate covalent compounds of carbon such as CF_4, C_2H_6, CHXYZ, etc.
	$M_m(CO)_n$	Metal carbonyl clusters with triply bridging CO groups (p. 351)
(see-saw, C_{2v})	$[Fe_4C(CO)_{13}]$	The μ_4-carbido C caps Fe_4 "butterfly"[7f]
5	Al_2Me_6	Alkyl-bridged organometallic compounds involving 3-centre, 2-electron bonds (pp. 289, etc.)
	$C_2B_4H_6$, etc.	Several extremely stable carboranes (p. 205)
	$[(\eta^5\text{-}C_5H_5)NiRu_3(CO)_9CCHBu^t]$	The terminal C atom is bonded to $CHBu^t$ and to all 4 metal atoms (which adopt a "butterfly" arrangement)[7g]
	$[Os_5C(CO)_{13}HL_2]$	The μ_5-carbido C bonds to all 5 Os[7h]
6	$C_2B_{10}H_{12}$, etc.	Numerous very stable carboranes (p. 207)
7	$(LiMe)_4$ crystals	See structure, p. 113
8	Be_2C (antifluorite), $[Co_8C(CO)_{18}]^{2-}$	See structure, p. 129 See structure, p. 358

8.3 Graphite Intercalation Compounds[8, 9, 9a, 9b]

The large interlayer distance between the parallel planes of C atoms in graphite implies that the interlayer bonding is relatively weak. This accounts for the ready cleavage along the basal plane and the remarkable softness of the crystals. It also enables a wide range of substances to intercalate between the planes under mild conditions to give lamellar compounds of variable composition. These reactions are often reversible (unlike those with O and F discussed above) and the graphitic nature of the host lattice is retained. The compounds have quite different structures and properties from those previously encountered in this book and so will be described in some detail. They may be compared with the materials formed by intercalation into certain sheet silicates (p. 406).

The first alkali–metal graphite compound was reported in 1926: bronze-coloured C_8K was formed by direct reaction of graphite with K vapour at 300°C. Rubidium and Cs behave similarly. When heated at $\sim360°$ under reduced pressure the metal is removed in

[7f] J. S. BRADLEY, G. B. ANSELL, M. E. LEONOWICZ, and E. W. HILL, Synthesis and molecular structure of μ_4-carbido-μ_2-carbonyl-dodecacarbonyltetrairon, a neutral iron butterfly cluster bearing an exposed carbon atom, *J. Am. Chem. Soc.* **103**, 4968–70 (1981).

[7g] E. SAPPA, A. TIRIPICCHIO, and M. T. CAMELLINI, A new mixed ruthenium–nickel cluster. Synthesis and X-ray crystal structure of $[(\eta^5\text{-}C_5H_5)NiRu_3(CO)_9(CCHBu^t)]$, *JCS Chem. Comm.* 1979, 154.

[7h] J. M. FERNANDEZ, B. F. G. JOHNSON, J. LEWIS, and P. RAITHBY, The molecular and crystal structure of $[Os_5C(CO)_{13}H\{OP(OMe_2)\}\{P(OMe)_3\}]$, *JCS Dalton*, 1981, 2250–57.

[8] A. K. HOLLIDAY, G. HUGHES, and S. M. WALKER, Carbon, Chap. 13 in *Comprehensive Inorganic Chemistry*, Vol. 1, pp. 1173–1294. Pergamon Press, Oxford, 1973.

[9] L. B. EBERT, Intercalation compounds of graphite, *A. Rev. Materials Sci.* **6**, 181–211 (1976).

[9a] A. HÉROLD, Crystallochemistry of carbon intercalation compounds, in F. LEVY (ed.), *Intercalated Layered Materials*, pp. 323–421, Reidel, 1979.

[9b] H. SELIG and L. B. EBERT, Graphite intercalation compounds, *Adv. Inorg. Chem. Radiochem.* **23**, 281–327 (1980); a review with ~350 references.

stages to give a series of intercalation compounds C_8M (bronze-red), $C_{24}M$ (steel-blue), $C_{36}M$ (dark blue), $C_{48}M$ (black), and $C_{60}M$ (black). The compounds can also be prepared by electrolysis of fused melts with graphite electrodes, by reaction of graphite with solutions of M in liquid ammonia or amines, and by exchange reactions using M/aromatic radical anions. Intercalation is more difficult to achieve with Li and Na though direct reaction with highly purified graphite at 500° yields C_6Li (brass coloured), $C_{12}Li$ (copper), and $C_{18}Li$ (steel), and reaction with Li/naphthalene in thf yields $C_{16}Li$ and $C_{40}Li$. Corresponding reaction of graphite and molten Na at 450° gives $C_{64}Na$ (deep violet) whereas Na/naphthalene gives $C_{32}Na$ and $C_{120}Na$.

The crystal structure of C_8K is shown in Fig. 8.9; the graphite layers remain intact but are stacked vertically above each other instead of in the sequence . . . ABAB . . . found in α-graphite itself. Each graphite layer is interleaved by a layer of K atoms having a commensurate lattice in which the spacing between each K is twice the spacing between the centres of the graphite hexagons (Fig. 8.10). The stoichiometries of the other stages can then be achieved by varying the frequency of occurrence of the intercalated M layers in the host lattice. An idealized representation of this model is shown in Fig. 8.11. Difficulties are encountered in devising a plausible mechanistic route to the formation of these compounds since the direct preparation of one stage from an adjacent stage apparently requires both the complete emptying and the complete filling of inserted layers. It may be that the situation is more complex, with distributions of stages rather than a single uniform arrangement for each stoichiometry.

The electrical resistance of graphite intercalation compounds is even lower than for graphite itself, resistance along the *a*-axis dropping by a factor of ~ 10 and that along the *c*-axis by ~ 100; moreover, in contrast to graphite, which is diamagnetic, the compounds have a temperature-independent (Pauli) paramagnetism, and also behave as true metals in having a resistivity that increases with increase in temperature. This is illustrated by the comparative data shown in Table 8.2.

TABLE 8.2   *Resistivity of graphite and its intercalates*

Material	ρ (90 K)/ohm cm	ρ (285 K)/ohm cm
α-graphite	37.7	28.4
C_8K	0.768	1.02
$C_{12}K$	0.932	1.15

These data, and the other properties of C_nM, suggest that bonding occurs by transfer of electrons from the alkali metal atoms to the conduction band of the host graphite. Consistent with this, metal intercalation has only been observed with the most electropositive elements (Group IA) though Ba, with a first-stage ionization energy intermediate between those of Li and Na, was recently (1974) found to give C_6Ba.

Alkali–metal graphites are extremely reactive in air and may explode with water. In general, reactivity decreases with ease of ionization of M in the sequence Li > Na > K > Rb > Cs. Under controlled conditions H_2O or ROH produce only H_2, MOH, and graphite, unlike the alkali–metal carbides M_2C_2 (p. 319) which produce hydrocarbons such as acetylene. In an important new reaction C_8K has been found to

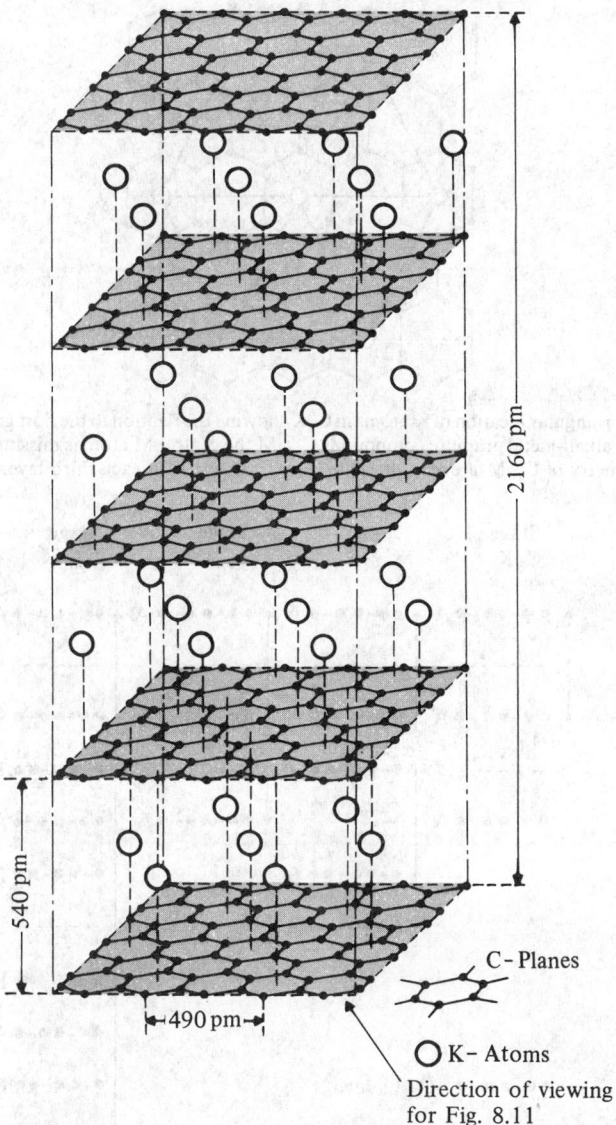

C- Planes

○ K- Atoms

Direction of viewing
for Fig. 8.11

FIG. 8.9 Crystal lattice of C_8K showing the vertical packing of graphitic layers. The C–C distance within layers almost identical to that in graphite itself but the interplanar spacing (540 pm) is much larger than for graphite (335 pm) due to the presence of K atoms. The spacing increases still further to 561 pm for C_8Rb and to 595 pm for C_8Cs.

react smoothly with transition metal salts in tetrahydrofuran at room temperature to give the corresponding transition metal lamellar compounds:[9c]

$$nC_8K + MX_n \xrightarrow{\text{thf}} C_{8n}M + nKX$$

[9c] D. BRAGA, A. RIPAMONTI, D. SAVOIA, C. TROMBINI, and A. UMANI-RONCHI, Graphite lamella compounds. Structure of transition-metal intercalates, *JCS Dalton* 1979, 2026–8.

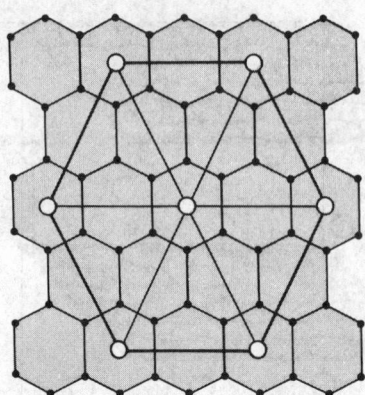

FIG. 8.10   Triangular location of K atoms in C_8K showing the relation to the host graphite layers. In the other alkali-metal graphite compounds $C_{12n}M$ the central M atom is missing, leading to a stoichiometry of $C_{12}M$ if every alternate layer is M, $C_{24}M$ if each third layer is M, etc.

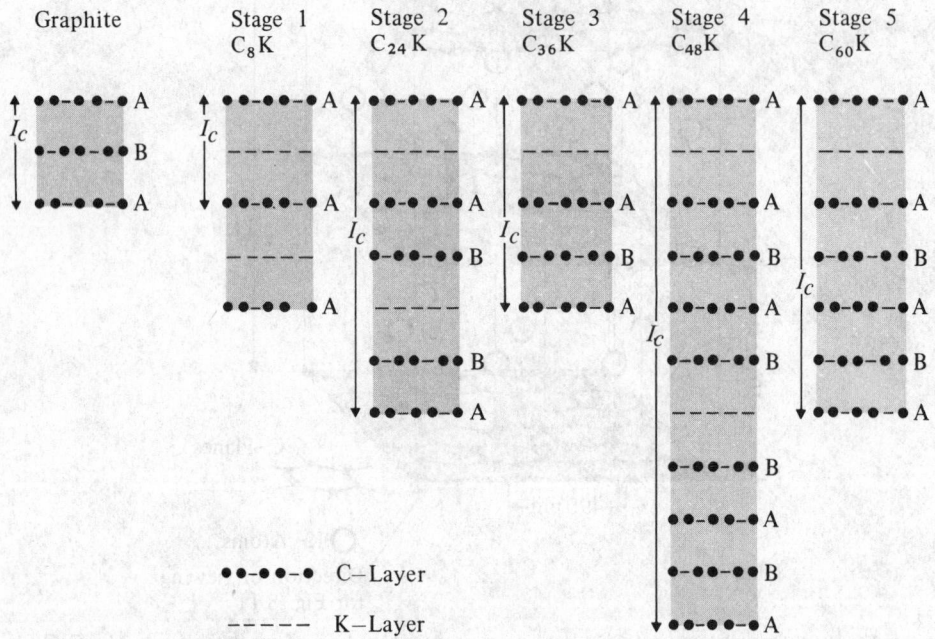

FIG. 8.11   Layer-plane sequence along the *c*-axis for graphite in various stage 1–5 of alkali-metal graphite intercalation compounds. Comparison with Fig. 8.10 shows that the horizontal planes are being viewed diagonally across the figure. I_c is the interlayer repeat distance along the *c*-axis.

Examples of MX_n include $Ti(OPr^i)_4$, $MnCl_2.4H_2O$, $FeCl_3$, $CoCl_2.6H_2O$, $CuCl_2.2H_2O$, and $ZnCl_2$.

A quite different sort of graphite intercalation compound is formed by the halides of many elements, particularly those halides which themselves have layer structures or weak intermolecular binding. The first such compound (1932) was with $FeCl_3$; chlorides, in general, have been the most studied, but fluoride and bromide intercalates are also known.

Halides which have been reported to intercalate include the following:

HF; ClF_3, BrF_3, IF_5; XeF_6, $XeOF_4$; CrO_2F_2, SbF_3Cl_2, TiF_4, MF_5 (M = As, Sb, Nb, Ta); UF_6

MCl_2: M = Be; Mn, Co, Ni, Cu; Zn, Cd, Hg

MCl_3: M = B, Al, Ga, In, Tl; Y; Sm, Eu, Gd, Tb, Dy; Cr, Fe, Co; Ru, Rh, Au; I

MCl_4: M = Zr, Hf; Re, Ir; Pd, Pt

MCl_5: M = Sb; Mo; U

MCl_6: M = W, U; also CrO_2Cl_2, UO_2Cl_2

Mixtures of $AlCl_3$ plus Br_2, I_2, ICl_3, $FeCl_3$, WCl_6

Bromides: $CuBr_2$; $AlBr_3$, $GaBr_3$; $AuBr_3$

The intercalates are usually prepared by heating a mixture of the reactants though sometimes the presence of free Cl_2 is also necessary, particularly for "non-oxidizing" chlorides such as $MnCl_2$, $NiCl_2$, $ZnCl_2$, $AlCl_3$, etc. Many of the compounds appear to show various stages of intercalation, the first stage usually exhibiting a typical blue colour. A common feature of many of the intercalated halides is their ability to act as electron-pair acceptors (Lewis acids). Low heat of sublimation is a further characteristic of most of the intercalating compounds. It may be that an important feature is an ability of the guest molecule to form a layer lattice commensurate with the host graphite. For example, in $C_{6.69}FeCl_3$ the intercalated $FeCl_3$ has a layer structure similar to that in $FeCl_3$ itself with Cl in approximately close-packed arrangement though with some distortion, and with extensive stacking disorder. The "first-stage" compound varies in composition in the range $C_{\sim 6-7}FeCl_3$; in addition a "second-stage" compound corresponding to $C_{\sim 12}FeCl_3$ is known, and also a "third-stage" with composition in the range $C_{\sim 24-30}FeCl_3$. Another well-characterized phase occurs with $MoCl_5$: layers of close-packed Mo_2Cl_{10} molecules alternate with sets of 4 graphite layers along the c-axis.

There appears to be a small but definite transfer of electron charge from the graphite to the guest species and this has led to formulations such as $C_{70}^+Cl^-.FeCl_2.5FeCl_3$. Similarly, the intercalate of $AlCl_3$ (which is formed in the presence of free Cl_2) has been formulated as $C_{27}^+Cl^-.3AlCl_3$ or $C_{27}^+AlCl_4^-.2AlCl_3$. This would explain the enhanced conductivity of the graphite–metal halide compounds, due to the formation of positive holes near the top of the valence band. However, despite extensive work by a variety of techniques, many structural problems remain unresolved and there is still no consensus on the detailed description of the bonding. (For further recent work see ref. 9d.)

The halogens themselves show a curious alternation of behaviour towards graphite. F_2 gives the compounds CF, C_2F, and C_4F (p. 309) whereas liquid Cl_2 reacts slowly to give C_8Cl, and I_2 appears not to intercalate at all. By contrast, Br_2 readily intercalates in several stages to give compounds of formula C_8Br, $C_{12}Br$, $C_{16}Br$, and $C_{20}Br$; the compounds $C_{14}Br$ and $C_{28}Br$ have also been well-characterized crystallographically but may be metastable phases. A notable feature of the Br_2 intercalation reaction is that it is completely prevented by prior coating of the *basal* plane of the sample of graphite with a layer impervious to Br_2. The lamellar character of blue C_8Br has been confirmed by X-ray diffraction and the intercalation of bromine, is accompanied by a marked decrease in the resistivity of the graphite—more than tenfold along the a-axis and twofold along the c-axis. C_8ICl and $C_{36}ICl$ have also been prepared.

[9d] E. M. McCarron and N. Bartlett, Composition and staging in the graphite–AsF_6 system and its relationship to graphite–AsF_5, *JCS Chem. Comm.* 1980, 404–6.

Numerous oxides, sulfides, and oxoacids have been found to intercalate into graphite. For example, lamellar compounds with SO_3, N_2O_5, and Cl_2O_7 are known (but not with SO_2, NO, or NO_2). CrO_3 and MoO_3 readily intercalate as do several sulfides such as V_2S_3, $Cr_2S_3(+S)$, WS_2, $PdS(+S)$, and Sb_2S_5. The reversible intercalation of various oxoacids under oxidizing conditions leads to lamellar graphite "salts" some of which have been known for over a century and are now particularly well characterized structurally. For example, the formation of the blue, "first-stage" compound with conc H_2SO_4 can be expressed by the idealized equation

$$24C + 3H_2SO_4 + \tfrac{1}{4}O_2 \longrightarrow C_{24}{}^{+}HSO_4{}^{-}.2H_2SO_4 + \tfrac{1}{2}H_2O$$

The overall stoichiometry is thus close to $C_8H_2SO_4$ and the structure is very similar to that of C_8K (p. 314) except for the detail of vertical alignment of the carbon atoms in the c direction which is . . . ABAB . . .. Several later stages (2, 3, 4, 5, 11) have been established and their properties studied. Intercalation is accompanied by a marked decrease in electrical resistance. A series of graphite nitrates can be prepared similarly, e.g. $C_{24}{}^{+}NO_3{}^{-}.2HNO_3$ (blue), $C_{48}{}^{+}NO_3{}^{-}.3HNO_3$ (black), etc. Other oxoacids which intercalate (particularly under electrolytic conditions) include $HClO_4$, HSO_3F, HSO_3Cl, H_2SeO_4, H_3PO_4, $H_4P_2O_7$, H_3AsO_4, CF_3CO_2H, CCl_3CO_2H, etc. The extent of intercalation depends both on the strength of the acid and its concentration, and the reactions are of considerable technological importance because they can lead to the swelling and eventual destruction of the graphite electrodes used in many electrochemical processes.

8.4   Carbides

Carbon forms binary compounds with most elements: those with metals are considered in this section whilst those with H, the halogens, O, and the chalcogens are discussed in subsequent sections. General methods of preparation of metal carbides are:[9e]

(1) Direct combination of the elements above $\sim 2000°C$.
(2) Reaction of the metal oxide with carbon at high temperature.
(3) Reaction of the heated metal with gaseous hydrocarbon.
(4) Reaction of acetylene with electropositive metals in liquid ammonia.

Attempts to classify carbides according to structure or bond type meet the same difficulties as were encountered with hydrides (p. 70) and borides (p. 162) and for the same reasons. The general trends in properties of the three groups of compounds are, however, broadly similar, being most polar (ionic) for the electropositive metals, most covalent (molecular) for the electronegative non-metals, and somewhat complex (interstitial) for the elements in the centre of the d block. There are also several elements with poorly characterized, unstable, or non-existent carbides, namely the later transition elements (Groups IB and IIB), the platinum metals, and the post transition-metal elements in Group IIIB.

Salt-like carbides containing individual C "anions" are sometimes called "methanides" since they yield predominantly CH_4 on hydrolysis. Be_2C and Al_4C_3 are the best-

---

[9e] Reference 4, pp. 476–535: Carbides (p. 476); Cemented carbides (p. 483); Industrial heavy metal carbides (p. 490); Calcium carbide (p. 505); Silicon carbide (p. 520).

characterized examples, indicating the importance of small compact cations. Be_2C is prepared from BeO and C at 1900–2000°C; it is brick-red, has the antifluorite structure (p. 129), and decomposes to graphite when heated above 2100°. Al_4C_3, prepared by direct union of the elements in an electric furnace, forms pale-yellow crystals, mp 2200°C. It has a complex structure in which {AlC_4} tetrahedra of two types are linked to form a layer lattice: this defines two types of C atom, one surrounded by a deformed octahedron of 6 Al at 217 pm and the other surrounded by 4 Al at 190–194 pm and a fifth Al at 221 pm. The closest C$\cdots$C approach is at the nonbonding distance of 316 pm. Although it is formally possible to describe the structure as ionic, $(Al^{3+})_4(C^{4-})_3$, such a gross separation of charges is unlikely to occur over the observed interatomic distances.

Carbides containing a C_2 unit are well known; they are exemplified by the acetylides (ethynides) of the alkali metals, $M_2^IC_2$, alkaline earth metals, $M^{II}C_2$, and the lanthanoids LnC_2 and Ln_2C_3 i.e. $Ln_4(C_2)_3$. The corresponding compounds of Group IB (Cu, Ag, Au) are explosive and those of Group IIB (Zn, Cd, Hg) are poorly characterized. $M_2^IC_2$ are best prepared by the action of C_2H_2 on a solution of alkali metal in liquid NH_3; they are colourless crystalline compounds which react violently with water and oxidize to the carbonate on being heated in air. $M^{II}C_2$ can be prepared by heating the alkaline earth metal with ethyne above 500°C. By far the most important compound in this group is CaC_2—it is manufactured on a huge scale, ~ 7 million tonnes worldwide in 1959. As such, it has in the past† been the major source of ethyne for the chemical industry and for oxyacetylene welding; petroleum is now an alternative source of this important gas especially when used as a chemical intermediate (~ 1 million tonnes pa in 1976). Industrially, CaC_2 is produced by the endothermic reaction of lime and coke:

$$CaO + 3C \xrightarrow{2200-2250°C} CaC_2 + CO; \quad \Delta H = 465.7 \text{ kJ mol}^{-1}$$

Subsequent hydrolysis is highly exothermic and must be carefully controlled:

$$CaC_2 + 2H_2O \longrightarrow C_2H_2 + Ca(OH)_2; \quad \Delta H = -120 \text{ kJ mol}^{-1}$$

Another industrially important reaction of CaC_2 is its ability to fix N_2 from the air by formation of calcium cyanamide:

$$CaC_2 + N_2 \xrightarrow{1000°-1200°} CaCN_2 + C; \quad \Delta H = -296 \text{ kJ mol}^{-1}$$

$CaCN_2$ is widely used as a fertilizer because of its ready hydrolysis to cyanamide, H_2NCN (p. 341).

Pure CaC_2 is a colourless solid, mp 2300°C. It can be prepared on the laboratory scale by passing ethyne into a solution of Ca in liquid NH_3, followed by decomposition of the complex so formed, under reduced pressure at $\sim 325°$:

$$Ca(\text{liq } NH_3) + 2C_2H_2 \xrightarrow{-80°} H_2 + CaC_2 \cdot C_2H_2 \xrightarrow{325°} CaC_2 + C_2H_2$$

It exists in at least four crystalline forms, the one stable at room temperature being a tetragonally distorted NaCl-type structure (Fig. 8.12) in which the C_2 units are aligned along the c-axis. The ethynides of Mg, Sr, and Ba have the same structure and also

---

† US production of CaC_2 reached a maximum of 0.9 million tonne pa in 1960 and had declined to 0.3 million tonne pa by 1975.

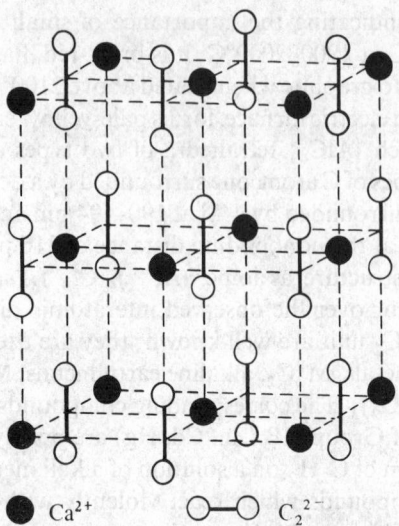

FIG. 8.12 Crystal structure of tetragonal CaC_2 showing the resemblance to NaCl (p. 273). Above 450°C the parallel alignment of the C_2 units breaks down and the structure transforms to a cubic phase.

hydrolyse to give ethyne. In addition, BaC_2 absorbs N_2 from the atmosphere to give $Ba(CN)_2$ (cf. CaC_2 above).

The carbides of the lanthanoids and actinoids can be prepared by heating M_2O_3 with C in an electric furnace or by arc-melting compressed pellets of the elements in an inert atmosphere. They contain the C_2 unit and have a stoichiometry MC_2 or $M_4(C_2)_3$. MC_2 have the CaC_2 structure or a related one of lower symmetry in which the C_2 units lie at right-angles to the c-axis of an orthogonal NaCl-type cell.[10] They are more reactive than the alkaline-earth metal carbides, combining readily with atmospheric oxygen and hydrolysing to a complex mixture of hydrocarbons. This derives from their more complicated electronic structure and, indeed, LnC_2 are metallic conductors (not insulators like CaC_2); they are best regarded as ethynides of Ln^{III} with the supernumerary electron partly delocalized in a conduction band of the crystal. This would explain the evolution of H_2 as well as C_2H_2 on hydrolysis, and the simultaneous production of the reduced species C_2H_4 and C_2H_6 together with various other hydrocarbons up to C_6H_{10}:

$$LnC_2 + 3H_2O \longrightarrow Ln(OH)_3 + C_2H_2 + [H]$$

An interesting feature of the ethynides MC_2 and $M_4(C_2)_3$ is the variation in the C–C distance as measured by neutron diffraction. Typical values (in pm) are:

CaC_2	YC_2	CeC_2	LaC_2	UC_2	$La_4(C_2)_3$	$Ce_4(C_2)_3$	$U_4(C_2)_3$
119.2	127.5	128.3	130.3	135.0	123.6	127.6	129.5

[10] A. F. WELLS, *Structural Inorganic Chemistry*, 4th edn., Oxford University Press, Oxford, 1975, 1095 pp.

The C–C distance in CaC_2 is close to that in ethyne (120.5 pm) and it has been suggested that the observed increase in the lanthanoid and actinoid carbides results from a partial localization of the supernumerary electron in the antibonding orbital of the ethynide ion $[C{\equiv}C]^{2-}$ (see p. 360). The effect is noticeably less in the sesquicarbides than in the dicarbides. The compounds EuC_2 and YbC_2 differ in their lattice parameters and hydrolysis behaviour from the other LnC_2 and this may be related to the relative stability of Eu^{II} and Yb^{II} (p. 1436).

The lanthanoids also form metal-rich carbides of stoichiometry M_3C in which individual C atoms occupy at random one-third of the octahedral Cl sites in a NaCl-like structure. Several of the actinoids (e.g. Th, U, Pu) form monocarbides, MC, in which all the octahedral Cl sites in the NaCl structure are occupied and this stoichiometry is also observed for several other carbides of the early transition elements, e.g. M = Ti, Zr, Hf; V, Nb, Ta; Mo, W. These are best considered as interstitial carbides and in this sense the lanthanoids and actinoids occupy an intermediate position in the classification of the carbides, as they did with the hydrides (p. 72).

Interstitial carbides are infusible, extremely hard, refractory materials that retain many of the characteristic properties of metals (lustre, metallic conductivity).[10a] Reported mps are frequently in the range 3000–4000°C. Interstitial carbides derive their name from the fact that the C atoms occupy octahedral interstices in a close-packed lattice of metal atoms, though the arrangement of metal atoms is not always the same as in the metal itself. The size of the metal atoms must be large enough to generate a site of sufficient size to accommodate C, and the critical radius of M seems to be ~ 135 pm: thus the transition metals mentioned in the preceding paragraph all have 12-coordinate radii >135 pm, whereas metals with smaller radii (e.g. Cr, Mn, Fe, Co, Ni) do not form MC and their interstitial carbides have a more complex structure (see below). If the close-packed arrangement of M atoms is hexagonal (h) rather than cubic (c) then the 2 octahedral interstices on either side of a close-packed M layer are located directly above one another and only one of these is ever occupied. This gives a stoichiometry M_2C as in V_2C, Nb_2C, Ta_2C, and W_2C. Intermediate stoichiometries are encountered when the M atom stacking sequence alternates, e.g. Mo_3C_2 (hcc) and V_4C_3 (hhcc). Ordered defect NaCl-type structures are also known, e.g. V_8C_7 and V_6C_5, thus illustrating the wide range of stoichiometries which occur among interstitial carbides. Unlike the "ionic" carbides, interstitial carbides do not react with water and are generally very inert, though some do react with air when heated above 1000° and most are degraded by conc HNO_3 or HF. The extreme hardness and inertness of WC and TaC have led to their extensive use as high-speed cutting tools.

The carbides of Cr, Mn, Fe, Co, and Ni are profuse in number, complicated in structure, and of great importance industrially. Cementite, Fe_3C, is an important constituent of steel (p. 1248). Typical stoichiometries are listed in Table 8.3 though it should be noted that several of the phases can exist over a range of composition.

The structures, particularly of the most metal-rich phases, are frequently related to those of the corresponding metal-rich borides (and silicides, germanides, phosphides, arsenides, sulfides, and selenides), in which the non-metal is surrounded by a trigonal prism of M atoms with 0, 1, 2, or 3 additional neighbours beyond the quadrilateral prism

[10a] H. H. JOHANSEN, Recent developments in the chemistry of transition metal carbides and nitrides, *Survey of Progress in Chemistry* **8**, 57–81 (1977).

TABLE 8.3.   *Stoichiometries of some transition element carbides*

$Cr_{23}C_6$	$Mn_{23}C_6$	Fe_3C	Co_3C	Ni_3C
Cr_7C_3	$Mn_{15}C_4$	Fe_2C	Co_2C	
Cr_3C_2	Mn_3C			
	Mn_5C_2			
	Mn_7C_3			

faces (p. 166), e.g. Fe_3C (cementite), Mn_3C, and Co_3B; Mn_5C_2 and Pd_5B_2; Cr_7C_3 and Re_7B_3. Numerous ternary carbides, carbonitrides, and oxocarbides are also known.

The carbides of Cr, Mn, Fe, Co, and Ni are much more reactive than the interstitial carbides of the earlier transition metals. They are rapidly hydrolysed by dilute acid and sometimes even by water to give H_2 and a mixture of hydrocarbons. For example, M_3C give H_2 (75%), CH_4 (15%), and C_2H_6 (8%) together with small amounts of higher hydrocarbons.

8.5   Hydrides, Halides, and Oxohalides

The ability of C to catenate (i.e. to form bonds to itself in compounds) is nowhere better illustrated than in the compounds it forms with H. Hydrocarbons occur in great variety in petroleum deposits and elsewhere, and form various homologous series in which the C atoms are linked into chains, branched chains, and rings. The study of these compounds and their derivatives forms the subject of organic chemistry and is fully discussed in the many textbooks and treatises on that subject. The matter is further considered on p. 436 in relation to the much smaller ability of other Group IV elements to form such catenated compounds. Methane, CH_4, is the archetype of tetrahedral coordination in molecular compounds; some of its properties are listed in Table 8.4 where they are compared with those of the corresponding halides. Unsaturated hydrocarbons such as ethene (C_2H_4), ethyne (C_2H_2), benzene (C_6H_6), cyclooctatetraene (C_8H_8), and homocyclic radicals such as cyclopentadienyl (C_5H_5) and cycloheptatrienyl (C_7H_7) are effective ligands to metals and form many organometallic complexes (pp. 357–378).

Tetrafluoromethane (CF_4) is an exceptionally stable gas with mp close to that of CH_4. It can be prepared on a laboratory scale by reacting SiC with F_2 or by fluorinating CO_2, CO, or $COCl_2$ with SF_4. Industrially it is prepared by the aggressive reaction of F_2 on CF_2Cl_2 or CF_3Cl, or by electrolysis of MF or MF_2 using a C anode. CF_4 was first

TABLE 8.4   *Some properties of methane and* CX_4

Property	CH_4	CF_4	CCl_4	CBr_4	CI_4
MP/°C	−182.5	−183.5	−22.9	90.1	171 (d)
BP/°C	−161.5	−128.5	76.7	189.5	∼130 (subl)
Density/g cm^{-3}	0.424	1.96	1.594	2.961	4.32
(at T°C)	(−164°)	(−184°)	(20°)	(100°)	(20°) (s)
$-\Delta H_f^\circ$/kJ mol^{-1}	74.87	679.9	106.7 (g)	160 (l)	—
			139.3 (l)		
D (X_3C–X)/kJ mol^{-1}	435	515	295	235	—

obtained pure in 1926; C_2F_6 was isolated in 1930 and C_2F_4 in 1933; but it was not until 1937 that the various homologous series of fluorocarbons were isolated and identified. Replacement of H by F greatly increases both thermal stability and chemical inertness because of the great strength of the C–F bond (Table 8.4). Accordingly, fluorocarbons are resistant to attack by acids, alkalis, oxidizing agents, reducing agents, and most chemicals up to 600°. They are immiscible with both water and hydrocarbon solvents, and when combined with other groups they confer water-repellance and stain-resistance to paper, textiles, and fabrics.[11] Tetrafluoroethene (C_2F_4) can be polymerized to a chemically inert, thermosetting plastic PTFE (polytetrafluoroethene); this has an extremely low coefficient of friction and is finding increasing use as a protective coating in non-stick kitchen utensils, razor blades, and bearings. PTFE is made by partial fluorination of chloroform using HF in the presence of $SbFCl_4$ as catalyst, followed by thermolysis to C_2F_4 and subsequent polymerization:

$$CCl_3H \longrightarrow CF_2ClH \overset{\Delta}{\longrightarrow} C_2F_4 \longrightarrow (C_2F_4)_n$$

As a ligand towards metals, C_2F_4 and other unsaturated fluorocarbons differ markedly from alkenes (p. 361).

CCl_4 is a common laboratory and industrial solvent with a distinctive smell, usually prepared by reaction of CS_2 or CH_4 with Cl_2. Its use as a solvent has declined somewhat because of its toxicity, but CCl_4 is still extensively used as an intermediate in preparing "Freons" such as $CFCl_3$, CF_2Cl_2, and CF_3Cl:[11]

$$CCl_4 + HF \overset{SbFCl_4}{\longrightarrow} CFCl_3 + HCl$$

$$CFCl_3 + HF \overset{SbFCl_4}{\longrightarrow} CF_2Cl_2 + HCl$$

The catalyst is formed by reaction of HF on $SbCl_5$. The Freons have a unique combination of properties which make them ideally suited for use as refrigerants and aerosol propellants. They have low bp, low viscosity, low surface tension, and high density, and are non-toxic, non-flammable, odourless, chemically inert, and thermally stable. The most commonly used is CF_2Cl_2, bp, $-29.8°$. The market for Freons and other fluorocarbons expanded rapidly in the sixties: production in the USA alone exceeded 200 000 tonnes in 1964, and by 1977 there was an annual production of 2.4×10^9 spray-cans. However, there has been growing concern that chlorofluorocarbons from spray-cans gradually work their way into the upper atmosphere where they may, through a complex chemical reaction, deplete the earth's ozone layer (p. 708). For this reason there was an enforced progressive elimination of this particular application in the USA during the 2-y period starting 15 October 1978.

CBr_4 is a pale-yellow solid which is markedly less stable than the lighter tetrahalides. Preparation involves bromination of CH_4 with HBr or Br_2 or, more conveniently, reaction of CCl_4 with Al_2Br_6 at 100°. The trend to diminishing thermal stability continues to CI_4 which is a bright-red crystalline solid with a smell reminiscent of I_2. It is prepared by the $AlCl_3$-catalysed halogen exchange reaction between CCl_4 and EtI.

[11] Ref. 4, Vol. 11, pp. 1–81 (1980), Fluorine compounds, organic.

Carbon oxohalides are reactive gases or volatile liquids which feature planar molecules of C_{2v} symmetry; they are isoelectronic with BX_3 (p. 220) and the bonding is best described in terms of molecular orbitals spanning all 4 atoms rather than in terms of localized orbitals as implied by the formulation $\overset{X}{\underset{X}{>}}C{=}O$. Some physical properties and molecular dimensions are in Table 8.5. The values call for little comment except to note that the XCX angle is significantly less (as expected) than the value of 120° found for the more symmetrical isoelectronic species BX_3 and CO_3^{2-}. The C–Br distance is unusually long; it comes from a very early diffraction measurement and could profitably be checked.

TABLE 8.5 *Some physical properties and molecular dimensions of* COX_2

Property	COF_2	$COCl_2$	$COBr_2$
MP/°C	−114°	−127.8°	—
BP/°C	−83.1°	7.6°	64.5°
Density (T°C)/g cm^{-3}	1.139 (−144°)	1.392 (19°)	—
Distance (C–O)/pm	117.4	116.6	113
Distance (C–X)/pm	131.2	174.6	(205)
Angle X–C–X	108.0°	111.3°	110±5°
Angle X–C–O	126.0°	124.3°	∼125°

Mixed oxohalides are also known and their volatilities are intermediate between those of the parent species, e.g. COFCl (bp −42°), COFBr (bp −20.6°). COI_2 is unknown but COFI has been prepared (mp −90°, bp 23.4°). Synthetic routes are as follows: COFCl from $COCl_2/HF$; COFBr from CO/BrF_3; COFI from CO/IF_3; and COClBr from CCl_3Br/H_2SO_4.

COF_2 can be made by fluorinating $COCl_2$ with standard fluorinating agents such as $NaF/MeCN$ or SbF_5/SbF_3; direct fluorination of CO with AgF_2 affords an alternative route. COF_2 is rapidly hydrolysed by water to CO_2 and HX, as are all the other COX_2. It is a useful laboratory reagent for producing a wide range of fluoroorganic compounds and the heavier alkali metal fluorides react in MeCN to give trifluoromethoxides $MOCF_3$.

$COCl_2$ (phosgene) is highly toxic and should be handled with great caution. It was first made in 1812 by John Davy (Sir Humphry Davy's brother) by the action of sunlight on $CO + Cl_2$, whence its otherwise surprising name (Greek φῶς *phos*, light; -γενής, *-genes*, born of). It is now a major industrial chemical and is made on the kilotonne scale by combining the two gases catalytically over activated C (p. 301). It was used briefly and rather ineffectively as a chemical warfare gas in 1916 but is now principally used to prepare isocyanates as intermediates to polyurethanes. It also acts as a ligand (Lewis base) towards $AlCl_3$, $SnCl_4$, $SbCl_5$, etc., forming adducts $Cl_2CO{\rightarrow}MCl_n$, and is a useful chlorinating agent, converting metal oxides into highly pure chlorides. It reacts with NH_3 to form mainly urea, $CO(NH_2)_2$, together with more highly condensed products such as guanidine, $C(NH)(NH_2)_2$; biuret, $NH_2CONHCONH_2$; and cyanuric acid, i.e. *cyclo*-$[CO(NH)]_3$ (p. 341).

8.6 Oxides and Carbonates

Carbon forms 2 extremely stable oxides, CO and CO_2, 3 oxides of considerably lower stability, C_3O_2, C_5O_2, and $C_{12}O_9$, and a number of unstable or poorly characterized oxides including C_2O, C_2O_3, and the nonstoichiometric graphite oxide (p. 308). Of these, CO and CO_2 are of outstanding importance and their chemistry will be discussed in subsequent paragraphs after a few brief remarks about some of the others.

Tricarbon dioxide, C_3O_2, often called "carbon suboxide" and ponderously referred to in *Chemical Abstracts* as 1,2-propadiene-1,3-dione, is a foul-smelling gas obtained by dehydrating malonic acid at reduced pressure over P_4O_{10} at 140°: it has mp $-112.5°$, bp 6.7°, is stable at $-78°$, and polymerizes at room temperature to a yellow solid. C_3O_2 forms linear molecules ($D_{\infty h}$ symmetry) which can be written as $O{=}C{=}C{=}C{=}O$, consistent with the short interatomic distances C–C 128 pm and C–O 116 pm. Above 100°, polymerization yields a ruby-red solid; at 400° the product is violet and at 500° the polymer decomposes to C. The basic structure of all the polymers appears to be a polycyclic 6-membered lactone (Fig. 8.13).

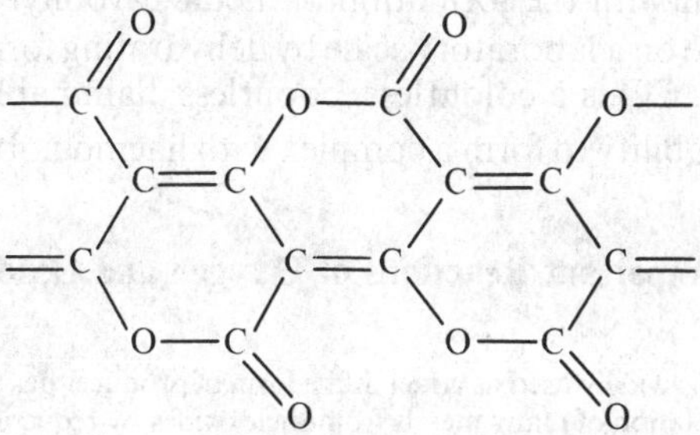

FIG. 8.13

C_3O_2 readily rehydrates to malonic acid, $CH_2(CO_2H)_2$, and reacts with NH_3 and HCl to give respectively the corresponding amide and acid chloride: $CH_2(CONH_2)_2$ and $CH_2(COCl)_2$. Thermolysis of C_3O_2 in a flow system has been reported to give a liquid product C_5O_2 though this has been disputed; if the compound is confirmed it would be the next member of the linear catenated series OC_nO with n odd as required by simple π-bond theory. The other moderately stable lower oxide is $C_{12}O_9$, a white sublimable solid which is the anhydride of mellitic acid (Fig. 8.14).

Direct oxidation of C in a limited supply of oxygen or air yields CO; in a free supply CO_2 results. Some properties of these familiar gases are in Table 8.6. The great strength of the

TABLE 8.6 *Some properties of CO, CO_2, and C_3O_2*

Property	CO	CO_2	C_3O_2
MP/°C	-205.1	-56.6 (5.2 atm)	-112.5
BP/°C	-191.5	-78.5 (subl)	6.7
ΔH_f°/kJ mol^{-1}	-110.5	-393.5	$+97.8$
Distance (C–O)/pm	112.8	116.3	116
Distance (C–C)/pm	—	—	128
D(C–O)/kJ mol^{-1}	1070.3	531.4	—

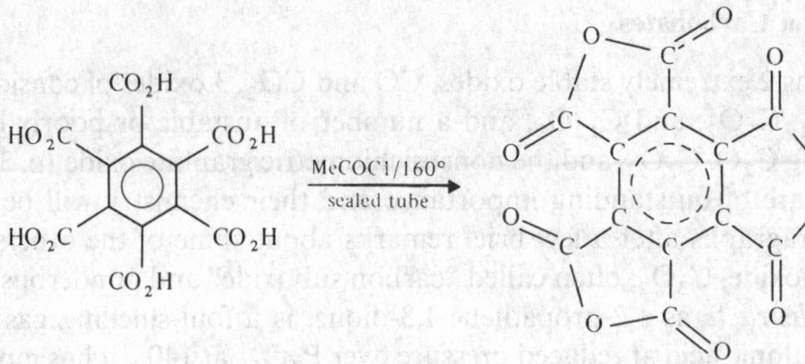

FIG. 8.14

C–O bond confers considerable thermal stability on these molecules but the compounds are also quite reactive chemically, and many of the reactions are of major industrial importance. Some of these are discussed more fully in the Panel.

The nature of the bonding, particularly in CO, has excited much attention because of the unusual coordination number (1) and oxidation state (+2) of carbon: it is discussed on p. 349 in connection with the formation of metal–carbonyl complexes.

Pure CO can be made on a laboratory scale by dehydrating formic acid (HCOOH) with conc H_2SO_4 at $\sim 140°$. CO is a colourless, odourless, flammable gas; it has a relatively high toxicity due to its ability to form a complex with haemoglobin that is some 300 times

---

Industrially Important Reactions of Oxygen and Oxides with Carbon

Carbon monoxide is widely used as a fuel in the form of producer gas or water gas and is also formed during the isolation of many metals from their oxides by reduction with coke. Producer gas is obtained by blowing air through incandescent coke and consists of about 25% CO, 4% CO_2, and 70% N_2, together with traces of H_2, CH_4, and O_2. The reactions occurring during production are:

$$2C + O_2 \longrightarrow 2CO; \quad \Delta H° = -221.0 \text{ kJ/mol } O_2; \quad \Delta S° + 179.4 \text{ J K}^{-1} \text{ mol}^{-1}$$

$$C + O_2 \longrightarrow CO_2; \quad \Delta H° = -393.5 \text{ kJ mol}^{-1}; \quad \Delta S° + 2.89 \text{ J K}^{-1} \text{ mol}^{-1}$$

Water gas is made by blowing steam through incandescent coke: it consists of about 50% H_2, 40% CO, 5% CO_2, and 5% $N_2 + CH_4$. The oxidation of C by H_2O is strongly endothermic:

$$C + H_2O \longrightarrow CO + H_2; \quad \Delta H° = +131.3 \text{ kJ mol}^{-1}; \quad \Delta S° + 133.7 \text{ J K}^{-1} \text{ mol}^{-1}$$

Consequently, the coke cools down and the steam must be intermittently replaced by a flow of air to reheat the coke.

At high temperatures, particularly in the presence of metal catalysts, CO undergoes reversible disproportionation:[†]

$$2CO \rightleftharpoons C + CO_2; \quad \Delta H° = -172.5 \text{ kJ/mol } CO_2; \quad \Delta S° = -176.5 \text{ J K}^{-1} \text{ mol}^{-1}$$

[†] Very recently it has been shown that, at all pressures, there is a fairly wide range of temperatures in which CO_2 dissociates directly into CO and O_2 without precipitation of carbon:[11a]

$$CO_2 \rightleftharpoons CO + \tfrac{1}{2}O_2$$

For example, the temperature range is 250–370°C at 10^{-2} atm, 320–480°C at 1 atm, and 405–630°C at 100 atm. At higher temperatures in each case, C is also formed, but always in the presence of some O_2.

[11a] M. H. LIETZKE and C. MULLINS, Thermal decomposition of carbon dioxide, *J. Inorg. Nucl. Chem.* **43**, 1769–71 (1981).

The equilibrium concentration of CO is 10% at 550°C and 99% at 1000°. As the forward reaction involves a reduction in the number of gaseous molecules in the system it is accompanied by a large decrease in entropy. Remembering that $\Delta G = \Delta H - T\Delta S$ this implies that the reverse reaction becomes progressively more favoured at higher temperatures. The thermodynamic data for the formation of CO and CO_2 can be represented diagrammatically on an Ellingham diagram (Fig. A) which plots standard free energy changes per mol of O_2 as a function of the absolute temperature.

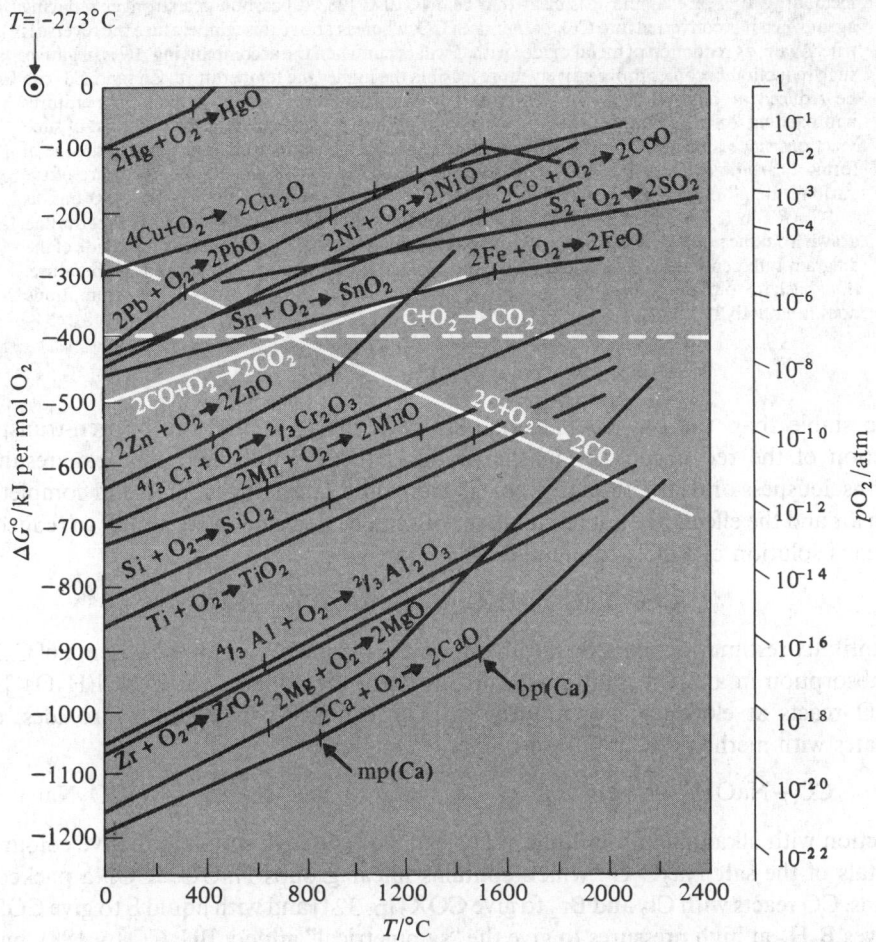

FIG. A Ellingham diagram for the free energy of formation of metallic oxides. (After F. D. Richardson and J. H. E. Jeffes, *J. Iron Steel Inst.* **160**, 261 (1948).) The oxygen dissociation pressure of a given M–MO system at a given temperature is obtained by joining ⊙ on the top left hand to the appropriate point on the M–MO free-energy line, and extrapolating to the scale on the right hand ordinate for p_{O_2} (atm).

The oxidation of C to CO results in an increase in the number of gaseous molecules; it is therefore accompanied by a large increase in entropy and is favoured at high temperature. By contrast, oxidation to CO_2 leaves the number of gaseous molecules unchanged; there is little change in entropy ($\Delta S° 2.93$ J K^{-1} mol^{-1}), and the free energy is almost independent of temperature. The two lines (and that for the oxidation of CO to CO_2) intersect at 983 K; it follows that ΔG for the disproportionation reaction is zero at this temperature. The diagram also includes the plots of ΔG (per mole of O_2) for the oxidation of several representative metals. On the left of the diagram (at

$T=0$ K) $\Delta G = \Delta H$ and the sequence of elements is approximately that of the electrochemical series. The slope of most of the lines is similar and corresponds to the loss of 1 mol of gaseous O_2; small changes of slope occur at the temperature of phase changes or the mp of the metal, and a more dramatic increase in slope signals the bp of the metal. For example, for MgO(s), the slope increases about three-fold at the bp of Mg since, above this temperature, reaction removes 3 gaseous species $(2Mg + O_2)$ rather than 1 (O_2).

Such diagrams are of great value in codefying a mass of information of use in extractive metallurgy.[12] For example, it is clear that below 710°C (983 K) carbon is a stronger reducing agent when it is converted into CO_2 rather than CO, whereas above this temperature the reverse is true. Again, as reduction of metal oxides with C will occur when the accompanying ΔG is negative, such reduction becomes progressively more feasible the higher the temperature: Zn (and Cd) can be reduced at relatively low temperatures but MgO can only be reduced at temperatures approaching 2000 K. Caution should be exercised, however, in predicting the outcome of such reactions since a number of otherwise reasonable reductions cannot be used because the metal forms a carbide (e.g. Cr, Ti). The temperature at which the oxygen dissociation pressure of the various metal oxides reaches a given value can also be obtained from the diagram: as $-\Delta G = RT \ln K_p [= 2.303RT \log\{p(O_2)/atm\}$ for the reactions considered] it follows that the line drawn from the point $\odot$ ($\Delta G = 0$, $T = 0$) to the appropriate scale mark on the right-hand side of the diagram intercepts the free-energy line for the element concerned at the required temperature. (Establish to your own satisfaction that this statement is approximately true—what assumptions does it embody?)

more stable than the oxygen–haemoglobin complex (p. 1276): the oxygen-transport function of the red corpuscles in the blood is thereby impeded. This can result in unconsciousness or death, though recovery from mild poisoning is rapid and complete in fresh air and the effects are not cumulative. CO can be detected by its ability to reduce an aqueous solution of $PdCl_2$ to metallic Pd:

$$CO + PdCl_2 + H_2O \longrightarrow Pd + CO_2 + 2HCl$$

Quantitative estimation relies on the liberation of I_2 from I_2O_5 or (in the absence of C_2H_2) on absorption in an acid solution of CuCl to form the adduct $[Cu(CO)Cl(H_2O)_2]$.

CO reacts at elevated temperatures to give formates with alkali hydroxides, and acetates with methoxides:

$$CO + NaOH \longrightarrow HCO_2Na; \qquad CO + MeONa \longrightarrow MeCO_2Na$$

Reaction with alkali metals in liquid NH_3 leads to reductive coupling to give colourless crystals of the salt $Na_2C_2O_2$ which contains linear groups $NaOC{\equiv}CONa$ packed in chains. CO reacts with Cl_2 and Br_2 to give COX_2 (p. 324) and with liquid S to give COS. It cleaves B_2H_6 at high pressures to give the "symmetrical" adduct BH_3CO (p. 188), but in the presence of $NaBH_4/thf$ the reaction takes a different course to yield *B*-trimethylboroxine:

[12] C. B. ALCOCK, *Principles of Pyrometallurgy*, Academic Press, London, 1976, 348 pp.

With BR_3, CO inserts in successive stages to give, ultimately, the corresponding trialkylmethylboroxine $(R_3CBO)_3$. Alternative products are obtained in the presence of other reagents, e.g. aqueous alkali yields R_3COH; water followed by alkaline peroxide yields R_2CO; and alkaline $NaBH_4$ yields RCH_2OH (p. 190). CO can also insert into M–C bonds (M=Mo, W; Mn, Fe, Co; Ni, Pd, Pt):

$$MeMn(CO)_5 + CO \longrightarrow MeC(O)Mn(CO)_5$$

A detailed discussion of CO as a ligand and the chemistry of metal carbonyls is on pp. 349–352. CO is a key intermediate in the catalytic production of a wide variety of organic compounds on an industrial scale.[13, 13a, 13b]

CO_2 is much less volatile than CO (p. 325). It is a major industrial chemical but its uses, though occasionally chemical, more frequently depend on its properties as a refrigerant, as an inert atmosphere, or as a carbonating (gasifying) agent in drinks and foam plastic (see Panel). Of more chemical interest is the synthesis of radioactive ^{14}C compounds from $^{14}CO_2$ which is conveniently stored as a carbonate. ^{14}C is generated by an (n, p) reaction on a nitride or nitrate in a nuclear reactor (see p. 1456). More than 500 compounds specifically labelled with ^{14}C are now available commercially, the starting point of many of the syntheses being one of the following reactions:

1. $NaH^{14}CO_3 + H_2/Pd/C \longrightarrow H^{14}CO_2H$

2. $^{14}CO_2 + RMgX \longrightarrow R^{14}CO_2H$

3. $^{14}CO_2 + LiAlH_4 \longrightarrow {}^{14}CH_3OH$

4. $Ba^{14}CO_3 + Ba \longrightarrow Ba^{14}C_2 \xrightarrow{H_2O} {}^{14}C_2H_2$

5. $Ba^{14}CO_3 + NH_3 \longrightarrow Ba^{14}CN_2 \longrightarrow {}^{14}C/N$ compounds

When CO_2 dissolves in water at 25° it is only partly hydrated to carbonic acid according to the equilibrium

$$H_2CO_3 \rightleftharpoons CO_2 + H_2O; \quad K = [CO_2]/[H_2CO_3] \approx 600$$

Interpretation of acid–base behaviour in this system is further complicated by the slowness of some of the reactions and their dependence on pH. The main reactions are:

$$\left.\begin{array}{l} CO_2 + H_2O \rightleftharpoons H_2CO_3 \text{ (slow)} \\ H_2CO_3 + OH^- \rightleftharpoons HCO_3^- + H_2O \text{ (fast)} \end{array}\right\} pH < 8$$

$$\left.\begin{array}{l} CO_2 + OH^- \rightleftharpoons HCO_3^- \text{ (slow)} \\ HCO_3^- + OH^- \rightleftharpoons CO_3^{2-} + H_2O \text{ (fast)} \end{array}\right\} pH > 10$$

In the range pH 8–10 both sets of equilibria are important. The apparent dissociation constant of carbonic acid is

$$K_1 = [H^+][HCO_3^-]/[CO_2 + H_2CO_3] = 4.45 \times 10^{-7} \text{ mol } l^{-1}$$

[13] Ref. 4, Vol. 4, Carbon monoxide, pp. 772–93; Carbonyls, pp. 794–814.
[13a] R. L. PRUETT, Hydroformylation, *Adv. Organometallic Chem.* **17**, 1–60 (1979).
[13b] C. MASTERS, The Fischer–Tropsch Reaction, *Adv. Organometallic Chem.* **17**, 61–103 (1979).

Production and Uses of CO_2

CO_2 can be readily obtained in small amounts by the action of acids on carbonates. On an industrial scale the main source is as a byproduct of the synthetic ammonia process in which the H_2 required is generated either by the catalytic reaction

$$CH_4 + 2H_2O \longrightarrow CO_2 + 4H_2$$

or by the water–gas shift reaction

$$CO + H_2O \rightleftharpoons CO_2 + H_2$$

CO_2 is also recovered economically from the flue gases resulting from combustion of carbonaceous fuels, from fermentation processes, and from the calcination of limestone: recovery is by reversible absorption either in aqueous Na_2CO_3 or aqueous ethanolamine (Girbotol process).

$$Na_2CO_3 + H_2O + CO_2 \underset{}{\overset{cool}{\rightleftharpoons}} 2NaHCO_3$$

$$2HOC_2H_4NH_2 + H_2O + CO_2 \underset{100-150°}{\overset{heat \atop 25-65°}{\rightleftharpoons}} (HOC_2H_4NH_3)_2CO_3$$

In certain places CO_2 can be obtained from natural gas wells. H_2S impurity is removed by oxidation using a buffered alkaline solution saturated with $KMnO_4$:

$$3H_2S + 2KMnO_4 + 2CO_2 \longrightarrow 3S + 2MnO_2 + 2KHCO_3 + 2H_2O$$

The scale of production has increased rapidly in recent years and in 1980 exceeded 33 million tonnes in the USA alone.

The most extensive application of CO_2 is as a refrigerant, some 52% of production being consumed in this way. CO_2 can be liquefied at any temperature between its triple point $-56.6°$ and its critical point $+31.0°$ (75.28 atm). The gas can either be pressurized to 75 atm and then water-cooled to room temperature, or precooled to about $-15°$ ($\pm 5°$) and then pressurized to 15.25 atm. Solid CO_2 is obtained by expanding liquid CO_2 from cylinders to give a "snow" which is then mechanically compressed into blocks of convenient size. Until about 20 y ago the bulk of CO_2 refrigerant was in the form of solid CO_2, but since 1960 production of liquid CO_2 has overtaken the solid form because of lower production costs and ease of transporting and metering the material. Some typical production figures are shown in Table 8.7:

TABLE 8.7   *USA production of* CO_2

CO_2 production/kilotonnes	1955	1960	1962	1977
Solid	520	426	406	340
Liquid and gas	185	432	522	1660
Total	705	858	928	2000

Solid CO_2 is used as a refrigerant for ice-cream, meat, and frozen foods, and as a convenient laboratory cooling agent and refrigerant. Liquid CO_2 is extensively used to improve the grindability of low-melting metals (and hamburger meat), and for the rapid cooling of loaded trucks and rail cars; it is also used for inflating life rafts, in fire extinguishers, and in blasting shells for coal mining. A related application of growing importance is as a replacement for chlorofluorocarbon aerosol propellants (p. 323) though this application will never consume large amounts of the gas since the amount in each tin is extremely small.

Gaseous CO_2 is extensively used to carbonate soft drinks and this use alone accounts for 25% of production.† Other quasi-chemical applications are its use as a gas purge, as an inert protective

---

† Some idea of the colossal consumption of carbonated beverages can be gained from US statistics; in 1975 the number of cases [240-ml (i.e. 8 oz) cans or bottles] produced was 3 916 000 000, i.e. 439 bottles per head of population.

gas for welding, and for the neutralization of caustic and alkaline waste waters. Small amounts are also used in the manufacture of sodium salicylate, basic lead carbonate ("white lead"), and various carbonates such as $M_2^ICO_3$ and M^IHCO_3 (M^I = Na, K, NH_4, etc.).

A substantial amount of CO_2 is now used to manufacture urea via ammonium carbamate:

$$CO_2 + 2NH_3 \xrightarrow[200\ atm]{185°} NH_2CO_2NH_4 \xrightarrow{-H_2O} CO(NH_2)_2$$

Urea is used to make urea-formaldehyde plastics and resins and, increasingly, as a nitrogenous fertilizer (46.7% N). Urea is readily soluble in H_2O (51.6 g per 100 g solution at 20°) and it hydrolyses very slowly firstly to ammonium carbamate and thence to $(NH_4)_2CO_3$; this is the basis for its use as a fertilizer. World production of urea was 6.5 million tonnes in 1970.

As $[CO_2]/[H_2CO_3] = K \approx 600$, it follows that the true dissociation constant is:

$$K_a = [H^+][HCO_3^-]/[H_2CO_3] = K_1(1+K) \approx 2.5 \times 10^{-4} \text{ mol } l^{-1}$$

This value is in the range expected from an acid of structure $(HO)_2CO$ (p. 54). The second dissociation constant is given by

$$K_2 = [H^+][CO_3^{2-}]/[HCO_3^-] = 4.84 \times 10^{-11} \text{ mol } l^{-1}$$

A hydrate $CO_2.8H_2O$ can be crystallized from aqueous solutions at 0° and $p(CO_2) \sim 45$ atm.

The coordination chemistry of CO_2 is by no means as extensive as that of CO (p. 349) but some exciting developments have recently been published.[14] The first transition metal complexes with CO_2 were claimed by M. E. Volpin's group in 1969: tertiary phosphine or N_2 ligands were displaced from Rh and Ni complexes to give binuclear products whose definitive structure has not yet been established. CO_2 also displaced N_2 from $[Co(N_2)(PPh_3)_3]$ to give $[Co(CO_2)(PPh_3)_3]$. The Ni^0 complexes $[Ni(PEt_3)_4]$ (violet) and $[Ni(PBu_3^n)_4]$ (red) react in toluene at room temperature with CO_2 (1 atm) to give the yellow complexes $[Ni(CO_2)L_3]$. The structure of the analogous complex with $P(C_6H_{11})_3$ was established by X-ray diffraction analysis; it features a pseudo-3-coordinate Ni atom μ-bonded to a bent CO_2 ligand as in Fig. 8.15. The isoelectronic Rh^I appears to

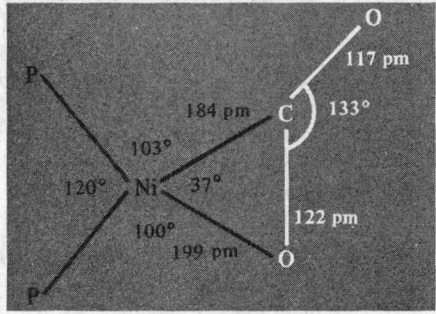

FIG. 8.15   Coordination about the Ni atom in the complex
$[Ni(CO_2)\{P(C_6H_{11})_3\}_2]0.75C_6H_5Me$.

[14] M. E. VOLPIN and I. S. KOLOMNIKOV, The reactions of organometallic compounds with carbon dioxide, *Organometallic Reactions* **5**, 313–86 (1975). Further references to isolable CO_2-transition metal adducts are given in R. L. HARLOW, J. B. KINNEY, and T. HERSKOVITZ, *JCS Chem. Comm.* 1980, 813–14. See also G. S. BRISTOW, P. B. HITCHCOCK, and M. F. LAPPERT, A novel carbon dioxide complex: Synthesis and crystal structure of $[Nb(\eta-C_5H_4Me)_2(CH_2SiMe_3)(\eta^2-CO_2)]$, *JCS Chem. Comm.* 1981, 1145–46.

form two types of complex: an orange-red series $[Rh(CO_2)ClL_2]$ (L = tertiary phosphine) with a μ-bonded bent CO_2 as in Fig. 8.15 and a somewhat less-stable yellow series $[Rh(CO_2)ClL_3]$ which is thought to contain the ligand configuration Rh—C$\lesssim^O_O$. A Pt compound which had earlier (1965) been thought to contain CO_2 as a ligand was subsequently found to require the presence of O_2 for its formation and to be, in fact, a novel bidentate carbonato complex (Fig. 8.16).[†]

$$[Pt(PPh_3)_3] + CO_2 + O_2 \xrightarrow{C_6H_6/25°} [Pt(CO_3)(PPh_3)_2] + Ph_3PO$$

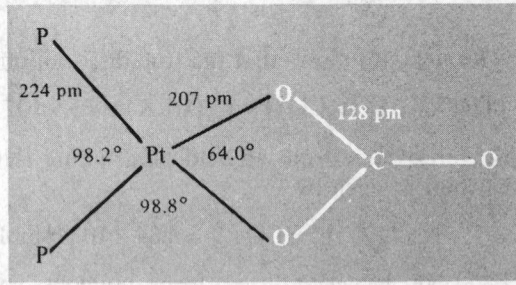

FIG. 8.16   Coordination about the Pt atom in the complex $[Pt(CO_3)(PPh_3)_2] \cdot C_6H_6$.

[†] The CO_3^{2-} ligand can be both bidentate and bridging as in the recently studied complex cation $[(CuL_2)_2(\mu\text{-}CO_3)]^{2+}$, where L is a tridentate macrocyclic triaza ligand (structure 1), [14a] and in the binuclear molecular complex molecule $[\{CuCl(Me_2NCH_2CH_2NMe_2\}_2(\mu\text{-}CO_3)]$ (structure 2).[14b] This mode of coordination confers some unusual properties including diamagnetism on these Cu^{II} complexes.

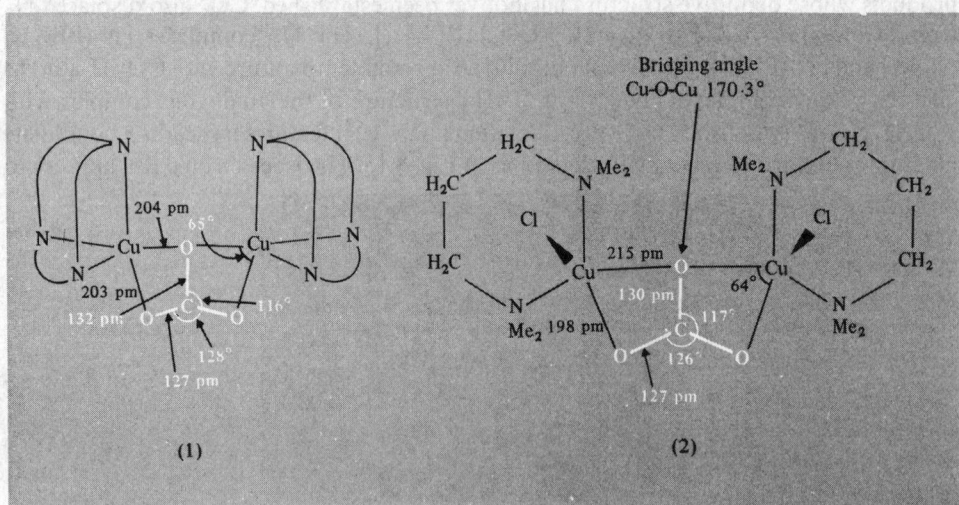

(1)   (2)

[14a] A. R. DAVIS, F. W. P. EINSTEIN, N. F. CURTIS, and J. W. L. MARTIN, A novel mode of carbonate binding. Structure of spin-paired μ-carbonato-bis(2,4,4,7-tetramethyl-1,5,9-triazacyclododec-1-ene)dicopper(II) perchlorate, *J. Am. Chem. Soc.* **100**, 6259–60 (1978).

[14b] M. R. CHURCHILL, G. DAVIES, M. A. EL-SAYED, M. F. EL-SHAZLY, J. P. HUTCHINSON, M. RUPICH, and K. O. WATKINS, Synthesis, physical properties, and structural characterization of μ-carbonato-dichlorobis(N,N,N',N'-tetramethyl-1,3-propanediamine)dicopper(II), [LCuCl(CO_3)ClCuL], a diamagnetic initiator for the oxidative coupling of phenols by dioxygen, *Inorg. Chem.* **18**, 2296–2300 (1979).

If the starting material contains M–H or M–C bonds a further complication can arise due to the possibility of a CO_2 insertion reaction. Thus, both $[Ru(H)_2(N_2)(PPh_3)_3]$ and $[Ru(H)_2(PPh_3)_4]$ react to give the formate $[Ru(H)(OOCH)(PPh_3)_3]$, and similar CO_2 insertions into M–H are known for M = Co, Fe, Os, Ir, Pt. These "normal" insertion reactions are consistent with the expected bond polarities $M^{\delta+}-H^{\delta-}$ and $O^{\delta-}=C^{\delta+}=O$, but occasionally "abnormal" insertion occurs to give metal carboxylic acids M–COOH. Likewise, normal insertion into M–C yields alkyl carboxylates M–OOCR, though metalloacid esters M–COOR are sometimes obtained. The reactions have obvious catalytic implications and are being actively studied at the present time by several groups.

CO_2 insertion into M–C bonds has, of course, been known since the first papers of V. Grignard in 1901 (p. 150). Organo-Li (and other M^I and M^{II}) also react extremely vigorously to give salts of carboxylic acids, RCO_2Li, $(RCO_2)_2Be$, etc. Zinc dialkyls are much less reactive towards CO_2, e.g.

$$ZnEt_2 + 2CO_2 \xrightarrow{150°} (EtCOO)_2Zn,$$

and organo-Cd and -Hg compounds are even less reactive. With AlR_3, one CO_2 inserts at room temperature and a second at 220° under pressure to give $R_2Al(OOCR)$ and $RAl(OOCR)_2$ respectively. B–C, Si–C, Ge–C, and Sn–C are rather inert to CO_2 but insertion readily occurs into bonds between these elements and N. A few examples are:

$$PhB(NHEt)_2 + CO_2 \xrightarrow{25°} PhB(OCONHEt)_2$$

$$Me_3SiNEt_2 + CO_2 \xrightarrow{Et_2NH/25°} Me_3SiOCONEt_2$$

$$Me_3SnNMe_2 + CO_2 \xrightarrow{20°} Me_3SnOCONMe_2$$

$$As(NMe_2)_3 + 1(3)CO_2 \xrightarrow{20°-40°} (Me_2N)_2AsOCONMe_2, As(OCONMe_2)_3$$

$$Ti(NMe_2)_4 + CO_2 \xrightarrow{20°} Ti(OCONMe_2)_4, \text{ etc.}$$

8.7   Chalcogenides and Related Compounds

Carbon forms several sulfides. CS (unlike CO) is an unstable reactive radical even at −196°: it reacts with the other chalcogens and with halogens to give CSSe, CSTe, and CSX_2. It is formed by action of a high-frequency discharge on CS_2 vapour. Passage of an electric discharge or arc through liquid or gaseous CS_2 yields C_3S_2, a red liquid mp −5° which polymerizes slowly at room temperature (cf. C_3O_2). By far the most important sulfide is CS_2, a colourless, volatile, flammable liquid (mp −111.6°, bp 46.25°, flash point −30°, auto-ignition temperature 100°, explosion limits in air 1.25–50%). Impure samples have a fetid almost nauseating stench due to organic impurities but the purified liquid has a rather pleasant ethereal smell; it is very poisonous and can have disastrous effects on the nervous system and brain. CS_2 was formerly manufactured by direct reaction of S vapour and coke in Fe or steel retorts at 750–1000°C but, since the early 1950s, the preferred synthesis has been the catalysed reaction between sulfur and natural gas:

$$CH_4 + 4S \xrightarrow[\text{SiO}_2 \text{ gel or Al}_2\text{O}_3]{\sim 600°} CS_2 + 2H_2S$$

World production in 1976 was over 1 million tonnes the principal industrial uses being in the manufacture of viscose rayon (35–50%), cellophane films (15%) (see below), and CCl_4 (15–30%) depending on country.

CS_2 reacts with aqueous alkali to give a mixture of M_2CO_3 and the trithiocarbonate M_2CS_3. NH_3 gives ammonium dithiocarbamate $NH_4[H_2NCS_2]$; under more forcing conditions in the presence of Al_2O_3 the product is NH_4CNS and this can be isomerized at 160° to thiourea, $(NH_2)_2CS$. Water itself reacts only reluctantly, yielding COS at 200° and $H_2S + CO_2$ at higher temperatures; many other oxocompounds also convert CS_2 to COS, e.g. MgO, SO_3, HSO_3Cl, and urea. With aqueous NaOH/EtOH carbon disulfide yields sodium ethyl dithiocarbonate (xanthate):

$$CS_2 + NaOH + EtOH \longrightarrow S{=}C\begin{smallmatrix} \diagup OEt \\ \diagdown SNa \end{smallmatrix}$$

When ethanol is replaced by cellulose, sodium cellulose xanthate is obtained; this dissolves in aqueous alkali to give a viscous solution (viscose) from which either viscose rayon or cellophane can be obtained by adding acid to regenerate the (reconstituted) cellulose. Trithiocarbonates (CS_3^{2-}), dithiocarbonates (COS_2^{2-}), xanthates (CS_2OR^-), dithiocarbamates ($CS_2NR_2^-$), and 1,2-dithiolates have an extensive coordination chemistry which has been reviewed.[15]

Chlorination of CS_2, when catalysed by $Fe/FeCl_3$, proceeds in two steps:

$$CS_2 + 3Cl_2 \longrightarrow CCl_4 + S_2Cl_2$$
$$CS_2 + 2S_2Cl_2 \longrightarrow CCl_4 + 6S$$

With I_2 as catalyst the main product is perchloromethylthiol (Cl_3CSCl). Reaction products with F_2 depend on the conditions used, typical products being SF_4, SF_6, S_2F_{10}, $F_2C(SF_3)_2$, $F_2C(SF_5)_2$, F_3CSF_5, and $F_3SCF_2SF_5$.

CS_2 is rather more reactive than CO_2 in forming complexes and in undergoing insertion reactions. The field was opened up by G. Wilkinson and his group in 1966 when they showed that $[Pt(PPh_3)_3]$ reacts rapidly and quantitatively with CS_2 at room temperature to give orange needles of $[Pt(CS_2)(PPh_3)_2]$, mp 170°. X-ray crystal diffraction analysis revealed the structure shown diagramatically in Fig. 8.17. The geometry of the bent CS_2 ligand is similar to that in the first excited state of the molecule and the CS_2 is almost coplanar with PtP_2 (dihedral angle 6°). Bonding is considered to involve a 1-electron transfer via the intermediary of Pt from the highest filled π MO of the ligand to its lowest

[15] G. D. THORN and R. A. LUDWIG, *The Dithiocarbamates and Related Compounds*, Elsevier 1962, 298 pp. J. A. MCCLEVERTY, Metal 1,2-dithiolene and related complexes, *Prog. Inorg. Chem.* **10**, 49–221 (1968) (188 refs.). D. COUCOUVANIS, The chemistry of the dithio acid and 1,1-dithiolate complexes, *Prog. Inorg. Chem.* **11**, 233–271 (1970) (516 refs.). R. E. EISENBERG, Structural systematics of 1,1- and 1,2-dithiolate chelates, *Prog. Inorg. Chem.* **12**, 295–369 (1971) (173 refs.).

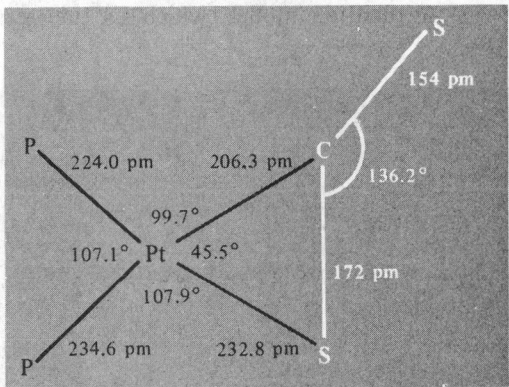

FIG. 8.17   Coordination about the Pt atom in $[Pt(CS_2)(PPh_3)_2]$.

antibonding MO, and the Pt can be thought of as being oxidized from Pt^0 to Pt^{II}. However, the substantial difference between the two Pt–P distances and the wide deviation of the angles of Pt from 90° emphasize the inadequacy of describing the bonding of such complicated species in terms of simple localized bonding theory. The orange complex $[Pd(CS_2)(PPh_3)_2]$ is isostructural and further work yielded deep-green $[V(\eta^5\text{-}C_5H_5)_2 (CS_2)]$, dimeric $[(Ph_3P)Ni(\mu\text{-}CS_2)_2Ni(PPh_3)]$, and various CS_2 complexes of Fe, Ru, Rh, and Ir. The deep-red complex $[Rh(CS_2)_2Cl(PPh_3)_2]$ probably involves pseudo-octahedral Rh^{III} with one of the CS_2 ligands η^2-bonded as above and the other one σ-bonded via a single S atom. The numerous η^1, η^2, and bridging modes of coordination now known for CS_2 are indicated schematically below:[15a]

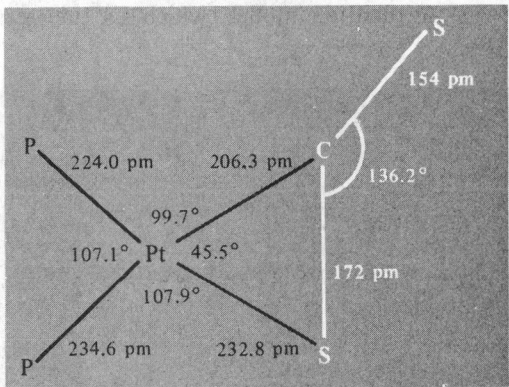

Insertion reactions of CS_2 are known for all the elements which undergo CO_2 insertion (p. 333) and also for M–N bonds involving Sb^{III}, Zr^{IV}, Nb^V, Ta^V, etc.

Stable thiocarbonyl complexes containing the elusive CS ligand are also now well

[15a] T. G. SOUTHERN, U. OEHMICHEN, J. Y. LE MAROUILLE, H. LE BOZEC, D. GRANDJEAN, and P. H. DIXNEUF, Use of organometallic ligands in the synthesis of CS_2-bridged heterodinuclear complexes. X-ray structure of $[(PhMe_2P)_2(CO)_2FeCS_2Mn(CO)_2(C_5H_5)]$, *Inorg. Chem.* **19**, 2976–80 (1980). Other references to recent key papers in this burgeoning field are: G. FACHINETTI, C. FLORIANI, A. CHIESI-VILLA, and C. GUESTINI, *JCS Dalton*, 1979, 1612–17. P. CONWAY, S. M. GRANT, and A. R. MANNING, *JCS Dalton* 1979, 1920–4. P. J. VERGAMINI and P. G. ELLER, *Inorg. Chim. Acta* **34**, L291–L292 (1979). C. BIANCHINI, A. MELI, A. ORLANDINI, and L. SACCONI, *Inorg. Chim. Acta* **35**, L375–L376 (1979). C. BIANCHINI, C. MEALLI, A. MELI, A. ORLANDINI, and L. SACCONI, *Angew. Chem. Int. Edn. (Engl.)*, **18**, 673–4 (1979). C. BIANCHINI, C. MEALLI, A. MELLI, A. ORLANDINI, and L. SACCONI, *Inorg. Chem.* **19**, 2968–75 (1980). W. P. FEHLHAMMER and H. STOLZENBERG, *Inorg. Chim. Acta* **44**, L151–L152 (1980), P. V. BROADHURST, B. F. G. JOHNSON, J. LEWIS, and P. R. RAITHBY, *JCS Chem. Comm.* 1982, 140–41, and references therein.

established and known coordination modes, which include terminal, bridging, and polyhapto, are as follows:[15b]

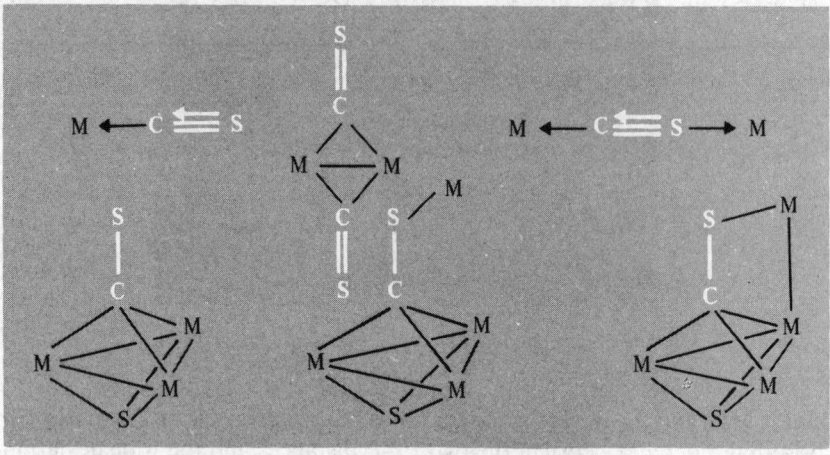

Likewise complexes of CSe and CTe have been characterized.[15c]

8.8 Cyanides and Other Carbon–Nitrogen Compounds

The chemistry of compounds containing the CN group is both extensive and varied. The types of compound to be discussed are listed in Table 8.8, which also summarizes some basic structural information. The names, cyanide, cyanogen, etc., refer to the property of forming deep-blue pigments such as Prussian blue with iron salts (Greek κύανος, *cyanos*, dark blue).

A useful theme for cohering much of the chemistry of compounds containing the CN group is the concept of pseudohalogens, a term introduced in 1925 for certain strongly bound, univalent radicals such as CN, OCN, SCN, SeCN, (and N_3, etc.). These groups can form anions X^-, hydracids HX, and sometimes neutral species X_2, XY, etc. It is also helpful to recognize that CN^- is isoelectronic with C_2^{2-} (p. 319) and with several notable ligands such as CO, N_2, and NO^+. Similarly, the cyanate ion OCN^- is isoelectronic with CO_2, N_3^-, fulminate (CNO^-), etc.

Cyanogen, $(CN)_2$, is a colourless poisonous gas (like HCN) mp $-27.9°$, bp $-21.2°$ (cf. Cl_2, Br_2). When pure it possesses considerable thermal stability (800°C) but trace

[15b] I. S. BUTLER, *Acc. Chem. Res.* **10**, 359–65 (1977). P. V. YANEFF, *Coord. Chem. Rev.* **23**, 183–220 (1977) (includes CS_2 complexes also). H. WERNER and K. LEONHARD, *Angew. Chem.*, Int. Edn. (Engl.), **18**, 627–8 (1979). H HERBERHOLD and P. D. SMITH, *Angew. Chem.*, Int. Edn. (Engl.) **18**, 631–2 (1979). W. W. GREAVES, R. J. ANGELICI, B. J. HELLAND, R. KLIMA, and R. A. JACOBSON, *J. Am. Chem. Soc.* **101**, 7618–20 (1979). F. FARONE, G. TRESOLDI, and G. A. LOPRETE, *JCS Dalton* 1979, 933–7; *JCS Dalton* 1979, 1053–6. P. V. BROADHURST, B. F. G. JOHNSON, J. LEWIS, and R. W. RAITHBY, *JCS Chem. Comm.* 1980, 812–13; *J. Am. Chem. Soc.* **103**, 3198–3200 (1981), and references therein.
[15c] G. R. CLARK, K. MARSDEN, W. R. ROPER, and L. J. WRIGHT, Carbonyl, thiocarbonyl, selenocarbonyl, and tellurocarbonyl complexes derived from a dichlorocarbene complex of osmium, *J. Am. Chem. Soc.* **102**, 1206–7 (1981), J.-P. BATTIONI, D. MANSUY, and J.-C. CHOTTARD, A new route to selenocarbonyl-transition-metal complexes, *Inorg. Chem.* **19**, 791–792 (1980).

TABLE 8.8   *Some compounds containing the CN group*

Name	Conventional formula	r (C–N)/pm	Remarks[a]
Cyanogen	N≡C—C≡N	115	Linear; r (C–C) 138 pm (short)
Paracyanogen	$(CN)_x$	—	Involatile polymer, see text
Hydrogen cyanide	H—C≡N	115.6	Linear; r (C–H) 106.5 pm
Cyanide ion	(C≡N)⁻	116	r_{eff} 192 pm when "freely rotating" in MCN
Cyanides (nitriles)	M—C≡N (R—C≡N)	115.8	Linear; r (C–C) 146.0 pm (for MeCN)
Isocyanides	R—N≡C	116.7	Linear, r (H_3C–N) 142.6 pm (for MeNC). Coordinated isocyanides are slightly bent, e.g. [M(←C≡N—C_6H_5)$_6$] angle CNC 173°, r (C≡N) 117.6 pm; bridging modes also known, e.g.:

Cyanogen halides (halogen cyanides)	X—C≡N	116	Linear
Cyanamide	H_2N—C≡N	115	Linear NCN; r (C–NH_2) 131 pm
Dicyandiamide	N≡C—N=C(NH_2)$_2$	122–136	

Cyanuric compounds	{—C(X)=N—}$_3$	134	Cyclic trimers; X = halogen, OH, NH_2
Cyanate ion	(O—C≡N)⁻	~121	Linear
Isocyanates	R—N=C=O	120	Linear NCO; ∠RNC ~126°
Fulminate ion	⟩(C=N—O)⁻	109	Linear; another form of AgCNO has r(C–N) 112 pm
Thiocyanate ion	(S—C≡N)⁻	115	Linear
Thiocyanates	R—S—C≡N (M—S—C≡N)	116	Linear NCS; ∠RSC 100° in MeSCN; ∠MSC variable (80–107°)
Isothiocyanates	R—N=C=S	122	Linear NCS; ∠HNC 135° in HNCS; ∠MNC variable (111–180°)
Selenocyanate ion	(Se—C≡N)⁻	~112	Linear NCSe

[a] Several groups can also act as bridging ligands in metal complexes, e.g. —CN—, ⟩NCO, —SCN—.

impurities normally facilitate polymerization at 300–500° to paracyanogen a dark-coloured solid which may have a condensed polycyclic structure.

The polymer reverts to $(CN)_2$ above 800° and to CN radicals above 850°. $(CN)_2$ can be prepared in 80% yield by mild oxidation of CN^- with aqueous Cu^{II}; the reaction is complex but can be idealized as

$$2CuSO_4 + 4KCN \xrightarrow{H_2O/60°} (CN)_2 + 2CuCN + 2K_2SO_4$$

CO_2 which is also formed (20%) can be removed by passage of the product gas over solid NaOH and the byproduct CuCN can be further oxidized with hot aqueous Fe^{III} to complete the conversion:

$$2CuCN + 2FeCl_3 \xrightarrow{H_2O/heat} (CN)_2 + 2CuCl + 2FeCl_2$$

Industrially it is now made by direct gas-phase oxidation of HCN with O_2 (over a silver catalyst), or with Cl_2 (over activated charcoal), or NO_2 (over CaO glass). $(CN)_2$ is fairly stable in H_2O, EtOH, and Et_2O but slowly decomposes in solution to give HCN, HNCO, $(H_2N)_2CO$, and $H_2NC(O)C(O)NH_2$ (oxamide). Alkaline solutions yield CN^- and $(OCN)^-$ (cf. halogens).

$$(CN)_2 + 2OH^- \longrightarrow CN^- + OCN^- + H_2O$$

Hydrogen cyanide, mp $-13.4°$ bp 25.6°, is an extremely poisonous compound of very high dielectric constant (p. 59). It is miscible with H_2O, EtOH, and Et_2O. In aqueous solution it is an even weaker acid than HF, the dissociation constant K_a being 2.1×10^{-9}. It was formerly produced industrially by acidifying NaCN or $Ca(CN)_2$ but the most modern catalytic processes are based on direct reaction between CH_4 and NH_3, e.g.:[15d]

Andrussow process: $CH_4 + NH_3 + 1\frac{1}{2}O_2 \xrightarrow[\text{2 atm/1000–1200°}]{\text{Pt/Rh or Pt/Ir}} HCN + 3H_2O$

Degussa process: $CH_4 + NH_3 \xrightarrow[\text{1200–1300°}]{\text{Pt}} HCN + 3H_2$

Both processes rely on a fast flow system and the rapid quenching of product gases; yields of up to 90% can be attained. It is salutory to note that US production of this highly toxic compound now approaches 300 000 tonnes pa (1980) and world production exceeds 500 000 tonnes pa. Of this, 60% is used to manufacture methyl methacrylate ($HCN + Me_2CO \longrightarrow$ acetone cyanohydrin $\xrightarrow{H_2SO_4}$ methacrylamide sulfate $\xrightarrow{MeOH}$ methyl methacrylate); this represents a major change from 15 y ago when 51% of HCN production was used to manufacture acrylonitrile (by addition to C_2H_2) and only 18% was used for methyl methacrylate. HCN is now also used to make $(ClCN)_3$ (15%), NaCN (10%), chelates (10%), hexacyanoferrates, etc.

As noted above, CN^-(aq) is fairly easily oxidized to $(CN)_2$ or OCN^-; $E°$ values calculated from free energy data (p. 498) are:

$$\frac{1}{2}(CN)_2 + H^+ + e^- \rightleftharpoons HCN; \qquad E° + 0.37 \text{ V}$$
$$OCN^- + 2H^+ + 2e^- \rightleftharpoons CN^- + H_2O; \quad E° -0.14 \text{ V}$$

[15d] Ref. 4, Vol. 7 (1979), Cyanides (including HCN, M^ICN, and $M^{II}(CN)_2$, pp. 307–19; Cyanamides (including CaNCN, H_2NCN, dicyandiamide, and melamine), pp. 291–306.

HCN can also be reduced to $MeNH_2$ by powerful reducing agents such as Pd/H_2 at $140°$.

The alkali metal cyanides MCN are produced by direct neutralization of HCN; they crystallize with the NaCl structure (M = Na, K, Rb) or the CsCl structure (M = Cs, Tl) consistent with "free" rotation of the CN^- group. The effective radius is ~ 190 pm, intermediate between those of Cl^- and Br^-. At lower temperatures the structures transform to lower symmetries as a result of alignment of the CN^- ions. LiCN differs in having a loosely packed 4-coordinate arrangement and this explains its low density (1.025 g cm^{-3}) and unusually low mp ($160°$, cf. NaCN $564°$). World production of alkali metal cyanides was $\sim 113\,000$ tonnes in 1977. NaCN readily complexes metallic Ag and Au under mildly oxidizing conditions and is much used in the extraction of these metals from their low-grade ores (first patented in 1888 by R. W. Forrest, W. Forrest, and J. S. McArthur):

$$8NaCN + 4M + 2H_2O + O_2 \longrightarrow 4Na[M(CN)_2] + 4NaOH$$

Until the 1960s, when HCN became widely available, NaCN was made by the Castner process via sodamide and sodium cyanamide:

$$2Na + C + 2NH_3 \xrightarrow{750°} 2NaCN + 3H_2$$

The CN^- ion can act either as a monodentate or bidentate ligand.[16] Because of the similarity of electron density at C and N it is not usually possible to decide from X-ray data whether C or N is the donor atom in monodentate complexes, but in those cases where the matter has been established by neutron diffraction C is always found to be the donor atom (as with CO). Very frequently CN^- acts as a bridging ligand —CN— as in AgCN, and AuCN (both of which are infinite linear chain polymers), and in Prussian-blue type compounds (p. 1271). The same tendency for a coordinated M–CN group to form a further donor–acceptor bond using the lone-pair of electrons on the N atom is illustrated by the mononuclear BF_3 complexes with tetracyanonickelates and hexacyanoferrates, e.g. $K_2[Ni(CN.BF_3)_4]$ and $K_4[Fe(CN.BF_3)_6]$.

The complex $CuCN.NH_3$ provides the only example so far of CN acting as a bridging ligand at C, a mode which is common in μ-CO complexes (p. 351); indeed, the complex is unique in featuring tridentate CN groups which link the metal atoms into plane nets via the grouping $\begin{smallmatrix} Cu \\ \vdots \\ Cu \end{smallmatrix}\!\!\!>\!\!C\!-\!N\!-\!Cu$ as shown in Fig. 8.18. Other cyanide complexes are discussed under the appropriate metals. In organic chemistry, both nitriles R–CN and isonitriles (isocyanides) R–NC are known. Isocyanides have been extensively studied as ligands (p. 349).[17]

Cyanogen halides, X–CN, are colourless, volatile, reactive compounds which can be regarded as pseudohalogen analogues of the interhalogen compounds, XY (p. 946) (Table 8.9). All tend to trimerize to give cyclic cyanuric halides (Fig. 8.19) especially in the presence of free HX. FCN is prepared by pyrolysis of $(FCN)_3$ which in turn is made by fluorinating $(ClCN)_3$ with NaF in tetramethylene sulfone. ClCN and BrCN are prepared by direct reaction of X_2 on MCN in water or CCl_4, and ICN is prepared by a dry route

[16] A. G. SHARPE, *The Chemistry of Cyano Complexes of the Transition Metals*, Academic Press, London, 1976, 302 pp.
[17] L. MALATESTA and F. BONATI, *Isocyanide Complexes of Metals*, Wiley, London, 1969, 199 pp.

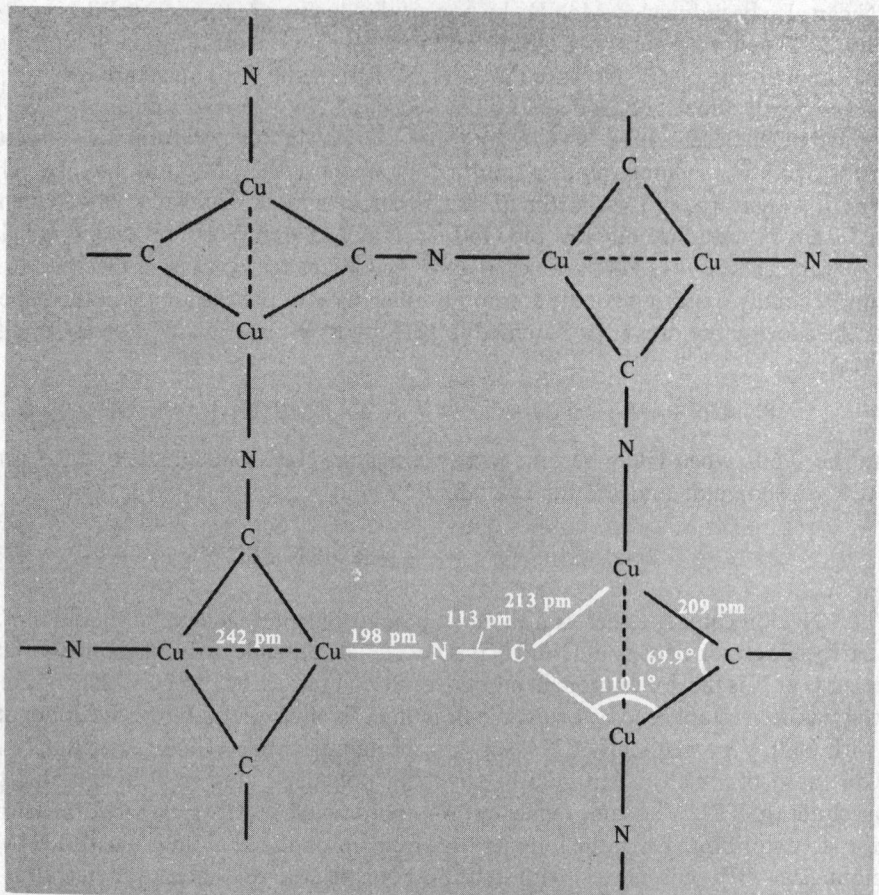

FIG. 8.18 Schematic diagram of the layer structure of CuCN . NH$_3$ showing the tridentate CN groups; each Cu is also bonded to 1 NH$_3$ molecule at 207 pm. Note also the unusual 5-coordination of Cu including one near neighbour Cu at 242 pm (13 pm closer than Cu–Cu in the metal). The lines in the diagram delineate the geometry and do not represent pairs of electrons.

TABLE 8.9 *Cyanogen halides*

Property	FCN	ClCN	BrCN	ICN
MP/°C	−82	−6.9	51.3	146
BP/°C	−46	13.0	61.3	146 (subl)

from Hg(CN)$_2$ and I$_2$. Similarly, colourless crystals of cyanamide (H$_2$NCN mp 46°) result from the reaction of NH$_3$ on ClCN and trimerize to melamine at 150° (Fig. 8.19). The industrial preparation is by acidifying CaNCN (see Panel). The "dimer", dicyandiamide, NCNC(NH$_2$)$_2$, can be made by boiling calcium cyanamide with water: the colourless crystals are composed of nonlinear molecules which feature three different C–N distances (see Table 8.8).

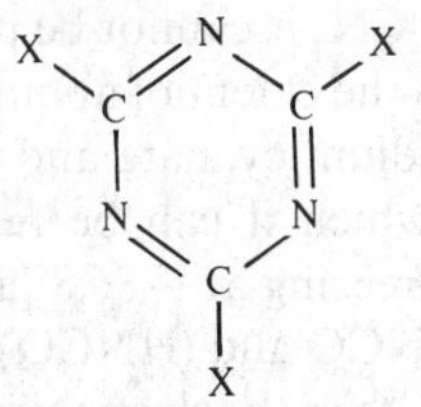

Cyanuric halides 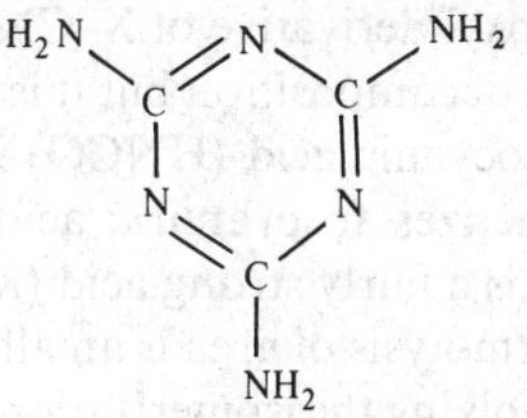   Cyanuric amide (melamine)

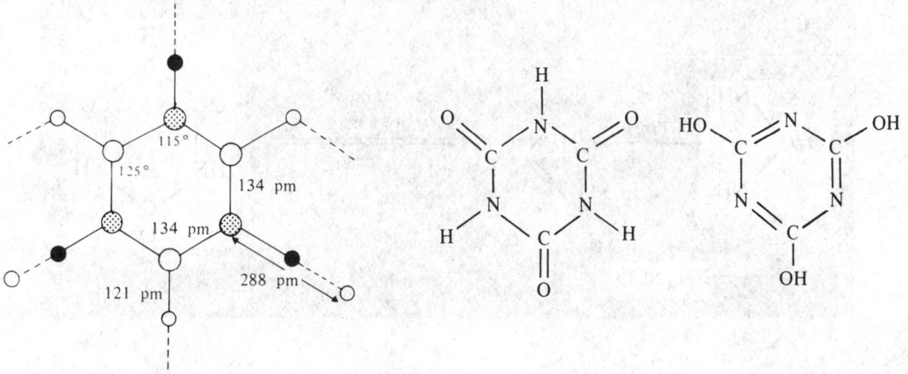

Cyanuric acid: H-bonded crystal structure and alternative valence-bond formulations

FIG. 8.19   The planar structure of various cyanuric compounds: all 6 C–N distances within the ring are equal.

The Cyanamide Industry[15d]

The basic chemical of the cyanamide industry is calcium cyanamide $CaNCN$, mp 1340°, obtained by nitrogenation of CaC_2.

$$CaC_2 + N_2 \xrightarrow{1000°C} CaNCN + C$$

$CaNCN$ is used as a direct application fertilizer, weed killer, and cotton defoliant; it is also used for producing cyanamide, dicyandiamide, and melamine plastics. Production formerly exceeded 1.3 million tonnes pa, but this has fallen considerably in the last few years, particularly in the USA where the use of $CaNCN$ as a nitrogenous fertilizer has gradually been replaced by other materials. In 1975 most of the world's supply was made in Japan, the Federal Republic of Germany, and Canada.

Acidification of $CaNCN$ yields free cyanamide, H_2NCN which reacts further to give differing products depending on pH: at pH $\leqslant 2$ or > 12 urea is formed, but at pH 7–9 dimerization to dicyandiamide $NCNC(NH_2)_2$ occurs. Solutions are most stable at pH ~ 5; accordingly commercial preparation of H_2NCN is by continuous carbonation of an aqueous slurry of $CaNCN$ in the presence of graphite: the overall reaction can be represented

$$CaNCN + CO_2 + H_2O \longrightarrow H_2NCN + CaCO_3$$

Reaction of H_2NCN with H_2S gives thiourea, $SC(NH_2)_2$.

Dicyandiamide forms white, non-hygroscopic crystals which melt with decomposition at 209°. Its most important reaction is conversion to melamine (Fig. 8.19) by pyrolysis above the mp under a pressure of NH_3 to counteract the tendency to deammonation. Melamine is mainly used for melamine–formaldehyde plastics. US annual production of both H_2NCN and $NCNC(NH_2)_2$ is on the 10^5 tonne scale.

The hydroxyl derivative of X–CN is cyanic acid HO–CN: it cannot be prepared pure due to rapid decomposition but it is probably present to the extent of about 3% when its tautomer, isocyanic acid (HNCO) is prepared from sodium cyanate and HCl. HNCO rapidly trimerizes to cyanuric acid (Fig. 8.19) from which it can be regenerated by pyrolysis. It is a fairly strong acid (K_a 1.2×10^{-4} at 0°) freezing at $-86.8°$ and boiling at 23.5°C. Thermolysis of urea is an alternative route to HNCO and $(HNCO)_3$; the reverse reaction, involving the isomerization of ammonium cyanate, is the classic synthesis of urea by F. Wöhler (1828):[17a]

Alkali metal cyanates are stable and readily obtained by mild oxidation of aqueous cyanide solutions using oxides of Pb^{II} or Pb^{IV}. The commercial preparation of NaNCO is by reaction of urea with Na_2CO_3.

$$Na_2CO_3 + 2OC(NH_2)_2 \xrightarrow[\text{(dry)}]{\text{heat}} 2NaOCN + CO_2 + 2NH_3 + H_2O$$

The pseudohalogen concept might lead one to expect a cyanate analogue of cyanogen but there is little evidence for NCO–OCN, consistent with the known reluctance of oxygen to catenate. By contrast, thiocyanogen $(SCN)_2$ is moderately stable; it can be prepared as white crystals by suspending AgSCN in Et_2O or SO_2 and oxidizing the anion at low temperatures with Br_2 or I_2. $(SCN)_2$ melts at $\sim -7°$ to an unstable orange suspension which rapidly polymerizes to the brick-red solid parathiocyanogen $(SCN)_x$. This ready polymerization hampers structural studies but it is probable that the molecular structure is N≡C–S–S–C≡N with a nonlinear central C–S–S–C group. $(SeCN)_2$ can be prepared similarly as a yellow powder which polymerizes to a red solid.

Thiocyanates and selenocyanates can be made by fusing the corresponding cyanide with S or Se. The SCN^- and $SeCN^-$ ions are both linear, like OCN^-. Treatment of KSCN with dry $KHSO_4$ produces free isothiocyanic acid HNCS, a white crystalline solid which is stable below 0° but which decomposes rapidly at room temperatures to HCN and a yellow solid $H_2C_2N_2S_3$. Thiocyanic acid, HSCN, (like HOCN) has not been prepared pure but compounds such as MeSCN and $Se(SCN)_2$ are known.

The thiocyanate ion has been much studied as an ambidentate ligand (in which either S or N is the donor atom); it can also act as a bidentate bridging ligand –SCN–, and even as

---

[17a] J. SHORTER, The conversion of ammonium cyanate into urea—a saga in reaction mechanisms, *Chem. Soc. Revs.* **7**, 1–14 (1978).

a tridentate ligand $>$SCN–.[18, 19] The ligands OCN$^-$ and SeCN$^-$ have been less studied but appear to be generally similar. A preliminary indication of the mode of coordination can sometimes be obtained from vibrational spectroscopy since N coordination raises both $v(CN)$ and $v(CS)$ relative to the values of the uncoordinated ion, where S coordination leaves $v(CN)$ unchanged and increases $v(CS)$ only somewhat. The bridging mode tends to increase both $v(CN)$ and $v(CS)$. Similar trends are noted for OCN$^-$ and SeCN$^-$ complexes. However, these "group vibrations" are in reality appreciably mixed with other modes both in the ligand itself and in the complex as a whole, and vibrational spectroscopy is therefore not always a reliable criterion. Increasing use is being made of ^{14}N and ^{13}C nmr data[20] but the most reliable data, at least for crystalline complexes, come from X-ray diffraction studies.[20a] The variety of coordination modes so revealed is illustrated in Table 8.10, which is based on one by A. H. Norbury.[18] Phenomenologically it is observed that class a metals tend to be N-bonded whereas class b tend to be S-bonded (see below), though it should be stressed that kinetic and solubility factors as well as relative thermodynamic stability are sometimes implicated, and so-called 'linkage isomerism" is well established, e.g. $[Co(NH_3)_5(NCS)]Cl_2$ and $[Co(NH_3)_5(SCN)]Cl_2$. In terms of the a and b (or "hard" and "soft") classification of ligands and acceptors it is noted that metals in Groups IIIA–VIII together with the lanthanoids and actinoids tend to form –NCS complexes; in the later transition groups Co, Ni, Cu and Zn also tend to form –NCS complexes whereas their heavier congeners Rh, Ir; Pd, Pt; Au; and Hg are predominantly S-bonded. Ag and Cd are intermediate and readily form both types of complex. The interpretation to be placed on these observations is less certain. Steric influences have been mentioned (N bonding, which is usually linear, requires less space than the bent M–S–CN mode). Electronic factors also play a role, though the detailed nature of the bonding is still a matter of debate and devotees of the various types of electronic influence have numerous interpretations to select from. Solvent effects (dielectric constant ε, coordinating power, etc.) have also been invoked and it is clear that these various explanations are not mutually exclusive but simply tend to emphasize differing aspects of an extremely complicated and delicately balanced situation. The interrelation of these various interpretations is summarized in Table 8.11.

Fewer data are available for SeCN$^-$ complexes but similar generalizations seem to hold. By contrast, OCN$^-$ complexes are not so readily discussed in these terms: in fact, very few cyanato (–OCN) complexes have been characterized and the ligand is usually N-bonded (isocyanato).[20a]

[18] A. H. NORBURY, Coordination chemistry of the cyanate, thiocyanate, and selenocyanate ions, *Adv. Inorg. Chem. Radiochem.* **17**, 231–402 (1975) (825 refs.).

[19] A. A. NEWMAN (ed.), *Chemistry and Biochemistry of Thiocyanic Acid and its Derivatives*, Academic Press, London, 1975, 351 pp.; a set of six reviews emphasizing inorganic aspects of general chemistry (M. N. HUGHES), Coordination chemistry (J. L. BURMEISTER), Molten thiocyanates (D. H. KERRIDGE), Biochemistry (J. L. WOOD), Technology and industrial applications (H. A. BEEKHUIS), and Analytical chemistry (M. R. F. ASHWORTH).

[20] J. A. KARGOL, R. W. CRECELY, and J. L. BURMEISTER, ^{13}C nuclear magnetic resonance spectra of coordinated thiocyanate, *Inorg. Chim. Acta* **25**, L109–L110 (1977), and references therein.

[20a] S. J. ANDERSON, D. S. BROWN, and K. J. FINNEY, Crystal and molecular structure of $[Ti(\eta^5\text{-}C_5H_5)_2(NCO)_2]$ and $[Zr(\eta^5\text{-}C_5H_5)_2(NCO_2]$, *JCS Dalton* 1979, 152–4. (The compounds, originally thought to be O-bonded on the basis of infrared and ^{14}N nmr spectroscopy, now shown by single-crystal X-ray diffractometry to be N-bonded.)

TABLE 8.10   *Modes of bonding established by X-ray crystallography*

Mode	Example	Comment
Ag—NCO	$[AsPh_4][Ag(NCO)_2]$	Linear anion
Mo—OCN	$[Mo(OCN)_6]^{3-}$, $[Rh(OCN)(PPh_3)_3]$	Based on infrared data only
Ag⎯NCO⎯Ag	AgNCO	Cf. fulminate in Table 8.8
Ni⎯OCN/NCO⎯Ni	$[Ni_2(NCO)_2\{N(CH_2CH_2NH_2)_3\}_2][BPh_4]_2$	Note bent Ni–N–C
Co—NCS	$[Co(NH_3)_5(NCS)]Cl_2$	Linkage isomerism
Co—SCN	$[Co(NH_3)_5(SCN)]Cl_2$	
Pd—NCS	$[Pd(NCS)(SCN)\{Ph_2P(CH_2)_3PPh_2\}]$	Both N and S monodentate in a single crystal
Pd⎯SCN⎯Pd	$K_2[Pd(SCN)_4]$	Weak S bridging to a second Pd
Re⎯SCN/NCS⎯Re	$[NBu_4^n]_3[Re(NCS)_{10}]$	N-bonded bridging (and terminal)[20b]
Co—NCS—Hg	$[Co(NCS)_4Hg]$	Bidentate, different metals
Pt⎯SCN/NCS⎯Pt	$[Pt_2(Cl)_2(PPr_3)_2(SCN)_2]$	Bidentate, same metal
Hg⎯SCN—Co⎯Hg	$[Co(NCS)_6Hg_2]\cdot C_6H_6$	Tridentate
Ni—NCSe	$[Ni(HCONMe_2)_4(NCSe)_2]$	N donor
Co—SeCN	$K[Co(Me_2glyoxime)_2(SeCN)_2]$	*Se* donor

[20b] F. A. Cotton, A. Davison, W. H. Isley, and H. S. Trop, Reformulation from X-ray crystallography of a dinuclear thiocyanate complex of rhenium. The first observation of a solely N-bonded bridging thiocyanate, *Inorg. Chem.* **18**, 2719–23 (1979).

TABLE 8.11   *Mode of bonding in thiocyanate complexes*

Metal type[a]	σ-Donor ligand	High-ε solvent	Low-ε solvent	π-Acceptor ligand
Class a	–NCS	–NCS	–SCN	–SCN
Class b	–SCN	–SCN	–NCS	–NCS

[a] Sometimes discussed in terms of "hard" and "soft" acids and bases.

8.9 Organometallic Compounds

This section gives a brief summary of the vast and burgeoning field of organometallic chemistry.† No area of chemistry has produced more surprises and challenges during the past 25 y and the field continues to be one of great excitement and activity. A rich harvest of new and previously undreamed of structure types is reaped each year, the rewards of elegant and skilful synthetic programmes being supplemented by an unusual number of chance discoveries and totally unsuspected reactions. Synthetic chemists can take either a buccaneering or an intellectual approach (or both); structural chemists are able to press their various techniques to the limit in elucidating the products formed; theoretical chemists and reaction kineticists, though badly outpaced in predictive work, provide an invaluable underlying rationale for various aspects of the continually evolving field and just occasionally run ahead of the experimentalists; industrial chemists can exploit and extend the results by developing numerous catalytic processes of immense importance. The field is not new, but was transformed in 1952 by the recognition of the "sandwich" structure of dicyclopentadienyliron (ferrocene).[21,22] Many books[23–25c] and extended

---

† The term *organometallic* is somewhat vague since definitions of *organo* and *metallic* are themselves necessarily imprecise. We shall use the term to refer to compounds that involve at least one close M–C interaction: this includes metal complexes with ligands such as CO, CO_2, CS_2, and CN⁻ but excludes "ionic" compounds such as NaCN or Na acetate; it also excludes metal alkoxides $M(OR)_n$ and metal complexes with organic ligands such as C_5H_5N, PPh_3, OEt_2, SMe_2, etc., where the donor atom is not carbon. A permissive view is often taken in the literature of what constitutes a "metal" and the elements B; Si, Ge; As, Sb; Se and Te are frequently included for convenience and to give added perspective. However, it is not helpful to include as metals all elements less electronegative than C since this includes I, S, and P. Metal carbides (p. 318) and graphite intercalation compounds (p. 313) are also normally excluded.

²¹ G. WILKINSON, M. ROSENBLUM, M. C. WHITING, and R. B. WOODWARD, The structure of iron *bis*-cyclopentadienyl, *J. Am. Chem. Soc.* **74**, 2125–6 (1952). For some personal recollections on the events leading up to this paper, see G. WILKINSON, The iron sandwich. A recollection of the first four months, *J. Organometallic Chem.* **100**, 273–8 (1975).

²² J. S. THAYER, Organometallic chemistry: a historical perspective, *Adv. Organometallic Chem.* **13**, 1–49 (1975).

²³ G. E. COATES, M. L. H. GREEN, and K. WADE, *Organometallic Compounds*, 3rd edn., Vol. I, *The Main Group Elements* (573 pp.); Vol. II, *The Transition Elements* (376 pp.), Methuen, London, 1967, 1968. An excellent systematic account. The fourth edition is beginning to appear: B. J. AYLETT, Vol. 1, Part 2, Groups IV and V. Chapman & Hall, London, 1979, 521 pp.

²⁴ G. E. COATES, M. L. H. GREEN, P. POWELL, and K. WADE, *Principles of Organometallic Chemistry*, Methuen, London, 1968, 259 pp. A useful general introduction.

²⁵ J. M. SWAN and D. ST. C. BLACK, *Organometallics in Organic Synthesis*, Chapman & Hall, London, 1974, 158 pp.

²⁵ᵃ H. ALPER (ed.), *Transition Metal Organometallics in Organic Synthesis*, Academic Press, London, Vol. 1, 1976, 258 pp.; Vol. 2, 1978, 192 pp.

²⁵ᵇ E. NEGISHI, *Organometallics in Organic Synthesis*, Vol. 1, *General Discussions and Organometallics of Main Group Metals in Organic Synthesis*, Wiley, Chichester, 1980, 532 pp.

²⁵ᶜ G. WILKINSON, F. G. A. STONE, and E. W. ABEL (eds.), *Comprehensive Organometallic Chemistry*, 9 Vols., Pergamon Press, Oxford, 1982, 9569 pp.

reviews[26-29b] are available on various aspects, and continued progress is summarized in annual volumes.[30-33]

The most satisfactory way of classifying the wide variety of organometallic compounds is by the number of C atoms attached to (or closely associated with) the metal atom. This essentially structural criterion can be established by several techniques and is more definite than other features such as the presumed number of electrons involved in the bonding. The number of attached carbon atoms is called the *hapticity* of the organic group (Greek ηαπτειν, *haptien*, to fasten) and hapticities from 1 to 8 have been observed. Monohapto groups are specified as η^1, dihapto as η^2, etc.† Occasionally an observed M–C interatomic distance is intermediate between values that are typical of bonding and nonbonding interactions, and to this extent the structural classification is uncertain, but this is not usually a problem and need not concern us further. Earlier classifications of organometallic compounds were based either on the formal number of bonding electrons thought to be contributed by the organic moiety, or on the "type" of bonding (σ, π, etc.). However, such classifications depend on the bonding model adopted and to this extent are indeterminate or ambiguous; they also change from time to time as the various bonding models evolve, and are inevitably formal in being restricted to integral members of electrons, whereas most modern theories of bonding envisage partial bond orders and the involvement of varying amounts of electron charge from both the organic moiety and the metal atoms. We shall, of course, frequently refer to various bonding theories since these are extremely important in gaining an understanding of organometallic compounds, in rationalizing their physical and chemical properties, and in planning further experiments. But they are unsatisfactory as the basis for a relatively permanent general classification of the types of compound.

The various classes of ligands and attached groups that occur in organometallic compounds are summarized in Table 8.12, and these will be briefly discussed in the

† Some authors use the notation h^1, h^2, etc., but this is being replaced and is not recommended by IUPAC.

[26] M. L. H. GREEN, An introduction to the organic chemistry of the metallic elements, Chap. 14 in *Comprehensive Inorganic Chemistry*, Vol. 1, pp. 1295–1321, Pergamon Press, Oxford, 1973.

[27] W. P. GRIFFITH, Carbonyls, cyanides, isocyanides, and nitrosyls, Chap. 46 in *Comprehensive Inorganic Chemistry*, Vol. 4, pp. 105–95, Pergamon Press, Oxford, 1973.

[28] B. F. G. JOHNSON, Transition metal chemistry, Chap. 52 in *Comprehensive Inorganic Chemistry*, Vol. 4, pp. 673–779, Pergamon Press, Oxford, 1973.

[29] B. L. SHAW and N. I. TUCKER, Organo transition metal compounds and related aspects of homogeneous catalysis, Chap. 53 in *Comprehensive Inorganic Chemistry*, Vol. 4, pp. 781–994, Pergamon Press, Oxford, 1973.

[29a] T. J. MARKS, Chemistry and spectroscopy of f element organometallics. Part 1, The lanthanides, *Prog. Inorg. Chem.* **24**, 51–107 (1978); Part 2, The Actinides, *Prog. Inorg. Chem.* **25**, 223–333 (1979).

[29b] F. A. COTTON and G. WILKINSON, *Advanced Inorganic Chemistry*, 4th edn., Wiley, New York, 1980, particularly Chap. 25, Metal carbonyls and other complexes with π-acceptor ligands, pp. 1049–79; Chap. 27, Transition metal compounds with bonds to hydrogen and carbon, pp. 1113–82; Chap. 29, Transition metal to carbon bonds in synthesis, pp. 1234–64; Chap. 30, Transition metal to carbon bonds in catalysis, pp. 1265–1309. These four chapters contain numerous references to relevant books, reviews, and recent literature.

[30] F. G. A. STONE and R. WEST (eds.), *Advances in Organometallic Chemistry*, Academic Press, New York, Vol. 1 (1964)–Vol. 19 (1981).

[31] *Organometallic Chemistry Reviews*, Vol. 1 (1966)–Vol. 10 (1972), Elsevier. Appeared as Series A (extended subject reviews) and Series B (annual surveys of individual groups of elements). Now incorporated as annual surveys in *J. Organometallic Chem.*

[32] *Organometallic Chemistry Reactions*, Wiley, Vol. 1, (1967)–Vol. 12 (1981).

[32a] *Topics in Inorganic and Organometallic Stereochemistry*, Wiley, Vol. 1, (1967)–Vol. 12 (1981).

[32b] J. J. EISCH, Organometallic Syntheses, Vol. 2, Academic Press, New York, 1981, 194 pp.

[33] Chemical Society Specialist Periodical Reports, *Organometallic Chemistry*, Vol. 1 (1971)–Vol. 9 (1979). Shorter, more selective reviews, appear in *Annual Reports* of the Chemical Society.

TABLE 8.12 *Classification of organometallic ligands according to the number of attached C atoms*[(a)]

Number	Examples
η^1, monohapto	Alkyl ($-R$), aryl ($-Ar$), perfluoro ($-R_f$), acyl ($-\overset{\displaystyle O}{\overset{\|}{C}}R$), σ-allyl ($-CH_2CH{=}CH_2$), σ-ethynyl ($-C{\equiv}CR$), CO, CO$_2$, CS$_2$. CN$^-$, isocyanide (RNC), carbene ($=CR_2, =C\overset{\displaystyle OR'}{\underset{\displaystyle R}{\big\langle}}$, $=C\overset{\displaystyle OR}{\underset{\displaystyle NHAr}{\big\langle}}$, $=$Ccyclo, etc.) carbyne ($\equiv CR, \equiv CAr$), carbido (C)
η^2, dihapto	Alkene ($\rangle C{=}C\langle$), perfluoroalkene (e.g. C$_2$F$_4$), alkyne ($-C{\equiv}C-$), etc. [non-conjugated dienes are bis-dihapto]
η^3, trihapto	π-allyl ($\rangle C{-}C{-}C\langle$)
η^4, tetrahapto	Conjugated diene (e.g. butadiene), cyclobutadiene derivatives
η^5, pentahapto	Dienyl (e.g. cyclopentadienyl derivatives, cycloheptadienyl derivatives)
η^6, hexahapto	Arene (e.g. benzene, substituted benzenes) cycloheptatriene, cycloocta-1,3,5-triene
η^7, heptahapto	Tropylium (cycloheptatrienyl)
η^8, octahapto	Cyclooctatetraene

[(a)] Many ligands can bond in more than one way:[(29)] e.g. allyl can be η^1 (σ-allyl) or η^3 (π-allyl); cyclooctatetraene can be η^4 (1,3-diene), η^4 (chelating, 1,5-diene), η^6 (1,3,5-triene), η^6 (bis-1,2,3,–5,6,7-π-allyl), η^8 (1,3,5,7-tetraene), etc.

following paragraphs. Aspects which concern the general chemistry of carbon will be emphasized in order to give coherence and added significance to the more detailed treatment of the organometallic chemistry of individual elements given in other sections, e.g. Li (p. 110), Be (p. 141), Mg (p. 146), etc.

8.9.1 *Monohapto ligands*

Alkyl and aryl derivatives of many main-group metals have already been discussed in previous chapters, and compounds such as PbMe$_4$ and PbEt$_4$ are made on a huge scale, larger than all other organometallics put together (p. 432). The alkyl and aryl groups are usually regarded as 1-electron donors but it is important to remember that even a monohapto 1-electron donor can bond simultaneously to more than 1 metal atom, e.g. to 2 in Al$_2$Me$_6$ (p. 289), 3 in Li$_4$Bu$_4^t$ (p. 112), and 4 in [Li$_4$Me$_4$]$_n$. Similarly, an η^1 ligand such as CO, which is often regarded as a 2-electron donor, can bond simultaneously to either 1, 2, or 3 metal atoms (p. 351). There is thus an important distinction to be drawn between (a) hapticity (the number of C atoms in the organic group that are closely associated with a metal atom), (b) metal connectivity (the number of M atoms simultaneously bonded to the organic group), and (c) the number of ligand electrons formally involved in bonding to the metal atom(s). The metal connectivity is also to be distinguished from the coordination number of the C atom, which also includes all other atoms or groups attached to it: e.g. the bridging C atoms in Al$_2$Me$_6$ are monohapto with a metal connectivity of 2 and a coordination number of 5.

Although zinc alkyls were first described by E. Frankland in 1849 and the alkyls and aryls of most main group elements had been prepared and often extensively studied during the subsequent 100 y, very few such compounds were known for the transition metals even as recently as the late 1960s. The great burst of current activity[34, 35] stems from the independent suggestion[36, 37] that M–C bonds involving transition elements are not inherently weak and that kinetically stable complexes can be made by a suitable choice of organic groups. In particular, the use of groups which have no β-hydrogen atom (e.g. –CH$_2$Ph, –CH$_2$CMe$_3$, or –CH$_2$SiMe$_3$) often leads to stable complexes since this prevents at least one facile decomposition route namely β-elimination.

The reverse reaction (formation of metal alkyls by addition of alkenes to M–H) is the basis of several important catalytic reactions such as alkene hydrogenation, hydro-formylation, hydroboration, and isomerization. A good example of decomposition by β-elimination is the first-order intramolecular reaction:

$$[Pt(Bu)_2(PPh_3)_2] \longrightarrow 1\text{-}C_4H_8 + [Pt(Bu)(H)(PPh_3)_2] \longrightarrow n\text{-}C_4H_{10} + [Pt(PPh_3)_2]$$

β-Elimination reactions have been much studied[34, 35] but should not be over emphasized since other decomposition routes must also be considered. Amongst these are:

homolytic fission, e.g. HgPh$_2 \rightarrow$ Hg + 2Ph

reductive elimination, e.g. $[Au^{III}Me_3(PPh_3)] \rightarrow [Au^IMe(PPh_3)] + C_2H_6$

binuclear elimination (or formation of Bu radicals) e.g.

$$2[Cu(Bu)(PBu_3)] \longrightarrow 2Cu + 2PBu_3 + n\text{-}C_4H_{10} + 1\text{-}C_4H_8$$

α-Elimination to give a carbene complex, e.g.

$$[Ta(CH_2CMe_3)_3Cl_2] + 2\,LiCH_2CMe_3 \longrightarrow 2\,LiCl + \text{``}[Ta(CH_2CMe_3)_5]\text{''}$$

$$\xrightarrow{\alpha\text{-elim}} CMe_4 + [Ta(CH_2CMe_3)_3(=C\stackrel{H}{\underset{CMe_3}{\diagdown}})]$$

Stabilization of η^1-alkyl and -aryl derivatives of transition metals can be enhanced by the judicious inclusion of various other stabilizing ligands in the complex, even though such ligands are now known not to be an essential prerequisite.[38] Particularly efficacious

[34] P. J. DAVIDSON, M. F. LAPPERT, and R. PEARCE, Metal σ-hydrocarbonyls, MR$_n$: stoichiometry, structures, stabilities, and thermal decomposition pathways, *Chem. Revs.* 76, 219–42 (1976)—a review with 329 refs.

[35] R. R. SCHROCK and G. W. PARSHALL, σ-Alkyl and σ-aryl complexes of Group IV–VII transition metals, *Chem. Revs.* 76, 243–68 (1976)—a review with 357 refs.

[36] M. R. COLLIER, M. F. LAPPERT, and M. M. TRUELOCK, μ-Methylene transition metal binuclear compounds: complexes with Me$_3$SiCH$_2$– and related ligands, *J. Organometallic Chem.* 25, C36–C38 (1970).

[37] G. YAGUPSKY, W. MOWAT, A. SHORTLAND, and G. WILKINSON, Trimethylsilylmethyl compounds of transition metals, *JCS Chem. Comm.* 1369–71 (1970).

[38] P. S. BRATERMAN and R. J. CROSS, Organo-transition-metal complexes: stability, reactivity, and orbital correlations, *Chem. Soc. Rev.* 2, 271–94 (1973).

are potential π acceptors (see below) such as $AsPh_3$, PPh_3, CO, or $\eta^5\text{-}C_5H_5$ in combination with the heavier transition metals since the firm occupation of coordination sites prevents their use for concerted decomposition routes. Steric protection may also be implicated. Similar arguments have been used to interpret the observed increase in stability of η^1 complexes in the sequence alkyl $<$ aryl $<$ o-substituted aryl $<$ ethynyl ($-C\equiv CH$).

The next group of η^1 ligands comprise the isoelectronic species, CO, CN^-, and RNC. They are closely related to other 14-electron (10 valence electron) ligands such as N_2 and NO^+ (and also to tertiary phosphines and arsines, and to organic sulfides, selenides, etc.), and it is merely the presence of C as the donor atom which classifies their complexes as organometallics. All have characteristic donor properties that distinguish them from simple electron-pair donors (Lewis bases, p. 223) and these have been successfully interpreted in terms of a synergic or mutually reinforcing interaction between σ donation from ligand to metal and π back donation from metal to ligand as elaborated below. CO is undoubtedly the most important and most widely studied of all organometallic ligands and it is the prototype for this group of so-called π-acceptor ligands.[38a] The currently accepted view of the bonding is represented diagramatically in Figs. 8.20 and 8.21. Figure 8.20 shows a schematic molecular orbital energy level diagram for the heteronuclear diatomic molecule CO. The AOs lie deeper in O than in C because of the higher effective nuclear charge on O; consequently O contributes more to bonding MOs and C contributes more to the antibonding MOs. It can be seen that all the bonding MOs are filled and, in this description, the CO molecule can be said to have a triple bond :C≡O: with the lone-pair on carbon weakly available for donation to an acceptor. The top part (a) of Fig. 8.21 shows the formation of a σ bond by donation of the lone-pair into a suitably directed hybrid orbital on M, and the lower part (b) shows the accompanying back donation from a filled metal d orbital into the vacant antibonding CO orbital having π symmetry (one node) with respect to the bonding axis. This at once interprets why CO, which is a very weak σ donor to Lewis acids such as BF_3 and $AlCl_3$, forms such strong complexes with transition elements, since the drift of π-electron density from M to C tends to make the ligand more negative and so enhances its σ-donor power. The pre-existing negative charge on CN^- increases its σ-donor propensity but weakens its effectiveness as a π acceptor. It is thus possible to rationalize many chemical observations by noting that effectiveness as a σ donor decreases in the sequence $CN^- > RNC > NO^+ \sim CO$ whereas effectiveness as a π acceptor follows the reverse sequence $NO^+ > CO \gg RNC > CN^-$. By implication, back donation into antibonding CO orbitals weakens the CO bond and this is manifest in the slight increase in interatomic distance from 112.8 pm in free CO to ~ 115 pm in many complexes. There is also a decrease in the C–O force constant, and the drop in the infrared stretching frequency from 2143 cm^{-1} in free CO to 2125–1850 cm^{-1} for terminal COs in neutral carbonyls has been interpreted in the same way.

The occurrence of stable neutral binary carbonyls is restricted to the central area of the d block (Table 8.13), where there are low-lying vacant metal orbitals to accept σ-donated lone-pairs and also filled d orbitals for π back donation. Outside this area carbonyls are either very unstable (e.g. Cu, Ag, p. 1392), or anionic, or require additional ligands besides CO for stabilization. As with boranes and carboranes (p. 203), CO can be replaced by isoelectronic equivalents such as $2e^-$, H^-, $2H^{\cdot}$, or L. Mean bond dissociation energies

[38a] *Topics in Current Chemistry*, Vol. 71, *Metal Carbonyl Chemistry*, Springer-Verlag, Berlin, 1977, 190 pp.

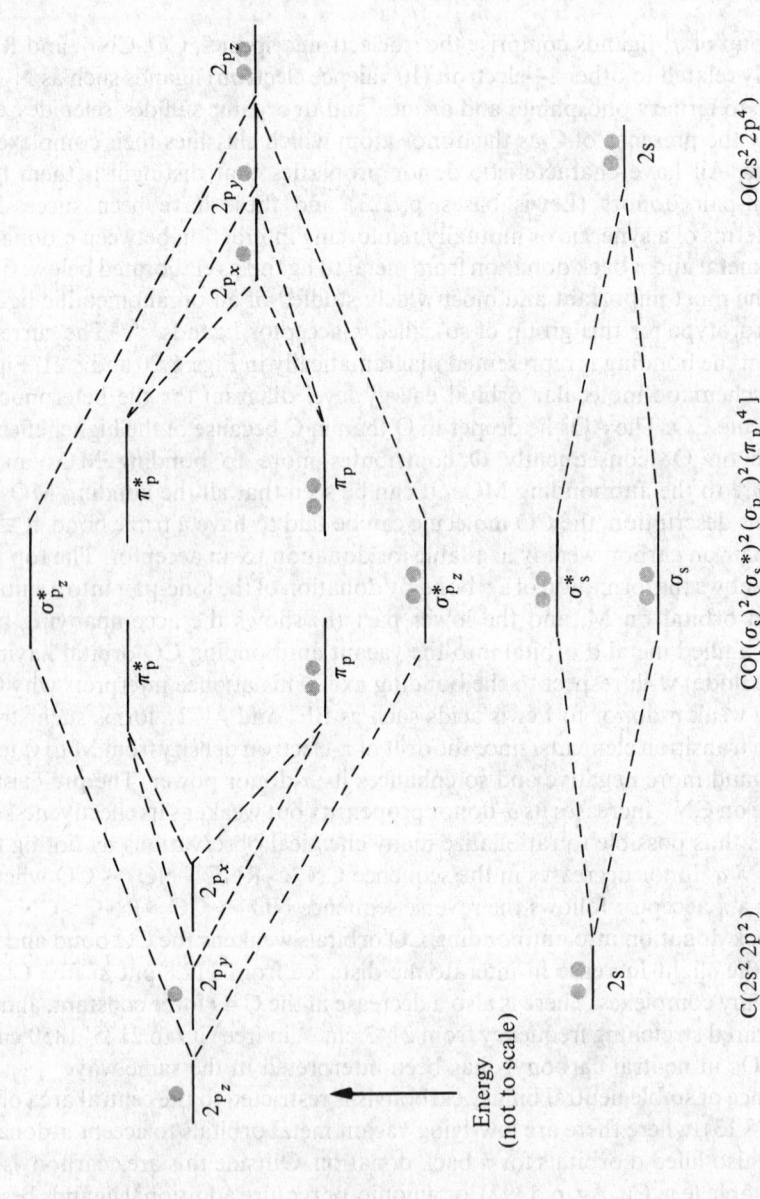

FIG. 8.20 Schematic molecular energy level diagram for CO. The 1s orbitals have been omitted as they contribute nothing to the bonding. A more sophisticated treatment would allow some mixing of the 2s and $2p_z$ orbitals in the bonding direction (z) as implied by the orbital diagram in Fig. 8.21.

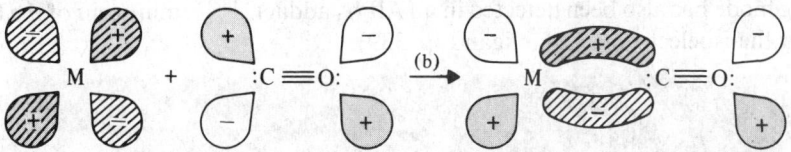

FIG. 8.21   Schematic representation of the orbital overlaps leading to M–CO bonding: (a) σ overlap and donation from the lone-pair on C into a vacant (hybrid) metal orbital to form a σ M←C bond, and (b) π overlap and the donation from a filled d_{xz} or d_{yz} orbital on M into a vacant antibonding π_p^* orbital on CO to form a π M→C bond.

TABLE 8.13   *Known neutral binary metal carbonyls. Osmium also forms $Os_5(CO)_{16}$, $Os_5(CO)_{19}$, $Os_6(CO)_{18}$, $Os_6(CO)_{20}$, $Os_7(CO)_{21}$, and $Os_8(CO)_{23}$. Carbonyls of elements in the shaded area are either very unstable, or anionic, or require additional ligands besides CO for stabilization*

IIIA	IVA	VA	VIA	VIIA	← ——— VIII ———— →		IB	IIB
		$V(CO)_6$	$Cr(CO)_6$	$Mn_2(CO)_{10}$	$Fe(CO)_5$	$Co_2(CO)_8$	$Ni(CO)_4$	
	Ti				$Fe_2(CO)_9$	$Co_4(CO)_{12}$		Cu
					$Fe_3(CO)_{12}$	$Co_6(CO)_{16}$		
			$Mo(CO)_6$	$Tc_2(CO)_{10}$	$Ru(CO)_5$	$Rh_2(CO)_8$		
	Zr	Nb		$Tc_3(CO)_{12}$	$Ru_2(CO)_9$	$Rh_4(CO)_{12}$	Pd	Ag
					$Ru_3(CO)_{12}$	$Rh_6(CO)_{16}$		
			$W(CO)_6$	$Re_2(CO)_{10}$	$Os(CO)_5$	$Ir_2(CO)_8$		
	Hf	Ta			$Os_2(CO)_9$	$Ir_4(CO)_{12}$	Pt	Au
					$Os_3(CO)_{12}$	$Ir_6(CO)_{16}$		

$\bar{D}$(M–CO)/kJ mol^{-1} increase in the sequence $Cr(CO)_6$ 109, $Mo(CO)_6$ 151, $W(CO)_6$ 176, and in the sequence $Mn_2(CO)_{10}$ 100, $Fe(CO)_5$ 121, $Co_2(CO)_8$ 138, $Ni(CO)_4$ 147.

CO can act as a terminal ligand, as an unsymmetrical or symmetrical bridging ligand (μ_2-CO) or as a triply bridging ligand (μ_3-CO):

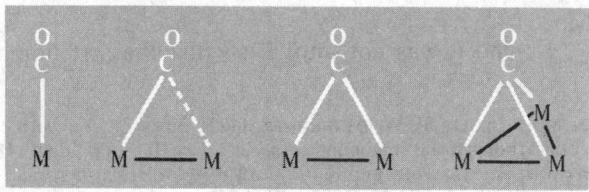

In all these cases CO is η^1 but the connectivity to metal increases from 1 to 3. It is notable that in the μ_2-bridging carbonyls the angle M–C(O)–M is usually very acute (77–80°), whereas in organic carbonyls the C–C(O)–C angle is typically 120–124°. This suggests a fundamentally differing bonding mode in the two cases and points to the likelihood of a 2-electron 3-centre bond (p. 179) for the bridging metal carbonyls. The hapticity can also rise, and recent structural determinations afford examples in which one or both of the π^* orbitals in CO are thought to contribute to η^2 bonding to 1 or 2 M atoms.[38b] A bis-η^1-bridging mode has also been detected in an $AlPh_3$ adduct,[38c] reminiscent of the bridging mode in the isoelectronic CN^- ligand (p. 339):

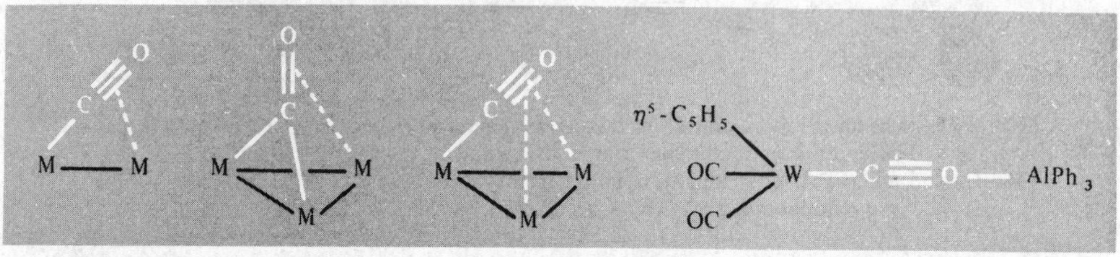

Numerous examples of metal carbonyls will be found in later chapters dealing with the chemistry of the individual transition metals. CO also has an unrivalled capacity for stabilizing metal clusters and for inserting into M–C bonds (p. 329). Synthetic routes include:

(a) direct reaction, e.g.: $Ni + 4CO \xrightarrow{30°/1\ atm} Ni(CO)_4$

$$Fe + 5CO \xrightarrow{200°/200\ atm} Fe(CO)_5$$

(b) reductive carbonylation, e.g.:

$$OsO_4 + 9CO \xrightarrow{250°/350\ atm} Os(CO)_5 + 4CO_2$$

$$RuI_3 + 5CO + 3Ag \xrightarrow{175°/250\ atm} Ru(CO)_5 + 3Ag$$

$$WCl_6 + 3Fe(CO)_5 \xrightarrow{100°} W(CO)_6 + 3FeCl_2 + 9CO$$

(c) photolysis or thermolysis, e.g.:

$$2Fe(CO)_5 \xrightarrow{h\nu} Fe_2(CO)_9 + CO$$

$$2Co_2(CO)_8 \xrightarrow{70°} Co_4(CO)_{12} + 4CO$$

The remaining classes of monohapto organic ligands listed in Table 8.12 are carbene ($=CR_2$), carbyne ($\equiv CR$), and carbido (C). Stable carbene complexes were first reported in 1964 by E. O. Fischer and A. Maasböl.[39] Initially they were of the type $[W(CO)_5(:C\underset{\textstyle R}{\overset{\textstyle OMe}{<}})]$; and it was not until 1968 that the first homonuclear carbene

---

[38b] W. A. HERRMANN, M. L. ZIEGLER, K. WEIDENHAMMER, and H. BIERSACK, A novel mode of coordination in the structural chemistry of carbonyl metal compounds, *Angew. Chem.*, Int. Edn. (Engl.) **18**, 960–61 (1979), and references therein. See also J. A. MARSELLA and K. G. CAULTON, Molecular-dynamics of the four-electron bridging carbonyl, *Organometallics* **1**, 274–79 (1982); W. A. HERRMANN, J. BIERSACK, M. L. ZIEGLER, K. WEIDENHAMMER, R. SIEGEL, and D. REHDER, Carbon monoxide—a six-electron ligand?, *J. Am. Chem. Soc.* **103**, 1692–99 (1981).

[38c] J. M. BURLICH, M. E. LEONOWICZ, R. B. PETERSEN, and R. E. HUGHES, Coordination of metal carbonyl anions to triphenyl-aluminium, -gallium, and -indium, *Inorg. Chem.* **18**, 1097–1105 (1979).

[39] E. O. FISCHER, On the way to carbene and carbyne complexes, *Adv. Organometallic Chem.* **14**, 1–32 (1976).

complex was reported $[Cr(CO)_5(:C{\overset{CPh}{\underset{CPh}{\|}}})]$, and isolation of a carbene containing the parent methylene group $:CH_2$ was not achieved until 1975:[40]

$$[Ta(\eta^5\text{-}C_5H_5)_2Me_3] \xrightarrow{Ph_3CBF_4} [Ta(\eta^5\text{-}C_5H_5)_2Me_2]^+BF_4^- \xrightarrow{base}$$

$$[Ta(\eta^5\text{-}C_5H_5)_2(Me)(:CH_2)]$$

Other preparative routes are:

$$[W(CO)_6]+LiR \xrightarrow{Et_2O} [W(CO)_5(C{\overset{OLi(OEt_2)_n}{\underset{R}{}}})] \xrightarrow{Me_3OBF_4} [(W(CO)_5(:C{\overset{OMe}{\underset{R}{}}})]$$

$$[cyclo\text{-}(PhC)_2CCl_2]+Na_2Cr(CO)_5 \xrightarrow[-20°]{thf} [Cr(CO)_5(:C{\overset{CPh}{\underset{CPh}{\|}}})]$$

The metal is in the formal oxidation state zero. As expected, the M–C bonds are somewhat shorter than M–R bonds to alkyls, but they are noticeably longer than M–CO bonds suggesting only limited double-bond character M=C, e.g.:

in $[Ta(\eta^5\text{-}C_5H_5)_2(Me)(CH_2)]$ Ta–CH_2 220.6 pm   Ta–CH_3 225 pm

in $[W(CO)_5\{C(OMe)Ph\}]$ W–C(OMe)Ph 205 pm   W–CO 189 pm

in $[Cr(CO)_4\{C(OMe)Me\}(PPh_3)]$ Cr–C(OMe)Me 204 pm   Cr–CO 186 pm

Carbene complexes are highly reactive species.[29, 39]

Carbyne complexes were first made in 1973 by the unexpected reaction of methoxycarbene complexes with boron trihalides:

$$[M(CO)_5\{C(OMe)R\}]+BX_3 \xrightarrow[low\ temp]{pentane} [M(CO)_4(\equiv CR)X]+CO+BX_2(OMe)$$

$$M = Cr, Mo, W; \ R = Me, Et, Ph; \ X = Cl, Br, I$$

Several other routes are now also available in which BX_3 is replaced by $AlCl_3$, $GaCl_3$, Al_2Br_6, or Ph_3PBr_2, e.g.:

$$[W(CO)_5\{C(OMe)Ph\}]+Al_2Br_6 \xrightarrow[-30°]{pentane} [WBr(CO)_4(\equiv CPh)]+CO+Al_2Br_5(OMe)$$

X-ray studies reveal the expected short M–CR distance, but the bond angle at the carbyne C atom is not always linear. Some structural data are annexed, see page 354. A compound which features all three types of η^1 ligand, alkyl, alkylidene, and alkylidyne, is the red, square-pyramidal tungsten(VI) complex $[W(\equiv CCMe_3)(=CHCMe_3)$ $(CH_2CMe_3)(Me_2PCH_2CH_2PMe_2)]$ in which the W–C distance is 226 pm to neopentyl,

[40] R. R. SCHROCK, The first isolable transition metal methylene complex and analogs. Characterization, mode of decomposition, and some simple reactions, *J. Am. Chem. Soc.* **97**, 6577–78 (1975); L. J. GUGGENBERGER and R. R. SCHROCK, Structure of bis(cyclopentadienyl)methylmethylenetantalum and the estimated barrier to rotation about the tantalum-methylene bond, ibid. 6578–79.

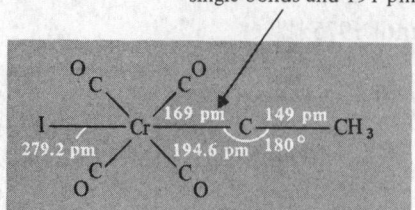

This is the shortest known Cr–C distance cf. 217–222 pm in Cr–C single bonds and 191 pm in Cr(CO$_6$)

Cf. 227–232 pm for W–C single bond and 206 pm in W(CO)$_6$

194 pm to neopentylidene, and 176 pm to the apical neopentylidyne ligand; the corresponding W–C–C angles are 125°, 150°, and 175° respectively.[40a]

A carbido-complex $[Fe_5C(CO)_{15}]$ was first obtained in 1962 as an unexpected side product in very low yield from a reaction whose mechanism still remains obscure.

$$[Fe_3(CO)_{12}] + PhC \equiv CMe \xrightarrow[90°]{\text{petrol ether}} [Fe_5C(CO)_{15}]; \quad \text{yield} \leqslant 0.5\%$$

More recently it has become apparent that the individual C atom fulfils predictable steric and electronic functions and a somewhat more systematic approach to the synthesis of such complexes is beginning to emerge.[41] Three routes hold promise. For Fe, Ru, and Os derivatives pyrolysis under carefully defined conditions produces respectable though unaccountable yields, e.g.:

$$[Ru_3(CO)_{12}] \xrightarrow[6 \text{ h, } H_2O]{Bu_2^n O/142°} Ru, \quad [Ru_4(CO)_{12}H_4], \quad [Ru_4(CO)_{13}H_2], \quad [Ru_6C(CO)_{17}]$$
$$\qquad\qquad\qquad\qquad\quad 53\% \qquad\qquad\quad 10\% \qquad\qquad\quad 3\% \qquad\qquad\quad 30\%$$

For Co and Rh carbidocarbonyl clusters, CCl$_4$ can be used as the source of C, e.g.:

$$6[NMe_3Bz] [Rh(CO)_4] + CCl_4 \xrightarrow[90\% \text{ yield}]{Pr^iOH/25°} [NMe_3Bz]_2[Rh_6C(CO)_{15}] + 4[NMe_3Bz]Cl$$

[40a] M. R. CHURCHILL and W. J. YOUNGS, The crystal structure and molecular geometry of [W(≡CCMe$_3$) (=CHCMe$_3$) (CH$_2$CMe$_3$) (dmpe)], a mononuclear tungsten(VI) complex with metal-alkylidyne, metal-alkylidene, and metal-alkyl linkages, *Inorg. Chem.* **18**, 2454–8 (1979).

[41] P. CHINI, G. LONGONI, and V. G. ALBANO, High nuclearity metal carbonyl clusters, *Adv. Organometallic Chem.* **14**, 285–344 (1976). M. TACHIKAWA and E. L. MUETTERTIES, Metal carbide clusters, *Prog. Inorg. Chem.* **28**, 203–38 (1981). D. H. FARRAR, P. F. JACKSON, B. F. G. JOHNSON, J. LEWIS, and J. N. NICHOLLS, A high-yield synthesis of Ru$_5$C(CO)$_{15}$ by the carbonylation of Ru$_6$C(CO)$_{17}$, *JCS Chem. Comm.* 1981, 415–16.

The analogous reaction cannot be carried out directly with $[Co(CO)_4]^-$ since it is oxidized by CCl_4 to $[Co_2(CO)_8]$; instead the reaction is carried out in two steps via the well-known triply-bridging chloromethynyl derivative $[Co_3(CO)_9(CCl)]$:

$$5[Co_2(CO)_8] + 3CCl_4 \xrightarrow[90\% \text{ yield}]{CCl_4/40°} 2[Co_3(CO)_9(CCl)] + 4CoCl_2 + 22CO + \tfrac{1}{2}C_2Cl_4$$

$$[Co_3(CO)_9(CCl)] + 3Na[Co(CO)_4] \xrightarrow{Pr_2^i O/25°} Na_2[Co_6C(CO)_{15}] + NaCl + 6CO$$

The second step can be viewed as a redox cocondensation reaction and can be used to synthesize even larger clusters, e.g.:

$$2[Co_6C(CO)_{15}]^{2-} + [Co_4(CO)_{12}] \xrightarrow[80\% \text{ yield}]{Pr_2^i O/60°} 2[Co_8C(CO)_{18}]^{2-} + 6CO$$

Known examples of these compounds are listed in Table 8.14 and Figs. 8.22–8.28 give representative structures.[41, 42] $[Fe_5C(CO)_{15}]$ comprises a square-based pyramid of $\{Fe(CO)_3\}$ groups with the unique C atom symmetrically disposed just below the basal plane; the similarity to the "isoelectronic" ferraborane structure $[Fe(B_4H_8)(CO)_3]$ has already been discussed (p. 198) and formal electron-counting rules indicate that the carbido C contributes 4 electrons to the cluster bonding. Octahedral M_6 clusters have been established for the 86-electron clusters $[Fe_6C(CO)_{16}]^{2-}$ and $[Ru_6C(CO)_{17}]$ (Fig. 8.23), whereas the 90-electron species $[Rh_6C(CO)_{15}]^{2-}$ features trigonal prismatic coordination of Rh_6 about the central C(Fig. 8.24). More complex geometries are found for $[Rh_8C(CO)_{19}]$ (Fig. 8.25) and $[Co_8C(CO)_{18}]^{2-}$ (Fig. 8.26): these two isoelectronic clusters are not isostructural though a slight distortion would (hypothetically) transform one into the other. The central carbido C in the square antiprismatic $[Co_8C(CO)_{18}]^{2-}$ is formally 8-coordinate, the Co–C distances being in the range 195–220 pm with a mean

TABLE 8.14 *Some carbidocarbonyl metal clusters†*

$[Fe_5C(CO)_{15}]$	Black	$[Co_6C(CO)_{14}]^-$	Dark brown, paramag
$[Fe_5C(CO)_{14}]^{2-}$	Red-black	$[Co_6C(CO)_{15}]^{2-}$	Brown-red
$[Fe_6C(CO)_{16}]^{2-}$	Black	$[Co_8C(CO)_{18}]^{2-}$	Brown
		$[Co_{13}(C)_2(CO)_{24}H]^{4-}$	Brown
$[Ru_5C(CO)_{15}]$	Red		
$[Ru_6C(CO)_{17}]$	Deep red	$[Rh_6C(CO)_{13}]^{2-}$	Red-brown
$[Ru_6C(CO)_{16}(L)]$	Orange	$[Rh_6C(CO)_{15}]^{2-}$	Yellow
$[Ru_6(arene)C(CO)_{14}]$	Dark red	$[Rh_8C(CO)_{19}]$	Black
		$[Rh_{12}(C)_2(CO)_{24}]^{2-}$	Black
$[Os_5C(CO)_{15}]$	Orange	$[Rh_{12}(C_2)(CO)_{25}]$	Black
$[Os_7C(CO)_{19}(H)_2]$	Brown	$[Rh_{15}(C)_2(CO)_{28}]^-$	Brown
$[Os_8C(CO)_{21}]$	Purple		

† Recently an example has also been synthesized with a Group VIIA metal[42a] namely the red species $[Re_7C(CO)_{21}]^{2-}$.

[42] V. G. ALBANO, P. CHINI, G. CHINI, M. SANSONI, D. STRUMOLO, B. T. HEATON, and S. MARTINENGO, Stabilization of carbonyl clusters by a carbide atom: synthesis and characterization of $[Co_8(CO)_{18}C]^{2-}$ and of paramagnetic $[Co_6(CO)_{14}C]^-$ anions, *J. Am. Chem. Soc.* **98**, 5027–8 (1976).

[42a] G. CIANI, G. D'ALFONSO, M. FRENI, P. REMITI, and A. SIRONI, A high nuclearity carbonyl cluster of rhenium stabilized by an interstitial carbide; synthesis and X-ray crystal structure of the anion $[Re_7C(CO)_{21}]^{2-}$, *JCS Chem. Comm.* **1982**, 339–40.

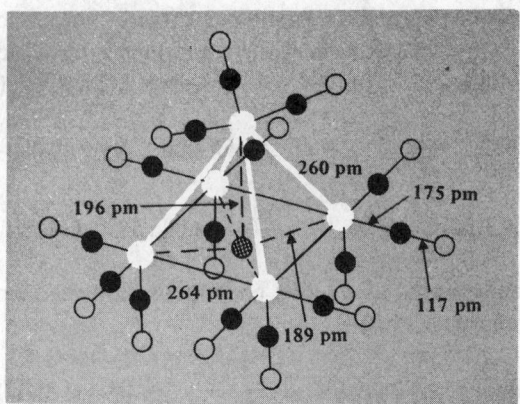

FIG. 8.22   Structure of $[Fe_5C(CO)_{15}]$; the 5-coordinate carbido C is 15 pm below the basal plane.

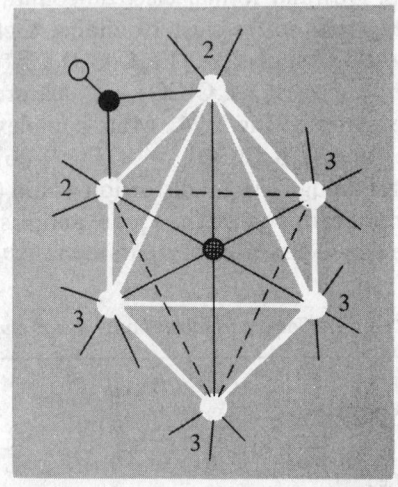

FIG. 8.23   Structure of $[Ru_6C(CO)_{17}]$ with one bridging CO group; the isoelectronic dianion $[Fe_6C(CO)_{16}]^{2-}$ has a similar octahedral M_6 cluster but has 3 μ_2-CO groups.

value of 207 pm. Even more complicated structures are found for the large Rh clusters containing 2 carbido C atoms: $[Rh_{12}(C_2)(CO)_{25}]$ (Fig. 8.27 has no symmetry elements but it is clear that the Rh_{12} cluster surrounds an ethanide unit C_2 in which the C–C distance is only 147 pm); the cluster also has 14 pendant terminal CO groups, 10 μ_2-CO groups and one μ_3-CO. In contrast, $[Rh_{15}(C)_2(CO)_{28}]^-$ (Fig. 8.28) has individual 6-coordinate (octahedral) carbido C atoms symmetrically placed on each side of a central Rh which itself has 12 Rh nearest neighbours in addition to the 2 C atoms. The approach to metal structures is notable and is one of the main interests in constructing large clusters and studying their chemical and catalytic activity.

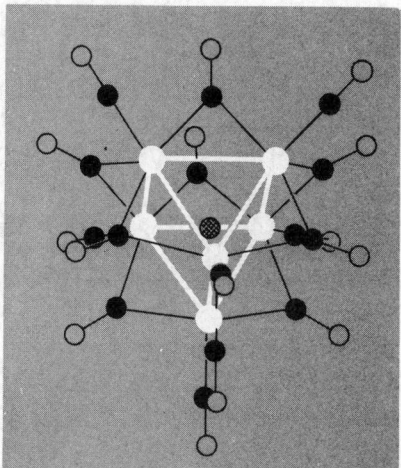

FIG. 8.24   Structure of $[Rh_6C(CO)_{15}]^{2-}$; the C is surrounded by a trigonal prism of Rh atoms each of which carries 1 terminal CO. The other 9 CO groups bridge the 9 edges of the prism.

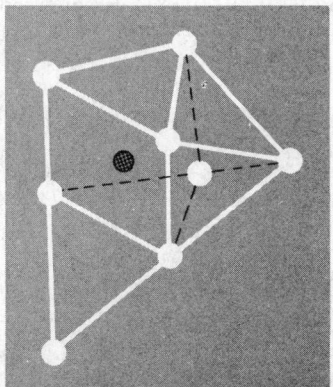

FIG. 8.25   Heavy-atom structure of $[Rh_8C(CO)_{19}]$ showing a trigonal prism of 6 Rh capped by a 7th Rh on one square face and bridged by the 8th Rh on one triangular edge. There are 11 terminal CO, 6 μ_2-CO, and 2 μ_3-CO.

8.9.2   *Dihapto ligands*

Reference to Table 8.12 places this section in context. The first complex between a hydrocarbon and a transition metal was isolated by the Danish chemist W. C. Zeise in 1825 and in the following years he characterized the pale-yellow compound now formulated as $K[Pt(\eta^2\text{-}C_2H_4)Cl_3] H_2O$.† Zeise's salt, and a few closely related complexes such as the chloro-bridged binuclear compound $[Pt_2(\eta^2\text{-}C_2H_4)(\mu_2\text{-}Cl)_2Cl_2]$, remained as chemical curiosities and a considerable theoretical embarrassment for over 100 y but are now seen as the archetypes of a large family of complexes based on the bonding of unsturated organic molecules to transition metals. The structure of the anion of Zeise's salt has been extensively studied and the most recent refinements from neutron

† The original reaction was obscure: Zeise heated a mixture of $PtCl_2$ and $PtCl_4$ in EtOH under reflux and then treated the resulting black solid with aqueous KCl and HCl to give ultimately the cream-yellow product. Subsequently the compound was isolated by direct reaction of C_2H_4 with $K_2[PtCl_4]$ in aqueous HCl.

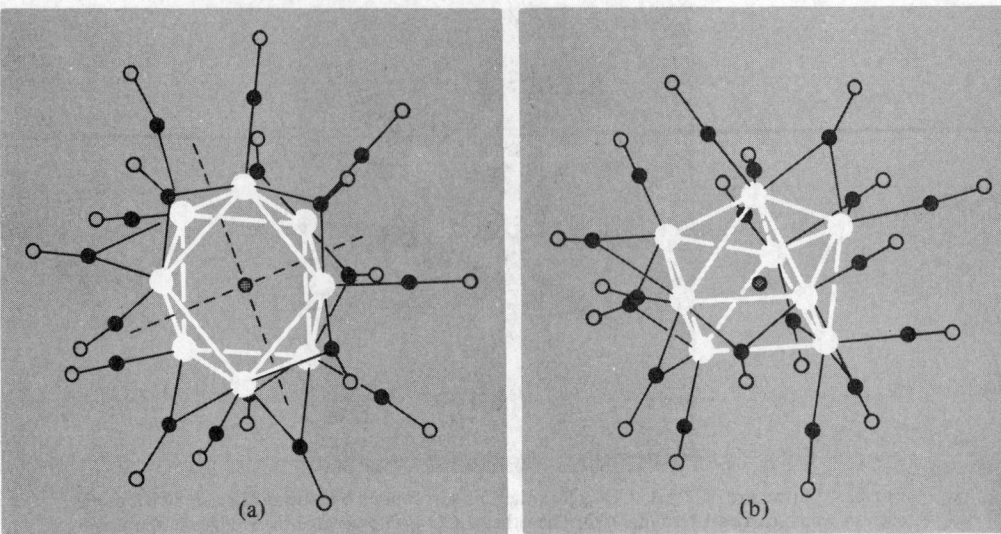

FIG. 8.26   Two views of $[Co_8C(CO)_{18}]^{2-}$; the 8 Cs define a distorted square antiprism (a) which can also be viewed as a distorted, bicapped trigonal prism (b) (cf. Fig. 8.25).

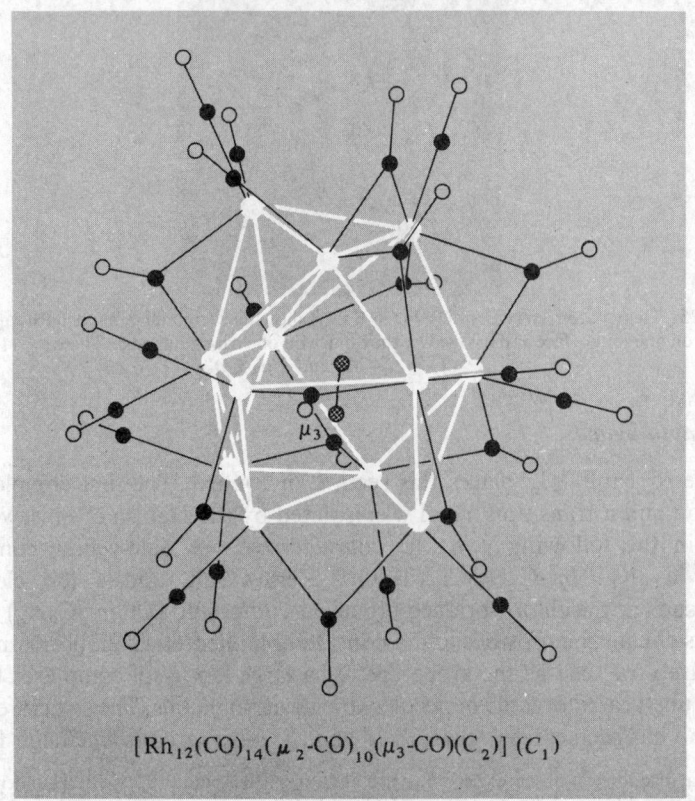

$[Rh_{12}(CO)_{14}(\mu_2\text{-CO})_{10}(\mu_3\text{-CO})(C_2)]\ (C_1)$

FIG. 8.27   Schematic representation of the molecular structure of $Rh_{12}(C_2)(CO)_{25}$.

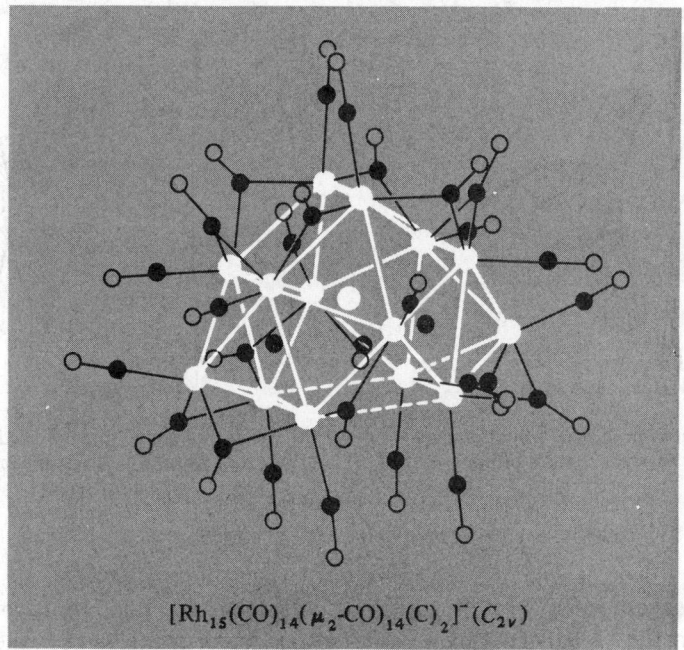

$$[Rh_{15}(CO)_{14}(\mu_2\text{-}CO)_{14}(C)_2]^-(C_{2v})$$

FIG. 8.28   Schematic representation of the molecular structure of the anion $[Rh_{15}(C)_2(CO)_{28}]^-$.

diffraction data[43] are in Fig. 8.29. Significant features are (a) the C=C bond is perpendicular to the $PtCl_3$ plane and is only 3.8 pm longer than in free C_2H_4, (b) the C_2H_4 group is significantly distorted from planarity, each C being 16.4 pm from the plane of 4H, (c) the angle between the normals to the CH_2 planes is 32.5°, and (d) there is an unambiguous *trans*- effect (p. 1352), i.e. the Pt–Cl distance *trans* to C_2H_4 is longer than the 2 *cis*-Pt–Cl distances by 3.8 pm (19 standard deviations). The key to our present understanding of the bonding in Zeise's salt and all other alkene complexes stems from the

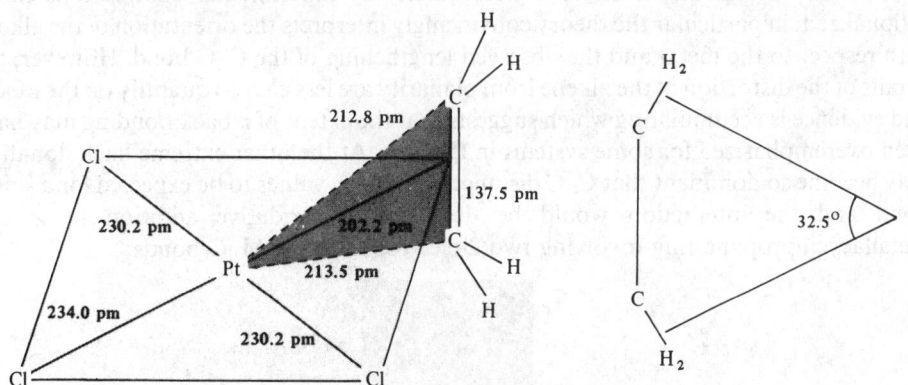

FIG. 8.29   Structure of the anion of Zeise's salt $[Pt(\eta^2\text{-}C_2H_2)Cl_3]^-$; standard deviations are
Pt–Cl 0.2 pm; Pt–C 0.3 pm, and C–C 0.4 pm.

[43] R. A. LOVE, T. F. KOETZLE, G. J. B. WILLIAMS, L. C. ANDREWS, and R. BAU, Neutron diffraction study of the structure of Zeise's salt, $KPtCl_3(C_2H_4) \cdot H_2O$, *Inorg. Chem.* **14**, 2653–7 (1975).

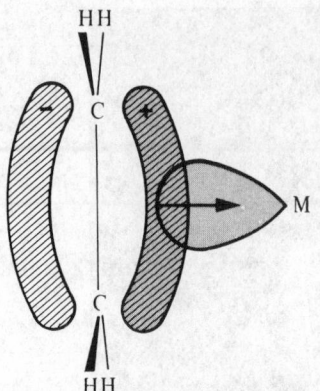

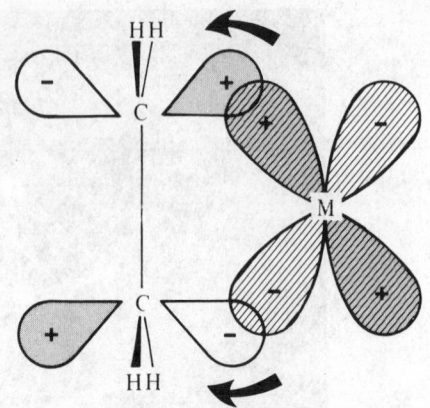

(a) σ donation from filled π orbital of alkene into vacant metal hybrid orbital.

(b) π-back donation from a filled metal d orbital (or hydrid) into the vacant antibonding orbital of alkene.

FIG. 8.30   Schematic representation of the two components of an η^2-alkene–metal bond.

perceptive suggestion by M. J. S. Dewar in 1951 that the bonding involves electron donation from the π bond of the alkene into a vacant metal orbital of σ symmetry; this idea was modified and elaborated by J. Chatt and L. A. Duncanson in a seminal paper in 1953 and the Dewar–Chatt–Duncanson theory forms the basis for most subsequent discussion.[23, 24, 29] The bonding is considered to arise from two interdependent components as illustrated schematically in Fig. 8.30. In the first part, σ overlap between the filled π orbital of ethene and a suitably directed vacant hybrid metal orbital forms the "electron-pair donor bond". This is reinforced by the second component which derives from overlap of a filled metal d orbital with the vacant antibonding orbital of ethene; these orbitals have π symmetry with respect to the bonding axis and allow $M \rightarrow C_2$ π back bonding to assist the $\sigma C_2 \rightarrow M$ bond synergically as for CO (p. 351). The flexible interplay of these two components allows a wide variety of experimental observations to be rationalized: in particular the theory convincingly interprets the orientation of the alkene with respect to the metal and the observed lengthening of the C–C bond. However, the details of the distortion of the alkene from planarity are less easy to quantify on the model and evidence is accumulating which suggests that the extent of π back bonding may have been overemphasized for some systems in the past. At the other extreme back donation may become so dominant that C–C distances approach values to be expected for a single bond and the interaction would be described as oxidative addition to give a metallacyclopropane ring involving two 2-electron 2-centre M–C bonds:

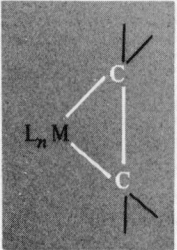

For example, tetracyanoethylene has a formal C=C double bond (133.9 pm) in the free ligand but in the complex $[Pt\{C_2(CN)_4\}(PPh_3)_2]$ the C–C distance (152 pm) is that of a single bond and the CN groups are bent away from the Pt and 2P atoms; moroever, the 2P and 2C that are bonded to Pt are nearly coplanar, as expected for Pt^{II} but not as in (tetrahedral) 4-coordinate Pt^0complexes. $[Rh(C_2F_4)Cl(PPh_3)_2]$ affords another example of the tendency to form a metallacyclopropane-type complex (C–C 141 pm) with pseudo-5-coordinate Rh^{III} rather than a pseudo-4-coordinate η^2-alkene complex of Rh^I. However, the two descriptions are not mutually exclusive and, in principle, there can be a continuous gradation between them.

Compounds containing $M–\eta^2$-alkene bonds are generally prepared by direct replacement of a less strongly bound ligand such as a halide ion (cf. Zeise's salt), a carbonyl, or another alkene.[29] Chelating dialkene complexes can be made similarly, e.g. with *cis-cis*-cycloocta-1,5-diene (COD):

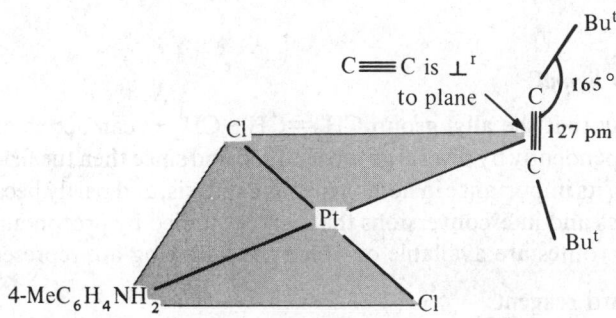

Numerous examples are given in later sections dealing with the chemistry of individual transition metals. Few, if any, η^2-alkene or -diene complexes have been reported for the first three transition-metal groups (why?), but all later groups are well represented, including Cu^I, Ag^I, and Au^I. Indeed, an industrial method for the separation of alkenes uses the differing stabilities of their complexes with CuCl. For many metals it is found that increasing alkyl substitution of the alkene lowers the stability of the complex and that *trans*-substituted alkenes give less stable complexes than do *cis*-substituted alkenes. For Rh^I complexes F substitution of the alkene enhances the stability of the complex and Cl substitution lowers it.

Alkyne complexes have been less studied than alkene complexes but are similar. Preparative routes are the same and bonding descriptions are also analogous. In some cases, e.g. the pseudo-4-coordinate complex $[Pt(\eta^2\text{-}C_2Bu^t_2)Cl_2(4\text{-toluidine})]$ (Fig. 8.31) the C≡C bond remains short and the alkyne group is normal to the plane of coordination; in others, e.g. the pseudo-3-coordinate complex $[Pt(\eta^2\text{-}C_2Ph_2)(PPh_3)_2]$

C≡C is $\perp^r$ to plane

Bu^t

165°

C

127 pm

C

Bu^t

Cl

Pt

4-MeC$_6$H$_4$NH$_2$

Cl

FIG. 8.31 Structure of $[Pt(\eta^2\text{-}C_2Bu^t_2)Cl_2(4\text{-toluidine})]$.

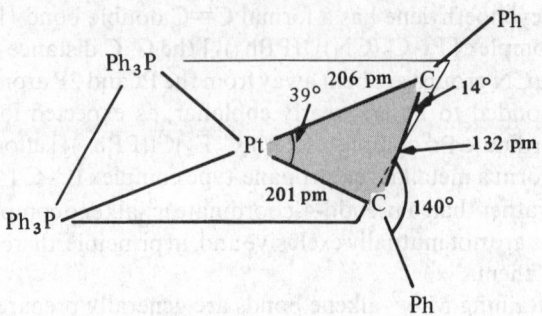

FIG. 8.32   Structure of $[Pt(\eta^2\text{-}C_2Ph_2)(PPh_3)_2]$.

(Fig. 8.32), the alkyne group is almost in the plane (14°) and the attached substituents are bent back to an angle of 140° suggesting a formulation intermediate between 3-coordinate Pt^0 and 4-coordinate Pt^{II}. One important difference between alkynes and alkenes is that the former have a triple bond which can be described in terms of a σ bond and two mutually perpendicular π bonds. The possibility thus arises that η^2-alkynes can function as bridging ligands and several such complexes have been characterized. The classic example is $[Co_2(CO)_6(C_2Ph_2)]$ which is formed by direct displacement of the 2 bridging carbonyls in $[Co_2(CO)_8]$ to give the structure sketched below:

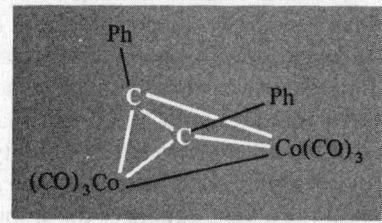

The C–C group lies above and at right angles to the Co–Co vector; the C–C distance is 146 pm (27 pm greater than in the free alkyne) and this has been taken to indicate extensive back donation from the 2 Co atoms. The Co–Co distance is 247 pm compared with 252 pm in $Co_2(CO)_8$.

8.9.3   *Trihapto ligands*

The possibility that the allyl group $CH_2=CH-CH_2-$ can act as an η^3 ligand was recognized independently by several groups in 1960 and since then the field has flourished, partly because of its importance in homogeneous catalysis, and partly because of the novel steric possibilities and interconversions that can be studied by proton nmr spectroscopy. Many synthetic routes are available of which the following are representative.

(a) Allyl Grignard reagent:

$$NiBr_2 + 2C_3H_5MgBr \longrightarrow 2MgBr_2 + [Ni(\eta^3\text{-}C_3H_5)_2] \text{ (also Pd, Pt)}$$

A mixture of *cis* and *trans* isomers is obtained

Tris-(η^3-allyl) complexes [M(C$_3$H$_5$)$_3$] can be prepared similarly for V, Cr, Fe, Co, Rh, Ir, and *tetrakis* complexes [M(η^3-C$_3$H$_5$)$_4$] for Zr, Th, Mo, and W.

(b) Conversion of η^1-allyl to η^3-allyl:

$$Na[Mn(CO)_5] \xrightarrow{C_3H_5Cl} [Mn(CO)_5(\eta^1\text{-}C_3H_5)] \xrightarrow[80°]{hv\ or} [Mn(CO)_4(\eta^3\text{-}C_3H_5)] + CO$$

Similarly many other η^1-allyl carbonyl complexes convert to η^3-allyl complexes with loss of 1 CO.

(c) From allylic halides (e.g. 2-methylallyl chloride):

$$Na_2PdCl_4 + C_4H_7Cl + CO + H_2O \xrightarrow[100\%\ yield]{MeOH} \tfrac{1}{2}[Pd_2(\eta^3\text{-}C_4H_7)_2(\mu_2\text{-}Cl_2)] + 2NaCl$$
$$+ 2HCl + CO_2$$

(d) Oxidative addition of allyl halides, e.g.:

$$[Fe^0(CO)_5] + C_3H_5I \longrightarrow [Fe^{II}(\eta^3\text{-}C_3H_5)(CO)_3I] + 2CO$$
$$[Co^I(\eta^5\text{-}C_5H_5)(CO)_2] + C_3H_5I \longrightarrow [Co^{III}(\eta^3\text{-}C_3H_5)(\eta^5\text{-}C_5H_5)I] + 2CO$$

(e) Elimination of HCl from an alkene metal halide complex, e.g.:

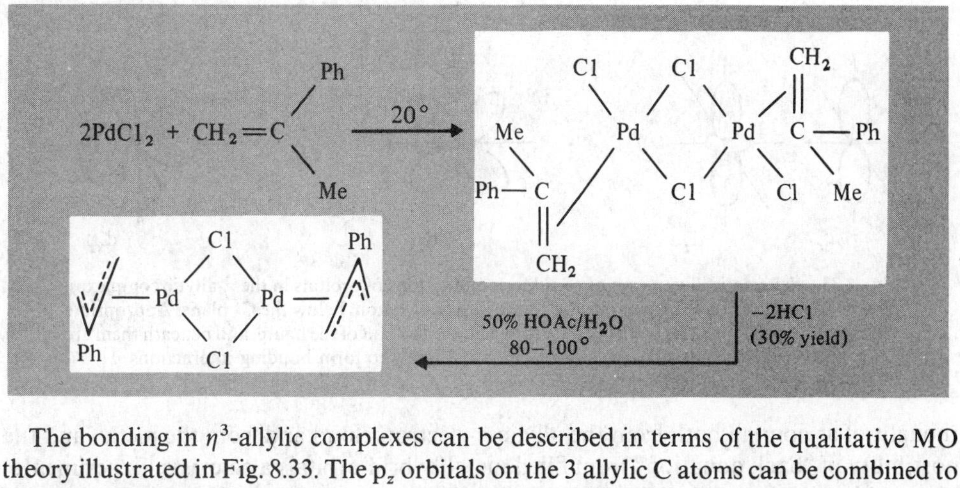

The bonding in η^3-allylic complexes can be described in terms of the qualitative MO theory illustrated in Fig. 8.33. The p$_z$ orbitals on the 3 allylic C atoms can be combined to give the 3 orbitals shown in the upper part of Fig. 8.33; each retains π symmetry with respect to the C$_3$ plane but has, in addition, 0, 1, or 2 nodes perpendicular to this plane. The metal orbitals of appropriate symmetry to form bonding MOs with these 3 combinations are shown in the lower part of Fig. 8.33. The extent to which these orbitals are, in fact, involved in bonding depends on their relative energies, their radial diffuseness, and the actual extent of orbital overlap. Electrons to fill these bonding MOs can be

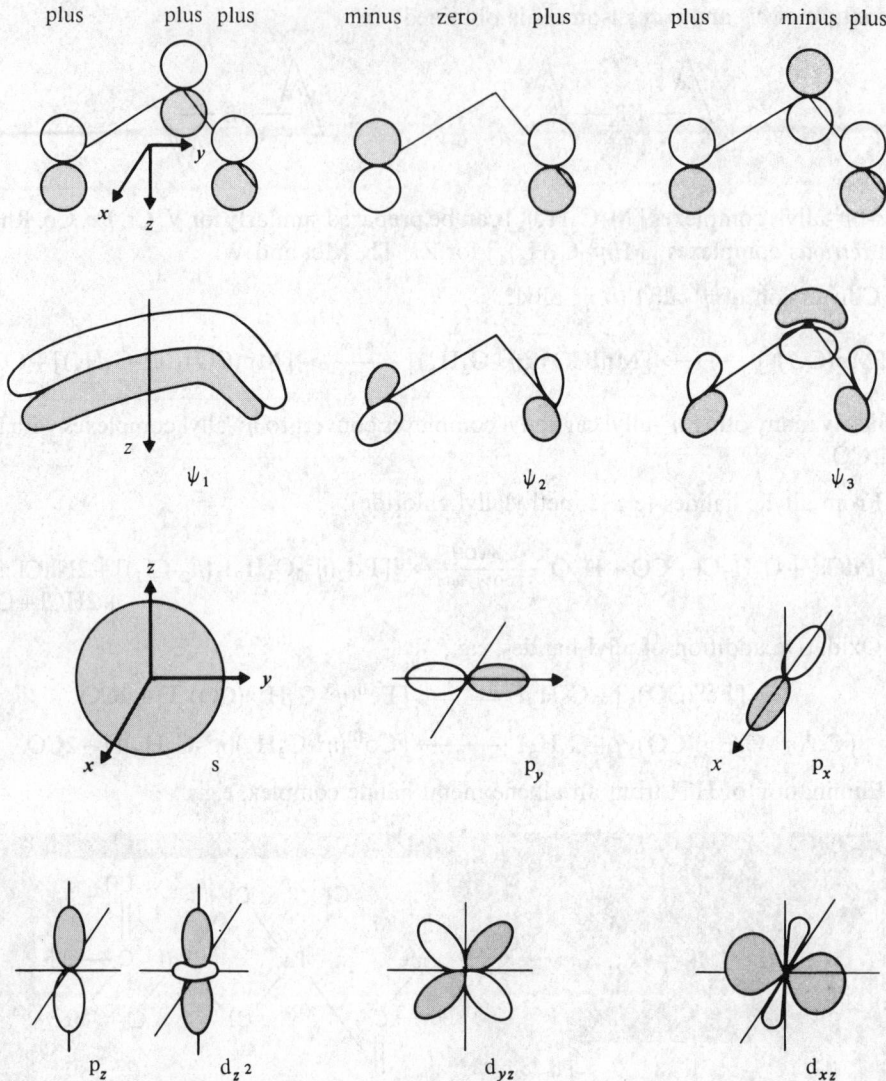

FIG. 8.33 Schematic illustration of possible combinations of orbitals in the π-allytic complexes. The bonding direction is taken to be the z-axis with the M atom below the C_3 plane. Appropriate combinations of p_π orbitals on the 3 C are shown in the top half of the figure, and beneath them are the metal orbitals with which they are most likely to form bonding interactions.

thought of as coming both from the allylic π-electron cloud and from the metal, and the possibility of "back donation" from filled metal hybrid orbitals also exists. Experimental observables which must be interpreted in any quantitative treatment are the variations (if any) in the M–C distances to the 3 C atoms and the tilt of the C_3 plane to the bonding plane of the metal atom.

In addition to acting as an η^1 and an η^3 ligand the allyl group can also act as a bridging ligand by η^1 bonding to one metal atom and η^2 bonding via the alkene function to a second metal atom. For example $[Pt_2(acac)_2(\eta^1,\eta^2\text{-}C_3H_5)_2]$ has the dimeric structure

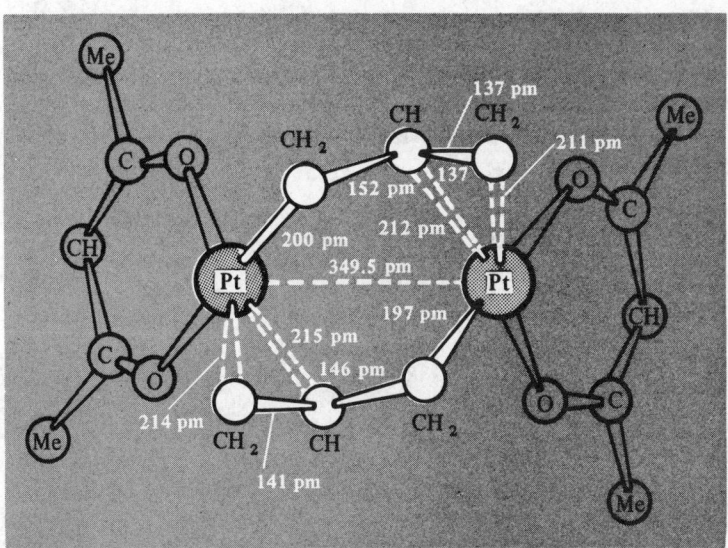

Fig. 8.34. Structure of $[Pt_2(acac)_2(\mu\text{-}C_3H_5)_2]$ showing the bridging allyl groups, each η^1 bonded to 1 Pt and η^2-bonded to the other. Interatomic distances are in pm with standard deviations of ~ 5 pm for Pt–C and ~ 7 for C–C. The distance of Pt to the centre of the η^2-C_2 group is 201 pm, very close to the η^1-Pt–C distance of 199 pm.

shown in Fig. 8.34. The compound was made from $[Pt(\eta^3\text{-}C_3H_5)_2]$ by treatment first with HCl to give polymeric $[Pt(C_3H_5)Cl]$ and then with thallium(I) acetylacetonate.

Many η^3-allyl complexes are fluxional (p. 1071) at room temperature or slightly above, and this property has been extensively studied by ^1H nmr spectroscopy.[29] Exceedingly complex patterns can emerge. The simplest interchange that can occur is between the *syn* and *anti* H atoms probably via a short-lived η^1-allyl metal intermediate. The fluxional behaviour can be slowed down at lower temperatures, and separate resonances from the various types of H atom are observed. Fluxionality can also sometimes be quenched by incorporating the allylic group in a ring system which restricts its mobility.

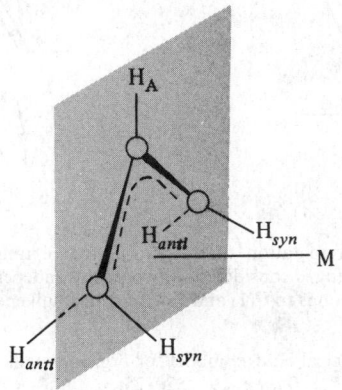

8.9.4   *Tetrahapto ligands*

Conjugated dienes such as butadiene and its open-chain analogues can act as η^4 ligands; the complexes are usually prepared from metal carbonyl complexes by direct replacement of 2CO by the diene. Isomerization or rearrangement of the diene may occur as indicated schematically below:

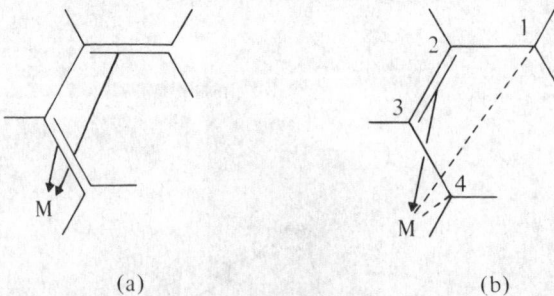

No new principles are involved in describing the bonding in these complexes and appropriate combinations of the $4p_\pi$ orbitals on the diene system can be used to construct MOs with the metal-based orbitals for donation and back donation of electron density.[43a] As with ethene, two limiting cases can be envisaged which can be represented schematically as in Fig. 8.35. Consistent with this view, the C–C distances in diene complexes vary and the central C(2)–C(3) distance is often less than the two outer C–C distances.

Cyclobutadiene complexes are also well established though they must be synthesized by

Fig. 8.35   Schematic representation of the two formal extremes of bonding in 1,3-diene complexes. In (a) the bonding is considered as two almost independent η^2-alkene–metal bonds, whereas in (b) there are σ bonds to C(1) and C(4) and an η^2-alkene–metal bond from C(2)–C(3).

[43a] D. M. P. MINGOS, A topological Hückel model for organometallic complexes. Part 1. Bond lengths in complexes of conjugated olefins, *JCS Dalton*, 1977, 20–25.

indirect routes since the parent dienes are either unstable or non-existent. Four general routes are available:

(a) Dehalogenation of dihalocyclobutenes, e.g.:

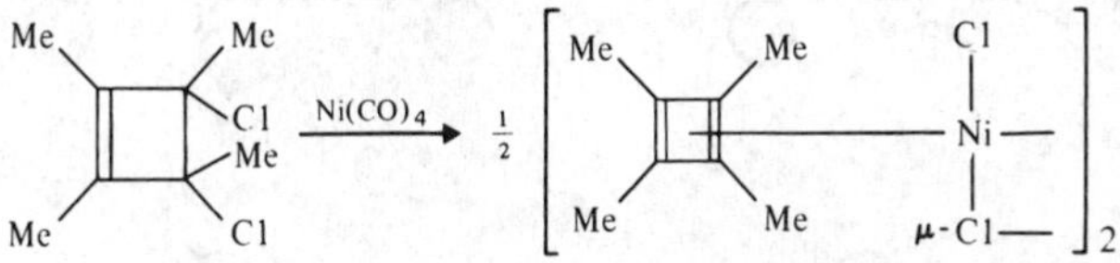

(b) Cyclodimerization of alkynes, e.g. with cyclopentadienyl-(cycloocta-1,5-diene) cobalt:

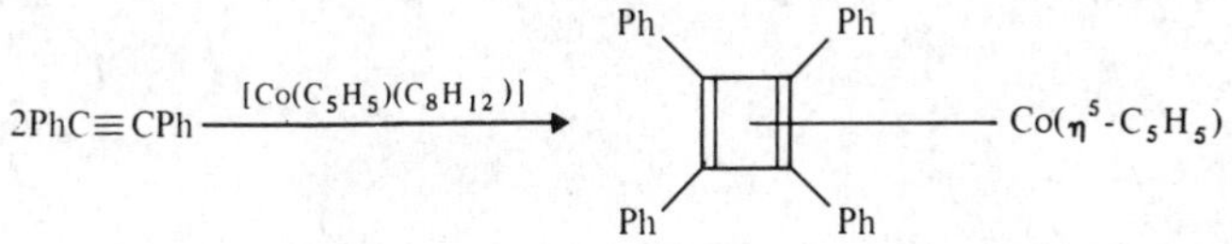

(c) From metallacyclopentadienes:

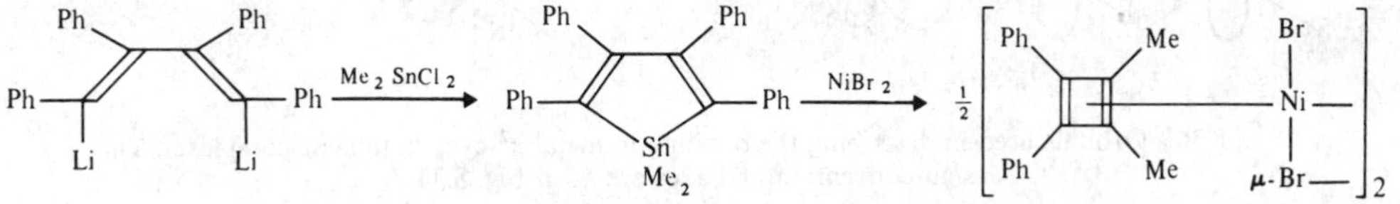

(d) Ligand exchange from other cyclobutadiene complexes, e.g.:

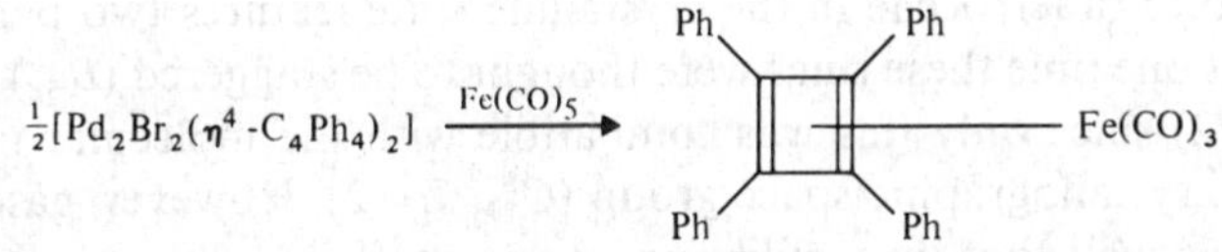

A schematic interpretation of the bonding in cyclobutadiene complexes can be given within the framework outlined in the preceding sections and this is illustrated in Fig. 8.36. Cyclobutadiene complexes afford a classic example of the stabilization of a ligand by coordination to a metal and, indeed, were predicted theoretically on this basis by H. C. Longuet-Higgins and L. E. Orgel (1956) some 3 y before the first examples were synthesized. In the (hypothetical) free cyclobutadiene molecule 2 of the 4 π-electrons would occupy ψ_1 and there would be an unpaired electron in each of the 2 degenerate orbitals ψ_2, ψ_3. Coordination to a metal provides further interactions and avoids this unstable configuration. See also the discussion on ferraboranes (p. 197).

8.9.5 *Pentahapto ligands*

The importance of bis(cyclopentadienyl)iron [Fe(η^5-C_5H_5)$_2$] in the development of organometallic chemistry has already been alluded to (p. 345). The compound, which forms orange crystals, mp 174°, has extraordinary thermal stability (>500°) and a remarkable structure which was unique when first established. It also has an extensive aromatic-type reaction chemistry which is reflected in its common name "ferrocene". The

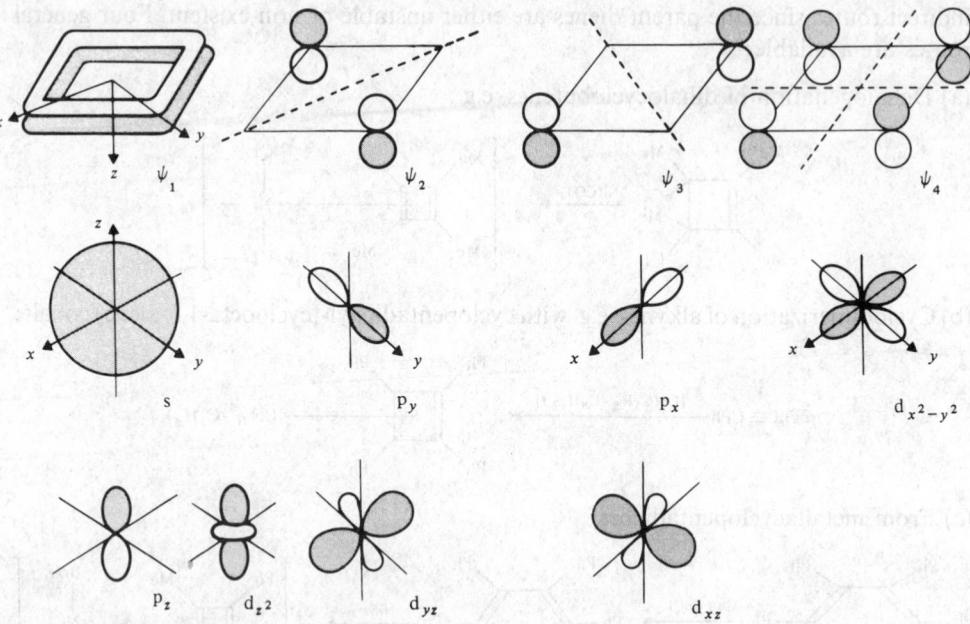

FIG. 8.36   Orbitals used in describing the bonding in metal-η^2-cyclobutadiene complexes. The
sign convention and axes are as in Fig. 8.34.

molecular structure of ferrocene in the crystalline state features two parallel cyclopentadienyl rings: at one time these rings were thought to be staggered (D_{5d}) as in Fig. 8.37a
and b (see p. 372) since only this was compatible with the molecular inversion centre
required by the crystallographic space group (C_{2h}^5, $Z = 2$). However, gas-phase electron
diffraction data suggest that the equilibrium structure of ferrocene is eclipsed (D_{5h}) as in
Fig. 8.37c rather than staggered, with a rather low barrier to internal rotation of
~ 4 kJ mol^{-1}. Recent X-ray crystallographic[43b] and neutron diffraction studies[43c]
confirm this general conclusion, the space-group symmetry requirement being met by a
disordered arrangement of nearly eclipsed molecules (rotation angle between the rings
$\sim 9°$ rather than 0° for precisely eclipsed or 36° for staggered conformation). Below 169 K
the molecules become ordered, the rotation angle remaining $\sim 9°$. The perpendicular
distance between the rings is 325 pm (cf. graphite 335 pm) and the mean interatomic
distances are Fe–C 203 $\pm$ 2 pm and C–C 139 $\pm$ 6 pm. The Ru and Os analogues [M(η^5-
$C_5H_5)_2$] have similar molecular structures with eclipsed parallel C_5 rings. A molecular-
orbital description of the bonding can be developed along the lines indicated in previous
sections and is elaborated in the Panel. Because of the importance of ferrocene, numerous
calculations have been made of the detailed sequence of energy levels in the molecule;
though these differ slightly depending on the assumptions made and the computational
methods adopted, there is now a general consensus concerning the bonding. One of the
most recent sets of results (1976) is included in the Panel.

[43b] P. SEILER and J. D. DUNITZ, A new interpretation of the disordered crystal structure of ferrocene, *Acta
Cryst.* **B35**, 1068–74 (1979).
[43c] F. TAKUSAGAWA and T. F. KOETZLE, A neutron diffraction study of the crystal structure of ferrocene, *Acta
Cryst.* **B35**, 1074–81 (1979).

A Molecular Orbital Description of the Bonding in $[Fe(\eta^5\text{-}C_5H_5)_2]$

The 5 p_π atomic orbitals on the planar C_5H_5 group can be combined to give 5 group MOs as shown in Fig. A; one combination has the full symmetry of the ring (a) and there are two doubly degenerate combinations (e_1 and e_2) having respectively 1 and 2 planar nodes at right angles to the plane of the ring. These 5 group MOs can themselves be combined in pairs with a similar set from the second C_5H_5 group. Figure B shows that one of these combinations is symmetric with respect to inversion (German *gerade*, even) whereas the other combination is antisymmetric to inversion (*ungerade*, odd). These combinations are labelled g and u respectively and can in turn be combined with metal orbitals having similar symmetry properties. Each of the combinations [(ligand orbitals) + (metal orbitals)] shown in Fig. B leads, in principle to a bonding MO of the molecule, providing that the energy of the two component sets is not very different. There are an equal number of antibonding combinations (not shown) with the sign [(ligand orbitals)–(metal orbital)]. Note, however, that there are no metal orbitals of appropriate symmetry to combine with the degenerate ligand sets labelled e_{2u}.

Calculation of the detailed sequence of energy levels arising from these combinations poses severe computational problems but a schematic indication of the sequence (not to scale) is shown in Fig. C. Thus, starting from the foot of the figure, the a_{1g} bonding MO is mainly ligand-based

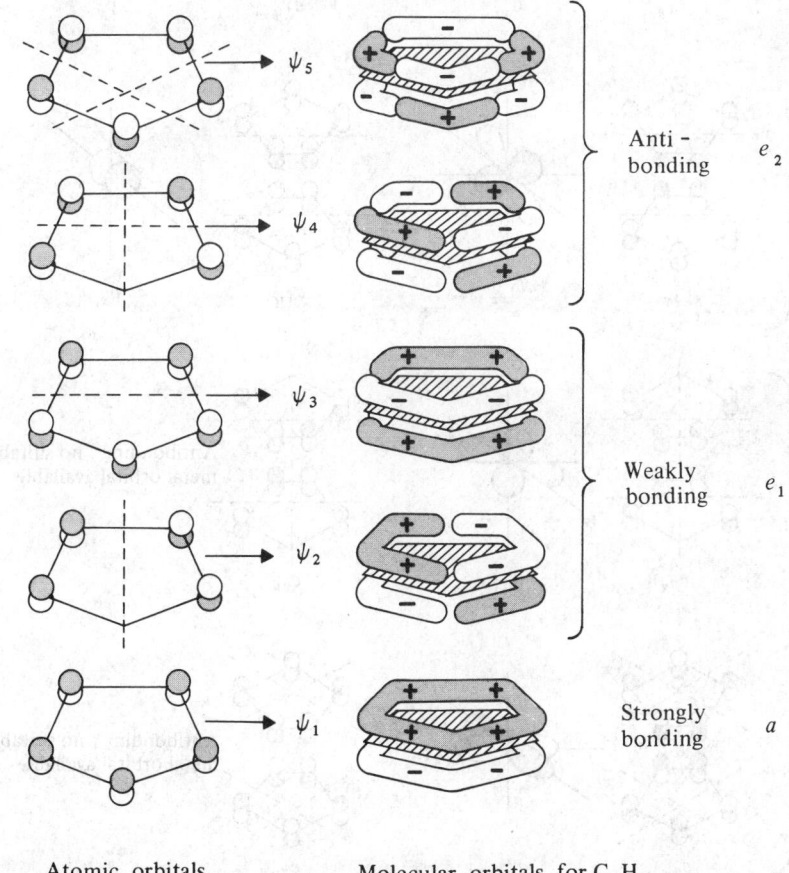

Atomic orbitals   Molecular orbitals for C_5H_5

FIG. A   The π molecular orbitals formed from the set of p_π orbitals of the C_5H_5 ring.

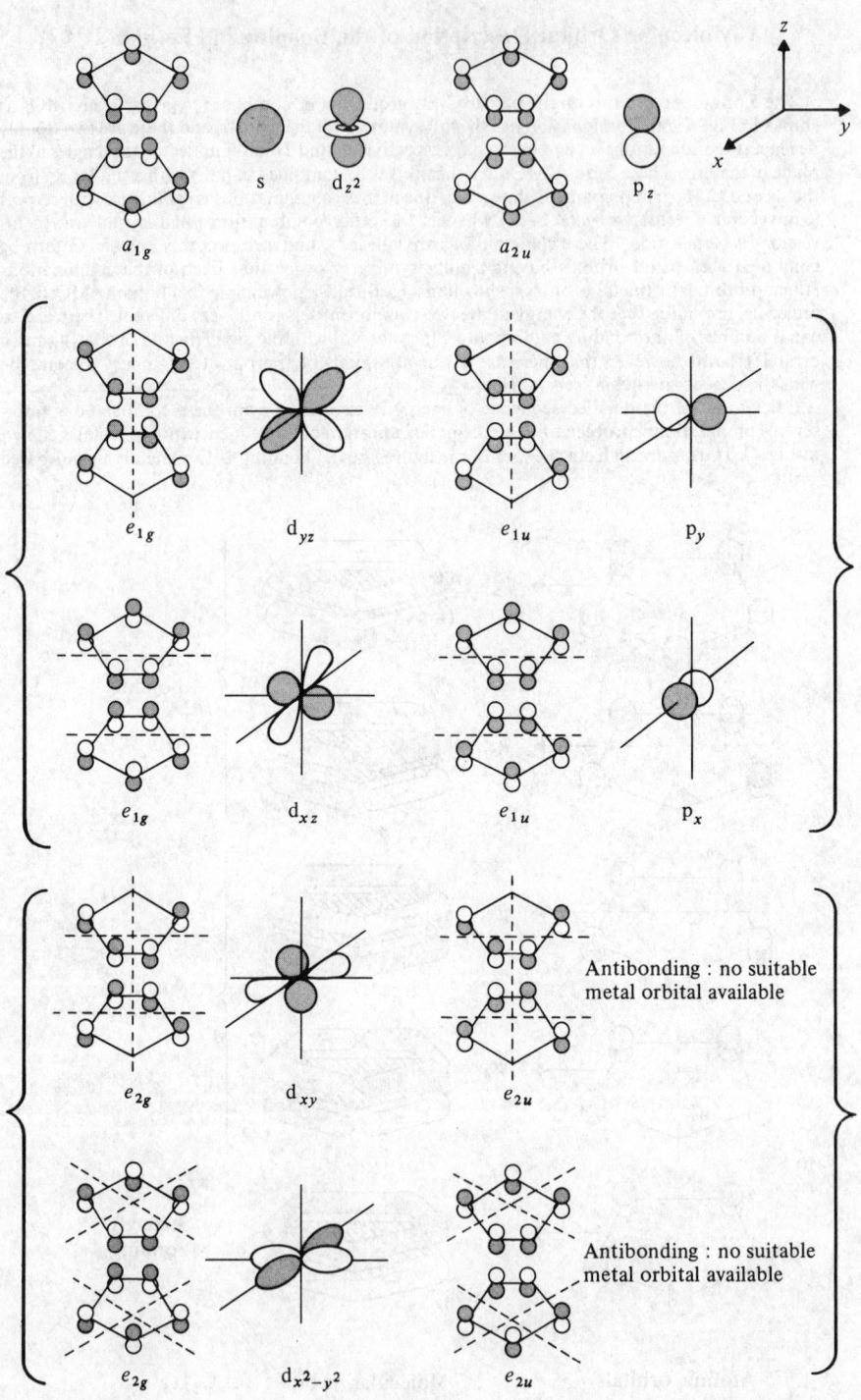

FIG. B

Continued

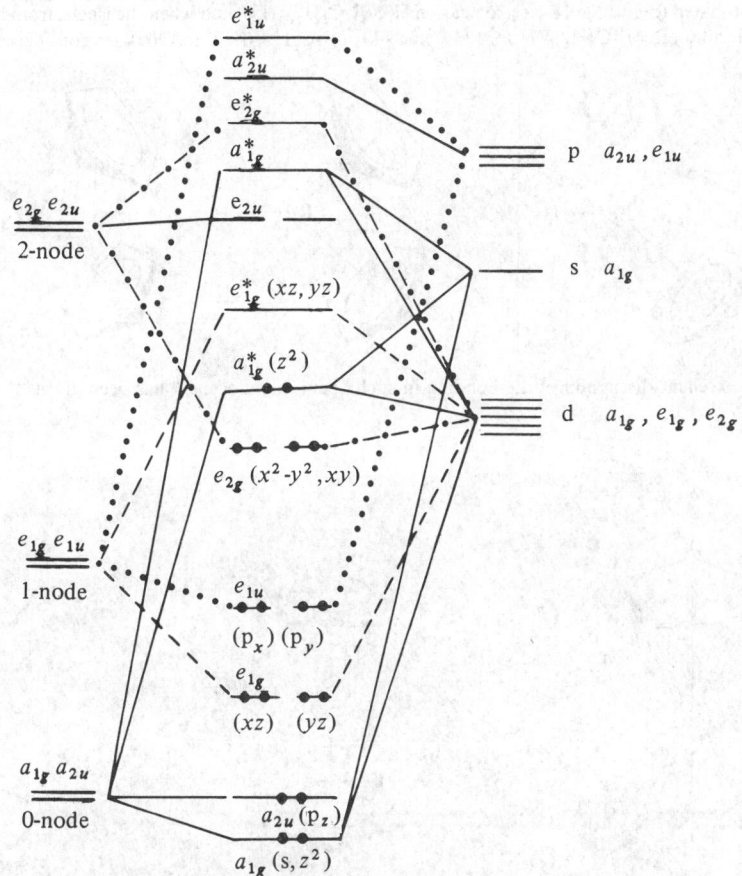

Fig. C   A qualitative molecular orbital diagram for ferrocene.

with only a slight admixture of the Fe 4s and $3d_{z^2}$ orbitals. Similarly, the a_{2u} level has little, if any, admixture of the even higher-lying Fe $4p_z$ orbital with which it is formally able to combine. The e_{1g} MO arises from the bonding combination of the ligand e_{1g} orbitals with Fe $3d_{xz}$ and $3d_{yz}$ and this is the main contribution to the stability of the complex; the corresponding antibonding $e_{1g}{}^*$ are unoccupied in the ground state but will be involved in optical transitions. The e_{1u} bonding MOs are again mainly ligand-based but with some contribution from Fe $4p_x$ and $4p_y$, etc. It can be seen that there is room for just 18 electrons in bonding and nonbonding MOs and that the antibonding MOs are unoccupied. In terms of electron counting the 18 electrons can be thought of as originating from the Fe atom (8e) and the two C_5H_5 groups (2 × 5e) or from an Fe^{II} ion (6e) and two $C_5H_5{}^-$ groups (2 × 6e).

The stability of $[Fe(\eta^5\text{-}C_5H_5)_2]$ compared with the 19 electron system $[Co(\eta^5\text{-}C_5H_5)_2]$ and the 20-electron system $[Ni(\eta^5\text{-}C_5H_5)_2]$ is readily interpreted on this bonding scheme since these latter species have 1 and 2 easily oxidizable electrons in the antibonding $e_{1g}{}^*$ orbitals. Similarly, $[Cr(\eta^5\text{-}C_5H_5)_2]$ (16e) and $[V(\eta^5\text{-}C_5H_5)_2]$ (15e) have unfilled bonding MOs and are highly reactive. However, attachment of additional groups or ligands destroys the D_{5d} (or D_{5h}) symmetry of the simple metallocene and this modifies the orbital diagram. This also happens when ferrocene is

Continued

protonated to give the 18-electron cation $[Fe(\eta^5\text{-}C_5H_5)_2H]^+$ and when the isoelectronic neutral molecules $[Re(\eta^5\text{-}C_5H_5)_2H]$ (p. 1239) and $[Mo(\eta^5\text{-}C_5H_5)_2H_2]$ (p. 1209) are considered:

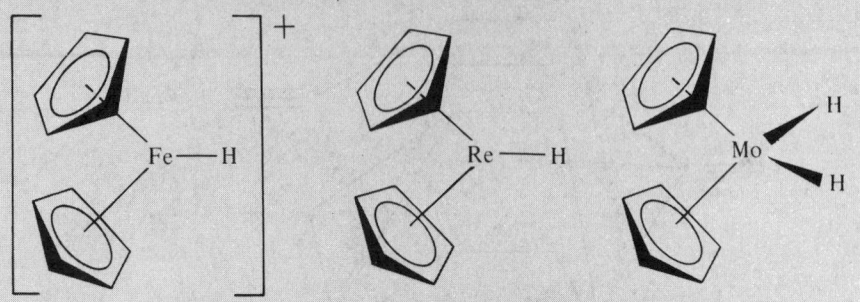

An excellent discussion of the bonding in such "bent metallocenes" has been given.[44]

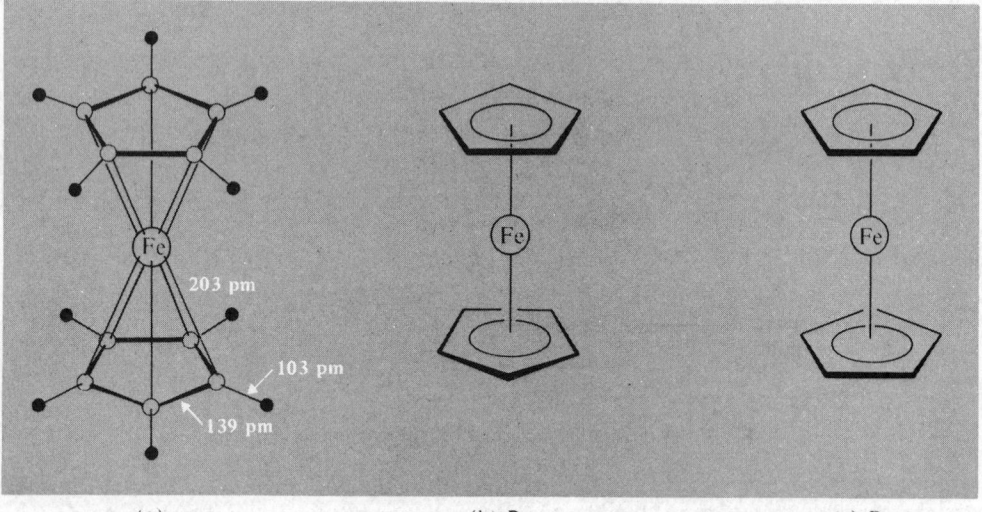

(a)   (b) D_{5d} (c) D_{5h}

FIG. 8.37 Structure of ferrocene, $[Fe(\eta^5\text{-}C_5H_5)_2]$, and a conventional "shorthand" representation (see text on p. 368).

A general preparative route to $\eta^5\text{-}C_5H_5$ compounds is the reaction of NaC_5H_5 with a metal halide or complex halide in a polar solvent such as thf, Me_2O (bp $-23°$), $(MeO)C_2H_4(OMe)$, or $HC(O)NMe_2$:

$$C_5H_6 + Na \xrightarrow{-\frac{1}{2}H_2} \{NaC_5H_5\} \xrightarrow{L_nMX} [M(\eta^5\text{-}C_5H_5)L_n] + NaX$$

A very convenient though somewhat less general method is to use a strong nitrogen base to deprotonate the C_5H_6:

$$2C_5H_6 + 2NEt_2H + FeCl_2 \xrightarrow[\text{amine}]{\text{excess of}} [Fe(\eta^5\text{-}C_5H_5)_2] + [NEt_2H_2]Cl$$

---

[44] J. W. LAUHER and R. HOFFMAN, Structure and chemistry of bis(cyclopentadienyl)-ML$_n$ complexes, *J. Am. Chem. Soc.* **98**, 1729–42 (1976), and references therein.

An enormous number of $\eta^5\text{-}C_5H_5$ complexes is now known. Thus the isoelectronic yellow Co^I species $[Co(\eta^5\text{-}C_5H_5)_2]^+$ is stable in aqueous solutions and its salts are thermally stable to $\sim 400°$. The bright-green paramagnetic complex $[Ni(\eta^5\text{-}C_5H_5)_2]$, mp $173°$(d), is fairly stable as a solid but is rapidly oxidized to $[Ni(\eta^5\text{-}C_5H_5)_2]^+$. In contrast, the scarlet, paramagnetic complex $[Cr(\eta^5\text{-}C_5H_5)_2]$, mp $173°$, is very air sensitive; it dissolves in aqueous HCl to give C_5H_6 and a blue cation which is probably $[Cr(\eta^5\text{-}C_5H_5)Cl(H_2O)_n]^+$. Other stoichiometries are exemplified by $[Ti(\eta^5\text{-}C_5H_5)_3]$ and $[M(\eta^5\text{-}C_5H_5)_4]$, where M is Zr, Hf, Th. Innumerable derivatives have been synthesized in which one or more $\eta^5\text{-}C_5H_5$ group is present in a mononuclear or polynuclear metal complex together with other ligands such as CO, NO, H, or X. It should also be borne in mind that C_5H_5 can act as an η^1-ligand by forming a σ M–C bond and mixed complexes are sometimes obtained, e.g.:

$$MoCl_5 \xrightarrow{\text{NaC}_5\text{H}_5} [Mo(\eta^1\text{-}C_5H_5)_3(\eta^5\text{-}C_5H_5)]$$

$$NbCl_5 \xrightarrow{\text{NaC}_5\text{H}_5} [Nb(\eta^1\text{-}C_5H_5)_2(\eta^5\text{-}C_5H_5)_2]$$

Such $\eta^1\text{-}C_5H_5$ complexes are often found to be fluxional in solution at room temperature, the 5 H atoms giving rise to a single sharp ^1H nmr resonance. At lower temperatures the spectrum usually broadens and finally resolves into the expected complex spectrum at temperatures which are sufficiently low to prevent interchange on the nmr time scale ($\sim 10^{-3}$ s). Numerous experiments have been devised to elucidate the mechanism by which the H atoms become equivalent and, at least in some systems, it seems likely that a non-dissociative (unimolecular) 1,2-shift occurs.

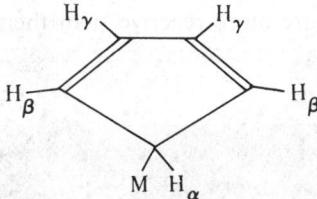

8.9.6 *Hexahapto ligands*

Arenes such as benzene and its derivatives can form complexes precisely analogous to ferrocene and related species. Though particularly exciting when first recognized as η^6 complexes in 1955 these compounds introduce no new principles and need only be briefly considered here. Curiously, the first such compounds were made as long ago as 1919 when F. Hein reacted $CrCl_3$ with PhMgBr to give compounds which he formulated as "polyphenylchromium" compounds $[CrPh_n]^{0, +1}$ ($n = 2$, 3, or 4); their true nature as η^6-arene complexes of benzene and diphenyl was not recognized until over 35 y later.[45] The best general method for making bis (η^6-arene) metal complexes is due to E. O. Fischer and W. Hafner (1955) who devised it originally for dibenzenechromium—the isoelectronic

^{45} H. Zeiss, P. J. Wheatley, and H. J. S. Winkler, *Benzenoid-Metal Complexes*, Ronald Press, New York, 1966, 101 pp.

analogue of ferrocene: $CrCl_3$ was reduced with Al metal in the presence of C_6H_6, using $AlCl_3$ as a catalyst:

$$3CrCl_3 + 2Al + AlCl_3 + 6C_6H_6 \xrightarrow{140°/\text{press}} 3[Cr(\eta^6\text{-}C_6H_6)_2]^+[AlCl_4]^-$$

The yield is almost quantitative and the orange-yellow Cr^I cation can be reduced to the neutral species with aqueous dithionite:

$$[Cr(\eta^6\text{-}C_6H_6)_2]^+ + \tfrac{1}{2}S_2O_4^{2-} + 2OH^- \longrightarrow [Cr(\eta^6\text{-}C_6H_6)_2] + SO_3^{2-} + H_2O$$

Dibenzenechromium(0) forms dark-brown crystals, mp 284°, and the molecular structure (Fig. 8.38) comprises plane parallel rings in eclipsed configuration above and below the Cr atom (D_{6h}); the C–H bonds are tilted slightly towards the metal and, most significantly, the C–C distances show no alternation around the rings. A bonding scheme can be constructed as for ferrocene (p. 369) using the six p_z orbitals on each benzene ring.[23]

Bis (η^6-arene) metal complexes have been made for many transition metals by the $Al/AlCl_3$ reduction method and cationic species $[M(\eta^6\text{-}Ar)_2]^{n+}$ are also well established for $n = 1$, 2, and 3. Numerous arenes besides benzene have been used, the next most common being $1,3,5\text{-}Me_3C_6H_3$ (mesitylene) and C_6Me_6. Reaction of arenes with metal carbonyls in high-boiling solvents or under the influence of ultraviolet light results in the displacement of 3CO and the formation of arene–metal carbonyls:

$$Ar + [M(CO)_6] \xrightarrow[\text{or heat}]{h\nu} [M(\eta^6\text{-}Ar)(CO)_3] + 3CO$$

For Cr, Mo, and W the benzenetricarbonyl complexes are yellow solids melting at 162°, 125°, and 140°, respectively. The structure of $[Cr(\eta^6\text{-}C_6H_6)(CO)_3]$ is in Fig. 8.39. In general, η^6-arene complexes are more reactive than their $\eta^5\text{-}C_5H_5$ analogues and are thermally less stable.

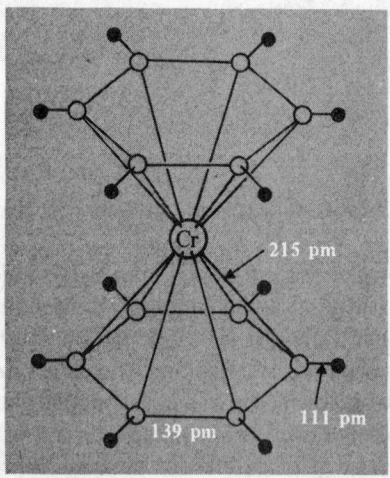

FIG. 8.38 The eclipsed (D_{6h}) structure of $[Cr(\eta^6\text{-}C_6H_6)_2]$ as revealed by X-ray diffraction, showing the two parallel rings 323 pm apart. Newton diffraction shows the H atoms are tilted slightly towards the Cr, and electron diffraction on the gaseous compound shows that the eclipsed configuration is retained without rotation.

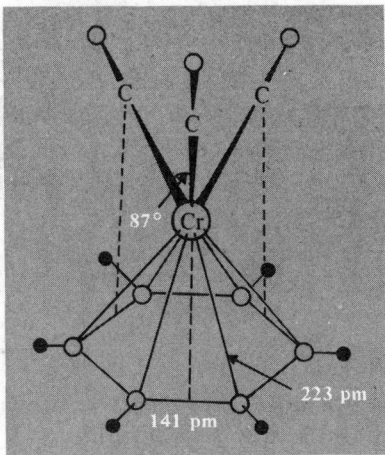

FIG. 8.39   The structure of $[Cr(\eta^6\text{-}C_6H_6)(CO)_3]$ showing the three CO groups in staggered configuration with respect to the benzene ring: the Cr–O distance is 295 pm and the plane of the 3 O atoms is parallel to the plane of the ring.

8.9.7   *Heptahapto and octahapto ligands*

Treatment of cycloheptatriene complexes of the type $[M(\eta^6\text{-}C_7H_8)(CO)_3]$ (M = Cr, Mo, W) with $Ph_3C^+BF_4^-$ results in hydride abstraction to give orange-coloured η^7-cycloheptatrienyl (or tropylium) complexes:

$$
\xrightarrow{\ Ph_3C^+BF_4^-\ } \qquad BF_4^- + Ph_3CH
$$

In some cases the loss of hydrogen may occur spontaneously, e.g.:

$$[V(\eta^5\text{-}C_5H_5)(CO)_4] + C_7H_8 \longrightarrow [V(\eta^5\text{-}C_5H_5)(\eta^7\text{-}C_7H_7)] + 4CO + \tfrac{1}{2}H_2$$

$$3[V(CO)_6] + 3C_7H_8 \longrightarrow [V(\eta^7\text{-}C_7H_7)(CO)_3] + [V(\eta^6\text{-}C_7H_8)(\eta^7\text{-}C_7H_7)]^+[V(CO)_6]^-$$
$$+ 9CO + H_2$$

The purple paramagnetic complex $[V(\eta^5\text{-}C_5H_5)(\eta^7\text{-}C_7H_7)]$ and the dark-brown diamagnetic complex $[V(\eta^7\text{-}C_7H_7)(CO)_3]$ both feature symmetrical planar C_7 rings as illustrated in Fig. 8.40. The bonding appears to be similar to that in $\eta^5\text{-}C_5H_5$ and $\eta^6\text{-}C_6H_6$ complexes but, as expected from the large number of bonding electrons formally provided by the ligand, its complexes are restricted to elements in the early part of the transition series, e.g. V, Cr, Mo, Mn[I]. For $[V(\eta^5\text{-}C_5H_5)(\eta^7\text{-}C_7H_7)]$ the rings are "eclipsed" as shown, and a notable feature of the structure is the substantially closer approach of the C_7H_7 ring to the V atom, suggesting that equality of V–C distances to the 2 rings is the controlling factor; consistent with this V–C(7 ring) is 225 pm and V–C(5 ring) is 223 pm. In addition to acting as an η^7 ligand, cycloheptatrienyl can also bond in the η^5, η^3, and even η^1 mode (see ref. 47 on p. 378).

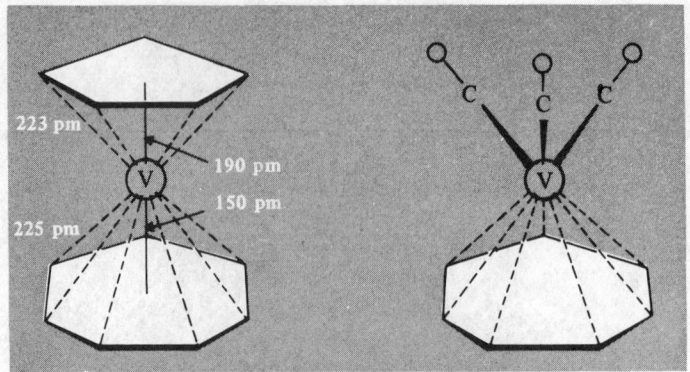

FIG. 8.40 Schematic representation of the structures of $[V(\eta^5\text{-}C_5H_5)(\eta^7\text{-}C_7H_7)]$ and $[V(\eta^7\text{-}C_7H_7)(CO)_3]$.

Octahapto ligands are rare but cyclooctatetraene fulfils this role in some of its complexes—the metal must clearly have an adequate number of unfilled orbitals and be large enough to bond effectively with such a large ring. Th, Pa, U, Np, and Pu satisfy these criteria and the complexes $[M(\eta^8\text{-}C_8H_8)_2]$ have been shown by X-ray crystallography to have eclipsed parallel planar rings (Fig. 8.41).[29a] The deep-green U complex can be made by reducing C_8H_8 with K in dry thf and then reacting the intense yellow solution of $K_2C_8H_8$ with UCl_4:

$$2C_8H_8 + 4K \xrightarrow[-30°]{\text{thf}} 2K_2C_8H_8 \xrightarrow[0°]{UCl_4/\text{thf}} [U(\eta^8\text{-}C_8H_8)_2] + 4KCl$$

The compound inflames in air but is stable in aqueous acid or alkali solutions. The colourless complex $[Th(\eta^8\text{-}C_8H_8)_2]$, yellow complexes $[Pa(\eta^8\text{-}C_8H_8)_2]$ and $[Np(\eta^8\text{-}$

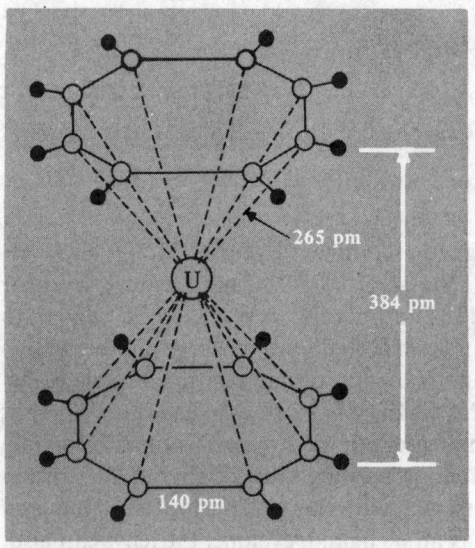

FIG. 8.41 The structure of $[U(\eta^8\text{-}C_8H_8)_2]$ showing D_{8h} symmetry.

228 pm
243 pm
236 pm
Ti
Ti
140 pm
254 pm
235 pm
∠C-Ti-C 34·5°

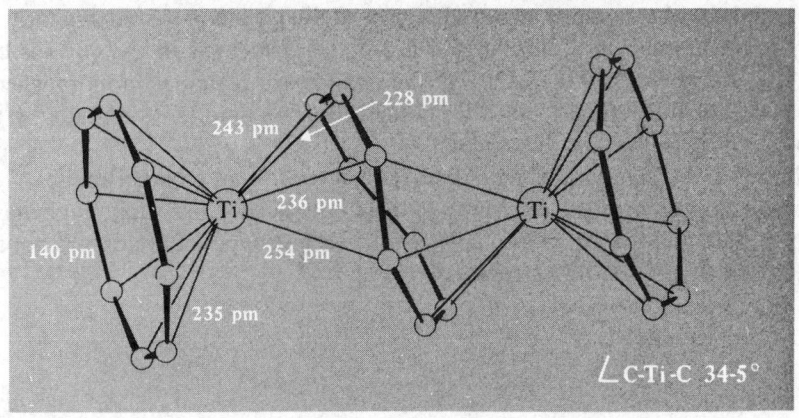

FIG. 8.42 Structure of $[Ti_2(C_8H_8)_3]$ showing it to be $[\{Ti(\eta^8\text{-}C_8H_8)\}_2\mu\text{-}(\eta^4,\eta^4\text{-}C_8H_8)]$. Ti–C to outer 16 C = 235 pm. H atoms are omitted for clarity.

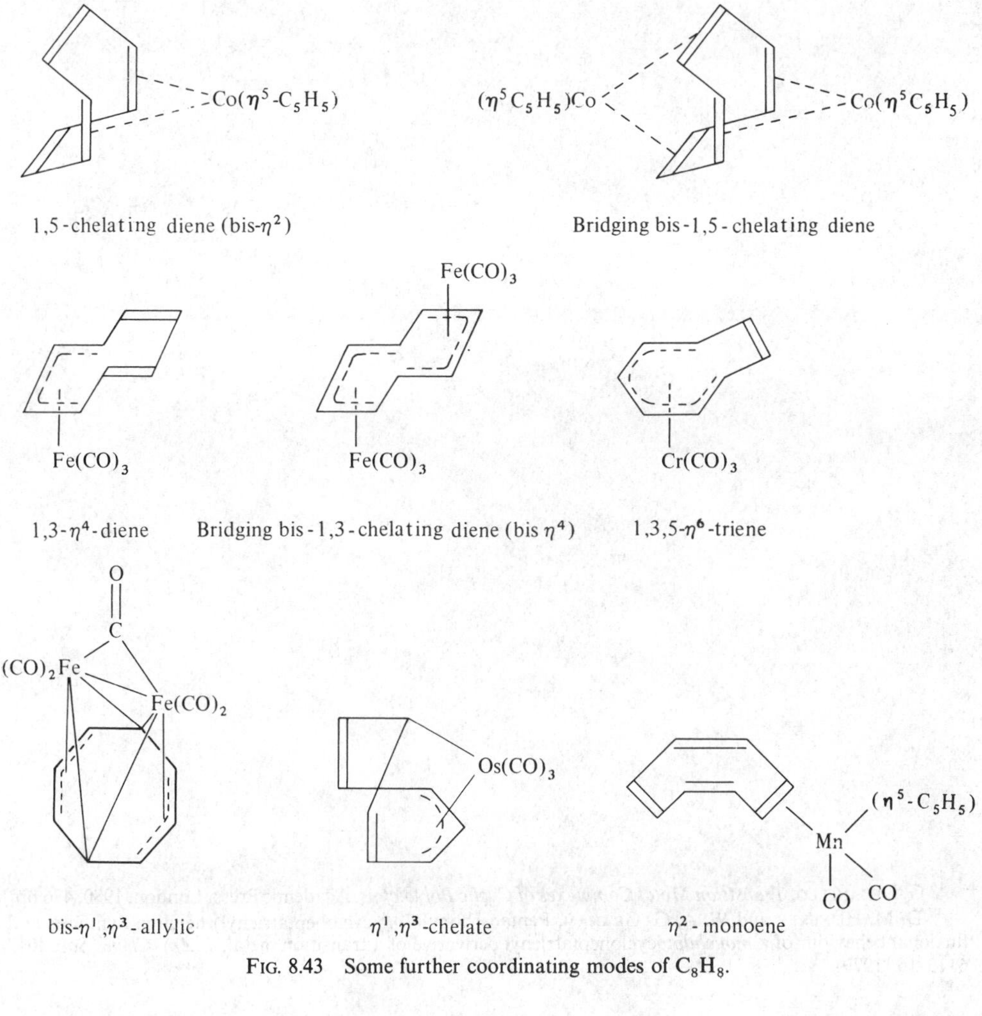

Co(η⁵-C₅H₅)
1,5-chelating diene (bis-η²)
(η⁵C₅H₅)Co
Co(η⁵C₅H₅)
Bridging bis-1,5-chelating diene
Fe(CO)₃
Fe(CO)₃
Fe(CO)₃
Cr(CO)₃
1,3-η⁴-diene
Bridging bis-1,3-chelating diene (bis η⁴)
1,3,5-η⁶-triene
O
C
(CO)₂Fe
Fe(CO)₂
Os(CO)₃
(η⁵-C₅H₅)
Mn
CO
CO
bis-η¹,η³-allylic
η¹,η³-chelate
η² - monoene

FIG. 8.43 Some further coordinating modes of C_8H_8.

$C_8H_8)_2$], and the cherry red compound [$Pu(\eta^8\text{-}C_8H_8)_2$] are prepared similarly. One of the very few examples of η^8 bonding to a d-block element is in the curious complex $Ti_2(C_8H_8)_3$. As shown in Fig. 8.42, 2 of the ligands are planar η^8 donors whereas the central puckered ring bridges the 2 Ti atoms in a bis-η^4-mode. It is made by treating $Ti(OBu^n)_4$ with C_8H_8 in the presence of $AlEt_3$.

In addition to acting as an η^8 ligand, C_8H_8 can coordinate in other modes,[46] some of which are illustrated in Fig. 8.43. Many of these complexes show fluxional behaviour in solution (p. 365) and the distinction between the various types of bonding is not as clear-cut as implied by the limiting structures in Fig. 8.43.

[46] G. DEGANELLO, *Transition Metal Complexes of Cyclic Polyolefins*, Academic Press, London, 1980, 476 pp.

[47] D. M. HEINEKEY and W. A. G. GRAHAM, Pentacarbonyl(7-η^1-cycloheptatrienyl)rhenium. Synthesis and fluxional behaviour of a *monohapto*cycloheptatrienyl derivative of a transition metal, *J. Am. Chem. Soc.* **101**, 6115–16 (1979).

		1 H	2 He														
3 Li	4 Be											5 B	6 C	7 N	8 O	9 F	10 Ne
11 Na	12 Mg											13 Al	14 Si	15 P	16 S	17 Cl	18 Ar
19 K	20 Ca	21 Sc	22 Ti	23 V	24 Cr	25 Mn	26 Fe	27 Co	28 Ni	29 Cu	30 Zn	31 Ga	32 Ge	33 As	34 Se	35 Br	36 Kr
37 Rb	38 Sr	39 Y	40 Zr	41 Nb	42 Mo	43 Tc	44 Ru	45 Rh	46 Pd	47 Ag	48 Cd	49 In	50 Sn	51 Sb	52 Te	53 I	54 Xe
55 Cs	56 Ba	57 La	72 Hf	73 Ta	74 W	75 Re	76 Os	77 Ir	78 Pt	79 Au	80 Hg	81 Tl	82 Pb	83 Bi	84 Po	85 At	86 Rn
87 Fr	88 Ra	89 Ac	104 Unq	105 Unp	106 Unh	107 Uns											

58 Ce	59 Pr	60 Nd	61 Pm	62 Sm	63 Eu	64 Gd	65 Tb	66 Dy	67 Ho	68 Er	69 Tm	70 Yb	71 Lu
90 Th	91 Pa	92 U	93 Np	94 Pu	95 Am	96 Cm	97 Bk	98 Cf	99 Es	100 Fm	101 Md	102 No	103 Lr

9

Silicon

9.1 Introduction

Silicon shows a rich variety of chemical properties and it lies at the heart of much modern technology.[1] Indeed, it ranges from such bulk commodities as concrete, clays, and ceramics, through more chemically modified systems such as soluble silicates, glasses, and glazes to the recent industries based on silicone polymers and solid-state electronics devices. The refined technology of ultrapure silicon itself, is perhaps the most elegant example of the close relation between chemistry and solid-state physics and has led to numerous developments such as the transistor, printed circuits, and microelectronics (p. 383).

In its chemistry, silicon is clearly a member of Group IV of the periodic classification but there are notable differences from carbon, on the one hand, and the heavier metals of the group on the other (p. 431). Perhaps the most obvious questions to be considered are why the vast covalent chemistry of carbon and its organic compounds finds such pallid reflection in the chemistry of silicon, and why the intricate and complex structural chemistry of the mineral silicates is not mirrored in the chemistry of carbon–oxygen compounds.†

Silica (SiO_2) and silicates have been intimately connected with the evolution of mankind from prehistoric times: the names derive from the Latin *silex*, gen. *silicis*, flint, and serve as a reminder of the simple tools developed in paleolithic times (~ 500 000 years ago) and the shaped flint knives and arrowheads of the neolithic age which began some 20 000 years ago. The name of the element, silicon, was proposed by Thomas Thomson in 1831, the ending *on* being intended to stress the analogy with carbon and boron.

The great affinity of silicon for oxygen delayed its isolation as the free element until 1823 when J. J. Berzelius succeeded in reducing K_2SiF_6 with molten potassium. He first made

---

† Throughout this chapter we will notice important differences between the chemical behaviour of carbon and silicon, and one is reminded of Grant Urry's memorable words: "It is perhaps appropriate to chide the polysilane chemist for milking the horse and riding the cow in attempting to adapt the success of organic chemistry in the study of polysilanes. A valid argument can be made for the point of view that the most effective chemistry of silicon arises from the differences with the chemistry of carbon compounds rather than the similarities" (see ref. 15 on p. 393).

1 *Kirk–Othmer Encyclopedia of Chemical Technology*, 2nd edn., Vol. 18, pp. 46–286, 1969 (Silica, silicon and silicides; Silicon compounds); 3rd edn., Vol. 5 (1979), Ceramics, pp. 234–90; Ceramics as electrical materials, pp. 290–314; Clays, Vol. 6, pp. 190–223 (1979).

$SiCl_4$ in the same year, SiF_4 having previously been made in 1771 by C. W. Scheele who dissolved SiO_2 in hydrofluoric acid. The first volatile hydrides were discovered by F. Wöhler who synthesized $SiHCl_3$ in 1857 and SiH_4 in 1858, but major advances in the chemistry of the silanes awaited the work of A. Stock during the first third of the twentieth century. Likewise, the first organosilicon compound $SiEt_4$ was synthesized by C. Friedel and J. M. Crafts in 1863, but the extensive development of the field was due to F. S. Kipping in the first decades of this century.[2] The unique properties and industrial potential of siloxanes escaped attention at that time and the dramatic development of silicone polymers, elastomers, and resins has occurred during the past three decades (p. 424).

The solid-state chemistry of silicon has shown similar phases. The bizarre compositions derived by analytical chemistry for the silicates only became intelligible following the pioneering X-ray structural work of W. L. Bragg in the 1920s[3] and the concurrent development of the principles of crystal chemistry by L. Pauling[4] and of geochemistry by V. M Goldschmidt.[5] More recently the complex crystal chemistry of the silicides has been elucidated and the solid-state chemistry of doped semiconductors has been developed to a level of sophistication that was undreamt of even in the 1960s.

9.2 Silicon

9.2.1 *Occurrence and distribution*

Silicon (27.2 wt%) is the most abundant element in the earth's crust after oxygen (45.5%), and together these 2 elements comprise 4 out of every 5 atoms available near the surface of the globe. This implies that there has been a substantial fractionation of the elements during the formation of the solar system since, in the universe as a whole, silicon is only seventh in order of abundance after H, He, C, N, O, and Ne (p. 3). Further fractionation must have occurred within the earth itself: the core, which has 31.5% of the earth's mass, is commonly considered to have a composition close to $Fe_{25}Ni_2Co_{0.1}S_3$; the mantle (68.1% of the mass) probably consists of dense oxides and silicates such as olivine $(Mg,Fe)_2SiO_4$, whereas the crust (0.4% of the mass) accumulates the lighter siliceous minerals which "float" to the surface. The crystallization of igneous rocks from magma (molten rock, e.g. lava) depends on several factors such as the overall composition, the lattice energy, mp, and crystalline complexity of individual minerals, the rate of cooling, etc. This has been summarized by N. L. Bowen in a reaction series which gives the approximate sequence of appearance of crystalline minerals as the magma is cooled: olivine $[M_2^{II}SiO_4]$, pyroxene $[M_2^{II}Si_2O_6]$, amphibole $[M_7^{II}\{(Al,Si)_4O_{11}\}(OH)_2]$, biotite mica $[(K,H)_2(Mg,Fe)_2(Al,Fe)_2(SiO_4)_3]$, orthoclase feldspar $[KAlSi_3O_8]$, muscovite mica $[KAl_2(AlSi_3O_{10}(OH)_2]$, quartz $[SiO_2]$, zeolites, and hydrothermal minerals. The structure of these mineral types is discussed later (p. 399), but it is clear that the reaction

[2] E. G. ROCHOW, Silicon, Chap. 15 in *Comprehensive Inorganic Chemistry*, Vol. 1, pp. 1323–467, Pergamon Press, Oxford, 1973.

[3] W. L. BRAGG, *The Atomic Structure of Minerals*, Oxford University Press, 1937, 292 pp.

[4] L. PAULING, *The Nature of the Chemical Bond*, 3rd edn., pp. 543–62, Cornell University Press, 1960, and references therein.

[5] V. M. GOLDSCHMIDT, Crystal structure and chemical constitution, *Trans. Faraday Soc.* **25**, 253–83 (1929); *Geochemistry*, Oxford University Press, Oxford, 1954, 730 pp.

series leads to progressively more complex silicate structural units and that the later part of the series is characterized by the introduction of OH (and F) into the structures. Extensive isomorphous substitution among the metals is also possible. Subsequent weathering, transport, and deposition leads to sedimentary rocks such as clays, shales, and sandstones. Metamorphism at high temperatures and pressures can effect further changes during which the presence or absence of water plays a vital role.[6, 6a, 7]

Silicon never occurs free: it invariably occurs combined with oxygen and, with trivial exceptions, is always 4-coordinate in nature. The $\{SiO_4\}$ unit may occur as an individual group or be linked into chains, ribbons, rings, sheets, or three-dimensional frameworks (pp. 399–416).

9.2.2 *Isolation, production, and industrial uses*

Silicon (96–99% pure) is now invariably made by the reduction of quartzite or sand with high purity coke in an electric arc furnace; the SiO_2 is kept in excess to prevent the accumulation of SiC (p. 386):

$$SiO_2 + 2C = Si + 2CO; \quad 2SiC + SiO_2 = 3Si + 2CO$$

The reaction is frequently carried out in the presence of scrap iron (with low P and S content) to produce ferrosilicon alloys: these are used in the metallurgical industry to deoxidize steel, to manufacture high-Si corrosion-resistant Fe, and Si/steel laminations for electric motors. The scale of operations can be gauged from the 1974 US production figures for various ranges:

% Si	22–55	56–70	71–80	81–95	96–99
Production/tonnes	426 500	57 500	14 400	90	127 500

World production is measured in millions of tonnes per year.

Silicon for the chemical industry is usually purified to ~98.5% by leaching the powdered 96–97% material with water. Very pure Si for semiconductor applications is obtained either from $SiCl_4$ (made from the chlorination of scrap Si) or from $SiHCl_3$ (a byproduct of the silicone industry, p. 389). These volatile compounds are purified by exhaustive fractional distillation and then reduced with exceedingly pure Zn or Mg; the resulting spongy Si is melted, grown into cylindrical single crystals, and then purified by zone refining. Alternative routes are the thermal decomposition of SiI_4/H_2 on a hot tungsten filament (cf. boron, p. 157), or the epitaxial growth of a single-crystal layer by thermal decomposition of SiH_4. Very recently (*Chem. & Eng. News*, 3 Sept., p. 8, 1979) a one-step process has been developed to produce high-purity Si for solar cells at one-tenth of the cost of rival methods. In this process Na_2SiF_6 (which is a plentiful waste product from the phosphate fertilizer industry) is reduced by metallic Na; the reaction is highly exothermic and is self-sustaining without the need for external fuel.

[6] B. MASON, *Principles of Geochemistry*, 3rd edn., Wiley, New York, 1966, 329 pp.
[6a] P. HENDERSON, *Inorganic Geochemistry*, Pergamon Press, Oxford, 1982, 372 pp.
[7] D. K. BAILEY and R. MACDONALD (eds.), *The Evolution of the Crystalline Rocks*, Academic Press, London, 1976, 484 pp.

Hyperfine Si is one of the purest materials ever made on an industrial scale: the production of transistors (p. 383) requires the routine preparation of crystals with impurity levels below 10^{-9} to 10^{-10}, and levels approaching 10^{-12} can be attained in special cases.

9.2.3 Atomic and physical properties

Silicon consists predominantly of ^{28}Si (92.23%) together with 4.67% ^{29}Si and 3.10% ^{30}Si. No other isotopes are stable. The ^{29}Si isotope (like the proton) has a nuclear spin $I = \frac{1}{2}$, and is being increasingly used in nmr spectroscopy.[7a] ^{31}Si, formed by neutron irradiation of ^{30}Si, has $t_{\frac{1}{2}}$ 2.62 h; it can be detected by its characteristic β^- activity (E_{max} 1.48 MeV) and is very useful for the quantitative analysis of Si by neutron activation. The radioisotope with the longest half-life (~ 650 y) is the soft β^- emitter ^{32}Si (E_{max} 0.2 MeV).

In its ground state, the free atom Si has the electronic configuration $[\text{Ne}]3s^2 3p^2$. Ionization energies and other properties are compared with those of the other members of Group IV on p. 431. Silicon crystallizes in the diamond lattice (p. 305) with a_0 541.99 pm at $25°$, corresponding to an Si–Si distance of 235.2 pm and a covalent atomic radius of 117.6 pm. There appear to be no allotropes though at high pressures Si (like Ge) changes to a denser form in which the Si–Si distance remains unaltered but the tetrahedral bond angles distort so that 3 are $99°$ and 3 are $108°$. Physical properties are summarized in Table 9.1 (see also p. 434). Silicon is notably more volatile than C and has a substantially lower

TABLE 9.1 *Some physical properties of silicon*

MP/°C	1420
BP/°C	~ 3280
Density (20°C)/g cm^{-3}	2.336
ΔH_{fus}/kJ mol^{-1}	50.6 ± 1.7
ΔH_{vap}/kJ mol^{-1}	383 ± 10
ΔH_f (monatomic gas)/kJ mol^{-1}	454 ± 12
a_0/pm	541.99
r (covalent)/pm	117.6
r ("ionic")/pm	26[a]
Pauling electronegativity	1.8

[a] This is the "effective ionic radius" for 4-coordinate Si^{IV} in silicates, obtained by subtracting r (O^{-II}) = 140 pm from the observed Si–O distance. The value for 6-coordinate Si^{IV} is 40 pm. [R. D. Shannon, *Acta Cryst.* **A32**, 751–67 (1976).]

energy of vaporization, thus reflecting the smaller Si–Si bond energy. The element is a semiconductor with a distinct shiny, blue–grey metallic lustre; the resistivity decreases with increase of temperature, as expected for a semiconductor. The actual value of the resistivity depends markedly on purity but is ~ 40 ohm cm at $25°$ for very pure material.

[7a] J.-P. KINTZINGER and H. MARSMANN, *Oxygen-17 and Silicon-29 NMR*, Vol. 17 of *NMR Basic Principles and Progress* (P. DIEHL, E. FLUCK, and R. KOSFIELD, eds.), Springer-Verlag, Berlin, 1980, 250 pp.

The immense importance of Si in transistor technology stems from the chance discovery of the effect in Ge at Bell Telephone Laboratories, New Jersey, in 1947, and the brilliant theoretical and practical development of the device by J. Bardeen, W. H. Brattain, and W. Shockley for which they were awarded the 1956 Nobel Prize in Physics. A brief description of the physics and chemistry underlying transistor action in Si is given in the Panel.

The Physics and Chemistry of Transistors

In ultrapure semiconductor grade Si there is an energy gap E_g between the highest occupied energy levels (the valence band) and the lowest unoccupied energy levels (the conduction band). This is shown diagrammatically in Fig. a: the valence band is completely filled, the conduction band is empty, the Fermi level (E_F)† lies approximately midway between these, and the material is an insulator at room temperature. If the Si is doped with a Group V element such as P, As, or Sb, each atom of dopant introduces a supernumerary electron and the impurity levels can act as a source of electrons which can be thermally or photolytically excited into the conduction band (Fig. b): the material is an n-type semiconductor with an activation energy ΔE_n (where n indicates negative current carriers, i.e. electrons). Conversely, doping with a Group III element such as B, Al, or Ga introduces acceptor levels that can act as traps for electrons excited from the filled valence band (Fig. c): the material is a p-type semiconductor and the current is carried by the positive holes in the valence band.

When an n-type sample of Si is joined to a p-type sample, a p–n junction is formed having a common Fermi level as in Fig. d: electrons will flow from n to p and holes from p to n thereby producing a voltage drop V_0 across the space charge region. A p–n junction can thus act as a diode for rectifying alternating current, the current passing more easily in one direction than the other. In practise a large p–n junction might cover 10 mm², whereas in integrated circuits such a device might cover no more than 10^{-4} mm² (i.e. a square of side 10 μm).

A transistor, or n–p–n junction is built up of two n-type regions of Si separated by a thin layer of weakly p-type (Fig. e). When the emitter is biased by a small voltage in the forward direction and the collector by a larger voltage in the reverse direction, this device acts as a triode amplifier. The relevant energy level diagram is shown schematically in Fig. f.

The large-scale reproducible manufacture of minute, electronically-stable, single-crystal transistor junctions is a triumph of the elegant techniques of solid-state chemical synthesis. The sequence of steps is illustrated in diagrams (i)–(v).

(i) A small wafer of single-crystal n-type Si is oxidized by heating it in O_2 or H_2O vapour to form a thin surface layer of SiO_2.

(ii) The oxide coating is covered by a photosensitive film called "photoresist".

(iii) A mask is placed over the photoresist to confine the exposure to the desired pattern and the chip is exposed to ultraviolet light; the exposed photoresist is then removed by treatment with acid, leaving a tough protective layer over the parts of the oxide coating that are to be retained.

(iv) The unprotected areas of Si are etched away with hydrofluoric acid and the remaining photoresist is also removed.

(v) The surface is exposed to the vapour of a Group III element and the impurity atoms diffuse into the unprotected area to form a layer of p-type Si.

(vi) Steps (i)–(v) are repeated, with a different mask, and then the newly exposed areas are treated with the vapour of a Group V element to produce a patterned layer of n-type Si.

(vii) Finally, again with a different mask, the surface is reoxidized and then re-etched to produce openings into which metal is deposited so as to connect the n- and p-regions into an integrated circuit.

Each individual p–n diode or n–p–n transistor can be made almost unbelievably minute by these techniques; for example computer memory units storing over 10^5 bits of information on a single small chip are routinely used. Further information can be obtained from textbooks of solid-state physics or electronic engineering.

† The Fermi level is the energy at which the chance of a state being occupied by an electron is $\frac{1}{2}$.

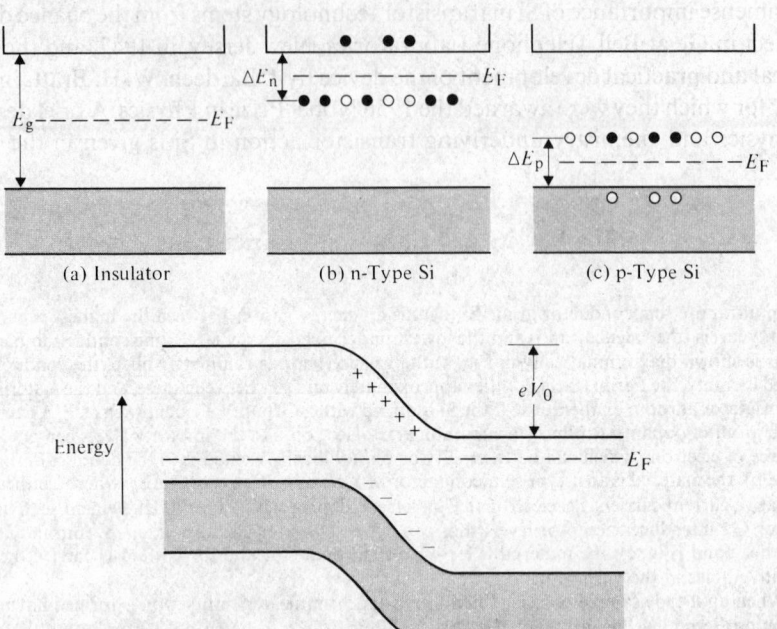

(a) Insulator   (b) n-Type Si   (c) p-Type Si

(d) p-n Junction

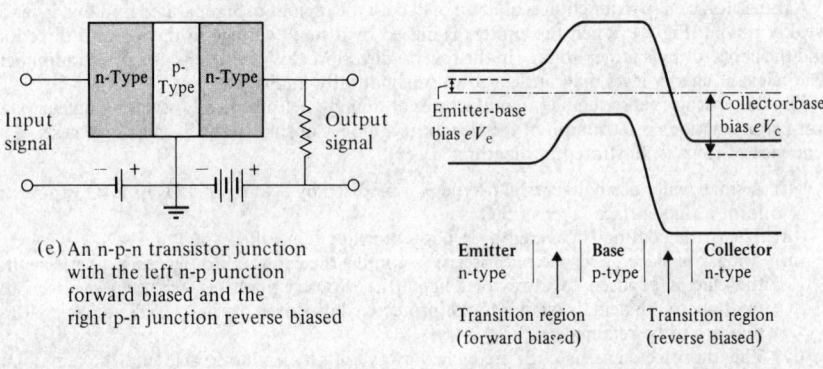

(e) An n-p-n transistor junction with the left n-p junction forward biased and the right p-n junction reverse biased

(f) Energy level diagrams

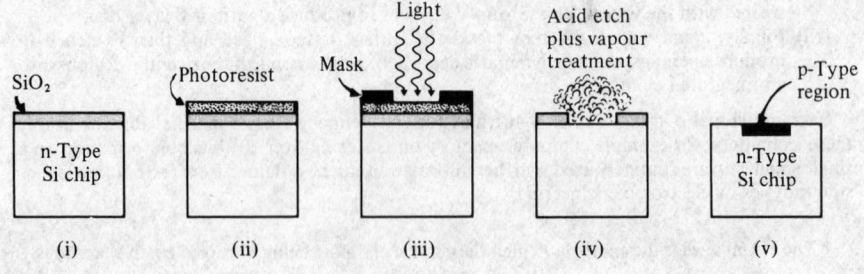

(i)   (ii)   (iii)   (iv)   (v)

9.2.4 *Chemical properties*

Silicon in the massive, crystalline form is relatively unreactive except at high temperatures. Oxygen, water, and steam all have little effect probably because of the formation of a very thin, continuous, protective surface layer of SiO_2 a few atoms thick (cf. Al, p. 253). Oxidation in air is not measurable below 900°; between 950° and 1160° the rate of formation of vitreous SiO_2 rapidly increases and at 1400° the N_2 in the air also reacts to give SiN and Si_3N_4. Sulfur vapour reacts at 600° and P vapour at 1000°. Silicon is also unreactive towards aqueous acids, though the aggressive mixture of conc HNO_3/HF oxidizes and fluorinates the element. Silicon dissolves readily in hot aqueous alkali due to reactions of the type $Si + 4OH^- = SiO_4^{4-} + 2H_2$. Likewise, the thin film of SiO_2 is no barrier to attack by halogens, F_2 reacting vigorously at room temperature, Cl_2 at $\sim 300°$, and Br_2, I_2 at $\sim 500°$. Even alkyl halides will react at elevated temperatures and, in the presence of Cu catalysts, this constitutes the preferred "direct" synthesis of organosilicon chlorides for the manufacture of silicones (p. 421).

In contrast to the relative inertness of solid Si to gaseous and liquid reagents, molten Si is an extremely reactive material: it forms alloys or silicides with most metals (see below) and rapidly reduces most metal oxides because of the very large heat of formation of SiO_2 (~ 900 kJ mol^{-1}). This presents problems of containment when working with molten Si, and crucibles must be made of refractories such as ZrO_2 or the borides of transition metals in Groups IVA–VIA (p. 163).

Chemical trends within Group IVB are discussed on p. 434. Silicon does not form binary compounds with the heavier members of the group (Ge, Sn, Pb) but its compound with carbon, SiC, is of outstanding academic and practical interest, and is manufactured on a huge scale industrially (see Panel). In the vast majority of its compounds Si is tetrahedrally coordinated but sixfold coordination also occurs and occasional examples of several other coordination geometries are known as indicated in Table 9.2.

TABLE 9.2 *Coordination geometries of silicon*

Coordination number	Examples
2 (bent)	SiF_2(g), $SiMe_2$ (matrix, 77 K)
3 (planar)	Silabenzene, SiC_5H_6; silatoluene, C_5H_5SiMe (see ref. 26)
3 (pyramidal)	(?) SiH_3^- in $KSiH_3$ (NaCl structure)
4 (tetrahedral)	SiH_4, SiX_4, SiX_nY_{4-n}, SiO_2, silicates, etc.
4 (planar)[a]	
5 (trigonal bipyramidal)	SiX_5^-(?), *cyclo*-$[Me_2NSiH_3]_5$, $PhSi(-OC_6H_4O-)_2$
5 (square pyramidal)[b]	$[SiF(1,2-O_2C_6H_4)_2]^-$
6 (octahedral)	SiF_6^{2-}, $[Si(acac)_3]^+$, $[L_2SiX_4]$, SiO_2(stishovite), SiP_2O_7
8 (cubic)	Mg_2Si(antifluorite)
10	TiS_2, $CrSi_2$, $MoSi_2$ (see ref. 10, p. 790)

[a] H. Meyer and G. Nagorsen, *Angew. Chem.*, Int. Edn. (Engl.) **18**, 551–3 (1979); E.-U. Würthwein and P. von R. Schleyer, ibid, 553–4. This conclusion, based on space-group arguments, has been challenged by J. Dunitz, *Angew. Chem.*, Int. Edn. (Engl.) **19**, 1034 (1980).

[b] J. J. Harland, R. O. Day, J. F. Vollano, A. C. Sau, and R. R. Holmes, *J. Am. Chem. Soc.* **103**, 5269–70 (1981).

Silicon Carbide, SiC

Silicon carbide was made accidently by E. G. Acheson in 1891; he recognized its abrasive power and coined the name "carborundum" from carbo(n) and (co)rundum (Al_2O_3) to indicate that its hardness on the Mohs scale (9.5) was intermediate between that of diamond (10) and Al_2O_3 (9). Within months he had formed the Carborundum Co. for its manufacture, and current world production is several hundred thousand tonnes annually.

Despite its simple formula, SiC exists in at least 70 crystalline modifications based on hexagonal α-SiC (wurtzite-type ZnS, p. 1405) or cubic β-SiC (diamond or zincblende-type, p. 1405). The complexity arises from the numerous stacking sequences of the *a* and *b* "layers" in the crystal.[10] The α-form is marginally the more stable thermodynamically. Industrially, α-SiC is obtained as black, dark green, or purplish iridescent crystals by reducing high-grade quartz sand with a slight excess of coke or anthracite in an electric furnace at 2000–2500°C:

$$SiO_2 + 2C \longrightarrow Si + 2CO; \quad Si + C \longrightarrow SiC$$

The dark colour is caused by impurities such as iron, and the irridescence is due to a very thin layer of SiO_2 formed by surface oxidation. Purer samples are pale yellow or colourless. Even higher temperatures (and vacuum conditions) are required to produce the β-forms. Alternatively, very pure β-SiC can be obtained by heating grains of ultrapure Si with graphite at 1500°. Lattice constants for α-SiC are *a* 307.39 pm, *c* 1006.1 pm, *c/a* 3.273; for cubic β-SiC, a_0 435.02 pm (cf. diamond 356.68, Si 541.99, mean 449.34 pm).

SiC has greater thermal stability than any other binary compound of Si and decomposition by loss of Si only becomes appreciable at $\sim 2700°$. It resists attack by most aqueous acids (including HF but not H_3PO_4) and is oxidized in air only above 1000° because of the protective layer of SiO_2; this can be removed by molten hydroxides or carbonates and oxidation is much more rapid under these conditions, e.g.:

$$SiC + 2NaOH + 2O_2 \longrightarrow Na_2SiO_3 + H_2O + CO_2$$

Cl_2 attacks SiC vigorously, yielding $SiCl_4 + C$ at 100° and $SiCl_4 + CCl_4$ at 1000°.

Technical interest in SiC originally stemmed from its excellence as an abrasive powder; this derives not only from the great intrinsic hardness of the compound but also from its peculiar fracture to give sharp cutting edges. As a refractory, α-SiC combines great strength and chemical stability, with an extremely low thermal expansion coefficient ($\sim 6 \times 10^{-6}$) which shows no sudden discontinuities due to phase transitions. Pure α-SiC is an intrinsic semiconductor with an energy band gap sufficiently large (1.90 ± 0.10 eV) to make it a very poor electrical conductor ($\sim 10^{-13}$ ohm^{-1} cm^{-1}). However, the presence of controlled amounts of impurities makes it a valuable extrinsic semiconductor (10^{-2}–3 ohm^{-1} cm^{-1}) with a positive temperature coefficient. This, combined with its mechanical and chemical stability, accounts for its extensive use in electrical heating elements. In recent years pure β-SiC has received much attention as a high-temperature semiconductor with applications in transistors, diode rectifiers, electroluminescent diodes, etc. (see p. 383).

9.3 Compounds

9.3.1 *Silicides*[8, 9]

As with borides (p. 162) and carbides (p. 318) the formulae of metal silicides cannot be rationalized by the application of simple valency rules, and the bonding varies from essentially metallic to ionic and covalent. Observed stoichiometries include M_6Si, M_5Si,

[8] A. S. Berezhoi, *Silicon and its Binary Systems*, Consultants Bureau, New York, 1960, 275 pp.

[9] B. Aronsson, T. Lundström, and S. Rundqvist, *Borides, Silicides, and Phosphides*, Methuen, London, 1965, 120 pp.

[10] A. F. Wells, *Structural Inorganic Chemistry*, 4th edn., Chap. 23, Silicon, pp. 784–832, Oxford University Press, Oxford, 1975.

M_4Si, $M_{15}Si_4$, M_3Si, M_5Si_2, M_2Si, M_5Si_3, M_3Si_2, MSi, M_2Si_3, MSi_2, MSi_3, and MSi_6. Silicon, like boron, is more electropositive than carbon, and structurally the silicides are more closely related to the borides than the carbides (cf. diagonal relation, p. 32). However, the covalent radius of Si (118 pm) is appreciably larger than for B (88 pm) and few silicides are actually isostructural with the corresponding borides. Silicides have been reported for virtually all A subgroup metals except Be, the greatest range of stoichiometries being shown by the transition metals in Groups IVA–VIII and uranium. No silicides are known for the B subgroup metals except Cu; most form simple eutectic mixtures, but the heaviest post-transition metals Hg, Tl, Pb, and Bi are completely immiscible with molten Si.

Some metal-rich silicides have isolated Si atoms and these occur either in typical metal-like structures or in more polar structures. With increasing Si content, there is an increasing tendency to catenate into isolated Si_2 or Si_4, or into chains, layers, or 3D networks of Si atoms. Examples are in Table 9.3 and further structural details are in refs. 8–10.

TABLE 9.3 *Structural units in metal silicides*[8–10]

Unit	Examples	
Isolated Si	Cu_5Si (β-Mn structure)	Metal structures (good electrical conductors)
	M_3Si (β-W structure) M = V, Cr, Mo	
	Fe_3Si (Fe_3Al superstructure)	
	Mn_3Si (random bcc)	
	M_2Si (anti-CaF_2); M = Mg, Ge, Sn, Pb	Non-metal structures (non-conductors)
	M_2Si (anti-$PbCl_2$); M = Ca, Ru, Ce, Rh, Ir, Ni	
Si_2 pairs	U_3Si_2 (Si–Si 230 pm), also for Hf and Th	
Si_4 tetrahedra	KSi (Si–Si 243 pm), i.e. $[M]_4^+[Si_4]^{4-}$ cf. isoelectronic P_4 (M = K, Rb, Cs; also for M_4Ge_4)	
Si chains	USi (FeB structure) (Si–Si 236 pm); also for Ti, Zr, Hf, Th, Ce, Pu	
	CaSi (CrB structure) (Si–Si 247 pm); also for Sr, Y	
Plane hexagonal Si nets	β-USi_2 (AlB_2 structure) (Si–Si 222–236 pm); also for other actinoids and lanthanoids	
Puckered hexagonal Si nets	$CaSi_2$ (Si–Si 248 pm)—as in "puckered graphite" layer	
Open 3D Si frameworks	$SrSi_2$, α-$ThSi_2$ (Si–Si 239 pm; closely related to AlB_2), α-USi_2	

Silicides are usually prepared by direct fusion of the elements but coreduction of SiO_2 and a metal oxide with C or Al is sometimes used. Heats of formation are similar to those of borides and carbides but mps are substantially less; e.g. TiC 3140°, TiB_2 2980°, $TiSi_2$ 1540°; and TaC 3800°, TaB_2 3100°, $TaSi_2$ 1560°C. Few silicides melt as high as 2000–2500°, and above this temperature only SiC is solid (decomp ~2700°C).

Silicides of groups IA and IIA are generally much more reactive than those of the transition elements (cf. borides and carbides). Hydrogen and/or silanes are typical products; e.g.:

$$Na_2Si + 3H_2O \xrightarrow[\text{complete}]{\text{rapid and}} Na_2SiO_3 + 3H_2$$

$$Mg_2Si + 2H_2SO_4(aq) \longrightarrow 2MgSO_4 + SiH_4$$

Products also depend on stoichiometry (i.e. structural type). For example, the polar, non-conducting Ca_2Si (anti-$PbCl_2$ structure with isolated Si atoms) reacts with water to give $Ca(OH)_2$, SiO_2 (hydrated), and H_2, whereas CaSi (which features zigzag Si chains) gives silanes and the polymeric SiH_2; $CaSi_2$, which has puckered layers of Si atoms yields H_2 and polysilene (SiH_2) but not silanes. Transition metal silicides are usually inert to aqueous reagents except HF, but yield to more aggressive reagents such as molten KOH, or F_2 (Cl_2) at red heat.

9.3.2 *Silicon hydrides (silanes)*

The great development which occurred in synthetic organic chemistry from the 1830s onward encouraged early speculations that a similar extensive chemistry might be generated based on Si. The first silanes were made in 1857 by F. Wöhler and H. Buff who reacted Al/Si alloys with aqueous HCl; the compounds prepared were shown to be SiH_4 and $SiHCl_3$ by C. Friedel and A. Ladenburg in 1867 but it was not until 1902 that the first homologue, Si_2H_6, was prepared by H. Moissan and S. Smiles from the protonolysis of magnesium silicide. The thermal instability and great chemical reactivity of the compounds precluded further advances until A. Stock developed his greaseless vacuum techniques and first began to study them as contaminants of his boron hydrides in 1916. He proposed the names silanes and boranes (p. 171) by analogy with the alkanes.

Silanes Si_nH_{2n+2} are now known as unbranched and branched chains (up to $n=8$) but no cyclic compounds or unsaturated analogues of alkenes and alkynes have been prepared.† Silanes are colourless gases or volatile liquids; they are extremely reactive and spontaneously ignite or explode in air. Thermal stability decreases with increasing chain length and only SiH_4 is stable indefinitely at room temperature. Some physical properties are in Table 9.4 from which it can be seen that silanes are less volatile than both the alkanes

TABLE 9.4 *Some properties of silanes*

Property	SiH_4	Si_2H_6	Si_3H_8	n-Si_4H_{10}	i-Si_4H_{10}
MP/°C	−185°	−132.5°	117.2°	−89.9°	−99.1°
BP/°C	−111.8°	−14.3°	53.1°	108°	101°
Density $(T°C)$/g cm^{-3}	0.68 (−186°)	0.686 (−25°)	0.725 (0°)	0.825 (0°)	—
Thermal decomposition at room temperature	Stable	Very slow[a]	Slow	Fairly rapid	Fairly rapid

[a] 2.5% in 8 months.

† Recent exciting developments require this statement to be modified. For example, hexakis(2,6-dimethylphenyl)-cyclotrisilane $\{(2,6\text{-}Me_2C_6H_3)_2Si\}_3$ can readily be prepared by reductive coupling in dimethoxyethane solution:

$$3(2,6\text{-}Me_2C_6H_3)_2SiCl_2 \xrightarrow[-78°]{Li/C_{10}H_8} \text{cyclo-}\{(2,6\text{-}Me_2C_6H_3)_2Si\}_3 \sim 10\% \text{ yield}$$

The compound is stable to air and moisture, and to heat up to its mp 272°. X-ray structure analysis shows the Si_3 ring is an isosceles triangle (bond angles 58.7°, 60.7°, 60.8°; Si–Si distances 237.5, 242.2, 242.5 pm). Irradiation of the cyclotrisilane in cyclohexane with a low-pressure Hg lamp rapidly gave the corresponding highly reactive disilene $Ar_2Si{=}SiAr_2$ in almost quantitative yield: yellow crystals, mp 215° (S. Masamune, Y. Hanzawa, S. Murakami, T. Bally, and J. F. Blount, *J. Am. Chem. Soc.* **104**, 1150–3 (1982) and references therein). Likewise, photolysis of Me_3Si-$Si(mesityl)_2$-$SiMe_3$ gives bright orange-yellow crystals of $(mesityl)_2Si{=}Si(mesityl)_2$, stable in the absence of air up to its mp 176° (mesityl = 2,4,6-trimethylphenyl): R. West, M. J. Fink, and J. Michl, *Science* **214**, 1343–4 (1981).

and boranes (p. 186) of similar formula, but more volatile than the corresponding germanes (p. 436).

There are three general types of preparative route to the silanes and their derivatives. Early methods (pre-1945) treated materials such as metal silicides which contained negatively charged $Si^{\delta-}$ with a protonic reagent such as an aqueous acid. Concurrent hydrolysis of the products limited the yield but considerable improvement resulted from the use of nonaqueous systems such as $NH_4Br/liq NH_3$ (1934). The second general preparative route involves treatment of compounds such as SiX_4 ($Si^{\delta+}$) with hydridic reagents such as LiH, NaH, $LiAlH_4$, etc., in ether solvents at low temperatures. This is now the preferred route: e.g. reaction of Si_nCl_{2n+2} ($n = 1, 2, 3$) with $LiAlH_4$ gives essentially quantitative yields of SiH_4, Si_2H_6, and Si_3H_8. Organosilanes can be prepared similarly, e.g. Me_2SiCl_2 gives Me_2SiH_2. The third general method for preparing Si–H compounds involves direct reaction of HX or RX with Si or a ferrosilicon alloy in the presence of a catalyst such as Cu when necessary (p. 421), e.g.:

$$Si + 3HCl \xrightarrow{350^\circ} SiHCl_3 + H_2$$

$$Si + 2MeCl \xrightarrow{Cu/300^\circ} MeSiHCl_2 + H_2 + C$$

Combination of these various methods has led to a vast number of derivatives in which H is progressively replaced by one or more monofunctional group such as F, Cl, Br, I, CN, R, Ar, OR, SH, SR, NH_2, NR_2, etc.[1]

Silanes are much more reactive than the corresponding C compounds.[1, 2, 11] This has been ascribed to several factors including: (a) the larger radius of Si which would facilitate attack by nucleophiles, (b) the great polarity of Si–X bonds, and (c) the presence of low-lying d orbitals which permit the formation of 1:1 and 1:2 adducts, thereby lowering the activation energy of the reaction. The relative magnitude of the various bond energies is also an important factor in deciding which bonds will survive and which will be formed. Thus, it can be seen in Table 9.5, Si–Si < Si–C < C–C, and Si–H < C–H, whereas for

TABLE 9.5   *Some typical bond energies/kJ mol^{-1}*

X =	C	Si	H	F	Cl	Br	I	O—	N⟨
C–X	368	360	435	453	351	293	216	~360	~305
Si–X	360	340	393	565	381	310	234	452	322

bonds for the other elements the energy C–X < Si–X. These data should be used only for broad comparisons since the estimated bond energies depend markedly on the particular compounds being studied and also on the experimental technique employed and the method of computation.

[11] E. WIBERG and E. AMBERGER, *Hydrides of the Elements of Main Groups I–IV*, Chap. 7, Silicon hydrides, pp. 462–638, Elsevier, Amsterdam, 1971. A comprehensive review of compounds containing Si–H bonds; over 700 references.

The pyrolysis of silanes leads to polymeric species and ultimately to Si and H_2; indeed, pyrolysis of SiH_4 is a commercial route to ultrapure Si. The reactions occurring have been less studied than those of alkenes (and boranes, p. 187), but it is clear that there are significant differences. Thus the initial step in the thermal decomposition of alkanes is the cleavage of a C–H or C–C bond with formation of radical intermediates $R_3C^•$. However, studies using deuterium-substituted compounds suggest that the initial step in the decomposition of polysilanes is the elimination of silenes $:SiH_2$ or $:SiHR$.[12]

$$RSiH_2SiH_3 \longrightarrow \{:SiH_2\} + RSiH_3 \quad \text{or} \quad \{:SiHR\} + SiH_4$$

Activation energies for this process ($\sim 210 \text{ kJ mol}^{-1}$) are substantially less than Si–Si and Si–H bond energies and the reaction appears to involve a 1,2-H shift with a 5-coordinate

transition state $\overset{\diagdown}{\underset{\diagup}{Si}} \cdots \overset{\diagup}{\underset{\diagdown}{Si}}-$; this is presumably energetically unstable for carbon.

Pure silanes do not react with pure water or dilute acids in silica vessels, but even traces of alkali dissolved out of glass apparatus catalyse the hydrolysis which is then rapid and complete ($SiO_2 . nH_2O + 4H_2$). Solvolysis with MeOH can be controlled to give several products $SiH_{4-n}(OMe)_n$ ($n = 2, 3, 4$). Si–H adds (with difficulty) to alkenes though the reaction occurs more readily with substituted silanes. Similarly, SiH_4 adds to Me_2CO at 450° to give $C_3H_7OSiH_3$, and it ring-opens ethylene oxide at the same temperature to give $EtOSiH_3$ and other products. Silanes explode in the presence of Cl_2 or Br_2 but the reaction with Br_2 can be moderated at $-80°$ to give good yields of SiH_3Br and SiH_2Br_2. More conveniently, halogenosilanes SiH_3X can be made by the catalysed reaction of SiH_4 and HX in the presence of Al_2X_6, or by the reaction with solid AgX in a heated flow reactor, e.g.:

$$SiH_4 + 2AgCl \xrightarrow{260°} SiH_3Cl + HCl + 2Ag$$

SiH_3I in particular is a valuable synthetic intermediate and some of its reactions are summarized in Table 9.6. SiH_3I is a dense, colourless, mobile liquid, mp $-57.0°$, bp $+45.4°$, $d(15°)$ 2.035 g cm^{-3}. Another valuable reagent is $KSiH_3$, a colourless crystalline compound with NaCl-type structure; it is stable up to $\sim 200°$ and is prepared by direct reaction of potassium on silane in monoglyme or diglyme:

$$SiH_4 + 2K \longrightarrow KSiH_3 + KH; \qquad SiH_4 + K \longrightarrow KSiH_3 + \tfrac{1}{2}H_2$$

When hexamethylphosphoramide, $(NMe_2)_3PO$, is used as solvent only the second reaction occurs. The synthetic utility of $KSiH_3$ can be gauged from Table 9.7 which summarizes some of its reactions.

[12] I. M. T. DAVIDSON and A. V. HOWARD, *JCS Faraday I*, **71**, 69–77 (1975) and references therein. C. H. HAAS and M. A. RING, *Inorg. Chem.* **14**, 2253–6 (1975). A. J. VANDERWIELEN, M. A. RING, and H. E. O'NEAL, *J. Am. Chem. Soc.* **97**, 993–8 (1975).

TABLE 9.6 *Some reactions of* $SiH_3I^{(a)}$

Reagent	Major Si product	Reagent	Major Si product
Na/Hg	Si_2H_6	N_2H_4	$(SiH_3)_2NN(SiH_3)_2$
H_2O	$O(SiH_3)_2$	$LiN(SiCl_3)_2$	$SiH_3N(SiCl_3)_2$
HgS	$S(SiH_3)_2$	P_4	$(SiH_3)_nPI_{3-n}$ $(n=1, 2, 3)$
Ag_2Se	$Se(SiH_3)_2$	$AgXCN$ (N_2 atm)	SiH_3NCX (X=O, Se)
Li_2Te	$Te(SiH_3)_2$	AgSCN	SiH_3NCS
$Si_2H_5Br + H_2O$	$SiH_3OSi_2H_5$	AgCN	SiH_3CN
$Hg(SCF_3)_2$	SiH_3SCF_3	Ag_2NCN	$(SiH_3)_2NCN$
$Hg(SeCF_3)_2$	SiH_3SeCF_3	$HC{\equiv}CMgBr$	$SiH_3C{\equiv}CH$
NH_3	$N(SiH_3)_3$	$NaMn(CO)_5$	$[Mn(CO)_5(SiH_3)]$
R_2NH	SiH_3NR_2	$Na_2Fe(CO)_4$	$[Fe(CO)_4(SiH_3)_2]$
$NMe_3^{(b)}$	$SiH_3I.NMe_3$ and $SiH_3I.2NMe_3$	$[Co(CO)_4]^-$	$[Co(CO)_4(SiH_3)]$

(a) Detailed references to conditions, yields, and other minor products are given in ref. 1 which also summarizes the extensive reaction chemistry of $O(SiH_3)_2$, $S(SiH_3)_2$, and $N(SiH_3)_3$.
(b) Many other ligands (L) also give 1:1 and 1:2 adducts.

TABLE 9.7 *Some reactions of* $KSiH_3^{(a)}$

Reagent	Major Si product	Reagent	Major Si product
H_2O	$SiO_2.nH_2O$	Me_3SiCl	$[SiMe_3(SiH_3)]$
MeOH	$Si(OMe)_4$	Me_3GeBr	$[GeMe_3(SiH_3)]$
HCl	SiH_4	Me_3SnBr	$[SnMe_3(SiH_3)]$
MeI	SiH_3Me	GeH_3Cl	GeH_3SiH_3
SiH_3Br	Si_2H_6, SiH_4	$MeOCH_2Cl$	$SiH_3(CH_2OMe)$
Si_2H_5Br	Si_3H_8, Si_2H_6		

(a) Detailed references to conditions, yields, and minor products are given in ref. 1.

9.3.3 *Silicon halides and related complexes*

Silicon and silicon carbide both react readily with all the halogens to form colourless volatile reactive products SiX_4. $SiCl_4$ is particularly important and is manufactured on the multikilotonne scale for producing boron-free transistor grade Si, fumed silica (p. 397), and various silicon esters. When two different tetrahalides are heated together they equilibrate to form an approximately random distribution of silicon halides which, on cooling, can be separated and characterized:[13]

$$nSiX_4 + (4-n)SiY_4 \rightleftharpoons 4SiX_nY_{4-n}$$

Mixed halides can also be made by halogen exchange reaction, e.g. by use of SbF_3 to successively fluorinate $SiCl_4$ or $SiBr_4$. The mps and bps of these numerous species are compared with those of the parent hydride and halides in Fig. 9.1. While there is a clear trend to higher mps and bps with increase in molecular weight, this is by no means always

[13] J. C. LOCKHART, *Redistribution Reactions*, Chap. 7.1, Silicon, pp. 107–19, Academic Press, London, 1970.

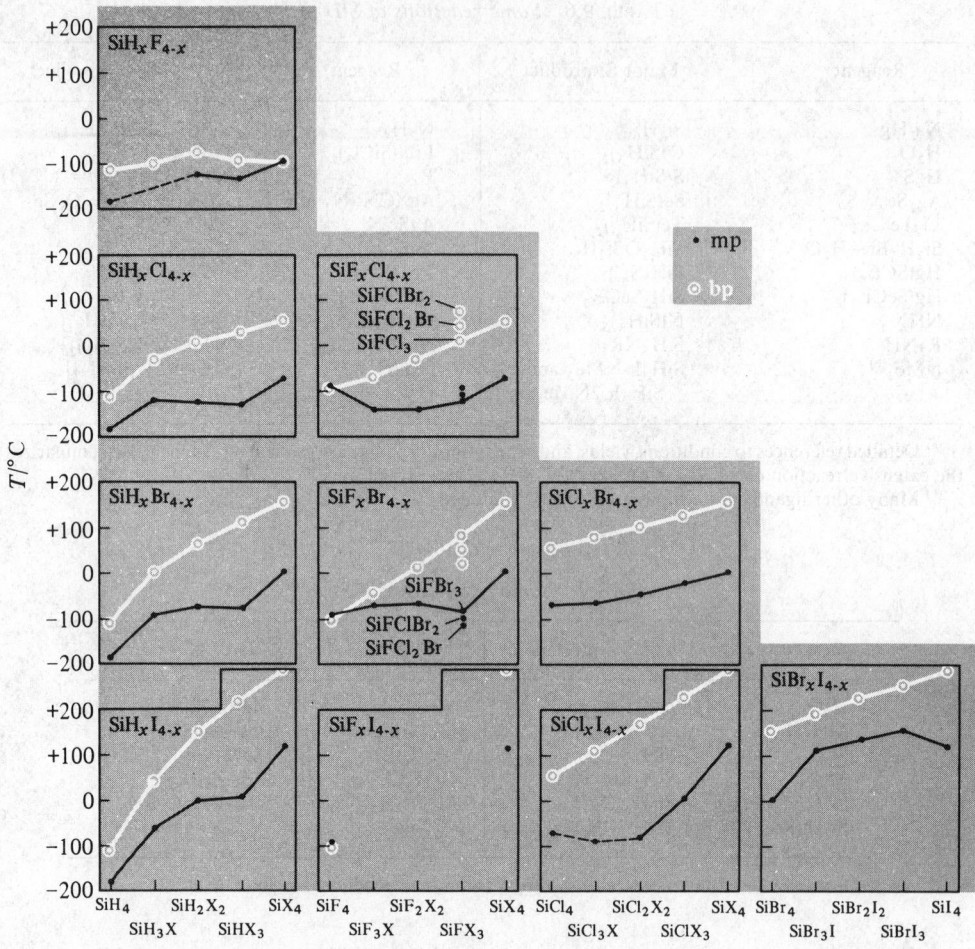

FIG. 9.1   Trends in the mp and bp of silicon hydride halides and mixed halides.

regular. More notable is the enormous drop in mp (bp) which occurs for the halides of Si when compared with Al and earlier elements in the same row of the periodic table, e.g.:

Compound	NaF	MgF$_2$	AlF$_3$	SiF$_4$	PF$_5$	SF$_6$
MP/°C	988	1266	1291 (subl)	−90	−94	−50

This is sometimes erroneously ascribed to a discontinuous change from "ionic" to "covalent" bonding, but the electronegativity and other bonding parameters of Al are fairly similar to those of Si and the difference is more convincingly seen merely as a consequence of the change from an infinite lattice structure (in which each Al is surrounded by 6 F) to a lattice of discrete SiF$_4$ molecules as dictated by stoichiometry and size. Several other examples of this effect will be noticed amongst compounds of the Group IVB elements.

The reactions of SiX_4 are straightforward and call for little comment.[1, 2]

Higher homologues Si_nX_{2n+2} are volatile liquids or solids and, contrary to the situation in carbon chemistry, catenation in Si compounds reaches its maximum in the halides rather than the hydrides. This has been ascribed to additional back-bonding from filled halogen p_π orbitals into the Si d_π orbitals which thus synergically compensates for electron loss from Si via σ bonding to the electronegative halogens (cf. CO, pp. 349–51). Fluoropolysilanes up to $Si_{16}F_{34}$ and other series up to at least Si_6Cl_{14} and Si_4Br_{10} are known. Preparative routes are exemplified by the following reactions:

$$Si + SiF_4 \xrightarrow[\text{system/1250°C}]{\text{low press. flow}} 2SiF_2(g) \xrightarrow[-196°]{\text{condense}} (SiF_2)_x \text{ polymer} \xrightarrow{\sim 200°} Si_nF_{2n+2} \text{ mixture}$$

$$Si + SiCl_4 \xrightarrow{\text{red heat}} Si_2Cl_6 \text{ plus higher homologues}$$

$$Si_2Cl_6 + ZnF_2 \longrightarrow Si_2F_6 + ZnCl_2$$

$$5Si_2Cl_6 \xrightarrow[\text{0.1 mol% NMe}_3]{\text{heat with}} Si_6Cl_{14} + 4SiCl_4$$

$$3Si_2Cl_6 \xrightarrow{\text{NMe}_3 \text{ catalyst}} Si_5Cl_{12} + 2Si_2Cl_6$$

These compounds show many unusual reactions and reviews of their chemistry make fascinating reading.[14, 15] Partial hydrolysis of $SiCl_4$ (or the reaction of $Cl_2 + O_2$ on Si at 700°) leads to a series of volatile chlorosiloxanes $Cl_3Si(OSiCl_2)_nOSiCl_3$ ($n = 0$–5) and to the cyclic $(SiOCl_2)_4$. The corresponding bromo compounds are prepared similarly, using Br_2 and O_2.

9.3.4 *Silica and silicic acids*

Silica has been more studied than any other chemical compound except water. More than 22 phases have been described and, although some of these may depend on the presence of impurities or defects, at least a dozen polymorphs of "pure" SiO_2 are known. This intriguing structural complexity, coupled with the great scientific and technical utility of silica, have ensured continued interest in the compound from the earliest times. The various forms of SiO_2 and their structural inter relations will be described in the following paragraphs. By far the most commonly occurring form of SiO_2 is α-quartz which is a major mineral constituent of many rocks such as granite and sandstone; it also occurs alone as rock crystal and in impure forms as rose quartz, smoky quartz (red brown), morion (dark brown), amethyst (violet), and citrine (yellow). Poorly crystalline forms of quartz include chalcedony (various colours), chrysoprase (leek green), carnelian (deep red), agate (banded), onyx (banded), jasper (various), heliotrope (bloodstone), and flint (often black due to inclusions of carbon). Less-common crystalline modifications of SiO_2

[14] J. L. MARGRAVE and P. W. WILSON, SiF$_2$, a carbene analogue. Its reactions and properties, *Acc. Chem. Res.* **4**, 145–52 (1971).
[15] G. URRY, Systematic synthesis in the polysilane series, *Acc. Chem. Res.* **3**, 306–12 (1970).

are tridymite, cristobalite, and the extremely rare minerals coesite and stishovite. Earthy forms are particularly prevalent as kieselguhr and diatomaceous earth.†

Vitreous SiO_2 occurs as tectites, obsidian, and the rare mineral lechatelierite. Synthetic forms include keatite, and W-silica. Opals are an exceedingly complex crystalline aggregate of partly hydrated silica.

The main crystalline modifications of SiO_2 consist of infinite arrays of corner-shared $\{SiO_4\}$ tetrahedra. In α-quartz, which is thermodynamically the most stable form at room temperature, the tetrahedra form interlinked helical chains; there are two slightly different Si–O distances (159.7 and 161.7 pm) and the angle Si–O–Si is 144°. The helices in any one crystal can be either right-handed or left-handed so that individual crystals have non-superimposable mirror images and can readily be separated by hand. This enantiomorphism also accounts for the pronounced optical activity of α-quartz (specific rotation of the Na D-line 27.71°/mm). At 573°C α-quartz transforms into β-quartz which has the same general structure but is somewhat less distorted (Si–O–Si 155°): only slight displacements of the atoms are required so the transition is readily reversible on cooling and the "handedness" of the crystal is preserved throughout. This is called a non-reconstructive transformation. A more drastic structural change occurs at 867° when β-quartz transforms into β-tridymite. This is a reconstructive transformation which requires the breaking of Si–O bonds to enable the $\{SiO_4\}$ tetrahedra to be rearranged into a simpler, more open hexagonal structure of lower density. For this reason the change is often sluggish and this enables tridymite to occur as a (metastable) mineral phase below the transition temperature. When β-tridymite is cooled to ~120° it undergoes a fast, reversible, non-reconstructive transition to (metastable) α-tridymite by slight displacements of the atoms. Conversely, when β-tridymite is heated to 1470° it undergoes a sluggish reconstructive transformation into β-cristobalite and this, in turn, can retain its structure as a metastable phase when cooled below the transition temperature; further slight displacements occur rapidly and reversibly in the temperature range 200–280° to give α-cristobalite (Si–O 161 pm, Si–O–Si 147°). These transitions are summarized below and further structural details are in ref. 10.

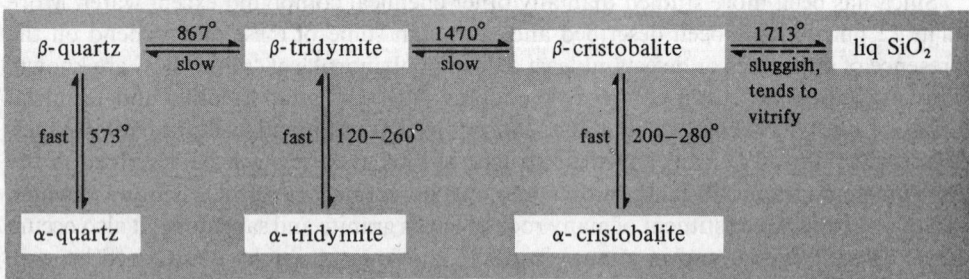

† The names of minerals often give a clue to their properties or discovery. Coesite, stishovite, and keatite are named after their discoverers (p. 395). Quartz derives from *kwardy*, a West Slav dialectal equivalent of the Polish *twardy*, hard. Tridymite was recognized as a new polymorph by von Rath in 1861 because of its typical occurrence as trillings or groups of 3 crystals (Greek τριδυμος, *tridymos*, threefold). Cristobalite was discovered by von Rath in 1884 on the slopes of Mt San Cristobal, Mexico, where tridymite had also first been discovered. Kieselguhr is a combination of the German *Kiesel*, flint, and *Guhr*, earthy deposit. Diatomaceous earth refers to its origin as the remains of minute unicellular algae called diatoms: these marine organisms (0.01–0.1 mm diam) have the astonishing property of accreting silica on their cell walls and this preserves the shape of the organism after death—enormous deposits occur in many places (see p. 397).

The α-form of each of the three minerals can thus be obtained at room temperature and, because of the sluggishness of the reconstructive interconversions of the β-forms, it is even possible to melt β-quartz (1550°) and β-tridymite (1703°) if they are heated sufficiently rapidly. The bp of SiO_2 is not accurately known but is about 2800°C.

Other forms of SiO_2 can be made at high pressure (Fig. 9.2). Coesite was first made by L.

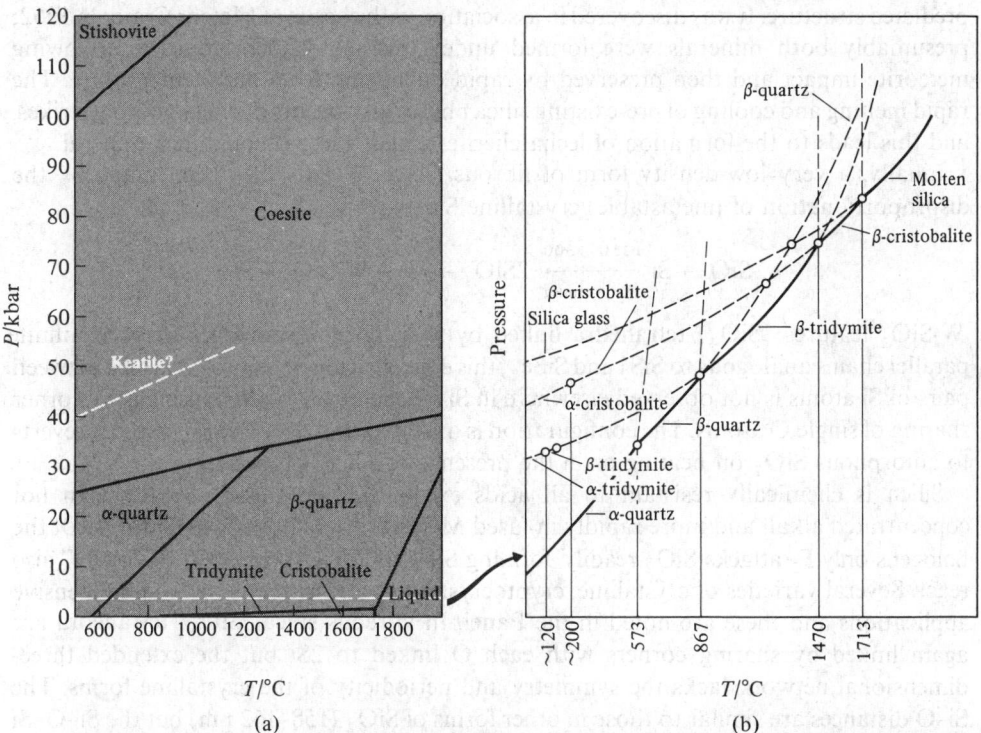

FIG. 9.2   (a) Pressure-temperature phase diagram for SiO_2 showing the stability regions for the various polymorphs. The low-pressure segment below the broken line is shown in (b) using an (arbitrary) expanded scale to illustrate the relationships described in the preceding paragraphs.

Coes in 1953 by heating dry Na_2SiO_3 and $(NH_4)_2HPO_4$ at 700° and 40 kbar, and was subsequently found in Nature at Meteor Crater, Arizona (1960). Its structure consists of 4-connected networks of $\{SiO_4\}$ in which the smallest rings are 4- and 8-membered, and this compact structure explains its high density (Table 9.8). On being heated it rapidly

TABLE 9.8   *Density of the main forms of SiO_2 (room temperature)*

	d/g cm^{-3}		d/g cm^{-3}
W (fibrous)	1.97	β-quartz (600°)	2.533
Lechatelierite	2.19	α-quartz	2.648
Vitreous	2.196	Coesite	2.911
Tridymite	2.265	Keatite	3.010
Cristobalite	2.334	Stishovite	4.287

converts to tridymite or cristobalite. At still higher pressures (40–120 kbar, 380–585°) keatite is formed under hydrothermal conditions from amorphous silica and dilute alkali (P. P. Keat, 1959); the $\{SiO_4\}$ are connected into 5-, 7-, and 8-membered rings as in ice(III) (p. 731). The highest density form of SiO_2 was predicted in 1952 by J. B. Thompson who visualized 6-coordinate Si in a rutile structure (p. 1120). It was first synthesized in S. M. Stishov's laboratory (1961) at 1200–1400° and 160–180 kbar, and found to have the predicted structure. It was discovered in association with coesite at Meteor Crater in 1962: presumably both minerals were formed under transient shock pressures following meteorite impact and then preserved by rapid quenching from high temperature. The rapid melting and cooling of pre-existing silica phases also occurs during lightning strikes, and this leads to the formation of lechatelierite, a glassy or vitreous silica mineral.

Finally, a very low-density form of fibrous silica, $W\text{-}SiO_2$ has been made by the disproportionation of (metastable) crystalline SiO:

$$SiO_2 + Si \xrightarrow[10^{-4}\text{ mmHg}]{1250-1300°} 2SiO \longrightarrow W\text{-}SiO_2 + Si$$

$W\text{-}SiO_2$ features $\{SiO_4\}$ tetrahedra linked by sharing opposite edges to form infinite parallel chains analogous to SiS_2 and $SiSe_2$; this edge-sharing of pairs of O atoms between pairs of Si atoms is not observed elsewhere in Si–O chemistry where linking is by corner sharing of single O atoms. The configuration is unstable, and fibrous SiO_2 rapidly reverts to amorphous SiO_2 on heating or in the presence of traces of moisture.

Silica is chemically resistant to all acids except HF but dissolves slowly in hot concentrated alkali and more rapidly in fused MOH or M_2CO_3 to give M_2SiO_3. Of the halogens only F_2 attacks SiO_2 readily, forming SiF_4 and O_2. Above 1000° H_2 and C also react. Several varieties of crystalline, cryptocrystalline, and vitreous SiO_2 find extensive applications and these are noted in the Panel. In vitreous silica $\{SiO_4\}$ tetrahedra are again linked by sharing corners with each O linked to 2Si but the extended three-dimensional network lacks the symmetry and periodicity of the crystalline forms. The Si–O distances are similar to those in other forms of SiO_2 (158–162 pm) but the Si–O–Si angles vary by as much as 15–20° on either side of the mean value of 153°.

The detailed reactions of SiO_2 with the oxides of the metals and semi-metals are of great importance in glass technology and ceramics but will not be treated here.[1,2,15a] Suffice it to say that, in addition to innumerable crystalline compounds and vitreous phases, many water-soluble compositions are known and many of these find extensive commercial application. Perhaps the best known are the soluble sodium (and potassium) silicates which are made by fusing sand with the appropriate carbonate in a glass-making furnace at ~1400°. The resulting soluble glass is dissolved in hot water under pressure and any insoluble glass or unreacted sand filtered off. The ternary phase diagram for $Na_2O\text{-}SiO_2\text{-}H_2O$ (Fig. 9.3) indicates that only certain limited regions are of commercial interest, e.g. the stable liquid materials (area 9) in the composition range 30–40% SiO_2, 10–20% Na_2O, 60–40% H_2O, i.e. $\sim Na_2Si_2O_5 \cdot 6H_2O$. These find extensive use in industrial and domestic liquid detergents because they maintain high pH by means of their buffering ability and can saponify animal and vegetable oils and fats; they also emulsify mineral oils, deflocculate dirt particles, and prevent redeposition of suspended dirt and soil. The more dilute solutions (area 10) are used in production of silica gels by

[15a] S. FRANK, *Glass and Archaeology*, Academic Press, London, 1982, 156 pp.

Some Uses of Silica[1]

The main types of SiO_2 used in industry are high-purity α-quartz, vitreous silica, silica gel, fumed silica, and diatomaceous earth. The most important application of quartz is as a piezoelectric material (p. 62); it is used in crystal oscillators and filters for frequency control and modulation, and in electromechanical devices such as transducers and pickups: tens of millions of such devices are made each year. There is insufficient natural quartz of adequate purity so it must be synthesized by hydrothermal growth of a seed crystal using dilute aqueous NaOH and vitreous SiO_2 at 400°C and 1.7 kbar. The technique was first successfully employed by G. R. Spezia in 1905. (Crystal growth from molten SiO_2 cannot be used—why?)

Vitreous silica combines exceptionally low thermal expansion† and high thermal shock resistance with high transparency to ultraviolet light, good refractory properties, and general chemical inertness. As a glass it is hard to work because of its very high softening point, high viscosity, short liquid range, and high volatility at forming temperatures. It is familiar in high-quality laboratory glassware, particularly for photolysis experiments and as sample cells in ultraviolet/visible spectroscopy; it is also much used as a protective sheath in the form of tubing or as thin films deposited from the vapour.

Silica gel is an amorphous form of SiO_2 with a very porous structure, formed by acidification of aqueous solutions of sodium silicate; the gelatinous precipitate is washed free of electrolytes and then dehydrated either by roasting or spray drying. The properties of the resulting microporous material depend critically on the conditions of preparation, but typical samples have a pore diameter of 2200–2600 pm, a surface area of 750–800 m^2 g^{-1}, and an apparent bulk density of 0.67–0.75 g cm^{-3}. Such material finds extensive use as a desiccant, selective absorbant, chromatographic support,[15b] catalyst substrate, and insulator (thermal and sound). It can absorb more than 40% of its own weight of water and, when stained with cobalt salts such as the nitrate or $(NH_4)_2CoCl_4$, is familiar as a self-indicating desiccant that can readily be regenerated by heating (anhydrous, blue; hydrated, pink). It is chemically inert, non-toxic, and dimensionally stable, and finds a growing application in the food industry as an anticaking agent in cocoa, fruit juice powders, $NaHCO_3$, and powdered sugar and spices. It is also used as a flatting agent to produce an attractive matte finish on lacquers, varnishes, and paints, and on the surface of vinyl plastics and synthetic fabrics.

Another manufactured form of ultrafine powdered SiO_2 is pyrogenic or fumed silica, formed by the high-temperature hydrolysis of $SiCl_4$ in an oxyhydrogen flame in specially designed burners; the SiO_2 is formed as a very fine white smoke which is collected on cooled rotating rollers. The bulk density is only 0.03–0.06 g cm^{-3} and the surface area 150–500 m^2 g^{-1}. Its main use is as a thixotropic thickening agent in the processing of epoxy and polyester resins and plastics, and as a reinforcing filler in silicone rubbers where, in contrast to carbon black fillers (p. 301), its chemical inertness does not interfere with the peroxide initiated cure (p. 424).

Diatomaceous earth or kieselguhr (p. 394) is mined by open-cast methods on a very substantial scale, particularly in Europe and North America, which respectively account for 59% and 39% of the world production (1.8 million tonnes in 1977). The principal use is in filtration plants, and this accounts for about 60% of the supply; a further 20% is used in abrasives, fillers, light-weight aggregates, and insulation material, and the remainder is used as an inert carrier, coating agent, or in the manufacture of pozzolan.

† The linear coefficient of thermal expansion of vitreous silica is $\sim 0.25 \times 10^{-6}$. This can be compared with a value of $\sim 100 \times 10^{-6}$ for ordinary soda-lime glasses ($\sim 79\%$ SiO_2, $\sim 12.5\%$ Na_2O, $\sim 8.5\%$ CaO). Addition of B_2O_3 (as in Pyrex) sharply reduces this value to 3×10^{-6} (typical laboratory glassware has a composition 83.9% SiO_2, 10.6% B_2O_3, 1.2% Al_2O_3, 3.9% Na_2O, 0.4% K_2O).

[15b] K. K. UNGER, *Porous Silica: Its Properties and Use as a Support in Column Liquid Chromatography,* Journal of Chromatography Library, Vol. 16, 1979, pp. 294.

acidification (pp. 397 and below). There are numerous other uses of soluble silicates including adhesives, glues, and binders, especially for corrugated cardboard boxes, and as refractory acid-resistant cements and sealants.[15c] World production in 1967 was ~2.7 million tonnes of which the sodium silicates formed the major part.

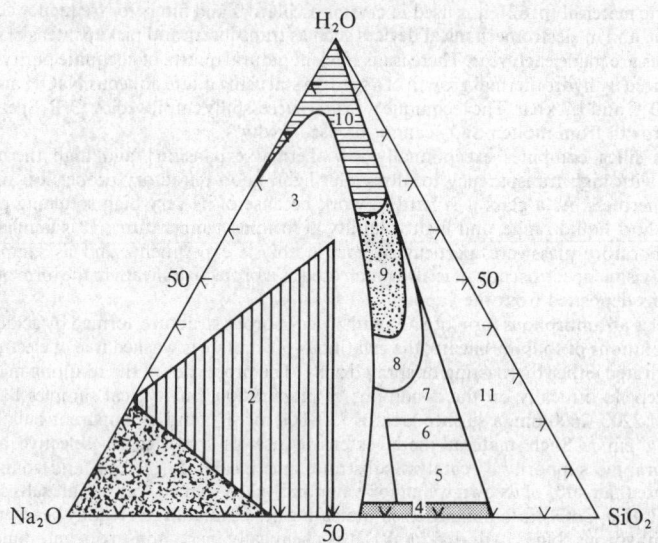

FIG. 9.3 Simplified schematic ternary phase diagram for the system $Na_2O-SiO_2-H_2O$. Commercially important areas are shaded. (1) anhydrous "Na_4SiO_4" and its granular mixtures with NaOH; (2) granular crystalline alkaline silicates such as Na_2SiO_3 and its hydrates; (3) uneconomic partially crystallized mixtures; (4) glasses; (5) uneconomic hydrated glasses; (6) dehydrated liquids; (7) uneconomic semi-solids and gels; (8) uneconomic, unstable viscous liquids; (9) ordinary commercial liquids; (10) dilute liquids; (11) unstable liquids and gels. (From J. G. Vail, *Soluble Silicates*, Reinhold, New York, 1952.)

The system SiO_2-H_2O, even in the absence of metal oxides, is particularly complex and of immense geochemical and industrial importance.[15d] The mp of pure SiO_2 decreases dramatically by as much as 800° on addition of 1–2% H_2O (at high pressure), presumably as a result of the structure-breaking effect of replacing Si–O–Si links by "terminal" Si–OH groups. With increasing concentrations of H_2O one obtains hydrated silica gels and colloidal dispersions of silica; there are also numerous hydrates and distinct silicic acids in very dilute aqueous solutions, but these tend to be rather insoluble and rapidly precipitate with further condensation when aqueous solutions of soluble silicates are acidified. Structural information is sparse, particularly for the solid state, but in solution evidence has been claimed for at least 5 species (Table 9.9). It is unlikely that any of these species exist in the solid state since precipitation is accompanied by further condensation and cross-linking to form "polysilicic acids" of indefinite and variable composition $[SiO_x(OH)_{4-2x}]_n$ (cf B, Al, Fe, etc.).

[15c] L. S. DENT GLASSER, Sodium silicates, *Chem. Br.* **18**, 33–9 (1982).
[15d] R. K. ILER, *The Chemistry of Silica: Solubility, Polymerization, Colloid and Surface Properties, and Biochemistry*, Wiley, New York, 1979, 866 pp.

TABLE 9.9 *Silicic acids in solution*

Formula	$n^{(a)}$	Name	Sol. (H_2O, 20°)/ mol l^{-1}
$H_{10}Si_2O_9$	2.5	Pentahydrosilicic acid	2.9×10^{-4}
H_4SiO_4	2	Orthosilicic acid	7×10^{-4}
$H_6Si_2O_7$	1.5	Pyrosilicic acid	9.6×10^{-4}
H_2SiO_3	1	Metasilicic acid	10×10^{-4}
$H_2Si_2O_5$	0.5	Disilicic acid	20×10^{-4}

(a) Number of mols H_2O per mol SiO_2, i.e. $SiO_2 . nH_2O$.

9.3.5 *Silicate minerals*[10, 16, 17, 17a]

The earth's crustal rocks and their breakdown products—the various soils, clays, and sands—are composed almost entirely ($\sim 95\%$) of silicate minerals and silica. This predominance of silicates and aluminosilicates is reflected in the abundance of O, Si, and Al, which are the commonest elements in the crust (p. 380). Despite the great profusion of structural types and the widely varying stoichiometries which are unmatched elsewhere in chemistry, it is possible to classify these structures on the basis of a few simple principles. Almost invariably Si is coordinated tetrahedrally by 4 oxygen atoms and these $\{SiO_4\}$ units can exist either as discrete structural entities or can combine by corner sharing of O atoms into larger units. The resulting O lattice is frequently close-packed, or approximately so, and charge balance is achieved by the presence of further cations in tetrahedral, octahedral, or other sites depending on their size. Typical examples are as follows (radii in pm):†

CN 4 : Li^I (59) Be^{II} (27) Al^{III} (39) Si^{IV} (26)
CN 6 : Na^I (102) Mg^{II} (72) Al^{III} (54) Ti^{IV} (61) Fe^{II} (78)
CN 8 : K^I (151) Ca^{II} (112)
CN 12: K^I (164)

The quoted radii, which in turn depend on the CN, are the empirical "effective ionic radii" deduced by R. D. Shannon (and C. T. Prewitt)[18] and do not imply full charge separation such as $\{Si^{4+}(O^{2-})_4\}$, etc. Note that Al^{III} can occupy either 4- or 6-coordinate sites so that it can replace either Si or M in the lattice—this is particularly important in discussing the structures of the aluminosilicates. Several other cations can occupy sites of differing CN, e.g. Li (4 and 6), Na (6 and 8), K (6–12), though they are most commonly observed in the CN shown.

---

† The metals which form silicate and aluminosilicate minerals are the more electropositive metals, i.e. those in Groups IA, IIA, and the 3d transition series (except Co), together with Y, La, and the lanthanoids, Zr, Hf, Th, U, and to a much lesser extent the post-transition elements Sn^{II}, Pb^{II}, and Bi^{III}.

[16] W. A. DEER, R. A. HOWIE, and J. ZUSSMAN, *An Introduction to the Rock-forming Minerals*, Longmans, London, 1966, 528 pp.

[17] B. MASON and L. G. BERRY, *Elements of Mineralogy*, W. H. Freeman, San Francisco, 1968, 550 pp.

[17a] F. LIEBAU, Silicon, element 14, in K. H. WEDEPOHL (ed.), *Handbook of Geochemistry*, Vol. II-2, Chap. 14,

[18] R. D. SHANNON, Revised effective ionic radii and systematic studies of interatomic distances, *Acta Cryst.* **A32**, 751–67 (1976).

As with the borates (p. 231) and to a lesser extent the phosphates (p. 616), the $\{SiO_4\}$ units can build up into chains, multiple chains (or ribbons), rings, sheets, and three-dimensional networks as summarized below and elaborated in the following paragraphs.

Neso-silicates	discrete $\{SiO_4\}$	no O atoms shared
soro-silicates	discrete $\{Si_2O_7\}$	1 O atom shared
cyclo-silicates	closed ring structures	
ino-silicates	continuous chains or ribbons	2 O atoms shared
phyllo-silicates	continuous sheets	3 O atoms shared
tecto-silicates	continuous 3D frameworks	all 4 O atoms shared

Silicates with discrete units

Discrete $\{SiO_4\}$ units occur in the orthosilicates $M_2^{II}SiO_4$ (M = Be, Mg, Mn, Fe, and Zn) and in $ZrSiO_4$. (Discrete $\{SiO_4\}$ has also recently been established in the synthetic orthosilicates Na_4SiO_4 and K_4SiO_4.)[18a] In phenacite, Be_2SiO_4, both Be and Si occupy sites of CN 4 and the structure could equally well be described as a 3D network M_3O_4. When octahedral sites are occupied, isomorphous replacement of M^{II} is often extensive as in olivine, $(Mg,Fe,Mn)_2SiO_4$ which derives its name from its olive-green colour (Fe^{II}). The structure is shown in Fig. 9.4. In zircon, $ZrSiO_4$, the stoichiometry of the crystal and the

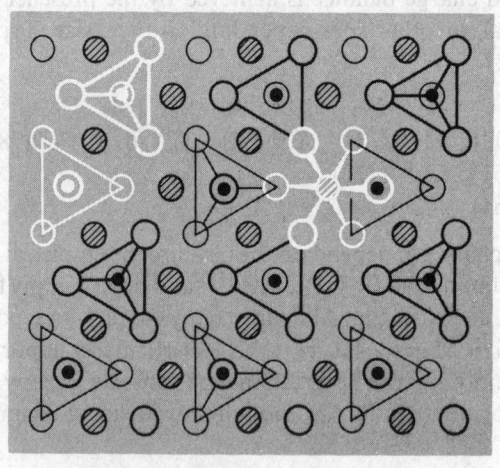

⬨ Mg;   ● Si;   ◯ and ◯ Oxygen

FIG. 9.4   Plan of the structure of Mg_2SiO_4; light and heavy lines are used to distinguish between $\{SiO_4\}$ tetrahedra at different levels and alternate tetrahedra point up and down. The octahedral coordination of one $\{MgO_6\}$ is also indicated.

[18a] M. G. BARKER and P. G. GOOD, The preparation and crystal structure of sodium orthosilicate, Na_4SiO_4, *J. Chem. Research (S)*, 1981, 274, and references therein.

larger radius of Zr (84 pm) dictate eightfold coordination of the cation. Another important group of orthosilicates is the garnets, $[M_3^{II}M_2^{III}(SiO_4)_3]$, in which M^{II} are 8-coordinate (e.g. Ca, Mg, Fe) and M^{III} are 6-coordinate (e.g. Al, Cr, Fe).[19] Orthosilicates are also vital components of Portland cement (p. 284): β-Ca_2SiO_4 has discrete $\{SiO_4\}$ with rather irregularly coordinated Ca in sixfold and eightfold environments (the α-form has the K_2SO_4 structure and the γ-form has the olivine structure). Again alite, Ca_3SiO_5, which is intimately involved in the "setting" process, has individual Ca, $\{SiO_4\}$, and O as the structural units.

Disilicates, containing the discrete $\{Si_2O_7^{6-}\}$ unit, are rare. One example is the mineral thortveitite, $Sc_2Si_2O_7$, which features octahedral Sc^{III} (r 75 pm) and a linear Si–O–Si bond between staggered tetrahedra (Fig. 9.5a). There is also a series of lanthanoid disilicates $Ln_2Si_2O_7$ in which the Si–O–Si angle decreases progressively from 180° to 133° and the CN of Ln increases from 6 through 7 to 8 as the size of Ln increases from 6-coordinated Lu^{III} (86 pm) to 8-coordinated Nd^{III} (111 pm). In the Zn mineral hemimorphite the angle is 150° but the conformation of the 2 tetrahedra is eclipsed (C_{2v}) rather than staggered (Fig. 9.5b); the mineral was originally formulated as $Zn_2SiO_4 \cdot H_2O$ or $H_2Zn_2SiO_5$, but

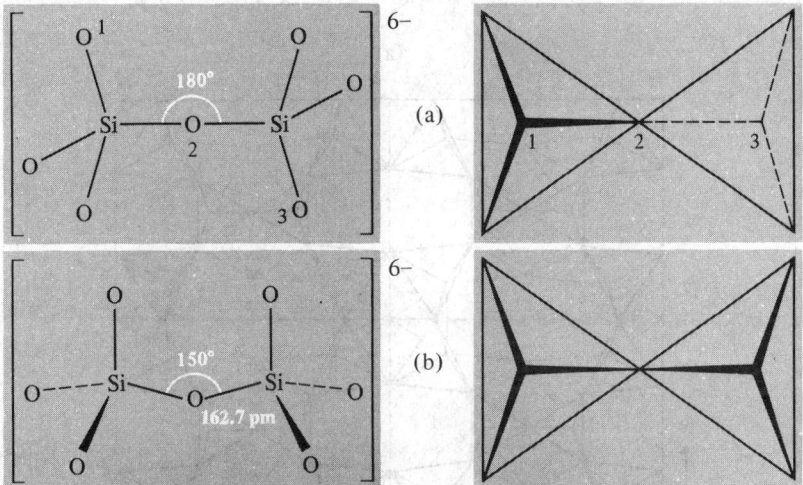

FIG. 9.5   (a) Two representations of the $\{Si_2O_7\}$ unit in $Sc_2Si_2O_7$ showing the linear Si–O–Si link between the two tetrahedra and the D_{3d} (staggered) conformation, and (b) eclipsed (C_{2v}) conformation of the $\{Si_2O_7\}$ unit in hemimorphite, $[Zn_4Si_2O_7(OH)_2] \cdot H_2O$.

X-ray studies showed that the correct formula was $[Zn_4(OH)_2Si_2O_7] \cdot H_2O$, i.e. "$2H_2Zn_2SiO_5$". Two further features of importance also emerged. The first was that there was no significant difference between the Si–O distances to "bridging" and "terminal" O atoms as would be expected for isolated $\{Si_2O_7^{6-}\}$ groups, and the structure is best considered as a 3D framework of $\{ZnO_3(OH)\}$ and $\{SiO_4\}$ tetrahedra linked in threes to form 6-atom rings Zn–O–Si–O–Zn–OH. The rings are linked into infinite sheets in the (010) planes as shown in Fig. 9.6a. Each O in the sheet is linked to 2 Zn and 1 Si atom and

---

[19] See ref. 10, p. 500, for a description of the garnet structure which is also adopted by many synthetic and non-silicate compounds; these have been much studied recently because of their important optical and magnetic properties, e.g. ferrimagnetic yttrium iron garnet (YIG), $Y_3^{III}Fe_2^{III}(Al^{III}O_4)_3$.

the fourth O atom in each tetrahedron is involved in bridging the sheets via Zn–O(H)–Zn or Si–O–Si bonds. Finally, the 3D framework so generated leaves large channels which open into large cavities that accommodate the removable H_2O molecules. The structure is thus very similar in principle to that of the framework aluminosilicates (p. 414) and its conventional description in terms of discrete $Si_2O_7^{6-}$ ions is rather misleading and uninformative.

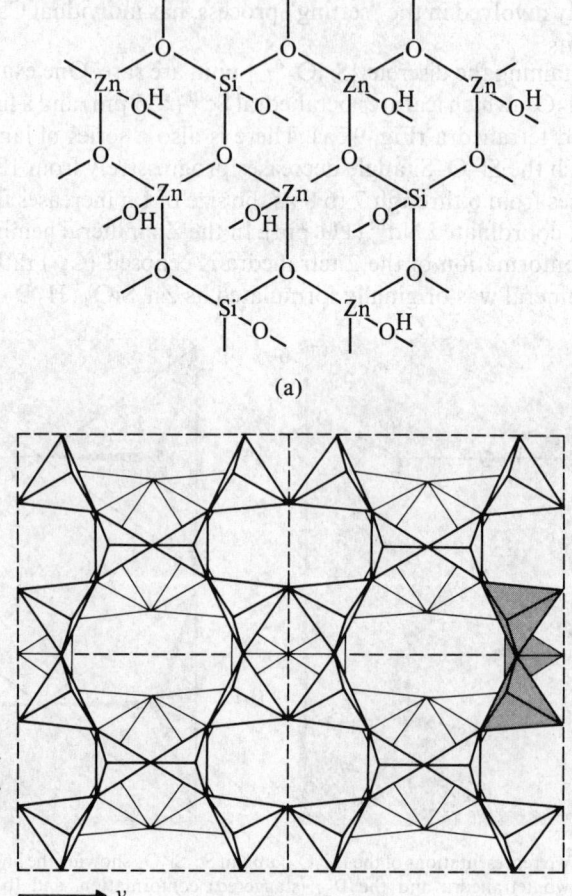

(a)

(b)

FIG. 9.6   (a) A section of the structure of hemimorphite, $[Zn_4(OH)_2Si_2O_7].H_2O$, showing the 6-membered Zn_2SiO_3 rings in the (010) plane (perpendicular to the *y*-axis). (b) The structure projected down the *z*-axis on to the (001) plane and showing the linking of tetrahedra to form channels; the larger tetrahedra represent $\{ZnO_3(OH)\}$ (Zn–O 195 pm) and the smaller tetrahedra represent $\{SiO_4\}$ (Si–O 163 pm); 4 unit cells are shown; one group of 3 fused tetrahedra forming a Zn_2SiO_3 ring at right angles to the *y*-axis is shaded for ease of recognition.

Structures having triple tetrahedral units are extremely rare but they exist in aminoffite, $Ca_3(BeOH)_2(Si_3O_{10})$, and kinoite, $Cu_2Ca_2(Si_3O_{10}).2H_2O$. The first tetrasilicate was synthesized as recently as 1979:[19a] $Ag_{10}Si_4O_{13}$ was prepared as stable vermilion crystals

[19a] M. JANSEN and H.-L. KELLER, $Ag_{10}Si_4O_{13}$—the first tetrasilicate, *Angew. Chem.*, Int. Edn. (Engl.) **18**, 464 (1979).

by heating AgO and SiO_2 for 1–3 days at 500–600°C under a pressure of 2–4.5 kbar of O_2.

When each $\{SiO_4\}$ shares 2 O with contiguous tetrahedra, metasilicates of empirical formula $SiO_3{}^{2-}$ are formed. Cyclic metasilicates $[(SiO_3)_n]^{2n-}$ having 3, 4, 6, or 8 linked tetrahedra are known, though 3 and 6 are the most common. These anions are shown schematically in Fig. 9.7 and are exemplified by the mineral benitoite $[BaTi\{Si_3O_9\}]$, the synthetic compound $[K_4\{Si_4O_8(OH)_4\}]$, and by beryl $[Be_3Al_2\{Si_6O_{18}\}]$ (p. 117) and murite $[Ba_{10}(Ca,Mn,Ti)_4\{Si_8O_{24}\}(Cl,OH,O)_{12}] \cdot 4H_2O$.

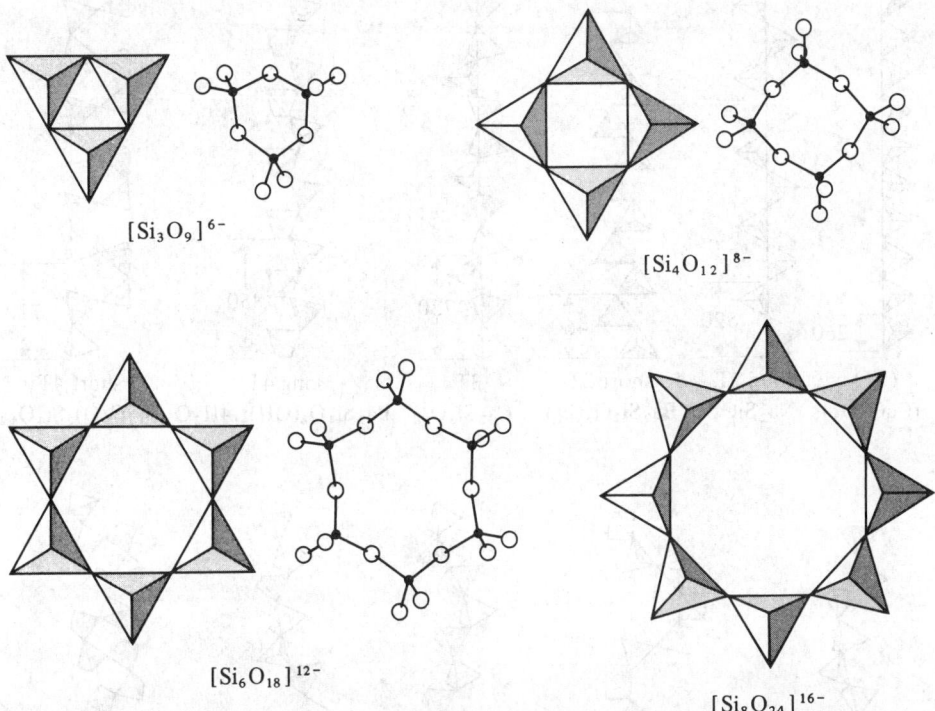

$[Si_3O_9]^{6-}$

$[Si_4O_{12}]^{8-}$

$[Si_6O_{18}]^{12-}$

$[Si_8O_{24}]^{16-}$

FIG. 9.7   Schematic representations of the structures of cyclic metasilicate anions with $n = 3, 4, 6,$ and 8.

Silicates with chain or ribbon structures

Chain metasilicates $\{SiO_3{}^{2-}\}_\infty$ formed by corner-sharing of $\{SiO_4\}$ tetrahedra are particularly prevalent in nature and many important minerals have this basic structural unit (cf. polyphosphates, p. 616). Despite the apparent simplicity of their structure motif and stoichiometry considerable structural diversity is encountered because of the differing conformations that can be adopted by the linked tetrahedra. As a result, the repeat distance along the *c*-axis can be (1), 2, 3···7, 9, or 12 tetrahedra (T), as illustrated schematically in Fig. 9.8. The most common conformation for metasilicates is a repeat after every second tetrahedron (2T) with the chains stacked parallel so as to provide sites of 6- or 8-coordination for the cations; e.g. the pyroxene minerals enstatite $[Mg_2Si_2O_6]$, diopside $[CaMgSi_2O_6]$, jadeite $[NaAlSi_2O_6]$, and spodumene $[LiAlSi_2O_6]$ (p. 76). The synthetic metasilicates Li_2SiO_3 and Na_2SiO_3 are similar; for the latter compound

Si–O–Si is 134° and the Si–O distance is 167 pm within the chain and 159 pm for the other 2 O. The minerals wollastonite [$Ca_3Si_3O_9$] and pectolite [$Ca_2NaHSi_3O_9$] have a 3T repeat unit, haradaite [$Sr_2(VO)Si_4O_{12}$] is 4T, rhodonite [$CaMn_4Si_5O_{15}$] has a 5T repeat, etc.[10, 17a]

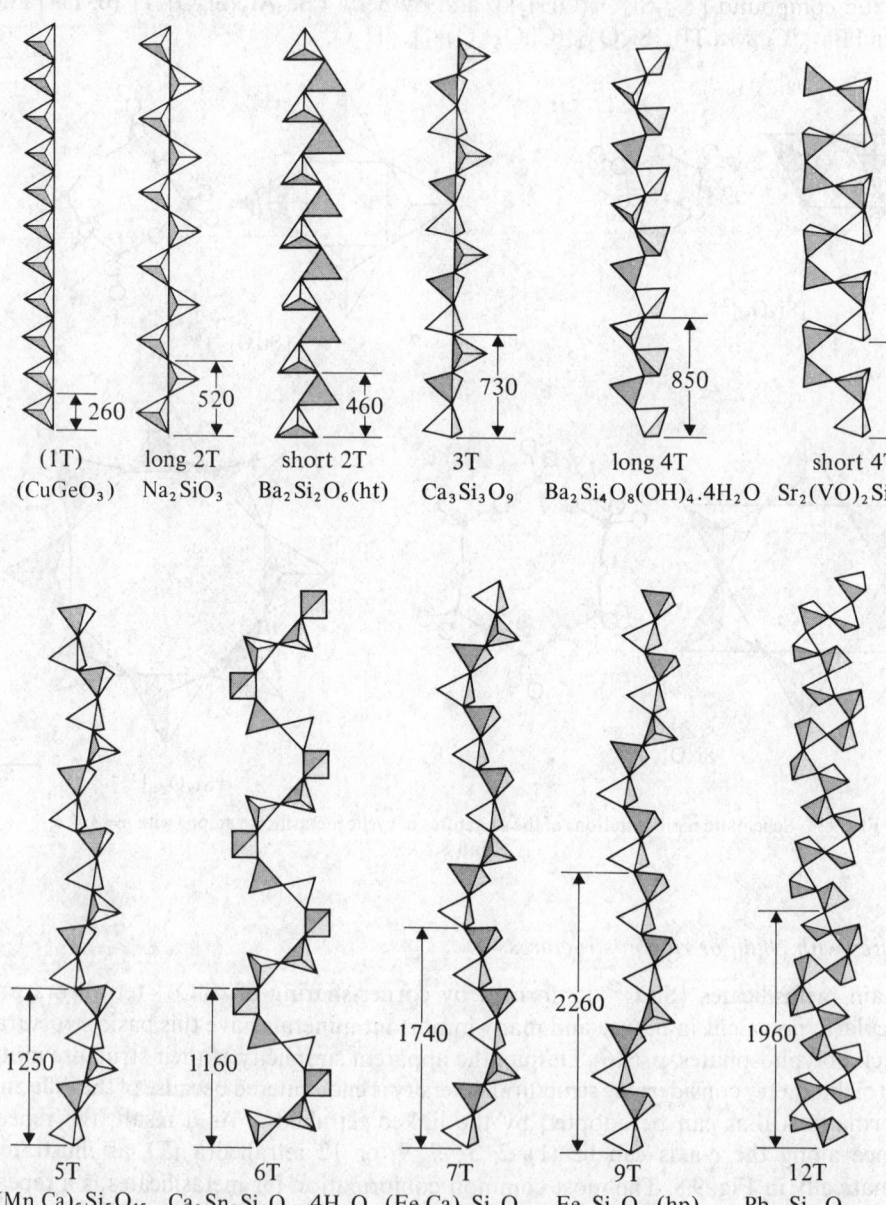

FIG. 9.8   Schematic representation and examples of various chain metasilicates $\{SiO_3{}^{2-}\}_\infty$ with repeat distances (in pm) after 1, 2 . . . 7, 9, or 12 tetrahedra (T). [(ht) high-temperature form; (hp) high-pressure form.]

In the next stage of structural complexity the single $\{SiO_3{}^{2-}\}_\infty$ chains can link laterally to form double chains or ribbons whose stoichiometry depends on the repeat unit of the single chain (Fig. 9.9). By far the most numerous are the amphiboles or asbestos minerals

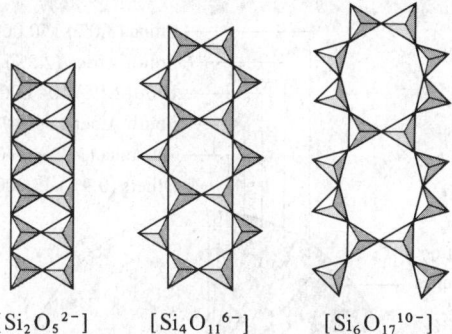

$[Si_2O_5{}^{2-}]$ $\qquad$ $[Si_4O_{11}{}^{6-}]$ $\qquad$ $[Si_6O_{17}{}^{10-}]$

FIG 9.9 Double chains of $\{SiO_4\}$ tetrahedra: (a) the double chain based on the 1T metasilicate structure, stoichiometry $\{Si_2O_5{}^{2-}\}$—it is found in the aluminosilicate sillimanite $[Al(AlSiO_5)]$; (b) $\{Si_4O_{11}{}^{6-}\}$ chain based on the 2T metasilicate occurs in the amphiboles (see text); and (c) the rare 3T double chain $Si_6O_{17}{}^{10-}$ occurs in xonotlite $[Ca_6Si_6O_{17}(OH)_2]$. More complex 3T, 4T, and 6T double chains are also known.[17a]

which adopt the $\{Si_4O_{11}{}^{6-}\}$ double chain, e.g. tremolite $[Ca_2Mg_5(Si_4O_{11})_2(OH)_2]$; the structure of this compound is very similar to that of diopside (above) except that the length of the *b*-axis of the unit cell is doubled. The fibrous nature of the asbestos minerals thus finds a ready interpretation on the basis of their crystal structures (see Panel). In addition to these well-established double chains of linked $\{SiO_4\}$ tetrahedra, examples of infinite one-dimensional structures consisting of linked triple, quadruple, and sextuple chains have recently been discovered in nephrite jade by means of electron microscopy,[20]

Production and Uses of Asbestos

The fibrous silicate minerals known collectively as asbestos (Greek ἄσβεστος, unquenchable) have been used both in Europe and the far east for thousands of years. In ancient Rome the wicks of the lamps of the vestal virgins were woven from asbestos, and Charlemagne astounded his barbarian guests by throwing the festive table cloth into the fire whence, being woven asbestos, it emerged cleansed and unburnt.[20a] Its use has accelerated during the past 100 y and it is now an important ingredient in over 3000 different products. Its desirable characteristics are high tensile strength, great flexibility, resistance both to heat and flame and also to corrosion by acids or alkalis, and its low cost.

Asbestos is derived from two large groups of rock-forming minerals—the serpentines and the amphiboles. Chrysotile, or white asbestos $[Mg_3(Si_2O_5)(OH)_4]$, is the sole representative of the serpentine layer silicate group (p. 408) but is by far the most abundant kind of asbestos and constitutes 95% of world production. The amphibole group includes the blue asbestos mineral crocidolite $[Na_2Fe_3^{II}Fe_2^{III}Si_8O_{22}(OH)_2]$ (3.5% of world production) and the grey–brown mineral amosite $[(Mg,Fe)_7Si_8O_{22}(OH)_2]$ (1.5%). Annual production in 1976 was 5.2 million tonnes. Of

[20] L. G. MALLINSON, J. L. HUTCHINSON, D. A. JEFFERSON, and J. M. THOMAS, Discovery of new types of chain silicates by high resolution electron microscopy, *JCS Chem. Comm.* 910–11 (1977).
[20a] W. J. SMITHER, Asbestos, *School Science Review* **60** (210), 59–69 (1978).

this, 4 927 000 tonnes was chrysotile, distributed as indicated in the Figure.[20b] In addition South Africa produced 259 000 tonnes of amphiboles—mainly crocidolite (178 000 tonnes) and amosite (79 000 tonnes).

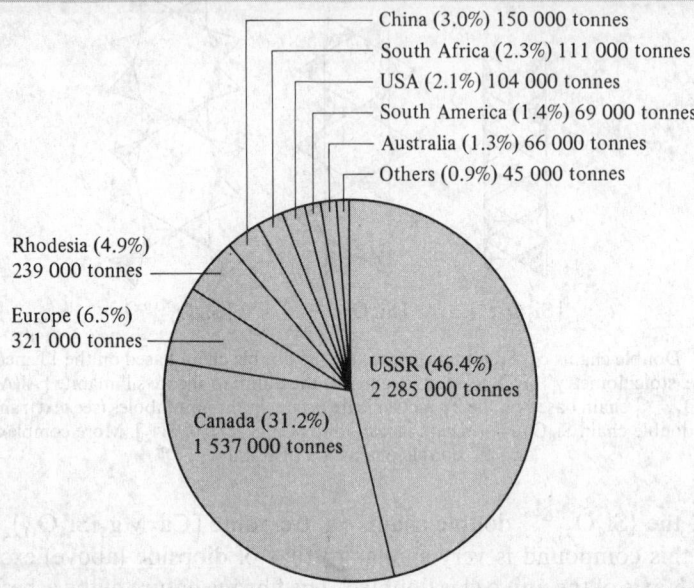

China (3.0%) 150 000 tonnes
South Africa (2.3%) 111 000 tonnes
USA (2.1%) 104 000 tonnes
South America (1.4%) 69 000 tonnes
Australia (1.3%) 66 000 tonnes
Others (0.9%) 45 000 tonnes

Rhodesia (4.9%)
239 000 tonnes

Europe (6.5%)
321 000 tonnes

USSR (46.4%)
2 285 000 tonnes

Canada (31.2%)
1 537 000 tonnes

Asbestos-reinforced cements ($\sim 12.5\%$ asbestos) absorb nearly two-thirds of the world's annual production of chrysotile: it is used in corrugated and flat roofing sheets, pressure pipes and ducts, and many other hard-wearing, weather-proof, long-lasting products. About 8% is used in asbestos papers and a further 7% is used for making vinyl floor tiles. Other important uses include composites for brake linings, clutch facings, and other friction products. Long-fibred chrysotile (fibre length >20 mm) is woven into asbestos textiles for fire-fighting garments and numerous fire-proofing and insulating applications.

Prolonged exposure to airborne suspensions of asbestos fibre dust can be very dangerous and there has been increasing concern at the incidence of asbestosis (non-malignant scarring of lung tissue) and lung carcinoma among certain workers in the industry. Unfortunately, there is an extended latent period (typically 20–30 y) before these diseases are manifest, and many of the cases diagnosed during the last 15 y may have been due to conditions in wartime factories in the years 1940–5 and thereafter. Incidence in the UK now appears to be dropping from a peak of about 150 cases in 1970 (representing about 5 per 1000 workers in the industry). Asbestosis is dose-related and the best form of control is to reduce the level of dust exposure in places where the mineral is mined, processed, or fabricated.[20c]

and these form a satisfying link between the pyroxenes and amphiboles, on the one hand, and the sheet silicates (to be described in the next paragraph), on the other.

Silicates with layer structures

Silicates with layer structures include some of the most familiar and important minerals known to man, particularly the clay minerals [such as kaolinite (china clay),

[20b] Reference 1, 3rd edn., Asbestos 3, 267–83 (1978).
[20c] L. MICHAELIS and S. S. CHISSICK (eds.), *Asbestos: Properties, Applications and Hazards*, Vol. 1, Wiley, New York, 1979, 553 pp.

montmorillonite (bentonite, fuller's earth), and vermiculite], the micas (e.g. muscovite, phlogopite, and biotite), and others such as chrysotile (white asbestos), talc, soapstone, and pyrophyllite. The physical and chemical properties of these minerals, which have made many of them so valued for domestic and industrial use for several milleniums (p. 379), can be directly related to the details of their crystal structure. The simplest silicate layer structure can be thought of as being formed either by the horizontal cross-linking of the 2T metasilicate chain $\{Si_2O_6^{4-}\}$ in Fig. 9.8 or by the planar condensation of the $\{Si_6O_{18}^{12-}\}$ unit in Fig. 9.7 to give a 6T network of composition $\{Si_2O_5^{2-}\}$ in which 3 of the 4 O atoms in each tetrahedron are shared; this is shown in both plan and elevation in Fig. 9.10. In fact, such a structure with a completely planar arrangement is extremely rare

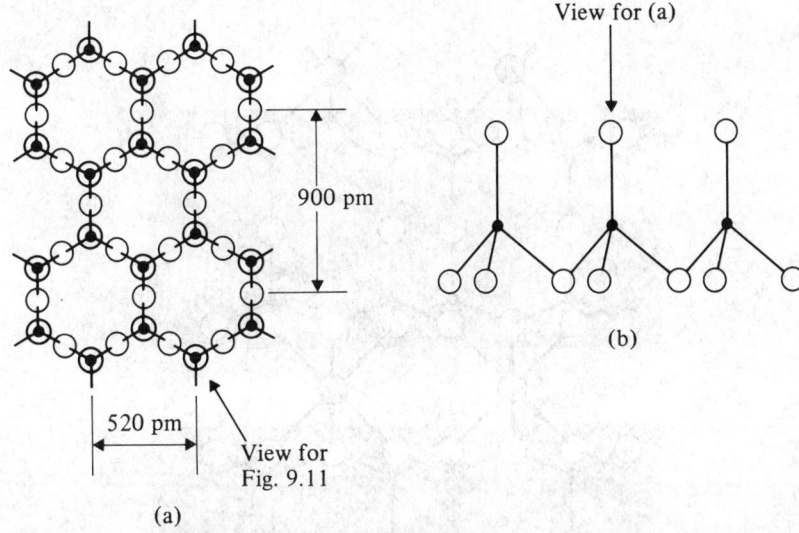

FIG. 9.10   Planar network formed by extended 2D condensation of rings of 6 $\{SiO_4\}$ tetrahedra to give $\{Si_2O_5^{2-}\}$. (a) Plan as seen looking down the O–Si direction, and (b) side elevation.

though closely related puckered 6T networks are found in $M_2Si_2O_5$ (M = Li, Na, Ag, H) and in petalite ($LiAlSi_4O_{10}$, p. 76). More complex arrangements are also found in which the 6T rings forming the network are replaced by alternate 4T and 8T rings, or by equal numbers of 4T, 6T, and 8T rings, or even by a network of 4T, 6T, and 12T rings.[10, 17a]

Double layers can be generated by sharing the fourth (apical) O atom between pairs of tetrahedra as in Fig. 9.11. This would give a stoichiometry SiO_2 (since each O atom is shared between 2 Si atoms) but if half the Si^{IV} were replaced by Al^{III} then the composition would be $\{Al_2Si_2O_8^{2-}\}$ as found in $Ca_2Al_2Si_2O_8$ and $Ba_2Al_2Si_2O_8$ (Fig. 9.12). Another way of building up double layers involves the interleaving of layers of the gibbsite $Al(OH)_3$ or brucite $Mg(OH)_2$ structure (pp. 274–77, 133) which happen to have closely similar dimensions and can thus share O atoms with the silicate network. This leads to the china-clay mineral kaolinite $[Al_2(OH)_4Si_2O_5]$ illustrated in Fig. 9.13.†

† Named in 1867 from "kaolin", a corruption of the Chinese *kauling*, or high-ridge, the name of the hill where this china clay was found.

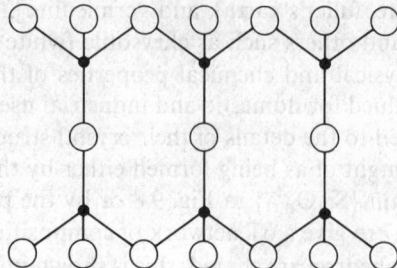

FIG. 9.11   Side elevation of double layers of formula $\{Al_2Si_2O_8{}^{2-}\}$ formed by sharing the fourth (apical) O in Fig. 9.10(b). Sites marked ● are occupied by equal numbers of Al and Si atoms.

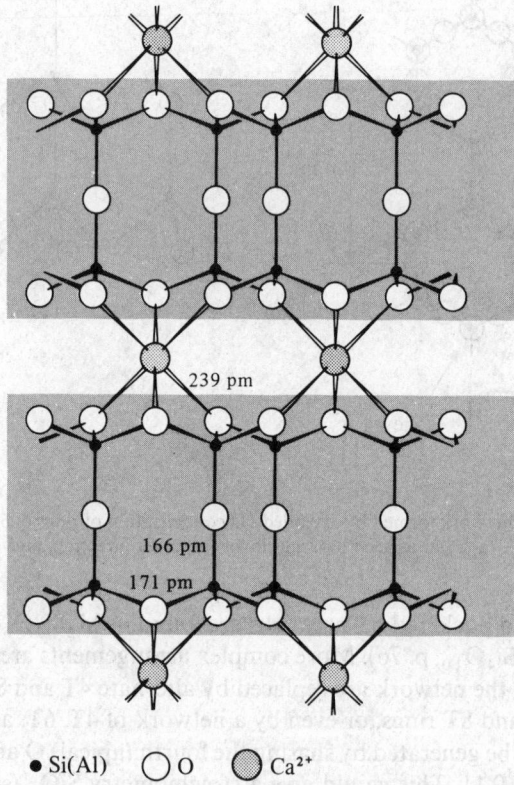

● Si(Al)   ◯ O   ◯ Ca²⁺

FIG. 9.12   Structure of $Ca_2Al_2Si_2O_8$ formed by interleaving 6-coordinated Ca^II atoms between the double layers depicted in Fig. 9.11.

Repetition of the process on the other side of the Al/O layer leads to the structure of pyrophyllite $[Al_2(OH)_2Si_4O_{10}]$ (Fig. 9.13b). Replacement of 2Al^III by 3Mg^II in kaolinite $[Al_2(OH)_4Si_2O_5]$ gives the serpentine asbestos mineral chrysotile $[Mg_3(OH)_4Si_2O_5]$ and a similar replacement in pyrophyllite gives talc $[Mg_3(OH)_2Si_4O_{10}]$. The gibbsite series is sometimes called dioctahedral and the brucite series trioctahedral in obvious reference to the number of octahedral sites occupied in the "non-silicate" layer.

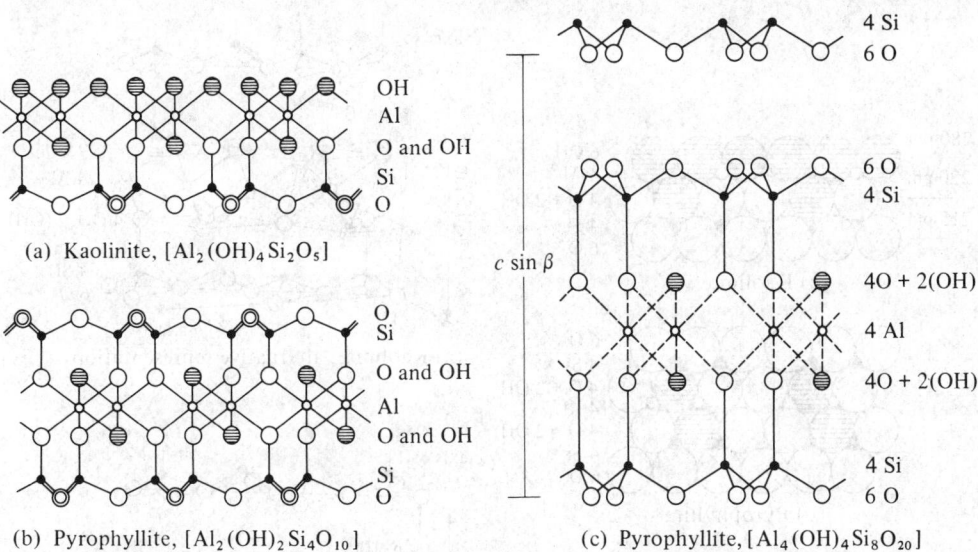

(a) Kaolinite, $[Al_2(OH)_4Si_2O_5]$

OH
Al
O and OH
Si
O

(b) Pyrophyllite, $[Al_2(OH)_2Si_4O_{10}]$

O
Si
O and OH
Al
O and OH
Si
O

$c \sin \beta$

4 Si
6 O

6 O
4 Si

4O + 2(OH)

4 Al

4O + 2(OH)

4 Si
6 O

(c) Pyrophyllite, $[Al_4(OH)_4Si_8O_{20}]$

FIG. 9.13   (a) Schematic representation of the structure of kaolinite (side elevation) showing $\{SiO_3O\}$ tetrahedra (bottom) sharing common O atoms with $\{Al(OH)_2O\}$ to give a composite layer of formula $[Al_2(OH)_4Si_2O_5]$. The double lines and double circles in the tetrahedra indicate bonds to 2 O atoms (one in front and one behind). (b) Similar representation of the structure of pyrophyllite, showing shared $\{SiO_3O\}$ tetrahedra above and below the $\{Al(OH)O_2\}$ layer to give a composite layer of formula $[Al_2(OH)_2Si_4O_{10}]$. (c) Alternative representation of pyrophyllite to be compared with (b), and showing the stoichiometry of each layer.

Alternative representations of the structures are given in Figs. 9.14 and 9.15, and it is well worth while looking carefully at these three sets of diagrams since they have the pleasing property of becoming simpler and easier to understand the longer they are contemplated. It should be stressed that the formulae given are ideal limiting compositions and that Al^{III} or Mg^{II} can be replaced by several other cations of appropriate size. The stoichiometry is further complicated by the possibility that Si^{IV} can be partly replaced by Al^{III} in the tetrahedral sites thereby giving rise to charged layers. These layers can be interleaved with M^I or M^{II} cations to give the micas or by layers of hydrated cations to give montmorillonite. Alternatively, charge balance can be achieved by interleaving positively

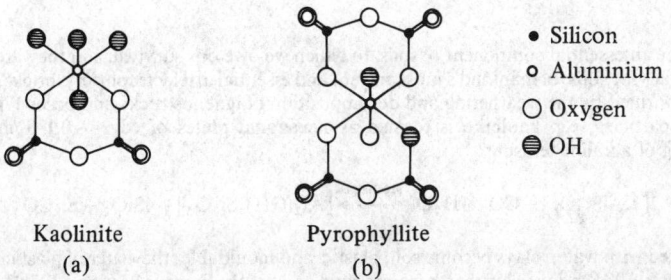

Kaolinite
(a)

Pyrophyllite
(b)

• Silicon
○ Aluminium
○ Oxygen
⊜ OH

FIG. 9.14   An alternative representation of the layer structures of (a) kaolinite, and (b) pyrophyllite.

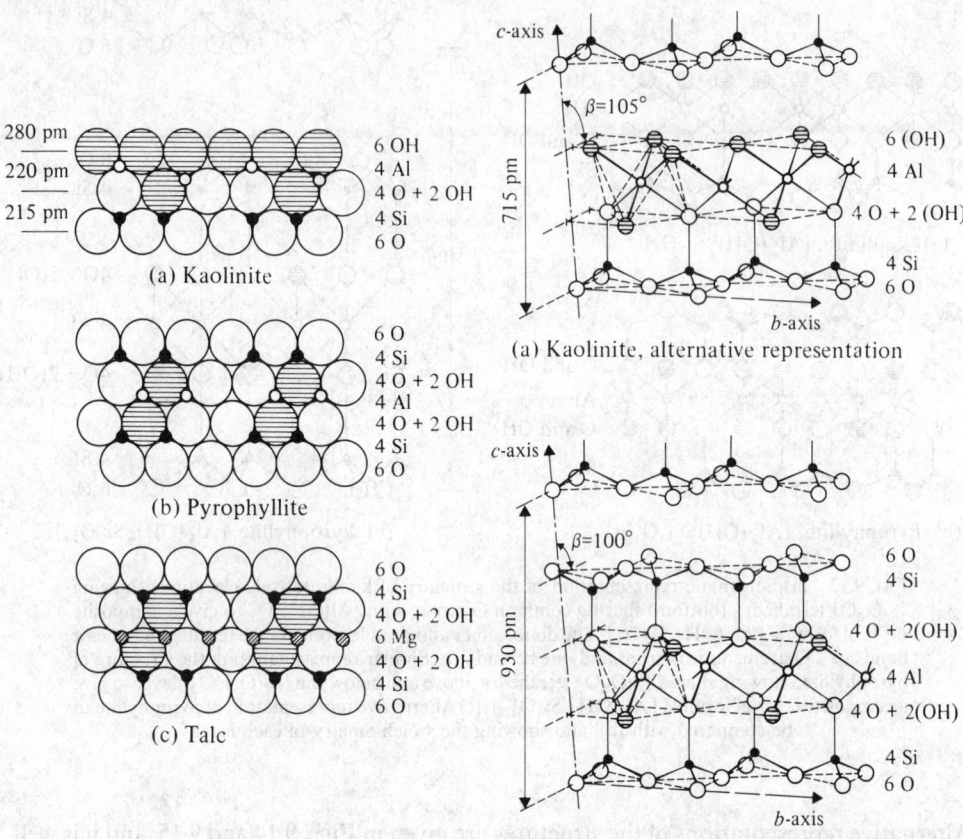

FIG. 9.15 Alternative representations of the layer structures of (a) kaolinite, (b) pyrophyllite, and (c) talc. (After H. J. Emeleus and J. S. Anderson, 1960 and B. Mason and L. G. Berry, 1968.)

charged $(Mg,Al)(OH)_2$ layers as in the chlorites. These possibilities are shown schematically in Fig. 9.16 and elaborated in the following paragraphs.

The technological importance of the clay minerals is outlined in the Panel.

Clay Minerals and Related Aluminosilicates

Clays are an essential component of soils, to which we owe our survival, and they are also the raw materials for some of mankind's most ancient and essential artefacts: pottery, bricks, tiles, etc. Clays are formed by the weathering and decomposition of igneous rocks and occur typically as very fine particles: e.g. kaolinite is formed as hexagonal plates of edge ~ 0.1–3 μm by the weathering of alkaline feldspar:

$$2[KAlSi_3O_8] + CO_2 + H_2O \xrightarrow{\text{idealized}} [Al_2(OH)_4Si_2O_5] + 4SiO_2 + K_2CO_3$$

When mixed with water, clays become soft, plastic, and mouldable; the water of plasticity can be removed at $\sim 100°$ and the clay then becomes rigid and brittle. Further heating ($\sim 500°$) removes structural water of crystallization and results in the oxidation of any carbonaceous material or Fe^{II}

(600–900°). Above about 950° mullite $(Al_6Si_2O_{13})$ begins to form and glassy phases appear.[1]

China clay or kaolin, which is predominantly kaolinite, is particularly valuable because it is essentially free from iron impurities (and therefore colourless). World production in 1974 was about 17 million tonnes, the main producers being the USA (36%), the UK (22%), and the USSR (13%). In the USA almost half of this vast tonnage was used for paper filling or paper coating and one-quarter was used for various fire-bricks and refractories. Only 82 000 tonnes was used for china, crockery, and earthenware, which is now usually made from ball clay, a particularly fine-grained, highly plastic material which is predominantly kaolinite together with clay-mica and quartz. Some 800 000 tonnes of ball clay is used annually in the USA for white ware, table ware, wall and floor tiles, sanitary ware, and electrical porcelain.

Fuller's earth† is a montmorillonite in which the principal exchangeable cation is calcium. It has a high absorptive and adsorptive capacity, and pronounced cation exchange properties which enable it to be converted to sodium-montmorillonite (bentonite). Nomenclature is confusing and, in American usage, the fibrous hydrated magnesium aluminosilicate attapulgite is also called fuller's earth. US production exceeds 1 million tonnes pa and is used mainly as an oil and grease absorbent (40%) and for what the government statisticians coyly call "pet absorbent" (30%).

Bentonite (sodium-montmorillonite) is extensively used as a drilling mud, but this apparently mundane application is based on the astonishing thixotropic properties of its aqueous suspensions. Thus, replacement of Ca by Na in the montmorillonite greatly enhances its ability to swell in one dimension by the reversible uptake of water; this effectively cleaves the clay particles causing a separation of the lamellar units to give a suspension of very finely divided, exceedingly thin plates. These plate-like particles have negative charges on the surface and positive charges on the edges and, even in a suspension of quite low solid content, the particles orient themselves negative to positive to give a jelly-like mass or gel; on agitation, however, the weak electrical bonds are broken and the dispersion becomes a fluid whose viscosity diminishes with the extent of agitation. This indefinitely reversible property is called thixotropy and is widely used in civil engineering applications, in oil-well drilling, and in non-drip paints. The plasticity of bentonite is also used in mortars, putties, and adhesives, in the pelletizing of iron ore, and in foundry sands. Some 3 million tonnes were produced in the USA in 1975 and world production was 4.3 million tonnes.

Micas occur as a late crystallization phase in igneous rocks. Usually the crystals are 1–5 mm on edge but in pegmatites (p. 119) they may considerably exceed this to give the valuable block mica: one spectacular find in Ontario gave single crystals of muscovite up to 4.3 m in diameter and weighing over 80 tonnes. Uses of muscovite mica depend on its perfect basal cleavage, toughness, elasticity, transparency, high dielectric strength, chemical inertness, and thermal stability to 500°. Phlogopite (Mg-mica) is less used except when stability to 850–1000° is required. Sheet mica is used for furnace windows, for electrical insulation (condensers, heating elements, etc.), and in vacuum tubes. Ground mica is used as a filler for rubber, plastics, and insulating board, for silver glitter paints, etc. World production (excluding China) was ~240 000 tonnes in 1974 of which 53% came from the USA, 20% from India, and 17% from the USSR.

Talc, unlike the micas, consists of electrically neutral layers without the interleaving cations. It is valued for its softness, smoothness, and dry lubricating properties, and for its whiteness, chemical inertness, and foliated structure. Its most important applications are in ceramics, insecticides, paints, and paper manufacture. The more familiar use in cosmetics and toilet preparations accounts for only 3% of world production which was 5.5 million tonnes in 1974 and 4.9 million tonnes in 1975. Half of this came from Japan and the USA, and other major producers are Korea, the USSR, France, and China. Talc and its more massive mineral form soapstone or steatite are widely distributed throughout the world and many countries produce it for domestic consumption either by open-cast or underground mining.

† The name derives from its early use (from Roman times) for the fulling or cleansing of woollen cloth from its natural oils and greases. It is now little used for this purpose since soaps and detergents are more effective.

Micas are formed when one-quarter of the Si^{IV} in pyrophyllite and talc is replaced by Al^{III} and the resulting negative charge is balanced by K^I:

pyrophyllite $[Al_2(OH)_2Si_4O_{10}] \longrightarrow [KAl_2(OH)_2(Si_3AlO_{10})]$ muscovite (white mica)

talc $[Mg_3(OH)_2Si_4O_{10}] \longrightarrow [KMg_3(OH)_2(Si_3AlO_{10})]$ phlogopite

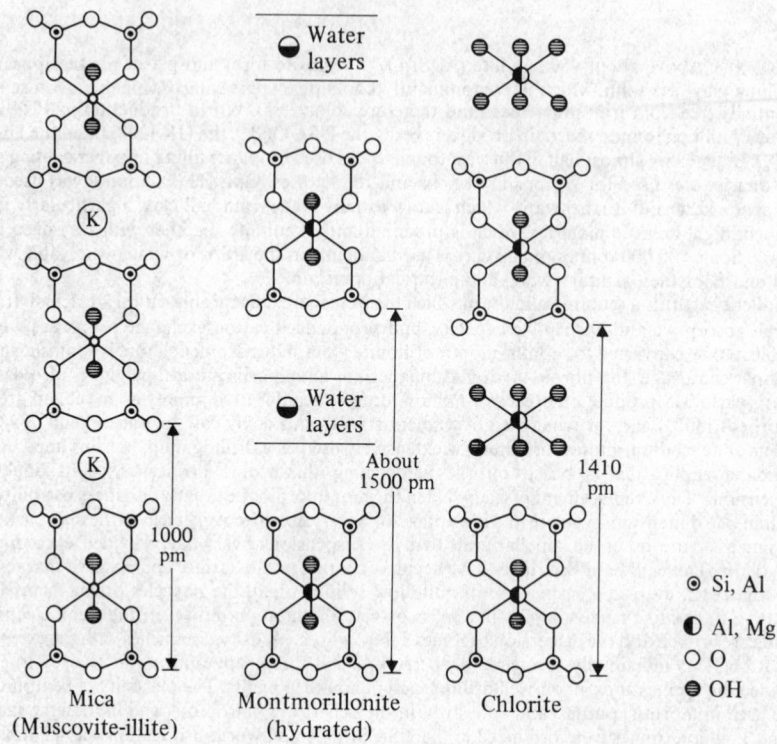

FIG. 9.16 Schematic representation of the structures of muscovite mica, $[K_2Al_4(Si_6Al_2)O_{20}(OH)_4]$, hydrated montmorillonite, $[Al_4Si_8O_{20}(OH)_4].xH_2O$ and chlorite, $[Mg_{10}Al_2(Si_6Al_2)O_{20}(OH)_{16}]$, see text.

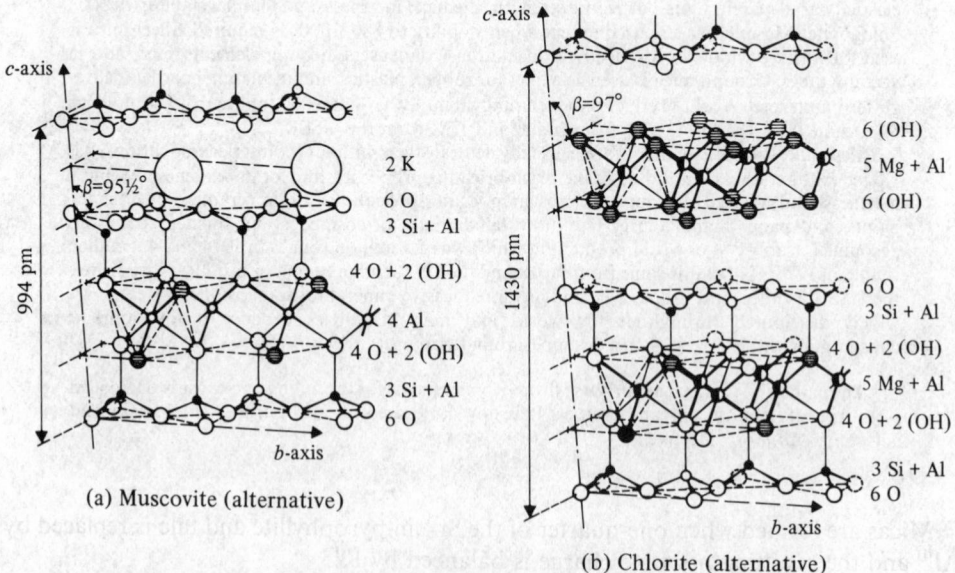

FIG. 9.16 Alternative representations of the structures of muscovite and chlorite (after B. Mason and L. G. Berry[17]).

The OH can be partly replaced by F and, in phlogopite, partial replacement of Mg^{II} by Fe^{II} gives biotite (black mica) $[K(Mg,Fe^{II})_3(OH,F)_2(Si_3AlO_{10})]$. The presence of K^I between the layers makes the micas appreciably harder than pyrophyllite and talc but the layers are still a source of weakness and micas show perfect cleavage parallel to the layers. With further substitution of up to half the Si by Al charge balance can be restored by the more highly charged Ca^{II} and brittle micas result, such as margarite $[CaAl_2(OH)_2(Si_2Al_2O_{10})]$ which is even harder than muscovite.

Another set of minerals, the montmorillonites, result if, instead of replacing tetrahedral Si^{IV} by Al^{III} in phlogopite, the octahedral Al^{III} is *partially* replaced by Mg^{II} (not *completely* as in talc). The resulting partial negative charge per unit formula can be balanced by incorporating hydrated M^I or M^{II} between the layers; this leads to the characteristic swelling, cation exchange, and thixotropy of these minerals (see Panel). A typical sodium montmorillonite might be formulated $Na_{0.33}[Mg_{0.33}Al_{1.67}(OH)_2(Si_4O_{10})].nH_2O$, but more generally they can be written as $M_x[(Mg,Al,Fe)_2(OH)_2(Si_4O_{10})].nH_2O$ where $M = H$, Na, K, $\frac{1}{2}Mg$, or $\frac{1}{2}Ca$. Simultaneous altervalent substitution in both the octahedral and tetrahedral sites in talc leads to the vermiculites of which a typical formula is

$$[Mg_{0.32}(H_2O)_{4.32}]^{0.64+}[(Mg_{2.36}Al_{0.16}Fe^{III}_{0.48})(OH)_2(Si_{2.72}Al_{1.20}O_{10})]^{0.64-}$$

$$\diagdown \diagup$$

total 3.00

When these minerals are heated they dehydrate in a remarkable way by extruding little worm-like structures as indicated by their name (Latin *vermiculus*, little worm); the resulting porous light-weight mass is much used for packing and insulation. The relationship between the various layer silicates is summarized with idealized formula in Table 9.10.

TABLE 9.10   *Summary of layer silicate structures (idealized formulae)*[17]

Dioctahedral (with gibbsite-type layers)	Trioctahedral (with brucite-type layers)
Two-layer structures	
Kaolinite, nacrite, dickite $[Al_4(OH)_8(Si_4O_{10})]$	Antigorite (platy serpentine) $[Mg_6(OH)_8(Si_4O_{10})]$
Halloysite $[Al_4(OH)_8(Si_4O_{10})]$	Chrysotile (fibrous serpentine) $[Mg_6(OH)_8(Si_4O_{10})]$
Three-layer structures	
Pyrophyllite $[Al_2(OH)_2(Si_4O_{10})]$	Talc $[Mg_3(OH)_2(Si_4O_{10})]$
Montmorillonite $[Al_2(OH)_2(Si_4O_{10})].xH_2O^{(a)}$	Vermiculite $[Mg_3(OH)_2(Si_4O_{10})].xH_2O^{(b)}$
Muscovite (mica) $[KAl_2(OH)_2(AlSi_3O_{10})]$	Phlogopite (mica) $[KMg_3(OH)_2(AlSi_3O_{10})]$
Margarite (brittle mica) $[CaAl_2(OH)_2(Al_2Si_2O_{10})]$	Clintonite $[CaMg_3(OH)_2(Al_2Si_2O_{10})]$
	Chlorite $[Mg_5Al(OH)_8(AlSi_3O_{10})]^{(c)}$

(a) With partial replacement of octahedral Al by Mg and with adsorbed cations and variable water content.

(b) With partial replacement of octahedral Mg by Al and with adsorbed cations and variable water content.

(c) That is, regularly alternating talc-like and brucite-like sheets.

Silicates with framework structures

The structural complexity of the 3D framework aluminosilicates precludes a detailed treatment here, but many of the minerals are of paramount importance. The group includes the feldspars (which are the most abundant of all minerals, and comprise $\sim 60\%$ of the earth's crust), the zeolites (which find major applications as molecular sieves, desiccants, ion exchangers, and water softeners), and the ultramarines which, as their name implies, often have an intense blue colour. All are constructed from SiO_4 units in which each O atom is shared by 2 tetrahedra (as in the various forms of SiO_2 itself), but up to one-half of the Si atoms have been replaced by Al, thus requiring the addition of further cations for charge-balance.

Most feldspars can be classified chemically as members of the ternary system $NaAlSi_3O_8$–$KAlSi_3O_8$–$CaAl_2Si_2O_8$. This is illustrated in Fig. 9.17, which also indicates

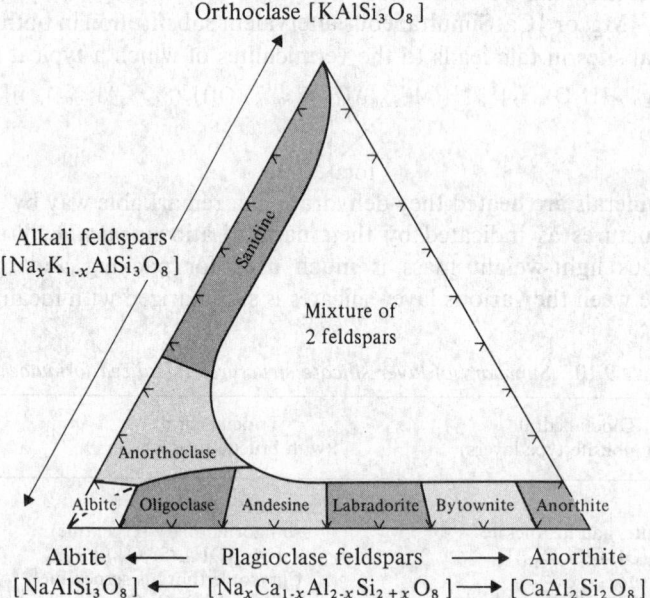

FIG. 9.17   Ternary phase diagram for feldspars. The precise positions of the various phase boundaries depend on the temperature of formation.

the names of the mineral phases. Particularly notable is the continuous plagioclase series in which Na^I (102 pm) is replaced by Ca^{II} (100 pm) on octahedral sites, the charge-balance being maintained by a simultaneous substitution of Al^{III} for Si^{IV} on the tetrahedral sites. K^I (138 pm) is too disparate in size to substitute for Ca^{II} and 2-phase mixtures result, though orthoclase does form a continuous series of solid solutions with the Ba feldspar celsian $[BaAl_2Si_2O_8]$ (Ba^{II} 136 pm). Likewise, most of the alkaline feldspars are not homogeneous but tend to contain separate K-rich and Na-rich phases unless they have crystallized rapidly from solid solutions at high temperatures (above $\sim 660°$). Feldspars have tightly constructed aluminosilicate frameworks that generate large interstices in which the large M^I or M^{II} are accommodated in irregular coordination.[16] Smaller cations, which are common in the chain and sheet silicates (e.g. Li^I, Mg^{II}, Fe^{III}), do not occur as major constituents in feldspars presumably because they are unable to fill the interstices adequately.

Pressure is another important variable in the formation of feldspars and at sufficiently high pressures there is a tendency for Al to increase its coordination number from 4 to 6 with consequent destruction of the feldspar lattice.† For example:

$$NaAl_tSi_3O_8 \xrightarrow{\text{pressure}} NaAl_oSi_2O_6 + SiO_2$$
$$\text{albite (feldspar)} \qquad \text{jadeite (clinopyroxene)} \quad \text{quartz}$$

$$NaAl_tSi_3O_8 + NaAl_tSiO_4 \xrightarrow{\text{pressure}} 2NaAl_oSi_2O_6$$
$$\text{albite} \qquad \text{nepheline} \qquad \text{jadeite}$$

$$3Ca(Al_t)_2Si_2O_8 \xrightarrow{\text{pressure}} Ca_3(Al_o)_2(SiO_4)_3 + 2(Al_o)_2SiO_5 + SiO_2$$
$$\text{anorthite (feldspar)} \qquad \text{grossular (Ca garnet)} \quad \text{kyanite} \quad \text{quartz}$$

$$Ca(Al_t)_2Si_2O_8 + Ca_2(Al_t)_2SiO_7 + 3CaSiO_3 \xrightarrow{\text{pressure}} 2Ca_3(Al_o)_2(SiO_4)_3$$
$$\text{anorthite} \qquad \text{gehlenite} \qquad \text{wollastonite} \qquad \text{grossular}$$

Such reactions marking the disappearance of plagioclase feldspars may be responsible for the Mohorovicic discontinuity between the earth's crust and mantle: this implies that the crust and mantle are isocompositional, the crustal rocks above having phases characteristic of gabbro rock (olivine, pyroxene, plagioclase) whilst the mantle rocks below are an eclogite-containing garnet, Al-rich pyroxene, and quartz. Not all geochemists agree, however.

Zeolites have much more open aluminosilicate frameworks than feldspars and this enables them to take up loosely bound water or other small molecules in their structure. Indeed, the name zeolite was coined by the mineralogist A. F. Cronstedt in 1756 (ζειν *zein, to boil;* λιθος *lithos, stone*) because the mineral appeared to boil when heated in the blow-pipe flame. Zeolite structures are characterized by the presence of tunnels or systems of interconnected cavities; these can be linked either in one direction giving fibrous crystals, or more usually in two or three directions to give lamellar and 3D structures respectively. Figure 9.18a shows the construction of a single cavity from 24 linked $\{SiO_4\}$ tetrahedra and Fig. 9.18b shows how this can be conventionally represented by a truncated cubo-octahedron formed by joining the Si atom positions. Several other types of polyhedron have also been observed.[10, 16] These are then linked in three dimensions to build the aluminosilicate framework. A typical structure is shown in Fig. 9.18c for the synthetic zeolite "Linde A" which has the formula $[Na_{12}(Al_{12}Si_{12}O_{48})].27H_2O.$[20d] Other cavity frameworks are found in other zeolites such as faujasite, which has the idealized formula $[NaCa_{0.5}(Al_2Si_5O_{14})].10H_2O$, and chabazite $[Ca(Al_2Si_4)O_{12}].6H_2O$. There is great current interest in this field since it offers scope for the reproducible synthesis of structures having cavities, tunnels, and pores of precisely defined dimensions on the atomic scale. By appropriate design such molecular sieves can be used to selectively

---

† In some compounds of course, octahedrally coordinated Al is stable at normal atmospheric pressure, e.g. in Al_2O_3, $Al(OH)_3$, and spinels such as $MgAl_2O_4$. Much higher pressures still are required to transform 4-coordinated Si to 6-coordinated (p. 396).

[20d] J. M. THOMAS, L. A. BURSILL, E. A. LODGE, A. K. CHEETHAM, and C. A. FYFE, A reassessment of Zeolite A: Evidence that the structure is rhombohedral with unexpected ordering in the aluminosilicate framework, *JCS Chem. Comm.* 1981, 276–7.

remove water or other small molecules, to separate normal from branched-chain paraffins, to generate highly dispersed metal catalysts, and to promote specific size-dependent chemical reactions.[21-23, 23a] Zeolites are made commercially by crystallizing aqueous gels of mixed alkaline silicates and aluminates at 60–100°. Zeolite-A is being increasingly used as a detergent builder to replace sodium tripolyphosphate (p. 609).

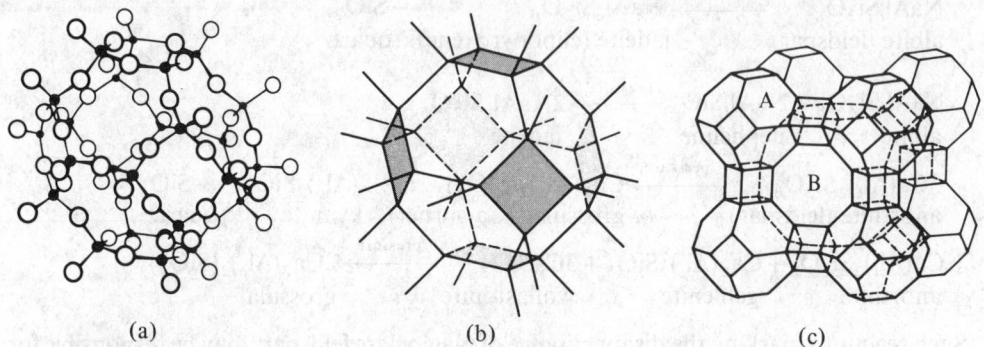

(a)   (b)   (c)

FIG. 9.18 (a) 24{SiO$_4$} tetrahedra linked by corner sharing to form a framework surrounding a truncated cubo-octahedral cavity; (b) conventional representation of the polyhedron in (a); and (c) space-filling arrangement of the polyhedra A which also generates larger cavities B.

The final group of framework aluminosilicates are the ultramarines which have alternate Si and Al atoms at the corners of the polyhedra shown in Fig. 18a and b and, in addition, contain substantial concentrations of anions such as Cl^-, SO_4^{2-}, or S_2^{2-}. These minerals tend to be anhydrous, like the feldspars, and in contrast to the even more open zeolites. Examples are sodalite $[Na_8Cl_2(Al_6Si_6O_{24})]$, noselite $[Na_8(SO_4)(Al_6Si_6O_{24})]$, and ultramarine $[Na_8(S_2)(Al_6Si_6O_{24})]$. Sodalite is colourless if the supernumerary anions are all chloride, but partial replacement by sulfide gives the brilliant blue mineral lapiz lazuli. Further replacement gives ultramarine which is now manufactured synthetically as an important blue pigment for oil-based paints and porcelain, and as a "blueing" agent to mask yellow tints in domestic washing, paper making, starch, etc.† The colour is due to the presence of the sulfur radical anions S_2^- and S_3^- and shifts from green to blue as the ratio S_3^-/S_2^- increases; in ultramarine red the predominant species may be the neutral S_4 molecule though S_3^- and S_2^- are also present.[23b]

† According to H. Remy the artificial production of ultramarine was first suggested by J. W. von Goethe in his *Italian Journey* (1786–8); it was first accomplished by L. Gmelin in 1828 and developed industrially by the Meissen porcelain works in the following year. It can be made by firing kaolin and sulfur with sodium carbonate; various treatments yield greens, reds, and violets, as well as the deep blue, the colours being reminiscent of the highly coloured species obtained in nonaqueous solutions of S, Se, and Te (pp. 785, 896).

[21] D. W. BRECK, *Zeolite Molecular Sieves (Structure, Chemistry, and Uses)*, Wiley, New York, 1974, 771 pp.

[22] F. SCHWOCHOW and L. PUPPE, Zeolites—their synthesis, structure and applications, *Angew. Chem.*, Int. Edn. (Engl.) **14**, 620–8 (1975).

[23] K. SEFF, Structural chemistry inside zeolite A, *Acc. Chem. Res.* **9**, 121–8 (1976).

[23a] R. M. BARRER, *Zeolites and Clay Minerals as Sorbents and Molecular Sieves*, Academic Press, London, 1978, 496 pp.

[23b] R. J. H. CLARK and D. G. COBBOLD, Characterization of sulfur radical anions in solutions of alkali polysulfides in dimethylformamide and hexamethylphosphoramide, and in the solid state in ultramarine blue, green, and red, *Inorg. Chem.* **17**, 3169–74 (1978).

9.3.6 *Other inorganic compounds of silicon*

This section briefly considers compounds in which Si is bonded to elements other than hydrogen, the halogens, or oxygen, especially compounds in which Si is bonded to S or N. Silicon burns in S vapour at $100°$ to give SiS_2 which can be sublimed in a stream of N_2 to give long, white, flexible, asbestos-like fibres, mp $1090°$, sublimation $1250°C$. The structure consists of infinite chains of edge-shared tetrahedra (like W-silica, p. 396) and these transform at high temperature and pressure to a (corner-shared) cristobalite modification. The structural complexity of SiO_2 is not repeated, however. SiS_2 hydrolyses rapidly to SiO_2 and H_2S and is completely ammonolysed by liquid NH_3 to the imide

$$SiS_2 + 4NH_3 \longrightarrow Si(NH)_2 + 2NH_4SH$$

Sulfides of Na, Mg, Al, and Fe convert SiS_2 into metal thiosilicates, and ethanol yields "ethylsilicate" $Si(OEt)_4$ and H_2S.† Volatile thiohalides have been reported from the

reaction of SiX_4 with H_2S at red heat; e.g. $SiCl_4$ yields $S(SiCl_3)_2$, cyclic $Cl_2Si\overset{\displaystyle S}{\underset{\displaystyle S}{\diamond}}SiCl_2$,

and crystalline $(SiSCl_2)_4$. The absence of Si=S bonds is notable and the structure can be compared with those of the oxohalides (p. 393).

The most important nitride of Si is Si_3N_4; this is formed by direct reaction of the elements above $1300°$ or, more economically by heating SiO_2 and coke in a stream of N_2/H_2 at $1500°$. The compound is of considerable interest as an engineering material since it is almost completely inert chemically, and retains its strength, shape, and resistance to corrosion and wear even above $1000°$. Its great hardness (Mohs 9), high dissociation temperature ($1900°$, 1 atm) and high density (3.185 g cm^{-3}) can all be related to its compact structure which resembles that of phenacite (Be_2SiO_4, p. 400). It is an insulator with a resistivity at room temperature $\sim 6.6 \times 10^{10}$ ohm cm. Another refractory, Si_2N_2O, is formed when $Si + SiO_2$ are heated to $1450°$ in a stream of Ar containing 5% N_2. The structure (Fig. 9.19) comprises puckered hexagonal nets of alternating Si and N atoms interlinked by nonlinear Si–O–Si bonds to similar nets on either side; the Si atoms are thus each 4-coordinate and the N atoms 3-coordinate.

Volatile silylamides are readily prepared by reacting a silyl halide with NH_3, RNH_2, or R_2NH in the vapour phase or in Et_2O, e.g.:

$$3SiH_3Cl + 4NH_3 \longrightarrow N(SiH_3)_3 + 3NH_4Cl$$
$$SiH_3Br + 2Me_2NH \longrightarrow SiH_3NMe_2 + Me_2NH_2Br$$
$$4SiH_3I + 5N_2H_4 \longrightarrow (SiH_3)_2NN(SiH_3)_2 + 4N_2H_5I$$

Silicon-substituted derivatives may require the use of lithio or sodio reagents, e.g.:

$$Me_3SiCl + NaN(SiMe_3)_2 \longrightarrow N(SiMe_3) + NaCl$$
$$Ph_3SiLi + R_2NH \longrightarrow Ph_3SiNR_2 + LiH$$

---

† $Si(OEt)_4$ is an important industrial chemical that is made on a large scale (~ 1400 tonnes pa in 1968) by the action of EtOH on $SiCl_4$. Though first made in this way by J. von Ebelman in 1845 it was first correctly characterized by D. I. Mendeleev in 1860. It has mp $-77°$, bp $168.5°$, and d_{20} 0.9346 g cm^{-3}. Uses almost all depend on its controlled hydrolysis to produce silica in an adhesive or film-producing form. It is also a source of metal-free silica for use in phosphors in fluorescent lamps and TV tubes. In partly hydrolysed form it is used as a paint vehicle, a protective coating for porous stone, and as a vehicle for zinc-containing galvanic corrosion-preventing coatings. Many other orthoesters $Si(OR)_4$ are known but none are commercially important.

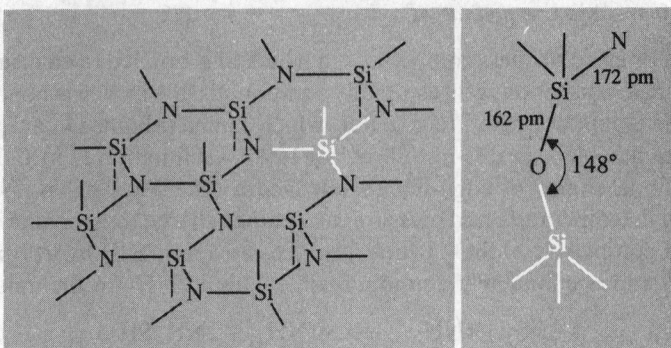

FIG. 9.19 (a) Schematic representation of the puckered hexagonal net of Si and N atoms in
Si_2N_2O; half of the Si atoms bond to the adjacent net above via bent Si–O–Si bonds as shown in
(b) and the other half bond likewise to the net below.

The N atom is always tertiary in these compounds and no species containing the SiH–NH
group is stable at room temperature. Apart from this restriction, innumerable such
compounds have been prepared including cyclic and polymeric analogues, e.g. [cyclo-
$\{Me_2SiN(SiMe_3)_2\}$] and [cyclo-$(Me_2SiNH)_4$]. Interest has focused on the stereochemistry of the N atom which is often planar, or nearly so.[24] Thus $N(SiH_3)_3$ features a planar N
atom and this has been ascribed to p_π–d_π interaction between the "nonbonding" pair of
electrons on N and the "vacant" d_π orbitals on Si as shown schematically in Fig. 9.20.†

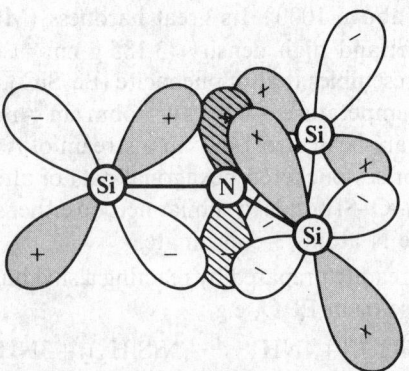

FIG. 9.20 Symmetry relation between p_π orbital on N and d_π orbitals on the 3 Si atoms in planar
$\{NSi_3\}$ compounds such as $N(SiH_3)_3$.

Consistent with this trisilylamines are notably weaker ligands than their tertiary amine
analogues though replacement of one or two SiH₃ by CH₃ enhances the donor power
again; e.g. $N(SiH_3)_3$ forms no adduct with BH_3 even at low temperature; $MeN(SiH_3)_2$
forms a 1:1 adduct with BH_3 at $-80°$ but this decomposes when warmed; $Me_2N(SiH_3)$
gives a similar adduct which decomposes at room temperature into Me_2NBH_2 and SiH_4
(cf. the stability of Me_3NBH_3, p. 188).

† The linear skeleton of H_3SiNCO and H_3SiNCS has also been interpreted in terms of N≡Si p_π–d_π bonding.

24 E. A. V. EBSWORTH, *Volatile Silicon Compounds*, Pergamon Press, Oxford, 1963, 179 pp.

9.3.7 *Organosilicon compounds and silicones*[2]

Tens of thousands of organosilicon compounds have been synthesized and, during the past three decades, silicone oils, elastomers, and resins have become major industrial products. Many organosilicon compounds have considerable thermal stability and chemical inertness; e.g. $SiPh_4$ can be distilled in air at its bp 428°, as can Ph_3SiCl (bp 378°) and Ph_2SiCl_2 (bp 305°). These, and many similar compounds, reflect the considerable strength of the Si–C bond which is, indeed, comparable with that of the C–C bond (p. 389). A further illustration is the compound SiC which closely resembles diamond in its properties (p. 386). However, attempts to replicate the vast domain of organic chemistry were thwarted from the start by the reluctance of Si to catenate except when joined to electronegative substituents such as F and Cl (p. 393). This can be explained in part by the comparative weakness of the Si–Si bond which limits the development of a polysilane chemistry comparable with that of the hydrocarbons. Another characteristic and limiting feature of Si is its reluctance to form stable compounds containing double bonds such as Si=Si, Si=C, Si=N, Si=O, etc. Thus, although the word "silicone" was coined by F. S. Kipping in 1901 to indicate the similarity in formula of Ph_2SiO with that of the ketone benzophenone, Ph_2CO, he stressed that there was no chemical resemblance between them and that Ph_2SiO was polymeric.[25] It is now recognized that the great thermal and chemical stability of the silicones derives from the strength both of the Si–C bonds and of the Si–O–Si linkages. However, despite the warnings of G. Urry (p. 379), a recurring theme in organosilicon chemistry has been the attempt to prepare unsaturated compounds, particularly those with Si=C double bonds. These efforts were finally crowned with success in 1976 when two groups independently isolated and characterized 1,1,2-trimethylsilaethene, Me_2Si=CHMe, in low temperature matrices.[26] Thus, irradiation of trimethylsilyldiazomethane in an argon matrix at 8 K sets up a photo-equilibrium with the isomeric trimethylsilyldiazirine (see scheme); both compounds lose N_2 to give the desired product via a carbene intermediate. Me_2Si=CHMe was identified by several spectroscopic techniques and by its dimerization above 45 K to the *cis*- and *trans*-isomers of the corresponding disilacyclobutane:

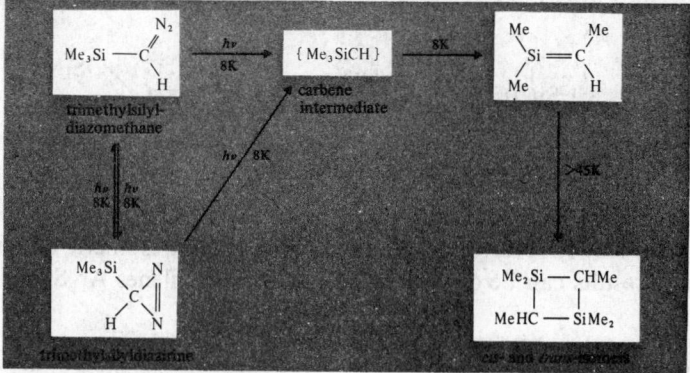

[25] F. S. KIPPING and L. L. LOYD, Organic derivatives of silicon. Triphenylsilicol and alkoxysilicon chlorides, *J. Chem. Soc. (Transactions)* **79**, 449–59 (1901).

[26] O. L. CHAPMAN, C.-C. CHANG, J. KOLE, M. E. JUNG, J. A. LOWE, T. J. BARTON, and M. L. TUMEY, 1,1,2-Trimethylsilaethylene, *J. Am. Chem. Soc.* **98**, 7844–6 (1976). M. R. CHEDEKEL, M. SKOGLUND, R. L. KREEGER, and H. SHECHTER, Solid state chemistry—discrete trimethylsilylmethylene, ibid., 7846–8 (1976). T. J. BARTON and G. T. BURNS, Silabenzene, ibid. **100**, 5246 (1978). C. L. KREIL, O. L. CHAPMAN, G. T. BURNS, and T. J. BARTON, Silatoluene, ibid. **102**, 841–2 (1980).

Transient reaction species containing Si═C bonds have been known since about 1966 and can be generated thermally, photolytically, or even chemically:[27]

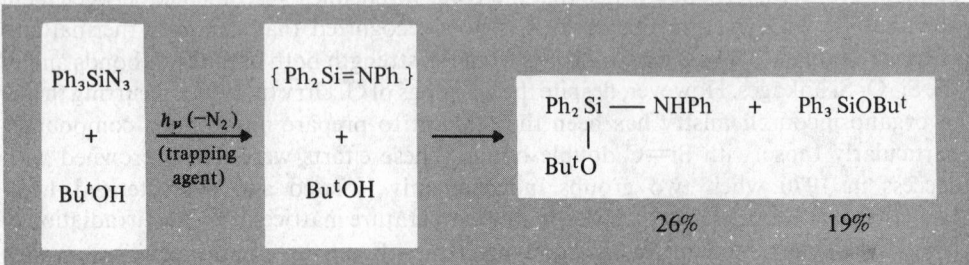

The gaseous product can be detected mass spectrometrically but rapidly dimerizes to a disilacyclobutane or undergoes Diels–Alder or other cycloaddition reactions with appropriate reagents. Similar methods can also be used to generate Si═N compounds as transient reaction intermediates, e.g. by the photolysis of several alkyl- and aryl-silyl azides:[28]

$$Ph_3SiN_3 + Bu^tOH \xrightarrow[\text{(trapping agent)}]{h\nu \ (-N_2)} \{Ph_2Si=NPh\} + Bu^tOH \longrightarrow$$

$$Ph_2Si\!\!-\!\!NHPh \quad + \quad Ph_3SiOBu^t$$
$$\underset{Bu^tO}{|}$$
$$26\% \qquad\qquad 19\%$$

However, despite concerted well-planned attempts over many years, it was not until 1981 that a stable compound containing the Si═C bond was reported.[28a] A. G. Brook and his group prepared 2-adamantyl-2-trimethylsiloxy-1,1-bis(trimethylsilyl)-1-silaethene as pale yellow needles mp 92°:

$$(Me_3Si)_3Si\overset{O}{\overset{\|}{C}}(C_{10}H_{15}) \underset{\text{warm solution}}{\overset{\text{ether}/h\nu}{\rightleftharpoons}} (Me_3Si)_2Si{=}C\overset{OSiMe_3}{\underset{C_{10}H_{15}}{\diagdown}}$$

mp 64° mp 92°

The solid silaethene was stable at room temperature in the absence of air or other reagents but in solution it slowly reverted (over several days) to the isomeric acylsilane starting material. These results can be compared with those in disilenes $Ar_2Si{=}SiAr_2$ (p. 388).

[27] L. E. GUSEL'NIKOV, N. NAMETKIN, and V. M. VDOVIN, Unstable silicon analogs of unsaturated compounds, *Acc. Chem. Res.* **8**, 18–25 (1975). N. WIBERG and G. PREINER, Synthesis and reactions of Me₂Si=C(SiMe₃)₂, *Angew. Chem.*, Int. Edn. (Engl.) **16**, 329–30 (1977).

[28] D. R. PARKER and L. H. SOMMER, Photolysis of silyl azides. Generation and reactions of silicon–nitrogen $(p_\pi{-}p_\pi)$ doubly bonded intermediates {R₂Si = NR}, *J. Am. Chem. Soc.* **98**, 618–20 (1976). See also N. WIBERG and G. PREINER, Preparation and reactions of dimethyl(trimethylsilylimino)silane, Me₂Si = NSiMe₃, *Angew. Chem.*, Int. Edn. (Engl.) **17**, 362–3 (1978).

[28a] A. G. BROOK, F. ABDESAKEN, B. GUTEKUNST, G. GUTEKUNST, and R. K. KALLURY, A solid silaethene: isolation and characterization, *JCS Chem. Comm.* 1981, 191–2.

More frequent success has attended the synthesis of stable homocyclic polysilane derivatives and many compounds have been described.[29] A dramatic advance was the recent one-step high-yield synthesis of a cyclohexasilane from readily available materials:[30]

$$6Me_2SiCl_2 + 12Li \xrightarrow[\substack{6h(90\% \\ yield)}]{thf/0°} [cyclo\text{-}(SiMe_2)_6] + 12LiCl$$
$$mp\ 250°$$

Photolysis of [cyclo-$(SiMe_2)_6$] in a rigid hydrocarbon glass at 77 K or in an argon matrix at 10 K produces the known $(SiMe_2)_5$ and a new bright-yellow species dimethylsilylene $SiMe_2$ which is stable indefinitely under these conditions.[30a]

There are three general methods of forming Si–C bonds. The most convenient laboratory method for small-scale preparations is by the reaction of $SiCl_4$ with organolithium, Grignard, or organoaluminium reagents.† A second attractive route is the hydrosilylation of alkenes, i.e. the catalytic addition of Si–H across C=C double bonds; this is widely applicable except for the crucially important methyl and phenyl silanes. Industrially, organosilanes are made by the direct reaction of RX or ArX with a fluidized bed of Si in the presence of about 10% by weight of metallic Cu as catalyst (cf. the direct preparation of organo compounds of Ge, Sn, and Pb, pp. 459ff.). The method was patented by E. G. Rochow in 1945 and ensured the commercial viability of the now extensive silicone industry.[30b]

$$2MeCl + Si \xrightarrow[\sim 300°]{Cu\ powder} Me_2SiCl_2 \quad (70\%\ yield)$$

By-products are $MeSiCl_3$ (12%) and Me_3SiCl (5%) together with 1–2% each of $SiCl_4$, $SiMe_4$, $MeSiHCl_2$, etc. Relative yields can readily be altered by modifying the reaction conditions or by adding HCl (which increases $MeSiHCl_2$ and drastically reduces Me_2SiCl_2). The overall reaction is exothermic and heat must be removed from the fluidized bed. It has been suggested that the role of Cu is to form {CuMe} and {CuCl} as reaction intermediates:

$$MeCl + 2Cu^0 \longrightarrow \{CuMe\} + \{CuCl\}$$
$$\{CuMe\} \longrightarrow Cu^0 + \{Me^•\}$$
$$\{CuCl\} + Si^0 \longrightarrow Cu^0 + \{SiCl\}$$
$$\{Me^•\} + \{SiCl\} \longrightarrow MeSiCl$$

and possibly also $\quad \{CuMe\} + \{SiCl\} \longrightarrow MeSiCl + Cu^0$

---

† A very recent alternative is the use of silyl anions which can readily be prepared by cleavage of Si–H or Si–Si bonds with KH in dimethoxyethane or similar solvents: e.g. $Et_3SiH \rightarrow Et_3SiK$, $Ph_3SiH \rightarrow Ph_3SiK$, Me_3Si-$SiMe_3 \rightarrow Me_3SiK + Me_3SiH$. Reaction with alkyl halides, benzyl halides, or α-enones then gives the corresponding tetraorganosilane in a 70–75% yield. Likewise Et_3SiK and Me_3SiCl yield Et_3Si-$SiMe_3$ (75%) and Me_3SiK plus Ph_3GeBr yield Me_3Si-$GePh_3$ (80%). See R. J. P. Corriu and C. Guerin, *JCS Chem. Comm.* 1980, 168–9.

[29] A. L. REINGOLD (ed.) *Homoatomic Rings, Chains, and Macromolecules of Main-Group Elements*, Elsevier, Amsterdam, 1977, 615 pp. A Symposium Proceedings containing three chapters on polysilanes.
[30] M. LAGUERRE, J. DUNOGUES, and R. CALAS, One-step synthesis of dodecamethylcyclohexasilane, *JCS Chem. Comm.* 1978, 272.
[30a] T. J. DRAHNAK, J. MICHL, and R. WEST, Dimethylsilylene, $SiMe_2$, *J. Am. Chem. Soc.* **101**, 5427–8 (1979).
[30b] M. G. VORONKOV, V. P. MILESHKEVICH, and YU. A. YUZHELEVSKII, *The Siloxane Bond*, Consultants Bureau, New York, 1978, 493 pp. A massive compendium on the physical and chemical properties of compounds with the Si–O bond; over 3400 references, including patents, up to mid-1974.

Me$^\bullet$ radicals are known not to react directly with Si itself. Because of their very similar bps, careful fractionation is necessary if pure products are required: Me_3SiCl 57.7°, Me_2SiCl_2 69.6°, $MeSiCl_3$ 66.4°. Mixtures of ethylchlorosilanes or phenylchlorosilanes (or their bromo analogues) can be made similarly. All these compounds are mobile, volatile liquids (except Ph_3SiCl, mp 89°, bp 378°).

Innumerable derivatives have been prepared by the standard techniques of organic chemistry.[2] The organosilanes tend to be much more reactive than their carbon analogues, particularly towards hydrolysis, ammonolysis, and alcoholysis. Further condensation to cyclic oligomers or linear polymers generally ensues, e.g.:

$$Ph_2SiCl_2 \xrightarrow{\;H_2O\;} Ph_2Si(OH)_2 \xrightarrow{\;>100°\;} \tfrac{1}{n}(Ph_2SiO)_n + H_2O$$

$$\begin{array}{ll} \text{white crystals} & n = 3\text{(cyclo), 4(cyclo), or } \infty \\ \text{mp} \sim 132° \text{ (d)} & \end{array}$$

$$Me_2SiCl_2 \xrightarrow{\;NH_3/-35°\;} \{Me_2Si(NH_2)_2\} \longrightarrow [\text{cyclo-}(Me_2SiNH)_3] + [\text{cyclo-}(Me_2SiNH)_4]$$
$$\text{not isolated}$$

For both economic and technical reasons, commercial production of such polymers is almost entirely restricted to the methyl derivatives (and to a lesser extent the phenyl derivatives) and hydrolysis of the various methylchlorosilanes has, accordingly, been much studied. Hydrolysis of Me_3SiCl yields trimethylsilanol as a volatile liquid (bp 99°); it is noticeably more acidic than the corresponding Bu^tOH and can be converted to its Na salt by aqueous NaOH (12M). Condensation gives hexamethyldisiloxane which has a very similar bp (100.8°):

$$2Me_3SiCl \xrightarrow[(-2HCl)]{+2H_2O} 2Me_3SiOH \xrightarrow{\;-H_2O\;} [O(SiMe_3)_2]$$

Hydrolysis of Me_2SiCl_2 usually gives high polymers, but under carefully controlled conditions leads to cyclic dimethylsiloxanes $[(Me_2SiO)_n]$ ($n = 3, 4, 5, 6$). Linear siloxanes have also been made by hydrolysing Me_2SiCl_2 in the presence of varying amounts of Me_3SiCl as a "chain-stopping" group, i.e. $[Me_3SiO(Me_2SiO)_xSiMe_3]$ ($x = 0, 1, 2, 3, 4$), etc. Cross-linking is achieved by hydrolysis and condensation in the presence of $MeSiCl_3$ since this generates a third Si–O function in addition to the two required for polymerization:

$$Me_3SiCl \xrightarrow{\;H_2O\;} Me_3Si-O- \quad \text{terminal group}$$

$$Me_2SiCl_2 \xrightarrow{\;H_2O\;} \begin{array}{c} Me \\ | \\ -O-Si-O- \\ | \\ Me \end{array} \quad \text{chain-forming group}$$

$$MeSiCl_3 \xrightarrow{\;H_2O\;} \begin{array}{c} Me \\ | \\ -O-Si-O- \\ | \\ O \\ | \end{array} \quad \text{branching and bridging group}$$

Comparison with the mineral silicates is instructive since there is a 1:1 correspondence between the two sets of compounds, the methyl groups in the silicones being replaced by the formally isoelectronic O^- in the silicates:

Orthosilicate

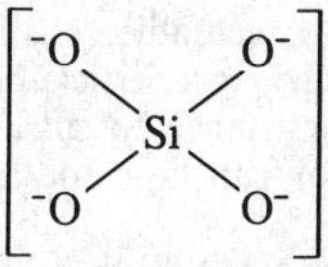

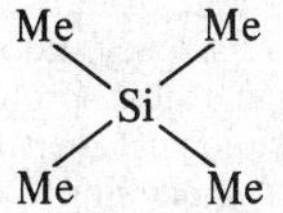

$$\begin{bmatrix} ^-O & & O^- \\ & Si & \\ ^-O & & O^- \end{bmatrix}$$

$$\begin{array}{cc} Me & Me \\ & Si & \\ Me & Me \end{array}$$

Tetramethylsilane

Disilicate

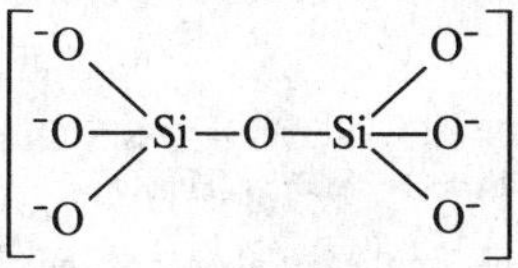

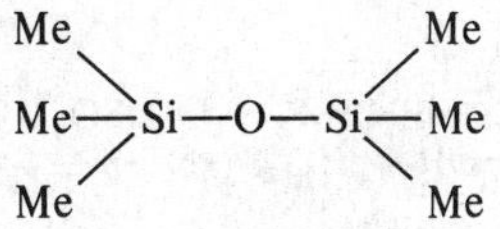

$$\begin{bmatrix} ^-O & & O^- \\ ^-O-Si-O-Si-O^- \\ ^-O & & O^- \end{bmatrix}$$

$$\begin{array}{cc} Me & Me \\ Me-Si-O-Si-Me \\ Me & Me \end{array}$$

Hexamethyl-disiloxane

Pyroxenes

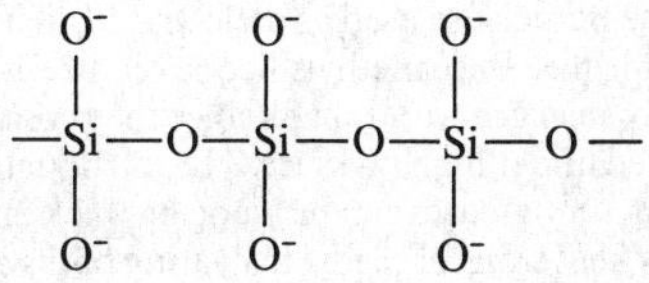

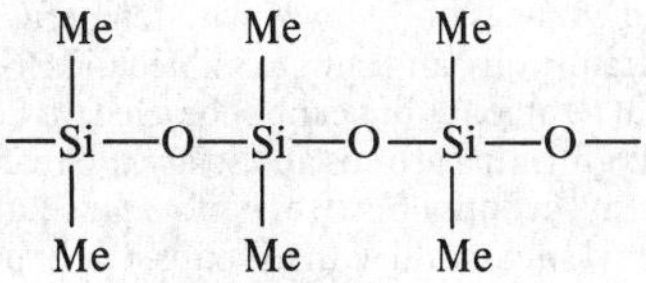

$$\begin{array}{ccc} O^- & O^- & O^- \\ | & | & | \\ -Si-O-Si-O-Si-O- \\ | & | & | \\ O^- & O^- & O^- \end{array}$$

$$\begin{array}{ccc} Me & Me & Me \\ | & | & | \\ -Si-O-Si-O-Si-O- \\ | & | & | \\ Me & Me & Me \end{array}$$

Polydimethyl-siloxane
(silicone oil)

Cyclic metasilicates

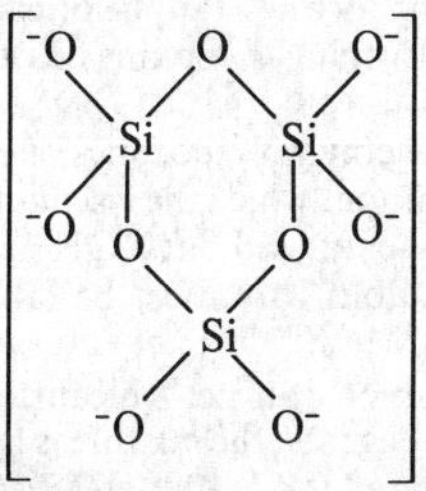

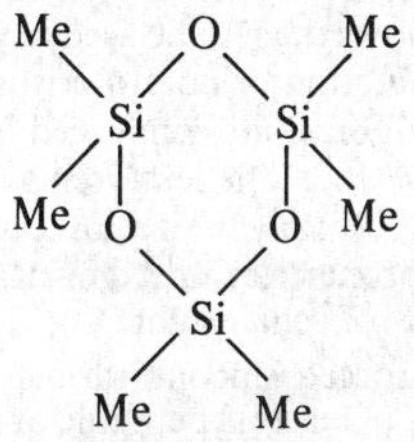

$$\begin{bmatrix} ^-O & & O & & O^- \\ & Si & & Si & \\ ^-O & O & & O & O^- \\ & & Si & & \\ & ^-O & & O^- & \end{bmatrix}$$

$$\begin{array}{ccccc} Me & & O & & Me \\ & Si & & Si & \\ Me & O & & O & Me \\ & & Si & & \\ & Me & & Me & \end{array}$$

Cyclic dimethylsiloxane

Amphiboles

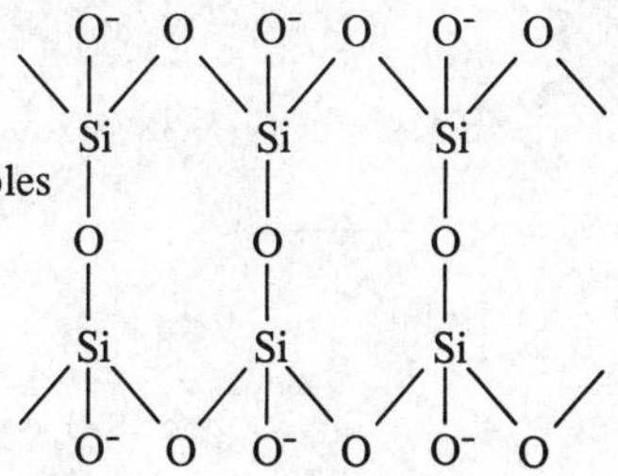

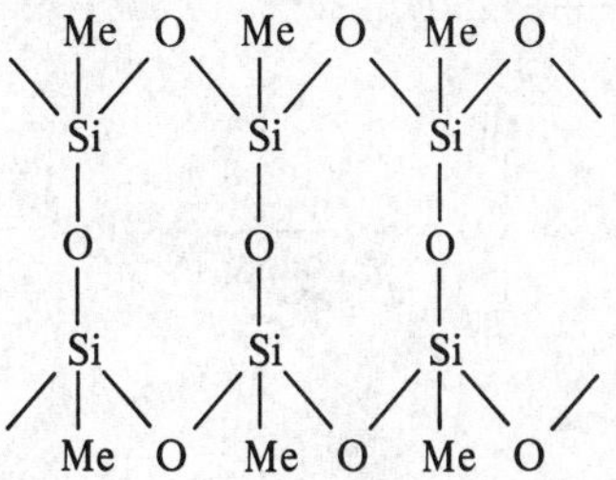

$$\begin{array}{ccc} O^- \; O & O^- \; O & O^- \; O \\ \diagdown | \diagup \diagdown | \diagup \diagdown | \diagdown \\ Si & Si & Si \\ | & | & | \\ O & O & O \\ | & | & | \\ Si & Si & Si \\ \diagup | \diagdown \diagup | \diagdown \diagup | \diagdown \diagup \\ O^- \; O & O^- \; O & O^- \; O \end{array}$$

$$\begin{array}{ccc} Me \; O & Me \; O & Me \; O \\ \diagdown | \diagup \diagdown | \diagup \diagdown | \diagdown \\ Si & Si & Si \\ | & | & | \\ O & O & O \\ | & | & | \\ Si & Si & Si \\ \diagup | \diagdown \diagup | \diagdown \diagup | \diagdown \diagup \\ Me \; O & Me \; O & Me \; O \end{array}$$

Methylsil-sesquioxane ladder polymer

Infinite sheet silicates   Siloxenes

Framework silicates   Silicone resins

This reminds us of the essentially covalent nature of the Si–O–Si linkage, but the analogy should not be taken to imply identity of structures in detail, particularly for the more highly condensed polymers. Some aspects of the technology of silicones are summarized in the concluding Panel.

Silicone Polymers[2, 31]

Silicones have good thermal and oxidative stability, valuable resistance to high and low temperatures, excellent water repellency, good dielectric properties, desirable antistick and antifoam properties, chemical inertness, prolonged resistance to ultraviolet irradiation and weathering, and complete physiological inertness. They can be made as fluids (oils), greases, emulsions, elastomers (rubbers), and resins.

Silicone oils are made by shaking suitable proportions of $[O(SiMe_3)_2]$ and $[cyclo\text{-}(Me_2SiO)_4]$ with a small quantity of 100% H_2SO_4; this randomizes the siloxane links by repeatedly cleaving the Si–O bonds to form HSO_4 esters and then reforming new Si–O bonds by hydrolysing the ester group:

$$\equiv Si-O-Si\equiv + H_2SO_4 \longrightarrow \equiv Si-O-SO_3H + \equiv Si-OH$$
$$\equiv Si-OH + HO_3S-O-Si\equiv \longrightarrow \equiv Si-O-Si\equiv + H_2SO_4$$

The molecular weight of the resulting polymer depends only on the initial proportion of the chain-ending groups (Me_3SiO- and Me_3Si-) and the chain-building groups ($-Me_2SiO-$) from the two components. Viscosity at room temperature is typically in the range 50–300 000 times that of water and it changes only slowly with temperature. These liquids are used as dielectric insulating media, hydraulic oils, and compressible fluids for liquid springs. Pure methylsilicone oils are good lubricants at light loads but cannot be used for heavy-duty steel gears and shafts since they contain no polar film-forming groups and so are too readily exuded under high pressure. The introduction of some phenyl groups improves performance, and satisfactory greases can be made by thickening methyl phenyl silicone oil with Li soaps. Other uses are as heat transfer media in heating baths and as components in car polish, sun-tan lotion, lipstick, and other cosmetic formulations. Their low surface tension leads to their extensive use as antifoams in textile dyeing, fermentation processes, and sewage disposal: about 10^{-2} to $10^{-4}\%$ is sufficient for these applications. Likewise their complete non-toxicity allows them to be used to prevent frothing in cooking oils, the processing of fruit juices, and the production of potato crisps.

Silicone elastomers (rubbers) are reinforced linear dimethylpolysiloxanes of exceedingly high molecular weight (5×10^5–10^7). The reinforcing agent, without which the viscous gum is useless, is usually fumed silica (p. 397). Polymerization can be acid-catalysed but KOH produces a rubber with superior physical properties; in either case scrupulous care must be taken to avoid the presence of precursors of chain-blocking groups $[Me_3Si-O-]$ or cross-linking groups $[MeSi(-O-)_3]$. The reinforced silicone rubber composition can be "vulcanized" by oxidative cross-linking using 1–3% of benzoyl peroxide or similar reagents; the mixture is heated to 150° for 10 min at the time of pressing or moulding and then cured for 1–10 h at 250°:

$$ROOR \xrightarrow{heat} 2RO^\bullet$$

$$2RO^\bullet + 2 \underset{CH_3}{\overset{Me}{\sim\!\!\sim\!\!\sim SiO \sim\!\!\sim\!\!\sim}} \longrightarrow 2ROH + 2 \underset{CH_2^\bullet}{\overset{Me}{\sim\!\!\sim\!\!\sim SiO \sim\!\!\sim\!\!\sim}} \longrightarrow \begin{array}{c} Me \\ | \\ \sim\!\!\sim SiO \sim\!\!\sim \\ | \\ CH_2 \\ | \\ CH_2 \\ | \\ \sim\!\!\sim SiO \sim\!\!\sim \\ | \\ Me \end{array}$$

Alternatively, and more elegantly, the process can be achieved at room temperature or slightly above by incorporating a small controlled concentration of Si–H groups which can be catalytically added across pre-introduced $Si-CH=CH_2$ groups in adjacent chains. Again, the

[31] C. A. PEARCE, *Silicon Chemistry and Applications*, Chemical Society Monograph for Teachers, No. 20, 1972, 74 pp.

cross-linking of 1-component silicone rubbers containing acetoxy groups can be readily effected at room temperature by exposure to moisture:

$$
\text{\textasciitilde\textasciitilde\textasciitilde O}-\underset{\underset{\text{(OCOMe)}_2}{|}}{\overset{\overset{\text{Me}}{|}}{\text{Si}}} \xrightarrow{\text{H}_2\text{O}} \text{\textasciitilde\textasciitilde\textasciitilde O}-\underset{\underset{\text{OCOMe}}{|}}{\overset{\overset{\text{Me}}{|}}{\text{Si}}}-\text{OH} \ + \ \text{MeCO}_2\text{H}
$$

$$
\text{\textasciitilde\textasciitilde O}-\underset{\underset{\text{O}}{|}}{\overset{\overset{\text{Me}}{|}}{\text{Si}}}-\text{O}\text{\textasciitilde\textasciitilde}
$$

$$
\text{\textasciitilde\textasciitilde O}-\underset{\underset{\text{O}}{|}}{\overset{\overset{\text{Me}}{|}}{\text{Si}}}-\text{O}-\underset{}{\text{Si}}-\text{O}\text{\textasciitilde\textasciitilde} \ + \ \text{MeCO}_2\text{H}
$$

$$
\text{\textasciitilde\textasciitilde O}-\underset{\underset{\text{Me}}{|}}{\text{Si}}-\text{O}\text{\textasciitilde\textasciitilde}
$$

Such rubbers generally have 1 cross-link for every 100–1000 Si atoms and are unmatched by any other synthetic or natural rubbers in retaining their inertness, flexibility, elasticity, and strength up to 250° and down to −100°. They find use in cable-insulation sleeving, static and rotary seals, gaskets, belting, rollers, diaphragms, industrial sealants and adhesives, electrical tape insulation, plug-and-socket connectors, oxygen masks, medical tubing, space suits, fabrication of heart-valve implants, etc. They are also much used for making accurate moulds and to give rapid, accurate, and flexible impressions for dentures and inlays.

Silicone resins are prepared by hydrolysing phenyl substituted dichloro- and trichloro-silanes in toluene. The Ph groups increase the heat stability, flexibility, and processability of the resins. The hydrolysed mixture is washed with water to remove HCl and then partly polymerized or "bodied" to a carefully controlled stage at which the resin is still soluble. It is in this form that the resins are normally applied, after which the final cross-linking to a 3D siloxane network is effected by heating to 200° in the presence of a heavy metal or quaternary ammonium catalyst to condense the silanol groups, e.g.:

$$
3\text{PhSiCl}_3 + \text{PhSiMeCl}_2 \xrightarrow{\text{H}_2\text{O/toluene}} \left[\text{HO}-\underset{\underset{\text{OH}}{|}}{\overset{\overset{\text{Ph}}{|}}{\text{Si}}}-\text{O}-\underset{\underset{\text{Me}}{|}}{\overset{\overset{\text{Ph}}{|}}{\text{Si}}}-\text{O}-\underset{\underset{\text{OH}}{|}}{\overset{\overset{\text{Ph}}{|}}{\text{Si}}}-\text{O}-\underset{\underset{\text{OH}}{|}}{\overset{\overset{\text{Ph}}{|}}{\text{Si}}}-\text{OH} \right]
$$

a typical intermediate species

$\xrightarrow{\ 200° \ / \text{catalyst}}$

[cross-linked resin] + $n\text{H}_2\text{O}$

Silicone resins are used in the insulation of electrical equipment and machinery, and in electronics as laminates for printed circuit boards; they are also used for the encapsulation of components such as resistors and integrated circuits by means of transfer moulding. Non-electrical uses include high-temperature paints and the resinous release coatings familiar on domestic cooking ware and industrial tyre moulds. When one recalls the very small quantities of silicones needed in many of these individual applications, the global production figures are particularly impressive: they have grown from a few tonnes in the mid-1940s to over 100 000 tonnes on 1969 and an estimated production of 280 000 tonnes in 1978. About half of this is in the USA, distributed so that some 62% is as fluid silicones, 25% as elastomers, and 13% as resins.

10

Germanium, Tin, and Lead

10.1 Introduction

Germanium was predicted as the missing element of a triad between silicon and tin by J. A. R. Newlands in 1864, and in 1871 D. I. Mendeleev specified the properties that "eka-silicon" would have (p. 34). The new element was discovered by C. A. Winkler in 1886 during the analysis of a new and rare mineral argyrodite, Ag_8GeS_6;[1] he named it in honour of his country, Germany.† By contrast, tin and lead are two of the oldest metals known to man and both are mentioned in early books of the Old Testament. The chemical symbols for the elements come from their Latin names *stannum* and *plumbum*. Lead was used in ancient Egypt for glazing pottery (7000–5000 BC); the Hanging Gardens of Babylon were floored with sheet lead to retain moisture, and the Romans used lead extensively for water-pipes and plumbing; they extracted some 6–8 million tonnes in four centuries with a peak annual production of 60 000 tonnes. Production of tin, though equally influential, has been on a more modest scale and dates back to 3500–3200 BC. Bronze weapons and tools containing 10–15% Sn alloyed with Cu have been found at Ur, and Pliny described solder as an alloy of Sn and Pb in AD 79.

Germanium and Sn are non-toxic (like C and Si). Lead is now recognized as a heavy-metal poison;[2, 2a, 2b] it acts by complexing with oxo-groups in enzymes and affects virtually all steps in the process of haeme synthesis and porphyrin metabolism. It also inhibits acetylcholine-esterase, acid phosphatase, ATPase, carbonic anhydrase, etc., and inhibits

---

† The astonishing correspondence between the predicted and observed properties of Ge (p. 34) has tempted later writers to overlook the fact that Winkler thought he had discovered a metalloid like As and Sb and he originally identified Ge with Mendeleev's (incorrectly) predicted "eka-stibium" between Sb and Bi; Mendeleev himself thought it was "eka-cadmium", which he had (again incorrectly) predicted as a missing element between Cd and Hg. H. T. von Richter thought it was "eka-silicon"; so did Lothar Meyer, and they proved to be correct. This illustrates the great difficulties encountered by chemists working only 90 y ago, yet three decades before the rationale which stemmed from the work of Moseley and Bohr.

[1] M. E. WEEKS, *Discovery of the Elements*, 6th edn., Journal of Chemical Education Publ. 1956, 910 pp. Germanium, pp. 683–93; Tin and lead, pp. 41–47.

[2] J. J. CHISHOLM, Lead poisoning, *Scientific American* **224**, 15–23 (1971). Reprinted as Chap. 36 in *Chemistry in the Environment*, Readings from *Scientific American*, pp. 335–43, W. H. Freeman, San Francisco, 1973.

[2a] D. TURNER, Lead in petrol. Part 2. Environmental health, *Chem. Br.*, **16**, 312–14 (1980); see also Part 1, preceding paper, pp. 308–10.

[2b] R. M. HARRISON and D. P. H. LAXEN, *Lead Pollution*, Chapman and Hall, London, 1981, 175 pp.

protein synthesis probably by modifying transfer-RNA. In addition to O complexation (in which it resembles Tl^I, Ba^{II}, and Ln^{III}), Pb^{II} also inhibits SH enzymes (though less strongly than Cd^{II} and Hg^{II}), especially by interaction with cysteine residues in proteins. Typical symptoms of lead poisoning are cholic, anaemia, headaches, convulsions, chronic nephritis of the kidneys, brain damage, and central nervous-system disorders. Treatment is by complexing and sequestering the Pb using a strong chelating agent such as EDTA $\{-CH_2N(CH_2CO_2H)_2\}_2$ or BAL, i.e. British anti-Lewisite, $HSCH_2CH(SH)CH_2OH$.

10.2 The Elements

10.2.1 *Terrestrial abundance and distribution*

Germanium and Sn appear about half-way down the list of elements in order of abundance together with several other elements in the region of 1–2 ppm:

Element	Br	U	**Sn**	Eu	Be	As	Ta	**Ge**	Ho	Mo	W	Tb
PPM	2.5	2.3	**2.1**	2.1	2	1.8	1.7	**1.5**	1.4	1.2	1.2	1.2
Order	46	47	**48**	=48	50	51	52	**53**	54	55	=55	=55

Germanium minerals are extremely rare but the element is widely distributed in trace amounts (like its neighbour Ga). Recovery has been achieved from coal ash but is now normally from the flue dusts of smelters processing Zn ores.

Tin occurs mainly as cassiterite, SnO_2, and this has been the only important source of the element from earliest times. Julius Caesar recorded the presence of tin in Britain, and Cornwall remained the predominant supplier for European needs until the present century (apart from a minor flourish from Bohemia between 1400 and 1550). Today (1977) the major producers are Malaysia (25%), the USSR (14%), Bolivia (14%), Indonesia (10%), Thailand (10%), and China (9%), and annual world production is about 210 000 tonnes. The major consumer is the USA but virtually all of this has to be imported since less than 200 tonnes are mined domestically.

Lead (13 ppm) is by far the most abundant of the heavy elements, being approached amongst these only by thallium (8.1 ppm) and uranium (2.3 ppm). This abundance is related to the fact that 3 of the 4 naturally occurring isotopes of lead (206, 207, and 208) arise primarily as the stable end products of the natural radioactive series. Only ^{204}Pb (1.4%) is non-radiogenic in origin. The variation in isotopic composition of Pb with its origin also accounts for the variability of atomic weight and the limited precision with which it can be quoted (p. 22). The most important Pb ore is the heavy black mineral galena, PbS. Other ore minerals are anglesite ($PbSO_4$), cerussite ($PbCO_3$), pyromorphite ($Pb_5(PO_4)_3Cl$), and mimetesite ($Pb_5(AsO_4)_3Cl$). Some 25 other minerals are known[3] but are not economically important; all contain Pb^{II} in contrast to tin minerals which are invariably Sn^{IV} compounds. Lead ores are widely distributed and commercial deposits are

[3] D. GRENINGER, V. KOLLONITSCH, C. H. KLINE, L. C. WILLEMSENS, and J. F. COLE, *Lead Chemicals*, International Lead Zinc Research Organization, Inc., New York, 1976, 343 pp. This is the most up to date and comprehensive compilation of data on lead and its compounds. See also J. O. NRIAGU (ed.), *The Biogeochemistry of Lead in the Environment*. Part *A. Ecological Cycles*, Elsevier/North-Holland, Amsterdam, 1978, 422 pp.

worked in over fifty countries. World production in 1977 approached 3.5 million tonnes of contained Pb, over half of which came from the four main producers: the USA (16%), the USSR (15%), Australia (13%), and Canada (9%).

10.2.2 *Production and uses of the elements*

Recovery of Ge from flue dusts is complicated, not only because of the small concentration of Ge but also because its amphoteric properties are similar to those of Zn from which it is being separated.[4] Leaching with H_2SO_4, followed by addition of aqueous NaOH, results in the coprecipitation of the 2 elements at pH ~ 5 and enrichment of Ge from ~ 2 to 10%.† The concentrate is heated with HCl/Cl_2 to drive off $GeCl_4$, bp 83.1° (cf. $ZnCl_2$, bp 756°). After further fractionation of $GeCl_4$, hydrolysis affords purified GeO_2, which can be slowly reduced to the element by H_2 at $\sim 530°$. Final purification for semiconductor-grade Ge is effected by zone refining. World production of Ge in 1977 was 90 000 kg (90 tonnes). The largest use is in transistor technology and, indeed, transistor action was first discovered in this element (p. 383). This use is now diminishing somewhat whilst that in optics is growing—Ge is transparent in the infrared and is used in infrared windows, prisms, and lenses. Magnesium germanate is a useful phosphor, and other small-scale applications are in special alloys, strain gauges, and superconductors. Despite its spectacular increase in availability during the past 30 y from a laboratory rarity to a general article of commerce Ge and its compounds are still relatively expensive. Zone-refined Ge was quoted at $316 per kg in 1977 and GeO_2 at $177 per kg.

The ready reduction of SnO_2 by glowing coals accounts for the knowledge of Sn and its alloys in the ancient world. Modern technology uses a reverberatory furnace at 1200–1300°.[5] The main chemical problem in reducing SnO_2 comes from the presence of Fe in the ores which leads to a hard product with unacceptable properties. Reference to Ellingham-type diagrams of the sort shown on p. 327 shows that $-\Delta G(SnO_2)$ is very close to that for FeO/Fe_3O_4 and only about 80 kJ mol^{-1} above the line for reducing FeO to Fe at 1000–1200°. It is therefore essential to reduce cassiterite/iron oxide ores at an oxygen pressure sufficiently high to prevent extensive reduction to Fe. This is achieved in a two-stage process, the impure molten Sn from the initial carbon reduction being stirred vigorously in contact with atmospheric O_2 to oxidize the iron—a process that can be effected by "poling" with long billets of green wood—or, alternatively, by use of steam or compressed air. The price of tin is regulated by the International Tin Council and the imbalance between supply and demand resulted in a sharp increase in price from less than $4000 per tonne in 1972 to more than $11 000 per tonne in 1978 (i.e. $11 per kg) since when it has been more stable. The many uses of metallic tin and its alloys are summarized in the Panel.

Lead is normally obtained from PbS. This is first concentrated from low-grade ores by

---

† GeO_2 begins to precipitate at pH 2.4, is 90% precipitated at pH 3, and 98% precipitated at pH 5. $Zn(OH)_2$ begins to precipitate at pH 4 and is completely precipitated at pH 5.5.

[4] *Kirk–Othmer Encyclopedia of Chemical Technology* 3rd edn. **11**, 791–802 (1980), Germanium and germanium compounds.
[5] *Kirk–Othmer Encyclopedia of Chemical Technology* 2nd edn. **20**, 273–304 (1969), Tin and tin alloys; 304–27, Tin compounds.

Uses of Metallic Tin and Its Alloys

Because of its low strength and high cost, Sn is seldom used by itself but its use as a coating, and as alloys, is familiar in a variety of domestic and technological applications. Tin-plate accounts for almost 40% of tin used—it provides a non-toxic corrosion-resistant cover for sheet steel and can be applied either by hot dipping in molten Sn or more elegantly and controllably by electrolytic tinning. The layer is typically 0.4–25 μm thick. In addition to extensive use in food packaging, tin-plate is used increasingly for distributing beer and other drinks. In the USA alone 27 000 million of the 40 000 million drink cans sold annually are tin-plate, the rest being Al: this is a staggering per capita consumption of 190 pa. The UK in 1977 claimed a more modest 3000 million tin cans for drinks annually—55 per head of population—plus a further 6000 million food cans.

The main alloys of tin together with an indication of the percentage of total Sn production for these alloys in the USA are:

Solder (24%)	(Sn/Pb) typically containing 33% Sn by weight but varying between 2–63% depending on use; sometimes Cd, Ga, In, or Bi are added for increased fusibility.
Bronze (15%)	(Cu/Sn) typically 5–10% Sn often with added P or Zn to aid casting and impart superior elasticity and strain resistance. Gun metal is ~85% Cu, 5% Sn, 5% Zn, and 5% Pb. Coinage metal and brass also often contain small amounts of Sn. World production of bronzes approaches 500 000 tonnes pa.
Babbitt (5%)	(heavy duty bearing metal introduced by I. Babbitt in 1839). The two main compositions are 80–90% Sn, 0–5% Pb, 5% Cu; and 75% Pb, 12% Sn, 13% Sb, 0–1% Cu. They have the characteristics of a hard compound embedded in a soft matrix and are used mainly in railway wagons, diesel locomotives, etc.
Pewter (3%)	(90–95% Sn, 1–8% Sb, 0.5–3% Cu); a decorative and servicable alloy that can be cast, bent, spun, or formed into any shape; it is much used for coffee and tea services, trays, plates, jugs, tankards, candelabra, bowls, and trophies. A related alloy of 90–95% Sn with Pb and other elements is highly prized and much used for organ pipes because of its tonal qualities, e.g. the Royal Albert Hall organ in London has 10 000 pipes containing some 150 tonnes Sn.
Type Metals (0.5%)	(Pb/Sb/Sn). These lead-based alloys contain Sb and Sn in approximately a 2:1 ratio by weight; they increase in hardness and resistance to wear with increasing Sb and Sn content, e.g. linotype (5% Sn, Brinell Hardness 21), stereotype (7% Sn, BH 23), monotype (8% Sn, BH 25), foundry type (12% Sn, BH 32).

Other specialized uses of Sn and its alloys are as the molten-metal bath in the manufacture of float glass and as the alloy Nb_3Sn in superconducting magnets. The many industrial and domestic uses of tin compounds are discussed in later sections; these compounds account for about 8% of the tin produced.

froth flotation then roasted in a limited supply of air to give PbO which is then mixed with coke and a flux such as limestone and reduced in a blast furnace:[6]

$$PbS + 1.5O_2 \longrightarrow PbO + SO_2$$

$$PbO + C \longrightarrow Pb(liq) + CO; \qquad PbO + CO \longrightarrow Pb(liq) + CO_2$$

Alternatively, the carbon reduction can be replaced by reduction of the roasted ore with fresh galena:

$$PbS + 2PbO \longrightarrow 3Pb(liq) + SO_2(g)$$

In either case the Pb contains numerous undesirable metal impurities, notably Cu, Ag, Au, Zn, Sn, As, and Sb, some of which are clearly valuable in themselves. Copper is first

---

[6] *Kirk–Othmer Encyclopedia of Chemical Technology* 3rd edn. **14**, 98–139 (1981), Lead; 140–160, Lead alloys; 160–200, Lead compounds.

removed by liquation: the Pb bullion is melted and held just above its freezing point when Cu rises to the surface as an insoluble solid which is skimmed off. Tin, As, and Sb are next removed by preferential oxidation in a reverberatory furnace and skimming off the oxides; alternatively, the molten bullion is churned with an oxidizing flux of molten $NaOH/NaNO_3$ (Harris process). The softened Pb may still contain Ag, Au, and perhaps Bi. Removal of the first two depends on their preferential solubility in Zn: the mixed metals are cooled slowly from 480° to below 420° when the Zn (now containing nearly all the Ag and Au) solidifies as a crust which is skimmed off; the excess of dissolved Zn is then removed either by oxidation in a reverberatory furnace, or by preferential reaction with gaseous Cl_2, or by vacuum distillation. Final purification (which also removes any Bi) is by electrolysis using massive cast Pb anodes and an electrolyte of acid $PbSiF_6$ or a sulfamate;[6a] this yields a cathode deposit of 99.99% Pb which can be further purified by zone refining to <1 ppm impurity if required. Cost, for comparison with Ge and Sn, was ~$475 per tonne ($0.47 per kg) in 1974. Almost half the Pb produced is used for storage batteries and the remainder is used in a variety of alloys and chemicals as summarized in the Panel.

10.2.3 *Properties of the elements*

The atomic properties of Ge, Sn, and Pb are compared with those of C and Si in Table 10.1. Trends noted in previous groups are again apparent. The pairwise similarity in the ionization energies of Si and Ge (which can be related to the filling of the $3d^{10}$ shell) and of Sn and Pb (which is likewise related to the filling of the $4f^{14}$ shell) are notable (Fig. 10.1). Tin has more stable isotopes than any other element (why?) and one of these, ^{119}Sn (nuclear spin $\frac{1}{2}$), is particularly valuable both for nmr experiments and for Mössbauer spectroscopy.[7]

TABLE 10.1 *Atomic properties of Group IV elements*

Property		C	Si	Ge	Sn	Pb
Atomic number		6	14	32	50	82
Electronic structure		$[He]2s^22p^2$	$[Ne]3s^23p^2$	$[Ar]3d^{10}$ $4s^24p^2$	$[Kr]4d^{10}$ $5s^25p^2$	$[Xe]4f^{14}5d^{10}$ $6s^26p^2$
Number of naturally occurring isotopes		2+1	3	5	10	4
Atomic weight		12.011	28.0855($\pm$3)	72.59($\pm$3)	118.69($\pm$3)	207.2
Ionization energy/kJ mol^{-1}	I	1086.1	786.3	761.2	708.4	715.4
	II	2351.9	1576.5	1537.0	1411.4	1450.0
	III	4618.8	3228.3	3301.2	2942.2	3080.7
	IV	6221.0	4354.4	4409.4	3929.3	4082.3
r^{IV} (covalent)/pm		77.2	117.6	122.3	140.5	146
r^{IV} ("ionic", 6-coordinate)/pm		(15) (CN 4)	40	53	69	78
r^{II} ("ionic", 6-coordinate)/pm		—	—	73	118	119
Pauling electronegativity		2.5	1.8	1.8	1.8	1.9

[6a] A. T. KUHN (ed.), *The Electrochemistry of Lead*, Academic Press, London, 1977, 467 pp.
[7] N. N. GREENWOOD and T. C. GIBB, *Mössbauer Spectroscopy*, Chapman & Hall, London, 1971, 659 pp. T. C. GIBB, *Principles of Mössbauer Spectroscopy*, Chapman & Hall, London, 1976, 254 pp.

Uses of Lead Alloys and Chemicals

Although much lead is used as an inert material in cast, rolled, or extruded form, a far greater tonnage is consumed as alloys. Its major application is in storage batteries where an alloy of 91% Pb, 9% Sb forms the supporting grid for the oxidizing agent (PbO_2) and the reducing agent (spongy Pb).[6a, 6b] Over 80% of this Pb is recovered and recycled. In addition, its use (with Sn) in solders, fusible alloys, bearing metals (babbitt), and type metals has been summarized on p. 430. Other mechanical as distinct from chemical applications are in ammunition, lead shot, lead weights, and ballast.

The pattern of chemical usage of Pb compounds in a particular country depends very much on whether organolead compounds are allowed as antiknock additives in petrol for cars (gasoline for automobiles) (Table A.) In enlightened countries such as Germany, the USSR, and Japan such additives are considered to be wasteful, dangerous, and unnecessary. There are encouraging signs that public opinion and environmental legislation† are gradually achieving the elimination of $PbEt_4$ and $PbMe_4$ as antiknocks in the USA, the UK, and elsewhere,[2a, 2b] but at present these chemicals account for one-fifth of the total lead used and more than 99.7% of Pb chemicals, excluding pigments (e.g. in the USA in 1974 227 250 tonnes of organolead was burned as antiknock whereas the total consumption of other (non-pigment) Pb chemicals involved only 642 tonnes of contained Pb. The presence of Pb additives in petrol also interferes with the catalytic converters that are being developed to reduce or eliminate CO, NO_x, and hydrocarbons from exhaust fumes, and this will likewise encourage the change to other antiknocks.

TABLE A   *Consumption of lead in four countries in 1974*

Application	USA 10^3 tonne (%)	Japan 10^3 tonne (%)	Fed. Repub. Germany 10^3 tonne (%)	UK 10^3 tonne (%)
Batteries	669 (60.2)	198 (64.3)	118 (38.8)	80 (26.6)
Cable sheathing	35 (3.2)	21 (6.8)	52 (17.1)	45 (15.0)
Sheet, pipe, foil, tubes	34 (3.1)	28 (9.1)	54 (17.0)	50 (16.7)
Solders, etc.	49 (4.4)	24 (7.8)	— (0)	32 (10.7)
$PbEt_4$	227 (20.5)	— (0)	— (0)	56 (18.7)
Pigments and chemicals	96 (8.6)	37 (12.0)	80 (26.3)	37 (12.3)
Total	1110 (100.0)	308 (100.0)	304 (100.0)	300 (100.0)

Lead pigments are widely used as rust-inhibiting priming paints for iron and steel. Red lead (Pb_3O_4) is the traditional primer but Ca_2PbO_4 is finding increasing use, particularly for galvanized steel. Lead chromate, $PbCrO_4$, is a strong yellow pigment extensively used in yellow paints for road markings and as an ingredient (with iron blues) in many green paints and coloured plastics. Other pigments include $PbMoO_4$ (red-orange), litharge PbO (canary yellow), and white lead, $\sim 2PbCO_3 . Pb(OH)_2$. Lead compounds are also used for ceramic glazes, e.g. $PbSi_2O_5$ (colourless), in crown glass manufacture, and as polyvinylchloride plastic stabilizers, e.g. "tribasic lead sulfate", $3PbO.PbSO_4.H_2O$. See also p. 449.

† For example the USA Environmental Protection Agency reduced the permitted amount of Pb in gasoline from a typical value of 3 g/US gal to 1.4 g/gal in 1976 and then successively to 0.8 and 0.5 g/gal in 1978 and 1979 (1 US gal = 0.8327 imperial gal = 3.785 l). (See *Chem. Eng. News* 6 Feb. 1978, pp. 12–16.)

6b H. BODE, *Lead–Acid Batteries*, Wiley, New York, 1977, 408 pp.

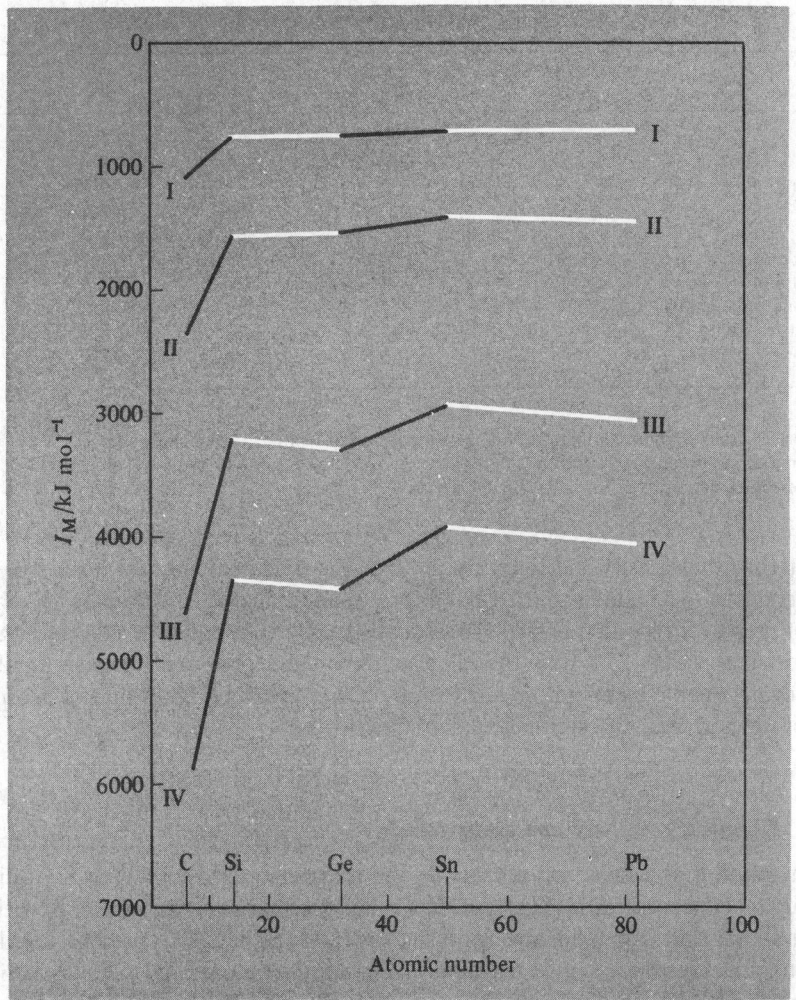

FIG. 10.1 Successive ionization energies for Group IVB elements showing the influence of the $3d^{10}$ shell between Si and Ge and the $4f^{14}$ shell between Sn and Pb.

Some physical properties of the elements are compared in Table 10.2. Germanium forms brittle, grey-white lustrous crystals with the diamond structure; it is a metalloid with a similar electrical resistivity to Si at room temperature but with a substantially smaller band gap. Its mp, bp, and associated enthalpy changes are also lower than for Si and this trend continues for Sn and Pb which are both very soft, low-melting metals.

Tin has two allotropes: at room temperature the stable modification is white, tetragonal, β-Sn, but at low temperatures this transforms into grey α-Sn which has the cubic diamond structure. The transition temperature is 13.2° but the transformation usually requires prolonged exposure at temperatures well below this. The reverse transition from $\alpha \rightarrow \beta$ involves a structural distortion along the c-axis and is remarkable for

TABLE 10.2　*Some physical properties of Group IV elements*

Property	C	Si	Ge	Sn	Pb
MP/°C	4100	1420	945	232	327
BP/°C	—	~3280	2850	2623	1751
Density (20°C)/g cm^{-3}	3.514	2.336 (β 2.905)[a]	5.323 (β 6.71)[a]	α 5.769 β 7.265[b]	11.342
a_0/pm	356.68[c]	541.99[c]	565.76[c]	α 648.9[b, c]	494.9[d]
ΔH_{fus}/kJ mol^{-1}	—	50.6	36.8	7.07	4.81
ΔH_{vap}/kJ mol^{-1}	—	383	328	296	178
ΔH_f (monatomic gas)/kJ mol^{-1}	716.7	454	283	300.7	195.0
Electrical resistivity (20°)/ohm cm	10^{14} 10^{16}	~48	~47	β 11 × 10^{-6}	20 × 10^{-6}
Band gap E_g/kJ mol^{-1}	~580	106.8	64.2	α 7.7, β 0	0

[a] See text.
[b] β-form (stable at room temperature) is tetragonal a_0 583.1 pm, c_0 318.1 pm.
[c] Diamond structure.
[d] Face-centred cubic.

the fact that the density increases by 26% in the high-temperature form.† A similar transformation to a metallic, tetragonal β-form can be effected in Si and Ge by subjecting them to pressures of ~200 and ~120 kbar respectively along the *c*-axis, and again the density increases by ~25% from the value at atmospheric pressure.[9] Lead is familiar as a blue-grey, malleable metal with a fairly high density (nearly 5 times that of Si and twice those of Ge and Sn, but only half that of Os).

10.2.4　*Chemical reactivity and group trends*

Germanium is somewhat more reactive and more electropositive than Si: it dissolves slowly in hot concentrated H_2SO_4 and HNO_3 but does not react with water or with dilute acids or alkalis unless an oxidizing agent such as H_2O_2 or NaOCl is present; fused alkalis react with incandescence to give germanates. Germanium is oxidized to GeO_2 in air at red heat, and both H_2S and gaseous S yield GeS_2; Cl_2 and Br_2 yield GeX_4 on moderate heating and HCl gives both $GeCl_4$ and $GeHCl_3$. Alkyl halides react with heated Ge (as with Si) to give the corresponding organogermanium halides.

† This arises because, although the Sn–Sn distances increase in the $\alpha \rightarrow \beta$ transition, the CN increases from 4 to 6 and the distortion also permits a closer approach of the 12 next-nearest neighbours:[8]

Modification	Bond angles	Nearest neighbours	Next nearest neighbours
α (grey, diamond)	6 at 109.5°	4 at 280 pm	12 at 459 pm
β (white, tetragonal)	4 at 94° 　　2 at 149.5°	4 at 302 pm 　　2 at 318 pm	4 at 377 pm 　　8 at 441 pm

[8] A. F. WELLS, *Structural Inorganic Chemistry*, 4th edn., Oxford University Press, Oxford, 1975, pp. 103, 1012–13.
[9] J. C. JAMIESON, Crystal structures at high pressures of metallic modifications of silicon and germanium, *Science* **139**, 762–4 (1963).

Tin is notably more reactive and electropositive than Ge though it is still markedly amphoteric in its aqueous chemistry. It is stable towards both water and air at ordinary temperatures but reacts with steam to give SnO_2 plus H_2 and with air or oxygen on heating to give SnO_2. Dilute HCl and H_2SO_4 show little, if any, reaction but dilute HNO_3 produces $Sn(NO_3)_2$ and NH_4NO_3. Hot concentrated HCl yields $SnCl_2$ and H_2 whereas hot concentrated H_2SO_4 forms $SnSO_4$ and SO_2. The occurrence of Sn^{II} compounds in these reactions is notable. By contrast, the action of hot aqueous alkali yields hydroxostannate(IV) compounds, e.g.:

$$Sn + 2KOH + 4H_2O \longrightarrow K_2[Sn(OH)_6] + 2H_2$$

Tin reacts readily with Cl_2 and Br_2 in the cold and with F_2 and I_2 on warming to give SnX_4. It reacts vigorously with heated S and Se, to form Sn^{II} and Sn^{IV} chalcogenides depending on the proportions used, and with Te to form SnTe.

Finely divided Pb powder is pyrophoric but the reactivity of the metal is usually greatly diminished by the formation of a thin, coherent, protective layer of insoluble product such as oxide, oxocarbonate, sulfate, or chloride. This inertness has been exploited as one of the main assets of the metal since early times: e.g. a temperature of $600-800°$ is needed to form PbO in air and Pb is widely used for handling hot concentrated H_2SO_4. Aqueous HCl does, in fact react slowly to give the sparingly-soluble $PbCl_2$ ($< 1\%$ at room temperature) and nitric acid reacts quite rapidly to liberate oxides of nitrogen and form the very soluble $Pb(NO_3)_2$ (~ 50 g per 100 cm^3, i.e. 1.5 M). Organic acids such as acetic acid also dissolve Pb in the presence of air to give $Pb(OAc)_2$, etc.; this precludes contact with the metal when processing or storing wine, fruit juices, and other drinks. Fluorine reacts at room temperatures to give PbF_2 and Cl_2 gives $PbCl_2$ on heating. Molten Pb reacts with the chalcogens to give PbS, PbSe, and PbTe.

The steady trend towards increasing stability of M^{II} rather than M^{IV} compounds in the sequence Ge, Sn, Pb is an example of the so-called "inert-pair effect" which is well established for the heavier B subgroup metals. The discussion on p. 255 is relevant here. A notable exception is the organometallic chemistry of Sn and Pb which is almost entirely confined to the M^{IV} state (pp. 459-65).

Catenation is also an important feature of the chemistry of Ge, Sn, and Pb though less so than for C and Si. The discussion on p. 393 can be extended by reference to the bond energies in Table 10.3 from which it can be seen that there is a steady decrease in the M–M bond strength. In general, with the exception of M–H bonds, the strength of other M–X bonds diminishes less noticeably, though the absence of Ge analogues of silicone polymers speaks for the lower stability of the Ge–O–Ge linkage.

The structural chemistry of the Group IVB elements affords abundant illustrations of the trends to be expected from increasing atomic size, increasing electropositivity, and increasing tendency to form M^{II} compounds, and these will become clear during the more detailed treatment of the chemistry in the succeeding sections. The often complicated stereochemistry of M^{II} compounds (which arises from the presence of a nonbonding electron-pair on the metal) is particularly revealing as also is the propensity of Sn^{IV} to become 5- and 6-coordinate.[9a] The ability of both Sn and Pb to form polyatomic cluster

[9a] J. A. ZUBIETA and J. J. ZUCKERMAN, Structural tin chemistry, *Prog. Inorg. Chem.* **24,** 251–475 (1978). An excellent comprehensive review with full structural diagrams and data, and more than 750 references.

TABLE 10.3   *Approximate average bond energies/kJ mol^{-1}*[a]

M–	–M	–C	–H	–F	–Cl	–Br	–I
C	368	368	435	452	351	293	216
Si	340	360	393	565	381	310	234
Ge	188	255	289	465	342	276	213
Sn	151	226	251	—	320	272	187
Pb	—	130	205	—	244	—	—

[a] These values often vary widely (by as much as 50–100 kJ mol^{-1}) depending on the particular compound considered and the method of computation used; e.g. Si–Si is quoted as 339, 337, and 368 kJ mol^{-1} in Si_2H_6, Si_2Me_6, and Si_2Ph_6, but is only 227 kJ mol^{-1} in crystalline Si. Individual values are thus less significant than general trends. The data represent a collation of values for typical compounds gleaned from refs. 10–12.

anions of very low formal oxidation state, (e.g. M_5^{2-}, M_9^{4-}, etc.) reflects the now well-established tendency of the heavier B subgroup elements to form chain, ring, or cluster homopolyatomic ions:[13] this was first established for the polyhalide anions and for Hg_2^{2+} but is also prevalent in Groups IVB, VB, and VIB, e.g. Pb_9^{4-} is isoelectronic with Bi_9^{5+} (see Section 10.3.6).

10.3   Compounds

10.3.1   *Hydrides and hydrohalides*

Germanes of general formula Ge_nH_{2n+2} are known as colourless gases or volatile liquids for $n=1$–5 and their preparation, physical properties, and chemical reactions are very similar to those of silanes (p. 388). Thus GeH_4 was formerly made by the inefficient hydrolysis of Mg/Ge alloys with aqueous acids but is now generally made by the reaction of $GeCl_4$ with $LiAlH_4$ in ether or even more conveniently by the reaction of GeO_2 with aqueous solutions of $NaBH_4$. The higher germanes are prepared by the action of a silent electric discharge on GeH_4; mixed hydrides such as SiH_3GeH_3 can be prepared similarly by circulating a mixture of SiH_4 and GeH_4 but no cyclic or unsaturated hydrides have yet been prepared. The germanes are all less volatile than the corresponding silanes (see Table) and, perhaps surprisingly, noticeably less reactive. Thus, in contrast to SiH_4 and

Property	GeH_4	Ge_2H_6	Ge_3H_8	Ge_4H_{10}	Ge_5H_{12}
MP/°C	− 164.8	− 109	− 105.6	—	—
BP/°C	− 88.1	29	110.5	176.9	234
Density (T°C)/g cm^{-3}	1.52 (− 142°)	1.98 (− 109°)	2.20 (− 105°)	—	—

[10] J. A. KERR and A. F. TROTMAN-DICKENSON, Bond strengths in polyatomic molecules, *CRC Handbook of Chemistry and Physics*, 57th edn., 1976–7, pp. F231–F237.
[11] W. E. DASENT, *Inorganic Energetics*, Penguin Books, Harmondsworth, 165 pp.
[12] C. F. SHAW and A. L. ALLRED, Nonbonded interactions in organometallic compounds of Group IVB, *Organometallic Chem. Rev.* **5A**, 95–142 (1970).
[13] J. D. CORBETT, Homopolyatomic ions of the post-transition elements—synthesis, structure, and bonding, *Prog. Inorg. Chem.* **21**, 129–55 (1976).

SnH_4, GeH_4 does not ignite in contact with air and is unaffected by aqueous acid or 30% aqueous NaOH. It acts as an acid in liquid NH_3 forming NH_4^+ and GeH_3^- ions and reacts with alkali metals in this solvent (or in $MeOC_2H_4OMe$) to give $MGeH_3$. Like the corresponding $MSiH_3$, these are white, crystalline compounds of considerable synthetic utility. X-ray diffraction analysis shows that $KGeH_3$ and $RbGeH_3$ have the NaCl-type structure, implying free rotation of GeH_3^-, and $CsGeH_3$ has the rare TlI structure (p. 273). The derived "ionic radius" of 229 pm emphasizes the similarity to SiH_3^- (226 pm) and this is reinforced by the bond angles deduced from broad-line nmr experiments: SiH_3^- $94\pm4°$ (cf. isoelectronic PH_3, 93.5°); GeH_3^- $92.5\pm4°$ (cf. isoelectronic AsH_3, 91.8°).[14]

The germanium hydrohalides GeH_xX_{4-x} (X = Cl, Br, I; x = 1, 2, 3) are colourless, volatile, reactive liquids. Preparative routes include reaction of Ge, GeX_2, or GeH_4 with HX. The compounds are valuable synthetic intermediates (cf. SiH_3I). For example, hydrolysis of GeH_3Cl yields $O(GeH_3)_2$, and various metatheses can be effected by use of the appropriate Ag salts or, more effectively, Pb^{II} salts, e.g. GeH_3Br with PbO, $Pb(OAc)_2$, and $Pb(NCS)_2$ affords $O(GeH_3)_2$, $GeH_3(OAc)$, and $GeH_3(SCN)$. Treatment of this latter compound with MeSH or $[Mn(CO)_5H]$ yields GeH_3SMe and $[Mn(CO)_5(GeH_3)]$ respectively. An extensive phosphinogermane chemistry is also known, e.g. $R_nGe(PH_2)_{4-n}$, R = alkyl or H.

Binary Sn hydrides are much less stable. Reduction of $SnCl_4$ with ethereal $LiAlH_4$ gives SnH_4 in 80–90% yield; $SnCl_2$ reacts similarly with aqueous $NaBH_4$. SnH_4 (mp $-146°$, bp $-52.5°$) decomposes slowly to Sn and H_2 at room temperature; it is unattacked by dilute aqueous acids or alkalis but is decomposed by more concentrated solutions. It is a potent reducing agent. Sn_2H_6 is even less stable, and higher homologues have not been obtained. By contrast, organotin hydrides are more stable, and catenation up to $H(SnPh_2)_6H$ has been achieved by thermolysis of Ph_2SnH_2. Preparation of R_nSnH_{4-n} is usually by $LiAlH_4$ reduction of the corresponding organotin chloride.

PbH_4 is the least well-characterized Group IVB hydride and it is unlikely that it has ever been prepared except perhaps in trace amounts at high dilution; methods which successfully yield MH_4 for the other Group IVB elements all fail even at low temperatures.[15] The alkyl derivatives R_2PbH_2 and R_3PbH can be prepared from the corresponding halides and $LiAlH_4$ at $-78°$ or by exchange reactions with Ph_3SnH, e.g.:

$$Bu_3^nPbX + Ph_3SnH \longrightarrow Bu_3^nPbH + Ph_3SnX$$

Me_3PbH (mp $\sim -106°$ decomp above $-30°$) and Et_3PbH (mp $\sim -145°$ decomp above $-20°$) readily add to alkenes and alkynes (hydroplumbation) to give stable tetra-organolead compounds.

10.3.2 *Halides and related complexes*

Germanium, Sn, and Pb form two series of halides: MX_2 and MX_4. PbX_2 are more stable than PbX_4, whereas the reverse is true for Ge, consistent with the steady increase in

[14] G. THIRASE, E. WEISS, H. J. HENNING, and H. LECHERT, Preparative, X-ray, and broad-line proton magnetic resonance studies on germylalkali compounds, $MGeH_3$, *Z. anorg. allgem. Chem.* **417**, 221–8 (1975).

[15] C. J. PORRIT, Plumbane, *Chem. Ind. (Lond.)* 1975, 398; E. WIBERG and E. AMBERGER *Hydrides of the Elements of Main Groups I–IV*, Chap. 10, Lead hydrides, pp. 757–64, Elsevier, Amsterdam 1971.

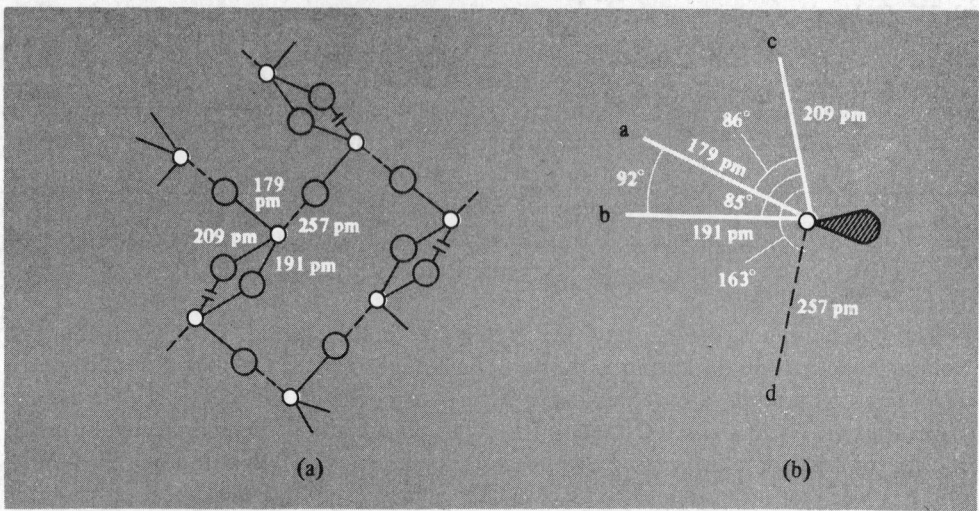

Fig. 10.2   Crystal structure of GeF$_2$: (a) projection along the chains, and (b) environment of Ge pseudo trigonal bipyramidal). The bond to the unshared F is appreciably shorter than those in the chain and there is a weaker interaction (257 pm) linking the chains into a 3D structure.

stability of the dihalides in the sequence $CX_2 \ll SiX_2 < GeX_2 < SnX_2 < PbX_2$. Numerous complex halides are also known for both oxidation states.

GeF$_2$ is formed as a volatile white solid (mp 110°) by the action of GeF$_4$ on powdered Ge at 150–300°; it has a unique structure in which trigonal pyramidal {GeF$_3$} units share 2 F atoms to form infinite spiral chains (Fig. 10.2). Pale-yellow GeCl$_2$ can be prepared similarly at 300° or by thermal decomposition of GeHCl$_3$ at 70°. Typical reactions are summarized in the scheme:

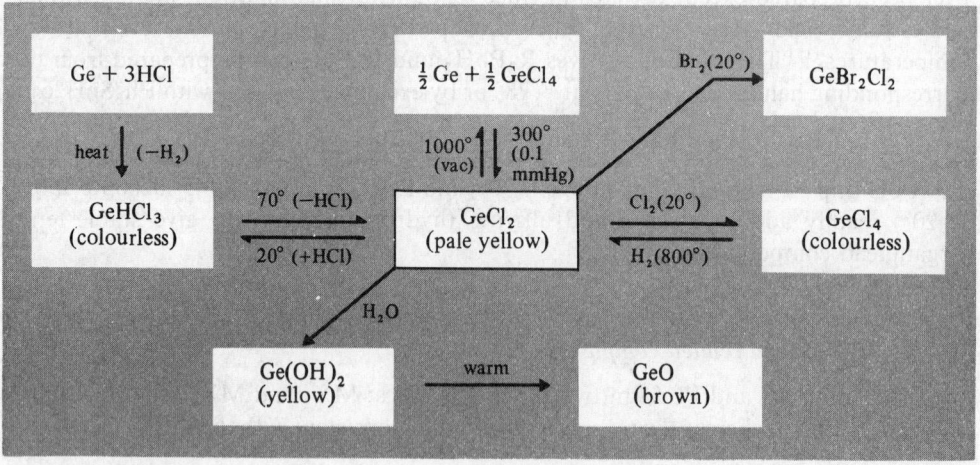

GeBr$_2$ is made by reduction of GeBr$_4$ or GeHBr$_3$ with Zn, or by the action of HBr on an excess of Ge at 400°; it is a yellow solid, mp 122°, which disproportionates to Ge and GeBr$_4$ at 150°, adds HBr at 40°, and hydrolyses to the unstable yellow Ge(OH)$_2$. GeI$_2$ is best prepared by reduction of GeI_4 with aqueous H_3PO_2 in the presence of HI to prevent hydrolysis; it sublimes to give bright orange-yellow crystals, is stable in dry air, and disproportionates only when heated above $\sim 550°$. The structure of the lemon-yellow, monomeric, 3-coordinate Ge^{II} complex [Ge(acac)I] (1) has recently been determined.[15a] GeI_2 is oxidized to GeI_4 in aqueous KI/HCl; it forms numerous adducts with nitrogen ligands, and reacts with C_2H_2 at $140°$ to give a compound that was originally formulated

as a 3-membered heterocycle $\overset{\displaystyle CH}{\underset{\displaystyle CH}{\|}} \!\!\!\!\searrow\!\!\! GeI_2$ but which has subsequently been shown by mass

spectroscopy to have the unusual dimeric structure (2).

Germanium tetrahalides are readily prepared by direct action of the elements or via the action of aqueous HX on GeO_2. The lighter members are colourless, volatile liquids, but GeI_4 is an orange solid (cf. CX_4, SiX_4). All hydrolyse readily and $GeCl_4$ in particular is an important intermediate in the preparation of organogermanium compounds via LiR or RMgX reagents. Many mixed halides and hydrohalides are also known, as are complexes

Property	GeF_4	$GeCl_4$	$GeBr_4$	GeI_4
MP/°C	-15 (4 atm)	-49.5	26	146
BP/°C	-36.5 (subl)	83.1	186	~ 400
Density $(T°C)$/g cm^{-3}	2.126 (0°)	1.844 (30°)	2.100 (30°)	4.322 (26°)

of the type $GeF_6{}^{2-}$, $GeCl_6{}^{2-}$, *trans*-L_2GeCl_4, and L_4GeCl_4 (L = tertiary amine or pyridine). The curious mixed-valency complex Ge_5F_{12}, i.e. $[(GeF_2)_4GeF_4]$ has recently been shown to feature distorted square pyramids of $\{:Ge^{II}F_4\}$ with the "lone-pair" of electrons pointing away from the 4 basal F atoms which are at 181, 195, 220, and 245 pm from the apical Ge^{II}; the Ge^{IV} atom is at the centre of a slightly distorted octahedron (Ge^{IV}–F 171–180 pm, angle F–Ge–F 87.5–92.5°) and the whole structure is held together by F bridges.[16]

The structural chemistry of Sn^{II} halides is particularly complex, partly because of the stereochemical activity (or non-activity) of the nonbonding pair of electrons and partly because of the propensity of Sn^{II} to increase its CN by polymerization into larger structural units such as rings or chains. Thus, the first and second ionization energies of Sn

[15a] S. R. STOBART, M. R. CHURCHILL, F. J. HOLLANDER, and W. J. YOUNGS, Characterization and X-ray crystal structure of a monomeric germylene derivative [Ge(acac)I] (Hacac = acetylacetone), *JCS Chem. Comm.* 1979, 911–12.

[16] J. C. TAYLOR and P. W. WILSON, The structure of the mixed valence fluoride Ge_5F_{12}, *J. Am. Chem. Soc.* **95**, 1834–8 (1973).

(p. 431) are very similar to those of Mg (p. 122), but Sn^{II} rarely adopts structures typical of spherically symmetrical ions because the nonbonding pair of electrons, which is $5s^2$ in the free gaseous ion, readily distorts in the condensed phase; this can be described in terms of ligand-field distortions or the adoption of some "p character". Again, the "nonbonding" pair can act as a donor to vacant orbitals, and the "vacant" third 5p orbital and 5d orbitals can act as acceptors in forming further covalent bonds. A particularly clear example of this occurs with the adducts $[SnX_2(NMe_3)]$ (X = Cl, Br, I): the Sn^{II} atom, which has accepted a pair of electrons from the ligand NMe_3, can itself donate its own lone-pair to a strong Lewis acid to form a double adduct of the type $[BF_3\{\leftarrow SnX_2(\leftarrow NMe_3)\}]$ (X = Cl, Br, I).[16a]

SnF_2 (which is obtained as colourless monoclinic crystals by evaporation of a solution of SnO in 40% aqueous HF) is composed of Sn_4F_8 tetramers interlinked by weaker Sn–F interactions;[17] the tetramers are puckered 8-membered rings of alternating Sn and F as shown in Fig. 10.3 and each Sn is surrounded by a highly distorted octahedron of F (1 Sn–F_t at ~205 pm, 2 Sn–F_μ at ~218 pm, and 3 much longer Sn···F in the range 240–329 pm, presumably due to the influence of the nonbonding pair of electrons. In aqueous solutions containing F^- the predominant species is the very stable pyramidal complex SnF_3^- but crystallization is attended by further condensation. For example, crystallization of SnF_2 from aqueous solutions containing NaF does not give $NaSnF_3$ as previously supposed but $NaSn_2F_5$ or $Na_4Sn_3F_{10}$ depending on conditions. The $\{Sn_2F_5\}$ unit in the first compound can be thought of as a discrete ion $[Sn_2F_5]^-$ or as an F^- ion coordinating to 2 SnF_2 molecules (Fig. 10.4a): each Sn is trigonal pyramidal with two

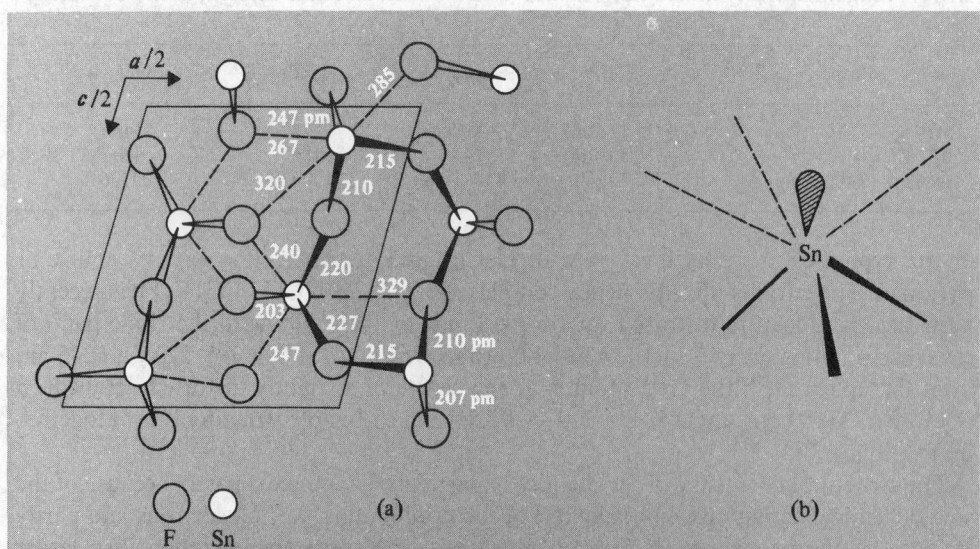

FIG. 10.3 Structure of SnF_2 showing (a) interconnected rings of $\{Sn_4F_4(F_4)\}$, and (b) the unsymmetrical 3 + 3 coordination around Sn.

[16a] C. C. HSU and R. A. GEANANGEL, Donor and acceptor behaviour of divalent tin compounds, *Inorg. Chem.* **19**, 110–19 (1980).

[17] R. C. MCDONALD, H. HO-KUEN HAU, and K. ERIKS, Crystal structure of monoclinic SnF_2, *Inorg. Chem.* **15**, 762–5 (1976).

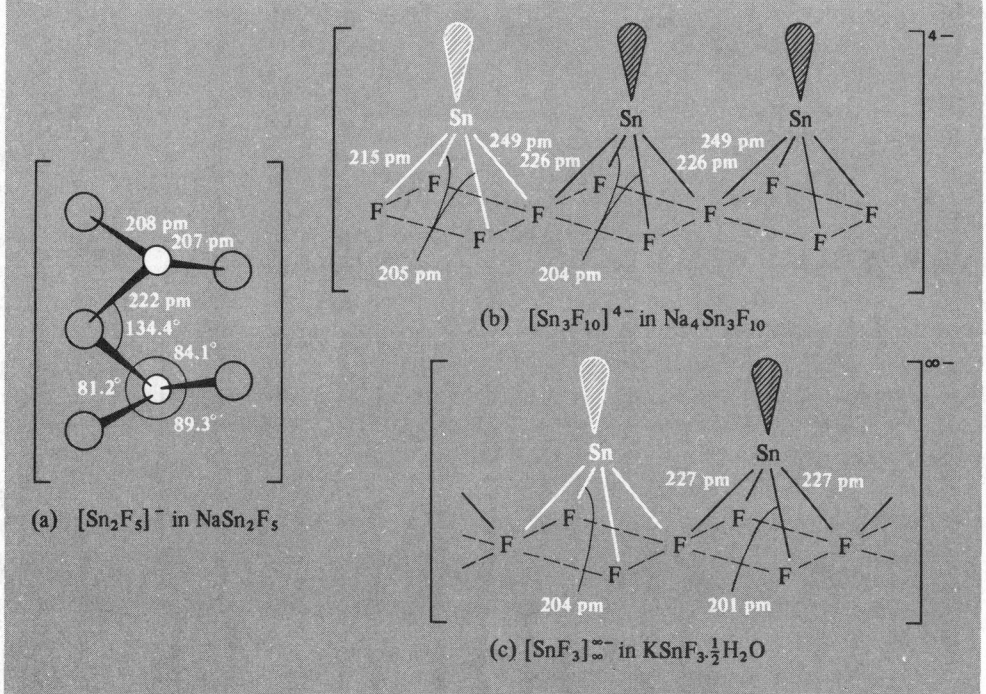

Fig. 10.4   Structure of some fluoro-complexes of Sn[II].

(a) $[Sn_2F_5]^-$ in $NaSn_2F_5$

(b) $[Sn_3F_{10}]^{4-}$ in $Na_4Sn_3F_{10}$

(c) $[SnF_3]_\infty^{-}$ in $KSnF_3 \cdot \frac{1}{2}H_2O$

close F_t, one intermediate $F_{t'}$, and 3 more distant F at 253, 298, and 301 pm. By contrast the compound $Na_4Sn_3F_{10}$ features 3 corner-shared square-pyramidal $\{SnF_4\}$ units (Fig. 10.4b) though the wide range of Sn–F distances could be taken to indicate incipient formation of a central $SnF_4{}^{2-}$ weakly bridged to two terminal $SnF_3{}^-$ groups. In the corresponding system with KF the compound that crystallizes is $KSnF_3 \cdot \frac{1}{2}H_2O$ in which the bridging of square pyramids is extended to give infinite chain polymers (Fig. 10.4c). There is also an intriguing mixed valence compound Sn_3F_8 which is formed when solutions of SnF_2 in anhydrous HF are oxidized at room temperature with F_2, O_2, or even SO_2; the structure features nearly regular $\{Sn^{IV}F_6\}$ octahedra *trans*-bridged to $\{Sn^{II}F_3\}$ pyramids which themselves form polymeric $Sn^{II}F$ chains: Sn^{IV}–F 196 pm and Sn^{II}–F 210, 217, 225 pm with weaker $Sn^{II} \cdots F$ interactions in the range 255–265 pm.[18] (The main commercial application of SnF_2 is in toothpaste and dental preparations where it is used to prevent demineralization of teeth and to lessen the development of dental caries.)

Tin(II) chlorides are similarly complex (Fig. 10.5). In the gas phase, $SnCl_2$ forms bent molecules, but the crystalline material (mp 246°, bp 623°) has a layer structure with chains of corner-shared trigonal pyramidal $\{SnCl_3\}$ groups. The dihydrate also has a 3-coordinated structure with only 1 of the H_2O molecules directly bonded to the Sn[II] (Fig. 10.5c); the neutral aquo complexes are arranged in double layers with the second H_2O molecules interleaved between them to form a two-dimensional H-bonded network with

[18] M. F. A. DOVE, R. KING, and T. J. KING, Preparation and X-ray crystal structure of Sn_3F_8, *JCS Chem. Comm.* 1973, 944–5.

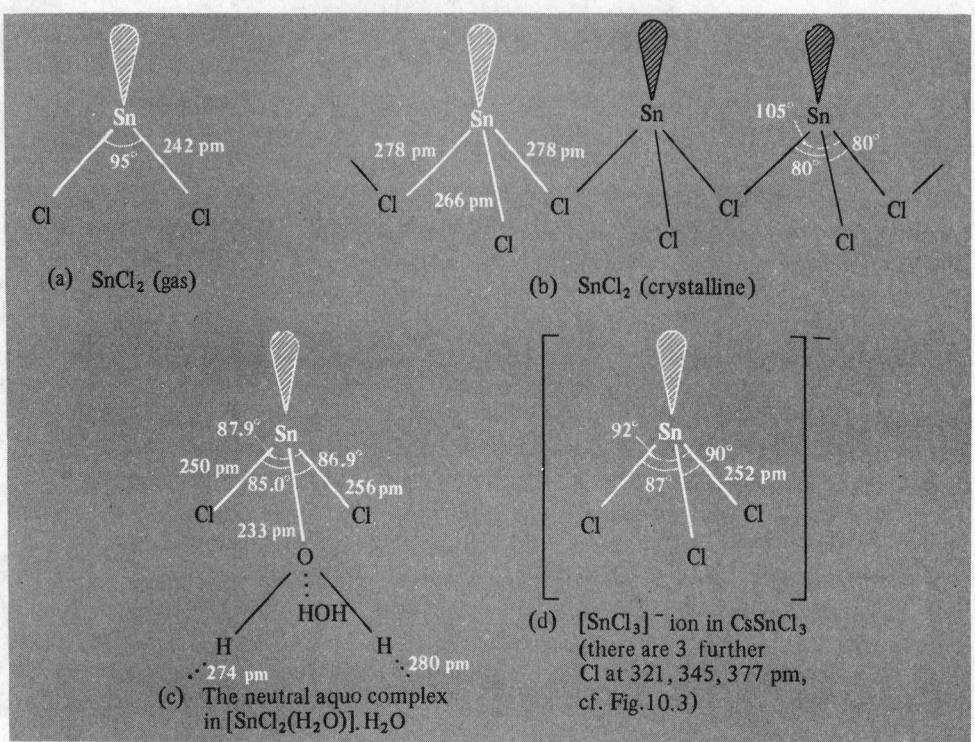

(a) SnCl$_2$ (gas)

(b) SnCl$_2$ (crystalline)

(c) The neutral aquo complex in [SnCl$_2$(H$_2$O)].H$_2$O

(d) [SnCl$_3$]$^-$ ion in CsSnCl$_3$ (there are 3 further Cl at 321, 345, 377 pm, cf. Fig.10.3)

Fig. 10.5 Structure of SnCl$_2$ and some chloro complexes of Sn^{II}.

the coordinated H$_2$O (O—H···O 274, 279, and 280 pm.[19] If the aquo ligand is replaced by Cl$^-$ the pyramidal SnCl$_3$$^-$ ion (isoelectronic with SbCl$_3$) is obtained, e.g. in CsSnCl$_3$ (Fig. 10.5d). There seems little tendency to add a second ligand:† e.g. the compound K$_2$SnCl$_4$.H$_2$O has been shown to contain pyramidal SnCl$_3$$^-$ and "isolated" Cl$^-$ ions, i.e. K$_2$[SnCl$_3$]Cl.H$_2$O with Sn–Cl 259 pm and the angle Cl–Sn–Cl ~85°. Apart from its structural interest, SnCl$_2$ is important as a widely used mild reducing agent in acid solution. The dihydrate is commercially available for use in electrolytic tin-plating baths, as a sensitizer in silvering mirrors and in the plating of plastics, and as a perfume stabilizer in toilet soaps. The anhydrous material can be obtained either by dehydration using acetic anhydride or directly by reacting heated Sn with dry HCl gas.

SnBr$_2$ is a white solid when pure (mp 216°, bp 620°); it has a layer–lattice structure but the details are unknown. It forms numerous hydrates (e.g. 3SnBr$_2$.H$_2$O, 2SnBr$_2$.H$_2$O, 6SnBr$_2$.5H$_2$O) all of which have a distorted trigonal prism of 6Br about the Sn^{II} with a further Br and H$_2$O capping one or two of the prism faces and leaving the third face

† Thus, [Co(en)$_3$][SnCl$_3$](Cl)$_2$, prepared from [Co(en)$_3$]Cl$_3$ and SnCl$_2$.2H$_2$O in excess HCl contains [SnCl$_3$]$^-$ and Cl$^-$ units but a similar reaction of [Co(NH$_3$)$_6$]Cl$_3$ with SnCl$_2$ in an HCl-solution saturated with NaCl has recently been shown to yield [Co(NH$_3$)$_6$][SnCl$_4$](Cl) in which [SnCl$_4$]$^{2-}$ has a pseudo trigonal bipyramidal structure with an equatorial lone-pair of electrons; even here, however, the fourth Sn–Cl distance (310 pm, axial) is substantially longer than the other axial Sn–Cl (267 pm) or the two equatorial distances (247, 253 pm). [H. J. Haupt, F. Huber, and H. Preut, *Z. anorg. Chem.* **422**, 97, 255 (1976).]

19 H. Kiriyama, K. Kitahama, O. Nakamura, and R. Kiriyama, An X-ray redetermination of the crystal structure of tin(II) chloride dihydrate, *Bull. Chem. Soc. Japan* **46**, 1389–95 (1973).

uncapped (presumably because of the presence of the nonbonding pair of electrons in that direction).[20] A similar pseudo-9-coordinate structure is adopted by $3PbBr_2.2H_2O$. By contrast, $NH_4SnBr_3.H_2O$ adopts a structure in which Sn^{II} is coordinated by a tetragonal pyramid of 5 Br atoms which form chains by edge sharing of the 4 basal Br; the Sn^{II} is slightly above the basal plane with Sn–Br 304–350 pm and Sn–Br$_{apex}$ 269 pm. The $NH_4{}^+$ and H_2O form rows between the chains.

SnI_2 forms as brilliant red needles (mp 316°, bp 720°) when Sn is heated with I_2 in 2 M hydrochloric acid. It has a unique structure in which one-third of the Sn atoms are in almost perfect octahedral coordination in rutile-like chains (2 Sn–I 314.7 pm, 4 Sn–I 317.4 pm, and no significant distortions of angles from 90°); these chains are in turn cross-linked by double chains containing the remaining Sn atoms which are themselves 7-coordinate (5 Sn–I all on one side at 300.4–325.1 and 2 more-distant I at 371.8 pm).[21] There is an indication here of reduced distortion in the octahedral site and this has been observed more generally for compounds with the heavier halides and chalcogenides in which the nonbonding electron pair on Sn^{II} can delocalize into a low-lying band of the crystal. Accordingly, SnTe is a metalloid with cubic NaCl structure. Likewise, $CsSn^{II}Br_3$ has the ideal cubic perovskite structure (p. 1122);[22] the compound forms black lustrous crystals with a semi-metallic conductivity of $\sim 10^3$ ohm^{-1} cm^{-1} at room temperature due, it is thought, to the population of a low-lying conduction band formed by the overlap of "empty" t_2 5d orbitals on Br. In this connection it is noteworthy that $Cs_2Sn^{IV}Br_6$ has a very similar structure to $CsSn^{II}Br_3$ (i.e. $Cs_2Sn_2^{II}Br_6$) but with only half the Sn sites occupied—it is white and non-conducting since there are no high-energy nonbonding electrons to populate the conduction band which must be present. Similarly, yellow $CsSn^{II}I_3$, $CsSn_2^{II}Br_5$, $Cs_4Sn^{II}Br_6$, and compositions in the system $CsSn_2^{II}X_5$ (X = Cl, Br) all transform to black metalloids on being warmed and even yellow monoclinic $CsSnCl_3$ (Fig. 10.5d) transforms at 90° to a dark-coloured cubic perovskite structure.

Tin(IV) halides are more straightforward. SnF_4 (prepared by the action of anhydrous HF on $SnCl_4$) is an extremely hygroscopic, white crystalline compound which sublimes above 700°. The structure (unlike that of CF_4, SiF_4, and GeF_4) is polymeric with octahedral coordination about Sn: the $\{SnF_6\}$ units are joined into planar layers by edge-sharing of 4 equatorial F atoms (Sn–F$_\mu$ 202 pm) leaving 2 further (terminal) F in *trans* positions above and below each Sn (Sn–F$_t$ 188 pm). The other SnX_4 can be made by direct action of the elements and are unremarkable volatile liquids or solids comprising

Property	SnF$_4$	SnCl$_4$	SnBr$_4$	SnI$_4$
Colour	White	Colourless	Colourless	Brown
MP/°C	—	−33.3	31	144
BP/°C	~705 (subl)	114	205	348
Density (T°C)/g cm^{-3}	4.78 (20°)	2.234 (20°)	3.340 (35°)	4.56 (20°)
Sn–X/pm	188, 202	231	244	264

[20] J. ANDERSON, The crystal structures of SnBr$_2$ hydrates, *Acta Chem. Scand.* **26**, 1730, 2543, 3813 (1973).

[21] R. A. HOWIE, W. MOSER, and I. C. TREVENA, The crystal structure of tin(II) iodide, *Acta Cryst.* **B28**, 2965–71 (1972).

[22] J. D. DONALDSON, J. SILVER, S. HADJIMINOLIS, and S. D. ROSS, Effects of the presence of valence-shell non-bonding electron pairs on the properties and structures of caesium tin(II) bromides and of related antimony and tellurium compounds, *JCS Dalton* 1975, 1500–6; see also J. D. DONALDSON *et al.*, *JCS Dalton* 1973, 666.

tetrahedral molecules. Similarities with the tetrahalides of Si and Ge are obvious. The compounds hydrolyse readily but definite hydrates can also be isolated from acid solution, e.g. $SnCl_4.5H_2O$, $SnBr_4.4H_2O$. Complexes with a wide range of organic and inorganic ligands are known, particularly the 6-coordinate *cis-* and *trans-*L_2SnX_4 and occasionally the 1:1 complexes $LSnX_4$. Stereochemistry has been deduced by infrared and Mössbauer spectroscopy and, when possible, by X-ray crystallography. The octahedral complexes SnX_6^{2-} (X = Cl, Br, I) are also well characterized for numerous cations. Five-coordinate trigonal bipyramidal complexes are less common but have been established for $SnCl_5^-$ and $Me_2SnCl_3^-$. A novel rectangular pyramidal geometry for Sn^{IV} has recently been revealed by X-ray analysis of the spirocyclic dithiolato complex anion $[(MeC_6H_3S_2)_2SnCl]^-$: the Cl atom occupies the apical position and the Sn atom is slightly above the plane of the 4 S atoms (mean angle Cl–Sn–Cl 103°).[22a] A similar stereochemistry has also very recently been established for Si^{IV} (p. 385) and for Ge^{IV} in $[(C_6H_4O_2)_2GeCl]^-$.[22b]

Lead continues the trends outlined in preceding sections, PbX_2 being much more stable thermally and chemically than PbX_4. Indeed, the only stable tetrahalide is the yellow PbF_4 (mp 600°); $PbCl_4$ is a yellow oil (mp −15°) stable below 0° but decomposing to $PbCl_2$ and Cl_2 above 50°; $PbBr_4$ is even less stable and PbI_4 is of doubtful existence (cf. discussion on TlI_3, p. 269). Stability can be markedly increased by coordination: e.g. direct chlorination of $PbCl_2$ in aqueous HCl followed by addition of an alkali metal chloride gives stable yellow salts M_2PbCl_6 (M = Na, K, Rb, Cs, NH_4) which can serve as a useful source of Pb^{IV}. By contrast, PbX_2 are stable crystalline compounds which can readily be prepared by treating any water-soluble Pb^{II} salt with HX or halide ions to precipitate the insoluble PbX_2. As with Sn, the first two ionization energies of Pb are very similar to those of Mg; moreover, the 6-coordinate radius of Pb^{II} (119 pm) is virtually identical with that of Sr^{II} (118 pm) and there is less evidence of the structurally distorting influence of the nonbonding pair of electrons. Thus α-PbF_2, $PbCl_2$, and $PbBr_2$ all form colourless orthorhombic crystals in which Pb^{II} is surrounded by 9 X at the corners of a tricapped trigonal prism.† The high-temperature β-form of PbF_2 has the cubic fluorite (CaF_2) structure with 8-coordinated Pb^{II}. PbI_2 (yellow) has the CdI_2 hexagonal layer lattice structure. Like many other heavy-metal halides, $PbCl_2$ and $PbBr_2$ are photo-sensitive and deposit metallic Pb on irradiation with ultraviolet or visible light. PbI_2 is a photoconductor and decomposes on exposure to green light (λ_{max} 494.9 nm). Many mixed halides have also been characterized by PbFCl, PbFBr, PbFI, $PbX_2.4PbF_2$, etc. Of these PbFCl is an important tetragonal layer–lattice structure type frequently adopted by large cations in the presence of 2 anions of differing size;[23] its sparing solubility in water (37 mg per 100 cm³ at 25°C) forms the basis of a gravimetric method of determining F.

---

† There are, in fact, never 9 equidistant X neighbours but a range of distances in which one can discern 7 closer and 2 more distant neighbours. This (7 + 2)-coordination is also a feature of the structures of BaX_2 (X = Cl, Br, I), $EuCl_2$, CaH_2, etc.—see p. 222 of ref. 8; see also hydrated tin(II) bromides, p. 442.

22a A. C. SAU, R. O. DAY, and R. R. HOLMES, A new geometrical form for tin. Synthesis and structure of the spirocyclic complex $[NMe_4][(C_7H_6S_2)_2SnCl]$ and the related monocyclic derivative $[NEt_4][(C_7H_6S_2)Ph_2SnCl]$, *Inorg. Chem.* **20**, 3076–81 (1981).
22b A. C. SAU, R. O. DAY, and R. R. HOLMES, A new geometrical form of germanium. Synthesis and structure of tetraethylammonium 2-chloro-2,2′-spirobis(1,3,2-benzodioxagermole), *J. Am. Chem. Soc.* **102**, 7972–3 (1980).
23 N. N. GREENWOOD, *Ionic Crystals, Lattice Defects, and Nonstoichiometry*, pp. 59–60, Butterworths, London 1968; see also ref. 8, pp. 408–9.

Property	PbF$_2$	PbCl$_2$	PbBr$_2$	PbI$_2$
MP/°C	818	500	367	400
BP/°C	1290	953	916	860–950 (decomp)
Density/g cm^{-3}	8.24 (α), 7.77 (β)	5.85	6.66	6.2
Solubility in H$_2$O (T°C)/	64 (20°)	670 (0°)	455 (0°)	44 (0°)
mg per 100 cm^3		3200 (100°)	4710 (100°)	410 (100°)

Pb^{II} apparently forms complexes with an astonishing range of stoichiometries,[24] but structural information is frequently lacking. Cs$_4$PbX$_6$ (X = Cl, Br, I) have the K$_4$CdCl$_6$ structure with discrete [Pb^{II}X$_6$]$^{4-}$ units. CsPb^{II}X$_3$ also feature octahedral coordination (in perovskite-like structures, cf. p. 1122) but there is sometimes appreciable distortion as in the yellow, low-temperature form of CsPbI$_3$ which adopts the NH$_4$CdCl$_3$ structure with three Pb–I distances, 301, 325, and 342 pm.

10.3.3 *Oxides and hydroxides*

GeO is obtained as a yellow sublimate when powdered Ge and GeO$_2$ are heated to 1000°, and dark-brown crystalline GeO is obtained on further heating at 650°. The compound can also be obtained by dehydrating Ge(OH)$_2$ (p. 438) but neither compound is particularly well characterized. Both are reducing agents and GeO disproportionates rapidly to Ge and GeO$_2$ above 700°. Much more is known about GeO$_2$ and, as pointed out by A. F. Wells (p. 929 of ref. 8), there is an impressive resemblance between the oxide chemistry of Ge^{IV} and Si^{IV}. Thus hexagonal GeO$_2$ has the 4-coordinated β-quartz structure (p. 394), tetragonal GeO$_2$ has the 6-coordinated rutile-like structure of stishovite (p. 396), and vitreous GeO$_2$ resembles fused silica. Similarly, Ge analogues of all the major types of silicates and aluminosilicates (pp. 399–416) have been prepared. Be$_2$GeO$_4$ and Zn$_2$GeO$_4$ have the phenacite and willemite structures with "isolated" {GeO$_4$} units; Sc$_2$Ge$_2$O$_7$ has the thortveitite structure; BaTiGe$_3$O$_9$ has the same type of cyclic ion as benitoite, and CaMgGe$_2$O$_6$ has a chain structure similar to diopside. Further, the two crystalline forms of Ca$_2$GeO$_4$ are isostructural with two forms of Ca$_2$SiO$_4$, and Ca$_3$GeO$_5$ crystallizes in no fewer than 4 of the known structures of Ca$_3$SiO$_5$. The reaction chemistry of the two sets of compounds is also very similar.

SnO exists in several modifications. The commonest is the blue-black tetragonal modification formed by the alkaline hydrolysis of Sn^{II} salts to the hydrous oxide and subsequent dehydration in the absence of air. The structure features square pyramids of {SnO$_4$} arranged in parallel layers with Sn^{II} at the apex and alternately above and below the layer of O atoms as shown in Fig. 10.6. The Sn–Sn distance between tin atoms in adjacent layers is 370 pm, very close to the values in β-Sn (p. 434). The structure can also be described as a fluorite lattice with alternate layers of anions missing. A metastable, red modification of SnO is obtained by heating the white hydrous oxide; this appears to have a similar structure and it can be transformed into the blue-black form by heating, by pressure, by treatment with strong alkali, or simply by contact with the stable form. Both

[24] E. W. ABEL, Lead, Chap. 18 in *Comprehensive Inorganic Chemistry*, Vol. 2, pp. 105–46, Pergamon Press, Oxford, 1973.

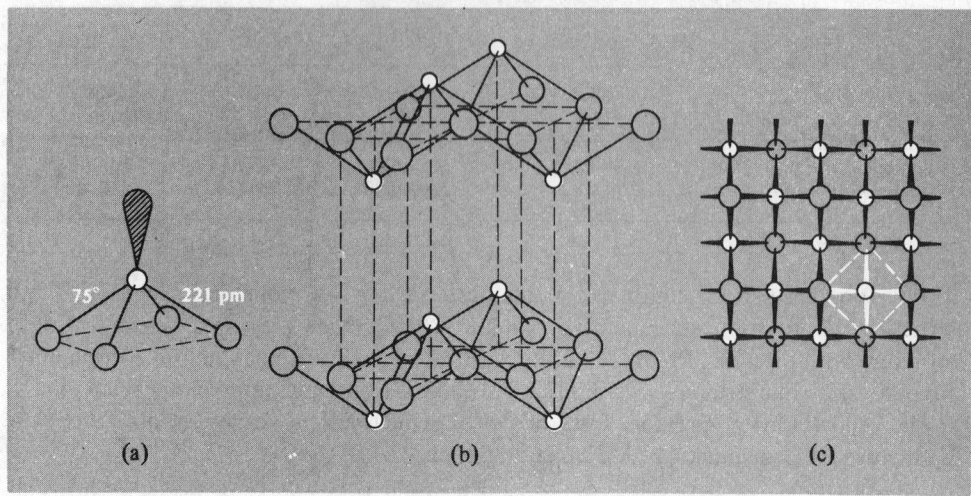

FIG. 10.6 Structure of tetragonal SnO (and PbO) showing (a) a single square-based pyramid
{:SnO₄}, (b) the arrangement of the pyramids in layers, and (c) a plane view of a single layer.

forms oxidize to SnO_2 with incandescence when heated in air to $\sim 300°$ but when heated
in the absence of O_2, the compound disproportionates like GeO. Various mixed-valence
oxides have also been reported of which the best characterized is Sn_3O_4, i.e. $Sn_2^{II}Sn^{IV}O_4$.

SnO and hydrous tin(II) oxide are amphoteric, dissolving readily in aqueous acids to
give Sn^{II} or its complexes, and in alkalis to give the pyramidal $Sn(OH)_3^-$; at intermediate values of pH, condensed basic oxide–hydroxide species form, e.g.
$[(OH)_2SnOSn(OH)_2]^{2-}$ and $[Sn_3(OH)_4]^{2+}$, etc. Analytically the hydrous oxide
frequently has a composition close to $3SnO.H_2O$ and an X-ray study shows it to contain
pseudo-cubic Sn_6O_8 clusters resembling $Mo_6Cl_8^{4+}$ (p. 1191) with 8 oxygen atoms centred
above the faces of an Sn_6 octahedron and joined in infinite array by H bonds, i.e.
$Sn_6O_8H_4$; the compound can be thought of as being formed by the deprotonation and
condensation of $2[Sn_3(OH)_4]^{2+}$ units as the pH is raised:

$$2[Sn_3(OH)_4]^{2+} + 4OH^- \longrightarrow [Sn_6O_8H_4] + 4H_2O$$

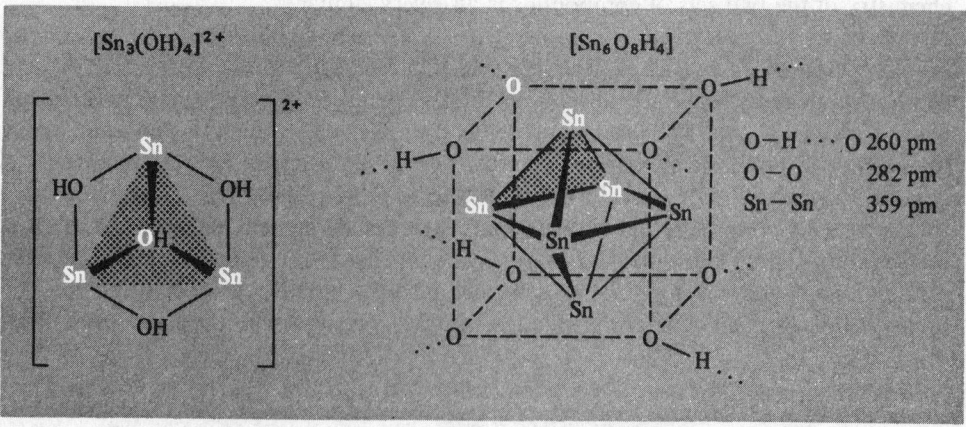

(Hydrolysis of Pb^{II} salts leads to different structures, p. 458.) It seems unlikely that pure $Sn(OH)_2$ itself has ever been prepared from aqueous solutions but recently it has been obtained as a white, amorphous solid by an anhydrous organometallic method:[25]

$$2Me_3SnOH + SnCl_2 \xrightarrow{\text{thf}} Sn(OH)_2 + 2Me_3SnCl$$

SnO_2, cassiterite, is the main ore of tin and it crystallizes with a rutile-type structure (p. 1120). It is insoluble in water and dilute acids or alkalis but dissolves readily in fused alkali hydroxides to form "stannates" $M_2^I Sn(OH)_6$.† Conversely, aqueous solutions of tin(IV) salts hydrolyse to give a white precipitate of hydrous tin(IV) oxide which is readily soluble in both acids and alkalis thereby demonstrating the amphoteric nature of tin(IV). $Sn(OH)_4$ itself is not known, but a reproducible product of empirical formula $SnO_2.H_2O$ can be obtained by drying the hydrous gel at 110°, and further dehydration at temperatures up to 600° eventually yields crystalline SnO_2. Similarly, thermal dehydration of $K_2[Sn(OH)_6]$, i.e. "$K_2SnO_3.3H_2O$", yields successively $K_2SnO_3.H_2O$, $3K_2SnO_3.2H_2O$, and, finally, anhydrous K_2SnO_3; this latter compound also results when K_2O is heated directly with SnO_2, and variations in the ratio of the two reactants yield K_4SnO_4 and $K_2Sn_3O_7$. The structure of K_2SnO_3 does not have 6-coordinate Sn^{IV} but chains of 5-coordinate Sn^{IV} of composition $\{SnO_3\}$ formed by the edge sharing of tetragonal pyramids of $\{SnO_5\}$ as shown in Fig. 10.7. Some industrial uses of tin((IV) oxide systems and other tin compounds are summarized in the Panel.

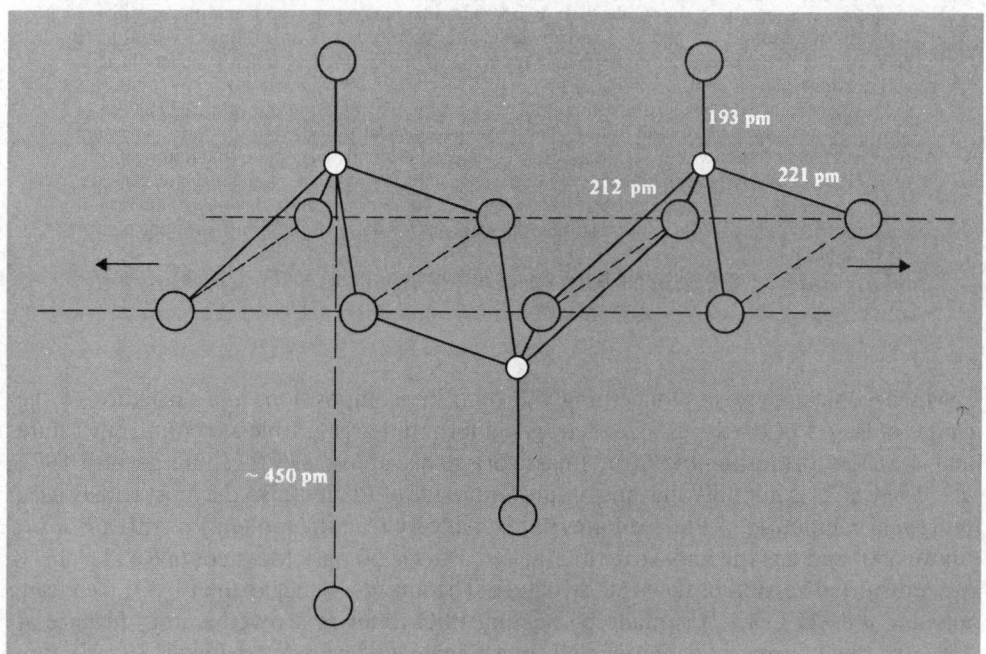

FIG. 10.7 $\{SnO_3\}$ chain in the structure of K_2SnO_3 (and K_2PbO_3).

† $K_2Sn(OH)_6$ has the brucite-like structure of $Mg(OH)_2$ but with each 3 Mg replaced regularly by 2K + Sn.

25 W. D. HONNICK and J. J. ZUCKERMAN, Tin(II) hydroxide, *Inorg. Chem.* **15**, 3034–7 (1976).

Some Industrial Uses of Tin Compounds

Tin(IV) oxide is much used an the ceramics industry as an opacifier for glazes and enamels. Because of its insolubility (or, rather, slow solubility) in glasses and glazes it also serves as a base for pigments, e.g. SnO_2/V_2O_5 yellows, SnO_2/Sb_2O_5 blue-greys, and SnO_2/Cr_2O_3 pinks. These latter, which can vary from a delicate pale pink to a dark maroon, probably involve substitutional incorporation of Cr^{III} for Sn^{IV} with concomitant oxide-ion vacancies, i.e. $[Sn^{IV}_{1-2x}Cr^{III}_{2x}O^{II}_{2-x}(\square_-)_x]$. The vanadium- and antimony–tin glazes, on the other hand, probably involve reductive substitution without vacant sites, e.g. $[Sn^{IV}_{1-3x}Sn^{II}_xSb^V_{2x}O^{II}_2]$. Some 3500 tonnes of SnO_2 is consumed annually for ceramic glazes. An example familiar to users of the London Underground is the dominant blue-grey colour of the tiles on the Victoria Line completed in 1969: these are based on the Sn/Sb oxide system with a small admixture ($\sim 1\%$) of V_2O_5 to control the blueness.

A related application is the use of $SnCl_4$ vapour to toughen freshly fabricated glass bottles by deposition of an invisible transparent film of SnO_2 ($<0.1\ \mu m$) which is then incorporated in the surface structure of the glass. This increases the strength of the glass and improves its abrasion resistance so that bottles so treated can be made considerably lighter without loss of robustness. When the thickness of the SnO_2 film is similar to the wavelength of visible light ($0.1–1.0\ \mu m$), then thin-film interference effects occur and the glass acquires an attractive iridescence. Still thicker films give electrically conducting layers which, after suitable doping with Sb or F ions, can be used as electrodes, electro-luminescent devices (for low-intensity light panels and display signs, in aircraft, cinemas, etc.), fluorescent lamps, antistatic cover-glasses, transparent tube furnaces, de-iceable windscreens (especially for aircraft), etc. Another property of these thicker films is their ability to reflect a high proportion of infrared (heat) radiation whilst remaining transparent to visible radiation—the application to heat insulation of windows is obvious.

Attention should be drawn to the use of tin oxide systems as heterogeneous catalysts. The oldest and most extensively patented systems are the mixed tin–vanadium oxide catalysts for the oxidation of aromatic compounds such as benzene, toluene, xylenes, and naphthalene in the synthesis of organic acids and acid anhydrides. More recently mixed tin–antimony oxides have been applied to the selective oxidation and ammoxidation of propylene to acrolein, acrylic acid, and acrylonitrile.

Homogeneous catalysis by tin compounds is also of great industrial importance. The use of $SnCl_4$ as a Friedel–Crafts catalyst for homogeneous acylation, alkylation, and cyclization reactions has been known for many decades. The most commonly used industrial homogeneous tin catalysts, however, are the Sn(II) salts of organic acids (e.g. acetate, oxalate, oleate, stearate, and octoate) for the curing of silicone elastomers, and, more importantly, for the production of polyurethane foams. World consumption of tin catalysts for these last applications alone is over 1000 tonnes pa.

For uses of organotin compounds (i.e. compounds having at least one Sn–C bond), see p. 460.

Much confusion exists concerning the number, composition, and structure of the oxides of lead. PbO exists as a red tetragonal form (litharge) stable at room temperature and a yellow orthorhombic form (massicot) stable above 488°C. Litharge (mp 897°, d 9.355 g cm^{-3}) is not only the most important oxide of Pb, it is also the most widely used inorganic compound of Pb (see Panel); it is made by reacting molten Pb with air or O_2 above 600° and has the SnO structure (p. 446, Pb–O 230 pm). Massicot (d 9.642 g cm^{-3}) has a distorted version of the same structure. The mixed-valency oxide Pb_3O_4 (red lead, minium, d 8.924 g cm^{-3}) is made by heating PbO in air in a reverberatory furnace at 450–500° and is important commercially as a pigment and primer (see Panel). Its structure (Fig. 10.8) consists of chains of $Pb^{IV}O_6$ octahedra (Pb–O 214 pm) sharing opposite edges, these chains being linked by the Pb^{II} atoms which themselves are pyramidally coordinated by 3 oxygen atoms (2 at 218 pm and 1 at 213 pm). The dioxide normally occurs as the maroon-coloured $PbO_2(I)$ which has the tetragonal, rutile structure (Pb^{IV}– O 218 pm,

The Oxides of Lead[3, 6, 6a]

PbO (red, orange, or yellow depending on the method of preparation) is amphoteric and dissolves readily in both acids and alkalis. It is much used in glass manufacture since a high Pb content leads to greater density, lower thermal conductivity, higher refractive index (greater brilliance), and greater stability and toughness. The replacement of the very mobile alkali ions by Pb also leads to high electrical capacities, comparable with mica. PbO is also used to form stable ceramic glazes and vitreous enamels (see SnO_2, p. 448). Electric storage batteries are the other major user of PbO (either as litharge or as "black oxide", i.e. PbO + Pb). The plates of the battery consist of an inactive grid or support onto which is applied a paste of PbO/H_2SO_4. The positive plates are activated by oxidizing PbO to PbO_2 and the negative plates by reducing PbO to Pb. Another use of PbO is in the production of pigments (p. 432). Production of litharge in the USA was 121 000 tonnes in 1975 and production of black oxide in the same year was 333 000 tonnes.

Red lead (Pb_3O_4) is manufactured on the 18 000-tonne scale annually and is used primarily as a surface coating to prevent corrosion of iron and steel (check oxidation–reduction potentials). It is also used in the production of leaded glasses and ceramic glazes and, very substantially, as an activator, vulcanizing agent, and pigment in natural and artificial rubbers and plastics.

PbO_2 is a strong oxidizing agent and, in addition to its *in situ* production in storage batteries, it is independently manufactured for use as an oxidant in the manufacture of chemicals, dyes, matches, and pyrotechnics. It is also used in considerable quantity as a curing agent for sulfide polymers and in high-voltage lightning arresters. Because of the instability of Pb^{IV}, PbO_2 tends to give salts of Pb^{II} with liberation of O_2 when treated with acids, e.g.

$$PbO_2 + H_2SO_4 \xrightarrow{\text{warm}} PbSO_4 + H_2O + \tfrac{1}{2}O_2$$

$$PbO_2 + 2HNO_3 \longrightarrow Pb(NO_3)_2 + H_2O + \tfrac{1}{2}O_2$$

Warm HCl reacts similarly but in the cold $PbCl_4$ is obtained. PbO_2 is produced commercially by the oxidation of Pb_3O_4 in alkaline slurry with Cl_2 and the technical product is marketed in 90-kg drums.

Mixed oxides of Pb^{IV} with other metals find numerous applications in technology and industry. They are usually made by heating PbO_2 (or PbO) in air with the appropriate oxide, hydroxide, or oxoacid salt, the product formed being dependent on the stoichiometry used, e.g. $M^{II}Pb^{IV}O_3$, $M_2^{II}Pb^{IV}O_4$ (M^{II} = Ca, Sr, Ba). $CaPbO_3$ in particular is increasingly replacing Pb_3O_4 as a priming pigment to protect steel against corrosion by salt water. Mixed oxides of Pb^{II} are also important. Ferrimagnetic oxides of general formula $PbO \cdot nFe_2O_3$ (n = 6, 5, 2.5, 1, 0.5) can be prepared by direct reaction but have not proved to be as attractive, commercially, as the hard ferrite $BaFe_{12}O_{19}$. By contrast, the ferroelectric behaviour (p. 61) of several mixed oxides with Pb^{II} has excited considerable interest. Many of these compounds have a distorted perovskite-type structure (p. 1122); e.g. yellow $PbTiO_3$ (ferroelectric below 490°C), colourless $PbZrO_3$ (230°), and $PbHfO_3$ (215°, antiferroelectric). Others have a tetragonal tungsten-bronze-type structure (p. 1185), e.g. $PbNb_2O_6$ (ferroelectric up to 560°), $PbTi_2O_6$ (~215°). The mode of action and uses of hard ferroelectrics has been discussed on p. 62, and the high Curie temperature of many Pb^{II} ferroelectrics makes them particularly useful for high-temperature applications.

d 9.643 g cm^{-3}), but there is also a high-pressure, black, orthorhombic polymorph, PbO_2(II) (d 9.773 g cm^{-3}).

When PbO_2 is heated in air it decomposes as follows:[26]

$$PbO_2 \xrightarrow{293°} Pb_{12}O_{19} \xrightarrow{351°} Pb_{12}O_{17} \xrightarrow{374°} Pb_3O_4 \xrightarrow{605°} PbO$$

In addition, a sesquioxide Pb_2O_3 can be obtained as vitreous black monoclinic crystals (d 10.046 g cm^{-3}) by decomposing PbO_2 (or PbO) at 580–620° under an oxygen pressure of

[26] W. B. WHITE and R. RAY, Phase relations in the system lead–oxygen, *J. Am. Ceram. Soc.* **47**, 242–7 (1964) and references therein.

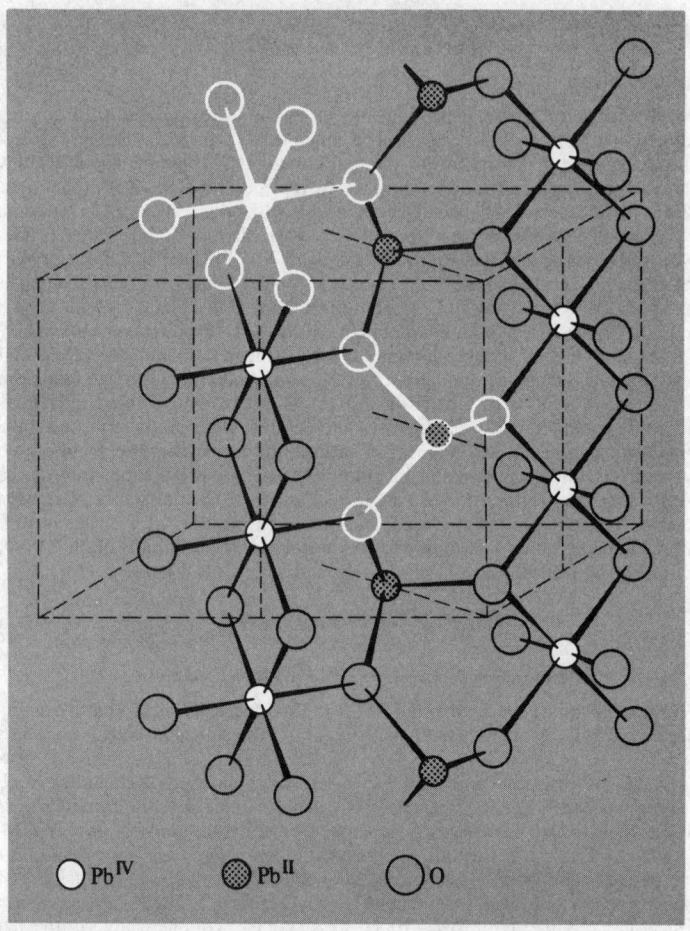

FIG. 10.8 Portion of the crystal structure of Pb_3O_4 showing chains of edge-shared $Pb^{IV}O_6$ octahedra joined by pyramids of $Pb^{II}O_3$; the mean $O–Pb^{II}–O$ angle is 76° as in PbO.

1.4 kbar: in this compound the Pb^{II} atoms are situated between layers of distorted $Pb^{IV}O_6$ octahedra (mean $Pb^{IV}–O$ 218 pm) with 3 $Pb^{II}–O$ in the range 231–246 pm and 3 in the range 264–300 pm. The monoclinic compound $Pb_{12}O_{19}$ (i.e. $PbO_{1.583}$) forms dark-brown or black crystals which have a pseudocubic defect-fluorite structure with 10 ordered anion vacancies according to the formulation $[Pb_{24}O_{38}(\square_-)_{10}]$ and no detectable variability of composition. It will be recalled that PbO can be considered as a defect fluorite structure in which each alternate layer of O atoms in the (001) direction is missing (p. 448), i.e. $[Pb_{24}O_{24}(\square_-)_{24}]$; it therefore seems reasonable to suppose that the anion vacancies in $[Pb_{24}O_{38}(\square_-)_{10}]$ are also confined to alternate layers, though it is not clear why this structure should show no variability in composition. Further heating above 350° (or careful oxidation of PbO) yields $Pb_{12}O_{17}$ (i.e. $PbO_{1.417}$) which is also a stoichiometric ordered defect fluorite structure $[Pb_{24}O_{34}(\square_-)_{14}]$. However, oxidation of this phase under increasing oxygen pressure leads to a bivariant nonstoichiometric phase of variable composition between $PbO_{1.42}$ and $PbO_{1.57}$ in which there appears to be a quasi-random array of anion vacancies.[27]

Lead does not appear to form a simple hydroxide, $Pb(OH)_2$, [cf. $Sn(OH)_2$, p. 447]. Instead, increasing the pH of solutions of Pb^{II} salts leads to hydrolysis and condensation, see $[Pb_6O(OH)_6]^{4+}$ (p. 458).

10.3.4 *Derivatives of oxoacids*

Oxoacid salts of Ge are usually unstable, generally uninteresting, and commercially unimportant. The tetraacetate $Ge(OAc)_4$ separates as white needles, mp 156°, when $GeCl_4$ is treated with TlOAc in acetic anhydride and the resulting solution is concentrated at low pressure and cooled. An unstable sulfate $Ge(SO_4)_2$ is formed in a curious reaction when $GeCl_4$ is heated with SO_3 in a sealed tube at 160°:

$$GeCl_4 + 6SO_3 \longrightarrow Ge(SO_4)_2 + 2S_2O_5Cl_2$$

Numerous oxoacid salts of Sn^{II} and Sn^{IV} have been reported and several basic salts are also known. Anhydrous $Sn(NO_3)_2$ has not been prepared but the basic salt $Sn_3(OH)_4(NO_3)_2$ can be made by reacting a paste of hydrous tin(II) oxide with aqueous HNO_3; the compound may well contain the oligomeric cation $[Sn_3(OH)_4]^{2+}$ illustrated on p. 446. $Sn(NO_3)_4$ can be obtained in anhydrous reactions of $SnCl_4$ with N_2O_5, $ClNO_3$, or $BrNO_3$; the compound readily oxidizes or nitrates organic compounds, probably by releasing reactive NO_3 radicals. Many phosphates and phosphato complexes have been described: typical examples for Sn^{II} are $Sn_3(PO_4)_2$, $SnHPO_4$, $Sn(H_2PO_4)_2$, $Sn_2P_2O_7$, and $Sn(PO_3)_2$. Examples with Sn^{IV} are $Sn_2O(PO_4)_2$, $Sn_2O(PO_4)_2.10H_2O$, SnP_4O_7, $KSn(PO_4)_3$, $KSnOPO_4$, and $Na_2Sn(PO_4)_2$. One remarkable compound is tin(IV) hypophosphite, $Sn(H_2PO_2)_4$ since it contains Sn^{IV} in the presence of the strongly reducing hypophosphorous anion; it has been suggested that the isolation of $Sn(H_2PO_2)_4$ (colourless crystals) by bubbling O_2 through a solution of SnO in hypophosphorous acid, $[H_2PO(OH)]$, may be due to a combination of kinetic effects and the low solubility of the product.

Treatment of SnO_2 with hot dilute H_2SO_4 yields the hygroscopic dihydrate $Sn(SO_4)_2.2H_2O$. In the Sn^{II} series $SnSO_4$ is a stable, colourless compound which is probably the most convenient laboratory source of Sn^{II} uncontaminated with Sn^{IV}; it is readily prepared by using metallic Sn to displace Cu from aqueous solutions of $CuSO_4$. $SnSO_4$ was at one time thought to be isostructural with $BaSO_4$ but this seemed unlikely in view of the very different sizes of the cations and the known propensity of Sn^{II} to form distorted structures; it is now known to consist of $\{SO_4\}$ groups linked into a framework by $O-Sn-O$ bonds in such a way that Sn is pyramidally coordinated by 3 O atoms at 226 pm ($O-Sn-O$ angles 77–79°); other $Sn-O$ distances are much larger and fall in the range 295–334 pm.[28] A basic sulfate and oxosulfate are also known:

$$3Sn^{II}SO_4(aq) \xrightarrow{\text{NH}_3(aq)} [Sn_3^{II}(OH)_2O(SO_4)] \xrightarrow{230°} [Sn_3^{II}O_2SO_4]$$

---

[27] J. S. ANDERSON and M. STERNS, The intermediate oxides of lead, *J. Inorg. Nucl. Chem.* **11**, 272–85 (1959).
[28] J. D. DONALDSON and D. C. PUXLEY, The crystal structure of tin(II) sulfate, *Acta Cryst.* **28B**, 864–7 (1972).

In general, the product obtained by the thermal decomposition of Sn^{II} oxoacid salts depends on the coordinating strength of the oxoacid anion. For strong ligands such as formate, acetate, and phosphite, other Sn^{II} compounds are formed (often SnO), whereas for less-strongly coordinating ligands such as the sulfate or nitrate internal oxidation to SnO_2 occurs, e.g.:

Strong ligands:

$$2Sn(HCO_2)_2 \xrightarrow{200°} 2SnO + H_2CO + CO_2$$

$$Sn(MeCO_2)_2 \xrightarrow{240°} SnO + Me_2CO + CO_2$$

$$5SnHPO_3 \xrightarrow{325°} Sn_2^{II}P_2O_7 + Sn_3^{II}(PO_4)_2 + PH_3 + H_2$$

$$2SnHPO_4 \xrightarrow{395°} Sn_2^{II}P_2O_7 + H_2O$$

Weak ligands:

$$2SnO \xrightarrow{350°} SnO_2 + Sn$$

$$SnSO_4 \xrightarrow{378°} SnO_2 + SO_2$$

$$Sn_3(OH)_4(NO_3)_2 \xrightarrow[\text{(explosive)}]{125°} 3SnO_2 + 2NO + 2H_2O$$

Most oxoacid derivatives of lead are Pb^{II} compounds, though $Pb(OAc)_4$ is well known and is extensively used as a selective oxidizing agent in organic chemistry.[29] It can be obtained as colourless, moisture-sensitive crystals by treating Pb_3O_4 with glacial acetic acid. $Pb(SO_4)_2$ is also stable when dry and can be made by the action of conc H_2SO_4 on $Pb(OAc)_4$ or by electrolysis of strong H_2SO_4 between Pb electrodes. $PbSO_4$ is familiar as a precipitate for the gravimetric determination of sulfate (solubility 4.25 mg per 100 cm³ at 25°C); $PbSeO_4$ is likewise insoluble. By contrast $Pb(NO_3)_2$ is very soluble in water (37.7 g per 100 cm³ at 0°, 127 g at 100°). The diacetate is similarly soluble (19.7 and 221 g per 100 cm³ at 0° and 50° respectively). Both compounds find wide use in the preparation of Pb chemicals by wet methods and are made simply by dissolving PbO in the appropriate aqueous acid. A large number of basic nitrates and acetates is also known. The thermal decomposition of anydrous $Pb(NO_3)_2$ above 400° affords a convenient source of N_2O_4 (see p. 524):

$$2Pb(NO_3)_2 \longrightarrow 2PbO + O_2 + 4NO_2 (\rightleftharpoons 2N_2O_4)$$

Other important Pb^{II} salts are the carbonate, basic carbonate, silicates, phosphates, and perchlorate, but little new chemistry is involved.[3] $PbCO_3$ occurs as cerussite; the compound is made as a dense white precipitate by treating the nitrate or acetate with CO_2 in the presence of $(NH_4)_2CO_3$ or Na_2CO_3, care being taken to keep the temperature low to avoid formation of the basic carbonate $\sim 2Pb(CO_3) \cdot Pb(OH)_2$. These compounds were

[29] R. N. BUTLER, Lead tetra-acetate, Chap. 4 in J. S. PIZEY (ed.), *Synthetic Reagents*, Vol. 3, pp. 278–419, Wiley, Chichester, 1977.

formerly much used as pigments (white lead) but are now largely replaced by other white pigments such as TiO_2 which has higher covering power and lower toxicity. For example, US production of the basic carbonate fell from 14 200 to 3070 tonnes pa between 1962 and 1975. The highly soluble perchlorate [and even more the tetrafluoroborate $Pb(BF_4)_2$] are much used as electrolytic plating baths for the deposition of Pb to impart corrosion resistance or lubricating properties to various metal parts. Throughout the chemistry of the oxoacid salts of Pb^{II} the close correlation between anionic charge and aqueous solubility is apparent.

10.3.5 *Other inorganic compounds*

Few of the many other inorganic compounds of Ge, Sn, and Pb call for special comment. Many pseudo-halogen derivatives of Sn^{IV}, Pb^{IV}, and Pb^{II} have been reported, e.g. cyanides, azides, isocyanates, isothiocyanates, isoselenocyanates, and alkoxides.[24, 30]

All 9 chalcogenides MX are known (X = S, Se, Te). GeS and SnS are interesting in having layer structures similar to that of the isoelectronic black-P (p. 557). The former is prepared by reducing a fresh precipitate of GeS_2 with excess H_3PO_2 and purifying the resulting amorphous red-brown powder by vacuum sublimation. SnS is usually made by sulfide precipitation from Sn^{II} salts. PbS occurs widely as the black opaque mineral galena, which is the principal ore of Pb (p. 428). In common with PbSe, PbTe, and SnTe, it has the cubic NaCl-type structure. Pure PbS can be made by direct reaction of the elements or by reaction of $Pb(OAc)_2$ with thiourea; the pure compound is an intrinsic semiconductor which, in the presence of impurities or stoichiometric imbalance, can develop either n-type or p-type semiconducting properties (p. 383). It is also a photoconductor (like PbSe and PbTe)† and is one of the most sensitive detectors of infrared radiation; the photovoltaic effect in these compounds is also widely used in photoelectric cells, e.g. PbS in photographic exposure meters.

Of the selenides, GeSe (mp 667°) forms as a dark-brown precipitate when H_2Se is passed into an aqueous solution of $GeCl_2$. SnSe (mp 861°) is a grey-blue solid made by direct reaction of the elements above 350°. PbSe (mp 1075°) can be obtained by volatilizing $PbCl_2$ with H_2Se, by reacting $PbEt_4$ with H_2Se in organic solvents, or by reducing $PbSeO_4$ with H_2 or C in an electric furnace; thin films for semiconductor devices are generally made by the reaction of $Pb(OAc)_2$ with selenourea, $(NH_2)_2CSe$. The tellurides are best made by heating Ge, Sn, or Pb with the stoichiometric amount of Te.

Other chalcogenides that have been described include GeS_2, $GeSe_2$, Sn_2S_3, and $SnSe_2$, but these introduce no novel chemistry or structural principles. Very recently the first sulfide halide of Ge was made by the apparently straightforward reaction:

$$4GeBr_4 + 6H_2S \xrightarrow{CS_2/Al_2Br_6} Ge_4S_6Br_4 + 12HBr$$

The unexpectedly complex product was isolated as an almost colourless air-stable powder, and a single-crystal X-ray analysis showed that the compound had the molecular

---

† These 3 compounds are unusual in that their colour diminishes with increasing molecular weight, PbS is black, PbSe grey, and PbTe white.

30 E. W. ABEL, Tin, Chap. 17 in *Comprehensive Inorganic Chemistry*, Vol. 2, pp. 43–104, Pergamon Press, Oxford, 1973.

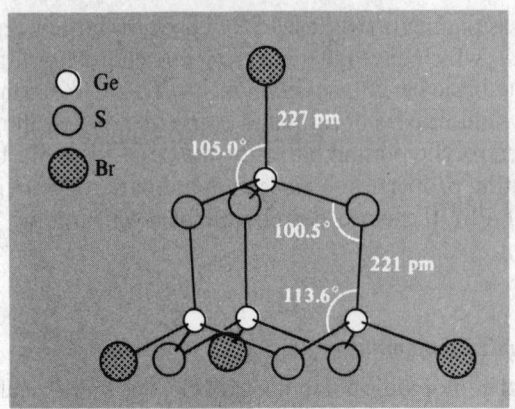

FIG. 10.9   Molecular structure and dimensions of $Ge_4S_6Br_4$.

adamantane-like structure shown in Fig. 10.9.[31] This is very similar to the structure of the "isoelectronic" compound P_4O_{10} (p. 579).

Another recent and exciting development has been the isolation of the first stable monomeric 2-coordinate compounds of bivalent Ge, Sn, and Pb. Thus, treatment of the corresponding chlorides MCl_2 with lithium di-*tert*-butyl phenoxide derivatives in thf afford a series of yellow (Ge^{II}, Sn^{II}) and red (Pb^{II}) compounds $M(OAr)_2$ in high yield.[31a] The O–M–O bond angle in $M(OC_6H_2Me\text{-}4\text{-}Bu^t_2\text{-}2,6)_2$ was 92° for Ge and 89° for Sn. Similar reactions of MCl_2 with $LiNBu^t_2$ yielded the (less stable) monomeric di-*tert*-butylamide, $Ge(NBu^t_2)_2$ (orange), and $Sn(NBu^t_2)_2$ (maroon);[31b] the more stable related compound $[Ge\{\overline{NCMe_2(CH_2)_3CMe_2}\}_2]$ was found to have a somewhat larger bond angle at Ge (N–Ge–N = 111°) and a rather long Ge–N bond (189 pm).

10.3.6   *Metal–metal bonds and clusters*

The catenation of Group IV elements has been discussed on pp. 393 and 435, and further examples are in Section 10.3.7. In addition, heteroatomic metal–metal bonds can be formed by a variety of synthetic routes as illustrated below for tin:

$$
\underline{\text{Insertion:}} \qquad SnCl_2 + Co_2(CO)_8 \longrightarrow (CO)_4Co\overset{\displaystyle Cl}{\underset{\displaystyle Cl}{\overset{|}{\underset{|}{-Sn-}}}}Co(CO)_4
$$

[31] S. POHL, The first sulfide halide of germanium, *Angew. Chem.*, Int. Edn. (Engl.) **15**, 162 (1976).

[31a] B. CETINKAYA, I. GÜMRÜKÇÜ, M. F. LAPPERT, J. L. ATWOOD, R. D. ROGERS, and M. J. ZAWOROTKO, Bivalent germanium, tin, and lead 2,6-di-*tert*-butylphenoxides and the crystal and molecular structures of $[M(OC_6H_2Me\text{-}4\text{-}Bu^t_2\text{-}2,6)_2]$ (M = Ge or Sn), *J. Am. Chem. Soc.* **102**, 2088–9 (1980).

[31b] M. F. LAPPERT, M. J. SLADE, J. L. ATWOOD, and M. J. ZAWOROTKO, Monomeric, coloured germanium(II) and tin(II)- di-*t*-butylamides, and the crystal and molecular structure of $[Ge\{\overline{NCMe_2(CH_2)_3CMe_3}\}_2]$, *JCS Chem. Comm.* 1980, 621–2.

Metathesis:

$$Me_2SnCl_2 + 2NaRe(CO)_5 \longrightarrow (CO)_5Re\!-\!\overset{\displaystyle Me}{\underset{\displaystyle Me}{\overset{|}{\underset{|}{Sn}}}}\!-\!Re(CO)_5 + 2NaCl$$

Elimination:

$$SnCl_2 + [Fe(\eta^5\text{-}C_5H_5)(CO)_2HgCl] \longrightarrow [(\eta^5\text{-}C_5H_5)\,\overset{\displaystyle CO}{\underset{\displaystyle CO}{\overset{|}{\underset{|}{Fe}}}}\!-\!SnCl_3] + Hg$$

Oxidative
addition:

$$SnCl_4 + [Ir(CO)Cl(PPh_3)_2] \longrightarrow [(PPh_3)_2\,\overset{\displaystyle CO}{\underset{\displaystyle SnCl_3}{\overset{|}{\underset{|}{IrCl_2}}}}]$$

Some representative examples, all featuring tetrahedral Sn, are in Fig. 10.10.[30] Several reactions are known in which the Sn–M bond remains intact, e.g.:

$$Ph_3SnMn(CO)_5 + 3Cl_2 \longrightarrow Cl_3SnMn(CO)_5 + 3PhCl$$

$$Cl_2Sn\{Co(CO)_4\}_2 + 2RMgX \longrightarrow R_2Sn\{Co(CO)_4\}_2 + 2MgClX$$

Others result in cleavage, e.g.:

$$Me_3SnCo(CO)_4 + I_2 \longrightarrow Me_3SnI + Co(CO)_4I$$

$$Me_3SnMn(CO)_5 + Ph_2PCl \longrightarrow Me_3SnCl + \tfrac{1}{2}[Ph_2PMn(CO)_4]_2 + CO$$

$$Me_3SnMn(CO)_5 + C_2F_4 \longrightarrow Me_3SnCF_2CF_2Mn(CO)_4$$

A similar though less extensive range of Pb–M compounds has been established;[24] e.g. $[Ph_2Pb\{Mn(CO)_5\}_2]$, $[Ph_3PbRe(CO)_5]$, $[Ph_2Pb\{Co(CO)_4\}_2]$, $[(PPh_3)_2Pt(PbPh_3)_2]$, $[(CO)_3Fe(PbEt_3)_2]$, and the cyclic dimer $[(CO)_4Fe\text{-}PbEt_2]_2$. Reaction of these compounds with halogens results in fission of the Pb–M bonds.

It has been known since the early 1930s that reduction of Ge, Sn, and Pb by Na in liquid ammonia gives polyatomic Group IV metal anions, and crystalline compounds can be isolated using ethylenediamine, e.g. $[Na_4(en)_5Ge_9]$ and $[Na_4(en)_7Sn_9]$. A dramatic advance was achieved recently[32] by means of the polydentate crypt ligand $[N\{(C_2H_4)O(C_2H_4)O(C_2H_4)\}_3N]$ (p. 109). Thus, reaction of crypt in ethylenediamine with the alloys $NaSn_{1-1.7}$ and $NaPb_{1.7-2}$ gave red crystalline salts $[Na(crypt)]_2^+[Sn_5]^{2-}$ and $[Na(crypt)]_2^+[Pb_5]^{2-}$ containing the D_{3h} cluster anions illustrated in Fig. 10.11. If each Sn or Pb atom is thought to have 1 nonbonding pair of electrons then the M_5^{2-} clusters have 12 framework bonding electrons as has $[B_5H_5]^{2-}$ (p. 185); the anions are

[32] P. A. EDWARDS and J. D. CORBETT, Synthesis and crystal structures of salts containing the pentaplumbide(2−) and pentastannide(2−) anions, *Inorg. Chem.* **16**, 903–7 (1977). J. D. CORBETT and P. A. EDWARDS, The nonastannide(4−) anion Sn_9^{4-}—a novel capped antiprismatic configuration (C_{4v}), *J. Am. Chem. Soc.* **99**, 3313–17 (1977).

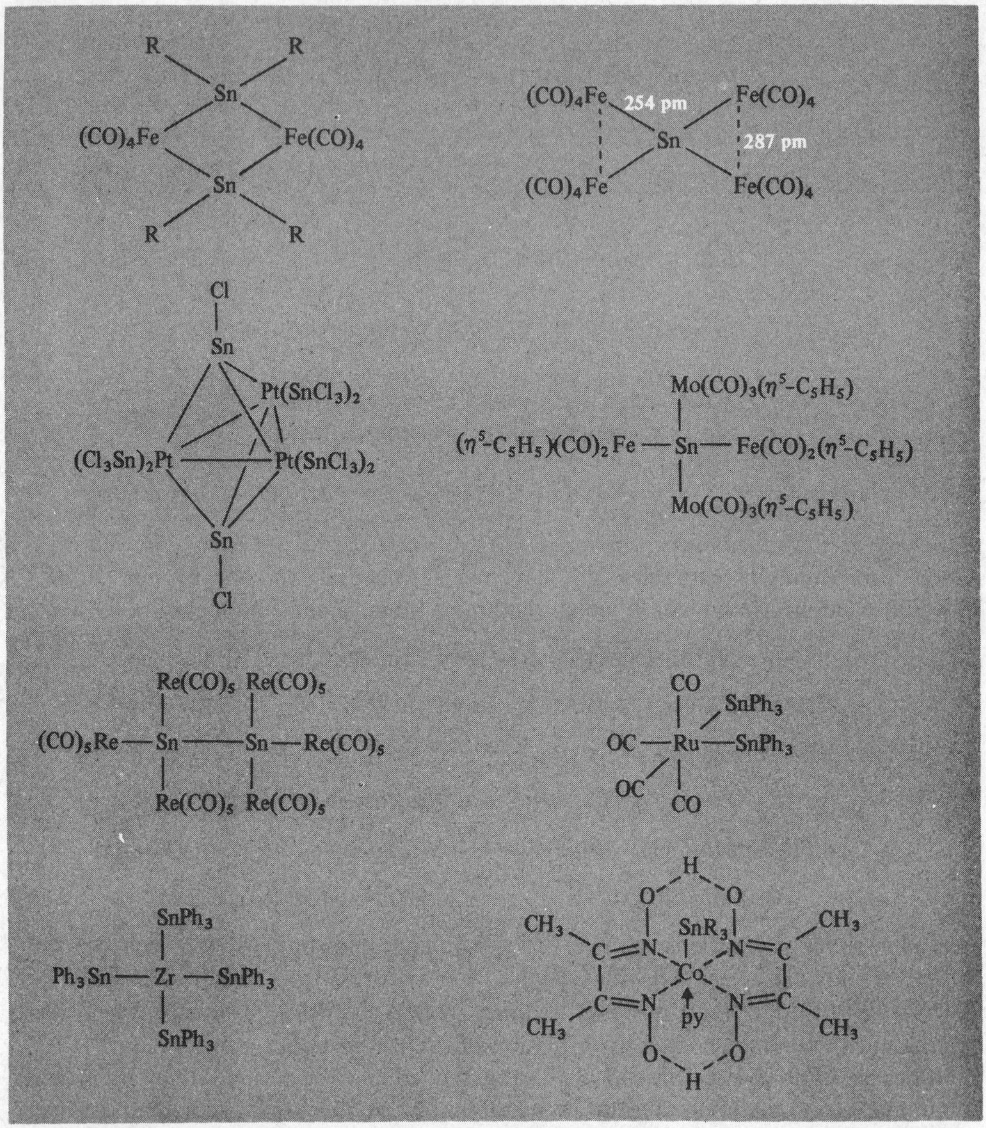

FIG. 10.10 Some examples of metal sequences and metal clusters containing tin–transitional metal bonds.

also isoelectronic with the well-known cation $[Bi_5]^{3+}$. Similarly, the alloy $NaSn_{\sim 2.25}$ reacts with crypt in ethylenediamine to give dark-red crystals of $[Na(crypt)]_4^+[Sn_9]^{4-}$; the anion is the first example of a C_{4v} unicapped Archimedian antiprism (Fig. 10.11) and differs from the D_{3h} structure of the isoelectronic cation $[Bi_9]^{5+}$ which, in the salt $Bi^+[Bi_9]^{5+}[HfCl_6]^{2-}_3$ (p. 688), features a tricapped trigonal prism, as in $[B_9H_9]^{2-}$ (p. 172). The emerald green species $[Pb_9]^{4-}$, which is stable in liquid NH_3 solution, has not so far proved amenable to isolation via crypt-complexed cations.

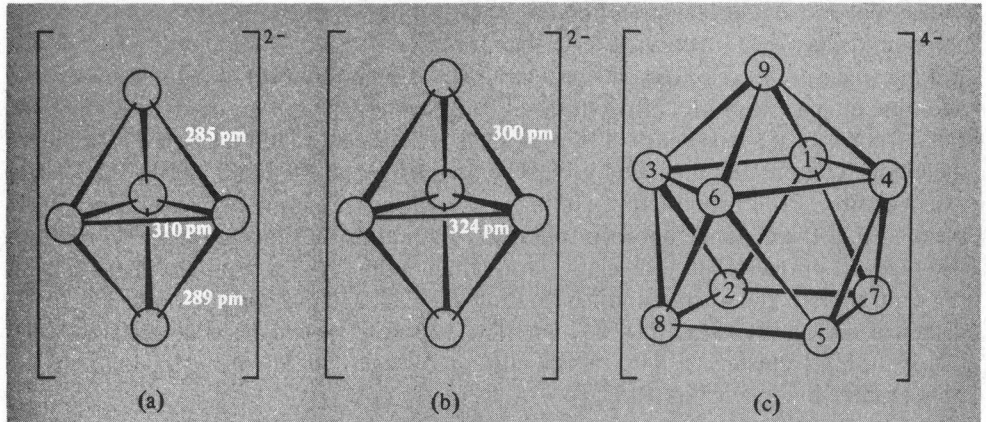

Fig. 10.11   The structure of polystannide and polyplumbide anions: (a) the slightly distorted D_{3h} structure of $[Sn_5]^{2-}$, (b) the D_{3h} structure of $[Pb_5]^{2-}$, and (c) the unique C_{4v} structure of $[Sn_9]^{4-}$: all Sn–Sn distances are in the range 295–302 pm except those in the slightly longer upper square (1,3,6,4) which are in the range 319–331 pm; the angles within the two parallel squares are all 90° (±0.8°).

The influence of electron-count on cluster geometry has been very elegantly shown by a crystallographic study of the deep-red compound $[K(\text{crypt})]_6^+[Ge_9]^{2-}[Ge_9]^{4-}.2.5\text{en}$, prepared by the reaction of KGe with crypt in ethylenediamine. $[Ge_9]^{4-}$ has the C_{4v} unicapped square-antiprismatic structure (10.11c) whereas $[Ge_9]^{2-}$, with 2 less electrons, adopts a distorted D_{3h} structure which clearly derives from the tricapped trigonal prism (p. 172).[32a] The field is one of great current interest and activity, as evidenced by papers describing the synthesis of and structural studies on tetrahedral Ge_4^{2-} and Sn_4^{2-},[32b, 32c] tricapped trigonal-prismatic $TlSn_8^{3-}$,[32d] bicapped square-antiprismatic $TlSn_9^{3-}$,[32d] and the two *nido*-series $Sn_{9-x}Ge_x^{4-}$ ($x = 0$–9) and $Sn_{9-x}Pb_x^{4-}$ ($x = 0$–9).[32e] Theoretical studies on many of these polymetallic-cluster anions have also been published.[32f]

[32a] C. H. E. BELIN, J. D. CORBETT, and A. CISAR, Homopolyatomic anions and configurational questions. Synthesis and structures of the nonagermanide(2−) and nonagermanide(4−) ions $[Ge_9]^{2-}$ and $[Ge_9]^{4-}$, *J. Am. Chem. Soc.* **99**, 7163–9 (1977).

[32b] S. C. CRITCHLOW and J. D. CORBETT, Stable homopolyatomic anions: the tetrastannide(2−) and tetragermanide(2−) anions, Sn_4^{2-} and Ge_4^{2-}. X-ray crystal structure of $[K^+(\text{crypt})]_2Sn_4^{2-}.\text{en}$, *JCS Chem. Comm.*, 1981, 236–7.

[32c] M. J. ROTHMAN, L. S. BARTELL, and L. L. LOHR, Prediction of fluxional behaviour for Sn_4^{2-} in solution, *J. Am. Chem. Soc.* **103**, 2482–3 (1981).

[32d] R. C. BURNS and J. D. CORBETT, Heteroatomic polyanions of the post transition metals. The synthesis and structure of a compound containing $TlSn_9^{3-}$ and $TlSn_8^{3-}$ with a novel structural disorder, *J. Am. Chem. Soc.* **104**, 2804–10 (1982).

[32e] R. W. RUDOLPH and W. L. WILSON, Naked-metal clusters in solution. 4. Indications of the variety of cluster species obtainable by extraction of Zintl phases: Sn_4^{2-}, $TlSn_8^{5-}$, $Sn_{9-x}Ge_x^{4-}$ ($x = 0$–9), and $SnTe_4^{4-}$, *J. Am. Chem. Soc.* **103**, 2480–1 (1981), and references therein.

[32f] L. L. LOHR, Relativistically parameterized extended Hückel calculations. 5. Charged polyhedral clusters of germanium, tin, lead, and bismuth atoms, *Inorg. Chem.* **20**, 4229–35 (1981); R. C. BURNS, R. J. GILLESPIE, J. A. BARNES, and M. J. McGLINCHEY, Molecular orbital investigation of the structure of some polyatomic cations and anions of the main-group elements, *Inorg. Chem.* **31**, 799–807 (1982).

The polymeric cluster compound $[Sn_6O_4(OH)_4]$ formed by hydrolysis of Sn^{II} compounds has been mentioned on p. 446. Hydrolysis of Pb^{II} compounds also leads to polymerized species; e.g. dissolution of PbO in aqueous $HClO_4$ followed by careful addition of base leads to $[Pb_6O(OH)_6]^{4+}[ClO_4]^-_4 \cdot H_2O$. The cluster cation (Fig. 10.12a) consists of 3 tetrahedra of Pb sharing faces; the central tetrahedron encompasses the unique O atom and the 6 OH groups lie on the faces of the 2 end tetrahedra.[33] The extent of direct Pb–Pb interaction within the overall cluster has not been established but it is noted that the distance between "adjacent" Pb atoms falls in the range 344–409 pm (average 381 pm) which is appreciably larger than in the Pb_5^{2-} anion. The distance from the central O to the 4 surrounding Pb atoms is 222–235 pm and the other Pb–O(H) distances are in the range 218–267 pm. The structure should be compared with the $[Sn_6O_4(OH)_4]$ cluster (p. 446), which also has larger Sn–Sn distances than in the polystannide anions in Fig. 10.11.

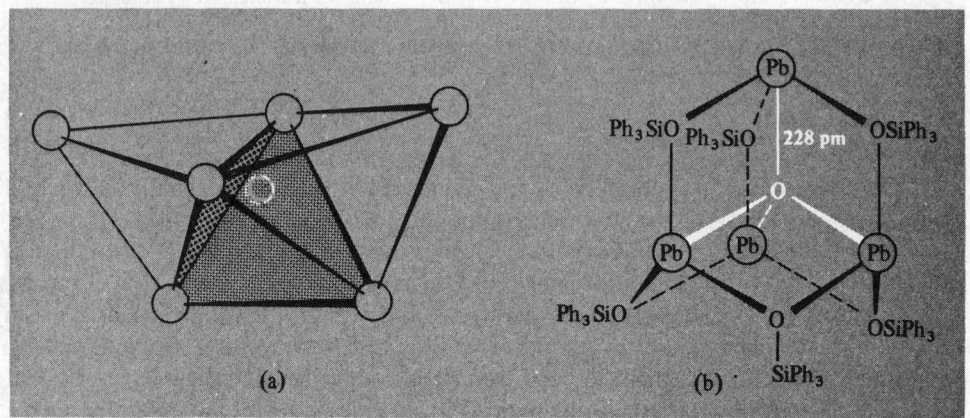

FIG. 10.12 (a) The three face-sharing tetrahedra of Pb atoms in the $Pb_6O(OH)_6^{4+}$ cluster; only the unique 4-coordinate O atom at the centre of the central tetrahedron is shown (in white). (b) The adamantane-like structure of $[Pb_4O(OSiPh_3)_6]$ showing the fourfold coordination about the central O atom.

Another polycyclic structure in which a unique O atom is surrounded tetrahedrally by 4 Pb^{II} atoms is the colourless adamantane-like complex $[Pb_4O(OSiPh_3)_6]$, obtained as a 1:1 benzene solvate by reaction of Ph_3SiOH with $[Pb(C_5H_5)_2]$ (p. 464). The local geometry about Pb^{II} is also noteworthy: it comprises pseudo-trigonal bipyramidal coordination in which the bridging $OSiPh_3$ groups occupy the equatorial sites whilst the apical sites are occupied by the unique O atom and axially directed lone pairs of electrons.[33a]

[33] T. G. SPIRO, D. H. TEMPLETON, and A. ZALKIN, The crystal structure of a hexanuclear basic lead(II) perchlorate hydrate: $Pb_6O(OH)_6(ClO_4)_4.H_2O$, *Inorg. Chem.* **8,** 856–61 (1969).

[33a] C. GAFFNEY, P. G. HARRISON, and T. J. KING, The crystal and molecular structure of *adamanta*-(μ_4-oxo-hexakis(μ-triphenylsiloxy)tetralead(II), *JCS Chem. Comm.* 1980, 1251–2.

10.3.7 *Organometallic compounds*[34, 35]

Germanium

Organogermanium chemistry closely resembles that of Si though the Ge compounds tend to be somewhat less thermally stable and rather more chemically reactive than their Si counterparts. The table of comparative bond energies on p. 436 indicates that the Ge–C and Ge–H bonds are weaker than the corresponding bonds involving Si but are nevertheless quite strong; Ge–Ge is noticeably weaker. There is little tendency to form double bonds by p_π–p_π interaction to C, O, N or any other element and (with the exceptions noted below and on p. 463) no organic compounds of Ge^{II} have been prepared. Preparative routes parallel those for organosilicon compounds (p. 421) and most of the several thousand known organogermanes can be considered as derivatives of R_nGeX_{4-n} or Ar_nGeX_{4-n} where X = hydrogen, halogen, pseudohalogen, OR, etc. The compounds are colourless, volatile liquids, or solids. Attempts to prepare $(-R_2GeO-)_x$ analogues of the silicones (p. 422) show that the system is different: hydrolysis of Me_2GeCl_2 is reversible and incomplete, but extraction of aqueous solutions of Me_2GeCl_2 with petrol leads to the cyclic tetramer $[Me_2GeO]_4$, mp 92°; the compound is monomeric in water. Organo-digermanes and -polygermanes have also been made by standard routes, e.g.:

$$2Me_3GeBr + 2K \xrightarrow{\text{reflux}} 2KBr + Ge_2Me_6 \ (\text{mp} -40°, \text{bp } 140°)$$

$$Et_3GeBr + NaGePh_3 \xrightarrow{\text{liq NH}_3} NaBr + Et_3GeGePh_3 \ (\text{mp } 90°)$$

$$2Ph_3GeBr + 2Na \xrightarrow[\text{xylene}]{\text{boiling}} 2NaBr + Ge_2Ph_6 \ (\text{mp } 340°)$$

$$Ph_2GeCl_2 + 2NaGePh_3 \longrightarrow 2NaCl + Ge_3Ph_8 \ (\text{mp } 248°)$$

Crystalline cyclic oligomers $(GePh_2)_n$ ($n = 4, 5, 6$) have also been prepared. In general, the Ge–Ge bond is readily cleaved by Br_2 either at ambient or elevated temperatures but the compounds are stable to thermal cleavage at moderate temperatures. Ge_2R_6 compounds can even be distilled unchanged in air (like Si_2R_6 but unlike the more reactive Sn_2R_6) and are stable towards hydrolysis and ammonolysis.

Tin

Organotin compounds have been much more extensively investigated than those of $Ge^{(34, 35)}$ and, as described in the Panel, many have important industrial applications.[36, 36a] Syntheses are by standard techniques (pp. 150, 289, 421) of which the following are typical:

---

[34] G. E. COATES and K. WADE, *Organometallic Compounds*, Vol. 1, 3rd edn., Chap. 4, Group IV: germanium, tin, and lead, pp. 375–509, Methuen, London, 1967.

[35] D. SEYFERTH (ed.), *Organometallic Chemistry Reviews: Annual Surveys: Silicon, Germanium, Tin, Lead*, Journal of Organometallic Chemistry Library **4**, 1977, 540 pp.; ibid. **6**, 1978, 550 pp.; ibid. **8**, 1979, 608 pp.

[36] J. J. ZUCKERMAN (ed.), *Organotin Compounds: New Chemistry and Applications*, Advances in Chemistry Series No. 157, 1976, 299 pp. (A symposium of 19 reviews and original papers.)

[36a] A. G. DAVIES and P. J. SMITH, Recent advances in organotin chemistry, *Adv. Inorg. Chem. Radiochem.* **23**, 1–77 (1980). A review, mainly of the last 10 years, with 566 references (over 1000 papers are published annually on organotin chemistry).

Grignard: $SnCl_4 + 4RMgCl \longrightarrow SnR_4 + 4MgCl_2$ (also with ArMgCl)

Organo Al: $3SnCl_4 + 4AlR_3 \longrightarrow 3SnR_4 + 4AlCl_3$ (alkyl only)

Direct (Rochow)† $Sn + 2RX \longrightarrow R_2SnX_2$ (and R_nSnX_{4-n}) (alkyl only).

All three routes are used on an industrial scale and the Grignard route (or the equivalent organo-Li reagent) is convenient for laboratory scale. Rather less used is the modified

Uses of Organotin Compounds

Tin is unsurpassed by any other metal in the multiplicity of applications of its organometallic compounds. The first organotin compound was made in 1849 but large-scale applications have developed only recently; indeed, world production figures for organotin compounds have increased more than 600-fold since 1950:

Year	1950	1960	1965	1970	1975	1978
Tonnes pa	< 50	2000	5000	15 000	25 000	30 000

The largest application for organotin compounds (66% by weight) is as stabilizers for PVC plastics; in their absence halogenated polymers are rapidly degraded by heat, light, or oxygen to give discoloured, brittle products. The most effective stabilizers are R_2SnX_2, where R is an alkyl residue (typically *n*-octyl) and X is laurate, maleate, etc. For food packaging the *cis*-butenedioate polymer, $[Oct_2^i Sn-OC(O)CH=CHC(O)O]_n$, and the *S,S'*-bis-(*iso*-octyl mercaptoethanoate), $Oct_2^i Sn\{SCH_2C(O)OOct^i\}_2$ have been approved and are used when colourless non-toxic materials with high transparency are required (Pl. 10.1). The compounds are thought to be such effective stabilizers because (i) they inhibit the onset of dehydrochlorination by exchanging their anionic groups X with reactive Cl sites in the polymer, (ii) they react with and hence scavenge the HCl which is produced and which would otherwise catalyse further elimination, and (iii) they act as antioxidants and thereby prevent breakdown of the polymer initiated by atmospheric O_2.

Another major use of organotin compounds is as curing agents for the room temperature "vulcanization" of silicones; the 3 most commonly used compounds are Bu_2SnX_2, where X is acetate, 2-ethylhexanoate, or laurate. The same compounds are also used to catalyse the addition of alcohols to isocyanates to produce polyurethanes.

The next major use of organotin compounds (30%) is as agricultural biocides and here triorganotins are the most active materials; the importance of this application can readily be appreciated since, at present, over one-third of the world's food crops are lost annually to pests such as fungi, bacteria, insects, or weeds. The great advantage of organotin compounds in these applications is that their toxic action is selective and there is little danger to higher (mammalian) life; furthermore, their inorganic degradation products are completely non-toxic. $Bu_3^n SnOH$ and Ph_3SnOAc control fungal growths such as potato blight and related infections of sugar-beet, peanuts, and rice. They also eradicate red spider mite from apples and pears. Other R_3SnX are effective in controlling insects, either by acting as chemosterilants or by killing the larvae. Again, $O(SnBu_3^n)_2$ is an excellent wood preserver, and derivatives of Ph_3Sn- and (cyclohexyl)$_3Sn-$ are also used for this. Related applications are as marine antifouling agents for timber-hulled boats: paints containing $Bu_3^n Sn-$ or Ph_3Sn- derivatives slowly release these groups and provide long-term protection against attachment of barnacles or attack by Teredo woodworm borers (Pl. 10.2). Cellulose and woollen fabrics are likewise protected against fungal attack or destruction by moths. R_3SnX are also used as bacteriostats to control slime in paper and wood-pulp manufacture.

Recently, Me_2SnCl_2 has been introduced in Japan as an alternative to $SnCl_4$ for coating glass with a thin film of SnO_2 since it is a non-corrosive solid which is easier to handle. The glass (or ceramic) surface is treated with Me_2SnCl_2 vapour at temperatures above 450° and, depending on the thickness of the oxide film produced, the glass is toughened and the surface can be rendered scratch-resistant, lustrous, or electroconductive (p. 448).

† For example, with MeCl at 175° in the presence of catalytic amounts of CH_3I and NEt_3, the yields were Me_2SnCl_2 (39%), $MeSnCl_3$ (6.6%), Me_3SnCl (4.6%).

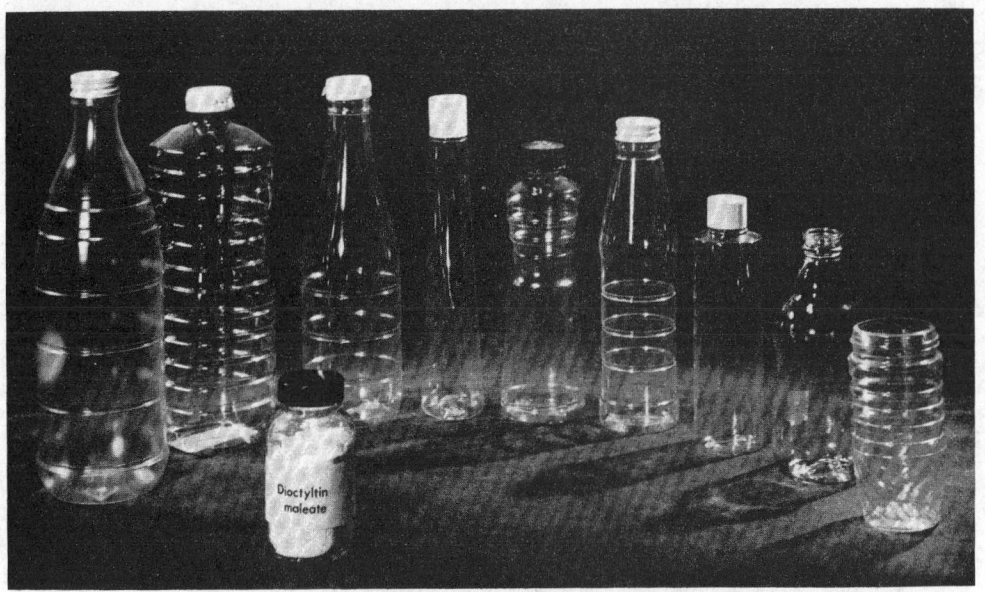

PLATE 10.1   These bottles and jars for food products illustrate the excellent clarity of rigid PVC stabilized with dioctyltin compounds. (Courtesy of International Tin Research Institute.)

PLATE 10.2   Sea water immersion tests show the effectiveness of antifouling paints in preventing the attachment of marine organisms—the untreated panels are heavily encrusted with marine growth. (International Red Hand Marine Coatings.)

Wurtz-type reaction $(SnCl_4 + 4RCl \xrightarrow{8\ Na} SnR_4 + 8NaCl)$. Conversion of SnR_4 to the partially halogenated species is readily achieved by scrambling reactions with $SnCl_4$. Reduction of R_nSnX_{4-n} with $LiAlH_4$ affords the corresponding hydrides and hydrostannation (addition of Sn–H) to C–C double and triple bonds is an attractive route to unsymmetrical or heterocyclic organotin compounds.

Most organotin compounds can be regarded as derivatives of $R_nSn^{IV}X_{4-n}$ $(n = 1\text{–}4)$ and even compounds such as SnR_2 or $SnAr_2$ are in fact cyclic oligomers $(Sn^{IV}R_2)_x$ (p. 462). The physical properties of tetraorganostannanes closely resemble those of the corresponding hydrocarbons or tetraorganosilanes but with higher densities, refractive indices, etc. They are colourless, monomeric, volatile liquids or solids. Chemically they resist hydrolysis or oxidation under normal conditions though when ignited they burn to SnO_2, CO_2, and H_2O. Ease of Sn–C cleavage by halogens or other reagents varies considerably with the nature of the organic group and generally increases in the sequence Bu (most stable) $<$ Pr $<$ Et $<$ Me $<$ vinyl $<$ Ph $<$ Bz $<$ allyl $<$ CH_2CN $<$ CH_2CO_2R (least stable). The lability of Sn–C bonds and the ease of redistribution in mixed organostannane systems frustrated early attempts to prepare optically active tin compounds and the first synthesis of a 4-coordinate Sn compound in which the metal is the sole chiral centre was only achieved in 1971 with the isolation and resolution of [MeSn(4-anisyl)(1-naphthyl) $\{CH_2CH_2C(OH)Me_2\}]$.[37]

The association of SnR_4 via bridging alkyl groups (which is such a notable feature of many organometallic compounds of Groups I–III) is not observed at all. However, many compounds of general formula R_3SnX or R_2SnX_2 are strongly associated via bridging X-groups which thereby raise the coordination number of Sn to 5, 6, or even 7. As expected, F is more effective in this role than the other halogens (why?). For example Me_3SnF has a chain structure with trigonal bipyramidal coordination about Sn (Fig. 10.13a† and Me_2SnF_2 has a layer structure with octahedral Sn and *trans*-Me groups above and below the F-bridged layers as in SnF_4 (p. 443). The weaker Cl bridging in Me_2SnCl_2 leads to the more distorted structure shown in Fig. 10.13b. By contrast the O atom is even more effective than F and, amongst the numerous compounds R_3SnOR' and $R_2Sn(OR')_2$ that have been studied by X-ray crystallography, the only ones with 4-coordinate tin (presumably because of the bulky ligands) are 1,4-$(Et_3SnO)_2C_6Cl_4$ and $[Mn(CO)_3\{\eta^5\text{-}C_5Ph_4(OSnPh_3)\}]$.

Catenation is well established in organotin chemistry and distannane derivatives are readily prepared by standard methods (see Ge, p. 459). The compounds are more reactive than organodigermanes; e.g. Sn_2Me_6 (mp 23°) inflames in air at its bp (182°) and absorbs oxygen slowly at room temperature to give $(Me_2Sn)_2O$. Typical routes to higher polystannanes are:

$$2Me_3SnBr + NaMe_2SnSnMe_2Na \xrightarrow{\text{liq } NH_3} Me_3Sn(SnMe_2)_2SnMe_3 \text{ (oil)}$$

$$3Ph_3SnLi + SnCl_2 \longrightarrow [(Ph_3Sn)_3SnLi] \xrightarrow{Ph_3SnCl} Sn(SnPh_3)_4 \text{ (mp } \sim 320°)$$

† The volatile compound Me_3SnCl (mp 39.5°, bp 154°) has a similar chain structure at low temperature[37a] whereas Ph_3SnCl and Ph_3SnBr are monomeric molecules in the crystal.

[37] M. GIELEN, From kinetics to the synthesis of chiral tetraorganotin compounds, *Acc. Chem. Res.* **6**, 198–202 (1973).

[37a] M. B. HOSSAIN, J. L. LEFFERTS, K. C. MOLLOY, D. VAN DER HELM, and J. J. ZUCKERMAN, The crystal and molecular structure of trimethyltin chloride at 135 K. A highly volatile organotin polymer, *Inorg. Chim. Acta* **36**, L409–L410 (1979).

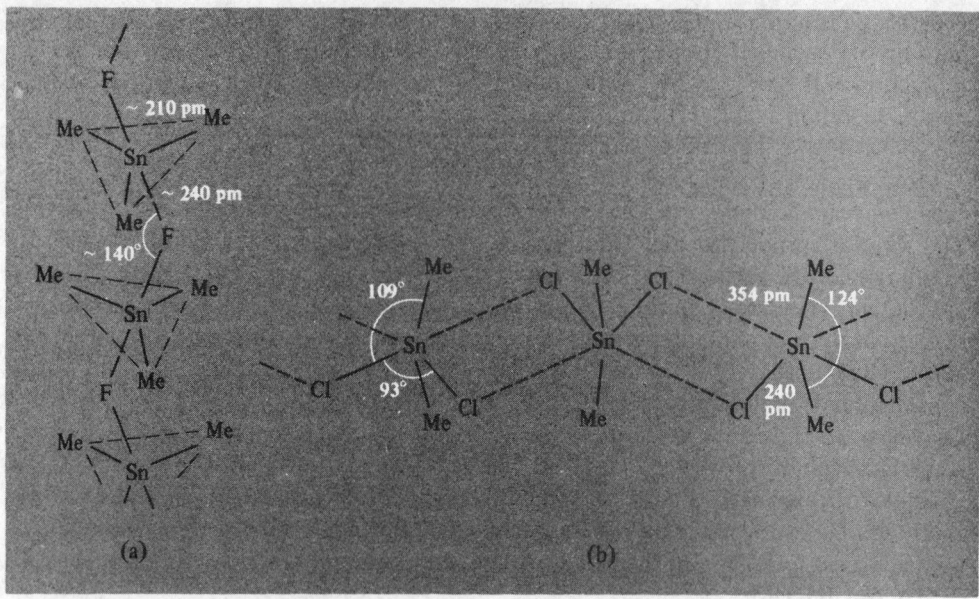

FIG. 10.13   Crystal structure of (a) Me₃SnF, and (b) Me₂SnCl₂, showing tendency to polymerize via Sn—X···Sn bonds.

Cyclo-dialkyl stannanes(IV) can also be readily prepared, e.g. reaction of Me₂SnCl₂ with Na/liq NH₃ yields cyclo-(SnMe₂)₆ together with acyclic X(SnMe₂)ₙX (n = 12–20). Yellow crystalline cyclo-(SnEt₂)₉ is obtained almost quantitatively when Et₂SnH₂, dissolved in toluene/pyridine, is catalytically dehydrogenated at 100° in the presence of a small amount of Et₂SnCl₂. Similarly, under differing conditions, the following have been prepared:[34] (SnEt₂)₆, (SnEt₂)₇, (SnBu₂ᵗ)₄, (SnBu₂ⁱ)₄, (SnBu₂ⁱ)₆, and (SnPh₂)₅. The compounds are highly reactive yellow or red oils or solids. A crystal structure of the colourless hexamer (SnPh₂)₆ shows that it exists in the chair conformation (I) with Sn–Sn distances very close to the value of 280 pm in α-Sn (p. 434).

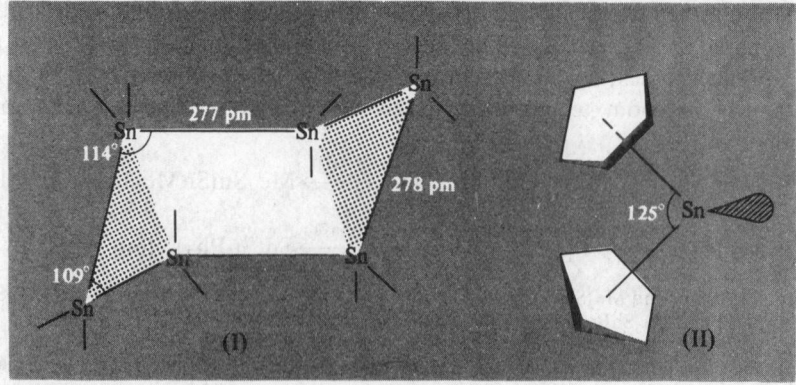

True monomeric organotin(II) compounds have proved exceedingly elusive. The cyclopentadienyl compound Sn(η⁵-C₅H₅)₂ (which is obtained as white crystals mp 105°

from the reaction of NaC_5H_5 and $SnCl_2$ in thf) probably has the angular structure II in the gas phase. Stabilization can also be achieved by the use of bulky ligands (see also p. 454), and this approach has recently led to the first preparation of 2- and 3-coordinate organo-derivatives of Ge^{II}, Sn^{II}, and Pb^{II}, e.g.:[38]

$$SnCl_2 + 2[LiCH(SiMe_3)_2] \xrightarrow{\text{Et}_2O} [Sn\{CH(SiMe_3)_2\}_2]$$

The compound was obtained as air-sensitive red crystals (mp 136°); it is monomeric in benzene and behaves chemically as a "stannylene", displacing CO from $M(CO)_6$ to give orange $[Cr(CO)_5(SnR_2)]$ and yellow $[Mo(CO)_5(SnR_2)]$. The corresponding lead compound $[Pb\{CH(SiMe_3)_2\}_2]$ is purple. An alternative route reacts $2[LiCH(SiMe_3)_2]$ with $[Sn\{N(SiMe_3)_2\}_2]$ and an analogous reaction with $[Ge\{N(SiMe_3)_2\}_2]$ yielded the yellow crystalline $[Ge\{CH(SiMe_3)_2\}_2]$, mp 180°. A crystal structure determination of the Sn compound shows that it dimerizes in the solid state, perhaps by donation of the lone-pair hybrid ($\sim sp_xp_y$) on each Sn into the "vacant" p_z orbital of its neighbour to give a weak, bent, double bond as illustrated schematically below; this would interpret the orientation of the $-CH(SiMe_3)_2$ groups attached terminally to the tin atoms.

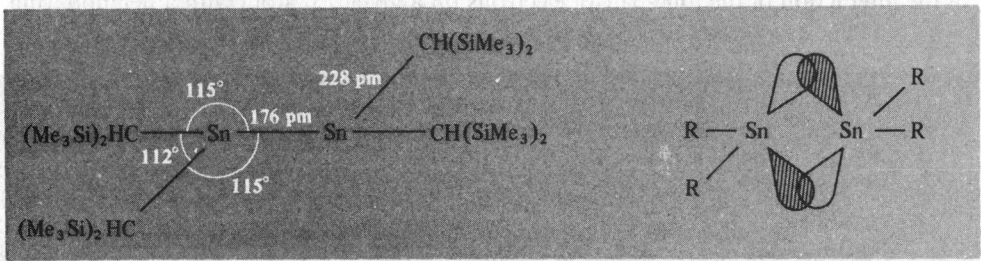

Lead[3]

The organic chemistry of Pb is much less extensive than that of Sn, though over 2000 organolead compounds are known and $PbEt_4$ is produced on a larger tonnage than any other single organometallic compound (p. 432). The most useful laboratory-scale routes to organoleads involve the use of LiR, RMgX, or AlR_3 on lead(II) compounds such as $PbCl_2$, or lead(IV) compounds such as $R_2'PbX_2$, $R_3'PbX$, or K_2PbCl_6. On the industrial scale the reaction of RX on a Pb/Na alloy is much used; an alternative is the electrolysis of RMgX, M^IBR_4, or M^IAlR_4 using a Pb anode. The simple tetraalkyls are volatile, monomeric molecular liquids which can be steam-distilled without decomposition; $PbPh_4$ (mp 227–228°) is even more stable thermally: it can be distilled at 240° (15–20 mmHg) but decomposes above 270°. Diplumbanes Pb_2R_6 are much less stable and higher polyplumbanes are unknown except for the thermally unstable, reactive red solid, $Pb(PbPh_3)_4$.

The decreasing thermal stability of Group IV organometallics with increasing atomic

[38] P. J. DAVIDSON and M. F. LAPPERT, Stabilization of metals in low coordinative environment using the bis(trimethylsilyl)methyl ligand; coloured Sn^{II} and Pb^{II} alkyls, $M[CH(SiMe_3)_2]_2$, *JCS Chem. Comm.* 1973, 317. See also M. F. LAPPERT *et al.*, *JCS Chem. Comm.* 1976, 261–2; *JCS Dalton* 1976, 2268–74, 2275–86, 2286–90.

number of M reflects the decreasing M–C and M–M bond energies. This in turn is related to the increasing size of M and the consequent increasing interatomic distance (see table).

M	C	Si	Ge	Sn	Pb
M–C distance in MR_4/pm	154	194	199	217	227

Parallel with these trends and related to them is the increase in chemical reactivity which is further enhanced by the increasing bond polarity and the increasing availability of low-lying vacant orbitals for energetically favourable reaction pathways.

It is notable that the preparation of alkyl and aryl derivatives from Pb^{II} starting materials always results in Pb^{IV} organometallic compounds. The only well-defined examples of Pb^{II} organometallics are purple compound $Pb[CH(SiMe_3)_3]_2$ and the cyclopentadienyl compound $Pb(\eta^5\text{-}C_5H_5)_2$ and its ring-methyl derivative. Like the Sn analogue (p. 462) $Pb(\eta^5\text{-}C_5H_5)_2$ features non-parallel cyclopentadienyl rings in the gas phase, the angle subtended at Pb being $135 \pm 15°$. Two crystalline forms are known and the orthorhombic polymorph has the unusual chain-like structure shown in Fig. 10.14:[39] 1 C_5H_5 is between 2 Pb and perpendicular to the Pb–Pb vector whilst the other C_5H_5 is bonded (more closely) to only 1 Pb. The chain polymer can be thought to arise as a result of the interaction of the lone-pair of electrons on a given Pb atom with a neighbouring

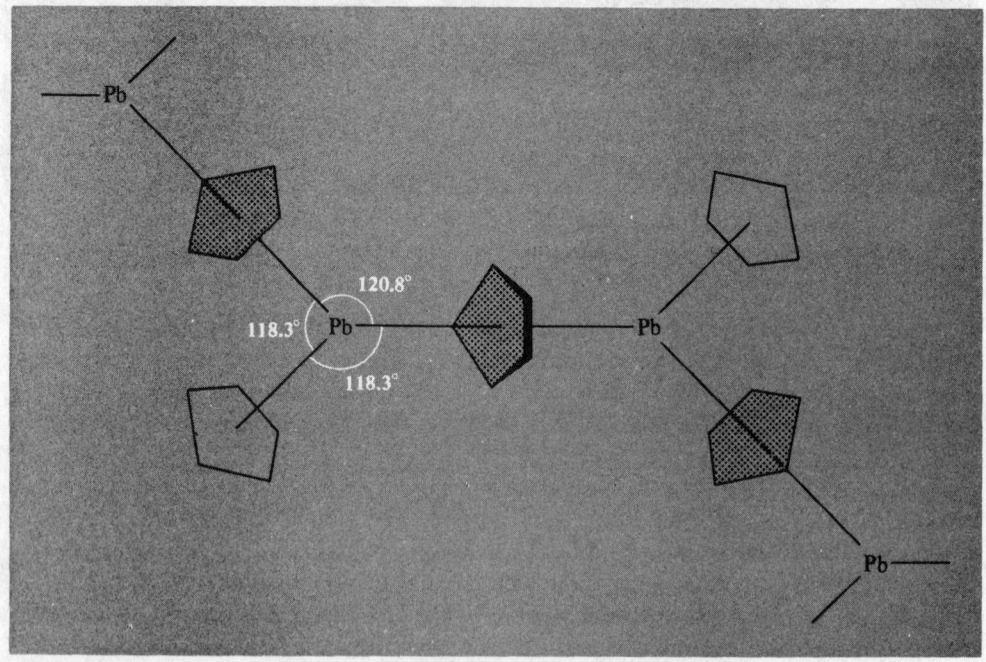

FIG. 10.14   Schematic diagram of the chain structure of orthorhombic $Pb(\eta^5\text{-}C_5H_5)_2$. For the doubly coordinate C_5H_5 ring (shaded) $Pb\text{-}C_{av}$ is 306 pm, and for the "terminal" C_5H_5 ring $Pb\text{-}C_{av}$ is 276 pm; the $Pb\cdots Pb$ distance within the chain is 564 pm.

[39] C. PANATTONI, G. BOMBIERI, and U. CROATO, Chain structure of the orthorhombic modification of dicyclopentadienyl-lead containing bridging π-cyclopentadienyl rings, *Acta Cryst.* **21**, 823–6 (1966).

(chain) C_5H_5 ring; a 3-centre bond is constructed by overlapping 2 opposite sp^2 hybrids on 2 successive Pb atoms in the chain with the σ MO (A''_2) of the C_5H_5 group: this forms 1 bonding, 1 nonbonding, and 1 antibonding MO of which the first 2 are filled and the third empty.

Another unusual organo-Pb^{II} compound is the η^6-benzene complex [Pb^{II}(AlCl$_4$)$_2$(η^6-C_6H_6)]$\cdot$C$_6$H$_6$ in which Pb^{II} is in a distorted pentagonal bipyramidal site with 1 axial Cl and the other axial site occupied by the centre of the benzene ring (Fig. 10.15). The other C_6H_6 is a molecule of solvation far removed from the metal. One {AlCl$_4$} group chelates the Pb in an axial-equatorial configuration and the other {AlCl$_4$} chelates and bridges neighbouring Pb atoms to form a chain. There is a similar Sn^{II} compound with the same structure. The original paper should be consulted for a discussion of the bonding.[40]

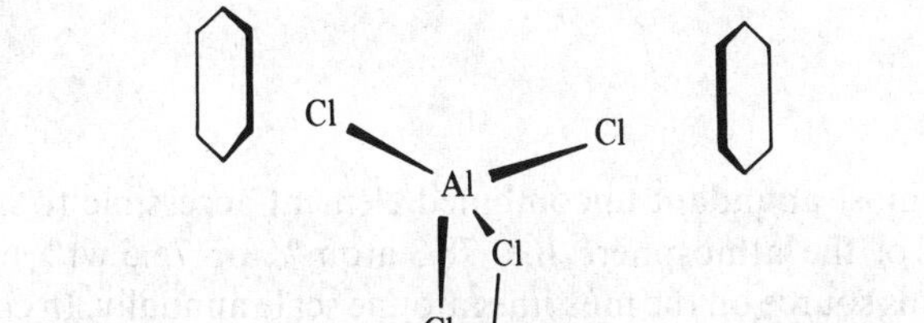

FIG. 10.15   Schematic diagram of the chain structure of [Pb^{II}(AlCl$_4$)$_2$(η^6-C$_6$H$_6$)]$\cdot$C$_6$H$_6$: Pb–Cl varies from 285–322 pm, Pb–C$_{av}$(bound) 311 pm, Pb–centre of C$_6$H$_6$(bound) 277 pm.

[40] A. G. GASH, P. F. RODESILER, and E. L. AMMA, Synthesis, structure, and bonding of π-C$_6$H$_6$Pb(AlCl$_4$)$_2$·C$_6$H$_6$, *Inorg. Chem.* **13**, 2429–24 (1974). See also J. L. LEFFERTS, M. B. HOSSAIN, K. C. MOLLOY, D. VAN DER HELM, and J. J. ZUCKERMAN, X-ray structure analysis of bis(*O,O'*-diphenyldithiophosphato)tin(II), a bicyclic dimer held together by η^6-C$_6$H$_6$/Sn^{II} interactions, *Angew. Chem.*, Int. Edn. (Engl.) **19**, 309–10 (1980).

Periodic table

1 H	2 He																
3 Li	4 Be											5 B	6 C	7 N	8 O	9 F	10 Ne
11 Na	12 Mg											13 Al	14 Si	15 P	16 S	17 Cl	18 Ar
19 K	20 Ca	21 Sc	22 Ti	23 V	24 Cr	25 Mn	26 Fe	27 Co	28 Ni	29 Cu	30 Zn	31 Ga	32 Ge	33 As	34 Se	35 Br	36 Kr
37 Rb	38 Sr	39 Y	40 Zr	41 Nb	42 Mo	43 Tc	44 Ru	45 Rh	46 Pd	47 Ag	48 Cd	49 In	50 Sn	51 Sb	52 Te	53 I	54 Xe
55 Cs	56 Ba	57 La	72 Hf	73 Ta	74 W	75 Re	76 Os	77 Ir	78 Pt	79 Au	80 Hg	81 Tl	82 Pb	83 Bi	84 Po	85 At	86 Rn
87 Fr	88 Ra	89 Ac	104 Unq	105 Unp	106 Unh	107 Uns											

58 Ce	59 Pr	60 Nd	61 Pm	62 Sm	63 Eu	64 Gd	65 Tb	66 Dy	67 Ho	68 Er	69 Tm	70 Yb	71 Lu
90 Th	91 Pa	92 U	93 Np	94 Pu	95 Am	96 Cm	97 Bk	98 Cf	99 Es	100 Fm	101 Md	102 No	103 Lr

11
Nitrogen

11.1 Introduction

Nitrogen is the most abundant uncombined element accessible to man. It comprises 78.1% by volume of the atmosphere (i.e. 78.3 atom% or 75.5 wt%) and is produced industrially from this source on the multimegatonne scale annually. In combined form it is essential to all forms of life, and constitutes, on average, about 15% by weight of proteins. The industrial fixation of nitrogen for agricultural fertilizers and other chemical products is now carried out on a vast scale in many countries, and the tonnage of anhydrous ammonia manufactured is exceeded only by those of sulfuric acid and lime. Indeed, of the top-twelve "high-volume" industrial chemicals produced in the USA, four contain nitrogen (Fig. 11.1).[1] This has important consequences, predominantly beneficial but occasionally detrimental, since of all man's recent interventions in the cycles of nature the industrial fixation of nitrogen is by far the most extensive. These aspects will be discussed further in later sections.

The "discovery" of nitrogen in 1772 is generally credited to Daniel Rutherford, though the gas was also isolated independently about the same time by both C. W. Scheele and H. Cavendish.[2] Rutherford (at the suggestion of his teacher Joseph Black who had earlier discovered CO_2, p. 297) was studying the properties of the residual "air" left after carbonaceous substances were burned in a limited supply of air; he removed the CO_2 by means of KOH and so obtained nitrogen which he thought was ordinary air that had taken up phlogiston from the combusted material. The elementary nature of nitrogen was disputed by some even as late as 1840 despite the work of A. L. Lavoisier. The name "nitrogen" was suggested by Jean-Antoine-Claude Chaptal in 1790 when it was realized that the element was a constituent of nitric acid and nitrates (Greek νίτρον, nitron; γεννᾶν, to form). Lavoisier preferred azote (Greek ἀζωτικός, no life) because of the asphyxiating properties of the gas, and this name is still used in the French language and in such forms as azo, diazo, azide, etc. The German name Stickstoff refers to the same property (sticken, to choke or suffocate).

---

[1] Top 50 Chemicals, Chem. Eng. News 9 June, 1980, 36. (See also ibid. 14 June, 1982, 33.)

[2] M. E. WEEKS, in H. M. LEICESTER (ed.), Discovery of the Elements, 6th edn., Journal of Chemical Education Publication, 1956: Nitrogen, pp. 205–8; Rutherford, discoverer of nitrogen, pp. 235–51; Old compounds of nitrogen, pp. 188–95.

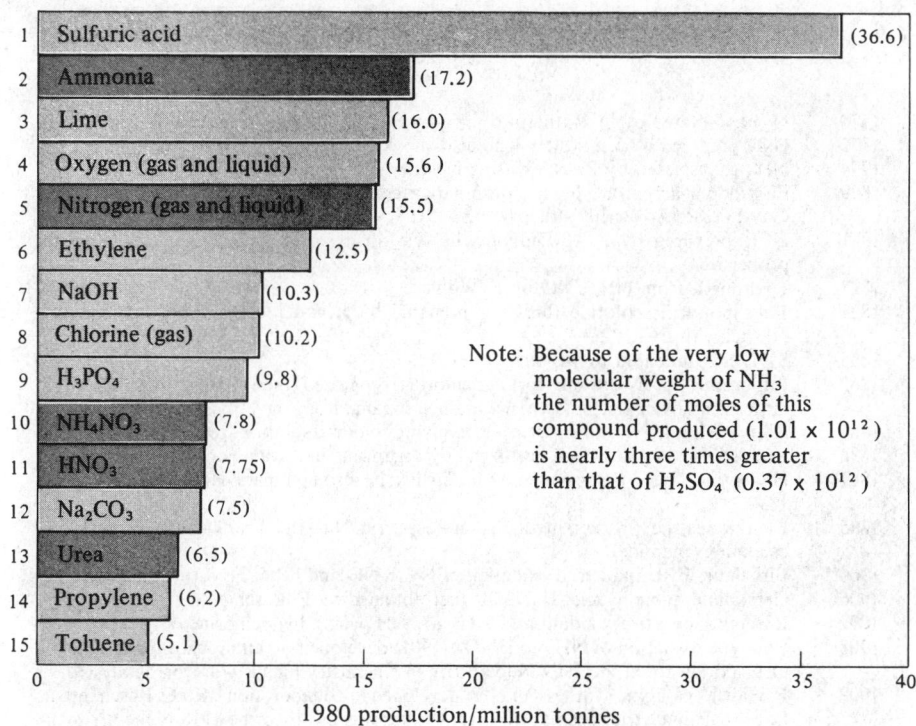

		1980 production/million tonnes
1	Sulfuric acid	(36.6)
2	Ammonia	(17.2)
3	Lime	(16.0)
4	Oxygen (gas and liquid)	(15.6)
5	Nitrogen (gas and liquid)	(15.5)
6	Ethylene	(12.5)
7	NaOH	(10.3)
8	Chlorine (gas)	(10.2)
9	H_3PO_4	(9.8)
10	NH_4NO_3	(7.8)
11	HNO_3	(7.75)
12	Na_2CO_3	(7.5)
13	Urea	(6.5)
14	Propylene	(6.2)
15	Toluene	(5.1)

Note: Because of the very low molecular weight of NH_3 the number of moles of this compound produced (1.01×10^{12}) is nearly three times greater than that of H_2SO_4 (0.37×10^{12})

FIG. 11.1 US production of industrial chemicals (1980).

Compounds of nitrogen have an impressive history. Ammonium chloride was first mentioned in the *Historia* of Herodotus (fifth century BC)[†] and ammonium salts, together with nitrates, nitric acid, and aqua regia, were well known to the early alchemists.[2] Some important dates in the subsequent development of the chemistry of nitrogen are given in the Panel. Exciting discoveries are still being made at the present time and, indeed, the mechanisms by which bacteria fix nitrogen at ambient temperatures and pressures has been termed "one of the most conspicuously challenging, unsolved problems in chemistry".[3] Several recent reviews, monographs, and proceedings of symposia have been published.[3a, 3b, 3c]

† "There are pieces of salt in large lumps on the hills of Libya and the Ammonians who live there worship the god Ammon in a temple resembling that of the Theban Jupiter." (Greek Αμμων, the name of the Egyptian diety Amun, whence *sal ammoniac* from ἀμμωνιακόν, belonging to Ammon.)

³ F. A. COTTON and G. WILKINSON, *Advanced Inorganic Chemistry*, 3rd edn., p. 345, Interscience, 1972; see also K. J. SKINNER, Nitrogen fixation, *Chem. Eng. News* 4 Oct. 1976, 22–35.
³ᵃ R. L. RICHARDS, Nitrogen fixation, *Educ. Chem.* **16**, 66–69 (1979).
³ᵇ R. W. F. HARDY, F. BOTTOMLEY, and R. C. BURNS (eds.), *A Treatise on Dinitrogen Fixation*, Sections 1 and 2, Wiley, New York 1979, 812 pp.
³ᶜ J. CHATT, L. M. DA C. PINA, and R. L. RICHARDS, *New Trends in the Chemistry of Nitrogen Fixation*, Academic Press, London, 1980, 284 pp. (See also ref. 17a.)

Time Chart for Nitrogen Chemistry

1772	N_2 gas isolated by D. Rutherford (also by C. W. Scheele and H. Cavendish).
1772	N_2O prepared by J. Priestley who also showed it supported combustion.
1774	NH_3 gas isolated by Priestley using mercury in a pneumatic trough.
1809	First donor-acceptor adduct (coordination compound) $NH_3.BF_3$ prepared by J. L. Gay Lussac (A. Werner's theory, 1891–5).
1811	NCl_3 prepared by P. L. Dulong who lost an eye and three fingers studying its properties.
1828	Urea made from NH_4CNO by F. Wöhler.
1832	Phosphonitrilic chloride $(NPCl_2)_x$ prepared by J. von Liebig by heating NH_3 or NH_4Cl with PCl_5.
1835	S_4N_4 first prepared by M. Gregory.
1862	Importance of N *in soil* for agriculture recognized (von Liebig, despite fierce opposition, having steadfastly maintained it came from the atmosphere directly).
1864	Ability of liquid NH_3 to dissolve metals giving coloured solutions reported by W. Weyl.
1886	Atmospheric N_2 shown to be "fixed" by organisms in certain root nodules.
1887	Hydrazine, N_2H_4, first isolated by T. Curtius; he also first made HN_3 (from N_2H_4) in 1890.
1895	First industrial process involving atmospheric N_2—the Frank–Caro process for calcium cyanamide.
1900	Birkeland–Eyde industrial oxidation of N_2 to NO and hence HNO_3 (now obsolete).
1906	Crystalline sulfamic acid, H_3NSO_3, first obtained by F. Raschig.
1907	Raschig's industrial oxidation of NH_3 to N_2H_4 using hypochlorite.
1908	Catalytic oxidation of NH_3 to HNO_3 (1901) developed on an industrial scale by W. Ostwald (awarded the 1909 Nobel Prize in Chemistry for his work on catalysis).
1909	F. Haber's catalytic synthesis of NH_3 developed in collaboration with C. Bosch into a large-scale industrial process by 1913. (Haber was awarded the 1918 Nobel Prize in Chemistry "for the synthesis of ammonia from its elements", and Bosch shared the 1931 Nobel Prize for his "contributions to the invention and development of chemical high-pressure methods", the Haber synthesis of NH_3 being the first high-pressure industrial process.)
1925	Borazine, $(HBNH)_3$, analogous to benzene prepared by A. Stock and E. Pohland.
1928	NF_3 first prepared by O. Ruff and E. Hanke, 117 y after NCl_3.
1925–35	Spectrum of atomic N gradually analysed.
1929	Discovery of a nitrogen isotope ^{15}N by S. M. Naudé following the discovery of isotopes of O and C by others earlier in the same year.
1934	Microwave absorption in NH_3 (due to molecular inversion) first observed—this marks the start of microwave spectroscopy.
1950	Nuclear magnetic resonance in compounds, containing ^{14}N and ^{15}N first observed by W. E. Proctor and F. C. Yu.
1957	N_2F_4 first made by C. B. Colburn and A. Kennedy and later (1961) shown to be in dissociative equilibrium with paramagnetic NF_2 above 100°C.
1958	$NH_3.BH_3$ isoelectronic with ethane prepared by S. G. Shore and R. W. Parry (direct reaction of NH_3 and B_2H_6 gives $[BH_2(NH_3)_2]^+[BH_4]^-$).
1962	First "bent" NO complex encountered, viz. $[Co(NO)(S_2CNMe_2)_2]$ (P. R. H. Alderman, P. G. Owston, and J. M. Rowe).
1965	First N_2 ligand complex prepared by A. D. Allan and C. V. Senoff.
1966	ONF_3 (isoelectronic with CF_4) discovered independently by two groups.
1968	N_2 recognized as a bridging ligand in $[(NH_3)_5RuN_2Ru(NH_3)_5]^{4+}$ (D. F. Harrison, E. Weissberger, and H. Taube).
1974	First thionitrosyl (NS) complex isolated by J. Chatt and J. R. Dilworth.
1975	$(SN)_x$ polymer, known since 1910, found to be metallic (and a superconductor at temperatures below 0.33 K).

11.2 The Element

11.2.1 *Abundance and distribution*

Despite its ready availability in the atmosphere, nitrogen is relatively unabundant in the crustal rocks and soils of the earth. At 19 ppm it is equal 33rd with Ga in the order of abundance, and similar to Nb (20 ppm) and Li (18 ppm). The only major minerals are KNO_3 (nitre, saltpetre) and $NaNO_3$ (sodanitre, Chile saltpetre). Both occur widespread, usually in small amounts as evaporites in arid regions, often as an efflorescence on soils or in caverns. $NaNO_3$ is isomorphous with calcite (p. 119) whereas KNO_3 is isomorphous with aragonite (p. 119), thus reflecting the similar size of NO_3^- and CO_3^{2-}, and the fact that K^+ is considerably larger than Na^+ and Ca^{2+}. Major deposits of KNO_3 occur in India and there are smaller amounts in Bolivia, Italy, Spain, and the USSR. There are vast deposits of $NaNO_3$ in the desert regions of northern Chile where it occurs with other evaporites such as $NaCl$, Na_2SO_4, and KNO_3 on the eastern slopes of the coastal ranges at an elevation of 1200–2500 m; the nitrates may have resulted from the oxidative decomposition products of vegetable and animal remains or alternatively from volcanic sources, though the detailed process by which they became concentrated in this highly localized belt 700 km long and 10–60 km broad remain unclear. The beds are usually about 1 m below the sandy overburden and vary both in thickness (0.2–4.3 m) and in concentration of $NaNO_3$ (10–30%). Because of the development of the synthetic ammonia and nitric acid industries these large deposits are no longer a major source of nitrates, though they played an important role in agriculture until the 1920s (as also did guano, the massive deposits of bird excreta on certain islands).

The continuous interchange of nitrogen between the atmosphere and the biosphere is called the nitrogen cycle. Global estimates are difficult to obtain and there are frequently regional and local impacts which vary greatly from the mean. However, some indication of the size of the various "reservoirs" of nitrogen in the atmosphere, on land, and in the seas is given by Fig. 11.2 together with the estimated annual rate of transfer between these various pools.[4–6] Estimates frequently vary by a factor of 3 or more. Atmospheric nitrogen is fixed by biological action (p. 1205), industrial processes (p. 482), and to a significant extent by fires, lightning, and other atmospheric discharges which produce NO_x. There is also a minute production (on the global scale) of NO_x from internal combustion engines and from coal-burning, though the local concentration in some urban environments can be very high and extremely unpleasant.[7,8] Absorption of fixed nitrogen by both terrestrial and aquatic plants leads to protein synthesis followed by death, decay, oxidation, and denitrification by bacterial and other action which eventually returns the nitrogen to the seas and the atmosphere as N_2. An alternative sequence involves the digestion of plants by animals, the synthesis of animal proteins, excretion of nitrogenous material, and, again, ultimate death, decay, and denitrification. Figure 11.2 indicates that the greatest anthropogenic impact on the cycle arises from the industrial

[4] C. C. DELWICHE, The nitrogen cycle, Chap. 5 in C. L. HAMILTON (ed.), *Chemistry in the Environment, Readings from Scientific American*, W. H. Freeman, San Francisco, 1973.

[5] B. H. SVENSSON and R. SÖDERLUND (eds.), *Nitrogen, Phosphorus, and Sulfur—Global Biogeochemical Cycles*, SCOPE Report No. 7, Sweden 1976, 170 pp. Also available from the Swedish National Science Research Center, Wenner-Gren Center, Box 23136, S-10435, Stockholm, Sweden.

[6] SCOPE Report No. 10, *Environmental Issues*, Wiley, New York, 1977, 220 pp.

[7] J. HEICKLEN, *Atmospheric Chemistry*, Academic Press, 1976, 406 pp.

[8] I. M. CAMPBELL, *Energy and the Atmosphere*, The nitrogen cycle, pp. 198–202, Wiley, London, 1971.

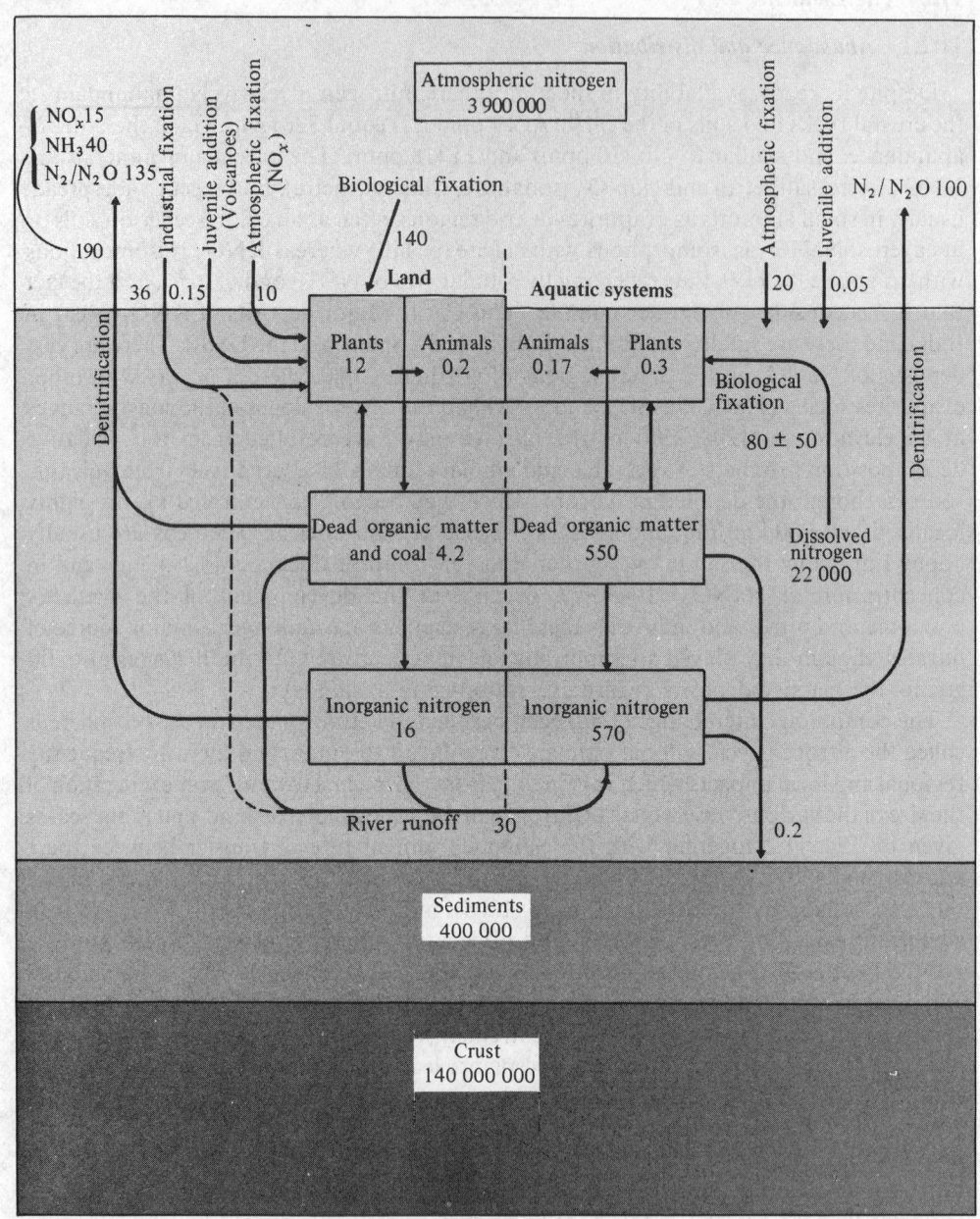

FIG. 11.2 Distribution of nitrogen in the biosphere and annual transfer rates can be estimated only within broad limits. The two quantities known with high confidence are the amount of nitrogen in the atmosphere and the rate of industrial fixation. The inventories (within the boxes) are expressed in terms of 10^9 tonnes of N; the transfers (indicated by arrows) are in 10^6 tonnes of N. Taken from ref. 4 with more recent data from ref. 5.

fixation of nitrogen by the Haber and other processes. Much of this is used beneficially as fertilizers but the leaching of excess nitrogenous material can lead to eutrophication in freshwater systems, and the increased nitrate concentration in waters used for human consumption can pose a health hazard.[8a] Nevertheless, there is no doubt that the high yields of agriculture necessary to maintain even the present human population of the world cannot be achieved without the judicious application of manufactured nitrogenous fertilizers. Concern has also been expressed that increasing levels of N_2O following denitrification may eventually impoverish the ozone layer in the stratosphere. Much more data are required and the subject is being actively pursued by several international agencies as well as by national and local governments and individual scientists.

11.2.2 *Production and uses of nitrogen*

The only important large-scale process for producing N_2 is the liquefaction and fractional distillation of air[9] (see Panel). Production has grown remarkably during the past few years, partly as a result of the increasing demand for its coproduct O_2 for steelmaking. For example, US domestic production has increased 250-fold in the past 25 y from 0.12 million tonnes in 1955 to 30 million tonnes in 1980.[10] In 1974 world production was 23.5 million tonnes (USA 34%; Eastern Europe 23%; Western Europe 20%; Asia 19%). Commercial N_2 is a highly purified product, typically containing less than 20 ppm O_2. Specially purified "oxygen-free" N_2, containing less than 2 ppm, is available commercially, and "ultrapure" N_2 (99.999%) containing less than 10 ppm Ar is also produced on the multitonne per day scale.

Laboratory routes to highly purified N_2 are seldom required. Thermal decomposition of sodium azide at 300°C under carefully controlled conditions is one possibility:

$$2NaN_3 \longrightarrow 2Na + 3N_2$$

Hot aqueous solutions of ammonium nitrite also decompose to give nitrogen though small amounts of NO and HNO_3 are also formed (p. 497) and must be removed by suitable absorbents such as dichromate in aqueous sulfuric acid:

$$NH_4NO_2 \xrightarrow{\text{aq}} N_2 + 2H_2O$$

Other routes are the thermal decomposition of $(NH_4)_2Cr_2O_7$, the reaction of NH_3 with bromine water, or the high-temperature reaction of NH_3 with CuO; overall equations can be written as:

$$(NH_4)_2Cr_2O_7 \longrightarrow N_2 + Cr_2O_3 + 4H_2O$$

$$8NH_3 + 3Br_2 \xrightarrow{\text{aq}} N_2 + 6NH_4Br(aq)$$

$$2NH_3 + 3CuO \longrightarrow N_2 + 3Cu + 3H_2O$$

[8a] PANEL ON NITRATES OF THE NATIONAL RESEARCH COUNCIL COORDINATING COMMITTEE FOR SCIENTIFIC AND TECHNICAL ASSESSMENTS OF ENVIRONMENTAL POLLUTANTS, *Nitrates: An Environmental Assessment*, National Academy of Sciences Printing and Publishing Office, Washington DC, 1978, 723 pp.

[9] W. J. GRANT and S. L. REDFEARN, Industrial gases, in R. THOMPSON (ed.), *The Modern Inorganic Chemicals Industry*, Chem. Soc. Special Publ. **31**, 273–301 (1977).

[10] W. F. KEYES, Nitrogen, in *Mineral Facts and Problems: Bicentennial Edition 1975*, US Bureau of Mines Bulletin No. 667, pp. 749–60; Nitrogen, *Chem. Eng. News* 4 Aug. 1980, 13.

Industrial Gases from Air

Air is the source of six industrial gases, N_2, O_2, Ne, Ar, Kr, and Xe. As the mass of the earth's atmosphere is approximately 5×10^9 million tonnes, the supply is unlimited and the annual industrial production, though vast, is insignificant by comparison. The composition of air at low altitudes is remarkably constant, the main variable component being water vapour which ranges from ~4% by volume in tropical jungles to very low values in cold or arid climates. Other minor local variations result from volcanism or human activity. The main invariant part of the air has the following composition (% by volume, bp in parentheses):

N_2	78.03 (77.2 K)	CO_2	0.033 (194.7 K)	He	0.0005 (4.2 K)
O_2	20.99 (90.1 K)	Ne	0.0015 (27.2 K)	Kr	0.0001 (119.6 K)
Ar	0.93 (87.2 K)	H_2	0.0010 (20.2 K)	Xe	0.000008 (165.1 K)

Details of the production and uses of O_2 (p. 702) and the noble gases (p. 1044) are in later chapters.

About two-thirds of the N_2 produced industrially is supplied as a gas, mainly in pipes but also in cylinders under pressure. The remaining one-third is supplied as liquid N_2 since this is also a very convenient source of the dry gas. The main use is as an inert atmosphere in the iron and steel industry and in many other metallurgical and chemical processes where the presence of air would involve fire or explosion hazards or unacceptable oxidation of products. Thus, it is extensively used as a purge in petrochemical reactors and other chemical equipment, as an inert diluent for chemicals, and in the float glass process to prevent oxidation of the molten tin (p. 430). It is also used as a blanketing gas in the electronics industry, in the packaging of processed foods and pharmaceuticals, and to pressurize electric cables, telephone wires, and inflatable rubber tyres, etc.

About 10% of the N_2 produced is used as a refrigerant. Typical of such applications are (a) freeze grinding of normally soft or rubbery materials, (b) low-temperature machining of rubbers, (c) shrink fitting and assembly of engineering components, (d) the preservation of biological specimens such as blood, semen, etc., and (e) as a constant low-temperature bath ($-196°C$). Liquid N_2 is also frequently used for convenience in applications where a very low temperature is not essential such as (a) food freezing (and hamburger meat grinding), (b) in-transit refrigeration, (c) freeze branding of cattle, (d) pipe-freezing for stopping flow in the absence of valves, and (e) soil-freezing for consolidating unstable ground in tunnelling or excavation.

The cost of N_2, like that of O_2, is particularly dependent on electricity costs, though plant maintenance and transport costs also obtrude. Typical prices in 1979 for N_2 in the USA were about \$40 per tonne for bulk liquid (exclusive of transportation and handling charges). Costs for small-scale users of N_2 from gas cylinders are proportionately much higher.

11.2.3 *Atomic and physical properties*

Nitrogen has two stable isotopes ^{14}N (relative atomic mass 14.003 07, abundance 99.634%) and ^{15}N (15.000 11, 0.366%); their relative abundance (272:1) is almost invariant in terrestrial sources and corresponds to an atomic weight of 14.0067. Both isotopes have a nuclear spin and can be used in nmr experiments.[11] though the sensitivity at constant field is only one-thousandth that of ^1H. The ^{14}N nucleus has a spin quantum number of 1 and, in consequence, the spectra are broadened by quadrupole effects. The ^{15}N nucleus with spin $\frac{1}{2}$ does not have this difficulty though its low abundance poses

[11] B. E. Mann, Nitrogen NMR, Chap. 4 in B. E. Mann and R. K. Harris (eds.), *NMR and the Periodic Table*, pp. 96–98, Academic Press, 1979; *Nitrogen NMR*, M. Witanowski and G. A. Webb, Plenum, New York, 1973. G. J. Martin, M. L. Martin, and J.-P. Gouesnard, *NMR Volume 18: ^{15}N NMR Spectroscopy*, Springer-Verlag, Berlin, 1981, 382 pp.

problems.[11a] Interestingly, the first chemical shift to be observed in nmr spectroscopy ("as an annoying ambiguity in the magnetic moment of ^{14}N") was in 1950 in an aqueous solution of NH_4NO_3.[11b]

Isotopic enrichment of ^{15}N is usually effected by chemical exchange, and samples containing up to 99.5% ^{15}N have been obtained from the 2-phase equilibrium

$$^{15}NO(g) + {}^{14}NO_3{}^-(aq) \rightleftharpoons {}^{14}NO(g) + {}^{15}NO_3{}^-(aq)$$

Other exchange reactions that have been used are:

$$^{15}NH_3(g) + {}^{14}NH_4{}^+(aq) \rightleftharpoons {}^{14}NH_3(g) + {}^{15}NH_4{}^+(aq)$$
$$^{15}NO(g) + {}^{14}NO_2(g) \rightleftharpoons {}^{14}NO(g) + {}^{15}NO_2(g)$$

Fractional distillation of NO provides another effective route and, as the heavier isotope of oxygen is simultaneously enriched, the product has a high concentration of $^{15}N^{18}O$. Many key nitrogen compounds are now commercially available with ^{15}N enriched to 5%, 30%, or 95%, e.g. N_2, NO, NO_2, NH_3, HNO_3, and several ammonium salts and nitrates. Fortunately the use of these compounds in tracer experiments is simplified by the absence of exchange with atmospheric N_2 under normal conditions, in marked contrast with labelled H, C, and O compounds where contact with atmospheric moisture and CO_2 must be avoided.[12]

The ground state electronic configuration of the N atom is $1s^2 2s^2 2p_x^1 2p_y^1 2p_z^1$ with three unpaired electrons (^4S). The electronegativity of N (~ 3.0) is exceeded only by those of F and O. Its "single-bond" covalent radius (~ 70 pm) is slightly smaller than those of B and C, as expected; the nitride ion, N^{3-}, is much larger and has been assigned a radius in the range 140–170 pm. Ionization energies and other properties are compared with those of the other Group V elements (P, As, Sb, and Bi) on p. 642.

Molecular N_2, i.e. dinitrogen (see p. 40), (mp $-210°C$, bp $-195.8°C$) is a colourless, odourless, tasteless, diamagnetic gas. The short interatomic distance (109.76 pm) and very high dissociation energy (945.41 kJ mol^{-1}) are both consistent with multiple bonding. The free-energy change for the equilibrium $N_2 \rightleftharpoons 2N$ is $\Delta G = 911.13$ kJ mol^{-1} from which it is clear that the dissociation constant $K_p = [N]^2/[N_2]$ atm is negligible under normal conditions; it is 1.6×10^{-24} atm at 2000 K and still only 1.3×10^{-12} atm at 4000 K. Detailed tabulations of other physical properties of nitrogen are available.[13]

11.2.4 *Chemical reactivity*

Gaseous N_2 is rather inert at room temperature presumably because of the great strength of the N≡N bond and the large energy gap between the highest occupied molecular orbitals (HOMO) and the lowest unoccupied molecular orbitals (LUMO). Further contributory factors are the very symmetrical electron distribution in the molecule and the absence of bond polarity—when these are modified, as in the

[11a] G. C. LEVY and R. L. LICHTER, *Nitrogen-15 Nuclear Magnetic Resonance Spectroscopy*, Wiley, New York, 1979, 221 pp.
[11b] W. G. PROCTOR and F. C. YU, The dependence of a nuclear magnetic resonance frequency upon chemical compound, *Phys. Rev.* **77**, 717 (1950).
[12] W. SPINDEL, in R. H. HERBER (ed.), *Inorganic Isotopic Syntheses*, pp. 74–118, Benjamin, New York, 1962.
[13] B. R. BROWN, Physical properties of nitrogen, in *Mellor's Comprehensive Treatise on Inorganic and Theoretical Chemistry*, Vol. 8, Suppl. 1, *Nitrogen*, Part 1, pp. 27–149, Longmans, London, 1964.

isoelectronic analogues CO, CN^-, and NO^+, the reactivity is considerably enhanced. Nitrogen reacts readily with Li at room temperature (p. 88) and with several transition-element complexes (p. 476).

Reactivity increases rapidly with rising temperature and the element combines directly with Be, the alkaline earth metals, and B, Al, Si, and Ge to give nitrides (p. 479); hydrogen yields ammonia (p. 483), and coke yields cyanogen, $(CN)_2$, when heated to incandescence (p. 336). Many finely divided transition metals also react directly at elevated temperatures to give nitrides of general formula MN (M = Sc, Y, lanthanoids; Zr, Hf; V; Cr, Mo, W; Th, U, Pu). Although not always directly preparable from $N_2(g)$, many other nitrides are known (p. 479) and, indeed, nitrides as a class include some of the most stable compounds in the whole of chemistry. Nitrogen forms bonds with almost all elements in the periodic table, the only exceptions apparently being the noble gases (other than Xe, p. 1058). A wide range of stereochemistries is observed and typical examples of coordination numbers 0, 1, 2, 3, 4, 5, 6, and 8 are given in Table 11.1.

A particularly reactive form of nitrogen can be obtained by passing an electric discharge through $N_2(g)$ at a pressure of 0.1–2 mmHg.[13, 14] Atomic N is formed, and the process is accompanied by a peach-yellow emission which persists as an afterglow, often for several minutes after the discharge has been stopped. Atoms of N in their ground state (^4S) have a relatively long lifetime since recombination involves either a 3-body collision on the surface of the vessel (first-order reaction in N at pressures below ~ 3 mmHg) or a termolecular homogeneous association reaction (second order in N at pressures above ~ 3 mmHg):

$$N(^4S) + N(^4S) \xrightarrow{M} N_2^* \longrightarrow N_2 + h\nu \text{ (yellow)} \tag{1}$$

The molecules of N_2^* so formed are in an excited state $(B^3\Pi_g)$ and give rise to the emission of the first positive band system of the spectrum of molecular N_2 in returning to the ground state $(A^3\Sigma_u^+)$. The N-atom content of "active nitrogen" can be determined by its very fast reaction with the NO radical in a flow system. As the flow rate of NO is increased the colour changes successively from yellow to purple, blue, darkness (equal concentrations of N and NO), and greenish yellow. The NO reacts very rapidly with $N(^4S)$ and, at equal concentrations, quenches all emission in the visible region of the spectrum:

$$N(^4S) + NO \longrightarrow N_2 + O(^3P) \tag{2}$$

With insufficient NO the O atoms so formed react with the excess of N atoms to produce excited NO molecules which emit the blue β and γ bands of NO.

$$N(^4S) + O(^3P) \longrightarrow NO^* \longrightarrow NO + h\nu \text{ (blue)} \tag{3}$$

This, together with the simultaneous peach-yellow emission from reaction (1), gives the purple coloration, though near the end point the blue will predominate. With an excess of NO all the N atoms are rapidly eliminated by reaction (2) and the $O(^3P)$ reacts with the excess NO to form excited NO_2^* molecules which emit the greenish-yellow "airglow":

$$NO + O(^3P) \longrightarrow NO_2^* \longrightarrow NO_2 + h\nu \text{ (greenish yellow)} \tag{4}$$

[14] A. N. WRIGHT and C. A. WINKLER, *Active Nitrogen*, Academic Press, New York, 1968.

TABLE 11.1 *Stereochemistry of nitrogen*[a]

CN		Examples
0		N(g) in "active nitrogen"
1		N_2, NO, NNO, $[NNN]^-$, HNNN, $RC\equiv N$, $XC\equiv N$ (X = Hal), $[OsO_3N]^-$
2	*Linear:*	$[NO_2]^+$, NNO, $[NNN]^-$, HNNN; η^1-N_2 complexes, e.g. $[Ru(N_2)(NH_3)_5]^{2+}$; η^1-NO complexes, e.g. $[Fe(CN)_5(NO)]^{2-}$; μ_2-N complexes, e.g. $[(H_2O)Cl_4RuNRuCl_4(OH_2)]^{3-}$ and $[Cl_5WNWCl_5]^{2-}$ (see ref. 14a)
	Bent:	NO_2, $[NO_2]^-$, $[NH_2]^-$, HNNN, HNCO, RNCO, XNCO, N_2F_2, *cyclo*-$\overline{CH_2NN}$, *cyclo*-$[NSF(O)]_3$
3	*Planar:*	$[NO_3]^-$, N_2O_4, XNO_2, $(HO)NO_2$, $K[ON(NO)(SO_3)]$, $K_2[ON(SO_3)_2]$ (Fremy's salt), $N(SiH_3)_3$, $N(GeH_3)_3$, $N(PF_2)_3$, Si_3N_4, and Ge_3N_4 (Be_2SiO_4 structure, p. 400), μ_3-N complexes, e.g. $[\{H_2O(SO_4)_2Ir\}_3N]^{4-}$
	Pyramidal:	NH_3, NF_3, NH_2F, NHF_2, $(HO)NH_2$, N_2H_4, N_2F_4, $[N_4(CH_2)_6]$
	T-shaped:	$[Mo_3(\mu_3\text{-}N)O(\eta^5\text{-}C_5H_5)_3(CO)_4]$ (see ref. 14b)
4	*Tetrahedral:*	$[NH_4]^+$, $[NH_3(OH)]^+$, $[NF_4]^+$, H_3NBF_3 and innumerable other coordination complexes of NH_3, NR_3, en, edta, etc., including Me_3NO and sulfamic acid (H_3NSO_3) BN (layer structure and Zn blende-type), AlN (wurtzite-type), $[PhAlNPh]_4$ (cubane-type)
	See-saw:	$[\{Fe(CO)_3\}_4(\mu_4\text{-}N)]^-$ (see refs. 14c, 14d)
5	*Square-pyram:*	$[Fe_5(CO)_{14}H(\mu_5\text{-}N)]$ (see ref. 14d)
6	*Octahedral:*	MN (interstitial nitrides with NaCl or hcp structure, e.g. M = Sc; La, Ce, Pr, Nd; Ti, Zr, Hf, V, Nb, Ta; Cr, Mo, W; Th, U), Ti_2N (anti-rutile TiO_2-type), Cu_3N (ReO_3-type), Ca_3N_2 (anti-Mn_2O_3)
	Trigonal prism:	$[NCo_6(CO)_{15}]^-$ (see ref. 14e)
8	*Cubic:*	Ternary nitrides with anti-CaF_2 structure, e.g. BeLiN, $AlLi_3N_2$, $TiLi_5N_3$, $NbLi_7N_4$, and $CrLi_9N_5$

[a] For coordination numbers 1, 2, and 3 the CN is sometimes increased in the condensed phase as a result of H bonding (p. 57), e.g. HCN, NH_2^-, NH_3, N_2H_4, $NH_2(OH)$, $NO_2(OH)$.

Several elements react with the N atoms in active nitrogen to form nitrides. The excited N_2 molecules are also highly reactive and can dissociate molecules that are normally stable to attack either by ordinary N_2 or even N atoms, e.g.:

$$N_2^* + CO_2 \longrightarrow N_2 + CO + O(^3P)$$
$$N_2^* + H_2O \longrightarrow N_2 + OH(^2\Pi) + H(^2S).$$

One of the most dramatic recent developments in the chemistry of N_2 was the discovery by A. D. Allen and C. V. Senoff in 1965 that dinitrogen complexes such as $[Ru(NH_3)_5(N_2)]^{2+}$ could readily be prepared from aqueous $RuCl_3$ using hydrazine hydrate in aqueous solution.[15] Since that time virtually all transition metals have been

14a F. WELLER, W. LIEBELT, and K. DEHNICKE, $[AsPh_4]_2[W_2NCl_{10}]$, a μ-nitrido complex containing tungsten(V) and tungsten (VI), *Angew. Chem.*, Int. Edn. (Engl.), **19**, 220 (1980). [The W–N–W linkage is linear and the interatomic distances are 166 pm (W^{VI}–N) and 207 pm (W^V–N).

14b N. D. FEASEY, S. A. R. KNOX, and A. G. ORPEN, A new coordination mode for nitrogen in a nitrido metal cluster: X-ray crystal structure of $[Mo_3(N)(O)(CO)_4(\eta\text{-}C_5H_5)_3]$. *JCS Chem. Comm.* 1982, 75-6.

14c D. E. FJARE and W. L. GLADFELTER, Synthesis and characterization of a nitrosyl, a nitrido, and an imido carbonyl cluster, *J. Am. Chem. Soc.* **103**, 1572–4 (1981).

14d M. TACHIKAWA, J. STEIN, E. L. MUETTERTIES, R. G. TELLER, M. A. BENO, E. GEBERT, and J. M. WILLIAMS, Metal clusters with exposed and low-coordinate nitride nitrogen atoms, *J. Am. Chem. Soc.* **102**, 6648–9 (1980).

14e S. MARTINENGO, G. CIANI, A. SIRONI, B. T. HEATON, and J. MASON, Synthesis, and X-ray characterization of the $[M_6N(\mu\text{-}CO)_9(CO)_6]^-$ (M = Co, Rh) anions. A new class of metal carbonyl cluster compounds containing an interstitial nitrogen atom *J. Am. Chem. Soc.* **101**, 7095–7 (1979).

15 A. D. ALLEN and C. V. SENOFF, Nitrogenopentamminetruthenium(II) complexes, *Chem. Comm.* 1965, 621–2. See also H. TAUBE, The researches of A. D. Allen—an appreciation, *Coord. Chem. Rev.* **26**, 1–5 (1978).

found to give dinitrogen complexes and well over 200 such compounds are now characterized.[16, 17, 17a] Three general preparative methods are available:

(a) Direct replacement of labile ligands in metal complexes by N_2: such reactions proceed under mild conditions and are often reversible, e.g.:

$$[Ru(NH_3)_5(H_2O)]^{2+} + N_2 \underset{H_2O}{\rightleftharpoons} H_2O + [Ru(NH_3)_5(N_2)]^{2+} \quad \text{pale yellow}$$

$$[CoH_3(PPh_3)_3] + N_2 \underset{EtOH}{\rightleftharpoons} H_2 + [CoH(N_2)(PPh_3)_3] \quad \text{orange-red}$$

$$2[Cr(C_6Me_6)(CO)_3] + N_2 \xrightarrow{thf/h\nu} 2CO + [\{Cr(C_6Me_6)(CO)_2\}_2N_2] \quad \text{red-brown}$$

(b) Reduction of a metal complex in the presence of an excess of a suitable coligand under N_2, e.g.:

$$[MoCl_4(PMe_2Ph)_2] + 2N_2 \xrightarrow{Na/toluene/PMe_2Ph} cis\text{-}[Mo(N_2)_2(PMe_2Ph)_4]$$

$$[WCl_4(PMe_2Ph)_2] + 2N_2 \xrightarrow{(Na/Hg)/PMe_2Ph} cis\text{-}[W(N_2)_2(PMe_2Ph)_4] \quad \text{yellow}$$

$$[FeClH(depe)_2] + N_2 \xrightarrow{NaBH_4/Me_2CO} NaCl + trans\text{-}[Fe(depe)_2H(N_2)] \quad \text{orange}$$

$$2[Ni(acac)_2] + 4PCy_3 + N_2 \xrightarrow{AlMe_3} [\{Ni(N_2)(PCy_3)_2\}_2]$$

where depe is $Et_2PCH_2CH_2PEt_2$, acac is 3,5-pentanedionate, and PCy_3 is tris(cyclohexyl)phosphine. In some systems Mg/thf is a better reducing agent than Na, e.g.:

$$MoCl_5 + 4PMe_2PH + 2N_2 \xrightarrow{Mg/thf} cis\text{-}[Mo(N_2)_2(PMe_2Ph)_4]$$

(c) Conversion of a ligand with N–N bonds into N_2; in the early development of N_2 complex chemistry this was the most successful and widely used route, e.g.:

$$[Mn(\eta^5\text{-}C_5H_5)(CO)_2(N_2H_4)] + 2H_2O_2 \xrightarrow{Cu^{II}/thf/-40°}$$

$$4H_2O + [Mn(\eta^5\text{-}C_5H_5)(CO)_2(N_2)] \quad \text{red-brown}$$

$$[(PPh_3)_2Cl_2\overset{\overset{\displaystyle O}{\frown}}{Re-N{=}N-CPh}] + 2PPh_3 \xrightarrow{MeOH}$$

$$HCl + PhCO_2Me + trans\text{-}[ReCl(N_2)(PPh_3)_4] \quad \text{yellow}$$

$$[RuCl(das)_2(N_3)]PF_6 + NOPF_6 \longrightarrow [RuCl(das)_2(N_2)] \quad \text{white,}$$

$$\text{plus byproducts}$$

$$(das = Ph_2AsCH_2CH_2AsPh_2)$$

$$trans\text{-}[Ir(CO)Cl(PPh_3)_2] + PhCON_3 \xrightarrow{CHCl_3/0} trans\text{-}[IrCl(N_2)(PPh_3)_2] \quad \text{yellow}$$

[16] A. D. ALLEN, R. O. HARRIS, B. R. LOESCHER, J. R. STEVENS, and R. N. WHITELEY, Dinitrogen complexes of transition metals, *Chem. Revs.* **73**, 11–20 (1973).

[17] D. SELLMANN, Dinitrogen–transition metal complexes: synthesis, properties, and significance, *Angew. Chem.*, Int. Edn. (Engl.) **13**, 639–49 (1974).

[17a] J. CHATT, J. R. DILWORTH, and R. L. RICHARDS, Recent advances in the chemistry of nitrogen fixation, *Chem. Rev.* **78**, 589–625 (1978).

Occasionally an $N\equiv N$ triple bond can be formed within a metal complex, e.g. by reaction of coordinated NH_3 with HNO_2, but this method is of limited application, e.g.:

$$[Os(N_2)(NH_3)_5]^{2+} + HNO_2 \longrightarrow 2H_2O + cis\text{-}[Os(N_2)_2(NH_3)_4]$$

Frequently dinitrogen complexes have colours in the range white–yellow–orange–red–brown but other colours are known, e.g. $[\{Ti(\eta^5\text{-}C_5H_5)_2\}_2(N_2)]$ is blue.

Dinitrogen might coordinate to metals in at least 4 ways,[18] but only the end-on modes, structures (1) and (2), are well established as common bonding modes by numerous well-defined examples:

The side-on structure (3) has been established in two dinickel complexes which have very complicated structures involving lithium atoms also in association with the bridging N_2.[18a] The "side-on" η^2 mode (structure 4) was at one time thought to be exemplified by the rhodium(I) complex $[RhCl(N_2)(PPr_3^i)_2]$ but a reinvestigation of the X-ray structure by another group [18b] showed conclusively that the N_2 ligand was coordinated in the "end-on" mode (1)—an instructive example of mistaken conclusions that can be drawn from this technique. The side on structure (4) has been postulated for the zirconium(III) complex $[Zr(\eta^5\text{-}C_5H_5)(N_2)R]$ on the basis of its ^{15}N nmr spectrum.[19] A unique triply-coordinated bridging mode ($\mu_3\text{-}N_2$) has also very recently been established by X-ray crystallography.[19a]

Complexes are known which feature more than 1 N_2 ligand, e.g. $cis\text{-}[W(N_2)_2(PMe_2Ph)_4]$ and $trans\text{-}[W(N_2)_2(diphos)_2]$ (where $diphos = Ph_2PCH_2CH_2PPh_2$) and some complexes feature more than 1 bonding mode, e.g.:[20]

$$[(\eta^5\text{-}C_5Me_5)_2(\eta^1\text{-}N_2)Zr\leftarrow N\equiv N\rightarrow Zr(\eta^1\text{-}N_2)(\eta^5\text{-}C_5Me_5)_2]$$

[18] K. JONAS, D. J. BRAUER, C. KRÜGER, P. J. ROBERTS, and Y.-H. TSAY, Side-on dinitrogen-transition metal complexes, *J. Am. Chem. Soc.* **98**, 74–81 (1976). P. R. HOFFMAN, T. YOSHIDA, T. OKANO, S. OTSUKA, and J. IBERS, Crystal and molecular structure of hydrido(dinitrogen)bis[phenyl(di-*tert*-butyl)phosphine]rhodium(I), *Inorg. Chem.* **15**, 2462–6 (1976).

[18a] K. KRÜGER and Y.-H. TSAY, Molecular structure of a μ-dinitrogen–nickel–lithium complex, *Angew. Chem.*, Int. Edn. (Engl.) **12**, 998–9 (1973).

[18b] D. L. THORN, T. H. TULIP, and J. A. IBERS, The structure of *trans*-chloro(dinitrogen)bis(tri-isopropylphosphine)rhodium(I): an X-ray study of the structure in the solid state and a nuclear magnetic resonance study of the structure in solution, *JCS Dalton*, 1979, 2022–5.

[19] M. J. S. GYNANE, J. JEFFREY, and M. F. LAPPERT, Organozirconium(III)–dinitrogen complexes: evidence for ($\eta^2\text{-}N_2$)-metal bonding in $[Zr(\eta^5\text{-}C_5H_5)_2(N_2)(R)]$ [R = (Me_3Si)_2CH]. *JCS Chem. Comm.* 1978, 34–36.

[19a] G. P. PEZ, P. APGAR, and R. K. CRISSEY, Reactivity of $[\mu\text{-}(\eta^1\text{:}\eta^5\text{-}C_5H_4)](\eta\text{-}C_5H_5)_3Ti_2$ with dinitrogen. Structure of a titanium complex with a triply-coordinated N_2 ligand, *J. Am. Chem. Soc.* **104**, 482–90 (1982).

[20] R. D. SANNER, J. M. MANRIQUEZ, R. E. MARSH, and J. E. BERCAW, Structure of μ-dinitrogen-bis(bispentamethylcyclopentadienyl)dinitrogenzirconium(II), $[\{(\eta^5\text{-}C_5Me_5)_2Zr(N_2)\}_2N_2]$, *J. Am. Chem. Soc.* **98**, 8351–7 (1976).

X-ray structural studies have shown that for N_2 complexes with structure (1), the M–N–N group is linear or nearly so (172–180°); the N–N internuclear distance is usually in the range 110–113 pm, only slightly longer than in gaseous N_2 (109.8 pm). Such complexes have a strong, sharp, infrared absorption in the range 1900–2200 cm^{-1}, corresponding to the Raman-active band at 2331 cm^{-1} in free N_2. Similarly, in complexes with structure (2), when both transition metals have a closed d-shell, the N–N distance falls in the range 112–120 pm and v(N–N) often occurs near 2100 cm^{-1}, i.e. little altered from that of the corresponding complexes of structure (1). On the other hand, if one of the M is a transition metal with a closed d-shell and the other is either a main-group metal such as Al in $AlMe_3$ or an open-shell transition metal such as Mo in $MoCl_4$, then the N–N bond is greatly lengthened and the N–N stretching frequency is lowered even to 1600 cm^{-1}. Compounds with structure (3) have N–N $\sim$134–136 pm, and this very substantial lengthening has been attributed to interaction with the Li atoms in the structure.[18a]

As implied above, N_2 is isoelectronic with both CO and C_2H_2, and the detailed description of the bonding in structures 1–4 follows closely along the lines already indicated (pp. 351 and 361) though there are some differences in the detailed sequences of orbital energies. Crystallographic and vibrational spectroscopic data have been taken to indicate that N_2 is weaker than CO in both its σ-donor and π-acceptor functions. Theoretical studies suggest that σ donation is more important for the formation of the M–N bond than is π back-donation, which mainly contributes to the weakening of the N–N bond, and end-on (η^1) donation is more favourable than side-on (η^2).[20a]

To conclude this section on the chemical reactivity of nitrogen it will be helpful to compare the element briefly with its horizontal neighbours C and O, and also with the heavier elements in Group VB, P, As, Sb, and Bi. The diagonal relationship with S is vestigial. Nitrogen resembles oxygen in its high electronegativity and in its ability to form H bonds (p. 57) and coordination complexes (p. 223) by use of its lone-pair of electrons. Catenation is more limited than for carbon, the longest chain so far reported being the N_8 unit in PhN=N—N(Ph)—N=N—N(Ph)—N=NPh.

Nitrogen shares with C and O the propensity for multiple bonding via p_π–p_π interactions both with another N atom or with a C or O atom. In this it differs sharply from its Group VB congeners which have no analogues of the oxides of nitrogen, nitrites, nitrates, nitro-, nitroso-, azo-, and diazo-compounds, azides, nitriles, cyanates, thiocyanates, or imino-derivatives. Conversely, there are no nitrogen analogues of the various oxoacids of phosphorus (p. 586).

A further reason why N is atypical of Group V is that its covalency is limited to a maximum of 4. Extended structures in which N has a coordination number of 6 or 8 are, of course, well known (p. 475) but in covalent molecules and finite complexes the absence of low-lying d orbitals precludes the formation of compounds with higher coordination numbers analogous to PF_5, PF_6^-, etc. The recently prepared cluster complexes $[NFe_5(CO)_{14}H]$[14d] and $[NCo_6(\mu\text{-}CO)_9(CO)_6]$[14e] should, however, be noted as outstanding exceptions.

[20a] T. YAMABE, K. HORI, T. MINATO, and K. FUKUI, Theoretical study on the bonding nature of transition-metal complexes of molecular nitrogen, *Inorg. Chem.* **19**, 2154–9 (1980).

11.3 Compounds

This section deals with the binary compounds that nitrogen forms with metals, and then describes the extensive chemistry of the hydrides, halides, pseudohalides, oxides, and oxoacids of the element. The chemistry of P–N compounds is deferred until Chapter 12 (p. 619) and S–N compounds are discussed in Chapter 15 (p. 854). Compounds with B (p. 234) and C (p. 336) have already been treated.

11.3.1 *Nitrides, azides, and nitrido complexes*

Nitrogen forms binary compounds with almost all elements of the periodic table and for many elements several stoichiometries are observed, e.g. MnN, Mn_6N_5, Mn_3N_2, Mn_2N, Mn_4N, and Mn_xN ($9.2 < x < 25.3$). Nitrides are frequently classified into 4 groups: "salt-like", covalent, "diamond-like", and metallic (or "interstitial"). The remarks on p. 70 concerning the limitations of such classifications are relevant here. The two main methods of preparation are by direct reaction of the metal with N_2 or NH_3 (often at high temperatures) and the thermal decomposition of metal amides, e.g.:

$$3Ca + N_2 \longrightarrow Ca_3N_2$$

$$3Mg + 2NH_3 \xrightarrow{\;900^\circ\;} Mg_3N_2 + 3H_2$$

$$3Zn(NH_2)_2 \longrightarrow Zn_3N_2 + 4NH_3$$

Common variants include reduction of a metal oxide or halide in the presence of N_2 and the formation of a metal amide as an intermediate in reactions in liquid NH_3:

$$Al_2O_3 + 3C + N_2 \longrightarrow 2AlN + 3CO$$

$$2ZrCl_4 + N_2 + 4H_2 \longrightarrow 2ZrN + 8HCl$$

$$3Ca + 6NH_3 \xrightarrow{\;-3H_2\;} \{3Ca(NH_2)_2\} \longrightarrow Ca_3N_2 + 4NH_3$$

Metal nitrides have also been prepared by adding KNH_2 to liquid-ammonia solutions of the appropriate metal salts in order to precipitate the nitride, e.g. Cu_3N, Hg_3N_2, AlN, Tl_3N, and BiN.

"Salt-like" nitrides are exemplified by Li_3N (mp 548°C, decomp) and M_3N_2 (M = Be, Mg, Ca, Sr, Ba). It is possible to write ionic formulations of these compounds using the species N^{3-} though charge separation is unlikely to be complete, particularly for the corresponding compounds of Group IB and IIB, i.e. Cu_3N, Ag_3N, and M_3N_2 (M = Zn, Cd, Hg). The N^{3-} ion has been assigned a radius of 146 pm, slightly larger than the value for the isoelectronic ions O^{2-} (140 pm) and F^- (133 pm), as expected. Stability varies widely; e.g. Be_3N_2 melts at 2200°C whereas Mg_3N_2 decomposes above 271°C. The existence of Na_3N is doubtful and the heavier alkali metals appear not to form analogous compounds, perhaps for steric reasons (p. 88). The azides NaN_3, KN_3, etc., are, however, well characterized as colourless crystalline salts which can be melted with little decomposition; they feature the symmetrical linear N_3^- group as do $Sr(N_3)_2$ and $Ba(N_3)_2$. The corresponding B subgroup metal azides such as AgN_3, $Cu(N_3)_2$, and $Pb(N_3)_2$ are shock-sensitive and detonate readily; they are far less ionic and have more complex structures.[21] Further discussion of azides is on p. 496. Other stoichiometries are also known, e.g. Ca_2N (anti-$CdCl_2$ layer structure), Ca_3N_4, and $Ca_{11}N_8$.

The covalent binary nitrides are more conveniently treated under the appropriate element. Examples include cyanogen $(CN)_2$ (p. 336), P_3N_5 (p. 619), As_4N_4 (p. 676), S_2N_2 (p. 859), and S_4N_4 (p. 856). The Group III nitrides MN (M = B, Al, Ga, In, Tl) are a special case since they are isoelectronic with graphite, diamond, SiC, etc., to which they are structurally related (p. 288). Their physical properties suggest a gradation of bond-type from covalent, through partially ionic, to essentially metallic as the atomic number increases. Si_3N_4 and Ge_3N_4 are also known and have the phenacite (Be_2SiO_4)-type structure. Si_3N_4, in particular, has excited considerable interest in recent years as a ceramic material with extremely desirable properties: high strength and wear resistance, high decomposition temperature and oxidation resistance, excellent thermal-shock properties and resistance to corrosive environments, low coefficient of friction, etc. Unfortunately it is extremely difficult to fabricate and sinter suitably shaped components, and considerable efforts have therefore been spent on developing related nitrogen ceramics by forming solid solutions between Si_3N_4 and Al_2O_3 to give the "sialons" (SiAlON) of general formula $Si_{6-0.75x}Al_{0.67x}O_xN_{8-x}$ $(0 < x < 6)$.[22, 23]

The most extensive group of nitrides are the metallic nitrides of general formulae MN, M_2N, and M_4N in which N atoms occupy some or all of the interstices in cubic or hcp metal lattices (examples are in Table 11.1, p. 475). These compounds are usually opaque, very hard, chemically inert, refractory materials with metallic lustre and conductivity and sometimes having variable composition. Similarities with borides (p. 162) and carbides (p. 318) are notable. Typical mps (°C) are:

TiN	ZrN	HfN		VN	NbN	TaN		CrN	ThN	UN
2950	2980	2700		2050	2300	3090		d1770	2630	2800

Hardness on the Mohs scale is often above 8 and sometimes approaches 10 (diamond). These properties commend nitrides for use as crucibles, high-temperature reaction vessels, thermocouple sheaths, and related applications. Several metal nitrides are also used as heterogeneous catalysts, notably the iron nitrides in the Fischer–Tropsch hydriding of carbonyls. Few chemical reactions of metal nitrides have been studied; the most characteristic (often extremely slow but occasionally rapid) is hydrolysis to give ammonia or nitrogen:

$$2AlN + (n+3)H_2O \longrightarrow Al_2O_3 . nH_2O + 2NH_3$$
$$2VN + 3H_2SO_4 \longrightarrow V_2(SO_4)_3 + N_2 + 3H_2$$

The nitride ion N^{3-} is an excellent ligand, particularly towards second- and third-row transition metals.[24] It is considered to be by far the strongest π donor known, the next strongest being the isoelectronic species O^{2-}. Nitrido complexes are usually prepared by the thermal decomposition of azides (e.g. those of phosphine complexes of V^V, Mo^{VI}, W^{VI},

[21] A. F. WELLS, *Structural Inorganic Chemistry*, 4th edn., Oxford University Press, Oxford, 1975, 649 pp.
[22] K. H. JACK, Nitrogen ceramics, *Trans. J. Br. Ceram. Soc.* **72**, 376–84 (1973).
[23] F. L. RILEY (ed.), *Nitrogen Ceramics*, Noordhoff-Leyden, 1977, 694 pp.
[24] W. P. GRIFFITH, Transition-metal nitrido-complexes, *Coord. Chem. Revs.* **8**, 369–96 (1972).

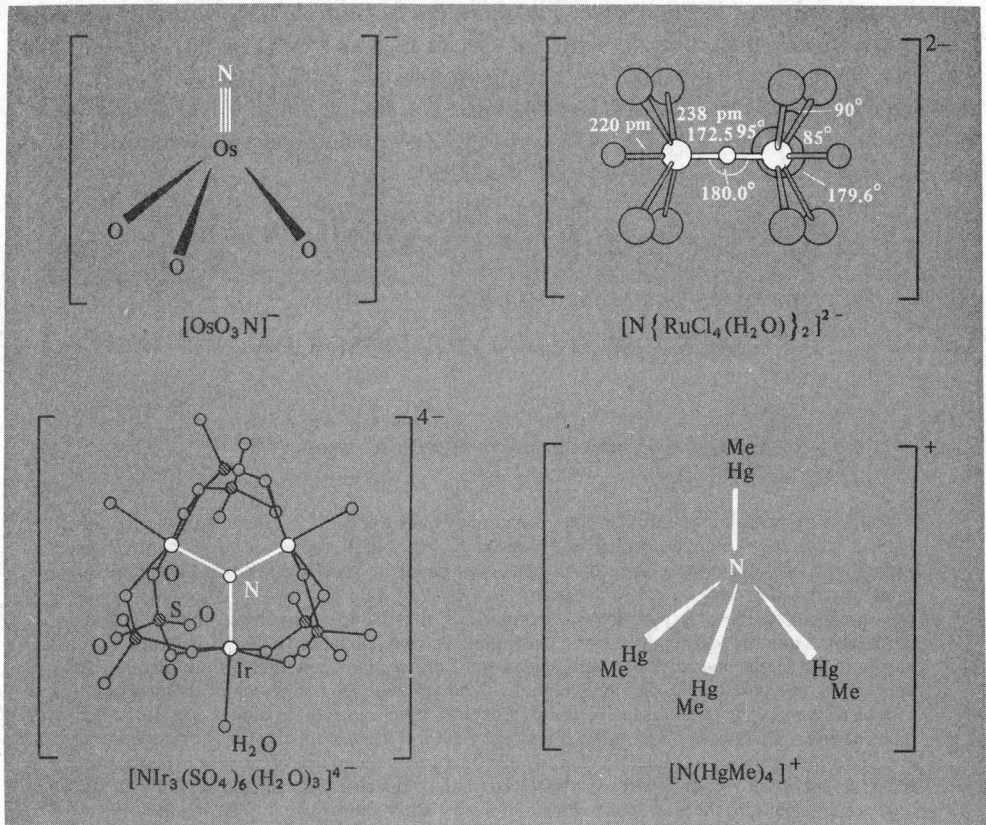

Fig. 11.3 Structures of some nitrido complexes.[24]

Ru^{VI}, Re^V) or by deprotonation of NH_3 (e.g. $[OsO_4 \rightarrow OsO_3N]^-$). Most involve a terminal $\{\equiv N\}^{3-}$ group as in $[VCl_3N]^-$, $[MoO_3N]^-$, $[WCl_5N]^{2-}$, $[ReN(PR_3)_3X_2]$, and $[RuN(OH_2)X_4]^-$. The M–N distance is much shorter (by 40–50 pm) than the "normal" σ-(M–N) distance, consistent with strong multiple bonding. Other bonding modes feature linear symmetrical bridging as in $[(H_2O)Cl_4Ru-N-RuCl_4(OH_2)]^{3-}$, trigonal planar μ_3 bridging as in $[\{(H_2O)(SO_4)_2Ir\}_3N]^{4-}$, and tetrahedral coordination as in $[(MeHg)_4N]^+$ (Fig. 11.3).

11.3.2 *Ammonia and ammonium salts*

NH_3 is a colourless, alkaline gas with a unique, penetrating odour that is first preceptible at concentrations of about 20–50 ppm. Noticeable irritation to eyes and the nasal passages begins at about 100–200 ppm, and higher concentrations can be dangerous.[25] NH_3 is prepared industrially in larger amounts (number of moles) than any other single compound (p. 467) and the production of synthetic ammonia is of major

[25] J. R. LeBlanc, S. Madhavan, and R. E. Porter, Ammonia, *Kirk–Othmer Encyclopedia of Chemical Technology*, 3rd edn., Vol. 2, pp. 470–516, Wiley, New York, 1978.

importance for several industries (see Panel). In the laboratory NH_3 is usually obtained from cylinders unless isotopically enriched species such as $^{15}NH_3$ or ND_3 are required. Pure dry $^{15}NH_3$ can be prepared by treating an enriched $^{15}NH_4^+$ salt with an excess of KOH and drying the product gas over metallic Na. Reduction of $^{15}NO_3^-$ or $^{15}NO_2^-$ with Devarda's alloy (50% Cu, 45% Al, 5% Zn) in alkaline solution provides an alternative route as does the hydrolysis of a nitride, e.g.:

$$3Ca + {}^{15}N_2 \longrightarrow Ca_3{}^{15}N_2 \xrightarrow{6H_2O} 2{}^{15}NH_3 + 3Ca(OH)_2$$

ND_3 can be prepared similarly using D_2O, e.g.:

$$Mg_3N_2 + 6D_2O \longrightarrow 2ND_3 + 3Mg(OD)_2$$

Industrial Production of Synthetic Ammonia[25-27]

The first industrial production of NH_3 began in 1913 at the BASF works in Ludwigshaven-Oppau, Germany. The plant, which had a design capacity of 30 tonnes per day, involved an entirely new concept in process technology; it was based on the Haber–Bosch high-pressure catalytic reduction of N_2 with H_2 obtained by electrolysis of water. Modern methods employ the same principles for the final synthesis but differ markedly in the source of hydrogen, the efficiency of the catalysts, and the scale of operations, many plants now having a capacity of 1650 tonnes per day or more. Great ingenuity has been shown not only in plant development but also in the application of fundamental thermodynamics to the selection of feasible chemical processes. Except where electricity is unusually cheap, reduction by electrolytic hydrogen has now been replaced either by coke/H_2O or, more recently, by natural gas (essentially CH_4) or naphtha (a volatile aliphatic petrol-like fraction of crude oil). The great advantages of modern hydrocarbon reduction methods over coal-based processes are that, comparing plant of similar capital costs, they occupy one-third the land area, use half the energy, and require one-tenth the manpower, yet produce 4 times the annual tonnage of NH_3.

The operation of a large synthetic ammonia plant based on natural gas (Pl. 11.1) involves a delicately balanced sequence of reactions. The gas is first *desulfurized* to remove compounds which will poison the metal catalysts, then compressed to ~ 30 atm and reacted with steam over a nickel catalyst at 750°C in the *primary steam reformer* to produce H_2 and oxides of carbon:

$$CH_4 + H_2O \underset{}{\overset{Ni/750°}{\rightleftharpoons}} CO + 3H_2$$

$$CH_4 + 2H_2O \underset{}{\overset{Ni/750°}{\rightleftharpoons}} CO_2 + 4H_2$$

Under these conditions the issuing gases contain some 9% of unreacted methane; sufficient air is injected via a compressor to give a final composition of $1:3$ $N_2:H_2$ and the air burns in the hydrogen thereby heating the gas to $\sim 1100°C$ in the *secondary reformer*:

$$\underset{air}{2H_2 + (O_2 + 4N_2)} \overset{1100°}{\rightleftharpoons} 2H_2O + 4N_2$$

$$CH_4 + H_2O \overset{Ni/1000°}{\rightleftharpoons} CO + 3H_2$$

[26] S. P. S. ANDREW, Modern processes for the production of ammonia, nitric acid, and ammonium nitrate, in R. THOMPSON (ed.), *The Modern Inorganic Chemicals Industry*, pp. 201–31, The Chemical Society, London, 1977. See also Ammonia in *Inorganic Chemicals*, pp. 3–13, ICI Educational Publication, Kynoch Press, London, 1978.

[27] S. D. LYON, Development of the modern ammonia industry, *Chem. Ind.* 1975, 731–9.

PLATE 11.1   A general view of three 1000-tonne-per-day synthetic ammonia plants. The CO_2 removal section is to the fore on each plant with the primary and secondary reformer structures and shift converters in the background (see text). (Photograph by courtesy of ICI Agricultural Division, Billingham, Teesside.)

The emerging gas, now containing only 0.25% CH_4, is cooled in heat exchangers which generate high-pressure steam for use first in the turbine compressors and then as a reactant in the primary steam reformer. Next, the CO is converted to CO_2 by the *shift reaction* which also produces more H_2:

$$CO + H_2O \xrightleftharpoons{(a)+(b)} CO_2 + H_2$$

Maximum conversion occurs by equilibration at the lowest possible temperature so the reaction is carried out sequentially on two beds of catalyst:

(a) iron oxide (400°C) which reduces the CO concentration from 11% to 3%;
(b) a copper catalyst (200°) which reduces the CO content to 0.3%.

Removal of CO_2 ($\sim 18\%$) is effected in a *scrubber* containing either a concentrated alkaline solution of K_2CO_3 or an amine such as ethanolamine:

$$CO_2 + H_2O + K_2CO_3 \xrightleftharpoons[\text{regeneration (heat)}]{\text{absorption}} 2KHCO_3$$

Remaining trace quantities of CO (which would poison the iron catalyst during ammonia synthesis) are converted back to CH_4 by passing the damp gas from the scrubbers over a Ni *methanation catalyst* at 325°:

$$CO + 3H_2 \xrightleftharpoons{Ni/325°} CH_4 + H_2O$$

This reaction is the reverse of that occurring in the primary steam reformer. The *synthesis gas* now emerging has the approximate composition

$$H_2 \ 74.3\%, \ N_2 \ 24.7\%, \ CH_4 \ 0.8\%, \ Ar \ 0.3\%, \ CO \ 1\text{--}2 \ ppm;$$

it is *compressed* in three stages from 25 atm to ~ 200 atm and then passed over a *promoted iron catalyst* at 380–450°C:

$$N_2 + 3H_2 \xrightleftharpoons{Fe/400°/200 \ atm} 2NH_3$$

The gas leaving the catalyst beds contains about 15% NH_3; this is condensed by refrigeration and the remaining gas mixed with more incoming synthesis gas and recycled. Variables in the final reaction are the synthesis pressure, synthesis temperature, gas composition, gas flow rate[†] and catalyst composition and particle size. Since the earliest days the "promoted" Fe catalysts have been prepared by fusing magnetite (Fe_3O_4) on a table with KOH in the presence of a small amount of mixed refractory oxides such as MgO, Al_2O_3 and SiO_2; the solidified sheet is broken up into chunks 5–10 mm in size. These chunks are then reduced inside the ammonia synthesis converter to give the active catalyst which consists of Fe crystallites separated by the amorphous refractory oxides and partly covered by the alkali promotor which increases its activity by at least an order of magnitude.

World production of synthetic ammonia has increased dramatically during the past 30 y and is still continuing to rise. Production in 1950 was little more than 1 million tonnes; though this was huge when compared with the production of most other compounds, it is dwarfed by today's rate of production which approaches 100 million tonnes pa.[‡] Anhydrous liquid NH_3 is now usually stored in refrigerated tanks at $-33.5°$ (104.7 kPa); such tanks were first installed (in the USA) in 1956 and now generally replace the smaller insulated Horton spheres (275 kPa). Tank capacity can be up to 36 000 tonnes. Anhydrous NH_3 is shipped in tank trucks (up to 30 m^3), tank cars (up to 130 m^3, i.e. 34 000 US gal), barges or pipelines, some of which are several thousand km long (e.g. from southern Louisiana to northern Indiana, Illinois, Iowa, and Nebraska).

† Flow rate is usually quoted as "space velocity", i.e. the ratio of volumetric rate of gas at STP to volume of catalyst; typical values are in the range 8000–60 000 h^{-1}.

‡ For example in 1975 world production capacity was estimated to be 84.5 million tonnes distributed as follows: Eastern Europe and the USSR 27.9%, Asia 25.7%, North America 21.4%, Western Europe 18.8%, Central America 3.9%, Africa 2.1%, Oceania 0.2%.

The applications of NH_3 are dominated (over 80%) by its use in various forms as a fertilizer. Of these, direct application is the most common (27.1%), followed by NH_4NO_3 (18.9%), urea (13.9%), ammonium phosphates (8.8%), N solutions and mixed fertilizers (8.2%), and $(NH_4)_2SO_4$ (3.5%). Industrial uses include (a) commercial explosives (5%)—such as NH_4NO_3, nitroglycerine, TNT, and nitrocellulose, which are produced from NH_3 via HNO_3—and (b) fibres/plastics (10%), e.g. in the manufacture of caprolactam for nylon-6, hexamethylenediamine for nylon-6,6, polyamides, rayon, and polyurethanes. Other uses ($\sim 5\%$) include a wide variety of applications in refrigeration, wood pulping, detinning of scrap-metal, and corrosion inhibition; it is also used as a rubber stabilizer, pH controller, in the manufacture of household detergents, in the food and beverage industry, pharmaceuticals, water purification, and the manufacture of numerous organic and inorganic chemicals. Indeed, synthetic ammonia is the key to the industrial production of most inorganic nitrogen compounds, as indicated in Scheme A.

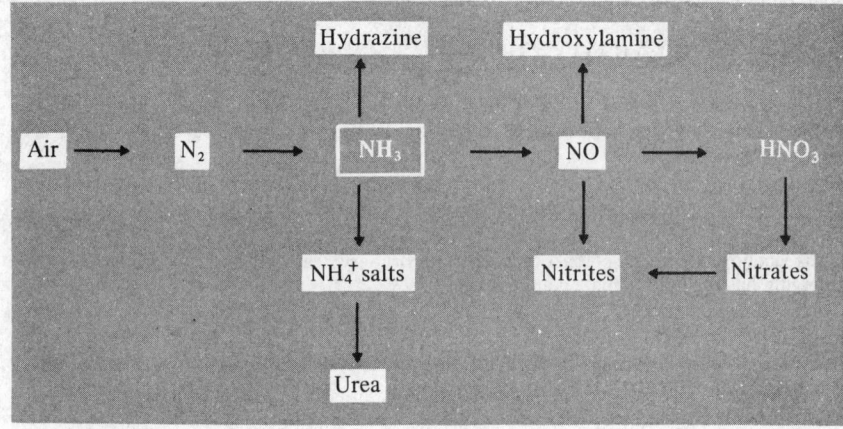

SCHEME A

The average price of NH_3 (USA) leaped from $100 per tonne in 1973 to $300 in 1975 (why?) but then settled back to about $150–200 from 1976 on. The 1982 price was $160–190 per tonne.

Some physical and molecular properties of NH_3 are given in Table 11.2. The influence of H bonding on the bp and other properties has already been noted (p. 57). It has been estimated that 26% of the H bonding in NH_3 breaks down on melting, 7% on warming from the mp to the bp, and the final 67% on transfer to the gas phase at the bp. The low density, viscosity, and electrical conductivity, and the high dielectric constant of liquid ammonia are also notable. Liquid NH_3 is an excellent solvent and a valuable medium for chemical reactions (p. 486); its high heat of vaporization (23.35 kJ mol^{-1} at the bp) makes it relatively easy to handle in simple vacuum flasks. The molecular properties call for little comment except to note that the rapid inversion frequency with which the N atom moves through the plane of the 3 H atoms has a marked effect on the vibrational spectrum of the molecule. The inversion itself occurs in the microwave region of the spectrum at 23.79 GHz ($\equiv 1.260$ cm $\equiv 0.793$ cm^{-1}) and was, in fact, the first microwave absorption spectrum to be detected (C. E. Cleeton and N. H. Williams, 1934). Inversion also occurs in ND_3 at a frequency of 1.591 GHz, i.e. less than for NH_3 by a factor of 14.95. The inversion can be stopped in NH_3 by increasing the pressure to ~ 2 atm. The corresponding figure for ND_3 is ~ 90 mmHg (i.e. again a factor of about 15).

TABLE 11.2   *Some properties of ammonia,* NH_3

Physical properties		Molecular properties	
MP/K	195.42	Symmetry	C_{3v} (pyramidal)
BP/K	239.74	Distance (N–H)/pm	101.7
Density (l; 239 K)/g cm^{-3}	0.6826	Angle H–N–H	107.8°
Density (g; rel. air = 1)	0.5963	Pyramid height/pm	36.7
η (239.5 K)/centipoise[a]	0.254	μ/Debye[b]	1.46
Dielectric constant ε (239 K)	22	Inversion barrier/kJ mol^{-1}	24.7
κ (234.3 K)/ohm^{-1} cm^{-1}	1.97×10^{-7}	Inversion frequency/GHz[c]	23.79
ΔH_f° (298 K)/kJ mol^{-1}	-46.1	D (H–NH$_2$)/kJ mol^{-1}	435
ΔG_f° (298 K)/kJ mol^{-1}	-16.5	Ionization energy/kJ mol^{-1}	979.7
S° (298 K)/J K^{-1} mol^{-1}	192.3	Proton affinity (gas)/kJ mol^{-1}	841

[a] 1 centipoise = 10^{-3} kg m^{-1} s^{-1}.
[b] 1 Debye = 10^{-18} esu = $3.335\,64 \times 10^{-30}$ C m.
[c] 1 GHz = 10^9 s^{-1}.

Ammonia is readily absorbed by H_2O with considerable evolution of heat (~ 37.1 kJ per mol of NH_3 gas). Aqueous solutions are weakly basic due to the equilibrium

$$NH_3(aq) + H_2O \underset{H_2O}{\rightleftharpoons} NH_4^+(aq) + OH^-(aq);$$

$$K_{298.2} = [NH_4^+][OH^-]/[NH_3] = 1.81 \times 10^{-5} \text{ mol l}^{-1}$$

The equilibrium constant at room temperature corresponds to $pK_b = 4.74$ and implies that a 1 molar aqueous solution of NH_3 contains only 4.25 mmol l^{-1} of NH_4^+ (or OH^-). Such solutions do not contain the undissociated "molecule" NH_4OH, though weakly bonded hydrates have been isolated at low temperature:

$$NH_3 \cdot H_2O \text{ (mp 194.15 K)} \quad \text{and} \quad 2NH_3 \cdot H_2O \text{ (mp 194.32 K)}$$

These hydrates are not ionically dissociated but contain chains of H_2O molecules cross-linked by NH_3 molecules into a three-dimensional H-bonded network.

Ammonia burns in air with difficulty, the flammable limits being 16–25 vol%. Normal combustion yields nitrogen but, in the presence of a Pt or Pt/Rh catalyst at 750–900°C, the reaction proceeds further to give the thermodynamically less-favoured products NO and NO_2:

$$4NH_3 + 3O_2 \xrightarrow{\text{burn}} 2N_2 + 6H_2O$$

$$4NH_3 + 5O_2 \xrightarrow{\text{Pt/800°}} 4NO + 6H_2O$$

$$2NO + O_2 \xrightarrow{\text{Pt/800°}} 2NO_2$$

These reactions are very important industrially in the production of HNO_3 (p. 537). See also the industrial production of HCN by the Andrussov process (p. 338): $2NH_3 + 3O_2 + 2CH_4 \rightarrow 2HCN + 6H_2O$.

Gaseous NH_3 burns with a greenish-yellow flame in F_2 (or ClF_3) to produce NF_3 (p. 503). Chlorine yields several products depending on conditions: NH_4Cl, NH_2Cl, $NHCl_2$,

NCl_3, $NCl_3.NH_3$, N_2, and even small amounts of N_2H_4. The reaction to give chloramine, NH_2Cl, is important in urban and domestic water purification systems.[25] Reactions with other non-metals and their halides or oxides are equally complex and lead to a variety of compounds, many of which are treated elsewhere (pp. 570, 580, 624, 856, etc.). At red heat carbon reacts with NH_3 to give $NH_4CN + H_2$, whereas phosphorus yields PH_3 and N_2, and sulfur gives H_2S and N_4S_4. Metals frequently react at higher temperature to give nitrides (p. 479). Of particular importance is the attack on Cu in the presence of oxygen (air) at room temperature since this precludes the use of this metal and its alloys in piping and valves for handling either liquid or gaseous NH_3. Corrosion of Cu and brass by moist NH_3/air mixtures and by air-saturated aqueous solutions of NH_3 is also rapid. Contact with Ni and with polyvinylchloride plastics should be avoided for the same reason.

Liquid ammonia as a solvent[28-30, 30a]

Liquid ammonia is the best-known and most widely studied non-aqueous ionizing solvent. Its most conspicuous property is its ability to dissolve alkali metals to form highly coloured, electrically conducting solutions containing solvated electrons, and the intriguing physical properties and synthetic utility of these solutions have already been discussed (p. 88). Apart from these remarkable solutions, much of the chemistry in liquid ammonia can be classified by analogy with related reactions in aqueous solutions. Accordingly, we briefly consider in turn, solubility relationships, metathesis reactions, acid–base reactions, amphoterism, solvates and solvolysis, redox reactions, and the preparation of compounds in unusual oxidation states. Comparison of the physical properties of liquid NH_3 (p. 485) with those of water (p. 730) shows that NH_3 has the lower mp, bp, density, viscosity, dielectric constant and electrical conductivity; this is due at least in part to the weaker H bonding in NH_3 and the fact that such bonding cannot form cross-linked networks since each NH_3 molecule has only 1 lone-pair of electrons compared with 2 for each H_2O molecule. The ionic self-dissociation constant of liquid NH_3 at $-50°C$ is $\sim 10^{-33}$ mol^2 l^{-2}.

Most ammonium salts are freely soluble in liquid NH_3 as are many nitrates, nitrites, cyanides, and thiocyanates. The solubilities of halides tend to increase from the fluoride to the iodide; solubilities of multivalent ions are generally low suggesting that (as in aqueous systems) lattice-energy and entropy effects outweigh solvation energies. The possibility of H-bond formation also influences solubility. Some typical solubilities at 25°C expressed as g per 100 g solvent are: NH_4OAc 253.2, NH_4NO_3 389.6, $LiNO_3$ 243.7, $NaNO_3$ 97.6, KNO_3 10.4, NaF 0.35, $NaCl$ 3.0, $NaBr$ 138.0, NaI 161.9, $NaSCN$ 205.5. Some of these solubilities are astonishingly high, particularly when expressed as the number of moles of

[28] W. L. JOLLY and C. J. HALLADA, Liquid ammonia, Chap. 1 in T. C. WADDINGTON (ed.), *Non-Aqueous Solvent Systems*, pp. 1–45, Academic Press, London, 1965.

[29] G. W. A. FOWLES, Inorganic reactions in liquid ammonia, Chap. 7, in C. B. COLBURN (ed.), *Developments in Inorganic Nitrogen Chemistry*, pp. 522–76, Elsevier, Amsterdam, 1966.

[30] J. J. LAGOWSKI and G. A. MOCZYGEMBA, Liquid ammonia, Chap. 7 in J. J. LAGOWSKI (ed.), *The Chemistry of Non-aqueous Solvents*, Vol. 2, pp. 320–71, Academic Press, 1967.

[30a] D. NICHOLLS, *Inorganic Chemistry in Liquid Ammonia: Topics in Inorganic and General Chemistry*, Monograph 17, Elsevier, Amsterdam, 1979, 238 pp.

solute per 10 mol NH_3, e.g.: NH_4NO_3 8.3, $LiNO_3$ 6.1, NaSCN 4.3. Further data at 25° and other temperatures are in ref. 31.

Metathesis reactions are sometimes the reverse of those in aqueous systems because of the differing solubility relations. For example because AgBr forms the complex ion $[Ag(NH_3)_2]^+$ in liquid NH_3 it is readily soluble, whereas $BaBr_2$ is not, and can be precipitated:

$$Ba(NO_3)_2 + 2AgBr \xrightarrow{\text{liq } NH_3} BaBr_2\downarrow + 2AgNO_3$$

Reactions analogous to the precipitation of AgOH and of insoluble oxides from aqueous solution are:

$$AgNO_3 + KNH_2 \xrightarrow{\text{liq } NH_3} AgNH_2\downarrow + KNO_3$$

$$3HgI_2 + 6KNH_2 \xrightarrow{\text{liq } NH_3} Hg_3N_2\downarrow + 6KI + 4NH_3$$

Acid-base reactions in many solvent systems can be thought of in terms of the characteristic cations and anions of the solvent (see also p. 972)

$$\text{solvent} \rightleftharpoons \text{characteristic cation (acid)} + \text{characteristic anion (base)}$$
$$2H_2O \rightleftharpoons H_3O^+ \qquad\qquad + OH^-$$
$$2NH_3 \rightleftharpoons NH_4^+ \qquad\qquad + NH_2^-$$

On this basis NH_4^+ salts can be considered as solvo-acids in liquid NH_3 and amides as solvo-bases. Neutralization reactions can be followed conductimetrically, potentiometrically, or even with coloured indicators such as phenolphthalein:

$$NH_4NO_3 + KNH_2 \xrightarrow{\text{liq } NH_3} KNO_3 + 2NH_3$$
$$\text{solvo-acid} \quad \text{solvo-base} \qquad\qquad \text{salt} \qquad \text{solvent}$$

Likewise, amphoteric behaviour can be observed. For example $Zn(NH_2)_2$ is insoluble in liquid NH_3 (as is $Zn(OH)_2$ in H_2O), but it dissolves on addition of the solvo-base KNH_2 due to the formation of $K_2[Zn(NH_2)_4]$; this in turn is decomposed by NH_4^+ salts (solvo-acids) with reprecipitation of the amide:

$$K_2[Zn(NH_2)_4] + 2NH_4NO_3 \xrightarrow{\text{liq } NH_3} Zn(NH_2)_2 + 2KNO_3 + 4NH_3$$

Solvates are perhaps less prevalent in compounds prepared from liquid ammonia solutions than are hydrates precipitated from aqueous systems, but large numbers of ammines are known, and their study formed the basis of Werner's theory of coordination compounds (1891–5). Frequently, however, solvolysis (ammonolysis) occurs (cf. hydrolysis).[29] Examples are:

$$M^IH + NH_3 \longrightarrow MNH_2 + H_2$$

$$M_2^IO + NH_3 \longrightarrow MNH_2 + MOH$$

$$SiCl_4 \xrightarrow[\text{temp}]{\text{low}} [Si(NH_2)_4] \xrightarrow{0°} Si(NH)(NH_2)_2 \xrightarrow{1200°} Si_3N_4$$

³¹ K. JONES, Nitrogen, Chap. 19 in *Comprehensive Inorganic Chemistry* Vol. 2, pp. 147–388, Pergamon Press, Oxford, 1973.

Amides are one of the most prolific classes of ligand and the subject of metal and metalloid amides has recently been extensively reviewed.[31a]

Redox reactions are particularly instructive. If all thermodynamically allowed reactions in liquid NH_3 were kinetically rapid, then no oxidizing agent more powerful than N_2 and no reducing agent more powerful than H_2 could exist in this solvent. Using data for solutions at $25°$:[28]

Acid solutions (1 M NH_4^+)

$$NH_4^+ + e^- = NH_3 + \tfrac{1}{2}H_2 \qquad E° = 0.0 \text{ V}$$
$$3NH_4^+ + \tfrac{1}{2}N_2 + 3e^- = 4NH_3 \qquad E° = -0.04 \text{ V}$$

Basic solutions (1 M NH_2^-)

$$NH_3 + e^- = NH_2^- + \tfrac{1}{2}H_2 \qquad E° = 1.59 \text{ V}$$
$$2NH_3 + \tfrac{1}{2}N_2 + 3e^- = 3NH_2^- \qquad E° = 1.55 \text{ V}$$

Obviously, with a range of only 0.04 V available very few species are thermodynamically stable. However, both the hydrogen couple and the nitrogen couple usually exhibit "overvoltages" of ~ 1 V, so that in acid solutions the practical range of potentials for solutes is from $+1.0$ to -1.0 V. Similarly in basic solutions the practical range extends from 2.6 to 0.6 V. It is thus possible to work in liquid ammonia with species which are extremely strong reducing agents (e.g. alkali metals) and also with extremely strong oxidizing agents (e.g. permanganates, superoxides, and ozonides; p. 710). For similar reasons the NO_3^- ion is effectively inert towards NH_3 in acid solution but in alkaline solutions N_2 is slowly evolved:

$$3K^+ + 3NH_2^- + 3NO_3^- \longrightarrow 3KOH\downarrow + N_2 + 3NO_2^- + NH_3$$

The use of liquid NH_3 to prepare compounds of elements in unusual (low) oxidation states is exemplified by the successive reduction of $K_2[Ni(CN)_4]$ with Na/Hg in the presence of an excess of CN^-: the dark-red dimeric Ni^I complex $K_4[Ni_2(CN)_6]$ is first formed and this can be further reduced to the yellow Ni^0 complex $K_4[Ni(CN)_4]$. The corresponding complexes $[Pd(CN)_4]^{4-}$ and $[Pt(CN)_4]^{4-}$ can be prepared similarly, though there is no evidence in these systems for the formation of the M^I dimer. A ditertiaryphosphine complex of Pd^0 has also been prepared:

$$[Pd\{1,2\text{-}(PEt_2)_2C_6H_4\}_2]Br_2 \xrightarrow{\text{Na/NH}_3} [Pd\{1,2\text{-}(PEt_2)_2C_6H_4\}_2] + 2NaBr$$

$[Co^{III}(CN)_6]^{3-}$ yields the pale-yellow complex $[Co^I(CN)_4]^{3-}$ and the brown-violet complex $[Co_2^{(0)}(CN)_8]^{8-}$ (cf. the dimeric carbonyl $[Co_2(CO)_8]$).

Liquid NH_3 is also extensively used as a preparative medium for compounds which are unstable in aqueous solutions, e.g.:

$$2Ph_3GeNa + Br(CH_2)_xBr \xrightarrow{\text{liq NH}_3} Ph_3Ge(CH_2)_xGePh_3 + 2NaBr$$

$$Me_3SnX + NaPEt_2 \xrightarrow{\text{liq NH}_3} Me_3SnPEt_2 + 2NaX$$

[31a] M. F. LAPPERT, P. P. POWER, A. R. SANGER, and R. C. SRIVASTAVA, *Metal and Metalloid Amides*, Ellis Horwood Ltd., Chichester, 1980, 847 pp. (approximately 3000 references).

Alkali metal acetylides M_2C_2, MCCH, and MCCR can readily be prepared by passing C_2H_2 or C_2HR into solutions of the alkali metal in liquid NH_3, and these can be used to synthesize a wide range of transition-element acetylides,[32] e.g.:

$$Ni(SCN)_2 . 6NH_3 + 5KC_2Ph \xrightarrow{\text{liq } NH_3} K_2[Ni(C_2Ph)_4] . 2NH_3 + 2KSCN + 4NH_3$$

$$K_2[Ni(C_2Ph)_4] . 2NH_3 \xrightarrow{\text{vac}} K_2[Ni^{II}(C_2Ph)_4] + 2NH_3$$
$$\text{yellow}$$

Other examples are orange-red $K_3[Cr^{III}(C_2H)_6]$, rose-pink $Na_2[Mn^{II}(C_2Me)_4]$, dark-green $Na_4[Co^{II}(C_2Me)_6]$, orange $K_4[Ni^{(0)}(C_2H)_4]$, yellow $K_6[Ni_2^I(C_2Ph)_6]$. Such compounds are often explosive, though the analogues of Cu^I and Zn^{II} are not, e.g. yellow $Na[Cu(C_2Me)_2]$, colourless $K_2[Cu(C_2H)_3]$, and colourless $K_2[Zn(C_2H)_4]$.

11.3.3 *Other hydrides of nitrogen*

Nitrogen forms at least seven binary compounds with hydrogen: ammonia (p. 481), hydrazine N_2H_4, hydrogen azide HN_3, ammonium azide NH_4N_3 (i.e. N_4H_4), hydrazinium azide $N_2H_5N_3$ (i.e. N_5H_5), and the unstable diazene (diimide) N_2H_2 and tetrazene ($H_2N-N{=}N-NH_2$, i.e. N_4H_4).† Ammonium hydride (NH_5), is unknown. Hydroxylamine, $NH_2(OH)$, is closely related in structure and properties to both ammonia, $NH_2(H)$, and hydrazine, $NH_2(NH_2)$. and it will be convenient to discuss this compound in the present section also.

Hydrazine

Anhydrous N_2H_4 is a fuming, colourless, liquid with a faint ammoniacal odour first detectable at a concentration of 70–80 ppm. Many of its physical properties (Table 11.3)

---

† These last 2 hydrides have only recently been characterized (see *Ann. Rept.* **73A**, 146–47, 1976 for refs.). N_2H_2 is an extremely unstable yellow solid at $-196°$ and was obtained as the *trans*-isomer HN=NH by thermal decomposition of substituted alkali metal hydrazides, e.g. $H_2NN(Ar)Li$. The *cis*-isomer has not been characterized but the labile isodiazene $H_2N{=}N$(singlet) has been made.[32a] Tetrazene is more stable, decomposing only above $0°$ either by disproportionation to N_2 and N_2H_4 (75%) or by isomerization to ammonium azide (15%); it was made by the following low-temperature reaction sequence:

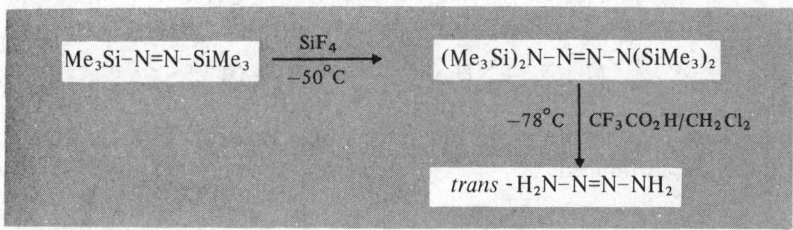

[32] R. NAST and coworkers; for summary of results and detailed refs., see pp. 568–71 of ref. 29.

[32a] N. WIBERG, G. FISCHER, and H. BACHHUBER, Diazene and other dinitrogen hydrides: heats of formation, dissociation energies, appearance potentials, and proton affinities, *Z. Naturforsch.* **34b**, 1385–90 (1979) and references therein.

TABLE 11.3　*Some physical and thermochemical properties of hydrazine*

MP/°C	2.0	Dielectric constant ε (25°)	51.7
BP/°C	113.5	κ (25°)/ohm^{-1} cm^{-1}	$\sim 2.5 \times 10^{-6}$
Density/(solid at $-5°$)/g cm^{-3}	1.146	$\Delta H_{combustion}$/kJ mol^{-1}	148.635
Density (liquid at 25°)/g cm^{-3}	1.00	ΔH_f° (25°)/kJ mol^{-1}	50.6
η (25°)/centipoise$^{(a)}$	0.9	ΔG_f° (25°)/kJ mol^{-1}	149.2
Refractive index n_D^{25}	1.470	S° (25°)/J K^{-1} mol^{-1}	121.2

$^{(a)}$ 1 centipoise $= 10^{-3}$ kg m^{-1} s^{-1}.

are remarkably similar to those of water (p. 730); comparisons with NH_3 (p. 485) H_2O_2 (p. 742) are also instructive, and the influence of H bonding is apparent. In the gas phase 4 conformational isomers are conceivable (Fig. 11.4) but the large dipole (1.85 D) clearly eliminates the staggered *trans*-conformation and electron diffraction data (and infrared) indicate the *gauche*-conformation with an angle of rotation of 90–95° from the eclipsed position.

The most effective preparative routes to hydrazine are still based on the process introduced by F. Raschig in 1907: this involves the reaction of ammonia with an alkaline solution of sodium hypochlorite in the presence of gelatin or glue. The overall reaction can be written as

$$2NH_3 + NaOCl \xrightarrow{\text{aqueous alkali}} N_2H_4 + NaCl + H_2O \tag{1}$$

but it proceeds in two main steps. First there is a rapid formation of chloramine which proceeds to completion even in the cold:

$$NH_3 + OCl^- \longrightarrow NH_2Cl + OH^- \tag{2}$$

The chloramine then reacts further to produce N_2H_4 either by slow nucleophilic attack of NH_3 (3) and subsequent rapid neutralization (4), or by preliminary rapid formation of the chloramide ion (5) followed by slow nucleophilic attack of NH_3 (6):

$$NH_2Cl + NH_3 \xrightarrow{\text{slow}} N_2H_5^+ + Cl^- \tag{3}$$

$$N_2H_5^+ + OH^- \xrightarrow{\text{fast}} N_2H_4 + H_2O \tag{4}$$

$$NH_2Cl + OH^- \xrightarrow{\text{fast}} NHCl^- + H_2O \tag{5}$$

$$NHCl^- + NH_3 \xrightarrow{\text{slow}} N_2H_4 + Cl^- \tag{6}$$

In addition there is a further rapid but undesirable reaction with chloramine which destroys the N_2H_4 produced:

$$N_2H_4 + 2NH_2Cl \xrightarrow{\text{fast}} 2NH_4Cl + N_2 \tag{7}$$

This reaction is catalysed by traces of heavy metal ions such as Cu^{II} and the purpose of the gelatin is to suppress reaction (7) by sequestering the metal ions; it is probable that gelatin also assists the hydrazine-forming reactions between ammonia and chloramine in a way

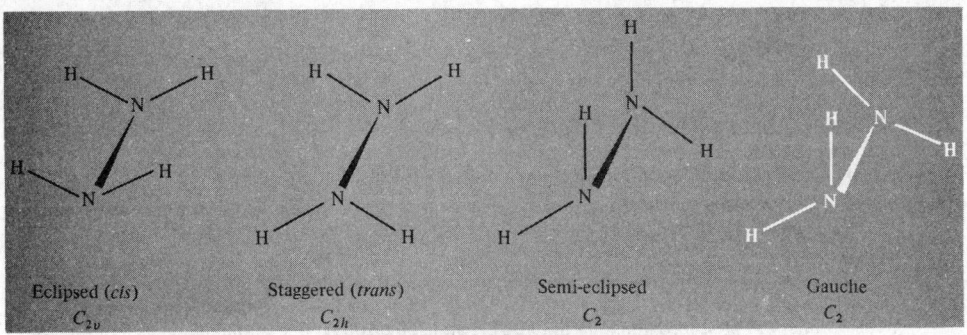

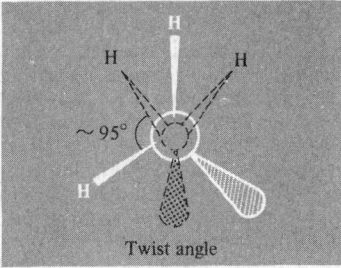

FIG. 11.4 Possible conformations of N_2H_4 with pyramidal N. Hydrazine adopts the gauche form with N–N 145 pm, H–N–H 108°, and the twist angle ~95° as illustrated schematically in the lower diagram.

that is not fully understood. The industrial preparation and uses of N_2H_4 are summarized in the Panel.

At room temperature, pure N_2H_4 and its aqueous solutions are kinetically stable with respect to decomposition despite the endothermic nature of the compound and its positive free energy of formation:

$$N_2(g) + 2H_2(g) \longrightarrow N_2H_4(l); \quad \Delta H_f^\circ = 50.6 \text{ kJ mol}^{-1}$$
$$\Delta G_f^\circ = 149.2 \text{ kJ mol}^{-1}$$

Industrial Production and Uses of Hydrazine

Hydrazine is usually prepared in a continuous process based on the Raschig reaction. Solutions of ammonia and sodium hypochlorite (30:1) are mixed in the cold with a gelatin solution and then passed rapidly under pressure through a reactor at 150° (residence time 1 s). This results in a 60% conversion based on hypochlorite and produces a solution of ~0.5% by weight of N_2H_4. The excess of NH_3 and steam are stripped off in stages and the solution finally distilled to give pure hydrazine hydrate $N_2H_4.H_2O$ (mp −51.7°, bp 118.5°, d 1.0305 g cm^{-3} at 21°). In the Olin Mathieson variation of this process, NH_2Cl is preformed from $NH_3 + NaOCl$ (3:1) and then anhydrous NH_3 is injected to a ratio of ~30:1; this simultaneously raises the temperature and pressure in the reactor.[33] An alternative industrial route, which is economical only for smaller plants, uses urea instead of ammonia in a process very similar to Raschig's:

$$(NH_2)_2CO + NaOCl + 2NaOH \xrightarrow[\text{protein inhibitor}]{\text{rapid heat}} N_2H_4 . H_2O + NaCl + Na_2CO_3$$

[33] L. A. RAPHAELIAN, Hydrazine and its derivatives, *Kirk–Othmer Encyclopedia of Chemical Technology*, 2nd edn., pp. 164–96, 1966.

Hydrazine hydrate contains 64.0% by weight of N_2H_4 and is frequently preferred to the pure compound not only because it is cheaper but also because its much lower mp avoids problems of solidification. Anhydrous N_2H_4 can be obtained from concentrated aqueous solutions by distillation in the presence of dehydrating agents such as solid NaOH or KOH. Alternatively, hydrazine sulfate can be precipitated from dilute aqueous solutions using dilute H_2SO_4 and the precipitate treated with liquid NH_3 to liberate the hydrazine:

$$N_2H_4(aq) + H_2SO_4(aq) \longrightarrow N_2H_6SO_4 \xrightarrow[\text{(anhydr)}]{2NH_3} N_2H_4 + (NH_4)_2SO_4$$

Production figures are hard to obtain, partly because of the extensive use of anhydrous N_2H_4 and its methyl derivatives as rocket fuels, but it seems likely that amounts in excess of 10 000 tonnes are produced annually in the USA and some 2000 tonnes annually in each of the UK, the FRG, and Japan.

The major use (non-commercial) of N_2H_4 and its methyl derivatives $MeNHNH_2$ and Me_2NNH_2 is as a rocket fuel in guided missiles, space shuttles, lunar missions, etc. For example the Apollo lunar modules were decelerated on landing and powered on blast-off for the return journey by the oxidation of a 1:1 mixture of $MeNHNH_2$ and Me_2NNH_2 with liquid N_2O_4; the landing required some 3 tonnes of fuel and 4.5 tonnes of oxidizer, and the relaunching about one-third of this amount. Other oxidants used are O_2, H_2O_2, HNO_3, or even F_2.

The major commercial applications (USA) are as blowing agents ($\sim10\%$), agricultural chemicals ($\sim5\%$), medicinals ($\sim2\%$), and—increasingly—in boiler water treatment. In other countries, particularly those with more modest missile and space programmes, these uses consume a much greater proportion of the hydrazine produced.

Aqueous solutions of N_2H_4 are versatile and attractive reducing agents. They have long been used to prepare silver (and copper) mirrors, to precipitate many elements (such as the platinum metals) from solutions of their compounds, and in other analytical applications. A major application is now in the treatment of high-pressure boiler water: this was first introduced in Germany about 1945 and in the USA in 1951. Most power stations and other large steam-boiler installations now use N_2H_4 rather than the previously favoured Na_2SO_3 because of the following advantages:

(a) N_2H_4 is completely miscible with H_2O and reacts with dissolved O_2 to give merely N_2 and H_2O:

$$N_2H_4 + O_2 \longrightarrow N_2 + 2H_2O$$

(b) N_2H_4 does not increase the dissolved solids (cf. Na_2SO_3) since N_2H_4 itself and all its reaction and decomposition products are volatile.

(c) These products are either alkaline (like N_2H_4) or neutral, but never acidic.

(d) N_2H_4 is also a corrosion inhibitor (by reducing Fe_2O_3 to hard, coherent Fe_3O_4) and it is therefore useful for stand-by and idle boilers.

As N_2H_4 and O_2 have the same molecular weight, 1 kg N_2H_4 will remove 1 kg dissolved O_2 (cf. 8 kg Na_2SO_3); the usual concentration of O_2 in boiler feed water is ~0.01 ppm so that, even allowing for a twofold excess, 1 kg N_2H_4 is sufficient to treat 50 000 tonnes of feed water (say ~4 days' supply at the rate of 500 tonnes per hour).

Hydrazine and its derivatives find considerable use in the synthesis of biologically active materials, dyestuff intermediates, and other organic derivatives. Thus isonicotinic hydrazine ($C_5H_4NCONHNH_2$) is an antituberculin drug, maleic hydrazide ($\overline{N}HCOCH{=}CHCO\overline{N}H$) and β-hydroxyethylhydrazine ($HOCH_2CH_2NHNH_2$) are plant growth regulators, 1,4-diphenyl-semicarbazide ($PhNHCONHNHPh$) is an insecticide, and other derivatives are algaecides and fungicides. Reactions of aldehydes to form hydrazides ($RCH{=}NNH_2$) and azines ($RCH{=}NN{=}CHR$) are well known in organic chemistry, as is the use of hydrazine and its derivatives in the synthesis of heterocyclic compounds.

When ignited, N_2H_4 burns rapidly and completely in air with considerable evolution of heat (see Panel):

$$N_2H_4(l) + O_2(g) \longrightarrow N_2(g) + 2H_2O; \quad \Delta H = -621.5 \text{ kJ mol}^{-1}$$

In solution, N_2H_4 is oxidized by a wide variety of oxidizing agents (including O_2) and it finds use as a versatile reducing agent because of the variety of reactions it can undergo. Thus the thermodynamic reducing strength of N_2H_4 depends on whether it undergoes a 1-, 2-, or 4-electron oxidation and whether this is in acid or alkaline solution. Typical examples in acid solution are as follows:[†]

1-electron change (e.g. using Fe^{III}, Ce^{IV}, or MnO_4^-):
$$NH_4^+ + \tfrac{1}{2}N_2 + H^+ + e^- = N_2H_5^+; \quad E^\circ = -1.74 \text{ V}$$

2-electron change (e.g. using H_2O_2 or HNO_2):
$$\tfrac{1}{2}NH_4^+ + \tfrac{1}{2}HN_3 + \tfrac{5}{2}H^+ + 2e^- = N_2H_5^+; \quad E^\circ = +0.11 \text{ V}$$

4-electron change (e.g. using IO_3^- or I_2):
$$N_2 + 5H^+ + 4e^- = N_2H_5^+; \quad E^\circ = -0.23 \text{ V}$$

For basic solutions the corresponding reduction potentials are:

$$NH_3 + \tfrac{1}{2}N_2 + H_2O + e^- = N_2H_4 + OH^-; \quad E^\circ = -2.42 \text{ V}$$
$$\tfrac{1}{2}NH_3 + \tfrac{1}{2}N_3^- + \tfrac{5}{2}H_2O + 2e^- = N_2H_4 + \tfrac{5}{2}OH^-; \quad E^\circ = -0.92 \text{ V}$$
$$N_2 + 4H_2O + 4e^- = N_2H_4 + 4OH^-; \quad E^\circ = -1.16 \text{ V}$$

In the 4-electron oxidation of acidified N_2H_4 to N_2, it has been shown by the use of N_2H_4 isotopically enriched in ^{15}N that both the N atoms of each molecule of N_2 originated in the same molecule of N_2H_4. This reaction is also the basis for the most commonly used method for the analytical determination of N_2H_4 in dilute aqueous solution:

$$N_2H_4 + KIO_3 + 2HCl \xrightarrow{H_2O/CCl_4} N_2 + KCl + ICl + 3H_2O$$

The IO_3^- is first reduced to I_2 which is subsequently oxidized to ICl by additional IO_3^-; the end-point is detected by the complete discharge of the iodine colour from the CCl_4 phase.

As expected N_2H_4 in aqueous solutions is somewhat weaker as a base than ammonia (p. 485):

$$N_2H_4(aq) + H_2O = N_2H_5^+ + OH^-; \quad K_{25} = 8.5 \times 10^{-7}$$
$$N_2H_5^+(aq) + H_2O = N_2H_6^{2+} + OH^-; \quad K_{25} = 8.9 \times 10^{-16}$$

The hydrate $N_2H_4 \cdot H_2O$ is an H-bonded molecular adduct and is not ionically dissociated. Two series of salts are known, e.g. N_2H_5Cl and $N_2H_6Cl_2$. (It will be noticed that $N_2H_6^{2+}$ is isoelectronic with ethane.) H bonding frequently influences the crystal structure and this is particularly noticeable in $N_2H_6F_2$ which features a layer lattice similar to CdI_2 though the structure is more open and the fluoride ions are not close-packed.[34] Sulfuric acid forms three salts, $N_2H_4 \cdot nH_2SO_4$ ($n = \tfrac{1}{2}, 1, 2$), i.e. $[N_2H_5]_2SO_4$, $[N_2H_6]SO_4$, and $[N_2H_6][HSO_4]_2$.

Hydrazido(2−)-complexes of Mo and W have been prepared by protonating dinitrogen complexes with concentrated solutions of HX and by ligand exchange.[34a] For

---

† See p. 498 for discussion of standard electrode potentials and their use. It is conventional to write the half-reactions as (oxidized form) + ne^- = (reduced form). Since $\Delta G = -nE^\circ F$ at unit activities, it follows that the reactions will occur spontaneously in the reverse direction to that written when E° is negative, i.e. hydrazine is oxidized by the reagents listed.

[34] Ref. 21, pp. 309–10, 645.

[34a] J. CHATT, A. J. PEARMAN, and R. L. RICHARDS, Hydrazido(2−)-complexes of molybdenum and tungsten formed from dinitrogen complexes by protonation and ligand exchange, *JCS Dalton* 1978, 1766–76.

example several dozen complexes of general formulae $[MX_2(NNH_2)L_3]$ and *trans*-$[MX(NNH_2)L_4]$ have been characterized for $M = Mo$; $X = halogen$; $L = phosphine$ or heterocyclic-N donor. Similarly, *cis*-$[W(N_2)_2(PMe_2Ph)_4]$ afforded *trans*-$[WF(NNH_2)(PMe_2Ph)_4][BF_4]$ when treated with HF/MeOH in a borosilicate glass vessel. Side-on coordination of a phenylhydrazido$(1-)$ ligand has also been established in compounds such as the dark-red $[W(\eta^5-C_5H_5)_2(\eta^2-H_2NNPh)][BF_4]$;[34b] these are synthesized by the ready isomerization of the first-formed yellow η^1-arylhydrazido$(2-)$ tungsten hydride complex above $-20°$ ($X = BF_4$, PF_6):

$$\left[[(\eta^5-C_5H_5)_2WH_2] \right] \xrightarrow[\text{MeOH/MePh}(-20°)]{\text{Arene diazonium X}} \left[(\eta^5-C_5H_5)_2W \underset{NN(H)R}{\overset{H}{<}} \right] X \xrightarrow{0°}$$

Yellow

$$\left[(\eta^5-C_5H_5)_2 W \underset{NR}{\overset{NH_2}{<}} \right] X$$

Dark red

In these reactions $R = Ph$, p-$MeOC_6H_4$, p-MeC_6H_4, or p-FC_6H_4. Further bonding modes are as an isodiazene (i.e. $M \leftarrow N = NMe_2$ rather than $M = N - NMe_2$)[34c] and as a bridging diimido group ($M = N - N = M$).[34d]

Hydroxylamine

Anhydrous NH_2OH is a colourless, thermally unstable hygroscopic compound which is usually handled as an aqueous solution or in the form of one of its salts. The pure compound (mp 32.05°C, d 1.204 g cm^{-3} at 33°C) has a very high dielectric constant (77.63–77.85) and a vapour pressure of 10 mmHg at 47.2°. It can be regarded as water in which 1 H has been replaced by the more electronegative NH_2 group or as NH_3 in which 1 H has been replaced by OH. Aqueous solutions are less basic than either ammonia or hydrazine:

$$NH_2OH(aq) + H_2O = NH_3OH^+ + OH^-; \quad K_{25°} \ 6.6 \times 10^{-9}$$

Hydroxylamine can be prepared by a variety of reactions involving the reduction of nitrites, nitric acid, or NO, or by the acid hydrolysis of nitroalkanes. In the conventional Raschig synthesis, an aqueous solution of NH_4NO_2 is reduced with HSO_4^-/SO_2 at 0° to give the hydroxylamido-*N,N*-disulfate anion which is then hydrolysed stepwise to

[34b] J. A. CARROLL, D. SUTTON, M. COWIE, and M. D. GAUTHIER, Side-on coordination of a phenylhydrazido ligand: synthesis and X-ray structure determination of [η^5-$C_5H_5)_2W(\eta^2-H_2NNPh)][BF_4]$, *JCS Chem. Comm.* 1979, 1058–9.

[34c] J. R. DILWORTH, J. ZUBIETA, and J. R. HYDE, Preparation and crystal structure of $[Mo_3S_8(NNMe_2)_2]^{2-}$, a trinuclear sulfido-bridged molybdenum anion with coordinated isodiazene ligands, *J. Am. Chem. Soc.* **104**, 365–7 (1982).

[34d] M. R. CHURCHILL and H. J. WASSERMAN, Crystal structure of $[Ta(=CHCMe_3)(CH_2CMe_3)(PMe_3)_2]_2(\mu-N_2)$, a molecule with a dinitrogen ligand behaving as a diimido group in a Ta=NN=Ta bridge, *Inorg. Chem.* **20**, 2899–904 (1981).

hydroxylammonium sulfate:

$$NH_4NO_2 + 2SO_2 + NH_3 + H_2O \rightarrow [NH_4]_2[N(OH)(OSO_2)_2]$$
$$[NH_4]^+{}_2[N(OH)(OSO_2)_2]^{2-} + H_2O \rightarrow [NH_4][NH(OH)(OSO_2)] + [NH_4][HSO_4]$$
$$2[NH_4]^+[NH(OH)(OSO_2)]^- + 2H_2O \rightarrow [NH_3(OH)]_2[SO_4] + [NH_4]_2[SO_4]$$

Aqueous solutions of NH_2OH can then be obtained by ion exchange, or the free compound can be prepared by ammonolysis with liquid NH_3; insoluble ammonium sulfate is filtered off and the excess of NH_3 removed under reduced pressure to leave solid NH_2OH.

Alternatively, hydroxylammonium salts can be made either (a) by the electrolytic reduction of aqueous nitric acid between amalgamated lead electrodes in the presence of H_2SO_4/HCl, or (b) by the hydrogenation of nitric oxide in acid solutions over a Pt/charcoal catalyst:

(a) $HNO_3(aq) + 6H^+(aq) \xrightarrow{6e^-} 2H_2O + NH_2OH \xrightarrow{HCl(g)} +[NH_3(OH)]Cl(s)$

(b) $2NO(g) + 3H_2(g) + H_2SO_4(aq) \xrightarrow{Pt/C} [NH_3(OH)]_2SO_4$

The vigorous hydrolysis of primary nitroalkanes by refluxing with strong mineral acids, though mechanistically obscure, is also commercially viable:

$$2RCH_2NO_2 + 2H_2O \xrightarrow{H_2SO_4} 2RCO_2H + [NH_3(OH)]_2SO_4$$

$$C_2H_4(g) + N_2O_4(g) \longrightarrow O_2NCH_2CH_2NO_2 \xrightarrow{H_2SO_4} CO_2 + CO + [NH_3(OH)]_2SO_4$$

A convenient laboratory route involves the reduction of an aqueous solution of nitrous acid or potassium nitrite with bisulfite under carefully controlled conditions: The hydroxylamidodisulfate first formed, though stable in alkaline solution, rapidly hydrolyses to the monosulfate in acid solution and this can then subsequently by hydrolysed to the hydroxylammonium ion by treatment with aqueous HCl at 100° for 1 h:

$$HNO_2 + 2HSO_3^- \longrightarrow [N(OH)(OSO_2)_2]^{2-} + H_2O \xrightarrow{fast}$$
$$[NH(OH)(OSO_2)]^- + [HSO_4]^-$$

$$[NH(OH)(OSO_2)]^- + H_3O^+ \xrightarrow{100°/1\ h} [NH_3(OH)]^+ + [HSO_4]^-$$

Anhydrous NH_3OH can be prepared by treating a suspension of hydroxylammonium chloride in butanol with NaOBu:

$$[NH_3(OH)]Cl + NaOBu \longrightarrow NH_2OH + NaCl + BuOH$$

The NaCl is removed by filtration and the NH_2OH precipitated by addition of Et_2O and cooling.

NH_2OH can exist as 2 configurational isomers (*cis* and *trans*) and in numerous intermediate *gauche* conformations as shown in Fig. 11.5. In the crystalline form, H bonding appears to favour packing in the *trans* conformation. The N–O distance is 147 pm consistent with its formulation as a single bond. Above room temperature the compound

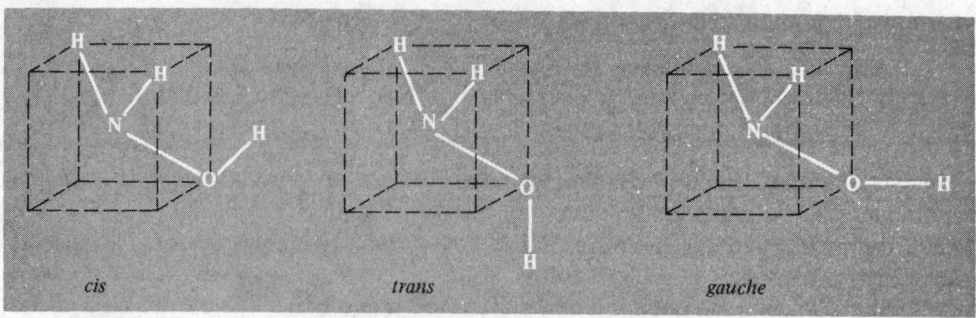

cis   trans   gauche

FIG 11.5   Configurations of NH_2OH.

decomposes (sometimes explosively) by internal oxidation–reduction reactions into a complex mixture of N_2, NH_3, N_2O, and H_2O. Aqueous solutions are much more stable, particularly acid solutions in which the compound is protonated, $[NH_3(OH)]^+$. Such solutions can act as oxidizing agents particularly when acidified but are more generally used as reducing agents, e.g. as antioxidants in photographic developers, stabilizers of monomers, and for reducing Cu^{II} to Cu^{I} in the dyeing of acrylic fibres. Comparisons with the redox chemistry of H_2O_2 and N_2H_4 are also instructive (see, for example, pp. 272–3 of ref. 31). The ability of NH_2OH to react with N_2O, NO, and N_2O_4 under suitable conditions (e.g. as the sulfate adsorbed on silica gel) makes it useful as an absorbent in combustion analysis. Another major use, which derives from its ability to form oximes with aldehydes and ketones, is in the manufacture of caprolactam, a key intermediate in the production of polyamide fibres such as nylon.

The extensive chemistry of the hydroxylamides of sulfuric acid is discussed later in the context of other H–N–O–S compounds (pp. 875–881).

Hydrogen azide

Aqueous solutions of HN_3 were first prepared in 1890 by T. Curtius who oxidized aqueous hydrazine with nitrous acid:

$$N_2H_5^+ + HNO_2 \longrightarrow HN_3 + H^+ + 2H_2O$$

Other oxidizing agents that can be used include nitric acid, hydrogen peroxide, peroxydisulfate, chlorate, and the pervanadyl ion. The anhydrous compound is extremely explosive and even dilute solutions should be treated as potentially hazardous. Pure HN_3 is best prepared by careful addition of H_2SO_4 to NaN_3; it is a colourless liquid or gas (mp $\sim -80°$, estimated bp 35.7°, d 1.126 g cm^{-3} at 0°). Its large positive enthalpy and free energy of formation emphasize its inherent instability: ΔH_f° (l, 298 K) 269.5, ΔG_f° (l, 298 K) 327.2 kJ mol^{-1}. It has a repulsive, intensely irritating odour and is a deadly (though non-cumulative) poison; even at concentrations less than 1 ppm in air it can be dangerous. In the gas phase the 3 N atoms are colinear, as expected for a 16 valence-electron species, and the angle HNN is 112°; the two N–N distances are appreciably different, $HN-N_2$ being 124 pm and HN_2-N 113 pm. Similar differences are found for organic azides (e.g. MeN_3). In ionic azides (p. 479) the N_3^- ion is both linear and symmetrical (both N–N distances being 116 pm) as befits a 16-electron species isoelectronic with CO_2 (cf. also the cyanamide ion NCN^{2-}, the cyanate ion NCO^-, the fulminate ion CNO^-, and the nitronium ion NO_2^+).

Aqueous solutions of HN_3 are about as strongly acidic as acetic acid:

$$HN_3(aq) = H^+(aq) + N_3^-(aq); \quad K_a \ 1.8 \times 10^{-5}, \ pK_a \ 4.77 \text{ at } 298 \text{ K}$$

Numerous metal azides have been characterized (p. 479) and covalent derivatives of non-metals are also readily preparable by simple metathesis using either NaN_3 or aqueous solutions of HN_3.[35, 36] In these compounds the N_3 group behaves as a pseudohalogen (p. 336) and, indeed, the unstable compounds FN_3, ClN_3, BrN_3, IN_3, and NCN_3 are known, though potential allotropes of nitrogen such as N_3-N_3 (analogous to Cl_2) and $N(N_3)_3$ (analogous to NCl_3) have not been isolated. More complex heterocyclic compounds are, however, well established, e.g. cyanuric azide $\{-NC(N)_3-\}_3$, B,B,B-triazidoborazine $\{-NB(N_3)-\}_3$, and even the azidophosphazene derivative $\{-NP(N_3)_2-\}_3$.

Most preparative routes to HN_3 and its derivatives involve the use of NaN_3 since this is reasonably stable and commercially available. NaN_3 can be made by adding powdered $NaNO_3$ to fused $NaNH_2$ at $175°$ or by passing N_2O into the same molten amide at $190°$:

$$NaNO_3 + 3NaNH_2 \longrightarrow NaN_3 + 3NaOH + NH_3$$
$$N_2O + 2NaNH_2 \longrightarrow NaN_3 + NaOH + NH_3$$

The latter reaction is carried out on an industrial scale using liquid NH_3 as solvent; a variant uses Na/NH_3 without isolation of the $NaNH_2$:

$$3N_2O + 4Na + NH_3 \longrightarrow NaN_3 + 3NaOH + 2N_2$$

The major use of inorganic azides depends on the explosive nature of heavy metal azides. $Pb(N_3)_2$ in particular is extensively used in detonators because of its reliability, especially in damp conditions; it is prepared by metathesis between $Pb(NO_3)_2$ and NaN_3 in aqueous solution.

11.3.4 *Thermodynamic relations between N-containing species*

The ability of N to exist in its compounds in at least 10 different oxidation states from -3 to $+5$ poses certain thermodynamic and mechanistic problems that invite systematic treatment. Thus, in several compounds N exists in more than one oxidation state, e.g. $[N^{-III}H_4]^+[N^{III}O_2]^-$, $[N^{-III}H_4]^+[N^VO_3]^-$, $[N_2^{-II}H_5]^+[N^VO_3]^-$, $[N^{-III}H_4]^+[N^{-\frac{1}{3}}_3]^-$, etc. Furthermore, we have seen (p. 485) that, under appropriate conditions, NH_3 can be oxidized by O_2 to yield N_2, NO, or NO_2, whereas oxidation by OCl^- yields N_2H_4 (p. 490). Likewise, using appropriate reagents, N_2H_4 can be oxidized either to N_2 or to HN_3 (in which the "average" oxidation number of N is $-\frac{1}{3}$). The thermodynamic relations between these various hydrido and oxo species containing N can be elegantly codified by means of their standard reduction potentials, and these can be displayed pictorially using the concept of the "volt equivalent" of each species (see Panel).

The standard reduction potentials in acidic aqueous solution are given in Table 11.4; these are shown diagrammatically in Fig. 11.6 which also includes the corresponding data

[35] Ref. 31, pp. 276–93.

[36] A. D. YOFFE, The inorganic azides, Chap. 2 in C. B. COLBURN (ed.), *Developments in Inorganic Nitrogen Chemistry*, Vol. 1, pp. 72–149, Elsevier, Amsterdam, 1966.

Standard Reduction Potentials and Volt Equivalents

Chemical reactions can often formally be expressed as the sum of two or more "half-reactions" in which electrons are transferred from one chemical species to another. Conventionally these are now almost always represented as equilibria in which the forward reaction is a reduction (addition of electrons):

$$(\text{oxidized form}) + ne^- \rightleftharpoons (\text{reduced form})$$

The electrochemical reduction potential (E volts) of such an equilibrium is given by

$$E = E^\circ - \frac{2.3026RT}{nF} \log_{10} \frac{a\,(\text{red})}{a\,(\text{ox})} \tag{1}$$

where E° is the "standard reduction potential" at unit activity a, R is the gas constant (8.3144 J mol^{-1} K^{-1}), T is the absolute temperature, F is the Faraday constant (96 485 C mol^{-1}), and 2.3026 is the constant $\ln_e 10$ required to convert from natural to decadic logarithms. At 298.15 K (25°C) the factor $2.3026RT/F$ has the value 0.059 16 V and, replacing activities by concentrations, one obtains the approximate expression

$$E \approx E^\circ - \frac{0.059\ 16}{n} \log_{10} \frac{[\text{red}]}{[\text{ox}]} \tag{2}$$

By convention, E° for the half-reaction (3) is taken as zero, i.e. $E^\circ(\text{H}^+/\tfrac{1}{2}\text{H}_2) = 0.0$ V:

$$\text{H}^+(\text{aq}, a=1) + e^- \rightleftharpoons \tfrac{1}{2}\text{H}_2 \text{ (g, 1 atm)} \tag{3}$$

Remembering that $\Delta G = -nEF$, it follows that the standard free energy change for the half reaction is $\Delta G^\circ = -nE^\circ F$, e.g.:

$$\text{Fe}^{3+}(\text{aq}) + e^- \rightleftharpoons \text{Fe}^{2+}(\text{aq}); \quad E^\circ(\text{Fe}^{3+}/\text{Fe}^{2+}) = 0.771 \text{ V} \tag{4}$$
$$\Delta G^\circ = -74.4 \text{ kJ mol}^{-1}$$

Coupling the half-reactions (3) and (4) gives the reaction (5) {i.e. (4)−(3)} which, because ΔG is negative, proceeds spontaneously from left to right as written:

$$\text{Fe}^{3+}(\text{aq}) + \tfrac{1}{2}\text{H}_2(\text{g}) \rightleftharpoons \text{Fe}^{2+}(\text{aq}) + \text{H}^+(\text{aq}); \quad E^\circ = 0.771 \tag{5}$$
$$\Delta G^\circ = -74.4 \text{ kJ mol}^{-1}$$

Again, $E^\circ(\text{Zn}^{2+}/\text{Zn}) = -0.763$ V, hence reaction (6) occurs spontaneously in the reverse direction:

$$\text{Zn}^{2+}(\text{aq}) + \text{H}_2(\text{g}) \rightleftharpoons \text{Zn}(\text{s}) + 2\text{H}^+(\text{aq}); \quad E^\circ = -0.763 \text{ V};$$
$$\Delta G^\circ = +147.5 \text{ kJ mol}^{-1} \text{ (note the factor of 2 for } n) \tag{6}$$

In summary, at pH 0 a reaction is spontaneous from left to right if $E^\circ > 0$ and spontaneous in the reverse direction of $E^\circ < 0$. At other H-ion concentrations eqn. (2) indicates that the potential of the H electrode (3) will be

$$E = -0.059\ 16 \log \frac{\{P_{\text{H}_2}/\text{atm}\}^{\frac{1}{2}}}{\{[\text{H}^+(\text{aq})]/\text{mol l}^{-1}\}} \text{ V}$$

and, in general, the potential of any half-reaction changes with the concentration of the species involved according to the Nernst equation (7):

$$E = E^\circ - \frac{0.059\ 16}{n} \log Q \tag{7}$$

where Q has the same form as the equilibrium constant but is a function of the actual activities of the reactants and products rather than those of the equilibrium state. Note also that the potential is independent of the coefficients of the half-reaction whereas the free energy is directly proportional to these, e.g.:

$$\tfrac{1}{2}\text{I}_2 + e^- \rightleftharpoons \text{I}^-; \quad E^\circ = 0.536 \text{ V}; \quad \Delta G^\circ = -51.5 \text{ kJ mol}^{-1}$$
$$\text{I}_2 + 2e^- \rightleftharpoons 2\text{I}^-; \quad E^\circ = 0.536 \text{ V}; \quad \Delta G^\circ = -103.0 \text{ kJ mol}^{-1}$$

Continued

It is vital to remember that, when half-reactions are added or subtracted, one should not add or subtract the corresponding $E°$ values but rather $nE°$. (We shall return to this point later.)

Lists of standard reduction potentials are given in many books[37] and are extensively quoted throughout this text. Almost all lists now use the IUPAC sign conventions employed above, though some earlier American books (including, unfortunately, the classic early text on the subject[38]) use the opposite sign convention. When standard reduction potentials are listed in sequence from the most negative to the most positive the strongest reducing agents are at the top of the list and a reducing agent should, in principle, be capable of reducing all oxidizing agents lying below it in the table. Conversely, the oxidizing agents are listed in order of *increasing* strength and a given oxidizing agent should be able to oxidize all reducing agents lying above it in the table. Such lists are an extremely compact way of summarizing a great deal of predictive information. For example a list of 100 independent reduction potentials enables the free energy change for $100 \times 99/2 = 4950$ reactions to be calculated and indicates the direction in which a hypothetical reaction would occur under appropriate conditions (which might involve the use of a catalyst).

When an element can exist in several oxidation states it is sometimes convenient to display the various reduction potentials diagramatically, the corresponding half-reactions under standard conditions being implied. Thus, in acidic aqueous solutions

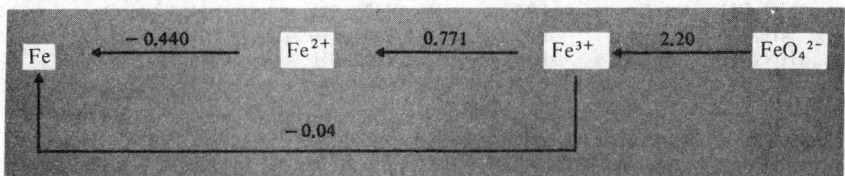

Note that the value of $E°(Fe^{3+}/Fe) = -0.04$ V is equivalent to $\{(2 \times -0.44) + 0.77\}/3$. Because the quantity $nE°$ is used in these calculations (rather than $E°$) it is convenient to define the "volt equivalent" of a species; the volt equivalent of a compound or ion is the reduction potential of the species relative to the element in its standard state multiplied by the oxidation state of the element in the compound (including its sign). The oxidation state is the number of electrons that must be added to an atom of an element to regain electroneutrality when all other atoms in the compound (ion) have been removed as their "normal" ions. For example the oxidation state of Fe is $+6$ in FeO_4^{2-}:

$$FeO_4^{2-} \xrightarrow{\ -4O^{2-}\ } Fe^{6+} \xrightarrow{\ +6e^-\ } Fe$$

It follows that, in the above example, the volt equivalents of Fe^{2+} and Fe^{3+} are -0.88 and -0.11 respectively and that of FeO_4^{2-} is $+5.49$ {i.e. $(2 \times -0.44) + 0.77 + (3 \times 2.20)$}. This leads to $E°(FeO_4^{2-}/Fe) = +0.91$ V.

The power of these various concepts in codifying and rationalizing the redox chemistry of the elements is illustrated for the case of nitrogen in the present section. Standard reduction potentials and plots of volt equivalents against oxidation state for other elements are presented in later chapters.

for alkaline solutions. The reduction potentials are readily converted to volt equivalents (by multiplying by the appropriate oxidation state) and these are plotted against oxidation state in Fig. 11.7. This latter diagram is particularly valuable in giving a visual representation of the redox chemistry of the element. Thus, it follows from the definition of "volt equivalent" that the reduction potential of any couple is the *slope* of the line joining

[37] G. CHARLOT, A. COLLUMEAU, and M. J. C. MARCHON, *Selected Constants: Oxidation–Reduction Potentials of Inorganic Substances in Aqueous Solution*, Butterworths, London, 1971, 73 pp. G. MILAZZO and S. CAROLI, *Tables of Standard Electrode Potentials*, Wiley, New York, 1978, 421 pp.

[38] W. M. LATIMER, *The Oxidation States of the Elements and their Potentials in Aqueous Solutions*, 2nd edn., Prentice-Hall, New York, 1952, 392 pp.

TABLE 11.4 *Standard reduction potentials for nitrogen species*[a] *in acidic aqueous solution (pH 0, 25°C)*

Couple	$E°/V$	Corresponding half-reaction
N_2/HN_3	-3.09	$\frac{3}{2}N_2 + H^+(aq) + e^- \rightarrow HN_3(aq)$
$N_2/N_2H_5^+$	-0.23	$N_2 + 5H^+ + 4e^- \rightarrow N_2H_5^+$
$H_2N_2O_2/NH_3OH^+$	$+0.387$	$H_2N_2O_2 + 6H^+ + 4e^- \rightarrow 2NH_3OH^+$
HN_3/NH_4^+	$+0.695$	$HN_3 + 11H^+ + 8e^- \rightarrow 3NH_4^+$
$NO/H_2N_2O_2$	$+0.712$	$2NO + 2H^+ + 2e^- \rightarrow H_2N_2O_2$
NO_3^-/N_2O_4	$+0.803$	$2NO_3^- + 4H^+ + 2e^- \rightarrow N_2O_4 + 2H_2O$
$HNO_2/H_2N_2O_2$	$+0.86$	$2HNO_2 + 4H^+ + 4e^- \rightarrow H_2N_2O_2 + 2H_2O$
NO_3^-/HNO_2	$+0.94$	$NO_3^- + 3H^+ + 2e^- \rightarrow HNO_2 + H_2O$
NO_3^-/NO	$+0.957$	$NO_3^- + 4H^+ + 3e^- \rightarrow NO + 2H_2O$
HNO_2/NO	$+0.983$	$HNO_2 + H^+ + e^- \rightarrow NO + H_2O$
N_2O_4/NO	$+1.035$	$N_2O_4 + 4H^+ + 4e^- \rightarrow 2NO + 2H_2O$
N_2O_4/HNO_2	$+1.065$	$N_2O_4 + 2H^+ + 2e^- \rightarrow 2HNO_2$
$N_2H_5^+/NH_4^+$	$+1.275$	$N_2H_5^+ + 3H^+ + 2e^- \rightarrow 2NH_4^+$
HNO_2/N_2O	$+1.29$	$2HNO_2 + 4H^+ + 4e^- \rightarrow N_2O + 3H_2O$
NH_3OH^+/NH_4^+	$+1.35$	$NH_3OH^+ + 2H^+ + 2e^- \rightarrow NH_4^+ + H_2O$
$NH_3OH^+/N_2H_5^+$	$+1.42$	$2NH_3OH^+ + H^+ + 2e^- \rightarrow N_2H_5^+ + 2H_2O$
HN_3/NH_4^+	$+1.96$	$HN_3 + 3H^+ + 2e^- \rightarrow NH_4^+ + N_2$
$H_2N_2O_2/N_2$	$+2.65$	$H_2N_2O_2 + 2H^+ + 2e^- \rightarrow N_2 + 2H_2O$

[a] All the half-reactions listed in this table have only (Ox), H^+, and e^- on the left-hand side of the half-reaction. Others, such as N_2/NH_3OH^+ -1.87 (i.e. $N_2 + 2H_2O + 4H^+ + 2e^- \rightarrow 2NH_3OH^+$) can readily be calculated by appropriate combinations (in this case, for example, $N_2/N_2H_5^+ - NH_3OH^+/N_2H_5^+$). There are also simple electron addition reactions, e.g. NO^+/NO, $E° + 1.46$ V (i.e. $NO^+ + e^- \rightarrow NO$) and more complex electron additions, e.g. $NO_3^-,NO/NO_2^-$, $E° + 0.49$ V (i.e. $NO_3^- + NO + e^- \rightarrow 2NO_2^-$), etc.

the two points: the greater the positive slope the stronger the oxidizing potential and the greater the negative slope the stronger the reducing power. Any pair of points can be joined. For example in acid solution N_2H_4 is a stronger reducing agent than H_2 (slope of tie-line -0.23 V) and NH_2OH is even stronger (slope -1.87 V). By contrast, the couple N_2O/NH_3OH^+ has virtually the same reducing power as H_2 (slope -0.05 V).

It also follows that, when three (or more) oxidation states lie approximately on a straight line in the volt-equivalent diagram, they tend to form an equilibrium mixture rather than a reaction going to completion (provided that the attainment of thermo-dynamic equilibrium is not hindered kinetically). This is because the slopes joining the several points are almost the same, so that $E°$ for the various couples (and hence $\Delta G°$) are the same; there is consequently approximately zero change in free energy and a balanced equilibrium is maintained between the several species. Indeed, the volt-equivalent diagram is essentially a plot of free energy versus oxidation state (as indicated by the right-hand ordinate of Fig. 11.7).

Two further points follow from these general considerations:

(a) a compound will tend to disproportionate into a higher and a lower oxidation state if it lies above the line joining the 2 compounds in these oxidation states, i.e. disproportionation is accompanied by a decrease in free energy and will tend to occur spontaneously if not kinetically hindered. Examples are the disproportionation of

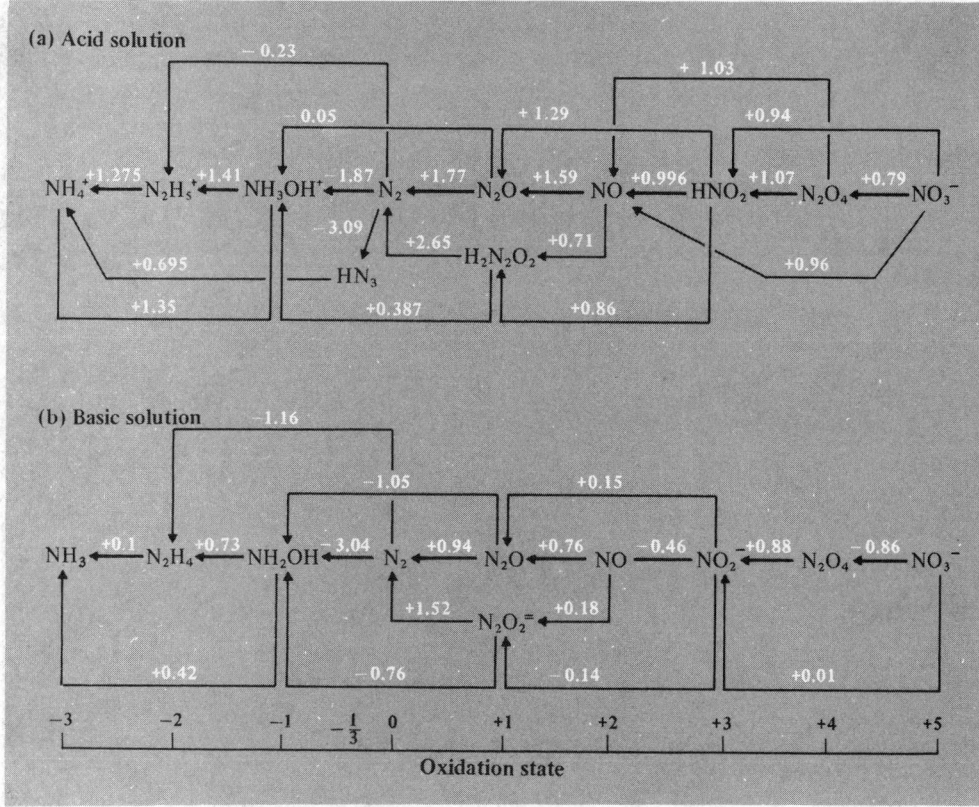

FIG. 11.6 Oxidation states of nitrogen showing standard reduction potentials in volts: (a) in acid solution at pH 0, and (b) in basic solution at pH 14.

hydroxylamine in acidic solutions (slow) and alkaline solutions (fast):

$$4NH_3OH^+ \longrightarrow N_2O + 2NH_4^+ + 3H_2O + 2H^+$$
$$3NH_2OH \longrightarrow N_2 + NH_3 + 3H_2O$$

(b) Conversely, a compound can be formed by conproportionation of compounds in which the element has a higher and lower oxidation state if it lies below the line joining these two states. A particularly important example is the synthesis of $HN_3^{-\frac{1}{3}}$ by reacting $N_2^{-II}H_5^+$ and $HN^{+III}O_2$ (p. 496). It will be noted that the reduction potential of HN_3 (-3.09 V) is more negative than that of any other reducing agent in acidic aqueous solution so it is thermodynamically impossible to synthesize HN_3 by reduction of N_2 or any of its compounds in such media unless the reducing agent itself contains N (as does hydrazine).

In basic solutions a different set of redox equilibria obtain and a different set of reduction potentials must be used, as shown on the following page.

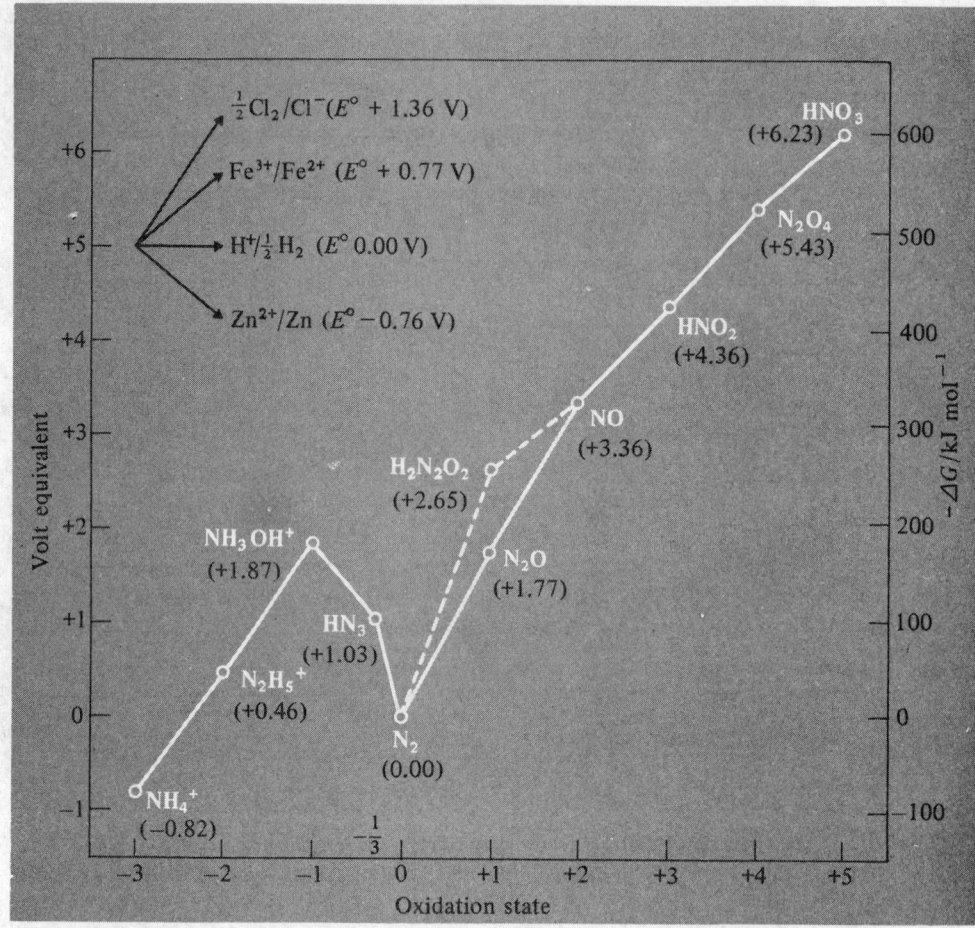

FIG. 11.7 Plot of volt equivalent against oxidation state for various compounds or ions containing N in acidic aqueous solution. Note that values of $-\Delta G$ refer to N_2 as standard (zero) but are quoted per mol of N atoms and per mol of N_2; they refer to reactions in the direction $(ox) + ne^- \rightarrow (red)$. Slopes corresponding to some common oxidizing and reducing agents are included for comparison.

For example:

$$
\begin{array}{ll}
 & E^\circ/V \\
N_2 + 4H_2O + 2e^- = 2NH_2OH + 2OH^- & -3.04 \\
N_2 + 4H_2O + 4e^- = N_2H_4 + 4OH^- & -1.16 \\
N_2O + 5H_2O + 4e^- = 2NH_2OH + 4OH^- & -1.05 \\
N_2O_2^{2-} + 6H_2O + 4e^- = 2NH_2OH + 6OH^- & -0.73 \\
NO_3^- + H_2O + 2e^- = NO_2^- + 2OH^- & +0.01 \\
N_2H_4 + 4H_2O + 2e^- = 2NH_4^+ + 4OH^- & +0.11 \\
2NH_2OH + 2e^- = N_2H_4 + 2OH^- & +0.73
\end{array}
$$

A more complete compilation is summarized in Fig. 11.6. It is instructive to use these data to derive a plot of volt equivalent versus oxidation state in basic solution and to compare this with Fig. 11.7 which refers to acidic solutions.

11.3.5 *Nitrogen halides and related compounds*[31]

It is a curious paradox that NF_3, the most stable binary halide of N, was not prepared until 1928, more than 115 y after the highly unstable NCl_3 was prepared in 1811 by P. L. Dulong (who lost three fingers and an eye studying its properties). Pure NBr_3 explodes even at $-100°$ and was not isolated until 1975[39] and NI_3 has not been prepared, though the explosive adduct $NI_3 . NH_3$ was first made by B. Courtois in 1813 and several other ammines are known. In all, there are now 5 binary fluorides of nitrogen (NF_3, N_2F_4, *cis*- and *trans*-N_2F_2, and N_3F) and these, together with various mixed halides, hydride halides, and oxohalides are discussed in this section.

NF_3 was first prepared by Otto Ruff's group in Germany by the electrolysis of molten NH_4F/HF and this process is still used commercially. An alternative is the controlled fluorination of NH_3 over a Cu metal catalyst.

$$4NH_3 + 3F_2 \xrightarrow{\text{Cu}} NF_3 + 3NH_4F$$

NF_3 is a colourless, odourless, thermodynamically stable, gas (mp $-206.8°$, bp $-129.0°$ ΔG_{298}° -83.3 kJ mol^{-1}). The molecule is pyramidal with an F–N–F angle of 102.5°, but the dipole moment (0.234 D) is only one-sixth of that of NH_3 (1.47 D) presumably because the N–F bond moments act in the opposite direction to that of the lone-pair moment:

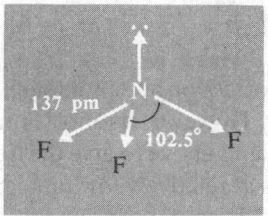

The gas is remarkably unreactive (like CF_4) being unaffected by water or dilute aqueous acid or alkali; at elevated temperatures it acts as a fluorinating agent and with Cu, As, Sb, or Bi in a flow reactor it yields N_2F_4 ($2NF_3 + 2Cu \rightarrow N_2F_4 + 2CuF$). As perhaps expected (p. 223) NF_3 shows little tendency to act as a ligand, though NF_4^+ is known[39a] and also the surprisingly stable isoelectronic species ONF_3 (mp $-160°$, bp $-87.6°$):

$$NF_3 + 2F_2 + SbF_3 \xrightarrow{200°/100 \text{ atm}} [NF_4]^+[SbF_6]^-$$

$$2NF_3 + O_2 \xrightarrow[-196°]{\text{electric discharge/}} 2ONF_3$$

$$3FNO + 2IrF_6 \xrightarrow{20°} ONF_3 + 2[NO]^+[IrF_6]^-$$

[39] J. LANDER, J. KNACKMUSS, and K.-U. THIEDEMANN, Preparation and isolation of NBr$_3$ and NBr$_2$I, *Z. Naturforsch.* **B30,** 464–5 (1975).

[39a] K. O. CHRISTE, C. H. SCHACK, and R. D. WILSON, Synthesis and characterization of (NF$_4$)$_2$SnF$_6$ and (NF$_4$)SnF$_5$, *Inorg. Chem.* **16,** 849–54 (1977), and references therein. See also K. O. CHRISTE, R. D. WILSON, and I. R. GOLDBERG, Formation and decomposition mechanism of NF$_4^+$ salts, *Inorg. Chem.* **18,** 2572–7 (1979). K. O. CHRISTE, R. D. WILSON, and C. J. SCHACK, Synthesis and properties of NF$_4^+$SO$_3$F$^-$, *Inorg. Chem.* **19,** 3046–9 (1980).

ONF_3 was discovered independently by two groups in 1966. Although isoelectronic with BF_4^-, CF_4, and NF_4^+ it has excited interest because of the short N–O distance (115.8 pm), which implies some multiple bonding, and the correspondingly long N–F distances (143.1 pm). Similar partial double bonding to O and highly polar bonds to F have also been postulated for the analogous ion $[OCF_3]^-$ in $Cs[OCF_3]$.[39b]

Dinitrogen tetrafluoride, N_2F_4, is the fluorine analogue of hydrazine and apparently exists in both the staggered (*trans*) C_{2h} and *gauche* C_2 conformations (p. 491). It was discovered in 1957 and is now made either by partial defluorination of NF_3 (see above) or by quantitative oxidation of NF_2H with alkaline hypochlorite:

$$(NH_2)_2CO(aq) \xrightarrow[\text{(70\% yield)}]{F_2/N_2} NH_2CONF_2 \xrightarrow[\text{(100\% yield)}]{\text{conc } H_2SO_4} NF_2H \xrightarrow[\text{(100\% yield)}]{\text{NaOCl (pH 12)}} \tfrac{1}{2}N_2F_4$$

N_2F_4 is a colourless reactive gas (mp $-164.5°$, bp $-73°$, $\Delta G_{298}° + 81.2$ kJ mol^{-1}) which acts as a strong fluorinating agent towards many substances, e.g.:

$$SiH_4 + N_2F_4 \xrightarrow{25°} SiF_4 + N_2 + 2H_2$$

$$10Li + N_2F_4 \xrightarrow{-80 \text{ to } +250°} 4LiF + 2Li_3N$$

$$S + N_2F_4 \xrightarrow{110-140°} SF_4 + SF_5NF_2 + \ldots$$

It forms adducts with strong fluoride-ion acceptors such as AsF_5 which can be formulated as salts, e.g. $[N_2F_3]^+[AsF_6]^-$. However, its most intriguing property is an ability to dissociate at room temperature and above to give the free radical NF_2. Thus, when N_2F_4 is frozen out from the warm gas at relatively low pressures the solid is dark blue whereas when it is frozen out from the cold gas at moderate pressures it is colourless. At 150°C the equilibrium constant for the dissociation $N_2F_4 \rightleftharpoons 2NF_2$ is $K = 0.03$ atm and the enthalpy of dissociation is 83.2 kJ mol^{-1}.[40] Such a dissociation, which interprets much of the reaction chemistry of N_2F_4,[41] is reminiscent of the behaviour of N_2O_4 (p. 522) but is unparalleled in the chemistry of N_2H_4:

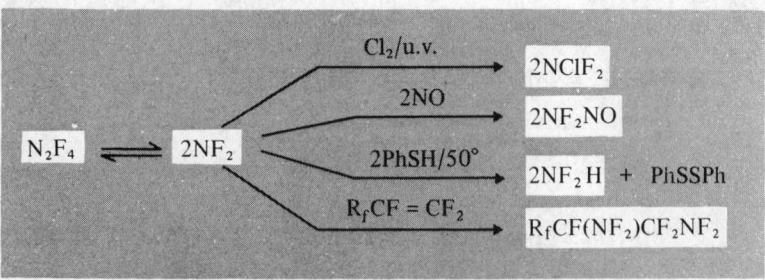

[39b] K. O. Christe, E. C. Curtis, and C. J. Schack, The CF_3O^- anion: vibrational spectrum of an unusual CF_3 compound, *Spectrochim. Acta* **31A**, 1035–8 (1975).
[40] F. H. Johnson and C. B. Colburn, The tetrafluorohydrazine–difluoroamino radical equilibrium, *J. Am. Chem. Soc.* **83**, 3043–7 (1961).
[41] C. L. Baumgardner and E. L. Lawton, Photofluoroamination and atomic fluorine reactions, *Acc. Chem. Res.* **7**, 14–20 (1974).

Dinitrogen difluoride, N_2F_2, was first identified in 1952 as a thermal decomposition product of the azide N_3F and it also occurs in small yield during the electrolysis of NH_4F/HF (p. 503), and in the reactions of NF_3 with Hg or with NF_3 in a Cu reactor (p. 503). Fluorination of NaN_3 gives good yields on a small scale but the compound is best prepared by the following reaction sequence:

$$KF + NF_2H \xrightarrow{-80°} KF \cdot NF_2H \xrightarrow[\sim 100\% \text{ yield}]{20°} N_2F_2 + KHF_2$$

All these methods give mixtures of the *cis-* and *trans-*isomers; these are thermally interconvertible but can be separated by low-temperature fractionation. The *trans-* form is thermodynamically more unstable than the *cis-* form but it can be stored in glass vessels whereas the *cis-* form reacts completely within 2 weeks to give SiF_4 and N_2O. *Trans-*N_2F_2 can be prepared free of the *cis-* form by the low-temperature reaction of N_2F_4 with $AlCl_3$ or MCl_2 (M = Mn, Fe, Co, Ni, Sn); thermal isomerization of *trans-*N_2F_2 at 70–100° yields an equilibrium mixture containing $\sim 90\%$ *cis-*N_2F_2 (ΔH_{isom} 12.5 kJ mol^{-1}). Pure *cis-*N_2F_2 can be obtained by selective complexation with AsF_5; only the *cis-* form reacts at room temperature to give $[N_2F]^+[AsF_6]^-$ and this, when treated with NaF/HF, yields pure *cis-*N_2F_2. Some characteristic properties are listed below.

*cis-*planar, C_{2v} *trans-*planar, C_{2h}

Isomer	MP/°C	BP/°C	$\Delta H_f°$/kJ mol^{-1}	μ/Debye
cis-N_2F_2	< -195	-105.7	69.5	0.18
trans-N_2F_2	-172	-111.4	82.0	0.00

Several mixed halides and hydrohalides of nitrogen are known but they tend to be unstable, difficult to isolate pure, and of little interest. Examples are[31] $NClF_2$, NCl_2F, $NBrF_2$, NF_2H, NCl_2H, and $NClH_2$.

The well known compound NCl_3 is a dense, volatile highly explosive liquid (mp $-40°$, bp $+71°$, $d(20°)$ 1.65 g cm^{-3}, μ 0.6 D) with physical properties which often closely resemble those of CCl_4 (p. 322). It is much less hazardous as a dilute gas and, indeed, is used industrially on a large scale for the bleaching and sterilizing of flour; for this purpose it is prepared by electrolysing an acidic solution of NH_4Cl at pH 4 and the product gas is swept out of the cell by means of a flow of air for immediate use. NCl_3 is rapidly hydrolysed by moisture and in alkaline solution can be used to prepare ClO_2:

$$NCl_3 + 3H_2O \longrightarrow NH_3 + 3HOCl \quad \text{(bleach, etc.)}$$
$$NCl_3 + 3H_2O + 6NaClO_2 \longrightarrow 6ClO_2 + 3NaCl + 3NaOH + NH_3$$
$$2NCl_3 + 6NaClO_2 \longrightarrow 6ClO_2 + 6NaCl + N_2$$

The elusive NBr$_3$ was finally prepared as a deep-red, very temperature-sensitive, volatile solid by the low-temperature bromination of bistrimethylsilylbromamine with BrCl:

$$(Me_3Si)_2NBr + 2BrCl \xrightarrow{\text{pentane}/-87°} NBr_3 + 2MeSiCl$$

It reacts instantly with NH$_3$ in CH$_2$Cl$_2$ solution at $-87°$ to give the dark-violet solid NBrH$_2$; under similar conditions I$_2$ yields the red-brown solid NBr$_2$I.

Pure NI$_3$ has not been isolated, but the structure of its well-known extremely shock-sensitive adduct with NH$_3$ has recently been elucidated—a feat of considerable technical virtuosity.[42] Unlike the volatile, soluble, molecular solid NCl$_3$, the involatile, insoluble compound [NI$_3$.NH$_3$]$_n$ has a polymeric structure in which tetrahedral NI$_4$ units are corner-linked into infinite chains of –N–I–N–I– (215 and 230 pm) as shown in Fig. 11.8. These in turn are linked into sheets by I–I interactions (336 pm) in the c-direction; in addition of one I of each NI$_4$ unit is also loosely attached to an NH$_3$ (253 pm) that projects into the space between the sheets of tetrahedra. The structure resembles that of the linked SiO$_4$ units in chain metasilicates (p. 403). A further interesting feature is the presence of linear or almost linear groupings N(1)–I(1')–N(1'), N(1)–I(2)–N(2), and N(1)–I(3)–I(2, neighbouring chain) which suggest the presence of 3-centre, 4-electron bonds (p. 69) characteristic of polyhalides and xenon halides (pp. 983, 1053).

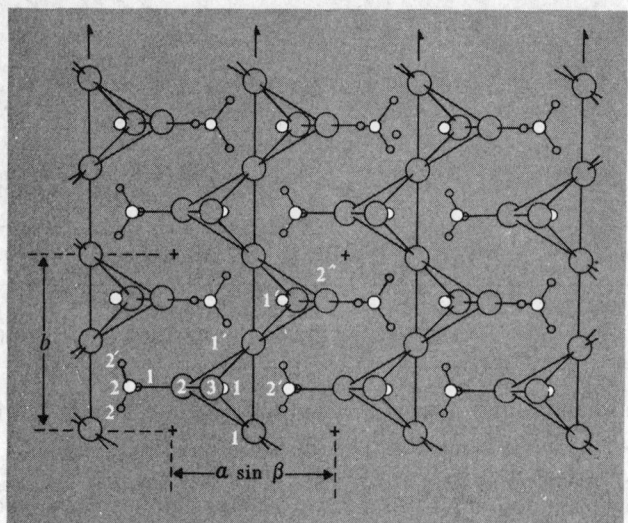

FIG. 11.8 The polymeric structure of NI$_3$.NH$_3$ shown projected on the (001) plane. ◯ I, ○ N, ○ H. Notes: (a) The chains of NI$_4$ tetrahedra involve I(1) (I1'). (b) The chains are built into sheets (perpendicular to paper) by I(3)–I(2, neighbouring chain) contacts (336 pm) between chains lying above one another along the c-axis. (c) {NI$_4$} are slightly distorted with N(I)–I(1) and N(1)–I(1') 230 pm, N(1)–I(2) and N(1)–I(3) ~215 pm. (d) The ammonia contact is I(2)–N(2) 253 pm. (e) Linear N–I–N groupings are N(1)–I(1')–N(1'), N(1)–I(2)–N(2), and N(1)–I(3)–N(2, neighbouring chain); of these, the first is symmetrical indicating that there is no difference between N–I contacts "within" and "between" tetrahedra.

[42] J. JANDER, Recent chemistry and structure investigation of NI$_3$, NBr$_3$, NCl$_3$ and related compounds, *Adv. Inorg. Chem. Radiochem.* **19**, 1–63 (1976).

Nitrogen forms two series of oxohalides—the nitrosyl halides XNO and the nitryl halides XNO_2. There are also two halogen nitrates $FONO_2$ (bp $-46°$) and $ClONO_2$ (bp $22.3°$), but these do not contain N–X bonds and can be considered as highly reactive derivatives of nitric acid, from which they can be prepared by direct halogenation:

$$HNO_3 + F_2 \longrightarrow FONO_2 + HF$$
$$HNO_3 + ClF \longrightarrow ClONO_2 + HF$$

The nitrosyl halides are reactive gases that feature bent molecules; they can be made by direct halogenation of NO with X_2, though fluorination of NO with AgF_2 has also been used and ClNO can be more conveniently made by passing N_2O_4 over moist KCl:

$$2NO + X_2 \longrightarrow 2XNO$$
$$NO + AgF_2 \longrightarrow FNO + AgF$$
$$N_2O_4 + KCl \longrightarrow ClNO + KNO_3$$

Some physical properties are in Table 11.5. FNO is colourless, ClNO orange-yellow and BrNO red. The compounds, though generally less reactive than the parent halogens, are

TABLE 11.5 *Some physical properties of XNO*[a]

Property	FNO	ClNO	BrNO
MP/°C	-132.5	-59.6	-56
BP/°C	-59.9	-6.4	~ 0
ΔH_f° (298 K)/kJ mol^{-1}	-66.5	$+51.7$	$+82.2$
ΔG_f° (298 K)/kJ mol^{-1}	-51.1	$+66.0$	$+82.4$
Angle X–N–O	$110°$	$113°$	$117°$
Distance N–O/pm	113	114	115
Distance N–X/pm	152	198	214
μ/D	1.81	0.42	—

[a] BrNO dissociates reversibly into NO and Br, the extent of dissociation being $\sim 7\%$ at room temperature and 1 atm pressure. A similar reversible dissociation occurs with ClNO at higher temperatures.

nevertheless extremely vigorous reagents. Thus FNO fluorinates many metals ($nFNO + M \rightarrow MF_n + nNO$) and also reacts with many fluorides to form salt-like adducts such as $NOAsF_6$, $NOVF_6$, and $NOBF_4$. ClNO acts similarly and has been used as an ionizing solvent to prepare complexes such as $NOAlCl_4$, $NOFeCl_4$, $NOSbCl_6$, and $(NO)_2SnCl_6$.[43] Aqueous solutions of XNO are particularly potent solvents for metals (like aqua regia, HNO_3/HCl) since the HNO_2 formed initially, reacts to give HNO_3:

$$XNO + H_2O \longrightarrow HNO_2 + HX; \quad 3HNO_2 \longrightarrow HNO_3 + 2NO + H_2O$$

Alkaline solutions contain a similar mixture:

$$4XNO + 3H_2O \longrightarrow HNO_3 + HNO_2 + 2NO + 4HX$$
$$4XNO + 6NaOH \longrightarrow NaNO_3 + NaNO_2 + 2NO + 4NaX + 3H_2O$$

[43] V. GUTMANN (ed.), in *Halogen Chemistry*, Vol. 2, p. 399, Academic Press, London, 1967; and V. GUTMANN, *Coordination Chemistry in Nonaqueous Solutions*, Springer-Verlag, New York, 1968.

With alcohols, however, the reaction stops at the nitrite stage:

$$XNO + ROH \longrightarrow RONO + HX$$

Nitryl fluoride and chloride, XNO_2, like their nitrosyl analogues, are reactive gases; they feature planar molecules, analogous to the isoelectronic nitrate anion, NO_3^-. Some physical properties are in Table 11.6. FNO_2 can be prepared by direct reaction of F_2 with

TABLE 11.6 *Some physical properties of* XNO_2

Property	FNO_2	$ClNO_2$	Property	FNO_2	$ClNO_2$
MP/°C	−166	−145	Angle X–N–O	118°	115°
BP/°C	−72.5	−15.9	Distance (N–O)/pm	123	120
ΔH_f° (298 K)/kJ mol^{-1}	−80	+13	Distance (N–X)/pm	135	184
ΔG_f° (298 K)/kJ mol^{-1}	−37.2	+54.4	μ/D	0.47	0.42

NO_2 or $NaNO_2$ or by fluorination of NO_2 using CoF_3 at 300°. $ClNO_2$ can not be made by direct chlorination of NO_2 but is conveniently synthesized in high yield by reacting anhydrous nitric acid with chlorosulfuric acid at 0°C:

$$HNO_3 + ClSO_3H \longrightarrow ClNO_2 + H_2SO_4$$

Reactions of XNO_2 often parallel those of XNO; e.g. FNO_2 readily fluorinates many metals and reacts with the fluorides of non-metals to give nitryl "salts" such as NO_2BF_4, NO_2PF_6, etc. Likewise, $ClNO_2$ reacts with many chlorides in liquid Cl_2 to give complexes such as NO_2SbCl_6. Hydrolysis yields aqueous solutions of nitric and hydrochloric acids, whereas ammonolysis in liquid ammonia yields chloramine and ammonium nitrite:

$$ClNO_2 + H_2O \longrightarrow \{HOCl + HNO_2\} \longrightarrow HNO_3 + HCl$$
$$ClNO_2 + 2NH_3 \longrightarrow ClNH_2 + NH_4NO_2$$

11.3.6 *Oxides of nitrogen*

Nitrogen is unique among the elements in forming no fewer than 7 molecular oxides, 3 of which are paramagnetic and all of which are thermodynamically unstable with respect to decomposition into N_2 and O_2. In addition there is evidence for fugitive species such as nitrosyl azide N_3NO and nitryl azide N_3NO_2, but these decompose rapidly below room temperature[44] and will not be considered further. Three of the oxides (N_2O, NO, and NO_2) have been known for over 200 y and were, in fact, amongst the very first gaseous compounds to be isolated and identified (J. Priestley and others in the 1770s). The physiological effects of N_2O (laughing gas, anaesthetic) and NO_2 (acrid, corrosive fumes) have been known from the earliest days, and the environmental problems of "NO_x" from automobile exhaust fumes and as a component in photochemical smog are well known in all industrial countries.[44a] NO and NO_2 are also important in the commercial

---

[44] M. P. DOYLE, J. J. MACIEJKO, and S. C. BUSMAN, Reaction between azide and nitronium ions. Formation and decomposition of nitryl azide, *J. Am. Chem. Soc.* **95**, 952–3 (1973).

[44a] S. D. LEE (ed.), *Nitrogen Oxides and their Effects on Health*, Ann Arbor Publishers, Michigan, 1980, 382 pp.

production of nitric acid (p. 537) and nitrate fertilizers. More recently N_2O_4 has been used extensively as the oxidizer in rocket fuels for space missions (p. 492).

The oxides of nitrogen played an important role in exemplifying Dalton's law of multiple proportions which led up to the formulation of his atomic theory in 1803–8, and they still pose some fascinating problems in bonding theory. Their formulae, molecular structure, and physical appearance are briefly summarized in Table 11.7 and each compound is discussed in turn in the following sections.

TABLE 11.7   *The oxides of nitrogen*

Formula	Name	Structure	Description
N_2O	Dinitrogen monoxide (nitrous oxide)	N——N——O linear $(C_{\infty v})$	Colourless gas (bp $-88.5°$) (cf. isoelectronic CO_2, NO_2^+, N_3^-)
NO	(Mono)nitrogen monoxide	N——O	Colourless paramagnetic gas (bp $-151.8°$) liquid and solid are also colourless when pure
N_2O_3	Dinitrogen trioxide	N——N planar (C_s)	Blue solid (mp $-100.7°$), dissociates reversibly in gas phase into NO and NO_2
NO_2	Nitrogen dioxide	N O ⋯ O planar (C_{2v})	Brown paramagnetic gas, dimerizes reversibly to N_2O_4
N_2O_4	Dinitrogen tetroxide	N——N planar D_{2h}	Colourless liquid (mp $-11.2°$) dissociates reversibly in gas phase to NO_2
N_2O_5	Dinitrogen pentoxide	$[NO_2]^+[NO_3]^-$ N——O——N planar C_{2v} $(\sim D_{2h})$	Colourless ionic solid; sublimes at $32.4°$ to unstable molecular gas (N–O–N $\sim 180°$)
NO_3	Nitrogen trioxide	O——N planar (D_{3h})	Unstable paramagnetic radical

Nitrous oxide, N_2O

Nitrous oxide can be made by the careful thermal decomposition of molten NH_4NO_3 at about 250°C:

$$NH_4NO_3 \xrightarrow{\Delta} N_2O + 2H_2O$$

Although the reaction has the overall stoichiometry of a dehydration it is more complex than this and involves a mutual redox reaction between N^{-III} and N^V. This is at once explicable in terms of the volt-equivalent diagram in Fig. 11.7 which also interprets why NO and N_2 are formed simultaneously as byproducts. It is probable that the mechanism involves dissociation of NH_4NO_3 into NH_3 and HNO_3, followed by autoprotolysis of HNO_3 to give $NO_2{}^+$, which is the key intermediate:

$$NH_4NO_3 \rightleftharpoons NH_3 + HNO_3; \quad 2HNO_3 \rightleftharpoons NO_2{}^+ + H_2O + NO_3{}^-$$
$$NH_3 + NO_2{}^+ \longrightarrow \{H_3NNO_2\}^+ \longrightarrow NNO + H_3O^+, \text{ etc.}$$

Consistent with this ^{15}NNO can be made from $^{15}NH_4NO_3$, and $N^{15}NO$ from $NH_4{}^{15}NO_3$. Alternative preparative routes (Fig. 11.7) are the reduction of aqueous nitrous acid with either hydroxylamine or hydrogen azide:

$$HNO_2 + NH_2OH \xrightarrow{aq} N_2O + 2H_2O$$

$$HNO_2 + HN_3 \xrightarrow{aq} N_2O + N_2 + H_2O$$

Thermal decomposition of nitramide, H_2NNO_2, or hyponitrous acid $H_2N_2O_2$ (both of which have the empirical formula $N_2O \cdot H_2O$) have also been used. The mechanisms of these and other reactions involving simple inorganic compounds of N have been reviewed.[44b] However, though N_2O can be made in this way it is not to be regarded as the anhydride of hyponitrous acid since $H_2N_2O_2$ is not formed when N_2O is dissolved in H_2O (a similar relation exists between CO and formic acid).

Nitrous oxide is a moderately unreactive gas comprised of linear unsymmetrical molecules, as expected for a 16-electron triatomic species (p. 496). The symmetrical structure N–O–N is precluded on the basis of orbital energetics. Some physical properties are in Table 11.8: it will be seen that the N–N and N–O distances are both short and the most recent calculations[45] give the bond orders as N–N 2.73 and N–O 1.61. N_2O is thermodynamically unstable and when heated above $\sim 600°C$ it dissociates by fission of the weaker bond ($N_2O \rightarrow N_2 + \frac{1}{2}O_2$). However, the reaction is much more complex than this simple equation might imply and the process involves a "forbidden" singlet–triplet transition in which electron spin is not conserved.[46] The activation energy for the process is high (~ 250 kJ mol^{-1}) and at room temperature N_2O is relatively inert: e.g. it does not react with the halogens, the alkali metals, or even ozone. At higher temperatures reactivity increases markedly: H_2 gives N_2 and H_2O; many other non-metals (and some metals) react to form oxides, and the gas supports combustion. Perhaps its most remarkable

[44b] G. STEDMAN, Reaction mechanisms of inorganic nitrogen compounds, *Adv. Inorg. Chem. Radiochem.* **22**, 114–70 (1979).

[45] K. JUG, Bond order orbitals and eigenvalues, *J. Am. Chem. Soc.* **100**, 6581–6 (1978).

[46] I. R. BEATTIE, Nitrous Oxide, Section 24 in *Mellor's Comprehensive Treatise on Inorganic and Theoretical Chemistry*, Vol. 8, pp. 189–215, Supplement 2, *Nitrogen* (Part 2), Longmans, London, 1967.

TABLE 11.8   *Some physical properties of* N_2O

MP/°C	−90.86	μ/D	0.166
BP/°C	−88.48	Distance (N–N)/pm	112.6
ΔH_f° (298 K)/kJ mol^{-1}	82.0	Distance (N–O)/pm	118.6
ΔG_f° (298 K)/kJ mol^{-1}	104.2		

reaction is with molten alkali metal amides to yield azides, the reaction with $NaNH_2$ being the commercial route to NaN_3 and hence all other azides (p. 497):

$$NaNH_2(1) + N_2O(g) \xrightarrow{200°} NaN_3 + H_2O(g)$$

$$NaNH_2 + H_2O \longrightarrow NaOH + NH_3(g)$$

It is also notable that N_2O (like N_2 itself) can act as a ligand by displacing H_2O from the aquo complex $[Ru(NH_3)_5(H_2O)]^{2+}$:[47]

$$[Ru(NH_3)_5(H_2O)]^{2+} + N_2O(aq) \longrightarrow [Ru(NH_3)_5(N_2O)]^{2+} + H_2O$$

The formation constant K is 7.0 mol^{-1} l for N_2O and 3.3×10^4 mol^{-1} l for N_2.

Notwithstanding the fascinating reaction chemistry of N_2O it is salutory to remember that its largest commercial use is as a propellant and aerating agent for "whipped" ice-cream—this depends on its solubility under pressure in vegetable fats coupled with its non-toxicity in small concentrations and its absence of taste. It has also been much used as an anaesthetic.

Nitric oxide, NO

Nitric oxide is the simplest thermally stable odd-electron molecule known and, accordingly, its electronic structure and reaction chemistry have been very extensively studied.[48] The compound is an intermediate in the production of nitric acid and is prepared industrially by the catalytic oxidation of ammonia (p. 537). On the laboratory scale it can be synthesized from aqueous solution by the mild reduction of acidified nitrites with iodide or ferrocyanide or by the disproportionation of nitrous acid in the presence of dilute sulfuric acid:

$$KNO_2 + KI + H_2SO_4 \xrightarrow{aq} NO + K_2SO_4 + H_2O + \tfrac{1}{2}I_2$$

$$KNO_2 + K_4[Fe(CN)_6] + 2MeCO_2H \longrightarrow NO + K_3[Fe(CN)_6] + H_2O + 2MeCO_2K$$

$$6NaNO_2 + 3H_2SO_4 \longrightarrow 4NO + 2HNO_3 + 2H_2O + 3Na_2SO_4$$

The dry gas has been made by direct reduction of a solid mixture of nitrite and nitrate with chromium(III) oxide $(3KNO_2 + KNO_3 + Cr_2O_3 \rightarrow 4NO + 2K_2CrO_4)$ but is now more conveniently obtained from a cylinder.

[47] J. N. ARMOR and H. TAUBE, Formation and reactions of $[Ru(NH_3)_5(N_2O)]^{2+}$, *J. Am. Chem. Soc.* **91**, 6874–6 (1969). A. A. DIAMANTIS and G. J. SPARROW, Nitrous oxide complexes: the isolation of pentammine-(dinitrogen oxide)ruthenium(II) tetrafluoroborate, *JCS Chem. Comm.* 1970, 819–20. J. N. ARMOR and H. TAUBE, Evidence of a binuclear nitrous oxide complex of ruthenium, *JCS Chem. Comm.* 1971, 287–8.
[48] Ref. 31, pp. 323–5.

TABLE 11.9 *Some physical properties of NO*

MP/°C	− 163.6	μ/D	0.15
BP/°C	− 151.8	Distance (N–O)/pm	115
ΔH_f° (298 K)/kJ mol^{-1}	90.2	Ionization energy/eV	9.23
ΔG_f° (298 K)/kJ mol^{-1}	86.6	Ionization energy/kJ mol^{-1}	890.6

Nitric oxide is a colourless, monomeric, paramagnetic gas with a low mp and bp (Table 11.9). It is thermodynamically unstable and decomposes into its elements at elevated temperatures (1100–1200°C), a fact which militates against its direct synthesis from N_2 and O_2. At high pressures and moderate temperatures ($\sim 50°$) it rapidly disproportionates:

$$3NO \longrightarrow N_2O + NO_2; \quad -\Delta H = 3 \times 51.8 \text{ kJ mol}^{-1}$$
$$-\Delta G = 3 \times 34.7 \text{ kJ mol}^{-1}$$

However, when the gas is occluded by zeolites the disproportionation takes a different course:

$$4NO \longrightarrow N_2O + N_2O_3; \quad -\Delta H = 4 \times 48.8 \text{ kJ mol}^{-1}$$
$$-\Delta G = 4 \times 25.7 \text{ kJ mol}^{-1}$$

The molecular orbital description of the bonding in NO is similar to that in N_2 or CO (p. 350) but with an extra electron in one of the π^* antibonding orbitals. This effectively reduces the bond order from 3 to ~ 2.5[†] and accounts for the fact that the interatomic N–O distance (115 pm) is intermediate between that in the triple-bonded NO$^+$ (106 pm) and values typical of double-bonded NO species (~ 120 pm). It also interprets the very low ionization energy of the molecule (9.25 eV, compared with 15.6 eV for N_2, 14.0 eV for CO, and 12.1 eV for O_2). Similarly, the notable reluctance of NO to dimerize can be related both to the geometrical distribution of the unpaired electron over the entire molecule and to the fact that dimerization to O=N—N=O leaves the total bond order unchanged ($2 \times 2.5 = 5$). When NO condenses to a liquid, partial dimerization occurs, the *cis*-form being more stable than the *trans*-. The pure liquid is colourless, not blue as sometimes stated: blue samples owe their colour to traces of the intensely coloured N_2O_3.[48a] Crystalline nitric oxide is also colourless (not blue) when pure,[48a] and X-ray diffraction data are best interpreted in terms of weak association into dimeric units. It

† For example a recent calculation gives:[45]

Species	Total bond order	σ_g	π_x	π_y
N_2	3.000	1.000	1.000	1.000
NO	2.417	0.982	0.966	0.469
NO$^+$	2.929	0.977	0.976	0.976
CO	2.760	0.908	0.926	0.926

[48a] J. MASON, Textbook errors, 122. The nitric oxide dimer—blue with rectangular molecules?, *J. Chem. Educ.* **52**, 445–7 (1975).

seems probable that the dimers adopt the cis-(C_{2v}) structure[49] rather than the rectangular C_{2h} structure which was at one time favoured,[49a] i.e.:

$$\text{rather than}$$

$$C_{2v} \qquad\qquad C_{2h}$$

In either case each dimer has two possible orientations, and random disorder between these accounts for the residual entropy of the crystal (6.3 J mol^{-1} of dimer). Very recently[50] an asymmetric dimer $O{\nearrow}^{N-O}{\searrow}N$ has been characterized; this forms as a red species when NO is condensed in the presence of polar molecules such as HCl or SO_2, or Lewis acids such as BX_3, SiF_4, $SnCl_4$, or $TiCl_4$.

The reactivity of NO towards atoms, free radicals, and other paramagnetic species has been much studied, and the chemiluminescent reactions with atomic N and atomic O have already been mentioned (p. 474). NO reacts rapidly with molecular O_2 to give brown NO_2, and this gas is the normal product of reactions which produce NO if these are carried out in air. The oxidation is unusual in following third-order reaction kinetics and, indeed, is the classic example of such a reaction (M. Bodenstein, 1918). The reaction is also unusual in having a negative temperature coefficient, i.e. the rate becomes progressively slower at higher temperatures. For example the rate drops by a factor of 2 between room temperature and 200°. This can be accounted for by postulating that the mechanism involves the initial equilibrium formation of the unstable dimer which then reacts with oxygen:

$$2NO \rightleftharpoons N_2O_2 \xrightarrow{\;O_2\;} 2NO_2$$

As the equilibrium concentration of N_2O_2 decreases rapidly with increase in temperature the decrease in rate is explained. However alternative mechanisms have also been suggested, e.g.:[48]

$$NO + O_2 \rightleftharpoons NO_3 \xrightarrow{\;NO\;} 2NO_2$$

$$NO_3 + NO_2 \longrightarrow N_2O_5 \xrightarrow{\;NO\;} 3NO_2$$

Nitric oxide reacts with the halogens to give XNO (p. 507). Some other facile reactions are listed below:-

$$ClNO_2 + NO \longrightarrow ClNO + NO_2 \text{ (?Cl transfer or O transfer)}$$

[49] W. N. LIPSCOMB, F. E. WANG, W. R. MAY, and E. L. LIPPERT, Comments on the structures of 1,2-dichloroethane and of N_2O_2, *Acta Cryst.* **14**, 1100–01 (1961).
[49a] W. J. DULMAGE, E. A. MEYERS, and W. N. LIPSCOMB, On the crystal and molecular structure of N_2O_2, *Acta Cryst.* **6**, 760–4 (1953).
[50] J. R. OLSEN and J. LAANE, Characterization of the asymmetric nitric oxide dimer O=N—O=N by resonance Raman and infrared spectroscopy, *J. Am. Chem. Soc.* **100**, 6948–55 (1978).

$$NCl_3 + 2NO \longrightarrow ClNO + N_2O + Cl_2 \text{ (stepwise at } -150°)$$
$$XeF_2 + 2NO \longrightarrow 2FNO + Xe \text{ (occurs stepwise; also with XeF}_4)$$
$$I_2O_5 + 5NO \longrightarrow \tfrac{5}{2}N_2O_4 + 2I_2 \text{ (N}_2O_5 \text{ also produced)}$$

Reactions with sulfides, polysulfides, sulfur oxides, and the oxoacids of sulfur are complex and the products depend markedly on reaction conditions (see also p. 880 for blue crystals in chamber acid). Some examples are:

$$SO_2 + 2NO \longrightarrow N_2O + SO_3$$

$$2SO_3 + NO \longrightarrow (SO_3)_2NO$$

$$2H_2SO_3 + 2NO \longrightarrow 2H_2SO_3NO \xrightarrow{-H_2SO_3} H_2SO_3(NO)_2 \longrightarrow N_2O + H_2SO_4$$

$$K_2SO_3(aq) + NO \xrightarrow{0°} K_2[ONSO_3] \xrightarrow{NO} K_2[O\underset{\underset{NO}{|}}{N}SO_3]$$
$$\qquad\qquad\qquad\qquad\qquad\text{radical anion}\qquad\qquad\qquad\qquad\qquad\text{white crystals}$$

Under alkaline conditions disproportionation reactions predominate. Thus with Na_2O the hydronitrite(II) first formed, disproportionates into the corresponding nitrite(III) and hyponitrite(I):

$$4Na_2O + 4N^{II}O \xrightarrow{100°C} 4Na_2N^{II}O_2 \longrightarrow 2Na_2O + 2NaN^{III}O_2 + Na_2N^I_2O_2$$

With alkali metal hydroxides, both N_2O and N_2 are formed in addition to the nitrite:

$$2MOH + 4N^{II}O \longrightarrow 2MN^{III}O_2 + N^I_2O + H_2O$$
$$4MOH + 6N^{II}O \longrightarrow 4Mn^{III}O_2 + N^{(0)}_2 + 2H_2O$$

Nitric oxide complexes. NO readily reacts with many transition metal compounds to give nitrosyl complexes and these are also frequently formed in reactions involving other oxo-nitrogen species. Classic examples are the "brown-ring" complex $[Fe(H_2O)_5NO]^{2+}$ formed during the qualitative test for nitrates, Roussin's red and black salts (p. 1272), and sodium nitroprusside, $Na_2[Fe(CN)_5NO] . 2H_2O$. The field has been extensively reviewed [51-55] and only the salient features need be summarized here. A variety of preparative routes is available (see Panel). Most nitrosyl complexes are highly coloured—deep reds, browns, purples, or even black. Apart from the intrinsic interest in the structure and bonding of these compounds there is much current interest in their potential use as homogeneous catalysts for a variety of chemical reactions.[55]

NO shows a wide variety of coordination geometries (linear, bent, doubly bridging, and triply bridging) and sometimes adopts more than one mode within the same complex. NO

[51] B. F. G. JOHNSON and J. A. McCLEVERTY, Nitric oxide compounds of transition metals, *Progr. Inorg. Chem.* **7**, 277–359 (1966).

[52] W. P. GRIFFITH, Organometallic nitrosyls, *Adv. Organometallic Chem.* **7**, 211–39 (1968).

[53] J. H. ENEMARK and R. D. FELTHAM, Principles of structure, bonding, and reactivity for metal nitrosyl complexes, *Coord. Chem. Revs.* **13**, 339–406 (1974).

[54] K. G. CAULTON, Synthetic methods in transition metal nitrosyl chemistry, *Coord. Chem. Revs.* **14**, 317–55 (1975).

[54a] J. A. McCLEVERTY, The reactions of nitric oxide coordinated to transition metals, *Chem. Rev.* **79**, 53–76 (1979).

[55] R. EISENBERG and C. D. MEYER, The coordination chemistry of nitric oxide, *Acc. Chem. Res.* **8**, 26–34 (1975).

Synthetic Routes to NO Complexes[54]

The coordination chemistry of NO is often compared to that of CO but, whereas carbonyls are frequently prepared by reactions involving CO at high pressures and temperatures, this route is less viable for nitrosyls because of the thermodynamic instability of NO and its propensity to disproportionate or decompose under such conditions (p. 512). Nitrosyl complexes can sometimes be made by transformations involving pre-existing NO complexes, e.g. by ligand replacement, oxidative addition, reductive elimination, or condensation reactions (reductive, thermal, or photolytic). Typical examples are:

$$[Mn(CO)_3(NO)(PPh_3)] + PPh_3 \longrightarrow [Mn(CO)_2(NO)(PPh_3)_2] + CO$$

$$2[Cr(\eta^5\text{-}C_5H_5)Cl(NO)_2] \xrightarrow{\ BH_4^-\ } [\{Cr(\eta^2\text{-}C_5H_5)(NO)_2\}_2]$$

$$2[Mn(CO)_4(NO)]' \xrightarrow{\ h\nu\ } [Mn_2(CO)_7(NO)_2]$$

$$[\{Mn(\eta^5\text{-}C_5H_5)(CO)(NO)\}_2] \xrightarrow{\ h\nu\ } [Mn_3(\eta^5\text{-}C_5H_5)_3(NO)_4]$$

Syntheses which *increase* the number of coordinated NO molecules can be classified into more than a dozen types, of which only the first three use free NO gas.

1. *Addition of NO to coordinatively unsaturated complexes:*

$$[CoCl_2L_2] + NO \longrightarrow [CoCl_2L_2(NO)]$$
$$[Co(OPPh_3)_2X_2] + 2NO \longrightarrow [Co(NO)_2(OPPh_3)_2X_2]$$

2. *Substitution (ligand replacement)*

Very frequently in these reactions 2NO replace 3CO. Alternatively, 1NO can replace 2CO with simultaneous formation of a metal–metal bond, or 1NO can replace CO + a halogen atom:

$$[Co(CO)_3(NO)] + 2NO \longrightarrow [Co(NO)_3] + 3CO$$
$$[Fe(CO)_5] + 2NO \longrightarrow [Fe(CO)_2(NO)_2] + 3CO$$
$$[Cr(CO)_6] + 4NO \longrightarrow [Cr(NO)_4] + 6CO$$
$$[Co(\eta^5\text{-}C_5H_5)(CO)_2] + NO \longrightarrow \tfrac{1}{2}[\{Co(\eta^5\text{-}C_5H_5)(NO)\}_2] + 2CO$$
$$[Mn(CO)_5I] + 3NO \longrightarrow [Mn(CO)(NO)_3] + 4CO + \{I\}$$
$$\tfrac{1}{2}[\{Mn(CO)_4I\}_2] + 3NO \longrightarrow [Mn(CO)(NO)_3] + 3CO + \{I\}$$

3. *Reductive nitrosylation* (cf. $MF_6 + NO \rightarrow NO^+MF_6^-$ for Mo, Tc, Re, Ru, Os, Ir, Pt)

$$CoCl_2 + 3NO + B + ROH \longrightarrow \tfrac{1}{2}[\{CoCl(NO)_2\}_2] + BH^+ + RONO$$

where B is a proton acceptor such as an alkoxide or amine.

4. *Addition of or substitution by* NO^+

This method uses $NOBF_4$, $NOPF_6$, or $NO[HSO_4]$ in MeOH or MeCN, e.g.:

$$[Rh(CNR)_4]^+ + NO^+ \longrightarrow [Rh(CNR)_4(NO)]^{2+}$$
$$[Ir(CO)ClL_2] + NO^+ \longrightarrow [Ir(CO)ClL_2(NO)]^+$$
$$[Ni(CO)_2L_2] + NO^+ \longrightarrow [Ir(CO)L_2(NO)]^+ + CO$$
$$[Cr(CO)_4(diphos)] + 2NO^+ \xrightarrow{\ MeCN\ } [Cr(NO)_2(MeCN)_4]^{2+} + 4CO + diphos$$

5. *Oxidative addition of XNO*

The reaction may occur with either coordinatively unsaturated or saturated complexes, e.g.:

$$[PtX_4]^{2-} + ClNO \longrightarrow [PtCl(NO)(X)_4]^{2-} \quad (X = Cl, CN, NO_2)$$
$$[Ni(PPh_3)_4] + ClNO \longrightarrow [Ni(Cl)(NO)(PPh_3)_2] + 2PPh_3$$

Continued

6. *Reaction of metal hydride complexes with N-nitrosoamides*, e.g. N-methyl-N-nitrosourea:

$$[Mn(CO)_5H] + MeN(NO)CONH_2 \longrightarrow Mn(CO)_4(NO)] + CO + MeNHCONH_2$$

7. *Transfer of coordinated NO* (especially from dimethylglyoximate complexes)

$$[Co(dmg)_2(NO)] + [MClL_n] \longrightarrow [CoCl(dmg)_2] + [M(NO)L_n]$$

$$[Ru(NO)_2(PPh_3)_2] + [RuCl_2(PPh_3)_3] \xrightarrow[\text{dust}]{\text{Zn}} 2[RuCl(NO)(PPh_3)_2] + PPh_3$$

8. *Use of NH_2OH in basic solution* (especially for cyano complexes)

The net transformation can be considered as the replacement of CN^- (or X^-) by NO^- and the reaction can be formally represented as

$$2NH_2OH \longrightarrow NH_3 + H_2O + \{NOH\} \longrightarrow NO^- + H^+ \text{ (removed by base)}$$

Examples are:

$$[Ni(CN)_4]^{2-} + 2NH_2OH \xrightarrow{\text{MOH}} [Ni(CN)_3(NO)]^{2-} + NH_3 + 2H_2O + MCN$$

$$[Cr(CN)_6]^{3-} + 2NH_2OH \xrightarrow{\text{MOH}} [Cr(CN)_5(NO)]^{3-} + NH_3 + 2H_2O + MCN$$

9. *Use of acidified nitrites* (i.e. $NO_2^- + 2H^+ \rightarrow NO^+ + H_2O$), e.g.:

$$K[Fe(CO)_3(NO)] + KNO_2 + CO_2 + H_2O \longrightarrow [Fe(CO)_2(NO)_2] + 2KHCO_3$$

$$Na[Fe(CO)_4H] + 2NaNO_2 + 3MeCO_2H \longrightarrow [Fe(CO)_2(NO)_2] + 2CO$$
$$+ 2H_2O + 3MeCO_2Na$$

10. *Use of (acidified) nitrites RONO*, i.e. $RONO + H^+ \rightleftharpoons NO^+ + ROH$, e.g.:

$$[Fe(CO)_3(PPh_3)_2] + RONO + H^+ \longrightarrow [Fe(CO)_2(NO)(PPh_3)_2]^+ + CO + ROH$$

Alternatively in aprotic solvents such as benzene:

$$[\{Mn(CO)_4(PPh_3)\}_2] \xrightarrow{\text{RONO}} 2[Mn(CO)_3(NO)(PPh_3)] + 2CO$$

11. *Use of concentrated nitric acid* (i.e. $2HNO_3 \rightleftharpoons NO^+ + NO_3^- + H_2O$)

Some of these reactions result, essentially, in the oxidative addition of $NO^+NO_3^-$ to coordinatively unsaturated metal centres whereas in others ligand replacement by NO^+ occurs—this is a favoured route for producing "nitroprusside", i.e. nitrosylpentacyanoferrate(II):

$$[Pt(en)_2]^{2+} \xrightarrow{\text{HNO}_3} [Pt(en)_2(NO)(NO_3)]^{2+}$$

$$[Fe(CN)_6]^{4-} \xrightarrow{\text{HNO}_3} [Fe(CN)_5(NO)]^{2-}$$

12. *Oxide ion abstraction from coordinated NO_2*, i.e.

$$[ML_x(NO_2)]^{n+} + H^+ \longrightarrow [ML_x(NO)]^{(n+2)+} + OH^-$$

e.g. $\quad cis\text{-}[Ru(bipy)_2(NO_2)X] \underset{\text{2OH}^-}{\overset{\text{2H}^+}{\rightleftharpoons}} cis\text{-}[Ru(bipy)_2(NO)X]^{2+} + H_2O$

$$[Fe(CN)_5(NO_2)]^{4-} \xrightarrow{\text{2H}^+} [Fe(CN)_5(NO)]^{2-} + H_2O$$

13. *Oxygen atom abstraction*

$$[Fe(CO)_5] + KNO_2 \longrightarrow K[Fe(CO)_3(NO)] + CO + CO_2$$

Many variations on these synthetic routes have been devised and the field is still being actively developed.

The reactions of NO coordinated to transition metals have recently been reviewed.[54a]

has one more electron than CO and often acts as a 3-electron donor—this is well illustrated by the following isoelectronic series of compounds in which successive replacement of CO by NO is compensated by a matching decrease in atomic number of the metal centre:

$[Ni(CO)_4]$	$[Co(CO)_3(NO)]$	$[Fe(CO)_2(NO)_2]$	$[Mn(CO)(NO)_3]$	$[Cr(NO)_4]$
mp $-25°$	$-11°$	$+18.4°$	$+27°$	decomp > rt
(colourless)	(red)	(deep red)	(dark green)	(red-black)

For the same reason 3CO can be replaced by 2NO; e.g.:

$$[Co(CO)_3(NO)] \longrightarrow [Co(NO)_3]; \quad [Mn(CO)_4(NO)] \longrightarrow [Mn(CO)(NO)_3];$$
$$[Fe(CO)_5] \longrightarrow [Fe(CO)_2(NO)_2]; \qquad [Cr(CO)_6] \longrightarrow [Cr(NO)_4]$$

In these and analogous compounds the M–N–O group is linear or nearly so, the M–N and N–O distances are short, and the N–O infrared stretching modes usually occur in the range 1650–1900 cm^{-1}. The bonding in such compounds is sometimes discussed in terms of the preliminary transfer of 1 electron from NO to the metal and the coordination of NO^+ to the reduced metal centre as a "2-electron σ donor, 2-electron π acceptor" analogous to CO (p. 351). This formal scheme, though useful in emphasizing similarities and trends in the coordination behaviour of NO^+, CO, and CN^-, is unnecessary even for the purpose of "book-keeping" of electrons; it is also misleading in implying an unacceptably large separation of electronic charge in these covalent complexes and in leading to uncomfortably low oxidation states for many metals, e.g. Cr($-$IV) in $[Cr(NO)_4]$, Mn($-$III) in $[Mn(CO)(NO)_3]$, etc. Many physical techniques (such as ESCA, Mössbauer spectroscopy, etc.) suggest a much more even distribution of charge and there is accordingly a growing trend to consider linear NO complexes in terms of molecular orbital energy level schemes in which an almost neutral NO contributes 3 electrons to the bonding system via orbitals of σ and π symmetry.[55a]

Compounds in which the {M–N–O} groups is nominally linear often feature a slightly bent coordination geometry and M–N–O bond angles in the range 165–180° are frequently encountered. However, another group of compounds is now known in which the angle M–N–O is close to 120°. The first example, $[Co(NO)(S_2CNMe_2)]$, appeared in 1962[56] though there were problems in refining the structure, and a second example was found in 1968[57] when the cationic complex $[Ir(CO)Cl(NO)(PPh_3)_2]^+$ was found to have a bond angle of 124° (Fig. 11.9); values in the range 120–140° have since been observed in several other compounds (Table 11.10). The related complex $[RuCl(NO)_2(PPh_3)_2]^+$, in which the CO ligand has been replaced by a second NO molecule, is interesting in having both linear and bent {M–NO} groups: as can be seen in Fig. 11.9 the nonlinear coordination is associated with a lengthening of the Ru–N and N–O distances. This is consistent with a weakening of these bonds and it is significant that

[55a] H. W. CHEN and W. L. JOLLY, An XPS study of the relative π-acceptor abilities of the nitrosyl and carbonyl ligands, *Inorg. Chem.* **18**, 2548–51 (1979).

[56] P. R. H. ALDERMAN, P. G. OWSTON, and J. M. ROWE, The crystal structure of $[Co(NO)(S_2CNMe_2)_2]$, *J. Chem. Soc.* 668–73 (1962).

[57] D. J. HODGSON and J. A. IBERS, The crystal and molecular structure of $[Ir(CO)Cl(NO)(PPh_3)_2]BF_4$, *Inorg. Chem.* **7**, 2345–52 (1968); see also *J. Am. Chem. Soc.* **90**, 4486–8 (1968).

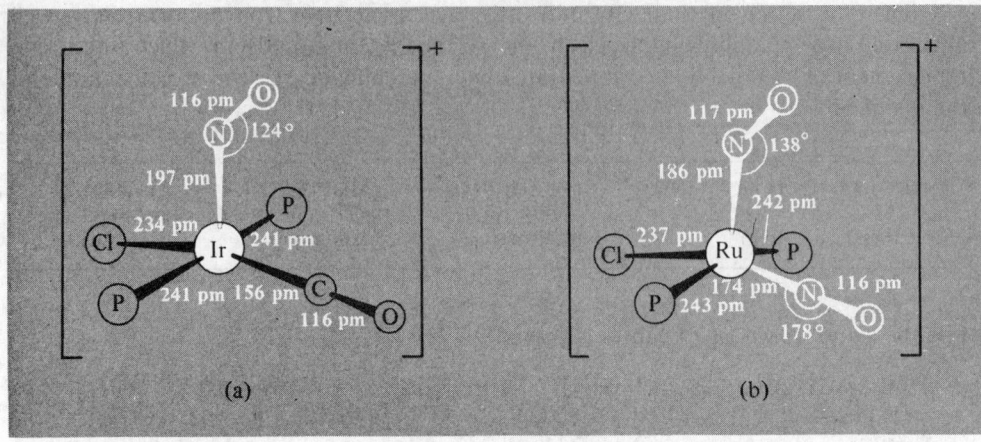

FIG. 11.9 Complexes containing bent NO groups: (a) $[Ir(CO)Cl(NO)(PPh_3)_2]^+$, and (b) $[RuCl(NO)_2(PPh_3)_2]^+$. This latter complex also has a linearly coordinated NO group. The diagrams show only the coordination geometry around the metal and omit the phenyl groups for clarity.

the N–O infrared stretching mode in such compounds tends to occur at lower wave numbers ($1525–1690$ cm^{-1}) than for linearly coordinated NO ($1650–1900$ cm^{-1}). In such systems neutral NO can be thought of as a 1-electron donor, as in the analogous (bent) nitrosyl halides, XNO (p. 507); it is unnecessary to consider the ligand as an NO$^-$ 2-electron donor. The implication is that the other pair of electrons on NO is placed in an essentially non-bonding orbital on N (which is thus approximately described as an sp^2 hybrid) rather than being donated to the metal as in the linear, 3-electron-donor mode (Fig. 11.10). Consistent with this, non-linear coordination is generally observed with the later transition elements in which the low-lying orbitals on the metal are already filled, whereas linear coordination tends to occur with earlier transition elements which can more readily accommodate the larger number of electrons supplied by the ligand. However, the energetics are frequently finely balanced and other factors must also be considered—a good example is supplied by the two "isoelectronic" complexes shown in Fig. 11.11: $[Co(diars)_2(NO)]^{2+}$ has a linear NO equatorially coordinated to a trigonal bipyramidal cobalt atoms whereas $[IrCl_2(NO)(PPh_3)_2]$ has a bent NO axially coordinated to a square-pyramidal iridium atom, even though both Co and Ir are in the same group in the periodic table. Indeed, the complex cation $[Ir(\eta^3\text{-}C_3H_5)(NO)(PPh_3)_2]^+$ shows a facile equilibrium (in CH$_2$Cl$_2$ or MeCN solutions) between the linear and the bent NO modes of coordination and, with appropriate counter anions, either the linear –NO (light brown) or bent –NO (red-brown) isomer can be crystallized.[57a] Some further examples of the two coordination geometries are in Table 11.10.

Like CO, nitric oxide can also act as a bridging ligand between 2 or 3 metals. Examples are the Cr and Mn complexes in Fig. 11.12. In $[\{Cr(\eta^5\text{-}C_5H_5)(NO)(\mu_2\text{-}NO)\}_2]$ the linear terminal NO has an infrared band at 1672 cm^{-1} whereas for the doubly bridging NO the

[57a] M. W. SCHOONOVER, E. C. BAKER, and R. EISENBERG, Ligand dynamics of $[Ir(\eta^3\text{-}C_3H_5)(NO)(PPh_3)_2]^+$. A facile linear–bent nitrosyl equilibrium, *J. Am. Chem. Soc.* **101**, 1880–2 (1979).

TABLE 11.10   *Some examples of "linear" and "bent" coordination of nitric oxide*

Compound	Angle M–N–O	ν (N–O)/cm^{-1}
Linear		
[Co(en)$_3$][Cr(CN)$_5$(NO)].2H$_2$O	176°	1630
[Cr(η^5-C$_5$H$_5$)Cl(NO)$_2$]	171°, 166°	1823, 1715
K$_3$[Mn(CN)$_5$(NO)].2H$_2$O	174°	1700
[Mn(CO)$_2$(NO)(PPh$_3$)$_2$]	178°	1661
[Fe(NO)(mnt)$_2$]$^-$	180°	1867
[Fe(NO)(mnt)$_2$]$^{2-}$	165°	1645
[Fe(NO)(S$_2$CNMe$_2$)$_2$]	170°	1690
Na$_2$[Fe(CN)$_5$(NO)].2H$_2$O	178°	1935
[Co(diars)(NO)]$^{2+}$	179°	1852
[Co(Cl)$_2$(NO)(PMePh$_2$)$_2$]	165°	1735, 1630
Na$_2$[Ru(NO)(NO$_2$)$_4$(OH)].2H$_2$O	180°	1893
[RuH(NO)(PPh$_3$)$_3$]	175°	1645
[Ru(diphos)$_2$(NO)]$^+$	174°	1673
[Os(CO)$_2$(NO)(PPh$_3$)$_2$]$^+$	177°	1750
[IrH(NO)(PPh$_3$)$_3$]$^+$	175°	1715
Bent		
[CoCl(en)$_2$(NO)]ClO$_4$	124°	1611
[Co(NH$_3$)$_5$NO]$^{2+}$	119°	1610
[Co(NO)(S$_2$CNMe$_2$)$_2$]$^{(a)}$	~135°	1626
[Rh(Cl)$_2$(NO)(PPh$_3$)$_2$]	125°	1620
[Ir(Cl)$_2$(NO)(PPh$_3$)$_2$]	123°	1560
[Ir(CO)Cl(NO)(PPh$_3$)$_2$]BF$_4$	124°	1680
[Ir(CO)I(NO)(PPh$_3$)$_2$]BF$_4$.C$_6$H$_6$	124°	1720
[Ir(CH$_3$)I(NO)(PPh$_3$)$_2$]	120°	1525
Both		
[RuCl(NO)$_2$(PPh$_3$)$_2$]$^+$	178°, 138°	1845, 1687
[Os(NO)$_2$(OH)(PPh$_3$)$_2$]$^+$	~180°, 127°	1842, 1632
[Ir(η^3-C$_3$H$_5$)(NO)(PPh$_3$)$_2$]$^+$	~180°, 129°	1763, 1631

mnt = maleonitriledithiolate.
diars = 1,2-bis(dimethylarsino)benzene.
diphos = 1,2-bis-diphenylphosphino)ethane.
[a] Value imprecise because of crystal twinning (see ref. 56).

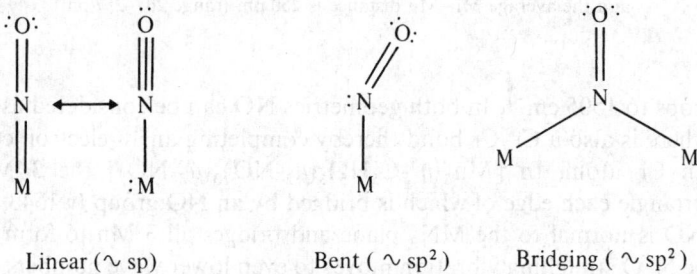

Linear (~ sp)   Bent (~ sp^2)   Bridging (~ sp^2)

FIG. 11.10   Schematic representation of the bonding in linear, bent, and bridging NO complexes. Note that bending would withdraw an electron-pair from the metal centre to the N atom thus creating a vacant coordination site: this may be a significant factor in the catalytic activity of such complexes.[55, 58]

[58] J. P. COLLMAN, N. W. HOFFMAN, and D. E. MORRIS, Oxidative addition and catalysis of olefin hydrogenation by [Rh(NO)(PPh$_3$)$_3$], *J. Am. Chem. Soc.* **91**, 5659–60 (1969).

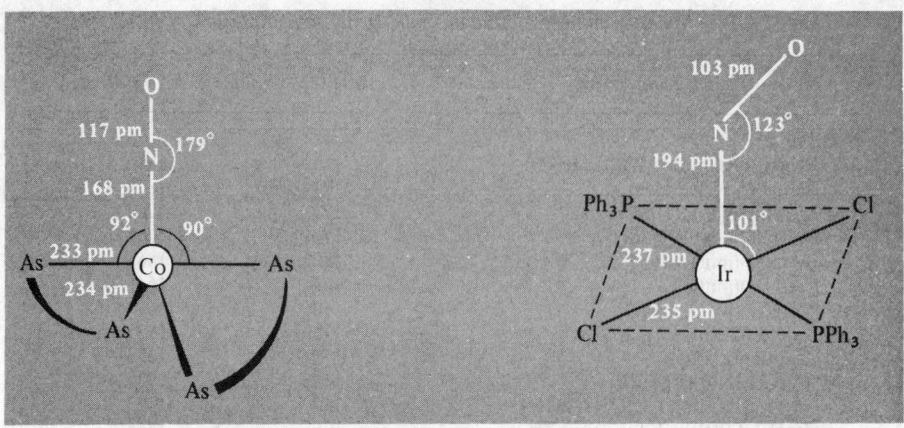

FIG. 11.11 Comparison of the coordination geometries of $[Co(diars)_2(NO)]^{2+}$ and $[IrCl_2(NO)(PPh_3)]$; diars = 1,2-bis(dimethylarsino)benzene.

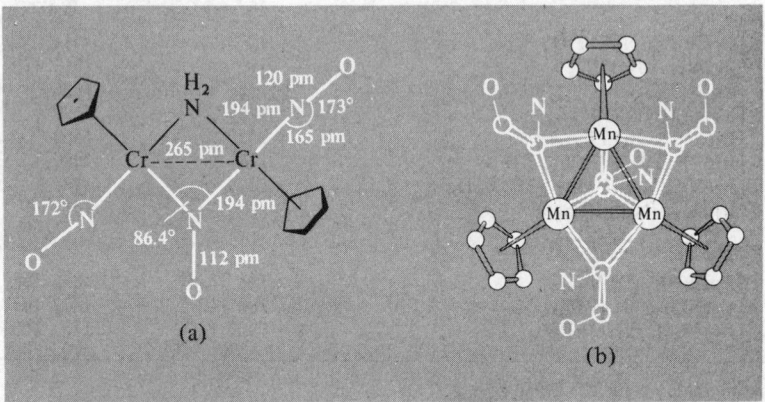

FIG. 11.12 Structures of polynuclear nitrosyl complexes: (a) $[\{Cr(\eta^5\text{-}C_5H_5)(NO)\}_2(\mu_2\text{-}NH_2)(\mu_2\text{-}NO)]$ showing linear-terminal and doubly bridging NO; and (b) $[Mn_3(\eta^5\text{-}C_5H_5)_3(\mu_2\text{-}NO)_3(\mu_3\text{-}NO)]$ showing double- and triply-bridging NO; the molecule has virtual C_{3v} symmetry and the average Mn–Mn distance is 250 pm (range 247–257 pm).

vibration drops to 1505 cm^{-1}. In both geometries NO can be considered as a 3-electron donor and there is also a Cr–Cr bond thereby completing an 18-electron configuration around each Cr atom. In $[Mn_3(\eta^5\text{-}C_5H_5)_3(\mu_2\text{-}NO)_3(\mu_3\text{-}NO)]$ the 3 Mn form an equilateral triangle each edge of which is bridged by an NO group (v 1543, 1481 cm^{-1}); the fourth NO is normal to the MN$_3$ plane and bridges all 3 Mn to form a triangular pyramid; the N–O stretching vibration moves to even lower wave numbers (1328 cm^{-1}). Again, each metal is associated with 18 valency electrons if each forms Mn–Mn bonds with its 2 neighbours and each NO is a 3-electron donor.

In contrast to the numerous complexes of NO which have been prepared and characterized, complexes of the thionitrosyl ligand (NS) are virtually unknown, as is the free ligand itself. The first such complex $[Mo(NS)(S_2CNMe_2)_3]$ was obtained as orange-red air-stable crystals by treating $[MoN(S_2CNMe_2)_3]$ with sulfur in refluxing MeCN, and was shown later to have an M–N–S angle of 172.1°.[58a] More recently [Cr(η^5-C$_5$H$_5$)(CO)$_2$(NS)] was made by reacting Na[Cr(η^5-C$_5$H$_5$)(CO)$_3$] with S$_3$N$_3$Cl$_3$ and again the NS group was found to adopt an essentially linear coordination with Cr–N–S 176.8°.[58b]

Dinitrogen trioxide, N$_2$O$_3$

Pure N$_2$O$_3$ can only be obtained at low temperatures because, above its mp (−100.1°C), it dissociates increasingly according to the equilibria:

$$N_2O_3 \rightleftharpoons NO + NO_2; \quad 2NO_2 \rightleftharpoons N_2O_4$$
$$\text{Blue} \qquad \text{Colourless} \quad \text{Brown} \qquad \text{Brown} \quad \text{Colourless}$$

The solid is pale blue; the liquid is an intense blue at low temperatures but the colour fades and becomes greenish due to the presence of NO$_2$ at higher temperatures. The dissociation also limits the precision with which physical properties of the compound can be determined. At 25°C the dissociative equilibrium in the gas phase is characterized by the following thermodynamic quantities:

$$N_2O_3(g) \rightleftharpoons NO(g) + NO_2(g); \quad \Delta H = 40.5 \text{ kJ mol}^{-1}; \quad \Delta G = -1.59 \text{ kJ mol}^{-1}$$

Hence $\Delta S = 139$ J K^{-1} mol^{-1} and the equilibrium constant $K(25°C) = 1.91$ atm. Molecules of N$_2$O$_3$ are planar with C_s symmetry.

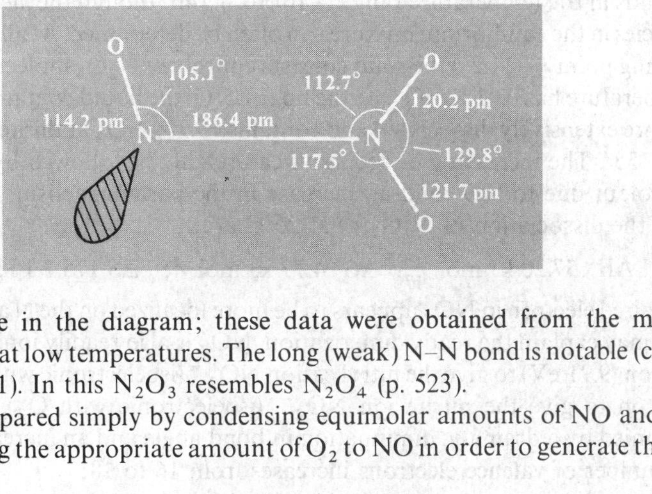

Structural data are in the diagram; these data were obtained from the microwave spectrum of the gas at low temperatures. The long (weak) N–N bond is notable (cf. 145 pm in hydrazine, p. 491). In this N$_2$O$_3$ resembles N$_2$O$_4$ (p. 523).

N$_2$O$_3$ is best prepared simply by condensing equimolar amounts of NO and NO$_2$ at −20°C or by adding the appropriate amount of O$_2$ to NO in order to generate the NO$_2$ *in situ*:

$$2NO + N_2O_4 \xrightarrow{\text{cool}} 2N_2O_3$$
$$4NO + O_2 \longrightarrow 2N_2O_3$$

[58a] J. CHATT and J. R. DILWORTH, Thionitrosyl complexes of molybdenum, *JCS Chem. Comm.* 1974, 508; crystal structure by M. B. HURSTHOUSE and M. MONTEVALLI quoted by J. CHATT in *Pure Appl. Chem.* **49**, 815–26 (1977). See also M. W. BISHOP, J. CHATT, and J. R. DILWORTH, Thionitrosyl complexes of Mo, Re, and Os, *JCS Dalton* 1979, 1–5.

[58b] T. J. GREENOUGH, B. W. S. KOLTHAMMER, P. LEGZDINS, and J. TROTTER, Thionitrosyl ligand; X-ray molecular structure of dicarbonyl(η^5-cyclopentadienyl)thionitrosylchromium, *JCS Chem. Comm.* 1978, 1036–7.

Alternative preparations involve the reduction of 1:1 nitric acid by As_2O_3 at 70°, or the reduction of fuming HNO_3 with SO_2 followed by hydrolysis:

$$2HNO_3 + 2H_2O + As_2O_3 \longrightarrow N_2O_3 + 2H_3AsO_4$$

$$2HNO_3 + 2SO_2 \longrightarrow 2NOHSO_4 \xrightarrow{2H_2O} N_2O_3 + H_2SO_4$$

However, these methods do not yield a completely anhydrous product and dehydration can prove difficult.

Studies of the chemical reactivity of N_2O_3 are complicated by its extensive dissociation into NO and NO_2 which are themselves reactive species. With water N_2O_3 acts as the formal anhydride of nitrous acid and in alkaline solution it is converted essentially quantitatively to nitrite:

$$N_2O_3 + H_2O \xrightarrow{aq} 2HNO_2$$

$$N_2O_3 + 2OH^- \longrightarrow 2NO_2^- + H_2O$$

Reaction with concentrated acids provides a preparative route to nitrosyl salts such as $NO[HSO_4]$, $NO[HSeO_4]$, $NO[ClO_4]$, and $NO[BF_4]$, e.g.:

$$N_2O_3 + 3H_2SO_4 \longrightarrow 2NO^+ + H_3O^+ + 3HSO_4^-$$

Nitrogen dioxide, NO_2, and dinitrogen tetroxide, N_2O_4

The facile equilibrium $N_2O_4 \rightleftharpoons 2NO_2$ makes it impossible to study the pure individual compounds in the temperature range $-10°$ to $+140°$ though the *molecular* properties of each species in the equilibrium mixture can often be determined. At all temperatures below the freezing point ($-11.2°$) the solid consists entirely of N_2O_4 molecules but the liquid at this temperature has 0.01% NO_2. At the bp ($21.5°C$) the liquid contains 0.1% NO_2 but the gas is more extensively dissociated and contains 15.9% NO_2 at this temperature and 99% NO_2 at 135°. The increasing dissociation can readily be followed by a deepening of the brown colour due to NO_2 and an increase in the paramagnetism; the thermodynamic data for the dissociation of $N_2O_4(g)$ at 25°C are:

$$\Delta H° \ 57.20 \text{ kJ mol}^{-1}; \quad \Delta G° \ 4.77 \text{ kJ mol}^{-1}; \quad \Delta S \ 175.7 \text{ J K}^{-1} \text{ mol}^{-1}$$

The unpaired electron in NO_2 appears to be more localized on the N atom than it is in NO and this may explain the ready dimerization. NO_2 is also readily ionized either by loss of an electron (9.91 eV) to give the nitryl cation NO_2^+ (isoelectronic with CO_2) or by gain of an electron to give the nitrite ion NO_2^- (isoelectronic with O_3). These changes are accompanied by a dramatic diminuation in bond angle and an increase in N–O distance as the number of valence electrons increases from 16 to 18:

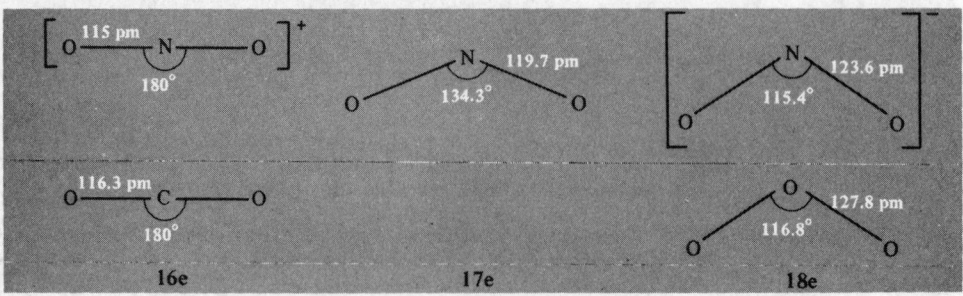

The structure of N_2O_4 in the gas phase is planar (D_{2h}) with a remarkably long N–N bond, and these features persist in both the monoclinic crystalline form near the mp and the more stable low-temperature cubic form. Data for the monoclinic form are in the diagram below† together with those for the isoelectronic species B_2F_4 and the oxalate ion $C_2O_4{}^{2-}$.

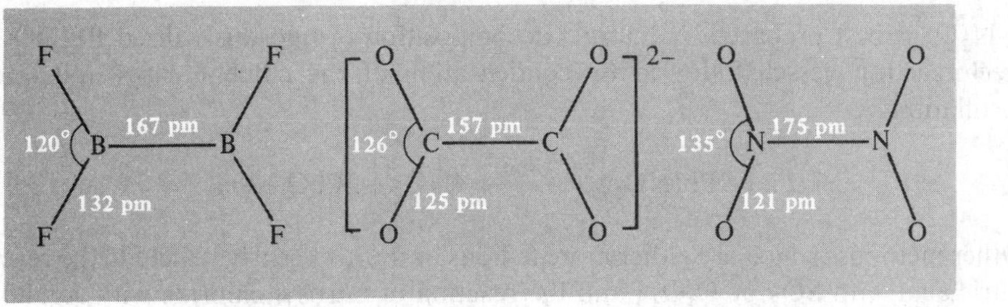

The trends in bond angles and terminal bond distances are clear but the long central bond in N_2O_4 is not paralleled in the other 2 molecules where the B–B distance (p. 166) and C–C distance (p. 311) are normal. However, the B–B bond in B_2Cl_4 is also long (175 pm).

In addition to the normal homolytic dissociation of N_2O_4 into $2NO_2$, the molecule sometimes reacts as if by heterolytic fission: $N_2O_4 \rightleftharpoons NO_2{}^+ + NO_2{}^-$ (see p. 526). Again, in media of high dielectric constant the compound often reacts as though dissociated according to the equilibrium $N_2O_4 \rightleftharpoons NO^+ + NO_3{}^-$ (see p. 525). This has sometimes been taken to imply the presence in liquid N_2O_4 of oxygen-bridged species such as $ONONO_2$

or even $ON{\overset{\displaystyle O}{\underset{\displaystyle O}{\big<}}}NO$ but there is no evidence for such species in solution and it seems

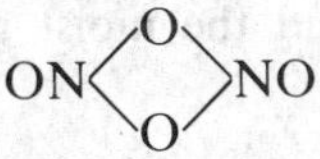

unnecessary to invoke them since similar reactions also occur with the oxalate ion:

Thus

$$N_2O_4 \longrightarrow NO^+ + NO_3{}^- \xrightarrow{2H^+} NO^+ + H_2NO_3{}^+ \xrightarrow{H^+} NO^+ + NO_2{}^+ + H_3O^+$$

Compare

$$C_2O_4{}^{2-} \longrightarrow CO + CO_3{}^{2-} \xrightarrow{2H^+} CO + H_2CO_3 \xrightarrow{H^+} CO + CO_2 + H_3O^+$$

There is no noticeable tendency for pure N_2O_4 to dissociate into ions and the electrical conductivity of the liquid is extremely low (1.3×10^{-13} ohm^{-1} cm^{-1} at 0°). The physical properties of N_2O_4 are summarized in Table 11.11.

---

† Values for the gas phase are similar but there is a noticeable contraction in the cubic crystalline form (in parentheses). N–N 175 pm (164 pm), N–O 118 pm (117 pm), angle O–N–O 133.7° (126°). In addition, infrared studies on N_2O_4 isolated in a low-temperature matrix at liquid nitrogen temperature ($-196°C$) have been interpreted in terms of a twisted (non-planar) molecule O_2N–NO_2, and similar experiments at liquid helium temperature ($-269°C$) have been interpreted in terms of the unstable oxygen-bridged species $ONONO_2$.

TABLE 11.11   *Some physical properties of* N_2O_4

MP/°C	−11.2	Density $(−195°C)/g\ cm^{−3}$	1.979 (s)
BP/°C	+21.15	Density $(0°C)/g\ cm^{−3}$	1.4927 (l)
$\Delta H_f^°$ (298 K)/kJ mol⁻¹	9.16	$\eta(0°C)/poise$	0.527
$\Delta G_f^°$ (298 K)/kJ mol⁻¹	97.83	$\kappa\ (0°C)/ohm^{−1}\ cm^{−1}$	$1.3 \times 10^{−13}$
$S^°$ (298 K)/J K⁻¹ mol⁻¹	304.2	Dielectric constant ε	2.42

N_2O_4 is best prepared by thermal decomposition of rigorously dried $Pb(NO_3)_2$ in a steel reaction vessel, followed by condensation of the effluent gases and fractional distillation:

$$2Pb(NO_3)_2 \xrightarrow{\sim 400°} 4NO_2 + 2PbO + O_2$$

Other methods (which are either more tedious or more expensive) include the reaction of nitric acid with SO_2 or P_4O_{10} and the reaction of nitrosyl chloride with $AgNO_3$:

$$2HNO_3 + SO_2 \longrightarrow N_2O_4 + H_2SO_4$$
$$4HNO_3 + P_4O_{10} \longrightarrow 2N_2O_4 + O_2 + 4HPO_3$$
$$NOCl + AgNO_3 \longrightarrow N_2O_4 + AgCl$$

The compound is also formed when NO reacts with oxygen:

$$2NO + O_2 \rightleftharpoons 2NO_2 \underset{warm}{\overset{cool}{\rightleftharpoons}} N_2O_4$$

These equilibria limit the temperature range in which reactions of N_2O_4 and NO_2 can be studied since dissociation of N_2O_4 into NO_2 is extensive above room temperature and is virtually complete by 140° whereas decomposition of NO_2 into NO and O_2 becomes significant above 150° and is complete at about 600°.

N_2O_4/NO_2 react with water to form nitric acid (p. 537) and the moist gases are therefore highly corrosive:

$$N_2O_4 + H_2O \longrightarrow HNO_3 + HNO_2; \quad 3HNO_2 \longrightarrow HNO_3 + 2NO + H_2O$$

The oxidizing action of NO_2 is illustrated by the following:

$$NO_2 + 2HCl \longrightarrow NOCl + H_2O + \tfrac{1}{2}Cl_2$$

$$NO_2 + 2HX \xrightarrow{heat} NO + H_2O + X_2 \quad (X = Cl,\ Br)$$

$$2NO_2 + F_2 \longrightarrow 2FNO_2$$

$$NO_2 + CO \longrightarrow NO + CO_2$$

N_2O_4 has been extensively studied as a nonaqueous solvent system[59] and it is uniquely useful for preparing anhydrous metal nitrates and nitrato complexes (p. 541). Much of the

[59] C. C. ADDISON, Chemistry in liquid dinitrogen tetroxide, in G. JANDER, H. SPANDAU, and C. C. ADDISON (eds.), *Chemistry in Non-aqueous Ionizing Solvents*, Vol. 3, Part 1, pp. 1–78, Pergamon Press, London, 1967. C. C. ADDISON, Dinitrogen tetroxide, nitric acid, and their mixtures as media for inorganic reactions, *Chem. Rev.* **80**, 21–39 (1980).

chemistry can be rationalized in terms of a self-ionization equilibrium similar to that observed for liquid ammonia (p. 487):

$$N_2O_4 \rightleftharpoons NO^+ + NO_3^-$$
$$\text{Solvent} \qquad \text{Solvo-acid} \quad \text{Solvo-base}$$

As noted above, there is no physical evidence for this equilibrium in pure N_2O_4, but the electrical conductivity is considerably enhanced when the liquid is mixed with a solvent of high dielectric constant such as nitromethane ($\varepsilon \approx 37$), or with donor solvents (D) such as $MeCO_2Et$, Et_2O, Me_2SO, or Et_2NNO (diethylnitrosamine):

$$N_2O_4 + nD \rightleftharpoons \{D_n \cdot N_2O_4\} \rightleftharpoons [D_nNO]^+ + NO_3^-$$

Typical solvent system reactions are summarized below together with the analogous reactions from the liquid ammonia solvent system:

"Neutralization"
$$NOCl + AgNO_3 \xrightarrow{N_2O_4} AgCl + N_2O_4$$

$$NH_4Cl + NaNH_2 \xrightarrow{NH_3} NaCl + 2NH_3$$

"Acid"
$$2NOCl + Sn \xrightarrow{N_2O_4} SnCl_2 + 2NO$$

$$2NH_4Cl + Sn \xrightarrow{NH_3} SnCl_2 + 2NH_3 + H_2$$

"Base/amphoterism"
$$2[EtNH_3][NO_3] + 2N_2O_4 + Zn \xrightarrow{N_2O_4} [EtNH_3]_2[Zn(NO_3)_4] + 2NO$$

$$2NaNH_2 + 2NH_3 + Zn \xrightarrow{NH_3} Na_2[Zn(NH_2)_4] + H_2$$

"Solvolysis"
$$CaO + 2N_2O_4 \xrightarrow{N_2O_4} Ca(NO_3)_2 + N_2O_3$$

$$Na_2O + NH_3 \xrightarrow{NH_3} NaNH_2 + NaOH$$

Similarly:
$$ZnCl_2 + N_2O_4 \xrightarrow{N_2O_4} Zn(NO_3)_2 + 2NOCl$$

Such reactions provide an excellent route to anhydrous metal nitrates, particularly when metal bromides or iodides are used, since then the nitrosyl halide decomposes and this prevents the possible formation of nitrosyl compounds, e.g.:

$$TiI_4 + 4N_2O_4 \rightleftharpoons Ti(NO_3)_4 + 4NO + 2I_2$$

Many carbonyls react similarly, e.g.:

$$[Mn_2(CO)_{10}] + N_2O_4 \longrightarrow [Mn(CO)_5(NO_3)] + [Mn(CO)_x(NO)_y]$$
$$[Fe(CO)_5] + 4N_2O_4 \longrightarrow [Fe(NO_3)_3 \cdot N_2O_4] + 5CO + 3NO$$

Solvates are frequently formed in these various reactions, e.g.:

$$Ca + 3N_2O_4 \xrightarrow{\text{MeNO}_2} [Cu(NO_3)_2 . N_2O_4] + 2NO$$

Some of these may contain undissociated solvent molecules N_2O_4 but structural studies have revealed that often such "solvates" are actually nitrosonium nitrato-complexes. For example it has been shown[60] that $[Sc(NO_3)_3 . 2N_2O_4]$ is, in fact, $[NO]_2^+[Sc(NO_3)_5]^{2-}$. Similarly, X-ray crystallography revealed[61] that $[Fe(NO_3)_3 . 1\frac{1}{2}N_2O_4]$ is $[NO]_3^+[Fe(NO_3)_4]_2^-[NO_3]^-$, in which there is a fairly close approach of 3 NO^+ groups to the "uncoordinated" nitrate ion to give a structural unit of stoichiometry $[N_4O_6]^{2+}$ (see also p. 544).

In contrast to the wealth of reactions in which N_2O_4 tends to behave as $NO^+NO_3^-$, reactions based on the alternative heterolytic dissociation $NO_2^+NO_2^-$ are less well documented. However, N_2O_4 reacts with the strong Lewis acid BF_3 to give stable white complexes of stoichiometry $N_2O_4 . BF_3$ and $N_2O_4 . 2BF_3$, and a variety of evidence points to their formulation as $[NO_2]^+[ONOBF_3]^-$ and $[NO_2]^+[F_3BONOBF_3]^-$ respectively.[62] The related adducts $N_2O_3 . 2BF_3$ and $N_2O_5 . BF_3$ can be formulated similarly as $[NO]^+[F_3BONOBF_3]^-$ and $[NO_2]^+[O_2NOBF_3]^-$.

Dinitrogen pentoxide, N_2O_5, and nitrogen trioxide, NO_3

N_2O_5 is the anhydride of nitric acid and is obtained as a highly reactive deliquescent, light-sensitive, colourless, crystalline solid by carefully dehydrating the concentrated acid with P_4O_{10} at low temperatures:

$$4HNO_3 + P_4O_{10} \xrightarrow{-10°} 2N_2O_5 + 4HPO_3$$

The solid has a vapour pressure of 100 mmHg at 7.5°C and sublimes (1 atm) at 32.4°C, but is thermally unstable both as a solid and as a gas above room temperature. Thermodynamic data at 25°C are:

	$\Delta H_f°/kJ\ mol^{-1}$	$\Delta G_f°/kJ\ mol^{-1}$	$S°/J\ K^{-1}mol^{-1}$
N_2O_5(cryst)	-43.1	113.8	178.2
N_2O_5(g)	11.3	115.1	355.6

X-ray diffraction studies show that solid N_2O_5 consists of an ionic array of linear NO_2^+ (N–O 115.4 pm) and planar NO_3^- (N–O 124 pm). In the gase phase and in solution (CCl_4, $CHCl_3$, $OPCl_3$) the compound is molecular; the structure is not well established but may be $O_2N–O–NO_2$ with a central N–O–N angle close to 180°. The molecular form

[60] C. C. ADDISON, A. J. GREENWOOD, M. J. HALEY, and N. LOGAN, Novel coordination numbers in scandium(III) and yttrium(III) nitrato complexes. X-ray crystal structures of $Sc(NO_3)_3 . 2N_2O_4$ and $Y(NO)_3)_3 . 2N_2O_4$, *JCS Chem. Comm.* 1978, 580–1.

[61] L. J. BLACKWELL, E. K. NUNN, and S. C. WALLWORK, Crystal structure of anhydrous nitrates and their complexes. Part VII. 1.5 dinitrogen tetroxide solvate of iron(III) nitrate, *JCS Dalton* 1975, 2068–72.

[62] R. W. SPRAGUE, A. B. GARRETT, and H. H. SISLER, Reactions of some oxides of nitrogen with boron trifluoride, *J. Am. Chem. Soc.* **82,** 1059–64 (1960).

can also be obtained in the solid phase by rapidly quenching the gas to $-180°$, but it rapidly reverts to the more stable ionic form on being warmed to $-70°$. [cf. ionic and covalent forms of $BF_3 . 2H_2O$ (p. 224), $AlCl_3$ (p. 263), PCl_5 (p. 573), PBr_5 (p. 574), etc.]

N_2O_5 is readily hydrated to nitric acid and reacts with H_2O_2 to give pernitric acid as a coproduct:

$$N_2O_5 + H_2O \longrightarrow 2HONO_2; \quad N_2O_5 + H_2O_2 \longrightarrow HONO_2 + HOONO_2$$

It reacts violently as an oxidizing agent towards many metals, non-metals, and organic substances, e.g.:

$$N_2O_5 + Na \longrightarrow NaNO_3 + NO_2$$
$$N_2O_5 + NaF \longrightarrow NaNO_3 + FNO_2$$
$$N_2O_5 + I_2 \longrightarrow I_2O_5 + N_2$$

Like N_2O_4 (p. 525) it dissociates ionically in strong anhydrous acids such as HNO_3, H_3PO_4, H_2SO_4, HSO_3F, and $HClO_4$, and this affords a convenient source of nitronium ions and hence a route to nitronium salts, e.g.:

$$N_2O_5 + 3H_2SO_4 \longrightarrow 2NO_2^+ + H_3O^+ + 3HSO_4^-$$
$$N_2O_5 + HSO_3F \longrightarrow [NO_2]^+[FSO_3]^- + HNO_3$$
$$N_2O_5 + 2SO_3 \longrightarrow [NO_2]_2^+[S_2O_7]^{2-}$$

In the gas phase, N_2O_5 decomposes according to a first-order rate law which can be explained by a dissociative equilibrium followed by rapid reaction according to the scheme

$$N_2O_5 \rightleftharpoons NO_2 + \{NO_3\} \longrightarrow NO_2 + O_2 + NO$$
$$N_2O_5 + NO \rightleftharpoons 3NO_2$$

The fugitive, paramagnetic species $\{NO_3\}$ is also implicated in several other gas-phase reactions involving the oxides of nitrogen and, in the N_2O_5-catalysed decomposition of ozone, its concentration is sufficiently high for its absorption spectrum to be recorded, thereby establishing its integrity as an independent chemical species. Such reactions are the subject of considerable current interest for environmental reasons. NO_3 probably has a symmetrical planar structure (like NO_3^-) but it has not been isolated as a pure compound.

11.3.7 *Oxoacids, oxoanions, and oxoacid salts of nitrogen*

Nitrogen forms numerous oxoacids, though several are unstable in the free state and are known only in aqueous solution or as their salts. The principal species are summarized in Table 11.12; of these by far the most stable is nitric acid and this compound, together with its salts the nitrates, are major products of the chemical industry (p. 537).

Hyponitrous acid and hyponitrites[63]

Hyponitrous acid crystallizes from ether solutions as colourless crystals which readily decompose (explosively when heated). Its structure has not been determined but the

---

[63] M. N. HUGHES, Hyponitrites, *Q. Rev.* **22**, 1–13 (1968).

TABLE 11.12 *Oxoacids of nitrogen and related species*

Formula	Name	Remarks
$H_2N_2O_2$	Hyponitrous acid	Weak acid HON=NOH, isomeric with nitramide, H_2N-NO_2;[a] salts are known (p. 529)
{HNO}	Nitroxyl	Reactive intermediate (p. 530),[b] salts are known
$H_2N_2O_3$	Hyponitric acid (oxyhyponitrous)	Known in solution and as salts, e.g. Angeli's salt $Na_2[ON=NO_2]$ (p. 530)
$H_4N_2O_4$	Nitroxylic acid (hydronitrous)	Explosive; sodium salt known $Na_4[O_2NNO_2]$[c]
HNO_2	Nitrous acid	Unstable weak acid, HONO (p. 531); stable salts (nitrites) are known
HOONO	Peroxonitrous acid	Unstable, isomeric with nitric acid; some salts are more stable[d]
HNO_3	Nitric acid	Stable strong acid $HONO_2$; many stable salts (nitrates) known (p. 536)
HNO_4	Peroxonitric acid	Unstable, explosive crystals, $HOONO_2$; no solid salts known. (For "orthonitrates", NO_4^{3-}, i.e. salts of the unknown orthonitric acid H_3NO_4, see p. 545)

[a] The structure of nitramide is as shown, the dihedral angle between NH_2 and NNO_2 is 52°. Nitramide is a weak acid pK_1 6.6 (K_1 2.6×10^{-7}) and it decomposes into N_2O and H_2O by a base-catalysed mechanism:

$$H_2NNO_2 + B \xrightarrow{\text{slow}} BH^+ + [HNNO_2]^- \xrightarrow{\text{fast}} N_2O + OH^-$$

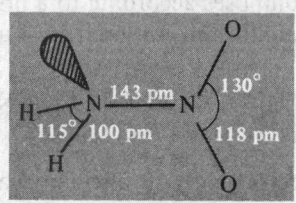

[b] A stable complex of HNO has recently been prepared by addition of HCl to $[Os^I(CO)Cl(NO)(PPh_3)_2]$; the ligand is *N*-bonded and essentially coplanar with the {$Os(CO)Cl_2$} group.[63a]

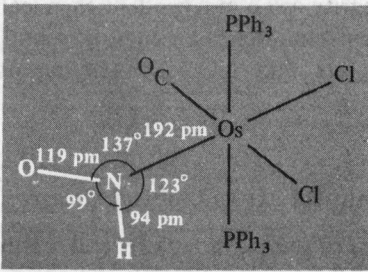

[c] Sodium nitroxylate can be prepared as a deep yellow solid by reduction of sodium nitrite with Na in liquid NH_3, i.e.

$$2NaNO_2 + 2Na \xrightarrow{NH_3} Na_4\left[\overset{O}{\underset{O}{>}}N-N\overset{O}{\underset{O}{<}}\right]$$

[d] Peroxonitrous acid is formed as an unstable intermediate during the oxidation of acidified aqueous solutions of nitrites using H_2O_2; such solutions are orange-red and are more highly oxidizing than either H_2O_2 or HNO_3 alone (e.g. they liberate Br_2 from Br^-). Alkaline solutions are more stable but the yellow peroxonitrites M[OONO] have not been isolated pure.

63a R. D. WILSON and J. A. IBERS, Coordinated nitrosyl hydride: structural and spectroscopic study of $[Os(CO)Cl_2(HNO)(PPh_3)_2]$, *Inorg. Chem.* **18**, 336–43 (1979).

molecular weight indicates a double formula $H_2N_2O_2$, i.e. $HON{=}NOH$; consistent with this the compound yields N_2O when decomposed by H_2SO_4, and hydrazine when reduced. The free acid is obtained by treating $Ag_2N_2O_2$ with anhydrous HCl in ethereal solution. It is a weak dibasic acid: pK_1 6.9, pK_2 11.6. Aqueous solutions are unstable between pH 4–14 due to base catalysed decomposition via the hydrogen-hyponitrite ion:

$$HONNOH \xrightarrow{\text{base}} [HONNO]^- \xrightarrow[\text{(minutes)}]{\text{fast}} N_2O + OH^-$$

At higher acidities (lower pH) decomposition is slower ($t_{\frac{1}{2}}$ days or weeks) and the pathways are more complex.

Hyponitrites can be prepared in variable (low) yields by several routes of which the commonest are reduction of aqueous nitrite solutions using sodium (or magnesium) amalgam, and condensation of organic nitrites with hydroxylamine in NaOEt/EtOH:

$$2NaNO_3 + 8Na/Hg + 4H_2O \longrightarrow Na_2N_2O_2 + 8NaOH + 8Hg$$
$$2AgNO_3 + 2NaNO_2 + 4Na/Hg + 2H_2O \longrightarrow Ag_2N_2O_2 + 2NaNO_3 + 4NaOH + 4Hg$$
$$Ca(NO_3)_2 + 4Mg/Hg + 4H_2O \longrightarrow CaN_2O_2 + 4Mg(OH)_2 + 4Hg$$

$$NH_2OH + RONO + 2NaOEt \xrightarrow{\text{EtOH}} Na_2N_2O_2 + ROH + 2EtOH$$

Vibrational spectroscopy indicates that the hyponitrite ion has the *trans-* (C_{2h}) configuration in the above salts.

As implied by the preparative methods employed, hyponitrites are usually stable towards reducing agents though under some conditions they can be reduced (p. 500). More frequently they themselves act as reducing agents and are thereby oxidized, e.g. the analytically useful reaction with iodine:

$$[ONNO]^{2-} + 3I_2 + 3H_2O \longrightarrow [NO_3]^- + [NO_2]^- + 6HI$$

There is also considerable current environmental interest in hyponitrite oxidation because it is implicated in the oxidation of ammonia to nitrite, an important step in the nitrogen cycle (p. 469). Specifically, it seems likely that the oxidation proceeds from ammonia through hydroxylamine and hyponitrous acid to nitrite (or N_2O).

With liquid N_2O_4 stepwise oxidation of hyponitrites occurs to give $Na_2N_2O_x$ ($x = 3–6$):

$$Na_2N_2O_2 \xrightarrow{\text{fast}} \beta\text{-}Na_2N_2O_3 \xrightarrow{\text{slow}} Na_2N_2O_5 \xrightarrow[100°]{\text{slow}} Na_2N_2O_6$$

$$\alpha\text{-}Na_2N_2O_3 \xrightarrow{\text{fast}} Na_2N_2O_4 \xrightarrow{\text{slow}} Na_2N_2O_6$$

The α isomer of $Na_2N_2O_3$ (i.e. sodium hyponitrate or trioxodinitrate) can be made by the following reaction in cooled methanol:

$$NH_2OH + BuONO_2 + 2NaOMe \xrightarrow{MeOH/0°} \alpha\text{-}Na_2[ONNO_2] + BuOH + 2MeOH$$

Its structure, shown below, was deduced by vibration spectroscopy and differs from that of the β isomer which is thought to feature an O–O linkage:

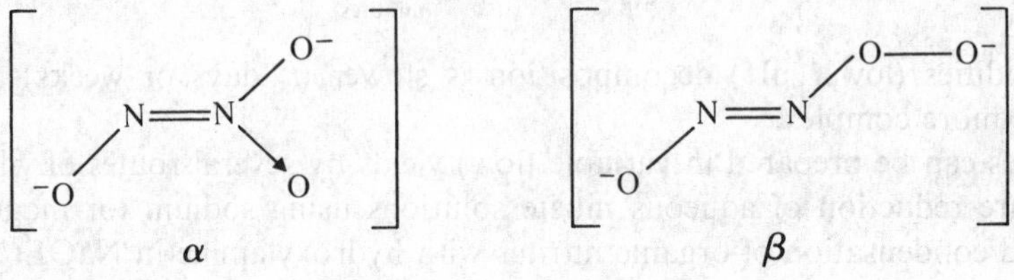

In contrast to the stepwise oxidation of sodium hyponitrite in liquid N_2O_4, the oxidation goes rapidly to the nitrate ion in an inert solvent of high dielectric constant such as nitromethane:

$$[ON=NO]^{2-} + 2N_2O_4 \xrightarrow{MeNO_2} 2[NO_3]^- + \{ONON=NONO\} \longrightarrow N_2 + 2NO_2$$

More recently it has been found that the hyponitrite ion can act as a bidentate ligand in either a bridging or a chelating mode. Thus, the controversy about the nature of the black and red isomers of nitrosyl pentammine cobalt(III) complexes has been resolved by X-ray crystallographic studies which show that the black chloride $[Co(NH_3)_5NO]Cl_2$ contains a mononuclear octahedral Co^{III} cation with a linear Co–N–O groups whereas the red isomer, in the form of a mixed nitrate-bromide, is dinuclear with a bridging *cis*-hyponitrite group as shown in Fig. 11.13.[64] The *cis*- configuration is probably adopted for steric reasons since this is the only configuration that allows the bridging of two $\{Co(NH_3)_5\}$ groups by an ONNO group without steric interference between them. The chelating mode was found in the air-sensitive yellow crystalline complex $[Pt(O_2N_2)(PPh_3)_2]$ which can be made by the action of NO on $[Pt(PPh_3)_3]$ or $[Pt(PPh_3)_4]$.[65] The compound was shown not to be a bisnitrosyl (20-electron Pt species) but to have the (18-electron) structure shown schematically in Fig. 11.14. The presence of the *cis*- configuration in both these complexes invites speculation as to whether *cis*-$[ON–NO]^{2-}$ can also exist in simple hyponitrites. Likely candidates appear to be the "alkali metal nitrosyls" MNO prepared by the action of NO on Na/NH_3; infrared data suggest they are not $M^+[NO]^-$ and might indeed contain the *cis*-hyponitrite ion. They would therefore not be salts of nitroxyl HNO which has often been postulated as an intermediate in reactions which give N_2O and which is well known in the gas phase. Nitroxyl can be prepared by the action of atomic H or HI on NO and decomposes to N_2O and H_2O. As expected, the molecule is bent (angle H–N=O 109°).

[64] B. F. HOSKINS, F. D. WHILLANS, D. H. DALE, and D. C. HODGKIN, The structure of the red nitrosylpenta-amminecobalt(III) cation, *JCS Chem. Comm.* 1969, 69–70.

[65] S. BHADURI, B. F. G. JOHNSON, A. PICKARD, P. R. RAITHBY, G. M. SHELDRICK, and C. I. ZUCCARO, Nitrosyl coupling in transition-metal complexes: the molecular structure of $[(PPh_3)_2Pt(N_2O_2)]$, *JCS Chem. Comm.* 1977, 354–5.

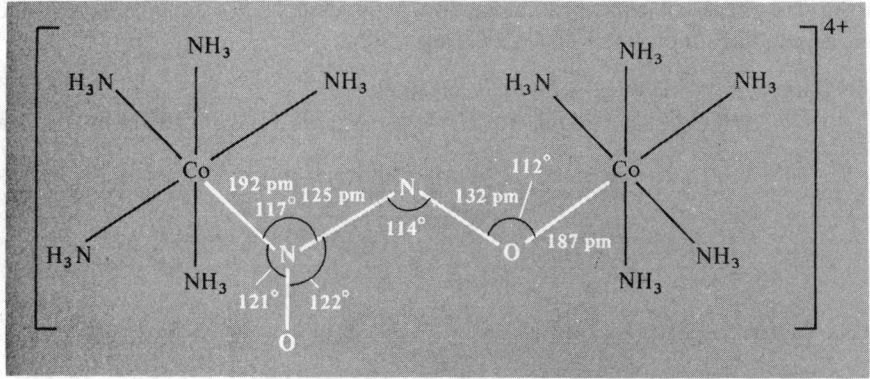

FIG. 11.13 Structure of the dinuclear cation in the red isomer $[\{Co(NH_3)_5NO\}_2](Br)_{2.5}$ $(NO_3)_{1.5}.2H_2O$; (mean Co–NH$_3$ 194$\pm$2 pm, mean angle 90$\pm$4°).

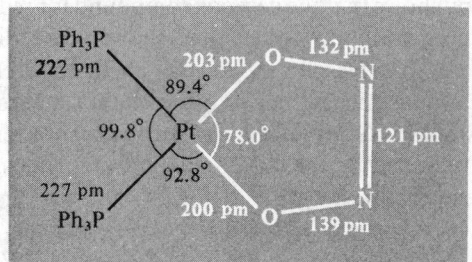

FIG. 11.14 Structure of the square-planar platinum complex $[Pt(O_2N_2)(PPh_3)_2]$ showing the cis chelating ONNO group and short N–N distance.

Nitrous acid and nitrites

Nitrous acid, HNO_2, has not been isolated as a pure compound but it is a well known and important reagent in aqueous solutions and has also been studied as a component in gas-phase equilibria. Solutions of the free acid can readily be obtained by acidification of cooled aqueous nitrite solutions but even at room temperature disproportionation is noticeable:

$$3HNO_2(aq) \rightleftharpoons H_3O^+ + NO_3^- + 2NO$$

It is a fairly weak acid with pK_a 3.35 at 18°C, i.e. intermediate in strength between acetic (4.75) and chloroacetic (2.85) acids at 25°, and very similar to formic (3.75) and sulfanilic (3.23) acids. Salt-free aqueous solutions can be made by choosing combinations of reagents which give insoluble salts, e.g.:

$$Ba(NO_2)_2 + H_2SO_4 \xrightarrow{aq} 2HNO_2 + BaSO_4$$

$$AgNO_2 + HCl \xrightarrow{aq} HNO_2 + AgCl$$

When the presence of salts in solution is unimportant, the more usual procedure is simply to acidify $NaNO_2$ with hydrochloric acid below 0°.

In the gas phase, an equilibrium reaction producing HNO_2 can be established by mixing equimolar amounts of H_2O, NO, and NO_2:

$$2HNO_2(g) \rightleftharpoons H_2O(g) + NO(g) + NO_2(g);$$
$$\Delta H°(298 \text{ K}) \ 38 \text{ kJ/2 mol } HNO_2; \quad K_p(298 \text{ K}) \ 8.0 \times 10^5 \text{ N m}^{-2} \text{ (7.9 atm)}$$

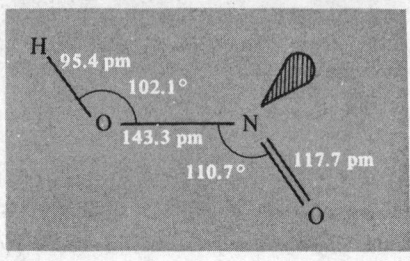

Microwave spectroscopy shows that the gaseous compound is predominantly in the *trans*-planar (C_s) configuration with the dimensions shown. The differences between the two N–O distances is notable. Despite the formal single-bond character of the central bond the barrier to rotation is 45.2 kJ mol^{-1}. Infrared data suggest that the *trans*-form is ~ 2.3 kJ mol^{-1} more stable $(\Delta G°)$ than the *cis*- form at room temperature.

Nitrites are usually obtained by the mild reduction of nitrates, using C, Fe, or Pb at moderately elevated temperatures, e.g.:

$$NaNO_3 + Pb \xrightarrow{\text{melt}} NaNO_2 + PbO$$

On the industrial scale, impure $NaNO_2$ is made by absorbing "nitrous fumes" in aqueous alkali or carbonate solutions and then recrystallizing the product:

$$NO + NO_2 + 2NaOH \text{ (or } Na_2CO_3) \xrightarrow{\text{aq}} 2NaNO_2 + H_2O \text{ (or } CO_2)$$

The sparingly soluble $AgNO_2$ can be obtained by metathesis, and simple variants yield the other stable nitrites, e.g.:

$$NaNO_2 + AgNO_3 \xrightarrow{\text{aq}} AgNO_2 + NaNO_3$$

$$NaNO_2 + KCl \xrightarrow{\text{aq}} KNO_2 + NaCl$$

$$2NH_3 + H_2O + N_2O_3 \longrightarrow 2NH_4NO_2$$

$$Ba(OH)_2 + NO + NO_2 \xrightarrow{\text{aq}} Ba(NO_2)_2 + H_2O$$

Many stable metal nitrites (Li, Na, K, Cs, Ag, Tl^I, NH$_4$, Ba) contain the bent $[O-N-O]^-$ anion (p. 475) with N–O in the range 113–123 pm and the angle 116–132°. Nitrites of less basic metals such as Co(II), Ni(II), and Hg(II) are often highly coloured and are probably essentially covalent assemblages. Solubility (g per 100 g H_2O at 25°) varies considerably, e.g. $AgNO_2$ 0.41, $NaNO_2$ (hygroscopic) 85.5, KNO_2 (deliquescent) 314. Thermal stability also varies widely: e.g. the alkali metal nitrites can be fused without decomposition (mp

NaNO$_2$ 284°, KNO$_2$ 441°C), whereas Ba(NO$_2$)$_2$ decomposes when heated above 220°, AgNO$_2$ above 140° and Hg(NO$_2$)$_2$ above 75°. Such trends are a general feature of oxoacid salts (pp. 540, 1011, 1017). NH$_4$NO$_2$ can decompose explosively.

The aqueous solution chemistry of nitrous acid and nitrites has been extensively studied. Some reduction potentials involving these species are given in Table 11.4 (p. 500) and these form a useful summary of their redox reactions. Nitrites are quantitatively oxidized to nitrate by permanganate and this reaction is used in titrimetric analysis. Nitrites (and HNO$_2$) are readily reduced to NO and N$_2$O with SO$_2$, to H$_2$N$_2$O$_2$ with Sn(II), and to NH$_3$ with H$_2$S. Hydrazinium salts yield azides (p. 496) which can then react with further HNO$_2$:

$$HNO_2 + N_2H_5{}^+ \longrightarrow HN_3 + H_2O + H_3O^+$$
$$HNO_2 + HN_3 \longrightarrow N_2O + N_2 + H_2O$$

This latter reaction is most unusual in that it simultaneously involves an element (N) in four different oxidation states. Use of ^{15}N-enriched reagents shows that all the N from HNO$_2$ goes quantitatively to the internal N of N$_2$O:[66]

$$HN_3 + HO^{15}NO \xrightarrow{\ -H_2O\ } \{NNN^{15}NO\} \longrightarrow NN + N^{15}NO$$

NaNO$_2$ is mildly toxic (tolerance limit ~ 100 mg/kg body weight per day, i.e. 4–8 g/day for humans). It has been much used for curing meat. NaNO$_2$ is used industrially on a large scale for the synthesis of hydroxylamine (p. 494), and in acid solution for the diazotization of primary *aromatic* amines:

$$ArNH_2 + HNO_2 \xrightarrow{\ HCl/aq\ } [ArNN]Cl + 2H_2O$$

The resulting diazo reagents undergo a wide variety of reactions including those of interest in the manufacture of azo dyes and pharmaceuticals. With primary *aliphatic* amines the course of the reaction is different: N$_2$ is quantitatively evolved and alcohols usually result:

$$RNH_2 + HNO_2 \xrightarrow{\ -H_2O\ } RNHNO \longrightarrow RN{=}NOH \xrightarrow{\ -OH^-\ } RN_2{}^+ \longrightarrow$$
$$N_2 + R^+ \longrightarrow \text{products}$$

The reaction is generally thought to involve carbonium-ion intermediates but several puzzling features remain.[67] Secondary aliphatic amines give nitrosamines without evolution of N$_2$:

$$R_2NH + HONO \longrightarrow R_2NNO + H_2O$$

Tertiary aliphatic amines react in the cold to give nitrite salts and these decompose on warming to give nitrosamines and alcohols:

$$R_3N + HNO_2 \xrightarrow{\ cold\ } [R_3NH][NO_2] \xrightarrow{\ warm\ } R_2NNO + ROH$$

---

[66] K. CLUSIUS and H. KNOPF, Reactions with ^{15}N. Part 22. The decomposition of azide with nitrous acid and the formation of azide from nitrous oxide and sodium amide, *Chem. Ber.* **89**, 681–5 (1956).

[67] C. J. COLLINS, Reactions of primary aliphatic amines with nitrous acid, *Acc. Chem. Res.* **4**, 315–22 (1971).

In addition to their general use in synthetic organic chemistry, these various reactions afford the major route for introducing ^{15}N into organic compounds by use of $Na^{15}NO_2$.

The nitrite ion, NO_2^-, is a versatile ligand and can coordinate in at least five different ways (i)–(v):

(i) Nitro   (ii) Nitrito   (iii) Chelating   (iv) Unsymmetrical bridging (N,O)   (v) η^1-O bridging

Nitro-nitrito isomerism (i), (ii), was discovered by S. M. Jörgensen in 1894–9 and was extensively studied during the classic experiments of A. Werner (p. 1068); the isomers usually have quite different colours, e.g. $[Co(NH_3)_5(NO_2)]^{2+}$, yellow, and $[Co(NH_3)_5(ONO)]^{2+}$, red. The nitrito form is usually less stable and tends to isomerize to the nitro form. The change can also be effected by increase in pressure since the nitro form has the higher density. For example application of 20 kbar pressure converts the violet nitrito complex $[Ni(en)_2(ONO)_2]$ to the red nitro complex $[Ni(en)_2(NO_2)_2]$ at 126°C, thereby reversing the change from nitro to nitrito which occurs on heating the complex from room temperature at atmospheric pressure.[68] An X-ray study of the thermally induced nitrito→nitro isomerization and the photochemically induced nitro→nitrito isomerization of Co(III) complexes has shown that both occur intramolecularly by rotation of the NO_2 group in its own plane, probably via a 7-coordinated cobalt intermediate.[68a] Similarly, the base-catalysed nitrito→nitro isomerization of $[M^{III}(NH_3)_5(ONO)]^{2+}$ (M = Co, Rh, Ir) is intramolecular and occurs without ^{18}O exchange of the coordinated ONO^- with $H_2^{18}O$, $^{18}OH^-$, or "free" $N^{18}O_2^-$.[68b] However, an elegant ^{17}O nmr study using specifically labelled $[Co(NH_3)_5(^{17}ONO)]^{2+}$ and $[Co(NH_3)_5(ON^{17}O)]^{2+}$ established that spontaneous intramolecular O-to-O exchange in the nitrite ligand occurs at a rate comparable to that of the spontaneous O-to-N isomerization.[68c]

A typical value for the N–O distance in nitro complexes is 124 pm whereas in nitrito complexes the terminal N–O (121 pm) is shorter than the internal N–O(M) ~129 pm. In the bidentate chelating mode (iii) the 2 M–O distances may be fairly similar as in

[68] J. R. FERRARO and L. FABBRIZZI, Effect of pressure on the nitro–nitrito linkage isomerism in solid $[Ni(en)_2(NO_2)_2]$, *Inorg. Chim. Acta* **26**, L15–L17 (1978).

[68a] I. GRENTHE and E. NORDIN, Nitrito–nitro linkage isomerization in the solid state. Part 1. X-ray crystallographic studies of *trans*-$[Co(en)_2(NCS)(ONO)]X$ and $[Co(en)_2(NCS)(NO_2)]X$ (X=ClO$_4$, I), *Inorg. Chem.* **18**, 1109–16 (1979). Part 2. A comparative study of the structures of nitrito- and nitropentaamminecobalt(III) dichloride, *Inorg. Chem.* **18**, 1869–74 (1979).

[68b] W. G. JACKSON, G. A. LAWRANCE, P. A. LAY, and A. N. SARGESON, Base-catalysed nitrito to nitro linkage isomerization of cobalt(III), rhodium(III), and iridium(III) pentaammine complexes, *Inorg. Chem.* **19**, 904–10 (1980).

[68c] W. G. JACKSON, G. A. LAWRENCE, P. A. LAY, and A. M. SARGESON, Oxygen-17 nmr study of linkage isomerization in nitritopentaamminecobalt(III): Evidence for intramolecular oxygen exchange, *JCS Chem. Comm.* 1982, 70–2.

$[Cu(bipy)_2(O_2N)]NO_3$ or quite different as in $[Cu(bipy)(O_2N)_2]$:

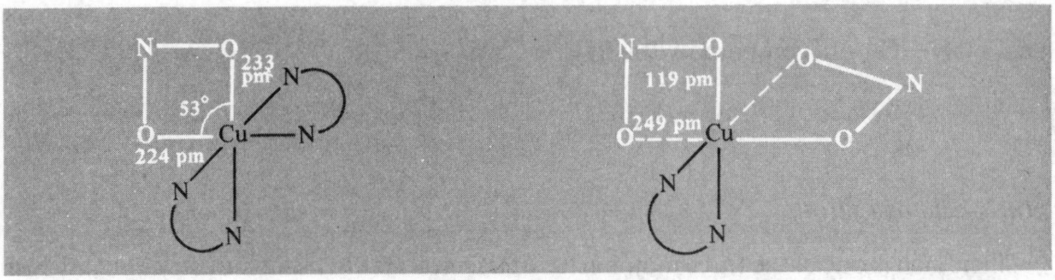

Examples of the unsymmetrical bridging mode (iv) are:

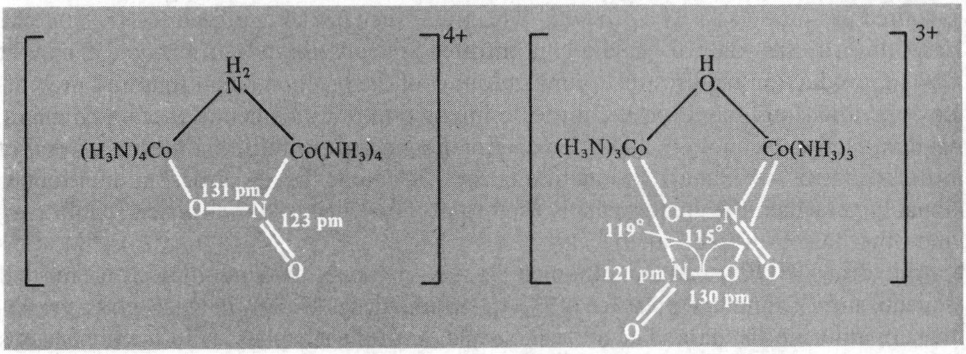

The oxygen-bridging mode (v) is less common but occurs together with modes (iii) and (iv) in the following centrosymmetrical trimeric Ni complex and related compounds.[69]

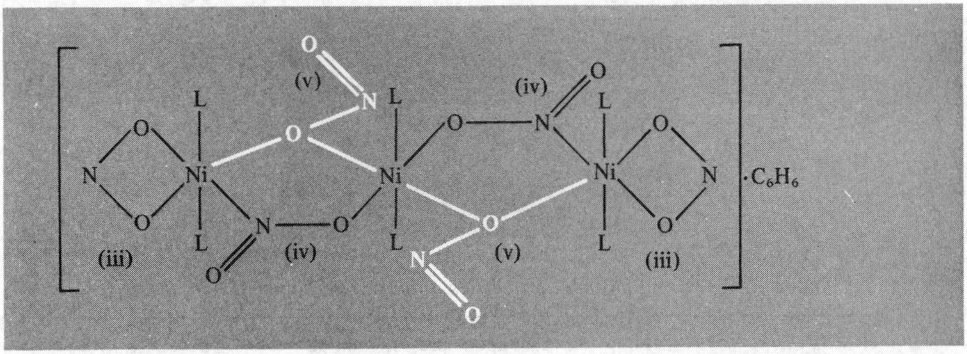

It is possible that a sixth (symmetrical bridging) mode M—O⟍ ⟋O—M occurs in some

complexes such as $Rb_3Ni(NO_2)_5$ but this has not definitely been established; an

[69] D. M. L. GOODGAME, M. A. HITCHMAN, D. F. MARSHAM, P. PHAVANANTHA, and D. ROGERS, The crystal structure of $[Ni(\beta\text{-picoline})_2(NO_2)_2]_3 \cdot C_6H_6$: a linear trimer with a novel nitrite bridge, *Chem. Comm.* 1969, 1383–4; see also *J. Chem. Soc. A*, 1971, 159–264.

unsymmetrical bridging mode with a *trans-* configuration of metal atoms is also possible,

i.e. $\begin{array}{c} O \quad\quad M \\ \diagdown \quad \diagup \\ N-O \\ \diagup \\ M \end{array}$ [compared with (iv)].

Nitric acid and nitrates

Nitric acid is one of the three major acids of the modern chemical industry and has been known as a corrosive solvent for metals since alchemical times in the thirteenth century.[70,71] It is now invariably made by the catalytic oxidation of ammonia under conditions which promote the formation of NO rather than the thermodynamically more favoured products N_2 or N_2O (p. 485). The NO is then further oxidized to NO_2 and the gases absorbed in water to yield a concentrated aqueous solution of the acid. The vast scale of production requires the optimization of all the reaction conditions and present-day operations are based on the intricate interaction of fundamental thermodynamics, modern catalyst technology, advanced reactor design, and chemical engineering aspects of process control (see Panel). Production in the USA alone now exceeds 7 million tonnes annually, of which the greater part is used to produce nitrates for fertilizers, explosives, and other purposes (see Panel).

Anhydrous HNO_3 can be obtained by low-pressure distillation of concentrated aqueous nitric acid in the presence of P_4O_{10} or anhydrous H_2SO_4 in an all-glass, grease-free apparatus in the dark. The molecule is planar in the gas phase with the dimensions shown (microwave). The difference in N–O distances, the slight but real tilt of the NO_2 group away from the H atom by 2°, and the absence of free rotation are notable features. The same general structure obtains in the solid state but detailed data are less reliable. Physical properties are shown in Table 11.13. Despite its great thermodynamic stability

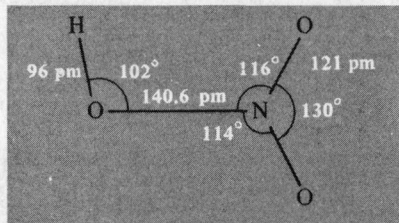

TABLE 11.13   *Some physical properties of anhydrous liquid* HNO_3 *at* 25°C

MP/°C	−41.6	Vapour pressure/mmHg	57
BP/°C	82.6	Density/g cm^{-3}	1.504
ΔH_f°/kJ mol^{-1}	−174.1	η/centipoise	7.46
ΔG_f°/kJ mol^{-1}	−80.8	κ/ohm^{-1} cm^{-1} (20°)	3.72×10^{-2}
S°/J K^{-1} mol^{-1}	155.6	Dielectric constant ε (14°)	50 ± 10

[70] J. W. MELLOR, Nitric acid—history and occurrence, in *A Comprehensive Treatise on Inorganic and Theoretical Chemistry*, Vol. 8, pp. 555–8, Longmans, Green, London, 1928.
[71] T. K. DERRY and T. I. WILLIAMS, *A Short History of Technology from the Earliest Times to AD 1900*, Oxford University Press, Oxford, 1960, 782 pp.

Production and uses of Nitric Acid[26, 72]

Before 1900 the large-scale production of nitric acid was based entirely on the reaction of concentrated sulfuric acid with $NaNO_3$ and KNO_3 (p. 469). The first successful process for making nitric acid directly from N_2 and O_2 was devised in 1903 by E. Birkeland and S. Eyde in Norway and represented the first industrial fixation of nitrogen:

$$\tfrac{1}{2}N_2 + 1\tfrac{1}{4}O_2 + \tfrac{1}{2}H_2O(l) \longrightarrow HNO_3(l); \quad \Delta H^\circ(298\ K)\ -30.3\ kJ\ mol^{-1}$$

The overall reaction is exothermic but required the use of an electric arc furnace which, even with relatively cheap hydroelectricity, made the process very expensive. The severe activation energy barrier, though economically regrettable, is in fact essential to life since, in its absence, all the oxygen in the air would be rapidly consumed and the oceans would be a dilute solution of nitric acid and its salts. [Dilution of $HNO_3(l)$ to $HNO_3(aq)$ evolves a further 33.3 kJ mol^{-1} at 25°C.]

The modern process for manufacturing nitric acid depends on the catalytic oxidation of NH_3 over heated Pt to give NO in preference to other thermodynamically more favoured products (p. 485). The reaction was first systematically studied in 1901 by W. Ostwald (Nobel Prize 1909) and by 1908 a commercial plant near Bochum, Germany, was producing 3 tonnes/day. However, significant expansion in production depended on the economical availability of synthetic ammonia by the Haber–Bosch process (p. 482). The reactions occurring, and the enthalpy changes per mole of N atoms at 25°C are:

$$NH_3 + 1\tfrac{1}{4}O_2 \longrightarrow NO + 1\tfrac{1}{2}H_2O(l); \quad \Delta H^\circ\ -292.5\ kJ\ mol^{-1}$$
$$NO + \tfrac{1}{2}O_2 \longrightarrow NO_2; \quad \Delta H^\circ\ -56.8\ kJ\ mol^{-1}$$
$$NO_2 + \tfrac{1}{3}H_2O(l) \longrightarrow \tfrac{2}{3}HNO_3(l) + \tfrac{1}{3}NO; \quad \Delta H^\circ\ -23.3\ kJ\ mol^{-1}$$

Whence, multiplying the second and third reactions by $\tfrac{3}{2}$ and adding:

$$NH_3(g) + 2O_2(g) \longrightarrow HNO_3(l) + H_2O(l); \quad \Delta H^\circ\ -412.6\ kJ\ mol^{-1}$$

In a typical industrial unit a mixture of air with 10% by volume of NH_3 is passed very rapidly over a series of gauzes (Pt, 10% Rh) at $\sim 850°C$ and 5 atm pressure; contact time with the catalyst is restricted to ~ 1 ms in order to minimize unwanted side reactions. Conversion efficiency is $\sim 96\%$ (one of the most efficient industrial catalytic reactions known) and the effluent gases are passed through an absorption column to yield 60% aqueous nitric acid at about 40°C. Plate 11.2a shows two such burner/boiler units in a plant which produces some 520-tonne-equivalent per day of 60% nitric acid, and Pl. 11.2b shows the layers of Pt/Rh catalyst gauze in another plant. Loss of platinum metal from the catalyst under operating conditions is reduced by alloying with Rh but tends to increase with pressure from about 50–100 mg/tonne of HNO_3 produced at atmospheric pressure to about 250 mg/tonne at 10 atm; though this is not a major part of the cost, the scale of operations means that about 0.5 tonne of Pt metals is lost annually in the UK from this cause and more than twice this amount in the USA.

Concentration by distillation of the 60% aqueous nitric acid produced in most modern ammonia-burning plants is limited by the formation of a maximum-boiling azeotrope (122°) at 68.5% by weight; further concentration to 98% can be effected by countercurrent dehydration using concentrated H_2SO_4, or by distillation from concentrated $Mg(NO_3)_2$ solutions. Alternatively 99% pure HNO_3 can be obtained directly from ammonia oxidation by incorporating a final oxidation of N_2O_4 with the theoretical amounts of air and water at 70°C and 50 atm over a period of 4 h:

$$N_2O_4 + \tfrac{1}{2}O_2 + H_2O \longrightarrow 2HNO_3$$

The largest use of nitric acid is in the manufacture of NH_4NO_3 for fertilizers and this accounts for some 80% of production. Many plants have a capacity of 2000 tonnes/day or more and great care must be taken to produce the NH_4NO_3 in a readily handleable form (e.g. prills of about 3 mm diameter); about 1% of a "conditioner" is usually added to improve storage and handling properties. NH_4NO_3 is thermally unstable (p. 541) and decomposition can become explosive. For this reason a temperature limit of 140°C is imposed on the neutralization step and pH is strictly controlled. The decomposition is catalysed by many inorganic materials including chloride, chromates, hypophosphites, thiosulfates, and powdered metals (e.g. Cu, Zn, Hg). Organic materials (oil, paper, string, sawdust, etc.) must also be rigorously excluded during neutralization

[72] S. STRELZOFF and D. J. NEWMAN, Nitric acid, in *Kirk–Othmer Encyclopedia of Chemical Technology*, 2nd edn., Vol. 13, pp. 796–814, 1967. (See also 3rd edn., Vol. 15, pp. 853–871, 1981.)

since their oxidation releases additional heat. Indeed, since the mid-1950s NH_4NO_3 prills mixed with fuel oil have been extensively used as a direct explosive in mining and quarrying operations (p. 541) and this use now accounts for up to 15% of the NH_4NO_3 produced.

Some 5–10% of HNO_3 goes to make cyclohexanone, the raw material for adipic acid and ε-caprolactam, which are the monomers for nylon-6,6 and nylon-6 respectively. A further 5–10% is used in other organic nitration reactions to give nitroglycerine, nitrocellulose, trinitrotoluene, and numerous other organic intermediates (p. 539). Minor uses (which still consume large quantities of the acid) include the pickling of stainless steel, the etching of metals, and its use as the oxidizer in rocket fuels. In Europe nitric acid is sometimes used to replace sulfuric acid in the treatment of phosphate rock to give nitrophosphate fertilizers according to the idealized equation:

$$Ca_{10}(PO_4)_6F_2 + 14HNO_3 \longrightarrow 3Ca(H_2PO_4)_2 + 7Ca(NO_3)_2 + 2HF$$

Another minority use is in the manufacture of nitrates (other than NH_4NO_3) for use in explosives, propellants, and pyrotechnics generally; typical examples are:

explosives: gun powder, KNO_3/S/powdered C (often reinforced with powdered Si)
white smokes: $ZnO/CaSi_2/KNO_3/C_2Cl_6$
incendiary agents: $Al/NaNO_3$/methylmethacrylate/benzene
local heat sources: $Al/Fe_3O_4/Ba(NO_3)_2$; $Mg/Sr(NO_3)_2/SrC_2O_4$/thiokol polysulfide
photoflashes: $Mg/NaNO_3$
flares (up to 10 min): $Mg/NaNO_3/CaC_2O_4$/polyvinyl chloride/varnish; $Ti/NaNO_3$/boiled linseed oil;
coloured flares: $Mg/Sr(NO_3)_2$/chlorinated rubber (red); $Mg(Ba(NO_3)_2)$/chlorinated rubber (green).

(with respect to the elements) pure HNO_3 can only be obtained in the solid state; in the gas and liquid phases the compound decomposes spontaneously to NO_2 and this occurs more rapidly in daylight (thereby accounting for the brownish colour which develops in the acid on standing):

$$2HNO_3 \rightleftharpoons 2NO_2 + H_2O + \tfrac{1}{2}O_2$$

In addition, the liquid undergoes self-ionic dissociation to a greater extent than any other nominally covalent pure liquid (cf. $BF_3 . 2H_2O$, p. 224); initial autoprotolysis is followed by rapid loss of water which can then react with a further molecule of HNO_3:

$$2HNO_3 \rightleftharpoons H_2NO_3^+ + NO_3^- \rightleftharpoons H_2O + [NO_2]^+ + [NO_3]^-$$
$$HNO_3 + H_2O \rightleftharpoons [H_3O]^+ + [NO_3]^-$$

These equilibria effect a rapid exchange of N atoms between the various species and only a single ^{15}N nmr signal is seen at the weighted average position of HNO_3, $[NO_2]^+$, and $[NO_3]^-$. They also account for the high electrical conductivity of the "pure" (stoichiometric) liquid (Table 11.13), and are an important factor in the chemical reactions of nitric acid and its non-aqueous solutions (p. 539).

The phase diagram HNO_3–H_2O shows the presence of two hydrates, $HNO_3.H_2O$ mp $-37.68°$, and $HNO_3.3H_2O$ mp $-18.47°$. A further hemihydrate, $2HNO_3.H_2O$, can be extracted into benzene or toluene from 6 to 16 M aqueous solutions of nitric acid, and a dimer hydrate, $2HNO_3.3H_2O$, is also known, though neither can be crystallized. The structure of the two crystalline hydrates is dominated by hydrogen bonding as expected; e.g. the monohydrate is $[H_3O]^+[NO_3]^-$ in which the puckered layers shown in Fig. 11.15 comprise pyramidal $[H_3O]^+$ hydrogen bonded to planar $[NO_3]^-$ so that there are 3 H bonds per ion. The trihydrate forms a more complex three-dimensional H-bonded framework. (See also p. 540 for the structure of hydrogen-nitrates.)

PLATE 11.2   (a) Two burner/boiler No. 10 Nitric Acid Plant at ICI Billingham, England.

PLATE 11.2 (b) Pt/Rh catalyst gauzes in the ammonia oxidation plant at Thames Nitrogen, Rainham, England; six gauzes have been rolled back to show the catchment gauze pack beneath them.

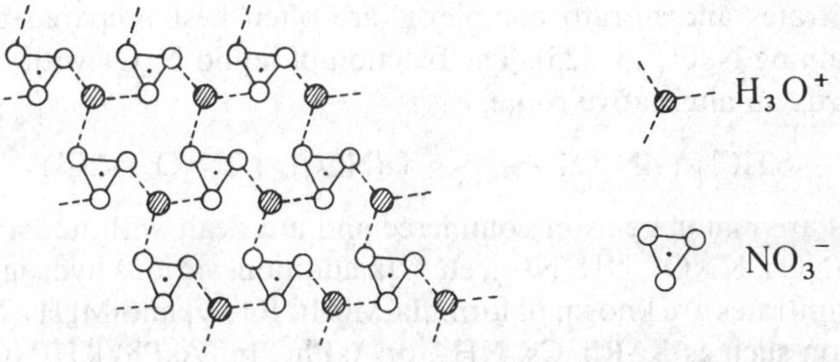

FIG. 11.15   Puckered layer structure of nitric acid monohydrate showing pyramidal H_3O^+ hydrogen bonded to planar NO_3^-.

The solution chemistry of nitric acid is extremely varied. Redox data are summarized in Table 11.4 and Fig. 11.7 (pp. 500–502). In dilute aqueous solutions (<2 M) nitric acid is extensively dissociated into ions and behaves as a typical strong acid in its reactions with metals, oxides, carbonates, etc. More concentrated aqueous solutions are strongly oxidizing and attack most metals except Au, Pt, Rh, and Ir, though some metals which react at lower concentrations are rendered passive, probably because of the formation of an oxide film (e.g. Al, Cr, Fe, Cu). Aqua regia (a mixture of concentrated hydrochloric and nitric acid in the ratio of $\sim 3:1$ by volume) is even more aggressive, due to the formation of free Cl_2 and ClNO and the superior complexing ability of the chloride ion; it has long been known to "dissolve" both gold and the platinum metals, hence its name. In concentrated H_2SO_4 the chemistry of nitric acid is dominated by the presence of the nitrosonium ion (p. 538):

$$HNO_3 + H_2SO_4 \rightleftharpoons NO_2^+ + H_3O^+ + 2HSO_4^- ; \quad K \sim 22 \text{ mol } l^{-1}$$

Such solutions are extensively used in aromatic nitration reactions in the heavy organic chemicals industry.

Anhydrous nitric acid has been studied as a nonaqueous ionizing solvent, though salts tend to be rather insoluble unless they produce NO_2^+ or NO_3^- ions.[73] Addition of water to nitric acid at first diminishes its electrical conductivity by repressing the autoprotolysis reactions mentioned above. For example, at $-10°$ the conductivity decreases from 3.67×10^{-2} ohm^{-1} cm^{-1} to a minimum of 1.08×10^{-2} ohm^{-1} cm^{-1} at 1.75 molal H_2O ($82.8\% \, N_2O_5$) before rising again due to the increasing formation of the hydroxonium ion according to the acid–base equilibrium

$$HNO_3 + H_2O \rightleftharpoons H_3O^+ + NO_3^-$$

By contrast, Raman spectroscopy and conductivity measurements show that N_2O_4 ionizes almost completely in anhydrous HNO_3 to give NO^+ and NO_3^- and such solutions show no evidence for the species N_2O_4, NO_2^+, or NO_2^-.[59] N_2O_5 is also extremely soluble in anhydrous nitric acid in which it is completely ionized as $NO_2^+NO_3^-$.

Nitrates, the salts of nitric acid, can readily be made by appropriate neutralization of the acid, though sometimes it is the hydrate which crystallizes from aqueous solution.

[73] W. H. LEE, Nitric acid, Chap. 4 in J. J. LAGOWSKI (ed.), *The Chemistry of Non-aqueous Solvents*, Vol. 2, pp. 151–89, Academic Press, New York, 1967.

Anhydrous nitrates and nitrato complexes are often best prepared by use of donor solvents containing N_2O_4 (p. 525). The reaction of liquid N_2O_5 with metal oxides and chlorides affords an alternative route, e.g.:

$$TiCl_4 + 4N_2O_5 \longrightarrow Ti(NO_3)_4 + 2N_2O_4 + 2Cl_2$$

Many nitrates are major items of commerce and are dealt with under the appropriate metal (e.g. $NaNO_3$, KNO_3, NH_4NO_3, etc.). In addition, various hydrogen dinitrates and dihydrogen trinitrates are known of formula $M[H(NO_3)_2]$ and $M[H_2(NO_3)_3]$ where M is a large cation such as K, Rb, Cs, NH_4, or $AsPh_4$. In $[AsPh_4][H(NO_3)_2]$ 2 coplanar NO_3^- ions are linked by a short H bond as shown in (a) whereas [*trans*-$RhBr_2(py)_4][H(NO_3)_2]$ features a slightly distorted tetrahedral group of 4 oxygen atoms in which the position of the H atom is not obvious [structure (b)]. In $[NH_4][H_2(NO_3)_3]$ there is a more extended system of H bonds in which 2 coplanar molecules of HNO_3 are symmetrically bridged by an NO_3^- ion at right angles as shown in (c).

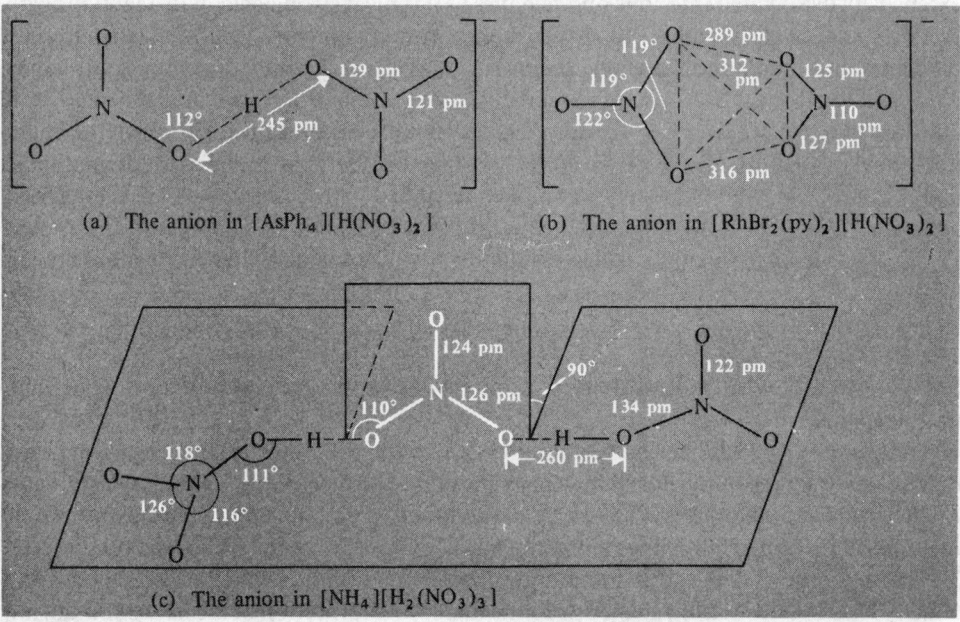

(a) The anion in $[AsPh_4][H(NO_3)_2]$ (b) The anion in $[RhBr_2(py)_2][H(NO_3)_2]$

(c) The anion in $[NH_4][H_2(NO_3)_3]$

As with the salts of other oxoacids, the thermal stability of nitrates varies markedly with the basicity of the metal, and the products of decomposition are equally varied.[74] Thus the nitrates of Group IA and IIA metals find use as molten salt baths because of their thermal stability and low mp (especially as mixtures). Representative values of mp and the temperature (T_d) at which the decomposition pressure of O_2 reaches 1 atm are:

M	Li	Na	K	Rb	Cs	Ag	Tl
MP of MNO_3/°C	255	307	333	310	414	212	206
T_d/°C	474	525	533	548	584	—	—

[74] B. O. FIELD and C. J. HARDY, Inorganic nitrates and nitrato-compounds, *Q. Rev.* **18**, 361–88 (1964).

The product of thermolysis is the nitrite or, if this is unstable at the temperature employed, the oxide (or even the metal if the oxide is also unstable):

$$2NaNO_3 \longrightarrow 2NaNO_2 + O_2 \qquad \text{(see p. 532)}$$
$$2KNO_3 \longrightarrow K_2O + N_2 + \tfrac{5}{2}O_2$$
$$Pb(NO_3)_2 \longrightarrow PbO + 2NO_2 + \tfrac{1}{2}O_2 \quad \text{(see p. 524)}$$
$$2AgNO_3 \longrightarrow Ag + 2NO_2 + O_2$$

As indicated in earlier sections, NH_4NO_3 can be exploded violently at high temperatures or by use of detonators (p. 538), but slow controlled thermolysis yields N_2O (p. 510):

$$2NH_4NO_3 \xrightarrow{>300°} 2N_2 + O_2 + 4H_2O$$

$$NH_4NO_3 \xrightarrow{200-260°} N_2O + 2H_2O$$

The presence of organic matter or other reducible material also markedly affects the thermal stability of nitrates and the use of KNO_3 in gunpowder has been known for centuries (p. 757).

The nitrate group, like the nitrite group, is a versatile ligand and numerous modes of coordination have been found in nitrato complexes.[75] The "uncoordinated" NO_3^- ion (isoelectronic with BF_3, BO_3^{3-}, CO_3^{2-}, etc.) is planar with N–O near 122 pm; this value increases to 126 pm in $AgNO_3$ and 127 pm in $Pb(NO_3)_2$. The most common mode of coordination is the symmetric bidentate mode Fig. 11.16a, though unsymmetrical bidentate coordination (b) also occurs and, in the limit, unidentate coordination (c). Bridging modes include the *syn–syn* conformation (d) (and the *anti–anti* analogue), and also geometries in which a single O atom bridges 2 or even 3 metal atoms (e), (f). Sometimes more than one mode occurs in the same compound.

Symmetrical bidentate coordination (a) has been observed in complexes with 1–6 nitrates coordinated to the central metal, e.g. $[Cu(NO_3)(PPh_3)_2]$; $[Cu(NO_3)_2]$, $[Co(NO_3)_2(OPMe_3)_2]$; $[Co(NO_3)_3]$ in which the 6 coordinating O atoms define an almost regular octahedron (Fig. 11.17a), $[La(NO_3)_3(bipy)_2]$; $[Ti(NO_3)_4]$, $[Mn(NO_3)_4]^{2-}$, $[Fe(NO_3)_4]^-$, and $[Sn(NO_3)_4]$, which feature dodecahedral coordination about the metal; $[Ce(NO_3)_5]^{2-}$ in which the 5 bidentate nitrate groups define a trigonal bipyramid leading to tenfold coordination of cerium (Fig. 11.17b); $[Ce(NO_3)_6]^{2-}$ and $[Th(NO_3)_6]^{2-}$, which feature nearly regular icosahedral (p. 158) coordination of the metal by 12 O atoms; and many lanthanide and uranyl $[UO_2]^{2+}$ complexes. It seems, therefore, that the size of the metal centre is not necessarily a dominant factor.

Unsymmetrical bidentate coordination (Fig. 11.16b) is observed in the high-spin d^7 complex $[Co(NO_3)_4]^{2-}$ (Fig. 11.17c), in $[SnMe_2(NO_3)_2]$, and also in several Cu^{II} complexes of formula $[CuL_2(NO_3)_2]$, where L is MeCN, H_2O, py, or 2-MeC_5H_4N (α-picoline). An example of unidentate coordination is furnished by $K[Au(NO_3)_4]$ as shown in Fig. 11.18, and further examples are in *cis*-$[Pd(NO_3)_2(OSMe_2)_2]$, $[Re(CO)_5(NO_3)]$, $[Ni(NO_3)_2(H_2O)_4]$, $[Zn(NO_3)_2(H_2O)_4]$, and several Cu^{II} complexes such as $[CuL_2(NO_3)_2]$ where L is pyridine *N*-oxide or 1,4-diazacycloheptane. It appears

[75] C. C. ADDISON, N. LOGAN, S. C. WALLWORK, and C. D. GARNER, Structural aspects of coordinated nitrate groups, *Q. Rev.* **25**, 289–322 (1971).

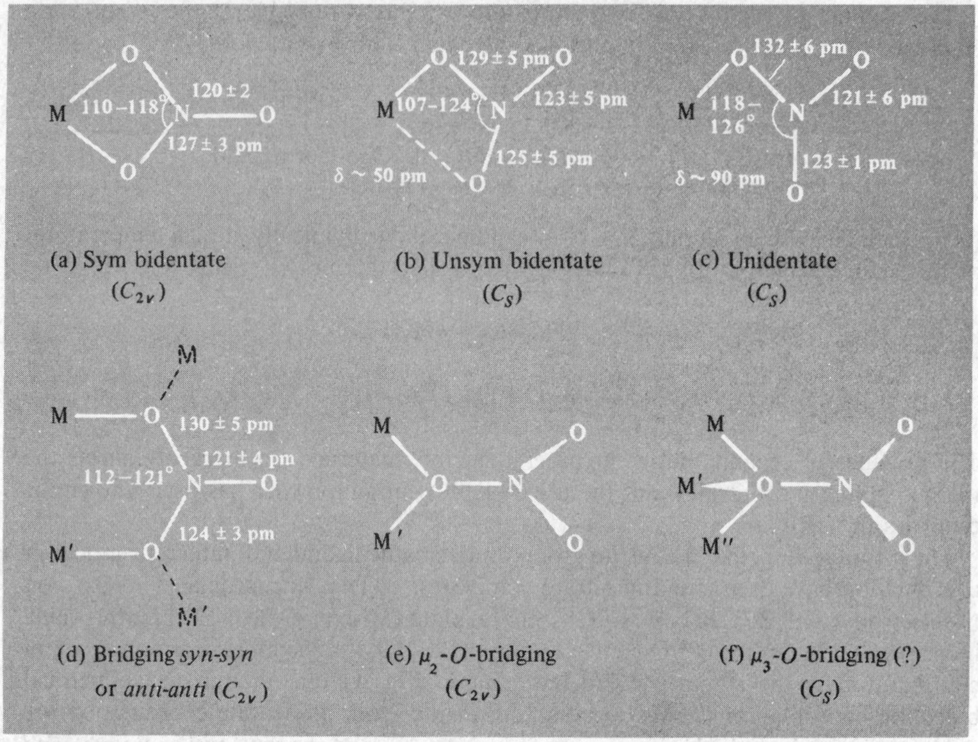

FIG. 11.16 Coordination geometries of the nitrate group showing typical values for the interatomic distances and angles. Further structural details are in ref. 75.

that a combination of steric effects and the limited availability of coordination sites in these already highly coordinated late-transition-metal complexes restricts each nitrate group to one coordination site. When more sites become available, as in $[Ni(NO_3)_2(H_2O)_2]$ and $[Zn(NO_3)_2(H_2O)_2]$, or when the co-ligands are less bulky, as in $[CuL_2(NO_3)_2]$, where L is H_2O, MeCN, or $MeNO_2$, then the nitrate groups become bidentate bridging (mode d in Fig. 11.16) and a further example of this is seen in $[\alpha\text{-}Cu(NO_3)_2]$, which forms a more extensive network of bridging nitrate groups, as shown in Fig. 11.19. The single oxygen atom bridging mode (e) occurs in $[Cu(NO_3)_2(py)_2]_2py$ (Fig. 11.20) and the triple-bridge (f) may occur in $[Cu_4(NO_3)_2(OH)_6]$ though there is some uncertainty about this structure and further refinement would be desirable. Finally, the structure of the unique yellow solvate of formula $[Fe(NO_3)_3.1\frac{1}{2}N_2O_4]$ (p. 526) has been shown[61] to be $[N_4O_6]^{2+}[Fe(NO_3)_4]^-{}_2$. Each anion has 4 symmetrically bidentate NO_3 groups in which the coordinating O atoms lie at the corners of a trigonal dodecahedron, as is commonly found in tetranitrato species ($N-O_t$ 120 pm, $N-O(Fe)$ 127 pm, angles $O-N-O$ 113.4° and $O-Fe-O$ 60.0°). The $[N_4O_6]^{2+}$ cation comprises a central planar nitrate group ($N-O$ 123 pm) surrounded by 3 NO groups at distances which vary from 241 to 278 pm (Fig. 11.21); the interatomic distance in the NO groups is very short (90–99 pm) implying NO^+ and the distances of these to the central NO_3 group are slightly less than the sum of the van der Waals radii for N and O.

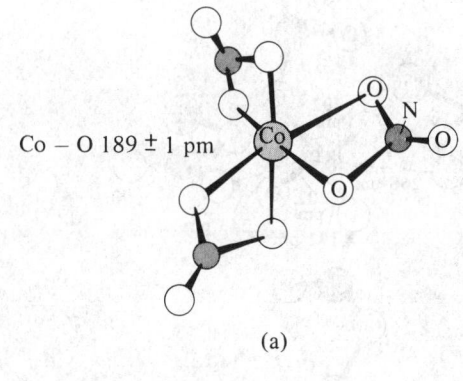

(a)

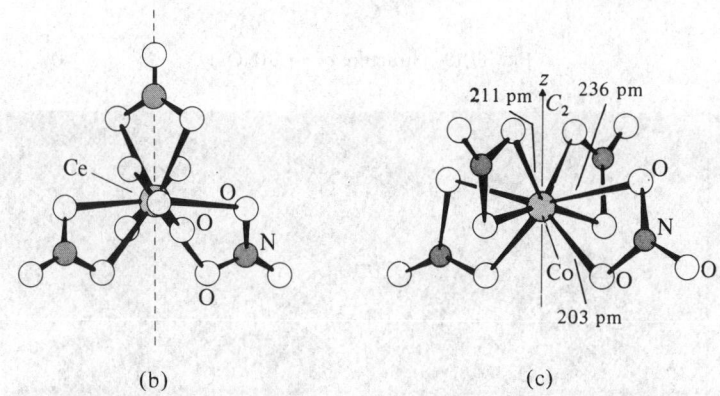

(b)   (c)

FIG. 11.17   Structures of (a) $Co(NO_3)_3$, (b) $[Ce(NO_3)_5]^{2-}$, and (c) $[Co(NO_3)_4]^{2-}$.

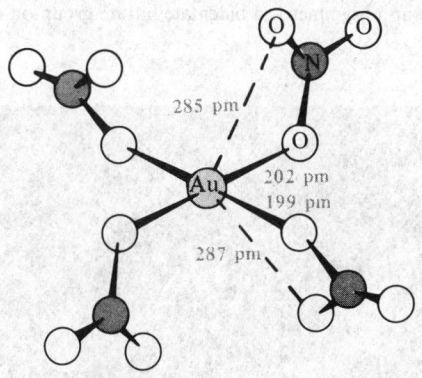

FIG. 11.18   Structure of $[Au(NO_3)_4]^-$.

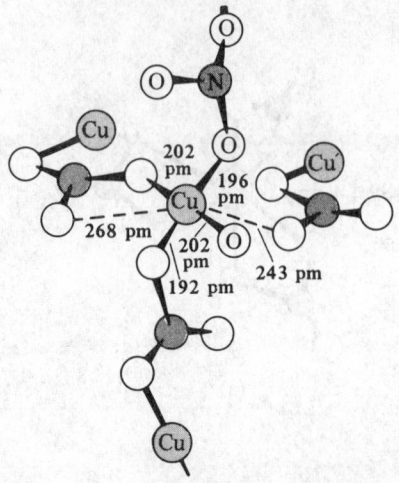

FIG. 11.19   Structure of α-Cu(NO₃)₂.

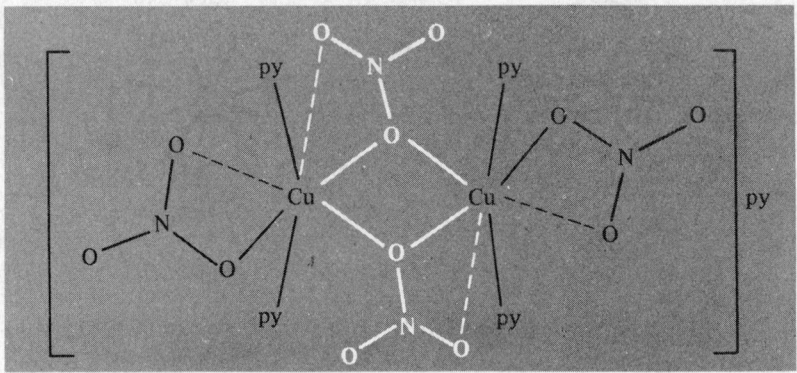

FIG. 11.20   Schematic diagram of the centrosymmetric dimer in [Cu₂(NO₃)₄(py)₄]py showing the two bridging nitrato groups each coordinated to the 2 Cu atoms by a single O atom; the dimer also has an unsymmetrical bidentate nitrate group on each Cu.

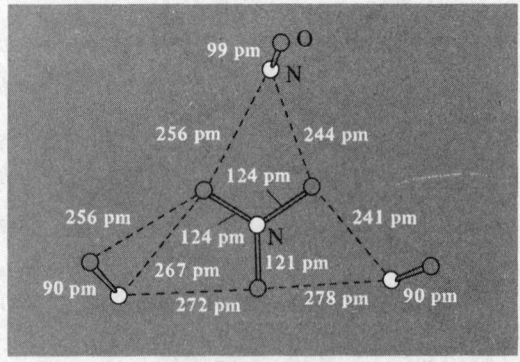

FIG. 11.21   The [N₄O₆]²⁺ cation.

Orthonitrates, $M_3^I NO_4$

There is no free acid H_3NO_4 analogous to orthophosphoric acid H_3PO_4 (p. 595), but the alkali metal orthonitrates Na_3NO_4 and K_3NO_4 have recently been synthesized by direct reaction at elevated temperatures, e.g.:[76]

$$NaNO_3 + Na_2O \xrightarrow[\text{300°C for 7 days}]{\text{Ag crucible}} Na_3NO_4$$

The compound forms white crystals that are very sensitive to atmospheric moisture and CO_2:

$$Na_3NO_4 + H_2O + CO_2 \longrightarrow NaNO_3 + NaOH + NaHCO_3$$

X-ray structural analysis showed that the NO_4^{3-} ion had regular T_d symmetry with the unexpectedly small N–O distance of 139 pm. This suggests that substantial polar interactions are superimposed on the N–O single bonds since the d_π orbitals on N are too high in energy to contribute significantly to multiple covalent bonding. It further implies that d_π–p_π interactions need not necessarily be invoked to explain the observed short interatomic distance in the isoelectronic oxoanions PO_4^{3-}, SO_4^{2-}, and ClO_4^-.

[76] M. JANSEN, Crystal structure of Na_3NO_4, *Angew. Chem.*, Int. Edn. (Engl.) **18**, 698 (1979). For prep. see M. JANSEN, ibid. **16**, 534 (1977).

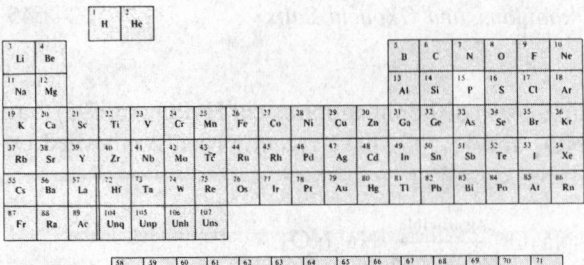

12

Phosphorus

12.1 Introduction

Phosphorus has an extensive and varied chemistry which transcends the traditional boundaries of inorganic chemistry not only because of its propensity to form innumerable covalent "organophosphorus" compounds, but also because of the numerous and crucial roles it plays in the biochemistry of all living things. It was first isolated by the alchemist Hennig Brandt in 1669 by the unsavoury process of allowing urine to putrify for several days before boiling it down to a paste which was then reductively distilled at high temperatures; the vapours were condensed under water to give the element as a white waxy substance that glowed in the dark when exposed to air.[1] Robert Boyle improved the process (1680) and in subsequent years made the oxide and phosphoric acid; he referred to the element as "aerial noctiluca", but the name phosphorus soon became generally accepted (Greek $\phi\omega\varsigma$ *phos*, light; Greek $\phi o \rho o \varsigma$ *phoros*, bringing). As shown in the Panel, phosphorus is probably unique among the elements in being isolated first from animal (human) excreta, then from plants, and only a century later being recognized in a mineral.

In much of its chemistry phosphorus stands in relation to nitrogen as sulfur does to oxygen. For example, whereas N_2 and O_2 are diatomic gases, P and S have many allotropic modifications which reflect the various modes of catenation adopted. Again, the ability of P and S to form multiple bonds to C, N, and O, though it exists, is less highly

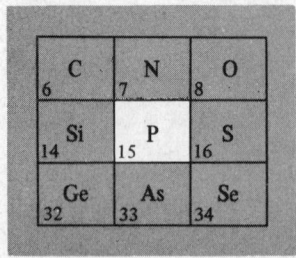

developed than for N (p. 478), whereas the ability to form extended networks of –P–O–P–O– and –S–O–S–O– bonds is greater; this is well illustrated by comparing the

---

[1] M. E. WEEKS, *Discovery of the Elements*, Journal of Chemical Education Publ., Easton, Pa., 1956; Phosphorus, pp. 109–39.

Time Chart for Phosphorus Chemistry

1669 Phosphorus isolated from urine by H. Brandt.

1680 Process improved by R. Boyle who also showed that air was necessary for the phosphorescence of P.

1688 Phosphorus first detected in the vegetable kingdom (by B. Albino).

1694 P_4O_{10} and H_3PO_4 first made by R. Boyle.

1769 Phosphorus shown by J. G. Gahn and C. W. Scheele to be an essential constituent in the bones of man and animals, thereby revealing a plentiful source of fertilizers.

1779 Phosphorus first discovered in a mineral by J. G. Gahn (pyromorphite, a lead phosphate); subsequently found in the much more abundant apatite by T. Bergman and J. L. Proust.

1783 PH_3 first prepared by P. Gengembre (and independently in 1786 by R. Kirwan).

1808 PCl_3 and PCl_5 made by J. L. Gay Lussac and L. J. Thenard (and by H. Davy).

1811 N. L. Vauquelin isolated the first organic P compound (lethicin) from brain fat; it was characterized as a phospholipid by Gobbley in 1850.

1816 P. L. Dulong first clearly demonstrated the existence of two oxides of P.

1820 First synthesis of an organo-P compounds by J. L. Lassaigne who made alkyl phosphites from H_3PO_4 + ROH.

1833 T. Graham, later to become the first President of the Chemical Society, classified phosphates into 3 groups—ortho, pyro, and meta, following J. J. Berzelius's preparation of pyrophosphoric acid by heat.

1834 $(PNCl_2)_n$ made by F. Wöhler and J. von Liebig (originally formulated as $P_3N_2Cl_5$).

1843 J. Murray patented his production of "superphosphate" fertilizer (a name coined by him for the product of H_2SO_4 on phosphate rock).

1844 A. Albright started the manufacture of elemental P in England (for matches); 0.75 tonne in 1844, 26.5 tonnes in 1851.

1845 Polyphosphoric acids made by T. Fleitmann and W. Henneberg.

1848 Red (amorphous) P discovered by A. Schrötter.

1850 First commercial production of "wet process" phosphoric acid.

1868 E. F. Hoppe-Seyler and F. Miescher isolated "nuclein", the first nucleic acid, from puss cells.

1880 Modern cyclic formulation of tetrametaphosphate anion suggested by A. Glatzel (Ring structure of metaphosphate definitely established by L. Pauling and J. Sherman 1937.)

1888 Electrothermal process for manufacturing P introduced by J. B. Readman (Edinburgh).

1898 "Strike-anywhere" matches devised by H. Sévène and E. D. Cahen in France; previously the brothers Lundström had exhibited "safety matches" in 1855, and the first P-containing striking match had been invented by F. Dérosne in 1812.

1929 C. H. Fiske and Y. Subbarow discovered adenosine triphosphate (ATP) in muscle fibre; synthesized some 20 y later by A. Todd *et al.* (Nobel Prize 1957).

1932 Elucidation of the glycolysis process (by G. Embden and by O. Meyerhof) followed by the glucose oxidation process (H. A. Krebs, 1937) established the intimate involvement of P compounds in many biochemical reactions.

1935 Radioactive ^{32}P made by (n,γ) reaction on ^{31}P.

~1940 Highly polymeric phosphate esters (nucleic acids) present in all cells and recognized as essential constituents of chromosomes.

1951 First ^{31}P nmr chemical shifts measured by W. C. Dickinson (for $POCl_3$, PCl_3, and PBr_3 relative to aqueous H_3PO_4).

1952 Detergents (using polyphosphates) overtake soap as main washing agent in the USA. (Heavy duty liquid detergents with polyphosphates introduced in 1955.)

1953 F. H. C. Crick, J. B. Watson, and M. H. F. Wilkins (with Rosalind Franklin) establish the double helix structure of nucleic acids (Nobel Prize 1962).

1960 Concept of "pseudorotation" introduced by R. S. Berry to interpret the stereochemical non-rigidity of trigonal bipyramidal PF_5 (and SF_4, ClF_3); the 5 F atoms are equivalent (1953) due to interconversion via a square pyramidal intermediate.

1961 First 2-coordinate compound of P prepared by A. B. Burg ($Me_3P=PCF_3$). First 1-coordinate P compound (HC≡P) made by T. E. Gier.

1966 First heterocyclic aromatic analogue of pyridine ($Ph_3C_5H_2P$) prepared by G. Märkl, followed by the parent compound C_5H_5P in 1971 (A. J. Ashe).

1979 G. Wittig shared the Chemistry Nobel Prize for his development of the Wittig reaction (first published with G. Geissler in 1953).

oxides and oxoanions of N and P. "Valency expansion" is another point of difference between the elements of the first and second periods of the periodic table for, although compounds in which N has a formal oxidation state of $+5$ are known, no simple "single-bonded" species such as NF_5 or NCl_6^- have been prepared, analogous to PF_5 and PCl_6^-. This finds interpretation in the availability of 3d orbitals for bonding in P (and S) but not for N (or O). The extremely important Wittig reaction for olefin synthesis (p. 635) is another manifestation of this property. Discussion of more extensive group trends in which N and P are compared with the other Group V elements As, Sb, and Bi, is deferred until the next chapter (pp. 642–646).

Because of the great importance of phosphorus and its compounds in the chemical industry, several books and reviews on their preparation and uses are available.[2-6d] Some of these applications reflect the fact that P is a vital element for the growth and development of all plants and animals and is therefore an important constituent in many fertilizers. Phosphorus compounds are involved in energy transfer processes (such as photosynthesis (p. 139), metabolism, nerve function, and muscle action), in heredity (via DNA, p. 611), and in the production of bones and teeth.[7, 7a]

12.2 The Element

12.2.1 *Abundance and distribution*

Phosphorus is the eleventh element in order of abundance in crustal rocks of the earth and it occurs there to the extent of ~ 1120 ppm (cf. H ~ 1520 ppm, Mn ~ 1060 ppm). All its known terrestrial minerals are orthophosphates though the reduced phosphide mineral schreibersite $(Fe,Ni)_3P$ occurs in most iron meteorites. Some 200 crystalline phosphate minerals have been described, but by far the major amount of P occurs in a single mineral family, the apatites, and these are the only ones of industrial importance, the others being

[2] J. EMSLEY and D. HALL, *The Chemistry of Phosphorus*, Harper & Row, London 1976, 534 pp.

[3] A. D. F. TOY, *Phosphorus Compounds in Everyday Living*, Am. Chem. Soc., Washington, DC, 1976, 238 pp.

[4] H. HARNISH, New developments in the field of the preparation and uses of inorganic phosphorus products, *Pure Appl. Chem.* **44**, 439–57 (1975) (in German).

[4a] V. I. KOSYAKOV and I. G. VASIL'EVA, Phosphorus rings, clusters, chains, and layers, *Russian Chem. Rev.* **48**, 153–61 (1979).

[5] D. E. C. CORBRIDGE, *Phosphorus: An Outline of its Chemistry, Biochemistry, and Technology*, Elsevier, Amsterdam, 1978, 464 pp.; 2nd edn., 560 pp., 1980.

[6] A. F. CHILDS, Phosphorus, phosphoric acid, and inorganic phosphates, in *The Modern Inorganic Chemicals Industry*, (R. THOMPSON, ed.), pp. 375–401, The Chemical Society, London, 1977.

[6a] *Proceedings of the First International Congress on Phosphorus Compounds and their Non-fertilizer Applications, 17–21 October 1977 Rabat, Morocco*, IMPHOS (Institut Mondial du Phosphat), Rabat, 1978, 767 pp.

[6b] L. D. QUIN and J. D. VERKADE (eds.), *Phosphorus Chemistry: Proceedings of the 1981 International Conference*, ACS Symposium Series No. 171, 1981, 640 pp.

[6c] E. C. ALYEA and D. W. MEEK (eds.), *Catalytic Aspects of Metal Phosphine Complexes*, ACS Symposium Series No. 196, 1982, 421 pp.

[6d] H. GOLDWHITE, *Introduction to Phosphorus Chemistry*, Cambridge University Press, Cambridge, 1981, 113 pp.

[7] J. R. VAN WAZER (ed.), *Phosphorus and its Compounds*, Vol. 2, *Technology, Biological Functions and Applications*, Interscience, New York, 1961, 2046 pp.

[7a] F. H. PORTUGAL and J. S. COHEN, *A Century of DNA. A History of the Discovery of the Structure and Function of the Genetic Substance*, MIT Press, Littleton, Mass., 1977, 384 pp.

rare curiosities. Apatites (p. 606) have the idealized general formula $3Ca_3(PO_4)_2 . CaX_2$, that is $Ca_{10}(PO_4)_6X_2$, and common members are fluorapatite $Ca_5(PO_4)_3F$, chloroapatite $Ca_5(PO_4)_3Cl$, and hydroxyapatite $Ca_5(PO_4)_3(OH)$. In addition, there are vast deposits of amorphous phosphate rock, phosphorite, which approximates in composition to fluoroapatite.[7] These deposits are widely spread throughout the world as indicated in Table 12.1 and reserves are adequate for several centuries. With present technology, the phosphate content of commercial phosphate rock generally falls in the range $(72 \pm 10)\%$ BPL [i.e. "bone phosphate of lime", $Ca_3(PO_4)_2$] corresponding to $(33 \pm 5)\%$ P_4O_{10} or 12–17% P. The USA is the principal producer, the mines of Florida alone having produced one-third of the total world output in 1975. Morocco is the largest exporter, mainly to the UK and continental Europe which no longer have significant reserves.† World production is a staggering 110 million tonnes of phosphate rock per annum, equivalent to some 15 million tonnes of contained phosphorus (p. 554).

TABLE 12.1 *Estimated reserves of phosphate rock (in gigatonnes of contained P)*[6]

Continent	Main areas	Reserves/10^9 tonnes P
Africa	Morocco, Senegal, Tunisia, Algeria, Sahara, Egypt, Togo, Angola, South Africa	3.5
North America	USA (Florida, Georgia, Carolina, Tennessee, Idaho, Montana, Utah, Wyoming) Mexico	1.4
South America	Peru, Brazil, Chile, Columbia	0.2
Asia/Middle East	USSR (Kola Peninsula, Kazakhstan, Siberia), Jordan, Israel, Saudi Arabia, India, Turkey	0.4
Australasia	Queensland, Nauru, Makatea	0.2
Total		5.7

Phosphorus also occurs in all living things and the phosphate cycle, including the massive use of phosphatic fertilizers, is of great current interest.[8–10, 10a] The movement of phosphorus through the environment differs from that of the other non-metals essential to life (H, C, N, O, and S) because it has no volatile compounds that can circulate via the atmosphere. Instead, it circulates via two rapid biological cycles (weeks and years) superimposed on a much slower primary inorganic cycle (millions of years) as shown schematically in Fig. 12.1.[9]

In the inorganic cycle, phosphates are slowly leached from the igneous or sedimentary rocks by weathering, and transported by rivers to the lakes and seas where they are

† The price of Moroccan phosphate rock which had been stable at $12 per tonne (70–72% BPL) for several years, rose sharply to $40 per tonne in October 1973; by 1975 it had reached $65 per tonne but subsequently fell considerably, thus improving the economics of phosphorus compounds in Europe.

[8] B. H. SVENSSON and R. SÖDERLUND (eds.), *Nitrogen, Phosphorus, and Sulfur–Global Biogeochemical Cycles*, SCOPE Report, No. 7, Sweden 1976, 170 pp.; also SCOPE Report No. 10 on Environmental Issues, Wiley, New York, 1977, 220 pp.

[9] J. EMSLEY, Phosphate Cycles, *Chem. Br.* **13**, 459–63, 1977.

[10] E. J. GRIFFITH, A. BEETON, J. M. SPENCER, and D. T. MITCHELL (eds.), *Environmental Phosphorus Handbook*, Wiley, New York, 1973, 718 pp.

[10a] Ciba Foundation Symposium 57 (New Series), *Phosphorus in the Environment: Its Chemistry and Biochemistry*, Elsevier, Amsterdam, 1978, 320 pp.

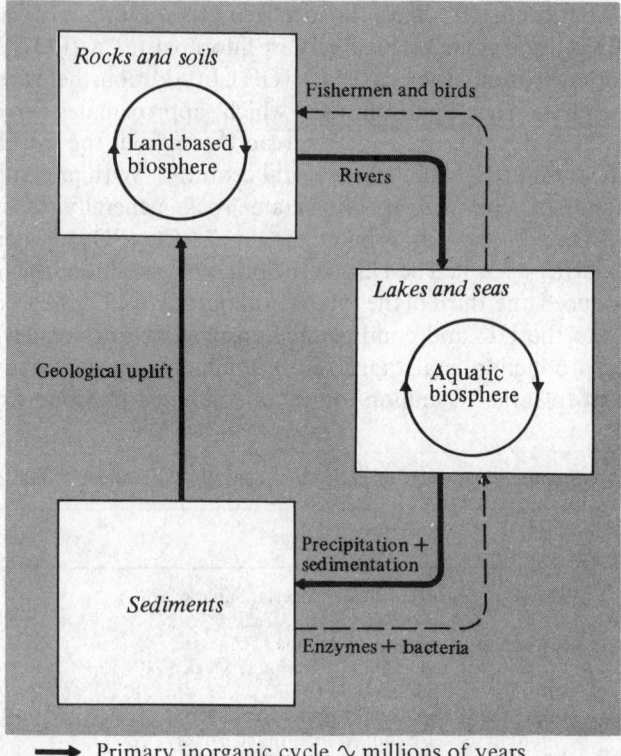

FIG. 12.1　The phosphate cycles.

precipitated as insoluble metal phosphates or incorporated into the aquatic food chain. The solubility of metal phosphates clearly depends on pH, salinity, temperature, etc., but in neutral solution $Ca_3(PO_4)_2$ (solubility product $\sim 10^{-29}$ $mol^5 l^{-5}$) may first precipitate and then gradually transform into the less soluble hydroxyapatite $[Ca_5(PO_4)_3(OH)]$, and, finally, into the least-soluble member, fluoropatite (solubility product $\sim 10^{-60}$ $mol^9 l^{-9}$). Sedimentation follows and eventually, on a geological time scale, uplift to form a new land mass. Some idea of actual concentrations of ions involved may be obtained from the fact that in sea water there is one phosphate group per million water molecules; at a salinity of 3.3%, pH 8, and 20°C, 87% of the inorganic phosphate exist as $[HPO_4]^{2-}$, 12% as $[PO_4]^{3-}$, and 1% as $[H_2PO_4]^-$. Of the $[PO_4]^{3-}$ species, 99.6% is complexed with cations other than Na^+.[11]

The secondary biological cycles stem from the crucial roles that phosphates and particularly organophosphates play in all life processes. Thus organophosphates are incorporated into the backbone structures of DNA and RNA (p. 612) which regulate the reproductive processes of cells, and they are also involved in many metabolic and energy-transfer processes either as adenosine triphosphate (ATP) (p. 611) or other such compounds. Another role, restricted to higher forms of life, is the structural use of calcium

[11] E. T. DEGENS, Molecular mechanisms on carbonate, phosphate, and silica deposition in the living cell, *Topics in Current Chem.* **64**, 1–112 (1976).

phosphates as bones and teeth. Tooth enamel is nearly pure hydroxyapatite and its resistance to dental caries is enhanced by replacement of OH^- by F^- (fluoridation) to give the tougher, less soluble $[Ca_5(PO_4)_3F]$. It is also commonly believed that the main inorganic phases in bone are hydroxyapatite and an amorphous phosphate, though many crystallographers favour an isomorphous solution of hydroxyapatite and the carbonate–apatite mineral dahlite, $[(Na,Ca)_5(PO_4,CO_3)_3(OH)]$, as the main crystalline phase with little or no amorphous material. Young bones also contain brushite, $[CaHPO_4 . 2H_2O]$, and the hydrated octacalcium phosphate $[Ca_8H_2(PO_4)_6 . 5H_2O]$ which is composed, essentially, of alternate layers of apatite and water oriented parallel to (001).[11]

The land-based phosphate cycle is shown in Fig. 12.2.[9] The amount of phosphate in untilled soil is normally quite small and remains fairly stable because it is present as the insoluble salts of Ca^{II}, Fe^{III}, and Al^{III}. To be used by plants, the phosphate must be released as the soluble $[H_2PO_4]^-$ anion, in which form it can be taken up by plant roots. Although acidic soil conditions will facilitate phosphate absorption, phosphorus is the nutrient which is often in shortest supply for the growing plant. Most mined phosphate is thus destined for use in fertilizers and this accounts for up to 75% of phosphate rock in technologically advanced countries and over 90% in less advanced (more agriculturally based) countries. Moderation in all things, however: excessive fertilization of natural waters due to detergents and untreated sewage in run-off water can lead to heavy

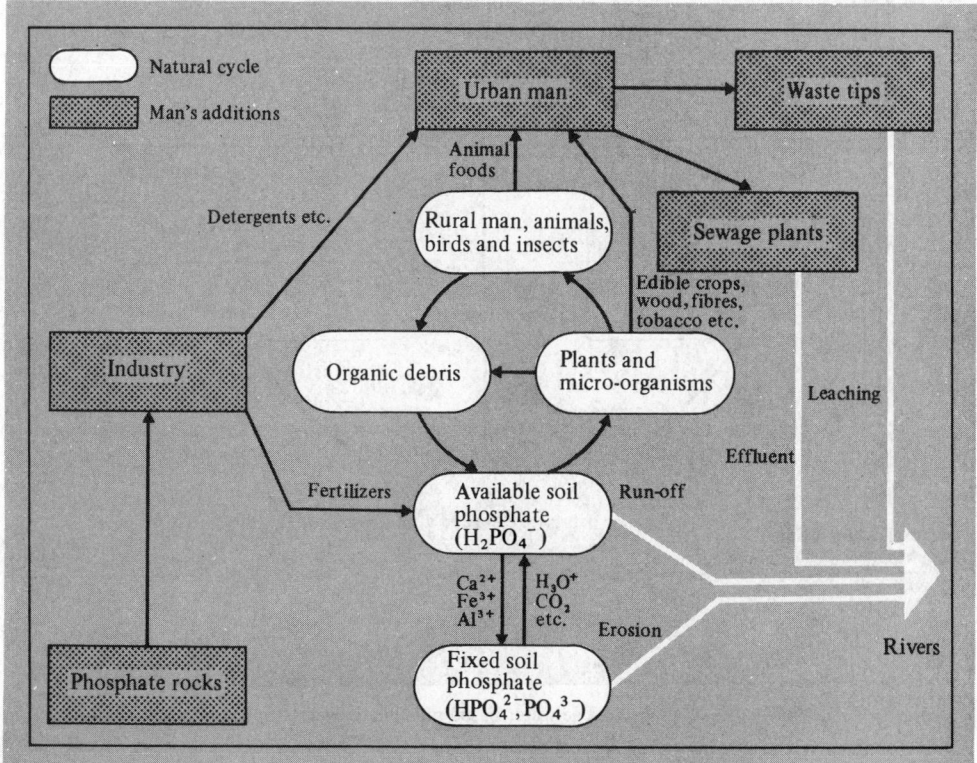

FIG. 12.2   The land-based phosphate cycle.

overgrowth of algae and higher plants, thus starving the water of dissolved oxygen, killing fish and other aquatic life, and preventing the use of lakes for recreation, etc. This unintended over-fertilization and its consequences has been termed eutrophication (Greek εὖ, *eu*, well; τρέφειν, *trephein*, to nourish) and is the subject of active environmental legislation in several countries.[12] Reclamation of eutrophied lakes can best be effected by addition of soluble Al^{III} salts to precipitate the phosphates.

As just implied, the land-based phosphorus cycle is connected to the water-based cycle via the rivers and sewers. It has been estimated that, on a global scale, about 2 million tonnes of phosphate are washed into the seas annually from natural processes and rather more than this amount is dumped from human activities. For example in the UK some 200 000 tonnes of phosphate enters the sewers each year: 100 000 tonnes from detergents, 75 000 from human excreta, and 25 000 tonnes from industrial processes. Details of the subsequent water-based phosphate cycle are shown schematically in Fig. 12.3. The water-based cycle is the most rapid of the three phosphate cycles and can be completed within weeks (or even days). The first members of the food chain are the algae and experiments with radioactive ^{32}P (p. 558) have shown that, within minutes of entering an aquatic

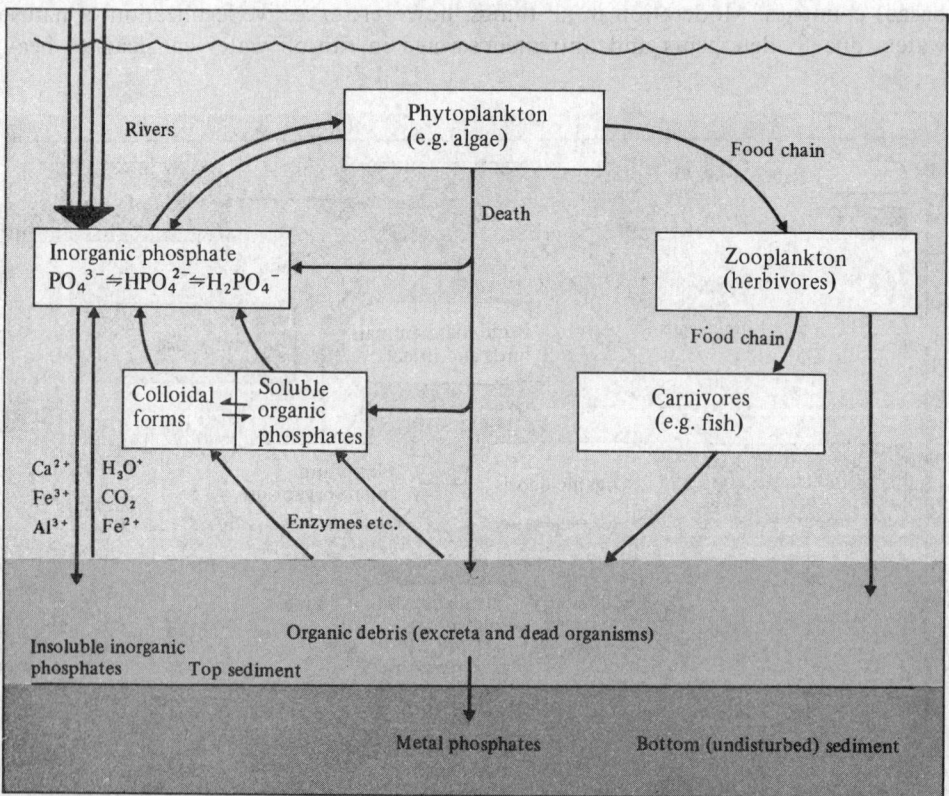

Fig. 12.3 The water-based phosphate cycle.[9]

[12] G. F. LEE, Eutrophication, in *Kirk–Othmer Encyclopedia of Chemical Technology*, 2nd edn., pp. 315–38, Supplement Volume, 1971.

environment, inorganic phosphate is absorbed by algae and bacteria (50% uptake in 1 min, 80% in 3 min). In the seas and oceans the various phosphate anions form insoluble inorganic phosphates which gradually sink to the sea bed. The concentration of phosphate therefore increases with depth (down to about 1000 m, below which it remains fairly constant); by contrast the sunlight, which is necessary for the primary photo-synthesis in the food chain, is greatest at the surface and rapidly diminishes with depth. It is significant that those regions of the sea where the deeper phosphate-rich waters come welling up to the surface support by far the greatest concentration of the world's fish population; such regions, which occur in the mid-Pacific, the Pacific coast of the Americas, Arabia, and Antartica, account for only 0.1% of the sea's surface but support 50% of the world's fish population.

12.2.2 *Production and uses of elemental phosphorus*

For a century after its discovery the only source of phosphorus was urine. The present process of heating phosphate rock with sand and coke was proposed by E. Aubertin and L. Boblique in 1867 and improved by J. B. Readman who introduced the use of an electric furnace. The reactions occurring are still not fully understood, but the overall process can be represented by the idealized equation:

$$2Ca_3(PO_4)_2 + 6SiO_2 + 10C \xrightarrow[1500°]{1400°} 6CaSiO_3 + 10CO + P_4; \quad \Delta H = -3060 \text{ kJ/mol } P_4$$

The presence of silica to form slag which is vital to large-scale production was perceptively introduced by Robert Boyle in his very early experiments. Two apparently acceptable mechanisms have been proposed and it is possible that both may be occurring. In the first, the rock is thought to react with molten silica to form slag and P_4O_{10} which is then reduced by the carbon:

$$2Ca_3(PO_4)_2 + 6SiO_2 \longrightarrow 6CaSiO_3 + P_4O_{10}$$

$$P_4O_{10} + 10C \longrightarrow 10CO + P_4$$

In the second possible mechanism, the rock is considered to be directly reduced by CO and the CaO so formed then reacts with the silica to form slag:

$$2Ca_3(PO_4)_2 + 10CO \longrightarrow 6CaO + 10CO_2 + P_4$$

$$6CaO + 6SiO_2 \longrightarrow 6CaSiO_3$$

$$10CO_2 + 10C \longrightarrow 20CO$$

Whatever the details, the process is clearly energy intensive and, even at 90% efficiency, requires ~15 MWh per tonne of phosphorus (see Panel).

12.2.3 *Allotropes of phosphorus*[13]

Phosphorus (like C and S) exists in many allotropic modifications which reflect the variety of ways of achieving catenation. At least five crystalline polymorphs are known

---

[13] D. E. C. CORBRIDGE, *The Structural Chemistry of Phosphorus*, Elsevier, Amsterdam, 1974, 542 pp.

Production of Phosphorus[6, 7]

A typical modern phosphorus furnace (12 m diameter) can produce some 4 tonnes per hour and is rated at 60–70 MW (i.e. 140 000 A at 500 V). Three electrodes, each weighing 60 tonnes, lead in the current. The amounts of raw material required to make 1 tonne of white phosphorus depend on their purity but are typically 8 tonnes of phosphate rock, 2 tonnes of silica, 1.5 tonnes of coke, and 0.4 tonnes of electrode carbon. The phosphorus vapour is driven off from the top of the furnace together with the CO and some H_2; it is passed through a hot electrostatic precipitator to remove dust and then condensed by water sprays at about 70° (P_4 melts at 44.1°). The byproduct CO is used for supplementary heating.

As most phosphate rock approximates in composition to fluoroapatite, $[Ca_5(PO_4)_3F]$, it contains 3–4 wt% F. This reacts to give the toxic and corrosive gas SiF_4 which must be removed from the effluent. The stoichiometry of phosphate rock might suggest that about 1 mole of SiF_4 is formed for each 3 moles of P_4, but only about 20% of the fluorine reacts in this way, the rest being retained in the slag. Nevertheless, since a typical furnace can produce over 30 000 tonnes of phosphorus per year this represents a substantial waste of a potentially useful byproduct ($\sim$ 5000 tonnes SiF_4 yearly per furnace). In some plants the SiF_4 is recovered by treatment with water and soda ash (Na_2CO_3) to give Na_2SiF_6 which can be used in the fluoridation of drinking water.

Another troublesome impurity in phosphate rock (1–5%) is Fe_2O_3 which is reduced in the furnace to "ferrophosphorus", an impure form of Fe_2P. This is a dense liquid at the reaction temperature; it sinks beneath the slag and can be drained away at intervals. As every tonne of ferrophosphorus contains $\sim$0.25 tonne of P, this is a major loss, but is unavoidable since the Fe_2O_3 cannot economically be removed beforehand. The few uses of ferrophosphorus depend on its high density ($\sim$6.6 g cm^{-3}). It can be mixed with dynamite for blasting or used as a filler in high-density concrete and in radiation shields for nuclear reactors. It is also used in the manufacture of special steels and cast-irons, especially for non-sparking railway brake-shoes. The other substantial byproduct, $CaSiO_3$ slag, has little economic use and is sold as hard core for road-fill or concrete aggregate; about 7–9 tonnes are formed per tonne of P produced.

World capacity for the production of elemental P is $\sim$1.2 million tonnes per year of which about half is in the USA. Some figures for 1976 are as follows:

Country	USA	USSR	Netherlands	Canada	Germany (FRG)	France
ktonne/y	528	250	90	85	80	39

Country	China	Japan	Germany (DDR)	India	South Africa
ktonne/y	35	20	15	6	6

About 80–90% of the elemental P produced is reoxidized to (pure) phosphoric acid (p. 599). The rest is used to make phosphorus sulfides (p. 585), phosphorus chlorides and oxochloride (p. 569), and organic P compounds. Bulk price (USA 1981) was ca. $1.50–1.75 per kg.

and there are also several "amorphous" or vitreous forms (see Fig. 12.4). All forms, however, melt to give the same liquid which consists of symmetrical P_4 tetrahedral molecules, P–P 225 pm. The same molecular form exists in the gas phase (P–P 221 pm), but at high temperatures (above $\sim$ 800°C) and low pressures P_4 is in equilibrium with the diatomic form P≡P (189.5 pm). At atmospheric pressure, dissociation of P_4 into $2P_2$ reaches 50% at $\sim$ 1800°C and dissociation of P_2 into 2P reaches 50% at $\sim$ 2800°.

The commonest form of phosphorus, and the one which is usually formed by condensation from the gaseous or liquid states, is the waxy, cubic, white form α-P_4 (d 1.8232 g cm^{-3} at 20°C). This, paradoxically, is also the most volatile and reactive solid form and thermodynamically the least stable. It is the slow phosphorescent oxidation of

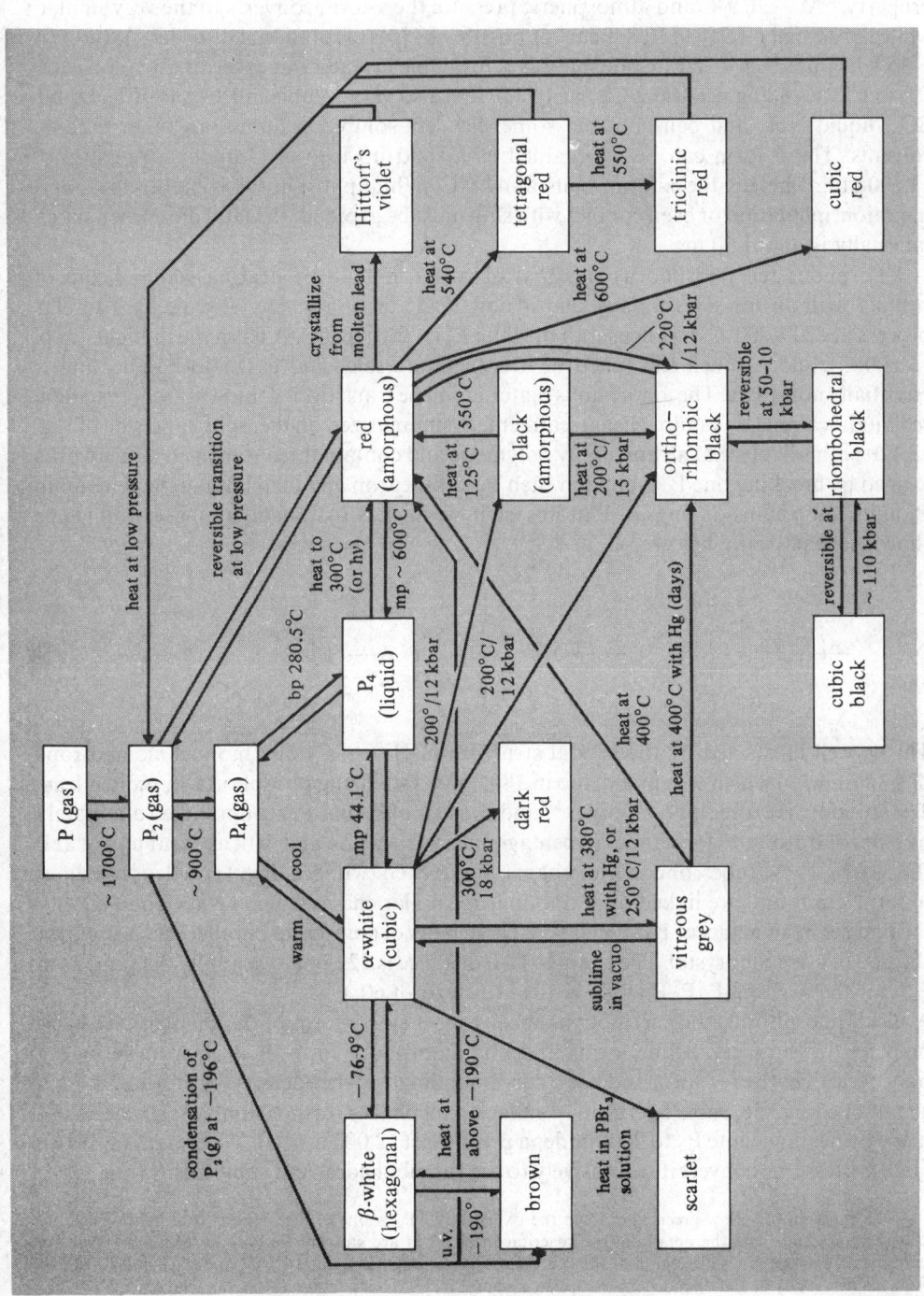

Fig. 12.4   Interconversion of the various forms of elemental phosphorus (1 kbar = 10^8 Pa = 987.2 atm).

the vapour above these crystals that gives white phosphorus its most characteristic property.† At $-76.9°C$ and atmospheric pressure the α-form converts to the very similar white hexagonal β-form (d 1.88 g cm^{-3}), possibly by loss of rotational disorder; $\Delta H(\alpha \rightarrow \beta)$ -15.9 kJ (mol P$_4$)$^{-1}$. White phosphorus is insoluble in water but exceedingly soluble (as P$_4$) in CS$_2$ ($\sim$880 g per 100 g CS$_2$ at 10°C). It is also very soluble in PCl$_3$, POCl$_3$, liquid SO$_2$, liquid NH$_3$ and benzene, and somewhat less soluble in numerous other organic solvents. The β form can be maintained as a solid up to 64.4°C under a pressure of 11 600 atm, whereas the α-form melts at 44.1°C. White phosphorus is highly toxic and ingestion, inhalation or even contact with skin must be avoided; the fatal dose when taken internally is about 50 mg.

Amorphous red phosphorus was first obtained in 1848 by heating white P$_4$ out of contact with air for several days, and is now made on a commercial scale by a similar process at 270°–300°C. It is denser than white P$_4$ ($\sim$2.16 g cm^{-3}), has a much higher m.p. ($\sim$600°C), and is much less reactive; it is therefore safer and easier to handle, and is essentially non-toxic. The amorphous material can be transformed into various crystalline red modifications by suitable heat treatment, as summarized on the right hand side of Fig. 12.4. It seems likely that all are highly polymeric and contain three-dimensional networks formed by breaking one P–P bond in each P$_4$ tetrahedron and then linking the remaining P$_4$ units into chains or rings of P atoms each of which is pyramidal and 3 coordinate as shown schematically below:

This is well illustrated by the crystal structure of Hittorf's violet monoclinic allotrope (d 2.35 g cm^{-3}) which was first made in 1865 by crystallizing phosphorus in molten lead. The structure is exceedingly complex[14] and consists of P$_8$ and P$_9$ groups linked alternately by pairs of P atoms to form tubes of pentagonal cross-section and with a repeat unit of 21 P (Fig. 12.5). These tubes, or complex chains, are stacked (without direct covalent bonding) to form sheets and are linked by P–P bonds to similar chains which lie at right angles to the first set in an adjacent parallel layer. These pairs of composite parallel sheets are then stacked to form the crystal. The average P–P distance is 222 pm (essentially the same as in P$_4$) but the average P–P–P angle is 101° (instead of 60°).

Black phosphorus, the thermodynamically most stable form of the element, has been prepared in three crystalline forms and one amorphous form. It is even more highly polymeric than the red form and has a correspondingly higher density (orthorhombic 2.69, rhombohedral 3.56, cubic 3.88 g cm^{-3}). Black phosphorus (orthorhombic) was originally made by heating white P$_4$ to 200° under a pressure of 12 000 atm (P. W. Bridgman, 1916). Higher pressures convert it successively to the rhombohedral and cubic forms (Fig. 12.4).

† The emission of yellow-green light from the oxidation of P$_4$ is one of the earliest recorded examples of chemiluminescence, but the details of the reaction mechanism are still not fully understood. The primary emitting species in the visible region of the spectrum are probably (PO)$_2$ and HPO; and ultraviolet emission from excited states of PO also occurs (R. J. van Zee and A. U. Khan, *J. Am. Chem. Soc.* **96**, 6805–6 (1974).

[14] VON H. THURN and H. KREBS, The crystal structure of Hittorf's phosphorus, *Acta Cryst.* **B25**, 125–35 (1969) (in German).

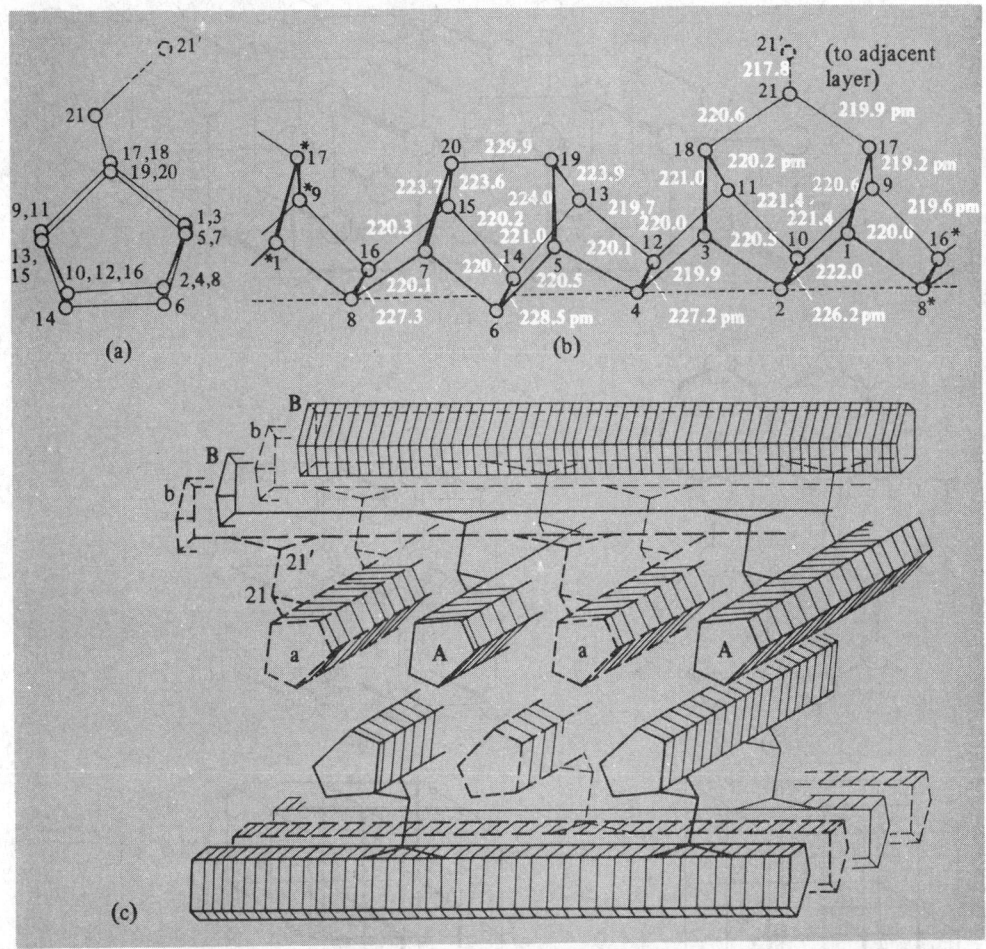

FIG. 12.5   Structure of Hittorf's violet monoclinic phosphorus showing (a) end view of one pentagonal tube, (b) the side view of a single tube (dimensions in pm), and (c) schematic representation of the stacking of the tubes in the crystal structure.

Orthorhombic black P (mp ∼610°) has a layer structure which is based on a puckered hexagonal net of 3-coordinate P atoms with 2 interatomic angles of 102° and 1 of 96.5° (P–P 223 pm). The relation of this form to the rhombohedral and cubic forms is shown in Fig. 12.6. Comparison with the rhombohedral forms of As, Sb, and Bi is also instructive in showing the increasing tendency towards octahedral coordination and metallic properties (p. 643). Black P is semiconducting but its electrical properties are probably significantly affected by impurities introduced during its preparation.

12.2.4   *Atomic and physical properties*[15]

Phosphorus has only 1 stable isotope, $^{31}_{15}P$, and accordingly (p. 20) its atomic weight is

[15] *Mellor's Comprehensive Treatise on Inorganic and Theoretical Chemistry*, Vol. 8, Suppl. 3, *Phosphorus*, Longman, London, 1971, 1467 pp.

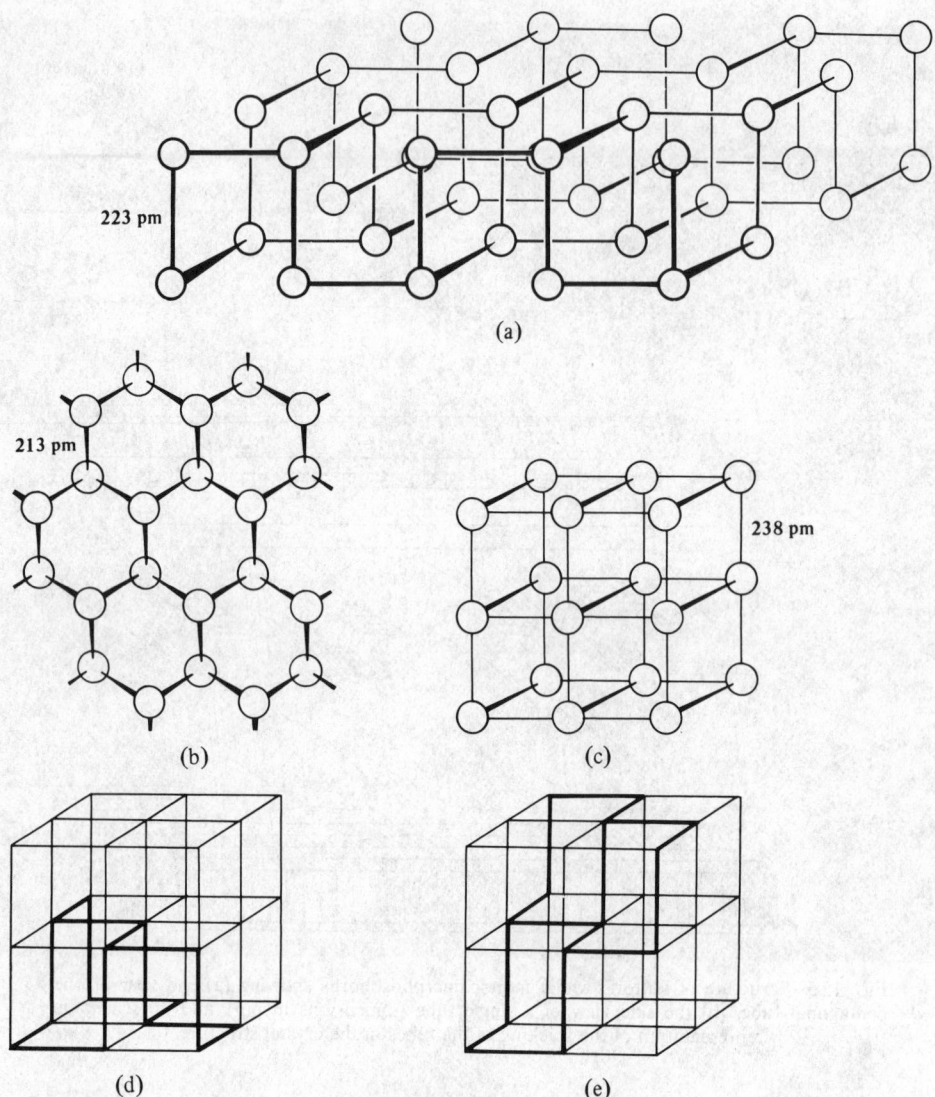

FIG. 12.6 The structures of black phosphorus: (a) portion of one layer of orthorhombic P
(idealized), (b) rhombohedral form, portion of one hexagonal layer, (c) cubic form, 4 unit cells, (d)
distortion of (a) to the cubic form, and (e) distortion of (b) to the cubic form.

known with extreme accuracy, 30.973 76. Six radioactive isotopes are known, of which ^{32}P
is by far the most important; it is made on the multikilogram scale by the neutron
irradiation of ^{32}S(n,p) or ^{31}P(n,γ) in a nuclear reactor, and is a pure β-emitter of half life
14.28 days, E_{max} 1.709 MeV, E_{mean} 0.69 MeV. It finds extensive use in tracer and
mechanistic studies. The stable isotope ^{31}P has a nuclear spin quantum number of $\frac{1}{2}$ and
this is much used in nmr spectroscopy.[16] Chemical shifts and coupling constants can
both be used diagnostically to determine structural information.

[16] P. S. PREGOSIN and L. M. VENANZI, Multinuclear nmr and the preparative coordination chemist, *Chem. in
Br.* **14**, 276–81 (1978).

In the ground state, P has the electronic configuration $[Ne]3s^2 3p_x^1 3p_y^1 3p_z^1$ with 3 unpaired electrons; this, together with the availability of low-lying vacant 3d orbitals, accounts for the predominant oxidation states III and V in phosphorus chemistry. Ionization energies, electronegativity, and atomic radii are compared with those of N, As, Sb, and Bi on p. 642. White phosphorus (α-P_4) has mp 44.1° (or 44.25° when ultrapure), bp 280.5°, and a vapour pressure of 0.122 mmHg at 40°C. It is an insulator with an electrical resistivity of $\sim 10^{11}$ ohm cm at 11°C, a dielectric constant of 4.1 (at 20°), and a refractive index $n_D(29.2°)$ 1.8244. The heat of combustion of P_4 to P_4O_{10} is -2971 kJ mol^{-1} and the heat of transition to amorphous red phosphorus is -29 kJ (mol $P_4)^{-1}$.

12.2.5 *Chemical reactivity*

The spontaneous chemiluminescent reaction of white phosphorus with moist air was the first property of the element to be observed and was the origin of its name (p. 546); its spontaneous ignition temperature in air is $\sim 35°$. We have already seen (p. 556) that the reactivity of phosphorus depends markedly upon which allotrope is being studied and that increasing catenation of the polymeric red and black forms notably diminishes both reactivity and solubility. The preference of phosphorus for these forms rather than for the gaseous form P_2, which is its most obvious distinction from nitrogen, can be rationalized in terms of the relative strengths of the triple and single bonds for the 2 elements. Reliable values are hard to obtain but generally accepted values are as follows:

E (N≡N)/kJ per mol of N	946		E (P≡P)/kJ per mol of P	490
E(> N—N <)/kJ per mol of N	159 (or 296)		E(> P—P <)/kJ per mol of P	200
Ratio	5.95		Ratio	2.45
	(or 3.20)			

It is clear that, for nitrogen, the triple bond is preferred since it has more than 3 times the energy of a single bond, whereas for phosphorus the triple-bond energy is less than 3 times the single-bond energy and so allotropes having 3 single bonds per P atom are more stable than that with a triple bond.

Phosphorus forms binary compounds with all elements except Sb, Bi, and the inert gases. It reacts spontaneously with O_2 and the halogens at room temperature, the mixtures rapidly reaching incandescence. Sulfur and the alkali metals also react vigorously with phosphorus on warming, and the element combines directly with all metals (except Bi, Hg, Pb) frequently with incandescence (e.g. Fe, Ni, Cu, Pt). White phosphorus (but not red) also reacts readily with heated aqueous solutions to give a variety of products (pp. 564 and 589 ff), and with many other aqueous and nonaqueous reagents.

The stereochemistry and bonding of P are very varied as will become apparent in later sections: the element is known in at least 14 coordination geometries with CN up to 9, though the most frequently met have CN 3, 4, 5, and 6. Some typical coordination geometries are summarized in Table 12.2 and illustrated in Fig. 12.7. Many of these compounds will be more fully discussed in later sections.

TABLE 12.2 *Stereochemistry of phosphorus*

CN	Geometry	Examples
0	—	P(g)—in equilibrium with $P_2(g)$ above $2200°C$
1	—	$P_2(g)$—in equilibrium with $P_4(g)$ above $800°C$; $HC\equiv P$; $FC\equiv P$; $MeC\equiv P$ (p. 633)
2	Bent[16a]	$HP{=}CH_2$,[17b] $[P(CN)_2]^-$,[17c] $[\{C_6H_4S(NR)C\}_2P]^+X^-$ (p. 635), $cyclo\text{-}C_5H_5P$, $2,4,6\text{-}Ph_3C_5H_2P$; $Me_3P{=}PCF_3$; $P_7{}^{3-}$ anion[17] (isoelectronic with P_4S_3) in Sr_3P_{14}; $P_{11}{}^{3-}$ anion in Na_3P_{11}; diazaphospholes[17d]
3	Planar	$[PhP\{Mn(\eta^5\text{-}C_5H_5)(CO)_2\}_2]$[17a]
	Pyramidal	P_4, PH_3, PX_3, P_4O_6, $[PhP\{Co(CO)_4\}_2]$[18]
4	Tetrahedral	$PH_4{}^+$, Cl_3PO, P_4O_{10}, $PO_4{}^{3-}$, polyphosphates, MP (zinc blende type, $M = B$, Al, Ga, In), $[Co_3(CO)_9(\mu_3\text{-}PPh)]$,[19] $[(P_4)Ni\{(Ph_2PCH_2CH_2)_3N\}]$;[17e] many complexes of PR_3, etc., with metal centres
	Local C_{2v}	$PBr_4{}^-$, $[PBr_2(CN)_2]^-$.[19a] $[\mu(\eta^3\text{-}P_3)\{Ni(triphos)\}_2]^{2+}$[20]
5	Trigonal bipyramidal	PF_5, PPh_5
	Square pyramidal	$[Co_4(CO)_8(\mu\text{-}CO)_2(\mu_4\text{-}PPh)_2]$, $[Os_5(CO)_{15}(\mu_4\text{-}POMe)]$[21]
6	Octahedral	$PF_6{}^-$, $PCl_6{}^-$, MP (NaCl-type, $M = $ La, Sm, Th, U, etc.)
	Trigonal primsatic	Rh_4P_3, Hf_3P_2 (also contains seven- and eight-fold coordination of P by M)
	Irregular $(4+2)$	$[Co_6(CO)_{14}(\mu\text{-}CO)_2P]^-$
7	Bicapped trigonal prismatic	Ta_2P, Hf_2P (contains P in seven-, eight-, and nine-fold coordination by M)
8	Cubic	M_2P (antifluorite type (p. 129), $M = $ Ir, Rh)
9	Tricapped trigonal prismatic	M_3P ($M = $ Ti, V, Cr, Mn, Fe, Ni, Zr, Nb, Ta) M_2P ($PbCl_2$-type, $M = $ Fe, Co, Ru)
	Monocapped square antiprismatic	$[Rh_9(CO)_{21}P]^{2-}$[22]

[16a] E. FLACK, Compounds of phosphorus with coordination number 2, *Topics in Phosphorus Chemistry* **10**, 193–284 (1980).

[17] W. DAHLMANN and H. G. VON SCHNERING, Sr_3P_{14}. A phosphide with isolated $P_7{}^{3-}$ groups, *Naturwissenschaften* **59**, 420 (1972). See also W. WICHELHAUS and H. G. VON SCHNERING, Na_3P_{11}. A phosphide with isolated $P_{11}{}^{3-}$ groups, ibid. **60**, 104 (1973).

[17a] G. HUTTNER, H.-D. MÜLLER, A. FRANK, and H. LORENZ, $[PhP\{Mn(C_5H_5)(CO)_2\}_2]$, a phosphinidene complex with trigonal planar coordinated phosphorus, *Angew. Chem.*, Int. Edn. (Engl.) **14**, 705–6 (1975).

[17b] H. W. KROTO, J. F. NIXON, K. OHNO, and N. P. C. SIMMONS, The microwave spectrum of phosphaethene, $HP{=}CH_2'$, *JCS Chem. Comm.* 1980, 709.

[17c] W. S. SHELDRICK, J. KRONER, F. ZWASCHKA, and A. SCHMIDPETER, Structure of the dicyanophosphide ion in a crown-ether sodium salt, *Angew. Chem.*, Int. Edn. (Engl.) **18**, 934–5 (1979).

[17d] J. H. WEINMAIER, A. SCHMIDPETER, *et al.*, Diazaphospholes, *Angew. Chem.*, Int. Edn. (Engl.) **18**, 412 (1979); *Chem. Ber.* **113**, 2278–90 (1980); *J. Organometallic Chem.* **185**, 53–68 (1980).

[17e] P. DAPPORTO, S. MIDOLLINI, and L. SACCONI, *Tetrahedro*-tetraphosphorus as a monodentate ligand in a nickel(0) complex, *Angew. Chem.*, Int. Edn. (Engl.) **18**, 469 (1979).

[18] J. C. BURT and G. SCHMID, Synthesis of μ-phenylphosphine-diyl-bis(tetracarbonylcobalt) and 1,1,1,2,2,2,3,3,3-nona-carbonyl-μ_3-phenylphosphinediyl-*triangulo*-dicobaltiron, *JCS Dalton* 1978, 1385–7.

[19] L. MARKÓ and B. MARKÓ, Paramagnetic phosphido cobalt carbonyl clusters, *Inorg. Chim. Acta* **14**, L39 (1975).

[19a] W. S. SHELDRICK, A. SCHMIDPETER, F. ZWASHKA, K. B. DILLON, A. W. G. PLATT, and T. C. WADDINGTON, The structures of hypervalent phosphorus(III) anions $P(CN)_{4-n}Br_n{}^-$. Transition from ψ-trigonal-bipyramidal to ψ-octahedral coordination and deviation from valence shell electron pair repulsion theory, *JCS Dalton* 1981, 413–18; see also *Angew. Chem.*, Int. Edn. (Engl.) **18**, 935–6 (1979).

[20] M. DI VAIRA, S. MIDOLLINI, and L. SACCONI, *cyclo*-Triphosphorus and *cyclo*-triarsenic as ligands in "double sandwich" complexes of cobalt and nickel, *J. Am. Chem. Soc.* **101**, 1757–63 (1979). For analogous complexes in which $\mu\text{-}(\eta^3\text{-}P_3)$ bridges RhCo, RhNi, IrCo, and RhRh, see C. BIANCHINI, M. DI VAIRA, A. MELI, and L. SACCONI, *Angew. Chem.*, Int. Edn. (Engl.) **19**, 405–6 (1980).

[21] J. M. FERNANDEZ, B. F. G. JOHNSON, J. LEWIS, and P. R. RAITHBY, Synthesis and molecular structure of $[Os_5(CO)_{15}(POMe)]$, *JCS Chem. Comm.* 1978, 1015–16.

[22] J. L. VIDAL, W. E. WALKER, R. L. PRUETT, and R. C. SCHOENING, $[Rh_9P(CO)_{21}]^{2-}$. Example of encapsulation of phosphorus by transition-metal-carbonyl clusters, *Inorg. Chem.* **18**, 129–36 (1979).

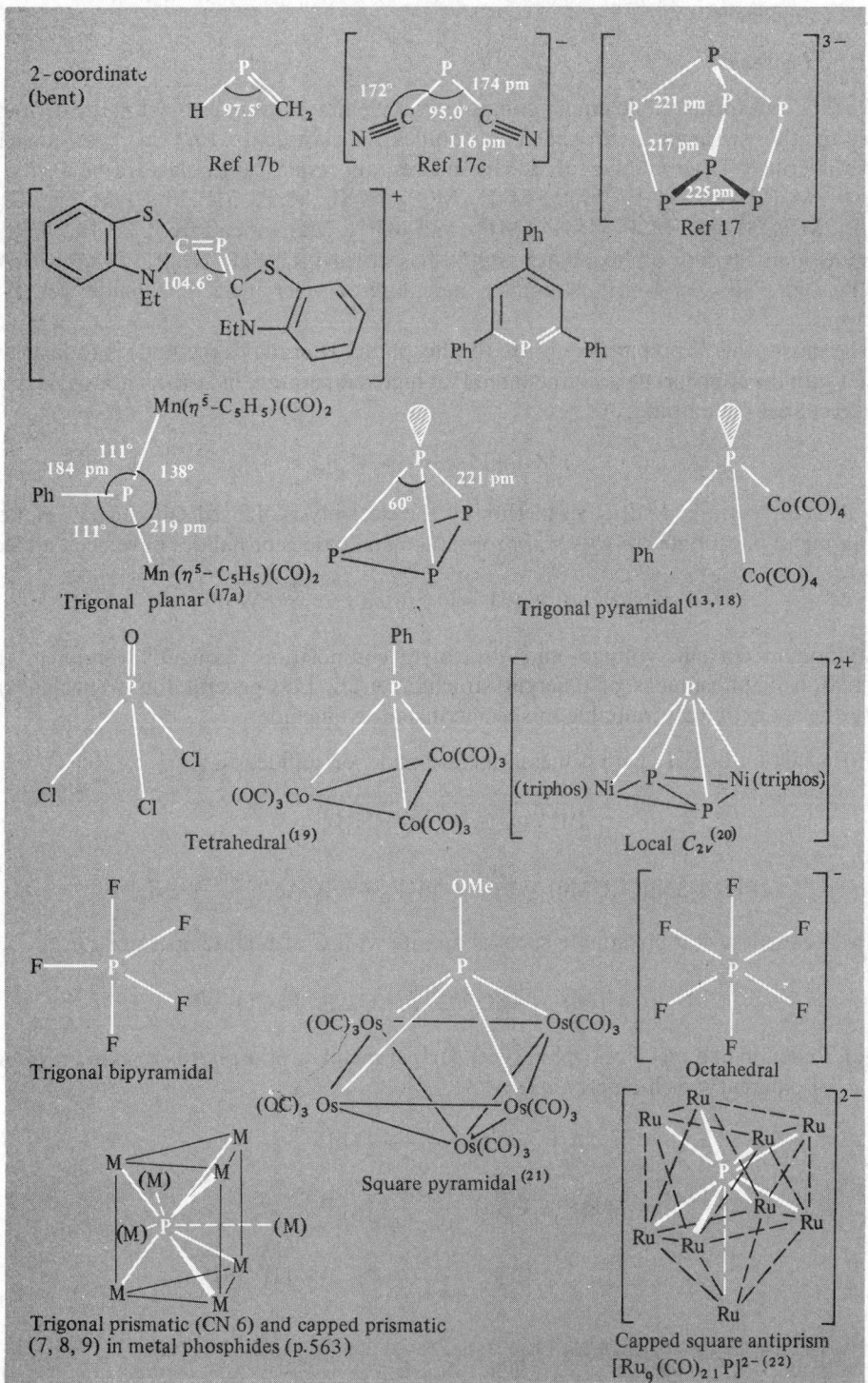

FIG. 12.7   Schematic representation of some of the coordination geometries of phosphorus.

12.3 Compounds

12.3.1 *Phosphides*[23-25]

Phosphorus forms stable binary compounds with almost every element in the periodic table and those with metals are called phosphides. Like borides (p. 162) they are known in a bewilderingly large number of stoichiometries, and typical formulae are M_4P, M_3P, $M_{12}P_5$, M_7P_3, M_2P, M_7P_4, M_5P_3, M_3P_2, M_4P_3, M_5P_4, M_6P_5, MP, M_3P_4, M_2P_3, MP_2, M_2P_5, MP_3, M_3P_{11}, M_3P_{14}, MP_5, MP_7, and MP_{15}. Many metals (e.g. Ti, Ta, W, Rh) form as many as 5 or 6 phosphides and Ni has at least 8 (Ni_3P, Ni_5P_2, $Ni_{12}P_5$, NiP_2, Ni_5P_4, NiP, NiP_2, and NiP_3). Ternary and more complex metal phosphides are also known.

The most general preparative route to phosphides (Faraday's method) is to heat the metal with the appropriate amount of red P at high temperature in an inert atmosphere or an evacuated sealed tube:

$$nM + mP \xrightarrow{\text{heat}} M_nP_m$$

An alternative route (Andrieux's method) is the electrolysis of fused salts such as molten alkali-metal phosphates to which appropriate metal oxides or halides have been added:

$$\{(NaPO_3)_n/NaCl/WO_3\}\text{fused} \xrightarrow{\text{electrol}} W_3P$$

Variation in current, voltage, and electrolyte composition frequently results in the formation of phosphides of different stoichiometries. Less-general routes (which are nevertheless extremely valuable in specific instances) include:

(a) Reaction of PH_3 with a metal, metal halide, or sulfide, e.g.:

$$PH_3 + 2Ti \xrightarrow{800°} Ti_2P$$

$$2PH_3 + 3Ni(O_2CMe)_2 \xrightarrow[\text{}]{H_2O} Ni_3P_2 + 6HOAc \xrightarrow[\text{reaction}]{\text{further}} Ni_5P_2$$

(b) Reduction of a phosphate such as apatite with C at high temperature, e.g.:

$$Ca_3(PO_4)_2 + 8C \xrightarrow{1200°} Ca_3P_2 + 8CO$$

(c) Reaction of a metal phosphide with further metal or phosphorus to give a product of different stoichiometry, e.g.:

$$Th_3P_4 + Th \xrightarrow{900°} 4ThP$$

$$4RuP + P_4(g) \xrightarrow{650°} 4RuP_2$$

$$4IrP_2 \xrightarrow[\text{(low press)}]{1150°} 2Ir_2P + 1\tfrac{1}{2}P_4(g)$$

[23] A. WILSON, The metal phosphides, Chap. 3 (pp. 289–363) in ref. 15; see also p. 256.
[24] A. D. F. TOY, "Phosphorus", Chap. 20 in *Comprehensive Inorganic Chemistry*, Vol. 2, pp. 389–545, Pergamon Press, Oxford, 1973 (Section 20.2, Phosphides, pp. 406–14).
[25] Ref. 13, Chap. 3, "Phosphides", pp. 25–64.

Phosphides resemble in many ways the metal borides (p. 162), carbides (p. 318), and nitrides (p. 479), and there are the same difficulties in classification and description of bonding. Perhaps the least-contentious procedure is to classify according to stoichiometry, i.e. (a) metal-rich phosphides $(M/P > 1)$, (b) monophosphides $(M/P = 1)$, and (c) phosphorus-rich phosphides $(M/P < 1)$:

(a) *Metal-rich phosphides* are usually hard, brittle, refractory materials with metallic lustre, high thermal and electrical conductivity, great thermal stability, and general chemical inertness. Phosphorus is often in trigonal prismatic coordination being surrounded by 6 M, or by 7, 8, or 9 M (see Fig. 6.7 on p. 166 and Fig. 12.7). The antifluorite structure of many M_2P also features eightfold (cubic) coordination of P by M. The details of the particular structure adopted in each case are influenced predominantly by size effects.

(b) *Monophosphides* adopt a variety of structures which appear to be influenced by both size and electronic effects. Thus the Group IIIA phosphides MP adopt the zinc-blende structure (p. 1405) with tetrahedral coordination of P, whereas SnP has the NaCl-type structure (p. 273) with octahedral coordination of P, VP has the hexagonal NiAs-type structure (p. 649) with trigonal prismatic coordination of isolated P atoms by V, and MoP has the hexagonal WC-type structure (p. 321) in which both Mo and P have a trigonal prismatic coordination by atoms of the other kind. More complicated arrangements are also encountered, e.g.:[25]

TiP, ZrP, HfP: half the P trigonal prismatic and half octahedral;

MP (M = Cr, Mn, Fe, Co, Ru, W): distorted trigonal prismatic coordination of P by M plus two rather short contacts to P atoms in adjacent trigonal prisms, thus building up a continuous chain of P atoms; NiP is a distortion of this in which the P atoms are grouped in pairs rather than in chains (or isolated as in VP).

(c) *Phosphorus-rich phosphides* are typified by lower mps and much lower thermal stabilities when compared with monophosphides or metal-rich phosphides. They are often semiconductors rather than metallic conductors and feature increasing catenation of the P atoms (cf. boron rich borides, p. 166). P_2 units occur in FeP_2, RuP_2, and OsP_2 (marcasite-type, p. 804) and in PtP_2 (pyrites type, p. 804) with P–P 217 pm. Planar P_4 rings (square or rectangular) occur in several MP_3 (M = Co, Ni, Rh, Pd, Ir) with P–P typically 223 pm in the square ring of RhP_3. Structures are also known in which the P atoms form chains (PdP_2, NiP_2, CdP_2, BaP_3), double chains ($ZnPbP_{14}$, $CdPbP_{14}$, $HgPbP_{14}$), or layers (CuP_2, AgP_2, CdP_4); in the last 3 phosphides the layers are made up by a regular fusion of puckered 10-membered rings of P atoms with the metal atoms in the interstices. The double-chained structure of $MPbP_{14}$ is closely related to that of violet phosphorus (p. 557).

In addition, phosphides of the electropositive elements in Groups IA, IIA, and the lanthanoids form phosphides with some degree of ionic bonding. The compounds Na_3P_{11} and Sr_3P_{14} have already been mentioned (p. 560) and other somewhat ionic phosphides are M_3P (M = Li, Na), M_3P_2 (M = Be, Mg, Zn, Cd), MP (M = La, Ce), and Th_3P_4. However, it would be misleading to consider these as fully ionized compounds of P^{3-} and there is extensive metallic or covalent interaction in the solids. Such compounds are characterized by their ready hydrolysis by water or dilute acid to give PH_3.

Few industrial uses have so far been found for phosphides. "Ferrophosphorus" is produced on a large scale as a byproduct of P_4 manufacture, and its uses have been noted

(p. 554). Phosphorus is also much used as an alloying element in iron and steel, and for improving the workability of Cu. Group IIIA monophosphides are valuable semi-conductors (p. 288) and Ca_3P_2 is an important ingredient in some navy sea-flares since its reaction with water releases spontaneously flammable phosphines. By contrast the phosphides of Nb, Ta, and W are valued for their chemical inertness, particularly their resistance to oxidation at very high temperatures, though they are susceptible to attack by oxidizing acids or peroxides.

12.3.2 *Phosphine and related compounds*

The most stable hydride of P is phosphine (phosphane), PH_3. It is the first of a homologous series P_nH_{n+2} ($n=1-6$) the members of which rapidly diminish in thermal stability, though P_2H_4 and P_3H_5 have been isolated pure. Other (unstable) homologous series are $P_nH_n(n=2-9)$, P_nH_{n-2} ($n=4-7$), and P_nH_{n-4} ($n=6-8$); in all of these there is a tendency to form cyclic and condensed polyphosphanes at the expense of open-chain structures. Phosphorane, PH_5, has not been prepared or even detected, despite numerous attempts.

PH_3 is an extremely poisonous, highly reactive, colourless gas which has a faint garlic odour at concentrations above about 2 ppm by volume. It is intermediate in thermal stability between NH_3 (p. 483) and AsH_3 (p. 650). Several convenient routes are available for its preparation:

1. Hydrolysis of a metal phosphide such as AlP or Ca_3P_2; the method is useful even up to the 10-mole scale and can be made almost quantitative

$$Ca_3P_2 + 6H_2O \longrightarrow 2PH_3 + 3Ca(OH)_2$$

2. Pyrolysis of phosphorous acid at 205–210°; under these conditions the yield of PH_3 is 97% though at higher temperatures the reaction can be more complex (p. 589)

$$4H_3PO_3 \xrightarrow{200°} PH_3 + 3H_3PO_4$$

3. Alkaline hydrolysis of PH_4I (for very pure phosphine):

$$P_4 + 2I_2 + 8H_2O \longrightarrow 2PH_4I + 2HI + 2H_3PO_4$$

$$PH_4I + KOH(aq) \longrightarrow PH_3 + KI + H_2O$$

4. Reduction of PCl_3 with $LiAlH_4$ or LiH:

$$PCl_3 + LiAlH_4 \xrightarrow{Et_2O/<0°} PH_3 + \ldots.$$

$$PCl_3 + 3LiH \xrightarrow{warm} PH_3 + 3LiCl$$

5. Alkaline hydrolysis of white P_4 (industrial process):

$$P_4 + 3KOH + 3H_2O \longrightarrow PH_3 + 3KH_2PO_2$$

Phosphine has a pyramidal structure, as expected, with P–H 142 pm and the H–P–H angle 93.6° (see p. 650). Other physical properties are mp −133.5°, bp −87.7°, dipole moment 0.58 D, heat of formation ΔH_f° -9.6 kJ mol^{-1} (uncertain), and mean P–H bond energy 320 kJ mol^{-1}. The free energy change (at 25°C) for the reaction $\frac{1}{4}P_4(\alpha\text{-white}) + \frac{3}{2}H_2(g) = PH_3(g)$ is -13.1 kJ mol^{-1}, implying a tendency for the elements to combine, though there is negligible reaction unless H_2 is energized photolytically or by a high-current arc. The inversion frequency of PH_3 is about 4000 times less than for NH_3 (p. 484); this reflects the substantially higher energy barrier to inversion for PH_3 which is calculated to be ~ 155 kJ mol^{-1} rather than 24.7 kJ mol^{-1} for NH_3.

Phosphine is rather insoluble in water at atmospheric pressure but is more soluble in organic liquids, and particularly so in CS_2 and CCl_3CO_2H. Some typical values are:

Solvent ($T°C$)	H_2O (17°)	CH_3CO_2H (20°)	C_6H_6 (22°)	CS_2 (21°)	CCl_3CO_2H
Solubility/ml PH_3(g) per 100 ml solvent	26	319	726	1025	1590

Aqueous solutions are neutral and there is little tendency for PH_3 to protonate or deprotonate:[26]

$$PH_3 + H_2O \rightleftharpoons PH_2^- + H_3O^+; \quad K_A = 1.6 \times 10^{-29}$$

$$PH_3 + H_2O \rightleftharpoons PH_4^+ + OH^-; \quad K_B = 4 \times 10^{-28}$$

In liquid ammonia, however, phosphine dissolves to give $NH_4^+PH_2^-$ and with potassium gives KPH_2 in the same solvent. Again, phosphine reacts with liquid HCl to give the sparingly soluble $PH_4^+Cl^-$ and this reacts further with BCl_3 to give PH_4BCl_4. The corresponding bromides and PH_4I are also known. More generally, phosphine readily acts as a ligand to numerous Lewis acids and typical coordination complexes are $[BH_3(PH_3)]$, $[BF_3(PH_3)]$, $[AlCl_3(PH_3)]$, $[Cr(CO)_2(PH_3)_4]$, $[Cr(CO)_3(PH_3)_3]$, $[Co(CO)_2(NO)(PH_3)]$, $[Ni(PF_3)_2(PH_3)_2]$, and $[CuCl(PH_3)]$. Further details are in the Panel and other aspects of the chemistry of PH_3 have been extensively reviewed.[26a]

Phosphine is also a strong reducing agent: many metal salts are reduced to the metal and PCl_5 yields PCl_3. The pure gas ignites in air at about 150° but when contaminated with traces of P_2H_4 it is spontaneously flammable:

$$PH_3 + 2O_2 \longrightarrow H_3PO_4$$

When heated with sulfur, PH_3 yields H_2S and a mixture of phosphorus sulfides. Probably the most important reaction industrially is its hydrophosphorylation of formic acid in aqueous hydrochloric acid solution:

$$PH_3 + 4HCHO + HCl \longrightarrow [P(CH_2OH)_4]Cl$$

The tetrakis(hydroxymethyl)phosphonium chloride so formed is the major ingredient with urea-formaldehyde or melamine-formaldehyde resins for the permanent flame-proofing of cotton cloth.

Of the other hydrides of phosphorus only diphosphane (diphosphine), P_2H_4, has been much studied. P_2H_4 is best made (30–50% yield) by passing PH_3 through an electric discharge at 5–10 kV. An older method, which involved passing products of hydrolysis of Ca_3P_2 through a cold trap, gave lower and rather unpredictable yields. P_2H_4 is a

[26] R. E. Weston and J. Bigeleisen, Kinetics of the exchange of hydrogen between phosphine and water: a kinetic estimate of the acid and base strengths of phosphine, *J. Am. Chem. Soc.* **76**, 3074–5 (1954).

[26a] E. Fluck, Chemistry of phosphine, *Topics in Current Chem.* **35**, 1–64 (1973). A review with 493 references.

Phosphine and its Derivatives as Ligands[6c, 27, 28, 28a, 28b]

A wide variety of 3-coordinate phosphorus(III) compounds are known and these have been extensively studied as ligands because of their significance in improving our understanding of the stability and reactivity of many coordination complexes. Among the most studied of these ligands are PH_3, PF_3 (p. 568), PCl_3 (p. 569), PR_3 (R = alkyl), PPh_3, and $P(OR)_3$, together with a large number of "mixed" ligands such as Me_2NPF_2, $PMePh_2$, etc., and many multidentate (chelating) ligands such as $Ph_2PCH_2CH_2PPh_2$, etc.

In many of their complexes PF_3 and PPh_3 (for example) resemble CO (p. 349) and this at one time encouraged the belief that their bonding capabilities were influenced not only by the factors (p. 223) which affect the stability of the σ P→M interaction which uses the lone-pair of electrons on P^{III} and a vacant orbital on M, but also by the possibility of synergic π back-donation from a "nonbonding" d_π pair of electrons on the metal into a "vacant" $3d_\pi$ orbital on P. It is, however, not clear to what extent, if any, the σ and π bonds reinforce each other,[29] and more recent descriptions are based on an MO approach which uses all (σ and π) orbitals of appropriate symmetry on both the phosphine and the metal-containing moiety. To the extent that σ and π bonding effects on the stability of metal–phosphorus bonds can be isolated from each other and from steric factors (see below) the accepted sequence of effects is as follows:

σ bonding: $PBu_3^t > P(OR)_3 > PR_3 \approx PPh_3 > PH_3 > PF_3 > P(OPh)_3$
π bonding: $PF_3 > P(OPh)_3 > PH_3 > P(OR)_3 > PPh_3 \approx PR_3 > PBu_3^t$
Steric interference: $PBu_3^t > PPh_3 > P(OPh)_3 > PMe_3 > P(OR)_3 > PF_3 > PH_3$

Steric factors are frequently dominant, particularly with bulky ligands, and their influence on the course of many reactions is crucial. One measure of the "size" of a ligand in so far as it affects bond formation is C. A. Tolman's cone angle (1970) which is the angle at the metal atom of the cone swept out by the van der Waals radii of the groups attached to P. This is illustrated in the diagram

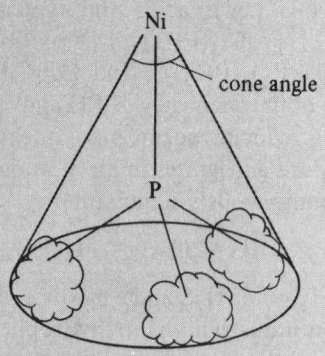

Continued

[27] Chapter 5 in ref. 2, Phosphorus(III) ligands in transition-metal complexes, pp. 177–207.

[28] C. A. McAuliffe and W. Levason, *Phosphine, Arsine, and Stibine Complexes of the Transition Elements*, Elsevier, Amsterdam, 1979, 546 pp. A review with over 2700 references, mainly to literature since 1971. Earlier work was comprehensively reviewed in C. A. McAuliffe (ed.), *Transition-Metal Complexes of Phosphorus, Arsenic, and Antimony Donor Ligands*, Macmillan, London, 1972.

[28a] O. Stelzer, Transition metal complexes of phosphorus ligands, *Topics in Phosphorus Chemistry* 9, 1–229 (1977). An extensive review with over 1700 references arranged element-by element and technique-by-technique but with no assessment or generalizations.

[28b] R. Mason and D. W. Meek, The versatility of tertiary phosphine ligands in coordination and organometallic chemistry, *Angew. Chem.*, Int. Edn. (Engl.), **17**, 183–94 (1978).

[29] A. J. Carty, N. J. Taylor, A. W. Coleman, and M. F. Lappert, The chromium–heavy Group 5 donor bond: a comparison of structural changes within the series [Cr(CO)$_5$(XPh$_3$)] (X = P, As, Sb, or Bi) via their X-ray crystal structures, *JCS Chem. Comm.* 1979, 639–40.

and will, of course, be dependent on the actual interatomic distance between M and P. For the particular case of Ni, for which a standard value of 228 pm was adopted for Ni–P, the calculated values for the cone angle are:

Ligand	PH_3	PF_3	$P(OMe)_3$	$P(OEt)_3$	PMe_3	$P(OPh)_3$
Cone angle	87°	104°	107°	109°	118°	121°

Ligand	PCl_3	PEt_3	PPh_3	PPr^i_3	PBu^t_3
Cone angle	125°	132°	145°	160°	182°

Bulky tertiary phosphine ligands exert both steric and electronic influences when they form complexes (since an increase in bulkiness of a substituent on P increases the inter-bond angles and this in turn can be thought of as an increase in "p-character" of the lone-pair of electrons on P). For example, the sterically demanding di-t-butylphosphines, PBu^t_2R (R = alkyl or aryl), promote spatially less-demanding features such as hydride formation, coordinative unsaturation at the metal centre, and even the stablization of unusual oxidation states, such as Ir^{II}. They also favour internal C– or O– metallation reactions for the same reasons. Indeed, the metallation of C–H and C–P bonds of coordinated tertiary phosphines can be considered as examples of intramolecular oxidative addition, and these have important mechanistic implications for homogeneous and heterogeneous catalysis.[29a] A possible mechanism would be by concerted 3-centre addition:

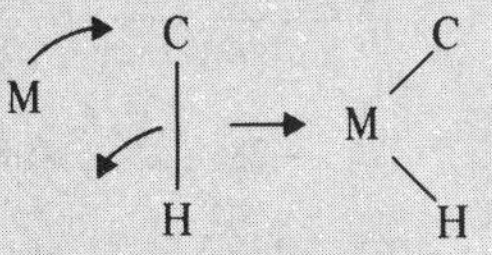

Other notable examples are the orthometallation (orthophenylation) reactions of many complexes of aryl phosphines (PAr_3) and aryl phosphites $P(OAr)_3$ with platinum metals in particular, e.g.:

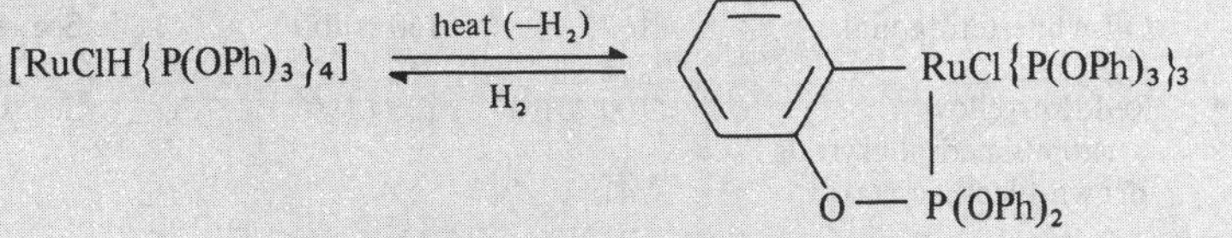

colourless, volatile liquid (mp $-99°$) which is thermally unstable above room temperature and is decomposed slowly by water. Its vapour pressure at 0°C is 70.2 mmHg but partial decomposition precludes precise determination of the bp (63.5° extrap). Electron-diffraction measurements on the gas establish the *gauche–C_2* configuration (p. 491) with P–P 222 pm, P–H 145 pm, and the angle H–P–H 91.3°, though vibration spectroscopy suggests a *trans–C_{2h}* configuration in the solid phase. These results can be compared with those for the halides P_2X_4 on p. 571.

12.3.3 *Phosphorus halides*

Phosphorus forms three series of halides P_2X_4, PX_3, and PX_5, and all 12 compounds are now known, the most recent additions being P_2Br_4 (1973) and PI_5 (1978). Numerous mixed halides PX_2Y and PX_2Y_3 are also known as well as various pseudohalides such as

[29a] G. PARSHALL, Homogeneous catalytic activation of C–H bonds, *Acc. Chem. Res.* **8**, 113–17 (1975).

$P(CN)_3$, $P(CNO)_3$, $P(CNS)_3$, and their mixed halogeno-counterparts. The compounds form an extremely useful extended series with which to follow the effect of progressive substitution on various properties, and the pentahalides are particularly significant in spanning the "ionic-covalent" border, so that they exist in various structural forms depending on the nature of the halogen, the phase of aggregation, or the polarity of the solvent. Some curious polyhalides such as PBr_7 and PBr_{11} have also been characterized. Some physical properties of the binary halides are summarized in Table 12.3. Ternary (mixed) halides tend to have properties intermediate between those of the parent binary halides.

TABLE 12.3   *Some physical properties of the binary phosphorus halides*

Compound	Physical State at 25°C	MP/°C	BP/°C	P–X/pm	Angle X–P–X
PF_3	Colourless gas	−151.5	−101.8	156	96.3°
PCl_3	Colourless liquid	−93.6	76.1	204	100°
PBr_3	Colourless liquid	−41.5	173.2	222	101°
PI_3	Red hexagonal crystals	61.2	decomp > 200	243	102°
P_2F_4	Colourless gas	−86.5	−6.2	159 (P–P 228)	99.1° (F–P–P 95.4°)
P_2Cl_4	Colourless oily liquid	−28	∼180 (d)	—	—
P_2Br_4	?	—	—	—	—
P_2I_4	Red triclinic needles	125.5	decomp	248 (P–P 221)	102.3° (I–P–P 94.0°)
PF_5	Colourless gas	−93.7	−84.5	153 (eq) 158 (ax)	120° (eq–eq) 90° (eq–ax)
PCl_5	Off-white tetragonal crystals	167	160 (subl)	See text	
PBr_5	Reddish-yellow rhombohedral crystals	<100 (d)	106 (d)	See text	
PI_5	Brown-black crystals	41	—	—	

Phosphorus trihalides

All 4 trihalides are volatile reactive compounds which feature pyramidal molecules. The fluoride is best made by the action of CaF_2, ZnF_2, or AsF_3 on PCl_3, but the others are formed by direct halogenation of the element. PF_3 is colourless, odourless, and does not fume in air, but is very hazardous due to the formation of a complex with blood haemoglobin (cf. CO, p. 1279). It is about as toxic as $COCl_2$. The similarity of PF_3 and CO as ligands was first noted by J. Chatt[30] and many complexes with transition elements are now known,[31] e.g. $[Ni(CO)_n(PF_3)_{4-n}]$ $(n=0-4)$, $[Pd(PF_3)_4]$, $[Pt(PF_3)_4]$, $[CoH(PF_3)_4]$, $[Co_2(\mu-PF_2)_2(PF_3)_6]$, etc. Such complexes can be prepared by ligand replacement reactions, by fluorination of PCl_3 complexes, by direct reaction of PF_3 with

[30] J. CHATT, The coordinate link in chemistry, *Nature* **165**, 637–8 (1950); J. CHATT and A. A. WILLIAMS, Complex formation by phosphorus trifluoride, *J. Chem. Soc.* 1951, 3061–7.

[31] T. KRUCK, Trifluorophosphine complexes of transition metals, *Angew. Chem.*, Int. Edn. (Engl.) **6**, 53–67 (1967); J. F. NIXON, Recent progress in the chemistry of fluorophosphines, *Adv. Inorg. Chem. Radiochem.* **13**, 363–469 (1970); R. J. CLARKE and M. A. BUSCH, Stereochemical studies of metal carbonyl–phosphorus trifluoride complexes, *Acc. Chem. Res.* **6**, 246–52 (1973).

metal salts, or even by direct reaction of PF_3 with metals at elevated temperatures and pressures.

PF_3, unlike the other trihalides of phosphorus, hydrolyses only slowly with water, the products being phosphorous acid and HF:

$$PF_3 + 3H_2O \longrightarrow H_3PO_3 + 3HF$$

The reaction is much more rapid in alkaline solutions, and in dilute aqueous $KHCO_3$ solutions the intermediate monofluorophosphorous acid is formed:

$$PF_3 + 2H_2O \xrightarrow{\ 2\% \text{ aq } KHCO_3\ } O{=\!=}P\!\!\begin{array}{c} H \\ \diagup \\ \!\!-\!F \\ \diagdown \\ OH \end{array} + 2HF$$

PCl_3 is the most important compound of the group and is made industrially on a large scale† by direct chlorination of phosphorus suspended in a precharge of PCl_3—the reaction is carried out under reflux with continuous take-off of the PCl_3 formed. PCl_3 undergoes many substitution reactions, as shown in the diagram, and is the main source of organophosphorus compounds. Particularly notable are PR_3, PR_nCl_{3-n}, $PR_n(OR)_{3-n}$, $(PhO)_3PO$, and $(RO)_3PS$. Many of these compounds are made on the 1000-tonne scale pa, and the major uses are as oil additives, plasticizers, flame retardants, fuel additives, and intermediates in the manufacture of insecticides.[32] PCl_3 is also readily oxidized to the important phosphorus(V) derivatives PCl_5, $POCl_3$, and $PSCl_3$. It is oxidized by As_2O_3 to P_2O_5 though this is not the commercial route to this compound (p. 578). It fumes in moist air and is more readily hydrolysed (and oxidized) by water than is PF_3. With cold N_2O_4 ($-10°$) it undergoes a curious oxidative coupling reaction to give $Cl_3P{=}N{-}POCl_2$, mp $35.5°$; this compound had previously been formulated as $P_4O_4Cl_{10}$, i.e.

$$\begin{array}{ccccccccc}
 & Cl & & Cl & Cl\ Cl & & Cl & & \\
 & | & & \diagdown\diagup & | & & | & & \\
O{=}P & {-}O{-} & P & {-} & P & {-}O{-} & P & {=}O \\
 & | & & | & \diagup\diagdown & & | & & \\
 & Cl & & Cl\ Cl & Cl & & Cl & &
\end{array}$$

but was shown to be

$$\begin{array}{ccccc}
Cl & & & & Cl \\
| & & & & | \\
Cl{-} & P & {=}N{-} & P & {=}O \\
| & & & & | \\
Cl & & & & Cl
\end{array}$$

with two different 4-coordinate $P^{(V)}$ atoms by ^{31}P nmr and other techniques.[33] Other

---

† World production exceeds a quarter of a million tonnes pa; of this the US produces ~150 000 tonnes, the UK and Europe ~90 000, and Japan ~33 000 tonnes pa.

[32] D. H. CHADWICK and R. S. WATT, Manufacture of phosphate esters and organic phosphorus compounds, Chap. 19 in ref. 7, pp. 1221–79.

[33] M. BECKE-GOEHRING, A. DEBO, E. FLUCK, and W. GOETZE, Concerning trichlorophosphornitrido-phosphoryl dichloride and the reaction between phosphorus trichloride and dinitrogen tetroxide, *Chem. Ber.* **94**, 1383–7 (1961).

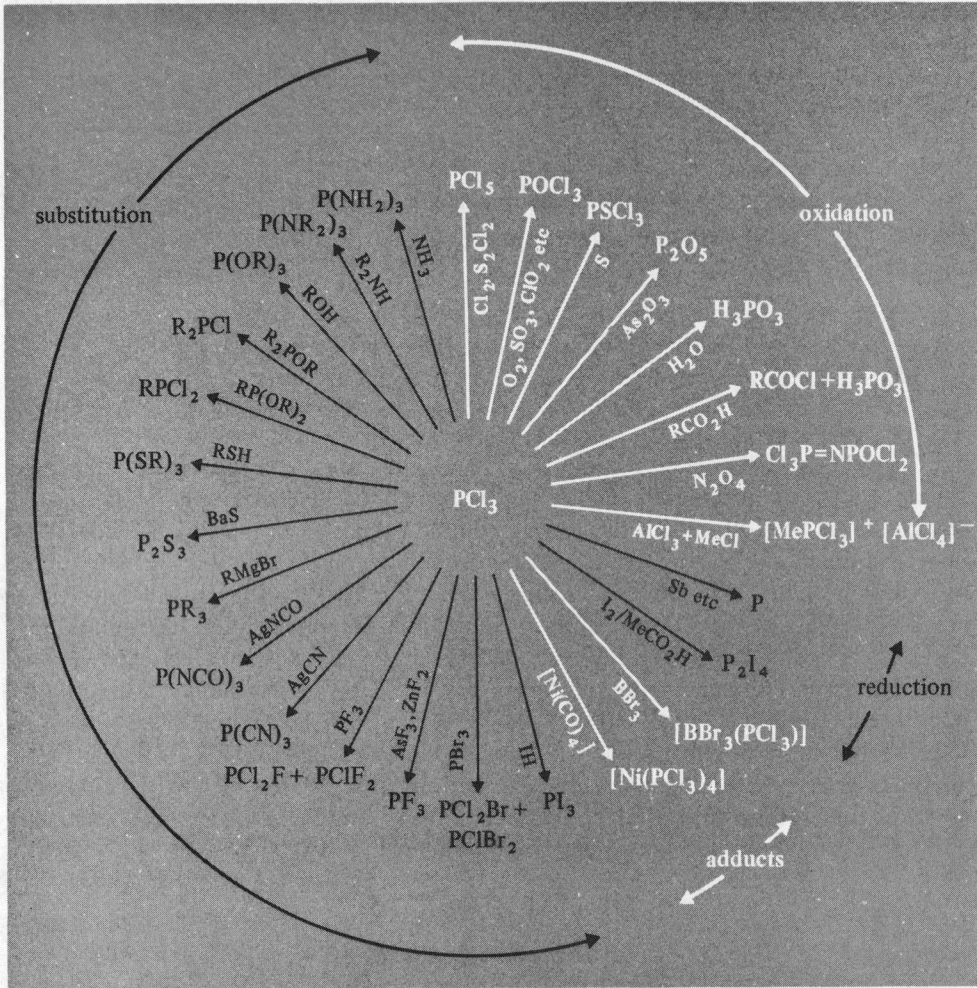

notable reactions of PCl_3 are its extensive use to convert alcohols to RCl and carboxylic acids to $RCOCl$, its reduction to P_2I_4 by iodine, and its ability to form coordination complexes with Lewis acids such as BX_3 and $Ni^{(0)}$.

PI_3 is emerging as a powerful and versatile deoxygenating agent.[33a] For example solutions of PI_3 in CH_2Cl_2 at or below room temperature convert sulfoxides ($RR'SO$) into diorganosulfides, selenoxides ($RR'SeO$) into selenides, aldehyde oximes ($RCH{=}NOH$) into nitriles, and primary nitroalkanes (RCH_2NO_2) into nitriles, all in high yield (75–95%). The formation of nitriles, RCN, in the last two reactions requires the presence of triethylamine in addition to the PI_3.

[33a] J. N. DENIS and A. KRIEF, Phosphorus triiodide (PI_3), a powerful deoxygenating agent, *JCS Chem. Comm.* 1980, 544–5.

Diphosphorus tetrahalides

The physical properties of P_2X_4, in so far as they are known, are summarized in Table 12.3. P_2F_4 was first made in other than trace amounts in 1966, using the very effective method of coupling two PF_2 groups at room temperature under reduced pressure:

$$2PF_2I + 2Hg \xrightarrow{86\% \text{ yield}} P_2F_4 + Hg_2I_2$$

The compound hydrolyses to F_2POPF_2 which can also be prepared directly in 67% yield by the reaction of O_2 on P_2F_4.

P_2Cl_4 can be made (in low yield) by passing an electric discharge through a mixture of PCl_3 and H_2 under reduced pressure or by microwave discharge through PCl_3 at 1–5 mmHg pressure. The compound decomposes slowly at room temperature to PCl_3 and an involatile solid, and can be hydrolysed in basic solution to give an equimolar mixture of P_2H_4 and $P_2(OH)_4$.

Little is known of P_2Br_4, said to be produced by an obscure reaction in the system C_2H_4–PBr_3–Al_2Br_6.[34] By contrast, P_2I_4 is the most stable and also the most readily made of the 4 tetrahalides; it is formed by direct reaction of I_2 and red P at 180° or by I_2 and white P_4 in CS_2 solution, and can also be made by reducing PI_3 with red P, or PCl_3 with iodine. Its X-ray crystal structure shows that the molecules of P_2I_4 adopt the *trans*-, centrosymmetric (C_{2h}) form (see N_2H_4, p. 491, N_2F_4, p. 504). Reaction of P_2I_4 with sulfur in CS_2 yields $P_2I_4S_2$, which probably has the symmetrical structure

$$
\begin{array}{ccc}
\text{I} & & \text{S} \\
\diagdown & & \diagup\diagup \\
\text{I}\!=\!\!\text{P} & \!\!-\!\!\text{P}\!\!- & \!\text{I} , \\
\diagup\diagup & & \diagdown \\
\text{S} & & \text{I}
\end{array}
$$

but most reactions of P_2I_4 result in cleavage of the P–P bond, e.g. Br_2 gives $PBrI_2$ in 90% yield. Hydrolysis yields various phosphines and oxoacids of P, together with a small amount of hypophosphoric acid, $(HO)_2(O)PP(O)(OH)_2$.

Phosphorus pentahalides

Considerable theoretical and stereochemical interest attaches to these compounds because of the variety of structures they adopt; PCl_5 is also an important chemical intermediate. Thus, PF_5 is molecular and stereochemically non-rigid (see below), PCl_5 is molecular in the gas phase, ionic in the crystalline phase, $[PCl_4]^+[PCl_6]^-$, and either molecular or ionically dissociated in solution, depending on the nature of the solvent. PBr_5 is also ionic in the solid state but exists as $[PBr_4]^+[Br]^-$ rather than $[PBr_4]^+[PBr_6]^-$. The iodide, which was first made only in 1978 (by the reaction of HI,

[34] R. I. PYRKIN, YA. A. LEVIN, and E. I. GOLDFARB, Reactions in the system ethylene-phosphorus tribromide-aluminium bromide, *J. Gen. Chem. USSR* **43**, 1690–6 (1973). See also A. HINKE, W. KUCHEN, and J. KUTTER, Stabilization of diphosphorus tetrabromide as the bis(pentacarbonylchromium) complex, *Angew. Chem.* Int. Ed. (Engl.) **20**, 1060 (1981).

LiI, NaI, or KI on PCl_5 dissolved in MeI), appears also to be $[PI_4]^+I^-$, at least in solution.[35]

PF_5 is a thermally stable, chemically reactive gas which can be made either by fluorinating PCl_5 with AsF_3 (or CaF_2), or by thermal decomposition of $NaPF_6$, $Ba(PF_6)_2$, or the corresponding diazonium salts. Electron diffraction, after early incorrect conclusions, firmly establishes the molecular structure as trigonal bipyramidal with the axial P–F bond distances (158 pm) longer than the equatorial (153 pm). This implies that the "static" structure, as obtained by techniques which monitor the structure on a very fast ("instantaneous") time scale, involves two geometrically different types of F atom. However, the ^{19}F nmr spectrum, as recorded down to $-100°C$, shows only a single fluorine resonance peak (split into a doublet by $^{31}P-^{19}F$ coupling) implying that on this longer time scale (milliseconds) all 5 F atoms are equivalent. This can be explained if the axial and equatorial F atoms interchange their positions more rapidly than this, a process termed "pseudorotation" by R. S. Berry (1960); indeed, PF_5 was the first compound to show this effect.[36] The proposed mechanism is illustrated in Fig. 12.8 and is discussed more fully in ref. 37.

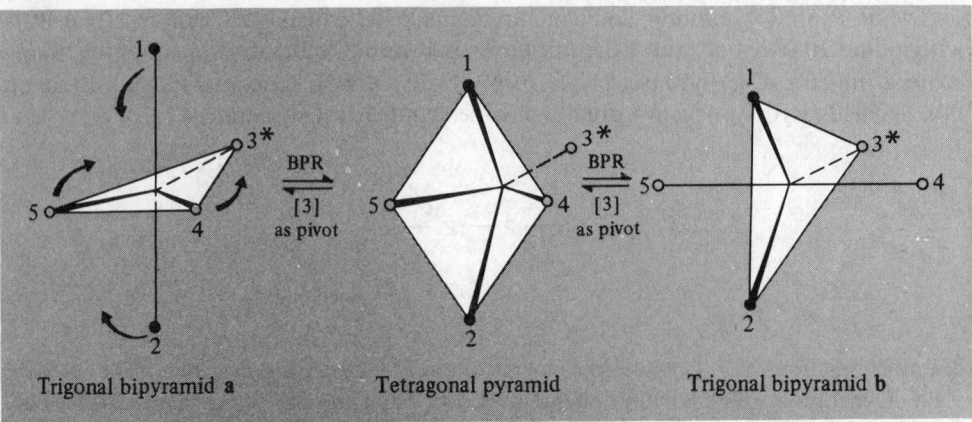

FIG. 12.8 Interchange of axial and equatorial positions by Berry pseudorotation (BPR).

The mixed chlorofluorides PCl_4F (mp $-59°$, bp $+67°$) and PCl_3F_2 (mp $-63°$) are also trigonal bipyramidal with axial F atoms; likewise PCl_2F_3 (mp $-125°$, bp $+7.1°$) has 2 axial and 1 equatorial F atoms, and $PClF_4$ (mp $-132°$, bp $-43.4°$) has both axial positions occupied by F atoms. These compounds are obtained by addition of halogen to the appropriate phosphorus(III) chlorofluoride, but if PCl_5 is fluorinated in a polar solvent, ionic isomers are formed, e.g. $[PCl_4]^+[PCl_4F_2]^-$ (colourless crystals, subl 175°) and $[PCl_4]^+[PF_6]^-$ (white crystals, subl 135° with decomposition). The crystalline hemifluoride $[PCl_4]^+[PCl_5F]^-$ has also been identified. The analogous parallel series of

[35] N. G. FESHCHENKO, V. G. KOSTINA, and A. V. KIRSANOV, Synthesis of PI_5, *J. Gen. Chem. USSR* **48**, 195–6 (1978).

[36] R. S. BERRY, Correlation of rates of intramolecular tunnelling processes, with application to some group V compounds, *J. Chem. Phys.* **32**, 933–8 (1960).

[37] R. LUCKENBACH, *Dynamic Stereochemistry of Pentacoordinate Phosphorus and Related Elements*, G. THIEME, Stuttgart, 1973, 259 pp.

covalent and ionic bromofluorides is less well characterized but PBr_2F_3 is known both as an unstable molecular liquid (decomp 15°) and as a white crystalline powder $[PBr_4]^+[PF_6]^-$ (subl 135° decomp).

PCl_5 is even closer to the ionic-covalent borderline than is PF_5, the ionic solid $[PCl_4]^+[PCl_6]^-$ melting (or subliming) to give a covalent molecular liquid (or gas). Again, when dissolved in non-polar solvents such as CCl_4 or benzene, PCl_5 is monomeric and molecular, whereas in ionizing solvents such as MeCN, $MeNO_2$, and $PhNO_2$ there are two competing ionizing equilibria:[38]

$$2PCl_5 \rightleftharpoons [PCl_4]^+ + [PCl_6]^-$$

$$PCl_5 \rightleftharpoons [PCl_4]^+ + Cl^-$$

As might be expected, the former equilibrium predominates at higher concentrations of PCl_5 (above about 0.03 mol l^{-1}) whilst the latter predominates below this concentration. The P–Cl distances (pm) in these various species are:

$$PCl_5 \ 214 \ (axial), \ 202 \ (equatorial); \quad [PCl_4]^+ \ 197; \quad [PCl_6]^- \ 208 \ pm$$

Ionic isomerism is also known and, in addition to $[PCl_4]^+[PCl_6]^-$, another (metastable) crystalline phase of constitution $[PCl_4]_2^+[PCl_6]^-Cl^-$ can be formed either by application of high pressure or by crystallizing PCl_5 from solutions of dichloromethane containing Br_2 or SCl_2.[38a] When gaseous PCl_5 (in equilibrium with $PCl_3 + Cl_2$) is quenched to 15 K the trigonal-bipyramidal molecular structure is retained; this forms an ordered molecular crystalline lattice on warming to ~ 130 K, but further warming towards room temperature results in chloride-ion transfer to give $[PCl_4]^+[PCl_6]^-$.[38b]

The delicate balance between ionic and covalent forms is influenced not only by the state of aggregation (solid, liquid, gas) or the nature of the solvent, but also by the effect of substituents. Thus $PhPCl_4$ is molecular with Ph equatorial whereas the corresponding methyl derivative is ionic, $[MePCl_3]^+Cl^-$. Despite this the $[PhPCl_3]^+$ is known and can readily be formed by reacting $PhPCl_4$ with a chlorine ion acceptor such as BCl_3, $SbCl_5$, or even PCl_5 itself:[39]

$$PhPCl_4 + PCl_5 \longrightarrow [PhPCl_4]^+[PCl_6]^-$$

Likewise crystalline Ph_2PCl_3 is molecular whereas the corresponding Me and Et derivatives are ionic $[R_2PCl_2]^+Cl^-$. However, all 3 triorganophosphorus dihalides are ionic $[R_3PCl]^+Cl^-$ (R = Ph, Me, Et). The pale-yellow, crystalline mixed halide P_2BrCl_9 appears to be $[PCl_4]_6^+[PCl_3Br]_2^+[PCl_6]_4^-[Br]_4^-$ (i.e. $P_{12}Br_6Cl_{54}$).[39a]

[38] R. W. SUTER, H. C. KNACHEL, V. P. PETRO, J. H. HOWATSON, and S. G. SHORE, Nature of phosphorus(V) chloride in ionizing and non-ionizing solvents, *J. Am. Chem. Soc.* **95**, 1474–9 (1973).

[38a] A. FINCH, P. N. GATES, H. D. B. JENKINS, and K. P. THAKUR, Ionic isomerism in phosphorus(V) chloride, *JCS Chem. Comm.* 1980, 579–80.

[38b] A. FINCH, P. N. GATES, and A. S. MUIR, Variable-temperature Raman spectra of phosphorus(V) chloride and bromide deposited at 15 K, *JCS Chem. Comm.* 1981, 812–4. See also H. D. B. JENKINS, K. P. THAKUR, A. FINCH, and P. N. GATES, Ionic isomerism. 2. Calculations of thermodynamic properties of phosphorus(V) chloride isomers: $\Delta H_f^\circ(PCl_4^+(g))$ and $\Delta H_f^\circ(PCl_6^-(g))$, *Inorg. Chem.* **21**, 423–6 (1982).

[39] K. B. DILLON, R. J. LYNCH, R. N. REEVE, and T. C. WADDINGTON, Solid-state ^{31}P nmr and ^{35}Cl nqr studies of some alkyl- and aryl-chlorophosphoranes and their addition compounds with Lewis acids, *J. Chem. Soc. (Dalton)* 1976, 1243–8.

[39a] F. F. BENTLEY, A. FINCH, P. N. GATES, F. J. RYAN, and K. B. DILLON, The structure of P_2Cl_9Br, *J. Inorg. Nucl. Chem.* **36**, 457–9 (1974). See also *JCS Dalton* 1973, 1863–6.

Phosphorus pentabromide is rather different. The crystalline solid is $[PBr_4]^+Br^-$ but this appears to dissociate completely to PBr_3 and Br_2 in the vapour phase; rapid cooling of this vapour to 15 K results in the formation of a disordered lattice of PBr_3 and PBr_7 (i.e. $[PBr_4]^+[Br_3]^-$) and this mixture reverts to $[PBr_4]^+Br^-$ on being warmed to 180 K :[38b]

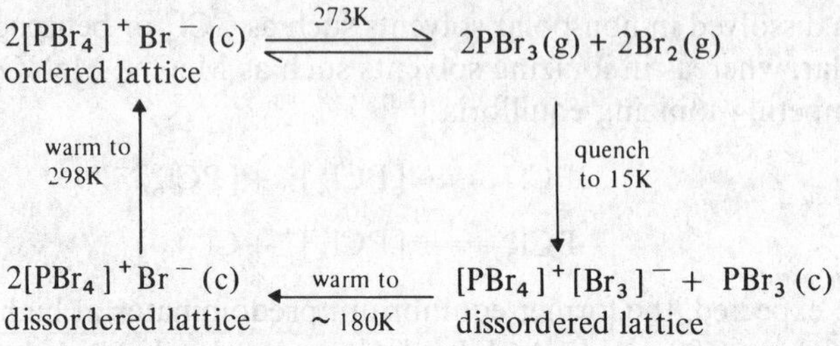

$$2[PBr_4]^+Br^-\text{(c)} \xrightleftharpoons{273K} 2PBr_3\text{(g)} + 2Br_2\text{(g)}$$

ordered lattice

warm to 298K

quench to 15K

$$2[PBr_4]^+Br^-\text{(c)} \xleftarrow{\text{warm to} \sim 180K} [PBr_4]^+[Br_3]^- + PBr_3\text{(c)}$$

dissordered lattice   dissordered lattice

PCl_5 is made on an industrial scale by the reaction of Cl_2 on PCl_3 dissolved in an equal volume of CCl_4. World production probably exceeds 20 000 tonnes pa. On the laboratory scale Cl_2 gas (or liquid) can be passed directly into PCl_3. PCl_5 reacts violently with water to give HCl and H_3PO_4 but in equimolar amounts the reaction can be moderated to give $POCl_3$:

$$PCl_5 + H_2O \longrightarrow POCl_3 + 2HCl$$

PCl_5 chlorinates alcohols to alkyl halides and carboxylic acids to the corresponding RCOCl. When heated with NH_4Cl the phosphonitrilic chlorides are obtained (p. 625). These and other reactions are summarized in the diagram.[5] The chlorination of phosphonic and phosphinic acids and esters are of considerable importance. PCl_5 can also act as a Lewis acid to give 6-coordinate P complexes, e.g. py PCl_5, and pyz PCl_5 where py $= C_5H_5N$ (pyridine) and pyz $= cyclo$-1,4-$C_4H_4N_2$ (pyrazine).[39b]

Pseudohalides of phosphorus(III)[40]

Paralleling the various phosphorus trihalides are numerous pseudohalides and mixed pseudohalide-halides of which the various isocyanates and isothiocyanates are perhaps the best known. Most are volatile liquids, e.g.

Compound	$P(NCO)_3$	$PF(NCO)_2$	$PF_2(NCO)$	$PCl(NCO)_2$
MP/°C	−2	−55	$\sim$ −108	−50
BP/°C	169.3	98.7	12.3	134.6

Compound	$PCl_2(NCO)$	$P(NCS)_3$	$PF_2(NCS)$	$PCl_2(NCS)$
MP/°C	−99	−4	−95	−76
BP/°C	104.5	$\sim$120/1 mmHg	90.3	148 (decomp)

[39b] B. N. MEYER, J. N. ISHLEY, A. V. FRATINI, and H. C. KNACHEL, Addition compounds of phosphorus(V) chloride and aromatic nitrogen bases. Crystal structure and ^1H nmr and Raman spectra of pyrazine-phosphorus(V) chloride, *Inorg. Chem.* **19**, 2324–7 (1980) and references therein.
[40] A. F. CHILDS, "Phosphorus pseudohalides", Section 19 of ref. 15, pp. 582–95.

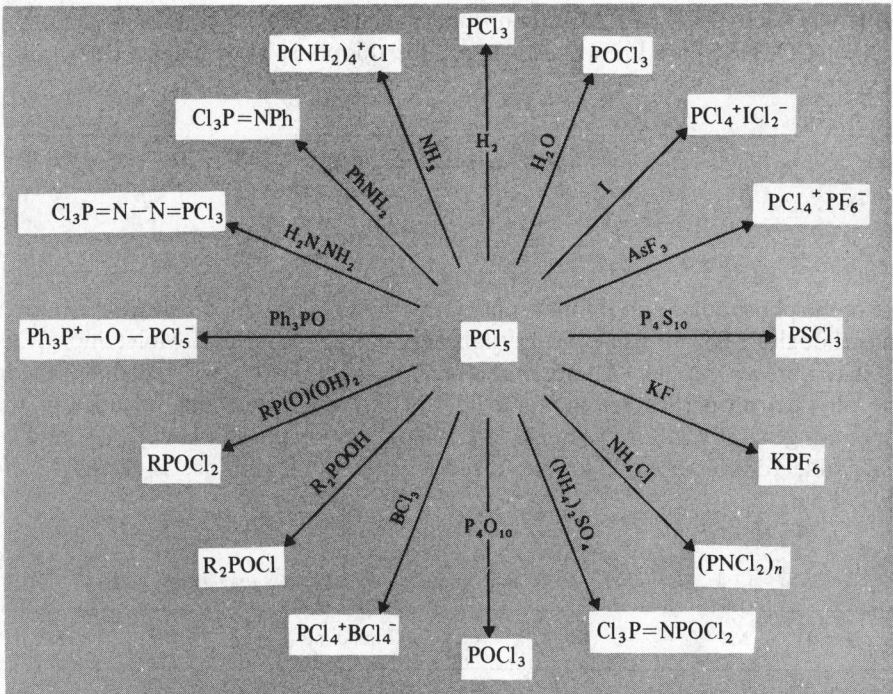

Some reactions of phosphorus pentachloride.

The corresponding phosphoryl and thiophosphoryl pseudohalides are also known, i.e. $PO(NCO)_3$, $PS(NCO)_3$, etc. Preparations are by standard procedures such as those on the diagram for PCl_3 (p. 570). As indicated there, $P(CN)_3$ has also been made: it is a highly reactive white crystalline solid mp 203° which reacts violently with water to give mainly phosphorous acid and HCN.

12.3.4 *Oxohalides and thiohalides of phosphorus*

The propensity of phosphorus(III) compounds to oxidize to phosphorus(V) by formation of an additional P=O bond is well illustrated by the ease with which the trihalides are converted to their phosphoryl analogues POX_3. Thus, PCl_3 reacts rapidly with pure O_2 (less rapidly with air) at room temperature or slightly above and this reaction is used on an industrial scale. Alternatively, a slurry of P_4O_{10} in PCl_3 can be chlorinated, the PCl_5 so formed reacting instantaneously with the P_4O_{10}:

$$P_4O_{10} + 6PCl_5 \longrightarrow 10POCl_3$$

$POBr_3$ can be made by similar methods, but POF_3 is usually made by fluorination of $POCl_3$ using a metal fluoride (e.g. M = Na, Mg, Zn, Pb, Ag, etc.). POI_3 was first made in 1973 by iodinating $POCl_3$ with LiI, or by reacting $ROPI_2$ with iodine

$(ROPI_2 + I_2 \rightarrow RI + POI_3)$.[41] Mixed phosphoryl halides, POX_nY_{3-n}, and pseudohalides (e.g. X = NCO, NCS) are known, as also are the thiophosphoryl halides PSX_3, e.g.:

$$P_2S_5 + 3PCl_5 \longrightarrow 5PSCl_3$$

$$PCl_3 + S \xrightarrow{\text{AlCl}_3} PSCl_3$$

$$PI_3 + S \xrightarrow{\text{CS}_2/\text{dark}} PSI_3$$

Most of the phosphoryl and thiophosphoryl compounds are colourless gases or volatile liquids though $PSBr_3$ forms yellow crystals, mp 37.8°, POI_3 is dark violet, mp 53°, and PSI_3 is red-brown, mp 48°. All are monomeric tetrahedral (C_{3v}) or pseudotetrahedral. Some physical properties are in Table 12.4. The P–O interatomic distance in these compounds generally falls in the range 154–158 pm, the small value being consistent with considerable "double-bond character". Likewise the P–S distance is relatively short (185–194 pm).

TABLE 12.4 *Some phosphoryl and thiophosphoryl halides and pseudohalides*

Compound	MP/°C	BP/°C	Compound	MP/°C	BP/°C
POF_3	−39.1	−39.7	POF_2Cl	−96.4	3.1
$POCl_3$	1.25	105.1	$POFCl_2$	−80.1	52.9
$POBr_3$	55	191.7	POF_2Br	−84.8	31.6
POI_3	53	—	$POFBr_2$	−117.2	110.1
$PO(NCO)_3$	5.0	193.1	$POCl_2Br$	11	52/3 mmHg
$PO(NCS)_3$	13.8	300.1	$POClBr_2$	31	49/12 mmHg
PSF_3	−148.8	−52.2	PSF_2Cl	−155.2	6.3
$PSCl_3$	−35	−125	$PSFCl_2$	−96.0	64.7
$PSBr_3$	37.8	212 (d)	PSF_2Br	−136.9	35.5
PSI_3	48	d	$PSFBr_2$	−75.2	125.3
$PS(NCO)_3$	8.8	215	$PO(NCO)FCl$	—	103
$PS(NCS)_3$	—	123/0.3 mmHg	$PS(NCS)F_2$	—	90

The phosphoryl and thiophosphoryl halides are reactive compounds that hydrolyse readily on contact with water. They form adducts with Lewis acids and undergo a variety of substitution reactions to form numerous organophosphorus derivatives and phosphate esters. Thus, alcohols give successively $(RO)POCl_2$, $(RO)_2POCl$, and $(RO)_3PO$; phenols react similarly but more slowly. Likewise, amines yield $(RNH)POCl_2$, $(RNH)_2POCl$, and $(RNH)_3PO$ whereas Grignard reagents yield R_nPOCl_{3-n} ($n = 1$–3). Many of these compounds find extensive use as oil additives, insecticides, plasticizers, surfactants, or flame retardants, and are manufactured on the multikilotonne scale.

In addition to the monophosphorus phosphoryl and thiophosphoryl compounds discussed above, several poly-phosphoryl and -thiophosphoryl halides have been characterized. Pyrophosphoryl fluoride, $O{=}PF_2{-}O{-}P({=}O)F_2$ (mp −0.1°, bp 72° extrap) and the white crystalline cyclic tetramer $[O{=}P(F){-}O]_4$ were obtained by

[41] A. V. KIRSANOV, ZH. K. GORBATENKO, and N. G. FESHCHENKO, Chemistry of phosphorus iodides, *Pure Appl. Chem.* **44**, 125–39 (1975).

subjecting equimolar mixtures of PF_3 and O_2 to a silent electric discharge at $-70°$. Pyrophosphoryl chloride, $O{=}PCl_2{-}O{-}P({=}O)Cl_2$, is conveniently prepared by passing Cl_2 into a boiling suspension of P_4O_{10} and PCl_3 diluted with CCl_4:

$$P_4O_{10} + 4PCl_3 + 4Cl_2 \longrightarrow 2P_2O_3Cl_4 + 4POCl_3$$

It is a colourless, odourless, non-fuming, oily liquid, mp $-16.5°$, bp $215°$ (decomp), with reactions similar to those of $POCl_3$. Sealed-tube reactions between P_4O_{10} and $POCl_3$ at 200–230° give more highly condensed cyclic and open-chain polyphosphoryl chlorides. A rather different structural motif occurs in $P_2S_4F_4$; this compound is obtained by fluorinating P_4S_{10} with an alkali-metal fluoride to give the anion $[S_2PF_2]^-$ which is then oxidized by bromine to $P_2S_4F_4$ (bp 60° at 10 mmHg). Vibrational and nmr spectra are consistent with the structure $F_2(S)PSSP(S)F_2$.

Bromination of P_4S_7 in cold CS_2 yields, in addition to PBr_3 and $PSBr_3$, two further thiobromides $P_2S_6Br_2$ (mp 118° decomp) and $P_2S_5Br_4$ (mp 90° decomp). The first of these has the cyclic structure shown in which the ring adopts a skew-boat configuration. An

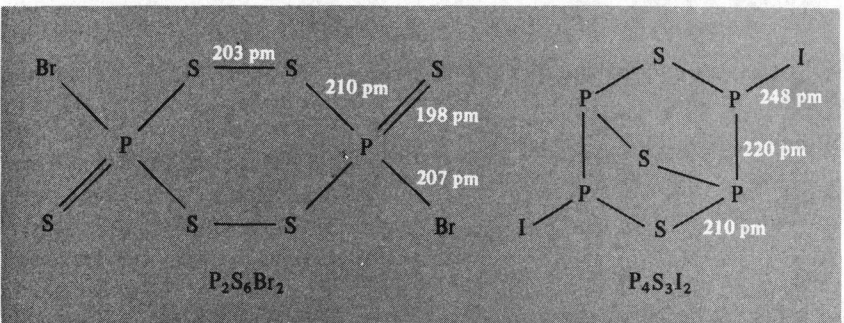

$$P_2S_6Br_2 \qquad\qquad P_4S_3I_2$$

even more complex, bicyclic arrangement is found in the orange-yellow compound $P_4S_3I_2$ (mp 120° decomp) which is formed (together with several other products) when equi-atomic amounts of P, S, and I are allowed to react. The P and S atoms are arranged in two 5-membered rings having a common P–S–P group as shown; in each there is a P–P group and the I atoms are bonded in *cis*- configuration to the P atoms not common to the two rings. The orange compound $P_2S_2I_4$ (mp 94°) was mentioned on p. 571.

12.3.5 *Phosphorus oxides, sulfides, and oxosulfides*

The oxides and sulfides of phosphorus are amongst the most important compounds of the element. At least 6 binary oxides and 6 well-defined sulfides are known, together with several oxosulfides. It will be convenient to discuss first the preparation and structure of each group of compounds and then mention the chemical reactions of the more important members in so far as they are known.

Oxides

P_4O_6 is obtained by controlled oxidation of P_4 in an atmosphere of 75% O_2 and 25% N_2 at 90 mmHg and ${\sim}50°$ followed by distillation of the product from the mixture.

Careful precautions are necessary if good yields are to be obtained.[42] It forms soft white crystals, mp 23.8°, bp 175.4°, and is soluble in many organic solvents. The molecular structure has tetrahedral symmetry and comprises 4 fused 6-membered P_3O_3 heterocycles each with the chair conformation as shown in Fig. 12.9.[42a] When P_4O_6 is heated to 200–400° in a sealed, evacuated tube it disproportionates into red phosphorus and a solid-solution series of composition P_4O_n depending on conditions. The α-phase has a composition in the range $P_4O_{8.1}$–$P_4O_{9.2}$ and comprises a solid solution of oxides in which one or two of the "external" O atoms in P_4O_{10} have been removed. The β-phase has a composition range $P_4O_{8.0}$–$P_4O_{7.7}$ and appears to be a solid solution of P_4O_8 and P_4O_7, the latter compound having only one O atom external to the P_4O_6 cluster (C_{3v} symmetry). P_4O_7 is now best prepared from P_4O_6 dissolved in thf, using Ph_3PO as a catalyst (not an oxidant) at room temperature. The molecular structure and dimensions of P_4O_7 are given in Fig. 12.9 from which it is apparent that there is a gradual lengthening of P–O distances in the sequence P^V–O_t < P^V–O_μ < P^{III}–O_μ. Similar trends are apparent in the dimensions of the other members of the series P_4O_{6+n} shown in Fig. 12.9.[42a] In addition, ring angles at P (96–103°) are always less than those at O (122–132°), as expected.

P_4O_6 hydrolyses in cold water to give H_3PO_3 (i.e. $H-\overset{\displaystyle O}{\underset{\displaystyle OH}{\overset{\|}{P}}}-OH$); this is interesting in view of the structure of P_4O_6 and implies an oxidative rearrangement of

$$P-OH \quad \text{to} \quad P\overset{\displaystyle O}{\underset{\displaystyle H}{\nwarrow}}$$

(p. 591). The oxide itself ignites and burns when heated in air; the progress of the reaction depends very much on the purity of the oxide and the conditions employed, and, when traces of elemental phosphorus are present in the oxide, the reaction is spontaneous even at room temperature. P_4O_6 reacts readily (often violently) with many simple inorganic and organic compounds but well-characterized products have rarely been isolated until recently.[42] It behaves as a ligand and successively displaces CO from $[Ni(CO)_4]$ to give compounds such as $[P_4O_6\{Ni(CO)_3\}_4]$, $[Ni(CO)_2(P_4O_6)_2]$, and $[Ni(CO)(P_4O_6)_3]$. With diborane adducts of formula $[P_4O_6(BH_3)_n]$ ($n = 1–3$) are obtained.

"Phosphorus pentoxide", P_4O_{10}, is the commonest and most important oxide of phosphorus. It is formed as a fine white smoke or powder when phosphorus burns in air and, when condensed rapidly from the vapour phase in this way, is obtained in the H (hexagonal) form comprising tetrahedral molecules as shown in Fig. 12.9. This compound and the other phosphorus oxides are the first we have considered that feature the $\{PO_4\}$ group as a structural unit; this group dominates most of phosphate chemistry and will recur repeatedly during the rest of this chapter. The common hexagonal form of P_4O_{10} is, in fact, metastable and can be transformed into several other modifications by suitable thermal or high-pressure treatment. A metastable orthorhombic (O) form is obtained by heating H for 2 h at 400° and the stable orthorhombic (O′) form is obtained after 24 h at

[42] D. HEINZE, Chemistry of phosphorus(III) oxide, *Pure Appl. Chem.* **44**, 141–72 (1975).
[42a] M. JANSEN and M. VOSS, Crystal structure of P_4O_7, *Angew. Chem.* Int. Ed. (Engl.) **20**, 100–01 (1981), and references therein to crystal structure determinations on the other members of the series P_4O_{6+n}.

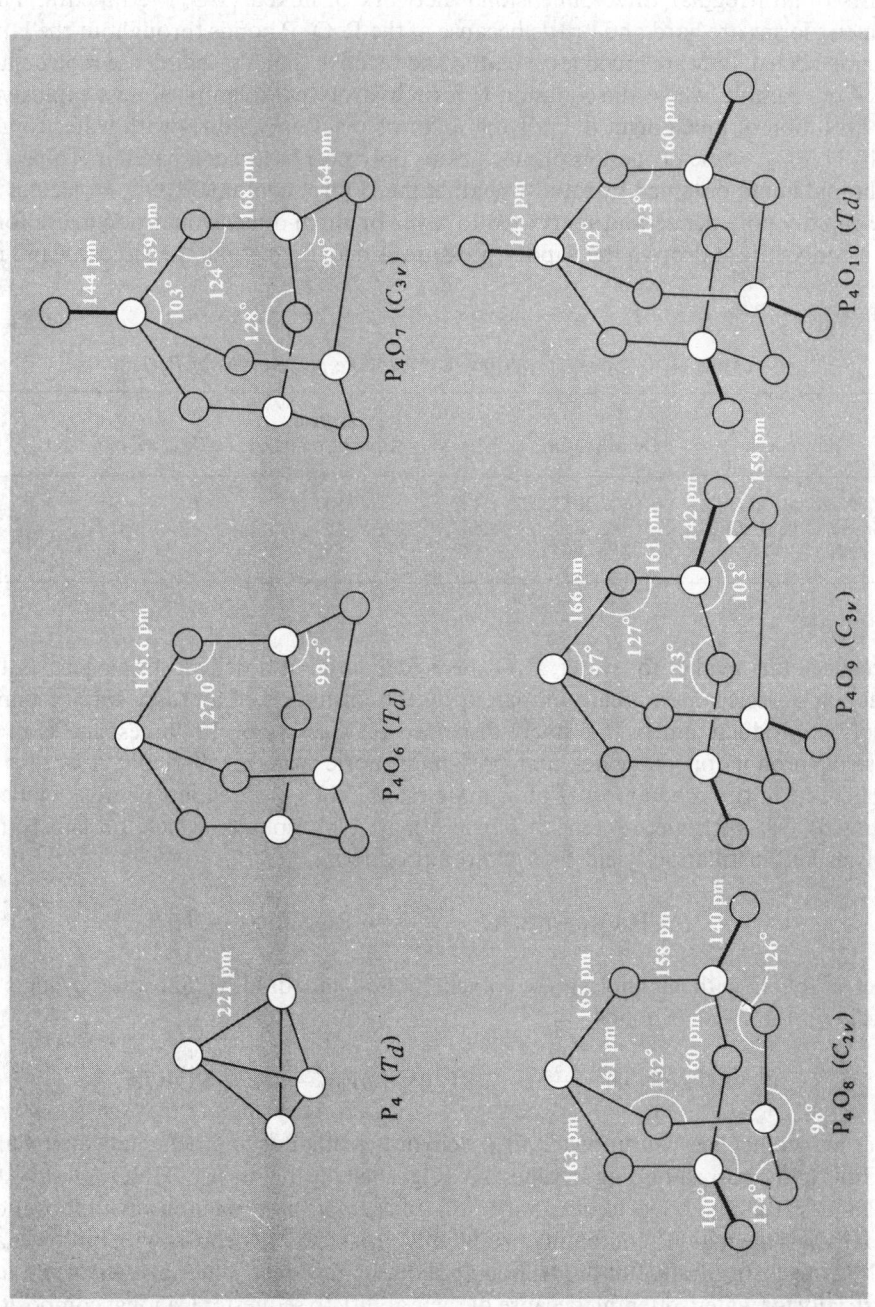

FIG. 12.9   Molecular structures, symmetries, and dimensions of the 5 oxides P_4O_{6+n} ($n = 0–4$) compared with α-P_4. The $P \cdots P$ distances in the oxides are $\sim 280–290$ pm, i.e. essentially nonbonding.

450°. Both consist of extensive sheet polymers of interlocking heterocyclic rings composed of fused {PO_4} groups. There is also a high-pressure form and a glass, which probably consists of an irregular three-dimensional network of linked {PO_4} tetrahedra. These polymeric forms are hard and brittle because of the P–O–P bonds throughout the lattice and, as expected, they are much less volatile and reactive than the less-dense molecular H form. For example, whilst the common H form hydrolyses violently, almost explosively, with evolution of much heat, the polymeric forms react only slowly with water to give, finally, H_3PO_4. Some properties of the various polymorphs are compared in Table 12.5. The limpid liquid obtained by rapidly heating the H form contains P_4O_{10} molecules but these rapidly polymerize and rearrange to layer or three-dimensional polymeric forms with a concomitant drop in the vapour pressure and an increase in the viscosity and mp.

TABLE 12.5   *Some properties of crystalline polymorphs of* P_2O_5

Polymorph	Density/g cm^{-3}	MP/°C	Pressure at triple pt/mmHg	ΔH_{subl}/kJ (mol P_4O_{10})$^{-1}$
H: hexagonal P_4O_{10}	2.30	420	3600	95
O: metastable $(P_2O_5)_n$	2.72	562	437	152
O′: stable $(P_2O_5)_n$	2.74–3.05	580	555	142

Because of its avidity for water, P_4O_{10} is widely used as a dehydrating agent, but its efficacy as a desiccant is greatly impaired by the formation of a crusty surface film of hydrolysis products unless it is finely dispersed on glass wool. Its largest use is in the industrial production of ortho- and poly-phosphoric acids (p. 600) but it is also an intermediate in the production of phosphate esters. Thus, triethylphosphate is made by reacting P_4O_{10} with diethyl ether to form ethylpolyphosphates which, on subsequent pyrolysis and distillation, yield the required product:

$$P_4O_{10} + 6Et_2O \xrightarrow{\text{heat}} 4\ PO(OEt)_3$$

Direct reaction with alcohols gives mixed mono- and di-alkyl phosphoric acids by cleavage of the P–O–P bonds:

$$P_4O_{10} + 6ROH \xrightarrow{65°} 2(RO)PO(OH)_2 + 2(RO)_2PO(OH)$$

Under less-controlled conditions P_4O_{10} dehydrates ethanol to ethene and methylaryl-carbinols to the corresponding styrenes. H_2SO_4 is dehydrated to SO_3, HNO_3 gives N_2O_5, and amides ($RCONH_2$) yield nitriles (RCN). In each of these reactions metaphosphoric acid HPO_3 is the main P-containing product. P_4O_{10} reacts vigorously with both wet and dry NH_3 to form a range of amorphous polymeric powdery materials which are used industrially for water softening because of their ability to sequester Ca ions; composition depends markedly on the preparative conditions employed but most of the commercial products appear to be condensed linear or cyclic amidopolyphosphates which can be represented by formulae as those at the top of the next page:

$$
\left[\begin{array}{c} \text{O} \\ \parallel \\ \text{HO} - \text{P} - \text{O} \\ \mid \\ \text{NH}_4\text{O} \end{array} \right.
\left[\begin{array}{ccc} \text{O} & \text{H} & \text{O} \\ \parallel & \mid & \parallel \\ \text{P} - \text{N} - \text{P} - \text{O} \\ \mid & & \mid \\ \text{ONH}_4 & \text{ONH}_4 \end{array} \right]_n
\left. \begin{array}{c} \text{O} \\ \parallel \\ \text{P} - \text{NH}_2 \\ \mid \\ \text{ONH}_4 \end{array} \right]
$$

$$
\begin{array}{ccccc}
& \text{O} & & \text{O} & \\
& \parallel & & \parallel & \\
\text{NH}_4\text{O} - & \text{P} & - \text{NH} - & \text{P} & - \text{ONH}_4 \\
& \mid & & \mid & \\
& \text{O} & & \text{O} & \\
& \mid & & \mid & \\
\text{NH}_4\text{O} - & \text{P} & - \text{NH} - & \text{P} & - \text{ONH}_4 \\
& \parallel & & \parallel & \\
& \text{O} & & \text{O} &
\end{array}
$$

Other oxides of phosphorus are less well characterized though the suboxide PO and the peroxide P_2O_6 seem to be definite compounds. PO was obtained as a brown cathodic deposit when a saturated solution of Et_3NHCl in anhydrous $POCl_3$ was electrolysed between Pt electrodes at 0°. Alternatively it can be made by the slow reaction of $POBr_3$ with Mg in Et_2O under reflux:

$$2POBr_3 + 3Mg \longrightarrow 2PO + 3MgBr_2$$

Its structure is unknown but is presumably based on a polymeric network of P–O–P links. It reacts with water to give PH_3 and is quantitatively oxidized to P_2O_5 by oxygen at 300°. The peroxide P_2O_6 is thought to be the active ingredient in the violet solid obtained when P_4O_{10} and O_2 are passed through a heated discharge tube at low pressure. The compound has not been obtained pure but liberates I_2 from aqueous KI, hydrolyses to a peroxophosphoric acid, and liberates O_2 when heated to 130° under reduced pressure. Its structure may be

$$
\begin{array}{ccc}
\text{O} & & \text{O} \\
\diagdown\diagdown & & \diagup\diagup \\
& \text{P} - \text{O} - \text{O} - \text{P} & \\
\diagup\diagup & & \diagdown\diagdown \\
\text{O} & & \text{O}
\end{array}
$$

or, in view of the variable composition of the product, it may be a mixture of P_4O_{11} and P_4O_{12} obtained by replacing P–O–P links by P–O–O–P in P_4O_{10}.

Sulfides[42b]

The sulfides of phosphorus form an intriguing series of compounds which continue to present puzzling structural features. The compounds P_4S_{10}, P_4S_9, P_4S_7, P_4S_5, α-P_4S_4, β-P_4S_4, and P_4S_3 are all based on the P_4 tetrahedron but only P_4S_{10} (and possibly P_4S_9) is structurally analogous to the oxide, and P_4S_6 is conspicuous by its absence. Structural data are summarized in Fig. 12.10 and some physical properties are in Table 12.6.

P_4S_3 is the most stable compound in the series and can be prepared by heating the required amounts of red P and sulfur above 180° in an inert atmosphere and then

---

[42b] H. HOFFMANN and M. BECKE-GOEHRING, Phosphorus sulfides, *Topics in Phosphorus Chemistry* **8**, 193–271 (1976); an excellent modern review of the history, phase equilibria, preparation, properties, and applications of the various phosphorus sulfides (466 refs.).

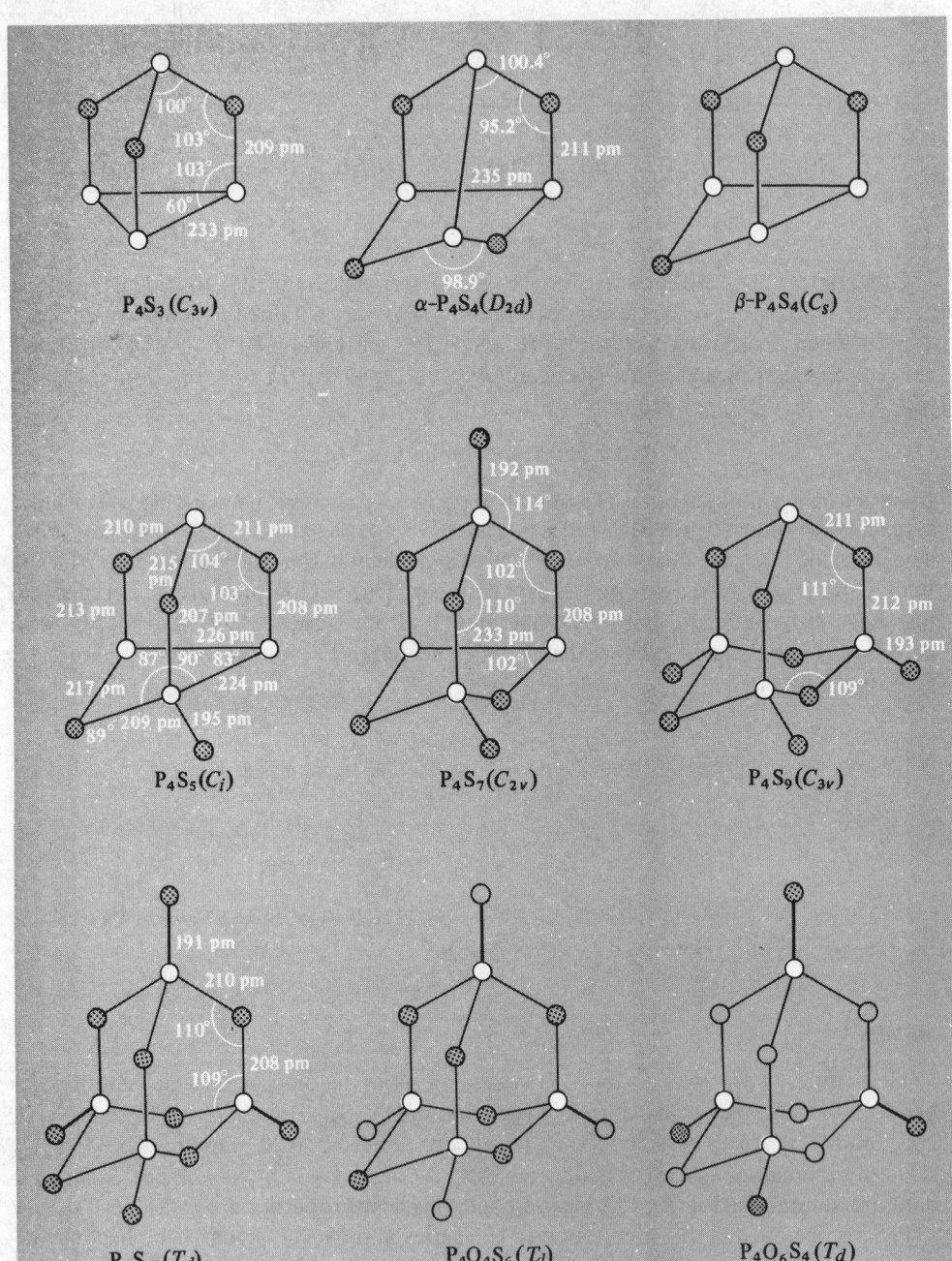

FIG. 12.10 Structures of phosphorus sulfides and oxosulfides (schematic).

TABLE 12.6 *Physical properties of some phosphorus sulfides*

Property	P_4S_3	α-P_4S_4	P_4S_5	P_4S_7	P_4S_{10}
Colour	Yellow	Pale yellow	Bright yellow	Very pale yellow	Yellow
MP/°C	174	230 (d)	170–220 (d)	308	288
BP/°C	408	—	—	523	514
Density/g cm^{-3}	2.03	2.22	2.17	2.19	2.09
Solubility in CS_2 (17°)/ g per 100 g CS_2	100	sol	0.5	0.029	0.222

purifying the product by distillation at 420° or by recrystallization from toluene.† The retention of a P_3 ring in the structure is notable. Its reactions and commercial application in match manufacture are discussed on p. 585.

P_4S_4 is the most recent sulfide to be isolated and characterized (1976) and it exists in two structurally distinct forms.[45, 46] Each can be made in quantitative yield by reacting the appropriate isomer of $P_4S_3I_2$ (p. 577) with [($Me_2Sn)_2S$] in CS_2 solution:

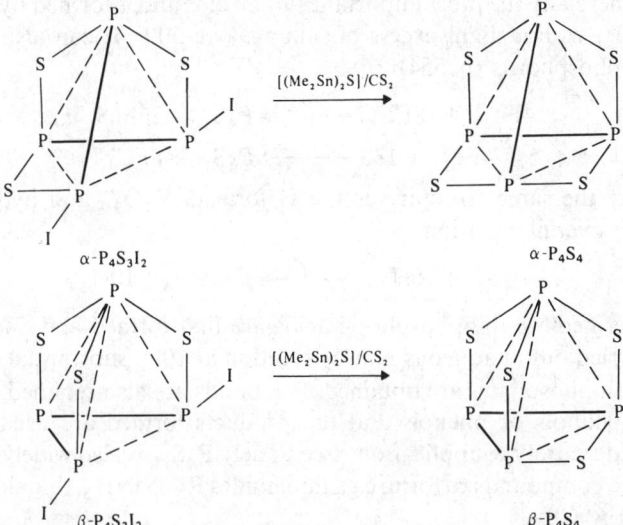

† The curious phase relations between phosphorus, sulfur, and their binary compounds are worth noting. Because both P_4 and S_8 are stable molecules the phase diagram, if studied below 100°, shows only solid solutions with a simple eutectic at 10° (75 atom % P). By contrast, when the mixtures are heated above 200° the elements react and an entirely different phase diagram is obtained; however, as only the most stable compounds P_4S_3, P_4S_7, and P_4S_{10} melt congruently, only these three appear as compounds in equilibrium with the melt. Careful work at lower temperatures is needed to detect peritectic equilibria involving P_4S_9, P_4S_5 (and possibly even P_4S_2),[43] and it is notable that these compounds are normally prepared by low-temperature reactions involving addition of 2S to P_4S_7 and P_4S_3 respectively. Likewise there is no sign of P_4S_4 on the phase diagram, and claims to have detected it in this way[44] have been shown to be erroneous.[45]

[43] H. VINCENT, Phosphorus–sulfur phase diagram, *Bull. Soc. Chim. France* 1972, 4517–21; R. FÖRTHMANN and A. SCHNEIDER, The system sulfur–phosphorus, *Z. Phys. Chem.* (NF) **49**, 22–37 (1966).

[44] H. VINCENT and C. VINCENT-FORAT, Phosphorus sulfides P_4S_9, P_4S_4, and P_4S_2, *Bull. Soc. Chim. France* 1973, 499–502.

[45] A. M. GRIFFIN, P. C. MINSHALL, and G. M. SHELDRICK, Two new molecular phosphorus sulfides: α-P_4S_4 and β-P_4S_4; X-ray crystal structure of α-P_4S_4, *JCS Chem. Comm.* 1976, 809–10.

[46] C.-C. CHANG, R. C. HALTIWANGER, and A. D. NORMAN, Synthesis of new phenylimido- and sulfido-tetraphosphorus ring and cage compounds, *Inorg. Chem.* **17**, 2056–62 (1978).

As seen from Fig. 12.10 the structure of α-P_4S_4 resembles that of As_4S_4 (p. 676) rather than N_4S_4 (p. 856). The 4 P atoms are in tetrahedral array and the 4 S atoms form a slightly distorted square. The 2 P–P bonds are long (as also in P_4S_3 and P_4S_7) when compared with corresponding distances in P_4S_5 (225 pm) and P_4 itself (221 pm). The structure of β-P_4S_4 has not been determined by X-ray crystallography but spectroscopic data indicate the absence of P=S groups and the C_s structure shown in Fig. 12.10 is the only other possible arrangement of 3 coordinate P for this composition.

P_4S_5 disproportionates below its mp ($2P_4S_5 \rightleftharpoons P_4S_3 + P_4S_7$) and so cannot be obtained directly from the melt. It is best prepared by irradiating a solution of P_4S_3 and S in CS_2 solution using a trace of iodine as catalyst. Its structure is quite unexpected and features a single exocyclic P=S group and 3 fused heterocycles containing, respectively, 4, 5, and 6 atoms; there are 2 short P–P bonds and the 4-membered P_3S ring is almost square planar.

P_4S_7 is the second most stable sulfide (after P_4S_3) and can be obtained by direct reaction of the elements. Perhaps surprisingly the structure retains a P–P bond and has two exocyclic P=S groups. P_4S_9 is formed reversibly by heating $P_4S_7 + 2P_4S_{10}$ and has the structure shown in Fig. 12.10.

P_4S_{10} is commercially the most important sulfide of P and is formed by direct reaction of liquid white P_4 with a slight excess of sulfur above 300°. It can also be made from byproduct ferrophosphorus (p. 554).

$$4Fe_2P + 18FeS_2 \xrightarrow{\text{heat}} P_4S_{10} + 26FeS$$

$$4Fe_2P + 18S \longrightarrow P_4S_{10} + 8FeS$$

It has essentially the same structure as the H form of P_4O_{10} and hydrolyses mainly according to the overall equation

$$P_4S_{10} + 16H_2O \longrightarrow 4H_3PO_4 + 10H_2S$$

Presumably intermediate thiophosphoric acids are first formed and, indeed, when the hydrolysis is carried out in aqueous NaOH solution at 100°, substantial amounts of the mono- and di-thiophosphates are obtained. P–S bonds are also retained during reaction of P_4S_{10} with alcohols or phenols and the products formed are used extensively in industry for a wide variety of applications (see Panel). P_4S_{10} is also widely used to replace O by S in organic compounds to form, e.g., thioamides $RC(S)NH_2$, thioaldehydes RCHS, and thioketones R_2CS.

A rather different series of cyclic thiophosphate(III) anions $[(PS_2)_n]^{n-}$ is emerging from a study of the reaction of elemental phosphorus with polysulfidic sulfur. Anhydrous compounds $M_5^I[cyclo\text{-}P_5S_{10}]$ and $M_6^I[cyclo\text{-}P_6S_{12}]$ were obtained using red phosphorus, whereas white P_4 yielded $[NH_4]_4[cyclo\text{-}P_4S_8] \cdot 2H_2O$ as shiny platelets. This unique $P_4S_8^{4-}$ anion is the first known homocycle of 4 tetracoordinated P atoms and X-ray studies reveal that the P atoms form a square with rather long P–P distances (228 pm).[46a]

Oxosulfides

When P_4O_{10} and P_4S_{10} are heated in appropriate proportions above 400°, $P_4O_6S_4$ is obtained as colourless hygroscopic crystals, mp 102°.

$$3P_4O_{10} + 2P_4S_{10} \longrightarrow 5P_4O_6S_4$$

46a H. FALIUS, W. KRAUSE, and W. S. SHELDRICK, $[NH_4]_4[P_4S_8] \cdot 2H_2O$, the salt of a "square" phosphoric acid, *Angew. Chem.* Int. Ed. (Engl.) **20**, 103–4 (1981).

Phosphorus Sulfides in Industry

The two compounds of importance are P_4S_3 and P_4S_{10}. The former is made on a large scale for use in "strike anywhere" matches according to a formula evolved by Sévène and Cahen in France in 1898. The ignition results from the violent reaction between P_4S_3 and $KClO_3$ which is initiated by friction of the match against glass paper (on the side of the box) or other abrasive material. A typical formulation for the match head is:

Reactants		Fillers (moderators)			Adhesives	
$KClO_3$	P_4S_3	Ground glass	Fe_2O_3	ZnO	Glue	Water
20%	9%	14%	11%	7%	10%	29%

Formulations of this type have completely replaced earlier "strike anywhere" matches based on (poisonous) white P_4, sulfur, and $KClO_3$, though "safety matches" still use a match head which is predominantly $KClO_3$ struck against the side of the match-box which has been covered with a paste of (non-toxic) red P (49.5%), antimony sulfide (27.6%), Fe_2O_3 (1.2%), and gum arabic (21.7%). About 10^{11} matches are used annually in the UK alone.

P_4S_{10} is made on an even larger scale than P_4S_3 and is the primary source of a very wide range of organic P–S compounds. World production of P_4S_{10} exceeds 250 000 tonnes annually of which about half is made in the USA, one-third in the UK/Europe, and the remaining 30 000 tonnes elsewhere (Japan, Romania, the USSR, Mexico, etc.). The most important reaction of P_4S_{10} is with alcohols or phenols to give dialkyl or diaryl dithiophosphoric acids:

$$P_4S_{10} + 8ROH \longrightarrow 4(RO)_2P(S)SH + 2H_2S$$

The zinc salts of these acids are extensively used as additives to lubricating oils to improve their extreme-pressure properties. The compounds also act as antioxidants, corrosion inhibitors, and detergents. Short-chain dialkyl dithiophosphates and their sodium and ammonium salts are used as flotation agents for zinc and lead sulfide ores. The methyl and ethyl derivatives $(RO)_2P(S)SH$ and $(RO)_2P(S)Cl$ are of particular interest in the large-scale manufacture of pesticides such as parathion, malathion, dimethylparathion, etc.[32] For example parathion, which first went into production as an insecticide in Germany in 1947, is made by the following reaction sequence:

$$(EtO)_2P(S)SH + Cl_2 \longrightarrow HCl + S + (EtO)_2P(S)Cl$$

$$(EtO)_2P(S)Cl + NaO\!-\!\langle\bigcirc\rangle\!-\!NO_2 \longrightarrow (EtO)_2P(S)O\!-\!\langle\bigcirc\rangle\!-\!NO_2$$

parathion: a monothiophosphate ester

Methylparathion is the corresponding dimethyl derivative. Later (1952) malathion found favour because of its decreased toxicity to mammals; it is readily made in 90% yield by the addition of dimethyldithiophosphate to diethylmaleate in the presence of NEt_3 as a catalyst and hydroquinone as a polymerization inhibitor:

$$(MeO)_2P(S)SH + \begin{matrix} CCHO_2Et \\ \| \\ CHCO_2Et \end{matrix} \xrightarrow[60\%]{NEt_3/HQ} \begin{matrix} (MeO)_2P(S)\,SCHCO_2Et \\ | \\ CH_2CO_2Et \end{matrix}$$

malathion: a dithiophosphate ester

The scale of manufacture of these organophosphorus pesticides can be guaged from data referring to the USA annual production in 1975 (tonnes): methylparathion 46 000, parathion 36 000, and malathion 16 000. In addition, some 15 other thioorganophosphorus insecticides are manufactured in the USA on a scale exceeding 2000 tonnes pa each.[6a] They act by inhibiting cholinesterase, thus preventing the natural hydrolysis of the neurotransmitter acetylcholine in the insect.[10a]

The structure is shown in Fig. 12.10. The related compound $P_4O_4S_6$ is said to be formed by the reaction of H_2S with $POCl_3$ at $0°$ (A. Besson, 1897) but has not been recently investigated. An amorphous yellow material of composition $P_4O_4S_3$ is obtained when a solution of P_4S_3 in CS_2 or organic solvents is oxidized by dry air or oxygen. Other oxosulfides of uncertain authenticity such as $P_6O_{10}S_5$ have been reported but their structural integrity has not been established and they may be mixtures. However, the following series can be prepared by appropriate redistribution reactions: $P_4O_6S_n(n=1-4)$, $P_4O_6Se_n$ $(n=1-3)$, P_4O_6SSe, $P_4O_7S_n$ $(n=1-3)$, P_4O_7Se, $P_4O_8S_n$ $(n=1, 2)$.[46b]

12.3.6 *Oxoacids of phosphorus and their salts*

The oxoacids of P are more numerous than those of any other element, and the number of oxoanions and oxo-salts is probably exceeded only by those of Si. Many are of great importance technologically and their derivatives are vitally involved in many biological processes (p. 611). Fortunately, the structural principles covering this extensive array of compounds are very simple and can be stated as follows:†

   (i) All P atoms in the oxoacids and oxoanions are 4-coordinate and contain at least one P=O unit.

   (ii) All P atoms in the oxoacids have at least one P—OH group and this often occurs in the anions also; all such groups are ionizable as proton donors.

   (iii) Some species also have one (or more) P–H group; such directly bonded H atoms are not ionizable.

---

† Heteropolyacids containing P fall outside this classification and are treated, together with the isopolyacids and their salts, on p. 1184. Organic esters such as $P(OR)_3$ are also excluded.

[46b] M. L. WALKER, D. E. PECKENPAUGH, and J. L. MILLS, Mixed phosphorus(III)-phosphorus(V) oxide chalcogenides, *Inorg. Chem.* **18**, 2792–6 (1979).

(iv) Catenation is by P–O–P links or via direct P–P bonds; with the former both open chain ("linear") and cyclic species are known but only corner sharing of tetrahedra occurs, never edge- or face-sharing.

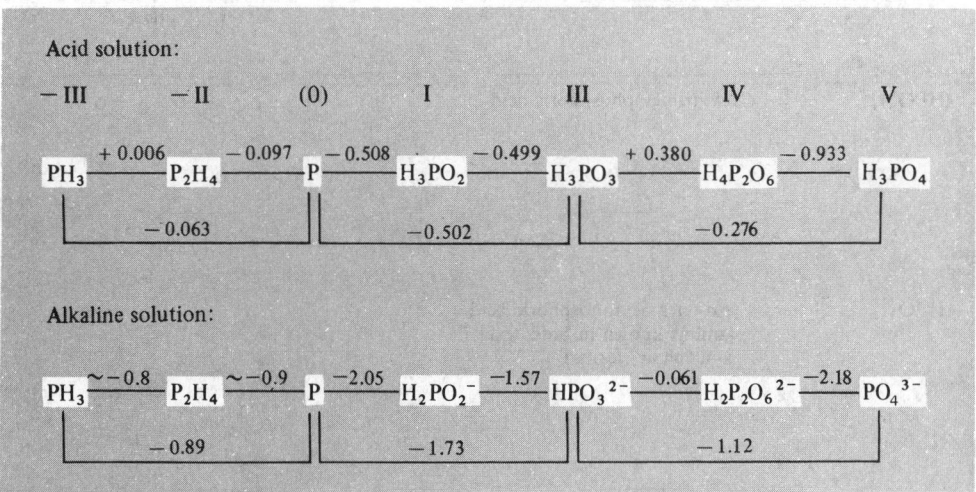

(v) Peroxo compounds feature either

    $\geqslant$P—OOH groups or $\geqslant$POOP$\leqslant$ links.

It follows from these structural principles that each P atom is 5-covalent. However, the oxidation state of P is 5 only when it is directly bound to 4 O atoms; the oxidation state is reduced by 1 each time a P–OH is replaced by a P–P bond and by 2 each time a P–OH is replaced by a P–H. Some examples of phosphorus oxoacids are listed in Table 12.7 together with their recommended and common names. It will be seen that the numerous structural types and the variability of oxidation state pose several problems of nomenclature which offer a rich source of confusion in the literature.

The oxoacids of P are clearly very different structurally from those of N (p. 528) and this difference is accentuated when the standard reduction potentials (p. 500) and oxidation-state diagrams (p. 501) for the two sets of compounds are compared. Some reduction potentials ($E°/V$) in acid solution are in Table 12.8[47] and these are shown schematically below, together with the corresponding data for alkaline solutions:

Acid solution:

$-$ III	$-$ II	(0)	I	III	IV	V

$$\text{PH}_3 \xrightarrow{+\,0.006} \text{P}_2\text{H}_4 \xrightarrow{-\,0.097} \text{P} \xrightarrow{-\,0.508} \text{H}_3\text{PO}_2 \xrightarrow{-\,0.499} \text{H}_3\text{PO}_3 \xrightarrow{+\,0.380} \text{H}_4\text{P}_2\text{O}_6 \xrightarrow{-\,0.933} \text{H}_3\text{PO}_4$$

$$\text{PH}_3 \underset{-\,0.063}{\rule{2cm}{0.4pt}} \text{P} \qquad \text{P} \underset{-\,0.502}{\rule{2cm}{0.4pt}} \text{H}_3\text{PO}_3 \qquad \text{H}_3\text{PO}_3 \underset{-\,0.276}{\rule{2cm}{0.4pt}} \text{H}_3\text{PO}_4$$

Alkaline solution:

$$\text{PH}_3 \xrightarrow{\sim\,-0.8} \text{P}_2\text{H}_4 \xrightarrow{\sim\,-0.9} \text{P} \xrightarrow{-\,2.05} \text{H}_2\text{PO}_2{}^- \xrightarrow{-\,1.57} \text{HPO}_3{}^{2-} \xrightarrow{-\,0.061} \text{H}_2\text{P}_2\text{O}_6{}^{2-} \xrightarrow{-\,2.18} \text{PO}_4{}^{3-}$$

$$\text{PH}_3 \underset{-\,0.89}{\rule{2cm}{0.4pt}} \text{P} \qquad \text{P} \underset{-\,1.73}{\rule{2cm}{0.4pt}} \qquad \underset{-\,1.12}{\rule{2cm}{0.4pt}}$$

The alternative presentation as an oxidation state diagram is in Fig. 12.11 which shows the dramatic difference to N (p. 502).

The fact that the element readily dissolves in aqueous media with disproportionation into PH$_3$ and an oxoacid is immediately clear from the fact that P lies above the line

[47] G. CHARLOT, A. COLLUMEAU, and M. J. C. MARCHON, *Selected Constants: Oxidation-Reduction Potentials of Inorganic Substances in Aqueous Solution*, Butterworths, London 1971, 73 pp. G. MILAZZO and S. CAROLI, *Tables of Standard Electrode Potentials*, Wiley, New York, 1978, 421 pp.

TABLE 12.7 *Some phosphorus oxoacids*[a]

Formula[b]	Name	Structure[a]
H_3PO_4	(Ortho)phosphoric acid	
$H_4P_2O_7$	Diphosphoric acid (pyrophosphoric acid)	
$H_5P_3O_{10}$	Triphosphoric acid	
$H_{n+2}P_nO_{3n+1}$	Polyphosphoric acid (*n* up to 17 isolated)	
$(HPO_3)_3$	*Cyclo*-trimetaphosphoric acid	
$(HPO_3)_4$	*Cyclo*-tetrametaphosphoric acid (anions known in both "boat" and "chair" forms)	
$(HPO_3)_n$	Polymetaphosphoric acid (see text for various conformations in salts)	

Continued

TABLE 12.7 *Some phosphorus oxoacids*[a]

Formula[b]	Name	Structure[a]
H_3PO_5	Peroxomonophosphoric acid	
$H_4P_2O_8$	Peroxodiphosphoric acid	
$H_4P_2O_6$	Hypophosphoric acid [diphosphoric(IV) acid]	
$H_4P_2O_6$	Isohypophosphoric acid [diphosphoric(III,V) acid]	
H_3PO_3 (2)[b]	Phosphonic acid (phosphorous acid)	
$H_4P_2O_5$ (2)[b]	Diphosphonic acid (diphosphorous or pyrophosphorous acid)	
H_3PO_2 (1)[b]	Phosphinic acid (hypophosphorous acid)	

[a] Some acids are known only as their salts in which one or more –OH group has been replaced by O^-.
[b] The number in parentheses after the formula indicates the maximum basicity, where this differs from the total number of H atoms in the formula.

TABLE 12.8 *Some reduction potentials in acid solution (pH 0)*[(a)]

Reaction	$E°/V$
$P + 3H^+ + 3e^- \rightleftharpoons PH_3(g)$	-0.063
$P + 2H^+ + 2e^- \rightleftharpoons \frac{1}{2}P_2H_4(g)$	-0.097
$\frac{1}{2}P_2H_4 + H^+ + e^- \rightleftharpoons PH_3$	$+0.006$
$H_3PO_2 + H^+ + e^- \rightleftharpoons P + 2H_2O$	-0.508
$H_3PO_3 + 3H^+ + 3e^- \rightleftharpoons P + 3H_2O$	-0.502
$H_3PO_4 + 5H^+ + 5e^- \rightleftharpoons P + 4H_2O$	-0.411
$H_3PO_3 + 2H^+ + 2e^- \rightleftharpoons H_3PO_2 + H_2O$	-0.499
$H_3PO_4 + 2H^+ + 2e^- \rightleftharpoons H_3PO_3 + H_2O$	-0.276
$H_3PO_4 + H^+ + e^- \rightleftharpoons \frac{1}{2}H_4P_2O_6 + H_2O$	-0.933
$\frac{1}{2}H_4P_2O_6 + H^+ + e^- \rightleftharpoons H_3PO_3$	$+0.380$

[(a)] P refers to white phosphorus, $\frac{1}{4}P_4(s)$.

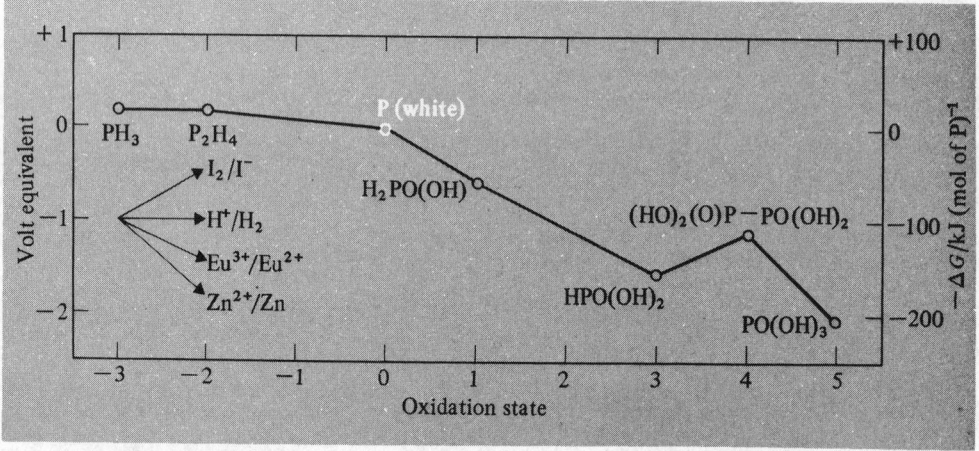

FIG. 12.11 Oxidation state diagram for phosphorus. (Note that all the oxoacids have a covalency of 5.)

joining PH_3 and either H_3PO_2 (hypophosphorous acid), H_3PO_3 (phosphorous acid), or H_3PO_4 (orthophosphoric acid). The reaction is even more effective in alkaline solution. Similarly, $H_4P_2O_6$ disproportionates into H_3PO_3 and H_3PO_4. Figure 12.11 also illustrates that H_3PO_2 and H_3PO_3 are both effective reducing agents, being readily oxidized to H_3PO_4, but this latter compound (unlike HNO_3) is not an oxidizing agent.

A comprehensive treatment of the oxoacids and oxoanions of P is inappropriate but selected examples have been chosen to illustrate interesting points of stereochemistry, reaction chemistry, or technological applications. The treatment begins with the lower oxoacids and their salts (in which P has an oxidation state less than $+5$) and then considers phosphoric acid, phosphates, and polyphosphates. The peroxoacids H_3PO_5 and $H_4P_2O_8$ and their salts will not be treated further,[(48)] nor will the peroxohydrates of orthophosphates obtained from aqueous H_2O_2 solutions.[(24)]

---

[48] I. I. CREASER and J. O. EDWARDS, Peroxophosphates, *Topics in Phosphorus Chemistry* **7**, 379–435 (1972).

Hypophosphorous acid and hypophosphites [$H_2PO(OH)$ and $H_2PO_2^-$]

The recommended names for these compounds (phosphinic acid and phosphinates) have not yet gained wide acceptance for inorganic compounds but are generally used for organophosphorus derivatives. Hypophosphites can be made by heating white phosphorus in aqueous alkali:

$$P_4 + 4OH^- + 4H_2O \xrightarrow[\text{[NaOH/Ca(OH)}_2]}{\text{warm}} 4H_2PO_2^- + 2H_2$$

Phosphite and phosphine are obtained as byproducts (p. 564) and the former can be removed via its insoluble calcium salt:

$$P_4 + 4OH^- + 2H_2O \xrightarrow{2\,Ca^{2+}} Ca(HPO_3)_2 + 2PH_3$$

Free hypophosphorous acid is obtained by acidifying aqueous solutions of hypophosphites but the pure acid cannot be isolated simply by evaporating such solutions because of its ready oxidation to phosphorous and phosphoric acids and disproportionation to phosphine and phosphorous acid (Fig. 12.11). Pure H_3PO_2 is obtained by continuous extraction from aqueous solutions into Et_2O; it forms white crystals mp 26.5° and is a monobasic acid pK 1.1.

During the past 10–20 y hydrated sodium hypophosphite, $NaH_2PO_2.H_2O$, has been increasingly used as an industrial reducing agent, particularly for the electroless plating of Ni onto both metals and non-metals.[49, 49a] This developed from an accidental discovery by A. Brenner and Grace E. Riddel at the National Bureau of Standards, Washington, in 1944. Acid solutions ($E \sim -0.40$ V at pH 4–6 and $T > 90°$) are used to plate thick Ni layers on to other metals, but more highly reducing alkaline solutions (pH 7–10; T 25–50°) are used to plate plastics and other non-conducting materials:

$$HPO_3^{2-} + 2H_2O + 2e^- \rightleftharpoons H_2PO_2^- + 3OH^-; \quad E \sim -1.57 \text{ V}$$

Typical plating solutions contain 10–30 g/l of nickel chloride or sulfate and 10–50 g/l NaH_2PO_2; with suitable pump capacities it is possible to plate up to 10 kg Ni per hour from such a bath (i.e. 45 m² surface to a thickness of 25 μm). Chemical plating is more expensive than normal electrolytic plating but is competitive when intricate shapes are being plated and is essential for non-conducting substrates.

Phosphorous acid and phosphites [$HPO(OH)_2$ and HPO_3^{2-}]

Again, the recommended names (phosphonic acid and phosphonates) have found more general acceptance for organic derivatives such as RPO_3^{2-}, and purely inorganic salts are still usually called phosphites. The free acid is readily made by direct hydrolysis of PCl_3 in cold CCl_4 solution:

$$PCl_3 + 3H_2O \longrightarrow HPO(OH)_2 + 3HCl$$

[49] H. NIEDERPRÜM, Chemical nickel plating, *Angew. Chem.* Int. Edn. (Engl.) **14**, 614–20 (1975); see also use of BH_4^- in this connection (p. 190).
[49a] G. A. KRULIK, Electroless plating, *Kirk–Othmer Encyclopedia of Chemical Technology*, 3rd edn., Vol. 8, pp. 738–50, Wiley, New York, 1979.

On an industrial scale PCl_3 is sprayed into steam at 190° and the product sparged of residual water and HCl using nitrogen at 165°. Phosphorous acid forms colourless, deliquescent crystals, mp 70.1°, in which the structural units shown are H bonded to give a complex network with $O{-}H\cdots\sim260$ pm. In aqueous solutions phosphorous acid is

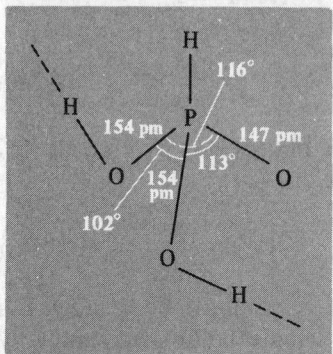

dibasic (pK_1 1.3, pK_2 6.7), and forms two series of salts: phosphites and hydrogen phosphites (acid phosphites), e.g.

"normal": $[NH_4]_2[HPO_3].H_2O$, $Li_2[HPO_3]$, $Na_2[HPO_3].5H_2O$, $K_2[HPO_3]$

"acid": $[NH_4][HPO_2(OH)]$, $Li[HPO_2(OH)]$, $Na[HPO_2(OH)].2\tfrac{1}{2}H_2O$, $K[HPO_2(OH)]$, and $M[HPO_2(OH)]_2$ (M = Mg, Ca, Sr).

Dehydration of these acid phosphites by warming under reduced pressure leads to the corresponding pyrophosphites $M_2^I[HP(O)_2\text{–}O\text{–}P(O)_2H]$ and $M^{II}[HP(O)_2\text{–}O\text{–}P(O)_2H]$.

Organic derivatives fall into 4 classes $RPO(OH)_2$, $HPO(OR)_2$, $R'PO(OR)_2$, and the phosphite esters $P(OR)_3$; this latter class has no purely inorganic analogues, though it is, of course, closely related to PCl_3. Some preparative routes have already been indicated. Reactions with alcohols depend on conditions:

$$PCl_3 + 3ROH \longrightarrow HPO(OR)_2 + RCl + 2HCl$$

$$PCl_3 + 3ROH + 3R_3'N \longrightarrow P(OR)_3 + 3R_3'NHCl$$

Phenols give triaryl phosphites $P(OAr)_3$ directly at $\sim160°$ and these react with phosphorous acid to give diaryl phosphonates:

$$2P(OAr)_3 + HPO(OH)_2 \longrightarrow 3HPO(OAr)_2$$

Trimethyl phosphite $P(OMe)_3$ spontaneously isomerizes to methyl dimethylphosphonate $MePO(OMe)_2$, whereas other trialkyl phosphites undergo the Michaelis–Arbusov reaction with alkyl halides via a phosphonium intermediate:

$$P(OR)_3 + R'X \rightarrow \{[R'P(OR)_3]X\} \longrightarrow R'PO(OR)_2 + RX$$

Further discussion of these fascinating series of reactions falls outside our present scope.[2]

Hypophosphoric acid ($H_4P_2O_6$) and hypophosphates

There has been much confusion over the structure of these compounds but their diamagnetism has long ruled out a monomeric formulation, H_2PO_3. In fact, as shown in Table 12.7, isomeric forms are known: (a) hypophosphoric acid and hypophosphates in which both P atoms are identical and there is a direct P–P bond; (b) isohypophosphoric acid and isohypophosphates in which 1 P has a direct P–H bond and the 2 different P atoms are joined by a P^{III}–O–P^V link.[15]

Hypophosphoric acid, $(HO)_2P(O)$–$P(O)(OH)_2$, is usually prepared by the controlled oxidation of red P with sodium chlorite solution at room temperature: the tetrasodium salt, $Na_4P_2O_6.10H_2O$, crystallizes at pH 10 and the disodium salt at pH 5.2:

$$2P + 2NaClO_2 + 8H_2O \longrightarrow Na_2H_2P_2O_6 . 6H_2O + 2HCl$$

Ion exchange on an acid column yields the crystalline "dihydrate" $H_4P_2O_6.2H_2O$ which is actually the hydroxonium salt of the dihydrogen hypophosphate anion $[H_3O]_2^+[(HO)P(O)_2–P(O)_2(OH)]^{2-}$; it is isostructural with the corresponding ammonium salt for which X-ray diffraction studies establish the staggered structure shown.

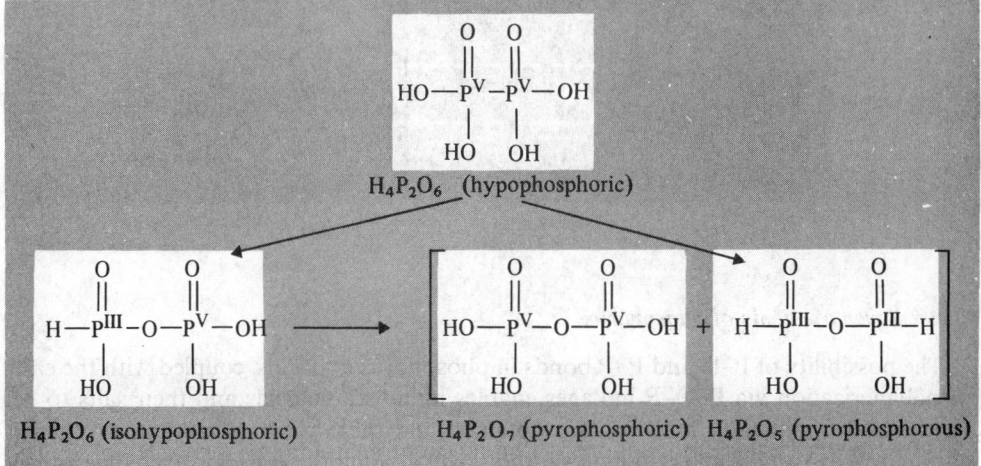

The anhydrous acid is obtained either by the vacuum dehydration of the dihydrate over P_4O_{10} or by the action of H_2S on the insoluble lead salt $Pb_2P_2O_6$. As implied above, the first proton on each $–PO(OH)_2$ unit is more readily removed than the second and the successive dissociation constants at $25°$ are pK_1 2.2, pK_2 2.8, pK_3 7.3, pK_4 10.0. Both $H_4P_2O_6$ and its dihydrate are stable at $0°$ in the absence of moisture. The acid begins to melt (with decomposition) at $73°$ but even at room temperature it undergoes rearrangement and disproportionation to give a mixture of isohypophosphoric, pyrophosphoric, and pyrophosphorous acids as represented schematically below:

Hypophosphoric acid is very stable towards alkali and does not decompose even when heated with 80% NaOH at 200°. However, in acid solution it is less stable and even at 25° hydrolyses at a rate dependent on pH (e.g. $t_{\frac{1}{2}}$ 180 days in 1 M HCl, $t_{\frac{1}{2}}$ <1 h in 4 M HCl):

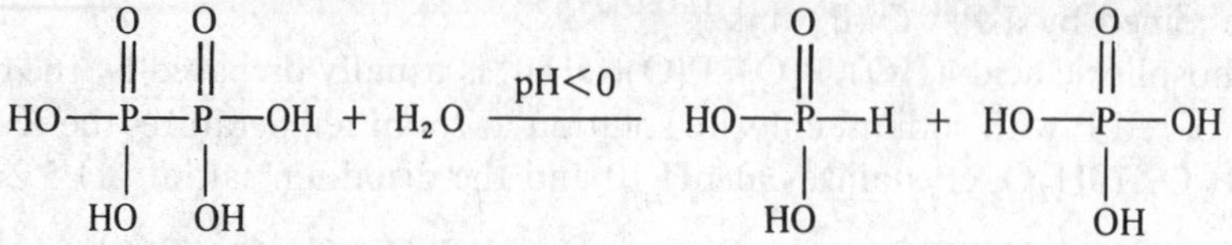

$$HO\!-\!\underset{\underset{HO}{|}}{\overset{\overset{O}{\|}}{P}}\!-\!\underset{\underset{OH}{|}}{\overset{\overset{O}{\|}}{P}}\!-\!OH + H_2O \xrightarrow{pH<0} HO\!-\!\underset{\underset{HO}{|}}{\overset{\overset{O}{\|}}{P}}\!-\!H \;+\; HO\!-\!\underset{\underset{OH}{|}}{\overset{\overset{O}{\|}}{P}}\!-\!OH$$

The presence of P–H groups amongst the products of these reactions was one of the earlier sources of confusion in the structures of hypophosphoric and isohypophosphoric acids.

The structure of isohypophosphoric acid and its salts can be deduced from ^{31}P nmr which shows the presence of 2 different 4-coordinate P atoms, the absence of a P–P bond, and the presence of a P–H group (also confirmed by Raman spectroscopy). It is made by the careful hydrolysis of PCl$_3$ with the stoichiometric amounts of phosphoric acid and water at 50°:

$$PCl_3 + H_3PO_4 + 2H_2O \xrightarrow{50°} iso\text{-}[H_3(HP_2O_6)] + 3HCl$$

The trisodium salt is best made by careful dehydration of an equimolar mixture of hydrated disodium hydrogen phosphate and sodium hydrogen phosphite at 180°:

$$Na_2HPO_4 . 12H_2O + NaH_2PO_3 . 2\tfrac{1}{2}H_2O \xrightarrow{180°} Na_3[HP_2O_6] + 15\tfrac{1}{2}H_2O$$

The structural relation between the reacting anions and the product is shown schematically below:

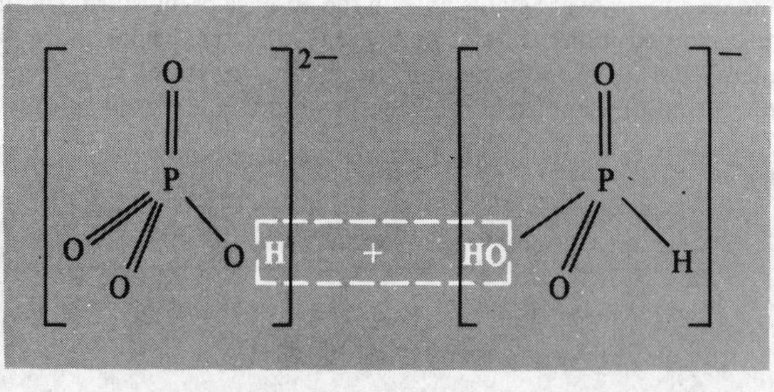

Other lower oxoacids of phosphorus

The possibility of P–H and P–P bonds in phosphorus oxoacids, coupled with the ease of polymerization via P–O–P linkages enables innumerable acids and their salts to be synthesized. Frequently mixtures are obtained and these can be separated by paper chromatography, paper electrophoresis, thin-layer chromatography, ion exchange, or gel chromatography.[50] Much ingenuity has been expended in designing appropriate syntheses but no new principles emerge. A few examples are listed in Table 12.9 to illustrate both the range of compounds available and also the abbreviated notation, which proves to be more convenient than formal systematic nomenclature in this area. In this notation the sequence of P–P and P–O–P links is indicated and the oxidation state of each P is shown as a superscript numeral which enables the full formula (including P–H groups) to be deduced.

The phosphoric acids

This section deals with orthophosphoric acid (H_3PO_4), pyrophosphoric acid ($H_4P_2O_7$), and the polyphosphoric acids ($H_{n+2}P_nO_{3n+1}$). Several of these compounds can be isolated pure but their facile interconversion renders this area of phosphorus chemistry far more complex than might otherwise appear. The corresponding phosphate salts are discussed in subsequent sections as also are the cyclic metaphosphoric acids (HPO_3)$_n$, the polymetaphosphoric acids (HPO_3)$_n$, and their salts.

Orthophosphoric acid is a remarkable substance: it can only be obtained pure in the crystalline state (mp 42.35°C) and when fused it slowly undergoes partial self-dehydration to diphosphoric acid:

$$2H_3PO_4 \rightleftharpoons H_2O + H_4P_2O_7$$

The sluggish equilibrium is obtained only after several weeks near the mp but is more rapid at higher temperatures. This process is accompanied by extremely rapid autoprotolysis (see below) which gives rise to several further (ionic) species in the melt. As the concentration of these various species builds up the mp slowly drops until at equilibrium it is 34.6°, corresponding to about 6.5 mole% of diphosphate.[51] Slow crystallization of stoichiometric molecular H_3PO_4 from this isocompositional melt gradually reverses the equilibria and the mp eventually rises to the initial value. Crystalline H_3PO_4 has a hydrogen-bonded layer structure in which each $PO(OH)_3$ molecule is linked to 6 others by H bonds which are of two lengths, 253 and 284 pm. The shorter bonds link OH and O=P groups whereas the longer H bonds are between 2 OH groups on adjacent molecules.

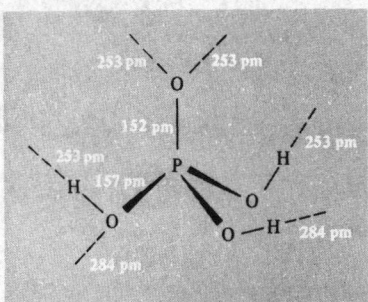

[50] S. OHASHI, Chromatography of phosphorus oxoacids, *Pure Appl. Chem.* **44**, 415–38 (1975).

[51] N. N. GREENWOOD and A. THOMPSON, The mechanism of electrical conduction in fused phosphoric and trideuterophosphoric acids, *J. Chem. Soc.* 1959, 3485–92; Anomalous conduction in phosphoric acid hemihydrate, *J. Chem. Soc.* 1959, 3864–7.

TABLE 12.9 *Some lower oxoacids of phosphorus*
(Superscript numerals in the abbreviated notation indicate oxidation states)

Formula (basicity)	Structure	Abbreviated notation
$H_4P_2O_4$ (2)		$\overset{2}{P}-\overset{2}{P}$
$H_4P_2O_5$ (3)		$\overset{2}{P}-\overset{4}{P}$
$H_4P_2O_6$ (3)		$\overset{3}{P}-O-\overset{5}{P}$
$H_5P_3O_7$ (4)		$\overset{3}{P}-O-\overset{4}{P}-\overset{4}{P}$
$H_5P_3O_8$ (5)		$\overset{4}{P}-\overset{3}{P}-\overset{4}{P}$
$H_5P_3O_9$ (5)		$\overset{5}{P}-O-\overset{4}{P}-\overset{4}{P}$
$H_6P_4O_{11}$ (6)		$\overset{4}{P}-\overset{4}{P}-O-\overset{4}{P}-\overset{4}{P}$
$H_4P_4O_{10}$ (4)		$(-\overset{4}{P}-\overset{4}{P}-O-)_2$ ring
$H_6P_6O_{12}$ (6)		$(-\overset{3}{P}-)_6$ ring

Extensive H bonding persists on fusion and phosphoric acid is a viscous syrupy liquid that readily supercools. At 45°C (just above the mp) the viscosity is 76.5 centipoise (cP) and this increases to 177.7 cP at 25°. These values can be compared with 1.00 cP for H_2O at 20° and 24.5 cP for anhydrous H_2SO_4 at 25°. As shown in Table 12.10, trideuterophosphoric acid has an even higher viscosity and deuteration also raises the mp and density.

TABLE 12.10 *Some physical properties of orthophosphoric acid, trideuterophosphoric acid, and the hemihydrate, $H_3PO_4 \cdot \frac{1}{2}H_2O$*[51]

Property	H_3PO_4	D_3PO_4	$H_3PO_4 \cdot \frac{1}{2}H_2O$
MP/°C	42.35	46.0	29.30
Density (25°C); supercooled/g cm^{-3}	1.8683	1.9083	1.7548
η (25°C)/centipoise	177.5	231.8	70.64
κ/ohm^{-1} cm^{-1}	4.68×10^{-2}	2.82×10^{-2}	7.01×10^{-2}

Despite this enormous viscosity, fused H_3PO_4 (and D_3PO_4) conduct electricity extremely well and this has been shown to arise from extensive self-ionization (autoprotolysis) coupled with a proton-switch conduction mechanism for the $H_2PO_4^-$ ion:[51, 52]

$$2H_3PO_4 \rightleftharpoons H_4PO_4^+ + H_2PO_4^- \qquad (1)$$

In addition, the diphosphate group is also deprotonated:

$$\left. \begin{array}{l} 2H_3PO_4 \rightleftharpoons H_2O + H_4P_2O_7 \rightleftharpoons H_3O^+ + H_3P_2O_7^- \\ H_3P_2O_7^- + H_3PO_4 \rightleftharpoons H_4PO_4^+ + H_2P_2O_7^{2-} \end{array} \right\}$$

i.e. $\qquad 3H_3PO_4 \rightleftharpoons H_3O^+ + H_4PO_4^+ + H_2P_2O_7^{2-} \qquad (2)$

At equilibrium, the concentration of H_3O^+ and $H_2P_2O_7^{2-}$ are each ~ 0.28 molal and $H_2PO_4^-$ is ~ 0.26 molal, thereby implying a concentration of 0.54 molal for $H_4PO_4^+$. These values are about 20–30 times greater than the concentrations of ions in molten H_2SO_4, namely $[HSO_4^-]$ 0.0178 molal, $[H_3SO_4^+]$ 0.0135 molal, and $[HS_2O_7^-]$ 0.0088 molal (see p. 843). Because of the very high viscosity of molten H_3PO_4 electrical conduction by normal ionic migration is negligible and the high conductivity is due almost entirely to a rapid proton-switch followed by a relatively slow reorientation involving the $H_2PO_4^-$ ion H-bonded to the solvent structure (Fig. 12.12).[51] Note that the tetrahedral $H_4PO_4^+$ ion, i.e. $[P(OH)_4]^+$, like the NH_4^+ ion in liquid NH_3, does not contribute to the proton-switch conduction mechanism in H_3PO_4 because, having no dipole moment, it does not orient preferentially in the applied electric field; accordingly any proton switching will occur randomly in all directions independently of the applied field and therefore will not contribute to the electrical conduction.

Addition of the appropriate amount of water to anhydrous H_3PO_4, or crystallization from a concentrated aqueous solution of syrupy phosphoric acid, yields the hemihydrate

[52] R. A. MUNSON, Self-dissociative equilibria in molten phosphoric acid, *J. Phys. Chem.* **68**, 3374–7 (1964).

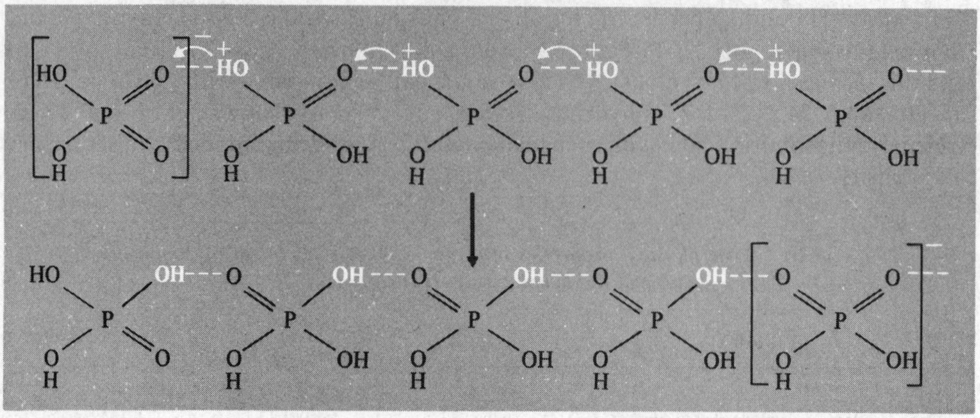

Fig. 12.12 Schematic representation of proton switch conduction mechanism involving $[H_2PO_4]^-$ in molten phosphoric acid.

$2H_3PO_4.H_2O$ as a congruently melting compound (mp 29.3°). The crystal structure[53] shows the presence of 2 similar H_3PO_4 molecules which, together with the H_2O molecule, are linked into a three-dimensional H-bonded network: each of the nine O atoms participates in at least 1 relatively strong O–H···O bond (255–272 pm) and the interatomic distances P=O (149 pm) and P–OH (155 pm) are both slightly shorter than the corresponding distances in H_3PO_4. Hydrogen bonding persists in the molten compound, and the proton-switch conductivity is even higher than in the anhydrous acid (Table 12.10).

In dilute aqueous solutions H_3PO_4 behaves as a strong acid but only one of the hydrogens is readily ionizable, the second and third ionization constants decreasing successively by factors of $\sim 10^5$ (see p. 54). Thus, at 25°:

$$H_3PO_4 + H_2O \rightleftharpoons H_3O^+ + H_2PO_4^-; \quad K_1 = 7.11 \times 10^{-3}; \quad pK_1 = 2.15$$

$$H_2PO_4^- + H_2O \rightleftharpoons H_3O^+ + HPO_4^{2-}; \quad K_2 = 6.31 \times 10^{-8}; \quad pK_2 = 7.20$$

$$HPO_4^{2-} + H_2O \rightleftharpoons H_3O^+ + PO_4^{3-}; \quad K_3 = 4.22 \times 10^{-13}; \quad pK_3 = 12.37$$

Accordingly, the acid gives three series of salts, e.g. NaH_2PO_4, Na_2HPO_4, and Na_3PO_4 (p. 603). A typical titration curve in this system is shown in Fig. 12.13: there are three steps with two inflexions at pH 4.5 and 9.5. The first inflexion, corresponding to the formation of NaH_2PO_4, can be detected by an indicator such as methyl orange (pK_i 3.5) the second, corresponding to Na_2HPO_4, is indicated by the phenolphthalein end point (pK_i 9.5). The third equivalence point cannot be detected directly by means of a coloured indicator. Between the two inflexions the pH changes relatively slowly with addition of NaOH and

---

[53] A. D. MIGHELL, J. P. SMITH, and W. E. BROWN, The crystal structure of phosphoric acid hemihydrate, $H_3PO_4 . \frac{1}{2}H_2O$, *Acta Cryst.* **B25**, 776–81 (1969).

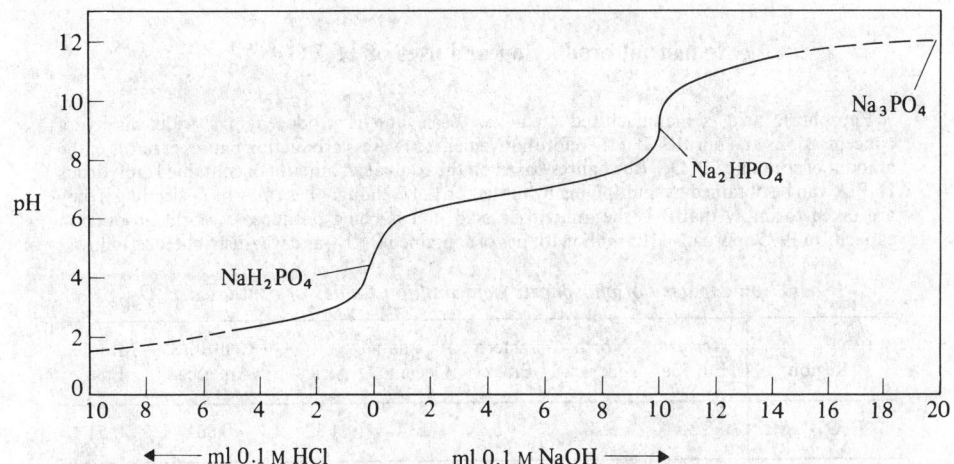

FIG. 12.13   Neutralization curve for aqueous orthophosphoric acid. For technical reasons the curve shown refers to 10 cm³ of 0.1 M NaH₂PO₄ titrated (to the left) with 0.1 M aqueous HCl and (to the right) with 0.1 M NaOH solutions. Extrapolations to points corresponding to 0.1 M H₃PO₄ (pH 1.5) and 0.1 M Na₃PO₄ (pH 12.0) are also shown.

this is an example of buffer action.† Indeed, one of the standard buffer solutions used in analytical chemistry comprises an equimolar mixture of Na_2HPO_4 and KH_2PO_4. Another important buffer, which has been designed to have a pH close to that of blood, consists of 0.030 43 M Na_2HPO_4 and 0.008 695 M KH_2PO_4, i.e. a mole ratio of 3.5:1 (pH 7.413 at 25°).

Concentrated H_3PO_4 is one of the major acids of the chemical industry and is manufactured on the multimillion-tonne scale for the production of phosphate fertilizers and for many other purposes (see Panel). Two main processes (the so-called "thermal" and "wet" processes) are used depending on the purity required. The "thermal" (or "furnace") process yields concentrated acid essentially free from impurities and is used in applications involving products destined for human consumption (see also p. 604); in this process a spray of molten phosphorus is burned in a mixture of air and steam in a stainless steel combustion chamber:

$$P_4 + 5O_2 \xrightarrow{\hspace{1.5cm}} P_4O_{10} \xrightarrow{6H_2O} 4H_3PO_4$$

Acid of any concentration up to 84 wt% P_4O_{10} can be prepared by this method (72.42% P_4O_{10} corresponds to anhydrous H_3PO_4) but the usual commercial grades are 75–85% (expressed as anhydrous H_3PO_4). The hemihydrate (p. 597) corresponds to 91.58% H_3PO_4 (66.33% P_4O_{10}). The somewhat older "wet" (or "gypsum") process involves the

† A buffer solution is one that resists changes in pH on dilution or on addition of acid or alkali. It consists of a solution of a weak acid (e.g. $H_2PO_4^-$) and its conjugate base (HPO_4^{2-}) and is most effective when the concentration of the two species are the same. For example at 25° an equimolar mixture of Na_2HPO_4 and KH_2PO_4 has pH 6.654 when each is 0.2 M and pH 6.888 when each is 0.01 M. The central section of Fig. 12.13 shows the variation in pH of an equimolar buffer of Na_2HPO_4 and NaH_2PO_4 at a concentration of 0.033 M (you should check this statement). Further discussion of buffer solutions is given in standard textbooks of volumetric analysis.

Industrial production and uses of H_3PO_4[3–7]

Phosphoric acid is manufactured on a vast scale and is produced in a wide variety of concentrations and purities. It is therefore convenient to express production figures in terms of the amount of contained P_4O_{10} (the figures based on the equivalent amount of contained anhydrous H_3PO_4 can be obtained by multiplying by the factor 1.380, though these may be misleading if they are taken to imply that it is the anhydrous acid that is being produced). World production capacity in 1975 approached 25 million tonnes of contained P_4O_{10} and was distributed as follows:

Production capacity of phosphoric acid (million tonnes of contained P_4O_{10})

Region	Western Europe	North America	Eastern Block	Asia and Oceania	Africa	Central/S. America	Middle East
"P_4O_{10}"/Mt	8.35	5.28	~5	2.34	1.32	0.86	0.54

Production is still increasing steadily in most countries, especially for fertilizer manufacture, and the 1981 figures for the USA alone were 9.3 million tonnes ("P_4O_{10}"). "Thermal" acid (made by oxidation of phosphorus in the presence of water vapour) is about 3 times as expensive as "wet" acid (made by treating rock phosphate with sulfuric acid). It costs ~£265 per tonne compared with ~£85 per tonne at 1976 UK prices and is being increasingly substituted by wet process acid in all but the most demanding applications. Corresponding 1978 US prices were $500 and $275 per tonne ("$P_4O_{10}$") respectively. The present approximate pattern of production and uses is shown in the following scheme:

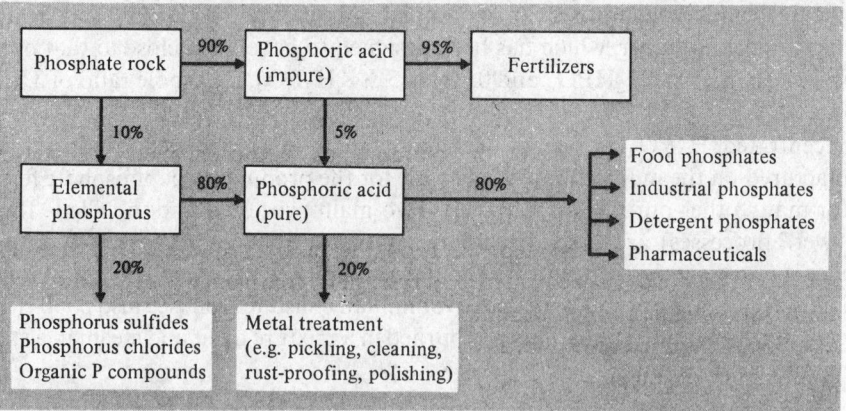

Many of these uses have already been discussed, or will be discussed further in later sections (pp. 603, 609).

Applications of phosphoric acid in metal treatment date from 1869 when a British patent was granted for the prevention of rusting of corset stays by damp air or perspiration; the process involved plunging the red-hot metal into H_3PO_4 solution. Improvements followed the incorporation of certain metal ions in the phosphatizing solution (notably Mn, Fe, and Zn), and today corrosion resistance is imparted in this way to innumerable metal objects such as nuts, bolts, screws, tools, car-engine parts, gears, etc. In addition, car-bodies, refrigerators, washing machines, and other electrical applicances with painted or enamelled surfaces all use phosphatized undercoatings to prevent the paint from blistering or peeling. The simple immersion process may take up to 2 h at 90°C but can be accelerated by adding small amounts of an oxidizing agent such as $NaNO_3$; addition of 0.002–0.4% $Cu(NO_3)_2$ accelerates the process even more by a mechanism that is not yet well understood, and under these conditions the coating time can be less than 5 min.

A zinc phosphatized coating is usually about 0.6 μm thick (i.e. 0.025 thousandths of an inch) corresponding to about 2.2 g m^{-2}.

Another important process is "bright dip" or chemical polishing of Al metal which is now replacing chrome plating for car trims and other uses: the metal is immersed at 91–99°C in a solution containing 95 parts by weight of 85% H_3PO_4, 4 parts of 68% HNO_3, and 0.01% $Cu(NO_3)_2$, followed by electrolytic anodization to give the mirror-like surface a protective coating of transparent Al_2O_3.

Polyphosphoric acid supported on diatomaceous earth (p. 394) is a petrochemicals catalyst for the polymerization, alkylation, dehydrogenation, and low-temperature isomerization of hydrocarbons. Phosphoric acid is also used in the production of activated carbon (p. 301). In addition to its massive use in the fertilizer industry (p. 605) free phosphoric acid can be used as a stabilizer for clay soils: small additions of H_3PO_4 under moist conditions gradually leach out Al and Fe from the clay and these form polymeric phosphates which bind the clay particles together. An allied, though more refined use is in the setting of dental cements.

By far the greatest consumption of pure aqueous phosphoric acid is in the preparation of various salts for use in the food, detergent, and tooth-paste industries (p. 603). When highly diluted the free acid is non-toxic and devoid of odour, and is extensively used to impart the sour or tart taste to many soft drinks ("carbonated beverages") such as the various colas ($\sim0.05\%$ H_3PO_4, pH 2.3), root beers ($\sim0.01\%$ H_3PO_4, pH 5.0), and sarsaparilla ($\sim0.01\%$ H_3PO_4, pH ~4.5).

treatment of rock phosphate (p. 549) with sulfuric acid, the idealized stoichiometry being:

$$Ca_5(PO_4)_3F + 5H_2SO_4 + 10H_2O \longrightarrow 3H_3PO_4 + 5CaSO_4 . 2H_2O + HF$$

The gypsum is filtered off together with other insoluble matter such as silica, and the fluorine is removed as insoluble Na_2SiF_6. The dilute phosphoric acid so obtained (containing 35–70% H_3PO_4 depending on the plant used) is then concentrated by evaporation. It is usually dark green or brown in colour and contains many metal impurities (e.g. Na, Mg, Ca, Al, Fe, etc.) as well as residual sulfate and fluoride, but is suitable for the manufacture of phosphatic fertilizers, metallurgical applications, etc. (see Panel).

Diphosphoric acid $H_4P_2O_7$ becomes an increasingly prevalent species as the stoichiometry in the system P_4O_{10}/H_2O becomes increasingly more concentrated; indeed, the phase diagram shows that, in addition to the hemihydrate (mp 29.30°) and orthophosphoric acid (mp 42.35°) the only other congruently melting phase in the system is $H_4P_2O_7$. The compound is dimorphic with a metastable modification mp 54.3° and a stable form mp 71.5°, but in the molten state it comprises an isocompositional mixture of various polyphosphoric acids and their autoprotolysis products. Equilibrium is reached only sluggishly and the actual constitution of the melt depends sensitively both on the precise stoichiometry and the temperature (Fig. 12.14).[54] For the nominal stoichiometry corresponding to $H_4P_2O_7$ typical concentrations of the species $H_{n+2}P_nO_{3n+1}$ from $n=1$ (i.e. H_3PO_4) to $n=8$ are as follows:†

n	1	2	3	4	5	6	7	8
mole%	35.0	42.5	14.6	5.0	1.8	0.7	0.3	0.1

† Note that this table indicates mole% of each molecular species present whereas the graphs in Fig. 12.14 plot weight percentage of P_2O_5 present as each acid shown.

54 R. F. JAMESON, The composition of the "strong" phosphoric acids, *J. Chem. Soc.* 1959, 752–9. The term "strong" phosphoric acids refers to those mixtures containing more P_2O_5 than required for the stoichiometry "H_3PO_4".

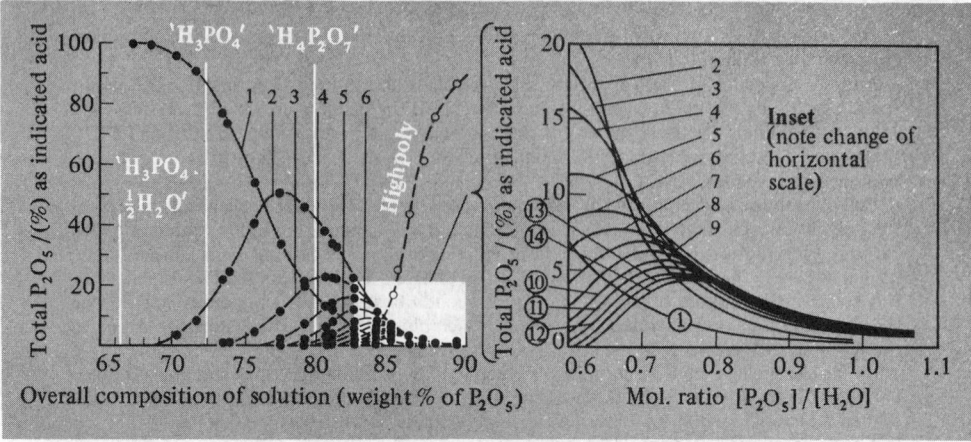

FIG. 12.14   The composition of the strong phosphoric acids shown as the weight per cent of P_2O_5 present in the form of each acid plotted against the overall stoichiometric composition of the mixture. The overall stoichiometries corresponding to the three congruently melting species $H_3PO_4 \cdot \frac{1}{2}H_2O$, H_3PO_4, and $H_4P_2O_7$ are indicated. Compositions above 82 wt% P_2O_5 are shown on an expanded scale in the inset using the mole ratio $[P_2O_5]/[H_2O]$ as the measure of stoichiometry. (For comparison, $H_4P_2O_7$ corresponds to a mole ratio of 0.500, $H_5P_3O_{10}$ to a ratio 0.600, $H_6P_4O_{13}$ to 0.667, etc.). In both diagrams the curves labelled 1, 2, 3, ... refer to ortho-, di-, tri- ... phosphoric acids, and "highpoly" refers to highly polymeric material hydrolysed from the column.

Thus, although $H_4P_2O_7$ is marginally the most abundant species present, there are substantial amounts of H_3PO_4, $H_5P_3O_{10}$, $H_6P_4O_{13}$, and higher polyphosphoric acids.

In dilute aqueous solution $H_4P_2O_7$ is a somewhat stronger acid than H_3PO_4: the 4 dissociation constants at 25° are: $K_1 \sim 10^{-1}$, $K_2 \sim 1.5 \times 10^{-2}$, K_3 2.7×10^{-7}, and K_4 2.4×10^{-10}, and the corresponding negative logarithms are: $pK_1 \sim 1.0$, $pK_2 \sim 1.8$, pK_3 6.57, and pK_4 9.62. The P–O–P linkage is kinetically stable towards hydrolysis in dilute neutral solutions at room temperature and the reaction half-life can be of the order of years. Such hydrolytic breakdown of polyphosphate is of considerable importance in certain biological systems (p. 611) and has been much studied. Some factors which affect the rate of degradation of polyphosphates are shown in Table 12.11.

TABLE 12.11   *Factors affecting the rate of polyphosphate degradation*

Factor	Approximate effect on rate
Temperature	10^5–10^6 faster from 0° to 100°
pH	10^3–10^4 slower from strong acid to base
Enzymes	Up to 10^5–10^6 faster
Colloidal gels	Up to 10^4–10^5 faster
Complexing cations	Often very many times faster
Concentration	Roughly proportional
Ionic environment in solution	Several-fold change

Orthophosphates[13, 24]

Phosphoric acid forms several series of salts in which the acidic H atoms are successively replaced by various cations; there is considerable commercial application for many of these compounds.

Lithium orthophosphates are unimportant and differ from the other alkali metal phosphates in being insoluble. At least 10 crystalline hydrated or anhydrous sodium orthophosphates are known and these can be grouped into three series:

$Na_3PO_4 . nH_2O$ ($n = 0, \frac{1}{2}, 6, 8, 12$)
$Na_2HPO_4 . nH_2O$ ($n = 0, 2, 7, 8, 12$)
$NaH_2PO_4 . nH_2O$ ($n = 0, 1, 2$), $NaH_2PO_4 . H_3PO_4$ [i.e. $NaH_5(PO_4)_2$]
   $NaH_2PO_4.Na_2HPO_4$ [i.e. $Na_3H_3(PO_4)_2$], and $2NaH_2PO_4.Na_2HPO_4.2H_2O$

Likewise, there are at least 10 well-characterized potassium orthophosphates and several ammonium analogues. The presence of extensive H bonding in many of these compounds leads to considerable structural complexity and frequently confers important properties (see later). The mono- and di-sodium phosphates are prepared industrially by neutralization of aqueous H_3PO_4 with soda ash (anhydrous Na_2CO_3, p. 102). However, preparation of the trisodium salts requires the use of the more expensive NaOH to replace the third H atom. Careful control of concentration and temperature are needed to avoid the simultaneous formation of pyrophosphates (diphosphates). Some indication of the structural complexity can be gained from the compound $Na_3PO_4 . 12H_2O$ which actually crystallizes with variable amounts of NaOH up to the limiting composition $4(Na_3PO_4 . 12H_2O) . NaOH$. The structure is built from octahedral $[Na(H_2O)_6]$ units which join to form "hexagonal" rings of 6 octahedra which in turn form a continuous two-dimensional network of overall composition $\{Na(H_2O)_4\}$; between the sheets lie $\{PO_4\}$ connected to them by H bonds (Fig. 12.15 on p. 607).[55] The positions (000) and ($00\frac{1}{2}$) are at the centres of large (octahedral) voids which are statistically occupied by Na^+ and OH^- up to the maximum allowed by the limiting formula. Some industrial, domestic, and scientific applications of Na, K, and NH_4 orthophosphates are given in the Panel.

Uses of Orthophosphates[3]

Phosphates are used in an astonishing variety of domestic and industrial applications but their ubiquitous presence and their substantial impact on everyday life is frequently overlooked. It will be convenient first to indicate the specific uses of individual compounds and the properties on which they are based, then to conclude with a brief summary of many different types of application and their interrelation. The most widely used compounds are the various phosphate salts of Na, K, NH_4, and Ca. The uses of diphosphates, triphosphates, and polyphosphates are mentioned on pp. 609 and 617.

Na_3PO_4 is strongly alkaline in aqueous solution and is thus a valuable constituent of scouring powders, paint strippers, and grease saponifiers. Its complex with NaOCl $[(Na_3PO_4 . 11H_2O)_4 . NaOCl]$, is also strongly alkaline (a 1% solution has pH 11.8) and, in addition, it releases active chlorine when wetted; this combination of scouring, bleaching, and bacterial action makes the adduct valuable in formulations of automatic dishwashing powders.

Na_2HPO_4 is widely used as a buffer component (p. 599). The use of the dihydrate ($\sim 2\%$ concentration) as an emulsifier in the manufacture of pasteurized processed cheese was patented

[55] E. TILLMANNS and W. H. BAUR, On the stoichiometry of trisodium orthophosphate dodecahydrate, *Inorg. Chem.* **9**, 1957–8 (1970).

by J. L. Kraft in 1916 and is still used, together with insoluble sodium metaphosphate or the mixed phosphate $Na_{15}Al_3(PO_4)_8$, to process cheese on the multikilotonne scale daily—despite much study, the reason why phosphate salts act as emulsifiers is still not well understood in detail. Na_2HPO_4 is also added ($\sim 0.1\%$) to evaporated milk to maintain the correct Ca/PO_4 balance and to prevent gelation of the milk powder to a mush. Its addition at the 5% level to brine (15–20% NaCl solution) for the pickling of ham makes the product more tender and juicy by preventing the exudation of juices during subsequent cooking. Another major use in the food industry is as a starch modifier: small additions enhance the ability to form stable cold-water gels (e.g. instant pudding mixes), and the addition of 1% to farinaceous products raises the pH to slightly above 7 and provides "quick-cooking" breakfast cereals.

NaH_2PO_4 is a solid, water-soluble acid, and this property finds use (with $NaHCO_3$) in effervescent laxative tablets and in the pH adjustment of boiler waters. It is also used as a mild phosphatizing agent for steel surfaces and as a constituent in the undercoat for metal paints.

K_3PO_4 (like Na_3PO_4) is strongly alkaline in aqueous solution and is used to absorb H_2S from gas streams; the solution can be regenerated simply by heating. K_3PO_4 is also used as a regulating electrolyte to control the stability of synthetic latex during the polymerization of styrene-butadiene rubbers. The buffering action of K_2HPO_4 has already been mentioned (p. 599) and this is the reason for its addition as a corrosion inhibitor to car-radiator coolants which otherwise tend to become acidic due to slow oxidation of the glycol antifreeze. KH_2PO_4 is a piezoelectric (pp. 60–2) and finds use in submarine sonar systems. For many applications, however, the cheaper sodium salts are preferred unless there is a specific advantage for the potassium salt; one example is the specialist balanced commercial fertilizer formulation $[KH_2PO_4 . (NH_4)_2HPO_4]$ which contains 10.5% N, 53% P_2O_5, and 17.2% K_2O (i.e. N–P–K 10–53–17).

$(NH_4)_2HPO_4$ and $(NH_4)H_2PO_4$ can be used interchangeably as specialist fertilizers and nutrients in fermentation broths; though expensive, their high concentration of active ingredients ameliorate this, particularly in localities where transportation costs are high. Indeed, $(NH_4)_2HPO_4$ in granulated or liquid form consumes more phosphate rock than any other single end-product (over 8 million tonnes pa in the USA alone in 1974). Ammonium phosphates are also much used as flame retardants for cellulosic materials, about 3–5% gain in dry weight being the optimum treatment. Their action probably depends on their ready dissociation into NH_3 and H_3PO_4 on heating; the H_3PO_4 then catalyses the decomposition of cellulose to a slow-burning char (carbon) and this, together with the suppression of flammable volatiles, smothers the flame. As the ammonium phosphates are soluble, they are used mainly for curtains, theatre scenery, and disposable paper dresses and costumes. The related compound urea phosphate $(NH_2CONH_2 . H_3PO_4)$ has also been used to flameproof cotton fabrics: the material is soaked in a concentrated aqueous solution, dried (15% weight gain), and cured at 160° to bond the retardant to the cellulose fibre. The advantage is that the retardant does not wash out, but the strength of the fabric is somewhat reduced by the process.

Calcium phosphates have a broad range of applications both in the food industry and as bulk fertilizers. The vast scale of the phosphate rock industry has already been indicated (p. 549) and this is further elaborated in Fig. A for the particular case of the USA.

The crucial importance of Ca and PO_4 as nutrient supplements for the healthy growth of bones, teeth, muscle, and nerve cells has long been recognized. The non-cellular bone structure of an average adult human consists of $\sim 60\%$ of some form of "tricalcium phosphate" $[Ca_5(PO_4)_3OH]$; teeth likewise comprise $\sim 70\%$ and the average person carries 3.5 kg of this material in his body. Phosphates in the body are replenished by a continuous cycle, and used P is carried by the blood to the kidneys and then excreted in urine, mainly as $Na(NH_4)HPO_4$. An average adult eliminates 3–4 g of PO_4 equivalent daily (cf. the discovery of P in urine by Brandt, p. 546).

Calcium phosphates are used in baking acids, toothpastes, mineral supplements, and stock feeds. $Ca(H_2PO_4)_2$ was introduced as a leavening acid in the late nineteenth century (to replace "cream of tartar" $KHC_4H_4O_6$) but the monohydrate (introduced in the 1930s) finds more use today. "Straight baking powder", a mixture of $Ca(H_2PO_4)_2 . H_2O$ and $NaHCO_3$ with some 40% starch coating, tends to produce CO_2 too quickly during dough mixing and so "combination baking powder", which also incorporates a slow-acting acid such as $NaAl(SO_4)_2$, is preferred. Nearly 90% of all US household baking powders now use such combinations, e.g.:

$$Ca(H_2PO_4)_2 . H_2O + 2NaHCO_3 \longrightarrow 2CO_2 + 3H_2O + \text{'}Na_2Ca(HPO_4)_2\text{'}$$

$$NaAl(SO_4)_2 + 3NaHCO_3 \xrightarrow{H_2O} 3CO_2 + Al(OH)_3 + 2Na_2SO_4$$

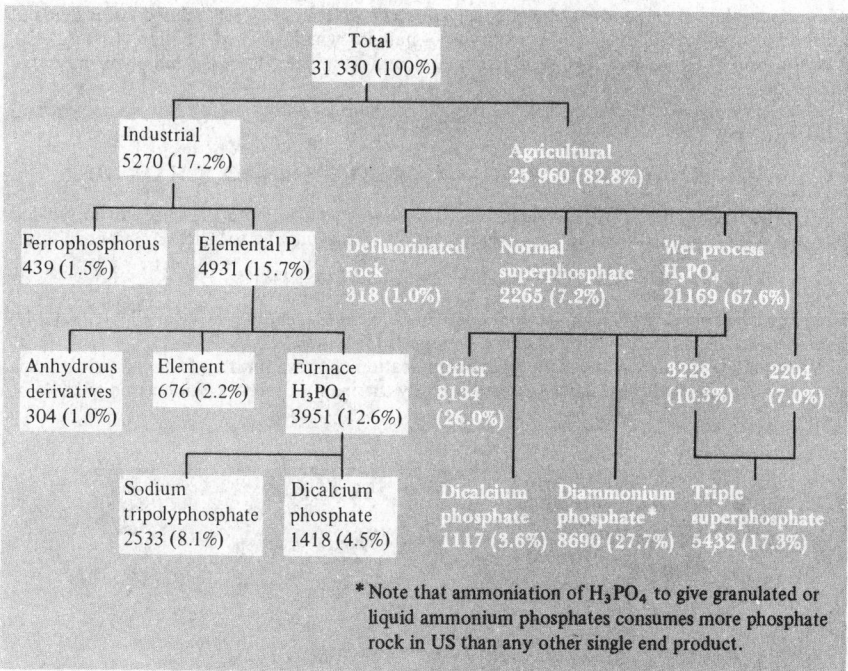

FIG. A   US Phosphate rock usage (1974)/kilotonnes pa (%) (excluding 12 610 kilotonnes exported).

A typical powder contains 28% $NaHCO_3$, 10.7% $Ca(H_2PO_4)_2.H_2O$, 21.4% $NaAl(SO_4)_2$, and 39.9% starch and the scale of manufacture approaches 10^5 tonnes pa.

In toothpastes, $CaHPO_4 . 2H_2O$ was first used to replace chalk as a mild abrasive and polishing agent in the early 1930s. It is still widely used provided the toothpaste does not also contain fluoride, since this would precipitate as CaF_2 and effectively eliminate the desired anion. Some 25 000 tonnes of $CaHPO_4 . 2H_2O$ are used in this way annually in the USA and the compound typically comprises 50% by weight of the paste. The first important fluoride toothpaste contained 39% of the diphosphate $Ca_2P_2O_7$ which is the most insoluble and inert of all the calcium phosphates. It is made by careful dehydration of $CaHPO_4 . 2H_2O$ at 150° and then above 400°. It was first used in Procter and Gamble's "Crest" which also contained 0.4% SnF_2 and 1% $Sn_2P_2O_7$.

Synthetic $Ca_5(PO_4)_3OH$ is added to table salt (1–2%) to impart free-flowing properties and it is likewise added to granulated sugar, baking powders, and even fertilizers. It is prepared by adding H_3PO_4 to a slurry of hydrated lime—this is the reverse order of addition to that used for making $Ca(H_2PO_4)_2$ and $CaHPO_4$ since the aim is to deprotonate all three OH groups. The compound is extremely insoluble and precipitates as very fine particles ($\sim$0.5–3 μm diameter).

The idea of converting insoluble "tricalcium phosphate" or phosphate rock into soluble "monocalcium phosphate" $Ca(H_2PO_2)_2$ dates back to the 1830s when J. von Liebig observed that acidulated bones made good fertilizer. The limited supply of bones (including those from old battlefields!) was soon replaced by Suffolk coprolites and apatites, though the vast North African deposits were still unknown. The phosphate fertilizer industry originated in England (Lawes, 1843); it grew rapidly as shown by the dramatic increase in world production of phosphate rock, which leapt from 500 tonnes to over 1 million tonnes in less than 40 y and now exceeds 110 million tonnes (p. 549):

Year	1847	1850	1860	1870	1880	1890	1900 . . .	1974
Kilotonnes	0.5	1.0	66	270	500	1600	3100 . . .	110 000

This unprecedented demand for phosphatic fertilizers is, of course, closely related to the demand for food from an exploding world population of humans which reached 1 billion (10^9) in 1830, 2 billion in 1930, 3 billion in 1960, 4 billion in 1974, and is predicted to be 6–8 billion by the end of the century.

"Superphosphate" is now made by the (highly exothermic) addition of H_2SO_4 to fine-ground phosphate rock:

$$2Ca_5(PO_4)_3F + 7H_2SO_4 + H_2O \longrightarrow 7CaSO_4 + 3Ca(H_2PO_4)_2 . H_2O + 2HF$$

The $CaSO_4$ or its hydrate (gypsum) acts only as an unwanted diluent. Its presence can be avoided by using H_3PO_4 instead of H_2SO_4 for the acidulation, thus giving rise to "triple superphosphate":

$$Ca_5(PO_4)_3F + 7H_3PO_4 + H_2O \longrightarrow 5Ca(H_2PO_4)_2 . H_2O + HF$$

Commercial triple superphosphate contains almost 3 times the amount of available (soluble) P_2O_5 as ordinary superphosphate, hence its name (45–50 wt% vs. 18–20 wt%).

A schematic summary of the main types of application of phosphoric acid and its derivatives is shown in Fig. B which also indicates very broadly the main properties being exploited.[6]

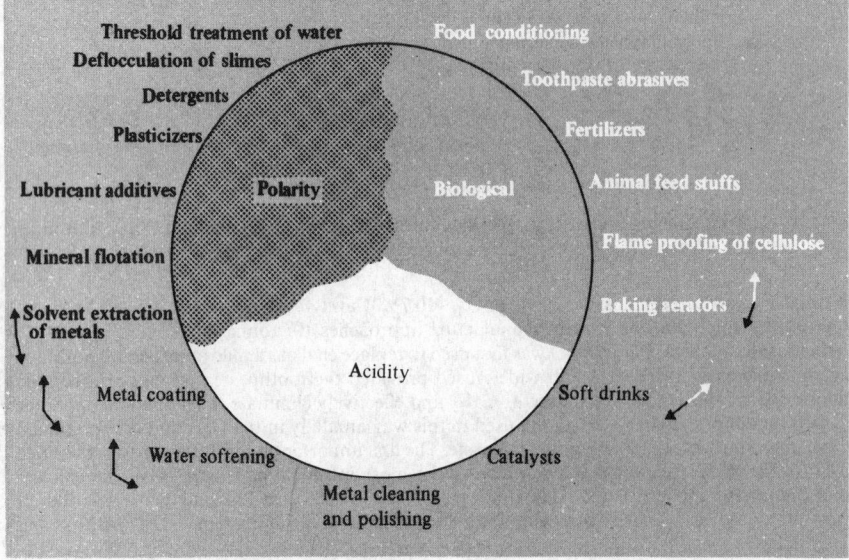

FIG. B. Applications of phosphoric acid and its derivatives.

Calcium orthophosphates are particularly important in fertilizer technology, in the chemistry of bones and teeth, and in innumerable industrial and domestic applications (see Panel). They are also the main source of phosphorus and phosphorus chemicals and occur in vast deposits as apatites and rock phosphate (p. 549). The main compounds occurring in the $CaO-H_2O-P_4O_{10}$ phase diagram are:

$$Ca(H_2PO_4)_2, \quad Ca(H_2PO_4)_2.H_2O, \quad Ca(HPO_4).nH_2O \quad (n=0, \tfrac{1}{2}, 2),$$

$$Ca_3(PO_4)_2, \quad Ca_2PO_4(OH).2H_2O, \quad Ca_5(PO_4)_3OH \text{ (i.e. apatite)},$$

$$Ca_4P_2O_9 \text{ [probably } Ca_3(PO_4)_2.CaO], \text{ and } Ca_8H_2(PO_4)_6.5H_2O$$

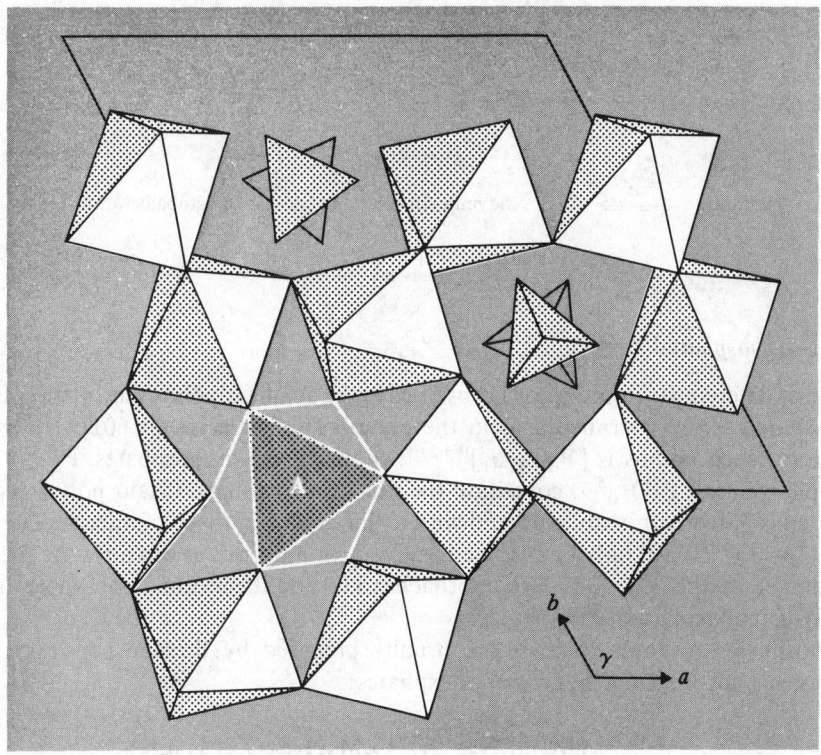

FIG. 12.15   The structure of $Na_3PO_4.12H_2O.\frac{1}{4}NaOH$ viewed down the c-axis perpendicular to the "hexagonal" sheets of fused $\{Na(H_2O)_6\}$ octahedra. Octahedral sites such as A—at (000)—are occupied statistically by Na^+ and OH^- up to a maximum allowed by the formula (and structure). The orientation of the PO_4 tetrahedra above and below the plane of corner-linked octahedra is also shown.

In all of these alkali-metal and alkaline earth-metal orthophosphates there are discrete, approximately regular tetrahedral PO_4 units in which P–O is usually in the range 153 ± 3 pm and the angle O–P–O is usually in the range $109 \pm 5°$. Extensive H-bonding and M–O interactions frequently induce substantial deviations from a purely ionic formulation (p. 94). This trend continues with the orthophosphates of tervalent elements $M^{III}PO_4$ (M = B, Al, Ga, Cr, Mn, Fe) which all adopt structures closely related to the polymorphs of silica (p. 393). $NaBePO_4$ is similar, and YPO_4 adopts the zircon ($ZrSiO_4$) structure. The most elaborate analogy so far revealed is for $AlPO_4$ which can adopt each of the 6 main polymorphs of silica as indicated in the scheme on p. 608. The analogy covers not only the structural relations between the phases but also the sequence of transformation temperatures (°C) and the fact that the α–β-transitions occur readily whilst the others are sluggish (p. 394). Similarly, the orthophosphates of B, Ga, and Mn are known in the β-quartz and the α- and β-cristobalite forms whereas $FePO_4$ adopts either the α- or β-quartz structure. Numerous hydrated forms are also known. The $Al–PO_4–H_2O$ system is used industrially as the basis for many adhesives, binders, and cements.[56]

[56] J. H. MORRIS, P. G. PERKINS, A. E. A. ROSE, and W. E. SMITH, The chemistry and binding properties of aluminium phosphates, *Chem. Soc. Revs.* **6**, 173–94 (1977).

$$\text{SiO}_2 \begin{cases} \text{quartz} \underset{867°}{\rightleftharpoons} \text{tridymite} \underset{1470°}{\rightleftharpoons} \text{cristobalite} \underset{1713°}{\rightleftharpoons} \text{melt} \\[2ex] \beta \underset{573°}{\rightleftharpoons} \alpha \qquad \beta \underset{117°}{\rightleftharpoons} \alpha_1 \underset{163°}{\rightleftharpoons} \alpha_2 \qquad \beta \underset{220°}{\rightleftharpoons} \alpha \end{cases}$$

$$\text{AlPO}_4 \begin{cases} \text{berlinite} \underset{705°}{\rightleftharpoons} \text{tridymite-form} \underset{1025°}{\rightleftharpoons} \text{cristobalite-form} \underset{>1600°}{\rightleftharpoons} \text{melt} \\[2ex] \beta \underset{586°}{\rightleftharpoons} \alpha \qquad \beta \underset{93°}{\rightleftharpoons} \alpha_1 \underset{130°}{\rightleftharpoons} \alpha_2 \qquad \beta \underset{210°}{\rightleftharpoons} \alpha \end{cases}$$

Chain polyphosphates[13, 24]

A rather different structure-motif is observed in the chain polyphosphates: these feature corner-shared $\{PO_4\}$ tetrahedra as in the polyphosphoric acids (p. 601). The general formula for such anions is $[P_nO_{3n+1}]^{(n+2)-}$, of which the diphosphates, $P_2O_7^{4-}$, and tripolyphosphates, $P_3O_{10}^{5-}$, constitute the first two members. Chain polyphosphates have been isolated with n up to 10 and with n "infinite", but those of intermediate chain length $(10 < n < 50)$ can only be obtained as glassy or amorphous mixtures. As the chain length increases, the ratio $(3n+1)/n$ approaches 3.00 and the formula approaches that of the polymetaphosphates $[PO_3^-]_\infty$.

Diphosphates (pyrophosphates) are usually prepared by thermal condensation of dihydrogen phosphates or hydrogen phosphates:

$$2MH_2PO_4 \xrightarrow{\Delta} M_2H_2P_2O_7 + H_2O$$

$$2M_2HPO_4 \xrightarrow{\Delta} M_4P_2O_7 + H_2O$$

They can also be prepared in specialized cases by (a) metathesis, (b) the action of H_3PO_4 on an oxide, (c) thermolysis of a metaphosphate, (d) thermolysis of an orthophosphate, or (e) reductive thermolysis, e.g.:

$$\text{(a)} \qquad Na_4P_2O_7 + 4AgNO_3 \longrightarrow Ag_4P_2O_7\downarrow + 4NaNO_3$$

$$\text{(b)} \qquad 2H_3PO_4 + PbO_2 \longrightarrow PbP_2O_7\downarrow + 3H_2O$$

$$\text{(c)} \qquad 4Cr(PO_3)_3 \xrightarrow{\Delta} Cr_4(P_2O_7)_3 + 3P_2O_5$$

$$\text{(d)} \qquad 2Hg_3(PO_4)_2 \xrightarrow{\Delta} 2Hg_2P_2O_7 + 2Hg + O_2$$

$$\text{(e)} \qquad 2FePO_4 + H_2 \longrightarrow Fe_2P_2O_7 + H_2O$$

Many diphosphates of formula $M^{IV}P_2O_7$, $M_2^{II}P_2O_7$, and hydrated $M_4^IP_2O_7$ are known and there has been considerable interest in the relative orientation of the two linked $\{PO_4\}$ groups and in the P–O–P angle between them.[56a] For small cations the 2

[56a] G. M. CLARK and R. MORLEY, Inorganic pyro compounds, $M_a[(X_2O_7)_b]$, *Chem. Soc. Revs.* **5**, 269–95 (1976).

{PO$_4$} are approximately staggered whereas for larger cations they tend to be nearly eclipsed. The P–O–P angle is large and variable, running from 130° in Na$_4$P$_2$O$_7$.10H$_2$O to 156° in α-Mg$_2$P$_2$O$_7$. The apparent colinearity in the higher-temperature (β) form of many diphosphates, which was previously ascribed to a P–O–P angle of 180°, is now generally attributed to positional disorder. Bridging P–O distances are invariably longer than terminal P–O distances, typical values (for Na$_4$P$_2$O$_7$.10H$_2$O) being P–O$_\mu$ 161 pm, P–O$_t$ 152 pm.

As diphosphoric acid is tetrabasic, four series of salts are possible though not all are always known, even for simple cations. The most studied are those of Na, K, NH$_4$, and Ca, e.g.:

$$Na_4P_2O_7.10H_2O \ (mp\ 79.5°), \quad Na_4P_2O_7 \ (mp\ 985°)$$

$$Na_3HP_2O_7.9H_2O \xrightarrow{30-35°} Na_3HP_2O_7.H_2O \xrightarrow{150°} Na_3HP_2O_7$$

$$Na_2H_2P_2O_7.6H_2O \xrightarrow{\sim 27°} Na_2H_2P_2O_7$$

$$NaH_3P_2O_7 \ (mp\ 185°)$$

Before the advent of synthetic detergents, Na$_4$P$_2$O$_7$ was much used as a dispersant for lime soap scum which formed in hard water, but it has since been replaced by the tripolyphosphate (see below). However, the ability of diphosphate ions to form a gel with soluble calcium salts has made Na$_4$P$_2$O$_7$ a useful ingredient for starch-type instant pudding which requires no cooking. The main application of Na$_2$H$_2$P$_2$O$_7$ is as a leavening acid in baking: it does not react with NaHCO$_3$ until heated, and so large batches of dough or batter can be made up and stored. Ca$_2$P$_2$O$_7$, because of its insolubility, inertness, and abrasive properties, is used as a toothpaste additive compatible with Sn^{II} and fluoride ions.

Of the tripolyphosphates only the sodium salt need be mentioned. It was introduced in the mid-1940s as a "builder" for synthetic detergents, and its production for this purpose is now measured in megatonnes per annum (see Panel). On the industrial scale Na$_5$P$_3$O$_{10}$ is usually made by heating an intimate mixture of powdered Na$_2$HPO$_4$ and NaH$_2$PO$_4$ of the required stoichiometry under carefully controlled conditions:

$$2Na_2HPO_4 + NaH_2PO_4 \longrightarrow Na_5P_3O_{10} + 2H_2O$$

Uses of Sodium Tripolyphosphate

Many synthetic detergents contain 25–45% Na$_5$P$_3$O$_{10}$ though the amount is lower in the USA than in Europe because of the problems of eutrophication in some areas (p. 552). It acts mainly as a water softener, by chelating and sequestering the Mg^{2+} and Ca^{2+} in hard water. Indeed, the formation constants of its complexes with these ions are nearly one million-fold greater than with Na$^+$: (NaP$_3$O$_{10}^{4-}$ pK $\sim$2.8; MgP$_3$O$_{10}^{3-}$ pK $\sim$8.6; CaP$_3$O$_{10}^{3-}$ pK $\sim$8.1). In addition, Na$_5$P$_3$O$_{10}$ increases the efficiency of the surfactant by lowering the critical micelle concentration, and by its ability to suspend and peptize dirt particles by building up a large negative charge on the particles by adsorption; it also furnishes a suitable alkalinity for cleansing action without irritating eyes or skin and it provides effective buffering action at these pHs. The dramatic growth of synthetic detergent powders during the 1950s is shown in Fig. A (p. 610).[7]

Na$_5$P$_3$O$_{10}$ is also used as a dispersing agent in clay suspensions used in oil-well drilling. Again, addition of $\lesssim$1% Na$_5$P$_3$O$_{10}$ to the slurries used in manufacturing cement and bricks enables much less water to be used to attain workability, and thus less to be removed during the setting or calcining processes.

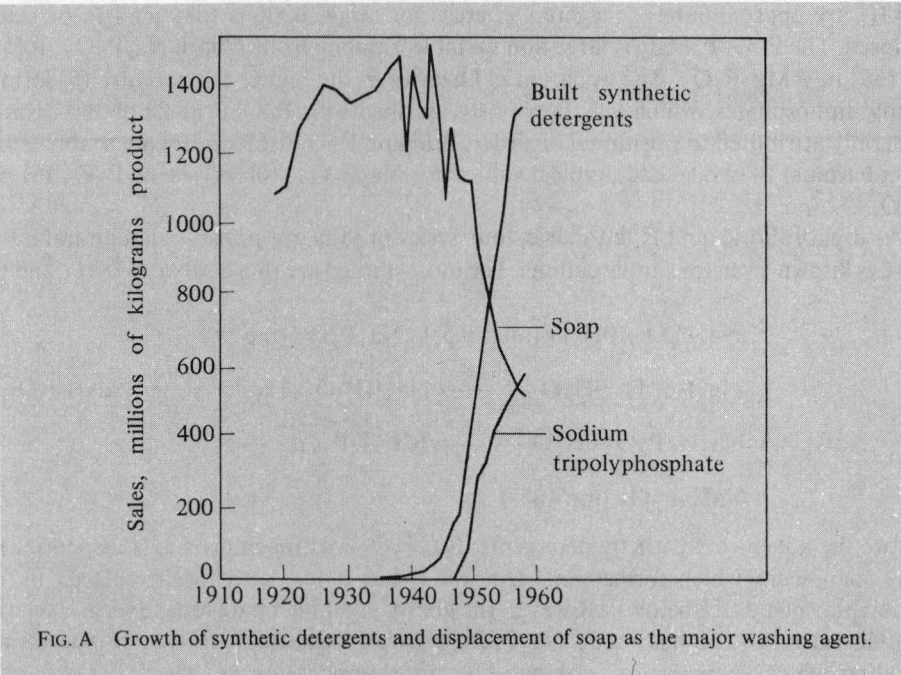

FIG. A   Growth of synthetic detergents and displacement of soap as the major washing agent.

The low-temperature form (I) converts to the high-temperature form (II) above 417°C and both forms react with water to give the crystalline hexahydrate. All three materials contain the tripolyphosphate ion $P_3O_{10}^{5-}$ with a *trans-* configuration of adjacent tetrahedra and a twofold symmetry axis; forms (I) and (II) differ mainly in the coordination of the sodium ions and the slight differences in the dimensions of the ion in the three crystals are probably within experimental error. Typical values are:

The complicated solubility relations, rates of hydrolysis, self-disproportionation, and interconversion with other phosphates depends sensitively on pH, concentration, temperature, and the presence of impurities. Though of great interest academically and of paramount importance industrially these aspects will not be further considered here.[7, 13, 24, 57] Triphosphates such as adenosine triphosphate (ATP) are also of vital importance in living organisms (see Panel).

[57] E. J. GRIFFITH, The chemical and physical properties of condensed phosphates, *Pure Appl. Chem.* **44**, 173–200 (1975).

Phosphates in Life Processes

Phosphate esters are crucial components of all living matter and they play a vital role in many life processes such as protein synthesis, genetic coding, photosynthesis, nitrogen fixation, and innumerable other metabolic pathways.[57a] The biophosphorus compound on which all life depends is adenosine triphosphate (ATP); its name is derived from adenine, β-D-ribose, and the tripolyphosphate ion $HP_3O_{10}^{4-}$ as shown below:

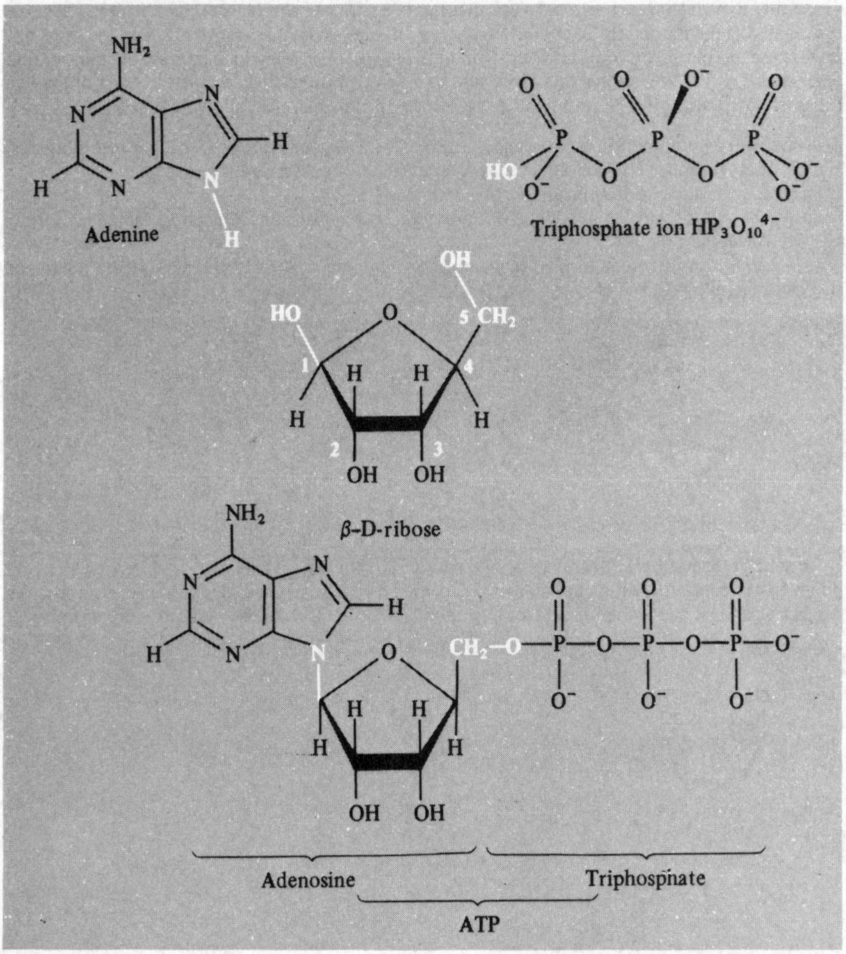

Adenine

Triphosphate ion $HP_3O_{10}^{4-}$

β-D-ribose

Adenosine   Triphosphate

ATP

The precise conformation of ATP will depend on whether the group is in a crystalline compound, whether it is bonded to metals in solution, interacting with enzymes, etc. Hydrolysis of this ionic triphosphate ester to adenosine diphosphate (ADP) or adenosine monophosphate (AMP), in which $\{P_3O_{10}^{5-}\}$ is replaced by $\{P_2O_7^{4-}\}$ or $\{PO_4^{3-}\}$, releases the energy required for many biochemical processes. The free-energy changes associated with these hydrolyses depend

[57a] I. S. KULAEV, *The Biochemistry of Inorganic Polyphosphates*, Wiley, Chichester, 1980, 225 pp.

sensitively on pH, temperature, and the presence of metal ions. Standard values† for pH 7.4, T 25°C, and $[Mg^{2+}]$ 10^{-4} M are:

$$ATP^{4-} + H_2O \longrightarrow ADP^{3-} + HPO_4^{2-} + H^+; \quad \Delta G^{\circ\prime} = -40.9 \text{ kJ mol}^{-1}$$
$$K' = 1.3 \times 10^7$$

$$ATP^{4-} + H_2O \longrightarrow AMP^{2-} + HP_2O_7^{3-} + H^+; \quad \Delta G^{\circ\prime} = -43.5 \text{ kJ mol}^{-1}$$
$$K' = 4.2 \times 10^7$$

Both reactions are highly favoured and release substantial amounts of energy.

It would be inappropriate within the confines of this brief panel to attempt a summary of the vast amount of elegant biochemical work that has been done to elucidate metabolic pathways and the details of enzyme action. This would require a full course on biochemistry, and many excellent texts are available, including ones which summarize aspects of immediate interest in bioinorganic chemistry.[2, 5, 57a] We confine ourselves here to a description of the structure of some of the key types of compound. It will be helpful to recall the meaning of the following terms:

nucleoside: a combination of an organic nitrogen base (e.g. adenine) and a sugar (e.g. ribose or 2'-deoxyribose); a typical nucleoside is adenosine (see scheme on p. 611).
nucleotide: a nucleoside phosphate ester (e.g. AMP).
nucleic acid: a nucleotide polymer, i.e. a polymeric diester of orthophosphoric acid (e.g. DNA, RNA, see below).
polypeptide: an oligomer formed from organic aminoacids, $RCH(NH_2)CO_2H$, by means of peptide links –CO–NH–, i.e.:

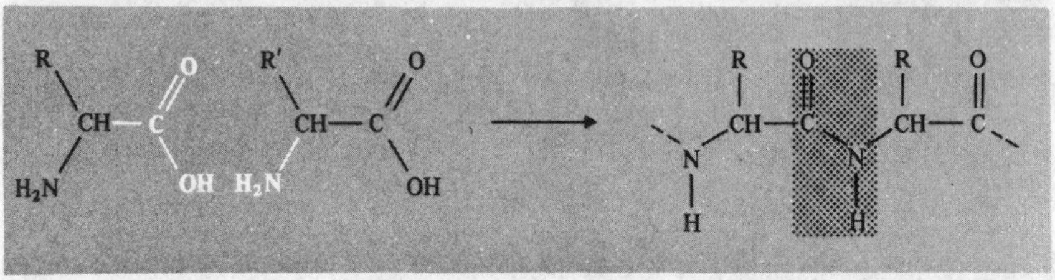

Genetic information is transferred by means of deoxyribonucleic acids, DNA, which are polymeric phosphate diesters of nucleosides formed from the sugar 2'-deoxyribose (cf. ribose on p. 611), and the 4 bases adenine, cytosine, guanine, and thymine whose structures are as follows:

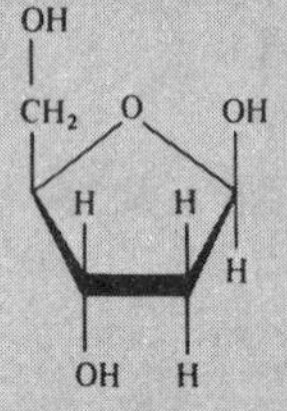

2'-Deoxyribose 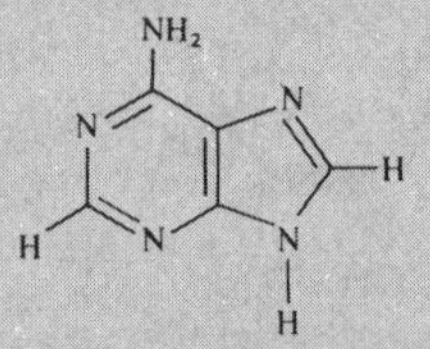   Adenine 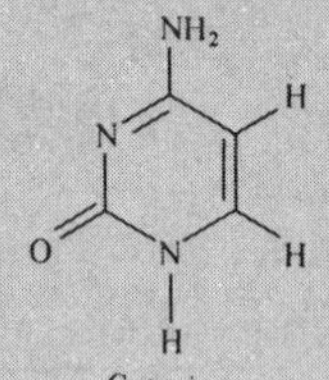   Cytosine

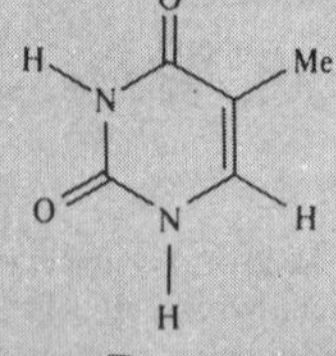

Thymine 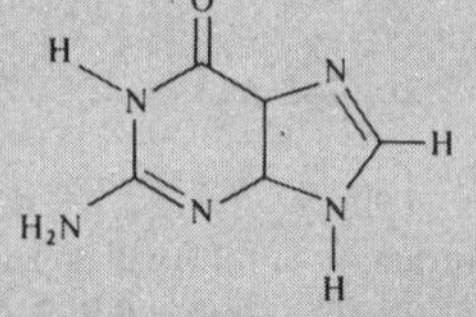   Guanine

† These values are designated $\Delta G^{\circ\prime}$ rather than ΔG° since the "standard state" refers to a biochemical standard state which resembles that *in vivo* rather than the normal thermochemical standard state of pH 0.0, T 25°C, unit activity, and no Mg^{2+} present.

The sequence of bases is specific for each nucleic acid which itself comprises two helical strands joined by H bonding between the base-pairs adenine–thymine and cytosine–guanine (pp. 66–7). A typical "representative" section of one of the DNA chains is in Fig. A and a schematic model of the double chain showing the coordination of the phosphate groups to hydrated sodium ions is shown

FIG. A A "representative" section of one of the polymeric phosphate diester chains of DNA showing the 4 constituent nucleosides formed by combination of deoxyribose with the 4 bases adenine, cytosine, guanine and thymine.

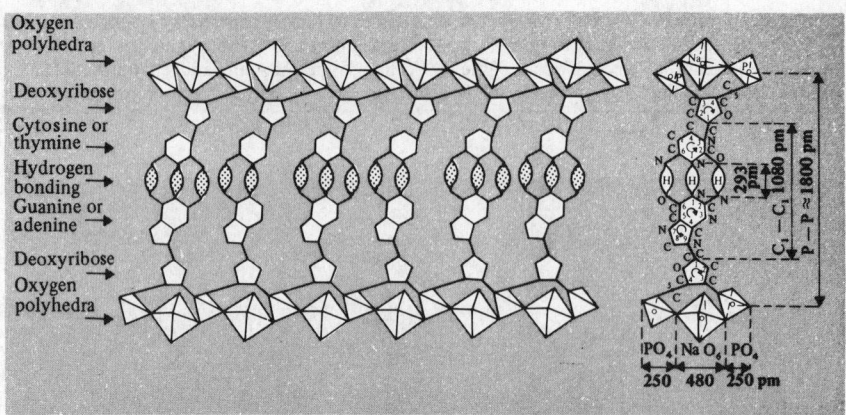

Fig. B   Model of DNA (schematic) showing metal-coordinated phosphate groups. A sodium ion
is bonded by ionic forces between 2 phosphate tetrahedra (After E. T. Degens[11].).

in Fig. B. Clearly the detailed conformation of the chain will depend on the nature of the cations present.

A related series of compounds, the ribonucleic acids (RNA), are based on the sugar ribose rather than 2'-deoxyribose as in DNA. The same four bases occur in the nucleosides which then form the polymeric phosphate diester chains. RNA molecules serve as a template for synthesizing protein molecules from amino acids. They are of three distinct types:

ribosomal RNA (rRNA): the combination of an RNA with a protein is called a ribosome; these occur in the cytoplasm of a cell surrounding the nucleus and are the site of protein synthesis.

messenger RNA (mRNA): this carries the code from the DNA which determines the amino-acid sequence in the polypeptide chain of the protein.

transfer RNA (tRNA): these are relatively small nucleotides containing about 80 {PO_4} units. Each tRNA carries 1 specific amino acid and 3 specific consecutive nucleotides with their bases; these bases serve as a code-word which fits exactly with the 3 consecutive complementary N-bases along the long mRNA strand.

As befits their function as templates, RNA molecules are mostly single helices. It should be noted, however, that protein structures are much more complex than the mere sequence of amino-acid residues in the polypeptide chain. Indeed, 4 elements of protein structure are now recognized. The *primary* structure refers to the sequence of amino acids; the *secondary* structure relates to the conformation of the polypeptide chain (random coil or α-helix, maintained by H bonds between the NH and CO groups); the *tertiary* structure involves the whole protein molecule and is maintained by various interactions including S–S covalent bonds between two cysteine residues, electrostatic forces, coordination to metal ions, H bonding, and even van der Waals' forces; finally, the *quaternary* structure concerns the subunit structure of the protein, e.g. haemoglobin has 4 subunits (2 α-chains and 2 β-chains).

Some other biochemically important phosphate esters are listed below together with their functions and structures (opposite):

phosphocreatine: an aminophosphate ester used to regenerate ATP;

uridine triphosphate, UTP: synthesis of the polysaccharide molecule glycogen from surplus carbohydrate;

nicotinamide-adenine dinucleotide (NAD$^+$): degradation of citric acid to succinic acid in the Krebs' cycle;

nicotinamide-adenine dinucleotide-2-phosphate (NADP$^+$): photosynthesis of carbohydrates from CO_2 in the presence of chlorophyll.

Long-chain polyphosphates, $M_{n+2}^I P_n O_{3n+1}$, approach the limiting composition $M^I PO_3$ as $n \to \infty$ and are often called linear metaphosphates to distinguish them from the cyclic metaphosphates of the same composition (p. 617). Their history extends back nearly 150 y to the time when Thomas Graham described the formation of a glassy sodium polyphosphate mixture known as Graham's salt. Various heat treatments converted this to crystalline compounds known as Kurrol's salt, Maddrell's salt, etc., and it is now appreciated, as a result of X-ray crystallographic studies, that these and many related substances all feature unbranched chains of corner-shared $\{PO_4\}$ units which differ only in the mutual orientations and repeat units of the constituent tetrahedra.[57b] These, in turn, are dictated by the size and coordination requirements of the counter cations present (including H). Some examples are shown schematically in Fig. 12.16 and the geometric resemblance between these and many of the chain metasilicates (p. 404) should be noted. In most of these polyphosphates P–O$_\mu$ 161±5 pm, P–O$_t$

[57b] J. MALING and F. HANIC, Phase chemistry of condensed phosphates, *Topics in Phosphorus Chemistry* **10**, 341–502 (1980). Contains many phase diagrams for metaphosphates and polyphosphates plus detailed structural diagrams of repeat units in condensed phosphate chains.

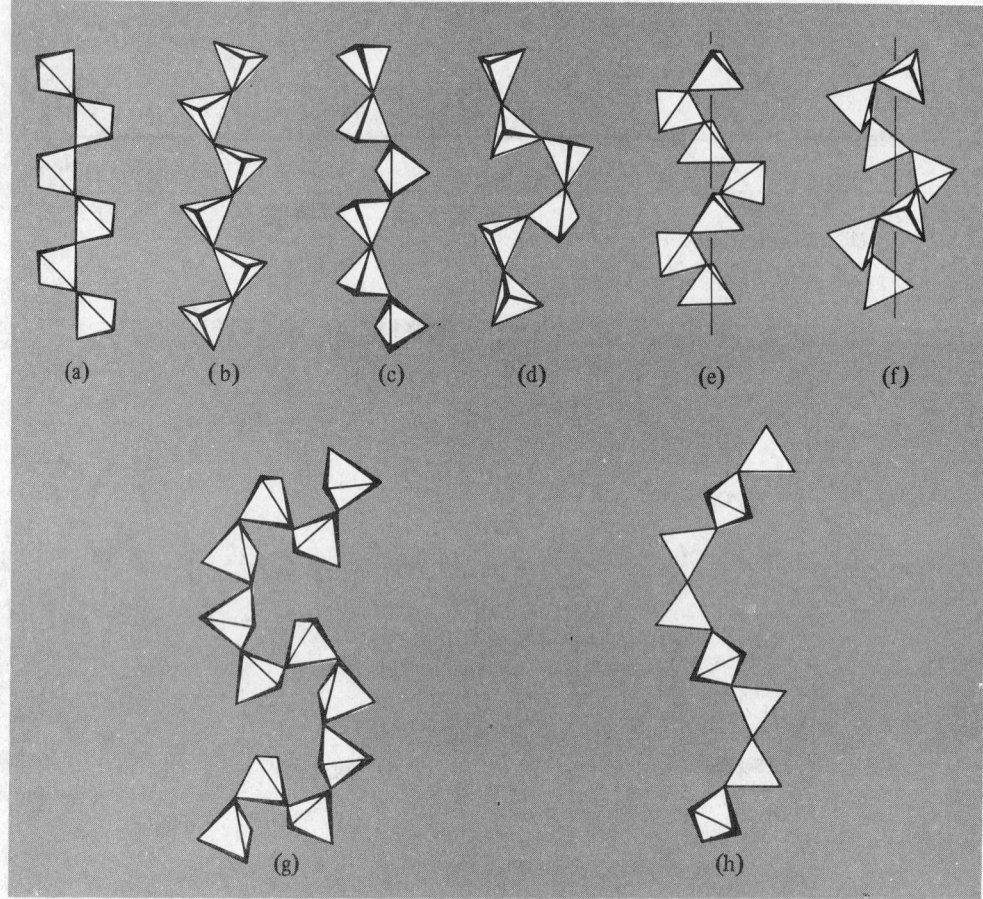

FIG. 12.16 Types of polyphosphate chain configuration. The diagrams indicate the relative orientations of adjacent PO_4 tetrahedra, extended along the chain axes. (a) $(RbPO_3)_n$ and $(CsPO_3)_n$, (b) $(LiPO_3)_n$ low temp, and $(KPO_3)_n$, (c) $(NaPO_3)_n$ high-temperature Maddrell salt and $[Na_2H(PO_3)_3]_n$, (d) $[Ca(PO_3)_2]_n$ and $[Pb(PO_3)_2]_n$, (e) $(NaPO_3)_n$, Kurrol A, and $(AgPO_3)_n$, (f) $(NaPO_3)_n$, Kurrol B, (g) $[CuNH_4(PO_3)_3]_n$, and isomorphous salts, (h) $[CuK_2(PO_3)_4]_n$ and isomorphous salts. Each crystalline form of Kurrol salt contains equal numbers of right-handed and left-handed spiralling chains.

150 ± 2 pm, $P-O_\mu-P$ 125–135° and O_t-P-O_t 115–120° (i.e. very similar to the dimensions and angles in the tripolyphosphate ion, p. 610).

The complex preparative interrelationships occurring in the sodium polyphosphate system are summarized in Fig. 12.17. Thus anhydrous NaH_2PO_4, when heated to 170° under conditions which allow the escape of water vapour, forms the diphosphate $Na_2H_2P_2O_7$, and further dehydration at 250° yields either Maddrell's salt (closed system) or the cyclic trimetaphosphate (water vapour pressure kept low). Maddrell's salt converts from the low-temperature to the high-temperature form above 300°, and above 400° reverts to the cyclic trimetaphosphate. The high-temperature form can also be obtained (via Graham's and Kurrol's salts) by fusing the cyclic trimetaphosphate (mp 526°C) and then quenching it from 625° (or from 580° to give Kurrol's salt directly). All these linear

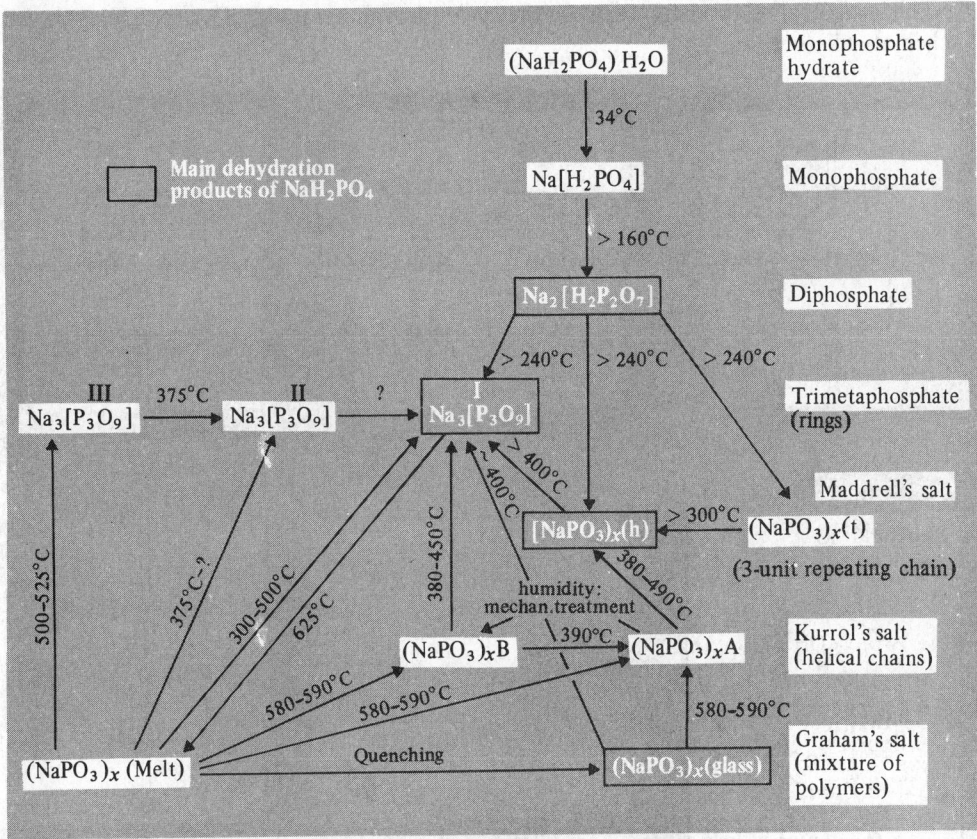

FIG. 12.17 Interrelationship of metaphosphates. (From CIC, Vol. 2, p. 521.)

polyphosphates of sodium revert to the cyclic trimetaphosphate on prolonged annealing at ~400°C.

Fuller treatments of the phase relations and structures of polyphosphates, and their uses as glasses, ceramics, refractories, cements, plasters, and abrasives, are available.[57b, 58]

Cyclo-*polyphosphoric acids and* cyclo-*polyphosphates*[59]

These compounds were formerly called metaphosphoric acids and metaphosphates but the IUPAC *cyclo*- nomenclature is preferred as being structurally more informative. The only two important acids in the series are *cyclo*-triphosphoric acid $H_3P_3O_9$ and *cyclo*-tetraphosphoric acid $H_4P_4O_{12}$, but well-characterized salts are known with heterocyclic anions $[cyclo-(PO_3)_n]^{n-}$ ($n = 3-8$), and larger rings are undoubtedly present in some mixtures.

[58] A. E. R. WESTMAN, Phosphate ceramics, *Topics in Phosphorus Chemistry* **9**, 231–405, 1977. A comprehensive account with 963 references.
[59] S. Y. KALLINEY, Cyclophosphates, *Topics in Phosphorus Chemistry* **7**, 255–309, 1972.

The structural relationship between the *cyclo*-phosphates and P_4O_{10} (p. 578) is shown schematically below.

In P_4O_{10} all 10 P–O(–P) bridges are equivalent and hydrolytic cleavage of any one leads to "$H_2P_4O_{11}$" in which P–O(–P) bridges are now of two types. Cleavage of "type a" leads to *cyclo*-tetraphosphoric acid or its salts (as shown in the upper line of the scheme), whereas cleavage of any of the other bridges leads to a *cyclo*-triphosphate ring with a pendant –OP(O)OH group which can subsequently be hydrolysed off to leave $(HPO_3)_3$ or its salts (lower line of scheme). *Cyclo*-$(HPO_3)_4$ can, indeed, be made by careful hydrolysis of hexagonal P_4O_{10} with ice-water, and similar treatment with iced NaOH or $NaHCO_3$ gives a 75% yield of the corresponding salt *cyclo*-$(NaPO_3)_4$. The preparation of *cyclo*-$(NaPO_3)_3$ by controlled thermolytic dehydration of NaH_2PO_4 was mentioned in the preceding section and acidification yields *cyclo*-triphosphoric acid. The *cyclo*-$(PO_3)_3^{3-}$ anion adopts the chair configuration with dimensions as shown; *cyclo*-$(PO_3)_4^{4-}$ is also known in this configuration though this can be modified by changing the cation. The crystal structure of the *cyclo*-hexaphosphate anion in $Na_6P_6O_{18} \cdot 6H_2O$ shows that all 6 P atoms are coplanar and that bond lengths are similar to those in the $P_3O_9^{3-}$ and $P_4O_{12}^{4-}$ anions.

Higher *cyclo*-metaphosphates can be isolated by chromatographic separation from Graham's salt in which they are present to the extent of $\sim 1\%$.

12.3.7 *Phosphorus–nitrogen compounds*

The P–N bond is one of the most intriguing in chemistry and many of its more subtle aspects still elude a detailed and satisfactory description. It occurs in innumerable compounds, frequently of great stability, and in many of these the strength of the bond and the shortness of the interatomic distance have been interpreted in terms of "partial double-bond character". In fact, the conventional symbols P—N and P=N are more an aid to electron counting than a description of the bond in any given compound (see p. 628).

Many compounds containing the P–N link can be considered formally as derivatives of the oxoacids of phosphorus and their salts (pp. 586–619) in which there has been isoelectronic replacement of:

$$\text{PH [or P(OH)]} \quad \text{by} \quad \text{P(NH}_2\text{)} \quad \text{or} \quad \text{P(NR}_2\text{)}$$

$$\text{P=O [or P=S]} \quad \text{by} \quad \text{P=NH} \quad \text{or} \quad \text{P=NR}$$

$$\text{P—O—P} \quad \text{by} \quad \text{P—NH—P} \quad \text{or} \quad \text{P—NR—P, etc.}^{[60, 60a]}$$

Examples are phosphoramidic acid, $H_2NP(O)(OH)_2$; Phosphordiamidic acid, $(H_2N)_2P(O)(OH)$; phosphoric triamide, $(H_2N)_3PO$; and their derivatives. There are an enormous number of compounds featuring the 4-coordinate group shown in structure (1) including the versatile nonaqueous solvent hexamethylphosphoramide $(Me_2N)_3PO$; this is readily made by reacting $POCl_3$ with $6Me_2NH$, and dissolves metallic Na to give paramagnetic blue solutions similar to those in liquid NH_3 (p. 88).

[60] D. A. PALGRAVE, Amido- and imido-phosphoric acids and their salts, Section 28, pp. 760–815, in ref. 15.

[60a] M. L. NIELSEN, Phosphorus–nitrogen compounds (excluding cyclic phosphonitrilic compounds), Chap. 5 in C. B. COLBURN (ed.), *Developments in Inorganic Nitrogen Chemistry*, Vol. 1, pp. 307–469, Elsevier, Amsterdam, 1966.

Another series includes the *cyclo*-metaphosphimic acids, which are tautomers of the *cyclo*-polyphosphazene hydroxides (p. 632). Similarly, halogen atoms in PX_3 or other P–X compounds can be successively replaced by the isoelectronic groups $-NH_2$, $-NHR$, $-NR_2$, etc., and sometimes a pair of halogens can be replaced by $=NH$ or $=NR$. These, in turn, can be used to prepare a large number of other derivatives as indicated schematically below for $P(NMe_2)_3$.[2]

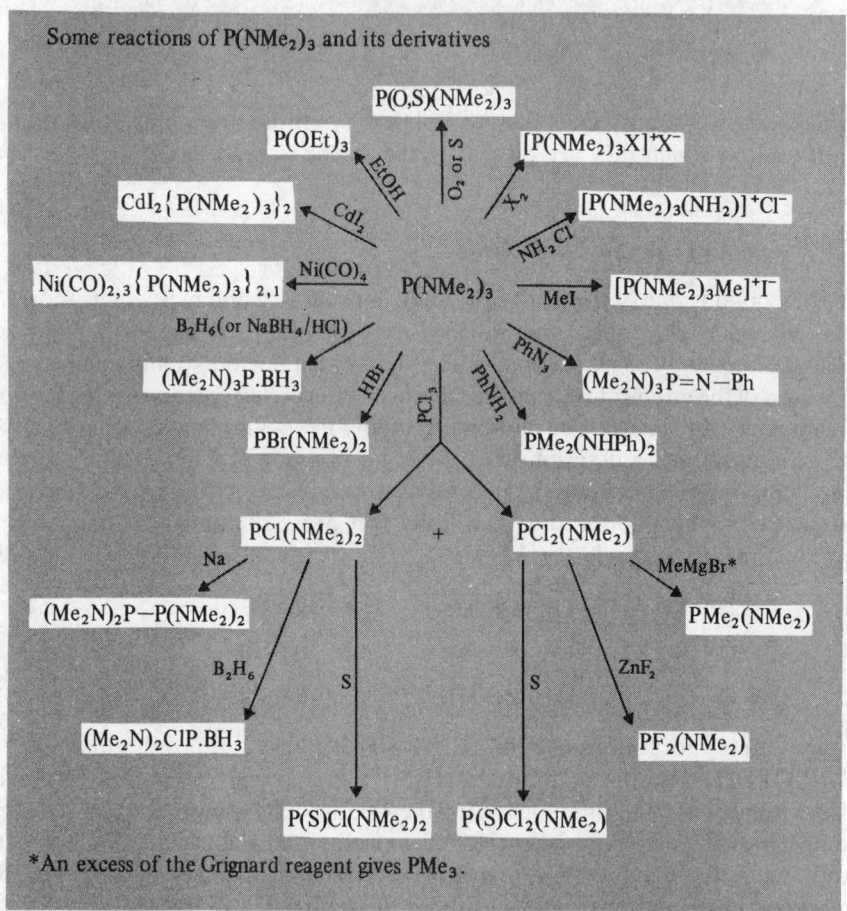

Some reactions of $P(NMe_2)_3$ and its derivatives

*An excess of the Grignard reagent gives PMe_3.

Although such compounds all formally contain P–N single bonds, they frequently display properties consistent with more extensive bonding. A particularly clear example is $PF_2(NMe_2)$ which features a short interatomic P–N distance and a planar N atom as indicated in the diagram opposite. (In the absence of this additional π bonding the P^{III}–N single-bond distance is close to 177 pm.) Again, the proton nmr of such compounds sometimes reveals restricted rotation about P–N at low temperatures and typical energy barriers to rotation (and coalescence temperatures of the non-equivalent methyl proton signals) are $PCl_2(NMe_2)$ 35 kJ mol^{-1} ($-120°$), $P(CF_3)_2(NMe_2)$ 38 kJ mol^{-1} ($-120°$), $PClPh(NMe_2)$ 50 kJ mol^{-1} ($-50°$).

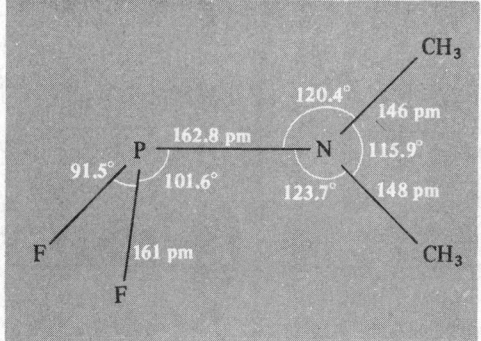

Cyclophosphazanes

Many heterocyclic compounds contain formally single-bonded P–N groups, the simplest being the *cyclo*-diphosphazanes $(X_3PNR)_2$ and $\{X(O,S)PNR\}_2$. These contain P^V and have the structures shown in Fig. 12.18. A few phosph(III)azane dimers are also known, e.g. $(RPNR')_2$. A more complex example, containing fused heterocycles of alternating P^{III} and N atoms, is the interesting hexamethyl derivative $P_4(NMe)_6$ mp 122°. This stable compound (Fig. 12.19a) is readily obtained by reacting PCl_3 with $6MeNH_2$; it is isoelectronic with and isostructural with P_4O_6 (p. 579) and undergoes many similar reactions. The stoichiometrically similar compound $P_4(NPr^i)_6$ can be prepared in the non-adamantane-type structure shown in Fig. 12.19b, though it converts to structure-type a on being heated at 157° for 12 days.[60b] A different sequence of atoms occurs in $P_2(NMe)_6$ (Fig. 12.19c) and many other "saturated" heterocycles featuring either P^{III} or P^V have been made. A recent example, made by slow addition of PCl_3 to $PhNH_2$ in toluene at 0°, is $[PhNHP_2(NPh)_2]_2NPh$; the crystal structure of the 1:1 solvate of this compound with CH_2Cl_2 (mp 250°) reveals that all N atoms are essentially planar with distances to P as indicated in the following diagram.[61]

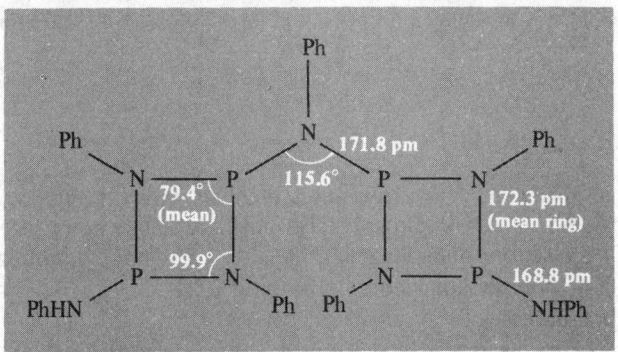

[60b] O. J. SCHERER, K. ANDRES, C. KRÜGER, Y.-H. TSAY, and G. WOLMERHÄUSER, $P_4(N\text{-}i\text{-}C_3H_7)_6$, a P_4X_6 molecule with and without adamantane structure, *Angew. Chem.*, Int. Edn. (Engl.) **19**, 571–2 (1980).

[61] M. L. THOMPSON, R. C. HALTIWANGER, and A. D. NORMAN, Synthesis and X-ray structure of a dinuclear cyclodiphosph(III)azane $[(PhNH)P_2(NPh)_2]_2NPh$, *JCS Chem. Comm.* 1979, 647–8.

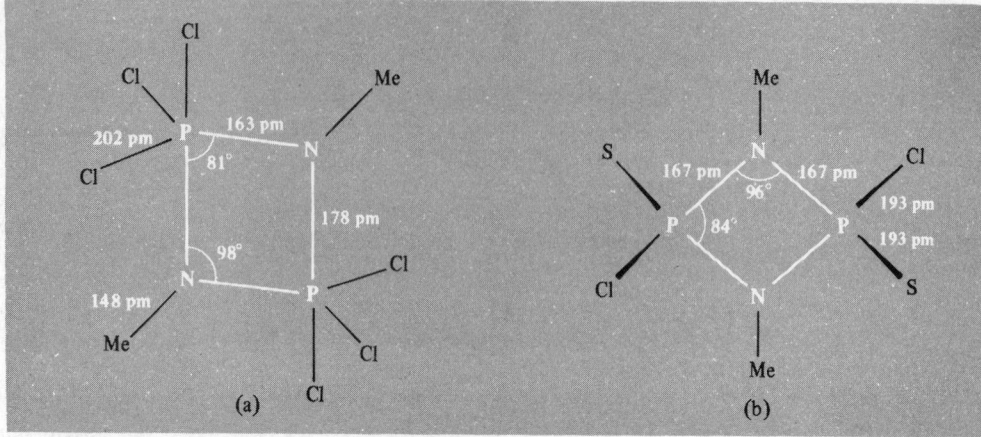

FIG. 12.18 Structures of (a) $(Cl_3PNMe)_2$, and (b) $\{Cl(S)PNMe\}_2$. Note the difference in length of the axial P–N and equatorial P–N bonds (and of the axial and equatorial P–Cl bonds) about the trigonal bipyramidal P atoms in (a).

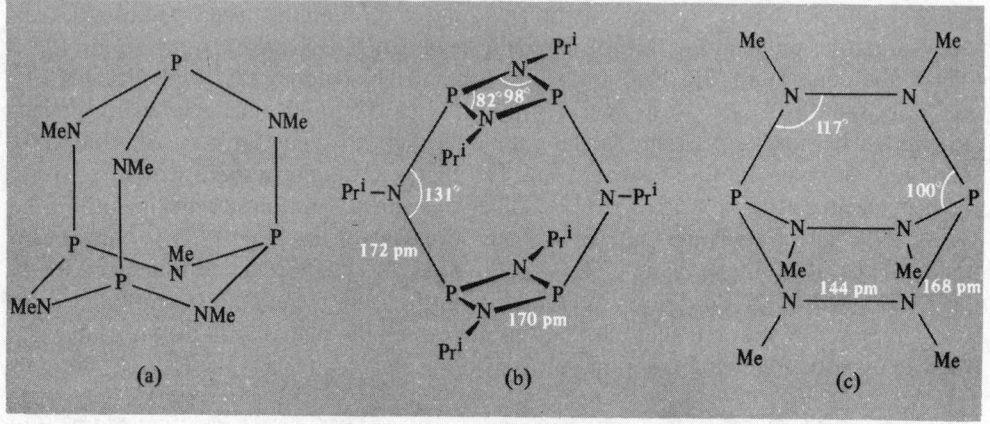

FIG. 12.19 Structures of (a) $P_4(NMe)_6$, (b) $P_4(NPr^i)_6$, and (c) $P_2(NMe)_6$.

Phosphazenes

Formally "unsaturated" PN compounds are called phosphazenes and contain P^V in the grouping $\geqslant P{=}N{-}$. A few phosph(III)azenes are also known. Phosphazenes can be classified into monophosphazenes (e.g. $X_3P{=}NR$), diphosphazenes (e.g. $X_3P{=}N{-}P(O)X_2$), polyphosphazenes containing $2, 3, 4, \ldots \infty$ $-X_2P{=}N-$ units, and the *cyclo*-polyphosphazenes $[-X_2P{=}N-]_n$, $n = 3, 4, 5 \ldots 17$.

Monophosphazenes, particularly those with organic substituents, $R_3P{=}NR'$, derive great interest from being the N analogues of phosphorus ylides $R_3P{=}CR_2$ (p. 635). They were first made by H. Staudinger in 1919 by reacting an organic azide such as PhN_3 with PR_3 ($R = Cl, OR, NR_2, Ar,$ etc.), e.g.:

$$PPh_3 + PhN_3 \longrightarrow N_2 + Ph_3P{=}NPh; \quad mp\ 132°$$

More recently they have been made via a reaction associated with the name of A. V. Kirsanov (1962), e.g.:

$$Ph_3PCl_2 + PhNH_2 \longrightarrow Ph_3P=NPh + 2HCl$$

As expected, the P–N distance is short and the angle at N is $\sim 120°$, e.g.:

Over 600 such compounds are now known, especially those with the $Cl_3P=N-$ group.[62]

Diphosphazenes can be made by reacting PCl_5 with NH_4Cl in a chlorohydrocarbon solvent under mild conditions:†

$$3PCl_5 + NH_4Cl \xrightarrow{\text{solvent}} 4HCl + [Cl_3P=N-PCl_3]^+ PCl_6^-; \quad \text{mp } 310°$$

$$\downarrow NH_4Cl$$

$$4HCl + [Cl_3P=N-PCl_2=N-PCl_3]^+ Cl^-$$

† The inverse of these compounds are the phosphadiazene cations, prepared by halide ion abstraction from diaminohalophosphoranes in CH_2Cl_2 or SO_2 solution, e.g.:

$$(R_2N)_2PCl + AlCl_3 \longrightarrow [(R_2N)_2P]^+ [AlCl_4]^-$$

An X-ray crystal structure of the Pr^i_2N-derivative shows the presence of a bent, 2-coordinate P atom, equal P–N distances, and accurately planar 3-coordinate N atoms (A. H. Cowley, M. C. Cashner, and J. S. Szobota, *J. Am. Chem. Soc.* **100**, 7784–6 (1978).)

[62] M. BERMANN, Compilation of physical data of phosphazo trihalides, *Topics in Phosphorus Chemistry* **7**, 311–78, 1972.

In liquid ammonia ammonolysis also occurs:

$$2PCl_5 + 16NH_3(liq.) \longrightarrow [(H_2N)_3P{=}N{-}P(NH_2)_3]^+Cl^- + 9NH_4Cl$$

The $P{=}N$ and $P{-}N$ bonds are equivalent in these compounds and they could perhaps better be written as $[X_3P{\cdots}N{\cdots}PX_3]^+$, etc. Like the parent phosphorus pentahalides (p. 571), these diphosphazenes can often exist in ionic and covalent forms and they are part of a more extended group of compounds which can be classified into several general series $Cl(Cl_2PN)_nPCl_4$, $[Cl(Cl_2PN)_nPCl_3]^+Cl^-$, $[Cl(Cl_2PN)_nPCl_3]^+PCl_6^-$, $Cl(Cl_2PN)_nPOCl_2$, etc., where $n = 0, 1, 2, 3 \ldots$. Some examples of the first series are PCl_5 (i.e. $n = 0$),

$$P_2NCl_7 \text{ (i.e. } \underset{\underset{Cl}{|}}{\overset{\overset{Cl}{|}}{Cl{-}P}}{=}N{-}\underset{\underset{Cl}{|}}{\overset{\overset{Cl}{|}}{PCl_2}}),$$

$$P_3N_2Cl_9 \text{ (i.e. } \underset{\underset{Cl}{|}}{\overset{\overset{Cl}{|}}{Cl{-}P}}{=}N{-}\underset{\underset{Cl}{|}}{\overset{\overset{Cl}{|}}{P}}{=}N{-}\underset{\underset{Cl}{|}}{\overset{\overset{Cl}{|}}{PCl_2}}),$$

and $P_4N_3Cl_{11}$ (i.e. $\underset{\underset{Cl}{|}}{\overset{\overset{Cl}{|}}{Cl{-}P}}{=}N{-}\underset{\underset{Cl}{|}}{\overset{\overset{Cl}{|}}{P}}{=}N{-}\underset{\underset{Cl}{|}}{\overset{\overset{Cl}{|}}{P}}{=}N{-}\underset{\underset{Cl}{|}}{\overset{\overset{Cl}{|}}{PCl_2}}),$

and some of these can exist in the ionic form represented by the second series

$$[\cdots\cdots\cdots\underset{\underset{Cl}{|}}{\overset{\overset{Cl}{|}}{P}}{-}Cl]^+Cl^-.$$

Likewise, the third series runs from $n = 0$ (i.e. $PCl_4^+PCl_6^-$) through P_3NCl_{12}, $P_4N_2Cl_{14}$, and $P_5N_3Cl_{16}$ to $P_6N_4Cl_{18}$

$$\text{(i.e. } [Cl({-}\underset{\underset{Cl}{|}}{\overset{\overset{Cl}{|}}{P}}{=}N{-})_4\underset{\underset{Cl}{|}}{\overset{\overset{Cl}{|}}{P}}{-}Cl]^+PCl_6^-).$$

In the limit, polymeric phosphazene dichlorides are formed $({-}NPCl_2{-})_n$, where n can exceed 10^4 and these polyphosphazenes and their *cyclo*-analogues form by far the most extensive range PN compounds.

Polyphosphazenes

The grouping

$$\underset{\underset{R}{|}}{\overset{\overset{R}{|}}{{-}N}}{=}P{-} \text{ is isoelectronic with the silicone grouping } \underset{\underset{R}{|}}{\overset{\overset{R}{|}}{{-}O{-}Si{-}}} \text{ (p. 422)}$$

and, after the silicones, the polyphosphazenes form the most extensive series of covalently bonded polymers with a non-carbon skeleton. This section will describe their preparation, structure, bonding, and potential applications.[2, 5, 63–65]

Preparation and structure. Polyphosphazenes have a venerable history. $(NPCl_2)_n$ oligomers were first made in 1834 by J. von Liebig and F. Wöhler who reacted PCl_5 with NH_3, but their stoichiometry and structure were not elucidated until much later. The fluoro analogues $(NPF_2)_n$ were first made in 1956 and the bromo compounds $(NPBr_2)_n$ in 1960. The synthesis of $(NPCl)_n$ was much improved by R. Schenk and G. Römer in 1924 and their method remains the basis for present-day production on both the laboratory and industrial scales:

$$nPCl_5 + nNH_4Cl \xrightarrow[120-150°]{\text{solvent}} (NPCl_2)_n + 4nHCl$$

Appropriate solvents are 1,1,2,2-tetrachloroethane (bp 146°), PhCl (bp 132°), and 1,2-dichlorobenzene (bp 179°). By varying the conditions, yields of the cyclic trimer or tetramer and other oligomers can be optimized and the compounds then separated by fractionation. Highly polymeric $(NPCl_2)_\infty$ can be made by heating *cyclo*-$(NPCl_2)_3$ to 150–300°, though heating to 350° induces depolymerization. Polycyclic compounds are rarely obtained in these preparations, one exception being $N_7P_6Cl_9$, mp 237.5°, which can be obtained in modest yield from the direct thermolytic reaction of PCl_5 and NH_4Cl. The tricyclic structure is strongly distorted from planarity though the central NP_3 group features an accurately planar 3-coordinate N atom with much longer N–P bonds than

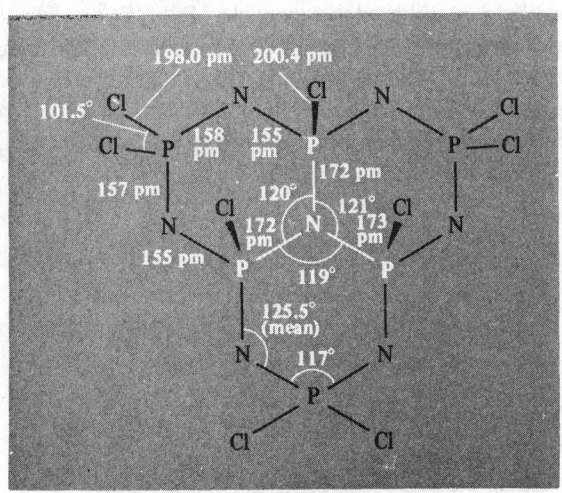

[63] H. R. ALLCOCK, *Phosphorus Nitrogen Compounds*, Academic Press, New York, 1972, 498 pp.; H. R. ALLCOCK, Recent advances in phosphazene (phosphonitrilic) chemistry, *Chem. Rev.* **72**, 315–56 (1972) (475 refs.).

[64] R. A. SHAW, Aspects of the structure and bonding in inorganic phosphorus compounds, *Pure Appl. Chem.* **44**, 317–41 (1975).

[65] S. S. KRISNAMURTHY, A. C. SAU, and M. WOODS, Cyclophosphazenes, *Adv. Inorg. Chem. Radiochem.* **21**, 41–112 (1978) (499 refs.).

those in the peripheral macrocycle. The 2 sorts of P–Cl bonds are also noticeably different in length and the 3 central Cl atoms are all on one side of the NP_3 plane with $\angle NPCl$ $104°$.

Many details of the preparative reaction mechanism remain unclear but it is thought that NH_4Cl partly dissociates into NH_3 and HCl, and that PCl_5 reacts in its ionic form $PCl_4^+PCl_6^-$ (p. 573). Nucleophilic attack by NH_3 on PCl_4^+ then occurs with elimination of HCl and the $\{HN=PCl_3\}$ attacks a second PCl_4^+ to give $[Cl_3P=N-PCl_3]^+$ and HCl. After 1 h the major (insoluble) intermediate product is $[Cl_3P=N-PCl_3]^+PCl_6^-$ (i.e. P_3NCl_{12}, p. 624) and this then slowly reacts with more NH_3 to give HCl and $\{Cl_3P=N-PCl_2=NH\}$, etc. It is probable that NH_4Br and $PBr_4^+Br^-$ react similarly to give $(NPBr_2)_n$ but NH_4F fluorinates PCl_5 to NH_4PF_4 and the fluoroanalogues $(NPF_2)_n$ are best prepared by fluorinating $(NPCl_2)_n$ with KSO_2F/SO_2 (i.e. KF in liquid SO_2). Similarly, standard substitution reactions lead to many derivatives in which all (or some) of the Cl atoms are replaced by OMe, OEt, OCH_2CF_3, OPh, NHPh, NMe_2, NR_2, R, Ar, etc. Partial replacement leads to geminal derivatives (in which both Cl on 1 P atom are replaced) and to non-geminal derivatives which, in turn, can exist as *cis-* or *trans-* isomers.

The cyclic trimer $(NPF_2)_3$, mp $28°$, has an accurately planar 6-membered ring (D_{3h} symmetry) in which all 6 P–N distances are equal (Fig. 12.20). Most other trimers are also more-or-less planar with equal P–N distances. Perhaps surprisingly, the cyclic tetramer $(NPF_2)_4$, mp $30.4°$, is also a planar heterocycle (D_{4h} symmetry) with even shorter P–N bonds (151 pm) and with ring angles of $122.7°$ and $147.4°$ at P and N respectively. However, other conformations are found in other derivatives, e.g. chair (C_{2h}), saddle (D_{2d}), boat (S_4), crown tetrameric (C_{4v}), and hybrid. Thus $(NPCl_2)_4$ exists in the metastable K form (in which it has the boat conformation) and the stable T form (chair configuration) as shown in Fig. 12.21. The remarkable diversity of molecular conformations observed for the 8-membered heterocycle $\{P_4N_4\}$ suggests that the particular structure adopted in each case results from a delicate balance of intra- and inter-molecular forces including the details of skeletal bonding, the orientation of substituents and their polar and steric nature, crystal-packing effects, etc. The mps for various series of *cyclo-*$(NPX_2)_n$ frequently shows an alternation, with values for n even being greater than those for adjacent n odd. Some examples are in Fig. 12.22.

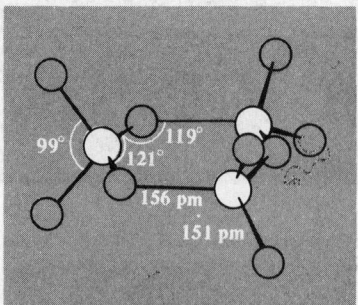

FIG. 12.20   Molecular structure and dimensions of $(NPF_2)_3$; the ring is accurately planar with all 6 P–N distances equal. $(NPCl_2)_3$ is almost planar (pseudo-chair) with P–N 158 pm, P–Cl 197 pm, $\angle NPN$ 118.4°, $\angle PNP$ 121.4°, $\angle ClPCl$ 102°.

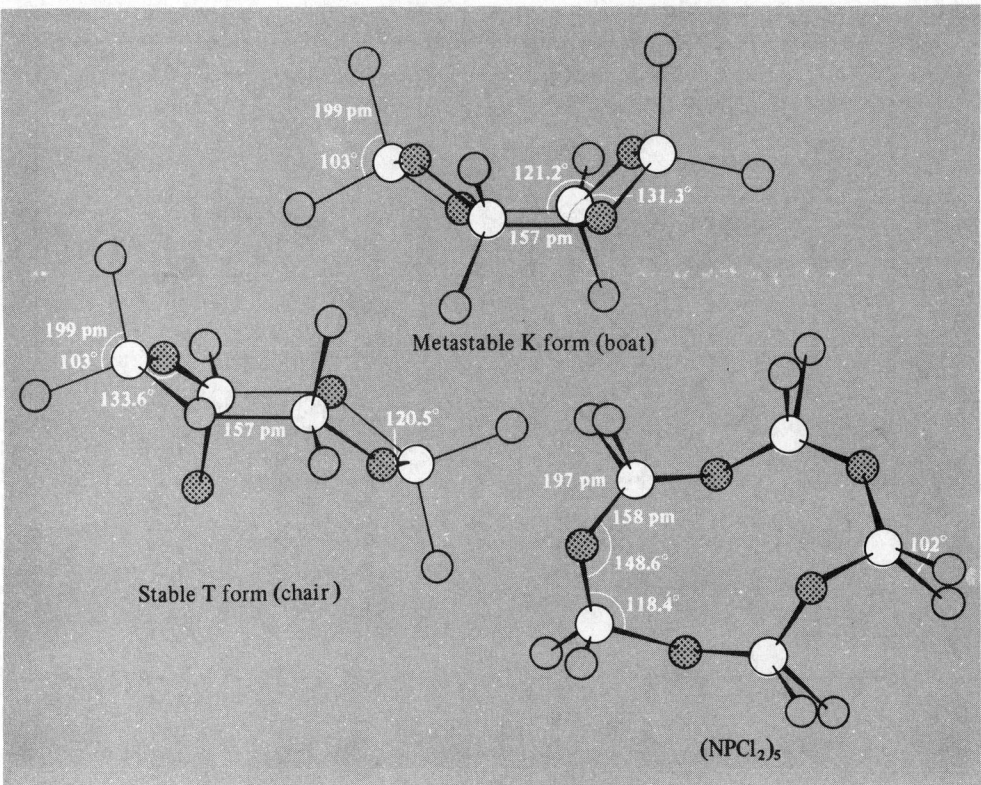

Metastable K form (boat)

Stable T form (chair)

(NPCl₂)₅

FIG. 12.21 Molecular structure and dimensions of the two forms of $(NPCl_2)_4$ and of $(NPCl_2)_5$.

Bonding. All phosphazenes, whether cyclic or chain, contain the formally unsaturated

group N⚌P with 2-coordinate N and 4-coordinate P. The experimental facts

that have to be interpreted by any acceptable theory of bonding are:

(i) the rings and chains are very stable;
(ii) the skeletal interatomic distances are equal around the ring (or along the chain) unless there is differing substitution at the various P atoms;
(iii) the P–N distances are shorter than expected for a covalent single bond (~ 177 pm) and are usually in the range 158 ± 2 pm (though bonds as short as 147 pm occur in some compounds);
(iv) the N–P–N angles are usually in the range $120 \pm 2°$ but the P–N–P angles in various compounds span the range from 120–148.6°;

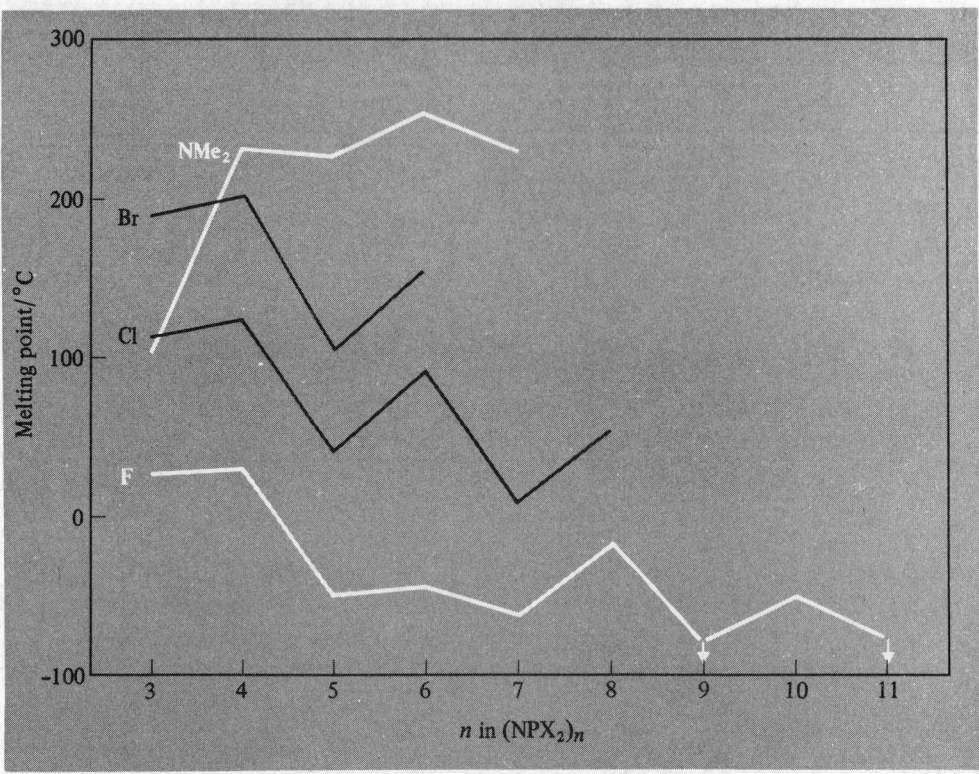

Fig. 12.22   Melting points of various series of *cyclo*-polyphosphazenes $(NPX_2)_n$ showing the higher values for *n* even.

(v) skeletal N atoms are weakly basic and can be protonated or form coordination complexes, especially when there are electron-releasing groups on P;

(vi) unlike many aromatic systems the phosphazene skeleton is hard to reduce electrochemically;

(vii) spectral effects associated with organic π-systems (such as the bathochromic ultraviolet shift that accompanies increased electron delocalization) are not found.

In short, the bonding in phosphazenes is not adequately represented by a sequence of alternating double and single bonds —N=P—N=P— yet it differs from aromatic σ-π systems in which there is extensive electron delocalization via p_π–p_π bonding. The possibility of p_π–d_π bonding in N–P systems has been considered by many authors since the mid-1950s but there is still no consensus, and for nearly every argument that can be mounted in favour of P(3d)-orbital contributions another can be raised against it.[63] It seems generally agreed that 2 electrons on N occupy an sp^2 lone-pair in the plane of the ring (or the plane of the local PNP triangle) as in Fig. 12.23a. The situation at P is less clear mainly because of uncertainties concerning the d-orbital energies and the radial extent (size) of these orbitals in the *bonding situation* (as distinct from the free atom). In so far as symmetry is concerned, the sp^2 lone-pair on each N can be involved in coordinate bonding in the xy plane to "vacant" $d_{x^2-y^2}$ and d_{xy} orbitals on the P (Fig. 12.23b); this is called π'-

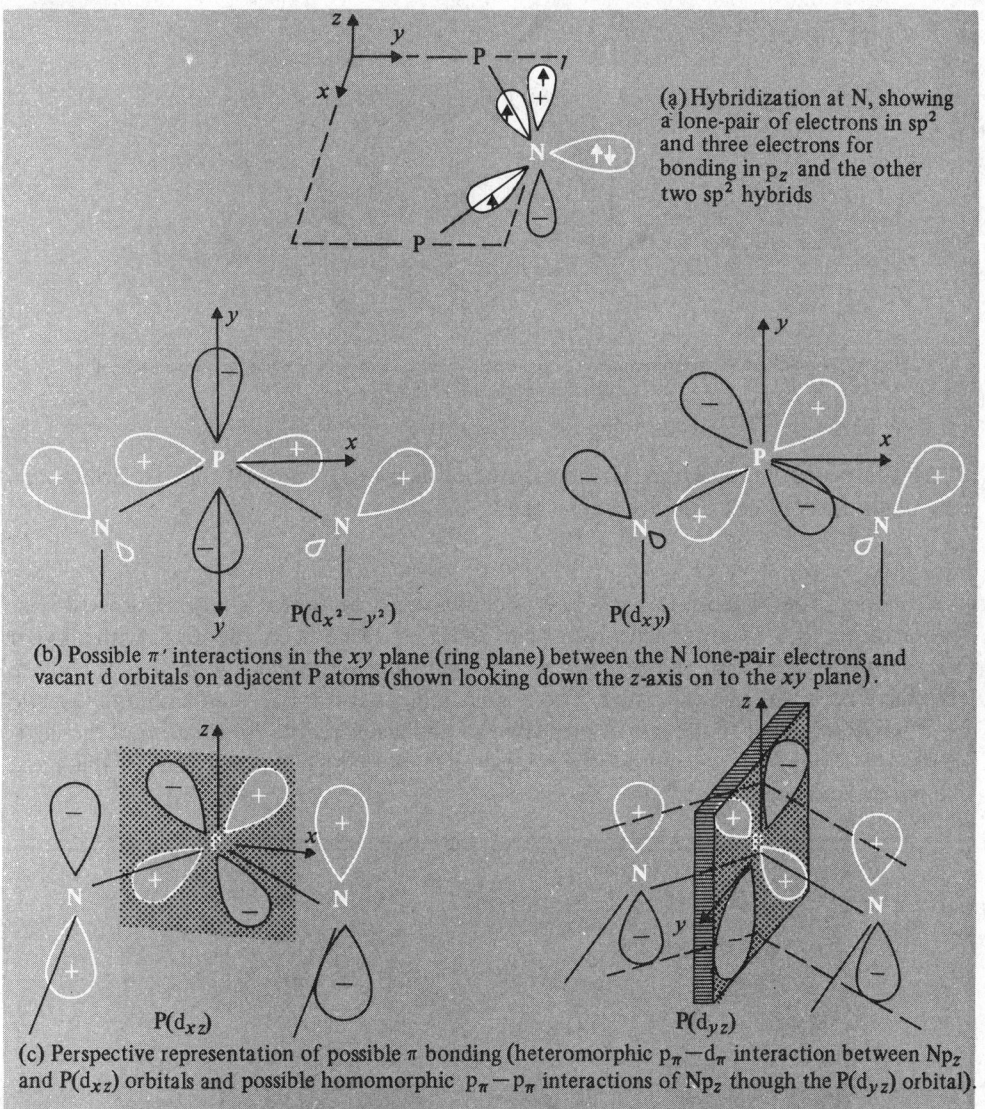

(a) Hybridization at N, showing a lone-pair of electrons in sp^2 and three electrons for bonding in p_z and the other two sp^2 hybrids

$P(d_{x^2-y^2})$ $P(d_{xy})$

(b) Possible π' interactions in the xy plane (ring plane) between the N lone-pair electrons and vacant d orbitals on adjacent P atoms (shown looking down the z-axis on to the xy plane).

$P(d_{xz})$ $P(d_{yz})$

(c) Perspective representation of possible π bonding (heteromorphic p_π–d_π interaction between Np_z and $P(d_{xz})$ orbitals and possible homomorphic p_π–p_π interactions of Np_z though the $P(d_{yz})$ orbital).

Fig. 12.23

bonding. Involvement of the out-of-plane d_{xz} and d_{yz} orbitals on the phosphorus with the singly occupied p_z orbital on N gives rise to the possibility of heteromorphic (N–P) "pseudoaromatic" p_π–d_π bonding (with d_{xz}), or homomorphic (N–N) p_π–p_π bonding (through d_{yz}) as in Fig. 12.23c. The controversy hinges in part on the relative contributions of the π' in-plane and of the two π out-of-plane interactions; approximately equal contributions from these latter two π systems would tend to separate the π orbitals into localized 3-centre islands of π character interrupted at each P atom, and broad delocalization effects would not then be expected. This is shown schematically in Fig.

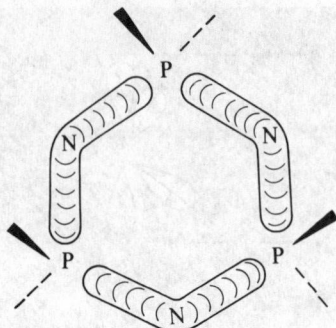

FIG. 12.24   Schematic representation of possible 3-centre islands of π bonding above and below
the ring plane for (NPX₂)₃.

12.24. The possibility of exocyclic π bonding between P(d$_{z^2}$) and appropriate orbitals on
the substituents X has also been envisaged.

Reactions. The N atom in *cyclo*-polyphosphazenes can act as a weak Brønsted base
(proton acceptor) towards such strong acids as HF and HClO₄; compounds with alkyl or
NR₂ substituents on P are more basic than the halides, as expected, and their adducts with
HCl have been well characterized. There is usually a substantial lengthening of the two
N–P bonds adjacent to the site of protonation and a noticeable contraction of the next-
nearest N–P bonds. For example, the relevant distances in [HN₃P₃Cl₂(NHPr^i)₄]Cl and
the parent compound are:[66]

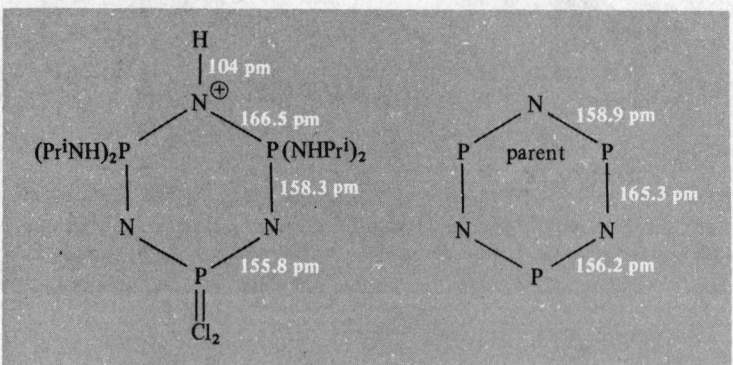

Typical basicities (pK_a' measured against HClO₄ in PhNO₂) for ring-*N* protonation are:

N₃P₃(NHMe)₆	N₃P₃(NEt₂)₆	N₃P₃Et₆	N₃P₃Ph₆	N₃P₃(OEt)₆	*trans*-N₃P₃Cl₃(NMe₂)₃
8.2	8.2	6.4	1.5	−0.2	−5.4

[66] N. V. MANI and A. J. WAGNER, The crystal structure of compounds with (N–P)$_n$ rings. Part 8.
Dichlorotetrakisisopropylamino-cyclotriphosphaza-triene hydrochloride, N₃P₃Cl₂(NHPr^i)₄·HCl, *Acta
Cryst.* **27B**, 51–58 (1971).

Cyclo-polyphosphazenes can also act as Lewis bases (*N* donor-ligands) to form complexes such as $[TiCl_4(N_3P_3Me_6)]$, $[SnCl_4N_3P_3Me_6)]$, $[AlBr_3(N_3P_3Br_6)]$, and $[2AlBr_3.(N_3P_3Br_6)]$. Not all such adducts are necessarily ring-*N* donors and the 1:1 adduct of $(NPCl_2)_3$ with $AlCl_3$ is thought to be a chloride ion donor, $[N_3P_3Cl_5]^+[AlCl_4]^-$. By contrast, the complex $[Pt^{II}Cl_2(\eta^2-N_4P_4Me_8)]$.MeCN features transannular bridging of 2 N atoms by the $PtCl_2$ moiety.[66a] An intriguing example of a *cyclo*-polyphosphazene acting as a multidentate macrocyclic ligand occurs in the bright orange complex formed when $N_6P_6(NMe_2)_{12}$ reacts with equal amounts of $CuCl_2$ and CuCl. The crystal structure of the resulting $[N_6P_6(NMe_2)_{12}CuCl]^+[CuCl_2]^-$ has been determined (Fig. 12.25b) and detailed comparison with the conformation and interatomic distances in the parent heterocycle (Fig. 12.25a) gives important clues as to the relative importance of the various π and π' bonding interactions involving N (and P) atoms.[67] Incidentally, the compound also affords the first example of the linear 2-coordinate Cu^I complex $[CuCl_2]^-$.

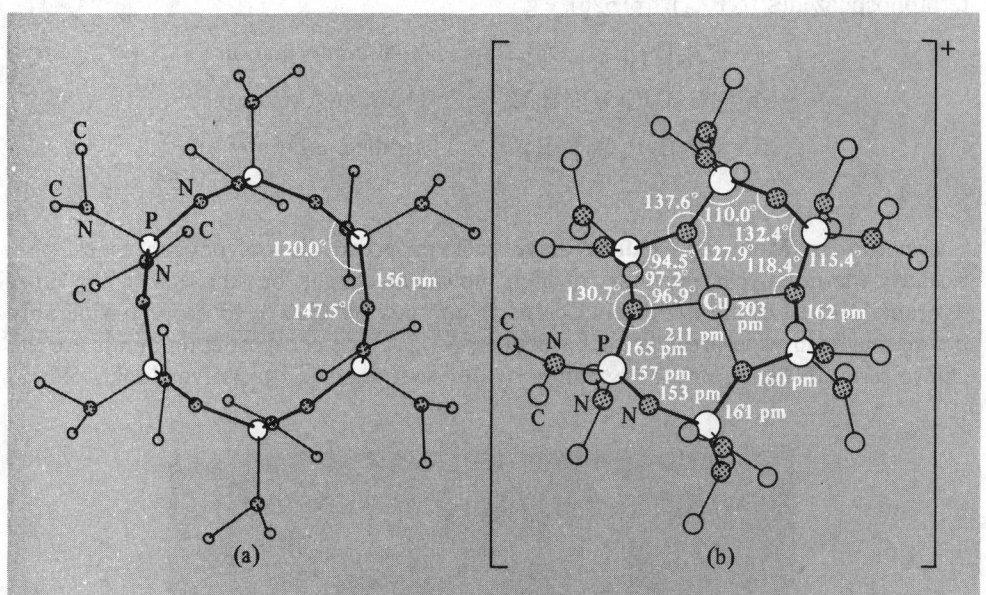

Fig. 12.25 Structure of (a) the free ligand $N_6P_6(NMe_2)_{12}$, and (b) the η^4 complex cation $[CuCl\{N_6P_6(NMe_2)_{12}\}]^+$ showing changes in conformation and interatomic distances in the phosphazene macrocycle. The Cl is obscured beneath the Cu and can be regarded as occupying either the apical position of a square pyramid or, since $\angle N(1)$-Cu-N(1') is large (160.9°), an equatorial position of a distorted trigonal bipyramid. Note that coordination tightens the ring, already somewhat crowded in the uncomplexed state, the mean angles at P being reduced from 120.0° to 107.5°, and the mean angles at N being reduced from 147.5° to 133.6°. The lengthening of the 8 P–N bonds contiguous to the 4 donor N atoms from 156 to 162 pm is significant, the other P–N distances (mean 156 pm) remaining similar to those in the free ligand.

[66a] J. P. O'BRIEN, R. W. ALLEN, and H. R. ALLCOCK, Crystal and molecular structures of two (cyclophosphazene)platinum compounds: $[N_4P_4Me_8]Pt^{II}Cl_2$.MeCN and $[H_2N_4P_4Me_8]^{2+}[PtCl_4]^{2-}$, *Inorg. Chem.* **18**, 2230–5 (1979).

[67] W. C. MARSH, N. L. PADDOCK, C. J. STEWART, and J. TROTTER, Dodeca(dimethylamido)cyclohexaphosphonitrile as a macrocyclic ligand, *JCS Chem. Comm.* 1970, 1190–1.

Many of the cyclic and chain dichloro derivatives $(NPCl_2)_n$ can be hydrolysed to *n*-basic acids and the lower members form well-defined salts frequently in the tautomeric metaphosphimic-acid form, e.g.:

$$
\left[\begin{array}{c} OH \qquad OH \\ \diagdown \qquad \diagup \\ N = P \\ \diagup \qquad \diagdown \end{array} \right]_{3,4} \rightleftharpoons \left[\begin{array}{c} H \quad O \quad\; OH \\ | \quad \| \quad \diagup \\ N - P \\ \diagup \qquad \diagdown \end{array} \right]_{3,4}
$$

The dihydrate of the tetramer is particularly stable and is, in fact, the bishydroxonium salt of tetrametaphosphimic acid $[H_3O]_2^+[(NH)_4P_4O_6(OH)_2]^{2-}$ the anion of which has a boat configuration and is linked by short H bonds (246 pm) into a two-dimensional sheet (Fig. 12.26). The related salts $M_4^I[NHPO_2)_4].nH_2O$ show considerable variation in conformation of the tetrametaphosphimate anion, as do the 8-membered heterocyclic tetraphosphazenes $(NPX_2)_4$ (p. 626), e.g.

$$[NH_4]_4[N_4H_4P_4O_8].2H_2O \quad \text{boat conformation}$$

$$K_4[N_4H_4P_4O_8].4H_2O \qquad\quad \text{chair conformation}$$

$$Cs_4[N_4H_4P_4O_8].6H_2O \qquad\; \text{saddle conformation.}$$

Applications. Many applications have been proposed for polyphosphazenes, particularly the non-cyclic polymers of high molecular weight, but those with the most desirable properties are extremely expensive and costs will have to drop considerably before they gain widespread use (cf. silicones, p. 424). The cheapest compounds are the chloro series $(NPCl_2)_n$ but these readily hydrolyse in moist air to polymetaphosphimic

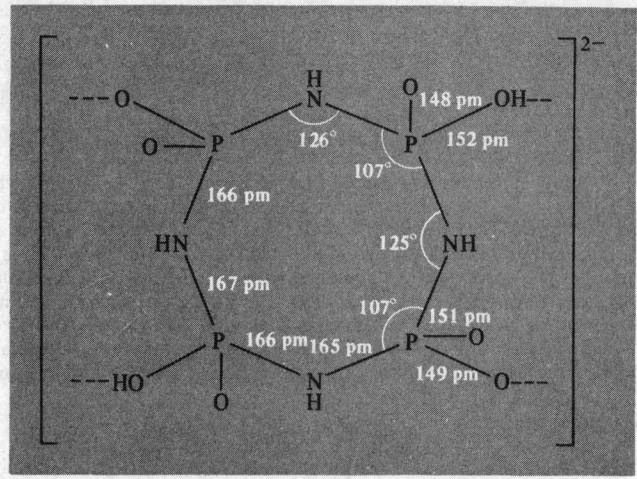

FIG. 12.26 Schematic representation of the boat-shaped anion $[(NH)_4P_4O_6(OH)_2]^{2-}$ showing important dimensions and the positions of H bonds.

acids. Greater stability is displayed by amino, alkoxy, phenoxy, and especially fluorinated derivatives, and these are attracting increasing interest as rigid plastics, elastomers, plastic films, extruded fibres, and expanded foams.[67a] Such materials are water-repellent, solvent-resistant, flame-resistant, and flexible at low temperatures (Fig. 12.27). Possible applications are as fuel hoses, gaskets, and O-ring seals for use in high-flying aircraft or for vehicles in Arctic climates. Their extraordinary dielectric strength makes them good candidates for metal coatings and wire insulation. Other applications of poly-phosphazenes include their use to improve the high-temperature properties of phenolic resins, and their use as composites with asbestos or glass for non-flammable insulating material. Some of the more reactive derivatives have been proposed as pesticides and even as ultra-high capacity fertilizers.

12.3.8 *Organophosphorus compounds*

A general treatment of the vast domain of organic compounds of phosphorus falls outside the scope of this book though several important classes of compound have already been briefly mentioned, e.g. tertiary phosphine ligands (p. 566), alkoxyphosphines and their derivatives (p. 569), organophosphorus halides (p. 573), phosphate esters in life processes (p. 611) and organic derivatives of PN compounds (preceding section). Within the general realm of organic compounds of phosphorus it is convenient to distinguish organophosphorus compounds as a particular group, i.e. those which contain one or more direct P–C bond. In such compounds the coordination number of P can be 1, 2, 3, 4, or 5 (p. 560).

Coordination number 1 is represented by the relatively unstable compounds HCP, FCP, and MeCP (cf. HCN, FCN, and MeCN). $HC\equiv P$ was first made in 1961 by subjecting PH_3 gas at 40 mmHg pressure to a low-intensity rotating arc struck between graphite electrodes;[68] it is a colourless, reactive gas, stable only below its triple point of $-124°$ (30 mmHg). Monomeric HCP slowly polymerizes at $-130°$ (more rapidly at $-78°$) to a black solid, and adds 2HCl at $-110°$ to give $MePCl_2$ as the sole product. Both monomer and polymer are pyrophoric in air even at room temperature. More recently[69] MeCP was made by pyrolysing $MeCH_2PCl_2$ at 930° in a low-pressure flow reactor and trapping the products at $-78°$. Dramatic stabilization of a phospha-alkyne has been achieved by complexation to a metal centre:[69a]

$$Bu^tC\equiv P + [Pt(C_2H_4)(PPh_3)_2] \xrightarrow[\text{room temp}]{C_6H_6} C_2H_4 + [Pt(\eta^2\text{-}Bu^tCP)(PPh_3)_2].C_6H_6$$

The translucent, cream-coloured benzene solvate was characterized by single-crystal X-ray analysis and by ^{31}P nmr spectroscopy.

[67a] H. R. ALLCOCK, Polyphosphazenes and the inorganic approach to polymer chemistry, *Sci. Progr. Oxf.* **66**, 355–69 (1980).

[68] T. E. GIER, HCP, a unique phosphorus compound, *J. Am. Chem. Soc.* **83**, 1769–70 (1961).

[69] N. P. C. WESTWOOD, H. W. KROTO, J. F. NIXON, and N. P. C. SIMMONS, Formation of 1-phosphapropyne, $CH_3C\equiv P$, by pyrolysis of dichloro(ethyl)phosphine: a He(I) photoelectron spectroscopic study, *JCS Dalton* 1979, 1405–8.

[69a] J. C. T. R. BURKETT-ST. LAURENT, P. B. HITCHCOCK, H. W. KROTO, and J. F. NIXON, Novel transition metal phospha-alkyne complexes; X-ray crystal and molecular structure of a side-bonded $Bu^tC\equiv P$ complex of zerovalent platinum, $[Pt(Bu^tCP)(PPh_3)_2]$, *JCS Chem. Comm.* 1981, 1141–3.

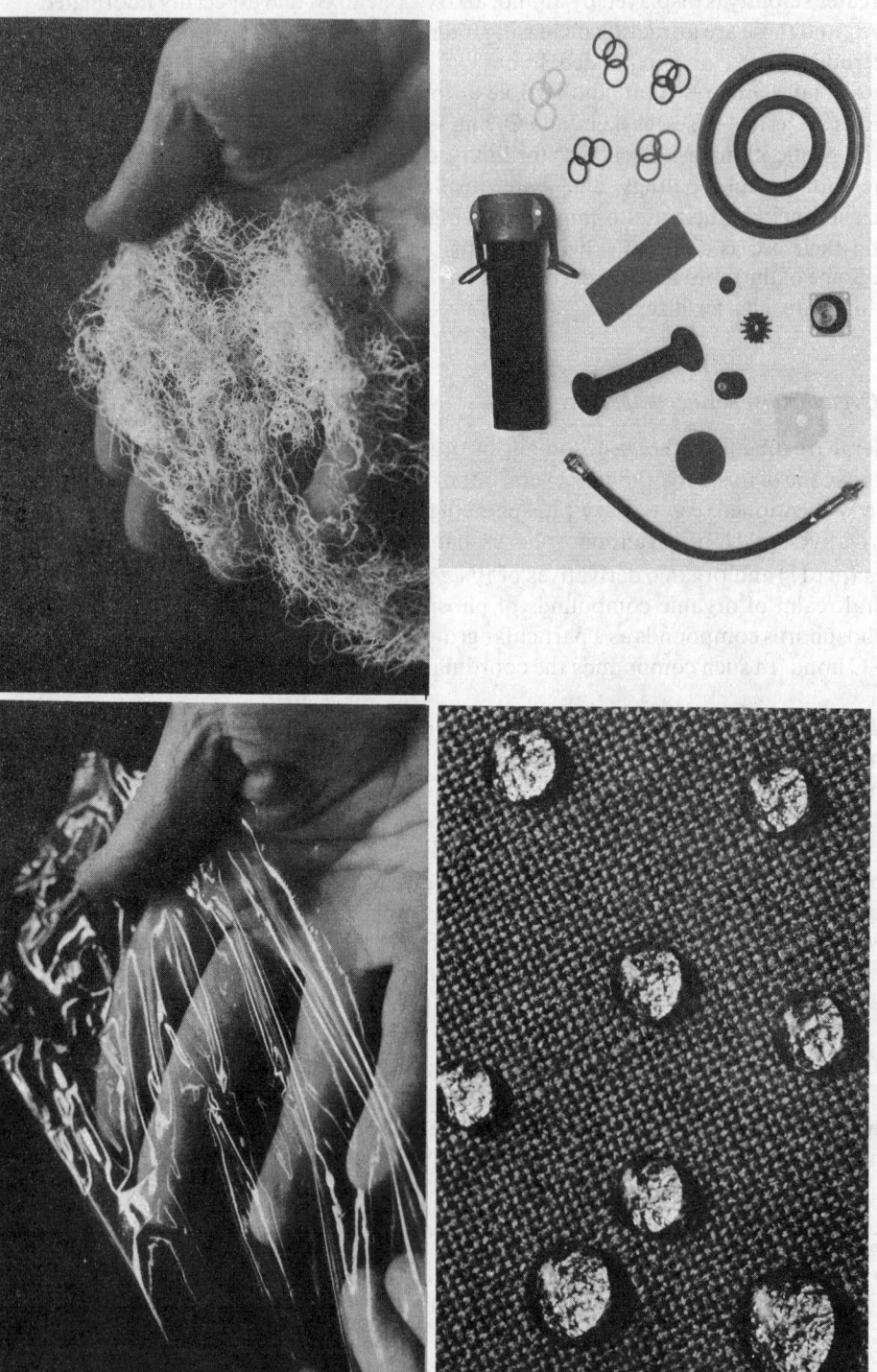

FIG. 12.27   Potential uses of polyphosphazenes: (a) A thin film of a poly(aminophosphazene); such materials are of interest for biomedical applications. (b) Fibres of poly[bis(trifluoroethoxy)phosphazene]; these fibres are water-repellant, resistant to hydrolysis or strong sunlight, and do not burn. (c) Cotton cloth treated with a poly(fluoroalkoxyphosphazene) showing the water repellancy conferred by the phosphazene. (d) Polyphosphazene elastomers are now being manufactured for use in fuel lines, gaskets, O-rings, shock absorbers, and carburettor components; they are impervious to oils and fuels, do not burn, and remain flexible at very low temperatures. Photographs by courtesy of H. R. Allcock (Pennsylvania State University) and the Firestone Tire and Rubber

The first 2-coordinate P compound also appeared in 1961:[70] $Me_3P{=}PCF_3$ was made as a white solid by cleaving *cyclo*-$[P(CF_3)]_{4 \text{ or } 5}$ with PMe_3; it is stable at low temperatures but readily dissociates into the starting materials above room temperature. More stable is the bent 2-coordinate phosphocation occurring in the orange salt[71]

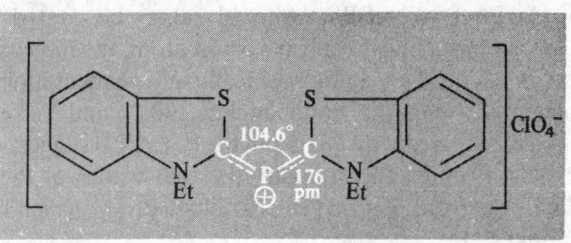

The aromatic heterocycle phosphabenzene C_5H_5P (analogous to pyridine) was reported in 1971,[72] some years after its triphenyl derivative 2,4,6-$Ph_3C_5H_2P$. See also $HP{=}CH_2$[17b] and $[P(CN)_2]^{-}$[17c] (p. 560). The growing field of phospha-alkynes and phospha-alkenes has been fully reviewed.[72a]

The most common coordination numbers for organophosphorus compounds are 3 and 4 as represented by tertiary phosphines and their complexes, and quaternary cations such as $[PMe_4]^+$ and $[PPh_4]^+$. Also of great significance are the 4-coordinate P ylides† $R_3P{=}CH_2$; indeed, few papers have created so much activity as the report by G. Wittig and G. Geissler in 1953 that methylene triphenylphosphorane reacts with benzophenone to give Ph_3PO and 1,1-diphenylethylene in excellent yield.[73]

$$Ph_3P{=}CH_2 + Ph_2CO \longrightarrow Ph_3PO + Ph_2C{=}CH_2$$

The ylide $Ph_3P{=}CH_2$ can readily be made by deprotonating a quaternary phosphonium halide with *n*-butyllithium and many such ylides are now known:

$$[Ph_3PCH_3]^+Br^- \xrightarrow{\text{LiBu}^n} Ph_3P{=}CH_2 + LiBr + Bu^nH$$

$$[PMe_4]^+Br^- + NaNH_2 \xrightarrow{\text{thf/0}°} Me_3P{=}CH_2 + NaBr + NH_3$$

† An ylide can be defined as a compound in which a carbanion is attached directly to a heteroatom carrying a high degree of positive charge:

$$\diagdown \!\!\! \diagup C^- \!\!—\, X \}^+ \longleftrightarrow \diagdown \!\!\! \diagup C = X \}$$

Thus $Ph_3P{=}CH_2$ is triphenylphosphonium methylide (see pp. 274–304 of reference 2, or textbooks of organic chemistry for a fuller treatment of the Wittig reaction).

[70] A. Burg and W. Mahler, Bicoordinate phosphorus: Base-PCF_3 adducts, *J. Am. Chem. Soc.* **83**, 2388–9 (1961).

[71] K. Dimroth and P. Hoffmann, Phosphamethincyanins, a new class of organic phosphorus compounds, *Chem. Ber.* **99**, 1325–31 (1966); R. Allmann, The crystal structure of bis(*N*-ethylbenzthiazol-(2)] phosphamethincyanin perchlorate, *Chem. Ber.* **99**, 1332–40 (1966).

[72] A. J. Ashe, Phosphabenzene and arsabenzene, *J. Am. Chem. Soc.* **93**, 3293–5 (1971).

[72a] R. Appel, F. Knoll, and I. Ruppert, Phospha-alkenes and phospha-alkynes: Genesis and properties of the (p–p)π-multiple bond, *Angew. Chem.* Int. Ed. (Engl.) **20**, 731–44 (1981).

[73] G. Wittig and G. Geissler, Reactions of pentaphenylphosphorus and several of its derivatives, *Annalen* **580**, 44–57 (1953).

The enormous scope of the Wittig reaction and its variants in affording a smooth, high-yield synthesis of C=C double bonds, etc., has been amply delineated by the work of Wittig and others and culminated in the award of the 1979 Nobel Prize for Chemistry (jointly with H. C. Brown for hydroboration, p. 192). The reaction of P ylides with many inorganic compounds has also led to some fascinating new chemistry.[74] The curious yellow compound $Ph_3P=C=PPh_3$ should also be noted:[75] unlike allene, $H_2C=C=CH_2$, which has a linear central carbon atom, the molecules are bent and the structure is unique in having 2 crystallographically independent molecules in the unit cell which have substantially differing bond angles, 130.1° and 143.8°. The short P=C distances (163 pm as compared with 183.5 pm for P–C(Ph)) suggest double bonding, but the nonlinear P=C=P unit and especially the two values of the angle, are hard to rationalize (cf. the isoelectronic cation $[Ph_3P=N=PPh_3]^+$ which has various angles in different compounds).

Pentaorgano derivatives of P are rare. The first to be made (by G. Wittig and M. Rieber in 1948) was PPh_5:

$$Ph_3PO \xrightarrow[\text{(2) HCl}]{\text{(1) LiPh}} [PPh_4]^+Cl^- \xrightarrow{\text{HI}} [PPh_4]^+I^- \xrightarrow{\text{LiPh}} PPh_5 \text{ (d. } 124°)$$

Unlike $SbPh_5$ (which has a square-pyramidal structure p. 697), PPh_5 adopts a trigonal bipyramidal coordination with the axial P–C distances (199 pm) being appreciably longer than the equatorial P–C distances (185 pm). More recently (1976) $P(CF_3)_3Me_2$ and $P(CF_3)_2Me_3$ were obtained by methylating the corresponding chlorides with $PbMe_4$. There are also many examples of 5-coordinate P in which not all the directly bonded atoms are carbon. One such is the dioxaphenylspiro-phosphorane shown in Fig. 12.28; the local symmetry about P is essentially square pyramidal, and the factors which affect the choice between this geometry and trigonal bipyramidal is a topic of active current interest.[76]

Many organophosphorus compounds are highly toxic and frequently lethal. They have been actively developed for herbicides, pesticides, and more sinister purposes such as nerve gases which disorient, harass, paralyse, or kill.[3]

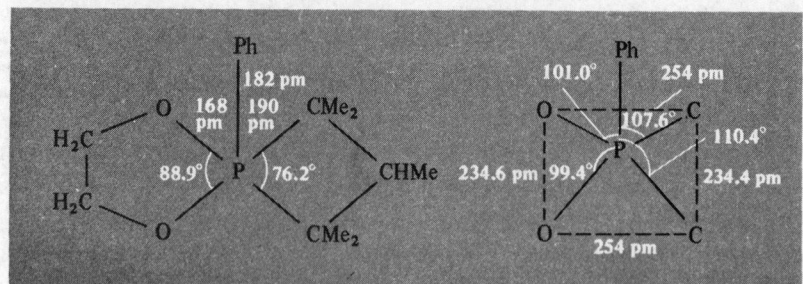

Fig. 12.28   Schematic representation of the molecular structure of $[P(C_3HMe_5)(O_2C_2H_4)Ph]$ showing the rectangular-based pyramidal disposition of the 5 atoms bonded to P; the P atom is 44 pm above the C_2O_2 plane.

[74] H. SCHMIDBAUR, Inorganic chemistry with ylides, *Acc. Chem. Res.* **8**, 62–70 (1975).

[75] A. T. VINCENT and P. J. WHEATLEY, Crystal structure of bis(triphenylphosphoranylidene)methane [hexaphenylcarbodi-phosphorane, $Ph_3P:C:PPh_3$], *J. Chem. Soc. (D), Chem. Comm.* 1971, 592.

[76] W. ALTHOFF, R. O. DAY, R. K. BROWN, and R. R. HOLMES, Molecular structure of 1,4-dioxa-5-phenyl-6,6,7,8,8-pentamethyl-$5\lambda^5$-phosphaspiro[3.4]octane. Structural preferences of four-membered rings in phosphoranes, *Inorg. Chem.* **17**, 3265–70 (1978); see also the immediately following two papers, pp. 3270–6 and 3276–85.

13

Arsenic, Antimony, and Bismuth

13.1 Introduction

The three elements arsenic, antimony, and bismuth, which complete Group V of the periodic table, were amongst the earliest elements to have been isolated and all were known before either nitrogen (1772) or phosphorus (1669) had been obtained as the free elements. The properties of arsenic sulfide and related compounds have been known to physicians and professional poisoners since the fifth century BC though their use is no longer recommended by either group of practitioners. Isolation of the element is sometimes credited to Albertus Magnus (AD 1193–1280) who heated orpiment (As_2S_3) with soap, and its name reflects its ancient lineage.† Antimony compounds were also known to the ancients and the black sulfide, stibnite, was used in early biblical times as a cosmetic to darken and beautify women's eyebrows; a rare Chaldean vase of cast antimony dates from 4000 BC and antimony-coated copper articles were used in Egypt 2500–2200 BC. Pliny (~ AD 50) gave it the name *stibium* and writings attributed to Jabir (~ AD 800) used the form *antimonium*; indeed, both names were used for both the element and its sulfide until the end of the eighteenth century (Lavoisier). The history of the element, like that of arsenic, is much obscured by the intentionally vague and misleading descriptions of the alchemists, though the elusive Benedictine monk Basil Valentine may have prepared it in 1492 (about the time of Columbus). N. Lémery published his famous *Treatise on Antimony* in 1707. Bismuth was known as the metal at least by 1480 though its previous history in the Middle Ages is difficult to unravel because the element was sometimes confused with Pb, Sn, Sb, or even Ag. The Gutenberg printing presses (1440 onwards) used type that had been cut from brass or cast from Pb, Sn, or Cu, but about 1450 a secret method of casting type from Bi alloys came into use and this particular use is still an important application of the element (p. 640). The name derives from the German *Wismut* (possibly white metal or meadow mines) and this was latinized to *bisemutum* by the sixteenth-century German scientist G. Bauer (Agricola) about 1530. Despite the difficulty of assigning precise dates to discoveries made by alchemists, miners, and metal workers (or indeed even discerning what those discoveries actually were), it seems clear

---

† Arsenic, Latin *arsenicum* from Greek ἀρσενικόν (*arsenicon*) which was itself derived (with addition of ον) from Persian *az-zarnīkh*, yellow orpiment (*zar* = gold).

that As, Sb, and Bi became increasingly recognized in their free form during the thirteenth–fifteenth centuries; they are therefore contemporary with Zn and Co, and predate all other elements except the 7 metals and 2 non-metallic elements known from ancient times (Au, Ag, Cu, Fe, Hg, Pb, Sn; C and S).[1]

Arsenic and antimony are classed as metalloids or semi-metals and bismuth is a typical B subgroup metal like tin and lead.

13.2 The Elements

13.2.1 *Abundance, distribution, and extraction*

None of the three elements is particularly abundant in the earth's crust though several minerals contain them as major constituents. As can be seen from Table 13.1, arsenic occurs about half-way down the elements in order of abundance, grouped with several others near 2 ppm. Antimony has only one-tenth of this abundance and Bi, down by a further factor of 20 or more, is about as unabundant as several of the commoner platinum metals and gold. In common with all the B subgroup metals, As, Sb, and Bi are chalcophiles, i.e. they occur in association with the chalcogens S, Se, and Te rather than as oxides and silicates.

TABLE 13.1   *Abundances of elements in crustal rocks (g tonne^{-1})*

Element	Sn	Eu	Be	**As**	Ta	Ge	In	**Sb**	Cd	Pd	Pt	**Bi**	Os	Au
PPM	2.1	2.1	2.0	**1.8**	1.7	1.5	0.24	**0.2**	0.16	0.015	0.01	**0.008**	0.005	0.004
Order	48	48	50	**51**	52	53	61	**62**	63	67	68	**69**	70	71

Arsenic minerals are widely distributed throughout the world and small amounts of the free element have also been found. Common minerals include the two sulfides realgar (As_4S_4) and orpiment (As_2S_3) and the oxidized form arsenolite (As_2O_3). The arsenides of Fe, Co, and Ni and the mixed sulfides with these metals form another set of minerals, e.g. loellingite ($FeAs_2$), safforlite (CoAs), niccolite (NiAs), rammelsbergite ($NiAs_2$), arseno-pyrite (FeAsS), cobaltite (CoAsS), enargite (Cu_3AsS_4), gerdsorfite (NiAsS), and the quaternary sulfide glaucodot [(Co,Fe)AsS]. Elemental As is obtained on an industrial scale by smelting $FeAs_2$ or FeAsS at 650–700°C in the absence of air and condensing the sublimed element:

$$FeAsS \xrightarrow{\ 700° \ } FeS + As(g) \longrightarrow As(s)$$

Residual As trapped in the sulfide residues can be released by roasting them in air and trapping the sublimed As_2O_3 in the flue system. The oxide can then either be used directly for chemical products or reduced with charcoal at 700–800° to give more As. As_2O_3 is also obtained in large quantities as flue dust from the smelting of Cu and Pb concentrates; because of the huge scale of these operations (pp. 1366, 432) this represents the most

[1] M. E. WEEKS, *Discovery of the Elements*, Chap. 3, pp. 91–119, Journal of Chemical Education, Easton, Pa, 1956.

important industrial source of As. Some production figures and major uses of As and its compounds are listed in the Panel.

Stibnite Sb_2S_3, is the most important ore of antimony and it occurs in large quantities in China, South Africa, Mexico, Bolivia, and Chile. Other sulfide ores include ullmanite (NiSbS), livingstonite ($HgSb_4S_8$), tetrahedrite (Cu_3SbS_3), wolfsbergite ($CuSbS_2$), and jamesonite ($FePb_4Sb_6S_{14}$). Indeed, complex ores containing Pb, Cu, Ag, and Hg are an

---

Production and Uses of Arsenic, Antimony, and Bismuth[2]

The major producers of "white arsenic", i.e. As_2O_3, are the USA, Sweden, France, the USSR, Mexico, and South-west Africa. World production (of which the USA accounted for some 70%) has been fairly steady at about 50 000 tonne pa during the past decade. Refined oxide (99.5%) was quoted at $230 per tonne (23 ¢ per kg in 1974 and commercial grade As (99%) cost about $4.00 per kg in 1976. Semiconductor grade As (>99.999%) obtained by zone refinement was considerably more expensive.

The main use of elemental As is in alloys with Pb and to a lesser extent Cu. Addition of small concentrations of As improves the properties of Pb/Sb for storage batteries (see below), up to 0.75% improves the hardness and castability of type metal, and 0.5–2.0% improves the sphericity of Pb ammunition. Automotive body solder is Pb (92%), Sb (5.0%), Sn (2.5%), and As (0.5%). Intermetallic compounds with Al, Ga, and In give the III–V semiconductors (p. 288) of which GaAs and InAs are of particular value for light-emitting diodes (LEDs), tunnel diodes, infrared emitters, laser windows, and Hall-effect devices (p. 290).

Arsenic compounds find extensive use in agriculture as herbicides for weed and pest control, e.g. MSMA (monosodium methylarsonate, $NaMeHAsO_3$), DSMA (the disodium salt Na_2MeAsO_3); and cacodylic acid (dimethylarsenic acid, $Me_2AsO(OH)$). Arsenic acid itself, $AsO(OH)_3$, is used as a desiccant for the defoliation of cotton bolls prior to harvesting, and as a wood preservative; sodium arsenite is used in sheep and cattle dips and for aquatic weed control. Medicinal and veterinary uses are declining: most arsenicals are to be considered as poisonous, though no toxic episodes have been reported for the element itself or its (very insoluble) sulfide, As_2S_3. The oxide is used to decolorize bottle-glass. The overall pattern of commercial uses in the USA 1974 is summarized in Table A, in terms of the amount of contained oxide.

TABLE A   *Some uses of arsenic compounds*

Use	Pesticides	Wood preservatives	Glass	Alloys and electronics	Miscellaneous
As_2O_3/tonnes	34 000	9500	5000	1500	2000
Percentage	65	18	10	3	4
		83			

The estimated world production of Sb and its compounds is about 80 000 tonnes pa of contained Sb and the price for the commercial grade metal was $3–5 per kg in 1975–6. The main supplies prior to 1934 came from China but production thereafter was disrupted for several decades and supplies are only now returning to significant proportions. More recent distribution of production (1974) is: South Africa (21.4%), Bolivia (18.5%), China (16.7%), the USSR (10.3%), Thailand (6.0%), Turkey (4.8%), Mexico (3.4%), and others (18.9%). US consumption in 1975 was 11 800 tonnes and this was dominated by the production of antimonial lead for use in storage batteries (4143 tonnes); the grid metal formerly contain 4–5% Sb to impart fluidity, creep resistance, fatigue strength, and electrochemical stability, but more recently alloys with 2.5–3% Sb and a trace of As have been preferred as this minimizes self-discharge, gassing, and poisoning of the

---

[2] *Kirk–Othmer Encyclopedia of Chemical Technology*, 3rd edn., Vol. 3, Wiley, New York, 1978; Arsenic and arsenic alloys (pp. 243–50); Arsenic compounds (251–66); Antimony and antimony alloys (96–105); Antimony compounds (105–28); Bismuth and bismuth alloys (912–21); Bismuth compounds (921–37).

negative electrode. Other typical uses of Sb alloys in the USA, 1975 (tonnes of contained Sb), are as shown in Table B.

TABLE B   *Some uses of antimony alloys*

Use	Sb/Pb batteries	Bearings	Ammunition	Solder	Type metal	Sheet pipe	Other metal	Non-metal products
Sb/tonnes	4143	365	216	121	68	55	144	6657
Percentage	35.2	3.5	1.8	1.0	0.6	0.5	1.2	56.5

As with arsenic, semiconductor grade Sb is prepared by chemical reduction of highly purified compounds. AlSb, GaSb, and InSb have applications in infrared devices, diodes, and Hall-effect devices. ZnSb has good thermoelectric properties. Applications of various *compounds* of Sb will be mentioned when the compounds themselves are discussed.

World annual production of bismuth and its compounds has hovered around 4000 tonnes of contained Bi for the past several years and production has been dominated by Japan (21.5%), Peru (17.2%), Bolivia (16.7%), Mexico (15.4%), and Australia (5.5%) which, between them, account for over 76% of all supplies. Prices have fluctuated wildly and quotes for the free element on the London Metal Exchange moved from \$14 per kg to \$31 and back to \$14 during 1974. Consumption of the metal and its compounds has also been unusual, usage in the USA dropping by a factor of 2 from 1973 to 1975, for example. This is shown in Table C which also indicates that the main uses are in pharmaceuticals, fusible alloys (including type metal, p. 637), and metallurgical additives.

No industrial poisoning by Bi metal has ever been reported but ingestion of compounds and inhalation of dust should be avoided.

TABLE C   *US use of bismuth (in tonnes)*

Product	1973	1975
Fusible alloys	423	182
Metallurgical additives	377	189
Other alloys	7	12
Pharmaceuticals and other chemicals	507	251
Others	4	4
Total	1318	638

important industrial source of Sb. Small amounts of oxide minerals formed by weathering are also known, e.g. valentinite (Sb_2O_3), cervantite (Sb_2O_4), and stibiconite ($Sb_2O_4.H_2O$), and minor finds of native Sb have occasionally been reported. Commercial ores have 5–60% Sb, and recovery methods depend on the grade. Low-grade sulfide ores (5–25% Sb) are volatilized as the oxide (any As_2O_3 being readily removed first by virtue of its greater volatility). The oxide can be reduced to the metal by heating it in a reverberatory furnace with charcoal in the presence of an alkali metal carbonate or sulfate as flux. Intermediate ores (25–40%) are smelted in a blast furnace and the oxide recovered from the flue system. Ores containing 40–60% Sb are liquated at 550–600° under reducing conditions to give Sb_2S_3 and then treated with scrap iron to remove the sulfide.

$$Sb_2S_3 + 3Fe \longrightarrow 2Sb + 3FeS$$

Some complex sulfide ores are treated by leaching and electro-winning, e.g. the electrolysis of alkaline solutions of the thioantimonate Na_3SbS_4, and the element is also recovered from the flue dusts of Pb smelters. Impure Sb contains Pb, As, S, Fe, and Cu; the latter two can be removed by stibnite treatment or heating with charcoal/Na_2SO_4 flux; the As and S can be removed by an oxidizing flux of $NaNO_3$ and NaOH (or Na_2CO_3); Pb is hard to remove but this is unnecessary if the Sb is to be used in Pb alloys (see below). Electrolysis yields $> 99.9\%$ purity and remaining impurities can be reduced to the ppm level by zone refining. The scale of production and the various uses of Sb and its compounds are summarized in the Panel.

Bismuth occurs mainly as bismite (α-Bi_2O_3), bismuthinite (Bi_2S_3), and bismutite $[(BiO)_2CO_3]$; very occasionally it occurs native, in association with Pb, Ag, or Co ores. The main commercial source of the element is as a byproduct from Pb/Zn and Cu plants, from which it is obtained by special processes dependent on the nature of the main product.[2] Sulfide ores are roasted to the oxide and then reduced by iron or charcoal. Because of its low mp, very low solubility in Fe, and fairly high oxidative stability in air, Bi can be melted and cast (like Pb) in iron and steel vessels. Like Sb, the metal is too brittle to roll, draw, or extrude at room temperature, but above 225°C Bi can be worked quite well.

13.2.2 *Atomic and physical properties*

Arsenic and Bi (like P) each have only 1 stable isotope and this occurs with 100% abundance in all natural sources of the elements. Accordingly (p. 20) their atomic weights are known with great precision (Table 13.2). Antimony has 2 stable isotopes (like N); however, unlike N, which has 1 predominantly abundant isotope, the 2 isotopes of Sb are approximately equal in abundance (^{121}Sb 57.25%, ^{123}Sb 42.75%) and consequently (p. 20) the atomic weight is known only to 1 part in 4000. It is also noteworthy that ^{209}Bi is the heaviest stable isotope of any element; all nuclides beyond $^{209}_{83}Bi$ are radioactive.

The ground-state electronic configuration of each element in the group is ns^2np^3 with an unpaired electron in each of the three p orbitals, and much of the chemistry of the group can be interpreted directly on this basis. However, smooth trends are sometimes modified (or even absent altogether), firstly, because of the lack of low-lying empty d orbitals in N, which differentiates it from its heavier congeners, and, secondly, because of the countervailing influence of the underlying filled d and f orbitals in As, Sb, and Bi. Such perturbations are apparent when the various ionization energies in Table 13.2 are plotted as a function of atomic number. Table 13.2 also contains approximate data on the conventional covalent single-bond radii for threefold coordination though these values vary by about ± 4 pm in various tabulations and should only be used as a rough guide. The 6-coordinate "effective ionic radii" for the $+3$ and $+5$ oxidation states are taken from R. D. Shannon's most recent tabulation,[3] but should not be taken to imply the presence of M^{3+} and M^{5+} cations in many of the compounds of these elements.

---

[3] R. D. SHANNON, Revised effective ionic radii and systematic studies of interatomic distances in halides and chalcogenides, *Acta Cryst.* **A32**, 751–67 (1976).

TABLE 13.2 *Atomic properties of Group V elements*

Property		N	P	As	Sb	Bi
Atomic number		7	15	33	51	83
Atomic weight (1981)		14.0067	30.97376	74.9216	121.75($\pm$3)	208.9804
Electronic configuration		$[He]2s^22p^3$	$[Ne]3s^23p^3$	$[Ar]3d^{10}4s^24p^3$	$[Kr]4d^{10}5s^25p^3$	$[Xe]4f^{14}5d^{10}6s^26p^3$
Ionization energies/MJ mol^{-1}	(I)	1.402	1.012	0.947	0.834	0.703
	(II)	2.856	1.903	1.798	1.595	1.610
	(III)	4.577	2.910	2.736	2.443	2.466
Sum (I+II+III)/MJ mol^{-1}		8.835	5.825	5.481	4.872	4.779
Sum (IV+V)/MJ mol^{-1}		16.920	11.220	10.880	9.636	9.776
Electronegativity χ		3.0	2.1	2.0	1.9	1.9
r_{cov} (M^{III} single bond)/pm		70	110	120	140	150
r_{ionic} (6-coordinate) (M^{III})/pm		(16)	44	58	76	103
(6-coordinate) (M^V)/pm		(13)	38	46	60	76

Arsenic, Sb, and Bi each exist in several allotropic forms[4] though the allotropy is not so extensive as in P (p. 555). There are three crystalline forms of As, of which the ordinary, grey, "metallic", rhombohedral, α-form is the most stable at room temperature. It consists of puckered sheets of covalently bonded As stacked in layers perpendicular to the hexagonal *c*-axis as shown in Fig. 13.1. Within each layer each As has 3 nearest neighbours at 251.7 pm and the angle As–As–As is 96.7°; each As also has a further 3 neighbours at 312 pm in an adjacent layer. The α-forms of Sb and Bi are isostructural with α-As and have the dimensions shown in Table 13.3. It can be seen that there is a progressive diminution in the difference between intra-layer and inter-layer distances though the inter-bond angles remain almost constant.

In the vapour phase As is known to exist as tetrahedral As_4 molecules (with As–As 243.5 pm) and when the element is sublimed, a yellow, cubic modification is obtained which probably also contains As_4 units though the structure has not yet been determined because the crystals decompose in the X-ray beam. The mineral arsenolamprite is another polymorph, ε-As; it is possibly isostructural with "metallic" orthorhombic P.

Antimony exists in 5 forms in addition to the ordinary α-form which has been discussed above. The yellow form is unstable above −90°; a black form can be obtained by cooling

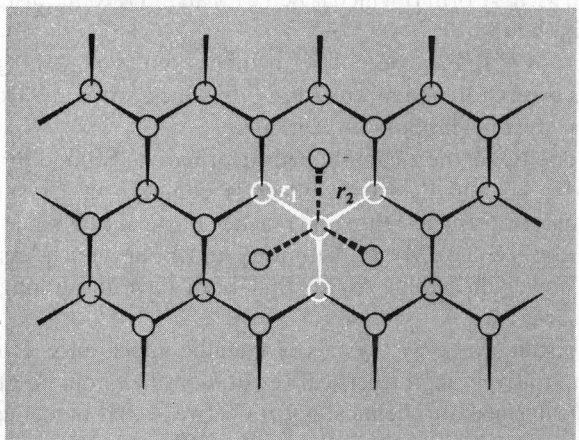

FIG. 13.1 Puckered layer structure of As showing pyramidal coordination of each As to 3 neighbours at a distance r_1 (252 pm). The disposition of As atoms in the next layer (r_2 312 pm) is shown by dashed lines.

TABLE 13.3 *Comparison of black P and α-rhombohedral arsenic, antimony, and bismuth*

	r_1/pm	r_2/pm	r_2/r_1	Angle M–M–M
Black P	223.1 (av)	332.4 (av)	1.490	2 at 96.3° (1 at 102.1°)
α-As	251.7	312.0	1.240	96.7°
α-Sb	290.8	335.5	1.153	96.6°
α-Bi	307.2	352.9	1.149	95.5°

[4] J. DONOHUE, *The Structure of the Elements*, Wiley, 1974, 436 pp.

TABLE 13.4   *Some physical properties of Group V elements*

Property	N_2	P_4	α-As	α-Sb	α-Bi
MP/°C	−210.0	44.1	816 (38.6 atm)	630.7	271.4
BP/°C	−195.8	280.5	615 (subl)	1587	1564
Density (25°C)/g cm^{-3}	0.879 (−210°)	1.823	5.778[a]	6.697	9.808
Hardness (Mohs)	—	—	3.5	3–3.5	7 (HB)
Electrical resistivity (20°C)/μohm cm	—	—	33.3	41.7	120
Contraction on freezing/%	—	—	10	0.8	−3.32

[a] Yellow As_4 has d_{25} 1.97 g cm^{-3}; cf. difference between the density of rhombohedral black P (3.56 g cm^{-3}) and white P_4 (1.823 g cm^{-3}) (p. 554).

gaseous Sb, and an explosive (impure?) form can be made electrolytically. The two remaining crystalline forms are made by high-pressure techniques: Form I has a primitive cubic lattice with a_0 296.6 pm: it is obtained from α-Sb at 50 kbar (5GPa, i.e. 5×10^9 N m^{-2}) by increasing the rhombohedral angle from 57.1° to 60.0° together with small shifts in atomic position so that each Sb has 6 equidistant neighbours. Further increase in pressure to 90 kbar yields Form II which is hcp with an interatomic distance of 328 pm for the 12 nearest neighbours.

Several polymorphs of Bi have been described but there is as yet no general agreement on their structures except for α-Bi (above) and ζ-Bi which forms at 90 kbar and has a bcc structure with 8 nearest neighbours at 329.1 pm.

The physical properties of the α-rhombohedral form of As, Sb, and Bi are summarized in Table 13.4. Data for N_2 and P_4 are included for comparison. Crystalline As is rather volatile and the vapour pressure of the solid reaches 1 atm at 615° some 200° below its mp of 816°C (at 38.6 atm, i.e. 3.91 MPa). Antimony and Bi are much less volatile and also have appreciably lower mps than As, so that both have quite long liquid ranges at atmospheric pressure.

Arsenic forms brittle steel-grey crystals of metallic appearance. However, its lack of ductility and comparatively high electrical resisitivity (33.3 μohm cm), coupled with its amphoterism and intermediate chemical nature between that of metals and non-metals, have lead to its being classified as a metalloid rather than a "true" metal. Antimony is also very brittle and forms bluish-white, flaky, lustrous crystals of high electrical resistivity (41.7 μohm cm). These values of resistivity can be compared with those for "good" metals such as Ag (1.59), Cu (1.72), and Al (2.82 μohm cm), and with "poor" metals such as Sn (11.5) and Pb (22 μohm cm). Bismuth has a still higher resistivity (120 μohm cm) which even exceeds that of commercial resistors such as Nichrome alloy (100 μohm cm). Bismuth is a brittle, white, crystalline metal with a pinkish tinge. It is the most diamagnetic of all metals (mass susceptibility 17.0×10^{-9} m^3 kg^{-1}—to convert this SI value to cgs multiply by $10^3/4\pi$, i.e. 1.35×10^{-6} cm^3 g^{-1}). It also has the highest Hall effect coefficient of any metal. It is also unusual in expanding on solidifying from the melt, a property which it holds uniquely with Ga and Ge among the elements.

13.2.3 *Chemical reactivity and group trends*

Arsenic is stable in dry air but the surface oxidizes in moist air to give a superficial golden bronze tarnish which deepens to a black surface coating on further exposure.

When heated in air it sublimes and oxidizes to As_4O_6 with a garlic like odour (poisonous). Above 250–300° the reaction is accompanied by phosphorescence (cf. P_4, p. 546). When ignited in oxygen, As burns brilliantly to give As_4O_6 and As_4O_{10}. Metals give arsenides (p. 646), fluorine enflames to give AsF_5 (p. 655), and the other halogens yield AsX_3 (p. 652). Arsenic is not readily attacked by water, alkaline solutions, or non-oxidizing acids, but dil HNO_3 gives arsenious acid (H_3AsO_3), hot conc HNO_3 yields arsenic acid (H_3AsO_4), and hot conc H_2SO_4 gives As_4O_6. Reaction with fused NaOH liberates H_2:

$$As + 3NaOH \longrightarrow Na_3AsO_3 + \tfrac{3}{2}H_2$$

One important property which As has in common with its neighbouring elements immediately following the 3d transition series (i.e. Ge, As, Se, Br) and which differentiates it from its Group VB neighbours P and Sb, is its notable reluctance to be oxidized to the group valence of $+5$. Consequently As_4O_{10} and H_3AsO_4 are oxidizing agents and arsenates are used for this purpose in titrimetric analysis (p. 674).

The ground-state electronic structure of As, as with all Group VB elements features 3 unpaired electrons ns^2np^3; there is a substantial electron affinity for the acquisition of 1 electron but further additions must be effected against considerable coulombic repulsion, and the formation of As^{3-} is highly endothermic. Consistent with this there are no "ionic" compounds containing the arsenide ion and compounds such as Na_3As are intermetallic or alloy-like. However, despite the metalloidal character of the free element, the ionization energies and electronegativity of As are similar to those of P (Table 13.2) and the element readily forms strong covalent bonds to most non-metals. Thus AsX_3 (X = H, hal, R, Ar, etc.) are covalent molecules like PX_3 and the tertiary arsines have been widely used as ligands to b-class transition elements (p. 1065).[5] Similarly, As_4O_6 and As_4O_{10} resemble their P analogues in structure; the sulfides are also covalent heterocyclic molecules though their stoichiometry and structure differ from those of P.

Antimony is in many ways similar to As, but it is somewhat less reactive. It is stable to air and moisture at room temperature, oxidizes on being heated under controlled conditions to give Sb_2O_3, Sb_2O_4, or Sb_2O_5, reacts vigorously with Cl_2 and more sedately with Br_2 and I_2 to give SbX_3, and also combines with S on being heated. H_2 is without direct reaction and SbH_3 (p. 650) is both very poisonous and thermally very unstable. Dilute acids have no effect on Sb; concentrated oxidizing acids react readily, e.g. conc HNO_3 gives hydrated Sb_2O_5, aqua regia gives a solution of $SbCl_5$, and hot conc H_2SO_4 gives the salt $Sb_2(SO_4)_3$.

Bismuth continues the trend to electropositive behaviour and Bi_2O_3 is definitely basic, compared with the amphoteric oxides of Sb and As and the acidic oxides of P and N. There is also a growing tendency to form salts of oxoacids by reaction of either the metal or its oxide with the acid, e.g. $Bi_2(SO_4)_3$ and $Bi(NO_3)_3$. Direct reaction of Bi with O_2, S, and X_2 at elevated temperatures yields Bi_2O_3, Bi_2S_3, and BiX_3 respectively, but the increasing size of the metal atom results in a steady decrease in the strength of covalent linkages in the sequence P > As > Sb > Bi. This is most noticeable in the instability of BiH_3 and of many organobismuth compounds (p. 697).

Most of the trends are qualitatively understandable in terms of the general atomic properties in Table 13.2 though they are not readily deducible from them in any

---

[5] C. A. McAuliffe (ed.), *Transition Metal Complexes of Phosphorus, Arsenic, and Antimony Ligands*, Macmillan, London, 1973, 428 pp.

quantitative sense. Again, the $+5$ oxidation state in Bi is less stable than in Sb for the reasons discussed on p. 250; not only is the sum of the 4th and 5th ionization energies for Bi greater than for Sb (9.78 vs. 9.63 MJ mol^{-1}) but the promotion energies of one of the ns^2 electrons to a vacant nd orbital is also greater for Bi (and As) than for Sb. The discussions on redox properties (p. 673) and the role of d orbitals (p. 250) are also relevant. Finally, Bi shows an interesting resemblance to La in the crystal structures of the chloride oxide, MOCl, and in the isomorphism of the sulfates and double nitrates; this undoubtedly stems from the very similar ionic radii of the 2 cations: Bi^{3+} 103, La^{3+} 103.2 pm.

The most frequently met coordination numbers are 3, 4, 5, and 6, though instances of coordination number 2 occur, for example, in the cluster anions As$_7^{3-}$, Sb$_7^{3-}$, and As$_{11}^{3-}$ (p. 686) and in organo heterocycles such as arsa-benzene, -naphthalene, and -anthracene (p. 690). Coordination number 1 is represented by the recently characterized "isolated" tetrahedral anions SiAs$_4^{8-}$ and GeAs$_4^{8-}$ in the lustrous, dark, metallic Zintl phases Ba$_4$SiAs$_4$ and Ba$_4$GeAs$_4$.[5a] High coordination numbers are exemplified by encapsulated As and Sb atoms in rhodium carbonyl cluster anions: for example As is surrounded by a bicapped square antiprism of 10 Rh atoms in [Rh$_{10}$As(CO)$_{22}$]$^{3-}$,[5b] and Sb is surrounded by an icosahedron of 12 Rh in [Rh$_{12}$Sb(CO)$_{27}$]$^{3-}$.[5c] In each case the anion is the first example of a complex in which As or Sb acts as a 5-electron donor (cf. P as a 5-electron donor in [Rh$_9$P(CO)$_{21}$]$^{3-}$): all these clusters then have precisely the appropriate number of valence electrons for *closo* structures on the basis of Wade's rules (pp. 185, 198).

13.3 Compounds of Arsenic, Antimony, and Bismuth

13.3.1 *Intermetallic compounds and alloys*[6, 7]

Most metals form arsenides, antimonides, and bismuthides, and many of these command attention because of their interesting structures or valuable physical properties. Like the borides (p. 162), carbides (p. 318), silicides (p. 386), nitrides (p. 479), and phosphides (p. 562), the multitude of stoichiometries, the complexities of the structures, and the intermediate nature of the bonding make classification difficult. The compounds are usually prepared by direct reaction of the elements in the required proportions, and typical compositions are M$_9$As, M$_5$As, M$_4$As, M$_3$As, M$_3$As$_2$, M$_2$As, M$_5$As$_3$, M$_3$As$_2$, M$_4$As$_3$, M$_5$As$_4$, MAs, M$_3$As$_4$, M$_2$As$_3$, MAs$_2$, and M$_3$As$_7$. Antimony and bismuth are similar. Many of these intermetallic compounds exist over a range of composition, and nonstoichiometry is rife.

The (electropositive) alkali metals of Group IA form compounds M$_3$E (E = As, Sb, Bi) and the metals of Groups IIA and IIB likewise form M$_3$E$_2$. These can formally be written as

[5a] B. EISENMANN, H. JORDAN, and H. SCHÄFER, Ba$_4$SiAs$_4$ and Ba$_4$GeAs$_4$, Zintl phases with isolated SiAs$_4^{8-}$ and GeAs$_4^{8-}$ anions, *Angew. Chem.* Int. Ed. (Engl.) **20**, 197–8 (1981).

[5b] J. L. VIDAL, [Rh$_{10}$As(CO)$_{22}$]$^{3-}$. Example of encapsulation of arsenic by transition metal carbonyl clusters, *Inorg. Chem.* **20**, 243–9 (1981).

[5c] J. L. VIDAL and J. M. TROUP, [Rh$_{12}$Sb(CO)$_{27}$]$^{3-}$. An example of encapsulation of antimony by a transition metal carbonyl cluster, *J. Organometallic Chem.* **213**, 351–63 (1981).

[6] J. D. SMITH, Arsenic, antimony, and bismuth, Chap. 21 in *Comprehensive Inorganic Chemistry*, Vol. 2, pp. 547–683, Pergamon Press, Oxford, 1973.

[7] F. HULLIGER, Crystal chemistry of chalcogenides and pnictides of the transition elements, *Struct. Bond.* **4**, 83–229 (1968). A comprehensive review with 532 references.

$M^+_3E^{3-}$ and $M^{2+}_3E^{3-}_2$ but the compounds are even less ionic than Li_3N (p. 88) and have many metallic properties. Moreover, other stoichiometries are found (e.g. LiBi, KBi_2, $CaBi_3$) which are not readily accounted for by the ionic model, and conversely, compounds M_3E are formed by many metals that are not usually thought of as univalent, e.g. Ti, Zr, Hf; V, Nb, Ta; Mn. Despite this, there are clearly strong additional interactions between unlike atoms as indicated by the structures adopted and the high mp of many of the compounds, e.g. Na_3Bi melts at 840°, compared with Na 98° and Bi 271°C. Many of the M_3E compounds have the hexagonal Na_3As (anti-LaF_3) structure (Fig. 13.2) in which equal numbers of Na and As form hexagonal nets as in boron nitride and the remaining Na atoms are arranged in layers on either side of these nets. Each As has 5 neighbours at the corners of a trigonal bipyramid (3 at 294 and 2 at 299) and 6 other Na atoms at 330 pm form a trigonal prism (i.e. 11-coordinate). The Na atoms are of two sorts: type a in the NaAs net has 3 As at 294 pm and 6 Na at 330 pm whereas type b has 4 As at the corners of a distorted tetrahedron (1 at 299 and 3 at 330) together with 6 Na at 328 or 330 pm. The structure thus involves high coordination numbers typical of metals and all the Na–Na distances are less than in Na metal (371.6 pm). The compounds show either metallic conductivity or are semiconductors. An even more compact metal structure (cubic) is adopted by β-Li_3Bi, β-Li_3Sb, and by M_3E, where M = Rb, Cs, and E = Sb, Bi.

Some of the alkali metal-group V element systems give compounds of stoichiometry ME. Of these LiBi and NaBi have typical alloy structures and are superconductors below 2.47 K and 2.22 K respectively. Others, like LiAs, NaSb, and KSb, have parallel infinite spirals of As or Sb atoms, and it is tempting to formulate them as $M^+_n(E_n)^{n-}$ in which the $(E_n)^{n-}$ spirals are isoelectronic with those of covalently catenated Se and Te (p. 887); however, their metallic lustre and electrical conductivity indicate at least some metallic

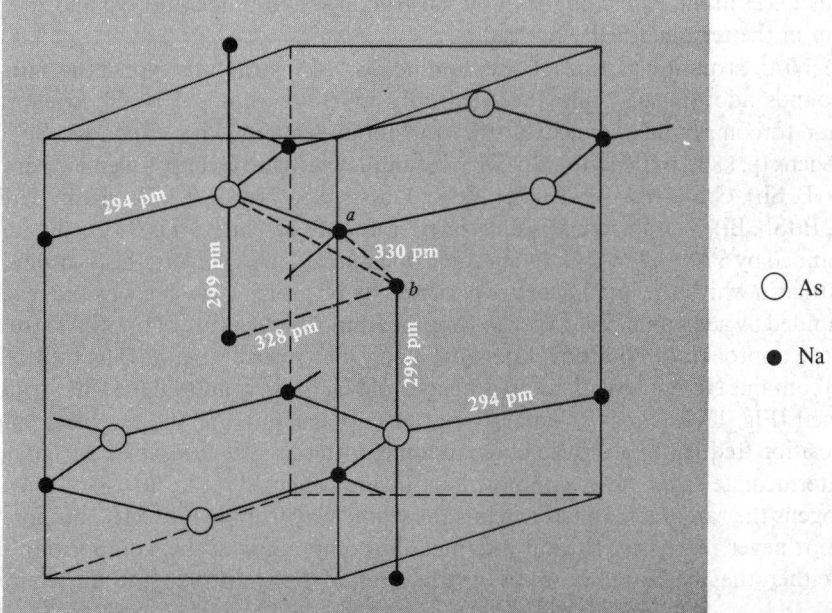

FIG. 13.2   The hexagonal structure of Na_3As (see text). This is adopted by M_3E where M = Li, Na, K and E = P, As, Sb, Bi, and by Rb_3As and Rb_3Sb.

bonding. Within the spiral chains As–As is 246 pm (cf. 252 pm in the element) and Sb–Sb is ~285 pm (cf. 291 pm in the element).

Compounds with Sc, Y, lanthanoids, and actinoids are of three types. Those with composition ME have the (6-coordinated) NaCl structure, whereas M_3E_4 (and sometimes M_4E_3) adopt the body-centred thorium phosphide structure (Th_3P_4) with 8-coordinated M, and ME_2 are like $ThAs_2$ in which each Th has 9 As neighbours. Most of these compounds are metallic and those of uranium are magnetically ordered. Full details of the structures and properties of the several hundred other transition metal Group VB element compounds fall outside the scope of this treatment, but three particularly important structure types should be mentioned because of their widespread occurrence and relation to other structure types, namely $CoAs_3$, NiAs, and structures related to those adopted by FeS_2 (marcasite, pyrites, loellingite, etc.).

$CoAs_3$ occurs in the mineral skutterudite; it is a diamagnetic semiconductor and has a cubic structure related to that of ReO_3 (p. 1219) but with a systematic distortion which results in the generation of well-defined *planar* rings of As_4. The same structure motif is found in MP_3 (M = Co, Ni, Rh, Pd), MAs_3 (M = Co, Rh, Ir), and MSb_3 (M = Co, Rh, Ir). The unit cell (Fig. 13.3) contains 8Co and 24As (i.e. $6As_4$), and it follows from the directions in which the various sets of atoms move, that 2 of the 8 original ReO_3 cells do not contain an As_4 group. Each As has a nearly regular tetrahedral arrangement of 2 Co and 2 As neighbours and each Co has a slightly distorted octahedral coordination group of 6 As. The planar As_4 groups are not quite square, the sides of the rectangle being 246 and 257 pm (cf. 244 pm in the tetrahedral As_4 molecule). The distortions from the ReO_3 structure (in which each As would have had 8 equidistant neighbours at about 330 pm) thus permit the closer approach of the As atoms in groups of 4 though this does not proceed so far as to form 6 equidistant As–As links as in the tetrahedral As_4 molecule. The P–P distances in the P_4 rectangles of the isostructural phosphides are 223 and 231 pm (cf. 225 pm in the tetrahedral P_4 molecule).

The NiAs structure is one of the commonest MX structure types, the number of compounds adopting it being exceeded only by those with the NaCl structure. It is peculiar to compounds formed by the transition elements with either As, Sb, Bi, the chalcogens (p. 883), or occasionally Sn.[8] Examples with the Group V elements are Ti(As, Sb), V(P, Sb), CrSb, Mn(As, Sb, Bi), FeSb, Co(As, Sb), Ni(As, Sb, Bi), RhBi, Ir(Sb, Bi), PdSb, Pt(Sb, Bi). The structure is illustrated in Fig. 13.4a: each Ni is 8-coordinate, being surrounded by 6 As and by 2 Ni (which are coplanar with 4 of the As); the As atoms form a hcp lattice in which the interstices are occupied by Ni atoms in such a way that each As is surrounded by a trigonal prism of 6 Ni. Another important feature of the NiAs structure is the close approach of Ni atoms in chains along the (vertical) c-axis. The unit cell (Fig. 13.4b) contains Ni_2As_2, and if the central layer of Ni atoms is omitted the CdI_2 structure is obtained (Fig. 13.4c). This structural relationship accounts for the extensive ranges of composition frequently observed in compounds with this structure, since partial filling of the intermediate layer gives compositions in the range $M_{1+x}X_2$ ($0 < x < 1$). With the chalcogens the range sometimes extends the whole way from ME to ME_2 but for As, Sb, and Bi it never reaches ME_2 and intermetallic compounds of this composition usually have either the marcasite or pyrites structures of FeS_2 (p. 804) or the compressed

[8] N. N. GREENWOOD, *Ionic Crystals, Lattice Defects, and Nonstoichiometry*, Butterworths, London, 1968, 194 pp.

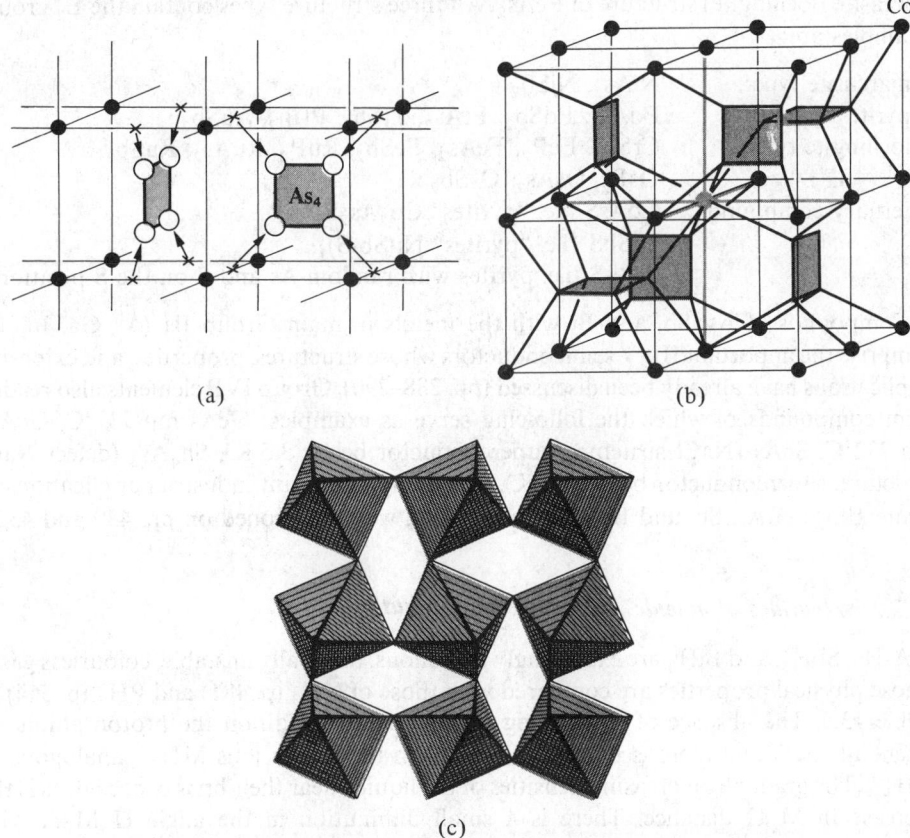

FIG. 13.3   The cubic structure of skutterudite (CoAs₃). (a) Relation to the ReO₃ structure; (b) unit cell (only sufficient Co–As bonds are drawn to show that there is a square group of As atoms in only 6 of the 8 octants of the cubic unit cell, the complete 6-coordination group of Co is shown only for the atom at the body-centre of the cell); and (c) section of the unit cell showing {CoAs₆} octahedra corner-linked to form As₄ squares.

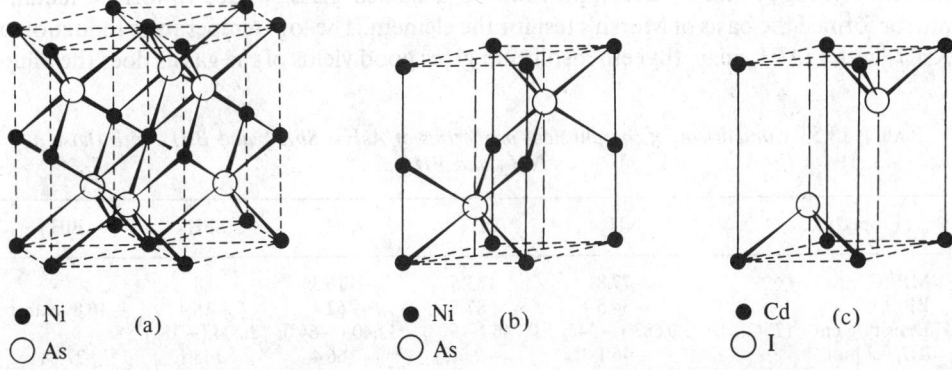

FIG. 13.4   Structure of nickel arsenide showing (a) 3 unit cells, (b) a single unit cell Ni₂As₂ and its relation to (c) the unit cell of the layer lattice compound CdI₂ (see text).

marcasite (loellingite) structure of $FeAs_2$. All three structure types contain the E_2 group. Examples are:

marcasite type:	$NiAs_2$, $NiSb_2$
pyrites type:	$PdAs_2$, $PdSb_2$, $PtAs_2$, $PtSb_2$, $PtBi_2$, $AuSb_2$
loellingite type:	$CrSb_2$, FeP_2, $FeAs_2$, $FeSb_2$, RuP_2, $RuAs_2$, $RuSb_2$, OsP_2, $OsAs_2$, $OsSb_2$
ternary compounds:	$CoAsS$ (i.e. "pyrites" $Co_2As_2S_2$), $NiSbS$ {i.e. "pyrites" $Ni(Sb\text{-}S)$}, $NiAsS$ (i.e. pyrites with random As and S on the S positions)

Compounds of As, Sb, and Bi with the metals in main Group III (Al, Ga, In, Tl) comprise the important III–V semiconductors whose structures, properties, and extensive applications have already been discussed (pp. 288–290). Group IVB elements also readily form compounds of which the following serve as examples: GeAs mp 737°C, $GeAs_2$ mp 732°C, SnAs (NaCl structure, superconductor below 3.5 K), Sn_4As_3 (defect NaCl structure, superconductor below 1.2 K). The many important industrial applications of dilute alloys of As, Sb, and Bi with tin and lead were mentioned on pp. 430 and 432.

13.3.2 *Hydrides of arsenic, antimony, and bismuth*

AsH_3, SbH_3, and BiH_3 are exceedingly poisonous, thermally unstable, colourless gases whose physical properties are compared with those of NH_3 (p. 481) and PH_3 (p. 564) in Table 13.5. The absence of H bonding is apparent; in addition the proton affinity is essentially zero and there is no tendency to form the onium ions MH_4^+ analogous to NH_4^+. The gradually increasing densities of the liquids near their bp is expected, as is the increase in M–H distance. There is a small diminution in the angle H–M–H with increasing molecular weight though the difference for AsH_3 and SbH_3 is similar to the experimental uncertainty. The rapid diminution in thermal stability is reflected in the standard heats of formation ΔH_f°; AsH_3 decomposes to the elements on being warmed to 250–300°, SbH_3 decomposes steadily at room temperature, and BiH_3 cannot be kept above $-45°$.

Arsine, AsH_3, is formed when many As-containing compounds are reduced with nascent hydrogen and its decomposition on a heated glass surface to form a metallic mirror formed the basis of Marsh's test for the element. The low-temperature reduction of $AsCl_3$ with $LiAlH_4$ in diethyl ether solution gives good yields of the gas as does the dilute

TABLE 13.5 *Comparison of the physical properties of AsH_3, SbH_3, and BiH_3 with those of NH_3 and PH_3*

Property	NH_3	PH_3	AsH_3	SbH_3	BiH_3
MP/°C	-77.8	-133.5	-116.3	-88	—
BP/°C	-34.5	-87.5	-62.4	-18.4	$+16.8$ (extrap)
Density/g cm^{-3}(T°C)	0.683 ($-34°$)	0.746 ($-90°$)	1.640 ($-64°$)	2.204 ($-18°$)	—
ΔH_f°/kJ mol^{-1}	-46.1	$-9.6(?)$	66.4	145.1	277.8
Distance (M–H)/pm	101.7	141.9	151.9	170.7	—
Angle H–M–H	$107.8°$	$93.6°$	$91.8°$	$91.3°$	—

acid hydrolysis of many arsenides of electropositive elements (Na, Mg, Zn, etc.). Similar reactions yield stibine, e.g.:

$$Zn_3Sb_2 + 6H_3O^+ \xrightarrow[14\% \text{ yield}]{\text{aq HCl}} 2SbH_3 + 3Zn^{2+} + 6H_2O$$

$$SbO_3^{3-} + 3Zn + 9H_3O^+ \longrightarrow SbH_3 + 3Zn^{2+} + 12H_2O$$

$$SbCl_3(\text{in aq NaCl}) + NaBH_4 \longrightarrow SbH_3 \text{ (high yield)}$$

Both AsH_3 and SbH_3 oxidize readily to the trioxide and water, and similar reactions occur with S and Se. AsH_3 and SbH_3 form arsenides and antimonides when heated with metals and this reaction also finds application in semiconductor technology; e.g. highly purified SbH_3 is used as a gaseous n-type dopant for Si (p. 383).

Bismuthine, BiH_3, is extremely unstable and was first detected in minute traces by F. Paneth using a radiochemical technique involving $^{212}Bi_2Mg_3$. These experiments, carried out in 1918, were one of the earliest applications of radiochemical tracer experiments in chemistry. Later work using BH_4^- to reduce $BiCl_3$ was unsuccessful in producing macroscopic amounts of the gas and the best preparation (1961) is the disproportionation of $MeBiH_2$ at $-45°$ for several hours; Me_2BiH can also be used:

$$Me_{3-n}BiH_n \xrightarrow{45°} \tfrac{n}{3} BiH_3 + \tfrac{3-n}{3} BiMe_3$$

Lower hydrides such as As_2H_4 have occasionally been reported as fugitive species but little is known of their properties (cf. N_2H_4, p. 489; P_2H_4, p. 565).

13.3.3 *Halides and related complexes*

The numerous halides of As, Sb, and Bi show highly significant gradations in physical properties, structure, bonding, and chemical reactivity. Distinctions between ionic, coordinate, and covalent (molecular) structures in the halides and their complexes frequently depend on purely arbitrary demarcations and are often more a hindrance than a help in discerning the underlying structural and bonding principles. Alternations in the stability of the $+5$ oxidation state are also illuminating. It will be convenient to divide the discussion into five subsections dealing in turn with the trihalides MX_3, the pentahalides MX_5, other halides, halide complexes of M^{III} and M^V, and oxohalides.

Trihalides, MX_3

All 12 compounds are well known and are available commercially; their physical properties are summarized in Table 13.6 Comparisons with the corresponding data for NX_3 (p. 503) and PX_3 (p. 568) are also instructive. Trends in mp, bp, and density are far from regular and reflect the differing structures and bond types. Thus AsF_3, $AsCl_3$, $AsBr_3$, $SbCl_3$, and $SbBr_3$ are clearly volatile molecular species, whereas AsI_3, SbF_3, and BiX_3 have more extended interactions in the solid state. Trends in the heats of formation from the elements are more regular being *ca.* -925 kJ mol^{-1} for MF_3, *ca.* -350 kJ mol^{-1} for MCl_3, *ca.* -245 kJ mol^{-1} for MBr_3, and *ca.* -100 kJ mol for MI_3. Within these average values, however, AsF_3 is noticeably more exothermic than SbF_3 and BiF_3, whereas the

TABLE 13.6 *Some physical properties of the trihalides of arsenic, antimony, and bismuth*

Compound	Colour and state at 25°C	MP/°C	BP/°C	$d/g\ cm^{-3}$ $(T°C)$	$\Delta H_f^\circ/$ kJ mol^{-1}
AsF$_3$	Colourless liquid	−6.0	62.8	2.666 (0°)	−956.5
AsCl$_3$	Colourless liquid	−16.2	130.2	2.205 (0°)	−305.0
AsBr$_3$	Pale-yellow crystals	+31.2	221°	3.66 (15°)	−197.0
AsI$_3$	Red crystals	140.4	∼400	4.39 (15°)	−58.2
SbF$_3$	Colourless crystals	290	∼345	4.38 (25°)	−915.5
SbCl$_3$	White, deliquescent crystals	73.4	223	3.14 (20°)	−382.2
SbBr$_3$	White, deliquescent crystals	96.0	288	4.15 (25°)	−259.4
SbI$_3$	Red crystals	170.5	401	4.92 (22°)	−100.4
BiF$_3$	Grey-white powder	649[a]	900	∼5.3	−900
BiCl$_3$	White, deliquescent crystals	233.5	441	4.75	−379
BiBr$_3$	Golden, deliquescent crystals	219	462	5.72	−276
BrI$_3$	Green-black crystals	408.6	∼542 (extrap)	5.64	−150

[a] BiF$_3$ is sometimes said to be "infusible" or to have mp at varying temperatures in the range 725–770°, but such materials are probably contaminated with the oxofluoride BiOF (p. 667).

reverse is true for the chlorides; there is also a regular trend towards increasing stability in the sequence As < Sb < Bi for the bromides and for the iodides of these elements.

The trifluorides are all readily prepared by the action of HF on the oxide M_2O_3 (direct fluorination of M or M_2O_3 with F_2 gives MF$_5$, p. 655). Because AsF$_3$ hydrolyses readily, the reaction is best done under anhydrous conditions using H_2SO_4/CaF_2 or HSO_3F/CaF_2, but aqueous HF can be used for the others. The trichlorides, tribromides, and triiodides of As and Sb can all be prepared by direct reaction of X_2 with M or M_2O_3, whereas the less readily hydrolysed BiX$_3$ can be obtained by treating Bi_2O_3 with the aqueous HX. Many variants of these reactions are possible: e.g., AsCl$_3$ can be made by chlorination of As_2O_3 with Cl_2, S_2Cl_2, conc HCl, or H_2SO_4/MCl.

The trihalides of As are all pyramidal molecular species in the gas phase with angle X–As–X in the range 96–100°. This structure persists in the solid state, and with AsI$_3$ the packing is such that each As is surrounded by an octahedron of six I with 3 short and 3 long As–I distances (256 and 350 pm; ratio 1.37, mean 303 pm). The I atoms form a regular hcp lattice. A similar layer structure is adopted by SbI$_3$ and BiI$_3$ but with the metal atoms progressively nearer to the centre of the I$_6$ octahedra:

3 Sb–I at 287 pm and 3 at 332 pm; ratio 1.16, mean 310 pm
all 6 Bi–I at 310 pm; "ratio" 1.00

This is sometimes described as a trend from covalent, molecular AsI$_3$ through intermediate SbI$_3$ to ionic BiI$_3$, but this exaggerates the difference in bond-type. Arsenic, Sb, and Bi have very similar electronegativities (p. 642) and it seems likely that the structural trend reflects more the way in which the octahedral interstices in the hcp iodine lattice are filled by atoms of gradually increasing size. The size of these interstices is about constant (see mean M–X distance) but only Bi is sufficiently large to fill them symmetrically.

Discrete molecules are apparent in the crystal structure of the higher trihalides of Sb, and, again, these pack to give 3 longer and 3 shorter interatomic distances (Table 13.7).

TABLE 13.7 *Structural data for antimony trihalides*

	SbF$_3$	SbCl$_3$	α-SbBr$_3$	β-SbBr$_3$	SbI$_3$
Sb–X in gas molecule/pm	?	233	251	251	272
Three short Sb–X in crystal/pm	192	236	250	249	287
Three long Sb–X in crystal/pm	261	$\geqslant$350	$\geqslant$375	$\geqslant$360	332
Ratio (long/short)	1.36	$\geqslant$1.48	$\geqslant$1.50	$\geqslant$1.44	1.16
Angle X–Sb–X in crystal	87°	95°	96°	95°	96°

The structure of BiF$_3$ is quite different: β-BiF$_3$ has the "ionic" YF$_3$ structure with tricapped trigonal prismatic coordination of Bi by 9 F. BiCl$_3$ has an essentially molecular structure (like SbX$_3$) but there is a significant distortion within the molecule itself, and the packing gives 5 (not 3) further Cl at 322–345 pm to complete a *bi*capped trigonal prism. As

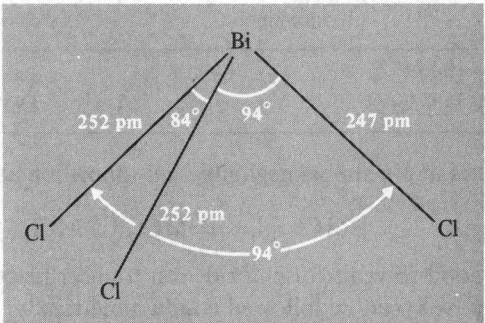

a consequence of this structure BiCl$_3$ has smaller unit cell dimensions than SbCl$_3$ despite the longer Bi–Cl bond (250 pm, as against 236 pm for Sb–Cl). The eightfold coordination has been rationalized by postulating that the ninth position is occupied by the stereochemically active lone-pair of electrons on Bi^{III}. On this basis, the 3 long and 3 short M–X distances in octahedrally coordinated structures can also be understood, the lone-pair being directed towards the centre of the more distant triangle of 3X. However, it is hard to quantify this suggestion, particularly as the X–M–X angles are fairly constant at $97 \pm 2°$ (rather than 109.5° for sp^3 hybrids), implying little variation in hybridization and a lone-pair with substantial s^2 character. The effect is less apparent in SbI$_3$ and absent BiI$_3$ (see above) and this parallels the diminishing steric influence of the lone-pair in some of the complexes of the heavier halides with Sn^{II} (p. 443) and Te^{IV} (p. 892).

Many of the trihalides of As, Sb, and Bi hydrolyse readily but can be handled without great difficulty under anhydrous conditions. AsF$_3$ and SbF$_3$ are important reagents for converting non-metal chlorides to fluorides. SbF$_3$ in particular is valuable for preparing organofluorine compounds (the Swarts reaction):

$$CCl_3CCl_3 + SbF_3 \longrightarrow CCl_2FCCl_2F$$
$$SiCl_4 + SbF_3 \longrightarrow SiCl_3F, \ SiCl_2F_2, \ SiClF_3$$
$$CF_3PCl_2 + SbF_3 \longrightarrow CF_3PF_2$$
$$R_3PS + SbF_3 \longrightarrow R_3PF_2$$

Sometimes the reagents simultaneously act as mild oxidants:

$$3PhPCl_2 + 4SbF_3 \longrightarrow 3PhPF_4 + 2Sb + 2SbCl_3$$
$$3Me_2P(S)P(S)Me_2 + 6SbF_3 \longrightarrow 6Me_2PF_3 + 2Sb + 2Sb_2S_3$$

AsF_3, though a weaker fluorinating agent than SbF_3, is preferred for the preparation of high-boiling fluorides since $AsCl_3$ (bp 130°) can be distilled off. SbF_3 is preferred for low-boiling fluorides, which can be readily fractionated from $SbCl_3$ (bp 223°). Selective fluorinations are also possible, e.g.:

$$[PCl_4]^+[PCl_6]^- + 2AsF_3 \longrightarrow [PCl_4]^+[PF_6]^- + 2AsCl_3$$

$AsCl_3$ and $SbCl_3$ have been used as non-aqueous solvent systems for a variety of reactions.[9, 10] They are readily available, have convenient liquid ranges (p. 652), are fairly easy to handle, have low viscosities η, moderately high dielectric constants ε, and good solvent properties (Table 13.8).

TABLE 13.8   *Some properties of liquid $AsCl_3$ and $SbCl_3$*

	η/centipoise	ε	κ/ohm^{-1}cm^{-1}
$AsCl_3$ at 20°C	1.23	12.8	1.4×10^{-7}
$SbCl_3$ at 75°C	2.58	33.2	1.4×10^{-6}

The low conductivities imply almost negligible self-ionization according to the formal scheme:

$$2MCl_3 \rightleftharpoons MCl_2^+ + MCl_4^-$$

Despite this, they are good solvents for chloride-ion transfer reactions, and solvo-acid–solvo-base reactions (p. 968) can be followed conductimetrically, voltametrically, or by use of coloured indicators. As expected from their constitution, the trihalides of As and Sb are only feeble electron-pair donors (p. 223) but they have marked acceptor properties, particularly towards halide ions (p. 659) and amines.

AsX_3 and SbX_3 react with alcohols (especially in the presence of bases) and with sodium alkoxide to give arsenite and antimonite esters, $M(OR)_3$ (cf. phosphorus, (p. 592):

$$AsCl_3 + 3PhOH \longrightarrow As(OPh)_3 + 3HCl$$

$$SbCl_3 + 3Bu^tOH + 3NH_3 \longrightarrow Sb(OBu^t)_3 + 3NH_4Cl$$

$$SbCl_3 + 3NaOSiEt_3 \longrightarrow Sb(OSiEt_3)_3 + 3NaCl$$

Halide esters $(RO)_2MX$ and $(RO)MX_2$ can be made similarly:

$$AsCl_3 + 2NaOEt \longrightarrow (EtO)_2AsCl + 2NaCl$$

$$AsCl_3 + EtOH \xrightarrow{CO_2} (EtO)AsCl_2 + HCl$$

$$As(OPr)_3 + MeCOCl \longrightarrow (PrO)_2AsCl + MeCO_2Pr$$

[9] D. S. PAYNE, Halides and oxyhalides of Group V elements as solvents, Chap. 8 in T. C. WADDINGTON (ed.), *Nonaqueous Solvent Systems*, pp. 301–25, Academic Press, London, 1965.

[10] E. C. BAUGHAN, Inorganic acid chlorides of high dielectric constant (with special reference to antimony trichloride), Chap. 5 in J. J. LAGOWSKI (ed.), *The Chemistry of Nonaqueous Solvents*, Vol. 4, pp. 129–65, Academic Press, London, 1976.

Amino derivatives are obtained by standard reactions with secondary amines, lithium amides, or by transaminations:

$$AsCl_3 + 6Me_2NH \longrightarrow As(NMe_2)_3 + 3[Me_2NH_2]Cl$$
$$SbCl_3 + 3LiNMe_2 \longrightarrow Sb(NMe_2)_3 + 3LiCl$$
$$As(NMe_2)_3 + 3Bu_2NH \longrightarrow As(NBu_2)_3 + 3Me_2NH$$

As with phosphorus (p. 620) there is an extensive derivative chemistry of these and related compounds.[6]

Pentahalides, MX_5

Until recently only the pentafluorides and $SbCl_5$ were known, but the exceedingly elusive $AsCl_5$ was finally prepared in 1976 by ultraviolet irradiation of $AsCl_3$ in liquid Cl_2 at $-105°C$.[11] Some properties of the 5 pentahalides are given in Table 13.9.

TABLE 13.9 *Some properties of the known pentahalides*

Property	AsF_5	SbF_5	BiF_5	$AsCl_5$	$SbCl_5$
MP/°C	-79.8	8.3	154.4	~ -50 (d)	4
BP/°C	-52.8	141	230	—	140 (d)
Density ($T°C$)/g cm^{-3}	2.33 ($-53°$)	3.11 (25°)	5.40 (25°)	—	2.35 (21°)

The pentafluorides are prepared by direct reaction of F_2 with the elements (As, Bi) or their oxides (As_2O_3, Sb_2O_3). $AsCl_5$, as noted above, has only a fugitive existence and decomposes to $AsCl_3$ and Cl_2 at about $-50°$. $SbCl_5$ is more stable and is made by reaction of Cl_2 on $SbCl_3$. No pentabromides or pentaiodides have been characterized, presumably because M^V is too highly oxidizing for these heavier halogens (cf. TlI_3, p. 269). The relative instability of $AsCl_5$ when compared with PCl_5 and $SbCl_5$ is a further example of the instability of the highest valency state of p-block elements following the completion of the first (3d) transition series (p. 645). This can be understood in terms of incomplete shielding of the nucleus which leads to a "d-block contraction" and a consequent lowering of the energy of the 4s orbital in As and $AsCl_3$, thereby making it more difficult to promote one of the $4s^2$ electrons for the formation of $AsCl_5$. There is no evidence that the As–Cl bond strength itself, in $AsCl_5$, is unduly weak. The non-existence of $BiCl_5$ likewise suggests that it is probably less stable than $SbCl_5$, due the analogous "f-block contraction" following the lanthanide elements (p. 1431).

Evidence from vibration spectroscopy suggests that gaseous AsF_5, solid $AsCl_5$, and liquid $SbCl_5$ are trigonal bipyramidal molecules like PF_5 (D_{3h}). By contrast SbF_5 is an extremely viscous, syrupy liquid with a viscosity approaching 850 centipoise at 20°: the liquid features polymeric chains of *cis*-bridged $\{SbF_6\}$ octahedra in which the 3 different types of F atom (a, b, c) can be distinguished by low-temperature ^{19}F nmr spectroscopy.[12] As shown in Fig. 13.5, F_a are the bridging atoms and are *cis* to each other in any one octahedron; F_b are also *cis* to each other and are, in addition, *cis* to 1 F_a and *trans* to the

[11] K. SEPPELT, Arsenic pentachloride, $AsCl_5$, *Angew. Chem.*, Int. Edn. (Engl.) **15**, 377–8 (1976).
[12] T. K. DAVIES and K. C. MOSS, Nuclear magnetic resonance studies on fluorine-containing compounds. Part 1. SbF_5 and its reaction with SbF_3, AsF_3, and NbF_5, *J. Chem. Soc.* (A) 1970, 1054–8.

other, whereas F_c are *trans* to each other and *cis* to both F_a. In the crystalline state the *cis* bridging persists but the structure has tetrameric molecular units (Fig. 13.6) rather than high polymers.[13] There are two different Sb–F–Sb bridging angles, 141° and 170°, and the terminal Sb–F_t distances (mean 182 ± 5 pm) are noticeably less than the bridging Sb–F_μ distances (mean 203 ± 5 pm). Yet another structure motif is adopted in BiF_5; this crystallizes in long white needles and has the α-UF_5 structure in which infinite linear chains of *trans*-bridged $\{BiF_6\}$ octahedra are stacked parallel to each other. The Bi–F–Bi bridging angle between adjacent octahedra in the chain is 180°.

The pentafluorides are extemely powerful fluorinating and oxidizing agents and they also have a strong tendency to form complexes with electron-pair donors. This latter property has already been presaged by the propensity of SbF_5 to polymerize and is discussed more fully on p. 662. Some typical reactions of SbF_5 and $SbCl_5$ are as follows:

$$ClCH_2PCl_2 + SbF_5 \longrightarrow ClCH_2PF_4$$

$$Me_3As + SbCl_5 \longrightarrow Me_3AsCl_2 + SbCl_3$$

$$R_3P + SbCl_5 \longrightarrow [R_3PCl]^+[SbCl_6]^- \quad (R = Ph, Et_2N, Cl)$$

$$SbCl_5 \xrightarrow{5\,NaOR} Sb(OR)_5 \xrightarrow{NaX} Na[Sb(OR)_5X] \quad (X = OR, Cl)$$

Perhaps the most reactive compound of the groups is BiF_5. It reacts explosively with H_2O to form O_3, OF_2, and a voluminous brown precipitate which is probably a hydrated bismuth(V) oxide fluoride. At room temperature BiF_5 reacts vigorously with iodine or sulfur; above 50° it converts paraffin oil to fluorocarbons; at 150° it fluorinates UF_4 to UF_6; and at 180° it converts Br_2 to BrF_3 and BrF_5, and Cl_2 to ClF.

Mixed halides and lower halides

Unlike phosphorus, which forms a large number of readily isolable mixed halides of both P^{III} and P^V, there is little tendency to form such compounds with As, Sb, and Bi, and few mixed halides have been characterized. AsF_3 and $AsCl_3$ are immiscible below 19°C, but at room temperature ^{19}F nmr indicates some halogen exchange; however equilibrium constants for the formation of AsF_2Cl and $AsFCl_2$ are rather small. Likewise, Raman spectra show the presence of $AsCl_2Br$ and $AsClBr_2$ in mixtures of the parent trihalides, though rapid equilibration prevents isolation of the mixed halides. It is said that $SbBrI_2$ (mp 88°) can be obtained by eliminating EtBr from $EtSbI_2Br_2$.

Mixed pentahalides are more readily isolated and are of at least two types: ionic and tetrameric. Thus, chlorination of a mixture of $AsF_3/AsCl_3$ with Cl_2, or fluorination of $AsCl_3$ with ClF_3 (p. 969) gives $[AsCl_4]^+[AsF_6]^-$ [mp 130°(d)]. Similarly, $AsCl_3 + SbCl_5 + Cl_2 \rightarrow [AsCl_4]^+[SbCl_6]^-$. Antimony chloride fluorides have been known since the turn of the century but the complexity of the system, the tendency to form mixtures of compounds, and their great reactivity have conspired against structural characterization until very recently.[14] It is now clear that fluorination of $SbCl_5$ depends crucially on the nature of the fluorinating agent. Thus, with AsF_3 it gives $SbCl_4F$ (mp 83°)

[13] A. J. Edwards and P. Taylor, Crystal structure of antimony pentafluoride, *JCS Chem. Comm.* 1971, 1376–7.

[14] J. G. Ballard, T. Birchall, and D. R. Slim, X-ray crystal structure of tetra-antimony(V) tridecachloride heptafluoride, $(Sb_2Cl_{6.5}F_{3.5})_2$, *JCS Dalton* 1979, 62–65, and references therein.

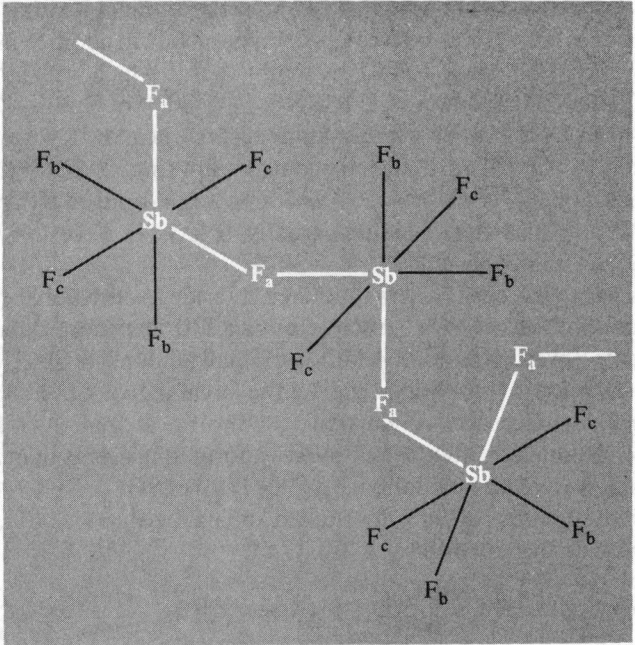

FIG. 13.5   Schematic representation of part of the *cis*-bridged polymer in liquid SbF$_5$ showing the three sorts of F atom.[12]

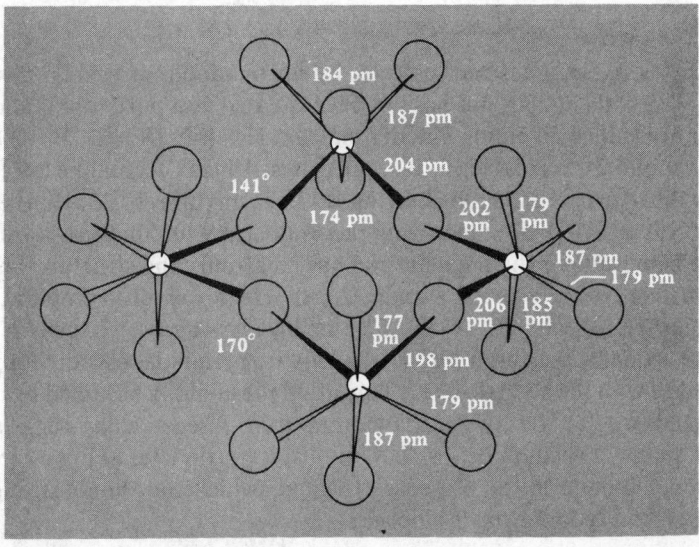

FIG. 13.6   Structure of the tetrameric molecular unit in (SbF$_5$)$_4$ showing the *cis*-bridging of 4 {SbF$_6$} octahedra.[13]

which is a *cis*-F-bridged tetramer as in Fig. 13.6 with the terminal F atoms replaced by Cl. Fluorination of $SbCl_5$ with HF also gives this compound but, in addition, $SbCl_3F_2$ mp 68° (*cis*-F-bridged tetramer) and $SbCl_2F_3$ mp 62°, which turns out to be $[SbCl_4]^+[Sb_2Cl_2F_9]^-$. The anion is F-bridged, i.e. $[ClF_4Sb-F-SbF_4Cl]^-$ with angle Sb–F–Sb 163°. Even more extensive fluorination occurs when $SbCl_5$ is reacted with SbF_5 and the product is $[SbCl_4]^+[Sb_2F_{11}]^-$. By contrast, fluorination of $(SbCl_4F)$ with SbF_5 in liquid SO_2 yields $Sb_4Cl_{13}F_7$ (mp $\sim$50°) which is a *cis*-F-bridged tetramer of $SbCl_3F_2$ with two of the Sb atoms having a Cl atom partially replaced by F, i.e. $(Sb_2Cl_{6.5}F_{3.5})_2$ and bridge angles Sb–F–Sb of 166–168°.

The attention which has been paid to the mixed chloride fluorides of Sb^V is due not only to the intellectual problem of their structure but also to their importance as industrial fluorinating agents (Swarts reaction). Addition of small amounts of $SbCl_5$ to SbF_5 results in a dramatic decrease in viscosity (due to the breaking of Sb–F–Sb links) and a substantial increase in electrical conductivity (due to the formation of fluoro-complex ions). Such mixed halides are often more effective fluorinating agents than SbF_3, provided that yields are not lowered by oxidation, e.g. $SOCl_2$ gives SOF_2; $POCl_3$ gives $POFCl_2$; and hexachlorobutadiene is partially fluorinated and oxidized to give $CF_3CCl=CClCF_3$ which can then be further oxidized to CF_3CO_2H:

$$CCl_2=CClCCl=CCl_2 \xrightarrow{\text{``SbF}_3Cl_2\text{''}} CF_3CCl=CClCF_3 \xrightarrow[\text{KOH}]{\text{KMnO}_4/} 2CF_3CO_2H$$

The use of SbF_5 in the preparation of "superacids" such as $(HSO_3F + SbF_5 + SO_3)$ is described in the following subsection.

The only well-established lower halide of As is As_2I_4 which is formed as red crystals (mp 137°) when stoichiometric amounts of the 2 elements are heated to 260° in a sealed tube in the presence of octahydrophenanthrene. The compound hydrolyses and oxidizes readily and disproportionates in warm CS_2 solution but is stable up to 150° in an inert atmosphere. Disproportionation is quantitative at 400°:

$$3As_2I_4 \longrightarrow 4AsI_3 + 2As$$

Sb_2I_4 is much less stable: it has been detected by emf or vapour pressure measurements on solutions of Sb in SbI_3 at 230° but has not been isolated as a pure compound. The lower halides of Bi are rather different. The diatomic species BiX (X = Cl, Br, I) occur in the equilibrium vapour above heated $Bi-BiX_3$ mixtures. A black crystalline lower chloride of composition $BiCl_{1.167}$ is obtained by heating $Bi-BCl_3$ mixtures to 325° and cooling them during 1–2 weeks to 270° before removing excess $BiCl_3$ by sublimation or extraction into benzene. The compound is diamagnetic and has an astonishing structure which involves cationic clusters of bismuth and 2 different chloro-complex anions:[15] $[(Bi_9^{5+})_2(BiCl_5^{2-})_4(Bi_2Cl_8^{2-})]$, i.e. $Bi_{24}Cl_{28}$ or Bi_6Cl_7. The Bi_9^{5+} cluster is a tricapped trigonal prism (p. 688); the anion $BiCl_5^{2-}$ has square pyramidal coordination of the 5 Cl atoms around Bi with the sixth octahedral position presumably occupied by the lone-pair of electrons, and $Bi_2Cl_8^{2-}$ has two such pyramids *trans*-fused at a basal edge (p. 660). The compound is stable in vacuum below 200° but disproportionates at higher temperatures. It also disproportionates in the presence of ligands which coordinate strongly to $BiCl_3$ and hydrolyses readily to the oxide chloride.

---

[15] A. HERSHAFT and J. D. CORBETT, the crystal structure of bismuth subchloride. Identification of the ion Bi_9^{5+}, *Inorg. Chem.* **2**, 979–85 (1963).

The first unambiguous identification of Bi^+ in the solid state came in 1971 when the structure of the complex halide $Bi_{10}Hf_3Cl_{18}$ was shown by X-ray diffraction analysis[16] to be $(Bi^+)(Bi_9^{5+})(HfCl_6^{2-})_3$. The compound was made by the oxidation of Bi with $HfCl_4/BiCl_3$.

Halide complexes of M^{III} *and* M^V

The trihalides of As, Sb, and Bi are strong halide-ion acceptors and numerous complexes have been isolated with a wide variety of compositions. They are usually prepared by direct reaction of the trihalide with the appropriate halide-ion donor. However, stoichiometry is not always a reliable guide to structure because of the possibility of oligomerization which depends both on the nature of M and X, and often also on the nature of the counter cation.[6, 17] Thus the tetra-alkylammonium salts of MCl_4^-, MBr_4^-, and MI_4^- may contain the monomeric C_{2v} ion as shown in Fig. 13.7a (cf. isoelectronic SeF_4, p. 905), whereas in $NaSbF_4$ there is a tendency to dimerize by formation of subsidiary $F \cdots Sb$ interactions (Fig. 13.7b) cf. $Bi_2Cl_8^{2-}$ in the preceding subsection. With $KSbF_4$ association proceeds even further to give tetrameric cyclic anions (Fig. 13.7c). In both $NaSbF_4$ and $KSbF_4$ the Sb atoms are 5-coordinate but coordination rises to 6 in the polymeric chain anions of the pyridinium and 2-methylpyridinium salts $pyHSbCl_4$, $(2\text{-}MeC_5H_4NH)BiBr_4$, and $(2\text{-}MeC_5H_4NH)BiI_4$. The structure of $(SbCl_4)_n^{n-}$ is shown schematically in Fig. 13.7d and the three differing Sb–Cl distances reflect, in part, the influence of the lone-pair of electrons on Sb^{III}. It will be noted that the shortest bonds are *cis* to each other, whereas the intermediate bonds are *trans* to each other; the longest bonds are *cis* to each other and *trans* to the short bonds. Corresponding distances in the Bi^{III} analogues are:

$(BiBr_4)_n^{n-}$: short (2 at 264 pm); intermediate (283, 297 pm); long (308, 327 pm)
$(BiI_4)_n^{n-}$: short (2 at 289 pm); intermediate (2 at 310 pm); long (331, 345 pm).

Complexes of stoichiometry MX_5^{2-} can feature either discrete 5-coordinate anions as in K_2SbF_5 and $(NH_4)_2SbCl_5$ (Fig. 13.8a), or 6-coordinate polymeric anions as in the piperidinium salt $(C_5H_{10}NH_2)_2BiBr_5$ (Fig. 13.8b). In the discrete anion $SbCl_5^{2-}$ the $Sb\text{-}Cl_{apex}$ distance (236 pm) is shorter than the $Sb\text{-}Cl_{base}$ distances (2 at 258 and 2 at 269 pm) and the Sb atom is slightly below the basal plane (by 22 pm). The same structure is observed in K_2SbCl_5.

In addition to the various complex fluoroantimonate(III) salts M^ISbF_4 and $M_2^ISbF_5$ mentioned above, the alkali metals form complexes of stoichiometry $M^ISb_2F_7$, $M^ISb_3F_{10}$, and $M^ISb_4F_{13}$, i.e. $[SbF_4^-(SbF_3)_n]$ ($n = 1, 2, 3$) but the mononuclear complexes $M_3^ISbF_6$ have not been found. The structure of $M^ISb_2F_7$ depends on the strength of the $Sb\text{-}F \cdots Sb$ bridge between the 2 units and this, in turn is influenced by the cation. Thus, in KSb_2F_7 there are distorted trigonal-bipyramidal SbF_4^- ions (Fig. 13.9a) and discrete pyramidal SbF_3 molecules (Sb–F 194 pm) with 2 (rather than 3) contacts between these and neighbouring SbF_4^- units of 241 and 257 pm (cf. SbF_3 itself, p. 653). By contrast $CsSb_2F_7$

[16] R. M. FRIEDMAN and J. D. CORBETT, Bismuth(I) in the solid state. The crystal structure of $(Bi^+)(Bi_9^{5+})(HfCl_6^{2-})_3$, *JCS Chem. Comm.* 1971, 422–3.
[17] A. F. WELLS, *Structural Inorganic Chemistry*, 4th edn., pp. 704–10 and 715–17, Oxford University Press, Oxford, 1975.

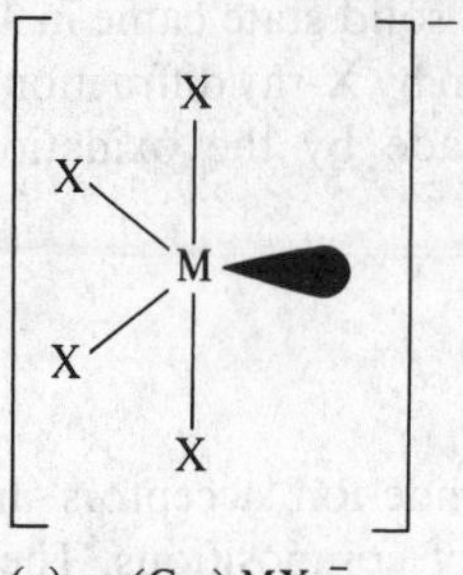

(a) $(C_{2v})\ MX_4^-$

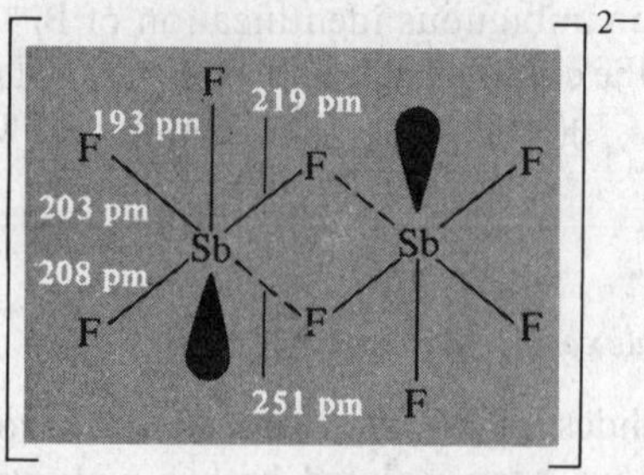

(b) $Sb_2F_8^{2-}$ in $NaSbF_4$

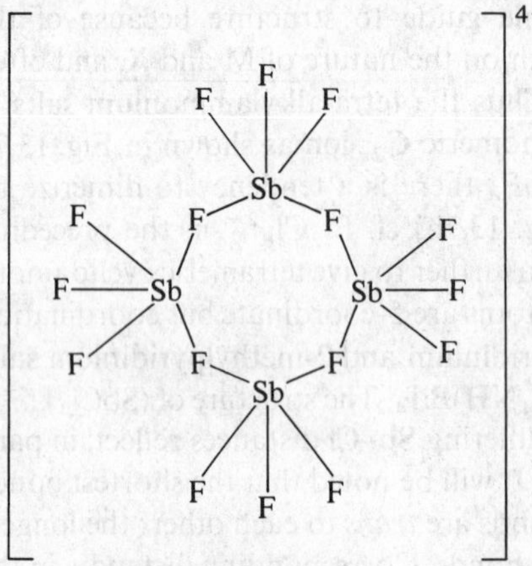

(c) $Sb_4F_{16}^{4-}$ in $KSbF_4$
(Sb–F 200–230 pm)

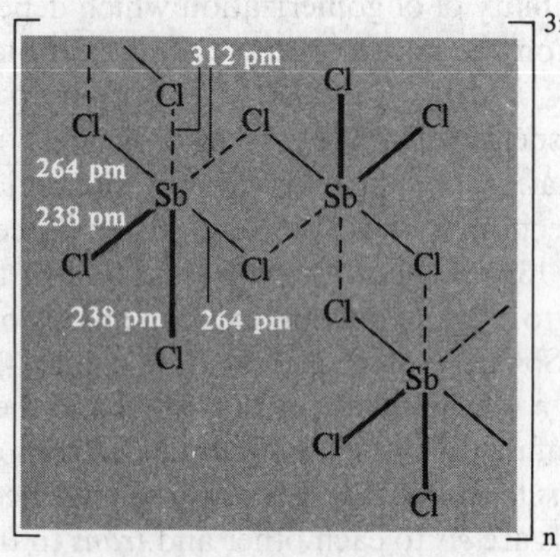

(d)   6-coordinate polymeric anion in
pyHSbCl$_4$, showing 2 short,
2 intermediate, and 2 long
Sb–Cl distances

FIG. 13.7   Structures of some complex halide anions of stoichiometry MX_4^-.

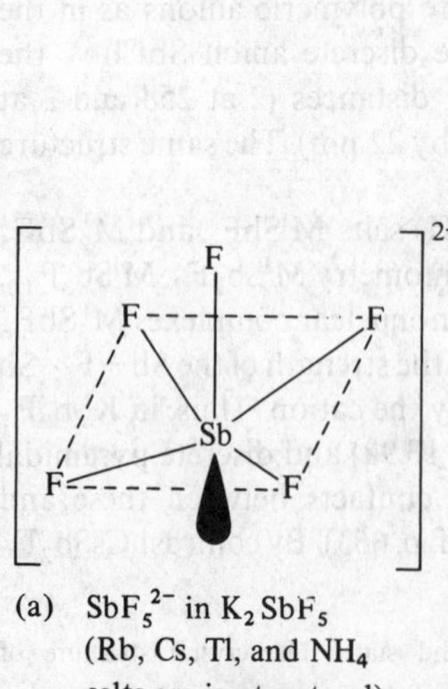

(a) SbF_5^{2-} in K_2SbF_5
(Rb, Cs, Tl, and NH$_4$
salts are isostructural)

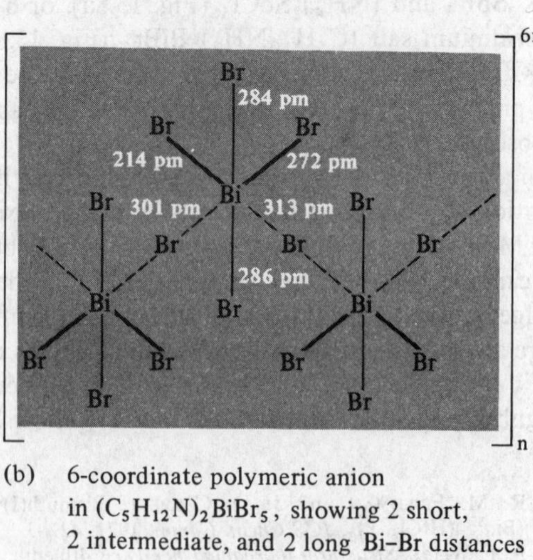

(b)   6-coordinate polymeric anion
in $(C_5H_{12}N)_2BiBr_5$, showing 2 short,
2 intermediate, and 2 long Bi–Br distances

FIG. 13.8   Structures of some complex halide anions of stoichiometry MX_5^{2-}.

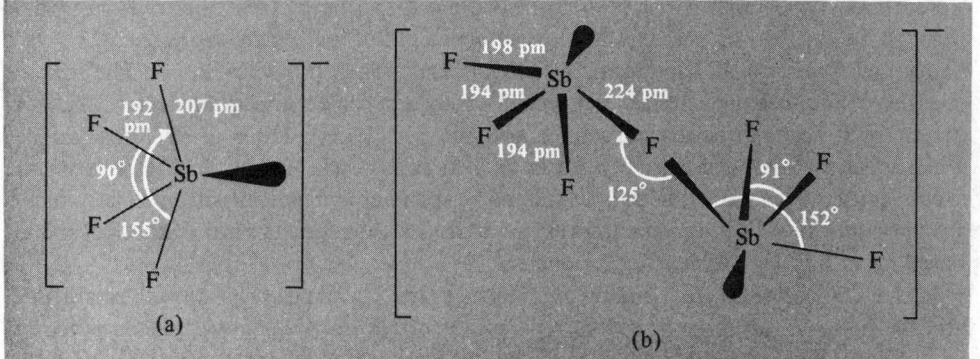

FIG. 13.9   Structures of SbF_4^- and $Sb_2F_7^-$ ions in $KSbF_4(SbF_3)$ and $CsSb_2F_7$ respectively.

has well-defined $Sb_2F_7^-$ anions (Fig. 13.9b) formed from 2 distorted trigonal bipyramidal $\{SbF_4\}$ groups sharing a common axial F atom with long bridge bonds.

Similar structural diversity characterizes the heavier halide complexes of the group. The $[MX_6]^{3-}$ group occurs in several compounds, and these frequently have a regular octahedral structure like the isoelectronic $[Te^{IV}X_6]^{2-}$ ions (p. 908), despite the formal 14-electron configuration on the central atom. For example the jet-black compound $(NH_4)_2SbBr_6$ is actually $[(NH_4^+)_4(Sb^{III}Br_6)^{3-}(Sb^VBr_6)^-]$ with alternating octahedral Sb^{III} and Sb^V ions. The undistorted nature of the $SbBr_6^{3-}$ octahedra suggests that the lone-pair is predominantly $5s^2$ but there is a sense in which this is still stereochemically active since the Sb–Br distance in $[Sb^{III}Br_6]^{3-}$ (279.5 pm) is substantially longer than in $[Sb^VBr_6]^-$ (256.4 pm). Similar dimensional changes are found in $(pyH)_6Sb_4Br_{24}$ which is $[(pyH^+)_6(Sb^{III}Br_6)^{3-}(Sb^VBr_6^-)_3]$. In $(Me_2NH_2)_3BiBr_6$ the $(Bi^{III}Br_6)^{3-}$ octahedron is only slightly distorted. Sixfold coordination also occurs in compounds such as $Cs_3Bi_2I_9$ and $[(pyH^+)_5(Sb_2Br_9)^{3-}(Br^-)_2]$ in which $M_2X_9^{3-}$ has the confacial bioctahedral structure of $Tl_2Cl_9^{3-}$ (p. 270) (Fig. 13.10). In β-$Cs_3Sb_2Cl_9$ and $Cs_3Bi_2Cl_9$, however, there are close-packed Cs^+ and Cl^- with Sb^{III} (or Bi^{III}) in octahedral interstices. In $Cs_3As_2Cl_9$ the $\{AsCl_6\}$ groups are highly distorted so that there are discrete $AsCl_3$ molecules (As–Cl 225 pm) embedded between Cs^+ and Cl^- ions (As–Cl 275 pm).

No completely general and quantitative theory of the stereochemical activity of the lone-pair of electrons in complex halides of tervalent As, Sb, and Bi has been developed but certain trends are discernible. The lone-pair becomes less decisive in modifying the

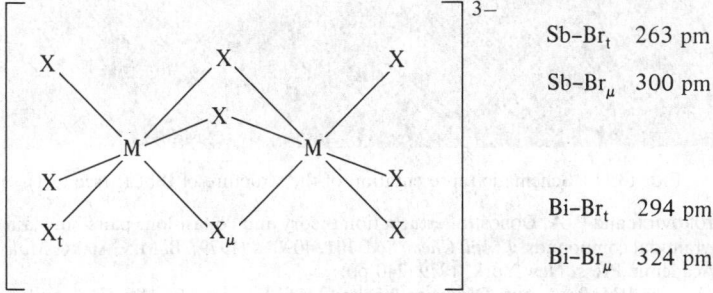

Sb–Br$_t$	263 pm
Sb–Br$_\mu$	300 pm
Bi–Br$_t$	294 pm
Bi–Br$_\mu$	324 pm

FIG. 13.10   Structure of $M_2X_9^{3-}$.

stereochemistry (a) with increase in the coordination number of the central atom from 4 through 5 to 6, (b) with increase in the atomic weight of the central atom (As > Sb > Bi), and (c) with increase in the atomic weight of the halogen (F > Cl > Br > I). The relative energies of the various valence-level orbitals may also be an important factor: the F(σ) orbital of F lies well below both the s and the p valence orbitals of Sb (for example) whereas the σ orbital energies of Cl, Br, and I lie between these two levels, at least in the free atoms. It follows that the lone pair is likely to be in a (stereochemically active) metal-based sp^x hybrid orbital in fluoro complexes of Sb but in a (stereochemically inactive) metal-based a_1 orbital for the heavier halogens.[17a]

In the +5 oxidation state, halide complexes of As, Sb, and Bi are also well established and the powerful acceptor properties of SbF$_5$ in particular have already been noted (p. 655). Such complexes are usually made by direct reaction of the pentahalide with the appropriate ligand. Thus KAsF$_6$ and NOAsF$_6$ have octahedral AsF$_6^-$ groups and salts of SbF$_6^-$ and SbCl$_6^-$ (as well as [Sb(OH)$_6$]$^-$) are also known. Frequently, however, there is strong residual interaction between the "cation" and the "complex anion" and the structure is better thought of as an extended three-dimensional network. For example the adduct SbCl$_5$.ICl$_3$ (i.e. ISbCl$_8$) comprises distorted octahedra of {SbCl$_6$} and angular {ICl$_2$} groups but, as shown in Fig. 13.11, there is additional interaction between the groups which links them into chains and the structure is intermediate between [ICl$_2$]$^+$[SbCl$_6$]$^-$ and [SbCl$_4$]$^+$[ICl$_4$]$^-$. Complexes are also formed by a variety of oxygen-donors, e.g. [SbCl$_5$(OPCl$_3$)] and [SbF$_5$(OSO)] as shown in Fig. 13.12. Fluoro-complexes in particular are favoured by large non-polarizing cations, and polynuclear complex anions sometimes then result as a consequence of fluorine bridging. For example irradiation of a mixture of SbF$_5$, F$_2$, and O$_2$ yields white crystals of O$_2$Sb$_2$F$_{11}$ which can be formulated[18] as O$_2^+$[Sb$_2$F$_{11}$]$^-$, and this complex, when heated under reduced pressure at 110°, loses SbF$_5$ to give O$_2^+$SbF$_6^-$. The dinuclear anion probably has a linear Sb–F–Sb bridge as in [BrF$_4$]$^+$[Sb$_2$F$_{11}$]$^-$ (p. 976), but in [XeF]$^+$[Sb$_2$F$_{11}$]$^-$ and [XeF$_3$]$^+$[Sb$_2$F$_{11}$]$^-$ (p. 1054) the bridging angle is reduced to 150°

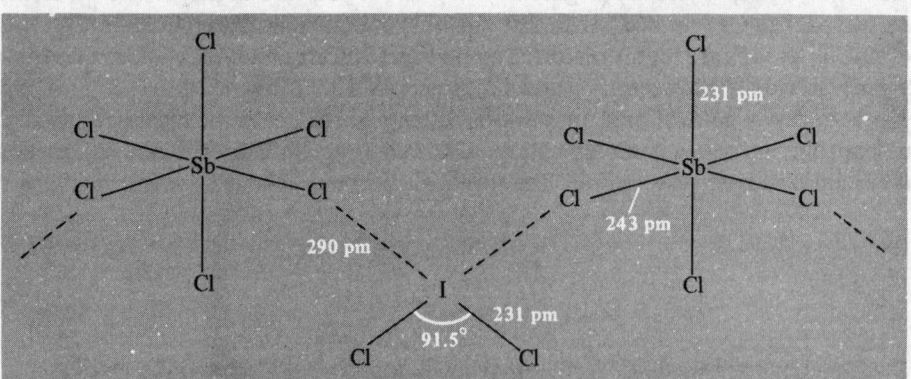

FIG. 13.11   Schematic representation of the structure of ISbCl$_8$ (see text).

[17a] E. SHUSTOROVICH and P. A. DOBOSH, Perturbation theory and "quasi-lone pairs" in main group angular and trigonal pyramidal compounds, *J. Am. Chem. Soc.* **101**, 4090–5 (1979). B. M. GIMARC, *Molecular Structure and Bonding*, Academic Press, New York, 1979, 240 pp.

[18] D. E. McKEE and N. BARTLETT, Dioxygenyl salts O$_2^+$SbF$_6^-$ and O$_2^+$Sb$_2$F$_{11}^-$ and their convenient laboratory synthesis, *Inorg. Chem.* **12**, 2738–40 (1973).

and 155° respectively. Even more extended coordination occurs in the 1:3 adduct $PF_5.3SbF_5$ which has been formulated as $[PF_4]^+[Sb_3F_{16}]^-$ on the basis of vibrational spectroscopy.[19] The same anion occurs in the scarlet paramagnetic complex $[Br_2]^+[Sb_3F_{16}]^-$ for which X-ray crystallography has established the *trans*-bridged octahedral structure shown in Fig. 13.13.[20] The compound (mp 69°) was prepared by adding a small amount of BrF_5 to a mixture of Br_2 and SbF_5. The structure of the compound $AsF_3.SbF_5$ can be described either as a molecular adduct, $F_2AsF{\rightarrow}SbF_5$, or as an ionic complex, $[AsF_2]^+[SbF_6]^-$; in both descriptions the alternating As and Sb units are joined into an infinite network by further F bonding.[21] Another mixed valency adduct, which can be formulated as $6SbF_3.5SbF_5$, has also recently been prepared by direct fluorination of Sb. This compound, $Sb_{11}F_{43}$, was obtained as a white, high-melting crystalline solid; it contains the polymeric chain cation $[Sb_6^{III}F_{13}^{5-}]_n$ and 5 $[Sb^VF_6]^-$ anions.[22] The 1:1 adduct $SbF_3.SbF_5$ has the pseudo-ionic structure $[Sb_2^{III}F_4]^{2+}[Sb^VF_6^-]_2$; however, the $[F_2Sb{-}F{\cdots}SbF]^{2+}$ cation features 5 different Sb–F distances (185, 187, 199, 201, and 215 pm) and can be regarded either as an SbF^{2+} cation coordinated by SbF_3, or as a fluorine-bridged dinuclear cation $[F_2Sb{-}F{-}SbF]^{2+}$, or as part of an infinite three-dimensional polymer $[(SbF_4)_4]_n$ when even longer Sb^{III}–F contacts are considered.[22a]

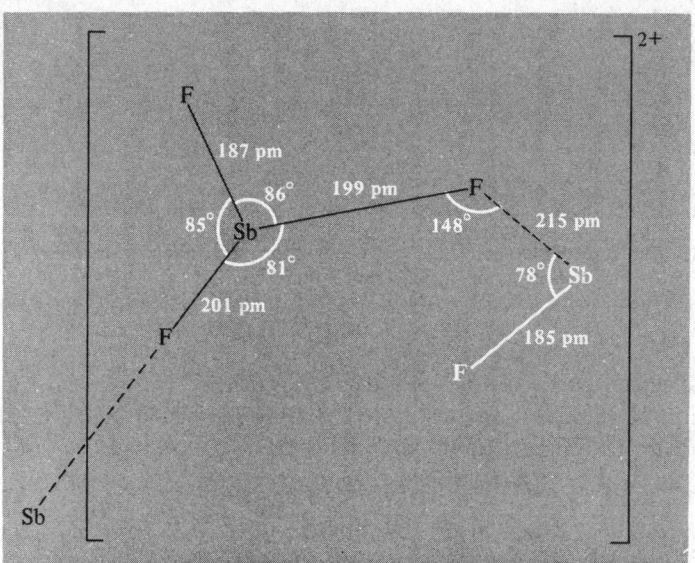

[19] G. S. H. CHEN and J. PASSMORE, Preparation and characterization of perfluorophosphonium hexadecafluoroantimonate(V) $[PF_4]^+[Sb_3F_{16}]^-$, *JCS Chem. Comm.* 1973, 559.

[20] A. J. EDWARDS and G. R. JONES, Fluoride crystal structures. Part XVI. Dibromine hexadecafluorotriantimonate(V) $[Br_2]^+[Sb_3F_{16}]^-$, *J. Chem. Soc.* A 1971, 2318–20.

[21] A. J. EDWARDS and R. J. C. SILLS, Fluoride crystal structures. Part XV. Arsenic trifluoride-antimony pentafluoride, *J. Chem. Soc.* A 1971, 942–5.

[22] A. J. EDWARDS and D. R. SLIM, Crystal structure of tridecafluorohexa-antimony(III) pentakis[hexafluoroantimonate(V)], *JCS Chem. Comm.* 1974, 178–9.

[22a] R. J. GILLESPIE, D. R. SLIM, and J. E. VEKRIS, The crystal structure of the 1:1 adduct of antimony trifluoride and antimony pentafluoride, *JCS Dalton* 1977, 971–4.

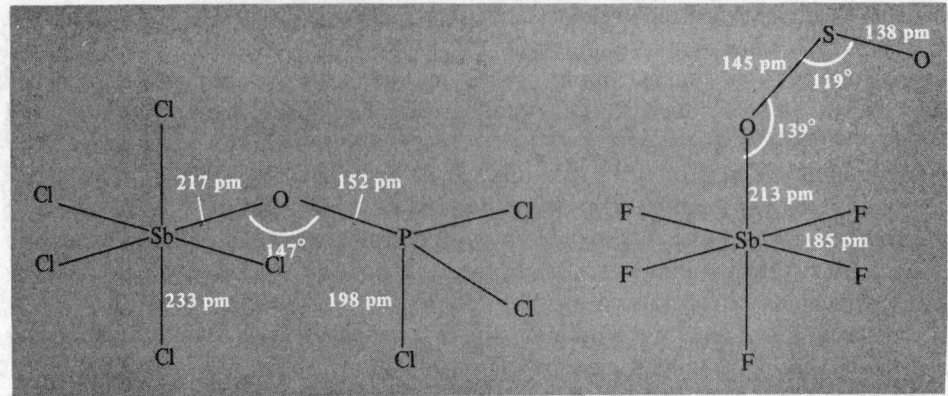

FIG. 13.12   Schematic representation of the pseudo-octahedral structures of [SbCl₅(OPCl₃)] and [SbF₅(OSO)].

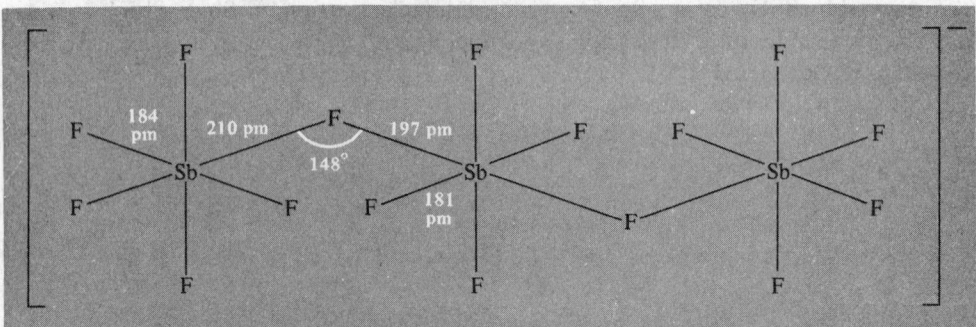

FIG. 13.13   Structure of the anion [Sb₃F₁₆]⁻ showing the *trans* arrangement of the 2 bridging F$_\mu$ atoms. The Sb–F$_\mu$ distances are significantly longer than Sb–F$_t$ and the asymmetry of the bridge is well outside the standard deviation (± 4 pm). The mean bridging distance (203 pm) is close to the value of 201 pm in [Sb₂F₁₁]⁻.

The great electron-pair acceptor capacity (Lewis acidity) of SbF₅ has been utilized in the production of extremely strong proton donors (Brønsted acids, p. 51). Thus the acidity of anhydrous HF is substantially increased in the presence of SbF₅:

$$2HF + SbF_5 \rightleftharpoons [H_2F]^+[SbF_6]^-$$

An even stronger acid results from the interaction of SbF₅ with an oxygen atom in fluorosulfuric acid HSO₃F (i.e. HF/SO₃):

$$SbF_5 + HSO_3F \rightleftharpoons$$

Many other equilibria occur in these solutions and the acidity is enhanced even further by the addition of 3 or more moles of SO_3.[23] Such solutions $SbF_5/HSO_3F/nSO_3$ have been termed "super-acid media" and they are, indeed, the strongest known proton donors at the present time. They have been extensively studied, particularly because they are able to protonate virtually all organic compounds.[24]

Oxide halides

The stable molecular nitrosyl halides NOX (p. 507) and phosphoryl halides POX_3 (p. 575) find few counterparts in the chemistry of As, Sb, and Bi. AsOF has been reported as a product of the reaction of As_4O_6 with AsF_3 in a sealed tube at 320° but has not been fully characterized. $AsOF_3$ is known only as a polymer. Again, just as $AsCl_5$ eluded preparation for over 140 y after Liebig's first attempt to make it in 1834, so $AsOCl_3$ defied synthesis until 1976 when it was made by ozonization of $AsCl_3$ in $CFCl_3/CH_2Cl_2$ at $-78°$: it is a white, monomeric, crystalline solid and is one of the few compounds that can be said to contain a "real" As=O double bond.[25] $AsOCl_3$ is thermally more stable than $AsCl_5$ (p. 655) but decomposes slowly at $-25°$ to give $As_2O_3Cl_4$:

$$3AsOCl_3 \xrightarrow[\text{(rapid)}]{0°} AsCl_3 + Cl_2 + As_2O_3Cl_4$$

The compound $As_2O_3Cl_4$ is polymeric and is thus not isostructural with $Cl_2P(O)OP(O)Cl_2$.

SbOF and SbOCl can be obtained as polymeric solids by controlled hydrolysis of SbX_3. Several other oxide chlorides can be obtained by varying the conditions, e.g.:

$$SbCl_3 \xrightarrow[\text{H}_2\text{O}]{\text{limited}} SbOCl \xrightarrow[\text{H}_2\text{O}]{\text{more}} Sb_4O_5Cl_2 \xrightarrow{460°/\text{Ar}} Sb_8O_{11}Cl_2$$

An alternative dry-way preparation which permits the growth of large, colourless, single crystals suitable for ferroelectric studies (p. 60) has recently been devised:[25a]

$$5Sb_2O_3 + 2SbCl_3 \xrightarrow[\text{vac}]{75°} 3Sb_4O_5Cl_2 \text{ (mp 590°)}$$

The compounds $Sb_4O_3(OH)_3Cl_2$ and Sb_8OCl_{22} have also been reported. SbOCl itself comprises polymeric sheets of composition $[Sb_6O_6Cl_4]^{2+}$ (formed by linking Sb atoms via O and Cl bridges) interleaved with layers of chloride ions. In addition to polymeric species, finite heterocyclic complexes can also be obtained. For example partial hydrolysis of the polymeric $[pyH]_3[Sb_2^{III}Cl_9]$ in ethanol leads to $[pyH^+]_2[Sb_2^{III}OCl_6]^{2-}$ in which the anion contains 2 pseudo-octahedral $\{:SbOCl_4\}$ units sharing a common face $\{\mu_3\text{-}OCl_2\}$ with the lone-pairs *trans* to the bridging oxygen atom (Fig. 13.14).[26] Another novel

[23] R. J. GILLESPIE, Fluorosulfuric acid and related superacid media, *Acc. Chem. Res.* **1**, 202–9 (1968).

[24] G. A. OLAH, A. M. WHITE, and D. H. O'BRIEN, Protonated heteroaliphatic compounds, *Chem. Rev.* **70**, 561–91 (1970).

[25] K. SEPPELT, Arsenic oxide trichloride, $AsOCl_3$, *Angew. Chem.*, Int. Edn. (Engl.) **15**, 766–7 (1976).

[25a] YA. P. KUTSENKO, Preparation and properties of $Sb_4O_5Cl_2$ single crystals, *Kristallografiya* (Engl. transl.) **24**, 349–51 (1979).

[26] M. HALL and D. B. SOWERBY, Preparation and crystal structure of bis(pyridinium-μ-dichloro-μ-oxo-tetrachlorodiantimonate(III); a discrete antimony(III) oxychloro anion, *JCS Chem. Comm.* 1979, 1134–5.

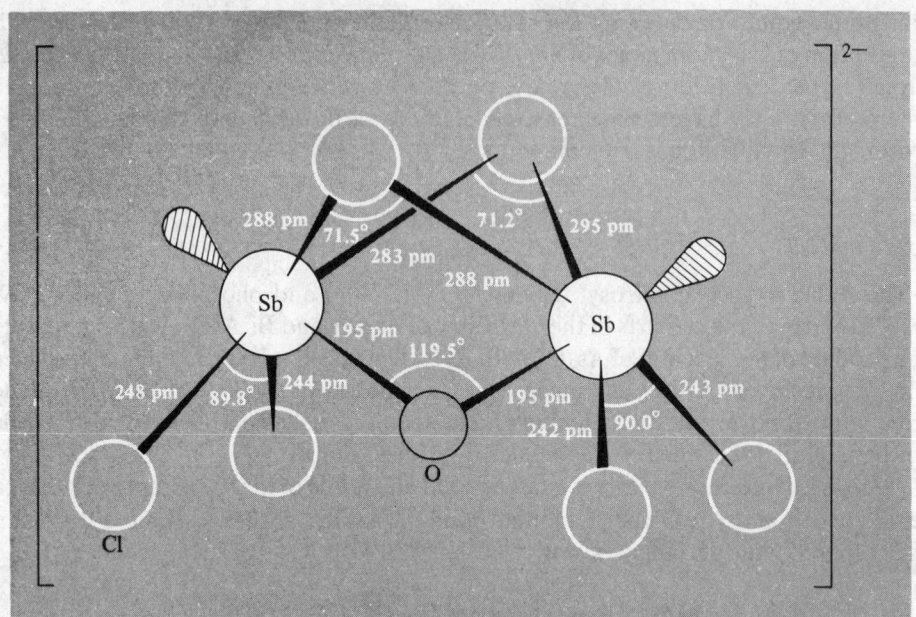

FIG. 13.14 Structure of the binuclear anion $[Sb_2^{III}OCl_6]^{2-}$ showing the bridging oxygen and chlorine atoms and the pseudooctahedral coordination about Sb; the O atom is at the common apex of the face-shared square pyramids and the lone-pairs are *trans*- to this below the $\{SbCl_4\}$ bases. The bridging distances Sb–Cl$_\mu$ are substantially longer than the terminal distances Sb–Cl$_t$.

polynuclear antimony oxide halide anion has very recently been established in the dark-blue ferrocenium complex $\{[Fe(\eta^5\text{-}C_5H_5)]_2[Sb_4Cl_{12}O]\}_2 \cdot 2C_6H_6$ which was made by photolysis of benzene solutions of ferrocene (p. 000) and $SbCl_3$ in the presence of oxygen:[27] the anion (Fig. 13.15) contains 2 square-pyramidal $\{Sb^{III}Cl_5\}$ units sharing a common edge and joined via a unique quadruply bridging Cl atom to 2 pseudo trigonal bipyramidal $\{Sb^{III}Cl_3O\}$ units which share a common bridging O atom and the unique Cl atom. The structure implies the presence of a lone-pair of electrons beneath the basal plane of the first 2 Sb atoms and in the equatorial plane (with O_μ and Cl_t) of the second 2 Sb atoms.

Other finite-complex anions occur in the oxyfluorides. For example the hydrated salts $K_2[As_2F_{10}O] \cdot H_2O$ and $Rb_2[As_2F_{10}O] \cdot H_2O$ contain the oxo-bridged binuclear anion $[F_5As\text{-}O\text{-}AsF_5]^{2-}$ as shown in Fig. 13.16[28] and the anhydrous salt $Rb_2[Sb_2F_{10}O]$ contains a similar anion with angle Sb–O–Sb 133°, Sb–F 188 pm, and Sb–O 191 pm.[29] The compound of empirical formula $CsSbF_4O$ is, in fact, trimeric with a 6-membered heterocyclic anion in the boat configuration, i.e. $Cs_3[Sb_3F_{12}O_3]$,[30] whereas the

[27] A. L. RHEINGOLD, A. G. LANDERS, P. DAHLSTROM, and J. ZUBIETA, Novel antimony cluster displaying a quadruply bridging chloride. X-ray crystal structure of $\{[Fe(\eta^5\text{-}C_5H_5)_2]_2[Sb_4Cl_{12}O]\}_2 \cdot 2C_6H_6$, *JCS Chem. Comm.* 1979, 143–4.

[28] W. HAASE, The crystal structure of dirubidium- and dipotassium-μ-oxo-bis(pentafluorodiarsenate) monohydrate, $M_2[As_2F_{10}O] \cdot H_2O$, *Acta Cryst.* **B30,** 1722–7 (1974).

[29] W. HAASE, Dirubidium-μ-oxo-bis(pentafluoroantimonate), *Acta Cryst.* **B30,** 2508–10 (1974).

[30] W. HAASE, Structure of tricaesium-cyclo-tri-μ-oxo-tris(tetrafluoroantimonate), $Cs_3[Sb_3F_{12}O_3]$, *Acta Cryst.* **B30,** 2465–9 (1974).

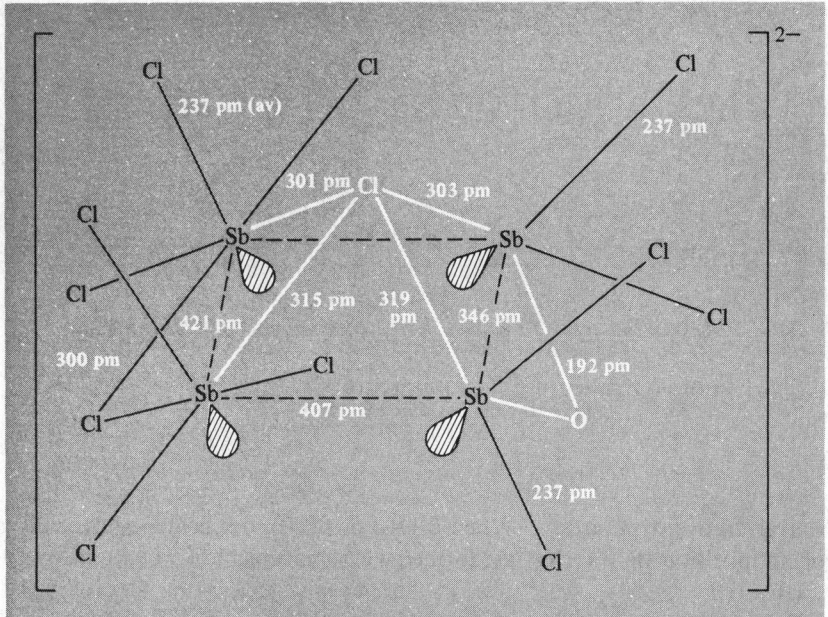

FIG. 13.15   Schematic representation of the structure of the complex anion $[Sb_4Cl_{12}O]^{2-}$ showing the two different coordination geometries about Sb and the unique quadruply bridging Cl atom.

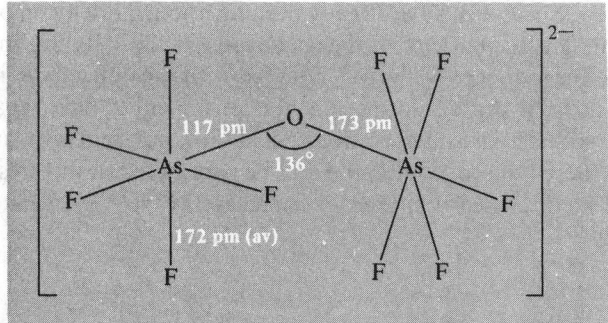

FIG. 13.16   Schematic representation of the anion structure in $M_2[As_2F_{10}O].H_2O$.

corresponding arsenic compound[31] has a dimeric anion $[As_2F_8O_2]^{2-}$ (Fig. 13.17). In both cases the Group VB element is octahedrally coordinated by 4 F and 2 O atoms in the *cis-* configuration.

Bismuth oxide halides BiOX are readily formed as insoluble precipitates by the partial hydrolysis of the trihalides (e.g. by dilution of solutions in concentrated aqueous HX). BiOF and BiOI can also be made by heating the corresponding BiX_3 in air. BiOI, which itself decomposes above 300°, is brick-red in colour; the other 3 BiOX are white. All have

[31] W. HAASE, The crystal and molecular structure of dicaesium-octafluoro-bis(μ-oxodiarsenate), $Cs_2[As_2F_8O_2]$, and a three-dimensional refinement of the structures of $K_2[As_2F_8O_2]$ and $Rb_2[As_2F_8O_2]$, *Chem. Ber.* **107**, 1009–18 (1974).

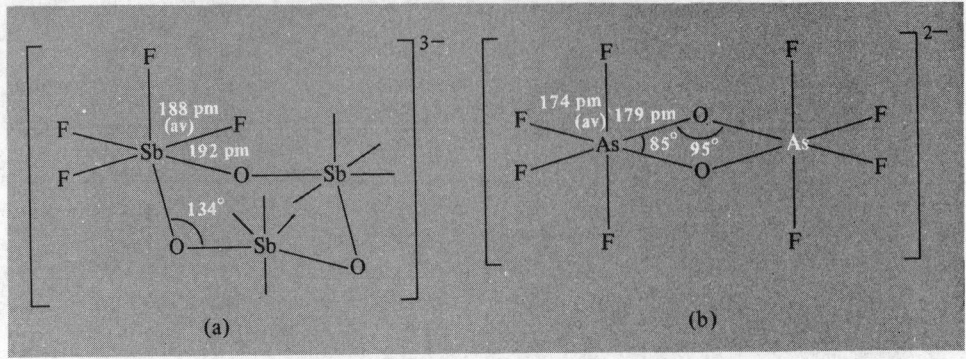

FIG. 13.17   Schematic representation of the structure of (a) the trimeric anion $[Sb_3F_{12}O_3]^{3-}$, and (b) the dimeric anion $[As_2F_8O_2]^{2-}$.

complex layer–lattice structures.[17] When BiOCl or BiOBr are heated above 600° oxide halides of composition $Bi_{24}O_{31}X_{10}$ are formed, i.e. replacement of 5 O atoms by 10 X in $Bi_{24}O_{36}$, (Bi_2O_3).

13.3.4   *Oxides and oxo compounds*

The amphoteric nature of As_2O_3 and the trends in properties of several of the oxides and oxoacids of As, Sb, and Bi have already been mentioned briefly on p. 645. Because of the trend towards greater basicity in the sequence As < Sb < Bi and the trend towards greater acidity in the sequence $M^{III} < M^V$, coupled with the difficulty of isolating some of the oxides from their "hydrated" forms, it is not convenient to have separate sections on oxides, hydrous oxides, hydroxides, acids, oxoacid salts, polyacid salts, and mixed oxides. Accordingly, all these types of compound will be considered in the present section: M^{III} compounds will be discussed first then intermediate M^{III}/M^V systems, and, finally, M^V oxo- compounds.

Oxo compounds of M^{III}

As_2O_3 (diarsenic trioxide) is the most important compound of As (Panel, p. 639). It is made (a) by burning As in air, (b) by hydrolysis of $AsCl_3$, or (c) industrially, by roasting sulfide ores such as arsenopyrite, FeAsS. Sb_2O_3 and Bi_2O_3 are made similarly. All 3 oxides exist in several modifications as shown in the subjoined scheme.[6] In the vapour phase As_2O_3 exists as As_4O_6 molecules isostructural with P_4O_6 (p. 579), and this unit also occurs in the cubic crystalline form. Above 800° gaseous As_4O_6 partially dissociates to an equilibrium mixture containing both As_4O_6 and As_2O_3 molecules. The less-volatile monoclinic form of As_2O_3 has a sheet-like structure of pyramidal $\{AsO_3\}$ groups sharing common O atoms. This transformation from molecular As_4O_6 units to polymeric As_2O_3 is accompanied by an 8.7% increase in density from 3.89 to 4.23 g cm^{-3}. A similar change from cubic, molecular Sb_4O_6 to polymeric Sb_2O_3 results in an 11.3% density increase from 5.20 to 5.79 g cm^{-3}.

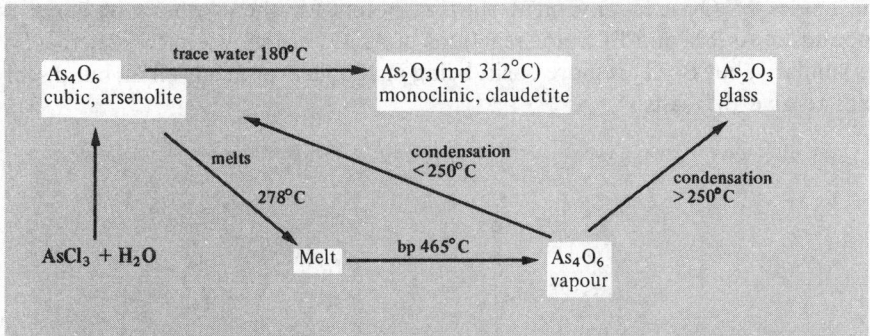

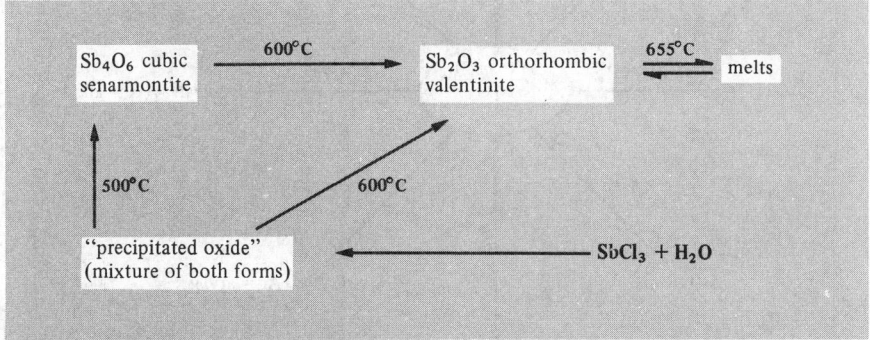

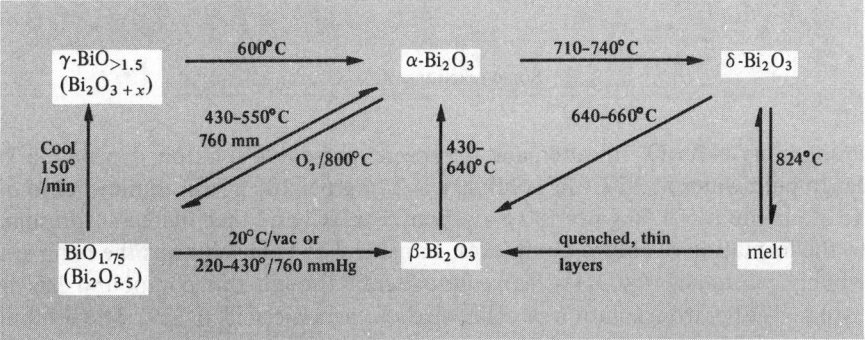

Polymorphs of M_2O_3 (M = As, Sb, Bi).

The structural relationships in Bi_2O_3 are more complex. At room temperature the stable form is monoclinic α-Bi_2O_3 which has a polymeric layer structure featuring distorted, 5-coordinate Bi in pseudo-octahedral $\{:BiO_5\}$ units. Above 717°C this transforms to the cubic δ-form which has a defect fluorite structure (CaF_2, p. 129) with randomly distributed oxygen vacancies, i.e. $[Bi_2O_3\square]$. The β-form and several oxygen-rich forms (in which some of the vacant sites are filled by O^{2-} with concomitant oxidation of some Bi^{III} to Bi^V) are related to the δ-Bi_2O_3 structure. There are also numerous double oxides $pMO_n.qBi_2O_3$, e.g. $Bi_{12}GeO_{20}$ (i.e. $GeO_2.6Bi_2O_3$), and other mixed oxides can be made by fusing Bi_2O_3 with oxides of Ca, Sr, Ba, Cd, or Pb; these latter have $(BiO)_n$ layers as in the oxide halides, interleaved with M^{II} cations.

The oxides M_2O_3 are convenient starting points for the synthesis of many other compounds of As, Sb, and Bi. Some reactions of As_2O_3 are shown in the scheme; Sb_2O_3 reacts similarly, but Bi_2O_3 is more basic, being insoluble in aqueous alkali but dissolving in acids to give Bi^{III} salts.

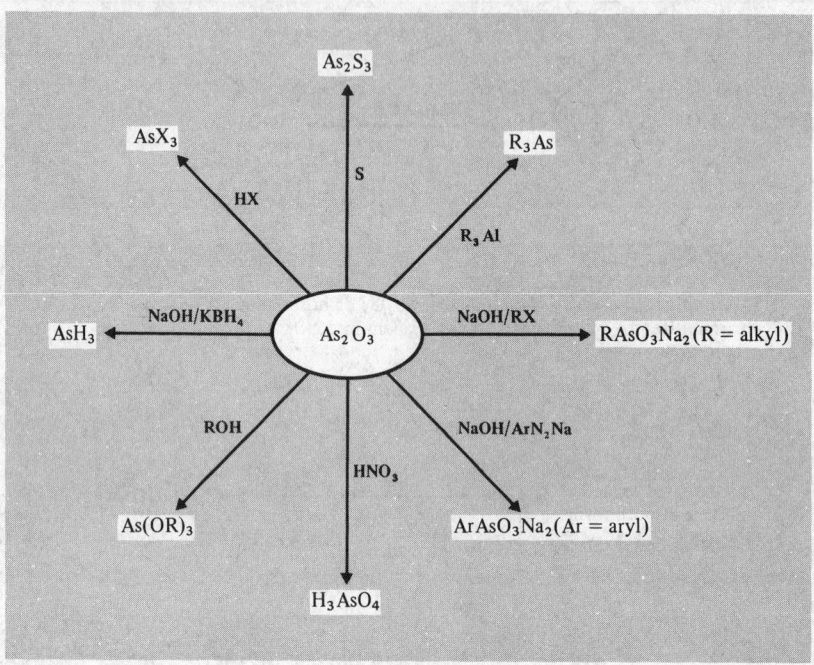

Some reactions of As_2O_3.

The solubility of As_2O_3 in water, and the species present in solution, depend markedly on pH. In pure water at 25°C the solubility is 2.16 g per 100 g; this diminishes in dilute HCl to a minimum of 1.56 g per 100 g at about 3 M HCl and then increases, presumably due to the formation of chloro-complexes. In neutral or acid solutions the main species is probably pyramidal $As(OH)_3$, "arsenious acid", though this compound has never been isolated either from solution or otherwise (cf. carbonic acid, p. 329). The solubility is much greater in basic solutions and spectroscopic evidence points to the presence of such anions as $[AsO(OH)_2]^-$, $[AsO_2(OH)]^{2-}$, and $[AsO_3]^{3-}$, corresponding to successive deprotonation of H_3AsO_3. The first stage dissociation constant at 25° is $K_a = [AsO(OH)_2{}^-][H^+]/[H_3AsO_3] \simeq 6 \times 10^{-10}$, pK_a 9.2; ortho-arsenious acid is therefore a very weak acid (as expected from Pauling's rules, p. 54) and is comparable in strength to boric acid (p. 230). Dissociation as a base is even weaker: $K_b = [As(OH)_2{}^+][OH^-]/[As(OH)_3] \simeq 10^{-14}$. There now seems to be less evidence for other species that were formerly considered to be present in solution, e.g. the monomeric meta-acid $HAsO_2$, i.e. $[AsO(OH)]$ (by loss of 1 H_2O) and the hexahydroxoacid $H_3[As(OH)_6]$ or its hydrate.

Arsenites of the alkali metals are very soluble in water, those of the alkaline earth metals less so, and those of the heavy metals are virtually insoluble. Many of the salts are obtained as meta-arsenites, e.g. $NaAsO_2$, which comprises polymeric chain anions formed by corner linkage of pyramidal {AsO₃} groups and held together by Na ions:

$$\left[\begin{array}{c} \text{(chain of corner-linked AsO}_3\text{ pyramids)} \end{array} \right]_\infty$$

The sparingly soluble yellow Ag_3AsO_3 is an example of an orthoarsenite. Copper(II) arsenites were formerly used as fine green pigments, e.g. Paris green, which is an acetate arsenite $[Cu_2(MeCO_2)(AsO_3)]$, and Scheele's green, which approximates to the hydrogen arsenite $CuHAsO_3$ or the dehydrated composition $Cu_2As_2O_5$.

Antimonious acid H_3SbO_3 and its salts are less well characterized but a few meta-antimonites and polyantimonites are known, e.g. $NaSbO_2$, $NaSb_3O_5 \cdot H_2O$, and $Na_2Sb_4O_7$. The oxide itself finds extensive use as a flame retardant in fabrics, paper, paints, plastics, epoxy resins, adhesives, and rubbers. The scale of industrial use can be gauged from the US statistics which indicate an annual consumption of Sb_2O_3 of some 10 000 tonnes in that country.

The corresponding Bi compound $Bi(OH)_3$ is definitely basic rather than acidic. It dissolves readily in acid giving solution of Bi^{III} ions but an increase in pH causes precipitation of oxo-salts. Before precipitation, however, polymeric oxocations can be detected in solution of which the best characterized is $[Bi_6(OH)_{12}]^{6+}$ in perchlorate solution. The species (Fig. 13.18) resembles $[Ta_6Cl_{12}]^{2+}$ and has 6 Bi at the corners of an octahedron with bridging OH groups above each of the 12 edges. The shortest Bi–O

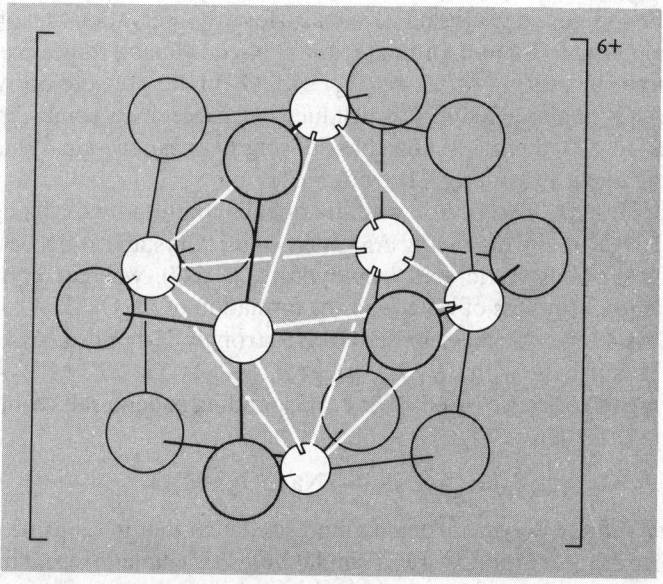

Fɪɢ. 13.18   The structure of the oxocation $[Bi_6(OH)_{12}]^{6+}$ obtained by hydrolysis of aqueous perchlorate solution.

distance is 233 pm and the (nonbonding) Bi$\cdots$Bi distance is 370 pm (307 and 353 in Bi metal). This contrasts with the bicapped tetrahedral distribution of metal atoms in $[Pb_6O(OH)_6]^{4+}$ (p. 458) where there is an O atom at the centre of the central tetrahedron and OH groups above the faces of the capping tetrahedra.

Mixed-valence oxides

The intermediate diamagnetic oxide α-Sb_2O_4 (i.e. $Sb^{III}Sb^VO_4$) has long been known as the massive, fine-grained, yellow, orthorhombic mineral cervantite and more recently a monoclinic β-form has been recognized. α-Sb_2O_4 can also be obtained by heating Sb_2O_3 in dry air at 460–540°C, and further heating in air or oxygen at 1130° produces β-Sb_2O_4. Both forms have similar structures with equal numbers of Sb^{III} and Sb^V. α-Sb_2O_4 is isostructural with $SbNbO_4$ and $SbTaO_4$ and consists of corrugated sheets of slightly distorted $\{Sb^VO_6\}$ octahedra sharing all their vertices (as in the plane layer in K_2NiF_4); the Sb^{III} lie between the layers in positions of irregular pyramidal fourfold coordination, all four O atoms lying on the same side of the Sb^{III}. Further oxidation to anhydrous Sb_2O_5 has not been achieved (see below). For oxygen-rich Bi_2O_{3+x} see p. 669.

Oxo compounds of M^V

Arsenic(V) oxide, As_2O_5, is one of the oldest-known oxides, but structural analysis has been thwarted until very recently because of poor thermal stability, ease of hydrolysis, and the difficulty of growing a single crystal. It is now known to consist of equal numbers of $\{AsO_6\}$ octahedra and $\{AsO_4\}$ tetrahedra completely linked by corner sharing to give cross-linked strands which define tubular cavities (cf. the corner sharing in ReO_3 octahedra, p. 1219, and SiO_2 tetrahedra, p. 394).[32] The structure accounts for the reluctance of the compound to crystallize and also for the observation that only half the As atoms can be replaced by Sb (6-coordinate) and P (4-coordinate) respectively. As_2O_5 can be prepared either by heating As (or As_2O_3) with O_2 under pressure or by dehydrating crystalline H_3AsO_4 at about 200°C. It is deliquescent, exceedingly soluble in water (230 g per 100 g H_2O at 20°), thermally unstable (loosing O_2 near the mp, ca. 300°C), and a strong oxidizing agent (liberating Cl_2 from HCl).

Arsenic acid, H_3AsO_4, can be obtained in aqueous solution by oxidizing As_2O_3 with concentrated HNO_3 or by dissolving As_2O_5 in water. Crystallization below 30° yields $2H_3AsO_4.H_2O$ (cf. phosphoric acid hemihydrate, p. 597), whereas crystallization at 100°C or above results in loss of water and the formation of $As_2O_5.\frac{5}{3}H_2O$, i.e. ribbon-like polymeric $(H_5As_3O_{10})_n$. All these materials are strongly H-bonded. Arsenic acid, like H_3PO_4 (p. 598), is tribasic with pK_1 2.2, pK_2 6.9, pK_3 11.5 at 25°. $M^IH_2AsO_4$ (M = K, Rb, Cs, NH_4) are ferroelectric (p. 60). The corresponding sodium salt readily dehydrates to give meta-arsenate $NaAs^VO_3$:

$$NaH_2AsO_4 \longrightarrow NaAsO_3 + H_2O$$

$NaAsO_3$ has an infinite polymeric chain anion similar to that in diopside (pp. 403, 616) but with a trimeric repeat unit; $LiAsO_3$ is similar but with a dimeric repeat unit whereas β-

---

[32] M. JANSEN, The crystal structure of As_2O_5, *Angew. Chem.*, Int. Edn. (Engl.) **16**, 214 (1977).

$KAsO_3$ appears to have a cyclic trimeric anion $As_3O_9^{3-}$ which resembles the *cyclo-trimetaphosphates* (p. 617). There is thus a certain structural similarity between arsenates and phosphates, though arsenic acid and the arsenates show less tendency to catenation (p. 608). The tetrahedral $\{As^VO_4\}$ group also resembles $\{PO_4\}$ in forming the central unit in several heteropolyacid anions (p. 1184).

One striking difference between arsenates and phosphates is the appreciable oxidizing tendency of the former. This is clear from the oxidation state diagram for the Group V elements shown in Fig. 13.19, which summarizes a great deal of relevant information (p. 498). Antimony is seen to resemble arsenic quite closely but Bi^V–Bi^{III} is a much more strongly oxidizing couple and, indeed (as is clear from Fig. 13.19), it is able to oxidize water to oxygen. It is also clear that the +3 oxidation states of As, Sb, and Bi do not disproportionate in solution. Nor do the elements themselves, so there are no reactions comparable to that of P_4 with alkali to give phosphine and hypophosphite (p. 589). Redox

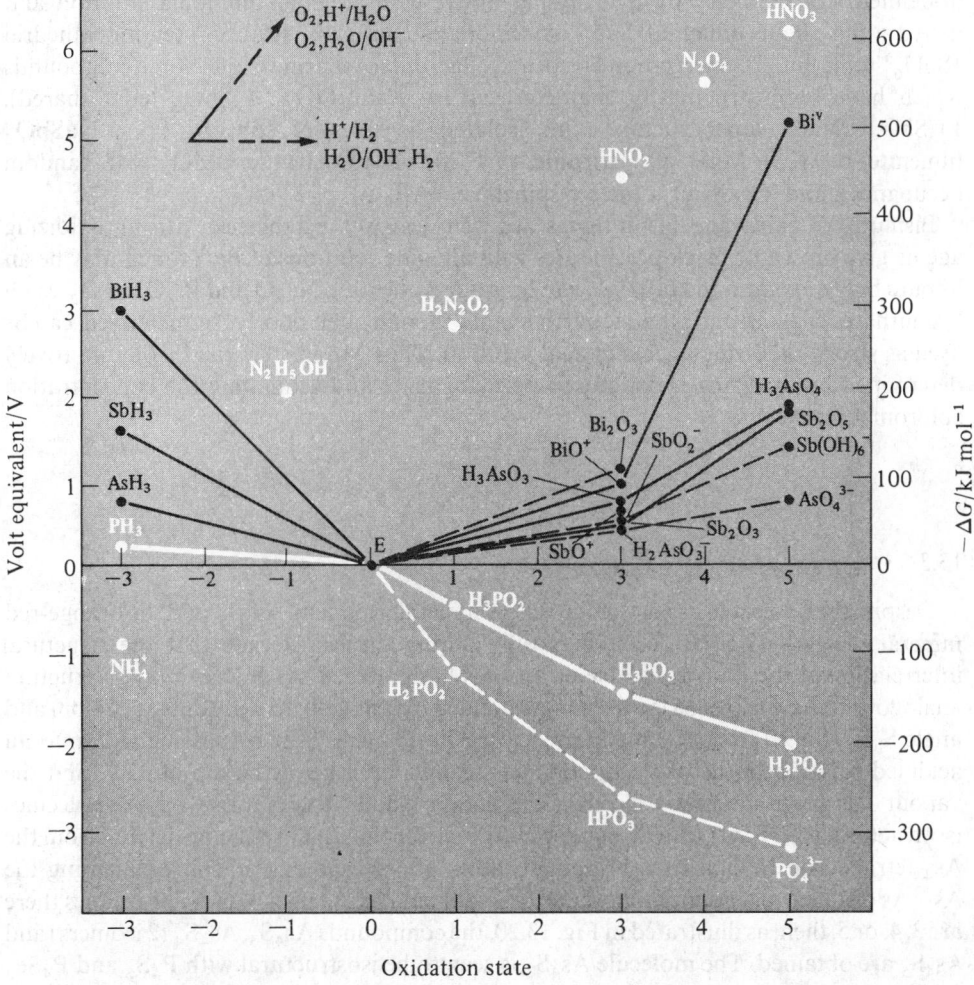

FIG. 13.19   Oxidation state diagram for As, Sb, and Bi in acid and alkaline solutions, together with selected data on N and P for comparison.

reactions have proved a useful volumetric method of analysis for both As and Sb. For example As^{III} is quantitatively oxidized in aqueous solution by I_2, or by potassium bromate, iodate, or permanganate. Such reactions can be formally represented as follows:

$$As^{III} + I_2 \longrightarrow As^V + 2I^-, \text{ etc.}$$

Thus, in an acid buffer such as borax–boric acid or Na_2HPO_4–NaH_2PO_4 (p. 599):

$$\tfrac{1}{2}As_2O_3(aq) + I_2 + H_2O \longrightarrow \tfrac{1}{2}As_2O_5(aq) + 2H^+(aq) + 2I^-$$

Such reactions are not available for Bi^{III} but this can readily be determined by complexometric titration using ethylenediaminetetraacetic acid or similar complexones:

$$Bi^{III} + H_4EDTA \xrightarrow{\ aq\ } [Bi(EDTA)]^- + 4H^+$$

Antimony(V) oxide has been obtained as a poorly characterized pale-yellow powder of ill-defined stoichiometry by hydrolysing $SbCl_5$ with aqueous ammonia solution and dehydrating the product at 275°. Antimonates generally feature pseudooctahedral $\{SbO_6\}$ units but polymerization by corner, edge, or face sharing is rife. Some compounds which have been structurally characterized are $NaSb(OH)_6$, $LiSbO_3$ (edge-shared), Li_3SbO_4 (NaCl superstructure with isolated lozenges of $\{Sb_4O_{16}\}^{12-}$), $NaSbO_3$ (ilmenite, p. 1122), $MgSb_2O_6$ (trirutile, p. 1120), $AlSbO_4$ (rutile, $2MO_2$ with random occupancy), and $Zn_7Sb_2O_{12}$ (defect spinel, i.e. $3AB_2O_4$, p. 279).

Bismuth(V) oxide and bismuthates are even less well established. Strong oxidizing agents give brown or black precipitates with alkaline solutions of Bi^{III}, which may be an impure higher oxide, and $NaBi^VO_3$ can be made by heating Na_2O and Bi_2O_3 in O_2. Such bismuthates of alkali and alkaline earth metals, though often poorly characterized, can be used as strong oxidizing agents in acid solution. Thus Mn in steel can be quantitatively determined by oxidizing it directly to permanganate and estimating the concentration colorometrically.

13.3.5 *Sulfides and related compounds*

Despite the venerable history of the yellow mineral orpiment, As_2S_3, and the orange-red mineral realgar, As_4S_4 (p. 637), it is only during the last decade that the structural interrelation of the numerous arsenic sulfides has emerged. As_2S_3 has a layer-structure analogous to As_2O_3 (p. 668) with each As bonded pyramidally to 3 S atoms at 224 pm and angle S–As–S 99°. It can be made by heating As_2O_3 with S or by passing H_2S into an acidified solution of the oxide. It sublimes readily, even below its mp of 320°, and the vapour has been shown by electron diffraction studies to comprise As_4S_6 molecules isostructural with P_4O_6 (p. 579). The structure can be thought of as being derived from the As_4 tetrahedron by placing a bridging S atom above each edge thereby extending the As$\cdots$As distance to a nonbonding value of ~ 290 pm. If instead of 6 As–S–As bridges there are 3, 4, or 5, then, as illustrated in Fig. 13.20, the compounds As_4S_3, As_4S_4 (2 isomers) and As_4S_5 are obtained. The molecule As_4S_3 is seen to be isostructural with P_4S_3 and P_4Se_3 (p. 581); it occurs in both the α- and the β-form of the orange-yellow mineral dimorphite (literally "two forms", discovered by A. Scacchi in volcanic fumaroles in Italy in 1849), the

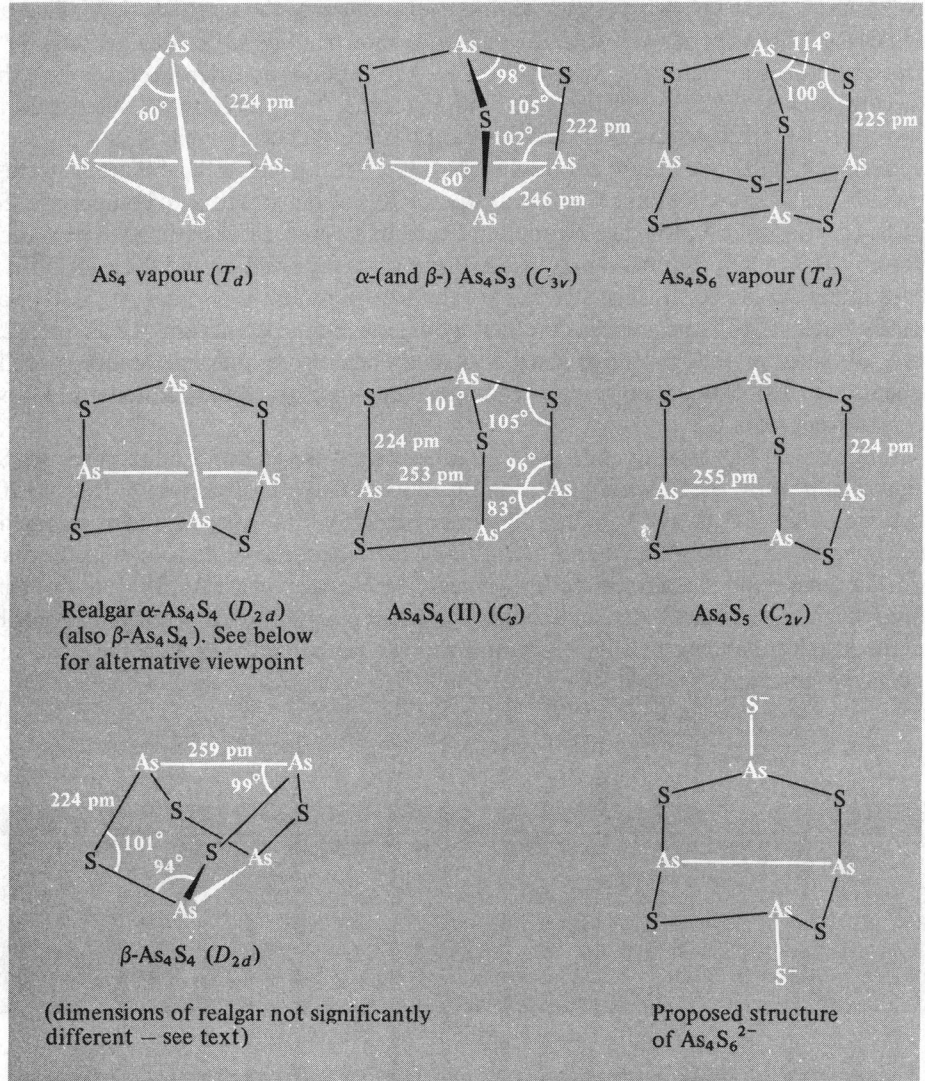

FIG. 13.20　Molecular structure of some sulfides of arsenic, stressing the relationship to the As_4 tetrahedron (point group symmetry in parentheses).

two forms differing only in the arrangement of the molecular units.[33] The compound can be synthesized by heating As and S in the required proportions and purifying the product by sublimation, the β-form being the stable modification at room temperature and the α-form above 130°. The same molecular form occurs in the recently synthesized isoelectronic cationic clusters $As_3S_4^+$ (yellow) and $As_3Se_4^+$ (orange)[33a] and in the isoelectronic clusters P_7^{3-}, As_7^{3-}, and Sb_7^{3-} (p. 686).

[33] H. J. WHITFIELD, The crystal structure of tetra-arsenic trisulfide, *J. Chem. Soc.* (A) 1970, 1800–3; Crystal structure of the β-form of tetra-arsenic trisulfide, ibid. 1973, 1737–8.

[33a] B. H. CHRISTIAN, R. J. GILLESPIE, and J. F. SAWYER, Preparation, X-ray crystal structures, and vibrational spectra of some salts of the $As_3S_4^+$ and $As_3Se_4^+$ cations, *Inorg. Chem.* **20**, 3410–20 (1981).

With As_4S_4 there are two possible geometrical isomers of the molecule depending on whether the 2 As–As bonds are skew or adjacent, as shown in Fig. 13.20. Realgar (mp 307°) adopts the more symmetric D_{2d} forms with skew As–As bonds and, depending on how the molecules pack in the crystal, either α- or β-As_4S_4 results.[34] In addition to the tetrahedral disposition of the 4 As atoms, note that the 4 S atoms are almost coplanar; this is precisely the inverse of the D_{2d} structure adopted by N_4S_4 (p. 856) in which the 4 S atoms form a tetrahedron and the 4 N atoms a coplanar square. It is also instructive to compare As_4S_4 with S_8 (p. 772): each S atom has 2 unpaired electrons available for bonding whereas each As atom has 3; As_4S_4 thus has 4 extra valency electrons for bonding and these form the 2 transannular As–As bonds. The structure of the second molecular isomer As_4S_4(II) was recently elucidated[35] and parallels the analogous geometrical isomerism of P_4S_4 (p. 582). It was obtained as yellow-orange platy crystals by heating equi-atomic amounts of the elements to 500–600°, then rapidly cooling the melt to room temperature and recrystallizing from CS_2.

Orange needle-like crystals of As_4S_5 occasionally form as a minor product when As_4S_4 is made by heating As_4S_3 with a solution of sulfur in CS_2. Its structure[36] (Fig. 13.20) differs from that of P_4S_5 and P_4Se_5 (p. 582) in having only 1 As–As bond and no exocyclic chalcogen As=S; this is a further illustration of the reluctance of As to oxidize beyond As^{III}. The compound can also be made by heterolytic cleavage of the $As_4S_6^{2-}$ anion. This anion, which is itself made by base cleavage of one of the As–As bonds in realgar, probably has the structure shown in Fig. 13.20 and this would certainly explain the observed sequence of reactions:[37]

$$As_4S_4 \xrightarrow[\text{in MeNHCH}_2\text{CH}_2\text{OH}]{\substack{\text{piperidine (or hexa-}\\ \text{methylenetetramine)}}} [pipH^+]_2[As_4S_6]^{2-} \xrightarrow{2\,HX} 2pipHX + H_2S + As_2S_5$$

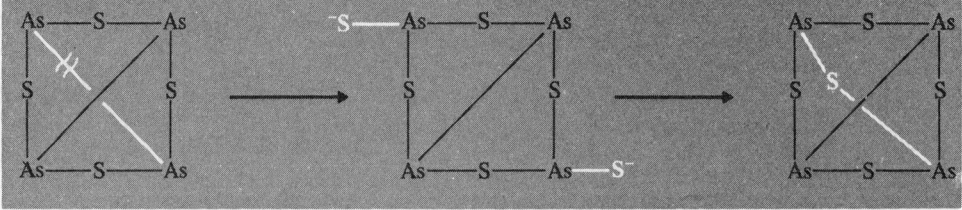

The structure of $As_2^VS_5$ is unknown. It is said to be formed as a yellow solid by passing a rapid stream of H_2S gas into an ice-cold solution of an arsenate in conc HCl; slower passage of H_2S at room temperature results in reduction of arsenate to arsenite and consequent precipitation of As_2S_3. It decomposes in air above 95° to give As_2S_3 and sulfur.

Reactions of the various sulfides of arsenic call for little further comment. As_2S_3 burns when heated in air to give As_2O_3 and SO_2. Chlorine converts it to $AsCl_3$ and S_2Cl_2. It is

[34] E. J. PORTER and G. M. SHELDRICK, The crystal structure of a new crystalline modification of As_4S_4, *JCS Dalton* 1972, 1347–9.

[35] A. KUTOGLU, Preparation and crystal structure of a new isomeric form of As_4S_4, *Z. anorg. Chem.* **419**, 176–84 (1976).

[36] H. J. WHITFIELD, Crystal and molecular structure of tetra-arsenic pentasulfide, *JCS Dalton* 1973, 1740–2.

[37] W. LAUER, M. BECKE-GOEHRING, and K. SOMMER, Studies on realgar. Part II, *Z. anorg. Chem.* **371**, 193–200 (1969).

insoluble in water but dissolved readily in aqueous alkali or alkali–metal sulfide solutions to give thioarsenites:

$$As_2S_3 + Na_2S \xrightarrow{\text{aq}} 2NaAs^{III}S_2$$

Reacidification reprecipitates As_2S_3 quantitatively. With alkali metal or ammonium polysulfides thioarsenates are formed which are virtually insoluble even in hot conc HCl:

$$As_2S_3 \xrightarrow{\text{aq (NH}_4)_2S_n} (NH_4)_3As^VS_4$$

When As_2S_3 is treated with boiling sodium carbonate solution it is converted to As_4S_4; this latter compound can also be made by fusing As_2O_3 with sulfur or (industrially) by heating iron pyrites with arsenical pyrites. As_4S_4 is scarcely attacked by water, inflames in Cl_2, and is used in pyrotechny as it violently enflames when heated with KNO_3. Above about 550° As_4S_4 begins to dissociate reversibly and at 1000° the molecular weight corresponds to As_2S_2 (of unknown structure).

Three selenides of arsenic are known: As_2Se_3, As_4Se_3, and As_4Se_4; each can be made by direct heating of the elements in appropriate proportions at about 500° followed by annealing at temperatures between 220–280°. As_2Se_3 is a stable, brown, semiconducting glass which crystallizes when annealed at 280°; it melts at 380° and is isomorphous with As_2S_3. α-As_4Se_3 forms fine, dark-red crystals isostructural with α-$As_4S_3(C_{3v})$ and the lighter-coloured β-form almost certainly contains the same molecular units.[38] Similarly, As_4Se_4 is isostructural with realgar, α-As_4S_4, and the directly linked As–As distances are very similar in the 2 molecules (257 and 259 pm respectively);[39] other dimensions are As–Se(av) 239 pm, angle Se–As–Se 95°, angle As–Se–As 97°, and angle As–As–Se 102° (cf. Fig. 13.20). The cationic cluster $As_3Se_4^+$ was mentioned on p. 675, and the heterocyclic anion $As_2Se_6^{2-}$ has very recently been isolated as its orange $[Na(crypt)]^+$ salt:[37a] the anion comprises a 6-membered heterocycle $\{As_2Se_4\}$ in the chain conformation and each As carries a further exocyclic Se atom to give overall C_{2h} symmetry, i.e.

$$\overset{Se}{\diagdown}As(\mu\text{-}Se_2)_2As\overset{}{\diagdown}_{Se}.$$

The chalcogenides of Sb and Bi are also readily prepared by direct reaction of the elements at 500–900°. They have rather complex ribbon or layer–lattice structures and have been much studied because of their semiconductor properties. Both *n*-type and *p*-type materials can be obtained by appropriate doping (pp. 290, 383) and for the compounds M_2X_3 the intrinsic band gap decreases in the sequence As > Sb > Bi for a given chalcogen, and in the sequence S > Se > Te for a given Group VB element. Some typical properties of these highly coloured compounds are in Table 13.10, but it should be mentioned that mp, density, and even colour are often dependent on crystalline form and purity. The large thermoelectric effect of the selenides and tellurides of Sb and Bi finds use

[37a] C. H. E. BELIN and M. M. CHARBONNEL, Heteroatomic polyanions of post transition elements. Synthesis and structure of a salt containing the diarsenichexaselenate(2−) anion, $As_2Se_6^{2-}$, *Inorg. Chem.* **21**, 2504-6 (1982).

[38] T. J. BASTOW and H. J. WHITFIELD, Crystal data and nuclear quadrupole resonance spectra of tetra-arsenic triselenide, *JCS Dalton* 1977, 959-61.

[39] T. J. BASTOW and H. J. WHITFIELD, Crystal structure of tetra-arsenic tetraselenide, *JCS Dalton* 1973, 1739-40.

in solid-state refrigerators.[6] Sb_2S_3 occurs as the black, or steely grey mineral stibnite and is made industrially on a moderately large scale for use in the manufacture of safety matches, military ammunition, explosives, and pyrotechnic products, and in the production of ruby-coloured glass. It reacts vigorously when heated with oxidizing agents but is also useful as a pigment in plastics such as polythene or polyvinylchloride because of its flame-retarding properties. Golden and crimson antimony sulfides (which comprise mixtures of Sb_2S_3, Sb_2S_4, and Sb_2OS_3) are likewise used as flame-retarding pigments in plastics and rubbers. A poorly characterized higher sulfide, sometimes said to be Sb_2S_5, can be obtained as a red solid by methods similar to those outlined for As_2S_5 (p. 676). It is used in fireworks, as a pigment, and to vulcanize red rubber.

Of the more complex chalcogenide derivatives of the Group VB elements a single example must suffice to indicate the great structural versatility of these elements, particularly in the $+3$ oxidation state where the nonbonding electron pair can play an important stereochemical role. Thus, the compound of unusual stoichiometry $Ba_4Sb_4^{III}Se_{11}$ was found to contain within 1 unit cell: one *trans*-$[Sb_2Se_4]^{2-}$ (1), two *cis*-$[Sb_2Se_4]^{2-}$ (2), two pyramidal $[SbSe_3]^{3-}$ (3), and two Se_2^{2-} (4) ions together with the requisite 8 Ba^{2+} cations.[39a]

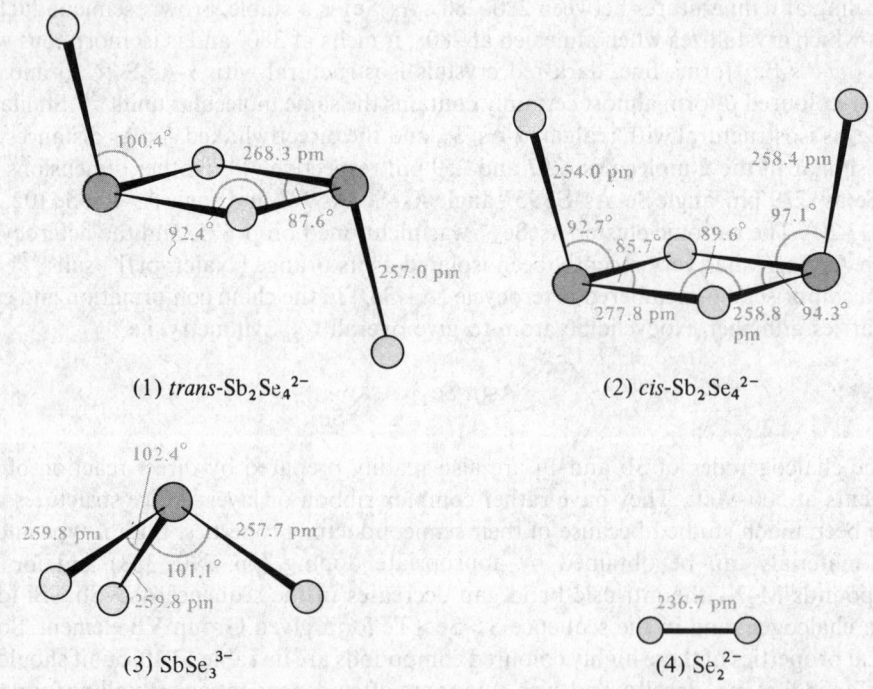

(1) *trans*-$Sb_2Se_4^{2-}$

(2) *cis*-$Sb_2Se_4^{2-}$

(3) $SbSe_3^{3-}$

(4) Se_2^{2-}

13.3.6 *Metal–metal bonds and clusters*

The limited tendency of N and P to catenate into homonuclear chains and rings has already been noted. The ability to form long chains is even less with As, Sb, and Bi, though numerous compounds containing one M–M bond are known. The Group V elements

[39a] G. CORDIER, R. COOK, and H. SCHÄFER, Novel Selenoantimonate(III) anions in $Ba_4Sb_4Se_{11}$, *Angew. Chem.*, Int. Edn. (Engl.) **19**, 324–5 (1980).

TABLE 13.10 Some properties of Group VB chalcogenides M_2X_3

Property	As_2S_3	Sb_2S_3	Bi_2S_3	As_2Se_3	Sb_2Se_3	Bi_2Se_3	As_2Te_3	Sb_2Te_3	Bi_2Te_3
Colour	Yellow	Black	Brown–black	Brown	Grey	Black	—	Grey	Grey
MP/°C	320	546	850	380	612	706	360	620	580
Density/ g cm^{-3}	3.49	4.61	6.78	4.80	5.81	7.50	6.25	6.50	7.74
E_g/eV[a]	2.5	1.7	1.3	2.1	1.3	0.35	~1	0.3	0.15

[a] 1 eV per atom = 96.485 kJ mol^{-1}.

therefore differ from C and the other Group IV elements, on the one hand (p. 436), and S and the Group VI elements, on the other (p. 887). However, As (like P) forms a well-defined set of *triangulo*-M_3 and *tetrahedro*-M_4 compounds, whilst Bi in particular has a propensity to form cluster cations Bi_m^{n+} reminiscent of Sn and Pb clusters (p. 457) and *closo*-borane anions (p. 172). Before discussing these various classes of compound it is convenient to recall that a particular grouping of atoms may well have strong interatomic bonds yet still be unstable because of disproportionation into even more stable groupings. A pertinent example concerns the bond dissociation energies of the diatomic molecules of the Group V elements themselves in the gas phase. Thus, the ground state electronic configuration of the atoms (ns^2np^3) allows the possibility of triple bonding between pairs of atoms M_2 (g), and it is notable that the bond dissociation energy of each of the Group V diatomic molecules is much greater than for those of neighbouring molecules in the same period (Fig. 13.21). Despite this, only N_2 is stable in the condensed phase because of the even greater stability of M_4 or M_{metal} for the heavier congeners (p.643). A notable advance has, however, recently been signalled in the isolation and X-ray structural characterization of Sb homologues of N_2 and azobenzene as complex ligands: the red compounds $[(\mu_3\eta^2\text{-}Sb\equiv Sb)\{W(CO)_5\}_3]$ and $[(\eta^1,\eta^1,(\mu,\eta^2)\text{-}(PhSb\equiv SbPh)\{W(CO)_5\}_3]$ are both stable at room temperature, even on exposure to air.[39b]

Diarsane, As_2H_4, is obtained in small yield as a byproduct of the formation of AsH_3 when an alkaline solution of arsenite is reduced by BH_4^- upon acidification:

$$2H_2AsO_3^- + BH_4^- + 3H^+ \longrightarrow As_2H_4 + B(OH)_3 + H_2O + H_2$$

Diarsane is a thermally unstable liquid with an extrapolated bp $\sim 100°$; it readily decomposes at room temperature to a mixture of AsH_3 and a polymeric hydride of approximate composition $(As_2H)_x$. Sb_2H_4 ($SbCl_3 + NaBH_4$/dil HCl) is even less stable. Both compounds can also be prepared by passing a silent electric discharge through MH_3 gas in an ozonizer at low temperature. Mass spectrometric measurements give the thermochemical bond energy E_{298}° (M–M) as 128 kJ mol^{-1} for Sb_2H_4 and 167 kJ mol^{-1} for As_2H_4, compared with 183 kJ mol^{-1} for P_2H_4. Of the halides, As_2I_4 is known (p. 658) but no corresponding compounds of Sb or Bi have yet been isolated (cf. P_2X_4, p. 571).

Organometallic derivatives M_2R_4 are rather more stable than the hydrides and, indeed, dicacodyl, $Me_2AsAsMe_2$, was one of the very first organometallic compounds to be made

[39b] G. HUTTNER, U. WEBER, B. SIGWARTH, and O. SCHEIDSTEGER, Antimony homologues of dinitrogen and azobenzene as complex ligands: preparation and structure of [Sb≡Sb{W(CO)₅}₃] and [PhSb=SbPh{W(CO)₅}₃], Angew. Chem., Int. Edn. (Engl.) 21, 215–6 (1982).

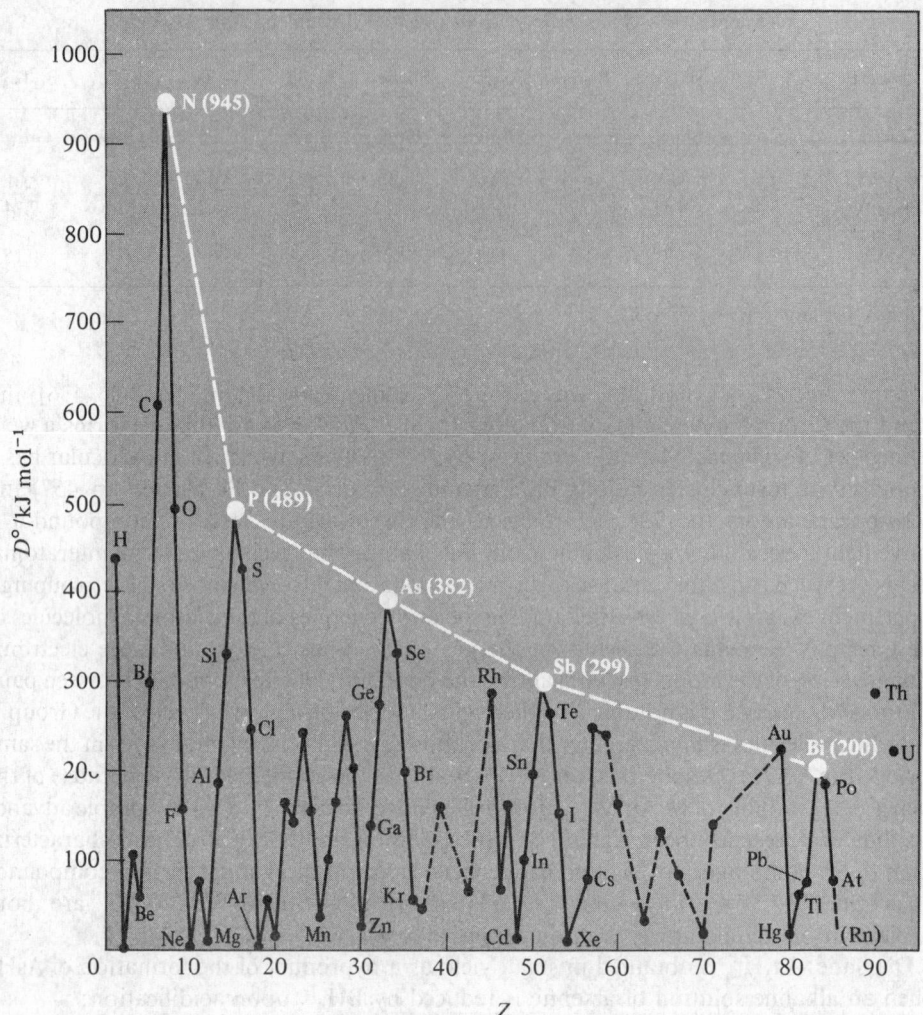

FIG. 13.21   Bond dissociation energies for gaseous, homonuclear diatomic molecules (data from
J. A. Kerr and A. F. Trotman-Dickenson, *Strengths of Chemical Bonds*, pp. F.220–F.229,
Handbook of Chemistry and Physics, 60th edn., 1979–80, CRC Press, Boca Raton, Florida). (Note
that in this compilation the published value for Pb, already in kJ mol⁻¹, has incorrectly been
multiplied by 4.184.)

(L. C. Cadet, 1760; R. Bunsen, 1837): it has mp −1°, bp 78°, is extremely poisonous, and
has a revolting smell, as indicated by its name (Greek κακωδία, *cacodia*, stink). It is now
readily made by the reaction

$$2Me_2AsI + 2Li \xrightarrow{\text{thf}} As_2Me_4 + 2LiI$$

Other preparative routes to As_2R_4 include reaction of R_2AsH with either R_2AsX or
R_2AsNH_2, and the reaction of R_2AsCl with $MAsR_2$ (M = Li, Na, K). In addition to alkyl
derivatives numerous other compounds are known, e.g. As_2Ph_4 mp 127°. $As_2(CF_3)_4$

bp 106° has the *trans* (C_{2h}) structure whereas As_2Me_4 has a temperature-dependent mixture of *trans* and *gauche* isomers (p. 491). Corresponding Sb compounds are of more recent lineage, the first to be made (1931) being the yellow crystalline Sb_2Ph_4 mp 122°. Other derivatives have R = Me, Bu^t, CF_3, cyclohexyl, *p*-tolyl, cyclopentadienyl, etc. Little is known of organodibismuthanes Bi_2R_4 despite sporadic attempts to prepare them.

More extensive catenation occurs in the *cyclo*-polyarsanes $(RAs)_n$ which can readily be prepared from organoarsenic dihalides or from arsonic acids as follows:

$$6PhAsCl_2 \xrightarrow{\text{Na/Et}_2\text{O}} (PhAs)_6 + 12NaCl$$

$$6PhAsI_2 \xrightarrow{\text{Hg (fast)}} 3PhIAsAsIPh \xrightarrow{\text{Hg (slow)}} (PhAs)_6$$

$$nPhAsO(OH)_2 \xrightarrow{\text{H}_3\text{PO}_2} (PhAs)_n \quad n = 5, 6$$

In addition to the 6-membered ring in $(PhAs)_6$, 5-membered rings have been obtained with R = Me, Et, Pr, Ph, CF_3, SiH_3, GeH_3, and 4-membered rings occur with R = CF_3, Ph. A 3-membered As_3 ring has recently been made and is the first *all-cis* organocyclotriarsane to be characterized.[40]

The factors influencing ring size and conformation have not yet become clear. Thus, the yellow $(MeAs)_5$ has a puckered As_5 ring with As–As 243 pm and angle As–As–As 102°; there is also a more stable red form. $(PhAs)_6$ has a puckered As_6 (chair form) with As–As 246 pm and angle As–As–As 91°.

In view of the excellent donor properties of tertiary arsines, it is of interest to inquire whether these *cyclo*-polyarsanes can also act as ligands. In this connection it has recently been shown that $(MeAs)_5$ can displace CO from metal carbonyls to form complexes in which it behaves as a uni-, bi-, or tridentate ligand. For example, direct reaction of $(MeAs)_5$ with $M(CO)_6$ in benzene at 170° (M = Cr, Mo, W) yielded red crystalline compounds $[M(CO)_3(\eta^3\text{-}As_5Me_5)]$ for which the structure in Fig. 13.22a has been proposed,[41] whereas reaction at room temperature with the ethanol derivative $[M(CO)_5(EtOH)]$ gave the yellow dinuclear product $[\{M(CO)_5\}_2\text{-}\mu\text{-}(\eta^1\eta^1\text{-}As_5Me_5)]$ for which a possible structure is given in Fig. 13.22b. Reaction can also lead to ring degradation; e.g. reaction with $Fe(CO)_5$ cleaves the ring to give dark-orange crystals of the *catena*-tetraarsane $[\{Fe(CO)_3\}_2(As_4Me_4)]$ whose structure (Fig. 13.23a) has been

[40] J. ELLERMANN and H. SCHÖSSNER, Synthesis of an all-*cis*-organocyclotriarsane derivative, *Angew. Chem., Int. Edn.* (Engl.) **13**, 601–2 (1974).

[41] P. S. ELMES and B. O. WEST, The coordinating properties of pentamethylcyclopenta-arsine, *Coord. Chem. Rev.* **3**, 279–91 (1968).

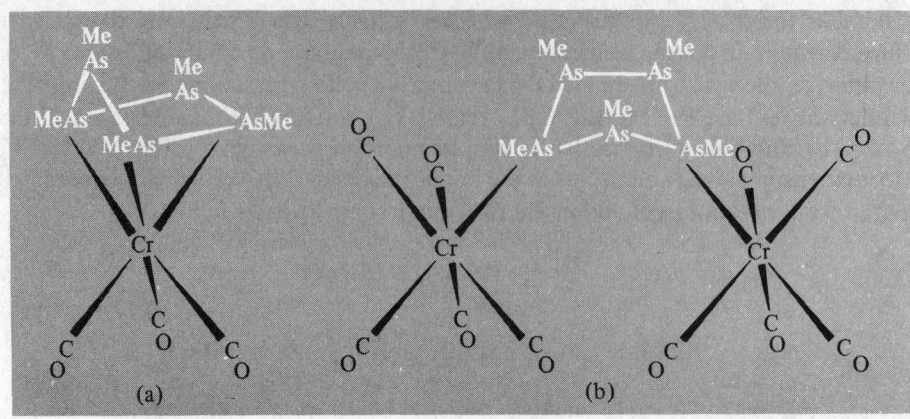

FIG. 13.22 Proposed structures for (a) the tridentate *cyclo*-polyarsine complex [Cr(CO)₃(As₅Me₅)], and (b) the bismonodentate binuclear complex [{Cr(CO)₅}₂(As₅Me₅)].

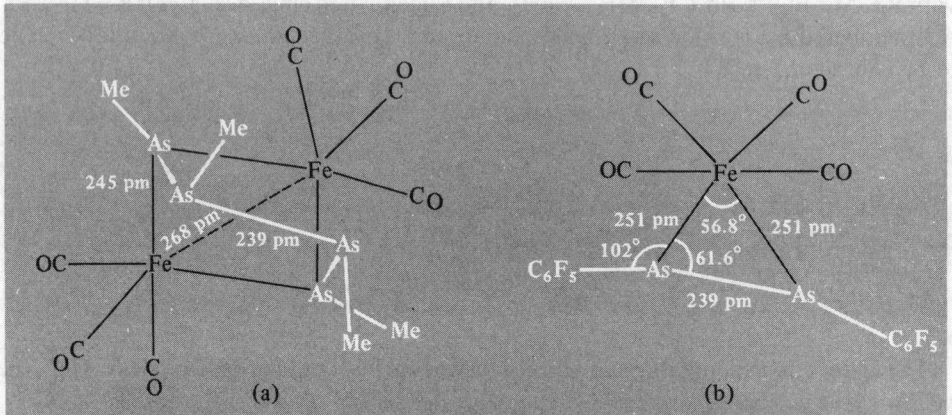

FIG. 13.23 Crystal structures of (a) [{Fe(CO)₃}₂{(AsMe)₄}], and (b) [Fe(CO)₄{(AsC₆F₅)₂}]. In (a) the distance between the 2 terminal As atoms is 189 pm, suggesting some "residual interaction" but no direct σ bond.

established by X-ray crystallography.[42] Even further degradation of the *cyclo*-polyarsane occurs when (C₆F₅As)₄ reacts with Fe(CO)₅ in benzene at 120° to give yellow plates of [Fe(CO)₄{(AsC₆F₅)₂}] mp 150° (Fig. 13.23b).[43] In other reactions homoatomic ring expansion or chain extension can occur. For example (AsMe)₅ when heated with Cr(CO)₆ in benzene at 150° gives crystals of [Cr₂(CO)₆-μ-{η⁶-*cyclo*(AsMe)₉}], whereas (AsPrⁿ)₅ and Mo(CO)₆ under similar conditions yield crystals of [Mo₂(CO)₆-μ-{η⁴-*catena*(AsPrⁿ)₈}]. The molecular structures were determined by X-ray analysis and are shown in Fig. 13.24.[43a] In the first, each Cr is 6-coordinate and the As₉ ring is hexahapto,

[42] B. M. GATEHOUSE, The crystal structure of [{Fe(CO)₃}₂(AsCH₃)₄]; a compound with a noncyclic tetramethyltetra-arsine ligand, *JCS Chem. Comm.* 1969, 948–9.

[43] P. S. ELMES, P. LEVERET, and B. O. WEST, X-ray determination of the structure of [Fe(CO)₄(AsC₆F₅)₂], a complex containing "decafluoroarsenobenzene", *JCS Chem. Comm.* 1971, 747–8.

[43a] P. S. ELMES, B. M. GATEHOUSE, D. J. LLOYD, and B. O. WEST, X-ray structures of [Cr₂(CO)₆(AsMe)₆] and [Mo₂(CO)₆(AsPrⁿ)₈]: the stabilization of an As₉ ring and an As₈ chain by coordination, *JCS Chem. Comm.* 1974, 953–4.

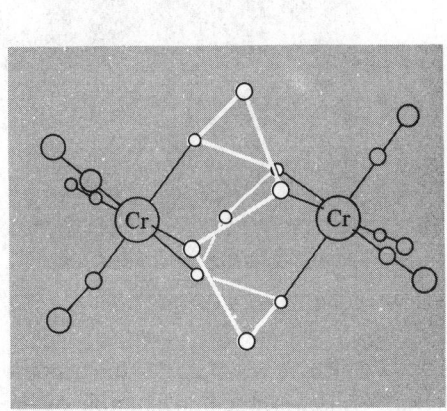
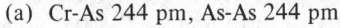
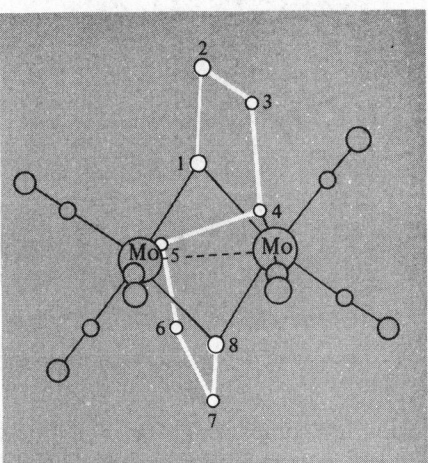

(a) Cr-As 244 pm, As-As 244 pm   (b) Mo-As(1) 255 pm, Mo-As(4) 262 pm
                                              Mo-Mo 310 pm
                                              As-As 243 pm

FIG. 13.24 Structures of (a) $[Cr_2(CO)_6-\mu\{\eta^6\text{-}cyclo(AsMe)_9\}]$, and (b) $[Mo_2(CO)_6-\mu\{\eta^4\text{-}catena(AsPr^n)_8\}]$. In both structures the alkyl group attached to each As atom has been omitted for clarity.

donating 3 pairs of electrons to each Cr atom. In the second the As atom at each end of the As_8 chain bridges the 2 Mo atoms whereas the 2 central As atoms each bond to 1 Mo atom only and there is an Mo–Mo bond.

Some of the compounds mentioned in the preceding paragraph can be thought of as heteronuclear cluster compounds and it is convenient to consider here other such heteronuclear cluster species before discussing compounds in which there are homonuclear clusters of Group VB atoms. Compounds structurally related to the As_4 cluster include the complete series $[As_{4-n}\{Co(CO)_3\}_n]$ $n = 0, 1, 2, 3, 4$. It will be noted that the atom As and the group $\{Co(CO)_3\}$ are "isoelectronic" in the sense that each requires 3 additional electrons to achieve a stable 8- or 18-electron configuration respectively. Yellow crystals of $[As_3Co(CO)_3]$ are obtained by heating $(MeAs)_5$ with $Co_2(CO)_8$ in hexane at 200° under a high pressure of CO.[44] The red air-sensitive liquid $[As_2\{Co(CO)_3\}_2]$ mp $-10°$ is obtained by the milder reaction of $AsCl_3$ with $Co_2(CO)_8$ in thf.[45] Substitution of some carbonyls by tertiary phosphines is also possible under ultraviolet irradiation. Typical structural details are in Fig. 13.25. In the first compound the η^3-triangulo-As_3 group can be thought of as a 3-electron donor to the cobalt atom; in the second, the very short As–As bond suggests multiple bonding and the structure closely resembles that of the "isoelectronic" acetylene complex $[\{Co(CO)_3\}_2PhC\equiv CPh]$ (p. 691). Phosphorus analogues are also known, e.g. the recently synthesized sand-coloured

[44] A. S. FOUST, M. F. FOSTER, and L. F. DAHL, Synthesis, structure, and bonding of new organometallic arsenic-metal atom clusters containing a metal-bridged multiply bonded As_2 ligand: $[(Co_2(CO)_6(As_2)]$ and $[Co_2(CO)_5(PPh_3)As_2)]$, J. Am. Chem. Soc. **91**, 5633–5 (1969); Preparation and structural characterization of the triarsenic-cobalt atom cluster system $[As_3Co(CO)_3]$. The first known X_3-transition metal analogue of Group V tetrahedral X_4 molecules, ibid. 5631–3.

[45] A. S. FOUST, C. F. CAMPANA, J. D. SINCLAIR, and L. F. DAHL, Preparation and structural characterization of the mono- and bis-(triphenylphosphine)-substituted derivatives of $[Co_2(CO)_6(\mu\text{-}As_2)]$, Inorg. Chem. **18**, 3047–54 (1979).

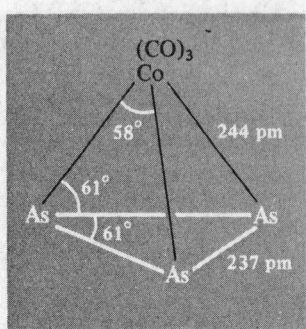
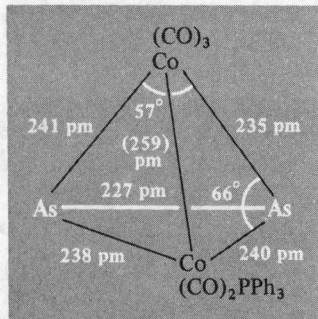

Fig. 13.25   Structures of $[As_3Co(CO)_3]$ and $[As_2\{Co(CO)_3\}\{Co(CO)_2(PPh_3)\}]$.

or colourless complexes $[M(\eta^3\text{-}P_3)L^*]$, where $M = Co$, Rh, or Ir, and L^* is the tripod-like tris(tertiary phosphine) ligand $MeC(CH_2PPh_2)_3$.[46] Likewise the first example of an η^2-P_2 ligand symmetrically bonded to 2 metal atoms to give a tetrahedral $\{P_2Co_2\}$ cluster was recently established by the X-ray structure determination of $[(\mu\text{-}P_2)$ $\{Co(CO)_3\}\{Co(CO)_2(PPh_3)\}]$.[47] If the $\mu\text{-}P_2$ (or $\mu\text{-}As_2$) ligand is replaced by $\mu\text{-}S_2$ (or $\mu\text{-}Se_2$), then isoelectronic and isostructural clusters can be obtained by replacing Co by Fe, as in $[(\mu\text{-}S_2)\{Fe(CO)_3\}_2]$ and $[(\mu\text{-}Se_2)\{Fe(CO)_3\}_2]$ (p. 894).

Even more intriguing are the "double sandwich" complexes which feature $\{\eta^3\text{-}P_3\}$ and $\{\eta^3\text{-}As_3\}$ as symmetrically bridging 3-electron donors. Thus As_4 reacts smoothly with Co^{II} or Ni^{II} aquo ions and the triphosphane ligand $L^* = MeC(CH_2PPh_2)_3$ in thf/ethanol/acetone mixtures to give the exceptionally air-stable dark-green paramagnetic cation $[L^*Co\text{-}\mu\text{-}(\eta^3\text{-}As_3)CoL^*]^{2+}$ with the dimensions shown in Fig. 13.26.[48] The structure of the related P_3 complex $[L^*\text{-}\mu\text{-}(\eta^3\text{-}P_3)NiL^*]^{2+}$ (prepared in the same way using white P_4) is closely similar[49] with P–P distances of 216 pm (smaller than for P_4 itself, 221 pm). Indeed, a whole series of complexes has now been established with the same structure-

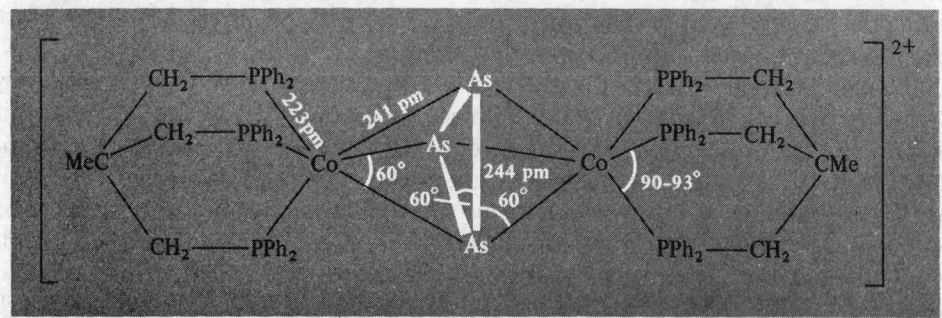

Fig. 13.26   Structure of the cation $[L^*Co\text{-}\mu\text{-}(\eta^3\text{-}As_3)CoL^*]^{2+}$.

[46] C. BIANCHINI, C. MEALLI, A. MELI, and L. SACCONI, Synthesis and crystal structure of two isomorphous Rh and Ir complexes with *cyclo*-triphosphorus and 1,1,1-tris(diphenylphosphinomethyl)ethane, *Inorg. Chim. Acta* **37**, L543–L544 (1979).

[47] C. F. CAMPANA, A. VIZI-OROSZ, G. PALYI, L. MARKÓ, and L. F. DAHL, Structural characterization of $[Co_2(CO)_5(PPh_3)(\mu\text{-}P_2)]$: a tricyclic complex containing a bridging P_2 ligand, *Inorg. Chem.* **18**, 3054–9 (1979).

[48] M. DI VAIRA, S. MIDOLLINI, L. SACCONI, and F. ZANOBINI, *cyclo*-Triarsenic as a μ,η-ligand in transition-metal complexes, *Angew. Chem.*, Int. Edn. (Engl.) **17**, 676–7 (1978).

[49] M. DI VAIRA, S. MIDOLLINI, and L. SACCONI, *cyclo*-Triphosphorus and *cyclo*-triarsenic as ligands in "double-sandwich" complexes of cobalt and nickel, *J. Am. Chem. Soc.* **101**, 1757–63 (1979).

TABLE 13.11 *Electronic configurations of the isostructural series of complexes containing bridging η^3-P_3 and η^3-As_3 ligands { L^* is the tridentate tertiary phosphine $MeC(CH_2PPh_2)_3$}*

(η^3-P_3) complex	Colour	Valence electrons	Unpaired electrons (e)	Electrons in highest orbital	Colour	(η^3-As_3) complex
$[L^*_2Co_2(P_3)]^{3+}$	Bright green	30	0	0		$[L^*_2Co_2(As_3)]^{3+}$
$[L^*_2Co_2(P_3)]^{2+}$		31	1	1	Dark green	$[L^*_2Co_2(As_3)]^{2+}$
$[L^*_2Co_2(P_3)]^{+}$		32	2	2		$[L^*_2Co_2(As_3)]^{+}$
$[L^*_2CoNi(P_3)]^{2+}$	Red-brown	32	2	2		—
$[L^*_2Ni_2(P_3)]^{2+}$		33	1	3		$[L^*_2Ni_2(As_3)]^{2+}$
$[L^*_2Ni_2(P_3)]^{+}$	Dark	34	0	4		$[L^*_2Ni_2(As_3)]^{+}$

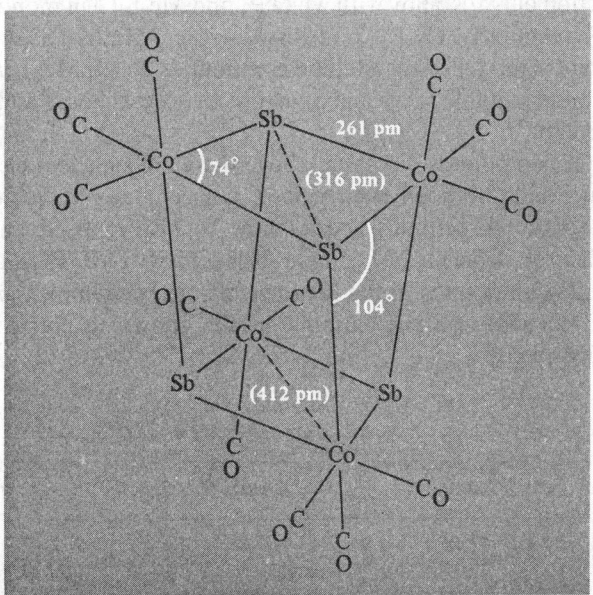

FIG. 13.27   Structure of the cubane-like mixed metal–metal cluster complex $[Sb_4\{Co(CO)_3\}_4]$.

motif and differing only in the number of valency electrons in the cluster; some of these are summarized in Table 13.11.[49, 50] The number of valence electrons in all these complexes falls in the range 30–34 as predicted by R. Hoffmann and his colleagues.[51]

With Sb even larger clusters can be obtained. For example reaction of $Co(OAc)_2.4H_2O$ and $SbCl_3$ in pentane at 150° under a pressure of H_2/CO gave black crystals of $[Sb_4\{Co(CO)_3\}_4]$ which was found to have a cubane like structure with Sb and Co at alternate vertices of a grossly distorted cube (Fig. 13.27).[52]

[50] F. FABBRIZZI and L. SACCONI, Electrochemical behaviour of the triple-decker sandwich cobalt and nickel complexes with *cyclo*-triphosphorus and *cyclo*-triarsenic, *Inorg. Chim. Acta*, **36**, L407–L408 (1979).

[51] J. W. LAUHER, M. ELIAN, R. H. SUMMERVILLE, and R. HOFFMANN, Triple-decker sandwiches, *J. Am. Chem. Soc.* **98**, 3219–24 (1976).

[52] A. S. FOUST and L. F. DAHL, Preparation, structure, and bonding of the tetrameric antimony-cobalt cluster system $Co_4(CO)_{12}Sb_4$: the first known (main group element)–(metal carbonyl) cubane-type structure, *J. Am. Chem. Soc.* **92**, 7337–41 (1970).

In addition to the heteronuclear clusters considered in the preceding paragraphs, As, Sb, and Bi also form homonuclear clusters. We have already seen that alkaline earth phosphides $M_3^{II}P_{14}$ contain the $[P_7]^{3-}$ cluster isoelectronic and isostructural with P_4S_3, and the analogous clusters $[As_7]^{3-}$ and $[Sb_7]^{3-}$ have also recently been synthesized. Thus, when As was heated with metallic Ba at 800°C, black, lustrous prisms of Ba_3As_{14} were obtained, isotypic with Ba_3P_{14}; these contained the $[As_7]^{3-}$ anion with dimensions as shown in Fig. 13.28.[53] Again, when powdered NaSb or NaSb$_3$ were treated with crypt, $[N(C_2H_4OC_2H_4OC_2H_4)_3N]$ (p. 109) in dry ethylenediamine, a deep-brown solution was obtained from which brown needles of $[Na(crypt)^+]_3[Sb_7]^{3-}$ were isolated with a C_{3v} anion like $[As_7]^{3-}$ and Sb–Sb distances 286 pm (base), 270 pm (side), and 278 pm (cap).[54] Isostructural, neutral molecular clusters can be obtained by replacing the 3 S or 3 Se atoms in P_4S_3 or As_4Se_3 by PR or AsR rather than by P$^-$ or As$^-$. For example reaction of Na/K alloy with white P_4 and Me_3SiCl in monoglyme gave P_7R_3, $P_{14}R_4$, and $P_{13}R_5$. Similarly, Cs_3P_{11} and Rb_3As_7 react with Me_3SiCl in toluene to give good yields of the bright-yellow crystalline compounds $P_{11}(SiMe_3)_3$ and $As_7(SiMe_3)_3$. This latter compound is stable to air and moisture for several hours and has the structure shown in Fig. 13.27b.[55]

In all the cluster compounds discussed above there are sufficient electrons to form 2-centre 2-electron bonds between each pair of adjacent atoms. Such is not the case, however, for the cationic bismuth species now to be discussed and these must be considered as "electron deficient". The unparalleled ability of Bi/BCl$_3$ to form numerous low oxidation-state compounds in the presence of suitable complex anions has already been mentioned (p. 658) and the cationic species shown in Table 13.12 have been unequivocally identified.

TABLE 13.12   *Cationic bismuth clusters*

Cation	Formal oxidation state	Cluster structure	Point group symmetry
Bi$^+$	1.00	—	—
Bi$_3^+$	0.33	Triangle	D_{3h}
Bi$_5^{3+}$	0.60	Trigonal bipyramid	D_{3h}
Bi$_8^{2+}$	0.25	Square antiprism	D_{4h}
Bi$_9^{5+}$	0.56	Tricapped trigonal prism	C_{3h} ($\sim D_{3h}$)

The structure of the last 3 cluster cations are shown in Fig. 13.29. In discussing the structure and bonding of these clusters it will be noted that Bi$^+$ ($6s^26p^2$) can contribute 2 p electrons to the framework bonding just as {BH} contributes 2 electrons to the cluster bonding in boranes (p. 181). Hence, using the theory developed for the boranes, it can be seen that $[B_nH_n]^{2-}$ is electronically equivalent to $(Bi^+)_n^{2-}$ i.e. $[Bi_n]^{n-2}$. This would account for the stoichiometries Bi$_3^+$ and Bi$_5^{3+}$ but would also lead one to expect Bi$_8^{6+}$

[53] W. SCHMETTOW and H. G. VON SCHNERING, Ba_3As_{14}, the first compound with the cluster anion As_7^{3-}, *Angew. Chem.*, Int. Edn. (Engl.) **16**, 857 (1977).

[54] J. D. CORBETT, D. G. ADOLPHSON, D. J. MERRIMAN, P. A. EDWARDS, and F. J. ARMATIS, Synthesis of stable homopolyatomic anions of Sb, Bi, Sn, and Pb. The crystal structure of a salt containing the heptaantimonide-(3−) anion, *J. Am. Chem. Soc.* **97**, 6267-8 (1975).

[55] H. G. VON SCHNERING, D. FENSKE, W. HÖNLE, M. BINNEWIES, and K. PETERS, Novel polycyclic phosphanes and arsanes: $P_{11}(SiMe_3)_3$ and $As_7(SiMe_3)_3$, *Angew. Chem.*, Int. Edn. (Engl.) **18**, 679 (1979).

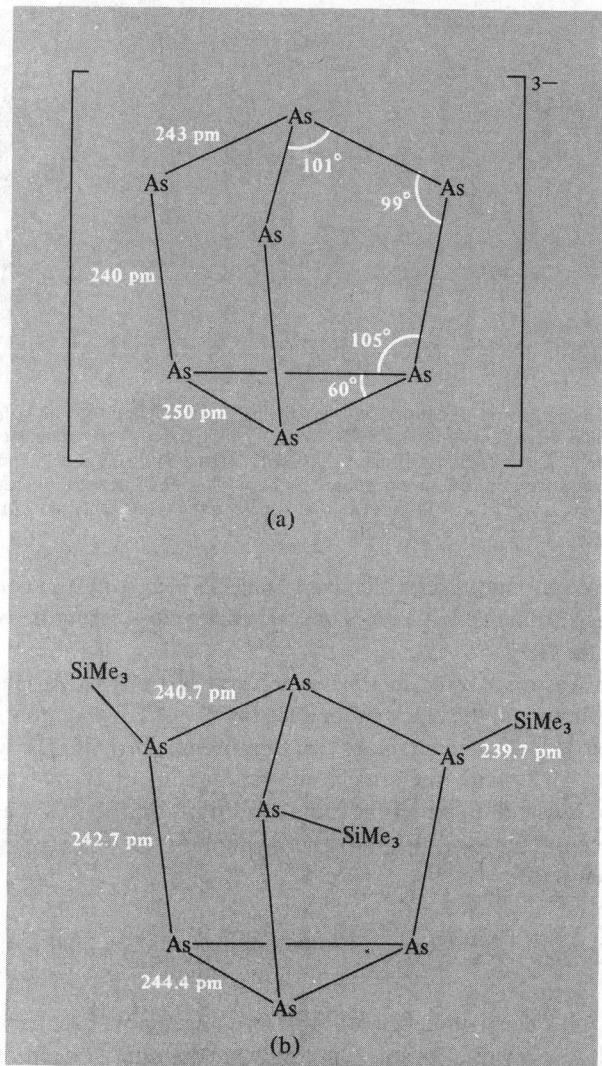

FIG. 13.28   (a) Structure of the anion As_7^{3-}, isoelectronic with As_4Se_3 (p. 677). The sequence of As–As distances (base > cap > side) is typical for such cluster anions but this alters to the sequence base > side > cap for neutral species such as $As_7(SiMe_3)_3$ shown in (b).

and Bi_9^{7+} for the larger clusters. However, these charges are very large and it seems likely that the lowest-lying nonbonding orbital would also be occupied in $(Bi^+)_n^{2-}$. For $(Bi^+)_8^{2-}$ this is an e_1 orbital which can accommodate 4 electrons, thereby reducing the charge from Bi_8^{6+} to Bi_8^{2+} as observed. In $(Bi^+)_9^{2-}$ the lowest nonbonding orbital is a_2'' which can accommodate 2 electrons, thus reducing the charge from Bi_9^{7+} to Bi_9^{5+} as observed.[56] It will also be noted that Bi_5^{3+} is isoelectronic with Sn_5^{2-} and Pb_5^{2-}

56 J. D. CORBETT, Homopolyatomic ions of the post-transition elements—Synthesis, structure, and bonding, *Prog. Inorg. Chem.* **21**, 129–58 (1976).

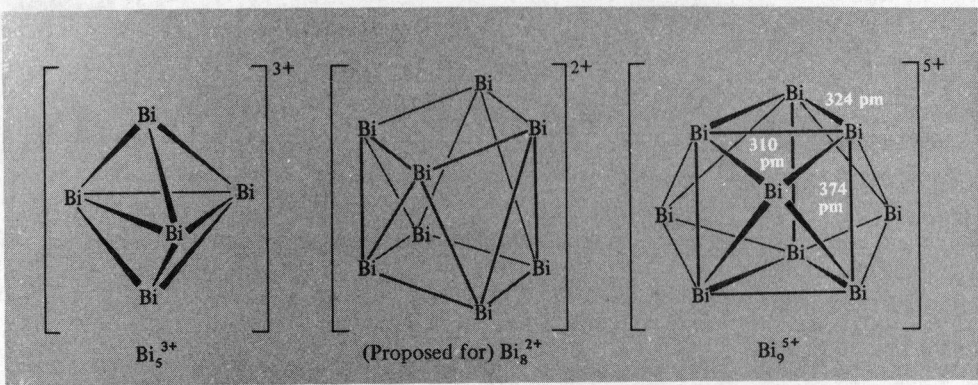

FIG. 13.29 The structures of cationic clusters of Bi_m^{n+}. The dimensions cited for Bi_9^{5+} were obtained from an X-ray study on $[(Bi_9^{5+})(Bi^+)(HfCl_6^{2-})_3]$; the corresponding average distances for Bi_9^{5+} in $BiCl_{1.167}$ i.e. $[(Bi_9^{5+})_2(BiCl_5^{2-})_4(Bi_2Cl_8^{2-})]$ are 310, 320, and 380 pm respectively. [The square antiprismatic structure proposed for Bi_8^{2+} has now been established by an X-ray study of $Bi_8[AlCl_4]_2$;[56a] The Bi–Bi distances are 309 pm (square ends) and 311 pm (sides).]

(p. 457); these penta-atomic species all have 12 valence electrons (not counting the "inert" s^2 electrons on each atom), i.e. $n + 1$ pairs ($n = 5$) hence a *closo*-structure would be expected by Wade's rules (p. 185).

The Bi_9^{5+} ion was discovered in 1963 as a result of work by A. Herschaft and J. D. Corbett on the structure of the black subhalide "BiCl" (p. 658) and more recently was also found in $Bi_{10}HfCl_{18}$.[16] The diamagnetic compound $Bi_5(AlCl_4)_3$ was prepared by reaction of $BiCl_3/AlCl_3$ with the stoichiometric amount of Bi in fused $NaAlCl_4$ (mp 151°).[57] With an excess of Bi under the same conditions $Bi_8(AlCl_4)_2$ was obtained. More recently it has been found that AsF_5 and other pentafluorides oxidize Bi in liquid SO_2 first to Bi_8^{2+} and then to Bi_5^{3+}.[58]

$$10Bi + 9AsF_5 \xrightarrow{SO_2} 2Bi_5(AsF_6)_3.2SO_2 + 3AsF_3$$
$$\text{(bright yellow)}$$

Very recently an "electron-deficient" polyarsenide anion has been synthesized by reaction of KAs_2 with crypt in ethylenediamine: deep-red crystals of $[K(crypt)^+]_3[As_{11}]^{3-}$ were obtained and X-ray diffractometry established the detailed structure of the As_{11}^{3-} anion as shown in Fig. 13.30a.[58a] This is very similar to the structure of the P_{11}^{3-} anion in Na_3P_{11}, and can be thought of as being derived from the 17-vertex polyhedron in Fig. 13.30b by removal of the 6 vertices shown as open circles; the resulting As_{11}^{3-} ion has approximate D_3 symmetry with eight 3-coordinate As atoms forming a bicapped twisted triangular prism with a "waist" of three 2-coordinate bridging

[56a] B. KREBS, M. HUCKE, and C. J. BRENDEL, Structure of the octabismuth(2+) cluster in crystalline $Bi_8(AlCl_4)_2$, *Angew. Chem.*, Int. Edn. (Engl.) **21**, 445–6 (1982).

[57] J. D. CORBETT, Homopolyatomic ions of the heavy post-transition elements. The preparation, properties and bonding of $Bi_5(AlCl_4)_3$ and $Bi_4(AlCl_4)$, *Inorg. Chem.* **7**, 198–208 (1968).

[58] R. C. BURNS, R. J. GILLESPIE, and WOON-CHUNG LUK, Preparation, spectroscopic properties, and structure of the pentabismuth(3+) cation Bi_5^{3+}, *Inorg. Chem.* **17**, 3596–604 (1978).

[58a] C. H. E. BELIN, Polyarsenide anions. Synthesis and structure of a salt containing the undecaarsenide(3−) ion, *J. Am. Chem. Soc.* **102**, 6036–40 (1980).

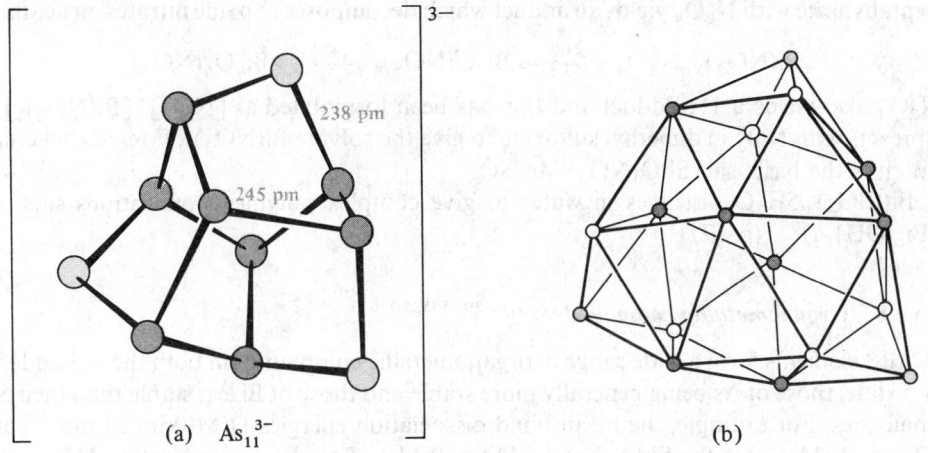

FIG. 13.30   (a) Structure of the anion $As_{11}{}^{3-}$; note that the As–As distances involving the 3 2-coordinate As atoms are significantly shorter than those between pairs of 3-coordinate As atoms. (b) A 17-vertex polyhedron having the same (D_3) symmetry as the anion $As_{11}{}^{3-}$ (see text).

atoms (Fig. 13.30a). In terms of Wade's rules (p. 185) the cluster can be rationalized by assuming that each As has 1 lone-pair of electrons and thus contributes 3 electrons to the cluster which thus has 36 bonding electrons $[(11 \times 3) + 3]$; these 18 bonding pairs imply a 17 vertex *closo*-polyhedron.

13.3.7 *Other inorganic compounds*

The ability to form stable oxoacid salts such as sulfates, nitrates, perchlorates, etc., increases in the order $As \ll Sb < Bi$. As^{III} is insufficiently basic to enable oxoacid salts to be isolated though species such as $[As(OH)(HSO_4)_2]$ and $[As(OH)(HSO_4)]^+$ have been postulated in anhydrous H_2SO_4 solutions of As_2O_3. In oleum, species such as $[As(HSO_4)_3]$, $[\{(HSO_4)_2As\}_2O]$, and $[\{(HSO_4)_2As\}_2SO_4]$ may be present. By contrast, $Sb_2(SO_4)_3$ can be isolated, as can the hydrates $Bi_2(SO_4)_3 \cdot nH_2O$ and the double sulfate $KBi(SO_4)_2$, though all are readily hydrolysed to basic salts.

The pentahydrate $Bi(NO_3)_3 \cdot 5H_2O$ can be crystallized from solutions of Bi^{III} oxide or carbonate in conc HNO_3. Dilution causes the basic salt $BiO(NO_3)$ to precipitate. Attempts at thermal dehydration yield complex oxocations by reactions which have been formulated as follows:

$$Bi(NO_3)_3 \cdot 5H_2O \xrightarrow{50-60°} [Bi_6O_6]_2(NO_3)_{11}(OH) \cdot 6H_2O \xrightarrow{77-130°}$$

$$[Bi_6O_6](NO_3)_6 \cdot 3H_2O$$

$$\alpha\text{-}Bi_2O_3 \xleftarrow{400-450°}$$

The $[Bi_6O_6]^{6+}$ ion is the dehydrated form of $[Bi_6(OH)_{12}]^{6+}$ (p. 671). Treatment of the pentahydrate with N_2O_4 yields an adduct which decomposes to oxide nitrates on heating:

$$Bi(NO_3)_3 . N_2O_4 \xrightarrow{200^\circ} Bi_2O(NO_3)_4 \xrightarrow{415^\circ} Bi_4O_5(NO_3)_2$$

N_2O_5 also yields a 1:1 adduct and this has been formulated as $[NO_2]^+[Bi(NO_3)_4]^-$. Bi reacts with NO_2 in dimethyl sulfoxide to give the solvate $Bi(NO_3)_3.3Me_2SO$, whereas Sb gives the basic salt $SbO(NO_3).Me_2SO$.

$Bi(ClO_4)_3.5H_2O$ dissolves in water to give complex polymeric oxocations such as $[Bi_6(OH)_{12}]^{6+}$ (p. 671).

13.3.8 *Organometallic compounds*[2, 5, 6, 59, 59a, 60, 61]

All 3 elements form a wide range of organometallic compounds in both the +3 and the +5 state, those of As being generally more stable and those of Bi less stable than their Sb analogues. For example, the mean bond dissociation energies $\bar{D}(M-Me)/kJ\ mol^{-1}$ are 238 for $AsMe_3$, 224 for $SbMe_3$ and 140 for $BiMe_3$. For the corresponding MPh_3, the values are 280, 267, and 200 kJ mol^{-1} respectively, showing again that the M–C bond becomes progressively weaker in the sequence As > Sb > Bi. Comparison with organophosphorus compounds (p. 633) is also apposite. In most of the compounds the metals are 3, 4, 5, or 6 coordinate though a few multiply-bonded compounds are known in which As has a coordination number of 2.[59a]

Organoarsenic(III) compounds

Some examples of 2-coordinate organoarsenic(III) compounds are:

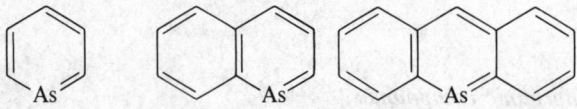

Arsabenzene   1-Arsanaphthalene   9-Arsa-anthracene

The first such compound to be prepared was the deep-yellow unstable compound 9-arsa-anthracene[62] but the thermally stable colourless arsabenzene (arsenin) can now conveniently be made by a general route from 1,4-pentadiyne:[63]

[59] G. E. COATES and K. WADE, Antimony and bismuth, Chap. 5 in *Organometallic Compounds*, Vol. 1, *The Main Group Elements*, 3rd edn., pp. 510–44, Methuen, London, 1967.

[59a] B. J. AYLETT, Arsenic, antimony, and bismuth in *Organometallic Compounds*, 4th edn., Vol. 1, *The Main Group Elements*, Part 2, Groups IV and V, pp. 387–521, Chapman & Hall, London, 1979.

[60] G. E. COATES, M. L. H. GREEN, P. POWELL, and K. WADE, Group VB elements, in *Principles of Organometallic Chemistry*, pp. 143–9, Methuen, London, 1968.

[61] F. G. MANN, *The Heterocyclic Derivatives of P, As, Sb, and Bi*, 2nd edn., Wiley, New York, 1970, 716 pp.

[62] P. JUZI and K. DEUCHERT, 9-Arsa-anthracene, *Angew. Chem.*, Int. Edn. (Engl.) **8**, 991 (1969). H. VERMEER and F. BICKELHAUPT, Dibenze[b,e]arsenin (9-arsa-anthracene), ibid. 992.

[63] A. J. ASHE, Phosphabenzene and arsabenzene, *J. Am. Chem. Soc.* **93**, 3293–5 (1971).

AsC_5H_5 is somewhat air sensitive but is distillable and stable to hydrolysis by mild acid or base. Using the same route, PBr_3 gave PC_5H_5 as a colourless volatile liquid (p. 635), $SbCl_3$ gave SbC_5H_5 as an isolable though rather labile substance which rapidly polymerized at room temperature, and $BiCl_3$ gave the even less-stable BiC_5H_5 which could only be detected spectroscopically by chemical trapping.[63, 64] Arsanaphthalene is an air-sensitive yellow oil.[65] No 1-coordinate organoarsenic compound $RC{\equiv}As$ (analogous to RCN and RCP, p. 633) has yet been isolated, though the compounds $[(MeCAs)Co_2(CO)_6]$ and $[(PhCAs)Co_2(CO)_6]$ both have tetrahedral cluster structures like those in Fig. 13.24 and can be regarded as acetylene-type complexes analogous to $[(RC{\equiv}CR)Co_2(CO)_6]$.[66] (See also p. 646 for 1-coordinate As in the "isolated" tetrahedral anions $SiAs_4^{8-}$ and $GeAs_4^{8-}$.)

Most organoarsenic(III) compounds are readily prepared by standard methods (p. 570) such as the treatment of $AsCl_3$ with Grignard reagents, organolithium reagents, organoaluminium compounds, or by sodium–alkyl halide (Wurtz) reactions. As_2O_3 can also be used as starting material as indicated in the scheme. AsR_3 and $AsAr_3$ are widely used as ligands in coordination chemistry.[5] Common examples are the 4 compounds $AsMe_{3-n}Ph_n$ ($n = 0, 1, 2, 3$). Multidentate ligands have also been extensively studied particularly the chelating ligand "*o*-phenylenebis(dimethylarsine)" i.e. 1,2-bis(dimethyl-arseno)benzene which can be prepared from cacodylic acid (dimethylarsinic acid) $Me_2AsO(OH)$ (itself prepared as indicated in the general scheme):

$$Me_2AsO(OH) \xrightarrow{Zn/HCl} Me_2AsH \xrightarrow{Na/thf} NaAsMe_2 \xrightarrow[thf]{1,2\text{-}Cl_2C_6H_4}$$

Arsine complexes are especially stable for b-class metals such as Rh, Pd, and Pt, and such complexes have found considerable industrial use in hydrogenation or hydroformylation of alkenes, oligomerization of isoprene, carbonylation of α-olefins, etc.

Haloarsines R_2AsX and dihaloarsines $RAsX_2$ are best prepared by reducing the corresponding arsinic acids $R_2AsO(OH)$ or arsonic acid $RAsO(OH)_2$ with SO_2 in the presence of HCl or HBr and a trace of KI. The actual reducing agent is I^- and the resulting I_2 is in turn reduced by the SO_2. Fluoro compounds are best prepared by metathesis of the chloro derivative with a metal fluoride, e.g. AgF.

Hydrolysis of R_2AsX yields arsinous acids R_2AsOH or their anhydrides $(R_2As)_2O$. An alternative route employs a Grignard reagent and As_2O_3, e.g. PhMgBr affords $(Ph_2As)_2O$. Hydrolysis of $RAsX_2$ yields either arsonous acids $RAs(OH)_2$ or their anhydrides $(RAsO)_n$. These latter are not arsenoso compounds $RAs{=}O$ analogous to nitroso compounds (p. 478) but are polymeric. Indeed, all these As^{III} compounds feature pyramidal 3-coordinate As as do the formally As^I compounds $(RAs)_n$ discussed on p. 681. Very recently a series of *planar* 3-coordinate arsenic(I) compounds have been prepared and these are discussed on p. 695.

[64] A. J. ASHE, The Group V heterobenzenes, *Acc. Chem. Res.* **11**, 153–7 (1978).

[65] A. J. ASHE, D. L. BELLVILLE, and H. S. FRIEDMAN, Synthesizing 1-arsanaphthalene, *JCS Chem. Comm.* 1979, 880–1.

[66] D. SEYFERTH and J. S. MEROLA, Arsa-acetylenes, $RC{\equiv}As$, as ligands in dicobalt hexacarbonyl complexes: novel main group element–transition metal hybrid cluster compounds, *J. Am. Chem. Soc.* **100**, 6783–4 (1978).

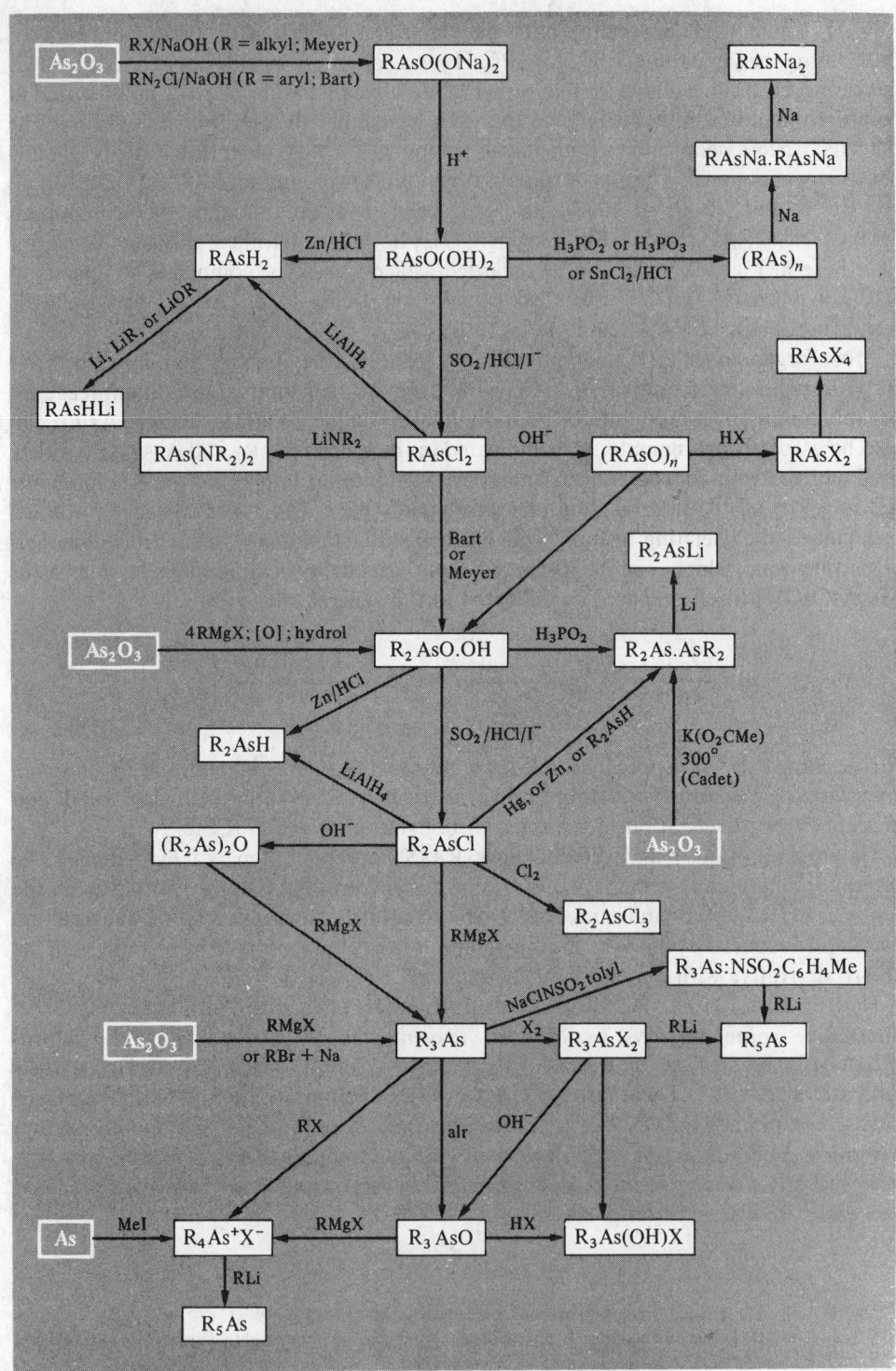

Some routes to organoarsenic compounds[60]

Organoarsenic(V) compounds

Among the compounds of As^V can be noted the complete series $R_{5-n}AsX_n$ ($n=0-5$) where R can be alkyl or aryl. Thus $AsPh_5$ (mp 150°) can be prepared by direct reaction of LiPh on either $[AsPh_4]I$, Ph_3AsCl_2, or $Ph_3As{=}O$. Similarly, $AsMe_5$ has recently been prepared as a colourless, volatile, mobile liquid (mp $-6°$):[67]

$$Me_3AsCl_2 \xrightarrow[(-LiCl)]{LiMe} [AsMe_4]Cl \xrightarrow[(-LiCl)]{LiMe} AsMe_5$$

The preparation is carried out in Me_2O at $-60°$ to avoid formation of the ylide $Me_3As{=}CH_2$ (mp 35°) by elimination of CH_4. $AsMe_5$ decomposes above 100° by one of two routes:

$$AsMe_5 \begin{cases} \xrightarrow{\Delta} C_2H_6 + AsMe_3 \\ \xrightarrow{\Delta} CH_4 + \{Me_3As{=}CH_2\} \longrightarrow AsMe_3 + (CH_2)_n \end{cases}$$

It is stable in air and hydrolyses only slowly:

$$AsMe_5 \begin{cases} \xrightarrow{H_2O} [AsMe_4]OH + CH_4 \\ \xrightarrow{HCl} [AsMe_4]Cl + CH_4 \end{cases}$$

The aryl analogues are rather more stable.

Of the quaternary arsonium compounds, methyltriaryl derivatives are important as precursors of arsonium ylides, e.g.

$$[Ph_3AsMe]Br + NaNH_2 \xrightarrow{thf} Ph_3As{=}CH_2 + NaBr + NH_3$$
$$\text{(mp 74°)}$$

Such ylides are unstable and react with carbonyl compounds to give both the Wittig product (p. 635) as well as $AsPh_3$ and an epoxide. However, this very reactivity is sometimes an advantage since As ylides often react with carbonyl compounds that are unresponsive to P ylides. Substituted quaternary arsonium compounds are also a useful source of heterocyclic organoarsanes, e.g. thermolysis of 4-(1,7-dibromoheptyl)trimethylarsonium bromide to 1-arsabicyclo[3.3.0]octane:

$$\left[Br(CH_2)_3{-}\overset{\overset{\displaystyle H}{|}}{\underset{\underset{\displaystyle AsMe_3}{|}}{C}}{-}(CH_2)_3Br \right]^+ Br^- \xrightarrow{\Delta} \text{[arsabicyclooctane]} + 3MeBr$$

Arsonic acids $RAsO(OH)$ are amongst the most important organoarsonium compounds. Alkyl arsonic acids are generally prepared by the Meyer reaction in which an alkaline solution of As_2O_3 is heated with an alkyl halide:

$$As(ONa)_3 + RX \xrightarrow{heat} NaX + RAsO(ONa)_2 \xrightarrow{acidify} RAsO(OH)_2$$

[67] K.-H. MITSCHKE and H. SCHMIDBAUR, Pentamethylarsenic, *Chem. Ber.* **106**, 3645–51 (1973).

Aryl arsonic acids can be made from a diazonium salt by the Bart reaction:

$$As(ONa)_3 + ArN_2X \longrightarrow NaX + N_2 + ArAsO(ONa)_2$$

Similar reactions on alkyl or aryl arsonites yield the arsinic acids $R_2AsO(OH)$ and $Ar_2AsO(OH)$. Arsine oxides are made by alkaline hydrolysis of R_3AsX_2 (or Ar_3AsX_2) or by oxidation of a tertiary arsine with $KMnO_4$, H_2O_2, or I_2.

Physiological activity of arsenicals

In general As^{III} organic derivatives are more toxic than As^V derivatives. The use of organoarsenicals in medicine dates from the discovery in 1905 by H. W. Thomas that "atoxyl" (first made by A. Béchamp in 1863) cured experimental trypanosomiasis (e.g. sleeping sickness). In 1907 P. Erlich and A. Bertheim showed that "atoxyl" was sodium hydrogen 4-aminophenylarsonate

and the field was systematically developed especially when some arsenicals proved effective against syphilis. Today such treatment is obsolete but arsenicals are still used against amoebic dysentery and are indispensable for treatment of the late neurological stages of African trypanosomiasis.

Organoantimony and organobismuth compounds

Organoantimony and organobismuth compounds are closely related to organoarsenic compounds but have not been so extensively investigated. Similar preparative routes are available and it will suffice to single out a few individual compounds for comment or comparison. MR_3 (and MAr_3) are colourless, volatile liquids or solids having the expected pyramidal molecular structure. Some properties are in Table 13.13. As expected (p. 223) tertiary stibines are much weaker ligands than phosphines or arsines.[5] Tertiary bismuthines are weaker still: among the very few coordination complexes that have been reported are $[Ag(BiPh_3)]ClO_4$, $Ph_3BiNbCl_5$, and $Ph_3BiM(CO)_5$ (M = Cr, Mo, W).

An intriguing 3-coordinate organoantimony compound, which is the first example of trigonal-planar Sb^I, has recently been characterized.[67a] The stibinidene complex

TABLE 13.13 *Some physical properties of* MMe_3 *and* MPh_3

Property	$AsMe_3$	$SbMe_3$	$BiMe_3$	$AsPh_3$	$SbPh_3$	$BiPh_3$
MP/°C	−87	−62	−86	61	55	78
BP/°C	50	80	109	—	—	—
Bond angle at M	96°	—	97°	102°	—	94°
Mean M–C bond energy/kJ mol^{-1}	229	215	143	267	244	177

[67a] J. VON SEYERL and G. HUTTNER, PhSb[Mn(CO)$_2$(η^5-C$_5$H$_5$)]$_2$—the first compound containing a trigonal-planar coordinated antimony(I), *Angew. Chem.*, Int. Edn. (Engl.) **17**, 843–4 (1978).

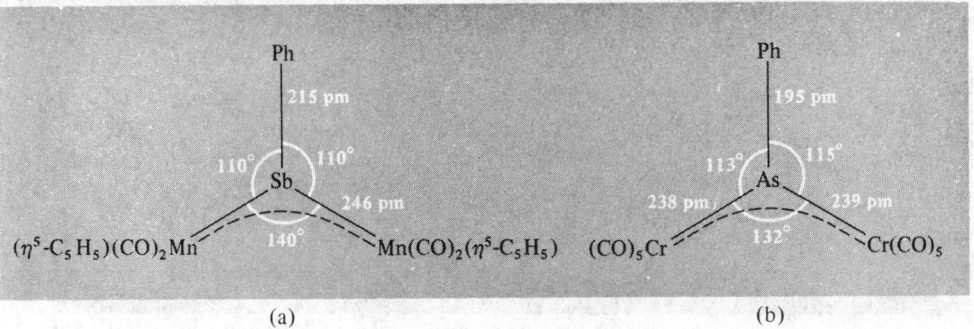

FIG. 13.31   Planar structure of (a) $[PhSb\{Mn(CO)_2(\eta^5\text{-}C_5H_5)_2\}]$, and (b) $[PhAs\{Cr(CO)_5\}_2]$. Note the relatively short Sb–Mn and As–Cr bonds.

$[PhSb\{Mn(CO)_2(\eta^5\text{-}C_5H_5)\}_2]$ has been isolated as shiny golden metallic crystals (mp 128°) from the crown-ether catalysed reaction:

$$[(\eta^5\text{-}C_5H_5)(CO)_2MnSbPhI_2] + [(\eta^5\text{-}C_5H_5)Mn(CO)_2] \cdot thf \xrightarrow[\text{[18]crown–6}]{\text{K/thf}}$$

$$[PhSb\{Mn(CO)_2(\eta^5\text{-}C_5H_5)\}_2] + 2KI + \cdots$$

The structure is shown in Fig. 13.31a: the interatomic angles and distances suggest that the bridging $\{PhSb^I\}$ group is stabilized by Sb–Mn π interactions. A similar route leads to 3-coordinate planar organoarsinidine complexes which can also be prepared by the following reaction sequence:

$$[Cr(CO)_6] \xrightarrow{PhAsH_2} [Cr(CO)_5(AsPhH_2)] \xrightarrow{LiBu} [Cr(CO)_5(AsPhLi_2)] \xrightarrow[\text{NCl}_2]{\text{cyclohexyl-}}$$

               yellow   orange

$$[\{Cr(CO)_5\}_2AsPh] \text{ dark violet (mp 104°)}$$

The chloro-derivative $[ClAs\{Mn(CO)_2(\eta^5\text{-}C_5H_5)\}_2]$ (shiny black crystals, mp 124°) can now be much more readily obtained by direct reaction of $AsCl_3$ with $[Mn(CO)_2(\eta^5\text{-}C_5H_5)] \cdot thf.^{[67b]}$

Halostibines R_2SbX and dihalostibines $RSbX_2$ (R = alkyl, aryl) can be prepared by standard methods. The former hydrolyse to the corresponding covalent molecular oxides $(R_2Sb)_2O$, whereas $RSbX_2$ yield highly polymeric "stiboso" compounds $(RSbO)_n$. The stibonic acids, $RSbO(OH)_2$, and stibinic acids, $R_2SbO(OH)$, differ in structure from phosphonic and phosphinic acids (p. 591) or arsonic and arsinic acids (p. 691) in being high molecular weight materials of unknown structure. They are probably best considered as oxide hydroxides of organoantimony(V) cations. Indeed, throughout its organometallic chemistry Sb shows a propensity to increase its coordination number by dimerization or polymerization. Thus Ph_2SbF consists of infinite chains of F-bridged pseudo trigonal-bipyramidal units as shown in Fig. 13.32.$^{[68]}$ The compound could not be prepared by the

^{67b} J. VON SEYERL, U. MOERING, A. WAGNER, A. FRANK, and G. HUTTNER, Facile new synthesis of chloroarsinidene complexes, *Angew Chem.*, Int. Edn. (Engl.) **17**, 844–5 (1978).

^{68} S. P. BONE and D. B. SOWERBY, Diphenylantimony(III) fluoride: preparation and crystal structure, *JCS Dalton* 1979, 1430–3.

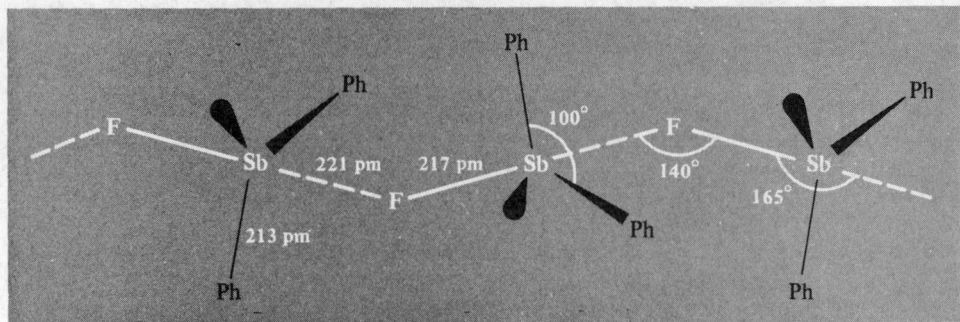

Fig. 13.32 Structure of Ph_2SbF_2 showing polymeric chains of apex-shared pseudo trigonal bipyramidal units $\{Ph_2\dot{S}\dot{b}\cdots F\}$.

normal methods of fluorinating Ph_2SbCl or phenylating SbF_3 but was recently made as a white, air-stable, crystalline solid mp 154° by the following sequence of steps:

$$PhSiCl_3 \xrightarrow[\text{(anhydr)}]{SbF_3/80°} PhSiF_3 \xrightarrow{aq\ NH_4F} [NH_4]_2[PhSiF_5] \xrightarrow{aq\ SbF_3} Ph_2SbF$$

Again, Me_2SbCl_3 is monomeric with equatorial methyl groups (C_{2v}) in solution (CH_2Cl_2, $CHCl_3$, or C_6H_6) but forms Cl-bridged dimers with *trans* methyl groups (D_{2h}) in the solid:[69]

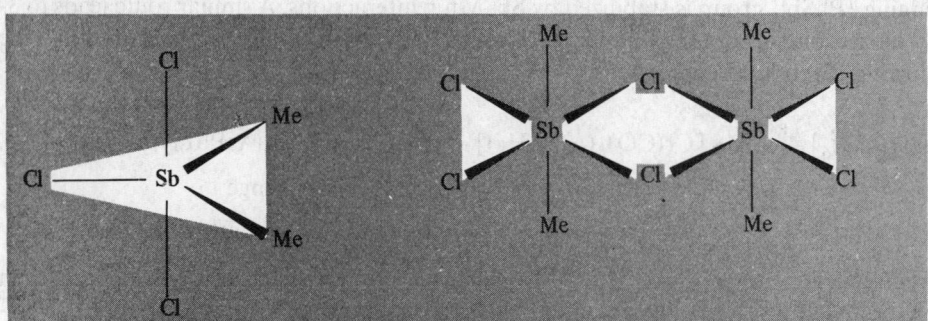

A similar Cl-bridged dimeric structure was established by X-ray analysis for Ph_2SbCl_3.[70]

Pentaphenylantimony, $SbPh_5$ (mp 171°), has attracted much attention as the first known example of a 10-valence-electron molecule of a main group element that has a square pyramidal rather than a trigonal bipyramidal structure.†[71] It can conveniently be prepared from $SbPh_3$ by chlorination to give Ph_3SbCl_2 and then reaction with LiPh:

$$Ph_3SbCl_2 + 3LiPh \longrightarrow 2LiCl + Li[SbPh_6] \xrightarrow{H_2O} LiOH + C_6H_6 + SbPh_5$$

† The *anion* $InCl_5^{2-}$ also has this arrangement (p. 268).

[69] N. Bertazzi, T. C. Gibb, and N. N. Greenwood, Antimony-121 Mössbauer spectra of some six-coordinate mono- and di-organoantimony(V) compounds, *JCS Dalton* 1976, 1153–7. K. Dehnicke and H. G. Nadler, Structural isomerism of dimethylantimony trichloride, *Chem. Ber.* **109**, 3034–8 (1976).

[70] J. Bordner, G. O. Doak, and J. R. Peters, Structure of the dimer of diphenylantimony trichloride, *J. Am. Chem. Soc.* **96**, 6763–5 (1974).

[71] P. J. Wheatley, An X-ray diffraction determination of the crystal and molecular structure of pentaphenylantimony, *J. Chem. Soc.* 1964, 3718–23.

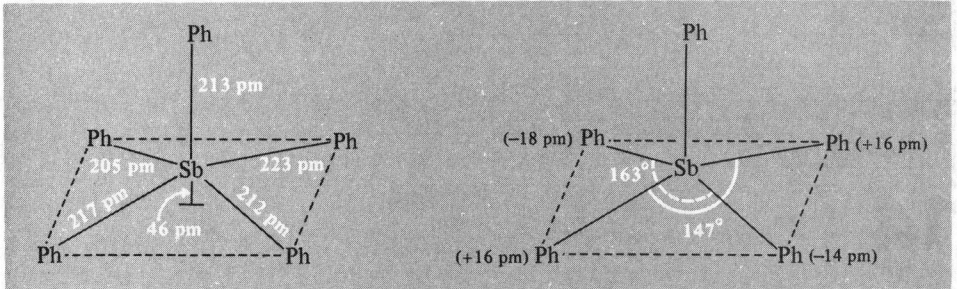

FIG. 13.33   Molecular geometry of SbPh$_5$ showing the distorted square-pyramidal structure. The Sb atom is 46 pm above the mean basal plane of the 4 C (phenyl) atoms which themselves alternate above and below this plane by some 16 pm.

The structure, shown in Fig. 13.33, is based on two-dimensional data and is somewhat imprecise, standard deviations on the Sb–C distances being 7 pm and on C–Sb–C angles 3°. The molecule appears to deviate slightly from true square-pyramidal geometry but has at least a twofold axis of symmetry. Thus the phenyl groups are bonded to antimony via C atoms that deviate alternately above and below the mean basal plane by some 16 pm and there are two basal diagonal C–Sb–C angles (147° and 163°). There is no significant difference between the Sb–C$_{axial}$ and mean Sb–C$_{basal}$ distances though more precise data may reveal such a difference. The reason why SbPh$_5$ does not adopt the trigonal bipyramidal structure found for PPh$_5$ and AsPh$_5$ is not clear. It may be related in part to crystal-packing forces since SbPh$_5 . \frac{1}{2}$C$_6$H$_{12}$ has an almost undistorted trigonal bipyramidal structure as does Sb(4-MeC$_6$H$_4$)$_5$.[72] However, vibrational spectroscopy suggests that SbPh$_5$ retains its square pyramidal configuration even in solution, and the same structure is apparently also adopted by the yellow *cyclo*-propyl derivative Sb(*cyclo*-C$_3$H$_5$)$_5$.[73] The pentamethyl compound, SbMe$_5$, is surprisingly stable in view of the difficulty of obtaining AsMe$_5$ and BiMe$_5$; it melts at $-19°$, boils at 127°, and does not inflame in air, though it oxidizes quickly and is hydrolysed by water. It resembles SbPh$_5$ in reacting with LiMe (LiPh) to give Li$^+$[SbR$_6$]$^-$ and in reacting with BPh$_3$ to give [SbR$_4$]$^+$[RBPh$_3$]$^-$.

Organobismuth(V) compounds are in general similar to their As and Sb analogues but are less stable and there are few examples known; e.g. [BiR$_4$]X and R$_3$BiX$_2$ are known but not R$_2$BiX$_3$ or RBiX$_4$, whereas all 4 classes of compound are known for P, As, and Sb. Similarly, no pentaalkylbismuth compound is known, though BiPh$_5$ has been prepared as a violet-coloured crystalline compound by treating [BiPh$_4$]Cl or Ph$_3$BiCl$_2$ with LiPh at $-75°$. It decomposes spontaneously over a period of days at room temperature and reacts readily with HX, X$_2$, or even BPh$_3$ by cleaving 1 phenyl to form quaternary bismuth compounds [BiPh$_4$]X and [BiPh$_4$][BPh$_4$]; this latter compound (mp 228°) is the most stable bismuthonium salt yet known.

[72] C. BRABANT, J. HUBERT, and A. L. BEAUCHAMP, The crystal and molecular structure of penta-*p*-tolylantimony (*p*-MeC$_6$H$_4$)$_5$Sb, *Can. J. Chem.* **51**, 2952–7 (1973).

[73] A. H. COWLEY, J. L. MILLS, T. M. LOEHR, and T. V. LONG, Synthesis and laser-Raman and infrared spectra of pentacyclopropyl-antimony(V)—a new square-pyramidal molecule, *J. Am. Chem. Soc.* **93**, 2150–3 (1971).

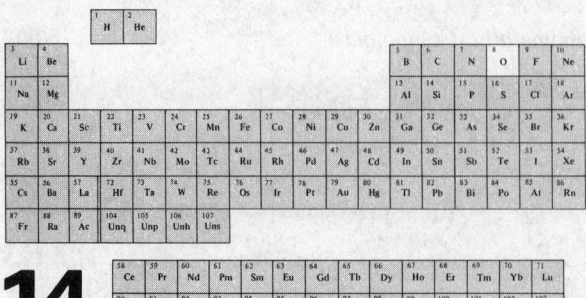

14

Oxygen

14.1 The Element

14.1.1 *Introduction*

Oxygen is the most abundant element on the earth's surface: it occurs both as the free element and combined in innumerable compounds, and comprises 23% of the atmosphere by weight, 46% of the lithosphere, and more than 85% of the hydrosphere ($\sim 85.8\%$ of the oceans and 88.81% of pure water). It is also, perhaps paradoxically, by far the most abundant element on the surface of the moon where, on average, 3 out of every 5 atoms are oxygen (44.6% by weight).

The "discovery" of oxygen is generally credited to C. W. Scheele and J. Priestley (independently) in 1773–4, though several earlier investigators had made pertinent observations without actually isolating and characterizing the gas.[1-4] Indeed, it is difficult to ascribe a precise meaning to the word "discovery" when applied to a substance so ubiquitously present as oxygen; particularly when (a) experiments on combustion and respiration were interpreted in terms of the phlogiston theory, (b) there was no clear consensus on what constituted "an element", and (c) the birth of Dalton's atomic theory was still far in the future. Moreover, the technical difficulties before the mid-eighteenth century of isolating and manipulating gases compounded the problem still further, and it seems certain that several investigators had previously prepared oxygen without actually collecting it or recognizing it as a constituent of "common air". Scheele, a pharmacist in Uppsala, Sweden, prepared oxygen at various times between 1771–3 by heating KNO_3, $Mg(NO_3)_2$, Ag_2CO_3, HgO, and a mixture of H_3AsO_4 and MnO_2. He called the gas "vitriol air" and reported that it was colourless, odourless, and tasteless, and supported combustion better than common air, but the results did not appear until 1777 because of his publisher's negligence. Priestley's classic experiment of using a "burning glass" to

[1] J. W. MELLOR, *A Comprehensive Treatise on Inorganic and Theoretical Chemistry*, Vol. 1, pp. 344–51, Longmans, Green, 1922. History of the discovery of oxygen.

[2] M. E. Weeks, *Discovery of the Elements*, 6th edn., pp. 209–23, Journal of Chemical Education, Easton, Pa, 1956. (Oxygen.)

[3] J. R. PARTINGTON, *A History of Chemistry*, Vol. 3, Macmillan, London, 1962; Scheele and the discovery of oxygen (pp. 219–22); Priestley and the discovery of oxygen (pp. 256–63); Lavoisier and the rediscovery of oxygen (pp. 402–10).

[4] *Gmelin's Handbuch der Anorganischen Chemie*, 8th edn., pp. 1–82. "Sauerstoff" System No. 3, Vol. 1, Verlag Chemie, 1943. (Historical.)

calcine HgO confined in a cylinder inverted over liquid mercury was first performed in Colne, England, on Monday, 1 August 1774; he related this to A. L. Lavoisier and others at a dinner party in Paris in October 1774 and published the results in 1775 after he had shown that the gas was different from nitrous oxide. Priestley's ingenious experiments undoubtedly established oxygen as a separate substance ("dephlogisticated air") but it was Lavoisier's deep insight which recognized the new gas as an element and as the key to our present understanding of the nature of combustion. This led to the overthrow of the phlogiston theory and laid the foundations of modern chemistry.[5] Lavoisier named the

Oxygen: Some Important Dates

15th century	Leonardo da Vinci noted that air has several constituents, one of which supports combustion.
1773–4	C. W. Scheele and J. Priestley independently discovered oxygen, prepared it by several routes, and studied its properties.
1775–7	A. L. Lavoisier recognized oxygen as an element, developed the modern theory of combustion, and demolished the phlogiston theory.
1777	A. L. Lavoisier coined the name "oxygen" (acid former).
1781	Composition of water as a compound of oxygen and hydrogen established by H. Cavendish.
1800	W. Nicholson and A. Carlisle decomposed water electrolytically into hydrogen and oxygen which they then recombined by explosion to resynthesize water.
1818	Hydrogen peroxide discovered by L.-J. Thenard.
1840	C. F. Schönbein detected and named ozone from its smell (see 1857).
1848	M. Faraday noted that oxygen was paramagnetic, correctly ascribed to the triplet $^3\Sigma_g^-$ ground state by R. S. Mulliken (in 1928).
1857	W. Siemens constructed the first machine to use the ozonator-discharge principle to generate ozone.
1877	Oxygen first liquefied by L. Cailletet and R. Pictet (independently).
1881	Oxygen gas first produced industrially (from BaO_2) by A. Brin and L. W. Brin's Oxygen Company.
1896	First production of liquid oxygen on a technical scale (C. von Linde).
1903	Ozonolysis of alkenes discovered and developed by C. D. Harries.
1921–3	The water molecule, previously thought to be linear, shown to be bent.
1929	Isotopes ^{17}O and ^{18}O discovered by W. F. Giauque and H. L. Johnston (see 1961).
1931	Singlet state of O_2, $^1\Sigma_g^+$, discovered by W. H. J. Childe and R. Mecke.
1934	A lower lying singlet state $^1\Delta_g$ discovered by G. Herzberg.
1931–9	H. Kautsky pointed out the significance of singlet O_2 in organic reactions; his views were discounted at the time time but the great importance of singlet O_2 was rediscovered in 1964 by (a) C. S. Foote and S. Wexler, and (b) E. J. Corey and W. C. Taylor.
1941	^{18}O-tracer experiments by S. Ruben and M. D. Kamen showed that the oxygen atoms in photosynthetically produced O_2 both come from H_2O and not CO_2; confirmed in 1975 by A. Stemler.
1951	First detection of ^{17}O nmr signal by H. E. Weaver, B. M. Tolbert and R. C. La Force.
1961	Dual atomic-weight scales based on oxygen = 16 (chemical) and $^{16}O = 16$ (physical) abandoned in favour of a unified scale based on $^{12}C = 12$.
1963	First successful launch of a rocket propelled by liquid H_2/liquid O_2 (Cape Kennedy).
1963	Reversible formation of a dioxygen complex by direct reaction of O_2 with *trans*-$[Ir(CO)Cl(PPh_3)_2]$ discovered by L. Vaska.
1967	Many crown ethers synthesized by C. J. Pederson who also studied their use as complexing agents for alkali metal and other cations.

[5] A. L. LAVOISIER, *La Traité Elémentaire de Chemie*, Paris, 1789, translated by R. Kerr, *Elements of Chemistry*, London, 1790; facsimile reprint by Dover Publications, Inc., New York, 1965.

element "oxygène" in 1777 in the erroneous belief that it was an essential constituent of all acids (Greek ὀξύς, *oxys*, sharp, sour; γείνομαι, *geinomai*, I produce; i.e. acid forming). Some other important dates in oxygen chemistry are in the Panel.

14.1.2 *Occurrence*

Oxygen occurs in the atmosphere in vast quantities as the free element O_2 (and O_3, p. 707) and there are also substantial amounts dissolved in the oceans and surface waters of the world. Virtually all of this oxygen is of biological origin having been generated by green-plant photosynthesis from water (and carbon dioxide).[6] The net reaction can be represented by:

$$H_2O + CO_2 + h\nu \xrightarrow[\text{enzymes}]{\text{chlorophyl}} O_2 + \{CH_2O\} \text{ (i.e. carbohydrates, etc.)}$$

However, this is misleading since isotope-tracer experiments using [18]O have shown that both the oxygen atoms in O_2 originate from H_2O, whereas those in the carbohydrates come from CO_2. The process is a complex multistage reaction involving many other species,[7] and requires 469 kJ mol^{-1} of energy (supplied by the light). The reverse process, combustion of organic materials with oxygen, releases this energy again. Indeed, except for very small amounts of energy generated from wind or water power, or from nuclear reactors, all the energy used by man comes ultimately from the combustion of wood or fossil fuels such as coal, peat, natural gas, and oil. Photosynthesis thus converts inorganic compounds into organic material, generates atmospheric oxygen, and converts light energy (from the sun) into chemical energy. The 1.5×10^9 km^3 of water on the earth is split by photosynthesis and reconstituted by respiration and combustion once every 2 million years or so.[8] The photosynthetically generated gas temporarily enters the atmosphere and is recycled about once every 2000 years at present rates. The carbon dioxide is partly recycled in the atmosphere and oceans after an average residence time of 300 years and is partly fixed by precipitation of $CaCO_3$, etc. (p. 302).

There was very little, if any, oxygen in the atmosphere 3000 million years ago. Green-plant photosynthesis probably began about 2500 My ago and O_2 first appeared in the atmosphere in geochemically significant amounts about 2000 My ago (this is signalled by the appearance of red beds of iron-containing minerals that have been weathered in an oxygen-containing atmosphere).[6, 8] The O_2 content of the atmosphere reached $\sim 2\%$ of the present level some 800 My ago and $\sim 20\%$ of the present level about 580 My ago. This can be compared with the era of rapid sea-floor spreading to give the separated continents which occurred 110–85 My ago. The concentration of O_2 in the atmosphere has probably remained fairly constant for the past 50 My, a period of time which is still extensive when compared with the presence of *homo sapiens*, < 1 My. The composition of the present atmosphere (excluding water vapour which is present in variable amounts depending on locality, season of the year, etc., is given in Table 14.1.[6] The oxygen content corresponds

[6] J. C. G. WALKER, *Evolution of the Atmosphere*, pp. 318, Macmillan, New York, 1977.

[7] R. Govindgee, Photosynthesis, *McGraw Hill Encyclopedia of Science and Technology*, 4th edn., Vol. 10, pp. 200–10, 1977. An excellent authoritative introductory article with bibliography for further reading.

[8] P. CLOUD and A. GIBOR, The oxygen cycle, Article 4 in *Chemistry in the Environment*, pp. 31–41, Readings from Scientific American, W. H. Freeman, San Francisco, 1973.

to 21.04 atom% and 23.15 wt%. The question of atmospheric ozone and pollution of the stratosphere is discussed on p. 708.

In addition to its presence as the free element in the atmosphere and dissolved in surface waters, oxygen occurs in combined form both as water, and a constituent of most rocks, minerals, and soils. The estimated abundance of oxygen in the crustal rocks of the earth is 455 000 ppm (i.e. 45.5% by weight): see silicates, p. 399; aluminosilicates, p. 399; carbonates, p. 119; phosphates, p. 548, etc.

TABLE 14.1 *Composition of the atmosphere*[a] *(excluding H_2O, variable)*

Constituent	Vol%	Total mass/tonnes	Constituent	Vol%	Total mass/tonnes
Dry air	100.0	$5.119(8) \times 10^{15}$	CH_4	$\sim 1.5 \times 10^{-4}$	$\sim 4.3 \times 10^9$
N_2	78.084(4)	$3.866(6) \times 10^{15}$	H_2	$\sim 5 \times 10^{-5}$	$\sim 1.8 \times 10^8$
O_2	**20.948(2)**	$\mathbf{1.185(2) \times 10^{15}}$	N_2O	$\sim 3 \times 10^{-5}$	$\sim 2.3 \times 10^9$
			CO	$\sim 1.2 \times 10^{-5}$	$\sim 5.9 \times 10^8$
Ar	0.934(1)	$6.59(1) \times 10^{13}$	NH_3	$\sim 1 \times 10^{-6}$	$\sim 3 \times 10^7$
CO_2	0.0315(10)	$2.45(8) \times 10^{12}$	NO_2	$\sim 1 \times 10^{-7}$	$\sim 8 \times 10^6$
Ne	$1.818(4) \times 10^{-3}$	$6.48(2) \times 10^{10}$	SO_2	$\sim 2 \times 10^{-8}$	$\sim 2 \times 10^6$
He	$5.24(5) \times 10^{-4}$	$3.71(4) \times 10^9$	H_2S	$\sim 2 \times 10^{-8}$	$\sim 1 \times 10^6$
Kr	$1.14(1) \times 10^{-4}$	$1.69(2) \times 10^{10}$	O_3	**Variable**	$\sim 3.3 \times 10^9$
Xe	$8.7(1) \times 10^{-6}$	$2.02(2) \times 10^9$			

[a] Total mass: $5.136(7) \times 10^{15}$ tonnes; H_2O $0.017(1) \times 10^{15}$ tonnes; dry atmosphere $5.119(8) \times 10^{15}$ tonnes. Figures in parentheses denote estimated uncertainty in last significant digit.

14.1.3 *Preparation*

Oxygen is now separated from air on a vast scale (see below) and is conveniently obtained for most laboratory purposes from high-pressure stainless steel cylinders. Small traces of N_2 and the rare gases, particularly argon, are the most persistent impurities. Occasionally, small-scale laboratory preparations are required and the method chosen depends on the amount and purity required and the availability of services. Electrolysis of degassed aqueous electrolytes produces wet O_2, the purest gas being obtained from 30% potassium hydroxide solution using nickel electrodes. Another source is the catalytic decomposition of 30% aqueous hydrogen peroxide on a platinized nickel foil.

Many oxoacid salts decompose to give oxygen when heated (p. 1011). A convenient source is $KClO_3$ which evolves oxygen when heated to 400–500° according to the simplified equation

$$2KClO_3 \xrightarrow{\Delta} 2KCl + 3O_2$$

The decomposition temperature is reduced to 150° in the presence of MnO_2 but then the product is contaminated with up to 3% of ClO_2 (p. 992). Small amounts of breathable oxygen for use in emergencies (e.g. failure of normal supply in aircraft or submarines) can be generated by decomposition of $NaClO_3$ in "oxygen candles". The best method for the controlled preparation of very pure O_2 is the thermal decomposition of recrystallized, predried, degassed $KMnO_4$ in a vacuum line. Mn^{VI} and Mn^{IV} are both formed and the reaction can formally be represented as:

$$2KMnO_4 \xrightarrow{215-235} K_2MnO_4 + MnO_2 + O_2$$

Oxygen gas and liquid oxygen are manufactured on a huge scale by the fractional distillation of liquid air at temperatures near $-183°C$. Although world production approaches 100 million tonnes pa this is still less than one ten-millionth part of the oxygen in the atmosphere; moreover, the oxygen is continuously being replenished by photosynthesis. Further information on the industrial production and uses of oxygen are in the Panel.

Industrial Production and Uses of Oxygen[9]

Air can be cooled and eventually liquified by compressing it isothermally and then allowing it to expand adiabatically to obtain cooling by the Joule–Thompson effect. Although this process was developed by C. von Linde (Germany) and W. Hampson (UK) at the end of the last century, it is thermodynamically inefficient and costly in energy. Most large industrial plants now use the method developed by G. Claude (France) in which air is expanded isentropically in an engine from which mechanical work can be obtained; this produces a much greater cooling effect than that obtained by the Joule–Thompson effect alone. Because N_2 (bp $-195.8°C$) is more volatile than O_2 (bp $-183.0°C$) there is a higher concentration of N_2 in the vapour phase above boiling liquid air than in the liquid phase, whilst O_2 becomes progressively enriched in the liquid phase. Fractional distillation of the liquified air is usually effected in an ingeniously designed double-column dual-pressure still which uses product oxygen from the upper column at a lower pressure (lower bp) to condense vapour for reflux at a higher pressure in the lower column. This is shown schematically in the Figure. The most volatile constituents of air (He, H_2, Ne) do not condense but accumulate as a high-pressure gaseous mixture with N_2 at the top of the lower column. Argon, which has a volatility between those of O_2 and N_2, concentrates in the upper column from which it can be withdrawn for further purification in a separate column, whilst the least-volatile constituents (Kr, Xe) accumulate in the oxygen boiler at the foot of the upper column. Typical operating pressures are 5 atm at the top of the lower column and 0.5 atm at the bottom of the upper column. A large plant might produce 700 tonnes per day of separated products. A rather different design is used if liquid (rather than gaseous) N_2 is required in addition to the liquid and/or gaseous O_2.

From modest beginnings at the turn of the century, oxygen has now become the third largest volume chemical produced in the USA (after H_2SO_4 and lime and ahead of NH_3 and N_2). Production in 1979 was 16.7 million tonnes (USA), 2.5 Mt (UK), and almost 100 Mt worldwide. About 20% of the USA production is as liquid O_2. This phenomenal growth derived mainly from the growing use of O_2 in steelmaking; the use of O_2 rather than air in the Bessemer process was introduced in the late 1950s and greatly increased the productivity by hastening the reactions. In many of the major industrial countries this use alone now accounts for 65–85% of the oxygen produced. Much of this is manufactured on site and is simply piped from the air-separation plant to the steel converter.

Oxygen is also used to an increasing extent in iron blast furnaces since enrichment of the blast enables heavy fuel oil to replace some of the more expensive metallurgical coke. Other furnace applications are in ferrous and non-ferrous metal smelting and in glass manufacture, where considerable benefits accrue from higher temperatures, greater productivity, and longer furnace life. Related, though smaller-scale applications, include steel cutting, oxy-gas welding, and oxygen lancing (concrete drilling).

In the chemical industry oxygen is used on a large scale in the production of TiO_2 by the chloride process (p. 1118), in the direct oxidation of ethene to ethylene oxide, and in the

[9] W. J. GRANT and S. L. REDFEARN, Industrial Gases, in R. Thompson (ed.), *The Modern Inorganic Chemicals Industry*, pp. 273–301, Chem. Soc. Special Publ. No. 31, 1978.

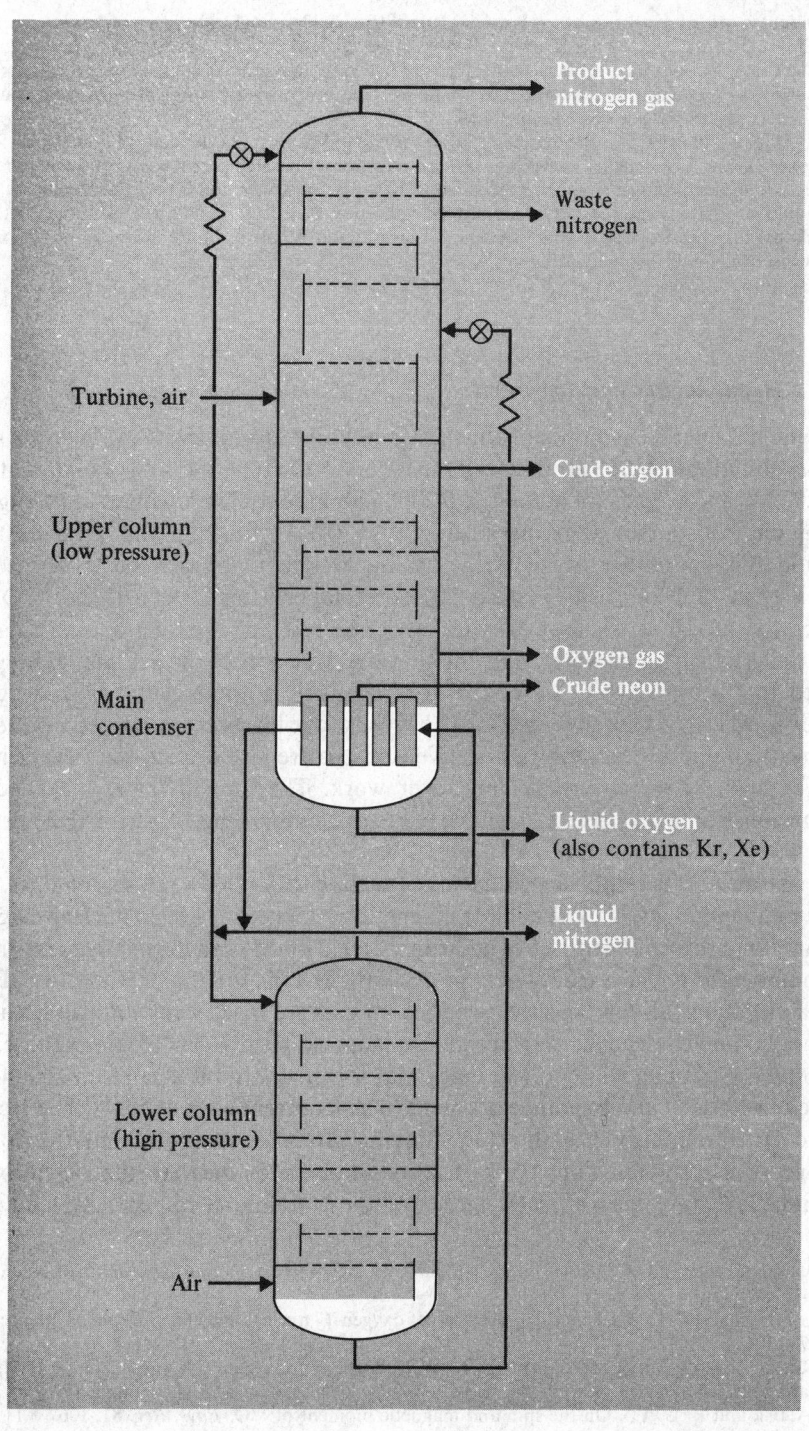

manufacture of synthesis gas ($H_2 + CO$), propylene oxide, vinyl chloride, vinyl acetate, etc. Environmental and biomedical uses embrace sewage treatment, river revival, paper-pulp bleaching, fish farming, artificial atmospheres for diving and submarine work, oxygen tents in hospitals, etc. Much of the oxygen for these applications is transported either in bulk liquid carriers or in high-pressure steel cylinders.

A final, somewhat variable outlet for large-scale liquid oxygen is as oxidant in rocket fuels for space exploration, satellite launching, and space shuttles. For example, in the Apollo mission to the moon, each Saturn 5 launch rocket used 1270 m^3 (i.e. 1.25 million litres or 1450 tonnes) of liquid oxygen in Stage 1, where it oxidized the kerosene fuel (195 000 l, or about 550 tonnes) in the almost unbelievably short time of 2.5 min. Stages 2 and 3 had 315 and 76.3 m^3 of liquid O_2 respectively, and the fuel was liquid H_2.

14.1.4 *Atomic and physical properties*

Oxygen has 3 stable isotopes of which ^{16}O (relative atomic mass 15.994 915) is by far the most abundant (99.763 atom%). Of the others, ^{17}O (16.999 134) has an abundance of only 0.037% and ^{18}O (17.999 160) is 0.200% abundant. These values vary slightly in differing natural sources (the ranges being 0.035–0.041% for ^{17}O and 0.188–0.209% for ^{18}O) and this variability prevents the atomic weight of oxygen being quoted more precisely than 15.9994 ± 0.0003 (see p. 20). Artificial enrichment of ^{17}O and ^{18}O can be achieved by several physical or chemical processes such as the fractional distillation of water, the electrolysis of water, and the thermal diffusion of oxygen gas. Heavy water enriched to 20 atom% ^{17}O or 98% ^{18}O is available commercially, as is oxygen gas enriched to 95% in ^{17}O or 99% in ^{18}O. The ^{18}O isotope has been much used in kinetic and mechanistic studies.[9a] Several radioactive isotopes are also known but their very short half-lives make them unsuitable for tracer work. The longest lived, ^{15}O, decays by positron emission with $t_{\frac{1}{2}}$ 122 s; it can be made by bombarding ^{16}O with ^3He particles— $^{16}O(^3He,\alpha)^{15}O$.

The isotope ^{17}O is important in having a nuclear spin ($I = \frac{5}{2}$) and this enables it to be used in nmr studies. [10,11] The nuclear magnetic moment is -1.8930 nuclear magnetons (very similar to the value for the free neutron, -1.9132 NM) and the relative sensitivity for equal numbers of nuclei is 0.0291, compared with ^1H 1.00, ^{11}B 0.17, ^{13}C 0.016, ^{31}P 0.066, etc. In addition to this low sensitivity, measurements are made more difficult because the quadrupolar nucleus leads to very broad resonances, typically 10^2–10^3 times those for ^1H. The observing frequency is ~ 0.136 times that for proton nmr. The resonance was first observed in 1951[12] and the range of chemical shifts extended in 1955.[13] The technique has proved particularly valuable for studying aqueous solutions and the solvation equilibria of electrolytes. Thus the hydration numbers for the diamagnetic cations Be^{II}, Al^{III}, and Ga^{III} have been directly measured as 4, 6, and 6 respectively, and several

[9a] I. D. DOSTROVSKY and D. SAMUEL, Oxygen-18, in R. H. Herber (ed.), *Inorganic Isotopic Syntheses*, Chap. 5, pp. 119–42, Benjamin, New York, 1962.

[10] B. L. SILVER and Z. LUZ, Chemical application of oxygen-17 nuclear and electron spin resonance, *Q. Rev.* **21**, 458–73 (1967).

[11] C. ROGER, N. SHEPPARD, C. MCFARLANE, and W. MCFARLANE, Oxygen-17, Chap. 12A in R. H. Harris and B. E. Mann (eds.). *NMR and the Periodic Table*, pp. 383–400, Academic Press, London, 1978.

[12] F. ALDER and F. C. YU, On the spin and magnetic moment of ^{17}O, *Phys. Rev.* **81**, 1067–8 (1951).

[13] H. E. WEAVER, B. M. TOLBERT, and R. C. LAFORCE, Observation of chemical shifts of ^{17}O nuclei in various chemical environments, *J. Chem. Phys.* **23**, 1956–7 (1955).

exchange reactions between "bound" and "free" water have been investigated. Chemical shifts for ^{17}O in a wide range of oxoanions $[XO_n]^{m-}$ have been studied and it has been found that the shifts for terminal and bridging O atoms in $[Cr_2O_7]^{2-}$ differ by as much as 760 ppm. The technique is proving increasingly valuable in the structure determination of complex polyanions in solution; for example all seven different types of O atoms in $[V_{10}O_{28}]^{6-}$ (p. 1149) have been detected.[14] The exchange of ^{17}O between $H_2^{17}O$ and various oxoanions has also been studied. Less work has been done so far on transition metal complexes of CO and NO though advances in techniques are now beginning to yield valuable structural and kinetic data.[14a] Attempts to obtain ^{17}O nmr spectra for dioxygen complexes (p. 718) have so far been unsuccessful, perhaps because the signals were too broad to be detected.

The electronic configuration of the free O atom is $1s^2 2s^2 2p^4$, leading to a $^3P_2$ ground state. The ionization energy of O is 1313.5 kJ mol^{-1} (cf. S on p. 782 and the other Group VI elements on p. 890). The electronegativity of O is 3.5; this is exceeded only by F and the high value is reflected in much of the chemistry of oxygen and the oxides. The single-bond atomic radius of O is usually quoted as 73–74 pm, i.e. slightly smaller than for C and N, and slightly larger than for F, as expected. The ionic radius of O^{2-} is assigned the standard value of 140 pm and all other ionic radii are derived from this.[15]

Molecular oxygen, O_2, is unique among gaseous diatomic species with an even number of electrons in being paramagnetic. This property, first observed by M. Faraday in 1848, receives a satisfying explanation in terms of molecular orbital theory. The schematic energy-level diagram is shown in Fig. 14.1; this indicates that the 2 least-strongly bound electrons in O_2 occupy degenerate orbitals of π symmetry and have parallel spins. This leads to a triplet ground state, $^3\Sigma_g^-$. As there are 4 more electrons in bonding MOs than in antibonding MOs, O_2 can be formally said to contain a double bond. If the 2 electrons, whilst remaining unpaired in separate orbitals, have opposite spin, then a singlet excited state of zero resultant spin results, $^1\Delta_g$. A singlet state also results if the 2 electrons occupy a single π^* orbital with opposed spins, $^1\Sigma_g^+$. These 2 singlet states lie 94.72 and 157.85 kJ mol^{-1} above the ground state and are extremely important in gas-phase oxidation reactions (p. 716). The excitation is accompanied by a slight but definite increase in the internuclear distance from 120.74 pm in the ground state to 121.55 and 122.77 pm in the excited states. The bond dissociation energy of O_2 is 493.4(2) kJ mol^{-1}; this is substantially less than for the triply bonded species N_2 (945.4 kJ mol^{-1}) but is much greater than for F_2 (158.8 kJ mol^{-1}).

Oxygen is a colourless, odourless, tasteless, highly reactive gas. It dissolves to the extent of 3.08 cm^3 (gas at STP) in 100 cm^3 H_2O at 20° and this drops to 2.08 cm^3 at 50°. Solubility in salt water is slightly less but is still sufficient for the vital support of marine and aquatic life. Solubility in many organic solvents is about 10 times that in water and necessitates careful degassing if these solvents are to be used in the preparation and handling of oxygen-sensitive compounds. Typical solubilities (expressed as gas volumes dissolved in 100 cm^3 of solvent at 25°C and 1 atm pressure) are Et_2O 45.0, CCl_4 30.2. Me_2CO 28.0 and C_6H_6 22.3 cm^3.

[14] W. G. KLEMPERER and W. SHUM, Charge distribution in large polyanions: determination of protonation sites in $[V_{10}O_{28}]^{6-}$ by ^{17}O nuclear magnetic resonance, *J. Am. Chem. Soc.* **99**, 3544–5 (1977).

[14a] R. L. KUMP and L. J. TODD, Oxygen-17 nmr studies of ^{17}O-enriched transition metal carbonyl complexes, *JCS Chem. Comm.* 1980, 292–3.

[15] R. D. SHANNON, revised effective ionic radii and systematic studies of interatomic distances in halides and chalcogenides, *Acta Cryst.* **A32**, 751–67 (1976).

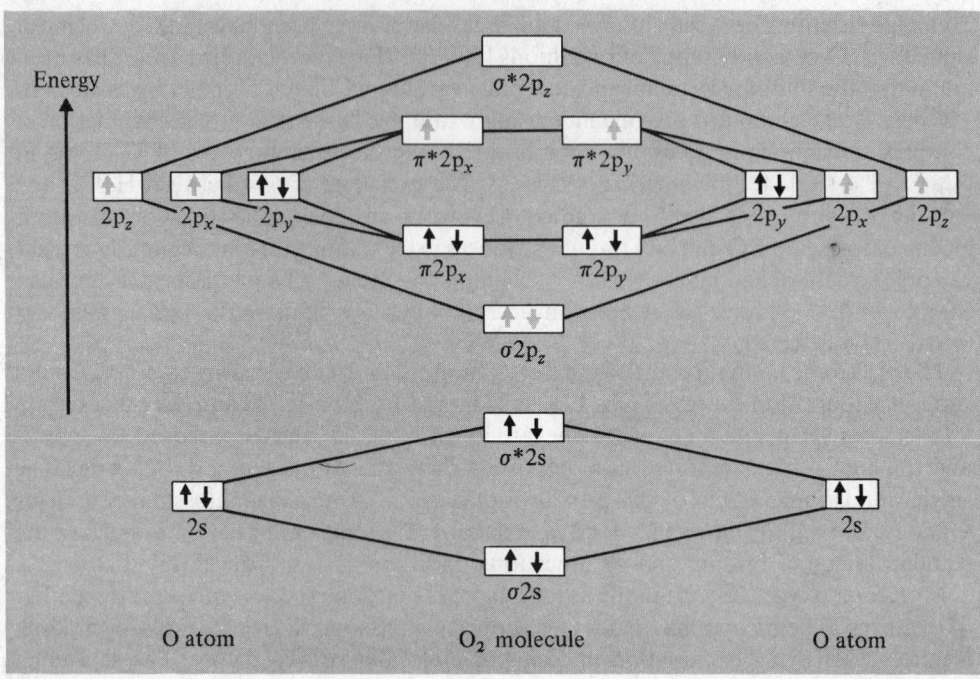

FIG. 14.1   Schematic molecular–orbital energy level diagram for the molecule O_2 in its ground
state, $^3\Sigma_g^-$. The internuclear vector is along the z-axis.

Oxygen condenses to a pale blue, mobile, paramagnetic liquid (bp $-183.0°C$ at 1 atm).
The viscosity (0.199 centipoise at $-183.5°$ and 10.6 atm) is about one-fifth that of water at
room temperature. The critical temperature, above which oxygen cannot be liquidified by
application of pressure alone, is $-118.4°C$ and the critical pressure is 50.15 atm. Solid
oxygen (pale blue, mp $-218.8°C$) also comprises paramagnetic O_2 molecules but, in the
cubic γ-phase just below the mp, these are rotationally disordered and the solid is soft,
transparent, and only slightly more dense than the liquid. There is a much greater increase
in density when the solid transforms to the rhombohedral β-phase at $-229.4°$ and there is
a further phase change to the monoclinic α-form at $-249.3°C$; these various changes and
the accompanying changes in molar volume ΔV_M are summarized in Table 14.2.

TABLE 14.2   *Densities and molar volumes of liquid and solid* O_2

Transition	bp/1 (atm)	mp (triple pt)	$\gamma \longleftrightarrow \beta$		$\beta \longleftrightarrow \alpha$	
T/K	90.18	54.35	43.80		23.89	
$d/g\ cm^{-3}$	1.1407(1)	1.3215(1)	1.334(γ)	1.495(β)		1.53(α)
ΔV_M /cm^3 mol^{-1}	3.84	0.23	2.58		0.49	

The blue colour of oxygen in the liquid and solid phases is due to electronic transitions
by which molecules in the triplet ground state are excited to the singlet states. These

transitions are normally forbidden in pure gaseous oxygen and, in any case, they occur in the infrared region of the spectrum at 7918 cm^{-1} ($^1\Delta_g$) and 13 195 cm^{-1} ($^1\Sigma_g^+$). However, in the condensed phases a single photon can elevate 2 colliding molecules simultaneously to excited states, thereby requiring absorption of energy in the visible (red–yellow–green) regions of the spectrum.[16] For example:

$$2O_2(^3\Sigma_g^-) + h\nu \longrightarrow 2O_2(^1\Delta_g); \quad \bar{\nu} = 15\ 800\ \text{cm}^{-1}, \text{ i.e. } \lambda = 631.2\ \text{nm}$$
$$2O_2(^3\Sigma_g^-) + h\nu \longrightarrow O_2(^1\Delta_g) + O_2(^1\Sigma_g^+); \quad \bar{\nu} \sim 21\ 100\ \text{cm}^{-1}, \text{ i.e. } \lambda = 473.7\ \text{nm}$$

The blue colour of the sky is, of course, due to Rayleigh scattering and not to electronic absorption by O_2 molecules.

14.1.5 *Other forms of oxygen*

Ozone

Ozone, O_3, is the triatomic allotrope of oxygen. It is an unstable, blue, diamagnetic gas with a characteristic pungent odour: indeed, it was first detected by means of its smell, as reflected by its name (Greek ὄζειν, *ozein*, to smell) coined by C. F. Schönbein in 1840. Ozone can be detected by its smell in concentrations as low as 0.01 ppm; the maximum permissible concentration for continuous exposure is 0.1 ppm but levels as high as 1 ppm are considered non-toxic if breathed for less than 10 min.

The molecule O_3 is bent, as are the isoelectronic species ONCl and ONO$^-$. Microwave measurements lead to a bond angle of $116.8 \pm 0.5°$ and an interatomic distance of 127.8 (± 0.3) pm between the central O and each of the 2 terminal O atoms as shown in Fig. 14.2a. This implies an O$\cdots$O distance of only 218 pm between the 2 terminal O atoms, compared with the normal van der Waals O$\cdots$O distance of 280 pm. A valence-bond description of the molecule is given by the resonance hybrids in Fig. 14.2b and a MO description of the bonding is indicated in Fig. 14.2c: in this, each O atom forms a σ bond to its neighbour using an sp^2-type orbital, and the 3 atomic p$_\pi$ orbitals can combine to give the 3 MOs shown. There are just sufficient electrons to fill the bonding and nonbonding MOs so that the π system can be termed a 4-electron 3-centre bond. The total bond order for each O–O bond is therefore approximately 1.5 (1 σ bond and half of 1 π-bonding MO). It is instructive to note that SO_2 has a similar structure (angle O–S–O 120°): the much greater stability of this molecule when compared with O_3 has been ascribed, in part, to the possible involvement of d$_\pi$ orbitals on the S atom which would allow the filled nonbonding orbital in O_3 to become bonding in SO_2. Other comparisons of O–O bond orders, interatomic distances and bond energies are in Table 14.4 (p. 720).

Ozone condenses to a deep blue liquid (bp $-111.9°$C) and to a violet-black solid (mp $-192.5°$C). The colour is due to an intense absorption band in the red region of the spectrum between 500–700 nm (λ_{max} 557.4 and 601.9 nm). Both the liquid and the solid are explosive due to decomposition into gaseous O_2. Gaseous ozone is also thermodynamically unstable with respect to decomposition into dioxygen though it decomposes only slowly, even at 200°, in the absence of catalysts or ultraviolet:

$$\tfrac{3}{2}O_2(g) \longrightarrow O_3(g); \quad \Delta H_f^\circ + 142.7\ \text{kJ mol}^{-1}; \quad \Delta G_f^\circ + 163.2\ \text{kJ mol}^{-1}$$

---

[16] E. A. OGRYZLO, Why liquid oxygen is blue, *J. Chem. Educ.* **42**, 647–8 (1965).

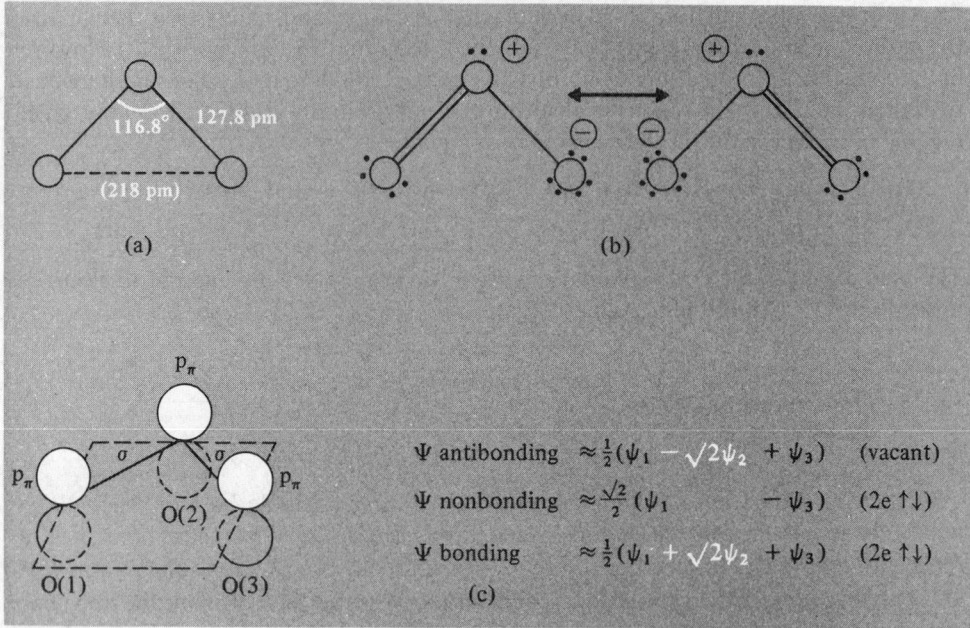

FIG. 14.2   (a) Geometry of the O_3 molecule, (b) valence-bond resonance description of the bonding in O_3, and (c) orbitals used in the MO description of the bonding in O_3, where ψ_1 is the $2p_\pi$ orbital of O(1), etc.

Other properties of ozone (which can be compared with those of dioxygen on p. 706) are: density at $-119.4°C$ 1.354 g cm^{-3} (liquid), density at $-195.8°C$ 1.728 g cm^{-3} (solid), viscosity at $-183°C$ 1.57 centipoise, dipole moment 0.54 D. Liquid ozone is miscible in all proportions with CH_4, CCl_2F_2, $CClF_3$, CO, NF_3, OF_2, and F_2 but forms two layers with liquid Ar, N_2, O_2, and CF_4.

A particularly important property of ozone is its strong absorption in the ultraviolet region of the spectrum between 220–290 nm (λ_{max} 255.3 nm); this protects the surface of the earth and its inhabitants from the intense ultraviolet radiation of the sun, and the possible depletion of this shield has been the subject of urgent study during the past decade.[17, 18, 18a, 18b] Possible chemical reactions with the oxides of nitrogen and with

[17] B. A. THRUSH, The chemistry of the stratosphere and its pollution, *Endeavour* (New Series) **1**, 3–6 (1977).

[18] R. J. DONOVAN, Chemistry and pollution of the stratosphere, *Educ. Chem.* **15**, 110–13 (1978).

[18a] Panel on Atmospheric Chemistry (Assembly of Mathematical and Physical Sciences, National Reasearch Council), *Halocarbons: Effects on Stratospheric Ozone*, National Academy of Sciences, Washington DC, 1976, 352 pp.

[18b] A. K. BISWAS (ed.), *Strategy for the Ozone Layer*, Pergamon Press, Oxford, 1980, 350 pp. A synthesis of papers based on the United Nations Environment Programme meeting on the ozone layer, Washington DC, and including a comprehensive survey of current research in 5 European countries and the USA, Canada, and Australia.

chlorofluoroalkanes have been identified but the extent to which they occur and whether the effects are irreversible have yet to be established.[19,20] See also pp. 323, 471, 904.

Ozone is best prepared by flowing O_2 at 1 atm and 25° through concentric metallized glass tubes to which low-frequency power at 50–500 Hz and 10–20 kV is applied to maintain a silent electric discharge (see also p. 712). The ozonizer tube, which becomes heated by dielectric loss, should be kept cooled to room temperature and the effluent gas, which contains up to 10% O_3 at moderate flow rates, can be used directly or fractionated if higher concentrations are required. Reaction proceeds via O atoms at the surface M, via excited O_2^* molecules, and by dissociative ion recombination:

$$O_2 + O + M \longrightarrow O_3 + M^*; \quad \Delta H^\circ_{298} - 109 \text{ kJ mol}^{-1}$$
$$O_2 + O_2^* \longrightarrow O_3 + O$$
$$O_2^+ + O_2^- \longrightarrow O_3 + O$$

However, the reverse reaction of ozone with atomic oxygen is highly exothermic and must be suppressed by trapping out the ozone if good yields are to be obtained:

$$O_3 + O \longrightarrow 2O_2; \quad \Delta H^\circ_{298} - 394 \text{ kJ mol}^{-1}$$

An alternative route to O_3 is by ultraviolet irradiation of O_2: this is useful for producing low concentrations of O_3 for sterilization of foodstuffs and disinfection, and also occurs during the generation of photochemical smog. The electrolysis of cold aqueous H_2SO_4 (or $HClO_4$) at very high anode current densities also affords modest concentrations of O_3, together with O_2 and $H_2S_2O_8$ (p. 846) as byproducts. Other reactions in which O_3 is formed are the reaction of elementary F_2 with H_2O (p. 939) and the thermal decomposition of periodic acid at 130° (p. 1022).

The concentration of ozone in O_2/O_3 mixtures can be determined by catalytic decomposition to O_2 in the gas phase and measurement of the expansion in volume. More conveniently it can be determined iodometrically by passing the gas mixture into an alkaline boric-acid-buffered aqueous solution of KI and determining the I_2 so formed by titration with sodium thiosulfate in acidified solution:

$$O_3 + 2I^- + H_2O \longrightarrow O_2 + I_2 + 2OH^-$$

The reaction illustrates the two most characteristic chemical properties of ozone: its strongly oxidizing nature and its tendency to transfer an O atom with coproduction of O_2. Standard reduction potentials in acid and in alkaline solution are:

$$O_3 + 2H^+ + 2e^- \rightleftharpoons O_2 + H_2O; \quad E^\circ + 2.076 \text{ V}$$
$$O_3 + H_2O + 2e^- \rightleftharpoons O_2 + 2OH^-; \quad E^\circ + 1.24 \text{ V}$$

The acid potential is exceeded only by fluorine (p. 939), perxenate (p. 1057), atomic O, the OH radical, and a few other such potent oxidants. Decomposition is rapid in acid solutions but the allotrope is much more stable in alkaline solution. At 25° the half-life of O_3 in 1 M NaOH is ~2 min; corresponding times for 5 M and 20 M NaOH are 40 min and 83 h respectively.

[19] L. DOTTO and H. SCHIFF, *The Ozone War*, Doubleday, Garden City, New York, 1978, 342 pp. A highly readable, somewhat partisan account of the debate in the USA reviewed by B. P. Block in The continuing controversy about ozone, *Chem. Eng. News* 16 July 1979, pp. 34 and 43.
[20] J. L. FOX, Atmospheric ozone issue looms again, *Chem. Eng. News* 15 October 1979, pp. 25–35.

The highly reactive nature of O_3 is further typified by the following reactions:

$$CN^- + O_3 \longrightarrow OCN^- + O_2$$

$$2NO_2 + O_3 \longrightarrow N_2O_5 + O_2$$

$$PbS + 4O_3 \longrightarrow PbSO_4 + 4O_2$$

$$3I^- + O_3 + 2H^+ \longrightarrow I_3^- + O_2 + H_2O$$

$$2Co^{2+} + O_3 + 2H^+ \longrightarrow 2Co^{3+} + O_2 + H_2O$$

An important reaction of ozone is the formation of ozonides MO_3. The formation of a red coloration when O_3 is passed into concentrated aqueous alkali was first noted by C. F. Schönbein in 1866, but the presence of O_3^- was not established until 1949.[21] The compounds are best prepared by action of gaseous O_3 on dry, powdered MOH below $-10°$, followed by extraction with liquid ammonia (which may also catalyse their formation). Mechanistic studies using ^{18}O suggest that the overall stoichiometry is:[22]

$$5O_3 + 2KOH \longrightarrow 2KO_3 + 5O_2 + H_2O$$

The compounds are red-brown paramagnetic solids ($\mu = 1.67$ BM) and they decrease in stability in the sequence $Cs > Rb > K > Na$; unsolvated LiO_3 has not been prepared but the ammine $LiO_3.4NH_3$ is known. Likewise the stability of $M^{II}(O_3)_2$ decreases in the sequence $Ba > Sr > Ca$. Above room temperature MO_3 decomposes to the superoxide MO_2 (p. 720) and the compounds are also hydrolytically unstable:

$$MO_3 \xrightarrow{\text{warm}} MO_2 + \tfrac{1}{2}O_2$$

$$KO_3 + H_2O \longrightarrow KOH + O_2 + \{OH\}$$

$$2\{OH\} \longrightarrow H_2O + \tfrac{1}{2}O_2$$

The ozonide ion O_3^- is expected to have C_{2v} symmetry like O_3 itself and the isoelectronic, paramagnetic molecule ClO_2 (p. 989). However, structural details are hard to obtain because of the thermal instability of the compounds and their great reactivity. So far, single crystals have not been obtained, but an X-ray powder diagram has shown that KO_3 is structurally related to KN_3.[21] This was previously (incorrectly) taken to imply that O_3^- was linear, but subsequent work has shown that both compounds are structurally related to the cubic CsCl structure type, the Cl^- being replaced either by a linear N_3^- or a bent O_3^- ion respectively in the now tetragonal unit cell.[23] The angle O–O–O was $100°$ and the O–O interatomic distance 119 pm; this, again, is surprising since the formal bond order in O_3^- is only 1.25 yet the interatomic distance is apparently less than that in O_3 itself (bond order 1.5, O–O 127.8 pm) or even in O_2 (bond order 2.0, O–O 121.0 pm). There is also the suggestion that a second, monoclinic form of KO_3

[21] I. A. KAZARNOVSKII, G. P. NIKOL'SKII and T. A. ABLETSOVA, A new oxide of potassium, *Dokl. Akad. Nauk SSSR* **64**, 69–72 (1949).

[22] I. I. VOL'NOV, V. N. CHAMOVA and E. I. LATYSHEVA, Mechanism of metal ozonide formation, *Izvest. Akad. Nauk SSSR*, Ser. Khim. 1967, 1183–7.

[23] L. V. AZÁROV and I. CORVIN, Crystal structure of potassium ozonide, *Proc. Natl. Acad. Sci. (US)* **49**, 1–5 (1963).

exists,[24] isostructural with KNO_2. The story illustrates the difficulty of obtaining reliable structural data on such compounds and a precise structural characterization of O_3^- has yet to be achieved.

Ozone adds readily to unsaturated organic compounds[24a] and can cause unwanted cross-linking in rubbers and other polymers with residual unsaturation, thereby leading to brittleness and fracture. Addition to alkenes yields "ozonides" which can be reductively cleaved by Zn/H_2O (or $I^-/MeOH$, etc.) to yield aldehydes or ketones. This smooth reaction, discovered by C. D. Harries in 1903, has long been used to determine the position of double bonds in organic molecules, e.g.:

butene-1: $EtCH{=}CH_2 \xrightarrow{O_3} EtCH{-}CH_2 \longrightarrow EtCHO + HCHO$

butene-2: $MeCH{=}CHMe \xrightarrow{O_3} MeCH{-}CHMe \longrightarrow 2MeCHO$

Until 1950 the reaction was thought to proceed as follows:

Primary ozonide   Final ozonide

However, the mechanism first suggested by R. Criegie in 1951 is now more generally preferred: this involves electrophilic attack by O_3 on a polarized double bond:

Primary ozonide

The primary ozonides are preparable below $-70°$ but in the temperature range $-70°$ to $-30°$ these rearrange to the final true ozonides. Normally, however, the ozonide is not isolated but is reductively cleaved to aldehydes and ketones in solution. Oxidative

[24] I. J. SOLOMON and A. J. KACMAREK, Sodium ozonide, *J. Phys. Chem.* **64,** 168–9 (1960). Also deals with KO_3.

[24a] P. S. BAILEY, *Ozonation in Organic Chemistry*, Vol. 1, Olefinic Compounds, Academic Press, New York, 1978, 272 pp.; Vol. 2, Nonolefinic Compounds, 1982, 496 pp.

cleavage (air or O_2) yields carboxylic acids and, indeed, the first large-scale application of the reaction was the commercial production of pelargonic and azelaic acids from oleic acid:

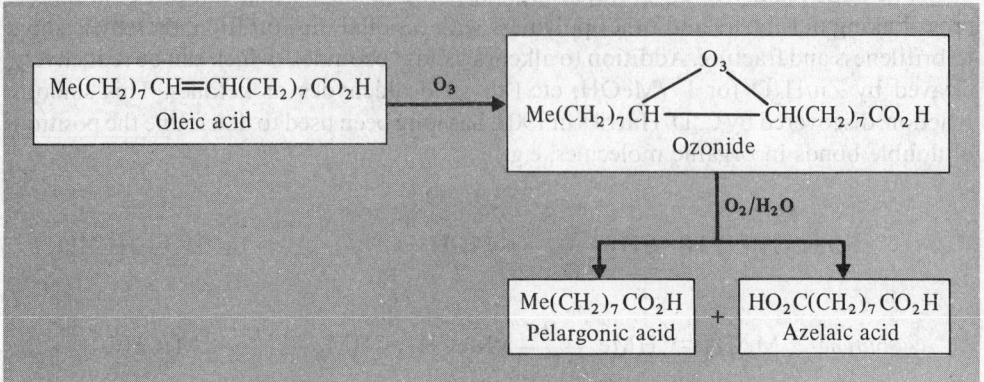

Esters of these acids are used as plasticizers for PVC (polyvinylchloride) and other plastics.

Because of the reactivity, instability, and hazardous nature of O_3 it is always generated on site. Typical industrial ozone generators operate at 1 or 2 atm, 15–20 kV, and 50 or 500 Hz. The concentration of O_3 in the effluent gas depends on the industrial use envisaged but yields of up to 10 kg per h or 150 kg per day from a single apparatus are not uncommon and some plants yield over 1 tonne per day. In addition to pelargonic and azelaic acid production, O_3 is used to make peroxoacetic acid from acetaldehyde and for various inorganic oxidations. At low concentrations it is used (particularly in Europe) to purify drinking water, since this avoids the undesirable taste and smell of chlorinated water, and residual ozone decomposes to O_2 soon after treatment.[24b] Of the 1039 plants operating in 1977 all but 40 were in Europe, with the greatest numbers in France (593), Switzerland (150), Germany (136), and Austria (42). Other industrial uses include the preservation of goods in cold storage, the treatment of industrial waste, and the deodorizing of air and sewage gases.[25]

Atomic oxygen

Atomic oxygen is an extremely reactive, fugitive species which cannot be isolated free from other substances. Many methods of preparing oxygen atoms also yield other reactive or electronically excited species, and this somewhat complicates the study of their properties. Passage of a microwave or electric discharge through purified O_2 gas diluted with argon produces O atoms in the ^3P ground state (2 unpaired electrons). Mercury-sensitized photolysis of N_2O is perhaps a more convenient route to ground state O atoms

[24b] J. KATZ (ed.), *Ozone and Chlorine Dioxide Technology for Disinfection of Drinking Water*, Noyes Data Corp., Park Ridge, New Jersey, 1980, 659 pp. R. G. RICE and M. E. BROWNING, *Ozone Treatment of Industrial Wastewater*, Noyes Data Corp., Park Ridge, New Jersey, 1981, 371 pp.
[25] C. NEBEL, Ozone, *Kirk–Othmer Encyclopedia of Chemical Technology*, 3rd edn. **16**, 683–713, Wiley, New York, 1981.

(plus inert N$_2$ molecules) though they can also be made by photolysis of O$_2$ or NO$_2$. Photolysis of N$_2$O in the absence of Hg gives O atoms in the spin-paired ^1D excited state, and this species can also be obtained by photolysis of O$_3$ or CO$_2$.

The best method for determining the concentration of O atoms is by their extremely rapid reaction with NO$_2$ in a flow system:

$$O + NO_2 \longrightarrow O_2 + NO$$

The NO thus formed reacts more slowly with any excess of O atoms to reform NO$_2$ and this reaction emits a yellow-green glow.

$$NO + O \longrightarrow NO_2{}^* \longrightarrow NO_2 + h\nu$$

The system is thus titrated with NO$_2$ until the glow is sharply extinguished (p. 474).

As expected, atomic O is a strong oxidizing agent and it is an important reactant in the chemistry of the upper atmosphere.[17, 18] Typical reactions are:

$$H_2 + O \longrightarrow OH + H$$

$$OH + O \longrightarrow O_2 + H$$

$$O_2 + O \longrightarrow O_3$$

$$H_2S + O \longrightarrow H_2O, SO_2, SO_3, H_2SO_4$$

$$Cl_2 + O \longrightarrow Cl_2O, ClO_2$$

$$NaCl + O \xrightarrow{\text{aq}} NaClO_3, Cl_2, NaOH$$

$$CO + O \longrightarrow CO_2$$

$$HCN + O \longrightarrow CO, CO_2, NO, H_2O$$

$$CH_4 + O \longrightarrow CO_2, H_2O$$

Many of these reactions are explosive and/or chemiluminescent.

14.1.6 *Chemical properties of dioxygen, O$_2$*

Oxygen is an extremely reactive gas which vigorously oxidizes many elements directly, either at room temperature or above. Despite the high bond dissociation energy of O$_2$ (493.4 kJ mol^{-1}) these reactions are frequently highly exothermic and, once initiated, can continue spontaneously (combustion) or even explosively. Familiar examples are its reactions with carbon (charcoal) and hydrogen. Some elements do not combine with oxygen *directly*, e.g. certain refractory or noble metals such as W, Pt, Au, and the noble gases, though oxo compounds of all elements are known except for He, Ne, Ar, and possibly Kr. This great range of compounds was one of the reasons why Mendeleev chose oxides to exemplify his periodic law (p. 24) and why oxygen was chosen as the standard element for the atomic weight scale in the early days when atomic weights were determined mainly by chemical stoichiometry (p. 18).

Many inorganic compounds and all organic compounds also react directly with O$_2$ under appropriate conditions. Reaction may be spontaneous, or may require initiation by heat, light, electric discharge, chemisorption, or various catalytic means. Oxygen is normally considered to be divalent, though the oxidation state can vary widely and includes the values of $+\frac{1}{2}, 0, -\frac{1}{3}, -\frac{1}{2}, -1$ and -2 in isolable compounds of such species as O_2^+, O_3, O_3^-, O_2^-, O_2^{2-} respectively. The coordination number of oxygen in its compounds also varies widely, as illustrated in Table 14.3 and numerous examples of

TABLE 14.3 *Coordination geometry of oxygen*

CN	Geometry	Examples
0	—	Atomic O
1	—	$O_2, CO, CO_2, NO, NO_2, SO_3(g), OsO_4$; terminal O_t in P_4O_{10}, $[VO(acac)_2]$, and many oxoanions $[MO_n]^{m-}$ (M = C, N, P, As, S, Se, Cl, Br, Cr, Mn, etc.)
2	Linear	Some silicates, e.g. $[O_3Si–O–SiO_3]^{6-}$ in $Sc_2Si_2O_7$; $[Cl_5Ru–O–RuCl_5]^{4-}$; $ReO_3(WO_3)$-type structures; coesite (SiO_2); $[O(SiPh_3)_2]$
2	Bent	O_3, H_2O, H_2O_2, F_2O; silica structures, GeO_2; P_4O_6 and many heterocyclic compounds with O_μ; complexes of ligands which have O_t as donor atom, e.g. $[BF_3(OSMe_2)]$, $[SnCl_4(OSeCl_2)_2]$, $[\{TiCl_4(OPCl_3)\}_2]$, $[HgCl_2(OAsPh_3)_2]$; complexes of O_2, e.g. $[Pt(O_2)(PPh_3)_2]$
3	Planar	$[O(HgCl)_3]^+Cl^-$, $Mg[OB_6O_6(OH)_6].4\frac{1}{2}H_2O$ (macallisterite); $Sr[OB_6O_8(OH)_2].3H_2O$, (tunellite); rutile-type structures, e.g. MO_2 (M = Ti; V, Nb, Ta; Cr, Mo, W; Mn, Tc, Re; Ru, Os; Rh, Ir; Pt; Ge, Sn, Pb; Te)
3	Pyramidal	$[H_3O]^+$; hydrato-complexes, e.g. $[M(H_2O)_6]^{n+}$; complexes of R_2O and crown ethers, etc.
4	Square planar	NbO (see text)
4	Tetrahedral	$[OBe_4(O_2CMe)_6]$; CuO, AgO, PdO; wurtzite structures, e.g. BeO, ZnO; corundum structures, e.g. M_2O_3 (M = Al, Ga, Ti, V, Cr, Fe, Rh); fluorite structures, e.g. MO_2 (M = Zr, Hf; Ce, Pr, Tb; Th, U, Np, Pu, Am, Cm; Po)
5		—
6	Octahedral	Central O in $[Mo_6O_{19}]^{2-}$; many oxides with NaCl-type structure, e.g. MO (M = Mg, Ca, Sr, Ba; Mn, Fe, Co, Ni; Cd; Eu)
7		—
8	Cubic	Anti-fluorite-type structure, e.g. M_2O (M = Li, Na, K, Rb)

stable compounds are known which exemplify each coordination number from 1 to 8 (with the possible exceptions of 5 and 7, for which unambiguous examples are more difficult to find). Most of these examples are straightforward and structural details will be found at appropriate points in the text. Linear 2-coordinate O occurs in the silyl ether $[O(SiPh_3)_2]$.[25a] Planar 3-coordinate O occurs in both cationic and anionic complexes (Fig. 14.3a, b), in two-dimensional layer lattices (Fig. 14.3c) and in the three-dimensional rutile structure (p. 1120). Planar 4-coordinate O occurs uniquely in NbO which can be considered as a defect-NaCl-type structure with O and Nb vacancies at (000) and $(\frac{1}{2}\frac{1}{2}\frac{1}{2})$ respectively, thereby having only 3 NbO (rather than 4) per unit cell as shown in Fig. 14.4 Tetrahedral, 4-coordinate O is featured in "basic beryllium acetate" (p. 135) and in

[25a] C. GLIDEWELL and D.C. LILES, A neutral silicon ether containing linear Si–O–Si, *JCS Chem. Comm.* 1977, 682.

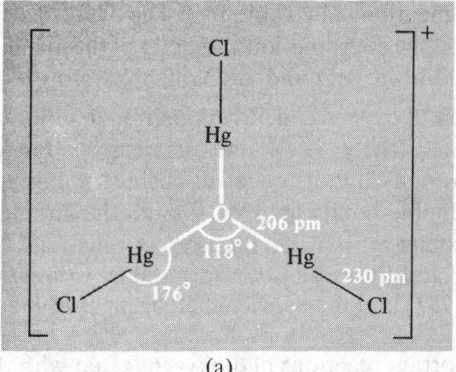

(a)

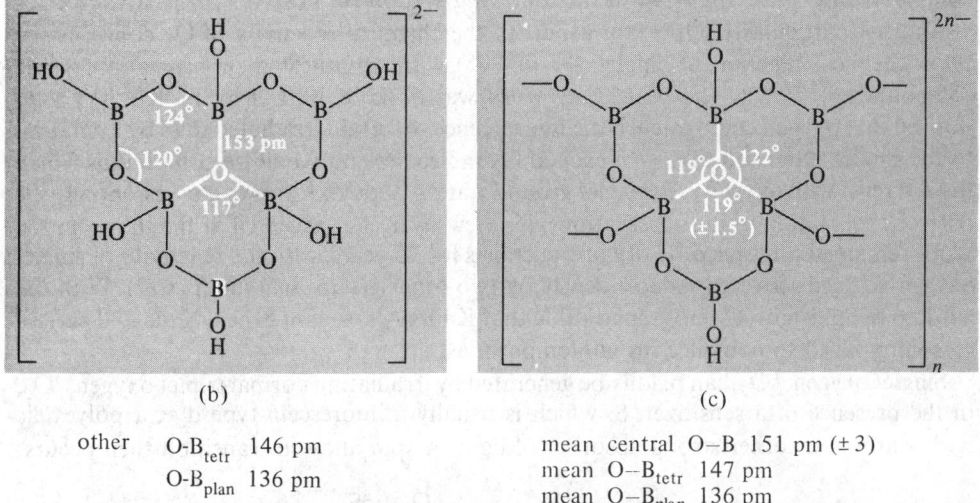

(b)   (c)

other O-B$_{\text{tetr}}$ 146 pm   mean central O–B 151 pm ($\pm$ 3)

     O-B$_{\text{plan}}$ 136 pm   mean O–B$_{\text{tetr}}$ 147 pm

                    mean O–B$_{\text{plan}}$ 136 pm

FIG. 14.3 Examples of planar 3-coordinate O: (a) the cation in [O(HgCl)$_3$]Cl, (b) the central O atom in the discrete borate anion [OB$_6$O$_6$(OH)$_6$]$^{2-}$ in macallisterite—the three heterocycles are coplanar but the 6 pendant OH groups lie out of the plane, and (c) the repeat unit in the polymeric anion [OB$_6$O$_8$(OH)$_2$]$_n^{2n-}$ in tunellite.

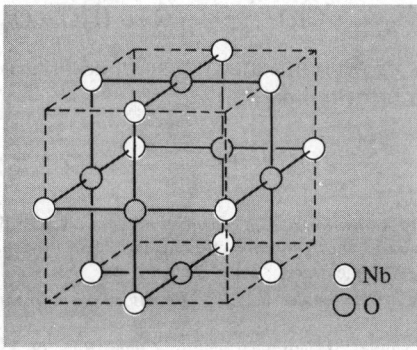

FIG. 14.4 Square-planar 4-coordinate O in the defect NaCl-type structure of NbO.

many binary oxides as mentioned in Table 14.3. The detailed structure depends both on the stoichiometry and on the coordination geometry of the metal, which is planar in CuO, AgO and PdO, tetrahedral in BeO and ZnO, octahedral in M_2O_3 and cubic in MO_2.

Much of the chemistry of oxygen can be rationalized in terms of its electronic structure $(2s^2 2p^4)$, high electronegativity (3.5), and small size. Thus, oxygen shows many similarities to nitrogen (p. 473) in its covalent chemistry, and its propensity to form H bonds (p. 57) and p_π double bonds (p. 478), though the anionic chemistry of O^{2-} and OH^- is much more extensive than for the isoelectronic ions N^{3-}, NH^{2-} and NH_2^-. Similarities to fluorine and fluorides are also notable. Comparisons with the chemical properties of sulfur (p. 782) and the heavier chalcogens (p. 890) are deferred to Chapters 15 and 16.

One of the most important reactions of dioxygen is that with the protein haemoglobin which forms the basis of oxygen transport in blood (p. 1276).[25b] Other coordination complexes of O_2 are discussed in the following section (p. 718).

Another particularly important aspect of the chemical reactivity of O_2 concerns the photochemical reaction of singlet O_2 (p. 705) with unsaturated or aromatic organic compounds.[26, 27, 27a] The pioneering work was done in 1931–9 by H. Kautsky who noticed that oxygen could quench the fluorescence of certain irradiated dyes by excitation to the singlet state, and that such excited O_2 molecules could oxidize compounds which did not react with oxygen in its triplet ground state. Although Kautsky gave essentially the correct explanation of his observations, his views were not accepted at the time and the work remained unnoticed by organic chemists for 25 years until the reactivity of singlet oxygen was rediscovered independently by two other groups in 1964 (p. 699). With the wisdom of hindsight it seems remarkable that Kautsky's elegant experiments and careful reasoning failed to convince his contemporaries.

Singlet oxygen, $^1O_2$, can readily be generated by irradiating normal triplet oxygen, $^3O_2$, in the presence of a sensitizer, S, which is usually a fluorescein-type dye, a polycyclic hydrocarbon, or other strong absorber of light. A spin-allowed transition then occurs:

$$^3O_2 + {}^1S \xrightarrow{\;h\nu\;} {}^1O_2 + {}^3S$$

Provided that the energy gap in the sensitizer is greater than 94.7 kJ mol^{-1}, the $^1\Delta_g$ singlet state of O_2 is generated (p. 705). Above 157.8 kJ mol^{-1} some $^1\Sigma_g^+$ O_2 is also produced and this species predominates above 200 kJ mol^{-1}. The $^1\Delta_g$ singlet state can also be conveniently generated chemically in alcoholic solution by the reaction

$$H_2O_2 + ClO^- \longrightarrow Cl + H_2O + O_2(^1\Delta_g)$$

Another chemical route is by decomposition of solid adducts of ozone with triaryl and other phosphites at subambient temperatures:

$$O_3 + P(OPh)_3 \xrightarrow{\;-78°\;} O_3P(OPh)_3 \xrightarrow{\;-15°\;} O_2(^1\Delta_g) + OP(OPh)_3$$

[25b] T. G. SPIRO (ed.), *Metal Ion Activation of Dioxygen*, Wiley, New York, 1980, 247 pp. Focuses on metal ions in biological systems including haem proteins, cytochrome-P450, dioxygenases, cytochrome c-oxidase, superoxide dismutases and oxygen toxicity.

[26] B. RÅNBY and J. F. RABEK (eds.) *Singlet Oxygen: Reactions with Organic Compounds and Polymers*, Wiley, Chichester, 1978, 331 pp.

[27] A. A. FRIMER, Reactions of singlet oxygen with olefins, *Chem. Rev.* **79**, 359–87 (1979).

[27a] H. H. WASSERMAN and R. W. MURRAY (eds.), *Singlet Oxygen*, Academic Press, New York, 1979, 688 pp.

Reactions of $^1O_2$ can be classified into three types: 1,2 addition, 1,3 addition and 1,4 addition. The 1,2 addition reaction converts sterically hindered alkenes into dioxetanes which can readily be cleaved either thermolytically or photolytically into two carbonyl-containing species:

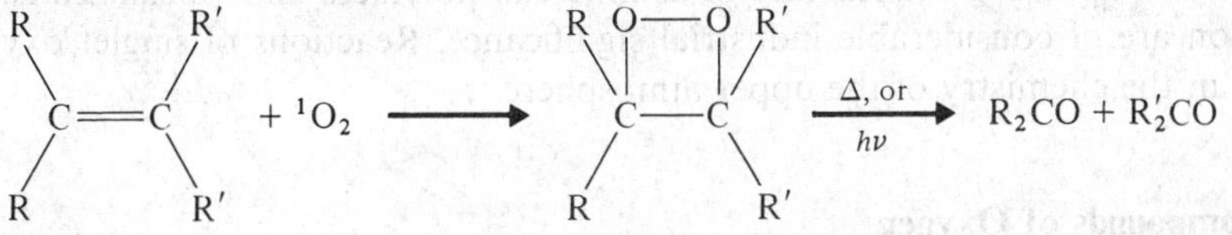

Other electron-rich compounds that undergo this [2+2] cycloaddition of singlet oxygen are strained alkynes, ketenes, allenes, sulfines, oximes, etc.

The second type of reaction (1,3 addition or "ene" reaction) occurs with alkenes having at least one allylic hydrogen: allylic hydroperoxides are formed in which the double bond has shifted to an adjacent position, e.g.:

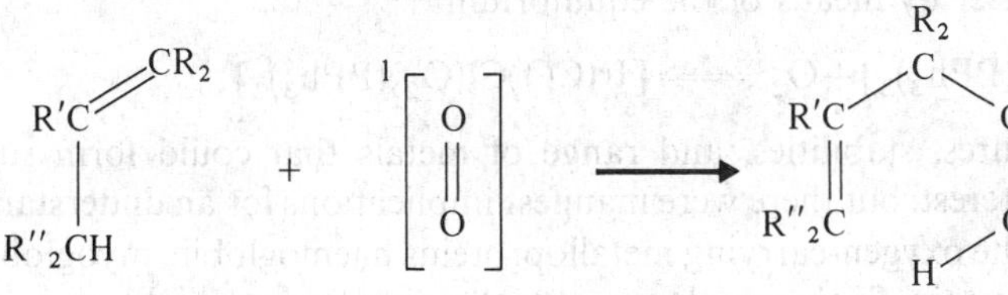

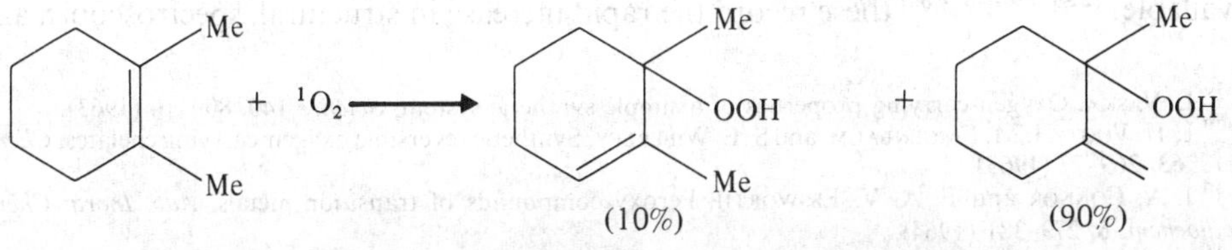

The third type of reaction involves 1,4 addition of $^1O_2$ to a *cisoid* 1,3-diene to produce an *endo*-peroxide, e.g.:

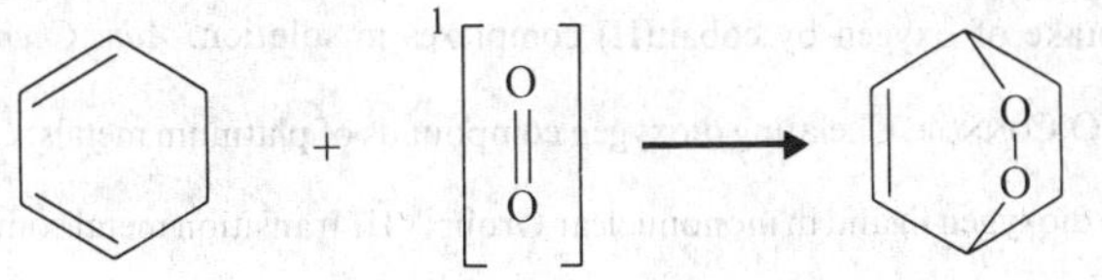

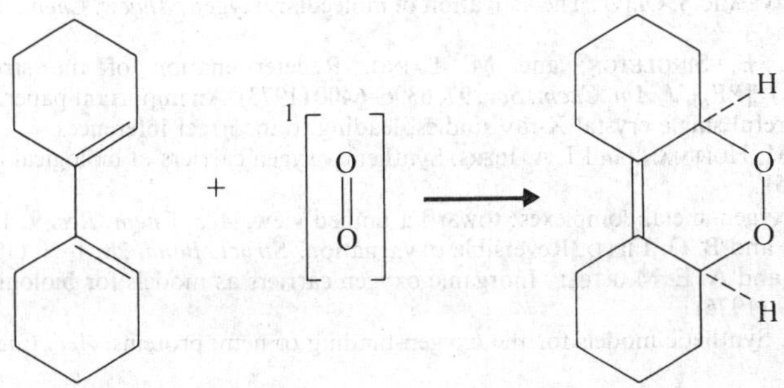

This reaction type is analogous to a photo-induced Diels–Alder reaction, with $^1O_2$ as the dienophile.

In addition to its importance in synthetic organic chemistry, singlet oxygen plays an important role in autoxidation (i.e. the photodegradation of polymers in air), and methods of improving the stability of commercial polymers and vulcanized rubbers to oxidation are of considerable industrial significance. Reactions of singlet oxygen also feature in the chemistry of the upper atmosphere.

14.2 Compounds of Oxygen

14.2.1 *Coordination chemistry: dioxygen as a ligand*

Few discoveries in synthetic chemistry during the past two decades have caused more excitement or had more influence on the direction of subsequent work than L. Vaska's observation in 1963 that the planar 16-electron complex *trans*-[Ir(CO)Cl(PPh$_3$)$_2$] can act as a reversible oxygen carrier by means of the equilibrium[28]

$$[Ir(CO)Cl(PPh_3)_2] + O_2 \rightleftharpoons [Ir(CO)Cl(O_2)(PPh_3)_2]$$

Not only were the structures, stabilities, and range of metals that could form such complexes of theoretical interest, but there were manifest implications for an understanding of the biochemistry of the oxygen-carrying metalloproteins haemoglobin, myoglobin, haemerythrin, and haemocyanin. Such complexes were also seen as potential keys to an understanding of the interactions occurring during homogeneous catalytic oxidations, heterogeneous catalysis, and the action of metalloenzymes. Several excellent reviews are available: [25b, 29–46, 46a] these record the rapid increase in structural, spectroscopic, and

[28] L. VASKA, Oxygen-carrying properties of a simple synthetic system, *Science* **140**, 809–10 (1963).

[29] L. H. VOGT, H. M. FAIGENBAUM, and S. E. WIBERLEY, Synthetic reversible oxygen-carrying chelates, *Chem. Revs.* **63**, 269–77 (1963).

[30] J. A. CONNOR and E. A. V. EBSWORTH, Peroxy compounds of transition metals, *Adv. Inorg. Chem. Radiochem.* **6**, 279–381 (1964).

[31] E. BAYER and P. SCHRETZMANN, Reversible oxygenation of metal complexes, *Struct. Bond.* **2**, 181–250 (1967). (in German).

[32] A. G. SYKES and J. A. WEIL, The formation, structure, and reactions of binuclear complexes of cobalt, *Prog. Inorg. Chem.* **13**, 1–106 (1970).

[33] R. G. WILKINS, Uptake of oxygen by cobalt(II) complexes in solution, *Adv. Chem. Ser.* **100**, 111–34 (1971).

[34] V. J. CHOY and C. J. O'CONNOR, Chelating dioxygen compounds of platinum metals, *Coord. Chem. Rev.* **9**, 145–70 (1972/3).

[35] J. S. VALENTINE, The dioxygen ligand in mononuclear Group VIII transition metal complexes, *Chem. Revs.* **73**, 235–45 (1973).

[36] G. HENRICI-OLIVÉ and S. OLIVÉ, The activation of molecular oxygen, *Angew. Chem.*, Int. Edn. (Engl.) **13**, 29–38 (1974).

[37] M. J. NOLTE, E. SINGLETON, and M. LAING, Redetermination of the structure of [Ir(O$_2$)(Ph$_2$PCH$_2$CH$_2$PPh$_2$)$_2$][PF$_6$], *J. Am. Chem. Soc.* **97**, 6396–6400 (1975). An important paper showing how errors can arise even in careful single crystal X-ray studies, leading to incorrect inferences.

[38] F. BASOLO, B. M. HOFFMAN, and J. A. IBERS, Synthetic oxygen carriers of biological interest, *Acc. Chem. Res.* **8**, 384–92 (1975).

[39] L. VASKA, Dioxygen–metal complexes: toward a unified view, *Acc. Chem. Res.* **9**, 175–83 (1976).

[40] R. W. ERSKINE and B. O. FIELD, Reversible oxygenation, *Struct. Bond.* **28**, 1–50 (1976).

[41] G. MOLENDON and A. E. MARTELL, Inorganic oxygen carriers as models for biological systems, *Coord. Chem. Rev.* **19**, 1–39 (1976).

[42] J. P. COLLMAN, Synthetic models for the oxygen-binding of hemoproteins, *Acc. Chem. Res.* **10**, 265–72 (1977).

mechanistic data available, and reveal the developing perceptions of the nature of the bonding involved. This extensive bibliography also affords an instructive case history of how at least one area of chemistry is at present advancing—the later reviews incorporate corrections to earlier data and, being based on a much greater range of information, present a more synoptic view of the field as at present delineated.

Dioxygen–metal complexes in which there is a 1:1 stoichiometry of O_2:M are of two main types, usually designated Ia (or superoxo) and IIa (or peroxo) for reasons which will shortly become apparent (Fig. 14.5). Dioxygen can also form 1:2 complexes in which O_2

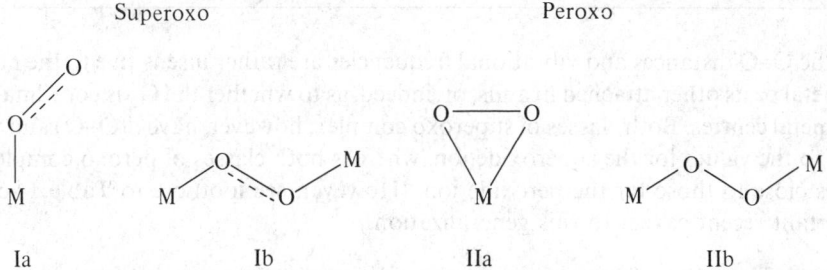

FIG. 14.5   The four main types of O_2–M geometry. The bridging modes Ib and IIb appear superficially similar but differ markedly in dihedral angles and other bonding properties. See also footnote to Table 14.5 for the recently established μ,η^1-superoxide bridging mode.

adopts a bidentate bridging geometry, labelled Ib and IIb in Fig. 14.5. Of these four classes of complex, the Vaska-type IIa peroxo complexes, are by far the most widespread amongst the transition metals, though many are not reversible oxygen carriers and some are formed by deprotonation of H_2O_2 (p. 747) rather than coordination of molecular O_2. By contrast, the bridging superoxo type Ib is known only for the green cobalt complexes formed by 1-electron oxidation of the corresponding IIb peroxo compounds. In all cases complex formation is accompanied by a significant increase in the O–O interatomic distances and a considerable decrease in the v(O–O) vibrational stretching frequency. Both effects are more marked for the peroxo (type II) complexes than for the superoxo (type I) complexes and have been interpreted in terms of a transfer of electrons from M into the antibonding orbitals of O_2 (p. 706) thereby weakening the O–O bond. The magnitude of the effects to be expected can be gauged from Table 14.4.[46] Comparative data for a wide range of dioxygen–metal complexes is in Table 14.5.[39, 46] It will be noted

[43] R. S. DRAGO, T. BEUGELSDIJK, J. A. BREESE and J. P. CANNADY, The relationship of thermodynamic data for base adduct formation with cobalt protoporphyrin IX dimethyl ester to the corresponding enthalpies of forming dioxygen adducts with implications to oxygen binding cooperativity, *J. Am. Chem. Soc.* **100**, 5374–82 (1978). (This discusses an alternative description of the Co–O_2 bonding in terms of the coupling of the spin of an unpaired electron in the d_{z^2} orbital of cobalt(II) with one electron in the π-antibonding orbital of O_2.)

[44] A. B. P. LEVER and H. B. GRAY, Electronic spectra of metal-dioxygen complexes, *Acc. Chem. Res.* **11**, 348–55 (1978).

[45] D. A. SUMMERVILLE, R. D. JONES, B. M. HOFFMAN and F. BASOLO, Assigning oxidation states to dioxygen complexes, *J. Chem. Educ.* **56**, 157–62 (1979).

[46] R. D. JONES, D. A. SUMMERVILLE and F. BASOLO, Synthetic oxygen carriers related to biological systems, *Chem. Revs.* **79**, 139–79 (1979).

[46a] A. B. P. LEVER, G. A. OZIN and H. B. GRAY, Electron transfer in metal–dioxygen adducts, *Inorg. Chem.* **19**, 1823–4 (1980).

TABLE 14.4 *Effect of electron configuration and charge on the bond properties of dioxygen species*

Species	Bond order	Compound	d(O–O)/pm	Bond energy/ kJ mol^{-1}	v(O–O)/cm^{-1}
O_2^+	2.5	$O_2[AsF_6]$	112.3	625.1	1858
$O_2(^3\Sigma_g^-)$	2	$O_2(g)$	120.7	490.4	1554.7
$O_2(^1\Delta_g)$	2	$O_2(g)$	121.6	396.2	1483.5
O_2^- (superoxide)	1.5	$K[O_2]$	128	—	1145
O_2^{2-} (peroxide)	1	$Na_2[O_2]$	149	204.2	842
–OO–	1	H_2O_2 (cryst)	145.3	213	882

that the O–O distances and vibrational frequencies are rather insensitive to the nature of the metal or its other attached ligands, or, indeed, as to whether the O_2 is coordinated to 1 or 2 metal centres. Both classes of superoxo complex, however, have d(O–O) and v(O–O) close to the values for the superoxide ion, whereas both classes of peroxo complex have values close to those for the peroxide ion. (However, see footnote to Table 14.5 for an important recent caveat to this generalization.)

TABLE 14.5 *Summary of properties of known dioxygen–metal complexes*[a]

Complex type	O_2 : M ratio	Structure	d(O–O)/pm (normal range)	v (O–O)/cm^{-1} (normal range)
superoxo Ia	1 : 1	O----O / M	125–135	1130–1195
superoxo Ib	1 : 2	M–O–O–M	126–136	1075–1122
peroxo IIa	1 : 1	O—O / M	130–155	800–932
peroxo IIb	1 : 2	M–O–O–M	144–149	790–884

[a] Reaction of K_2O with Al_2Me_6 in the presence of dibenzo-18-crown-6 (p. 107) yields the surprisingly stable anion $[(\mu,\eta^1-O_2)(AlMe_3)_2]^-$ in which one O of the superoxo ion bridges the 2 Al atoms (angle Al–O–Al 128°):

$$
\begin{array}{c}
O \\
| \\
O \\
Al \diagdown \quad \diagup Al
\end{array}
$$

In this new type of coordination mode d(O–O) is long (147 pm) and the weakness of the O–O linkage is also shown by the very low value of 851 cm^{-1} for v(O–O), both values being more characteristic of peroxo than of superoxo complexes.[46b]

Superoxo complexes having a nonlinear M–O–O configuration are known at present only for Fe, Co, Rh and perhaps a few other transition metals, whereas the Vaska-type

$$
M \diagdown\hspace{-0.3em}\diagup \begin{array}{c} O \\ | \\ O \end{array}
$$

complexes are known for almost all the transition metals except those in the Sc and Zn

[46b] D. C. HRNCIR, R. D. ROGERS, and J. L. ATWOOD, New bonding mode for a bridging dioxygen ligand: crystal and molecular structure of $[K(dibenzo\text{-}18\text{-}crown\text{-}6)][Al_2Me_6(O_2)].1.5C_6H_6$, *J. Am. Chem. Soc.* **103**, 4277–8 (1981).

groups and possibly Mn, Cr, Fe, and Cu. The two modes of formation are illustrated in Fig. 14.6 for the two reactions:

$$[\text{Co}(3\text{-Bu}^t\text{Salen})_2] \xrightarrow{\text{O}_2/\text{py}} [\text{Co}(3\text{-Bu}^t\text{Salen})_2(\text{O}_2)\text{py}]$$

$$[\text{Ir}(\text{CO})\text{Cl}(\text{PPh}_3)_2] \xrightarrow{\text{O}_2/\text{C}_6\text{H}_6} [\text{Ir}(\text{CO})\text{Cl}(\text{O}_2)(\text{PPh}_3)_2]$$

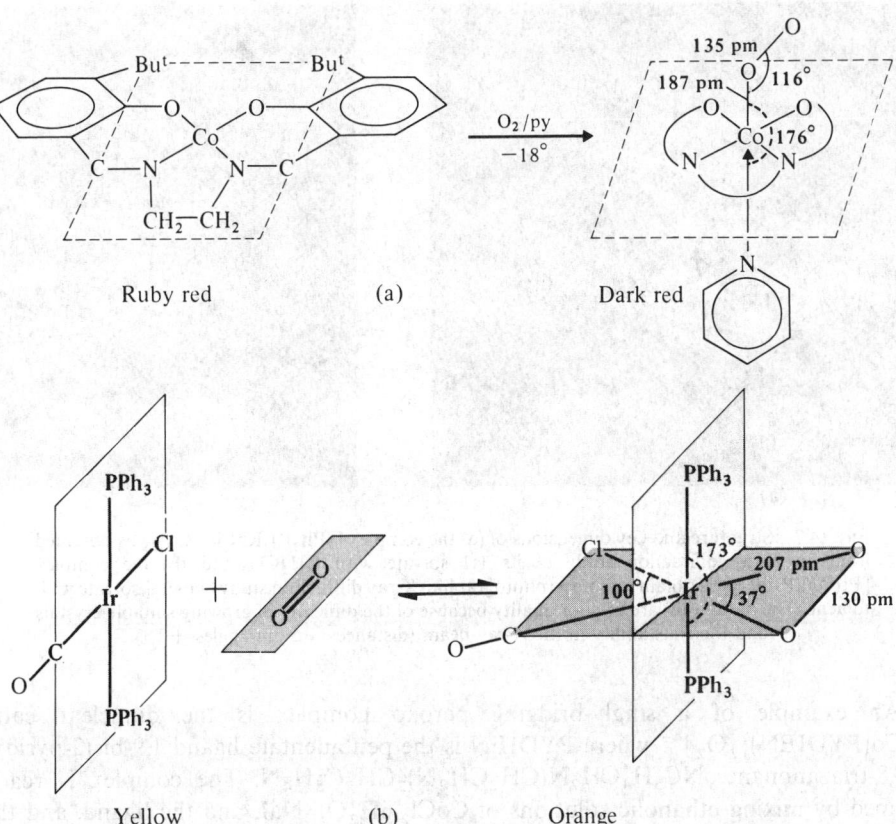

Ruby red   (a)   Dark red

Yellow   (b)   Orange

FIG. 14.6   (a) Reaction of *N,N'*-ethylenebis(3-*tert*-salicylideniminato)cobalt(II) with dioxygen and pyridine to form the superoxo complex [Co(3-Bu^{t}Salen)$_2$(O$_2$)py]; the py ligand is almost coplanar with the Co–O–O plane, the angle between the two being 18°.[47] (b) Reversible formation of the peroxo complex [Ir(CO)Cl(O$_2$)(PPh$_3$)$_2$]. The more densely shaded part of the complex is accurately coplanar.[48]

The sensitivity of the reaction type to the detailed nature of the bonding in the metal complex can be gauged from the fact that neither the PMe$_3$ analogue of Vaska's iridium complex, nor the corresponding rhodium complex [Rh(CO)Cl(PPh$_3$)$_2$] react with

[47] W. P. SCHAEFFER, B. T. HUIE, M. G. KURILLA and S. E. EALICK, Oxygen-carrying cobalt complexes. 10. Structures of *N,N'*-ethylenebis(3-*tert*-butylsalicylideniminato)cobalt(II) and its monomeric dioxygen adduct. *Inorg. Chem.* **19**, 340–4 (1980).
[48] S. J. LAPLACA and J. A. IBERS, Structure of [Ir(CO)Cl(O$_2$)(PPh$_3$)$_2$]. The oxygen adduct of a synthetic reversible molecular oxygen carrier, *J. Am. Chem. Soc.* **87**, 2581–6 (1965).

dioxygen in this way. By contrast, the closely related red complex $[RhCl(PPh_3)_3]$ reacts readily with O_2 in CH_2Cl_2 solution with elimination of PPh_3 to give the brown dinuclear doubly bridging complex $[(Ph_3P)_2RhCl(\mu-O_2)]_2 \cdot CH_2Cl_2$ the structure of which is shown in Fig. 14.7a.[49] With $[Pt(PPh_3)_4]$ reaction also occurs with elimination of PPh_3 but the product is the yellow, planar, mononuclear complex $[Pt(O_2)(PPh_3)_2]$ (Fig. 14.7b).[50]

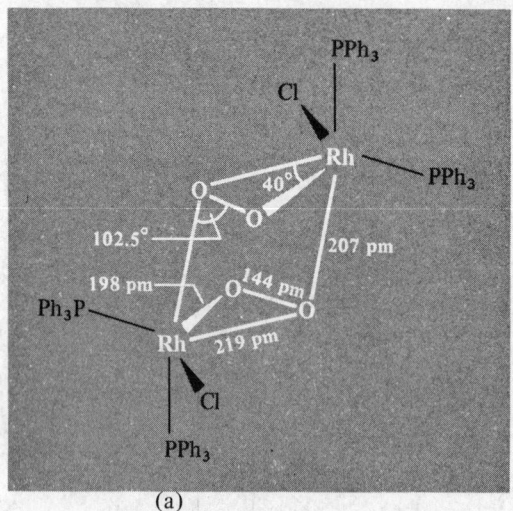

(a)

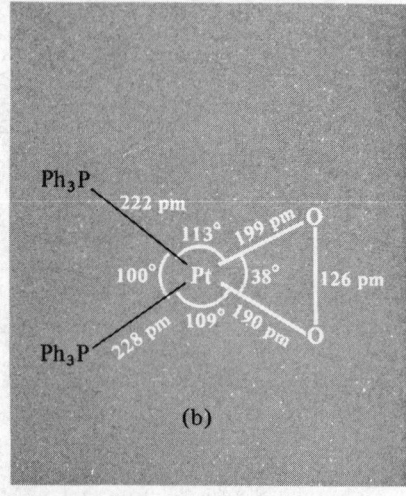

(b)

FIG. 14.7   Structure and key dimensions of (a) the complex $[(Ph_3P)_2RhCl(\mu-O_2)]_2$ as obtained from an X-ray diffraction study of its 1:1 solvate with CH_2Cl_2, and (b) the complex $[Pt(O_2)(PPh_3)_2]$ as obtained from a (photographic) X-ray diffraction study of its 1:1 solvate with toluene. The data in (b) are of poor quality because of the difficulty of growing suitable crystals and their instability in the X-ray beam (distances ± 5 pm angles $\pm 2°$).

An example of a singly-bridging peroxo complex is the dinuclear cation $[\{Co(PYDIEN)\}_2O_2]^{4+}$ where PYDIEN is the pentadentate ligand 1,9-bis(2-pyridyl)-2,5,8-triazanonane, $NC_5H_4CH_2N(CH_2CH_2N)_2CH_2C_5H_4N$. The complex is readily formed by mixing ethanolic solutions of $CoCl_2.6H_2O$, NaI, and the ligand, and then exposing the resulting solution to oxygen.[51] Some structural details are in Fig. 14.8. Such $\mu-O_2$ complexes are formed generally among the "Group VIII" metals (Fe), Ru, Os; Co, Rh, Ir; Ni, Pd, Pt.

Many mono- and di-nuclear peroxo-type dioxygen complexes can also be made by an alternative route involving direct reaction of transition metal compounds with H_2O_2 and it is, in fact, quite arbitrary to distinguish these complexes from those made directly from O_2. Many such compounds are discussed further on p. 745 and under the chemistry of

[49] M. J. BENNETT and P. B. DONALDSON, Molecular oxygen as a bridging ligand in a transition metal complex, *J. Am. Chem. Soc.* **93**, 3307–8 (1971).

[50] C. D. COOK, P.-T. CHENG and S. C. NYBURG, Molecular oxygen complexes of bis(triphenylphosphine-platinum(0), *J. Am. Chem. Soc.* **91**, 2123 (1969).

[51] J. H. TIMMONS, R. H. NISWANDER, A. CLEARFIELD and A. E. MARTELL, Crystal and molecular structure of μ-peroxo-bis[(1,9-bis(2-pyridyl)-2,5,8-triazanonane)cobalt(III) tetraiodide. Effect of chelate ring size on the structures and stabilities of dioxygen complexes, *Inorg. Chem.* **18**, 2977–82 (1979).

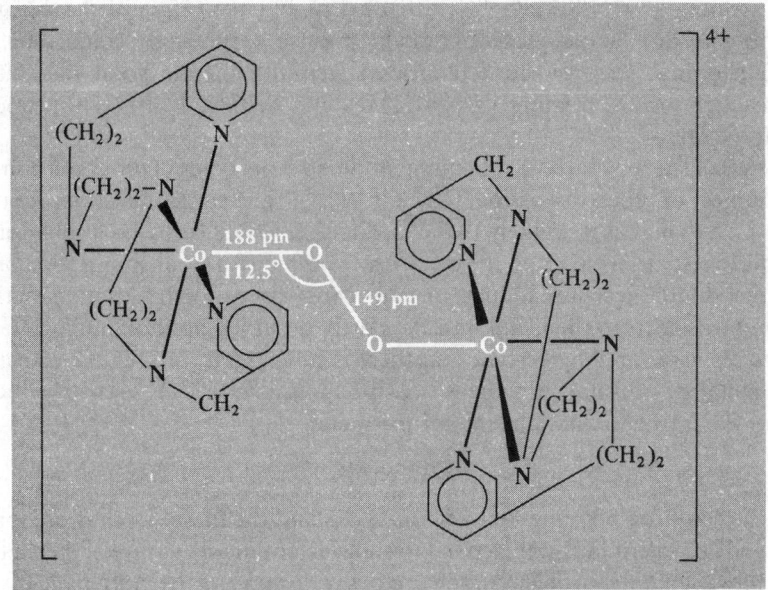

FIG. 14.8 Schematic representation of the structure of the dinuclear cation in [{Co(PYDIEN)}$_2$O$_2$]I$_4$ showing some important dimensions.

individual transition metals, but one very recent example calls for special mention since it is the first structurally characterized peroxo derivative to feature a symmetrical, doubly bidentate (side on) bridge linking two metal centres.[52] The local coordination geometry and dimensions of the central planar {LaO$_2$La} group are shown in Fig. 14.9; the very long O–O distance is particularly notable, being substantially longer than in the O$_2^{2-}$ ion itself (p. 720). The compound [La{N(SiMe$_3$)$_2$}$_2$(OPPh$_3$)]$_2$O$_2$, which is colourless, was

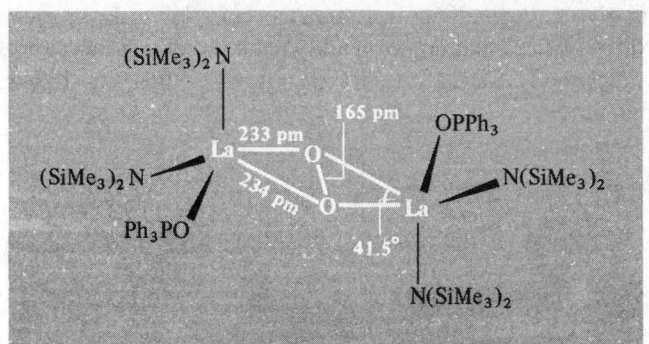

FIG. 14.9 Schematic representation of the planar central portion of the μ-peroxo complex [La{N(SiMe$_3$)$_2$}$_2$(OPPh$_3$)]$_2$O$_2$.

[52] D. C. BRADLEY, J. S. GHOTRA, F. A. HART, M. B. HURSTHOUSE and P. R. RAITHBY, Low coordination numbers in lanthanoid and actinoid compounds. Part 2. Synthesis, properties, and crystal and molecular structures of triphenylphosphine oxide and peroxo-derivatives of [bis(trimethylsilyl)amido] lanthanoids, *JCS Dalton* 1977, 1166–72.

made by treating $[La\{N(SiMe_3)_2\}_3]$ with Ph_3PO, but the origin of the peroxo group remains obscure. Similar complexes of Pr (which is also, surprisingly, colourless), Sm (pale yellow), Eu (orange red), and Lu (colourless), were obtained in good yield either by a similar reaction or by treating $[Ln\{N(SiMe_3)_2\}_3]$ with a half-molar proportion of $(Ph_3PO)_2 \cdot H_2O_2$.

The nature of the metal–oxygen bonding in the various types of dioxygen complex has been the subject of much discussion. [37, 39, 42–46, 46a] The electronic structure of the O_2 molecule (p. 706) makes it unlikely that coordination would be by the usual donation of an "onium" lone-pair (from O_2 to the metal centre) which forms an important component of most other donor–acceptor adducts (p. 223). Most discussion has centred on the extent of electron transfer from the metal into the partly occupied antibonding orbitals of O_2. There now seems general agreement that there is substantial transfer of electron density from the metal d_{z^2} orbital into the π^* antibonding orbitals of O_2 with concomitant increase in the formal oxidation state of the metal, e.g.:

$$\{Co^{II}\} + O_2 \longrightarrow \{Co^{III}(O_2^-)\}$$

Whether the resulting bonding between dioxygen and the metal atom is predominantly ionic or partly covalent may well depend to some extent on the nature of the metal centre and is largely a semantic problem which gradually disappears the more precisely one can define the detailed MOs or the actual electron distribution.[53] The difficulty is reminiscent of the discussion in Section 4.3.1 (p. 91). For those who like MO diagrams, the formal schematic energy level scheme for Fe^{II}–O_2 and Co^{II}–O_2 complexes in Fig. 14.10 should suffice; the overall sequence of energy levels is probably correct, but the detailed energy levels and the precise coefficients that describe the actual distribution of electron density are not known with any certainty. However, the drift of electron density from metal-based orbitals to oxygen-based orbitals seems well established.

In addition to their great importance for structural and bonding studies, dioxygen complexes undergo many reactions. As already indicated, some of these reactions are of unique importance in biological chemistry[27a, 54] and in catalytic systems. Some of the simpler inorganic reactions can be summarized as follows: aqueous acids yield H_2O_2 and reducing agents give coordinatively unsaturated complexes. Frequently the dioxygen complex can oxidize species that do not readily react directly with free molecular O_2, e.g. CO, CO_2, CS_2, NO, NO_2, SO_2, RNC, $RCHO$, R_2CO, PPh_3, etc. Illustrative examples of these reactions are:

$$[Pt(O_2)(PPh_3)_2] \xrightarrow{CO} [(Ph_3P)_2Pt\begin{smallmatrix}O\\ \diagup \ \diagdown\\ O\end{smallmatrix}C{=}O]$$

$$[Pt(O_2)(PPh_3)_2] \xrightarrow{CO_2} [(Ph_3O)_2Pt\begin{smallmatrix}O{-}O\\ \diagup \quad \diagdown\\ O\end{smallmatrix}C{=}O]$$

[53] S. Sakaki, K. Hori, and A. Ohyoshi, Electronic structures and stereochemistry of some side-on dioxygen complexes, *Inorg. Chem.* **17**, 3183–8 (1978).

[54] E.-I. Ochiai, Bioinorganic chemistry of oxygen, *J. Inorg. Nucl. Chem.* **37**, 1503–9 (1975). See also *Oxygen and Life: Second BOC Priestley Conference*, Roy. Soc. Chem. Special Publ. No. 39, London, 1981, 224 pp.

$$[Pt(O_2)(PPh_3)_2] \xrightarrow{\text{CS}_2} [Pt(PPh_3)_2(S_2CO)]$$

$$[Pd(O_2)(PPh_3)_2] \xrightarrow{\text{NO}} [Pt(NO_2)_2(PPh_3)_2]$$

$$[Ir(CO)Cl(O_2)(PPh_3)_2] \xrightarrow{\text{2NO}_2} [Ir(CO)Cl(NO_3)_2(PPh_3)_2]$$

$$[Ir(CO)Cl(O_2)(PPh_3)_2] \xrightarrow{\text{SO}_2} [Ir(CO)Cl(PPh_3)_2(SO_4)]$$

$$[RuCl(NO)(O_2)(PPh_3)_2] \xrightarrow{\text{SO}_2} [RuCl(NO)(PPh_3)_2(SO_4)]$$

$$[Ni(CNBu^t)_2(O_2)] \xrightarrow{\text{4Bu}^t\text{NC}} [Ni(CNBu^t)_4] + 2RNCO$$

$$[Pt(O_2)(PPh_3)_2] \xrightarrow{\text{MeCHO}} [(Ph_3P)_2Pt\underset{O}{\overset{O-O}{\diagup}}CHMe]$$

$$[Pt(O_2)(PPh_3)_2] \xrightarrow{\text{Me}_2\text{CO}} [(Ph_3P)_2Pt\underset{O}{\overset{O-O}{\diagup}}CMe_2]$$

$$[Ni(CNBu^t)_2(O_2)] \xrightarrow{\text{4PPh}_3} [Ni(CNBu^t)_2(PPh_3)_2] + 2Ph_3PO$$

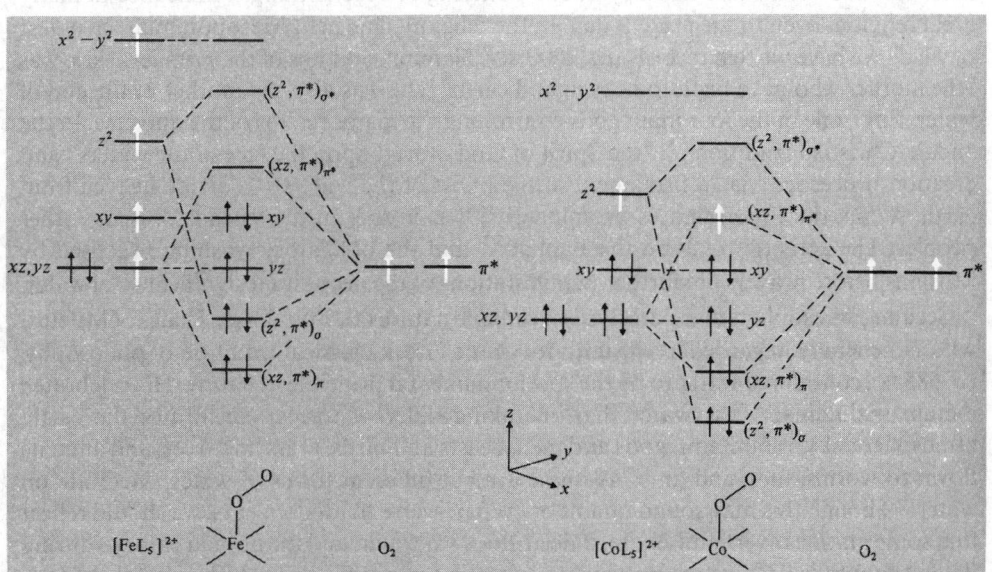

FIG. 14.10 Schematic representation of the energy levels in {Fe^{II}–O$_2$} and {Co^{II}–O$_2$} complexes. Note that the higher nuclear charge on Co^{II} tends to lower its energy levels and thus favour Co(d_{z^2})–O$_2(\pi^*)$ σ interaction. Electron spin resonance studies suggest that the unpaired electron in ($xz, \pi^*)_{\pi^*}$ resides almost completely on O$_2$; as the xy, yz, and ($xz, \pi^*)_\pi$ orbitals in the complex are predominantly localized on Co whilst ($z^2, \pi^*)_\sigma$ and ($xz, \pi^*)_{\pi^*}$ are predominantly associated with O$_2$, the coordinative interaction is sometimes said to involve oxidation of Co^{II}(d^7) to Co^{III}(d^6) and reduction of O$_2$ to O$_2^-$.

Explanations that have been advanced to explain the enhanced reactivity of coordinated dioxygen include:

(1) The diamagnetic nature of most O_2 complexes might facilitate reactions to form diamagnetic products which would otherwise be hindered by the requirement of spin conservation;

(2) the metal may hold O_2 and the reactant in *cis* positions thereby lowering the activation energy for oxidation, particularly with coordinatively unsaturated complexes;

(3) coordinated O_2 is usually partially reduced (towards O_2^- or O_2^{2-}) and this increased electron density on the molecule might activate it.

Detailed kinetic and mechanistic studies will be required to assess the relative importance of these and other possible factors in specific instances.

14.2.2 *Water*

Introduction

Water is without doubt the most abundant, the most accessible, and the most studied of all chemical compounds. Its omnipresence, its crucial importance for man's survival, and its ability to transform so readily from the liquid to the solid and gaseous states has ensured its prominence in man's thinking from the earliest times. Water plays a prominent role in most creation myths and has a symbolic purifying or regenerating significance in many great religions even to the present day. In the religion of ancient Mesopotamia, the oldest of which we have written records (*ca.* 2000 BC), Nammu, goddess of the primaeval sea, was "the mother who gave birth to heaven and earth"; she was also the mother of the god of water, Enki, one of the four main gods controlling the major realms of the universe. In the Judaic-Christian tradition[55] "the Spirit of God moved upon the face of the waters" and creation proceeded via "a firmament in the midst of the waters" to divide heaven from earth. Again, the Flood figures prominently[56] as it does in the legends of many other peoples. The activities of John the Baptist[57] and the obligatory washing practised by Muslims before prayers are further manifestations of the deep ritual significance of water.

Secular philosophers also perceived the unique nature of water. Thus, Thales of Miletus, who is generally regarded as the initiator of the Greek classical tradition of philosophy, *ca.* 585 BC, considered water to be the sole fundamental principle in nature. His celebrated dictum maintains: "It is water that, in taking different forms, constitutes the earth, atmosphere, sky, mountains, gods and men, beasts and birds, grass and trees, and animals down to worms, flies, and ants. All these are but different forms of water. Meditate on water!" Though this may sound quaint or even perverse to modern ears, we should reflect that some marine invertebrates are, indeed, 96–97% water, and the human embryo during its first month is 93% water by weight. Aristotle considered water to be one of the four elements, alongside earth, air, and fire, and this belief in the fundamental and elementary

---

[55] Holy Bible, Genesis, Chap. 1, verses 1–10.

[56] Holy Bible, Genesis, Chaps. 6–8.

[57] Holy Bible, Gospels according to St. Matthew, Chap. 3; St. Mark, Chap. 1; St. Luke, Chap. 3, St. John, Chap. 1.

nature of water persisted until the epoch-making experiments of H. Cavendish and others in the second half of the eighteenth century (pp. 39, 699) showed water to be a compound of hydrogen and oxygen.[58]

Distribution and availability

Water is distributed very unevenly and with very variable purity over the surface of the earth (Table 14.6). Desert regions have little rainfall and no permanent surface waters,

TABLE 14.6   *Estimated world water supply*

Source	Volume/10^3 km^3	% of total
Salt water		
Oceans	1 348 000	97.33
Saline lakes and inland seas	105[a]	0.008
Fresh water		
Polar ice and glaciers	28 200	2.04
Ground water	8 450	0.61
Lakes	125[b]	0.009
Soil moisture	69	0.005
Atmospheric water vapour	13.5	0.001
Rivers	1.5	0.0001
Total	1 385 000	100.0

[a] The Caspian Sea accounts for 75% of this.
[b] More than half of this is in the four largest lakes: Baikal 26 000; Tanganyika 20 000; Nyassa 13 000; and Superior 12 000 km^3.

whereas oceans, containing many dissolved salts, cover vast tracts of the globe; they comprise 97% of the available water and cover an area of 3.61×10^8 km^2 (i.e. 70.8% of the surface of the earth). Less than 2.7% of the total surface water is fresh and most of this is locked up in the Antarctic ice cap and to a much lesser extent the Arctic. The Antarctic ice cap covers some 1.5×10^7 km^2, i.e. larger than Continental Europe to the Urals (1.01×10^7 km^2), the USA including Alaska and Hawaii (0.94×10^7 km^2), or Australia (0.77×10^7 km^2): it comprises some 2.5–2.9×10^7 km^3 of fresh water which, if melted, would supply all the rivers of the earth for more than 800 years. Every year some 5000 icebergs, totalling 10^{12} m^3 of ice (i.e. 10^{12} tonnes), are calved from the glaciers and ice shelves of Antarctica. Each iceberg consists (on average) of ~ 200 Mtonnes of pure fresh water and, if towed at 1–2 km h^{-1}, could arrive 30% intact in Australia to provide water at one-tenth of the cost of current desalination procedures.[59] Transportation of crushed ice by ship from the polar regions is an alternative that was used intermittently towards the end of the last century.

[58] J. W. MELLOR, *A Comprehensive Treatise on Inorganic and Theoretical Chemistry*, Vol. 1, Chap. 3. Hydrogen and the composition of water, pp. 122–46, Longmans Green, London, 1922.
[59] F. FRANKS, *Introduction—Water, the Unique Chemical*, Vol. 1, Chap. 1, of F. FRANKS (ed.), *Water, a Comprehensive Treatise in 7 Volumes*, Plenum Press, New York, 1972–82.

Surface freshwater lakes contain 1.25×10^5 km^3 of water, more than half of which is in the four largest lakes. Though these huge lacustrine sources dwarf the innumerable smaller lakes, springs, and rivers of the earth, human habitation depends more on these widely distributed smaller sources which, in total, still far exceed the needs of man and the animal and plant kingdoms. Despite this, severe local problems can arise due to prolonged drought, the pollution of surface waters, or the extension of settlements into more arid regions. Indeed, droughts have been endemic since ancient times, and even pollution of local sources has been a cause of concern and the subject of legislation since at least 1847 (UK). Michael Faraday's letter to *The Times* of 7 July 1855, and the ensuing cartoon in *Punch* (Fig. 14.11) remind us of how long it can be before effective action is taken on urgent environmental problems. Fortunately, it now appears that the quality of water supplies and amenities is rising steadily in most communities since the nadir of 20–30 years ago, and public concern is now increasingly ensuring that funds are available on an appropriate scale to deal with the massive problems of water pollution.[60–65] (See also p. 551.)

Water purification and recycling is now a major industry.[66] The method of treatment depends on the source of the water, the use envisaged, and the volume required. Luckily the human body is very tolerant to changes in the composition of drinking water, and in many communities this may contain 0.5 g l^{-1} or more of dissolved solids (Table 14.7).

TABLE 14.7   *World Health Organization drinking-water standards*

Material	Maximum desirable conc/mg l^{-1}	Maximum permissible conc/mg l^{-1}
Total dissolved solids	500	1500
Mg	30	150
Ca	75	200
Chlorides	20	60
Sulfates	200	400

Prior treatment may consist of coagulation (by addition of alum or chlorinated FeSO$_4$ to produce flocs of Al(OH)$_3$ or Fe(OH)$_3$), filtration, softening (removal of Mg^{II} and Ca^{II} by ion exchange), and disinfection (by chlorination, p. 924, or addition of ozone, p. 712). In

[60] H. B. N. HYNES, *The Biology of Polluted Waters*, Liverpool Univ. Press, 4th impression 1973, 202 pp. (Probably the best textbook at present available on water pollution.)

[61] A. D. McKNIGHT, P. K. MARSTRAND and T. C. SINCLAIR (eds.), *Environmental Pollution Control*, Chap. 5: Pollution of inland waters; Chap. 6: The Law relating to pollution of inland waters; George, Allen and Unwin, 1974.

[62] C. E. WARREN, *Biology and Water Pollution Control*, Saunders, Philadelphia, 1971, 434 pp. (A good introductory book, encyclopaedic but readable.)

[63] B. COMMONER, The killing of a great lake, in *The 1968 World Book Year Book*, Field Enterprises Educ. Corp., 1968; Lake Erie water, Chap. 5 in *The Closing Circle*, London, Jonathan Cape, 1972.

[64] A. NISBETT, The myths of Lake Erie, *New Scientist*, 23 March 1972, pp. 650–52, Argues that the views in ref. 63 and elsewhere are unfounded: Lake Erie is not dead but it is damaged.

[65] M. WORBOYS, P. K. MARSTRAND and P. D. LOWE, *Science and the Environment-Unit Two*, Vol. 15 in the SISCON (Science in a Social Context) Project, 1975, 21 pp. Contains useful documentation on sources of water pollution and its treatment, and an annotated bibliography.

[66] T. V. ARDEN, Water purification and recycling, in R. THOMPSON (ed.), *The Modern Inorganic Chemicals Industry*, pp. 69–105, Chemical Society Special Publication, No. 31, 1977.

No. IX.

Observations on the Filth of the Thames, contained in a Letter addressed to the Editor of "The Times" Newspaper, by Professor Faraday.

Sir,

I TRAVERSED this day by steam-boat the space between London and Hungerford Bridges between half-past one and two o'clock; it was low water, and I think the tide must have been near the turn. The appearance and the smell of the water forced themselves at once on my attention. The whole of the river was an opaque pale brown fluid. In order to test the degree of opacity, I tore up some white cards into pieces, moistened them so as to make them sink easily below the surface, and then dropped some of these pieces into the water at every pier the boat came to; before they had sunk an inch below the surface they were indistinguishable, though the sun shone brightly at the time; and when the pieces fell edgeways the lower part was hidden from sight before the upper part was under water. This happened at St. Paul's Wharf, Blackfriars Bridge, Temple Wharf, Southwark Bridge, and Hungerford; and I have no doubt would have occurred further up and down the river. Near the bridges the feculence rolled up in clouds so dense that they were visible at the surface, even in water of this kind.

The smell was very bad, and common to the whole of the water; it was the same as that which now comes up from the gully-holes in the streets; the whole river was for the time a real sewer. Having just returned from out of the country air, I was, perhaps, more affected by it than others; but I do not think I could have gone on to Lambeth or Chelsea, and I was glad to enter the streets for an atmosphere which, except near the sink-holes, I found much sweeter than that on the river.

I have thought it a duty to record these facts, that they may be brought to the attention of those who exercise power or have responsibility in relation to the condition of our river; there is nothing figurative in the words I have employed, or any approach to exaggeration; they are the simple truth. If there be sufficient authority to remove a putrescent pond from the neighbourhood of a few simple dwellings, surely the river which flows for so many miles through London ought not to be allowed to become a fermenting sewer. The condition in which I saw the Thames may perhaps be considered as exceptional, but it ought to be an impossible state, instead of which I fear it is rapidly becoming the general condition. If we neglect this subject, we cannot expect to do so with impunity; nor ought we to be surprised if, ere many years are over, a hot season give us sad proof of the folly of our carelessness.

I am, Sir,

Your obedient servant,

Royal Institution, July 7.　　　　　　　M. FARADAY.

FARADAY GIVING HIS CARD TO FATHER THAMES;
And we hope the Dirty Fellow will consult the learned Professor.

FIG. 14.11.

most developed countries industrial needs for water are at least 10 times the volume used domestically. Moreover, some industrial processes require much purer water than that for human consumption, and for high-pressure boiler feedwater in particular the purity standard is 99.999 998%, i.e. no more than 0.02 ppm impurities. This is far purer than for reactor grade uranium, the finest refined gold, or the best analytical reagents, and is probably exceeded only by semiconductor grade germanium and silicon. In contrast to Ge and Si, however, water is processed on a megatonne-per-day scale at a cost of only about £1 per tonne.

The beneficiation of sea water and other saline sources to produce fresh water is also of increasing importance. Normal freshwater supplies from precipitation cannot meet the needs of the increasing world population, particularly in the semi-arid regions of the world, and desalination is being used increasingly to augment normal water supplies, or even to provide all the fresh water in some places such as the arid parts of the Arabian Peninsula. The most commonly used methods are distillation (e.g. multistage flash distillation processes) and ion-exchange techniques, including electrodialysis and reverse osmosis (hyperfiltration). The enormous importance of the field can be gauged from the fact that Gmelin's volume on *Water Desalting*,[67] which reviewed 14 000 papers published up to 1973/4, has already had to be supplemented by a further 360-page volume[68] dealing with the 4000 papers appearing during the following 4 years. An excellent example of the scale of modern desalination schemes is the Yuma Plant being built on the California–Arizona–Mexico border. Irrigation waters from the Imperial Dam across the Colorado River above Yuma have been percolating through porous soil, leaching out saline components and then collecting in a shallow aquifer below the irrigation region. By 1961 the water-level had begun to reach the root zone of the fields; wells were dug and the saline water (3200 mg l^{-1} dissolved solids) pumped back into the river, but this merely transferred the problem across the border where the Mexicans were unable to use the now brackish water for their own irrigation projects. The 1973 US-Mexico treaty set a limit of no more than 115 mg l^{-1} salinity above that of the Imperial Dam waters, and this has necessitated the construction of a huge reverse osmosis desalting plant that will process nearly 100 million gallons (US) per day and deliver over 90 000 acre-feet of product water annually for blending with the raw Colorado River water in Mexico. When completed in the mid-1980s the Yuma Plant will be the largest single desalting installation in the world, built at a cost of $190 million and with an annual operating cost of $12.4 million at 1977 prices.[69] A far cry from the first recorded use of desalination techniques in biblical times.[70]

Physical properties and structure

Water is a volatile, mobile liquid with many curious properties, most of which can be ascribed to extensive H bonding (p. 57). In the gas phase the H_2O molecule has a bond

[67] *Gmelin Handbook of Inorganic Chemistry*, 8th edn. (in English), *O: Water Desalting*, 1974, 339 pp.
[68] *Gmelin Handbook of Inorganic Chemistry*, 8th edn. (in English), *O: Water Desalting*, Supplement Vol. 1, 1979, 360 pp.
[69] Water desalination gets another look, *Chem. Eng. News*, 4 February 1980, p. 26–30.
[70] Holy Bible, Exodus, Chap. 15, verses 22–25: ". . . so Moses brought the sons of Israel from the Red Sea and they went into the desert of Sur. And they marched three days in the wilderness and found no water to drink. And then they arrived at Merra and they could not drink from the waters of Merra because they were bitter. . . . And the people murmured against Moses saying: What shall we drink? And Moses cried unto the Lord. And the Lord showed him a wood and he put it into the water and the water became sweet".

angle of 104.5° (close to tetrahedral) and an interatomic distance of 95.7 pm. The dipole moment is 1.84 D. Some properties of liquid water are summarized in Table 14.8 together

TABLE 14.8   *Some physical properties of H_2O, D_2O, and T_2O (at 25°C unless otherwise stated)*[a]

Property	H_2O	D_2O	T_2O
Molecular weight	18.0151	20.0276	22.0315
MP/°C	0.00	3.81	4.48
BP/°C	100.00	101.42	101.51
Temperature of maximum density/°C	3.98	11.23	13.4
Maximum density/g cm^{-3}	1.0000	1.1059	1.2150
Density (25°)/g cm^{-3}	0.997 01	1.1044	1.2138
Vapour pressure/mmHg	23.75	20.51	~19.8
Viscosity/centipoise	0.8903	1.107	—
Dielectric constant ε	78.39	78.06	—
Electrical conductivity (20°C)/ ohm^{-1} cm^{-1}	5.7×10^{-8}	—	—
Ionization constant $[H^+][OH^-]$/mol^2 l^{-2}	1.008×10^{-14}	1.95×10^{-15}	$\sim 6 \times 10^{-16}$
Ionic dissociation constant $K=[H^+][OH^-]/[H_2O]$/mol l^{-1}	1.821×10^{-16}	3.54×10^{-17}	$\sim 1.1 \times 10^{-17}$
Heat of ionization/kJ mol^{-1}	56.27	60.33	—
ΔH_f°/kJ mol^{-1}	-285.85	-294.6	—
ΔG_f°/kJ mol^{-1}	-237.19	-243.5	—

[a] Heavy water (p. 47) is now manufactured on the multikilotonne scale for use both as a coolant and neutron-moderator in nuclear reactors: its absorption cross-section for neutrons is much less than for normal water: σ_H 332, σ_D 0.46 mb (1 millibarn = 10^{-21} cm^2).

with those of heavy water D_2O and the tritium analogue T_2O (p. 48). The high bp is notable (cf. H_2S, etc.) as is the temperature of maximum density and its marked dependence on the isotopic composition of water. The high dielectric constant and measurable ionic dissociation equilibrium are also unusual and important properties. The ionic mobilities of $[H_3O]^+$ and $[OH]^-$ in water are abnormally high (350×10^{-4} and 192×10^{-4} cm s^{-1} per V cm^{-1} at 25° compared with 50–75×10^{-4} cm^2 V^{-1} s^{-1} for most other ions). This has been ascribed to a proton switch and reorientation mechanism involving the ions and chains of H-bonded solvent molecules. Other properties which show the influence of H bonding are the high heat and entropy of vaporization (ΔH_{vap} 44.02 kJ mol^{-1}, ΔS_{vap} 118.8 J deg^{-1} mol^{-1}), high surface tension (71.97 dyne cm^{-1}, i.e. 71.97 mN m^{-1}) and relatively high viscosity. The strength of the H bonds has been variously estimated at between 5–50 kJ per mol of H bonds and is most probably close to 20 kJ mol^{-1}. The structured nature of liquid water in which the molecules are linked to a small number of neighbours (2–3) by H bonds also accounts for its anomalously low density compared with a value of ~ 1.84 g cm^{-3} calculated for a normal close-packed liquid with molecules of similar size and mass. Details of the structure of liquid water have been probed for nearly five decades since the classic paper of J. D. Bernal and R. H. Fowler proposed the first plausible model.[71] Despite extensive work by X-ray and neutron

[71] J. D. BERNAL and R. H. FOWLER, A theory of water and ionic solution, with particular reference to hydrogen and hydroxyl ions, *J. Chem. Phys.* **1**, 515–48 (1933).

diffraction, Raman and infrared spectroscopy, and the theoretical calculation of thermodynamic properties based on various models, details are still controversial and there does not even appear to be general agreement on whether water consists of a mixture of two or more species of varying degrees of polymerization, or whether it is better described on a continuous model of highly bent H-bond configurations.[72]

When water freezes the crystalline form adopted depends upon the detailed conditions employed. At least nine structurally distinct forms of ice are known and the phase relations between them are summarized in Fig. 14.12. Thus, when liquid or gaseous water

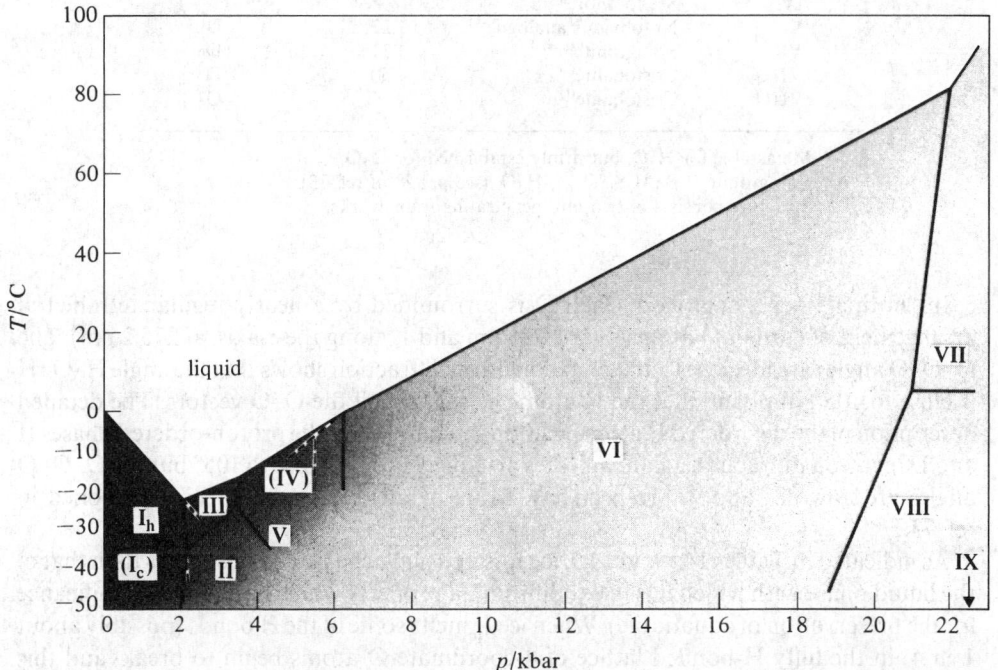

FIG. 14.12   Partial phase diagram for ice (metastable equilibrium shown by broken lines).

crystallizes at atmospheric pressure normal hexagonal ice I_h forms, but at very low temperatures $(-120°$ to $-140°)$ the vapour condenses to the cubic form, ice I_c. The relation between these structures is the same as that between the tridymite and cristobalite forms of SiO_2 (p. 394), though in both forms of ice the protons are disordered. Many of the high-pressure forms of ice are also based on silica structures (Table 14.9) and in ice II, VIII, and IX the protons are ordered, the last 2 being low-temperature forms of ice VII and III respectively in which the protons are disordered. Note also that the high-pressure polymorphs VI and VII can exist at temperatures as high as 80°C and that, as expected, the high-pressure forms have substantially greater densities than that for ice I. A vitreous form of ice can be obtained by condensing water vapour at temperatures of $-160°$C or below.

[72] P. KRINDEL and I. ELIEZER, Water structure models, *Coord. Chem. Rev.* **6**, 217–46 (1971).

TABLE 14.9   *Structural relations in the polymorphs of ice*[73]

Polymorph	Analogous silica polymorph	$d/(\text{g cm}^{-3})$	Ordered (O) or disordered (D) positions
I_h	Tridymite	0.92	D
I_c	Cristobalite	0.92	D
II	—	1.17	O
III ⎫	Keatite	1.16	D ⎫
IX ⎭	Keatite		O ⎭
IV	See footnote[a]	—	—
V	No obvious analogue	1.23	D
VI	Edingtonite[b, c]	1.31	D
VII ⎫	Cristobalite[c]	1.50	D ⎫
VIII ⎭	Cristobalite[c]		O ⎭

[a] Metastable for H_2O, but firmly established for D_2O.
[b] Edingtonite is $BaAl_2Si_3O_{10} \cdot 4H_2O$ (see p. 828 of ref. 73).
[c] Structure consists of two interpenetrating frameworks.

In "normal" hexagonal ice I_h each O is surrounded by a nearly regular tetrahedral arrangement of 4 other O atoms (3 at 276.5 pm and 1, along the *c*-axis, at 275.2 pm). The O–O–O angles are all close to 109.5° and neutron diffraction shows that the angle H–O–H is close to 105°, implying that the H atoms lie slightly off the O–O vectors. The detailed description of the disordered H atom positions is complex. In the proton-ordered phases II and IX neutron diffraction again indicates an angle H–O–H close to 105° but the O–O–O angles are now 88° and 99° respectively. More details are in the papers mentioned in ref. 73.

As indicated in Tables 14.8 and 14.9, ice I_h is unusual in having a density less than that of the liquid phase with which it is in equilibrium (a property which is of crucial significance for the preservation of aquatic life). When ice I_h melts some of the H bonds (possibly about 1 in 4) in the fully H-bonded lattice of 4-coordinate O atoms begin to break, and this process continues as the liquid is warmed, thereby enabling the molecules to pack progressively more closely with a consequent *increase* in density. This effect is opposed by the thermal motion of the molecules which tends to expand the liquid, and the net result is a maximum in the density at 3.98°C. Further heating reduces the density, though only slowly, presumably because the effects of thermal motion begin to outweigh the countervailing influence of breaking more H bonds. Again the qualitative explanation is clear but quantitative calculations of the density, viscosity, dielectric constant, etc., of H_2O, D_2O, and their mixtures remain formidable.

Until recently it was thought that pure ice had a low but measurable electrical conductivity of about 1×10^{-10} ohm^{-1} cm^{-1} at $-10°C$. This conductivity is now thought to arise almost exclusively from surface defects, and when these have been removed ice is essentially an insulator with an immeasurably small conductivity.[74]

[73] A. F. WELLS, Water and hydrates, Chap. 15 in *Structural Inorganic Chemistry*, 4th edn., pp. 537–69, Oxford University Press, Oxford, 1975.
[74] A. VON HIPPEL, From atoms towards living systems, *Mat. Res. Bull.* 14, 273–99 (1979).

Water of crystallization, aquo complexes, and solid hydrates

Many salts crystallize from aqueous solution not as the anhydrous compound but as a well-defined hydrate. Still other solid phases have variable quantities of water associated with them, and there is an almost continuous gradation in the degree of association or "bonding" between the molecules of water and the other components of the crystal. It is convenient to recognise five limiting types of interaction though the boundaries between them are vague and undefined, and many compounds incorporate more than one type.

(a) *H_2O coordinated in a cationic complex.* This is perhaps the most familiar class and can be exemplified by complexes such as $[Be(OH_2)_4]SO_4$, $[Mg(OH_2)_6]Cl_2$, $[Ni(OH_2)_6](NO_3)_2$, etc.; the metal ion is frequently in the $+2$ or $+3$ oxidation state and tends to be small and with high coordination power. Sometimes there is further interaction via H bonding between the aquocation and the anion, particularly if this derives from an oxoacid, e.g. the alums $\{[M(OH_2)_6]^+[Al(OH_2)_6]^{3+}[SO_4]_2^{2-}\}$ and related salts of Cr^{3+}, Fe^{3+}, etc. The species H_3O^+, $H_5O_2^+$, $H_7O_3^+$, and $H_9O_4^+$ are a special case in which the cation is a proton, i.e. $[H(OH_2)_n]^+$, and are discussed on p. 738.

(b) *H_2O coordinated by H bonding to oxoanions.* This mode is relatively uncommon but occurs in the classic case of $CuSO_4 . 5H_2O$ and probably also in $ZnSO_4 . 7H_2O$. Thus, in hydrated Cu sulfate, 1 of the H_2O molecules is held much more tenaciously than the other 4 (which can all be removed over P_4O_{10} or by warming under reduced pressure); the fifth can only be removed by heating the compound above 350°C (or to 250° *in vacuo*). The crystal structure shows that each Cu atom is coordinated by $4H_2O$ and $2SO_4$ groups in a *trans* octahedral configuration (Fig. 14.13) and that the fifth H_2O molecule is not bound to

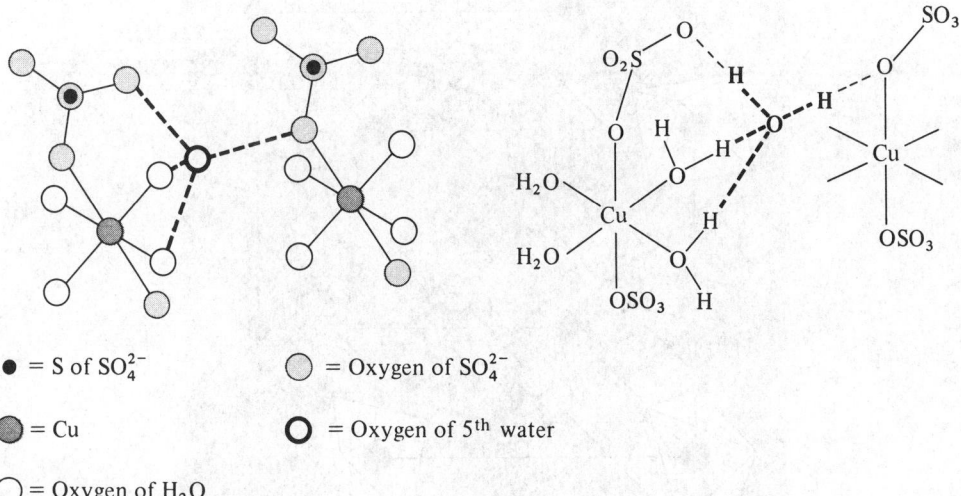

● = S of SO_4^{2-} ◯ = Oxygen of SO_4^{2-}

◓ = Cu   **O** = Oxygen of 5th water

◯ = Oxygen of H_2O

FIG. 14.13   Two representations of the repeating structural unit in $CuSO_4.5H_2O$ showing the geometrical distribution of ligands about Cu and the connectivity of the unique H_2O molecule.

Cu but forms H (donor) bonds to 2 SO_4 groups on neighbouring Cu atoms and 2 further H (acceptor) bonds with *cis*-H_2O molecules on 1 of the Cu atoms. It therefore plays a cohesive role in binding the various units of the structure into a continuous lattice.

(c) *Lattice water.* Sometimes hydration of either the cation or the anion is required to improve the size compatibility of the units comprising the lattice, and sometimes voids in the lattice so formed can be filled by additional molecules of water. Thus, although LiF and NaF are anhydrous, the larger alkali metal fluorides can form definite hydrates $MF \cdot nH_2O$ ($n=2$ and 4 for K; $1\frac{1}{2}$ for Rb; $\frac{2}{3}$ and $1\frac{1}{2}$ for Cs). Conversely, for the chlorides: KCl, RbCl, and CsCl are always anhydrous whereas LiCl can form hydrates with 1, 2, 3, and $5H_2O$, and $NaCl \cdot 2H_2O$ is also known. The space-filling role of water molecules is even more evident with very large anions such as those of the heteropoly acids (p. 1184), e.g. $H_3[PW_{12}O_{40}] \cdot 29H_2O$.

(d) *Zeoltic water.* The large cavities of the framework silicates (p. 414) can readily accommodate water molecules, and the lack of specific strong interactions enables the "degree of hydration" to vary continuously over very wide ranges. The swelling of ion-exchange resins and clay minerals (p. 407) are further examples of non-specific hydrates of variable composition.

(e) *Clathrate hydrates.* The structure motif of zeolite "hosts" accommodating "guest" molecules of water can be inverted in an intriguing way: just as the various forms of ice (p. 732) are formally related to those of silica (p. 394), so $(H_2O)_n$ can be induced to generate various cage-like structures with large cavities, thereby enabling the water structure itself to act as host to various guest molecules. Thus, polyhedral frameworks, sometimes with cavities of more than one size, can be generated from unit cells containing $12H_2O$, $46H_2O$, $136H_2O$, etc. In the first of these (Fig. 14.14) there is a cubic array of 24-cornered cavities,

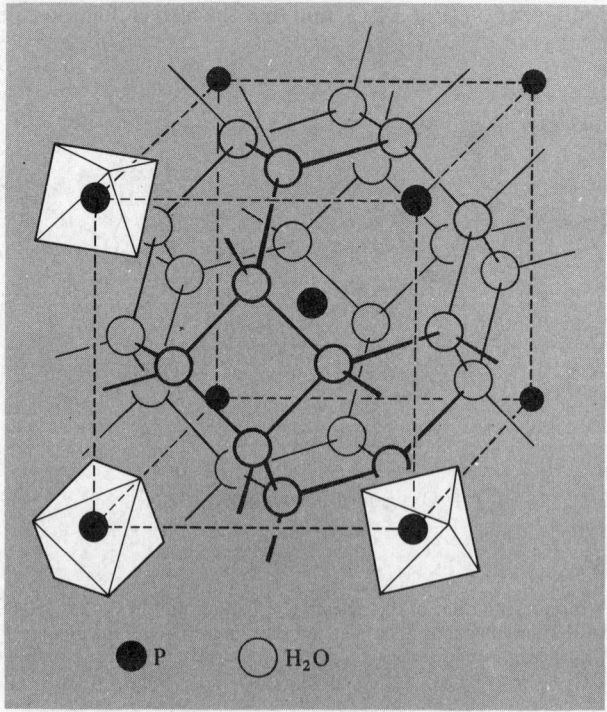

FIG. 14.14. Crystal structure of $HPF_6 \cdot 6H_2O$ showing the cavity formed by 24 H_2O molecules disposed with their O atoms at the vertices of a truncated octahedron. The PF_6 octahedra occupy centre and corners of the cubic unit cell, i.e. one PF_6 at the centre of each cavity.[73]

each cavity being a truncated octahedron with square faces of O atoms and each H_2O being common to 2 adjacent cavities (i.e. $24/2 = 12H_2O$). There is space for a guest molecule G at the centre of each cavity, i.e. at the centre of the cube and at each corner resulting in a stoichiometry $G(8G)_{1/8}.12H_2O$, i.e. $G.6H_2O$ as in $HPF_6.6H_2O$. The structure should be compared with the aluminosilicate framework in ultramarine (p. 416).

With the more complicated framework of $46H_2O$ there are 6 cavities of one size and 2 slightly smaller. If all are filled one has $46/8H_2O$ per guest molecule, i.e. $G.5\frac{3}{4}H_2O$ as in the high-pressure clathrates with $G = Ar$, Kr, CH_4, and H_2S. If only the larger cavities are filled, the stoichiometry rises to $G.7\frac{2}{3}H_2O$: this is approximated by the classic chlorine hydrate phase discovered by Humphry Davy and studied by Michael Faraday. The compound is now known to be $Cl_2.7\frac{1}{4}H_2O$, implying that up to 20% of the smaller guest sites are also occupied.

With the $136H_2O$ polyhedron there are 8 larger and 16 smaller voids. If only the former are filled, then $G.17H_2O$ results ($17 = 136/8$) as in $CHCl_3.17H_2O$ and $CHI_3.17H_2O$, whereas if both sets are filled with molecules of different sizes, compounds such as $CHCl_3.2H_2S.17H_2O$ result. Many more complicated arrays are possible, resulting from partial filling of the voids or partial replacement of H_2O in the framework by other species capable of being H-bonded into the network, e.g. $[NMe_4]F.4H_2O$, $[NMe_4]OH.5H_2O$, $Bu_3^nSF.20H_2O$, and $[N(i\text{-}C_5H_{11})_4]F.38H_2O$. Further structural details are in ref. 73.

Chemical properties

Water is an excellent solvent because of its high dielectric constant and very strong solvating power. Many compounds, whether hydrated or anhydrous, dissolve to give electrolytic solutions of hydrated cations and anions. However, detailed treatments of solubility relations, free energies and enthalpies of ionic hydration, temperature dependence of solubility, and the influence of dissolved ions on the H-bonded structure of the solvent, fall outside the scope of the present treatment. Even predominantly covalent compounds such as EtOH, $MeCO_2H$, Me_2CO, $(CH_2)_4O$, etc. can have high solubility or even complete miscibility with water due to H-bonded interaction with the solvent. Again, covalent compounds such as HCl can dissolve to give ionic solutions by heterolytic cleavage (e.g. to aquated $H_3O^+Cl^-$), and the process of dissolution sometimes also results in ionic cleavage of the solvent itself, e.g. $[H_3O]^+[BF_3(OH)]^-$ (p. 223). Because of the great affinity that many elements have for oxygen, solvolytic cleavage (hydrolysis) of "covalent" or "ionic" bonds frequently ensues, e.g.:

$$P_4O_{10}(s) + xH_2O \longrightarrow 4H_3PO_4(aq) \qquad \text{(p. 580)}$$

$$AlCl_3(s) + xH_2O \longrightarrow [Al(OH_2)_6]^{3+}(aq) + 3Cl^-(aq) \qquad \text{(p. 254)}$$

Such reactions are discussed at appropriate points throughout the book as each individual compound is being considered. A particularly important set of reactions in this category is the synthesis of element hydrides by hydrolysis of certain sulfides (to give H_2S), nitrides (to give NH_3), phosphides (PH_3), carbides (C_nH_m), borides (B_nH_m), etc.

Another important reaction (between H_2O, I_2, and SO_2) forms the basis of the quantitative determination of water when present in small amounts. The reaction, originally investigated by R. Bunsen in 1835, was introduced in 1935 as an analytical reagent by Karl Fischer who believed, incorrectly, that each mole of I_2 was equivalent to 2 moles of H_2O:

$$2H_2O + I_2 + SO_2 \xrightleftharpoons{\text{MeOH/py}} 2HI + H_2SO_4$$

In fact, the reaction is only quantitative in the presence of pyridine, and the methanol solvent is also involved leading to a 1:1 stoichiometry between I_2 and H_2O:

$$H_2O + I_2 + SO_2 + 3py \longrightarrow 2pyHI + C_5H_5N\!\!\begin{array}{c}SO_2\\ \diagdown\!\!\diagup\\ O\end{array} \xrightarrow{\text{MeOH}} [\,pyH]\,[MeOSO_3]$$

The stability of the reagent is much improved by replacing MeOH with $MeOCH_2CH_2OH$, and this forms the basis of the present-day Karl Fischer reagent.

In addition to simple dissolution, ionic dissociation, and solvolysis, two further classes of reaction are of pre-eminent importance in aqueous solution chemistry, namely acid–base reactions (p. 51) and oxidation–reduction reactions. In water, the oxygen atom is in its lowest oxidation state (-2). Standard reduction potentials (p. 498) of oxygen in acid and alkaline solution are listed in Table 14.10[75] and shown diagramatically in the

TABLE 14.10   *Standard reduction potentials of oxygen*

Acid solution (pH 0)	$E°/V$	Alkaline solution (pH 14)	$E°/V$
$O_2 + 4H^+ + 4e^- \rightleftharpoons 2H_2O$	1.229	$O_2 + 2H_2O + 4e^- \rightleftharpoons 4OH^-$	0.401
$O_2 + 2H^+ + 2e^- \rightleftharpoons H_2O_2$	0.695	$O_2 + H_2O + 2e^- \rightleftharpoons HO_2^- + OH^-$	-0.076
$O_2 + H^+ + e^- \rightleftharpoons HO_2$	-0.105	$O_2 + e^- \rightleftharpoons O_2^-$	-0.563
$HO_2 + H^+ + e^- \rightleftharpoons H_2O_2$	1.495	$O_2^- + H_2O + e^- \rightleftharpoons HO_2^- + OH^-$	0.413
$H_2O_2 + 2H^+ + 2e^- \rightleftharpoons 2H_2O$	1.776	$HO_2^- + H_2O + 2e^- \rightleftharpoons 3OH^-$	0.878
$H_2O_2 + H^+ + e^- \rightleftharpoons OH + H_2O$	0.71	$HO_2^- + H_2O + e^- \rightleftharpoons OH + 2OH^-$	-0.245
$OH + H^+ + e^- \rightleftharpoons H_2O$	2.85	$OH + e^- \rightleftharpoons OH^-$	2.02

scheme opposite. It is important to remember that if H^+ or OH^- appear in the electrode half-reaction, then the electrode potential will change markedly with the pH. Thus for the first reaction in Table 14.10: $O_2 + 4H^+ + 4e^- \rightleftharpoons 2H_2O$, although $E° = 1.229$ V, the actual potential at 25°C will be given by

$$E/\text{volt} = 1.229 + 0.05916 \log \{[H^+]/\text{mol } l^{-1}\}\,\{P_{O_2}/\text{atm}\}^{\frac{1}{4}}$$

which diminishes to 0.401 V at pH 14 (Fig. 14.15). Likewise, for the half-reaction $H^+ + e^- \rightleftharpoons \frac{1}{2}H_2$, $E°$ is zero by definition at pH 0, and at other concentrations

$$E/\text{volt} = -0.05916 \log\{P_{H_2}/\text{atm}\}^{\frac{1}{2}}/\{[H^+]/\text{mol } l^{-1}\}$$

[75] G. MILAZZO and S. CAROLI, *Tables of Standard Electrode Potentials*, p. 229, Wiley-Interscience, New York, 1978.

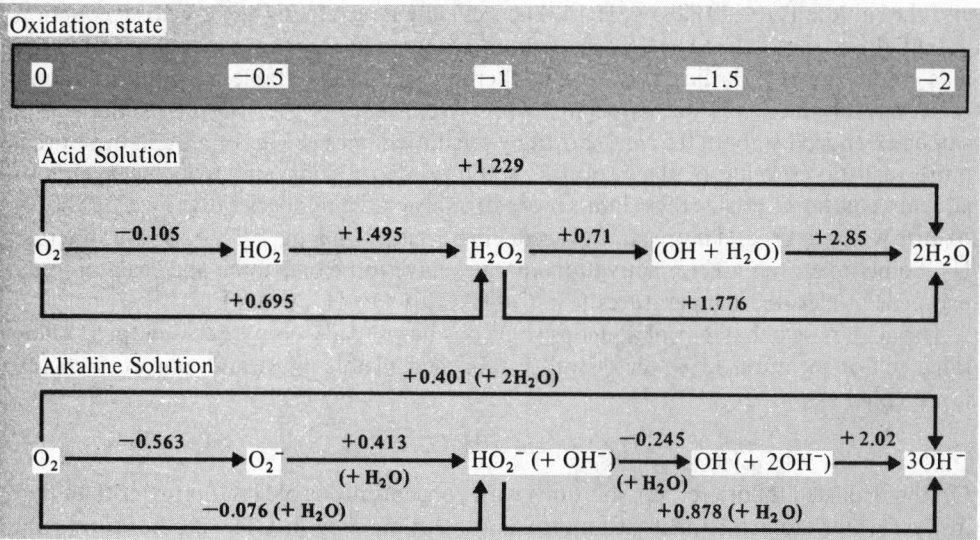

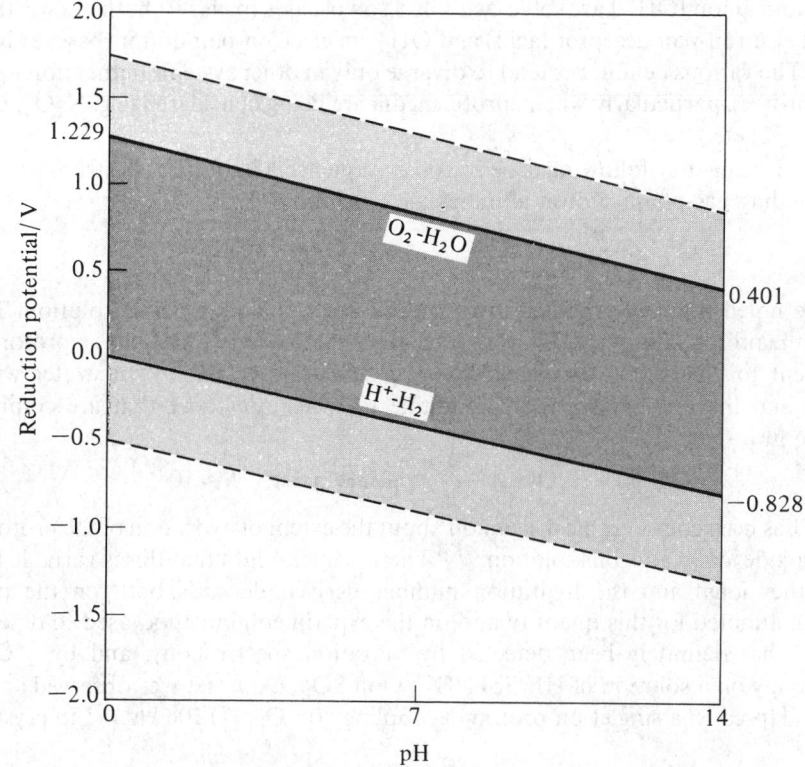

FIG. 14.15 Variation of the reduction potentials of the couples O_2/H_2O and H^+/H_2 as a function of pH (full lines). The broken lines lie 0.5 V above and below these full lines and give the approximate practical limits of oxidants and reductants in aqueous solution beyond which the solvent itself is oxidized to $O_2(g)$ or reduced to $H_2(g)$.

and the value falls to -0.828 at pH 14. Theoretically no oxidizing agent whose reduction potential lies above the O_2/H_2O line and no reducing agent whose reduction potential falls below the H^+/H_2 line can exist in thermodynamically stable aqueous solutions. However, for kinetic reasons associated with the existence of over-potentials, these lines can be extended by about 0.5 V as shown by the dotted lines in Fig. 14.15, and these are a more realistic estimate of the region of stability of oxidizing and reducing agents in aqueous solution. Outside these limits more strongly oxidizing species (e.g. F_2, $E°$ 2.866 V) oxidize water to O_2 and more strongly reducing agents (e.g. K_{metal}, $E°$ -2.931 V) liberate H_2. Sometimes even greater activation energies have to be overcome and reaction only proceeds at elevated temperatures (e.g. $C + H_2O \rightarrow CO + H_2$; p. 326).

The acid–base behaviour of aqueous solutions has already been discussed (p. 51). The ionic self-dissociation of water is well established (Table 14.8) and can be formally represented as

$$2H_2O \rightleftharpoons H_3O^+ + OH^-$$

On the Brønsted theory (p. 52), solutions with concentrations of H_3O^+ greater than that in pure water are acids (proton donors), and solutions rich in OH^- are bases (proton acceptors). The same classifications follow from the solvent-system theory of acids and bases since compounds enhancing the concentrations of the characteristic solvent cation (H_3O^+) and anion (OH^-) are solvo-acids and solvo-bases (p. 487). On the Lewis theory, H^+ is an electron-pair acceptor (acid) and OH^- an electron-pair donor (base, or ligand) (p. 223). The various definitions tend to diverge only in other systems (either nonaqueous or solvent-free), particularly when aprotic media are being considered (e.g. N_2O_4, p. 523; BrF_3, p. 972; etc.).

In considering the following isoelectronic sequence (8 valence electrons) and the corresponding gas-phase proton affinities ($A_{H+}/kJ\ mol^{-1}$):[76]

$$O^{2-} \xrightarrow[(-2860)]{H^+} OH^- \xrightarrow[(-1650)]{H^+} H_2O \xrightarrow[(-695)]{H^+} H_3O^+$$

it will be noted that only the last three species are stable in aqueous solution. This is because the proton affinity of O^{2-} is so huge that it immediately abstracts a proton from the solvent to give OH^-; as a consequence, oxides never dissolve in water without reaction and the only oxides that are stable to water are those that are completely insoluble in it:

$$O^{2-}(s) + H_2O(aq) \longrightarrow 2OH^-(aq); \quad K > 10^{22}$$

There has been considerable discussion about the extent of hydration of the proton and the hydroxide ion in aqueous solution.[76a] There is little doubt that this is variable (as for many other ions) and the hydration number derived depends both on the precise definition adopted for this quantity and on the experimental method used to determine it. H_3O^+ has definitely been detected by vibration spectroscopy, and by ^{17}O nmr spectroscopy on a solution of $HF/SbF_5/H_2^{17}O$ in SO_2; a quartet was observed at $-15°$ which collapsed to a singlet on proton decoupling, $J(^{17}O-^1H)$ 106 Hz.[77] In crystalline

[76] R. E. KARI and I. G. CSIZMADIA, A systematic study of the ionization potentials and electron, proton, hydrogen and hydride affinities of OH_n molecules and ions, *J. Am. Chem. Soc.* **99**, 4539–45 (1977).

[76a] P. A. GIGUÈRE, The great fallacy of the H^+ ion, *J. Chem. Educ.* **56**, 571–5 (1979).

[77] G. D. METEESCU and G. M. BENEDIKT, The hydronium ion (H_3O^+). Preparation and characterization by high resolution oxygen-17 nuclear magnetic resonance, *J. Am. Chem. Soc.* **101**, 3959–60 (1979).

hydrates there are a growing number of well-characterized hydrates of the series H_3O^+, $H_5O_2^+$, $H_7O_3^+$, $H_9O_4^+$, and $H_{13}O_6^+$, i.e. $[H(OH_2)_n]^+$ $n = 1-4, 6$. Thus X-ray studies have established the presence of H_3O^+ in the monohydrates of HCl, HNO_3, and $HClO_4$, and in the mono- and di-hydrates of sulfuric acid, $[H_3O][HSO_4]$ and $[H_3O]_2[SO_4]$. As expected, H_3O^+ is pyramidal like the isoelectronic molecule NH_3, but the values of the angles H–O–H vary considerably due to extensive H bonding throughout the crystal, e.g. 117° in the chloride, 112° in the nitrate, and 101°, 106°, and 126° in $[H_3O][HSO_4]$.[73] Likewise the H-bonded distance O–H$\cdots$O varies: it is 266 pm in the nitrate, 254–265 in $[H_3O][HSO_4]$, and 252–259 in $[H_3O]_2[SO_4]$. The most stable hydroxonium salt yet known is the white crystalline complex $[H_3O]^+[SbF_6]^-$, prepared by adding the stoichiometric amount of H_2O to a solution of SbF_5 in anhydrous HF;[78] it decomposes without melting when heated to 357°C. The analogous compound $[H_3O]^+[AsF_6]^-$ decomposes at 193°C.

The dihydrated proton $[H_5O_2]^+$ was first established in HCl.$2H_2O$ (1967) and $HClO_4$.$2H_2O$ (1968), and is now known in over a dozen compounds. The structure is shown schematically in the diagram though the conformation varies from staggered in the

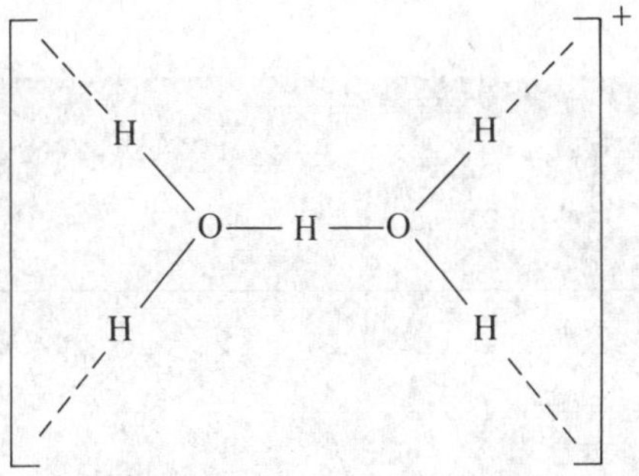

perchlorate, through an intermediate orientation for the chloride to almost eclipsed for $[H_5O_2]Cl.H_2O$. In the case of $[H_5O_2]^+{}_3[PW_{12}O_{40}]^{3-}$, an apparently planar arrangement of all 7 atoms in the cation is an artefact of disorder in the crystal. The O–H$\cdots$O distance is usually in the range 240–245 pm though in the deep-yellow crystalline compound $[NEt_4]_3[H_5O_2][Mo_2Cl_8H][MoCl_4O(OH_2)]$ it is only 234 pm, one of the shortest O–H$\cdots$O bonds known.[79]

The ions $[H_7O_3]^+$ and $[H_9O_4]^+$ are both featured in the compound HBr.$4H_2O$ which has the unexpectedly complicated formulation $[H_9O_4]^+[H_7O_3]^+[Br]^-{}_2.H_2O$. The structure of the cations are shown schematically in Fig. 14.16 which indicates that a bromide ion has essentially displaced the fourth water molecule of the second cation to give an effectively neutral H-bonded unit $[(H_2O)_3H^+Br^-]$. The discrete $[H_7O_3]^+$ ion is now known in about half-a-dozen complexes of which the most recent example is the deep-green complex $[NEt_4]_2[H_7O_3]_2[Ru_3Cl_{12}]$ in which the 2 O–H$\cdots$O distances are 245 and 255 pm and the O–O$\cdots$O angle is 115.9°.[80]

[78] K. O. CHRISTE, C. J. SCHACK and R. D. WILSON, Novel onium salts. Synthesis and characterization of $OH_3{}^+SbF_6{}^-$ and $OH_3{}^+AsF_6{}^-$, *Inorg. Chem.* **14**, 2224–30 (1975).

[79] A. BINO and F. A. COTTON, Structural characterization of the hydrido-bridged $[Mo_2Cl_8H]^{3-}$ ion, the $[MoCl_4O(OH_2)]$ ion and an $[H_5O_2]^+$ ion with an exceptionally short hydrogen bond, *J. Am. Chem. Soc.* **101**, 4150–4 (1979).

[80] A. BINO and F. A. COTTON, A linear, trinuclear, mixed valence chloro complex of ruthenium $[Ru_3Cl_{12}]^{4-}$, *J. Am. Chem. Soc.* **102**, 608–11 (1980).

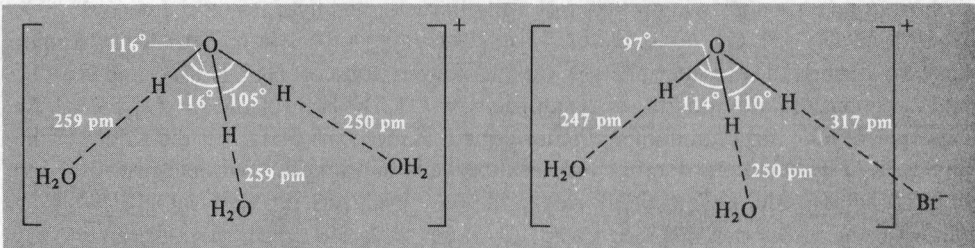

Fig. 14.16   Schematic representation of the structures of the $[H_9O_4]^+$ and $[H_7O_3]^+\cdots Br^-$ units
in HBr.4H$_2$O, showing bond angles and O–H$\cdots$O (O–H$\cdots$Br) distances.

The largest protonated cluster of water molecules yet characterized is the discrete unit
$[H_{13}O_6]^+$ formed serendipidously when the cage compound $[(C_9H_{18})_3(NH)_2Cl]^+Cl^-$ was
crystallized from a 10% aqueous hydrochloric acid solution.[81] The structure of the cage
cation is shown in Fig. 14.17 and the unit cell contains $4\{[C_9H_{18})_3(NH)_2Cl]Cl-$
$[H_{13}O_6]Cl\}$. The hydrated proton features a short symmetrical O–H–O bond at the

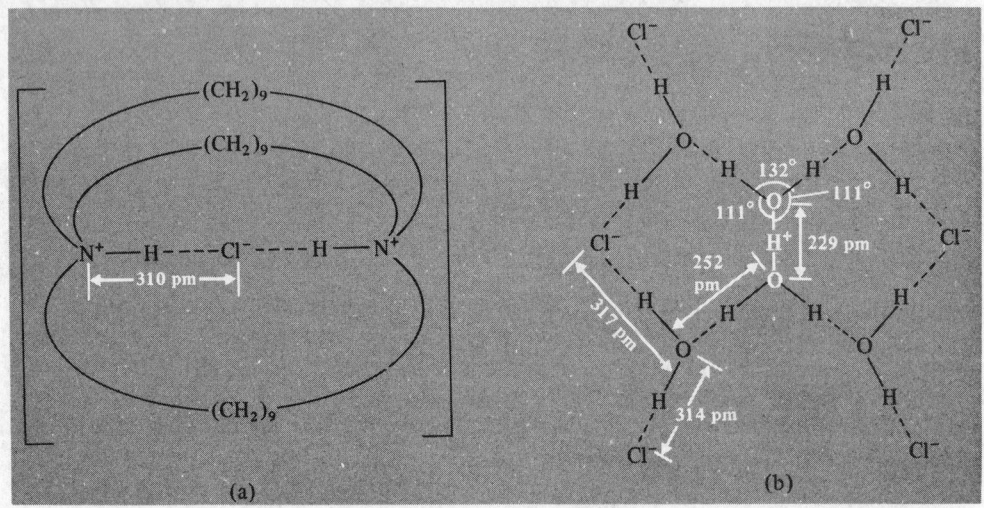

Fig.   14.17   (a)   Schematic   representation   of   the   structure   of   the   cage   cation
$[(C_9H_{18})_3(NH)_2Cl]^+$, and (b) detailed structure of the $[H_{13}O_6]^+$ ion showing its H bonding to
surrounding Cl$^-$ anions. The ion has C_{2h} symmetry with the very short central O–H–O lying
across the centre of symmetry.

centre of symmetry and 4 longer unsymmetrical O–H$\cdots$O bonds to 4 further H$_2$O
molecules, the whole $[H_{13}O_6]^+$ unit being connected to the rest of the lattice by H bonds
of normal length to surrounding chloride ions. It is clear from these various examples that
the stability of the larger proton hydrates is enhanced by the presence of large co-cations
and/or counter-anions in the lattice.

[81] R. A. BELL, G. G. CHRISTOPH, F. R. FRONCZEK and R. E. MARSH, The cation $[H_{13}O_6]^+$: a short symmetric
hydrogen bond, *Science* **190**, 151–2 (1975).

Hydrated forms of the hydroxide ion have been much less well characterized though the monohydrate $[H_3O_2]^-$ has recently been discovered in the mixed salt $Na_2[NEt_3Me]$-$[Cr\{PhC(S)=N(O)\}_3] . \frac{1}{2}NaH_3O_2 . 18H_2O$ which formed when $[NEt_3Me]I$ was added to a solution of tris(thiobenzohydroximato)chromate(III) in aqueous NaOH.[82] The compound tended to loose water at room temperature but an X-ray study identified the centro-symmetric $[HO-H-OH]^-$ anion shown in Fig. 14.18. The central O–H–O bond is

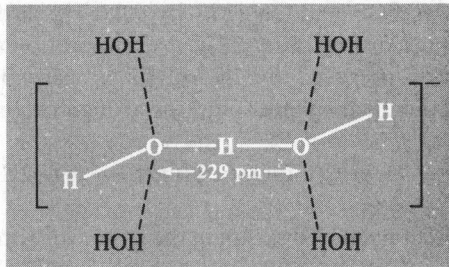

FIG. 14.18   Structure of the centrosymmetric $[H_3O_2]^-$ ion showing the disposition of longer H bonds to neighbouring water molecules.

very short indeed (229 pm) and is probably symmetrical, though the central H was not located on the electron density map. It will be noted that $[H_3O_2]^-$ is isoelectronic with the bifluoride ion $[F-H-F]^-$ which also features a very short, symmetrical H bond with $F\cdots F$ 227 pm (p. 64).

Polywater

The saga of polywater forms a fascinating and informative case history of the massive amount of work that can be done, even in modern times, on the preparation and characterization of a compound which was eventually found not to exist. Between 1966 and 1973 over 500 scientific papers were published on polywater following B. V. Deryagin's description of work done in the USSR during the preceding years.[83] The supposed compound, variously called anomalous water, orthowater, polywater, super-water, cyclimetric water, superdense water, water II, and water-X, was prepared in minute amounts by condensing purified "ordinary water" into fine, freshly drawn glass capillaries of diameter 1–3 μm. The thermodynamic difficulties inherent in the very existence of such a compound were soon apparent and it was proposed that polywater was, in fact, a dispersion of a silica gel leached from the glass capillaries,[84] despite the specific rejection of this possibility by several groups of earlier workers. The full panoply of physicochemical techniques was brought to bear on the problem, and it was finally conceded that the

[82] J. ABU-DARI, K. N. RAYMOND, and D. P. FREYBERG, The bihydroxide $[H_3O_2]^-$ anion. A very short symmetric hydrogen bond, *J. Am. Chem. Soc.* **101**, 3688–9 (1979).

[83] B. V. DERYAGIN, Effect of lyophile surfaces on the properties of boundary liquid films, *Discussions Faraday Soc.* **42**, 109–19 (1966).

[84] A. CHERKIN, Anomalous water—a silica dispersion?, *Nature* **224**, 1293 (1969). (See also *Nature* **222**, 159–61 (1969)).

anomalous properties were caused by a mixture of colloidal silicic acid and dissolved compounds of Na, K, Ca, B, Si, N (nitrate), O (sulfate), and Cl leached from the glass by the aggressive action of freshly condensed water.[85] A very informative annotated bibliography is available which traces the course of this controversy and analyses the reasons why it took so long to resolve.[86]

14.2.3 *Hydrogen peroxide*

Hydrogen peroxide was first made in 1818 by J. L. Thenard who acidified barium peroxide (p. 132) and then removed excess H_2O by evaporation under reduced pressure. Later the compound was prepared by hydrolysis of peroxodisulfates obtained by electrolytic oxidation of acidified sulfate solutions at high current densities:

$$2HSO_4^-(aq) \xrightarrow{-2e^-} HO_3SOOSO_3H(aq) \xrightarrow{2H_2O} 2HSO_4^- + H_2O_2$$

Such processes are now no longer used except in the laboratory preparation of D_2O_2, e.g.:

$$K_2S_2O_8 + 2D_2O \longrightarrow 2KDSO_4 + D_2O_2$$

On an industrial scale H_2O_2 is now almost exclusively prepared by the autoxidation of 2-alkylanthraquinols (see Panel).

Physical properties

Hydrogen peroxide, when pure, is an almost colourless (very pale blue) liquid, less volatile than water and somewhat more dense and viscous. Its more important physical properties are in Table 14.11 (cf. H_2O, p. 730). The compound is miscible with water in all

TABLE 14.11   *Some physical properties of hydrogen peroxide*[a]

Property	Value
MP/°C	−0.43
BP/°C (extrap)	150.2
Vapour pressure (25°)/mmHg	1.9
Density (solid at −4.5°)/g cm^{-3}	1.6434
Density (liquid at 25°)/g cm^{-3}	1.4425
Viscosity (20°)/centipoise	1.249
Dielectric constant ε (25°)	70.7
Electric conductivity (25°)/ohm^{-1} cm^{-1}	5.1×10^{-8}
ΔH_f°/kJ mol^{-1}	−187.6
ΔG_f°/kJ mol^{-1}	−118.0

(a) For D_2O_2: mp $+1.5°$; d_{20} 1.5348 g cm^{-3}; η_{20} 1.358 centipoise.

[85] B. V. DERYAGIN and N. V. CHURAEV, Nature of anomalous water, *Nature* **244**, 430–31 (1973); B. V. DERYAGIN, The state-of-the-art in liquids' modification by condensation, *Recent Advances in Adhesion*, 1973, 23–31.

[86] F. PERCIVAL and A. H. JOHNSTONE, *Polywater—A Library Exercise for Chemistry Degree Students*, The Chemical Society, London, 1978, 24 pp. [See also B. F. POWELL, Anomalous water—fact or figment? *J. Chem. Educ.* **48**, 663–7 (1971). H. FREIZER, Polywater and analytical chemistry—a lesson for the future, *J. Chem. Educ.* **49**, 445 (1972). F. FRANKS, *Polywater*, MIT Press, Cambridge, Mass., 1981, 208 pp.]

Preparation and Uses of Hydrogen Peroxide[87]

Hydrogen peroxide is a major industrial chemical manufactured on a multikilotonne scale by an ingenious cycle of reactions introduced by I. G. Farbenindustrie about 40 years ago. Since the value of the solvents and organic substrates used are several hundred times that of the H_2O_2 produced, the economic viability of the process depends on keeping losses very small indeed. The basic process consists of dissolving 2-ethylanthaquinone in a mixed ester/hydrocarbon or alcohol/hydrocarbon solvent and reducing it by a Raney nickel or supported palladium catalyst to the corresponding quinol. The catalyst is then separated and the quinol non-catalytically reoxidized in a stream of air:

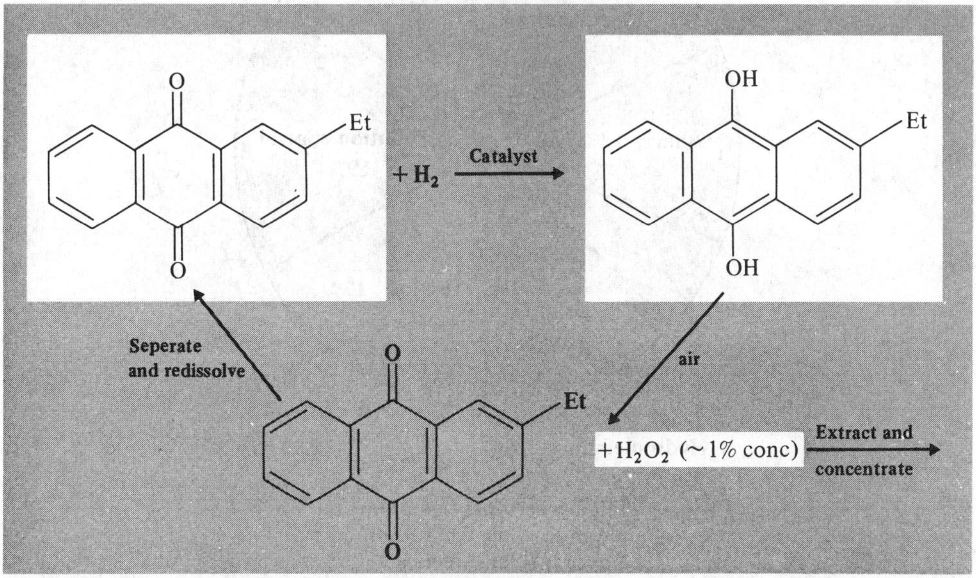

The H_2O_2 is extracted by water and concentrated to $\sim 30\%$ (by weight) by distillation under reduced pressure. Further low-pressure distillation to concentrations up to 85% are not uncommon.

World production (excluding the USSR) expressed as 100% H_2O_2 was some 520 000 tonnes in 1975 of which two-thirds was in Europe, one-fifth in the USA, and one-seventh in Japan. The earliest and still the largest industrial use for H_2O_2 is as a bleach for textiles, paper pulp, straw, leather, oils and fats, etc. Domestic use as a hair bleach and a mild disinfectant has diminished somewhat. Hydrogen peroxide is also extensively used to manufacture chemicals, notably sodium perborate (p. 234) and percarbonate (p. 102), which are major constituents of most domestic detergents at least in the UK and Europe. Normal formulations include $15–25\%$ of such peroxoacid salts, though the practice is much less widespread in the USA, and the concentrations, when included at all, are usually less than 10%.

In the organic chemicals industry, H_2O_2 is used in the production of epoxides, propylene oxide, and caprolactones for PVC stabilizers and polyurethanes, in the manufacture of organic peroxy compounds for use as polymerization initiators and curing agents, and in the synthesis of fine chemicals such as hydroquinone, pharmaceuticals (e.g. cephalosporin), and food products (e.g. tartaric acid).

Continued

[87.] C. A. CRAMPTON, G. FABER, R. JONES, J. P. LEAVER and S. SCHELLE, The manufacture, properties, and uses of hydrogen peroxide and other inorganic peroxy compounds, in R. THOMPSON (ed.), *The Modern Inorganic Chemicals Industry*, pp. 232–66. Chemical Society Special Publications, No. 31, 1977.

One of the rapidly growing uses of H_2O_2 is in environmental applications such as control of pollution by treatment of domestic and industrial effluents, e.g. oxidation of cyanides and obnoxious malodorous sulfides, and the restoration of aerobic conditions to sewage waters. An indication of the proportion of H_2O_2 production used for these various applications in the USA during 1979 is shown in the diagram.

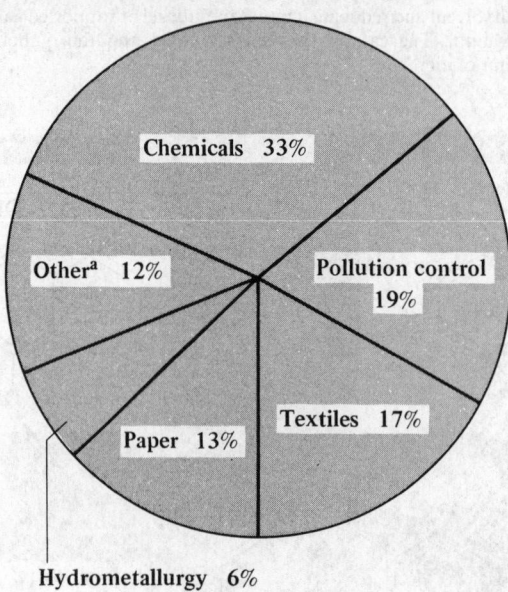

Hydrometallurgy 6%

[a]Includes personal care products,
food processing, and furniture
Source: Interox America

proportions and forms a hydrate $H_2O_2.H_2O$, mp $-52°$. Addition of water increases the already high dielectric constant of H_2O_2 (70.7) to a maximum value of 121 at $\sim 35\%$ H_2O_2, i.e. substantially higher than the value of water itself (78.4 at 25°).

In the gas phase the molecule adopts a skew configuration with a dihedral angle of 111.5° as shown in Fig. 14.19. This is due to repulsive interaction of the O–H bonds with the lone-pairs of electrons on each O atom. Indeed, H_2O_2 is the smallest molecule known to show hindered rotation about a single bond, the rotational barriers being 4.62 and 29.45 kJ mol^{-1} for the *trans* and *cis* conformations respectively. The skew form persists in the liquid phase, no doubt modified by H bonding, and in the crystalline state at $-163°C$ the most recent neutron diffraction study[88] gives the dimensions shown in Fig. 14.19b. The dihedral angle is particularly sensitive to H bonding, decreasing from 111.5° in the gas phase to 90.2° in crystalline H_2O_2; in fact, values spanning the complete range from 90° to 180° (i.e. *trans* planar) are known for various solid phases containing molecular H_2O_2

[88] J.-M. SAVARIAULT and M. S. LEHMANN, Experimental determination of the deformation electron density in hydrogen peroxide by combination of X-ray and neutron diffraction measurements, *J. Am. Chem. Soc.* **102**, 1298–1303 (1980).

(Table 14.12). The O–O distance in H_2O_2 corresponds to the value expected for a single bond (p. 720).

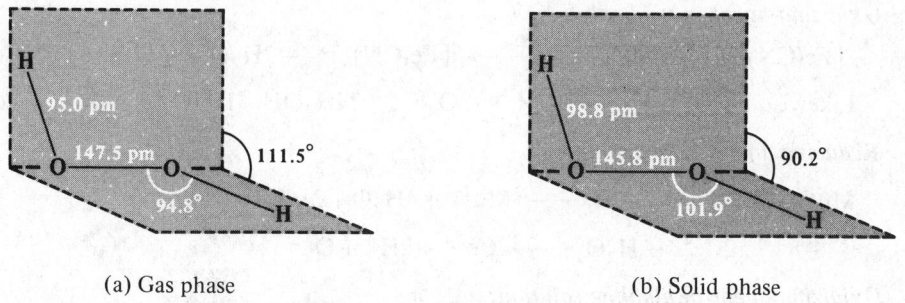

(a) Gas phase   (b) Solid phase

FIG. 14.19   Structure of the H_2O_2 molecule (a) in the gas phase, and (b) in the crystalline state. In the crystal the dihedral angle diminishes appreciably due to H bonding, and the O–O–H bond angle expands somewhat; changes in the interatomic distances are much smaller but are probably real.

TABLE 14.12   *Dihedral angle of H_2O_2 in some crystalline phases*

Compound	Dihedral angle	Compound	Dihedral angle
$H_2O_2(s)$	90.2°	$Li_2C_2O_4 . H_2O_2$	180°
$K_2C_2O_4 . H_2O_2$	101.6°	$Na_2C_2O_4 . H_2O_2$	180°
$Rb_2C_2O_4 . H_2O_2$	103.4°	$NH_4F . H_2O_2$ [89]	180°
$H_2O_2 . 2H_2O$	129°		

Chemical properties

In H_2O_2 the oxidation state of oxygen is -1, intermediate between the values for O_2 and H_2O, and, as indicated by the reduction potentials on p. 736, aqueous solutions of H_2O_2 should spontaneously disproportionate. For the pure liquid: $H_2O_2(l) \rightarrow H_2O(l) + \frac{1}{2}O_2(g)$; $\Delta H° = -98.2 \text{ kJ mol}^{-1}$, $\Delta G° = -119.2 \text{ kJ mol}^{-1}$. In fact, in the absence of catalysts, the compound decomposes negligibly slowly but the reaction is strongly catalysed by metal surfaces (Pt, Ag), by MnO_2, or by traces of alkali (dissolved from glass), and for this reason H_2O_2 is generally stored in wax-coated or plastic vessels with stabilizers such as urea; even a speck of dust can initiate explosive decomposition, and all handling of the anhydrous compound or its concentrated solutions must be carried out in dust-free conditions and in the absence of metal ions. A useful "carrier" for H_2O_2 in some reactions is the adduct $(Ph_3PO)_2 . H_2O_2$.

Hydrogen peroxide has a rich and varied chemistry which arises from (i) its ability to act either as an oxidizing or a reducing agent in both acid and alkaline solution, (ii) its ability to undergo proton acid/base reactions to form peroxonium salts $(H_2OOH)^+$, hydroperoxides $(OOH)^-$, and peroxides $(O_2)^{2-}$, and (iii) its reactions to give peroxometal complexes and peroxoacid anions.

[89] V. A. SARIN, V. YA. DUDAREV, T. A. DOBRYNINA and V. E. ZAVODNIK, An X-ray structure investigation of $NH_4F . H_2O_2$ crystals, *Soviet Phys. Crystallogr.* **24**, 472–3 (1979), and references therein.

The ability of H_2O_2 to act both as an oxidizing and a reducing agent is well known in analytical chemistry. Typical examples (not necessarily of analytical utility) are:

Oxidizing agent in acid solution:

$$2[Fe(CN)_6]^{4-} + H_2O_2 + 2H^+ \longrightarrow 2[Fe(CN)_6]^{3-} + 2H_2O$$

Likewise $Fe^{2+} \rightarrow Fe^{3+}$, $SO_3^{2-} \rightarrow SO_4^{2-}$, $NH_2OH \rightarrow HNO_3$ etc.

Reducing agent in acid solution:

$$MnO_4^- + 2\tfrac{1}{2}H_2O_2 + 3H^+ \longrightarrow Mn^{2+} + 4H_2O + 2\tfrac{1}{2}O_2$$

$$2Ce^{4+} + H_2O_2 \longrightarrow 2Ce^{3+} + 2H^+ + O_2$$

Oxidizing agent in alkaline solution:

$$Mn^{2+} + H_2O_2 \longrightarrow Mn^{4+} + 2OH^-$$

Reducing agent in alkaline solution:

$$2[Fe(CN)_6]^{3-} + H_2O_2 + 2OH^- \longrightarrow 2[Fe(CN)_6]^{4-} + 2H_2O + O_2$$

$$2Fe^{3+} + H_2O_2 + 2OH^- \longrightarrow 2Fe^{2+} + 2H_2O + O_2$$

$$KIO_4 + H_2O_2 \longrightarrow KIO_3 + H_2O + O_2$$

It will be noted that O_2 is always evolved when H_2O_2 acts as a reducing agent, and sometimes this gives rise to a red chemiluminescence if the dioxygen molecule is produced in a singlet state (p. 705), e.g.:

Acid solution:

$$HOCl + H_2O_2 \longrightarrow H_3O^+ + Cl^- + {}^1O_2{}^* \longrightarrow h\nu$$

Alkaline solution:

$$Cl_2 + H_2O_2 + 2OH^- \longrightarrow 2Cl^- + 2H_2O + {}^1O_2{}^* \longrightarrow h\nu$$

The catalytic decomposition of aqueous solutions H_2O_2 alluded to on p. 745 can also be viewed as an oxidation–reduction process and, indeed, most homogeneous catalysts for this reaction are oxidation–reduction couples of which the oxidizing agent can oxidize (be reduced by) H_2O_2 and the reducing agent can reduce (be oxidized by) H_2O_2. Thus, using the data on p. 736, any complex with a reduction potential between $+0.695$ and $+1.776$ V in acid solution should catalyse the reaction. For example:

$$\underline{Fe^{3+}/Fe^{2+}, E° +0.771\ V} \qquad\qquad \underline{Br_2/2Br^-, E° +1.078}$$

$$2Fe^{3+} + H_2O_2 \xrightarrow{-2H^+} 2Fe^{2+} + O_2 \qquad Br_2 + H_2O_2 \xrightarrow{-2H^+} 2Br^- + O_2$$

$$2Fe^{2+} + H_2O_2 \xrightarrow{+2H^+} 2Fe^{3+} + 2H_2O \quad 2Br^- + H_2O_2 \xrightarrow{+2H^+} Br_2 + 2H_2O$$

$$Net: 2H_2O_2 \longrightarrow 2H_2O + O_2 \qquad Net: 2H_2O_2 \longrightarrow 2H_2O + O_2$$

In many such reactions, experiments using ^{18}O show negligible exchange between H_2O_2 and H_2O, and all the O_2 formed when H_2O_2 is used as a reducing agent comes from the

H_2O_2, implying that oxidizing agents do not break the O–O bond but simply remove electrons. Not all reactions are heterolytic, however, and free radicals are sometimes involved, e.g. Ti^{3+}/H_2O_2 and Fenton's reagent (Fe^{2+}/H_2O_2). The most important free radicals are OH and O_2H.

Hydrogen peroxide is a somewhat stronger acid than water, and in dilute aqueous solutions has $pK_a(20°) = 11.75 \pm 0.02$, i.e. comparable with the third dissociation constant of H_3PO_4 (p. 598):

$$H_2O_2 + H_2O \rightleftharpoons H_3O^+ + OOH^-; \quad K_a = \frac{[H_3O^+][OOH^-]}{[H_2O_2]} = 1.78 \times 10^{-12} \text{ mol } l^{-1}$$

Conversely, H_2O_2 is a much weaker base than H_2O (perhaps by a factor of 10^6), and the following equilibrium lies far to the right:

$$H_3O_2{}^+ + H_2O \rightleftharpoons H_2O_2 + H_3O^+$$

As a consequence, salts of $H_3O_2{}^+$ cannot be prepared from aqueous solution but they have very recently (1979) been obtained as white solids from the strongly acid solvent systems anhydrous HF/SbF_5 and HF/AsF_5, e.g.:[90]

$$H_2O_2 + HF + MF_5 \longrightarrow [H_3O_2]^+[MF_6]^-$$

$$H_2O_2 + HF + 2SbF_5 \longrightarrow [H_3O_2]^+[Sb_2F_{11}]^-$$

The salts decompose quantitatively at or slightly above room temperature, e.g.:

$$2[H_3O_2][SbF_6] \xrightarrow{45°} 2[H_3O][SbF_6] + O_2$$

The ion $[H_2OOH]^+$ is isoelectronic with H_2NOH and vibrational spectroscopy shows it to have the same (C_s) symmetry.

Deprotonation of H_2O_2 yields OOH^-, and hydroperoxides of the alkali metals are known in solution. Liquid ammonia can also effect deprotonation and NH_4OOH is a white solid, mp 25°; infrared spectroscopy shows the presence of $NH_4{}^+$ and OOH^- ions in the solid phase but the melt appears to contain only the H-bonded species NH_3 and H_2O_2.[91] Double deprotonation yields the peroxide ion $O_2{}^{2-}$, and this is a standard route to transition metal peroxides.[30] Many such compounds are noted under the individual transition elements and it is only necessary here to note that the chemical identity of the products obtained is often very sensitive to the conditions employed because of the combination of acid–base and redox reactions in the system. For example, treatment of alkaline aqueous solutions of chromate(VI) with H_2O_2 yields the stable red paramagnetic tetraperoxochromate(V) compounds $[Cr^V(O_2)_4]^{3-}$ (μ 1.80 BM), whereas treatment of chromate(VI) with H_2O_2 in acid solution followed by extraction with ether and coordination with pyridine yields the neutral peroxochromate(VI) complex $[CrO(O_2)_2py]$ which has a small temperature-independent paramagnetism of about 0.5 BM. The structure of these two species is in Fig. 14.20 which also includes the structure

[90] K. O. CHRISTE, W. W. WILSON, and E. C. CURTIS, Novel onium salts. Synthesis and characterization of the peroxonium cation H_2OOH^+, *Inorg. Chem.* **18**, 2578–86 (1979).

[91] O. KNOP and P. A. GIGUÈRE, Infrared spectrum of ammonium hydroperoxide, *Canad. J. Chem.* **37**, 1794–7 (1959).

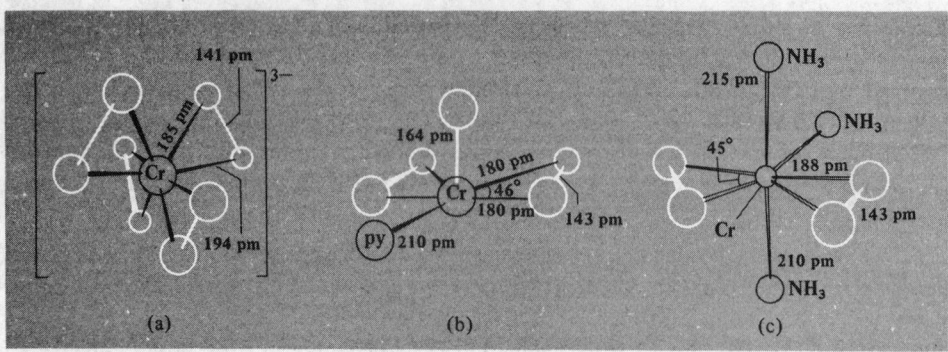

Fig. 14.20   Structures of (a) the tetraperoxochromate(V) ion $[Cr^V(O_2)_4]^{3-}$, (b) the pyridine oxodiperoxochromium(VI) complex $[Cr^{VI}O(O_2)_2py]$, and (c) the triamminodiperoxo-chromium(IV) complex $[Cr^{IV}(NH_3)_3(O_2)_2]$ showing important interatomic distances and angles. (This last compound was originally described as a chromium(II) superoxo complex $[Cr^{II}(NH_3)_3(O_2)_2]$ on the basis of an apparent O–O distance of 131 pm,[92] and is a salutory example of the factual and interpretative errors that can arise even in quite recent X-ray diffraction studies.[93]

of the brown diperoxochromium(IV) complex $[Cr^{IV}(NH_3)_3(O_2)_2]$ (μ 2.8 BM) prepared by treating either of the other two complexes with an excess of aqueous ammonia or more directly by treating an aqueous ammonical solution of $[NH_4]_2[Cr_2O_7]$ with H_2O_2. Besides deprotonation of H_2O_2, other routes to metal peroxides include the direct reduction of O_2 by combustion of the electropositive alkali and alkaline earth metals in oxygen (pp. 98, 132) or by reaction of O_2 with transition metal complexes in solution (p. 719).[94]

Peroxoanions are described under the appropriate element, e.g. peroxoborates (p. 232), peroxonitrates (p. 528), peroxophosphates (p. 588), peroxosulfates (p. 846), and peroxo-disulfates (p. 846).

14.2.4   *Oxygen fluorides*[95]

Oxygen forms several binary fluorides of which the most stable is OF_2. This was first made in 1929 by the electrolysis of slightly moist molten KF/HF but is now generally made by reacting F_2 gas with 2% aqueous NaOH solution:

$$2F_2 + 2NaOH \longrightarrow OF_2 + 2NaF + H_2O$$

Conditions must be controlled so as to minimize loss of the product by the secondary reaction:

$$OF_2 + 2OH^- \longrightarrow O_2 + 2F^- + H_2O$$

[92] E. H. McLaren and L. Helmholz, The crystal and molecular structure of triamminochromium tetroxide, *J. Chem. Phys.* **63**, 1279–83 (1959).

[93] R. Stromberg, The crystal and molecular structure of diperoxotriamminechromium(IV) $[Cr(O_2)_2(NH_3)_3]$, *Arkiv Kemi* **22**, 49–64 (1974).

[94] N.-G. Vannerberg, Peroxides, superoxides, and ozonides of the metals of Groups IA, IIA, and IIB, *Prog. Inorg. Chem.* **4**, 125–97 (1962).

[95] E. A. V. Ebsworth, J. A. Connor, and J. J. Turner, Oxygen fluorides, in J. C. Bailar, H. J. Eméleus, R. S. Nyholm, and A. F. Trotman-Dickenson (eds.), *Comprehensive Inorganic Chemistry*, Vol. 2, Chap. 22, Section 5, pp. 747–71. Pergamon Press, Oxford, 1973.

Oxygen fluoride is a colourless, very poisonous gas that condenses to a pale-yellow liquid (mp $-223.8°$, bp $-145.3°C$). When pure it is stable to $200°$ in glass vessels but above this temperature decomposes by a radical mechanism to the elements. Molecular dimensions (microwave) are in Fig. 14.21, where they are compared with those of related molecules.

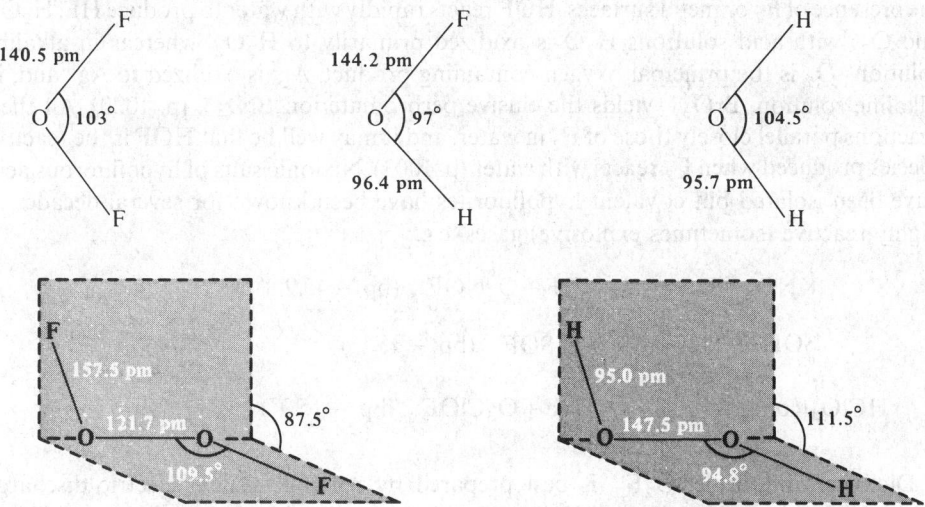

FIG. 14.21   Comparison of the molecular dimensions of various gaseous molecules having O–F and O–H bonds.

The heat of formation has been given as $\Delta H_f°$ 24.5 kJ mol^{-1}, leading to an average O–F bond energy of 187 kJ mol^{-1}. Though less reactive than elementary fluorine, OF_2 is a powerful oxidizing and fluorinating agent. Many metals give oxides and fluorides, phosphorus yields PF_5 plus POF_3, sulfur SO_2 plus SF_4, and xenon gives XeF_4 and oxofluorides (p. 1056). H_2S explodes on being mixed with OF_2 at room temperature. OF_2 is formally the anhydride of hypofluorous acid, HOF, but there is no evidence that it reacts with water to form this compound. Indeed, HOF had been sought for many decades but has only recently been prepared and fully characterized.[96]

HOF was first identified by P. N. Noble and G. C. Pimentel in 1968 using matrix isolation techniques: F_2/H_2O mixtures were frozen in solid N_2 and photolysed at 14–20 K:

$$F_2 + H_2O \rightleftharpoons HOF + HF$$

A more convenient larger-scale preparation was devised in 1971 by M. H. Studier and E. H. Appleman, who circulated F_2 rapidly through a Kel-F U-tube filled with Räschig rings of polytetrafluoroethylene (Teflon) which had been moistened with water and cooled to $-40°$. An essential further condition was the presence of traps at $-50°$ and $-79°$ to remove H_2O and HF (both of which react with HOF), and the product was retained in a trap at $-183°$. HOF is a white solid, melting at $-117°$ to a pale yellow liquid which boils

96 E. H. APPELMAN, Nonexistent compounds: two case histories, *Acc. Chem. Res.* **6**, 113–17 (1973).

below room temperature. Molecular dimensions are in Fig. 14.21; the small bond angle is particularly notable, being the smallest yet recorded for 2-coordinate O in an open chain. HOF is stable with respect to its elements: $\Delta H_f^\circ(298) = -98.2$, $\Delta G_f^\circ(298) = -85.7$ kJ mol^{-1}. However, HOF decomposes fairly rapidly to HF and O_2 at room temperature ($t_{1/2} \sim 30$ min at 100 mmHg in Kel-F or Teflon). Decomposition is accelerated by light and by the presence of F_2 or metal surfaces. HOF reacts rapidly with water to produce HF, H_2O_2, and O_2; with acid solutions H_2O is oxidized primarily to H_2O_2, whereas in alkaline solutions O_2 is the principal oxygen-containing product. Ag^I is oxidized to Ag^{II} and, in alkaline solution, BrO_3^- yields the elusive perbromate ion BrO_4^- (p. 1020). All these reactions parallel closely those of F_2 in water, and it may well be that HOF is the reactive species produced when F_2 reacts with water (p. 1003). No ionic salts of hypofluorous acid have been isolated but covalent hypofluorites have been known for several decades as highly reactive (sometimes explosive) gases, e.g.:

$$KNO_3 + F_2 \longrightarrow KF + O_2NOF \quad (\text{bp } -45.9°)$$

$$SOF_2 + 2F_2 \xrightarrow{\text{CsF}} F_5SOF \quad (\text{bp } -35.1°)$$

$$HClO_4(\text{conc}) + F_2 \longrightarrow HF + O_3ClOF \quad (\text{bp } -15.9°)$$

Dioxygen difluoride, O_2F_2, is best prepared by passing a silent electric discharge through a low-pressure mixture of F_2 and O_2: the products obtained depend markedly on conditions, and the yield of O_2F_2 is optimized by using a 1:1 mixture at 7–17 mmHg and a discharge of 25–30 mA at 2.1–2.4 kV. Alternatively, pure O_2F_2 can be synthesized by subjecting a mixture of liquid O_2 and F_2 in a stainless steel reactor at $-196°$ to 3 MeV bremstrahlung radiation for 1–4 h. O_2F_2 is a yellow solid and liquid, mp $-154°$, bp $-57°$ (extrapolated). It is much less stable than OF_2 and even at $-160°$ decomposes at a rate of some 4% per day. Decomposition by a radical mechanism is rapid above $-100°$. The structure of O_2F_2 (Fig. 14.21) resembles that of H_2O_2 but the remarkably short O–O distance is a notable difference in detail (cf. O_2 gas 120.7 pm). The O–F distance is, conversely, unusually long. These features are paralleled by the bond dissociation energies:

$$D(\text{FO–OF}) \ 430 \text{ kJ mol}^{-1}, \ D(\text{F–OOF}) \sim 75 \text{ kJ mol}^{-1}.$$

Accordingly, mass spectrometric, infrared, and electron spin resonance studies confirm dissociation into F and OOF radicals, and low-temperature studies have also established the presence of the dimer O_4F_2, which is a dark red-brown solid, mp $-191°C$. Impure O_4F_2 can also be prepared by silent electric discharge but the material previously thought to be O_3F_2 is probably a mixture of O_4F_2 and O_2F_2. Dioxygen difluoride, as expected, is a very vigorous and powerful oxidizing and fluorinating agent even at very low temperatures ($-150°$). It converts ClF to ClF_3, BrF_3 to BrF_5, and SF_4 to SF_6. Similar products are obtained from HCl, HBr, and H_2S, e.g.:

$$H_2S + 4O_2F_2 \longrightarrow SF_6 + 2HF + 4O_2$$

Interest in the production of high-energy oxidizers for use in rocket motors has stimulated the study of peroxo compounds bound to highly electronegative groups during the past few decades. Although such applications have not yet materialized, numerous new compounds of this type have been synthesized and characterized, e.g.:

$$2SO_3 + F_2 \xrightarrow[\text{90\% yield}]{160°/AgF_2 \text{ catalyst}} FO_2SOOSO_2F$$

$$2SF_5Cl + O_2 \xrightarrow{h\nu} F_5SOOSF_5 + Cl_2$$

$$2COF_2 + OF_2 \xrightarrow{CsF} F_3COOOCF_3$$

Such compounds are volatile liquids or gases (Table 14.13) and their extensive reaction chemistry has been very fully reviewed.[97]

TABLE 14.13 *Properties of some fluorinated peroxides*

Compound	MP/°C	BP/°C	Compound	MP/°C	BP/°C
FO$_2$SOOSO$_2$F	−55.4	67.1	F$_3$COONO$_2$	—	0.7
FO$_2$SOOF	—	0	F$_3$COOP(O)F$_2$	−88.6	15.5
FO$_2$SOOSF$_5$	—	54.1	F$_3$COOCl	−132	−22
F$_5$SOOSF$_5$	−95.4	49.4	(F$_3$C)$_3$COOC(CF$_3$)$_3$	12	98.6
F$_5$SOOCF$_3$	−136	7.7	F$_3$COOOCF$_3$	−138	−16

14.2.5 *Oxides*

Various methods of classification

Oxides are known for all elements of the periodic table except the lighter noble gases and, indeed, most elements form more than one binary compound with oxygen. Their properties span the full range of volatility from difficultly condensible gases such as CO (bp $-191.5°C$) to refractory oxides such as ZrO_2 (mp 3265°C, bp $\sim$4850°C). Likewise, their electrical properties vary from being excellent insulators (e.g. MgO), through semi-conductors (e.g. NiO), to good metallic conductors (e.g. ReO_3). They may be precisely stoichiometric or show stoichiometric variability over a narrow or a wide range of composition. They may be thermodynamically stable or unstable with respect to their elements, thermally stable or unstable, highly reactive to common reagents or almost completely inert even at very high temperatures. With such a vast array of compounds and such a broad spectrum of properties any classification of oxides is likely to be either too simplified to be reliable or too complicated to be useful. One classification that is both convenient and helpful at an elementary level stresses the acid–base properties of oxides; this can be complemented and supplemented by classifications which stress the structural relationships between oxides. General classifications based on redox properties or on presumed bonding models have proved to be less helpful, though they are sometimes of use when a more restricted group of compounds is being considered.

[97] R. A. DE MARCO and J. M. SHREEVE, Fluorinated peroxides, *Adv. Inorg. Chem. Radiochem.*, **16**, 109–76 (1974); J. M. SHREEVE, Fluorinated peroxides, *Endeavour* xxxv, No. 125, 79–82 (1976).

The acid–base classification[98] turns essentially on the thermodynamic properties of hydroxides in aqueous solution, since oxides themselves are not soluble as such (p. 738). Oxides may be:

> *acidic:* e.g. most oxides of non-metallic elements (CO_2, NO_2, P_4O_{10}, SO_3, etc.);
> *basic:* e.g. oxides of electropositive elements (Na_2O, CaO, Tl_2O, La_2O_3, etc.);
> *amphoteric:* oxides of less electropositive elements (BeO, Al_2O_3, Bi_2O_3, ZnO, etc.);
> *neutral:* oxides that do not interact with water or aqueous acids or bases (CO, NO, etc.).

Periodic trends in these properties are well documented (p. 31). Thus, in a given period, oxides progress from strongly basic, through weakly basic, amphoteric, and weakly acidic, to strongly acidic (e.g. Na_2O, MgO, Al_2O_3, SiO_2, P_4O_{10}, SO_3, ClO_2). Acidity also increases with increasing oxidation state (e.g. $MnO < Mn_2O_3 < MnO_2 < Mn_2O_7$). A similar trend is the decrease in basicity of the lanthanide oxides with increase in atomic number from La to Lu. In the main groups, basicity of the oxides increases with increase in atomic number down a group (e.g. $BeO < MgO < CaO < SrO < BaO$), though the reverse tends to occur in the later transition element groups.

The thermodynamic and other physical properties of binary oxides (e.g. ΔH_f°, ΔG_f°, mp, etc.) show characteristic trends and variations when plotted as a function of atomic number, and the preparation of such plots using readily available compilations of data[99] can be a revealing and rewarding exercise.[100]

Structural classifications of oxides recognize discrete molecular species and structures which are polymeric in one or more dimensions leading to chains, layers, and ultimately, to three-dimensional networks. Some typical examples are in Table 14.14; structural details

TABLE 14.14   *Structure types for binary oxides in the solid state*

Structure type	Examples
Molecular structures	CO, CO_2, OsO_4, Tc_2O_7, Sb_2O_6, P_4O_{10}
Chain structures	HgO, SeO_2, CrO_3, Sb_2O_3
Layer structures	SnO, MoO_3, As_2O_3. Re_2O_7
Three-dimensional structures	See text

are given elsewhere under each individual element. The type of structure adopted in any particular case depends (obviously) not only on the stoichiometry but also on the relative sizes of the atoms involved and the propensity to form p_π double bonds to oxygen. In structures which are conventionally described as "ionic", the 6-coordinate radius of O^{2-} (140 pm) is larger than all 6-coordinate cation radii except for Rb^I, Cs^I, Fr^I, Ra^{II}, and Tl^I

[98] C. S. G. PHILLIPS and R. J. P. WILLIAMS, *Inorganic Chemistry*, Vol. 1, Oxford University Press, Oxford, 1965; Section 14.1, Acid, base, and amphoteric character, pp. 516–23; see also ref. 95, pp. 722–29, Acid–base character of simple oxides.

[99] M. C. BALL and A. H. NORBURY, *Physical Data for Inorganic Chemists*, Longmans, London, 1974, 175 pp. G. H. AYLWARD and T. J. V. FINDLAY, *SI Chemical Data*, 2nd edn., Wiley, Sydney, 1975, 136 pp.

[100] R. V. PARISH, *The Metallic Elements*, Longmans, London 1977, 254 pp. (see particularly pp. 25–28, 40–44, 66–74, 128–33, 148–50, 168–77, 188–98.

though it is approached by K^I (138 pm) and Ba^{II} (135 pm).[101] Accordingly, many oxides are found to adopt structures in which there is a close-packed oxygen lattice with cations in the interstices (frequently octahedral). For "cations", which have very small effective ionic radii (say < 50 pm), particularly if they carry a high formal charge, the structure type and bonding are usually better described in covalent terms, particularly when π interactions enhance the stability of terminal M=O bonds (M = C, N, P^V, S^{VI}, etc.). Thus, for oxides of formula MO, a coordination number of 1 (molecular) is found for CO and NO though the latter tends towards a coordination number of 2 (dimers, p. 513). With the somewhat larger Be^{II} and Zn^{II} the wurtzite (4:4) structure is adopted, and monoxides of still larger divalent cations tend to adopt the sodium chloride (6:6) structure (e.g. M^{II} = Mg, Ca, Sr, Ba, Co, Ni, Cd, Eu, etc.).

A similar trend is observed for oxides of $M^{IV}O_2$ in Group IV of the periodic table. The small C atom, with its propensity to form p_π–p_π bonds to oxygen, adopts a linear, molecular structure O=C=O. Silicon, being somewhat larger and less prone to double bonding (p. 419), is surrounded by 4 essentially single-bonded O in most forms of SiO_2 (p. 393) and the coordination geometry is thus 4:2. Similarly, GeO_2 adopts the quartz structure; in addition a rutile form (p. 1120) is known in which the coordination is 6:3. SnO_2 and PbO_2 also have rutile structures as has TiO_2, but the largest Group IVA cations Zr and Hf adopt the fluorite (8:4) structure (p. 129) in their dioxides. Other large cations with a fluorite structure for MO_2 are Po; Ce, Pr, Tb; Th, U, Np, Pu, Am, and Cm. Conversely, the antifluorite structure is found for the alkali metal monoxides M_2O (p. 97). Such simple ideas are capable of considerable further elaboration.[102]

Nonstoichiometry

Transition elements, for which variable valency is energetically feasible, frequently show nonstoichiometric behaviour (variable composition) in their oxides, sulfides, and related binary compounds. For small deviations from stoichiometry a thermodynamic approach is instructive, but for larger deviations structural considerations supervene, and the possibility of thermodynamically unstable but kinetically isolable phases must be considered. These ideas will be expanded in the following paragraphs but more detailed treatment must be sought elsewhere.[103, 104, 104a, 104b]

Any crystal is contact with the vapour of one of its constituents is potentially a nonstoichiometric compound since, for true thermodynamic equilibrium, the composition of the solid phase must depend on the concentration (pressure) of this constituent in the vapour phase. If the solid and vapour are in equilibrium with each other ($\Delta G = 0$) at

[101] R. D. SHANNON, Revised effective ionic radii and systematic studies of interatomic distances in halides and chalcogenides, *Acta Cryst.* **A32**, 751–67 (1976).

[102] A. F. WELLS, *Structural Inorganic Chemistry*, 4th edn., Oxford University Press, Oxford, 1975; Chap. 12, Binary metal oxides, pp. 439–75; Chap. 13, Complex oxides, pp. 476–515.

[103] N. N. GREENWOOD, *Ionic Crystals, Lattice Defects, and Nonstoichiometry*, Chaps. 6 and 7, pp. 111–81, Butterworths, London, 1968.

[104] D. J. M. BEVAN, Nonstoichiometric compounds: an introductory essay, Chap. 49 in J. C. BAILAR, H. J. EMELÉUS, R. S. NYHOLM, and A. F. TROTMAN-DICKENSON (eds.), *Comprehensive Inorganic Chemistry*, Vol. 4, pp. 453–40, Pergamon Press, Oxford, 1973.

[104a] T. SØRENSEN, *Nonstoichiometric Oxides*, Academic Press, New York, 1981, 441 pp.

[104b] S. TRASATTI, *Electrodes of Conductive Metallic Oxides*, Elsevier, Amsterdam, Part A, 1980, 366 pp.; Part B, 1981, 336 pp.

a given temperature and pressure, then a change in this pressure will lead to a change (however minute) in the composition of the solid, provided that the activation energy for the reaction is not too high at the temperature being used. Such deviations from ideal stoichiometry imply a change in valency of at least some of the ions in the crystal and are readily detected for many oxides using a range of techniques such as pressure–composition isotherms, X-ray diffraction, neutron diffraction, electrical conductivity (semi-conductivity), visible and ultraviolet absorption spectroscopy (colour centres),[103] and Mössbauer (γ-ray resonance) spectroscopy.[105]

If the pressure of O_2 above a crystalline oxide is increased, the oxide-ion activity in the solid can be increased by placing the supernumerary O^{2-} ions in interstitial positions, e.g.:

$$UO_2 + \tfrac{x}{2}O_2 \xrightarrow{1150°C} UO_{2+x} \quad 0 < x < 0.25$$

The electrons required to reduce $\tfrac{1}{2}O_2$ to O^{2-} come from individual cations which are thereby oxidized to a higher oxidation state. Alternatively, if suitable interstitial sites are not available, the excess O^{2-} ions can build on to normal lattice sites thereby creating cation vacancies which diffuse into the crystal, e.g.:

$$(1 - \tfrac{x}{2})Cu_2O + \tfrac{x}{2}O_2 \longrightarrow Cu_{2-x}O$$

In this case the requisite electrons are provided by $2Cu^I$ becoming oxidized to $2Cu^{II}$.

Conversely, if the pressure of O_2 above a crystalline oxide is decreased below the equilibrium value appropriate for the stoichiometric composition, oxygen "boils out" of the lattice leaving supernumerary metal atoms or lower-valent ions in interstitial positions, e.g.:

$$(1 + x)ZnO \longrightarrow Zn_{1+x}O + \tfrac{x}{2}O_2$$

The absorption spectrum of this nonstoichiometric phase forms the basis for the formerly much-used qualitative test for zinc oxide: "yellow when hot, white when cold". Alternatively, anion sites can be left vacant, e.g.:

$$TiO \longrightarrow TiO_{1-x} + \tfrac{x}{2}O_2$$

In both cases the average oxidation state of the metal is reduced. It is important to appreciate that, in all such examples, the resulting nonstoichiometric compound is a homogeneous phase which is thermodynamically stable under the prevailing ambient conditions.

Sometimes the lattice defects form clusters amongst themselves rather than being randomly distributed throughout the lattice. A classic example is "ferrous oxide", which is unstable as FeO at room temperature but exists as $Fe_{1-x}O$ ($0.05 < x < 0.12$): the NaCl-type lattice has a substantial number of vacant Fe^{II} sites and these tend to cluster so that Fe^{III} can occupy tetrahedral sites within the lattice as shown schematically in Fig. 14.22. Such clustering can sometimes nucleate a new phase in which "vacant sites" are eliminated by being ordered in a new structure type. For example, PrO_{2-x} forms a disordered

---

[105] N. N. GREENWOOD and T. C. GIBB, *Mössbauer Spectroscopy*, Chapman & Hall, London, 1971, 659 pp.

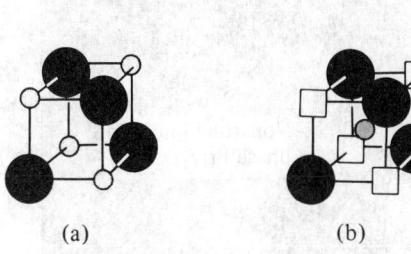
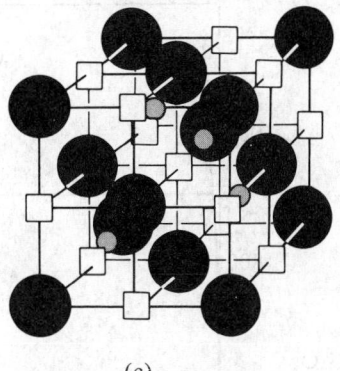

(a)　　　　　　　　　　(b)

(c)

FIG. 14.22　Schematic representation of defect clusters in $Fe_{1-x}O$. The normal NaCl-type structure (a) has Fe^{II} (small open circles) and O^{-II} (large dark circles) at alternate corners of a cube. In the 4:1 cluster (b), four octahedral Fe^{II} sites are left vacant and an Fe^{III} ion (grey) occupies the cube centre, thus being tetrahedrally coordinated by the $4O^{-II}$. In (c) a more extended 13:4 cluster is shown in which, again, all anion sites are occupied but the 13 octahedral Fe^{II} sites are vacant and four Fe^{III} occupy a tetrahedral array of cube centres.

nonstoichiometric phase $(0 < x < 0.25)$ at 1000°C but at lower temperatures (400–700°C) this is replaced by a succession of intermediate phases with only very narrow (and non-overlapping) composition ranges of general formula Pr_nO_{2n-2} with $n = 4, 7, 9, 10, 11, 12$ and ∞ as shown in Fig. 14.23 and Table 14.15.

TABLE 14.15　*Intermediate phases formed by ordering of defects in the praseodymium–oxygen system*

n	Formula Pr_nO_{2n-2}	y in PrO_y	Nonstoichiometric limits of x at $T°C$	$T°C$
4	Pr_2O_3	1.500	1.500–1.503	1000
7	Pr_7O_{12}	1.714	1.713–1.719	700
9	Pr_9O_{16}	1.778	1.776–1.778	500
10	Pr_5O_9	1.800	1.799 1.801	450
11	$Pr_{11}O_{20}$	1.818	1.817–1.820	430
12	Pr_6O_{11}	1.833	1.831–1.836	400
∞	PrO_2	2.000	1.999–2.000	400
			1.75 –2.00	1000

Oxygen (oxide ions) in crystal lattices can be progressively removed by systematically replacing corner-shared $\{MO_6\}$ octahedra with edge-shared octahedra. The geometrical principles involved in the conceptual generation of such successions of phases (chemical-shear structures) are now well understood, but many mechanistic details of their formation remain unresolved. Typical examples are the rutile series Ti_nO_{2n-1} ($n = 4, 5, 6, 7, 8, 9, 10, \infty$) between $TiO_{1.75}$ and TiO_2, and the ReO_3 series M_nO_{3n-1} which leads to a succession of 6 phases with $n = 8, 9, 10, 11, 12,$ and 14 in the narrow composition range $MO_{2.875}$ to $MO_{2.929}$ (M = Mo or W).

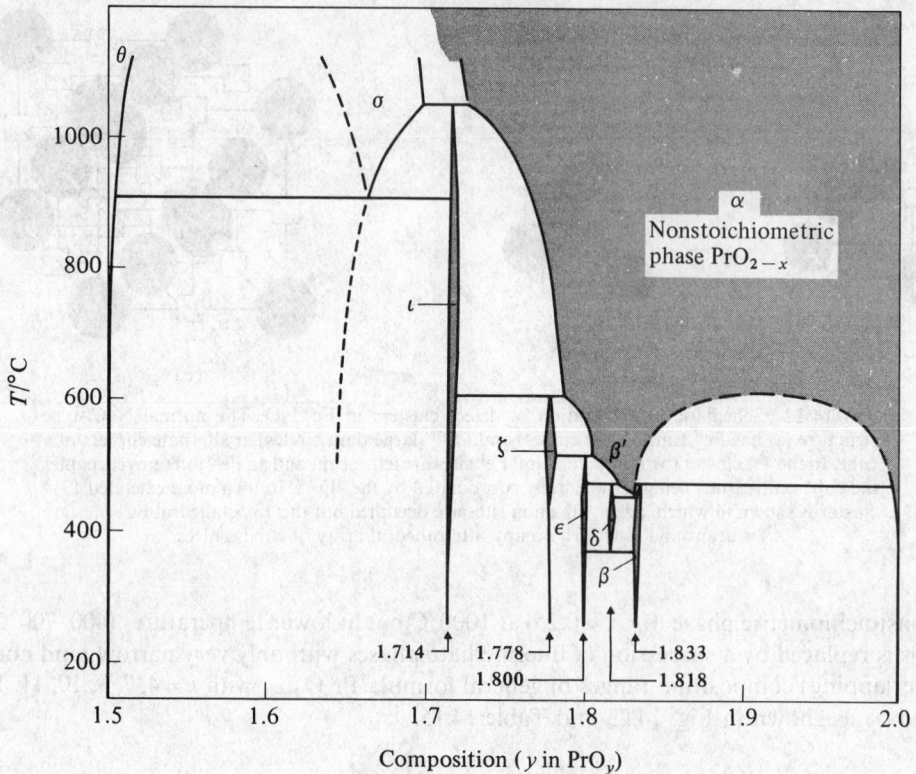

FIG. 14.23   Part of the Pr–O phase diagram showing the extended nonstoichiometric α phase PrO$_{2-x}$ at high temperatures (shaded) and the succession of phases Pr$_n$O$_{2n-2}$ at lower temperatures.

Nonstoichiometric oxide phases are of great importance in semiconductor devices, in heterogeneous catalysis, and in understanding photoelectric, thermoelectric, magnetic, and diffusional properties of solids. They have been used in thermistors, photoelectric cells, rectifiers, transistors, phosphors, luminescent materials, and computer components (ferrites, etc.). They are crucially implicated in reactions at electrode surfaces, the performance of batteries, the tarnishing and corrosion of metals, and many other reactions of significance in catalysis.[103, 104, 104a, 104b]

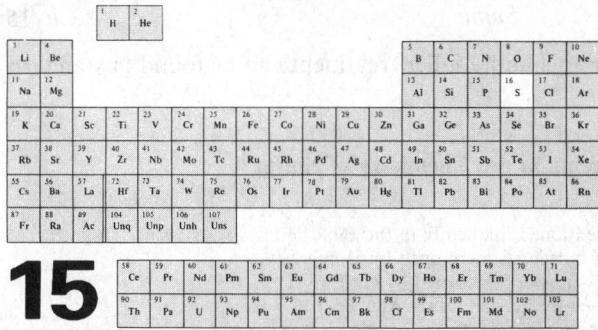

15

Sulfur

15.1 The Element

15.1.1 *Introduction*

Sulfur occurs uncombined in many parts of the world and has therefore been known since prehistoric times. Indeed, sulfur and carbon were the only two non-metallic elements known to the ancients. References to sulfur occur throughout recorded history from the legendary destruction of Sodom and Gomorrah by brimstone[1] to its recent discovery (together with H_2SO_4) as a major component in the atmosphere of the planet Venus. The element was certainly known to the Egyptians as far back as the sixteenth century BC and Homer refers to its use as a fumigant.[2] Pliny the Elder[3] mentioned the occurrence of sulfur in volcanic islands and other Mediterranean locations, spoke of its use in religious ceremonies and in the fumigation of houses, described its use by fullers, cotton-bleachers, and match-makers, and indicated fourteen supposed medicinal virtues of the element.

Gunpowder, which revolutionized military tactics in the thirteenth century, was the sole known propellant for ammunition until the mid-nineteenth century when smokeless powders based on guncotton (1846), nitroglycerine (1846), and cordite (1889) were discovered. Gunpowder, an intimate mixture of saltpeter ($NaNO_3$), powdered charcoal, and sulfur in the approximate ratios 75:15:10 by weight, was discovered by Friar Roger Bacon about 1245 though the Chinese and Arabs had produced incendiary mixtures somewhat earlier. It was first used in action at the Battle of Crécy (26 August 1346), but the guns lacked all power of manouevre and the devastating victory of Edward III was due chiefly to the long-bow men whom the French were also encountering for the first time. By 1415, however, gunpowder was decisive in Henry V's siege of Harfleur, and its increasing use in mobile field guns, naval artillery, and hand-held firearms was a dominant feature of world history for the next 500 y. Parallel with these activities, but largely independent of them, was the growing development of the alchemy and chemistry of sulfur and the growth of the emerging chemical industry based on sulfuric acid (p. 838). Some of the key points in

---

[1] Genesis *19*, 24: "Then the Lord rained upon Sodom and Gomorrah brimstone and fire from the Lord out of heaven." Other biblical references to brimstone are in Deuteronomy *29*, 23; Job *18*, 15; Psalm *11*, 6; Isaiah *30*, 33; Ezekiel *38*, 22; Revelation *19*, 20; etc.

[2] HOMER, *Odyssey*, Book 22, 481: "Bring me sulfur, old nurse, that cleanses all pollution and bring me fire, that I may purify the house with sulfur."

[3] G. PLINY (the Elder), AD 23–79, mentions sulfur in several of the many books of his posthumously published major work, *Naturalis Historia*.

this story are summarized in the Panel and a fuller treatment can be found in standard references.[4–7]

Developments in the Chemistry of Sulfur

Prehistory	Sulfur (brimstone) mentioned frequently in the Bible.[1]
∼800 BC	Fumigating power of burning S mentioned by Homer.[2]
∼AD 79	Occurrence and many uses of S recorded by G. Pliny.[3]
AD 940	Sulfuric acid mentioned by Persian writer Abu Bekr al Rases.
∼1245	Gunpowder discovered independently in Europe by Roger Bacon (England) and Berthold Swartz.
1661	Effects of SO_2 pollution in London dramatically described to Charles II by John Evelyn (p. 825).
1746	Lead chamber process for H_2SO_4 introduced by John Roebuck (Birmingham, UK); this immediately superseded the cumbersome small-scale glass bell-jar process (p. 838).
1777	Elemental character of S proposed by A.-L. Lavoisier though even in 1809 experiments (presumably on impure samples) led Humphry Davy to contend that oxygen and hydrogen were also essential constituents of S.
1781	Sulfur first detected in plants by N. Deyeux (roots of the dock, horse-radish, and cochlearia).
1809	Sulfur firmly established as an element by J. L. Gay Lussac and L. J. Thenard.
1813	Sulfur detected in the bile and blood of animals by H. A. Vogel.
1822	Xanthates (e.g. EtOCSSK) discovered by W. C. Zeise who also prepared the first mercaptan (EtSH) in 1834 (see also p. 357).
1831	Contact process for SO_3/H_2SO_4 patented by P. Philips of Bristol, UK (the original platinum catalyst was subsequently replaced by ones based on V_2O_5).
1835	S_4N_4 first made by M. Gregory ($S_2Cl_2 + NH_3$); X-ray structure by M. J. Bueger, 1936.
1839	Vulcanization of natural rubber latex by heating it with S discovered by Charles Goodyear (USA).
1865	Prospectors boring for petroleum in Louisiana discovered a great S deposit beneath a 150-m thick layer of quicksand.
1891–4	H. Frasch developed commercial recovery of S by superheated water process.
1912	E. Beckmann showed that rhombohedral sulfur was S_8 by cryoscopy in molten iodine.
1923	V. B. Goldschmid's geochemical classification includes "chalcophiles" (p. 760)
1926	Isotopes ^{33}S and ^{34}S discovered by F. W. Aston who previously (1920) had only detected ^{32}S in his mass spectrometer.
1935	Molecular structure of *cyclo*-S_8 established by X-ray methods (B. E. Warren and J. T. Burwell).
1944	Sulfur first produced from sour natural gas; by 1971 this source, together with crude oil, accounted for nearly one-third of world production.
1950	SF_4 first isolated by G. A. Silvery and G. H. Cady.
1951	Sulfur nmr signals (from ^{33}S) first detected by S. S. Dharmatti and H. E. Weaver.
1972	Sulfur and H_2SO_4 detected in the atmosphere of the planet Venus by USSR Venera 8 (subsequently confirmed in 1978 by US Venus Pioneer 2).
1973	S_{18} and S_{20} synthesized and characterized by M. Schmidt, A. Kutoglu, and their coworkers.
1975	The metallic and superconducting properties of polymeric $(SN)_x$ discovered independently by two groups in the USA (p. 860).

[4] J. W. MELLOR, *A Comprehensive Treatise on Inorganic and Theoretical Chemistry*, Vol. 10, Chap. 57, pp. 1–692, Longmans, Green, London, 1930.

[5] *Gmelins Handbuch der Anorganischen Chemie*, System Number 9A *Schwefel*, pp. 1–60, Verlag Chemie, Weinheim/Bergstrasse, 1953.

[6] M. E. WEEKS, *Discovery of the Elements*, Sulfur, pp. 52–73, Journal of Chemical Education, Easton, 1956.

[7] T. K. DERRY and T. I. WILLIAMS, *A Short History of Technology*, Oxford University Press, Oxford, 1960 (consult index).

15.1.2 *Abundance and distribution*

Sulfur occurs, mainly in combined form, to the extent of about 340 ppm in the crustal rocks of the earth. It is the sixteenth element in order or abundance, closely following barium (390 ppm) and strontium (384 ppm), and being about twice as abundant as the next element carbon (180 ppm). Earlier estimates placed its global abundance in the range 300–1000 ppm. Sulfur is widely distributed in nature but only rarely is it sufficiently concentrated to justify economic mining. Its ubiquity is probably related to its occurrence in nature in both inorganic and organic compounds, and to the fact that it can occur in at least five oxidation states: -2 (sulfides, H_2S, and organosulfur compounds), -1 (disulfides, S_2^{2-}), 0 (elemental S), $+4$ (SO_2), and $+6$ (sulfates). The three most important commercial sources are:

(1) elemental sulfur in the caprock salt domes in the USA and Mexico, and the sedimentary evaporite deposits in south-eastern Poland;
(2) H_2S in natural gas and crude oil, and organosulfur compounds in tar sands, oil shales, and coal (the latter two also contain pyrites inclusions);
(3) pyrites (FeS_2) and other metal-sulfide minerals.

Volcanic sources of the free element are also widespread; they have been of great economic importance until this century but are now little used. They occur throughout the mountain ranges bordering the Pacific, particularly in South and Central America, New Zealand, and the Philippines, Japan, Taiwan, the Kamchatka Peninsula, and the Kurile Islands. They also occur in Iceland and in the Mediterranean region, notably in Turkey, Italy, and formerly also in Sicily and Spain.

Elemental sulfur in the caprock of salt domes was almost certainly produced by the anaerobic bacterial reduction of sedimentary sulfate deposits (mainly anhydrite or gypsum, p. 761). The strata are also associated with hydrocarbons; these are consumed as a source of energy by the anaerobic bacteria, which use sulfur instead of O_2 as a hydrogen acceptor to produce $CaCO_3$, H_2O, and H_2S. The H_2S may then be oxidized to colloidal sulfur, or may form calcium hydrosulfide and polysulfide, which reacts with CO_2 generated by the bacteria to precipitate crystalline sulfur and secondary calcite. Alternatively, H_2S may escape from the system and the limestone caprock will then be free of sulfur. Indeed, of over 400 salt-dome structures known to exist in the coastal and off-shore area of the Gulf of Mexico, only about 12 contain commercial deposits of sulfur (5 in Louisiana, 5 in Texas, and 2 in Mexico). The mining operations are described in the Section 15.1.3.

The great evaporite basin deposits of elemental sulfur in Poland were discovered only in 1953 but have already had a dramatic impact on the economy of that country which is now third after the USA and Canada in the production of elemental sulfur. The deposits occur in the upper valley of the Vistula River near Tarnobrzeg in association with secondary limestone, gypsum, and anhydrite, and extend into the Ukraine where it is now also mined. The origin of the sulfur is believed to be derived from hydrocarbon reduction of sulfates assisted by bacterial action. The H_2S so formed is consumed by other bacteria to produce sulfur as waste—this accumulates in the bodies of the bacteria until death, when the sulfur remains.

The next great natural occurrence of sulfur is as H_2S in sour natural gas and as organosulfur compounds in crude oil. Again, distribution is widespread. Although commercial production of elemental sulfur from such sources was first effected only in 1944 (in the USA) it now represents a major source of the element in the USA, Canada, and France, and this growth has been one of the most significant trends in world sulfur production during the past two decades. Sulfur, of course, also occurs in many plant and animal proteins, and three of the principal amino-acid residues contain sulfur: cysteine, $HSCH_2CH(NH_2)CO_2H$; cystine $\{-SCH_2CH(NH_2)CO_2H\}_2$; and methionine, $MeSCH_2CH_2CH(NH_2)CO_2H$.

Oil shales represent a further source of sulfur though here (unlike the tar sands which yield crude oil and H_2S) the sulfur is predominantly in the form of pyrites. US oil shales contain about 0.7% S of which about 80% is pyritic; other major reserves are in Brazil, the USSR, China, and Africa, though these do not at present seem to be used as an industrial source of sulfur. Coal also contains about 1–2% S and is thus as huge a potential source of the element as it is an actual present source of air pollution (p. 825). From over 3×10^9 tonnes of coal mined annually, only some 500 000 tonnes of sulfur are recovered (as H_2SO_4) from a potential 50 million tonnes.

The third great source of sulfur and its compounds is from the mineral sulfides. V. M. Goldschmid's geochemical classification of the elements (1923), which has formed the basis of all subsequent developments in the field, proposed four main groups of elements: chalcophile, siderophile, lithophile, and atmophile.[8] Of these the chalcophiles (Greek χαλκος, *chalcos*, copper; φιλος, *philos*, loving) are associated with copper, specifically as sulfides. Elements which occur mainly as sulfide minerals are predominantly from Groups IB–VIB of the periodic table (together with iron, molybdenum, and to a lesser extent, some of the platinum metals as shown in Fig. 15.1. Some examples of the more important sulfide minerals are listed in Table 15.1, and a further discussion of the structural chemistry and reactivity of metal sulfides is on p. 798. Pyrites (fool's gold, FeS_2) is one of the most abundant of all sulfur minerals and is a major source of the element (see above). It often occurs in massive lenses but may also appear in veins or in disseminated zones. The

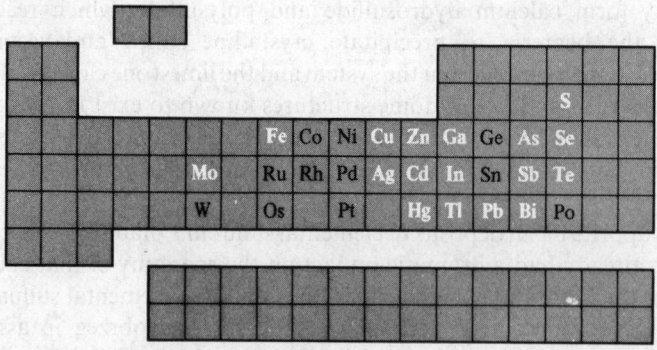

FIG. 15.1   Position of the chalcophilic elements in the periodic table: these elements (particularly those in white) tend to occur in nature as sulfide minerals; the tendency is much less pronounced for the elements in normal black type.

[8] R. W. FAIRBRIDGE, *Encyclopedia of Geochemistry and Environmental Sciences*, Van Nostrand, New York, 1972. See sections on Geochemical Classification of the Elements; Sulfates; Sulfate Reduction–Microbial; Sulfides; Sulfosalts; Sulfur; Sulfur Cycle; Sulfur Isotope Fractionation in Biological Processes, etc., pp. 1123–58.

largest commercial deposits extend from Seville (Spain) westward into Portugal, and, at the Rio Tinto mines in Huelva Province, one of the lenses is 1.5 km long and 240 m wide with a sulfur content of 48% (pure FeS_2 has 53.4% S). Other major deposits are in the USSR (the world's largest producer of pyrites), Japan, Italy, Cyprus, and Scandinavia. The most important non-ferrous metal sulfides are those of Cu, Ni, Zn, Pb, and As.

Finally, sulfur occurs in many localities as the sulfates of electropositive elements (see Chapters 4 and 5) and to a lesser extent as sulfates of Al, Fe, Cu, and Pb, etc. Gypsum ($CaSO_4 . 2H_2O$) and anhydrite ($CaSO_4$) are particularly notable but are little used as a source of sulfur because of high capital and operating costs. Similarly, by far the largest untapped source of sulfur is in the oceans as the dissolved sulfates of Mg, Ca, and K. It has been calculated that there are some 1.5×10^9 cubic km of water in the oceans of the world and that 1 cubic km of sea-water contains approximately 1 million tonnes of sulfur combined as sulfate.

TABLE 15.1   *Some sulfide minerals (those in bold are the more prevalent or important)*

Name	Idealized formula	Name	Idealized formula
Molybdenite	MoS_2	**Galena** (Pb glance)	**PbS**
Tungstenite	WS_2	**Realgar**	As_4S_4
Alabandite	MnS	**Orpiment**	As_2S_3
Pyrite (fool's gold)	FeS_2	Dimorphite	As_4S_3
Marcasite	FeS_2	**Stibnite**	Sb_2S_3
Pyrrhotite	$Fe_{1-x}S$	Bismuthinite	Bi_2S_3
Laurite	RuS_2	Pentlandite	$(Fe,Ni)_9S_8$
Linnaeite	Co_3S_4	**Chalcopyrite**	$CuFeS_2$
Millerite	NiS	**Bornite**	Cu_5FeS_4
Cooperite	PtS	**Arsenopyrite**	FeAsS
Chalcocite (Cu glance)	Cu_2S	Cobaltite	CoAsS
Argentite (Ag glance)	Ag_2S	Enargite	Cu_3AsS_4
Sphalerite (Zn blende)	**ZnS**	Bournoite	$CuPbSbS_3$
Wurtzite	ZnS	Proustite	Ag_3AsS_3
Greenockite	CdS	Pyrargyrite	Ag_3SbS_3
Cinnabar (vermillion)	**HgS**	**Tetrahedrite**[a]	$Cu_{12}As_4S_{13}$ [a]

[a] There is a second series in which As is replaced by Sb; in both series Cu is often substituted in part by Fe, Ag, Zn, Hg, or Pb.

15.1.3   *Production and uses of elemental sulfur*

Sulfur is produced commercially from one or more sources in over seventy countries of the world, and production of all forms in 1974 amounted to 51.67 million tonnes. The main producers are shown in Table 15.2. Until the beginning of this century, sulfur was obtained mainly by mining volcanic deposits of the element, but this only continues to be a major source in Japan, Turkey, Mexico, Italy, and the Andean countries of South America. During the first half of this century the prime method of production was the process developed by H. Frasch in 1891–4. This involves forcing superheated water into submerged sulfur-bearing strata and then forcing the molten element to the surface by compressed air (see Panel). This is the method used to obtain sulfur from the caprock of

TABLE 15.2   *Main producers of sulfur in 1974 (expressed in millions of tonnes of contained S)*

USA	USSR	Canada	Poland	Japan	Mexico	France	FRG	Spain	China
11.60	8.26	7.95	4.37	2.81	2.39	1.95	1.34	1.32	1.17

salt domes in the Gulf Coast region of the USA and Mexico, and from the evaporite basin deposits in west Texas, Poland, the USSR, and Iraq.

Recovery from sour natural gas and from crude oil was first developed in the USA in 1944, and by 1970 these sources exceeded the total volume of Frasch-mined sulfur for the first time. Canada (Alberta) and France are the principal producers from sour natural gas, which contains 15–20% H_2S. The USA and Japan are the largest producers from petroleum refineries. The phenomenal growth of these sources is clear from the following figures (in 10^6 tonnes): <0.5 (1950); 2.5 (1960); >15 (1972). Recovery from sour natural

The Frasch Process for Mining Elemental Sulfur[9, 10]

The ingeneous process of melting subterranean sulfur with superheated water and forcing it to the surface with compressed air was devised and perfected by Herman Frasch in the period 1891–4. Originally designed to overcome the problems of recovering sulfur from the caprock of salt domes far below the swamps and quicksands of Louisiana, the method is now also extensively used in Poland and elsewhere to extract native sulfur.

The caprock typically occurs some 150–750 m beneath the surface and the sulfur-bearing zone is typically about 30 m thick and contains 20–40% S. A diagrammatic representation of the strata is in Fig. 15.2. Using oil-well drilling techniques a cased 200 mm (8-inch) pipe is sunk through the caprock to the bottom of the S-bearing layer. Its lower end is perforated with small holes. Inside this pipe a 100 mm (4-inch) pipe is lowered to within a short distance of the bottom and, finally, a concentric 25-mm (1-inch) compressed-air pipe is lowered to a point rather more than half-way down to the bottom of the well as shown in Fig. 15.3a. Superheated water at 165°C is forced down the two outer pipes and melts the surrounding sulfur (mp 119°C). As liquid sulfur is about twice as dense as water under these conditions, it flows to the bottom of the well; the pumping of water down the 100-mm pipe is discontinued, but the static pressure of the hot water being pumped down the outer 200-mm pipe forces the liquid sulfur some 100 m up the 100-mm pipe as shown in Fig. 15.3b. Compressed air is then forced down the central 25-mm pipe to aerate the molten sulfur and carry it to the surface where it emerges from the 100-mm annulus. One well can extract sulfur (~35 000 tonnes) from an area of about 2000 m² (0.5 acre) and new wells must continually be sunk. Bleed-water wells must also be sunk to remove the excess of water pumped into the strata.

A Frasch mine can produce as much as 2.5 million tonnes of sulfur per annum. Such massive operations clearly require huge quantities of mining water (up to 5 million gallons daily) and abundant power supplies for the drilling, pumping, and superheating operations.

The sulfur can be piped long distances in liquid form or transported molten in ships, barges, or rail cars. Alternatively it can be prilled or handled as nuggets or chunks. The photos in Fig. 15.4 give some idea of these handling techniques.[10]

Despite the vast bulk of liquid sulfur mined by the Frasch process it is obtained in very pure form. There is virtually no selenium, tellurium, or arsenic impurity, and the product is usually 99.5–99.9% pure.

[9] W. HAYNES, *Brimstone: The Stone that Burns*, Van Nostrand, Princeton, 1959, 308 pp. (The story of the Frasch sulfur industry.)
[10] Freeport Sulfur Company, *Sulfur, Ally of Agriculture and Industry*, an informative 20-page colour brochure obtainable from the Company, PO Box 61520, New Orleans, Louisiana 70160, USA.

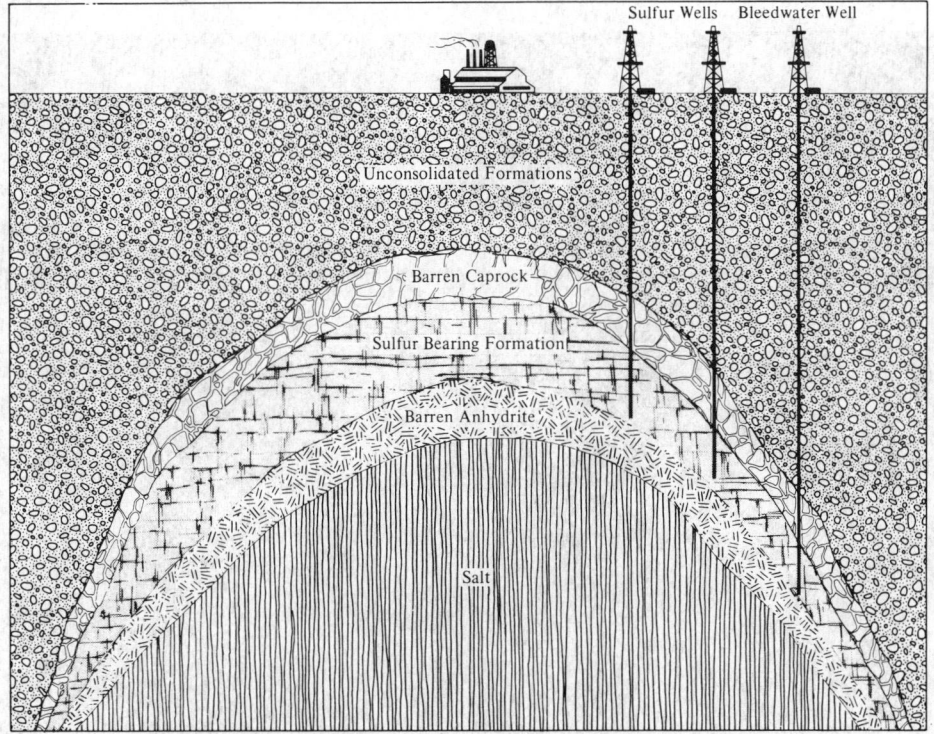

FIG. 15.2   Diagram of caprock strata. (From Freeport Sulfur Co. brochure.[10])

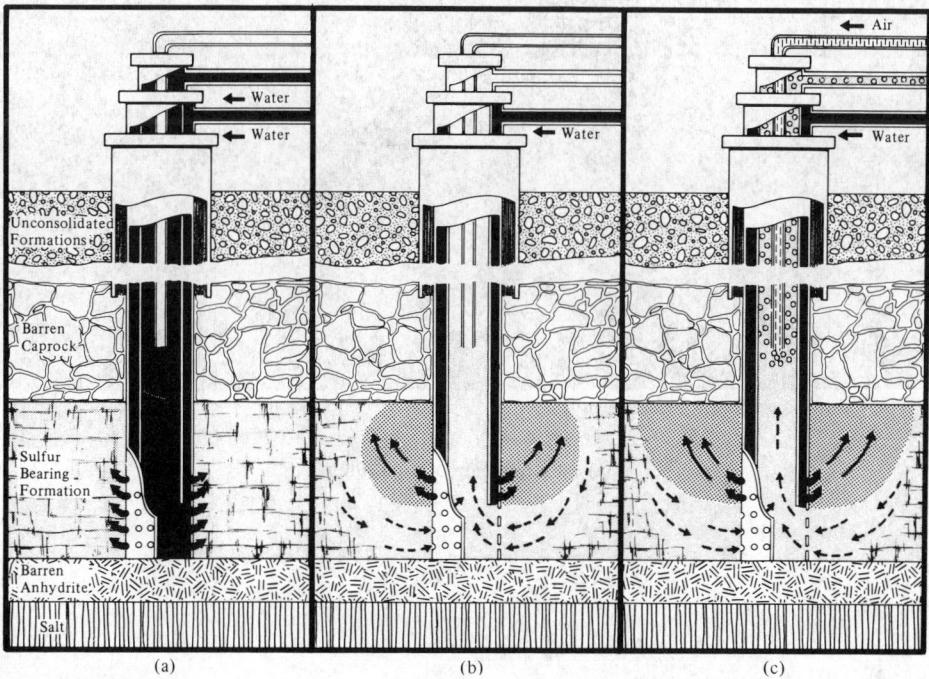

FIG. 15.3   (a)–(c) The Frasch Process.[10]

FIG. 15.4

(a) Bulk handling of solid sulfur.

(b) The world's first offshore sulfur mine (Frasch) at Grans Isle, Louisiana, USA.

(c) Ships, barges, and rail-cars used to transport sulfur.

(d) "Sulfur nuggets" (produced by spraying molten sulfur into a water bath) have low dusting properties and other advantages when compared with slates, lumps, or prills.

gas involves first separating out the H_2S by absorption in mono-ethanolamine and then converting it to sulfur by a process first developed by C. F. Claus in Germany about 1880. In this process one-third of the H_2S is burned to produce SO_2, water vapour, and sulfur vapour; the SO_2 then reacts with the remaining H_2S in the presence of oxide catalysts such as Fe_2O_3 or Al_2O_3 to produce more H_2O and S vapour:

$$H_2S + \tfrac{1}{2}O_2 \xrightarrow[\text{combustion}]{\text{low-temp}} \tfrac{1}{8}S_8 + H_2O$$

$$H_2S + 1\tfrac{1}{2}O_2 \longrightarrow SO_2 + H_2O$$

$$2H_2S + SO_2 \xrightarrow[\text{catalyst}/300°]{\text{oxide}} \tfrac{3}{8}S_8 + 2H_2O$$

Multiple reactors achieve 95–96% conversion and recovery, and stringent air pollution legislation has now pushed this to 99%. A similar sequence of reactions is used for sulfur production from crude oil except that the organosulfur compounds must first be removed from the refinery feed and converted to H_2S by a hydrogenation process before the sulfur can be recovered.

The third major source of sulfur is pyrite and related sulfide minerals. The ore is roasted to secure SO_2 gas which is then usually used directly for the manufacture of H_2SO_4 (p. 838). Again air pollution by SO_2 gas emissions has been the subject of increasing legislation and control during the past two decades (p. 825).

The proportion of sulfur and S-containing compounds recovered by these various methods has been changing rapidly and frequently depends on the nature of local sources available. The comparative figures for 1971 in Table 15.3 give some indication of the global scene at that time. Estimated reserves on the basis of present technology and prices are summarized in Table 15.4; these can increase more than tenfold if coal, gypsum,

TABLE 15.3   *World production of sulfur in 1971 by source (expressed in million tonnes of contained S)*

Source		10^6 tonnes	Percentage
Native deposits		13.2	31
Salt dome (Frasch)	7.1		
Evaporite (Frasch)	3.5		
Evaporite (mined)	2.3		
Volcanic	0.3		
Natural gas and crude oil		10.7	25
Sour natural gas	7.8		
Crude oil and tar sands	2.9		
Metal sulfides		17.2	41
Ferrous	12.3		
Nonferrous	4.9		
Miscellaneous		1.2	3
Sulfates	0.7		
Other	0.5		
Total, all sources		42.3	100

TABLE 15.4   *Estimated world reserves of sulfur*

Source	Natural gas	Petroleum	Native ore	Pyrite	Sulfide ore	Dome	**Total**
S/10^6 tonne	690	450	560	380	270	150	**2500**

anhydrite, and sea-water are included. At present these latter sources are economic only under special conditions though, as we have already seen (p. 760), vast quantities of SO_2 are lost from industrial coal each year. Recovery of useful sulfur compounds from anhydrite (and gypsum) can be achieved by two main routes. The Müller–Kühne process used in the UK and Austria involves the roasting of anhydrite with clay, sand, and coke in a rotary kiln at 1200–1400°:

$$2CaSO_4 + C \longrightarrow 2CaO + 2SO_2 + CO_2$$

The emergent SO_2 is then fed into a contact process for H_2SO_4 (p. 838). Alternatively, ammonia and CO_2 can be passed into a gypsum slurry to give ammonium sulfate for use in fertilizers:

$$CaSO_4 + (NH_4)_2CO_3 \longrightarrow CaCO_3 + (NH_4)_2SO_4$$

This double decomposition route was developed in Germany and has been used in the UK since 1971.

The pattern of uses of sulfur and its compounds in the chemical industry is illustrated in the flow chart on p. 771. Most sulfur is converted via SO_2/SO_3 into sulfuric acid which accounts, for example, for some 88% of the contained sulfur used in the USA. The proportion of sulfur used in making the extensive number of end products is shown in Fig. 15.5 Indeed, the uses of sulfur and its principal compounds are so widely spread throughout industry that a nation's consumption of sulfur is often used as a reliable measure of its economic development. Thus, the USA, the USSR, Japan, and Germany lead the world in industrial production and rank similarly in the consumption of sulfur. Further details of industrial uses will be found in subsequent sections dealing with specific compounds of sulfur, and two recent review books have appeared on the subject.[11, 12]

15.1.4 *Allotropes of sulfur*[13–15]

The allotropy of sulfur is far more extensive and complex than for any other element. This arises partly because of the great variety of molecular forms that can be achieved by –S–S– catenation and partly because of the numerous ways in which the molecules so formed can be arranged within the crystal. In fact, S–S bonds are very variable and

[11] J. R. WEST (ed.), *New Uses of Sulfur*, Advances in Chemistry Series No. 140, Am. Chem. Soc., Washington, DC, 1975, 230 pp.
[12] D. J. BOURNE (ed.), *New Uses of Sulfur—II*, Advances in Chemistry Series No. 165, Am. Chem. Soc., Washington, DC 1978, 282 pp.
[13] J. DONOHUE, *The Structures of the Elements*, Sulfur, pp. 324–69, Wiley, New York, 1974.
[14] B. MEYER, Elemental sulfur, *Chem. Revs.* **76**, 367–88, (1976).
[15] M. SCHMIDT, The scientific basis for the practical applications of elemental sulfur, Chap. 1, pp. 1–12, in ref. 12.

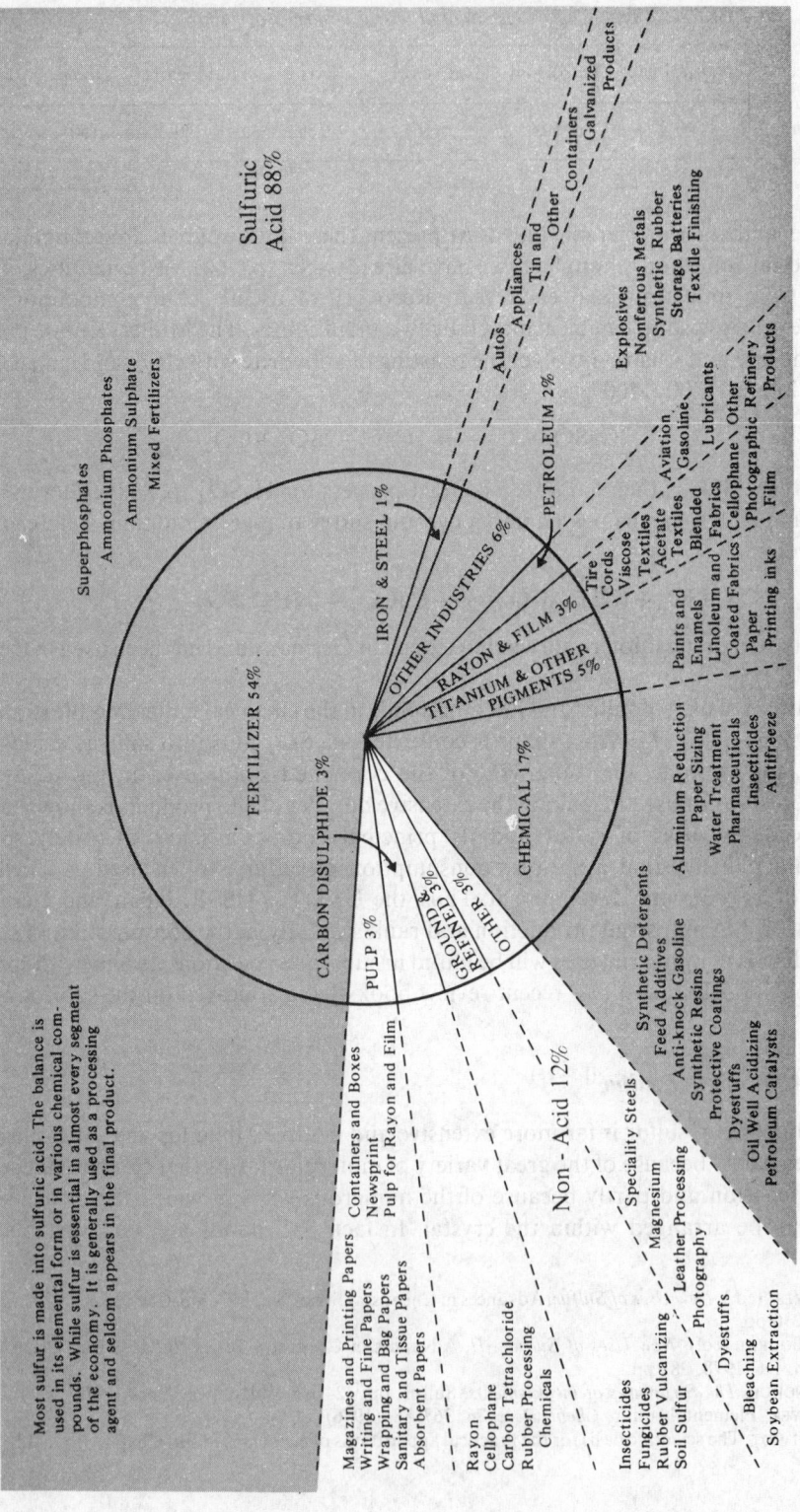

Most sulfur is made into sulfuric acid. The balance is used in its elemental form or in various chemical compounds. While sulfur is essential in almost every segment of the economy, it is generally used as a processing agent and seldom appears in the final product.

FIG. 15.5 Sulfur's uses as acid and as non-acid.

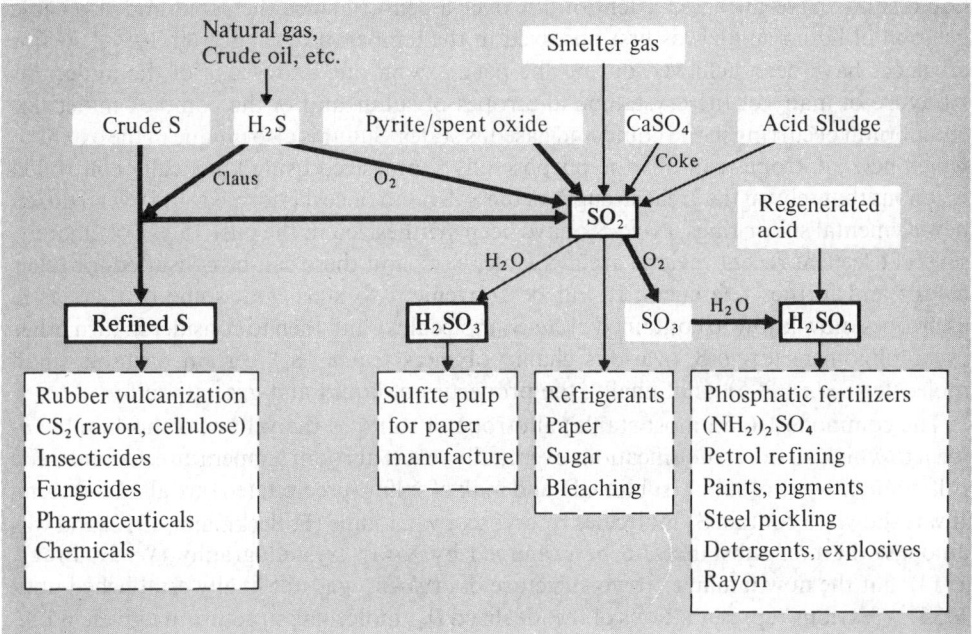

flexible: interatomic distances cover the enormous range 180–260 pm (depending to some extent on the amount of multiple bonding), whilst bond angles S–S–S vary from 90° to 180° and dihedral angles S–S–S–S from 0° to 180° (Fig. 15.6). Estimated S–S bond energies may be as high as 430 kJ mol^{-1} and the unrestrained –S–S– *single-bond* energy of 265 kJ mol^{-1} is exceeded amongst homonuclear single bonds only by those of H_2 (435 kJ mol^{-1}) and C–C (330 kJ mol^{-1}). Again, the amazing temperature dependence of the properties of

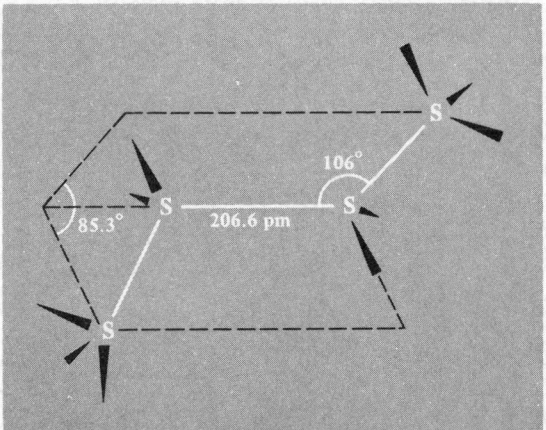

FIG. 15.6   Portion of an unrestrained –S$_n$– chain showing typical values for the S–S–S bond angle (106°) and S–S–S–S dihedral angle (85.3°). Possible alternative orientations of the bonds from the 2 inner S atoms, and possible directions for extensions of the chain from the 2 outer S atoms are indicated by the black lines. (See also p. 774.)

liquid sulfur have attracted attention for over a century since the rapid and reversible gelation of liquid sulfur was first observed in the temperature range 160–195°C. Major advances have been achieved during the past 15 y in our knowledge of the molecular structure of many of the crystalline allotropes of sulfur and of the complex molecular equilibrium occurring in the liquid and gaseous states. Sulfur is also unique in the extent to which new allotropes can now be purposefully synthesized using kinetically controlled reactions that rely on the great strength of the S–S bond once it is formed, and over a dozen new elemental sulfur rings, *cyclo*-S_n, have been synthesized in the past 15 y. Fortunately, several excellent recent reviews are available,[13–15] and these can be consulted for fuller details and further references. It will be convenient to start with some of the classic allotropes (now know to contained *cyclo*-S_8 molecules), and then to consider in turn other cyclic oligomers (*cyclo*-S_n) various chain polymers (*catena*-S_n), certain unstable small molecules S_n ($n = 2$–5), and, finally, the properties of liquid and gaseous sulfur.

The commonest (and most stable) allotrope of sulfur is the yellow, orthorhombic α-form to which all other modifications eventually revert at room temperature. Commercial roll sulfur, flowers of sulfur (sublimed), and milk of sulfur (precipitated) are all of this form. It was shown to contain S_8 molecules by cryoscopy in iodine (E. Beckmann, 1912) and was amongst the first substances to be examined by X-ray crystallography (W. H. Bragg, 1914), but the now familiar crown structure of *cyclo*-S_8 was not finally established until 1935.[13] Various representations of the idealized D_{4d} molecular structure are given in Fig. 15.7. The packing of the molecules within the crystal has been likened to a crankshaft arrangement extending in two different directions and leads to a structure which is very complex.[13] Orthorhombic α-S_8 has a density of 2.069 g cm^{-3}, is a good electrical insulator when pure, and is an excellent thermal insulator, the extremely low thermal

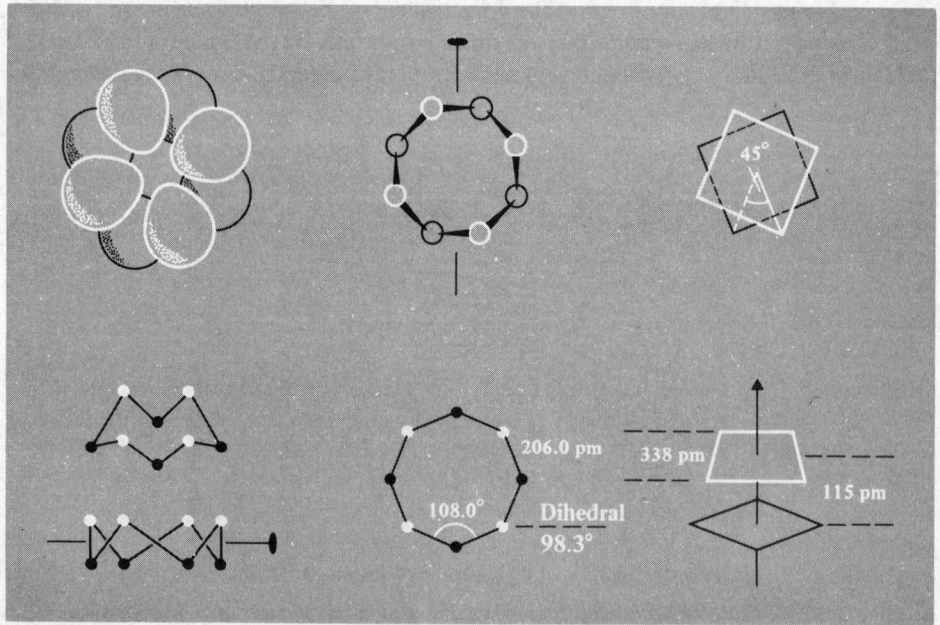

FIG. 15.7 Various representations of the molecule *cyclo*-S_8 found in α-rhombohedral, β-monoclinic, and γ-monoclinic sulfur.

conductivity being similar to those of the very best insulators such as mica (p. 411) and wood. Some solubilities in common solvents are in Table 15.5.

At about 95.3° α-S_8 becomes unstable with respect to β-monoclinic sulfur in which the packing of the S_8 molecules is altered and their orientation becomes partly disordered.[16] This results in a lower density (1.94–2.01 g cm^{-3}), but the dimensions of the S_8 rings in the two allotropes are very similar. The transition is somewhat sluggish even above 100°, and this enables a mp of metastable single crystals of α-S_8 to be obtained: a value of 112.8° is often quoted but microcrystals may melt as high as 115.1°. Monoclinic β-S_8 has a "mp" which is usually quoted as 119.6° but this can rise to 120.4° in microcrystals or may be as low as 114.6°. The uncertainty arises because the S_8 ring is unstable above ~119° and begins to form other species which progressively depress the mp. The situation is reminiscent of the equilibria accompanying the melting of anhydrous phosphoric acid (p. 595). Monoclinic β-S_8 is best prepared by crystallizing liquid sulfur at about 100° and then cooling it rapidly to room temperature to retard the formation of orthorhombic α-S_8; under these conditions β-S_8 can be kept for several weeks at room temperature before reverting to the more stable α-form.

TABLE 15.5   *Solubilities of α-orthorhombic sulfur (at 25°C unless otherwise shown)*

Solvent	CS_2	S_2Cl_2	Me_2CO	C_6H_6	CCl_4	Et_2O	C_6H_{14}	EtOH
g S per 100 g solvent (T°C)	35.5[a]	17[b] (21°)	2.5	2.1	0.86[c]	0.283 (23°)	0.25 (20°)	0.065

[a] 55.6 at 60°.   [b] 97 at 110°.   [c] 1.94 at 60°.

A third crystalline modification, γ-monoclinic sulfur, was first obtained by W. Muthmann in 1890. It is also called nacreous or mother-of-pearl sulfur and can be made by slowly cooling a sulfur melt that has been heated above 150°, or by chilling hot concentrated solutions of sulfur in EtOH, CS_2, or hydrocarbons. However, it is best prepared as pale-yellow needles by the mechanistically obscure reaction of pyridine with copper(I) ethyl xanthate, CuSSCOEt. Like α- and β-sulfur, γ-monoclinic sulfur comprises *cyclo*-S_8 molecules but the packing is more efficient and leads to a higher density (2.19 g cm^{-3}). It reverts slowly to α-S_8 at room temperature but rapid heating leads to a mp of 106.8°.

We now consider other homocyclic polymorphs of sulfur containing 6–20 S atoms per ring. A rhombohedral form, ε-sulfur, was first prepared by M. R. Engel in 1891 by the reaction of concentrated HCl on a saturated solution of thiosulfate $HS_2O_3^-$ at 0°. It was shown to be hexameric in 1914 but its structure as *cyclo*-S_6 was not established until 1958–61.[13] The allotrope is best prepared (1966) by the reaction

$$H_2S_4 + S_2Cl_2 \xrightarrow[\text{87\% yield}]{\text{dil solns in Et}_2\text{O}} cyclo\text{-}S_6 + 2HCl$$

The ring adopts the chair form shown in Fig. 15.8 and its dimensions are compared with those of other polymorphs in Table 15.6. Note that *cyclo*-S_6 has the smallest bond angles

[16] L. K. TEMPLETON, D. H. TEMPLETON, and A. ZALKIN, Crystal structure of monoclinic sulfur, *Inorg. Chem.* **15**, 1999–2001 (1976).

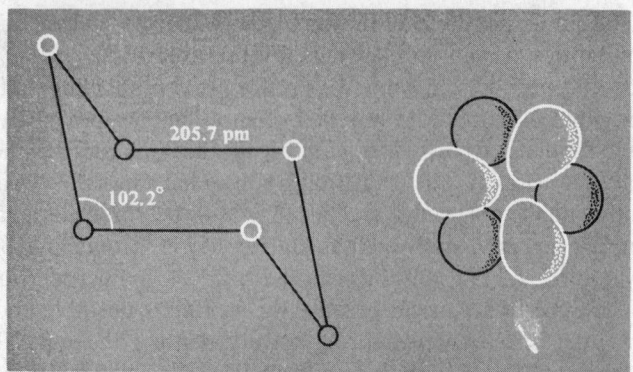

FIG. 15.8   The chair configuration of *cyclo*-S$_6$ found in rhombohedral ε-sulfur, and a "space-filling" representation of the molecule viewed down the *c*-axis of the crystal.

and dihedral angles of all poly-sulfur species for which data are available and this, together with the small "hole" at the centre of the molecule and the efficient packing within the crystal, lead to the highest density of any known polymorph of sulfur (Table 15.7).

In *cyclo*-S$_6$ and *cyclo*-S$_8$ all the S atoms are equivalent with essentially equal interatomic distances, angles, and conformations. This is not necessarily so for all homocyclic molecules. Thus, in building up cumulated –S$_n$– bonds, addition of S atoms to an S$_3$ unit can occur in three ways: *cis* (c), d-*trans* (dt), and l-*trans* (lt):

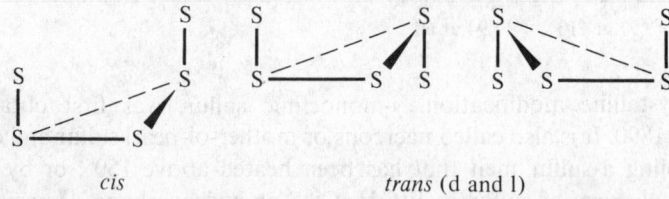

cis   *trans* (d and l)

Both S$_6$ (chair) and S$_8$ (crown) are all -*cis* conformations, but larger rings have more complex motifs.

TABLE 15.6   *Dimensions of some sulfur molecules. Average values are given except for S$_7$ where deviations from the mean are more substantial (see text)*

Molecule	Interatomic distance/pm	Bond angle	Dihedral angle
S$_2$ (matrix at 20 K)	188.9	—	
cyclo-S$_6$	205.7	102.2°	74.5°
cyclo-S$_7$	199.3–218.1	101.5°–107.5°	0.3°–107.6°
cyclo-S$_8$(α)	203.7	107.8°	98.3°
cyclo-S$_8$(β)	204.5	107.9°	—
cyclo-S$_{10}$	205.6	106.2°	−77° and +123°
cyclo-S$_{12}$	205.3	106.5°	86.1°
cyclo-S$_{18}$	205.9	106.3°	84.4°
cyclo-S$_{20}$	204.7	106.5°	83.0°
catena-S$_γ$	206.6	106.0°	85.3°

TABLE 15.7 *Some properties of sulfur allotropes*

Allotrope	Colour	Density/g cm^{-3}	mp or decomposition point/°C
S_2(g) or matrix at 20 K	Blue-violet	—	Very stable at high temp
S_3(g)	Cherry red	—	Stable at high temp
S_6	Orange red	2.209	d > 50
S_7	Yellow	2.182 (−110°)	d 39
α-S_8	Yellow	2.069	112.8 (see text)
β-S_8	Yellow	1.94–2.01	119.6 (see text)
γ-S_8	Light yellow	2.19	106.8 (see text)
S_9	Intense yellow	—	Stable below rt
S_{10}	Pale yellow green	2.103 (−110°C)	d > 0
S_{11}	—	—	—
S_{12}	Pale yellow	2.036	148
S_{18}	Lemon yellow	2.090	m 128(d)
S_{20}	Pale yellow	2.016	m 124(d)
S_x	Yellow	2.01	104(d)

At least eight further cyclic modifications of sulfur have been synthesized during the past 15 y by the elegant work of M. Schmidt and his group. The method is to couple two compounds which have the desired combined number of S atoms and appropriate terminal groups, e.g.:

$$S_{12-n}Cl_2 + H_2S_n \longrightarrow cyclo\text{-}S_{12} + 2HCl \quad [\text{also for } S_6, S_{10}, S_{18}, S_{20}]$$

$$S_xCl_2 + [Ti(\eta^5\text{-}C_5H_5)_2(S_5)] \longrightarrow cyclo\text{-}S_{x+5} + [Ti(\eta^5\text{-}C_5H_5)_2Cl_2]$$
$$[\text{for } S_7, S_9, S_{10}, S_{11}]$$

A variant is the ligand displacement and coupling reaction:

$$2[Ti(\eta^5\text{-}C_5H_5)_2(S_5)] + 2SO_2Cl_2 \xrightarrow[35\% \text{ yield}]{-78°} cyclo\text{-}S_{10} + 2[Ti(\eta^5\text{-}C_5H_5)_2Cl_2] + 2SO_2$$

The preparation and structures of the reactants are on p. 807 (H_2S_n), p. 815 (S_xCl_2), and p. 791 [$Ti(\eta^5\text{-}C_5H_5)_2(S_5)$].

S_7 is known in four crystalline modifications; one of these, obtained by crystallization from CS_2 at −78°, rapidly disintegrates to a powder at room temperature, but an X-ray study at −110° showed it to consist of *cyclo*-S_7 molecules with the dimensions shown in Fig. 15.9.[17] Notable features are the very large interatomic distance S(6)–S(7) (218.1 pm) which probably arises from the almost zero dihedral angle between the virtually coplanar atoms S(4) S(6) S(7) S(5), thus leading to maximum repulsion between nonbonding lone-pairs of electrons on adjacent S atoms. As a result of this weakening of S(6)–S(7), the adjacent bonds are strengthened (199.5 pm) and there are further alternations of bond lengths (210.2 and 205.2 pm) throughout the molecule.

The structures of S_9 and S_{11} have not yet been determined but the spectroscopic and other properties of these allotropes indicate that they almost certainly also contain cyclic molecules. *Cyclo*-decasulfur has been prepared as thermally unstable, light-sensitive yellow crystals and an X-ray study at −110°C leads to the molecular structure shown in

[17] R. Steudel, R. Reinhardt, and F. Schuster, Crystal and molecular structure of *cyclo*-heptasulfur (δ-S_7), *Angew. Chem.*, Int. Edn. (Engl.) **16**, 715 (1977).

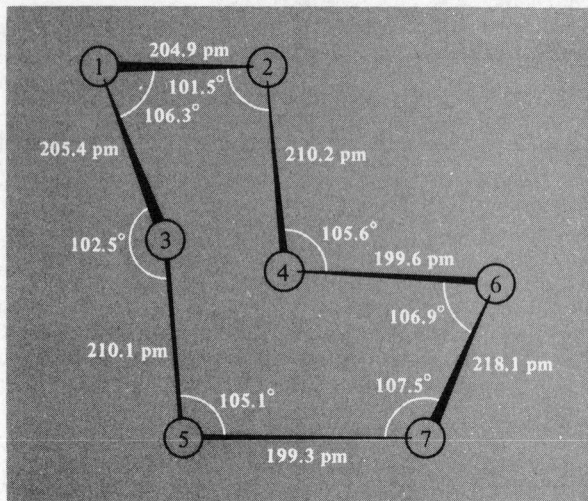

FIG. 15.9 Molecular structure of *cyclo*-S$_7$ showing the large distance S(6)–(7) and alternating interatomic distances away from this bond. The point group symmetry is approximately C$_s$.

Fig. 15.10.[18] The molecule belongs to the very rare point group symmetry D_2 (three orthogonal twofold axes of rotation as the only symmetry elements). The mean interatomic distance and bond angle are close to those in *cyclo*-S$_{12}$ (Table 15.6) and the molecule can be regarded as composed of two identical S$_5$ units obtained from the S$_{12}$ molecule (Fig. 15.11).

Cyclo-S$_{12}$ occupies an important place amongst the cyclic oligomers of sulfur. In a classic paper by L. Pauling[19] the molecule had been predicted to be unstable, though

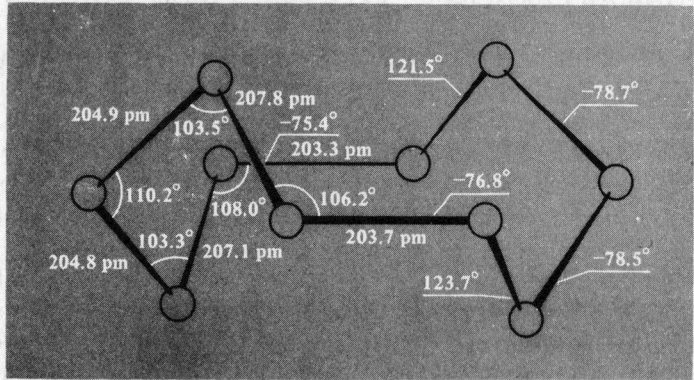

FIG. 15.10 Molecular structure of *cyclo*-S$_{10}$ showing interatomic distances (in pm), bond angles, and dihedral angles. The relationship between this structure and that of *cyclo*-S$_{12}$ (Fig. 15.11) can be seen by omitting the 2 "central" S atoms in S$_{12}$ and joining the 2 fragments so formed by the "horizontal" bonds in S$_{10}$ (203.5 pm). The distance between these 2 "horizontal" bonds is 541 pm.

[18] R. REINHARDT, R. STEUDEL, and F. SCHUSTER, X-ray structure analysis of *cyclo*-decasulfur, S$_{10}$, *Angew. Chem.*, Int. Edn. (Engl.) **17**, 57–58 (1978).

[19] L. PAULING, On the stability of the S$_8$ molecule and the structure of fibrous sulfur, *Proc. Natl. Acad. Sci. USA* **35**, 495–9 (1949).

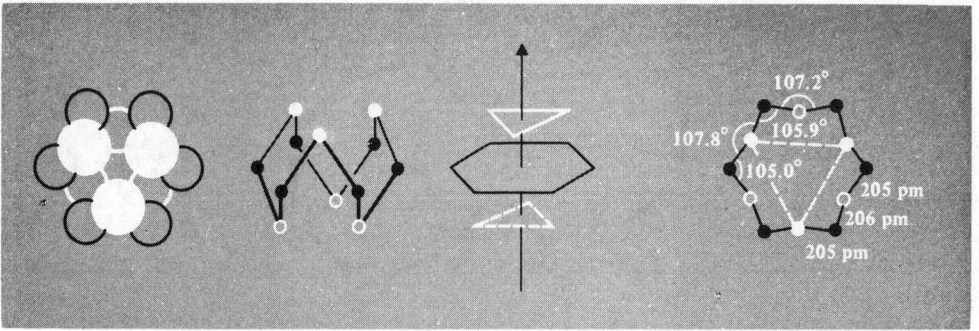

FIG. 15.11   Various representations of the molecular structure of *cyclo*-S_{12} showing S atoms in three parallel planes. The idealized point group symmetry is D_{3d} and the mean dihedral angle is $86.1 \pm 5.5°$. In the crystal the symmetry is slightly distorted to C_{2h} and the central group of 6 S atoms deviate from coplanarity by ± 14 pm.

subsequent synthesis showed it to be second only to *cyclo*-S_8 in stability. In fact, the basic principles underlying Pauling's prediction remain valid but he erroneously applied them to two sets of S atoms in two parallel planes whereas the configuration adopted has S atoms in three parallel planes. Several representations of the structure are in Fig. 15.11. Using the nomenclature of p. 774 it can be seen that, unlike S_6 and S_8, the conformation of all S atoms is not *cis*: the S atoms in upper and lower planes do indeed have this conformation but the 6 atoms in the central plane are alternately d-*trans* and l-*trans* leading to the sequence:

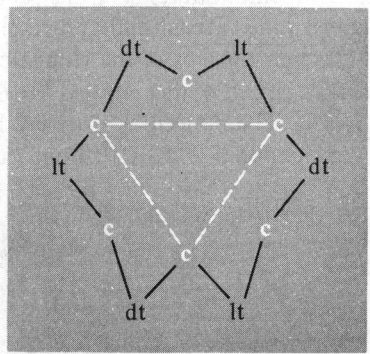

Cyclo-S_{12} was first prepared in 1966 in 3% yield by reacting H_2S_4 with S_2Cl_2 but a better route is the reaction between dilute solutions of H_2S_8 and S_4Cl_2 in Et_2O (18% yield). It can also be extracted from liquid sulfur. The stability of the allotrope can be gauged from its mp (148°), which is higher than that of any other allotrope and nearly 30° above the temperature at which the S_8 ring begins to decompose.

Two allotropes of *cyclo*-S_{18} are known. The structure of the first is shown in Fig. 15.12: if we take 3 successive atoms out of the S_{12} ring then the 9-atom fragment combines with a second one to generate the structure. Alternatively, the structure can be viewed as two parallel 9-atom helices (see below), one right-handed and one left-handed, mutually joined at each end by the *cis* atoms S(5) and S(14). Interatomic distances vary between

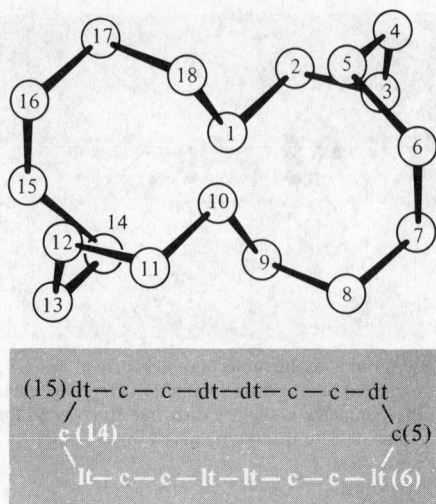

FIG. 15.12   The molecular structure of one form of *cyclo*-S_{18}, together with the conformational sequence of the two helical subunits[20]

204–211 pm (mean 206 pm), bond angles between 103.7–108.3° (mean 106.3°), and dihedral angles between 79.1–90.0° (mean 84.5°).

Cyclo-S_{18} is formed by the reaction between H_2S_8 and $S_{10}Cl_2$ and forms lemon-coloured crystals, mp 128°, which can be stored in the dark for several days without apparent change. A second form of *cyclo*-S_{18} has the molecular structure shown in Fig. 15.13: this has twice the 8-atom repeat motif *cis-cis-trans-cis-trans-cis-trans-cis* (one with d-*trans* and the other with l-*trans*) joined at each end by further bridging single *trans*-sulfur atoms which constitute the 2 extreme atoms of the elongated ring.

Pale yellow crystals of *cyclo*-S_{20}, mp 124° (decomp), d 2.016 g cm^{-3}, have been made by the reaction of H_2S_{10} and $S_{10}Cl_2$. The molecular structure is shown in Fig. 15.14. The interatomic S–S distances vary between 202.3–210.4 pm (mean 204.2 pm), the angles

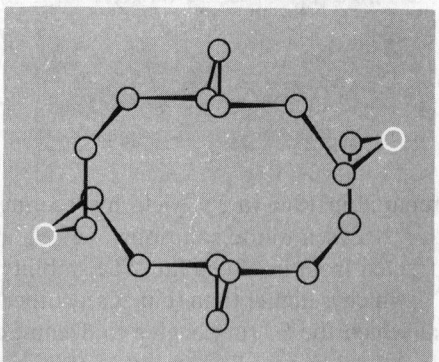

FIG. 15.13   The molecular structure of the second form of *cyclo*-S_{18} showing the *trans*-configuration of the S atoms at the extreme ends of the elongated rings.[15]

[20] T. DEBAERDEMAEKER and A. KUTOGLU, The crystal and molecular structure of a new elemental sulfur, S_{18}, *Naturwissenschaften* **60**, 49 (1973).

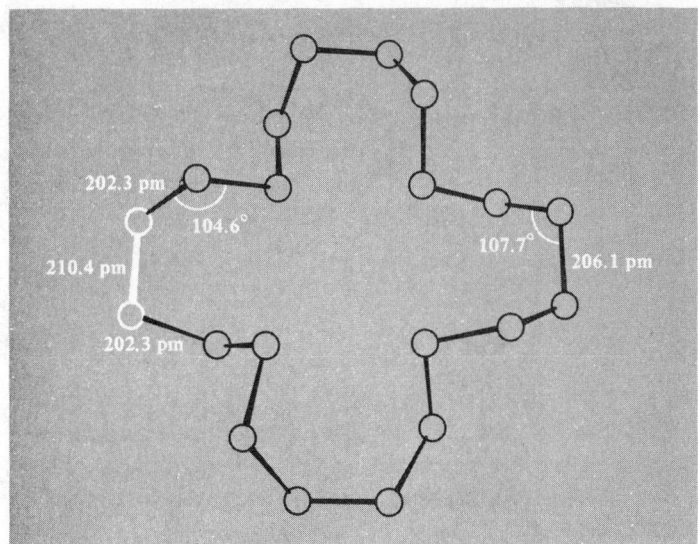

Fig. 15.14   Molecular structure of *cyclo*-S_{20} viewed along the [001] direction.[21] The 2 adjacent
S atoms with the longest interatomic distance are shown in white.

S–S–S between 104.6–107.7° (mean 106.4°), and the dihedral angles between 66.3–89.9°
(mean 84.7°). In this case the conformation motif is -c-lt-lt-lt-c- repeated 4 times and the
abnormally long bond required to achieve ring closure is notable; it is also this section of
the molecule which has the smallest dihedral angles thereby incurring increased repulsion
between adjacent nonbonding lone-pairs of electrons. Consistent with this the adjacent
bonds are the shortest in the molecule.

Solid poly*catena*sulfur comes in many forms: it is present in rubbery S, plastic (χ)S,
lamina S, fibrous (ψ, φ), polymeric (μ), and insoluble (ω)S, supersublimation S, white S,
and the commercial product Crystex. All these are metastable mixtures of allotropes
containing more or less defined concentrations of helices (S_x), *cyclo*-S_8, and other
molecular forms. The various modifications are prepared by precipitating S from solution
or by quenching *hot* liquid S (say from 400°C). The best-defined forms are fibrous (*d*
~2.01 g cm^{-3}) in which the helices are mainly parallel, and lamella S in which they are
partly criss-crossed. When carefully prepared by drawing filaments from hot liquid sulfur,
fibrous rubbery or plastic S can be repeatedly stretched to as much as 15 times its normal
length without substantially impairing its elasticity. All these forms revert to *cyclo*-S_8(α) at
room temperature and this has caused considerable difficulty in obtaining their X-ray
structures.[13] However, it is now established that fibrous S consists of infinite chains of S
atoms arranged in parallel helices whose axes are arranged on a close-packed (hexagonal)
net 463 pm apart. The structure contains both left-handed and right-handed helices of
radius 95 pm and features a repeat distance of 1380 pm comprising 10 S atoms in three
turns as shown in Fig. 15.15. Within each helix the interatomic distance S–S is 206.6 pm,
the bond angle S–S–S is 106.0°, and the dihedral angle S–S–S–S is 85.3°.

^{21} T. DEBAERDEMAEKER, E. HELLNER, A. KUTOGLU, M. SCHMIDT, and E. WILHELM, Synthesis, crystal- and
molecular-structure of *cyclo*-icosasulfur, S_{20}, *Naturwissenschaften* **60**, 300 (1973).

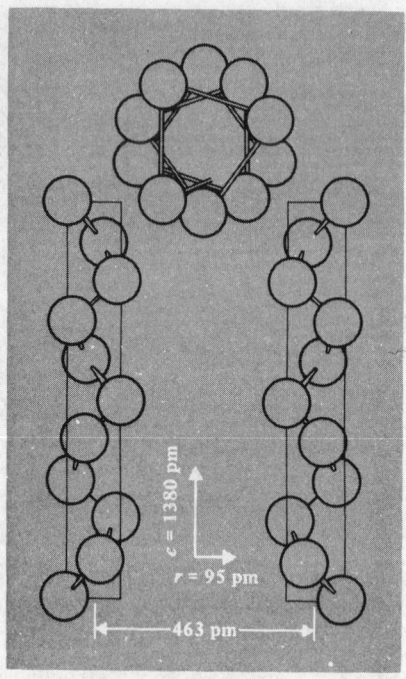

FIG. 15.15 The structure of right-handed and left-handed S_∞ helices in fibrous sulfur (see text).

The constitution of liquid sulfur has been extensively investigated, particularly in the region just above the remarkable transition at 159.4°. At this temperature virtually all properties of liquid sulfur change discontinuously, e.g. specific heat (λ point) density, velocity of sound, polarizability, compressibility, colour, electrical conductivity, surface tension, and, most strikingly, viscosity, which increases over 10 000-fold within the temperature range 160–195°C before gradually decreasing again. The phenomena can now be interpreted at least semi-quantitatively by a 2-step polymerization theory involving initiation and propagation:

$$cyclo\text{-}S_8 \rightleftharpoons catena\text{-}S_8$$
$$catena\text{-}S_8 + cyclo\text{-}S_8 \longrightarrow catena\text{-}S_{16}, \text{ etc.}$$

The polymerization is photosensitive, involves diradicals, and leads to chain lengths that exceed 200 000 S atoms at $\sim 180°C$ before dropping slowly to ~ 1000 S at 400° and ~ 100 S at 600°. Polymeric S_∞ is dark yellow with an absorption edge at 350 nm (cf. H_2S_n, p. 807) but the colour is often obscured either by the presence of trace organic impurities or, in pure S, by the presence of other highly coloured species such as the dark cherry-red trimer S_3 or the deeper-coloured diradicals S_4 and S_5.

The saturated vapour pressure above solid and liquid sulfur is given in Table 15.8. The molecular composition of the vapour has long been in contention but, mainly as a result of the work of J. Birkowitz and others[14] is now known to contain all molecules S_n with $2 \leqslant n \leqslant 10$ including odd-numbered species. The actual concentration of each species depends on both temperature and pressure. In the saturated vapour up to 600°C S_8 is the most common species followed by S_6 and S_7, and the vapour is green. Between 620–720°C

TABLE 15.8   *Vapour pressure of crystalline* cyclo-$S_8(\alpha)$ *and liquid sulfur*

p/mmHg	10^{-5}	10^{-3}	10^{-1}	1	10	100	760
T/°C	39.0	81.1	141	186	244.9	328	444.61

p/atm	1	2	5	10	50	100	200
T/°C	444.61	495	574	644	833	936	1035

S_7 and S_6 are slightly more prevalent than S_8 but the concentration of all three species falls rapidly with respect to those of S_2, S_3, and S_4, and above 720°C S_2 is the predominant species. At lower pressures S_2 is even more prominent, accounting for more than 80% of all vapour species at 530°C and 100 mmHg, and 99% at 730°C and 1 mmHg. This vapour is violet. The vapour above FeS_2 at 850°C is also S_2.

The best conditions for observing S_3 are 440°C and 10 mmHg when 10–20% of vapour species comprise this deep cherry-red bent triatomic species; like ozone, p. 707, it has a singlet ground state. The best conditions for S_4 are 450°C and 20 mmHg (concentration ~20%) but the structure is still not definitely established and may, in fact be a strained ring, an unbranched diradical chain, or a branched-chain isostructural with $SO_3(g)$ (p. 832).

The great stability of S_2 in the gas phase at high temperature is presumably due to the essentially double-bond character of the molecule and to the increase in entropy ($T\Delta S$) consequent on the breaking up of the single-bonded S_n oligomers. As with O_2 (p. 706) the ground state is a triplet level $^3\Sigma_g^-$ but the splitting within the triplet state is far larger than with O_2 and the violet colour is due to the transition $B^3\Sigma_u^- \leftarrow X^3\Sigma_g^-$ at 31 689 cm^{-1}. The corresponding B→X emission is observed whenever S compounds are burned in a reducing flame and the transition can be used for the quantitative analytical determination of the concentration of S compounds. There is also a singlet $^1\Delta$ excited state as for O_2. The dissociation energy $D_0^\circ(S_2)$ is 421.3 kJ mol^{-1}, and the interatomic distance in the gas phase 188.7 pm (cf. Table 15.6).

15.1.5   *Atomic and physical properties*

Several physical properties of sulfur have been mentioned in the preceding section; they vary markedly with the particular allotrope and its physical state.

Sulfur ($Z = 16$) has 4 stable isotopes of which ^{32}S is by far the most abundant in nature (95.02%). The others are ^{33}S (0.75%), ^{34}S (4.21%), and ^{36}S (0.02%). These abundances vary somewhat depending on the source of the sulfur, and this prevents the atomic weight of sulfur being quoted for general use more precisely than 32.06 ± 0.01 (p. 20). The variability is a valuable geochemical indication of the source of the sulfur and the isotope ratios of sulfur-containing impurities can even be used to identify the probable source of petroleum samples.[22] In such work it is convenient to define the abundance ratio of the 2 most important isotopes ($R = {}^{32}S/{}^{34}S$) and to take as standard the value of 22.22 for

[22] H. NIELSEN, Sulfur isotopes, in E. JÄGER and J. C. HUNZIKER (eds.), *Lectures in Isotope Geology*, pp. 283–312, Springer-Verlag, Berlin, 1979.

meteoritic troilite (FeS). Deviations from this standard ratio are then expressed in parts per thousand (sometimes confusingly called "per mil" or ‰):

$$\delta^{34}S = 1000(R_{sample} - R_{std})/R_{std}$$

On this definition, $\delta^{34}S$ is zero for meteoritic troilite; dissolved sulfate in ocean water is enriched $+20‰$ in ^{34}S, as are contemporary evaporite sulfates, whereas sedimentary sulfides are depleted in ^{34}S by as much as $-50‰$ due to fractionation during bacterial reduction to H_2S.

In addition to the 4 stable isotopes sulfur has 6 radioactive isotopes, the one with the longest half-life being ^{35}S which decays by β^- activity (E_{max} 0.167 MeV, $t_{\frac{1}{2}}$ 87.4 d). ^{35}S can be prepared by $^{35}Cl(n,p)$, $^{34}S(n,\gamma)$, or $^{34}S(d,p)$ and is commercially available as $S_{element}$, H_2S, $SOCl_2$, and KSCN. The β^- radiation has a similar energy to that of ^{14}C (E_{max} 0.155 MeV) and similar counting techniques can be used (p. 306). The maximum range is 300 mm in air and 0.28 mm in water, and effective shielding is provided by a perspex screen 3–10 mm thick. The preparation of many ^{35}S-containing compounds has been reviewed[23] and many of these have been used for mechanistic studies, e.g. the reactions of the specifically labelled thiosulfate ions $^{35}SSO_3^{2-}$ and $S^{35}SO_3^{2-}$. Another ingenious application, which won Barbara B. Askins the US Inventor of the Year award for 1978, is the use of ^{35}S for intensifying under-exposed photographic images: prints or films are immersed in dilute aqueous alkaline solutions of ^{35}S-thiourea, which complexes all the silver in the image (including invisibly small amounts), and the alkaline medium converts this to immobile, insoluble $Ag^{35}S$; the film so treated is then overlayed with unexposed film which reproduces the image with heightened intensity as a result of exposure to the β^- activity.[24]

The isotope ^{33}S has a nuclear spin quantum number $I = \frac{3}{2}$ and so is potentially useful in nmr experiments (receptivity to nmr detection 17×10^{-6} that of the proton). The resonance was first observed in 1951 but the low natural abundance of ^{33}S (0.75%) and the quadrupolar broadening of many of the signals has so far restricted the amount of chemically significant work appearing on this resonance.[25] However, more results are expected now that pulsed fourier-transform techniques are becoming widely available.

The S atom in the ground state has the electronic configuration $[Ne]3s^23p^4$ with 2 unpaired p electrons ($^3P_1$). Other atomic properties are: ionization energy 999.30 kJ mol^{-1}, electron affinities $+200$ and -414 kJ mol^{-1} for the addition of the first and second electrons respectively, electronegativity (Pauling) 2.5, covalent radius 103 pm, and ionic radius of S^{2-} 184 pm. These properties can be compared with those of the other elements in Group VIB on p. 890.

15.1.6 *Chemical reactivity*

Sulfur is a very reactive element especially at slightly elevated temperatures (which presumably facilitates cleavage of S–S bonds). It unites directly with all elements except

[23] R. H. HERBER, Sulfur-35, in R. H. HERBER (ed.), *Inorganic Isotopic Syntheses*, pp. 193–214, Benjamin, New York, 1962.

[24] ANON., NASA chemist named inventor of the year, *Chem. Eng. News*, 5 Feb. 1979, p. 8; see also US Patent 4 101 780 (1978).

[25] C. RODGER, N. SHEPPARD, C. McFARLANE, and W. McFARLANE, Sulfur-33, in R. H. HARRIS and B. E. MANN (eds.), *NMR and the Periodic Table*, pp. 401–2, Academic Press, London, 1978.

the noble gases, nitrogen, tellurium, iodine, iridium, platinum, and gold, though even here compounds containing S bonded directly to N, Te, I, Ir, Pt, and Au are known. Sulfur reacts slowly with H_2 at 120°, more rapidly above 200°, and is in reversible thermodynamic equilibrium with H_2 and H_2S at higher temperatures. It ignites in F_2 and burns with a livid flame to give SF_6; reaction with chlorine is more sedate at room temperature but rapidly accelerates above this to give (initially) S_2Cl_2 (p. 815). Sulfur dissolves in liquid Br_2 to form S_2Br_2, which readily dissociates into its elements; iodine has been used as a cryoscopic solvent for sulfur (p. 772) and no binary compound is formed even at elevated temperatures. Oxidation of sulfur by (moist?) air is very slow at room temperature though traces of SO_2 are formed; the ignition temperature of S in air is 250–260°. Pure dry O_2 does not react at room temperature though O_3 does. Likewise direct reaction with N_2 has not been observed but, in a discharge tube, activated N reacts. All other non-metals (B, C, Si, Ge; P, As, Sb; Se) react at elevated temperatures. Of the metals, sulfur reacts in the cold with all the main group representatives of Groups I, II, III, Sn, Pb, and Bi, and also Cu, Ag, and Hg (which even tarnishes at liquid-air temperatures). The transition metals (except Ir, Pt, and Au) and the lanthanides and actinides react more or less vigorously on being heated with sulfur to form binary metal sulfides (p. 798).

The reactivity of sulfur clearly depends sensitively on the molecular complexity of the reacting species. Little systematic work has been done. *Cyclo*-S_8 is obviously less reactive than the diradical *catena*-S_8, and smaller oligomers in the liquid or vapour phase also complicate the picture. In the limit atomic sulfur, which can readily be generated photolytically, is an extremely reactive species. As with atomic oxygen and the various methylenes, both singlet and triplet states are possible and these have different reactivities. The ground state is $^3P_2$, and the singlet state $^1D_2$ lies 110.52 kJ mol^{-1} above this. Triplet state S atoms (with 2 unpaired electrons) can be generated by the Hg-photosensitized irradiation of COS:

$$Hg + hv \text{ (253.7 nm)} \longrightarrow Hg(^3P_1)$$
$$Hg(^3P_1) + COS \longrightarrow Hg + CO + S(^3P)$$

Triplet S can also be generated by direct photolysis of CS_2 ($hv < 210$ nm) or ethylene episulfide

$$CH_2 \overset{S}{\diagup\!\!\!\diagdown} CH_2$$

(hv 220–260 nm). Photolysis of SPF_3 (hv 210–230 pm) generates singlet state S atoms (with no unpaired electrons) but the best syntheses of these is the direct primary photolysis of COS in the absence of Hg; this generates mainly singlet S (75%) with the rest being in the triplet state (^3P):

$$COS + hv \longrightarrow CO + S(^1D_2)$$

Generation of (excited state) singlet S in the presence of paraffins yields the corresponding mercaptan by a concerted single-step insertion:

$$RH + S(^1D_2) \longrightarrow RSH$$

By contrast, paraffins are inert to triplet (ground state) S atoms. Singlet S undergoes analogous insertion reactions with $MeSiH_3$, $SiMe_4$, and B_2H_6. Olefins can undergo insertion of singlet S atoms on stereospecific addition of triplet S atoms; according to experimental conditions, the products are alkenyl mercaptans, vinylic mercaptans, or

TABLE 15.9   *Coordination geometries of sulfur*

CN	Examples
1	$S_2(g)$, CS_2, HNCS, K[SCN] and "covalent" isothiocyanates, $P_4O_6S_4$, P_4S_n (terminal S), SSF_2, SSO_3^{2-}, $Na_3SbS_4.9H_2O$, Tl_3VS_4, M_2MoS_4, $(NH_4)_2WS_4$, $S=WCl_4^{(a)}$
2 (linear)	$[(\eta^5\text{-}C_5H_5)(CO)_2Cr\equiv S\equiv Cr(CO)_2(\eta^5\text{-}C_5H_5)]^{(b)}$
2 (bent)	S_n, H_2S, H_2S_n, Me_2S_n, S_nX_2 (Cl, Br), SO_2, P_4S_n (bridging S), $Se(SCN)_2$ and "covalent" thiocyanates
3 (planar, D_{3h})	SO_3 (g)

3 (T-shaped planar)

3 (pyramidal)	SSF_2, $OSCl_2$, $S_8O(1\ S)$, SO_3^{2-}, $S_2O_4^{2-}$, $S_2O_5^{2-}(1\ S)$, Me_3S^+, SF_3^+
4 (tetrahedral)	SO_3 (s) [i.e. cyclic S_3O_9 or fibrous $(SO_3)_x$], SO_2Cl_2, SO_4^{2-}, $S_2O_6^{2-}$ ($O_3SSO_3^{2-}$), $S_2O_7^{2-}$ ($O_3SOSO_3^{2-}$), $S_3O_{10}^{2-}$, $S_5O_{16}^{2-}$, ZnS (blende, and M=Be, Cd, Hg), ZnS(wurtzite, and M=Cd, Mn)
4 (ψ-tbp)	SF_4
5 (square pyramidal) (ψ-octahedral)	SF_5^-, SOF_4 NiS (millerite structure)
6 (octahedral)	SF_6, S_2F_{10}, MS(NaCl-type, M = Mg, Ca, Sr, Ba, Mn, Pb, Ln, Th, U, Pu)
6 (trigonal prismatic)	MS(NiAs-type), (M = Ti, V, Fe, Co, Ni), Hf_2S
7 (mono-capped trigonal prismatic)	Ta_6S,$^{(d)}$ $Ti_2S^{(e)}$
8 (cubic)	M_2S (antifluorite-type, M = Li, Na, K, Rb)
9 (mono-capped square antiprismatic)	$[Rh_{17}(CO)_{32}(S)_2]^{3-(f)}$
10 (bicapped square antiprismatic)	$[Rh_{10}(CO)_{10}(\mu\text{-}CO)_{12}S]^{2-(g)}$

(a) Ref. 26.   (b) Ref. 27.   (c) Ref. 28.   (d) Ref. 29.   (e) Ref. 30.   (f) Ref. 30a.   (g) Ref. 31.

[26] M. G. B. DREW and R. MANDYCZEWSKY, Crystal and molecular structures of tungsten(VI) sulfide tetrachloride and tungsten(VI) sulfide tetrabromide, *J. Chem. Soc.* A 1970, 2815–18.

[27] T. J. GREENHOUGH, B. W. S. KOLTHAMMER, P. LEGZDINS, and J. TROTTER, Crystal and molecular structure of bis[(η^5-cyclopentadienyl)dicarbonylchromium] sulfide, a novel organometallic complex possessing a $Cr\equiv S\equiv Cr$ linkage, *Inorg. Chem.* **18**, 3543–8 (1979).

[28] P. H. W. LAU and J. C. MARTIN, Tricoordinate hypervalent sulfur species. Sulfuranide anions, *J. Am. Chem. Soc.* **100**, 7077–9 (1978).

[29] H. F. FRANZEN and J. G. SMEGGIL, The crystal structure of Ta_6S, *Acta Cryst.* **B26**, 125–9 (1970).

[30] J. P. OWENS, B. R. CONARD, and H. F. FRANZEN, The crystal structure of Ti_2S, *Acta Cryst.* **23**, 77–82 (1967).

[30a] J. L. VIDAL, R. A. FIATO, L. A. CROSBY, and R. L. PRUETT, $[Rh_{17}(S)_2(CO)_{32}]^{3-}$. 1. An example of encapsulation of chalcogen atoms by transition-metal-carbonyl clusters, *Inorg. Chem.* **17**, 2574–82 (1978).

[31] G. CIANI, L. GARLASCHELLI, A. SIRONI, and S. MARTINENGO, Synthesis and X-ray characterization of the novel $[Rh_{10}S(CO)_{10}(\mu\text{-}CO)_{12}]^{2-}$ anion; a bicapped square-antiprismatic cluster containing an interstitial sulfur atom. *JCS Chem. Comm.* 1981, 563–5.

episulfides. Analogous reactions with inorganic compounds appear to be a very promising field for future research.

Sulfur compounds exhibit a rich and multifarious variety which derives not only from the numerous possible oxidation states of the element (from -2 to $+6$) but also from the range of bond types utilized (covalent, coordinate, ionic, and even metallic) and the multiplicity of coordination geometries adopted by the element. Oxidation states and their interrelationships as codified by oxidation state diagrams are dealt with more fully in the section on oxoacids of sulfur (p. 834) though the existence of several other series of compounds, notably the halides, also illustrates the element's versatility. The range of bond types, as reflected in the physical and chemical properties of the various compounds of the element, will become increasingly apparent throughout the rest of the chapter. The multiplicity of coordination geometries is amply demonstrated by the examples in Table 15.9. Most of these can be readily rationalized by the numerous variants of elementary bonding theory.

Polyatomic sulfur cations

As long ago as 1804 C. F. Bucholz observed that sulfur dissolves in oleum to give clear, brightly coloured solutions which could be yellow, deep blue, or red (or intermediate colours) depending on the strength of the oleum and the time of the reaction. These solutions are now known to contain S_n^{2+} cations, the structure of which has been elucidated during the past decade mainly by the elegant synthetic, Raman spectroscopic, and crystallographic studies of R. J. Gillespie and his coworkers.[31a] Selenium and tellurium behave similarly (p. 896). Sulfur can most conveniently be quantitatively oxidized using SbF_5 or AsF_5 in an inert solvent such as SO_2, e.g.:

$$S_8 + 2AsF_5 \xrightarrow{\text{SO}_2} [S_8]^{2+}[AsF_6]^-_2 + AsF_3$$

The bright-yellow solutions contain S_4^{2+}, a square-planar ring whose structure has been confirmed by an X-ray study on the unusual crystalline compound $As_6F_{36}I_4S_{32}$, i.e. $[S_4]^{2+}[S_7I]^+_4[AsF_6]^-_6$ (p. 896). The S–S interatomic distance is 198 pm compared with 204 pm for a single-bonded species. Note also that S_4^{2+} is isoelectronic with the known heterocyclic compound S_2N_2 (p. 859). The pale-yellow compound $[S_4]^{2+}[SbF_6]^-_2$ has also been isolated.

The deep-blue solutions contain S_8^{2+}, and the X-ray structure of $[S_8]^{2+}[AsF_6]^-_2$ reveals that the cation has an *exo-endo* cyclic structure with a long transannular bond as shown in Fig. 15.16 (see also p. 857). The bright-red solutions were originally thought to contain the S_{16}^{2+} cation and a compound thought to be $S_{16}(AsF_6)_2$ was isolated; however, a very recent crystallographic study has shown[31b] that the compound has the totally unexpected formulation $[S_{19}]^{2+}[AsF_6]^-_2$ which could not have been distinguished from the earlier stoichiometry on the basis of the original analytical data. This

[31a] R. J. GILLESPIE, Ring, cage, and cluster compounds of main group elements, *Chem. Soc. Rev.* **8**, 315–52 (1979).
[31b] R. C. BURNS, R. J. GILLESPIE, and J. F. SAWYER, Preparation and crystal structure of $S_{19}(AsF_6)_2$ and an esr and absorption spectral study of solutions containing the S_{19}^{2+} cation and related species, *Inorg. Chem.* **19**, 1423–32 (1980).

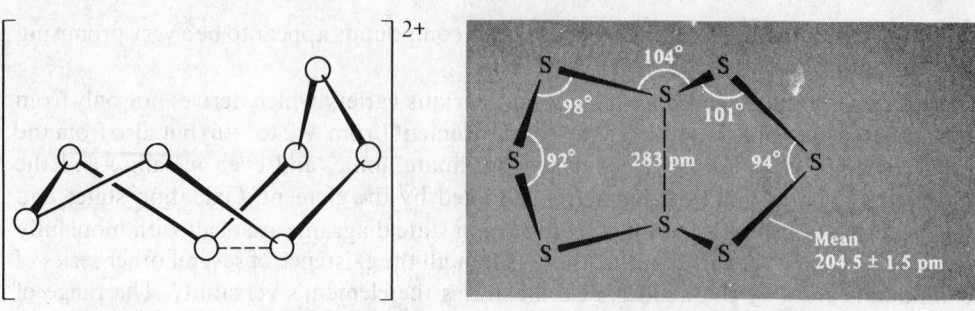

FIG. 15.16   The structure and dimensions of the S_8^{2+} cation in $[S_8]^{2+}[AsF_6]_2^{-}$.

astonishing cation consists of two 7-membered rings joined by a 5-atom chain. As shown in Fig. 15.17, one of the rings has a boat conformation whilst the other is disordered, existing as a 4:1 mixture of chain and boat conformations. S–S distances vary greatly from 187 to 239 pm and S–S–S angles vary from 91.9° to 127.6°.

Solutions of sulfur in oleum also give rise to paramagnetic species, probably S_n^{+}, but the nature of these has not yet been fully established. For polysulfur anions S_n^{2-}, see p. 805.

Sulfur as a ligand

The S atom can act either as a terminal or a bridging ligand. The dianion S_2^{2-} is also an effective ligand, and chelating polysulfides $-S_n-$ are well established. These various sulfur ligands will be briefly considered before dealing with the broad range of compounds in which S acts as the donor atom, e.g. H_2S, R_2S, dithiocarbamates and related anions, 1,2-dithiolenes etc. Ligands in which S acts as a donor atom are usually classified as class-b ligands ("soft" Lewis bases), in contrast to oxygen donor-atom ligands which tend to be class-a or hard (p. 1065). The larger size of the S atom and the consequent greater deformability of its electron cloud give a qualitative rationalization of this difference and

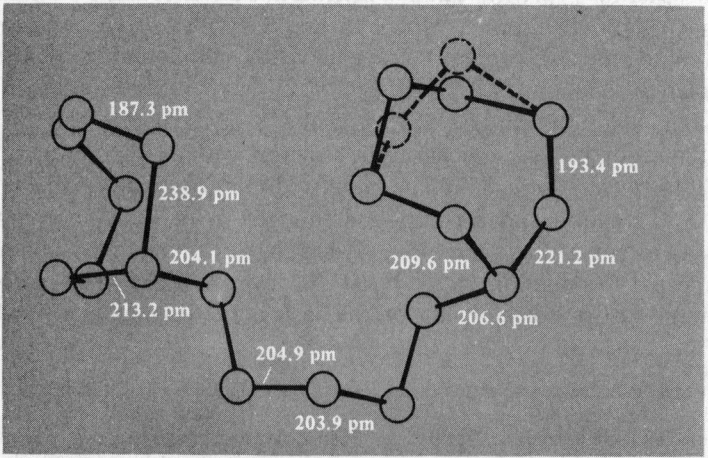

FIG. 15.17   The structure and some of the dimensions of the disordered cation S_{19}^{2+} (see text).

the possible participation of d_π orbitals in bonding to sulfur has also been invoked (see comparison of N and P, p. 478).

Some examples of the S atom as a bridging ligand are given in Fig. 15.18. In the μ_2 bridging mode S is usually regarded as a 2-electron donor, though in the recently characterized linear bridge $[\{(C_5H_5)(CO)_2Cr\}_2S]$ it is probably best regarded as a 6-electron donor.[27] In the μ_3 triply bridging mode S can be regarded as a 4-electron donor, using both its unpaired electrons and one lone-pair.[32] If the 3 bridged metal atoms are different then a chiral tetrahedrane molecule results and this has permitted the very recent (1980) resolution of the enantiomers of the first optically active metal cluster compound, the red complex $[\{Co(CO)_3\}\{Fe(CO)_3\}\{Mo(\eta^5\text{-}C_5H_5)(CO)_2\}S]$.[33] The pseudo-cubane structure adopted by some of the μ_3-S compounds is assuming added significance as a crucial structural unit in many biologically important systems, e.g. the $\{(RS)FeS\}_4$ units which cross-link the polypeptide chains in ferredoxins (p. 1280). In the μ_4-mode 6-electrons are involved, if the bonding is considered to be predominantly covalent, though metal-sulfides are sometimes treated as compounds of S^{2-}. No *molecular* compounds are known in which S bridges 6 or 8 metal atoms though, again, these coordinations are prevalent in solid-state compounds, many of which have interatomic bonding which is far from being purely ionic.

The disulfur ligand S_2 (sometimes more helpfully considered as S_2^{2-}) is attracting increasing attention since no other simple ligand is as versatile in the variety of its modes of coordination. Moreover, in one particular mode (see Type III below) it is particularly effective in stabilizing metal clusters. Many of the complexes of S_2 were first obtained accidentally, and their seemingly bizarre stoichiometries only became intelligible after structural elucidation by X-ray crystallography. The complexes can be prepared by reacting metals or their compounds with:

(a) a positive S_2 group as in S_2Cl_2;[34]
(b) a neutral S_2 group, usually derived from S_8;
(c) a negative S_2^{2-} group such as an alkaline polysulfide solution.

Examples are

$$Nb + S_2Cl_2 \xrightarrow{\text{heat}} NbS_2Cl_2 \quad \text{(see Fig. 15.20a), (p. 793)}$$

$$Nb + \tfrac{1}{4}S_8 + X_2 \xrightarrow{500°C} NbS_2X_2 \quad (X = Cl, Br, I)$$

$$(NH_4)_6Mo_7O_{24}\cdot4H_2O + H_2S + (NH_4)_2S_n \xrightarrow{\text{aq NH}_3} (NH_4)_2[Mo_2(S_2)_6]$$

(see Fig. 15.19g)

The S-S bond can also be formed by a direct coupling reaction, e.g.:

$$2[(H_2O)_5Cr(SH)]^{2+} + I_2 \longrightarrow [(H_2O)_5CrSSCr(OH_2)_5]^{4+} + 2HI$$

[32] H. VAHRENKAMP, Sulfur atoms as ligands in metal complexes, *Angew. Chem.*, Int. Edn. (Engl.) **14**, 322–9 (1975). A valuable review of the structure and bonding, with 121 references.

[33] F. RICHTER and H. VAHRENKAMP, The first optically active cluster: resolution of enantiomers and absolute configuration of $[SFeCoMo(C_5H_5)(CO)_8]$, *Angew. Chem.*, Int. Edn. (Engl.) **19**, 65 (1980).

[34] M. J. ATHERTON and J. H. HOLLOWAY, Thio-, seleno-, and telluro-halides of the transition metals, *Adv. Inorg. Chem. Radiochem.* **22**, 171–98 (1979).

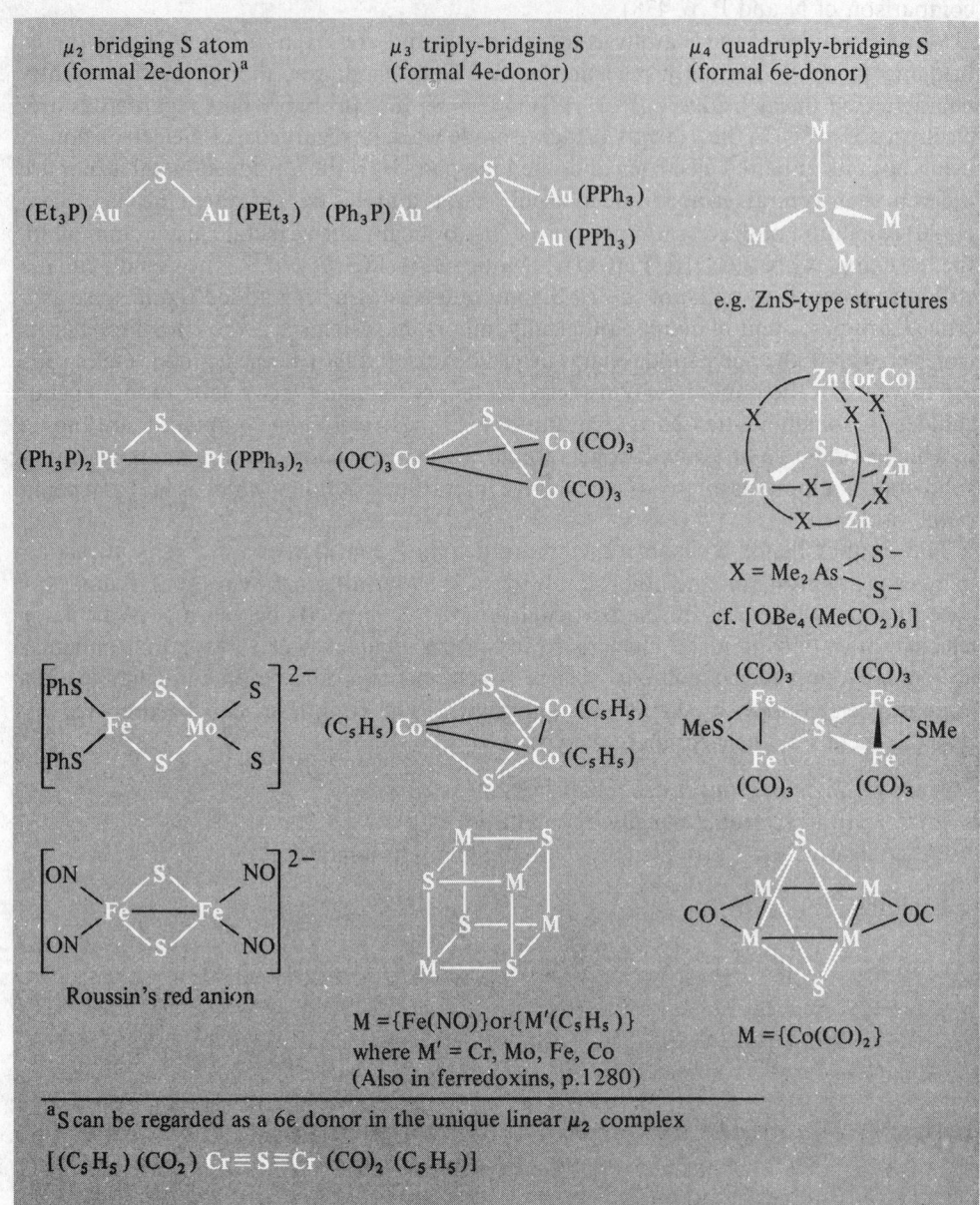

FIG. 15.18 The S atom as a bridging ligand.

At least 8 modes of coordination are known (Table 15.10);[35] they are all based on either side-on S_2 or bridging –S–S– with possible further ligation via one or two lone-pairs as shown schematically below:

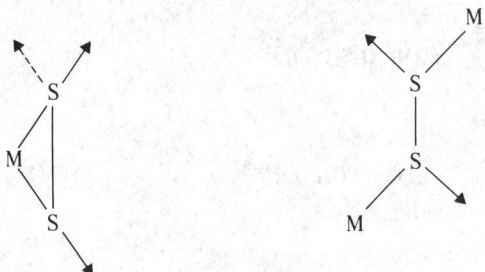

Frequently, more than one type of coordination occurs in a given complex, e.g. Figs. 15.19b, c, and g. Interestingly, there appear to be no known example of terminal "end-on" coordination, M–S–S (see dioxygen complexes, p. 718). Detailed descriptions of all the structures and their bonding are beyond the scope of this treatment but it will be noted from Table 15.10 that the S–S interatomic distances in disulfide complexes range from 201 to 209 pm. The following specific points of interest may also be mentioned. The orange-red anion $[Mo_4(NO)_4S_{13}]^{4-}$ (Fig. 15.19b) features two triangular arrays of Mo atoms joined by a common edge and with an angle of 127.6° between the two Mo_3 planes; each plane has a μ_3-bonded S atom above it (Mo–S 250.1 pm) and there is a further unique μ_4-bonded S atom which is 261.6 pm from each of the 4 Mo atoms. Four of the 5 S_2^{2-} ligands are simultaneously bonded both end on (Mo–S 246.5 pm) and side on (Mo–S 249.2 pm) whilst the fifth is side-on only. The complex therefore has sulfur in five different bonding states. In the red complex $[Mn_4(CO)_{15}(S_2)_2]$ (Fig. 15.19c) the 2 S_2^{2-} ligands are different (Types Ic and Id); the 4 Mn atoms are bonded, respectively, to 3, 3, 4, and 5 carbonyl ligands, but each achieves a distorted octahedral coordination by being bonded also to 3, 3, 2, and 1 S atoms respectively. There seems no reason to suppose that the diamagnetic bridged dinuclear anion $[(NC)_5Co^{III}SSCo^{III}(CN)_5]^{6-}$ is not a formal Type IIa disulfido S_2^{2-} complex, but there is evidence[39] that the superficially analogous paramagnetic dinuclear ruthenium cation in Fig. 15.19d is, in fact, a mixed-valence supersulfido S_2^- complex: $[H_3N)_5Ru^{II}SSRu^{III}[NH_3)_5]^{4+}$. The bridged dinuclear cobalt anion undergoes a remarkable aerial oxidation in aqueous ethanol solutions at $-15°C$; one of the bridging S atoms only is oxidized and this results in the formation of a bridging thiosulfito group $[(NC)_5CoSSO_2 Co(CN)_5]^{6-}$ coordinated through the two S atoms to the two Co atoms.[35a]

Not all disulfide complexes are discrete molecular or ionic species and several solid-state compounds of S_2^{2-} are known in addition to the familiar pyrites and marcasite-type disulfides (p. 804). Examples are the chlorine-bridged polymeric NbS_2Cl_2 mentioned on p. 787 (Fig. 15.20a) and the curious series of brown and red compounds formed by heating Mo or MoS_3 with S_2Cl_2, e.g.[34] MoS_2Cl_2, MoS_2Cl_3 (Fig. 15.20b), $Mo_2S_4Cl_5$ (Fig. 15.20c), $Mo_2S_5Cl_3$ and $Mo_3S_7Cl_4$.

[35] A. MULLER and W. JAEGERMANN, Disulfur ligands in transition-metal coordination and cluster compounds, *Inorg. Chem.* **18**, 2631–3 (1979).

[35a] F. R. FRONCZEK, R. E. MARSH, and W. P. SCHAEFER, A new oxygen-sulfur anion: Preparation and structure of potassium μ-(thiosulfito-*S,S'*)-decacyanodicobaltate(III), *J. Am. Chem. Soc.* **104**, 3382–5 (1982).

TABLE 15.10　*Types of metal-disulfide complex*

Type	Example	$d(S-S)$/pm	Structure
Ia　M$\langle$S$\vert$S	$[Mo_2O_2S_2(S_2)_2]^{2-}$	208(1)	Figure 15.19a (36)
Ib　M$\langle$S$\vert$S (–M)	$[Mo_4(NO)_4S_{13}]^{4-}$	204.8(7)	Figure 15.19b (37)
Ic　M$\langle$S$\vert$S (M above and below)	$[Mn_4(CO)_{15}(S_2)_2]$	207	Figure 15.19c (38)
Id　M–S (M,M above; S–M below)	$[Mn_4(CO)_{15}(S_2)_2]$	209	Figure 15.19c (38)
IIa　M–S–S–M	$[Ru_2(NH_3)_{10}S_2]^{4+}$	201.4(1)	Figure 15.19d (39)
IIb　M–S–S–M (bridged)	$[Co_4(\eta^5\text{-}C_5H_5)_4(\mu_3\text{-}S)_2(\mu_3\text{-}S_2)_2]$	201(3)	Figure 15.19e (40)
IIc　M–S–S–M (bridged)	$[\{SCo_3(CO)_7\}_2S_2]$	204.2(14)	Figure 15.19f (41)
III　S$\vert$S triangle M–M	$[Mo_2(S_2)_6]^{2-}$	204.3(5)	Figure 15.19g (42)

[36] W. CLEGG, N. MOHAN, A. MÜLLER, A. NEUMAN, W. RITTNER, and G. M. SHELDRICK, Crystal and molecular structure of [NMe$_4$]$_2$[Mo$_2$O$_2$S$_2$(S$_2$)$_2$]: a compound with two S$_2^{2-}$ ligands, *Inorg. Chem.* **19**, 2066–9 (1980).

[37] A. MÜLLER, W. ELTZNER, and N. MOHAN, S$_2^{2-}$ as a simultaneous end-on and side-on bonded ligand in the novel transition-metal complex [Mo$_4$(NO)$_4$S$_{13}$]$^{4-}$, *Angew. Chem.*, Int. Edn. (Engl.) **18**, 168–9 (1979).

[38] V. KÜLLMER, E. RÖTTINGER, and H. VAHRENKAMP, Preparation and crystal structure of [Mn$_4$S$_4$(CO)$_{15}$]; oxidation at sulfur of [{(CO)$_4$Mn–S–SnMe$_3$}$_2$], *JCS Chem. Comm.* 1977, 782–3.

Complexes with chelating polysulfide ligands can be made either by reacting complex metal halides with solutions of polysulfides or by reacting hydrido complexes with elemental sulfur, e.g.:

$$H_2PtCl_6 + (NH_4)_2S_x(aq) \xrightarrow{\text{boil}} (NH_4)_2[Pt^{IV}(S_5)_3]$$

$$[Ti(\eta^5\text{-}C_5H_5)_2Cl_2] + Na_2S_5 \longrightarrow [Ti^{IV}(\eta^5\text{-}C_5H_5)_2(S_5)] + 2NaCl$$

$$[W(\eta^5\text{-}C_5H_5)_2H_2] + \tfrac{5}{8}S_8 \longrightarrow [W^{IV}(\eta^5\text{-}C_5H_5)_2(S_4)] + H_2S$$

The red dianion $[PtS_{15}]^{2-}$ was first made in 1903 but its structure as a chiral tris chelating pentasulfido complex (Fig. 15.21a) was not established until 1969.[43] It is a rare example of a "purely inorganic" (carbon-free) optically active species.[44] [Other examples are S. Heřmánek and J. Plešek's resolution of the main group element cluster compound $i\text{-}B_{18}H_{22}$,[45] A. Werner's first-row transition-metal complex cation $[Co\{(\mu\text{-}OH)_2Co(NH_3)_4\}_3]^{6+}$,[46] and F. G. Mann's second-row complex anion $cis\text{-}[Rh\{\eta^2\text{-}(NH)_2SO_2\}_2(OH_2)_2]^-$.[47]] The structure of the complex $[Ti(\eta^5\text{-}C_5H_5)_2(S_5)]$ is in Fig. 15.21b; it has previously been mentioned in connection with the synthesis of *cyclo*-polysulfur allotropes (p. 775). The chair conformation of the 6-membered TiS_5 ring undergoes chair-to-chair inversion above room temperature with an activation energy of about 69 kJ mol^{-1}.[48] A similar ring inversion in $[Pt(S_5)_3]^{2-}$ is even more facile and ^{195}Pt n.m.r. studies lead to a value of 50.5 ± 1.3 kJ mol^{-1} for $\Delta G^\ddagger$ at 0°C.[48a] Other

[39] R. C. ELDER and M. TRKULA, Crystal structure of $[(NH_3)_5RuSSRu(NH_3)_5]Cl_4 \cdot 2H_2O$. A structural trans effect and evidence for a supersulfide S_2^- bridge, *Inorg. Chem.* **16**, 1048–51 (1977).

[40] V. A. UCHTMAN and L. F. DAHL, Preparation and structural characterization of $[Co_4(\eta^5\text{-}C_5H_5)_4S_6]$. A new mode of transition-metal bonding for a disulfide group, *J. Am. Chem. Soc.* **91**, 3756–63 (1969).

[41] D. L. STEVENSON, V. R. MAGNUSON, and L. F. DAHL, Structure of a hexanuclear cobalt sulfur complex, $[\{SCo_3(CO)_7\}_2S_2]$, containing a tetra metal-coordinated bridging disulfide group, *J. Am. Chem. Soc.* **89**, 3727–32 (1967).

[42] A. MÜLLER, W.-O. NOLTE, and B. KREBS, $[(S_2)_2Mo(S_2)_2Mo(S_2)_2]^{2-}$, a novel complex containing only S_2^{2-} ligands and a Mo–Mo bond, *Angew. Chem.*, Int. Edn. (Engl.) **17**, 279 (1978); A. MÜLLER, W.-O. NOLTE, and B. KREBS. $[NH_4]_2[(S_2)_2Mo(S_2)_2Mo(S_2)_2] \cdot 2H_2O$, a novel sulfur-rich coordination compound with two nonequivalent complex anions having the same point group but different structures: crystal and molecular structures. *Inorg. Chem.* **19**, 2835–6 (1980).

[43] P. E. JONES and L. KATZ, The crystal structure of ammonium tris(pentasulfido)platinum(IV) dihydrate, *Acta Cryst.* **B25**, 745–52 (1969).

[44] R. D. GILLARD and F. L. WIMMER, Inorganic optical activity, *JCS Chem. Comm.* 1978, 936–7.

[45] S. HEŘMÁNEK and J. PLEŠEK, Resolution of iso-octadecaborane(22) into optical enantiomers, *Coll. Czech. Chem. Comm.* **35**, 2488–93 (1970).

[46] A. WERNER, Study of asymmetric cobalt atoms. Part 12. Optical activity of carbon-free compounds, *Ber.* **47**, 3057–94 (1914).

[47] F. G. MANN, The complex metallic salts of sulfamide: an optically active inorganic salt, *J. Chem. Soc.* 1933, 412–19.

[48] E. W. ABEL, M. BOOTH, and K. G. ORRELL, The inversion barrier of the TiS_5 ring in di(cyclopentadienyl)titanium pentasulfide, *J. Organometall. Chem.* **160**, 75–79 (1978).

[48a] F. G. RIDDELL, R. D. GILLARD, and F. L. WIMMER, Inorganic six-membered ring inversion. The inversion barrier in the tris(pentasulfane-1,5-diyl)platinate(IV) anion, *JCS Chem. Comm.* 1982, 332–3.

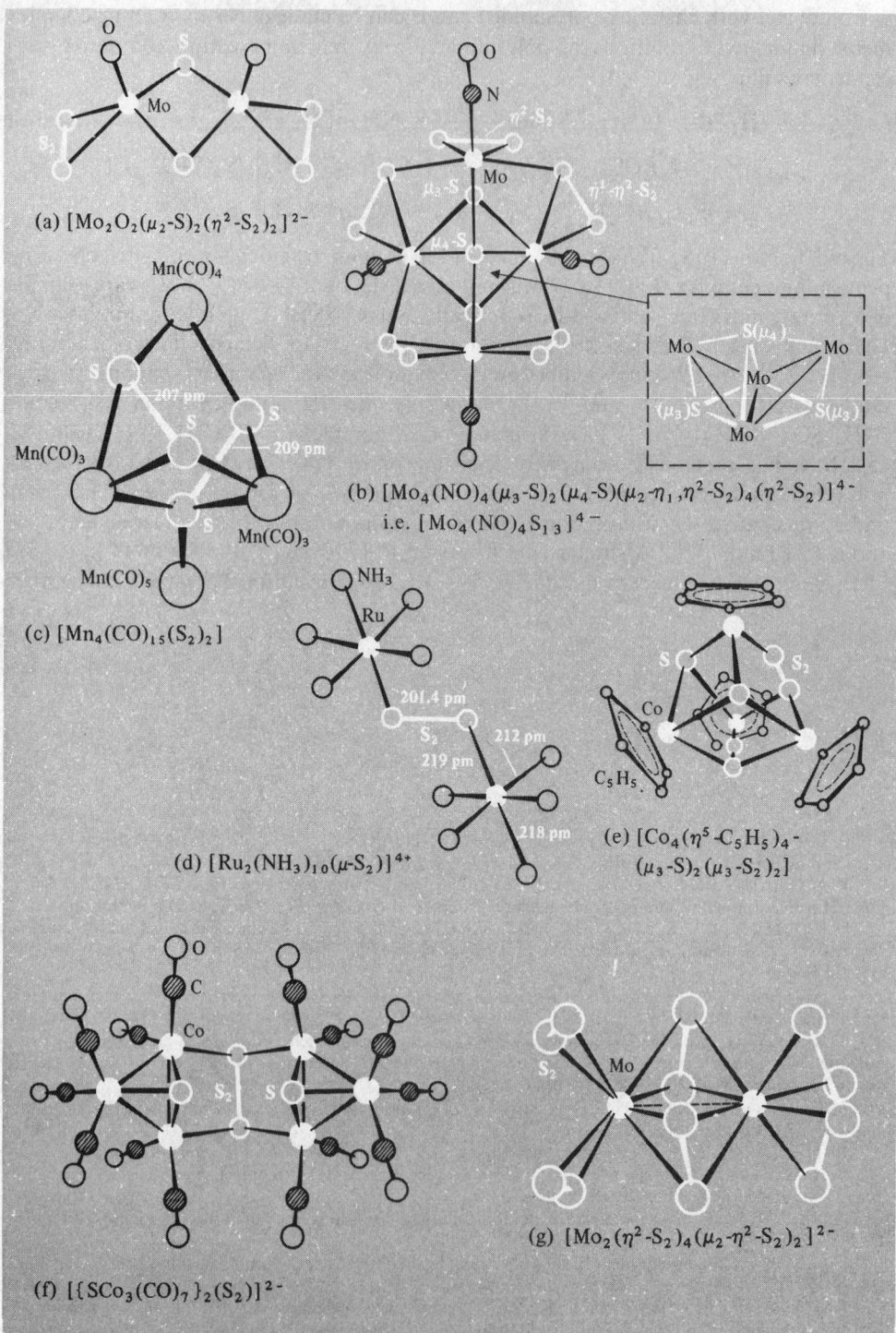

(a) $[Mo_2O_2(\mu_2\text{-}S)_2(\eta^2\text{-}S_2)_2]^{2-}$

(b) $[Mo_4(NO)_4(\mu_3\text{-}S)_2(\mu_4\text{-}S)(\mu_2\text{-}\eta_1,\eta^2\text{-}S_2)_4(\eta^2\text{-}S_2)]^{4-}$
i.e. $[Mo_4(NO)_4S_{13}]^{4-}$

(c) $[Mn_4(CO)_{15}(S_2)_2]$

(d) $[Ru_2(NH_3)_{10}(\mu\text{-}S_2)]^{4+}$

(e) $[Co_4(\eta^5\text{-}C_5H_5)_4\text{-}(\mu_3\text{-}S)_2(\mu_3\text{-}S_2)_2]$

(f) $[\{SCo_3(CO)_7\}_2(S_2)]^{2-}$

(g) $[Mo_2(\eta^2\text{-}S_2)_4(\mu_2\text{-}\eta^2\text{-}S_2)_2]^{2-}$

FIG. 15.19   Structures of some disulfide complexes.

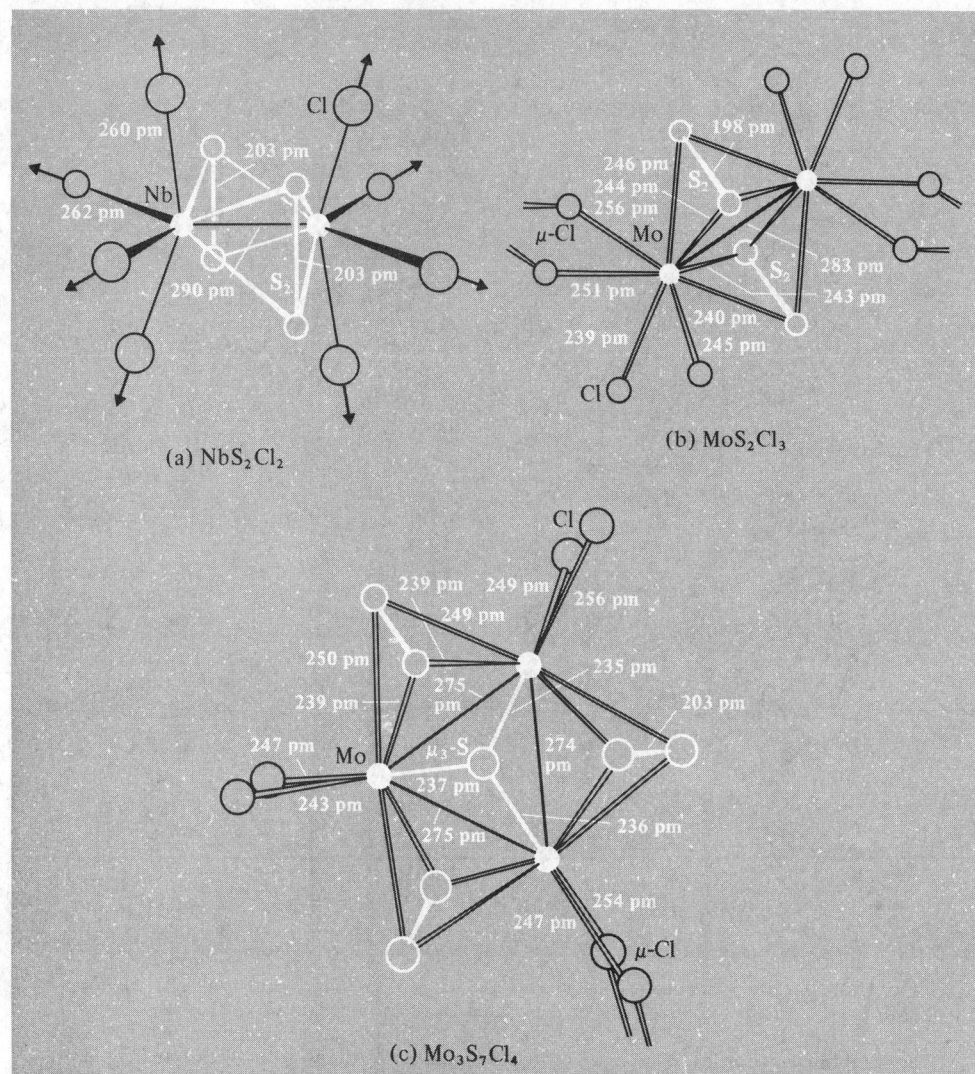

Fig. 15.20   Chlorine bridged polymeric structures of (a) NbS_2Cl_2, (b) MoS_2Cl_3, and (c) $Mo_3S_7Cl_4$.

recent examples of chelating S_n^{2-} ligands occur in the dark red-brown dianion[49]

$$[(\eta^2\text{-}S_5)Fe \underset{S}{\overset{S}{<}} Fe\,(\eta^2\text{-}S_5)]^{2-}$$

and in the intriguing black dianion $[Mo_2S_{10}]^{2-}$ which features 4 different sorts of sulfur ligand and at least 6 different S-atom environments (Fig. 15.21c).[50]

[49] D. COUCOUVANIS, D. SWENSON, P. STREMPLE, and N. C. BAENZIGER, Reactions of $[Fe(SPh)_4]^{2-}$ with organic trisulfides and implications concerning the biosynthesis of ferredoxins. Synthesis and structure of the $[PPh_4]_2[Fe_2S_{12}]$ complex, *J. Am. Chem. Soc.* **101**, 3392–4 (1979).

[50] W. CLEGG, G. CHRISTOU, C. D. GARNER, and G. M. SHELDRICK, $[Mo_2S_{10}]^{2-}$; a complex with terminal sulfido, bridging sulfido, persulfido and tetrasulfido groups, *Inorg. Chem.* **20**, 1562–6 (1981).

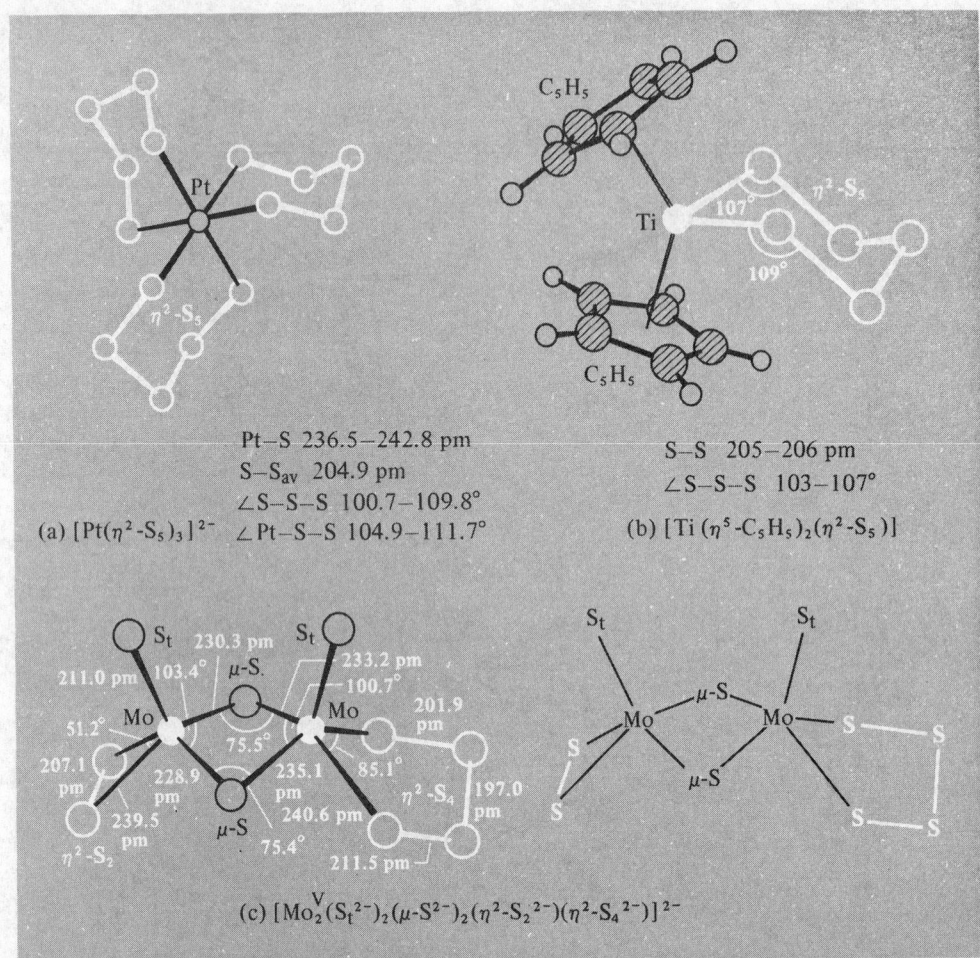

Pt–S 236.5–242.8 pm
S–S$_{av}$ 204.9 pm
∠S–S–S 100.7–109.8°
(a) $[Pt(\eta^2-S_5)_3]^{2-}$ ∠Pt–S–S 104.9–111.7°

S–S 205–206 pm
∠S–S–S 103–107°
(b) $[Ti(\eta^5-C_5H_5)_2(\eta^2-S_5)]$

(c) $[Mo_2^V(S_t^{2-})_2(\mu-S^{2-})_2(\eta^2-S_2^{2-})(\eta^2-S_4^{2-})]^{2-}$

FIG. 15.21 Structure and dimensions of (a) $[Pt(\eta^2-S_5)_3]^{2-}$, (b) $[Ti(\eta^5-C_5H_5)_2(\eta^2-S_5)]$, and (c) $[Mo_2S_{10}]^{2-}$: this last complex can be considered as an Mo^V derivative on the basis of the formulation $[Mo_2^V(S_t^{2-})_2(\mu-S^{2-})_2(\eta^2-S_2^{2-})(\eta^2-S_4^{2-})]^{2-}$. Note that the angles subtended by S atoms at Mo vary from 51.2° through 85.1° to 100.7° and 103.4°, the M–S distances from 211 pm through 229 and 235 pm to 241 pm, and the S–S distances from 197 to 211.5 pm with the S_2^{2-} group being 207 pm.

Other ligands containing sulfur as donor atom

H$_2$S, the simplest compound of sulfur, differs markedly from its homologue H$_2$O in complex-forming ability: whereas aquo complexes are extremely numerous and frequently very stable (p. 733), H$_2$S rarely forms simple adducts due to its ready oxidation to sulfur or its facile deprotonation to SH$^-$ or S^{2-}. [AlBr$_3$(SH$_2$)] has long been known as a stable compound of tetrahedral Al[50a] but the few transition metal complexes having some degree of stability at room temperature are of more recent vintage: examples include

[50a] A. WEISS, R. PLASS, and AL. WEISS, The crystal structure and acidity of AlBr$_3$.SH$_2$, *Z. anorg. allgem. Chem.* **283**, 390–400 (1956).

$[Mn(\eta^5\text{-}C_5H_5)(CO)_2(SH_2)]$, $[W(CO)_5(SH_2)]$, and the *triangulo* cluster complexes $[Ru_3(CO)_9(SH_2)]$ and $[Os_3(CO)_9(SH_2)]$.[32, 51] Action of H_2S on acidic aqueous solutions frequently precipitates the metal sulfide (cf. qualitative analysis separation schemes) but, in the presence of a reducing agent such as Eu^{II}, H_2S can displace H_2O from the pale-yellow aquopentammine ruthenium(II) ion:

$$[Ru(NH_3)_5(OH_2)]^{2+} + H_2S \rightleftharpoons [Ru(NH_3)_5(SH_2)]^{2+}; \quad K_{298} = 1.5 \times 10^{-3}\, l\, mol^{-1}$$

In the absence of Eu^{II}, oxidative deprotonation of the pale-yellow H_2S complex occurs to give the orange ruthenium(III) complex $[Ru(NH_3)_5(SH)]^{2+}$. Other examples of complexes containing the SH^- ligand are $[Cr(OH_2)_5(SH)]^{2+}$, $[W(\eta^5\text{-}C_5H_5)(CO)_3(SH)]$, $[Ni(\eta^5\text{-}C_5H_5)(PBu_3^n)(SH)]$, *trans*-$[PtH(PEt_3)_2(SH)]$, and *trans*-$[Pt(PEt_3)_2(SH)_2]$.[32, 52, 53]

The S-donor ligands SO, S_2O_2, and SO_2 are mentioned in Section 15.2.5 and S–N ligands in Section 15.2.7. Thiocyanate (SCN^-) is ambidentate, but towards heavier metals it tends to be S-bonded rather than N-bonded. The bridging mode is also known (e.g. p. 344).

Organic thio ligands are well established, examples being the thioethers SMe_2, SEt_2, tetrahydrothiophene, etc., the chelating dithioethers, e.g. $MeS(CH_2)_2SMe$, and macrocyclic ligands such as $\{-(CH_2)_3S-\}_n$ with $n = 3, 4$. Thiourea, $(H_2N)_2C{=}S$, affords a further example. Factors affecting the stability of the resulting complexes have already been reviewed (p. 223). It is also notable that when $B_{10}H_{14}$ reacts with solutions of thioethers in OEt_2, tetrahydrofuran, etc., it is the thio ligand rather than the oxygen-containing species which forms the stable *arachno*-bis adducts $[B_{10}H_{12}(SR_2)_2]$ (p. 200).

Another large class of S-donor ligands comprises the dithiocarbamates $R_2NCS_2^{2-}$ and related anions YCS_2^-, e.g. dithiocarboxylates RCS_2^-, xanthates $ROCS_2^-$, thioxanthates $RSCS_2^-$, dithiocarbonate OCS_2^{2-}, trithiocarbonate SCS_2^{2-}, and dithiophosphinates $R_2PS_2^-$ (see p. 585 for applications). Dithiocarbamates can function either as unidentate or bidentate (chelating) ligands:

In the chelating mode they frequently stabilize the metal centre in an unusually high apparent formal oxidation state, e.g. $[Fe^{IV}(S_2CNR_2)_3)]^+$ and $[Ni^{IV}(S_2CNR_2)_3]^+$. They also have a propensity for stabilizing novel stereochemical configurations, unusual mixed oxidation states (e.g. of Cu), intermediate spin states (e.g. Fe^{III}, $S = \frac{3}{2}$), and for forming a variety of tris chelated complexes of Fe^{III} which lie at the $^2T_2\text{-}^6A_1$ spin crossover.[53a]

---

[51] C. G. KUEHN and H. TAUBE, Ammine ruthenium complexes of hydrogen sulfide and related ligands, *J. Am. Chem. Soc.* **98**, 689–702 (1976).

[52] T. RAMASAMI and A. G. SYKES, Further characterization and aquation of the thiolopentaaquochromium(III) complex, $CrSH^{2+}$, and its equilibrium with thiocyanate, *Inorg. Chem.* **15**, 1010–14 (1976).

[53] I. M. BLACKLAWS, E. A. V. EBSWORTH, D. W. H. RANKIN, and H. E. ROBERTSON, Hydrides of platinum(II) and platinum(IV) incorporating hydrogensulfide and hydrogenselenide ligands, *JCS Dalton* 753–8 (1978).

[53a] R. L. MARTIN, Recent studies in the synthetic and structural chemistry of the transition metals, in D. BANERJEA (ed.), *Coordination Chemistry—20*, (International Conf. Calcutta, 1979) pp. 255–65, Pergamon Press, Oxford, 1980.

Dithiocarbamates and their analogues have 2 potential S-donor atoms joined to a single C atom and their complexes are sometimes called 1,1-dithiolato complexes. If the 2 S atoms are joined to adjacent C atoms then the equally numerous class of 1,2-dithiolato complexes results. Examples of chelating dithiolene ligands (drawn for convenience with localized valence bonds and ionic charges) are:

R = alkyl, aryl, CF$_3$, H   R = Me, F, Cl, H

Complexes of these ligands have been extensively studied during the past two decades not only because of the intrinsically interesting structural and bonding problems that they pose but also because of their varied industrial applications.[54, 55] These include their use as highly specific analytical reagents, chromatographic supports, polarizers in sun-glasses, mode-locking additives in neodymium lasers, semiconductors, fungicides, pesticides, vulcanization accelerators, high-temperature wear-inhibiting additives in lubricants, polymerization and oxidation catalysts, and even finger-print developers in forensic investigations.

Complexes in which dithiolenes are the only ligands present can be classified according to six structural types as shown schematically in Fig. 15.22. For bis(dithiolato) complexes the planar structure (a) with D_{2h} local symmetry about the metal is the commonest mode but occasionally 5-coordinate dimers (b) are observed. The very rare metal–metal bonded 5-coordinate dimeric bis(dithiolato) structure (c) has been found for the palladium and platinum complexes [{M(S$_2$C$_2$H$_2$)$_2$}$_2$] with Pd–Pd 279 pm and Pt–Pt 275 pm. For tris(dithiolato) complexes two limiting geometries are possible: trigonal prismatic (Fig. 15.22d) and octahedral (Fig. 15.22f). The two geometries are related by a 30° twist of one triangular S$_3$ face with respect to the other, and intermediate twists are also known (Fig. 15.22e). As a rough generalization, the less-common trigonal prismatic geometry (local D_{3h} symmetry) is adopted by "ligand-controlled" complexes which are often neutral or highly oxidized [e.g. M(S$_2$C$_2$R$_2$)$_3$, where M = V, Cr, Mo, W, Re], whereas the more usual octahedral (D_3) geometry tends to be formed when the central metal dominates the stereochemistry as in the reduced anionic complexes. Thus reduction of the trigonal prismatic [V{S$_2$C$_2$(CN)$_2$}$_3$] to the dianion [V{S$_2$C$_2$(CN)$_2$}$_3$]$^{2-}$ results in distortion to an intermediate geometry, whereas the iron analogue [Fe{S$_2$C$_2$(CN)$_2$}$_3$]$^{2-}$ has the chelated octahedral D_3 structure. Intermediate geometries (Fig. 15.22e) have also been found for [Mo{S$_2$C$_2$(CN)$_2$}$_3$]$^{2-}$ and its W analogue.

[54] R. EISENBERG, Structural systematics of 1,1- and 1,2-dithiolato chelates, *Prog. Inorg. Chem.* **12**, 295–369 (1970).
[55] R. P. BURNS and C. A. MCAULIFFE, 1,2-Dithiolene complexes of transition metals, *Adv. Inorg. Chem. Radiochem.* **22**, 303–48 (1979); R. P. BURNS, F. P. MCCULLOUGH, and C. A. MCAULIFFE, 1,1-Dithiolato complexes of the transition elements, *Adv. Inorg. Chem. Radiochem.* **23**, 211–80 (1980). A review with 442 references.

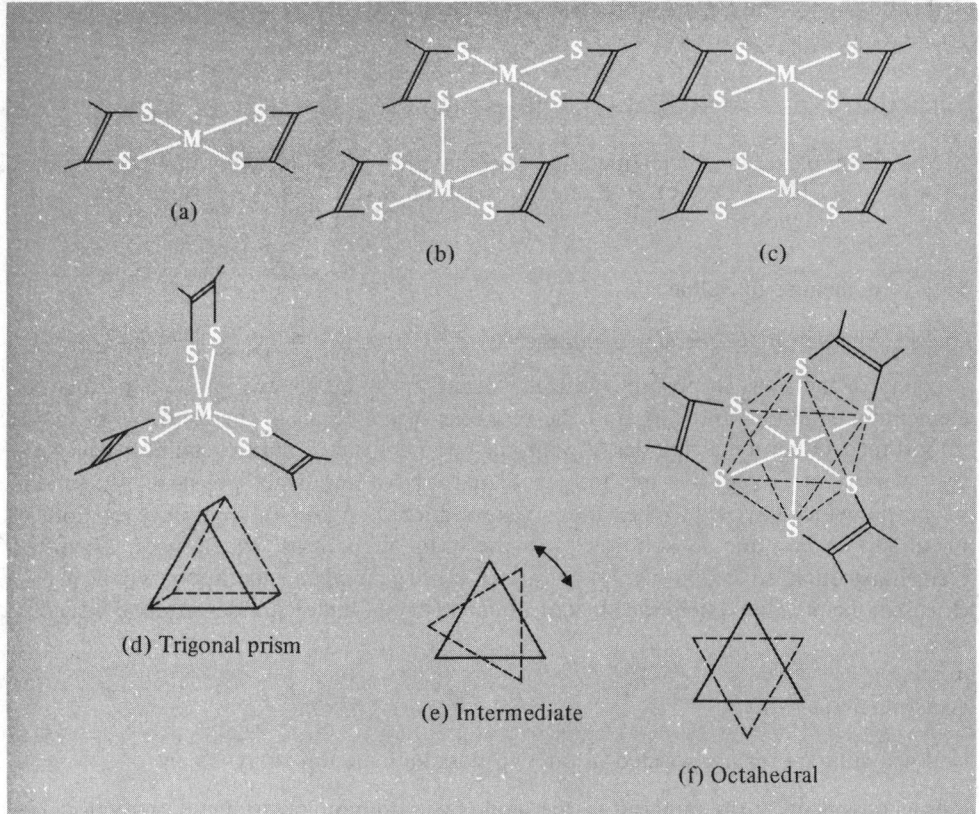

(a)

(b)   (c)

(d) Trigonal prism

(e) Intermediate

(f) Octahedral

FIG. 15.22   Coordination geometries of bis- and tris-1,2-dithiolene complexes (see text).

There has been much discussion about the detailed bonding in 1,2-dithiolene complexes because of the alternative ways that the ring system can be described, e.g.:

The formal oxidation state of the metal differs by 2 in these two limiting formulations (or by 6 in a tris complex). On this basis it is unclear whether the complex $[V\{S_2C_2(CN)_2\}_3]$ mentioned in the preceding paragraph should be formulated as V^{VI} (!) or V^0: it seems probable that an intermediate value would be more likely, but the example emphasizes the difficulty of assigning meaningful oxidation numbers to metal atoms in a redox series when the electronic configuration of the ligands themselves may also be undergoing change during reduction. Such reversible oxidation–reduction sequences are a characteristic feature of many 1,2-dithiolene complexes, e.g. for $L = \{S_2C_2(CN)_2\}$:

$$[CrL_3]^0 \underset{-e}{\overset{+e}{\rightleftharpoons}} [CrL_3]^{1-} \underset{-e}{\overset{+e}{\rightleftharpoons}} [CrL_3]^{2-} \underset{-e}{\overset{+e}{\rightleftharpoons}} [CrL_3]^{3-}$$

$$[NiL_2]^0 \underset{-e}{\overset{+e}{\rightleftharpoons}} [NiL_2]^{1-} \underset{-e}{\overset{+e}{\rightleftharpoons}} [NiL_2]^{2-} \underset{-e}{\overset{+e}{\rightleftharpoons}} [NiL_2]^{3-}$$

and similarly for the Pd, Pt, and other analogues.[56] Likewise for dimeric species with $L = \{S_2C_2(CF_3)_2\}$:

$$[\{CoL_2\}_2]^0 \underset{-e}{\overset{+e}{\rightleftharpoons}} [\{CoL_2\}_2]^{1-} \underset{-e}{\overset{+e}{\rightleftharpoons}} [\{CoL_2\}_2]^{2-} \underset{-2e}{\overset{+2e}{\rightleftharpoons}} 2[CoL_2]^{2-}$$

Mixed complexes in which a metal is coordinated by a dithiolene and by other ligands such as $(\eta^5\text{-}C_5H_5)$, CO, NO, R_3P, etc., are also known.

15.2   Compounds of Sulfur

15.2.1   *Sulfides of the metallic elements*[56a,b]

Many of the most important naturally occurring minerals and ores of the metallic elements are sulfides (p. 760), and the recovery of metals from these ores is of major importance. Other metal sulfides, though they do no occur in nature, can be synthesized by a variety of preparative methods, and many have important physical or chemical properties which have led to their industrial production. Again, the solubility relations of metal sulfides in aqueous solution form the basis of the most widely used scheme of elementary qualitative analysis. These various more general considerations will be briefly discussed before the systematic structural chemistry of metal sulfides is summarized.

General considerations

When sulfide ores are roasted in air two possible reactions may occur:

(a) conversion of the material to the oxide (as a preliminary to metal extraction, e.g. lead sulfide roasting);
(b) formation of water-soluble sulfates which can then be used in hydrometallurgical processes.

The operating conditions (temperature, oxygen pressure, etc.) required to achieve each of these results depend on the thermodynamics of the system and the duration of the roast is determined by the kinetics of the gas–solid reactions.[57] According to the Gibbs' phase rule:

$$F + P = C + 2$$

where F is the number of degrees of freedom (pressure, temperature, etc.), P is the number of phases in equilibrium, and C is the number of components (independently variable

[56] W. E. GIEGER, T. E. MINES, and F. E. SENFTLEBER, Vacuum electrochemical studies of Ni(I) products and intermediates produced in the reduction of nickel dithiolenes, *Inorg. Chem.* **14**, 2141–7 (1975); W. E. GEIGER, C. S. ALLEN, T. E. MINES, and F. C. SENFTLEBER, Paramagnetic M(I) complexes generated by the electrochemical reduction of M(II) nickel, palladium, and platinum 1,2-dithiolates, *Inorg. Chem.* **16**, 2003–8 (1977).

[56a] F. JELLINEK, Sulfides, Chap. 19 in G. NICKLESS (ed.), *Inorganic Sulfur Chemistry*, pp. 669–747, Elsevier, Amsterdam, 1968. A comprehensive review with 631 references.

[56b] D. J. VAUGHAN and J. R. CRAIG, *Mineral Chemistry of Metal Sulfides*, Cambridge University Press, Cambridge, 1978, 493 pp. A comprehensive account of the structure bonding, and properties of mineral sulfides.

[57] C. B. ALCOCK, *Principles of Pyrometallurgy*, Chap. 2, pp. 15 ff., Academic Press, London, 1967.

chemical entities) in the system. It follows that, for a 3-component system (metal–sulfur–oxygen) at a given temperature and total pressure of the gas phase, a maximum of *three* condensed phases can coexist in equilibrium. The ranges of stability of the various solid phases at a fixed temperature can be shown on a stability diagram which plots the equilibrium pressure of SO_2 against the pressure of oxygen on a log–log graph. An idealized stability diagram for a divalent metal M is shown in Fig. 15.23a, and actual stability diagrams for copper at 950 K and lead at 1175 K are in Fig. 15.23b, and c. Note that, ideally, all boundaries are straight lines: those between M/MO and MS/MSO$_4$ are vertical whereas the others have slopes of 1.0 (M/MS), 1.5 (MS/MO), and -0.5 (MO/MSO$_4$).†

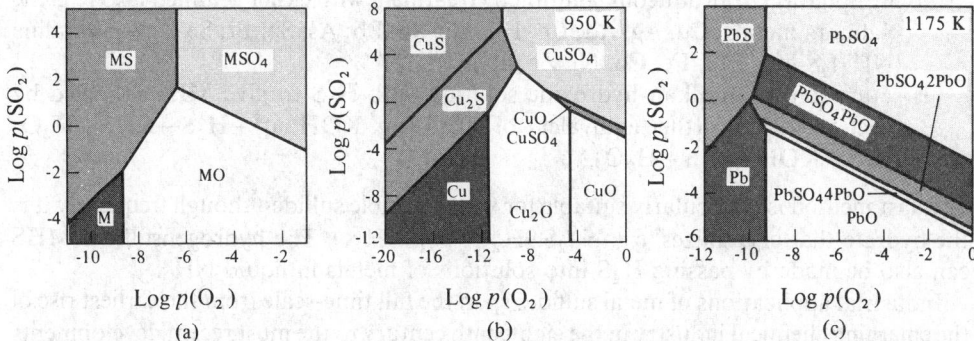

Fɪɢ. 15.23 Stability diagrams for the systems (a) metal (M)–sulfur–oxygen (idealized), (b) Cu–S–O, and (c) Pb–S–O.

The application of these generalizations to the extractive metallurgy of individual metals is illustrated at appropriate points in the text dealing with the chemistry of the various elements.

As noted above, the roasting of most metal sulfides yields either the oxide or sulfate. However, a few metals can be obtained directly by oxidation of their sulfides, and these all have the characteristic property that their oxides are much less stable than SO_2. Examples are Cu, Ag, Hg, and the platinum metals. In addition, metallic Pb can be extracted by partial oxidation of galena to form a sulfate (the "Scotch hearth" or Newnham process, p. 430). The oversimplified reaction is:

$$PbS + PbSO_4 \longrightarrow 2Pb + 2SO_2$$

---

† These simple relations can readily be deduced from the equilibria being represented. Thus at constant temperature:

M/MO boundary: $MO = M + \frac{1}{2}O_2(g)$; $K = p^{\frac{1}{2}}(O_2)$. Hence $\log p(O_2) = 2 \log K =$ constant [i.e. independent of $p(SO_2)$].

MS/MSO$_4$ boundary: $MSO_4 = MS + 2O_2(g)$; $K = p^2(O_2)$. Hence $\log p(O_2) = \frac{1}{2} \log K =$ constant.

M/MS boundary: $MS + O_2(g) = M + SO_2(g)$; $K = p(SO_2)/p(O_2)$. Hence $\log p(SO_2) = \log K + \log p(O_2)$, i.e. slope = 1.0.

MS/MO boundary: $MS + \frac{3}{2}O_2(g) = MO + SO_2(g)$; $K = p(SO_2)/p^{\frac{3}{2}}(O_2)$. Hence $\log p(SO_2) = \log K + \frac{3}{2} \log p(O_2)$, i.e. slope = 1.5.

MO/MSO$_4$ boundary: $MSO_4 = MO + SO_2(g) + \frac{1}{2}O_2(g)$: $K = p(SO_2).p^{\frac{1}{2}}(O_2)$. Hence $\log p(SO_2) = \log K - \frac{1}{2} \log p(O_2)$, i.e. slope = -0.5.

However, as indicated in Fig. 15.23c, the system is complicated by the presence of several stable "basic sulfates" $PbSO_4 . nPbO$ ($n = 1, 2, 4$), and these can react with gaseous PbS at lower metal-making temperatures,[58] e.g.:

$$PbSO_4 . 2PbO(s) + 2PbS(g) \longrightarrow 5Pb(l) + 3SO_2(g)$$

Metal sulfides can be prepared in the laboratory or on an industrial scale by a number of reactions; pure products are rarely obtained without considerable refinement and nonstoichiometric phases abound (p. 804). The more important preparative routes include:

(a) direct combination of the elements (e.g. $Fe + S \rightarrow FeS$);
(b) reduction of a sulfate with carbon (e.g. $Na_2SO_4 + 4C \rightarrow Na_2S + 4CO$);
(c) precipitation from aqueous solution by treatment with either acidified H_2S (e.g. the platinum metals; Cu, Ag, Au; Cd, Hg; Ge, Sn, Pb; As, Sb, Bi; Se, Te) or alkaline $(NH_4)_2S$ (e.g. Mn, Fe, Co, Ni, Zn; In, Tl);
(d) saturation of an alkali hydroxide solution with H_2S to give MHS followed by reaction with a further equivalent of alkali (e.g. $KOH(aq) + H_2S \rightarrow KHS + H_2O$; $KHS + KOH \rightarrow K_2S + H_2O$).

This last method is particularly suitable for water-soluble sulfides, though frequently it is the hydrate that crystallizes, e.g. $Na_2S.9H_2O$, $K_2S.5H_2O$. The hydrogensulfides MHS can also be made by passing H_2S into solutions of metals in liquid NH_3.

Industrial applications of metal sulfides span the full time-scale from the earliest rise of the emerging chemical industry in the eighteenth century to the most recent developments of Li/S and Na/S power battery systems (see Panel). Reduction of Na_2SO_4 by C was the first step in the now defunct Leblanc process (1791) for making Na_2CO_3 (p. 79). Na_2S (or NaHS) is still used extensively in the leather industry for removal of hair from hides prior to tanning, for making organo-sulfur dyes, as a reducing agent for organic nitro compounds in the production of amines, and as a flotation agent for copper ores. It is readily oxidized by atmospheric O_2 to give thiosulfate:

$$2Na_2S + 2O_2 + H_2O \longrightarrow Na_2S_2O_3 + 2NaOH$$

Industrial production in the USA exceeds 50 000 tonnes pa. Barium sulfide (from $BaSO_4 + C$) is the largest volume Ba compound manufactured but little of it is sold; almost all commercial Ba compounds are made by first making BaS and then converting it to the required compound.

Metal sulfides vary enormously in their solubility in water. As expected, the (predominantly ionic) alkali metal sulfides and alkaline earth metal sulfides are quite soluble though there is appreciable hydrolysis which results in strongly alkaline solutions ($M_2S + H_2O \rightarrow MSH + MOH$). Accordingly, solubilities depend sensitively not only on temperature but also on pH and partial pressure of H_2S. Thus, by varying the acidity, As can be separated from Pb, Pb from Zn, Zn from Ni, and Mn from Mg. In pure water the solubility of Na_2S is said to be 18.06 g per 100 g H_2O and for Ba_2S it is 7.28 g. In the case of some less-basic elements (e.g. Al_2S_3, Cr_2S_3) hydrolysis is complete and action of H_2S on solutions of the metal cation results in the precipitation of the hydroxide; likewise these sulfides (and SiS_2, etc.) react rapidly with water with evolution of H_2S.

---

[58] See ref. 57, Chap. 10, pp. 153-7.

Sodium–Sulfur Batteries

Alternatives to coal and hydrocarbon fuels as a source of power have been sought with increasing determination over the past two decades. One possibility is the Hydrogen Economy (p. 46). Another possibility, particularly for secondary, mobile sources of power, is the use of storage batteries. Indeed, electric vehicles were developed simultaneously with the first internal-combustion-engined vehicles, the first being made in 1888. In those days, nearly a century ago, electric vehicles were popular and sold well compared with the then noisy, inconvenient, and rather unreliable petrol-engined vehicles. In 1899 an electric car held the world land-speed record at 105 km per hour. In the early years of this century, taxis in New York, Boston, and Berlin were mainly electric; there were over 20 000 electric vehicles in the USA and some 10 000 cars and commercial vehicles in London. Even today battery-powered milk delivery vehicles are extensively used in the UK and there are now more than 50 000 in daily use. These use the traditional lead–sulfuric acid battery (p. 432), but this is extremely heavy and rather expensive.

The Na/S system has the potential to store 5-times as much energy (for the same weight) as the conventional lead battery and, in addition, shares with it the advantages of being silent, cheap to run, and essentially pollution-free; in general it is also reliable, has a long life, and has extremely low maintenance costs. However, until very recently it lacked the mileage range between successive chargings when compared with the highly developed petrol- or diesel-powered vehicles and it has a rather low performance (top speed and acceleration). A further disadvantage is the very long time taken to recharge the batteries (15–20 h) compared with the average time required to refill a petrol tank (1–2 min). The first work on an Na/S battery system was announced in 1966 by J. T. Kummer and N. Weber of the Ford Motor Company, USA, and active development has proceeded in several other countries since that time. Such power sources are likely to be extended from milk-floats to commuter buses and private cars for urban use in the very near future. Mixed power sources (petrol/electric battery) seem to be emerging as the favoured mode for development.

Conventional batteries consist of a liquid electrolyte separating two solid electrodes. In the Na/S battery this is inverted: a solid electrolyte separates two liquid electrodes (see figures): a

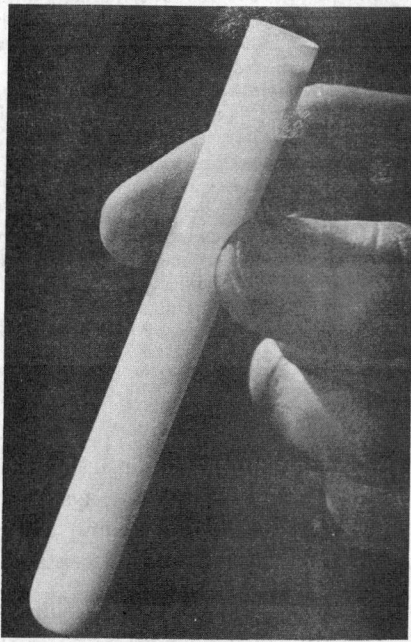

Fig. A   Beta-alumina ceramic electrolyte tube. (From the Electricity Council Research Centre, *Sodium Sulfur Batteries for Traction*, 1974.)

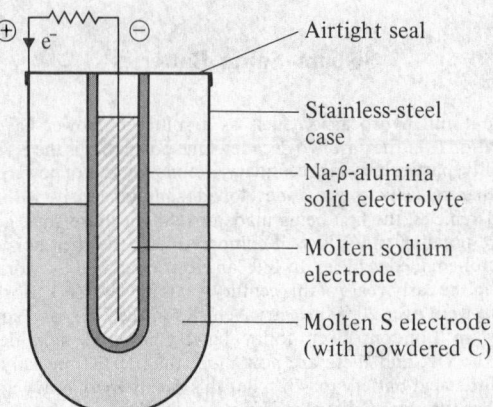

FIG. B   Schematic diagram of sodium–sulfur battery.

ceramic tube made from the solid electrolyte sodium β-alumina (p. 281) separates an inner pool of molten sodium (mp 98°) from an outer bath of molten sulfur (mp 119°) and allows Na^+ ions to pass through. The whole system is sealed and is encased in a stainless steel canister which also serves as the sulfur-electrode current collector. Within the battery, the current is passed by Na^+ ions which pass through the solid electrolyte and react with the sulfur. The cell reaction can be written formally as

$$2Na(l) + \tfrac{n}{8}S_8(l) \longrightarrow Na_2S_n(l)$$

In the central compartment molten Na gives up electrons which pass through the external circuit and reduce the molten S_8 to polysulfide ions S_n^{2-} (p. 805). The open circuit voltage is 2.08 V at 350°C. Since sulfur is an insulator the outer compartment is packed with porous carbon to provide efficient electrical conduction: the electrode volume is partially filled with sulfur when fully charged and is completely filled with sodium sulfide when fully discharged. To recharge, the polarity of the electrodes is changed and the passage of current forces the Na^+ ions back into the central compartment where they are discharged as Na atoms.

Typical dimensions for the β-alumina electrolyte tube are 380 mm long, with an outer diameter of 28 mm, and a wall thickness of 1.5 mm. A typical battery for automotive power might contain 980 of such cells (20 modules each of 49 cells) and have an open-circuit voltage of 100 V. Capacity exceeds 50 kWh. The cells operate at an optimum temperature of 300–350°C (to ensure that the sodium polysulfides remain molten and that the β-alumina solid electrolyte has an adequate Na^+ ion conductivity). This means that the cells must be thermally insulated to reduce wasteful loss of heat and to maintain the electrodes molten even when not in operation. Such a system is about one-fifth of the weight of an equivalent lead–acid traction battery and has a longer life (>1000 cycles). Further information on this and many other potential battery systems and materials is contained in the Proceedings of an International Conference on *Fast Ion Transport in Solids: Electrodes and Electrolytes*, P. Vashista, J. N. Mundy, and G. K. Shenoy (eds.), North-Holland, New York, 1979, 744 pp.

By contrast with the water-soluble sulfides of Groups IA and IIA, the corresponding heavy metal sulfides of Groups IB and IIB are amongst the least-soluble compounds known. Literature values are often wildly discordant, and care should be taken in interpreting the data. Thus, for black HgS the most acceptable value[59] of the solubility product $[Hg^{2+}][S^{2-}]$ is $10^{-51.8}$ mol^2 l^{-2}, i.e.

$$HgS(s) \rightleftharpoons Hg^{2+}(aq) + S^{2-}(aq); \quad pK = 51.8 \pm 0.5$$

[59] A. RINGBOM, Solubilities of sulfides—a preliminary report to the Analytical Division of IUPAC July 1953.

However, this should not be taken to imply a concentration of only $10^{-25.9}$ mol l^{-1} for mercury in solution (i.e. less than 10^{-2} of 1 atom of Hg per litre!) since complex formation can simultaneously occur to give species such as $[Hg(SH)_2]$ in weakly acid solutions and $[HgS_2]^{2-}$ in alkaline solutions:

$$HgS(s) + H_2S(1\ atm) \rightleftharpoons [Hg(SH)_2](aq); \quad pK = 6.2$$
$$HgS(s) + S^{2-}(aq) \rightleftharpoons [HgS_2]^{2-}(aq); \quad pK = 1.5$$

Hydrolysis also sometimes obtrudes.

Structural chemistry of metal sulfides

The predominantly ionic alkali metal sulfides M_2S (Li, Na, K, Rb, Cs) adopt the antifluorite structure (p. 129) in which each S atom is surrounded by a cube of 8 M and each M by a tetrahedron of S. The alkaline earth sulfides MS (Mg, Ca, Sr, Ba) adopt the NaCl-type 6:6 structure (p. 273) as do many other monosulfides of rather less basic metals (M = Pb, Mn, La, Ce, Pr, Nd, Sm, Eu, Tb, Ho, Th, U, Pu). However, many metals in the later transition element groups show substantial trends to increasing covalency leading either to lower coordination numbers or to layer-lattice structures.[60] Thus MS (Be, Zn, Cd, Hg) adopt the 4:4 zinc blende structure (p. 1405) and ZnS, CdS, and MnS also crystallize in the 4:4 wurtzite modification (p. 1405). In both of these structures both M and S are tetrahedrally coordinated, whereas PtS, which also has 4:4 coordination, features a square-planar array of 4 S atoms about each Pt, thus emphasizing its covalent rather than ionic bonding. Group IIIB sulfides M_2S_3 (p. 285) have defect ZnS structures with various patterns of vacant lattice sites.

The final major structure type found amongst monosulfides is the NiAs (nickel arsenide) structure (Fig. 15.24a). Each S atom is surrounded by a trigonal prism of 6 M

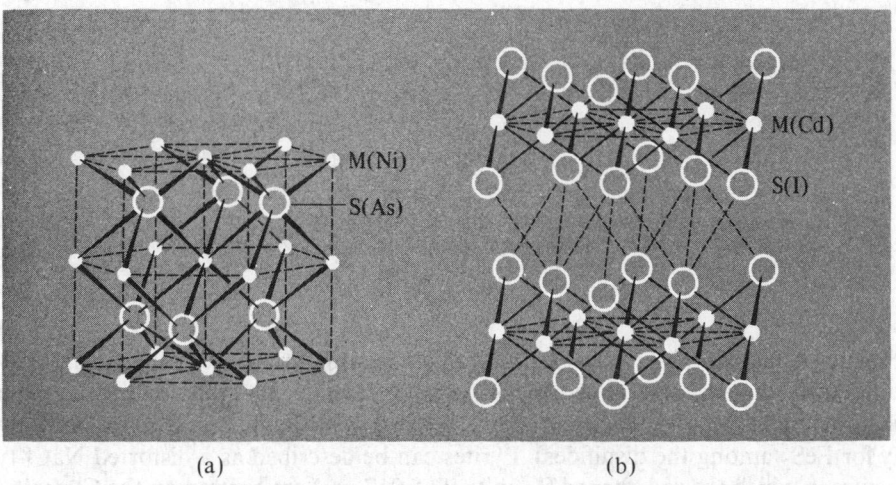

(a)　　　　　　　　　　　　　　　　　　(b)

FIG. 15.24　Comparison of the nickel arsenide structure (a) adopted by many monosulfides MS with the cadmium iodide structure (b) adopted by some disulfides MS_2. The structures are related simply by removing alternate layers of M from MS to give MS_2.

[60] N. N. GREENWOOD, *Ionic Crystals, Lattice Defects, and Nonstoichiometry*, Chap. 3, pp. 37–61; also pp. 153–55, Butterworths, London, 1968.

atoms whilst each M has eightfold coordination, being surrounded octahedrally by 6 S atoms and by 2 additional M atoms which are coplanar with 4 of the S atoms. A significant feature of the structure is the close approach of the M atoms in chains along the (vertical) c-axis (e.g. 260 pm in FeS) and the structure can be regarded as transitional between the 6:6 NaCl structure and the more highly coordinated structures typical of metals. The NiAs structure is adopted by most first row transition-metal monosulfides MS (M = Ti, V, Cr, Fe, Co, Ni) as well as by many selenides and tellurides of these elements.

The NiAs structure is closely related to the hexagonal layer-lattice CdI_2 structure shown in Fig. 15.24b, this stoichiometry being achieved simply by leaving alternate M layers of the NiAs structure vacant. Disulfides MS_2 adopting this structure include those of Ti, Zr, Hf, Ta, Pt, and Sn; conversely, Tl_2S has the anti-CdI_2 structure. Progressive partial filling of the alternate metal layers leads to phases of intermediate composition as exemplified by the Cr/S system (Table 15.11). For some elements these intermediate phases have quite extensive ranges of composition, the limits depending on the temperature of the system. For example, at 1000°C there is a succession of non-stoichiometric titanium sulfides $TiS_{0.97}$–$TiS_{1.06}$, $TiS_{1.204}$–$TiS_{1.333}$, $TiS_{1.377}$–$TiS_{1.594}$, $TiS_{1.810}$–$TiS_{1.919}$.[60] Many diselenides and ditellurides also adopt the CdI_2 structure and in some there is an almost continuous nonstoichiometric variation in composition, e.g. CoTe→$CoTe_2$. A related 6:3 layer structure is the $CdCl_2$-type adopted by TaS_2, and the layer structures of MoS_2 and WS_2 are mentioned on p. 1186.

TABLE 15.11 *Some sulfides of chromium (see text)*

Nominal formula	Ratio Cr/S calculated	Ratio Cr/S observed	Proportion of sites occupied in alternate layers	Random or ordered vacancies[a]
CrS[b]	1.000	≈0.97	1:1	None
Cr_7S_8	0.875	0.88–0.87	$1:\frac{3}{4}$	Random
Cr_5S_6	0.833	0.85	$1:\frac{2}{3}$	Ordered
Cr_3S_4	0.750	0.79–0.76	$1:\frac{1}{2}$	Ordered
Cr_2S_3	0.667	0.69–0.67	$1:\frac{1}{3}$	Ordered
(CrS_2)	0.500	Not observed	1:0	—

[a] Refers to the vacancies in the alternate metal layers.
[b] CrS has a unique monoclinic structure intermediate between NiAs and PtS types.

Finally, many disulfides have a quite different structure motif, being composed of infinite three-dimensional networks of M and discrete S_2 units. The predominate structural types are pyrites, FeS_2 (also for M = Mn, Co, Ni, Ru, Os), and marcasite (known only for FeS_2 among the disulfides). Pyrites can be described as a distorted NaCl-type structure in which the rod-shaped S_2 units (S–S 217 pm) are centred on the Cl positions but are oriented so that they are inclined away from the cubic axes. The marcasite structure is a variant of the rutile structure (TiO_2, p. 1120) in which the columns of edge-shared octahedra are rotated to give close approaches between pairs of S atoms in adjacent columns (S–S 221 pm).

Many metal sulfides have important physical properties.[56a, 61] They range from insulators, through semiconductors to metallic conductors of electricity, and some are even superconductors, e.g. NbS_2 (<6.2 K), TaS_2 (<2.1 K), $Rh_{17}S_{15}$ (<5.8 K), CuS (<1.62 K), and CuS_2 (<1.56 K). Likewise they can be diamagnetic, paramagnetic, temperature-independent paramagnetic, ferromagnetic, antiferromagnetic, or ferrimagnetic.

The structure of more complex ternary metal sulfides such as $BaZrS_3$ (perovskite-type, p. 1122), $ZnAl_2S_4$ (spinel type, p. 279), and $NaCrS_2$ (NaCl superstructure) introduce no new principles. Likewise, thiosalts, which may feature finite anions (e.g. $Tl_3[VS_4]$), vertex-shared chains (e.g. Ba_2MnS_3), edge-shared chains (e.g. $KFeS_2$), double chains (e.g. Ba_2ZnS_3), double layers (e.g. KCu_4S_3), or three-dimensional frameworks (e.g. $NH_4Cu_7S_4$).[62]

Anionic polysulfides

The pyrites and marcasite structures can be thought of as containing $S_2{}^{2-}$ units though the variability of the interatomic distance and other properties suggest substantial deviation from a purely ionic description. Numerous higher polysulfides $S_n{}^{2-}$ have been characterized, particularly for the more electropositive elements Na, K, Ba, etc. They are yellow at room temperature, turn dark red on being heated, and may be thought of as salts of the polysulfanes (p. 807). Typical examples are M_2S_n ($n = 2-5$ for Na, 2–6 for K, 6 for Cs), BaS_2, BaS_3, BaS_4, etc. The polysulfides, unlike the monosulfides, are low melting solids: published values for mps vary somewhat but representative values (°C) are:

Na_2S	Na_2S_2	Na_2S_4	Na_2S_5	K_2S_3	K_2S_4	K_2S_5	K_2S_6	BaS_3
1180°	484°	294°	255°	292°	~145°	211°	196°	554°

Structures are in Fig. 15.25.[63-67] The $S_3{}^{2-}$ ion is bent (C_{2v}) and is isoelectronic with SCl_2 (p. 815). The $S_4{}^{2-}$ ion has twofold symmetry, essentially tetrahedral bond angles, and a dihedral angle of 97.8° (see p. 771). The $S_5{}^{2-}$ ion also has approximately twofold symmetry (about the central S atom); it is a contorted but unbranched chain with bond angles close to tetrahedral and a small but significant difference between the terminal and internal S–S distances. The $S_6{}^{2-}$ ion has alternating S–S distances, and bond angles in the range 106.4–110.0° (mean 108.8°). Several of the references in Fig. 15.25 give preparative details: these can involve direct reaction of stoichiometric amounts of the elements in

[61] F. HULLIGER, Crystal chemistry of chalcogenides and pnictides of the transition elements, *Struct. Bonding* (Berlin) **4**, 83–229 (1968). A comprehensive review with 532 references, 65 structural diagrams, and a 34-page appendix tabulating the known phases and their physical properties.

[62] A. F. WELLS, *Structural Inorganic Chemistry*, 4th edn., Chap. 17, pp. 605–35, Oxford University Press, 1975.

[63] H. FOPPL, E. BUSMANN, and F.-K. FRORATH, The crystal structures of α-Na_2S_2 and K_2S_2, β-Na_2S_2 and Na_2Se_2, *Z. anorg. allgem. Chem.* **314**, 12–30 (1962). An early somewhat imprecise structure determination.

[64] H. G. VON SCHNERING and N.-K. GOH, The structure of the polysulfides BaS_3, SrS_3, BaS_2, and SrS_2, *Naturwissenschaften* **61**, 272 (1974).

[65] R. TEGMAN, The crystal structure of sodium tetrasulfide, *Acta Cryst.* **B29**, 1463–9 (1973).

[66] B. KELLY and P. WOODWARD, Crystal structure of dipotassium pentasulfide, *JCS Dalton*, 1976, 1314–16.

[67] S. C. ABRAHAMS and E. GRISON, The crystal structure of caesium hexasulfide, *Acta Cryst.* **6**, 206–13 (1953).

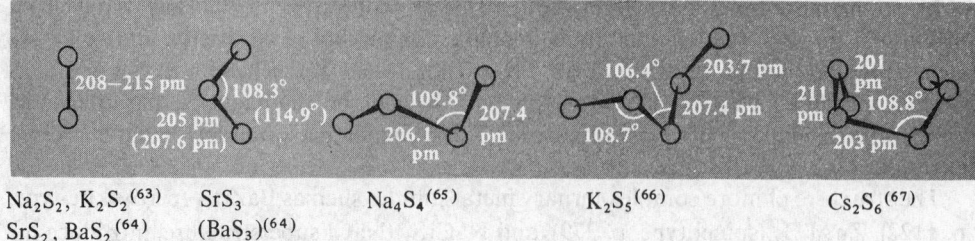

Na$_2$S$_2$, K$_2$S$_2$[63]   SrS$_3$ Na$_4$S$_4$[65]   K$_2$S$_5$[66]   Cs$_2$S$_6$[67]
SrS$_2$, BaS$_2$[64]   (BaS$_3$)[64]

FIG. 15.25 Structures of polysulfide anions S$_n^{2-}$ in M$_2^I$S$_n$ and BaS$_n$.

sealed tubes or reaction of MSH with S in ethanol.[68] It is interesting that, despite the
unequivocal presence of the S$_3^{2-}$ ion in K$_2$S$_3$, BaS$_3$, etc., a Raman spectroscopic study of
molten "Na$_2$S$_3$" showed that the ion had disproportionated into S$_2^{2-}$ and S$_4^{2-}$.[69]

15.2.2 *Hydrides of sulfur (sulfanes)*

Hydrogen sulfide is the only thermodynamically stable sulfane; it occurs widely in
nature as a result of volcanic or bacterial action and is, indeed, a prime source of elemental
S (p. 759). It has been known since earliest times and its classical chemistry has been
extensively studied since the seventeenth century.[70] H$_2$S is a foul smelling, very
poisonous gas familiar to all students of chemistry. Its smell is noticeable at 0.02 ppm but
the gas tends to anaesthetize the olefactory senses and the intensity of the smell is therefore
a dangerously unreliable guide to its concentration. H$_2$S causes irritation at 5 ppm,
headaches and nausea at 10 ppm, and immediate paralysis and death at 100 ppm; it is
therefore as toxic and as dangerous as HCN.

H$_2$S is readily prepared in the laboratory by treating FeS with dilute HCl in a Kipp
apparatus. Purer samples can be made by hydrolysing CaS, BaS, or Al$_2$S$_3$, and the purest
gas is prepared by direct reaction of the elements at 600°C. Some physical properties are in
Table 15.12:[71] comparison with the properties of water (p. 730) shows the absence of any
appreciable H bonding in H$_2$S. Comparisons with H$_2$Se, H$_2$Te, and H$_2$Po are on p. 900.

TABLE 15.12 *Some molecular and physical properties of H$_2$S*

Distance (S–H)/pm	133.6(g)	ΔH_f°/kJ mol^{-1}	20.1(g)
Angle H–S–H	92.1°(g)	Density (s)/g cm^{-3}	1.12 (−85.6°)
MP/°C	−85.6	Density (l)/g cm^{-3}	0.993 (−85.6°)
BP/°C	−60.3	Viscosity/centipoise	0.547 (−82°)
Critical temperature/°C	100.4	Dielectric constant ε	8.99 (−78°)
Critical pressure/atm	84	Electrical conductivity/ohm^{-1} cm^{-1}	3.7 × 10^{-11} (−78°)

[68] G. WEDDIDEN, H. KLEINSCHMAGER, and S. HOPPE, Synthesis of sodium polysulfides, *J. Chem. Res. (S)*,
1978, 96; *(M)*, 1978, 1101–12.

[69] G. J. JANZ *et al.*, Raman studies of sulfur-containing anions in inorganic polysulfides, *Inorg. Chem.* **15**,
1751–4, 1755–9, 1759–63 (1976).

[70] J. W. MELLOR, *A Comprehensive Treatise on Inorganic and Theoretical Chemistry*, Vol. 10, Hydrogen
sulfide, and The polysulfides of hydrogen, pp. 114–61, Longmans, London, 1930.

[71] F. FEHÉR, Liquid hydrogen sulfide, Chap. 4 in J. J. LAGOWSKI (ed.), *The Chemistry of Nonaqueous Solvents*,
Vol. 3, pp. 219–40, Academic Press, New York, 1970.

H_2S is readily soluble in both acidic and alkaline aqueous solutions. Pure water dissolves 4.65 volumes of the gas at $0°$ and 2.61 volumes at $20°$; in other units a saturated solution is 0.1 M at atmospheric pressure and $25°$, i.e.

$$H_2S(g) \rightleftharpoons H_2S(aq); \quad K = 0.1023 \text{ mol } l^{-1} \text{ atm}^{-1}; \quad pK = 0.99$$

In aqueous solution H_2S is a weak acid (p. 52). At $20°$:[72]

$$H_2S(aq) \rightleftharpoons H^+(aq) + SH^-(aq); \quad pK_{a_1} = 6.88 \pm 0.02$$
$$SH^-(aq) \rightleftharpoons H^+(aq) + S^{2-}(aq); \quad pK_{a_2} = 14.15 \pm 0.05$$

The chemistry of such solutions has been alluded to on p. 800. At low temperatures a hydrate $H_2S \cdot 5\frac{3}{4}H_2O$ crystallizes. In acid solution H_2S is also a mild reducing agent; e.g. even on standing in air solutions slowly precipitate sulfur. The gas burns with a bluish flame in air to give H_2O and SO_2 (or H_2O and S if the air supply is restricted). For adducts, see p. 794.

In very strongly acidic nonaqueous solutions (such as HF/SbF_5) H_2S acts as a base (proton acceptor) and the white crystalline solid $[SH_3]^+[SbF_6]^-$ has been isolated from such solutions.[73] The compound, which is the first known example of a stable salt of SH_3^+, can be stored at room temperature in Teflon or Kel-F containers but attacks quartz. Vibrational spectroscopy confirms the pyramidal C_{3v} structure expected for a species isoelectronic with PH_3 (p. 564).

Polysulfanes, H_2S_n, with $n = 2$–8 have been prepared and isolated pure, and many higher homologues have been obtained as mixtures with variable n. Our modern knowledge of these numerous compounds stems mainly from the elegant work of F. Fehér and his group in the 1950s. All polysulfanes have unbranched chains of n sulfur atoms thus reflecting the well-established propensity of this element towards catenation (p. 769). The polysulfanes are reactive liquids whose density d, viscosity, η, and bp increase with increasing chain length. H_2S_2, the analogue of H_2O_2, is colourless but the others are yellow, the colour deepening with increasing chain length.

The polysulfanes were at one time made by fusing crude $Na_2S \cdot 9H_2O$ with various amounts of sulfur and pouring the resulting polysulfide solution into an excess of dilute hydrochloric acid at $-10°C$. The resulting crude yellow oil is a mixture mainly of H_2S_n ($n = 4$–7). Polysulfanes can now also be readily prepared by a variety of other reactions, e.g.:

$$Na_2S_n(aq) + 2HCl(aq) \longrightarrow 2NaCl(aq) + H_2S_n \ (n = 4\text{–}6)$$
$$S_nCl_2(l) + 2H_2S(l) \longrightarrow 2HCl(g) + H_2S_{n+2}(l)$$
$$S_nCl_2(l) + 2H_2S_m(l) \longrightarrow 2HCl(g) + H_2S_{n+2m}(H_2S_6\text{–}H_2S_{18})$$

Purification is by low-pressure distillation. Some physical properties are in Table 15.13. Polysulfanes are readily oxidized and all are thermodynamically unstable with respect to disproportionation:

$$H_2S_n(l) \longrightarrow H_2S(g) + \tfrac{n-1}{8} S_8(s)$$

[72] M. WIDMER and G. SCHWARZENBACH, The acidity of the hydrogensulfide ion HS^-, *Helv. Chim. Acta* **47**, 266–71 (1964).

[73] K. O. CHRISTE, Novel onium salts. Synthesis and characterization of $SH_3^+ SbF_6^-$, *Inorg. Chem.* **14**, 2230–3 (1975).

TABLE 15.13   *Some physical properties of polysulfanes*[74]

Compound	d_{20}/g cm^{-3}	P_{20}/mmHg	BP/°C (extrap)
H_2S_2	1.334	87.7	70
H_2S_3	1.491	1.4	170
H_2S_4	1.582	0.035	240
H_2S_5	1.644	0.0012	285
H_2S_6	1.688	—	—
H_2S_7	1.721	—	—
H_2S_8	1.747	—	—

This disproportionation is catalysed by alkali, and even traces dissolved from the surface of glass containers is sufficient to effect deposition of sulfur. They are also degraded by sulfite and by cyanide ions:

$$H_2S_n + (n-1)SO_3{}^{2-} \longrightarrow H_2S + (n-1)S_2O_3{}^{2-}$$

$$H_2S_n + (n-1)CN^- \longrightarrow H_2S + (n-1)SCN^-$$

The former reaction, in particular, affords a convenient means of quantitative analysis by determination of the H_2S (precipitated as CdS) and iodometric determination of the thiosulfate produced.

15.2.3   *Halides of sulfur*

Sulfur fluorides

The seven known sulfur fluorides are quite different from the other halides of sulfur in their stability, reactivity, and to some extent even in their stoichiometries; it is therefore convenient to consider them separately. Moreover, they have proved a rich field for both structural and theoretical studies since they form an unusually extensive and graded series of covalent molecular compounds in which S has the oxidation states 1, 2, 3, 4, 5, and 6, and in which it also exhibits all coordination numbers from 1 to 6 (if $SF_5{}^-$ is also included). The compounds feature a rare example of structural isomerism amongst simple molecular inorganic compounds (FSSF and SSF_2) and also a monomer–dimer pair (SF_2 and F_3SSF). The structures and physical properties will be described first, before discussing the preparative routes and chemical reactions.

Structures and physical properties. The molecular structure, point group symmetries, and dimensions of the sulfur fluorides are summarized in Fig. 15.26.[75] S_2F_2 resembles H_2O_2, H_2S_2, O_2F_2, and S_2X_2, and detailed comparisons of bond distances, bond angles, and dihedral angles are instructive. The isomer SSF_2 (thiothionylfluoride) features 3-coordinate S^{IV} and 1-coordinate S^{II} and it is notable that the formally double-bonded S–S distance is very close to that in the singly bonded isomer. The fugitive species SF_2 has the expected bent configuration in the gas phase but is unique in readily undergoing dimerization by insertion of a second SF_2 into an S–F bond. The structure of the resulting

[74] M. SCHMIDT and W. SIEBERT, Sulfanes, Section 2.1 in *Comprehensive Inorganic Chemistry*, Vol. 2, Chap. 23, pp. 826–42, Pergamon Press, Oxford, 1973.
[75] F. SEEL, Lower sulfur fluorides, *Adv. Inorg. Chem. Radiochem.* **16**, 297–333 (1974).

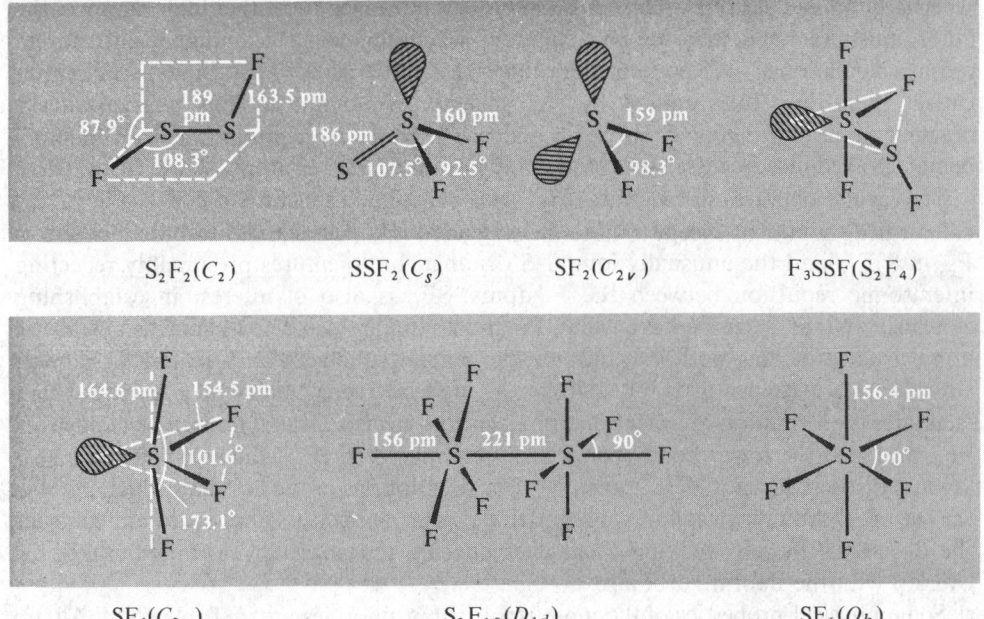

Fig. 15.26   Molecular structures of the sulfur fluorides.

molecule F_3SSF is, in a sense, intermediate between those of S_2F_2 and SF_4, being based on a trigonal bipyramid with the equatorial F atom replaced by an SF group. The fact that the ^{19}F nmr spectrum at $-100°$ shows four distinct F resonances indicates that the 2 axial F atoms are non-equivalent, implying restricted rotation about the S–S bond.

The structure of SF_4 is particularly significant. It is based on a trigonal bipyramid with one equatorial position occupied by the lone-pair; this distorts the structure by reducing the equatorial F–S–F bond angle from $120°$ to $101.6°$ and by repelling the axial F_{ax} atoms towards F_{eq}. There is also a significant difference between the (long) S–F_{ax} and (short) S–F_{eq} distances. Again, the low-temperature ^{19}F nmr spectrum is precisely diagnostic of the C_{2v} structure, since the observed doublet of 1:2:1 triplets is consistent only with the two sets of 2 equivalent F atoms in this point group symmetry (^{19}F, like ^1H, has nuclear spin $\frac{1}{2}$).[76] Thus, an axial lone-pair (C_{3v}) would lead to a doublet and a quartet of integrated relative intensity 3:1, whereas all other conceivable symmetries (T_d, C_{4v}, D_{4h}, D_{2d}, D_{2h}) would give a sharp singlet from the 4 equivalent F atoms. Above $-98°$ the 30 MHz ^{19}F nmr spectrum of SF_4 gradually broadens and it coalesces at $-47°$ into a single broad resonance which gradually sharpens again to a narrow singlet at higher temperatures; this is due to molecular fluxionality which permits intramolecular interchange of the axial and equatorial F atoms.

The structure of SF_4 can be rationalized on most of the simple bonding theories; the environment of S has 10 valency electrons and this leads to the observed structure in both

[76] F. A. COTTON, J. W. GEORGE, and J. S. WAUGH, Nuclear resonance spectrum and structure of SF_4, *J. Chem. Phys.* **28**, 944–95 (1958); E. MUETTERTIES and W. D. PHILLIPS, Structure and exchange processes in some inorganic fluorides by nuclear magnetic resonance, *J. Am. Chem. Soc.* **81**, 1084–8 (1959).

valence-bond and electron-pair repulsion models. However, the rather high energy of the 3d orbitals on S make their full participation in bonding via $sp^3d_{z^2}$ unlikely and, indeed, recent calculations[77] show that there may be as little as 12% d-orbital participation rather than the 50% implied by the scheme $sp_xp_y + p_zd_{z^2}$. Thus charge-transfer configurations or bonding via $sp_xp_y + p_z$ seem to be better descriptions, the p_z orbital on S being involved in a 3-centre 4-electron bond with the 2 axial F atoms (cf. XeF_2, p. 1053).

The regular octahedral structure of SF_6 and the related structure of S_2F_{10} (Fig. 15.26) call for little comment except to note the staggered (D_{4d}) arrangement of the two sets of F_{eq} in S_2F_{10} and the unusually long S–S distance, both features presumably reflecting interatomic repulsion between the F atoms. SF_6 is also of interest in establishing conclusively that S can be hexavalent. Its great stability (see below) contrasts with the non-existence of SH_4 and SH_6 despite the general similarity in S–F and S–H bond strengths; its existence probably reflects the high electronegativity of F (p. 30), which facilitates the formation of either polar or 3-centre 4-electron bonds as discussed above for SF_4, and also the lower bond energy of F_2 compared to H_2, which for SH_4 and SH_6 favours dissociation into $H_2S + nH_2$.[77a] For descriptions of the bonding which involve the use of 3d orbitals on sulfur, a net positive charge on the central atom would contract the d orbitals thereby making them energetically and spatially more favourable for overlap with the fluorine orbitals.

Some physical properties of the more stable sulfur fluorides are in Table 15.14. All are colourless gases or volatile liquids at room temperature. SF_6 sublimes at $-63.8°$ (1 atm) and can only be melted under pressure ($-50.8°$). It is notable both for its extreme thermal and chemical stability (see below), and also for having a higher gas density than any other substance that boils below room temperature (5.107 times as dense as air).

Synthesis and chemical reactions. Disulfur difluoride, S_2F_2, can be prepared by the mild fluorination of sulfur with AgF in a rigorously dried apparatus at 125°. It is best handled in the gas phase at low pressures and readily isomerizes to thiothionylfluoride, SSF_2, in the presence of alkali metal fluorides, SSF_2 can be made either by isomerizing S_2F_2 or directly by the fluorination of S_2Cl_2 using KF in SO_2:

$$2KSO_2F + S_2Cl_2 \longrightarrow SSF_2 + 2KCl + 2SO_2$$

SSF_2 can be heated to 250° but is, in fact, thermodynamically unstable with respect to disproportionation, being immediately transformed to SF_4 in the presence of acid catalysts such as BF_3 or HF:

$$2SSF_2 \longrightarrow \tfrac{3}{8}S_8 + SF_4$$

TABLE 15.14 *Physical properties of some sulfur fluorides*

	FSSF	S=SF$_2$	SF$_4$	SF$_6$	S$_2$F$_{10}$
MP/°C	−133	−164.6	−121	−50.5	−52.7
BP/°C	+15	−10.6	−38	−63.8 (subl)	+30
Density (T°C)/g cm^{-3}	—	—	1.919 (−73°)	1.88 (−50°)	2.08 (0°)

[77] P. J. Hay, Generalized valence bond studies of the electronic structure of SF_2, SF_4, and SF_6, *J. Am. Chem. Soc.* **99**, 1003–12 (1977).

[77a] G. M. Schwenzer and H. F. Schaeffer, The hypervalent molecules sulfurane (SH_4) and persulfurane (SH_6), *J. Am. Chem. Soc.* **97**, 1393–7 (1975).

Both S_2F_2 and SSF_2 are rapidly hydrolysed by pure water to give S_8, HF, and a mixture of polythionic acids $H_2S_nO_6$ (n 4–6), e.g.:

$$5S_2F_2 + 6H_2O \longrightarrow \tfrac{3}{4}S_8 + 10HF + H_2S_4O_6$$

Alkaline hydrolysis yields predominantly thiosulfate. SSF_2 burns with a pale-blue flame when ignited, to yield SO_2, SOF_2, and SO_2F_2.

Sulfur difluoride, SF_2, is a surprisingly fugitive species in view of its stoichiometric similarity to the stable compounds H_2S and SCl_2 (p. 815). It is best made by fluorinating gaseous SCl_2 with activated KF (from KSO_2F) or with HgF_2 at 150°, followed by a tedious fractionation from the other sulfur fluorides ($FSSF$, SSF_2, and SF_4) which form the predominant products. The chlorofluorides ClSSF and $ClSSF_3$ are also formed. The compound can only be handled as a dilute gas under rigorously anhydrous conditions or at very low temperatures in a matrix of solid argon, and it rapidly dimerizes to give F_3SSF.

Sulfur tetrafluoride, SF_4, though extremely reactive (and valuable) as a selective fluorinating agent, is much more stable than the lower fluorides. It is formed, together with SF_6, when a cooled film of sulfur is reacted with F_2, but is best prepared by fluorinating SCl_2 with NaF in warm acetonitrile solution:

$$3SCl_2 + 4NaF \xrightarrow[75°]{\text{MeCN}} S_2Cl_2 + SF_4 + 4NaCl$$

SF_4 is unusual in apparently acting both as an electron-pair acceptor and an electron-pair donor (amphoteric Lewis acid–base). Thus pyridine forms a stable 1:1 adduct $C_5H_5NSF_4$ which presumably has a pseudooctahedral (square-pyramidal) geometry. Likewise CsF (at 125°) and Me_4NF (at −20°) form $CsSF_5$ and $[NMe_4]^+[SF_5]^-$ (Fig. 15.27a). By contrast, SF_4 behaves as a donor to form 1:1 adducts with many Lewis acids;

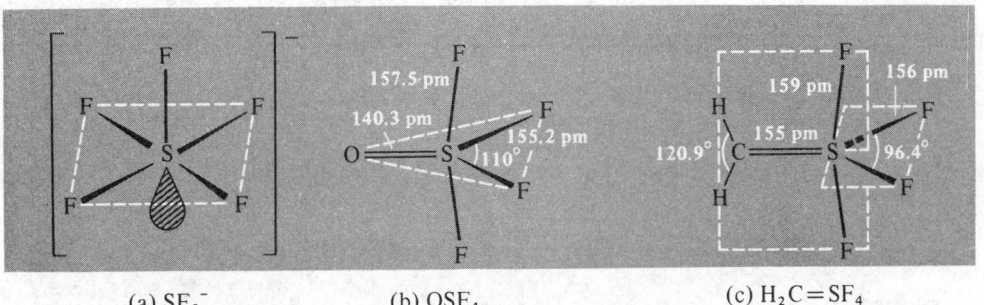

(a) SF_5^- (b) OSF_4 (c) $H_2C{=}SF_4$

FIG. 15.27   Comparison of the structures of three species in which S has 12 valence electrons: (a) the SF_5^- ion in $CsSF_5$, as deduced from vibrational spectroscopy,[78] (b) OSF_4 as deduced from gas-phase electron diffraction[79] (note the wider angle $F_{eq}SF_{eq}$ when compared with SF_4 (Fig. 15.26) and the shorter distance S–F_{ax}; the angle $F_{ax}SF_{ax}$ is 178.6±0.8°), and (c) H_2CSF_4 (X-ray crystal structure at −160°).[80] The angle $F_{eq}SF_{eq}$ is significantly smaller than in SF_4 and OSF_4 as is the angle $F_{ax}SF_{ax}$ (170.4°); the methylene group is coplanar with the axial SF_2 group as expected for p_π–d_π C=S overlap and, unlike SF_4, the molecule is non-fluxional.

[78] K. O. CHRISTE, E. C. CURTIS, C. J. SCHACK, and D. PILIPOVICH, Vibrational spectra and force constants of the square-pyramidal anions SF_5^-, SeF_5^-, and TeF_5^{-1}, *Inorg. Chem.* **11**, 1679–82 (1972).

[79] G. GUNDERSEN and K. HEDBERG, Molecular structure of thionyltetrafluoride, SOF_4, *J. Chem. Phys.* **51**, 2500–7 (1969).

[80] H. BOCK, J. E. BOGGS, G. KLEEMANN, D. LENTZ, H. OBERHAMMER, E. M. PETERS, K. SEPELT, A. SIMON, and B. SOLOUKI, Structure and reactions of methylenesulfur tetrafluoride, *Angew. Chem.*, Int. Edn. (Engl.) **18**, 944–5 (1979).

the stability decreases in the sequence $SbF_5 > AsF_5 > IrF_5 > BF_3 > PF_5 > AsF_3$. In view of the discussion on p. 223 it seems likely that SF_4 is acting here not as an S lone-pair donor but as a fluoride ion lone-pair donor and there is, indeed, infrared evidence to suggest that $SF_4 . BF_3$ is predominantly $[SF_3]^+[BF_4]^-$.

SF_4 rapidly decomposes in the presence of moisture, being instantly hydrolysed to HF and SO_2. Despite this it has been increasingly used as a powerful and highly selective fluorinating agent for both inorganic and organic compounds. In particular it is useful for converting ketonic and aldehyde $=C=O$ groups to $=CF_2$, and carboxylic acid groups $-COOH$ to $-CF_3$. Similarly, $\equiv P=O$ groups are smoothly converted to $\equiv PF_2$, and $=P(O)OH$ groups to $=PF_3$. It also undergoes numerous oxidative addition reactions to give derivatives of S^{VI}. The simplest of these are direct oxidation of SF_4 with F_2 or ClF (at 380°) to give SF_6 and $SClF_5$ respectively. Analogous reactions with $N_2F_4(h\nu)$ and F_5SOOSF_5 yield SF_5NF_2 and cis-$SF_4(OSF_5)_2$ respectively; likewise F_5SOF (p. 814) yields F_5SOSF_5. Direct oxidation of SF_4 with O_2, however, proceeds only slowly unless catalysed by NO_2: the product is OSF_4, which has a trigonal bipyramidal structure like SF_4 itself, but with the equatorial lone-pair replaced by the oxygen atom (Fig. 15.27b). A similar structure is adopted by the recently prepared methylene compound $H_2C=SF_4$ (Fig. 15.27c);[80] this is made by treating $SF_3–CH_2Br$ with $LiBu^n$ at $-110°$ and is more stable than the isoelectronic P or S ylides or metal carbene complexes, being stable in the gas phase up to 650° at low pressures.

Some other reactions of SF_4 are:

$$Cl_2 + CsF + SF_4 \xrightarrow{110°} SClF_5 + CsCl$$

$$I_2O_5 + 5SF_4 \longrightarrow 2IF_5 + 5OSF_2$$

$$4BCl_3 + 3SF_4 \longrightarrow 4BF_3 + 3SCl_2 + 3Cl_2$$

$$RCN + SF_4 \longrightarrow RCF_2N=SF_2$$

$$NaOCN + SF_4 \longrightarrow CF_3N=SF_2 + \ldots$$

$$CF_3CF=CF_2 + SF_4 \xrightarrow{CsF/150°} (CF_3)_2CFSF_3 \text{ (bp 46°)}$$

$$2CF_3CF=CF_2 + SF_4 \xrightarrow{CsF/150°} \{(CF_3)_2CF\}_2SF_2 \text{ (bp } \sim 111°)$$

Disulfur decafluoride, S_2F_{10}, is obtained as a byproduct of the direct fluorination of sulfur to SF_6 but is somewhat tedious to separate and is more conveniently made by the photolytic reduction of $SClF_5$ (prepared as above):

$$2SClF_5 + H_2 \xrightarrow{h\nu} S_2F_{10} + 2HCl$$

It is intermediate in reactivity between SF_4 and the very inert SF_6. Unlike SF_4 it is not hydrolysed by water or even by dilute acids alkalis and, unlike SF_6, it is extremely toxic. It disproportionates readily at 150° probably by a free radical mechanism involving $SF_5^•$ (note the long, weak S–S bond; Fig. 15.26):

$$2S_2F_{10} \xrightarrow{150°} SF_4 + SF_6$$

Similarly it reacts readily with Cl_2 and Br_2 to give $SClF_5$ and $SBrF_5$. It oxidizes KI (and I_3^-) in acetone solution to give iodine (note SF_4 converts acetone to Me_2CF_2). S_2F_{10} reacts with SO_2 to give F_5SSO_2F and with NH_3 to give $N{\equiv}SF_3$.

Sulfur hexafluoride is unique in its stability and chemical inertness: it is a colourless, odourless, tasteless, unreactive, non-flammable, non-toxic, insoluble gas prepared by burning sulfur in an atmosphere of fluorine. Because of its extraordinary stability and excellent dielectric properties it is extensively used as an insulating gas for high-voltage generators and switch gear: at a pressure of 2–3 bars it withstands 1.0–1.4 MV across electrodes 50 mm apart without breakdown, and at 10 bars it is used for high-power underground electrical transmission systems at 400 V and above. SF_6 can be heated to 500° without decomposition, and is unattacked by most metals, P, As, etc., even when heated. It is also unreactive towards high-pressure steam presumably as a result of kinetic factors since the gas-phase reaction $SF_6 + 3H_2O \rightarrow SO_3 + 6HF$ should release some 460 kJ mol^{-1} ($\Delta G° \sim 200$ kJ mol^{-1}). By contrast, H_2S yields sulfur and HF. Hot HCl and molten KOH at 500° are without effect. Boiling Na attacks SF_6 to yield Na_2S and NaF; indeed, this reaction can be induced to go rapidly even at room temperature or below in the presence of biphenyl dissolved in glyme (1,2-dimethoxyethane). It is also reduced by Na/liq NH_3 and, more slowly, by $LiAlH_4/Et_2O$. Al_2Cl_6 at 200° yields AlF_3, Cl_2, and sulfur chlorides. Recent experiments[80a] indicate that SF_6 becomes much more reactive at higher temperatures and pressures; for example PF_3 is quantitatively oxidized to PF_5 at 500° and 300 bars.

Derivatives of SF_6 are rather more reactive: S_2F_{10} and $SClF_5$ have already been mentioned. Further synthetically useful reactions of this latter compound are:

$$RC{\equiv}N \xrightarrow{h\nu} RCCl{=}NSF_5$$

$$HC{\equiv}CH \longrightarrow HCCl{=}CHSF_5$$

$$RCH{=}CH_2 \longrightarrow RCHClCH_2SF_5$$

$$CH_2{=}C{=}O \xrightarrow{SClF_5} SF_5CH_2COCl \xrightarrow{H_2O/Ag^+} SF_5CH_2CO_2Ag \xrightarrow[-CO_2]{Br_2} SF_5CH_2Br$$

$$SF_5CH_2CO_2Ag \text{ branch up: } Zn/HCl \to SF_5Me$$

$$SF_5CH_2Br \xrightarrow{LiBu^n(-LiF)} SF_4{=}CH_2$$
(p. 812)

$$SClF_5 + O_2 \xrightarrow{h\nu} F_5SOSF_5 + F_5SOOSF_5$$

$SClF_5$ is readily attacked by other nucleophiles, e.g. OH^- but is inert to acids. SF_5OH and SF_5OOH are known.

[80a] A. P. HAGEN and D. L. TERRELL, High-pressure reactions of small covalent molecules. 12. Interaction of PF_3 and SF_6, *Inorg. Chem.* **20**, 1325–6 (1981).

The very reactive yellow SF_5OF, which is one of the few known hypofluorites, can be made by the catalytic reaction:

$$SOF_2 + 2F_2 \xrightarrow{\text{CsF}/25°} SF_5OF$$

In the absence of CsF the product is SOF_4 (p. 812) and this can then be isomerized in the presence of CsF to give a second hypofluorite, SF_3OF. Derivatives of $-SF_5$ are usually reactive volatile liquids or gases, e.g.:

Compound	F_5SCl	F_5SBr	$(F_5S)_2O$	$(F_5SO)_2$	$F_5SNF_2^{(a)}$	$(F_5S)_2$	F_5SOF
MP/°C	−64	−79	−118	−95.4	—	−52.7	−86
BP/°C	−21	+3.1	+31	+49.4	−18	+30.0	−35.1

(a) See reference 80b for F_5SNClF, F_5SNHF, and $F_4S{=}NF$.

Of these, $(F_5SO-)_2$ is an amusing example of a compound accidentally prepared as a byproduct of SF_6 and S_2F_{10} due to the fortuitous presence of traces of molecular oxygen in the gaseous fluorine used to fluorinate sulfur. A small amount of material boiling somewhat above S_2F_{10} and having a molecular weight some 32 units higher was isolated. [How would you show that it was not S_3F_{10}, and that its structure was F_5SOOSF_5 rather than one of the 8 possible isomers of $F_4S(OF)-SF_4(OF)$ or $F_4S(OF)-OSF_5$?][81]

Numerous other highly reactive oxofluoro-sulfur compounds have been prepared but their chemistry, though sometimes hazardous because of a tendency to explosion, introduces no new principles. Some examples are:

Thionyl fluorides: OSF_2, $OSFCl$, $OSFBr$, $OSF(OM)$.
Sulfuryl fluorides: O_2SF_2, FSO_2-O-SO_2F, $FSO_2-O-SO_2-O-SO_2F$, $FSO_2-OO-SO_2F$, $FSO_2-OO-SF_5$.
Other peroxo compounds:[82] $SF_5OOC(O)F$, $SF_5OSF_4OOSF_5$, $SF_5OSF_4OOSF_4OSF_5$, $CF_3OSF_4OOSF_5$, $CF_3OSF_4OOSF_4OCF_3$, $(CF_3SO_2)_2O_2$, $HOSO_2OOCF_3$, $CF_3OOSO_2OCF_3$.
Fluorosulfuric acid:[83] $FSO_2(OH)$, FSO_3^-.

Of these the most extensively studied is fluorosulfuric acid, made by direct reaction of SO_3 and HF. Its importance derives from its use as a solvent system and from the fact that its mixtures with SbF_5 and SO_3 are amongst the strongest known acids (superacids, p. 665). Anhydrous HSO_3F is a colourless, dense, mobile liquid which fumes in moist air: mp −89.0°, bp 162.7°; d_{25} 1.726 g cm^{-3}, η_{25} 1.56 centipoise, κ_{25} 1.085 × 10^{-4} ohm^{-1} cm^{-1}.

80b D. D. DesMarteau, H. H. Eysel, H. Oberhammer, and H. Günther, Novel sulfur–nitrogen–fluorine compounds. Synthesis and properties of SF_5NClF, SF_5NHF, and $FN{=}SF_4$ and the molecular structure and vibrational analysis of $FN{=}SF_4$, *Inorg. Chem.* **21**, 1607–16 (1982).

81 R. B. Harvey and S. H. Bauer, An electron-diffraction study of disulfur decafluoridedioxide, *J. Am. Chem. Soc.* **76**, 859–64 (1954).

82 R. A. De Marco and J. M. Shreeve, Fluorinated peroxides, *Adv. Inorg. Chem. Radiochem.* **16**, 109–76 (1974).

83 A. W. Jache, Fluorosulfuric acid, its salts, and derivatives, *Adv. Inorg. Chem. Radiochem.* **16**, 177–200 (1974).

Chlorides, bromides, and iodides of sulfur

Sulfur is readily chlorinated by direct reaction with Cl_2 but the simplicity of the products obtained belies the complexity of the mechanisms involved. The reaction was first investigated by C. W. Scheele in 1774 and has been extensively studied since because of its economic importance (see below) and its intrinsic physicochemical interest. Direct chlorination of molten S followed by fractional distillation yields disulfur dichloride (S_2Cl_2) a toxic, golden-yellow liquid of revolting smell: mp $-76°$, bp $138°$, $d(20°)$ $1.677\,\mathrm{g\,cm^{-3}}$. The molecule has the expected C_2 structure (like S_2F_2, H_2O_2, etc.) with S–S 195 pm, S–Cl 206 pm, angle Cl–S–S $107.7°$, and a dihedral angle of $85.2°$.[84] Further chlorination of S_2Cl_2, preferably in the presence of a trace of catalyst such as $FeCl_3$, yields the more-volatile, cherry-red liquid sulfur dichloride, SCl_2: mp $-122°$, bp $59°$, $d(20°)$ $1.621\,\mathrm{g\,cm^{-3}}$. SCl_2 resembles S_2Cl_2 in being foul-smelling and toxic, but is rather unstable when pure due to the decomposition equilibrium $2SCl_2 \rightleftharpoons S_2Cl_2 + Cl_2$. However, it can be stabilized by the presence of as little as 0.01% PCl_5 and can be purified by distillation at atmospheric pressure in the presence of 0.1% PCl_5.[85] The sulfur dichloride molecule is nonlinear (C_{2v}) as expected, with S–Cl 201 pm and angle Cl–S–Cl $103°$.

S_2Cl_2 and SCl_2 both react readily with H_2O to give a variety of products such as H_2S, SO_2, H_2SO_3, H_2SO_4, and the polythionic acids $H_2S_xO_6$. Oxidation of SCl_2 yields thionyl chloride $(OSCl_2)$ and sulfuryl chloride (O_2SCl_2) (see Section 15.2.4). Reaction with F_2 produces SF_4 and SF_6 (p. 811), whereas fluorination with NaF is accompanied by some disproportionation:

$$3SCl_2 + 5NaF \longrightarrow SF_4 + S_2Cl_2 + 4NaCl$$

As indicated on p. 810, fluorination of S_2Cl_2 with KF/SO_2 occurs with concurrent isomerization to SSF_2. Both S_2Cl_2 and SCl_2 react with atomic N (p. 474) to give NSCl as the first step, and this can then react further with S_2Cl_2 to give the ionic heterocyclic compound $S_3N_2Cl^+Cl^-$ (p. 874). By contrast, reaction of S_2Cl_2 with NH_4Cl at $160°$ (or with $NH_3 + Cl_2$ in boiling CCl_4) yields S_4N_4 (p. 856).

S_2Cl_2 and SCl_2 are important industrial chemicals. The main use for S_2Cl_2 is in the vapour-phase vulcanization of certain rubbers, but other uses include its chlorinating action in the preparation of mono- and di-chlorohydrins, and the opening of some minerals in extractive metallurgy. Some idea of the scale of production can be gauged from the fact that S_2Cl_2 is shipped in 50-tonne tank cars; smaller quantities are transported in drums containing 300 or 60 kg of the liquid. Its less-stable homologue SCl_2 is notable for its ready addition across olefinic double bonds: e.g., thiochlorination of ethene yields the notorious vesicant, mustard gas:

$$SCl_2 + 2CH_2{=}CH_2 \longrightarrow S(CH_2CH_2Cl)_2$$

The compounds SCl_2 and S_2Cl_2 can be thought of as the first two members of an extended series of dichlorosulfanes S_nCl_2. The lower electronegativity of Cl (compared with F) and the lower S–Cl bond energy (compared with S–F) enable the natural

[84] C. J. MARSDEN, R. D. BROWN, and P. D. GODFREY, Microwave spectrum and molecular structure of disulfur dichloride, S_2Cl_2, *JCS Chem. Comm.* 1979, 399–401.

[85] R. J. ROSSEN and F. R. WHITT, Development of a large-scale distillation process for purifying crude sulfur dichloride. 1. Laboratory scale investigations, *J. Appl. Chem.* **10**, 229–37 (1960); see also the following paper (pp. 237–46) for large-scale distillation unit.

catenating propensity of S to have full reign and a series of dichlorosulfanes can be prepared in which S–S bonds in sulfur chains (and rings) can be broken and the resulting $-S_n-$ oligomers stabilized by the formation of chain-terminating S–Cl bonds. The first eight members with $n=1-8$ have been isolated as pure compounds, and mixtures up to perhaps $S_{100}Cl_2$ are known.† Specific compounds have been made by F. Fehér's group using the polysulfanes as starting materials (p. 807):[85a]

$$H_2S_x + 2S_2Cl_2 \longrightarrow 2HCl + S_{4+x}Cl_2$$

$$H_2S_x + 2SCl_2 \xrightarrow{-80°} 2HCl + S_{2+x}Cl_2$$

The dichlorosulfanes are yellow to orange-yellow viscous liquids with an irritating odour. They are thermally and hydrolytically unstable. S_3Cl_2 boils at $31°$ (10^{-4} mmHg) and has a density of 1.744 g cm^{-3} at $20°$. Higher homologues have even higher densities:

n in S_nCl_2	1	2	3	4	5	6	7	8
Density (20°)/g cm^{-3}	1.621	1.677	1.744	1.777	1.802	1.822	1.84	1.85

The higher chlorides of S (unlike the higher fluorides) are very unstable and poorly characterized. There is no evidence for the chloro analogues of SF_4, S_2F_{10}, and SF_6, though $SClF_5$ is known (p. 812). Chlorination of SCl_2 by liquid Cl_2 at $-78°$ yields a powdery off-white solid which begins to decompose when warmed above $-30°$. It analyses as SCl_4 and is generally formulated as $SCl_3^+Cl^-$, but little reliable structural work has been done on it. Consistent with this ionic formulation, reaction of SCl_4 with Lewis acids results in the formation stable adducts; e.g. $AlCl_3$ yields the white solid $SCl_4 . AlCl_3$ which has been shown by vibrational spectroscopy on both the solid and the melt (125°) to be $[SCl_3]^+[AlCl_4]^-$.[86] The compound $[SCl_3]^+[ICl_4^-]$ is also known (p. 819).[87] As expected from a species that is isoelectronic with PCl_3 the cation is pyramidal; dimensions are: S–Cl (average) 198.5 pm, angle Cl–S–Cl 101.3° (cf. PCl_3: P–Cl 204.3 pm, angle Cl–P–Cl 100.1°).

Sulfur bromides are but poorly characterized and there are few reliable data on them. SBr_2 probably does not exist at room temperature but has been claimed as a matrix-isolated product when a mixture of $S_2Cl_2/SCl_2:Br_2:Ar$ in the ratio 1:1:150 is passed through an 80-W microwave discharge and the product condensed on a CsI window at 9 K.[88] The dibromosulfanes S_nBr_2 ($n=2-8$) are formed by the action of anhydrous HBr

---

† Several related series of compounds are also known in which Cl is replaced by a pseudohalogen such as $-CF_3$ or $-C_2F_5$, e.g. $S_n(CF_3)_2$ ($n=1-4$), $CF_3S_nC_2F_5$ ($n=2-4$), and $S_n(C_2F_5)_2$ ($n=2-4$). These can be prepared by the reaction of CF_3I and S vapour in a glow discharge followed by fractionation and glc separation; other routes include reaction of CS_2 with IF_5 at 60–200°, reaction of CF_3I with sulfur at 310°, and fluorination of $SCCl_2$ or related compounds with NaF or KF at 150–250°. (See, for example, T. Yasumura and R. J. Lagow, *Inorg. Chem.* **17**, 3108–3110 (1978).)

[85a] F. Fehér, Dichloropolysulfanes and dibromopolysulfanes, pp. 370–9 in G. Brauer (ed.), *Handbook of Preparative Inorganic Chemistry*, 2nd edn., Vol. 1, Academic Press, New York, 1963.

[86] G. Mamantov, R. Marassi, F. W. Poulson, S. E. Springer, J. P. Wiaux, R. Huglen, and N. R. Smynl, $[SCl_3]^+[AlCl_4]^-$: improved synthesis and characterization, *J. Inorg. Nuclear Chem.* **41**, 260–1 (1979).

[87] A. J. Edwards, Crystal structure of trichlorosulfonium(IV) tetrachloroiodate(III), *JCS Dalton* 1978, 1723–5.

[88] M Feuerhan and G. Vahl, Infrared spectra of matrix-isolated sulfur dibromide and sulfur diiodide, *Inorg. Nuclear Chem. Lett.* **16**, 5–8 (1980).

on the corresponding chlorides.[85a] The best characterized compound (which can also be made directly from the elements at $100°C$) is the garnet-red oily liquid S_2Br_2 isostructural with S_2Cl_2 (S–S 198 pm, S–Br 224 pm, angle Br–S–S 105°, dihedral angle $84 \pm 11°$). It has mp $-46°$, bp (0.18 mmHg) 54°, and $d(20°)$ 2.629 g cm^{-3}, but even at room temperature S_2Br_2 tends to dissociate into its elements. Interestingly, the higher homologues have progressively lower densities (cf. S_nCl_2).

n in S_nBr_2	2	3	4	5	6	7	8
Density (20°)/g cm^{-3}	2.629	2.52	2.47	2.41	2.36	2.33	2.30

Sulfur iodides are a topic of considerable current interest. Until recently no authentic binary compounds of sulfur and iodine had been established though a 1980 report[88] suggests that SI_2 may have been made from SCl_2 and I_2 in a matrix-isolation experiment. S_2I_2 has been made as a reddish-brown solid by reacting HI/N_2 with an 0.1 M solution of S_2Cl_2 in pentane at $-90°C$ (S_8, I_2, and Cl_2 are also formed):[89] the vibrational spectrum of S_2I_2 was assigned on the basis of C_2 symmetry. The failure to prepare sulfur iodides by direct reaction of the elements probably reflects the weakness of the S–I bond: an experimental value is not available but extrapolation from representative values for the bond energies of other S–X bonds leads to a value of ~ 170 kJ mol^{-1}:

Bond	S–F	S–Cl	S–Br	S–I	S–S	I–I
Energy/kJ mol^{-1}	327	271	218	(~ 170)	225	150

The data indicate that formation of SI_2 from $\frac{1}{8}S_8 + I_2$ and the formation of S_2I_2 from $\frac{1}{4}S_8 + I_2$ are both endothermic to the extent of ~ 35 kJ mol^{-1}, implying that successful synthesis of these compounds must employ kinetically controlled routes to obviate decomposition back to the free elements.

It is probable that the first species unambiguously known to contain even one S–I bond was the curious and unexpected cation $[S_7I]^+$: this was found (1976) in the dark-orange compound $[S_7I]^+[SbF_6]^-$ formed when iodine and sulfur react in SbF_5 solution.[90] The structure of the cation is shown in Fig. 15.28a and features an S_7 ring with alternating S–S distances and a pendant iodine atom; the conformation of the ring is the same as in S_7, S_8, and S_8O (p. 822). Subsequently (1980) the same cation was found in $[S_7I]^+_4[S_4]^{2+}[AsF_6]^-_6$[91] and a similar motif forms part of the iodo-bridged species

[89] V. G. VAHL and R. MINKWITZ, Infrared spectroscopic investigation of solid disulfur diiodide, *Inorg. Nuclear Chem. Lett.* **13**, 213–15 (1977) and references therein.

[90] J. PASSMORE, P. TAYLOR, T. K. WHIDDEN, and P. WHITE, Preparation and X-ray structure of iodocycloheptasulfur hexafluoroantimonate(V) $[S_7I]^+[SbF_6]^-$, *JCS Chem. Comm.* 1976, 689. J. PASSMORE, G. SUTHERLAND, P. TAYLOR, T. K. WHIDDEN, and P. S. WHITE, Preparations and X-ray crystal structures of iodo-*cyclo*-heptasulfur hexafluoroantimonate(V) and hexafluoroarsenate(V), S_7ISbF_6 and S_7IAsF_6, *Inorg. Chem.* **20**, 3839–45 (1981). The cation is also one of the products formed when an excess of S reacts with $[I_3]^+[AsF_6]^-$ or $[I_3]^+[As_2F_{11}]^-$ or AsF_5/I_2, or when $[S_{16}]^{2+}[SbF_6]_2^-$ is iodinated with an excess of iodine.

[91] J. PASSMORE, G. SUTHERLAND, and P. S. WHITE, Preparation and crystal structures of $[S_7I]_4[S_4][AsF_6]_6$ and $[S_4][AsF_6]_2 . 0.6SO_2$; a convenient synthesis of hexafluoroarsenate salts of chalcogen homoatomic cations, *JCS Chem. Comm.* 330–1 (1980). (See also *Inorg. Chem.* **21**, 2717–23 (1982).)

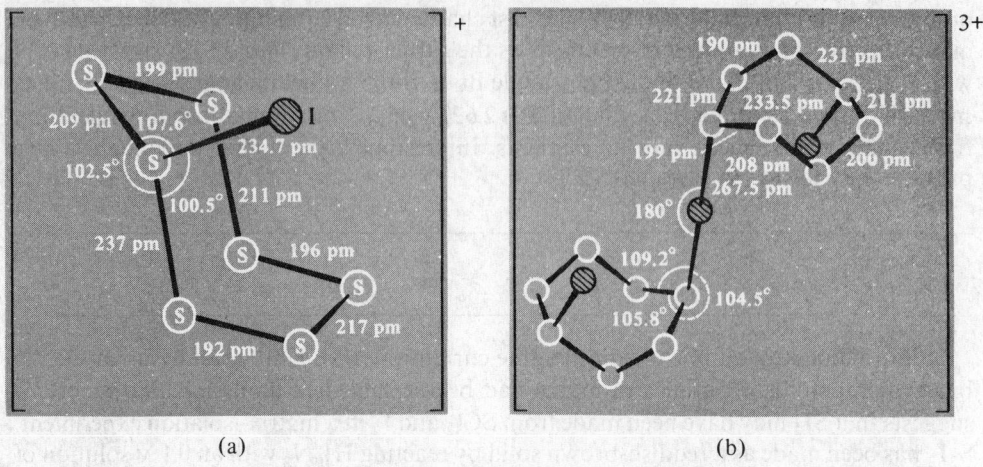

(a)   (b)

Fig. 15.28   (a) Structure of the iodocycloheptasulfur cation in $[S_7I]^+[SbF_6]^-$. The S–S–S angles in the S_7 ring are in the range 102.5–108.4° (mean 105.6°).[90] (b) Structure of the centrosymmetric cation $[(S_7I)_2I]^{3+}$ showing similar dimensions to those in $[S_7I]^+$; note also the long S–I distances in the linear bridging S–I–S group.[92]

$[(S_7I)_2I]^{3+}$ (Fig. 15.28b);[92] this latter cation was formed during the reaction of S_8 and I_2 with SbF_5 in the presence of AsF_3 according to the reaction stoichiometry:

$$1\tfrac{3}{4}S_8 + 1\tfrac{1}{2}I_2 + 6SbF_5 \xrightarrow{\;2\,AsF_3\;} [S_{14}I_3]^{3+}[SbF_6]^-_3 . 2AsF_3 + (SbF_3)_2SbF_5$$

The very long $S–I_\mu$ bonds in the linear S–I–S bridge (267.5 pm) are notable and have been interpreted in terms of an S–I bond order of $\tfrac{1}{2}$. Even weaker S···I interactions occur in the cation $[S_2I_4]^{2+}$ which could, indeed, alternatively be regarded as an S_2^{2+} cation coordinated side-on by two I_2 molecules (Fig. 15.29).[93] This curious right triangular prismatic conformation (notably at variance with that in the isoelectronic P_2I_4 molecule) is associated with a very short S–S bond (bond order $2\tfrac{1}{3}$) and rather short I–I distances (bond order $1\tfrac{1}{3}$). The cation is formed in AsF_5/SO_2 solution according to the equation:

$$\tfrac{1}{4}S_8 + 2I_2 + 3AsF_5 \xrightarrow{\;SO_2\;} [S_2I_4]^{2+}[AsF_6]^-_2 + AsF_3$$

Other species containing S–I bonds that have recently been characterized are the pseudopolyhalide anions $[I(SCN)_2]^-$ and $[I_2(SCN)]^{-}$ [93a]

We conclude this section with an amusing cautionary tale which illustrates the type of blunder that can still appear in the pages of a refereed journal (1975) when scientists (in this case physicists) attempt to deduce the structure of a compound by spectroscopic

[92] J. PASSMORE, G. SUTHERLAND, and P. S. WHITE, Preparation and X-ray crystal structure of μ-iodo-bis(4-iodo-*cyclo*-heptasulfur)-tris(hexafluoroantimonate)-bis(arsenic trifluoride), $[(S_7I)_2I]^{3+}[SbF_6]_3^-$. 2AsF₃, *JCS Chem. Comm.* 1979, 901–2. (See also *Inorg. Chem.* **21**, 2717–23 (1982).)

[93] J. PASSMORE, G. SUTHERLAND, T. WHIDDEN, and P. S. WHITE, Preparation and crystal structure of $[S_2I_4]^{2+}[AsF_6]_2^-$ containing the distorted right triangular prismatic disulfur tetraiodide(2+) cation and a disulfur unit of bond order greater than two, *JCS Chem. Comm.* 1980, 289–90.

[93a] G. A. BOWMAKER and D. A. ROGERS, Synthesis and vibrational spectroscopic study of compounds containing the $I(SCN)_2^-$, $I_2(SCN)^-$, and $I(SeCN)_2^-$ ions, *JCS Dalton*, 1981, 1146–51.

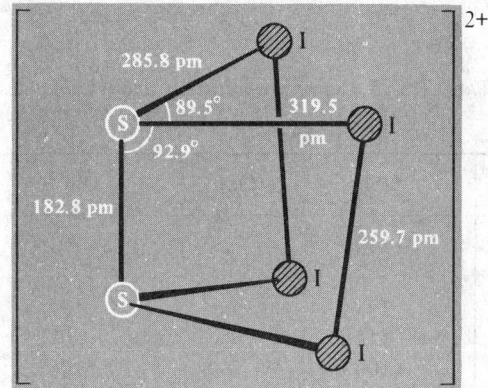

FIG. 15.29 Structure of the $[S_2I_4]^{2+}$ cation of C_2 symmetry, showing the very short S–S distance and the rather short I–I distances; note also the S–I distances which are even longer than in the weak charge transfer complex $[(H_2N_2CS)_2I]^+$ (262.9 pm). The nonbonding I···I distance is 426.7 pm.

techniques alone, without ever analysing the substance being investigated. The work[94] purported to establish the presence of a new molecule Cl_3SI in solid solution with an ionic complex $[SCl_3]^+[ICl_2]^-$, thus leading to an overall formula for the crystals of $S_2Cl_8I_2$. The mixed compound had apparently been made originally by M. Jaillard in 1860: he obtained it as beautiful transparent yellow-orange prismatic crystals by treating a mixture of sulfur and iodine with a stream of dry Cl_2. R. Weber obtained the same material in 1866 by passing Cl_2 into a solution of I_2 in CS_2 but he reported a composition of S_2Cl_7I rather than Jaillard's SCl_4I ($S_2Cl_8I_2$). The implausibility of forming a stable compound containing an S–I bond in this way, coupled with the perceptive recognition that the published Raman spectrum had bands that could be assigned to $[ICl_4]^-$ rather than $[ICl_2]^-$, led P. N. Gates and A. Finch to reinvestigate the compound.[95] It transpired that the nineteenth-century workers had used $S = 16$ as the atomic weight of sulfur so the true chemical composition of the crystals was, in fact, SCl_7I. The previous spectroscopic interpretation[94] was therefore totally incorrect and the compound was shown to be $[SCl_3]^+[ICl_4]^-$. This was later confirmed by a single-crystal X-ray diffraction study (p. 816).[87] In short, far from containing the new iodo-derivative Cl_3SI, the compound did not even contain an S–I bond.

15.2.4 *Oxohalides of sulfur*

Sulfur forms two main series of oxohalides, the thionyl dihalides $OS^{IV}X_2$ and the sulfuryl dihalides $O_2S^{VI}X_2$. In addition, various other oxofluorides and peroxofluorides are known (p. 814). Thionyl fluorides and chlorides are colourless volatile liquids (Table 15.15); $OSBr_2$ is rather less volatile and is orange-coloured. All have pyramidal molecules

[94] Y. TAVARES-FORNERIS and R. FORNERIS, Raman spectrum and structure of trichlorosulfonium iodide (Cl_3SI), *J. Mol. Structure* **24**, 205–13 (1975).

[95] A. FINCH, P. N. GATES, and T. H. PAGE, Trichlorosulfonium tetrachloroiodate and the constitution of Jaillard's compound, *Inorg. Chim. Acta* **25**, L49–L50 (1977).

TABLE 15.15 *Some properties of thionyl dihalides,* $O{=}S(_X^X)X$

Property	OSF$_2$	OSFCl	OSCl$_2$	OSBr$_2$
mp/°C	−110	−120	−101	−50
bp/°C	−44	12	76	140
d(O−S)/pm	141.2	−	145	145 (assumed)
d(S−X)/pm	158.5	−	207	227
angle O−S−X	106.8°	−	106°	108°
angle X−S−X	92.8°	−	114°(?)	96°

(C_s point group for OSX$_2$), and OSFCl is chiral though stereochemically labile. Dimensions are in Table 15.15: the short O–S distance is notable.

The most important thionyl compound is OSCl$_2$—it is readily prepared by chlorination of SO$_2$ with PCl$_5$ or, on an industrial scale, by oxygen-atom transfer from SO$_3$ to SCl$_2$:

$$SO_2 + PCl_5 \longrightarrow OSCl_2 + OPCl_3$$

$$SO_3 + SCl_2 \longrightarrow OSCl_2 + SO_2$$

OSCl$_2$ reacts vigorously with water and is particularly valuable for drying or dehydrating readily hydrolysable inorganic halides:

$$MX_n \cdot mH_2O + mOSCl_2 \longrightarrow MX_n + mSO_2 + 2mHCl$$

Examples are MgCl$_2$.6H$_2$O, AlCl$_3$.6H$_2$O, FeCl$_3$.6H$_2$O, etc. Thionyl chloride begins to decompose above its bp (76°) into S$_2$Cl$_2$, SO$_2$, and Cl$_2$; it is therefore much used as an oxidizing and chlorinating agent in organic chemistry. Fluorination with SbF$_3$/SbF$_5$ gives OSF$_2$; use of NaF/MeCN gives OSFCl or OSF$_2$ according to conditions. Thionyl chloride also finds some use as a nonaqueous ionizing solvent as does SO$_2$ (p. 828) and the formally related dimethylsulfoxide (DMSO), Me$_2$SO (mp 18.6°, bp 189°, viscosity η_{25} 1.996 centipoise, dielectric constant ε_{25} 46.7).

Sulfuryl halides, like their thionyl analogues, are also reactive, colourless, volatile liquids or gases (Table 15.16). The most important compound is O$_2$SCl$_2$, which is made on an industrial scale by direct chlorination of SO$_2$ in the presence of a catalyst such as activated charcoal (p. 301) or FeCl$_3$. It is stable to 300° but begins to dissociate into SO$_2$ and Cl$_2$ above this: it is a useful reagent for introducing Cl or O$_2$SCl into organic compounds. O$_2$SCl$_2$ can be regarded as the acid chloride of H$_2$SO$_4$ and, accordingly, slow hydrolysis (or ammonolysis) yields O$_2$S(OH)$_2$ or O$_2$S(NH$_2$)$_2$. Fluorination yields O$_2$SF$_2$ (also prepared by SO$_2$ + F$_2$) and comproportionation of this with O$_2$SCl$_2$ and O$_2$SBr$_2$ yield the corresponding O$_2$SFX species. All these compounds have (distorted) tetrahedral molecules, those of formula O$_2$SX$_2$ having C_{2v} symmetry and the others C_s. Dimensions are in Table 15.16: the remarkably short O–S and S–F distances in O$_2$SF$_2$ should be noted (cf. above). Indeed, the implied strength of bonding in this molecule is reflected by the fact

TABLE 15.16 *Some properties of sulfuryl dihalides*, $O{=}S(O)X_2$

Property	O_2SF_2	O_2SFCl	O_2SCl_2	O_2SFBr
mp/°C	−120	−125	−54	−86
bp/°C	−55	7	69	41
d(O–S)/pm	140.5	–	143	–
d(S–X)/pm	153.0	–	199	–
angle O–S–O	124°	–	120°	
angle X–S–X	96°	–	111°	

that it can be made by reacting the normally extremely inert compound SF_6 (p. 813) with the fluoro-acceptor SO_3:

$$SF_6 + 2SO_3 \longrightarrow 3O_2SF_2; \quad \Delta G^\circ_{298} = -202 \text{ kJ mol}^{-1}$$

A 20% conversion can be effected by heating the two compounds at 250° for 24 h.

15.2.5 *Oxides of sulfur*

At least thirteen proven oxides of sulfur are known to exist[96] though this profusion should not obscure the fact that SO_2 and SO_3 remain by far the most stable and unquestionably the most important economically. The six homocyclic polysulfur monoxides S_nO $(5 < n < 10)$ are made by oxidizing the appropriate cyclo-S_n (p. 774) with trifluoroperoxoacetic acid, $CF_3C(O)OOH$, at $-30°$. The dioxides S_7O_2 and S_6O_2 are also known. In addition there are the thermally unstable acyclic oxides S_2O, S_2O_2, SO, and the fugitive species SOO and SO_4. Several other compounds were described in the older literature (pre-1950s) but these reports are now known to be in error. For example, the blue substance of composition "S_2O_3" prepared from liquid SO_3 and sulfur now appears to be a mixture of salts of the cations S_4^{2+} and S_8^{2+} (p. 785) with polysulfate anions. Likewise a "sulfur monoxide" prepared by P. W. Schenk in 1933 was shown by D. J. Meschi and R. J. Meyers in 1956 to be a mixture of S_2O and SO_2. The well-established lower oxides of S will be briefly reviewed before SO_2 and SO_3 are discussed in more detail.

Lower oxides[96]

Elegant work by R. Steudel and his group in Berlin during the last few years has shown that, when cyclo-S_{10}, -S_9, and -S_8 are dissolved in CS_2 and oxidized by freshly prepared $CF_3C(O)O_2H$ at temperatures below $-10°$, modest yields (10–20%) of the corresponding

96 *Gmelin Handbuch der Anorganischen Chemie*, 8th edn., Schwefel Oxide, Engänzungsband 3, 1980, 344 pp.

crystalline monoxides S_nO are obtained. Similar oxidation of *cyclo*-S_7, and α- and β-S_6 in CH_2Cl_2 solution yields crystalline S_7O, S_7O_2, and α- and β-S_6O. Crystals of S_6O_2 and S_5O ($d > -50°$) have not yet been isolated but the compounds have been made in solution by the same technique. S_8O had previously been made (1972) by the reaction of $OSCl_2$ and H_2S_7 in CS_2 at $-40°$: it is one of the most stable compounds in the series and melts (with decomposition) at $78°$. All the compounds are orange or dark yellow and decompose with liberation of SO_2 and sulfur when warmed to room temperature or slightly above. Structures are in Fig. 15.30. It will be noted that S_7O is isoelectronic and isostructural with $[S_7I]^+$. This invites the question as to whether S_7S can be prepared as a new structural isomer of *cyclo*-S_8.

S_8O reacts with $SbCl_5$ in CS_2 over a period of 9 days at $-50°$ to give a 71% yield of the

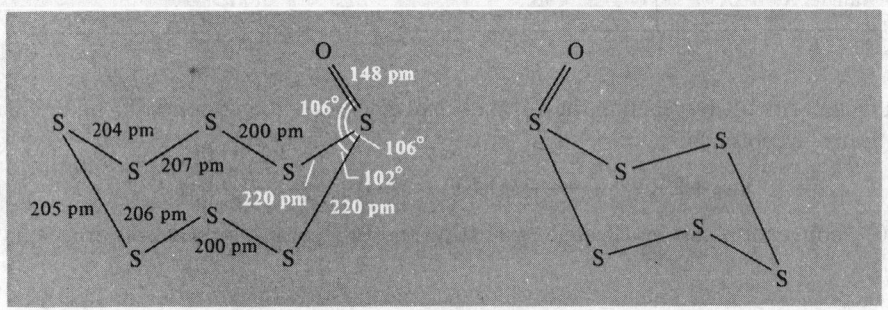

S₈O: orange-yellow crystals, mp 78° (decomp)

α-S₆O: orange crystals, mp 39° (d)
β-S₆O: dark orange, mp 34° (d)

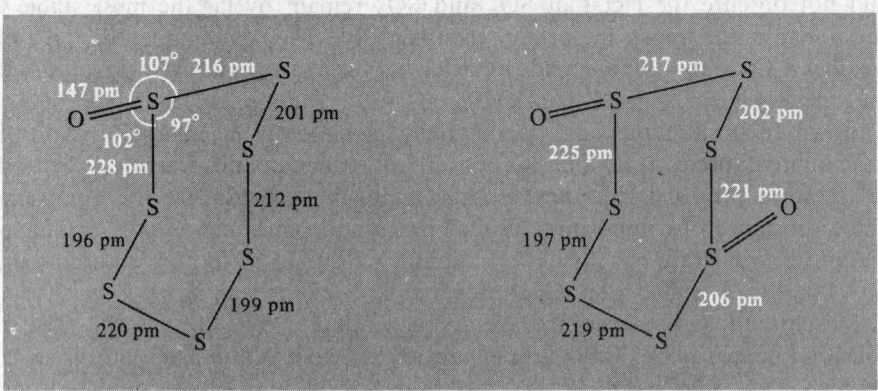

S₇O: orange crystals, mp 55° (d)

S₇O₂: dark orange crystals, d > room temp

FIG. 15.30 Structures of S_8O, S_7O, S_7O_2, and S_6O; in each case the O atom adopts an axial conformation. For S_8O there is an alternation of S–S distances, the longest being adjacent to the exocyclic O atoms; S–S–S angles are in the range 102–108° and torsion angles (p. 771) vary from 95° to 112° (+ and −). For S_7O there is again an alternation in S–S distances; ring angles are in the range 97–106° the smallest angle again being at the S atom carrying the pendant O. The structure of S_7O_2 was deduced from its Raman spectrum, the interatomic distances (d/pm) being computed from the relation $\log (d/\text{pm}) = 2.881 - 0.213 \log (\nu/\text{cm}^{-1})$. The two modifications α- and β-S_6O have the same Raman spectrum in solution.

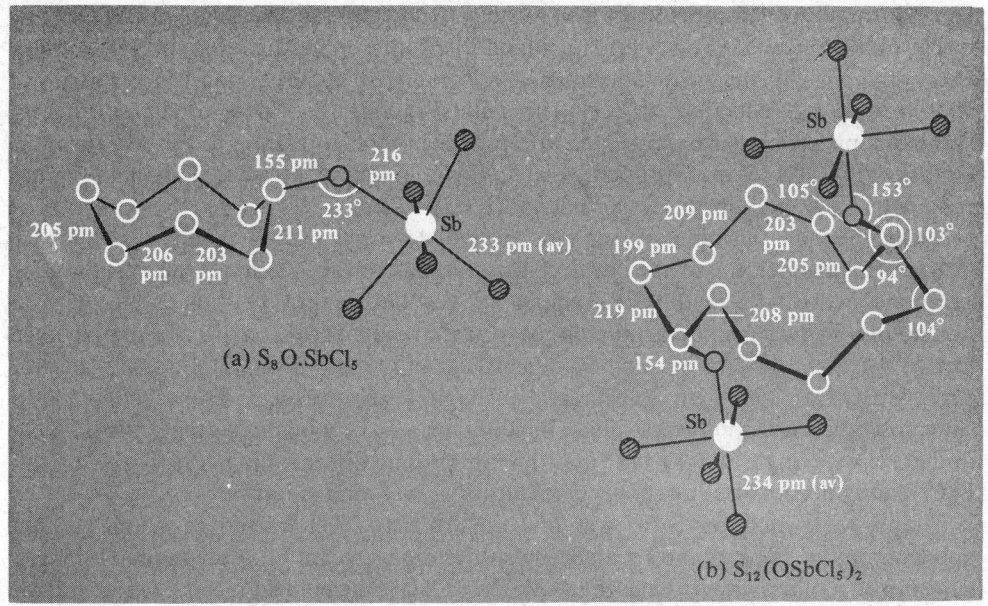

FIG. 15.31   Molecular structure and dimensions of (a) the adduct $S_8O.SbCl_5$ at $-100°$, and (b) the dimeric unit $Sb_{12}O_2.2SbCl_5$ in $Sb_{12}O_2.2SbCl_5.3CS_2$ at $-115°C$.

unstable orange adduct $S_8O . SbCl_5$:[97] its structure and dimensions are in Fig. 15.31a. It will be noted that the S_8O unit differs from molecular S_8O in having an equatorially bonded O atom and significantly different S–O and S–S interatomic distances. The X-ray crystal structure was determined at $-100°C$ as the adduct decomposes within 5 min at $25°$ to give $OSCl_2$, $SbCl_3$, and S_8. When a similar reaction was attempted with β-S_6O, the novel dimer $S_{12}O_2 . 2SbCl_5 . 3CS_2$ was obtained as orange crystals in 10% yield after 1 week at $-50°$[98] (Fig. 15.31b). Formation of the centrosymmetric $S_{12}O_2$ molecule, which is still unknown in the uncoordinated state, can be explained in terms of a dipolar addition reaction (Fig. 15.32):

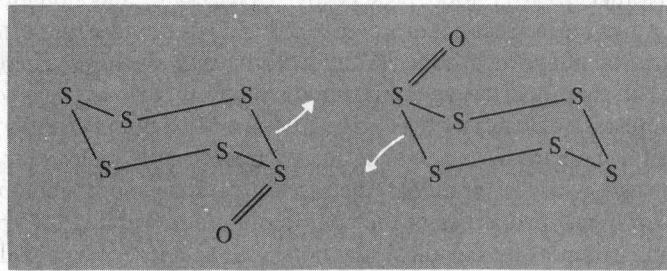

FIG. 15.32   Possible reaction to form $S_{12}O_2$ by dipolar addition of $2S_6O$.

[97] R. STEUDEL, T. SANDOW, and J. STEIDEL, Reversible isomerization of *cyclo*-octasulfur monoxide; preparation and X-ray crystal structure of $S_8O . SbCl_5$, *JCS Chem. Comm.* 180–1 (1980).
[98] R. STEUDEL, J. STEIDEL, and J. PICKARDT, Dimerization of S_6O to $S_{12}O_2$: crystal structure analysis of $S_{12}O_2 . 2SbCl_5 . 3CS_2$, *Angew. Chem.*, Int. Edn. (Engl.) **19**, 325–6 (1980).

Its conformation differs drastically from the D_{3d} symmetry of the parent *cyclo*-S_{12} (p. 777).

The fugitive species SO was first identified by its ultraviolet spectrum in 1929 but it is thermodynamically unstable and decomposes completely in the gas phase in less than 1 s. It is formed by reduction of SO_2 with sulfur vapour in a glow discharge and its spectroscopic properties have excited interest because of its relation to O_2 ($^3\Sigma^-$ ground state, p. 705). Molecular properties include internuclear distance 148.1 pm, dipole moment 1.55 D, equilibrium bond energy D_e 524 kJ mol^{-1}.

S_2O is also an unstable species but survives for several days in the gas phase at <1 mmHg pressure. It is formed by decomposition of SO (above) and by numerous other reactions between S- and O-containing species but cannot be isolated as a pure compound. Typical recipes include: (a) passing a stream of $OSCl_2$ at 0.1–0.5 mmHg over heated Ag_2S at 160°, (b) burning S_8 in a stream of O_2 at ~ 8 mmHg pressure, and (c) passing SO_2 at 120° and <1 mmHg through a high-voltage discharge (~ 5 kV). Spectroscopic studies in the gas phase have shown it to be a nonlinear molecule (like O_3 and SO_2) with angle S–S–O 118° and the interatomic distances S–S 188, S–O 146 pm. S_2O readily decomposes at room temperature to SO_2 and sulfur.

The unstable molecule S_2O_2 was first unambiguously characterized by microwave spectroscopy in 1974. It can be made by subjecting a stream of SO_2 gas at 0.1 mmHg pressure to a microwave discharge (80 W, 2.45 GHz): the effluent gas is predominantly SO_2 but also contains 20–30% SO, 5% S_2O, and 5% S_2O_2. This latter species has a planar C_{2v} structure with r (S–S) 202 pm, r (S–O) 146 pm, and angle S–S–O 113°; it decomposes directly into SO with a half-life of several seconds at 0.1 mmHg.

(ii) Sulfur dioxide, SO_2

Sulfur dioxide is made commercially on a very large scale either by the combustion of sulfur or H_2S or by roasting sulfide ores (particularly pyrite, FeS_2) in air (p. 768). It is also produced as a noxious and undesirable byproduct during the combustion of coal and fuel oil. The ensuing environmental problems and the urgent need to control this pollution are matters of considerable concern and activity (see Panel). Most of the technically produced SO_2 is used in the manufacture of sulfuric acid (p. 838) but it also finds use as a bleach, disinfectant (Homer, p. 757), food preservative, refrigerant, and nonaqueous solvent. Other chemical uses are in the preparation of sulfites and dithionites (p. 849) and, with Cl_2, in the derivatization of hydrocarbons via sulfochlorination reactions. There is also much current interest in its properties as a multimode ligand (p. 829).

SO_2 is a colourless, toxic gas with a choking odour. Maximum permitted atmospheric concentration for humans is 5 ppm but many green plants suffer severe distress in concentrations as low as 1–2 ppm. SO_2 neither burns nor supports combustion. Some molecular and physical properties of the compound are in Table 15.17 (p. 827).

By far the most important chemical reaction of SO_2 is its further oxidation to SO_3 according to the equilibrium:

$$SO_2 + \tfrac{1}{2}O_2 \Longrightarrow SO_3; \quad \Delta H° = -95.6 \text{ kJ mol}^{-1}$$

The equilibrium constant, $K_p = p(SO_3)/[p(SO_2) \cdot p^{\frac{1}{2}}(O_2)]$, decreases rapidly with increasing temperature; for example: log $K_p = 3.49$ at 800°C and -0.52 at 1100°C. Thus for maximum oxidation during the manufacture of H_2SO_4 it is necessary to work at lower

Atmospheric SO$_2$ and Environmental Pollution[98a-e]

The pollution of air by smoke and sulfurous fumes is no new problem† but the quickening pace of industrial development during the nineteenth century, and the growing concern for both personal health and protection of the environment generally since the 1950s, has given added impetus to measures required to eliminate or at least minimize the hazard.

As indicated on p. 759, there are vast amounts of volatile sulfur compounds in the environment as a result of natural processes. Geothermal activity (especially volcanic) releases large amounts of SO$_2$ together with smaller quantities of H$_2$S, SO$_3$, elemental S, and particulate sulfates. From a global viewpoint, however, this accounts for less than 1% of the naturally formed volatile S compounds (Fig. A). By far the most important source is the biological reduction of S compounds which occurs most readily in the presence of organic matter and under oxygen-deficient conditions. Much of this is released as H$_2$S but other compounds such as Me$_2$S are probably also implicated. The final natural source of atmospheric S compounds is sea-spray (sulfate is the second most abundant anion in sea-water being about one-seventh the concentration of chloride). Though much sulfur is transported as sulfate by wind-driven sea spray and by river run off, its environmental impact is not severe.

Much more serious is the effect of volatile S compounds (mainly SO$_2$) released into the atmosphere as a result of man's domestic and industrial activities. This has been estimated to be some 200 million tonnes pa, and is comparable in amount to all the sulfur released by natural processes ($\sim 310 \times 10^6$ tonnes pa). Unfortunately, by the very nature of its origin, this SO$_2$ is released in the heart of densely populated areas and does great damage to the respiratory organs of man and animals, to buildings, and perhaps most seriously to plants, lake-waters, and aquatic life as a result of "acid rain". Dispersal by means of high chimney stacks is inadequate since this merely transfers the problem to neighbouring regions. For example, only one-tenth of the serious SO$_2$/H$_2$SO$_4$ pollution of lakes and streams in Sweden is as a result of atmospheric SO$_2$ emissions in Sweden itself; one-tenth is due to emissions from the UK, and the remaining four-fifths is from industrial regions in northern Europe.[98f] The extent of the problem is shown by the map (Fig. B), which indicates the amount of dry-plus-wet-SO$_2$ deposited on Europe annually (1974). Similar problems afflict the North American continent.[98g]

In Europe and the USA (and presumably elsewhere) the major source of SO$_2$ pollution is in coal-based power generation; this, together with other coal consumption and coking operations accounts for some 60% of the emissions. A further 25% arises from oil-refinery operations, oil-fired power generation, and other oil consumption. Copper-smelting (together with much smaller amounts from zinc and lead ore processing) accounts for some 12% of the annual release of SO$_2$. The sulfuric acid manufacturing industry, which is the only one designed actually to make SO$_2$ on

---

† One of the earliest tracts on the matter was John Evelyn's *Fumifugium, or the Inconvenience of the Aer and Smoke of London Dissipated* which he submitted (with little effect) to Charles II in 1661. Evelyn, a noted diarist and a founder Fellow of the Royal Society, outlined the problem as follows: "For when in all other places the Aer is most Serene and Pure, it is here [in London] Ecclipsed with such a Cloud of Sulphure, as the Sun itself, which gives day to all the World besides, is hardly able to penetrate and impart it here; and the weary Traveller, at many Miles distance, sooner smells than sees the City to which he repairs. This is that pernicious Smoake which sullyes all her Glory, superinducing a sooty Crust or Fur upon all that it lights, spoyling the moveables, tarnishing the Plate, Gildings, and Furniture, and corroding the very Iron-bars and hardest Stones with these piercing and acrimonious Spirits which accompany its Sulphure; and executing more in one year than exposed to the pure Aer of the Country it could effect in some hundreds."

[98a] I. M. CAMPBELL, *Energy and the Atmosphere*, pp. 202–9, Wiley, London, 1977.

[98b] J. HEICKLEN, *Atmospheric Chemistry*, Academic Press, New York, 1976, 406 pp.

[98c] B. MEYER, *Sulfur, Energy, and the Environment*, Elsevier, Amsterdam, 1977, 448 pp.

[98d] R. B. HUSAR, J. P. LODGE, and D. J. MOORE (eds.), *Sulfur in the Atmosphere*, Pergamon Press, Oxford, 1978, 816 pp. Proceedings of the International Symposium at Dubrovnik, September 1977.

[98e] J. O. NRIAGU (ed.), *Sulfur in the Environment*. Part 2. *Ecological Impacts*, Wiley, Chichester, 1979, 494 pp.

[98f] B. E. A. FISHER, The long range transport of sulfur dioxide, *Atmospheric Environment* **9**, 1063–70 (1975).

[98g] L. R. EMBER, Acid pollutants: hitchhikers ride the winds, *Chem. and Eng. News*, 14 Sept., 1981, pp. 20–31.

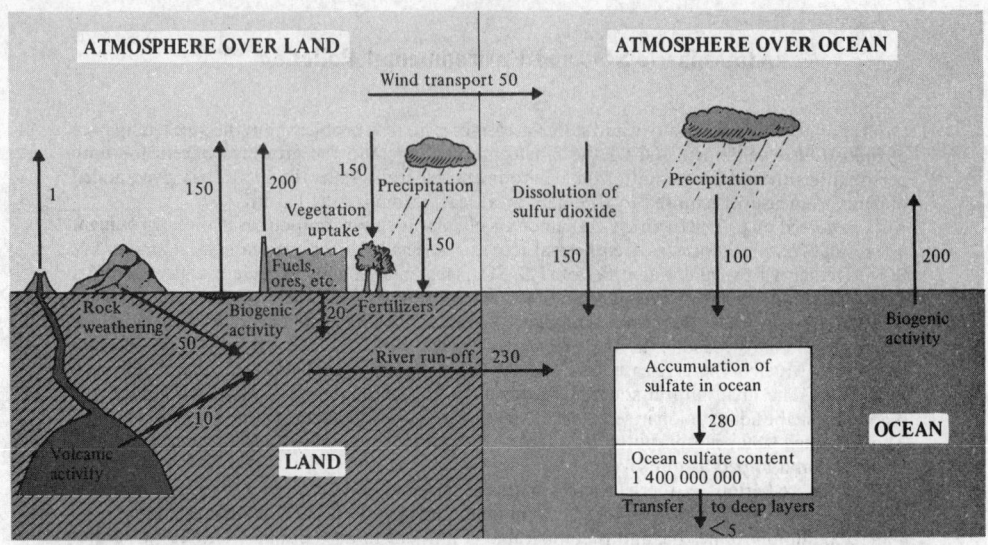

FIG. A   The sulfur budget for the land–atmosphere–ocean system. Annual turnover rates are indicated in units of 10^6 tonnes (as estimated for 1977).[98a]

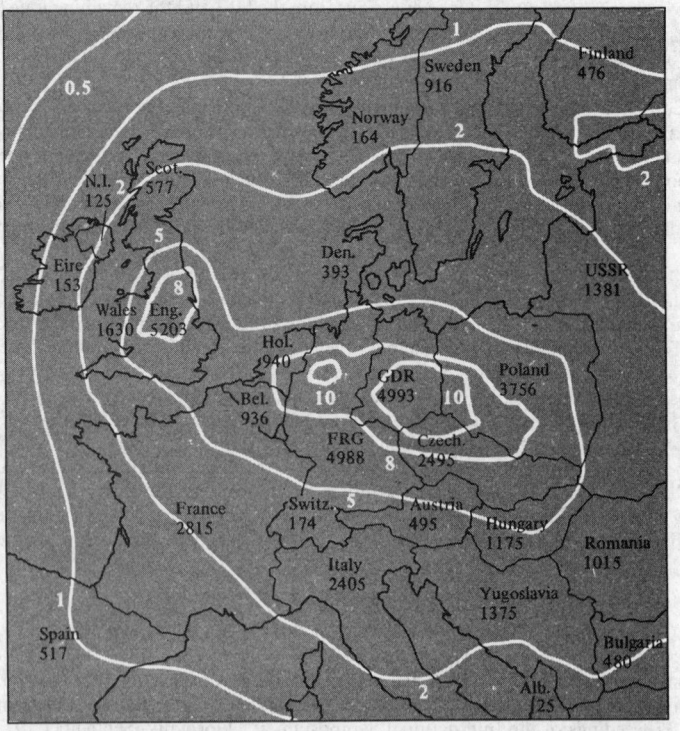

FIG. B   Dry plus wet deposition of SO_2 from all the European sources shown, expressed as contour lines in g (SO_2) m^{-2} y^{-1}. The emission strengths for each country are shown in units of 10^3 tonnes of SO_2 per year.[98f]

a large scale, only contributes <2% to the total, probably because of the efficient design of the process.

Ultimately, pollution can only be avoided by complete removal of SO_2 from the effluent gases, but this council of perfection is both technologically and economically unattainable. Many processes are available to reduce the SO_2 concentration to very low figures, but the vast scale of power generation and domestic heating by coal and oil still results in substantial emission. SO_2 can be removed by scrubbing with a slurry of "milk of lime", $Ca(OH)_2$. Alternatively, partial reduction to H_2S using natural gas (CH_4), naphtha, or coal, followed by catalytic conversion to elemental sulfur by the Claus process can be used:

$$2H_2S + SO_2 \xrightarrow[\text{Al}_2\text{O}_3]{\text{activated}} 3S\downarrow + H_2O$$

The detection of SO_2 in the atmosphere has become a refined analytical procedure. Several techniques are available:

(a) Absorption in aqueous H_2O_2 and titration (or conductimetric determination) of the resulting H_2SO_4:

$$H_2O_2 + SO_2 \longrightarrow H_2SO_4$$

(b) Reaction with $Na_2[HgCl_4]$ or $K_2[HgCl_4]$:

$$[HgCl_4]^{2-} + 2SO_2 + 2H_2O \longrightarrow [Hg(SO_3)_2]^{2-} + 4Cl^- + 4H^+$$

The resulting disulfitomercurate is determined colorimetrically after addition of acidic para-rosaniline and formaldehyde (P. W. West and G. C Gaeke, 1956).

(c) Flame-photometric monitoring of the gas stream using a reducing H_2/air flame and emission of S_2 at 394 nm. This method is sensitive down to 1 part in 10^9 by volume.

(d) Pulsed fluorescent analyser using radiation in the region of 214 nm; this is specific for SO_2 and response is linear over wide ranges down to 1 in 10^9. Commercial instruments are available.

TABLE 15.17 *Some molecular and physical properties of SO_2*

Property	Value	Property	Value
MP/°C	−75.5	Electrical conductivity κ/ohm^{-1} cm^{-1}	$<10^{-8}$
BP/°C	−10.0	Dielectric constant ε (0°)	15.4
Critical temperature/°C	157.5°	Dipole moment μ/D	1.62
Critical pressure/atm	77.7	Angle O–S–O	119°
Density ($-10°$)/g cm^{-3}	1.46	Distance r (S–O)/pm	143.1
Viscosity η (0°)/centipoise	0.403	ΔH_f° (g)/kJ mol^{-1}	−296.9

temperatures and to increase the rate of reaction by use of catalysts. Typical conditions would be to pass a mixture of SO_2 and air over Pt gauze or more commonly a V_2O_5/K_2O contact catalyst supported on Kieselguhr or zeolite.

Gaseous SO_2 is readily soluble in water (3927 cm^3 SO_2 in 100 g H_2O at 20°). Numerous species are present in this aqueous solution of "sulfurous acid" (p. 850). At 0° a cubic clathrate hydrate also forms with a composition $\sim SO_2 \cdot 6H_2O$; its dissociation pressure reaches 1 atm at 7.1°. The ideal composition would be $SO_2 \cdot 5\frac{3}{4}H_2O$ (p. 735).

Liquid SO_2 has been much studied as a nonaqueous solvent.[99, 100] Some of the early work (particularly on the physical properties of the solutions) is now known to be in error but the solvent is especially useful for carrying out a range of inorganic reactions. It is also an excellent solvent for proton nmr studies. In general, covalent compounds are very soluble: e.g. Br_2, ICl, OSX_2, BCl_3, CS_2, PCl_3, $OPCl_3$, and $AsCl_3$ are completely miscible, and most organic amines, ethers, esters, alcohols, mercaptans, and acids are readily soluble. Many uni-univalent salts are moderately soluble, and those with ions such as the tetramethylammonium halides and the alkali metal iodides are freely so. The low dielectric constant of liquid SO_2 leads to extensive ion-pair and ion-triplet formation but the solutions have limiting molar conductances in the range 190–250 ohm^{-1} cm^2 mol^{-1} at 0°. Solvate formation is exemplified by compounds such as $SnBr_4 . SO_2$ and $2TiCl_4 . SO_2$ (see p. 829 for SO_2 as a ligand). Solvolysis reactions are also documented, e.g.:

$$NbCl_5 + SO_2 \longrightarrow NbOCl_3 + OSCl_2$$

$$WCl_6 + SO_2 \longrightarrow WOCl_4 + OSCl_2$$

$$UCl_5 + 2SO_2 \longrightarrow UO_2Cl_2 + 2OSCl_2$$

Several other reaction types have also appeared in the literature but are sometimes purely formal schemes dating from the time when the solvent was (incorrectly) thought to undergo self-ionic dissociation into SO^{2+} and $SO_3{}^{2-}$ or SO^{2+} and $S_2O_5{}^{2-}$.[99] More recently it has been shown that, whereas neither SO_2 nor $OSMe_2$ (DMSO) react with first-row transition metals, the mixed solvent smoothly effects dissolution of the metals with simultaneous oxidation of S^{IV} to S^{VI}, thereby enabling the production of crystalline solvated metal disulfates $(S_2O_7{}^{2-})$ in high yield.[101] Examples are: colourless $[Ti^{IV}(OSMe_2)_6]$ $[S_2O_7]_2$, green $[V^{III}(OSMe_2)_6]_2[S_2O_7]_3$, and the salts $[M^{II}(OSMe_2)_6]$ $[S_2O_7]$, where $M = $ Mn (yellow), Fe (pale green), Co (pale pink), Ni (green), Cu (pale blue), Zn (white), and Cd (white). This is by far the most convenient way to prepare pure disulfates. Dissolution of metals in SO_2 mixed with other solvents such as DMF, DMA, or HMPA also occurs, but in these cases there is no oxidation of the S^{IV}, and the product is usually the metal dithionite, $M^{II}[S_2O_4]$.

Sulfur oxide ligands

Although the lower oxides SO, S_2O and S_2O_2 are unstable as free molecules, all three are known as ligands in stable complexes. Thus, a bridging SO group is formed in the violet-black complex $[\{Mn(\eta^5\text{-}C_5H_5)(CO)_2\}_2SO]$ which can be considered as a derivative

[99] T. C. WADDINGTON, Liquid sulfur dioxide, Chap. 6 in T. C. WADDINGTON (ed.), *Nonaqueous Solvent Systems*, pp. 253–84, Academic Press, London, 1965.

[100] W. KARCHER and H. HECHT, *Chemie in Flussigem Schwefeldioxid*, Vol. 3, Part 2, of G. JANDER, H. SPAUNDAU, and C. C. ADDISON, *Chemistry in Nonaqueous Ionizing Solvents*, pp. 79–193, Pergamon Press, Oxford, 1967. See also D. F. BUROW, Liquid sulfur dioxide, in J. J. LAGOWSKI (ed.), *Nonaqueous Solvents*, Vol. 3, pp. 138–85, Academic Press, New York, 1970.

[101] W. D. HARRISON, J. B. GILL, and D. C. GOODALL, Reactions in mixed nonaqueous systems containing sulfur dioxide. Part 2. The dissolution of transition metals in the binary mixture dimethyl sulfoxide-sulfur dioxide, and ion-pair formation involving the sulfoxylate radical ion in mixed solvents containing sulfur dioxide, *JCS Dalton* 1979, 847–50.

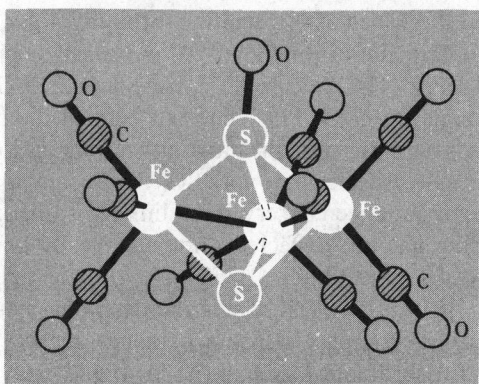

FIG. 15.33   Molecular structure of the complex [Fe$_3$(CO)$_9$S(SO)]. The S–O distance (147 pm) is close to that in free SO (148 pm). The Fe–SO distances (214–222 pm) are significantly shorter than the other three Fe–S distances (223–227 pm), and the S$\cdots$S distance of 272 pm indicates an incipient tendency to form an S$_2$O group. The two bonding Fe–Fe contacts are 262 and 265 pm respectively, and the fact that the third Fe$\cdots$Fe distance is nonbonding implies that both SO and S are acting as 4-electron donors (cf. B$_5$H$_9$ and *nido*-ferraboranes, p. 197).

of Cl$_2$SO.[102] SO can also act as a triply-bridging ligand, e.g. in the dark red-brown cluster compound [Fe$_2$(CO)$_9$S(SO)] (Fig. 15.33), which is formed, together with the known cluster [Fe$_3$(CO)$_9$S$_2$], when aqueous solutions of H$_2$O$_2$ and Na$_2$SO$_3$ are successively added to a solution of Fe(CO)$_5$ in aqueous methanolic NaOH and the mixture then acidified with dilute HCl.[103] Iridium complexes of S$_2$O and SO$_2$ can be made by successive periodate oxidations of the corresponding S$_2$ complex:[104]

$$\left[(dppe)_2\,Ir\begin{smallmatrix}S\\|\\S\end{smallmatrix} \right]^+ \xrightarrow{\ IO_4^-\ } \left[(dppe)_2\,Ir\begin{smallmatrix}S=O\\|\\S\end{smallmatrix} \right]^+ \underset{PPh_3}{\overset{IO_4^-}{\rightleftharpoons}} \left[(dppe)_2\,Ir\begin{smallmatrix}S=O\\|\\S=O\end{smallmatrix} \right]^+$$

All the complexes were isolated as chlorides and the yellow S$_2$O-compound could also be made by monodeoxygenation of the S$_2$O$_2$-complex using PPh$_3$. (See p. 789 for the *S,S*-bridging SSO$_2$ ligand.)

The coordination chemistry of SO$_2$ as a ligand has been extensively studied during the past decade and at least 6 different bonding modes have been established. These are illustrated schematically in Fig. 15.34 and typical examples are given in Table 15.18.[105]

[102] M. Höfler and A. Baitz, Zur Darstellung von [Mn(η^5-C$_5$H$_5$)(CO)$_2$(SOF$_2$)] und [Mn(η^5-C$_5$H$_5$)(CO)$_2$]$_2$SO, *Chem. Ber.* **109**, 3147–50 (1976).

[103] L. Markó, B. Markó-Monostory, T. Madach, and H. Vahrenkamp, First fixation of the unstable species sulfur monoxide in a cluster: synthesis and structure of [Fe$_3$(CO)$_9$S(SO)], *Angew. Chem.*, Int. Edn. (Engl.) **19**, 226–7 (1980).

[104] G. Schmidt and G. Ritter, Disulfur monoxide as complex ligand, *Angew. Chem.*, Int. Edn. (Engl.) **14**, 645–6 (1975).

[105] G. J. Kubas, Diagnostic features of transition-metal-SO$_2$ coordination geometries, *Inorg. Chem.* **18**, 182–8 (1979) and references therein. See also subsequent papers in *Inorg. Chem.* **18**, 223–7, 227–9 (1980) and **19**, 3003–7 (1980) and references 107–9 below. A comprehensive review is given in R. R. Ryan, G. J. Kubas, D. C. Moody, and P. G. Eller, Structure and bonding of transition metal-sulfur dioxide complexes, *Structure and Bonding*, **46**, 47–100 (1981).

The sole example so far established of SO_2 ligating through oxygen is the colourless main-group element complex $SbF_5.OSO$ (mp 66°);[106] dimensions, which can be compared where appropriate with those of free SO_2 (p. 827), are: $Sb-O_\mu$ 213, $O_\mu-S$ 145, $S-O_t$ 138 pm; angle $Sb-O_\mu-S$ 139°, angle $O_\mu-S-O_t$ 119°. The existence of the both *O*-bonded and *S*-bonded SO_2 complexes is reminiscent of the nitro-nitrito-linkage isomerism (p. 534) of the NO_2^- ion (which is formally isoelectronic with SO_2). Interconversion between *O*- and *S*-bonded SO_2 complexes has not been observed but linkage isomerization from pyramidal η^1 to side-on η^2 bonding has been effected photochemically in the solid state at low temperatures: ultra-violet irradiation (365 nm) of the red-brown compound $[Rh^{II}Cl(NH_3)_4(\eta^1-SO_2)]Cl$ at 25–195 K yields the deep-blue η^2-SO_2 isomer which, in turn, reverts to the original complex at room temperature.[107] The existence of both pyramidal and planar η^1 coordination geometries for the $M-SO_2$ group is analogous to the bent and linear geometries for the nitrosyl complexes, $M-NO$ (p. 514), though it has not yet been possible to define with certainty the stereochemical and electronic factors which determine the particular bonding mode adopted.[108] It is clear that nearly all the transition-metal complexes involve the metals in oxidation state zero or +1. Moreover, SO_2 in the pyramidal η^1-clusters tends to be reversibly bound (being eliminated when the complex is heated to <200° and recombining when the system is cooled to room temperature) whereas this tends not to be the case for the other bonding modes. Facile oxidation of the SO_2 by molecular O_2 to give coordinated sulfato complexes (SO_4^{2-}) is also a characteristic of pyramidal η^1-SO_2 which is not shared by the other types. Vibrational spectroscopy is a less-reliable criterion but, having identified pyramidal η^1-SO_2 by reversible dissociation or facile oxidation, the other main bonding modes can often be recognized by the position of the two $\nu(SO)$ stretching modes (cm^{-1}):[105]

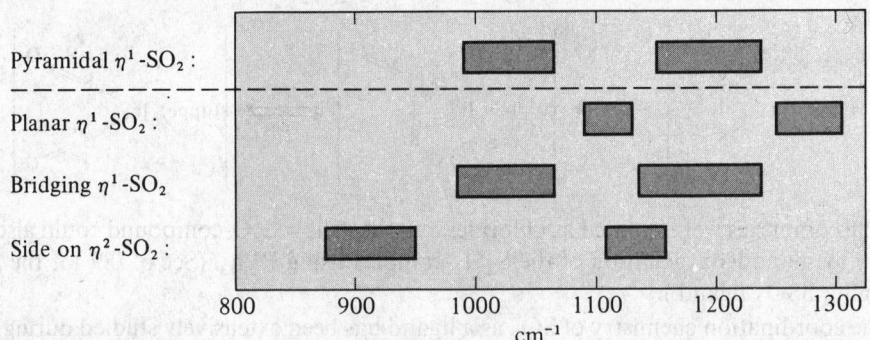

Sometimes coordination of SO_2 to an organometallic complex is followed by intramolecular insertion of SO_2 into the $M-C$ σ bond, e.g.

$$trans\text{-}[PtClPh(PEt_3)_2] + SO_2 \longrightarrow [PtClPh(PEt_3)_2(SO_2)] \quad \text{(5-coordinate)}$$
$$\text{(4 coordinate Pt)} \qquad\qquad\qquad\qquad \downarrow$$
$$trans\text{-}[PtCl(PEt_3)_2\{S(O)_2Ph\}] \quad \text{(4-coordinate)}$$

[106] J. W. MOORE, H. W. BAIRD, and H. B. MILLAR, The crystal and molecular structure of the 1:1 adduct of antimony(V) fluoride and sulfur dioxide, *J. Am. Chem. Soc.* **90**, 1358–9 (1968).

[107] D. A. JOHNSON and V. C. DREW. Photochemical linkage isomerization in coordinated SO_2, *Inorg. Chem.* **18**, 3273–4 (1979).

[108] P. G. ELLER and R. R. RYAN, Crystal and molecular structure of two square-pyramidal rhodium(I)-sulfur dioxide complexes. Bonding effects in pyramidal-SO_2 complexes, *Inorg. Chem.* **19**, 142–7 (1980).

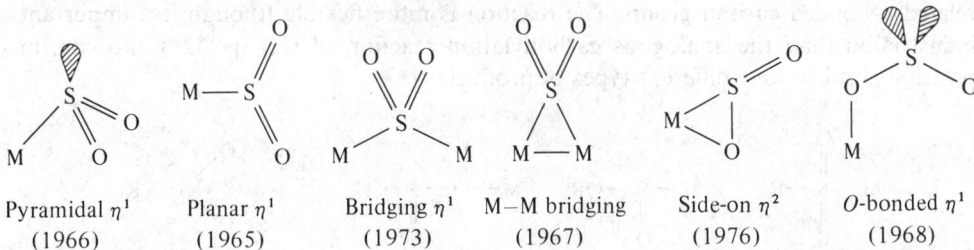

FIG. 15.34 Various bonding modes of SO_2 as a ligand (the date in parentheses refers to the first unambiguous example for each mode).

TABLE 15.18 *Examples of structurally characterized complexes containing SO_2*[a]

Pyramidal η^1	Planar η^1	Bridging η^1
[RhCl(CO)(PPh$_3$)$_2$(SO$_2$)]	[Mn(C$_5$H$_5$)(CO)$_2$(SO$_2$)]	[{Fe(C$_5$H$_5$)(CO)$_2$}$_2$SO$_2$]
[{RhCl(PPh$_3$)$_2$(SO$_2$)}$_2$]	[RuCl(NH$_3$)$_4$(SO$_2$)]Cl	[{IrH(CO)$_2$(PPh$_3$)}$_2$SO$_2$]
[IrCl(CO)(PPh$_3$)$_2$(SO$_2$)]	[Os(CO)ClH(PCy$_3$)$_2$(SO$_2$)]	[{IrI$_2$(CO)(PPh$_3$)}$_2$SO$_2$]
[Ir(SPh)(CO)	[Co(NO)(PPh$_3$)$_2$(SO$_2$)]	
(PPh$_3$)$_2$(SO$_2$)]	[Rh(C$_5$H$_5$)(C$_2$H$_4$)(SO$_2$)]	
[Pt(PPh$_3$)$_3$(SO$_2$)]	[Ni(PPh$_3$)$_3$(SO$_2$)]	
[Pt(PPh$_3$)$_2$(SO$_2$)$_2$]	[Ni(PPh$_3$)$_2$(SO$_2$)$_2$]	

M–M bridging	Side-on η^2
[Fe$_2$(CO)$_8$(μ-SO$_2$)]	[Mo(CO)$_3$(phen)$_3$(η^2-SO$_2$)]
[Fe$_2$(C$_5$H$_5$)$_2$(CO)$_3$(μ-SO$_2$)]	[RuCl(NO)(PPh$_3$)(η^2-SO$_2$)]
[Pd$_2$Cl$_2$(dpm)$_2$(μ-SO$_2$)]	[Rh(NO)(PPh$_3$)$_2$(η^2-SO$_2$)]
[Pd$_3$(CNBu^t)$_5$(μ-SO$_2$)$_2$]	
[Pt$_3$(PPh$_3$)$_3$(μ-SO$_2$)$_3$]	
[Pt(PPh$_3$)$_3$(μ, η^1-Ph)(μ-PPh$_2$)(μ-SO$_2$)]	

[a] See refs. 105–9. In addition there is the O-bonded complex [SbF$_5$(OSO)] (see text), and also some recent examples of SO_2 acting as a polyhapto bridging ligand via both S and O atoms.[109a, 109b]

Ligand abbreviations: PCy$_3$ = tricyclohexylphosphine, dmp = bis(diphenylphosphino)methane, (Ph$_2$P)$_2$CH$_2$; phen = 1,10-phenanthroline.

Intermolecular insertion of SO_2 can also occur (without prior formation of an insolable complex) and the general reaction can be represented by the equation:[110]

$$\{M\text{–}R + SO_2 \longrightarrow \{M\text{–}(SO_2)\text{–}R$$

where $\{M$ represents a metal atom and its pendant ligands and R is an alkyl, aryl, or

[109] D. G. EVANS, G. R. HUGHES, D. M. P. MINGOS, J.-M. BASSETT, and A. J. WELCH, Reactions of zero-valent platinum-sulfur dioxide complexes with dienes; isolation and X-ray crystal structure of a novel, μ-phenyl-, μ-phosphido-, μ-sulfur dioxide *triangulo*-triplatinum cluster, *JCS Chem. Comm.* 1980, 1255–7.

[109a] G. D. JARVINEN, G. J. KUBAS, and R. R. RYAN, A new bridging geometry for sulfur dioxide in [Mo(CO)$_2$(PPh$_3$)(pyridine)(μ-SO$_2$)]$_2$.2CH$_2$Cl$_2$; X-ray crystal structure, *JCS Chem. Comm.* 1981, 305–6.

[109b] C. E. BRIANT, B. R. C. THEOBALD, and D. M. P. MINGOS, Synthesis and X-ray structural characterization of [Rh$_4$(μ-CO)(μ-SO$_2$)$_3${P(OPh)$_3$}$_4$].$\frac{1}{2}$C$_6$H$_6$: A "butterfly" cluster with an unusual bridging mode for sulfur dioxide, *JCS Chem. Comm.* 1981, 963–5.

[110] A. WOJCICKI, Insertion reactions of transition-metal-carbon σ-bonded compounds. Part 2. Sulfur dioxide and other molecules, *Adv. Organomet. Chem.* 12, 31–81 (1974).

related σ-bonded carbon group. The reaction is more flexible (though less important industrially) than the analogous carbonylation reaction of CO (p. 329) and can, in principle, lead to four different types of product:

$$
\underset{\substack{\| \\ O}}{\overset{\substack{O \\ \|}}{M-S-R}} \qquad \underset{\substack{\| \\ O}}{M-S-OR} \qquad \underset{\substack{\| \\ O}}{M-O-S-R} \qquad M\overset{O}{\underset{O}{\diagup\diagdown}}S-R
$$

   S-sulfinate   *O*-alkyl-*S*-sulfoxylate   *O*-sulfinate   *O*-*O*'-sulfinate

Examples of all except the second mode are known.

Sulfur trioxide

SO$_3$ is made on a huge scale by the catalytic oxidation of SO$_2$ (p. 824): it is not usually isolated but is immediately converted to H$_2$SO$_4$ (p. 838). It can also be obtained by the thermolysis of sulfates though rather high temperatures are required (p. 844). SO$_3$ is available commercially as a liquid: such samples contain small amounts (0.03–1.5%) of additives to inhibit polymerization. Typical additives are simple compounds of boron (e.g. B$_2$O$_3$, B(OH)$_3$, HBO$_2$, BX$_3$, MBF$_4$, Na$_2$B$_4$O$_7$), silica, siloxanes, SOCl$_2$, sulfonic acids, etc. The detailed mode of action of these additives remains obscure. SO$_3$ is also readily available as fuming sulfuric acid (or oleum) which is a solution of 25–65% SO$_3$ in H$_2$SO$_4$ (p. 837). Because of its extremely aggressive reaction with most materials, pure anhydrous SO$_3$ is difficult to handle. It can be obtained on a laboratory scale by the double distillation of oleum in an evacuated all-glass apparatus; a small amount of KMnO$_4$ is sometimes used to oxidize any traces of SO$_2$.

In the gas phase, monomeric SO$_3$ has a planar (D_{3h}) structure with S–O 142 pm. This species is in equilibrium with the cyclic trimer S$_3$O$_9$ in both the gaseous and liquid phases: $K_p \approx 1$ atm^{-2} at 25°, $\Delta H° \approx 125$ kJ (mole S$_3$O$_9$)$^{-1}$. Bulk properties therefore often refer to this equilibrium mixture, e.g. bp 44.6°C, $d(25°)$ 1.903 g cm^{-3}, $\eta(25°)$ 1.820 centipoise. Below the mp (16.86°), colourless crystals of ice-like orthorhombic γ-SO$_3$ separate and structural studies reveal that the only species present is the trimer S$_3$O$_9$ (Fig. 15.35). Traces of water (10^{-3} mole%) lead to the rapid formation of glistening, white, needle-like crystals of β-SO$_3$ which is actually a mixture of fibrous, polymeric polysulfuric acids HO(SO$_2$O)$_x$H, where x is very large ($\approx 10^5$). The helical chain structure of β-SO$_3$ is shown in Fig. 15.35 (cf. polyphosphates, p. 615). A third and still more stable form, α-SO$_3$, also requires traces of moisture or other polymerizing agent for its formation but involves some cross-linking between the chains to give a complex layer structure (mp 62°). The standard enthalpies of formation ($\Delta H_f°$/kJ mol^{-1}) of the various forms of SO$_3$ at 25°C are: gas −395.2, liquid −437.9, γ-crystals −447.4, β-crystals −449.6, α-solid −462.4.

SO$_3$ reacts vigorously and extremely exothermically with water to give H$_2$SO$_4$. Substoichiometric amounts yield oleums and mixtures of various polysulfuric acids (p. 845). Hydrogen halides give the corresponding halogenosulfuric acids HSO$_3$X. SO$_3$ extracts the elements of H$_2$O from carbohydrates and other organic matter leaving a carbonaceous char. It acts as a strong Lewis acid towards a wide variety of inorganic and

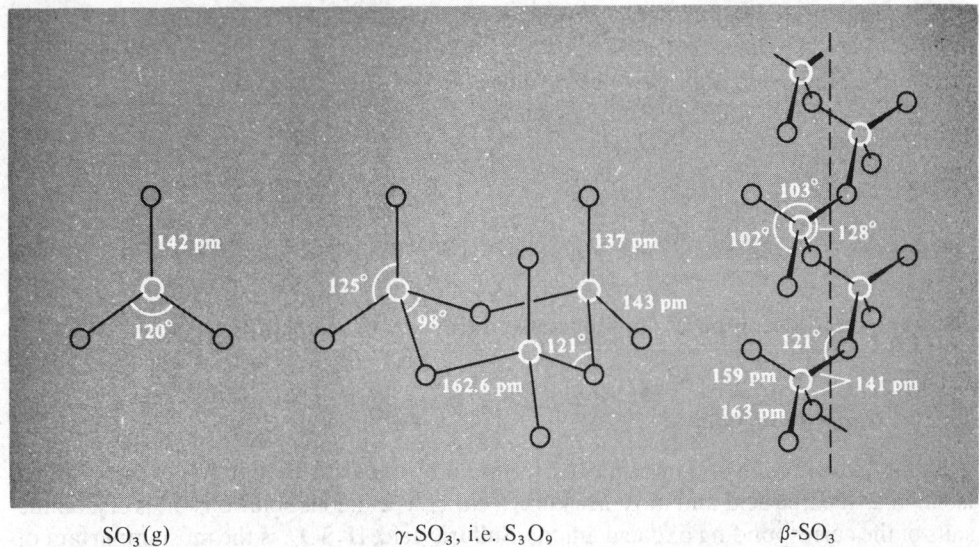

SO$_3$(g) γ-SO$_3$, i.e. S$_3$O$_9$ β-SO$_3$

FIG. 15.35   Structure of the monomeric, trimeric, and chain-polymeric forms of sulfur trioxide.

organic ligands to give adducts: e.g. oxides give SO$_4^{2-}$, Ph$_3$P gives Ph$_3$P.SO$_3$,[110a] Ph$_3$AsO gives Ph$_3$AsO.SO$_3$ etc. Frequently further reaction ensues: thus, under various conditions reaction with NH$_3$ yields H$_2$NSO$_3$H, HN(SO$_3$H)$_2$, HN(SO$_3$NH$_4$)$_2$, NH$_4$N(SO$_3$NH$_4$)$_2$, etc. SO$_3$ can also act as a ligand towards strong electron-pair acceptors such as AsF$_3$, SbF$_3$, and SbCl$_3$. It is reduced to SO$_2$ by activated charcoal or by metal sulfides. The reaction with metal oxides (particularly Fe$_3$O$_4$) to give sulfates is used industrially to rid stack-gases of unwanted byproduct SO$_3$.

Higher oxides

The reaction of gaseous SO$_2$ or SO$_3$ with O$_2$ in a silent electric discharge gives colourless polymeric condensates of composition SO$_{3+x}$ ($0 < x < 1$). These materials are derived from β-SO$_3$ by random substitution of oxo-bridges by peroxo-bridges:

$$
\begin{array}{c}
\quad\quad\; O \\
\quad\quad\; \| \\
-O-S-O-O-S-O- \\
\quad\quad\; \| \\
\quad\quad\; O
\end{array}
$$

Hydrolysis of the polymers yields H$_2$SO$_4$ and H$_2$SO$_5$ (p. 846), with H$_2$O$_2$ and O$_2$ as secondary products.

Monomeric neutral SO$_4$ can be obtained by reaction of SO$_3$ and atomic oxygen; photolysis of SO$_3$/ozone mixtures also yields monomeric SO$_4$, which can be isolated by inert-gas matrix techniques at low temperatures (15–78 K). Vibration spectroscopy

---

[110a] R. L. BEDDOES and O. S. MILLS, The structure of the 1:1 complex formed between triphenylphosphine and sulfur trioxide, *J. Chem. Research* (M) 1981, 2772–89; (S) 1981, 233; see also *JCS Chem. Comm.* 1981, 789–90.

indicates either an open peroxo C_s structure or a closed peroxo C_{2v} structure, the former being preferred by the most recent study, on the basis of agreement between observed and calculated frequencies and reasonable values for the force constants:[111]

C_s C_{2v}

The compound decomposes spontaneously below room temperature.

15.2.6 *Oxoacids of sulfur*

Sulfur, like nitrogen and phosphorus, forms many oxoacids though few of these can be isolated as the free acid and most are known either as aqueous solutions or as crystalline salts of the corresponding oxoacid anions. Sulfuric acid, H_2SO_4, is the most important of all industrial chemicals and is manufactured on an enormous scale, greater than for any other compound of any element (p. 467). Other compounds, such as thiosulfates, sulfites, disulfites, and dithionites, are valuable reducing agents with a wide variety of applications. Nomenclature is somewhat confusing but is summarized in Table 15.19 which also gives an indication of the various oxidation states of S and a schematic representation of the structures. Previously claimed species such as "sulfoxylic acid" (H_2SO_2), "thiosulfurous acid" $(H_2S_2O_2)$, and their salts are now thought not to exist.

Many of the sulfur oxoacids and their salts are connected by oxidation–reduction equilibria: some of the more important standard reduction potentials are summarized in Table 15.20 and displayed in graphic form as a volt-equivalent diagram (p. 498) in Fig. 15.36. By use of the couples in Table 15.20 data for many other oxidation–reduction equilibria can readily be calculated.† Several important points emerge which are immediately apparent from inspection of Fig. 15.36. For example, it is clear that, in acid solutions, the gradient between H_2S and S_8 is less than between S_8 and any positive oxidation state, so that H_2S is thermodynamically able to reduce any oxoacid of sulfur to the element. Again, as all the intermediate oxoacids lie above the line joining HSO_4^- and S_8, it follows that all can ultimately disproportionate into sulfuric acid and the element. Similarly, any moderately powerful oxidizing agent should be capable of oxidizing the intermediate oxoacids to sulfuric acid (sometimes with concurrent precipitation of sulfur) though by suitable choice of conditions it is often possible to obtain kinetically stable intermediate oxidation states (e.g. the polythionates with the stable S–S linkages). It follows that all the oxoacids except H_2SO_4 are moderately strong reducing agents (see below).

---

† It is an instructive exercise to check the derivation of the numerical data given in parentheses in Fig. 15.36 from the data given in Table 15.20 and to calculate the standard reduction potentials of other couples, e.g. HSO_4^-/H_2S 0.289 V, HSO_4^-/H_2SO_3 0.119 V, $H_2SO_3/S_2O_3^{2-}$ 0.433 V, etc.

[111] P. La Bonville, R. Kugel, and J. R. Ferraro, Normal coordinate analysis of the neutral CO_3 and SO_4 molecules, *J. Chem. Phys.* **67**, 1477–81 (1977).

TABLE 15.19　*Oxoacids of sulfur*

Formula	Name	Ox. states	Schematic structure*	Salt
H_2SO_4	sulfuric	VI		sulfate, $SO_4{}^{2-}$ H-sulfate, $HOSO_3{}^-$
$H_2S_2O_7$	disulfuric	VI		disulfate, $O_3SOSO_3{}^{2-}$
$H_2S_2O_3$	thiosulfuric	IV, 0, (or VI, −II)		thiosulfate, $SSO_3{}^{2-}$
H_2SO_5	peroxomonosulfuric	VI		peroxomonosulfate, $OOSO_3{}^{2-}$
$H_2S_2O_8$	peroxodisulfuric	VI		peroxodisulfate, $O_3SOOSO_3{}^{2-}$
$H_2S_2O_6$	dithionic*	V		dithionate, $O_3SSO_3{}^{2-}$
$H_2S_{n+2}O_6$	polythionic	V, 0		polythionate, $O_3S(S)_nSO_3{}^{2-}$
H_2SO_3	sulfurous*	IV		sulfite, $SO_3{}^{2-}$ H-sulfite, $HOSO_2{}^-$
$H_2S_2O_5$	disulfurous*	V, III		disulfite, $O_3SSO_2{}^{2-}$
$H_2S_2O_4$	dithionous*	III		dithionite, $O_2SSO_2{}^{2-}$

*Acids marked with an asterisk do not exist in the free state but are known as salts.

TABLE 15.20   *Some standard reduction potentials of*
sulfur species (25°, pH 0)

Couple	$E°/V$
$2H_2SO_3 + H^+ + 2e^- \rightleftharpoons HS_2O_4^- + 2H_2O$	-0.082
$S + 2H^+ + 2e^- \rightleftharpoons H_2S$	$+0.142$
$HSO_4^- + 7H^+ + 6e^- \rightleftharpoons S + 4H_2O$	0.339
$H_2SO_3 + 4H^+ + 4e^- \rightleftharpoons S + 3H_2O$	0.449
$S_2O_3^{2-} + 6H^+ + 4e^- \rightleftharpoons 2S + 3H_2O$	0.465
$4H_2SO_3 + 4H^+ + 6e^- \rightleftharpoons S_4O_6^{2-} + 6H_2O$	0.509
$S_2O_6^{2-} + 4H^+ + 2e^- \rightleftharpoons 2H_2SO_3$	0.564
$S_2O_8^{2-} + 2H^+ + 2e^- \rightleftharpoons 2HSO_4^-$	2.123

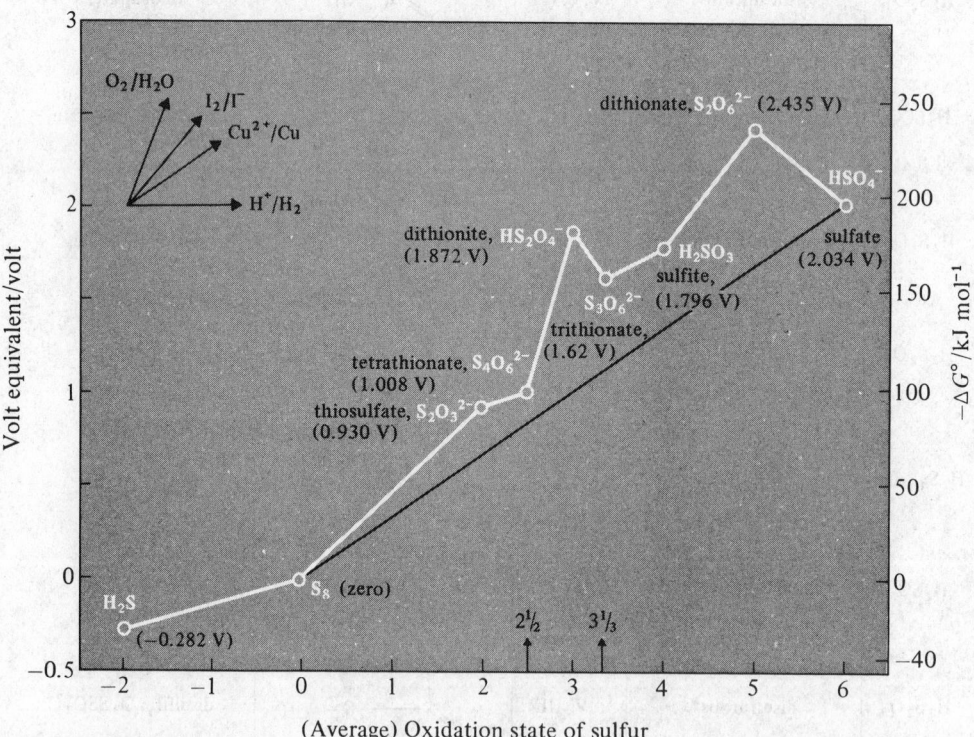

FIG. 15.36   Volt-equivalent diagram for sulfur-containing species in acid solution.

The formal interrelationship between the various oxoacids of sulfur can also be illustrated in a scheme[112] which places less emphasis on oxidation–reduction reactions but which is useful in suggesting possible alternative synthetic routes to these oxoacids.

[112] M. SCHMIDT and W. SIEBERT, Oxyacids of sulfur, Section 2.4 in *Comprehensive Inorganic Chemistry*, Vol. 2, Chapter 23, pp. 868–98.

Thus successive addition of SO_3 or SO_2 to H_2O can be represented by the scheme:

$$H_2O \xrightarrow{\text{SO}_3} H_2SO_4 \xrightarrow{\text{SO}_3} H_2S_2O_7$$
$$\text{sulfuric} \qquad\qquad \text{disulfuric}$$
$$\downarrow \text{SO}_2$$
$$H_2S_2O_6$$
$$\text{dithionic}$$
$$\uparrow \text{SO}_3$$
$$H_2O \xrightarrow{\text{SO}_2} H_2SO_3 \xrightarrow{\text{SO}_2} H_2S_2O_5$$
$$\text{sulfurous} \qquad\qquad \text{disulfurous}$$

Likewise addition of SO_3 to H_2O_2, H_2S, and H_2S_n generates the formulae of the other oxoacids as follows:

$$H_2O_2 \xrightarrow{\text{SO}_3} H_2SO_5 \xrightarrow{\text{SO}_3} H_2S_2O_8$$
$$\text{peroxomonosulfuric} \quad \text{peroxodisulfuric}$$

$$H_2S \xrightarrow{\text{SO}_3} H_2S_2O_3 \xrightarrow{\text{SO}_3} H_2S_3O_6$$
$$\text{thiosulfuric} \qquad\qquad \text{trithionic}$$

$$H_2S_n \xrightarrow{\text{SO}_3} H_2S_{n+1}O_3 \xrightarrow{\text{SO}_3} H_2S_{n+2}O_6$$
$$\text{polythionic}$$

It should be emphasized that not all the processes in these schemes represent viable syntheses, and other routes are frequently preferred. The following sections give a fuller discussion of the individual oxoacids and their salts.

Sulfuric acid, H_2SO_4

Anhydrous sulfuric acid is a dense, viscous liquid which is readily miscible with water in all proportions: the reaction is extremely exothermic (~ 880 kJ mol^{-1} at infinite dilution) and can result in explosive spattering of the mixture if the water is added to the acid; it is therefore important always to reverse the order and add the acid to the water slowly and with stirring. The large-scale preparation of sulfuric acid is a major industry in most countries and is described in the Panel.

TABLE 15.21   *Some physical properties of anhydrous H_2SO_4 and D_2SO_4*[a]

Property	H_2SO_4	D_2SO_4
MP/°C	10.371	14.35
BP/°C	~ 300 (decomp)	—
Density (25°)/g cm^{-3}	1.8267	1.8572
Viscosity (25°)/centipoise	24.55	24.88
Dielectric constant ε	100	—
Specific conductivity κ (25°)/ohm^{-1} cm^{-1}	1.0439×10^{-2}	0.2832×10^{-2}

[a] In the gas phase H_2SO_4 and D_2SO_4 adopt the C_2 conformation with r(O–H) 97 pm, r(S–OH) 157.4 pm, r(S–O) 142.2 pm; the various interatomic and dihedral angles were also determined and the molecular dipole moment calculated to be 2.73 D.[112a]

[112a] R. L. KUCZKOWSKI, R. D. SUENRAM, and F. J. LOVAS, Microwave spectrum, structure, and dipole moment of sulfuric acid, *J. Am. Chem. Soc.* **103**, 2561–6 (1981).

Industrial Manufacture of Sulfuric Acid

Sulfuric acid is the world's most important industrial chemical and is the cheapest bulk acid available in every country of the world. It was one of the first chemicals to be produced commercially in the USA (by John Harrison, Philadelphia, 1793); in Europe the history of its manufacture goes back even further—by at least two centuries.[120, 121] Concentrated sulfuric acid ("oil of vitriol") was first made by the distillation of "green vitriol", $FeSO_4 . nH_2O$ and was needed in quantity to make Na_2SO_4 from NaCl for use in the Leblanc Process (p. 82). This expensive method was replaced in the early eighteenth century by the burning of sulfur and saltpetre (p. 103) in the necks of large glass vessels containing a little water. The process was patented in 1749 by Joshua Ward (the Quack of Hogarth's *Harlot's Progress*) though it had been in use for several decades previously in Germany, France, and England. The price plumeted from £2 to 2 shillings per pound (say $5.00 to $0.50 per kg). It dropped by a further factor of 10 by 1830 following firstly John Roebuck's replacement (*ca.* 1755) of the fragile glass jars by lead-chambers of 200 ft^3 (5.7 m^3) capacity, and, secondly, the discovery (by N. Clement and C. B. Désormes in 1793) that the amount of $NaNO_3$ could be substantially reduced by admitting air for the combustion of sulfur. By 1860 James Muspratt (UK) was using lead chambers of 56 000 ft^3 capacity (1585 m^3) and the process was continuous. The maximum concentration of acid that could be produced by this method was about 78% and until 1870 virtually the only source of oleum was the Nordhausen works (distillation of $FeSO_4 . nH_2O$). Today both processes have been almost entirely replaced by the modern contact process. This derives originally from Peregrine Philips' observation (patented in 1831) that SO_2 can be oxidized to SO_3 by air in the presence of a platinum catalyst.

The modern process uses a potassium-sulfate-promoted vanadium(V) oxide catalyst on a silica or kieselguhr support.[122] The SO_2 is obtained either by burning pure sulfur or by roasting sulfide minerals (p. 768) notably iron pyrite, or ores of Cu, Ni, and Zn during the production of these metals. On a worldwide basis about 60% of the SO_2 comes from the burning of sulfur and some 40% by the roasting of sulfide ores but in some countries (e.g. the UK) over 95% comes from the former.

The oxidation of SO_2 to SO_3 is exothermic and reversible:

$$SO_2 + \tfrac{1}{2}O_2 \rightleftharpoons SO_3; \quad \Delta H° - 98 \text{ kJ mol}^{-1}$$

According to le Chatelier's principle the *yield* of SO_3 will increase with increase in pressure, increase in excess O_2 concentration, and removal of SO_3 from the reaction zone; each of these factors will also increase the *rate* of conversion somewhat (by the law of mass action). Reaction rate will also increase substantially with increase in temperature but this will simultaneously decrease the yield of the exothermic forward reaction. Accordingly, a catalyst is required to accelerate the reaction without diminishing the yield. Optimum conditions involve an equimolar feed of O_2/SO_2 (i.e. air/SO_2:5/1) and a 4-stage catalytic converter operating at the temperatures shown in Fig. A.[123] (The V_2O_5 catalyst is inactive below 400°C and breaks down above 620°C; it is dispersed as a thin film of molten salt on the catalyst support.) Such a converter may be 13 m high, 9 m in diameter, and contain 80 tonnes of catalyst pellets. A labelled model of a typical interpass absorption (IPA) sulfuric acid plant and a photograph of two identical 500-tonnes-per-day plants at ICI Billingham are shown in Figs. B and C.[123] The gas temperature rises during passage through the catalyst bed and is recooled by passage through external heat-exchanger loops between the first three stages. In the most modern "double-absorption" plants (IPA) the SO_3 is removed at this stage before the residual SO_2/O_2 is passed through a fourth catalyst bed for final conversion. The SO_3 gas cannot be absorbed directly in water because it would first come into contact with the water-vapour above the absorbant and so produce a stable mist of fine droplets of H_2SO_4 which would then pass right through the absorber and out into the

[120] T. K. DERRY and T. I. WILLIAMS, *A Short History of Technology from the Earliest Times to AD 1900*, pp. 268, and 534–5, Oxford University Press, Oxford, 1960.

[121] L. F. HABER, *The Chemical Industry During the Nineteenth Century*, Oxford University Press, Oxford, 1958, 292 pp; L. F. HABER, *The Chemical Industry 1900–1930*, Oxford University Press, Oxford, 1971, 452 pp.

[122] A. PHILLIPS, The modern sulfuric acid industry, in R. THOMPSON (ed.), *The Modern Inorganic Chemicals Industry*, pp. 183–200, The Chemical Society, London, 1977.

[123] R. FINCH, S. NEAL, and E. MOORE (eds.), *Inorganic Chemicals*, pp. 33–42, ICI Educational Publications, The Kynock Press, London, 1978.

Feed gas 10% SO$_2$, 11% O$_2$

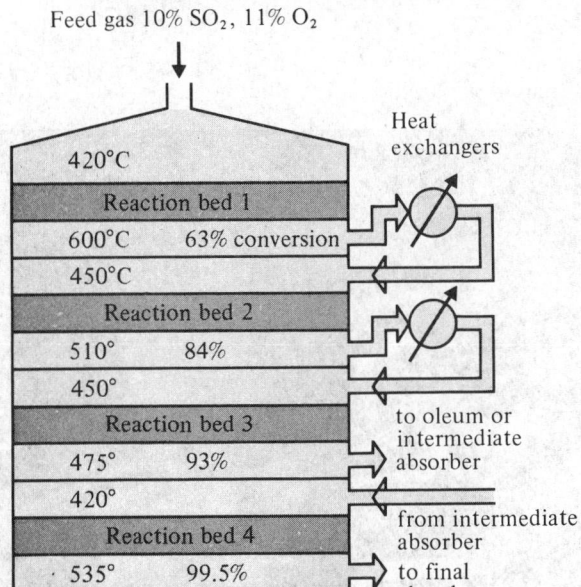

Schematic diagram of converter.

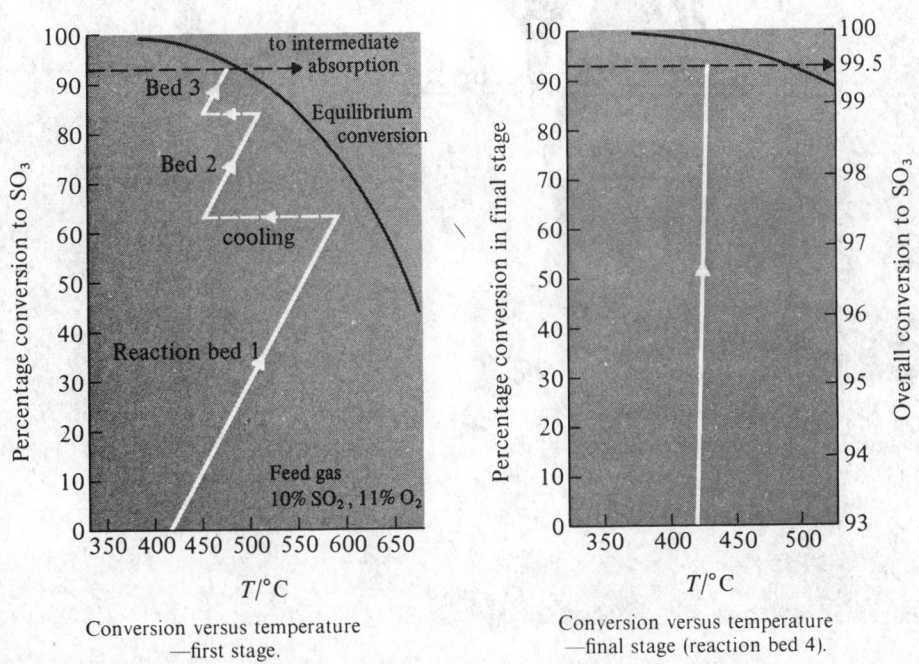

Conversion versus temperature —first stage.

Conversion versus temperature —final stage (reaction bed 4).

Fig. A   Double absorption (IPA) sulfuric acid plant.

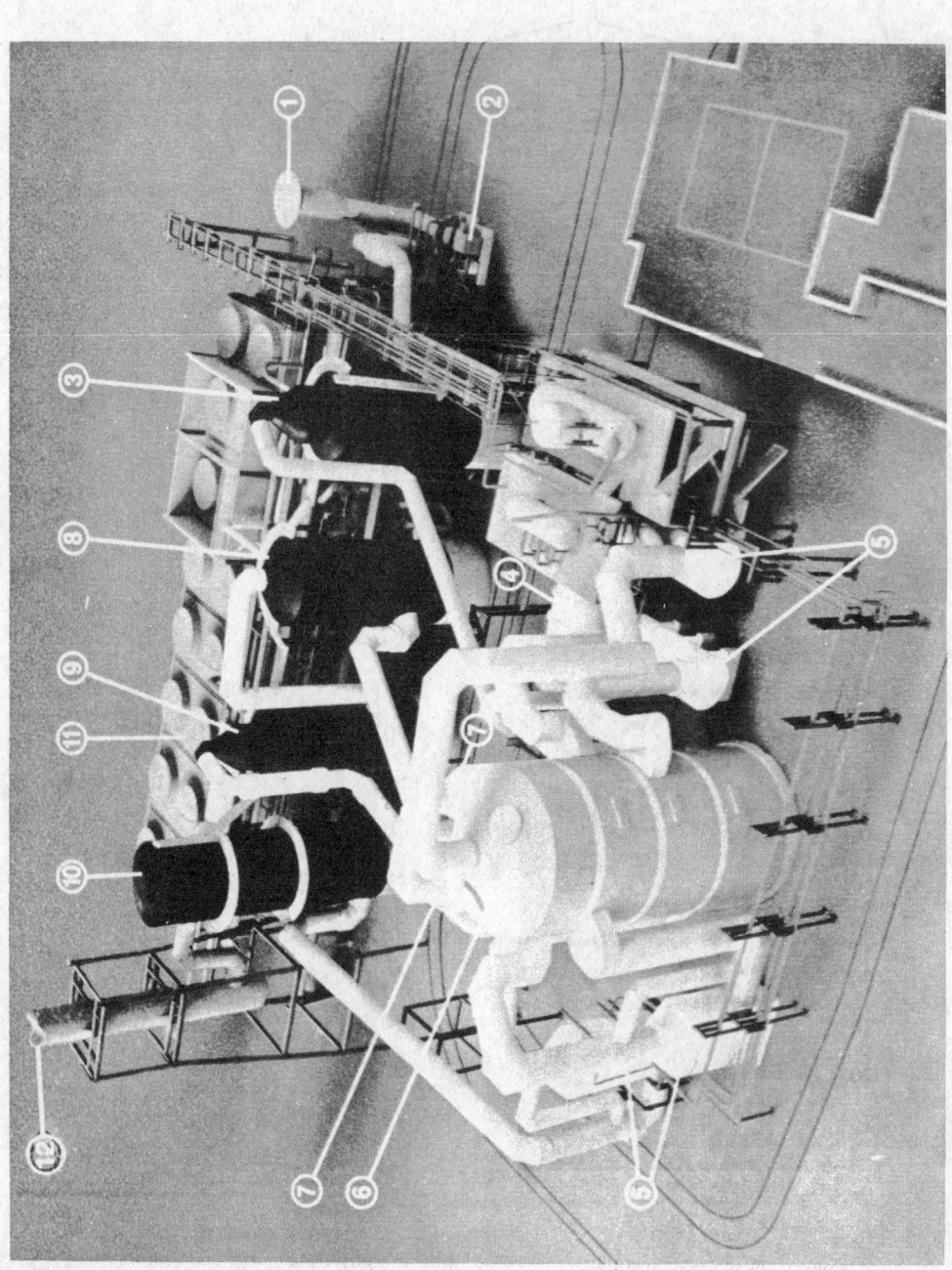

Fig. B   Model of double absorption sulfuric acid plant. 1, air intake; 2. air pump; 3, drying tower; 4, sulfur furnace (980°C); 5, boilers; 6, converter (420–600°C); 7, heat exchangers; 8, oleum absorber (60–180°C); 9, intermediate absorber; 10, final absorber (80°C); 11, acid coolers; 12, vent stack.

Fig. C Two identical 500-tonnes-per-day sulfuric acid plants at ICI Billingham.

atmosphere. Instead, absorption is effected by 98% H_2SO_4 in ceramic-packed towers and sufficient water is added to the circulating acid to maintain the required concentration. Commercial conc H_2SO_4 is generally 96–98% to prevent undesirable solidification of the product. The main construction materials of the sulfur burner, catalytic converter, absorption towers, and ducting are mild steel and stainless steel, and the major impurity in the acid is therefore Fe^{II} (10 ppm) together with traces of SO_2 and NO_x.

Some idea of the accelerating demand for sulfuric acid can be gained from the following UK production figures:

Year	1860	1870	1880	1890	1900	1917	1960	1980
10^3 tonnes	260	560	900	870	1100	1400	2750	4750

Figures for France, Germany, and the USA were lower than these until the turn of the century, but then the USA began to outstrip the rest. At about the same time superphosphate manufacture overtook the Leblanc soda process as the main user of H_2SO_4. H_2SO_4 production is now often taken as a reliable measure of a nation's industrial strength because it enters into so many industrial and manufacturing processes. Thus in 1900 production was equivalent to 4.05 million tonnes of 100% H_2SO_4 distributed as follows (%):

UK	USA	Germany	France	Austria	Belgium	Russia	Japan
25.9	23.2	21.0	15.5	4.9	4.0	3.1	1.2

By 1976 world production was 113 million tonnes and the distribution changed to the following (%):

USA	USSR	Japan	Germany	France	Poland	UK	Canada	Spain	Italy	Others
25.6	17.7	5.4	4.1	3.5	3.2	2.9	2.8	2.5	2.4	29.9

US production increased still further from 29.5 to 37.2 million tonnes between 1976 and 1981. Such vast quantities require huge plants: these frequently have a capacity in excess of 2000 tonnes per day in the USA but are more commonly in the range 300–750 tonnes per day in Europe and smaller still in less industrialized countries. Even so, the energy flows are enormous as can be appreciated by scaling up the following reactions:

$$S + O_2 \longrightarrow SO_2, \qquad \Delta H° \ -297 \text{ kJ mol}^{-1}$$
$$SO_2 + \tfrac{1}{2}O_2 \longrightarrow SO_3, \qquad \Delta H° \ -9.8 \text{ kJ mol}^{-1}$$
$$SO_3 + H_2O \text{ (in 98\% } H_2SO_4) \longrightarrow H_2SO_4, \qquad \Delta H° \ -130 \text{ kJ mol}^{-1}$$

For example, the oxidation of S to SO_3 liberates nearly 4×10^9 J per tonne of H_2SO_4 of which ~3 GJ can be sold as energy in the form of steam and much of the rest used to pump materials around the plant, etc. A plant producing 750 tonnes per day of H_2SO_4 produces ~25 MW of byproduct thermal energy, equivalent to ~7 MW of electricity if the steam is used to drive generators. Effective utilization of this energy is an important factor in minimizing the cost of sulfuric acid which remains a remarkably cheap commodity despite inflation ($50 per tonne in 1976, $90 per tonne in 1980).

Environmental legislation in the USA requires that sulfur emitted from the stack (SO_2 and persistent H_2SO_4 mist) must not exceed 0.3% of the sulfur burned (0.5% in the UK). Despite this, because of the vast scale of the industry, large quantities of unconverted SO_2 are vented to the atmosphere each year (say 0.3% of 113×10^6 tonnes $\times \frac{64}{98} = 220\,000$ tonnes SO_2 pa). It is a testament to the efficiency of the process that this represents a global impact of only some 600 tonnes SO_2 per day, which is minute compared with other sources of this pollution (p. 825).

The pattern of use of H_2SO_4 varies from country to country and from decade to decade. Current US usage is dominated by fertilizer production (65%) followed by chemical manufacture, metallurgical uses, and petroleum refining (~5% each). In the UK the distribution of uses is more even: only 30% of the H_2SO_4 manufactured is used in the fertilizer industry but 18% goes on paints, pigments, and dyestuff intermediates, 16% on chemicals manufacture, 12% on soaps and detergents, 10% on natural and manmade fibres, and 2.5% on metallurgical applications. Petroleum refining accounts for only 1% of the H_2SO_4 in the UK.

Some physical properties of anhydrous H_2SO_4 (and D_2SO_4) are in Table 15.21.[113, 114] In addition, several congruently melting hydrates, $H_2SO_4 . nH_2O$, are known with $n = 1, 2, 3, 4$ (mps 8.5°, $-39.5°$, $-36.4°$, and $-28.3°$, respectively). Other compounds in the

[113] R. J. GILLESPIE and E. A. ROBINSON, Sulfuric acid, Chap. 4 in T. C. WADDINGTON (ed.), *Nonaqueous Solvent Systems*, pp. 117–210, Academic Press, London, 1965. A definitive review with some 250 references.

[114] N. N. GREENWOOD and A. THOMPSON, Comparison of some properties of anhydrous sulfuric acid and dideuterosulfuric acid over a range of temperatures, *J. Chem. Soc.* 1959, 3474–84.

H_2O/SO_3 system are $H_2S_2O_7$ (mp 36°) and $H_2S_4O_{13}$ (mp 4°). Anhydrous H_2SO_4 is a remarkable compound with an unusually high dielectric constant, and a very high electrical conductivity which results from the ionic self-dissociation (autoprotolysis) of the compound coupled with a proton-switch mechanism for the rapid conduction of current through the viscous H-bonded liquid.† It thus has many features in common with anhydrous H_3PO_4 (p. 595) but the equilibria are reached much more rapidly (almost instantaneously) in H_2SO_4:

$$2H_2SO_4 \rightleftharpoons H_3SO_4^+ + HSO_4^-; \quad K_{ap} (25°) = [H_3SO_4^+][HSO_4^-] = 2.7 \times 10^{-4}$$

This value is compared with those for other acids and protonic liquids in Table 15.22:[113] the extent of autoprotolysis in H_2SO_4 is greater than that in water by a factor of more than 10^{10} and is exceeded only by anhydrous H_3PO_4 and $[HBF_3(OH)]$ (p. 223). In addition to autoprotolysis, H_2SO_4 undergoes ionic self-dehydration:

$$2H_2SO_4 \rightleftharpoons H_3O^+ + HS_2O_7^-; \quad K_{id} (25°)\ 5.1 \times 10^{-5}$$

TABLE 15.22 *Autoprotolysis constants at 25°*

Compound	$-\log K_{ap}$	Compound	$-\log K_{ap}$	Compound	$-\log K_{ap}$
$HBF_3(OH)$	~ −1	HCO_2H	6.2	H_2O_2	12
H_3PO_4	~2	HF	9.7	H_2O	14.0
H_2SO_4	3.6	$MeCO_2H$	12.6	D_2O	14.8
D_2SO_4	4.3	EtOH	18.9	NH_3	29.8

This arises from the primary dissociation of H_2SO_4 into H_2O and SO_3 which then react with further H_2SO_4 as follows:

$$H_2O + H_2SO_4 \rightleftharpoons H_3O^+ + HSO_4^-; \quad K_{H_2O} (25°) = [H_3O^+][HSO_4^-]/[H_2O] \sim 1$$
$$SO_3 + H_2SO_4 \rightleftharpoons H_2S_2O_7$$
$$H_2S_2O_7 + H_2SO_4 \rightleftharpoons H_3SO_4^+ + HS_2O_7^-;$$
$$K_{H_2S_2O_7} (25°) = [H_3SO_4^+][HS_2O_7^-]/[H_2S_2O_7] = 1.4 \times 10^{-2}$$

It is clear that "pure" anhydrous sulfuric acid, far from being a single substance in the bulk liquid phase, comprises a dynamic equilibrium involving at least seven well-defined species. The concentration of the self-dissociation products in H_2SO_4 and D_2SO_4 at 25° (expressed in millimoles of solute per kg solvent) are:

HSO_4^-	$H_3SO_4^+$	H_3O^+	$HS_2O_7^-$	$H_2S_2O_7$	H_2O	Total
15.0	11.3	8.0	4.4	3.6	0.1	42.4

DSO_4^-	$D_3SO_4^+$	D_3O^+	$DS_2O_7^-$	$D_2S_2O_7$	D_2O	Total
11.2	4.1	11.2	4.9	7.1	0.6	39.1

† For example, at 25° the single-ion conductances for $H_3SO_4^+$ and HSO_4^- are 220 and 150 respectively, whereas those for Na^+ and K^+ which are viscosity-controlled are only 3–5.

As the molecular weight of H_2SO_4 is 98.078 it follows that 1 kg contains 10.196 mol; hence the predominant ions are present to the extent of about 1 millimole per mole of H_2SO_4 and the total concentration of species in equilibrium with the parent acid is 4.16 millimole per mole. Many of the physical and chemical properties of anhydrous H_2SO_4 as a nonaqueous solvent stem from these equilibria.[113]

In the sulfuric acid solvent system, compounds that enhance the concentration of the solvo-cation HSO_4^- will behave as bases and those that give rise to $H_3SO_4^+$ will behave as acids (p. 487). Basic solutions can be formed in several ways of which the following examples are typical:

(a) Dissolution of metal hydrogen sulfates:

$$KHSO_4 \xrightarrow{H_2SO_4} K^+ + HSO_4^-$$

(b) Solvolysis of salts of acids that are weaker than H_2SO_4:

$$KNO_3 + H_2SO_4 \longrightarrow K^+ + HSO_4^- + HNO_3$$
$$NH_4ClO_4 + H_2SO_4 \longrightarrow NH_4^+ + HSO_4^- + HClO_4$$

(c) Protonation of compounds with lone-pairs of electrons:

$$H_2O + H_2SO_4 \longrightarrow H_3O^+ + HSO_4^-$$
$$Me_2CO + H_2SO_4 \longrightarrow Me_2COH^+ + HSO_4^-$$
$$MeCOOH + H_2SO_4 \longrightarrow MeC(OH)_2^+ + HSO_4^-$$

(d) Dehydration reactions:

$$HNO_3 + 2H_2SO_4 \longrightarrow NO_2^+ + H_3O^+ + 2HSO_4^-$$
$$N_2O_5 + 3H_2SO_4 \longrightarrow 2NO_2^+ + H_3O^+ + 3HSO_4^-$$

The reaction with HNO_3 is quantitative, and the presence of large concentrations of the nitronium ion, NO_2^+, in solutions of HNO_3, MNO_3, and N_2O_5 in H_2SO_4 enable a detailed interpretation to be given of the nitration of aromatic hydrocarbons by these solutions.

Because of the high acidity of H_2SO_4 itself, bases form the largest class of electrolytes and only few acids (proton donors) are known in this solvent system. As noted above, $H_2S_2O_7$ acts as a proton donor to H_2SO_4 and HSO_3F is also a weak acid:

$$HSO_3F + H_2SO_4 \rightleftharpoons H_3SO_4^+ + SO_3F^-$$

One of the few strong acids is tetra(hydrogen sulfato)boric acid $HB(HSO_4)_4$; solutions of this can be obtained by dissolving boric acid in oleum:

$$B(OH)_3 + 3H_2S_2O_7 \longrightarrow H_3SO_4^+ + [B(HSO_4)_4]^- + H_2SO_4$$

Other strong acids are $H_2Sn(HSO_4)_6$ and $H_2Pb(HSO_4)_6$.

Sulfuric acid forms salts (sulfates and hydrogen sulfates) with many metals. These are frequently very stable and, indeed, they are the most important mineral compounds of several of the more electropositive elements. They have been discussed in detail under the appropriate elements. Sulfates can be prepared by:

(a) dissolution of metals in aqueous H_2SO_4 (e.g. Fe);
(b) neutralization of aqueous H_2SO_4 with metal oxides or hydroxides (e.g. MOH);
(c) decomposition of salts of volatile acids (e.g. carbonates) with aqueous H_2SO_4;

(d) metathesis between a soluble sulfate and a soluble salt of the metal whose (insoluble) sulfate is required (e.g. $BaSO_4$);

(e) oxidation of metal sulfides or sulfites.

The sulfate ion is tetrahedral (S–O 149 pm) and can act as a monodentate, bidentate (chelating), or bridging ligand. Examples are in Fig. 15.37. Vibrational spectroscopy is a useful diagnostic, as the progressive reduction in local symmetry of the SO_4 group from T_d to C_{3v} and eventually C_{2v} increases the number of infrared active modes from 2 to 6 and 8 respectively, and the number of Raman active modes from 4 to 6 and 9.[115] (The effects of crystal symmetry and the overlapping of bands complicates the analysis but correct assignments are frequently still possible.)

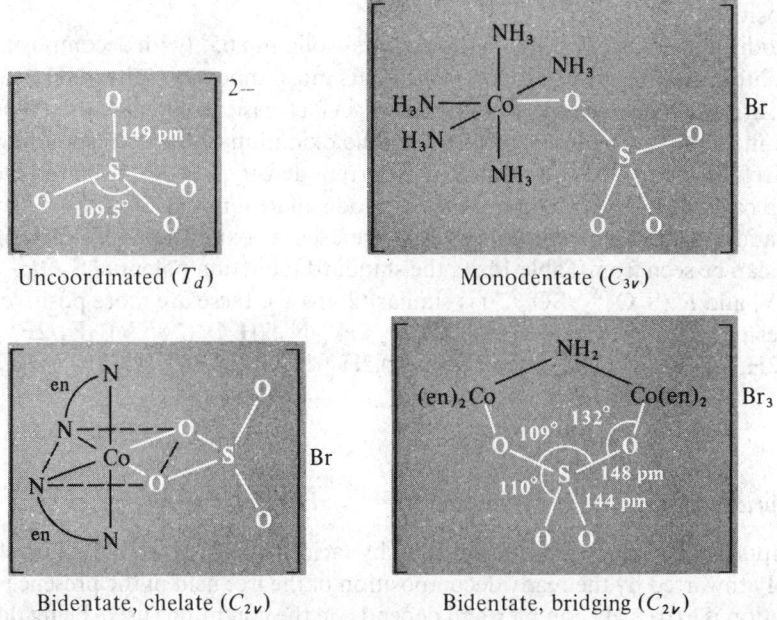

FIG. 15.37   Examples of SO_4^{2-} as a ligand.

Pairs of corner-shared SO_4 tetrahedra are found in the disulfates, $S_2O_7^{2-}$ (S–O_μ–S 124°, S–O_μ 164.5 pm, S–O_t 144 pm); they are made by thermal dehydration of $MHSO_4$. Likewise the trisulfate ion $S_3O_{10}^{2-}$ is known and also the pentasulfate ion, $S_5O_{16}^{2-}$ whose structure indicates an alternation of S–O interatomic distances and very long O–S distances to the almost planar terminal SO_3 groups:

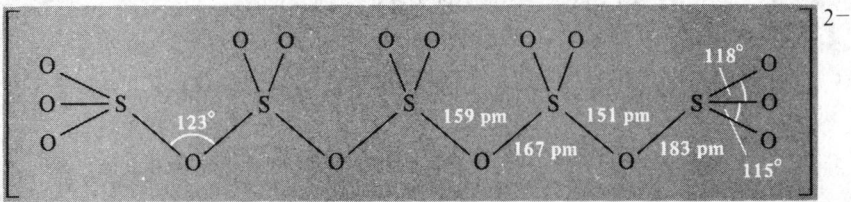

[115] K. NAKAMOTO, *Infrared Spectra of Inorganic and Coordination Compounds*, 2nd edn., Wiley, New York, 1970, 338 pp. (See also *J. Am. Chem. Soc.* **79**, 4904–8 (1957) for detailed correlation table.)

Peroxosulfuric acids, H_2SO_5 *and* $H_2S_2O_8$

Anhydrous peroxomonosulfuric acid (Caro's acid) can be prepared by reacting chlorosulfuric acid with anhydrous H_2O_2

$$HOOH + ClSO_2(OH) \longrightarrow HOOSO_2(OH) + HCl$$

It is colourless, beautifully crystalline, and melts at 45°, but should be handled carefully because of the danger of explosions. It can also be made by the action of conc H_2SO_4 on peroxodisulfates and is formed as a byproduct during the preparation of $H_2S_2O_8$ by electrolysis of aqueous H_2SO_4 (N. Caro, 1898). Its salts are unstable and the compound has few uses except those dependent on the formation of the H_2O_2 during its decomposition.

Peroxodisulfuric acid, $H_2S_2O_8$, is a colourless solid mp 65° (with decomposition). The acid is soluble in water in all proportions and its most important salts, $(NH_4)_2S_2O_8$ and $K_2S_2O_8$, are also freely soluble. These salts are, in fact, easier to prepare than the acid and both are made on an industrial scale by anodic oxidation of the corresponding sulfates under carefully controlled conditions (high current density, $T < 30°$, bright Pt electrodes, protected cathode). The structure of the peroxodisulfate ion is $O_3SOOSO_3{}^{2-}$ with O–O 131 pm and S–O 150 pm. The compounds are used as oxidizing and bleaching agents. Thus, as can be seen from Table 15.20, the standard reduction potential $S_2O_8{}^{2-}/HSO_4{}^-$ is 2.123 V, and $E°(S_2O_8{}^{2-}/SO_4{}^{2-})$ is similar (2.010 V); these are more positive than for any other aqueous couples except $H_2N_2O_2,2H^+/N_2,2H_2O$ (2.85 V), $F_2/2F^-$ (2.87 V) and $F_2,2H^+/2HF(aq)$ (3.06)—see also $O(g),2H^+/H_2O$ (2.42 V), $OH,H^+/H_2O$ (2.8 V).

Thiosulfuric acid, $H_2S_2O_3$

Attempts to prepare thiosulfuric acid by acidification of stable thiosulfates are invariably thwarted by the ready decomposition of the free acid in the presence of water. The reaction is extremely complex and depends on the conditions used, being dominated by numerous redox interconversions amongst the products: these can include sulfur (partly as cyclo-S_6), SO_2, H_2S, H_2S_n, H_2SO_4 and various polythionates. In the absence of water, however, these reactions are avoided and the parent acid is more stable: it decomposes quantitatively below 0° according to the reaction $H_2S_2O_3 \rightarrow H_2S + SO_3$ (cf. the analogous decomposition of H_2SO_4 to H_2O and SO_3 above its bp $\sim 300°$). Successful anhydrous syntheses have been devised by M. Schmidt and his group (1959–61), e.g.:

$$H_2S + SO_3 \xrightarrow{\text{Et}_2O/-78°} H_2S_2O_3 \cdot n\text{Et}_2O$$

$$Na_2S_2O_3 + 2HCl \xrightarrow{\text{Et}_2O/-78°} 2NaCl + H_2S_2O_3 \cdot 2\text{Et}_2O$$

$$HSO_3Cl + H_2S \xrightarrow[\text{low temp}]{\text{no solvent}} HCl + H_2S_2O_3 \text{ (solvent-free acid)}$$

Combination of stoichiometric amounts of H_2S and SO_3 at low temperature yields the white crystalline adduct $H_2S \cdot SO_3$ which is isomeric with thiosulfuric acid.

In contrast to the free acid, stable thiosulfate salts can readily be prepared by reaction of H_2S on aqueous solutions of sulfites:

$$2HS^- + 4HSO_3^- \longrightarrow 3S_2O_3^{2-} + 3H_2O$$

The reaction appears to proceed first by the formation of elemental sulfur which then equilibrates with more HSO_3^- to form the product:[116]

$$2HS^- + HSO_3^- \longrightarrow 3S + 3OH^-$$

$$3S + 3HSO_3^- \longrightarrow 3S_2O_3^{2-} + 3H^+$$

Consistent with this, experiments using HS^- labelled with radioactive ^{35}S (p. 782) show that acid hydrolysis of the $S_2O_3^{2-}$ produces elemental sulfur in which two-thirds of the ^{35}S activity is concentrated. Thiosulfates can also be made by boiling aqueous solutions of metal sulfites (or hydrogen sulfites) with elemental sulfur according to the stoichiometry

$$Na_2SO_3 + \tfrac{1}{8}S_8 \xrightarrow{\;H_2O/100°\;} Na_2S_2O_3$$

Aerial oxidation of polysulfides offers an alternative industrial route:

$$Na_2S_5 + \tfrac{3}{2}O_2 \longrightarrow Na_2S_2O_3 + \tfrac{3}{x}S_x$$

$$CaS_2 + \tfrac{3}{2}O_2 \longrightarrow CaS_2O_3$$

The thiosulfate ion closely resembles the SO_4^{2-} ion in structure and can act as monodentate η^1-S ligand, a monhapto bidentate bridging ligand $(\mu,\eta^1$-$S)$, or a dihapto chelating η^2-S,O ligand as illustrated in Fig. 15.38.[117] Hydrated sodium thiosulfate $Na_2S_2O_3 \cdot 5H_2O$ ("hypo") forms large, colourless, transparent crystals, mp 48.5°; it is readily soluble in water and is used as a "fixer" in photography to dissolve unreacted AgBr from the emulsion by complexation:

$$AgBr(cryst) + 3Na_2S_2O_3(aq) \longrightarrow Na_5[Ag(S_2O_3)_3](aq) + NaBr(aq)$$

The thiosulfate ion is a moderately strong reducing agent as indicated by the couple

$$S_4O_6^{2-} + 2e^- \rightleftharpoons 2S_2O_3^{2-}; \quad E° = 0.169 \text{ V}$$

Thus the quantitative oxidation of $S_2O_3^{2-}$ by I_2 to form tetrathionate and iodide is the basis for the iodometric titrations in volumetric analysis

$$2S_2O_3^{2-} + I_2 \longrightarrow S_4O_6^{2-} + 2I^-$$

Stronger oxidizing agents take the reaction through to sulfate, e.g.:

$$S_2O_3^{2-} + 4Cl_2 + 5H_2O \longrightarrow 2HSO_4^- + 8H^+ + 8Cl^-$$

This reaction is the basis for the use of thiosulfates as "antichlorine" in the bleaching industry where they are used to destroy any excess of Cl_2 in the fibres. Bromine, being intermediate between iodine and chlorine, can cause $S_2O_3^{2-}$ to act either as a 1-electron or an 8-electron reducer according to conditions. For example, in an amusing and

---

[116] G. W. HEUNISH, Stoichiometry of the reaction of sulfides with the hydrogen sulfide ion, *Inorg. Chem.* **16**, 1411–13 (1979) and references therein.

[117] See p. 589 of ref. 62 for detailed references.

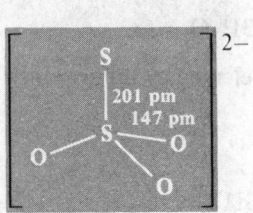

(a) Uncoordinated

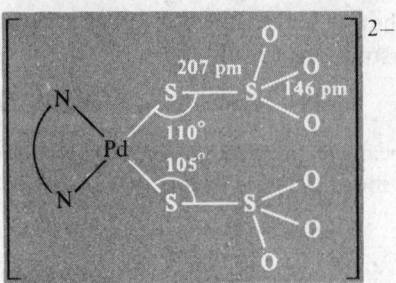

(b) Monodentate (η^1-S): the S^{VI} atoms are not
coplanar with the {PdN_2S_2} group

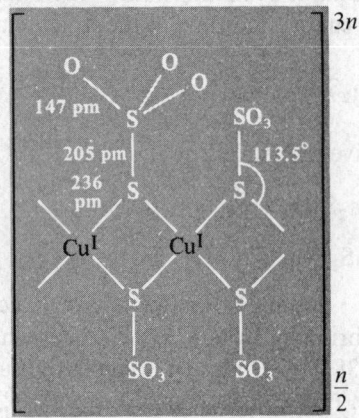

(c) Monohapto bidentate bridging
(μ,η^1-S)

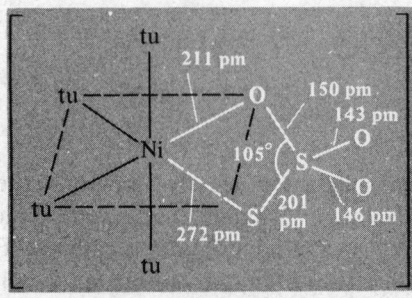

(d) Dihapto bidentate chelating (η^2-S,O)

Fig. 15.38 Structure of the thiosulfate ion and its various modes of coordination: (a) uncoordinated $S_2O_3^{2-}$; (b) monodentate (η^1-S) in the anion of the orange complex [$Pd^{II}(en)_2$][$Pd^{II}(en)(S_2O_3)_2$]; (c) monohapto bidentate bridging (μ,η^1-S) in the polymeric anion of the pale-violet mixed valence copper complex $Na_4[Cu^{II}(NH_3)_4][Cu^I(S_2O_3)_2]_2$; and (d) dihapto chelating (η^2-S,O) in the thiourea nickel complex [$Ni(S_2O_3)(tu)_4$].H_2O.

instructive experiment, if concentrated aqueous solutions of $S_2O_3^{2-}$ and Br_2 are titrated, and the titration is then repeated after having diluted both the $S_2O_3^{2-}$ and Br_2 solutions 100-fold, then the titre will be found to have increased by a factor of exactly 8.

Dithionic acid, $H_2S_2O_6$

In dithionic acid and dithionates, $S_2O_6^{2-}$, the oxidation state of the 2 S atoms has been reduced from VI to V by the formation of an S–S bond (Table 15.19, p. 835). The free acid has not been obtained pure, but quite concentrated aqueous solutions can be prepared by treatment of the barium salt with the stoichiometric amount of H_2SO_4:

$$BaS_2O_6(aq) + H_2SO_4(aq) \longrightarrow H_2S_2O_6(aq) + BaSO_4\downarrow$$

Crystalline dithionates are thermally stable above room temperature (e.g. $K_2S_2O_6$ decomp $258°$ to $K_2SO_4 + SO_2$). They are commonly made by oxidizing the corresponding sulfite. On a technical scale aqueous solutions of SO_2 are oxidized by a suspension of hydrated MnO_2 or Fe_2O_3:

$$2MnO_2 + 3SO_2 \xrightarrow{aq/0 \, °C} MnSO_4 + MnS_2O_6$$

$$Fe_2O_3 + 3SO_2 \xrightarrow{aq} \{Fe_2^{III}(SO_3)_3\} \longrightarrow Fe^{II}SO_3 + Fe^{II}S_2O_6$$

All the dithionates are readily soluble in water and can be made by standard metathesis reactions. For example, addition of an excess of Ba^{II} ions to the Mn^{II} solution above precipitates $BaSO_4$, after which $BaS_2O_6 . 2H_2O$ can be crystallized. The $[O_3SSO_3]^{2-}$ ion is centrosymmetric (staggered) D_{3d} in $Na_2S_2O_6 . 2H_2O$ but in the anhydrous potassium salt some of the $S_2O_6^{2-}$ ions have an almost eclipsed configuration for the two SO_3 groups (D_{3h}). Dimensions are unremarkable: S–S 215 pm, S–O 143 pm, and angle S–S–O $103°$. Dithionates are relatively stable towards oxidation in solution though strong oxidants such as the halogens, dichromate, and permanganate oxidize them to sulfate. Powerful reductants (e.g. Na/Hg) reduce dithionates to sulfites and dithionites ($S_2O_4^{2-}$).

Polythionic acids, $H_2S_nO_6$

The numerous acids and salts in this group have a venerable history and the chemistry of systems in which they occur goes back to John Dalton's studies (1808) of the effect of H_2S on aqueous solutions of SO_2. Such solutions are now named after H. W. F. Wackenroder (1846) who subjected them to systematic study. Work during the following 60–80 y indicated the presence of numerous species including, in particular, the tetrathionate $S_4O_6^{2-}$ and pentathionate $S_5O_6^{2-}$ ions. New perceptions have emerged during the past 25 y as a result of the work of H. Schmidt and his group in Germany: just as H_2S can react with SO_3 or HSO_3Cl to yield thiosulfuric acid, $H_2S_2O_3$ (p. 846), so reaction with H_2S_2 yields "disulfane monosulfonic acid", HS_2SO_3H; likewise polysulfanes H_2S_n $(n=2-6)$ yield HS_nSO_3H. Reaction at both ends of the polysulfane chain would yield "polysulfane disulfonic acids" $HO_3SS_nSO_3H$ which are more commonly called polythionic acids $(H_2S_{n+2}O_6)$. many synthetic routes are available, though mechanistic details are frequently obscure because of the numerous simultaneous and competing redox, catenation, and disproportionation reactions that occur. Typical examples include:

(a) Interaction of H_2S and SO_2 in Wackenroder's solution.
(b) Reaction of chlorosulfanes with HSO_3^- or $HS_2O_3^-$, e.g.:

$$SCl_2 + 2HSO_3^- \longrightarrow [O_3SSSO_3]^{2-} + 2HCl$$
$$S_2Cl_2 + 2HSO_3^- \longrightarrow [O_3SS_2SO_3]^{2-} + 2HCl$$
$$SCl_2 + 2HS_2O_3^- \longrightarrow [O_3SS_3SO_3]^{2-} + 2HCl, \text{ etc.}$$

(c) Oxidation of thiosulfates with mild oxidants (p. 847) such as I_2, Cu^{II}, $S_2O_8^{2-}$, H_2O_2.
(d) Specific syntheses as noted below.

Sodium trithionate, $Na_2S_3O_6$, can be made by oxidizing sodium thiosulfate with cooled hydrogen peroxide solution

$$2Na_2S_2O_3 + 4H_2O_2 \longrightarrow Na_2S_3O_6 + Na_2SO_4 + 4H_2O$$

The potassium (but not the sodium) salt is obtained by the obscure reaction of SO_2 on aqueous thiosulfate. Aqueous solutions of the acid $H_2S_3O_6$ can then be obtained from $K_2S_3O_6$ by treatment with tartaric acid or perchloric acid.

Sodium (and potassium) tetrathionate, $M_2S_4O_6$, can be made by oxidation of thiosulfate by I_2 (p. 847) and the free acid liberated (in aqueous solution) by addition of the stoichiometric amount of tartaric acid.

Potassium pentathionate, $K_2S_5O_6$, can be made by adding potassium acetate to Wackenroder's solution and solutions of the free acid $H_2S_5O_6$ can then be obtained by subsequent addition of tartaric acid.

Potassium hexathionate, $K_2S_6O_6$, is best synthesized by the action of KNO_2 on $K_2S_2O_3$ in conc HCl at low temperatures, though the ion is also a constituent of Wackenroder's solution.

Anhydrous polythionic acids can be made in ether solution by three general routes:

$$HS_nSO_3H + SO_3 \longrightarrow H_2S_{n+2}O_6 \quad (n+2 = 3, 4, 5, 6, 7, 8)$$
$$H_2S_n + 2SO_3 \longrightarrow H_2S_{n+2}O_6 \quad (n+2 = 3, 4, 5, 6, 7, 8)$$
$$2HS_nSO_3H + I_2 \longrightarrow H_2S_{2n+2}O_6 + 2HI \quad (2n+2 = 4, 6, 8, 10, 12, 14)$$

The structure of the trithionate ion (in $K_2S_3O_6$) is shown in Fig. 15.39a and calls for little comment (cf. the disulfate ion $O_3SOSO_3^{2-}$, p. 845). The tetrathionate ion (in $BaS_4O_6 \cdot 2H_2O$ and $Na_2S_4O_6 \cdot 2H_2O$) has the configuration shown in Fig. 15.39b with dihedral angles close to $90°$ and a small, but definite, alternation in S–S distances. The pentathionate ion in $BaS_5O_6 \cdot 2H_2O$ has the *cis* configuration in which the S_5 unit can be regarded as part of an S_8 ring (p. 772) from which 3 adjacent S atoms have been removed (Fig. 15.39c). By contrast, in the potassium salt $K_2S_5O_6 \cdot 1\frac{1}{2}H_2O$ the pentathionate ion adopts the *trans* configuration in which the two terminal SO_3 groups are on opposite sides of the central S_3 plane (Fig. 15.39d). These structural differences persist in the seleno- and telluro-analogues $O_3SSSeSSO_3^{2-}$ and $O_3SSTeSSO_3^{2-}$, the dihydrated Ba salts being *cis* and the potassium hemihydrates being *trans*.[118] There are three possible rotameric forms of the hexathionate ion $S_6O_6^{2-}$: the extended *trans-trans* form analogous to spiral chains of fibrous sulfur (p. 779) occurs in the *trans*$[Co^{III}(en)_2Cl_2]^+$ salt (Fig. 15.39e), whereas the *cis-cis* form (analogous to *cyclo*-S_8) occurs in the potassium barium salt (Fig. 15.39f); the *cis-trans* form of $S_6O_6^{2-}$ has not yet been observed in crystals but presumably occurs in equilibrium with the other two forms in solution since the energy barrier to rotation about the S–S bonds is only some 40 kJ mol^{-1}.

Sulfurous acid, H_2SO_3

Sulfurous acid has never been isolated, and exists in only minute concentrations (if at all) in aqueous solutions of SO_2. However, its salts, the sulfites, are quite stable and many

---

[118] O. Foss, Aspects of the structural chemistry of polythionates, *IUPAC Additional Publication* (24th International Congress, Hamburg, 1973), Vol. 4, *Compounds of Non-Metals*, pp. 103–13, Butterworths, London, 1974, and references therein.

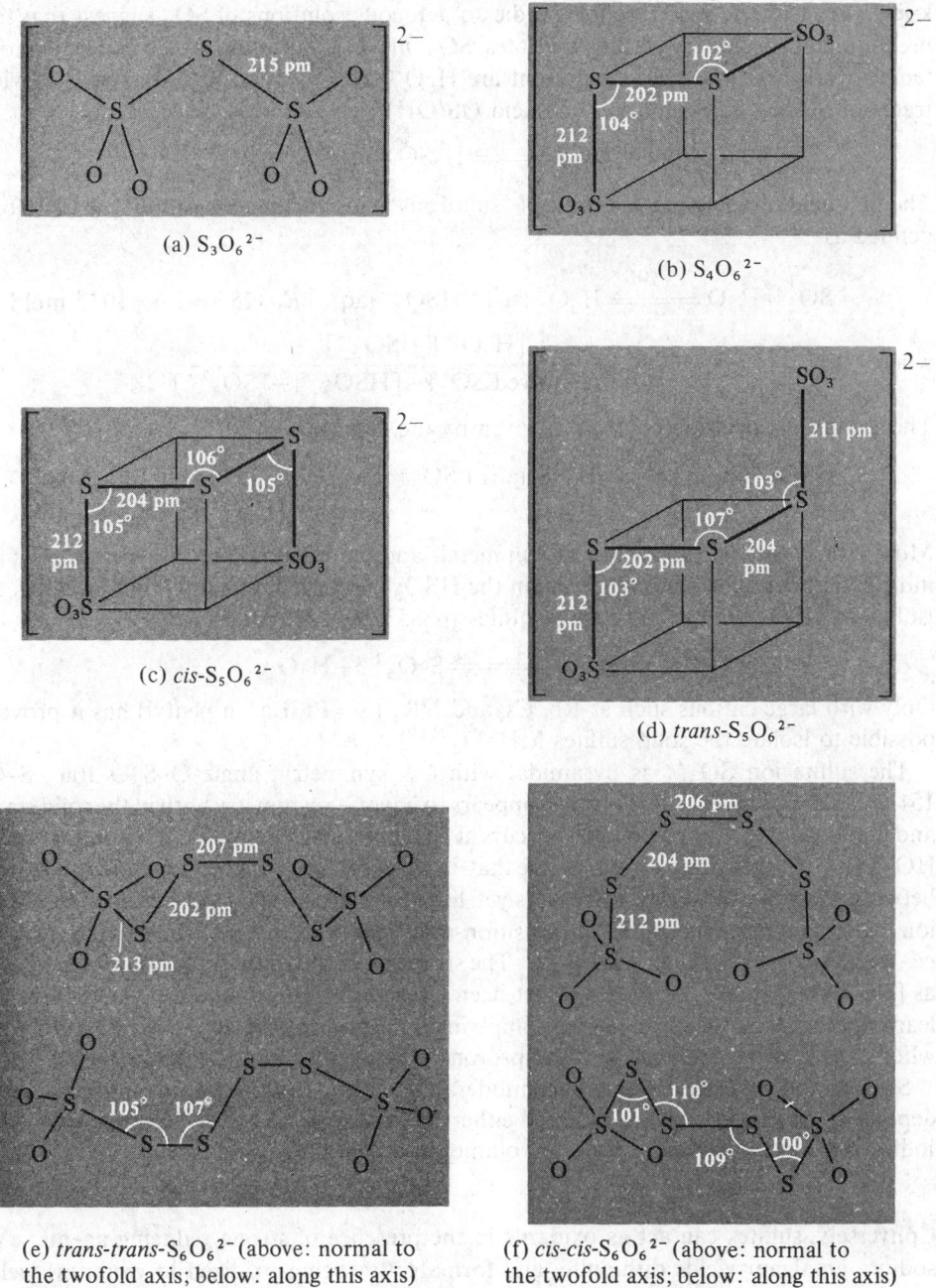

(a) $S_3O_6^{2-}$

(b) $S_4O_6^{2-}$

(c) *cis*-$S_5O_6^{2-}$

(d) *trans*-$S_5O_6^{2-}$

(e) *trans-trans*-$S_6O_6^{2-}$ (above: normal to the twofold axis; below: along this axis)

(f) *cis-cis*-$S_6O_6^{2-}$ (above: normal to the twofold axis; below: along this axis)

FIG. 15.39   Structures of some polythionate ions.[118]

are known in crystalline form; a second series of salts, the hydrogen sulfites HSO_3^-, are known in solution. Spectroscopic studies of aqueous solutions of SO_2 suggest that the predominant species are various hydrates, $SO_2 . nH_2O$; depending on the concentration, temperature, and pH, the ions present are H_3O^+, HSO_3^-, and $S_2O_5^{2-}$ together with traces of SO_3^{2-}. The undissociated acid $OS(OH)_2$ has not been detected:

$$SO_2 . nH_2O \rightleftharpoons H_2SO_3(aq); \quad K \ll 10^{-9}$$

The first acid dissociation constant of "sulfurous acid" in aqueous solution is therefore defined as:

$$SO_2 . nH_2O \overset{H_2O}{\rightleftharpoons} H_3O^+(aq) + HSO_3^-(aq); \quad K_1 \ (25°) = 1.6 \times 10^{-2} \ mol \ l^{-1}$$

where
$$K_1 = \frac{[H_3O^+][HSO_3^-]}{[total\ dissolved\ SO_2] - [HSO_3^-] - [SO_3^{2-}]}$$

The second dissociation constant is given by the equation

$$HSO_3^-(aq) \rightleftharpoons H_3O^+(aq) + SO_3^{2-}(aq); \quad K_2(25°) = 1.0 \times 10^{-7} \ mol \ l^{-1}$$
$$K_2 = [H_3O^+][SO_3^{2-}]/[HSO_3^-]$$

Most sulfites (except those of the alkali metals and ammonium) are rather insoluble; as indicated above such solutions contain the HSO_3^- ion predominantly, but attempts to isolate $M^I HSO_3$ tend to produce disulfites (p. 853) by "dehydration":

$$2HSO_3^- \rightleftharpoons S_2O_5^{2-} + H_2O$$

Only with large cations such as Rb, Cs, and NR_4 (R = Et, Bu^n, n-pentyl) has it proved possible to isolate the solid sulfites $MHSO_3$.[119]

The sulfite ion SO_3^{2-} is pyramidal with C_{3v} symmetry: angle O–S–O 106°, S–O 151 pm. The hydrogen sulfite ion also appears to have C_{3v} symmetry both in the solid state and in solution,[119] i.e. protonation occurs at S rather than O to give $H–SO_3^-$ rather than $HO–SO_2^-$ (C_s symmetry). It is possible that, in solution, there is a tautomeric equilibrium between these two forms but there is, as yet, little evidence for the C_s structure. The sulfite ion also coordinates through S in transition-metal complexes, e.g. $[Pd(NH_3)_3(\eta^1\text{-}SO_3)]$, cis- and trans-$[Pt(NH_3)_2(\eta^1\text{-}SO_3)_2]^{2-}$. The structure of hydrogen-sulfito complexes such as $[Ru^{II}(NH_3)_4(HSO_3)_2]$ have not yet been determined and it would be interesting to learn whether they, too, are S-bonded, implying a 1, 2 proton shift to give $M\{SO_2(OH)\}$ or whether they are O-bonded with the proton remaining attached to the S atom.

Sulfites and hydrogen sulfites are moderately strong reducing agents (p. 836) and, depending on conditions, are oxidized either to dithionate or sulfate. The reaction with iodine is quantitative and is used in volumetric analysis:

$$HSO_3^- + I_2 + H_2O \longrightarrow HSO_4^- + 2H^+ + 2I^-$$

Conversely, sulfites can act as oxidants in the presence of strong reducing agents; e.g. sodium amalgam yields dithionite, and formates (in being oxidized to oxalates) yield thiosulfate:

$$2SO_3^{2-} + 2H_2O + 2Na/Hg \longrightarrow S_2O_4^{2-} + 4OH^- + 2Na^+$$
$$2SO_3^{2-} + 4HCO_2^- \longrightarrow SSO_3^{2-} + 2C_2O_4^{2-} + 2OH^- + H_2O$$

[119] R. MAYLOR, J. B. GILL, and D. C. GOODALL, Tetra-n-alkylammonium bisulfites: a new example of the existence of the bisulfite ion in solid compounds, *JCS Dalton* 1972, 2001–3, and references therein.

Thiosulfates also result from reduction of $SO_3{}^{2-}$ or $HSO_3{}^-$ with elemental sulfur (p. 847), whereas reduction with H_2S in Wackenroder's solution (p. 849) yields polythionates.

On a technical scale, solutions of sodium hydrogen sulfite are prepared by passing SO_2 into aqueous Na_2CO_3; addition of a further equivalent of Na_2CO_3 allows the normal sulfite to be crystallized, whereas addition of more SO_2 yields the disulfite (see subsection *vii* below).

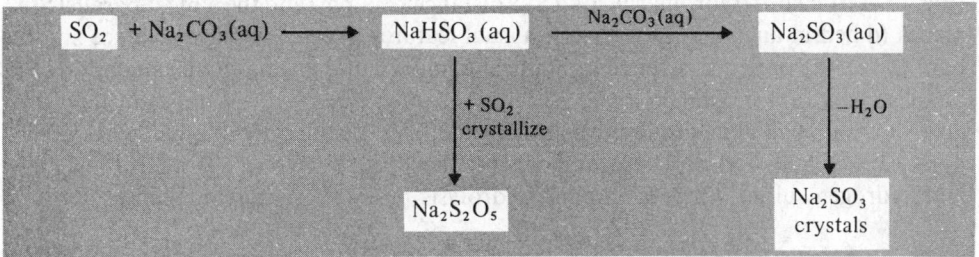

Crystallization of Na_2SO_3 above 37° gives the anhydrous salt; below this temperature $Na_2SO_3 \cdot 7H_2O$ is obtained. World production of the anhydrous salt exceeds 1 million tonnes pa; most is used in the paper pulp industry, but other applications are as an O_2 scavenger in boiler-water treatment, and as a reducing agent in photography. Similarly, $K_2SO_3 \cdot 2H_2O$ is obtained by passing SO_2 into aqueous KOH until samples of the solution are neutral to phenolphthalein.

Disulfurous acid, $H_2S_2O_5$

Like "sulfurous acid", disulfurous acid is unknown either in the free state or in solution. However, as indicated in the preceding section, its salts, the disulfites, are readily obtained from concentrated solutions of hydrogen sulfite: $2HSO_3{}^- \rightleftharpoons S_2O_5{}^{2-} + H_2O$. Unlike disulfates (p. 845), diphosphates (p. 602), etc., disulfites condense by forming an S–S bond. As indicated in Fig. 15.40a (on p. 855) this S–S bond is rather long, but the S–O distances are unexceptional.

Acidification of solutions of disulfites regenerates $HSO_3{}^-$ and SO_2 again, and the solution chemistry of $S_2O_5{}^{2-}$ is essentially that of the normal sulfites and hydrogen sulfites, despite the formal presence of S^V and S^{III} (rather than S^{IV}) in the solid state.

Dithionous acid, $H_2S_2O_4$

Dithionites, $S_2O_4{}^{2-}$ are quite stable when anhydrous, but in the presence of water they disproportionate (slowly at pH $\geqslant 7$, rapidly in acid solution):

$$2\overset{\text{III}}{S_2O_4{}^{2-}} + H_2O \longrightarrow 2\overset{\text{IV}}{HSO_3{}^-} + \overset{-\text{II/VI}}{SSO_3{}^{2-}}$$

The parent acid has no independent existence and has not been detected in aqueous solution either. Sodium dithionite is widely used as an industrial reducing agent and can be prepared by reduction of sulfite using Zn dust, Na/Hg, or electrolytically, e.g.:

$$2\overset{\text{IV}}{HSO_3{}^-} + \overset{\text{IV}}{SO_2} \cdot nH_2O + 2Zn \longrightarrow \overset{\text{IV}}{ZnSO_3} + \overset{\text{III}}{ZnS_2O_4} + (n+2)H_2O$$

The dihydrate $Na_2S_2O_4 . 2H_2O$ can be precipitated by "salting out" with NaCl. Air and oxygen must be excluded at all stages in the process to avoid reoxidation. The dithionite ion can also be produced *in situ* on an industrial scale by reaction between $NaHSO_3$ and $NaBH_4$ (p. 189). Its main use is as a reducing agent in dyeing, bleaching of paper pulp, straw, clay, soaps, etc., and in chemical reductions (see below).

The dithionite ion has a remarkable eclipsed structure of approximate C_{2v} symmetry (Fig. 15.40b). The extraordinarily long S–S distance (239 pm) and the almost parallel SO_2 planes (dihedral angle 30°) are other unusual features. Electron-spin-resonance studies have shown the presence of the $SO_2^{\cdot -}$ radical ion in solution (~ 300 ppm), suggesting the establishment of a monomer–dimer equilibrium $S_2O_4^{2-} \rightleftharpoons 2SO_2^{\cdot -}$. Consistent with this, air-oxidation of alkaline dithionite solutions at 30–60° are of order one-half with respect to $[S_2O_4^{2-}]$. Acid hydrolysis (second order with respect to $[S_2O_4^{2-}]$) yields thiosulfate and hydrogen sulfite, whereas alkaline hydrolysis produces sulfite and sulfide:

$$2S_2O_4^{2-} + H_2O \longrightarrow S_2O_3^{2-} + 2HSO_3^-$$
$$2Na_2S_2O_4 + 6NaOH \longrightarrow 5Na_2SO_3 + Na_2S + 3H_2O$$

Hydrated dithionites can be dehydrated by gentle warming, but the anhydrous salts themselves decompose on further heating. For example, $Na_2S_2O_4$ decomposes rapidly at 150° and violently at 190°:

$$2Na_2S_2O_4 \longrightarrow Na_2S_2O_3 + Na_2SO_3 + SO_2$$

Dithionites are strong reducing agents and will reduce dissolved O_2, H_2O_2, I_2, IO_3^-, and MnO_4^-. Likewise Cr^{VI} is reduced to Cr^{III} and TiO^{2+} to Ti^{III}. Heavy metal ions such as Cu^I, Ag^I, Pb^{II}, Sb^{III}, and Bi^{III} are reduced to the metal. Many of these reactions are useful in water-treatment and pollution control.

15.2.7 *Sulfur–nitrogen compounds*[124-7]

The study of S–N compounds is one of the most active areas of current inorganic research: many novel cyclic and acyclic compounds are being prepared which have unusual structures and which pose considerable problems in terms of simple bonding theory. The recent discovery that the polymer $(SN)_x$ is a metal whose conductivity *increases* with decrease in temperature and which becomes superconducting below 0.33 K has aroused tremendous additional interest and has stimulated still further the already

[124] M. BECKE-GOEHRING and E. FLUCK, Developments in the inorganic chemistry of compounds containing the sulfur–nitrogen bond, Chap. 3 in C. B. COLBURN (ed.), *Developments in Inorganic Nitrogen Chemistry*, Vol. 1, pp. 150–240, Elsevier, Amsterdam, 1966. A substantial general review with 277 references.

[125] I. HAIDUC, *The Chemistry of Inorganic Ring Systems*, Part 2, (sulfur–nitrogen heterocycles), pp. 909–83, Wiley, London, 1970.

[126] H. G. HEAL, The sulfur nitrides, *Adv. Inorg. Chem. Radiochem.* **15**, 374–412 (1972).

[126a] H. G. HEAL, *The Inorganic Heterocyclic Chemistry of Sulfur, Nitrogen, and Phosphorus*, Academic Press, London, 1981, 271 pp.

[127] H. W. ROESKY, Cyclic sulfur–nitrogen compounds, *Adv. Inorg. Chem. Radiochem.* **22**, 239–301 (1979). This is the most recent review on this aspect of S–N compounds—279 references. A shortened version appears in H. W. ROESKY, Structure and bonding in cyclic sulfur-nitrogen compounds, *Angew. Chem.*, Int. Edn. (Engl.) **18**, 91–97 (1979). See also R. J. GILLESPIE, J. P. KENT and J. F. SAWYER, Monomeric and dimeric thiodithiazyl cations, $S_3N_2^+$ and $S_6N_4^{2+}$. Preparation and crystal structures of $(S_3N_2)(AsF_6)$, $(S_6N_4)(S_2O_2F)_2$, and $(S_6N_4)(SO_3F)_2$, *Inorg. Chem.* **20**, 3784–99 (1981).

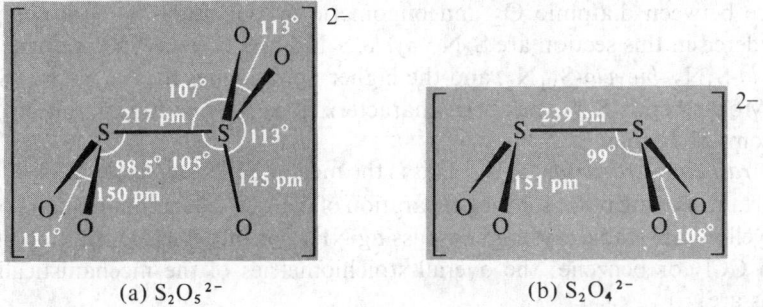

(a) $S_2O_5{}^{2-}$ (b) $S_2O_4{}^{2-}$

FIG. 15.40   Structure of (a) the disulfite ion $S_2O_5{}^{2-}$ in $(NH_4)_2S_2O_5$, and (b) the dithionite ion $S_2O_4{}^{2-}$ in $Na_2S_2O_4.2H_2O$.

substantial activity in this area of synthetic and structural chemistry. The field is not new. S_4N_4 was first prepared in an impure form by W. Gregory in 1835,† though the stoichiometry and tetrameric nature of the pure compound were not established until 1851 and 1896 respectively, and its cyclic, pseudo-cluster structure was not revealed until 1944.[128] Other important compounds containing S–N bonds that date from the first half of the nineteenth century included sulfamic acid $HSO_3.NH_2$, imidosulfonic acid $HSO_3N{=}NH$, sulfamide $SO_2(NH_2)_2$, nitrilotrisulfonic acid $N(HSO_3)_3$, hydroxy nitrilo-sulfonic acids $HSO_3NH(OH)$ and $(HSO_3)_2N(OH)$, and their many derivatives (p. 878).

It will be convenient to describe first the binary sulfur nitrides S_xN_y and then the related cationic and anionic species, $S_xN_y{}^{n\pm}$. The sulfur imides and other cyclic S–N compounds will then be discussed and this will be followed by sections on S–N– halogen and S–N–O compounds. Several compounds which feature isolated S←N, S–N, S=N, and S≡N bonds have already been mentioned in the section on SF_4; e.g. $F_4S{\leftarrow}NC_5H_5$, $F_5S{-}NF_2$, $F_2S{=}NCF_3$, and $F_3S{\equiv}N$ (p. 811). However, many SN compounds do not lend themselves to simple bond diagrams, and formal oxidation states are often unhelpful or even misleading.

Nitrogen and sulfur are diagonally related in the periodic table and might therefore be expected to have similar electronic charge densities for similar coordination numbers (p. 87). Likewise, they have similar electronegativities (N 3.0, S 2.5) and these become even more similar when additional electron-withdrawing groups are bonded to the S atoms. Extensive covalent bonding into acyclic, cyclic, and polycyclic molecular structures is thus not unexpected.

Binary sulfur nitrides

There is little structural similarity between the sulfur nitrides and the oxides of nitrogen (p. 508). The instability of NS when compared with the great stability of NO, and the paucity of thionitrosyl complexes have already been mentioned (p. 520), as has the

---

† Disulfur dichloride was added to an aqueous solution of ammonia to give a yellow precipitate of sulfur contaminated with S_4N_4 (*J. Pharm. Chim.* **21**, 315 (1835).)

[128] CHIA-SI LU and J. DONOHUE, An electron diffraction investigation of sulfur nitride, arsenic disulfide (realgar), arsenic trisulfide (orpiment), and sulfur, *J. Am. Chem. Soc.* **66**, 818–27 (1944). D. CLARK, The structure of sulfur nitride (by X-ray diffraction), *J. Chem. Soc.* 1615–20 (1952).

difference between diatomic O_2 and oligomeric or polymeric S_n. The compounds to be considered in this section are S_4N_4, *cyclo*-S_2N_2, and *catena*-$(SN)_x$ polymer, together with *cyclo*-S_4N_2, *bicyclo*-$S_{11}N_2$, and the higher homologues $S_{15}N_2$, $S_{16}N_2$, $S_{17}N_2$, and $S_{19}N_2$. Most recently S_5N_6 has been characterized as the first binary sulfur nitride with more atoms of N than S.

(a) *Tetrasulfur tetranitride*, S_4N_4. This is the most readily prepared sulfur nitride and is an important starting point for the preparation of many S–N compounds. It is obtained as orange-yellow, air-stable crystals† by passing NH_3 gas into a warm solution of S_2Cl_2 (or SCl_2) in CCl_4 or benzene; the overall stoichiometries of the mechanistically obscure reactions are:

$$6S_2Cl_2 + 16NH_3 \xrightarrow{50°} S_4N_4 + 8S + 12NH_4Cl$$

$$6SCl_2 + 16NH_3 \longrightarrow S_4N_4 + 2S + 14NH_4Cl$$

Alternatively, NH_4Cl can be heated with S_2Cl_2 at 160°:

$$6S_2Cl_2 + 4NH_4Cl \xrightarrow{26\% \text{ yield}} S_4N_4 + 8S + 16HCl$$

The compound also results from the reversible equilibrium reaction of sulfur with anhydrous liquid ammonia:

$$10S + 4NH_3 \rightleftharpoons S_4N_4 + 6H_2S$$

The H_2S, of course, reacts with further ammonia to form ammonium sulfides but the reaction can be made to proceed in the forward direction as written by addition of (soluble) AgI to precipitate AgS and form NH_4I.

S_4N_4 is kinetically stable in air but is endothermic with respect to its elements (ΔH_f° 460 ± 8 kJ mol^{-1}) and may detonate when struck or when heated rapidly. This is due more to the stability of elementary sulfur and the great bond strength of N_2 rather than to any inherent weakness in the S–N bonds. On careful heating S_4N_4 melts at 178.2°. The structure (Fig. 15.41a) is an 8-membered heterocycle in the extreme cradle configuration; it has D_{2d} symmetry and resembles that of As_4S_4 (p. 676) but with the sites of the Group V and Group VI elements interchanged. The S–N distance of 162 pm is rather short when compared with the sum of the covalent radii (178 pm) and this, coupled with the equality of all the S–N bond distances in the molecule, has been attributed to some electron delocalization in the heterocycle. The trans-annular S···S distances (258 pm) are intermediate between bonding S–S (208 pm) and nonbonding van der Waals (330 pm) distances; this suggests a weak but structurally significant bonding interaction between the pairs of S atoms. It is not possible to write down a single, satisfactory, classical bonding diagram for S_4N_4 and, in valence-bond theory, numerous resonance hybrids must be considered of which the following are typical:

† Crystalline S_4N_4 is thermochromic, being pale yellow below about $-30°$: the colour deepens to orange at room temperature and to a deep red at 100° (cf. sulfur, p. 775).

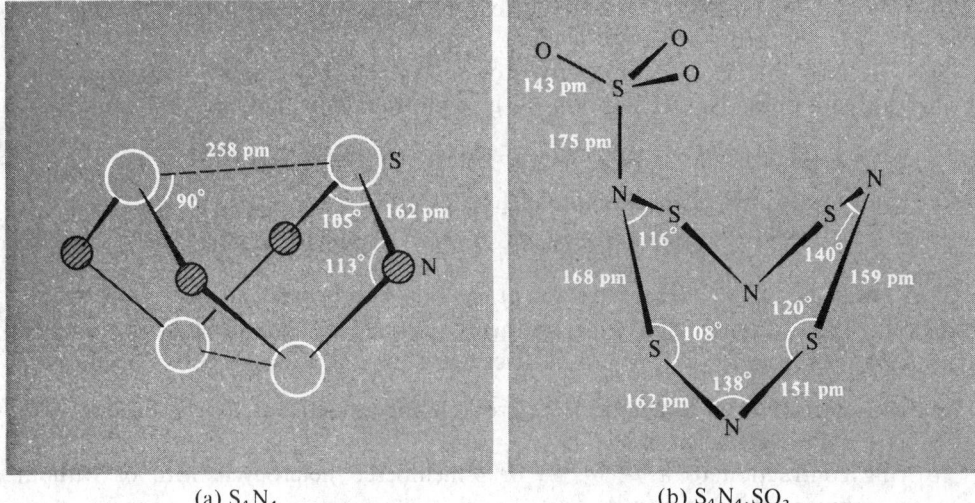

(a) S_4N_4 (b) $S_4N_4.SO_3$

FIG. 15.41   Structure of (a) S_4N_4, and (b) S_4N_4-SO_3.

The extent to which each hybrid is incorporated into the full bonding description of the molecule will depend on the extent to which 3d orbitals on S are involved and the extent of trans-annular S–S bonding. More recent MO-calculations lead to semiquantitative estimates of these features and to electron charge densities on the individual atoms.[129] It is also instructive to compare the structure of the 44-(valence)electron species S_4N_4 with those of the 46-electron species $S_8{}^{2+}$ (p. 785) and the 48-electron species S_8 (p. 772): successive formal addition of 2 and then 4 electrons results in the progressive opening of the S_4N_4 pseudocluster first to the *bicyclic*-$S_8{}^{2+}$ with a single weak trans-annular S–S bond and then to the open-crown structure of S_8 with no trans-annular bonding at all.

Interestingly, in the N-donor adducts $S_4N_4.BF_3$ and $S_4N_4.SbCl_5$ the S_4N_4 ring adopts the alternative D_{2d} configuration of As_4S_4, with the 4 S atoms now coplanar instead of the 4 N atoms; the mean S–N distance increases slightly to 168 pm but the (nonbonding) trans-annular S···S distances are 380 pm. The same interchange occurs in $S_4N_4.SO_3$ and Fig. 15.41b shows the substantial alternations in S–N distances and angles that are concurrently introduced into the ring. By contrast, in $S_4N_4.CuCl$ the heterocycle acts as a bridging ligand between zigzag chains of $(-Cu-Cl-)_\infty$; the S_4N_4 retains the same conformation and almost the same dimensions as in the free molecule, with 2 of the 4 planar N atoms acting as a *cisoid* bridge and the 2 trans-annular S···S distances remaining

[129] R. C. HADDON, S. R. WASSERMAN, F. WUDL, and G. R. J. WILLIAMS, Molecular orbital study of sulfur–nitrogen and sulfur–carbon conjugation: mode of bonding in (SN)$_x$ and related compounds, *J. Am. Chem. Soc.* **102**, 6687–93 (1980), and references therein. See also R. D. HARCOURT and H. M. HUGEL, Increased-valence structures and the bonding for some cyclic sulfur-nitrogen compounds, *J. Inorg. Nuclear Chem.* **43**, 239–52 (1981), for a less rigorous treatment. Other recent references include R. GLEITER, Structure and bonding in cyclic sulfur-nitrogen compounds—molecular orbital considerations, *Angew. Chem. Int. Edn.* (Engl.) **20**, 444–52 (1981); A. A. BHATTACHARYYA, A. BHATTACHARYYA, R. R. ADKINS, and A. G. TURNER, Localized molecular orbital studies of three-, four-, five-, and six-membered ring molecules and ions formed from sulfur and nitrogen, *J. Am. Chem. Soc.* **103**, 7458–65 (1981).

short (259 and 263 pm).[130] It is not yet clear in detail what factors determine the ring conformation adopted (see also p. 774).

S_4N_4 is insoluble in and unreactive towards water but readily undergoes base hydrolysis with dilute NaOH solutions to give thiosulfate, trithionate, and ammonia:

$$2S_4N_4 + 6OH^- + 9H_2O \longrightarrow S_2O_3{}^{2-} + 2S_3O_6{}^{2-} + 8NH_3$$

More concentrated alkali yields sulfite instead of trithionate:

$$S_4N_4 + 6OH^- + 3H_2O \longrightarrow S_2O_3{}^{2-} + 2SO_3{}^{2-} + 4NH_3$$

Milder bases such as Et_2NH leave some of the S–N bonds intact to yield, for example, $S(NEt_2)_2$. The value of S_4N_4 as a synthetic intermediate can be gauged from the representative reactions in Table 15.23. It can be seen that these reactions embrace:

(a) conservation of the 8-membered heterocycle and attachment of substituents to S or N (or subrogation of N by S);
(b) ring contraction to a 7-, 6-, 5-, or 4-membered heterocycle with or without attachment of substituents;
(c) ring fragmentation into non-cyclic S–N groups (which sometimes then coordinate to metal centres);
(d) complete cleavage of all S–N bonds;
(e) formation of more complex heterocycles with 3 (or more) different heteroatoms.

The molecular structures of the products are described as indicated at appropriate points in the text. $S_3N_2O_2$ was at one time thought to be cyclic but X-ray diffraction analysis has revealed an open chain structure (Fig. 15.42a). The structure of $[Pt(S_2N_2H)_2]$ (Fig.

TABLE 15.23   *Some further reactions of S_4N_4*[124-6]

Reagents and conditions	Products	Ref. for structure, etc.
Vacuum thermolysis (Ag wool 300°)	S_2N_2, $(SN)_x$	p. 859
$SnCl_2$ (boiling $C_6H_6 + EtOH$)	$S_4(NH)_4$	p. 868
NH_3	$S_2N_2 \cdot NH_3$	
$N_2H_4/SiO_2(C_6H_6, 46°)$	$S_{8-n}(NH)_n$, $n = 1$–4	p. 868
S/CS_2 (heat in autoclave)	S_4N_2	
S_2Cl_2	$[S_4N_3]^+Cl^-$	
AgF_2 (cold CCl_4)	$N_4(SF)_4$	
AgF_2 (hot CCl_4)	NSF, NSF_3	
$Cl_2(CCl_4)$	$N_3(SCl)_3$	
Br_2 (neat, heat in sealed tube)	$[S_4N_3]^+Br_3^-$	100% yield[131]
$HX(CCl_4)$ X = F, Cl, Br	$[S_4N_3]^+X^-$	
HI	H_2S, NH_3, I_2	
$OSCl_2$	$S_3N_2O_2$	Fig. 15.42a
$NiCl_2/MeOH$	$[Ni(S_2N_2H)_2]$ (also Co, Pd)	Fig. 15.42b
H_2PtCl_6	$[Pt(S_2N_2H)_2]$	Fig. 15.42b
PbI_2/NH_3	$[Pb(NSNS)(NH_3)]$	Fig. 15.42c

[130] U. THEWALT, S_4N_4 bridge ligand: structure of $CuCl \cdot S_4N_4$, *Angew. Chem.*, Int. Edn. (Engl.) **15**, 765–6 (1976).

[131] G. WOLMERSHÄUSER and G. B. STREET, Reaction of S_4N_4 with liquid Br_2 and ICl, *Inorg. Chem.* **17**, 2685–6 (1978).

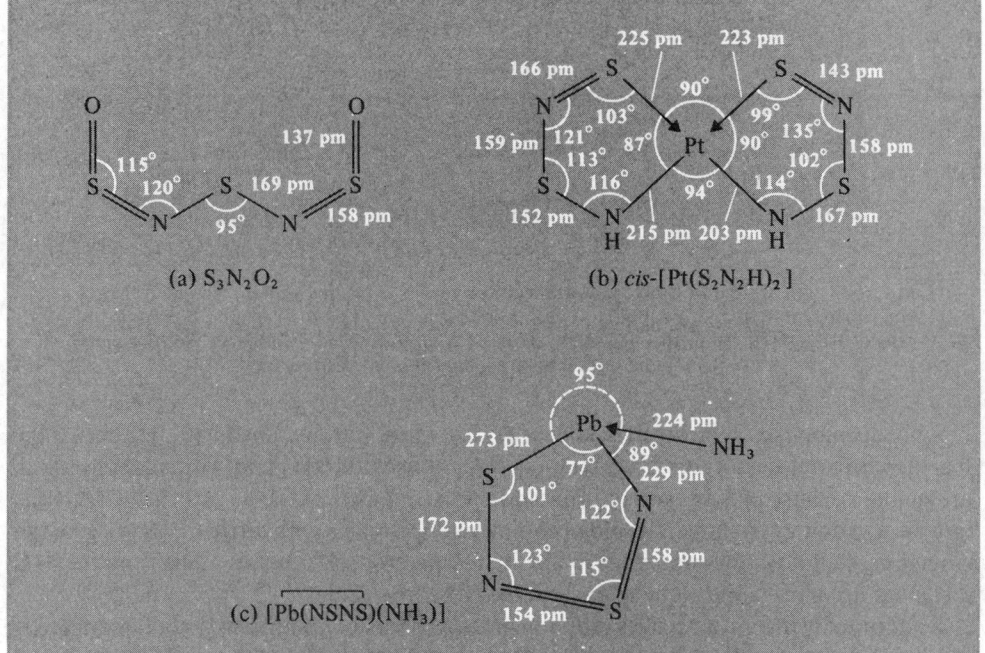

FIG. 15.42   Structure of some SN compounds mentioned in Table 15.23.[132]

15.42b) is typical of several such compounds. When S_4N_4 reacts with metal carbonyls in aprotic media, the products are the structurally similar $[M(S_2N_2)_2]$ (M = Fe, Co, Ni). The pyramidal Pb^{II} complex (Fig. 15.42c) is also notable, and features unequal S–N distances consistent with the bonding indicated.

(b) *Disulfur dinitrogen*, S_2N_2. When S_4N_4 is carefully depolymerized by passing the heated vapour over Ag wool at 250–300° and 0.1–1.0 mmHg, the unstable cyclic dimer S_2N_2 is obtained. The main purpose of the silver is to remove sulfur generated by the thermal decomposition of S_4N_4; the Ag_2S so formed then catalyses the depolymerization of further S_4N_4:

$$S_4N_4 + 8Ag \longrightarrow 4Ag_2S + 2N_2$$

$$S_4N_4 \xrightarrow{\;Ag_2S\;} 2S_2N_2$$

In the absence of Ag/Ag_2S the product is contaminated with S_4N_2 (p. 861) formed by the reaction of the excess sulfur with either S_4N_4 or S_2N_2. S_2N_2 forms large colourless crystals which are insoluble in water but soluble in many organic solvents. The molecular structure is a square-planar ring (D_{2h}) analogous to the isoelectronic cation S_4^{2+} (D_{4h}, p. 785). Figure 15.43 shows the structure obtained by X-ray diffraction at $-130°$ together with typical valence-bond representations:[129]

[132] J. WEISS, Crystal and molecular structure of trisulfurdinitrogendioxide, $S_3N_2O_2$, *Z. Naturforsch.* **16b**, 477 (1961); J. WEISS, Metal–sulfur–nitrogen compounds, *Fortsch. Chem. Forsch.* **5**, 635–62 (1966).

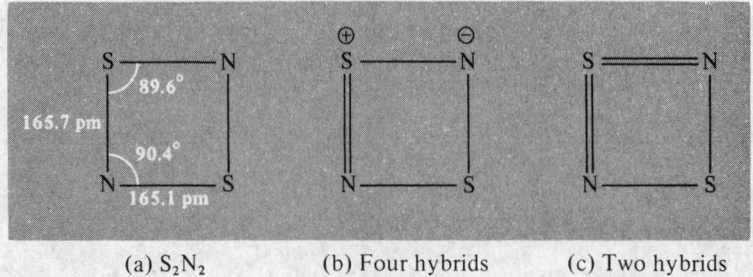

(a) S_2N_2 (b) Four hybrids   (c) Two hybrids

FIG. 15.43   (a) Molecular structure and dimensions of S_2N_2, together with (b) minimal valence-bond representation and (c) addition valence-bond representation involving 3d S orbitals for extended basis set. (Note that the molecule has 6π electrons and 4 unshared electron-pairs superimposed on the square-planar σ-bonded structure.)

S_2N_2 decomposes explosively when struck or when warmed above 30°. Its chemistry has therefore not been extensively studied. Reactions with NH_3 and with aqueous alkali are similar to those of S_4N_4 and it forms adducts with Lewis bases, e.g. $S_2N_2(SbCl_5)_2$; this latter is a yellow crystalline N-bonded complex which reacts with further S_2N_2 to give the orange crystalline monoadduct $S_2N_2.SbCl_5$. The heterocycle remains planar and the S–N distances are almost the same as in the free S_2N_2 molecule.

Undoubtedly the most exciting reaction of S_2N_2 is its slow spontaneous polymerization in the solid state at room temperature to give crystalline $(SN)_x$. Crystals up to several millimetres in length can be grown. Not only is this an unusually facile topochemical reaction for a solid at low temperature but it results in an unprecedented metallic superconducting polymer, as discussed in the following subsection.

(c) *Polythiazyl*, $(SN)_x$.[133] Polymeric sulfur nitride, also known as polythiazyl, was first prepared by F. B. Burt in 1910 using essentially the same method as is still used today—the solid-state polymerization of crystalline S_2N_2 at room temperature (or preferably at 0°C over several days).† Despite the bronze colour and metallic lustre of the polymer, over 50 y were to elapse before its metallic electrical conductivity, thermal conductivity, and thermoelectric effect were investigated. By 1973 it had been established that $(SN)_x$ was indeed a metal down to liquid helium temperatures, and in 1975 the polymer was shown to be a superconductor below 0.26 K. (For higher-quality crystals the transition temperature rises to 0.33 K.) Values of the conductivity σ depend on the purity and crystallinity of the polymer and on the direction of measurement, being much greater along the fibres (b-axis) than across them. At room temperature typical values of $\sigma_\parallel$ are 1000–4000 ohm^{-1} cm^{-1}, and this increases by as much as 1000-fold on cooling to 4.2 K. Typical values of the anisotropy ratio $\sigma_\parallel/\sigma_\perp$ are ~50 at room temperature and ~1000 at 40 K.

$(SN)_x$ is much more stable than its precursor S_2N_2. When heated in air it decomposes explosively at about 240°C but it sublimes readily in vacuum at about 135°. The crystal structure reveals an almost planar chain polymer with the dimensions shown in Fig. 15.44.

---

† Alternative high-yield syntheses have very recently been devised,[133a] e.g. $(SN)_x$ can be made in 65% yield by the reaction of $SiMe_3N_3$ with $S_3N_3Cl_3$, $S_3N_2Cl_2$, or S_3N_2Cl (p. 872) at $-15°C$ in MeCN solution, or by the reaction of $S_3N_3Cl_3$ with an excess of NaN_3.

133 M. M. LABES, P. LOVE, and L. F. NICHOLS, Polysulfur nitride—a metallic, superconducting polymer, *Chem. Revs.* **79**, 1–15 (1979). A definitive review with 150 references.

133a F. A. KENNETT, G. K. MACLEAN, J. PASSMORE, and M. N. S. RAO, Chemical synthesis of poly(sulfur nitride), $(SN)_x$, *JCS Dalton Trans.* 1982, 851–7.

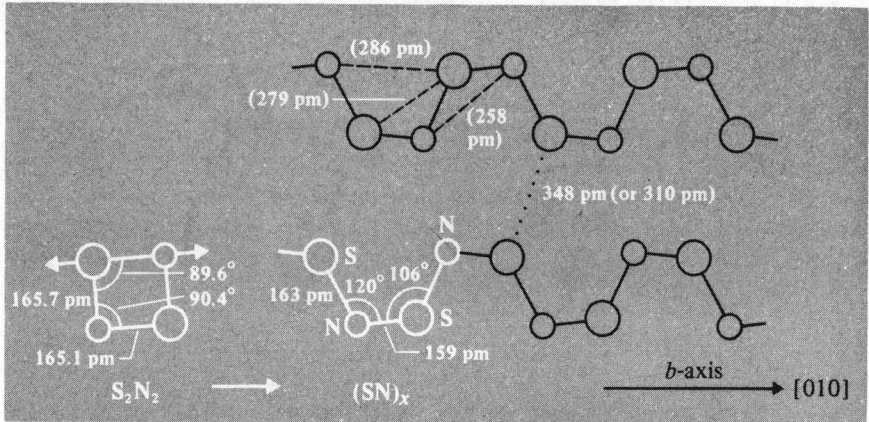

FIG. 15.44   Structure of fibrous $(SN)_x$ and its relation to S_2N_2.

The S and N atoms deviate by about 17 pm from the mean plane. The structure should be compared with that of helical S_∞ (p. 779), the (formal) replacement of alternate S atoms by N resulting both in a conformational change in the position of the atoms and an electronic change whereby 1 valence electron is removed for each SN unit in the chain. Polymerization is thought to occur by a one-point ring cleavage of each S_2N_2 molecule followed by the formation of the *cis-trans*-polymer along the *a*-axis of the S_2N_2 crystal which thereby transforms to the *b*-axis of the $(SN)_x$ polymer.

There is intense current interest in these one-dimensional metals and several related partially halogenated derivatives have also been made, some of which have an even higher metallic conductivity, e.g. partial bromination of $(SN)_x$ with Br_2 vapour yields blue-black single crystals of $(SNBr_{0.4})_x$ having a room-temperature conductivity of 2×10^4 ohm^{-1} cm^{-1} i.e. an order of magnitude greater than for the parent $(SN)_x$ polymer. An even more facile preparation involves direct bromination of S_4N_4 crystals ($\sigma \sim 10^{-14}$ ohm^{-1} cm^{-1} at 25°) with Br_2 vapour at 180 mmHg over a period of hours; subsequent pumping at room temperature gives stoichiometries in the range $(SNBr_{1.5})_x$ to $(SNBr_{0.4})_x$ and further pumping at 80°C for 4 h reduces the halogen content to $(SNBr_{0.25})_x$. Similar highly conducting nonstoichiometric polymers can be obtained by treating S_4N_4 with ICl, IBr, and I_2, the increase in conductivity being more than 16 orders of magnitude.

(d) *Other binary sulfur nitrides.* Six further sulfur nitrides can be briefly mentioned: S_4N_2, $S_{11}N_2$, and $(S_7N)_2S_x$ ($x = 1, 2, 3, 5$); as can be seen from Fig. 15.45, these belong to three distinct structural classes. (For a fourth structure class, exemplified by S_5N_6, see p. 863).

S_4N_2 is usually prepared by heating S_4N_4 with a solution of sulfur in CS_2 under pressure at 100–120°, though a more convenient laboratory preparation is now available by the reaction of activated Zn on S_4N_3Cl.[133b] The compound also results from the thermolytic loss of N_2 from S_4N_4 which occurs when S_4N_4 is heated under reflux in xylene for some hours. An alternative preparation (42% yield), which involves neither high

[133b] R. W. H. SMALL, A. J. BANISTER, and Z. V. HAUPTMAN, Tetrasulfur dinitride; its preparation, crystal structure, and solid state decomposition to give poly(sulfur nitride), *JCS Dalton Trans.* 1981, 2188–91. T. CHIVERS, P. W. CODDING, and R. T. OAKLEY, The X-ray crystal and molecular structure of tetrasulfur dinitride, *JCS Chem. Comm.* 1981, 584–5.

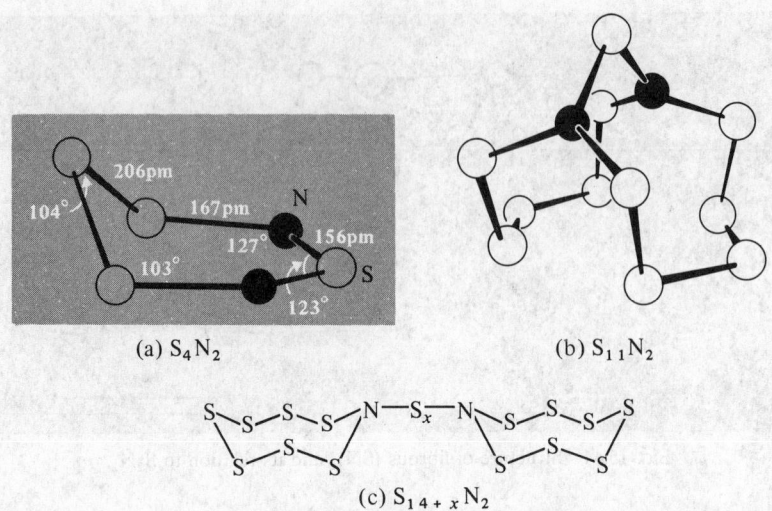

(a) S_4N_2 (b) $S_{11}N_2$

(c) $S_{14+x}N_2$

FIG. 15.45   Structures of (a) S_4N_2[133b] showing the "half-chair" conformation with the central S of the S_3 unit tilted out of the plane of the SNSNS group by 55°; (b) $S_{11}N_2$[134] showing the two planar N atoms; (c) $S_{14+x}N_2$ ($x = 1, 2, 3, 5$)—for $x = 2$ the linking S–S distance is 190 pm and S–N is 170 pm; for $x = 3$ the linking S–S is 204 pm and S–N 171 pm[135].

pressure or high temperature, is the smooth reaction of solutions of $Hg_5(NS)_8$ and S_2Cl_2 in CS_2:

$$Hg_5(NS)_8 + 4S_2Cl_2 \xrightarrow{CS_2/20°} Hg_2Cl_2 + 3HgCl_2 + 4S_4N_2$$

In all these reactions only the 1,3-diazaheterocycle (Fig. 15.45a) is obtained: the 1,1- and 1,4-heterocycles and acyclic isomers are unknown (cf. N_2O_4, p. 523). S_4N_2 forms opaque red-grey needles or transparent dark red prisms which melt at 25° to a dark-red liquid resembling Br_2. It decomposes explosively above 100°. S_4N_2 appears to be a weaker ligand than either S_4N_4 or S_2N_2: it does not react with BCl_3 in CS_2 solution, and $SbCl_5$ gives a complex reaction mixture which contains $S_4N_4 \cdot SbCl_5$ and $[S_4N_3]^+[SbCl_6]^-$ in addition to a poorly defined 1:1 adduct.

$S_{11}N_2$ is obtained as pale amber-coloured crystals by the double condensation of 1,3-$S_6(NH)_2$ with an equimolar amount of S_5Cl_2 in the presence of pyridine:

[134] H. GARCIA-FERNANDEZ, H. G. HEAL, and G. TESTE DE SAGEY, Molecular structure of a new bicyclic sulfur nitride, $S_{11}N_2$, *Compt. Rend.* **C275**, 323–6 (1972).
[135] H. GARCIA-FERNANDEZ, H. G. HEAL, and G. TESTE DE SAGEY, Molecular structure of the bicyclic sulfur nitrides $S_{16}N_2$ and $S_{17}N_2$ having linked cycles and of $S_{11}N_2$ having condensed cycles, *Compt. Rend.* **C282**, 241–3 (1976).

Some polymer is also formed but this can be converted into the bicyclic $S_{11}N_2$ by refluxing in CS_2. The X-ray crystal structure (Fig. 15.45b) shows that the 2 N atoms are planar.[134] This has been interpreted in terms of sp^2 hybridization at N, with some delocalization of the p_π lone-pair of electrons into S-based orbitals, thus explaining the considerably diminished donor power of the molecule. $S_{11}N_2$ is stable at room temperature but begins to decompose when heated above 145°.

The sulfur nitrides $S_{15}N_2$ and $S_{16}N_2$ are (formally) derived from *cyclo*-S_8 (or S_7NH) and can be prepared by reacting S_7NH with SCl_2 and S_2Cl_2 respectively:

$$2S_7NH + S_xCl_2 \longrightarrow S_7N-S_x-NS_7 + 2HCl$$

Both are yellow crystalline materials, stable at room temperature, and readily soluble in CS_2 (Fig. 15.45c).[135] Compounds with $x = 3$ and 5 can be prepared similarly.

Finally, in this subsection we mention the recent discovery of S_5N_6 which is best prepared (73% yield) by the reaction of $S_4N_5^-$ (p. 866) with Br_2 in CH_2Cl_2 at 0°C for several hours[136] Iodine reacts similarly but chlorine affords S_4N_5Cl (p. 864). S_5N_6 forms orange crystals which are stable for prolonged periods at room temperature in an inert atmosphere, though they immediately blacken in air. It can be sublimed unchanged at 45° (10^{-2} mmHg) and decomposes above 130°. The structure (Fig. 15.46) features a molecular basket in which an $-N{=}S{-}N-$ group bridges 2 S atoms of an S_4N_4 cradle. Comparison with S_4N_4 itself (p. 856) shows little change in the S–N distances in the cradle (161 pm) but the trans-annular S···S distances are markedly different: one is opened up from 258 pm to 394 pm (nonbonding) whereas the other contracts to 243 pm suggesting stronger trans-annular bonding between these 2 S atoms and the incipient formation of 2 fused 5-membered S_3N_2 rings.

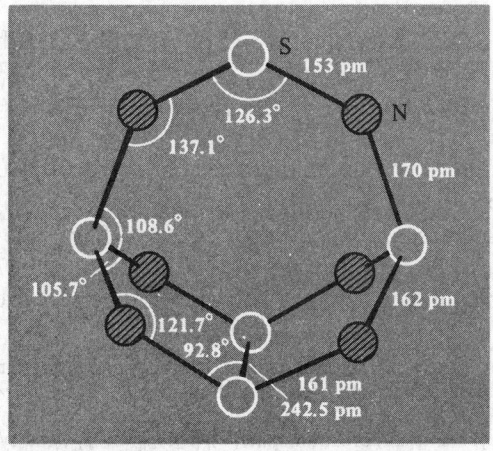

FIG. 15.46   Structure of S_5N_6.

[136] T. Chivers and J. Proctor, Preparation and crystal structure of a new sulfur nitride S_5N_6; a molecular basket, *JCS Chem. Comm.* 1978, 642–3. A fuller version is in *Can. J. Chem.* **57**, 1286–93 (1979). See also W. S. Sheldrick, M. N. S. Rao, and H. W. Roesky, Bicyclic sulfur–nitrogen compounds, etc., *Inorg. Chem.* **19**, 538–43 (1980).

(ii) Sulfur–nitrogen cations and anions

Numerous charged sulfur–nitrogen species have been synthesized in recent years, particularly those having an odd number of N atoms which would otherwise be paramagnetic. However, thio analogues of nitrites (NO_2^-, p. 531) and nitrates (NO_3^-, p. 536) are unknown.

The simplest cationic species is S_2N^+, prepared by oxidation of S_7NH, S_7NBCl_2, or 1,4-$S_6(NH)_2$ (p. 868) with $SbCl_3$.[137] An X-ray structure determination on $[S_2N]^+[SbCl_6]^-$ showed the cation to be linear ($D_{\infty h}$) as expected for a species isoelectronic with CS_2 and NO_2^+. The rather short N–S distance of 146.4 pm is consistent with the formulation $[S{=}N{=}S]^+$.

The radical cation $S_3N_2^+$ is formed in high yield from the oxidation of S_4N_4 with the anhydride $(CF_3SO_2)_2O$:[127]

$$(CF_3SO_2)_2O + S_4N_4 \longrightarrow [S_3N_2]^+[CF_3SO_3]^- + CF_3SO_2S_3N_3$$

The product is a black-brown solid that is very sensitive to oxygen. The same cation can be obtained by oxidation of S_4N_4 with AsF_5 and is unusual in being the only sulfur–nitrogen (paramagnetic) radical that has been obtained as a stable crystalline salt. X-ray diffraction analysis shows the structure to be a planar 5-membered ring with approximate C_{2v} symmetry (Fig. 15.47a). The corresponding diamagnetic dimer $S_6N_4^{2+}$ was obtained in low yield by oxidation of S_3N_2Cl with $ClSO_3H$: its structure (Fig. 15.47b) consists of 2 symmetry-related planar $S_3N_2^+$ units linked by 2 very long S–S bonds. Alternatively, the central S_4 unit can be thought of as being bound by a 4-centre 6-electron bond.

Cations containing 4 S atoms include $S_4N_3^+$, $S_4N_4^{2+}$, and $S_4N_5^+$. The structures are in Fig. 15.48 and typical preparative routes are:[127, 138, 139]

$$S_3N_3Cl_3 + S_2Cl_2 \longrightarrow [S_4N_3]^+Cl^- + SCl_2 + Cl_2$$

$$3S_3N_2Cl_2 + S_2Cl_2 \longrightarrow 2[S_4N_3]^+Cl^- + 3SCl_2$$

$$S_4N_4 + 4SbF_5 \xrightarrow{SO_2} [S_4N_4]^{2+}[SbF_6]^-[Sb_3F_{14}]^-$$

$$S_3N_3Cl_3 + (Me_3SiN)_2S \xrightarrow{CCl_4} [S_4N_5]^+Cl^- + \ldots$$

These compounds contain some fascinating and subtle structural and bonding problems and the original papers should be consulted for further details.

An interesting structural problem also emerges from the study of the final sulfur–nitrogen cation to be considered, $S_5N_5^+$. First made in 1972, this was originally thought to contain a planar, heart shaped 10-membered heterocycle on the basis of X-ray

[137] R. FAGGIANI, R. J. GILLESPIE, C. J. L. LOCK, and J. D. TYRER, Preparation, vibrational spectra, and crystal structure of dithionitronium hexachloroantimonate(V), $[S_2N][SbCl_6]$, *Inorg. Chem.* **17**, 2975–8 (1978).

[138] R. J. GILLESPIE, D. R. SLIM, and J. D. TYRER, A new cationic S–N ring system $S_4N_4^{2+}$. The crystal structure of cyclotetrathiazyl bis(hexafluoroantimonate(V), $[S_4N_4][SbCl_6]_2$, and cyclotetrathiazyl hexafluoro-antimonate(V), tetradecafluorotriantimonate, $[S_4N_4][SbF_6][Sb_3F_{14}]$, *JCS Chem. Comm.* 1977, 253–5. R. J. GILLESPIE, J. P. KENT, J. F. SAWYER, D. R. SLIM, and J. D. TYRER, Reactions of S_4N_4 with $SbCl_5$, SbF_5, AsF_5, PF_5, and HSO_3F. Preparation and crystal structures of salts of the $S_4N_4^{2+}$ cation, *Inorg. Chem.* **20**, 3799–3812 (1981).

[139] T. CHIVERS, L. FIELDING, W. G. LAIDLAW, and M. TRSIC, Synthesis and structure of salts of the bicyclic sulfur-nitrogen cation $S_4N_5^+$ and a comparison of the electronic structures of the tetrasulfur pentanitride(1+) and -(1−) ions, *Inorg. Chem.* **18**, 3379–87 (1979).

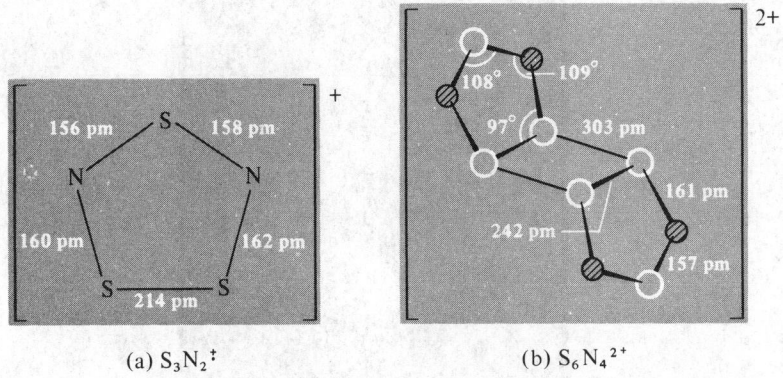

(a) $S_3N_2^{+}$

(b) $S_6N_4^{2+}$

FIG. 15.47   Structures of the radical cation $S_3N_2^{+}$ and its dimer $S_6N_4^{2+}$.

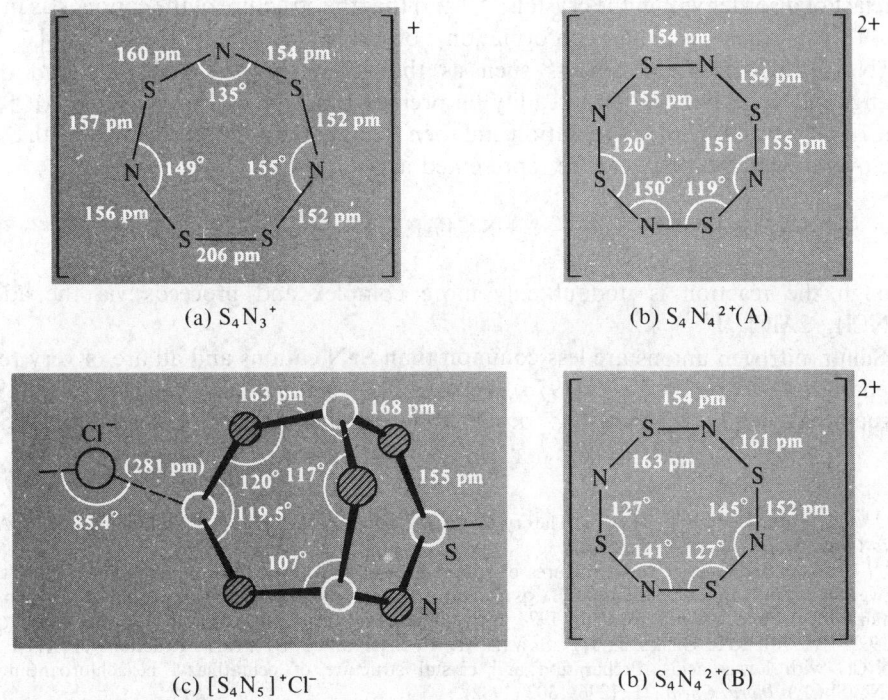

(a) $S_4N_3^{+}$

(b) $S_4N_4^{2+}$(A)

(c) $[S_4N_5]^{+}Cl^{-}$

(b) $S_4N_4^{2+}$(B)

FIG. 15.48 Structure of (a) planar $S_4N_3^{+}$; (b) the two types of planar ring in $[S_4N_4]^{2+}[SbF_6]^{-}[Sb_3F_{11}]^{-}$, ring A featuring equal interatomic distances around the hetero-cycle and ring B alternating distances; and (c) a portion of the polymeric structure of $[S_4N_5]^{+}Cl^{-}$ showing the trans-annular bridging N atom.

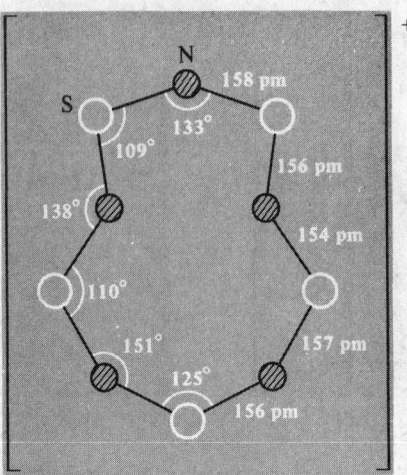

FIG. 15.49   Structure of $S_5N_5^+$.

diffraction studies on $[S_5N_5]^+[AlCl_4]^-$;[140] however, it now seems likely that this is an artefact of disorder within the crystals[141] and that the structure of the cation is as in Fig. 15.49[142] which is the conformation observed in $[S_5N_5]^+[S_3N_3O_4]^-$ and $[S_5N_5]^+[SnCl_5(POCl_3)]^-$. Salts such as the yellow $[S_5N_5]^+[AlCl_4]^-$ and dark-orange $[S_5N_5]^+[FeCl_4]^-$ can readily be prepared in high yield by adding $AlCl_3$ (or $FeCl_3$) to $S_3N_3Cl_3$ in $SOCl_2$ solution and then treating the adduct so formed with S_4N_4; the overall stoichiometry can be represented as:

$$\tfrac{1}{3}(SNCl)_3 + AlCl_3 \longrightarrow \text{``}[NS]^+[AlCl_4]^-\text{''} \xrightarrow{S_4N_4} [S_5N_5]^+[AlCl_4]^-$$

though the reaction is undoubtedly more complex and proceeds via the adduct $(SNCl)_3 \cdot 2AlCl_3$.[143]

Sulfur–nitrogen anions are less common than S–N cations and all are of very recent preparation: *bicyclo*-$S_4N_5^-$ (1976), *cyclo*-$S_3N_3^-$ (1977), and *catena*-S_4N^- (1979). Structures are in Fig. 15.50. $S_4N_5^-$ occurs as the product in a variety of reactions of S_4N_4

[140] A. C. HAZELL and R. G. HAZELL, The crystal structure of pentathiazyl tetrachloroaluminate, $S_5N_5AlCl_4$, *Acta Chim. Scand.* **26**, 1987–95 (1972).

[141] I. RAYMENT, Some crystal structures of sulfur nitrogen and carbon boron compounds, PhD thesis, University of Durham, UK, 1975; A. J. BANISTER and J. A. DURRANT, Sulfur bond angle-sulfur nitrogen bond distance correlations, *J. Chem. Res.* (*M*) 1978, 1928–36. See also R. BARTETZKO and R. GLEITER, *Inorg. Chem.* **17**, 995–7 (1978); R. J. GILLESPIE, J. F. SAWYER, D. R. SLIM, and J. D. TYRER, Reactions of $S_3N_3F_3$ and $S_3N_3Cl_3$ with Lewis acids. Preparation and crystal structure of pentathiazyl hexachloroantimonate, $(S_5N_5)(SbCl_6)$, *Inorg. Chem.* **21**, 1296–1302 (1982).

[142] H. W. ROESKY, W. G. BÖWING, I. RAYMENT, and H. M. M. SHEARER, Preparation and X-ray structure of sulfur-nitrogen oxides, *JCS Chem. Comm.* 1975, 735–6; A. J. BANISTER, J. A. DURRANT, I. RAYMENT, and H. M. M. SHEARER, The preparation and structure of cyclopentathiazenium pentachloro(phosphoryl chloride)stannate(IV), *JCS Dalton* 1976, 928–30.

[143] A. J. BANISTER and H. G. CLARKE, The preparation of cyclopentathiazenium ($S_5N_5^+$) salts and some observations on the structure of the cyclopentathiazenium cations, *JCS Dalton* 1972, 2661–3. See also A. J. BANISTER, A. J. FIELDER, R. G. HEY, and N. R. M. SMITH, ibid., 1457–60.

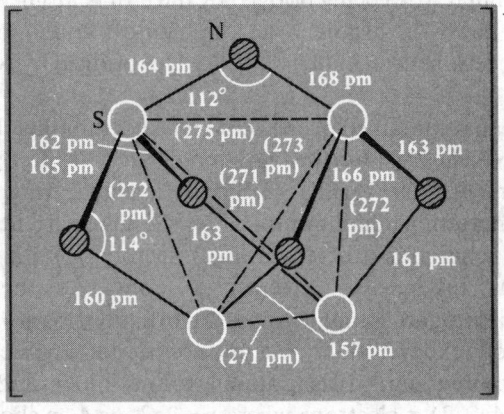

(a) $S_4N_5^-$

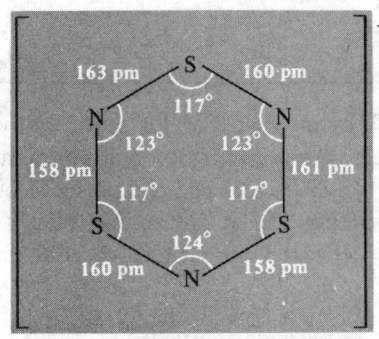

(b) $S_3N_3^-$

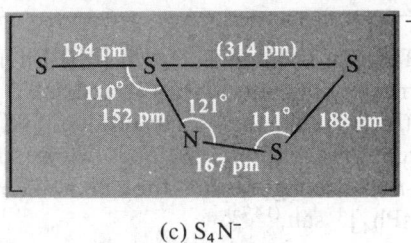

(c) S_4N^-

FIG. 15.50   Structure of sulfur–nitrogen anions.

with nucleophiles:[144] e.g. liquid NH_3, or ethanolic solutions of R_2NH, MN_3 (M = Li, Na, K, Rb), KCN, or even Na_2S. The course of these reactions suggests the initial formation of $S_3N_3^-$ which then reacts with further S_4N_4 to give $S_4N_5^-$. The ammonium salt $[NH_4]^+[S_4N_5]^-$ is a ubiquitous product of the reaction of ammonia with S_4N_4, $(SNCl)_3$, S_2Cl_2, SCl_2, or SCl_4.[145] Yet another route is the methanolysis of $(Me_3SiN)_2S$:

$$Me_3Si-N{=}S{=}N-SiMe_3 \xrightarrow{\text{MeOH}} [NH_4]^+[S_4N_5]^-$$

Subsequent metathesis with Bu_4^nNOH yielded yellow crystals suitable for X-ray structure analysis. The structure of $[S_4N_5]^-$ (Fig. 15.50a) is closely related to that of S_4N_4 (and

[144] J. BOJES, T. CHIVERS, I. DRUMMOND, and G. MACLEAN, Central role of the $S_3N_3^-$ and $S_4N_5^-$ ions in the deprotonation of $S_4(NH)_4$ and in the reductive or nucleophilic degradation of S_4N_4, *Inorg. Chem.* **17**, 3668–72 (1978).

[145] O. J. SCHERER and G. WOLMERSHÄUSER, Ubiquitous ammonium tetrasulfur pentanitride, $[NH_4]^+[S_4N_5]^-$, *Chem. Ber.* **110**, 3241–4 (1977).

$S_4N_5^+$), one trans-annular $S\cdots S$ being bridged by the fifth N atom.[146] One feature of the structure is that all the $S\cdots S$ distances become almost equal so that an alternative description is of an S_4 tetrahedron with 5 of the 6 edges bridged by N atoms, angle S–N–S 112–114°.

The anion $S_3N_3^-$ can be obtained by the action of azides (or metallic K) on S_4N_4 or the reaction of KH on $S_4(NH)_4$.[147] Further reaction of $S_3N_3^-$ with S_4N_4 yields $S_4N_5^-$ (as above). The structure of $S_3N_3^-$ (Fig. 15.50b) is a planar ring of approximate D_{3h} symmetry. This has interesting bonding implications. Thus each S in a heterocycle forms a σ bond to each of its neighbours (thereby using 2 electrons) and it also has an exocyclic lone-pair of electrons: this leaves 2 electrons to contribute to the π system of the heterocycle (which might or might not involve S 3d orbitals). Likewise, each N atom has 2 electrons in σ bonds, one exocyclic lone pair, and contributes one electron to the π system. Planar S–N heterocycles having 4–10 ring atoms are now known and all except the radical cation $S_3N_2^{\ddagger}$ have $(4n+2)$ π electrons where $n=2$, 3, or 4 as shown below:

Ringe/size	4	5	6	7	8	10
Species	S_2N_2	$S_3N_2^{\ddagger}$	$S_3N_3^-$	$S_4N_3^+$	$S_4N_4^{2+}$	$S_5N_5^+$
Number of π electrons	6	[7]	10	10	10	14

Thermal decomposition of $[N(PPh_3)_2]^+[S_4N_5]^-$ in MeCN yields sequentially the corresponding salts of $S_3N_3^-$ and S_4N^- (50% yield). An X-ray crystallographic analysis of the dark-blue air-stable product $[N(PPh_3)_2]^+[S_4N]^-$ revealed the presence of the unique acyclic anion $[SSNSS]^-$ whose structure is in Fig. 15.50c. The anion is planar with *cis-trans* configuration, though a different geometrical configuration occurs in the $[AsPh_4]^+$ salt.[147a]

Sulfur imides, $S_{8-n}(NH)_n$ [124]

The NH group is "isoelectronic" with S and so can successively subrogate S in *cyclo*-S_8. Thus we have already seen that reduction of S_4N_4 with dithionite or with $SnCl_2$ in boiling ethanol/benzene yields $S_4(NH)_4$. Again, whereas reaction of S_2Cl_2 or SCl_2 with NH_3 in non-polar solvents yields S_4N_4, heating these 2 reactants in polar solvents such as dimethylformamide affords a range of sulfur imides. In a typical reaction 170 g S_2Cl_2 and the corresponding amount of NH_3 yielded:

S_8 (32 g)	1,3-$S_6(NH)_2$	(0.98 g)	1,3,5-$S_5(NH)_3$	(0.08 g)
S_7NH (15.4 g)	1,4-$S_6(NH)_2$	(2.3 g)	1,3,6-$S_5(NH)_3$	(0.32 g)
	1,5-$S_6(NH)_2$	(0.82 g)		

[146] W. FLUES, O. J. SCHERER, J. WEISS, and G. WOLMERSHÄUSER, Crystal and molecular structure of the tetrasulfur pentanitride anion, *Angew. Chem.*, Int. Edn. (Engl.) **15**, 379–80 (1976).

[147] J. BOJES, T. CHIVERS, W. G. LAIDLAW, and M. TRSIC, Crystal and molecular structure of [[Bu$_4^n$N]$^+$[S$_3$N$_3$]$^-$ and the vibrational assignments and electronic structure of the planar six-membered ring of the trisulfur trinitride anion, *J. Am. Chem. Soc.* **101**, 4517–22 (1979), and references therein.

[147a] N. BURFORD, T. CHIVERS, A. W. CORDES, R. T. OAKLEY, W. T. PENNINGTON, and P. N. SWEPSTON, Variable geometry of the S_4N^- anion: crystal and molecular structure of Ph$_4$As$^+$S$_4$N$^-$ and a refinement of the structure of PPN$^+$S$_4$N$^-$, *Inorg. Chem.* **20**, 4430–2 (1981). See also T. CHIVERS and C. LAU, Raman spectroscopic identification of the S_4N^- and S_3^- ions in blue solutions of sulfur in liquid ammonia, *Inorg. Chem.* **21**, 453–5 (1982).

In no case have adjacent NH groups been observed.

S₇NH is a stable pale-yellow compound, mp 113.5°; the structure is closely related to that of *cyclo*-S₈ as shown in Fig. 15.51a. The proton is acidic and undergoes many reactions of which the following are typical (see also p. 863):

$$BX_3 \longrightarrow S_7N{-}BX_2 + HX \quad (X = Cl, Br)$$
$$NaCPh_3 \longrightarrow S_7NNa + Ph_3CH$$
$$(Me_3Si)_2NH \longrightarrow S_7N{-}SiMe_3 + NH_3$$
$$Hg(MeCO_2)_2 \longrightarrow Hg(NS_7)_2 + MeCO_2H$$

The 3 isomeric compounds S₆(NH)₂ form stable colourless crystals and have the structures illustrated in Fig. 15.51b, c, and d.[126a, 148] The 1,3-, 1,4-, and 1,5-isomers melt at 130°, 133°, and 155° respectively. The 1,3,5- and 1,3,6-triimides melt with decomposition at 128° and 133° (Fig. 15.51e and f). The tetraimide, S₄(NH)₄ (mp 145°) is structurally very similar (Fig. 15.51g):[149] the N atoms are each essentially trigonal planar and the heterocycle is somewhat flattened, the distance between the planes of the 4 N atoms and 4 S atoms being only 57 pm. Alkyl derivatives such as 1,4-S₆(NR)₂ and S₄(NR)₄ can be synthesized by reacting S₂Cl₂ with primary amines RNH₂ in an inert solvent. Very recently the bis-adduct [Ag(S₄N₄H₄)₂]⁺ has been isolated as its perchlorate; this has a sandwich-like structure and is unique in being S-bonded rather than N-bonded to the metal ion.[149a]

Other cyclic sulfur–nitrogen compounds[125, 127]

Incorporation of a third heteroatom into S–N compounds is now well established, e.g. for C, Si; P, As; O; Sn and Pb, together with the S₂N₂ chelates of Fe, Co, Ni, Pd, and Pt mentioned on p. 858. The field is very extensive but introduces no new concepts into the general scheme of covalent heterocyclic molecular chemistry. Illustrative examples are in Fig. 15.52 and fuller details including X-ray structures for many of the compounds are in the references cited above. A selenium analogue of S₆N₄²⁺ (p. 864) has also recently been prepared and structurally characterized, *viz.* [SN₂Se₂Se₂N₂S]²⁺.[149b]

Sulfur–nitrogen halogen compounds[150-2]

As with sulfur–halogen compounds (pp. 808–819) the stability of N–S–X compounds decreases with increase in atomic weight of the halogen. There are numerous fluoro and

[148] J. C. VAN DE GRAMPEL and A. VOS, The structure of the three isomers S₆(NH)₂, *Acta Cryst.* **B25,** 611–17 (1969), and references therein. See also H. J. POSTMA, F. VAN BOLHUIS, and A. VOS, The molecular and crystal structure of cyclohexasulfur-1,3-diimide, S₆(NH)₂ isomer (III), *Acta Cryst.* **B27,** 2480–6 (1971).

[149] T. M. SABINE and G. W. COX, A neutron structure analysis of S₄N₄H₄, *Acta Cryst.* **28,** 574–7 (1967).

[149a] M. B. HURSTHOUSE, K. M. A. MALIK, and S. N. NABI, Crystal and molecular structure of bis(tetrasulfur tetraimide)silver(I) perchlorate sesquihydrate, *JCS Dalton* 1980, 355–9.

[149b] R. J. GILLESPIE, J. P. KENT, and J. F. SAWYER, Preparation and crystal structure of the hexafluoroarsonate and hexafluoroantimonate salts of the dimeric thiodiselenazyl cation Se₄S₂N₄²⁺, *Inorg. Chem.* **20,** 4053–60 (1981).

[150] O. GLEMSER and M. FILD, Sulfur–nitrogen–halogen compounds, in V. GUTMANN (ed.), *Halogen Chemistry,* Vol. 2, pp. 1–30, Academic Press, London, 1967.

[151] R, MEWS, Nitrogen–sulfur–fluorine ions, *Adv. Inorg. Chem. Radiochem.* **19,** 185–237 (1976).

[152] O. GLEMSER and R. MEWS, Chemistry of thiazyl fluoride (NSF) and thiazyl trifluoride (NSF₃): a quarter century of sulfur–nitrogen–fluorine chemistry, *Angew. Chem.,* Int. Edn. (Engl.) **19,** 883–99 (1980).

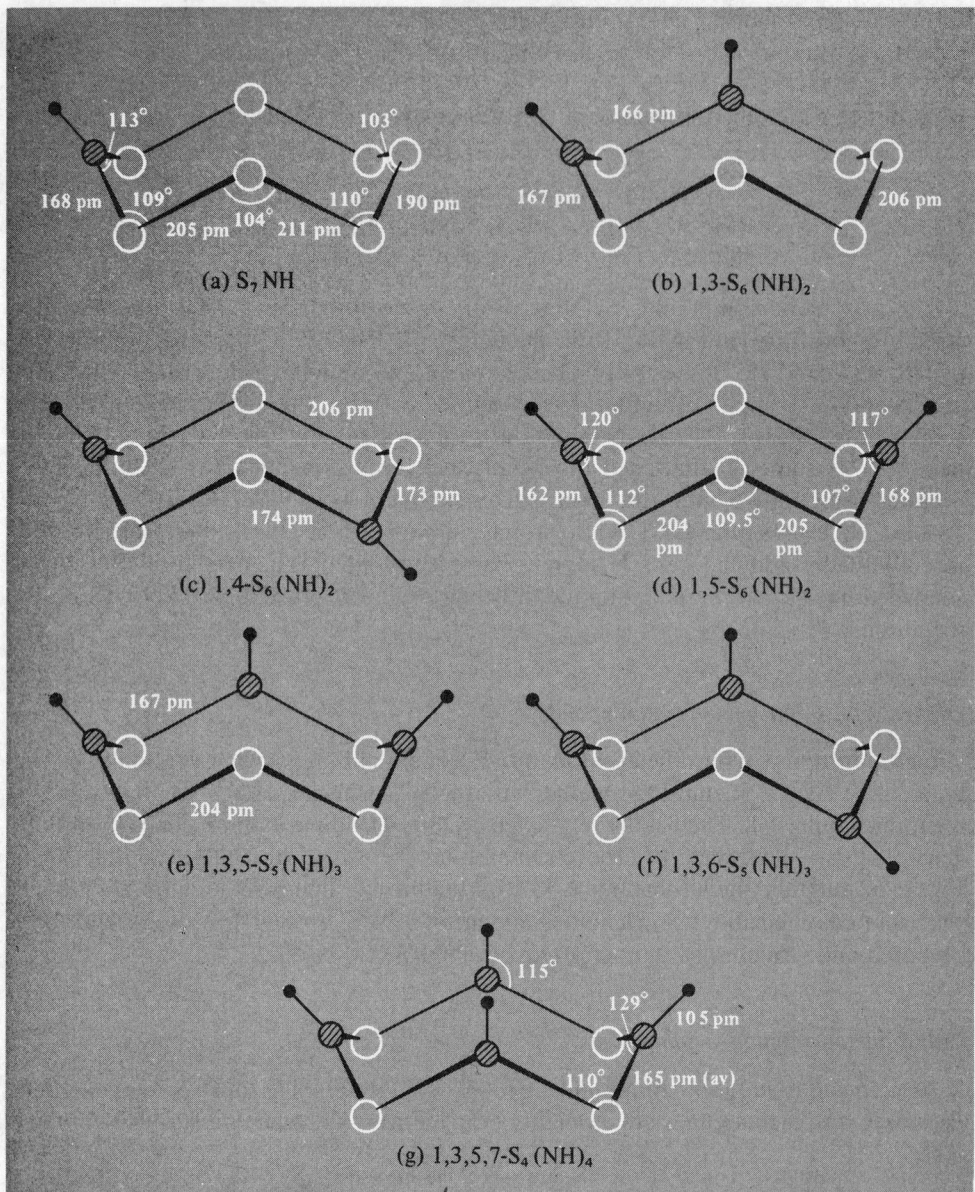

(a) S$_7$NH

(b) 1,3-S$_6$(NH)$_2$

(c) 1,4-S$_6$(NH)$_2$

(d) 1,5-S$_6$(NH)$_2$

(e) 1,3,5-S$_5$(NH)$_3$

(f) 1,3,6-S$_5$(NH)$_3$

(g) 1,3,5,7-S$_4$(NH)$_4$

FIG. 15.51   Structures of the various cyclo sulfur imides.

chloro derivatives but bromo and iodo derivatives are virtually unknown except for the nonstoichiometric (SNX$_x$)$_\infty$ polymers (p. 861). Unlike the H atoms in the sulfur imides (p. 868) the halogen atoms are attached to S rather than N. Fluoro derivatives have been known since 1965 but some of the chloro compounds have been known for over a century. The simplest compounds are the nonlinear thiazyl halides N≡S—F and N≡S—Cl: these form a noteworthy contrast to the nonlinear nitrosyl halides O=N—X. In all cases, the

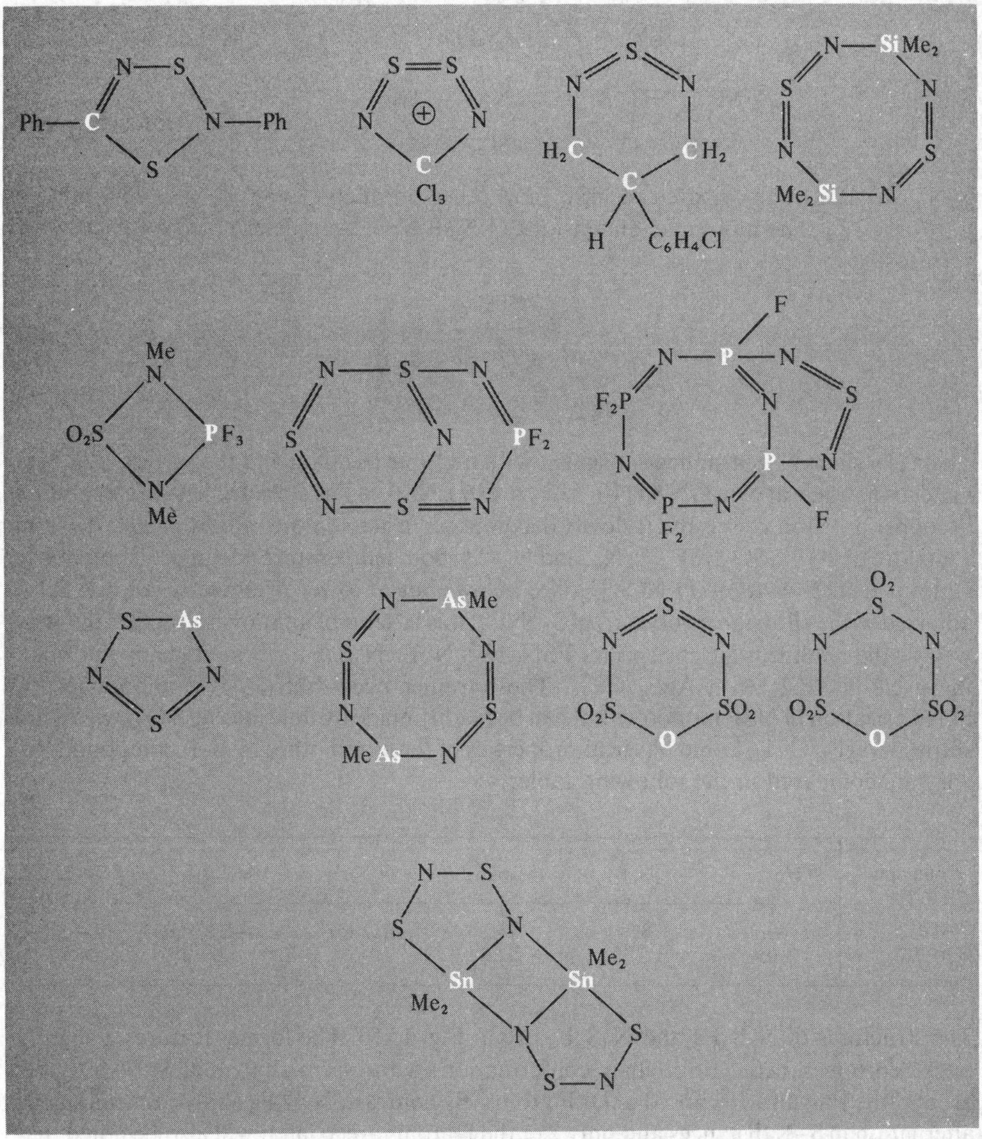

FIG. 15.52   Some heterocyclic S–N compounds incorporating a third heteroelement.

pairs of elements directly bonded are consistent with rule that the most electronegative atom of the trio bonds to the least electronegative, i.e. $\{S(NH)\}_4$, $\{N(SF)\}_{1,3,4}$, $\{N(SCl)\}_{1,3}$, $O(NF)$, $O(NCl)$ (formal Pauling electronegativities: H 2.1, S 2.5, N 3.0, Cl 3.0, O 3.5, F 4.0).

Thiazyl fluoride, NSF, is a colourless, reactive, pungent gas (mp $-89°$, bp $+0.4°$). It is best prepared by the action of HgF_2 on a slurry of S_4N_4 and CCl_4 but it can also be made by a variety of other reactions:[150]

$$S_4N_4 + 4HgF_2 \text{ (or } AgF_2) \xrightarrow{CCl_4} 4NSF + 2Hg_2F_2$$

$$S_4N_4 \xrightarrow{\text{IF}_5} \{S_4N_4(NSF)_4\} \xrightarrow{50°} S_4N_4 + 4NSF$$

$$SF_4 + NH_3 \longrightarrow NSF + 3HF$$

$$NF_3 + 3S \longrightarrow NSF + SSF_2$$

S_4N_4 can also be fluorinated to NSF (and other products) using F_2 at $-75°$, SeF_4 at $-10°$, or SF_4. The molecular dimensions of NSF have been determined by microwave spectroscopy:

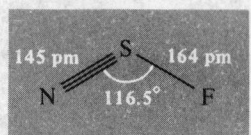

There is a short N–S distance consistent with multiple bonding and the angle at S is very close to the angle at N in ONX (110–117°, p. 507). NSF can be stored at room temperature in copper or teflon vessels but it slowly decomposes in glass (more rapidly at 200°) to form a mixture of OSF_2, SO_2, SiF_4, S_4N_4, and N_2. At room temperature and at pressures above 1 atm it trimerises to $cyclo$-$N_3S_3F_3$ (see below) but at lower pressures it affords S_4N_4 admixed with yellow-green crystals of $S_3N_2F_2$; this latter is of unknown structure but may well be the nonlinear acyclic species $FSN=S=NSF$. $N_3S_3F_3$ is best made by fluorinating $cyclo$-$N_3S_3Cl_3$ with AgF_2/CCl_4. The tetramer $cyclo$-$N_4S_4F_4$ is not obtained by polymerization of NSF monomer but can be readily made by fluorinating S_4N_4 with a hot slurry of AgF_2/CCl_4. Some physical properties of these and other N–S–F compounds (p. 858) are compared in the following table:

Compound	N≡S—F	$S_3N_2F_2$	$N_3S_3F_3$	$N_4S_4F_4$	N≡SF$_3$	FN=SF$_2$
MP/°C	−89	83	74.2	153 (d)	−72	—
BP/°C	+0.4	—	92.5	—	27.1	−6.7

The structures of $N_3S_3F_3$ and $N_4S_4F_4$ are in Fig. 15.53. The former features a slightly puckered 6-membered ring (chair conformation) with essentially equal S–N distances around the ring and 3 eclipsed axial F atoms. By contrast, $N_4S_4F_4$ shows a pronounced alternation in S–N distances and only 2 of the F atoms are axial; it will also be noted that the conformation of the N_4S_4 ring is very different to that in S_4N_4 (p. 856) or $S_4(NH)_4$ (p. 870). The chemistry of these various NSF oligomers has not been extensively studied. $N_3S_3F_3$ is stable in dry air but is hydrolysed by dilute aqueous NaOH to give NH_4F and sulfate. $N_4S_4F_4$ is reported to form an N-bonded 1:1 adduct with BF_3 whereas with AsF_5 or SbF_5 fluoride ion transfer occurs (accompanied by dethiazylation of the ring) to give $[N_3S_3F_2]^+[MF_6]^-$ and $[NS]^+[MF_6^-]$.

In the chloro series, the compounds to be considered are N≡S—Cl, $cyclo$-$N_3S_3Cl_3$, $cyclo$-$N_3S_3Cl_3O_3$, the ionic compounds $[S_4N_3]^+Cl^-$, $[cyclo$-$N_2S_3Cl]^+Cl^-$, and $[catena$-$N(SCl)_2]^+[BCl_4]^-$, together with various isomeric oxo- and fluoro-chloro derivatives. Thiazyl chloride, NSCl, is best obtained by vacuum pyrolysis of the trimer in vacuum at 100°. It can also be made by the reaction of Cl_2 on NSF (note that $NSF + F_2 \rightarrow NSF_3$)

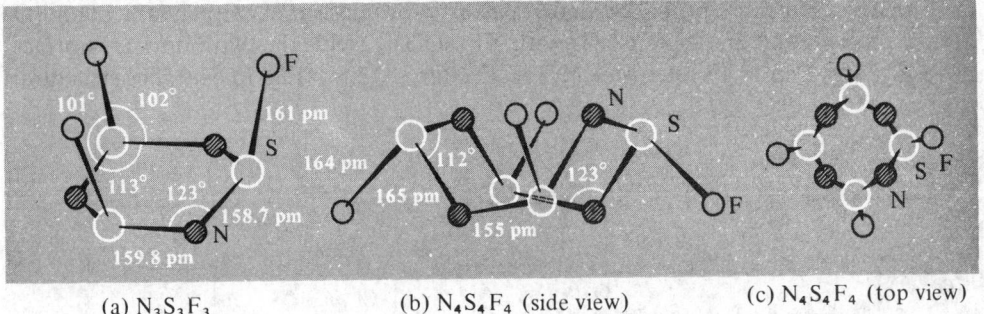

(a) $N_3S_3F_3$ (b) $N_4S_4F_4$ (side view)   (c) $N_4S_4F_4$ (top view)

FIG. 15.53   Molecular structures of (a) $N_3S_3F_3$, (b) $N_4S_4F_4$ (side view), and (c) $N_4S_4F_4$ (top view).

and by numerous other reactions.[150] It is a yellow-green gas that rapidly trimerizes at room temperature, and is isostructural with NSF.

By far the most common compound in the series is $N_3S_3Cl_3$ (yellow needles, mp 168°) which can be prepared by the direct action of Cl_2 (or $SOCl_2$) on S_4N_4 in CCl_4, and which is also obtained in all reactions leading to NSCl. The structure (Fig. 15.54) is very similar to that of $N_3S_3F_3$ and comprises a slightly puckered ring with equal S–N distances of 160.5 pm and the N atoms only 18 pm above and below the plane of the 3 S atoms. $N_3S_3Cl_3$ is sensitive to moisture and is oxidized by SO_3 above 100° to $N_3S_3Cl_3O_3$; at lower temperatures the adduct $N_3S_3Cl_3.6SO_3$ is formed and this dissociates at 100° to $N_3S_3Cl_3.3SO_3$. A more efficient preparation of $N_3S_3Cl_3O_3$ is by thermal decomposition of the product obtained by the reaction of amidosulfuric acid with PCl_5:

$$H_2NSO_3H + 2PCl_5 \longrightarrow 3HCl + OPCl_3 + ClSO_2-N{=}PCl_3$$

$$3ClSO_2-N{=}PCl_3 \xrightarrow{\text{heat}} 3OPCl_3 + (NSClO)_3$$

The compound is obtained in two isomeric forms from this reaction: α, mp 145° and β, mp 43°. The structure of the α-form is in Fig. 15.54b and is closely related to that of $(NSCl)_3$ with uniform S–N distances around the ring. The β-form may have a different ring

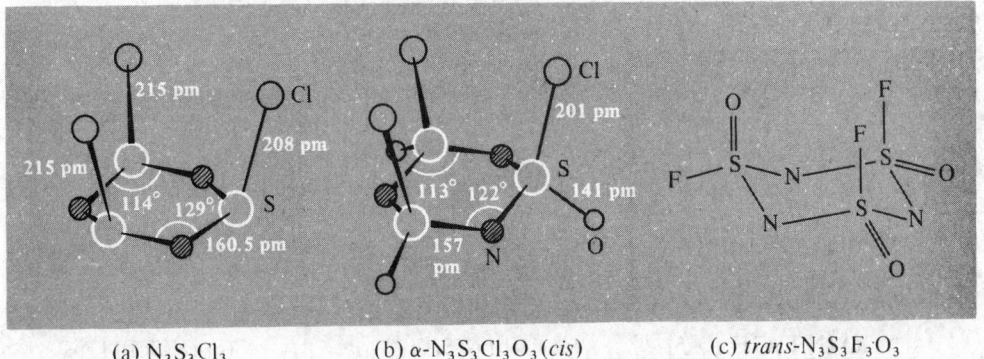

(a) $N_3S_3Cl_3$ (b) α-$N_3S_3Cl_3O_3$ (*cis*)   (c) *trans*-$N_3S_3F_3O_3$

FIG. 15.54   Molecular structure of (a) $N_3S_3Cl_3$, (b) α-$N_3S_3Cl_3O_3$ (*cis*), and (c) *trans*-$N_3S_3F_3O_3$.

conformation but more probably involves *cis-trans* isomerism of the pendant Cl and O atoms. Fluorination of α-$N_3S_3Cl_3O_3$ with KF in CCl_4 yields the two isomeric fluorides *cis*-$N_3S_3F_3O_3$ (mp 17.4°) and *trans*-$N_3S_3F_3O_3$ (mp $-12.5°$) (Fig. 15.54c). The structural

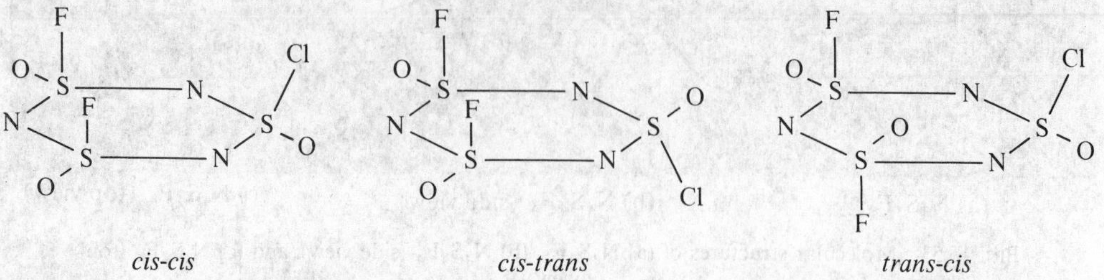

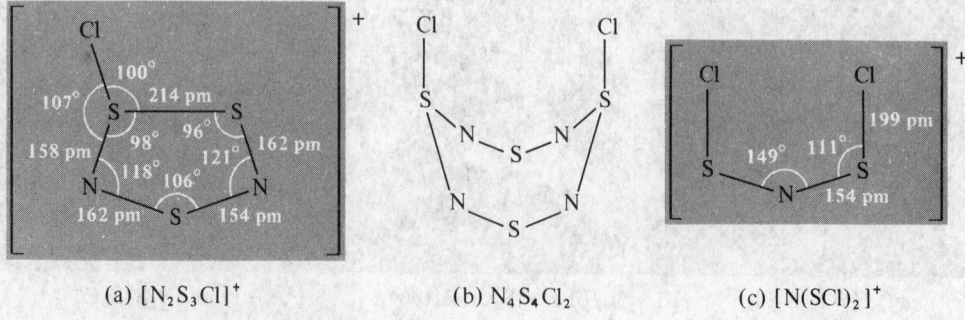

cis-cis   *cis-trans*   *trans-cis*

FIG. 15.54A   Schematic representation of the three geometric isomers of $N_3S_3ClF_2O_3$. The three isomers of the monofluoro derivative are similar but with Cl and F interchanged.

assignment of the 2 isomers was made on the basis of ^{19}F nmr. Fluorination with SbF_3 under reduced pressure yields both the monofluoro and difluoro derivatives $N_3S_3Cl_2FO_3$ and $N_3S_3ClF_2O_3$, each having 3 isomers which can be separated chromatographically and assigned by ^{19}F nmr as indicated schematically in Fig. 15.54A. Numerous other derivatives are known in which one or more halogen atom is replaced by $-NH_2$, $-N{=}SF_2$, $-N{=}PCl_3$, $-N{=}CHPh$, $-OSiMe_3$, etc.

A different structure motif occurs in S_4N_3Cl. This very stable yellow compound features the $S_4N_3^+$ cation (p. 865) and is obtained by many reactions, e.g.:

$$3S_4N_4 + 2S_2Cl_2 \xrightarrow{CCl_4/heat} 4[S_4N_3]^+Cl^-$$

The chloride ion is readily replaced by other anions to give, for example, the orange-yellow $[S_4N_3]Br$, bronze-coloured $[S_4N_3]SCN$, $[S_4N_3]NO_3$, $[S_4N_3]HSO_4$, etc.

Chlorination of S_4N_4 with NOCl or $SOCl_2$ in a polar solvent yields $S_3N_2Cl_2$:

$$S_4N_4 + 2NOCl \longrightarrow S_3N_2Cl_2 + \tfrac{1}{2}S_2Cl_2 + 2N_2O$$

The crystal structure again reveals an ionic formulation, $[N_2S_3Cl]^+Cl^-$, this time with a slightly puckered 5-membered ring carrying a single pendant Cl atom as shown in Fig. 15.55a; the alternation of S–N distances and the rather small angles at the 2 directly linked S atoms are notable features. Yet a further product can be obtained by the partial

(a) $[N_2S_3Cl]^+$ (b) $N_4S_4Cl_2$ (c) $[N(SCl)_2]^+$

FIG. 15.55   Structure of (a) the cation in $[N_2S_3Cl]^+Cl^-$, (b) $N_4S_4Cl_2$, and (c) $[N(SCl)_2]^+$.

chlorination of S_4N_4 with Cl_2 in CS_2 solution below room temperature: one of the transannular $S\cdots S$ "bonds" is opened to give yellow crystals of $N_4S_4Cl_2$ (Fig. 15.55b) and this derivatized heterocycle can be used to prepare several other compounds.[153]

Reaction of NSF_3 with BCl_3 yields the acyclic cation $[N(SCl)_2]^+$ as its BCl_4^- salt (Fig. 15.55c); the compound is very hygroscopic and readily decomposes to BCl_3, SCl_2, S_2Cl_2, and N_2.

The formation of highly conducting nonstoichiometric bromo and iodo derivatives of polythiazyl has already been mentioned (p. 861). It has recently been found that, whereas bromination of solid S_4N_4 with gaseous Br_2 yields conducting $(SNBr_{0.4})_x$, reaction with liquid bromine leads to the stable tribromide $[S_4N_3]^+[Br_3]^-$.[154] In contrast, the reaction of S_4N_4 with Br_2 in CS_2 solution results in a (separable) mixture of $[S_4N_3]^+[Br_3]^-$, $[S_4N_3]^+Br^-$, and the novel ionic compound $CS_3N_2Br_2$ which may be

$$[\overline{S{=}C{-}S{=}N{-}S{=}N}]^{2+}[Br^-]_2 \text{ or } [\overline{S{=}C{-}S{=}N{-}S(Br){=}N}]^+Br^-.$$

Sulfur–nitrogen–oxygen compounds[124]

This is a classic area of inorganic chemistry dating back to the middle of the last century and only a brief outline will be possible. It will be convenient first to treat the sulfur nitrogen oxides and then the amides, imides, and nitrides of sulfuric acid. Hydrazides and hydroxylamides of sulfuric acid will also be considered. Some of these compounds have remarkable properties and some are implicated in the lead-chamber process for the manufacture of H_2SO_4 (p. 838). The field is closely associated with the names of the great German chemists E. Frémy ($\sim$1845), A. Claus ($\sim$1870), F. Raschig ($\sim$1885–1925), W. Traube ($\sim$1890–1920), F. Ephraim ($\sim$1910), P. Baumgarten ($\sim$1925), and, in more recent years, M. Becke-Goehring ($\sim$1955) and F. Seel ($\sim$1955–65).

(a) *Sulfur–nitrogen oxides.* Trisulfur dinitrogen dioxide, $S_3N_2O_2$, is best made by treating S_4N_4 with boiling $OSCl_2$ under a stream of SO_2:

$$S_4N_4 + 2OSCl_2 \longrightarrow S_3N_2O_2 + 2Cl_2 + S_2N_2 + S$$

It is a yellow solid with an acyclic structure (Fig. 15.56a), cf N_2O_5 (p. 526). Moist air converts $S_3N_2O_2$ to SO_2 and S_4N_4 whereas SO_3 oxidizes it smoothly to $S_3N_2O_5$:

$$S_3N_2O_2 + 3SO_3 \longrightarrow S_3N_2O_5 + 3SO_2$$

The pentoxide $S_3N_2O_5$ can also be made directly from S_4N_4 and SO_3. It forms colourless, strongly refracting crystals which readily hydrolyse to sulfamic acid:

$$S_3N_2O_5 + 3H_2O \longrightarrow 2H_2NSO_3H + SO_2$$

It has a cyclic structure and may be regarded as a substituted diamide of disulfuric acid, $H_2S_2O_7$ (Fig. 15.56b).

[153] H. W. ROESKY, C. GRAF, M. N. S. RAO, B. KREBS, and G. HENKEL, $S_5N_6(CH_2)_4$—the first spirocycle 1´,6-thiacyclopentane derivative of a sulfur–nitrogen compound, *Angew. Chem.*, Int. Edn. (Engl.) **18**, 780–1 (1979), and references therein. H. W. ROESKY, M. N. S. RAO, C. GRAF, A. GIEREN, and E. HÄDICKE, 1,5-Bis(dimethylamino(tetrasulfur) tetranitride—a cage molecule with a non-symmetric nitrogen bridge, *Angew. Chem.* Int. Edn. (Engl.) **20**, 592–3 (1981).

[154] G. WOLMERSHÄUSER, G. B. STREET, and R. D. SMITH, Reactions of tetrasulfide tetranitride with bromine. Reaction in carbon disulfide solution to give $CS_3N_2Br_2$, *Inorg. Chem.* **18**, 383–5 (1979).

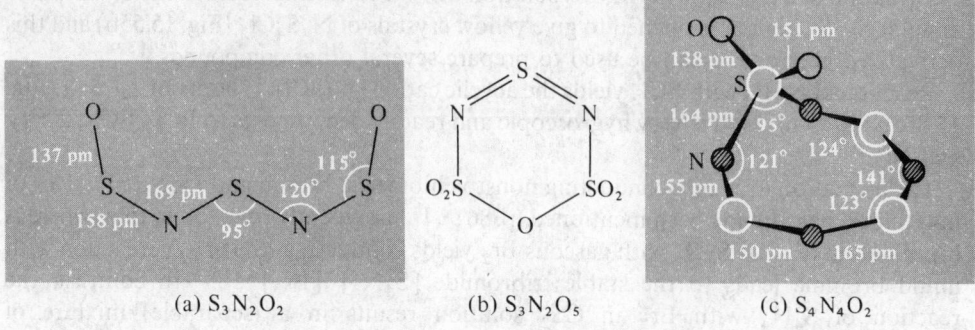

(a) S$_3$N$_2$O$_2$ (b) S$_3$N$_2$O$_5$ (c) S$_4$N$_4$O$_2$

FIG. 15.56   Structures of sulfur–nitrogen oxides.

An alternative synthetic strategy for sulfur–nitrogen oxides is exemplified by the recent reaction:[155]

$$S_3N_2Cl_2 + SO_2(NH_2)_2 \xrightarrow[\text{reflux}]{CCl_4} S_4N_4O_2$$

The product forms orange-yellow crystals, mp 166 (d), having a structure in which 1 S atom of an S$_4$N$_4$ ring carries both O atoms. X-ray diffractometry shows substantial deviation from the parent S$_4$N$_4$ structure, a notable feature being the coplanarity of the S$_3$N$_2$ moiety furthest removed from the SO$_2$ group (Fig. 15.56c).

(b) *Amides of sulfuric acid*. Amidosulfuric acid (better known as sulfamic acid, H[H$_2$NSO$_3$]), is a classical inorganic compound and an important industrial chemical. Formal replacement of both hydroxyl groups in sulfuric acid leads to sulfamide (H$_2$N)$_2$SO$_2$ (p. 878) which is also clearly related structurally to the sulfuryl halides X$_2$SO$_2$ (p. 820).

Sulfamic acid can be made by many routes, including addition of hydroxylamine to SO$_2$ and addition of NH$_3$ to SO$_3$:

$$H_2NOH + SO_2 \longrightarrow H[H_2NSO_3]$$
$$NH_3 + SO_3 \longrightarrow H[H_2NSO_3]$$

The industrial synthesis uses the strongly exothermic reaction between urea and anhydrous H$_2$SO$_4$ (or dilute oleum):

$$(H_2N)_2CO + 2H_2SO_4 \longrightarrow CO_2 + H[H_2NSO_3] + NH_4[HSO_4]$$

Salts are obtained by direct neutralization of the acid with appropriate oxides, hydroxides, or carbonates. Sulfamic acid is a dry, non-volatile, non-hygroscopic, colourless, white, crystalline solid of considerable stability. It melts at 205°, begins to decompose at 210°, and at 260° rapidly gives a mixture of SO$_2$, SO$_3$, N$_2$, H$_2$O, etc. It is a strong acid (dissociation constant 1.01×10^{-1} at 25° solubility $\sim$25 g per 100 g H$_2$O) and, because of its physical form and stability, is a convenient standard for acidimetry. Several thousand tonnes are

[155] H. W. ROESKY, W. SCHAPER, O. PETERSEN, and T. MÜLLER, Simple syntheses of sulfur–nitrogen compounds, *Chem. Ber.* **110**, 2695–8 (1977).

manufactured annually and its principal applications are in formulations for metal cleaners, scale removers, detergents, and stabilizers for chlorine in aqueous solution.[156] Its salts are used in flame retardants, weed killers, and for electroplating.

In the solid state sulfamic acid forms a strongly H-bonded network which is best described in terms of zwitterion units $^+H_3NSO_3^-$ rather than the more obvious formulation as aminosulfuric acid, $H_2NSO_2(OH)$. The zwitterion has the staggered configuration shown in Fig. 15.57a and the S–N distance is notably longer than in the sulfamate ion or sulfamide.

Dilute aqueous solutions of sulfamic acid are stable for many months at room temperature but at higher temperatures hydrolysis to $NH_4[HSO_4]$ sets in. Alkali metal salts are stable in neutral and alkaline solutions even at the bp. Sulfamic acid is a monobasic acid in water (see Fig. 15.57b for structure of the sulfamate ion). In liquid ammonia solutions it is dibasic and, with Na for example, it forms $NaNH.SO_3Na$. Sulfamic acid is oxidized to nitrogen and sulfate by Cl_2, Br_2, and ClO_3^-, e.g.:

$$2H[H_2NSO_3] + KClO_3 \longrightarrow N_2 + 2H_2SO_4 + KCl + H_2O$$

Concentrated HNO_3 yields pure N_2O whilst aqueous HNO_2 reacts quantitatively to give N_2:

$$H[H_2NSO_3] + HNO_3 \longrightarrow H_2SO_4 + H_2O + N_2O$$

$$H[H_2NSO_3] + NaNO_2 \xrightarrow{\text{acidify}} NaHSO_4 + H_2O + N_2$$

This last reaction finds use in volumetric analysis. The use of sulfamic acid to stabilize chlorinated water depends on the equilibrium formation of *N*-chlorosulfamic acid, which reduces loss of chlorine by evaporation, and slowly re-releases hypochlorous acid by the reverse hydrolysis:

$$Cl_2 + H_2O \rightleftharpoons HOCl + HCl$$

$$H[H_2NSO_3] + HOCl \rightleftharpoons HN(Cl)SO_3H + H_2O$$

$$HOCl \longrightarrow HCl + \tfrac{1}{2}O_2$$

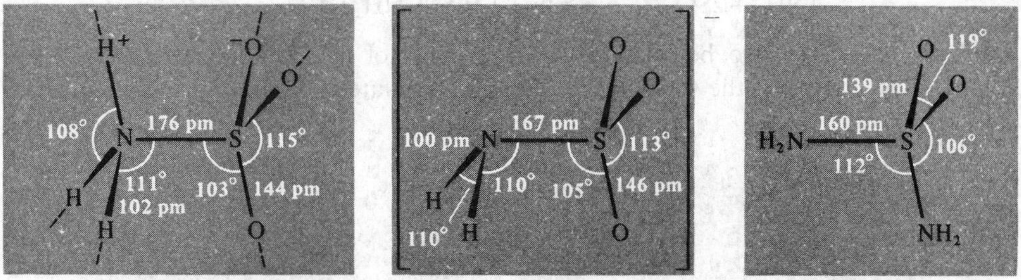

(a) $^+H_3NSO_3^-$ (b) $[H_2NSO_3]^-$ (c) $(H_2N)_2SO_2$
  sulfamic acid   sulfamate ion   sulfamide

FIG. 15.57   The structures of (a) sulfuric acid, (b) the sulfamate ion, and (c) sulfamide.

[156] D. SANTMEYER and R. AARONS, Sulfamic acid and sulfamates, *Kirk–Othmer Encyclopedia of Chemical Technology*, 2nd edn., Vol. 19, pp. 242–9, Wiley, New York, 1969.

Sulfamide, $(H_2N)_2SO_2$, can be made by ammonolysis of SO_3 or O_2SCl_2. It is a colourless crystalline material, mp 93°, which begins to decompose above this temperature. It is soluble in water to give a neutral non-electrolytic solution but in boiling water it decomposes to ammonia and sulfuric acid. The structure (Fig. 15.57c) can be compared with those of sulfuric acid, $(HO)_2SO_2$ (p. 837) and the sulfuryl halides X_2SO_2 (p. 820).

(c) *Imido and nitrido derivatives of sulfuric acid.* In the preceding section the sulfamate ion and related species were regarded as being formed by replacement of an OH group in $(HO)SO_3^-$ or $(HO)_2SO_3$ by an NH_2 group. They could equally well be regarded as sulfonates of ammonia in which each H atom is successively replaced by SO_3^- (or SO_3H):

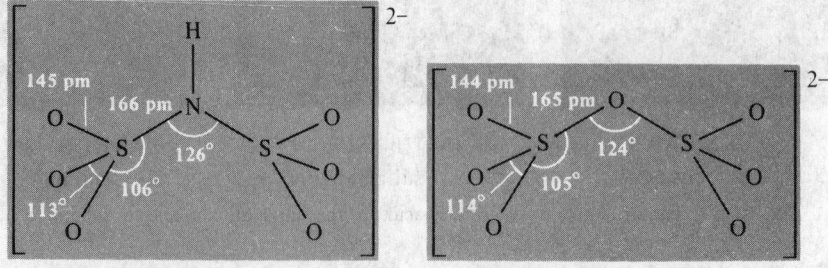

| Ammonia | (Sulfamate) Amidosulfate Aminosulfonate | Imidodisulfate Iminodisulfonate | Nitridotrisulfate Nitrilotrisulfonate |

Both sets of names are used in the literature. Free imidodisulfuric acid $HN(SO_3H)_2$ (which is isoelectronic with disulfuric acid $H_2S_2O_7$, p. 835) and free nitridotrisulfuric acid $N(SO_3H)_3$ are unstable, but their salts are well characterized and have been extensively studied.

Imidodisulfuric acid derivatives can be prepared from urea by using less sulfuric acid than required for sulfamic acid (p. 876):

$$4(H_2N)_2CO + 5H_2SO_4 \xrightarrow{\text{warm}} 4CO_2 + 2HN(SO_3NH_4)_2 + (NH_4)_2SO_4$$

Addition of aqueous KOH liberates NH_3 and affords crystalline $HN(SO_3K)_2$ on evaporation. All 3 H atoms in $HN(SO_3H)_2$ can be replaced by NH_4 or M^I, e.g. the direct reaction of NH_3 and SO_3 yields the triammonium salt:

$$4NH_3 + 2SO_3 \longrightarrow NH_4[N(SO_3NH_4)_2]$$

Imidodisulfates can also be obtained by hydrolysis of nitridotrisulfates (see below). Figure 15.58 compares the structure of the imidodisulfate and parent disulfate ions, as

Fig. 15.58 Comparison of the structures of the imidodisulfate and disulfate ions in their potassium salts.

determined from the potassium salts. Comparison with the hydroxylamine derivative $K[HN(OH)SO_3]$ (below) is also instructive.

Fluoro and chloro derivatives of imidodisulfuric acid can be made by reacting HSO_3F or HSO_3Cl (rather than H_2SO_4) with urea:

$$(H_2N)_2CO + 3HSO_3F \longrightarrow CO_2 + HN(SO_2F)_2 + [NH_4][HSO_4] + HF$$

$HN(SO_2F)_2$ melts at 17°, boils at 170° and can be further fluorinated with elementary F_2 at room temperatures to give $FN(SO_2F)_2$, mp $-79.9°$, bp 60°. The chloro derivative $HN(SO_2Cl)_2$ is a white crystalline compound, mp 37°: it is made in better yield from sulfamic acid by the following reaction sequence:

$$2PCl_5 + H_2NHSO_3 \longrightarrow Cl_3P{=}NSO_2Cl + OPCl_3 + HCl$$
$$Cl_3P{=}NSO_2Cl + ClSO_3H \longrightarrow HN(SO_2Cl)_2 + OPCl_3$$

Salts of nitridotrisulfuric acid, $N(SO_3M^I)_3$, are readily obtained by the exothermic reaction of nitrites with sulfites or hydrogen sulfites in hot aqueous solution:

$$KNO_2 + 4KHSO_3 \longrightarrow N(SO_3K)_3 + K_2SO_3 + 2H_2O$$

The dihydrate crystallizes as the solution cools. Such salts are stable in alkaline solution but hydrolyse in acid solution to imidodisulfate (and then more slowly to sulfamic acid):

$$[N(SO_3)_3]^{3-} + H_3O^+ \longrightarrow [HN(SO_3)_2]^{2-} + H_2SO_4$$

(d) *Hydrazine and hydroxylamine derivatives of sulfuric acid.* Hydrazine sulfonic acid, $H_2NNH.HSO_3$ is obtained as its hydrazinium salt by reacting anhydrous N_2H_4 with diluted gaseous SO_3 or its pyridine adduct:

$$C_5H_5NSO_3 + 2N_2H_4 \longrightarrow C_5H_5N + [N_2H_5]^+[H_2NNHSO_3]^-$$

The free acid is monobasic, pK 3.85; it is much more easily hydrolysed than sulfamic acid and has reducing properties comparable with those of hydrazine. Like sulfamic acid it exists as a zwitterion in the solid state: $^+H_3NNHSO_3^-$.

Symmetrical hydrazine disulfonic acid can be made by reacting a hydrazine sulfonate with a chlorosulfate:

$$H_2NNHSO_3^- + ClSO_3^- \longrightarrow HCl + [O_3SNHNHSO_3]^{2-}$$

Oxidation of the dipotassium salt with HOCl yields the azodisulfonate $KO_3SN{=}NSO_3K$. Numerous other symmetrical and unsymmetrical hydrazine polysulfonate derivatives are known.

With hydroxylamine, $HONH_2$, 4 of the 5 possible sulfonate derivatives have been prepared as anions of the following acids:

$HONHSO_3H$:   hydroxylamine *N*-sulfonic acid
$HON(SO_3H)_2$:   hydroxylamine *N,N*-disulfonic acid
$(HSO_3)ONHSO_3H$:   hydroxylamine *O,N*-disulfonic acid
$(HSO_3)ON(SO_3H)_2$:   hydroxylamine trisulfonic acid

The first of these can be made by careful hydrolysis of the *N,N*-disulfonate which is itself made by the reaction of SO_2 and a nitrite in cold alkaline solution:

$$KNO_2 + KHSO_3 + SO_2 \longrightarrow HON(SO_3K)_2$$

The potassium salt readily crystallizes from the cold solution thus preventing further reaction with the hydrogen sulfate to give nitridotrisulfate (p. 879). The structure of the hydroxylamine *N*-sulfonate ion is shown in Fig. 15.59a. The closely related *N*-nitrosohydroxylamine *N*-sulfonate ion (Fig. 15.59b) can be made directly by absorbing NO in alkaline K_2SO_3 solution: the 6 atoms ONN(O)SO all lie in one plane and the interatomic distances suggest an S–N single bond but considerable additional π bonding in the N–N bond.

Oxidation of hydroxylamine *N,N*-disulfonate with permanganate or PbO_2 yields the intriguing nitrosodisulfonate $K_2[ON(SO_3)_2]$: this was first isolated by Frémy as a yellow solid which was subsequently shown to be dimeric and diamagnetic due to the formation of long N···O bonds in the crystal (Fig. 15.59c). However, in aqueous solution the anion dissociates reversibly into the deep violet, paramagnetic monomer $[ON(SO_3)_2]^{2-}$.

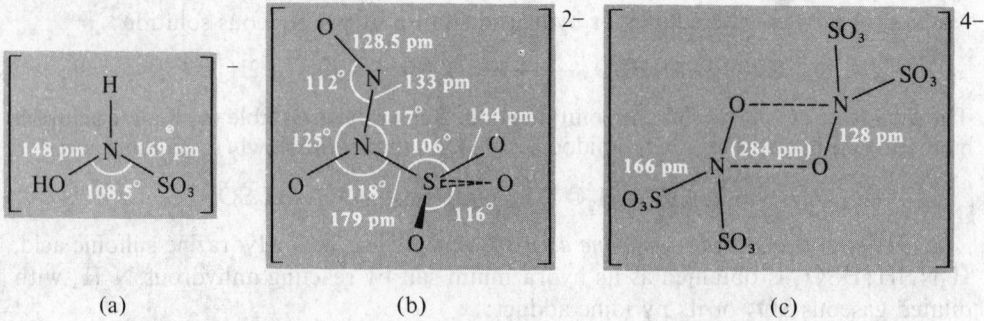

FIG. 15.59 Structures of various S–N oxoanions: (a) hydroxylamine-*N*-sulfonate, (b) *N*-nitrosohydroxylamine *N*-sulfonate, and (c) the dimeric anion in Frémy's salt $\{K_2[ON(SO_3)_2]\}_2$.

Hydroxylamine trisulfonates, e.g. $(KO_3S)ON(SO_3K)_2$ are made by the reaction of K_2SO_3 with potassium nitrosodisulfonate (Frémy's salt). Acidification of the product results in rapid hydrolysis to the *O,N*-disulfonate which can be isolated as the exclusive product:

$$(KO_3S)ON(SO_3K)_2 + H_2O \longrightarrow (KO_3S)ONH(SO_3K) + KHSO_4$$

Sulfonic acids containing nitrogen have long been implicated as essential intermediates in the synthesis of H_2SO_4 by the lead-chamber process (p. 838) and, as shown by F. Seel and his group, the crucial stage is the oxidation of sulfite ions by the nitrosyl ion NO^+:

$$SO_3{}^{2-} + NO^+ \longrightarrow [ONSO_3]^- \xrightarrow{NO^+} 2NO + SO_3$$

The NO^+ ions are thought to be generated by the following sequence of reactions:

$$NO + \tfrac{1}{2}O_2 \longrightarrow NO_2 \quad \text{(gas phase)}$$
$$\left.\begin{array}{r} NO_2 + NO + H_2O \longrightarrow 2\{ONOH\} \\ SO_2 + H_2O \longrightarrow H_2SO_3 \\ H_2SO_3 \longrightarrow H^+ + HSO_3{}^- \\ \{ONOH\} + H^+ \longrightarrow NO^+ + H_2O \end{array}\right\} \text{(surface reactions)}$$

The nitrososulfonate intermediate $[ONSO_3]^-$ can also react with SO_3^{2-} to give the hydroxylamine disulfonate ion which can likewise be oxidized by NO^+:

$$[ON(SO_3)_2]^{3-} + NO^+ \longrightarrow N_2O + SO_3 + SO_4^{2-}$$

In a parallel reaction the $[ONSO_3]^-$ intermediate can react with SO_2 to form nitrilotrisulfonate:

$$[ON(SO_3)_2]^{3-} + SO_2 \longrightarrow [N(SO_3)_3]^{3-}$$

This then reacts with NO^+ to form N_2, SO_3, and SO_4^{2-}.

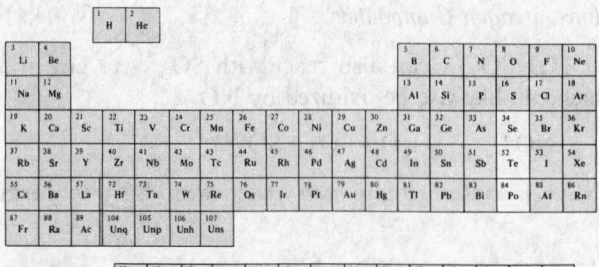

16

Selenium, Tellurium, and Polonium

16.1 The Elements[1-4]

16.1.1 *Introduction: history, abundance, distribution*

Tellurium was the first of these three elements to be discovered. It was isolated by the Austrian chemist F. J. Müller von Reichenstein in 1782 a few years after the discovery of oxygen by J. Priestley and C. W. Scheele (p. 698), though the periodic group relationship between the elements was not apparent until nearly a century later (p. 24). Tellurium was first observed in ores mined in the gold districts of Transylvania; Müller called it *metallum problematicum* or *aurum paradoxum* because it showed none of the properties of the expected antimony.[5] The name tellurium (Latin *tellus*, earth) is due to another Austrian chemist, M. H. Klaproth, the discoverer of zirconium and uranium.

Selenium was isolated some 35 y after tellurium and, since the new element resembled tellurium, it was named from the Greek σελήνη, *selene*, the moon. The discovery was made in 1817 by the Swedish chemist J. J. Berzelius (discoverer of Si, Ce, and Th) and J. G. Gahn (discoverer of Mn);[5] they observed a reddish-brown deposit during the burning of sulfur obtained from Fahlun copper pyrites, and showed it to be volatile and readily reducible to the new element.

The discovery of polonium by Marie Curie in 1898 is a story that has been told many times.[6] The immense feat of processing huge quantities of uranium ore and of following

[1] K. W. BAGNALL, *The Chemistry of Selenium, Tellurium, and Polonium*, Elsevier, Amsterdam, 1966, 200 pp. K. W. BAGNALL, Selenium, tellurium, and polonium, Chap. 24 in *Comprehensive Inorganic Chemistry*, Vol. 2, pp. 935–1008, Pergamon Press, Oxford, 1973.

[2] R. A. ZINGARO and W. C. COOPER (eds.), *Selenium*, Van Nostrand, Reinhold, New York, 1974, 835 pp.

[3] W. C. COOPER (ed.), *Tellurium*, Van Nostrand, Reinhold, New York, 1971, 437 pp.

[4] N. B. MIKEEV, Polonium, *Chemiker Zeitung* **102**, 277–86 (1978). See also K. W. BAGNALL, The chemistry of polonium, *Radiochim. Acta* **31**, in press (1982).

[5] M. E. WEEKS, *Discovery of the Elements*, 6th edn., Journal of Chemical Education, Easton, Pa., 1956: Selenium and tellurium, pp. 303–319; Klaproth–Kitaibel letters on tellurium, pp. 320–37.

[6] Ref. 4, Chap. 29, The natural radioactive elements, pp. 803–43. See also E. FARBER, *Nobel Prize Winners in Chemistry 1901–1961*, Abelard-Schuman, London, Marie Sklodowska Curie, pp. 45–8. F. C. WOOD, Marie Curie, in E. FARBER (ed.), *Great Chemists*, pp. 1263–75, Interscience, New York, 1961.

the progress of separation by the newly discovered phenomenon of radioactivity (together with her parallel isolation of radium by similar techniques, p. 118), earned her the Nobel Prize for Chemistry in 1911. She had already shared the 1902 Nobel Prize for Physics with H. A. Becquerel and her husband P. Curie for their joint researches on radioactivity. Indeed, this was the first time, though by no means the last, that invisible quantities of a new element had been identified, separated, and investigated solely by means of its radioactivity. The element was named after Marie Curie's home country, Poland.

Selenium and tellurium are comparatively rare elements, being sixty-sixth and seventy-third respectively in order of crustal abundance; polonium, on account of its radioactive decay, is exceedingly unabundant. Selenium comprises some 0.05 ppm of the earth's crust and is therefore similar to Ag and Hg, which are each about 0.08 ppm, and Pd (0.015 ppm). Tellurium, at about 0.002 ppm can be compared with Au (0.004 ppm) and Ir (0.001 ppm). Both elements are occasionally found native, in association with sulfur, and many of their minerals occur together with the sulfides of chalcophilic metals (p. 760),[2, 3] e.g. Cu, Ag, Au; Zn, Cd, Hg; Fe, Co, Ni; Pb, As, Bi. Sometimes the minerals are partly oxidized, e.g. $MSeO_3.2H_2O$ (M = Ni, Cu, Pb); $PbTeO_3$, $Fe_2(TeO_3)_3.2H_2O$, $FeTeO_4$, Hg_2TeO_4, $Bi_2TeO_4(OH)_4$, etc. Selenolite, SeO_2, and tellurite, TeO_2, have also been found.

Polonium has no stable isotopes, all 27 isotopes being radioactive, of these only ^{210}Po occurs naturally, as the penultimate member of the radium decay series:

$$^{210}_{82}Pb \xrightarrow[22.3\,y]{\beta^-} {}^{210}_{83}Bi \xrightarrow[5.01\,d]{\beta^-} {}^{210}_{84}Po \xrightarrow[138.38\,d]{\alpha} {}^{206}_{82}Pb$$

$$\text{RaD} \qquad\qquad \text{RaE} \qquad\qquad \text{RaF} \qquad\qquad \text{RaG}$$

Because of the fugitive nature of ^{210}Po, uranium ores contain only about 0.1 mg Po per tonne of ore (i.e. 10^{-4} ppm). The overall abundance of Po in crustal rocks of the earth is thus of the order of 3×10^{-10} ppm.

16.1.2   *Production and uses of the elements*[2–4, 7]

The main source of Se and Te is the anode slime deposited during the electrolytic refining of Cu (p. 1366); this mud also contains commercial quantities of Ag, Au, and the platinum metals. Direct recovery from minerals is not usually economically viable because of their rarity. Selenium is also recovered from the sludge accumulating in sulfuric acid plants and from electrostatic precipitator dust collected during the processing of Cu and Pb. Detailed procedures for isolation and purification depend on the relative concentrations of Se, Te, and other impurities, but a typical sequence involves oxidation by roasting in air with soda ash followed by leaching:

$$Ag_2Se + Na_2CO_3 + O_2 \xrightarrow{650°} 2Ag + Na_2SeO_3 + CO_2$$

$$Cu_2Se + Na_2CO_3 + 2O_2 \longrightarrow 2CuO + Na_2SeO_3 + CO_2$$

$$Cu_2Te + Na_2CO_3 + 2O_2 \longrightarrow 2CuO + Na_2TeO_3 + CO_2$$

---

[7] *Kirk–Othmer Encyclopedia of Chemical Technology*, 2nd edn., 1968, Selenium, Vol. 17, pp. 809–33 (by E. M. ELKIN and J. L. MARGRAVE), Tellurium and tellurium compounds, Vol. 19, pp. 756–74, 1969 (by E. M. ELKIN).

In the absence of soda ash, SeO_2 can be volatilized directly from the roast:

$$Cu_2Se + \tfrac{3}{2}O_2 \xrightarrow{300} CuO + CuSeO_3 \xrightarrow{650} 2CuO + SeO_2$$

$$Ag_2SeO_3 \xrightarrow{700} 2Ag + SeO_2 + \tfrac{1}{2}O_2$$

Separation of Se and Te can also be achieved by neutralizing the alkaline selenite and tellurite leach with H_2SO_4; this precipitates the tellurium as a hydrous dioxide and leaves the more acidic selenous acid, H_2SeO_3, in solution from which 99.5% pure Se can be precipitated by SO_2:†

$$H_2SeO_3 + 2SO_2 + H_2O \longrightarrow Se + 2H_2SO_4$$

Tellurium is obtained by dissolving the dioxide in aqueous NaOH followed by electrolytic reduction:

$$Na_2TeO_3 + H_2O \longrightarrow Te + 2NaOH + O_2$$

The NaOH is regenerated and only make-up quantities are required. However, the detailed processes adopted industrially to produce Se and Te are much more complex and sophisticated than this outline implies.[2, 3]

World production of Se (and its compounds) in 1975 was ~1500 tonnes of contained Se, the largest producers being Japan (415 t), Canada (305 t), and the USA (160 t). The pattern of use no doubt varies somewhat from country to country, but in the USA the largest single use of the element is as a decolorizor of glass (0.01–0.15 kg/tonne). Higher concentrations (1–2 kg/tonne) yield delicate pink glasses. The glorious selenium ruby glasses, which are the most brilliant reds known to glass-makers, are obtained by incorporating solid particles of cadmium sulfoselenide in the glass; the deepest ruby colour is obtained when Cd(S,Se) has about 10% CdS, but as the relative concentration of CdS increases the colour moderates to red (40% CdS), orange (75%), and yellow (100%). Cadmium sulfoselenides are also widely used as heat-resistant red pigments in plastics, paints, inks, and enamels. Another very important application of elemental Se is in xerography, which has developed during the past three decades into the pre-eminent process for document copying, as witnessed by the ubiquitous presence of xerox machines in offices and libraries (see Panel). Related uses are as a photoconductor (selenium photoelectric cells) and as a rectifier in semiconductor devices (p. 290). Small amounts of ferroselenium are used to improve the casting, forging, and machinability of stainless steels, and the dithiocarbamate $[Se(S_2CNEt_2)_4]$ finds some use in the processing of natural and synthetic rubbers. Selenium pharmaceuticals comprise a further small outlet. In addition to Se, Fe/Se, Cd(S,Se), and $[Se(S_2CNEt_2)_4]$ the main commercially available compounds of Se are SeO_2, Na_2SeO_3, Na_2SeO_4, H_2SeO_4, and $SeOCl_2$ (q.v.).

Production of Te is on a much smaller scale: The known amount in 1975 was 150 tonnes pa (USA 60 t, Canada 36 t, Peru 32 t, Japan 21 t) but this does not include statistics from the USSR and several other producing countries. More than 90% of the Te is used in iron and steel production and in non-ferrous metals and alloys. A small amount of TeO_2 is used in tinting glass, and Te compounds find some use as catalysts and as curing agents in the rubber industry. In addition to Te, Fe/Te, and TeO_2, commercially important compounds include Na_2TeO_4 and $[Te(S_2CNEt_2)_4]$.

---

† Very pure Se can be obtained by heating the crude material in H_2 at 650° and then decomposing the H_2Se so formed by passing the gas through a silica tube at 1000°. Any H_2S present, being more stable than H_2Se, passes through the tube unchanged, whereas hydrides which are less stable than H_2Se, such as those of Te, P, As, Sb, are not formed in the initial reaction at 650°.

Xerography

The invention of xerography by C. F. Carlson (USA) in the period 1934–42 was the culmination of a prolonged and concerted attack on the problem of devising a rapid, cheap, and dry process for direct document copying without the need for the intermediate formation of a permanent photographic "negative", or even the use of specially prepared photographic paper for the "print". The discovery that vacuum-deposited amorphous or vitreous selenium was the almost ideal photoconductor for xerography was made in the Battelle Memorial Institute (Ohio, USA) in 1948. The dramatic success of these twin developments is witnessed by the vast number of xerox machines in daily use throughout the world today. However, early xerox equipment was not automatic. Models introduced in 1951 became popular for making offset masters, and rotary xerographic machines were introduced in 1959, but it was only after the introduction of the Xerox 914 copier in the early 1960s that electrophotography came of age. the word "xerography" derives from the Greek ξηρό, *xero* dry, γραφή, *graphy*, writing.

The sequential steps involved in commercial machines which employ reusable photoreceptors for generating xerox copies are shown in the figure[2] and further elucidated below.

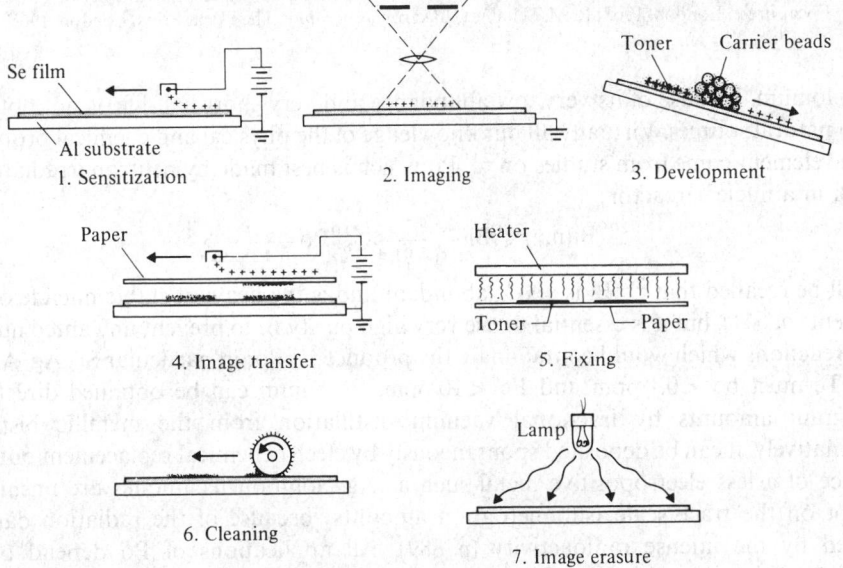

Process steps in reuseable xerography.

1. *Sensitization of the photoreceptor.* The photoreceptor consists of a vacuum-deposited film of amorphous Se, ~50μm thick, on an Al substrate; this is sensitized by electrostatic charging from a corona discharge using a field of ~10^5 V cm^{-1}.

2. *Exposure and latent image formation.* The sensitized photoreceptor is exposed to a light and dark image pattern; in the light areas the surface potential of the photoconductor is reduced due to a photoconductive discharge. Since current can only flow perpendicular to the surface, this step produces an electrostatic-potential distribution which replicates the light–dark pattern of the image.

3. *Development of the image.* This is done using a mixture of black (or coloured) toner particles, typically 10 μm in diameter, and spherical carrier beads (~100 μm diameter). The toner particles become charged triboelectrically (i.e. by friction) and are preferentially attracted either by the surface fringe field at light–dark boundaries or (in systems with a developing electrode) by the absolute potential in the dark areas; they adhere to the photoreceptor, thus forming a visible image corresponding to the latent electrostatic image.

Continued

4. *Image transfer.* This is best done electrostatically by charging the print paper to attract the toner particles.

5. *Print fixing.* The powder image is made permanent by fusing or melting the toner particles into the surface of the paper, either by heat, by heat and pressure, or by solvent vapours.

6. *Cleaning.* Any toner still left on the photoreceptor after the transfer process is removed mechanically with a cloth web or brush, or by a combination of electrostatic and mechanical means.

7. *Image erasure.* The potential differences due to latent image formation are removed by flooding the photoreceptors with a sufficiently intense light source to drive the surface potential to some uniformly low value (typically ~ 100 V corresponding to fields of $\sim 10^4$ V cm^{-1}); the photoreceptor is then ready for another print cycle.

The elegance, cheapness, and convenience of xerography for document copying has led to rapid commercial development. It would be impossible to form a precise inventory of its daily use throughout the world but there are probably more than 100 000 machines. Xerox or related photocopying machines are in use producing perhaps 100 million xerox copies daily. Even in 1967 the electrostatic copying industry exceeded $10^9 in sales and the current figure probably exceeds this one thousandfold worldwide.

Further reading. G. Lucovsky and M. D. Tabak, Selenium in electrophotography, Chapter 16 in reference 2, pp. 788–807; J. H. Dessauer and H. E. Clark, *Xerography and Related Processes*, The Focal Press, London, 1965; R. M. Schaffert, *Electrophotography*, The Focal Press, London, 1965.

Polonium, because of its very low abundance and very short half-life, is not obtained from natural sources. Virtually all our knowledge of the physical and chemical properties of the element come from studies on ^{210}Po which is best made by neutron irradiation of ^{209}Bi in a nuclear reactor:

$$^{209}_{83}\text{Bi}(n,\gamma)^{210}_{83}\text{Bi} \xrightarrow[t_{\frac{1}{2}} 5.01\,d]{\beta^-} {}^{210}_{84}\text{Po} \xrightarrow[t_{\frac{1}{2}} 138.38\,d]{\alpha} \cdots$$

It will be recalled that ^{209}Bi is 100% abundant and is the heaviest stable nuclide of any element (p. 641), but it is essential to use very high purity Bi to prevent unwanted nuclear side-reactions which would contaminate the product ^{210}Po; in particular Sc, Ag, As, Sb, and Te must be <0.1 ppm and Fe <10 ppm. Polonium can be obtained directly in milligram amounts by fractional vacuum distillation from the metallic bismuth. Alternatively, it can be deposited spontaneously by electrochemical replacement onto the surface of a less electropositive metal such as Ag. Solution techniques are unsuitable except on the trace scale (submicrogram amounts) because of the radiation damage caused by the intense radioactivity (p. 889). All applications of Po depend on its radioactivity: it is an almost pure α-emitter (E_α 5.30 MeV) and only 0.0011% of the activity is due to γ-rays (E_{max} 0.803 MeV). Because of its short half-life (138.38 d) this entails a tremendous energy output of ~ 140 W per gram of metal: in consequence, there is considerable self-heating of Po and its compounds. The element can therefore be used as a convenient light-weight heat source, or to generate spontaneous and reliable thermo-electric power for space satellites and lunar stations, since no moving parts are involved. Polonium also finds limited use as a neutron generator when combined with a light element of high α,n cross-section such as beryllium: $^9_4\text{Be}(\alpha,n)^{12}_6\text{C}$. The best yield (93 neutrons per 10^6 α-particles) is obtained with a BeO target.

16.1.3 *Allotropy*

At least six structurally distinct forms of Se are known: the three red monoclinic polymorphs (α, β, and γ) consist of Se$_8$ rings and differ only in the intermolecular packing of the rings in the crystals; the grey, "metallic", hexagonal crystalline form features helical polymeric chains and these also occur, somewhat deformed, in amorphous red Se. Finally, vitreous black Se, the ordinary commercial form of the element, comprises an extremely complex and irregular structure of large polymeric rings having up to 1000 atoms per ring.

The α- and β-forms of red crystalline Se_8 are obtained respectively by the slow and rapid evaporation of CS_2 or benzene solutions of black vitreous Se; very recently a third (γ) form of red crystalline Se_8 was obtained from the reaction of dipiperidinotetraselane with solvent CS_2:[8]

$$[Se_4(NC_5H_{10})_2] \xrightarrow{2CS_2} [Se(S_2CNC_5H_{10})_2] + \tfrac{3}{8}Se_8$$

All three allotropes consist of almost identical puckered Se_8 rings similar to those found in *cyclo*-S_8 (p. 772) and of average dimensions Se–Se 233.5 pm, angle Se–Se–Se 105.7°, dihedral angle 101.3° (Fig. 16.1a). The intermolecular packing is most efficient for the

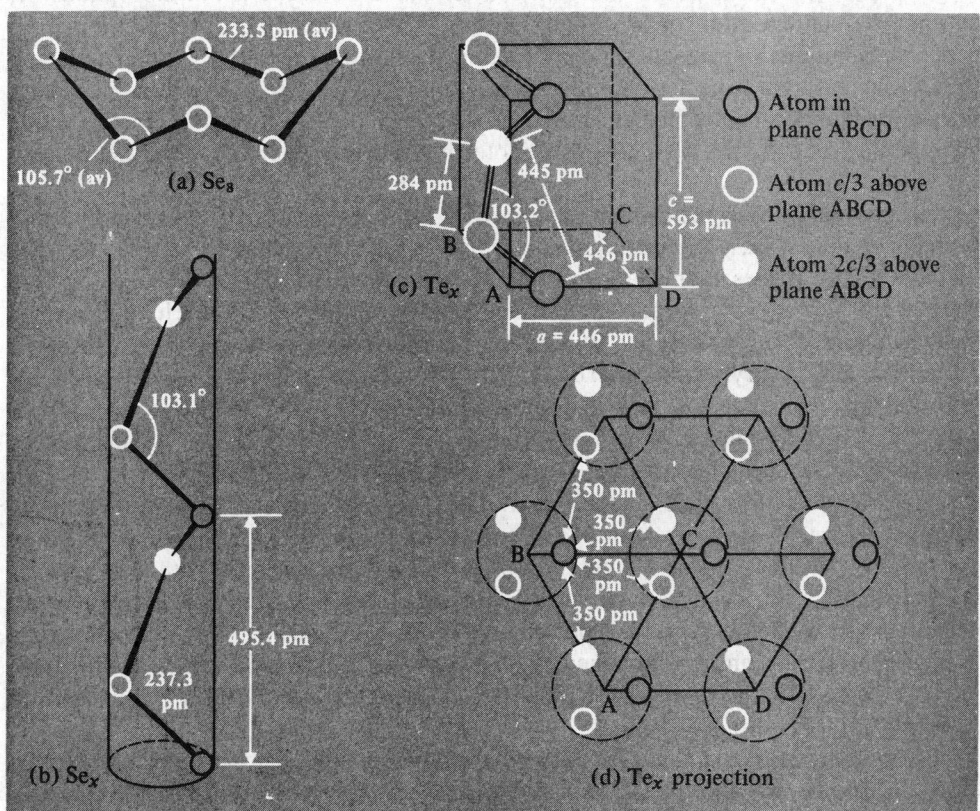

FIG. 16.1   Structures of various allotropes of selenium and the structure of crystalline tellurium:
(a) the Se_8 unit in α- β- and γ-red selenium; (b) the helical Se chain along the *c*-axis in hexagonal grey selenium; (c) the similar helical chain in crystalline tellurium shown in perspective; and (d) projection of the tellurium structure on a plane perpendicular to the *c*-axis.

[8] O. Foss and V. Janickis, X-ray crystal structure of a new red, monoclinic form of cyclo-octaselenium Se_8, *JCS Chem. Comm.* 1977, 834–5.

α-form, the volumes per Se_8 unit being α 0.2385 nm^3, β 0.2410 nm^3, and γ 0.2423 nm^3. [It is interesting to note that β-Se_8 was at one time thought, on the basis of an X-ray crystal structure determination, to be an 8-membered *chain* with the configuration of a puckered ring in which 1 Se–Se bond had been broken; the error was corrected in a very perceptive paper by L. Pauling and his co-workers.[9]] Both α- and β-Se_8 (and presumably also γ-Se_8) are appreciably soluble in CS_2 to give red solutions.

Grey, hexagonal, "metallic" selenium is thermodynamically the most stable form of the element and can be formed by warming any other modification; it can also be obtained by slowly cooling molten Se or by condensing Se vapour at a temperature just below the mp (220.5°). It is a photoconductor (p. 885) and is the only modification which conducts electricity. The structure (Fig. 16.1b) consists of unbranched helical chains with Se–Se 237.3 pm, angle Se–Se–Se 103.1°, and a repeat unit every 3 atoms (cf. fibrous sulfur, p. 779). The closest Se···Se distance between chains is 343.6 pm, which is very close to that in Te_x (350 pm) (see below). Grey Se_x is insoluble in CS_2 and its density, 4.82 g cm^{-3}, is the highest of any modification of the element. A related allotrope is red amorphous Se, formed by condensation of Se vapour onto a cold surface or by precipitation from aqueous solutions of selenous acid by treatment with SO_2 (p. 891) or other reducing agents such as hydrazine hydrate. It is slightly soluble in CS_2, and has a deformed chain structure but does not conduct electricity. The heat of transformation to the stable hexagonal grey form has been variously quoted but is in the region of 5–10 kJ per mole of Se atoms.

Vitreous, black Se is the ordinary commercial form of the element, obtained by rapid cooling of molten Se; it is a brittle, opaque, bluish-black lustrous solid which is somewhat soluble in CS_2. It does not melt sharply but softens at about 50° and rapidly transforms to hexagonal grey Se when heated to 180° (or at lower temperatures when catalysed by halogens, amines, etc.). There has been much discussion about the structure but it seems to comprise rings of varying size up to quite high molecular weights. Presumably these rings cleave and polymerize into helical chains under the influence of thermal soaking or catalysts. The great interest in the various allotropes of selenium and their stabilization or interconversion, stems from its use in photocells, rectifiers, and xerography (p. 885).[2]

Tellurium has only one crystalline form and this is composed of a network of spiral chains similar to those in hexagonal Se (Fig. 16.1c and d). Although the intra-chain Te–Te distance of 284 pm and the *c* dimension of the crystal (593 pm) are both substantially greater than for Se_x (as expected), nevertheless the closest interatomic distance between chains is almost identical for the 2 elements. Accordingly the elements form a continuous range of solid solutions in which there is a random alternation of Se and Te atoms in the helical chains.[10] The rapid diminution in allotropic complexity from sulfur through selenium to tellurium is notable.

Polonium is unique in being the only element known to crystallize in the simple cubic form (6 nearest neighbours at 335 pm). This α-form distorts at about 36° to a simple rhombohedral modification in which each Po also has 6 nearest neighbours at 335 pm. The precise temperature of the phase change is difficult to determine because of the self-

[9] R. E. MARSH, L. PAULING, and J. D. McCULLOUGH, The crystal structure of β selenium, *Acta Cryst.* 6, 71–5 (1953).

[10] A. A. KUDRYAVTSEV, *The Chemistry and Technology of Selenium and Tellurium*, Collet's Publishers, London, 1974, 278 pp.

heating of crystalline Po (p. 886) and it appears that both modifications can coexist from about 18° to 54°. Both are silvery-white metallic crystals with substantially higher electrical conductivity than Te.

16.1.4 *Atomic and physical properties*

Selenium, Te, and Po are the three heaviest members of Group VIB and, like their congenors O and S, have two p electrons less than the next following noble gases. Selenium is normally said to have 6 stable isotopes though the heaviest of these (^{82}Se, 9.4% abundant) is actually an extremely long-lived β^- emitter, $t_{\frac{1}{2}}$ 1.4 × 10^{20} y. The most abundant isotope is ^{80}Se (49.6%), and all have zero nuclear spin except the 7.6% abundant ^{77}Se ($I = \frac{1}{2}$), which is finding increasing use in nmr experiments.[11] Because of the plethora of isotopes the atomic weight is only known to about 1 part in 2600 (p. 19). Tellurium, with 8 naturally occurring stable isotopes, likewise suffers some imprecision in its atomic weight (1 part in 4300). The most abundant isotopes are ^{130}Te (33.8%) and ^{128}Te (31.7%), and again all have zero nuclear spin except the nmr active isotopes ^{123}Te (0.91%) and ^{125}Te (7.14%), which have spin $\frac{1}{2}$.[11a] ^{125}Te also has a low-lying nuclear isomer ^{125m}Te which decays by pure γ emission (E_γ 35.48 keV, $t_{\frac{1}{2}}$ 58 d)— this has found much use in Mössbauer spectroscopy.[12] Polonium, as we have seen (p. 883), has no stable isotopes. The 3 longest lived, together with their modes of production and other properties, are as shown in Table 16.1.

TABLE 16.1   *Production and properties of long-lived Po isotopes*

Isotope	Production	$t_{\frac{1}{2}}$	E /MeV	A_r (relative atomic mass)
^{208}Po	^{209}Bi(d,3n) or (p,2n)	2.90 y	5.11	207.981
^{209}Po	^{209}Bi(d,2n) or (p,n)	102 y	4.88	208.982
^{210}Po	^{209}Bi(n,γ)	138.38 d	5.305	209.983

Several atomic and physical properties of the elements are given in Table 16.2. The trends to larger size, lower ionization energy, and lower electronegativity are as expected. The trend to metallic conductivity is also noteworthy; indeed, Po resembles its horizontal neighbours Bi, Pb, and Tl not only in this but in its moderately high density and notably low mp and bp.

[11] C. RODGER, N. SHEPPARD, H. C. E. McFARLANE, and W. McFARLANE, Selenium-77 and Tellurium-125, in R. K. HARRIS and B. R. MANN (eds.), *NMR and the Periodic Table*, pp. 402–19, Academic Press, London, 1978.
[11a] For recent examples of the use of ^{123}Te and ^{125}Te nmr spectroscopy, see refs. 22 and 23 below, and also W. TÖTSCH and F. SLADKY, Tetrafluorotellurates(VI): *cis-* and *trans-*(OH)$_2$TeF$_4$, HOTeF$_4$OMe, and (MeO)$_2$TeF$_4$, *JCS Chem. Comm.* 1980, 927–8.
[12] N. N. GREENWOOD and T. C. GIBB, Tellurium-125 in *Mössbauer Spectroscopy*, pp. 452–62, Chapman & Hall, London, 1971.

TABLE 16.2 *Some atomic and physical properties of selenium, tellurium, and polonium*

Property	Se	Te	Po
Atomic number	34	52	84
Number of stable isotopes	6	8	0
Electronic structure	$[Ar]3d^{10}4s^24p^4$	$[Kr]4d^{10}5s^25p^4$	$[Xe]4f^{14}5d^{10}6s^26p^4$
Atomic weight	78.96($\pm$0.03)	127.60($\pm$0.03)	(210)
Atomic radius (12-coordinate)/pm[a]	140[a]	160[a]	164[a]
Ionic radius/pm (M^{2-})	198	221	(230?)
(M^{4+})	50	97	94
(M^{6+})	42	56	67
Ionization energy/kJ mol^{-1}	940.7	869.0	813.0
Pauling electronegativity	2.4	2.1	2.0
Density (25°)/g cm^{-3})	Hexag 4.189	6.25	α9.142
	α-monoclinic 4.389		β9.352
	Vitreous 4.285		
MP/°C	217	452	246–254
BP/°C	685	990	962
$H_{atomization}$/kJ mol^{-1}	206.7	192	—
Electrical resistivity (25°)/ohm cm	10^{10}[b]	1	α4.2 × 10^{-5}
			β4.4 × 10^{-5}
Band energy gap E_g/kJ mol^{-1}	178	32.2	0

[a] The 2-coordinate covalent radius is 119 pm for elemental Se and 142 pm for Te; the 6-coordinate metallic radius of Po is 168 pm.

[b] Depends markedly on purity, temperature, and photon flux; resistivity of liquid Se at 400° is 1.3×10^5 ohm cm.

16.1.5 *Chemical reactivity and trends*

The elements in Group VI share with the preceding main-group elements the tendency towards increasing metallic character as the atomic weight increases within the group. Thus O and S are insulators, Se and Te are semiconductors, and Po is a metal. Parallel with this trend is the gradual emergence of cationic (basic) properties with Te, and these are even more pronounced with Po. For example, Se is not appreciably attacked by dilute HCl whereas Te dissolves to some extent in the presence of air; Po dissolves readily to yield pink solutions of Po^{II} which are then rapidly oxidized further to yellow Po^{IV} by the products of radiolytic decomposition of the solvent. Likewise, the structure and bonding of the halides of these elements depends markedly on both the electronegativity of the halogen and on the oxidation state of the central element, thereby paralleling the "ionic-covalent" transition which has already been discussed for the halides of P (p. 573), As and Sb (p. 651), and S (p. 816).

Selenium, Te, and Po combine directly with most elements, though less readily than do O and S. The most stable compounds are (a) the selenides, tellurides, and polonides (M^{2-}) formed with the strongly positive elements of Groups IA, IIA, and the lanthanides, and (b) the compounds with the electronegative elements O, F, and Cl in which the oxidation states are +2, +4, and +6. The compounds tend to be less stable than the corresponding compounds of S (or O), and there are few analogues of the extensive range of sulfur–nitrogen compounds (p. 854). A similar trend (also noted in the preceding groups) is the decreasing thermal stability of the hydrides: $H_2O > H_2S > H_2Se > H_2Te > H_2Po$. Selenium and tellurium share to a limited extent sulfur's great propensity for catenation (see allotropy of the elements, polysulfanes, halides, etc.).

As found in preceding groups, there is a marked diminution in the stability of multiple bonds (e.g. to C, N, O) and a corresponding decrease in their occurrence as the atomic number of the group element increases. Thus $O=C=O$ and (to a lesser extent) $S=C=S$ are stable, whereas $Se=C=Se$ polymerizes readily, $Se=C=Te$ is unstable, and $Te=C=Te$ unknown. Again, SO_2 is a (nonlinear) gaseous molecule, $O^{\diagup S}\diagdown_O$, whereas SeO_2 is a chain polymer

$$-O-Se-$$
$$\|$$
$$O$$

(p. 911) and TeO_2 features 4-coordinate pseudo-trigonal-bipyramidal units $\{:TeO_4\}$ which are singly-bonded into extended layer or 3D structures (p. 912); in PoO_2 the coordination number increases still further to 8 and the compound adopts the typical "ionic" fluorite structure. It can be seen that double bonds are less readily formed between 2 elements the greater the electronegativity difference between them and the smaller the sum of their individual electronegativities; this is paralleled by a diminution in double-bond formation with increasing size of the more electropositive element and the consequent decrease in bond energy.

The redox properties of the elements also show interesting trends. In common with several elements immediately following the first (3d) transition series (especially Ge, As, Se, Br) selenium shows a marked resistance to oxidation up to its group valency, i.e. Se^{VI}. For example, whereas HNO_3 readily oxidizes S to H_2SO_4, selenium gives H_2SeO_3. Again dehydration of H_2SO_4 with P_2O_5 yields SO_3 whereas H_2SeO_4 gives $SeO_2 + \frac{1}{2}O_2$. Likewise S forms a wide range of sulfones, R_2SO_2, but very few selenones are known: Ph_2SeO is not oxidized either by HNO_3 or by acidified $K_2Cr_2O_7$, and alkaline $KMnO_4$ is required to produce Ph_2SeO_2 (mp 155°). As noted in the isolation of the element (p. 888), SO_2 precipitates Se from acidified solutions of Se^{IV}.

The standard reduction potentials of the elements in acid and alkaline solutions are summarized in the scheme below.[13] Similar data (for the acid solutions only) are shown in Fig. 16.2 which plots the volt equivalents (p. 499) for each oxidation state. The trends are obvious: e.g.

(i) the decreasing stability of H_2M from H_2S to H_2Po;
(ii) the greater stability of M^{IV} relative to M^0 and M^{VI} for Se, Te, and Po (but not for S, p. 836), as shown by the concavity of the graph;
(iii) the anomalous position of Se in its higher oxidation states, as mentioned in the preceding paragraph.

The known coordination geometries of Se, Te, and Po are summarized in Table 16.3 together with typical examples. Most of the common geometries are observed for Se and Te, though twofold (linear) and fivefold (trigonal bipyramidal) are conspicuous by their absence. The smaller range of established geometries for compounds of Po undoubtedly reflects the paucity of structural data occasioned by the rarity of this element and the extreme difficulty of obtaining X-ray crystallographic or other structural information. There appears, however, to be a clear preference for higher coordination numbers, as expected from the larger size of the Po atom. The various examples will be discussed more

[13] A. J. Bard (ed.), *Encyclopedia of Electrochemistry of the Elements* 4, 445–54 (1975).

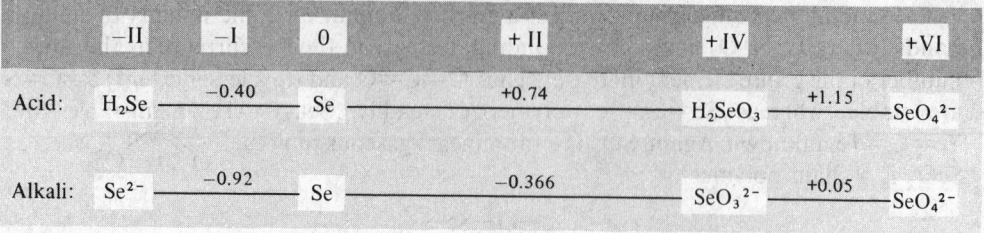

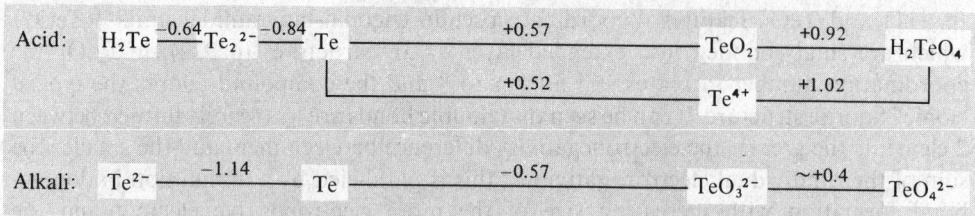

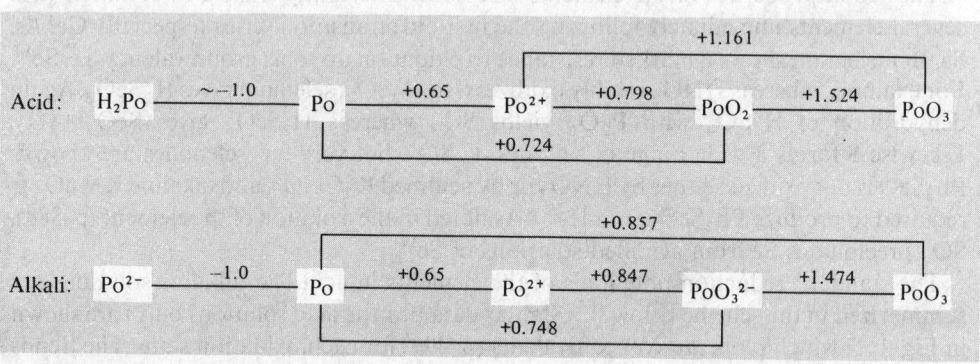

Standard reduction potentials of Se, Te, and Po.[13]

fully in subsequent sections but the rare pentagonal planar coordination formed in the ethyl xanthato complex $[Te(\eta^2\text{-}S_2COEt)_2(\eta^1\text{-}S_2COEt)]^-$ should be noted (Fig. 16.3);[14] this structure is consistent with a pentagonal bipyramidal set of orbitals on Te^{II}, 2 of which are occupied by stereochemically active lone-pairs directed above and below the TeS_5 plane. By contrast, the single lone-pairs in $Se^{IV}X_6^{2-}$, $Te^{IV}X_6^{2-}$, and $Po^{IV}I_6^{2-}$ are sterically inactive and the 14-(valence)electron anions are accurately octahedral, as in molecular $Se^{VI}F_6$.

Other less-symmetrical coordination geometries for Se and Te occur in the μ-Se$_2$ complexes and the polyatomic cluster cations Se_{10}^{2+} and Te_6^{4+}, as mentioned below.

The coordination chemistry of complexes in which Se is the donor atom has been extensively studied.[2, 15] Ligands with Te as donor atom have been less widely

[14] B. F. HOSKINS and C. D. PANNAN, A novel 5-coordinate pentagonal-planar complex: X-ray structure of the tris(*O*-ethyl xanthato)tellurium(II) anion, *JCS Chem. Comm.* 1975, 408–9.

[15] S. E. LIVINGSTONE, Metal complexes of ligands containing sulfur, selenium, or tellurium donor atoms. *Q. Rev.* **19**, 386–425 (1965).

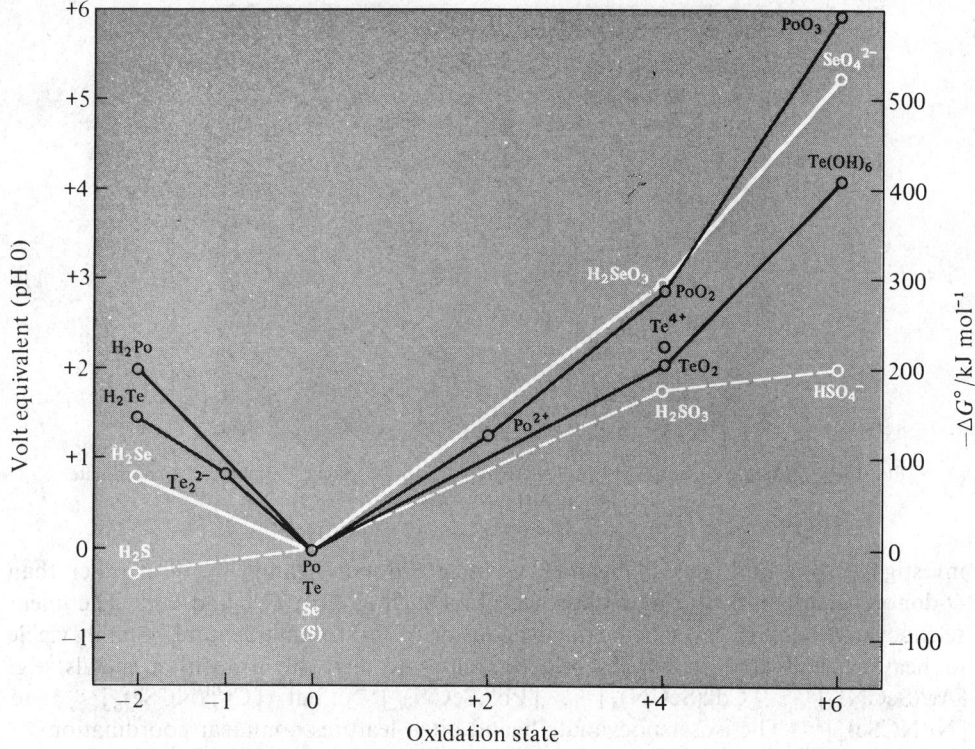

FIG. 16.2   Volt-equivalent diagram for S, Se, Te, and Po.

TABLE 16.3   *Coordination geometries of selenium, tellurium, and polonium*

Coordination number	Se	Te	Po
1	COSe, CSe$_2$, NCSe$^-$, MoSe$_4^{2-}$, WSe$_4^{2-}$	COTe, CSTe Te$_3^{2-}$ [(a)]	
2 (bent)	Se$_x$, H$_2$Se, R$_2$Se cyclo-Se$_4^{2+}$	Te$_x$, H$_2$Te, R$_2$Te, TeBr$_2$ cyclo-Te$_4^{2+}$	
3 (trigonal planar)	SeO$_3$(?)	TeO$_3$(g)	
3 (pyramidal)	(SeO$_2$)$_x$, SeOX$_2$, SeMe$_3^+$	TeO$_3^{2-}$, TeMe$_3^+$	
4 (planar)	—	[TeBr$_2${SC(NH$_2$)$_2$}$_2$]	
4 (tetrahedral)	SeO$_4^{2-}$, SeO$_2$Cl$_2$	—	CdPo (ZnS)
4 (pseudo-trigonal bipyramidal)	R$_2$SeX$_2$	TeO$_2$, Me$_2$TeCl$_2$	
5 (square pyramidal)	[SeOCl$_2$py$_2$]	TeF$_5^-$, [TeI$_4$Me]$^+$	
5 (pentagonal planar)	—	[Te(S$_2$COEt)$_3$]$^-$ (Fig. 16.3)	
6 (octahedral)	SeF$_6$, SeBr$_6^{2-}$	Te(OH)$_6$, TeBr$_6^{2-}$	PoI$_6^{2-}$, Po metal CaPo (NaCl)
6 (trigonal prismatic)	VSe, CrSe, MnSe (NiAs)	ScTe, VTe, MnTe, (NiAs)	MgPo (NiAs)
8 (cubic)	—	TeF$_8^{2-}$ (?)	Na$_2$Po, PoO$_2$ (CaF$_2$)

[(a)] Te$_3^{2-}$ (Te–Te 270 pm, angle 113°) has been isolated as the deep-red complex [K(crypt)]$^+_2$Te$_3^{2-}$. (A. Cisar and J. D. Corbett, *Inorg. Chem.* **16**, 632 (1977).)

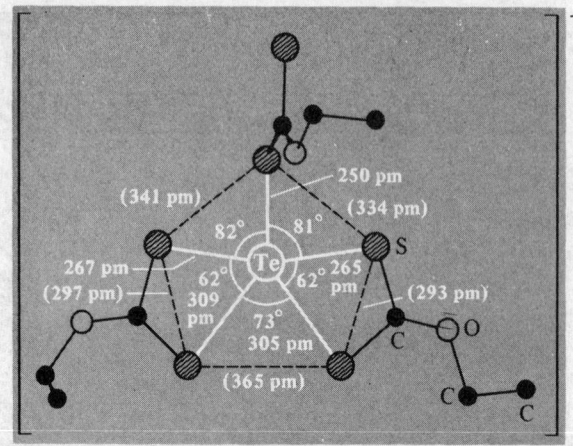

FIG. 16.3   Structure of the anion $[Te(S_2COEt)_3]^-$, the first authentic example of 5-coordinate pentagonal planar geometry.

investigated but both sets of ligands resemble *S*-donor ligands (p. 794) rather than *O*-donor ligands in favouring b-class acceptors such as Pd^{II}, Pt^{II}, and Hg^{II}. The linear selenocyanate ion $SeCN^-$, like the thiocyanate ion (p. 795) is ambidentate, bonding via Se to heavy metals and via N (isoselenocyanate) to first-row transition metals, e.g. $[Ag^I(SeCN)_3]^{2-}$, $[Cd^{II}(SeCN)_4]^{2-}$, $[Pb^{II}(SeCN)_6]^{4-}$, but $[Cr^{III}(NCSe)_6]^{3-}$ and $[Ni(NCSe)_4]^{2-}$. The isoselenocyanate ligand often features nonlinear coordination

$$M{\diagdown}N{=}C{=}Se$$

but in the presence of bulky ligands it tends to become linear $M-\overset{+}{N}{\equiv}C-\overset{-}{Se}$. A bidentate bridging mode is also well established, e.g. {Cd–Se–C–N–Cd} and {Ag–Se–C–N–Cr}. Monodentate organoselenium ligands include R_2Se, Ar_2Se, $R_3P{=}Se$, and selenourea $(H_2N)_2C{=}Se$, all of which bond well to heavy metal acceptors. Tellurium appears to be analogous:[3] e.g. $Me_2Te\cdot HgX_2$, $C_4H_8Te\cdot HgCl_2$, $Ph_2Te\cdot HgX_2$, etc.

The structure of complexes containing the η^2-Se_2 ligand have recently been determined and, where appropriate, compared with analogous η^2-S_2, η^2-P_2, and η^2-As_2 complexes (p. 684). Examples are in Fig. 16.4 and the original papers should be consulted for further details.[16–18]

The compounds of Se, Te, and Po should all be treated as potentially toxic. Volatile compounds such as H_2Se, H_2Te, and organo derivatives are particularly dangerous and maximum permissible limits for air-borne concentrations are 0.1 mg m^{-3} (cf. 10 mg m^{-3} for HCN). The elements are taken up by the kidneys, spleen, and liver, and even in minute concentrations cause headache, nausea, and irritation of mucous membrane.

[16] C. F. CAMPANA, F. Y.-K. LO, and L. F. DAHL, Stereochemical analysis of $[Fe_2(CO)_6(\mu$-$Se_2)]$: a diselenium analogue of $[Fe_2(CO)_6(\mu$-$S_2)]$, *Inorg. Chem.* **18**, 3060–4 (1979); see also pp. 3047 and 3054.

[17] D. H. FARRAR, K. R. GRUNDY, N. C. PAYNE, W. R. ROPER, and A. WALKER, Preparation and reactivity of some η^2-S_2 and η^2-Se_2 complexes of osmium and the X-ray crystal structure of $[Os(CO)_2(PPh_3)_2$-$(\eta^2$-$Se_2)]$, *J. Am. Chem. Soc.* **101**, 6577–82 (1979).

[18] M. G. B. DREW, G. W. A. FOWLES, E. M. PAGE, and D. A. RICE, A unique triple atom bridge: X-ray structure of the μ-selenido-μ-diselenido-bis{tetrachlorotungstate(V)} ion, *J. Am. Chem. Soc.* **101**, 5827–8 (1979).

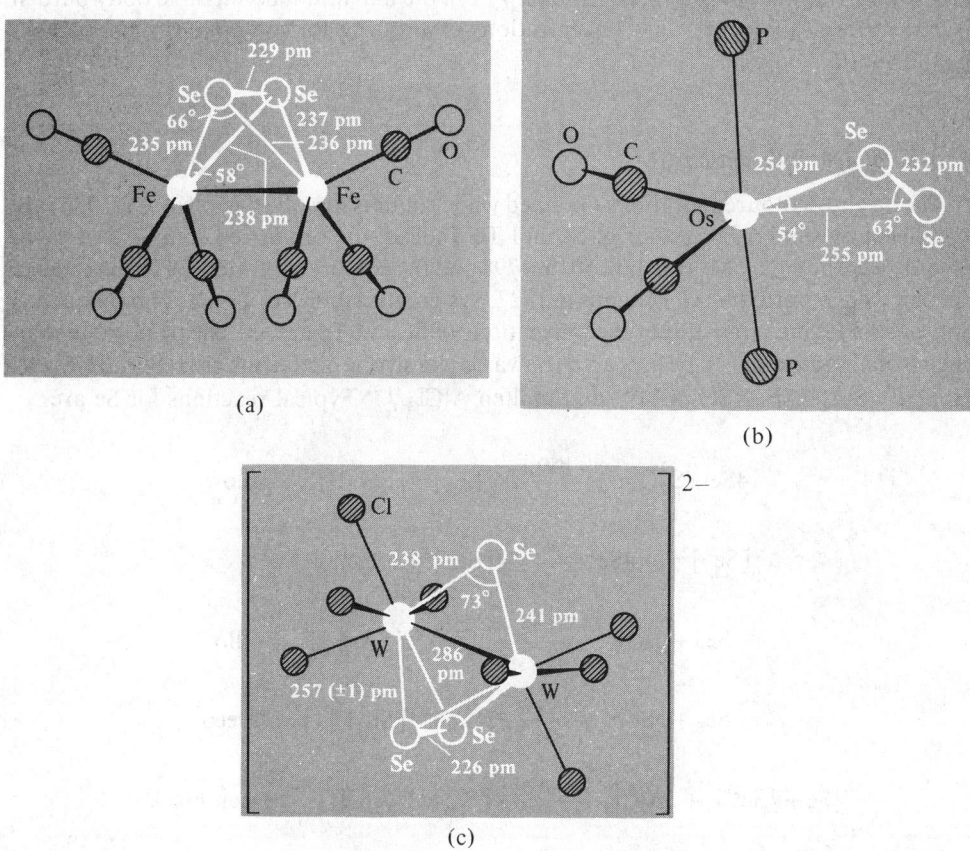

Fig. 16.4 Structures of some η^2-Se$_2$ complexes. (a) red [Fe$_2$(CO)$_6$(μ,η^2-Se$_2$)],[16] (b) reddish-purple [Os(CO)$_2$(PPh$_3$)$_2$(η^2-Se$_2$)],[17] and (c) brown [W$_2$Cl$_8$(μ-Se)(μ-Se$_2$)]$^{2-}$.[18]

Organoselenium compounds in particular, once ingested, are slowly released over prolonged periods and result in foul-smelling breath and perspiration. The element is also highly toxic towards grazing sheep, cattle, and other animals, and, at concentrations above about 5 ppm, causes severe disorders. Despite this Se was found (in 1957) to play an essential dietary role in animals and also in humans—it is required in the formation of the enzyme glutathione peroxidase which is involved in fat metabolism. It has also been found that the incidence of kwashiorkor (severe protein malnutrition) in children is associated with inadequate uptake of Se, and it may well be involved in protection against certain cancers. The average dietary intake of Se in the USA is said to be ~ 150 μg daily, usually in meat and sea food. Considerable caution should be taken in handling compounds of Se and Te, but the hazards should also be kept in perspective—no human fatalities directly attributable to either Se or Te poisoning have ever been recorded.

Polonium is extremely toxic at all concentrations and is never beneficial. Severe radiation damage of vital organs follows ingestion of even the minutest concentrations and, for the most commonly used isotope, ^{210}Po, the maximum permissible body burden is 0.03 μCi, i.e. $\sim 7 \times 10^{-12}$ g. Concentrations of airborne Po compounds must be kept below 4×10^{-11} mg m^{-3}.

16.1.6 *Polyatomic cations,* M_x^{n+}

The brightly coloured solutions obtained when sulfur is dissolved in oleums (p. 785) are paralleled by similar behaviour of Se and Te. Indeed, the bright-red solutions of Te in H_2SO_4 were noted by M. H. Klaproth in 1798 and the coloured solutions of Se in the same solvent were reported by G. Magnus in 1827. Systematic studies in a range of nonaqueous solvents have since shown that the polycations of Se and Te are less electropositive than their S analogues and can be prepared in a variety of strong acids such as H_2SO_4, $H_2S_2O_7$, HSO_3F, SO_2/AsF_5, SO_2/SbF_5, and molten $AlCl_3$.[19] Typical reactions for Se are:

$$4Se + S_2O_6F_2 \xrightarrow{\text{HSO}_3\text{F}} [Se_4]^{2+}[SO_3F]^-_2, \text{ yellow}$$

$$[Se_4]^{2+} + 4Se \xrightarrow{\text{HSO}_3\text{F}} [Se_8]^{2+}, \text{ green}$$

$$Se_8 + 6AsF_5 \xrightarrow[(-2AsF_3)]{\text{SO}_2/80°} 2[Se_4]^{2+}[AsF_6]^-_2, \text{ yellow}$$

$$Se_8 + 5SbF_5 \xrightarrow[(-SbF_3)]{\text{SO}_2/-23°} [Se_8]^{2+}[Sb_2F_{11}]^-_2, \text{ green}$$

$$7\tfrac{1}{2}Se + \tfrac{1}{2}SeCl_4 + 2AlCl_3 \xrightarrow{\text{fuse at 250°}} [Se_8]^{2+}[AlCl_4]^-_2, \text{ green-black}$$

X-ray crystal structure studies on $[Se_4]^{2+}[HS_2O_7]^-_2$ show that the cation is square planar (like S_4^{2+}, p. 785) as in Fig. 16.5a. The Se–Se distance of 228 pm is significantly less

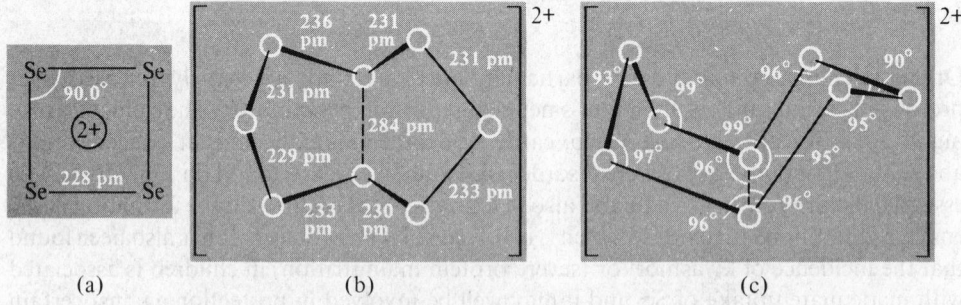

FIG. 16.5   (a) Structure of $[Se_4]^{2+}$; (b) and (c) views of $[Se_8]^{2+}$.

[19] R. J. GILLESPIE and J. PASSMORE, Homopolyatomic cations of the elements, *Adv. Inorg. Chem. Radiochem.* **17**, 49–87 (1975). M. J. TAYLOR, *Metal–Metal Bonded States in Main Group Elements*, Academic Press, London, 1975, 211 pp. J. D. CORBETT, Homopolyatomic ions of the post-transition elements: synthesis, structure, and bonding, *Prog. Inorg. Chem.* **21**, 121–58 (1976).

than the value of 234 pm in Se_8 and 237 pm in Se_∞, consistent with some multiple bonding. The structure of $[Se_8]^{2+}$ in the salt $[Se_8]^{2+}[AlCl_4]^-{}_2$ is in Fig. 16.5b and c: it comprises a bicyclo C_s structure with the *endo-exo* configuration with a long trans-annular link of 284 pm. Other Se–Se distances are very similar to those in Se_8 itself, but the Se–Se–Se angles are significantly smaller in the cation, being $\sim 96°$ rather than $106°$. Very recently [20] the deep-red crystalline compound $Se_{10}(SbF_6)_2$ has been isolated from the reaction of SbF_5 with an excess of Se in SO_2 under pressure at $\sim 50°$. Two views of the bicyclic cation are shown in Fig. 16.6; it features a 6-membered boat-shaped ring linked across the middle

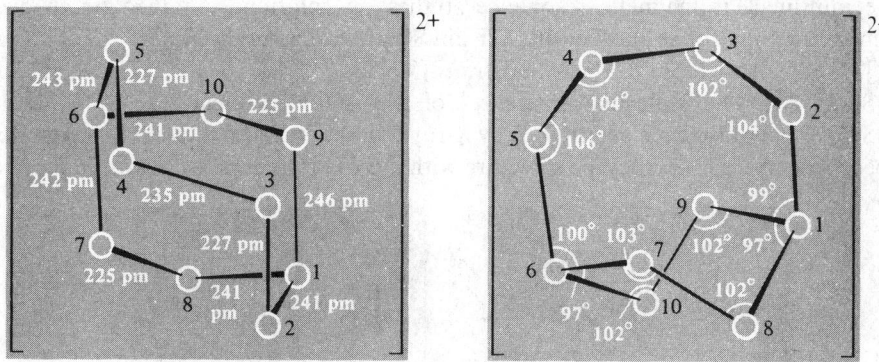

FIG. 16.6   Structure of the $[Se_{10}]^{2+}$ cation $Se_{10}(SbF_6)_2$ along the *b*- and *c*-axes of the crystal; angles Se(2)-Se(1)-Se(9) and Se(5)-Se(6)-Se(10) are each 101.7°.

of a zigzag chain of 4 further Se atoms. The Se–Se distances vary from 225 to 240 pm and Se–Se–Se angles range from 97° to 106°, with 6 angles at the bridgehead atoms Se(1) and Se(6) being significantly smaller than the other 8 in the linking chains.

Polyatomic tellurium cations can be prepared by similar routes. The bright-red species Te_4^{2+}, like S_4^{2+} and Se_4^{2+}, is square planar with the Te–Te distance (266 pm) somewhat less than in the element (284 pm) (Fig. 16.7a). No analogue of Se_8^{2+} has yet been identified

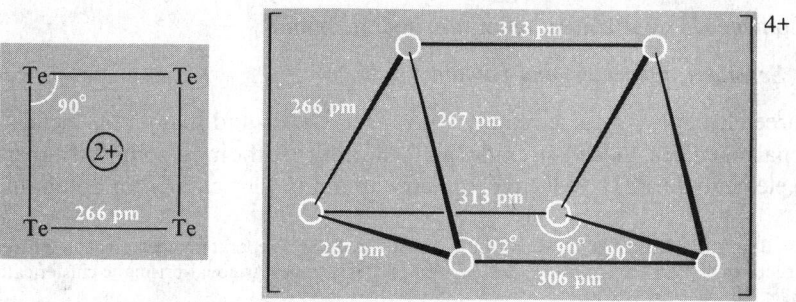

FIG. 16.7   Structure of the cations $[Te_4]^{2+}$ and $[Te_6]^{4+}$.

[20] R. C. BURNS, W.-L. CHAN, R. J. GILLESPIE, W.-C. LUK, J. F. SAWYER, and D. R. SLIM, Preparation and characterization of $Se_{10}(AsF_6)_2$, $Se_{10}(SbF_6)_2$, and $Se_{10}(AlCl_4)_2$ and crystal structure of $Se_{10}(SbF_6)_2$; *Inorg. Chem.* **19**, 1432–9 (1980).

but oxidation of Te with AsF_5 in AsF_3 as solvent yields the brown crystalline compound $Te_6(AsF_6)_4.2AsF_3$. X-ray studies reveal the presence of $[Te_6]^{4+}$ which is the first example of a simple trigonal prismatic cluster cation (Fig. 16.7b). The Te–Te distances between the triangular faces (313 pm) are substantially larger than those within the triangle (267 pm).[21]

Mixed Se/Te polatomic cations are also known. For example, when Se and Te are dissolved in 65% oleum at room temperature the resulting orange-brown solutions were shown by ^{125}Te and ^{123}Te nmr spectroscopy to contain the four species $[Te_nSe_{4-n}]^{2+}$ ($n = 1$–4) and the species $[Se_4]^{2+}$ was also presumably present.[22] Likewise ^{77}Se and ^{125}Te multinuclear magnetic resonance studies on solutions obtained by oxidizing equimolar mixtures of Se and Te with AsF_5 in SO_2 reveal not only $[Se_4]^{2+}$, $[Te_4]^{2+}$, and $[Te_6]^{4+}$ but also $[TeSe_3]^{2+}$, *cis-* and *trans-*$[Te_2Se_2]^{2+}$, $[Te_3Se]^{2+}$, $[Te_2Se_4]^{2+}$, and $[Te_3Se_3]^{2+}$.[23] The molecular structures of the sulfur analogue $[Te_3S_3]^{2+}$ and of $[Te_2Se_4]^{2+}$ have also been determined by X-ray diffractometry and found to have a boat-shaped 6-membered heterocyclic structure with a cross-ring bond as shown in Fig. 16.8.

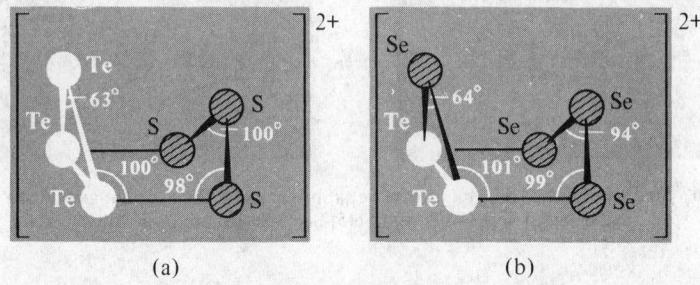

FIG. 16.8   Structures of the heteroatomic cluster cations (a) $[Te_3S_3]^{2+}$, and (b) $[Te_2Se_4]^{2+}$.

As expected, these M_6^{2+} species are more open than the corresponding Te_6^{4+} cluster because of the presence of 2 extra valency-shell electrons (p. 857).

The mixed *anionic* species $[Tl_2Te_2]^{2-}$ (20 valence electrons) is butterfly-shaped with Tl_2 at the "hinge" and 2Te at the "wing tips",[23a] in contrast to the 22 valence-electron *cationic* species Te_4^{2+} and Se_4^{2+} which are square planar.

16.2   Compounds of selenium, tellurium, and polonium

16.2.1   *Selenides, tellurides, and polonides*

All three elements combine readily with most metals and many non-metals to form binary chalcogenides. Indeed, selenides and tellurides are the most common mineral forms of these elements (p. 883). Nonstoichiometry abounds, particularly for compounds with

[21] R. C. BURNS, R. J. GILLESPIE, W.-C. LUK, and D. R. SLIM, Preparation spectroscopic properties, and crystal structures of $Te_6(AsF_6)_4.2AsF_3$ and $Te_6(AsF_6)_4.2SO_2$: a new trigonal-prismatic cluster cation Te_6^{4+}, *Inorg. Chem.* **18**, 3086–94 (1979).

[22] C. R. LASSIGNE and E. J. WELLS, Identification of the species $[Te_nSe_{4-n}]^{2+}$ by high resolution ^{125}Te and ^{123}Te fourier transform nmr spectroscopy, *JCS Chem. Comm.* 1978, 956–7.

[23] G. J. SCHROBILGEN, R. C. BURNS, and P. GRANGER, Tellurium-125 and selenium-77 multinuclear magnetic resonance studies of $[Te_6]^{4+}$, $[Te_{4-n}Se_n]^{2+}$, $[Te_2Se_4]^{2+}$ and $[Te_3Se_3]^{2+}$ polyatomic cations, *JCS Chem. Comm.* 1978, 957–60.

[23a] R. C. BURNS and J. D. CORBETT, Heteroatomic polyanions of post-transition metals. The synthesis and structure of $Tl_2Te_2^{2-}$. Skeletal requirements for bonding, *J. Am. Chem. Soc.* **103**, 2627–32 (1981).

the transition elements (where electronegativity differences are minimal and variable valency is favoured), and many of the chalcogenides can be considered as metallic alloys. Many such compounds have important technological potentialities for solid-state optical, electrical, and thermoelectric devices and have been extensively studied. For the more electropositive elements (e.g. Groups IA and IIA), the chalcogenides can be considered as "salts" of the acids, H_2Se, H_2Te, and H_2Po (p. 900).

The alkali metal selenides and tellurides can be prepared by direct reaction of the elements at moderate temperatures in the absence of air, or more conveniently in liquid ammonia solution. They are colourless, water soluble, and readily oxidized by air to the element. The structures adopted are not unexpected from general crystallochemical principles. Thus Li_2Se, Na_2Se, and K_2Se have the antifluorite structure (p. 129); MgSe, CaSe, SrSe, BaSe, ScSe, YSe, LuSe, etc., have the rock-salt structure (p. 273); BeSe, ZnSe, and HgSe have the zinc-blende structure (p. 1405); and CdSe has the wurtzite structure (p. 1405). The corresponding tellurides are similar, though there is not a complete 1:1 correspondence. Polonides can also be prepared by direct reaction and are amongst the stablest compounds of this element: Na_2Po has the antifluorite structure; the NaCl structure is adopted by the polonides of Ca, Ba, Hg, Pb, and the lanthanide elements; BePo and CdPo have the ZnS structure and MgPo the nickel arsenide structure (p. 649). Decomposition temperatures of these polonides are about $600 \pm 50°C$ except for the less-stable HgPo (d 300°) and the extremely stable lanthanide derivatives which do not decompose even at 1000° (e.g. PrPo mp 1253°, TmPo mp 2200°C).

Transition-element chalcogenides are also best prepared by direct reaction of the elements at 400–1000°C in the absence of air. They tend to be metallic nonstoichiometric alloys though intermetallic compounds also occur, e.g. $Ti_{\sim 2}Se$, $Ti_{\sim 3}Se$, $TiSe_{0.95}$, $TiSe_{1.05}$, $Ti_{0.9}Se$, Ti_3Se_4, $Ti_{0.7}Se$, Ti_5Se_8, $TiSe_2$, $TiSe_3$, etc.[24, 25] Fuller details of these many compounds are in the references cited.

Most selenides and tellurides are decomposed by water or dilute acid to form H_2Se or H_2Te but the yields, particularly of the latter, are poor.

Polychalcogenides are less stable than polysulfides (p. 805). Reaction of alkali metals with Se in liquid ammonia affords M_2Se_2, M_2Se_3, and M_2Se_4, and analogous polytellurides have also been reported (see p. 893 for structure of $Te_3{}^{2-}$). However, all these compounds are rather unstable thermally and tend to be oxidized in air.

Reaction of GaSe with Cs results in the formation of $Cs_{10}Ga_6Se_{14}$ which contains the extraordinary, discrete, rectilinear, C_{2h} anion $[Ga_6Se_{14}]^{10-}$; this comprises 6 edge-fused $\{GaSe_4\}$ tetrahedra, $[Se_2\{Ga(\mu\text{-}Se)_2\}_5GaSe_2]^{10-}$ and is 1900 pm long (Ga–Se$_t$ 236 pm, Ga–Se$_\mu$ 239–250 pm).[25a] It therefore forms a unique link between finite edge-fused dimers such as Al_2Cl_6 and the infinite chains of edge-linked tetrahedra such as occur in SiS_2.

16.2.2 Hydrides

H_2Se (like H_2O and H_2S) can be made by direct combination of the elements (above 350°), but H_2Te and H_2Po cannot be made in this way because of their thermal instablity.

[24] D. M. CHIZHIKOV and V. P. SHCHASTLIVYI, *Selenium and Selenides*, Collet's, London, 1968, 403 pp.
[25] F. HULLIGER, Crystal chemistry of chalcogenides and pnictides of the transition elements, *Struct. Bonding* (*Berlin*), **4**, 83–229 (1968).
[25a] H.-J. DIESEROTH and H. FU-SON, $[Ga_6Se_{14}]^{10-}$: A 1900 pm long, hexameric anion, *Angew. Chem.* Int. Edn. (Engl.) **20**, 962–3 (1981).

H_2Se is a colourless, offensive-smelling, poisonous gas which can be made by hydrolysis of Al_2Se_3, the action of dilute mineral acids on FeSe, or the surface-catalysed reaction of gaseous Se and H_2:

$$Al_2Se_3 + 6H_2O \longrightarrow 3H_2Se + 2Al(OH)_3$$

$$FeSe + 2HCl \longrightarrow H_2Se + FeCl_2$$

$$Se + H_2 \rightleftharpoons H_2Se$$

In this last reaction, conversion at first rises with increase in temperature and then falls because of increasing thermolysis of the product: conversion exceeds $\sim 40\%$ between 350–650° and is optimum (64%) at 520°.

H_2Te is also a colourless, foul-smelling, toxic gas which is best made by electrolysis of 15–50% aqueous H_2SO_4 at a Te cathode at $-20°$, 4.5 A, and 75–110 V. It can also be made by hydrolysis of Al_2Te_3, the action of hydrochloric acid on the tellurides of Mg, Zn, or Al, or by reduction of Na_2TeO_3 with $TiCl_3$ in a buffered solution. The compound is unstable above 0° and decomposes in moist air and on exposure to light. H_2Po is even less stable and has only been made in trace amounts ($\sim 10^{-10}$ g scale) by reduction of Po using Mg foil/dilute HCl and the reaction followed by radioactive tracer techniques.

Physical properties of the three gases are compared with those of H_2O and H_2S in Table 16.4. The trends are obvious, as is the "anomalous" position of water (p. 729). The densities

TABLE 16.4 *Some physical properties of H_2O, H_2S, H_2Se, H_2Te, and H_2Po*

Property	H_2O	H_2S	H_2Se	H_2Te	H_2Po
MP/°C	0.0	−85.6	−65.7	−51	−36(?)
BP/°C	100.0	−60.3	−41.3	−4	+37(?)
$\Delta H_f°$/kJ mol^{-1}	−285.9	+20.1	+73.0	+99.6	—
Bond length (M–H)/pm	95.7	133.6	146	169	—
Bond angle (H–M–H) (g)	104.5°	92.1°	91°	90°	—
Dissociation constant:					
HM$^-$, K_1	1.8×10^{-16}	1.3×10^{-7}	1.3×10^{-4}	2.3×10^{-3}	—
M^{2-}, K_2	—	7.1×10^{-15}	$\sim 10^{-11}$	1.6×10^{-11}	—

of liquid and solid H_2Se are 2.12 and 2.45 g cm^{-3}. H_2Te condenses to a colourless liquid (d 4.4 g cm^{-3}) and then to lemon-yellow crystals. Both gases are soluble in water to about the same extent as H_2S, yielding increasingly acidic solutions (cf. acetic acid $K_1 \sim 2 \times 10^{-5}$). Such solutions precipitate the selenides and tellurides of many metals from aqueous solutions of their salts but, since both H_2Se and H_2Te are readily oxidized (e.g. by air), elementary Se and Te are often formed simultaneously.

H_2Se and H_2Te burn in air with a blue flame to give the dioxide (p. 911). Halogens and other oxidizing agents (e.g. HNO_3, $KMnO_4$) also rapidly react with aqueous solutions to precipitate the elements. Reaction of H_2Se with aqueous SO_2 is complex, the products formed depending critically on conditions (cf. Wackenroder's solution, p. 849): addition of the selenide to aqueous SO_2 yields a 2:1 mixture of S and Se together with oxoacids of sulfur, whereas addition of SO_2 to aqueous H_2Se yields mainly Se:

$$H_2Se + 5SO_2 + 2H_2O \longrightarrow 2S + Se + 3H_2SO_4$$

$$H_2Se + 6SO_2 + 2H_2O \longrightarrow 2S + Se + H_2S_2O_6 + 2H_2SO_4$$

$$H_2Se + 6SO_2 + 2H_2O \longrightarrow Se + H_2S_4O_6 + 2H_2SO_4$$

16.2.3 **Halides**

As with sulfur, there is a definite pattern to the stoichiometries of the known halides of the heavier chalcogens. Selenium forms no binary iodides whereas the more electropositive Te and Po do. Numerous chlorides and bromides are known for all 3 elements, particularly in oxidation states $+1$, $+2$, and $+4$. In the highest oxidation state, $+6$, only the fluorides MF_6 are known for the 3 elements; in addition SeF_4 and TeF_4 have been characterized but no fluorides of lower oxidation states except the fugitive $FSeSeF$, $Se=SeF_2$, and SeF_2 which can be trapped out at low temperature.[25b, 25c] The compound previously thought to be Te_2F_{10} is now known to be $O(TeF_5)_2$[25c, 25d] (p. 910). Finally, Te forms a range of curious lower halides which are structurally related to the Te_x chains in elementary tellurium.

The known compounds are summarized in Table 16.5 which also lists their colour, mp, bp, and decomposition temperature where these have been reported. It will be convenient to discuss the preparation, structure, and chemical properties of these various compounds in ascending order of formal oxidation state. For comparable information on the halides of S, see pp. 808–819.

Lower halides

The phase relations in the tellurium–halogen systems have only recently been elucidated and the results show a series of subhalides with various structural motifs based on the helical-chain structure of Te itself. These are summarized in Fig. 16.9. Thus, reaction of Te and Cl_2 under carefully controlled conditions in a sealed tube[26] results in Te_3Cl_2 (Fig. 16.9b) in which every third Te atom in the chain is oxidized by addition of 2 Cl atoms, thereby forming a series of 4-coordinate pseudo-trigonal-bipyramidal groups with axial Cl atoms linked by pairs of unmodified Te atoms $-Te-Te-TeCl_2-Te-Te-TeCl_2-$.[27] Te_2Br and Te_2I consist of zigzag chains of Te in planar arrangement (Fig. 16.9c); along the chain is an alternation of trigonal pyramidal (pseudo-tetrahedral) and square-planar (pseudo-octahedral) Te atoms. These chains are joined in pairs by cross-linking at the trigonal pyramidal Te atoms, thereby forming a ribbon of fused 6-membered Te rings in the boat configuration.[27] A similar motif occurs in β-TeI (Fig. 16.9d) which is formed by rapidly cooling partially melted α-TeI (see below) from 190°: in this case the third bond from the trigonal pyramidal Te atoms carries an I atom instead of being cross-linked to a similar chain.[28] The second, more stable modification, α-TeI, features tetrameric molecules

[25b] B. COHEN and R. D. PEACOCK, Fluorine compounds of selenium and tellurium, *Adv. Fluorine Chem.* **6**, 343–85 (1970).

[25c] E. ENGELBRECHT and F. SLADKY, Selenium and tellurium fluorides, *Adv. Inorg. Chem. Radiochem.* **24**, 189–223 (1981). This review also includes oxofluorides of Se and Te, and related anions.

[25d] P. M. WATKINS, Ditellurium decafluoride: a continuing myth, *J. Chem. Educ.*, **51**, 520–21 (1974).

[26] A. RABENAU and H. RAU, The systems Te–TeCl₄ and Te–TeBr₄, *Z. anorg. allgem. Chem.* **395**, 273–9 (1973).

[27] R. KNIEP, D. MOOTZ, and A. RABENAU, Structural investigations of tellurium subhalides: modified tellurium structures, *Angew. Chem.*, Int. Ed. (Engl.) **12**, 499–500 (1973). M. TAKEDA and N. N. GREENWOOD, Tellurium-125 Mössbauer spectra of some tellurium subhalides, *JCS Dalton* 1976, 631–6.

[28] R. KNIEP, D. MOOTZ, and A. RABENAU, Subhalides of tellurium: crystal structures of α-TeI and β-TeI, *Angew. Chem.*, Int. Edn. (Engl.) **13**, 403–4 (1973). Modified chain and ribbon structures are also observed for the ternary compounds α-AsSeI, β-AsSeI, α-AsTeI, and β-AsTeI, all of which are isoelectronic with Se_x and Te_x but which have more complex chain structures (R. KNIEP and H. D. RESKI, Chalcogenide iodides of arsenic, *Angew. Chem.* Int. Edn. (Engl.) **20**, 212–4 (1981)).

TABLE 16.5   *Halides of selenium, tellurium, and polonium*

Oxidation state	Fluorides	Chlorides	Bromides	Iodides
< 1		Te_2Cl Te_3Cl_2 silver grey mp 238 (peritectic)	Te_2Br grey needles mp 224 (peritectic)	Te_2I silver grey
+ 1	(FSeSeF) and (Se=SeF$_2$) trapped at low temperature	Se_2Cl_2 yellow-brown liquid mp −85, bp 130 (d)	(β-)Se$_2$Br$_2$ blood-red liquid bp 225 (d) (α-SeBr, mp +5)	α-Te$_4$I$_4$ black mp 185 (peritectic) β-TeI black
+ 2	(SeF$_2$) trapped at low temperature	(SeCl$_2$) d in vapour ("TeCl$_2$") black eutectic PoCl$_2$ dark ruby red mp 355, subl 130	(SeBr$_2$) d in vapour ("TeBr$_2$") brown d (see text) PoBr$_2$ purple-brown mp 270 (d)	(PoI$_2$) impure (from decomp of PoI$_4$ at 200)
+ 4	SeF$_4$ colourless liquid mp −10, bp 101 TeF$_4$ colourless mp 129, d >194 PoF$_4$(?) solid from decomp of PoF$_6$	Se$_4$Cl$_{16}$ colourless mp 305, subl 196 Te$_4$Cl$_{16}$ pale-yellow solid, maroon liquid mp 228, bp 390 PoCl$_4$ yellow d >200 to PoCl$_2$ mp 300, bp ∼390 extrapolated	α-Se$_4$Br$_{16}$ orange-red mp 123 (also β-Se$_4$Br$_{16}$) Te$_4$Br$_{16}$ yellow mp 363 (under Br$_2$) bp 414 (under Br$_2$) PoBr$_4$ bright red mp 330, bp 360 200 mmHg	Te$_4$I$_{16}$ Black mp 280, d 100 PoI$_4$ black d >200

			Mixed halides
			TeBr$_2$Cl$_2$ yellow solid, ruby-red liquid
			       mp 292, bp 415
			TeBr$_2$I$_2$ garnet-red crystals
			       mp 325, d 420
			PoBr$_2$Cl$_2$ salmon pink (PoCl$_2$ + Br$_2$ vap)	

+ 6	SeF$_6$ colourless gas mp −35 (2 atm), subl −47 TeF$_6$ colourless gas mp −38°, subl −39°			

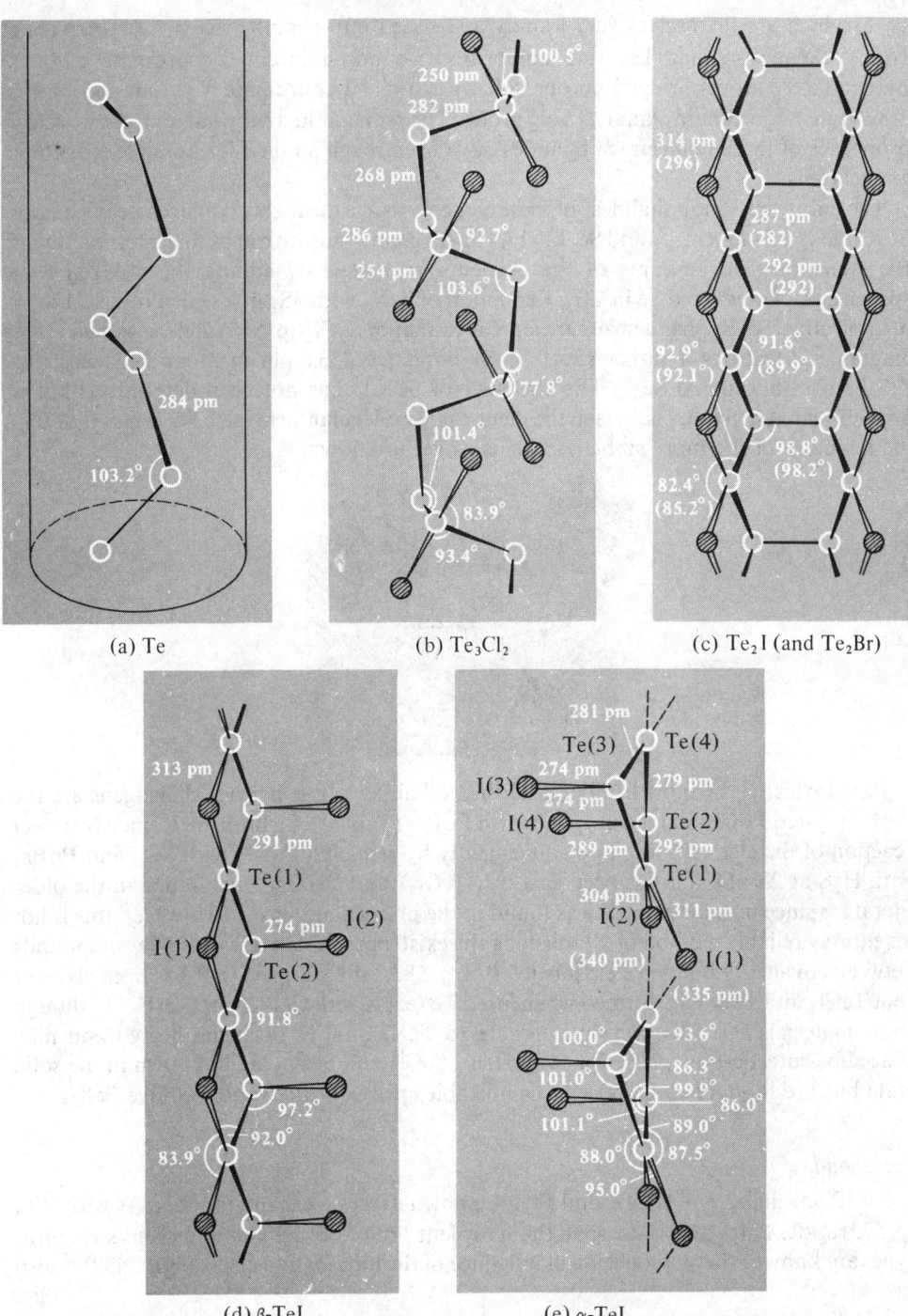

Fig. 16.9   Structural relations between tellurium and its subhalides: (a) tellurium, (b) Te_3Cl_2, (c) Te_2Br and Te_2I, (d) β-TeI, and (e) α-TeI.

Te_4I_4 which are themselves very loosely associated into chains by Te–I···Te links (Fig. 16.9e); the non-planar Te_4 ring comprises two non-adjacent 3-coordinate trigonal pyramidal Te atoms bridged on one side by a single 2-coordinate Te atom and on the other by a 4-coordinate planar $>TeI_2$ group. The semiconductivity and nonlinear optical properties of these tellurium subhalides have been much studied for possible electronic applications.

The only other "monohalides" of these chalcogens are the highly coloured heavy liquids Se_2Cl_2 (d_{25} 2.774 g cm^{-3}) and Se_2Br_2 (d_{15} 3.604 g cm^{-3}). Both can be made by reaction of the stoichiometric amounts of the elements or better, by adding the halogen to a suspension of powdered Se in CS_2. Reduction of SeX_4 with 3Se in a sealed tube at 120° is also effective. Se_2Br_2 has a structure similar to that of S_2Cl_2 (p. 815) and, as shown in the diagram,[29] features a rather short Se–Se bond (cf. 233.5 pm in monoclinic Se_8 and 237.3 pm in hexagonal Se_∞). The structure of Se_2Cl_2 has not been determined but is probably similar. Se_2Br_2 is, in fact, the metastable molecular form (also known as β-SeBr); the structure of the more stable α-SeBr is as yet unknown.

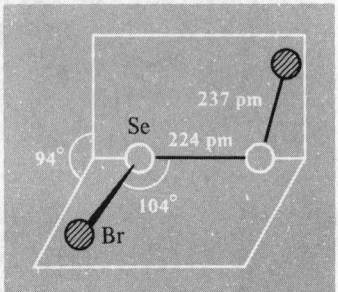

Paradoxically, the most firmly established dihalides of the heavier chalcogens are the dark ruby-red $PoCl_2$ and the purple-brown $PoBr_2$ (Table 16.5). Both are formed by direct reaction of the elements or more conveniently by reducing $PoCl_4$ with SO_2 and $PoBr_4$ with H_2S at 25°. Doubt has been cast on "$TeCl_2$" and "$TeBr_2$" mentioned in the older literature since no sign of these was found in the phase diagrams.[26] However, this is not an entirely reliable method of establishing the existence of relatively unstable compounds between covalently bonded elements (cf. P/S, p. 583, and S/I, p. 817). It has been claimed that $TeCl_2$ and $TeBr_2$ are formed when fused Te reacts with CCl_2F_2 or $CBrF_3$,[30] though these materials certainly disproportionate to TeX_4 and Te on being heated and may indeed be eutectic-type phases in the system. $SeCl_2$ and $SeBr_2$ are unknown in the solid state but are thought to be present as unstable species in the vapour above SeX_4.

Tetrahalides

All 12 tetrahalides of Se, Te, and Po are known except, perhaps, for SeI_4. As with PX_5 (p. 573) and SX_4 (p. 816) these span the "covalent-ionic" border and numerous structural types are known; the stereochemical influence of the lone-pair of electrons (p. 439) is also

[29] D. KATRYNIOK and R. KNIEP, The molecular and crystal structure of β-SeBr, *Angew. Chem.*, Int. Edn. (Engl.) **19**, 645 (1980).

[30] E. E. AYNSLEY, The preparation and properties of tellurium dichloride, *J. Chem. Soc.* 1953, 3016–19. E. E. AYNSLEY and R. H. WATSON, The preparation and properties of tellurium dibromide, *J. Chem. Soc.* 1955, 2603–6.

prominent. SeF_4 is a colourless reactive liquid which fumes in air and crystallizes to a white hygroscopic solid (Table 16.5). It can be made by the controlled fluorination of Se (using F_2 at 0°, or AgF) or by reaction of SF_4 with SeO_2 above 100°. SeF_4 can be handled in scrupulously dried borosilicate glassware and is a useful fluorinating agent. Its structure in the gas phase, like that of SF_4 (p. 809), is pseudo-trigonal-bipyramidal with C_{2v} symmetry; the dimensions shown in Fig. 16.10a were obtained by microwave spectroscopy.

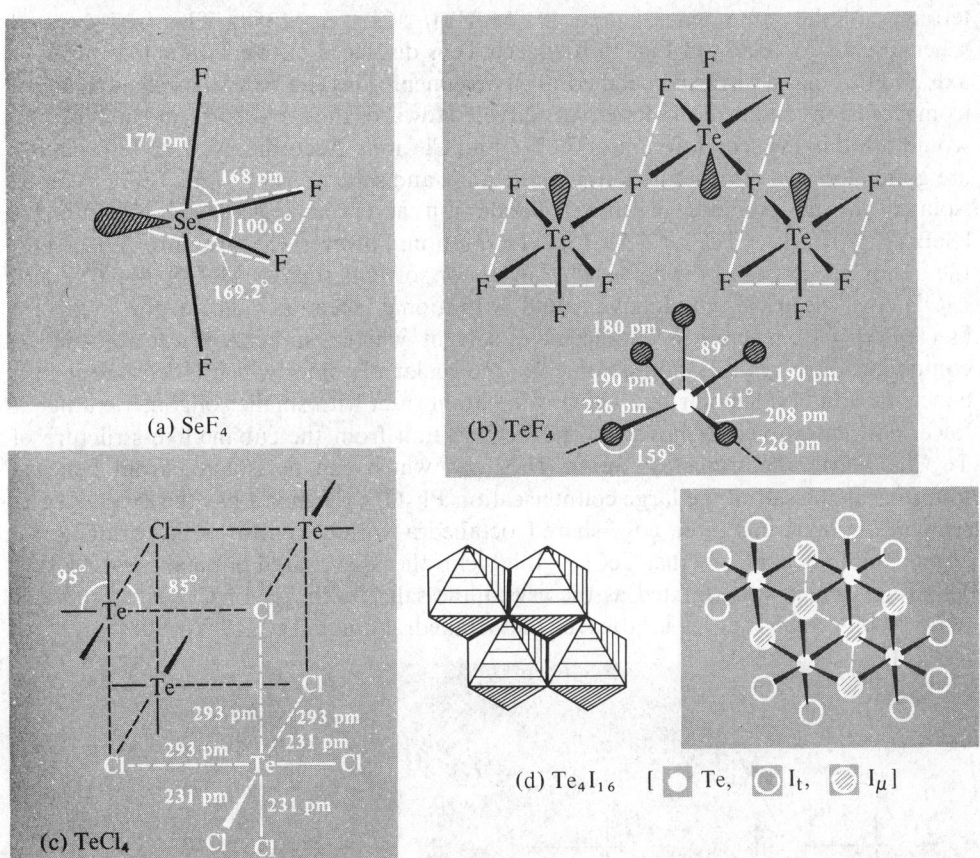

(a) SeF_4

(b) TeF_4

(c) $TeCl_4$

(d) Te_4I_{16} [⬜ Te, ◧ I_t, ▨ I_μ]

FIG. 16.10   Structures of some tetrahalides of Se and Te: (a) SeF_4, (b) schematic representation of the $(TeF_4)_\infty$ chains in crystalline TeF_4 and the dimensions of one TeF_5 unit, (c) the tetrameric unit in crystalline $(TeCl_4)_4$, and (d) two representations of the tetrameric molecules in Te_4I_{16} showing the shared edges of the $\{TeI_6\}$ octahedral subunits.

TeF_4 can be obtained as colourless, hygroscopic, sublimable crystals by controlled fluorination of Te or TeX_2 with F_2/N_2 at 0°, or more conveniently by reaction of SeF_4 with TeO_2 at 80°. It decomposes above 190° with formation of TeF_6 and is much more reactive than SeF_4. For example, it readily fluorinates SiO_2 above room temperature and reacts with Cu, Ag, Au, and Ni at 185° to give the metal tellurides and fluorides. Adducts with BF_3, AsF_5, and SbF_5 are known. Although probably monomeric in the gas phase, crystalline TeF_4 comprises chains of *cis*-linked square-pyramidal TeF_5 groups (Fig. 16.10b) similar to those in the isoelectronic $(SbF_4^-)_x$ chains in $NaSbF_4$ (p. 660). The lone-pair is alternately above and below the mean basal plane and each Te atom is displaced some 30 pm in the same direction. However, the local Te environment is somewhat less symmetrical than implied by this idealized description, and the Te–F distances span the range 180–226 pm.

The other tetrahalides can all readily be made by direct reactions of the elements. Crystalline $SeCl_4$, $TeCl_4$, and β-$SeBr_4$ are isotypic and the structural unit is a cubane-like tetramer of the same general type as $[Me_3Pt(\mu_3\text{-}Cl)]_4$ (p. 1358). This is illustrated schematically for $TeCl_4$ in Fig. 16.10c: each Te is displaced outwards along a threefold axis and thus has a distorted octahedral environment. This can be visualized as resulting from repulsions due to the Te lone-pairs directed towards the cube centre and, in the limit, would result in the separation into $TeCl_3^+$ and Cl^- ions. Accordingly, the 3 tetrahalides are good electrical conductors in the fused state, and salts of SeX_3^+ and $TeCl_3^+$ can be isolated in the presence of strong halide ion acceptors, e.g. $[SeCl_3]^+[GaCl_4]^-$, $[SeBr_3]^+[AlBr_4]^-$, $[TeCl_3]^+[AlCl_4]^-$. In solution, however, the structure depends on the donor properties of the solvent:[31] in donor solvents such as MeCN, Me_2CO, and EtOH the electrical conductivity and vibrational spectra indicate the structure $[L_2TeCl_3]^+Cl^-$, where L is a molecule of solvent, whereas in benzene and toluene the compound dissolves as a non-conducting molecular oligomer which is tetrameric at a concentration of 0.1 molar but which is in equilibrium with smaller oligomeric units at lower concentrations. Removal of one $TeCl_3^+$ unit from the cubane-like structure of Te_4Cl_{16} leaves the trinuclear anion $Te_3Cl_{13}^-$ which can be isolated from benzene solutions as the salt of the large counter-cation Ph_3C^+; the anion has the expected C_{3v} structure comprising three edge-shared octahedra with a central triply bridging Cl atom.[32] Removal of a further $TeCl_3^+$ unit yields the edge-shared bi-octahedral dianion $Te_2Cl_{10}^{2-}$ which was isolated as the crystalline salt $[AsPh_4]_2^+[Te_2Cl_{10}]^{2-}$. Notional removal of a final $\{TeCl_4\}$ unit leaves the octahedral anion $TeCl_6^{2-}$ (p. 908):

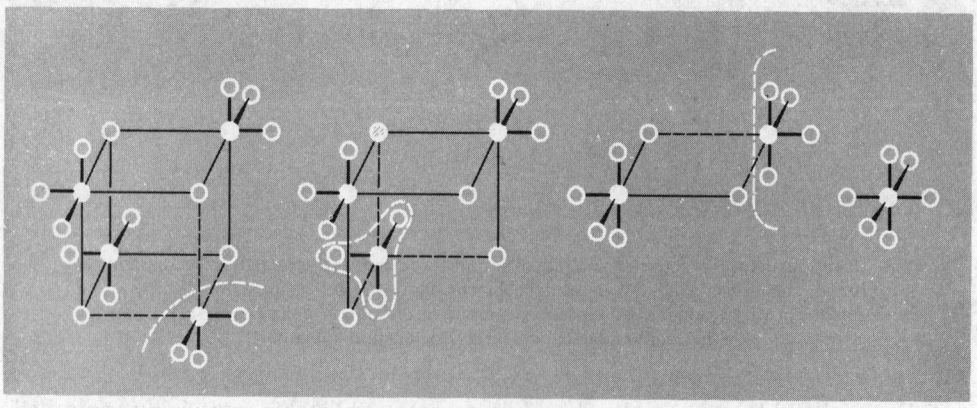

$$Te_4Cl_{16} \xrightarrow{-TeCl_3^+} [Te_3Cl_{13}]^- \xrightarrow{-TeCl_3^+} [Te_2Cl_{10}]^{2-} \xrightarrow{-TeCl_4} [TeCl_6]^{2-}$$

[31] N. N. GREENWOOD, B. P. STRAUGHAN, and A. E. WILSON, Behaviour of tellurium(IV) chloride, bromide, and iodide in organic solvents, and the structure of the species present, *J. Chem. Soc.* (A) 1968, 2209–12.

[32] B. KREBS and V. PAULAT, Preparation and properties of trimeric chlorotellurates(IV). Crystal structure of $Ph_3CTe_3Cl_{13}$, *Z. Naturforsch.* **34b**, 900–905 (1979), and references therein.

SeBr$_4$ is dimorphic: the α-form, like β-SeBr$_4$ mentioned in the preceding paragraph, has a cubane-like tetrameric unit (Se–Br$_t$ 237 pm, Se–Br$_\mu$ 297 pm) but the two forms differ in the spacial arrangement of the tetramers.[33] TeI$_4$ has yet another structure which involves a tetrameric arrangement of edge-shared {TeI$_6$} octahedra not previously encountered in binary inorganic compounds (Fig. 16.10d).[34] The molecule is close to idealized C_{2h} symmetry with each terminal octahedron sharing 2 edges with the 2 neighbouring central octahedra and each central octahedron sharing 3 edges with its 3 neighbours (Te–I$_t$ 277 pm, Te–I$_{\mu_2}$ 311 pm, Te–I$_{\mu_3}$ 323 pm). There is no significant intermolecular I···I bonding. Comparison of the structures and bond data for the homologous series TeF$_4$, TeCl$_4$(TeBr$_4$), TeI$_4$ reveals an increasing delocalization of the Te^{IV} lone-pair. This effect is also observed in the compounds of other ns^2 elements (e.g. Sn^{II}, Pb^{II}, As^{III}, Sb^{III}, Bi^{III}, I^V; see pp. 439, 445, 661) and correlates with the gradation of electronegativities and the polarizing power of the halogens.

The detailed structures of PoX$_4$ are unknown. Some properties are in Table 16.5. PoF$_4$ is not well characterized. PoCl$_4$ forms bright-yellow monoclinic crystals which can be melted under an atmosphere of chlorine, and PoBr$_4$ has a fcc lattice with $a_o = 560$ pm. These compounds and PoI$_4$ can be made by direct combination of the elements or indirectly, e.g. by the chlorination of PoO$_2$ with HCl, PCl$_5$, or SOCl$_2$, or by the reaction of PoO$_2$ with HI at 200°. Similar methods are used to prepare the tetrahalides of Se and Te, e.g.:

TeCl$_4$:   Cl$_2$ + Te;   SeCl$_2$ on Te, TeO$_2$, or TeCl$_2$;   CCl$_4$ + TeO$_2$ at 500°

TeBr$_4$:   Te + Br$_2$ at room temp;   aq HBr on Te or TeO$_2$

TeI$_4$:   Heat Te + I$_2$;   Te + MeI;   TeBr$_4$ + EtI

The two mixed tellurium(IV) halides listed in Table 16.5 were prepared by the action of liquid Br$_2$ on TeCl$_2$ to give the yellow solid TeBr$_2$Cl$_2$, and by the action of I$_2$ on TeBr$_2$ in ether solution to give the red crystalline TeBr$_2$I$_2$; their structures are as yet unknown.

Hexahalides

The only hexahalides known are the colourless gaseous fluorides SeF$_6$ and TeF$_6$ and the volatile liquids TeClF$_5$ and TeBrF$_5$. The hexafluorides are prepared by direct fluorination of the elements or by reaction of BrF$_3$ on the dioxides. Both are octahedral with Se–F 167–170 pm and Te–F 184 pm. SeF$_6$ resembles SF$_6$ in being inert to water but it is decomposed by aqueous solutions of KI or thiosulfate. TeF$_6$ hydrolyses completely within 1 day at room temperature.

The mixed halides TeClF$_5$ and TeBrF$_5$ are made by oxidative fluorination of TeCl$_4$ or TeBr$_4$ in a stream of F$_2$ diluted with N$_2$ at 25°. Under similar conditions TeI$_4$ gave only

[33] P. Born, R. Kniep, and D. Mootz, Phase relations in the system Se–Br and the crystal structures of dimorphic SeBr$_4$, *Z. anorg. allgem. Chem.* **451**, 12–24 (1979).

[34] V. Paulat and B. Krebs, Tellurium(IV) iodide: tetrameric Te$_4$I$_{16}$ molecules in the solid, *Angew. Chem., Int. Edn.* (Engl.) **15**, 39–40 (1976).

TeF_6 and IF_5. $TeClF_5$ can also be made by the action of ClF on TeF_4, $TeCl_4$, or TeO_2 below room temperature; it is a colourless liquid, mp $-28°$, bp $13.5°$, which does not react with Hg, dry metals, or glass at room temperature.

Halide complexes

It is convenient to include halide complexes in this section on the halides of Se, Te, and Po and, indeed, some have already been alluded to above. In addition, pentafluoroselenates(IV) can be obtained as rather unstable white solids $MSeF_5$ by dissolving alkali metal fluorides or TlF in SeF_4. The tellurium analogues are best prepared by dissolving MF and TeO_2 in aqueous HF or SeF_4; they are white crystalline solids. The TeF_5^- ion (see diagram) has a distorted square-based pyramidal structure (C_{4v}) in which the Te atom (and pendant lone-pair of electrons) is about 40 pm below the basal plane (cf. TeF_4, Fig. 16.10b). The resemblance to other isoelectronic $MF_5^{n\pm}$ species is illustrated in

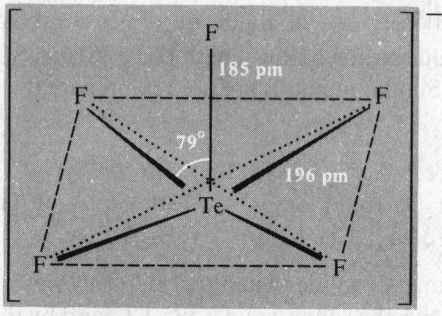

Table 16.6; in each case, the fact that the distance $M-F_{base}$ is greater than $M-F_{apex}$ and that the angle $F_{apex}-M-F_{base}$ is less than $90°$ can be ascribed to repulsive interaction of the basal M–F bonds with the lone-pair of electrons. Attempts to prepare compounds containing the TeF_6^{2-} ion have not been successful though numerous routes have been

TABLE 16.6   *Dimensions of some isoelectronic square-pyramidal species*

Species	$M-F_{apex}$/pm	$M-F_{base}$/pm	Angle $F_{apex}-M-F_{base}$
SbF_5^{2-}	200	204	$83°$
TeF_5^-	185	196	$79°$
BrF_5	168	181	$84°$
XeF_5^+	181	188	$79°$

tried. By contrast, compounds of the complex anions SeX_6^{2-} and TeX_6^{2-} (X = Cl, Br, I) are readily prepared in crystalline form by direct reaction (e.g. $TeX_4 + 2MX$) or by precipitating the complex from a solution of SeO_2 or TeO_2 in aqueous HX. Their most notable feature is a regular octahedral structure despite the fact that they are formally 14-electron species; it appears that with large monatomic ligands of moderate electronegativity the stereochemistry is dominated by inter-ligand repulsions and the lone-pair then either resides in an ns^2 orbital for isolated ions or is delocalized in a low-energy solid-state band.[35] Similar results were noted for octahedral Sn^{II} (p. 443) and Sb^{III} (p. 661).

16.2.4   *Oxohalides and pseudohalides*[1]

Numerous oxohalides of Se^{IV} and Se^{VI} are known. $SeOF_2$ and $SeOCl_2$ are colourless, fuming, volatile liquids, whereas $SeOBr_2$ is a rather less-stable orange solid which decomposes in air above 50° (Table 16.7). The compounds can be conveniently made by

TABLE 16.7   *Some physical properties of selenium oxohalides*

Property	$SeOF_2$	$SeOCl_2$	$SeOBr_2$	SeO_2F_2	$(SeOF_4)_2$	F_5SeOF	$F_5SeOOSF_5$
MP/°C	15	10.9	41.6	−99.5	−12	−54	−62.8
BP/°C	125	177.2	~220(d)	−8.4	65	−29	76.3
Density/g cm^{-3} (T°C)	2.80 (21.5°)	2.445 (16°)	3.38 (50°)	—	—	—	—

reacting SeO_2 with the appropriate tetrahalide and their molecular structure is probably pyramidal (like SOX_2, p. 820). $SeOF_2$ is an aggressive reagent which attacks glass, reacts violently with red phosphorus and with powdered SiO_2 and slowly with Si. In the solid state, X-ray studies have revealed that the pyramidal $SeOF_2$ units are linked by O and F bridges into layers thereby building a distorted octahedral environment around each Se with 3 close contacts (to O and 2F) and 3 (longer) bridging contacts grouped around the lone-pair to neighbouring units.[36] This contrasts with the discrete molecular structure of SOF_2 and affords yet another example of the influence of preferred coordination number on the structure and physical properties of isovalent compounds, e.g. molecular BF_3 and 6-coordinate AlF_3, molecular GeF_4, and the 6-coordinate layer lattice of SnF_4, and, to a less extent, molecular AsF_3 and F-bridged SbF_3. (See also the Group IV dioxides, etc.)

$SeOCl_2$ (Table 16.7) is a useful solvent: it has a high dielectric constant (46.2 at 20°), a high dipole moment (2.62 D in benzene), and an appreciable electrical conductivity (2×10^{-5} ohm^{-1} cm^{-1} at 25°). This last has been ascribed to self-ionic dissociation resulting from chloride-ion transfer:

$$2SeOCl_2 \rightleftharpoons SeOCl^+ + SeOCl_3^-$$

[35] For experimental results and theoretical discussion see I. D. BROWN, The crystal structure of K_2TeBr_6, *Can. J. Chem.* **42**, 2758–67 (1964). D. S. URCH, The stereochemically inert lone pair? A speculation on the bonding in $SbCl_6{}^{3-}$, $SeBr_6{}^{2-}$, $TeBr_6{}^{2-}$, $IF_6{}^-$, and XeF_6, *J. Chem. Soc.* 1964, 5775–81. N. N. GREENWOOD and B. P. STRAUGHAN, Metal–halogen vibrations of simple octahedral ions containing tin, selenium, and tellurium, *J. Chem. Soc.* (A) 1966, 962–4. T. C. GIBB, R. GREATREX, N. N. GREENWOOD, and A. C. SARMA, Tellurium-125 Mössbauer spectra of some tellurium complexes, *J. Chem. Soc.* (A) 1970, 212–17. J. D. DONALDSON, S. D. ROSS, J. SILVER, and P. WATKISS, Solid-state effects in the vibrational spectra of hexahalogeno-stannates(IV) and -tellurates(IV), *J. Chem. Soc. Dalton* 1975, 1980–83, and references therein.

[36] J. C. DEWAN and A. J. EDWARDS, Crystal structure of selenyl difluoride at −35 C, *J. Chem. Soc. Dalton* 1976, 2433–5.

Oxohalides of Se^{VI} are known only for fluorine (Table 16.7). SeO_2F_2 is a readily hydrolysable colourless gas which can be made by fluorinating SeO_3 with SeF_4 (or KBF_4 at 70°) or by reacting $BaSeO_4$ with HSO_3F under reflux at 50°. Its vibrational spectra imply a tetrahedral structure with C_{2v} symmetry as expected. By contrast $SeOF_4$ is a dimer $[F_4Se(\mu\text{-}O)_2SeF_4]$ in which each Se achieves octahedral coordination via the 2 bridging O atoms: the planar central Se_2O_2 ring has Se–O 178 pm and angle Se–O–Se 97.5°, and Se–F_{eq} and Se–F_{ax} are 167 and 170 pm respectively.[37]

Two further oxofluorides of Se^{VI} can be prepared by reaction of SeO_2 with a mixture of F_2/N_2: at 80° the main product is the "hypofluorite" F_5SeOF whilst at 120° the peroxide $F_5SeOOSeF_5$ predominates. The compounds (Table 16.7) can be purified by fractional sublimation and are reactive, volatile, colourless solids. The analogous sulfur compounds were discussed on p. 814. The colourless liquid $F_5SeOSeF_5$ (mp $-85°$, bp 53°) is made by a somewhat more esoteric route as follows:[38]

$$Xe(OSeF_5)_2 \xrightarrow{130} Xe + \tfrac{1}{2}O_2 + F_5SeOSeF_5$$

The corresponding tellurium analogue is made by fluorinating TeO_2 in a copper vessel at 60° using a stream of F_2/N_2 (1:10). It is a colourless, mobile, unreactive liquid, mp $-36.6°$, bp 59.8°. The Se–O–Se angle in $F_5SeOSeF_5$ is 142.4° ($\pm1.9°$) as in the sulfur analogue, and the Te–O–Te angle is very similar (145.5 $\pm$ 2.1°). Both the $-OSeF_5$ group and the $-OTeF_5$ group have very high electronegativities as can be seen, for example, by reactions of the ligand-transfer reagent $[B(OTeF_5)_3]$:[39]

$$IF_5 + B(OTeF_5)_3 \xrightarrow[(-BF_3)]{20} FI(OTeF_5)_4$$

$$XeF_4 + B(OTeF_5)_3 \xrightarrow[(-BF_3)]{0} Xe(OTeF_5)_4 \quad \text{(see also p. 1055)}$$

The oxohalides of Te^{IV} are less well characterized.[3] Representatives are $TeOBr_2$, and the white sublimate crystalline $Te_6^{IV}O_{11}Cl_2$ formed by heating TeO_2 and $TeCl_4$ at 500°. The fluorination of Te in the presence of oxygen yields (in addition to Te_2F_{10}, p. 907) the dense colourless liquids $Te_3^{VI}O_2F_{14}$ and $Te_6^{VI}O_5F_{26}$. Very recently more purposeful synthetic routes have been devised, leading to the isolation and structural characterization of the 6-coordinate Te^{VI} oxofluorides *cis-* and *trans-*$F_4Te(OTeF_5)_2$, *cis-* and *trans-*$F_2Te(OTeF_5)_4$, $FTe(OTeF_5)_5$, and even $Te(OTeF_5)_6$.[40] Similarly, thermolysis of $B(OTeF_5)_3$ at 600° in a flow system yields the oxygen-bridged dimer $Te_2O_2F_8$ analogous to $Se_2O_2F_8$ above. $Te_2O_2F_8$ is a colourless liquid with a garlic-like smell, mp 28°, bp 77.5°. The planar central Te_2O_2 ring has Te–O 192 pm and angle Te–O–Te 99.5°, and again the equatorial Te–F distances (180 pm) are shorter than the axial ones (185 pm).[37]

Pseudohalides of Se in which the role of halogen is played by cyanide, thiocyanate, or selenocyanate are known and, in the case of Se^{II} are much more stable with respect to

[37] H. OBERHAMMER and K. SEPPELT, Molecular structures of $Se_2O_2F_8$ and $Te_2O_2F_8$, *Inorg. Chem.* **18**, 2226–9 (1979).

[38] H. OBERHAMMER and K. SEPPELT, Electron diffraction study of bis(pentafluorosulfur), bis(pentafluoroselenium), and bis(pentafluorotellurium) oxides, *Inorg. Chem.* **17**, 1435–9 (1978).

[39] D. LENZ and K. SEPPELT, Extremely high electronegativities, *Angew. Chem.*, Int. Edn. (Engl.) **17**, 355–6 (1978); Xenon tetrakis(pentafluoroorthotellurate), ibid., pp. 356–61.

[40] D. LENTZ, H. PRITZKOW, and K. SEPPELT, Novel tellurium oxide fluorides, *Inorg. Chem.* **17**, 1926–31 (1978).

disproportionation than are the halides themselves. Examples are $Se(CN)_2$, $Se_2(CN)_2$, $Se(SeCN)_2$, $Se(SCN)_2$, $Se_2(SCN)_2$. Tellurium and polonium analogues include $Te(CN)_2$ and $Po(CN)_4$ but have been much less studied than their Se counterparts. The long-sought tellurocyanate ion $TeCN^-$ has recently been made, and isolated in crystalline form by the use of large counter-cations;[41] as expected, the anion is essentially linear (angle Te–C–N 175°), and the distances Te–C and C–N are 202 and 107 pm respectively.

The selenohalides and tellurohalides of both main-group elements and transition metals have been compared with the corresponding thiohalides in two extensive reviews.[41a]

16.2.5 *Oxides*

The monoxides SeO and TeO have transient existence in flames but can not be isolated as stable solids. PoO has been obtained as a black, easily oxidized solid by the spontaneous radiolytic decomposition of the sulfoxide $PoSO_3$.

The dioxides of all 3 elements are well established and can be obtained by direct combination of the elements. SeO_2 is a white solid which melts in a sealed tube to a yellow liquid at 340° (sublimes at 315°/760 mmHg). It is very soluble in water to give selenous acid H_2SeO_3 from which it can be recovered by dehydration. It is also very soluble (as a trimer) in $SeOCl_2$ and in H_2SO_4 in which it behaves as a weak base. SeO_2 is thermodynamically less stable than either SO_2 or TeO_2 and is readily reduced to the elements by NH_3, N_2H_4, or aqueous SO_2 (but not gaseous SO_2). It also finds use as an oxidizing agent in organic chemistry. In the solid state SeO_2 has a polymeric structure of corner-linked flattened $\{SeO_3\}$ pyramids each carrying a pendant terminal O atom:

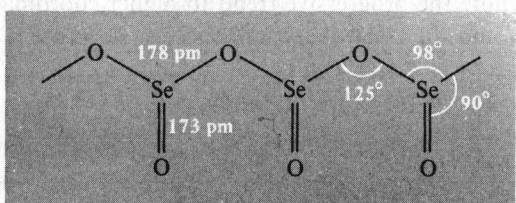

TeO_2 is dimorphic: the yellow, orthorhombic mineral tellurite (β-TeO_2) has a layer structure in which pseudo-trigonal bipyramidal $\{TeO_4\}$ groups form edge-sharing pairs (Fig. 16.11a) which then further aggregate into layers (Fig. 16.11b) by sharing the remaining vertices. By contrast, synthetic α-TeO_2 ("paratellurite") forms colourless tetragonal crystals in which very similar $\{TeO_4\}$ units (Fig. 16.11c) share all vertices (angle Te–O–Te 140°) to form a rutile-like (p. 1120) three-dimensional structure. TeO_2 melts to a red liquid at 733° and is much less volatile than SeO_2. It can be prepared by the action of O_2 on Te, by dehydrating H_2TeO_3, or by thermal decomposition of the basic nitrate above 400°. TeO_2 is virtually insoluble in water and, being amphoteric, shows a minimum in solubility (at pH ~ 4.0). It is very soluble in $SeOCl_2$.

[41] A. S. Foust, Tellurocyanate: X-ray crystal structure of the bis(triphenylphosphoranylid)ammonium salt, *JCS Chem. Comm.* 1979, 414–15.

[41a] M. J. Atherton and J. H. Holloway, Thio-, seleno-, and telluro-halides of the transition metals, *Adv. Inorg. Chem. Radiochem.* **22**, 171–98 (1979). J. Fenner, A. Rabenau, and G. Trageser, Solid-state chemistry of thio-, seleno-, and telluro-halides of representative and transition elements, ibid. **23**, 329–425 (1980).

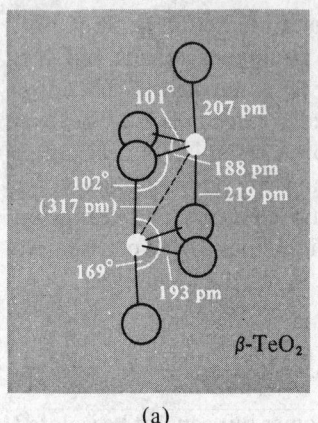 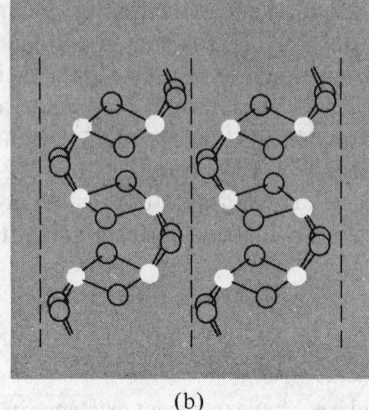 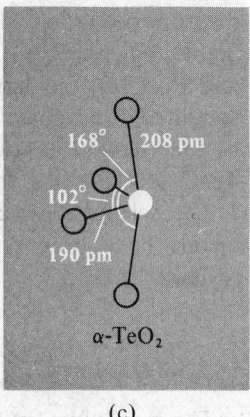

(a)   (b)   (c)

FIG. 16.11 Structural units in crystalline TeO_2: (a) pair of edge-sharing pseudo-trigonal bipyramidal $\{TeO_4\}$ groups in tellurite (β-TeO_2) which aggregate into layers as shown in (b) by sharing the remaining vertices with neighbouring pairs, and (c) the $\{TeO_4\}$ unit in paratellurite (α-TeO_2).

PoO_2 is obtained by direct combination of the elements at 250° or by thermal decomposition of polonium(IV) hydroxide, nitrate, sulfate, or selenate. The yellow (low-temperature) fcc form has a fluorite lattice; it becomes brown when heated and can be sublimed in a stream of O_2 at 885°. However, under reduced pressure it decomposes into the elements at almost 500°. There is also a high-temperature, red, tetragonal form. PoO_2 is amphoteric, though appreciably more basic than TeO_2: e.g. it forms the disulfate $Po(SO_4)_2$ for which no Te analogue is known.

It is instructive to note the progressive trend to higher coordination numbers in the Group VIA dioxides, and the consequent influence on structure:

Compound	SO_2	SeO_2	TeO_2	PoO_2
Coordination number	2	3	4	8
Structure	molecule	chain polymers	layer or 3D	3D "fluorite"

The difficulty of oxidizing Se to the $+6$ state has already been mentioned (p. 891). Indeed, unlike SO_3 and TeO_3, SeO_3 is thermodynamically unstable with respect to the dioxide:

$$SeO_3 \longrightarrow SeO_2 + \tfrac{1}{2}O_2; \quad \Delta H° = -46 \text{ kJ mol}^{-1}$$

Some comparative figures for the standard heats of formation $\Delta H_f°$ are in Table 16.8.

TABLE 16.8 $\Delta H_f°$ (298)/kJ mol^{-1} for MO_n from elements in standard states

SO_2	297	SeO_2	230	TeO_2	325
SO_3	432	SeO_3	184	TeO_3	348

Accordingly, SeO_3 can not be made by direct oxidation of Se or SeO_2 and is even hard to make by the dehydration of H_2SeO_4 with P_2O_5; a better route is to treat anhydrous K_2SeO_3 with SO_3 under reflux, followed by vacuum sublimation at 120°. SeO_3 is a white, hygroscopic solid which melts at 118°, sublimes readily above 100° (40 mmHg), and decomposes above 165°. The crystal structure is built up from cyclic tetramers, Se_4O_{12}, which have a configuration very similar to that of $(PNCl_2)_4$ (p. 626). In the vapour phase, however, there is some dissociation into the monomer. In the molten state SeO_3 is probably polymeric like the isoelectronic polymetaphosphate ions (p. 615).

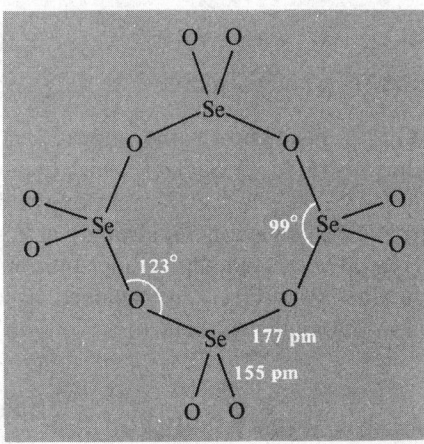

TeO$_3$ exists in two modifications. The yellow-orange α-form and the more stable, less reactive, grey β-form. The α-TeO$_3$ is made by dehydrating Te(OH)$_6$ (p. 915) at 300–360°; the β-TeO$_3$ is made by heating α-TeO$_3$ or Te(OH)$_6$ in a sealed tube in the presence of H$_2$SO$_4$ and O$_2$ for 12 h at 350°. α-TeO$_3$ has a structure like that of FeF$_3$, in which TeO$_6$ octahedra share all vertices to give a 3D lattice. It is unattacked by water, but is a powerful oxidizing agent when heated with a variety of metals or non-metals. It is also soluble in hot concentrated alkalis to form tellurates (p. 916). The β-form is even less reactive but can be cleaved with fused KOH.

PoO$_3$ may have been detected on a tracer scale but has not been characterized with weighable amounts of the element.

16.2.6 *Hydroxides and oxoacids*

The rich oxoacid chemistry of sulfur (pp. 834–854) is not paralleled by the heavier elements of the group. The redox relationships have already been summarized (p. 891). Apart from the dark-brown hydrated monoxide "Po(OH)$_2$", which precipitates when alkali is added to a freshly prepared solution of Po(II), only compounds in the +4 and +6 oxidation states are known.

Selenous acid, O=Se(OH)$_2$, i.e. H$_2$SeO$_3$, and tellurous acid, H$_2$TeO$_3$, are white solids which can readily be dehydrated to the dioxide (e.g. in a stream of dry air). H$_2$SeO$_3$ is best prepared by slow crystallization of an aqueous solution of SeO$_2$ or by oxidation of powdered Se with dilute nitric acid:

$$3Se + 4HNO_3 + H_2O \longrightarrow 3H_2SeO_3 + 4NO$$

The less-stable H$_2$TeO$_3$ is obtained by hydrolysis of a tetrahalide or acidification of a cooled aqueous solution of a telluride. Crystalline H$_2$SeO$_3$ is built up of pyramidal SeO$_3$

groups (Se–O 174 pm) which are hydrogen-bonded to give an orthorhombic layer lattice. The detailed structure of H_2TeO_3 is unknown. Both acids form acid salts $MHSeO_3$ and $MHTeO_3$ by reaction of the appropriate aqueous alkali. The neutral salts M_2SeO_3 and M_2TeO_3 can be obtained similarly or by heating the metal oxide with the appropriate dioxide. Dissociation constants have not been precisely determined but approximate values are:

$$H_2SeO_3: \quad K_1 \sim 3.5 \times 10^{-3}; \quad K_2 \sim 5 \times 10^{-8}$$

$$H_2TeO_3: \quad K_1 \sim 3 \times 10^{-3} \quad\quad K_2 \sim 2 \times 10^{-8}$$

Alkali diselenites $M_2^I Se_2O_5$ are also known and appear (on the basis of vibrational spectroscopy) to contain the ion $[O_2Se\text{–}O\text{–}SeO_2]^{2-}$, with C_{2v} symmetry and a nonlinear Se–O–Se bridge (cf. disulfite $O_3S\text{–}SO_2{}^{2-}$, p. 853). Selenous acid, in contrast to H_2TeO_3, can readily be oxidized to H_2SeO_4 by ozone in strongly acid solution; it is reduced to elementary selenium by H_2S, SO_2, or aqueous iodide solution.

Hydrated polonium dioxide, $PoO(OH)_2$, is obtained as a pale-yellow flocculent precipitate by addition of dilute aqueous alkali to a solution containing Po(IV). It is appreciably acidic, e.g.:

$$PoO(OH)_2 + 2KOH \underset{}{\overset{22°}{\rightleftharpoons}} K_2PoO_3 + 2H_2O; \quad K_a = \frac{[PoO_3{}^{2-}]}{[OH^-]^2} = 8.2 \times 10^{-5}$$

In the $+6$ oxidation state the oxoacids of Se and Te show little resemblance to each other. H_2SeO_4 resembles H_2SO_4 (p. 837) whereas orthotelluric acid $Te(OH)_6$ and polymetatelluric acid $(H_2TeO_4)_n$ are quite different.

Anhydrous H_2SeO_4 is a viscous liquid which crystallizes to a white deliquescent solid (mp 62°). It loses water on being heated and combines readily with SeO_3 to give "pyroselenic acid", $H_2Se_2O_7$ (mp 19°), and triselenic acid, $H_4Se_3O_{11}$ (mp 25°). It also resembles H_2SO_4 in forming several hydrates: $H_2SeO_4 . H_2O$ (mp 26°) and $H_2SeO_4 . 4H_2O$ (52°). Crystalline H_2SeO_4 (d 2.961 g cm^{-3}) comprises tetrahedral SeO_4 groups strongly H-bonded into layers through all 4 O atoms (Se–O 161 pm, O—H$\cdots$O 261–268 pm). H_2SeO_4 can be prepared by several routes:

(i) Oxidation of H_2SeO_3 with H_2O_2, $KMnO_4$, or $HClO_3$, which can be formally represented by the equations:

$$H_2SeO_3 + H_2O_2 \longrightarrow H_2SeO_4 + H_2O$$

$$8H_2SeO_3 + 2KMnO_4 \longrightarrow 5H_2SeO_4 + K_2SeO_3 + 2MnSeO_3 + 3H_2O$$

$$5H_2SeO_3 + 2HClO_3 \longrightarrow 5H_2SeO_4 + Cl_2 + H_2O$$

(ii) Oxidation of Se with chlorine or bromine water, e.g.:

$$Se + 3Cl_2 + 4H_2O \longrightarrow H_2SeO_4 + 6HCl$$

(iii) Action of bromine water on a suspension of silver selenite:

$$Ag_2SeO_3 + Br_2 + H_2O \longrightarrow H_2SeO_4 + 2AgBr$$

The acid dissociation constants of H_2SeO_4 are close to those of H_2SO_4, e.g. $K_2 (H_2SeO_4)$ 1.2×10^{-2}. Selenates resemble sulfates and both acids form a series of alums (p. 88). Selenic acid differs from H_2SO_4, however, in being a strong oxidizing agent: this is perhaps most dramatically shown by its ability to dissolve not only Ag (as does H_2SO_4) but also Au, Pd (and even Pt in the presence of Cl^-):

$$2Au + 6H_2SeO_4 \longrightarrow Au_2(SeO_4)_3 + 3H_2SeO_3 + 3H_2O$$

It oxidizes halide ions (except F^-) to free halogen. Solutions of S, Se, Te, and Po in H_2SeO_4 are brightly coloured (cf. p. 785).

By contrast, the two main forms of telluric acid do not resemble H_2SO_4 and H_2SeO_4 and tellurates are not isomorphous with sulfates and selenates. Orthotelluric acid is a white solid, mp 136°, whose crystal structure is built up of regular octahedral molecules, $Te(OH)_6$. This structure, which persists in solution (Raman spectrum), is also reflected in its chemistry; e.g. breaks occur in the neutralization curve at points corresponding to NaH_5TeO_6, $Na_2H_4TeO_6$, $Na_4H_2TeO_6$, and Na_6TeO_6. Similar salts include Ag_6TeO_6 and Hg_3TeO_6. Moreover diazomethane converts it to the hexamethyl ester $Te(OMe)_6$. In this respect Te resembles its horizontal neighbours in the periodic table Sn, Sb, and I which form the isoelectronic species $[Sn(OH)_6]^{2-}$, $[Sb(OH)_6]^-$, and $IO(OH)_5$. Orthotelluric acid can be prepared by oxidation of powdered Te with chloric acid solution or oxidation of TeO_2 with permanganate in nitric acid:

$$5Te + 6HClO_3 + 12H_2O \longrightarrow 5H_6TeO_6 + 3Cl_2$$

$$5TeO_2 + 2KMnO_4 + 6HNO_3 + 12H_2O \longrightarrow 5H_6TeO_6 + 2KNO_3 + 2Mn(NO_3)_2$$

Alternatively, Te or TeO_2 can be oxidized by CrO_3/HNO_3 or by 30% H_2O_2 under reflux. Acidification of a tellurate with an appropriate precipitating acid offers a further convenient route:

$$BaTeO_4 + H_2SO_4 + 2H_2O \longrightarrow BaSO_4\downarrow + H_6TeO_6$$

$$Ag_2TeO_4 + 2HCl + 2H_2O \longrightarrow 2AgCl\downarrow + H_6TeO_6$$

Crystallization from aqueous solutions below 10° gives the tetrahydrate $H_6TeO_6 \cdot 4H_2O$. The anhydrous acid is stable in air at 100° but above 120° gradually looses water to give polymetatelluric acid and allotelluric acid (see below). Unlike H_2SO_4 and H_2SeO_4, H_6TeO_6 is a weak acid, approximate values of its successive dissociation constants being $K_1 \sim 2 \times 10^{-8}$, $K_2 \sim 10^{-11}$, $K_3 \sim 3 \times 10^{-15}$. It is a fairly strong oxidant, being reduced to the element by SO_2 and to H_2TeO_3 in hot HCl:

$$H_6TeO_6 + 3SO_2 \longrightarrow Te + 3H_2SO_4$$

$$H_6TeO_6 + 2HCl \longrightarrow H_2TeO_3 + 3H_2O + Cl_2$$

Polymetatelluric acid $(H_2TeO_4)_{\sim 10}$ is a white, amorphous hygroscopic powder formed by incomplete dehydration of H_6TeO_6 in air at 160°. Alternatively, in aqueous solution the equilibrium $nH_6TeO_6 \rightleftharpoons (H_2TeO_4)_n + 2nH_2O$ can be shifted to the right by increasing the temperature; rapid cooling then precipitates the sparingly soluble polymetatelluric acid. The structure is unknown but appears to contain 6-coordinate Te. Allotelluric acid "$(H_2TeO_4)_3(H_2O)_4$" is an acid syrup obtained by heating $Te(OH)_6$ in a sealed tube at 305°: the compound has not been obtained pure but tends to revert to H_6TeO_6 at room temperature or to $(H_2TeO_4)_n$ when heated in air; indeed, it may well be a mixture of these two substances.

Tellurates are prepared by fusing a tellurite with a corresponding nitrate, by oxidizing a tellurite with chlorine, or neutralizing telluric acid with a hydroxide.[42]

Numerous peroxoacid or thioacid derivatives of Se and Te have been reported[1] but these add little to the discussion of the reaction chemistry or the structure types already described. Examples are peroxoselenous acid $HOSeO(OOH)$ (stable at $-10°$) and potassium peroxo-orthotellurate $K_2H_4TeO_7$ which also loses oxygen at room temperature. Isomeric selenosulfates, $M_2^ISO_3Se$, and thioselenates, $M_2^ISeO_3S$, are known and can be made by the obvious routes of $[SO_3^{2-}(aq)+Se]$ and $[SeO_3^{2-}(aq)+S]$. Likewise, colourless or yellow-green crystalline selenopolythionates $M_2Se_xS_yO_6$ ($x=1, 2$; $y=2, 4$) and orange-yellow telluropentathionates $M_2^ITeS_4O_6$ are known. X-ray structure analysis reveals unbranched chains with various conformations as found for the polythionates themselves (p. 851).[43] Typical examples are in Fig. 16.12. It will be seen

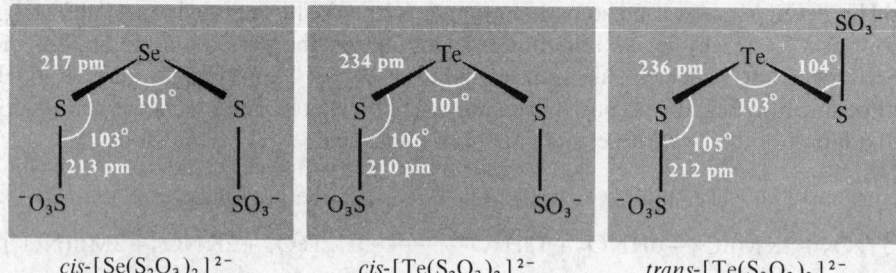

cis-$[Se(S_2O_3)_2]^{2-}$ cis-$[Te(S_2O_3)_2]^{2-}$ $trans$-$[Te(S_2O_3)_2]^{2-}$

FIG. 16.12 Structures and conformations of unbranched chain anions in (a) $Ba[Se(S_2O_3)_2].2H_2O$, (b) $Ba[Te(S_2O_3)_2].2H_2O$, and (c) $(NH_4)_2[Te(S_2O_3)_2]$.

that these compounds contain Se and Te bonded to S rather than O and they therefore form a natural link with the Group VI sulfides to be described in the next section.

16.2.7 *Other inorganic compounds*

The red compound Se_4S_4, obtained by fusing equimolar amounts of the elements, is a covalent molecular species which can be crystallized from benzene. Similar procedures yield Se_2S_6, SeS_7, and TeS_7, all of which are structurally related to S_8 (p. 772).

PoS forms as a black precipitate when H_2S is added to acidic solutions of polonium compounds. Its solubility product is $\sim 5 \times 10^{-29}$. The action of aqueous ammonium sulfide on polonium(IV) hydroxide gives the same compound. It decomposes to the elements when heated to $275°$ under reduced pressure and is of unknown structure.

Se_4N_4 is an orange, shock-sensitive crystalline compound which decomposes violently at $160°$. It resembles its sulfur analogue (p. 856) in being thermochroic (yellow-orange at $-195°$ and red at $+100°$) and in having the same D_{2d} molecular structure. Se_4N_4 can be made by reacting anhydrous NH_3 with $SeBr_4$ (or with SeO_2 at $70°$ under pressure). Tellurium nitride can be prepared similarly; it is a lemon-yellow, violently explosive compound with a formula that appears to be Te_3N_4 rather than Te_4N_4; its structure is unknown.

[42] Ref. 10, pp. 94–7.
[43] A. F. WELLS, *Structural Inorganic Chemistry*, 4th edn., pp. 596–604, Oxford University Press, Oxford, 1975. For more recent examples see *J. Chem. Soc. Dalton* 1978, 1528–32 ($Pb_2Te_3O_8$). *Inorg. Chem.* **19**, 1040–3, 1044–8, 1063–4 (1980) ($SeS_3O_6^{2-}$, $Se_2S_2O_6^{2-}$, $SeS_2O_6^{2-}$).

The increasing basicity of the heavier members of Group VI is reflected in the increasing incidence of oxoacid salts. Thus polonium forms $Po(NO_3)_4.xN_2O_4$, $Po(SO_4)_2.xH_2O$, and a basic sulfate and selenate $2PoO_2 . SO_3$ and $2PoO_2 . SeO_3$ all of which are white, and a hydrated yellow chromate $Po(CrO_4)_2 . xH_2O$. There is also fragmentary information on the precipitation of an insoluble polonium(IV) carbonate, iodate, phosphate, and vanadate.[4] Tellurium(IV) forms a white basic nitrate $2TeO_2 . HNO_3$ and a basic sulfate and selenate $2TeO_2 . XO_3$, and there are indications of a white, hygroscopic basic sulfate of selenium(IV), $SeO_2 . SO_3$ or $SeOSO_4$. Most of these compounds have been prepared by evaporation of aqueous solutions of the oxide or hydrated oxide in the appropriate acid. There is no doubt that more imaginative nonaqueous synthetic routes could be devised, but the likely products seem rather uninteresting and the field has attracted little recent attention.

16.2.8   *Organo-compounds*[(44-47)]

Organoselenium and organotellurium chemistry is a large and expanding field which parallels but is distinct from organosulfur chemistry. The biochemistry of organoselenium compounds has also been much studied (p. 894). Organopolonium chemistry is almost entirely restricted to trace-level experiments because of the charring and decomposition of the compounds by the intense α activity of polonium (p. 886).

The principal classes of organoselenium compound are summarized in the scheme below which indicates the central synthetic role of the selenides R_2Se and diselenides R_2Se_2.[(1)] Detailed discussion of these and related tellurium compounds falls outside the

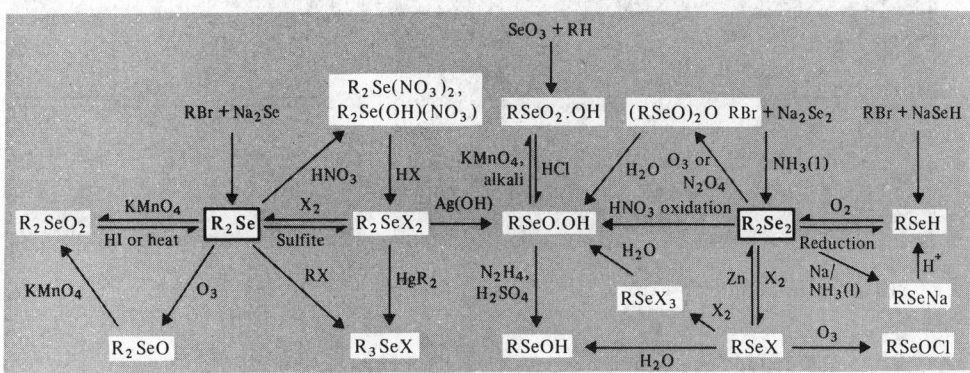

Reaction scheme for the formation of organo-selenium compounds (X = halogen).

scope of the present treatment. Other compounds such as the cyano derivatives (p. 911) and CSe_2, $COSe$, $COTe$, and $CSTe$ (p. 891) have already been briefly mentioned.

[44] K. J. IRGOLIC and M. V. KUDCHADKER, The organic chemistry of selenium, Chap. 8 in ref. 2, pp. 408–545. H. E. GANTHER, Biochemistry of selenium, Chap. 9 in ref. 2, pp. 546–614. W. C. COOPER and J. R. GLOVER, The toxicology of selenium and its compounds, Chap. 11 in ref. 2, pp. 654–74.

[45] R. A. ZINGARO and K. IRGOLIC, Organic compounds of tellurium, Chap. 5 in ref. 3, pp. 184–280. W. C. COOPER, Toxicology of tellurium and its compounds, Chap. 7 in ref. 3, pp. 313–72.

[46] P. D MAGNUS, Organic selenium and tellurium compounds, in D. BARTON and W. D. OLLIS (eds.), *Comprehensive Organic Chemistry*, Vol. 3, Chap. 12, pp. 491–538, Pergamon Press, Oxford, 1979.

[47] Specialist Periodical Reports of the Chemical Society (London), *Organic Compounds of Sulfur, Selenium, and Tellurium*, Vols. 1–5 . . ., 1970–9 . . . .

Tellurocarbonyl derivatives $R^1\!-\!C\!-\!OR^2$ and telluroamides, e.g. $Ph\!-\!C\!-\!NMe_2$

(mp 73°) have recently been prepared[48] and shown to be similar to, though more reactive than, the corresponding seleno derivatives.

Stoichiometry is frequently an inadequate guide to structure in organo-derivatives of Se and Te particularly when other elements (such as halogens) are also present. This arises from the incipient tendency of many of the compounds to undergo ionic dissociation or, conversely, to increase the coordination number of the central atom by dimerization or other oligomeric interactions. Thus Me_3SeI features pyramidal ions $[SeMe_3]^+$ but these are each associated rather closely with 1 iodide which is colinear with 1 Me–Se bond to give a distorted pseudotrigonal bipyramidal configuration (Fig. 16.13a).[43] A regular

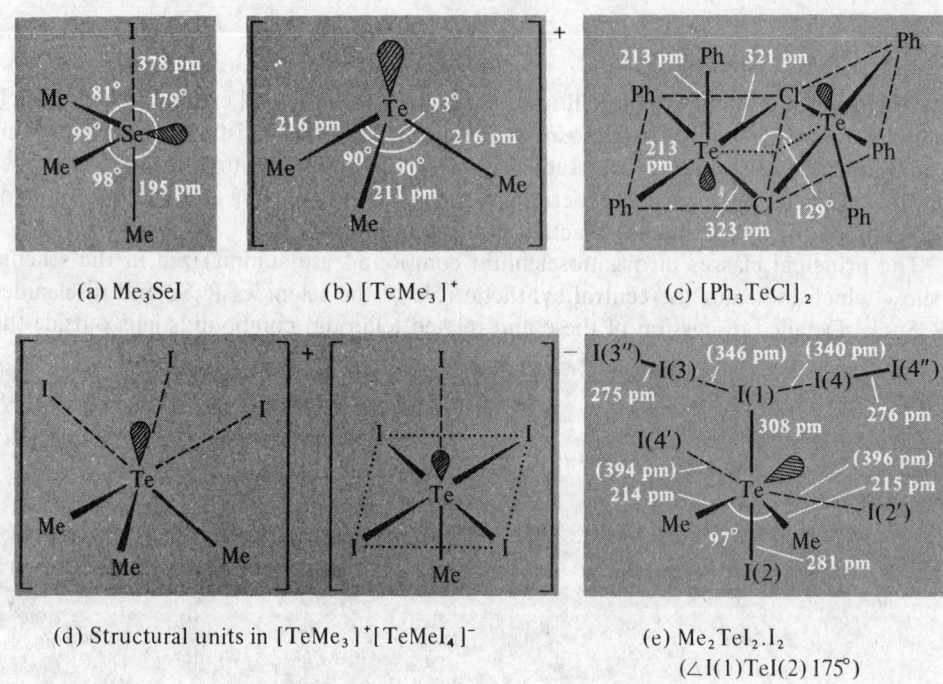

(a) Me_3SeI　　　　　　　(b) $[TeMe_3]^+$　　　　　　　(c) $[Ph_3TeCl]_2$

(d) Structural units in $[TeMe_3]^+[TeMeI_4]^-$　　　　　　(e) $Me_2TeI_2.I_2$
　　　　　　　　　　　　　　　　　　　　　　　　　　　　　($\angle I(1)TeI(2)$ 175°)

Fig. 16.13 Some coordination environments of Se and Te in their organohalides.

pyramidal cation can, however, be obtained by use of a large non-coordinating counter-anion, as in $[TeMe_3]^+[BPh_4]^-$ (Fig. 16.13b).[49] By contrast, Ph_3TeCl is a chloride-bridged dimer with 5-coordinate square-pyramidal Te (Fig. 16.13c).[50] The possibility of isomerism also exists: e.g. 4-coordinate, monomeric molecular Me_2TeI_2 and its ionic counterpart $[TeMe_3]^+[TeMeI_4]^-$ in which interionic interactions make both the cation

[48] K. A. LERSTRUP and L. HENRIKSEN, Preparation of telluroamides and tellurohydrazides, *JCS Chem. Comm.* 1979, 1102–3, and references therein.

[49] R. F. ZIOLO and J. M. TROUP, Threefold configuration of tellurium(IV). Crystal structure of trimethyltellurium tetraphenylborate, *Inorg. Chem.* **18**, 2271–4 (1979).

[50] R. F. ZIOLO and M. EXTINE, Crystal and molecular structure of unsolvated Ph_3TeCl, *Inorg. Chem.* **19**, 2964–7 (1980).

and the anion pseudo-6-coordinate (Fig. 16.13d).[43] Further complications obtrude when the halogen itself is capable of forming polyhalide units in the crystal. Thus reaction of molecular Me_2TeI_2 with iodine readily affords Me_2TeI_4 but the chemical behaviour and spectra of the product give no evidence for oxidation to Te(VI), and X-ray analysis indicates the formation of an adduct $Me_2TeI_2.I_2$ in which the axially disposed iodine atoms of the pseudo-trigonal-bipyramidal Me_2TeI_2 are weakly bonded to molecules of iodine to form a network as shown in Fig. 16.13e[51] (cf. TlI_3, p. 269).

[51] H. PRITZKOW, Crystal and molecular structure of dimethyltellurium tetraiodide, Me_2TeI_4, *Inorg. Chem.* **18**, 311–13 (1979).

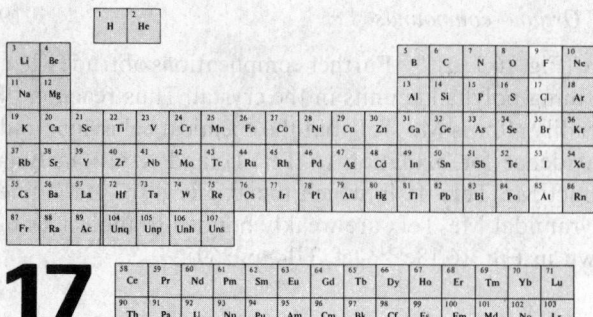

17

The Halogens:
Fluorine, Chlorine,
Bromine, Iodine,
and Astatine

17.1 The Elements

17.1.1 *Introduction*

Compounds of the halogens have been known from earliest times and the elements have played a particularly important role during the past two hundred years in the development of both experimental and theoretical chemistry.[1] Some of this early history is summarized in Table 17.1. The name "halogen" was introduced by J. S. C. Schweigger in 1811 to describe the property of chlorine, at that time unique among the elements, of combining directly with metals to give salts (Greek ἅλς, sea salt, plus the root -γεν, produce). The name has since been extended to cover all five members of Group VIIB of the periodic table.

Fluorine

Fluorine derives its name from the early use of fluorspar (CaF_2) as a flux (Latin *fluor*, flowing). The name was suggested to Sir Humphry Davy by A.-M. Ampère in 1812. The corrosive nature of hydrofluoric acid and the curious property that fluorspar has of emitting light when heated ("fluorescence") were discovered in the seventeenth century. However, all attempts to isolate the element either by chemical reactions or by electrolysis were foiled by the extreme reactivity of free fluorine. Success was finally achieved on 26 June 1886 by H. Moissan who electrolysed a cooled solution of KHF_2 in anhydrous liquid HF, using Pt/Ir electrodes sealed into a platinum U-tube sealed with fluorspar caps: the gas evolved immediately caused crystalline silicon to burst into flames, and Moissan reported the results to the Academy two days later in the following cautious words: "One

---

[1] M. E. WEEKS, *Discovery of the Elements*, 6th edn., Journal of Chemical Education, Easton, 1956, Chap. 27, 'The halogen family', pp. 729–77.

TABLE 17.1 *Early history of the halogens and their compounds*

3000 BC	Archaeological evidence for the use of rock-salt
~400 BC	Written records on salt (ascribed to Herodot)
~200 BC	Use of salt as part payment for services (salary)
~21 AD	Strabo described dyeworks for obtaining tyrian purple (dibromoindigo) in his *Geographica*
~100	Use of salt to purify noble metals
~900	Dilute hydrochloric acid prepared by Arabian alchemist Rhazes
~1200	Development of *aqua regia* (HCl/HNO_3) to dissolve gold—presumably Cl_2 was formed but not studied
1529	Georgius Agricola described use of fluorspar as a flux
~1630	Chlorine recognized as a gas by Belgian physician J. B. van Helmont (see Scheele, 1774)
1648	Concentrated HCl prepared by J. L. Glauber (by heating hydrated $ZnCl_2$ and sand)
1670	H. Schwanhard (Nürnberg) found that CaF_2 + strong acid gave acid vapours (HF) that etched glass (used decoratively)
1678	J. S. Elsholtz described emission of bluish-white light when fluorspar was heated. Also described by J. G. Wallerius, 1750; see 1852 "fluorescence"
1768	First chemical study of fluorite undertaken by A. S. Marggraf
1771	Crude hydrofluoric acid prepared by C. W. Scheele
1772	Gaseous HCl prepared over mercury by J. Priestley
1774	C. W. Scheele prepared and studied gaseous chlorine (MnO_2 + HCl) but thought it a compound
1785	Chemical bleaching (eau de Javel: aqueous $KOH + Cl_2$) introduced by C.-L. Berthollet
1787	N. Leblanc devised a technical process for obtaining NaOH from NaCl (beginnings of the chemical industry)
1798	Bleaching powder patented by C. Tennant (Cl_2 + slaked lime) following preparation of bleaching liquors from Cl_2 and lime solutions by T. Henry (1788)
1801	W. Cruickshank recommended use of Cl_2 as a disinfectant (widely installed in hospitals by 1823 and notably effective in the European cholera epidemic of 1831 and the outbreak of puerperal fever in Vienna, 1845)
1802	Fluoride found in fossil ivory and teeth by D. P. Morichini (soon confirmed by J. J. Berzelius who found it also in bones)
1810	H. Davy announced proof of elementary nature of chlorine to the Royal Society (15 November) and suggested the name "chlorine" (1811)
1811	B. Courtois isolated iodine by sublimation (H_2SO_4 + seaweed ash)
1811	The term "halogen" introduced by J. S. C. Schweigger to denote the (then) unique property of the element chlorine to combine directly with metals to give salts
1812	A.-M. Ampère wrote to H. Davy (12 August) suggesting the name *le fluore* (fluorine) for the presumed new element in CaF_2 and HF (by analogy with *le chlore*, chlorine). Adopted by Davy 1813
1814	Starch/iodine blue colour-reaction described by J.-J. Colin and H.-F. Gaultier de Claubry; developed by F. Stromeyer in same year as an analytical test sensitive to 2–3 ppm iodine
1814	First interhalogen compound (ICl) prepared by J. L. Gay Lussac
1819	Potassium iodide introduced as a remedy for goitre by J.-F. Coindet (Switzerland), the efficacy of extracts from kelp having been known in China and Europe since the sixteenth century
1823	M. Faraday showed that "solid chlorine" was chlorine hydrate (analysis corresponded to $Cl_2. \sim 10H_2O$ with present-day nomenclature). He also liquefied Cl_2 (5 March) by warming the hydrate in a sealed tube
1825	First iodine containing mineral (AgI) identified by A. M. del Río (Mexico) and N.-L. Vauquelin (Paris)
1826	Bromine isolated by A.-J. Balard (aged 23 y)
1835	L. J. M. Daguerre's photographic process (silver plate sensitized by exposure to iodine vapour)
~1840	Introduction of (light sensitive) AgBr into photography
1840	Iodine (as iodate) found in Chilean saltpetre by A. A. Hayes
1841	First mineral bromide (bromyrite, AgBr) discovered in Mexico by P. Berthier—later also found in Chile and France
1851	Diaphragm cell for electrolytic generation of Cl_2 invented by C. Watt (London) but lack of practical electric generators delayed commercial exploitation until 1886–90 (Matthes and Weber of Duisberg)
1852	The term "fluorescence" coined by G. G. Stokes for light emission from fluorspar
1857	Bromide therapy introduced by Lacock as a sedative and anticonvulsant for treatment of epilepsy
1858	Discovery of Stassfurt salt deposits opened the way for bromine production (for photography and medicine) as a by-product of potash

1863	Alkali Act (UK) prohibited atmospheric pollution and enforced the condensation of by-product HCl from the Leblanc process
1886	H. Moissan isolated F_2 by electrolysis of KHF_2/HF (26 June) after over 70 y of unsuccessful attempts by others (Nobel Prize for Chemistry 1906—died 2 months later)
1892–5	H. Y. Castner (US/UK) and C. Kelner (Vienna) independently developed commercial mercury-cathode cell for chlor-alkali production

can indeed make various hypotheses on the nature of the liberated gas; the simplest would be that *we are in the presence of fluorine*, but it would be possible, of course, that it might be a perfluoride of hydrogen or even a mixture of hydrofluoric acid and ozone. . . ." For this achievement, which had eluded some of the finest experimental chemists of the nineteenth century [including H. Davy (1813–14), G. Aimé (1833), M. Faraday (1834), C. J. and T. Knox (1836), P. Louyet (1846), E. Frémy (1854), H. Kammerer (1862), and G. Gore (1870)], and for his development of the electric furnace, Moissan was awarded the Nobel Prize for Chemistry in 1906.

Fluorine technology and the applications of fluorine-containing compounds have developed dramatically during the twentieth century.[2] Some highlights are included in Table 17.2 and will be discussed more fully in later sections. Noteworthy events are the

TABLE 17.2   *Halogens in the twentieth century*

~1900	First manufacture of inorganic fluorides for aluminium industry
1902	J. C. Downs (of E. I. du Pont de Nemours, Delaware) patented the first practical molten-salt cell for Cl_2 and Na metal
1908	HCl shown to be present in gastric juices of animals by P. Sommerfeld
1909	H. Friedlander showed that Tyrian Purple from *Murex brandaris* was 6,6'-dibromoindigo (synthesized by F. Sachs in 1904)
1920	Bromine detected in blood and organs of humans and other animals and birds by A. Damiens
1928	T. Midgley, A. L. Henne, and R. R. McNary synthesized Freon (CCl_2F_2) as a non-flammable, non-toxic gas for refrigeration
1928	ClF made by O. Ruff *et al.* ($Cl_2 + F_2$ at 250°)
1930	IF_7 made by O. Ruff and R. Keim (IF_5 having been made in 1871 by G. Gore)
1930+	H. T. Deans *et al.* put the correlation between decreased incidence of dental caries and the presence of F^- in drinking water on a quantitative basis
1931	First bulk shipment of commercial anhydrous HF (USA)
1938	R. J. Plunket discovered Teflon (polytetrafluoroethylene, PTFE)
1940	Astatine made via $^{209}Bi(\alpha,2n)$ by D. R. Corson, K. R. Mackenzie, and E. Segré
1940/1	Industrial production of $F_2(g)$ begun (in the UK and the USA for manufacture of UF_6 and in Germany for ClF_3)
1950	Chemical shifts for ^{19}F and nmr signals for ^{35}Cl and ^{37}Cl first observed
1962	ClF_5 (the last halogen fluoride to be made) synthesized by W. Maya
1965	LaF_3 crystals developed by J. W. Ross and M. S. Frant as the first non-glass membrane electrode (for ion-selective determination of F^-)
1965	Perchlorate ion established as a monodentate ligand (to Co) by X-ray crystallography, following earlier spectrosopic and conductimetric indications of coordination (1961)
1968	Perbromates first prepared by E. H. Appelman
1967	First example of $\mu(\eta^1,\eta^1)$-ClO_4^- as a bidentate bridging ligand (to Ag^+); chelating η^2-ClO_4^- identified in 1974
1971	HOF first isolated in weighable amounts (p. 1003)

[2] *Kirk-Othmer Encyclopedia of Chemical Technology*, 3rd edn., 1980: Fluorine; inorganic fluorine compounds, Vol. 10, pp. 630–828; Organic fluorine compounds, Vol. 10, pp. 829–962 and Vol. 11, pp. 1–81.

development of inert fluorinated oils, greases, and polymers: Freon gases such as CCl_2F_2 (1928) were specifically developed for refrigeration engineering; others were used as propellants in pressurized dispensers and aerosols; and the non-stick plastic polytetra-fluoroethylene (PTFE or Teflon) was made in 1938. Inorganic fluorides, especially for the aluminium industry (p. 247) have been increasingly exploited from about 1900, and from 1940 UF_6 has been used in gaseous diffusion plants for the separation of uranium isotopes for nuclear reactor technology. The great oxidizing strength of F_2 and many of its compounds with N and O have attracted the attention of rocket engineers and there have been growing large-scale industrial applications of anhydrous HF (p. 945).

The aggressive nature of HF fumes and solutions has been known since Schwanhard of Nürnberg used them for the decorative etching of glass. Hydrofluoric acid inflicts excruciatingly painful skin burns (p. 946) and any compound that might hydrolyse to form HF should be treated with great caution.[3] Maximum allowable concentration for continuous exposure to HF gas is 2–3 ppm (cf. HCN 10 ppm). The free element itself is even more toxic, maximum allowable concentration for a daily 8-h exposure being 0.1 ppm. Low concentrations of fluoride ion in drinking water have been known to provide excellent protection against dental caries since the classical work of H. T. Dean and his colleagues in the early 1930s; as there are no deleterious effects, even over many years, providing the concentration is kept at or below 1 ppm, fluoridation has been a recommended procedure in several countries for many years (p. 945). However, at 2–3 ppm a brown mottling of teeth can occur and at 50 ppm harmful toxic effects are noted. Ingestion of 150 mg of NaF can cause nausea, vomiting, diarrhoea, and acute abdominal pains though complete recovery is rapid following intravenous or intra-muscular injection of calcium ions.

Chlorine

Chlorine was the first of the halogens to be isolated and common salt (NaCl) has been known from earliest times (see Table 17.1). Its efficacy in human diet was well recognized in classical antiquity and there are numerous references to its importance in the Bible. On occasion salt was used as part payment for the services of Roman generals and military tribunes (salary) and, indeed, it is an essential ingredient in mammalian diets (p. 75). The alchemical use of *aqua regia* (HCl/HNO_3) to dissolve gold is also well documented from the thirteenth century onwards. Concentrated hydrochloric acid was prepared by J. L. Glauber in 1648 by heating hydrated $ZnCl_2$ and sand in a retort and the pure gas, free of water, was collected over mercury by J. Priestley in 1772. This was closely followed by the isolation of gaseous chlorine by C. W. Scheele in 1774: he obtained the gas by oxidizing nascent HCl with MnO_2 in a reaction which could now formally be written as:

$$4NaCl + 2H_2SO_4 + MnO_2 \xrightarrow{\text{heat}} 2Na_2SO_4 + MnCl_2 + 2H_2O + Cl_2$$

However, Scheele believed he had prepared a *compound* (dephlogisticated marine acid air) and the misconception was compounded by C.-L. Berthollet who showed in 1785 that the action of chlorine on water releases oxygen: $[Cl_2(g) + H_2O \rightarrow 2HCl(soln) + \frac{1}{2}O_2(g)]$; he concluded that chlorine was a loose compound of HCl and oxygen and called it

---

[3] A. J. FINKEL, Treatment of hydrogen fluoride injuries, *Adv. Fluorine Chem.* **7**, 199–203 (1973).

oxymuriatic acid.† The two decades from 1790 to 1810 were characterized by two major advances in chemical theory: Lavoisier's demolition of the phlogiston theory of combustion, and Davy's refutation of Lavoisier's contention that oxygen is a necessary constituent of all acids. Only when both these transformations had been achieved could the elementary nature of chlorine and the true composition of hydrochloric acid be appreciated, though some further time was to elapse (Dalton, Avogadro, Cannizaro) before gaseous chlorine was universally recognized to consist of diatomic molecules, Cl_2, rather than single atoms, Cl. The name, proposed by Davy in 1811, refers to the colour of the gas (Greek χλωρός, *chloros*, yellowish or light green—cf. chlorophyl).

The bleaching power of Cl_2 was discovered by Scheele in his early work (1774) and was put to technical use by Berthollet in 1785. This was a major advance on the previous time-consuming, labour-intensive, weather-dependent method of solar bleaching, and numerous patents followed (see Table 17.1). Indeed, the use of chlorine as a bleach remains one of its principal industrial applications (bleaching powder, elemental chlorine, hypochlorite solutions, chlorine dioxide, chloramines, etc.).[4] Another all-pervading use of chlorine, as a disinfectant and germicide, also dates from this period (1801), and the chlorination of domestic water supplies is now almost universal in developed countries. Again, as with fluoride, higher concentrations are toxic to humans: the gas is detectable by smell at 3 ppm, causes throat irritation at 15 ppm, coughing at 30 ppm, and rapid death at 1000 ppm. Prolonged exposure to concentrations above 1 ppm should be avoided.

Sodium chloride, by far the most abundant compound of chlorine, occurs in extensive evaporite deposits, saline lakes and brines, and in the ocean (p. 927). It has played a dominant role in the chemical industry since its inception in the late eighteenth century (p. 79). The now defunct Leblanc process for obtaining NaOH from NaCl signalled the beginnings of large-scale chemical manufacture, and NaCl remains virtually the sole source of chlorine and hydrochloric acid for the vast present-day chlorine-chemicals industry.[4] This embraces not only the large-scale production and distribution of Cl_2 and HCl, but also the manufacture of chlorinated methanes and ethanes, vinyl chloride, innumerable chlorinated organics, aluminium trichloride catalysts, and the chlorides of Mg, Ti, Zr, Hf, etc., for production of the metals. Details of many of these processes are to be found either in other chapters or in later sections of the present chapter.

Bromine

The magnificent purple pigment referred to in the Bible[5] and known to the Romans as Tyrian purple after the Phoenician port of Tyre (Lebanon), was shown by H. Friedländer

---

† *Muriatic* acid and *marine* acid were synonymous terms for what is now called hydrochloric acid, thus signifying its relation to the sodium chloride contained in brine (Latin *muria*) or sea water (Latin *mare*). Both names were strongly criticized by H. Davy in a scathing paper entitled "Some reflections on the nomenclature of oxymuriatic compounds" in *Phil. Trans. R. Soc.* for 1811: "To call a body which is not known to contain oxygen, and which cannot contain muriatic acid, oxymuriatic acid, is contrary to the principles of that nomenclature in which it is adopted; and an alteration of its seems necessary to assist the progress of the discussion, and to diffuse just ideas on the subject. If the great discoverer of this substance (i.e. Scheele) had signified it by any simple name it would have been proper to have referred to it; but 'dephlogisticated marine acid' is a term which can hardly be adopted in the present advanced area of the science. After consulting some of the most eminent chemical philosophers in the country, it has been judged most proper to suggest a name founded upon one of its most obvious and characteristic properties—its colour, and to call it *Chlorine*."

• ⁴ J. S. Sconce, *Chlorine: Its Manufacture, Properties and Uses*, Reinhold, New York, 1962, 901 pp.
⁵ Holy Bible, Ezekiel **27**:7, 16.

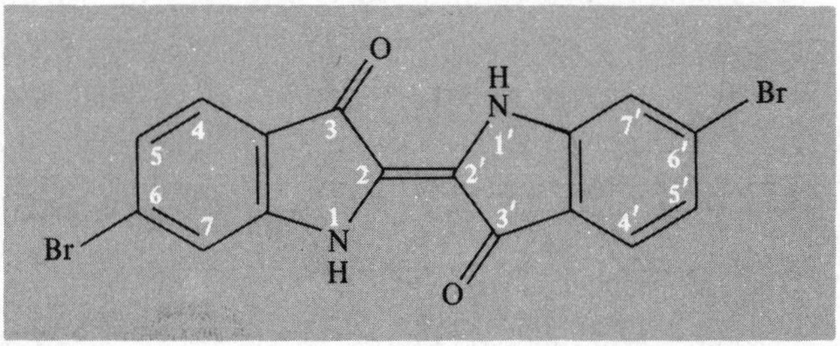

in 1909 to be 6,6'-dibromoindigo. This precious dye was extracted in the early days from the small purple snail *Murex brandaris*, as many as 12 000 snails being required to prepare 1.5 g of dye. The element was isolated by A.-J. Balard in 1826 from the mother liquors remaining after the crystallization of sodium chloride and sulfate from the waters of the Montpellier salt marshes; the liquor is rich in $MgBr_2$, and the young Balard, then 23 y of age, noticed the deep yellow coloration that developed on addition of chlorine water. Extraction with ether and KOH, followed by treatment of the resulting KBr with H_2SO_4/MnO_2, yielded the element as a red liquid. Astonishingly rapid progress was possible in establishing the chemistry of bromine and in recognizing its elemental nature because of its similarity to chlorine and iodine (which had been isolated 15 y earlier). Indeed, J. von Liebig had missed discovering the element several years previously by misidentifying a sample of it as iodine monochloride.[1] Balard had proposed the name *muride*, but this was not accepted by the French Academy, and the element was named bromine (Greek βρῶμος, stink) because of its unpleasant, penetrating odour. It is perhaps ironic that the name fluorine had already been pre-empted for the element in CaF_2 and HF (p. 920) since bromine, as the only non-metallic element that is liquid at room temperature, would pre-eminently have deserved the name.

The first mineral found to contain bromine (bromyrite, AgBr) was discovered in Mexico in 1841, and industrial production of bromides followed the discovery of the giant Stassfurt potash deposits in 1858. The major use at that time was in photography and medicine: AgBr had been introduced as the light-sensitive agent in photography about 1840, and the use of KBr as a sedative and anti-convulsant in the treatment of epilepsy was begun in 1857. The scale of the present-day production of bromine and bromine chemicals will become clear in later sections of this chapter.

Iodine

The lustrous, purple-black metallic sheen of resublimed crystalline iodine was first observed by the industrial chemist B. Courtois in 1811, and the name, proposed by J. L. Gay Lussac in 1813, reflects this most characteristic property (Greek ἰώδης, violet-coloured). Courtois obtained the element by treating the ash of seaweed (which had been calcined to extract saltpetre and potash) with concentrated sulfuric acid. Extracts of the brown kelps and seaweeds *Fucus* and *Laminaria* had long been known to be effective for the treatment of goitre and it was not long before J. F. Coindet and others introduced pure KI as a remedy in 1819.[6] It is now known that the thyroid gland produces the growth-regulating hormone thyroxine, an iodinated aminoacid:

⁶ E. BOOTH, The history of the seaweed industry. Part 3. The iodine industry, *Chem. Ind.* (*Lond.*) 1979, 31 and 52–55.

If the necessary iodine input is insufficient the thyroid gland enlarges in an attempt to garner more iodine: addition of 0.01% NaI to table salt (iodized salt) prevents this condition. Tincture of iodine is a useful antiseptic.

The first iodine-containing mineral (AgI) was discovered in Mexico in 1825 but the discovery of iodate as an impurity in Chilean saltpetre in 1840 proved to be more significant industrially. The Chilean nitrate deposits provided the largest proportion of the world's iodine until overtaken in the later 1960s by Japanese production from natural brines (pp. 927, 933).

In addition to its uses in photography and medicine, iodine and its compounds have been much exploited in volumetric analysis (iodometry and iodimetry, p. 1011). Organoiodine compounds have also played a notable part in the development of synthetic organic chemistry, being the first compounds used in A. W. von Hofmann's alkylation of amines (1850), A. W. Williamson's synthesis of ethers (1851), A. Wurtz's coupling reactions (1855), and V. Grignard's reagents (1900).

Astatine

From its position in the periodic table, all isotopes of element 85 would be expected to be radioactive. Those isotopes that occur in the natural radioactive series all have half-lives of less than 1 min and thus occur in negligible amounts in nature (p. 928). Astatine (Greek ἄστατ-ος, unstable) was first made and characterized by D. R. Corson, K. R. Mackenzie, and E. Segré in 1940: they synthesized the isotope ^{211}At ($t_{\frac{1}{2}}$ 7.21 h) by bombarding ^{209}Bi with α-particles in a large cyclotron:

$$^{209}_{83}Bi + {}^{4}_{2}He \longrightarrow {}^{211}_{85}At + 2{}^{1}_{0}n$$

In all, some 24 isotopes from ^{196}At to ^{219}At have now been prepared by various routes but all are short-lived. The only ones besides ^{211}At having half-lives longer than 1 h are ^{207}At (1.8 h), ^{208}At (1.63 h), ^{209}At (5.4 h), and ^{210}At (8.3 h): this means that weighable amounts of astatine or its compounds cannot be isolated, and nothing is known of the bulk physical properties of the element. For example, the least-unstable isotope (^{210}At) has a specific activity corresponding to 2 curies per μg. The largest preparations of astatine to date have involved about 0.05 μg and our knowledge of the chemistry of this element comes from extremely elegant tracer experiments, typically in the concentration range 10^{-11}–10^{-15} M. The most concentrated aqueous solutions of the element or its compounds ever investigated were only ~10^{-8} M.

17.1.2 **Abundance and distribution**

Because of their reactivity, the halogens do not occur in the free elemental state but they are both widespread and abundant in the form of their ions, X^-. Iodine also occurs as iodate (see below). In addition to large halide mineral deposits, particularly of NaCl and KCl, there are vast quantities of chloride and bromide in ocean waters and brines.

Fluorine is the thirteenth element in order of abundance in crustal rocks of the earth, occurring to the extent of 544 ppm (cf. twelfth Mn, 1060 ppm; fourteenth Ba, 390 ppm; fifteenth Sr, 384 ppm). The three most important minerals are fluorite CaF_2, cryolite Na_3AlF_6, and fluorapatite $Ca_5(PO_4)_3F$. Of these, however, only fluorite is extensively processed for recovery of fluorine and its compounds (p. 944). Cryolite is a rare mineral, the only commercial deposit being in Greenland, and most of the Na_3AlF_6 needed for the huge aluminium industry (p. 246) is now synthetic. By far the largest amount of fluorine in the earth's crust is in the form of fluorapatite, but this contains only about 3.5% by weight of fluorine and the mineral is processed almost exclusively for its phosphate content. Despite this, about 7% of the domestic requirement for fluorine compounds in the USA was obtained from fluorosilicic acid recovered as a by-product of the huge phosphate industry (pp. 549, 600). Minor occurrences of fluorine are in the rare minerals topaz $Al_2SiO_4(OH,F)_2$, sellaite MgF_2, villiaumite NaF, and bastnaesite $(Ce,La)(CO_3)F$ (but see p. 1425). The insolubility of alkaline-earth and other fluorides precludes their occurrence at commercially useful concentration in ocean water (1.2 ppm) and brines.

Chlorine is the twentieth most abundant element in crustal rocks where it occurs to the extent of 126 ppm (cf. nineteenth V, 136 ppm, and twenty-first Cr, 122 ppm). The vast evaporite deposits of NaCl and other chloride minerals have already been described (pp. 77, 83). Dwarfing these, however, are the inconceivably vast reserves in ocean waters (p. 79) where more than half the total average salinity of 3.4 wt% is due to chloride ions (1.9 wt%). Smaller quantities, though at higher concentrations, occur in certain inland seas and in subterranean brine wells, e.g. Great Salt Lake (23% NaCl), Dead Sea (8.0% NaCl, 13.0% $MgCl_2$, 3.5% $CaCl_2$).

Bromine is substantially less abundant in crustal rocks than either fluorine or chlorine; at 2.5 ppm it is forty-sixth in order of abundance being similar to Hf 2.8, Cs 2.6, U 2.3, Eu 2.1, and Sn 2.1 ppm. Like chlorine, the largest natural source of bromine is the oceans, which contain $\sim 6.5 \times 10^{-3}\%$, i.e. 65 ppm or 65 mg/l. The mass ratio Cl:Br is $\sim 300:1$ in the oceans, corresponding to an atomic ratio of $\sim 660:1$. Salt lakes and brine wells are also rich sources of bromine, and these are usually proportionately richer in bromine than are the oceans: the atom ratio Cl:Br spans the range $\sim 200-700$. Typical bromide-ion concentrations in such waters are: Dead Sea 0.4% (4 g/l), Sakskoe Ozoro (Crimea) 0.28%, Searle's Lake (California) 0.085%.

Iodine is considerably less abundant than the lighter halogens both in the earth's crust and in the hydrosphere. It comprises 0.46 ppm of the crustal rocks and is sixtieth in order of abundance (cf. Tl 0.7, Tm 0.5, In 0.24, Sb 0.2). It occurs but rarely as iodide minerals, and commercial deposits are usually as iodates, e.g. lautarite, $Ca(IO_3)_2$, and dietzeite, $7Ca(IO_3)_2.8CaCrO_4$. Thus the caliche nitrate beds of Chile contain iodine in this form ($\sim 0.02-1$ wt% I). These mine workings soon replaced calcined seaweeds as the main source of iodine during the last century, but have recently been themselves overtaken by iodine recovered from brines. Brines associated with oil-well drillings in Louisiana and California were found to contain 30–40 ppm iodine in the 1920s, and independent subterranean brines were located at Midland, Michigan, in the 1960s, and in Oklahoma (1977), which is now the main US source. Natural brine wells in Japan (up to 100 ppm I) were discovered after the Second World War, and exploitation of these now ensures Japan first place among the world's iodine producers. The concentration of iodine in ocean waters is only 0.05 ppm, too low for commercial recovery, though brown seaweeds of the *Laminaria* family (and to a lesser extent *Fucus*) can concentrate this up to 0.45% of their dry weight (see above).

The fugitive radioactive element astatine can hardly be said to exist in nature though the punctillious would rightly point to its temporary participation in the natural radioactive series. Thus ^{219}At ($t_{\frac{1}{2}}$ 54 s) occurs as a rare and inconspicuous branch (4×10^{-3}%) of another minor branch (1.2%) of the ^{235}U series (see scheme). Another branch

(5×10^{-4}%) at ^{215}Po yields ^{215}At by β emission before itself decaying by α emission ($t_{\frac{1}{2}}$ 1.0×10^{-4} s), likewise ^{218}At($t_{\frac{1}{2}}$ ~2 s) is a descendant of ^{238}U, and traces have been detected of ^{217}At ($t_{\frac{1}{2}}$ 0.0323 s) and ^{216}At ($t_{\frac{1}{2}}$ 3.0×10^{-4} s). Estimates suggest that the outermost kilometre of the earth's crust contains no more than 44 mg of astatine compared with 15 g of francium (p. 76) or the relatively abundant polonium (2500 tonnes) and actinium (7000 tonnes). Astatine can therefore be regarded as the rarest naturally occurring terrestrial element.

17.1.3 *Production and uses of the elements*

The only practicable method of preparing F_2 gas is Moissan's original procedure based on the electrolysis of KF dissolved in anhydrous HF. Moissan used a mole ratio KF:HF of about 1:13, but this has a high vapour pressure of HF and had to be operated at $-24°$. Electrolyte systems having mole ratios of 1:2 and 1:1 melt at ~72° and ~240°C respectively and have much lower vapour pressures of HF; accordingly these compositions are now favoured and are used in the so-called medium-temperature and high-temperature generators. Indeed, the medium-temperature cells are now almost universally employed, being preferred over the high-temperature cells because (a) they have a lower pressure of HF gas above the cell, (b) there are fewer corrosion problems, (c) the anode has a longer life, and (d) the composition of the electrolyte can vary within fairly wide limits without impairing the operating conditions or efficiency. The highly corrosive nature of the electrolyte, coupled with the aggressive oxidizing power of F_2, pose considerable problems of handling, and these are exacerbated by the explosive reaction of F_2 with its co-product H_2, so that accidental mixing of the gases must be prevented at all costs. Scrupulous absence of grease and other flamable contaminants must also be ensured since they can lead to spectacular fires which puncture the protective fluoride coating of the metal containers and cause the whole system to enflame. Another hazard in early generators was the formation of explosive graphite-fluorine compounds at the anode (p. 309). All these problems have now been overcome and F_2 can be routinely generated with safety both in the laboratory and on a large industrial scale.[2, 7] A typical generator (Fig. 17.1a) consists of a mild-steel pot (cathode) containing the electrolyte KF.2HF which is kept at 80–100°C either by a heating jacket when the cell is quiescent or by a cooling system when the cell is working. The anode consists of a central rod of compacted, ungraphitized carbon, and the product gases are kept separate by a skirt or diaphragm dipping below the electrolyte surface. The temperature is automatically controlled, as is the level of the electrolyte by controlled addition of make-up anhydrous HF. Laboratory generators usually operate at about 10–50 A whereas industrial production, employing banks of cells, may operate at 4000–6000 A and 8–12 V (Fig. 17.1b). An individual cell in such a bank might typically be $3.0 \times 0.8 \times 0.6$ m and hold 1 tonne of electrolyte; it might have 12 anode assemblies each holding two anode blocks and produce 3–4 kg F_2 per hour. A large-scale plant can produce *ca.* 9 tonnes of liquefied F_2 per day. The total annual production in the USA and Canada is about 5000 tonnes, and a similar though somewhat smaller amount is produced in the UK annually.

Cylinders of F_2 are now commercially available in various sizes from 230-g to 2.7-kg capacity; 1979 price ~$50 per kg. Liquid F_2 is shipped in tank trucks of 2.27 tonnes capacity, the container being itself cooled by a jacket of liquid N_2 which boils 8° below F_2. Alternatively, it can be converted to ClF_3, bp 11.7°C (p. 969), which is easier to handle and transport than F_2. In fact, some 70–80% of the elemental F_2 produced is used captively for

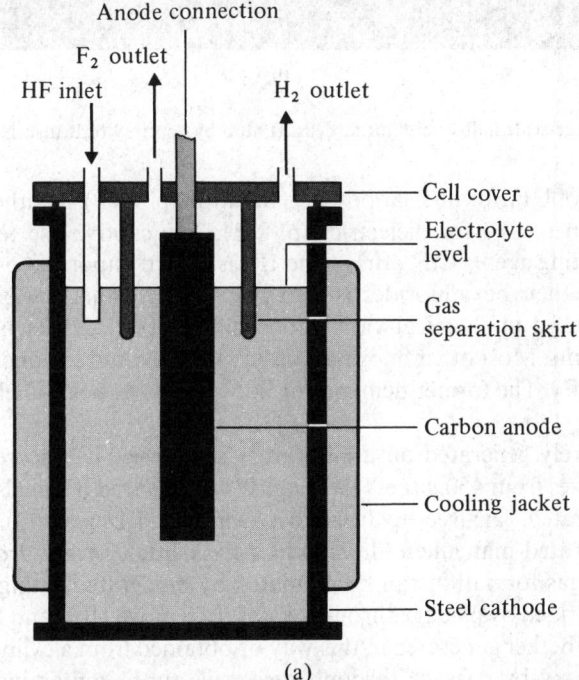

(a)

FIG. 17.1(a)   Schematic diagram of an electrolytic fluorine-generating cell.

[7] H. C. FIELDING and B. E. LEE, Hydrofluoric acid, inorganic fluorides, and fluorine, in R. THOMPSON (ed.), *The Modern Inorganic Chemicals Industry*, pp. 149–67, Chemical Society Special Publication No. 31, 1977.

(b)

Fig. 17.1 (b) Large-scale industrial fluorine cells. (Photo by courtesy of British Nuclear Fuels Ltd.)

the manufacture of UF_6 for nuclear power generation (p. 1460). Another important use is in the production of SF_6 for dielectrics (p. 813); The captive use to manufacture the versatile fluorinating agents ClF_3, BrF_3, and IF_5 is a third important outlet. Fluorination of W and Re to their hexafluorides is also industrially important since these volatile compounds are used in chemical vapour deposition of W and Re films on intricately shaped components. Most other fluorinations of inorganic and organic compounds avoid the direct use of F_2. The former demand for liquid F_2 as a rocket-fuel oxidizer has now virtually ceased.

Chlorine is rarely generated on a laboratory scale since it is so readily available in cylinders of all sizes from 450 g (net) to 70 kg. When required it can also be generated by adding concentrated, air-free hydrochloric acid (d 1.16 g cm^{-3}) dropwise on to precipitated hydrated manganese dioxide in a flask fitted with a dropping funnel and outlet tube: the gas formation can be regulated by moderate heating and the Cl_2 thus formed can be purified by passage through water (to remove HCl) and H_2SO_4 (to remove H_2O). The gas, whether generated in this way or obtained from a cylinder, can be further purified if necessary by passage through successive tubes containing CaO and P_2O_5, followed by condensation in a bath cooled by solid CO_2 and fractionation in a vacuum line.

Industrial production of Cl_2 and chlorine chemicals is on a vast scale and comprises a major section of the heavy chemical industry.[4, 8, 9] Some aspects have already been discussed on p. 924, and further details are in the Panel.

Industrial Production and Uses of Chlorine[4, 8, 9, 10]

The large-scale production of Cl_2 is invariably achieved by the electrolytic oxidation of the chloride ion. Natural brines or aqueous solutions of NaCl can be electrolysed in an asbestos diaphragm cell or a mercury cathode cell, though these latter are being phased out for environmental and other reasons (p. 1422). Electrolysis of molten NaCl is also carried out on a large scale: in this case the co-product is Na rather than NaOH. Electrolysis of by-product HCl is also used where this is cheaply available. World production of Cl_2 in 1979 exceeded 35 million tonnes. Production is dominated by the USA, but large tonnages are produced in all industrial countries: USA 11.0, FRG 6.4, USSR 3.2, Japan 3.1, France 2.6, UK 2.0, Italy 1.4, Canada 1.0, others ~5 million tonne. Cl_2 was ranked eighth among the large-volume chemicals manufactured in the USA during 1979. Diaphragm cells predominated though there is a growing interest in membrane cells in which the anolyte and catholyte are separated by a porous Nafion membrane (Nafion is a copolymer of tetrafluoroethylene and a perfluorosulfonylethoxy ether and the membrane is reinforced with a Teflon mesh).[11] In addition to cylinders of varying capacity up to 70 kg, chlorine can be transported in drums (865 kg), tank wagons (road: 15 tonnes; rail 27–82 tonnes), or barges (600–1100 tonnes).

The three main categories of use for Cl_2 are:

(a) Production of organic compounds by chlorination and/or oxychlorination using a fluidized bed of copper chloride catalyst (e.g. chloromethanes, chloroethanes, chloroethenes, etc.; pre-eminent amongst these are ethylenedichloride and vinyl chloride monomer, which in the USA alone are produced on a scale of 5.5 and 3.5 million tonnes respectively). For example:

$$CH_2{=}CH_2 + Cl_2 \xrightarrow[\text{100 ppm}]{FeCl_3} CH_2ClCH_2Cl \qquad \text{ethylene dichloride}$$

$$CH_2ClCH_2Cl \xrightarrow{450-500^\circ} HCl + CH_2{=}CHCl \qquad \text{vinyl chloride}$$

$$\left.\begin{array}{l} 2CuCl + 2HCl + \tfrac{1}{2}O_2 \xrightarrow{250^\circ} 2CuCl_2 + H_2O \\[2mm] CH_2{=}CH_2 + 2CuCl_2 \xrightarrow{250^\circ} CH_2ClCH_2Cl + 2CuCl \end{array}\right\} \text{catalytic cycle}$$

Production of chlorinated organic compounds accounts for about 70% of the Cl_2 produced.

(b) Bleaches (for paper, pulp, and textiles) sanitation and disinfection of municipal water supplies and swimming pools, sewage treatment and control. These uses account for about 20% of the Cl_2 produced.

(c) Production of inorganic compounds, notably HCl, Cl_2O, HOCl, $NaClO_3$, chlorinated isocyanurates, $AlCl_3$, $SiCl_4$, $SnCl_4$, PCl_3, PCl_5, $POCl_3$, $AsCl_3$, $SbCl_3$, $SbCl_5$, $BiCl_3$, S_2Cl_2, SCl_2, $SOCl_2$, ClF_3, ICl, ICl_3, $TiCl_3$, $TiCl_4$, $MoCl_5$, $FeCl_3$, $ZnCl_2$, Hg_2Cl_2, $HgCl_2$, etc. (see index for page references to production and uses). About 10% of Cl_2 production is used to manufacture inorganic chemicals.

[8] R. W. PURCELL, The chor-alkali industry, in ref. 7, pp. 106–3.   CAMPBELL, Chlorine and chlorination, ibid., pp. 134–48.

[9] Ref. 2, Vol. 1 (1978), pp. 799–883, Alkali and chlorine products.

[10] J. S. ROBINSON (ed.), *Chlorine Production Processes*, Noyes Data Corp., Park Ridge, New Jersey, 1981, 388 pp.

[11] Anon., Chloralkali membrane set for market, *Chem. Eng. News* 20 March 1978, 20–22.

Bromine is invariably made on an industrial scale by oxidation of bromide ion with Cl_2. The four main sources of Br^- are Arkansas brines (4000–5000 ppm) which account for 75% of US production, Michigan brines (~ 2000 ppm) which account for the remaining US production, the Dead Sea (4000–5000 ppm), and ocean waters (65 ppm). Following the oxidation of Br^- the Br_2 is removed from the solution either by passage of steam ("steaming out") or air ("blowing out"), and then condensed and purified. Although apparently simple, these unit operations must deal with highly reactive and corrosive materials, and the industrial processes have been ingeniously developed and refined to give optimum yields at the lowest possible operating costs.[12, 13]

World production of Br_2 is about 270 000 tonnes pa, i.e. about one-hundredth of the scale of the chlorine industry. The main countries for which production figures (1975) are known are (tonnes): USA 185 000, UK 28 000, Israel 18 000, France 15 000, Japan 11 000, FRG 4300, Italy 3500, Spain 450, India 270. However, this does not include several large producers such as the USSR for which data are not available. The production capacity of Israel has recently increased almost threefold because of expanded facilities on the Dead Sea. Historically, bromine was shipped in individual 3-kg (net) bottles to minimize damage due to breakage, but during the 1960s bulk transport in monel metal drums (100-kg capacity) or lead-lined tanks (6–27 tonnes) was developed and these are now used for transport by road, rail, and ship.

The industrial usage of bromine has been dominated by the single compound ethylene dibromide which has been (with ethylene dichloride) a valuable gasoline (petrol) additive where it acts as a scavenger for lead from the anti-knock additive $PbEt_4$. Environmental legislation has dramatically reduced the amount of leaded petrol produced and, accordingly, ethylene dibromide, which accounted for 90% of US bromine production in 1955, declined to 75% a decade later and now represents less than 50% of the total bromine consumption in the USA. Fortunately this decline has been matched by a steady increase in other applications, and the industry worldwide has shown a modest growth. Most of these large-volume applications involve organic compounds, notably MeBr, which is one of the most effective nematocides known (i.e. kills worms) and is also used as a general pesticide (herbicide, fungicide, and insecticide). Ethylene dibromide and dibromochloropropane are also much used as pesticides. Bromine compounds are also extensively used as fire retardants, especially for fibres, carpets, rugs, and plastics; they are about 3–4 times as effective (weight for weight) as chlorocompounds which gives them a substantial cost advantage. Bromo compounds can be added to the fibre either during the spinning operation or after fabrication: one of the most widely used retardants is tris(dibromopropyl)phosphate, $(Br_2C_3H_5O)_3PO$. Alternatively, the flame retardant can be included as a co-monomer before polymerization, e.g. vinyl bromide for acrylic fibres and the bis(2-hydroxyethyl ether) of tetrabromobisphenol A for polyester fibres.

Other uses of bromo-organics include high-density drilling fluids, dyestuffs, and pharmaceuticals. Bromine is also used in water sanitation and to synthesize a wide range of inorganic compounds, e.g. AgBr for photography, HBr, alkali metal bromides, bromates, etc. (see later sections). A composite estimate of the scale of various end uses of bromine worldwide in 1975 is shown in Table 17.3[12]

[12] R. B. McDonald and W. R. Merriman, The bromine and bromine chemicals industry, ref. 7, pp. 168–82.
[13] Ref. 2, Vol. 4 (1978), pp. 226–63, Bromine and bromine compounds.

TABLE 17.3   *End-use pattern for compounds manufactured with* Br_2

Use	10^3 tonnes	%	Use	10^3 tonnes	%
Fuel additive	150	53.6	Dyestuffs	10	3.6
Agriculture	50	17.9	Drilling fluids	5	1.8
Flame retardants	25	8.9	Photography	5	1.8
Cleaning agents	10	3.6	Pharmaceuticals	5	1.8
Water sanitation	10	3.6	Other organic and inorganic syntheses	10	3.6

The commercial recovery of iodine on an industrial scale depends on the particular source of the element. From natural brines, such as those at Midland (Michigan) or in Japan, chlorine oxidation followed by air blow-out as for bromine (above) is much used, the final purification being by resublimation. Alternatively the brine, after clarification, can be treated with just sufficient $AgNO_3$ to precipitate the AgI which is then treated with clean scrap iron or steel to form metallic Ag and a solution of FeI_2; the Ag is redissolved in HNO_3 for recycling and the solution is treated with Cl_2 to liberate the I_2:

$$I^-(\text{brine}) \xrightarrow{\;AgNO_3\;} AgI \downarrow \xrightarrow{\;Fe\;} Ag \downarrow + FeI_2(\text{soln}) \xrightarrow{\;Cl_2\;} FeCl_2(\text{soln}) + I_2$$

$$\text{(recycle via } HNO_3\text{)}$$

The newest process to be developed oxidizes the brine with Cl_2 and then treats the solution with an ion-exchange resin: the iodine is adsorbed in the form of polyiodide which can be eluted with alkali followed by NaCl to regenerate the column.

Recovery of iodine from Chilean saltpetre differs entirely from its recovery from brine since it is present as iodate. $NaIO_3$ is extracted from the caliche and is allowed to accumulate in the mother liquors from the crystallization of $NaNO_3$ until its concentration is about 6 g/l. Part is then drawn off and treated with the stoichiometric amount of sodium hydrogen sulfite required to reduce it to iodide:

$$2IO_3^- + 6HSO_3^- \longrightarrow 2I^- + 6SO_4^{2-} + 6H^+$$

The resulting acidic mixture is treated with just sufficient fresh mother liquor to liberate all the contained iodine:

$$5I^- + IO_3^- + 6H^+ \longrightarrow 3I_2 \downarrow + 3H_2O$$

The precipitated I_2 is filtered off and the iodine-free filtrate returned to the nitrate-leaching cycle after neutralization of any excess acid with Na_2CO_3.

World production of I_2 in 1977 was 10 900 tonnes, the dominant producers being Japan 56%, USSR 16%, Chile 15%, and USA 13%. Crude iodine is packed in double polythene-lined fibre drums of 45–90-kg capacity. Resublimed iodine is transported in lined fibre drums (11.3 kg) or in bottles containing 0.11, 0.45, or 2.26 kg. Unlike Cl_2 and Br_2, iodine has no predominant commercial outlet. About 50% is incorporated into a wide variety of organic compounds and about 15% each is accounted for as resublimed iodine, KI, and other inorganics. The end uses include catalysts for synthetic rubber manufacture, animal-

and fowl-feed supplements, stabilizers, dyestuffs, colourants, and pigments for inks, pharmaceuticals, sanitary uses (tincture of iodine, etc.), and photographic chemicals for high-speed negatives. Uses of iodine compounds as smog inhibitors and cloud-seeding agents are small. In analytical chemistry $KHgI_3$ forms the basis for Nessler's reagent for the detection of NH_3, and Cu_2HgI_4 was used in Mayer's reagent for alkaloids. Iodides and iodates are standard reagents in quantitative volumetric analysis (p. 1011). $AgHgI_4$ has the highest ionic electrical conductivity of any known solid at room temperature but this has not yet been exploited on a large scale in any solid-state device.

17.1.4 *Atomic and physical properties*

The halogens are volatile, diatomic elements whose colour increases steadily with increase in atomic number. Fluorine is a pale yellow gas which condenses to a canary yellow liquid, bp $-188.1°C$ (intermediate between N_2, bp $-195.8°$, and O_2, bp $-183.0°C$). Chlorine is a greenish-yellow gas, bp $-34.0°$, and bromine a dark-red mobile liquid, bp $59.5°$: interestingly the colour of both elements diminishes with decrease in temperature and at $-195°$ Cl_2 is almost colourless and Br_2 pale yellow. Iodine is a lustrous, black, crystalline solid, mp $113.6°$, which sublimes readily and boils at $185.2°C$.

Atomic properties are summarized in Table 17.4 and some physical properties are in Table 17.5. As befits their odd atomic numbers, the halogens have few naturally occurring

TABLE 17.4   *Atomic properties of the halogens*

Property	F	Cl	Br	I	At
Atomic number	9	17	35	53	85
Number of stable isotopes	1	2	2	1	0
Atomic weight	18.998 403	35.453	79.904	126.9045	(210)
Electronic configuration	$[He]2s^22p^5$	$[Ne]3s^23p^5$	$[Ar]3d^{10}4s^24p^5$	$[Kr]4d^{10}5s^25p^5$	$[Xe]4f^{14}5d^{10}6s^26p^5$
Ionization energy/kJ mol^{-1}	1680.6	1255.7	1142.7	1008.7	[926]
Electron affinity/kJ mol^{-1}	332.6	348.5	324.7	295.5	[270]
ΔH_{dissoc}/kJ mol$(X_2)^{-1}$	158.8	242.58	192.77	151.10	—
Ionic radius, X^-/pm	133	184	196	220	—
van der Waals radius/pm	135	180	195	215	—
Distance X–X in X_2/pm	143	199	228	266	—

TABLE 17.5   *Physical properties of the halogens*

Property	F_2	Cl_2	Br_2	I_2
MP/°C	-218.6	-101.0	-7.25	$113.6^{(a)}$
BP/°C	-188.1	-34.0	59.5	$185.2^{(a)}$
d (liquid, $T°C$)/g cm^{-3}	1.513 ($-188°$)	1.655 ($-70°$)	3.187 ($0°$)	3.960$^{(b)}$ ($120°$)
ΔH_{fusion}/kJ mol$(X_2)^{-1}$	0.51	6.41	10.57	15.52
ΔH_{vap}/kJ mol$(X_2)^{-1}$	6.54	20.41	29.56	41.95
Temperature (°C) for 1% dissoc at 1 atm	765	975	775	575

(a) Solid iodine has a vapour pressure of 0.31 mmHg at 25°C and 90.5 mmHg at the mp (113.6°).
(b) Solid iodine has a density of 4.940 g cm^{-3} at 20°C.

isotopes (p. 4). Only one isotope each of F and I occurs in nature and the atomic weights of these elements are therefore known very accurately indeed (p. 20). Chlorine has two naturally occurring isotopes (^{35}Cl 75.77%, ^{37}Cl 24.23%) as also does bromine (^{79}Br 50.69%, ^{81}Br 49.31%). All isotopes of At are radioactive. The ionization energies of the halogen atoms show the expected trend to lower values with increase in atomic number. The electronic configuration of each atom (ns^2np^5) is one p electron less than that of the next succeeding noble gas, and energy is evolved in the reaction $X(g) + e^- \rightarrow X^-(g)$. The electron affinity, which traditionally (though misleadingly) is given a positive sign despite the negative enthalpy change in the above reaction, is maximum for Cl, the value for F being intermediate between those for Cl and Br. Even more noticeable is the small enthalpy of dissociation for F_2 which is similar to that of I_2 and less than two-thirds of the value for Cl_2.[14] In this connection it can be noted that N–N single bonds in hydrazines are weaker than the corresponding P–P bonds and that O–O single bonds in peroxides are weaker than the corresponding S–S bonds. This was explained (R. S. Mulliken and others, 1955) by postulating that partial pd hybridization imparts some double-bond character to the formal P–P, S–S, and Cl–Cl single bonds thereby making them stronger than their first-row counterparts. However, following C. A. Coulson and others (1962), it seems unnecessary to invoke substantial d-orbital participation and the weakness of the F–F single bond is then ascribed to decreased overlap of bonding orbitals, appreciable internuclear repulsion, and the relatively large electron–electron repulsions of the lone-pairs which are much closer together in F_2 than in Cl_2.[14a] The rapid diminution of bond-dissociation energies in the sequence $N_2 \gg O_2 \gg F_2$ is, of course, due to successive filling of the antibonding orbitals (p. 706), thus reducing the formal bond order from triple in N≡N to double and single in O=O and F–F respectively.

Radioactive isotopes of the lighter halogens have found use in the study of isotope-exchange reactions and the mechanisms of various other reactions.[15, 16] The properties of some of the most used isotopes are in Table 17.6. Many of these isotopes are available commercially. A fuller treatment with detailed references of the use of radioactive isotopes of the halogens, including exchange reactions, tracer studies of other reactions, studies of diffusion phenomena, radiochemical methods of analysis, physiological and biochemical applications, and uses in technology and industry is available.[17] Excited states of ^{127}I and ^{129}I have also been used extensively in Mössbauer spectroscopy.[18]

The nuclear spin of the stable isotopes of the halogens has been exploited in nmr spectroscopy. The use of ^{19}F in particular, with its 100% abundance, convenient spin of $\frac{1}{2}$, and excellent sensitivity, has resulted in a vast and continually expanding literature since

[14] J. BERKOWITZ and A. C. WAHL, The dissociation energy of fluorine, *Adv. Fluorine Chem.* **7**, 147–74 (1973). A. A. WOOLF, Thermochemistry of inorganic fluorine compounds, *Adv. Inorg. Chem. Radiochem.* **24**, 1–55 (1981). J. J. TURNER, Physical and spectroscopic properties of the halogens, *MTP International Review of Science: Inorganic Chemistry Series* 1, Vol. 3, pp. 253–91, Butterworths, London, 1972.

[14a] P. POLITZER, Anomalous properties of fluorine, *J. Am. Chem. Soc.* **91**, 6235–7 (1969); Some anomalous properties of oxygen and nitrogen, *Inorg. Chem.* **16**, 3350–1 (1977).

[15] M. F. A. DOVE and D. B. SOWERBY, Isotopic halogen exchange reactions, in V. GUTMANN (ed.), *Halogen Chemistry*, Vol. 1, pp. 41–132, Academic Press, London, 1967.

[16] R. H. HERBER (ed.), *Inorganic Isotopic Syntheses*, W. H. Benjamin, New York, 1962; Radio-chlorine (B. J. MASTERS), pp. 215–26; Iodine-131 (M. KAHN), pp. 227–42.

[17] A. J. DOWNS and C. J. ADAMS, Chlorine, bromine, iodine, and astatine, in J. C. BAILAR, H. J. EMELÉUS, R. S. NYHOLM, and A. F. TROTMAN-DICKENSON, *Comprehensive Inorganic Chemistry*, Vol. 2, pp. 1148–61 (Isotopes), Pergamon Press, Oxford, 1973.

[18] N. N. GREENWOOD and T. C. GIBB, *Mössbauer Spectroscopy*, pp. 462–82, Chapman & Hall, London, 1971.

TABLE 17.6 *Some radioactive isotopes of the halogens*

Isotope	Nuclear spin and parity	Half-life	Principal mode of decay (E/MeV)	Principal source
^{18}F	1+	109.8 min	β^+ (0.649)	^{19}F(n,2n)
^{36}Cl	2+	3.00×10^5 y	β^- (0.714)	^{35}Cl(n,γ)
^{38}Cl	2−	37.3 min	β^- (4.81, 1.11, 2.77)	^{37}Cl(n,γ)
^{80m}Br	5−	4.42 h	γ (internal trans) (0.086)	^{79}Br(n,γ)
^{80}Br	1+	17.6 min	β^- (2.02, 1.35)	^{80m}Br(IT)
^{82}Br	5−	35.34 h	β^- (0.44)	^{81}Br(n,γ)
^{125}I	$\frac{5}{2}$+	60.2 d	Electron capture (0.035)	^{123}Sb(α,2n), ^{124}Te(d,n), or ^{125}Xe(β^-)
^{128}I	1+	24.99 min	β^- (2.12, 1.66)	^{127}I(n,γ)
^{129}I	$\frac{7}{2}$+	1.6×10^7 y	β^- (0.189)	U fission
^{131}I	$\frac{7}{2}$+	8.04 d	β^- (0.806)	^{130}Te(n,γ), U or Pu fission

^{19}F chemical shifts were first observed in 1950.[19] The resonances for ^{35}Cl and ^{37}Cl were also first observed in 1950.[20] Appropriate nuclear parameters are in Table 17.7. From this it is clear that the ^{19}F resonance can be observed with high receptivity at a frequency fairly close to that for ^{1}H. Furthermore, since $I < 1$ there is no nuclear quadrupole moment and hence no quadrupolar broadening of the resonance. The observed range of ^{19}F chemical shifts is more than an order of magnitude greater than for ^{1}H and spans more than 800 ppm of the resonance frequency. The signal moves to higher frequency with increasing electronegativity and oxidation state of the attached atom thus following the usual trends.[21] Results are regularly reviewed, the most recent report on ^{19}F (for 1976–8) occupying over 500 pages and including over 750 references.[22] For other halogens, as seen from Table 17.7, the nuclear spin I is greater than $\frac{1}{2}$ which means that the nuclear charge distribution is non-spherical; this results in a nuclear quadrupole moment, and resonance broadening due to quadrupolar relaxation severely restricts the use of the technique except for the halide ions X^- or for tetrahedral species such as ClO_4^- which have zero electric field gradient at the halogen nucleus. The receptivity is also much less than for ^{1}H or ^{19}F which accordingly renders observation difficult. Despite these technical problems, much useful information has been obtained, especially in physicochemical and biological investigations.[23] The quadrupole moments of Cl, Br, and I have also been

[19] W. C. DICKENSON, Dependence of the ^{19}F nuclear resonance position on chemical compound, *Phys. Rev.* **77**, 736–7 (1950). H. S. GUTOWSKY and C. J. HOFFMAN, Chemical shifts in the magnetic resonance of ^{19}F, *Phys. Rev.* **80**, 110–11 (1950).

[20] W. G. PROCTOR and F. C. YU, On the magnetic moments of ^{14}N, ^{15}N, ^{37}Cl, ^{59}Co, and ^{55}Mn, *Phys. Rev.* **77**, 716–17 (1950).

[21] J. W. EMSLEY, J. FEENEY, and L. H. SUTCLIFFE, *High Resolution Nuclear Magnetic Resonance Spectroscopy*, Vols. 1 and 2, Pergamon Press, Oxford, 1966, Chap. 11, Fluorine-19, pp. 871–968.

[22] *Annual Reports on NMR Spectroscopy*, Vol. 1 (1968)–Vol. 10b (1980) (Fluorine).

[23] B. LINDMAN and S. FORSEN, The halogens—chlorine, bromine, and iodine, Chap. 13 in R. K. HARRIS and B. E. MANN (eds.), *NMR and the Periodic Table*, pp. 421–38, Academic Press, London, 1978. B. LINDMAN and S. FORSEN, Chlorine, bromine, and iodine nmr. Physicochemical and biological applications, Vol. 12 of P. DIEHL, E. FLUCK, and R. KOSFELD (eds.), *NMR Basic Principles and Progress*, Springer-Verlag, Berlin, 1976, 365 pp.

TABLE 17.7   *Nuclear magnetic resonance parameters for the halogen isotopes*

Isotope	Nuclear spin quantum no. I	NMR frequency rel to $^1H(SiMe_4)$ $= 100.000$	Relative receptivity $D_p{}^{(a)}$	Nuclear quadrupole moment $Q/$ $(e\ 10^{-28}\ m^2)$
^1H	1/2	100.000	1.000	0
^{19}F	1/2	94.094	0.8328	0
^{35}Cl	3/2	9.798	3.55×10^{-3}	-7.89×10^{-2}
^{37}Cl	3/2	8.156	6.44×10^{-4}	-6.21×10^{-2}
$(^{79}Br)^{(b)}$	3/2	25.054	3.97×10^{-2}	0.33
^{81}Br	3/2	27.006	4.87×10^{-2}	0.28
^{127}I	5/2	20.007	9.34×10^{-2}	-0.69

(a) Receptivity D is proportional to $\gamma^3 N I(I+1)$ where γ is the magnetogyric ratio, N the natural abundance of the isotope, and I the nuclear spin quantum number; D_p is the receptivity relative to that of the proton taken as 1.000.

(b) Less-favourable isotope.

exploited successfully in nuclear quadrupole resonance studies of halogen-containing compounds in the solid state.[24]

The molecular and bulk properties of the halogens, as distinct from their atomic and nuclear properties, were summarized in Table 17.5 and have to some extent already been briefly discussed. The high volatility and relatively low enthalpy of vaporization reflect the diatomic molecular structure of these elements. In the solid state the molecules align to give a layer lattice: F_2 has two modifications (a low-temperature, α-form and a higher-temperature, β-form) neither of which resembles the orthorhombic layer lattice of the isostructural Cl_2, Br_2, and I_2. The layer lattice is illustrated for I_2 in Fig. 17.2: the I-I distance of 271.5 pm is appreciably longer than in gaseous I_2 (266.6 pm) and the closest interatomic approach between the molecules is 350 pm within the layer and 427 pm

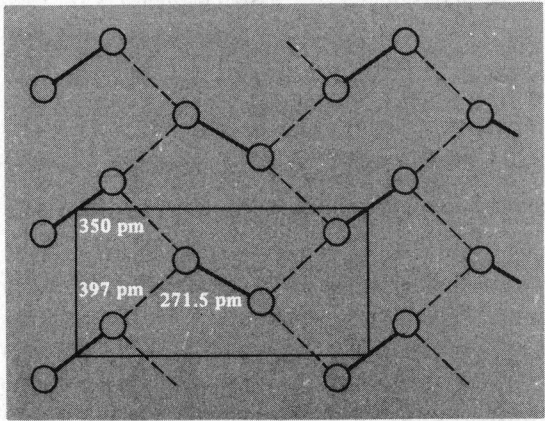

FIG. 17.2   Two-dimensional network of I_2 molecules in the (100) or *bc* plane of solid iodine.

24 T. P. DAS and E. L. HAHN, *Nuclear Quadrupole Resonance Spectroscopy*, Academic Press, New York, 1958, 223 pp; E. A. C. LUCKEN, *Nuclear Quadrupole Coupling Constants*, Academic Press, London, 1969, 360 pp.

between layers. These values are compared with similar data for the other halogens in Table 17.8 from which two further features of interest emerge: (a) the intralayer intermolecular distances Cl⋯Cl and Br⋯Br are almost identical, and (b) the differences between intra- and inter-layer X⋯X distances decreases with increase in atomic number. (Fluorine is not directly comparable because of its differing structure.)

TABLE 17.8   *Interatomic distances in crystalline halogens (pm)*

X	X–X	X⋯X Within layer	X⋯X Between layers	Ratio X⋯X / X–X
F	149	324	284	(1.91)
Cl	198	332, 382	374	1.68
Br	227	331, 379	399	1.46
I	272	350, 397	427	1.29

As expected from their structures, the elements are poor conductors of electricity: solid F_2 and Cl_2 have negligible conductivity and Br_2 has a value of $\sim 5 \times 10^{-13}$ ohm^{-1} cm^{-1} just below the mp. Iodine single crystals at room temperature have a conductivity of 5×10^{-12} ohm^{-1} cm^{-1} perpendicular to the *bc* plane but this increases to 1.7×10^{-8} ohm^{-1} cm^{-1} within the plane; indeed, the element is a two-dimensional semiconductor with a band gap $E_g \sim 1.3$ eV (125 kJ mol^{-1}). Even more remarkably, when crystals of iodine are compressed they become metallic, and at 350 kbar have a conductivity of $\sim 10^4$ ohm^{-1} cm^{-1}.[25] The metallic nature of the conductivity is confirmed by its negative temperature coefficient.

The ease of dissociation of the X_2 molecules follows closely the values of the enthalpy of dissociation since the entropy change for the reaction is almost independent of X. Thus F_2 at 1 atm pressure is 1% dissociated into atoms at 765°C but a temperature of 975°C is required to achieve the same degree of dissociation for Cl_2; thereafter, the required temperature drops to 775°C for Br_2 and 575°C for I_2 (see also next section for atomic halogens).

17.1.5   *Chemical reactivity and trends*

General reactivity and stereochemistry

Fluorine is the most reactive of all elements. It forms compounds, under appropriate conditions, with every other element in the periodic table except He, Ar, and Ne, frequently combining directly and with such vigour that the reaction becomes explosive. Some elements such as O_2 and N_2 react less readily with fluorine (pp. 750, 503) and some bulk metals (e.g. Al, Fe, Ni, Cu) acquire a protective fluoride coating, though all metals react exothermically when powdered and/or heated. For example, powdered Fe (0.84 mm size, 20 mesh) is not attacked by liquid F_2 whereas at 0.14 mm size (100 mesh) it ignites and burns violently. Perhaps the most striking example of the reactivity of F_2 is the ease

[25] A. S. BALCHIN and H. G. DRICKAMER, Effect of pressure on the resistance of iodine and selenium, *J. Chem. Phys.* **34**, 1948–9 (1961).

with which it reacts directly with Xe under mild conditions to produce crystalline xenon fluorides (p. 1049). This great reactivity of F_2 can be related to its small dissociation energy (p. 935) (which leads to low activation energies of reaction), and to the great strength of the bonds that fluorine forms with other elements. Both factors in turn can be related to the small size of the F atom and ensure that enthalpies of fluorination are much greater than those of other halogenations. Some typical average bond energies (kJ mol^{-1}) illustrating these points are:

X	XX	HX	BX_3	AlX_3	CX_4
F	158	574	645	582	456
Cl	243	428	444	427	327
Br	193	363	368	360	272
I	151	294	272	285	239

The tendency for F_2 to give F^- ions in solution is also much greater than for the other halogens as indicated by the steady decrease in oxidation potential ($E°$) for the reaction $X_2(soln) + 2e^- \rightleftharpoons 2X^-(aq)$:

X_2	F_2	Cl_2	Br_2	I_2	At_2
$E°/V$	2.866	1.395	1.087	0.615	~0.3

The corresponding free energy changes can be calculated from the relation $\Delta G = -nE°F$ where $n = 2$ and $F = 96.485$ kJ mol^{-1}. Note that $E°(F_2/2F^-)$ is greater than the decomposition potential for water (p. 737). Note also the different sequence of values for $E°(X_2/2X^-)$ and for the electron affinities of X^-(g) (p. 934). A similar "anomaly" was observed (p. 86) for $E°(Li^+/Li)$ and the ionization energy of Li(g), and in both cases the reason is the same, namely the enhanced enthalpy of hydration of the smaller ions. Other redox properties of the halogens are compared on p. 1000.

It follows from the preceding paragraph that F_2 is an extremely strong oxidizing element that can engender unusually high oxidation states in the elements with which it reacts, e.g. IF_7, PtF_6, PuF_6, BiF_5, TbF_4, CmF_4, $KAg^{III}F_4$, and AgF_2. Indeed, fluorine (like the other first-row elements Li, Be, B, C, N, and O) is atypical of the elements in its group and for the same reasons. For all 7 elements deviations from extrapolated trends can be explained in terms of three factors:

(1) their atoms are small;
(2) their electrons are tightly held and not so readily ionized or distorted (polarized) as in later members of the group;
(3) they have no low-lying d orbitals available for bonding.

Thus the ionization energy I_M is much greater for F than for the other halogens, thereby making formal positive oxidation states virtually impossible to attain. Accordingly, fluorine is exclusively univalent and its compounds are formed either by gain of 1 electron to give F^- ($2s^2 2p^6$) or by sharing 1 electron in a covalent single bond. Note, however, that the presence of lone-pairs permits both the fluoride ion itself and also certain molecular fluorides to act as Lewis bases in which the coordination number of F is greater than 1, e.g.

it is 2 for the bridging F atoms in $As_2F_{11}^-$, $Sb_3F_{16}^-$, Nb_4F_{20}, $(HF)_n$, and $(BeF_2)_x$. The coordination number of F^- can rise to 3 (planar) in compounds with the rutile structure (e.g. MgF_2, MnF_2, FeF_2, CoF_2, NiF_2, ZnF_2, and PdF_2). Likewise, fourfold coordination (tetrahedral) is found in the zinc-blende-type structure of CuF and in the fluorite structure of CaF_2, SrF_2, BaF_2, RaF_2, CdF_2, HgF_2, and PbF_2. A coordination number of 6 occurs in the alkali metal fluorides MF (NaCl type). In many of these compounds F^- resembles O^{2-} stereochemically rather than the other halides, and the radii of the 2 ions are very similar (F^- 133, O^{2-} 140 pm, cf. Cl^- 184, Br^- 196 pm).

The heavier halogens, though markedly less reactive than fluorine, are still amongst the most reactive of the elements. Their reactivity diminishes in the sequence $Cl_2 > Br_2 > I_2$. For example, Cl_2 reacts with CO, NO, and SO_2 to give $COCl_2$, NOCl, and SO_2Cl_2, whereas iodine does not react with these compounds. Again, in the direct halogenation of metals, Cl_2 and Br_2 sometimes produce a higher metal oxidation state than does I_2, e.g. Re yields $ReCl_6$, $ReBr_5$, and ReI_4 respectively. Conversely, the decreasing ionization energies and increasing ease of oxidation of the elements results in the readier formation of iodine cations (p. 986) and compounds in which iodine has a higher stable oxidation state than the other halogens (e.g. IF_7). The general reactivity of the individual halogens with other elements (both metals and non-metals) is treated under the particular element concerned. Reaction between the halogens themselves is discussed on p. 964. In general, reaction of X_2 with compounds containing M–M, M–H, or M–C bonds results in the formation of M–X bonds (M = metal or non-metal). Reaction with metal oxides sometimes requires the presence of C and the use of elevated temperatures.

The stereochemistry of the halogens in their various compounds is summarized in Table 17.9 and will be elucidated in more detail in subsequent sections.

Reactivity is enhanced in conditions which promote the generation of halogen atoms, though this does not imply that all reactions proceed via the intermediacy of X atoms. The reversible thermal dissociation of gaseous $I_2 \rightleftharpoons 2I$ was first demonstrated by Victor Meyer in 1880 and has since been observed for the other halogens as well (p. 938). Atomic Cl and Br are more conveniently produced by electric discharge though, curiously, this particular method is not successful for I. Microwave and radiofrequency discharges have also been used as well as optical dissociation by ultraviolet light. At room temperature and at pressures below 1 mmHg, up to 40% atomization can be achieved, the mean lives of the Cl and Br atoms in glass apparatus being of the order of a few milliseconds. The reason for the slow and relatively inefficient reversion to X_2 is the need for a 3-body collision in order to dissipate the energy of combination:

$$X^\bullet + X^\bullet + M \longrightarrow X_2 + M^*$$

A fuller account of the production, detection, and chemical reactions of atomic Cl, Br, and I is in reference 17 (pages 1141–8 and 1165–72).

Solutions and charge-transfer complexes[26]

The halogens are soluble to varying extents in numerous solvents though their great reactivity sometimes results in solvolysis or in halogenation of the solvent. Reactions with water are discussed on p. 1000. Iodine is only slightly soluble in water (0.340 g/kg at 25°,

[26] Ref. 17, pp. 1196–1220.

TABLE 17.9   *Stereochemistry of the halogens*

CN	Geometry	F	Cl	Br	I
0	—	$F^{\bullet}$(g), F^-(soln)	$Cl^{\bullet}$(g), Cl^-(soln)	$Br^{\bullet}$(g), Br^-(soln)	$I^{\bullet}$(g), I^-(soln)
1	—	F_2, ClF, BrF_3, BF_3, RF	Cl_2, ICl, BCl_3, RCl	Br_2, IBr, BBr_3, RBr	I_2, IX, PI_3, RI
2	Linear	Nb_4F_{20} NbF_3(ReO$_3$-type)	ClF_2^- YCl_3(ReO$_3$-type)	Br_3^-, $(MeCN)_2Br_2$ $CrBr_3$(ReO$_3$-type)	I_3^-, ICl_2^-, $BrICl^-$, Me_3NI_2 BiI_3(ReO$_3$-type)
	Bent	$(BeF_2)_x$, $(HF)_n$, Sn_4F_8	ClO_2, ClO_2^-, Al_2Cl_6, $[Nb_6Cl_{12}]^{2+}$, ClF_2^+ $BeCl_2$(polym), $PdCl_2$	BrF_2^+, Al_2Br_6	IR_2^+, Al_2I_6, AuI(polymeric)
3	Trigonal pyramidal		ClO_3^-, $CdCl_2$, $[Mo_6Cl_8]^{4+}$	BrO_3^-, $MgBr_2$	HIO_3, IO_3^-, CdI_2
	T-shaped		ClF_3	BrF_3	$RICl_2$
	Planar	MgF_2(rutile)			
4	Tetrahedral	CaF_2(fluorite) CuF(blende)	$SrCl_2$(fluorite), ClO_4^-, $FClO_3$, $CuCl$	BrO_4^-, $FBrO_3$, $CuBr$	IO_4^- CuI
	Square planar			BrF_4^-	ICl_4^-, I_2Cl_6
	See-saw (C_{2v} or C_s)		F_3ClO, $[F_2ClO_2]^-$	F_3BrO, $[F_2BrO_2]^-$	$[F_2IO_2]^-$, IF_4^+
5	Square pyramidal		ClF_5, $[F_4ClO]^-$	BrF_5, $[F_4BrO]^-$	IF_5
	Trigonal bipyramidal		F_3ClO_2		IO_5^{3-}(?)
6	Octahedral	NaF	$NaCl$	$NaBr$	IO_6^{5-}, F_5IO, NaI, IF_6^+
	Distorted octahedral			BrF_6^-	IF_6^- (?)
7	Pentagonal bipyramidal				IF_7
	Hexagonal pyramidal		$C_6H_6 . Cl_2$	$C_6H_6 . Br_2$	
8	Cubic		$CsCl$, $TlCl$	$CsBr$, $TlBr$	CsI, TlI

4.48 g/kg at 100°). It is more soluble in aqueous iodide solutions due to the formation of polyiodides (p. 978) and these can achieve astonishing concentrations; e.g. the solution in equilibrium with solid iodine and $KI_7.H_2O$ at 25° contains 67.8 wt% of iodine, 25.6% KI, and 6.6% H_2O. Iodine is also readily soluble in many organic solvents, typical values of its solubility at 25°C being (gI/kg solvent): Et_2O 337.3, EtOH 271.7, mesitylene 253.1, *p*-xylene 198.3, CS_2 197.0, toluene 182.5, benzene 164.0, ethyl acetate 157, EtBr 146, EtCN 141, $C_2H_4Br_2$ 115.1, Bu^{i}OH 97, $CHBr_3$ 65.9, $CHCl_3$ 49.7, cyclohexane 27.9, CCl_4 19.2, *n*-hexane 13.2, perfluoroheptane 0.12.

The most notable feature of such solutions is the dramatic dependence of their colour on the nature of the solvent chosen. Thus, solutions in aliphatic hydrocarbons or CCl_4 are bright violet (λ_{max} 520–540 nm), those in aromatic hydrocarbons are pink or reddish brown, and those in stronger donors such as alcohols, ethers, or amines are deep brown (λ_{max} 460–480 nm). This variation can be understood in terms of a weak donor–acceptor interaction leading to complex formation between the solvent (donor) and I_2 (acceptor) which alters the optical transition energy. Thus, referring to the conventional molecular orbital energy diagram for I_2 (or other X_2) as shown in Fig. 17.3, the violet colour of I_2 vapour can be seen to arise as a result of the excitation of an electron from the highest occupied MO (the antibonding π_g level) into the lowest unoccupied MO (the antibonding

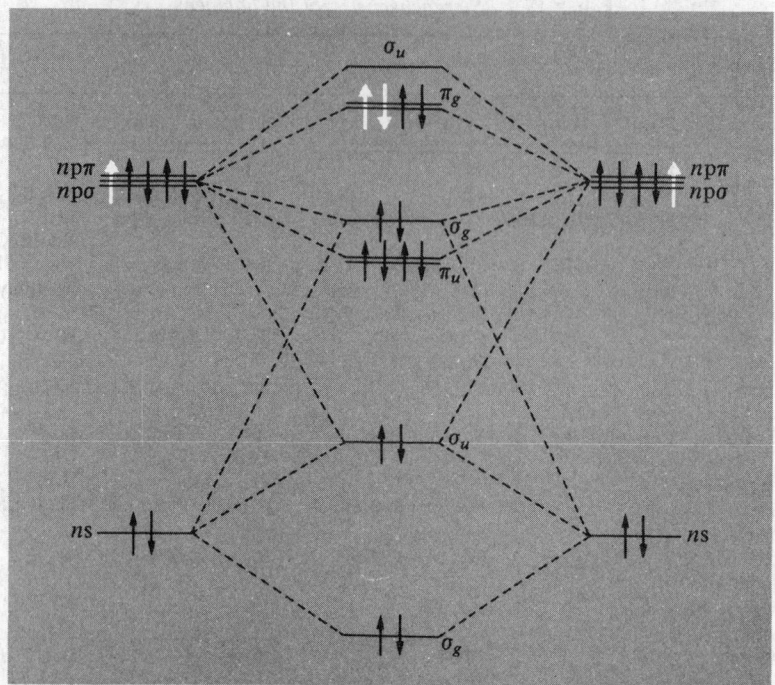

FIG. 17.3 Schematic molecular orbital energy diagram for diatomic halogen molecules. (For F_2 the order of the upper σ_g and π_u bonding MOs is inverted.)

σ_u level). In non-coordinating solvents such as aliphatic hydrocarbons or their fluoro- or chloro-derivatives the transition energy (and hence the colour) remains essentially unmodified. However, in electron-donor solvents, L, the vacant antibonding σ_u orbital of I_2 acts as an electron acceptor thus weakening the I–I bond and altering the energy of the electronic transitions:

$$\sigma_g^2\pi_u^4\pi_g^3\sigma_u^1(^3\Pi_u \text{ and } ^1\Pi_u)\leftarrow\sigma_g^2\pi_u^4\pi_g^4(^1\Sigma_g^+)$$

Consistent with this: (a) the solubility of iodine in the donor solvents tends to be greater than in the non-donor solvents (see list of solubilities), (b) brown solutions frequently turn violet on heating, and brown again on cooling, due to the ready dissociation and reformation of the complex, and (c) addition of a small amount of a donor solvent to a violet solution turns the colour brown. Such donor solvents can be classified as (i) weak π donors (e.g. the aromatic hydrocarbons and alkenes), (ii) stronger σ donors such as nitrogen bases (amines, pyridines, nitriles), oxygen bases (alcohols, ethers, carbonyls), and organic sulfides and selenides.

The most direct evidence for the formation of a complex $L\rightarrow I_2$ in solution comes from the appearance if an intense new charge-transfer band in the near ultraviolet spectrum. Such a band occurs in the region 230–330 nm with a molar extinction coefficient ε of the order of 5×10^3–5×10^4 l mol^{-1} cm^{-1} and a half-width typically of 4000–8000 cm^{-1}. Detailed physicochemical studies further establish that the formation constants of such complexes span the range 10^{-1}–10^4 l mol^{-1} with enthalpies of formation 5–50 kJ mol^{-1}. Some typical examples are in Table 17.10. The donor strength of the various solvents

TABLE 17.10 *Some iodine complexes in solution*

Donor solvent	Formation constant K (20°C)/1 mol^{-1}	$-\Delta H_f/$ kJ mol^{-1}	Charge-transfer band		
			λ_{max}/nm	ε_{max}	$\Delta v_{\frac{1}{2}}$/cm^{-1}
Benzene	0.15	5.9	292	16 000	5100
Ethanol	0.26	18.8	230	12 700	6800
Diethyl ether	0.97	18.0	249	5 700	6900
Diethyl sulfide	210	32.7	302	29 800	5400
Methylamine	530	29.7	245	21 200	6400
Dimethylamine	6800	41.0	256	26 800	6450
Trimethylamine	12 100	50.6	266	31 300	8100
Pyridine	269	32.6	235	50 000	5200

(ligands) is rather independent of the particular halogen (or interhalogen) solute and follows the approximate sequence benzene < alkenes < polyalkylbenzenes ≈ alkyl iodides ≈ alcohols ≈ ethers ≈ ketones < organic sulfides < organic selenides < amines. Conversely, for a given solvent the relative acceptor strength of the halogens increases in the sequence $Cl_2 < Br_2 < I_2 < IBr < ICl$, i.e. they are class b or "soft" acceptors (p. 1065). Further interactions may also occur in polar solvents leading to ionic dissociation which renders the solutions electrically conducting, e.g.:

$$2pyI_2 \rightleftharpoons [py_2I]^+ + I_3^-$$

Numerous solid complexes have been crystallized from brown solutions of iodine and extensive X-ray structural data are available. Complexes of the type L→I—X and L→I—X←L (L = Me$_3$N, py, etc.; X = I, Br, Cl, CN) feature a linear configuration as expected from the involvement of the σ_u antibonding orbital of IX (Fig. 17.4a, b, c). When the ligand has 2 donor atoms (as in dioxan) or the donor atom has more than 1 lone-pair of electrons (as in acetone) the complexes can associate into infinite chains (Fig. 17.4d, e), whereas with methanol, the additional possibility of hydrogen bonding permits further association into layers (Fig. 17.4f). The structure of $C_6H_6.Br_2$ is also included in Fig. 17.4(g). In all these examples, the lengthening of the X—X bond from that in the free halogen molecule is notable.

The intense blue colour of starch-iodine was mentioned on p. 921.

17.2 Compounds of Fluorine, Chlorine, Bromine, and Iodine

17.2.1 *Hydrogen halides*, HX

It is common practice to refer to the molecular species HX and also the pure (anhydrous) compounds as hydrogen halides, and to call their aqueous solutions hydrohalic acids. Both the anhydrous compounds and their aqueous solutions will be considered in this section. HCl and hydrochloric acid are major industrial chemicals and there is also a substantial production of HF and hydrofluoric acid. HBr and hydrobromic acid are made on a much smaller scale and there seems to be little industrial demand for HI and hydriodic acid. It will be convenient to discuss first the preparation and industrial uses of the compounds and then to consider their molecular and bulk physical properties.

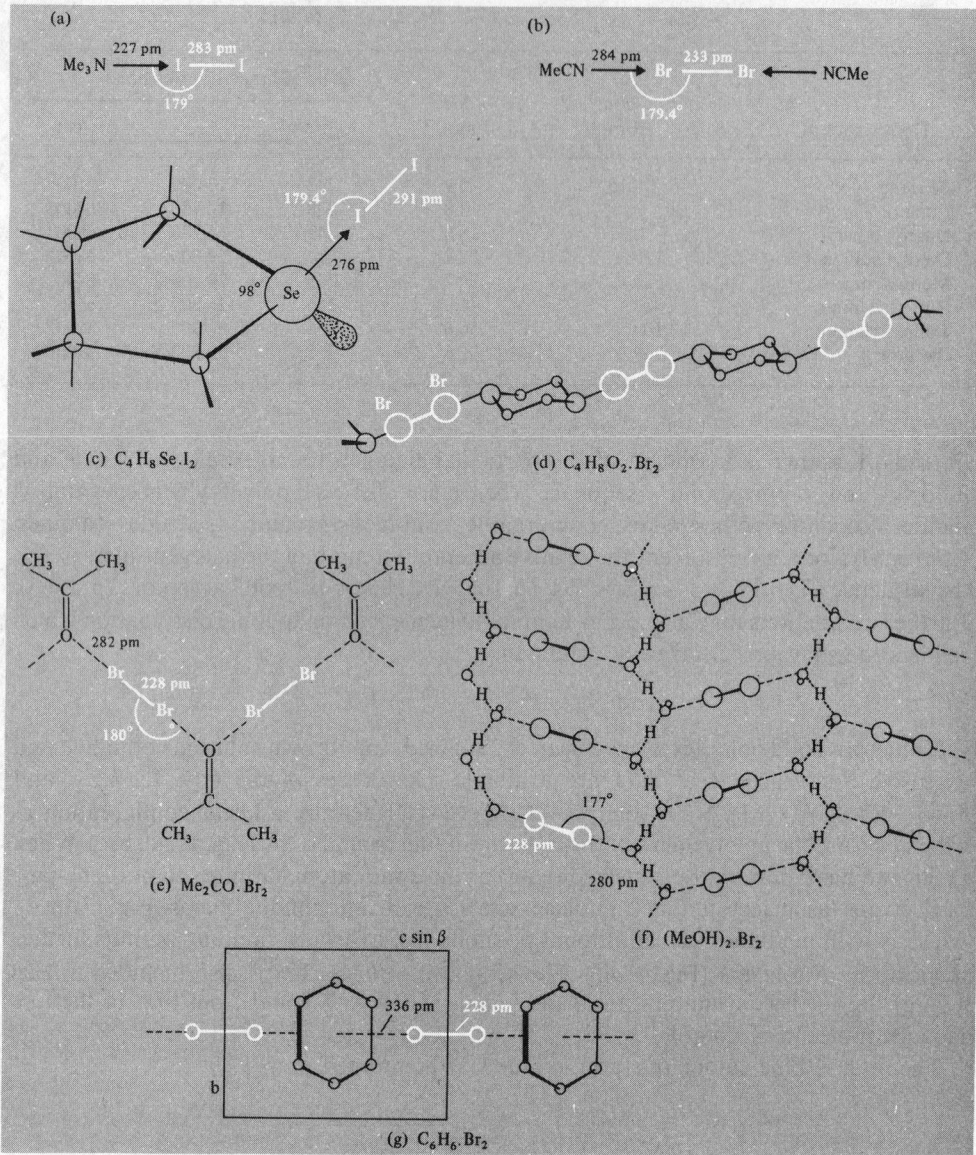

FIG. 17.4   Structures of some molecular complexes of the halogens.

The chemical reactivity of the anhydrous compounds and their acidic aqueous solutions will then be reviewed, and the section concludes with a discussion of the anhydrous compounds as nonaqueous solvents.

Preparation and uses

Anhydrous HF is almost invariably made by the action of conc H_2SO_4 ($\geqslant 95\%$) on "acid grade" fluorspar ($\geqslant 98\%$ CaF_2):

$$CaF_2(s) + H_2SO_4(l) \longrightarrow CaSO_4(s) + 2HF(g)$$

As the reaction is endothermic heat must be supplied to obtain good yields in reasonable time (e.g. 30–60 min at 200–250°C). Silica is a particularly undesirable impurity in the fluorspar since it consumes up to 6 moles of HF per mole of SiO_2 by reacting to form SiF_4 and then H_2SiF_6. A typical unit, producing up to 20 000 tonnes of HF pa, consists of an externally heated, horizontal steel kiln about 30 m long rotating at 1 revolution per minute. The product gas emerges at 100–150°C and, after appropriate treatment to remove solid, liquid, and gaseous impurities, is condensed to give a 99% pure product which is then redistilled to give a final product of 99.9% purity. The technical requirements to enable the safe manufacture and handling of so corrosive a product are considerable.[2, 7] In principle, HF could also be obtained from the wet-processing of fluorapatite to give phosphoric acid (p. 599) but the presence of SiO_2 preferentially yields SiF_4 and H_2SiF_6 from which HF can only be recovered uneconomically.

$$Ca_5(PO_4)_3F + 5H_2SO_4 \longrightarrow 5CaSO_4 + 3H_3PO_4 + 2HF$$

$$SiO_2 + 4HF \longrightarrow SiF_4 + 2H_2O \xrightarrow{\text{aq HF}} \text{aq } H_2SiF_6$$

Some of the H_2SiF_6 so produced finds commercial outlets (p. 946), but it has been estimated that $\sim 500\,000$ tonnes of H_2SiF_6 is discarded annually by the US phosphoric acid industry, equivalent to ~ 1 million tonnes of fluorspar—enough to supply that nation's entire requirements for HF. Production figures and major uses are in the Panel.

Production and Uses of Hydrogen Fluoride

Anhydrous HF was first produced commercially in the USA in 1931 and in the UK from about 1942. Now, some sixteen countries each produce at least 1000 tonnes pa with the USA accounting for some 300 000 tonnes of the estimated annual world production of about 1 million tonnes. In 1979 the US price was about $1.00 per kg for anhydrous HF and about $0.75 per kg for 70% acid. The primary suppliers ship HF in tank-cars of 20–91-tonne capacity and the product is also repackaged in steel cylinders holding 8.0–900 kg (2.7–635 kg in the UK). Lecture bottles contain 340 g HF. The 70% acid is shipped in tank-cars of 32–80-tonne capacity, tank trucks of 20-tonne capacity, and in polyethylene-lined drums holding 114 or 208 l.

The early need for HF was in the production of chlorofluorocarbons for refrigeration units and pressurizing gases. The large increase in aluminium production in 1935–40 brought an equivalent requirement for HF (for synthetic cryolite, p. 247) and these two uses still account for the bulk of HF produced in the USA (37% and 40% respectively). Other major outlets are in uranium processing (7%),† petroleum alkylation catalysts (5%), and stainless steel pickling (4%). The remaining 7% is distributed amongst traditional uses (such as glass etching and the frosting of light bulbs and television tubes, and the manufacture of fluoride salts), and newer applications such as rocket-propellant stabilizers, preparation of microelectronic circuits, laundry sours, and stain removers.

Probably about 40 000 tonnes of HF are used worldwide annually to make inorganic compounds other than UF_4/UF_6. Prominent amongst these products are:

NaF: for water fluoridation, wood preservatives, the formulation of insecticides and fungicides, and use as a fluxing agent. It is also used to remove HF from gaseous F_2 in the manufacture and purification of F_2.

Continued

† This includes both the use of HF to make UF_4 and its use to manufacture F_2 (p. 928) to convert UF_4 to UF_6.

SnF$_2$: in toothpastes to prevent dental caries.

HBF$_4$(aq) and metal fluoroborates: electroplating of metals, catalysts, fluxing in metal processing and surface treatment.

H$_2$SiF$_6$ and its salts: fluoridation of water, glass and ceramics manufacture, metal-ore treatment.

The highly corrosive nature of HF and aqueous hydrofluoric acid solutions have already been alluded to (pp. 923, 928) and great caution must be exercised in their handling. The salient feature of HF burns is the delayed onset of discomfit and the development of a characteristic white lesion that is excruciatingly painful. The progressive action of HF on skin is due to dehydration, low pH, and the specific toxic effect of high concentrations of fluoride ions: these remove Ca^{2+} from tissues as insoluble CaF$_2$ and thereby delay healing; in addition the immobilization of Ca^{2+} results in a relative excess of K$^+$ within the tissue, so that nerve stimulation ensues. Treatment of HF burns involves copious sluicing with water for at least 15 min followed either by (a) immersion in (or application of wet packs of) cold MgSO$_4$, or (b) subcutaneous injection of a 10% solution of calcium gluconate (which gives rapid relief from pain), or (c) surgical excision of the burn lesion.[3] Medical attention is essential, even if the initial effects appear slight, because of the slow onset of the more serious symptoms.

Hydrogen chloride is a major industrial chemical and is manufactured on a huge scale. It is also a familiar laboratory reagent both as a gas and as an aqueous acid. The industrial production and uses of HCl are summarized in the Panel. One important method for synthesis on a large scale is the burning of H$_2$ in Cl$_2$: no catalyst is needed but economic sources of the two elements are obviously required. Another major source of HCl is as a by-product of the chlorination of hydrocarbons (p. 931). The traditional "salt-cake" process of treating NaCl with conc H$_2$SO$_4$ also remains an important industrial source of the acid. On a small laboratory scale, gaseous HCl can be made by treating concentrated aqueous hydrochloric acid with conc H$_2$SO$_4$. Preparation of DCl is best effected by the action of D$_2$O on PhCOCl or a similar organic acid chloride; PCl$_3$, PCl$_5$, SiCl$_4$, AlCl$_3$, etc., have also been used.

Industrial Production and Uses of Hydrogen Chloride

World production of HCl is well in excess of 5 million tonnes pa, thus making it one of the largest volume chemicals to be manufactured. Production for some of the leading industrial countries in 1979 was as follows:[27]

Country	USA	France	Japan	Canada	UK
HCl (1979)/tonnes	2 697 000	710 000	528 000	154 000	135 000

Four major processes account for the bulk of HCl produced, the choice of method invariably being dictated by the ready availability of the particular starting materials, the need for the co-products, or simply the availability of by-product HCl which can be recovered as part of an integrated process.[28]

1. The classic salt-cake process was introduced with the Leblanc process towards the end of the

[27] Anon., *Chem. Eng. News* 9 June 1980, 35–90.
[28] D. S. ROSENBERG, Hydrogen chloride, *Kirk–Othmer's Encyclopedia of Chemical Technology*, 3rd Edn., Vol. 12, pp. 983–1015 (1980).

eighteenth century and is still used to produce HCl where rock-salt mineral is cheaply available (as in the UK Cheshire deposits). The process is endothermic and takes place in two stages:

$$NaCl + H_2SO_4 \xrightarrow{\sim 150} NaHSO_4 + HCl$$

$$NaCl + NaHSO_4 \xrightarrow{540-600} Na_2SO_4 + HCl$$

2. The Hargreaves process (late nineteenth century) is a variant of the salt-cake process in which NaCl is reacted with a gas-phase mixture of SO_2, air, and H_2O (i.e. "H_2SO_4") in a self-sustaining exothermic reaction:

$$2NaCl + SO_2 + \tfrac{1}{2}O_2 + H_2O \xrightarrow{430-450} Na_2SO_4 + 2HCl$$

Again the economic operation of the process depends on abundant rock-salt or the need for the by-product Na_2SO_4 for the paper and glass industries. It is probably this latter need which accounts for the vestigial survival of the processs in countries such as the USA where other processes are more economic (see Table below).

3. Direct synthesis of HCl by the burning of hydrogen in chlorine is the favoured process when high-purity HCl is required. The reaction is highly exothermic and requires specially designed burners and absorption systems.

4. By-product HCl from the heavy organic-chemicals industry (p. 931) accounted for nearly 90% of the HCl produced in the USA in 1978. Where such petrochemical industries are less extensive this source of HCl becomes correspondingly smaller. The crude HCl so produced may be contaminated with unreacted Cl_2, organics, chloro-organics, or entrained solids (catalyst supports, etc.), all of which must be removed. The growth of this source of HCl in the USA, and the consequent relative decline of the other methods during the past 30 y, is shown in the Table. The 1978 US bulk price was $66 per tonne, indicating the extraordinary cost-effectiveness of the major recovery processes.

Production of HCl in USA expressed as 100% $HCl/10^3$ tonnes

Year	Total	$NaCl/H_2SO_4$	$H_2 + Cl_2$	by-product HCl
1950	562	149.7 (26.6%)	101.6 (18.1%)	310 (55.2%)
1960	880	82.1	134.5	663 (75.3%)
1970	1826	113.7	86.2	1627 (89.1%)
1978	2478	87.1 (3.5%)	162.4 (6.6%)	2228 (89.9%)[a]

[a] Comprising 30% from vinyl chloride, 15% from chlorinated ethanes and ethenes, 15% from chlorinated fluorocarbons, 14% from other chlorinated organics, 10% from isocyanates and 6% from miscellaneous industrial processes, making 90% in all.

Most of the HCl produced from salt or by direct combination of the elements (at least in the USA) is destined for sale, whereas much of the by-product HCl is used captively, primarily in oxyhydrochlorination processes for making vinyl chloride and chlorinated solvents or for Mg processing (p. 120). HCl gas for industrial use can be transmitted without difficulty over moderate distances in mild-steel piping. It is also available in cylinders of varying size down to laboratory scale lecture bottles containing 225 g.

Industrial use of HCl gas for the manufacture of inorganic chemicals includes the preparation of anhydrous NH_4Cl by direct reaction with NH_3, and the synthesis of anhydrous metal chlorides by reaction with appropriate carbides, nitrides, oxides, or even the free metals themselves, e.g.:

$$SiC \xrightarrow[\text{(NiCl}_2\text{catal)}]{HCl/700} SiCl_4$$

$$MN_y \xrightarrow{HCl/heat} MCl_x(pure) \qquad (Ti, Zr, Hf; Nb, Ta; Cr, Mo, W, etc.)$$

Continued

$$MO + 2HCl \longrightarrow MCl_2 + H_2O \quad \text{(especially for removal of impurities or waste recovery)}$$

$$Al + 3HCl \longrightarrow AlCl_3 + \tfrac{3}{2}H_2$$

HCl is also used in the industrial synthesis of ClO_2 (p. 992):

$$NaClO_3 + 2HCl \longrightarrow ClO_2 + \tfrac{1}{2}Cl_2 + NaCl + H_2O$$

The reaction is catalysed by various salts of Ti, Mn, Pd, and Ag which promote the formation of ClO_2 rather than the competing reaction which otherwise occurs:

$$NaClO_3 + 6HCl \longrightarrow 3Cl_2 + NaCl + 3H_2O$$

HCl is also used in the production of Al_2O_3 (p. 273) and TiO_2 (p. 1118), the isolation of Mg from sea water (p. 120), and in many extractive metallurgical processes for isolating or refining metals, e.g. Ge, Sn, V, Mn, Ta, W, and Ra.

The largest single use of aqueous HCl is in the pickling of steel and other metals to remove the adhering oxide scale. Some $\tfrac{3}{4}$ million tonnes of hydrochloric acid (calculated as 32% HCl) is used annually for this purpose in the USA alone. It is also used in pH control (effluent neutralization, etc.), the desliming of hides and chrome tanning, the desulfurizing of petroleum, ore beneficiation, the coagulation of latex, and the production of aniline from $PhNO_2$ for dyestuffs intermediates. The manufacture of gelatine requires large quantities of hydrochloric acid to decompose the bones used as raw materials—high purity acid must be used since much of the gelatine is used in foodstuffs for human consumption. Another food-related application is the hydrolysis of starch to glucose under pressure: this process is catalysed by small concentrations of HCl and is extensively used to produce "maple syrup" from maize (corn) starch. At higher concentrations of HCl wood (lignin) can be converted to glucose.

Other uses of HCl are legion[28] and range from the purification of fine silica for the ceramics industry, and the refining of oils, fats, and waxes, to the manufacture of chloroprene rubbers, PVC plastics, industrial solvents and organic intermediates, the production of viscose rayon yarn and staple fibre, and the wet processing of textiles (where hydrochloric acid is used as a sour to neutralize residual alkali and remove metallic and other impurities).

Similar routes are available for the production of HBr and HI. The catalysed combination of H_2 and Br_2 at elevated temperatures (200–400°C in the presence of Pt/asbestos, etc.) is the principal industrial route for HBr, and is also used, though on a relatively small scale, for the energetically less-favoured combination of H_2 and I_2 (Pt catalyst above 300°C). Commercially HI is more often prepared by the reaction of I_2 with H_2S or hydrazine, e.g.:

$$2I_2 + N_2H_4 \xrightarrow{\text{H}_2\text{O}} 4HI + N_2 \quad \text{(quantitative)}$$

Reduction of the parent halogen with red phosphorus and water provides a convenient laboratory preparation of both HBr and HI:

$$2P + 6H_2O + 3X_2 \longrightarrow 6HX + 2H_3PO_3$$

$$H_3PO_3 + H_2O + Br_2 \longrightarrow 2HBr + H_3PO_4$$

The rapid reaction of tetrahydronaphthalene (tetralin) with Br_2 at 20° affords an alternative small-scale preparation though only half the Br_2 is converted, the other half being lost in brominating the tetralin:

The action of conc H_2SO_4 on metal bromides or iodides (analogous to the "salt-cake" process of HCl) causes considerable oxidation of the product HX but conc H_3PO_4 is satisfactory. Dehydration of the aqueous acids with P_2O_5 is a viable alternative. DBr and DI are obtained by reaction of D_2O on PBr_3 and PI_3 respectively.

Anhydrous HBr is available in cylinders (6.8-kg and 68-kg capacity) under its own vapour pressure (24 atm at 25°C) and in lecture bottles (450-g capacity). Its main industrial use is in the manufacture of inorganic bromides and the synthesis of alkyl bromides either from alcohols or by direct addition to alkenes. HBr also catalyses numerous organic reactions. There seem to be no large-scale uses for HI outside the laboratory, where it is used in various iodination reactions (lecture bottles containing 400 g HI are available). Commercial solutions contain 40–55 wt% of HI (cf. azeotrope at 56.9% HI, p. 953) and these solutions are thermodynamically much more stable than pure HI as indicated by the large negative free energy of solution.

Physical properties of the hydrogen halides

HF is a colourless volatile liquid and an oligomeric H-bonded gas $(HF)_x$, whereas the heavier HX are colourless diatomic gases at room temperature. Some molecular and bulk physical properties are summarized in Table 17.11. The influence of H bonding on the

TABLE 17.11 *Physical properties of the hydrogen halides*

Property	HF	HCl	HBr	HI
MP/°C	−83.4	−114.7	−88.6	−51.0
BP/°C	19.5[(a)]	−84.2	−67.1	−35.1
Liquid range (1 atm)/°C	102.9	30.5	21.5	15.9
Density $(T°C)/g\ cm^{-3}$	1.002(0°)[(b)]	1.187 (−114°)	2.603 (−84°)	2.85 (−47°)
Viscosity $(T°C)$/centipoise	0.256 (0°)	0.51 (−95°)	0.83 (−67°)	1.35 (−35.4°)
Dielectric constant, $\varepsilon^{(c)}$	84 (0°)	9.28 (−95°)	7.0 (−85°)	3.39 (−50°)
Electrical conductivity $(T°C)$/ohm^{-1} cm^{-1}	$\sim 10^{-6}$ (0°)	$\sim 10^{-9}(-85°)$	$\sim 10^{-9}$ (−85°)	$\sim 10^{-10}$ (−50°)
ΔH_f° (298°)/kJ mol^{-1}	−271.12	−92.31	−36.40	26.48
$\Delta G_f^{\ast}$ (298°)/kJ mol^{-1}	−273.22	−95.30	−53.45	1.72
$S°$ (298°)/J mol^{-1} K^{-1}	173.67	186.80	198.59	206.48
ΔH_{dissoc}(H–X)/kJ mol^{-1}	573.98	428.13	362.50	294.58
r_e(H–X)/pm	91.7	127.4	141.4	160.9
Vibrational frequency ω_e/cm^{-1}	4138.33	2990.94(H^{35}Cl) 2988.48(H^{37}Cl)	2649.65	2309.53
Dipole moment μ/D	1.74	1.07	0.788	0.382

(a) Vapour pressure of HF 363.8 mmHg at 0°.
(b) Density of liquid HF 1.23 g cm^{-3} near melting point.
(c) Dielectric constant ε(HF) 175 at −73°C.

(low) vapour pressure, (long) liquid range, and (high) dielectric constant of HF have already been discussed (pp. 57–9). Note also that the viscosity of liquid HF is lower than that of water (or indeed of the other HX) and this has been taken to imply the absence of a three-dimensional network of H bonds such as occurs in H_2O, H_2SO_4, H_3PO_4, etc. However, it should be remembered that the viscosity of HF is quoted for 0°C, i.e. some 80° above its mp and only 20° below its bp; a more relevant comparison might be its value of 0.772 centipoise at $-62.5°$ (i.e. 19° above its mp) compared with a value of 1.00 centipoise for water at 20°. Hydrogen bonding is also responsible for the association of HF molecules in the vapour phase: the vapour density of the gas over liquid HF reaches a maximum value of ~ 86 at $-34°$. At atmospheric pressure the value drops from 58 at 25° to 20.6 at 80° (the limiting vapour density of monomeric HF is $\frac{20.0063}{2.0158} = 9.925$). These results, together with infrared and electron diffraction studies, indicate that gaseous HF comprises an equilibrium mixture of monomers and cyclic hexamers, though chain dimers may also occur under some conditions of temperature and pressure:

$$6HF \rightleftharpoons (HF)_6; \quad 2HF \rightleftharpoons (HF)_2$$

The crystal structure of HF shows it to consist of planar zigzag chain polymers with an $F\cdots F$ distance of 249 pm and an angle of 120.1°:

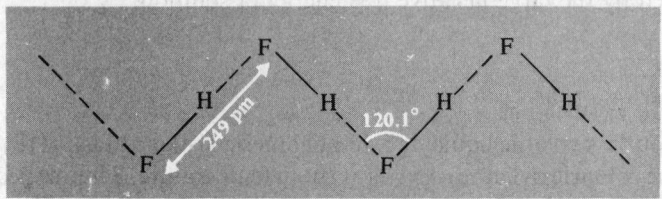

The other HX are not associated in the gaseous or liquid phases but the low-temperature forms of crystalline HCl and HBr both feature weakly H-bonded zigzag chains similar to those in solid HF. At higher temperatures substantial disorder sets in.

The standard heats of formation $\Delta H_f°$ of gaseous HX diminish rapidly with increase in molecular weight and HI is endothermic. The very small (and positive) value for the standard free energy of formation $\Delta G_f°$ of HI indicates that (under equilibrium conditions) this species is substantially dissociated at room temperature and pressure. However, dissociation is slow in the absence of a catalyst. The bond dissociation energies of HX show a similar trend from the very large value of 574 kJ mol^{-1} for HF to little more than half this (295 kJ mol^{-1}) for HI.

Chemical reactivity of the hydrogen halides

Anhydrous HX are versatile and vigorous reagents for the halogenation of metals, non-metals, hydrides, oxides, and many other classes of compound, though reactions that are thermodynamically permissible do not always occur in the absence of catalysts, thermal initiation or photolytic encouragement, because of kinetic factors. For example,[29]

[29] T. C. WADDINGTON, The hydrogen halides, Chap. 3 in V. GUTMANN (ed.), *Main Group Elements: Group VII and Noble Gases*, MTP International Review of Science: Inorganic Chemistry Series 1, Vol. 3, pp. 85–125, Butterworths, London, 1972.

reaction of HX(g) with elements (M) can thermodynamically proceed according to the equation

$$M + n\text{HX} = MX_n + \tfrac{1}{2}n\text{H}_2$$

providing that ΔG for the reaction [i.e. $\Delta G_f^\circ(MX_n) - n\Delta G_f^\circ(\text{HX, g})$] is negative. From the data in Table 17.11 this means that M could be oxidized to the n-valent halide MX_n if:

for the fluoride $\Delta G_f^\circ(MF_n)$ is $< -274n$ kJ mol^{-1}
for the chloride $\Delta G_f^\circ(MCl_n)$ is $< -96n$ kJ mol^{-1}
for the bromide $\Delta G_f^\circ(MBr_n)$ is $< -54n$ kJ mol^{-1}
for the iodide $\Delta G_f^\circ(MI_n)$ is $< \sim 0$ kJ mol^{-1}

Using tables of free energies of formation it is clear that most metals will react with most HX. Moreover, in many cases, e.g. with the alkali metals, alkaline earth metals, Zn, Al, and the lanthanide elements, such reactions are extremely exothermic. It is also clear that Ag should react with HCl, HBr, and HI but not with HF, and Cu should form CuF_2 with HF but not CuX_2 with the other HX. Iron should give $FeCl_3$ but in practice the reaction only proceeds to $FeCl_2$. TiX_4 can be made, but only at high temperatures. Reactions of Si to form SiX_4 are very favourable for X = F, Cl, Br, but only HF reacts at room temperature. With As, reaction with HF to give AsF_3 is thermodynamically favourable but reactions with the other HX are not. Similar, though more complicated, schemes can be worked out for the reactions of HX with oxides, other halides, hydrides, etc.

HF is miscible with water in all proportions and the phase diagram (Fig. 17.5a) shows the presence of three compounds: $H_2O.HF$ (mp $-35.5°$), $H_2O.2HF$ (mp $-75.5°$), and $H_2O.4HF$ (mp $-100.4°$, i.e. $17°$ below the mp of pure HF). Recent X-ray studies have confirmed earlier conjectures that these compounds are best formulated as H-bonded

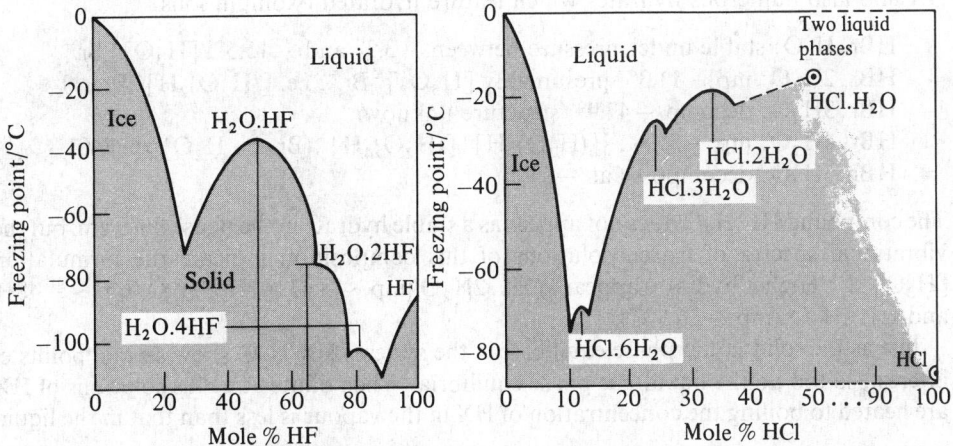

FIG. 17.5   The phase diagram of the systems HF/H$_2$O and HCl/H$_2$O. Note that for hydrofluoric acid all the solvates contain $\geqslant 1$HF per H$_2$O, whereas for hydrochloric acid they contain $\leqslant 1$HCl per H$_2$O. This is because the H bonds F—H$\cdots$F and F—H$\cdots$O are *stronger* than O—H$\cdots$O, whereas Cl—H$\cdots$Cl and Cl—H$\cdots$O are *weaker* than O—H$\cdots$O. Accordingly the solvates in the former system have the crystal structures $[H_3O]^+F^-$, $[H_3O]^+[HF_2]^-$, and $[H_3O]^+[H_3F_4]^-$, whereas the latter are $[H_3O]^+Cl^-$, $[H_5O_2]^+Cl^-$, and $[H_5O_2]^+Cl^-.H_2O$. The structures of HCl.6H$_2$O and the metastable HCl.4H$_2$O are not known.

oxonium salts $[H_3O]F$, $[H_3O][HF_2]$, and $[H_3O][H_3F_4]$ with three very strong H bonds per oxonium ion and average $O\cdots F$ distances of 246.7, 250.2, and 253.6 pm respectively.[30] Such H bonds are very relevant to the otherwise surprising observation that, unlike the other aqueous hydrohalic acids which are extremely strong, hydrofluoric acid is a very weak acid in aqueous solution. Indeed, the behaviour of such solutions is remarkable in showing a dissociation constant (as calculated from electrical conductivity measurements) that *diminishes* continuously on dilution. Detailed studies reveal the presence of two predominant equilibria:[31]

$$H_2O + HF \longrightarrow [(H_3O)^+F^-] \rightleftharpoons [H_3O]^+(aq) + F^-(aq); \quad pK_a \ 2.95$$

$$F^-(aq) + HF \rightleftharpoons HF_2^-(aq); \quad K_2 = \frac{[HF_2^-]}{[HF][F^-]}$$

The dissociation constant for the first process is only 1.1×10^{-3} l mol^{-1} at 25°C; this corresponds to pK_a 2.95 and indicates a rather small free hydrogen-ion concentration (cf. $ClCH_2CO_2H$, pK_a 2.85) as a result of the strongly H-bonded, undissociated ion-pair $[(H_3O)^+F^-]$. By contrast, $K_2 = 2.6 \times 10^{-1}$ l mol^{-1} (pK_2 0.58), indicating that an appreciable number of the fluoride ions in the solution are coordinated by HF to give HF_2^- rather than by H_2O despite the very much higher concentration of H_2O molecules.

Numerous hydrates also occur in the HCl/H_2O system (Fig. 17.5b), e.g. $HCl \cdot H_2O$ (mp $-15.4°$), $HCl \cdot 2H_2O$ (mp $-17.7°$), $HCl \cdot 3H_2O$ (mp $-24.9°$), $HCl \cdot 4H_2O$, and $HCl \cdot 6H_2O$ (mp $-70°$). The system differs from HF/H_2O not only in the stoichiometry of the hydrates but also in separating into two liquid phases at HCl concentrations higher than 1:1. The weakness of the $O-H\cdots Cl$ hydrogen bond also ensures that there is very little impediment to complete ionic dissociation and aqueous solutions of HCl (and also of HBr and HI) are strong acids; approximate values of pK_a are HCl -7, HBr -9, HI -10. The systems HBr/H_2O and HI/H_2O also show a miscibility gap at high concentrations of HX and also numerous hydrates which feature hydrated oxonium ions:

HBr . H_2O: stable under pressure between $-3.3°$ and $-15.5°$; $[H_3O]^+Br^-$
HBr . $2H_2O$: mp $-11.3°$; presumably $[H_5O_2]^+Br^-$, i.e. $[(H_2O)_2H]^+Br^-$
HBr . $3H_2O$: decomp $-47.9°$; structure unknown
HBr . $4H_2O$: mp $-55.8°$; $\{[(H_2O)_3H]^+[(H_2O)_4H]^+(Br^-)_2 . H_2O\}$ (p. 740)
HBr . $6H_2O$: decomposes at $-88.2°$

The compound HI . H_2O does not appear as a stable hydrate in the phase diagram, but the vibrational spectra of frozen solutions of this composition indicate the formulation $[H_3O]^+I^-$. Higher hydrates appear at HI . $2H_2O$ (mp $\sim -43°$), HI . $3H_2O$ (mp $\sim -48°$), and HI . $4H_2O$ (mp $-36.5°C$).

Just as the solid/liquid phase equilibria in the systems HX/H_2O show several points of interest, so too do the liquid/gas phase equilibria. When dilute aqueous solutions of HX are heated to boiling the concentration of HX in the vapour is less than that in the liquid

[30] D. Mootz, Crystallochemical correlate to the anomaly of hydrofluoric acid, *Angew. Chem.*, Int. Edn. (Engl.) **20**, 791 (1981).
[31] L. G. Sillén and A. E. Martell, *Stability Constants of Metal–Ion Complexes*, Special Publication No. 17, pp. 256–7, The Chemical Society, London, 1964; *Supplement No. 1* (Special Publication No. 17), pp. 152–3 (1971). These tables summarize over 50 papers on these equilibria; the results are sometimes inconsistent but the most recent values tend towards those selected in the text.

phase, so that the liquid becomes progressively more concentrated and the bp progressively rises until a point is reached at which the liquid has the same composition as the gas phase so that it boils without change in composition and at constant temperature. This mixture is called an azeotrope (Greek ἀ, without; ζεῖη, *zein*, to boil; τροπή, *trope*, change). The phenomenon is illustrated for HF and HCl in Fig. 17.6. Conversely, when more concentrated aqueous solutions are boiled, the concentration of HX in the vapour is greater than that in the liquid phase which thereby becomes progressively diluted by distillation until the azeotropic mixture is again reached, whereupon distillation continues without change of composition and at constant temperature. The bps and azeotropic compositions at atmospheric pressure are listed below, together with the densities of the azeotropic acids at 25°C:

Azeotrope	HF	HCl	HBr	HI
BP (1 atm)/°C	112	108.58	124.3	126.7
g (HX)/100 g soln	38	20.22	47.63	56.7
Density (25°)/g cm^{-3}	1.138	1.096	1.482	1.708

Of course, the bp and composition of the azeotrope both vary with pressure, as illustrated below for the case of hydrochloric acid (1 mmHg = 0.1333 kPa):

P/mmHg	50	250	500	700	**760**	800	1000	1200
BP/°C	48.72	81.21	97.58	106.42	**108.58**	110.01	116.19	122.98
g(HX)/100 g soln	23.42	21.88	20.92	20.36	**20.222**	20.16	19.73	19.36
Density(25°)/g cm^{-3}	1.112	1.104	1.099	1.097	**1.0959**	1.095$_5$	1.093	1.091$_5$

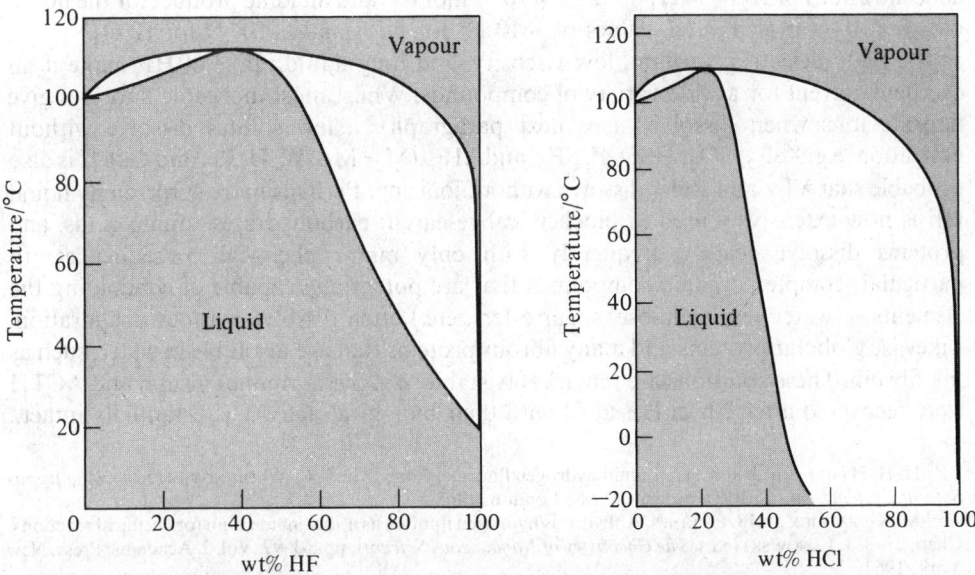

FIG. 17.6 Liquid/gas phase equilibria for the systems HF/H$_2$O and HCl/H$_2$O showing the formation of maximum boiling azeotropes as described in the text.

The occurrence of such azeotropes clearly restricts the degree to which aqueous solutions of HX can be concentrated by evaporation. However, they do afford a ready means of obtaining solutions of precisely known concentration: in the case of hydrochloric acid, its azeotrope is particularly stable over long periods of time and has found much use in analytical chemistry.

The hydrogen halides as nonaqueous solvents

The great synthetic value of liquid NH_3 as a nonaqueous solvent (p. 486) has encouraged the extensive study of the other neighbour of H_2O in the periodic table, namely, HF.[29, 32-34] Early studies were hampered by the aggressive nature of anhydrous HF towards glass and quartz, but the pure acid can now be safely handled without contamination using fluorinated plastics such as polytetrafluoroethylene. The self-ionic dissociation of the solvent, as evidenced by the residual electrical conductivity of highly purified HF, can be represented as $HF \rightleftharpoons H^+ + F^-$; however, since both ions will be solvated it is more usual to represent the equilibrium as

$$3HF \rightleftharpoons H_2F^+ + HF_2^-$$

The fluoride ion has an anomalously high conductance, λ_α, as shown by the following values obtained at $0°$:

Ion	Na^+	K^+	H_2F^+	BF_4^-	SbF_6^-	HF_2^-
λ_α/ohm^{-1} cm^2 mol^{-1}	117	117	**79**	183	196	**273**

As the specific conductivity of pure HF is $\sim 10^{-6}$ ohm^{-1} cm^2 at $0°$, these values imply concentrations of $H_2F^+ = HF_2^- \simeq 2.9 \times 10^{-6}$ mol l^{-1} and an ionic product for the liquid of $\sim 8 \times 10^{-12}$ mol^2 l^{-1} (cf. values of $\sim 10^{-33}$ for NH_3 and $\sim 10^{-14}$ for H_2O).

The high dielectric constant, low viscosity, and long liquid range of HF make it an excellent solvent for a wide variety of compounds. Whilst most inorganic fluorides give fluoride ions when dissolved (see next paragraph), a few solutes dissolve without ionization, e.g. XeF_2, SO_2, HSO_3F, SF_6, and MF_6 (M = Mo, W, U, Re, and Os). It is also probable that VF_5 and ReF_7 dissolve without ionizing. Perhaps more surprisingly liquid HF is now extensively used in biochemical research: carbohydrates, amino acids, and proteins dissolve readily, frequently with only minor chemical consequences. In particular, complex organic compounds that are potentially capable of eliminating the elements of water (e.g. cellulose, sugar esters, etc.) often dissolve without dehydration. Likewise globular proteins and many fibrous proteins that are insoluble in water, such as silk fibroin. These solutions are remarkably stable: e.g. the hormones insulin and ACTH were recovered after 2 h in HF at $0°$ with their biological activity substantially intact.

[32] H. H. HYMAN and J. J. KATZ, Liquid hydrogen fluoride, Chap. 2 in T. C. WADDINGTON (ed.), *Nonaqueous Solvent Systems*, pp. 47–81, Academic Press, London, 1965.
[33] M. KILPATRICK and J. G. JONES, Anhydrous hydrogen fluoride as a solvent medium for chemical reactions, Chap. 2 in J. J. LAGOWSKI (ed.), *The Chemistry of Nonaqueous Solvents*, pp. 43–99, Vol. 2, Academic Press, New York, 1967.
[34] T. A. O'DONNELL, Fluorine, Chap. 25 in *Comprehensive Inorganic Chemistry*, Vol. 2, pp. 1009–1106, Pergamon Press, Oxford, 1973.

Many of the ionic fluorides of M^I, M^{II}, and M^{III} dissolve to give highly conducting solutions due to ready dissociation. Some typical values of the solubility of fluorides in HF are in Table 17.12: the data show the expected trend towards greater solubility with increase in ionic radius within the alkali metals and alkaline earth metals, and the expected decrease in solubility with increase in ionic charge so that $MF > MF_2 > MF_3$. This is dramatically illustrated by AgF which is 155 times more soluble than AgF_2 and TlF which is over 7000 times more soluble than TlF_3.

TABLE 17.12 *Solubility of some metal fluorides in anhydrous HF (in g/100 g HF and at 12°C unless otherwise stated)*

LiF	NaF(11°)	NH$_4$F(17°)	KF(8°)	RbF(20°)	CsF(10°)	**AgF**	**TlF**
10.3	30.1	32.6	36.5	110	199	**83.2**	**580**

Hg$_2$F$_2$	BeF$_2$(11°)	MgF$_2$	CaF$_2$	SrF$_2$	BaF$_2$	**AgF$_2$**	CaF$_2$
0.87	0.015	0.025	0.817	14.83	5.60	**0.54**	0.010

HgF$_2$	CdF$_2$(14°)	ZnF$_2$(14°)	CrF$_2$(14°)	FeF$_2$	NiF$_2$	PbF$_2$
0.54	0.201	0.024	0.036	0.006	0.037	2.62

AlF$_3$	CeF$_3$	**TlF$_3$**	MnF$_3$	FeF$_3$	CoF$_3$	SbF$_3$	BrF$_3$
0.002	0.043	**0.081**	0.164	0.008	0.257	0.536	0.010

With inorganic solutes other than fluorides, solvolysis usually occurs. Thus chlorides, bromides, and iodides give the corresponding fluorides with evolution of HX, and fluorides are also formed from oxides, hydroxides, carbonates, and sulfites. Indeed, this is an excellent synthetic route for the preparation of anhydrous metal fluorides and has been used with good effect for TiF_4, ZrF_4, UF_4, SnF_4, VOF_3, VF_3, NbF_5, TaF_5, SbF_5, MoO_2F_2, etc. (Note, however, that AgCl, $PdCl_2$, $PtCl_4$, Au_2Cl_6, and ICl are apparently exceptions.[33]) Less-extensive solvolysis occurs with sulfates, phosphates, and certain other oxoanions. For example, a careful cryoscopic study of solutions of K_2SO_4 in HF (at $\sim -84°C$) gave a value of $v = 5$ for the number of solute species in solution, but this increased to about 6 when determined by vapour-pressure depressions at 0°. These observations can be rationalized if unionized H_2SO_4 is formed at the lower temperature and if solvolysis of this species to unionized HSO_3F sets in at the higher temperatures:

$$K_2SO_4 + 4HF \xrightarrow{-80} 2K^+ + 2HF_2^- + H_2SO_4$$

$$H_2SO_4 + 3HF \xrightarrow{0°} H_3O^+ + HSO_3F + HF_2^-$$

Consistent with this, the ^{19}F nmr spectra of solutions at 0° showed the presence of HSO_3F, and separate cryoscopic experiments with pure H_2SO_4 as the sole solute gave a value of v close to unity.

Solvolysis of phosphoric acids in the system $HF/P_2O_5/H_2O$ gave successively H_2PO_3F, HPO_2F_2, and $H_3O^+PF_6^-$, as shown by ^{19}F and ^{31}P nmr spectroscopy. Raman studies show that KNO_3 solvolyses according to the reaction

$$KNO_3 + 6HF \longrightarrow K^+ + NO_2^+ + H_3O^+ + 3HF_2^-$$

Permanganates and chromates are solvolysed by HF to oxide fluorides such as MnO_3F and CrO_2F_2.

Acid–base reactions in anhydrous HF are well documented. Within the Brønsted formalism, few if any acids would be expected to be sufficiently strong proton donors to be able to protonate the very strong proton-donor HF (p. 56), and this is borne out by observation. Conversely, HF can protonate many Brønsted bases, notably water, alcohols carboxylic acids, and other organic compounds having one or more lone-pairs on O, N, etc.:

$$H_2O + 2HF \rightleftharpoons H_3O^+ + HF_2^-$$

$$RCH_2OH + 2HF \longrightarrow RCH_2OH_2^+ + HF_2^-$$

$$RCO_2H + 2HF \longrightarrow RC(OH)_2^+ + HF_2^-$$

Alternatively, within the Lewis formalism, acids are fluoride-ion acceptors. The prime examples are AsF_5 and SbF_5 (which give MF_6^-) and to a lesser extent BF_3 which yields BF_4^-. A greater diversity is found amongst Lewis bases (fluoride-ion donors), typical examples being XeF_6, SF_4, ClF_3, and BrF_3:

$$MF_n + HF \longrightarrow MF_{n-1}^+ + HF_2^-$$

Such solutions can frequently be "neutralized" by titration with an appropriate Lewis acid, e.g.:

$$BrF_2^+HF_2^- + H_2F^+SbF_6^- \longrightarrow BrF_2^+SbF_6^- + 3HF$$

Oxidation-reduction reactions in HF form a particularly important group of reactions with considerable industrial application. The standard electrode potentials $E°(M^{n+}/M)$ in HF follow the same sequence as for H_2O though individual values in the two series may differ by up to ± 0.2 V. Early examples showed that CrF_2 and UF_4 reduced HF to H_2 whereas VCl_2 gave VF_3, 2HCl, and H_2. Of more significance is the very high potential needed for the anodic oxidation of F^- in HF:

$$F^- \rightleftharpoons \tfrac{1}{2}F_2 + e^-; \quad E°(F_2/2F^-) = 2.71 \text{ V at } 0°C$$

This enables a wide variety of inorganic and organic fluorinations to be effected by the electrochemical insertion of fluorine. For example, the production of NFH_2, NF_2H, and NF_3 by electrolysis of NH_4F in liquid HF represents the only convenient route to these compounds. Again, CF_3CO_2H is most readily obtained by electrolysis of CH_3CO_2H in HF. Other examples of anodic oxidations in HF are as follows:

Reactant	Products	Reactant	Products
NH_4F	NF_3, NF_2H, NFH_2	NMe_3	$(CF_3)_3N$
H_2O	OF_2	$(MeCO)_2O$	CF_3COF
SCl_2, SF_4	SF_6	SMe_2, CS_2	CF_3SF_5, $(CF_3)_2SF_4$
$NaClO_4$	ClO_3F	$MeCN$	CF_3CN, $C_2F_5NF_2$

The other hydrogen halides are less tractable as solvents, as might be expected from their physical properties (p. 949), especially their low bps, short liquid ranges, low dielectric constants, and negligible self-dissociation into ions. Nevertheless, they have received some attention, both for comparison with HF, and as preparative media with their own special advantages.[29, 35, 36] In particular, because of their low bp and consequent ease of removal, the liquid HX solvent systems have provided convenient routes to BX_4^-, BF_3Cl^-, $B_2Cl_6^{2-}$, NO_2Cl, $Al_2Cl_7^-$, R_2SCl^+, $RSCl_2^+$, PCl_3Br^+, $Ni_2Cl_4(CO)_3$ (from nickel tetracarbonyl and Cl_2), and $Ni(NO)_2Cl_2$ (from nickel tetracarbonyl and NOCl). Solubilities in liquid HX are generally much smaller than in HF and tend to be restricted to molecular compounds (e.g. NOCl, PhOH, etc.) or salts with small lattice energies, e.g. the tetraalkylammonium halides. Concentrations rarely attain $0.5 \, mol \, l^{-1}$ (i.e. $0.05 \, mol/100 \, g \, HF$). Ready protonation of compounds containing lone-pairs or π bonds is observed, e.g. amines, phosphines, ethers, sulfides, aromatic olefins, and compounds containing $-C\equiv N$, $-N=N-$, $>C=O$, $>P=O$, etc. Of particular interest is the protonation of phosphine in the presence of BX_3 to give $PH_4^+BCl_4^-$, $PH_4^+BF_3Cl^-$, and $PH_4^+BBr_4^-$. $Fe(CO)_5$ affords $[Fe(CO)_5H]^+$ and $[Fe(\eta^5-C_5H_5)(CO)_2]_2$ yields $[Fe(\eta^5-C_5H_5)(CO)_2]_2H^+$. Solvolysis is also well established:

$$Ph_3SnCl + HCl \longrightarrow Ph_2SnCl_2 + PhH$$

$$Ph_3COH + 3HCl \longrightarrow Ph_3C^+HCl_2^- + H_3O^+Cl^-$$

Likewise ligand replacement reactions and oxidations, e.g.:

$$Me_4N^+HCl_2^- + BCl_3 \longrightarrow Me_4N^+BCl_4^- + HCl$$

$$PCl_3 + Cl_2 + HCl \longrightarrow PCl_4^+HCl_2^-$$

The preparation and structural characterization of the ions HX_2^- has been an important feature of such work.[29] As expected, these H-bonded ions are much less stable than HF_2^- though crystalline salts of all three anions and of the mixed anions HXY^- (except $HBrI^-$) have been isolated by use of large counter cations, typically Cs^+ and NR_4^+ (R = Me, Et, Bu^n)—see pp. 1313–21, of ref. 17 for further details. Neutron and X-ray diffraction studies suggest that $[Cl-H\cdots Cl]^-$ can be either centrosymmetric or non-centrosymmetric depending on the crystalline environment. A recent example of the latter mode involves interatomic distances of 145 and 178 pm respectively and a bond angle of 168° ($Cl\cdots Cl$ 321.2 pm).[37]

17.2.2 *Halides of the elements*

The binary halides of the elements span a wide range of stoichiometries, structure types, and properties which defy any but the most grossly oversimplified attempt at a unified

[35] M. E. PEACH and T. C. WADDINGTON, The higher hydrogen halides as ionizing solvents, Chap. 3 in T. C. WADDINGTON (ed.), *Nonaqueous Solvent Systems*, pp. 83–115, Academic Press, London, 1965.

[36] F. KLANBERG, Liquid HCl, HBr, and HI, Chap. 1 in J. J. LAGOWSKI (ed.), *The Chemistry of Nonaqueous Solvents*, Vol. 2, pp. 1–41, Academic Press, New York, 1967.

[37] W. KUCHEN, D. MOOTZ, H. SOMBERG, H. WUNDERLICH, and H.-G. WUSSOW, Diorganodichlorophosphonium hydrogen dichlorides, a novel class of compounds containing [ClHCl]⁻ ions, *Angew. Chem.*, Int. Edn. (Engl.) **17**, 869–70 (1978).

classification. Indeed, interest in the halides as a class of compound derives in no small measure from this very diversity, and from the fact that, being so numerous, there are many examples of well-developed and well-graded trends between the limiting cases. Thus the fluorides alone include OF_2, one of the most volatile molecular compounds known (bp $-145°$), and CaF_2, which is one of the least-volatile "ionic" compounds (bp $2513°C$). Between these extremes of discrete molecules on the one hand, and 3D lattices on the other, is a continuous sequence of oligomers, polymers, and extended layer lattices which may be either predominantly covalent [e.g. ClF, $(MoF_5)_4$, $(CF_2)_\infty$, $(CF)_\infty$, p. 309] or substantially ionic [e.g. Na^+F^- (g), $(SnF_2)_4$, $(BeF_2)_\infty$ (quartz type), SnF_4, NaF(cryst)], or intermediate in bond type with secondary interactions also complicating the picture. The problems of classifying binary compounds according to presumed bond types or limiting structural characteristics have already been alluded to for the hydrides (p. 69), borides (p. 162), oxides, sulfides, etc. Such diversity and gradations are further compounded by the existence of four different halogens (F, Cl, Br, I) and by the possibility of numerous oxidation states of the element being considered, e.g. CrF_2, Cr_2F_5, CrF_3, CrF_4, CrF_5, and CrF_6 or S_2F_2, SF_2, SF_4, S_2F_{10}, SF_6.

A detailed discussion of individual halides is given under the chemistry of each particular element. This section deals with more general aspects of the halides as a class of compound and will consider, in turn, general preparative routes, structure, and bonding. For reasons outlined on p. 939, fluorides tend to differ from the other halides either in their method of synthesis, their structure, or their bond-type. For example, the fluoride ion is the smallest and least polarizable of all anions and fluorides frequently adopt 3D "ionic" structures typical of oxides. By contrast, chlorides, bromides, and iodides are larger and more polarizable and frequently adopt mutually similar layer-lattices or chain structures (cf. sulfides). Numerous examples of this dichotomy can be found in other chapters and in several general references.[38–41] Because of this it is convenient to discuss fluorides as a group first, and then the other halides.

Fluorides

Binary fluorides are known with stoichiometries that span the range from C_4F to IF_7 (or even, possibly, XeF_8). Methods of synthesis turn on the properties of the desired products.[40, 42–44] If hydrolysis poses no problem, fluorides can be prepared by halide metathesis in aqueous solution or by the reactions of aqueous hydrofluoric acid with an

[38] V. GUTMANN (ed.), *Halogen Chemistry*, Academic Press, London, 1967: Vol. 1, 473 pp.; Vol. 2, 481 pp; Vol. 3, 471 pp.

[39] R. COLTON and J. H. CANTERFORD, *Halides of the First Row Transition Elements*, Wiley, London, 1969, 579 pp.; *Halides of the Second and Third Row Transition Elements*, Wiley, London, 1968, 409 pp.

[40] Ref. 34, pp. 1062–1106; ref. 17, pp. 1232–80.

[41] A. F. WELLS, Simple halides, in *Structural Inorganic Chemistry*, pp. 345–76, Oxford University Press, Oxford, 1975.

[42] E. L. MUETTERTIES and C. W. TULLOCK, Binary fluorides, Chap. 7 in W. L. JOLLY (ed.), *Preparative Inorganic Reactions*, Vol. 2, pp. 237–99 (1965). R. J. LAGOW and L. J. MARGRAVE, Direct fluorination: a "new" approach to fluorine chemistry, *Prog. Inorg. Chem.* **26**, 161–210 (1979). M. R. C. GERSTENBERGER and A. HAAS, Methods of fluorination in organic chemistry, *Angew. Chem.*, Int. Edn. (Engl.) **20**, 647–67 (1981).

[43] J. PORTIER, Solid state chemistry of ionic fluorides, *Angew. Chem.*, Int. Edn. (Engl.) **15**, 475–86 (1976).

[44] R. D. PEACOCK, Transition metal pentafluorides and related compounds, *Adv. Fluorine Chem.* **7**, 113–45 (1973).

appropriate oxide, hydroxide, carbonate, or the metal itself. The following non-hydrated fluorides precipitate as easily filterable solids:

LiF	MgF_2	SnF_2	SbF_3
NaF	CaF_2	PbF_2	
NH_4F	SrF_2		
	BaF_2		

Gaseous SiF_4 and GeF_4 can also be prepared from aqueous HF. Furthermore, the following fluorides separate as hydrates that can readily be dehydrated thermally, though an atmosphere of HF is required to suppress hydrolysis except in the case of the univalent metal fluorides:

$KF.2H_2O$	$CuF_2.4H_2O$	$AlF_3.H_2O$
$RbF.3H_2O$	$ZnF_2.4H_2O$	$GaF_3.3H_2O$
$CsF.1\frac{1}{2}H_2O$	$CdF_2.4H_2O$	$InF_3.3H_2O$
$TlF.2HF.\frac{1}{2}H_2O$	$HgF_2.2H_2O$	$LnF_3.xH_2O$
$AgF.4H_2O$	$MF_2.6H_2O$	(Ln = lanthanide metal)
	(M = Fe, Co, Ni)	

By contrast $BeF_2.xH_2O$, $TiF_4.2H_2O$, and $ThF_4.4H_2O$ cannot be dehydrated without hydrolysis.

When hydrolysis is a problem then the action anhydrous HF on the metal (or chloride) may prove successful (e.g. the difluorides of Zn, Cd, Ge, Sn, Mn, Fe, Co, Ni; the trifluorides of Ga, In, Ti, and the lanthanides; the tetrafluorides of Ti, Zr, Hf, Th, and U; and the pentafluorides of Nb and Ta). However, many higher fluorides require the use of a more aggressive fluorinating agent or even F_2 itself. Typical of the fluorides prepared by oxidative fluorination with F_2 are:

difluorides:	Ag, Xe
trifluorides:	Cl, Br, Mn, Co
tetrafluorides:	Sn, Pb, Kr, Xe, Mo, Mn, Ce, Am, Cm
pentafluorides:	As, Sb, Bi, Br, I, V, Nb, Ta, Mo
hexafluorides:	S, Se, Te, Xe, Mo, W, Tc, Ru, Os, Rh, Ir, Pt, U, Np, Pu
heptafluorides:	I, Re
octafluorides:	Xe(?)

Wherever possible the use of elementary F_2 is avoided because of its cost and the difficulty of handling it; instead one of a graded series of halogen fluorides can often be used, the fluorinating power steadily diminishing in the sequence: $ClF_3 > BrF_5 > IF_7 > ClF > BrF_3 > IF_5$. Other "hard" oxidizing fluorinating agents are AgF_2, CoF_3, MnF_3, PbF_4, CeF_4, BiF_5, and UF_6. When selective fluorination of certain groups in organic compounds is required, then "moderate" fluorinating agents are employed, e.g. HgF_2, SbF_5, $SbF_3/SbCl_5$, AsF_3, CaF_2, or KSO_2F. Such nucleophilic reagents may replace other halogens in halohydrocarbons by F but rarely substitute F for H. An electrophilic variant is ClO_3F. Most recently XeF_2, which is available commercially, has been used to effect fluorinations via radical cations: it can oxidatively fluorinate CC double bonds and can replace either aliphatic or aromatic H atoms with F. Even gentler are the "soft" fluorinating agents which do not cause fragmentation of functional groups, do not saturate double bonds, and do not oxidize metals to their highest oxidation states; typical of such mild fluorinating agents are the monofluorides of H, Li, Na, K, Rb, Cs, Ag, and Tl and compounds such as SF_4, SeF_4, COF_2, SiF_4, and $NaSiF_6$.

The fluorination reactions considered so far can be categorized as metathesis, oxidation, or substitution. Occasionally reductive fluorination is the preferred route to a lower fluoride. Examples are:

$$2PdF_3 + SeF_4 \xrightarrow{\text{warm}} 2PdF_2 + SeF_6$$

$$6ReF_6 + W(CO)_6 \xrightarrow{\text{room temp}} 6ReF_5 + WF_6 + 6CO$$

$$2RuF_5 + \tfrac{1}{x}I_2 \xrightarrow{50°} 2RuF_4 + \tfrac{2}{x}IF_x$$

$$2EuF_3 + H_2 \xrightarrow{1100°} 2EuF_2 + 2HF$$

$$6ReF_7 + Re \xrightarrow{400°} 7ReF_6$$

Further examples of this last type of reductive fluorination in which the element itself is used to reduce its higher fluoride are:

Product	ClF	CrF_2	GeF_2	MoF_3	UF_3	IrF_4	TeF_4
Reactants	Cl_2/ClF_3	Cr/CrF_3	Ge/GeF_4	Mo/MoF_5	U/UF_4	Ir/IrF_6	Te/TeF_6
$T/°C$	350	1000	300	400	1050	170	180

The final route to fluorine compounds is electrofluorination (anodic fluorination) usually in anhydrous or aqueous HF. The preparation of NF_xH_{3-x} ($x = 1, 2, 3$) has already been described (p. 956). Likewise the only reliable route to OF_2 is the electrolysis of 80% HF in the presence of dissolved MF (p. 748). Perchloryl fluoride has been made by electrolysing $NaClO_4$ in HF but a simpler route (p. 1031) is the direct reaction of a perchlorate with fluorosulfuric acid:

$$KClO_4 + HSO_3F \longrightarrow KHSO_4 + ClO_3F$$

Electrolysis of organic sulfides in HF affords a variety of fluorocarbon derivatives:

$$Me_2S \text{ or } CS_2 \longrightarrow CF_3SF_5 \text{ and } (CF_3)_2SF_4$$

$$(-CH_2S-)_3 \longrightarrow (-CF_2SF_4-)_3, \ CF_3SF_5 \text{ and } SF_5CF_2SF_5$$

$$R_2S \longrightarrow R_fSF_5 \text{ and } (R_f)_2SF_4$$

where R_f is a perfluoroalkyl group.

The application of the foregoing routes has led to the preparation and characterization of fluorides of virtually every element in the periodic table except the three lightest noble gases, He, Ne, and Ar. The structures, bonding, reactivity, and industrial applications of these compounds will be found in the treatment of the individual elements and it is an instructive exercise to gather this information together in the form of comparative tables.[2, 40, 42–44]

Chlorides, bromides, and iodides

A similar set of preparative routes is available as were outlined above for the fluorides, though the range of applicability of each method and the products obtained sometimes vary from halogen to halogen. When hydrolysis is not a problem or when hydrated halides are sought, then wet methods are available, e.g. dissolution of a metal or its oxide, hydroxide, or carbonate in aqueous hydrohalic acid followed by evaporative crystallization:

$$Fe + 2HCl(aq) \longrightarrow [Fe(H_2O)_6]Cl_2 + H_2$$

$$CoCO_3 + 2HI(aq) \longrightarrow [Co(H_2O)_6]I_2 + H_2O + CO_2$$

Dehydration can sometimes be effected by controlled removal of water using a judicious combination of gentle warming and either reduced pressure or the presence of anhydrous HX:

$$[M(H_2O)_6]Br_3 \xrightarrow[\text{reduced press}]{70-170°} MBr_3 + 6H_2O \ (M = Ln \text{ or actinide})$$

$$CuCl_2 . 2H_2O \xrightarrow{HCl(g)/150°} CuCl_2 + 2H_2O$$

Hydrated chlorides that are susceptible to hydrolysis above room temperature can often be dehydrated by treating them with $SOCl_2$ under reflux:

$$[Cr(H_2O)_6]Cl_3 + 6SOCl_2 \xrightarrow{79°} CrCl_3 + 12HCl + 6SO_2$$

Alternative wet routes to hydrolytically stable halides are metathetical precipitation and reductive precipitation reactions, e.g.:

$$Ag^+(aq) + Cl^-(aq) \longrightarrow AgCl$$

$$Cu^{2+}(aq) + 2I^-(aq) \longrightarrow CuI + \tfrac{1}{2}I_2$$

More complex is the hydrolytic disproportionation of the molecular halogens themselves in aqueous alkali which is a commercial route to several alkali-metal halides:

$$3X_2 + 6OH^- \longrightarrow 5X^- + XO_3^- + 3H_2O$$

When the desired halide is hydrolytically unstable then dry methods must be used, often at elevated temperatures. Pre-eminent amongst these methods is the oxidative halogenation of metals (or non-metals) with X_2 or HX; when more than one oxidation state is available X_2 sometimes gives the higher and HX the lower, e.g.:

$$Cr + \tfrac{3}{2}Cl_2 \xrightarrow{600°} CrCl_3$$

$$Cr + 2HCl(g) \xrightarrow{\text{red heat}} CrCl_2 + H_2$$

Similarly, Cl_2 sometimes yields a higher and Br_2 a lower oxidation state, e.g. $MoCl_5$ and $MoBr_3$.

Other routes include the high-temperature halogenation of metal oxides, sometimes in the presence of carbon, to assist removal of oxygen; the source of halogen can be X_2, a volatile metal halide, CX_4, or another organic halide. A few examples of the many reactions that have been used industrially or for laboratory scale preparations are:

$$ZrO_2 \xrightarrow{\text{Cl}_2} ZrCl_4$$

$$Ta_2O_5 \xrightarrow[>460]{\text{C + Br}_2} TaBr_5$$

$$Nb_2O_5 \xrightarrow[370]{\text{CBr}_4} NbBr_5$$

$$UO_3 \xrightarrow[\text{reflux}]{\text{CCl}_2\,=\,\text{CClCCl}_3} UCl_4$$

$$MoO_2 \xrightarrow[230]{\text{AlI}_3} MoI_2$$

The last two of these reactions also feature a reduction in oxidation state. A closely related route is halogen exchange usually in the presence of an excess of the "halogenating reagent", e.g.:

$$FeCl_3 + BBr_3 \text{ (excess)} \longrightarrow FeBr_3 + BCl_3$$

$$MCl_3 + 3HBr \text{ (excess)} \xrightarrow{400-600} MBr_3 + 3HCl \text{ (M = Ln or Pu)}$$

$$3TaCl_5 + 5AlI_3 \text{ (excess)} \xrightarrow{400} 3TaI_5 + 5AlCl_3$$

Reductive halogenation can be achieved by reducing a higher halide with the parent metal, another metal, or hydrogen:

$$TaI_5 + Ta \xrightarrow[630\,\rightarrow 575]{\text{thermal gradient}} Ta_6I_{14}$$

$$3WBr_5 + Al \xrightarrow[475\,\rightarrow 240]{\text{thermal gradient}} 3WBr_4 + AlBr_3$$

$$MX_3 + \tfrac{1}{2}H_2 \longrightarrow MX_2 + HX \text{ (M = Sm, Eu, Yb, etc.,}$$
$$X = Cl, Br, I)$$

Alternatively, thermal decomposition or disproportionation can yield the lower halide:

$$ReCl_5 \xrightarrow{\text{at "bp"}} ReCl_3 + Cl_2$$

$$MoI_3 \xrightarrow{100^\circ} MoI_2 + \tfrac{1}{2}I_2$$

$$AuCl_3 \xrightarrow{160^\circ} AuCl + Cl_2$$

$$2TaBr_4 \xrightarrow{500^\circ} TaBr_3 + TaBr_5$$

Many significant trends are apparent in the structures of the halides and in their physical and chemical properties. The nature of the element concerned, its position in the periodic table, the particular oxidation state, and, of course, the particular halogen involved, all play a role. The majority of pre-transition metals (Groups IA, IIA) together with Group IIIA, the lanthanides, and the actinides in the +2 and +3 oxidation states form halides that are predominantly ionic in character, whereas the non-metals and metals in higher oxidation states ($\geqslant +3$) tend to form covalent molecular halides. The "ionic-covalent transition" in the halides of Group VB (P, As, Sb, Bi) and VIB (S, Se, Te, Po) has already been discussed at length (pp. 568, 652, 904) as has the tendency of the refractory transition metals to form cluster halides (pp. 1155, 1190, etc.). The problems associated with the ionic bond model and its range of validity were considered in Chapter 4 (p. 91). Presumed bond types tend to show gradual rather than abrupt changes within series in which the central element, the oxidation state, or the halogen are systematically varied. For example, in a sequence of chlorides of isoelectronic metals such as KCl, $CaCl_2$, $ScCl_3$, and $TiCl_4$ the first member is predominantly ionic with a 3D lattice of octahedrally coordinated potassium ions; $CaCl_2$ has a framework structure (distorted rutile) in which Ca is surrounded by a distorted octahedron of 6Cl; $ScCl_3$ has a layer structure and $TiCl_4$ is a covalent molecular liquid. The sudden discontinuity in physical properties at $TiCl_4$ is more a function of stoichiometry and coordination number than a sign of any discontinuous or catastrophic change in bond type. Numerous other examples can be found amongst the transition metal halides and the halides of the post-transition elements. In general, the greater the difference in electronegativity between the element and the halogen the greater will be the tendency to charge separation and the more satisfactory will be the ionic bond model. With increasing formal charge on the central atom or with decreasing electronegativity difference the more satisfactory will be the various covalent bond models. The complexities of the situation can be illustrated by reference to the bp (and mp) of the halides: for the more ionic halides these generally follow the sequence $MF_n > MCl_n > MBr_n > MI_n$, being dominated by coulombic interactions which are greatest for the small F^- and least for the large I^-, whereas for molecular halides the sequence is usually the reverse, viz. $MI_n > MBr_n > MCl_n > MF_n$ being dictated rather by polarizability and London dispersion forces which are greatest for I and least for F. As expected, intermediate halides are less regular as the first sequence yields to the reverse, and no general pattern can be discerned.

Similar observations hold for solubility. Predominantly ionic halides tend to dissolve in polar, coordinating solvents of high dielectric constant, the precise solubility being dictated by the balance between lattice energies and solvation energies of the ions, on the one hand, and on entropy changes involved in dissolution of the crystal lattice, solvation of the ions, and modification of the solvent structure, on the other: [ΔG (cryst→saturated soln)$= 0 = \Delta H - T\Delta S$]. For a given cation (e.g. K^+, Ca^{2+}) solubility in water typically follows the sequence $MF_n < MCl_n < MBr_n < MI_n$. By contrast for less-ionic halides with significant non-coulombic lattice forces (e.g. Ag) solubility in water follows the reverse sequence $MI_n < MBr_n < MCl_n < MF_n$. For molecular halides solubility is determined principally by weak intermolecular van der Waals' and dipolar forces, and dissolution is commonly favoured by less-polar solvents such as benzene, CCl_4, or CS_2.

Trends in chemical reactivity are also apparent, e.g. ease of hydrolysis tends to increase from the non-hydrolysing predominantly ionic halides, through the intermediate halides to the readily hydrolysable molecular halides. Reactivity depends both on the relative energies of M–X and M–O bonds and also, frequently, on kinetic factors which may hinder or even prevent the occurrence of thermodynamically favourable reactions. Further trends become apparent within the various groups of halides and are discussed at appropriate points throughout the text.

17.2.3 *Interhalogen compounds*[45-47]

The halogens combine exothermically with each other to form interhalogen compounds of four stoichiometries: XY, XY_3, XY_5, and XY_7 where X is the heavier halogen. A few ternary compounds are also known, e.g. $IFCl_2$ and IF_2Cl. For the hexatomic series, only the fluorides are known (ClF_5, BrF_5, IF_5), and IF_7 is the sole example of the octatomic series. All the interhalogen compounds are diamagnetic and contain an even number of halogen atoms. Similarly, the closely related polyhalide anions XY_{2n}^- and polyhalonium cations XY_{2n}^+ ($n = 1, 2, 3$) each have an odd number of halogen atoms: these ions will be considered in subsequent sections (pp. 978, 983).

Related to the interhalogens chemically, are compounds formed between a halogen atom and a pseudohalogen group such as CN, SCN, N_3. Examples are the linear molecules ClCN, BrCN, ICN, and the corresponding compounds XSCN and XN_3. Some of these compounds have already been discussed (p. 336) and need not be considered further. A recent microwave study[48] shows that chlorine thiocyanate is ClSCN (angle Cl–S–C 99.8°) rather than ClNCS, in contrast to the cyanate which is ClNCO. The chemistry of iodine azide has been reviewed[49]—it is obtained as volatile, golden yellow, shock-sensitive needles by reaction of I_2 with AgN_3 in non-oxygen-containing solvents such as CH_2Cl_2, CCl_4, or benzene: the structure in the gas phase (as with FN_3, ClN_3, and BrN_3 also) comprises a linear N_3 group joined at an obtuse angle to the pendant X atom, thereby giving a molecule of C_s symmetry.

Diatomic interhalogens, XY

All six possible diatomic compounds between F, Cl, Br, and I are known. Indeed, ICl was first made (independently) by J. L. Gay Lussac and H. Davy in 1813–14 soon after the isolation of the parent halogens themselves, and its existence led J. von Liebig to miss the discovery of the new element bromine, which has similar properties (p. 925). The compounds vary considerably in thermal stability: ClF is extremely robust; ICl and IBr are moderately stable and can be obtained in very pure crystalline form at room temperature; BrCl readily dissociates reversibly into its elements; BrF and IF disproportionate rapidly and irreversibly to a higher fluoride and Br_2 (or I_2). Thus,

[45] Ref. 17, pp. 1476–1563, Interhalogens and polyhalide anions; see also D. M. MARTIN, R. ROUSSON, and J. M. WEULERSSE, The Interhalogens, in J. J. LAGOWSKI (ed.), *The Chemistry of Nonaqueous Solvents*, Chap. 3, pp. 157–95, Academic Press, New York, 1978.

[46] A. I. POPOV, Interhalogen compounds and polyhalide anions, Chap. 2, in V. GUTMANN (ed.), *MTP International Review of Science: Inorganic Chemistry Series* 1, Vol. 3, pp. 53–84, Butterworths, London, 1972.

[47] K. O. CHRISTE, Halogen fluorides, *IUPAC Additional Publication 24th Int. Congr. Pure Appl. Chem.*, Hamburg, 1973, Vol. 4. *Compounds of Non-Metals*, pp. 115–41, Butterworths, London, 1974.

[48] R. J. RICHARDS, R. W. DAVIS, and M. C. L. GERRY, The microwave spectrum and structure of chlorine thiocyanate, *JCS Chem. Comm.* 1980, 915–16.

[49] K. DEHNICKE, The chemistry of iodine azide, *Angew. Chem.*, Int. Edn. (Engl.) **18**, 507–14 (1979).

although all six compounds can be formed by direct, controlled reaction of the appropriate elements, not all can be obtained in pure form by this route. Typical preparative routes (with comments) are as follows:

$$Cl_2 + F_2 \xrightarrow{225°} 2ClF; \quad \text{must be purified from } ClF_3 \text{ and reactants}$$

$$Cl_2 + ClF_3 \xrightarrow{300°} 3ClF; \quad \text{must be purified from excess } ClF_3$$

$$Br_2 + F_2 \xrightarrow{\text{gas phase}} 2BrF; \quad \text{disproportionates to } Br_2 + BrF_3 \text{ (and } BrF_5 \text{) at room temp}$$

$$Br_2 + BrF_3 \longrightarrow 3BrF; \quad \text{BrF favoured at high temp}$$

$$I_2 + F_2 \xrightarrow[\text{at } -45°]{\text{in } CCl_3F} 2IF; \quad \text{disproportionates rapidly to } I_2 + IF_5 \text{ at room temp}$$

$$I_2 + IF_3 \xrightarrow[\text{at } -78°]{\text{in } CCl_3F} 3IF$$

$$I_2 + AgF \xrightarrow{0°} IF + AgI$$

$$Br_2 + Cl_2 \underset{\text{or in } CCl_4}{\overset{\text{gas phase}}{\rightleftharpoons}} 2BrCl; \quad \text{compound cannot be isolated free from } Br_2 \text{ and } Cl_2$$

$$I_2 + X_2 \xrightarrow{\text{room temp}} 2IX \quad \text{purify by fractional crystallization of the molten}$$
$$(X = Cl, Br); \text{ compound}$$

In general the compounds have properties intermediate between those of the parent halogens, though a combination of aggressive chemical reactivity and/or thermal instability militates against the determination of physical properties such as mp, bp, etc., in some instances. However, even for such highly dissociated species as BrCl, precise molecular (as distinct from bulk) properties can be determined by spectroscopic techniques. Table 17.13 summarizes some of the more important physical properties of the diatomic interhalogens. The most volatile compound, ClF, is a colourless gas which condenses to a very pale yellow liquid below $-100°$. The least volatile is IBr; it forms black crystals in which the IBr molecules pack in a herring-bone pattern similar to that in I_2 (p. 937) and in which the internuclear distance $r(I–Br)$ is 252 pm, i.e. slightly longer than in the gas phase (248.5 pm). ICl is unusual in forming two crystalline modifications: the stable (α) form crystallizes as large, transparent ruby-red needles from the melt and features zigzag chains of molecules with two different ICl units and appreciable inter-chain intermolecular bonding as in Fig. 17.7a. The packing is somewhat different in the yellow, metastable (β) form (Fig. 17.7b) which can be obtained as brownish-red crystals from strongly supercooled melts.

The chemical reactions of XY can be conveniently classified as halogenation reactions, donor-acceptor interactions, and use as solvent systems. Reactions frequently parallel those of the parent halogens but with subtle and revealing differences. ClF is an effective

TABLE 17.13   Physical properties of interhalogen compounds XY

Property	ClF	BrF	IF	BrCl	ICl	IBr
Form at room temperature	Colourless gas	Pale brown (Br$_2$)	Unstable	Red brown gas	Ruby red crystals	Black crystals
MP/°C	−155.6	ca. −33 Disprop[a]	— Disprop[a]	ca. −66 Dissoc[a]	27.2(α) 13.9(β)	41 Some dissoc
BP/°C	−100.1	ca. 20	—	ca. 5	97–100[b]	~116[b]
ΔH_f°(298 K)/kJ mol^{-1}	−56.5	−58.6	−95.4	+14.6	−35.3(α)	−10.5 (cryst)
ΔG_f°(298 K)/kJ mol^{-1}	−57.7	−73.6	−117.6	−1.0	−13.95(α)	+3.7(gas)
Dissociation energy/ kJ mol^{-1}	252.5	248.6	~277	215.1	207.7	175.4
d(liq, T°C)/g cm^{-3}	1.62(−100°)	—	—	—	3.095(30°)	3.762(42°)
r(X–Y)/pm	162.81	175.6	190.9	213.8	232.07	248.5
Dipole moment/D	0.881	1.29	—	0.57	0.65	1.21
κ(liq, T°C)/ ohm^{-1} cm^{-1}	1.9×10^{-7} (−128°)	—	—	—	5.50×10^{-3}	3.4×10^{-4}

[a] Substantial disproportionation or dissociation prevents meaningful determination of mp and bp; the figures merely indicate the approximate temperature range over which the (impure) compound is liquid at atmospheric pressure.

[b] Fused ICl and IBr both dissociate into the free halogens to some extent: ICl 0.4% at 25° (supercooled) and 1.1% at 100°C; IBr 8.8% at 25° (supercooled) and 13.4% at 100°C.

fluorinating agent (p. 959) and will react with many metals and non-metals either at room temperature or above, converting them to fluorides and liberating chlorine, e.g.:

$$W + 6ClF \longrightarrow WF_6 + 3Cl_2$$

$$Se + 4ClF \longrightarrow SeF_4 + 2Cl_2$$

It can also act as a chlorofluorinating agent by addition across a multiple bond and/or by oxidation, e.g.:

$$(CF_3)_2CO + ClF \xrightarrow{\text{MF}} (CF_3)_2CFOCl \quad (M = K, Rb, Cs)$$

$$CO + ClF \longrightarrow COFCl$$

$$RCN + 2ClF \longrightarrow RCF_2NCl_2$$

$$SO_3 + ClF \longrightarrow ClOSO_2F$$

$$SO_2 + ClF \longrightarrow ClSO_2F$$

$$SF_4 + ClF \xrightarrow{\text{CsF}} SF_5Cl$$

$$N{\equiv}SF_3 + 2ClF \longrightarrow Cl_2NSF_5$$

Reaction with OH groups or NH groups results in the exothermic elimination of HF and the (often violent) chlorination of the substrate, e.g.:

$$HOH + 2ClF \longrightarrow 2HF + Cl_2O$$

$$HONO_2 + ClF \longrightarrow HF + ClONO_2$$

$$HNF_2 + ClF \longrightarrow HF + NF_2Cl$$

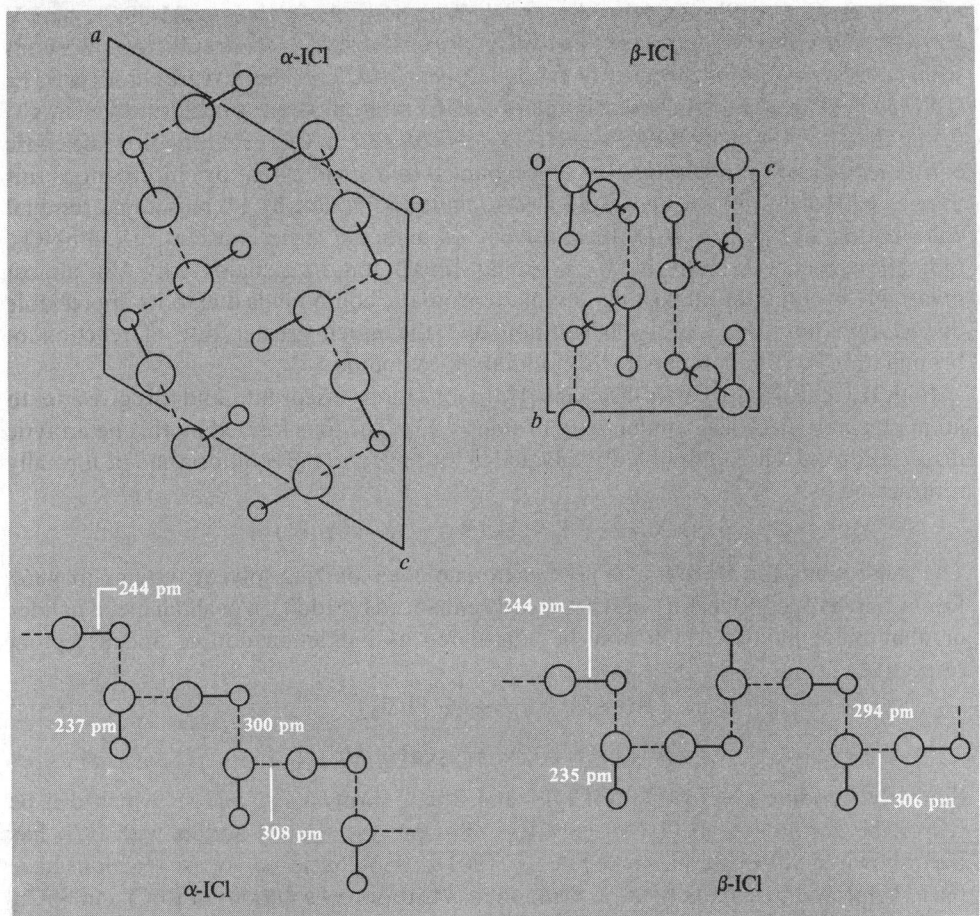

FIG. 17.7   Structures of α- and β-forms of crystalline ICl.

Lewis acid (fluoride ion acceptor) behaviour is exemplified by reactions with NOF and MF to give $[NO]^+[ClF_2]^-$ and $M^+[ClF_2]^-$ respectively (M = alkali metal or NH_4). Lewis base (fluoride ion donor) activity includes reactions with BF_3 and AsF_5:

$$BF_3 + 2ClF \longrightarrow [Cl_2F]^+[BF_4]^-$$

$$AsF_5 + 2ClF \longrightarrow [Cl_2F]^+[AsF_6]^-$$

The linear polyhalide anion $[F{-}Cl{-}F]^-$ and the angular polyhalonium cation $[F{\diagdown}Cl]^+$ are members of a more extensive set of ions to be treated on p. 978. ClF is commercially available in steel lecture bottles of 500-g capacity but must be handled with extreme circumspection in scrupulously dried and degreased apparatus constructed in steel, copper, Monel, or nickel; fluorocarbon polymers such as Teflon can also be used, but not at elevated temperatures.

The reactivity of ICl and IBr, though milder than that of ClF is nevertheless still extremely vigorous and the compounds react with most metals including Pt and Au, but not with B, C, Cd, Pb, Zr, Nb, Mo, or W. With ICl, phosphorus yields PCl_5 and V conveniently yields VCl_3 (rather than VCl_4). Reaction with organic substrates depends subtly on the conditions chosen. For example, phenol and salicylic acid are chlorinated by ICl *vapour*, since homolytic dissociation of the ICl molecule leads to chlorination by Cl_2 rather than iodination by the less-reactive I_2. By contrast, in CCl_4 solution (low dielectric constant) iodination predominates, accompanied to a small extent by chlorination: this implies heterolytic fission and rapid electrophilic iodination by I^+ plus some residual chlorination by Cl_2 (or ICl). In a solvent of high dielectric constant, e.g. $PhNO_2$, iodination occurs exclusively.[50] A similar interpretation explains why IBr almost invariably brominates rather than iodinates aromatic compounds due to its appreciable dissociation into Br_2 and I_2 in solution and the much greater rate of reaction of bromination by Br_2 compared with iodination by iodine.

Both ICl and IBr are partly dissociated into ions in the fused state, and this gives rise to an appreciable electrical conductivity (Table 17.13). The ions formed by this heterolytic dissociation of IX are undoubtedly solvated in the melt and the equilibria can be formally represented as

$$3IX \rightleftharpoons I_2X^+ + IX_2^- \quad (X = Cl, Br)$$

The compounds can therefore be used as nonaqueous ionizing solvent systems (p. 486). For example the conductivity of ICl is greatly enhanced by addition of alkali metal halides or aluminium halides which may be considered as halide-ion donors and acceptors respectively:

$$ICl + MCl \longrightarrow M^+[ICl_2]^-$$

$$2ICl + AlCl_3 \longrightarrow [I_2Cl]^+[AlCl_4]^-$$

Similarly pyridine gives $[pyI]^+[ICl_2]^-$ and $SbCl_5$ forms a 2:1 adduct which can be reasonably formulated as $[I_2Cl]^+[SbCl_6]^-$. By contrast, the 1:1 adduct with PCl_5 has been shown by X-ray studies to be $[PCl_4]^+[ICl_2]^-$. Solvoacid–solvobase reactions have been monitored by conductimetric titration; e.g. titration of solutions of RbCl and $SbCl_5$ in ICl (or of KCl and $NbCl_5$) shows a break at 1:1 molar proportions, whereas titration of NH_4Cl with $SnCl_4$ shows a break at the 2:1 mole ratio:

$$Rb^+[ICl_2]^- + [I_2Cl]^+[SbCl_6]^- \longrightarrow Rb^+[SbCl_6]^- + 3ICl$$

$$K^+[ICl_2]^- + [I_2Cl]^+[NbCl_6]^- \longrightarrow K^+[NbCl_6]^- + 3ICl$$

$$2NH_4{}^+[ICl_2]^- + [I_2Cl]^+{}_2[SnCl_6]^{2-} \longrightarrow [NH_4]^+{}_2[SnCl_6]^{2-} + 6ICl$$

The preparative utility of such reactions is, however, rather limited, and neither ICl or IBr has been much used except to form various mixed polyhalide species. Compounds must frequently be isolated by extraction rather than by precipitation, and solvolysis is a further complicating factor.

Tetra-atomic interhalogens, XY_3

The compounds to be considered are ClF_3, BrF_3, IF_3, and ICl_3 (I_2Cl_6). All can be prepared by direct reaction of the elements, but conditions must be chosen so as to avoid

[50] F. W. BENNETT and A. G. SHARP, Reactions of iodine monohalides in different solvents, *J. Chem. Soc.* 1950, 1383–4.

formation of mixtures of interhalogens of different stoichiometries. ClF_3 is best formed by direct fluorination of Cl_2 or ClF in the gas phase at 200–300° in Cu, Ni, or Monel metal apparatus. BrF_3 is formed similarly at or near room temperature and can be purified by distillation to give a pale straw-coloured liquid. With IF_3, which is only stable below $-30°$ the problem is to avoid the more facile formation of IF_5; this can be achieved either by the action of F_2 on I_2 suspended in CCl_3F at $-45°$ or more elegantly by the low-temperature fluorination of I_2 with XeF_2:

$$I_2 + 3XeF_2 \longrightarrow 2IF_3 + 3Xe$$

I_2Cl_6 is readily made as a bright-yellow solid by reaction of I_2 with an excess of liquid chlorine at $-80°$ followed by the low-temperature evaporation of the Cl_2; care must be taken with this latter operation, however, because of the very ready dissociation of I_2Cl_6 into ICl and Cl_2.

Physical properties are summarized in Table 17.14. Little is known of the unstable IF_3 but ClF_3 and BrF_3 are well-characterized volatile molecular liquids. Both have an unusual T-shaped structure of C_{2v} symmetry, consistent with the presence of 10 electrons in the valency shell of the central atom (Fig. 17.8). A notable feature of both structures is the slight deviation from colinearity of the apical F–X–F bonds, the angle being 175.0° for ClF_3 and 172.4° for BrF_3; this reflects the greater electrostatic repulsion of the nonbonding pair of electrons in the equatorial plane of the molecule. For each molecule the X–F$_{apical}$ distance is some 5–6% greater than the X–F$_{equatorial}$ distance but the mean X–F distance is very similar to that in the corresponding monofluoride. The structure of crystalline ICl_3 is quite different, being built up of planar I_2Cl_6 molecules separated by normal van der Waals' distances between the Cl atoms (Fig. 17.9). The terminal I–Cl distances are similar to those in ICl but the bridging I–Cl distances are appreciably longer.

ClF_3 is one of the most reactive chemical compounds known[51] and reacts violently with many substances generally thought of as inert. Thus it spontaneously ignites

TABLE 17.14 *Physical properties of interhalogen compounds* XY_3

Property	ClF_3	BrF_3	IF_3	I_2Cl_6
Form at room temperature	Colourless gas/liquid	Straw-coloured liquid	Yellow solid (decomp above $-28°$)	Bright yellow solid
MP/°C	-76.3	8.8	—	101 (16 atm)
BP/°C	11.8	125.8	—	—
ΔH_f°(298 K)/kJ mol^{-1}	-164 (g)	-301 (l)	ca. -485 (g) calc	-89.3 (s)
ΔG_f°(298 K)/kJ mol^{-1}	-124 (g)	-241 (l)	ca. -460 (g) calc	-21.5 (s)
Mean X–Y bond energy of XY_3/kJ mol^{-1}	174	202	ca. 275 (calc)	—
Density(T°C)/g cm^{-3}	1.885 (0°)	2.803 (25°)	—	3.111 (15°)
Dipole moment/D	0.557	1.19	—	—
Dielectric constant $\varepsilon(T°)$	4.75 (0°)	—	—	—
κ(liq, T°C)/ ohm^{-1} cm^{-1}	6.5×10^{-9} (0°)	8.0×10^{-3} (25°)	—	8.6×10^{-3} (102°)

[51] L. STEIN, Physical and chemical properties of halogen fluorides, in V. GUTMANN (ed.), *Halogen Chemistry*, Vol. 1, pp. 133–224, Academic Press, London, 1967.

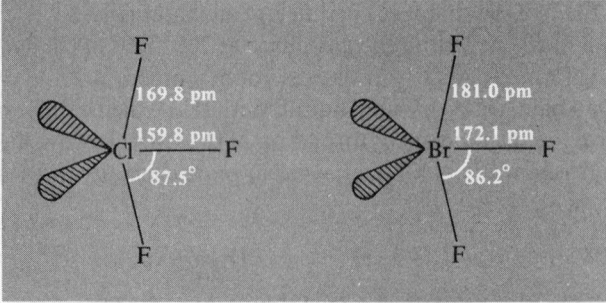

FIG. 17.8   Molecular structures of ClF_3 and BrF_3 as determined by microwave spectroscopy. An X-ray study of crystalline ClF_3 gave slightly longer distances (171.6 and 162.1 pm) and a slightly smaller angle (87.0°).

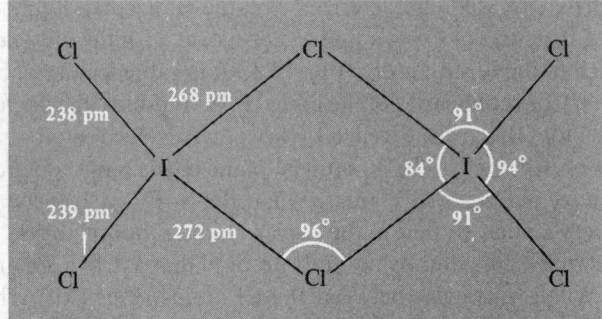

FIG. 17.9   Structure of I_2Cl_6 showing planar molecules of approximate D_{2h} symmetry.

asbestos, wood, and other building materials and was used in incendiary bomb attacks on UK cities during the Second World War. It reacts explosively with water and with most organic substances, though reaction can sometimes be moderated by dilution of ClF_3 with an inert gas, by dissolution of the organic compound in an inert fluorocarbon solvent, or by the use of low temperatures. Spontaneous ignition occurs with H_2, K, P, As, Sb, S, Se, Te, and powdered Mo, W, Rh, Ir, and Fe. Likewise, Br_2 and I_2 enflame and produce higher fluorides. Some metals (e.g. Na, Mg, Al, Zn, Sn, Ag) react at room temperature until a fluoride coating is established; when heated they continue to react vigorously. Palladium Pt, and Au are also attacked at elevated temperatures and even Xe and Rn are fluorinated. Mild steel can be used as a container at room temperature and Cu is only slightly attacked below 300° but the most resistant are Ni and Monel metal. Very pure ClF_3 has no effect on Pyrex or quartz but traces of HF, which are normally present, cause slow etching.

ClF_3 converts most chlorides to fluorides and reacts even with refractory oxides such as MgO, CaO, Al_2O_3, MnO_2, Ta_2O_5, and MoO_3 to form higher fluorides, e.g.:

$$AgCl + ClF_3 \longrightarrow AgF_2 + \tfrac{1}{2}Cl_2 + ClF$$

$$NiO + \tfrac{2}{3}ClF_3 \longrightarrow NiF_2 + \tfrac{1}{3}Cl_2 + \tfrac{1}{2}O_2$$

$$Co_3O_4 + 3ClF_3 \longrightarrow 3CoF_3 + \tfrac{3}{2}Cl_2 + 2O_2$$

With suitable dilution to moderate the otherwise violent reactions NH_3 gas and N_2H_4 yield HF and the elements:

$$NH_3 + ClF_3 \longrightarrow 3HF + \tfrac{1}{2}N_2 + \tfrac{1}{2}Cl_2$$

$$N_2H_4 + \tfrac{4}{3}ClF_3 \longrightarrow 4HF + N_2 + \tfrac{2}{3}Cl_2$$

At one time this latter reaction was used in experimental rocket motors, the ClF_3 oxidizer reacting spontaneously with the fuel (N_2H_4 or $Me_2N_2H_2$). At low temperatures NH_4F and NH_4HF_2 react with liquid ClF_3 when allowed to warm from -196 to $-5°$ but the reaction is hazardous and may explode above $-5°$:

$$NH_4F + \tfrac{5}{3}ClF_3 \longrightarrow NF_2Cl + 4HF + \tfrac{1}{3}Cl_2$$

The same products are obtained more safely by reacting gaseous ClF_3 with a suspension of NH_4F or NH_4HF_2 in a fluorocarbon oil.

ClF_3 is manufactured on a moderately large scale, considering its extraordinarily aggressive properties which necessitate major precautions during handling and transport. Production plant in Germany had a capacity of ~ 5 tonnes/day in 1940 (~ 1500 tonnes pa). It is now used in the USA, the UK, France, and the USSR primarily for nuclear fuel processing, production in the USA being several hundred tonnes pa (in 1979). ClF_3 is used to produce $UF_6(g)$:

$$U(s) + 3ClF_3(l) \xrightarrow{50-90} UF_6(l) + 3ClF(g)$$

It is also invaluable in separating U from Pu and other fission products during nuclear fuel reprocessing, since Pu reacts only to give the (involatile) PuF_4 and most fission products (except Te, I, and Mo) also yield involatile fluorides from which the UF_6 can readily be separated. ClF_3 is available in steel cylinders of up to 82 kg capacity and the price in 1979 was \$30 per kg.

Liquid ClF_3 can act both as a fluoride donor (Lewis base) or fluoride ion acceptor (Lewis acid) to give difluorochloronium compounds and tetrafluorochlorides respectively, e.g.:

$$MF_5 + ClF_3 \longrightarrow [ClF_2]^+[MF_6]^-; \quad \text{colourless solids: } M = As, Sb$$

$$PtF_5 + ClF_3 \longrightarrow [ClF_2]^+[PtF_6]^-; \quad \text{orange, paramagnetic solid, mp } 171°$$

$$BF_3 + ClF_3 \longrightarrow [ClF_2]^+[BF_4]^-; \quad \text{colourless solid, mp } 30°$$

$$MF + ClF_3 \longrightarrow M^+[ClF_4]^-; \quad \text{white or pink solids, decomp } \sim 350°: M = K, Rb, Cs$$

$$NOF + ClF_3 \longrightarrow [NO]^+[ClF_4]^-; \quad \text{white solid, dissociates below } 25°$$

Despite these reaction products there is little evidence for an ionic self-dissociation equilibrium in liquid ClF_3 such as may be formally represented by $2ClF_3 \rightleftharpoons ClF_2^+ + ClF_4^-$, and the electrical conductivity of the pure liquid (p. 969) is only of the order of 10^{-9} ohm^{-1} cm^{-1}. The structures of these ions is discussed more fully in subsequent sections.

Bromine trifluoride, though it reacts explosively with water and hydrocarbon tap greases, is somewhat less violent and vigorous a fluorinating agent than is ClF_3. The sequence of reactivity usually quoted for the halogen fluorides is:

$$ClF_3 > BrF_5 > IF_7 > ClF > BrF_3 > IF_5 > BrF > IF_3 > IF$$

It can be seen that, for a given stoichiometry of XF_n, the sequence follows the order $Cl > Br > I$ and for a given halogen the reactivity of XF_n diminishes with decrease in n, i.e. $XF_5 > XF_3 > XF$. (A possible exception is ClF_5; this is not included in the above sequence but, from the fragmentary data available, it seems likely that it should be placed near the beginning—perhaps between ClF_3 and BrF_5.) BrF_3 reacts vigorously with B, C, Si, As, Sb, I, and S to form fluorides. It has also been used to prepare simple fluorides from metals, oxides, and other compounds: volatile fluorides such as MoF_6, WF_6, and UF_6 distil readily from solutions in which they are formed whereas less-volatile fluorides such as AuF_3, PdF_3, RhF_4, PtF_4, and BiF_5 are obtained as residues on removal of BrF_3 under reduced pressure. Reaction with oxides often evolves O_2 quantitatively (e.g. B_2O_3, Tl_2O_3, SiO_2, GeO_2, As_2O_3, Sb_2O_3, SeO_3, I_2O_5, CuO, TiO_2, UO_3):

$$B_2O_3 + 2BrF_3 \longrightarrow 2BF_3 + Br_2 + \tfrac{3}{2}O_2$$

$$SiO_2 + \tfrac{4}{3}BrF_3 \longrightarrow SiF_4 + \tfrac{2}{3}Br_2 + O_2$$

The reaction can be used as a method of analysis and also as a procedure for determining small amounts of O (or N) in metals and alloys of Li, Ti, U, etc. In cases when BrF_3 itself only partially fluorinates the refractory oxides, the related reagents $KBrF_4$ and BrF_2SbF_6 have been found to be effective (e.g. for MgO, CaO, Al_2O_3, MnO_2, Fe_2O_3, NiO, CeO_2, Nd_2O_3, ZrO_2, ThO_2). Oxygen in carbonates and phosphates can also be determined by reaction with BrF_3. Sometimes partial fluorination yields new compounds, e.g. perrhenates afford tetrafluoroperrhenates:

$$MReO_4 + \tfrac{4}{3}BrF_3 \xrightarrow{\text{liq}} MReO_2F_4 + \tfrac{2}{3}Br_2 + O_2$$

$$M = K, Rb, Cs, Ag, \tfrac{1}{2}Ca, \tfrac{1}{2}Sr, \tfrac{1}{2}Ba$$

Likewise, $K_2Cr_2O_7$ and $Ag_2Cr_2O_7$ yield the corresponding $MCrOF_4$ (i.e. reduction from Cr^{VI} to Cr^V). Other similar reactions, which nevertheless differ slightly in their overall stoichiometry, are:

$$KClO_3 + \tfrac{5}{3}BrF_3 \longrightarrow KBrF_4 + \tfrac{2}{3}Br_2 + \tfrac{3}{2}O_2 + ClO_2F$$

$$ClO_2 + \tfrac{1}{3}BrF_3 \longrightarrow ClO_2F + \tfrac{1}{6}Br_2$$

$$N_2O_5 + \tfrac{1}{3}BrF_3 \longrightarrow \tfrac{1}{3}Br(NO_3)_3 + NO_2F$$

$$IO_2F + \tfrac{4}{3}BrF_3 \longrightarrow IF_5 + \tfrac{2}{3}Br_2 + O_2$$

As with ClF_3, BrF_3 is used to fluorinate U to UF_6 in the processing and reprocessing of nuclear fuel. It is manufactured commercially on the multitonne pa scale and is available as a liquid in steel cylinders of varying size up to 91 kg capacity. The US price in 1979 was $\sim\$30$ per kg.

In addition to its use as a straight fluorinating agent, BrF_3 has been extensively investigated and exploited as a preparative nonaqueous ionizing solvent. The appreciable electrical conductivity of the pure liquid (p. 969) can be interpreted in terms of the dissociative equilibrium

$$2BrF_3 \rightleftharpoons BrF_2^+ + BrF_4^-$$

Electrolysis gives a brown coloration at the cathode but no visible change at the anode:

$$2BrF_2^+ + 2e^- \longrightarrow BrF_3 + BrF \text{ (brown)}$$

$$2BrF_4^- \longrightarrow BrF_3 + BrF_5 + 2e^- \text{ (colourless)}$$

The specific conductivity decreases from 8.1×10^{-3} ohm^{-1} cm^{-1} at $10°$ to 7.1×10^{-3} ohm^{-1} cm^{-1} at $55°$ and this unusual behaviour has been attributed to the thermal instability of the BrF_2^+ and BrF_4^- ions at higher temperatures. Consistent with the above scheme KF, BaF_2 and numerous other fluorides (such as NaF, RbF, AgF, NOF) dissolve in BrF_3 with enhancement of the electrical conductivity due to the formation of the solvobases $KBrF_4$, $Ba(BrF_4)_2$, etc. Likewise, Sb and Sn give solutions of the solvoacids BrF_2SbF_6 and $(BrF_2)_2SnF_6$. Conductimetric titrations between these various species can be carried out, the end point being indicated by a sharp minimum in the conductivity:

$$BrF_2^+SbF_6^- + Ag^+BrF_4^- \longrightarrow Ag^+SbF_6^- + 2BrF_3$$

$$(BrF_2^+)_2SnF_6^{2-} + 2Ag^+BrF_4^- \longrightarrow (Ag^+)_2SnF_6^{2-} + 4BrF_3$$

Other solvoacids that have been isolated include the BrF_2^+ compounds of AuF_4^-, BiF_6^-, NbF_6^-, TaF_6^-, RuF_6^-, and PdF_6^{2-} and reactions of BrF_3 solutions have led to the isolation of large numbers of such anhydrous complex fluorides with a variety of cations.[51] Solvolysis sometimes complicates the isolation of a complex by evaporation of BrF_3 and solvates are also known, e.g. $K_2TiF_6 . BrF_3$ and $K_2PtF_6 . BrF_3$. It is frequently unnecessary to isolate the presumed reaction intermediates and the required complex can be obtained by the action of BrF_3 on an appropriate mixture of starting materials:

$$Ag + Au \xrightarrow{BrF_3} \{AgBrF_4 + BrF_2AuF_4\} \xrightarrow{-2BrF_3} Ag[AuF_4]$$

$$N_2O_4 + Sb_2O_3 \xrightarrow{BrF_3} [NO_2][SbF_6]$$

$$Ru + KCl \xrightarrow{BrF_3} K[RuF_6]$$

In these reactions BrF_3 serves both as a fluorinating agent and as a nonaqueous solvent reaction medium.

Molten I_2Cl_6 has been much less studied as an ionizing solvent because of the high dissociation pressure of Cl_2 above the melt. The appreciable electrical conductivity may well indicate an ionic self-dissociation equilibrium such as

$$I_2Cl_6 \rightleftharpoons ICl_2^+ ICl_4^-$$

Such ions are known from various crystal-structure determinations, e.g. $K[ICl_2].H_2O$, $[ICl_2][AlCl_4]$, and $[ICl_2][SbCl_6]$ (p. 983). I_2Cl_6 is a vigorous chlorinating agent, no doubt due at least in part to its ready dissociation into ICl and Cl_2. Aromatic compounds, including thiophen, C_4H_4S, give chlorosubstituted products with very little if any iodination. By contrast, reaction of I_2Cl_6 with aryl–tin or aryl–mercury compounds yield the corresponding diaryliodonium derivatives, e.g.:

$$2PhSnCl_3 + ICl_3 \longrightarrow Ph_2ICl + 2SnCl_4$$

Hexa-atomic and octa-atomic interhalogens, MF_5 and IF_7

The three fluorides ClF_5, BrF_5, and IF_5 are the only known hexa-atomic interhalogens, and IF_7 is the sole representative of the octa-atomic class. The first to be made (1871) was IF_5 which is the most readily formed of the iodine fluorides, whereas the more vigorous conditions required for the others delayed the synthesis of BrF_5 and IF_7 until 1930/1 and ClF_5 until 1962. The preferred method of preparing all four compounds on a large scale is by direct fluorination of the element or a lower fluoride:

$$Cl_2 + 5F_2 \xrightarrow{\text{excess } F_2, 350°C, 250 \text{ atm}} 2ClF_5$$

$$ClF_3 + F_2 \xrightarrow{hv, \text{room temp, 1 atm}} ClF_5$$

$$Br_2 + 5F_2 \xrightarrow{\text{excess } F_2, \text{ above } 150} BrF_5$$

$$I_2(s) + 5F_2 \xrightarrow{\text{room temp}} IF_5$$

$$I_2(g) + 7F_2 \xrightarrow{250-300} IF_7$$

Small-scale preparations can conveniently be effected as follows:

$$MCl(s) + 3F_2 \xrightarrow{100-300°} MF(s) + ClF_5$$

$$KBr + 3F_2 \xrightarrow{25°} KF(s) + BrF_5$$

$$I_2 \xrightarrow{\text{AgF, ClF}_3, \text{ or BrF}_3} IF_5$$

$$I_2O_5 \xrightarrow{ClF_3, \text{ BrF}_3, \text{ or SF}_4} IF_5$$

$$KI + 4F_2 \xrightarrow{250°} KF(s) + IF_7$$

$$PdI_2 + 8F_2 \longrightarrow PdF_2 + 2IF_7$$

This last reaction is preferred for IF_7 because of the difficulty of drying I_2. (IF_7 reacts with SiO_2, I_2O_5, or traces of water to give OIF_5 from which it can be separated only with difficulty.)

ClF_5, BrF_5, and IF_7 are extremely vigorous fluorinating reagents, being excelled in this only by ClF_3. IF_5 is (relatively) a much milder fluorinating agent and can be handled in glass apparatus: it is manufactured in the USA on a scale of several hundred tonnes pa and production may soon reach 1000 tpa. It is available as a liquid in steel cylinders up to 1350 kg capacity (i.e. $1\frac{1}{3}$ tonnes) and the price (1979) is *ca.* $25 per kg. All four compounds are colourless, volatile molecular liquids or gases at room temperature and their physical properties are given in Table 17.15. It will be seen that the liquid range of IF_5 resembles that of BrF_3 and that BrF_5 is similar to ClF_3. The free energies of formation of these and

TABLE 17.15 *Physical properties of the higher halogen fluorides*

Property	ClF_5	BrF_5	IF_5	IF_7
MP/°C	−103	−60.5	9.4	6.5 (triple point)
BP/°C	−13.1	41.3	104.5	4.8 (subl 1 atm)
ΔH_f°(gas, 298 K)/kJ mol^{-1}	−255	−429[a]	−843[b]	−962
ΔG_f°(gas, 298 K)/kJ mol^{-1}	−165	−351[a]	−775[b]	−842
Mean X–F bond energy/ kJ mol^{-1}	154	187	269	232
$d_{liq}(T^\circ C)$/g cm^{-3}	2.105 (−80°)	2.4716 (25°)	3.207 (25°)	2.669 (25°)
Dipole moment/D	—	1.51	2.18	0
Dielectric constant ε $(T^\circ C)$	4.28 (−80°)	7.91 (25°)	36.14 (25°)	1.75 (25°)
κ(liq at $T^\circ C$)/ohm^{-1} cm^{-1}	3.7×10^{-8} (−80°)	9.9×10^{-8} (25°)	5.4×10^{-6} (25°)	$<10^{-9}$ (25°)

[a] For liquid BrF_5: ΔH_f° (298 K) −458.6 kJ mol^{-1}, ΔG_f° (298 K) −351.9 kJ mol^{-1}.
[b] For liquid IF_5: ΔH_f° (298 K) −885 kJ mol^{-1}, ΔG_f° (298 K) −784 kJ mol^{-1}.

the other halogen fluorides in the gas phase are compared in Fig. 17.10. The trends are obvious; it is also clear from the convexity (or concavity) of the lines that BrF and IF might be expected to disproportionate into the trifluoride and the parent halogen, whereas ClF_3, BrF_3, and IF_5 are thermodynamically the most stable fluorides of Cl, Br, and I respectively. Plots of average bond energies are in Fig. 17.11: for a given value of n in XF_n the sequence of energies is $ClF_n < BrF_n < IF_n$, reflecting the increasing difference in electronegativity between X and F. ClF is an exception. As expected, for a given halogen, the mean bond energy decreases as n increases in XF_n, the effect being most marked for Cl and least for I. Note that high bond energy (as in BrF and IF) does not necessarily confer stability on a compound (why?).

The molecular structure of XF_5 has been shown to be square pyramidal (C_{4v}) with the central atom slightly below the plane of the four basal F atoms (Fig. 17.12). The structure

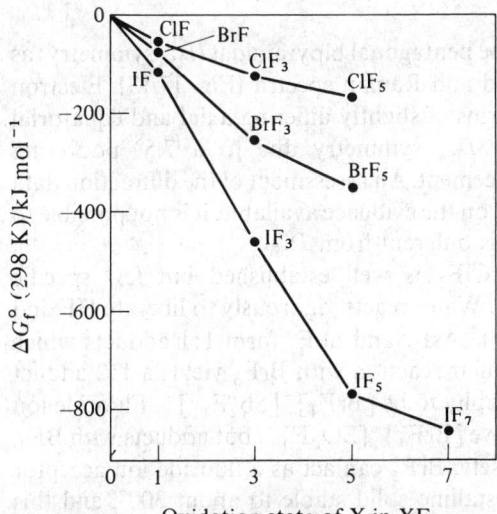

FIG. 17.10 Free energies of formation of gaseous halogen fluorides at 298 K.

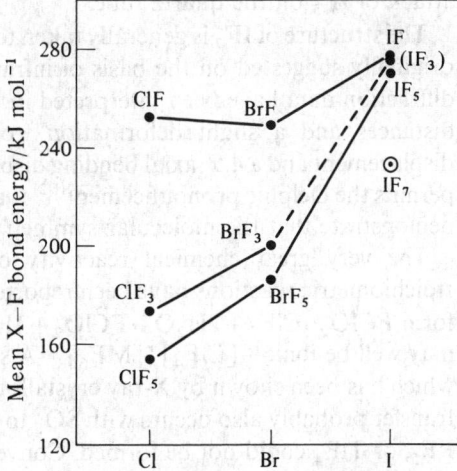

FIG. 17.11 Mean bond energies of halogen fluorides.

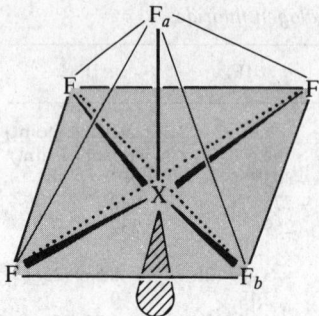

	ClF$_5$ (gas)	BrF$_5$ (gas)	BrF$_5$ (cryst)	IF$_5$ (gas)	IF$_5$ (cryst)
X–F$_b$/pm	~172	177.4	178	186.9	189
X–F$_a$/pm	~162	168.9	168	184.4	186
<F$_a$–X–F$_b$	~90° (assumed)	84.8°	84.5°	81.9°	80.9°

FIG. 17.12   The structure of XF$_5$ (X = Cl, Br, I) showing X slightly below the basal plane of the four F$_b$.

is essentially the same in the gaseous, liquid, and crystalline phases and has been established by some (or all) of the following techniques: electron diffraction, microwave spectroscopy, infrared and Raman spectroscopy, ^{19}F nmr spectroscopy and X-ray diffractometry. This structure immediately explains the existence of a small permanent dipole moment, which would be absent if the structure were trigonal bipyramidal (C_{3v}), and is consistent with the presence of 12 valence-shell electrons on the central atom X. Electrostatic effects account for the slight displacement of the 4 F$_b$ away from the lone-pair of electrons and also the fact that X–F$_b$ > X–F$_a$. The ^{19}F nmr spectra of both BrF$_5$ and IF$_5$ consist of a highfield doublet (integrated relative area 4) and a 1:4:6:4:1 quintet of integrated area 1: these multiplets can immediately be assigned on the basis of ^{19}F–^{19}F coupling and relative area to the 4 basal and the unique apical F atom respectively. The molecules are fluxional at higher temperatures: e.g. spin–spin coupling disappears in IF$_5$ at 115° and further heating leads to broadening and coalescence of the two signals, but a sharp singlet could not be attained at still higher temperatures because of accelerated attack of IF$_5$ on the quartz tube.

The structure of IF$_7$ is generally taken to be pentagonal bipyramidal (D_{5h} symmetry) as originally suggested on the basis of infrared and Raman spectra (Fig. 17.13). Electron diffraction data have been interpreted in terms of slightly differing axial and equatorial distances and a slight deformation from D_{5h} symmetry due to a 7.5° puckering displacement and a 4.5° axial bending displacement. An assessment of the diffraction data permits the Delphic pronouncement[52] that, on the evidence available, it is not possible to demonstrate that the molecular symmetry is different from D_{5h}.

The very great chemical reactivity of ClF$_5$ is well established but few specific stoichiometric reactions have been reported. Water reacts vigorously to liberate HF and form FClO$_2$ (ClF$_5$ + 2H$_2$O → FClO$_2$ + 4HF). AsF$_5$ and SbF$_5$ form 1:1 adducts which may well be ionic—[ClF$_4$]$^+$[MF$_6$]$^-$. A similar reaction with BrF$_5$ yields a 1:2 adduct which has been shown by X-ray crystallography to be [BrF$_4$]$^+$[Sb$_2$F$_{11}$]$^-$. Fluoride ion transfer probably also occurs with SO$_3$ to give [BrF$_4$]$^+$[SO$_3$F]$^-$, but adducts with BF$_3$, PF$_5$, or TiF$_4$ could not be formed. Conversely, BrF$_5$ can act as a fluoride ion acceptor (from CsF) to give CsBrF$_6$ as a white, crystalline solid stable to about 300°, and this

^{52} J. D. DONOHUE, Concerning the evidence for the molecular symmetry of IF$_7$, *Acta Cryst.* **18**, 1018–21 (1965).

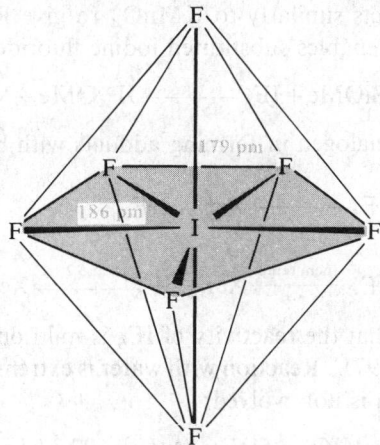

FIG. 17.13　Approximate structure of IF_7 (see text).

solvobase can be titrated with the solvoacid $[BrF_4]^+[Sb_2F_{11}]^-$ according to the following stoichiometry:

$$[BrF_4]^+[Sb_2F_{11}]^- + 2Cs^+[BrF_6]^- \longrightarrow 3BrF_5 + 2CsSbF_6$$

BrF_5 reacts explosively with water but when moderated by dilution with MeCN gives bromic and hydrofluoric acids:

$$BrF_5 + 3H_2O \longrightarrow HBrO_3 + 5HF$$

The vigorous fluorinating activity of BrF_5 is demonstrated by its reaction with silicates, e.g.:

$$KAlSi_3O_8 + 8BrF_5 \xrightarrow{450°C} KF + AlF_3 + 3SiF_4 + 4O_2 + 8BrF_3$$

The chemical reactions of IF_5 have been more extensively and systematically studied because the compound can be handled in glass apparatus and is much less vigorous a reagent than the other pentafluorides. The (very low) electrical conductivity of the pure liquid has been ascribed to slight ionic dissociation according to the equilibrium

$$2IF_5 \rightleftharpoons IF_4^+ + IF_6^-$$

Consistent with this, dissolution of KF increases the conductivity and KIF_6 can be isolated on removal of the solvent. Likewise NOF affords $[NO]^+[IF_6]^-$. Antimony compounds yield $ISbF_{10}$, i.e. $[IF_4]^+[SbF_6]^-$, which can be titrated with $KSbF_6$. However, the milder fluorinating power of IF_5 frequently enables partially fluorinated adducts to be isolated and in some of these the iodine is partly oxygenated. Complete structural identification of the products has not yet been established in all cases but typical stoichiometries are as follows:

$CrO_3 \longrightarrow CrO_2F_2$	$V_2O_5 \longrightarrow 2VOF_3 \cdot 3IOF_3$
$MoO_3 \longrightarrow 2MoO_3 \cdot 3IF_5$	$Sb_2O_5 \longrightarrow SbF_5 \cdot 3IO_2F$
$WO_3 \longrightarrow WO_3 \cdot 2IF_5$	$KMnO_4 \longrightarrow MnO_3 + IOF_3 + KF$

Potassium perrhenate reacts similarly to $KMnO_4$ to give ReO_3F. Similarly, the mild fluorinating action of IF_5 enables substituted iodine fluorides to be synthesized, e.g.:

$$Me_3SiOMe + IF_5 \longrightarrow IF_4OMe + Me_3SiF$$

IF_5 is unusual as an interhalogen in forming adducts with both XeF_2 and XeF_4:

$$XeF_2 + 2IF_5 \xrightarrow{5°} XeF_2 . 2IF_5$$

$$XeF_4 + IF_5 \xrightarrow{\text{room temp}} XeF_4 . IF_5 \xrightarrow{>92} XeF_4 + IF_5$$

It should be emphasized that the reactivity of IF_5 is mild only in comparison with the other halogen fluorides (p. 971). Reaction with water is extremely vigorous but the iodine is not reduced and oxygen is not evolved:

$$IF_5 + 3H_2O \longrightarrow HIO_3 + 5HF; \quad \Delta H = -92.3 \text{ kJ mol}^{-1}$$

$$IF_5 + 6KOH(aq) \longrightarrow 5KF(aq) + KIO_3(aq) + 3H_2O; \quad \Delta H = -497.5 \text{ kJ mol}^{-1}$$

Boron enflames in contact with IF_5; so do P, As, and Sb. Molybdenum and W enflame when heated and the alkali metals react violently. KH and CaC_2 become incandescent in hot IF_5. However, reaction is more sedate with many other metals and non-metals, and compounds such as $CaCO_3$ and $Ca_3(PO_4)_2$ appear not to react with the liquid.

IF_7 is a stronger fluorinating agent that IF_5 and reacts with most elements either in the cold or on warming. CO enflames in IF_7 vapour but NO reacts smoothly and SO_2 only when warmed. IF_7 vapour hydrolyses without violence to HIO_4 and HF; with small amounts of water at room temperature the oxyfluoride can be isolated:

$$IF_7 + H_2O \longrightarrow IOF_5 + 2HF$$

The same compound is formed by action of IF_7 on silica (at 100°) and Pyrex glass:

$$2IF_7 + SiO_2 \longrightarrow 2IOF_5 + SiF_4$$

IF_7 acts as a fluoride ion donor towards AsF_5 and SbF_5 and the compounds $[IF_6]^+[MF_6]^-$ have been isolated. Until recently no complexes with alkali metal fluorides had been isolated but it now appears that CsF and NOF can form adducts which have been characterized by X-ray powder data, and formulated on the basis of Raman spectroscopy as $Cs^+[IF_8]^-$ and $[NO]^+[IF_8]^-$.[53]

17.2.4 *Polyhalide anions*

Polyhalides anions of general formula XY_{2n}^- ($n = 1, 2, 3, 4$) have been mentioned several times in the preceding section. They can be made by addition of a halide ion to an interhalogen compound, or by reactions which result in halide-ion transfer between molecular species. Ternary polyhalide anions $X_mY_nZ_p^-$ ($m + n + p$ odd) are also known as are numerous polyiodides I_n^-. Stability is often enhanced by use of a large counter-cation, e.g. Rb^+, Cs^+, NR_4^+, PCl_4^+, etc.; likewise, for a given cation, thermal stability is enhanced the more symmetrical the polyhalide ion and the larger the central atom (i.e.

[53] C. J. ADAMS, Acceptor properties of iodine heptafluoride, *Inorg. Nuclear Chem. Letters* **10**, 831–5 (1974).

stability decreases in the sequence $I_3^- > IBr_2^- > ICl_2^- > I_2Br^- > Br_3^- > BrCl_2^- > Br_2Cl^-$). The structures of many of these polyhalide anions have been established by X-ray diffractometry or inferred from vibrational spectroscopic data and in all cases the gross stereochemistry is consistent with the expectations of simple bond theories (p. 1053); however, subtle deviations from the highest expected symmetry sometimes occur, probably due to crystal-packing forces and residual interactions between the various ions in the condensed phase.

Typical examples of linear (or nearly linear) triatomic polyhalides are in Table 17.16;[47, 54] the structures are characterized by considerable variability of interatomic distances and these distances are individually always substantially greater than for the corresponding diatomic interhalogen (p. 966). Note also that for $[Cl-I-Br]^-$ the I–Cl distance is greater than the I–Br distance, and in $[Br-I-I]^-$ I–Br is greater than I–I. On dissociation, the polyhalide yields the solid monohalide corresponding to the smaller of the halogens present, e.g. $CsICl_2$ gives CsCl and ICl rather than $CsI + Cl_2$. Likewise for CsIBrCl the favoured products are $CsCl(s) + IBr(g)$ rather than $CsBr(s) + ICl(g)$ or $CsI(s) + BrCl(g)$. Thermochemical cycles have been developed to interpret these results.[54]

Penta-atomic polyhalide anions $[XY_4]^-$ favour the square-planar geometry (D_{4h}) as expected for species with 12 valence-shell electrons on the central atom. Examples are the

TABLE 17.16   *Triatomic polyhalides* $[X-Y-Z]^-$

Polyhalide	Cations	Structure	Dimensions x/pm, y/pm		Angle
ClF_2^-	NO^+	$[F^xCl^yF]^-$	$x = y$		$\sim 180°$
	Rb^+, Cs^+	$[F-Cl-F]^-$	$x \neq y$		
Cl_3^-	NEt_4^+, NPr_4^{n+}, NBu_4^{n+}	$[Cl-Cl-Cl]^-$	$x = y$		$\sim 180°$
BrF_2^-	Cs^+	$[F-Br-F]^-$			
$BrCl_2^-$	Cs^+, NR_4^+ (R = Me, Et, Pr^n, Bu^n)	$[Cl-Br-Cl]^-$	$x = y$		$\sim 180°$
Br_2Cl^-		$[Br-Br-Cl]^-$	$x \neq y$		
Br_3^-	$Me_3NH^{+(a)}$	$[Br-Br-Br]^-$	$x = y = 254$		$171°$
	Cs^+ (and PBr_4^+)	$[Br-Br-Br]^-$	244 (239)	270 (291)	$177.5°$ ($177.3°$)
IF_2^-	NEt_4^+	$[F-I-F]^-$			
$IBrF^-$		$[F-I-Br]^-$			
$IBrCl^-$	NH_4^+	$[Cl-I-Br]^-$	291	251	$179°$
ICl_2^-	NMe_4^+ (and PCl_4^+)	$[Cl-I-Cl]^-$	$x = y = 255$		$180°$
	piperazinium[b]	$[Cl-I-Cl]^-$	247	269	$180°$
	triethylenediammonium[c]	$[Cl-I-Cl]^-$	254 (253)	267 (263)	$180°$ ($180°$)
IBr_2^-	Cs^+	$[Br-I-Br]^-$	262	278	$178°$
I_2Cl^-		$[Cl-I-I]^-$			
I_2Br^-	Cs^+	$[Br-I-I]^-$	291	278	$178°$
I_3^-	$AsPh_4^+$	$[I-I-I]^-$	$x = y = 290$		$176°$
	$[PhCONH_2]_2H^+$	$[I-I-I]^-$	291	295	$177°$
	NEt_4^+ (form I)	$[I-I-I]^-$	293	294	$180°$
	(form II)		291 (& 289), 296 (& 298)		$180°$ (& $178°$)
	Cs^+ (and NH_4^+)	$[I-I\cdots I]^-$	283 (282)	303 (310)	$176°$ ($177°$)

(a) In the compound $[Me_3NH]^+{}_2Br^-Br_3^-$; same dimensions for Br_3^- in PhN_2Br_3 and in $[C_6H_7NH]_2[SbBr_6][Br_3]$.

(b) piperazinium, $[H_2NC_4H_8NH_2]^{2+}$.

(c) triethylenediammonium, $[HN(C_2H_4)_3NH]^{2+}$: compound contains 2 non-equivalent ICl_2^- ions.

54 Ref. 17, pp. 1534–63, Polyhalide anions, and references therein.

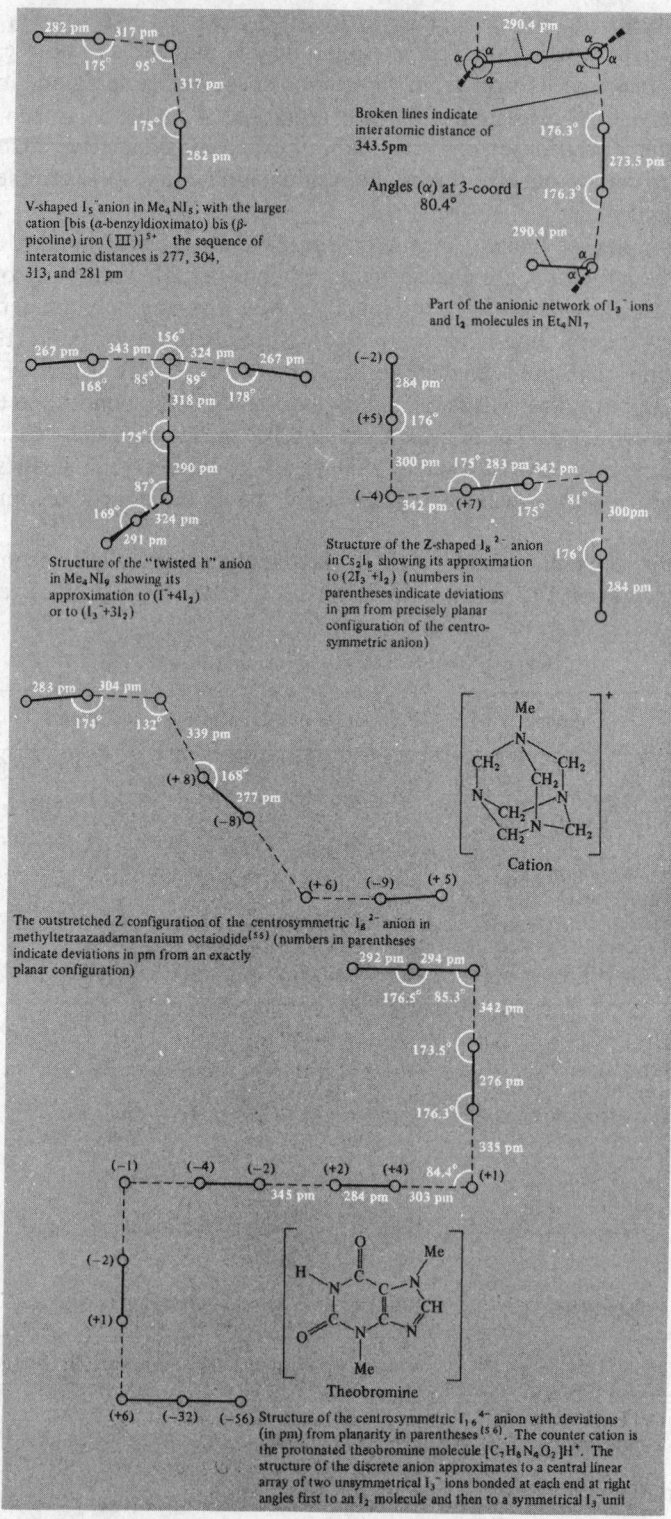

V-shaped I_5^- anion in Me_4NI_5; with the larger cation [bis (α-benzyldioximato) bis (β-picoline) iron (III)]$^{5+}$ the sequence of interatomic distances is 277, 304, 313, and 281 pm

Broken lines indicate interatomic distance of 343.5pm

Angles (α) at 3-coord I 80.4°

Part of the anionic network of I_3^- ions and I_2 molecules in Et_4NI_7

Structure of the "twisted h" anion in Me_4NI_9 showing its approximation to $(I^- + 4I_2)$ or to $(I_3^- + 3I_2)$

Structure of the Z-shaped I_8^{2-} anion in Cs_2I_8 showing its approximation to $(2I_3^- + I_2)$ (numbers in parentheses indicate deviations in pm from precisely planar configuration of the centro-symmetric anion)

The outstretched Z configuration of the centrosymmetric I_8^{2-} anion in methyltetraazaadamantanium octaiodide[55] (numbers in parentheses indicate deviations in pm from an exactly planar configuration)

Cation

Theobromine

Structure of the centrosymmetric I_{16}^{4-} anion with deviations (in pm) from planarity in parentheses[56]. The counter cation is the protonated theobromine molecule $[C_7H_8N_4O_2]H^+$. The structure of the discrete anion approximates to a central linear array of two unsymmetrical I_3^- ions bonded at each end at right angles first to an I_2 molecule and then to a symmetrical I_3^- unit

Fig. 17.14   Structure of some polyiodides.

Rb^+ and Cs^+ salts of $[ClF_4]^-$, and $KBrF_4$ (in which Br–F is 189 pm and adjacent angles F–Br–F are 90° ($\pm 2°$). The symmetry of the anion is slightly lowered in $CsIF_4$ (C_{2v}) and also in $KICl_4 \cdot H_2O$ (in which I–Cl is 242, 247, 253, and 260 pm and the adjacent angles Cl–I–Cl are 90.6°, 90.7°, 89.2°, and 89.5°. Other penta-atomic polyhalide anions for which the structure has not yet been determined are $[ICl_3F]^-$, $[IBrCl_3]^-$, $[I_2Cl_3]^-$, $[I_2BrCl_2]^-$, $[I_2Br_2Cl]^-$, $[I_2Br_3]^-$, $[I_4Br]^-$, and $[I_4Cl]^-$. Some of these may be "square planar" but the polyiodo species might well be more closely related to I_5^-: the tetramethylammonium salt of this anion features a planar V-shaped array in which two I_2 units are bonded to a single iodide ion, i.e. $[I(I_2)_2]^-$ as in Fig. 17.14. The V-shaped ions are arranged in a planar array which bear an interesting relation to a (hypothetical) array of planar IX_4^- ions. This is illustrated in Fig. 17.15.

Hepta-atomic polyhalide anions are exemplified by BrF_6^- (K^+, Rb^+, and Cs^+ salts) and IF_6^- (K^+, Cs^+, NMe_4^+, and NEt_4^+ salts). The anions have 14 valence-shell electrons on the central atom and spectroscopic studies indicate non-octahedral geometry (D_{3d} for BrF_6^-). Other possible examples are Br_6Cl^- and I_6Br^- but these have not been shown to contain discrete hepta-atomic species and may be extended anionic networks such as that found in Et_4NI_7 (Fig. 17.14).

IF_7 has been shown to act as a weak Lewis acid towards CsF and NOF, and the compounds $CsIF_8$ and $NOIF_8$ have been characterized by X-ray powder patterns and by Raman spectroscopy; they are believed to contain the IF_8^- anion.[53] A rather different structure motif occurs in the polyiodide Me_4NI_9; this consists of discrete units with a "twisted h" configuration (Fig. 17.14). Interatomic distances within these units vary from 267 to 343 pm implying varying strengths of bonding, and the anions can be thought of as being built up either from $I^- + 4I_2$ or from a central unsymmetrical I_3^- and $3I_2$. (The rather arbitrary recognition of discrete I_9^- anions is emphasized by the fact that the closest interionic I···I contact is 349 pm which is only slightly greater than the 343 pm separating one I_2 from the remaining I_7^- in the structure.)

The propensity for iodine to catenate is well illustrated by the numerous polyiodides which crystallize from solutions containing iodide ions and iodine. The symmetrical and unsymmetrical I_3^- ions (Table 17.16) have already been mentioned as have the I_5^- and I_9^- anions and the extended networks of stoichiometry I_7^- (Fig. 17.14). The stoichiometry of the crystals and the detailed geometry of the polyhalide depend sensitively on the relative concentrations of the components and the nature of the cation. For example, the linear I_4^{2-} ion may have the following dimensions:

$$\underset{334\ pm}{\qquad}\underset{280\ pm}{\qquad}\underset{334\ pm}{\qquad}$$
$$[I\text{--------}I\text{------}I\text{--------}I]^{2-} \quad \text{in } [Cu(NH_3)_4]I_4^{(57)}$$

$$\underset{318\ pm}{\qquad}\underset{314\ pm}{\qquad}\underset{318\ pm}{\qquad}$$
$$[I\text{-------}I\text{-------}I\text{-------}I]^{2-} \quad \text{in } Tl_6PbI_6^{(57a)}$$

[55] P. K. HON, T. C. M. MAK, and J. TROTTER, Synthesis and structure of 1-methyl-1,3,5,7-tetraazaadamantan-l-ium octaiodide, $[(CH_2)_6N_4Me]_2I_8$. A new outstretched Z configuration for the polyiodide ion I_8^{2-}, *Inorg. Chem.* **18**, 2916–7 (1979) and references therein.

[56] F. H. HERBSTEIN and M. KAPON, I_{16}^{4-} ions in crystalline (theobromine)$_2$.H_2I_8; X-ray structure of (theobromine)$_2$.H_2I_8, *JCS Chem. Comm.* 1975, 677–8.

[57] E. DUBLER and L. LINOWSKY, Proof of the existence of a linear centrosymmetric polyiodide ion I_4^{2-}: the crystal structure of $Cu(NH_3)_4I_4$, *Helv. Chim. Acta* **58**, 2604–9 (1978).

[57a] A RABENAU, H. SCHULZ, and W. STOEGER, A polyiodide with regular I_4^{2-} chains: Tl_6PbI_{10}, *Naturwissenschaften* **63**, 245 (1976).

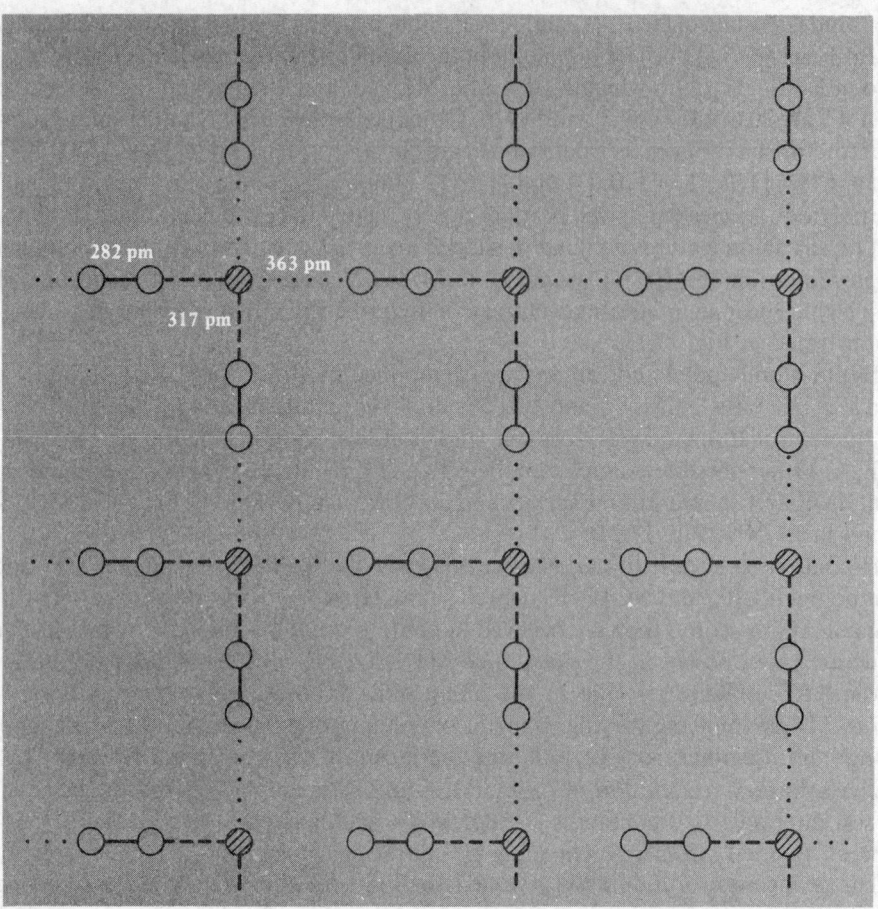

FIG. 17.15   Planar array of I_5^- ions showing distorted square coordination about the central I atom (shaded). Contraction of the 317 pm bond by, say, 17 pm to 300 pm and of the 363 pm distances by 63 pm to 300 pm would given an array of square-planar II_4^- ions separated along the I···I vectors by 362 pm.

(Note, however, that the overall length of the two I_4^{2-} ions is virtually identical.) Again, the I_8^{2-} anion is found with an acute-angled planar Z configuration in its Cs^+ salt but with an outstretched configuration in the black methyltetraazaadamantanium salt (Fig. 17.14). The largest discrete polyiodide ion so far encountered is the planar centro-symmetric I_{16}^{4-} anion; this was shown by X-ray diffractometry[56] to be present in the dark-blue needle-shaped crystals of (theobromine)$_2$.H_2I_8 which had first been prepared over a century earlier by S. M. Jorgensen in 1869.

The bonding in these various polyiodides as in the other polyhalides and neutral interhalogens has been the subject of much speculation, computation, and altercation. The detailed nature of the bonds probably depends on whether F is one of the terminal atoms or whether only the heavier halogens are involved. There is now less tendency than formerly to invoke much d-orbital participation (because of the large promotion energies required) and Mössbauer spectroscopic studies in iodine-containing species[58] also suggest rather scant s-orbital participation. The bonding appears predominantly to involve p orbitals only, and multicentred (partially delocalized) bonds such as are invoked in discussions of the isoelectronic xenon halides (p. 1053) are currently favoured. However, no bonding model yet comes close to reproducing the range of interatomic distances and angles observed in the crystalline polyhalides.[54]

17.2.5 *Polyhalonium cations* XY_{2n}^+

Numerous polyhalonium cations have already been mentioned in Section 17.2.3 during the discussion of the self-ionization of interhalogen compounds and their ability to act as halide-ion donors. The known species are summarized in Table 17.17.[59] Preparations are usually by addition of the appropriate interhalogen and halide-ion acceptor, or by straightforward modification of this general procedure in which the interhalogen or halogen is also used as an oxidant. For example Au dissolves in BrF_3 to give $[BrF_2][AuF_4]$, BrF_3 fluorinates and oxidizes $PdCl_2$ and $PdBr_2$ to $[BrF_2][PdF_4]$; ClF_3 converts $AsCl_3$ to $[ClF_2][AsF_6]$; stoichiometric amounts of I_2, Cl_2, and $2SbCl_5$ yield $[ICl_2][SbCl_6]$. The fluorocations tend to be colourless or pale yellow but the colour deepens with increasing atomic weight so that compounds of ICl_2^+ are wine-red or bright orange whilst I_2Cl^+ compounds are dark brown or purplish black.

Structures are as expected from simple valency theory and the isoelectronic principle

TABLE 17.17 *Polyhalonium cations,* XY_{2n}^+

Cation	(date)[a]	Examples of co-anions (mp of compound in parentheses)
ClF_2^+	(1950)	BF_4^- (30°), PF_6^-, AsF_6^-, SbF_6^- (78°), PtF_6^- (171°), SnF_6^{2-}
Cl_2F^+	(1969)	BF_4^-, AsF_6^-
BrF_2^+	(1949)	PdF_4^-, AuF_4^-, AsF_6^-, SbF_6^- (130°), $Sb_2F_{11}^-$ (33.5°), BiF_6^-, NbF_6^-, TaF_6^-, GeF_6^{2-} (subl 20°), SnF_6^{2-}, PtF_6^{2-} (136°), SO_3F^-
IF_2^+	(1968)	BF_4^-, AsF_6^- (d−22°), SbF_6^- (d 45°)
ICl_2^+	(1959)	$AlCl_4^-$ (105°), $SbCl_6^-$ (83.5°), $Sb_2F_{11}^-$ (62°), SO_3F^- (42°), SO_3Cl^- (8°)
I_2Cl^+	(1972)	$AlCl_4^-$ (53°), $SbCl_6^-$ (70°), $TaCl_6^-$ (102°), SO_3F^- (40°)
IBr_2^+	(1971)	$Sb_2F_{11}^-$ (65°), SO_3F^- (97°), $SO_3CF_3^-$ (75°)
I_2Br^+	(1974)	SO_3F^- (70°)
$IBrCl^+$	(1973)	$SbCl_6^-$, SO_3F^- (65°)
ClF_4^+	(1967)	AsF_6^-, SbF_6^- (88°), $Sb_2F_{11}^-$ (64°), PtF_6^-
BrF_4^+	(1957)	AsF_6^-, $Sb_2F_{11}^-$ (60°), SnF_6^{2-}
IF_4^+	(1950)	SbF_6^- (103°), $Sb_2F_{11}^-$, PtF_6^-, SO_3F^-, SnF_6^{2-}
ClF_6^+	(1972)	PtF_6^- (d140°)
BrF_6^+	(1973)	AsF_6^-, $Sb_2F_{11}^-$
IF_6^+	(1958)	BF_4^-, AsF_6^- (subl 120°), SbF_6^- (175°), $Sb_2F_{11}^-$, $[(SbF_5)_3F]^-$ (94°), AuF_6^-

[a] The date given refers to the first isolation of a compound containing the cation, or the characterization of the cation in solution.

[58] N. N. GREENWOOD and T. C. GIBB, *Mössbauer Spectroscopy*, pp. 462–82, Chapman & Hall, London, 1971.

[59] J. SHAMIR, Polyhalogen cations, *Struct. Bonding* **37**, 141–210 (1979).

(20 valency electrons). Thus the triatomic species are bent, rather than linear, as illustrated in Fig. 17.16 for ClF_2^+, BrF_2^+, and ICl_2^+; there is frequently some residual interionic interaction due to close approach of the cation and anion and this sometimes complicates the interpretation of vibrational spectroscopic data. In the case of $[ICl_2][SbF_6]$ (Fig. 17.16c) the very short I···F distance implies one of the strongest secondary interactions known between these two elements and the Sb—F···I angle deviates appreciably from linearity.[60] The ion $[Cl_2F]^+$ was originally thought to have the symmetrical bent C_{2v} structure $[Cl–F–Cl]^+$ but later Raman spectroscopic studies were interpreted on the basis of the unsymmetrical bent structure $[Cl–Cl–F]^+$. Recent calculations[61] suggest that the

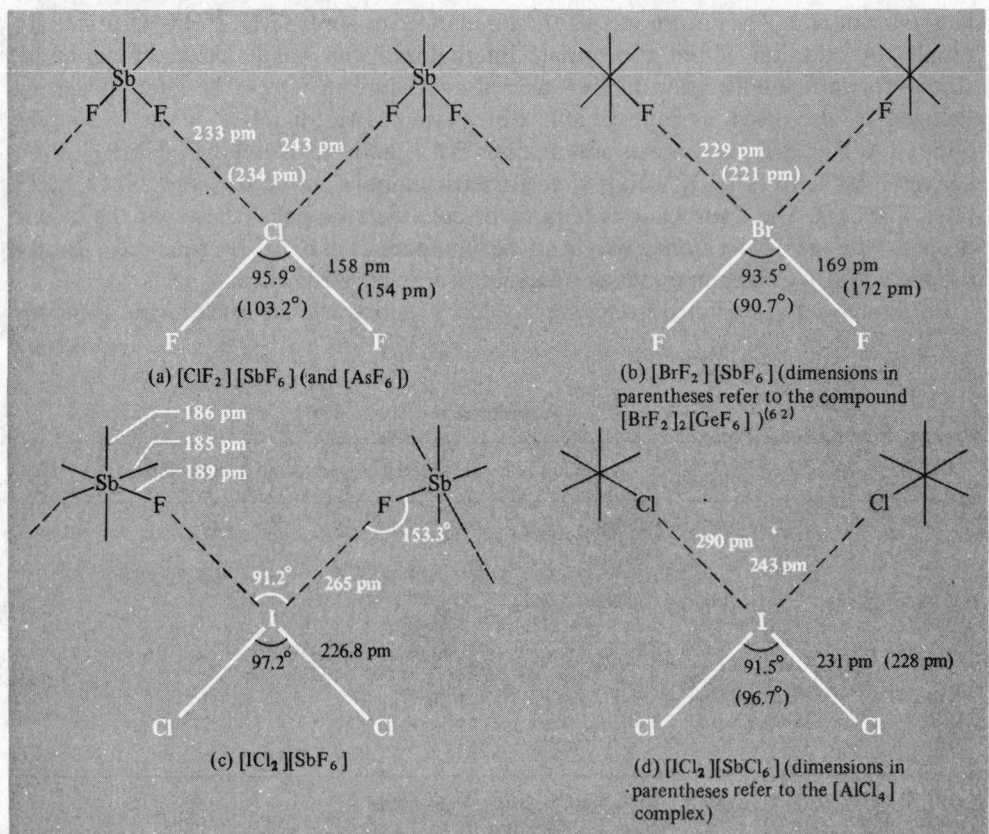

FIG. 17.16 Chain structures of compounds containing the triatomic cations XY_2^+: (a) $[ClF_2][SbF_6]$ (with dimensions for $[ClF_2][AsF_6]$ in parentheses); (b) $[BrF_2][SbF_6]$; (c) $[ICl_2][SbF_6]$ indicating slightly bent Sb—F···I configuration and very short I···F distance; and (d) $[ICl_2][SbCl_6]$ (with dimensions for the $[AlCl_4]^-$ salt in parentheses).

[60] T. Birchall and R. D. Myers, Preparation and crystal structure of $[ICl_2]^+[SbF_6]^-$ and iodine-127 Mössbauer spectra of some compounds containing the $[ICl_2]^+$ cation, *Inorg. Chem.* **20**, 2207–10 (1981).

[61] B. D. Joshi and K. Morokuma, A theoretical investigation of the structure of Cl_2F^+ and protonated ClF, *J. Am. Chem. Soc.* **101**, 1714–17 (1979), and references therein.

[62] A. J. Edwards and K. O. Christe, Fluoride crystal structures. Part 26. Bis[difluorobromonium(III)] hexafluorogermanate(IV), *JCS Dalton* 1976, 175–7.

symmetrical C_{2v} structure is indeed the more stable form at least for the isolated cation and the question must be regarded as still open: it may well be that the configuration adopted is determined by residual interactions in the solid state or in solution. In fact the ion is rather unstable in solution and disproportionates completely in SbF_5/HF even at $-76°$:

$$2Cl_2F^+ \longrightarrow ClF_2^+ + Cl_3^+$$

The pentaatomic cations ClF_4^+, BrF_4^+, and IF_4^+ are precisely isoelectronic with SF_4, SeF_4, and TeF_4 and adopt the same T-shaped (C_{2v}) configuration. This is illustrated in Fig. 17.17 for the case of $[BrF_4][Sb_2F_{11}]$: again there are strong subsidiary interactions, the coordination about Br being pseudooctahedral with four short Br–F distances and two longer Br···F distances which are no doubt influenced by the presence of the stereochemically active nonbonding pair of electrons on the Br atom. In addition, the mean Sb–F distance in the central SbF_6 unit is substantially longer than the mean of the five "terminal" Sb–F distances in the second unit and the structure can be described approximately as $[BrF_4^+ \cdots SbF_6^- \cdots SbF_5]$.

Of the heptaatomic cations, IF_6^+ has been known for some time since it can be made by fluoride-ion transfer from IF_7. Because ClF_7 and BrF_7 do not exist, alternative preparative procedures must be devised and compounds of ClF_6^+ and BrF_6^+ are of more recent vintage (Table 17.17). The cations have been made by oxidation of the pentafluorides with extremely strong oxidizers such as PtF_6, KrF^+, or KrF_3^+, e.g.:[59]

$$ClF_5 \text{ (excess)} + PtF_6 \text{ (red gas)} \xrightarrow[\text{room temp}]{\text{sapphire reactor}} ClF_6^+PtF_6^- + ClF_4^+PtF_6^-$$
$$\text{(bright yellow solids)}$$

$$BrF_5 \text{ (excess)} + KrF^+AsF_6^- \longrightarrow Kr + BrF_6^+AsF_6^-$$

Vibrational spectra and ^{19}F nmr studies on all three cations XF_6^+ and the ^{129}I Mössbauer spectrum of $[IF_6][AsF_6]$ establish octahedral (O_h) symmetry as expected for species isoelectronic with SF_6, SeF_6, and TeF_6 respectively.

Attempts to prepare ClF_7 and BrF_7 by reacting the appropriate cation with NOF failed; instead the following reactions occurred:

$$ClF_6^+PtF_6^- + NOF \longrightarrow NO^+PtF_6^- + ClF_5 + F_2$$

$$BrF_6^+AsF_6^- + 2NOF \xrightarrow{-78°} NO^+AsF_6^- + NO^+BrF_6^- + F_2$$

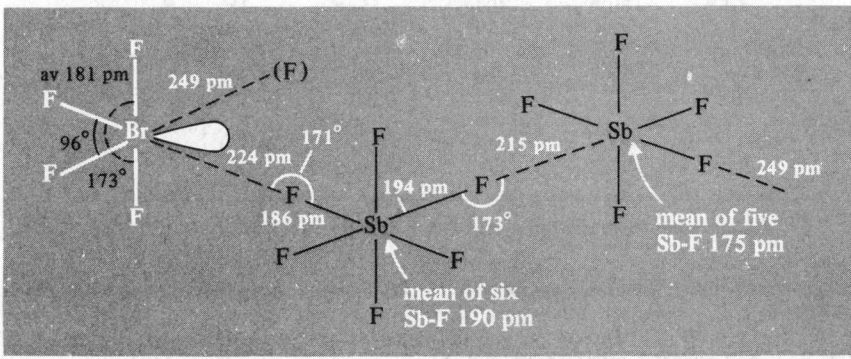

FIG. 17.17   Structure of $[BrF_4][Sb_2F_{11}]$ (see text).

As expected, the cations are extremely powerful oxidants, e.g.:

$$O_2 + BrF_6^+AsF_6^- \longrightarrow O_2^+AsF_6^- + BrF_5 + \tfrac{1}{2}F_2$$

$$Xe + BrF_6^+AsF_6^- \longrightarrow XeF^+AsF_6^- + BrF_5$$

$$Rn + IF_6^+SbF_6^- \longrightarrow RnF^+SbF_6^- + IF_5$$

17.2.6 *Halogen cations*[59, 63]

It has been known for many years that iodine dissolves in strongly oxidizing solvents such as oleum to give bright blue paramagnetic solutions, but only in 1966 was this behaviour unambiguously shown to be due to the formation of the diiodine cation I_2^+. (The production of similar brightly coloured solutions of S, Se, and Te has already been discussed on p. 785.) The ionization energies of Br_2 and Cl_2, whilst greater than that for I_2 (Table 17.18), are nevertheless smaller than for O_2, which can likewise be oxidized to O_2^+ (p. 720). Accordingly, compounds of the bright-red cationic species Br_2^+ are now well established, but Cl_2^+ is known only from its electronic band spectrum obtained in a low-pressure discharge tube. Some properties of the three diatomic cations X_2^+ are compared with those of the parent halogen molecules X_2 in Table 17.18; as expected, ionization reduces the interatomic distance and increases the vibration frequency ($v \text{ cm}^{-1}$) and force constant ($k \text{ Nm}^{-1}$). The principal synthetic routes to crystalline compounds of Br_2^+ and I_2^+ have been either (a) the comproportionation of BrF_3, BrF_5, or IF_5 with the stoichiometric amount of halogen in the presence of SbF_5, or (b) the direct oxidation of the halogen by an excess of SbF_5 or by SbF_5 dissolved in SO_2, e.g.:

$$2I_2 + 5SbF_5 \xrightarrow{\;SO_2/20°\;} [I_2]^+[Sb_2F_{11}]^- + SbF_3$$

More recently[64] a simpler route has been devised which involves oxidation of Br_2 or I_2

TABLE 17.18   *Comparison of diatomic halogens X_2 and their cations X_2^+*

Species	$I/\text{kJ mol}^{-1}$	r/pm	v/cm^{-1}	$k/\text{Nm}^{-1(a)}$	λ_{max}/nm
Cl_2	1110	199	554	316	330
Cl_2^+	—	189	645	429	—
Br_2	1014	228	319	238	410
Br_2^+	—	213	360	305	510
I_2	900	267	215	170	520
I_2^+	—	256	238	212	640

(a) Force constant k in newton/metre: 1 millidyne/Å $= 100$ N m^{-1}.

[63] R. J. GILLESPIE and J. PASSMORE, Homopolyatomic cations of the elements, *Adv. Inorg. Chem. Radiochem.* **17**, 49–87 (1975).

[64] W. W. WILSON, R. C. THOMPSON, and F. AUBKE, Improved synthesis of $[I_2]^+[Sb_2F_{11}]^-$ and $[Br_2]^+[Sb_3F_{16}]^-$ and magnetic, spectroscopic, and some chemical properties of the dihalogen cations I_2^+ and Br_2^+, *Inorg. Chem.* **19**, 1489–93 (1980).

with the peroxide $S_2O_6F_2$ (p. 751) followed by solvolysis using an excess of SbF_5, e.g.:

$$Br_2 + \tfrac{1}{2}S_2O_6F_2 \xrightarrow{\text{rt}} (\tfrac{1}{2}Br_2 \text{ dissolved in } BrSO_3F) \xrightarrow{3SbF_5} [Br_2]^+[Sb_3F_{16}]^-$$

The bright-red crystals of $[Br_2]^+[Sb_3F_{16}]^-$ melt at 85.5°C to a cherry-red liquid. Dark-blue crystals of $[I_2]^+[Sb_2F_{11}]^-$ melt sharply at 127°C and the corresponding blue solid $[I_2]^+[Ta_2F_{11}]^-$ melts at 120°C. When solutions of I_2^+ in HSO_3F are cooled below -60°C. When solutions of I_2^+ in HSO_3F are cooled below -60°C there is a dramatic colour change from deep blue to red as the cation dimerizes: $2I_2^+ \rightleftharpoons I_4^{2+}$. There is a simultaneous drop in the paramagnetic susceptibility of the solution and in its electrical conductivity. The changes are rapid and reversible, the blue colour appearing again on warming.

During the past 15 y numerous other highly coloured halogen cations have been characterized by Raman spectroscopy, X-ray crystallography, and other techniques, as summarized in Table 17.19. Typical preparative routes involve direct oxidation of the halogen (a) in the absence of solvent, (b) in a solvent which is itself the oxidant (e.g. AsF_5), or (c) in a non-reactive solvent (e.g. SO_2). Some examples are listed below:

$$Cl_2 + ClF + AsF_5 \xrightarrow{-78°} Cl_3^+AsF_6^-$$

$$\tfrac{3}{2}Br_2 + O_2^+AsF_6^- \longrightarrow Br_3^+AsF_6^-$$

$$Br_2 + BrF_3 + AsF_5 \longrightarrow Br_3^+AsF_6^- \text{ (subl 50°, decomp 70°)}$$

$$3I_2 + 3AsF_5 \xrightarrow{\text{in } AsF_5 \text{ or } SO_2} 2I_3^+AsF_6^- + AsF_3$$

$$3I_2 + S_2O_6F_2 \longrightarrow 2I_3^+SO_3F^- \text{ (mp 101.5°)}$$

$$I_2 + ICl + AlCl_3 \longrightarrow I_3^+AlCl_4^- \text{ (mp 45°)}$$

$$2I_2 + ICl + AlCl_3 \longrightarrow I_5^+AlCl_4^- \text{ (mp 50°)}$$

$$7I_2 + S_2O_6F_2 \longrightarrow 2I_7SO_3F \text{ (mp 90.5°)}$$

Other compounds that have recently been prepared[65] include the dark-brown gold(III) complexes $Br_3[Au(SO_3F)_4]$ (decomp ~105°C) and $Br_5[Au(SO_3F)_4]$ (mp 65°).

The triatomic cations X_3^+ are nonlinear and thus isostructural with other 20-electron

TABLE 17.19 *Summary of known halogen cations*

(Cl_2^+)	Br_2^+ cherry red	I_2^+ bright blue
Cl_3^+ yellow	Br_3^+ brown	I_3^+ dark brown/black
—		I_4^{2+} red-brown
	Br_5^+ dark brown	I_5^+ green/black[a]
—		(I_7^+) black

[a] $[I_5][AlCl_4]$ is described as greenish-black needles, dark brown-red in thin sections.

[65] K. C. LEE and F. AUBKE, Gold(III) fluorosulfate as fluorosulfate ion acceptor. Part 2. Compounds containing halogen and halogeno(fluorosulfato) cations, *Inorg. Chem.* **19**, 119–22 (1980).

species such as XY_2^+ (p. 984) and SCl_2 (p. 815). The contrast in bond lengths and angles between I_3^+ (Fig. 17.18)[66] and the linear 22-electron anion I_3^- (p. 979) is notable, as is its similarity with the isoelectronic Te_3^{2-} anion (p. 893). The structure of the pentaatomic cations Br_5^+ and I_5^+ have not been unambiguously established. The black compound I_7SO_3F (mp 90.5°) was established[67] as a local mp maximum in the phase diagram of the system $I_2/S_2O_6F_2$, together with the known compounds I_3SO_3F (mp 101.5°), ISO_3F (mp 50.2°), and $I(SO_3F)_3$ (mp 33.7°), but its structure has not been determined and there is at present no evidence for the presence of the discrete heptaatomic cation I_7^+ in the crystals.

17.2.7 *Oxides of chlorine, bromine, and iodine*

Perhaps nowhere else are the chemical differences between the halogens so pronounced as in their binary compounds with oxygen. This stems partly from the factors that distinguish F from its heavier congenors (p. 939) and partly from the fact that oxygen is less electronegative than F but more electronegative than Cl, Br, and I. The varying relative strengths of O–X bonds and the detailed redox properties of the halogens also ensure considerable diversity in stoichiometry, structure, thermal stability, and chemical reactivity of the various species. The binary compounds between O and F have already been described (p. 748). About 25 further binary halogen oxide species are known, which vary from shock-sensitive liquids and short-lived free radicals to rather stable solids. It will be convenient to treat the 3 halogens separately though intercomparison of corresponding

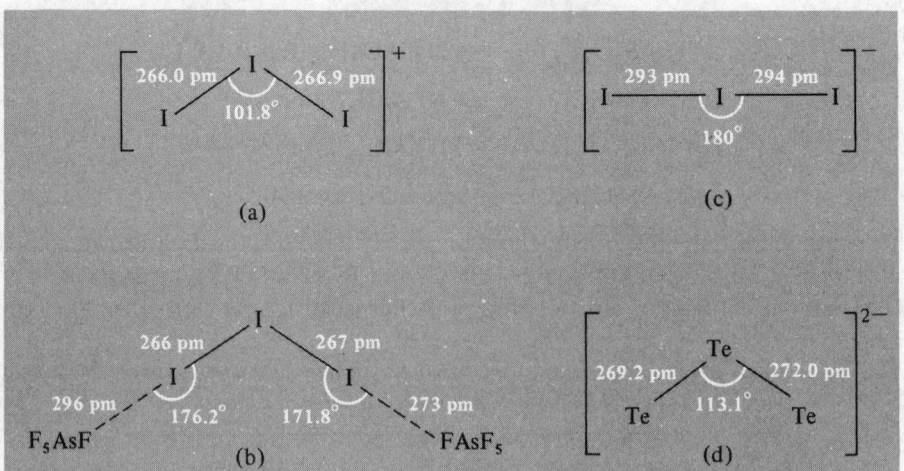

FIG. 17.18 The structure of (a) the nonlinear I_3^+ cation in I_3AsF_6 and (b) the weaker cation–anion interactions along the chain (cf. Fig. 17.16). For comparison, the dimensions of (c) the linear 22-electron cation I_3^- and (d) the nonlinear 20-electron cation Te_3^{2-} are given. The data for this latter species refer to the compound $[K(crypt)]_2Te_3.en$; in K_2Te_3 itself, where there are stronger cation–anion interactions, the dimensions are $r = 280$ pm and angle $= 104.4°$.

[66] J. PASSMORE, G. SUTHERLAND, and P. S. WHITE, The crystal structure of triiodine(1+) hexafluoroarsenate(V), I_3AsF_6, *Inorg. Chem.* **20**, 2169–71 (1981).

[67] C. CHUNG and G. H. CADY, The iodine-peroxydisulfuryl difluorite system, *Inorg. Chem.* **11**, 2528–31 (1972).

species is instructive and the chemistry is also, at times, related to that of the oxoacids (p. 999) and the halogen oxide fluorides (p. 1026).

Oxides of chlorine[68, 69]

Despite their instability (or perhaps because of it) the oxides of chlorine have been much studied and some (such as Cl_2O and particularly ClO_2) find extensive industrial use. They have also assumed considerable importance in studies of the upper atmosphere because of the vulnerability of ozone in the stratosphere to destruction by the photolysis products of chlorofluorocarbons (p. 994). The compounds to be discussed are:

Cl_2O a brownish-yellow gas at room temperature (or red-brown liquid and solid at lower temperatures) discovered in 1834; it explodes when heated or sparked.

Cl_2O_3 a dark-brown solid (1967) which explodes even below 0°.

ClO_2 a yellow paramagnetic gas (deep-red paramagnetic liquid and solid) discovered in 1811 by H. Davy; the liquid explodes above $-40°$ and the gas at room temperature may explode at pressures greater than 50 mmHg (6.7 kPa); despite this more than 10^5 tonnes are made for industrial use each year in the USA alone.

Cl_2O_4 a pale-yellow liquid (1970), $ClOClO_3$, which readily decomposes at room temperature into Cl_2, O_2, ClO_2, and Cl_2O_6.

Cl_2O_6 a dark-red liquid (1843) which is in equilibrium with its monomer ClO_3 in the gas phase; it decomposes to ClO_2 and O_2.

Cl_2O_7 a colourless oily liquid (1900) which can be distilled under reduced pressure.

In addition, there are the short-lived radical ClO, the chlorine peroxide radical ClOO (cf. OClO above), and the tetroxide radical ClO_4 (p. 990).

Some physical and molecular properties are summarized in Table 17.20. All the compounds are endothermic, having large positive enthalpies and free energies of formation. Structural data are in Fig. 17.19. Cl_2O has C_{2v} symmetry, as expected for a molecule with 20 valency-shell electrons; the dimensions indicate normal single bonds, and the bond angle can be compared with those for similar molecules such as OF_2, H_2O,

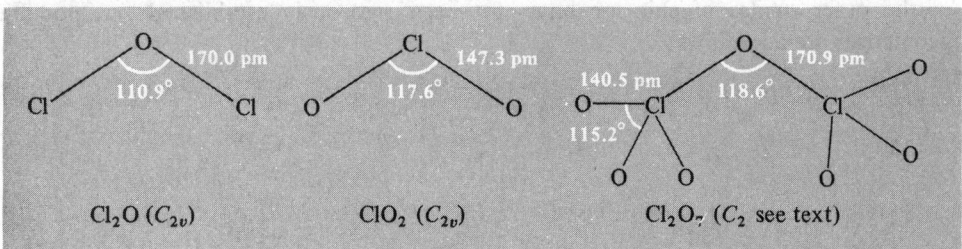

$Cl_2O\ (C_{2v})$ $ClO_2\ (C_{2v})$ $Cl_2O_7\ (C_2$ see text)

Fig. 17.19 Molecular structure and dimensions of gaseous molecules of chlorine oxides as determined by microwave spectroscopy (Cl_2O and ClO_2) or electron diffraction (Cl_2O_7).

[68] Ref. 17, pp. 1361–86, The oxides of the halogens.
[69] J. A. WOJTOWICZ, Chlorine monoxide, hypochlorous acid, and hypochlorites, *Kirk–Othmer Encyclopedia of Chemical Technology*, 3rd edn., Wiley, New York, 1979, Vol. 5, pp. 580–611. M. G. NOACK and R. L. DOERR, Chlorine dioxide, chlorous acid, and chlorites, ibid., pp. 612–32.

TABLE 17.20   *Physical and molecular properties of the oxides of chlorine*

Property	Cl_2O	ClO_2	$ClOClO_3$	$Cl_2O_6(l)$ ($\rightleftharpoons 2ClO_3(g)$)	Cl_2O_7
Colour and form at room temperature	Yellow-brown gas	Yellow-green gas	Pale yellow liquid	Dark red liquid	Colourless liquid
Oxidation states of Cl	+1	+4	+1, +7	+6	+7
MP/°C	−120.6	−59	−117	3.5	−91.5
BP/°C	2.0	11	44.5 (extrap)	203 (extrap)	81
d (liq, 0°C)/g cm^{-3}	—	1.64	1.806	—	2.02
ΔH_f° (gas, 25°C)/ kJ mol^{-1}	80.3	102.6	∼180	(155)	272
ΔG_f° (gas, 25°C)/ kJ mol^{-1}	97.9	120.6	—	—	—
S° (gas, 25°C)/ J K^{-1} mol^{-1}	265.9	256.7	327.2	—	—
Dipole moment μ/D$^{(a)}$	0.78 ± 0.08	1.78 ± 0.01	—	—	0.72 ± 0.02

$^{(a)}$ 1 D ≡ 3.3356 × 10^{-30} C m.

SCl_2, etc. Chlorine dioxide, ClO_2, also has C_{2v} symmetry but there are only 19 valency-shell electrons and this is reflected in the considerable shortening of the Cl–O bonds and the increase in the bond angle, which is only 1.7° less than in the 18-electron species SO_2 (p. 827). Cl_2 is an interesting example of an odd-electron molecule which is stable towards dimerization (cf. NO, p. 511); calculations suggest that the odd electron is delocalized throughout the molecule and this probably explains the reluctance to dimerize. Indeed, there is no evidence of dimerization even in the liquid or solid phases, or in solution. This contrasts with the precisely isoelectronic thionite ion SO_2^- which exists as dithionite, $S_2O_4^{2-}$, albeit with a rather long S–S bond (p. 854). The trioxide ClO_3 is also predominantly dimeric in the condensed phase (see below) as probably is BrO_2 (p. 996).

The gaseous molecule of Cl_2O_7 has C_2 symmetry (Fig. 17.19) the ClO_3 groups being twisted 15° from the staggered (C_{2v}) configuration; the Cl–O$_\mu$ bonds are also inclined at an angle of 4.7° to the three-fold axis of the ClO_3 groups and there is a substantial decrease from the (single-bonded) Cl–O$_\mu$ to the (multiple-bonded) Cl–O$_t$ distance. The structures of the other oxides of chlorine have not been rigorously established but possible geometries are as follows:

Cl_2O_3 (p 994) Cl_2O_4 (p 995) Cl_2O_6 (p 995) ClO_4 (p 995)

We next consider the synthesis and chemical reactions of the oxides of chlorine. Because the compounds are strongly endothermic and have large positive free energies of formation it is not possible to prepare them by direct reaction of Cl_2 and O_2. Dichlorine monoxide, Cl_2O, is best obtained by treating freshly prepared yellow HgO and Cl_2 gas (diluted with dry air or by dissolution in CCl_4):

$$2Cl_2 + 2HgO \longrightarrow HgCl_2 \cdot HgO + Cl_2O(g)$$

The reaction is convenient for both laboratory scale and industrial preparations. Another large-scale process is the reaction of Cl_2 gas on moist Na_2CO_3 in a tower or rotary tube reactor:

$$2Cl_2 + 2Na_2CO_3 + H_2O \longrightarrow 2NaHCO_3 + 2NaCl + Cl_2O(g)$$

Cl_2O is very soluble in water, a saturated solution at $-9.4°C$ containing 143.6 g Cl_2O per 100 g H_2O; in fact the gas is the anhydride of hypochlorous acid, with which it is in equilibrium in aqueous solutions:

$$Cl_2O + H_2O \rightleftharpoons 2HOCl$$

Much of the Cl_2O manufactured industrially is used to make hypochlorites, particularly $Ca(OCl)_2$, and it is an effective bleach for wood-pulp and textiles. Cl_2O is also used to prepare chloroisocyanurates (p. 339) and chlorinated solvents (via mixed chain reactions in which Cl and OCl are the chain-propagating species).[70] Its reactions with inorganic reagents are summarized in the subjoined scheme.

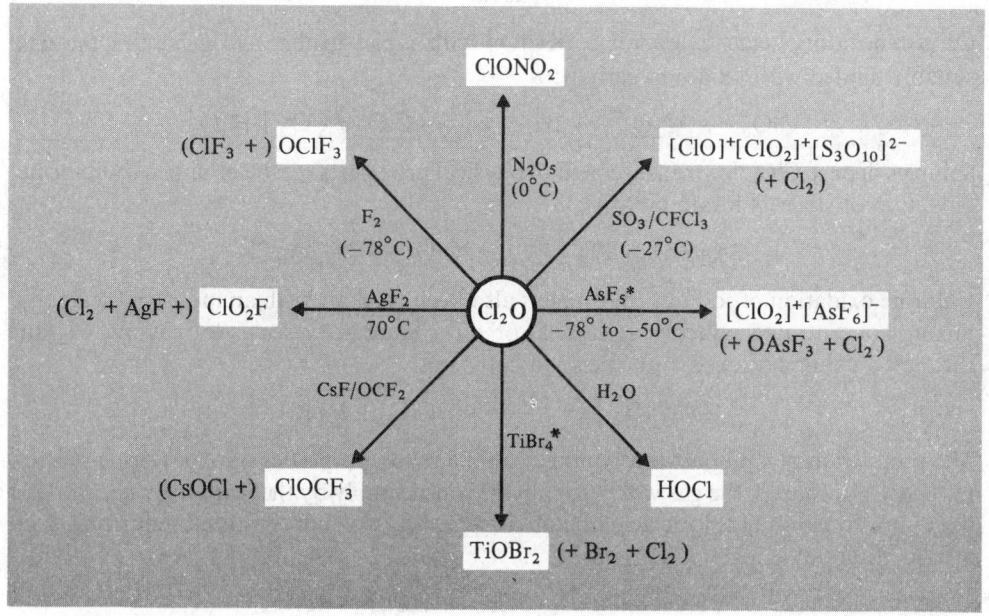

SCHEME Some reactions of dichlorine monoxide. *[In addition $AsCl_3 \rightarrow AsO_2Cl$; $SbCl_5 \rightarrow SbO_2Cl$; $P(NCl_2) \rightarrow PO_2Cl$; $VOCl_3 \rightarrow VO_2Cl$; $TiCl_4 \rightarrow TiOCl_2$.][70]

[70] J. J. RENARD and H. I. BOLKER, The chemistry of dichlorine monoxide, *Chem. Revs.* **76**, 487–505 (1976).

Gaseous mixtures of Cl_2O and NH_3 explode violently: the overall stoichiometry of the reaction can be represented as

$$3Cl_2O + 10NH_3 \longrightarrow 2N_2 + 6NH_4Cl + 3H_2O$$

Chlorine dioxide, ClO_2, was the first oxide of chlorine to be discovered and is now manufactured on a massive scale for the bleaching of wood-pulp and for water treatment;[69, 71] however, because of its explosive character as a liquid or concentrated gas, it must be made at low concentrations where it is to be used. For this reason, production statistics can only be estimated, but it is known that its use in the US wood-pulp and paper industry increased tenfold from 7800 tonnes in 1955 to 78 800 tonnes in 1970; thereafter captive production for this purpose levelled off at about 80 000 tonnes pa and the total US production of this gas for all purposes is now about 10^5 tonnes pa. Usually ClO_2 is prepared by reducing $NaClO_3$ with NaCl, HCl, SO_2, or MeOH in strongly acid solution; other reducing agents that have been used on a laboratory scale include oxalic acid, N_2O, EtOH, and sugar. With Cl^- as reducing agent the formal reaction can be written:

$$ClO_3^- + Cl^- + 2H^+ \longrightarrow ClO_2 + \tfrac{1}{2}Cl_2 + H_2O$$

Contamination of the product with Cl_2 gas is not always undesirable but can be avoided by using SO_2:

$$2ClO_3^- + SO_2 \xrightarrow[\text{solution}]{\text{acid}} 2ClO_2 + SO_4^{2-}$$

On a laboratory scale reduction of $KClO_3$ with moist oxalic acid generates the gas suitably diluted with oxides of carbon:

$$ClO_3^- + \tfrac{1}{2}C_2O_4^{2-} + 2H^+ \longrightarrow ClO_2 + CO_2 + H_2O$$

Samples of pure ClO_2 for measurement of physical properties can be obtained by chlorine reduction of silver chlorate at 90°C:

$$2AgClO_3 + Cl_2 \longrightarrow 2ClO_2 + O_2 + 2AgCl$$

Chlorine oxidation of sodium chlorite has also been used on both an industrial scale (by mixing concentrated aqueous solutions) or on a laboratory scale (by passing Cl_2/air through a column packed with the solid chlorite):

$$2NaClO_2 + Cl_2 \longrightarrow 2ClO_2 + 2NaCl$$

The production of ClO_2 obviously hinges on the redox properties of oxochlorine species (p. 1000) and, indeed, the gas was originally obtained simply by the (extremely hazardous) disproportionation of chloric acid liberated by the action of concentrated sulfuric acid on a solid chlorate:

$$3HClO_3 \longrightarrow 2ClO_2 + HClO_4 + H_2O$$

ClO_2 is a strong oxidizing agent towards both organic and inorganic materials and it reacts readily with S, P, PX_3, and KBH_4. Some further reactions are in the scheme:[68]

[71] W. J. MASSCHELEIN, *Chlorine Dioxide: Chemistry and Environmental Impact of Oxychlorine Compounds*, Ann Arbor Science Publishers, Ann Arbor, 1979, 190 pp. J. KATZ (ed.), *Ozone and Chlorine Dioxide Technology for Disinfection of Drinking Water*, Noyes Data Corp., Park Ridge, New Jersey, 1980, 659 pp.

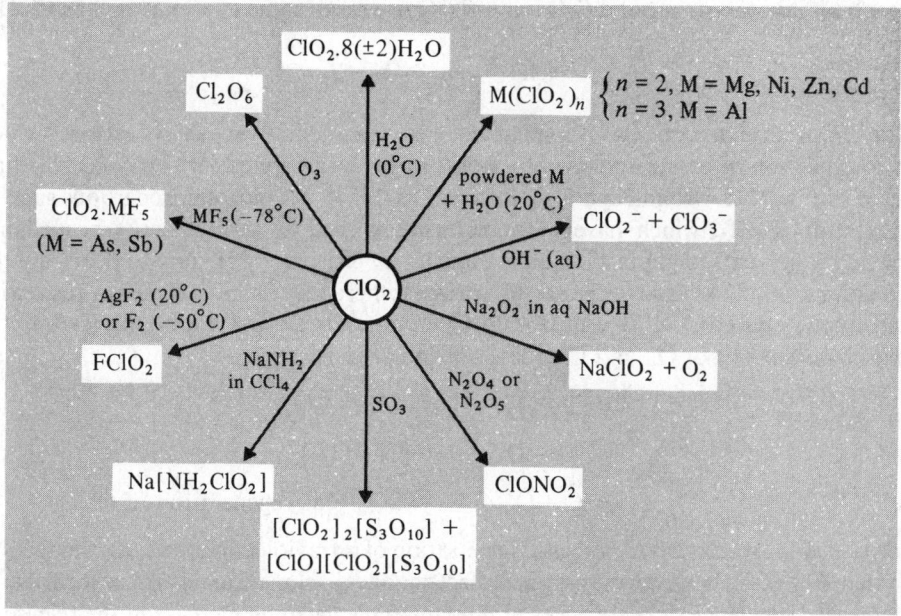

SCHEME   Some reactions of chlorine dioxide.

ClO_2 dissolves exothermically in water and the dark-green solutions, containing up to 8 g/l, decompose only very slowly in the dark. At low temperatures crystalline clathrate hydrates, $ClO_2 \cdot nH_2O$, separate ($n \approx 6$–10). Illumination of neutral aqueous solutions initiates rapid photodecomposition to a mixture of chloric and hydrochloric acids:

$$ClO_2 \xrightarrow{h\nu} ClO + O$$

$$ClO + H_2O \longrightarrow H_2ClO_2 \xrightarrow{ClO} HCl + HClO_3$$

By contrast, alkaline solutions hydrolyse vigorously to a mixture of chlorite and chlorate (see scheme).

The photochemical and thermal decomposition of ClO_2 both begin by homolytic scission of a Cl–O bond:

$$ClO_2 \xrightarrow{\Delta \ or \ h\nu} ClO + O; \quad \Delta H^\circ_{298} = 278 \text{ kJ mol}^{-1}$$

Subsequent reactions depend on conditions. Ultraviolet photolysis of isolated molecules in an inert matrix yields the radicals ClO and ClOO. At room temperature, photolysis of dry gaseous ClO_2 yields Cl_2, O_2, and some ClO_3 which either dimerizes or is further photolysed to Cl_2 and O_2:

$$ClO_2 + O \longrightarrow ClO_3$$

$$ClO + ClO \longrightarrow Cl_2 + O_2$$

$$2ClO_3 \longrightarrow Cl_2O_6$$

$$2ClO_3 \longrightarrow Cl_2 + 3O_2$$

By contrast, photolysis of solid ClO_2 at $-78°C$ produces some Cl_2O_3 as well as Cl_2O_6:

$$ClO_2 + ClO \rightleftharpoons O-Cl\cdots Cl\begin{matrix} O \\ \diagup \\ \diagdown \\ O \end{matrix}$$

The ClO radical in particular is implicated in environmentally sensitive reactions which lead to depletion of ozone and oxygen atoms in the stratosphere.[72] Thus (as was first pointed out by M. J. Molina and F. S. Rowland in 1974[73]) chlorofluorocarbons such as $CFCl_3$ and CF_2Cl_2, which have been increasingly used as aerosol spray propellants, refrigerants, solvents, and plastic foaming agents (p. 823), are beginning to penetrate the stratosphere (10–50 km above the earth's surface) where they are photolysed or react with electronically excited $O(^1D)$ atoms to yield Cl atoms and chlorine oxides: this leads to the continuous removal of O_3 and O atoms via such reactions as:

$$Cl + O_3 \longrightarrow ClO + O_2$$

$$ClO + O \longrightarrow Cl + O_2$$

$$\text{i.e. } O + O_3 \longrightarrow 2O_2 \text{ plus regeneration of Cl}$$

Depletion of O_3 results in an increased penetration of ultraviolet light with wavelengths in the range 290–320 nm which may in time effect changes in climate and perhaps lead also to an increased incidence of skin cancer in humans. The alarming increase in the release of chlorofluorocarbons into the atmosphere is illustrated by the following figures:

$$CFCl_3: \text{ 2270 tonnes in 1948} \longrightarrow \text{302 000 tonnes in 1973}$$

$$CF_2Cl_2: \text{ 2220 tonnes in 1948} \longrightarrow \text{383 000 tonnes in 1973}$$

Calculations suggested that this might eventually produce a 7% depletion in the steady-state concentration of O_3, though more recent studies taking into account the formation of chlorine nitrate ($ClO + NO_2 \rightarrow ClONO_2$) suggest this is an overestimate. Environmental legislation restricting the use of chlorofluorocarbons has also halted and, indeed, reversed the rapid increase in atmospheric concentration of these materials. The use of Al powder/NH_4ClO_4 as a propellent in the US Space Shuttle will not remove significant amounts of ozone from the stratosphere.

The great importance of the short-lived ClO radical has stimulated numerous investigations of its synthesis and molecular properties. Several routes are now available to this species (some of which have already been indicated above):

(a) thermal decomposition of ClO_2 or ClO_3;
(b) decomposition of $FClO_3$ in an electric discharge;
(c) passage of a microwave or radio-frequency discharge through mixtures of Cl_2 and O_2;
(d) reactions of Cl atoms with ClO or O_3 at 300 K;
(e) gas-phase photolysis of Cl_2O, ClO_2, or mixtures of Cl_2 and O_2.

It is an endothermic species with $\Delta H_f^\circ(298 \text{ K})$ 101.8 kJ mol^{-1}, $\Delta G_f^\circ(298 \text{ K})$

---

[72] R. J. DONOVAN, Chemistry and pollution of the stratosphere, *Educ. in Chem.* **15**, 110–13 (1978). B. A. THRUSH, The chemistry of the stratosphere and its pollution, *Endeavour* (New Series) **1**, 3–6 (1977), and references therein.

[73] M. J. MOLINA and F. S. ROWLAND, Stratospheric sink for chlorofluoromethanes: chlorine-atom catalysed destruction of ozone, *Nature* **249**, 810–12 (1974).

98.1 kJ mol^{-1}, $S°$(298 K) 226.5 J K^{-1} mol^{-1}. The interatomic distance Cl–O is 156.9 pm, its dipole moment is 1.24 D, and the bond dissociation energy D_0 is 264.9 kJ mol^{-1} (cf. BrO p. 997, IO p. 999).

Chlorine perchlorate ClOClO$_3$ is made by the following low-temperature reaction:

$$MClO_4 + ClOSO_2F \xrightarrow{-45°} MSO_3F + ClOClO_3 \quad (M = Cs, NO_2)$$

Little is known of its structure and properties; it is even less stable than ClO$_2$ and decomposes at room temperature to Cl$_2$, O$_2$, and Cl$_2$O$_6$.

Dichlorine hexoxide, Cl$_2$O$_6$, is best made by ozonolysis of ClO$_2$:

$$2ClO_2 + 2O_3 \longrightarrow Cl_2O_6 + 2O_2$$

The dark-red liquid freezes to a solid which is yellow at $-180°$C. The structure in neither phase is known but two possibilities that have been considered are shown on p. 990. The Cl–Cl linked structure is superficially attractive as the product of dimerization of the paramagnetic gaseous species ClO$_3$, but magnetic susceptibility studies of the equilibrium Cl$_2$O$_6 \rightleftharpoons 2$ClO$_3$ in the liquid phase were flawed by the subsequent finding that there was no esr signal from ClO$_3$ and that ClO$_2$ (as an impurity) was the sole paramagnetic species present. Accordingly, the much-quoted value of 7.24 kJ mol^{-1} for the derived heat of dimerization is without foundation. The alternative oxygen-bridged dimer, though requiring more electronic and geometric rearrangement of the presumed pyramidal ˙ClO$_3$ monomers, is rather closer to the ionic formulation [ClO$_2$]$^+$[ClO$_4$]$^-$ which has been proposed from the vibrational spectrum of the solid. Cl$_2$O$_6$ does, in fact, frequently behave as chloryl perchlorate in its reactions though experience with N$_2$O$_4$ as "nitrosyl nitrate" (p. 523) engenders caution in attempting to deduce a geometrical structure from chemical reactions (cf. however, diborane, p. 188).

Hydrolysis of Cl$_2$O$_6$ gives a mixture of chloric and perchloric acids, whereas anhydrous HF sets up an equilibrium:

$$Cl_2O_6 + H_2O \longrightarrow HOClO_2 + HClO_4$$

$$Cl_2O_6 + HF \rightleftharpoons FClO_2 + HClO_4$$

Nitrogen oxides and their derivatives displace ClO$_2$ to form nitrosyl and nitryl perchlorates. These and other reactions are summarized in the scheme on p. 996.

Dichlorine heptoxide, Cl$_2$O$_7$, is the anhydride of perchloric acid (p. 1013) and is conveniently obtained by careful dehydration of HClO$_4$ with H$_3$PO$_4$ at $-10°$C followed by cautious low-pressure distillation at $-35°$C and 1 mmHg. The compound is a shock-sensitive oily liquid with physical properties (p. 990) and structure (p. 989) as already described. Cl$_2$O$_7$ is less reactive than the lower oxides of chlorine and does not ignite organic materials at room temperature. Dissolution in water or aqueous alkalis regenerates perchloric acid and perchlorates respectively. Thermal decomposition (which can be explosive) is initiated by rupture of a Cl–O$_\mu$ bond, the activation energy being ~ 135 kJ mol^{-1}:

$$Cl_2O_7 \xrightarrow{\Delta} ClO_3 + ClO_4$$

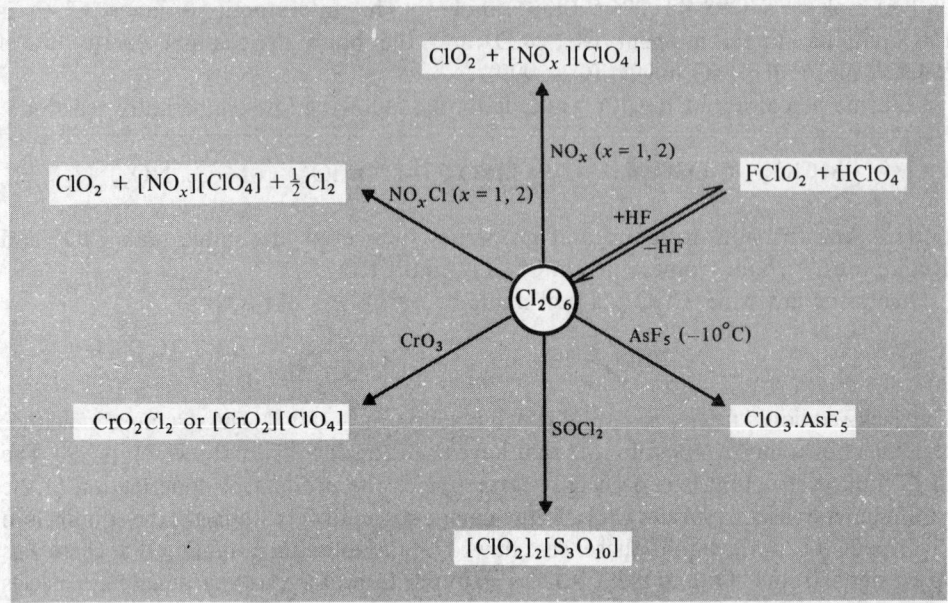

SCHEME   Some reactions of dichlorine hexoxide.

Oxides of bromine

The oxides of Br are less numerous, far less studied, and much less well characterized than the ten oxide species of chlorine discussed in the preceding section. The reasonably well established compounds are listed below.

Br_2O: a dark-brown solid moderately stable at $-60°$ (mp $-17.5°$ with decomposition), prepared by reaction of Br_2 vapour on HgO (cf. Cl_2O p. 991) or better, by low-temperature vacuum decomposition of BrO_2. The molecule has C_{2v} symmetry. It oxidizes I_2 to I_2O_5, benzene to 1,4-quinone, and yields OBr^- in alkaline solution.

BrO_2: a pale yellow crystalline solid formed quantitatively by low-temperature ozonolysis of Br_2:†

$$Br_2 + 4O_3 \xrightarrow{CF_3Cl/-78°C} 2BrO_2 + 4O_2$$

It is thermally unstable above $-40°C$ and decomposes violently to the elements at $0°C$; slower warming yields BrO_2 (see above). Alkaline hydrolysis leads to disproportionation:

$$6BrO_2 + 6OH^- \longrightarrow 5BrO_3^- + Br^- + 3H_2O$$

Reaction with F_2 yields $FBrO_2$ and with N_2O_4 yields $[NO_2]^+[Br(NO_3)_2]^-$.

In addition to these compounds the unstable momeric radicals BrO, BrO_2, and BrO_3 have

---

† Ozonolysis of Br_2 at $0°C$ yields white, poorly characterized solids which, depending on the conditions used, have compositions close to Br_2O_5, Br_3O_8, and BrO_3; no structural data are available.

been made by γ-radiolysis or flash photolysis of the anions OBr^-, BrO_2^-, and BrO_3^-. For BrO the interatomic distance is 172.1 pm, the dipole moment 1.55 D, and the thermodynamic properties $\Delta H_f^\circ(298\ K)$ 125.8 kJ mol^{-1}, $\Delta G_f^\circ(298\ K)$ 108.2 kJ mol^{-1}, and $S^\circ(298\ K)$ 237.4 J K^{-1} mol^{-1}.

Oxides of iodine

Iodine forms the most stable oxides of the halogens and I_2O_5 was made (independently) by J. L. Gay Lussac and H. Davy in 1813. However, despite this venerable history the structure of the compound was not determined unambiguously until 1970. It is most conveniently prepared by dehydrating iodic acid (p. 1010) at 200°C in a stream of dry air but it also results from the direct oxidation of I_2 with oxygen in a glow discharge. The structure (Fig. 17.20a) features molecular units of O_2IOIO_2 formed by joining two pyramidal IO_3 groups at a common oxygen. The bridging I–O distances correspond to single bonds, whereas the terminal I–O distances are substantially shorter.[74] There are also appreciable intermolecular interactions which join the molecular units into cross-linked chains (Fig. 17.20b); this gives each iodine pseudo-fivefold coordination, the sixth position of the distorted octahedron presumably being occupied by the lone-pair of electrons on the iodine atom.

I_2O_5 forms white, hygroscopic, thermodynamically stable crystals: ΔH_f° -158.1 kJ mol^{-1}, d 4.980 g cm^{-3}. The compound is very soluble in water, reforming the parent acid HIO_3. So great is the affinity for water that commercial "I_2O_5" consists almost entirely of HI_3O_8, i.e. $I_2O_5 \cdot HIO_3$. The interrelations between these compounds and the rather less stable oxides I_4O_9 and I_2O_4 are shown in the scheme. I_4O_9 is a hygroscopic

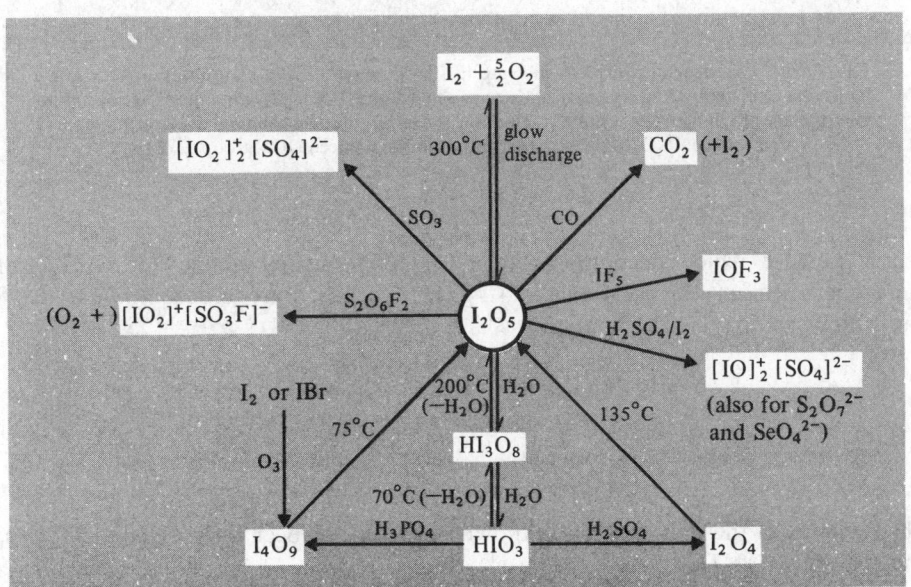

SCHEME   Preparation and reactions of iodine oxides.

[74] K. SELTE and A. KJEKSHUS, Iodine oxides. Part 3. The crystal structure of I_2O_5, *Acta Chem. Scand.* **24,** 1912–24 (1970).

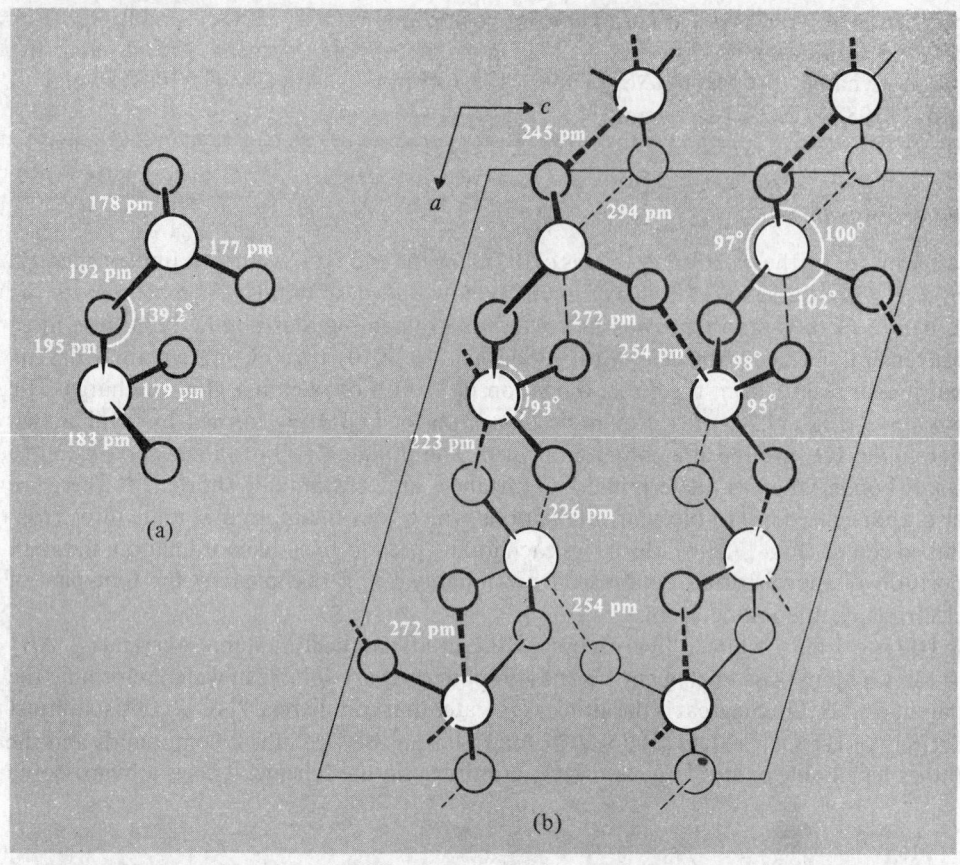

FIG. 17.20   The structure of I_2O_5 showing (a) the dimensions and conformation of a single molecular unit, and (b) the crystal structure projected along [010]. Note that the molecule has no mirror plane of symmetry so is not C_{2v}; also note the intramolecular interaction of one terminal O atom in each I_2O_5 unit with the I atom in the adjacent IO_3 group (272 pm).

yellow powder which decomposes to I_2O_5 when heated above 75°; I_2O_4 forms diamagnetic lemon-yellow crystals (d 4.2 g cm^{-3}) which start to decompose above 85° and which rapidly yield I_2O_5 at 135°:

$$2I_4O_9 \xrightarrow{75} 3I_2O_5 + I_2 + \tfrac{3}{2}O_2$$

$$5I_2O_4 \xrightarrow{135} 4I_2O_5 + I_2$$

The structure of these oxides are unknown but I_4O_9 has been formulated as $I^{III}(I^VO_3)_3$ and I_2O_4 as $[IO]^+[IO_3]^-$.

I_2O_5 is notable in being one of the few chemicals that will oxidize CO rapidly and completely at room temperature:

$$5CO + I_2O_5 \xrightarrow{} I_2 + 5CO_2$$

The reaction forms the basis of a useful analytical method for determining the concentration of CO in the atmosphere or in other gaseous mixtures. I_2O_5 also oxidizes NO, C_2H_4, and H_2S. SO_3 and $S_2O_6F_2$ yield iodyl salts, $[IO_2]^+$, whereas concentrated H_2SO_4 and related acids reduce I_2O_5 to iodosyl derivatives, $[IO]^+$. Fluorination of I_2O_5 with F_2, BrF_3, SF_4, or $FClO_2$ yields IF_5 which itself reacts with the oxide to give IOF_3.

In addition to the stable I_2O_5 and moderately stable I_4O_9 and I_2O_4, several short-lived radicals have been detected and characterized during γ-radiolysis and flash photolysis of iodates in aqueous alkali:

$$IO_3^- \xrightarrow{h\nu} IO_2 + O^-$$

$$IO_3^- + O^- \xrightarrow{H_2O} IO_3 + 2OH^-$$

$$IO_3^- + IO_2 \longrightarrow IO + IO_4^-$$

The endothermic radical IO has also been studied in the gas phase: the interatomic distance is 186.7 pm and the bond dissociation energy $\sim 175 \pm 20$ kJ mol^{-1}. It thus appears that, although the higher oxides of iodine are much more stable than any oxide of Cl or Br, nevertheless, IO is much less stable than ClO (p. 995) or BrO (p. 997). Its enthalpy of formation and other thermodynamic properties are: $\Delta H_f^\circ(298$ K$)$ 175.1 kJ mol^{-1}, $\Delta G_f^\circ(298$ K$)$ 149.8 kJ mol^{-1}, $S^\circ(298$ K$)$ 245.5 J K^{-1} mol^{-1}.

17.2.8 *Oxoacids and oxoacid salts*

General considerations[75]

The preparative chemistry and technical applications of the halogen oxoacids and their salts have been actively pursued and developed for over two centuries (p. 921) and can now be very satisfactorily systematized in terms of general thermodynamic principles. The thermodynamic data are codified in the form of reduction potentials and equilibrium constants and these, coupled with the relative rates of competing reactions, allow a vast range of aqueous solution chemistry of the halogens to be interrelated. Thus, although all the halogens are to some extent soluble in water, extensive disproportionation reactions and/or mutual redox reactions with the solvent can occur to an extent that depends crucially on conditions such as pH and concentration (which influence the thermo-dynamic variables) and the presence of catalysts or light quanta (which can overcome kinetic activation barriers). Fluorine is again exceptional and, because of its very high standard reduction potential, $E^\circ(\frac{1}{2}F_2/F^-) + 2.866$ V, reacts very strongly with water at all values of pH (p. 736). Its inability to achieve formal oxidation states higher than $+1$ also limits the available oxoacids to hypofluorous acid HOF (p. 1003). Numerous other oxoacids are known for the heavier halogens (Table 17.21) though most cannot be isolated pure and are stable only in aqueous solution or in the form of their salts. Anhydrous perchloric acid ($HClO_4$), iodic acid (HIO_3), paraperiodic acid (H_5IO_6), and metaperiodic acid (HIO_4) have been isolated as pure compounds.

The standard reduction potentials for Cl, Br, and I species in acid and in alkaline

75 Ref. 17, Chemical properties of the halogens—redox properties: aqueous solutions, pp. 1188–95; Oxoacids and oxoacid salts of the halogens, pp. 1396–1465.

TABLE 17.21   *Oxoacids of the halogens*

Generic name	Chlorine	Bromine	Iodine	Salts
Hypohalous acids[b]	HOCl[a]	HOBr[a]	HOI[a]	Hypohalites
Halous acids	HOClO[a]	(HOBrO?)[a]	—	Halites
Halic acids	HOClO$_2$[a]	HOBrO$_2$[a]	HOIO$_2$	Halates
Perhalic acids	HOClO$_3$	HOBrO$_3$[a]	HOIO$_3$, (HO)$_5$IO, H$_4$I$_2$O$_7$	Perhalates

[a] Stable only in aqueous solution.
[b] HOF also known (p. 1003).

aqueous solutions are summarized in Fig. 17.21. The couples $\frac{1}{2}X_2/X^-$ are independent of pH and, together with the value for F_2, indicate a steadily decreasing oxidizing power of the halogens in the sequence $F_2(+2.866 \text{ V}) > Cl_2(+1.358 \text{ V}) > Br_2(+1.066 \text{ V}) > I_2(+0.536 \text{ V})$. Remembering that $E°(\frac{1}{2}O_2/H_2O) = 1.229$ V these values indicate that the potentials for the reaction

$$X_2 + H_2O \longrightarrow \tfrac{1}{2}O_2 + 2H^+ + 2X^-$$

decrease in the sequence $F_2(+1.637 \text{ V}) > Cl_2(+0.129 \text{ V}) > Br_2(-0.163 \text{ V}) > I_2(-0.693 \text{ V})$. As already mentioned, this implies that F_2 will oxidize water to O_2 and the same should happen with chlorine in the absence of sluggish kinetic factors. In fact, were it not for the further fortunate circumstance of an appreciably higher overvoltage for oxygen, chlorine would not be evolved during the electrolysis of aqueous chloride solutions at low current densities: the phenomenon is clearly of great technical importance for the industrial preparation of chlorine by electrolysis of brines (p. 931).

For all other couples in Fig. 17.21 (i.e. for all couples involving oxygenated species) an increase in pH causes a dramatic reduction in $E°$ as expected (p. 498). For example, in acid solution the couple $BrO_3^-/\frac{1}{2}Br_2(l)$ refers to the equilibrium reaction

$$BrO_3^- + 6H^+ + 5e^- \rightleftharpoons \tfrac{1}{2}Br_2 + 3H_2O; \quad (E° \ 1.495 \text{ V})$$

The equilibrium constant clearly depends on the sixth power of the hydrogen-ion concentration and, when this is reduced (say to 10^{-14} in 1 M alkali), the potential is likewise diminished by an amount $\sim (RT/nF)\ln[H^+]^6$, i.e. by ca. $(0.0592/5) \times 14 \times 6 \simeq 0.99$ V. In agreement with this (Fig. 17.21) the potential at pH 14 is 0.519 V (calc ~ 0.51 V) for the reaction

$$BrO_3^- + 3H_2O + 5e^- \rightleftharpoons \tfrac{1}{2}Br_2(l) + 6OH^-$$

The data in Fig. 17.21 are presented in graphical form in Fig. 17.22 which shows the volt-equivalent diagrams (p. 499) for acid and alkaline solutions. It is clear from these that Cl_2 and Br_2 are much more stable towards disproportionation in acid solution (concave angle at X_2) than in alkaline solutions (convex angle). In terms of equilibrium constants:

$$\text{acid solution: } X_2 + H_2O \rightleftharpoons HOX + H^+ + X^-; \quad K_{ac} = \frac{[HOX][H^+][X^-]}{[X_2]}$$

$$\text{alkaline solution: } X_2 + 2OH^- \rightleftharpoons OX^- + H_2O + X^-; \quad K_{alk} = \frac{[OX^-][X^-]}{[X_2][OH^-]^2}$$

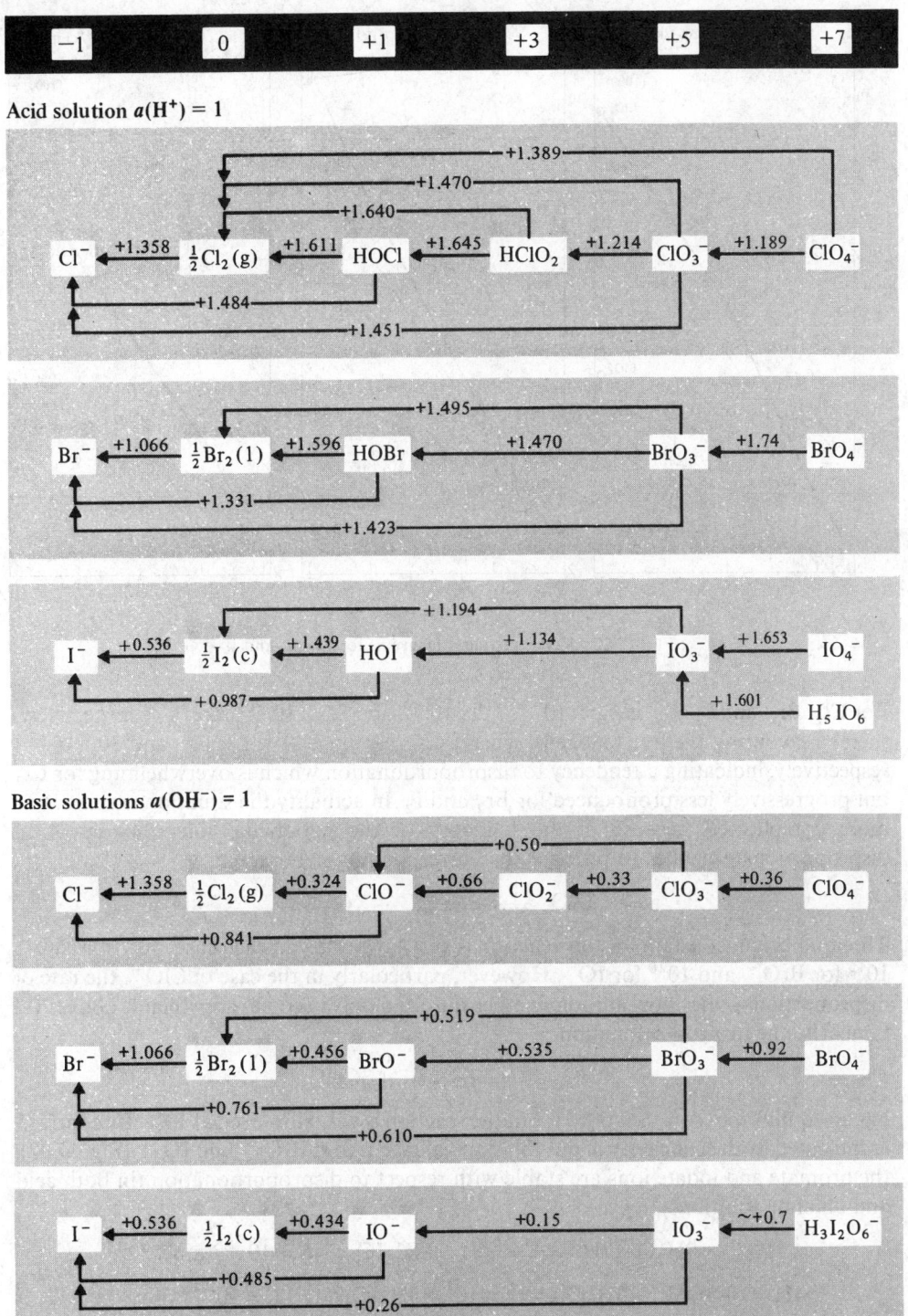

FIG. 17.21   Standard reduction potentials for Cl, Br, and I species in acid and alkaline aqueous solutions.

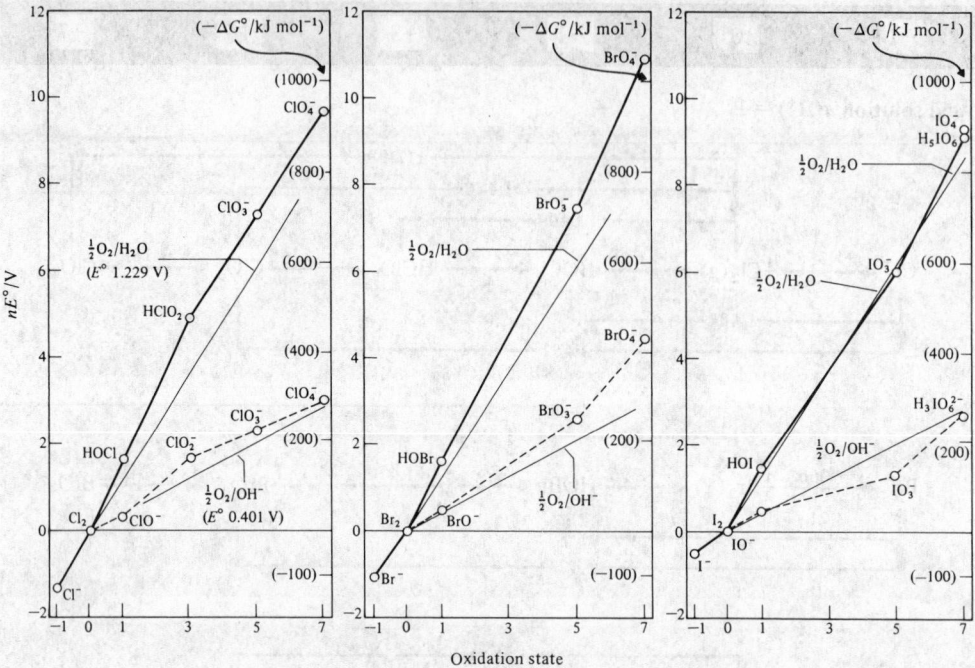

FIG. 17.22   Volt-equivalent diagrams for Cl, Br, and I.

For Cl_2, Br_2, and I_2, K_{ac} is 4.2×10^{-4}, 7.2×10^{-9}, and 2.0×10^{-13} mol^2 l^{-2} respectively, thereby favouring the free halogens, whereas K_{alk} is 7.5×10^{15}, 2×10^8, and 30 mol^{-1} l respectively, indicating a tendency to disproportionation which is overwhelming for Cl_2 but progressively less pronounced for Br_2 and I_2. In actuality the situation is somewhat more complicated because of the tendency of the hypohalite ions themselves to disproportionate further to produce the corresponding halite ions:

$$3XO^- \rightleftharpoons 2X^- + XO_3^-$$

The equilibrium constant for this reaction is very favourable in each case: 10^{27} for ClO^-, 10^{15} for BrO^-, and 10^{20} for IO^-. However, particularly in the case of ClO^-, the rate of disproportionation is slow at room temperature and only becomes appreciable above 70°. Similarly, the disproportionation

$$4ClO_3^- \rightleftharpoons Cl^- + 3ClO_4^-$$

has an equilibrium constant of 10^{20} but the reaction is very slow even at 100°. By contrast, as indicated by the concavity of the volt-equivalent curve at BrO_3^- and IO_3^- (Fig. 17.22), the bromate and iodate ions are stable with respect to disproportionation (in both acid and alkaline solutions), e.g.:

$$7IO_3^- + 9H_2O + 7H^+ \rightleftharpoons I_2 + 5H_5IO_6; \quad K = 10^{-85} \text{ mol}^{-8} \text{ l}^8$$

$$4IO_3^- + 3OH^- + 3H_2O \rightleftharpoons I^- + 3H_3IO_6^{2-}; \quad K = 10^{-44} \text{ mol}^{-3} \text{ l}^3$$

More detailed consideration of these various equilibria and other redox reactions of the halogen oxoacids will be found under the separate headings below.

The strengths of the monobasic acids increase rapidly with increase in oxidation state of the halogen in accordance with Pauling's rules (p. 54). For example, approximate values of pK_a are: HOCl 7.52, HOClO 1.94, HOClO$_2$ −3, HOClO$_3$ −10. The pK_a values of related acids increase in the sequence Cl < Br < I.

Hypohalous acids, HOX, *and hypohalites*, XO$^-$ [69, 75]

Hypofluorous acid is the most recent of the halogen oxoacids to be prepared.[76] Traces were obtained in 1968 by photolysis of a mixture of F$_2$ and H$_2$O in a matrix of solid N$_2$ at 14–20 K but weighable amounts of the compound were first obtained by M. H. Studier and E. H. Appelman in 1971 by the fluorination of ice:

$$F_2 + H_2O \underset{-40°C}{\rightleftharpoons} HOF + HF$$

The isolation of HOF depends on removing it rapidly from the reaction zone so that it is prevented from reacting further with HF, F$_2$, or H$_2$O (see below). The method used was to recirculate F$_2$ at ∼ 100 mmHg through a Kel-F U-tube filled with moistened Räschig rings cut from Teflon "spaghetti" tubing (Kel-F is polymerized chlorotrifluoroethene; Teflon is polymerized tetrafluoroethene). The U-tube was held at about −40°C and the effluent was passed through U-tubes cooled to −50° and −79° to remove water and HF, and, finally, through a U-tube at −183° to trap the HOF. The use of the −50° trap was found to be critical because without it all of the HOF was caught in the −79° with the H$_2$O, from which it could not be isolated because of subsequent reaction.

HOF is a white solid which melts at −117° to a pale-yellow liquid. Its bp (extrap) is somewhat below room temperature and its volatility is thus comparable to that of HF with which it is always slightly contaminated. Spectroscopic data establish a nonlinear structure with H–O 96.4 pm, O–F 144.2 pm, and bond angle H–O–F 97.2°: this is the smallest known bond angle at an unrestricted O atom (cf. H–O–H 104.7°, F–O–F 103.2°). It has been suggested that this arises in part from electrostatic attraction of the 2 terminal atoms, since nmr data lead to a charge of ∼ +0.5e on H and ∼ −0.5e on F. The negative charge on F is intermediate between those estimated for F in HF and OF$_2$ and this emphasizes the strictly formal nature of the +1 oxidation state for F in HOF.

The most prominent chemical property of HOF is its instability. It decomposes spontaneously to HF and O$_2$ with a half-life of *ca.* 30 min in a Teflon apparatus at room temperature and 100 mmHg. It reacts rapidly with water to produce HF, H$_2$O$_2$, and O$_2$; in dilute aqueous acid H$_2$O$_2$ is the predominant product whereas in alkaline solution O$_2$ is the principal O-containing product. It reacts with HF to reverse the equilibrium used in its preparation. It does not dehydrate to its formal anhydride OF$_2$ but in the presence of H$_2$O it reacts with F$_2$ to form this species.

$$F_2 + HOF \xrightarrow{H_2O} OF_2 + HF$$

This reaction does not occur in the gas phase, however, in the absence of H$_2$O.

By contrast with the elusive though isolable HOF, the history of HOCl goes back over

---

[76] E. H. APPELMAN, Nonexistent compounds: two case histories, *Acc. Chem. Res.* **6**, 113–17 (1973).

two centuries to the earliest experiments of C. W. Scheele with Cl_2 in 1774 (p. 923), and the bleaching and sterilizing action of hypochlorites have long been used both industrially and domestically. HOCl, HOBr, and HOI are all highly reactive, relatively unstable compounds that are known primarily in aqueous solutions. The most convenient preparation of such solutions is by perturbing the hydrolytic disproportionation equilibrium (p. 1000):

$$X_2 + H_2O \rightleftharpoons H^+ + X^- + HOX$$

by addition of HgO or Ag_2O so as to remove the halide ions. On an industrial scale, aqueous solutions of HOCl (containing Cl^-) are readily prepared by reacting Cl_2 with aqueous alkali. With strong bases {NaOH, $Ca(OH)_2$} the reaction proceeds via the intermediate formation of hypochlorite, but this intermediate product is not formed with weaker bases such as $NaHCO_3$ or $CaCO_3$:

$$Cl_2 + 2OH^- \xrightarrow{\text{aq}} \{ClO^- + Cl^- + H_2O\} \xrightarrow{Cl_2} 2HOCl + 2Cl^-$$

Chloride-free solutions (up to 5 M concentration) can be made by treating Cl_2O with water at 0° or industrially by passing Cl_2O gas into water. In fact, concentrated solutions of HOCl also contain appreciable amounts of Cl_2O which can form a separate layer and which is probably the source of the yellow colour of such solutions:

$$2HOCl(aq) \rightleftharpoons Cl_2O(aq) + H_2O(l); \quad K\ (0°C) = 3.55 \times 10^{-3}\ mol^{-1}\ l$$

Organic solutions can be obtained in high yield by extracting HOCl from Cl^--containing aqueous solutions into polar solvents such as ketones, nitriles, or esters. Electrodialysis using semipermeable membranes affords an alternative route.

Solutions of the corresponding hypohalites can be made by the rapid disproportionation of the individual halogens in cold alkaline solutions (p. 1000):

$$X_2 + 2OH^- \rightleftharpoons X^- + OX^- + H_2O$$

Such solutions are necessarily contaminated with halide ions and with the products of any subsequent decomposition of the hypohalite anions themselves. Alternative routes are the electrochemical oxidation of halides in cold dilute solutions or the chemical oxidation of bromides and iodides:

$$X^- + OCl^- \longrightarrow OX^- + Cl^- \quad (X = Br, I)$$

$$I^- + OBr^- \longrightarrow OI^- + Br^-$$

Hypochlorites can also be made by careful neutralization of aqueous solutions of hypochlorous acid or Cl_2O.

The most stable solid hypochlorites are those of Li, Ca, Sr, and Ba (see below). NaOCl has only poor stability and cannot be isolated pure; KOCl is known only in solution, Mg yields a basic hypochlorite, and impure Ag and Zn hypochlorites have been reported. Hydrated salts are also known. Solid, yellow, hydrated hypobromites $NaOBr.xH_2O$ ($x = 5, 7$) and $KOBr.3H_2O$ can be crystallized from solutions obtained by adding Br_2 to cold conc solutions of MOH but the compounds decompose above 0°C. No solid metal hypoiodites have yet been isolated.

HOCl is more stable than HOBr and HOI and its microwave spectrum in the gas phase confirms the expected nonlinear geometry with H–O 97 pm, O–Cl 169.3 pm, and angle

H–O–Cl $103 \pm 3°$ (cf. HOF, p. 1003). All three hypohalous acids are weak and solutions of their salts are therefore alkaline since the equilibrium

$$OX^- + H_2O \rightleftharpoons HOX + OH^-$$

lies well to the right. Except at high pH, hypohalite solutions contain significant amounts of the undissociated acid. Approximate values for the acid dissociation constants K_a at room temperature are HOCl 2.9×10^{-8}, HOBr 5×10^{-9}, HOI $\sim 10^{-11}$: these values are close to those of many α-aminoacids and may also be compared with carbonic acid K_a 4.3×10^{-7}, which is some 10 times stronger than HOCl, and phenol, which has K_a 1.3×10^{-10}.

The manner and rate of decomposition of hypohalous acids (and hypohalite ions) in solution are much influenced by the concentration, pH, and temperature of the solutions, by the presence or absence of salts which can act as catalysts, promotors, activators, and by light quanta. The main competing modes of decomposition are:

$$2HOX \longrightarrow 2H^+ + 2X^- + O_2 \quad (\text{or } 2OX^- \longrightarrow 2X^- + O_2)$$

$$\text{and } 3HOX \longrightarrow 3H^+ + 2X^- + XO_3^- \quad (\text{or } 3OX^- \longrightarrow 2X^- + XO_3^-)$$

The acids decompose more readily than the anions so hypohalites are stabilized in basic solutions. The stability of the anions diminishes in the sequence $ClO^- > BrO^- > IO^-$.

Hypochlorites are amongst the strongest of the more common oxidizing agents and they react with inorganic species, usually by the net transfer of an O atom. Kinetic studies suggest that the oxidizing agent can be either HOCl or OCl$^-$ in a given reaction, but rarely both simultaneously. Some typical examples are in Table 17.22. Hypochlorites react with

TABLE 17.22   *Oxidation of inorganic substrates with HOCl or OCl$^-$*

HOCl		OCl$^-$	
Substrate	Products	Substrate	Products
HCO_2^-	CO_3^{2-}	ClO^-	ClO_2^-
$HC_2O_4^-$	CO_2	ClO_2^-	ClO_3^-
OCN^-	CO_3^{2-}, N_2, NO_3^-	CN^-	OCN^-
NH_3	NCl_3	NH_3	NH_2Cl
NO_2^-	NO_3^-	SO_3^{2-}	SO_4^{2-}
H_2O_2	O_2	IO_3^-	IO_4^-
S	SO_4^{2-}	Mn^{2+}	MnO_4^-
Br^-	Br_2 (acid)	Br^-	OBr^-, BrO_3^- (alkaline)
I^-	I_2 (acid)	I^-	OI^-, IO_3^- (alkaline)

ammonia and organic amino compounds to form chloramines. The characteristic "chlorine" odour of water that has been sterilized with hypochlorite is, in fact, due to chloramines produced from attack on bacteria. By contrast, hypobromites oxidize amines quantitatively to N_2, a reaction that is exploited in the analysis of urea (p. 342):

$$(NH_2)_2CO + 3OBr^- + 2OH^- \longrightarrow N_2 + CO_3^{2-} + 3Br^- + 3H_2O$$

Other uses of hypohalous acids and hypohalites are described in the Panel.

Some Uses of Hypohalous Acids and Hypohalites

In addition to the applications just indicated, hypohalous acids are useful halogenating agents for both aromatic and aliphatic compounds. HOBr and HOI are usually generated *in situ*. The ease of aromatic halogenation increases in the sequence $OCl^- < OBr^- < OI^-$ and is facilitated by salts of Pb or Ag. Another well-known reaction of hypohalites is their cleavage of methyl ketones to form carboxylates and haloform:

$$RCOCH_3 + 3OX^- \longrightarrow RCO_2^- + 2OH^- + CHX_3$$

This is the basis of the iodoform test for the CH_3CO group. In addition to these reactions there is considerable industrial use for HOCl and hypochlorites in the manufacture of hydrazine (p. 490), chlorhydrins, and α-glycols:

By far the largest tonnage of hypochlorites is used for bleaching and sterilizing. "Liquid bleach" is an alkaline solution of NaOCl (pH $\geqslant$ 11); domestic bleaches have about 5% "available chlorine" content† whereas small-scale commercial installations such as laundries use $\sim$10% concentration. Chlorinated trisodium phosphate, which is a crystalline efflorescent product of approximate empirical composition $(Na_3PO_4 \cdot 11H_2O)_4 \cdot NaOCl$, has 3.5–4.5% available Cl and is used in automatic dishwasher detergents, scouring powders, and acid metal cleaners for dairy equipment. Paper and pulp bleaching is effected by "bleach liquor", a solution of $Ca(OCl)_2$ and $CaCl_2$, yielding $\sim$85 g l^{-1} of "available chlorine". Powdered calcium hypochlorite, $Ca(OCl)_2 \cdot 2H_2O$ (70% available Cl), is used for swimming-pool sanitation whereas "bleaching powder", $Ca(OCl)_2 \cdot CaCl_2 \cdot Ca(OH)_2 \cdot 2H_2O$ (obtained by the action of Cl_2 gas on slaked lime) contains 35% available Cl and is used for general bleaching and sanitation:

$$3Ca(OH)_2 + 2Cl_2 \longrightarrow Ca(OCl)_2 \cdot CaCl_2 \cdot Ca(OH)_2 \cdot 2H_2O$$

The speciality chemical LiOCl (40% "Cl") is used when calcium is contra-indicated, such as in the sanitation of hard water and in some dairy applications. Some idea of the scale of these applications can be gained from the following production figures which relate to the USA:[69]

LiOCl $\sim$4500 tonnes pa.

NaOCl $>$ 180 000 tpa (on a dry basis) of which about 250 tonnes per day (say 63 000 tpa) is for household liquid bleach and the rest for laundries, disinfection of swimming pools, municipal water supplies and sewage, and the industrial manufacture of N_2H_4 and organic chemicals.

NaOCl . $(Na_3PO_4 \cdot 11H_2O)_4$ was commercialized in 1930 and demand rose to 81 000 tonnes in 1973.

$Ca(OCl)_2$ $\sim$67 000 tpa plus production facilities in numerous other countries (e.g. the USSR $\sim$26 000 tonnes).

Bleaching powder is now much less used than formerly in highly industrialized countries but is still manufactured on a large scale in less-developed regions. In the USA its production peaked at 133 000 tonnes in 1923 but had fallen to 23 600 tonnes by 1955.

† "Available chlorine" content is defined as the weight of Cl_2 which liberates the same amount of I_2 from HI as does a given weight of the compound; it is often expressed as a percentage. For example, from the two (possibly hypothetical) stoichiometric equations $Cl_2 + 2HI \rightarrow I_2 + 2HCl$ and $LiOCl + 2HI \rightarrow I_2 + LiCl + H_2O$ it can be seen that 1 mol of I_2 is liberated by 70.92 g Cl_2 or by 58.4 g LiOCl. Whence the "available chlorine" content of pure LiOCl is $(70.92/58.4) \times 100 = 121\%$. The commercial product is usually diluted by sulfates to about one-third of this strength (see below).

This section concludes with a reminder that, in addition to the hypohalous acids HOX and metal hypohalites $M(OX)_n$, various covalent (molecular) hypohalites are known. Hypochlorites are summarized in Table 17.23. All are volatile liquids or gases at room temperature and are discussed elsewhere (see index). Organic hypohalites are unstable and rapidly expel HX or RX to form the corresponding aldehyde or ketone:

$$ROH + HOX \longrightarrow ROX + H_2O$$

$$RCH_2OX \longrightarrow RCHO + HX$$

$$RR'CHOX \longrightarrow RR'C{=}O + HX$$

$$RR'R''COX \xrightarrow{h\nu} RR'C{=}O + R''X$$

TABLE 17.23   *Physical properties of some molecular hypochlorites*

Compound	MP/°C	BP/°C	Compound	MP/°C	BP/°C
$ClONO_2$	−107	18	$ClOSeF_5$	−115	31.5
$ClOClO_3$	−117	44.5	$ClOTeF_5$	−121	38.5
$ClOSO_2F$	−84.3	45.1	$ClOOSF_5$	−130	26.4
$ClOSF_5$	—	8.9	$ClOOCF_3$	−132	−22

Halous acids, HOXO, and halites, XO_2^- [69, 75, 77, 78]

Chlorous acid is the least stable of the oxoacids of chlorine; it cannot be isolated but is known in dilute aqueous solution. HOBrO and HOIO are even less stable, and, if they exist at all, have only a fleeting presence in aqueous solutions. Several chlorites have been isolated and $NaClO_2$ is sufficiently stable to be manufactured as an article of commerce on the kilotonne pa scale. Little reliable information is available on bromites and still less is established for iodites which are essentially non-existent.

$HClO_2$ is formed (together with $HClO_3$) during the decomposition of aqueous solutions of ClO_2 (p. 993) but the best preparation is to treat an aqueous suspension of $Ba(ClO_2)_2$ with dilute sulfuric acid:

$$Ba(OH)_2(aq) + H_2O_2 + ClO_2 \longrightarrow Ba(ClO_2)_2 + 2H_2O + O_2$$

$$Ba(ClO_2)_2(\text{suspension}) + \text{dil } H_2SO_4 \longrightarrow BaSO_4\downarrow + 2HClO_2$$

Evidence for the undissociated acid comes from spectroscopic data but the solutions cannot be concentrated without decomposition. $HClO_2$ is a moderately strong acid K_a (25°C) 1.1×10^{-2} (cf H_2SeO_4 K_a 1.2×10^{-2}, $H_4P_2O_7$ K_a 2.6×10^{-2}).

The decomposition of chlorous acid depends sensitively on its concentration, pH, and the presence of catalytically active ions such as Cl^- which is itself produced during the

---

[77] G. GORDON, R. G. KIEFFER, and D. H. ROSENBLATT, The chemistry of chlorine dioxide, *Progr. Inorg. Chem.* **15**, 201–86 (1972). The first half of this review deals with the aqueous solution chemistry of chlorous acid and chlorites.

[78] F. SOLYMOSI, *Structure and Stability of Salts of the Halogen Oxyacids in the Solid Phase*, Wiley, UK, 1978, 468 pp.

decomposition. The main mode of decomposition (particularly if Cl^- is present) is to form ClO_2:

$$5HClO_2 \longrightarrow 4ClO_2 + Cl^- + H^+ + 2H_2O; \quad \Delta G° \ -144 \ kJ \ mol^{-1}$$

Competing modes produce ClO_3^- or evolve O_2:

$$3HClO_2 \longrightarrow 2ClO_3^- + Cl^- + 3H^+; \quad \Delta G° \ -139 \ kJ \ mol^{-1}$$

$$HClO_2 \longrightarrow Cl^- + O_2 + H^+; \quad \Delta G° \ -123 \ kJ \ mol^{-1}$$

Metal chlorites are normally made by reduction of aqueous solutions of ClO_2 in the presence of the metal hydroxide or carbonate. As with the preparation of $Ba(ClO_2)_2$ above, the reducing agent is usually a peroxide since this adds no contaminant to the resulting chlorite solution:

$$2ClO_2 + O_2^{2-} \longrightarrow 2ClO_2^- + O_2$$

The ClO_2^- ion is nonlinear, as expected, and X-ray studies of NH_4ClO_2 (at $-35°$) and $AgClO_2$ lead to the dimensions Cl–O 156 pm, angle O–Cl–O 111°. The chlorites of the alkali metals and alkaline earth metals are colourless or pale yellow. Heavy metal chlorites tend to explode or detonate when heated or struck (e.g. those of Ag^+, Hg^+, Tl^+, Pb^{2+}, and also those of Cu^{2+} and NH_4^+). Sodium chlorite is the only one to have sufficient stability and to be sufficiently inexpensive to be a major article of commerce.

Anhydrous $NaClO_2$ crystallizes from aqueous solutions above 37.4° but below this temperature the trihydrate is obtained. The commercial product contains about 80% $NaClO_2$. The anhydrous salt forms colourless deliquescent crystals which decompose when heated to 175–200°: the reaction is predominantly a disproportionation to ClO_3^- and Cl^- but about 5% of molecular O_2 is also released (based on the ClO_2^- consumed). Neutral and alkaline aqueous solutions of $NaClO_2$ are stable at room temperature (despite their thermodynamic instability towards disproportionation as evidenced by the reduction potentials on p. 1001). This is a kinetic activation-energy effect and, when the solutions are heated near to boiling, slow disproportionation occurs:

$$3ClO_2^- \longrightarrow 2ClO_3^- + Cl^-$$

Photochemical decomposition is rapid and the products obtained depend on the pH of the solution;

at pH 8.4: $$6ClO_2^- \xrightarrow{h\nu} 2ClO_3^- + 4Cl^- + 3O_2$$

at pH 4.0: $$10ClO_2^- \xrightarrow{h\nu} 2ClO_3^- + 6Cl^- + 2ClO_4^- + 3O_2$$

The stoichiometry in acid solution implies that, in addition to the more usual disproportionation into ClO_3^- and Cl^-, the following disproportionation also occurs:

$$2ClO_2^- \longrightarrow Cl^- + ClO_4^-$$

The mechanisms of these various reactions have been the object of many studies.[69, 75, 77]

The main commercial applications of $NaClO_2$ are in the bleaching and stripping of textiles, and as a source of ClO_2 where required volumes are comparatively small. It is also used as an oxidant for removal of nitrogen oxide pollutants from industrial off-gases. The specific oxidizing properties of $NaClO_2$ towards certain malodorous or toxic compounds such as unsaturated aldehydes, mercaptans, thioethers, H_2S, and HCN have likewise led to its use for scrubbing the off-gases of processes where these noxious pollutants are formed. Production statistics are rather sparse but the main non-communist production plants are in the EEC, which produced some 18 000 tonnes pa in 1977 (mostly in France); Japan, whose production rate was about 6000 tpa in the mid 1970s; and the USA, where the 1976 price for technical grade (80%) $NaClO_2$ was *ca.* \$1.30 per kg.

Crystalline barium bromite $Ba(BrO_2)_2 . H_2O$ was first isolated in 1959; it can be made by treating the hypobromite with Br_2 at pH 11.2 and 0°C, followed by slow evaporation. $Sr(BrO_2)_2 . 2H_2O$ was obtained similarly.

Halic acids, $HOXO_2$, and halates, XO_3^{-} [79, 80]

Disproportionation of X_2 in hot alkaline solution has long been used to synthesize chlorates and bromates (see oxidation state diagrams, p. 1002):

$$3X_2 + 6OH^- \longrightarrow XO_3^- + 5X^- + 3H_2O$$

For example, J. von Liebig developed the technical preparation of $KClO_3$ by passing Cl_2 into a warm suspension of $Ca(OH)_2$ and then adding KCl to enable the less-soluble chlorate to crystallize on cooling:

$$6Ca(OH)_2 + 6Cl_2 \longrightarrow Ca(ClO_3)_2 + 5CaCl_2 + 6H_2O$$

$$Ca(ClO_3)_2 + 2KCl \longrightarrow 2KClO_3 + CaCl_2$$

However, only one-sixth of the halogen present is oxidized and alternative routes are more generally preferred for large-scale manufacture. Thus, the most important halate, $NaClO_3$, is manufactured on a huge scale[†] by the electrolysis of brine in a diaphragmless cell which promotes efficient mixing. Under these conditions, the Cl_2 produced by anodic oxidation of Cl^- reacts with cathodic OH^- to give hypochlorite which then either disproportionates or is itself further anodically oxidized to ClO_3^-:

anode: $Cl^- \rightarrow \frac{1}{2}Cl_2 + e^-$;   *cathode:* $H_2O + e^- \rightarrow \frac{1}{2}H_2 + OH^-$
mixing: $Cl_2 + 2OH^- \rightarrow Cl^- + OCl^- + H_2O$
further disproportionation: $3OCl^- \rightarrow ClO_3^- + 2Cl^-$
further anodic oxidation: $OCl^- + 2H_2O \rightarrow ClO_3^- + 2H_2$

[†] $NaClO_3$ production approaches 400 000 tonnes pa in the USA alone. Its major use (78%) is for conversion to ClO_2 for bleaching paper pulp—this gives superior whiteness without degrading the cellulose fibre. A second important use (12%) is as an intermediate in making other chlorates and perchlorates. Some 4500 tonnes of $NaClO_3$ are used annually in the USA as a herbicide and it also finds large-scale application as a defoliant for cotton and a desiccant for soyabeans (to remove the leaves before mechanical picking). The use of $NaClO_3$ in pyrotechnic formulations is hampered by its hygroscopicity. $KClO_3$ does not suffer this disadvantage and is unexcelled as an oxidizer in fireworks and flares, the colours being obtained by admixture with salts of Sr (red), Ba (green), Cu (blue), etc. In addition $KClO_3$ is a crucial component in head of "safety matches" ($KClO_3$, S, Sb_2S_3, powdered glass, and dextrin paste). The US match industry consumes some 11 500 tonnes of $KClO_3$ annually. Representative prices (1977) for bulk lots of these commodities are \$0.33 per kg for $NaClO_3$ and \$0.32 per kg for $KClO_3$.

[79] Ref. 17, pp. 1418–35, Halic acids and halates.
[80] T. W. CLAPPER, Chloric acid and chlorates, *Kirk–Othmer Encyclopedia of Chemical Technology*, 3rd edn., Vol. 5, pp. 633–45, Wiley, New York, 1979.

Modern cells employ arrays of anodes (TiO_2 coated with a noble metal) and cathodes (mild steel) spaced 3 mm apart and carrying current at 2700 A m^{-2} into brine (80–100 g l^{-1}) at 60–80°C. Under these conditions current efficiency can reach 93% and 1 tonne of $NaClO_3$ can be obtained from 565 kg NaCl and 4535 kWh of electricity. The off-gas H_2 is also collected.

Bromates and iodates are prepared on a much smaller scale, usually by chemical oxidation. For example, Br$^-$ is oxidized to BrO_3^- by aqueous hypochlorite (conveniently effected by passing Cl_2 into alkaline solutions of Br$^-$). Iodates can be prepared either by direct high-pressure oxidation alkali metal iodides with oxygen at 600° or by oxidation of I_2 with chlorates:

$$I_2 + NaClO_3 \longrightarrow 2NaIO_3 + Cl_2$$

Salts of other metals are obtained by metathesis, and aqueous solutions of the corresponding acids are obtained by controlled addition of sulfuric acid to the barium salts:

$$Ba(XO_3)_2 + H_2SO_4 \longrightarrow 2HXO_3 + BaSO_4$$

Chloric acid, $HClO_3$ is fairly stable in cold water up to about 30% concentration but, on being warmed, such solutions evolve Cl_2 and ClO_2. Evaporation under reduced pressure can increase the concentration up to about 40% ($\sim HClO_3 . 7H_2O$) but thereafter it is accompanied by decomposition to $HClO_4$ and the evolution of Cl_2, O_2, and ClO_2:

$$8HClO_3 \longrightarrow 4HClO_4 + 2H_2O + 2Cl_2 + 3O_2$$
$$3HClO_3 \longrightarrow HClO_4 + H_2O + 2ClO_2$$

Likewise, aqueous $HBrO_3$ can be concentrated under reduced pressure to about 50% concentration ($\sim HBrO_3 . 7H_2O$) before decomposition obtrudes:

$$4HBrO_3 \longrightarrow 2H_2O + 2Br_2 + 5O_2$$

Both chloric and bromic acids are strong acids in aqueous solution ($pK_a \lesssim 0$) whereas iodic acid is slightly weaker, with pK_a 0.804, i.e. K_a 0.157.

Iodic acid is more conveniently synthesized by oxidation of an aqueous suspension of I_2 either electrolytically or with fuming HNO_3. Crystallization from acid solution yields colourless, orthorhombic crystals of α-HIO_3 which feature H-bonded pyramidal molecules of $HOIO_2$: $r(I–O)$ 181 pm, $r(I–OH)$ 189 pm, angle O–I–O 101.4°, angle O–I–(OH) 97°. When heated to $\sim$100°C iodic acid partly dehydrates to HI_3O_8 (p. 997); this comprises an H-bonded array of composition $HOIO_2 . I_2O_5$ in which the HIO_3 has almost identical dimensions to those in α-HIO_3. Further heating to 200° results in complete dehydration to I_2O_5. In concentrated aqueous solutions of HIO_3, the iodate ions formed by deprotonation react with undissociated acid according to the equilibrium

$$IO_3^- + HIO_3 \rightleftharpoons [H(IO_3)_2]^-; \quad K \approx 41 \, mol^{-1}$$

Accordingly, crystallization of iodates from solutions containing an excess of HIO_3 sometimes results in the formation of hydrogen biiodates, $M^IH(IO_3)_2$, or even dihydrogen triiodates, $M^IH_2(IO_3)_3$.

Chlorates and bromates feature the expected pyramidal ions XO_3^- with angles close to the tetrahedral (106–107°). With iodates the interatomic angles at iodine are rather less (97–105°) and there are three short I–O distances (177–190 pm) and three somewhat longer distances (251–300 pm) leading to distorted perovskite structures (p. 1122) with pseudo-sixfold coordination of iodine and piezoelectric properties (p. 62). In $Sr(IO_3)_2 . H_2O$ the coordination number of iodine rises to 7 and this increases still further to 8 (square antiprism) in $Ce(IO_3)_4$ and $Zr(IO_3)_4$.

The modes of thermal decomposition of the halates and their complex oxidation-reduction chemistry reflect the interplay of both thermodynamic and kinetic factors. On the one hand, thermodynamically feasible reactions may be sluggish, whilst, on the other, traces of catalyst may radically alter the course of the reaction. In general, for a given cation, thermal stability decreases in the sequence iodate > chlorate > bromate, but the mode and ease of decomposition can be substantially modified. For example, alkali metal chlorates decompose by disproportionation when fused:

$$4ClO_3^- \longrightarrow Cl^- + 3ClO_4^-$$

e.g. $LiClO_3$, mp 125° (d 270°); $NaClO_3$, mp 248° (d 265°); $KClO_3$, mp 368° (d 400°).† However, in the presence of a transition-metal catalyst such as MnO_2 decomposition of $KClO_3$ to KCl and oxygen begins at about 70° and is vigorous at 100°

$$2ClO_3^- \longrightarrow 2Cl^- + 3O_2$$

This is, indeed, a classic laboratory method for preparing small amounts of oxygen (p. 701). For bromates and iodates, disproportionation to halide and perhalate is not thermodynamically feasible and decomposition occurs either with formation of halide and liberation of O_2 (as in the catalysed decomposition of ClO_3^- just considered), or by formation of the oxide:

$$4XO_3^- \longrightarrow 2O^{2-} + 2X_2 + 5O_2$$

For all three halates (in the absence of disproportionation) the preferred mode of decomposition depends, again, on both thermodynamic and kinetic considerations. Oxide formation tends to be favoured by the presence of a strongly polarizing cation (e.g. magnesium, transition-metal, and lanthanide halates), whereas halide formation is observed for alkali-metal, alkaline-earth, and silver halates.

The oxidizing power of the halate ions in aqueous solution, as measured by their standard reduction potentials (p. 1001), decreases in the sequence bromate $\gtrsim$ chlorate > iodate but the rates of reaction follow the sequence iodate > bromate > chlorate. In addition, both the thermodynamic oxidizing power and the rate of reaction depend markedly on the hydrogen-ion concentration of the solution, being substantially greater in acid than in alkaline conditions (p. 1000).

An important series of reactions, which illustrates the diversity of behaviour to be expected, is the comproportionation of halates and halides. Bromides are oxidized quantitatively to bromine and iodides to iodine, this latter reaction being much used in volumetric analysis:

$$XO_3^- + 5X^- + 6H^+ \longrightarrow 3X_2 + 3H_2O \quad (X = Cl, Br, I)$$

† Note, however, that thermal decomposition of NH_4ClO_3 begins at 50°C and the compound explodes on further heating; this much lower decomposition temperature may result from prior proton transfer to give the less-stable acid

$$NH_4ClO_3 \longrightarrow NH_3 + HClO_3$$

A similar thermal instability afflicts NH_4BrO_3 (d $-5°$) and NH_4IO_3 (d $\sim 100°$).

Numerous variants are possible, e.g.:

$$ClO_3^- + 6Br^- \text{ (or } I^-) + 6H^+ \longrightarrow Cl^- + 3Br_2 \text{ (or } I_2) + 3H_2O$$
$$BrO_3^- + 6I^- + 6H^+ \longrightarrow Br^- + 3I_2 + 3H_2O$$
$$IO_3^- + 5Br^- + 6H^+ \longrightarrow 2Br_2 + IBr + 3H_2O$$
$$2ClO_3^- + 2Cl^- + 4H^+ \longrightarrow Cl_2 + 2ClO_2 + 2H_2O \quad \text{(p. 992)}$$
$$2BrO_3^- + 2Cl^- + 12H^+ \longrightarrow Br_2 + Cl_2 + 6H_2O$$
$$IO_3^- + 3Cl^- + 6H^+ \longrightarrow ICl + Cl_2 + 3H_2O$$

The greater thermodynamic stability of iodates enables iodine to displace Cl_2 and Br_2 from their halates:

$$I_2 + 2XO_3^- \longrightarrow X_2 + 2IO_3^- \quad (X = Cl, Br)$$

With bromate at pH 1.5–2.5 the reaction occurs in four stages:

(1) an induction period in which a catalyst (probably HOBr) is produced;
(2) $I_2 + BrO_3^- \longrightarrow IBr + IO_3^-$;
(3) $3IBr + 2BrO_3^- + 3H_2O \longrightarrow 5Br^- + 3IO_3^- + 6H^+$;
(4) $5Br^- + BrO_3^- + 6H^+ \longrightarrow 3Br_2 + 3H_2O$.

The dependence of reaction rates on pH and on the relative and absolute concentrations of reacting species, coupled with the possibility of autocatalysis and induction periods, has led to the discovery of some spectacular kinetic effects such as H. Landolt's "chemical clock" (1885): an acidified solution of Na_2SO_3 is reacted with an excess of iodic acid solution in the presence of starch indicator—the induction period before the appearance of the deep-blue starch-iodine colour can be increased systematically from seconds to minutes by appropriate dilution of the solutions before mixing. With an excess of sulfite, free iodine periodically appears and disappears as a result of the following sequence of reactions:

(1) $IO_3^- + 3SO_3^{2-} \longrightarrow I^- + 3SO_4^{2-}$
(2) $5I^- + IO_3^- + 6H^+ \longrightarrow 3I_2 + 3H_2O$
(3) $3I_2 + 3SO_3^{2-} + 3H_2O \longrightarrow 6I^- + 6H^+ + 3SO_4^{2-}$

Another periodic reaction, discovered by W. C. Bray in 1921, concerns the reduction of iodic acid to I_2 by H_2O_2 followed by the reoxidation of I_2 to HIO_3:

$$2HIO_3 + 5H_2O_2 \longrightarrow 5O_2 + I_2 + 6H_2O$$
$$I_2 + 5H_2O_2 \longrightarrow 2HIO_3 + 4H_2O$$

The net reaction is the disproportionation of H_2O_2 to $H_2O + \frac{1}{2}O_2$ and the starch indicator oscillates between deep blue and colourless as the iodine concentration pulsates.

Even more intriguing is the Belousov–Zhabotinskii class of oscillating reactions some of which can continue for hours. Such a reaction was first observed in 1959 by B. P. Belousov who noticed that, in stirred sulfuric acid solutions containing initially $KBrO_3$, cerium(IV) sulfate and malonic acid, $CH_2(CO_2H)_2$, the concentrations of Br^- and Ce^{4+} underwent repeated oscillations of major proportions (e.g. tenfold changes on a time-scale which was constant but which could be varied from a few seconds to a few minutes depending on concentrations and temperature). These observations were extended by A. M Zhabotinskii in 1964 to the bromate oxidation of several other organic substrates containing a reactive methylene group catalysed either by Ce^{IV}/Ce^{III} or Mn^{III}/Mn^{II}. Not surprisingly these reactions have attracted considerable attention, but detailed studies of their mechanisms are beyond the scope of this chapter.[81-83]

The various reactions of bromates and iodates are summarized in the following schemes on p. 1014.[79]

The oxidation of halates to perhalates is considered further in the next section.

Perhalic acid and perhalates

Because of their differing structures, chemical reactions, and applications, perchloric acid and the perchlorates are best considered separately from the various periodic acids and their salts; the curious history of perbromates also argues for their individual treatment.

Perchloric acid and perchlorates[84-87]

The most stable compounds of chlorine are those in which the element is in either its lowest oxidation state $(-I)$ or its highest (VII): accordingly perchlorates are the most stable oxo-compounds of chlorine (see oxidation-state diagram, (p. 1002) and most are extremely stable both as solids and as solutions at room temperature. When heated they tend to decompose by loss of O_2 (e.g. $KClO_4$ above 400°). Aqueous solutions of perchloric acid and perchlorates are not notable oxidizing agents at room temperature but when heated they become vigorous, even violent, oxidants. Considerable CAUTION should therefore be exercised when handling these materials, and it is crucial to avoid the presence of readily oxidizable organic (or inorganic) matter since this can initiate reactions of explosive intensity.

On an industrial scale, perchlorates are now invariably produced by the electrolytic oxidation of $NaClO_3$ (see Panel). Alternative routes have historical importance but are now only rarely used, even for small-scale laboratory syntheses. For example, perchlorates were first discovered in 1816–18 by F. von Stadion who treated fused $KClO_3$ with H_2SO_4: ClO_2 was evolved and $KClO_4$ crystallized. Likewise, von Stadion isolated perchloric acid by distilling mixtures of $KClO_3/H_2SO_4$. As previously noted (p. 1002) chlorates can be thermally disproportionated to give perchlorates, and they can also be chemically oxidized to perchlorates by ozone, peroxodisulfates, or PbO_2.

Perchloric acid is best made by treating anhydrous $NaClO_4$ or $Ba(ClO_4)_2$ with

[81] C. VIDAL, J. C. ROUX, and A. ROSSI, Quantitative measurements of intermediate species in sustained Belousov–Zhabotinskii oscillations, *J. Am. Chem. Soc.* **102**, 1241–5 (1980), and references therein.
[82] R. J. FIELD, E. KÖRÖS, and R. M. NOYES, Oscillations in chemical systems. II. Thorough analysis of temporal oscillations in the bromate–cerium–malonic acid system, *J. Am. Chem. Soc.* **94**, 8649–64 (1972).
[83] D. O. COOKE, Homogeneous oscillating reactions, *Educ. in Chem.* **14**, 53–56 (1977); see also *J. Chem. Educ.* **49**, 302–7, 308–11, 312–14 (1972); **50**, 357–8, 496 (1973).
[84] Ref. 17, pp. 1435–60, Perhalic acids and perhalates.
[85] F. SOLYMOSI, *Structure and Stability of Salts of Halogen Oxyacids in the Solid Phase*, Wiley, New York, 1978, 468 pp.
[86] A. A. SCHILT, *Perchloric Acid and Perchlorates*, Northern Illinois University Press, 1979, 189 pp.
[87] R. C. LEES, Perchloric acid and perchlorates, *Kirk–Othmer Encyclopedia of Chemical Technology*, 3rd edn., Vol. 5, pp. 646–67, Wiley, New York, 1979.

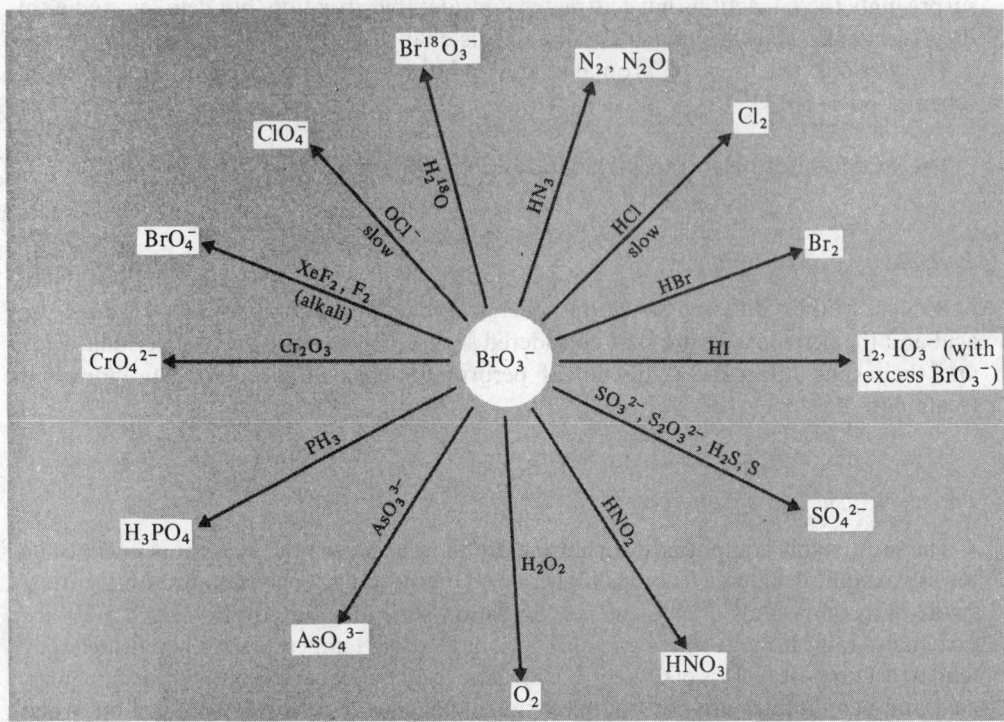

SCHEME   Some reactions of aqueous bromates.

Benzil may be fully oxidised ... (partially illegible) the reaction should
then be assessed when building the ... material, applying criteria laid out in prescribed
... multi-oxidisable organic (or inorganic) ... since they are rather ... (illegible)
application to industry.

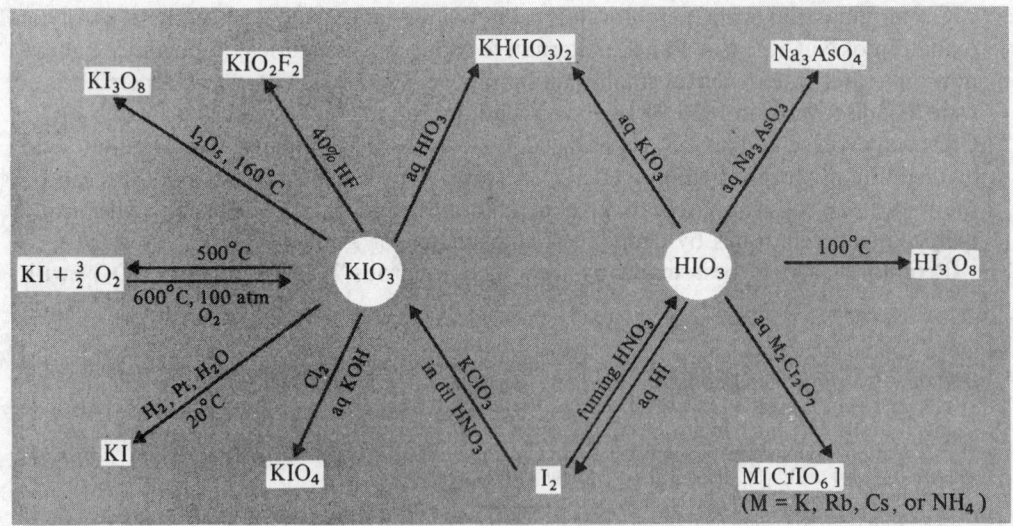

SCHEME   Some reactions of iodates.

Production and Uses of Perchlorates

$NaClO_4$ is made by the electrolytic oxidation of aqueous $NaClO_3$ using smooth Pt or PbO_2 anodes and a steel cathode which also acts as the container. All other perchlorates, including $HClO_4$, are made either directly or indirectly from this $NaClO_4$. In a typical cell $NaClO_3$ (600 g/l pH 6.5) is oxidized at 30–50°C with 90% current efficiency at 5000 A and 6.0 V with an anode current density of 3100 A m^{-2} and an electrode separation of ~ 5 mm. The process can be either batch or continuous and energy consumption is ~ 2.5 kWh/kg. A small concentration of $Na_2Cr_2O_7$ (1–5 g/l) is found to be extremely beneficial in inhibiting cathodic reduction of ClO_4^-.

World production of perchlorates was less than 1800 tonnes pa until 1940 when wartime missile and rocket requirements boosted this tenfold. World production capacity peaked at around 40 000 tpa in 1963 and is now about 30 000 tpa. More than half of this is converted to NH_4ClO_4 for use as a propellent:

$$NaClO_4 + NH_4Cl \longrightarrow NH_4ClO_4 + NaCl$$

A typical conversion plant can produce about 45 tonnes NH_4ClO_4 daily: its bulk price is $\sim$\$1.00 per kg. Ultrapure NH_4ClO_4 for physical measurements and research purposes can be made by direct neutralization of aqueous solutions of NH_3 and $HClO_4$. One of the main current uses of NH_4ClO_4 is in the Space Shuttle Programme: the two booster rockets use a solid propellent containing 70% by weight of NH_4ClO_4, this being the oxidizer for the "fuel" (powdered Al metal) which comprises most of the rest of the weight. Each shuttle launch requires almost 700 tonnes of NH_4ClO_4.

The annual consumption of 70% $HClO_4$ is about 450 tonnes mainly for making other perchlorates. Most of the $NaClO_4$ produced is used captively to make NH_4ClO_4 and $HClO_4$, but about 725 tpa is used for explosives, particularly in slurry blasting formulations: its bulk price is $\sim$\$0.41 per kg.

The two other perchlorates manufactured on a fairly large scale industrially are $Mg(ClO_4)_2$ and $KClO_4$. The former is used as the electrolyte in "dry cells" (batteries), whereas $KClO_4$ is a major constituent in pyrotechnic devices such as fireworks, flares, etc. Thus the white flash and thundering boom in fireworks displays are achieved by incorporating a compartment containing $KClO_4/S/Al$, whereas the flash powder commonly used in rock concerts and theatricals comprises $KClO_4/Mg$. Vivid blues, perhaps the most difficult pyrotechnic colour to achieve, are best obtained from the low temperature (< 1200°C) flame emission of CuCl in the 420–460 nm region: because of the instability of copper chlorate and perchlorate this colour is generated by ignition of a mixture containing 38% $KClO_4$, 29% NH_4ClO_4, and 14% $CuCO_3$ bound with red gum (14%) and dextrin (5%).

concentrated HCl, filtering off the precipitated chloride and concentrating the filtrate by distillation. The azeotrope (p. 953) boils at 203°C and contains 71.6% $HClO_4$ (i.e. $HClO_4 \cdot 2H_2O$). The anhydrous acid is obtained by low-pressure distillation of the azeotrope ($p < 1$ mmHg $= 0.13$ kPa) in an all-glass apparatus in the presence of fuming sulfuric acid. Commercially available perchloric acid is usually 60–62% (~ 3.5 H_2O) or 70–72% (~ 2 H_2O); more concentrated solutions are hygroscopic and are also unstable towards loss of Cl_2O_7 or violent decomposition by accidental impurities.

Pure $HClO_4$ is a colourless mobile, shock-sensitive, liquid: $d(25°)$ 1.761 g cm^{-3}. At least 6 hydrates are known (Table 17.24). The structure of $HClO_4$, as determined by electron diffraction in the gas phase, is as shown in Fig. 17.23. This molecular structure persists in the liquid phase, with some H bonding. The (low) electrical conductivity and other physical properties of anhydrous $HClO_4$ have been interpreted on the basis of slight dissociation according to the overall equilibrium:

$$3HClO_4 \rightleftharpoons Cl_2O_7 + H_3O^+ + ClO_4^-; \quad K(25°)\ 0.68 \times 10^{-6}$$

TABLE 17.24   *Perchloric acid and its hydrates*

n in $HClO_4 \cdot nH_2O$	Structure	MP/°C	BP/°C	$\Delta H_f^\circ / kJ\ mol^{-1}$
0	$HOClO_3$	−112	110 (expl)	−40.6 (liq)
0.25	$(HClO_4)_4 \cdot H_2O$	d −73.1	—	—
1	$[H_3O]^+[ClO_4]^-$	49.9	decomp	−382.2 (cryst)
2	$[H_5O_2]^+[ClO_4]^-$	−20.7	203	−688 (liq)
2.5	—	−33.1	—	—
3	$[H_7O_3]^+[ClO_4]^-$	−40.2	—	—
3.5	—	−45.9	—	—

(cf. H_2SO_4, p. 843; H_3PO_4, p. 597, etc.). The monohydrate forms an H-bonded crystalline lattice $[H_3O]^+[ClO_4]^-$ that undergoes a phase transition with rotational disorder above −30°; it melts to a viscous, highly ionized liquid at 49.9°. The other hydrates also feature hydroxonium ions $[(H_2O)_nH]^+$ as described more fully on p. 739. It is particularly notable that hydration does not increase the coordination number of Cl and in this perchloric acid differs markedly from periodic acid (p. 1022). This parallels the difference between sulfuric and telluric acids in the preceding group (p. 915).

Anhydrous $HClO_4$ is an extremely powerful oxidizing agent. It reacts explosively with most organic materials, ignites HI and $SOCl_2$, and rapidly oxidizes Ag and Au. Thermal decomposition in the gas phase yields a mixture of HCl, Cl_2, Cl_2O, ClO_2, and O_2 depending on the conditions. Above 310° the decomposition is first order and homogeneous, the rate-determining step being homolytic fission of the Cl–OH bond:

$$HOClO_3 \longrightarrow HO^\bullet + ClO_3^\bullet$$

The hydroxyl radical rapidly abstracts an H atom from a second molecule of $HClO_4$ to

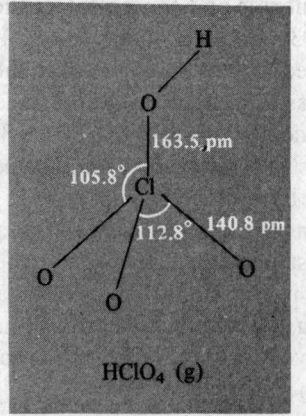
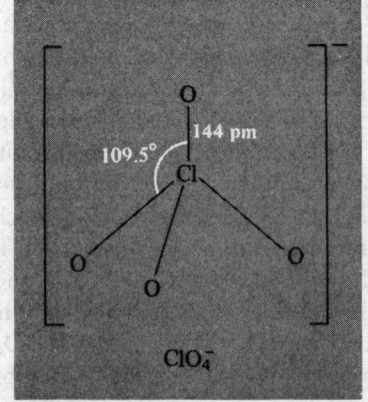

FIG. 17.23   Structure of $HClO_4$ and ClO_4^-.

give H_2O plus $ClO_4^{\bullet}$ and the 2 radicals $ClO_3^{\bullet}$ and $ClO_4^{\bullet}$ then decompose to the elements via the intermediate oxides. Above 450° the Cl_2 produced reacts with H_2O to give $2HCl$ plus $\frac{1}{2}O_2$ whilst in the low-temperature range (150–310°) the decomposition is heterogeneous and second order in $HClO_4$.

Aqueous perchloric acid solutions exhibit very little oxidizing power at room temperature, presumably because of kinetic activation barriers, though some strongly reducing species slowly react, e.g. Sn^{II}, Ti^{III}, V^{II} and V^{III}, and dithionite. Others do not, e.g. H_2S, SO_2, HNO_2, HI, and, surprisingly, Cr^{II} and Eu^{II}. Electropositive metals dissolve with liberation of H_2 and oxides of less basic metals also yield perchlorates, e.g. with 72% acid:

$$Mg + 2HClO_4 \xrightarrow{20°} [Mg(H_2O)_6](ClO_4)_2 + H_2$$

$$Ag_2O + HClO_4 \longrightarrow AgClO_4 + H_2O$$

NO and NO_2 react to give $NO^+ClO_4^-$ and F_2 yields $FOClO_3$ (p. 750). P_2O_5 dehydrates the acid to Cl_2O_7 (p. 995).

Perchlorates are known for most metals in the periodic table.[86, 87] The alkali-metal perchlorates are thermally stable to several hundred degrees above room temperature but NH_4ClO_4 deflagrates with a yellow flame when heated to 200°:

$$2NH_4ClO_4 \longrightarrow N_2 + Cl_2 + 2O_2 + 4H_2O$$

NH_4ClO_4 has a solubility in water of 20.2 g per 100 g solution at 25° and 135 g per 100 g liquid NH_3 at the same temperature. Aqueous solubilities decrease in the sequence $Na > Li > NH_4 > K > Rb > Cs$; indeed, the low solubility of the last 3 perchlorates in this series has been used for separatory purposes and even for gravimetric analysis (e.g. $KClO_4$ 1.99 g per 100 g H_2O at 20°). Many of these perchlorates and those of M^{II} can also be obtained as hydrates. $AgClO_4$ has the astonishing solubility of 557 g per 100 g H_2O at 25° and even in toluene its solubility is 101 g per 100 g PhMe at 25°. This has great advantages in the metathetic preparation of other perchlorates, particularly organic perchlorates, e.g. RI yields $ROClO_3$, Ph_3CCl yields $Ph_3C^+ClO_4^-$, and CCl_4 affords CCl_3OClO_3.

Oxidation-reduction reactions involving perchlorates have been mentioned in several of the preceding sections and the reactivity of aqueous solutions is similar to that of aqueous solutions of perchloric acid.

The perchlorate ion has for long been considered to be a non-coordinating ligand and has frequently been used to prepare "inert" ionic solutions of constant ionic strength for physicochemical measurements. Though it is true that ClO_4^- is a weaker ligand than H_2O it is not entirely toothless and, as shown schematically in Fig. 17.24, examples are known in which the perchlorate acts as a monodentate (η^1), bidentate chelating (η^2), and bidentate bridging (μ,η^2) ligand. The first unambiguous structural evidence for coordinated ClO_4^- was obtained in 1965 for the 5 coordinate cobalt(II) complex $[Co(OAsMePh_2)_4(\eta^1\text{-}OClO_3)_2]^{[88]}$ and this was quickly followed by a second example, the red 6-coordinate *trans* complex $[Co(\eta^2\text{-}MeSCH_2CH_2SMe)_2(\eta^1\text{-}OClO_3)_2]$.[89] The two structures are shown in Fig. 17.25. The perchlorate ion has now been established as a

[88] P. PAULING, G. B. ROBERTSON, and G. A. RODLEY, Coordination of transition metals by the perchlorate ion: the crystal structure of $[(Co(Ph_2MeAsO)_4(ClO_4)_2]$, *Nature* **207**, 73–74 (1965).

[89] F. A. COTTON and D. L. WEAVER, An authenticated perchlorate complex, *J. Am. Chem. Soc.* **87**, 4189–90 (1965).

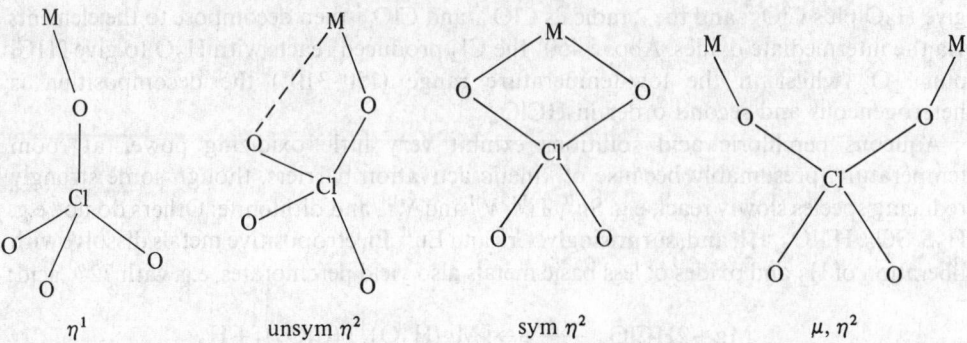

FIG. 17.24   Coordination modes of ClO_4^- as determined by X-ray crystallography.

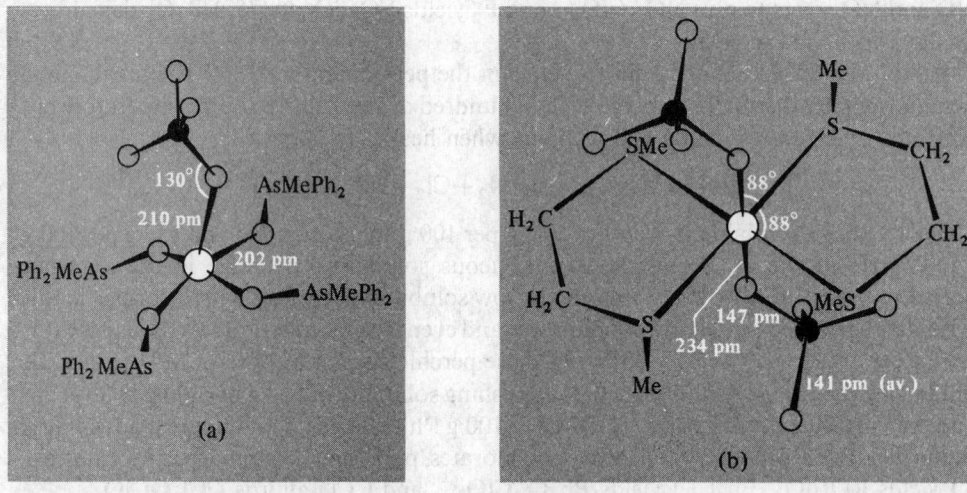

FIG. 17.25 The structures of monodentate perchlorate complexes: (a) the cation in $[Co(OAsMePh_2)_4(\eta^1\text{-}OClO_3)]ClO_4$, and (b) *trans*-bis(perchlorato)bis(2,5-dithiahexane)cobalt(II).

monodentate ligand towards an s-block element (Ba),[90] a p-block element (Sn^{II} and Sn^{IV}),[91] and an f-block element (Sm^{III})[92] as well as to the d-block elements Co^{II}, Ni^{II}, Cu^{II}, and Ag^I.[89, 93] It is also known to function as a bidentate ligand towards Na,[94] Sn^{IV},[91] Sm^{III},[92] and perhaps Ba,[90] sometimes both η^1 and η^2 modes occurring in the

[90] D. L. HUGHES, C. L. MORTIMER, and M. R. TRUTER, *Acta Cryst.* **B34,** 800–7 (1978) [Ba(dibenzo-24-crown-8) $(\eta^1\text{-}OClO_3)$ $(\eta^{1+1}\text{-}O_2ClO_2)$]; *Inorg. Chim. Acta* **29,** 43–55 (1978) [Ba(benzo-18-crown-6) $(\eta^1\text{-}OClO_3)_2(H_2O)_2$].

[91] R. C. ELDER, M. J. HEEG, and E. DEUTSCH, *Inorg. Chem.* **17,** 427–31 (1978) [$Cl_3Sn^{II}(\eta^1\text{-}OClO_3)]^{2-}$. C. BELIN, M. CHAABOUNI, J.-L. PASCAL, J. POTIER, and J. ROZIERE, *JCS Chem. Comm.* 1980. 105–6 [$Sn_3^{IV}O_2Cl_4(\eta^1\text{-}OClO_3)$ $(\eta^2\text{-}OClO_3)_3]_2$.

[92] M. CIAMPOLINI, N. NARDI, R. CINI, S. MANGANI, and P. ORIOLI, *JCS Dalton* 1979 1983–6 [Sm(dibenzo-18-crown-6) $(\eta^1\text{-}OClO_3)_2(\eta^2\text{-}O_2ClO_2)$].

[93] F. MADAULE-AUBRY and G. M. BROWN, *Acta Cryst.* **B24,** 745–53 (1968) [Ni(3,5-dimethylpyridine)$_4(\eta^1\text{-}OClO_3)_2$]. F. BIGOLI, M. A. PELLINGHELLI, and A. TIRIPICCHIO, *Cryst. Struct. Comm.* **4,** 123–6 (1976) [$Cu\{S\!=\!C(NHNH_2)_2\}(\eta^1\text{-}OClO_3)_2$]. E. A. HALL GRIFFITH and E. L. AMMA, *J. Am. Chem. Soc.* **93,** 3167–72 (1971) [Ag(cyclohexylbenzene)$_2(\eta^1\text{-}OClO_3)$].

[94] H. MILBURN, M. R. TRUTER, and B. L. VICKERY, *JCS Dalton* 1974, 841–6 [Na($\eta^2\text{-}O_2ClO_2$) $\{Cu^{II}(salen)\}_2$].

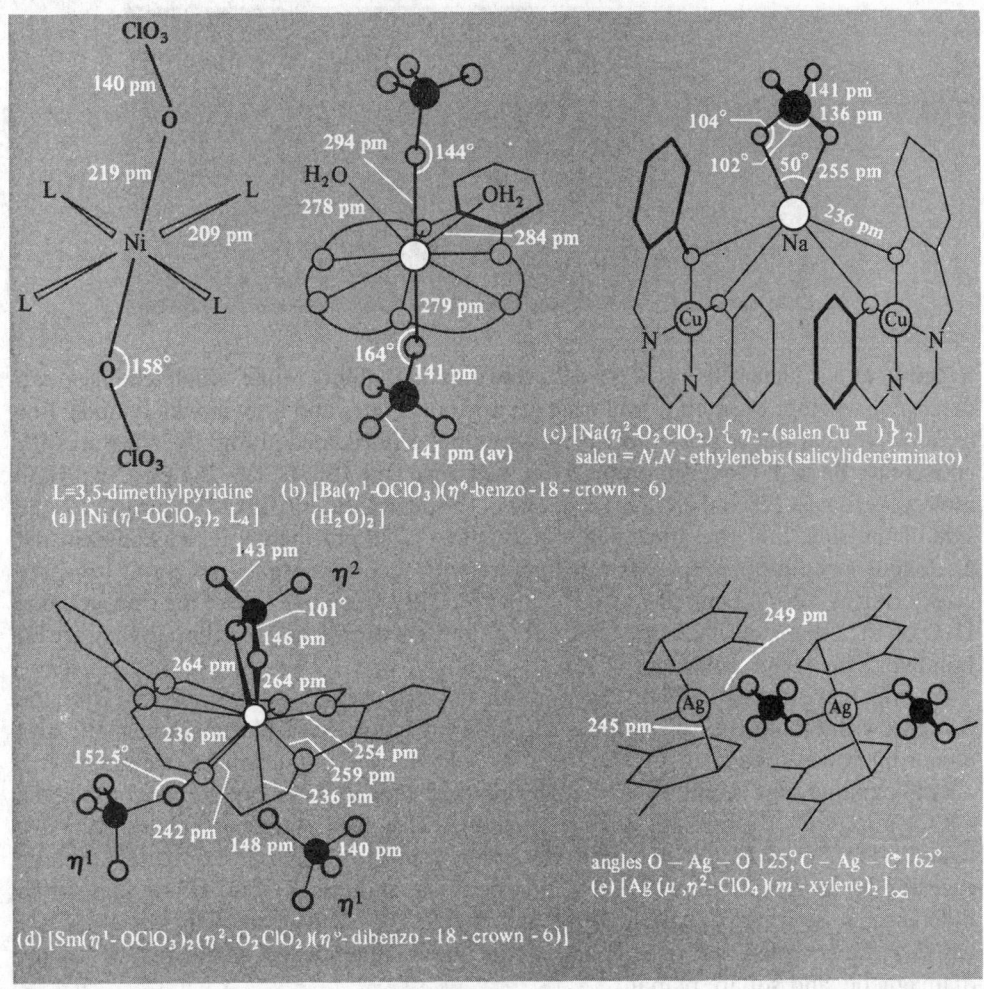

FIG. 17.26   Structures of compounds having monodentate, chelating, or bridging perchlorate
ligands.

same compound. The bidentate bridging mode occurs in the silver complex $[Ag\{\mu,\eta^2\text{-}OCl(O)_2O\text{-}\}(m\text{-}xylene)_2].^{(95)}$ The structure of appropriate segments of some of these compounds are in Fig. 17.26. The distinction between coordinated and non-coordinated ("ionic") perchlorate is sometimes hard to make and there is an almost continuous gradation between the two extremes. Similarly it is sometimes difficult to distinguish unambiguously between η^1 and unsymmetrical η^2 and, in the colourless complex $[Ag(cyclohexylbenzene)_2(ClO_4)]$, the η^1 bonding between Ag and $OClO_3$ (Ag–O 266 pm) is accompanied by a further weak symmetrical η^2 bonding from each ClO_4 to the neighbouring Ag (2Ag–O 284 pm) thereby generating a weakly-bridged chain-like structure involving pseudo-η^3 coordination of the perchlorate group.$^{(93)}$

^{95} I. F. TAYLOR, E. A HALL, and E. L. AMMA, *J. Am. Chem. Soc.* **91**, 5745–9 (1969) [Ag(m-xylene)$_2$ {μ,η^2-OCl(O)$_2$O-}].

Because of its generally rather weak coordinating ability quite small changes can determine whether or not a perchlorate group coordinates and if so, in which mode. For example, the barium crown-ether dihydrate complex illustrated in Fig. 17.26 features 10-coordinate Ba with 6 oxygen atoms from the crown ring (Ba–O 280–285 pm), two H_2O molecules (Ba–O 278 and 284 pm), and one of the perchlorates (Ba–O 294 pm) all on one side of the ring, and the other perchlorate (Ba–O 279 pm) below it. By contrast, the analogous strontium complex is a trihydrate with 9-coordinate Sr (six Sr–O from the crown ring at 266–272 pm, plus two H_2O at 257, 259 pm, on one side of the ring, and one H_2O on the other side at 255 pm); the ClO_4^- ions are uncoordinated though they are H-bonded to the water molecules.

An even more dramatic change occurs with nickel(II) perchlorate complexes. Thus, the complex with 4 molecules of 3,5-dimethylpyridine (Fig. 17.26a) is blue, paramagnetic, and 6-coordinate with $trans$-$(\eta^1$-$OClO_3)$ ligands, whereas the corresponding complex with 3,4-dimethylpyridine is yellow and diamagnetic with square-planar Ni^{II} and uncoordinated ClO_4^- ions.[93, 96] There is no steric feature of the structure which prevents the four 3,4-ligands from adopting the propeller-like configuration of the four 3,5-ligands thereby enabling Ni to accept two η^1-$OClO_3$, or vice versa, and one must conclude that subtle differences in secondary valency forces and energies of packing are sufficient to dictate whether the complex that crystallizes is blue, paramagnetic, and octahedral or yellow, diamagnetic, and square planar.

Perbromic acid and perbromates

The quest for perbromic acid and perbromates and the various reasons adduced for their apparent non-existence make fascinating and salutory reading.[97] The esoteric radiochemical synthesis of BrO_4^- in 1968 using the β-decay of radioactive ^{83}Se, whilst not providing a viable route to macroscopic quantities of perbromate, proved that this previously elusive species could exist, and this stimulated the search for chemical syntheses:

$$^{83}SeO_4^{2-} \xrightarrow[t_{\frac{1}{2}}\ 22.5\ min]{-\beta^-} {}^{83}BrO_4^- \xrightarrow[t_{\frac{1}{2}}\ 2.39\ h]{-\beta^-} \{^{83}Kr + 2O_2\}$$

---

[96] F. MADAULE-AUBRY, W. R. BUSING, and G. M. BROWN, Crystal structure of tetrakis(3,4-dimethyl-pyridine)nickel(II) perchlorate, *Acta Cryst.* **B24**, 754–60 (1968).

[97] E. H. APPELMAN, Nonexistent compounds: two case histories, *Acc. Chem. Res.* **6**, 113–17 (1973).

Electrolytic oxidation of aqueous $LiBrO_3$ produced a 1% yield of perbromate, but the first isolation of a solid perbromate salt ($RbBrO_4$) was achieved by oxidation of BrO_3^- with aqueous XeF_2:[98]

$$BrO_3^- + XeF_2 + H_2O \xrightarrow{10\% \text{ yield}} BrO_4^- + Xe + 2HF$$

The best synthesis is now by oxidation of alkaline solutions of BrO_3^- using F_2 gas under rather specific conditions:[99]

$$BrO_3^- + F_2 + 2OH^- \xrightarrow{20\% \text{ yield}} BrO_4^- + 2F^- + H_2O$$

In practice, F_2 is bubbled in until the solution is neutral, at which point excess bromate and fluoride are precipitated as $AgBrO_3$ and CaF_2; the solution is then passed through a cation exchange column to yield a dilute solution of $HBrO_4$. Several hundred grams at a time can be made by this route. The acid can be concentrated up to 6 M (55%) without decomposition and such solutions are stable for prolonged periods even at 100°. More concentrated solutions of $HBrO_4$ can be obtained but they are unstable; a white solid, possibly $HBrO_4 . 2H_2O$, can be crystallized.

Pure $KBrO_4$ is isomorphous with $KClO_4$ and contains tetrahedral BrO_4^- anions (Br–O 161 pm, cf. Cl–O 145 pm in ClO_4^- and I–O 179 pm in IO_4^-). Oxygen-18 exchange between 0.14 M $KBrO_4$ and H_2O proceeds to less than 7% completion during 19 days at 94° in either acid or basic solutions and there is no sign of any increase in coordination number of Br; in this BrO_4^- resembles ClO_4^- rather than IO_4^-. $KBrO_4$ is stable to 275–280° at which temperature it begins to dissociate into $KBrO_3$ and O_2. Even NH_4BrO_4 is stable to 170°. Dilute solutions of BrO_4^- show little oxidizing power at 25°; they slowly oxidize I^- and Br^- but not Cl^-. More concentrated $HBrO_4$ (3 M) readily oxidizes stainless steel and 12 M acid rapidly oxidizes Cl^-. The general inertness of BrO_4^- at room temperature stands in sharp contrast to its high thermodynamic oxidizing power, which is greater than that of any other oxohalogen ion that persists in aqueous solution. The oxidation potential calculated from thermochemical data is

$$BrO_4^- + 2H^+ + 2e^- \longrightarrow BrO_3^- + H_2O; \quad E° + 1.74 \text{ V}$$

(cf. 1.23 V and 1.64 V for ClO_4^- and IO_4^- respectively). Accordingly, only the strongest oxidants would be expected to convert bromates to perbromates. As seen above, F_2/H_2O ($E° \sim 2.87$ V) and XeF_2/H_2O ($E° \sim 2.64$ V) are effective, but ozone ($E°$ 2.07 V) and $S_2O_8^{2-}$ ($E°$ 2.01 V) are not, presumably for kinetic reasons. Thermochemical measurements[100] further show that $KBrO_4$ is thermodynamically stable with respect to its elements, but less so than the corresponding $KClO_4$ and KIO_4: this is not due to any significant difference in entropy effects or lattice energies and implies that the Br–O bond in BrO_4^- is substantially weaker than the X–O bond in the other perhalates. Some

[98] E. H. APPELMAN, The synthesis of perbromates, *J. Am. Chem. Soc.* **90**, 1900–1 (1968); *Inorg. Chem.* **8**, 223–7 (1969).

[99] E. H. APPELMAN, Perbromic acid and potassium perbromate, *Inorg. Synth.* **13**, 1–9 (1972).

[100] F. SCHREINER, D. W. OSBORNE, A. V. POCIUS, and E. H. APPELMAN, The heat capacity of potassium perbromate, $KBrO_4$, between 5 and 350 K, *Inorg. Chem.* **9**, 2320–4 (1970).

comparative data (298.15 K) are:

	$KClO_4$	$KBrO_4$	KIO_4
$\Delta H_f^\circ/kJ\ mol^{-1}$	-431.9	-287.6	-460.6
$\Delta G_f^\circ/kJ\ mol^{-1}$	-302.1	-174.1	-349.3

No entirely satisfactory explanation of these observations has been devised, though they are paralleled by the similar reluctance of other elements following the completion of the 3d subshell to achieve their highest oxidation states—see particularly Se (p. 891) and As (p. 645) immediately preceding Br in the periodic table. The detailed kinetics of several oxidation reactions involving aqueous solutions of BrO_4^- have been studied.[101] In general, the reactivity of perbromates lies between that of the chlorates and perchlorates which means that, after the perchlorates, perbromates are the least reactive of the known oxohalogen compounds. It has even been suggested[97] that earlier investigators may actually have made perbromates, but not realized this because they were expecting a highly reactive product rather than an inert one.

Periodic acids and periodates[84]

At least four series of periodates are known, interconnected in aqueous solutions by a complex series of equilibria involving deprotonation, dehydration, and aggregation of the parent acid H_5IO_6—cf. telluric acids (p. 915) and antimonic acids (p. 674) in the immediately preceding groups. Nomenclature is summarized in Table 17.25, though not all of the fully protonated acids have been isolated in the free state. The structural relationship between these acids, obtained mainly from X-ray studies on their salts, are shown in Fig. 17.27. H_5IO_6 itself (mp 128.5° decomp) consists of molecules of $(HO)_5IO$ linked into a three-dimensional array by $O-H\cdots O$ bonds (10 for each molecule, 260–278 pm).

Periodates can be made by oxidation of I^-, I_2, or IO_3^- in aqueous solution. Industrial processes involve oxidation of alkaline $NaIO_3$ either electrochemically (using a PbO_2 anode) or with Cl_2:

$$IO_3^- + 6OH^- - 2e^- \longrightarrow IO_6^{5-} + 3H_2O$$
$$IO_3^- + 6OH^- + Cl_2 \longrightarrow IO_6^{5-} + 2Cl^- + 3H_2O$$

TABLE 17.25   Nomenclature of periodic acids

Formula	Name	Alternative	Formal relation to H_5IO_6
H_5IO_6	Orthoperiodic	Paraperiodic	Parent
HIO_4	Periodic	Metaperiodic	$H_5IO_6-2H_2O$
"H_3IO_5"	Mesoperiodic	Diperiodic	$\begin{cases}2H_5IO_6-2H_2O\\2H_5IO_6-3H_2O\end{cases}$
$H_7I_3O_{14}$	Triperiodic		$3H_5IO_6-4H_2O$

[101] E. H. APPELMAN, U. K. KLÄNING, and R. C. THOMPSON, Some reactions of the perbromate ion in aqueous solution, *J. Am. Chem. Soc.* **101**, 929–34 (1979).

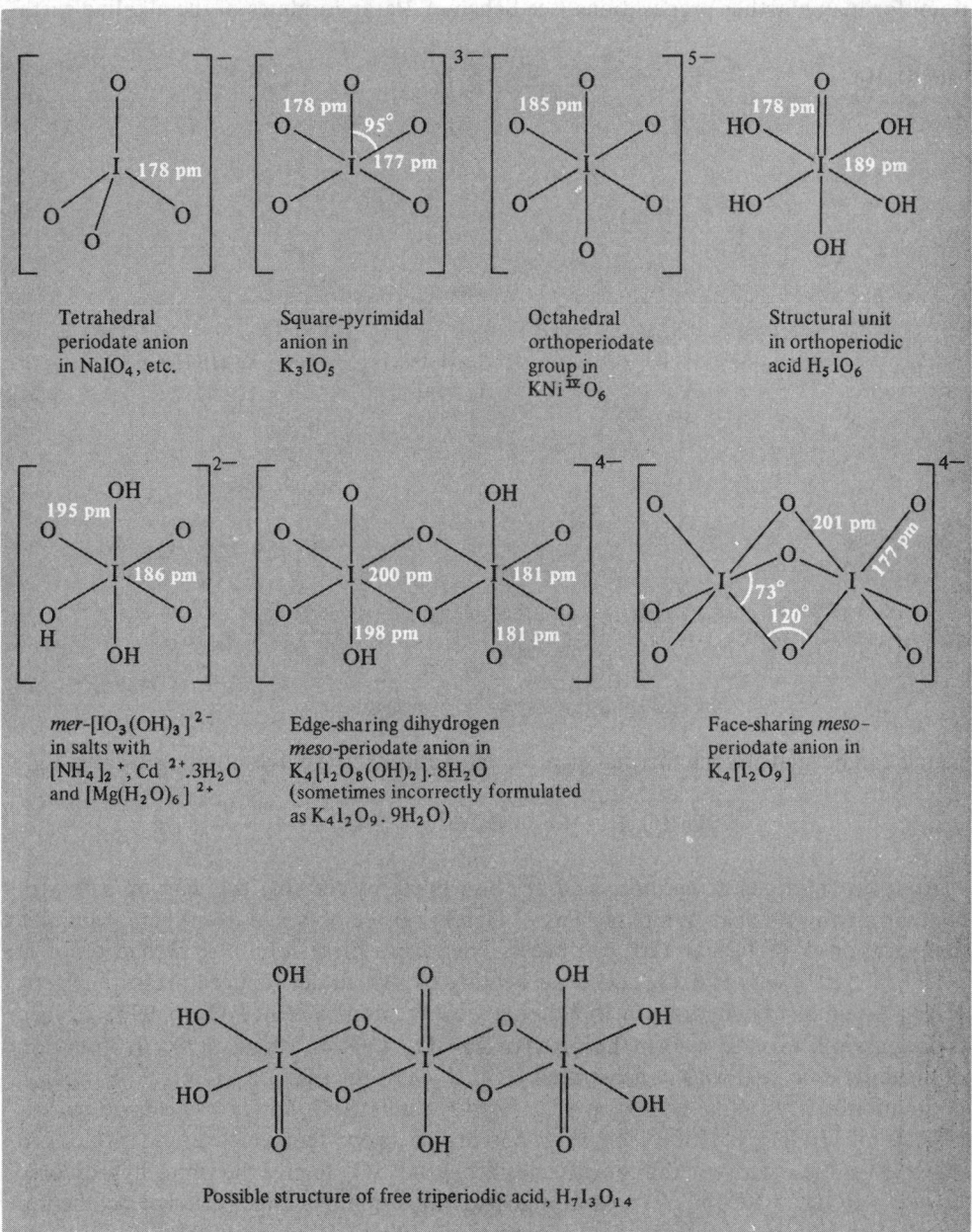

Fig. 17.27   Structures of periodic acids and periodate anions.

The product is the dihydrogen orthoperiodate $Na_3H_2IO_6$, which is a convenient starting point for many further preparations (see Scheme). Paraperiodates of the alkaline earth

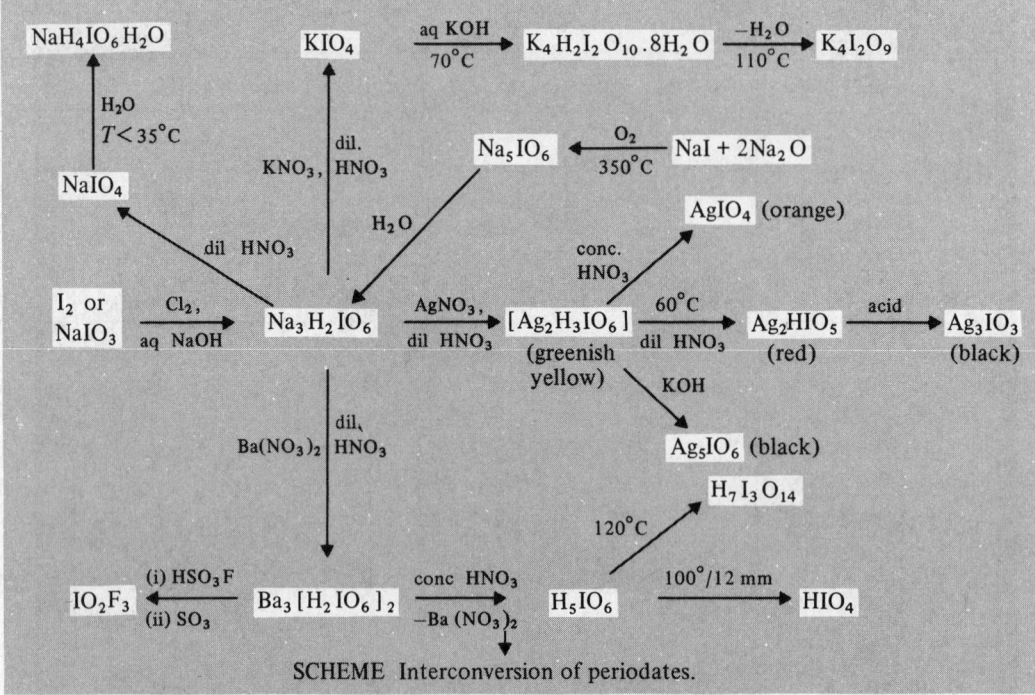

SCHEME Interconversion of periodates.

metals can be made by the thermal disproportionation of the corresponding iodates, e.g.:

$$5Ba(IO_3)_2 \xrightarrow{\Delta} Ba_5(IO_6)_2 + 4I_2 + 9O_2$$

Aqueous solutions of periodic acid are best made by treating this barium salt with concentrated nitric acid. White crystals of H_5IO_6 can be obtained from these solutions. Dehydration of H_5IO_6 at 120° can be obtained from these solutions. Dehydration of H_5IO_6 at 120° yields $H_7I_3O_{14}$, whereas heating to 100° under reduced pressure affords HIO_4. Attempts to dehydrate further do not yield the non-existent I_2O_7 (p. 997); oxygen is progressively evolved to form the mixed oxide $I_2O_5 . I_2O_7$ and finally I_2O_5. Protonation of orthoperiodic acid with concentrated $HClO_4$ yields the cation $[I(OH)_6]^+$. Similarly, dissolution of crystalline H_5IO_6 in 95% H_2SO_4 (or H_2SeO_4) at 120° yields colourless crystals of $[I(OH)_6][HSO_4]$ on slow cooling to room temperature and prolonged digestion of these with trichloroacetic acid extracts H_2SO_4 to give the white, hygroscopic powder $[I(OH)_6]_2SO_4$.[102] These compounds thus complete the series of octahedral hexahydroxo species $[Sn(OH)_6]^{2-}$, $[Sb(OH)_6]^-$, $[Te(OH)_6]$, and $[I(OH)_6]^+$.

In aqueous solution increase in pH results in progressive deprotonation, dehydration, and dimerization, the principal species being $[(HO)_4IO_2]^-$, $[(HO)_3IO_3]^{2-}$, $[(HO)_2IO_4]^{3-}$, $[IO_4]^-$, and $[(HO)_2I_2O_8]^{4-}$. The various equilibrium constants are:

[102] H. SIEBERT and U. WOERNER, Hexahydroxoiodine(VII) salts, *Z. anorg. allgem. Chem.* **398**, 193–7 (1973).

	$K(25°C)$	pK
$H_6IO_6^+ \rightleftharpoons H_5IO_6 + H^+$	6.3	-0.80
$H_5IO_6 \rightleftharpoons H_4IO_6^- + H^+$	5.1×10^{-4}	3.29
$H_4IO_6^- \rightleftharpoons H_3IO_6^{2-} + H^+$	4.9×10^{-9}	8.31
$H_3IO_6^{2-} \rightleftharpoons H_2IO_6^{3-} + H^+$	2.5×10^{-12}	11.60
$H_4IO_6^- \rightleftharpoons IO_4^- + 2H_2O$	29	-1.46
$2H_4IO_6^{2-} \rightleftharpoons H_2I_2O_{10}^{4-} + 2H_2O$	~ 820	-2.91

Periodates are both thermodynamically potent and kinetically facile oxidants. The oxidation potential is greatest in acid solution (p. 1000) and can be progressively diminished by increasing the pH of the solution. In acid solution it is one of the few reagents that can rapidly and quantitatively convert Mn^{II} to $Mn^{VII}O_4^-$. In organic chemistry it specifically cleaves 1,2-diols (glycols) and related compounds such as α-diketones, α-ketols, α-aminoalcohols, and α-diamines, e.g.:

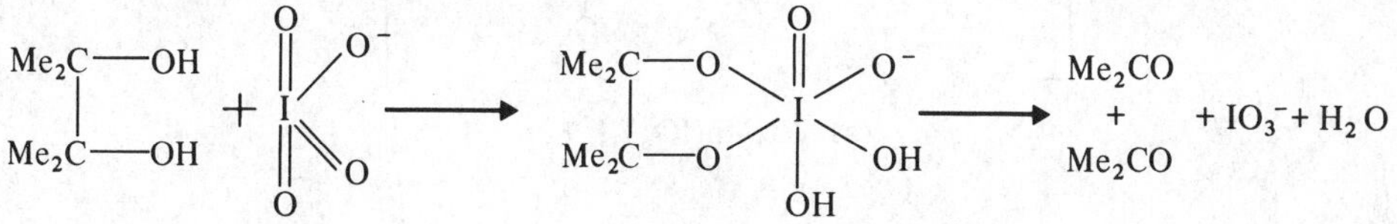

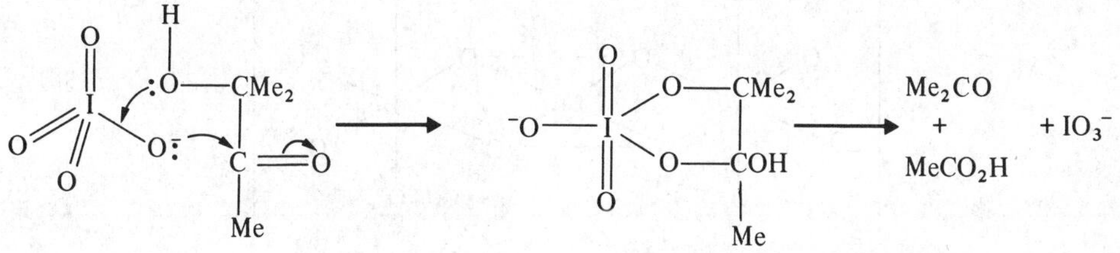

In rigid systems only *cis*-difunctional groups are oxidized, the specificity arising from the formation of the cyclic intermediate. Such reactions have been widely used in carbohydrate and nucleic acid chemistry.

Periodates form numerous complexes with transition metals in which the octahedral IO_6^{5-} unit acts as a bidentate chelate. Examples are:

$$[Mn^{IV}(IO_6)]^-, \quad [Ni^{IV}(IO_6)]^-, \quad [Fe^{III}(IO_6)]^{2-}, \quad [Co^{III}(IO_6)]^{2-}$$
$$[M^{IV}(IO_6)_2]^{6-} \ (M^{IV} = Pd, Pd, Ce); \quad [M^{III}(IO_6)_2]^{7-} \ (M^{III} = Fe, Co, Cu, Ag, Au)$$
$$[Mn^{IV}(IO_6)_3]^{11-}; \quad [Fe_4^{III}(IO_6)_3]^{3-}, \quad [Co_4^{III}(IO_6)_3]^{3-}$$

The stabilization of Ni^{IV}, Cu^{III}, and Ag^{III} is notable and many of the complexes have very high formation constants, e.g. $[Cu(IO_6)_2]^{7-} \sim 10^{10}$, $[Co(IO_6)_2]^{7-} \sim 10^{18}$. The high formal charge on the anion is frequently reduced by protonation of the $\{I(\mu\text{-}O)_2O_4\}$ moiety, as in orthoperiodic acid itself. For example $H_{11}[Mn(IO_6)_3]$ is a heptabasic acid with pK_1 and $pK_2 < 0$, pK_3 2.75, pK_4 4.35, pK_5 5.45, pK_6 9.55, and pK_7 10.45. The crystal structure of $Na_7[H_4Mn(IO_6)_3] \cdot 17H_2O$ features a 6-coordinate paramagnetic Mn^{IV} anion (Fig. 17.28a) whereas the diamagnetic compound $Na_3K[H_3Cu(IO_6)_2] \cdot 14H_2O$ has square-planar Cu^{III} (Fig. 17.28b).

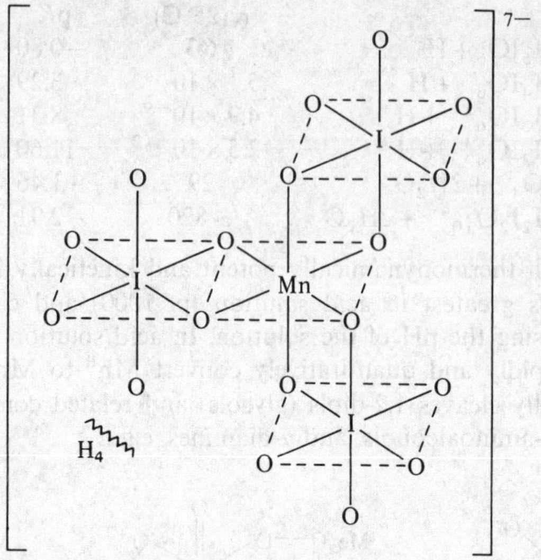

(a) $[H_4Mn(IO_6)_3]^{7-}$

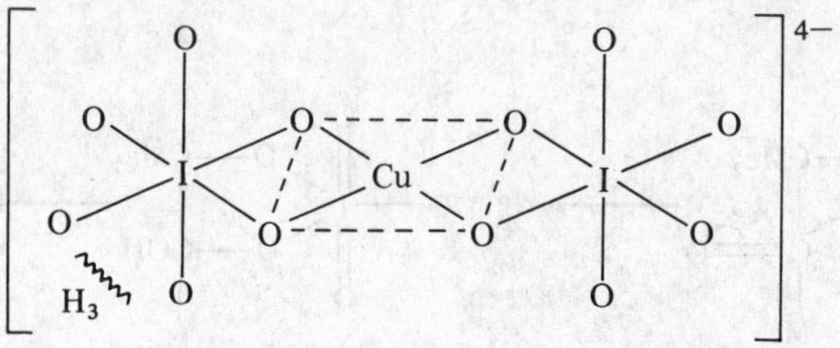

(b) $[H_3Cu(IO_6)_2]^{4-}$

Fig.   17.28 Structure   of   the   anions   in   (a) $Na_7[H_4Mn(IO_6)_3].17H_2O$,   and
(b) $Na_3K[H_3Cu(IO_6)_2].14H_2O$.

17.2.9 *Halogen oxide fluorides and related compounds*[103]

This section considers compounds in which X (Cl, Br, or I) is bonded to both O and F, i.e. F_nXO_m. Oxofluorides –OF and peroxofluorides –OOF have already been discussed (p. 748) and halogen derivatives of oxoacids, containing –OX bonds are treated in the following section (p. 1037).

Chlorine oxide fluorides[104]

Of the 6 possible oxide fluorides of Cl, 5 have been characterized: they range in stability from the thermally unstable $FCl^{III}O$ to the chemically rather inert perchloryl fluoride $FCl^{VII}O_3$. The others are FCl^VO_2, F_3Cl^VO, and $F_3Cl^{VII}O_2$. The remaining compound

103 Ref. 17, pp. 1386–96, The oxyfluorides of the halogens.
104 K. O. CHRISTE and C. J. SCHACK, Chlorine oxyfluorides, *Adv. Inorg. Chem. Radiochem.* **18**, 319–98 (1976).

$F_5Cl^{VII}O$ has been claimed but the report could not be confirmed. Fewer bromine oxide fluorides are known, only $FBrO_2$, F_3BrO, and possibly $FBrO_3$ being characterized. The compounds of iodine include the I^V derivatives FIO_2 and F_3IO, and the I^{VII} derivatives FIO_3, F_3IO_2, and F_5IO. All the halogen oxide fluorides resemble the halogen fluorides (p. 964), to which they are closely related both structurally and chemically. Thus they tend to be very reactive oxidizing and fluorinating agents and several can act as Lewis acids or bases (or both) by gain or loss of fluoride ions, respectively.

The structures of the chlorine oxide fluorides are summarized in Fig. 17.29, together with those of related cationic and anionic species formed from the neutral molecules by gain or loss or F^-. The first conclusive evidence for free FClO in the gas phase came in 1972 during a study of the hydrolysis of ClF_3 with substoichiometric amounts of H_2O in a flow reactor:

$$ClF_3 + H_2O \longrightarrow FClO + 2HF$$

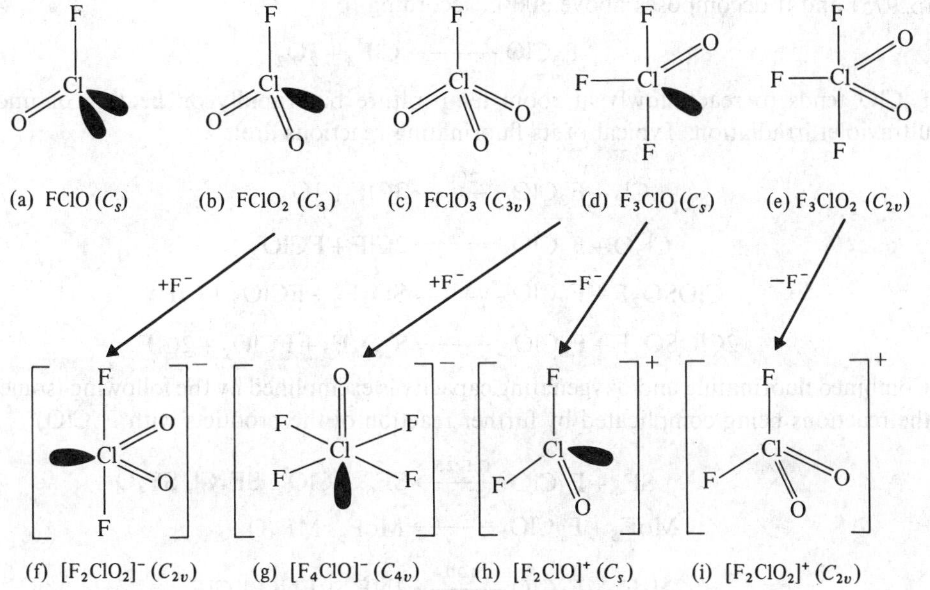

(a) FClO (C_s)   (b) $FClO_2$ (C_3)   (c) $FClO_3$ (C_{3v})   (d) F_3ClO (C_s)   (e) F_3ClO_2 (C_{2v})

(f) $[F_2ClO_2]^-$ (C_{2v})   (g) $[F_4ClO]^-$ (C_{4v})   (h) $[F_2ClO]^+$ (C_s)   (i) $[F_2ClO_2]^+$ (C_{2v})

FIG. 17.29   Structures of chlorine oxide fluorides and related cations and anions.

The compound is thermally unstable, and decomposes with a half-life of about 25 s at room temperature:

$$2FClO \xrightarrow{rt/t_{\frac{1}{2}} 25 s} FClO_2 + ClF$$

The compound can also be made by photolysis of a mixture of ClF and O_3 in Ar at 4–15 K; evidence for the expected nonlinear by structure comes from vibration spectroscopy (Fig. 17.29a).

F_3ClO was discovered in 1965 but not published until 1972 because of US security classification. It has low kinetic stability and is an extremely powerful fluorinating and oxidizing agent. It can be made in yields of up to 80% by fluorination of Cl_2O in the presence of metal fluorides, e.g. NaF:

$$Cl_2O + 2F_2 \xrightarrow[-78°]{MF} F_3ClO + ClF$$

However, the unpredictably explosive nature of Cl_2O in the liquid state renders this process somewhat hazardous and the best large-scale preparation is the low-temperature fluorination of $ClONO_2$ (p. 1038):

$$ClONO_2 + 2F_2 \xrightarrow[(>80\% \text{ yield})]{-35°} F_3ClO + FNO_2$$

F_3ClO is a colourless gas or liquid: mp $-43°$, bp $28°$ $d(l, 20°)$ 1.865 g cm^{-3}. The compound is stable at room temperature: $\Delta H_f°(g)$ -148 kJ mol^{-1}, $\Delta H_f°(l)$ -179 kJ mol^{-1}. It can be handled in well-passivated metal, Teflon, or Kel-F but reacts rapidly with glass or quartz. Its thermal stability is intermediate between those of ClF_3 and ClF_5 (p. 975) and it decomposes above 300°C according to

$$F_3ClO \longrightarrow ClF_3 + \tfrac{1}{2}O_2$$

F_3ClO tends to react slowly at room temperature but rapidly on heating or under ultraviolet irradiation. Typical of its fluorinating reactions are:

$$Cl_2 + F_3ClO \xrightarrow{200°} 3ClF + \tfrac{1}{2}O_2$$

$$Cl_2O + F_3ClO \xrightarrow{rt} 2ClF + FClO_2$$

$$ClOSO_2F + F_3ClO \longrightarrow SO_2F_2 + FClO_2 + ClF$$

$$2ClOSO_2F + F_3ClO \longrightarrow S_2O_5F_2 + FClO_2 + 2ClF$$

Combined fluorinating and oxygenating capacity is exemplified by the following (some of the reactions being complicated by further reaction of the products with F_3ClO):

$$SF_4 + F_3ClO \xrightarrow{CsF/25°} SF_6, FClO_2, SF_5Cl, SF_4O$$

$$MoF_5 + F_3ClO \longrightarrow MoF_6, MF_4O$$

$$2N_2F_4 + F_3ClO \xrightarrow{100°} 3NF_3 + FNO + ClF$$

$$HNF_2 + F_3ClO \xrightarrow{\text{low temp}} NF_3O, NF_2Cl, N_2F_4, FClO_2, HF$$

$$F_2NC(O)F + F_3ClO \longrightarrow NF_3O, ClNF_2, N_2F_4$$

It reacts as a reducing agent towards the extremely strong oxidant PtF_6:

$$F_3ClO + PtF_6 \longrightarrow [F_2ClO]^+[PtF_6]^- + \tfrac{1}{2}F_2$$

Hydrolysis with small amounts of water yields HF but this can react further by fluoride ion abstraction:

$$F_3ClO + H_2O \longrightarrow FClO_2 + 2HF$$
$$F_3ClO + HF \longrightarrow [F_2ClO]^+[HF_2]^-$$

This last reaction is typical of many in which F_3ClO can act as a Lewis base by fluoride ion donation to acceptors such as MF_5 (M = P, As, Sb, Bi, V, Nb, Ta, Pt, U), MoF_4O, SiF_4, BF_3, etc. These complexes are all white, stable, crystalline solids (except the canary yellow PtF_6^-) and contain the $[F_2ClO]^+$ cation (see Fig. 17.29h) which is isostructural with the isoelectronic F_2SO. Chlorine trifluoride oxide can also act as a Lewis acid (fluoride ion acceptor) and is therefore to be considered as amphoteric (p. 253). For example KF, RbF, and CsF yield $M^+[F_4ClO]^-$ as white solids whose stabilities increase with increasing size of M^+. Vibration spectroscopy establishes the C_{4v} structure of the anion (Fig. 17.29g).

The other Cl^V oxide fluoride $FClO_2$ (1942) can be made by the low-temperature fluorination of ClO_2 but is best prepared by the reaction:

$$6NaClO_3 + 4ClF_3 \xrightarrow[\text{(high yield)}]{\text{rt/1 day}} 6FClO_2 + 6NaF + 2Cl_2 + 3O_2$$

The C_s structure and dimensions (Fig. 17.29b) were established by microwave spectroscopy which also yielded a value for the molecular dipole moment μ 1.72 D. Other physical properties of this colourless gas are mp $-115°$ (or $-123°$), bp $\sim -6°$, ΔH_f° (g, 298 K) -34 ± 10 kJ mol^{-1} [or -273 kJ mol^{-1} when corrected for ΔH_f° (HF, g)!]. $FClO_2$ is thermally stable at room temperature in dry passivated metal containers and quartz. Thermal decomposition of the gas (first-order kinetics) only becomes measurable above 300° in quartz and above 200° in monel:

$$FClO_2(g) \xrightarrow{300°} ClF(g) + O_2(g)$$

It is far more chemically reactive than $FClO_3$ (p. 1031) despite the lower oxidation state of Cl. Hydrolysis is slow at room temperature and the corresponding reaction with anhydrous HNO_3 results in dehydration to the parent N_2O_5:

$$2FClO_2 + H_2O \longrightarrow 2HF + 2ClO_2 + \tfrac{1}{2}O_2$$
$$2FClO_2 + 2HONO_2 \longrightarrow 2HF + 2ClO_2 + \tfrac{1}{2}O_2 + N_2O_5$$

Other reactions with protonic reagents are:

$$2OH^-(aq) + FClO_2 \longrightarrow ClO_3^- + F^- + H_2O$$

$$NH_3(l) + FClO_2 \xrightarrow[\text{ignites}]{-78°} NH_4Cl, NH_4F$$

$$HCl(l) + FClO_2 \xrightarrow{-110°} HF + ClO_2 + \tfrac{1}{2}Cl_2$$

$$HOSO_2F + FClO_2 \xrightarrow{-78°} HF + ClO_2OSO_2F$$

$$HOClO_3(\text{anhydrous}) + FClO_2 \longrightarrow HF + ClO_2OClO_3$$

$$SO_3 + FClO_2 \xrightarrow{-10°} ClO_2OSO_2F(\text{insertion})$$

$FClO_2$ explodes with the strong reducing agent SO_2 even at $-40°$ and HBr likewise explodes at $-110°$.

Chlorine dioxide fluoride is a good fluorinating agent and a moderately strong oxidant:

SF_4 is oxidized to SF_6, SF_4O, and SF_2O_2 above 50°, whereas N_2F_4 yields NF_3, FNO_2, and FNO at 30°. UF_4 is oxidized to UF_5 at room temperature and to UF_6 at 100°. Chlorides (and some oxides) are fluorinated and the products can react further to form fluoro complexes. Thus, whereas $AlCl_3$ yields AlF_3, B_2O_3 affords $[ClO_2]^+BF_4^-$, and the Lewis acid chlorides $SbCl_5$, $SnCl_4$, and $TiCl_4$ yield $[ClO_2]^+[SbF_6]^-$, $[ClO_2]^+_2[SnF_6]^{2-}$, and $[ClO_2]^+_2[TiF_6]^{2-}$. Such complexes, and many others can, of course, be prepared directly from the corresponding fluorides either with or without concurrent oxidation, e.g.:

$$AsF_3 + 3FClO_2 \longrightarrow [ClO_2]^+[AsF_6]^- + 2ClO_2$$
$$SbF_5 + FClO_2 \longrightarrow [ClO_2]^+[SbF_6]^- \text{ (mp 220°)}$$
$$2SbF_5 + FClO_2 \longrightarrow [ClO_2]^+[Sb_2F_{11}]^-$$

An X-ray study on this last compound showed the chloryl cation to have the expected nonlinear structure, with angle OClO 122° and Cl–O 131 pm. $FClO_2$ can also act as a fluoride ion acceptor, though not so readily as F_3ClO above. For example CsF reacts at room temperature to give the white solid $Cs[F_2ClO_2]$; this is stable at room temperature but dissociates reversibly into its components above 100°. The C_{2v} structure of $[F_2ClO_2]^-$ (Fig. 17.29f) is deduced from its vibration spectrum.

The two remaining Cl^{VII} oxide fluorides are F_3ClO_2 and $FClO_3$. At one time F_3ClO_2 was thought to exist in isomeric forms but the so-called violet form, previously thought to be the peroxo compound F_2ClOOF has now been discounted.[104] The well-defined compound F_3ClO_2 was first made in 1972 as an extremely reactive colourless gas: mp −81.2°, bp −21.6°. It is a very strong oxidant and fluorinating agent and, because of its corrosive action, must be handled in Teflon or sapphire apparatus. It thus resembles the higher chlorine fluorides. The synthesis of F_3ClO_2 is complicated and depends on an ingenious sequence of fluorine-transfer reactions as outlined below:

$$2FClO_2 + 2PtF_6 \longrightarrow \underbrace{[F_2ClO_2]^+[PtF_6]^- + [ClO_2]^+[PtF_6]^-}$$
$$F_3ClO_2 + FClO_2 + 2[NO_2]^+[PtF_6]^- \longleftarrow \overset{2FNO_2}{\qquad}$$

Fractional condensation at −112° removes most of the $FClO_2$, which is slightly less volatile than F_3ClO_2. The remaining $FClO_2$ is removed by complexing with BF_3 and then relying on the greater stability of the F_3ClO_2 complex:

$$\left.\begin{array}{l} FClO_2 \\ \text{(mixture)} \\ F_3ClO_2 \end{array}\right\} + 2BF_3 \begin{array}{l} \longrightarrow [ClO_2]^+[BF_4]^- \text{ (dissoc press 1 atm at 44°C)} \\ \\ \longrightarrow [F_2ClO_2]^+[BF_4]^- \text{ stable at room temperature} \end{array}$$

Pumping at 20° removes $[ClO_2]^+[BF_4]^-$ as its component gases, leaving $[F_2ClO_2]^+$ $[BF_4]^-$ which, on treatment with FNO_2, releases the desired product:

$$[F_2ClO_2]^+[BF_4]^- + FNO_2 \longrightarrow [NO_2]^+[BF_4]^- + F_3ClO_2$$

The whole sequence of reactions represents a *tour de force* in the elegant manipulation of extremely reactive compounds. F_3ClO_2 is a violent oxidizing reagent but forms stable adducts by fluoride ion transfer to Lewis acids such as BF_3, AsF_5, and PtF_6. The structures of F_3ClO_2 and $[F_2ClO_2]^-$ have C_{2v} symmetry as expected (Fig. 17.27e and i).

In dramatic contrast to F_3ClO_2, perchloryl fluoride ($FClO_3$) is notably inert, particularly at room temperature. This colourless tetrahedral molecular gas (Fig. 17.29c) was first synthesized in 1951 by fluorination of $KClO_3$ at $-40°$ and it can also be made (in 50% yield) by the action of F_2 on an aqueous solution of $NaClO_3$. Electrolysis of $NaClO_4$ in anhydrous HF has also been used but the most convenient route for industrial scale manufacture is the fluorination of a perchlorate with SbF_5, SbF_5/HF, $HOSO_2F$, or perhaps best of all $HOSO_2F/SbF_5$:

$$KClO_4 \xrightarrow[\text{rt (or above)}]{HOSO_2F/SbF_5} FClO_3 \quad (97\% \text{ yield})$$

Because of its remarkably low reactivity at room temperature and its very high specific impulse, the gas has been much studied as a rocket propellent oxidizer (e.g. it compares favourably with N_2O_4 and with ClF_3 as an oxidizer for fuels such as N_2H_4, Me_2NNH_2, and LiH). $FClO_3$ has mp $-147.8°$, bp $-46.7°$, $d(l, -73°C)$ 1.782 g cm^{-3}, viscosity η ($-73°$) 0.55 centipoise. The extremely low dipole moment ($\mu = 0.023$ D) is particularly noteworthy. $FClO_3$ has high kinetic stability despite its modest thermodynamic instability: $\Delta H_f°(g, 298 \text{ K})$ -23.8 kJ mol^{-1}, $\Delta G_f°(g, 298 \text{ K})$ $+48.1$ kJ mol^{-1}. $FClO_3$ offers the highest known resistance to dielectric breakdown for any gas (30% greater than for SF_6, p. 813) and has been used as an insulator in high-voltage systems.

Perchloryl fluoride is thermally stable up to about 400°. Above 465° it undergoes decomposition with first-order kinetics and an activation energy of 244 kJ mol^{-1}. Hydrolysis is slow even at 250–300° and quantitative reaction is only achieved with concentrated aqueous hydroxide in a sealed tube under high pressure at 300°C:

$$FClO_3 + 2NaOH \longrightarrow NaClO_4 + NaF + H_2O$$

However, alcoholic KOH effects a similar quantitative reaction at 25°C. Reaction with liquid NH_3 is also smooth particularly in the presence of a strong nucleophile such as $NaNH_2$:

$$FClO_3 + 3NH_3 \longrightarrow [NH_4]^+[HNClO_3]^- + NH_4F$$

Metallic Na and K react only above 300°.

$FClO_3$ shows no tendency to form adducts with either Lewis acids or bases. This is in sharp contrast to most of the other oxide fluorides of chlorine discussed above and has been related to the preferred tetrahedral (C_{3v}) geometry as compared with the planar (D_{3h}) and trigonal bipyramidal (D_{3h}) geometries expected for $[ClO_3]^+$ and $[F_2ClO_3]^-$ respectively. Conversely the pseudo-trigonal bipyramidal C_s structure F_3ClO gains stability when converted to the pseudo-tetrahedral $[F_2ClO]^+$ or pseudo-octahedral $[F_4ClO]^-$ (see Fig. 17.29).

In reactions with organic compounds $FClO_3$ acts either as an oxidant or as a 1- or 2-centre electrophile which can therefore be used to introduce either F, a $-ClO_3$ group, or both F and O into the molecule. As $FClO_3$ is highly susceptible to nucleophilic attack at Cl it reacts readily with organic anions:

$$FClO_3 + Li^+Ph^- \longrightarrow PhClO_3 + LiF$$

$$FClO_3 + Li^+[C_4H_3S]^- \longrightarrow \text{[thiophene]}-F + LiClO_3$$

Compounds having a cyclic double bond conjugated to an aromatic ring (e.g. indene) undergo oxofluorination, with $FClO_3$ acting as a 2-centre electrophile:

$FClO_3$ also acts as a mild fluorinating agent for compounds possessing a reactive methylene group, e.g.:

$$CH_2(CO_2R)_2 \xrightarrow{FClO_3} CF_2(CO_2R)_2$$

It is particularly useful for selective fluorination of steroids.

Bromine oxide fluorides[105]

These compounds are less numerous and rather less studied than their chlorine analogues; indeed, until 15 y ago only $FBrO_2$ was well characterized. The known species are:

Oxidation state of Br	Cations	**Neutral species**	Anions
V	$[BrO_2]^+$	**$FBrO_2$** (1955)	$[F_2BrO_2]^-$
	$[F_2BrO]^+$	**F_3BrO** (1976)	$[F_4BrO]^-$
VII		**$FBrO_3$** (1969)	

Despite several attempts at synthesis, there is little or no evidence for the existence of $FBrO$, F_3BrO_2, or F_5BrO. The bromine oxide fluorides are somewhat less thermally stable than their chlorine analogues and somewhat more reactive chemically. The structures are as already described for the chlorine oxide fluorides (Fig. 17.29).

Bromyl fluoride, $FBrO_2$ is a colourless liquid, mp $-9°$, which attacks glass at room temperature and which undergoes rapid decomposition above $55°$:

$$3FBrO_2 \xrightarrow{\Delta} BrF_3 + Br_2 + 3O_2$$

It is best prepared by fluorine transfer reactions such as

$$K[F_2BrO_2] + HF(l) \longrightarrow KHF_2 + FBrO_2$$

The $K[F_2BrO_2]$ can be prepared by fluorination of $KBrO_3$ with BrF_5 in the presence of a trace of HF:

$$KBrO_3 + BrF_5 \longrightarrow FBrO_2 + K[F_4BrO]$$
$$K[F_4BrO] + KBrO_3 \longrightarrow 2K[F_2BrO_2]$$

[105] R. J. GILLESPIE and P. H. SPEKKENS, Oxyfluoro compounds of bromine, *Israel J. Chem.* **17**, 11–19 (1978). R. BOUGON, T. B. HUY, P. CHARPIN, R. J. GILLESPIE, and P. H. SPEKKENS, Preparation and characterization of some salts of the difluoro-oxo-bromine(V) cation, *JCS Dalton* 1979, 6–12.

However, the most convenient method of preparation of $K[F_2BrO_2]$ is by reaction of $KBrO_3$ with $KBrF_6$ in MeCN:

$$KBrO_3 + KBrF_6 \xrightarrow{\text{MeCN}} K[F_2BrO_2]\downarrow + K[F_4BrO] \text{ (sol)}$$

Bromyl fluoride is also produced by fluorine–oxygen exchange between BrF_5 and oxoiodine compounds (p. 1034), e.g.:

$$FIO_2 + BrF_5 \longrightarrow FBrO_2 + IF_5$$
$$2F_3IO + BrF_5 \longrightarrow FBrO_2 + 2IF_5$$
$$2I_2O_5 + 5BrF_5 \longrightarrow 5FBrO_2 + 4IF_5$$

As with $FClO_2$ and FIO_2, hydrolysis regenerates the halate ion, the reaction with $FBrO_2$ being of explosive violence. Hydrolysis in basic solution at $0°$ can be represented as

$$FBrO_2 + 2OH^- \longrightarrow BrO_3^- + F^- + H_2O$$

Organic substances react vigorously, often enflaming. Co-condensation of $FBrO_2$ with the Lewis acid AsF_5 produced $[BrO_2]^+[AsF_6]^-$. Vibrational spectra establish the expected nonlinear structure of the cation (3 bands active in both Raman and infrared). $FBrO_2$ can also react as a fluoride ion acceptor (from KF).

Bromine oxide trifluoride, F_3BrO, is made by reaction of $K[F_4BrO]$ with a weak Lewis acid:

$$K[F_4BrO] + [O_2]^+[AsF_6]^- \longrightarrow F_3BrO + KAsF_6 + O_2 + \tfrac{1}{2}F_2$$

$$K[F_4BrO] + HF(\text{anhydr}) \xrightarrow{-72°} F_3BrO + KHF_2$$

The product is a white solid which melts to a clear liquid at about $-5°$; it is only marginally stable at room temperature and slowly decomposes with loss of oxygen:

$$F_3BrO \longrightarrow BrF_3 + \tfrac{1}{2}O_2$$

The molecular symmetry is C_s (like F_3ClO; Fig. 17.29d) and there is some evidence for weak intermolecular association via F_{ax}—Br$\cdots F_{ax}$ bonding. Fluoride ion transfer reactions have been established and yield compounds such as $[F_2BrO]^+[AsF_6]^-$, $[F_2BrO]^+[BF_4]^-$, and $K[F_4BrO]$, though this last compound is more conveniently made independently, e.g. by the reaction of $KBrO_3$ with $KBrF_6$ mentioned above, or by direct fluorination of $K[F_2BrO_2]$:

$$K[F_2BrO_2] + F_2 \longrightarrow K[F_4BrO] + \tfrac{1}{2}O_2$$

Perbromyl fluoride, $FBrO_3$, is made by fluorinating the corresponding perbromate ion with AsF_5, SbF_5, BrF_5, or $[BrF_6]^+[AsF_6]^-$ in HF solutions. The reactions are smooth and quantitative at room temperature:

$$KBrO_4 + 2AsF_5 + 3HF \longrightarrow FBrO_3 + [H_3O]^+[AsF_6]^- + KAsF_6$$
$$2KBrO_4 + BrF_5 + 2HF \longrightarrow 2FBrO_3 + FBrO_2 + 2KHF_2$$
$$KBrO_4 + [BrF_6]^+[AsF_6]^- \longrightarrow FBrO_3 + BrF_5 + \tfrac{1}{2}O_2 + KAsF_6$$

Perbromyl fluoride is a reactive gas which condenses to a colourless liquid (bp $2.4°$) and then solidifies to a white solid (mp *ca.* $-110°$). It has the expected C_{3v} symmetry Fig. 17.30

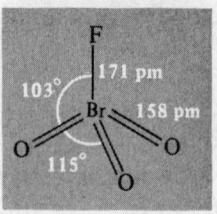

FIG. 17.30   Structure of $FBrO_3$ as determined by gas-phase electron diffraction.

and decomposes slowly at room temperature; it is more reactive than $FClO_3$ and, unlike that compound, it reacts rapidly with water, aqueous base, and even glass:

$$FBrO_3 + H_2O \longrightarrow BrO_4^- + HF + H^+$$
$$FBrO_3 + 2OH^- \longrightarrow BrO_4^- + F^- + H_2O$$

Fluoride ion transfer reactions have not been established for $FBrO_3$ and may be unlikely, (see p. 1031).

Iodine oxide fluorides

The compounds to be considered are the I^V derivatives FIO_2 and F_3IO, and the I^{VII} derivatives FIO_3, F_3IO_2, and F_5IO. Note that, unlike Cl, no I^{III} compound FIO has been reported and that conversely, F_5IO (but not F_5ClO) has been characterized.

FIO_2 has been prepared both by direct fluorination of I_2O_5 in anhydrous HF at room temperature and by thermal dismutation of F_3IO:

$$I_2O_5 + F_2 \xrightarrow{\text{HF}/20°} 2FIO_2 + \tfrac{1}{2}O_2$$

$$2F_3IO \xrightarrow{110°} FIO_2 + IF_5$$

Unlike gaseous molecular $FClO_2$, it is a colourless polymeric solid which decomposes without melting when heated above 200°. Like the other halyl fluorides it readily undergoes alkaline hydrolysis and also forms a complex with F^-:

$$FIO_2 + 2OH^- \longrightarrow IO_3^- + F^- + H_2O$$

$$FIO_2 + KF \xrightarrow{\text{HF}} K^+[F_2IO_2]^-$$

An X-ray study of this latter complex reveals a C_{2v} anion as in the chlorine analogue (Fig. 17.31a). This is closely related to the C_s structure of the neutral molecule F_3IO (Fig. 17.31b).

F_3IO is prepared as colourless crystals by dissolving I_2O_5 in boiling IF_5 and then cooling the mixture:

$$I_2O_5 + 3IF_5 \xrightarrow{105°} 5F_3IO$$

Above 110° it dismutates into FIO_2 and IF_5 as mentioned above.

Of the I^{VII} oxide fluorides FIO_3 has been prepared by the action of F_2/liquid HF on

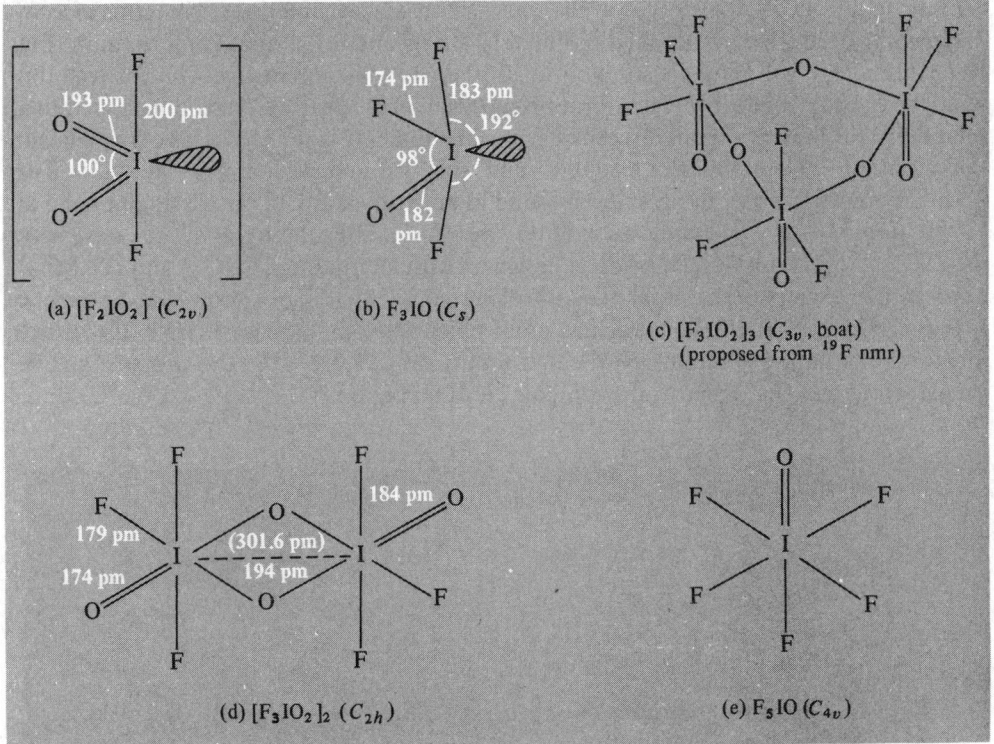

FIG. 17.31   Structures of iodine oxide fluorides.

HIO_4. It is a white, crystalline solid, stable in glass but decomposing with loss of oxygen on being heated:

$$FIO_3 \xrightarrow{\ 100°\ } FIO_2 + \tfrac{1}{2}O_2$$

Unlike its analogue $FClO_3$ it forms adducts with BF_3 and AsF_5, possibly by F^- donation to give $[IO_3]^+[BF_4^-]$ and $[IO_3]^+[AsF_6]^-$, though the structures have not yet been determined. Alternatively, the coordination number of the central I atom might be increased. SO_3 reduces FIO_3 to iodyl fluorosulfate:

$$FIO_3 + SO_3 \longrightarrow IO_2SO_2F + O_2$$

Like $FClO_3$ it reacts with NH_3 but the products has not been fully characterized.

F_3IO_2, first made in 1969, has posed an interesting structural problem. The yellow solid, mp 41°, can be prepared by partial fluorination of a periodate with fluorosulfuric acid:

$$Ba_3H_4(IO_6)_2 \xrightarrow{\ HSO_3F\ } [HIO_2F_4] \xrightarrow{\ SO_3\ } F_3IO_2$$

Unlike monomeric F_3ClO_2 (p. 1030) the structure is oligomeric not only in the solid state but also in the gaseous and solution phases. This arises from the familiar tendency of iodine to increase its coordination number to 6. Fluorine-19 nmr and Raman spectroscopy of F_3IO_2 dissolved in BrF_3 at $-48°$ have been interpreted in terms of a *cis*-oxygen-bridged trimer with axial terminal O atoms and a C_{3v} boat conformation (Fig. 17.31c).[106] On warming the solution to 50° there is a fast interconversion between this and the C_s chair conformer. The vibration spectrum of the gas phase at room temperature has been interpreted in terms of a centrosymmetric dimer (Fig. 17.31d). There is significant dissociation into monomers at 100° and this is almost complete at 185°. The centrosymmetric dimer has also been found in an X-ray study of the crystalline solid at $-80°$ (Fig. 31.7d).[107] Complexes of F_3IO_2 with AsF_5, SbF_5, NbF_5, and TaF_5 have been studied:[108] they are oxygen-bridged polymers with alternating $\{F_4IO_2\}$ and $\{O_2MF_4\}$ groups. For example, the crystal structure of the complex with SbF_5 shows it to be dimeric (Fig. 17.32a).[109] A similar structure motif is found in the adduct $F_3IO.F_3IO_2$ which features alternating 5- and 6-coordinate I atoms (Fig. 17.32b);[110] the structure can be regarded as a cyclic dimer of the ion pair $[F_2IO]^+[F_4IO_2]^-$.

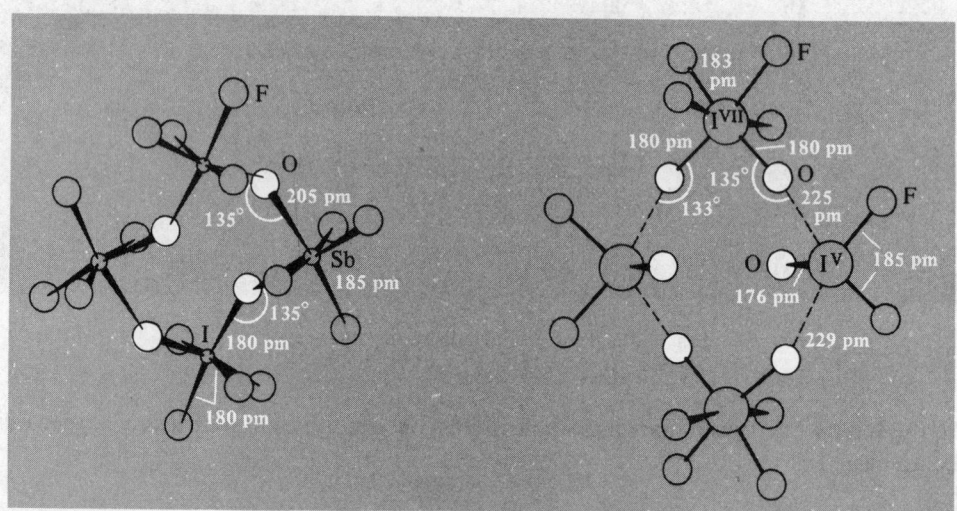

FIG. 17.32   Structures of dimeric adducts of F_3IO_2.

Finally in this section we mention iodine oxide pentafluoride, F_5IO, obtained as a colourless liquid, mp 45°, when IF_7 is allowed to react with water, silica, glass, or I_2O_5. As implied by its preparation from water, F_5IO is not readily hydrolysed. Vibrational spectroscopy and ^{19}F nmr studies point to the 6-coordinate C_{4v} geometry in Fig. 17.31e (i.e. I^{VII}) rather than the alternative 5-coordinate structure F_4I^VOF. Microwave spectroscopy yields a value of 1.08 D for the molecular dipole moment.

[106] R. J. GILLESPIE and J. P. KRASZNAI, A fluorine-19 nmr and Raman spectroscopic investigation of the structure of F_3IO_2, *Inorg. Chem.* **15**, 1251–6 (1976).

[107] L. E. SMART, X-ray crystal structure of iodine trifluoride dioxide, *JCS Chem. Comm.* 1977, 519–20.

[108] R. J. GILLESPIE and J. P. KRASZNAI, Lewis acid–base properties of iodine(VII) dioxide trifluoride, *Inorg. Chem.* **16**, 1384–92 (1977).

[109] A. J. EDWARDS and A. A. K. HANA, Fluoride crystal structures. Part 34. Antimony pentafluoride–iodine trifluoride dioxide, *JCS Dalton* 1980, 1734–6.

[110] R. J. GILLESPIE, J. P. KRASZNAI, and D. R. SLIM, Crystal structure of $(IOF_3 . IO_2F_3)_2$, *JCS Dalton* 1980, 481–3.

17.2.10   *Halogen derivatives of oxoacids*

Numerous compounds are known in which the H atom of an oxoacid has been replaced by a halogen atom. Examples are:

halogen(I) perchlorates $XOClO_3$ (X = F, Cl, Br, ?I)
halogen(I) fluorosulfates $XOSO_2F$ (X = F, Cl, Br, I)
halogen(I) nitrates $XONO_2$ (X = F, Cl, Br, I)

In addition, halogen(III) derivatives such as $Br(ONO_2)_3$, $I(ONO_2)_3$, $Br(OSO_2F)_3$, and $I(OSO_2F)_3$ are known, as well as complexes $M^I[X^I(ONO_2)_2]$, $M^I[I^{III}(ONO_2)_4]$, $M^I[X^{III}(OSO_2F)_4]$ (X = Br, I). In general, thermal stability decreases with increase in atomic number of the halogen.

The properties of halogen(I) perchlorates are in Table 17.26. $FOClO_3$ was originally prepared by the action of F_2 on concentrated $HOClO_3$, but the product had a pronounced

TABLE 17.26   *Properties of halogen(I) perchlorates*

Property	$FOClO_3$	$ClOClO_3$	$BrOClO_3$	$IOClO_3$
Colour	Colourless	Pale yellow	Red	Not obtained pure
MP/°C	−167.3	−117	< −78	
BP/°C	−15.9	44.5	—	
Decomp temp/°C	∼100	20	−20	

tendency to explode on freezing. More recently,[111] extremely pure $FOClO_3$ has been obtained by thermal decomposition of NF_4ClO_4 and such samples can be manipulated and repeatedly frozen without mishap. Thermal decomposition occurs via two routes:

$$FOClO_3 \begin{cases} \longrightarrow ClF + 2O_2 \\ \longrightarrow FClO_2 + O_2 \end{cases}$$

It readily oxidizes iodide ions:

$$FOClO_3 + 2I^- \longrightarrow ClO_4^- + F^- + I_2$$

$FOClO_3$ also adds to C=C double bonds in fluorocarbons to give perfluoroalkyl perchlorates:

$$CF_2{=}CF_2 + FOClO_3 \xrightarrow{-45°} CF_3CF_2OClO_3$$

$$CF_3CF{=}CF_2 + FOCl_3 \xrightarrow{-45°} CF_3CF_2CF_2OClO_3 + CF_3CF(OClO_3)CF_3$$
$$\qquad\qquad\qquad\qquad\qquad\qquad 68\% \qquad\qquad\qquad\qquad 32\%$$

[111] C. J. SCHACK and K. O. CHRISTE, Reactions of fluorine perchlorate with fluorocarbons and the polarity of the O–F bond in covalent hypofluorites, *Inorg. Chem.* **18**, 2619–20 (1979). For vibrational spectra, thermodynamic properties, and confirmation of C_s structure see K. O. CHRISTE and E. C. CURTIS, *Inorg. Chem.* **21**, 2938–45 (1982).

The formation of isomers in this last reaction implies a low bond polarity of FO– in $FOClO_3$.

Chlorine perchlorate, $ClOClO_3$, is made by low-temperature metathesis:

$$MClO_4 + ClOSO_2F \xrightarrow{\,-45°\,} ClOClO_3 + MSO_3F \quad (M = Cs, NO_2)$$

The bromine analogue can be made similarly using $BrOSO_2F$ at $-20°$ or by direct bromination of $ClOClO_3$ with Br_2 at $-45°$. Both compounds are thermally unstable and shock sensitive; e.g. $ClOClO_3$ decomposes predominantly to Cl_2O_6 with smaller amounts of ClO_2, Cl_2, and O_2 on gentle warming. Direct iodination of $ClOClO_3$ at $-50°$ yields the polymeric white solid $I(OClO_3)_3$ rather than $IOClO_3$; this latter compound has never been obtained pure but is among the products of the reaction of I_2 with $AgClO_4$ at $-85°$, the other products being $I(OClO_3)_3$, $Ag[I(OClO_3)_2]$, and AgI.

Halogen nitrates are even less thermally stable than the perchlorates: they are made by the action of $AgNO_3$ on an alcoholic solution of the halogen at low temperature. With an excess of $AgNO_3$, bromine and iodine yield $X(ONO_2)_3$. Numerous other routes are available; e.g., the reaction of ClF on $HONO_2$ gives a 90% yield of $ClONO_2$ and the best preparation of this compound is probably the reaction

$$Cl_2O + N_2O_5 \xrightarrow{\,0°\,} 2ClONO_2$$

Some physical properties are in Table 17.27. Both $FONO_2$ and $ClONO_2$ feature planar NO_3 groups with the halogen atom out of the plane. $ClONO_2$ has been used to convert metal chlorides to anhydrous metal nitrates, e.g. $Ti(NO_3)_4$. Likewise ICl_3 at $-30°$ yields $I(ONO_2)_3$. $ClONO_2$ and $IONO_2$ add across C=C double bonds, e.g.:

$$CH_2{=}CMe_2 + Cl{-}ONO_2 \xrightarrow{\,-78°\,} ClCH_2C(Me_2)ONO_2$$

TABLE 17.27   *Some properties of halogen(I) nitrates*

Property	$FONO_2$	$ClONO_2$	$BrONO_2$	$IONO_2$
Colour	Colourless	Colourless	Yellow	Yellow
MP/°C	-175	-107	-42	—
BP/°C	-45.9	18	—	—
Decomp temp/°C	Ambient	Ambient	<0	<0
$\Delta H_f^\circ(g,\ 298\ \text{K})/\text{kJ mol}^{-1}$	$+10.5$	$+29.2$	—	—
$\Delta G_f^\circ(g,\ 298\ \text{K})/\text{kJ mol}^{-1}$	$+73.5$	$+92.4$	—	—

Several other reactions have been studied but the overall picture is one of thermal instability, hazardous explosions, and vigorous chemical reactivity leading to complex mixtures of products.

The halogen fluorosulfates are amongst the most stable of the oxoacid derivatives of the halogens. $FOSO_2F$ is made by direct addition of F_2 to SO_3 and the others are made by direct combination of the halogen with an equimolar quantity of peroxodisulfuryl difluoride, $S_2O_6F_2$ (p. 751). With an excess of $S_2O_6F_2$, bromine and iodine yield $X(OSO_2F)_3$. An alternative route to $ClOSO_2F$ is the direct addition of ClF to SO_3, whilst $BrOSO_2F$ and $IOSO_2F$ can be made by thermal decomposition of the corresponding

TABLE 17.28 *Some physical properties of halogen fluorosulfates*[a]

Property	$FOSO_2F$	$ClOSO_2F$	$BrOSO_2F$	$IOSO_2F$
Colour	Colourless	Yellow	Red-brown	Black
State at room temp	Gas	Liquid	Liquid	Solid
MP/°C	-158.5	-84.3	-31.5	51.5
BP/°C	-31.3	45.1	117.3	—

[a] $Br(OSO_2F)_3$ is a pale yellow solid, mp 59°; $I(OSO_2F)_3$ is a pale yellow solid, mp 32°.

$X(OSO_2F)_3$. The halogen fluorosulfates are thermally unstable, moisture sensitive, highly reactive compounds. Some physical properties are summarized in Table 17.28. The vibrational spectra of $FOSO_2$ and $ClOSO_2F$ are consistent with C_s molecular symmetry as in $HOSO_2F$:

$$
\begin{array}{c}
F \\
\diagdown \\
O \\
\| \\
S \\
\diagup \diagdown \diagdown \\
O \qquad \quad F \\
\| \\
O
\end{array}
$$

Much of the chemistry of the halogen fluorosulfates resembles that of the interhalogens (p. 964) and in many respects the fluorosulfate group can be regarded as a pseudohalogen (p. 336). There is some evidence of ionic self-dissociation and reactions can be classified as exchange, addition, displacement, and complexation. This is illustrated for the iodine fluorosulfates in the following scheme:[112]

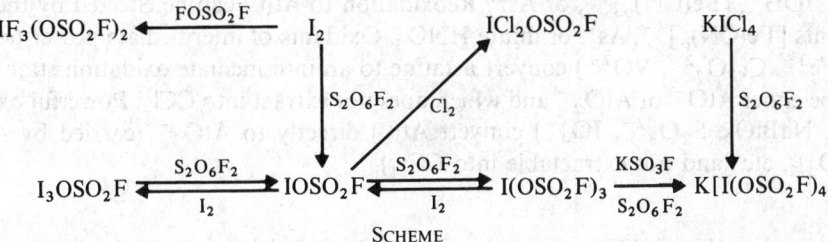

SCHEME

17.3 The Chemistry of Astatine[113]

All isotopes of element 85, astatine, are intensely radioactive with very short half-lives (p. 926). As a consequence weighable amounts of the element or its compounds cannot be

112 Ref. 17, pp. 1466–75, Halogen derivatives of oxyacids.
113 E. H. APPELMAN, Astatine, Chap. 6 in *MTP International Review of Science, Inorganic Chemistry*, Series 1. Vol. 3, *Main Group Elements Group VII and Noble Gases*, pp. 181–98, Butterworths, London, 1972; see also ref. 17, pp. 1573–94, Astatine.

prepared and no bulk properties are known. The chemistry of the element must, of necessity, be studied by tracer techniques on extremely dilute solutions, and this introduces the risk of experimental errors and the consequent possibility of erroneous conclusions. Nevertheless, a picture of the element is gradually emerging as outlined below. The synthesis of the element (p. 926), its natural occurrence in rare branches of the ^{235}U decay series (p. 928), and its atomic properties (p. 934) have already been mentioned.

The chemistry of At is most conveniently studied using ^{211}At ($t_{\frac{1}{2}}$ 7.21 h). This isotope is prepared by α-particle bombardment of ^{209}Bi using acceleration energies in the range 26–29 MeV. Higher energies result in the concurrent formation of ^{210}At and ^{209}At which complicate the subsequent radiochemical assays. The Bi is irradiated either as the metal or its oxide and the target must be cooled to avoid volatilization of the At produced. Astatine is then removed by heating the target to 300–600° (i.e. above the mp of Bi, 217°) in a stream of N_2 and depositing the sublimed element on a glass cold finger or cooled Pt disc. Aqueous solutions of the element can be prepared by washing the cold finger or disc with dilute HNO_3 or HCl. Alternatively, the irradiated target can be dissolved in perchloric acid containing a little iodine as carrier for the astatine; the Bi is precipitated as phosphate and the aqueous solution of AtI used as it is or the activity can be extracted into CCl_4 or $CHCl_3$.

Three oxidation states of At have been definitely established ($-I$, 0, V) and two others have been tentatively identified [$+I$(or III?) and VII]. The standard oxidation potentials connecting these states in 0.1 M acid solution are $E°/V$):

$$At^- \xleftarrow{\ +0.3\ } At(0) \xleftarrow{\ +1.0\ } HOAt(?) \xleftarrow{\ +1.5\ } AtO_3^- \xleftarrow{\ >+1.6\ } H_5AtO_6(?)$$

These values should be compared with those for the other halogens (in 1 M acid) (p. 1001). Noteworthy features are that At is the only halogen with an oxidation state between 0 and V that is thermodynamically stable towards disproportionation, and that the smooth trends in the values of $E°(\frac{1}{2}X_2/X^-)$ and $E°(HOX/\frac{1}{2}X_2)$ continue to At.

The astatide ion At^- (which coprecipitates with Ag, I, or PdI_2) can be obtained from At(0) or AtI using moderately powerful reducing agents, e.g. Zn/H^+, SO_2, SO_3^{2-}/OH^-, $[Fe(CN)_6]^{4-}$, or As^{III}. Reoxidation to At(0) can be effected by the weak oxidants $[Fe(CN)_6]^{3-}$, As^V, or dilute HNO_3. Oxidants of intermediate power (e.g. Cl_2, Br_2, Fe^{3+}, $Cr_2O_7^{2-}$, VO^{2+}) convert astatine to an intermediate oxidation state which may be either AtO^- or AtO_2^- and which does not extract into CCl_4. Powerful oxidants (Ce^{IV}, $NaBiO_3$, $S_2O_8^{2-}$, IO_4^-) convert At(0) directly to AtO_3^- (carried by $AgIO_3$, $Ba(IO_3)_2$, etc., and not extractable into CCl_4).

TABLE 17.29   *Formation constants for trihalide ions at 25°C*

Reaction	$K/l\ mol^{-1}$	Reaction	$K/l\ mol^{-1}$
$Cl_2 + Cl^- \rightleftharpoons Cl_3^-$	0.12	$AtI + Br^- \rightleftharpoons AtIBr^-$	120
$Br_2 + Cl^- \rightleftharpoons Br_2Cl^-$	1.4	$ICl + Cl^- \rightleftharpoons ICl_2^-$	170
$I_2 + Cl^- \rightleftharpoons I_2Cl^-$	3	$AtBr + Br^- \rightleftharpoons AtBr_2^-$	320
$AtI \geqslant Cl^- \rightleftharpoons AtICl^-$	9	$IBr + Br^- \rightleftharpoons IBr_2^-$	440
$Br_2 + Br^- \rightleftharpoons Br_3^-$	17	$I_2 + I^- \rightleftharpoons I_3^-$	800
$IBr + Cl^- \rightleftharpoons IBrCl^-$	43	$AtI + I^- \rightleftharpoons AtI_2^-$	2000

At(0) reacts with halogens X_2 to produce interhalogen species AtX, which can be extracted into CCl_4, whereas halide ions X^- yield polyhalide ions AtX_2^- which are not extracted by CCl_4 but can be extracted into Pr_2^iO. The equilibrium formation constants of the various trihalide ions are intercompared in Table 17.29.

A rudimentary chemistry of organic derivatives of astatine is emerging, but the problems of radiation damage, product separation, and tracer identification, already severe for inorganic compounds of astatine, are even worse with organic derivatives. Various compounds of the type RAt, $RAtCl_2$, R_2AtCl, and $RAtO_2$ (R = phenyl or *p*-tolyl) have been synthesized using astatine-labelled iodine reagents, e.g.:

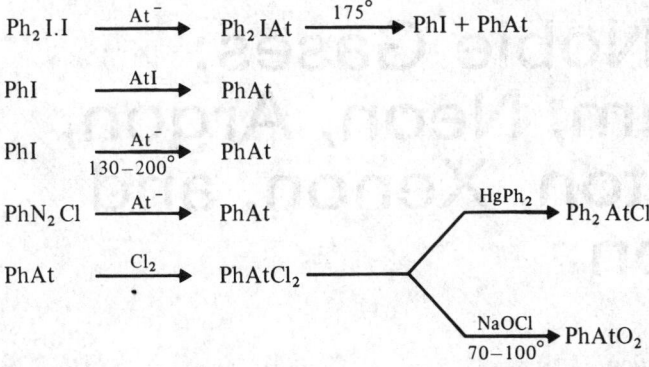

Astatine has been shown to be superior to radio-iodine for the destruction of abnormal thyroid tissue (p. 925) because of the localized action of the emitted α-particles which dissipate 5.9 MeV within a range of 70 μm of tissue, whereas the much less energetic β-rays of radio-iodine have a maximum range of *ca.* 2000 μm. However, its general inaccessibility and high cost render its extensive application unlikely.

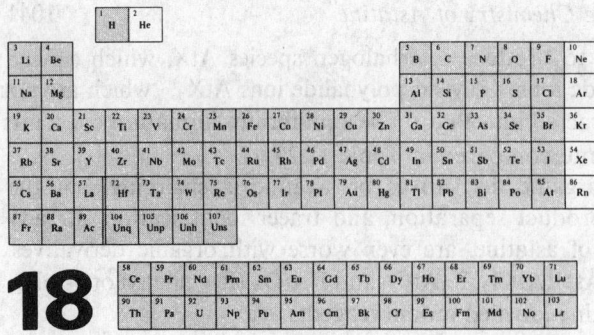

18

The Noble Gases: Helium, Neon, Argon, Krypton, Xenon, and Radon

18.1 Introduction

In 1785 H. Cavendish in his classic work on the composition of air (p. 466) noted that, after repeatedly sparking a sample of air with an excess of O_2, there was a small residue of gas which he was unable to remove by chemical means and which he estimated with astonishing accuracy to be "not more than $\frac{1}{120}$th part of the whole". He could not further characterize this component of air, and its identification as argon had to wait for more than a century. But first came the discovery of helium, which is unique in being the only element discovered extraterrestrially before being found on earth. During the solar eclipse of 18 August 1868, a new yellow line was observed close to the sodium D lines in the spectrum of the sun's chromosphere. This led J. N. Lockyer (founder in 1869 of the journal *Nature*) and E. Frankland to suggest the existence of a new element which, appropriately, they named helium (Greek ἥλιος, the sun). The same line was observed by L. Palmieri in 1881 in the spectrum of volcanic gas from Mount Vesuvius, and the terrestrial existence of helium was finally confirmed by W. Ramsay[1] in the course of his intensive study of atmospheric gases which led to the recognition of a new group in the periodic table. This work was initiated by the physicist, Lord Rayleigh, and was recognized in 1904 by the award of the Nobel Prizes for Chemistry and Physics to Ramsay and Rayleigh respectively.

In order to test Prout's hypothesis (that the atomic weights of all elements are multiples of that of hydrogen) Rayleigh made accurate measurements of the densities of common gases and found, to his surprise, that the density of nitrogen obtained from air by the removal of O_2, CO_2, and H_2O was consistently about 0.5% higher than that of nitrogen obtained chemically from ammonia. Ramsay then treated "atmospheric nitrogen" with heated magnesium ($3Mg + N_2 \rightarrow Mg_3N_2$), and was left with a small amount of a much

---

[1] M. W. TRAVERS, *Life of Sir William Ramsay*, E. Arnold, London, 1956.

denser, monatomic gas† which, in a joint paper (*Proc. R. Soc.* **57**, 265 (1895), was identified as a new element which was named *argon* (Greek ἀργόν, idle or lazy) because of its inert nature. Unfortunately there was no space for a new and unreactive, gaseous, element in the periodic table (p. 24), which led to Ramsay's audacious suggestion that a whole new group might be accommodated. By 1898 Ramsay and M. W. Travers had isolated by the low-temperature distillation of liquid air (which had only recently become available) and characterized by spectroscopic analysis, the further new elements: krypton (Greek κρυπτόν, hidden, concealed), neon (Greek νέον, new), and xenon (Greek ξένον, strange).

In 1895 Ramsay also identified helium as the gas previously found occluded in uranium minerals and mistakenly reported as nitrogen. Five years later he and Travers isolated helium from samples of atmospheric neon.

Element 86, the final member of the group, is a short-lived, radioactive element, formerly known as radium-emanation or niton or, depending on which radioactive series it originates in (i.e. which isotope) as radon, thoron, or actinon. It was first isolated and studied in 1902 by E. Rutherford and F. Soddy and is now universally known as radon (from radium and the termination -on adopted for the noble gases; Latin *radius*, ray).

Once the existence of the new group has been established it was apparent that it did not merely fit into the periodic table but actually improved it by providing a bridge between the strongly electronegative halogens and strongly electropositive alkali metals. The elements became known as "inert gases" comprising Group 0, though A. von Antropoff suggested that a maximum valency of eight might be attainable and designated them as Group VIIIB. They have also been described as the "rare gases" but, since the lighter members are by no means rare and the heavier ones are not entirely inert, "noble" gases seems a more appropriate name and has come into general use during the past two decades.

The apparent inertness of the noble gases gave them a key position in the electronic theories of valency as developed by G. N. Lewis (1916) and W. Kossel (1916) and the attainment of a "stable octet" was regarded as a prime criterion for bond formation between atoms (p. 25). Their monatomic, non-polar nature makes them the most nearly "perfect" gases known, and has led to continuous interest in their physical properties.

18.2 The Elements

18.2.1 *Distribution, production, and uses*[2, 3]

Helium is the second most abundant element in the universe (76% H, 23% He) as a result of its synthesis from hydrogen (p. 11) but, being too light to be retained by the earth's gravitational field, all primordial helium has been lost and terrestrial helium, like

---

† The molecular weight (mean relative molecular mass) was obtained by determination of density but, in order to determine that the gas was monatomic and its atomic and molecular weights identical, it was necessary to measure the velocity of sound in the gas and to derive from this the ratio of its specific heats: kinetic theory predicts that $C_p/C_v = 1.67$ for a monatomic and 1.40 for a diatomic gas.

² Helium group gases, in *Kirk–Othmer Encyclopedia of Chemical Technology*, 3rd edn, Vol. 12, pp. 249–87, Interscience, New York, 1980.
³ W. J. GRANT and S. L. REDFEARN, Industrial gases, in *The Modern Inorganic Chemicals Industry*, pp. 273–301. The Chemical Society, London, 1977.

argon, is the result of radioactive decay (^{4}He from α-decay of heavier elements, ^{40}Ar from electron capture by ^{40}K (p. 21)).

The noble gases make up about 1% of the earth's atmosphere in which their major component is Ar. Smaller concentrations are occluded in igneous rocks, but the atmosphere is the principal commercial source of Ne, Ar, Kr, and Xe, which are obtained as by-products of the liquefaction and separation of air (p. 702). Some Ar is also obtained from synthetic ammonia plants in which it accumulates after entering as impurity in the N_2 and H_2 feeds. World production of Ar in 1975 was 700 000 tonnes for use mainly as an inert atmosphere in high-temperature metallurgical processes and, in smaller amounts, for filling incandescent lamps. Along with Ne, Kr, and Xe, which are produced on a much smaller scale, Ar is also used in discharge tubes—the so-called neon lights for advertisements—(the colour produced depending on the particular mixture of gases used). They are also used in fluorescent tubes, though here the colour produced depends not on the gas but on the phosphor which is coated on the inside walls of the tube.

Although the concentration of He in the atmosphere is five times that of Kr and sixty times that of Xe (see Table 18.1), its recovery from this source is uneconomical compared to that from natural gas if more than 0.4% He is present. This concentration is attained in a number of gases in the USA (concentrations as high as 7% are known) and in eastern Europe (mainly Poland), and, by the liquefaction of the hydrocarbon and other gases present, about 20×10^6 m^3 (3400 tonnes) and 6×10^6 m^3 (1020 tonnes), respectively, are produced each year. The former use of He as a non-flammable gas (it has a lifting power of approximately 1 kg per m^3) in airships is no longer important, though it is still employed in meteorological balloons, and, as with Ar, its main use is in metallurgical processes such as arc welding. The choice between Ar and He for these purposes is determined by cost and, except in the USA, this generally favours Ar. Smaller, but important, uses are:

(a) as a substitute for N_2 in synthetic breathing gas for deep-sea diving (its low solubility in blood minimizes the degassing which occurs with N_2 when divers are depressurized and which produces the sometimes fatal "bends");

(b) as a cryogen (He has the lowest bp of any substance and liquid He provides the only practicable means of studying and applying such low-temperature phenomena as superconductivity);

(c) as a coolant in HTR nuclear reactors (p. 1459);

(d) as a flow-gas in gas–liquid chromatography;

(e) for deaeration of solutions and as a general inert diluent or inert atmosphere.

Radon has been used in the treatment of cancer and as a radioactive source in testing metal castings but, because of its short half-life (3.824 days) it has been superseded by more convenient materials. Such small quantities as are required are obtained as a decay product of ^{226}Ra (1 g of which yields 0.64 cm^3 in 30 days).

18.2.2 *Atomic and physical properties of the elements*[2-4]

Some of the important properties of the elements are given in Table 18.1. The

[4] A. H. COCKETT and K. C. SMITH, The monatomic gases: physical properties and production, Chap. 5 in *Comprehensive Inorganic Chemistry*, Vol. 1, pp. 139–211, Pergamon Press, Oxford, 1973. G. A. COOK (ed.), *Argon, Helium and the Rare Gases*, 2 vols, Interscience, New York, 1961, 818 pp.

TABLE 18.1 *Some properties of the noble gases*

Property	He	Ne	Ar	Kr	Xe	Rn
Atomic number	2	10	18	36	54	86
Number of naturally occurring isotopes	2	3[a]	3	6	9	(1)
Atomic weight	4.002 60	20.179	39.948	83.80	131.29($\pm$3)	(222)[b]
Abundance in dry air/ppm by vol	5.24	18.18	93.40	1.14	0.087	Variable traces
Abundance in igneous rocks/ppm by wt	3×10^{-3}	7×10^{-5}	4×10^{-2}	—	—	1.7×10^{-10}
Outer shell electronic configuration	$1s^2$	$2s^2 2p^6$	$3s^2 3p^6$	$4s^2 4p^6$	$5s^2 5p^6$	$6s^2 6p^6$
First ionization energy/kJ mol^{-1}	2372	2080	1520	1351	1170	1037
BP/K	4.215	27.07	87.29	119.7	165.04	211
/°C	−268.93	−246.06	−185.86	−153.35	−108.13	−62
MP/K	—[c]	24.55	83.78	115.90	161.30	202
/°C	—	−248.61	−189.37	−157.20	−111.80	−71
ΔH_{vap}/kJ mol^{-1}	0.08	1.74	6.52	9.05	12.65	18.1
Density at STP/mg cm^{-3}	0.178 47	0.899 94	1.784 03	3.7493	5.8971	9.73
Thermal conductivity at 0°C/J s^{-1} m^{-1} K^{-1}	0.1430	0.0461	0.0165	0.008 54	0.005 40	
Solubility in water at 20°C/cm^3 kg^{-1}	8.61	10.5	33.6	59.4	108.1	230

[a] In the pioneering work of J. J. Thomson and F. W. Aston on mass-spectrometry, neon was the first non-radioactive element shown to exist in different isotopic forms.

[b] The relative atomic mass of this nuclide is 222.018.

[c] Helium is the only liquid which cannot be frozen by the reduction of temperature alone. Pressure must also be applied. It is also the only substance lacking a "triple point", i.e. a combination of temperature and pressure at which solid, liquid, and gas coexist in equilibrium.

imprecision of the atomic weights of Kr and Xe reflects the natural occurrence of several isotopes of these elements. For He, however, and to a lesser extent Ar, a single isotope predominates (^{4}He, 99.999 86%; ^{40}Ar, 99.600%) and much greater precision is possible. The natural preponderance of ^{40}Ar is indeed responsible for the well-known inversion of atomic weight order of Ar and K in the periodic table (p. 36), and the position of Ar in front of K was only finally accepted when it was shown that the atomic weight of He placed it in front of Li. The second isotope of helium, ^{3}He, has only been available in significant amounts since the 1950s when it began to accumulate as a β-decay product of tritium stored for thermonuclear weapons.

All the elements have stable electronic configurations ($1s^2$ or ns^2np^6) and, under normal circumstances are colourless, odourless, and tasteless monatomic gases. The non-polar, spherical nature of the atoms which this implies, leads to physical properties which vary regularly with atomic number. The only interatomic interactions are weak van der Waals forces. These increase in magnitude as the polarizabilities of the atoms increase and the ionization energies decrease, the effect of both factors therefore being to increase the interactions as the sizes of the atoms increase. This is shown most directly by the enthalpy of vaporization, which is a measure of the energy required to overcome the interactions, and increases from He to Rn by a factor of over 200. However, ΔH_{vap} is in all cases small and bps are correspondingly low, that of He being the lowest of any substance.

The stability of the electronic configuration is indicated by the fact that each element has the highest ionization energy in its period, though the value decreases down the group as a result of increasing size of the atoms. For the heavier elements is it actually smaller than for first-row elements such as O and F with consequences for the chemical reactivities of the noble gases which will be considered in the next section.

As the first member of this unusual group He has, of course, a number of unique properties. Among these is the astonishing transition from so-called HeI to HeII which occurs around 2.2 K (the λ-point temperature) when liquid He (^{4}He to be precise, since ^{3}He does not behave in this way) is cooled by continuous pumping. The transition is clearly seen as the sudden cessation of turbulent boiling, even though evaporation continues. HeI is a normal liquid but at the transition the specific heat increases abruptly by a factor of 10, the thermal conductivity by the order of 10^6, and the viscosity, as measured by its flow through a fine capillary, becomes effectively zero (hence its description as a "superfluid"). HeII also has the curious ability to cover, with a film a few hundred atoms thick, all solid surfaces which are connected to it and are below the λ point. This can be spectacularly demonstrated by dipping the bottom of a suitable container into a bath of HeII. Once the vessel has cooled, liquid He flows, apparently without friction, up and over the edge of the container until the levels inside and outside are equal. These phenomena are evidently the result of quantum effects on a macroscopic scale, and HeII is believed to consist of two components: a true superfluid with zero viscosity and entropy, together with a normal fluid, the fraction of the former increasing to 1 at absolute zero. No completely satisfactory explanation of these phenomena is yet available.

Finally, a property of practical importance which may be noted is the ability of noble gases, especially He, to diffuse through many materials commonly used in laboratories. Rubber and PVC are cases in point, and He will even diffuse through most glasses so that glass Dewar vessels cannot be used in cryoscopic work involving liquid He.

18.3 Chemistry of the Noble Gases[5, 5a, 6, 6a, 6b]

The discovery of the noble gases was a direct result of their unreactive nature, and early unsuccessful attempts to induce chemical reactions reinforced the belief in their inertness. Nevertheless, attempts were made to make the heavier gases react, and in 1933 Linus Pauling, from a consideration of ionic radii, suggested that KrF_6 and XeF_6 should be preparable. D. M. Yost and A. L. Kaye attempted to prepare the latter by passing an electric discharge through a mixture of Xe and F_2 but failed† and, until "$XePtF_6$" was prepared in 1962, the only compounds of the noble gases which could be prepared were clathrates.

While investigating the chemistry of PtF_6, N. Bartlett noticed that its accidental exposure to air produced a change in colour, and with D. H. Lohmann he later showed this to be $O_2^+[PtF_6]^-$.[7] Recognizing that PtF_6 must therefore be an oxidizing agent of unprecedented power, he noted that Rn and Xe should similarly be oxidizable by this reagent since the first ionization energy of Rn is less than, and that of Xe is comparable to, that of molecular oxygen (1175 kJ mol^{-1} for $O_2 \rightarrow O_2^+ + e^-$). He quickly proceeded to show that deep-red PtF_6 vapour spontaneously oxidized Xe to produce an orange-yellow solid and announced this in a brief note.[7a] Within a few months XeF_4 and XeF_2 had been synthesized in other laboratories. Noble-gas chemistry had begun.

Isolable compounds are obtained only with the heavier noble gases Kr and Xe; no doubt Rn would be included as well, were it not for its intense radioactivity which is not only hazardous but also decomposes the reagents involved. The compounds almost always involve bonds to F or O, in most cases exclusively so. A few compounds involving bonds to Cl, N, and even C are, however, known (p. 1058). Chemical combinations involving the lighter noble gases have been observed but are very unstable, and frequently occur only as transient species.

18.3.1 *Clathrates*

Probably the most familiar of all clathrates are those formed by Ar, Kr, and Xe with quinol, $1,4-C_6H_4(OH)_2$, and with water. The former are obtained by crystallizing quinol from aqueous or other convenient solution in the presence of the noble gas at a pressure of 10–40 atm. The quinol crystallizes in the less-common β-form, the lattice of which is held together by hydrogen bonds in such a way as to produce cavities in the ratio 1 cavity : 3 molecules of quinol. Molecules of gas (G) are physically trapped in these cavities, there being only weak van der Waals interactions between "guest" and "host" molecules. The

---

† By what must have seemed to these workers a cruel irony, essentially the same method, but using sunlight instead of a discharge, when tried 30 years later produced XeF_2.

[5] N. BARTLETT and F. E. SLADKY, The chemistry of krypton, xenon and radon, Chap. 6, in *Comprehensive Inorganic Chemistry*, Vol. 1, pp. 213–330, Pergamon Press, Oxford, 1973.

[5a] D. T. HAWKINS, W. E. FALCONER and N. BARTLETT, *Noble Gas Compounds, A Bibliography 1962–1976*, Plenum Press, New York, 1978.

[6] J. H. HOLLOWAY, *Noble-gas Chemistry*, Methuen, London, 1968, 213 pp.

[6a] Helium group gases, compounds, pp. 288–97 of ref. 2.

[6b] K. SEPPELT and D. LENTZ, Novel developments in noble gas chemistry, *Progr. Inorg. Chem.* **29**, 167–202 (1982).

[7] N. BARTLETT and D. H. LOHMANN, Dioxygenyl hexafluoroplatinate(V), $O_2^+[PtF_6]^-$, *Proc. Chem. Soc.* 1962, 115–16.

[7a] N. BARTLETT, Xenon hexafluoroplatinate(V), $Xe^+[PtF_6]^-$, *Proc. Chem. Soc.* 1962, 218.

clathrates are therefore nonstoichiometric but have an "ideal" or "limiting" composition of $[G\{C_6H_4(OH)_2\}_3]$. Once formed they have considerable stability but the gas is released on dissolution or melting. Similar clathrates are obtained with numerous other gases of comparable size, such as O_2, N_2, CO, and SO_2 (the first clathrate to be fully characterized, by H. M. Powell in 1947) but not He or Ne, which are too small or insufficiently polarizable to be retained.

Noble gas hydrates are formed similarly when water is frozen under a high pressure of gas (p. 735). They have the ideal composition, $[G_8(H_2O)_{46}]$, and again are formed by Ar, Kr, and Xe but not by He or Ne. A comparable phenomenon occurs when synthetic zeolites (molecular sieves) are cooled under a high pressure of gas, and Ar and Kr have been encapsulated in this way (p. 415). Samples containing up to 20% by weight of Ar have been obtained.

Clathrates provide a means of storing noble gases and of handling the various radioactive isotopes of Kr and Xe which are produced in nuclear reactors.

18.3.2 *Compounds of xenon*

The chemistry of Xe is much the most extensive in this group and the known oxidation states of Xe range from $+2$ to $+8$. Details of some of the more important compounds are given in Table 18.2. There is clearly a rich variety of stereochemistries, though the description of these depends on whether only nearest-neighbour atoms are considered or whether the supposed disposition of lone-pairs of electrons is also included. Weaker secondary interactions in crystalline compounds also tend to increase the number of atoms surrounding a central Xe atom. For example, $[XeF_5]^+[AsF_6]^-$ has 5 F at 179–182 pm and *three* further F at 265–281 pm, whereas $[XeF_5]^+[RuF_6]^-$ has 5 F at 179–184 pm

TABLE 18.2   *Some compounds of xenon with fluorine and oxygen*

Oxidation State	Compound	MP/°C	Stereochemistry of Xe Actual	Pseudo (i.e. electron lone-pairs included, in parentheses)
$+2$	XeF_2	129	$D_{\infty h}$, linear	Trigonal bipyramidal (3)
$+4$	XeF_4	117.1	D_{4h}, square planar	Octahedral (2)
$+6$	XeF_6	49.5	Distorted octahedral (fluxional)	Pentagonal bipyramidal or capped octahedral (1)
	$[XeF_5]^+[AsF_6]^-$	130.5	C_{4v}, square pyramidal	Octahedral (1)
	$CsXeF_7$	Dec > 50		
	$[NO]^+{}_2[XeF_8]^{2-}$		D_{4d}, square antiprismatic	(Lone-pair inactive)
	$XeOF_4$	(-46)	C_{4v}, square pyramidal	Octahedral (1)
	XeO_2F_2	30.8	C_{2v}, "seee-saw"	Trigonal bipyramidal (1)
	$CsXeOF_5$		Distorted octahedral	Capped octahedral (1)
	$KXeO_3F$		Square pyramidal (chain)	Octahedral (1)
	XeO_3	Explodes	C_{3v}, pyramidal	Tetrahedral (1)
$+8$	XeO_4	-35.9	T_d, tetrahedral	(No lone-pair)
	XeO_3F_2	-54.1	D_{3h}, trigonal bipyramidal	Trigonal bipyramidal
	Ba_2XeO_6	Dec > 300	O_h, octahedral	(No lone-pair)

TABLE 18.3 *Stereochemistry of xenon*

CN	Stereochemistry	Examples	Structure
0	—	Xe(g)	Xe
1	—	$[XeF]^+$, $[XeOTeF_5]^-$	Xe—
2	Linear	XeF_2, $[FXeFXeF]^+$, $FXeOSO_2F$	—Xe—
3	Pyramidal	XeO_3	
	T-shaped	$[XeF_3]^+$, $XeOF_2$	—Xe—
4	Tetrahedral	XeO_4	
	Square	XeF_4	—Xe—
	C_{2v}, "see-saw"	XeO_2F_2	—Xe—
5	Trigonal bipyramidal	XeO_3F_2	—Xe—
	Square pyramidal	$XeOF_4$, $[XeF_5]^+$	
6	Octahedral	$[XeO_6]^{4-}$	
	Distorted octahedral	$XeF_6(g)$, $[XeOF_5]^-$	
7	(?)	$CsXeF_7$	
8	Square antiprismatic	$[XeF_8]^{2-}$	

and *four* further F at 255–292 pm. If only the most closely bonded atoms are counted, then Xe is known with all coordination numbers from 0 to 8 as shown schematically in Table 18.3.

The three fluorides of Xe can be obtained by direct reaction but conditions need to be carefully controlled if these are to be produced individually in pure form. XeF_2 can be prepared by heating F_2 with an excess of Xe to 400°C in a sealed nickel vessel or by irradiating mixtures of Xe and F_2 with sunlight. The product is a white, crystalline solid consisting of parallel linear XeF_2 units (Fig. 18.1). It is sublimable and its infrared and Raman spectra show that the linear molecular structure is retained in the vapour. XeF_2 is a mild fluorinating agent and will, for instance, difluorinate olefins (alkenes). It dissolves in water to the extent of 25 g dm^{-3} at 0°C, the solution being fairly stable (half-life ~7 h at 0°C) unless base is present, in which case almost instantaneous decomposition takes place:

$$2XeF_2 + 2H_2O \longrightarrow 2Xe + 4HF + O_2$$

The aqueous solutions are powerful oxidizing agents, converting $2Cl^-$ to Cl_2, Ce^{III} to Ce^{IV}, Cr^{III} to Cr^{VI}, Ag^I to Ag^{II}, and even BrO_3^- to BrO_4^- (p. 1021).

XeF_4 is best prepared by heating a 1:5 volume mixture of Xe and F_2 to 400°C under 6 atm pressure in a nickel vessel. It also is a white, crystalline, easily sublimed solid; the molecular shape is square planar (Xe–F 195.2 pm) and is essentially the same in both the solid and gaseous phases. Its properties are similar to those of XeF_2 except that it is a rather stronger fluorinating agent, as shown by the reactions:

$$2Hg + XeF_4 \longrightarrow Xe + 2HgF_2$$

$$Pt + XeF_4 \longrightarrow Xe + PtF_4$$

$$2SF_4 + XeF_4 \longrightarrow Xe + 2SF_6$$

It is also hydrolysed instantly by water, yielding a variety of products which include XeO_3:

$$XeF_4 + 2H_2O \longrightarrow \tfrac{1}{3}XeO_3 + \tfrac{2}{3}Xe + \tfrac{1}{2}O_2 + 4HF$$

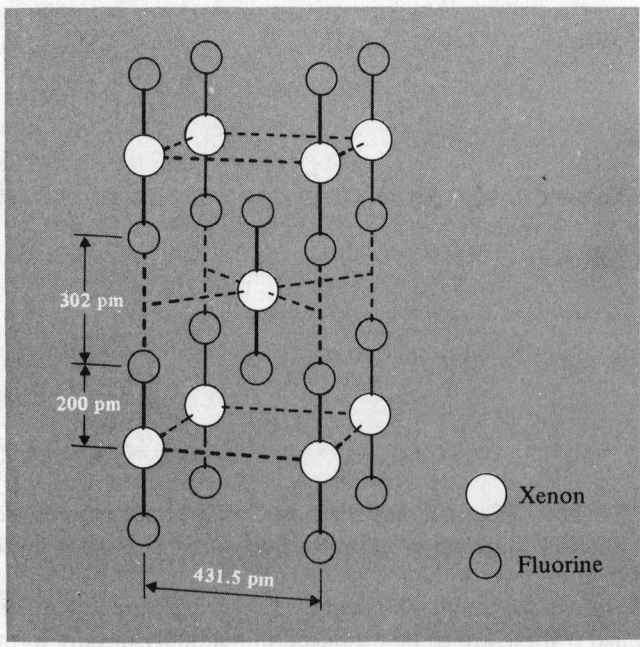

302 pm

200 pm

431.5 pm

Xenon

Fluorine

FIG. 18.1   The unit cell of crystalline XeF_2.

This reaction is indeed a major hazard in Xe/F chemistry, since XeO_3 is highly explosive, and the complete exclusion of moisture is therefore essential (see p. 165 of ref. 6). Interestingly, the maximum yield of XeO_3 is 33% rather than the 50% that would be expected from a simple disproportionation of $2Xe^{IV} \rightarrow Xe^{VI} + Xe^{II}$, and the following reaction sequence has been suggested to explain this:

$$3Xe^{IV}F_4 + 6H_2O \longrightarrow 2\{Xe^{II}O\} + \{Xe^{V\,III}O_4\} + 12HF$$

Decomposition in solution $\longrightarrow 2Xe^0 + O_2$

Decomposition $\longrightarrow Xe^{VI}O_3 + \tfrac{1}{2}O_2$

The stoichiometry of the reaction also depends sensitively on the precise conditions of hydrolysis.[8]

XeF_6 is produced by the prolonged heating of 1:20 volume mixtures of Xe and F_2 at 250–300°C under 50–60 atm pressure in a nickel vessel. It is a crystalline solid, even more volatile than XeF_2 and XeF_4, and although colourless in the solid it is yellow in the liquid and gaseous phases. It is also more reactive than the other fluorides, being both a stronger oxidizing and a stronger fluorinating agent. Hydrolysis occurs with great vigour and the compound cannot be handled in glass or quartz apparatus because of a stepwise reaction which finally produces the dangerous XeO_3:

$$2XeF_6 + SiO_2 \longrightarrow 2XeOF_4 + SiF_4$$
$$2XeOF_4 + SiO_2 \longrightarrow 2XeO_2F_2 + SiF_4$$
$$2XeO_2F_2 + SiO_2 \longrightarrow 2XeO_3 + SiF_4$$

The structure of XeF_6 has been the source of continuing controversy since its discovery in 1963. This is partly a result of the obvious problems associated with a substance which attacks most of the materials used to construct apparatus for structural determinations. It is now clear that in the gaseous phase this seemingly simple molecule is not a regular octahedron; it appears to be a non-rigid, distorted octahedron although, in spite of numerous theoretical studies, the precise nature of the distortion is uncertain (see, for instance, p. 299 of ref. 5). In the crystalline state at least four different forms of XeF_6 are known comprising square-pyramidal XeF_5^+ ions bridged by F^- ions. Three of these forms are tetramers, $[(XeF_5^+)F^-]_4$, while in the fourth and best-characterized cubic form,[8a] the unit cell comprises 24 tetramers and 8 hexamers, $[(XeF_5^+)F^-]_6$ (Fig. 18.2).

The nature of the bonding in these xenon fluorides is discussed in the Panel on p. 1053.

Apart from XeF, which has been produced as an unstable free radical by γ-irradiation of XeF_4 at 77 K†, there is no evidence for the existence of any odd-valent fluorides. Reports of XeF_8 have not been confirmed. Of the other halides, $XeCl_2$, $XeBr_2$, and $XeCl_4$ have been

---

† Excited XeF is the light-emitting species in certain lasers where it is produced by the action of a beam of 400 keV electrons on Xe/F_2 mixtures.

[8] J. L. HUSTON, Chemical and physical properties of some xenon compounds, *Inorg. Chem.* **21**, 685–8 (1982).
[8a] R. D. BURBANK and G. R. JONES, Structure of the cubic phase of xenon hexafluoride at 193 K, *J. Am. Chem. Soc.* **96**, 43–48 (1974).

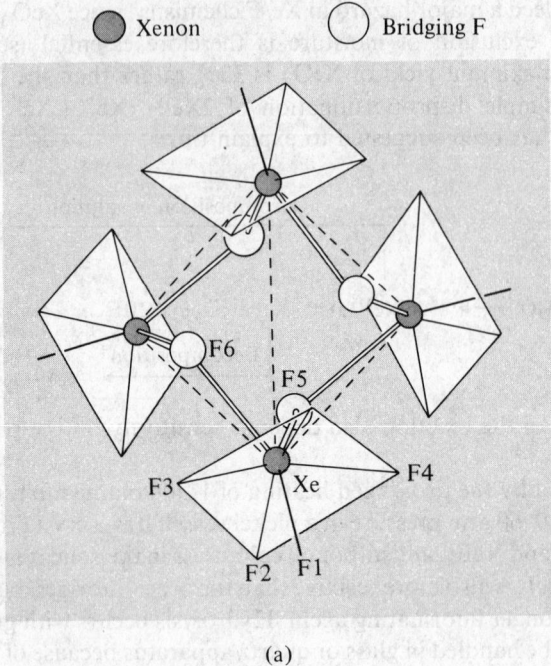

(a)

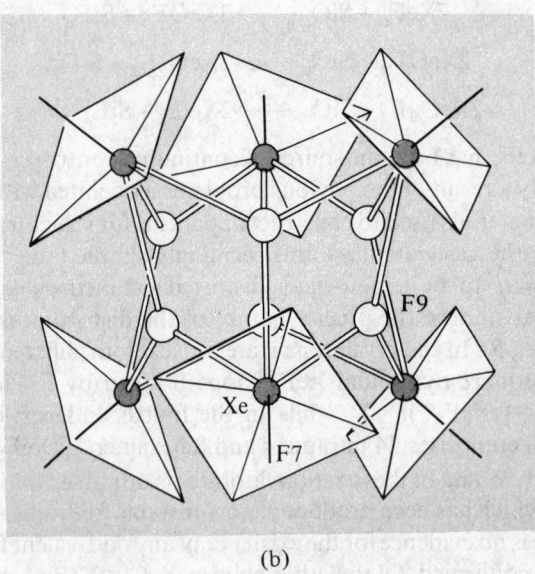

(b)

Fig. 18.2　(a) Tetrameric, and (b) hexameric units in the cubic crystalline form of XeF_6. In (a) the Xe atoms form a tetrahedron, with the apical F atoms of the square-pyramidal XeF_5^+ ions pointing outwards, approximately from the centre, and the bridging F^- ions near four of the six edges of the tetrahedron: Xe–F(1–5) 184 pm, Xe–F(6), 223 pm and 260 pm, angle Xe–F(6)–Xe 120.7°. In (b) the Xe atoms form an octahedron with the apical F atoms of the XeF_5^+ ions pointing outwards from the centre, and the briding F^- ions over six of the eight faces of the octahedron: Xe–F(7) 175 pm, Xe–F(8) 188 pm, Xe–F(9) 256 pm, angle Xe–F(9)–Xe 118.8°. XeF_5^+ ions are shown in skeletal form for clarity.

Bonding in Noble Gas Compounds

As it was widely believed, prior to 1962, that the noble gases were chemically inert because of the stability, if not inviolability, of their electronic configurations, the discovery that compounds could in fact be prepared, immediately necessitated a description of the bonding involved. A variety of approaches has been suggested,[9] none of which is universally applicable. The simplest molecular-orbital description is that of the 3-centre, 4-electron σ bond in XeF_2, which involves only valence shell p orbitals and eschews the use of higher energy d orbitals. The orbitals involved are the colinear set comprising the $5p_x$ orbital of Xe, which contains 2 electrons, and the $2p_x$ orbitals from each of the F atoms, each containing 1 electron. The possible combinations of these orbitals are shown in Fig. A and yield 1 bonding, 1 nonbonding, and 1 antibonding orbital. A single bonding pair of electrons is responsible for binding all 3 atoms, and the occupation of the nonbonding orbital, situated largely on the F atoms, implies significant ionic character. The scheme should be compared with the 3-centre, 2-electron bonding proposed for boron hydrides (p. 180).

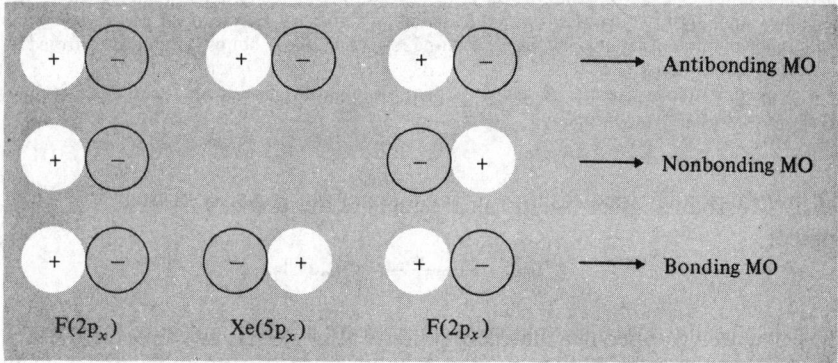

(a)

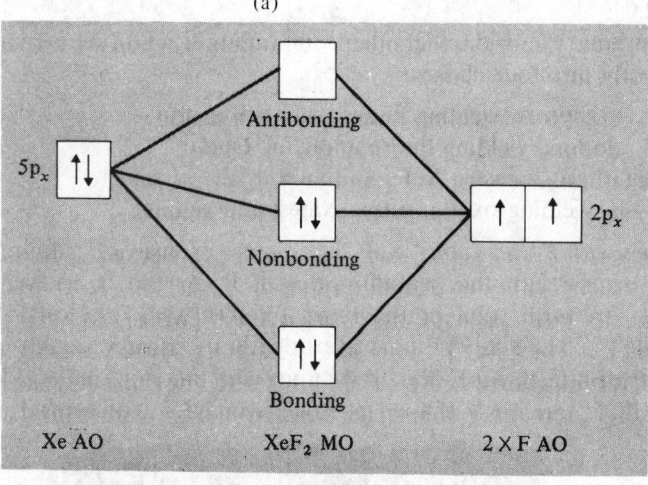

(b)

Fɪɢ. A   Molecular–orbital representation of the 3-centre F–Xe–F bond. (a) The possible combinations of colinear p_x atomic orbitals, and (b) the energies of the resulting MOs.

Continued

[9] C. A. Cᴏᴜʟsᴏɴ, The nature of the bonding in xenon fluorides and related molecules, *J. Chem. Soc.* 1964, 1442–54. J. G. Mᴀʟᴍ, H. Sᴇʟɪɢ, J. Jᴏʀᴛɴᴇʀ and S. A. Rɪᴄᴇ, The chemistry of xenon, *Chem. Revs.* **65**, 199–236 (1965).

A similar treatment, involving two 3-centre bonds accounts satisfactorily for the planar structure of XeF_4 but fails when applied to XeF_6 since three 3-centre bonds would produce a regular octahedron instead of the distorted structure actually found. An improvement is possible if involvement of the Xe 5d orbitals is invoked,[10] since this produces a triplet level which would be subject to a Jahn–Teller distortion (p. 1087). However, the approach which has most consistently rationalized the stereochemistries of noble-gas compounds, and provides the most readily visualized description of their shapes (as distinct from their bonding) is the electron-pair repulsion theory of Gillespie and Nyholm.[11] This assumes that stereochemistry is determined by the repulsions between valence-shell electron-pairs, both nonbonding and bonding, and that the former exert the stronger effect. Thus, in XeF_2 the Xe is surrounded by 10 electrons (8 from Xe and 1 from each F) distributed in 5 pairs; 2 bonding and 3 nonbonding. The 5 pairs are directed to the corners of a trigonal bipyramid and, because of their greater mutual repulsions, the 3 nonbonding pairs are situated in the equatorial plane at 120° to each other, leaving the 2 bonding pairs perpendicular to the plane and so producing a linear F–Xe–F molecule.

In the same way XeF_4, with 6 electron-pairs, is considered as pseudo-octahedral with its 2 nonbonding pairs trans to each other, leaving the 4 F atoms in a plane around the Xe. More distinctively, the 7 electron-pairs of XeF_6 suggest the possibility of a non-regular octahedral geometry and imply a distorted structure based on either a monocapped octahedral or a pentagonal pyramidal arrangement of electron-pairs, with the Xe-F bonds bending away from the projecting nonbonding pair.

It is an instructive exercise to devise similar rationalizations for the xenon oxides and oxofluorides listed in Table 18.3.

detected by Mössbauer spectroscopy as products of the β-decay of their $^{129}_{53}I$ analogues, for instance:

$$^{129}_{53}ICl_2{}^- \xrightarrow{\ -\beta^-\ } {}^{129}_{54}XeCl_2$$

$XeCl_2$ has also been trapped in a matrix of solid Xe after Xe/Cl_2 mixtures had been passed through a microwave discharge, but these halides are too unstable to be chemically characterized.

It is from the binary fluorides that other compounds of xenon are prepared, by reactions which fall mostly into four classes:

(a) with F^- acceptors, yielding fluorocations of xenon;
(b) with F^- donors, yielding fluoroanions of xenon;
(c) F/H metathesis between XeF_2 and an anhydrous acid;
(d) hydrolysis, yielding oxofluorides, oxides, and xenates.

(a) *Reactions with F^- acceptors.* XeF_2 has a more extensive F^- donor chemistry than has XeF_4; it reacts with the pentafluorides of P, As, Sb, I, as well as with metal pentafluorides, to form salts of the types $[XeF]^+[MF_6]^-$, $[XeF]^+[M_2F_{11}]^-$, and $[Xe_2F_3]^+[MF_6]^-$. The $[XeF]^+$ ions are apparently always weakly attached to the counter-anion forming linear F–Xe···F–M units with one short and one long Xe–F bond, while the $[Xe_2F_3]^+$ ions are V-shaped (cf. isoelectronic I_5^- with central angle 95°, p. 980).

[10] G. L. GOODMAN, Electronic states and molecular geometry of xenon hexafluoride: case of electronic isomerism, *J. Chem. Phys.* **56,** 5038–41 (1972).
[11] R. J. GILLESPIE, *Molecular Geometry*, van Nostrand Rheinhold, London, 1972, 228 pp.

With SbF_5 the bright-green paramagnetic Xe_2^+ cation has recently been identified as a further product.[12] MOF_4 (M = W, Mo) are also weak F acceptors and form $[XeF]^+[MOF_5]^-$, which again contain linear F–Xe$\cdots$F–M units.[13]

Although the orange-yellow solid prepared by Bartlett (p. 1047) was originally formulated as $Xe^+[PtF_6]^-$, it was subsequently found to have the variable composition $Xe(PtF_6)_x$, x lying between 1 and 2. The material has still not been fully characterized but probably contains both $[XeF]^+[PtF_6]^-$ and $[XeF]^+[Pt_2F_{11}]^-$.

XeF_4 forms comparable complexes only with the strongest F^- acceptors such as SbF_5 and BiF_5, but XeF_6 combines with a variety of pentafluorides to yield 1:1 adducts. In view of the structure of XeF_6 (see Fig. 18.2) it is not surprising that these adducts contain XeF_5^+ cations, as for instance in $[XeF_5]^+[AsF_6]^-$ and $[XeF_5]^+[PtF_6]^-$. In a similar manner, reactions with FeF_3 and CoF_3 yield $[XeF_5][MF_4]$ in which layers of corner-sharing FeF_6 octahedra are separated by $[XeF_5]^+$ ions.[14]

(b) *Reactions with F^- donors.* F^- acceptor behaviour of xenon fluorides is evidently confined to XeF_6 which reacts with alkali metal fluorides to form $MXeF_7$ (M = Rb, Cs) and M_2XeF_8 (M = Na, K, Rb, Cs). These compounds lose XeF_6 when heated:

$$2MXeF_7 \longrightarrow M_2XeF_8 + XeF_6$$

$$M_2XeF_8 \longrightarrow 2MF + XeF_6$$

Their thermal stability increases with molecular weight. Thus the Cs and Rb octafluoro complexes only decompose above 400°C, whereas the Na complex decomposes below 100°C. NaF can therefore conveniently be used to separate XeF_6 from XeF_2 and XeF_4, with which it does not react, the purified XeF_6 being regenerated on heating.

A similar product, $[NO]^+{}_2[XeF_8]^{2-}$, is formed with NOF and its anion has been shown by X-ray crystallography to be a slightly distorted square antiprism[15] (probably due to weak $F\cdots NO^+$ interactions). The absence of any clearly defined ninth coordination position for the lone-pair of valence electrons which is present, implies that this must be stereochemically inactive.

(c) *F/H metathesis between XeF_2 and an anhydrous acid:*

$$XeF_2 + HL \longrightarrow F\text{–}Xe\text{–}L + HF$$

where L = OSO_2F, $OClO_3$, $OTeF_5$, or $OC(O)CF_3$ (see also p. 1058 for an analogous reaction with $HN(SO_2F)_2$). The fluorosulfate and perchlorate are colourless and the others are pale yellow. Many of these compounds are thermodynamically unstable and the perchlorate (mp 16.5°) is dangerously explosive. The fluorosulfate (mp 36.6°) can be stored for many weeks at 0° but decomposes with a half-life of a few days at 20°:

$$2FXeOSO_2F \longrightarrow XeF_2 + Xe + S_2O_6F_2$$

The molecular structure of $FXeOSO_2F$ is in Fig. 18.3.

[12] L. STEIN and W. H. HENDERSON, Production of dixenon cation by reversible oxidation of xenon, *J. Am. Chem. Soc.* **102**, 2856–7 (1980).

[13] J. H. HOLLOWAY and G. J. SCHROBILGEN, Fluorine-19 and xenon-129 NMR studies of the $XeF_2.nWOF_4$ and $XeF_2.nMoOF_4$ (n = 1–4) adducts: examples of nonlabile xenon-fluorine-metal bridges in solution, *Inorg. Chem.* **19**, 2632–40 (1980).

[14] J. SLIVNIK, B. ZEMVA, M. BOHINC, D. HANZEL, J. GRANNEC and P. HAGENMULLER, On the synthesis and some properties of xenon(VI) fluoroferrate(III) and fluorocobaltate(III), *J. Inorg. Nucl. Chem.* **38**, 997–1000 (1976).

[15] S. W. PETERSON, J. H. HOLLOWAY, B. A. COYLE and J. M. WILLIAMS, Antiprismatic coordination about xenon. Structure of nitrosonium octafluoroxenate(VI), *Science*, **173**, 1238–9 (1971).

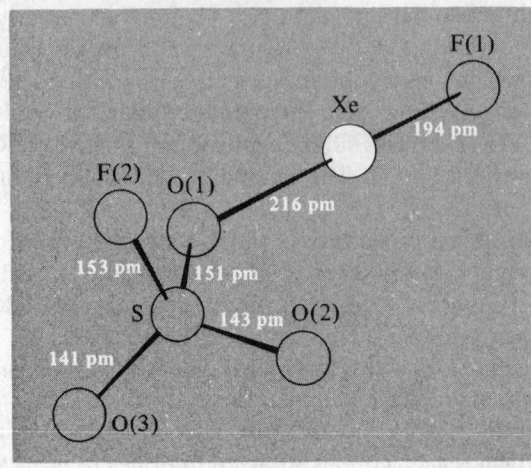

FIG. 18.3   The molecular structure of FXeOSO$_2$F. Precision of bond lengths is *ca.* 1 pm (uncorrected for thermal motion). The angles F(1)–Xe–O(1) and Xe–O(1)–S are 177.5$\pm$0.4° and 123.4$\pm$0.6°.

(d) *Hydrolysis and related reactions.* Two Xe^{VI} oxofluorides, XeOF$_4$ and XeO$_2$F$_2$, have been characterized, and the Xe^{VIII} derivative XeO$_3$F$_2$ is also known (see below). XeOF$_4$ is a colourless volatile liquid with a square-pyramidal molecular structure, the O atom being at the apex. It can be prepared by the controlled hydrolysis of XeF$_6$:

$$XeF_6 + H_2O \longrightarrow XeOF_4 + 2HF$$

Its most pronounced chemical characteristic is its propensity to hydrolyse further to XeO$_2$F$_2$ and then XeO$_3$ (p. 1057). This reaction is difficult to control, and the low-melting, colourless solid, XeO$_2$F$_2$, is more reliably obtained by the reaction:

$$XeO_3 + XeOF_4 \longrightarrow 2XeO_2F_2$$

An analogous reaction with XeO$_4$ (see p. 1057) is:

$$XeO_4 + XeF_6 \longrightarrow XeO_3F_2 + XeOF_4$$

Indeed, many of the reactions of the xenon oxides, fluorides, and oxofluorides can be systematized in terms of generalized acid–base theory in which any acid (here defined as an oxide acceptor) can react with any base (oxide donor) lying beneath it in the sequence of descending acidity: XeF$_6$ > XeO$_2$F$_4$ > XeO$_3$F$_2$ > XeO$_4$ > XeOF$_4$ > XeF$_4$ > XeO$_2$F$_2$ > XeO$_3 \approx$ XeF$_2$.[8]

In addition, oxofluoro anions may be produced by treating hydrolysis products with F$^-$. Thus aqueous XeO$_3$ and MF (M = K, Cs) yield the stable white solids M[XeO$_3$F] in which the anion consists of chains of pseudo-octahedral Xe atoms (the lone-pair of valence electrons occupying one of the six positions) linked by angular F bridges. More recently,[16] the reaction of XeOF$_4$ and dry CsF has been shown to yield the labile Cs[(XeOF$_4$)$_3$F] in which the anion consists of three equivalent XeOF$_4$ groups attached to

---

[16] G. J. SCHROBILGEN, D. M. ROVET, P. CHARPIN, and M. LANCE, XeOF$_5^-$ and [(XeOF$_4$)$_3$F]$^-$ Anions, *JCS Chem. Comm.* 894–7 (1980).

a central F^- ion. The ready loss of $2XeOF_4$ produces the more stable $CsXeOF_5$, the anion of which has a distorted octahedral geometry, the lone-pair of electrons again being stereochemically active and apparently occupying an octahedral face.

Complete hydrolysis of XeF_6 is the route to XeO_3. The most effective control of this potentially violent reaction is achieved by using a current of dry N_2 to sweep XeF_6 vapour into water:[17]

$$XeF_6 + 3H_2O \longrightarrow XeO_3 + 6HF$$

The HF may then be removed by adding MgO to precipitate MgF_2 and the colourless deliquescent solid XeO_3 obtained by evaporation. The aqueous solution known as "xenic acid" is quite stable if all oxidizable material is excluded, but the solid is a most dangerous explosive (reported to be comparable to TNT) which is easily detonated. The X-ray analysis, made even more difficult by the tendency of the crystals to disintegrate in an X-ray beam, shows the solid to consist of trigonal pyramidal XeO_3 units, with the xenon atom at the apex[18] (cf. the isoelectronic iodate ion IO_3^-; p. 1010).

In aqueous solution XeO_3 is an extremely strong oxidizing agent (for $XeO_3 + 6H^+ + 6e^- \rightleftharpoons Xe + 3H_2O$; $E° = 2.10$ V), but may be kinetically slow; the oxidation of Mn^{II} takes hours to produce MnO_2 and days before MnO_4^- is obtained. Treatment of aqueous XeO_3 with alkali produces xenate ions:

$$XeO_3 + OH^- \rightleftharpoons HXeO_4^-; \quad K = 1.5 \times 10^3$$

However, although some salts have been isolated, alkaline solutions are not stable and immediately, if slowly, begin to disproportionate into Xe^{VIII} (perxenates) and Xe gas by routes such as:

$$2HXeO_4^- + 2OH^- \longrightarrow XeO_6^{4-} + Xe + O_2 + 2H_2O$$

Similar results are obtained by the alkaline hydrolysis of XeF_6:

$$2XeF_6 + 16OH^- \longrightarrow XeO_6^{4-} + Xe + O_2 + 12F^- + 8H_2O$$

The most efficient production of perxenate is the treatment of XeO_3 in aqueous NaOH with ozone, when $Na_4XeO_6.2\frac{1}{3}H_2O$ precipitates almost quantitatively. The crystal structures of $Na_4XeO_6.6H_2O$ and $Na_4XeO_6.8H_2O$ show them to contain octahedral XeO_6^{4-} units with Xe–O 184 pm and 186.4 pm respectively. Perxenates of other alkali metals (Li^+, K^+) and of several divalent and trivalent cations (e.g. Ba^{2+}, Am^{3+}) have also been prepared. They are colourless solids, thermally stable to over 200°C, and contain octahedral XeO_6^{4-} ions. They are powerful oxidizing agents, the reduction of Xe^{VIII} to Xe^{VI} in aqueous acid solution being very rapid. The oxidation of Mn^{II} to MnO_4^- by perxenates, unlike that by XeO_3, is thus immediate and is accompanied by evolution of O_2:

$$2H_2XeO_6^{2-} + 2H^+ \longrightarrow 2HXeO_4^- + O_2 + 2H_2O$$

The addition of solid Ba_2XeO_6 to cold conc H_2SO_4 produces the second known oxide of xenon, XeO_4. This is an explosively unstable gas which may be condensed in a liquid

[17] B. Jaselskis, T. M. Spittler, and J. L. Huston, Preparation and properties of monocesium xenate, *J. Am. Chem. Soc.* **88**, 2149–50 (1966).

[18] D. H. Templeton, A. Zalkin, J. D. Forrester and S. M. Williamson, Crystal and molecular structure of xenon trioxide, *J. Am. Chem. Soc.* **85**, 817 (1963).

nitrogen trap. The solid tends to detonate when melted but small sublimed crystals have been shown to melt sharply at $-35.9°C$.[8] XeO_4 has only been incompletely studied, but electron diffraction and infrared evidence show the molecule to be tetrahedral.

Whilst the great bulk of noble-gas chemistry concerns Xe–F or Xe–O bonds, attempts to bond Xe to other atoms have not been entirely unsuccessful. Compounds containing Xe–N bonds have been produced by the replacement of F atoms by $-N(SO_2F)_2$ groups.[19] The relevant reactions may be represented as:

$$2XeF_2 + 2HN(SO_2F)_2 \xrightarrow[\text{0°C, 4 days}]{\text{in } CF_2Cl_2} 2F\text{–}Xe\text{–}N(SO_2F)_2$$
white solid

$$\downarrow {\scriptstyle -10\,C \quad 2AsF_5}$$

$$[F\{Xe\text{–}N(SO_2F)_2\}_2]^+ AsF_6^- \xleftarrow[\text{22 C, vacuum}]{-AsF_5} 2[FXe(NSO_2F)_2 \cdot AsF_5]$$
pale yellow solid   unstable, bright yellow solid

The first (white) product has been characterized by X-ray diffraction at $-55°$ and features a linear F–Xe–N group and a planar N atom (Fig. 18.4).[20] On the basis of Raman and ^{19}F nmr data, the cation of the final (pale yellow) product is believed to be essentially like the V-shaped $[Xe_2F_3]^+$ cation but with the 2 terminal F atoms replaced by $-N(SO_2F)_2$ groups.[19]

The Xe–C bond too has apparently been synthesized recently: reaction of XeF_2 vapour

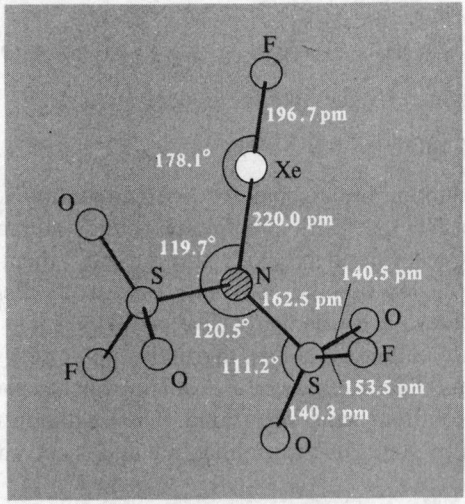

FIG. 18.4   The structure of $FXeN(SO_2F)_2$ (point group symmetry C_2) showing essentially linear Xe and planar N. Other bond angles are OSO 122.6°, OSF 106.3°, NSO 107.2° and 111.2°, NSF 101.2°.

[19] D. D. DesMarteau, An unusual xenon cation containing xenon–nitrogen bonds, *J. Am. Chem. Soc.* **100**, 6270–1 (1978). D. D. DesMarteau, R. D. LeBlond, S. F. Hossain, and D. Nothe, *J. Am. Chem. Soc.* **103**, 7734–9 (1981).
[20] J. F. Sawyer, G. J. Schrobilgen, and S. J. Sutherland, X-ray crystal and multinuclear NMR study of $FXeN(SO_2F)_2$; the first example of a xenon–nitrogen bond, *JCS Chem. Comm.* 1982, 210–11.

with $CF_3\cdot$ radicals yields a volatile, waxy, white solid thought to be $Xe(CF_3)_2$ with a half-life at room temperature of about 30 min. [21]

$$XeF_2 + C_2F_6 \xrightarrow{\text{plasma}} Xe(CF_3)_2 + F_2$$

$$Xe(CF_3)_2 \xrightarrow[t_{\frac{1}{2}} \, 30 \, \text{min}]{20°} XeF_2 + C_nF_m$$

18.3.3 *Compounds of other noble gases*

No stable compounds of He, Ne, or Ar are known. Radon apparently forms a difluoride and some complexes, but the evidence is based solely on radiochemical tracer techniques since Rn has no stable isotopes. The remaining noble gas, Kr, has an emerging chemistry though this is far less extensive than that of Xe.

Apart from the violet free radical KrF, which has been generated in minute amounts by γ-radiation of KrF_2 and exists only below $-153°C$, the chemistry of Kr is confined to the difluoride and its derivatives. An early claim for KrF_4 remains unsubstantiated. The volatile, colourless solid, KrF_2, is produced when mixtures of Kr and F_2 are cooled to temperatures near $-196°C$ and then subjected to electric discharge, or irradiated with high-energy electrons or X-rays. It is a thermally (and thermodynamically) unstable compound which slowly decomposes even at room temperature. It has the same linear molecular structure as XeF_2 (Kr–F 188.9 pm) but, consistent with its lower stability, is a stronger fluorinating agent and is rapidly decomposed by water without requiring the addition of a base.

Complexes of KrF_2 are analogous to those of XeF_2 and are confined to cationic species formed with F^- acceptors. Thus, such compounds as $[KrF]^+[MF_6]^-$, $[Kr_2F_3]^+[MF_6]^+$ (M = As, Sb) are known, and more recently $[KrF]^+[MoOF_5]^-$ and $[KrF]^+[WOF_5]^-$ have been prepared and characterized by ^{19}F nmr and Raman spectroscopy.[22]

[21] L. J. TURBINI, R. E. AIKMAN and R. J. LAGOW, Evidence for the synthesis of a "stable" σ-bonded xenon–carbon compound: bis(trifluoromethyl)xenon, *J. Am. Chem. Soc.* **101**, 5833–4 (1979).

[22] J. H. HOLLOWAY and G. J. SCHROBILGEN, Preparation and study by Raman spectroscopy of $KrF_2.MOF_4$, $XeF_2.MOF_4$ and $XeF_2.2MOF_4$ (M = Mo, W) and a solution ^{19}F NMR study of $KrF_2.nMoOF_2$ (n = 1–3) and $KrF_2.WOF_4$, *Inorg. Chem.* **20**, 3363–8 (1981).

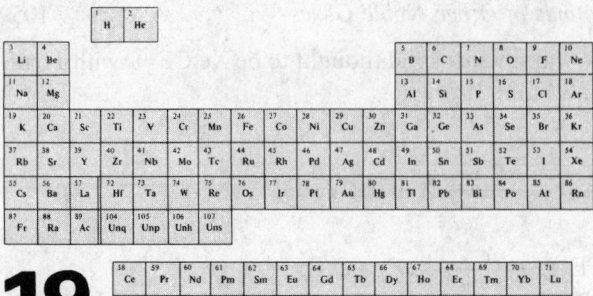

19

Coordination Compounds

19.1 Introduction

The three series of elements arising from the filling of the 3d, 4d, and 5d shells, and situated in the periodic table following the alkaline earth metals, are commonly described as "transition elements", though this term is sometimes also extended to include the lanthanide and actinide (or inner transition) elements. They exhibit a number of characteristic properties which together distinguish them from other groups of elements:

(i) They are all metals and as such are lustrous and deformable and have high electrical and thermal conductivities. In addition, their melting and boiling points tend to be high and they are generally hard and strong.

(ii) Most of them display numerous oxidation states which vary by steps of 1 rather than 2 as is usually the case with those main-group elements which exhibit more than one oxidation state.

(iii) They have an unparalleled propensity for forming coordination compounds with Lewis bases.

(i) and (ii) will be dealt with more fully in later chapters but it is the purpose of the present chapter to expand the theme of (iii).

A coordination compound, or complex, is formed when a Lewis base (ligand)[1] is attached to a Lewis acid (acceptor) by means of a "lone-pair" of electrons. Where the ligand is composed of a number of atoms, the one which is directly attached to the acceptor is called the "donor atom". This type of bonding has already been discussed (p. 223) and is exemplified by the addition compounds formed by the trihalides of the elements of Group IIIB (p. 267); it is also the basis of much of the chemistry of the transition elements. The precise nature of the bond between a transition metal ion and a ligand varies enormously and, though inevitably the line of demarcation is rather ill-defined, it is conventional to distinguish two extremes. On the one hand, are those cases in which the bond may be considered profitably as a single σ bond and in which the metal has an oxidation number of $+2$ or higher. On the other hand, are those cases where the

---

[1] W. H. BROCK, K. A. JENSEN, C. K. JØRGENSEN, and G. B. KAUFFMAN, The origin and dissemination of the term "ligand" in Chemistry, *Ambix* **27**, 171–83 (1981).

bonding is multiple, the ligand acting simultaneously as both a σ donor and a π acceptor (p. 1100). This situation is typified by the carbonyls (p. 349) and other organometallic compounds (p. 345) and often involves formal metal oxidation numbers of +1 or less, though the significance of these values is often unclear. It is convenient to restrict the term "complex" to the former type of compound, and it was through the investigation of such materials that A. Werner in the period 1893–1913 laid the foundations of coordination chemistry[1a] (see also p. 1068).

19.2   Types of Ligand

Ligands are most conveniently classified according to the number of donor atoms which they contain and are known as uni-, bi-, ter-, quadri-, quinqi-, and sexa-dentate accordingly as the number is 1, 2, 3, 4, 5, or 6. Unidentate ligands may be simple monatomic ions such as halide ions, or polyatomic ions or molecules which contain a donor atom from Groups VI or V, or even IV (e.g. CN^-). Bidentate ligands are frequently chelating ligands (from Greek χηλή, crab's claw) and, with the metal ion, produce chelate rings[1b] which in the case of the most commonly occurring bidentate ligands are 5- or 6-membered, e.g.: see p. 1062. Terdentate ligands produce 2 ring systems when coordinated to a single metal ion and in consequence may impose structural limitations on the complex, particularly where rigidity is introduced by the incorporation of conjugated double bonds within the rings. Thus diethylenetriamine, dien (1), being flexible is stereochemically relatively undemanding, whereas terpyridine, terpy (2), can only coordinate when the 3 donor nitrogen atoms and the metal ion are in the same plane.

| (1) | (2) | (3) | (4) |

Quadridentate ligands produce 3, and in some cases 4, rings on coordination, and so even greater restrictions on the stereochemistry of the complex may be imposed by an appropriate choice of ligand. The open-chain ligand triethylenetetramine, trien (3), is, like dien, flexible and undemanding, whereas triethylaminetriamine, tren, i.e. $N(CH_2CH_2NH_2)_3$, is one of the so-called "tripod" ligands which are quite unable to give planar coordination but instead favour trigonal bipyramidal structures (4). By contrast, the highly conjugated phthalocyanine (5), see p. 1063, which is an example of the class of macrocyclic ligands of which the crown ethers have already been mentioned (p. 107), forces the complex to adopt a virtually planar structure and has proved to be a valuable model for the naturally occurring porphyrins which, for instance, are involved in haem (p. 1277), B_{12} (p. 1322), and the chlorophylls (p. 139). Another well-known ligand, which

[1a] G. B. KAUFFMAN, *Alfred Werner Founder of Coordination Theory*, Springer, Berlin, 1966, 127 pp.

[1b] C. F. BELL, *Principles and Applications of Metal Chelation*, Oxford University Press, Oxford, 1977, 147 pp.

[1c] G. A. MELSON (ed.), *Coordination Chemistry of Macrocyclic Compounds*, Plenum Press, New York, 1979, 664 pp.

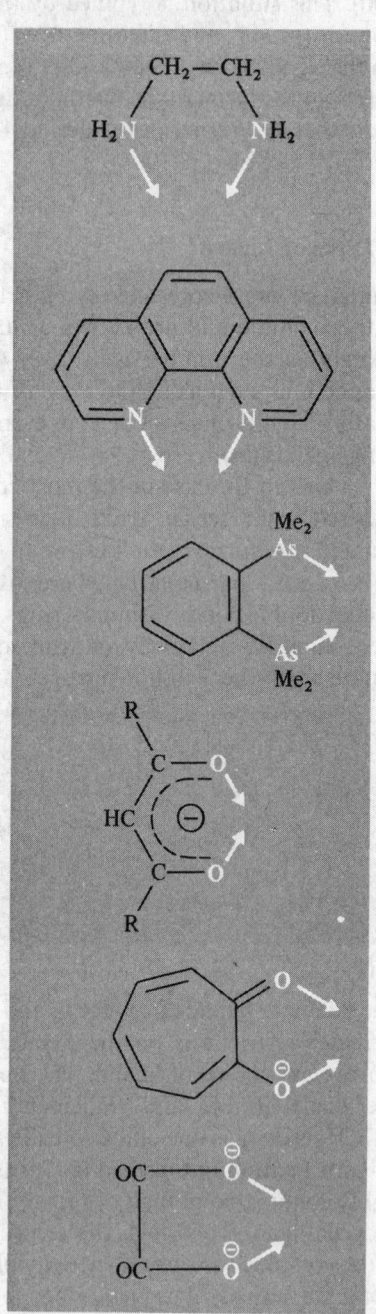

Ethylenediamine, en:

1,10-phenanthroline, phen:

o-phenylenebis(dimethylarsine), diars:
[1,2-bis(dimethylarsino)benzene]

β-diketonates (e.g. R = Me:
acetylacetonate, acac)

Tropolonate:

Oxalate:

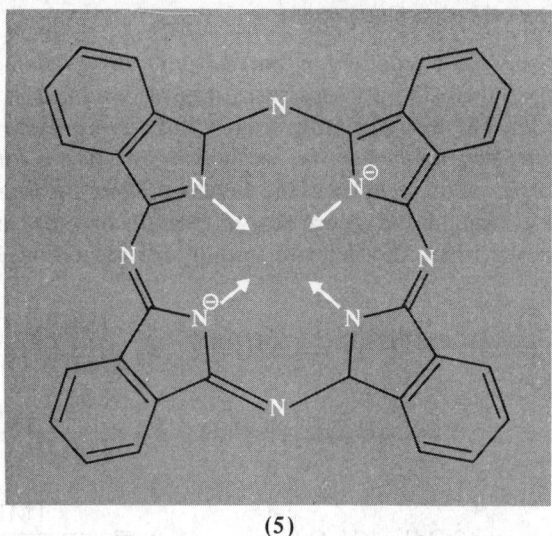

(5)

has been used to synthesize oxygen-carrying molecules, is bis(salicylaldehyde)ethylene-diimine, salen (6). Quinquidentate and sexadentate ligands are most familiarly exemplified by the anions derived from ethylenediaminetetraacetic acid, $EDTAH_4$ i.e.:

$$(HO_2CCH_2)_2N(CH_2)_2N(CH_2CO_2H)_2$$

which is used with remarkable versatility in the volumetric analysis of metal ions. As the fully ionized anion, $EDTA^{4-}$, it has 4 oxygen and 2 nitrogen donor atoms and has the flexibility to wrap itself around a variety of metal ions to produce a pseudo-octahedral complex involving five 5-membered rings as in (7):

(6)   (7)

In the incompletely ionized form, $EDTAH^{3-}$, one of the oxygen atoms is no longer able to coordinate to the metal and the anion is quinquidentate.

Ambidentate ligands possess more than 1 donor atom and can coordinate through either one or the other. This leads to the possibility of "linkage" isomerism (p. 1081). The commonest examples are the ions NO_2^- (p. 534) and SCN^- (p. 342). Such ligands can also coordinate via both donor sites simultaneously, thereby acting as bridging ligands.

19.3 Stability of Coordination Compounds

Because complexes are not generally prepared from their components in the gaseous phase, measurements of their stability necessarily imply a comparison with the stability of some starting material. The overwhelming majority of quantitative measurements have been made in aqueous solutions when the complex in question is formed by the ligand displacing water from the aquo complex of the metal ion. If, for simplicity, we take the case where L is a unidentate ligand and ignore charge, then the process can be represented as a succession of steps for which the stepwise stability (or formation) constants K are as shown:[†]

$$M(H_2O)_n + L \rightleftharpoons ML(H_2O)_{n-1} + H_2O; \quad K_1 = \frac{[ML(H_2O)_{n-1}]}{[M(H_2O)_n][L]}$$

$$ML(H_2O)_{n-1} + L \rightleftharpoons ML_2(H_2O)_{n-2} + H_2O; \quad K_2 = \frac{[ML_2(H_2O)_{n-2}]}{[ML(H_2O)_{n-1}][L]}$$

$$ML_{n-1}H_2O + L \rightleftharpoons ML_n + H_2O; \quad K_n = \frac{[ML_n]}{[ML_{n-1}H_2O][L]}$$

$$M(H_2O)_n + nL \rightleftharpoons ML_n + nH_2O; \quad \beta_n = \frac{[ML_n]}{[M(H_2O)_n][L]^n}$$

By convention the displaced water is ignored since its concentration can be assumed to be constant. The overall stability (or formation) constant β_n can clearly be expressed in terms of the stepwise constants:

$$\beta_n = K_1 \times K_2 \times \ldots \times K_n$$

These are thermodynamic constants which relate to the system when it has reached equilibrium, and must be distinguished from any considerations of kinetic lability or inertness which refer to the speed with which that equilibrium is attained.

A vast amount of data[(2)] has been accumulated from which a number of generalizations can be inferred concerning the factors which determine the stabilities of such complexes. Some of these are as follows:

(i) *The metal ion and its charge.* For a given metal and ligand the stability is generally greater if the oxidation state of the metal is $+3$ rather than $+2$. Furthermore, the stabilities of corresponding complexes of the bivalent ions of the first transition series,

---

† These constants are expressed here in terms of concentrations which means that the activity coefficients have been assumed to be unity. When pure water is the solvent this will only be true at infinite dilution, and so stability constants should be obtained by taking measurements over a range of concentrations and extrapolating to zero concentration. In practice, however, it is more usual to make measurements in the presence of a relatively high concentration of an inert electrolyte (e.g. 3 M NaClO$_4$) so as to maintain a constant ionic strength, thereby ensuring that the activity coefficients remain essentially constant. Stability constants obtained in this way (sometimes referred to as "concentration quotients" or "stoichiometric stability constants") are true thermodynamic stability constants referred to the standard state of solution in 3 M NaClO$_4$(aq), but they will, of course, differ from stability constants referred to solution in the pure solvent as standard state.

² L. G. SILLÉN and A. E. MARTELL, *Stability Constants of Metal-ion Complexes*, The Chemical Society, London, Special Publications No. 17, 1964, 754 pp., and No. 25, 1971, 865 pp. *Stability Constants of Metal-Ion Complexes, Part A. Inorganic Ligands* (E. Högfeldt ed.), 1982, pp. 310, *Part B. Organic Ligands* (D. Perrin ed.), 1979, pp. 1263. Pergamon Press, Oxford.

irrespective of the particular ligand involved, usually vary in the Irving–Williams[3] order (1953):

$$Mn^{II} < Fe^{II} < Co^{II} < Ni^{II} < Cu^{II} > Zn^{II}$$

which is the reverse of the order for the cation radii (p. 1497). These observations are consistent with the view that, at least for metals in oxidation states $+2$ and $+3$, the coordinate bond is largely electrostatic. This was a major factor in the acceptance of crystal field theory which will be discussed shortly (p. 1084).

(ii) *The relationship between metal and donor atom.* Some metal ions (known as class-a acceptors or alternatively as "hard" acids) form their most stable complexes with ligands containing N, O, or F donor atoms. Others (known as class-b acceptors or alternatively as "soft" acids) form their most stable complexes with ligands whose donor atoms are the heavier elements of the N, O, or F groups. The metals of Groups IA and IIA along with the inner transition elements and the early members of the transition series (Groups IIIA→VA) fall into class-a. The transition elements Rh, Pd, Ag, and Ir, Pt, Au, Hg comprise class-b, while the remaining transition elements may be regarded as borderline (Fig. 19.1). The difference between the class-a elements of Group IIA and the borderline class-b elements of Group IIB is elegantly and colourfully illustrated by the equilibrium

$$[CoCl_4]^{2-} + 6H_2O \underset{Ca^{II}}{\overset{Zn^{II}}{\rightleftharpoons}} [Co(H_2O)_6]^{2+} + 4Cl^-$$
$$\text{Blue} \qquad\qquad\qquad\qquad \text{Pink}$$

1 H																	2 He
3 Li	4 Be											5 B	6 C	7 N	8 O	9 F	10 Ne
11 Na	12 Mg											13 Al	14 Si	15 P	16 S	17 Cl	18 Ar
19 K	20 Ca	21 Sc	22 Ti	23 V	24 Cr	25 Mn	26 Fe	27 Co	28 Ni	29 Cu	30 Zn	31 Ga	32 Ge	33 As	34 Se	35 Br	36 Kr
37 Rb	38 Sr	39 Y	40 Zr	41 Nb	42 Mo	43 Tc	44 Ru	45 Rh	46 Pd	47 Ag	48 Cd	49 In	50 Sn	51 Sb	52 Te	53 I	54 Xe
55 Cs	56 Ba	57 La	72 Hf	73 Ta	74 W	75 Re	76 Os	77 Ir	78 Pt	79 Au	80 Hg	81 Tl	82 Pb	83 Bi	84 Po	85 At	86 Rn
87 Fr	88 Ra	89 Ac	104 Unq	105 Unp	106 Unh	107 Uns											

58 Ce	59 Pr	60 Nd	61 Pm	62 Sm	63 Eu	64 Gd	65 Tb	66 Dy	67 Ho	68 Er	69 Tm	70 Yb	71 Lu
90 Th	91 Pa	92 U	93 Np	94 Pu	95 Am	96 Cm	97 Bk	98 Cf	99 Es	100 Fm	101 Md	102 No	103 Lr

□ Class a   ▨ Class b   ▨ Borderline

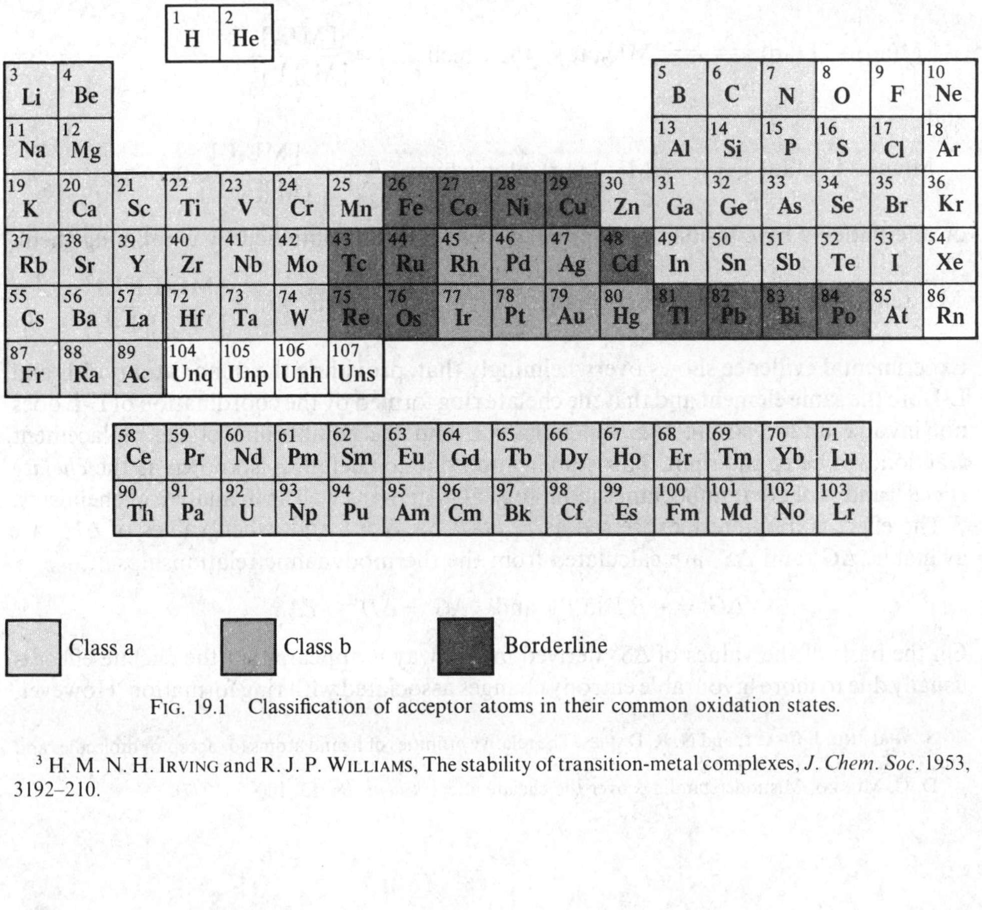

FIG. 19.1 Classification of acceptor atoms in their common oxidation states.

³ H. M. N. H. IRVING and R. J. P. WILLIAMS, The stability of transition-metal complexes, *J. Chem. Soc.* 1953, 3192–210.

If Ca^{II} is added it pushes the equilibrium to the left by bonding preferentially to H_2O, whereas Zn^{II}, with its partial b character, prefers the heavier Cl^- and so pushes the equilibrium to the right.

It seems that, as suggested by Ahrland *et al.*[4] in 1958, this distinction can be explained at least partly on the basis that class-a acceptors are the more electropositive elements which tend to form their most stable complexes with ligands favouring electrostatic bonding, so that, for instance, the stabilities of their complexes with halide ions should decrease in the order

$$F^- > Cl^- > Br^- > I^-$$

Class-b acceptors, on the other hand, are less electropositive, have relatively full d orbitals, and form their most stable complexes with ligands which, in addition to possessing lone-pairs of electrons, have empty π orbitals available to accommodate some charge from the d orbitals of the metal. The order of stability will now be the reverse of that for class-a acceptors, the increasing accessibility of empty d orbitals in the heavier halide ions for instance, favouring an increase in stability of the complexes in the sequence

$$F^- < Cl^- < Br^- < I^-$$

(iii) *The type of ligand.* In comparing the stabilities of complexes formed by different ligands, one of the most important factors is the possible formation of chelate rings. If L is a unidentate ligand, L-L a bidentate ligand, the simplest illustration of this point is provided by comparing the two reactions:

$$M(aq) + 2L(aq) \rightleftharpoons ML_2(aq) \quad \text{for which} \quad \beta_L = \frac{[ML_2]}{[M][L]^2}$$

and

$$M(aq) + L\text{-}L(aq) \rightleftharpoons ML\text{-}L(aq) \quad \text{for which} \quad \beta_{L\text{-}L} = \frac{[ML\text{-}L]}{[M][L\text{-}L]}$$

or alternatively by considering the replacement reaction obtained by combining them:

$$ML_2(aq) + L\text{-}L(aq) \rightleftharpoons ML\text{-}L(aq) + 2L(aq) \quad \text{for which} \quad K = \frac{[ML\text{-}L][L]^2}{[ML_2][L\text{-}L]} = \frac{\beta_{L\text{-}L}}{\beta_L}$$

Experimental evidence shows overwhelmingly that, providing the donor atoms of L and L-L are the same element and that the chelate ring formed by the coordination of L-L does not involve undue strain, L-L will replace L and the equilibrium of the replacement reaction will be to the right. This stabilization due to chelation is known as the *chelate effect*[5] and is of great importance in biological systems as well as in analytical chemistry.

The effect is frequently expressed as $\beta_{L\text{-}L} > \beta_L$ or $K > 1$ and, when values of $\Delta H°$ are available, $\Delta G°$ and $\Delta S°$ are calculated from the thermodynamic relationships

$$\Delta G° = -RT \ln \beta \quad \text{and} \quad \Delta G° = \Delta H° - T\Delta S°$$

On the basis of the values of $\Delta S°$ derived in this way it appears that the chelate effect is usually due to more favourable entropy changes associated with ring formation. However,

[4] S. AHRLAND, J. CHATT, and N. R. DAVIES, The relative affinities of ligand atoms for acceptor molecules and ions, *Q. Revs.* **12**, 265–76 (1958).

[5] D. C. MUNRO, Misunderstandings over the chelate effect, *Chem. Br.* **13**, 100–5 (1977).

the objection can be made that β_L and $\beta_{L \cdot L}$ as just defined have different dimensions and so are not directly comparable. It has been suggested that to surmount this objection concentrations should be expressed in the dimensionless unit "mole fraction" instead of the more usual mol dm^{-3}. Since the concentration of pure water at 25°C is approximately 55.5 mol dm^{-3}, the value of concentration expressed in mole fractions = conc in mol dm^{-3}/55.5. Thus, while β_L is thereby increased by the factor $(55.5)^2$, $\mathbf{B_{L \cdot L}}$ is increased by the factor (55.5) so that the derived values of $\Delta G°$ and $\Delta S°$ will be quite different. The effect of this change in units is shown in Table 19.1 for the Cd^{II} complexes of L = methylamine and L-L = ethylenediamine. It appears that the entropy advantage of the chelate, and with it the chelate effect itself, virtually disappears when mole fractions replace mol dm^{-3}.

TABLE 19.1   *Stability constants and thermodynamic functions for some complexes of Cd^{II} at 25°C*

Complex	log β	$\Delta H°$ (kJ mol^{-1})	$\Delta G°$ (kJ mol^{-1})	$T\Delta S°$ (kJ mol^{-1})
(a) [Cd(NH$_2$Me)$_4$]$^{2+}$	6.55	-57.32	-37.41	-19.91
	13.53		*-72.20*	*$+19.98$*
(b) [Cd(en)$_2$]$^{2+}$	10.62	-56.48	-60.67	$+4.19$
	14.11		*-80.51*	*$+24.04$*
Difference (b)−(a)	4.07	$+0.84$	-23.26	$+24.1$
	0.58		*-3.31*	*$+4.06$*

Values in roman type are based on concentrations measured in mol dm^{-3}.
Values in italics are based on concentrations measured in mole fractions.
The difference (b)−(a) corresponds to the replacement reaction

$$[Cd(NH_2Me)_4]^{2+}(aq) + 2en(aq) \rightleftharpoons [Cd(en)_2]^{2+}(aq) + 4NH_2Me(aq)$$

The resolution of this paradox lies in the assumptions about standard, or reference, states which are unavoidably involved in the above definitions of β_L and $\beta_{L \cdot L}$. In order to ensure that β_L and $\beta_{L \cdot L}$ are dimensionless (as they have to be if their logarithms are to be used) when concentrations are expressed in units which have dimensions, it is necessary to use the ratios of the actual concentrations to the concentrations of some standard state. Accordingly, the expression for any β should incorporate an additional factor composed of standard state concentrations, and the expression $\Delta G° = -RT \ln \beta$ should have an additional term involving the logarithm of this factor. Not to include this factor and this term inevitably implies the choice of standard states of concentration = 1 in whatever units are being used. Only in this way can the factor associated with β be 1 and its logarithm zero. It should be stressed, however, that irrespective of these definitional niceties, it remains true as stated above that chelating ligands which form unstrained complexes always tend to displace their monodentate counterparts under normally attainable experimental conditions.

Probably the most satisfactory model with which to explain the chelate effect is that proposed by G. Schwarzenbach[6] in 1952. If L and L-L are present in similar concentrations and are competing for two coordination sites on the metal, the probability of either of them coordinating to the first site may be taken as equal. However, once one end of L-L has become attached it is much more likely that the second site will be won by

[6] G. SCHWARZENBACH, Der Chelateffekt, *Helv. Chim. Acta* **35**, 2344–59 (1952).

its other end than by L, simply because its other end must be held close to the second site and its effective concentration where it matters is therefore much higher than the concentration of L. Because $\Delta G°$ refers to the transfer of the separate reactants at concentrations = 1 to the products, also at concentrations = 1, it is clear from this model that the advantage of L-L over L, as denoted by $\Delta G°$ or β, will be greatest when the units of concentration are such that a value of 1 corresponds to a dilute solution. Conversely, where a value of 1 corresponds to an exceedingly high concentration, the advantage will be much less and may even disappear. In normal practice even a concentration of 1 mol dm^{-3} is regarded as high, and a concentration of 1 mole fraction is so high as to be of only hypothetical significance, so it need cause no surprise that the choice of the latter unit should lead to rather bizarre results.

The chelate effect is usually most pronounced for 5- and 6-membered rings. Smaller rings generally involve excessive strain while increasingly large rings offer a rapidly decreasing advantage for coordination to the second site. Naturally the more rings there are in a complex the greater the total increase in stability.

19.4 The Various Coordination Numbers[7]

In 1893 at the age of 26, Alfred Werner† produced his classic coordination theory.[1a, 8] It is said that, after a dream which crystallized his ideas, he set down his views and by midday had written the paper which was the starting point for work which culminated in the award of the Nobel Prize for Chemistry in 1913. The main thesis of his argument was that metals possess two types of valency: (i) the primary, or ionizable, valency which must be satisfied by negative ions and is what is now referred to as the "oxidation state"; and (ii) the secondary valency which has fixed directions with respect to the central metal and can be satisfied by either negative ions or neutral molecules. This is the basis for the various stereochemistries found amongst coordination compounds. Without the armoury of physical methods available to the modern chemist, in particular X-ray crystallography, the early workers were obliged to rely on purely chemical methods to identify the more important of these stereochemistries. They did this during the next 20 y or so, mainly by preparing vast numbers of complexes of various metals of such stoichiometry that the number of isomers which could be produced would distinguish between alternative stereochemistries.

The term "secondary valency" has now been superseded by the term "coordination number". This may be defined as the number of donor atoms associated with the central metal atom or ion. For many years a distinction was made between coordination number in this sense and in the crystallographic sense, where it is the number of nearest-neighbour ions of opposite charge in an ionic crystal. Though the former definition applies to species

---

† Born in Mulhouse, Alsace, in 1866, he was French by birth, German in upbringing, and, working in Zürich, he became a Swiss citizen in 1894.

[7] B. F. G. JOHNSON, Transition metal chemistry, Chap. 52 in *Comprehensive Inorganic Chemistry*, Vol. 4, pp. 673–779, Pergamon Press, Oxford, 1973; E. L. MUETTERTIES and C. M. WRIGHT, Molecular polyhedra of high co-ordination number, *Q. Revs.* **21**, 109–94 (1967).

[8] J. C. BAILAR (ed.), *Chemistry of the Coordination Compounds*, Reinhold, New York, 1956, 834 pp.; *Werner Centennial*, Am. Chem. Soc., Adv. in Chemistry Series No. 62 (ed. R. F. GOULD), Washington, 1967, 661 pp. G. B. KAUFFMAN, Alfred Werner's research on structural isomerism, *Coord. Chem. Revs.* **11**, 161–88 (1973); Alfred Werner's research on optically active coordination compounds, *Coord. Chem. Revs.* **12**, 105–49 (1974).

which can exist independently in the solid or in solution, while the latter applies to extended lattice systems, the distinction is rather artificial, particularly in view of the fact that crystal field theory (one of the theories of bonding most commonly applied to coordination compounds) assumes that the coordinate bond is entirely ionic! Indeed, the concept can be extended to all molecules. $TiCl_4$, for instance, can be regarded as a complex of Ti^{4+} with $4\,Cl^-$ ions in which one lone-pair of electrons on each of the latter is completely shared with the Ti^{4+} to give essentially covalent bonds.

The most commonly occurring coordination numbers for transition elements are 4 and 6, but all values from 2 to 9 are known and a few examples of even higher ones have been established. The stereochemistries found for each of these coordination numbers, together with representative examples, will be discussed shortly.

The more important factors determining the most favourable coordination number for a particular metal and ligand are summarized below. However it is important to realize that, with so many factors involved, it is not difficult to provide facile explanations of varying degrees of plausibility for most experimental observations, and it is therefore prudent to treat such explanations with caution.

(i) If electrostatic forces are dominant the attractions between the metal and the ligands should exceed the destabilizing repulsions between the ligands. The attractions are proportional to the product of the charges on the metal and the ligand whereas the repulsions are proportional to the square of the ligand charge. High cation charge and low ligand charge should consequently favour high coordination numbers, e.g. halide ions usually favour higher coordination numbers than does O^{2-}.

(ii) There must be an upper limit to the number of molecules (atoms) of a particular ligand which can physically be fitted around a particular cation. For monatomic ligands this limit will be dependent on the radius ratio of cation and anion, just as is the case with extended crystal lattices.

(iii) Where covalency is important the distribution of charge is equalized by the transference of charge in the form of lone-pairs of electrons from ligands to cation. The more polarizable the ligand the lower the coordination number required to satisfy the particular cation though, if back-donation of charge from cation to ligand via suitable π orbitals is possible, then more ligands can be accommodated. Thus the species most readily formed with Fe^{III} in aqueous solutions are $[FeF_5(H_2O)]^{2-}$ for the non-polarizable F^-, $[FeCl_4]^-$ for the more polarizable Cl^-, but $[Fe(CN)_6]^{3-}$ for CN^- which is even more polarizable but which also possesses empty antibonding π orbitals suitable for back-donation.

(iv) The availability of empty metal orbitals of suitable symmetries and energies to accommodate electron-pairs from the ligands must also be important in covalent compounds. This is probably one of the main reasons why the lowest coordination numbers (2 and 3) are to be found in the Ag, Au, Hg region of the periodic table where the d shell has been filled. However, it would be unwise to draw the converse conclusion that the highest coordinations are found amongst the early members of the transition and inner-transition series because of the availability of empty d or f orbitals. It seems more likely that these high coordination numbers are achieved by electrostatic attractions between highly charged but rather large cations and a large number of relatively non-polarizable ligands.

Coordination number 2

Examples of this coordination number are virtually confined to linear $D_{\infty h}$ complexes of Cu^I, Ag^I, Au^I, and Hg^{II} of which a well-known instance is the ammine formed when ammonia is added to an aqueous solution of Ag^+:

$$[H_3N-Ag-NH_3]^+$$

Coordination number 3[8a]

This is rather rare and even in $[HgI_3]^-$, the example usually cited, the coordination number is dependent on the counter cation. In $[SMe_3][HgI_3]$ the Hg^{II} lies at the centre of an almost equilateral triangle of iodide ions (D_{3h}) whereas in $[NMe_4][HgI_3]$ the anion apparently polymerizes into loosely linked chains of 4-coordinate Hg^{II}. Other examples feature bulky ligands, e.g. $[Fe\{N(SiMe_3)_2\}_3]$, $[Cu\{(SC(NH_2)_2)_3\}]Cl$, and $[Cu(SPPh_3)_3]ClO_4$.

Coordination number 4

This is very common and usually gives rise to stereochemistries which may be regarded essentially as either tetrahedral T_d or (square) planar D_{4h}. Where a complex may be thought to have been formed from a central cation with a spherically symmetrical electron configuration, the ligands will lie as far from each other as possible, that is they will be tetrahedrally disposed around the cation. This has already been seen in complex anions such as BF_4^- and is also common amongst complexes of transition metals in their group oxidation states and of d^5 and d^{10} ions. $[MnO_4]^-$, $[Ni(CO)_4]$, and $[Cu(py)_4]^+$ exemplify these types. Central cations with other d configurations, in particular d^8, may give rise to a square-planar stereochemistry and the complexes of Pd^{II} and Pt^{II} are predominantly of this type. Then again, the difference in energy between tetrahedral and square-planar forms may be only slight, in which case both forms may be known or, indeed, interconversions may be possible as happens with a number of Ni^{II} complexes (p. 1347). In the $M_2^I CuX_4$ series of complexes of Cu^{II}, variation of M^I and X gives complex anions with stereochemistries ranging from square planar, e.g. $(NH_4)_2CuCl_4$, to almost tetrahedral, e.g. (Cs_2CuBr_4). Figure 19.2 shows that the change from square planar to tetrahedral requires a $90°$ rotation of one L,L pair and a $19\frac{1}{2}°$ change in the LML angles, and a continuous range of distortions from one extreme to the other would appear to be feasible.

Four-coordinate complexes provide good examples of the early use of preparative methods for establishing stereochemistry. For complexes of the type $[Ma_2b_2]$, where a and b are unidentate ligands, a tetrahedral structure cannot produce isomerism whereas a planar structure leads to *cis* and *trans* isomers (see opposite). The preparation of 2 isomers of $[PtCl_2(NH_3)_2]$, for instance, was taken as good evidence for their planarity.†

---

† On the basis of this evidence alone it is logically possible that one isomer could be tetrahedral. Early coordination chemists, however, assumed that the directions of the "secondary valencies" were fixed, which would preclude this possibility. X-ray structural analysis shows that, in the case of Pt^{II} complexes, they were correct.

[8a] P. G. ELLER, D. C. BRADLEY, M. B. HURSTHOUSE, and D. W. MEEK, Three coordination in metal complexes, *Coord. Chem. Revs.* **24**, 1–95 (1977).

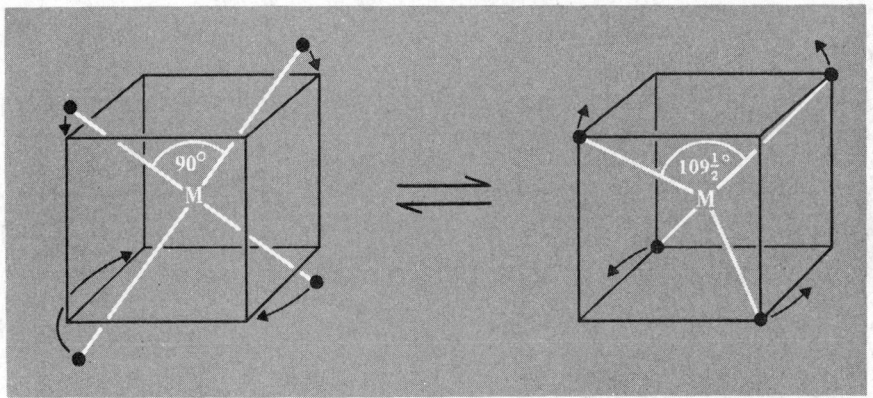

FIG. 19.2   Schematic interconversion of square planar and tetrahedral geometries.

Coordination number 5[8b]

Five-coordinate complexes are far more common than was once supposed and are now known for all configurations from d^1 to d^9. Two limiting stereochemistries may be distinguished (Fig. 19.3). One of the first authenticated examples of 5-coordination was [VO(acac)$_2$] which has the square-pyramidal C_{4v} structure with the $=O$ occupying the unique apical site. However, many of the complexes with this coordination number have structures intermediate between the two extremes and it appears that the energy required for their interconversion is frequently rather small.[9] Because of this stereochemical non-rigidity a number of 5-coordinate compounds behave in a manner described as "fluxional". That is, they exist in two or more configurations which are chemically equivalent and which interconvert at such a rate that some physical measurement (commonly nmr) is unable to distinguish the separate configurations and instead "sees" only their time-average. If ML$_5$ has a trigonal bipyramidal D_{3h} structure then 2 ligands must be "axial" and 3 "equatorial", but interchange via a square-pyramidal intermediate is possible (Fig. 19.4). This mechanism has been suggested as the reason why the ^{13}C nmr spectrum of trigonal bipyramidal Fe(CO)$_5$ (p. 1282) fails to distinguish two different kinds of carbon nuclei. See also the discussion of PF$_5$ on p. 572.

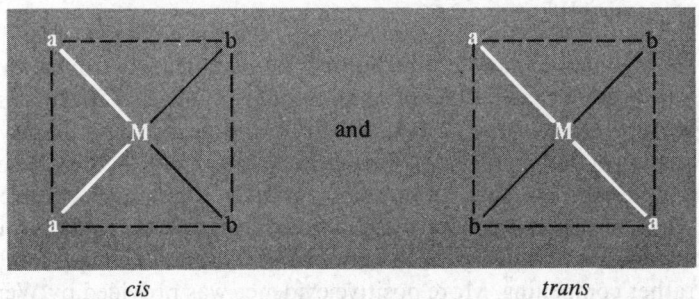

cis   and   trans

[8b] J. S. WOOD, Stereochemical and electronic structural aspects of five coordination, *Prog. Inorg. Chem.* **16**, 227–486 (1972).

[9] E. L. MUETTERTIES and R. A. SCHUNN, Pentaco-ordination, *Q. Revs.* **20**, 245–99 (1966).

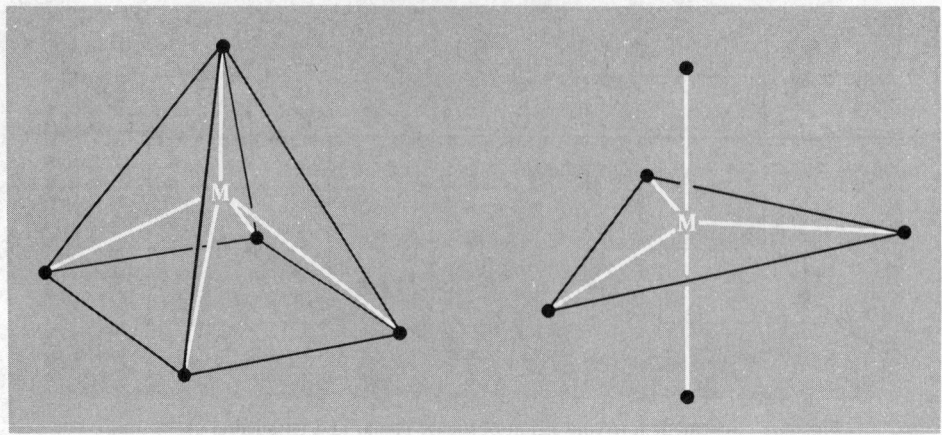

Square pyramidal   Trigonal bipyramidal

FIG. 19.3   Limiting stereochemistries for 5-coordination.

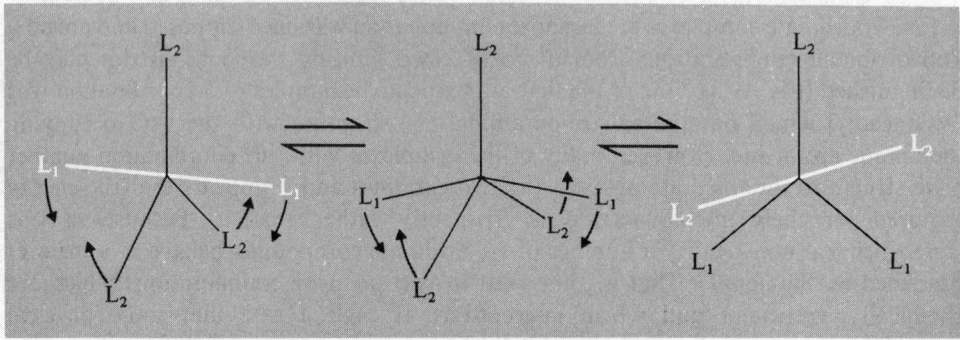

FIG. 19.4   The interconversion of trigonal bipyramidal configurations via a square-pyramidal
intermediate. Notice that the L_1 ligands, which in the left-hand tbp are axial, become equatorial in
the right-hand tbp and simultaneously 2 of the L_2 ligands change from equatorial to axial.

Coordination number 6[9a]

This is the most common coordination number for complexes of transition elements. It
can be seen by inspection that, for compounds of the type (Ma_4b_2), the three symmetrical
structures (Fig. 19.5) can give rise to 3, 3, and 2 isomers respectively. Exactly the same is
true for compounds of the type [Ma_3b_3]. In order to determine the stereochemistry of 6-
coordinate complexes very many examples of such compounds, particularly with
$M = Cr^{III}$ and Co^{III}, were prepared, and in no case were more than 2 isomers found. This, of
course, was only negative evidence for the octahedral structure, though the sheer volume
of it made it rather compelling. More positive evidence was provided by Werner, who in

[9a] D. L. KEPERT, Aspects of the stereochemistry of six-coordination, *Prog. Inorg. Chem.* **23**, 1–66 (1977).

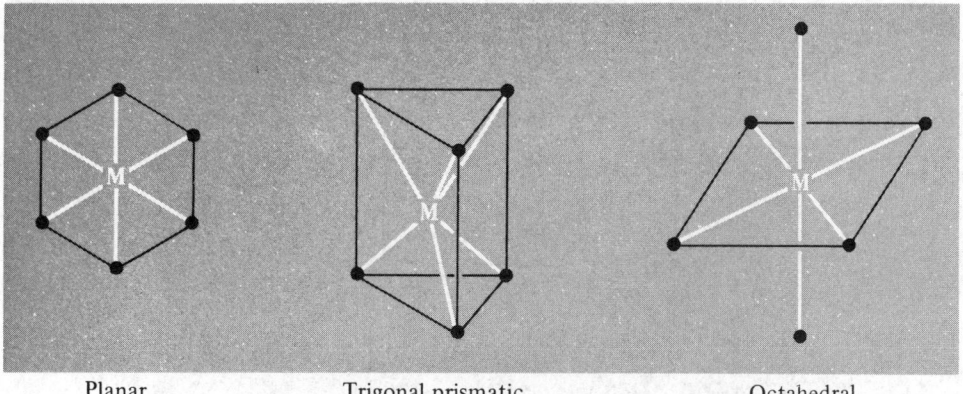

Planar   Trigonal prismatic   Octahedral

FIG. 19.5   Possible stereochemistries for 6-coordination.

1914 achieved the first resolution into optical isomers of an entirely inorganic compound,

$$\left[Co \left(\begin{array}{c} OH \\ \diagup \quad \diagdown \\ \diagdown \quad \diagup \\ OH \end{array} Co(NH_3)_4 \right)_3 \right]^{6+}$$

since neither the planar nor trigonal prismatic structures can give rise to such optical isomers.

Nevertheless, it cannot be assumed that every 6-coordinate complex is octahedral.[10] In 1923 the first example of trigonal prismatic coordination was reported for the infinite layer lattices of MoS_2 and WS_2. A limited number of further examples are now known following the report in 1965[11] of the structure of $[Re(S_2C_2Ph_2)_3]$ (Fig. 19.6). Intermediate structures also occur and can be defined by the "twist angle" which is the angle through which one face of an octahedron has been rotated with respect to the opposite face as "viewed along" a threefold axis of the octahedron. A twist angle of 60° suffices to convert an octahedron into a trigonal prism:

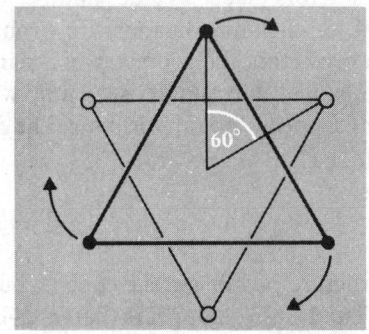

[10] R. A. D. WENTWORTH, Trigonal prismatic vs. octahedral stereochemistry in complexes derived from innocent ligands, *Coord. Chem. Revs.* **9**, 171–87 (1972–3).

[11] R. EISENBERG and J. A. IBERS, Trigonal prismatic coordination: the molecular structure of tris(*cis*-1,2-diphenylethene-1,2-dithiolato)rhenium, *J. Am. Chem. Soc.* **87**, 3776–8 (1965).

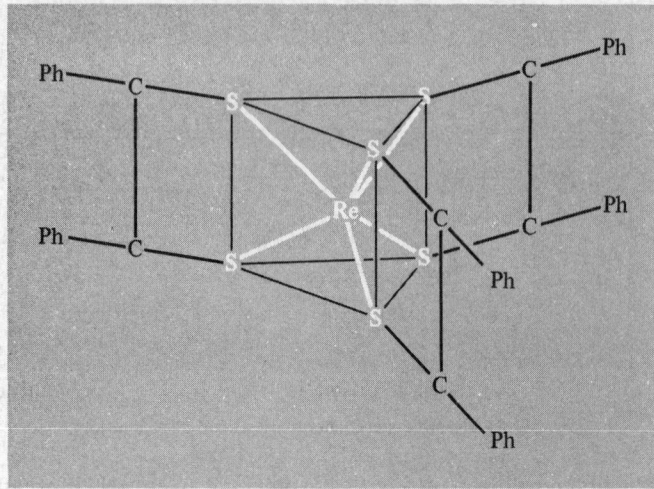

FIG. 19.6   Trigonal prismatic structure of $[Re(S_2C_2Ph_2)_3]$.

In fact the vast majority of 6-coordinate complexes are indeed octahedral or distorted octahedral. In addition to the twist distortion just considered distortions can be of two other types: trigonal and tetragonal distortions which mean compression or elongation along a threefold and a fourfold axis of the octahedron respectively (Fig. 19.7).

Coordination number 7[11a]

There appear to be three main stereochemistries for complexes of this coordination number: pentagonal bipyramidal D_{5h}, capped trigonal prismatic C_{2v}, and capped octahedral C_{3v}, the last two being obtained by the addition of a seventh ligand above one of the rectangular faces of a trigonal prism or above a triangular face of an octahedron respectively. These structures may conveniently be visualized within circumscribed spheres (Fig. 19.8).

As with other high coordination numbers, there seems to be little difference in energy between these structures. Factors such as the number of counter ions and the stereochemical requirements of chelating ligands are probably decisive and *a priori* arguments are unreliable in predicting the geometry of a particular complex. $[ZrF_7]^{3-}$ and $[HfF_7]^{3-}$ have the pentagonal bipyramidal structure, whereas the bivalent anions, $[NbF_7]^{2-}$ and $[TaF_7]^{2-}$ are capped trigonal prismatic. The capped octahedral structure is exemplified by $[NbOF_6]$.

Coordination number 8[11b]

The most symmetrical structure possible is the cube O_h but, except in extended ionic lattices such as those of CsCl and CaF_2, it appears that inter-ligand repulsions are nearly

[11a] D. L. KEPERT, Aspects of the stereochemistry of seven-coordination, *Prog. Inorg. Chem.* **25**, 41–144 (1979); M. G. B. DREW, Seven-coordination chemistry, *ibid.* **23**, 67–210 (1977).
[11b] D. L. KEPERT, "Aspects of the stereochemistry of eight-coordination, *Prog. Inorg. Chem.* **24**, 179–249 (1978).

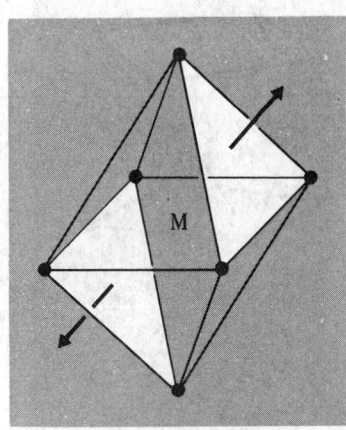

Trigonal elongation

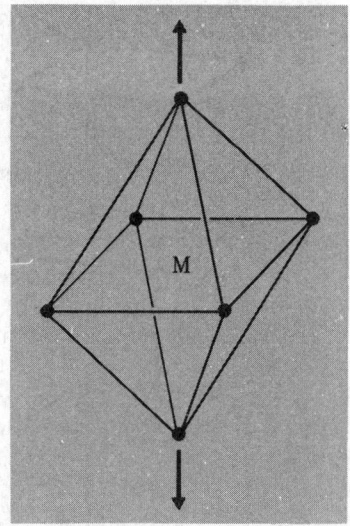

Tetragonal elongation

Fig. 19.7   Distortions of octahedral geometry.

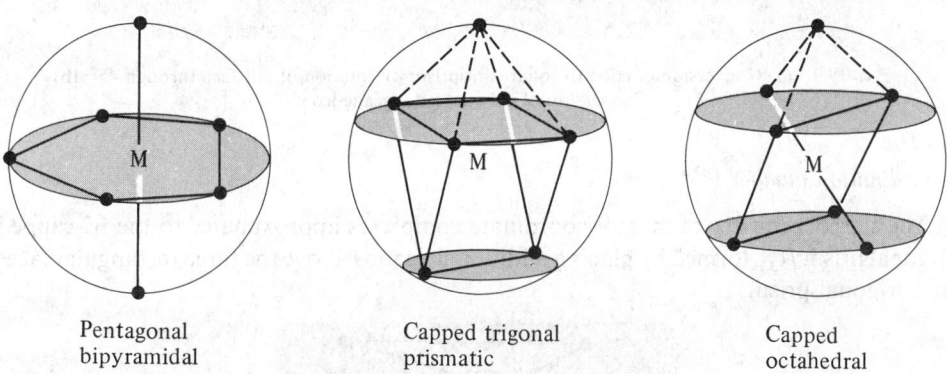

| Pentagonal bipyramidal | Capped trigonal prismatic | Capped octahedral |

Fig. 19.8   The three main stereochemistries for 7-coordination.

always (but see p. 1480) reduced by distorting the cube, the two most important resultant structures being the square antiprism D_{4h} and the dodecahedron D_{2d} (Fig. 19.9).

Again, these forms are energetically very similar; distortions from the idealized structures make it difficult to specify one or other, and the particular structure actually found must result from the interplay of many factors. $[TaF_8]^{3-}$, $[ReF_8]^{2-}$, and $[Zr(acac)_4]$ are square antiprismatic, whereas $[ZrF_8]^{4-}$ and $[Mo(CN)_8]^{4-}$ are dodecahedral. The nitrates $[Co(NO_3)_4]^{2-}$ and $Ti(NO_3)_4$ may both be regarded as dodecahedral, the former with some distortion.[11c] Each nitrate ion is bidentate but the 2 oxygen atoms are necessarily close together so that the structure of the complexes is probably more easily visualized from the point of view of the 4 nitrogen atoms which form a flattened tetrahedron around the metal (p. 1126).

[11c] C. C. ADDISON, N. LOGAN, S. C. WALLWORK, and C. D. GARNER, Structural aspects of co-ordinated nitrate groups, *Q. Revs.* **25**, 289–322 (1971).

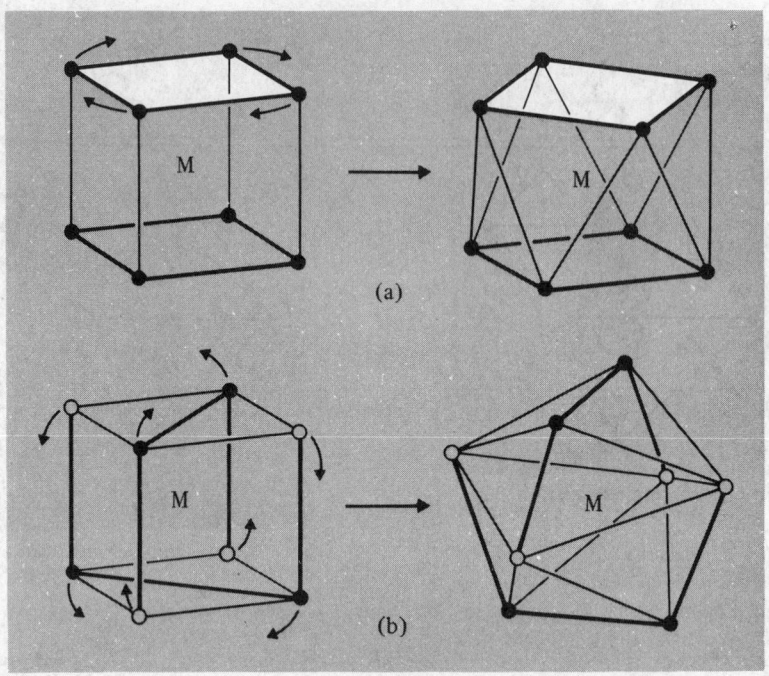

FIG. 19.9   (a) Conversion of cube to square antiprism by rotation of one face through 45°. (b) Conversion of cube into dodecahedron.

Coordination number 9[12]

The stereochemistry of most 9-coordinate complexes approximates to the tri-capped trigonal prism D_{3h}, formed by placing additional ligands above the three rectangular faces of a trigonal prism:

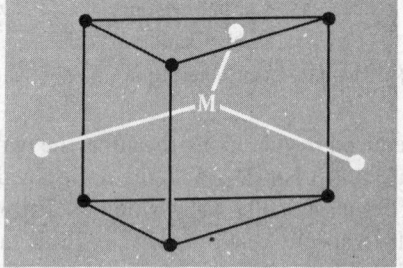

Amongst the known examples of this arrangement are a number of $[M(H_2O)_9]^{3+}$ hydrates of lanthanide salts and $[ReH_9]^{2-}$. The latter is interesting in that it is presumably only the small size of the H ligand which allows such a high coordination number for rhenium. Very occasionally 9-coordination results in a capped square antiprismatic C_{4v} arrangement in which the ninth ligand lies above one of the square faces, e.g. the Cl-bridged $[\{LaCl(H_2O)_7\}_2]^{4+}$.

[12] M. C. FAVAS and D. L. KEPERT, Aspects of the stereochemistry of nine-coordination, ten-coordination, and twelve-coordination, *Prog. Inorg. Chem.* **28**, 309–67 (1981).

Coordination numbers above $9^{(12)}$

Such high coordination numbers are not common and it is difficult to generalize about their structures since so few have been accurately determined. They are found mainly with ions of the early lanthanide and actinide elements and it is therefore tempting to assume that the availability of empty and accessible f orbitals is necessary for their formation. However, it appears that the bonding is predominantly ionic and that the really important point is that these are the elements which provide stable cations with charges high enough to attract a large number of anions and yet are large enough to ensure that the inter-ligand repulsions are not unacceptably high. $K_4[Th(O_2CCO_2)_4(H_2O)_2] \cdot 2H_2O$ (bicap-ped square antiprism D_{4d}) and $[La(EDTA)(H_2O)_4]$ afford examples of 10-coordination. Higher coordination numbers are reached only by chelating ligands such as NO_3^-, SO_4^{2-}, and 1,8-naphthyridine with donor atoms close together (i.e. ligands with only a small "bite").[11c] $[La(DAPBAH)(NO_3)_3]$,† is a good example[12a] (see also p. 1482). In it the 5 donor atoms of the DAPBAH are situated in a plane, with the N atoms (but not the donor oxygens) of the 3 bidentate nitrates in a second plane at right angles to the first. $Ce_2Mg_3(NO_3)_{12} \cdot 24H_2O$ contains 12-coordinate Ce in the complex ion $[Ce(NO_3)_6]^{3-}$. This has a distorted icosahedral stereochemistry, though it is more easily visualized as an octahedral arrangement of the nitrogen atoms around the Ce^{III}. Another example is $[Pr(naph)_6]^{3+}$ where naph is 1,8-naphthyridine,

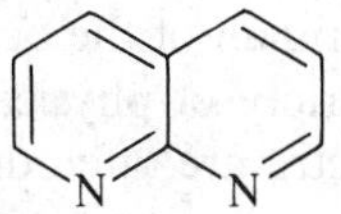

Higher coordination numbers (up to 16) are known, particularly among organometal-lic compounds (pp. 374–78) and metal borohydrides (p. 189).

19.5   Isomerism[13]

Isomers are compounds with the same chemical composition but different structures, and the possibility of their occurrence in coordination compounds is manifest. Their importance in the early elucidation of the stereochemistries of complexes has already been

† DAPBAH = 2,6-diacetylpyridinebis(benzoic acid hydrazone):

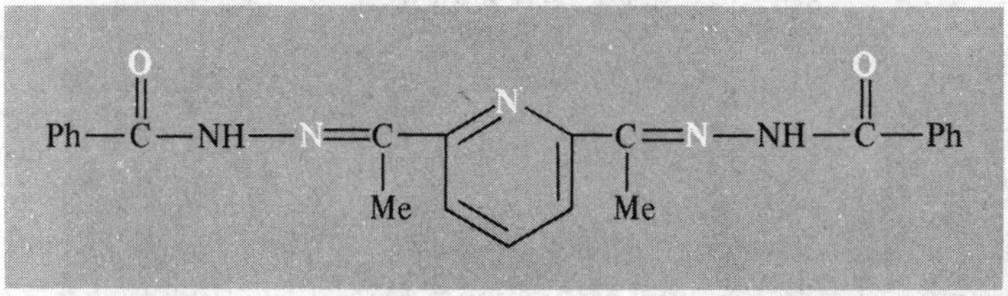

[12a] J. E. THOMAS, R. C. PALENIK, and G. J. PALENIK, A decahexahedral lanthanum complex: synthesis and characterization of the eleven coordinate trisnitrato[2,6-diacetylpyridinebis(benzoic acid hydrazone)]-lanthanum(III), *Inorg. Chim. Acta* **37**, L459–L460 (1979).

[13] J. C. BAILAR JR. (ed.), *Chemistry of the Coordination Compounds*, Chaps. 7 and 8, pp. 261–353, Reinhold, New York, 1956.

referred to and, though the purposeful preparation of isomers is no longer common, the preparative chemist must still be aware of the diversity of the compounds which he may produce. The more important of the various types of isomerism are listed below.

Conformational isomerism

In principle this type of isomerism (also known as "polytopal" isomerism) is possible with any coordination number for which there is more than one known stereochemistry. However, to actually occur the isomers need to be of comparable stability, and to be separable there must be a significant energy barrier preventing their interconversion. This behaviour is confined primarily to 4-coordinate nickel(II), an example being $[NiCl_2(Ph_2PCH_2Ph)_2]$ which is known in both planar and tetrahedral forms (p. 1347).

Geometrical isomerism

This is of most importance in square-planar and octahedral compounds where ligands, or more specifically donor atoms, can occupy positions next to one another (*cis*) or opposite each other (*trans*) (Fig. 19.10).

A similar type of isomerism occurs for $[Ma_3b_3]$ octahedral complexes since each trio of donor atoms can occupy either adjacent positions at the corners of an octahedral face (*facial*) or positions around the meridian of the octahedron (*meridional*). (Fig. 19.10A.) Geometrical isomers differ in a variety of physical properties, amongst which dipole moment and visible/ultraviolet spectra are often diagnostically important.

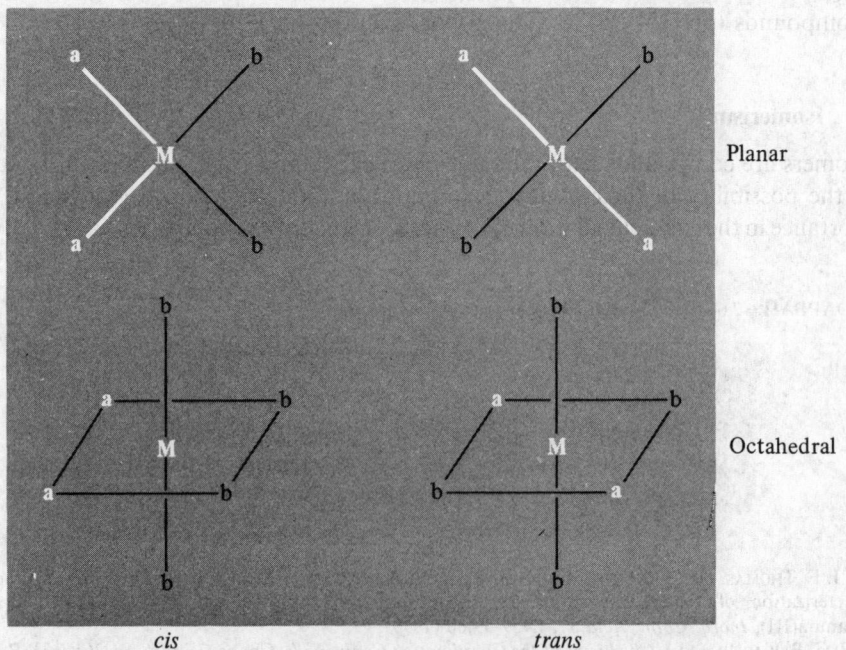

FIG. 19.10 *Cis* and *trans* isomerism.

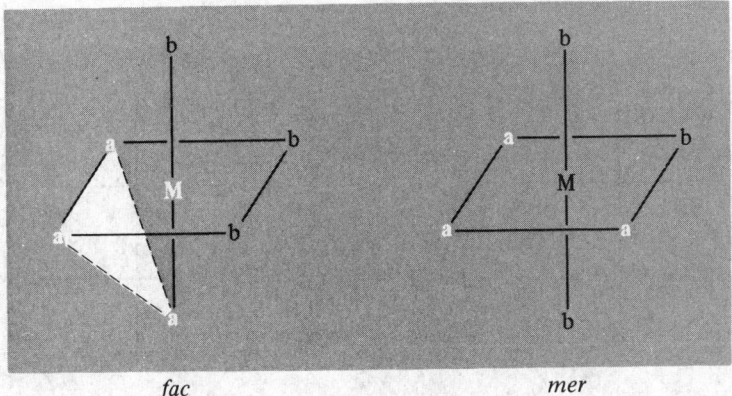

fac   *mer*

FIG. 19.10A   Facial and meridional isomers.

Optical isomerism

Optical isomers, enantiomorphs or enantiomers, as they are also known, are pairs of molecules which are non-superimposable mirror images of each other. Such isomers have the property of chirality (from Greek χειρ, hand), i.e. handedness, and virtually the only physical or chemical difference between them is that they rotate the plane of polarized light, one of them to the left and the other to the right. They are consequently designated as laevo (*l* or −) and dextro (*d* or +) isomers.

A few cases of optical isomerism are known for planar and tetrahedral complexes involving unsymmetrical bidentate ligands, but by far the most numerous examples are afforded by octahedral compounds of chelating ligands, e.g. $[Cr(oxalate)_3]^{2-}$ and $[Co(EDTA)]^-$ (Fig. 19.11).

Where unidentate ligands are present, the ability to effect the resolution of an octahedral complex (i.e. to separate 2 optical isomers) is proof that the 2 ligands are *cis* to each other. Resolution of $[PtCl_2(en)_2]^{2-}$ therefore shows it to be *cis* while of the 2 known geometrical isomers of $[CrCl_2en(NH_3)_2]^+$ the one which can be resolved must have the *cis-cis* structure since the *trans* forms would give superimposable, and therefore identical, mirror images:

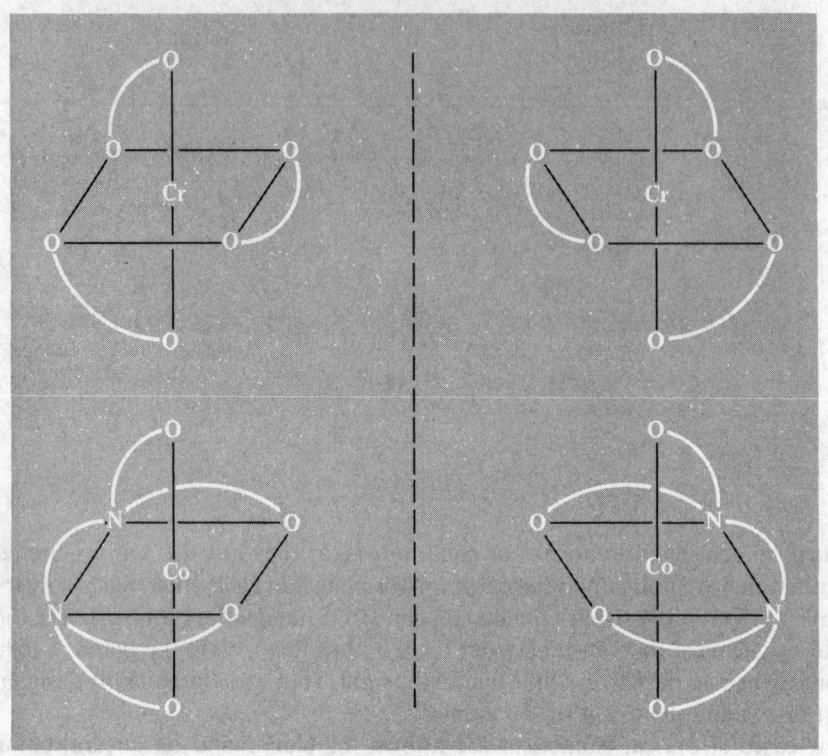

FIG. 19.11   Non-superimposable mirror images

Ionization isomerism

This type of isomerism occurs when isomers produce different ions in solution, and is possible in compounds which consists of a complex ion with a counter ion which is itself a potential ligand. The pairs: $[Co(NH_3)_5NO_3]SO_4$, $[Co(NH_3)_5SO_4]NO_3$ and $[PtCl_2(NH_3)_4]Br_2$, $[PtBr_2(NH_3)_4]Cl_2$, and the series $[CoCl(en)_2(NO_2)]SCN$, $[CoCl(en)_2(SCN)]NO_2$, $[Co(en)_2(NO_2)(SCN)]Cl$ are examples of ionization isomers.

A subdivision of this type of isomerism, known as "hydrate isomerism", occurs when water may be inside or outside the coordination sphere. It is typified by $CrCl_3.6H_2O$ which exists in the three distinct forms $[Cr(H_2O)_6]Cl_3$ (violet), $[CrCl(H_2O)_5]Cl_2.H_2O$ (pale green), and $[CrCl_2(H_2O)_4]Cl.2H_2O$ (dark green). These are readily distinguished by the action of $AgNO_3$ in aqueous solution which immediately precipitates 3, 2, 1 chloride ions respectively.

Linkage isomerism

This is in principle possible in any compound containing an ambidentate ligand. However, that such a ligand can *under different circumstances* coordinate through either of the 2 different donor atoms is by no means a guarantee that it will form isolable linkage isomers with the same cation. In fact, in only a very small proportion of the complexes of ambidentate ligands can linkage isomers actually be isolated, and these are confined largely to complexes of NO_2^- (p. 534) and, to a lesser extent, SCN^- (p. 342). Examples are:

$$[Co(en)_2(NO_2)_2]^-, [Co(en)_2(ONO)_2]^- \text{ and } [Pd(PPh_3)_2(NCS)_2], [Pd(PPh_3)_2(SCN)_2]$$

It should be noted that, by convention, the ambidentate ligand is always written with its donor atom first, i.e. NO_2 for the nitro, ONO for the nitrito, NCS for the *N*-thiocyanato and SCN for the *S*-thiocyanato complex. Differences in infrared spectra arising from the differences in bonding are often used to distinguish between such isomers.

Coordination isomerism

In compounds made up of both anionic and cationic complexes it is possible for the distribution of ligands between the ions to vary and so lead to isomers such as:

$$[Co(en)_3][Cr(CN)_6] \text{ and } [Cr(en)_3][Co(CN)_6]$$
$$[Cu(NH_3)_4][PtCl_4] \text{ and } [Pt(NH_3)_4][CuCl_4]$$
$$[Pt^{II}(NH_3)_4][Pt^{IV}Cl_6] \text{ and } [Pt^{IV}(NH_3)_4Cl_2][Pt^{II}Cl_4]$$

It can be seen that other intermediate isomers are feasible but in the above cases they have not been isolated. Substantial differences in both physical and chemical properties are to be expected between coordination isomers.

When the two coordinating centres are not in separate ions but are joined by bridging groups, the isomers are often distinguished as "coordination position isomers" as is the case for:

$$[(NH_3)_4Co \overset{\overset{\text{H}}{\overset{\text{O}}{\diagup \diagdown}}}{\underset{\underset{\text{H}}{\underset{\text{O}}{\diagdown \diagup}}}{}} Co(NH_3)_2Cl_2]^{2+} \text{ and } [Cl(NH_3)_3Co \overset{\overset{\text{H}}{\overset{\text{O}}{\diagup \diagdown}}}{\underset{\underset{\text{H}}{\underset{\text{O}}{\diagdown \diagup}}}{}} Co(NH_3)_3Cl]^{2+}$$

Polymerization isomerism

Compounds whose molecular compositions are multiples of a simple stoichiometry are polymers, strictly, only if they are formed by repetition of the simplest unit. However, the name "polymerization isomerism" is applied rather loosely to cases where the same stoichiometry is retained but where the molecular arrangements are different. The stoichiometry $PtCl_2(NH_3)_2$ applies to the 3 known compounds, $[Pt(NH_3)_4][PtCl_4]$, $[Pt(NH_3)_4][PtCl_3(NH_3)]_2$, and $[PtCl(NH_3)_3]_2[PtCl_4]$ (in addition to the *cis* and *trans* isomers of monomeric $[PtCl_2(NH_3)_2]$). There are actually 7 known compounds with the stoichiometry $Co(NH_3)_3(NO_2)_3$. Again it is clear that considerable differences are to be expected in the chemical properties and in physical properties such as conductivity.

Ligand isomerism

Should a ligand exist in different isomeric forms then of course the corresponding complexes will also be isomers, often described as "ligand isomers". In $[CoCl(en)_2(NH_2C_6H_4Me)]Cl_2$, for instance, the toluidine may be of the *o*-, *m*-, or *p*- form.

19.6   The Coordinate Bond (see also p. 223)

The concept of the coordinate bond as an interaction between a cation and an ion or molecule possessing a lone-pair of electrons can be accepted before specifying the nature of that interaction. Indeed, it is now evident that in different complexes the bond can span the whole range from electrostatic to covalent character. This is why the various theories which have been accorded popular favour at different times have been acceptable and useful even though based on apparently incompatible assumptions; a dichotomy reflected in the now obsolete adjectives "dative-covalent", "semi-polar", and "co-ionic", which have been used to describe the coordinate bond. The first of these descriptions arises from the idea advanced by N. V. Sidgwick in 1927, that the coordinate bond is a covalent bond formed by the donation of a lone-pair of electrons from the donor atom to the central metal. Since noble gases are extremely unreactive, and compounds in which atoms have attained the electronic configuration of a noble gas either by sharing or transferring electrons also tend to be stable, Sidgwick further suggested that, in complexes, the metal would tend to surround itself with sufficient ligands to ensure that the number of electrons around it (its "effective atomic number" or EAN) would be the same as that of the next noble gas. If this were true then a metal would have a unique coordination number for each oxidation state, which is certainly not always the case. However, the EAN rule is still of use in rationalizing the coordination numbers and structures of simple metal carbonyls.

In his *valence bond theory* (VB), L. Pauling extended the idea of electron-pair donation by considering the orbitals of the metal which would be needed to accommodate them, and the stereochemical consequences of their hybridization (1931–3). He was thereby able to account for much that was known in the 1930s about the stereochemistry and kinetic behaviour of complexes, and demonstrated the diagnostic value of measuring their magnetic properties. Unfortunately the theory offers no satisfactory explanation of spectroscopic properties and so was eventually superseded by *crystal field theory* (CF).

About the same time that VB theory was being developed, CF theory was also being used by H. Bethe, J. H. van Vleck, and other physicists to account for the colours and magnetic properties of hydrated salts of transition metals (1933–6). It is based on what, to chemists, appeared to be the outrageous assumption that the coordinate bond is entirely electrostatic. Nevertheless, in the 1950s a number of theoretical chemists used it to interpret the electronic spectra of transition metal complexes. It has since been remarkably successful in explaining the properties of M^{II} and M^{III} ions of the first transition series, especially when modifications have been incorporated to include the possiblity of some covalency. (The theory is then often described as *ligand field theory*, but there is no general agreement on this terminology.)

In order to take full account of both ionic and covalent character, recourse must be made to *molecular orbital theory* (MO) which, like the VB and CF theories, originated in the 1930s. The reason why it has not totally displaced the other bonding theories is that quantitative calculations are difficult and it is not as pictorially intelligible as CF theory.

Valence bond theory[14]

In this theory the formation of a coordination compound is described by the hypothetical sequence:

(i) removal of electrons from the metal to give the appropriate cation;

(ii) hybridization of those metal atomic orbitals which will provide a set of equivalent hybrid orbitals directed towards the ligands;

(iii) where necessary, rearrangement of the metal's electrons in order to ensure that the hybrid orbitals are empty;

(iv) formation of σ covalent bonds by overlap of the hybrid orbitals with the ligand orbitals containing the lone-pairs of electrons.

The common types of hybridization and the spatial distribution of the hybrid orbitals are given in Table 19.2 and examples of the way in which these are involved in the VB description of complexes are illustrated by the following diagrams of the electronic configurations of Co^{II} in 4- and 6-coordinate complexes:

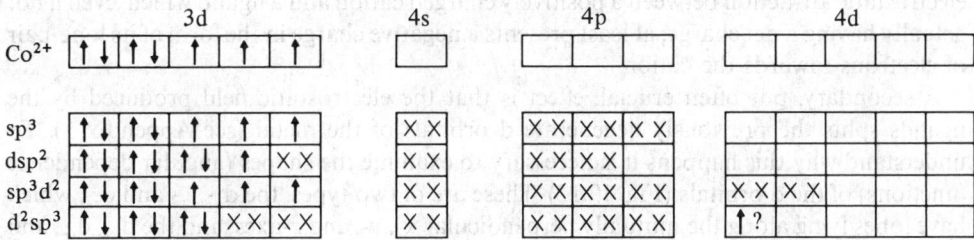

TABLE 19.2   *The spatial arrangements of different types of hybrid orbitals*[a]

CN	Atomic orbitals	Hybrid orbitals	Spatial arrangement
2	$s\ p_x$	sp	Linear
3	$s\ p_x\ p_y$	sp^2	Trigonal planar
4	$s\ p_x\ p_y\ p_z$	sp^3	Tetrahedral
4	$s\ p_x\ p_y\ d_{x^2-y^2}$	dsp^2 or sp^2d	Square planar
5	$s\ p_x\ p_y\ p_z\ d_{z^2}$	dsp^3 or sp^3d	Trigonal bipyramidal
5	$s\ p_x\ p_y\ p_z\ d_{x^2-y^2}$	dsp^3 or sp^3d	Square pyramidal
6	$s\ p_x\ p_y\ p_z\ d_{x^2-y^2}\ d_{z^2}$	d^2sp^3 or sp^3d^2	Octahedral

[a] Despite the rather definite language used in this section for clarity of presentation, it is important to remember that hybridization is not a phenomenon but a mathematical manipulation, and that hybrid orbitals have no physical existence in reality.

Two kinds of octahedral complex are apparent, one which uses only 3d orbitals and the other which makes use of 4d orbitals for bonding. These have been described as "covalent" or "inner orbital" and "ionic" or "outer orbital", respectively. In the former case the necessary promotion of a single electron to an unspecified outer orbital is consistent with

[14] L. PAULING, *The Nature of the Chemical Bond*, 3rd edn., Chap. 5, pp. 145–82, Cornell University Press, New York, 1960.

the fact that Co^{II} complexes are indeed easily oxidized to Co^{III}. It can also be seen that the two forms of 6-coordinate and also of 4-coordinate complexes differ in the number of unpaired, nonbonding electrons present. Since the magnetic moment of a complex is dependent on the number of unpaired electrons, it follows that magnetic measurements can provide useful indications of structure and bond type. Unfortunately, VB theory provides no means of explaining other aspects of magnetic behaviour, such as the variation of magnetic moment with temperature, and provides no explanation whatever for spectroscopic properties. In view of the fact that the variety of colours found amongst complexes of transition metals is one of the most immediately obvious contrasts with compounds of main group elements, this is a serious deficiency. It is remedied to a large extent by crystal field theory, as indicated in the next section.

19.7 Crystal Field Theory

In this approximation the coordinate bond is assumed to consist entirely of the electrostatic attraction between a positively charged cation and a ligand which, even if not actually having a net charge, at least presents a negative charge in the form of its lone-pair of electrons towards the cation.

A secondary, but often crucial, effect is that the electrostatic field produced by the ligands splits the previously degenerate d orbitals of the metal (see Appendix 1). To understand why this happens it is necessary to examine the shapes (angular dependence functions) of the d orbitals (Fig. 19.12). These are of two types, the $d_{x^2-y^2}$ and d_{z^2}, which have lobes lying along the mutually perpendicular x-, y-, and z-axes, and the d_{xy}, d_{yz}, and d_{xz}, whose lobes point between the axes. An octahedral complex, for instance, can now be imagined to be formed by 6 ligands (represented by negative point charges) approaching

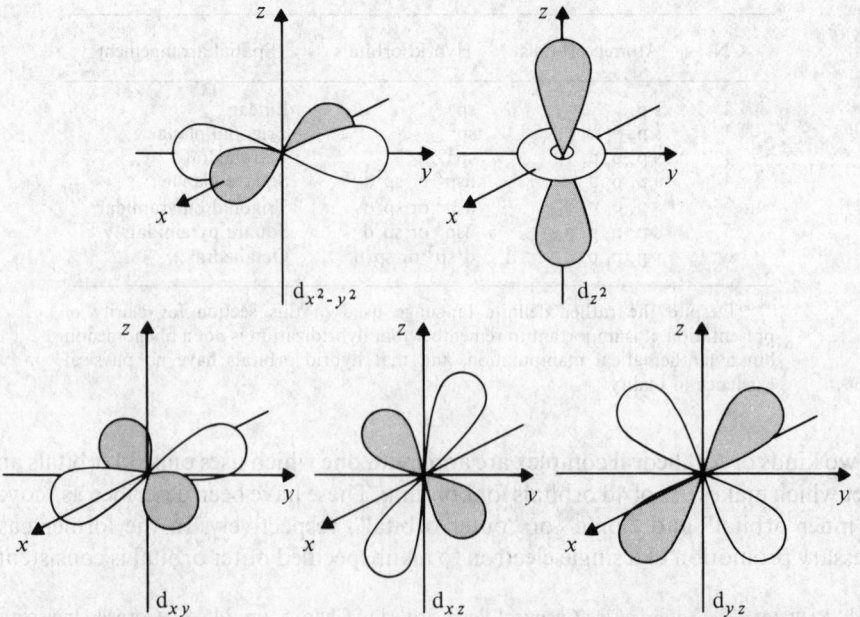

FIG. 19.12 Directional properties of d orbitals; see also Appendix I, p. 1487.

the cation along the x-, y-, and z-axes until they reach their final equilibrium positions. As they approach, the electrons in the metal d orbitals will be repelled, that is to say the energy of the d orbitals will be increased—but not equally. The $d_{x^2-y^2}$ and d_{z^2} orbitals which point directly towards the ligands will be affected more than the d_{xy}, d_{yz}, d_{xz}, which point between the approaching ligands. The effect can be considered in two parts as shown in Fig. 19.13. First, there is an average increase in the energy of the d orbitals which is what

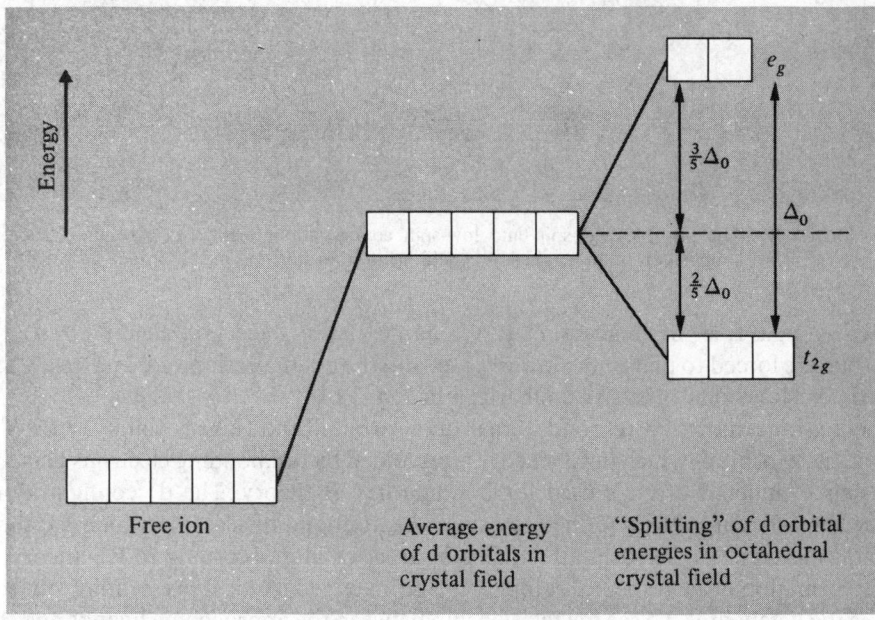

FIG. 19.13   The effect of imposing an octahedral crystal field on the d orbitals of a free ion.

would be observed if the d orbitals were individually spherically symmetrical and so were each affected identically. Secondly, the energies of the d orbitals split into two sets because two are axial and three are non-axial and so are affected by the ligands differently. Relative to the "average" energy the axial orbitals (described as the e_g set) increase in energy while the non-axial orbitals (described as the t_{2g} set) decrease in energy. Their separation is given the symbol Δ_o or $10Dq$. The major change in energy occurs during the first stage, but the second stage is important because it introduces the possibility of rearranging the metal d electrons. Since energy is required to force 2 electrons to occupy a single orbital, these d electrons will tend to remain unpaired in accordance with Hund's rules (p. 1090). However, they will also tend to occupy the orbitals of lowest energy. If Δ_o is large enough then electrons which would otherwise remain unpaired in the e_g orbitals may instead be forced to pair in the lower t_{2g} orbitals. This poses no problem for d^1, d^2, and d^3 configurations (since the t_{2g} set remain degenerate), nor for d^8 and d^9 configurations in which electrons are necessarily paired in the t_{2g} set. But for the d^4, d^5, d^6, and d^7 configurations two possibilities arise depending on the magnitude of Δ_o. If Δ_o is small (compared with electron–electron repulsion energies within one orbital), i.e. the crystal field is *weak*, then the maximum possible number of electrons remain unpaired and the configurations are

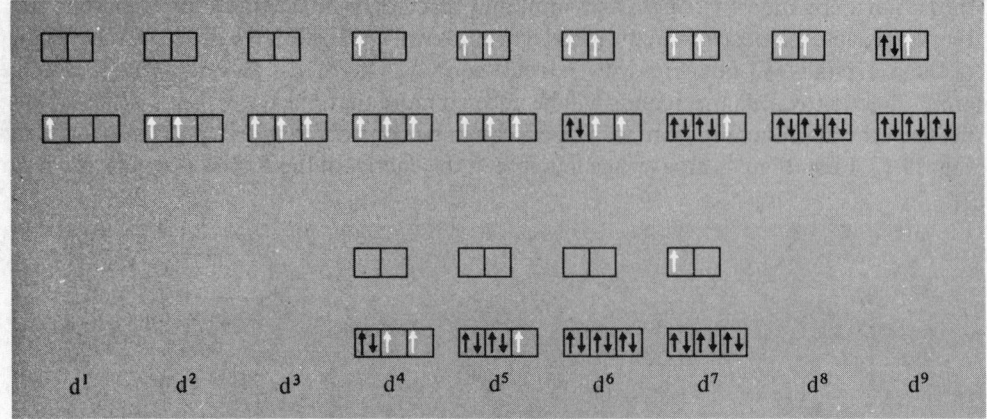

FIG. 19.14 The possible high-spin and low-spin configurations arising as a result of the imposition of an octahedral crystal field on a transition metal ion.

known as "spin-free" or "high-spin". If Δ_o is large, that is the crystal field is *strong*, then electrons are forced to pair in the lower t_{2g} set and the configurations are known as "spin-paired" or "low-spin". This is summarized in Fig. 19.14.

These arrangements correspond to the "outer orbital" and "inner orbital" types of VB theory, the e_g orbitals which in CF theory are avoided by nonbonding electrons being just that pair of inner d orbitals used for bonding in VB theory. The d^7 configuration is somewhat different in the two theories, since the nonbonding electron which VB theory promotes to some higher orbital, remains in an e_g orbital according to CF theory.

The formation of a tetrahedral complex can be imagined to occur in essentially the same way as the octahedral. To see the relationship between the approaching ligands and the d orbitals of the metal it is helpful to define the tetrahedron by reference to a cube. Now the axial *e* orbitals point towards the centres of the 6 faces of this cube, while the non-axial t_2 orbitals point towards the centres of the edges. (By convention, the *g* subscripts are omitted in the tetrahedral case.) The ligands then approach towards four corners of the cube (Fig. 19.15) and, though they do not approach any orbital along the direction of maximum electron density, it is quite clear that electrons in t_2 will be more strongly repelled than those in *e*. The d orbital splitting (Fig. 19.16) is therefore inverted as compared to the octahedral case and it can be shown that for the same metal, ligands and metal–ligand distances, $\Delta_t \simeq -\frac{4}{9}\Delta_o$. The occurrence of high- and low-spin configuration, this time for d^3, d^4, d^5, and d^6 ions, is again possible in principle, but in practice examples of low spin are very rare because of the smaller crystal field splittings involved.

Four-coordinate complexes with the alternative, square planar, stereochemistry can be described as distorted octahedral complexes in which the 2 *trans* ligands on the z-axis have been removed. This produces a great stabilization of the d_{z^2} orbital relative to the octahedral case, a smaller stabilization of the d_{xz} and d_{yz} pair and, because the loss of 2 ligands allows the remaining 4 to move closer to the metal, a destabilization of the $d_{x^2-y^2}$ and d_{xy} orbitals (Fig. 19.17). In square-planar complexes of the d^7 ion, Co^{II}, and the d^8 ions Ni^{II}, Pd^{II}, and Pt^{II}, electrons are forced to pair in the lower 4 orbitals leaving the top $d_{x^2-y^2}$ orbital empty. This corresponds closely to the VB result where this orbital is used for bonding.

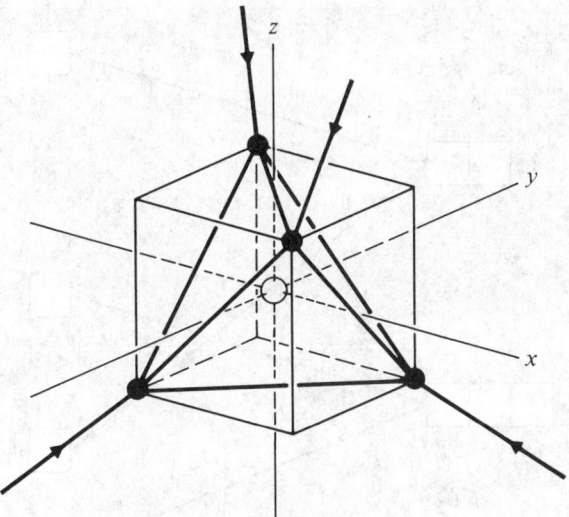

Fig. 19.15   Hypothetical formation of a tetrahedral complex.

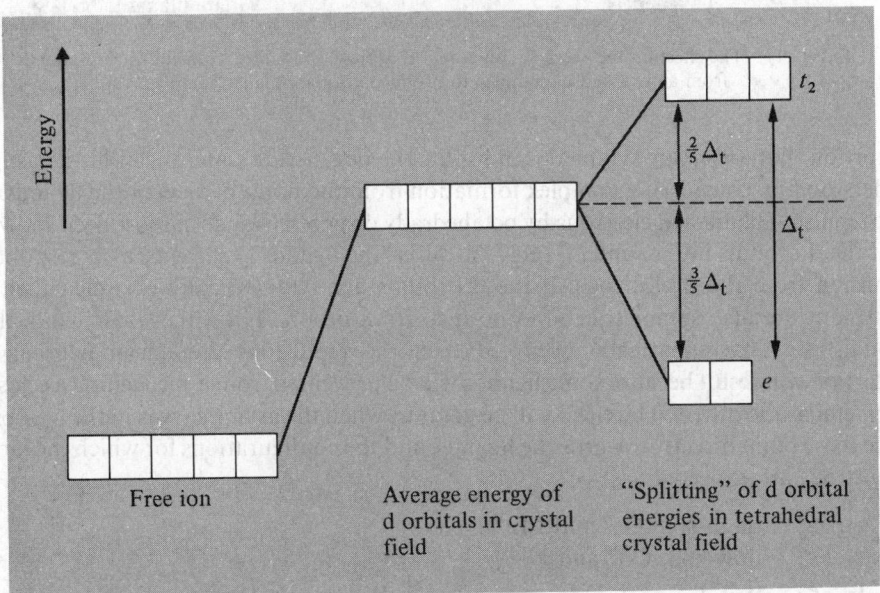

Fig. 19.16   The effect of imposing a tetrahedral field on the d orbitals of a free ion.

Tetragonally distorted octahedral complexes are intermediate between the octahedral and square-planar extremes and may arise, for instance, because the 6 ligands are not identical or because of the effect of ions outside the coordination sphere. Another important reason for such distortions is believed to be the *Jahn–Teller effect*. This is a manifestation of the theorem due to H. A. Jahn and E. Teller (1937) which states that a molecule in a degenerate electronic state will be unstable and will undergo a geometrical

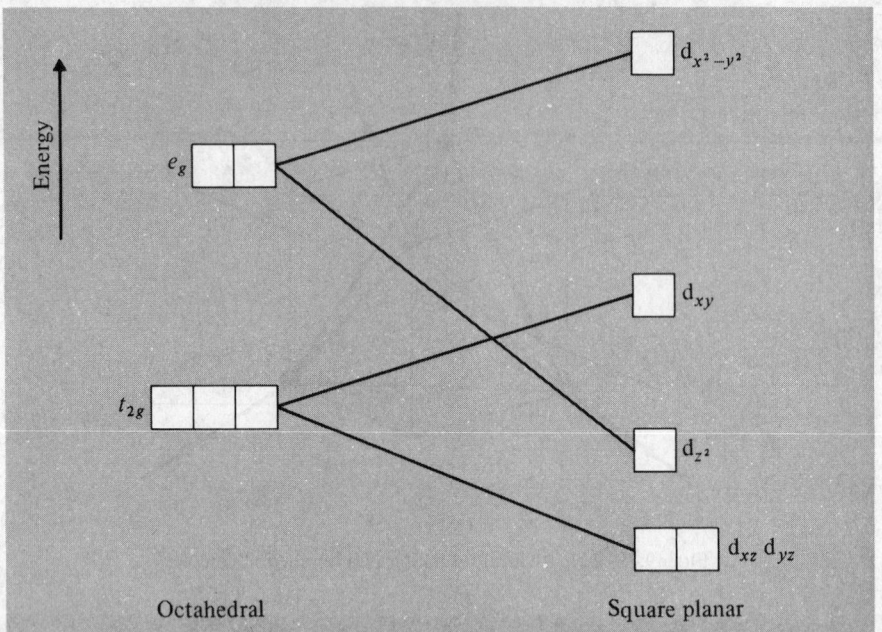

FIG. 19.17   The effect on the energy of an atom's d orbitals caused by removing 2 *trans*-ligands from an octahedral complex to produce a square-planar complex.

distortion that lowers its symmetry and splits the degenerate state. It can most easily be understood by considering complex formation from the point of view of the ligands.[14a] The repulsion of metal d electrons by octahedrally disposed ligands has just been shown to split the d orbitals into e_g and t_{2g} sets—however, the ligands experience a corresponding repulsion from the d electrons. If the d orbitals are symmetrically occupied then the repulsions are also symmetrical and no distortion occurs. If, on the other hand, the d orbitals are unsymmetrically occupied then the repulsions are unsymmetrical and distortion will result because some ligands will be prevented from approaching as close to the metal as the others. The effect will be greatest when the asymmetry is in the e_g orbitals since these point directly towards the ligands, and the configurations for which the largest distortions are expected are:

$$t_{2g}^{\,3}\,e_g^{\,1} \quad \text{(high-spin Cr}^{\text{II}} \text{ and Mn}^{\text{III}}\text{)}$$
$$t_{2g}^{\,6}\,e_g^{\,1} \quad \text{(low-spin Co}^{\text{II}} \text{ and Ni}^{\text{III}}\text{)}$$
$$t_{2g}^{\,6}\,e_g^{\,3} \quad \text{(Cu}^{\text{II}}\text{)}$$

If it is the d_{z^2} orbital which contains 1 electron more than the $d_{x^2-y^2}$, then the distortion should consist of an elongation of the octahedron along the z-axis. If, however, it is the $d_{x^2-y^2}$ which contains the extra electron, then elongations along the x- and y-axes (equivalent to compression along the z-axis) are expected. It is not possible to predict which type of distortion will occur but, in fact, elongation along one axis is the more commonly observed.

[14a] I. B. BERSUKER, Jahn–Teller effects in crystal chemistry and spectroscopy, *Coord. Chem. Rev.* **14**, 357–412 (1975).

The similarity of the predictions of VB and CF theories regarding the number of unpaired electrons shows that the use of magnetic measurements to distinguish different oxidation states and stereochemistries is still valid. In addition, however, CF theory offers the possibility of interpreting the colours of transition metal complexes.

19.8   Colours of Complexes[15]

In general a substance is coloured if it is capable of some internal energy change requiring an amount of energy E which can be supplied by a quantum of visible light. The frequency v of this light is defined by $E = hv$, or $E = hc/\lambda$, where h is Planck's constant and c is the velocity of light. The values of E corresponding to the frequency range of visible light are such as occur in electronic transitions (Fig. 19.18), and it is the possible electronic transitions which must therefore be examined if the colours of transition metal compounds are to be explained.

The simplest situation is the d^1 configuration, exemplified by the Ti^{III} ion, which in aqueous solution is reddish violet. This information is given more fully by its absorption spectrum (see Fig. 19.20, p. 1095) which shows that the observed colour arises because of absorption of the green portion of the incident white light. This can be satisfactorily explained if we assume that the aqueous solution of Ti^{III} contains octahedral

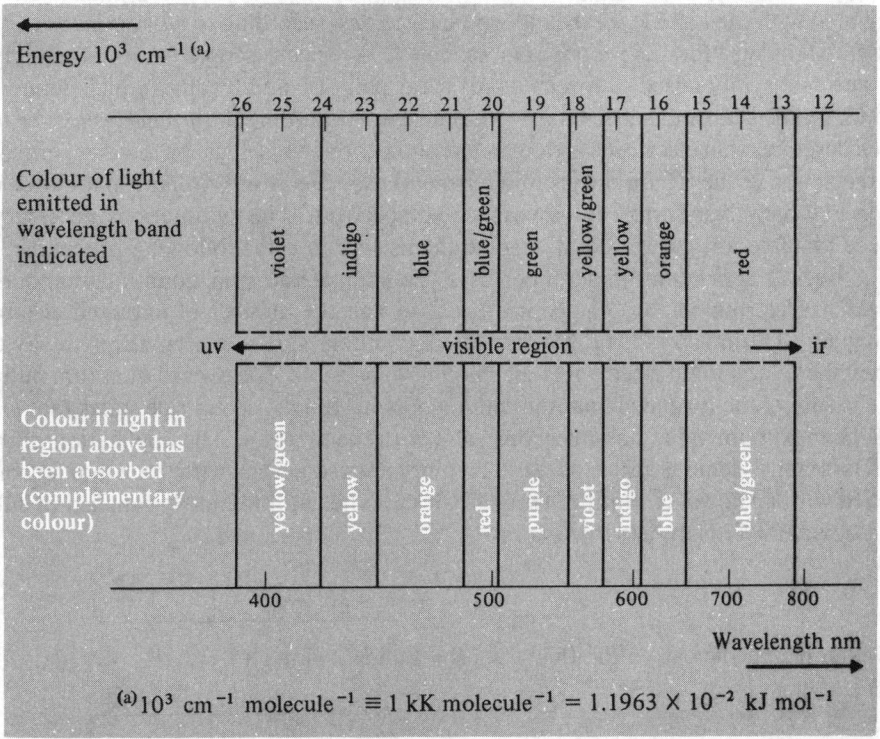

FIG. 19.18   Colours of the visible spectral region.

[15] B. N. FIGGIS, *Introduction to Ligand Fields*, Chap. 9, pp. 203–47, Interscience, New York, 1966.

$[Ti(H_2O_6]^{3+}$, and that absorption of light at $\sim 20\,000$ cm^{-1} (or ~ 500 nm) causes the promotion of the single d electron from a lower t_{2g} to a higher e_g orbital. Furthermore it indicates the value of Δ_o ($= hcN_o/\lambda$): substituting the values for $h = 6.626 \times 10^{-37}$ kJ s, $c = 2.998 \times 10^8$ m s^{-1}, Avogadro's constant $= 6.022 \times 10^{23}$ mol^{-1}, and 500 nm $= 500 \times 10^{-9}$ m gives $\Delta_o \simeq 239$ kJ mol^{-1}. [In practice it is usual simply to say $\Delta_o = 20\,000$ cm^{-1}, though of course cm^{-1} is not a real unit of energy.]

Values of Δ_o are usually of the same order of magnitude as bond energies, and in fact this one is very similar to that of the Cl–Cl bond in Cl$_2$ (242 kJ mol^{-1}).

The d^9 system, being 1 electron short of a filled d shell, is very similar to the d^1 because its single "positive hole" is the inverse of an electron and requires an energy Δ_o to move it from an e_g to a t_{2g} orbital.

Configurations d^2 to d^8 are not so simple and require a rather more complicated treatment. The proper CF theory explanation of their spectra requires an understanding of the way an electrostatic field splits the *spectroscopic terms* of a metal ion, and the details of this are beyond the scope of this account. Accepting some loss of precision, however, it is possible to relate these terms to electron configurations and so obtain a more readily understandable account.

Just as the angular momentum of an individual electron is characterized by quantum numbers l and m_s (see Chap. 2), so the angular momentum of a multi-electron atom can be characterized by the analogous quantum numbers L and S. The angular momenta associated with L and S are derived by the vectorial summation of the angular momenta associated with the l and m_s of the individual electrons, according to what is known as the Russell–Saunders (or *LS*) coupling scheme.† A spectroscopic term is simply a conventional symbol used to specify a particular pair of L and S values which defines the angular momentum and hence (in part) the energy of the atom. The *ground term* is the term which describes that electronic arrangement of the atom which has the lowest energy, and it is the action of the CF on the ground term of the isolated transition metal ion (free ion) which is largely responsible for the observed spectrum. The summations necessary to deduce this free-ion ground term, resolve themselves into the following procedure.

The overall spin quantum number S is the sum of the spin quantum numbers of individual electrons ($m_s = \pm\frac{1}{2}$), i.e. S is equal to half the number of unpaired electrons. According to Hund's first rule, S takes the maximum value it can, which is to say that the number of unpaired electrons is the maximum possible. The overall quantum number L is the sum of the magnetic quantum numbers (m_l). Hund's second rule states that L will take the maximum value possible, consistent with the first rule. The application of these rules is readily demonstrated using a common box diagram with each box (orbital) labelled with an m_l value. Both of Hund's rules are then automatically satisfied by filling these boxes singly from left to right.

$$m_l = \quad +2 \quad +1 \quad 0 \quad -1 \quad -2$$

For example, in the case of d^3 this gives $S = \frac{3}{2}$ and $L = 2+1+0 = 3$

$$m_l = \quad +2 \quad +1 \quad 0 \quad -1 \quad -2$$
$$\uparrow \quad \uparrow \quad \uparrow$$

† This coupling scheme works well for the 3d transition elements but is less satisfactory for the 4d and 5d transition elements.

These values of L and S define the spectroscopic ground term of a d^3 ion and the conventional symbolism describes it as ^4F. The superscript is the "spin multiplicity", $2S+1$ and the letter arises from the value of L in exactly the same way that the symbols for orbitals arise from the value of l, viz.:

$$S, P, D, F, \ldots, \text{ correspond to } L = 0, 1, 2, 3, \ldots$$

In this way the free-ion ground terms of d^x ions can readily be shown to be:

$x =$	1	2	3	4	5	6	7	8	9
Ground term $=$	^2D	^3F	^4F	^5D	^6S	^5D	^4F	^3F	^2D

Besides the spin multiplicity already noted, each of these terms has an *orbital multiplicity* of $2L+1$. This is analogous to the multiplicity $2l+1$, which tells us how many degenerate orbitals there are of a particular kind (s, p, d, or f) in a free-ion. Just as CF lifts the degeneracy of atomic orbitals, so it lifts the orbital degeneracy of the ground terms and produces a number of component terms as shown in Fig. 19.19.

The symbols used for the component terms and the subscripts appended are related to the symmetries of the atomic wave functions and g should be added to the subscript in the octahedral case. It is sufficient for present purposes if the subscripts are ignored and it is simply understood that A is an orbitally non-degenerate term, E is doubly degenerate, and

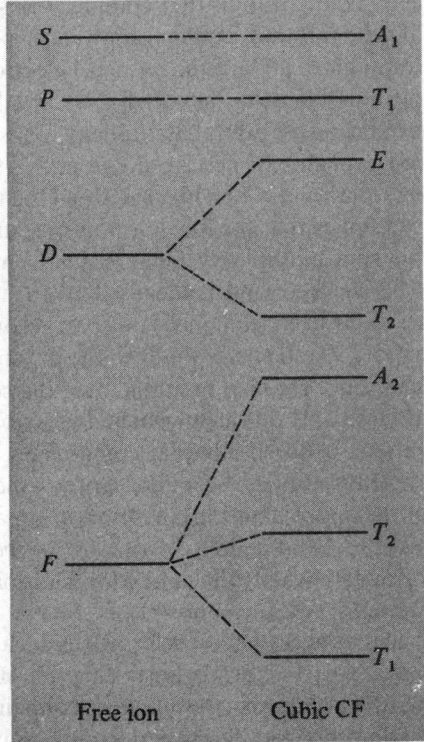

FIG. 19.19   The splitting of free-ion spectroscopic terms by a cubic crystal field.

T is triply degenerate. It is then plain that the orbital multiplicity of the free-ion terms is always retained in the aggregate of the component terms into which it splits. The spin multiplicity, which is the same for the component terms as for the parent term, is added as a superscript as before.

The splitting of the *D* and *F* terms under the action of a cubic CF is not always as shown above; it may be inverted and it is important to know when this will be. The problem is that, whereas for the splitting of d orbitals it is only necessary to know whether the cubic CF is octahedral or tetrahedral in order to decide whether t_{2g} or e orbitals lie lowest, for *D* and *F* terms it is also necessary to know the number of electrons involved before the splitting can be described.

For a d^1 ion in an octahedral field, the ground configuration is t_{2g}^1 and the single electron can occupy any of the 3 degenerate t_{2g} orbitals. It therefore has an orbital degeneracy of 3, corresponding to a ground *T* term. If the electron is excited to an e_g orbital it then has an orbital degeneracy of only 2, so that the excited e_g^1 configuration corresponds to an *E* term at an energy Δ above the *T*. Conversely, a d^9 ion in an octahedral field has the ground configuration $t_{2g}^6 e_g^3$, when the "positive hole" can occupy either of the 2 degenerate e_g orbitals. This is an orbital degeneracy of 2, corresponding to an *E* term. If a t_{2g} electron is excited to the e_g set, then the "hole" is effectively transferred to a t_{2g} orbital and has an orbital degeneracy of 3. The excited $t_{2g}^5 e_g^4$ configuration therefore corresponds to a *T* term at an energy of Δ above the *E*. In other words, the ground term of d^9 is inverted with respect to that of d^1. The d^1 and d^9 configurations, however, are just those whose spectra can most easily be accounted for on the basis of a single electron "jump", and it is with the d^2 configuration that complications arise.

In an octahedral field, the ground configuration t_{2g}^2 permits three degenerate arrangements and so corresponds to a *T* ground term. If 1 electron is excited it will be able to occupy either of the e_g orbitals while the unexcited electron will be able to occupy any of the three t_{2g} orbitals. This indicates an orbital multiplicity of $2 \times 3 = 6$, i.e. enough for two *T* terms. With the *T* ground term already described, this makes three *T* terms with a total orbital multiplicity of 9, yet we started with a free-ion *F* ground term which has an orbital multiplicity of only 7. This implies that the free ion *F* term is in fact accompanied by an excited *P* term of the same spin multiplicity. That this is true for every ion with an *F* ground term can readily be checked, and it means that a total orbital multiplicity of $7 + 3 = 10$ (of which 9 have so far been explained) must be accounted for. The remaining term must be a non-degenerate *A* and corresponds to the promotion of *both* electrons to the excited e_g orbitals where, since the spin multiplicity of the component term does not change, they remain unpaired so only one arrangement is possible. On this basis it is to be expected that both separations between adjacent components of the *F* term will be Δ, corresponding to the promotion of first one, and then the other electron from t_{2g} to e_g orbitals. It is a measure of the approximations involved in equating terms with electron configurations that, whereas the *A* and middle *T* terms are indeed separated by Δ, the two *T* terms are only $4\Delta/5$ ($8Dq$) apart. Exactly the same consideration can be applied to all the d^x ions and Table 19.3 summarizes the results.

Before continuing further it is necessary to refer briefly to a rule, known as the *spin selection rule* which simplifies matters considerably. This rule states that only electronic transitions which do not involve a change in the number of unpaired electrons are allowed. This means that only transitions between terms of the same (maximum) spin multiplicity as the ground term need be considered and other (excited) terms can be ignored. Now,

TABLE 19.3   *Correlation of spectroscopic ground terms with electron configuration for d^x ions in octahedral CF[(a)] (for simplicity, g subscripts are omitted)*

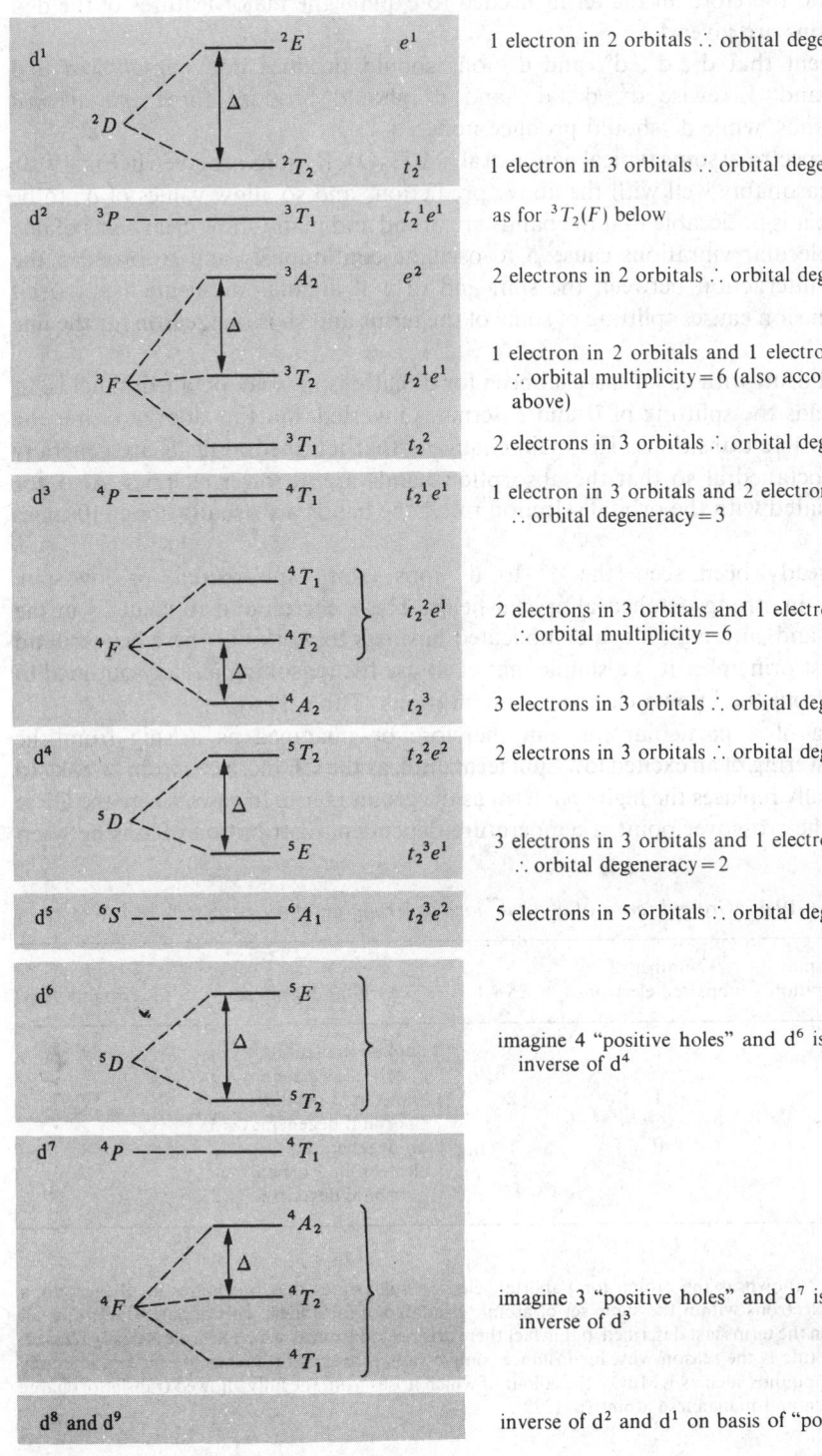

^2E e^1	1 electron in 2 orbitals ∴ orbital degeneracy $= 2$
$^2T_2$ t_2^1	1 electron in 3 orbitals ∴ orbital degeneracy $= 3$
$^3T_1$ $t_2^1e^1$	as for $^3T_2(F)$ below
$^3A_2$ e^2	2 electrons in 2 orbitals ∴ orbital degeneracy $= 1$
$^3T_2$ $t_2^1e^1$	1 electron in 2 orbitals and 1 electron in 3 orbitals ∴ orbital multiplicity $= 6$ (also accounts for $^3T_1(P)$ above)
$^3T_1$ t_2^2	2 electrons in 3 orbitals ∴ orbital degeneracy $= 3$
$^4T_1$ $t_2^1e^1$	1 electron in 3 orbitals and 2 electrons in 2 orbitals ∴ orbital degeneracy $= 3$
$^4T_1$ $^4T_2$ $t_2^2e^1$	2 electrons in 3 orbitals and 1 electron in 2 orbitals ∴ orbital multiplicity $= 6$
$^4A_2$ t_2^3	3 electrons in 3 orbitals ∴ orbital degeneracy $= 1$
$^5T_2$ $t_2^2e^2$	2 electrons in 3 orbitals ∴ orbital degeneracy $= 3$
^5E $t_2^3e^1$	3 electrons in 3 orbitals and 1 electron in 2 orbitals ∴ orbital degeneracy $= 2$
$^6A_1$ $t_2^3e^2$	5 electrons in 5 orbitals ∴ orbital degeneracy $= 1$
^5E $^5T_2$	imagine 4 "positive holes" and d^6 is then the exact inverse of d^4
$^4A_2$ $^4T_2$ $^4T_1$	imagine 3 "positive holes" and d^7 is then the exact inverse of d^3
d⁸ and d⁹	inverse of d^2 and d^1 on basis of "positive holes"

[(a)] In every case the splitting is reversed for tetrahedral CFs.

careful inspection of Table 19.3 reveals that for each d^x ion all possible electron arrangements involving the maximum number of unpaired electrons have been considered, and therefore all the terms needed to explain the major features of the d–d spectra of 3 ions are given.†

It is apparent that d^1, d^4, d^6, and d^9 ions should produce one *spin-allowed* d–d absorption band. Likewise d^2, d^3, d^7, and d^8 should produce three *spin-allowed* absorption bands, while d^5 should produce none.

The actual spectra of some typical octahedral $[M(H_2O)_6]^{n+}$ ions are given in Fig. 19.20. They agree reasonably well with the above predictions and so allow values of Δ_o to be estimated, but it is noticeable that the bands are broad and many show clear signs of fine structure. Molecular vibrations cause Δ to oscillate continuously and so broaden the bands, while interaction between the spin and orbital angular momenta (spin–orbit coupling) of the ion causes splitting of some of the terms and so is one reason for the fine structure.

The spectra dealt with so far have all been for d^x cations in weak octahedral fields. In tetrahedral fields the splitting of D and F terms is inverted, but this does not alter the number of possible transitions. The main change is that tetrahedral fields are generally smaller than octahedral so that the absorption bands are at lower energies. Also, for reasons associated with the orbital selection rule,† the bands are usually about 10 times more intense.

As has already been seen, the d^4 to d^7 ions adopt spin-paired, or low-spin, configurations in strong octahedral crystal fields. These correspond to changes in the ground terms and, although it is a complicated business to work out these new ground terms from first principles, it is a simple matter to use the reasoning already outlined to deduce them from the known electron configurations (Table 19.4).

Spin-pairing of a particular ion can therefore be imagined as arising from the progressive lowering of an excited low-spin term until, as the CF increases from "weak" to "strong", it finally replaces the high-spin term as the ground term. In cases where the CF is very close to the crossover point, a temperature-dependent distribution of ions between

TABLE 19.4   *Ground terms of d^4 to d^7 ions in strong octahedral crystal fields*

Ion	Low-spin configuration	Number of unpaired electrons	$\therefore 2S+1$	Orbital degeneracy	$\therefore$ Ground term
d^4	t_{2g}^4	2	3	2 "holes" in 3 orbitals $\therefore$ orbital degeneracy $= 3$	^3T
d^5	t_{2g}^5	1	2	1 "hole" in 3 orbitals $\therefore$ orbital degeneracy $= 3$	^2T
d^6	t_{2g}^6	0	1	Non-degenerate	^1A
d^7	$t_{2g}^6 e_e^1$	1	2	1 electron in 2 orbitals $\therefore$ orbital degeneracy $= 2$	^2E

† A second rule, known as the orbital (or Laporte) selection rule, states that transitions involving only a redistribution of electrons within the same set of atomic orbitals are forbidden. This apparently forbids all transitions between the terms just described, but in fact there are mechanisms by which the rule is partly relaxed. Nevertheless, this rule is the reason why, for instance, simple salts of transition metals are far less intensely coloured than compounds such as $KMnO_4$, the colour of which arises from the fully allowed transfer of charge between the oxygen and manganese atoms (p. 1222).

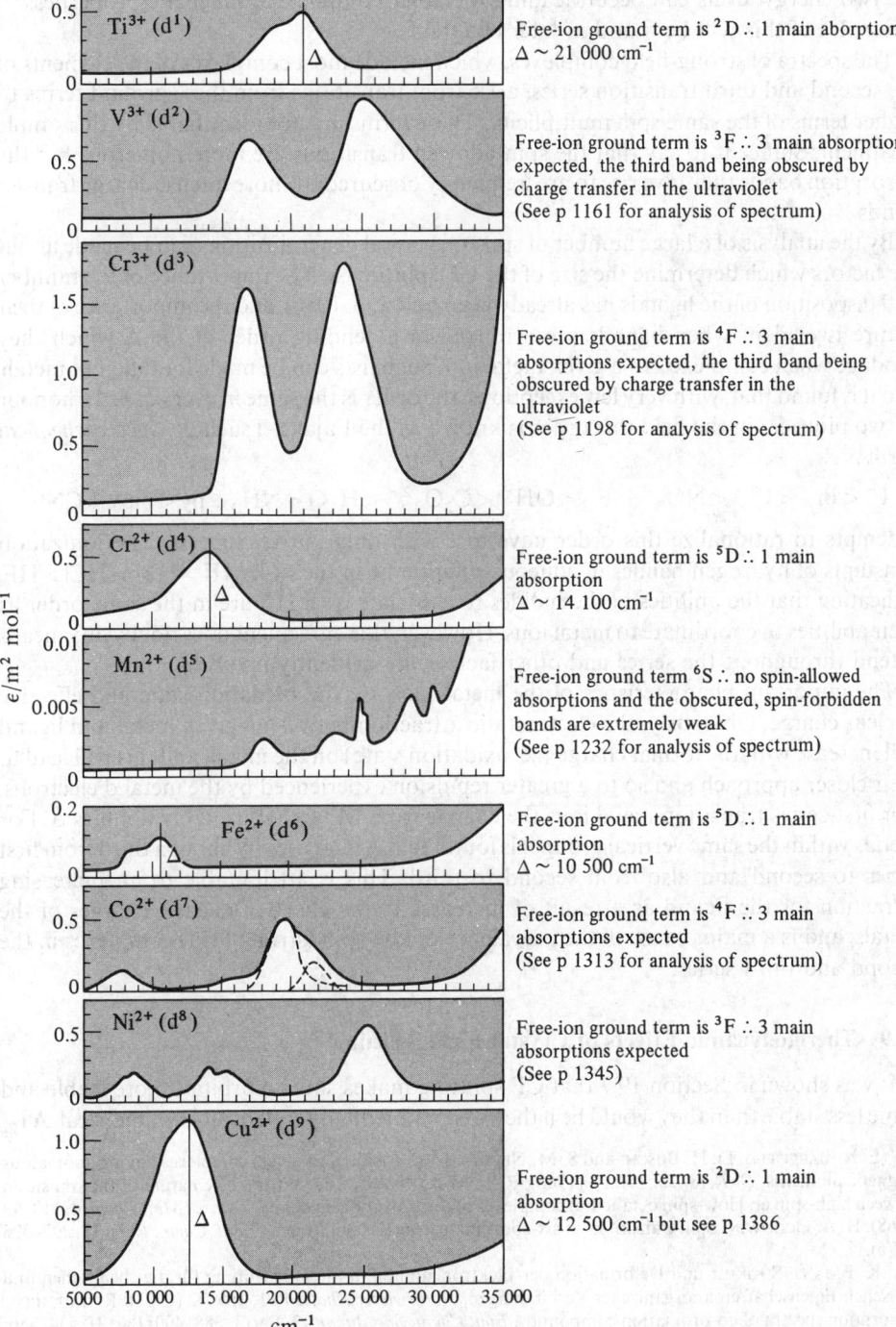

Free-ion ground term is ^2D ∴ 1 main aborption
$\Delta \sim 21\,000\ cm^{-1}$

Free-ion ground term is ^3F ∴ 3 main absorptions
expected, the third band being obscured by
charge transfer in the ultraviolet
(See p 1161 for analysis of spectrum)

Free-ion ground term is ^4F ∴ 3 main
absorptions expected, the third band being
obscured by charge transfer in the
ultraviolet
(See p 1198 for analysis of spectrum)

Free-ion ground term is ^5D ∴ 1 main
absorption
$\Delta \sim 14\,100\ cm^{-1}$

Free-ion ground term ^6S ∴ no spin-allowed
absorptions and the obscured, spin-forbidden
bands are extremely weak
(See p 1232 for analysis of spectrum)

Free-ion ground term is ^5D ∴ 1 main
absorption
$\Delta \sim 10\,500\ cm^{-1}$

Free-ion ground term is ^4F ∴ 3 main
absorptions expected
(See p 1313 for analysis of spectrum)

Free-ion ground term is ^3F ∴ 3 main
absorptions expected
(See p 1345)

Free-ion ground term is ^2D ∴ 1 main
absorption
$\Delta \sim 12\,500\ cm^{-1}$ but see p 1386

FIG. 19.20   The absorption spectra of aqueous solutions of first transition series ions. (From *Introduction to Ligand Fields*, B. N. Figgis, Interscience, 1966.)

the two energy levels can occur, leading to rather complicated magnetic properties.[16] Examples of this will be found in later chapters.

The spectra of strong-field complexes, which include most complexes of the elements of the second and third transition series, arise from transitions from these ground terms to higher terms of the same spin multiplicity. These terms are not identifiable by this simple treatment. Suffice it to say that the spin-allowed transitions are more numerous but the absorption bands they give rise to are frequently obscured by more intense charge transfer bands.

By the analysis of a large number of spectra, several generalizations can be made about the factors which determine the size of the CF splitting Δ. The importance of the number and disposition of the ligands has already been noted (p. 1086), and the importance of their nature is evident when ligands are arranged in ascending order of the Δ which they produce when complexed to a given metal ion. Such lists can be made for different metals and it is found that, with very few exceptions, the order is the same in every case. In honour of two pioneers in the field this order is known as the Fajans–Tsuchida *Spectrochemical Series*,[17]

$$I^- < Br^- < Cl^- < NO_3^- < F^- < OH^- < C_2O_4^{2-} < H_2O < NH_3 \simeq py < bipy \ll CN^-$$

Attempts to rationalize this order have met with only partial success. The ionization constants of hydrogen halides in aqueous solution lie in the order $HI > HBr > HCl > HF$, indicating that the abilities of the halides to associate with H^+ are in the same order as their abilities to coordinate to metal ions. However, this agreement does not by any means extend throughout the series and other factors are evidently involved.

The important characteristics of the metal ions are the oxidation state and effective nuclear charge. Obviously, the electrostatic attraction between a given metal and ligand will increase with the formal charge (i.e. oxidation state) on the metal, and this will lead to their closer approach and so to a greater repulsion experienced by the metal d electrons. For first-series transition metal ions the change from M^{II} to M^{III} roughly doubles Δ. For metals within the same vertical group it is found that Δ increases by about a third from first series to second and also from second to third. This is attributable to an increasing attraction for the ligand as a result of increases in the effective nuclear charges of the metals, and is a major cause of the prevalence of low-spin (strong field) complexes in the second and third series.

19.9 Thermodynamic Effects of Crystal Field Splitting[18]

It was shown in Section 19.7 that CF splitting makes some d orbitals more stable and some less stable than they would be if they were individually spherically symmetrical. A t_{2g}

---

[16] E. K. BAREFIELD, D. H. BUSCH, and S. M. NELSON, Iron, cobalt, and nickel complexes having anomalous magnetic moments, *Q. Revs.* **22**, 457–98 (1968). R. L. MARTIN and A. H. WHITE, The nature of the transition between high-spin and low-spin octahedral complexes of the transition elements, *Transit. Met. Chem.* **4**, 113–98 (1968). H. A. GOODWIN, Spin transitions in six-coordinate iron(II) complexes, *Coord. Chem. Revs.* **18**, 293–325 (1976).

[17] K. FAJANS, Struktur und Deformation der Elektronenhüllin in ihrer Bedeutung für die chemischen und optischen Eigenschaften anorganischer Verbindungen, *Naturwissenschaften* **11**, 165–72 (1923). R. TSUCHIDA, Absorption spectra of co-ordination compounds, *Bull. Chem. Soc. Japan* **13**, Part I, 388–400; Part II, 434–450; Part III, 471–80 (1938). C. K. JØRGENSEN, *Absorption Spectra and Chemical Bonding in Complexes*, Chap. 7, pp. 107–33, Pergamon Press, Oxford, 1962.

[18] P. GEORGE and D. S. McCLURE, The effect of inner orbital splitting on the thermodynamic properties of transition metal compounds and coordination complexes, *Prog. Inorg. Chem.* **1**, 381–463 (1959).

electron increases the stability of an octahedral complex by $\frac{2}{5}\Delta_o$ and an e_g electron decreases it by $\frac{3}{5}\Delta_o$. The net effect of all the d electrons represents the additional stability which may be thought to accrue because the CF splits the d orbitals and is known as the crystal field stabilization energy (CFSE).

No. of d electrons	0	1	2	3	4	5	6	7	8	9	10
CFSE/Δ_o (for high-spin octahedral ions)	0	$\frac{2}{5}$	$\frac{4}{5}$	$\frac{6}{5}$	$\frac{3}{5}$	0	$\frac{2}{5}$	$\frac{4}{5}$	$\frac{6}{5}$	$\frac{3}{5}$	0

This extra stability manifests itself in the experimental values of such thermodynamic quantities as ligation energies, lattice energies, and standard reduction potentials. The most complete set of data is the hydration energies of bivalent ions

$$M^{2+}(g) + \infty H_2O \longrightarrow M^{2+}aq + \Delta H^\circ$$

As shown in Fig. 19.21, the experimental values of ΔH° lie on an irregular double-humped curve.

If, however, the CFSE is in each case subtracted from ΔH°, then the "corrected" values fall very nearly on a smooth curve passing through the points for the spherically symmetrical d^0, d^5, and d^{10} ions. CFSE therefore affords a ready explanation[19] of the Irving–Williams order of stability (p. 1065) and was taken by many to indicate the essential correctness of the electrostatic CF model. It was, indeed, one of the major reasons why chemists accepted it.

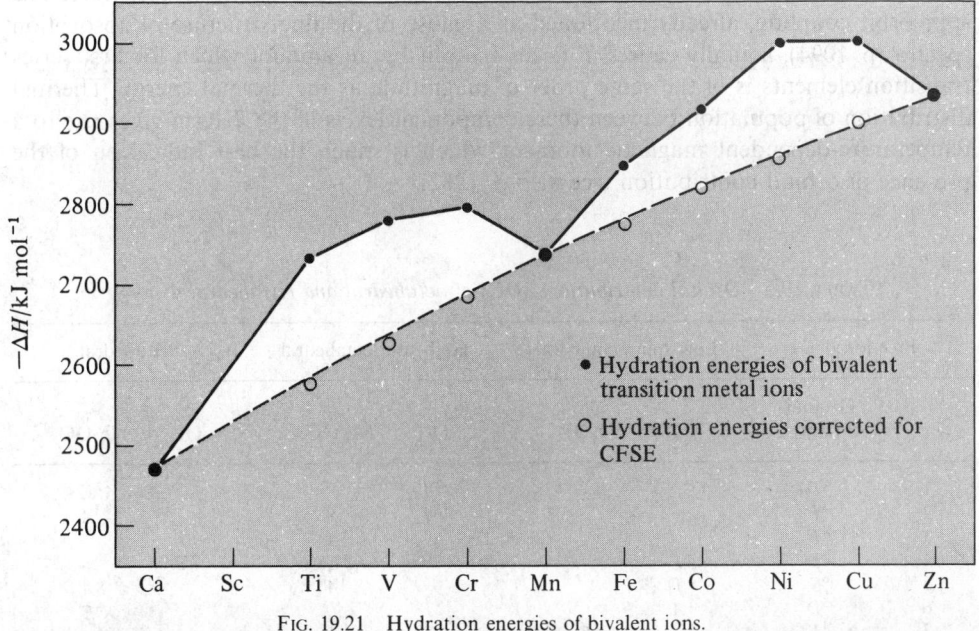

FIG. 19.21   Hydration energies of bivalent ions.

[19] J. BJERRUM and C. K. JØRGENSEN, Crystal field stabilization and tendency towards complex formation of first transition group elements, *Rec. Trav. Chim.* **75**, 658–64 (1956).

19.10 Magnetic Properties

Measurements of effective magnetic moment μ_e can be used to estimate the number of unpaired electrons n and hence the stereochemistry, bond type, or oxidation state of a coordinated metal ion by means of either the VB or CF theory. At its simplest this has meant using the so-called "spin-only" formula:

$$\mu_e = \sqrt{n(n+2)} \ \text{BM} \quad \text{(i.e. Bohr magnetons)}$$

This equality, as its name implies, is derived by assuming that the spin angular momentum alone is effective in producing a magnetic moment. In fact the orbital angular momentum also contributes to a greater or lesser extent, and experimental values of μ_e differ from the spin-only value accordingly. It is a distinct advantage of CF over VB theory that it offers an explanation for this.

Orbital angular momentum (i.e. the rotation of electronic charge around the nucleus) can be generated and so produce an orbital contribution to μ_e if at least 2 degenerate orbitals of the same shape are unequally occupied. Under these circumstances an electron is able to rotate about a particular axis by interconversion of the degenerate orbitals. These are actually the very same circumstances which produce a T ground term, whereas A and E ground terms are produced when orbital contribution is excluded. It has already been remarked that octahedral and tetrahedral CFs produce mutually inverted term-splittings so that for every d^x ion, one of these stereochemistries produces a T ground term and the other an A or E ground term (Table 19.5). If an orbital contribution can be detected, then the stereochemistries can be distinguished. In some cases values of μ_e in excess of the spin-only value are sufficient to indicate orbital contribution. However, this is rarely reliable and a better method is to measure μ_e over a range of temperature. The spin–orbit coupling, already mentioned as a cause of the fine structure of absorption spectra (p. 1094), actually causes T terms to split by an amount which for first series transition elements is of the same order of magnitude as the thermal energy. Thermal distribution of population between these component levels of the T term gives rise to a temperature-dependent magnetic moment which is much the best indication of the presence of orbital contribution (see also p. 1262).

TABLE 19.5   *Orbital contributions (OC) of octahedral and tetrahedral* d^x *ions*

		Low-spin octahedral		High-spin octahedral		Tetrahedral	
x	Ground term	OC	No OC	OC	No OC	OC	No OC
1	^2D			^2T			^2E
2	^3F			^3T			^3A
3	^4F				^4A	^4T	
4	^5D	^3T			^5E	^5T	
5	^6S	^2T			^6A		^6A
6	^5D	^1A		^5T			^5E
7	^4F		^2E	^4T			^4A
8	^3F				^3A	^3T	
9	^2D				^2E		^2T

19.11 Ligand Field Theory

In spite of the not inconsiderable success of the electrostatic CF model in interpreting physical properties of a number of transition metal compounds, there is much evidence to suggest that covalency also plays an essential part in coordinate bonding. Some of the most direct evidence is provided by electron spin resonance and nuclear magnetic resonance measurements, showing that appreciable unpaired electron density lies on the ligand and not the metal; the order of ligands in the spectrochemical series cannot be explained on purely electrostatic grounds, and it is clear that any satisfactory explanation must involve a variety of factors of which covalency is one; the variation of heats of hydration of M^{II} ions is very well explained by the CFSE but, for M^{III} ions with which a higher degree of covalency is to be expected, this is not adequate to explain the observed variation.

In order to make allowance for a degree of covalency within a basically electrostatic description, a number of adjustments have been made. In magnetic theory the "orbital reduction factor" or "electron delocalization factor" k is introduced to account for the reduction in orbital contribution due to the delocalization of d-electron density from metal to ligand. The Racah "interelectron repulsion parameter" B is similarly introduced in the analysis of absorption spectra. The further the value of B in a complex is reduced below that in the free ion, the more the d-electron charge is assumed to have expanded. In 1958 C. E. Schäffer and C. K. Jørgensen[20] arranged a list of ligands in order of decreasing B and called it the nephelauxetic (cloud expanding) series:

$$F^-, H_2O, NH_3, C_2O_4^{2-}, en, Cl^-, CN^-, Br^-, I^-$$

Like the spectrochemical series it is more or less independent of metal ion but, unlike the spectrochemical series, it is much more nearly the order (left to right) to be expected for increasing covalency.

19.12 Molecular Orbital Theory

This is undoubtedly the most complete theory, encompassing as it does all possibilities from the electrostatic to the covalent extreme. Unfortunately, it is also the most complicated, and quantitative calculations are made only with difficulty. However, modern computing techniques have gone a long way to overcoming this disadvantage and the results of various MO calculations are becoming increasingly available.

The fundamental assumption of MO theory is that metal and ligand orbitals will overlap and combine, providing they are of the correct symmetries to do so. In one approximation the appropriate AOs of the metal and atomic or molecular orbitals of the ligand, are used to produce the MOs by the linear combination of atomic orbitals (LCAO) method. Since combination of metal and ligand orbitals of widely differing energies can be neglected, only valence orbitals are considered.

In the case of an octahedral complex ML_6, the metal has 6 σ orbitals, i.e. the e_g pair of the nd set, together with the $(n+1)s$ and the three $(n+1)p$. The ligands each have one σ orbital (containing the lone-pair of electrons) and these are combined to give orbitals with the correct symmetry to overlap with the metal σ orbitals (Fig. 19.22). The 6 electron pairs

[20] C. E. SCHÄFFER and C. K. JØRGENSEN, The nephelauxetic series of ligands corresponding to increasing tendency of partly covalent bonding, *J. Inorg. Nuclear Chem.* **8**, 143–8 (1958).

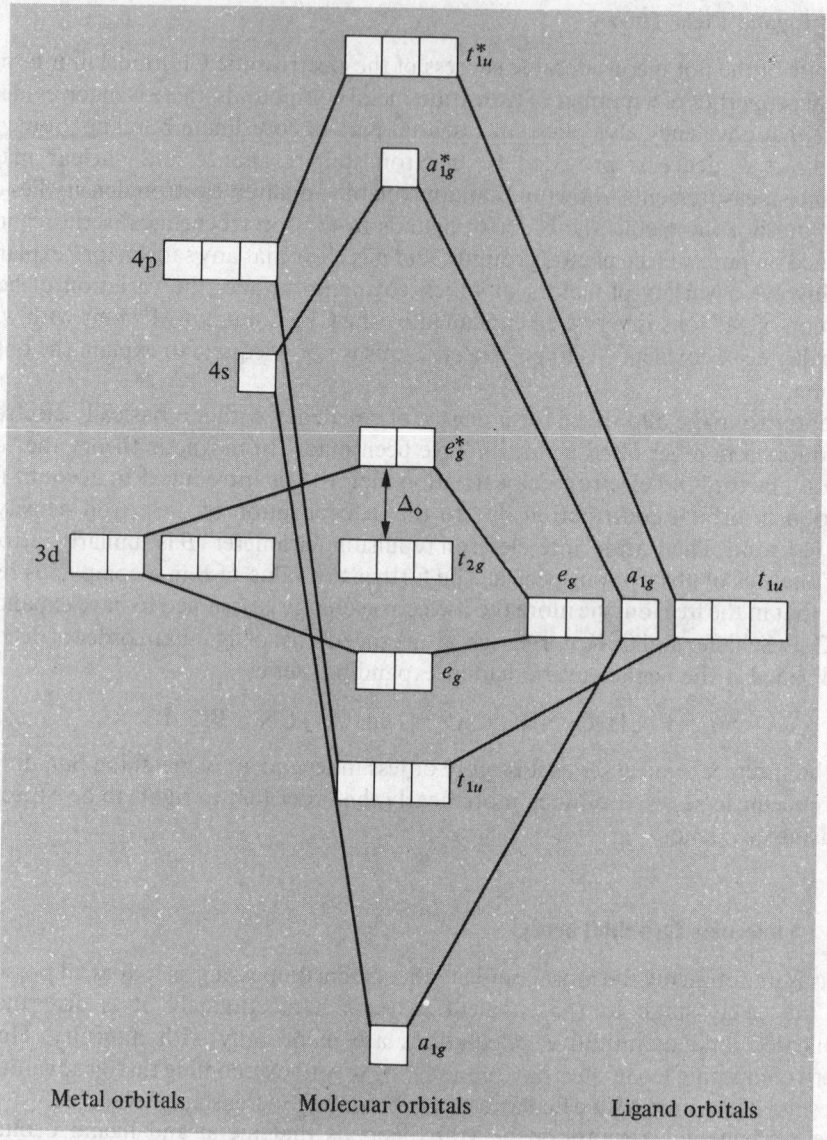

FIG. 19.22   Molecular orbital diagram for an octahedral complex of a first series transition metal (only σ interactions are considered in this simplified diagram).

from the ligands are placed in the 6 lowest MOs, leaving the nonbonding metal t_{2g} and the antibonding $e_g{}^*$ orbitals to accommodate the electrons originally on the metal. This central portion of the figure is indeed much the same as the e_g/t_{2g} splitting described in CF theory, with the difference that the $e_g{}^*$ orbitals now have some ligand character which implies covalency. The lower in energy the ligand orbitals are with respect to the AOs of the metal the nearer is the bonding to the electrostatic extreme.

If the ligand possesses orbitals of π as well as σ symmetry the situation is drastically changed because of the overlap of these orbitals with the t_{2g} orbitals of the metal. Two

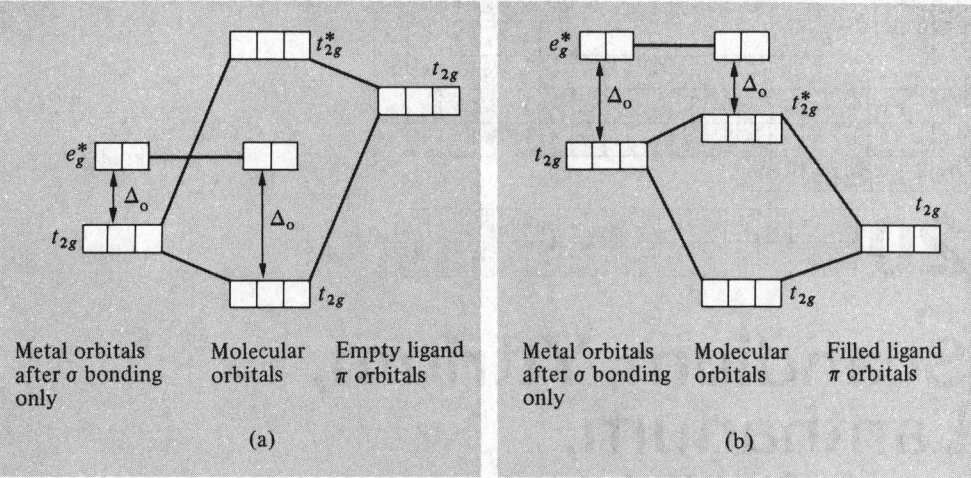

FIG. 19.23   Possible effects of π bonding on Δ_o: (a) when ligand π orbitals are empty, and (b) when ligand π orbitals are filled.

situations may arise. Either the ligand π orbitals are empty and of higher energy than the metal t_{2g}, or they are filled and of lower energy than the metal t_{2g} orbitals (Fig. 19.23). The former in effect increases Δ_o and is the most important case, including ligands such as CO, NO^+, and CN^-. This type of covalency, called π bonding or back bonding, provides a plausible explanation of the position of these ligands in the spectrochemical series and, more importantly, for the stability of such compounds as the metal carbonyls discussed in Chapter 8 (p. 349).

Similar MO treatments are possible for tetrahedral and square planar complexes but are increasingly complicated.

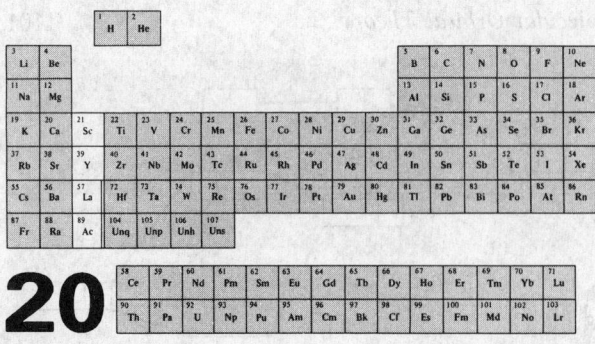

20

Scandium, Yttrium, Lanthanum, and Actinium

20.1 Introduction

In 1794 the Finnish chemist J. Gadolin, while examining a mineral recently discovered in a quarry at Ytterby, near Stockholm, isolated what he thought was a new oxide (or "earth") which A. G. Ekeberg in 1797 named yttria. In fact this was a mixture of a number of metal oxides from which yttrium oxide was separated by C. G. Mosander in 1843. This is actually part of the fascinating story of the "rare earths" to which we shall return in Chapter 30. The first sample of the metal, albeit very impure, was obtained by F. Wöhler in 1828 by the reduction of the trichloride by potassium.

Four years before isolating yttria, Mosander extracted lanthanum oxide as an impurity from cerium nitrate (hence the name from Greek λανθάνειν, to hide), but it was not until 1923 that metallic lanthanum in a relatively pure form was obtained by electrolysis of fused halides.

Scandium is also present in the Swedish ores from which yttrium and lanthanum had been extracted, but in only very small amounts and, probably for this reason, its discovery was delayed until 1879 when L. F. Nilsen isolated a new oxide and named it scandia. A few years later and with larger amounts at his disposal, P. T. Cleve prepared a large number of salts from this oxide and was able to show that it was the oxide of a new element whose properties tallied very closely indeed with those predicted by D. I. Mendeleev for eka-boron, an element missing from his classification (p. 32). It was only in 1937 that the metal itself was prepared by the electrolysis of molten chlorides of potassium, lithium, and scandium, and only in 1960 that the first pound of 99% pure metal was produced.

The final member of the group, actinium, was identified in uranium minerals by A. Debierne in 1899, the year after P. and M. Curie had discovered polonium and radium in the same minerals. However, the naturally occurring isotope, ^{227}Ac, is a β^- emitter with a half-life of 21.77 y and the intense γ activity of its decay products makes it difficult to study.

20.2 The Elements[1]

20.2.1 *Terrestrial abundance and distribution*

With the exception of actinium, which is found naturally only in traces in uranium ores, these elements are by no means rare though they were once thought to be so: Sc 25, Y 31, La 35 ppm of the earth's crustal rocks, (cf. Co 29 ppm). This was, no doubt, at least partly because of the considerable difficulty experienced in separating them from other constituent rare earths. As might be expected for class-a metals, in most of their minerals they are associated with oxoanions such as phosphate, silicate, and to a lesser extent carbonate.

Scandium is very widely but thinly distributed and its only rich mineral is the rare thortveitite, $Sc_2Si_2O_7$ (p. 401), found in Norway, but since scandium has no significant commercial use, this poses no great problem. In any case, considerable amounts of Sc_2O_3 are now available as a byproduct of the extraction of uranium. Yttrium and lanthanum are invariably associated with lanthanide elements, the former (Y) with the heavier or "yttrium group" lanthanides in minerals such as xenotime, $M^{III}PO_4$, and gadolinite, $M_2^{III}M_3^{II}Si_2O_{10}$ (M^{II} = Fe, Be), and the latter (La) with the lighter or "cerium group" lanthanides in minerals such as monazite, $M^{III}PO_4$, and bastnaesite, $M^{III}CO_3F$. This association of similar metals is a reflection of their ionic radii. While La^{III} is similar in size to the early lanthanides which immediately follow it in the periodic table, Y^{III}, because of the steady fall in ionic radius along the lanthanide series (p. 1431), is more akin to the later lanthanides.

20.2.2 *Preparation and uses of the metals*

Because of its very slight technological importance, little scandium is produced. Some is obtained from thortveitite, which contains 35–40% Sc_2O_3, but a major source is as a byproduct in the processing of uranium ores which contain only about 0.02% Sc_2O_3. Although a number of uses have been suggested they are all apparently such as can adequately be performed by cheaper alternatives.

Yttrium and lanthanum are both obtained from lanthanide minerals and the method of extraction depends on the particular mineral involved. Digestions with hydrochloric acid, sulfuric acid, or caustic soda are all used to extract the mixture of metal salts. Prior to the Second World War the separation of these mixtures was effected by fractional crystallizations, sometimes numbered in their thousands. However, during the period 1940–45 the main interest in separating these elements was in order to purify and characterize them more fully. The realization that they are also major constituents of the products of nuclear fission effected a dramatic sharpening of interest in the USA. As a result, ion-exchange techniques were developed and, together with selective complexation and solvent extraction, have now completely supplanted the older methods of separation (p. 1425). In cases where the free metals are required, reduction of the trifluorides with metallic calcium can be used.

---

[1] R. C. VICKERY, Scandium, yttrium and lanthanum, Chap. 31 in *Comprehensive Inorganic Chemistry*, Vol. 3, pp. 329–53, Pergamon Press, Oxford, 1973, and references therein. C. T. HOROVITZ (ed.), *Scandium: Its Occurrence, Chemistry, Physics, Metallurgy, Biology and Technology*, Academic Press, London, 1975, 598 pp. The most recent and comprehensive account of scandium.

In recent years yttrium has achieved some importance in the field of electronics. It is the basis of the phosphors used to produce the red colour on television screens, and yttrium garnets such as $Y_3Fe_5O_{12}$ are employed as microwave filters in radar. The related $Y_3Al_5O_{12}$ is used as a gemstone and has been used as a substitute for the famous Cartier diamond. Because of its low neutron cross-section, yttrium has considerable potential as a moderator in nuclear reactors though this use has yet to be developed.

Lanthanum has also found modest uses. Its oxide is an additive in high-quality optical glasses to which it imparts a high refractive index (sparkle). "Mischmetal", an unseparated mixture of lanthanide metals containing about 25% La, is used in making lighter flints, and some mixed oxide systems have been suggested as cheaper alternatives to platinum as catalysts in the pollution control of exhaust emissions from cars.

Actinium occurs naturally as a decay product of ^{235}U:

$$^{235}_{92}U \xrightarrow[7.04 \times 10^8 \text{ y}]{\alpha} {}^{231}_{90}Th \xrightarrow[25.52 \text{ h}]{\beta^-} {}^{231}_{91}Pa \xrightarrow[3.28 \times 10^4 \text{ y}]{\alpha} {}^{227}_{89}Ac \xrightarrow[21.77 \text{ y}]{\beta^-} {}^{227}_{90}Th - - - - \rightarrow$$

but the half-lives are such that one tonne of the naturally occurring uranium ore contains on average only about 0.2 mg of Ac. An alternative source is the neutron irradiation of ^{226}Ra in a nuclear reactor:

$$^{226}_{88}Ra + {}^1_0n \longrightarrow {}^{227}_{88}Ra \xrightarrow[42.2 \text{ min}]{\beta^-} {}^{227}_{89}Ac - - - - \rightarrow$$

In either case, ion-exchange or solvent extraction techniques are needed to separate the element and, at best, it can be produced in no more than milligram quantities. Large-scale use is therefore impossible even if desired.

20.2.3 *Properties of the elements*

A number of the properties of Group IIIA elements are summarized in Table 20.1. Each of the elements has an odd atomic number and so has few stable isotopes. All are rather soft, silvery-white metals, and they display the gradation in properties that might be expected for elements immediately following the strongly electropositive alkaline-earth metals and preceding the transition elements proper. Each is less electropositive than its predecessor in Group IIA but more electropositive than its successors in transition series, while the increasingly electropositive character of the heavier elements of the group is in keeping with the increase in size. The inverse trends in electronegativity are illustrated in Fig. 20.1.

As is the case for boron and aluminium, the underlying electron cores are those of the preceding noble gases and indeed, as was pointed out in Chapter 7, a much more regular variation in atomic properties occurs in passing from B and Al to Group IIIA than to Group IIIB (p. 251). However, the presence of a d electron on each of the atoms of this group (in contrast to the p electron in the atoms of B, Al, and Group IIIB) has consequences which can be seen in some of the bulk properties of the metals. For instance, the mps and bps (Fig. 20.2), along with the enthalpies associated with these transitions, all show discontinuous increases in passing from Al to Sc rather than to Ga, indicating that the d electron has a more cohesive effect than the p electron. It appears that this is due to d electrons forming more localized bonds within the metals. Thus, although Sc, Y, and La have typically metallic (hcp) structures (with other metallic modifications at higher

TABLE 20.1   *Some properties of Group IIIA elements*

Property	Sc	Y	La	Ac
Atomic number	21	39	57	89
Number of naturally occurring isotopes	1	1	2	(2)
Atomic weight	44.9559	88.9059	138.9055[a]	227.0278[b]
Electronic configuration	[Ar]$3d^14s^2$	[Kr]$4d^15s^2$	[Xe]$5d^16s^2$	[Rn]$6d^17s^2$
Electronegativity	1.3	1.2	1.1	1.1
Metal radius (12-coordinate)/pm	162	180	187	—
Ionic radius (6-coordinate)/pm	74.5	90.0	103.2	112
$E°$ $(M^{3+} + 3e^- = M(s))$/V	-2.077	-2.372	-2.522	-2.6
MP/°C	1539	1530	920	817
BP/°C	2748	3264	3420	2470
ΔH_{fus}/kJ mol^{-1}	15.77	11.5	8.5	(10.5)
ΔH_{vap}/kJ mol^{-1}	332.71	367	402	(293)
ΔH_f (monatomic gas)/kJ mol^{-1}	376 (±20)	425 (±8)	423 (±6)	—
Density (20°C)/g cm^{-3}	3.0	4.5	6.17	—
Electrical resistivity (20°C)/μohm cm	50–61	57–70	57–80	—

[a] Reliable to ±3 in last digit; other atomic weights are reliable to ±1 in the last digit.
[b] This value is for the radioisotope with the longest half-life (^{227}Ac).

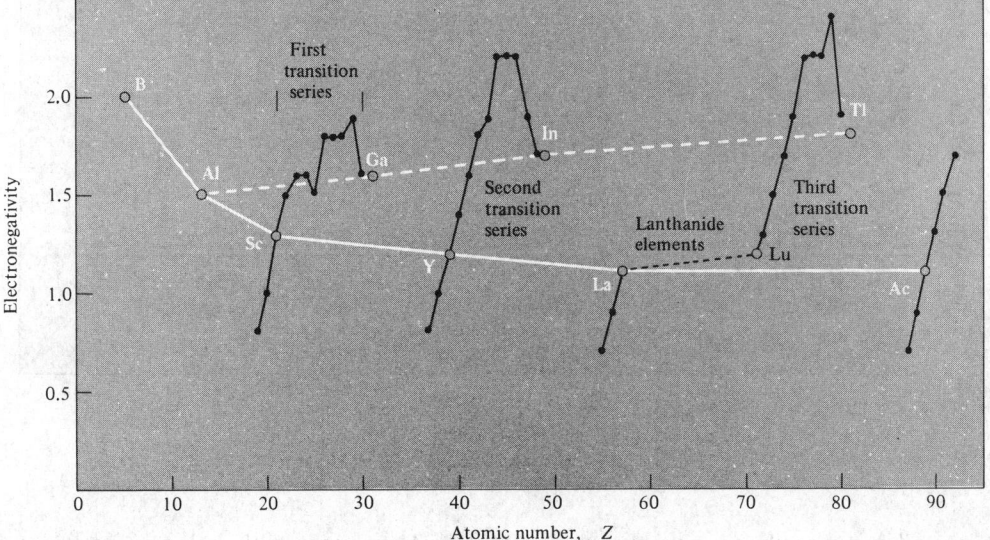

FIG. 20.1   Electronegativity of Group III elements.

temperatures), their electrical resistivities are much higher than that of Al (Fig. 20.3). Admittedly, resistivity is a function of thermal vibrations of the crystal lattice as well as of the degree of localization of valence electrons, but even so the marked changes between Al and Sc seem to indicate a marked reduction in the mobility of the d electron of the latter.

20.2.4   *Chemical reactivity and trends*

The general reactivity of the metals increases down the group. They tarnish in air—La rapidly, but Y much more slowly because of the formation of a protective oxide coating—

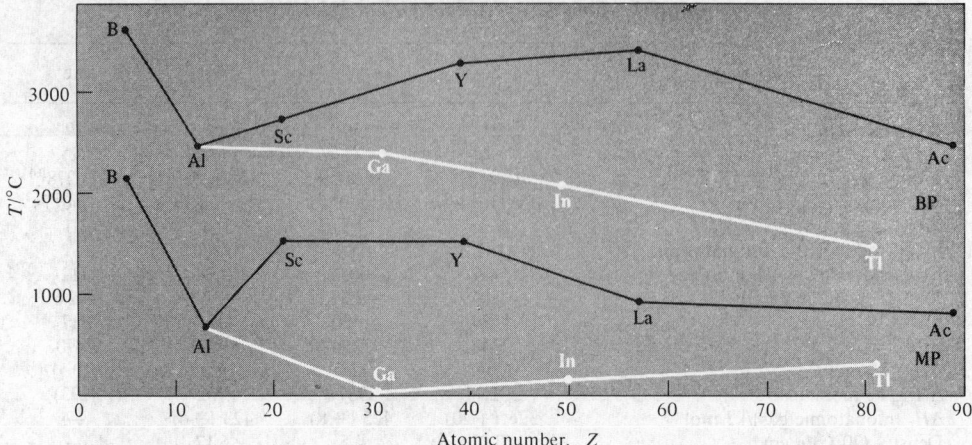

FIG. 20.2   Mps and bps of Group III elements.

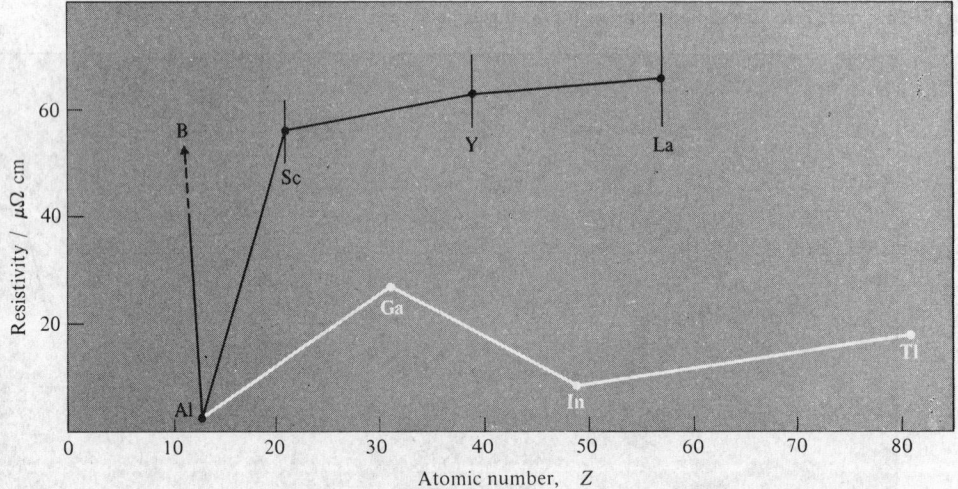

FIG. 20.3   Resistivities of Group III elements.

and all burn easily to give the oxides M_2O_3. They react with halogens at room temperature and with most non-metals on warming. They reduce water with evolution of hydrogen, particularly if finely divided or heated, and all dissolve in dilute acid. Strong acids produce soluble salts whereas weak acids such as HF, H_3PO_4, $H_2C_2O_4$, produce sparingly soluble or insoluble salts.

In the main, the chemistry of these elements concerns the formation of a predominantly ionic +3 oxidation state arising from the loss of all 3 valence electrons and giving a well-defined cationic aqueous chemistry. Because of this, although each member of this group is the first member of a transition series, its chemistry is largely atypical of the transition elements. The variable oxidation states and the marked ability to form coordination compounds with a wide variety of ligands are barely hinted at in this group. Differences in chemical behaviour within the group are largely a consequence of the differing sizes of the

M^{III} ions. Scandium, the lightest of these elements, with the smallest ionic radius, is the least basic and the strongest complexing agent, with properties not unlike those of aluminium. Its aqueous solutions are appreciably hydrolysed and its oxide has some acidic properties. On the other hand, lanthanum and actinium (in so far as its properties have been examined) show basic properties approaching those of calcium.

20.3 Compounds of Scandium, Yttrium, Lanthanum, and Actinium

20.3.1 *Simple compounds*

The oxides, M_2O_3, are white solids which can be prepared directly from the elements. In Sc_2O_3 and Y_2O_3 the metals are 6-coordinate but the larger La^{III} ion adopts this structure only at elevated temperatures, a 7-coordinate structure being normally more stable. When water is added to La_2O_3 it "slakes" like lime with evolution of much heat and a hissing sound. The hydroxides, $M(OH)_3$, (or in the case of scandium possibly the hydrated oxide) are obtained as gelatinous precipitates from aqueous solutions of the metal salts by addition of alkali hydroxide. In the case of scandium only, this precipitate can be dissolved in an excess of conc NaOH to give anionic species such as $[Sc(OH)_6]^{3-}$. Yttrium and lanthanum hydroxides possess only basic properties, and the latter especially will absorb atmospheric CO_2 to form basic carbonates.

Dissolution of the oxide or hydroxide in the appropriate acid provides the most convenient method for producing the salts of the colourless, diamagnetic M^{III} ions. Such solutions, especially those of Sc^{III}, are significantly hydrolysed with the formation of polymeric hydroxy species.

With the exception of the fluorides, the halides are all very water-soluble and deliquescent. Precipitation of the insoluble fluorides can be used as a qualitative test for these elements. The distinctive ability of Sc^{III} to form complexes is illustrated by the fact that an excess of F^- causes the first-precipitated ScF_3 to redissolve as $[ScF_6]^{3-}$; indeed, $M_3[ScF_6]$, $M = NH_4$, Na, K, were isolated as long ago as 1914. The anhydrous halides are best prepared by direct reaction of the elements rather than by heating the hydrates which causes hydrolysis. Heating the hydrated chlorides, for instance, gives Sc_2O_3, YOCl, and LaOCl respectively, though to produce AcOCl it is necessary to use superheated steam.

Sulfates and nitrates are known and in all cases they decompose to the oxides on heating. Double sulfates of the type $M_2^{III}(SO_4)_3.3Na_2SO_4.12H_2O$ can be prepared, and La (unlike Sc and Y) forms a double nitrate, $La(NO_3)_3.2NH_4NO_3.4H_2O$, which is of the type once used extensively in fractional crystallization procedures for separating individual lanthanides.

Reaction of the metals with hydrogen[2] produces highly conducting materials with the composition MH_2, similar to the metallic nonstoichiometric hydrides of the subsequent transition elements (pp. 72–3). Except in the case of ScH_2, further H_2 can then be absorbed causing loss of electrical conductivity until materials similar to the ionic hydrides of the alkaline-earth metals, and with the limiting composition MH_3, are produced. The dihydrides, though ostensibly containing the divalent metals, are probably best considered as pseudo-ionic compounds of M^{3+} and $2H^-$ with the extra electron in a conducting band. However, the question of the type of bonding is highly controversial, as was explained more fully in Chapter 3 (p. 73).

[2] W. G. Bos, Bonding in Group IIIb hydrides: an empirical approach, *Acc. Chem. Res.* **5**, 342–7 (1972).

Another example of a "divalent" metal of this group, but which in fact is probably entirely analogous to the dihydrides, is LaI_2. More recently,[3] the blue-black compound $CsScCl_3$, prepared by prolonged heating of $3CsCl + 2ScCl_3$ with an excess of scandium metal at 700°C, has been claimed to contain divalent Sc^{II} in linear $[ScCl_3^-]_n$ chains. Indeed, on heating Sc and $ScCl_3$ at temperatures up to 900°C in a sealed tantalum container, no less than five different reduced phases have been recognized, containing scandium in formal oxidation states down to $+1$ and being sensitive to oxygen and moisture. The structures of two of these phases[4] have been examined in some detail by X-ray diffraction and it is clear that extensive metal–metal bonding is present. ScCl is a grey-black graphite-like substance made up of close-packed layers of Sc and Cl atoms in the sequence, Cl–Sc–Sc–Cl. Each Sc atom has 6 near neighbours in the same layer and 3 in the next layer (at 347 and 322 pm respectively compared to 336 and 326 pm in scandium metal). $ScCl_{1.43}$ (or Sc_7Cl_{10}) is composed of two parallel chains of Sc octahedra, sharing a common edge and with even shorter Sc–Sc distances (314.7 and 315.3 pm) than in ScCl or the metal. However, the precise electronic structures of these materials are still unclear and the assignment of low oxidation numbers is therefore only formal. Evidence of similar chemistry in Y/YCl_3 and $La/LaCl_3$ systems is also emerging; e.g. the compounds YCl, YBr, Y_2Cl_3, Y_2Br_3, LaBr, and La_7I_{12} have recently been prepared and structurally characterized.[4a]

20.3.2 *Complexes*[5]

The complex anion $[ScF_6]^{3-}$ has already been mentioned and other complexes such as $[Sc(bipy)_3]^{3+}$, $[Sc(bipy)_2(NCS)_2]^+$, $[Sc(bipy)_2Cl_2]^+$, and $[Sc(DMSO)_6]^{3+}$ (where DMSO is dimethylsulfoxide, Me_2SO) exhibit scandium's usual coordination number of 6. However, compared to later elements in their respective transition series, scandium, yttrium, and lanthanum have rather poorly developed coordination chemistries and form weaker coordinate bonds, lanthanum generally being even less inclined to form strong coordinate bonds than scandium. This is reflected in the stability constants of a number of relevant 1:1 metal, EDTA complexes:

Metal ion	Sc^{III}	Y^{III}	La^{III}	Fe^{III}	Co^{III}
$\log_{10} K_1$	23.1	18.1	15.5	25.5	36.0

This may seem somewhat surprising in view of the charge of $+3$ ions, but this is coupled with appreciably larger ionic radii and also with greater electropositive character which

[3] K. R. POEPPELMEIER and J. D. CORBETT, $CsScCl_3$: A new linear metal chain compound containing divalent scandium. A paper presented at the ACS meeting, Chicago, 1977, INOR 08. K. R. POEPPELMEIER, J. D. CORBETT, T. P. MCMULLIN, D. R. TORGESON, and R. G. BARNES, Study of the crystal structures and nonstoichiometry in the system $Cs_3Sc_2Cl_9–CsScCl_3$, *Inorg. Chem.* **19**, 129–34 (1980).

[4] K. R. POEPPELMEIER and J. D. CORBETT, Metal–metal bonding in reduced scandium halides: synthesis and crystal structure of scandium monochloride, *Inorg. Chem.* **16**, 294–7 (1977); Synthesis and characterization of heptascandium decachloride (Sc_7Cl_{10}): a novel metal chain structure, ibid. **16**, 1107–11 (1977).

[4a] H. J. MATTAUSCH, J. B. HENDRICKS, R. EIGER, J. D. CORBETT, and A. SIMON, Reduced halides of yttrium with strong metal–metal bonding: yttrium monochloride, monobromide, sesquichloride, and sesquibromide, *Inorg. Chem.* **19**, 2128–32 (1980), and references therein.

[5] G. A. MELSON and R. W. STOTZ, The coordination chemistry of scandium, *Coord. Chem. Revs.* **7**, 133–60 (1971).

inhibits covalent contribution to their bonding. Lanthanum of course exhibits these characteristics more clearly than Sc, and, while La and Y closely resemble the lanthanide elements, Sc has more similarity with Al.

The gradation of properties within this group is well illustrated by the oxalates and β-diketonates which are formed. On addition of alkali-metal oxalate to aqueous solutions of M^{III}, oxalate precipitates form but their solubilities in an excess of the alkali-metal oxalate decrease very markedly down the group. Scandium oxalate dissolves readily with evidence of such anionic species as $[Sc(C_2O_4)_2]^-$. Yttrium oxalate also dissolves to some extent but lanthanum oxalate dissolves only slightly. All three elements form acetylacetonates: that of scandium is usually anhydrous, $[Sc(acac)_3]$, and presumably pseudo-octahedral: $[Y(acac)_3H_2O]$ is 7-coordinate with a capped trigonal prismatic structure (p. 1074); $[Y(acac)_3(H_2O)_2]H_2O$ and $[La(acac)_3(H_2O)_2]$ are 8-coordinate with distorted square-antiprismatic structures (p. 1075); the scandium compound can be sublimed without decomposition whereas the yttrium and lanthanum compounds decompose at about 500°C and dehydration without decomposition or polymerization is difficult.

By far the most common donor atom is oxygen, particularly in chelating ligands, and moving down the group the complexes are found to be increasingly ionic, and coordination numbers greater than 6 become the rule rather than the exception as used to be thought. Indeed, $[Y(NCS)_6]^{3-}$ and $[La(NCS)_6]^{3-}$ are amongst the few well-authenticated, 6-coordinate complexes of Y and La, and it seems likely[5a] that in aqueous solutions, in the absence of other preferred ligands, Y^{III} is directly coordinated to 8 water molecules and La^{III} to 9. Similarly high coordination numbers are common in many simple binary compounds; in LaX_3 (X = F, Cl) and in $M(OH)_3$, (M = Y, La) the metal ion is 9-coordinate with a stereochemistry approximating to tri-capped trigonal prismatic.

A coordination number of 8 is probably the most characteristic of La and possibly even of Y, with the square antiprism and the dodecahedron being the preferred stereochemistries. The acac complexes referred to above are good examples of the former type, while $Cs[Y(CF_3COCHCOCF_3)_4]$ typifies the latter. On the basis of ligand–ligand repulsions the cubic arrangement is expected to be much less favoured in discrete complexes, but, nonetheless, the complex $[La(bipyO_2)_4]ClO_4$, in which $bipyO_2$ is 2,2'-bipyridine dioxide,

has been shown to be very nearly cubic.[6]

EDTA complexes of La and the lanthanides have been examined.[7] $K[La(EDTA)(H_2O)_3].5H_2O$ is a 9-coordinate complex but steric constraints imposed by the EDTA produce deviations from a tricapped trigonal prismatic structure. $[La(EDTAH)(H_2O)_4].3H_2O$ is 10-coordinate and its structure is probably best regarded as being based on the same structure but with an extra water "squeezed" between the three coordinated water molecules.

[5a] J. BURGESS, *Metal Ions in Solution*, Ellis Horwood, Chichester, 1977, 481 pp.
[6] A. R. AL-KARAGHOULI, R. O. DAY, and J. S. WOOD, Crystal structure of [La(2,2-bipyridine-dioxide)](ClO_4)_3: an example of cubic eight co-ordination, *Inorg. Chem.* **17**, 3702–6 (1978).
[7] J. L. HOARD, B. LEE, and M. D. LIND, Structure and bonding in a ten-coordinate lanthanum(III) chelate of ethylenediaminetetraacetic acid, *J. Am. Chem. Soc.* **87**, 1611–12 (1965).

The highest coordination numbers of all are attained only with the aid of chelating ligands, such as SO_4^{2-} and NO_3^-, with very small "bites" (p. 1077). In $La_2(SO_4)_3.9H_2O$ there are actually two types of La^{III}, one being coordinated to 12 oxygens in SO_4^{2-} ions while the other is coordinated to 6 water molecules and 3 oxygens in SO_4^{2-} ions. In $[Y(NO_3)_5]^{2-}$ the Y^{III} is 10-coordinate and in $[Sc(NO_3)_5]^{2-}$, even though one of the nitrate ions is only unidentate (p. 541), the coordination number of 9 is extraordinarily high for scandium.

The low symmetries of many of the above highly coordinated species, which appear to be determined largely by the stereochemical requirements of the ligands, together with the fact that these high coordination numbers are attained almost exclusively with oxygen-donor ligands, are consistent with the belief that the bonding is essentially of an electrostatic rather than a directional covalent character.

20.3.3 *Organometallic compounds*[8]

In view of the electronic structures of the elements of this group it is not surprising that π bonding is of no importance in their chemistry and that consequently they form comparatively few organometallic compounds. The best characterized are the ionic cyclopentadienides, $M(C_5H_5)_3$, prepared by the reactions of anhydrous MCl_3 with NaC_5H_5 in tetrahydrofuran. They are crystalline solids, stable in dry air up to $\sim 400°C$ but sensitive to water. Indenyl compounds, $M(C_9H_7)_3$, have also been prepared. Treatment of the anhydrous chlorides with LiPh yields polymeric phenyl compounds, highly unstable to air and moisture and with the stoichiometries MPh_3 (M = Sc, Y), and $LiLaPh_4$. Monomeric alkyl compounds of the form MR_3 have also been obtained[9] for Sc and Y, where the alkyl groups are of the types Me_3SiCH_2 and Me_3CCH_2 which are bulky and contain no β hydrogen atoms (p. 348).

[8] H. GYSLING and M. TSUTSUI, Organolanthanides and organoactinides, *Adv. Organometallic Chem.* **9**, 361–95 (1970).

[9] P. J. DAVIDSON, M. F. LAPPERT and R. PEARCE, Stable homoleptic metal alkyls, *Acc. Chem. Res.* **7**, 209–17 (1974).

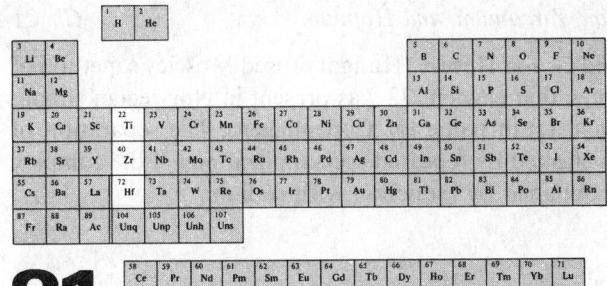

21

Titanium, Zirconium, and Hafnium

21.1 Introduction

In 1791 William Gregor,[1] a Cornish vicar and amateur chemist, examined sand from the local river Helford. Using a magnet he extracted a black material (now called ilmenite) from which he removed iron by treatment with hydrochloric acid. The residue, which dissolved only with difficulty in concentrated sulfuric acid, was the impure oxide of a new element, and Gregor proceeded to discover the reactions which were to form the basis of the production of virtually all TiO_2 up to about 1960. Four years later the German chemist M. H. Klaproth independently discovered the same oxide (or "earth"), in a sample of ore now known to be rutile, and named the element titanium after the Titans who, in Greek mythology, were the children of Heaven and Earth condemned to live amongst the hidden fires of the earth. Klaproth had previously (1789) isolated the oxide of zirconium from a sample of zircon, $ZrSiO_4$. Various forms of zircon (Arabic *zargun*) have been known as gemstones since ancient times. Impure samples of the two metals were prepared by J. J. Berzelius (Sweden) in 1824 (Zr) and 1825 (Ti) but samples of high purity were not obtained until much later. M. A. Hunter (USA) reduced $TiCl_4$ with sodium in 1910 to obtain titanium, and A. E. van Arkel and J. H. de Boer (Netherlands) produced zirconium in 1925 by their iodide-decomposition process (see below).

The discovery of hafnium was one of chemistry's more controversial episodes.[2] In 1911 G. Urbain, the French chemist and authority on "rare earths", claimed to have isolated the element of atomic number 72 from a sample of rare-earth residues, and named it celtium. With hindsight, and more especially with an understanding of the consequences of H. G. J. Moseley's and N. Bohr's work on atomic structure, it now seems very unlikely that element 72 could have been found in the necessary concentrations along with rare earths. But this knowledge was lacking in the early part of the century and, indeed, in 1922 Urbain and A. Dauvillier claimed to have X-ray evidence to support the discovery. However, by that time Niels Bohr had developed his atomic theory and so was confident that element 72 would be a member of Group IVA and was more likely to be found along with zirconium than with the rare earths. Working in Bohr's laboratory in Copenhagen in

[1] F. R. WILLIAMS, William Gregor, 1761–1817, *Educ. Chem.* **11**, 115 (1974).
[2] R. T. ALLSOP, Element 72—the great controversy, *Educ. Chem.* **10**, 222–3 (1973).

1922/3, D. Coster (Netherlands) and G. von Hevesy (Hungary) used Moseley's method of X-ray spectroscopic analysis to show that element 72 was present in Norwegian zircon, and it was named hafnium (*Hafnia*, Latin name for Copenhagen). The separation of hafnium from zirconium was then effected by repeated recrystallizations of the complex fluorides and hafnium metal was obtained by reduction with sodium.

21.2 The Elements

21.2.1 *Terrestrial abundance and distribution*

Titanium, which comprises 0.63% (i.e. 6320 ppm) of the earth's crustal rocks, is a very abundant element (ninth of all elements, second of the transition elements), and of the transition elements only Fe, Ti, and Mn are more abundant than zirconium (0.016%, 162 ppm). Even hafnium (2.8 ppm) is as common as Cs and Br.

That these elements have in the past been considered unfamiliar has been due largely to the difficulties involved in preparing the pure metals and also to their rather diffuse occurrence. Like their predecessors in Group IIIA, they are classified as type-a metals and are found as silicates and oxides in many silicaceous materials. These are frequently resistant to weathering and so often accumulate in beach deposits which can be profitably exploited.

The two most important minerals of titanium are ilmenite ($FeTiO_3$) and rutile (TiO_2). The former is a black sandy material mined in Canada, the USA, Australia, Scandinavia, and Malaysia, while the latter is mined principally in Australia. Zirconium's main minerals are zircon ($ZrSiO_4$) and baddeleyite (ZrO_2) found in the USA, Australia, and Brazil, and invariably containing hafnium, most commonly in quantities around 2% of the zirconium content. Only in a few minerals, such as alvite, $MSiO_4.xH_2O$ (M = Hf, Th, Zr), does the hafnium content occasionally exceed that of zirconium. As a result of the lanthanide contraction (p. 1431) the ionic radii of Zr and Hf are virtually identical and their association in nature parallels their very close chemical similarity.

21.2.2 *Preparation and uses of the metals*

Viable methods of producing the metals from oxide ores have to surmount two problems. In the first place, reduction with carbon is not possible because of the formation of intractable carbides (p. 321), and even reduction with Na, Ca, or Mg is unlikely to remove all the oxygen. In addition, the metals are extremely reactive at high temperatures and, unless prepared in the absence of air, will certainly be contaminated with oxygen and nitrogen.

In 1932 Wilhelm Kroll of Luxembourg produced titanium by reducing $TiCl_4$ with calcium and then later (1940) with magnesium and even sodium. The expense of this process was a severe deterrent to any commercial use of titanium. However, the metal has a very low density ($\sim 57\%$ that of steel) combined with good mechanical strength and, in fact, when alloyed with small quantities of such metals as Al and Sn, has the highest strength:weight ratio of any of the engineering metals. Accordingly, about 1950, a demand developed for titanium for the manufacture of gas-turbine engines, and this demand has rapidly increased as production and fabrication problems have been overcome. Its major uses are still in the aircraft industry for the production of both engines and airframes, but it is also widely used in chemical processing and marine equipment. Its world production now approaches 10^5 tonnes pa (45 000 tonnes in the USA pa). The Kroll method[3] still dominates the industry: in this ilmenite or rutile is heated with chlorine and carbon, e.g.:

$$2FeTiO_3 + 7Cl_2 + 6C \xrightarrow{900\,°C} 2TiCl_4 + 2FeCl_3 + 6CO$$

The $TiCl_4$ is fractionally distilled from $FeCl_3$ and other impurities and then reduced with molten magnesium in a sealed furnace under Ar,

$$TiCl_4 + 2Mg \xrightarrow{950–1150\,°C} Ti + 2MgCl_2$$

The $MgCl_2$ and any excess of magnesium are removed by leaching with water and dilute hydrochloric acid or by distillation, leaving titanium "sponge" which, after grinding and cleaning with aqua regia (1:3 mixture of concentrated nitric and hydrochloric acids), is melted under argon or vacuum and cast into ingots. The use of sodium instead of magnesium requires little change in the basic process but gives a more readily leached product. This yields titanium metal in a granular form which is fabricated by somewhat different techniques and is preferred by some users.

Zirconium, too, is produced commercially by the Kroll process,[4] but the van Arkel–de Boer process is also useful when it is especially important to remove all oxygen and nitrogen. In this latter method the crude zirconium is heated in an evacuated vessel with a little iodine, to a temperature of about 200°C when ZrI_4 volatilizes. A tungsten or zirconium filament is simultaneously electrically heated to about 1300°C. This decomposes the ZrI_4 and pure zirconium is deposited on the filament. As the deposit grows the current is steadily increased so as to maintain the temperature. The method is applicable to many metals by judicious adjustment of the temperatures. Zirconium has a high corrosion resistance and in certain chemical plants is preferred to alternatives such as stainless steel, titanium, and tantalum. It is also used in a variety of alloy steels and, when added to niobium, forms a superconducting alloy which retains its superconductivity in strong magnetic fields. The small percentage of hafnium normally present in zirconium is of no detriment in these cases and may even improve its properties, but zirconium's major use is as a cladding for uranium dioxide fuel rods in water-cooled nuclear reactors. When alloyed with $\sim 1.5\%$ tin, its corrosion resistance and mechanical properties, which are stable under irradiation, coupled with its extremely low absorption of "thermal" neutrons, make it an ideal material for this purpose. Unfortunately, hafnium is a powerful absorber of thermal neutrons (600 times more so than Zr) and its removal, though difficult,[5] is therefore necessary. Solvent extraction methods, taking advantage of the different solubilities of, for instance, the 2 nitrates in tri-*n*-butyl phosphate or the thiocyanates in hexone (methyl isobutyl ketone) have been developed and reduce the hafnium content to

[3] G. BRAUER, *Handbook of Preparative Inorganic Chemistry*, 2nd edn., pp. 1161–72, Academic Press, New York, 1965.
[4] G. L. MILLER, *Zirconium*, 2nd edn., Chaps. 3–7, pp. 31–142, Butterworths, London, 1957.
[5] W. FISCHER, B. DEIERLING, H. HEITSCH, G. OTTO, H.-P. POHLMANN, and K. REINHARDT, The separation of zirconium and hafnium by liquid–liquid partition of their thiocyanates, *Angew. Chem.* Int. Edn. (Engl.) **5**, 15–23 (1966).

less than 100 ppm. The neutron absorbing ability of hafnium is not always disadvantageous, however, since it is the reason for hafnium's use for reactor control rods in nuclear submarines. Hafnium is produced in the same ways as zirconium but on a much smaller scale—about 200 tonnes pa worldwide in 1978.

21.2.3 *Properties of the elements*

Table 21.1 summarizes a number of properties of these elements. The difficulties in attaining high purity has led to frequent revision of the estimates of several of these properties. Each element has a number of naturally occurring isotopes and, in the case of zirconium and hafnium, the least abundant of these is radioactive, though with a very long half-life ($^{96}_{40}$Zr, 2.76%, 3.6×10^{17} y; $^{174}_{72}$Hf, 0.18%, 2.0×10^{15} y).

TABLE 21.1   *Some properties of Group IVA elements*

Property		Ti	Zr	Hf
Atomic number		22	40	72
Number of naturally occurring isotopes		5	5	6
Atomic weight		47.88[a]	91.22	178.49[a]
Electronic configuration		[Ar]3d^{2}4s^2	[Kr]4d^{2}5s^2	[Xe]4f^{14}5d^{2}6s^2
Electronegativity		1.5	1.4	1.3
Metal radius/pm		147	160	159
Ionic radius (6-coordinate)/pm	M(IV)	60.5	72	71
	M(III)	67.0	—	—
	M(II)	86	—	—
MP/°C		1667	1857	2222 (or 2467)
BP/°C		3285	4200	4450
ΔH_{fus}/kJ mol^{-1}		18.8	19.2	(25)
ΔH_{vap}/kJ mol^{-1}		425 ($\pm$11)	567	571 ($\pm$25)
ΔH_f (monatomic gas)/kJ mol^{-1}		469 ($\pm$4)	612 ($\pm$11)	611 ($\pm$17)
Density (25°C)/g cm^{-1}		4.50	6.51	13.28
Electrical resistivity (20°C)/μohm cm		42.0	40.0	35.1

[a] Atomic weight reliable to $\pm$3 in the last digit.

The elements are all lustrous, silvery metals with high mps and they have typically metallic hcp structures which transform to bcc at high temperatures (882°, 870°, and 1760°C for Ti, Zr, and Hf). They are better conductors of heat and electricity than their predecessors in Group IIIA but are not to be regarded as "good" conductors in comparison with most other metals. The enthalpies of fusion, vaporization, and atomization have also increased, indicating that the additional d electron has in each case contributed to stronger metal bonding. As was noticed for Group III, the d electrons of the A subgroup contribute more effectively to the metal–metal bonding in the bulk materials than do the p electrons of the B subgroup (Ge, Sn, Pb). Figure 21.1 illustrates the consequent discontinuous increases in mp, bp, and enthalpy of atomization in passing from C and Si to Ti, Zr, and Hf, rather than to Ge, Sn, and Pb.

The mechanical properties of these metals are markedly affected by traces of impurities such as O, N, and C which have an embrittling effect on the metals, making them difficult to fabricate.

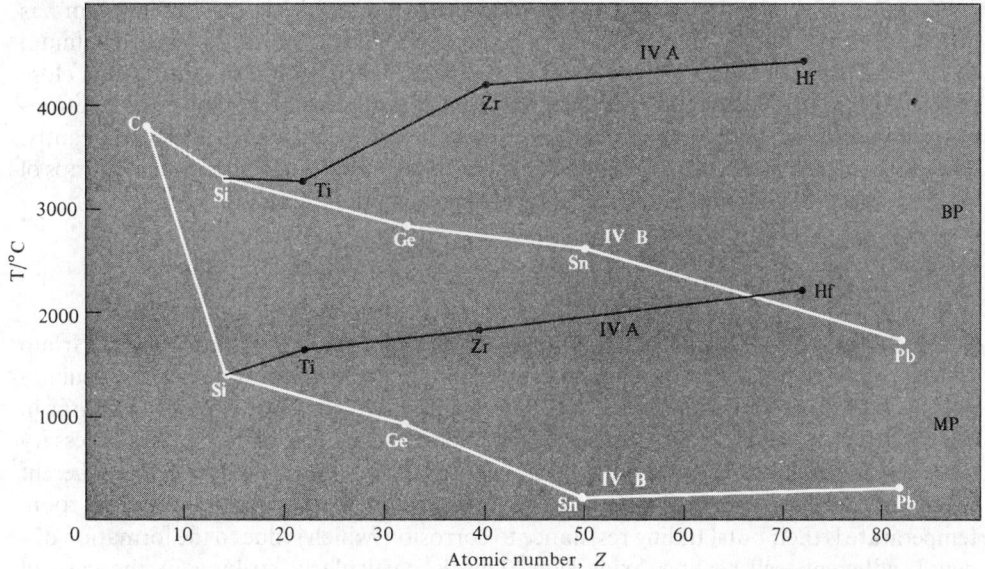

(a) Melting points and boiling points.

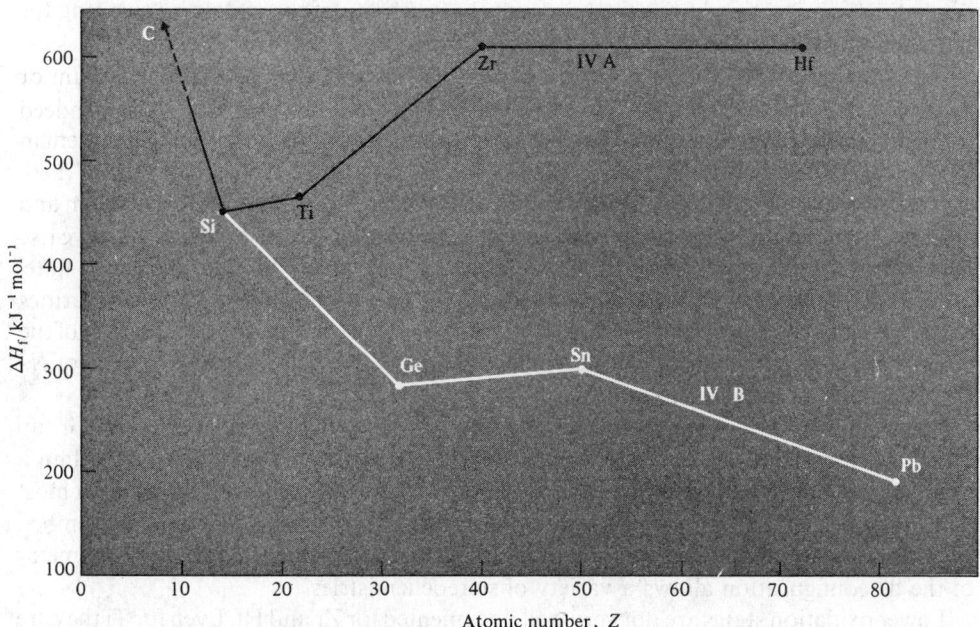

(b) Enthalpies of atomization.

FIG. 21.1   Trends in some properties of Group IV elements.

The effect of the lanthanide contraction on the metal and ionic radii of hafnium has already been mentioned. That these radii are virtually identical for zirconium and hafnium has the result that the ratio of their densities, like that of their atomic weights, is very close to Zr: Hf = 1:2.0. Indeed, the densities, the transition temperatures, and the neutron-absorbing abilities are the only properties of these two elements which differ significantly. This close similarity of second and third members is noticeable in all subsequent groups of the transition elements but is never more pronounced than here.

21.2.4 *Chemical reactivity and trends*

The elements of this group are relatively electropositive but less so than those of Group IIIA. If heated to high temperatures they react directly with most non-metals, particularly oxygen, hydrogen (reversibly), and, in the case of titanium, nitrogen (Ti actually burns in N_2). When finely divided the metals are pyrophoric and for this reason care is necessary when machining them to avoid the production of fine waste chips. In spite of this inherent reactivity, the most noticeable feature of these metals in the massive form at room temperature is their outstanding resistance to corrosion, which is due to the formation of a dense, adherent, self-healing oxide film. This is particularly striking in the case of zirconium. With the exception of hydrofluoric acid (which is the best solvent, probably because of the formation of soluble fluoro complexes) mineral acids have little effect unless hot. Even when hot, aqueous alkalis do not attack the metals. The presence of oxidizing agents such as nitric acid frequently reduces the reactivity of the metals by ensuring the retention of the protective oxide film.

The chemistry of hafnium has not received the same attention as that of titanium or zirconium, but it is clear that its behaviour follows that of zirconium very closely indeed with only minor differences in such properties as solubility and volatility being apparent in most of their compounds. The most important oxidation state in the chemistry of these elements is the group oxidation state of +4. This is too high to be ionic, but zirconium and hafnium, being larger, have oxides which are more basic than that of titanium and give rise to a more extensive and less-hydrolysed aqueous chemistry. In this oxidation state, particularly in the case of the dioxide and tetrachloride, titanium shows many similarities with tin which is of much the same size. A large number of coordination compounds of the M^{IV} metals have been studied[6] and complexes such as $[MF_6]^{2-}$ and those with *O*- or *N*-donor ligands are especially stable.

The M^{IV} ions, though much smaller than their triply charged predecessors in Group IIIA, are, nonetheless, sufficiently large, bearing in mind their high charge, to attain a coordination number of 8 or more, which is certainly higher than is usually found for most transition elements. Eight is not a common coordination number for the first member, titanium, but is very well known for zirconium and hafnium, and the spherical symmetry of the d^0 configuration allows a variety of stereochemistries.

Lower oxidation states are not very well documented for Zr and Hf. Even for Ti they are readily oxidized to +4 but they are undoubtedly well defined and, whatever arguments may be advanced against applying the description to Sc, there is no doubt that Ti is a "transition metal". In aqueous solution Ti^{III} can be prepared by reduction of Ti^{IV}, either with Zn and dilute acid or electrolytically, and it exists in dilute acids as the violet,

[6] D. L. KEPERT, *The Early Transition Metals*, Chap. 2, pp. 62–141, Academic Press, London, 1972.

octahedral $[Ti(H_2O)_6]^{3+}$ ion (p. 1131). Although this is subject to a certain amount of hydrolysis, normal salts such as halides and sulfates can be separated. Zr^{III} and Hf^{III} are only known as the trihalides or their derivatives and have no aqueous chemistry since they reduce water. Table 21.2 gives the oxidation states and stereochemistries found in the complexes of Ti, Zr, and Hf along with illustrative examples.

M–C σ bonds are not strong and, as might be expected for metals with so few d electrons, little help is available from synergic π bonding: for instance, of the simple carbonyls only $Ti(CO)_6$ has been reported, and that only on the basis of spectroscopic evidence. However, as will be seen on p. 1134, the discovery that titanium compounds can be used to catalyse the polymerization of alkenes (olefins) turned organo-titanium chemistry into a topic of major commercial importance.

TABLE 21.2   *Oxidation states and stereochemistries of titanium, zirconium, and hafnium*

Oxidation state	Coordination number	Stereochemistry	Ti	Zr/Hf
−1 (d⁵)	6	Octahedral	$[Ti(bipy)_3]^-$	—
0 (d⁴)	6	Octahedral	$[Ti(bipy)_3]$	$[Zr(bipy)_3]$
2 (d²)	6	Octahedral	$TiCl_2$	Layer structures and clusters
	12	—	$[Ti(\eta^5\text{-}C_5H_5)_2(CO)_2]$	
3 (d¹)	3	Planar	$[Ti\{N(SiMe_3)_2\}_3]$	
	5	Trigonal bipyramidal	$[TiBr_3(NMe_3)_2]$	
	6	Octahedral	$[Ti(urea)_6]^{3+}$	ZrX_3 (Cl, Br, I), HfI_3
4 (d⁰)	4	Tetrahedral	$TiCl_4$	$ZrCl_4$(g) (solid is octahedral)
	5	Trigonal bipyramidal	$[TiOCl_2(NMe_3)_2]$	—
		Square pyramidal	$[TiOCl_4]^{2-}$	
	6	Octahedral	$[TiF_6]^{2-}$	$[ZrF_6]^{2-}$, $ZrCl_4$(s)
	7	Pentagonal bipyramidal	$[TiCl(S_2CNMe_2)_3]$	$[NH_4]^+_3[ZrF_7]^{3-}$
		Capped trigonal prismatic	$[TiF_5(O_2)]^{3-}$	$[Zr_2F_{13}]^{5-}$
	8	Dodecahedral	$[Ti(\eta^2\text{-}NO_3)_4]$	$[Zr(C_2O_4)_4]^{4-}$
		Square antiprismatic	—	$[Zr(acac)_4]$
	11	—	$[Ti(\eta^5\text{-}C_5H_5)(S_2CNMe_2)_3]$	
	12	—	—	$[M(\eta^3\text{-}BH_4)_4]$

21.3   Compounds of Titanium, Zirconium, and Hafnium

The binary hydrides (p. 69), borides (p. 162), carbides (p. 321), and nitrides (p. 480) are hard, refractory, nonstoichiometric materials with metallic conductivities. They have already been discussed in relation to comparable compounds of other metals in earlier chapters.

21.3.1   *Oxides*[7] *and sulfides*

The main oxides are the dioxides, and TiO_2 is by far the most important compound

[7] C. N. R. RAO and G. V. S. RAO, *Transition Metal Oxides*, National Standard Reference Data System NSRDS-NBS49, Washington, 1974, 130 pp.

formed by the elements of this group, its importance arising predominantly from its use as a white pigment (see Panel). It exists at room temperature in three forms—rutile, anatase, and brookite, each of which occurs naturally. Each contains 6-coordinate titanium but rutile is the most common form, both in nature and as produced commercially, and the others transform into it on heating. The rutile structure is based on a slightly distorted hcp of oxygen atoms with half the octahedral interstices being occupied by titanium atoms. The octahedral coordination of the titanium atoms and trigonal planar coordination of the oxygen can be seen in Fig. 21.2. This is a structure commonly adopted by ionic

Titanium Dioxide as a Pigment

Of all white pigments, TiO_2 is now the most widely used: the impressive growth in demand is shown in Table A:[8]

TABLE A   *Annual world production of* TiO_2

Year	1925	1937	1975	1980
TiO_2/tonnes	5000	100 000	2 000 000	3 000 000

Well over half of this is used in the manufacture of paint, and other major uses are as a surface coating on paper (25%) and as a filler in rubber and plastics (15%).

The value of TiO_2 as a pigment is due to its exceptionally high refractive index in the visible region of the spectrum. Thus although large crystals are transparent, fine particles scatter light so strongly that they can be used to produce films of high opacity. Table B gives the refractive indices of a number of relevant materials. In the manufacture of TiO_2 either the anatase or the rutile form is produced depending on modifications in the process employed. Because of its slightly higher refractive index, rutile has a somewhat greater opacity and most of the TiO_2 currently produced is of this form.

In addition to these optical properties, TiO_2 is chemically inert which is why it displaced "white lead", $2PbCO_3 \cdot Pb(OH)_2$: in industrial atmospheres this formed PbS (black) during the production of or weathering of the paint and was also a toxic hazard. Unfortunately the naturally occurring forms of TiO_2 are invariably coloured, sometimes intensely, by impurities, and expensive processing is required to produce pigments of acceptable quality. The two main processes in use are the *sulfate process* and the *chloride process* (Fig. A), which account for approximately 70% and 30% respectively of total world production. The principal reactions of the chloride process are:

$$2TiO_2 + 3C + 4Cl_2 \xrightarrow{950^\circ C} 2TiCl_4 + CO_2 + 2CO$$

and

$$TiCl_4 + O_2 \xrightarrow{1000-1400^\circ} TiO_2 + 2Cl_2$$

It is operated mainly in the USA where it is responsible for about 60% of the production and is most economical when high-grade ores are used, becoming less economical with poorer feed materials containing iron, because of the production of chloride wastes from which the chlorine cannot be recovered. By contrast the sulfate process cannot make use of rutile and other high-grade ores which do not dissolve in sulfuric acid, but it has the advantage of being able to use the more plentiful ilmenite, as well as slags from which much of the iron has already been recovered.

The physical properties of the base pigments produced from both processes are further improved by slurrying in water and selectively precipitating on the finely divided particles a surface coating of SiO_2, Al_2O_3, or TiO_2 itself.

---

[8] R. S. DARBY and J. LEIGHTON, Titanium dioxide pigments, in *The Modern Inorganic Chemicals Industry*, pp. 354–74, Special Publication No. 31, (1977), The Chemical Society, London. *World Mineral Statistics 1976–80*, Inst. of Geological Sciences, HMSO, London, 1982.

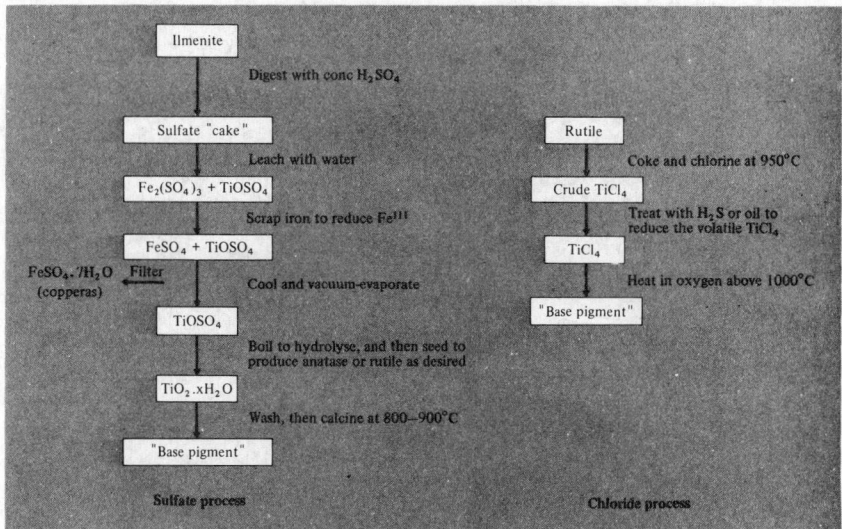

FIG. A Flow diagrams for the manufacture of TiO_2 pigments.

TABLE B *Refractive indices of some pigments and other materials*

Substance	Refractive index	Substance	Refractive index	Substance	Refractive index
NaCl	1.54	BaSO$_4$	1.64–1.65	Diamond	2.42
CaCO$_3$	1.53–1.68	ZnO	2.0	TiO$_2$ (anatase)	2.49–2.55
SiO$_2$	1.54–1.56	ZnS	2.36–2.38	TiO$_2$ (rutile)	2.61–2.90

dioxides and difluorides where the relative sizes of the ions are such as to favour 6-coordination (i.e. when the ratio of cation : anion lies in the range 0.73 to 0.41).[9] Anatase and brookite are both based on cubic rather than hexagonal close packing of oxygen atoms, but again the titanium atoms occupy half the octahedral interstices. TiO_2 melts at $1892 \pm 30°C$ when heated in an atmosphere of O_2; when heated in air the compound tends to loose oxygen and then melts at $1843 \pm 15°C$ ($TiO_{1.985}$).

Though it is unreactive, rutile can be reduced with difficulty to give numerous nonstoichiometric oxide phases, the more important of which are the Magnéli-type phases Ti_nO_{2n-1} ($4 \leqslant n \leqslant 9$), the lower oxides Ti_3O_5 and Ti_2O_3, and the broad, nonstoichiometric phase TiO_x ($0.70 \leqslant x \leqslant 1.30$). The Magnéli phases Ti_nO_{2n-1} are built up of slabs of rutile-type structure with a width of $nTiO_6$ octahedra and with adjacent slabs mutually related by a crystallographic shear which conserves oxygen atoms by an increased sharing between adjacent octahedra. Ti_4O_7 is metallic at room temperature but the other members of the series tend to be semiconductors.

[9] A. F. WELLS, *Structural Inorganic Chemistry*, 4th edn., Chap. III, pp. 65–92, Oxford University Press, Oxford, 1975.

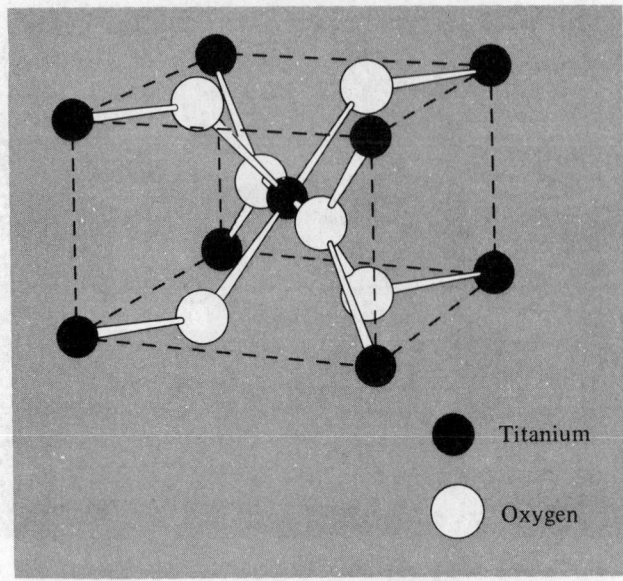

Titanium

Oxygen

FIG. 21.2   The tetragonal unit cell of rutile, TiO_2.

Of the lower oxides Ti_3O_5 is a blue-black material prepared by the reduction TiO_2 with H_2 at 900°C; it shows a transition from semiconductor to metal at 175°C. Ti_2O_3 is a dark-violet material with the corundum structure (p. 274); it is prepared by reacting TiO_2 and Ti metal at 1600°C and is generally inert, being resistant to most reagents except oxidizing acids. It has a narrow composition range ($x = 1.49$–1.51 in TiO_x) and undergoes a semiconductor-to-metal transition above ~ 200°C.

TiO, a bronze coloured, readily oxidized material, is again prepared by the reaction of TiO_2 and Ti metal. It has a defect rock-salt structure which tolerates a high proportion of vacancies (Schottky defects) in both Ti and O sites and so is highly nonstoichiometric with a composition range at 1700°C of $TiO_{0.75}$ to $TiO_{1.25}$. This range diminishes somewhat at lower temperatures and, at equilibrium below about 900°C, various ordered phases separate with smaller ranges of composition-variation, e.g. $TiO_{0.9}$–$TiO_{1.1}$ and $TiO_{1.25}$ (i.e. Ti_4O_5). In this latter compound the tetragonal unit cell can be thought of as being related to the NaCl-type structure: there are 10 Ti sites and 10 oxygen sites but 2 of the Ti sites are vacant in a regular or ordered way to generate the structure of Ti_4O_5. Further details of these various oxide structures and their interrelations are given in refs. 7 and 9a.

Finally, oxygen is soluble in metallic titanium up to a composition of $TiO_{0.5}$ with the oxygen atoms occupying octahedral sites in the hcp metal lattice: distinct phases that have been crystallographically characterized are Ti_6O, Ti_3O, and Ti_2O. It seems likely that in all these reduced oxide phases there is extensive metal–metal bonding.

In the case of zirconium and hafnium there is little evidence of stable phases other than MO_2, and at room temperature ZrO_2 (baddeleyite) and the isomorphous HfO_2 have a structure in which the metal is 7-coordinate (Fig. 21.3). ZrO_2 has at least two more high-temperature modifications (tetragonal above 1100°C and cubic, fluorite-type, above

[9a] D. J. M. BEVAN, Nonstoichiometric compounds: an introductory essay, Chap. 49, in *Comprehensive Inorganic Chemistry*, Vol. 3, Pergamon Press, Oxford, 1973; see especially pp. 485–6, 511–20, 538–40.

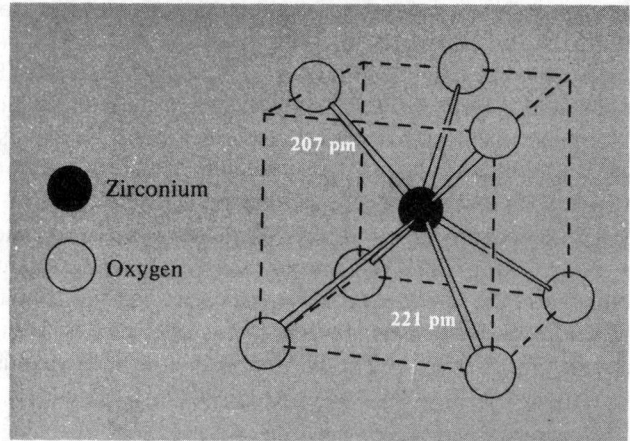

FIG. 21.3   The coordination of Zr^{IV} in baddeleyite ZrO_2; the 3 O atoms in the upper plane are each coordinated by 3 Zr atoms in a plane, whereas the 4 lower O atoms are each tetrahedrally coordinated by 4 Zr atoms.

2300°C) but it is notable that, presumably because of the greater size of Zr compared to Ti, neither of them has the 6-coordinate rutile structure. ZrO_2 is unreactive, has a low coefficient of thermal expansion, and a very high melting point (2710±25°C) and is therefore a useful refractory material, being used in the manufacture of crucibles and furnace cores. However, the phase change at 1100° severely restricts the use of pure ZrO_2 as a refractory because repeated thermal cycling through this temperature causes cracking and disintegration—the problem is avoided by using solid solutions of CaO or MgO in the ZrO_2 since these retain the cubic fluorite structure throughout the temperature range. ZrO_2 has also recently been produced in fibrous form suitable for weaving into fabrics, as already mentioned for Al_2O_3 (p. 275), and its chemical inertness and refractivity—coupled with an apparent lack of toxicity—can be expected to lead to increasing applications as an insulator and for the filtration of corrosive liquids. Production of ZrO_2 concentrates in 1980 exceeded 600 000 tonnes, of which more than 75% was mined in Australia.

The sulfides have been less thoroughly examined than the oxides but it is clear that a number of stable phases can be produced and nonstoichiometry is again prevalent (p. 804). The most important are the disulfides, which are semiconductors with metallic lustre. TiS_2 and ZrS_2 have the CdI_2 structure (p. 1407) in which the cations occupy the octahedral sites between alternate layers of hcp anions.

21.3.2 *Mixed, or complex, oxides*

Although the dioxides, MO_2, are notable for their inertness, particularly if they have been heated, fusion or firing at high temperatures (sometimes up to 2500°C) with the stoichiometric amounts of appropriate oxides produces a number of "titanates", "zirconates", and "hafnates". The titanates are of two main types: the orthotitanates $M_2^{II}TiO_4$ and the metatitanates $M^{II}TiO_3$. The names are misleading since the compounds almost never contain the discrete ions $[TiO_4]^{4-}$ and $[TiO_3]^{2-}$ analogous to phosphates or sulfites. Rather, the structures comprise three-dimensional networks of ions which are of particular interest and importance because two of the metatitanates are the archetypes of common mixed metal oxide structures.

When M^{II} is approximately the same size as Ti^{IV} (i.e. M = Mg, Mn, Fe, Co, Ni) the structure is that of *ilmenite*, $FeTiO_3$, which consists of hcp oxygens with one-third of the octahedral interstices occupied by M^{II} and another third by Ti^{IV}. This is essentially the same structure as corundum (Al_2O_3, p. 274) except that in that case there is only one type of cation which occupies two-thirds of the octahedral sites.

If, however, M^{II} is significantly larger than Ti^{IV} (e.g. M = Ca, Sr, Ba), then the preferred structure is that of *perovskite*, $CaTiO_3$. This can be envisaged as a ccp array of calcium and oxygen atoms, with the former regularly disposed, and the titanium atoms then occupying octahedral sites formed by oxygen atoms only and so being as remote as possible from the calciums (Fig. 21.4). The Ba^{II} ion is so large and expands the perovskite

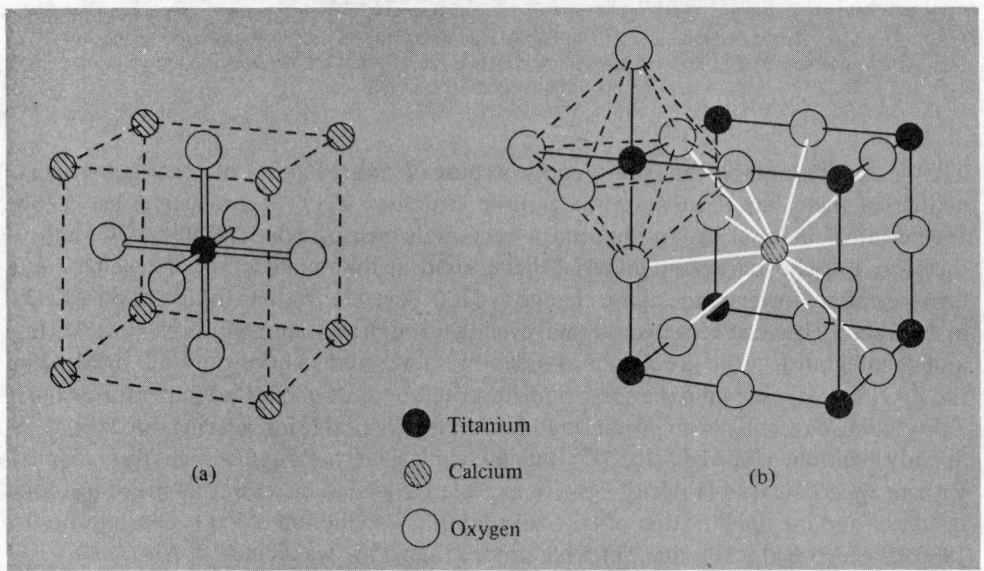

Titanium

Calcium

Oxygen

FIG. 21.4   Two representations of the structure of perovskite, $CaTiO_3$, showing (a) the octahedral coordination of Ti, and (b) the twelve-fold coordination of Ca by oxygen. Note the relation of (b) to the cubic structure of ReO_3 (p. 1219).

lattice to such an extent that the titanium is too small to fill the octahedral interstice which accommodates it. This leads to ferroelectric and piezoelectric behaviour as discussed in Chapter 3 (p. 60). In consequence, $BaTiO_3$ has found important applications in the production of compact capacitors (because of its high permittivity) and as a ceramic transducer in devices such as microphones and gramophone pick-ups. For such purposes it compares favourably with Rochelle salt (sodium potassium tartrate, $NaKC_4H_4O_6$) in terms of thermal stability, and with quartz in terms of the strength of the effect.

$M_2^{II}TiO_4$ (M = Mg, Zn, Mn, Fe, Co) have the *spinel* structure ($MgAl_2O_4$, p. 279) which is the third important structure type adopted by many mixed metal oxides; in this the cations occupy both octahedral and tetrahedral sites in a ccp array of oxide ions. Ba_2TiO_4, although having the same stoichiometry, is unique amongst titanates in that it contains discrete $[TiO_4]^{4-}$ ions[10] which have a somewhat distorted tetrahedral structure.

High-temperature reduction of Na_2TiO_3 with hydrogen produces nonstoichiometric materials, Na_xTiO_2 ($x = 0.20-0.25$), called titanium "bronzes" by analogy with the better-known tungsten bronzes (p. 1185). They have a blue-black, metallic appearance with high electrical conductivity and are chemically inert (even hydrofluoric acid does not attack them).

"Zirconates" and "hafnates" can be prepared by firing appropriate mixtures of oxides, carbonates, or nitrates. None of them are known to contain discrete $[MO_4]^{4-}$ or $[MO_3]^{2-}$ ions. Compounds $M^{II}ZrO_3$ usually have the perovskite structure whereas $M_2^{II}ZrO_4$ frequently adopt the spinel structure.

21.3.3　*Halides*[11, 12]

The most important of these are the tetrahalides, all 12 of which are well characterized. The titanium compounds (Table 21.3) show an interesting gradation in colour, the charge-transfer band moving steadily to lower energies (i.e. absorbing increasingly in the visible region of the spectrum) as the anion becomes more easily oxidized (F^- to I^-) by the small, highly polarizing titanium cation. The larger Zr^{IV} and Hf^{IV}, however, do not have the same polarizing effect and their tetrahalides are all white solids; the fluorides are involatile but the other tetrahalides sublime readily at temperatures in the range 320–430°C.

TABLE 21.3 *Some physical properties of titanium tetrahalides*

Compound	Colour	MP/°C	BP/°C
TiF_4	White	284	—
$TiCl_4$	Colourless	−24	136.5
$TiBr_4$	Orange	38	233.5
TiI_4	Dark brown	155	377

Though numerous preparative methods are possible, convenient general procedures are as follows:

tetrafluorides by the action of anhydrous HF on the tetrachloride;

tetrachlorides and tetrabromides by passing the halogen over the heated dioxide in the presence of a reducing agent such as carbon (this reaction is central to the chloride process for manufacturing TiO_2, p. 1118);

tetraiodides by the iodination of the dioxide with aluminium triiodide at a temperature of 130–400° depending on the metal ($3MO_2 + 4AlI_3 \rightarrow 3MI_4 + 2Al_2O_3$).

[10] J. A. BLAND, The crystal structure of barium orthotitanate, Ba_2TiO_4, *Acta Cryst.* **14**, 875–81 (1961).

[11] R. J. H. CLARK, *The Chemistry of Titanium and Vanadium*, Chaps. 2, 3, and 4, pp. 63–131, Elsevier, Amsterdam, 1968.

[12] R. COLTON and J. H. CANTERFORD, *Halides of the First Row Transition Metals*, Chap. 2, pp. 37–106, Wiley, London, 1969; and *Halides of the Second and Third Row Transition Metals*, Chap. 4, pp. 110–45, Wiley, London, 1968.

Not all the structures have been determined but in the vapour phase all the tetrahalides of titanium and probably all those of zirconium and hafnium have monomeric, tetrahedral structures. In the solid, TiF_4 is probably a fluoride-bridged polymer with 6-coordinate titanium, but the other tetrahalides of titanium retain the tetrahedral configuration around the metal even in the solids. In those cases where data are available, the larger zirconium (and presumably hafnium) exhibit higher coordination numbers. Thus solid ZrF_4 contains 8-coordinate (square antiprismatic) zirconium while $ZrCl_4$ is a polymer consisting of zigzag chains of $[ZrCl_6]^{2-}$ octahedra.

All the tetrahalides, but especially the chlorides and bromides, behave as Lewis acids dissolving in polar solvents to give rise to series of addition compounds; they also form complex anions with halides. They are all hygroscopic and hydrolysis follows the same pattern as complex formation, with the chlorides and bromides being more vulnerable than the fluorides and iodides. $TiCl_4$ fumes in and is completely hydrolysed by moist air $(TiCl_4 + 2H_2O \rightarrow TiO_2 + 4HCl)$; a variety of intermediate hydrolysis products, such as the oxochlorides $MOCl_2$, can be formed with aqueous HCl of varying concentration. Even in concHCl, $ZrCl_4$ gives $ZrOCl_2.8H_2O$. This contains the tetrameric cation $[Zr_4(OH)_8(H_2O)_{16}]^{8+}$ in which the 4 zirconium atoms are connected in a ring by 4 pairs of OH^- bridges and each zirconium atom is dodecahedrally coordinated to 8 oxygen atoms. The fluorides are less susceptible to hydrolysis and, though aqueous HF produces the oxofluorides, MOF_2, the hydrates $TiF_4.2H_2O$, $MF_4.H_2O$, and $MF_4.3H_2O$ (M = Zr, Hf) can be produced. Rather curiously the trihydrates of ZrF_4 and HfF_4 actually have different structures,[13] though both contain 8-coordinated metal atoms. The zirconium compound is essentially dimeric

$$[(H_2O)_3F_3Zr <\genfrac{}{}{0pt}{}{F}{F}> ZrF_3(H_2O)_3]$$

with dodecahedral Zr, whereas the hafnium compound consists of infinite chains of

$$[>HfF_2(H_2O)_2 <\genfrac{}{}{0pt}{}{F}{F}>]$$

with the third water molecule held in the lattice.

Besides being important as an intermediate in one of the processes for making TiO_2, $TiCl_4$ is also used to produce Ziegler–Natta catalysts (p. 1134) for the polymerization of ethylene (ethene) and is the starting point for the production of most of the commercially important organic titanium compounds (in most cases these are actually titanium alkoxides rather than true organometallic compounds). The iodides MI_4 are all utilized in the van Arkel–de Boer process for producing pure metals (p. 1113).

All the trihalides except HfF_3 have been prepared,[14] the most general method being the high-temperature reduction of the tetrahalide with the metal, though a variety of other methods have been used especially for the titanium compounds. Since the tetrahalides are quite stable to reduction, lower halides are not easily prepared in a pure state, incomplete reactions and the presence of excess metal being common. Apart from TiF_3, which, as expected for a d^1 ion has a magnetic moment of 1.75 BM at room temperature, all the

[13] D. Hall, C. E. F. Rickard, and T. N. Waters, The crystal structure of catena-di-μ-fluorodifluorodi-aquohafnium(IV) monohydrate, $HfF_4.3H_2O$, *J. Inorg. Nuclear Chem.* **33**, 2395–401 (1971).

[14] D. A. Miller and R. D. Bereman, The chemistry of d^1 complexes of niobium, tantalum, zirconium, and hafnium, *Coord. Chem. Revs.* **9**, 107–43 (1972).

titanium compounds have low magnetic moments, indicative of appreciable M–M bonding, and this is even more the case with the zirconium and hafnium compounds. All are coloured, halogen-bridged polymers with 6-coordinate metal and, apart from TiF_3 which is stable in air unless heated, all show reducing properties; indeed, ZrX_3 and HfX_3 reduce water and so have no aqueous chemistry, but aqueous solutions of TiX_3 are stable if kept under an inert atmosphere. Hexahydrates of $TiX_3 . 6H_2O$ are well known and the chloride is notable in that, like its chromium(III) analogue, it exhibits hydrate isomerism, existing as violet $[Ti(H_2O)_6]^{3+}Cl^-_3$ and green $[TiCl_2(H_2O)_4]^+Cl^-.2H_2O$.

TiX_2 (X = Cl, Br, I) have been characterized and are black solids with the CdI_2 structure (p. 1407) but their low magnetic moments again indicate extensive M–M bonding. They are very strongly reducing and decompose water. There have been numerous reports of similar black, solid dihalides of zirconium and hafnium but difficulties in preparing them and a tendency for them to disproportionate at high temperatures have made their characterization difficult. The chlorides and bromides Ti_7X_{16}, which are the first halides of titanium with a stoichiometry between TiX_2 and TiX_3, have recently been reported.[15] They are black crystalline solids sensitive to hydrolysis and oxidation and can be regarded as being composed of octahedrally coordinated Ti^{IV} and Ti^{II} in the ratio of 1:6 (i.e. $TiCl_4 . 6TiCl_2$) with the bivalent metal ions arranged in triangular groups involving Ti–Ti bonds. A novel ZrCl phase has been prepared[16] by reducing $ZrCl_4$ with Zr foil in a sealed tantalum container at 600–800°C. It has a layer structure, and strong metallic bonding is evidenced by Zr–Zr separations of 303 and 342 pm (compared to 319 pm in the metal itself) and metallic conduction in the plane of the sheets.

21.3.4 *Compounds with oxoanions*

Because of the high ratio of ionic charge to radius, normal salts of Ti^{IV} cannot be prepared from aqueous solutions, which only yield basic, hydrolysed species. Even with Zr^{IV} and Hf^{IV}, normal salts such as $Zr(NO_3)_4.5H_2O$ and $Zr(SO_4)_2.4H_2O$ can only be isolated if the solution is sufficiently acidic. Several oxometal(IV) compounds (i.e. "titanyl", "zirconyl") have been isolated and were originally thought to contain discrete MO^{2+} ions. This, however, now seems to be incorrect[17] and the compounds are actually polymeric in the solid state. Thus, $TiOSO_4.H_2O$ contains chains of –Ti–O–Ti–O– with each Ti being approximately octahedrally coordinated to 2 bridging oxygen atoms, 1 water molecule and an oxygen atom from each of 3 sulfates; $ZrO(NO_3)_2$ is also an oxygen-bridged chain, though hydroxy bridging, as in $ZrOCl_2.8H_2O$ mentioned above, is more common. By contrast, ion-exchange studies on aqueous solutions of Ti^{IV} in 2 M $HClO_4$ are consistent with the presence of monomeric doubly-charged cationic species rather than polymers, though it is not clear whether the predominent species is $[TiO]^{2+}$ or $[Ti(OH)_2]^{2+}$.[17a]

[15] H. Schäfer, R. Laumanns, B. Krebs, and G. Henkel, Novel titanium halides containing metal–metal bonds: Ti_7Cl_{16} and Ti_7Br_{16}, *Angew. Chem.*, Int. Edn. (Engl.) **18**, 325–6 (1979).

[16] D. G. Adolphson and J. D. Corbett, Crystal structure of zirconium monochloride: a novel phase containing metal–metal bonded sheets, *Inorg. Chem.* **15**, 1820–3 (1976).

[17] T. E. MacDermott, The structural chemistry of zirconium compounds, *Coord. Chem. Revs.* **11**, 1–20 (1973).

[17a] J. D. Ellis and A. G. Sykes, Kinetic studies on the vanadium(II)–titanium(IV) and titanium (III)–vanadium(IV) redox reactions in aqueous solutions, *JCS Dalton* 1973, 537–43.

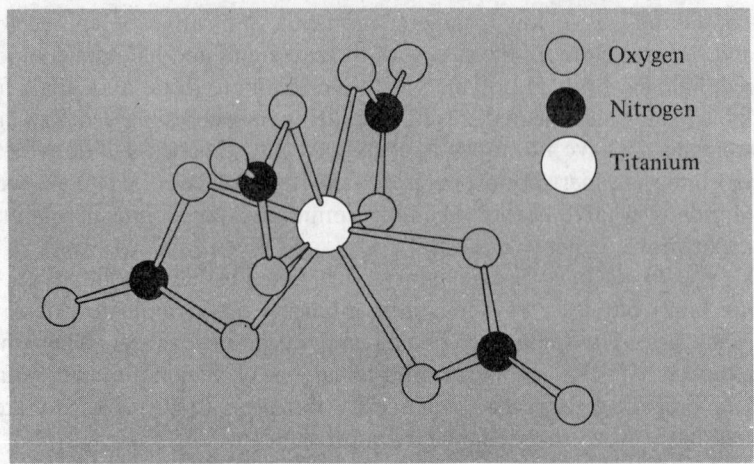

FIG. 21.5   The molecular structure of $Ti(NO_3)_4$. Eight O atoms form a dodecahedron around the Ti and the 4 N atoms form a flattened tetrahedron.

The $SO_4{}^{2-}$ ion has a strong affinity for Zr^{IV} and, besides the normal sulfate, forms many anionic complexes and basic salts, all of which appear to contain 7- or 8-coordinate zirconium.[18]

The anhydrous nitrates can be prepared by the action of N_2O_5 on MCl_4. $Ti(NO_3)_4$ is a white sublimable and highly reactive compound (mp 58°C) in which the bidentate nitrate ions are disposed tetrahedrally around the titanium which thereby attain a coordination number of 8 (Fig. 21.5). Infrared evidence suggests that $Zr(NO_3)_4$ is isostructural but hafnium nitrate sublimes under vacuum at 100°C as the adduct $Hf(NO_3)_4.N_2O_5$.

In oxidation states lower than $+4$, only Ti^{III} forms a sulfate and this gives rise to the alums, $MTi(SO_4)_2.12H_2O$ (M = Rh, Cs), containing the octahedral hexaquotitanium(III) ion.

21.3.5   *Complexes*

Oxidation state IV (d^0)

A very large number of these complexes, particularly of titanium, have been prepared and, as is to be expected for the d^0 configuration, they are invariably diamagnetic. Hydrolysis, resulting in polymeric species with –OH– or –O– bridges, is common especially with titanium and is still a preparative problem with zirconium and hafnium. A coordination number of 6 is the most usual for Ti^{IV} but 7- and even 8-coordination is possible. However, these high coordination numbers are much more characteristic of Zr^{IV} and Hf^{IV}, whose complexes are more labile (consistent with greater electrostatic character in the bonding). Furthermore, because changes in geometry entail smaller changes in energy for higher coordination numbers, these include a greater variety of stereochemistries.

[18] I. J. BEAR and W. G. MUMME, Normal sulphates of zirconium and hafnium, *Rev. Pure Appl. Chem. (Austral.)* **21,** 189–211 (1971).

The neutral and anionic adducts of the halides constitute a large proportion of the complexes of Ti^{IV}, and the alkoxides (also prepared from TiCl$_4$) are of commercial importance (p. 1128).

TiF$_4$ forms 6-coordinate adducts mainly with O- and N-donor ligands, and complexes of the type TiF$_4$L have all the appearances of fluorine-bridged polymers. TiCl$_4$ and TiBr$_4$ are especially prolific and are clearly "softer" acceptors than the fluoride. They form mainly yellow to red adducts of the types [MX$_4$L$_2$] and [MX$_4$(L-L)] with ligands such as ethers, ketones, OPCl$_3$, amines,[19] imines, nitriles, thiols, and thioethers. Apparently similar zirconium and hafnium analogues are known but are often less well characterized because of insolubility and also difficulty in preparing samples suitable for X-ray analysis. Phosphorus- and As-donor ligands also complex readily with the chlorides of all three metals, particularly as chelates, and are of interest in that they produce coordination numbers which, for titanium, are unusually high. Thus o-phenylenebis(dimethyldiarsine), diars, and its phosphorus analogue form not only the 6-coordinate [MX$_4$(L-L)] but also the 8-coordinate [MX$_4$(L-L)$_2$]. [TiCl$_4$(diars)$_2$] (Fig. 21.6) was in fact one of the first examples of an 8-coordinate complex of a first-row transition element.[20] The terdentate arsine, MeC(CH$_2$AsMe$_2$)$_3$, forms a 1:1 adduct with TiCl$_4$ which is monomeric and so presumably 7-coordinate. [TiCl$_4$L] adducts are usually 6-coordinate dimers with double chloride bridges.

Octahedral, anionic complexes, [MX$_6$]$^{2-}$, show a marked increase in susceptibility to hydrolysis and consequent difficulty in preparation, in passing from the stable fluorides to the heavier halides, with the result that the hexaiodo complexes cannot be isolated. The fluorozirconates and fluorohafnates display considerable variety,[12, 21] complexes of the types [MF$_7$]$^{3-}$, [M$_2$F$_{14}$]$^{6-}$, and [MF$_8$]$^{4-}$ having been prepared, often by fusion of the appropriate fluorides. In Na$_3$ZrF$_7$ the anion has the 7-coordinate pentagonal bipyramidal structure; in Li$_6$[BeF$_4$][ZrF$_8$] the zirconium anion is 8-coordinate, dodecahedral (distorted); in Cu$_6$[ZrF$_8$].12H$_2$O it is 8-coordinate, square antiprismatic, and in Cu$_3$[Zr$_2$F$_{14}$].18H$_2$O dimerization by the edge-sharing of two square antiprisms maintains 8-coordination. Stoichiometry does not, however, define coordination type, and this is very well illustrated by the ostensibly [MF$_6$]$^{2-}$ complexes which may contain 6-, 7-, or 8-coordinate Zr^{IV} or Hf^{IV} depending on the counter anion. In Rb$_2$MF$_6$, M is indeed octahedrally coordinated, but in (NH$_4$)$_2$MF$_6$ and K$_2$MF$_6$ polymerization occurs to give respectively 7- and 8-coordinate species.

Alkoxides of all 3 metals are well characterized but it is those of titanium which are of particular importance. The solvolysis of TiCl$_4$ with an alcohol yields a dialkoxide:

$$TiCl_4 + 2ROH \longrightarrow TiCl_2(OR)_2 + 2HCl$$

If dry ammonia is added to remove the HCl, then the tetraalkoxides can be produced:

$$TiCl_4 + 4ROH + 4NH_3 \longrightarrow Ti(OR)_4 + 4NH_4Cl$$

These alkoxides are liquids or sublimable solids and, unless the steric effects of the alkyl chain prevent it, apparently attain octahedral coordination of the titanium by

[19] G. W. A. Fowles, Reaction by metal halides with ammonia and aliphatic amines, *Prog. Inorg. Chem.* **6**, 1–36 (1964).

[20] R. J. H. Clark, J. Lewis, R. S. Nyholm, P. Pauling, and G. B. Robertson, Eight-coordinate diarsine complexes of quadrivalent metal halides, *Nature* **192**, 222–3 (1961).

[21] S. A. Cotton and F. A. Hart, *The Heavy Transition Elements*, Chap. 1, pp. 3–4, Macmillan, London, 1975.

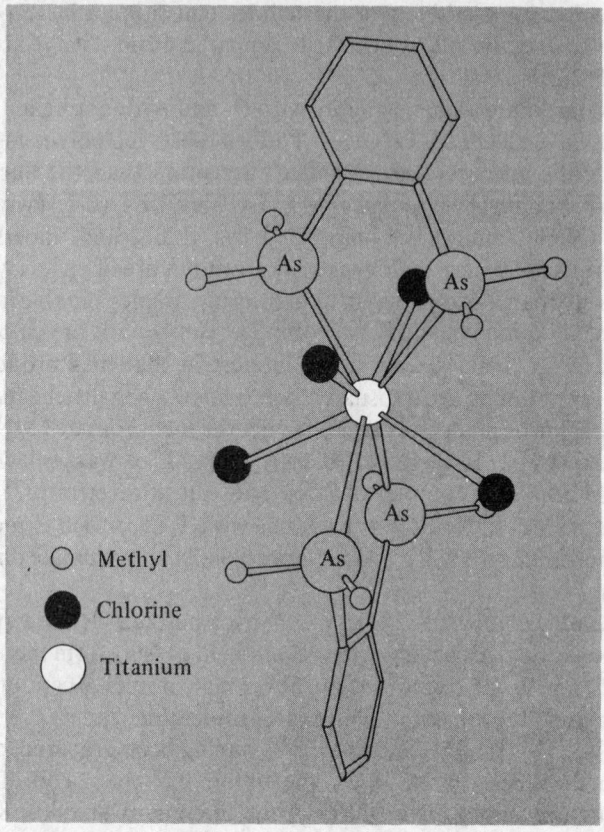

○ Methyl

● Chlorine

○ Titanium

Fig. 21.6   Molecular structures of [TiCl$_4$(diars)$_2$]. The dodecahedral coordination is produced
by two interpenetrating tetrahedra (slightly distorted) of chlorine and arsenic atoms.

polymerization (Fig. 21.7). The lower alkoxides are especially sensitive to moisture, hydrolysing to the dioxide. Application of these "organic titanates" (as they are frequently described) can therefore give a thin, transparent, and adherent coating of TiO$_2$ to a variety of materials merely by exposure to the atmosphere. In this way they are used to waterproof fabrics and also in heat-resistant paints. They are also used on glass and enamels which after firing retain a coating of TiO$_2$ which confers a resistance to scratching and often enhances the appearance. However, the most important commercial application is in the production of "thixotropic" paints which do not "drip" or "run". For this the Ti(OR)$_4$ is chelated with ligands such as β-diketonates to give products of the type [Ti(OR)$_2$(L-L)$_2$] which are water soluble and more resistant to hydrolysis. In concentrations of 1% or less, they form gels with the cellulose ether colloids used to thicken latex paints and so produce the desired characteristics.

One of the most sensitive methods for estimating titanium (or, conversely, for estimating H$_2$O$_2$) is to measure the intensity of the orange colour produced when H$_2$O$_2$ is added to acidic solutions of titanium(IV). The colour is probably due to the peroxo complex, [Ti(O$_2$)(OH)(H$_2$O)$_x$]$^+$, though alkaline solutions are needed before crystalline solids such as M$_3^I$[Ti(O$_2$)F$_5$] or M$_2^I$[Ti(O$_2$)(SO$_4$)$_2$] can be isolated. The peroxo ligand is

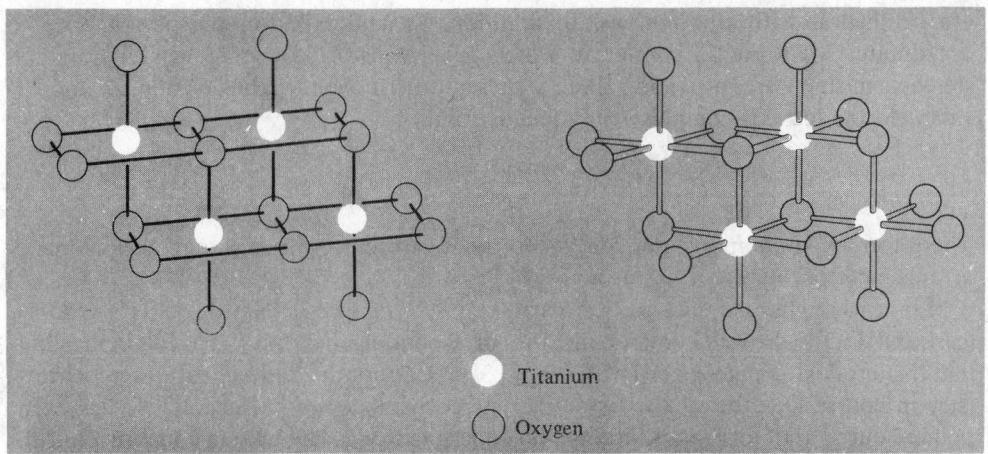

Titanium

Oxygen

FIG. 21.7   Two representations of the tetrameric structure of $[Ti(OEt)_4]_4$.

apparently bidentate, the 2 oxygen atoms being equidistant from the metal (see also p. 719). Bent σ bonds have been suggested,[22] as shown in Fig. 21.8.

Not surprisingly, in view of their greater size, zirconium and hafnium show a greater preference than titanium for *O*-donor ligands as well as for high coordination numbers, and this is shown by the greater variety of β-diketonates, carboxylates, and sulfato complexes which they form. Bis-β-diketonates such as $[MCl_2(acac)_2]$ of all 3 metals are made by the reaction of MCl_4 and the β-diketone in inert solvents such as benzene. They

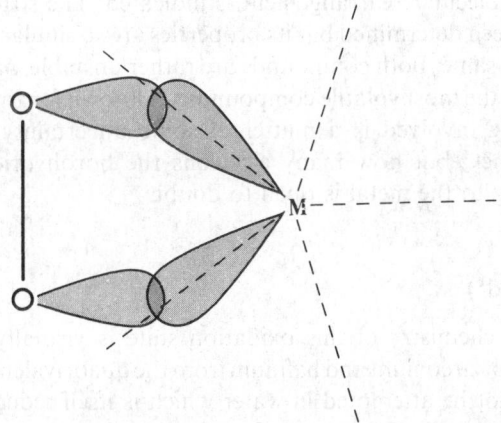

FIG. 21.8   Diagrammatic representation of bent σ bonds in complexes of the bidentate peroxo ligand. In many peroxo complexes the geometry is pentagonal bipyramidal, with the peroxo ligand situated in the pentagonal plane, for which the undistorted L–M–L angle (i.e. the angle between the metal sp^3d^3 hybrid orbitals) is 72° as shown here. The same situation occurs in 5-coordinate pentagonal pyramidal complexes (e.g. $[Cr(O_2)_5py]$); likewise in 8-coordinate, dodecahedral complexes (e.g. $[M(O_2)_4]^{3-}$, (M = V, Nb, Ta, Cr)) virtually identical undistorted angles of 71° are found.

[22] W. P. GRIFFITH, Studies on transition-metal peroxy-complexes. Pt. III. Peroxy complexes of Groups IVA, VA, VIA, *J. Chem. Soc.* 1964, 5248–53.

are octahedral with *cis*-chlorides. In addition, Zr and Hf form the monomeric, 7-coordinate [MCl(acac)$_3$] complexes which have a distorted pentagonal bipyramidal stereochemistry. Also, providing alkali is present to remove the labile proton, Zr and Hf will yield the tetrakis complexes in aqueous solution:

$$MOCl_2 \xrightarrow[+ Na_2CO_3]{4 acacH} [M(acac)_4]$$

These too are monomeric, and the 8-coordinate structure has a square-antiprismatic arrangement of oxygen atoms around M.

Monocarboxylates of the types [Zr(carbox)$_4$], [ZrO(carbox)$_3$(H$_2$O)$_x$] and [ZrO(OH)(carbox)(H$_2$O)$_x$] are well known, as are the corresponding dicarboxylates. It is interesting that the tetrakis(oxalates), Na$_4$[M(C$_2$O$_4$)$_4$].3H$_2$O, adopt the dodecahedral stereochemistry in contrast to the square-pyramidal stereochemistry of [M(acac)$_4$]. It has been pointed out[23] that for 8-coordination the dodecahedron is favoured in comparison with the square pyramid by ligands with a smaller "bite". It may therefore be significant that the 5-membered chelate ring formed by the oxalate ion features a smaller "bite" than does the 6-membered ring formed by acac. It may also be noted that, although optical and geometrical isomerism is conceivable for these stereochemistries, intramolecular rearrangement of the ligands is too rapid for sets of isomers of the above compounds (or, indeed, of any compound of Zr^{IV} of Hf^{IV}) to have been isolated.[17]

Intramolecular rearrangement evidently also occurs in the borohydride, [Zr(BH$_4$)$_4$] (p. 189). X-ray analysis of a single crystal at $-160°$C, at which temperature thermal vibrations are sufficiently reduced to allow the positions of the hydrogen atoms to be determined, showed it to have T_d symmetry[24] (Fig. 21.9), with triple-hydrogen bridges, implying two types of hydrogen. Yet the proton nmr distinguishes only one type of proton, so that rapid intramolecular rearrangement is indicated. The structure of the hafnium compound has not been determined but its properties are so similar that its structure may be assumed to be the same. Both compounds are rather unstable, have virtually the same mp ($\sim 29°$C) and are the most volatile compounds yet known for zirconium and hafnium. The type of bonding involved is a matter of some uncertainty.[25] The volatility is indicative of covalency, but how many electrons the borohydride groups should be regarded as donating to the metal is open to doubt.

Oxidation state III (d^1)

The coordination chemistry of this oxidation state is virtually confined to that of titanium. Reduction of zirconium and hafnium from the quadrivalent to the tervalent state is not easy and cannot be attempted in water which is itself reduced by Zr^{III} and Hf^{III}. Only a handful of complexes of these two elements therefore are known in this oxidation state[14] and these are confined almost entirely to adducts of the trihalides with *N*-donor

[23] D. G. BLIGHT and D. L. KEPERT, The stereochemistry of eight coordination: the effect of bidentate ligands, *Inorg. Chem.* **11**, 1556–61 (1972).

[24] P. H. BIRD and M. R. CHURCHILL, The crystal structure of zirconium(IV) borohydride (at $-160°$), *JCS Chem. Comm.* 1967, 403.

[25] A. DAVISON and S. S. WREFORD, Some comments on the bonding of tetrahydroborate ion to transition metals, *Inorg. Chem.* **14**, 703 (1975). T. A. KEIDERLING, W. T. WOZNIAK, R. S. GAY, D. JARKOWITZ, E. R. BERNSTEIN, S. J. LIPPARD, and T. G. SPIRO, Raman and infrared spectra of Hf(BH$_4$)$_4$ and Hf(BD$_4$)$_4$: evidence for Hf–B bonding, *Inorg. Chem.* **14**, 576–9 (1975).

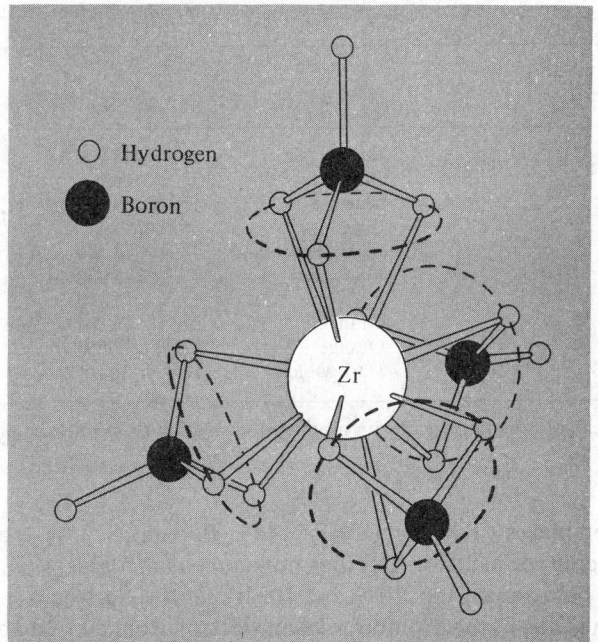

Fig. 21.9   Molecular structure of $[Zr(BH_4)_4]$ showing 4 trihapto BH_4 groups.

ligands such as pyridine, 2,2′-bipyridyl and 1,10-phenanthroline. Structural data are sparse. Even Ti^{III} is prone to aerial oxidation, and its compounds must be prepared under anaerobic conditions. However, because of the simplicity of its electronic configuration, it has been the focus of considerable attention, and the interpretation[26] of the electronic spectrum of aqueous Ti^{III} was an early landmark in the development of crystal field theory (p. 1084).

Most of the complexes of titanium(III) are octahedral and are produced by reacting $TiCl_3$ with an excess of the ligand, giving rise to stoichiometries such as $[TiL_6]X_3$, $[TiL_4X_2]X$, $[TiL_3X_3]$, and $M^I_3[TiX_6]$ (L = neutral unidentate ligand, X = singly charged anion) (Table 21.4) together with corresponding complexes involving multidentate ligands.

The first of these types is most familiarly represented by the hexaquo ion which is present in acidic aqueous solutions and, in the solid state, in the alum $CsTi(SO_4)_2.12H_2O$. In fact few other neutral ligands besides water form a $[TiL_6]^{3+}$ complex. Urea is one of these few and $[Ti(OCN_2H_4)_6]I_3$, in which the urea ligands coordinate to the titanium via their oxygen atoms, is one of the compounds of titanium(III) most resistant to oxidation.

Hydrate isomerism of $TiCl_3.6H_2O$, yielding $[TiCl_2(H_2O)_4]^+Cl^-$ as one of the isomers, has already been referred to (p. 1125) and analogous complexes are formed by a variety of alcohols. Neutral complexes, $[TiL_3X_3]$ have been characterized for a variety of ligands such as tetrahydrofuran (C_4H_8O), dioxan $(C_4H_8O_2)$, acetonitrile, pyridine, and picoline,

[26] F. E. ILSE and H. HARTMAN, Termsysteme elektrostatischer Komplexionen der Übergangsmetalle mit einem d-Elektron, *Z. Phys. Chem.* **197**, 239–46 (1951).

TABLE 21.4   *Spectroscopic and magnetic properties of some complexes of titanium(III)*

Complex	Colour	$^2E_g \leftarrow {}^2T_{2g}/(cm^{-1})$	μ (room temperature)/ BM
$[Cs(H_2O)_6][Ti(H_2O)_6][SO_4]_2$	Red-purple	19 900, 18 000	1.79
$[Ti(urea)_6]I_3$	Blue	17 550, 16 000	1.77
$[TiCl_3(NCMe)_3]$	Blue	17 100, 14 700	1.68
$[TiCl_3(NC_5H_5)_3]$	Green	16 600, Asym[a]	1.63
$[TiCl_3(thf)_3]$	Blue-green	14 700, 13 500	1.70
$[TiCl_3(dioxan)_3]$	Blue-green	15 150, 13 400	1.69
$[NH_4]_3[TiF_6]$	Purple	19 000, 15 100	1.78
$[C_5H_5NH]_3[TiCl_6]$	Orange	12 750, 10 800	1.78
$[C_5H_5NH]_3[TiBr_6]$	Orange	11 400, 9 650	1.81
$[NBu^n_4]_3[Ti(NCS)_6]$	Dark violet	18 400, Asym[a]	1.81

[a] The band "envelope" is asymmetrical with insufficient resolution to identify the position of the weaker component.

while anionic complexes $[TiX_6]^{3-}$ ($X = F^-$, Cl^-, Br^-, NCS^-) have been prepared by electrolytic reduction of melts or by other nonaqueous methods.

It has already been shown (pp. 1089 and 1092) that the electronic spectrum of Ti^{III} in aqueous solution is due to the promotion of an electron from a t_{2g} to an e_g orbital which can alternatively be described as the transition $^2E_g \leftarrow {}^2T_{2g}$. However, the absorption band actually observed for this and for other octahedral complexes of Ti^{III} is never of the symmetrical shape expected for a single transition, but is rather an asymmetrical peak with a (usually) distinct shoulder on the low-energy side. The whole absorption "envelope" is apparently made up of two superimposed bands whose positions are indicated in Table 21.4 and which are generally assumed to be a consequence of the Jahn–Teller effect—not acting on the ground term, as was discussed in Chapter 19, but acting in this case on the excited term. Although this is actually rather complicated, the result is that the excited ^2E_g term splits into two components,[27] corresponding to the $(d_{z^2})^1$ and $(d_{x^2-y^2})^1$ configurations, and so engenders two distinct excitations. The value of $10Dq$ is usually identified with the energy of the stronger of the two bands rather than an average, and the results in Table 21.4 indicate that this varies with the ligand in the order:

$$Br^- < Cl^- < urea < NCS^- < F^- < H_2O$$

which agrees with the spectrochemical series established for other metals (p. 1096).

Magnetically, the t_{2g}^1 ground configuration in an octahedral crystal field is expected (p. 1098) to produce an "orbital contribution" to the magnetic moment which varies with temperature. The observed magnetic moments do indeed vary with temperature, but by no means as much as would be expected for precisely octahedral fields, and, as is borne out by Table 21.4, only rarely do the magnetic moments at room temperature exceed the spin-only value of 1.73 BM. These discrepancies have been attributed[28] to distortions which split the $^2T_{2g}$ ground term by a few hundred cm^{-1} along with some delocalization of the single electron from the metal, due to partial covalent character of the metal–ligand bond.

[27] S. F. A. KETTLE, *Coordination Compounds*, pp. 109–14, Nelson, London, 1969.

[28] B. N. FIGGIS, The magnetic properties of transition metal ions in asymmetric ligand fields, Pt. 2. Cubic field $^2T_2$ terms, *Trans. Faraday Soc.* **57**, 198–203 (1961); but see also M. GERLOCH, *J. Chem. Soc.* (A) 1968, 2023–8, for details of the over-simplifications sometimes involved in this treatment.

Amongst the few complexes of Ti^{III} which have been shown to be non-octahedral are $[TiBr_3(NMe_3)_2]$ and $[Ti\{N(SiMe_3)_2\}_3]$. The former has a 5-coordinate, trigonal bipyramidal structure while the latter is one of a series of complexes of tervalent metals which have a 3-coordinate, planar structure.[29] It appears that the silylamide ligands are simply too bulky for the Ti^{III} ion to accommodate more than three of them, and this consideration overrides any preference which the metal might have for a higher coordination number.

Lower oxidation states

With the exception of TiO and the lower halides already mentioned, very little chemistry has yet been established for any of these metals in oxidation states lower than 3. $TiCl_2$ has a polymeric structure involving extensive metal–metal bonding and, although addition compounds of the type $[TiCl_2L_2]$ can be formed with difficulty with ligands such as dimethylformamide and acetonitrile,[30] magnetic moments in the range 1.0 to 1.2 BM at room temperature (compared to a spin-only value of 2.83 BM for 2 unpaired electrons) suggest that they also are polymeric with appreciable metal–metal bonding. However, the electronic spectra of Ti^{II} in $TiCl_2/AlCl_3$ melts[31] and also of Ti^{II} incorporated in NaCl crystals[32] (prepared by the reaction of $CdCl_2$ and titanium in molten NaCl and subsequent sublimation of Cd metal) have been shown to be as expected for a d^2 ion in an octahedral field.

Perhaps the most interesting of the complexes of these metals in low oxidation states are those formed when MCl_4 is reduced in tetrahydrofuran by lithium in the presence of bipyridyl. Compounds of the type $[M(bipy)_3]$, $Li[M(bipy)_3]$, and $Li_2[M(bipy)_3]$ with varying amounts of solvent of crystallization have been isolated, implying oxidation states of 0, -1, and -2. Such magnetic evidence as is available suggests that, if these oxidation states are realistic, then the consequent t_{2g}^4, t_{2g}^5, and t_{2g}^6 configurations are all low-spin. However, it seems[33] that there is very considerable delocalization of charge in the π^* orbitals of the ligands, so that reduction of the ligands may well be more important than reduction of the metal, and assigning oxidation states to the metals is a purely formal exercise. A more "realistic" claim to zero valency in Zr and Hf compounds is provided by $[M(\eta\text{-PhMe})_2(PMe_3)]$. Metal vapour was produced from an "electron-gun furnace" and condensed with an excess of toluene and trimethylphosphine at $-196°C$. On warming up, a dark-green solution was produced from which the pure solids were isolated.[34]

21.3.6　*Organometallic compounds*

Until the 1950s this was an unexplored area of chemistry, but then two events occurred:

[29] D. C. BRADLEY, Steric control of metal coordination, *Chem. Br.* **11**, 393–7 (1975).

[30] G. W. A. FOWLES, T. E. LESTER, and R. A. WALTON, Co-ordination complexes of titanium(II), *J. Chem. Soc.* (A) 1968, 1081–5.

[31] H. A. ØYE and D. M. GRUEN, Octahedral absorption spectra of the dipositive 3d metal ions in molten aluminium chloride, *Inorg. Chem.* **3**, 836–41 (1964).

[32] W. E. SMITH, The spectrum of Ti^{2+} ions isolated in a sodium chloride crystal, *JCS Chem. Comm.* 1972, 1121–2.

[33] L. E. ORGEL, Double bonding in chelated metal complexes, *J. Chem. Soc.* 1961, 3683–6.

[34] F. G. N. CLARKE and M. L. H. GREEN, Synthesis of zerovalent (η-arene)(PMe$_3$)-hafnium and -zirconium compounds using metal vapours, *JCS Chem. Comm.* 1979, 127–8.

ferrocene was discovered (pp. 345, 1286) and K. Ziegler[35] catalysed the polymerization of ethylene using an organo-titanium derivative. The first event initiated a systematic study of cyclopentadienyl compounds, and so led to the preparation of the most stable of the organometallic compounds of this group, while the second event provided a strong commercial incentive for the investigation of this field (see Panel).

Ziegler–Natta Catalysts[35a]

The original ICI process for producing polythene involved the use of high temperatures and pressures but K. Ziegler discovered[35] that, in the presence of a mixture of $TiCl_4$ and $AlEt_3$ in a hydrocarbon solvent, the polymerization will take place at room temperature and atmospheric pressure. G. Natta[36] then showed that by suitable modification of the catalyst stereoregular polymers of almost any alkene (olefin), $CH_2{=}CHR$, can be produced. In general, these catalysts can be formed from an alkyl of Li, Be, or Al together with a halide of one of the metals of Groups IVA to VIA in an oxidation state less than its maximum. As a result of their work, Ziegler and Natta were jointly awarded the 1963 Nobel Prize for Chemistry. Because of its commercially sensitive nature, much of the voluminous literature on this subject is in the form of patents, but a great deal of work has also been directed at ascertaining the mechanism of the catalyst.[37] The initial reaction of $TiCl_4$ and $AlEt_3$ produces insoluble $TiCl_3$ (alternatively, preformed $TiCl_3$ can be used). The most plausible sequence of events on the surface of this catalyst is then as illustrated in Fig. A:

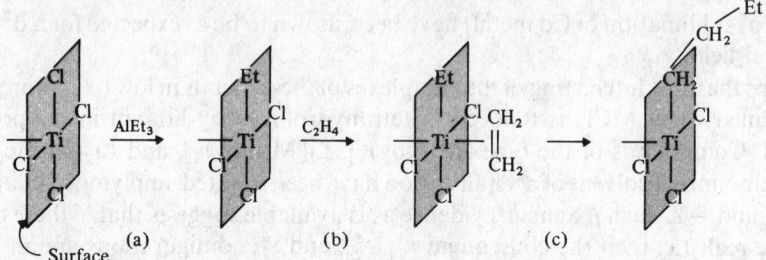

FIG. A   Possible mechanism of Ziegler–Natta catalyst.

(a) one of the chlorine atoms coordinated to a titanium atom is replaced by an ethyl group from $AlEt_3$;
(b) then, because the titanium atom on the surface of the solid has a vacant coordination site, a molecule of ethylene (ethene) can attach itself;
(c) migration of the ethyl group to the ethylene by a well-known process known as "cis-insertion" occurs.

The result of this cis-insertion is that a vacant site is left behind, and this can be occupied by another ethylene molecule and steps (a) and (b) repeated indefinitely.

The efficacy of the catalyst seems to lie in the fact that in the case of propylene $(CH_2{=}CH{-}CH_3)$, for instance, the steric hindrance inherent in the surface coordination sites ensures that the polymer which is produced is stereoregular. Such a stereoregular polymer is stronger and has a higher mp than the non-regular (so-called "atactic") polymer. Furthermore, while the titanium provides bonds sufficiently strong to be able to hold the olefin and the alkyl in the correct orientations for reaction, they are not so strong as to prevent the migration which is essential to the reaction. For an alternative suggestion for the mechanism of catalysis, see p. 293.

[35] K. Ziegler, E. Holzkamp, H. Breiland and H. Martin, Das Mülheimer Normaldruck-Polyäthylen-Verfahren, *Angew. Chem.* **67**, 541–7 (1955).

[35a] J. Boor, *Ziegler-Natta Catalysts and Polymerizations*, Academic Press, New York, 1979, 670 pp.

[36] G. Natta, Une nouvelle classe de polymeres d[1] α-olefines ayant une Regularité de Structure Exceptionnelle, *J. Polymer Sci.* **16**, 143–54 (1955).

[37] R. Feld and P. L. Cowe, *The Organic Chemistry of Titanium*, Chap. 11, pp. 149–67, Butterworths, London, 1965.

In sharp contrast to the Group IVB elements Ge, Sn, Pb, Group IVA metals form few alkyl and aryl compounds and those which are known are very unstable to both air and water. MMe_4 can be prepared by the reactions of LiMe and MCl_4 in ether at low temperatures, but the yellow titanium and the red zirconium compounds decompose to the metals at temperatures above -20 and $-15°C$ respectively. Phenyltitanium triisopropoxide, $(C_6H_5)Ti(OPr^i)_3$, containing only one Ti–C bond, is thermally stable up to 10°C and was the compound with the first authenticated Ti–C bond.

Perhaps because of inadequate or non-existent back-bonding (p. 1101), the only binary carbonyl so far reported is $Ti(CO)_6$ which has been produced by condensation of titanium metal vapour with CO in a matrix of inert gases at 10–15 K, and identified spectroscopically.[38] Cyclopentadienyls, however, are much more stable and common, and are known for metal oxidation states of IV, III, and II. The compounds $M(C_5H_5)_4$ are prepared from MCl_4 and NaC_5H_5 and the structure of the green-black titanium compound is shown in Fig. 21.10. It is therefore formulated as $[Ti(\eta^1\text{-}C_5H_5)_2(\eta^5\text{-}C_5H_5)_2]$ (p. 373). Rather surprisingly, the ^1H nmr distinguishes only one type of proton at room temperature. Studies at lower temperatures suggest that two fluxional processes occur. Firstly, the mono- and pentahapto- (i.e. σ- and π-bonded) rings rapidly interchange their roles, and secondly, the point of attachment of each of the monohapto rings changes continually between the 5 carbons of each ring ("ring whizzing"). The result of these time-averaging processes is that all 20 protons become indistinguishable. Reports of the structures of the yellow-orange Zr, and yellow Hf compounds have been somewhat confused but it is now known that the Hf compound, like that of Ti, contains 2 monohapto- and 2 pentahapto-rings[39] whereas that of Zr contains 1 monohapto- and 3 pentahapto-rings, $[Zr(\eta^1\text{-}C_5H_5)(\eta^5\text{-}C_5H_5)_3]$.[39a] This formulation is unexpected since it entails a formally 20-electron configuration; the two compounds also provide the first authenticated example of a structural difference in the organometallic chemistries of Zr and Hf.

Best known of all are the bis(cyclopentadienyls) of the type $[M(\eta^5\text{-}C_5H_5)_2X_2]$, the halides being prepared again by the action of NaC_5H_5 on MCl_4. Replacement of X by SCN^-, N_3^-, $-NR_2$, $-OR$, or $-SR$ is also possible and $[M(\eta^5\text{-}C_5H_5)_2(CO)_2]$ are the only well-characterized carbonyls of Group IVA[40] (Fig. 21.11). In all cases the structures are distorted tetrahedral with both rings pentahapto-. Interesting derivatives of the type $[(C_5H_5)_2Ti(CH_2)_4]$ have also been produced.[41] Amongst the many other reactions of the dihalides, ring replacement to give compounds such as $[Ti(C_5H_5)X_3]$ and reductions to $[Ti(C_5H_5)_2X]$ and $[Ti(C_5H_5)_2]$ may be noted. The last of these is of interest as a potential analogue of ferrocene. Several preparative routes have been suggested,[42] and the usual

[38] R. BUSBY, W. KLOTZBÜCHER, and G. A. OZIN, Titanium hexacarbonyl $Ti(CO)_6$ and titanium hexadinitrogen, $Ti(N_2)_6$: synthesis using titanium atoms and characterization by matrix infrared and ultraviolet-visible spectroscopy, *Inorg. Chem.* **16**, 822–8 (1977).

[39] R. D. ROGERS, R. V. BYNUM, and J. L. ATWOOD, First authentic example of a difference in the structural organometallic chemistry of zirconium and hafnium: crystal and molecular structure of $(\eta^5\text{-}C_5H_5)_2Hf(\eta^1\text{-}C_5H_5)_2$, *J. Am. Chem. Soc.* **103**, 692–3 (1981).

[39a] R. D. ROGERS, R. V. BYNUM, and J. L. ATWOOD, Crystal and molecular structure of tetra(cyclopentadienyl)zirconium, *J. Am. Chem. Soc.* **100**, 5238–9 (1978).

[40] D. J. SIKORA, M. D. RUASCH, R. D. ROGERS, and J. L. ATWOOD, Structure and reactivity of the first hafnium carbonyl, $[(\eta^5\text{-}C_5H_5)_2Hf(CO)_2]$, *J. Am. Chem. Soc.* **101**, 5079–81 (1979).

[41] J. S. MCDERMOTT, M. E. WILSON, and G. M. WHITESIDE, Synthesis and reactions of bis(cyclopentadienyl) titanium(IV) metallocycles, *J. Am. Chem. Soc.* **98**, 6529–36 (1976).

[42] G. W. WATT, L. J. BAYE, and F. O. DRUMMOND, Concerning the status of bis(cyclopentadienyl)titanium, *J. Am. Chem. Soc.* **88**, 1138–40 (1966).

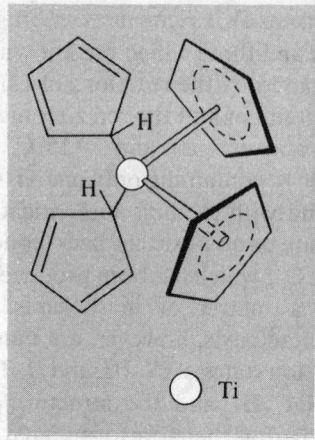

FIG. 21.10   Molecular structure of $Ti(C_5H_5)_4$.

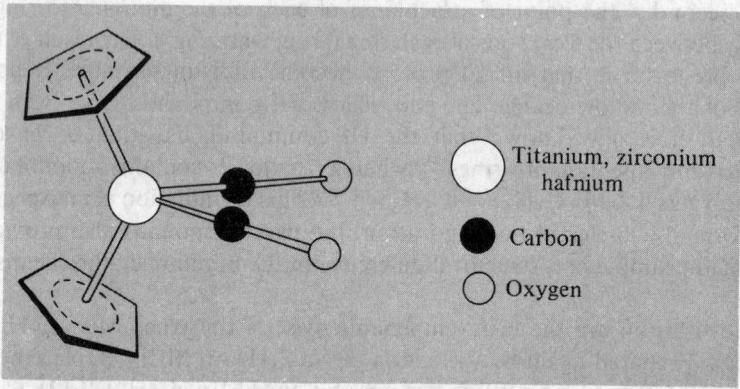

FIG. 21.11   Molecular structure of $[M(\eta^5\text{-}C_5H_5)_2(CO)_2]$. For M = Ti the C_5H_5 rings are "eclipsed" as shown here, but for M = Hf they are "staggered".

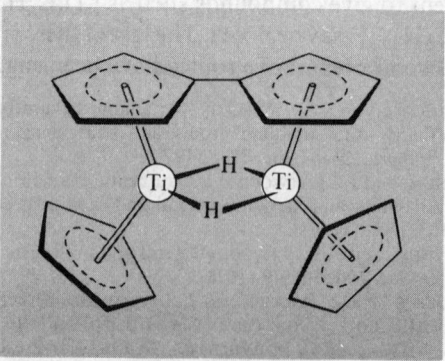

FIG. 21.12   Dimeric $Ti(C_5H_5)_2$. Actually (μ-(η^5,η^5-fulvalene(-di-(μ-hydrido)-bis(η^5-cyclopenta-dienyltitanium).

product is a dark-green, pyrophoric, diamagnetic dimer, though a monomeric isomer may be produced in some cases as an intermediate.[43] The structure of the dimer has only recently been shown by ^{13}C nmr to be that in Fig. 21.12.[44] The black, pyrophoric and diamagnetic zirconium compound is prepared similarly and is isomorphous with the titanium compound; it may therefore be presumed to be isostructural. A number of cyclopentadienyl and related compounds of titanium and zirconium have been found to absorb molecular nitrogen which in some cases can be recovered in a reduced form (i.e. as ammonia or hydrazine) upon hydrolysis. While this has obvious interest as a potential route to nitrogen fixation, a compound which can be regenerated and so act catalytically has so far proved elusive.

[43] R. H. MARVICH and H. H. BRINTZINGER, A metastable form of titanocene. Formation from a hydride complex and reaction with hydrogen, nitrogen, and carbon monoxide, *J. Am. Chem. Soc.* **93**, 2046–8 (1971).

[44] A. DAVISON and S. S. WREFORD, The structure of "titanocene": clarification by carbon-13 nuclear magnetic resonance, *J. Am. Chem. Soc.* **96**, 3017–8 (1974).

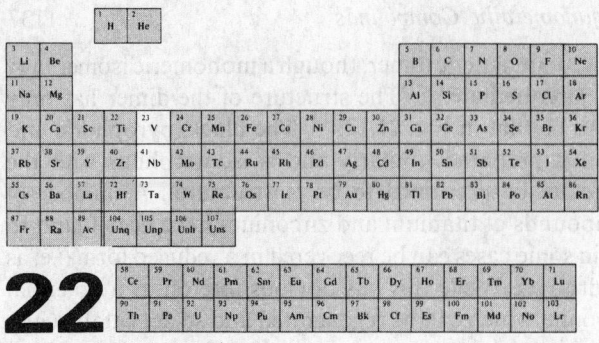

22

Vanadium, Niobium, and Tantalum

22.1 Introduction

The discoveries of all three of these elements were made at the beginning of the nineteenth century and were marked by initial uncertainty and confusion due, in the case of the heavier pair of elements, to the overriding similarity of their chemistries.

A. M. del Rio in 1801 claimed to have discovered the previously unknown element 23 in a sample of Mexican lead ore and, because of the red colour of the salts produced by acidification, he called it erythronium. Unfortunately he withdrew his claim when, 4 years later, it was (incorrectly) suggested by the Frenchman, H. V. Collett-Desotils, that the mineral was actually basic lead chromate. In 1830 the element was "rediscovered" by N. G. Sefström in some Swedish iron ore. Because of the richness and variety of colours found in its compounds he called it vanadium after Vanadis, the Scandinavian goddess of beauty. One year later F. Wöhler established the identity of vanadium and erythronium. The metal itself was isolated in a reasonably pure form in 1867 by H. E. Roscoe who reduced the chloride with hydrogen and was then responsible for much of the early work on the element.

In the same year that del Rio found his erythronium, C. Hatchett examined a mineral which had been sent to England from Massachusetts and had lain in the British Museum since 1753. From it he isolated the oxide of a new element which he named columbium, and the mineral columbite, in honour of its country of origin. Meanwhile in Sweden A. G. Ekeberg was studying some Finnish minerals and in 1802 claimed to have identified a new element which he named tantalum because of the difficulty he had had in dissolving the mineral in acids.† It was subsequently thought that the two elements were one and the same, and this view persisted until at least 1844 when H. Rose examined a columbite sample and showed that two distinct elements were involved. One was Ekeberg's tantalum and the other he called niobium (Niobe was the daughter of Tantalus). Despite the chronological precedence of the name columbium, IUPAC adopted niobium in 1950,

---

† The classical allusion refers to Tantalus, the mythical king of Phrygia, son of Zeus and a nymph, who was condemned for revealing the secrets of the gods to man: one of his punishments was being made to stand in Tartarus up to his chin in water, which constantly receded as he stooped to drink. As Ekeberg wrote (1802): "This metal I call *tantalum* . . . partly in allusion to its incapacity, when immersed in acid, to absorb any and be saturated."

though the former is still used widely in US industry. Impure niobium metal was first isolated by C. W. Bloomstrand in 1866 by the reduction of the chloride with hydrogen, but the first pure samples of metallic niobium and tantalum were not prepared until 1907 when W. von Bolton reduced the fluorometallates with sodium.

22.2 The Elements

22.2.1 *Terrestrial abundance and distribution*

The abundances of these elements decrease by approximately an order of magnitude from V to Nb and again from Nb to Ta. Vanadium has been estimated to comprise about 136 ppm (i.e. 0.0136%) of the earth's crustal rocks, which makes it the nineteenth element in order of abundance (between Zr, 162 ppm, and Cl, 126 ppm); it is the fifth most abundant transition metal after Fe, Ti, Mn, and Zr. It is widely, though sparsely, distributed; thus although more than 60 different minerals of vanadium have been characterized, there are few concentrated deposits and most of it is obtained as a coproduct along with other materials. One of its important minerals is the polysulfide, patronite, VS_4, but, being a class-a metal, it is more generally associated with oxygen. Thus vanadinate approximates to lead chloride vanadate, $PbCl_2 \cdot 3Pb_3(VO_4)_2$, and carnotite to potassium uranyl vanadate, $K(UO_2)(VO_4) \cdot 1.5H_2O$. Vanadium is also found in some crude oils, in particular those from Venezuala and Canada, and can be recovered from the oil residues and from flue dusts after burning. Of considerable biochemical interest is the astonishing ability of certain invertebrates to accumulate vanadium in their blood. For example, the ascidian seaworm *Phallusia mammilata* has a blood concentration of V up to 1900 ppm, which represents more than a millionfold concentration with respect to the sea-water in which it lives. The related organism *Ascidia nigra* has an even more spectacular accumulation with concentrations up to 1.45% V (i.e. 14 500 ppm) in its blood cells, which also contain considerable concentrations of sulfuric acid (pH ~0). Vanadium also occurs in the protein haemovanadin (which is not in fact a haem protein). However, the earlier suggestion that V plays a part in the oxygen transport cycle is now less accepted, and definite evidence concerning its role is still lacking. One possibility that has been mooted is that the ascidia accumulates vanadate and polyvanadate ions in mistake for phosphate and polyphosphates (p. 611).

The crustal abundances of niobium and tantalum are 20 ppm and 1.7 ppm, comparable to N (19 ppm), Ga (19 ppm), and Li (18 ppm), on the one hand, and to As (1.8 ppm) and Ge (1.5 ppm), on the other. Of course, in view of their chemical similarities, Nb and Ta invariably occur together, and their chief mineral, $(Fe,Mn)M_2O_6$ (M = Nb, Ta), is known as columbite or tantalite, depending on which metal preponderates. Deposits are widespread but rarely very concentrated, and significant amounts of the metals are obtained as a byproduct in the extraction of tin in Malaysia and Nigeria.

22.2.2 *Preparation and uses of the metals*[1]

Because it is usually produced along with other metals, the availability of vanadium and the economics of its production are intimately connected with the particular coproduct

[1] *Kirk–Othmer Encyclopedia of Chemical Technology*, 2nd edn., Vol. 21, pp. 157–66. Interscience, New York, 1970.

involved. This is dramatically illustrated by its extraction in the USA. Up to the 1950s the USA was the world's chief importer of vanadium ores, but then the US Atomic Energy Commission initiated an intensive programme of uranium production using domestically available carnotite containing 1–2% vanadium. This provided large quantities of vanadium concentrates with the result that about 1956 the USA became a net exporter. The main sources of the metal are South Africa, the USSR, China and the USA.

The usual extraction procedure is to roast the crushed ore, or vanadium residue, with NaCl or Na_2CO_3 at 850°C. This produces sodium vanadate, $NaVO_3$, which is leached out with water. Acidification with sulfuric acid to pH 2–3 precipitates "red cake", a polyvanadate which, on fusing at 700°C, gives a black, technical grade vanadium pentoxide. Reduction is then necessary to obtain the metal, but, since about 80% of vanadium produced is used as an additive to steel, it is usual to effect the reduction in an electric furnace in the presence of iron or iron ore to produce ferrovanadium, which can then be used without further refinement. Carbon was formerly used as the reductant, but it is difficult to avoid the formation of an intractable carbide, and so it has been superseded by aluminium or, more commonly, ferrosilicon (p. 381) in which case lime is also added to remove the silica as a slag of calcium silicate. If pure vanadium metal is required it can be obtained[2] by reduction of VCl_5 with H_2 or Mg, by reduction of V_2O_5 with Ca, or by electrolysis of partially refined vanadium in fused alkali metal chloride or bromide.

The benefit of vanadium as an additive in steel is that it forms V_4C_3 with any carbon present, and this disperses to produce a fine-grained steel which has increased resistance to wear and is stronger at high temperatures. Such steels are widely used in the manufacture of springs and high-speed tools. Annual world production of vanadium metal, alloys and concentrates exceeded 35,000 tonnes of contained vanadium in 1980.

Production of niobium and tantalum is on a smaller scale and the processes involved are varied and complicated.[3] Alkali fusion, or digestion of the ore with acids can be used to solubilize the metals, which can then be separated from each other. The process originally developed by M. C. Marignac in 1866 and in use for a century utilized the fact that in dil HF tantalum tends to form the sparingly soluble K_2TaF_7, whereas niobium forms the soluble $K_3NbOF_5.2H_2O$. Nowadays it is more usual to employ a solvent extraction technique. For instance, tantalum can be extracted from dilute aqueous HF solutions by methyl isobutyl ketone, and increasing the acidity of the aqueous phase allows niobium to be extracted into a fresh batch of the organic phase. The metals can then be obtained, after conversion to the pentoxides, by reduction with Na or C, or by the electrolysis of fused fluorides. Annual world production of Nb/Ta concentrates in 1980 (80% of which came from Brazil) incorporated 14 000 tonnes of contained Nb and 500 tonnes of contained Ta.

Niobium finds use in the production of numerous stainless steels for use at high temperatures, and Nb/Zr wires are used in superconducting magnets. The extreme corrosion-resistance of tantalum at normal temperatures (due to the presence of an exceptionally tenacious film of oxide) leads to its application in the construction of chemical plant, especially where it can be used as a liner inside cheaper metals. Its complete inertness to body fluids makes it the ideal material for surgical use in bone repair and internal suturing. It is widely used by the electronics industry in the manufacture of

---

[2] G. BRAUER, *Handbook of Preparative Inorganic Chemistry*, 2nd edn., pp. 1252–5, Academic Press, New York, 1965.

[3] F. FAIRBROTHER, *The Chemistry of Niobium and Tantalum*, pp. 7–15, Elsevier, Amsterdam, 1967.

capacitors, where the oxide film is an efficient insulator, and as a filament or filament support. Indeed, it was for a while widely used to replace carbon as the filament in incandescent light bulbs but, by about 1911, was, itself superseded by tungsten.

22.2.3 *Atomic and physical properties of the elements*

Some of the important properties of Group VA elements are summarized in Table 22.1. Having odd atomic numbers, they have few naturally occurring isotopes; Nb only 1 and V and Ta 2 each, though the second ones are present only in very low abundance (^{50}V 0.250%, ^{180}Ta 0.012%). As a consequence (p. 20) their atomic weights have been determined with considerable precision. On the other hand, because of difficulties in removing all impurities, reported values of their bulk properties have often required revision.

TABLE 22.1   *Some properties of Group VA elements*

Property		V	Nb	Ta
Atomic number		23	41	73
Number of naturally occurring isotopes		2	1	2
Atomic weight		50.9415	92.9064	180.9479
Electronic configuration		$[Ar]3d^34s^2$	$[Kr]4d^35s^2$	$[Xe]4f^{14}5d^36s^2$
Electronegativity		1.6	1.6	1.5
Metal radius (12-coordinate)/pm		134	146	146
Ionic radius (6-coordinate)/pm	V	54	64	64
	IV	58	68	68
	III	64	72	72
	II	79	—	—
MP/°C		1915	2468	2980
BP/°C		3350	4758	5534
ΔH_{fus}/kJ mol^{-1}		17.5	26.8	24.7
ΔH_{vap}/kJ mol^{-1}		459.7	680.2	758.2
ΔH_f (monatomic gas)/kJ mol^{-1}		510 ($\pm$29)	724	782 ($\pm$6)
Density (20°C)/g cm^{-3}		6.11	8.57	16.65
Electrical resistivity (20°C)/μ ohm cm		~25	~12.5	(12.4)

All three elements are shiny, silvery metals with typically metallic bcc structures. When very pure they are comparatively soft and ductile but impurities usually have a hardening and embrittling effect. When compared to the elements of Group IVA the expected trends are apparent. These elements are slightly less electropositive and are smaller than their predecessors, and the heavier pair Nb and Ta are virtually identical in size as a consequence of the lanthanide contraction. The extra d electron again appears to contribute to stronger metal–metal bonding in the bulk metals, leading in each case to a higher mp, bp, and enthalpy of atomization. Indeed, these quantities reach their maximum values in this and the following group. In the first transition series, vanadium is the last element before some of the $(n-1)d$ electrons begin to enter the inert electron-core of the atom and are therefore not available for bonding. As a result, not only is its mp the highest in the series but it is the last element whose compounds in the group oxidation state (i.e. involving all $(n-1)d$ and ns electrons) are not strongly oxidizing. In the second and third series the entry of $(n-1)d$ electrons into the electron core is delayed somewhat and it is molybdenum and tungsten in Group VIA whose mps are the highest.

22.2.4 *Chemical reactivity and trends*

The elements of Group VA are in many ways similar to their predecessors in Group IVA. They react with most non-metals, giving products which are frequently interstitial and nonstoichiometric, but they require high temperatures to do so. Their general resistance to corrosion is largely due to the formation of surface films of oxides which are particularly effective in the case of tantalum. Unless heated, tantalum is appreciably attacked only by oleum, hydrofluoric acid or, more particularly, a hydrofluoric/nitric acid mixture. Fused alkalis will also attack it. In addition to these reagents, vanadium and niobium are attacked by other hot concentrated mineral acids but are resistant to fused alkali.

The most obvious factor in comparing the chemistry of the three elements is again the very close similarity of the second and third members although, in this group, slight differences can be discerned as will be discussed shortly. The stability of the lower oxidation states decreases as the group is descended. As a result, although each element shows all the formal oxidation states from $+5$ down to -1, the most stable one in the case of vanadium under normal conditions is the $+4$, and even the $+3$ and $+2$ oxidation states (which are admittedly strongly reducing) have well-characterized cationic aqueous chemistries; by contrast most of the chemistries of niobium and tantalum are confined to the group oxidation state $+5$. Of the halogens, only the strongly oxidizing fluorine produces a pentahalide of vanadium, and the other vanadium(V) compounds are based on the oxohalides and the pentoxide. The pentoxide also gives rise to the complicated but characteristic aqueous chemistry of the polymerized vanadates (isopolyvanadates) which anticipates the even more extensive chemistry of the polymolybdates and polytungstates; this is only incompletely mirrored by niobium and tantalum.

The $+4$ oxidation state, which for Nb and Ta is best represented by their halides, is most notable for the uniquely stable VO^{2+} (vanadyl) ion which retains its identity throughout a wide variety of reactions and forms many complexes. Indeed it is probably the most stable diatomic ion known. The M^{IV} ions have only slightly smaller radii than those of Group IVA, and, again, coordination numbers as high as 8 are found. In the $+5$ state, however, only Nb and Ta are sufficiently large to achieve this coordination number with ligands other than bidentate ones with very small "bites", such as the peroxo group. Table 22.2 illustrates the various oxidation states and stereochemistries of compounds of V, Nb, and Ta.

Niobium and tantalum provide no counterpart to the cationic chemistry of vanadium in the $+3$ and $+2$ oxidation states. Instead, they form a series of "cluster" compounds based on octahedral M_6X_{12} units. The occurrence of such compounds is largely a consequence of the strength of metal–metal bonding in this part of the periodic table (as reflected in high enthalpies of atomization), and similar cluster compounds are found also for molybdenum and tungsten (see also discussion on p. 1145).

Compounds containing M–C σ-bonds are frequently unstable and do not give rise to an extensive chemistry (p. 1163). Vanadium forms a neutral (paramagnetic) hexacarbonyl which, though not very stable, contrasts with that of titanium in that it can at least be prepared in quantity. All three elements give a number of η^5-cyclopentadienyl derivatives. Nevertheless, lacking the commercial incentives which have been so important in the case of titanium (p. 1134), the organometallic chemistry of this group has not been so actively studied, and is not yet so extensive.

TABLE 22.2 *Oxidation states and stereochemistries of compounds of vanadium, niobium, and tantalum*

Oxidation state	Coordination number	Stereochemistry	V	Nb/Ta
$-3\ (d^8)$	5	—	$[V(CO)_5]^{3-}$	—
$-1\ (d^6)$	6	Octahedral	$[V(CO)_6]^-$	$[M(CO)_6]^-$
$0\ (d^5)$	6	Octahedral	$[V(CO)_6]$	—
$1\ (d^4)$	6	Octahedral	$[V(bipy)_3]^+$	—
	7	Capped octahedral	—	$[TaH(CO)_2(diphos)_2]$
$2\ (d^3)$	4	Square planar	—	NbO
	6	Octahedral	$[V(CN)_6]^{4-}$	TaO(?)
		Trigonal prismatic	VS	NbS
$3\ (d^2)$	3	Planar	$[V\{N(SiMe_3)_2\}_3]$	—
	4	Tetrahedral	$[VCl_4]^-$	
	5	Trigonal bipyramidal	$[VCl_3(NMe_3)_2]$	
	6	Octahedral	$[V(C_2O_4)_3]^{3-}$	$[Nb_2Cl_9]^{3-}$
		Trigonal prismatic	—	$LiNbO_2$
	7	Complex	—	$[Ta(CO)Cl_3(PMe_2Ph)_3].EtOH$
	8	Dodecahedral	—	$[Nb(CN)_8]^{5-}$
$4\ (d^1)$	4	Tetrahedral	VCl_4	$[Nb(NEt_2)_4]$ (not Ta)
	5	Trigonal bipyramidal	$[VOCl_2(NMe_3)_2]$	
		Square pyramidal	$[VO(acac)_2]$	
	6	Octahedral	$[VCl_4(bipy)]$	$[MCl_6]^{2-}$
	7	Pentagonal bipyramidal	—	$[NbF_7]^{3-}$
	8	Dodecahedral	$[VCl_4(diars)_2]$	$[NbCl_4(diars)_2]$ (not Ta)
		Square antiprismatic	—	$[Nb(\beta\text{-diketonate})_4]$
$5\ (d^0)$	4	Tetrahedral	$VOCl_3$	$ScNbO_4$
	5	Trigonal bipyramidal	$VCl_5(g)$	$MF_5(g)$
		Square pyramidal	$[VOF_4]^-$	$[M(NMe_2)_5]$
	6	Octahedral	$[VF_6]^-$	$[MF_6]^-$
		Trigonal prismatic	—	$[M(S_2C_6H_4)_3]^-$
	7	Pentagonal bipyramidal	$[VO(S_2CNEt_2)_3]$	$[TaS(S_2CNEt_2)_3]$
		Capped trigonal prismatic	—	$[MF_7]^{2-}$
	8	Dodecahedral	$[V(O_2)_4]^{3-}$	$[M(O_2)_4]^{3-}$
				$[Ta(S_2CNMe_2)_4]^+$
		Square antiprismatic	—	$[MF_8]^{3-}$

22.3 Compounds of Vanadium, Niobium, and Tantalum[4, 5]

The binary hydrides (p. 73), borides (p. 165), carbides (p. 321), and nitrides (p. 480) of these metals have already been discussed and will not be described further except to note that, as with the analogous compounds of Group IVA, they are hard, refractory, and nonstoichiometric materials with high conductivities. The intriguing cryo-compound $[V(N_2)_6]$ has been isolated by cocondensing V atoms and N_2 molecules at 20–25 K; it has an infrared absorption at $2100\ cm^{-1}$ and its d–d and charge-transfer spectra are strikingly similar to those of the isoelectronic 17-electron species $[V(CO)_6]$.[5a]

[4] D. L. KEPERT, *The Early Transition Metals*, Chap. 3, V, Nb, Ta, pp. 142–254, Academic Press, London, 1972.

[5] R. J. H. CLARK, Vanadium, Chap. 34, pp. 491–551, and D. BROWN, The chemistry of niobium and tantalum, Chap. 35, pp. 553–622, in *Comprehensive Inorganic Chemistry*, Vol. 3, Pergamon Press, Oxford, 1973.

[5a] H. HUBER, T. A. FORD, W. KLOTZBÜCHER, and G. A. OZIN, Direct synthesis with vanadium atoms. Part 2. Synthesis and characterization of vanadium hexadinitrogen, $[V(N_2)_6]$, in low-temperature matrices, *J. Am. Chem. Soc.* **98**, 3176–8 (1976).

22.3.1 *Oxides*[6]

Table 22.3 gives the principal oxides formed by the elements of this group. Besides the 4 oxides of vanadium shown, a number of other phases of intermediate composition have been identified and the lower oxides in particular have wide ranges of homogeneity. V_2O_5 is orange yellow when pure (due to charge transfer) and is the final product when the metal is heated in an excess of oxygen, but contamination with lower oxides is then common and a better method is to heat ammonium "metavanadate":

$$2NH_4VO_3 \longrightarrow V_2O_5 + 2NH_3 + H_2O$$

TABLE 22.3 *Oxides of Group VA metals*

Oxidation state:	+5	+4	+3	+2
V	V_2O_5	VO_2	V_2O_3	VO
Nb	Nb_2O_5	NbO_2	—	NbO
Ta	Ta_2O_5	TaO_2	—	(TaO)

On the basis of simple radius ratio arguments, vanadium(V) is expected to be rather large for tetrahedral coordination to oxygen, but rather small for octahedral coordination.[7] It is perhaps not surprising therefore that, though the structure of V_2O_5 is somewhat complicated, it consists essentially of distorted trigonal bipyramids of VO_5 sharing edges to form zigzag double chains. It is homogeneous over only a small range of compositions but loses oxygen reversibly on heating, which is probably why it is such a versatile catalyst. For instance, it catalyses the oxidation of numerous organic compounds by air or hydrogen peroxide, and the reduction of olefins (alkenes) and of aromatic hydrocarbons by hydrogen, but most importantly it catalyses the oxidation of SO_2 to SO_3 in the contact process for the manufacture of sulfuric acid (p. 838). For this purpose it replaced metallic platinum which, besides being far more expensive, was also prone to "poisoning" by impurities such as arsenic. V_2O_5 is amphoteric. It is slightly soluble in water, giving a pale yellow, acidic solution. It dissolves in acids producing salts of the pale-yellow dioxovanadium(V) ion, $[VO_2]^+$, and in alkalis producing colourless solutions which, at high pH, contain the orthovanadate ion, VO_4^{3-}. At intermediate pHs a series of hydrolysis-polymerization reactions occur yielding the isopolyvanadates to be discussed in the next section. It is also a mild oxidizing agent and in aqueous solution is reduced by, for instance, hydrohalic acids to vanadium(IV). In the solid, mild reduction with CO, SO_2, or fusion with oxalic acid gives the deep-blue VO_2.

At room temperature VO_2 has a rutile-like structure (p. 1118) distorted by the presence of pairs of vanadium atoms bonded together. Above 70°C, however, an undistorted rutile structure is adopted as the atoms in each pair separate, breaking the localized V–V bonds and releasing the bonding electrons, so causing a sharp increase in electrical conductivity and magnetic susceptibility. It is again amphoteric, dissolving in non-oxidizing acids to give salts of the blue oxovanadium(IV) (vanadyl) ion $[VO]^{2+}$, and in alkali to give the

[6] C. N. R. RAO and G. V. S. RAO, *Transition Metal Oxides*, National Standard Reference Data System NSRDS-NBS49, Washington, 1964, 130 pp.

[7] L. E. ORGEL, Ferroelectricity and the structure of transition-metal oxides, *Faraday Soc. Dis.* **26**, 138–44 (1958).

yellow to brown vanadate(IV) (hypovanadate) ion $[V_4O_9]^{2-}$, or at high pH $[VO_4]^{4-}$. Between V_2O_5 and VO_2 is a succession of phases V_nO_{2n+1} of which V_3O_7, V_4O_9, and V_6O_{13} have been characterized.

Further reduction with H_2, C, or CO produces a series of discrete chemical-shear phases (Magnéli phases) of general formula V_nO_{2n-1} based on a rutile structure with periodic defects (p. 1119), before the black, refractory sesquioxide V_2O_3 is reached. Examples are V_4O_7, V_5O_9, V_6O_{11}, V_7O_{13}, and V_8O_{15}. The oxides VO, V_2O_3, and V_3O_5 also conform to the general formula V_nO_{2n-1}, but this is a purely formal relation and their structures are not related by chemical-shear to those of the Magnéli phases.

V_2O_3 has a corundum structure (p. 273) and is notable for the transition occurring as it is cooled below about 170 K when its electrical conductivity changes from metallic to insulating in character. Chemically it is entirely basic, dissolving in aqueous acids to give blue or green vanadium(III) solutions which are strongly reducing. On still further reducing the oxide system, the corundum structure is retained down to compositions as low as $VO_{1.35}$, after which the grey metallic monoxide VO, with a defect rock-salt structure, is formed. This too is markedly nonstoichiometric with a composition range from $VO_{0.8}$ to $VO_{1.3}$. In all, therefore, at least 13 distinct oxide phases of vanadium have been identified between $VO_{\sim 1}$ and V_2O_5.

Niobium and tantalum also form different oxide phases but they are not so extensive or well characterized as those of vanadium. Their pentoxides are relatively much more stable and difficult to reduce. As they are attacked by conc HF and will dissolve in fused alkali, they may perhaps be described as amphoteric, but inertness is the more obvious characteristic. Their structures are extremely complicated and Nb_2O_5 in particular displays extensive polymorphism. It is interesting to note that the polymorphs of Nb_2O_5 and Ta_2O_5 are by no means all analogous.

High temperature reduction of Nb_2O_5 with hydrogen gives the bluish-black dioxide NbO_2 which has a distorted rutile structure. As in VO_2 the distortion is caused by pairs of metal atoms evidently bonded together, but the distortion is in a different direction. Between Nb_2O_5 and NbO_2 there is a homologous series of structurally related phases of general formula $Nb_{3n+1}O_{8n-2}$ with $n = 5$, 6, 7, 8 (i.e. Nb_8O_{19}, $Nb_{19}O_{46}$, $Nb_{11}O_{27}$, and $Nb_{25}O_{62}$). In addition, oxides of formula $Nb_{12}O_{29}$ and $Nb_{47}O_{116}$ have been reported: the numerical relationship to Nb_2O_5 is clear since $Nb_{12}O_{29}$ is $(12Nb_2O_5-2O)$ and $2Nb_{47}O_{116}$ (or $Nb_{94}O_{232}$) is $(47 Nb_2O_5-3O)$. Further reduction produces the grey monoxide NbO which has a cubic structure and metallic conductivity but differs markedly from its vanadium analogue in that its composition range is only $NbO_{0.982}$ to $NbO_{1.008}$. The structure is a unique variant of the rock-salt NaCl structure (p. 273) in which there are vacancies (Nb) at the eight corners of the unit cell and an O vacancy at its centre (see structure). The structure could therefore be described as a vacancy-defect NaCl structure $Nb_{0.75}\boxplus_{0.25}O_{0.75}\boxminus_{0.25}$, but as all the vacancies are ordered it is better to consider it as a new structure type in which both Nb and O form 4 coplanar bonds. The central feature is a 3D framework of Nb_6 octahedral clusters (Nb–Nb 298 pm, cf. Nb–Nb 285 pm in Nb metal) and this accounts for the metallic conductivity of the compound: $\kappa \sim 10^6$ ohm^{-1} cm^{-1} at 77 K and $\sim 10^5$ ohm^{-1} cm^{-1} at room temperature. The structure is reminiscent of the structure-motif of the lower halides of Nb and Ta (p. 1155). It is also instructive to compare the structures adopted by NbO and MgO (NaCl-type) in terms of the Born–Haber model for ionic compounds (p. 91). The ionic radius for Nb^{II} would be expected (p. 1141) to be only slightly larger than that for Mg^{II} (72 pm). The sum of the first

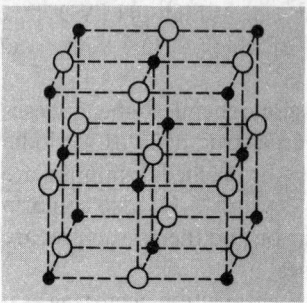

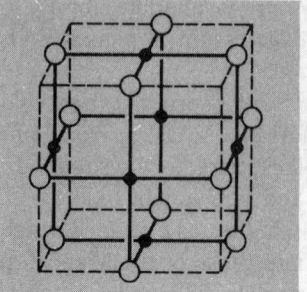

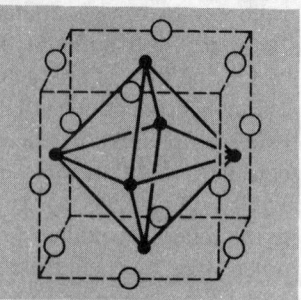

NaCl (MgO) showing all sites oc-cupied by M($\bullet$) and O($\bigcirc$)	NbO showing planar coordination of Nb (and O) and vacancies at the cube corners (Nb) and centre (O)	NbO showing octahedral Nb_6 cluster (joined by corner sharing to neighbouring unit cells)

two ionization energies is more favourable for Nb by 142.6 kJ mol^{-1} (Mg^I 737.6, Mg^{II} 1450.5; Nb^I 664.0, Nb^{II} 1381.5 kJ mol^{-1}). All other terms in the lattice energy equation are equal for the two compounds except for the heat of sublimation of the metal, which is dramatically less favourable for Nb [ΔH_f(monatomic vapour): Mg 146.4, Nb 724 kJ mol^{-1}]. This difference of more than 577 kJ mol^{-1} can thus be taken as the dominant reason for the retention of Nb$_6$ clusters in the structure rather than the adoption of the true NaCl-type structure Nb^{II}O^{-II}.

The heavier metal tantalum is distinctly less inclined than niobium to form oxides in lower oxidation states. The rutile phase TaO$_2$ is known but has not been studied, and a cubic rock-salt-type phase TaO with a narrow homogeneity range has also been reported but not yet confirmed. Ta$_2$O$_5$ has two well-established polymorphs which have a reversible transition temperature at 1355°C but the detailed structure of these phases is too complex to be discussed here.[6]

22.3.2 *Isopolymetallates*[8, 9]

The amphoteric nature of V$_2$O$_5$ has already been noted. In fact, if the colourless solution produced by dissolving V$_2$O$_5$ in strong aqueous alkali such as NaOH is gradually acidified, it first deepens in colour, becoming orange to red as the neutral point is passed; it then darkens further and, around pH = 2, a brown precipitate of hydrated V$_2$O$_5$ separates and redissolves at still lower pHs to give a pale-yellow solution. As a result of spectrophotometric studies there is general agreement that the predominant species in the initial colourless solution is the tetrahedral VO$_4^{3-}$ ion and, in the final pale-yellow solution, the angular VO$_2^+$ ion. In the intervening orange to red solutions a complicated series of hydrolysis-polymerization reactions occur, which have direct counterparts in the chemistries of Mo and W and to a lesser extent Nb, Ta, and Cr. The polymerized species involved are collectively known as isopolymetallates or isopolyanions. The determination of the equilibria involved in their formation, as well as their stoichiometries and structures, has been a confused and disputed area, some aspects of which are by no means settled even

[8] D. L. KEPERT, Isopolyanions and heteropolyanions, Chap. 51 in *Comprehensive Inorganic Chemistry*, Vol. 4, pp. 607–72, Pergamon Press, Oxford, 1973.

[9] M. T. POPE and B. W. DALE, Isopoly-vanadates, -niobates, and -tantalates, *Q. Revs.* **22**, 527–48 (1968).

now. That this is so is perfectly understandable because:

(i) Some of the equilibria are reached only slowly (possibly months in some cases) and it is likely that much of the reported work has been done under non-equilibrium conditions.

(ii) Often in early work, solid species were crystallized from solution and their stoichiometries, quite unjustifiably as it turns out, were used to infer the stoichiometries of species in solution.

(iii) When a series of experimental measurements has been made it is usual to see what combination of plausible ionic species will best account for the observed data. However, the greater the complexity of the system, the greater the number of apparently acceptable models there will be, and the greater the accuracy required if the measurements are to distinguish reliably and unambiguously between them.

Of the many experimental techniques which have been used in this field, the more important are: pH measurements, cryoscopy, ion-exchange and ultraviolet/visible spectroscopy for studying the stoichiometry of the equilibria, and infrared/Raman and nmr spectroscopy for studying the structures of the ions in solution. Oxygen-17 and metal atom nmr spectroscopy are also beginning to play an increasingly important role. Probably the best summary of our current understanding of the vanadate system is given by Fig. 22.1. This shows how the existence of the various vanadate species depends on the pH and on the total concentration of vanadium. Their occurrence can be accounted for by protonation and condensation equilibria such as the following:

In alkaline solution:

$$[VO_4]^{3-} + H^+ \rightleftharpoons [HVO_4]^{2-}$$

$$2[HVO_4]^{2-} \rightleftharpoons [V_2O_7]^{4-} + H_2O$$

$$[HVO_4]^{2-} + H^+ \rightleftharpoons [H_2VO_4]^-$$

$$3[H_2VO_4]^- \rightleftharpoons [V_3O_9]^{3-} + 3H_2O$$

$$4[H_2VO_4]^- \rightleftharpoons [V_4O_{12}]^{4-} + 4H_2O$$

In acid solution:

$$10[V_3O_9]^{3-} + 15H^+ \rightleftharpoons 3[HV_{10}O_{28}]^{5-} + 6H_2O$$

$$[H_2VO_4]^- + H^+ \rightleftharpoons H_3VO_4$$

$$[HV_{10}O_{28}]^{5-} + H^+ \rightleftharpoons [H_2V_{10}O_{28}]^{4-}$$

$$H_3VO_4 + H^+ \rightleftharpoons VO_2^+ + 2H_2O$$

$$[H_2V_{10}O_{28}]^{4-} + 14H^+ \rightleftharpoons 10VO_2^+ + 8H_2O$$

In these equilibria the site of protonation in the species $[HVO_4]^{2-}$, $[H_2VO_4]^-$, etc., is an oxygen atom (not vanadium); a more precise representation would therefore be $[VO_3(OH)]^{2-}$, $[VO_2(OH)_2]^-$, etc. However, the customary formulation is retained for convenience (cf. HNO_3, HSO_4^-, H_2SO_4, etc.).

It is evident from Fig. 22.1 that only in very dilute solutions are monomeric vanadium ions found and any increase in concentrations, particularly if the solution is acidic, leads to

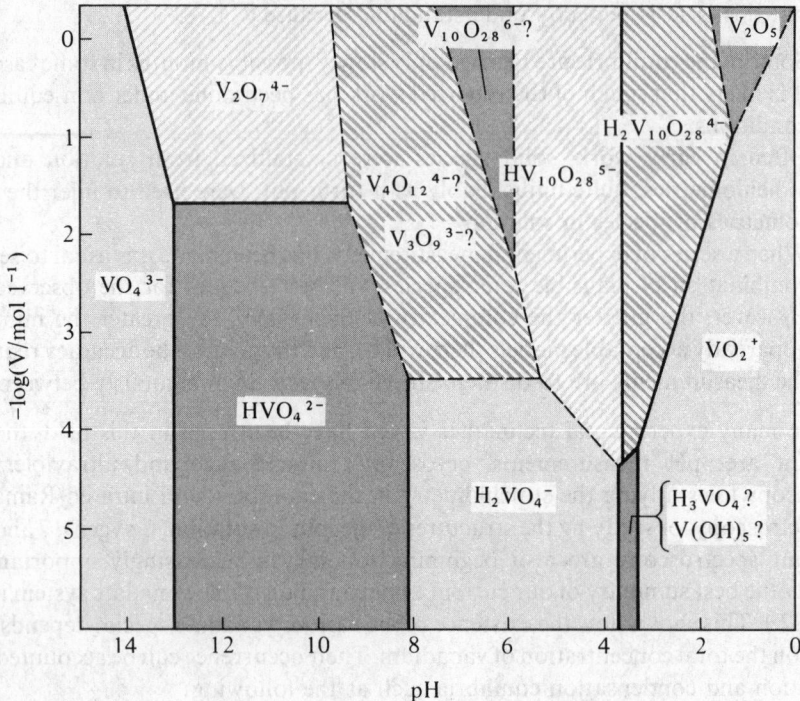

FIG. 22.1   Occurrence of various vanadate and polyvanadate species as a function of pH and total
concentration of vanadium.

polymerization. [10] ^{51}V nmr work indicates[10] that, starting from the alkaline side, the
various ionic species are all based on 4-coordinate vanadium(V) in the form of linked VO_4
tetrahedra until the decavanadates appear. These evidently involve a higher coordination
number, but whether or not it is the same in solution as in the solids which can be
separated is uncertain. However, it is interesting to note that similarities between the
vanadate and chromate systems cease with the appearance of the decavanadates which
have no counterpart in chromate chemistry. The smaller chromium(VI) is apparently
limited to tetrahedral coordination with oxygen, whereas vanadium(V) is not.

More information is of course available on the structures of the various crystalline
vanadates which can be separated from solution.[11] Traditionally, the colourless salts
obtained from alkaline solution were called ortho-, pyro-, or meta-vanadates by analogy
with the phosphates of corresponding stoichiometry. "Ortho"-vanadates, $M_3^IVO_4(aq)$,
apparently contain discrete, tetrahedral, VO_4^{3-} ions; "pyro"-vanadates, $M_4^IV_2O_7(aq)$,
contain dinuclear $[V_2O_7]^{4-}$ ions consisting of 2 VO_4 tetrahedra sharing a corner; the
structure of "meta"-vanadates depends on the state of hydration (Fig. 22.2) but in no cases
do they involve discrete VO_3^- ions. Anhydrous metavanadates such as NH_4VO_3 contain
infinite chains of corner-linked VO_4 tetrahedra, while hydrated metavanadates, such as
$KVO_3.H_2O$, contain infinite chains of approximately trigonal bipyramidal VO_5 units,

    [10] O. W. HOWARTH and R. E. RICHARDS, Nuclear magnetic resonance study of polyvanadate equilibria by use
of vanadium-51, *J. Chem. Soc.* 1965, 864–70.
    [11] H. T. EVANS and S. BLOCK, The crystal structures of potassium and cesium trivanadates, *Inorg. Chem.* **5**,
1808–14 (1966).

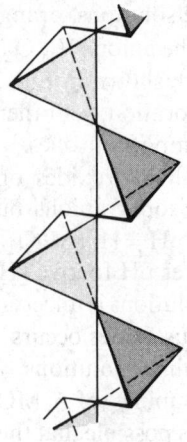

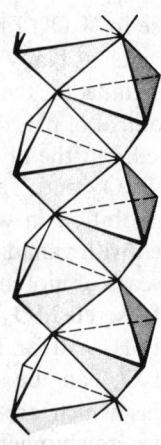

(a) Anhydrous metavanadates, consisting of infinte chains of corner-shared VO_4 tetrahedra.

(b) Hydrated metavanadates, consisting of infinite chains of edge-shared VO_5 trigonal bipyramids.

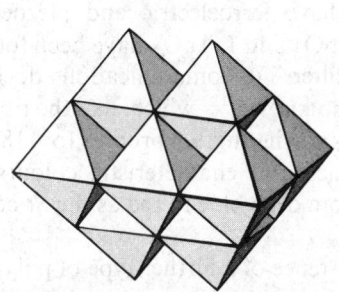

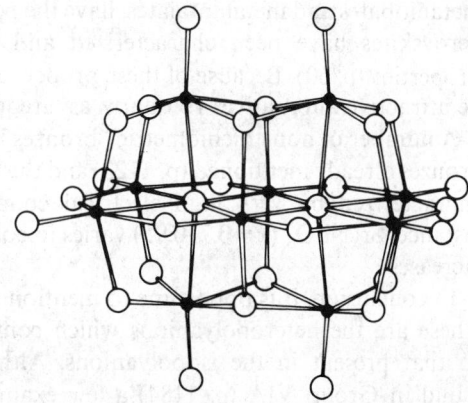

(c) The decavanadate, $[V_{10}O_{28}]^{6-}$, ion made up of 10 VO_6 octahedra (2 are obscured).

(d) An alternative representation of $[V_{10}O_{28}]^{6-}$ emphasizing the V–O bonds.

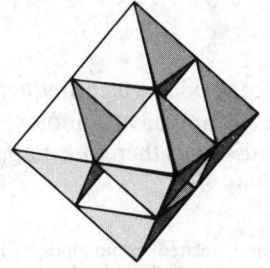

(e) The $[M_6O_{19}]^{8-}$ (M = Nb, Ta) ions made up of 6 MO_6 octahedra (one is obscured).

FIG. 22.2 The structures of some isopoly-anions in the solid state using the conventional representation in which each polyhedron contains a metal atom and each vertex of a polyhedron represents an oxygen atom.

not unlike those in V_2O_5. From the bright orange, acidic solutions, orange, crystalline decavanadates such as $Na_6V_{10}O_{28}.18H_2O$ are obtained: the anion $[V_{10}O_{28}]^{6-}$ is made up of 10 VO_6 octahedra, two representations of which are shown in Fig. 22.2.

By careful control of pH, temperature, and rate of evaporation, yet other compounds may be obtained, but the above examples are the more important ones.

Fusion of Nb_2O_5 and Ta_2O_5 with an excess of alkali hydroxides or carbonates, followed by dissolution in water, produces solutions of isopolyanions but not in the variety produced with vanadium. It appears that, down to pH $= 11$, $[M_6O_{19}]^{8-}$ ions are present; in the case of niobium protonation occurs at lower pH to give $[HNb_6O_{19}]^{7-}$. The presence of discrete $MO_4{}^{3-}$ ions in strongly alkaline solutions is uncertain. Below pH ~ 7 for Nb and pH ~ 10 for Ta, precipitation of the hydrous oxides occurs. Salts such as $K_8M_6O_{19}.16H_2O$ can be crystallized from the alkaline solutions and contain $[M_6O_{19}]^{8-}$ ions which are made up of octahedral groupings of 6 MO_6 octahedra (Fig. 22.2). Other stoichiometries have been claimed but it is possible that they were based on samples merely contaminated with alkali.

Most niobates and tantalates, however, are insoluble and may be regarded as mixed oxides in which the Nb or Ta is octahedrally coordinated and with no discrete anion present. Thus KMO_3, known inaccurately (since they have no discrete $MO_3{}^-$ anions) as metaniobates and metatantalates, have the perovskite (p. 1122) structure. Several of these perovskites have been characterized and some have ferroelectric and piezoelectric properties (p. 60). Because of these properties, $LiNbO_3$ and $LiTaO_3$ have been found to be attractive alternatives to quartz as "frequency filters" in communications devices.

A number of nonstoichiometric "bronzes" are also known[12] which, like the titanium bronzes already mentioned (p. 1123) and the better known tungsten bronzes (p. 1185), are characterized by very high electrical conductivities and characteristic colours. For instance, Sr_xNbO_3 ($x = 0.7$–0.95) varies in colour from deep blue to red as the Sr content increases.

In conclusion, it is opportune to mention the existence of a further type of polyanion. These are the heteropolyanions which contain a further cationic element in addition to that present in the isopolyanions. Although the greatest variety of such ions is found in Group VIA (p. 1184) a few examples, such as $Na_7[PV_{12}O_{36}].38H_2O$ and $Na_{12}[MnNb_{12}O_{38}].50H_2O$, are known for vanadium and niobium and it seems likely that others could be obtained.

22.3.3 *Sulfides, selenides, and tellurides*[5, 12a]

All three metals form a wide variety of binary chalcogenides which frequently differ both in stoichiometry and in structure from the oxides. Many have complex structures which are not easily described,[12a, 12b] and detailed discussion is therefore inappropriate.

[12] P HAGENMULLER, Tungsten bronzes, vanadium bronzes and related compounds, Chap. 50 in *Comprehensive Inorganic Chemistry*, Vol. 4, pp. 541–605, Pergamon Press, Oxford, 1973.

[12a] F. HULLINGER, Crystal chemistry of chalcogenides and pnictides of the transition elements, *Structure and Bonding* **4**, 83–229 (1968). Includes a valuable appendix summarizing crystal structures, electrical and magnetic properties.

[12b] A. F. WELLS, *Structural Inorganic Chemistry*, 4th edn., pp. 187, 606–26, Oxford University Press, Oxford, 1975.

The various sulfide phases are listed in Table 22.4: phases approximating to the stoichiometry MS have the NiAs-type structure (p. 648) whereas MS_2 have layer lattices related to MoS_2 (p. 1186), CdI_2, or $CdCl_2$ (p. 1407). Sometimes complex layer-sequences occur in which the 6-coordinate metal atom is alternately octahedral and trigonal prismatic. Most of the phases exhibit metallic conductivity and magnetic properties range from diamagnetic (e.g. VS_4), through paramagnetic (VS, V_2S_3), to antiferromagnetic (V_7S_8). Selenides and tellurides show a similar profusion of stoichiometries and structural types (Table 22.5).

TABLE 22.4   *Sulfides of vanadium, niobium, and tantalum*

V_3S	$Nb_{21}S_8$	Ta_6S	V_2S_3	—	—
—	—	Ta_2S	V_5S_8	$Nb_{1+x}S_2$	$Ta_{1+x}S_2$
V_5S_4	—	—	—	NbS_2	TaS_2
VS	NbS_{1-x}	TaS	—	NbS_3	TaS_3
V_7S_8	—	—	VS_4	—	—
V_3S_4	Nb_3S_4	—			

TABLE 22.5   *Selenides and tellurides of vanadium, niobium and tantalum*

V_2Se	—	—	—	—	—
V_5Se_4	Nb_5Se_4	—	V_5Te_4	Nb_5Te_4	—
VSe	$NbSe$	—	VTe_{1+x}	—	$TaTe$
V_7Se_8	—	—	—	—	—
V_3Se_4	Nb_3Se_4	—	V_3Te_4	Nb_3Te_4	—
(V_2Se_3)	Nb_2Se_3	Ta_2Se_3	V_2Te_3	—	—
V_5Se_8	—	—	V_5Te_8	—	—
—	$Nb_{1+x}Se_2$	$Ta_{1+x}Se_2$	$V_{1+x}Te_2$	$Nb_{1+x}Te_2$	$Ta_{1-x}Te_2$
VSe_2	$NbSe_2$	$TaSe_2$	VTe_2	$NbTe_2$	$TaTe_2$
—	—	$TaSe_3$	—	—	—
—	$NbSe_4$	—	—	$NbTe_4$	$TaTe_4$

In addition to these binary chalcogenides, many of which exist over wide ranges of composition because of the structural relation between the NiAs and CdI_2 structure types (p. 648), several ternary phases have been studied. Some, like $BaVS_3$ and $BaTaS_3$ have three-dimensional structures in which the Ba and V(Ta) are coordinated by 12 and 6 S atoms respectively. In other compounds, such as Tl_3VS_4, there are discrete tetrahedral $[VS_4]^{3-}$ anions. Very recently a thiovanadyl complex containing the coordinated $(V{=}S)^{2+}$ unit has been prepared by the reaction of solid B_2S_3 on a solution of the corresponding vanadyl complex in CH_2Cl_2.[12c] Thus, using the ligands salen [H_2salen $= N,N'$-ethylenebis(salicylideneamine), p. 1063] and acen [H_2acen $= N,N'$-ethylenebis(acetylacetonylideneamine)], deep magenta crystals of the square pyramidal V^{IV} complexes [VS(salen)] and [VS(acen)] were obtained; these were stable to the atmosphere in the solid state but acutely sensitive to oxygen and moisture in solution.

[12c] K. P. CALLAHAN, P. J. DURAND, and P. H. RIEGER, Synthesis and characterization of thiovanadyl $(V{=}S)^{2+}$ complexes, *JCS Chem. Comm.* 1980, 75–76.

22.3.4 *Halides and oxohalides*[13, 14]

The known halides of vanadium, niobium, and tantalum, are listed in Table 22.6. These are illustrative of the trends within this group which have already been alluded to. Vanadium(V) is only represented at present by the fluoride, and even vanadium(IV) does not form the iodide, though all the halides of vanadium(III) and vanadium(II) are known. Niobium and tantalum on the other hand, form all the halides in the high oxidation state, and are in fact unique (apart only from protoactinium) in forming pentaiodides but, even in the +4 state, tantalum fails to form a fluoride and neither metal produces a trifluoride. In still lower oxidation states, niobium and tantalum give a number

TABLE 22.6   *Halides of vanadium, niobium, and tantalum*[a]

Oxidation state	Fluorides	Chlorides	Bromides	Iodides
+5	VF_5 colourless mp 19.5°, bp 48.3°	—	—	—
	NbF_5 white mp 79°, bp 234°	$NbCl_5$ yellow mp 203°, bp 247°	$NbBr_5$ orange mp 254°, bp 360°	NbI_5 brass coloured
	TaF_5 white mp 97°, bp 229°	$TaCl_5$ white mp 210°, bp 233°	$TaBr_5$ pale yellow mp 280°, bp 345°	TaI_5 black mp 496°, bp 543°
+4	VF_4 lime green (subl >150°)	VCl_4 red-brown mp = −26°, bp 148°	VBr_4 magenta (d −23°)	—
	NbF_4 black (d >350°)	$NbCl_4$ violet-black	$NbBr_4$ dark brown	NbI_4 dark grey mp 503°
	—	$TaCl_4$ black	$TaBr_4$ dark blue	TaI_4
+3	VF_3 yellow-green mp 800°	VCl_3 red-violet	VBr_3 grey-brown	VI_3 brown-black
	NbF_3(?) blue	$NbCl_3$ black	$NbBr_3$ dark brown	NbI_3
	TaF_3(?) blue	$TaCl_3$ black	$TaBr_3$	—
+2	VF_2 blue	VCl_2 pale green (subl 910°)	VBr_2 orange-brown (subl 800°)	VI_2 red-violet

[a] Niobium and Ta form a number of polynuclear halides in which the metal has non-integral oxidation states (see text).

[13] R. A. WALTON, Halides and oxyhalides of the early transition series and their stability and reactivity in nonaqueous media, *Prog. Inorg. Chem.* **16**, 1–226 (1972).
[14] R. COLTON and J. H. CANTERFORD, *Halides of the First Row Transition Metals*, Chap. 3, pp. 107–60, Wiley, London 1969; and *Halides of the Second and Third Row Transition Metals*, Chap. 5, pp. 145–206, Wiley, London, 1968.

of (frequently nonstoichiometric) cluster compounds which can be considered to involve fragments of the metal lattice.

VF_5 and all pentahalides of Nb and Ta can be prepared conveniently by direct action of the appropriate halogen on the heated metal. They are all relatively volatile, hydrolysable solids (indicative of the covalency to be anticipated in such a high oxidation state) in which the metals attain octahedral coordination by means of halide bridges (Fig. 22.3). VF_5 is an infinite chain polymer, whereas NbF_5 and TaF_5 are tetramers, and the chlorides and bromides are dimers. The colours vary from white fluorides, yellow chlorides, and orange bromides, to brown iodides. The decreasing energy of the charge-transfer bands responsible for these colours is a reflection of the increasing polarizability of the anions from F^- to I^-, and for each anion it is the least readily reduced Ta which produces the palest colour. All the pentahalides can be sublimed in an atmosphere of the appropriate halogen and they are then monomeric, probably trigonal bipyramidal. Potentially they are all Lewis acids but their ability to form adducts (LMX_5) diminishes and the iodides rarely do so.

The tetrahalides can be prepared by direct action of the elements. However, whereas VF_4 tends to disproportionate into $VF_5 + VF_3$ and must be sublimed from them, VCl_4 and VBr_4 tend to dissociate into $VX_3 + \frac{1}{2}X_2$ and so require the presence of an excess of halogen. Even so, VBr_4 has only been isolated by quenching the mixed vapours at $-78°C$.

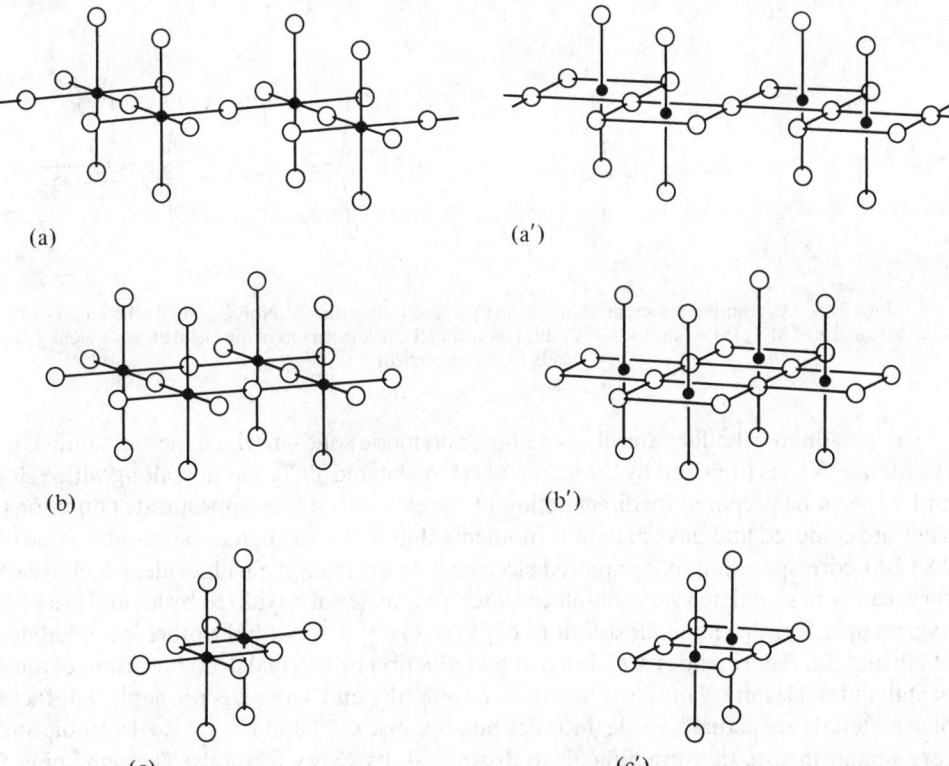

(a)   (a')

(b)   (b')

(c)   (c')

Fig. 22.3 Alternative representations of: (a) infinite chains of vanadium atoms in VF_5, (b) tetrameric structures of NbF_5 and TaF_5, and (c) dimeric structure of MX_5 (M=Nb, Ta; X=Cl, Br).

VF_4 is a bright-green hygroscopic solid, probably consisting of fluorine-bridged VF_6 octahedra. VCl_4 is a red-brown oil, rapidly hydrolysed by water to give solutions of oxovanadium(IV) chloride, and magnetic and spectroscopic evidence indicate that it consists of unassociated tetrahedral molecules. As far as its properties are known, the magneta-coloured VBr_4 is similar.

The Nb and Ta tetrahalides (except TaF_4 which is unknown, and NbI_4 which is prepared by thermal decomposition of NbI_5) are generally prepared by reduction of the corresponding pentahalide and are all readily hydrolysed. NbF_4 is a black involatile solid and its low magnetic moment suggests extensive metal–metal interaction, presumably via the intervening F^- ions since it consists of infinite sheets of NbF_6 octahedra (Fig. 22.4a). The chlorides, bromides, and iodides are brown to black solids with a chain structure (Fig. 22.4b) in which pairs of metal atoms are displaced towards each other, so facilitating the interaction which leads to their diamagnetism.

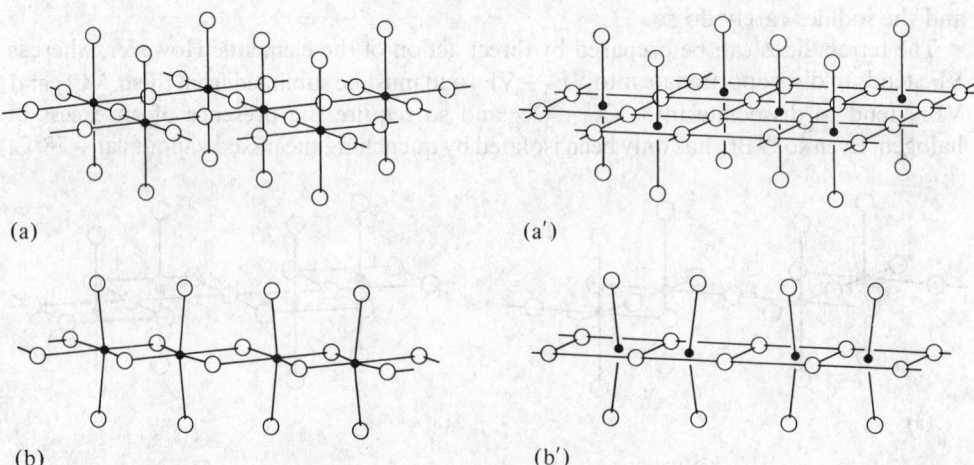

(a)   (a′)

(b)   (b′)

FIG. 22.4   Alternative representations of: (a) the sheet structure of NbF_4, and (b) the chain structure of MX_4 ($M = Nb, Ta; X = Cl, Br, I$) showing the displacement of the metal atoms which leads to diamagnetism.

The vanadium trihalides are all crystalline, polymeric solids in which the vanadium is 6-coordinate. VF_3 is prepared by the action of HF on heated VCl_3 and this, along with VBr_3 and VI_3, can be prepared by direct action of the elements under appropriate conditions. They are coloured and have magnetic moments slightly lower than the spin-only value of 2.83 BM corresponding to 2 unpaired electrons. Apart from the trifluoride, which is not very readily oxidized nor very soluble in water, they are easily oxidized by air and are very hygroscopic, forming aqueous solutions of $[V(H_2O)_6]^{3+}$. As with the other lower halides of Nb and Ta, the trihalides are obtained by reduction or thermal decomposition of their pentahalides. Despite claims for the existence of NbF_3 and TaF_3 it is probable that these blue materials are actually oxide fluorides but, because O^{2-} and F^- are isoelectronic and very similar in size, they are difficult to distinguish by X-ray methods. The remaining 5 known trihalides of this pair of metals are dark coloured, rather unreactive materials. The Nb–Cl system has been the most thoroughly studied but the others appear to be entirely analogous. They are nonstoichiometric and the composition "MX_3" is best considered as a single unexceptional point within a broad homogeneous phase based on hcp halide ions. At one extreme is M_3X_8 i.e. $MX_{2.67}$) in which one-quarter of the octahedral sites are empty and the others occupied by triangular groups of metal atoms. Of the 15 valence electrons provided by the 3 metal atoms, 8 are lost by ionization and transfer to the 8 Cl atoms and, of the remaining 7 available for metal–metal bonding, 6 are considered to be in bonding orbitals and 1 in a nonbonding orbital.[15] This accounts for the magnetic moment of 1.86 BM for each trinuclear cluster in Nb_3Cl_8. Metal deficiency then produces stoichiometries to somewhat beyond MX_3 (i.e. $M_{2.67-x}X_8$) after which the MX_4 phase separates, containing pairs of interacting metal atoms, as already mentioned (i.e. M_2X_8).

In the still lower oxidation state, $+2$, the halides of vanadium on the one hand, and niobium and tantalum, on the other, diverge still further. The dihalides of V are prepared by reduction of the corresponding trihalides and have simple structures based on the close-packing of halide ions: the rutile structure (p. 1118) for VF_2, and the CdI_2 structure (p. 1407) for the others. They are strongly reducing and hygroscopic, dissolving in water to give lavender-coloured solutions of $[V(H_2O)_6]^{2+}$. By contrast, high-temperature reductions of NbX_5 or TaX_5 with the metals (or Na or Al) have been shown to yield a series of phases based on $[M_6X_{12}]^{n+}$ units consisting of octahedral clusters of metal atoms with the halogen atoms situated above each edge of the octahedra (Fig. 22.5). These may

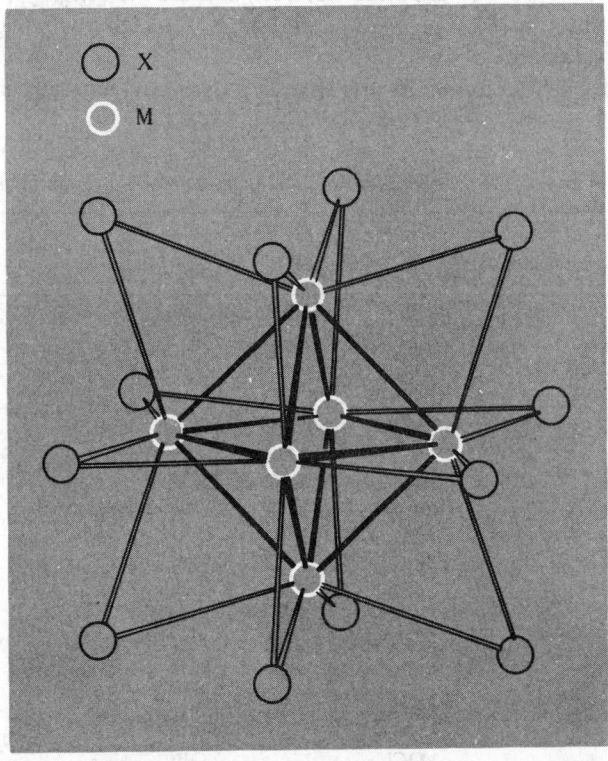

Fig. 22.5 $[M_6X_{12}]^{n+}$ cluster with X bridges over each edge of the octahedron of metal ions.

[15] F. A. COTTON, Metal atom clusters in oxide systems, *Inorg. Chem.* **3**, 1217–20 (1964).

be surrounded by:

(a) Four similar units, with each of which a halogen atom is shared, producing a sheet structure with the composition $[M_6X_{12}]X_{4/2} = M_6X_{14}$ (i.e. $MX_{2.3}$). These compounds are diamagnetic as a result of the metal–metal bonding.

(b) Six similar units, with each of which a halogen atom is shared, producing a three-dimensional array with the composition $[M_6X_{12}]X_{6/2} = M_6X_{15}$ (i.e. $MX_{2.5}$). These have magnetic moments corresponding to 1 unpaired electron per hexamer and so indicate the same metal–metal bonding within the cluster as in (a).

In one other case (Nb_6I_{11}) the octahedron of metal atoms has iodine atoms above each *face* and is then joined to 6 similar units by shared iodines giving a three-dimensional array with the composition $[Nb_6I_8]I_{6/2} = Nb_6I_{11}$. The magnetic behaviour is complicated but suggests the presence of 1 unpaired electron at low temperatures and 3 at higher temperatures. The formation of these Nb and Ta clusters, in contrast to the more ionic structures adopted by VX_2, can be related to the much higher heats of sublimation of the heavier metals (Table 22.1) and the effect of this term in the Born–Haber cycle (p. 94) for simple mononuclear halides.

Many of these cluster compounds are water-soluble and yield solutions in which the clusters are retained throughout chemical reactions. Thus, $[M_6X_{12}]^{2+}$ (diamagnetic) can be oxidized to $[M_6X_{12}]^{3+}$ (1 unpaired electron) and then to $[M_6X_{12}]^{4+}$ (diamagnetic), and compounds such as M_6X_{14}, M_6X_{15}, and M_6X_{16}, usually with 7 or 8 molecules of H_2O, can be crystallized.

The known oxohalides are listed in Table 22.7. They are generally prepared from the

TABLE 22.7   *Oxohalides of vanadium, niobium, and tantalum*

Oxidation State	Fluorides		Chlorides		Bromides	Iodides	
+5	VOF_3 yellow mp 300° bp 480°	VO_2F brown	$VOCl_3$ yellow mp −77° bp 127°	VO_2Cl orange	$VOBr_3$ deep red (d 180°)	—	
		NbO_2F white	$NbOCl_3$ white	NbO_2Cl white	$NbOBr_3$ yellow-brown	$NbOI_3$ black	NbO_2I red
	$TaOF_3$	TaO_2F	$TaOCl_3$ white	TaO_2Cl white	$TaOBr_3$ pale yellow	—	TaO_2I
+4	VOF_2 yellow		$VOCl_2$ green		$VOBr_2$ yellow-brown (d 180°)		
			$NbOCl_2$ black			$NbOI_2$ black	
			$TaOCl_2$				
+3	—		$VOCl$ yellow-brown bp 127°		$VOBr$ violet (d 480°)		

oxides but are not particularly well known and, as can be seen, are limited almost entirely to the oxidation states of +4 and +5. Those in the former oxidation state are relatively stable but those in the latter are notably hygroscopic and hydrolyse vigorously to the hydrous pentoxides. The Nb(V) and Ta(V) compounds are rather volatile, though less so than the pentahalides. $NbOCl_3$ is the best known, mainly because of its propensity for occurring as an unwanted impurity in the preparation of VCl_5 if O_2 is not rigorously excluded or, more specially, if V_2O_5 is used.

22.3.5 *Compounds with oxoanions*

The group oxidation state of +5 is too high to allow the formation of simple ionic salts even for Nb and Ta, and in lower oxidation states the higher sublimation energies of these heavier metals, coupled with their ease of oxidation, again militates against the formation of simple salts of the oxoacids. As a consequence the only simple oxoanion salts are the sulfates of vanadium in the oxidation states +3 and +2. These can be crystallized from aqueous solutions as hydrates and are both strongly reducing. They give rise to blue-violet alums, $MV(SO_4)_2.12H_2O$, the ammonium alum being air-stable when dry, and to the reddish-violet Tutton's salts, $M_2V(SO_4)_2.6H_2O$, the ammonium analogue of which is again relatively more stable to oxidation.

In the higher oxidation states partially hydrolysed species dominate the aqueous chemistry, the most important being the oxovanadium(IV), or vanadyl, ion VO^{2+}. This gives the sulphate $VOSO_4.5H_2O$, containing monodentate sulphate and octahedrally coordinated vanadium, and the polymeric $VOSO_4$. Oxovanadium(V) species are not well characterized, outside the oxohalides VOX_3, but in strongly acid solutions VO_2^+ is formed and reportedly gives the nitrate $VO_2(NO_3)$. The VO_2^+ ion is also found in anionic complexes such as $[VO_2(oxalate)_2]^{3-}$ and in all cases the oxygens are mutually *cis* as they are in the isoelectronic MoO_2^{2+} (p. 1192). Niobium and tantalum produce a variety of complicated and ill-defined, but probably polymeric, species which include the nitrates, $MO(NO_3)_3$, sulfates such as $Nb_2O(SO_4)$, and double sulfates such as $(NH_4)_6Nb_2O(SO_4)_7$, all of which are extremely readily hydrolysed.

22.3.6 *Complexes*

Oxidation State V (d⁰)

Vanadium(V) forms few complexes and probably the best known are the white, diamagnetic hexafluorovanadates, MVF_6; however, even these are extremely sensitive to moisture. In aqueous solutions of vanadium(V), various peroxocomplexes can be formed by the addition of H_2O_2. In neutral or alkaline solutions the yellow diperoxo-orthovanadate ion $[VO_2(O_2)_2]^{3-}$ is formed whereas in strongly acid solution the red-brown peroxovanadium cation $[V(O_2)]^{3+}$ predominates. The yellow "acid salts" $KH_2[VO_2(O_2)_2].H_2O$ and $(NH_4)_2H[VO_2(O_2)_2].nH_2O$ have been isolated and, at 0°, strongly alkaline solutions containing an excess of H_2O_2 deposit blue-violet needles of $M_3^I[V(O_2)_4].nH_2O$ (M^I = Li, Na, K, NH_4). $K_3[V(O_2)_4]$ is isomorphous with the corresponding Cr compound, which is known to be 8-coordinate and dodecahedral, (p. 748) but such a high coordination number is not common for vanadium. Niobium and tantalum produce similar peroxo-compounds, e.g. pale yellow $K_3[Nb(O_2)_4]$ and white

$K_3[Ta(O_2)_4]$. However, most of the currently known complexes are derived from the pentahalides, especially the pentafluorides, which act as Lewis acids and form complexes of the type MX_5L with O, S, N, P, and As donor ligands. Though most of these and other 6-coordinate complexes are no doubt octahedral, $[AsPh_4][Ta(benzenedithiolate)_3]$ has been shown to possess a trigonal prismatic stereochemistry. [16]

NbF_5 and TaF_5 dissolve in aqueous solutions of HF to give $[MOF_5]^{2-}$ and, if the concentration of HF is increased, $[MF_6]^-$. This is normally the highest coordination number attained in solution though some $[NbF_7]^-$ may form, and $[TaF_7]^-$ definitely does form, in very high concentrations of HF. However, by suitably regulating the concentration of metal, fluoride ion, and HF, octahedral $[MF_6]^-$, capped trigonal prismatic $[MF_7]^{2-}$, and even square-antiprismatic $[MF_8]^{3-}$ salts can all be isolated.

By contrast with the fluorides, aqueous solutions of MCl_5 and MBr_5 (M = Nb, Ta) yield only oxochloro- and oxobromo-complexes. Niobium(V) is generally considered to be a class-a metal, but a recent nmr study using ^{93}Nb has shown that the SCN^- ligand yields a series of both N-bonded thiocyanato and S-bonded isothiocyanato complexes, e.g. $[Nb-(NCS)_n(SCN)_{6-n}]^-$ ($n = 0, 2, 4, 5, 6$).[16a]

Oxidation State IV (d¹)

In this oxidation state there are considerable similarities between this group and the preceding group but, by and large, the complexes of this group have until recently received much less attention. Again, the tetrahalides are Lewis acids and produce a number of adducts with a variety of donor atoms, the most common coordination number being 6, though this and the precise geometry have often not been determined. Thus $[VF_4L]$ (L = NH_3, py) are insoluble in common organic solvents, have magnetic moments of about 1.8 BM, and are thought to be fluorine-bridged polymers. $[VCl_4 2L]$ (L = py, MeCN, aldehydes, etc.) and $VCl_4(L-L)$ (L-L = bipy, phen, diars) are brown paramagnetic, readily hydrolysed compounds assumed to be 6-coordinate monomers. Similar compounds of Nb and Ta are also paramagnetic and the metal–metal bonding which led to the diamagnetism of the parent tetrahalides is presumed to have been broken to give adducts which again are 6-coordinate monomers. Hexahalo-complexes $[MX_6]^{2-}$ (M = V, X = F, Cl; M = Nb, Ta, X = Cl, Br) are known, the vanadium compounds being especially sensitive to moisture though stable to air.

Higher coordination numbers are also found. Vanadium and Nb produce the dodecahedral $[MCl_4(diars)_2]^{[17]}$ just like the Group IVA metals. This is probably the most common stereochemistry for this coordination number but others are possible; differences in energy are slight and this facilitates non-rigidity. For example the yellow-coloured solid $K_4[Nb(CN)_8].2H_2O$ has recently been shown by X-ray analysis to contain dodecahedral niobium(IV) (like its molybdenum isomorph), whereas esr and infrared data suggest that, in solution, the anion has the square-antiprismatic

[16] J. L. MARTIN and J. TAKATS, Crystal and molecular structure of tetraphenylarsonium tris(benzenedithiolato)tantalate(V), *Inorg. Chem.* **14**, 1358–64 (1975).

[16a] R. G. KIDD and H. G. SPINNEY, The ambidentate thiocyanate ligand. Niobium-93 nuclear magnetic resonance detection of thiocyanato- and isothiocyanato-niobium(V) complexes, *J. Am. Chem. soc.* **103**, 4759–63 (1981).

[17] R. J. H. CLARK, J. LEWIS, and R. S. NYHOLM, Diarsine complexes of quadrivalent-metal halides, *J. Chem. Soc.* 1962, 2460–5; and J. C. DEWAN, D. L. KEPERT, C. L. RALSTON, and A. H. WHITE, Addition complexes of niobium and tantalum pentachlorides with *o*-phenylenebis(dimethylarsine), *JCS Dalton*, 1975, 2031–38.

configuration.[18] Likewise, the deep-red niobium(III) complex $K_5[Nb(CN)_8]$ adopts a dodecahedral (D_{2d}) configuration for the anion in the crystal whereas the single ^{13}C nmr signal in aqueous solution implies either a square-pyramidal (D_{4d}) or fluxional (D_{2d}) structure. [18a]

The major contrast with the Group IVA metals is the stability of VO^{2+} complexes which are the most important and the most widely studied of the vanadium(IV) complexes, and are the usual products of the hydrolysis of other vanadium(IV) complexes. VO^{2+} behaves are a class-a cation, forming stable compounds with F (especially), Cl, O, and N donor ligands. These "vanadyl" complexes are generally green or blue-green and can be cationic, neutral, or anionic.[19] They are very frequently 5-coordinate in which case the stereochemistry is almost invariably square pyramidal. $[VO(acac)_5]$ (Fig. 22.6) is the prime example[20] of this geometry in coordination compounds. In this and in similar

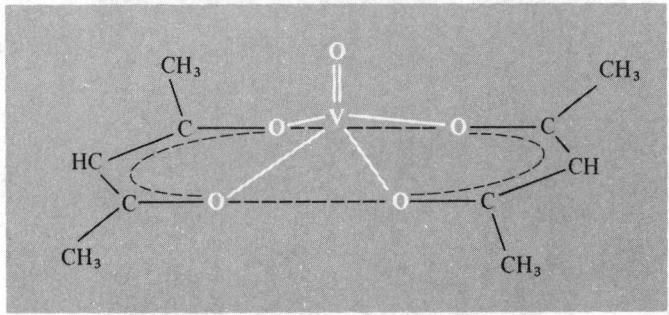

FIG. 22.6   The square-pyramidal structure of $[VO(acac)_2]$.

compounds the V=O bond length is ~ 157–168 pm which is about 50 pm shorter than the 4 equatorial V–O bonds. This, as well as spectroscopic evidence, is consistent with the formulation of the bond as double. A sixth ligand may be weakly bonded *trans* to the V=O to produce a distorted octahedral structure; the concomitant reduction in the stretching frequency of the V=O bond[21] has been interpreted in terms of electron donation from this sixth ligand, thereby making the vanadium atom less able to accept charge from the oxygen and so reducing the bond order. Tetradentate Schiff bases, produced for instance by the condensation of salicylaldehyde with primary diamines, in most cases give entirely analogous compounds, but some are yellow and may be polymeric with the vanadium attaining 6-coordination by "stacking" so that the sixth position of each vanadium is occupied by the oxygen from the V=O beneath. The corresponding thiovanadyl complexes were mentioned on p. 1151.

In spite of the evident proclivity of VO^{2+} to form square pyramidal or distorted octahedral complexes, it must not be assumed that 5-coordination inevitably results in the

^{18} M. LAING, G. GAFNER, W. P. GRIFFITH and P. M. KIERNAN, The X-ray crystal structure of an eight co-ordinate cyanide complex of niobium, $K_4[Nb(CN)_8]2H_2O$, *Inorg. Chim. Acta* **33**, L119 (1979).

^{18a} M. B. HURSTHOUSE, A. M. GALAS, A. M. SOARES, and W. P. GRIFFITH, X-ray crystal structure of $K_5[Nb(CN)_8]$, and a change in the structure of $[Nb(CN)_8]^{5-}$ in solution, *JCS Chem. Comm.* 1980, 1167–8.

^{19} J. SELBIN, Oxovanadium(IV) complexes, *Coord. Chem. Revs.* **1**, 293–314 (1966).

^{20} P. K. HON, R. L. BELFORD, and C. E. PFLUGER, Vanadyl bond-length variations, *J. Chem. Phys.* **43**, 3111–15 (1965).

^{21} C. J. POPP, J. H. NELSON, and R. O. RAGSDALE, Thermodynamic and infrared studies of tertiary amine oxides with bis(2,4-pentanedionato)oxovanadium(IV), *J. Am. Chem. Soc.* **91**, 610–14 (1969).

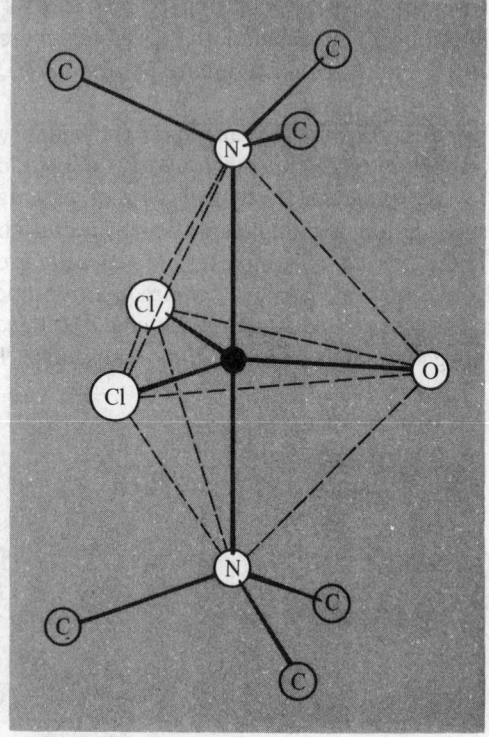

FIG. 22.7   The trigonal bipyramidal structure of [VOCl$_2$(NMe$_3$)$_2$].

former shape. [VOCl$_2$(NMe$_3$)$_2$] is in fact trigonal bipyramidal[22] (Fig. 22.7), no doubt because of a dominant steric effect of the bulky trimethylamine ligands rather than any electronic effect.

Most oxovanadium(IV) complexes are magnetically simple, having virtually "spin-only" moments of 1.73 BM corresponding to 1 unpaired electron, but their electronic spectra are less easily understood. This is primarily due to the presence of a strong π contribution to the bond between the vanadium and the oxygen which makes it difficult to assign an unequivocal sequence to the molecular orbitals involved.

Oxidation State III (d^2)

Of the members of this group, only vanadium provides a significant coordination chemistry: even so the majority of vanadium(III) compounds are easily oxidized and must be prepared with air rigorously excluded. The usual methods are to use VCl$_3$ as the starting material, or to reduce solutions of vanadium(V) or (IV) electrolytically.

The chemistry of vanadium(III) closely parallels that of titanium(III) and it likewise favours octahedral coordination. The interpretation of the electronic spectra of its complexes, as the prime examples of d^2 ions in an octahedral field, has provided the

[22] J. E. DRAKE, J. VEKRIS, and J. S. WOOD, The crystal and molecular structure of bis(trimethylamine) oxovanadium(IV) dichloride, *J. Chem. Soc.* (A) 1968, 1000–5.

stimulus for much of the preparative work in this area. In general, the spectra are characterized by two bands in the visible region with a further much more intense absorption in the ultraviolet. The two former bands are believed to arise from d–d transitions and others from charge-transfer. Since the d^2 configuration in a cubic field is expected to give rise to three spin-allowed transitions (p. 1094) it is assumed that the most energetic of these is obscured by the charge-transfer band. Table 22.8 gives data for some octahedral vanadium(III) complexes. It turns out, on examination of data such as these, that a coherent interpretation of the spectra is only possible if the bands are assigned as:

$$\nu_1 = {}^3T_{2g}(F) \leftarrow {}^3T_{1g}(F)$$

$$\nu_2 = {}^3T_{1g}(P) \leftarrow {}^3T_{1g}(F)$$

and the third, obscured one, therefore as:

$$\nu_3 = {}^3A_{2g}(F) \leftarrow {}^3T_{1g}(F)$$

TABLE 22.8   *Typical octahedral complexes of vanadium(III)*

Complex	Colour	ν_1/cm^{-1}	ν_2/cm^{-1}	$10Dq$/cm^{-1}	B/cm^{-1}	μ/BM (room temperature)
$[NH_4][V(H_2O_6][SO_4]_2$·6H$_2$O	Blue-violet	17 800	25 700	19 200	620	2.80
$[VCl_3(MeCN)_3]$	Green	14 400	21 400	15 500	540	2.79
$[VCl_3(thf)_3]$	Orange	13 300	19 900	14 000	553	2.80
$K_3[VF_6]$	Green	14 800	23 250	16 100	649	2.79
$[pyH]_3[VCl_6]$	Purple-pink	16 650	18 350	12 650	513	2.71

Now in the absence of this last band the simple and direct method of deriving the crystal field splitting, Δ_o or $10Dq$, by measuring the difference between ν_3 and ν_1, is not possible. Instead, Δ_0 can be obtained by fitting the two observed bands to the appropriate "Tanabe–Sugano" diagram[23] (Fig. 22.8). In these diagrams the calculated energy difference between each term and the ground term $T_{1g}(F)$ is plotted as a function of the strength of the crystal field. B is the "interelectronic repulsion parameter" already introduced in connection with the nephelauxetic series (p. 1099). It is included in Fig. 22.8 in order to retain generality and obviate the necessity of drawing separate diagrams for each d^2 metal ion. If we take as an example the $[V(H_2O)_6]^{3+}$ ion in the ammonium alum, we have:

$$\frac{\nu_1}{\nu_2} = \frac{17\ 800}{25\ 700} = 0.693.$$

This can only be fitted to Fig. 22.8 at $Dq/B \sim 3.1$ when $\nu_1/B = 28.7$ and $\nu_2 = 41.7$. It follows that $\nu_1/B = 28.7 = 17\ 800/B$, where $B = 620$ cm^{-1} and $10Dq = 10 \times 3.1 \times 620 = 19\ 200$ cm^{-1}. This value of B should be compared to that of 860 cm^{-1} for the free V^{III} ion, the difference being a measure of the extent to which the d-electron charge of the metal is assumed to have expanded on complexation.

[23] B. N. FIGGIS, *Introduction to Ligand Fields*, Chap. 7, pp. 145–72, Chap. 9, pp. 203–48, Wiley, New York, 1966.

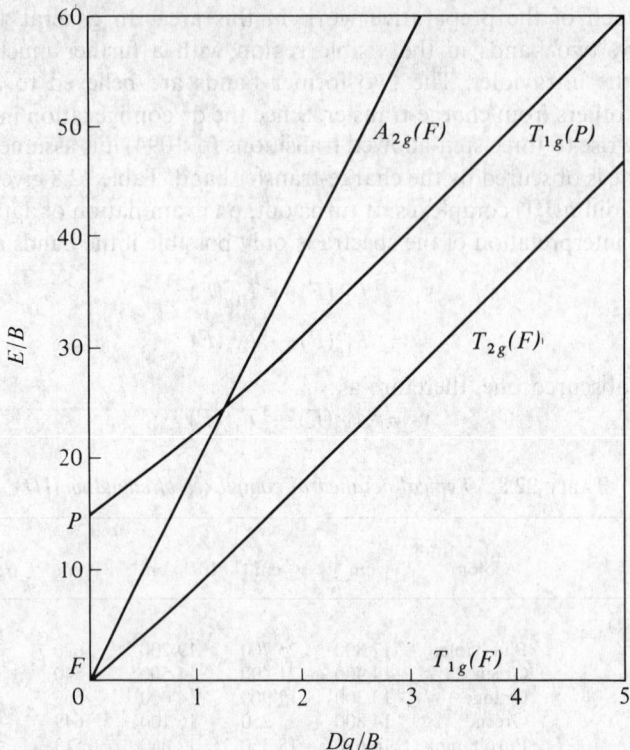

FIG. 22.8   Tanabe-Sugano diagram for a d^2 ion in an octahedral crystal field.

Cationic complexes of the type $[VL_6]^{3+}$ of which $[V(H_2O_6]^{3+}$ is the best-known example are actually rather rare. Anionic $[VX_6]^{3-}$ and neutral $[VCl_3L_3]$ are more common. In general the electronic spectra of these 6-coordinate complexes are accounted for moderately well on the assumption of basically octahedral crystal fields, but the inclusion of trigonal distortions gives more satisfactory results. The magnetic moments of d^2 ions in perfectly octahedral fields are expected to involve "orbital contribution" which varies with temperature. In practice the moments at room temperature rarely exceed the spin-only value and their variation with temperature is less than anticipated for a T ground term. This also is in accord with the presence of some distortion which splits the $^3T_{2g}$ ground term and so reduces the temperature dependence of the magnetic moment.

In spite of the preponderance of 6-coordinate complexes, other coordination numbers are known: the ions $[VCl_4]^-$ and $[VBr_4]^-$ are tetrahedral and are notable in that 4-coordination with ligands other than O-donors is common only later in the transition series. Their spectra exhibit two bands in the regions of $9000\,cm^{-1}$ and $15\,000\,cm^{-1}$ which are assigned to $^3T_1(F) \leftarrow ^3A_2$ and $^3T_1(P) \leftarrow ^3A_2$ transitions respectively, corresponding quite reasonably to values of Δ_t of about 5000 to $5500\,cm^{-1}$. Their magnetic moments too are about 2.7 BM and independent of temperature, as expected.

Neutral complexes of the type $[VX_3(NMe_3)_2]$ (X=Cl, Br) are trigonal bipyramidal with the trimethylamines occupying the axial positions.[24] By contrast $[V\{N(SiMe_3)_2\}_3]$

[24] M. W. DUCKWORTH, G. W. A. FOWLES, and P. T. GREENE, Five-coordinate complexes, *J. Chem. Soc.* (A) 1967, 1592–7.

has a 3-coordinate, planar structure,[25] presumably because the bis(trimethylsilyl)amido ligands are too big for the V^{III} to accommodate more.

Finally, the 7-coordinate $K_4[V(CN)_7] \cdot 2H_2O$ has been shown to have a pentagonal bipyramidal structure[26] and is probably the first known example of a 7-coordinated transition metal complex which persists in solution and in which the ligand is not F^-.

Oxidation State II (d^3)

Again, vanadium is the only member of the group to provide any significant coordination chemistry in this oxidation state, and even this is somewhat unfamiliar. Vanadium(II) complexes are usually prepared by electrolytic or zinc reduction of acidic solutions of vanadium in one of its higher oxidation states. The resulting blue-purple solutions are strongly reducing, and reduction of the water is, in general, only prevented by the presence of acid. Several salts and double sulfates have been prepared[27] and contain the $[V(H_2O)_6]^{2+}$ ion. Adducts of VCl_2 of the type $[VCl_2L_4]$, where L is one of a number of *O*- or *N*-donor ligands, have also been prepared.[28] The spectroscopic and magnetic properties of these compounds are typical of a d^3 ion and their interpretation follows closely that for the chromium(III) ion which, being far more familiar, will be discussed more fully (p. 1197). Also typical of a d^3 ion is the fact that vanadium(II) is kinetically inert and undergoes substitution reactions only slowly.

Other complexes of the type $[VCl_2L_2]$ are distinguished by their colour (green) and magnetic moment (~ 3.2 BM), well below the spin-only value for 3 unpaired electrons. They are therefore thought to be halogen-bridged polymers of V^{II}.

Organometallic compounds apart, oxidation states below $+2$ are best represented by complexes with tris-bidentate nitrogen-donor ligands such as 2,2'-bipyridyl. Reduction by $LiAlH_4$ in thf yields tris(bipyridyl) complexes in which the formal oxidation state of vanadium is $+2$ to -1. Magnetic moments are compatible with low-spin configurations of the metal but, as with the analogous compounds of titanium, it may well be that they would be better regarded as complexes with reduced, i.e. anionic, ligands.

22.3.7 *Organometallic compounds*

In spite of increasing interest recently, the organometallic chemistry of this group has not received a great deal of attention and most of it can be divided into the chemistry of the carbonyls and cyanides, on the one hand, and of the cyclopentadienyls, on the other. The chemistry of σ alkyls or aryls is less well developed than for many other elements but $[V^{III}\{CH(SiMe_3)_2\}_3]$, $[V^{IV}(CH_2SiMe_3)_4]$, and $[V^V O(CH_2SiMe_3)_3]$ have been isolated.[29] In these compounds the possibility of decomposition by alkene elimination or

[25] D. C. BRADLEY, Steric control of metal coordination, *Chem. Br.* **11**, 393–7 (1975).

[26] R. L. R. TOWNS and R. A. LEVENSON, The structure of the seven-coordinate cyano complex of vanadium(III), *J. Am. Chem. Soc.* **94**, 4345–6 (1972).

[27] L. F. LARKWORTHY, K. C. PATEL and D. J. PHILLIPS, Vanadium(II) chemistry. Part III. Spectroscopic and magnetic properties of the double chlorides, *J. Chem. Soc.* (A), 1971, 1347–50.

[28] H. J. SEIFERT and T. AUEL, Elektrolytische reduktion höherer metallchloride in nichwässerigen medien-II, *J. Inorg. Nuclear Chem.* **30**, 2081–6 (1968).

[29] G. YAGUPSKY, W. MOWAT, A. SHORTLAND and G. WILKINSON, Trimethylsilylmethyl compounds of transition metals, *JCS Chem. Comm.* 1970, 1369–70.

other routes is circumvented by the absence of β hydrogen atoms (p. 348) and the bulkiness of the trimethylsilylmethyl groups. Complexes such as $[MMe_5(dmpe)]$ ($M = Nb$, Ta; dmpe $= Me_2PCH_2CH_2PMe_2$) decompose spontaneously above room temperature[30, 30a] and, although free $TaMe_5$ has been isolated, it can explode spontaneously at room temperature even in the absence of air.[30b] Despite this instability, the Ta–Me bond itself is rather strong: thermochemical studies have shown that the mean bond dissociation energy $D(Ta–Me)$ in $TaMe_5$ is $261 \pm 6 \, kJ \, mol^{-1}$, which is substantially greater than, for example, the mean dissociation energy $D(W–CO)$ of $178 \pm 3 \, kJ \, mol^{-1}$ in the kinetically much more stable $W(CO)_6$.[30c]

Reduction of VCl_3 by Na in pyridine or diglyme under CO at 200 atm yields a salt of $[V(CO)_6]^-$ which on acidification and extraction with petroleum ether yields volatile, blue-green, pyrophoric crystals of $V(CO)_6$. Unlike other, formally odd-electron, transition metal carbonyls, this does not attain the noble gas configuration by dimerization and the formation of a M–M bond. It is in fact monomeric and isomorphous with Group VIA octahedral hexacarbonyls (p. 1208); it undergoes substitution reactions typical of metal carbonyls, but is unique amongst simple carbonyls in being paramagnetic with a moment at room temperature of 1.81 BM. The noble gas configuration is, however, attained in the isolable anionic intermediate $[V(CO)_6]^-$. Further reduction of $[Na(diglyme)_2][V(CO)_6]$ with Na metal in liquid NH_3 yields the super-reduced 18-electron species $[V(CO)_5]^{3-}$ which contains V in its lowest known formal oxidation state (-3).[30d] Although the Li, Na, and various onium salts of $[V(CO)_5]^{3-}$ decompose above $0°C$, the trianion can be isolated as the thermally rather stable Rb and Cs salts. $K_3[V(CO)_5]$ is treacherously shock sensitive. Reaction of $[V(CO)_5]^{3-}$ with $[AuCl(PPh_3)]$ in thf gives good yields of the novel, robust, tetrahedral cluster compound $[(Ph_3PAu)_3 V(CO)_5]$.[30e] Significantly, the Nb and Ta hexacarbonyl anions are also known, though these metals do not afford simple neutral carbonyls. A direct synthesis of $V(CO)_6$ and possibly also of the M–M bonded dimer, $V_2(CO)_{12}$, by condensation of vanadium vapour with CO in a matrix of noble gases has recently been reported.[31] The same technique has also been used[32] to prepare the hexakis(dinitrogen) compound, $[V(N_2)_6]$ (p. 1143) which is probably isoelectronic and isostructural with the hexacarbonyl.

Vanadium forms the bis(cyclopentadienyl) compounds: $[(V(\eta^5\text{-}C_5H_5)_2Cl_2]$,

[30] R. R. SCHROCK and P. MEAKIN, Pentamethyl complexes of niobium and tantalum, *J. Am. Chem. Soc.* **96**, 5288–90 (1974).

[30a] R. R. SCHROCK, The preparation of $NbH_5(Me_2PCH_2CH_2Me_2)_2$ and $NbHL_2(Me_2PCH_2CH_2PMe_2)_2$ (L=CO or C_2H_4), *J. Organometallic Chem.* **121**, 373–9 (1976).

[30b] K. MERTIS, L. GALYER and G. WILKINSON, Permethyls of tantalum, tungsten, and rhenium: a warning, *J. Organometallic Chem.* **97**, C65 (1975).

[30c] F. A. ADEDEJI, J. A. CONNOR, H. A. SKINNER, L. GALYER, and G. WILKINSON, Heat of formation of pentamethyltantalum and hexamethyltungsten, *JCS Chem. Comm.* 1976, 159–60.

[30d] J. E. ELLIS, K. L. FJARE, and T. G. HAYES, Highly reduced organometallics. 4. Synthesis and chemistry of pentacarbonylvanadate(3-) ion, $[V(CO)_5]^{3-}$, *J. Am. Chem. Soc.* **103**, 6100–106 (1981).

[30e] J. E. ELLIS, Highly reduced organometallics. 5. Synthesis, properties and the molecular structure of $[(Ph_3PAu)_3 V(CO)_5]$, a gold–vanadium cluster, *J. Am. Chem. Soc.* **103**, 6106–10, (1981).

[31] T. A. FORD, M. HUBER, W. K. KLOTZBÜCHER, M. MOSKOVITS, and G. A. OZIN, Direct synthesis with vanadium atoms. 1. Synthesis of hexacarbonylvanadium and dodecacarbonyldivanadium, *Inorg. Chem.* **15**, 1666–9 (1976).

[32] H. HUBER, T. A. FORD, W. KLOTZBÜCHER, and G. A. OZIN, Direct synthesis with vanadium atoms. II. Synthesis and characterization of vanadium hexadinitrogen, $V(N_2)_6$, in low-temperature matrices, *J. Am. Chem. Soc.* **98**, 3176–8 (1976).

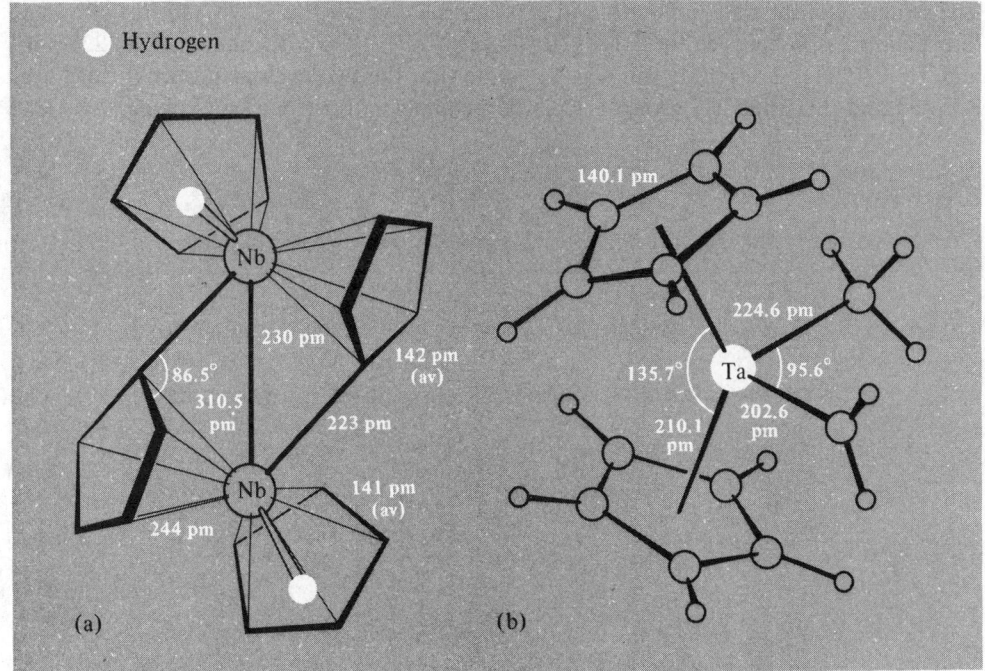

FIG. 22.9   (a) The structure of dimeric $[Nb(\eta^5\text{-}C_5H_5)H\text{-}\mu\text{-}(\eta^5, \eta^1\text{-}C_5H_4)]_2$. The observed diamagnetism of this compound is consistent with the Nb–Nb bond shown. Each of the two bridging rings is η^5-bonded to one Nb and η^1-bonded to the other. (b) The structure of $[Ta(\eta^5\text{-}C_5H_5)_2(CH_3)(=CH_2)]$.

$[V(\eta^5\text{-}C_5H_5)_2Cl]$, and $[V(\eta^5\text{-}C_5H_5)_2]$. The last of these is the dark-violet, paramagnetic and extremely air-sensitive "vanadocene". Unlike the titanium compound (p. 1135), this is a simple "sandwich" compound (p. 367). Niobium and tantalum form $[M(C_5H_5)_4]$ (p. 373), in which two rings are η^5- and two η^1-bonded, and also $[M(\eta^5\text{-}C_5H_5)_2X_3]$ and $[M(\eta^5\text{-}C_5H_5)_2X_2]$ (X = Cl, Br). They do not, though, form simple sandwich compounds. "Niobocene" is actually a dimer and a hydride[33] (Fig. 22.9a), while $[Ta(\eta^5\text{-}C_5H_5)_2H_3]$ is shown by its nmr spectrum to be one of an interesting class of bis(cyclopentadienyl) compounds in which the two rings are not parallel.[34] Another important compound is the mixed methylmethylene derivative of bis(cyclopentadienyl)tantalum(V) prepared by the following sequence of high-yield reactions:[35]

$$[Ta(\eta^5\text{-}C_5H_5)_2Me_3] \xrightarrow{\ Ph_3C^+BF_4^-\ } [Ta(\eta^5\text{-}C_5H_5)_2Me_2] \xrightarrow{\ +\ Me_3P=CH_2\ }$$
$$[Ta(\eta^5\text{-}C_5H_5)_2(CH_3)(=CH_2)]$$

[33] L. J. GUGGENBERGER and F. N. TEBBE, Structure of the dimer of niobocene, *J. Am. Chem. Soc.* **93**, 5924–5 (1971).

[34] J. W. LAUHER and R. HOFFMAN, Structure and chemistry of bis(cyclopentadienyl)-ML_n complexes, *J. Am. Chem. Soc.* **98**, 1729–42 (1976).

[35] R. R. SCHROCK, The first isolable transition metal methylene complex and analogues: characterization mode of decomposition, and some simple reactions, *J. Am. Chem. Soc.* **97**, 6577–8 (1978).

The structure of the pale buff-coloured product is shown in Fig. 22.9b and this allows a direct comparison between the 3 Ta–C distances:[36] Ta=CH_2 203 pm, Ta–CH_3 225 pm, and Ta–$C(C_5H_5)$ 216 pm. It will also be noted that the two cyclopentadienyl rings are eclipsed and that the CH_2 group orients perpendicular to the C–Ta–C plane.

[36] L. J. GUGGENBERGER and R. R. SCHROCK, Structure of bis(cyclopentadienyl)methylmethylenetantalum and the estimated barrier to rotation about the tantalum–methylene bond, *J. Am. Chem. Soc.* **97**, 6578–9 (1975).

1 H																	2 He
3 Li	4 Be											5 B	6 C	7 N	8 O	9 F	10 Ne
11 Na	12 Mg											13 Al	14 Si	15 P	16 S	17 Cl	18 Ar
19 K	20 Ca	21 Sc	22 Ti	23 V	24 Cr	25 Mn	26 Fe	27 Co	28 Ni	29 Cu	30 Zn	31 Ga	32 Ge	33 As	34 Se	35 Br	36 Kr
37 Rb	38 Sr	39 Y	40 Zr	41 Nb	42 Mo	43 Tc	44 Ru	45 Rh	46 Pd	47 Ag	48 Cd	49 In	50 Sn	51 Sb	52 Te	53 I	54 Xe
55 Cs	56 Ba	57 La	72 Hf	73 Ta	74 W	75 Re	76 Os	77 Ir	78 Pt	79 Au	80 Hg	81 Tl	82 Pb	83 Bi	84 Po	85 At	86 Rn
87 Fr	88 Ra	89 Ac	104 Unq	105 Unp	106 Unh	107 Uns											

58 Ce	59 Pr	60 Nd	61 Pm	62 Sm	63 Eu	64 Gd	65 Tb	66 Dy	67 Ho	68 Er	69 Tm	70 Yb	71 Lu
90 Th	91 Pa	92 U	93 Np	94 Pu	95 Am	96 Cm	97 Bk	98 Cf	99 Es	100 Fm	101 Md	102 No	103 Lr

23

Chromium, Molybdenum, and Tungsten

23.1 Introduction

The discoveries of these elements span a period of about 20 y at the end of the eighteenth century. In 1778 the famous Swedish chemist C. W. Scheele produced from the mineral molybdenite (MoS_2) the oxide of a new element, thereby distinguishing the mineral from graphite with which it had hitherto been thought to be identical. Molybdenum metal was isolated 3 or 4 y later by P. J. Hjelm by heating the oxide with charcoal. The name is derived from the Greek word for lead (μόλυβδο ς, *molybdos*), owing to the ancient confusion between any soft black minerals which could be used for writing (this is further illustrated by the use of the names "plumbago" and "black lead" for graphite).

In 1781 Scheele, and also T. Bergmann, isolated another new oxide, this time from the mineral now known as scheelite ($CaWO_4$) but then called "tungsten" (Swedish *tung sten*, heavy stone). Two years later the Spanish brothers J. J. and F. d'Elhuyar showed that the same oxide was a constituent of the mineral wolframite and reduced it to the metal by heating with charcoal. The name "wolfram", from which the symbol of the element is derived, is still widely used in the German literature and is recommended by IUPAC, but the allowed alternative "tungsten" is used in the English-speaking world.

Finally, in 1797, the Frenchman L. N. Vauquelin discovered the oxide of a new element in a Siberian mineral, now known as crocoite ($PbCrO_4$), and in the following year isolated the metal itself by charcoal reduction. This was subsequently named chromium (Greek χρωμα, *chroma*, colour) because of the variety of colours found in its compounds. Since their discoveries the metals and their compounds have become vitally important in many industries and, as one of the biologically active transition elements, molybdenum has been the subject of a great deal of attention in recent years, especially in the field of nitrogen fixation (p. 1205).

23.2 The Elements

23.2.1 *Terrestrial abundance and distribution*

Chromium, 122 ppm of the earth's crustal rocks, is comparable in abundance with vanadium (136 ppm) and chlorine (126 ppm), but molybdenum and tungsten (both

~ 1.2 ppm) are much rarer, and the concentration in their ores is low (cf. Ho 1.4 ppm, Tb 1.2 ppm). The only ore of chromium of any commercial importance is chromite, $FeCr_2O_4$, which is produced principally in the USSR, southern Africa (where 96% of the known reserves are located), and the Philippines. Other less plentiful sources are crocoite, $PbCrO_4$, and chrome ochre, Cr_2O_3, while the gemstones emerald and ruby owe their colours to traces of chromium (pp. 117, 273).

The most important ore of molybdenum is the sulphide molybdenite, MoS_2, of which the largest known deposit is in Colorado, USA, but it is also found in Canada and Chile. Less important ores are wulfenite, $PbMoO_4$, and powellite, $Ca(Mo,W)O_4$.

Tungsten occurs in the form of the tungstates scheelite, $CaWO_4$, and wolframite, $(Fe,Mn)WO_4$, which are found in China (thought to have perhaps 75% of the world's reserves), the USA, South Korea, Bolivia, the USSR, and Portugal.

23.2.2 *Preparation and uses of the metals*

Chromium is produced in two forms:[1]

(a) Ferrochrome by the reduction of chromite with coke in an electric arc furnace. A low-carbon ferrochrome can be produced by using ferrosilicon (p. 381) instead of coke as the reductant. This iron/chromium alloy is used directly as an additive to produce chromium-steels which are "stainless" and hard.

(b) Chromium metal by the reduction of Cr_2O_3. This is obtained by aerial oxidation of chromite in molten alkali to give sodium chromate, Na_2CrO_4, which is leached out with water, precipitated, and then reduced to the Cr(III) oxide by carbon. The oxide can be reduced by aluminium (aluminothermic process) or silicon:

$$Cr_2O_3 + 2Al \longrightarrow 2Cr + Al_2O_3$$

$$2Cr_2O_3 + 3Si \longrightarrow 4Cr + 3SiO_2$$

The main use of the chromium metal so produced is in the production of non-ferrous alloys, the use of pure chromium being limited because of its low ductility at ordinary temperatures. Alternatively, the Cr_2O_3 can be dissolved in sulphuric acid to give the electrolyte used to produce the ubiquitous chromium-plating which is at once both protective and decorative.

The sodium chromate produced in the isolation of chromium is itself the basis for the manufacture of all industrially important chromium chemicals. World production of chromite ores approached $9\frac{1}{2}$ million tonnes in 1980.

Molybdenum is obtained both as a primary product and as a byproduct in the production of copper. In either case MoS_2 is separated by flotation and then roasted to MoO_3. In the manufacture of stainless steel and high-speed tools, which account for about 85% of molybdenum consumption, the MoO_3 may be used directly or after conversion to ferromolybdenum by the aluminothermic process. Otherwise, further purification is possible by dissolution in aqueous ammonia and crystallization of ammonium molybdate (sometimes as the dimolybdate, $[NH_4]_2[Mo_2O_7]$, sometimes as the paramolybdate, $[NH_4]_6[Mo_7O_{24}] \cdot 4H_2O$, depending on conditions), which is the starting material for

---

[1] *Kirk–Othmer, Encyclopedia of Chemical Technology*, 3rd edn., Vol. 6, pp. 54–120, Interscience, New York, 1979.

the manufacture of molybdenum chemicals. Pure molybdenum, which finds important applications as a catalyst in a variety of petrochemical processes and as an electrode material, can be obtained by hydrogen reduction of ammonium molybdate. In 1980 the world production of molybdenum ores was equivalent to 108 000 tonnes of contained Mo.

The isolation of tungsten[2] is effected by the formation of "tungstic acid" (hydrous WO_3), but the chemical route chosen depends on the ore being used. After pulverization and concentration of the ore:

(a) Wolframite is converted to soluble alkali tungstate either by fusing with NaOH and leaching the cooled product with water, or by protracted boiling with aqueous alkali. Acidification with hydrochloric acid then precipitates the tungstic acid.
(b) Scheelite is converted to insoluble tungstic acid by direct treatment with hydrochloric acid and separated from the soluble salts of other metals.

Tungstic acid is then roasted to WO_3 which is reduced to the metal by heating with hydrogen at 850°C. Half of the tungsten produced is used as the carbide, WC, which is extremely hard and wear-resistant and so ideal as a tool-tip. Other major uses are in the production of numerous heat-resistant alloys, but the most important use of the *pure* metal is still as a filament in electric light bulbs, in which role it has never been bettered since it was first used in 1908. In 1980, world production of tungsten ores contained 50 000 tonnes of tungsten.

Both molybdenum and tungsten are obtained initially in the form of powders and, since fusion is impracticable because of their high mps, they are converted to the massive state by compression and sintering under H_2 at high temperatures.

23.2.3 *Properties of the elements*

As can be seen from Table 23.1, which summarizes some of the important properties of Group VIA, each of these elements has several naturally occurring isotopes which imposes limits on the precision with which their atomic weights have been determined, especially for Mo and W.

The elements all have typically metallic bcc structures and in the massive state are lustrous, silvery, and (when pure) fairly soft. However, the most obvious characteristic at least of molybdenum and tungsten, is their refractive nature, and tungsten has the highest mp of all metals—indeed, of all elements except carbon. For this reason, metallic Mo and W are fabricated by the techniques of powder metallurgy, and, in consequence, many of their bulk physical properties depend critically on the nature of their mechanical history.

As in the preceding transition-metal groups, the refractory behaviour and the relative stabilities of the different oxidation states can be explained by the role of the $(n-1)d$ electrons. Compared to vanadium, chromium has a lower mp, bp, and enthalpy of atomization which implies that the 3d electrons are now just beginning to enter the inert electron core of the atom, and so are less readily delocalized by the formation of metal bonds. This is reflected too in the fact that the most stable oxidation state has dropped to $+3$, while chromium(VI) is strongly oxidizing:

$$\tfrac{1}{2}Cr_2O_7^{2-} + 7H^+ + 3e^- \rightleftharpoons Cr^{3+} + 3\tfrac{1}{2}H_2O; \quad E° = 1.33 \text{ V}$$

[2] S. W. H. YIH and C. T. WANG, *Tungsten: Sources, Metallurgy, Properties and Applications*, Plenum Press, New York, 1979, 500 pp. M. HUDSON, Tungsten: its sources, extraction and uses, *Chem. Br.* **18**, 438–42 (1982).

TABLE 23.1   *Some properties of Group VIA elements*

Property		Cr	Mo	W
Atomic number		24	42	74
Number of naturally occurring isotopes		4	7	5
Atomic weight		51.996	95.94	183.85 ± 0.03
Electronic configuration		$[Ar]3d^54s^1$	$[Kr]4d^55s^1$	$[Xe]4f^{14}5d^46s^2$
Electronegativity		1.6	1.8	1.7
Metal radius (12-coordinate)/pm		128	139	139
Ionic radius (6-coordinate)/pm	VI	44	59	60
	V	49	61	62
	IV	55	65	66
	III	61.5	69	—
	II[a]	73 (ls), 80 (hs)	—	—
MP/°C		1900	1620	(3380)
BP/°C		2690	4650	(5500)
ΔH_{fus}/kJ mol^{-1}		21($\pm$2)	28($\pm$3)	(35)
ΔH_{vap}/kJ mol^{-1}		342($\pm$6)	590($\pm$21)	824($\pm$21)
ΔH_f (monatomic gas)/kJ mol^{-1}		397($\pm$3)	664($\pm$13)	849($\pm$13)
Density (20°C)/g cm^{-3}		7.14	10.28	19.3
Electrical resistivity (20°C)/μohm cm		13	~5	~5

[a] Radius depends on whether Cr(II) is low-spin (ls) or high-spin (hs).

For the heavier congenors, tungsten in the group oxidation state is much more stable to reduction, and it is apparently the last element in the third transition series in which all the 5d electrons participate in metal bonding.

23.2.4 *Chemical reactivity and trends*

At ambient temperatures all three elements resist atmospheric attack, which is why chromium is so widely used to protect other more reactive metals. They become more susceptible to attack at high temperatures, when they react with many non-metals giving frequently interstitial and nonstoichiometric products. Chromium reacts more readily with acids than does either molybdenum or tungsten though its reactivity depends on its purity and it can easily be rendered passive. Thus, it dissolves readily in dil HCl but, if very pure, will often resist dil H_2SO_4; again, HNO_3, whether dilute or concentrated, and aqua regia will render it passive for reasons which are by no means clear. In the presence of oxidizing agents such as KNO_3 or $KClO_3$, alkali melts rapidly attack the metals producing MO_4^{2-}.

Once again the two heavier elements are closely similar to each other and show marked differences from the lightest element. This is reflected particularly in the relative stabilities of the oxidation states, all of which are known from +6 down to -2.

The stability of the group oxidation state +6 was referred to above and it may be further noted that, while chromium(VI) tends to form poly oxoanions, the diversity of these is but a pale shadow of that of the polymolybdates and polytungstates (p. 1175). Oxidation states +5 and +4 are represented by chromium largely as unstable intermediates, whereas molybdenum and tungsten provide a significant aqueous chemistry. For chromium, +3 is much the most stable oxidation state, the symmetrical

t_{2g}^3 configuration leading to a coordination chemistry, the fecundity of which is exceeded only by that of cobalt(III), and which has no counterpart in the chemistries of molybdenum or tungsten. Chromium(II) is strongly reducing (Cr^{3+}/Cr^{2+}, $E° -0.41$ V) but it still has an extensive cationic chemistry while molybdenum(II) and tungsten(II) are stabilized only by the retention of appreciable metal–metal bonding, as in the cluster compounds based on the $[M_6X_8]^{4+}$ unit (p. 1190). However, the formation of strong M–M multiple bonds occurs with each of the divalent elements (though less extensively with tungsten than the other two). In the still lower oxidation states, found in compounds with π-acceptor ligands, the metals are quite similar.

Table 23.2 lists the oxidation states of the elements along with representative examples of their compounds. Coordination numbers as high as 12 can be attained, but those over 7 in the case of Cr and 9 in the cases of Mo and W involve the presence of the peroxo ligand or π-bonded aromatic rings systems such as η^5-$C_5H_5^-$ or η^6-C_6H_6.

23.3 Compounds[3, 4, 5]

The binary borides (p. 162), carbides (p. 321), and nitrides (p. 480) have already been discussed. Suffice it to note here that the chromium atom is too small to allow the ready insertion of carbon into its lattice, and its carbide is consequently more reactive than those of its predecessors. As for the hydrides, only CrH is known which is consistent with the general trend in this part of the periodic table that hydrides become less stable across the d block and down each group.

23.3.1 *Oxides of chromium, molybdenum, and tungsten*[6]

The principal oxides formed by the elements of this group are given in Table 23.3.

CrO_3, as is to be expected with such a small cation, is a strongly acidic and rather covalent oxide with a mp of only 197°C. Its deep-red crystals are made up of chains of corner-shared CrO_4 tetrahedra. It is commonly called "chromic acid" and is generally prepared by the addition of conc H_2SO_4 to a saturated aqueous solution of a dichromate. Its strong oxidizing properties are widely used in organic chemistry. CrO_3 melts with some decomposition and, if heated above 220–250°, it loses oxygen to give a succession of lower oxides until the green Cr_2O_3 is formed.

Like the analogous oxides of Ti, V, and Fe, Cr_2O_3 has the corundum structure (p. 274), and it finds wide applications as a green pigment. It is a semiconductor and is antiferromagnetic below 35°C. Cr_2O_3 is the most stable oxide of chromium and is the final product of combustion of the metal, though it is more conveniently obtained by heating ammonium dichromate:

$$(NH_4)_2Cr_2O_7 \longrightarrow Cr_2O_3 + N_2 + 4H_2O$$

[3] D. L. KEPERT, Chromium, molybdenum, tungsten, Chap. 4 in *The Early Transition Metals*, pp. 255–374, Academic Press, London, 1972.

[4] C. L. ROLLINSON, Chromium, molybdenum and tungsten, Chap. 36 in *Comprehensive Inorganic Chemistry*, Vol. 3, pp. 623–769, Pergamon Press, Oxford, 1973.

[5] Z. DORI, The coordination chemistry of tungsten, *Progr. Inorg. Chem.* **18**, 239–307 (1981).

[6] C. N. R. RAO and G. V. S. RAO, *Transition Metal Oxides*, National Standard Reference Data System NSRDS-NBS49, Washington, 1964, 130 pp.

TABLE 23.2   *Oxidation states and stereochemistries of compounds of chromium, molybdenum, and tungsten*

Oxidation state	Coordination number	Stereochemistry	Cr	Mo/W
-2 (d^8)	5	Trigonal bipyramidal?	$[Cr(CO)_5]^{2-}$	$[M(CO)_5]^{2-}$
-1 (d^7)	6	Octahedral	$[Cr_2(CO)_{10}]^{2-}$	$[M_2(CO)_{10}]^{2-}$
0 (d^6)	6	Octahedral	$[Cr(bipy)_3]$	$[M(CO)_6]$
	9	—	$[Cr(\eta^6\text{-}C_6H_6)(CO)_3]$	—
	12	—	$[Cr(\eta^6\text{-}C_6H_6)_2]$	—
1 (d^5)	6	Octahedral	$[Cr(CNR)_6]^+$	$[MoCl(N_2)(diphos)_2]$
	8	—	—	$[Mo(\eta^5\text{-}C_5H_5)(CO)_3]$
	11	—	—	$[Mo(\eta^5\text{-}C_5H_5)(\eta^6\text{-}C_6H_6)]$
	12	—	—	$[Mo(\eta^6\text{-}C_6H_6)_2]^+$
2 (d^4)	4	Tetrahedral	$[CrI_2(OPPh_3)_2]$	—
	5	Trigonal bipyramidal	$[CrBr\{N(C_2H_4NMe_2)_3\}]^+$	—
		Square pyramidal	—	$[Mo_2Cl_8]^{4-}$, $[W_2Me_8]^{4-}$
	6	Octahedral	$[Cr(en)_3]^{2+}$	$[M(diars)_2I_2]$
	7	Capped trigonal prismatic	$[Cr(CO)_2(diars)_2X]^+$	$[Mo(CNR)_7]^{2+}$†
		Pentagonal bipyramidal	—	$[MoH(\eta^2\text{-}O_2CCF_3)\{P(OMe)_3\}_4]$
	8	—	$[Cr(\eta^5\text{-}C_5H_5)Cl(NO)_2]$	—
	9	—	—	$[W(\eta^5\text{-}C_5H_5)(CO)_3Cl]$, M_6Cl_{12} clusters
	10	—	$[Cr(\eta^5\text{-}C_5H_5)_2]$	—
3 (d^3)	3	Planar	$[Cr(NPr_2)_3]$	—
	4	Tetrahedral	$[CrCl_4]^-$	$[(RO)_3Mo\equiv Mo(OR)_3]$, $[(R_2N)_3W\equiv W(NR_2)_3]$
	5	Trigonal bipyramidal	$[CrCl_3(NMe_3)_2]$	—
	6	Octahedral	$[Cr(NH_3)_6]^{3+}$	$[M_2Cl_9]^{3-}$
	7	?	—	$[WBr_2(CO)_3(diars)]^+$
	8	Dodecahedral?	—	$[Mo(CN)_7(H_2O)]^{4-}$
	8 or 12	—	—	$[Mo(\eta^1\text{-}C_5H_5)(\eta^x\text{-}C_5H_5)_2(NO)]$, $x = 3$ or 5
4 (d^2)	4	Tetrahedral	$[Cr(CO)_4]^{4-}$	$[Mo(NMe_2)_4]$
	6	Octahedral	$[CrF_6]^{2-}$	$[MCl_6]^{2-}$
		Trigonal prismatic	—	MS_2
	8	Dodecahedral	—	$[M(CN)_8]^{4-}$
		Square antiprismatic(?)	—	$Mo(S_2CNMe_2)_4]$, $[M(picolinate)_4]$
	12	—	—	$[M(\eta^5\text{-}C_5H_5)_2X_2]$
5 (d^1)	4	Tetrahedral	$[CrO_4]^{3-}$	—
	5	Square pyramidal	$[CrOCl_4]^-$	—
		Trigonal bipyramidal	—	$MoCl_5$(g)
	6	Octahedral	$[CrOCl_5]^{2-}$	$[MF_6]^-$
	8	Dodecahedral	$[Cr(O_2)_4]^{3-}$	$[M(CN)_8]^{3-}$
	13	—	—	$[W(\eta^5\text{-}C_5H_5)_2H_3]$
6 (d^0)	4	Tetrahedral	$[CrO_4]^{2-}$	$[MO_4]^{2-}$
	5	?	—	$[MOX_4]$
		Square pyramidal	—	$[W(\equiv CCMe_3)(=CHCMe_3)(CH_2CMe_3)\{(PMe_2CH_2^-)_2\}]$
	6	Octahedral	—	$\{MO_6\}$ in polymetallates
		Trigonal prismatic	—	$[M(S_2C_2H_2)_3]$ *Continued*

† The structure of these complexes is not regular and has recently[2(a)] been described as "4:3 (C_s) piano stool", which is obtained by slight distortion of a capped trigonal prism (C_{2v}).

[2a] J. C. DEWAN and S. J. LIPPARD, Structure of a homoleptic seven-coordinate molybdenum(II) aryl isocyanide complex [Mo(CNPh)$_7$](PF$_6$)$_2$, *Inorg. Chem.* **21**, 1682–4 (1982). See also *Inorg. Chem.* **20**, 3851–57 (1981) and *J. Am. Chem. Soc.* **104**, 133–6 (1982).

TABLE 23.2—*continued*

Oxidation state	Coordination number	Stereochemistry	Cr	Mo/W
	7	Pentagonal bipyramidal	—	$[WOCl_4(diars)]$
	8	?	—	$[MF_8]^{2-}$
	9	Tricapped trigonal prismatic (C_{2v})	—	$[WH_6(PPhPr^i_2)_3]$

TABLE 23.3　*Oxides of Group VIA*

Oxidation state:	+6	Intermediate				+4	+3
Cr	CrO_3	Cr_3O_8, Cr_2O_5, Cr_5O_{12}, etc.				CrO_2	Cr_2O_3
Mo	MoO_3	Mo_9O_{26}, Mo_8O_{23}, Mo_5O_{14}, $Mo_{17}O_{47}$, Mo_4O_{11}				MoO_2	—
W	WO_3	$W_{40}O_{119}$, $W_{50}O_{148}$, $W_{20}O_{58}$, $W_{18}O_{49}$				WO_2	—

When produced by such dry methods it is frequently unreactive but, if precipitated as the hydrous oxide (or "hydroxide") from aqueous chromium(III) solutions it is amphoteric. It dissolves readily in aqueous acids to give an extensive cationic chemistry based on the $[Cr(H_2O)_6]^{3+}$ ion, and in alkalis to produce complicated, extensively hydrolysed chromate(III) species ("chromites").

The third major oxide of chromium is the brown-black, CrO_2, which is an intermediate product in the decomposition of CrO_3 to Cr_2O_3 and has a rutile structure (p. 1120). It has metallic conductivity and its ferromagnetic properties lead to its commercial importance in the manufacture of magnetic recording tapes which are claimed to give better resolution and high-frequency response than those made from iron oxide. Other more- or less-stable phases with compositions between CrO_2 and CrO_3 have been identified but are of little importance. An example is Cr_5O_{12}, which can be considered as $Cr_2^{III}Cr_3^{VI}O_{12}$: its structure comprises pairs of edge-shared octahedra (presumably involving the larger Cr^{III} ions) linked by corner sharing with tetrahedra (presumably $\{Cr^{VI}O_4\}$) into a three-dimensional framework.

The trioxides of molybdenum and tungsten differ from CrO_3 in that, though they are acidic and dissolve in aqueous alkali to give salts of the MO_4^{2-} ions, they are insoluble in water and have no appreciable oxidizing properties, being the final products of the combustion of the metals. MoO_3 and WO_3 have mps of 795 and 1473°C respectively (i.e. much higher than for CrO_3) and their crystal structures are different. The white MoO_3 has an unusual layer structure composed of distorted MoO_6 octahedra while the yellow WO_3 (like ReO_3) consists of a three-dimensional array of corner-linked WO_6 octahedra. In fact, WO_3 is known in at least seven polymorphic forms and is unique in being the only oxide of any element that can undergo numerous facile crystallographic transitions near room temperature. Thus the monoclinic ReO_3-type phase (which is slightly distorted from cubic by W–W interactions) transforms to a ferroelectric monoclinic phase when cooled to $-43°C$, and transforms to another monoclinic variety above $+20°C$; there are further transitions to an orthorhombic phase at 325° and to a succession of tetragonal phases at 725°, 900°, and 1225°C.

If either MoO_3 or WO_3 is heated *in vacuo* or is heated with the powdered metal, reduction occurs until eventually MO_2 with a distorted rutile structure (p. 1120) is formed. In between these extremes, however, lie a variety of intensely coloured (usually violet or blue) phases whose structural complexity has excited great interest over many years.[7,8] Following the pioneer work of the Swedish chemist A. Magnéli in the late 1940s these materials, which were originally thought to consist of a comparatively small number of rather grossly nonstoichiometric phases, are now known to be composed of a much larger number of distinct and accurately stoichiometric phases with formulae such as Mo_4O_{11}, $Mo_{17}O_{47}$, Mo_8O_{23}, $W_{18}O_{49}$, and $W_{20}O_{58}$. As oxygen is progressively eliminated, a whole series of M_nO_{3n-1} stoichiometries are feasible between the MO_3 structure containing *corner*-shared MO_6 octahedra and the rutile structure consisting of edge-shared MO_6 octahedra. These are produced as slabs of corner-shared octahedra move so as to share edges with the octahedra of identical adjacent slabs (Fig. 23.1). This is the phenomenon of crystallographic shear and occurs in an ordered fashion throughout the solid.[9] The situation is further complicated by the formation of structures involving (a) 7-coordinate, and (b) 4-coordinate, alongside the more prevalent 6-coordinate, metal atoms. The reasons for the formation of these intermediate phases is by no means fully

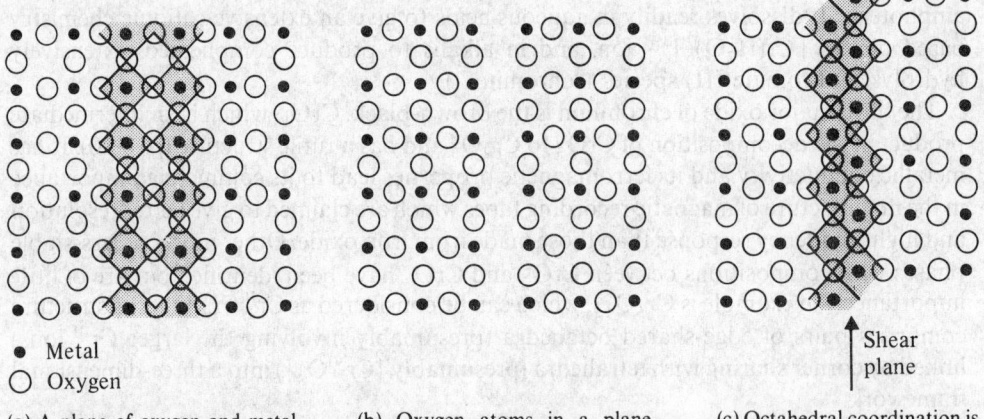

● Metal
○ Oxygen

            Shear plane

(a) A plane of oxygen and metal atoms. Planes of oxygen atoms above and below this, complete the octahedral coordination of each metal atom. Note the corner sharing of adjacent MO_6 octahedra (for clarity some are shaded)

(b) Oxygen atoms in a plane perpendicular to the page are removed, temporarily producing a plane of vacant sites, and destroying the octahedral coordination of metal atoms adjacent to it

(c) Octahedral coordination is retrieved by the movement of a slab of the crystal so as to produce edge sharing between adjacent MO_6 octahedra (shaded) along the shear plane

FIG 23.1   Formation of a crystallographic shear plane.

[7] D. J. M. BEVAN, *Non-stoichiometric compounds*, Chap. 49 in *Comprehensive Inorganic Chemistry*, Vol. 4, pp. 491–7, Pergamon Press, Oxford, 1973.
[8] N. N. GREENWOOD, *Ionic Crystals, Lattice Defects, and Nonstoichiometry*, pp. 140–7, Butterworths, London, 1968.
[9] R. V. PARISH, *The Metallic Elements*, pp. 223–5, Longman, London, 1977.

understood but, although their "nonstoichiometric" $M:O$ ratios imply mixed valence compounds, their largely metallic conductivities suggest that the electrons released as oxygen is removed are in fact delocalized within a conduction band permeating the whole lattice.

Reduction of a solution of a molybdate(VI), or of a suspension of MoO_3, in water or acid by a variety of reagents including Sn^{II}, SO_2, N_2H_4, Cu/acid, or Sn/acid, leads to the production of intense blue, sometimes transient, and probably colloidal products, referred to rather imprecisely as *molybdenum blues*.[10] They appear to be oxide/hydroxide species of mixed valence, forming a series between the extremes of $Mo^{VI}O_3$ and $Mo^VO(OH)_3$, but a precise explanation of their colour is lacking. Their formation can be used as a sensitive test for the presence of reducing agents. The behaviour of tungsten is entirely analogous to that of molybdenum and, as will be seen presently, the reduction of heteropolyanions of these metals produces similar coloured products which may be distinguished from the above "blues" as "heteropoly blues" (though this is not always done).

The dioxides of molybdenum (violet) and tungsten (brown) are the final oxide phases produced by reduction of the trioxides with hydrogen; they have rutile structures sufficiently distorted to allow the formation of M–M bonds and concomitant metallic conductivity and diamagnetism. Strong heating causes disproportionation:

$$3MO_2 \longrightarrow M + 2MO_3$$

No other oxide phases below MO_2 have been established but a yellow "hydroxide", precipitated by alkali from aqueous solutions of chromium(II), spontaneously evolves H_2 and forms a chromium(III) species of uncertain composition. The sulfides, selenides, and tellurides of this triad are considered on p. 1186.

23.3.2 *Isopolymetallates*[11, 12]

Acidification of aqueous solutions of the yellow, tetrahedral chromate ion, CrO_4^{2-}, initiates a series of labile equilibria involving the formation of the orange-red dichromate ion, $Cr_2O_7^{2-}$:

$$HCrO_4^- \rightleftharpoons CrO_4^{2-} + H^+; \quad K\ 10^{-5.9}$$
$$H_2CrO_4 \rightleftharpoons HCrO_4^- + H^+; \quad K\ 10^{+0.26}$$
$$Cr_2O_7^{2-} + H_2O \rightleftharpoons 2HCrO_4^-; \quad K\ 10^{-2.2}$$
$$HCr_2O_7^- \rightleftharpoons Cr_2O_7^{2-} + H^+; \quad K\ 10^{0.85}$$
$$H_2CrO_4 \rightleftharpoons HCr_2O_7^- + H^+; \quad K\ \text{large}$$

Above pH 8 only CrO_4^{2-} ions exist in appreciable concentration but, as the pH is lowered, the equilibria shift and between pH 2–6 the $HCrO_4^-$ and $Cr_2O_7^{2-}$ ions are in equilibrium. Because of the lability of these equilibria the addition of the cations Ag^I, Ba^{II}, or Pb^{II} to aqueous dichromate solutions causes their immediate precipitation as insoluble chromates rather than their more soluble dichromates. Polymerization beyond the

[10] N. V. SIDGWICK, *The Chemical Elements and their Compounds*, pp. 1046–7, Oxford University Press, Oxford, 1950.
[11] K.-H. TYTKO and O. GLEMSER, Isopolymolybdates and isopolytungstates, *Adv. Inorg. Chem. Radiochem.* **19**, 239–315 (1976).
[12] D. L. KEPERT, Isopolyanions and heteropolyanions, Chap. 51 in *Comprehensive Inorganic Chemistry*, Vol. 4, pp. 607–72, Pergamon Press, Oxford, 1973.

dichromate ion is apparently limited to the formation of tri- and tetra-chromates ($Cr_3O_{10}{}^{2-}$ and $Cr_4O_{13}{}^{2-}$), which can be crystallized as alkali-metal salts from very strongly acid solutions. These anions, as well as the dichromate ion, are formed by the corner sharing of CrO_4 tetrahedra, giving Cr–O–Cr angles very roughly in the region of 120° (Fig. 23.2). The simplicity of this anionic polymerization of chromium, as compared to that shown by the elements of the preceding groups and the heavier elements of the present triad, is probably due to the small size of Cr^{VI}. This evidently limits it to tetrahedral rather than octahedral coordination with oxygen, whilst simultaneously favouring Cr–O double bonds and so inhibiting the sharing of attached oxygens.

Sodium dichromate, $Na_2Cr_2O_7 \cdot 2H_2O$, produced from the chromate is commercially much the most important compound of chromium. It yields a wide variety of pigments used in the manufacture of paints, inks, rubber, and ceramics, and from it are formed a host of other chromates used as corrosion inhibitors and fungicides, etc. It is also the oxidant in many organic chemical processes; likewise, acidified dichromate solutions are used as strong oxidants in volumetric analysis:

$$Cr_2O_7{}^{2-} + 14H^+ + 6e^- \longrightarrow 2Cr^{3+} + 7H_2O; \quad E° = 1.33 \text{ V}$$

For this purpose the potassium salt $K_2Cr_2O_7$ is preferred since it lacks the hygroscopic character of the sodium salt and may therefore be used as a primary standard.

The polymerization of acidified solutions of molybdenum(VI) or tungsten(VI) yields the most complicated of all the polyanion systems and, in spite of the fact that the tungsten system has been the most intensively studied, it is still probably the least well understood. This arises from the problem inevitably associated with studies of such equilibria, and which were noted (p. 1147) in the discussion of the Group VA isopolyanions. It must also be admitted that, whilst the observed structures of individual polyanions are reasonable, it is often difficult to explain why, under given circumstances, a particular degree of aggregation or a particular structure is preferred over other alternatives.

When the trioxides of molybdenum and tungsten are dissolved in aqueous alkali, the resulting solutions contain tetrahedral $MO_4{}^{2-}$ ions and simple, or "normal", molybdates and tungstates such as Na_2MO_4 can be crystallized from them. If these solutions are made strongly acid, precipitates of yellow "molybdic acid", $MoO_3 \cdot 2H_2O$, or white

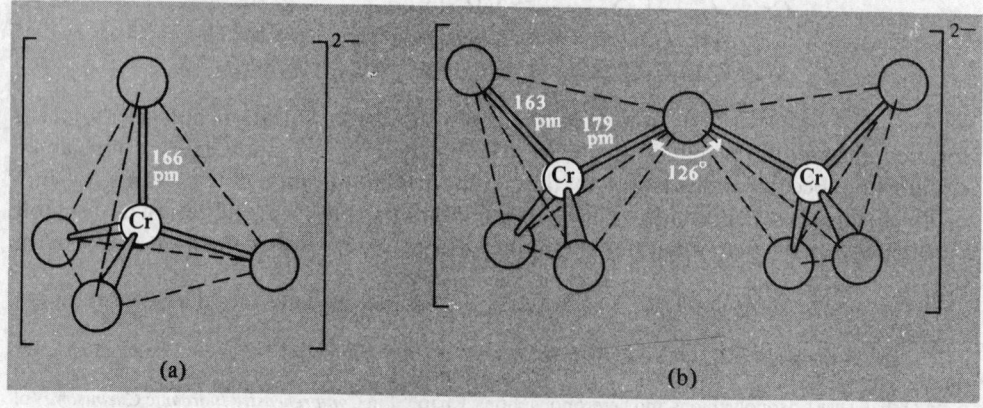

Fig. 23.2   (a) $CrO_4{}^{2-}$ ion, and (b) $Cr_2O_7{}^{2-}$ ion.

"tungstic acid", $WO_3 . 2H_2O$ are obtained which convert to the monohydrates if warmed. At pHs between these two extremes, however, polymeric anions, made up almost invariably of MO_6 octahedra, are formed and their salts can be crystallized. A plethora of physical techniques[13] has been used to characterize these species and unravel the complexity of their structures. Examination of the alkali metal (or ammonium) and alkaline earth salts, particularly by X-ray analysis, forms the basis of classical studies of the isopoly-molybdates and -tungstates. Providing that it can be shown that the anion in a solid is also the predominant species in the solution from which the solid crystallizes (and this is not necessarily so as evidenced by the very simple example of the precipitation of $PbCrO_4$, etc., from dichromate solutions), then experimental results obtained on the solid are taken to be applicable also to the species in solution.

Important differences distinguish the molybdenum and tungsten systems. In aqueous solution, equilibration of the molybdenum species is complete within a matter of minutes whereas for tungsten this may take several weeks; it also transpires that there is scarcely a polyanion in one system which has a precise counterpart in the other. The two must therefore be considered separately.

Undoubtedly the first major polyanion formed when the pH of an aqueous molybdate solution is reduced below about 6 is the heptamolybdate $[Mo_7O_{24}]^{6-}$, traditionally known as the paramolybdate. Anions with 8 and 36 Mo atoms are also formed before the increasing acidity suffices to precipitate the hydrous oxide. The formation of these isopolyanions may be represented by the net equations:

$$7[MoO_4]^{2-} + 8H^+ \rightleftharpoons [Mo_7O_{24}]^{6-} + 4H_2O$$
$$8[MoO_4]^{2-} + 12H^+ \rightleftharpoons [Mo_8O_{26}]^{4-} + 6H_2O$$
$$36[MoO_4]^{2-} + 64H^+ \rightleftharpoons [Mo_{36}O_{112}]^{8-} + 32H_2O$$

Of course many intermediate reactions occur: in particular, hydration (which raises the coordination number of the molybdenum from 4 to 6) and protonation (which reduces the high charges on some of the ions). Of the ions mentioned, $[MoO_4]^{2-}$, $[Mo_7O_{24}]^{6-}$, and $[Mo_{36}O_{112}]^{8-}$ are known to have the same structures in solids and in solution, whereas $[Mo_8O_{26}]^{4-}$ forms two different isomers in solids and its structure in solution is uncertain.[14] Careful adjustment of acidity, concentration, and temperature, often coupled with slow crystallization, can produce solids containing many other ions which are apparently not present in solution. Mixtures abound, but amongst the distinct species which have been characterized are: the dimolybdate,[15] $[Mo_2O_7]^{2-}$; the hexa-molybdate,[16] $[Mo_6O_{19}]^{2-}$; and the decamolybdate,[16] $[Mo_{10}O_{34}]^{8-}$. Figure 23.3 depicts the structures of some of these ions and it can be seen that the basic units are MoO_6 octahedra which are joined by shared corners or shared edges, but not by shared faces. MoO_4 tetrahedra are also involved in some of the ions. The structure of $[Mo_{36}O_{112}(H_2O)_{16}]^{8-}$, the largest isopolyanion yet known, has recently been reported[17] and consists predominantly of MoO_6 octahedra but includes, uniquely in the

[13] Ref. 11, pp. 245–65.

[14] W. G. KLEMPERER and W. SHUM, Synthesis and interconversion of the isomeric α- and β-Mo_8O_{26} ions, *J. Am. Chem. Soc.* **98**, 8291–3 (1976).

[15] A. W. ARMOUR, M. G. B. DREW, and P. C. H. MITCHELL, Crystal and molecular structure and properties of ammonium dimolybdate, *JCS Dalton* 1975, 1493–6.

[16] J. FUCHS, H. HARTH, W. D. HUNNIUS, and S. MAHJOUR, Anion structure of ammonium decamolybdate $(NH_4)_8Mo_{10}O_{34}$, *Angew. Chem.*, Int. Edn. (Engl.) **14**, 644 (1975).

[17] I. P. BÖSCHEN, X-ray crystallographic determination of the structure of the isopolyanion $[Mo_{36}O_{112}(H_2O)_{16}]^{8-}$ in the compound $K_8[Mo_{36}O_{112}(H_2O)_{16}].36H_2O$, *JCS Chem. Comm.* 1979, 780–2.

isopolymolybdates, MoO_7 pentagonal bipyramids. As noted in the equations above, the condensation of MoO_6 and MoO_4 polyhedra to produce these large anions requires large quantities of strong acid, as the supernumerary oxygen atoms are removed in the form of water molecules.

The most important species produced by the progressive acidification of normal tungstate solutions are the paratungstates which, indeed, were the only ones reported prior to the mid-1940s. They are generally less soluble than the normal tungstates and can

$[Mo_2O_7^{2-}]_n$

Polymeric chain of pairs of MoO_6 octahedra linked by double bridges of MoO_4 tetrahedra (note difference from $Cr_2O_7^{2-}$, p. 000)

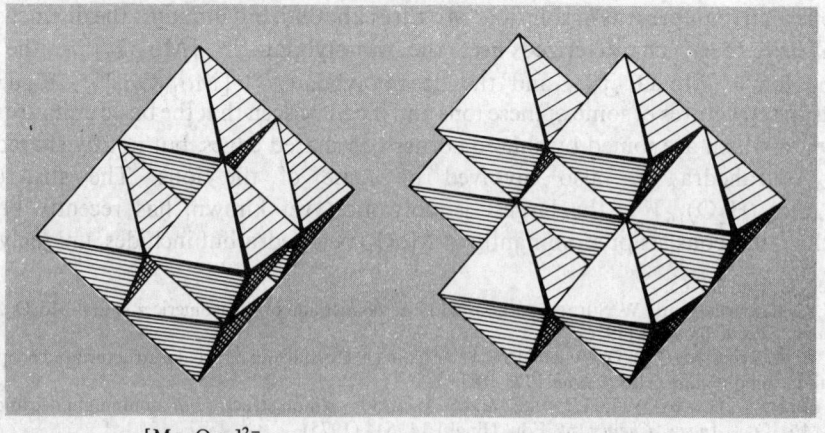

$[Mo_6O_{19}]^{2-}$
(The sixth octahedron is obscured)

Paramolybdate, $[Mo_7O_{24}]^{6-}$

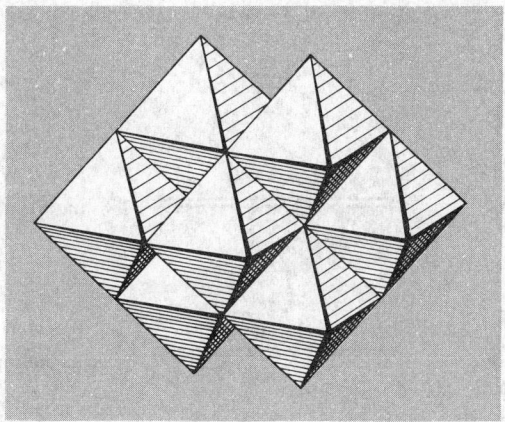

β-[Mo_8O_{26}]$^{4-}$ (one MoO_6 unit is obscured)

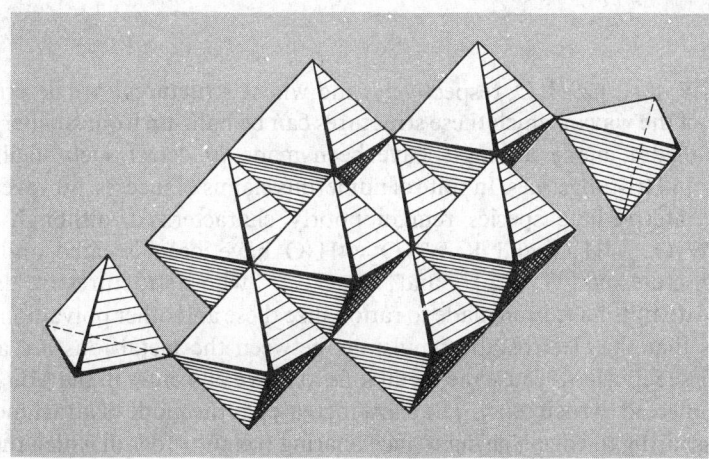

[$Mo_{10}O_{34}$]$^{8-}$

FIG. 23.3   Idealized structures of isopolymolybdate ions. (In general the polyhedra are distorted
so that the Mo atoms lie closest to oxygen atoms which are unshared).

be crystallized over a period of several days. Further acidification produces metatungstates which are rather more soluble but will crystallize either on standing for some
months or on prolonged heating of the solution. It seems that comparatively rapid
condensation produces relatively soluble species which, if left, will very slowly condense
further into less-soluble species. The much-simplified reaction scheme below contains
only those species whose existence is generally accepted; even so, it must be noted that the
degree of hydration of these ions is uncertain and the hexatungstates have not been
characterized with certainty. The best characterized are the dodeca-, para-, and metatungstates which have been crystallized as salts such as $(NH_4)_{10}$[$H_2W_{12}O_{42}$]. $10H_2O$

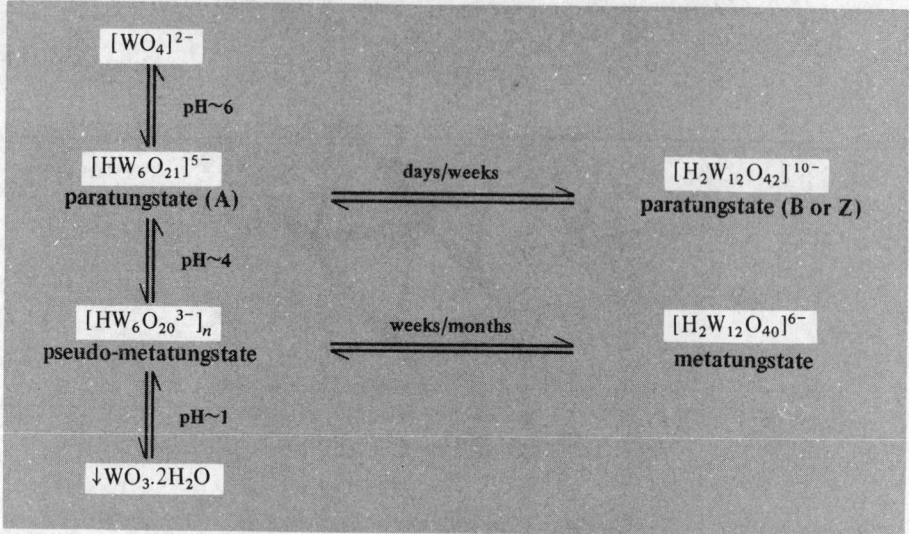

Reaction scheme for the condensation of tungstate ions in aqueous solution.

and $Na_6[H_2W_{12}O_{40}].29H_2O$, respectively, and whose structures† are described in Fig. 23.4. In view of the way in which these structures can be built-up from smaller aggregates of WO_6 octahedra, many attempts have been made to detect such smaller species, especially di- and tri-tungstates. In spite of numerous claims of success, however, these and many other intermediate species remain poorly characterized, although the solids, $Li_{14}(WO_4)(W_4O_{16}).4H_2O$ and $K_4W_{10}O_{32}.4H_2O$ have been isolated and shown to contain the discrete ions $[W_4O_{16}]^{8-}$ and $[W_{10}O_{32}]^{4-}$ whose structures are also included in Fig. 23.4. Attempts have been made to rationalize these and other polyanion structures, and it seems that the electrostatic repulsions between the metal ions are a dominant factor.[18] These repulsions cause the metal ions to move off-centre in the MO_6 octahedra which are connected to each other. The effect increases as the mode of attachment changes from corner-sharing to edge-sharing to face-sharing (i.e. the order in which the centres of adjacent octahedra move closer together) though the last of these does not occur in the polyanions. Thus, while the avoidance of unfavourably high, overall anionic charge favours edge-sharing as opposed to corner-sharing (thereby reducing the number of O^{2-} ions), the consequent distortion arising from electrostatic repulsions between the metal atoms becomes increasingly difficult to accommodate as the size of the polyanion increases. Ultimately edge-sharing is no longer possible, and this stage is reached by W^{VI} before it occurs with the smaller Mo^{VI}. Inspection of Figs. 23.3 and 23.4 shows the greater incidence of corner-sharing in the higher polytungstates than in the polymolybdates. Also,

---

† It is instructive to recall a particular problem facing early workers in establishing these structures by X-ray diffraction. Large scattering by the heavy tungsten atoms made it extremely difficult to locate the positions of the lighter oxygen atoms and this sometimes led to ambiguity in the assignment of precise structures (relative scattering $O/W = (8/74)^2 = 1/85$, cf. $H/C = (1/6)^2 = 1/36$). This is no longer a problem because of the greater precision of modern techniques of X-ray data acquisition and processing, but good-quality crystals are still necessary and these may be very difficult to produce.

[18] Ref. 3, pp. 46–60.

except in $[Mo_7O_{24}]^{6-}$, linear sets of 3 MO_6 octahedra on which the distortions are most difficult to accommodate are not found, triangular $M\text{-----}M$ sets being preferred.

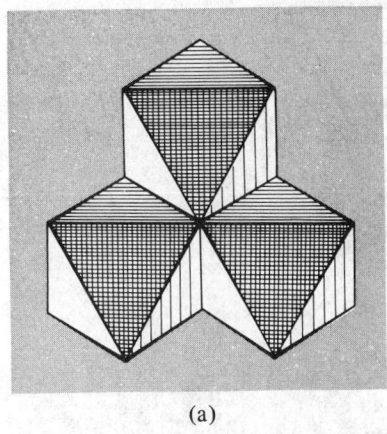

(a)

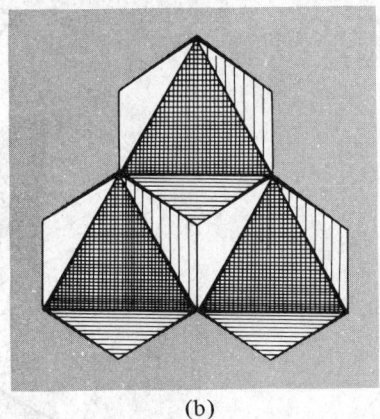

(b)

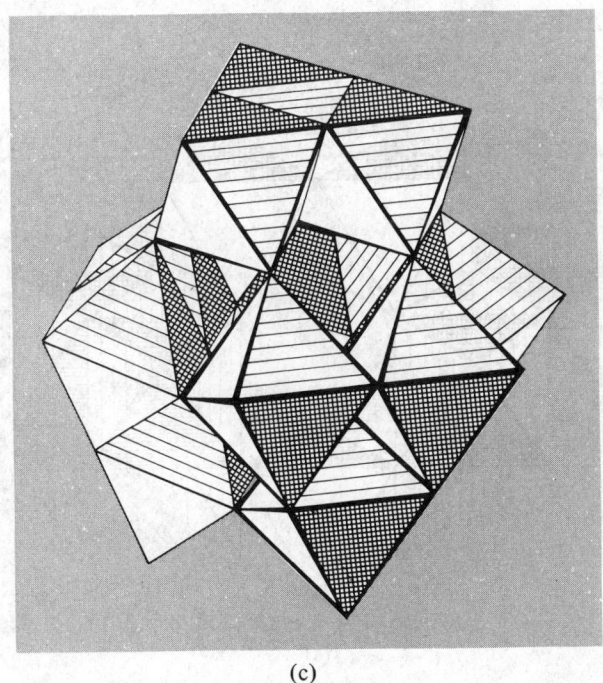

(c)

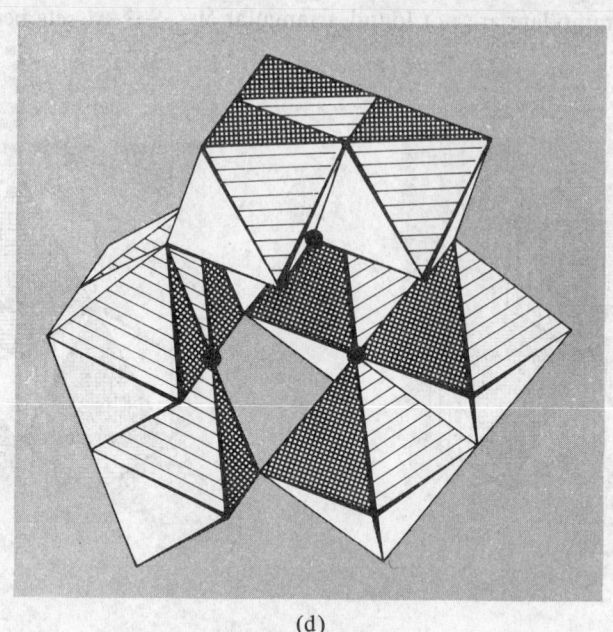

(d)

(e)

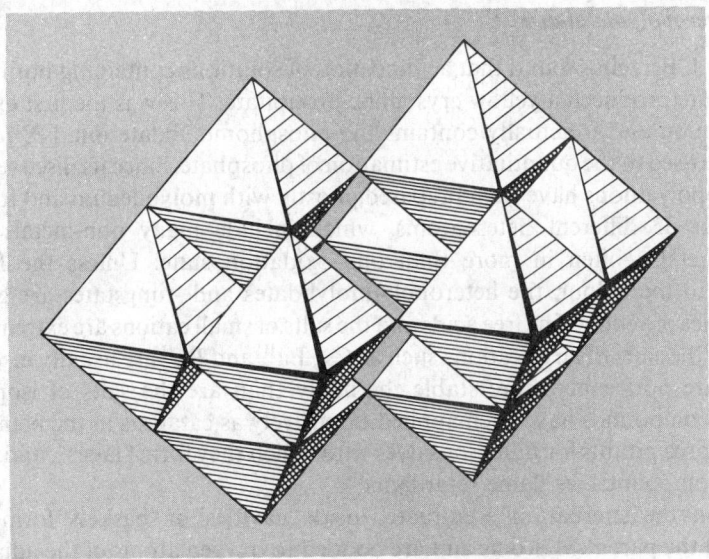

(f) $[W_{10}O_{32}]^{4-}$, composed of two identical W_5O_{16} groups.

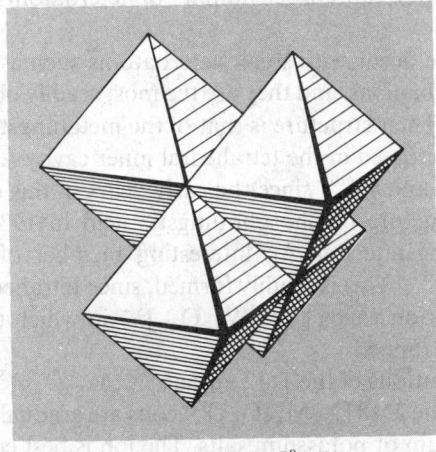

(g) $[W_4O_{16}]^{8-}$.

FIG. 23.4 The structures of some isopolytungstate ions. The metatungstate ion $[H_2W_{12}O_{40}]^{6-}$, is composed of four identical "tritungstate" groups. Each of these "tritungstate" groups is made up of 3 WO_6 octahedra joined by shared edges in such a way as to display, on opposite sides, the characteristic composite faces shown in (a) and (b). These four "tritungstate" groups are then attached to each other by corner sharing with their (a) faces directed inwards, and their (b) faces directed outwards. As can be seen in (c), an inner cavity is thereby produced and this is almost certainly where the 2 protons of the metatungstate ion are situated. The cavity is more clearly seen in (d) where one "tritungstate" group has been removed. The structure is generally known as the "Keggin structure" after its discoverer (J. F. Keggin, *Proc. R. Soc. A* **144**, 75–100 (1934).

The 1:12 tetrahedral heteropolyanions are isomorphous with the metatungstates, the hetero-atom being simply incorporated inside the cavity. The nearest neighbours to the hetero-atom are those oxygen atoms (marked by dots in (d)) which lie at the centres of (a) faces of the "tritungstate" groups, and they are tetrahedrally disposed around the hetero-atom.

The paratungstate (B or Z), $[H_2W_{12}O_{42}]^{10-}$, ion is also composed of four "tritungstate" groups, but these are now of two different types. The top and bottom groups in (e) are the same as those in the metatungstate ion, but the two groups making up the central layer are different in that the 3 WO_6 octahedra in each of these groups form a chain, so that there is no longer any 1 oxygen atom common to all 3.

23.3.3 *Heteropolymetallates*[12]

In 1826 J. J. Berzelius found that acidification of solutions containing both molybdate and phosphate produced a yellow crystalline precipitate. This was the first example of a heteropolyanion and it actually contains the phosphomolybdate ion, $[PMo_{12}O_{40}]^{3-}$, which can be used in the quantitative estimation of phosphate. Since its discovery a host of other heteropolyanions have been prepared, mostly with molybdenum and tungsten but with at least 35 different heteroatoms, which include many non-metals and most transition metals—often in more than one oxidation state. Unless the heteroatom contributes to the colour, the heteropoly-molybdates and -tungstates are generally of varying shades of yellow. The free acids and the salts of small cations are extremely soluble in water but the salts of large cations such as Cs^I, Ba^{II}, and Pb^{II} are usually insoluble. The solid salts are noticeably more stable thermally than are the salts of isopolyanions. Heteropoly compounds have been applied extensively as catalysts in the petrochemicals industry, as precipitants for numerous dyes with which they form "lakes", and, in the case of the Mo compounds, as flame retardants.

In these ions the heteroatoms are situated inside "cavities" or "baskets" formed by MO_6 octahedra of the parent M atoms and are bonded to oxygen atoms of the adjacent MO_6 octahedra. The stereochemistry of the heteroatom is determined by the shape of the cavity which in turn depends on the ratio of the number of heteroatoms to parent atoms. Four major classes are found.

1:12, tetrahedral. These occur with small heteroatoms such as P^V, As^V, Si^{IV}, and Ti^{IV} which yield tetrahedral oxoanions, and they are the most readily obtained and best known of the heteropolyanions. Their structure is that of the metatungstate ion (Fig. 23.4(c)) in which the heteroatom is situated in the tetrahedral inner cavity. Their isomorphism with metatungstates had been known[19] since the early years of this century, and it was the determination of the structure of phosphotungstic acid in 1934 which indicated the structure of the metatungstate ion. An interesting member of this class. is the Co^{II} derivative, $[Co^{II}W_{12}O_{40}]^{6-}$. This is readily formed, since tetrahedrally coordinated Co^{II} is not unusual, but oxidation yields $[Co^{III}W_{12}O_{40}]^{5-}$ in which the very unusual, high-spin, tetrahedral Co^{III} is trapped.

2:18, tetrahedral. If solutions of the 1:12 anions $[X^VM_{12}O_{40}]^{3-}$ (X = P, As; M = Mo, W) are allowed to stand, the 2:18 $[X_2M_{18}O_{62}]^{6-}$ ions are gradually produced and can be isolated as their ammonium or potassium salts. The ion is best considered to be formed from two 1:12 anions, each of which loses 3 MO_6 octahedra (the 3 "basal" octahedra in Fig. 23.4(c)) before fusing together.

1:6, octahedral. These are formed with larger heteroatoms such as Te^{VI}, I^{VII}, Co^{III}, and Al^{III} which coordinate to 6 edge-sharing MO_6 octahedra in the form of a hexagon around the central XO_6 octahedron. It is noticeable that tungsten forms this type of ion less frequently than does molybdenum, which again probably reflects a greater readiness of molybdenum to form large structures based solely on edge-sharing, rather than corner-sharing, of octahedra. A comparison of the structures shown in Figs. 23.3 and 23.4 tends to reinforce this conclusion.

1:9, octahedral. The structures of these ions are also based solely on edge-sharing MO_6 octahedra and tungsten apparently forms none. The best characterized examples are $[Mn^{IV}Mo_9O_{32}]^{6-}$ and $[Ni^{IV}Mo_9O_{32}]^{6-}$.

[19] Ref. 10, p. 1042.

Mild reduction of $1:12$ and $2:18$ heteropoly-molybdates and -tungstates produces characteristic and very intense blue colours ("heteropoly blues") which find application in the quantitative determinations of Si, Ge, P, and As, and commercially as dyes and pigments. The reductions are most commonly of 2 electron equivalents but may be up to 6 electron equivalents. Many of the reduced anions can be isolated as solid salts in which the unreduced structure apparently remains essentially unchanged. The reduction evidently occurs on individual M atoms, producing a proportion of M^V ions. Transfer of electrons from M^V to M^{VI} ions is then responsible for the intense "charge-transfer" absorption.

23.3.4 *Tungsten and molybdenum bronzes*

These materials owe their name to their metallic lustre and are used in the production of "bronze" paints. They provide a further example of the formation of intense and characteristic colours by the reduction of oxo-species of Mo and W. The tungsten bronzes[20] were the first to be discovered when, in 1823, F. Wöhler reduced a mixture of Na_2WO_4 and WO_3 with H_2 at red heat. The product was the precursor of a whole series of nonstoichiometric materials of general formula $M_x^IWO_3$ ($x < 1$) in which M^I is an alkali metal cation and W has an oxidation state between $+5$ and $+6$. Corresponding materials can also be obtained in which M is an alkaline earth or lanthanide metal. The alkali-metal molybdenum bronzes are analogous to, but less well-known than, those of tungsten, being less stable and requiring high pressure for their formation; they were not produced until the 1960s. The lower stability of the molybdenum bronzes may be a consequence of the greater tendency of Mo^V to disproportionate as compared to W^V.

Tungsten bronzes can be prepared by a variety of reductive techniques but probably the most general method consists of heating the normal tungstate with tungsten metal. They are extremely inert chemically, being resistant both to alkalis and to acids, even when hot and concentrated. Their colours depend in the proportion of M and W present. In the case of sodium tungsten bronze the colour varies from golden yellow, when $x \sim 0.9$, through shades of orange and red to bluish-black when $x \sim 0.3$. Within this range of x-values the structure consists of corner-shared WO_6 octahedra† as in WO_3 (p. 1173), with Na^I ions in the interstices—in other words an M-deficient perovskite lattice (p. 1122). The observed electrical conductivities are metallic in magnitude and decrease linearly with increase in temperature, suggesting the existence of a conduction band of delocalized electrons.[21] Measurements of the Hall effect (used to measure free electron concentrations) indicate that the concentration of free electrons equals the concentration of sodium atoms, implying that the conduction electrons arise from the complete ionization of sodium atoms. Several mechanisms have been suggested for the formation of this conduction band but it seems most likely that the t_{2g} orbitals of the tungsten overlap, not directly (since adjacent W atoms are generally more than 500 pm apart) but via oxygen $p\pi$ orbitals, so

---

† This corner-sharing in tungsten bronzes is to be compared with a mixture of corner and edge-sharing in molybdenum bronzes which presumably occurs, as in the case of the polymetallates, because the increased electrostatic repulsion entailed in edge-sharing is less disruptive when the smaller Mo is involved. The prevalence of edge-sharing is still more marked in the vanadium and titanium bronzes (pp. 1150, 1123) where the smaller charges on the metal ions produce correspondingly smaller repulsions.

[20] P. HAGENMULLER, Tungsten bronzes, vanadium bronzes, and related compounds, Chap. 50 in *Comprehensive Inorganic Chemistry*, Vol. 4, pp. 541–605, Pergamon Press, Oxford, 1973.
[21] Ref. 8, pp. 177–9.

forming a partly filled π^* band permeating the whole WO_3 framework. If the value of x is reduced below about 0.3 the resulting electrical properties are semiconducting rather than metallic. This change coincides with structural distortions which probably disrupt the mechanism by which the conduction band is formed and instead cause localization of electrons in t_{2g} orbitals of specific tungsten atoms.

23.3.5 *Sulfides, selenides, and tellurides*

The sulfides of this triad, though showing some similarities in stoichiometry to the principal oxides (p. 1171), tend to be more stable in the lower oxidation states of the metals. Thus Cr forms no trisulfide and it is the di- rather than the tri-sulfides of Mo and W which are the more stable. However, tungsten (unlike Cr and Mo) does not form M_2S_3. Many of the compounds are nonstoichiometric, most are metallic (or at least semiconducting), and they exhibit a wide variety of magnetic behaviour encompassing diamagnetic, para-magnetic, antiferro-, ferri-, and ferro-magnetic.[21a]

Cr_2S_3 is formed by heating powered Cr with sulfur, or by the action of $H_2S(g)$ on Cr_2O_3, $CrCl_3$, or Cr. It decomposes to CrS on being heated, via a number of intermediate phases which approximate in composition to Cr_3S_4, Cr_5S_6, and Cr_7S_8. The structural relationship between these various phases can be elegantly related to the CdI_2–NiAs structure motif (pp. 648 and 1407). Removal of M atoms from alternate layers of the NiAs structure yields the CdI_2 layer lattice. Thus, in Cr_7S_8 one-quarter of the Cr atoms in every second metal layer are randomly removed: the composition range is $Cr_{0.88}S$ to $Cr_{0.87}S$. With Cr_5S_6 one-third of the Cr atoms in every second metal layer are absent in an ordered way and with Cr_4S_6 (i.e. Cr_2S_3) two-thirds of the Cr atoms in every second metal layer are missing in an ordered way. Because of the stringent geometric criteria for this ordering, these two phases are truly stoichiometric and show no detectable range of composition variation. By contrast, in Cr_3S_4 every second Cr atom in alternate layers is missing but, because it is not possible for every *second* atom in a *trigonal* array to be missing in a regular pattern and still preserve trigonal symmetry, there is some disordering, the symmetry is lowered, and a small range of composition variation is permitted ($Cr_{0.79}S$–$Cr_{0.76}S$). Of these various phases Cr_2S_3 and CrS are semiconductors, whereas Cr_7S_8, Cr_5S_6, and Cr_3S_4 are metallic, and all exhibit magnetic ordering. The corresponding selenides CrSe, Cr_7Se_8, Cr_3Se_4, Cr_2Se_3, Cr_5Se_8, and Cr_7Se_{12} are broadly similar, as are the tellurides CrTe, Cr_7Te_8, Cr_5Te_6, Cr_3Te_4, Cr_2Te_3, Cr_5Te_8, and $CrTe_{\sim 2}$.

Of the many molybdenum sulfides which have been reported, only MoS, MoS_2, and Mo_2S_3 are well established. A hydrated form of the trisulfide of somewhat variable composition is precipitated from aqueous molybdate solutions by H_2S in classical analytical separations of molybdenum, but it is best prepared by thermal decomposition of the thiomolybdate, $(NH_4)_2MoS_4$. MoS is formed by heating the calculated amounts of Mo and S in an evacuated tube. The black MoS_2, however, is the most stable sulfide and, besides being the principal ore of Mo, is much the most important Mo compound commercially. In 1923 its structure was shown by R. G. Dickinson and L. Pauling (in the latter's first research paper) to consist of layers of MoS_2 in which the molybdenum atoms are each coordinated to 6 sulfides, but forming a trigonal prism rather than the more usual

[21a] F. HULLIGER, Crystal chemistry of chalcogenides and pnictides of the transition elements, *Struct. Bonding* **4**, 83–229 (1968).

octahedron. This layer structure promotes easy cleavage and graphite-like lubricating properties, which have led to its widespread use as a lubricant both dry and in suspensions in oils and greases. It also has applications as a catalyst[21b] in many hydrogenation reactions and, even when the original catalyst takes the form of an oxide, it is likely that impurities (which often "poison" other catalysts) quickly produce a sulfide catalytic system.

WS_3 and WS_2 are similar to their molybdenum analogues and all 4 compounds are diamagnetic semiconductors.

Selenides and tellurides are, again, broadly similar to the sulfides in structure and properties.[21a]

23.3.6 *Halides and oxohalides*[22]

The known halides of chromium, molybdenum, and tungsten are listed in Table 23.4. The observed trends are as expected. The group oxidation state of $+6$ is attained by chromium only with the strongly oxidizing fluorine, and even tungsten is unable to form a hexaiodide. Precisely the same is true in the $+5$ oxidation state, and in the $+4$ oxidation state the iodides have a doubtful or unstable existence. In the lower oxidation states all the chromium halides are known, but molybdenum has not yet been induced to form a difluoride nor tungsten a di- or tri-fluoride. Similarly, in the oxohalides (which are largely confined to the $+6$ and $+5$ oxidation states, see p. 1192) tungsten alone forms an oxoiodide, while only chromium (as yet) forms an oxofluoride in the lower of these oxidation states.

All the known hexahalides can be prepared by the direct action of the halogen on the metal and all are readily hydrolysed. The yellow CrF_6, however, requires a temperature of 400°C and a pressure of 200–300 atms for its formation, and reduction of the pressure causes it to dissociate into CrF_5 and F_2 even at temperatures as low as -100°C. The monomeric and octahedral hexafluorides MoF_6 and WF_6 are colourless liquids and the former is strongly oxidizing. Only tungsten is known with certainty to produce other hexahalides and these are the dark-blue solids WCl_6 and WBr_6, the latter in particular being susceptible to reduction.

Of the pentahalides, chromium again forms only the fluoride which is a strongly oxidizing, bright red, volatile solid prepared from the elements using less severe conditions than for CrF_6. MoF_5 and WF_5 can be prepared by reduction of the hexahalides with the metal but the latter dissociates into WF_6 and WF_4 if heated above about 80°C. They are yellow volatile solids, isostructural with the tetrameric $(NbF_5)_4$ and $(TaF_5)_4$ (Fig. 22.3b, p. 1153). Similarity with Group VA is again evident in the pentachlorides of Mo and W, $MoCl_5$ being the most extensively studied of the pentahalides. These, respectively, black and dark-green solids are obtained by direct reaction of the elements under carefully controlled conditions and have the same dimeric structure as their Nb and Ta analogues (Fig. 22.3c, p. 1153). WBr_5 can be prepared similarly but is not yet well characterized.

[21b] *J. Less Common Metals* **36**, 277–404 (1974). A series of papers on catalysis presented to the 1st International Conference on the Chemistry and Uses of Molybdenum at Reading, 1973.

[22] R. COLTON and J. H. CANTERFORD, *Halides of the First Row Transition Metals*, Chap. 4, pp. 161–211, Wiley, London, 1969; and *Halides of the Second and Third Row Transition Metals*, Chap. 6, pp. 206–72, Wiley, London, 1968.

TABLE 23.4　*Halides of Group VIA (mp/°C)*

Oxidation state	Fluorides	Chlorides	Bromides	Iodides
+6	CrF_6 yellow (d > -100°)			
	MoF_6 colourless (17.4°) bp 34°	$(MoCl_6)$ black		
	WF_6 colourless (1.9°) bp 17.1°	WCl_6 dark blue (275°) bp 346°	WBr_6 dark blue (309°)	
+5	CrF_5 red (34°) bp 117°			
	MoF_5 yellow (67°) bp 213°	$MoCl_5$ black (194°) bp 268°		
	WF_5 yellow	WCl_5 dark green (242°) bp 286°	WBr_5 black	
+4	CrF_4 green (277°)	$CrCl_4$ (d > 600°, gas phase)	$CrBr_4$?	CrI_4
	MoF_4 pale green	$MoCl_4$ black	$MoBr_4$ black	MoI_4?
	WF_4 red-brown	WCl_4 black	WBr_4 black	WI_4?
+3	CrF_3 green (1404°)	$CrCl_3$ red-violet (1150°)	$CrBr_3$ very dark green (1130°)	CrI_3 very dark green
	MoF_3 brown (>600°)	$MoCl_3$ very dark red (1027°)	$MoBr_3$ green (977°)	MoI_3 black (927°)
		WCl_3 red	WBr_3 black (d > 80°)	WI_3
+2	CrF_2 green (894°)	$CrCl_2$ white (820°)	$CrBr_2$ white (842°)	CrI_2 red-brown (868°)
		$MoCl_2$ yellow (d > 530°)	$MoBr_2$ yellow-red (d > 900°)	MoI_2
		WCl_2 yellow	WBr_2 yellow	WI_2 brown

The tetrahalides are scarcely more numerous or familiar than the hexa- and penta-halides, the 3 tetraiodides together with $CrBr_4$ and $CrCl_4$ being either of uncertain existence or occurring only at high temperatures in the gaseous phase. The most stable representatives are the fluorides: CrF_4 is an unreactive solid which melts without decomposition at 277°; MoF_4 is an involatile green solid; and WF_4 begins to decompose only when heated above 800°. $MoCl_4$ exists in two crystalline modifications: α-$MoCl_4$ is probably made up of linear chains of edge-shared octahedra, whereas β-$MoCl_4$ has recently been found to have a unique structure comprised of hexameric cyclic molecules $(MoCl_4)_6$ generated by edge-shared $\{MoCl_6\}$ octahedra with Mo-Cl$_t$ 220 pm, Mo-Cl$_\mu$ 243 and 251 pm and Mo$\cdots$Mo 367 pm.[22a] General preparative methods include controlled reaction of the elements, reduction of higher halides, and halogenation of lower halides. The tetrahalides of Mo and W are readily oxidized and hydrolysed and produce some adducts of the form MX_4L_2.

The trihalides show major differences between the 3 metals. All 4 of the chromium trihalides are known, this being much the most stable oxidation state for chromium; they can be prepared by reacting the halogen and the metal, though CrF_3 is better obtained from HF and $CrCl_3$ at 500°C. The fluoride is green, the chloride red-violet, and the bromide and iodide dark green to black. In all cases layer structures lead to octahedral coordination of the metal. $CrCl_3$ consists of a ccp lattice of chloride ions with Cr^{III} ions occupying two-thirds of the octahedral sites of alternate layers. The other alternate layers of octahedral sites are empty and, without the cohesive effect of the cations, easy cleavage in these planes is possible and this accounts for the flaky appearance. Stable, hydrated forms of CrX_3 can also be readily obtained from aqueous solutions, and $CrCl_3 . 6H_2O$ provides a well-known example of hydrate isomerism, mentioned on p. 1080. In view of this clear ability of Cr^{III} to aquate it may seem surprising that anhydrous $CrCl_3$ is quite insoluble in pure water (though it dissolves rapidly on the addition of even a trace of a reducing agent). It appears that the reducing agent produces at least some Cr^{II} ions. Solubilization then follows as a result of electron transfer from $[Cr(aq)]^{2+}$ in solution via a chloride bridge to Cr^{III} in the solid, which leaves $[Cr(aq)]^{3+}$ in solution and Cr^{II} in the solid. The latter is kinetically far more labile than Cr^{III} and can readily leave the solid and aquate, so starting the cycle again and rapidly dissolving the solid.

The Mo trihalides are obtained by reducing a higher halide with the metal (except for the triiodide which, being the highest stable iodide, is best prepared directly). They are insoluble in water and generally inert. $MoCl_3$ is structurally similar to $CrCl_3$ but is distorted so that pairs of Mo atoms lie only 276 pm apart which, in view of the low and temperature-dependent magnetic moment, is evidently close enough to permit appreciable Mo–Mo interaction. Electrolytic reduction of a solution of MoO_3 in aqueous HCl changes the colour to green, then brown, and finally red, when complexes of the octahedral $[MoCl_6]^{3-}$, $[MoCl_5(H_2O)]^{2-}$, and $[Mo_2Cl_9]^{3-}$ can be isolated using suitable cations. The diversity of the coordination chemistry of molybdenum(III) is, however, in no way comparable to that of chromium(III).

By contrast, the tungsten trihalides (the trifluoride is not known) are "cluster" compounds similar to those of Nb and Ta. The trichloride and tribromide are prepared by halogenation of the dihalides. The structure of the former is based on the $[M_6X_{12}]^{n+}$ cluster (Fig. 22.5) with a further 6 Cl atoms situated above the apical W atoms. WBr_3, on the other hand, has a structure based on the $[M_6X_8]^{n+}$ cluster (see Fig. 23.6), but as it is formed by only a 2-electron oxidation of $[W_6Br_8]^{4+}$ it does not contain tungsten(III) and is best formulated as $[W_6Br_8]^{6+} (Br_4{}^{2-})(Br^-)_2$, where $(Br_4{}^{2-})$ represents a bridging polybromide group. Electrolytic reduction of WO_3 in aqueous HCl fails to produce the mononuclear complexes obtained with molybdenum, but forms the green $[W_2Cl_9]^{3-}$ ion. This and its Cr and Mo analogues provide an interesting reflection of the increasing tendency to form cluster compounds in the order $Cr^{III} < Mo^{III} < W^{III}$. The structure

[22a] U. MÜLLER, Hexameric molybdenum tetrachloride, *Angew. Chem.* Int. Edn. (Engl.) **20**, 692–3 (1981).

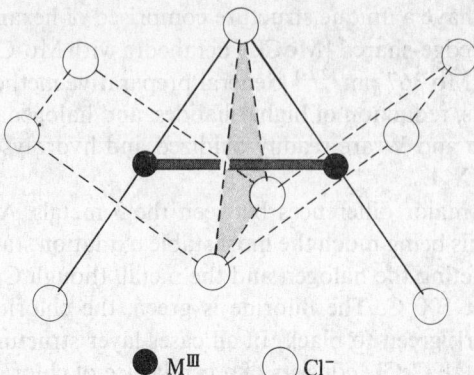

$\bullet$ M^{III} $\bigcirc$ Cl$^-$

FIG. 23.5   Structure of $[M_2Cl_9]^{3-}$ showing the M–M bond through the shared face of two inclined MCl$_6$ octahedra. See also Fig. 7.8, p. 270, for an alternative representation of the confacial bioctahedral structure.

consists of 2 MCl$_6$ octahedra sharing a common face (Fig. 23.5) which allows the possibility of direct M–M bonding. In the Cr ion the Cr atoms are 312 pm apart, being actually displaced in their CrO$_6$ octahedra away from each other. The magnetic moment of $[Cr_2Cl_9]^{3-}$ is normal for a metal ion with 3 unpaired electrons and indicates the absence of Cr–Cr bonding. In $[Mo_2Cl_9]^{3-}$ the Mo atoms are 267 pm apart and the magnetic moment is low and temperature dependent, indicating appreciable Mo–Mo bonding. Finally, $[W_2Cl_9]^{3-}$ is diamagnetic: the metal atoms are displaced towards each other, being only 242 pm apart (compared to 274 pm in the metal itself), clearly indicating stronger W–W bonding.

Anhydrous chromium dihalides are conveniently prepared by reduction of the trihalides with H$_2$ at 300–500°C, or by the action of HX (or I$_2$ for the diiodide) on the metal at temperatures of the order of 1000°C. They are all deliquescent and the hydrates can be obtained by reduction of the trihalides using pure chromium metal and aqueous HX. All have distorted octahedral structures as anticipated for a metal ion with the d^4 configuration which is particularly susceptible to Jahn–Teller distortion. This is typified by CrF$_2$, which adopts a distorted rutile structure in which 4 fluoride ions are 200 pm from the chromium atom while the remaining 2 are 243 pm away. The strongly reducing properties of chromium(II) halides contrast, at first sight surprisingly, with the redox stability of the molybdenum(II) halides. Even the tungsten(II) halides, which admittedly are also strong reducing agents (being oxidized to their trihalides), may by their very existence be thought to depart from the expected trend.

Of the various preparative methods available for the dihalides of Mo and W, thermal decomposition or reduction of higher halides is the most general. The reason for their enhanced stability lies in the prevalence of metal–atom clusters, stabilized by M–M bonding. All 6 of these dihalides (Mo and W do not form difluorides) are isomorphous, with a structure based on the $[M_6X_8]^{4+}$ unit briefly mentioned above for WBr$_3$. It can be seen (Fig. 23.6) that in this cluster each metal atom has a free coordination position. These 6 positions can be occupied by the counter ions X$^-$, providing that for each $[M_6X_8]^{4+}$ unit 4 of them bridge to the other such units. The precise details of the bonding scheme are not settled but, since all these compounds are diamagnetic, it is clear that in each cluster

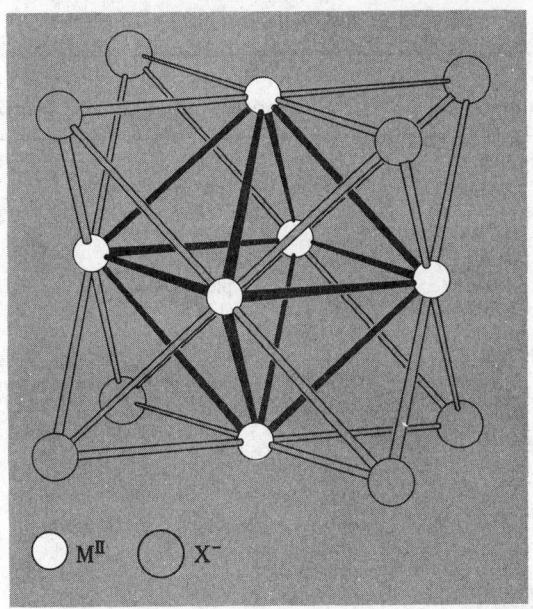

FIG. 23.6 $[M_6X_8]^{4+}$ clusters with X bridges over each face of the octahedron of metal ions.

the 6 metals contribute $6 \times 6 = 36$ valence electrons of which 4 are transferred to the counter anions, so producing the net charge, and 8 are used in bonding to the 8 chlorines of the cluster. Twenty-four electrons remain which can provide M–M bonds along each of the 12 edges of the octahedron of metal atoms. The observed chemical reactions, through a variety of which the $[M_6X_8]^{4+}$ persists as a stable hexanuclear centre, are entirely consistent with this.

The oxohalides of all three elements (Table 23.5) are very susceptible to hydrolysis and their oxidizing properties decrease in the order $Cr > Mo > W$. They are yellow to red liquids or volatile solids; probably the best known is the deep-red liquid, chromyl chloride, CrO_2Cl_2. It is most commonly encountered as the distillate in qualitative tests for chromium or chloride and can be obtained by heating a dichromate and chloride in conc H_2SO_4; it is an extremely aggressive oxidizing agent. The Mo and W oxohalides are prepared by a variety of oxygenation and halogenation reactions which frequently produce mixtures, and many specific preparations have therefore been devised.[23] They are possibly best known as impurities in preparations of the halides from which air or moisture have been inadequately excluded, and their formation is indicative of the readiness with which metal–oxygen bonds are formed by these elements in high oxidation states.

23.3.7   *Complexes of chromium, molybdenum, and tungsten* [4, 5]

Oxidation state VI (d⁰)

No halo complexes of the type $[MX_{6+x}]^{x-}$ are known and, apart from adducts of the oxohalides such as $[MoO_2F_4]^{2-}$ and $[WOCl_5]^-$, the coordination chemistry of this

²³ Ref. 3, pp. 275–81.

TABLE 23.5 *Oxohalides of Group VIA (mp/°C)*

Oxidation state	Fluorides		Chlorides		Bromides		Iodides
+6	CrOF$_4$ red (55°)	CrO$_2$F$_2$ violet (32°)		CrO$_2$Cl$_2$ red (−96.5°) bp 117°		CrO$_2$Br$_2$ red (d < rt)	
	MoOF$_4$ white (97°) bp 186°	MoO$_2$F$_2$ white (subl 270°)	MoOCl$_4$ green (101°) bp 159°	MoO$_2$Cl$_2$ pale yellow (175°) bp 250°		MoO$_2$Br$_2$ purple-brown	
	WOF$_4$ white (101°) bp 186°	WO$_2$F$_2$ white	WOCl$_4$ red (209°) bp 224°	WO$_2$Cl$_2$ pale yellow (265°)	WOBr$_4$ dark brown or black (321°)	WO$_2$Br$_2$ red (277°)	WO$_2$I$_2$
+5	(CrOF$_3$)		CrOCl$_3$ dark red				
			MoOCl$_3$ black (d > 200°)	MoO$_2$Cl	MoOBr$_3$ black (subl 270° vac)		
			WOCl$_3$ olive green		WOBr$_3$ dark brown		WO$_2$I
+3			CrOCl green		CrOBr		

oxidation state is centred mainly on oxomolybdenum compounds and the peroxo complexes of all three metals. The former class, along with similar complexes of Mo^V and Mo^{IV}, are of interest because of their possible role in the catalytic behaviour of molybdenum. Most are apparently octahedral chelates of the type [Mo^{VI}O$_2$(L-L$^-$)$_2$] with the oxygens mutually *cis*, thereby maximizing O(p$_\pi$)→Mo(d$_\pi$) bonding, and in which the MoO$_2$$^{2+}$ group is reminiscent of the uranyl UO$_2$$^{2+}$ ion (p. 1478), though its chemistry is by no means as extensive and the latter is a linear ion. The best-known example of this type of compound is [MoO$_2$(oxinate)$_2$] used for the gravimetric determination of molybdenum; oxine is 8-hydroxyquinoline, i.e.

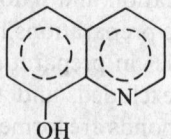

The peroxo-complexes provide further examples of the ability of oxygen to coordinate to the metals in their high oxidation states. The production of blue solutions when acidified dichromates are treated with H$_2$O$_2$ is a qualitative test for chromium.† The colour arises from the unstable CrO$_5$ which can, however, be stabilized by extraction into ether, and blue solid adducts such as [CrO$_5$(py)] can be isolated. This is more correctly formulated as [CrO(O$_2$)$_2$py] and has an approximately pentagonal pyramidal structure

† The acidity is important. In alkaline solution [Cr^V(O$_2$)$_4$]$^{3-}$ is produced, but from neutral solutions explosive violet salts, probably containing [Cr^{VI}O(O$_2$)$_2$OH]$^-$, are produced.

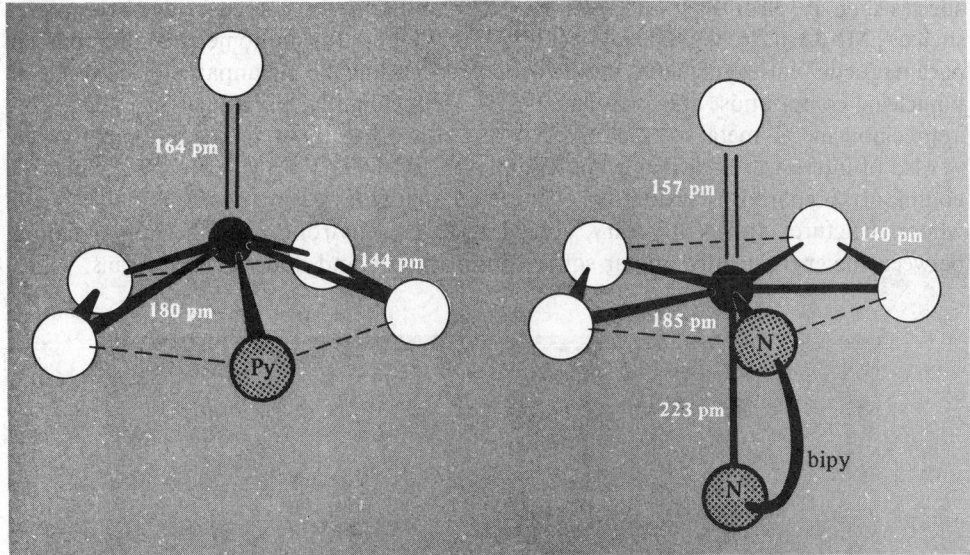

FIG. 23.7   Molecular structures of (a) [CrO(O$_2$)$_2$py] and (b) [CrO(O$_2$)$_2$(bipy)].

(Fig. 23.7a). Bidentate ligands, such as phenanthroline and bipyridyl produce pentagonal bipyramidal complexes in which the second N-donor atom is loosely bonded *trans* to the =O (Fig. 23.7b). This 7-coordinate structure is favoured in numerous peroxo-complexes of Mo and W, and the dark-red peroxo anion [Mo(O$_2$)$_4$]$^{2-}$ is 8-coordinate, with Mo–O 197 pm and O–O 155 pm.

Oxidation state V (d^1)

This is an unstable state for chromium and, apart from the fluoride and oxohalides already mentioned, it is represented primarily by the blue to black chromates of the alkali and alkaline earth metals and the red-brown tetraperoxochromate(V). The former contain the tetrahedral [CrO$_4$]$^{3-}$ ion and hydrolyse with disproportionation to Cr(III) and Cr(VI). The latter can be isolated as rather more stable salts from alkaline solutions of dichromate treated with H$_2$O$_2$. These red salts contain the paramagnetic 8-coordinate, dodecahedral, [Cr(O$_2$)$_4$]$^{3-}$ ion, which is isomorphous with the corresponding complex ions of the Group VA metals (p. 1157). The η^2-O$_2$ groups are unsymmetrically coordinated, with Cr–O 185 and 195 pm and the O–O distance 141 pm.

The heavier elements have a much more extensive +5 chemistry including, in the case of molybdenum, a number of compounds of considerable biological interest which will be discussed separately (p. 1205). A variety of reactions involving fusion and nonaqueous solvents have been used to produce octahedral hexahalo complexes. These are very susceptible to hydrolysis, and the affinity of Mo^V for oxygen is further demonstrated by the propensity of MoCl$_5$ to produce green oxomolybdenum(V) compounds by oxygen-abstraction from appropriate oxygen-containing materials. This leads to a number of well-characterized complexes of the type [MoOCl$_3$L] and [MoOCl$_3$L$_2$]. Oxomolyb-denum(V) compounds are also obtained from aqueous solution and include monomeric species such as $[MoOX_5]^{2-}$ (X = Cl, Br, NCS) and dimeric, oxygen-bridged complexes such as $[Mo_2O_4(C_2O_4)_2(H_2O)_2]^{2-}$ (Fig. 23.8). Whereas the monomeric compounds are paramagnetic with magnetic moments corresponding to 1 unpaired electron, the binuclear compounds are diamagnetic, or only slightly paramagnetic, suggesting appreciable metal–metal interaction occurring either directly or via the bridging oxygen.

Also of interest are the octacyano complexes, $[M(CN)_8]^{3-}$ (M = Mo, W), which are commonly prepared by oxidation of the M^{IV} analogues (using MnO_4^- or Ce^{IV}) and whose structures apparently vary, according to the environment and counter cation, between the energetically similar square-antiprismatic and dodecahedral forms.

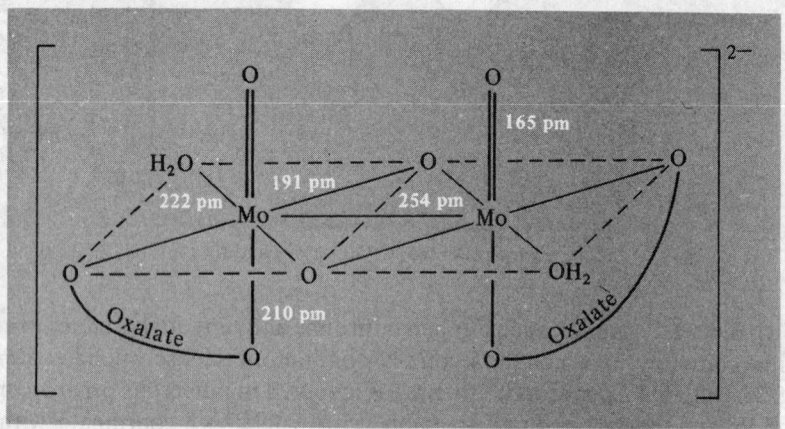

FIG. 23.8   The dimeric, oxygen bridged, $[Mo_2O_4(C_2O_4)_2(H_2O)_2]^{2-}$ showing the close approach of the 2 Mo atoms and unusually large range of Mo–O distances from 165 to 222 pm.

Oxidation state IV (d²)

This is not dissimilar to the previous oxidation state. Chromium is again represented by peroxo- and fluoro-complexes, while octacyano complexes are important for Mo and W which also yield some oxo complexes.

Treatment of the red Cr^V peroxo complex, $[Cr(O_2)_4]^{3-}$, with warm aqueous ammonia gives the Cr^{IV} complex, $[Cr(O_2)_2(NH_3)_3]$, which has a dark red-brown metallic lustre. The compound can also be made by the action of H_2O_2 on ammoniacal solutions of $(NH_4)_2CrO_4$ and the NH_3 molecules can be replaced by other ligands. The structures of these complexes are pentagonal bipyramidal with the peroxo-groups filling 4 of the planar positions. Green Ba_3CrO_5 is also a compound of Cr^{IV}.

Direct fluorination of anhydrous $CrCl_3$ and an alkali-metal chloride produces the very hydrolysable salts of $[CrF_6]^{2-}$. Other, more or less hydrolysable, hexahalo salts of $[MX_6]^{2-}$ (M = Mo, X = F, Cl, Br; M = W, X = Cl, Br) are also known, but without doubt the most extensively studied compounds of this oxidation state are the yellow octacyano complexes. These are prepared by aerial oxidation of aqueous Mo^{III} and W^{III} solutions in the presence of KCN, and are of interest because of their structures. The classical work of J. L. Hoard[24] in 1939 on $K_4[Mo(CN)_8] \cdot 2H_2O$ was the first authentication of an 8-

[24] J. L. HOARD and H. H. NORDSIECK, The structure of potassium molybdocyanide dihydrate, *J. Am. Chem. Soc.* **61**, 2853–63 (1939).

coordinate complex. It was shown to have a dodecahedral structure and is diamagnetic. It is not a simple matter to predict the precise splitting of d orbitals caused by this arrangement of ligands but it is agreed that it will be such as to stabilize one of the d orbitals (probably the d_{xy}) more than the others. This facilitates the pairing of the two d electrons and accounts for the diamagnetism. In aqueous solution, infrared and Raman data have been interpreted on the basis of a square-antiprismatic structure, and the ^{13}C nmr spectrum shows only a single band which is indicative of 8 equivalent CN groups. This could, however, arise from rapid intramolecular (fluxional) rearrangement of a dodecahedral structure which, in view of the very slight energy barrier between the two structures, is a possibility which cannot be dismissed. Photolysis of the otherwise stable $[M(CN)_8]^{4-}$ solutions causes loss of four CN^- ions to give octahedral oxo compounds such as $K_4[MO_2(CN)_4].6H_2O$ (M = Mo, W).

Yellow, air-stable carboxylate complexes of W^{IV} have recently been prepared by the reaction of $W(CO)_6$ with carboxylic acids. They have been shown[25] to be based on a central W_3O_2 unit in which 3 tungsten atoms are singly bonded to each other to form a triangle, with 1 oxygen above and 1 below the triangle (Fig. 23.9). This appears to be a

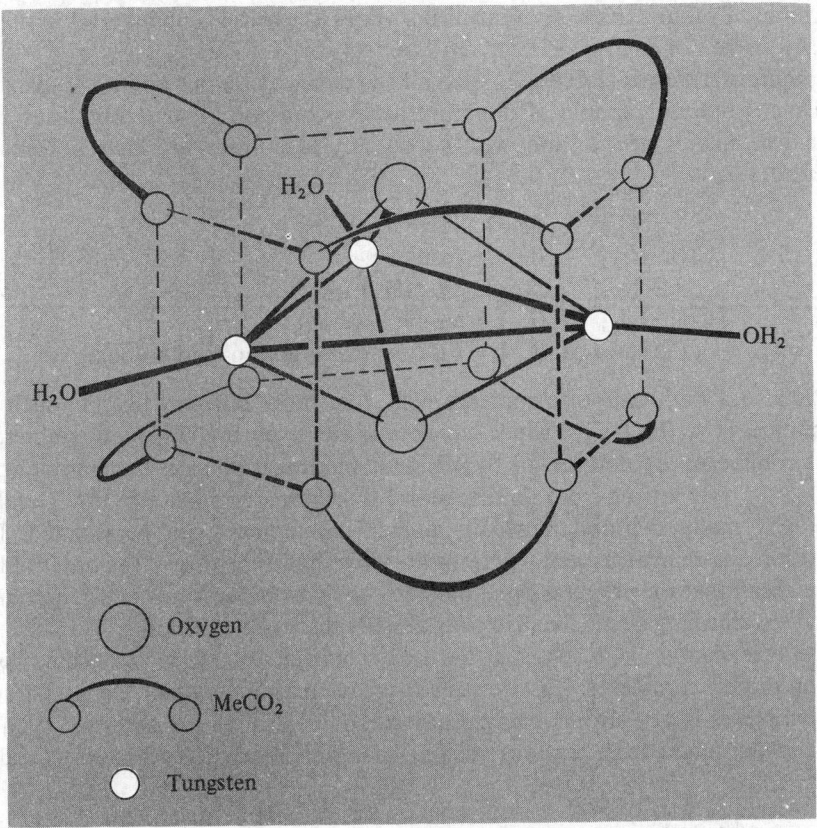

Fig. 23.9 $[W_3O_2(O_2CMe)_6(H_2O)_3]^{2+}$

[25] A. BINO, F. A. COTTON, Z. DORI, S. KOCH, H. KÜPPERS, M. MILLAR, and J. C. SEKUTOWSKI, A new class of trinuclear tungsten(IV) cluster compounds with W–W single bonds, *Inorg. Chem.* **17**, 3245–53 (1978).

unique arrangement and contrasts with the trinuclear M^{III} and dinuclear M^{II} carboxylates discussed below, which tungsten is not with certainty known to form.

Oxidation state III (d^3)

This is by far the most stable and best-known oxidation state for chromium[26] and is characterized by thousands of compounds, most of them prepared from aqueous solutions.[27] By contrast molybdenum(III) compounds are sparse and hardly any are known for tungsten(III). Thus Mo, but not W, gives rise to complexes $[MoX_6]^{3-}$ ($X = F$, Cl, Br, NCS), and direct action of acetylacetone on the hexachloromolybdate(III) ion produces the sublimable $[Mo(acac)_3]$ which, however, unlike its chromium analogue, is oxidized by air to Mo^V products. A black Mo^{III} cyanide, $K_4Mo(CN)_7 . 2H_2O$, has been precipitated from aqueous solution by the addition of ethanol. Its magnetic moment (~ 1.75 BM) is consistent with 7-coordinate Mo^{III} in which the loss of degeneracy of the t_{2g} orbitals has caused pairing of 2 of the three d electrons. The purple, and unusually air-stable, $Cs_2[Mo_2(HPO_4)_4(H_2O)_2]$ has recently been prepared[28] by the action of $K_4MoCl_8 . 2H_2O$ and CsCl in aqueous H_3PO_4. The cation has the dinuclear structure more commonly found in the divalent carboxylates (see below) and involves in this case a $Mo \equiv Mo$ triple bond.

Chromium(III) forms stable salts with all the common anions and it complexes with virtually any species capable of donating an electron-pair. These complexes may be anionic, cationic, or neutral and, with hardly any exceptions, are hexacoordinate and octahedral, e.g.:

$$[CrX_6]^{3-} \quad (X = \text{halide, CN, SCN, N}_3)$$
$$[Cr(L-L)_3]^{3-} \quad (L-L = \text{oxalate})$$
$$[CrX_6]^{3+} \quad (X = H_2O, NH_3)$$
$$[Cr(L-L)_3]^{3+} \quad (L-L = \text{en, bipy, phen})$$
$$[Cr(L-L)_3] \quad (L-L = \beta\text{-diketonates, amino-acid anions})$$

There is also a multitude of complexes with 2 or more different ligands, such as the pentammines $[Cr(NH_3)_5X]^{n+}$ which have been extensively used in kinetic studies. These various complexes are notable for their kinetic inertness, which is compatible with the half-filled t_{2g} level arising from an octahedral d^3 configuration (see p. 1097), and is the reason why many thermodynamically unstable complexes can be isolated. Ligand substitution and rearrangement reactions are slow (half-times are of the order of hours), with the result that the preparation of different, solid, isomeric forms of a compound as a means of establishing stereochemistry (as already discussed in Chapter 19) is frequently possible. It is mainly for this reason that early coordination chemists devoted so much attention to Cr^{III} complexes. For precisely the same reason, however, the preparation of these complexes is not always straightforward. Salts such as the hydrated sulfate and halides, which might seem obvious starting materials, themselves contain coordinated

[26] J. E. EARLEY and R. D. CANNON, Aqueous chemistry of chromium(III), *Transition Metal Chem.* **1**, 33–109 (1966).

[27] I. D. BROWN, Stereochemistry of chromium complexes: a bibliographic history, *Coord Chem. Revs.* **26**, 161–206 (1978). A non-critical list of Cr compounds whose structures have been determined by X-ray diffraction.

[28] A. BINO and F. A. COTTON, An easily prepared, air-stable compound with a triple metal-to-metal (MoMo) bond, *Angew. Chem.*, Int. Edn. (Engl.) **18**, 462–3 (1979).

water or anions and these are not always easily displaced. Simple addition of the appropriate ligand to an aqueous solution of a Cr^{III} salt is therefore not a usual preparative method, though in the presence of charcoal it is feasible in the case, for instance, of $[Cr(en)_3]^{3+}$. Some alternative routes, which avoid these pre-formed inert complexes, are:

(i) *Anhydrous methods*: ammine and amine complexes can be prepared by the reaction of CrX_3 with NH_3 or amine, and salts of $[CrX_6]^{3-}$ anions are best obtained by fusion of CrX_3 with the alkali metal salt.

(ii) *Oxidation of Cr(II)*: ammine and amine complexes can also be prepared by the aerial oxidation of mixtures of aqueous $[Cr(H_2O)_6]^{2+}$ (which is kinetically labile) and the appropriate ligands.

(iii) *Reduction of Cr(VI)*: CrO_3 and dichromates are commonly used to prepare such complexes as $K_3[Cr(C_2O_4)_3]$ and $NH_4[Cr(NH_3)_2(NCS)_4].H_2O$ (Reinecke's salt).

The violet hexaquo ion, $[Cr(H_2O)_6]^{3+}$, occurs in the chrome alums, $Cr_2(SO_4)_3$-$M_2^I SO_4.24H_2O$ (e.g. $[K(H_2O)_6][Cr(H_2O)_6][SO_4]_2$), but in hydrated salts and aqueous solutions, green species, produced by the replacement of some of the water molecules by other ligands, are more usual. So, the common form of the hydrated chloride is the dark-green *trans*-$[CrCl_2(H_2O)_4]Cl.2H_2O$, and other isomers are known (see p. 1080).

Chromium(III) is the archetypal d^3 ion and the electronic spectra and magnetic properties of its complexes have therefore been exhaustively studied. In an octahedral field, the splitting of the free-ion ground F term, along with the presence of the excited P term of the same multiplicity, provides the possibility of 3 spin-allowed d–d transitions (see Table 19.3, p. 1093). This is much the same situation as that found for the vanadium(III) ion, except that the splitting of the F term now produces a ground A instead of a T term. This has the simplifying effect that the crystal field splitting, Δ or $10Dq$, is given directly by the energy of the lowest transition. Furthermore, the magnetic moment is expected to be very close to the spin-only value for 3 unpaired electrons (3.87 BM) and also, because of the absence of any orbital contribution, to be independent of temperature. In practice, providing the compounds are mononuclear, these expectations are realized remarkably well apart from the fact that, as was noted for octahedral complexes of vanadium(III), the third high-energy band in the spectrum is usually wholly or partially obscured by more intense charge-transfer absorption. Data for a representative sample of complexes are given in Table 23.6.

TABLE 23.6   *Spectroscopic data for typical octahedral complexes of chromium(III)*

Complex	Colour	v_1/cm^{-1}	v_2/cm^{-1}	v_3/cm^{-1}	$10Dq/cm^{-1}$	B/cm^{-1}	$\mu_{rt}/BM^{(a)}$
$K[Cr(H_2O)_6]$ $[SO_4]_2.6H_2O$	Violet	17 400	24 500	37 800	17 400	725	3.84
$K_3[Cr(C_2O_4)_3].3H_2O$	Reddish-violet	17 500	23 900		17 500	620	3.84
$K_3[Cr(NCS)_6].4H_2O$	Purple	17 800	23 800		17 800	570	3.77
$[Cr(NH_3)_6]Br_3$	Yellow	21 550	28 500		21 550	650	3.77
$[Cr(en)_3]I_3.H_2O$	Yellow	21 600	28 500		21 600	650	3.84
$K_3[Cr(CN)_6]$	Yellow	26 700	32 200		26 700	530	3.87

(a) Room temperature value of μ_e.

The observed bands are assigned as:

$$\nu_1 = {}^4T_{2g}(F) \longleftarrow {}^4A_{2g}(F) = 10Dq$$
$$\nu_2 = {}^4T_{1g}(F) \longleftarrow {}^4A_{2g}(F)$$
$$\nu_3 = {}^4T_{1g}(P) \longleftarrow {}^4A_{2g}(F)$$

The internal consistency of this interpretation can be checked and a value obtained for B, the interelectronic repulsion parameter, by fitting the ratio ν_1/ν_2 to the appropriate "Tanabe–Sugano" diagram (Fig. 23.10).

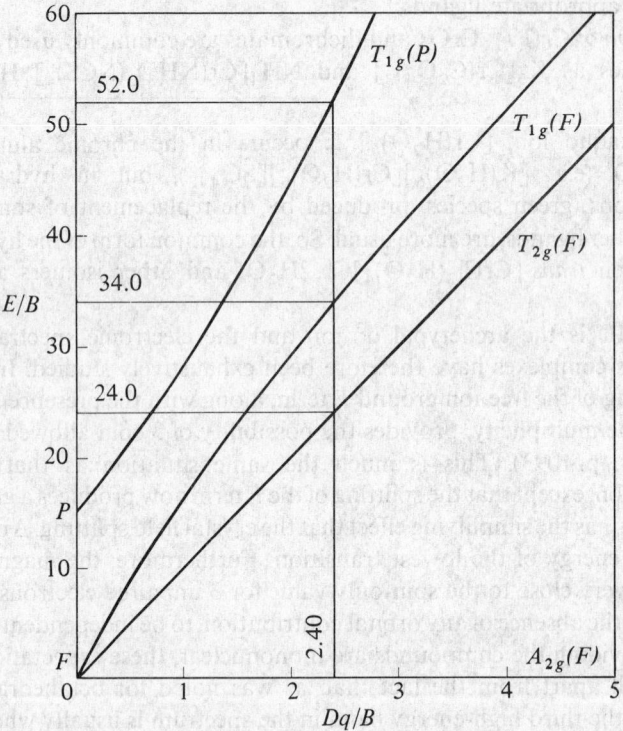

FIG. 23.10   Tanabe–Sugano diagram for d^3 ions in an octahedral CF.

Using $[Cr(H_2O)_6]^{3+}$ as an example, we have:

$$\frac{\nu_1}{\nu_2} = \frac{17\ 400}{24\ 500} = 0.710$$

This fits the T–S diagram at $Dq/B = 2.40$ when $\nu_1/B = 24.0$, $\nu_2/B = 34.0$, and $\nu_3/B = 52.0$. It follows that

$$10Dq = \nu_1 = 17\ 400\ \text{cm}^{-1}$$
$$B = \frac{17\ 400}{24.0} = 725\ \text{cm}^{-1}$$
$$\nu_3 = 52.0 \times 725 = 37\ 700\ \text{cm}^{-1}$$

There is indeed a shoulder on the high-energy charge transfer band at about 37 800 cm^{-1} which is in excellent agreement with the value calculated above. The value of B is below that of the free-ion (1030 cm^{-1}) because the expansion of d-electron charge on complexation reduces the interelectronic repulsions.

One of the most obvious characteristics of Cr^{III} is its tendency to hydrolyse and form polynuclear complexes containing OH$^-$ bridges. This is thought to occur by the loss of a proton from coordinated water, followed by coordination of the OH$^-$ so formed to a second cation:

$$[Cr(H_2O)_6]^{3+} \underset{+H^+}{\overset{-H^+}{\rightleftharpoons}} [Cr(H_2O)_5OH]^{2+}$$

$$[Cr(H_2O)_5OH]^{2+}: \quad [(H_2O)_4Cr \overset{OH}{\underset{OH}{<>}} Cr(H_2O)_4]^{4+} + 2H_2O$$

$$[Cr(H_2O)_6]^{3+}: \quad [(H_2O)_5Cr \overset{OH}{<>} Cr(H_2O)_5]^{5+} + H_2O$$

The ease with which the proton is removed can be judged by the fact that the hexaquo ion (p$K_a \sim 4$) is almost as strong an acid as formic acid. Further deprotonation and polymerization can occur and, as the pH is raised, the final product is hydrated chromium(III) oxide or "chromic hydroxide". Formation of this is the reason why amine complexes are not prepared by simple addition of the amine base to an aqueous solution of Cr^{III}. By methods which commonly start with Cr^{II}, binuclear compounds such as

$$[en_2Cr \overset{OH}{\underset{OH}{<>}} Cren_2]X_4$$

and $[(NH_3)_5Cr-OH-Cr(NH_3)_5]X_5$ are obtained. These have temperature-dependent magnetic moments, somewhat lower than those usual for octahedral Cr^{III}, but explicable on the assumption of partial spin-pairing via the bridging OH groups. However, the latter of these compounds can be converted to an oxo-bridged complex whose magnetic moment at room temperature is reduced still further to 1.3 BM (and virtually to zero below 100 K):[29]

$$[(NH_3)_5Cr-OH-Cr(NH_3)_5]^{5-} \underset{H^+}{\overset{OH^-}{\rightleftharpoons}} [(NH_3)_5Cr-O-Cr(NH_3)_5]^{4-}$$
$$\text{red} \qquad\qquad\qquad\qquad\qquad\qquad\qquad\qquad \text{blue}$$

The linear Cr–O–Cr bridge evidently permits pairing of the d electrons of the 2 metal atoms via d_π–p_π bonds, much more readily than the bent Cr–OH–Cr bridge.

Other interesting examples of an O atom providing a π pathway for the pairing of metal d electrons (but less effectively since the magnetic moments at room temperature are reduced only to about 2 BM) is found in a series of basic carboxylates of the general type $[Cr_3O(RCOO)_6L_3]^+$. These have the structure found in the carboxylates of other M^{III} ions and shown in Fig. 23.11.

[29] E. PEDERSON, Magnetic properties and molecular structure of the μ-oxo-bis{pentamminechromium(III)}ion, *Acta Chem. Scand.* **26**, 333–42 (1972).

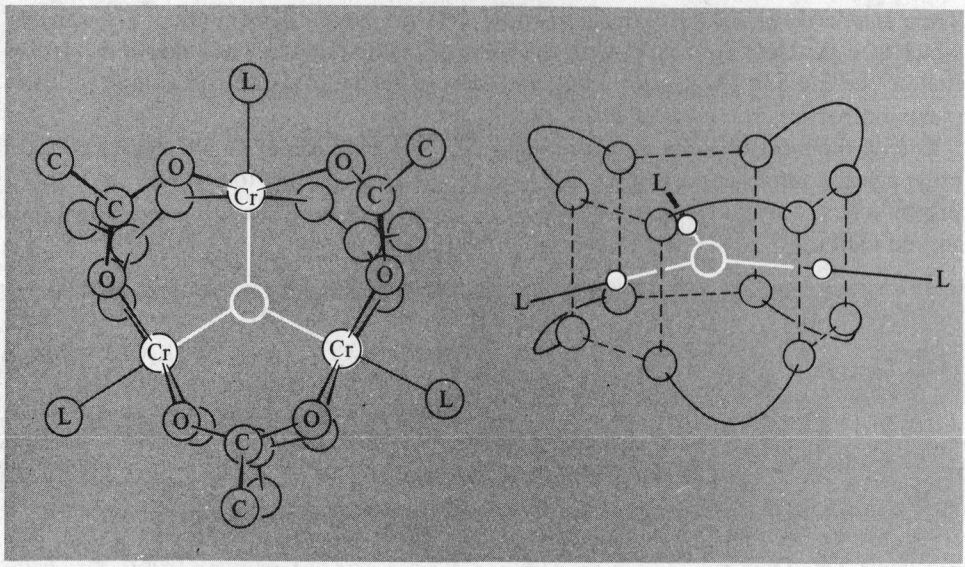

FIG. 23.11   Alternative representations of the trinuclear structure of $[Cr_3O(O_2CMe)_6L_3]^+$.

Hydrolysed, polynuclear Cr^{III} complexes are of considerable commercial importance in the dying and tanning industries. In the former the role is that of a mordant to the dye. After soaking in aqueous chrome alum or acetate, the fabric is steamed to precipitate colloidal hydrolysis products which will subsequently "fix" the dye. In leather production it is necessary to treat animal hides to prevent putrefaction and to render them supple when dry. This is done by using substances which have the ability to combine with the proteins in the gelatinous fibres of the hides. Traditionally, tannin was used, hence the name of the process, but this was superseded towards the end of the nineteenth century by solutions of chromium(III) sulfate. After soaking in sulfuric acid the hides are impregnated with the Cr^{III} solution. This is subsequently made alkaline, when the polynuclear complexes form and bridge neighbouring chains of proteins, presumably by coordinating to the carboxyl groups of the proteins.

Oxidation state II (d^4)

For chromium, this oxidation state is characterized by the aqueous chemistry of the strongly reducing Cr^{II} cation, and a noticeable tendency to form dinuclear compounds with multiple metal–metal bonds. The tendency to form such bonds is also very marked in the case of molybdenum but, perhaps surprisingly, is much less so in the case of tungsten,†

---

† A few compounds containing W–W quadruple bonds have been prepared recently[29a] and it may be that a major reason for their paucity is that the dinuclear acetate, which in the case of Mo is the most common starting material in the preparation of quadruply bonded dimeric complexes, is unknown for W.

[29a] P. R. SHARP and R. R. SCHROCK, Synthesis of a class of complexes containing tungsten–tungsten quadruple bonds, *J. Am. Chem. Soc.* **102**, 1430–1, (1980); F. A. COTTON, T. R. FELTHOUSE, and D. G. LAY, Structural characterization of four quadruply bonded ditungsten compounds and a new trinuclear tungsten cluster, *J. Am. Chem. Soc.* **102**, 1431–3 (1980).

though single M–M bonds are present in the $[M_6X_8]^{4+}$ clusters of the dihalides of both Mo and W (p. 1190). Due not only to the intrinsically interesting chemistry of the cluster and multiple-bonded species, but also to their potential value in the field of catalysis, there has been an effusion of publications in this field. This may be thought to have started in the late 1950s when the separate and stable identity of the $[Mo_6Cl_8]^{4+}$ species was recognized,[30] but has gained its main momentum in the 1970s.[31]

With the exception of the nitrate, which has not been prepared because of internal oxidation–reduction, the simple hydrated, sky-blue, salts of chromium(II) are best obtained by the reaction of the appropriate dilute acid with pure chromium metal, air being rigorously excluded. A variety of complexes are formed, especially with *N*-donor chelating ligands which commonly produce stoichiometries such as $[Cr(L–L)_3]^{2+}$, $[Cr(L–L)_2X_2]$, and $[Cr(L–L)X_2]$. They (and other complexes of Cr^{II}) are generally extremely sensitive to atmospheric oxidation if moist, but are considerably more stable when dry, probably the most air-stable of all being the pale-blue hydrazinium sulfate, $(N_2H_5^+)_2Cr^{II}(SO_4^{2-})_2$. In the solid state this consists[31a] of linear chains of Cr^{II} ions, bridged by SO_4^{2-} ions:

The majority of Cr^{II} complexes are octahedral and can be either high-spin or low-spin. The former are characterized by magnetic moments close to 4.90 BM and visible/ultraviolet spectra consisting typically of a broad band in the region of 16 000 cm^{-1} with another band around 10 000 cm^{-1}. Since a d^4 ion in a perfectly octahedral field can give rise to only one d–d transition (p. 1093) it is clear that some lowering of symmetry has occurred. Indeed, this is expected as a consequence of the Jahn–Teller effect (p. 1087), even when the metal is surrounded by 6 equivalent donor atoms. The splitting of the free-ion ^5D term is shown in Fig. 23.12 and the two observed bands are assigned to superimposed $^5B_{2g} \leftarrow {}^5B_{1g}$ and $^5E_g \leftarrow {}^5B_{1g}$ transitions and to the $^5A_{1g} \leftarrow {}^5B_{1g}$ transition respectively. The low-spin, intensely coloured compounds such as $K_4[Cr(CN)_6] \cdot 3H_2O$ and $[Cr(L–L)_3]X_2 \cdot nH_2O$ (L–L = bipy, phen; X = Cl, Br, I) have magnetic moments in the range 2.74–3.40 BM and electronic spectra showing clear evidence of extensive π bonding, as is to be expected with such ligands.

[30] J. C. SHELDON, Polynuclear complexes of molybdenum(II), *Nature* **184**, 1210–13 (1959).

[31] F. A. COTTON, Quadruple bonds and other multiple metal to metal bonds, *Chem. Soc. Rev.* **4**, 27–53 (1975); Discovering and understanding multiple metal–metal bonds, *Acc. Chem. Res.* **11**, 225–32 (1978); and (with M. H. CHISHOLM) Chemistry of compounds containing metal-to-metal triple bonds between molybdenum and tungsten, *Acc. Chem. Res.* **11**, 356–62 (1978). J. L. TEMPLETON, Metal-metal bonds of order four, *Prog. Inorg. Chem.* **26**, 211–300, (1979).

[31a] D. W. HAND and C. K. PROUT, The unit-cell dimensions of the metal hydrazinium sulphates $(N_2H_5)_2M^{II}(SO_4)_2$, *J. Chem. Soc. A*, 1966, 168–70.

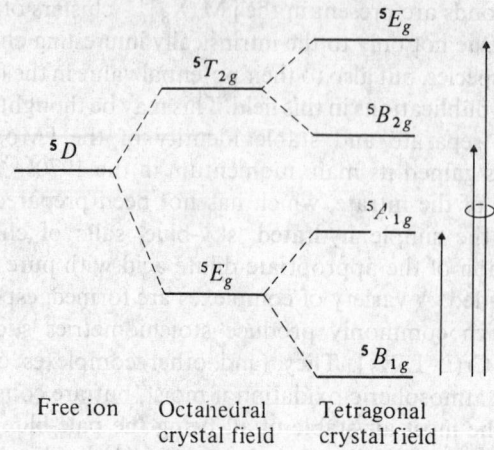

Fig. 23.12 Crystal field splitting of the ^5D term of a d^4 ion.

Although octahedral stereochemistry is certainly the most usual for Cr^{II} it is not the only one known. [CrI$_2$(OPPh$_3$)$_2$] is tetrahedral, and the quadridentate, "tripod" ligand N(CH$_2$CH$_2$NMe$_2$)$_3$ imposes trigonal bipyramidal geometry in [CrBr{N(CH$_2$CH$_2$NMe$_2$)$_3$}]Br as it does in analogous compounds of other divalent metal ions.

One of the best known of Cr^{II} compounds, and one which has often been used as the starting material in preparations of other Cr^{II} compounds, is the acetate. The red colour of the hydrated acetate is in sharp distinction to the blue of the simple salts—a contrast reflected in its dinuclear structure (Fig. 23.13a). The hydrated and anhydrous chromium(II) acetates are very nearly, but not quite, diamagnetic, which is indicative of appreciable metal–metal bonding. Stronger M–M bonding is evident in the completely diamagnetic [Mo$_2$(η^2-O$_2$CMe)$_4$] which is obtained by the action of acetic acid on [Mo(CO)$_6$]. Other carboxylates of Cr and Mo are similar. The same dinuclear structure, but with SO$_4^{2-}$ rather than acetate bridges, was suggested[32] for the violet, nearly diamagnetic, double sulfate, Cs$_2$SO$_4$.CrSO$_4$.2H$_2$O, in which case the formulation Cs$_4$[Cr$_2$(SO$_4$)$_4$(H$_2$O)$_2$].2H$_2$O would be appropriate. This compound is obtained by partial dehydration of the hexahydrate which, like the other double sulfates of Cr^{II}, is blue and high spin with a magnetic moment close to 4.9 BM. The plausibility of the dinuclear structure involving bridging SO$_4^{2-}$ groups was confirmed by the X-ray determination[33] of the structure of the pink compound, K$_4$[Mo$_2$(SO$_4$)$_4$].2H$_2$O, which is diamagnetic and does indeed have the structure of Fig. 23.13a. Attempted recrystallization produces K$_3$[Mo$_2$(SO$_4$)$_4$].3.5H$_2$O which retains the dinuclear skeleton but is paramagnetic and contains Mo in the formal oxidation +2.5. In view of these examples, and of the compound Cs$_2$[Mo$_2$(HPO$_4$)$_4$(H$_2$O)$_2$] previously noted,[28] it may well transpire that simple oxoanions produce the dinuclear structure more readily than once seemed likely.

[32] A. EARNSHAW, L. F. LARKWORTHY, K. C. PATEL, and G. BEECH, Chromium(II) chemistry. Part IV. Double sulphates, *J. Chem. Soc.* A 1969, 1334–9.

[33] F. A. COTTON, B. A. FRENZ, E. PEDERSON, and T. R. WEBB, Magnetic and electrochemical properties of transition metal complexes with multiple metal-to-metal bonds. III. Characterization of tetrapotassium and tripotassium tetrasulfatodimolybdates, *Inorg. Chem.* **14**, 391–8 (1975).

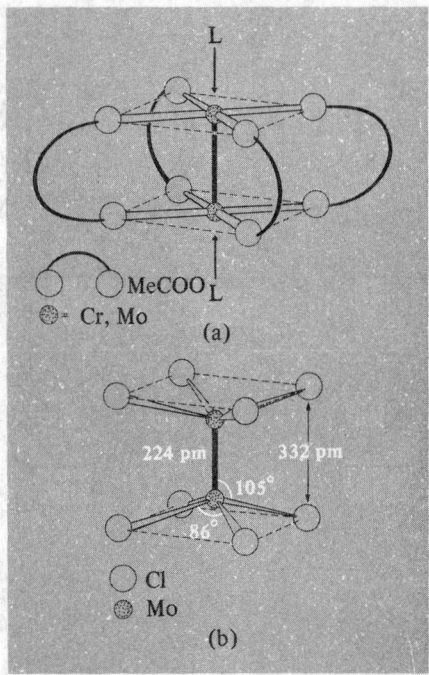

FIG. 23.13   (a) $[M_2(\eta^2\text{-}O_2CMe)_4]$, M = Cr. Mo. In the case of Cr, but not Mo, the hydrate and other adducts can be formed by attachment of H_2O (or in general, L) molecules as arrowed. (b) $[Mo_2Cl_8]^{4-}$.

The reaction of conc HCl and molybdenum acetate at 0°C produces the diamagnetic red anion $[Mo_2Cl_8]^{4-}$ (Fig. 23.13b) in which the 2 $MoCl_4$ are in the "eclipsed" orientation relative to each other and are held together solely by the Mo–Mo bond. At somewhat higher temperatures ($\sim 50°C$) the above reactants also produce the $[Mo_2Cl_8H]^{3-}$ ion[33a] which has the $[M_2^{III}Cl_9]^{3-}$ structure (Fig. 23.5) but with 1 of the bridging Cl atoms replaced by a H atom.

By now a large number of dinuclear compounds have been produced with a wide range of bridging groups, or like $[Mo_2Cl_8]^{4-}$ with no bridging groups at all, and not only for Cr^{II} and Mo^{II} but for the isoelectronic Re^{III} and Tc^{III} as well. Considerable attention has been devoted to the problem of how best to describe the M–M bonds, which are a central feature of these compounds, since the usual criteria by which bond order and oxidation state are judged are no longer wholly applicable. The best simple description of the d^4 systems is that shown in Fig. 23.14. The $d_{x^2-y^2}$ orbital is assumed to have been used in σ bonding to the ligands and the four d electrons on each metal atom are then used to form a M–M quadruple bond ($\sigma + 2\pi + \delta$). The eclipsed orientation noted in $[Mo_2Cl_8]^{4-}$ provides strong confirmation of this bonding scheme since, without the δ bond, this configuration would be sterically unfavourable. A variety of experimental techniques,

[33a] A. BINO and F. A. COTTON, A complex reaction product of dimolybdenum tetraacetate with aqueous hydrochloric acid. Structural characterization of the hydrido-bridged $[Mo_2Cl_8H]^{3-}$ ion, the $[Mo(O)Cl_4(H_2O)]^-$ ion, and an $H_5O_2^-$ ion with an exceptionally short hydrogen bond, *J. Am. Chem. Soc.* **101**, 4150–4 (1979).

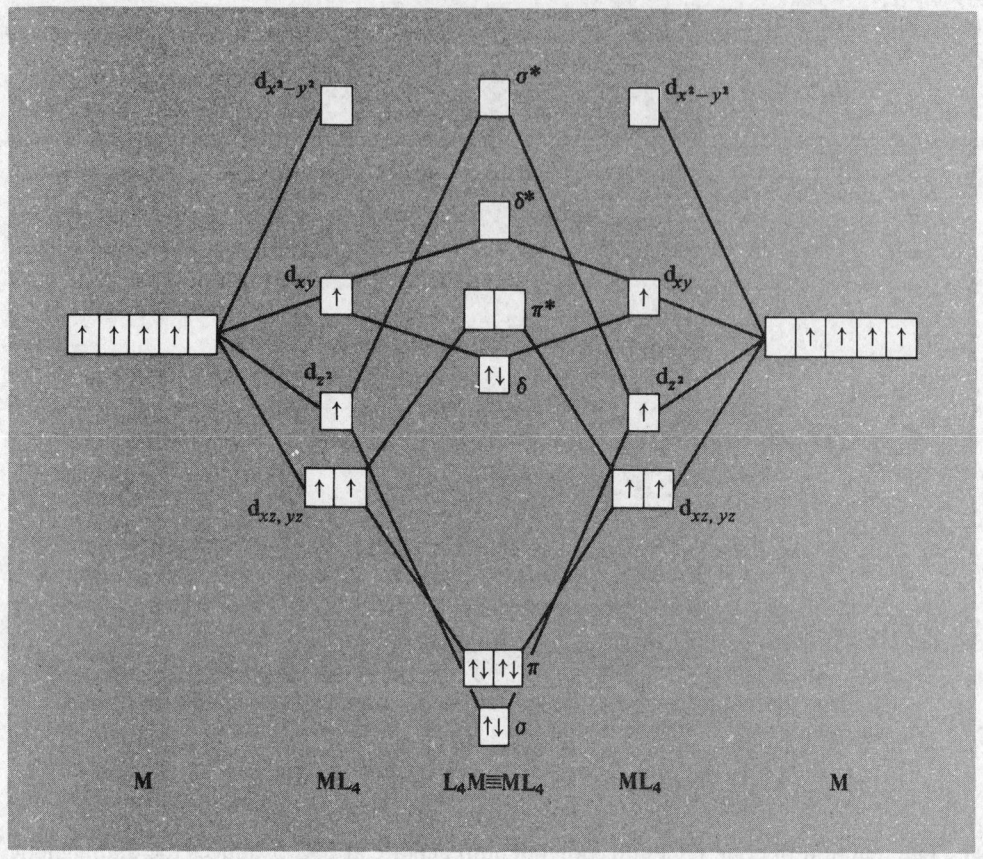

Fig. 23.14   Simplified MO diagram showing the formation of an M–M quadruple bond in M_2L_8 systems of d^4 metal ions (the d_{x2-y2}, along with p_x, p_y and s, orbitals of the metal ions are assumed to be used in the formation of M-L σ bonds).

which includes polarized single-crystal spectroscopy, photoelectron spectroscopy, and X-ray emission spectroscopy,[34] has been used to further substantiate this view of the bonding. Accurately determined M–M distances provide the most readily available indication of bond strength. Dichromium(II) compounds show the widest range of M–M distances (183–253 pm), whereas dimolybdenum(II) compounds cover only the range 203.7–218.3 pm.[34a] The shortest, and therefore presumably the strongest, bonds are found in those compounds with no axial ligands (i.e. like Fig. 23.13b, or a without the Ls). It is significant that $[Mo_2(\eta^2\text{-}O_2CMe)_4]$, which has an Mo–Mo distance of 211 pm, forms no adducts, whereas $[Cr_2(\eta^2\text{-}O_2CMe)_4]$, which in spite of the smaller size of the metal has a larger Cr–Cr distance of 228.8 pm, forms a hydrate and many other adducts. In order to

[34] D. E. HAYCOCK, D. S. URCH, C. D. GARNER, I. H. HILLIER, and G. R. MITCHESON, Direct evidence for the quadruple metal–metal bond in the octachlorodimolybdenum(II) ion, $[Mo_2Cl_8]^{4-}$, using X-ray emission spectroscopy, *JCS Chem. Comm.* 1978, 262–3.

[34a] F. A. COTTON, B. E. HANSON, W. H. ILSLEY, and G. W. RICE, Tetrakis(dimethylphosphonium dimethylide)-dichromium and -dimolybdenum. I. Crystal and molecular structures, *Inorg. Chem.* **18,** 2713–17 (1979).

effect acceptable comparisons between compounds of different metals, F. A. Cotton has introduced[35] the concept of the "formal shortness" (FS) ratio:

$$FS\ ratio = \frac{M\text{–}M\ distance\ in\ compound}{M\text{–}M\ distance\ in\ metal}$$

The compound with the shortest Cr–Cr distance, and indeed the shortest of all M–M distances, yet discovered (183.0 pm) is the yellow-orange, $Li_6[Cr_2(o\text{-}C_6H_4O)_4]$ $Br_2.Et_2O$. The central feature of this is the anion $[Cr_2(o\text{-}C_6H_4O)_4]^{4-}$ in which the bridging groups are:

The FS ratio† is the lowest yet found for any homonuclear bond. It is perhaps premature to regard this and similar Cr≡Cr bonds as being the strongest known but they are undoubtedly very strong. The M–M bonding discussed above may be regarded as "intramolecular antiferromagnetism". In contradistinction to this, weak ferromagnetism has been observed in a number of chloro and bromo complexes[36] of the type $M_2[CrX_4]$ (M = a variety of protonated amines and alkali metal cations, X = Cl, Br), which are analogous to previously known copper(II) complexes (p. 1384). They have magnetic moments at room temperature in the region of 6 BM (compared to 4.9 BM expected for magnetically dilute Cr^{II}) and these increase markedly as the temperature is lowered, the ferromagnetic interactions evidently being transmitted via Cr–Cl–Cr bridges. The electronic spectra consist of the usual absorptions expected for tetragonally distorted octahedral complexes of Cr^{II} but with two sharp and intense bands characteristically superimposed at higher energies (around 15 500 and 18 500 cm^{-1}). These are ascribed to spin-forbidden transitions intensified by the magnetic exchange.

Complexes in which the metal exhibits still lower oxidation states (such as I, 0, −I, −II) occur amongst the organometallic compounds (pp. 1172 and 1207).

23.3.8 *Biological activity and nitrogen fixation*

Tungsten has no known biological role but it appears that chromium(III) is an essential trace element in mammalian metabolism and, together with insulin, is responsible for the clearance of glucose from the blood-stream. However, from the point of view of biological activity, the focus of interest in this group is unquestionably on the role of molybdenum in

---

† 183.0/256 = 0.715 compared to 110/140 = 0.786 and 120.6/154 = 0.783 for N≡N and C≡C, respectively, which are the strongest homonuclear bonds for which bond energies are known.

35 F. A. COTTON, S. KOCH, and M. MILLAR, Exceedingly short metal-to-metal multiple bonds, *J. Am. Chem. Soc.* **99**, 7372–4 (1977).

36 L. F. LARKWORTHY and A. YAVARI, Ferromagnetic monoalkylammonium tetrabromochromates(II), *JCS Chem. Comm.* 1977, 172.

nitrogen fixation.[37] Estimates of the amount of nitrogen fixed biologically[38] range from 90×10^6 to 175×10^6 tonnes pa compared to a world total of about 85×10^6 tonnes pa fixed industrially by the Haber process (p. 482). In order to break the very stable $N\equiv N$ triple bond and convert dinitrogen to ammonia it is necessary in the Haber process to use energetically expensive high temperatures and pressures, whereas the biological processes occur under ambient conditions. The economic incentive to achieve an understanding of the mechanism of natural nitrogen fixation is therefore enormous.

Nitrogen fixation takes place in a wide variety of bacteria, the best known of which is *rhizobium* which is found in nodules on the roots of leguminous plants such as peas, beans, soya, and clover. It has been known since 1930 that traces of molybdenum are necessary for the growth of these bacteria, and pasture land deficient in Mo requires the application of suitable fertilizer containing Mo. More recently it has been recognized that essential constituents of all nitrogen-fixing bacteria are:

(i) adenosine triphosphate (ATP) which is a highly active energy transfer agent (p. 611), operating by means of its hydrolysis which requires the presence of Mg^{2+};

(ii) ferredoxin, $Fe_4S_4(SR)_4$ (p. 1280), which is an efficient electron-transfer agent that can be replaced in artificial systems by reducing agents such as dithionite, $[S_2O_4]^{2-}$;

(iii) a metallo-enzyme.

The best known of these metallo-enzymes is "nitrogenase" which has been isolated in an active form from about twenty different bacteria and in a pure form from a handful of these. Its activity is not specific to the reduction of N_2 to NH_3. It also effects the reduction of other species such as CN^- and N_3^- containing a triple bond, and reduces acetylenes to olefins.

Nitrogenase consists of two distinct proteins. One of these contains both Mo and Fe and is therefore known as the "MoFe protein" or "molybdoferredoxin". It is brown, air-sensitive, and has a molecular weight in the approximate range of 220 000 to 230 000; it involves 2 atoms of Mo, 24–36 atoms of Fe, and about the same number of S atoms. The other protein contains Fe but no Mo and is known as "Fe protein" or "azoferredoxin"; it is yellow and extremely air sensitive. Its molecular weight lies in the range 50 000 to 70 000 and the structure involves the Fe_4S_4 ferredoxin core. Unfortunately, investigation of the structures of these proteins is hampered by the extreme sensitivity of nitrogenase to oxygen and the inherent difficulty of obtaining pure crystalline derivatives from biological materials. (Bacteria evidently protect nitrogenase from oxygen by a process of respiration, $O_2 \rightarrow CO_2$, but if too much oxygen is present the system cannot cope and nitrogen fixation ceases.) Physical techniques which have been used to study the system include esr, Mössbauer spectroscopy, and most recently X-ray absorption spectroscopy (analysis of the "extended X-ray absorption fine structure" or EXAFS).

In addition to work on nitrogenase itself, attention has been directed to the preparation and characterization of transition metal/dinitrogen complexes in general and to the

[37] E. I. STIEFEL, The coordination and bioinorganic chemistry of molybdenum, *Prog. Inorg. Chem.* **22**, 1–223 (1977); J. CHATT, J. R. DILWORTH, and R. L. RICHARDS, Recent advances in the chemistry of nitrogen fixation, *Chem. Revs.* **78**, 589–625 (1978). M. COUGHLIN (ed.), *Molybdenum and Molybdenum-containing Enzymes.* Pergamon Press, Oxford, 1980, 577 pp.
[38] R. L. RICHARDS, Nitrogen fixation, *Educ. Chem.* **16**, 66–69 (1979); K. J. SKINNER, *Chem. Eng. News* Oct. 1976, 22–35.

preparation of model compounds which might simulate, if only partly, the action of nitrogenase. As a result it is concluded that the essential steps in nitrogen fixation are:

(i) Binding of N_2 to the MoFe protein. There is no evidence of any direct interaction between Fe and N_2 and it is generally assumed that the N_2 interacts with the Mo. If this is the case then the Mo must be in a sufficiently reduced state to have electrons of adequate energy to donate to the high-energy, antibonding orbitals of the N_2.

(ii) Reduction of the Fe protein by ferredoxin (or artifically by $[S_2O_4]^{2-}$).

(iii) Electron transfer from the reduced Fe protein to the MoFe protein/N_2 complex, where the electron passes to the Mo, possibly via the Fe, and then on to the N_2. This facilitates the protonation of the N_2 and its eventual reduction to ammonia, when H_2 is also produced as a byproduct. The simultaneous hydrolysis of ATP provides the energy for the process, by transferring it from the ultimate energy source, which is the oxidation of sugars to CO_2. That the natural process is unlikely to yield a viable method for the industrial fixation of nitrogen is evident from the fact that the fixation of 14 g of N_2 requires the oxidation of no less than 1 kg of glucose.

Since reduction of N_2 to $2NH_3$ involves 6 electrons it is tempting to infer that the 2 Mo atoms in the MoFe protein are adjacent and produce these electrons by the process $2Mo^{III} \rightarrow 2Mo^{VI} + 6e^-$. However, EXAFS studies[39] suggest that the Mo atoms are in fact associated not with each other but each with an Fe atom, and that they are surrounded by atoms of sulfur. Compounds incorporating Mo in a "ferredoxin cube", as shown in Fig. 23.15, have been prepared and may possibly mirror the situation in nitrogenase. However, there is as yet no evidence that these synthetic materials interact in any way with dinitrogen.

The most intensively studied "model compounds" of Mo are of the type *trans*-$[Mo(N_2)_2(dppe)_2]$ (dppe $= Ph_2PCH_2CH_2PPh_2$), but these contain Mo in the zero oxidation state. Whilst the molybdenum in the MoFe protein could be situated in some kind of protein "pocket" protecting it against oxidation by water, it seems unlikely that an oxidation state as low as zero is feasible. Alternatively, dinuclear, mixed-metal complexes such as $[(MeO)Cl_4MoN_2Re(PR_3)_4Cl]$, which involves N_2 with an elongated and hence weakened N–N bond, have been prepared and are believed to involve Mo in a high oxidation state.

Obviously, ideas about the precise function of molybdenum in nitrogenase are still at a very speculative stage. Even if it is Mo to which the N_2 attaches itself, the mode of attachment (linear or side-on), and the oxidation state of the metal, remain controversial. Continued work on the enzymes themselves and on more realistic models, will no doubt provide answers to these questions, but whether it will ever provide a viable alternative to the Haber process is far less clear.

23.3.9 *Organometallic compounds*

In this group a not-insignificant number of M–C σ-bonded compounds are known of

[39] S. P. CRAMER, K. O. HODGSON, W. O. GILLUM, and L. E. MORTENSON, The molybdenum site of nitrogenase, preliminary structural evidence from X-ray absorption spectroscopy, *J. Am. Chem. Soc.* **100**, 3398–407 (1978).

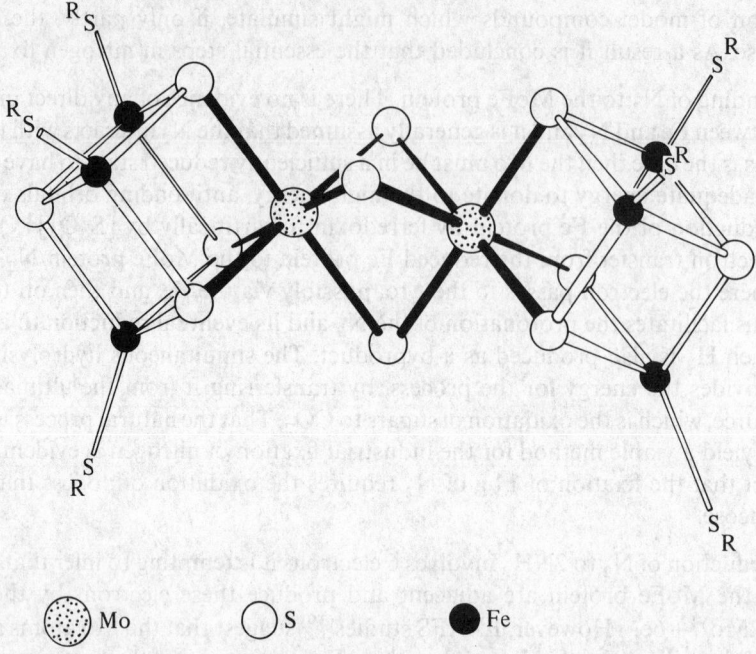

FIG. 23.15 Idealized structure of model compounds incorporating Mo in "ferredoxin-like" cubes. Note the octahedral disposition of S atoms around each Mo.

which hexamethyltungsten(VI) is conceptually the simplest. It is made by the reaction

$$WCl_6 + 6MeLi \longrightarrow WMe_6 + 6LiCl$$

or still better by the reaction

$$WCl_6 + 3Al_2Me_6 \longrightarrow WMe_6 + 6AlClMe_2$$

It is a dark-red, crystalline solid which explodes in air and can detonate in a vacuum. Trimethylsilylmethyl ($-CH_2SiMe_3$) derivatives are also known for all three metals but, whereas chromium forms the mononuclear [$Cr(TMS)_4$], the two heavier elements form dimers, [$(TMS)_3M\equiv M(TMS)_3$], involving $M\equiv M$ triple bonds. A similar dimeric structure is presumed to occur in octamethylditungsten(II).[40] As with the preceding group, however, the bulk of organometallic chemistry is concerned with carbonyl, cyclopentadienyl, and related derivatives of the metals in low oxidation states stabilized by π bonding. Cyanides have been discussed on pp. 1194–1201.

Stable, crystalline hexacarbonyls, $M(CO)_6$, are prepared by reductive carbonylation of compounds (often halides) in higher oxidation states and are octahedral and diamagnetic as anticipated from the 18-electron rule. Replacement of the carbonyl groups by either π-donor or σ-donor ligands is possible, giving a host of materials of the form [$M(CO)_{6-x}L_x$] or [$M(CO)_{6-2x}(L-L)_x$] (e.g. L = NO, NH_3, CN, PF_3; L–L = bipy, butadiene).

⁴⁰ F. A. COTTON, S. KOCH, K. MERTIS, M. MILLAR, and G. WILKINSON, Preparation and characterization of compounds containing the octamethylditungsten(II) anion and partially chlorinated analogues, *J. Am. Chem. Soc.* **99**, 4989–92 (1977).

$[M(CO)_5X]^-$ ions (X=halogen, CN, or SCN) are formed in this way. The low-temperature reaction $(-78°)$ of the halogens with $[Mo(CO)_6]$ or $[W(CO)_6]$ (but not with $[Cr(CO)_6]$) produces the M^{II} carbonyl halides, $[M(CO)_4X_2]$ from which many adducts, $[M(CO)_3L_2X_2]$, are obtained. Although these have not been fully characterized, they are diamagnetic and presumably involve 7-coordination. Reduction of the hexacarbonyls with a borohydride in liquid ammonia forms dimeric $[M_2(CO)_{10}]^{2-}$ which are isostructural with the isoelectronic $[Mn_2(CO)_{10}]$ (p. 1234). Hydrolysis of these dimers produces the hydrides $[HM_2(CO)_{10}]^-$. The chromium compound is $[(CO)_5Cr-H-Cr(CO)_5]^-$ with eclipsed carbonyls and it maintains the 18-valence electron configuration by means of a linear, 3-centre, electron-pair Cr–H–Cr bond. There is, however, evidence suggesting that M–H–M bridges are relatively weak and may be bent by crystal-packing forces arising from different counter-cations. A related compound, $[NEt_4]^+_2[H_2W_2(CO)_8]^{2-}$ is of interest[41] because it has 2 hydrogen bridges and a W–W distance indicative of a W–W double bond (301.6 pm compared to 322.2 pm for a W–W single bond) (Fig. 23.16). The compound also illustrates the improved refinement now possible with modern X-ray methods (p. 1180) and it was, in fact, the first case of the successful location of hydrogens bridging third-row transition metals.

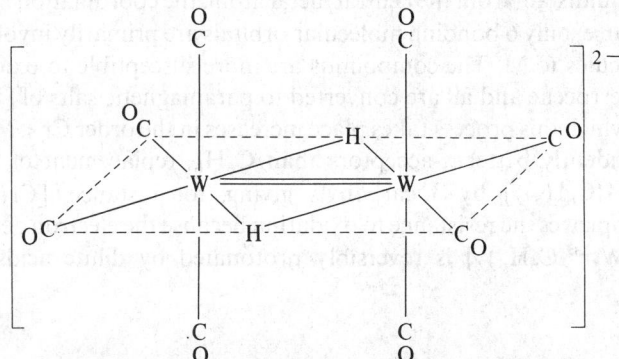

FIG. 23.16   Structure of $[H_2W_2(CO)_8]^{2-}$.

Metallocenes, $[M^{II}(\eta^5-C_5H_5)_2]$, analogous to ferrocene would have only 16 valence electrons and could therefore be considered "electron deficient". Chromocene can be formed by the action of sodium cyclopentadienide on $[Cr(CO)_6]$. It is isomorphous with ferrocene but paramagnetic and much more reactive. Under similar conditions Mo and W hexacarbonyls form only $[(\eta^5-C_5H_5)M^I(CO)_3]_2$, in which dimerization achieves the 18-valence-electron configuration by means of an M–M bond. The chromium analogue of these dimers has one of the longest M–M bonds found in dinuclear transition metal compounds (328.1 pm). Monomeric molybdenocene and tungstenocene are unknown but a red-brown polymeric solid $[M^{II}(C_5H_5)_2]_n$ can be obtained and, with CO, it produces the "bent" molecule $[Mo(\eta^5-C_5H_5)_2(CO)]$ (Fig. 23.17).

Of greater stability than the metallocenes in this group are the dibenzene sandwich

[41] M. R. CHURCHILL and S. W. Y. CHANG, X-ray crystallographic determination of the molecular structure of $(Et_4N^+)_2[H_2W_2(CO)_8{}^{2-}]$, *Inorg. Chem.* **13**, 2413–19 (1974).

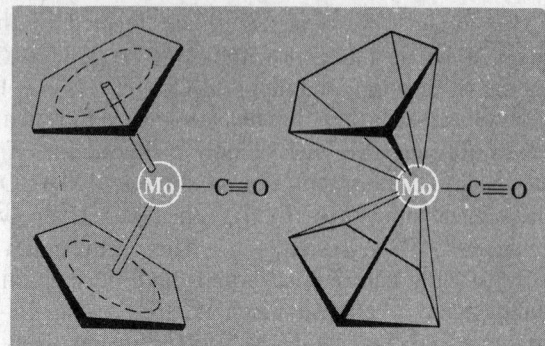

FIG. 23.17 Alternative representations of the structure of Mo(η^5-C$_5$H$_5$)$_2$CO.

compounds which are isoelectronic with ferrocene and of which the dark brown [Cr(η^6-C$_6$H$_6$)$_2$] was the first to be prepared (p. 374) and remains the best known. The green [Mo(η^6-C$_6$H$_6$)$_2$] and yellow-green [W(η^6-C$_6$H$_6$)$_2$] are also well characterized and all contain the metal in the formal oxidation state of zero. As the 12 C atoms in [M(η^6-C$_6$H$_6$)$_2$] are equidistant from the central metal atom, the coordination number of M is 12, though, of course, only 6 bonding molecular orbitals are primarily involved in linking the 2 ligand molecules to M. The compounds are more susceptible to oxidation than is the isoelectronic ferrocene and all are converted to paramagnetic salts of [M^I(η^6-C$_6$H$_6$)$_2$]$^+$: the ease with which this process takes place increases in the order Cr < Mo < W. Since CO groups are evidently better π-acceptors than C$_6$H$_6$, replacement of 1 of the benzene ligands in [M(C$_6$H$_6$)$_2$] by 3 carbonyls giving, for instance, [Cr(η^6-C$_6$H$_6$)(CO)$_3$], appreciably improves the resistance to oxidation because the electron density on the metal is lowered. [W(η^6-C$_6$H$_6$)$_2$] is reversibly protonated by dilute acids to give [W(η^6-C$_6$H$_6$)$_2$H]$^+$.

3	4											5	6	7	8	9	10
Li	Be											B	C	N	O	F	Ne
11	12											13	14	15	16	17	18
Na	Mg											Al	Si	P	S	Cl	Ar
19	20	21	22	23	24	25	26	27	28	29	30	31	32	33	34	35	36
K	Ca	Sc	Ti	V	Cr	Mn	Fe	Co	Ni	Cu	Zn	Ga	Ge	As	Se	Br	Kr
37	38	39	40	41	42	43	44	45	46	47	48	49	50	51	52	53	54
Rb	Sr	Y	Zr	Nb	Mo	Tc	Ru	Rh	Pd	Ag	Cd	In	Sn	Sb	Te	I	Xe
55	56	57	72	73	74	75	76	77	78	79	80	81	82	83	84	85	86
Cs	Ba	La	Hf	Ta	W	Re	Os	Ir	Pt	Au	Hg	Tl	Pb	Bi	Po	At	Rn
87	88	89	104	105	106	107											
Fr	Ra	Ac	Unq	Unp	Unh	Uns											

24

58	59	60	61	62	63	64	65	66	67	68	69	70	71
Ce	Pr	Nd	Pm	Sm	Eu	Gd	Tb	Dy	Ho	Er	Tm	Yb	Lu
90	91	92	93	94	95	96	97	98	99	100	101	102	103
Th	Pa	U	Np	Pu	Am	Cm	Bk	Cf	Es	Fm	Md	No	Lr

Manganese, Technetium, and Rhenium

24.1 Introduction

In terms of history, abundance, and availability, it is difficult to imagine a greater contrast than exists in this group between manganese and its congeners, technetium and rhenium. Millions of tonnes of manganese are used annually, and its most common mineral, pyrolusite, has been used in glassmaking since the time of the Pharaohs. On the other hand, technetium and rhenium are exceedingly rare and were only discovered comparatively recently, the former being the first new element to have been produced artificially and the latter being the last naturally occurring element to be discovered.

Metallic manganese was first isolated in 1774 when C. W. Scheele recognized that pyrolusite contained a new element, and his fellow Swede, J. G. Gahn, heated the MnO_2 with a mixture of charcoal and oil. The purity of this metal was low, and high-purity (99.9%) manganese was only produced in the 1930s when electrolysis of Mn^{II} solutions was used.

In Mendeleev's table, this group was completed by the then undiscovered eka-manganese ($Z = 43$) and dvi-manganese ($Z = 75$). Confirmation of the existence of these missing elements was not obtained until H. G. J. Moseley had introduced the method of X-ray spectroscopic analysis. Then in 1925 W. Noddack, I. Tacke (later Frau Noddack), and O. Berg discovered element 75 in a sample of gadolinite (a basic silicate of beryllium, iron, and lanthanides) and named it rhenium after the river Rhine. The element was also discovered, independently by F. H. Loring and J. F. G. Druce, in manganese compounds, but is now most usually recovered from the flue dusts produced in the roasting of CuMo ores.

The Noddacks also claimed to have detected element 43 and named it masurium after Masuren in Prussia. This claim proved to be incorrect, however, and the element was actually detected in 1937 in Italy by C. Perrier and E. Segré in a sample of molybdenum which had been bombarded with deuterons in the cyclotron of E. O. Lawrence in California. It was present in the form of the β^- emitters ^{95}Tc and ^{97}Tc with half-lives of 61 and 90 days respectively. The name technetium (from Greek τεχνικός, artificial) is clearly appropriate even though minute traces of the more stable ^{99}Tc (half-life $= 2.14 \times 10^5$ y) do occur naturally as a result of spontaneous fission of uranium. As this isotope is also produced in the fission products of nuclear reactors, it is currently isolated from this source in multikilogram quantities and its chemistry is therefore more familiar than might perhaps have been expected.

24.2 The Elements

24.2.1 *Terrestrial abundance and distribution*

The natural abundance of technetium is, as just indicated, negligibly small. The concentration of rhenium in the earth's crust is extremely low (of the order of $7 \times 10^{-8}\%$, i.e. 0.0007 ppm) and it is also very diffuse. Being chemically akin to molybdenum it is in molybdenites that its highest concentrations (0.2%) are found. By contrast, manganese (0.106%, i.e. 1060 ppm of the earth's crustal rocks) is the twelfth most abundant element and the third most abundant transition element (exceeded only by iron and titanium). It is found in over 300 different and widely distributed minerals of which about twelve are commercially important. As a class-a metal it occurs in primary deposits as the silicate. Of more commercial importance are the secondary deposits of oxides and carbonates such as pyrolusite (MnO_2), which is the most common, hausmannite (Mn_3O_4), and rhodochrosite ($MnCO_3$). These have been formed by weathering of the primary silicate deposits and are found in the USSR, Gabon, South Africa, Brazil, Australia, India, and China.

A further consequence of this weathering is that colloidal particles of the oxides of manganese, iron, and other metals are continuously being washed into the sea where they agglomerate and are eventually compacted into the "manganese nodules" (so called because Mn is the chief constituent), first noted during the voyage of HMS *Challenger* (1872–6). Following a search in the Pacific organized by the University of California during the International Geophysical Year (1957), the magnitude and potential value of manganese nodules has become apparent.[1] More than 10^{12} tonnes are estimated to cover vast areas of the ocean beds and a further 10^7 tonnes are deposited annually. The composition varies but the dried nodules generally contain between 15 and 30% of Mn. This is less than the 35% normally regarded as the lower limit required for present-day commercial exploitation but, since the Mn is accompanied not only by Fe but more importantly by smaller amounts of Ni, Cu, and Co, the combined recovery of Ni, Cu, and Co, with Mn effectively as a byproduct could well be economical if performed on a sufficient scale. The technical, legal, and political problems involved are enormous, but exploration and research is currently being undertaken by several multinational groups.

24.2.2 *Preparation and uses of the metals*

Ninety-five per cent of all the manganese ores produced are used in steel manufacture, mostly in the form of ferromanganese.[1a] This contains about 80% Mn and is made by reducing appropriate amounts of MnO_2 and Fe_2O_3 with coke in a blast furnace

[1] G. P. GLASBY (ed.), *Marine Manganese Deposits*, Elsevier, Amsterdam, 1977, 523 pp. A. J. MONHEMIUS, The extractive metallurgy of deep-sea manganese nodules, pp. 42–69, in *Critical Reports on Applied Chemistry*, Vol. 1. (A. R. BURKIN, ed.), Blackwell, Oxford, 1980.

[1a] *Kirk–Othmer Encyclopedia of Chemical Technology*, 3rd edn., Vol. 14, pp. 824–43, Interscience, New York, 1981.

or, if cheap electricity is available, in an electric-arc furnace. Dolomite or limestone is also added to remove silica as a slag. Where the Mn content is lower (because of the particular ores used) the product is known as silicomanganese (65–70% Mn, 15–20% Si) or spiegeleisen (5–20% Mn). Where pure manganese metal is required it is now prepared by the electrolysis of aqueous managanese(II) sulfate. Ore with an Mn content of nearly 10 million tonnes is produced annually (1980), of which 3.4 million tonnes are from the USSR, 2.13 from South Africa, about 1.0 from each of Gabon, Brazil, and Australia, 0.58 from India and 0.5 million tonnes from China.

All steels contain some Mn, and its addition in 1856 by R. Mushet ensured the success of the Bessemer process. It serves two main purposes. As a "scavenger" it combines with sulfur to form MnS which passes into the slag and prevents the formation of FeS which would induce brittleness, and it also combines with oxygen to form MnO, so preventing the formation of bubbles and pinholes in the cold steel. Secondly, the presence of Mn as an alloying metal increases the hardness of the steel. The hard, non-magnetic Hadfield steel containing about 13% Mn and 1.25% C, is the best known, and is used when resistance to severe mechanical shock and wear is required, e.g. for excavators, dredgers, rail crossings, etc.

Important, but less extensive, uses are found in the production of non-ferrous alloys. It is a scavenger in several Al and Cu alloys, while "manganin" is a well-known alloy (84% Cu, 12% Mn, 4% Ni) which is used in electrical instruments because the temperature coefficient of its resistivity is almost zero. A variety of other major uses have been found for Mn in the form of its compounds and these will be dealt with later under the appropriate headings.

Technetium[2] is obtained from nuclear power stations where it makes up about 6% of uranium fission products and is recovered from these after storage for several years to allow the highly radioactive, short-lived fission products to decay. The original process used the precipitation of $[AsPh_4]^+[ClO_4]^-$ to carry with it $[AsPh_4]^+[TcO_4]^-$ and so separate the Tc from other fission products, but solvent extraction and ion-exchange techniques are now used. The metal itself can be obtained by the high-temperature reduction of either NH_4TcO_4 or Tc_2S_7 with hydrogen. There is at present no significant use for Tc although ^{99}Tc, due to its long half-life, provides a practically invariant source of β^- radiation, and ammonium pertechnetate has been suggested as a good corrosion inhibitor.

In the roasting of molybdenum sulfide ores, any rhenium which might be present is oxidized to volatile Re_2O_7 which collects in the flue dusts and is the usual source of the metal via conversion to $(NH_4)ReO_4$ and reduction by H_2 at elevated temperatures.[3] Being highly refractory and corrosion-resistant, rhenium metal would no doubt find widespread use were it not for its scarcity and consequent high cost. As it is, uses are essentially small scale. Mass spectrometer filaments, furnace heater windings, thermocouples (Pt–Re, etc.) and catalysts for hydrogenation and dehydrogenation reactions are such uses. In the USA over 95% of the Re production is for bimetallic Pt/Re catalysts used to produce low-Pb, high-octane gasoline. World production is about 7000 kg annually.

[2] R. D. PEACOCK, Technetium, Chap. 38 in *Comprehensive Inorganic Chemistry*, Vol. 3, pp. 877–903, Pergamon Press, Oxford, 1973.
[3] R. D. PEACOCK, Rhenium, Chap. 39 in *Comprehensive Inorganic Chemistry*, Vol. 3, pp. 905–78, Pergamon Press, Oxford, 1973.

24.2.3 *Properties of the elements*

Some of the important properties of Group VIIA elements are summarized in Table 24.1. Technetium is an artificial element, so its atomic weight depends on which isotope has been produced. The atomic weights of Mn and Re, however, are known with considerable accuracy. In the case of the former this is because it has only 1 naturally occurring isotope and, in the case of the latter, because it has only 2 and the relative proportions of these in terrestrial samples are essentially constant (^{185}Re 37.4%, ^{187}Re 62.6%).

TABLE 24.1   *Some properties of Group VIIA elements*

Property		Mn	Tc	Re
Atomic number		25	43	75
Number of naturally occurring isotopes		1	—	2
Atomic weight		54.9380	98.906[a]	186.207
Electronic configuration		[Ar]3d^{5}4s^2	[Kr]4d^{6}5s^1	[Xe]4f^{14}5d^{5}6s^2
Electronegativity		1.5	1.9	1.9
Metal radius (12-coordinate)/pm		127	136	137
Ionic radius/pm	VII	46	56	53
(4-coordinate	VI	25.5*	—	55
if marked *;	V	33*	60	58
otherwise	IV	53	64.5	63
6-coordinate)	III	58 (ls), 64.5 (hs)	—	—
	II	67	—	—
MP/°C		1244	2200	3180
BP/°C		2060	4567	(5650)
ΔH_{fus}/kJ mol^{-1}		(13.4)	23.8	34($\pm$4)
ΔH_{vap}/kJ mol^{-1}		221($\pm$8)	585	704
ΔH_f (monatomic gas)/kJ mol^{-1}		281($\pm$6)	—	779($\pm$8)
Density (25°C)/g cm^{-3}		7.43	11.5	21.0
Electrical resistivity (20°C)/μohm cm		185.0	—	19.3

[a] This refers to ^{99}Tc ($t_\frac{1}{2}$ 2.13 × 10^5 y). For ^{97}Tc ($t_\frac{1}{2}$ 2.6 × 10^6 y) and ^{98}Tc ($t_\frac{1}{2}$ 4.2 × 10^6 y) the values are 96.906 and 97.907 respectively.

In the solid state all three elements have typically metallic structures. Technetium and Re are isostructural with hcp lattices, but there are 4 allotropes of Mn of which the α-form is the one stable at room temperature. This has a bcc structure in which, for reasons which are not clear, there are 4 distinct types of Mn atom. It is hard and brittle, and noticeably less refractory than its predecessors in the first transition series.

In continuance of the trends already noticed, the most stable oxidation state of manganese is +2, and in the group oxidation state of +7 it is even more strongly oxidizing than Cr(VI). Evidently the 3d electrons are more tightly held by the Mn atomic nucleus and this reduced delocalization produces weaker metallic bonding than in Cr. The same trends are also starting in the second and third series with Tc and Re, but are less marked, and Re in particular is very refractory, having a mp which is second only to that of tungsten amongst transition elements.

24.2.4 *Chemical reactivity and trends*

Manganese is more electropositive than any of its neighbours in the periodic table and the metal is more reactive, especially when somewhat impure. In the massive state it is superficially oxidized on exposure to air but will burn if finely divided. It liberates hydrogen from water and dissolves readily in dilute aqueous acids to form manganese(II) salts. With non-metals it is not very reactive at ambient temperatures but frequently reacts vigorously when heated. Thus it burns in oxygen, nitrogen, chlorine, and fluorine giving Mn_3O_4, Mn_3N_2, $MnCl_2$, and $MnF_2 + MnF_3$ respectively, and it combines directly with B, C, Si, P, As, and S.

Technetium and rhenium metals are less reactive than manganese and, as is to be expected for the two heavier elements, they are closely similar to each other. In the massive form they resist oxidation and are only tarnished slowly in moist air. However, they are normally produced as sponges or powders in which case they are more reactive. Heated in oxygen they burn to give volatile heptaoxides (M_2O_7), and with fluorine they give $TcF_5 + TcF_6$ and $ReF_6 + ReF_7$ respectively. MS_2 can also be produced by direct action. Although insoluble in hydrofluoric and hydrochloric acids, the metals dissolve readily in oxidizing acids such as HNO_3 and conc H_2SO_4 and also in bromine water, when "pertechnetic" and "perrhenic" acids (HMO_4) are formed.

Detailed comparisons of the stabilities of the different oxidation states of the three elements are not always possible owing to incomplete knowledge of the chemistry of technetium arising from its scarcity and the problems associated with the handling of radioactive materials. However, the activity consists only of weak β^- emission and, providing the appropriate precautions are taken, presents no major difficulty. It is therefore no doubt only a matter of time, as the element becomes more readily available, before these deficiencies are remedied. Already many of the similarities and contrasts expected in the chemistry of transition elements are evident in this triad. The relative stabilities of different oxidation states in aqueous, acidic solutions is summarized in Table 24.2 and Fig. 24.1.

TABLE 24.2 *$E°$ for some manganese, technetium, and rhenium couples in acid solution at 25°C*

Couple	$E°$/V
$Mn^{2+}(aq) + 2e^- \rightleftharpoons Mn(s)$	-1.185
$Mn^{3+}(aq) + 3e^- \rightleftharpoons Mn(s)$	-0.283
$MnO_2 + 4H^+ + 4e^- \rightleftharpoons Mn(s) + 2H_2O$	0.024
$MnO_4^{2-} + 8H^+ + 4e^- \rightleftharpoons Mn^{2+}(aq) + 4H_2O$	1.742
$MnO_4^- + 8H^+ + 5e^- \rightleftharpoons Mn^{2+}(aq) + 4H_2O$	1.507
$Tc^{2+}(aq) + 2e^- \rightleftharpoons Tc(s)$	0.400
$TcO_2 + 4H^+ + 4e^- \rightleftharpoons Tc(s) + 2H_2O$	0.272
$TcO_3 + 2H^+ + 2e^- \rightleftharpoons TcO_2 + H_2O$	0.757
$TcO_4^- + 8H^+ + 5e^- \rightleftharpoons Tc^{2+}(aq) + 4H_2O$	0.500
$Re^{3+}(aq) + 3e^- \rightleftharpoons Re(s)$	0.300
$ReO_2 + 4H^+ + 4e^- \rightleftharpoons Re(s) + 2H_2O$	0.251
$ReO_3 + 6H^+ + 3e^- \rightleftharpoons Re^{3+}(aq) + 3H_2O$	0.318
$ReO_4^{2-} + 8H^+ + 3e^- \rightleftharpoons Re^{3+}(aq) + 4H_2O$	0.795
$ReO_4^- + 8H^+ + 4e^- \rightleftharpoons Re^{3+}(aq) + 4H_2O$	0.422

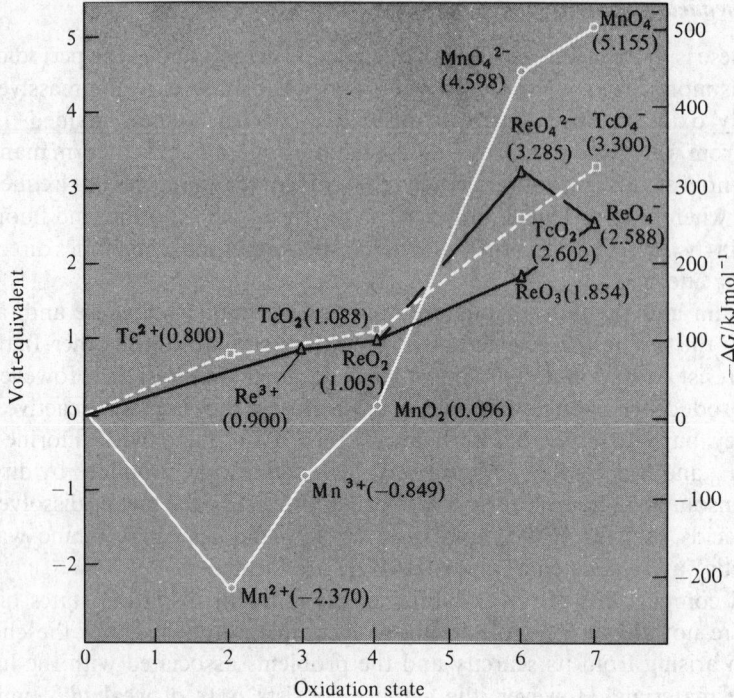

FIG. 24.1 Plot of volt-equivalent versus oxidation state for Mn, Tc, and Re.

The most obvious features of Fig. 24.1 are the relative positions of the +2 oxidation states. For manganese this state, represented by the high-spin Mn^{II} cation, is much the most stable. This may be taken as an indication of the stability of the symmetrical d^5 electron configuration. However, like the mp, bp, and enthalpy of atomization, it also reflects the weaker cohesive forces in the metallic lattice since for Tc and Re, which have much stronger metallic bonding, the +2 state is of little importance and the occurrence of cluster compounds with M–M bonds is a dominant feature of rhenium(III) chemistry.

Another marked contrast is evident in the +7 oxidation state where the manganate(VII) (permanganate) ion is an extremely strong oxidizing agent but $(TcO_4)^-$ and $(ReO_4)^-$ show only mild oxidizing properties. Indeed, the greater stability of Tc and Re compared to Mn in any oxidation state higher than +2 is apparent, as will be seen more fully in the following account of individual compounds.

Table 24.3 lists representative examples of the compounds of these elements in their various oxidation states. The wide range of the oxidation states is particularly noteworthy. It arises from the fact that, in moving across the transition series, the number of d electrons has increased and, in this mid-region, the d orbitals have not yet sunk energetically into the inert electron core. The number of d electrons available for bonding is consequently maximized, and not only are high oxidation states possible, but back donation of electrons from metal to ligand is also facilitated with resulting stabilization of low oxidation states.

A further point of interest is the noticeably greater tendency of rhenium, as compared to either manganese or technetium to form compounds with high coordination numbers.

TABLE 24.3   *Oxidation states and stereochemistries of manganese, technetium, and rhenium compounds*

Oxidation state	Coordination number	Stereochemistry	Mn	Tc/Re
-3 (d^{10})	4	Tetrahedral	$[Mn(NO)_3(CO)]$	$[M(CO)_4]^{3-}$
-2 (d^9)	4	Square planar	$[Mn(phthalocyanine)]^{2-}$	—
-1 (d^8)	5	Trigonal bipyramidal	$[Mn(CO)_5]^-$	$[M(CO)_5]^-$
	4	Square planar	$[Mn(phthalocyanine)]^-$	—
0 (d^7)	6	Octahedral	$[Mn_2(CO)_{10}]$	$[M_2(CO)_{10}]$
1 (d^6)	6	Octahedral	$[Mn(CN)_6]^-$	$[M(CN)_6]^-$
2 (d^5)	4	Tetrahedral	$[MnBr_4]^{2-}$	
		Square planar	$[Mn(phthalocyanine)]$	
	5	Trigonal bipyramidal	$[MnBr\{N(C_2H_4NMe_2)_3\}]^+$	
	6	Octahedral	$[Mn(H_2O)_6]^{2+}$	$[M(diars)_2Cl_2]$
	7	Capped trigonal prismatic	$[Mn(EDTA)(H_2O)]^{2-}$	
	8	Dodecahedral	$[Mn(NO_3)_4]^{2-}$	
3 (d^4)	5	Square pyramidal	$[MnCl_5]^{2-}$	$[Re_2Cl_8]^{2-}$
	6	Octahedral	$K_3[Mn(CN)_6]$	$[M(diars)_2Cl_2]^+$
	7	Pentagonal bipyramidal	$[Mn(NO_3)_3(bipy)]$	$[M(CN)_7]^{4-}$
	11	See Fig. 24.8	—	$[Re(\eta^5\text{-}C_5H_5)_2H]$
4 (d^3)	5	—	—	$[(Me_3SiCH_2)_4Re(N_2)$ $Re(CH_2SiMe_3)_4]$
	6	Octahedral	$[MnF_6]^{2-}$	$[MI_6]^{2-}$
5 (d^2)	4	Tetrahedral	$[MnO_4]^{3-}$	—
	5	Trigonal bipyramidal (?)	—	ReF_5
		Square pyramidal	—	$[MOCl_4]^-$
	6	Octahedral	—	$[Tc(NCS)_6]^-$, $[ReNCl_2(PEt_2Ph)_3]$
	8	Dodecahedral	—	$[M(diars)_2Cl_4]^+$
6 (d^1)	4	Tetrahedral	$[MnO_4]^{2-}$	—
	5	Square pyramidal		$ReOCl_4$
	6	Trigonal prismatic		$[Re(S_2C_2Ph_2)_3]$ (see p. 1227)
		Octahedral		ReF_6
	8	Dodecahedral		$[ReMe_8]^{2-}$
		Square antiprismatic		$[ReF_8]^{2-}$
7 (d^0)	4	Tetrahedral	$[MnO_4]^-$	$[MO_4]^-$
	5	Trigonal bipyramidal		cis-$[ReO_2Me_3]$
	6	Octahedral		$[ReO_3Cl_3]^{2-}$
	7	Pentagonal bipyramidal		ReF_7
	9	Tricapped trigonal prismatic		$[ReH_9]^{2-}$

24.3 Compounds of Manganese, Technetium, and Rhenium[2–5b]

Binary borides (p. 162), carbides (p. 318), and nitrides (p. 479) have already been mentioned. Manganese, like chromium (and also the succeeding elements in the first transition series), is too small to accommodate interstitial carbon without significant distortion of the metal lattice. As a consequence it forms a number of often readily hydrolysed carbides with rather complicated structures.

[4] R. D. W. KEMMITT, Manganese, Chap. 37 in *Comprehensive Inorganic Chemistry*, Vol. 3, pp. 771–876, Pergamon Press, Oxford, 1973.

[5] R. D. PEACOCK, *The Chemistry of Technetium and Rhenium*, Elsevier, Amsterdam, 1966, 137 pp.

[5a] K. SCHWOCHAU, The chemistry of technetium, *Topics in Current Chemistry*, **96**, 109–147 (1981).

[5b] M. J. CLARKE and P. H. FACKLER, The chemistry of technetium: toward improved diagnostic agents, *Structure and Bonding* **50**, 57–78 (1982).

Hydrido complexes are well-known but simple binary hydrides are not, which is in keeping with the position of these metals in the "hydrogen gap" portion of the periodic table (p. 74).

24.3.1 *Oxides and chalcogenides*

All three metals form heptoxides but, whereas Tc_2O_7 and Re_2O_7 are the final products formed when the metals are burned in an excess of oxygen, Mn_2O_7 requires prior oxidation of the manganese to the $+7$ state. It is a green explosive oil (mp 5.9°) produced by the action of conc H_2SO_4 on a manganate(VII) salt. On standing, it slowly loses oxygen to form MnO_2 but detonates around 95°C and will explosively oxidize most organic materials. The molecule is composed of 2 corner-sharing MnO_4 tetrahedra with a bent Mn–O–Mn bridge. The other 2 heptoxides are yellow solids whose volatility provides a

TABLE 24.4 *Oxides of Group VIIA*

Oxidation state	$+7$	$+6$	$+5$	$+4$	$+3$	$+2$
Mn	Mn_2O_7			MnO_2	Mn_2O_3 Mn_3O_4 MnO	
Tc	Tc_2O_7	$TcO_3(?)$		TcO_2		
Re	Re_2O_7	ReO_3	Re_2O_5	ReO_2		

useful means of purifying the elements and, as has been pointed out, is a crucial factor in the commercial production of rhenium (Tc_2O_7: mp 119.5°, bp 310.6°; Re_2O_7: mp 300.3°, bp 360.3°). In the vapour phase both consist of corner-sharing MO_4 tetrahedra but, whereas this structure is retained in the solid phase by Tc_2O_7, solid Re_2O_7 has an unusual structure consisting of polymeric double layers of corner-sharing ReO_4 tetrahedra alternating with ReO_6 octahedra. The same basic unit, though this time discrete, is found in the dihydrate (perrhenic acid), which is therefore best formulated as O_3Re–O–$ReO_3(H_2O)_2$.

Only rhenium forms a stable trioxide. It is a red solid with a metallic lustre and is obtained by the reduction of Re_2O_7 with CO. ReO_3 has a structure in which each Re is octahedrally surrounded by oxygens (Fig. 24.2). It has an extremely low electrical resistivity which decreases with decrease in temperature like a true metal: $\rho_{300\,K}$ 10 μohm cm, ρ_{100} 0.6 μohm cm. It is clear that the single valency electron on each Re atom is delocalized in a conduction band of the crystal. ReO_3 is unreactive towards water and aqueous acids and alkalis, but when boiled with conc alkali it disproportionates into ReO_4^- and ReO_2. A blue pentoxide Re_2O_5 has been reported but is also prone to disproportionation into $+7$ and $+4$ species.

The $+4$ oxidation state is the only one in which all three elements form stable oxides, but only in the case of technetium is this the most stable oxide. TcO_2 is the final product when any Tc/O system is heated to high temperatures, but ReO_2 disproportionates at 900°C into Re_2O_7 and the metal. Hydrated TcO_2 and ReO_2 may be conveniently prepared by reduction of aqueous solutions of MO_4^- with zinc and hydrochloric acid and are easily dehydrated. TcO_2 is dark brown and ReO_2 is blue-black: both solids have distorted rutile structures like MoO_2 (p. 1174).

It is MnO_2, however, which is by far the most important oxide in this group, though it

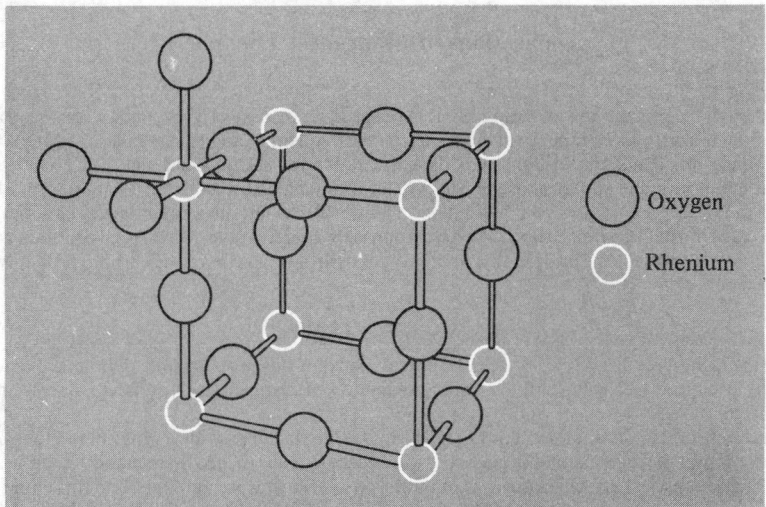

Fɪɢ. 24.2   The structure of ReO$_3$. Note the similarity to perovskite (p. 1122) which can be understood as follows: if the Re atom, which is shown here with its full complement of 6 surrounding O atoms, is imagined to be the small cation at the centre of Fig. 21.5(a), then the perovskite structure is obtained by placing the large cations (Ca^{II}) into the centre of the cube drawn above and in the 7 other equivalent positions around the Re.

is not the most stable oxide of manganese, decomposing to Mn$_2$O$_3$ above about 530°C and being a useful oxidizing agent. Hot concentrated sulfuric and hydrochloric acids reduce it to manganese(II):

$$2MnO_2 + 2H_2SO_4 \longrightarrow 2MnSO_4 + O_2 + 2H_2O$$
$$MnO_2 + 4HCl \longrightarrow MnCl_2 + Cl_2 + 2H_2O$$

the latter reaction being formerly the basis of the manufacture of chlorine. It is, however, extremely insoluble and, as a consequence, often unreactive. As pyrolusite it is the most plentiful ore of manganese and it finds many industrial uses (see Panel).

The structural history of MnO$_2$ is complex and confused due largely to the facts that nonstoichiometry is prevalent and in its hydrated forms it behaves as a cation-exchanger. Many of the various polymorphs which have been reported are probably therefore simply impure forms. The only stoichiometric form is the so-called β-MnO$_2$, which is that of pyrolusite and possesses the rutile structure (p. 1118), but even here a range of composition from MnO$_{1.93}$ to MnO$_{2.0}$ is possible. β-MnO$_2$ can be prepared by careful decomposition of manganese(II) nitrate[4] but, when precipitated from aqueous solutions, for instance by reduction of alkaline MnO$_4$$^-$, the hydrated MnO$_2$ has a more open structure which exhibits cation-exchange properties and cannot be fully dehydrated without some loss of oxygen.

Apart from the black Re$_2$O$_3$.2H$_2$O (which is readily oxidized to the dioxide and is prepared by boiling ReCl$_3$ in air-free water) oxides of oxidation states below +4 are known only for manganese. Mn$_3$O$_4$ is formed when any oxide of manganese is heated to about 1000°C in air and is the black mineral, hausmannite. It has the spinel structure (p. 279) and as such is appropriately formulated as Mn^{II}Mn$_2$^{III}O$_4$, with Mn^{II} and Mn^{III} occupying tetrahedral and octahedral sites respectively within a ccp lattice of oxide ions;

Applications of Manganese Dioxide[6]

Although the primary use of manganese is in the production of steel it also finds widespread and important uses in non-metallurgical industries. These frequently use the manganese as MnO_2 but even where this is not the case the dioxide is invariably the starting material.

The largest non-metallurgical use of MnO_2 is in the manufacture of dry-cell batteries (p. 1397) which in 1976 accounted for about half a million tonnes of ore. The most common dry batteries are of the carbon–zinc Leclanché type in which carbon is the positive pole. MnO_2 is incorporated as a depolarizer to prevent the undesirable liberation of hydrogen gas on to the carbon, probably by the reaction

$$MnO_2 + H^+ + e^- \longrightarrow MnO(OH)$$

Only the highest quality MnO_2 ore can be used directly for this purpose, and "synthetic dioxide", usually produced electrolytically by anodic oxidation of manganese(II) sulfate, is increasingly employed.

The brick industry is another major user of MnO_2 since it can provide a range of red to brown or grey tints. In the manufacture of glass its use as a decolourizer (hence "glassmaker's soap") is its most ancient application. Glass always contains iron at least in trace amounts, and this imparts a greenish colour; the addition of MnO_2 to the molten glass produces red-brown Mn^{III} which equalizes the absorption across the visible spectrum so giving a "colourless", i.e. grey glass. In recent times selenium compounds have replaced MnO_2 for this application, but in larger proportions the latter is still used to make pink to purple glass.

The oxidizing properties of MnO_2 are utilized in the oxidation of aniline for the preparation of hydroquinone which is important as a photographic developer and also in the production of dyes and paints.

In the electronics industry the advantages of higher electrical resistivity and lower cost of ceramic ferrites ($M^{II}Fe_2O_4$) (p. 1256) over metallic magnets have been recognized since the 1950s and the "soft" ferrites ($M^{II} = Mn, Zn$) are the most common of these. Between 0.3–1 kg of these are used on the sweep transformer and deflection yoke of every television set and, of course, MnO_2, either natural or synthetic, is required in their production.

there is, however, a tetragonal distortion due to a Jahn–Teller effect on Mn^{II}. A related structure, but with fewer cation sites occupied, is found in the black γ-Mn_2O_3 which can be prepared by aerial oxidation and subsequent dehydration of the hydroxide precipitated from aqueous Mn^{II} solutions. If MnO_2 is heated less strongly (say $<800°C$) than is required to produce Mn_3O_4, then the more stable α-form of Mn_2O_3 results which has a structure involving 6-coordinate Mn but with 2 Mn–O bonds longer than the other 4. This is no doubt a further manifestation of the Jahn–Teller effect (p. 1087) expected for the high-spin d^4 Mn^{III} ion and is presumably the reason why Mn_2O_3, alone among the oxides of transition metal M^{III} ions, does not have the corundum (p. 274) structure.

Reduction with hydrogen of any oxide of manganese produces the lowest oxide, the grey to green MnO. This is an entirely basic oxide, dissolving in acids and giving rise to the aqueous Mn^{II} cationic chemistry. It has a rock-salt structure and is subject to nonstoichiometric variation ($MnO_{1.00}$ to $MnO_{1.045}$), but its main interest is that it is a classic example of an antiferromagnetic compound. If the temperature is reduced below about 118 K (its Néel point), a rapid fall in magnetic moment takes place as the electron spins on adjacent Mn atoms pair-up. This is believed to take place by the process of "superexchange" by which the interaction is transferred through intervening, non-magnetic, oxide ions. (MnO_2 is also antiferromagnetic below 92 K whereas the alignment in Mn_3O_4 results in ferrimagnetism below 43 K.)

[6] S. A. WEISS, *Manganese. The other uses*, Metal Bulletin Books, London, 1977, 360 pp.

The sulfides are fewer and less familiar than the oxides but, as is to be expected, favour lower oxidation states of the metals. Thus manganese forms MnS_2 which has the pyrite structure (p. 804) with discrete Mn^{II} and S_2^{-II} ions and is converted on heating to MnS and sulfur. This green MnS is the most stable manganese sulfide and, like MnO, has a rock-salt structure and is strongly antiferromagnetic ($T_N - 121°C$). Less-stable red forms are also known and the pale-pink precipitate produced when H_2S is bubbled through aqueous Mn^{II} solutions is a hydrated form which passes very slowly into the green variety. The corresponding selenides are very similar: $Mn^{II}Se_2$ (pyrite-type), and MnSe (NaCl-type), antiferromagnetic with $T_N - 100°C$.

Technetium and rhenium favour higher oxidation states in their binary chalcogenides. Both form diamagnetic heptasulfides, M_2S_7, which are isomorphous and which decompose to $M^{IV}S_2$ and sulfur on being heated. These disulfides, unlike the pyrite-type $Mn^{II}S_2$, contain monatomic S^{-II} units. The diselenides are similar. TcS_2, $TcSe_2$, and ReS_2 feature trigonal prismatic coordination of M^{IV} by S (or Se) in a layer-lattice structure which is isomorphous with a rhombohedral polymorph of MoS_2. $ReSe_2$ also has a layer structure but the Re^{IV} atoms are octahedrally coordinated.

24.3.2 *Oxoanions*

The lower oxides of manganese are basic and react with aqueous acids to give salts of Mn^{II} and Mn^{III} cations. The higher oxides, on the other hand, are acidic and react with alkalis to yield oxoanion salts, but the polymerization which was such a feature of the chemistry of the preceding group is absent here.

Fusion of MnO_2 with an alkali metal hydroxide and an oxidizing agent such as KNO_3 produces very dark-green manganate(VI) salts (manganates) which are stable in strongly alkaline solution but which disproportionate readily in neutral or acid solution (see Fig. 24.1):

$$3MnO_4^{2-} + 4H^+ \longrightarrow 2MnO_4^- + MnO_2 + 2H_2O$$

The deep-purple manganate(VII) salts (permanganates) may be prepared in aqueous solution by oxidation of managanese(II) salts with very strong oxidizing agents such as PbO_2 or $NaBiO_3$, and are manufactured commercially by the electrolytic oxidation of manganate(VI):

$$2K_2MnO_4 + 2H_2O \longrightarrow 2KMnO_4 + 2KOH + H_2$$

An older and less-efficient method was to induce disproportionation of the alkaline K_2MnO_4 by reducing the pH with CO_2.

The most important manganate(VII) is $KMnO_4$ of which several tens of thousands of tonnes are produced annually. It is a well-known oxidizing agent, used analytically:

in acid solution: $MnO_4^- + 8H^+ + 5e^- \rightleftharpoons Mn^{2+} + 4H_2O$; $E° = +1.51$ V
in alkaline solution: $MnO_4^- + 2H_2O + 3e^- \rightleftharpoons MnO_2 + 4OH^-$; $E° = +1.23$ V

It is also important as an oxidizing agent in the industrial production of saccharin and benzoic acid, and medically as a disinfectant. It is increasingly being used also for purifying water, since it has the dual advantage over chlorine that it does not affect the taste, and the MnO_2 produced acts as a coagulant for colloidal impurities.

Reduction of $KMnO_4$ with aqueous Na_2SO_3 produces the bright-blue manganate(V) (hypomanganate), MnO_4^{3-}, which has also been postulated as a reaction intermediate in some organic oxidations; it is not stable, being prone to disproportionation.

All $[MnO_4]^{n-}$ ions are tetrahedral with Mn–O 162.9 pm in MnO_4^{-} and 165.9 pm in MnO_4^{2-}. K_2MnO_4 is isomorphous with K_2SO_4 and K_2CrO_4. By contrast, the only tetrahedral oxoanions of Tc and Re are the technetate(VII) (pertechnetate) and rhenate(VII) (perrhenate) ions. $HTcO_4$ and $HReO_4$ are strong acids like $HMnO_4$ and are formed when the heptoxides are dissolved in water. Careful evaporation of such solutions produces dark-red crystals with the composition $HTcO_4$ in the case of technetium and, in the case of rhenium, yellowish crystals of $Re_2O_7.2H_2O$. This latter compound has the structure $O_3Re–O–ReO_3(H_2O)_2$ already mentioned and is the nearest approach in this group to an isopolyacid.

$[TcO_4]^{-}$ and $[ReO_4]^{-}$ are produced whenever compounds of Tc and Re are treated with oxidizing agents such as nitric acid or hydrogen peroxide and, although reduced in aqueous solution by, for instance, Sn^{II}, Fe^{II}, Ti^{III}, and I^{-}, they are much weaker oxidizing agents than $[MnO_4]^{-}$. In further contrast to $[MnO_4]^{-}$ they are also stable in alkaline solution and are colourless whereas $[MnO_4]^{-}$ is an intense purple. In fact, the absorption spectra of the 3 $[MO_4]^{-}$ ions are very similar, arising in each case from charge transfer transitions between O^{2-} and M^{VII}, but the energies of these transitions reflect the relative oxidizing properties of M^{VII}. Thus the intense colour of $[MnO_4]^{-}$ arises because the absorption occurs in the visible region, whereas for $[ReO_4]^{-}$ it has shifted to the more energetic ultraviolet, and the ion is therefore colourless. $[TcO_4]^{-}$ is also normally colourless but the absorption starts on the very edge of the visible region and it may be that the red colour of crystalline $HTcO_4$, and other transient red colours which have been reported in some of its reactions, are due to slight distortions of the ion from tetrahedral symmetry causing the absorption to move sufficiently for it to "tail" into the blue end of the visible, thereby imparting a red coloration.

Fusion of rhenates(VII) with a basic oxide yield so-called ortho- and meso-perrhenates (M_5ReO_6 and M_3ReO_5, $M = Na$, $\frac{1}{2}Ca$, etc.) while addition of rhenium metal to the fusion (and exclusion of oxygen) produces rhenate(VI) (e.g. Ca_3ReO_6). There is evidence suggesting the existence of $[ReO_6]^{5-}$ and $[ReO_6]^{6-}$ but it may be better to regard all these compounds as mixed oxides. In any case, it is clear that the coordination sphere of the metal has expanded compared to that of the smaller Mn in the tetrahedral $[MnO_4]^{n-}$ ions. Comparable technetium compounds have also been prepared.

24.3.3 *Halides and oxohalides*[7]

The known halides and oxohalides of this group are listed in Tables 24.5 and 24.6 respectively.

The highest halide of each metal is of course a fluoride: ReF_7 (the only thermally stable heptahalide of a transition metal), TcF_6, and MnF_4 (indicating the diminished ability of manganese to attain high oxidation states when compared not only to Tc and Re but also to chromium, which forms CrF_5 and CrF_6). The most interesting of the lower halides are

---

[7] R. COLTON and J. H. CANTERFORD, *Halides of the First Row Transition Metals*, Chap. 5, pp. 212–70, Wiley, London, 1969; and *Halides of the Second and Third Row Transition Metals*, Chap. 7, pp. 272–322, Wiley, London, 1968.

TABLE 24.5 *Halides of Group VIIA*

Oxidation state	Fluorides	Chlorides	Bromides	Iodides
+7	ReF$_7$ yellow mp 48.3°, bp 73.7°			
+6	TcF$_6$ yellow mp 37.4°, bp 55.3°			
	ReF$_6$ yellow mp 18.5°, bp 33.7°	ReCl$_6$ red-green mp 29° (dichroic)		
+5	TcF$_5$ yellow mp 50°, bp (d)			
	ReF$_5$ yellow-green mp 48°, bp(extrap) 221°	ReCl$_5$ brown-black mp 220°	ReBr$_5$ dark brown (d 110°)	
+4	MnF$_4$ blue (d above rt)			
	—	TcCl$_4$ red (subl > 300°)	(?TcBr$_4$) (red-brown)	
	ReF$_4$ blue (subl > 300°)	ReCl$_4$ purple-black (d 300°)	ReBr$_4$	ReI$_4$ d above rt
+3	MnF$_3$ red-purple			
	—	[ReCl$_3$]$_3$ dark red (subl 500°) (d)	[ReBr$_3$]$_3$ red-brown	[ReI$_3$]$_3$ lustrous black (d on warming)
+2	MnF$_2$ pale pink mp 920°	MnCl$_2$ pink mp 652°, bp ~1200°	MnBr$_2$ rose mp 695°	MnI$_2$ pink mp 613°

the rhenium trihalides which exist as trimeric clusters which persist throughout much of the chemistry of Re^{III}.

Apart from ReF$_5$, which is produced when ReF$_6$ is reduced by tungsten wire at 600°C, all the known penta-, hexa-, and hepta- halides of Re and Tc can be prepared directly from the elements by suitably adjusting the temperature and pressure, although various specific methods have been suggested. They are volatile solids varying in colour from pale yellow (ReF$_7$) to dark brown (ReBr$_5$), and are readily hydrolysed by water with accompanying disproportionation into the comparatively more stable [MO$_4$]$^-$ and MO$_2$, e.g.:

$$3ReCl_5 + 8H_2O \longrightarrow HReO_4 + 2ReO_2 + 15HCl$$

Because of the tendency to produce mixtures of the halides, and the facile formation of

TABLE 24.6   *Oxohalides of Group VIIA*

Oxidation state	Fluorides			Chlorides	Bromides
+7	—	—	MnO$_3$F dark green mp −78°, bp(extrap) 60°	MnO$_3$Cl vol liq	
	—	—	TcO$_3$F yellow mp 18.3°, bp ∼100°	TcO$_3$Cl colourless	
	ReOF$_5$ cream mp 43.8°, bp 73.0°	ReO$_2$F$_2$ yellow mp 90°, bp 185°	ReO$_3$F yellow mp 147°, bp 164°	ReO$_3$Cl colourless mp 4.5°, bp 130°	ReO$_3$Br colourless
+6	—			MnO$_2$Cl$_2$ vol liq	
	TcOF$_4$ blue mp 134° bp (extrap) 165°			TcOCl$_4$ blue	
	ReOF$_4$ blue mp 108°, bp 171°			ReOCl$_4$ brown mp 30°, bp(extrap) 228°	ReOBr$_4$ blue
+5	—			MnOCl$_3$ vol liq	—
	—			TcOCl$_3$	TcOBr$_3$ black
	ReOF$_3$ black				—

oxohalides if air and moisture are not rigorously excluded (or even, in some cases, also by attacking glass), not all of these halides have been characterized as well as might be desired. There is spectroscopic evidence that ReF$_7$ has a pentagonal bipyramidal structure, and ReX$_6$ are probably octahedral. ReCl$_5$ is actually a dimer, Cl$_4$ReCl$_2$ReCl$_4$, in which the rhenium is octahedrally coordinated.

The tetrahalides are made by a variety of methods. MnF$_4$, being the highest halide formed by Mn, can be prepared directly from the elements, as can TcCl$_4$, which is the only thermally stable chloride of Tc. TcCl$_4$ is a red sublimable solid consisting of infinite chains of edge-sharing TcCl$_6$ octahedra. By contrast the black ReCl$_4$, which is prepared by heating ReCl$_3$ and ReCl$_5$ in a sealed tube at 300°C, is made up of pairs of ReCl$_6$ octahedra which share faces (as in [W$_2$Cl$_9$]$^{3-}$, p. 1190), these dimeric units then being linked in chains by corner-sharing. The closeness of the Re atoms in each pair (273 pm) is indicative of a metal–metal bond though not so pronounced as the more extensive metal–metal bonding found in Re^{III} chemistry.

MnF$_3$ is a red-purple, reactive, but thermally stable solid; it is prepared by fluorinating any of the Mn^{II} halides and its crystal lattice consists of MnF$_6$ octahedra which are distorted, presumably because of the Jahn–Teller effect expected for d^4 ions. The Re^{III} halides are obtained by thermal decomposition of $ReCl_5$, $ReBr_5$, and ReI_4. The dark-red chloride is composed of triangular clusters of chloride-bridged Re atoms with 1 of the 2 out-of-plane Cl on each Re bridging to adjacent trimeric clusters (Fig. 24.3). After allowing for the Re–Cl bonds, each Re^{III} has a d^4 configuration and the observed diamagnetism can be accounted for by assuming that these four d electrons on each Re are used in forming double bonds $(\sigma + \pi)$ to its 2 Re neighbours. The Re–Re distance of 249 pm is consistent with this (cf. 275 pm in Re metal). Re_3Cl_9 can be sublimed under vacuum but the green colour of the vapour probably indicates breakdown of the cluster in the vapour phase. The compound dissolves in water to give a red solution which slowly hydrolyses to hydrated Re_2O_3, and in conc hydrochloric acid it gives a red solution which is stable to oxidation and from which can be precipitated a number of complex chlorides in which the trimeric clusters persist.

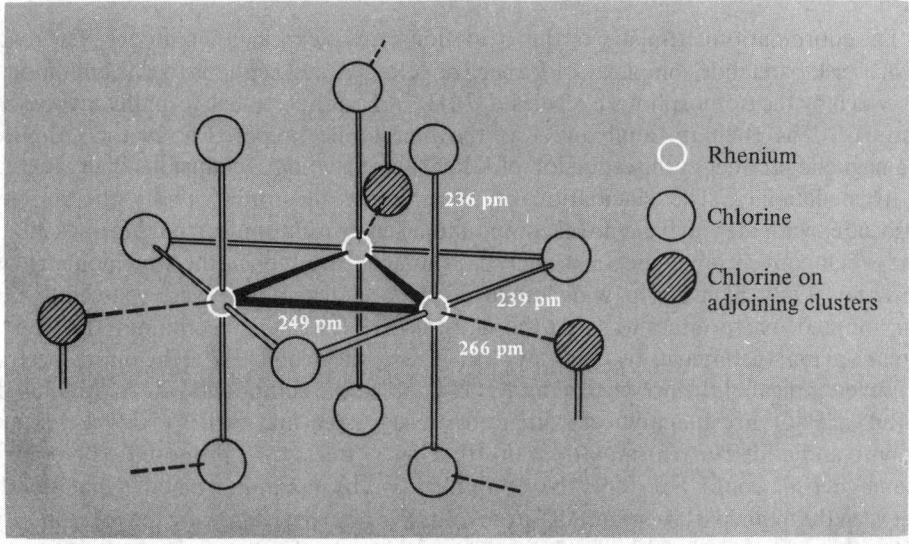

FIG. 24.3   Idealized structure of Re_3Cl_9; in crystalline $ReCl_3$ the trimeric units are linked into planar hexagonal networks.

Re_3Br_9 is similar to Re_3Cl_9 but the iodide, which is a black solid and is similarly trinuclear, differs in that it is thermally less stable and only 2 Re atoms in each cluster are linked to adjacent clusters, thereby forming infinite chains of trimeric units rather than planar networks.

Except for the possible existence of ReI_2, the only simple dihalides of this group that are known (so far) are those of manganese. They are pale-pink salts obtained by simply dissolving the metal or carbonate in aqueous HX. MnF_2 is insoluble in water and forms no hydrate, but the others form a variety of very water-soluble hydrates of which the tetrahydrates are the most common.

The oxohalides of manganese are green liquids (except MnO_2Cl_2 which is brown); they are notable for their explosive instability. MnO_3F can be prepared by treating $KMnO_4$

with fluorosulfuric acid, HSO_3F, whereas reaction of Mn_2O_7 with chlorosulfuric acid yields $MnO_3Cl + MnO_2Cl_2 + MnOCl_3$.

The oxohalides of technetium and rhenium are more numerous than those of manganese and are not so unstable, although all of them readily hydrolyse (with disproportionation to $[MO_4]^-$ and MO_2 in the case of oxidation states $+5$ and $+6$). In this respect they may be regarded as being intermediate between the halides and the oxides which, in the higher oxidation states, are the more stable. Treatment of the oxides with the halogens, or the halides with oxygen are common preparative methods. The structures are not all known with certainty, but $ReOCl_4$ may be noted as an example of a square-pyramidal structure.

24.3.4 *Complexes of manganese, technetium, and rhenium*[8]

Oxidation state VII (d^0)

The coordination chemistry of this oxidation state is confined mainly to a few readily hydrolysed oxohalide complexes of Re such as $KReO_2F_4$. Exceptions to this limitation are provided by the isomorphous hydrides K_2MH_9 of Tc and Re which formally involve M^{VII} and H^-. The rhenium analogue was the first to be prepared,[9] as the colourless, diamagnetic product of the reduction of $KReO_4$ by potassium in aqueous diaminoethane (ethylenediamine). The elucidation of its structure illustrates vividly the problems associated with identifying a novel compound when its isolation in a pure form is difficult and when conventional chemical analysis is unable to establish the stoichiometry with accuracy. Only by using a wide range of physical techniques were these difficulties surmounted. The product was first thought to be $K[Re(H_2O)_4]$ containing Re^{-1} with a square-planar geometry, by analogy with the isoelectronic Pt^{II}. The nmr spectrum, however, indicated the presence of an Re–H bond so the compound was reformulated as $KReH_4.2H_2O$. Fresh analytical evidence then suggested that earlier products had been impure and a further reformulation, this time as K_2ReH_8, was proposed. The observed diamagnetism could then only be accounted for by assuming metal–metal bonding between the implied d^1 rhenium(VI) atoms, but X-ray analysis showing the Re atoms to be 550 pm apart, precluded this possibility. The problem was finally resolved when a neutron diffraction study established[10] the formula as K_2ReH_9 and the structure is tricapped trigonal prismatic (Fig. 24.4). The nmr spectrum actually shows only one proton signal in spite of the existence of distinct capping and prismatic protons, and this is thought to be due to rapid exchange between the sites.

Oxidation state VI (d^1)

Again, fluoro and oxo complexes of rhenium predominate. The reaction of KF and ReF_6 in an inert PTFE vessel yields pink $K_2[ReF_8]$, the anion of which has a square-prismatic structure; hydrolysis converts it to $K[ReOF_5]$.

[8] G. ROUSCHIAS, Recent advances in the chemistry of rhenium, *Chem. Rev.* **74**, 531–66 (1974).

[9] Ref. 3, p. 912.

[10] S. C. ABRAHAMS, A. P. GINSBERG, and K. KNOX, The crystal and molecular structure of potassium rhenium hydride K_2ReH_9, *Inorg. Chem.* **3**, 558–67 (1964).

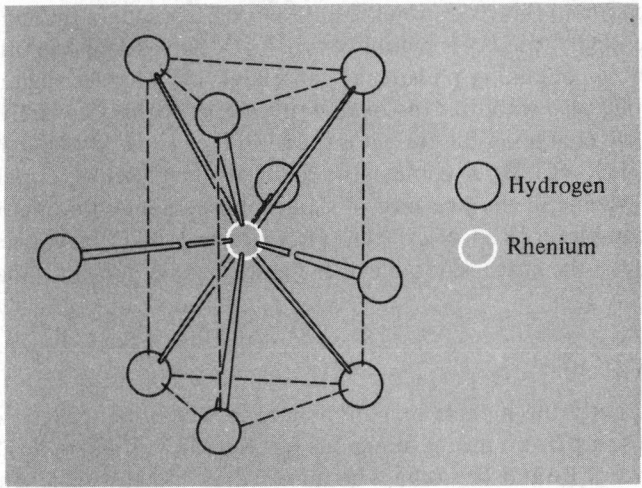

FIG. 24.4   The tricapped prismatic structure of the $[ReH_9]^{2-}$ anion.

An interesting compound which is usually included in discussions of Re^{VI} chemistry is the green crystalline dithiolate, $[Re(S_2C_2Ph_2)_3]$. This was the first authenticated example of a trigonal prismatic complex[11] (Fig. 19.6, p. 1074) but besides its structural interest it has, along with other complexes of such ligands, posed problems regarding the oxidation state of the metal. The ligand may be thought to coordinate in either of two extreme ways (or some intermediate state between them):

$$\left(\begin{array}{c} Ph-C=S \\ | \\ Ph-C=S \end{array}\right)_3 Re^0 \quad \text{or} \quad \left(\begin{array}{c} Ph-C-S \\ \| \\ Ph-C-S \end{array}\right)_3 Re^{VI}$$

The difference between the two extremes is essentially that, in the former, the rhenium Re retains its valence electrons in its d orbitals whereas in the latter it loses 6 of them to delocalized ligand orbitals. In either case paramagnetism is anticipated since rhenium has an odd number of valence electrons. The magnetic moment of 1.79 BM corresponding to 1 unpaired electron, and esr evidence showing that this electron is situated predominantly on the ligands, indicates that an intermediate oxidation state is involved but does not specify which one. Because of this uncertainty, dithiolate ligands, and others like them, have been expressively termed "non-innocent" ligands by C. K. Jørgensen.[11a]

Oxidation state V (d²)

Some fluoride complexes such as the salts of $[ReF_6]^-$ are known, but in oxo compounds such as $[ReOCl_5]^{2-}$ and $[ReOX_4]^-$ (X = Cl, Br, I) other halides are also able to co-

[11] R. EISENBERG and J. A. IBERS, Trigonal prismatic coordination: the molecular structure of tris(*cis*-1,2-diphenylethene-1,2-dithiolato)rhenium, *J. Am. Chem. Soc.* **87**, 3776–8 (1965).
[11a] C. K. JØRGENSEN, *Oxidation Numbers and Oxidation States*, Springer-Verlag, Berlin, 1969, 291 pp.

ordinate to rhenium in this oxidation state. $[ReOX_4]^-$ is square pyramidal with apical $Re=O$ and the ReO^{3+} moiety is reminiscent of VO^{2+}, being found in other compounds (particularly those containing phosphines) and labilizing whatever ligand is *trans* to it. The $Re\equiv N$ group also stabilizes the oxidation state, probably because the π bonds are able to reduce the charge on the Re^V; it is found in the yellow $[ReNX_2(PR_3)_3]$ and red $[ReNX_2(PR_3)_2]$ (X = Cl, Br, I) complexes produced when $[ReO_4]^-$ in alcoholic HCl is reduced by hydrazine in the presence of a phosphine. Eight-coordinate and probably dodecahedral complexes $[Re(CN)_8]^{3-}$ and $[ReCl_4(diars)_2]ClO_4$ have been prepared and the Tc analogue of the latter is notable as the first example of 8-coordinate technetium.

Oxidation state IV (d^3)

This is apparently the highest oxidation state in which manganese is able to form complexes. These are by no means numerous but $K_2[MnX_6]$, where X = F, Cl, IO_3, and CN, are known. For Re and Tc, perhaps because of the symmetrical t_{2g}^3 configuration, this is one of the stable oxidation states, giving rise to the oxides and halides already mentioned. Complexes are not especially numerous, however, the most important being the salts of $[MX_6]^{2-}$ [M = Tc, Re; X = F, Cl, Br, I]. The fluoro complexes are obtained by the reaction of HF on one of the other halogeno complexes and these in turn are obtained by reducing $[MO_4]^-$ (commonly by using I^-) in aqueous HX. The corresponding Tc and Re complexes are closely similar, but an interesting difference between Tc^{IV} and Re^{IV} is found in their behaviour with CN^-. The reaction of KCN and K_2ReI_6 in methanol causes oxidation of Re^{IV} to Re^V with the formation of brown diamagnetic $K_3[Re(CN)_8]$. Assuming that each CN^- ion donates an electron pair, the octacyanorhenium(IV) ion would contain 19 outer electrons and the loss of 1 electron is therefore easily understood. However, this high coordination number, and with it the stable 18-electron configuration, is evidently not so easily attained by technetium since the analogous reaction of KCN and K_2TcI_6 produces a reddish-brown, paramagnetic precipitate, thought to be $K_2[Tc(CN)_6]$.

Oxidation state III (d^4)

Manganese(III) is strongly oxidizing in aqueous solution with a marked tendency to disproportionate into Mn^{IV} (i.e. MnO_2) and Mn^{II} (see Fig. 24.1). It is, however, stabilized by O-donor ligands, as evidenced by the way in which the virtually white $Mn(OH)_2$ rapidly darkens in air as it oxidizes to hydrous Mn_2O_3 or $MnO(OH)$, and by the preparation of $[Mn(acac)_3]$ via the aerial oxidation of aqueous Mn^{II} in the presence of acetylacetonate. $K_3[Mn(C_2O_4)_3].3H_2O$ is also known, while the complexing oxoanions, phosphate and sulfate, have a stabilizing effect on aqueous solutions. The main preparative routes to Mn^{III} are by reduction of $KMnO_4$ or oxidation of Mn^{II}. The latter may be effected electrolytically but a common method is by way of the red-brown acetate. This is similar to the basic acetate of chromium(III) and so involves the $[Mn_3O(MeCOO)_6]^+$ unit (see Fig. 23.11, p. 1200). The hydrate is prepared by oxidation of manganese(II) acetate with $KMnO_4$ in glacial acetic acid, and the anhydrous salt by the action of acetic hydride on hydrated manganese(II) nitrate.

Nearly all manganese(III) complexes are octahedral and high-spin with magnetic moments close to the spin-only value of 4.90 BM expected for 4 unpaired electrons. A good deal of the interest in them has been occasioned by the expectation that they might be subject to Jahn–Teller distortions. Complexes involving 6 equivalent donor atoms provide the most telling evidence and the alums, of which $CsMn(SO_4)_2.12H_2O$ is the most stable, are of interest because they contain the $[Mn(H_2O)_6]^{3+}$ ion. For reasons which are not obvious, this ion apparently shows no appreciable distortion from precise octahedral symmetry although distortions in solid MnF_3, the octahedral Mn^{III} sites in Mn_3O_4, and more recently[12] in tris(tropolonato)manganese(III), have been ascribed to the Jahn–Teller effect. Earlier claims that $[Mn(acac)_3]$ was undistorted have since been convincingly refuted.[13]

The most important low-spin octahedral complex of Mn^{III} is the dark-red cyano complex, $[Mn(CN)_6]^{3-}$, which is produced when air is bubbled through an aqueous solution of Mn^{II} and CN^-. $[MnX_5]^{2-}$ $(X = F, Cl)$ are also known; the chloro ion, at least when combined with the cation $[bipyH_2]^{2+}$, is notable as an example of a square pyramidal manganese complex.[14]

Technetium(III) complexes are rather sparse: $[TcCl_2(diars)_2]ClO_4$, prepared by the reaction of *o*-phenylenebisdimethylarsine and HCl with $HTcO_4$ in aqueous alcohol, is probably the best known. The rhenium(III) analogue is isomorphous but requires the help of a reducing agent such as H_3PO_2 to effect the reduction from $[ReO_4]^-$. In general Re^{III} is readily oxidized to Re^{IV} or Re^{VII} unless it is stabilized by metal–metal bonding[15] as in the case of the trihalides already discussed. Rhenium(III) complexes with Cl^- and Br^- have been characterized and are of two types, $[Re_3X_{12}]^{3-}$ and $[Re_2X_8]^{2-}$, both of which involve multiple Re–Re bonds. If Re_3Cl_9 or Re_3Br_9 are dissolved in conc HCl or conc HBr respectively, stable red, diamagnetic salts may be precipitated by adding a suitable monovalent cation. Their stoichiometry is M^IReX_4 and they were formerly thought to be unique examples of low-spin, tetrahedral complexes. X-ray analysis, however, showed that the anions are trimeric with the same structure as the halides (Fig. 24.3) and likewise incorporating Re=Re double bonds. Their chemistry reflects the structure since 3 halide ions per trimeric unit can be replaced by ligands such as MeCN, Me_2SO, Ph_3PO, and PEt_2Ph yielding neutral complexes $[Re_3X_9L_3]$.

The blue diamagnetic complexes $[Re_2X_8]^{2-}$ are produced when $[ReO_4]^-$ in aqueous HCl or HBr is reduced by H_3PO_2 and they can then be precipitated by the addition of a suitable cation. The structure of $[Re_2Cl_8]^{2-}$ is shown in Fig. 24.5 and, as in $[Mo_2Cl_8]^{4-}$ (Fig. 23.13, p. 1203), the chlorine atoms are eclipsed. In both ions the metal has a d^4 configuration which is to be expected if a δ-bond is present. $[Re_2Cl_8]^{2-}$ can be reduced polarographically to unstable $[Re_2Cl_8]^{3-}$ and $[Re_2Cl_8]^{4-}$, and also undergoes a variety of substitution reactions (Fig. 24.5b, c, d).

One Cl on each Re may be replaced by phosphines, while $MeSCH_2CH_2SMe$ (DTH) takes up 4 coordination positions on 1 Re to give $[Re_2Cl_5(DTH)_2]$. In this case the average oxidation state of the rhenium has been reduced to $+2.5$ and the reduction in

[12] A. AVDEEF, J. A. COSTAMAGNA, and J. E. GUERCHAIS, Crystal and molecular structure of tris(tropolonato)-manganese(III), $Mn(O_2C_7H_5)_3$, a high-spin complex having structural features consistent with Jahn–Teller behavior for two distinct MnO_6 centers, *Inorg. Chem.* **13**, 1854–63 (1974).

[13] J. P. FACKLER and A. AVDEEF, Crystal structure of tris(2,4-pentanedionato)manganese(III), $Mn(O_2C_5H_7)_3$, a distorted complex as predicted by Jahn–Teller arguments, *Inorg. Chem.* **13**, 1864–75 (1974).

[14] I. BERNAL, N. ELLIOTT, and R. LALANCETTO, Molecular configuration of the anion $MnCl_5{}^{2-}$: a square pyramidal pentahalide of the 3d transition series, *Chem. Comm.* 1971, 803–5.

[15] F. A. COTTON, Quadruple bonds and other multiple metal to metal bonds, *Chem. Soc. Rev.* **4**, 27–53 (1975).

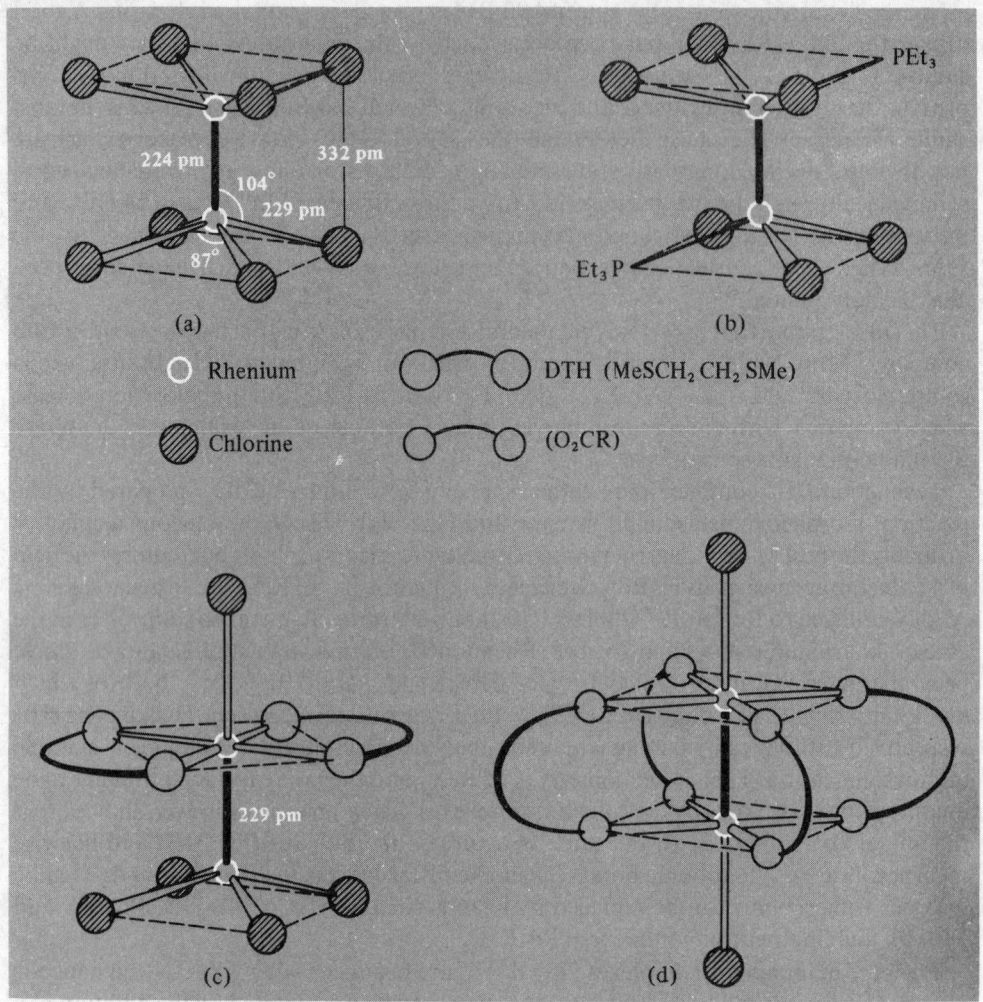

FIG. 24.5 Some complexes of Re with multiple Re-Re bonds: (a) $[Re_2Cl_8]^{2-}$. (b) $[Re_2Cl_6(PEt_3)_2]$. (c) $[Re_2Cl_5(DTH)_2]$: note the staggered configuration which is now possible because of the loss of the δ bond as the Re is reduced from $+3$ to $+2.5$ and the bond order decreased from 4 to 3. (d) $[Re_2(O_2CR)_4Cl_2]$.

bond order from 4 to 3.5, which the addition of a δ^* electron implies, causes some lengthening of the Re-Re distance and the configuration is staggered. Carboxylates are able to bridge the metal atoms forming complexes of the type $[Re_2Cl_2(O_2CR)_4]$ which are clearly analogous to the dimeric carboxylates found in the previous group.

In the octachloro technetium system, by contrast, the paramagnetic $[Tc_2Cl_8]^{3-}$ is the most readily obtained species. The Tc has a formal oxidation state of $+2.5$ with a $d(Tc-Tc)$ 212 pm and the configuration is eclipsed. The pale green $[NBu^n_4]^+{}_2[Tc^{III}Cl_8]^{2-}$ has recently also been isolated from the products of reduction of $[TcCl_6]^{2-}$ with $Zn/HCl(aq)$.[15a] The compound is strictly isomorphous with $[NBu^n_4]_2[Re_2Cl_8]$ and has

[15a] F. A. COTTON, L. DANIELS, A. DAVISON, and C. ORVIG, Structural characterization of the octachloroditechnetate(III) ion in its tetra-*n*-butylammonium salt, *Inorg. Chem.* **20**, 3051–5 (1981).

d(Tc–Tc) 215 pm. The reason for the increase in Tc–Tc distance on removal of the δ^* electron, and the consequent increase in the presumed bond order from 3.5 to 4, is not clear.

Oxidation state II (d⁵)

Oxidation state II (d^5)

The chemistry of technetium(II) and rhenium(II) is meagre and mainly confined to arsine and phosphine complexes. The best known of these are [MCl$_2$(diars)$_2$], obtained by reduction with hypophosphite and Sn^{II} respectively from the corresponding Tc^{III} and Re^{III} complexes, and in which the low oxidation state is presumably stabilized by π donation to the ligands. This oxidation state, however, is really best typified by manganese for which it is the most thoroughly studied and, in aqueous solution, by far the most stable; accordingly it provides the most extensive cationic chemistry in this group.

Salts of manganese(II) are formed with all the common anions and most are water-soluble hydrates. The most important of these commercially and hence the most widely produced is the sulfate, which forms several hydrates of which MnSO$_4$.5H$_2$O is the one commonly formed. It is manufactured either by treating pyrolusite with sulfuric acid and a reducing agent, or as a byproduct in the production of hydroquinone (MnO$_2$ is used in the conversion of aniline to quinone:

$$PhNH_2 + 2MnO_2 + 2\tfrac{1}{2}H_2SO_4 \longrightarrow OC_6H_4O + 2MnSO_4 + \tfrac{1}{2}(NH_4)_2SO_4 + 2H_2O)$$

It is the starting material for the preparation of nearly all manganese chemicals and is used in fertilizers in areas of the world where there is a deficiency of Mn in the soil, since Mn is an essential trace element in plant growth. The anhydrous salt has a surprising thermal stability; it remains unchanged even at red heat, whereas the sulfates of Fe^{II}, Co^{II}, and Ni^{II} all decompose under these conditions.

Aqueous solutions of salts with non-coordinating anions contain the pale-pink, [Mn(H$_2$O)$_6$]$^{2+}$, ion which is one of a variety of high-spin octahedral complexes which have been prepared more especially with chelating ligands such as en, EDTA, and oxalate. As is expected, most of these have magnetic moments close to the spin-only value of 5.92 BM. The absorption spectrum of [Mn(H$_2$O)$_6$]$^{2+}$ is described in the Panel.

The resistance of Mn^{II} to both oxidation and reduction is generally attributed to the effect of the symmetrical d^5 configuration, and there is no doubt that the steady increase in resistance of M^{II} ions to oxidation found with increasing atomic number across the first transition series suffers a discontinuity at Mn^{II}, which is more resistant to oxidation than either Cr^{II} to the left or Fe^{II} to the right. However, the high-spin configuration of the Mn^{II} ion provides no CFSE (p. 1096) and the stability constants of its high-spin complexes are consequently lower than those of corresponding complexes of neighbouring M^{II} ions and are kinetically labile. In addition, a zero CFSE confers no advantage on any particular stereochemistry which must be one of the reasons for the occurrence of a wider range of stereochemistries than is normally found for M^{II} ions.

Green-yellow salts of the tetrahedral [MX$_4$]$^{2-}$ (X = Cl, Br, I) ions can be obtained from ethanolic solutions and are well characterized. Furthermore, a whole series of adducts [MnX$_2$L$_2$] (X = Cl, Br, I) are known where L is an *N*-, *P*-, or *As*-donor ligand, and both octahedral and tetrahedral stereochemistries are found. Of interest because of the possible role of manganese porphyrins in photosynthesis is [Mn^{II}(phthalocyanine)] which is

Absorption Spectrum of $[Mn(H_2O)_6]^{2+}$

This spectrum (Fig. A) differs from those of other transition metal cations (see Fig. 19.20) in two important ways:

(a) the bands (apart from those in the ultraviolet region due to charge transfer) are of much lower intensity;

(b) there are more bands.

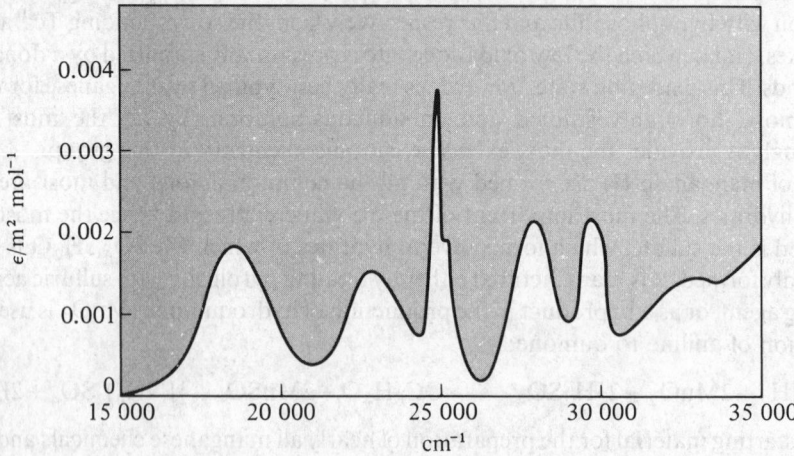

FIG. A   The electronic absorption spectrum of $[Mn(H_2O)_6]^{2+}$ (approximately 2 M $Mn(ClO_4)_2$ in approximately 2 M $HClO_4$).

The explanations of both these features stem from the fact that any electronic d–d transition from a high-spin d^5 configuration must of necessity involve the pairing of some electron spins. It follows immediately that such transitions are both spin forbidden and orbitally forbidden (p. 1094) and that the absorptions should be of exceedingly low intensity. That they occur at all is due to weak spin–orbit coupling. Furthermore, although there is only one possible way of arranging 5 electrons with parallel spins in 5 orbitals, there are many more ways if 2 electrons are paired. This is seen more quantitatively in the Tanabe–Sugano diagram (Fig. B) in which the ground state ^6S term corresponds to the configuration with 5 unpaired electrons and the excited "spin quadruplet" terms correspond to configurations with 2 paired electrons.† There are 4 of these in the free ion and in a cubic crystal field they split into no less than 10 component terms. The key to the assignment[16] of the possible transitions to the observed bands lies in the fact that the band near 25 000 cm^{-1} is unusually sharp, and has a shoulder indicating that there are actually two absorptions very close together. It has already been pointed out in Chapter 19 that absorption bands are broadened by molecular vibrations which cause Δ to oscillate, and by spin–orbit coupling which splits T terms. Now, inspection of Fig. B shows that the degenerate $^4E_g(G)$ and $^4A_{1g}(G)$ terms are parallel to the ground $^6A_{1g}$ term. The energy of the transition from the ground term will therefore not be affected either by molecular vibrations or, since no T term is involved, by

† "Doublet" terms corresponding to configurations with only 1 unpaired electron are generally of higher energy but, since transitions to them from the ^6S term would be doubly spin-forbidden the absorptions are too weak to be observed.

[16] L. J. HEIDT, G. F. KOSTER, and A. M. JOHNSON, Experimental and crystal field study of the absorption spectrum at 2000 to 8000 Å of manganous perchlorate in aqueous perchloric acid, *J. Am. Chem. Soc.* **80**, 6471–7 (1958).

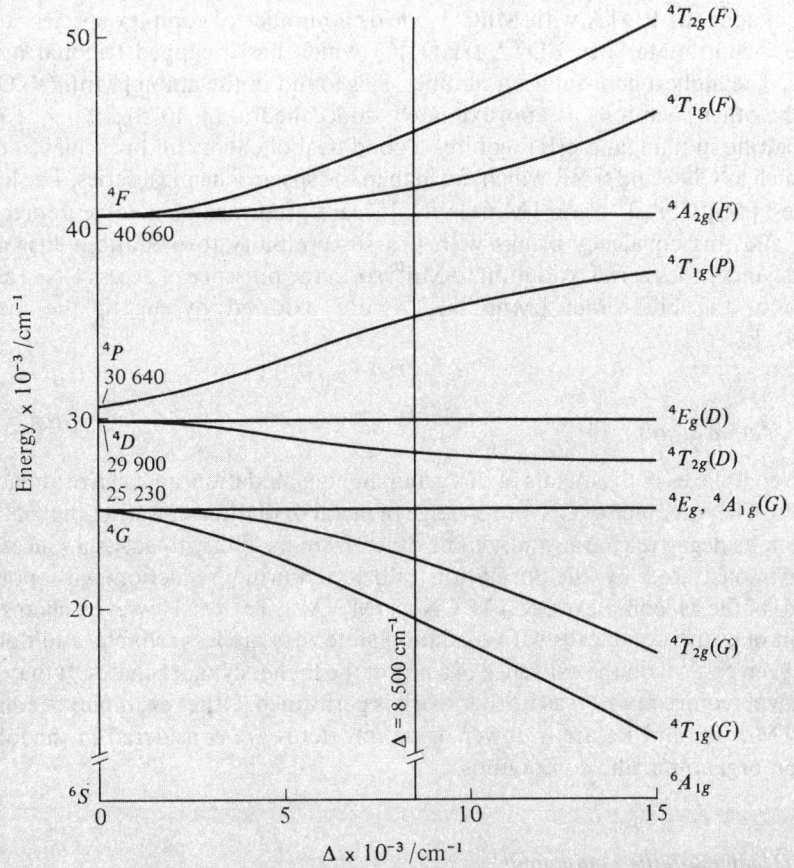

FIG. B   Tanabe–Sugano diagram for the Mn^{II} ion in an octahedral crystal field. All spin doublet terms are omitted.

spin–orbit coupling. A sharp band is consequently to be expected and the band and shoulder near 25 000 cm^{-1} may, with some confidence, be assigned to the $^4E_g(G){\leftarrow}^6A_{1g}$ and $^4A_{1g}(G){\leftarrow}^6A_{1g}$ transitions. This indicates a value of 8500 cm^{-1} for Δ and the assignments of the other bands as shown in Table A. The agreement between prediction and observation indicates the reliability of these assignments.

TABLE A   *The absorption bands of* $[Mn(H_2O)_6]^{2+}$ *in aqueous solution*

Band position/cm^{-1}		
Observed	Estimated	Assignment
18 600	19 400	$^4T_{1g}(G){\leftarrow}^6A_{1g}$
22 900	22 600	$^4T_{2g}(G){\leftarrow}^6A_{1g}$
24 900		$\Big\{\ ^4E_g(G){\leftarrow}^6A_{1g}$
25 150	25 200	$\ \ ^4A_{1g}(G){\leftarrow}^6A_{1g}$
27 900	28 200	$^4T_{2g}(D){\leftarrow}^6A_{1g}$
29 700	29 900	$^4E_g(D){\leftarrow}^6A_{1g}$

square planar, as is the bright-yellow dithiocarbamato complex, $[Mn(S_2CNEt_2)_2]$. The reaction of aqueous EDTA with $MnCO_3$ yields a number of complex species, amongst them the 7-coordinate $[Mn(EDTA)(H_2O)]^{2-}$ which has a capped trigonal prismatic structure. The highest coordination number, 8, is found in the anion $[Mn(\eta^2\text{-}NO_3)_4]^{2-}$ which, like other such ions, is approximately dodecahedral (p. 1075).

Spin-pairing in manganese(II) requires a good deal of energy and is achieved only by ligands such as CN^- and CNR which are high in the spectrochemical series. The low-spin complexes, $[Mn(CN)_6]^{4-}$ and $[Mn(CNR)_6]^{2+}$ are presumed to involve appreciable π bonding and this covalency brings with it a susceptibility to oxidation. Just as Mn^{II} hydroxide undergoes aerial oxidation to Mn^{III} so, in the presence of excess CN^-, aqueous solutions of the blue-violet $[Mn(CN)_6]^{4-}$ are oxidized by air to the dark red $[Mn(CN)_6]^{3-}$.

Lower oxidation states

Cyano complexes of the metals of this group in high oxidation states have already been referred to. The tolerance of CN^- to a range of metal oxidation states, arising, on the one hand, from its negative charge and, on the other, from its ability to act as a π-acceptor, is further demonstrated by the formation (albeit requiring reduction with potassium amalgam) of the M^I complexes, $K_5[M(CN)_6]$ ($M = Mn, Tc, Re$). However, claims for the formation of cyano complexes with oxidation state zero are less reliable, and doubt has recently been cast[17] on the existence of some of the higher cyanorhenates. It may be that this area will require revision as further work is performed. Other examples of complexes in which Mn, Tc, and Re are in lower oxidation states are considered in the following section on organometallic compounds.

24.3.5 *Organometallic compounds*

Carbonyls,[18] cyclopentadienyls, and their derivatives occupy a central position in the chemistry of this as of preceding groups, the bonding involved and even their stoichiometries having in some cases posed difficult problems. Increasingly, however, interest has focused on the chemistry of compounds involving M–C σ bonds, of which rhenium probably provides the richest variety of any transition metal.

Only one well-characterized binary carbonyl is formed by each of the elements of this group.† That of manganese is best prepared by reducing MnI_2 (e.g. with $LiAlH_4$) in the presence of CO under pressure. Those of technetium and rhenium are made by heating their heptoxides with CO under pressure. They are sublimable, isomorphous, crystalline solids: golden-yellow for $[Mn_2(CO)_{10}]$, mp 154°, and colourless for $[Tc_2(CO)_{10}]$, mp 160°, and $[Re_2(CO)_{10}]$, mp 177°. Their stabilities in air show a regular gradation: manganese carbonyl is quite stable below 110°C, technetium carbonyl decomposes slowly

---

† A trinuclear carbonyl, $[Tc_3(CO)_{12}]$, has been reported but not characterized.

[17] W. P. Griffith, P. M. Kiernan, and J.-M. Bregeault, Studies on transition-metal cyano-complexes. Part 2. Unsubstituted cyanorhenates, $[Re(CN)_x]^{n-}$, and cyanorhenates with thio-, seleno-, and nitrosyl ligands, *JCS Dalton* 1978, 1411–17.

[18] W. P. Griffith, Carbonyls, cyanides, isocyanides, and nitrosyls, Chap. 46 in *Comprehensive Inorganic Chemistry*, Vol. 4, pp. 105–95, Pergamon Press, Oxford, 1973.

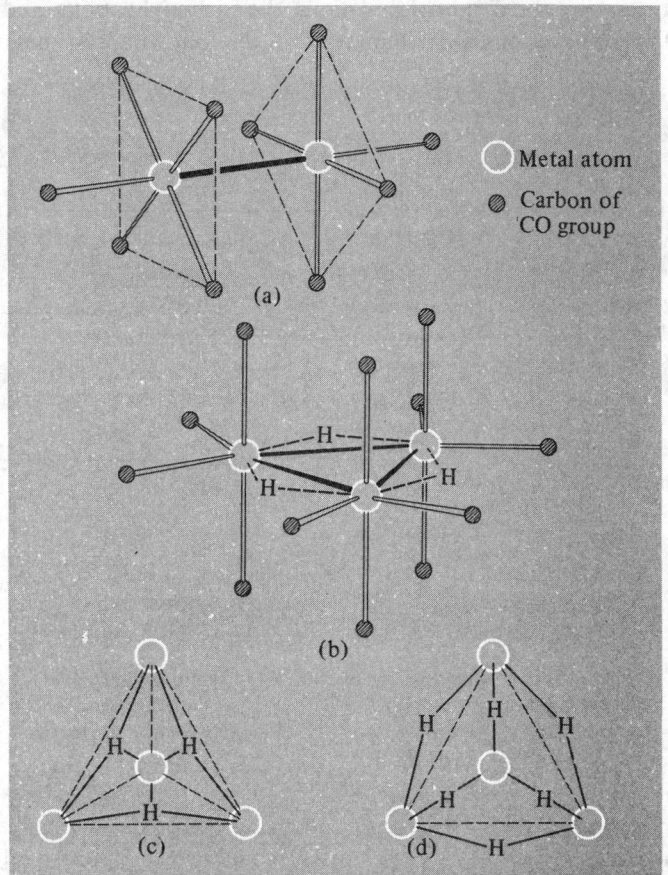

FIG. 24.6   Some carbonyls and carbonyl hydrides of Group VIIA metals. (a) $[M_2(CO)_{10}]$, M = Mn, Tc. Re (Mn–Mn 293 pm, Tc–Tc 304 pm, Re–Re 302 pm). (b) $[H_3Mn_3(CO)_{12}]$. Mn–Mn 311 pm. (c) $[H_4Re_4(CO)_{12}]$, Re–Re 289.6–294.5 pm. (In this structure the 4 Re atoms lie at the corners of a tetrahedron, the faces of which are bridged by 4 H atoms; the molecule is viewed from above 1 Re which obscures the fourth H. For clarity the CO groups are not shown but 3 are attached to each Re so as to "eclipse" the edges of the tetrahedron.) (d) $[H_6Re_4(CO)_{12}]^{2-}$, Re–Re 314.2–317.2 pm. (As in (c) the 4 Re atoms lie at the corners of a tetrahedron and the CO groups have been omitted for clarity. This time, however, the 3 CO groups attached to each Re are "staggered" wrt the edges of the tetrahedron, whilst the H atoms (6) are presumed to bridge these edges.)

and rhenium carbonyl may ignite spontaneously. The empirical stoichiometry $M(CO)_5$ would imply a paramagnetic molecule with 17 valence electrons, but the observed diamagnetism (for Mn and Re) suggests at least a dimeric structure. In fact, X-ray analysis reveals the structure shown in Fig. 24.6a in which two $M(CO)_5$ groups in staggered configuration are held together by an M–M bond, unsupported by bridging ligands (cf. S_2F_{10}, p. 809).

Very many derivatives of the carbonyls of Mn, Tc, and Re have been prepared since the parent carbonyls were first synthesized in 1949, 1961, and 1941 respectively:[2–4] among the more important are the carbonylate anions, the carbonyl cations, and the carbonyl hydrides.[19] Typical reactions are summarized in the following schemes.

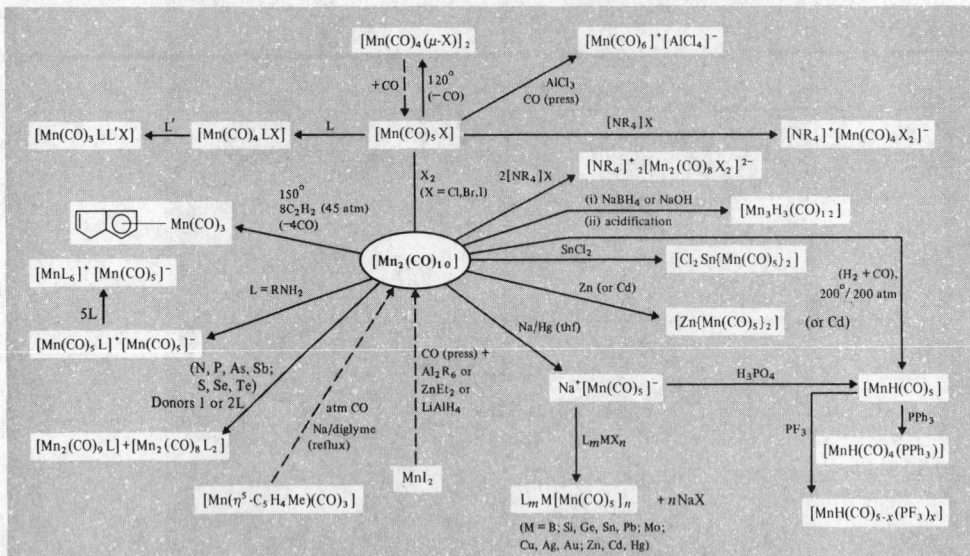

SCHEME A   Some reactions of $[Mn_2(CO)_{10}]$ and its derivatives.

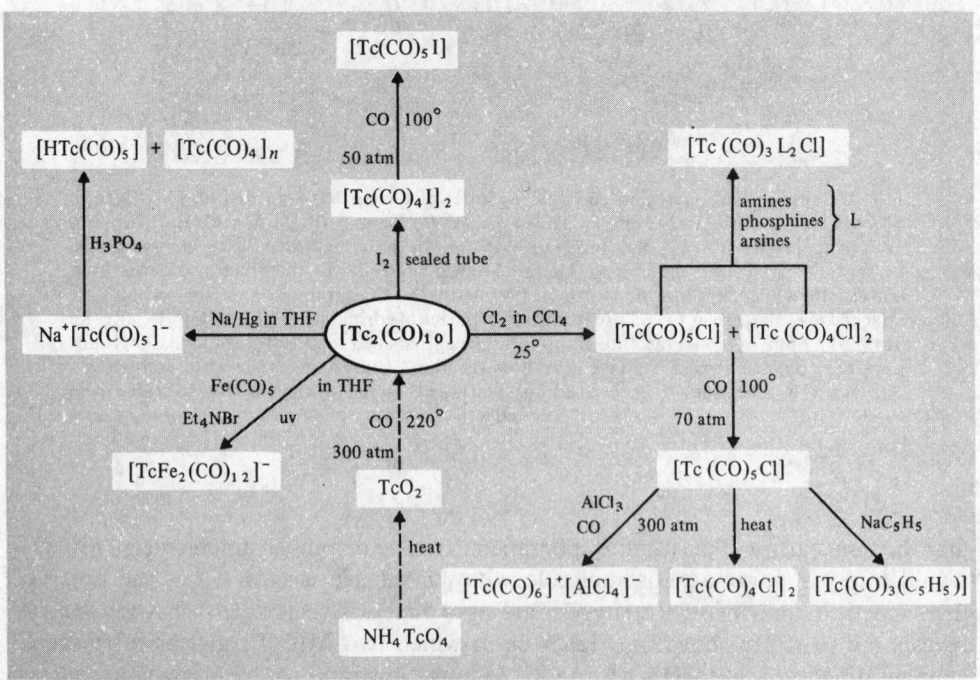

SCHEME B   Some reactions of $[Tc_2(CO)_{10}]$ and its derivatives.[2]

[19] B. F. G. JOHNSON (ed.), *Transition Metal Clusters*, Wiley, Chichester, 1980, 681 pp.

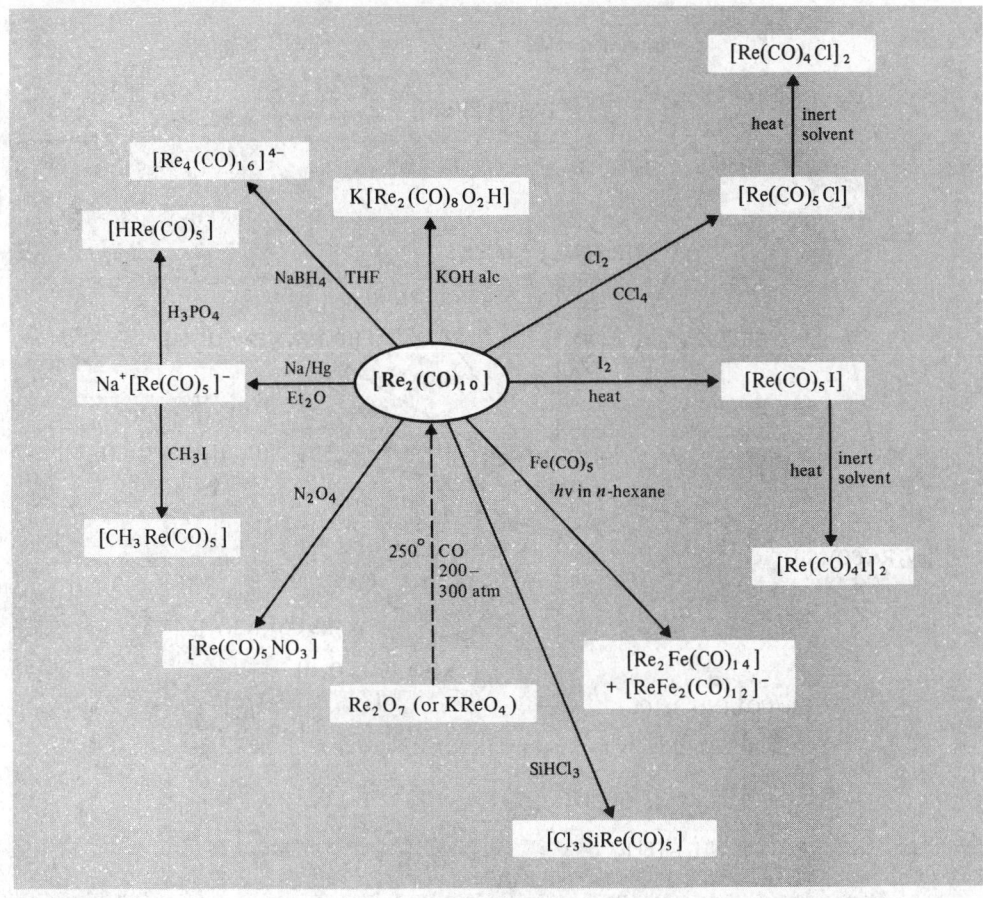

SCHEME C   Some reactions of rhenium carbonyl[3]

Sodium amalgam reductions of $M_2(CO)_{10}$ give $Na^+[M(CO)_5]^-$ and, indeed, further reduction[20] leads to the "super reduced" species $[M(CO)_4]^{3-}$ in which the metals exhibit their lowest known formal oxidation state of -3. On the other hand, treatment of $[M(CO)_5Cl]$ with $AlCl_3$ and CO under pressure produces $[M(CO)_6]^+AlCl_4^-$ from which other salts of the cation can be obtained. Acidification of $[M(CO)_5]^-$ produces the octahedral and monomeric, $[MH(CO)_5]$, and a number of polymeric carbonyls have been obtained by reduction of $[M_2(CO)_{10}]$, including interesting hydrogen-bridged complexes such as $[H_3Mn_3(CO_{12})]$ (the first transition metal cluster in which the H atoms were located[21]), $[H_4Re_4(CO)_{12}]$, and $[H_6Re_4(CO)_{12}]^{2-}$ (Fig. 24.6).[21a]

[20] J. E. ELLIS and R. A. FALTYNEK, Highly reduced organometallic anions. 1. Syntheses and properties of tetracarbonyl-metalate(3−) anions of manganese and rhenium, *J. Am. Chem. Soc.* **99**, 1801–8 (1977).

[21] S. W. KIRTLEY, J. P. OLSEN, and R. BARR, Location of the hydrogen atoms in $H_3Mn_3(CO)_{12}$: a crystal structure determination, *J. Am. Chem. Soc.* **95**, 4532–6 (1973).

[21a] A. P. HUMPHRIES and H. D. KAESZ, The hydrido-transition metal cluster complexes, *Prog. Inorg. Chem.* **25**, 145–222 (1979); R. HOFFMANN, B. E. R. SCHILLING, R. BAU, H. D. KAESZ, and D. M. P. MINGOS, Electronic structure of $M_4(CO)_{12}H_n$ and $M_4Cp_4H_n$ complexes, *J. Am. Chem. Soc.* **100**, 6088–93 (1978).

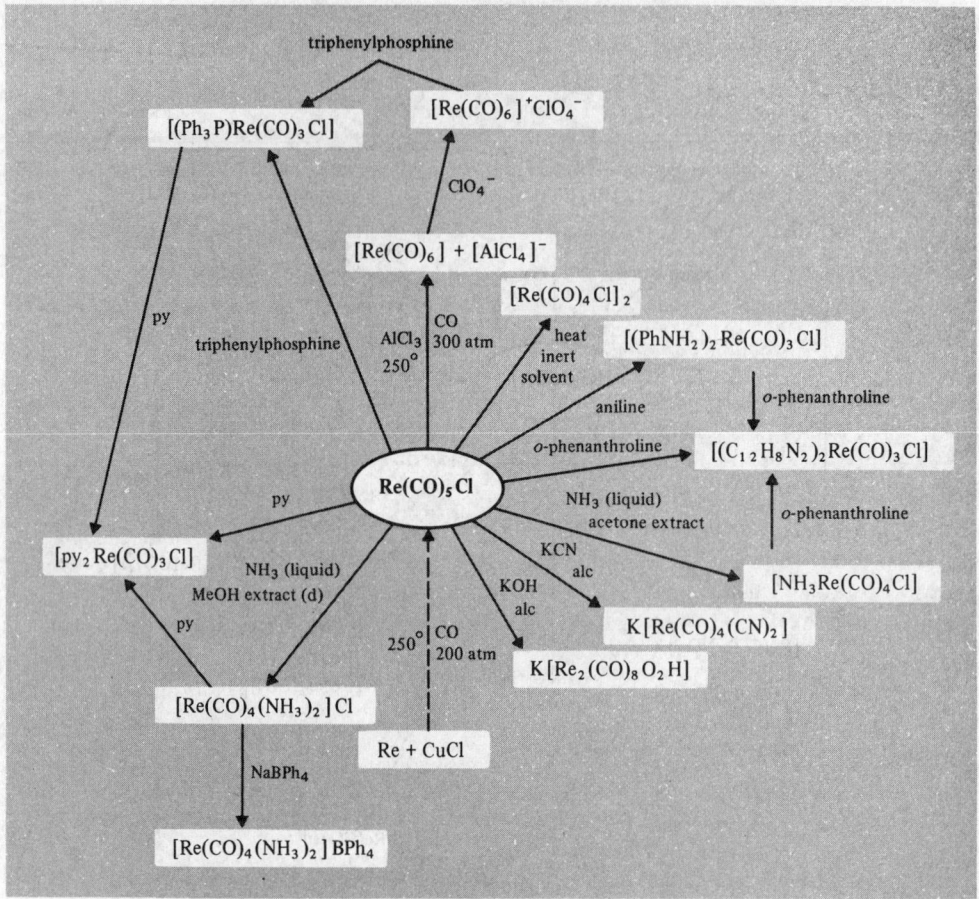

SCHEME D   Some reactions of rhenium carbonyl chloride.

When $MnCl_2$ in thf is treated with C_5H_5Na, amber-coloured crystals of manganocene, $[Mn(C_5H_5)_2]$, mp 172°, are produced. It is very sensitive to both air and water and has most unusual magnetic properties. It is antiferromagnetic with a Néel temperature of 134°C but, in dilute hydrocarbon solutions and when "doped" with solid $Mg^{II}(C_5H_5^-)_2$, it has a magnetic moment of 5.86 BM, corresponding to 5 unpaired electrons. This was taken as evidence for a high-spin, ionic configuration with intermolecular interactions leading to the antiferromagnetism. On the other hand, the similarity of the infrared spectrum and melting point to those of ferrocene itself, and the low electrical conductivity of ethereal solutions are suggestive of covalency; the sandwich structure of $[Mn(\eta^5$-$C_5H_5)_2]$, as revealed by X-ray studies, is consistent with either type of bonding.[22] However, $[Mn(C_5H_4Me)_2]$, which retains a subnormal magnetic moment (for 5 unpaired electrons) even in dilute solutions, has more recently been cited[23] as the first example of

[22] M. L. H. GREEN, *Organometallic Compounds*, 3rd edn., Vol. II, Methuen, London, 1967 (see p. 110).

[23] M. E. SWITZER, R. WANG, M. F. RETTIG, and A. H. MAKI, On the electronic ground state of manganocene and 1,1′-dimethylmanganocene, *J. Am. Chem. Soc.* **96**, 7669–74 (1974); A. ALMENNINGEN, S. SAMDAL, and A. HAALAND, Molecular structure of high- and low-spin 1,1′-dimethylmanganocenes determined by gas phase electron diffraction, *JCS Chem. Comm.* 1977, 14–15.

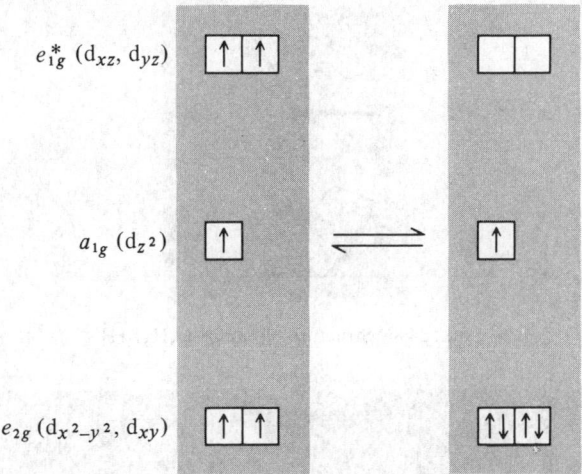

e_{1g}^* (d_{xz}, d_{yz})

a_{1g} (d_{z^2})

e_{2g} $(d_{x^2-y^2}, d_{xy})$

Fig. 24.7 Spin equilibrium in $[Mn(\eta^5\text{-}C_5H_4Me)_2]$: the orbitals shown here are the mainly metal-based orbitals in the centre of the MO diagram for metallocenes (see Fig. C, p. 371).

an Mn^{II} compound showing a high-spin⇌low-spin equilibrium (Fig. 24.7). In view of the similarity of the ligands C_5H_5 and C_5H_4Me, this suggests that manganocene itself is electronically probably very close to the high-spin/low-spin crossover, and the bonding may be regarded as similar to that in ferrocene but with the MOs sufficiently close together to permit a high-spin configuration which is unique among metallocenes. This would provide another example of the known resistance of Mn^{II} to spin-pairing. The precise origin of the antiferromagnetism is still undecided.

Technetium and Re analogues of manganocene are not known. Instead, when $TcCl_4$ or $ReCl_5$ are treated with C_5H_5Na in thf, yellow crystalline products are obtained which are diamagnetic and so cannot be the monomeric d^5 compounds $[M(C_5H_5)_2]$, though this formulation is consistent with the elemental analyses. It turns out that the rhenium compound (and the technetium compound is presumed to be the same) is a hydride of Re^{III}, in which the two C_5H_5 rings are tilted with respect to each other[24] (Fig. 24.8). In spite of this lack of symmetry, the protons on the cyclopentadienyl rings give rise to only one nmr signal. This is considered to be due to rapid rotation of the rings about the metal–ring axes, thereby making the protons indistinguishable.

The simplest of the σ-bonded Re–C compounds is the green, paramagnetic, crystalline, thermally unstable $ReMe_6$, which, after WMe_6, is only the second hexamethyl transition metal compound to have been synthesized (1976).[25] It reacts with LiMe to give the unstable, pyrophoric, $Li_2[ReMe_8]$, which has a square-antiprismatic structure, and incorporation of oxygen into the coordination sphere greatly increases the stability, witness $Re^{VI}OMe_4$, which is thermally stable up to 200°C, and $Re^{VII}O_3Me$, which is stable to air.[26]

[24] J. W. Lauher and R. Hoffmann, Structure and chemistry of bis(cyclopentadienyl)-ML_n complexes, *J. Am. Chem. Soc.* **98**, 1728–42 (1976).

[25] K. Mertis and G. Wilkinson, The chemistry of rhenium alkyls. Pt. III. The synthesis and reactions of hexamethylrhenium(VI), *cis*-trimethyldioxorhenium(VII), and the octamethylrhenium(VI) ion, *JCS Dalton* 1976, 1488–92.

[26] I. R. Beattie and P. J. Jones, Methyltrioxorhenium: an air stable compound containing a carbon–rhenium bond, *Inorg. Chem.* **18**, 2318–19 (1979).

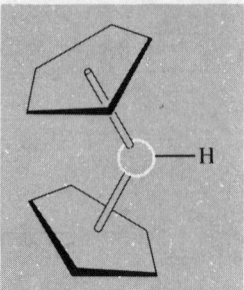

FIG. 24.8   Structure of $[Re(\eta^5\text{-}C_5H_5)_2H]$.

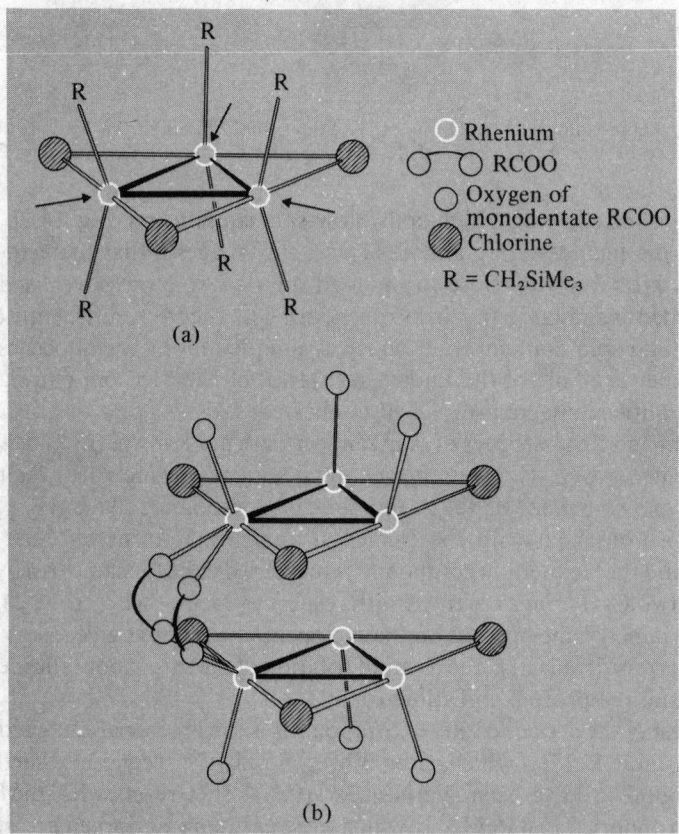

FIG. 24.9   Rhenium clusters: (a) $[Re_3Cl_3R_6]$: arrows indicate vacant coordination sites where further ligands can be attached. (b) A possible structure of $[Re_6Cl_6(O_2CR)_{12}]$: for clarity only 2 of the 6 bridging carboxylate groups are shown.

A whole series of rhenium alkyl cluster compounds (especially where the alkyl is CH_2SiMe_3) have been prepared starting from the blue trimeric $Re_3Cl_3R_6$ whose structure (Fig. 24.9a) is based on that of Re_3Cl_9 (see Fig. 24.3). Replacement of the 3 Cl bridges by 3 R groups gives $Re_3^{III}R_9$ to which up to 3 more ligands (e.g. phosphines) may be added in the Re_3 plane. If the additional groups are alkyls, then new Re^{IV} clusters Re_3R_{12} are obtained.

Alternatively, the use of bridging groups such as carboxylates[27] produce Re_6 species with structures which are possibly of the type shown in Fig. 24.9b though this has yet to be verified. The field of rhenium alkyl and cluster chemistry is currently expanding rapidly and compounds are being produced whose existence were quite unsuspected only a few years ago.

[27] P. G. EDWARDS, F. FELIX, K. MERTIS, and G. WILKINSON, Interaction of trirhenium(III) cluster alkyls with carboxylic acids, β-diketones and diphenyltriazine, *JCS Dalton* 1979, 361–6.

			1 H	2 He													

25

Iron, Ruthenium, and Osmium

25.1 Introduction

The nine elements, Fe, Ru, Os; Co, Rh, Ir; Ni, Pd, and Pt, together form Group VIII of Mendeleev's periodic table. They will be treated here, like the other transition elements, in "vertical" triads, but because of the marked "horizontal" similarities it is not uncommon for Fe, Co, and Ni to be distinguished from the other six elements (known collectively as the "platinum" metals) and the two sets of elements considered separately.

The triad Fe, Ru, and Os is dominated, as indeed is the whole block of transition elements, by the immense importance of iron. This element has been known since prehistoric times and no other element has played a more important role in man's material progress. Iron beads dating from around 4000 BC were no doubt of meteoric origin, and later samples, produced by reducing iron ore with charcoal, were not cast because adequate temperatures were not attainable without the use of some form of bellows. Instead, the spongy material produced by low-temperature reduction would have had to be shaped by prolonged hammering. It seems that iron was first smelted by the Hittites in Asia Minor sometime in the third millennium BC, but the value of the process was so great that its secret was carefully guarded and it was only with the eventual fall of the Hittite empire around 1200 BC that the knowledge was dispersed and the "Iron Age" began.[1] In more recent times the introduction of coke as the reductant had far-reaching effects, and was one of the major factors in the initiation of the Industrial Revolution. The name "iron" is Anglo-Saxon in origin (iren, cf. German Eisen). The symbol Fe and words such as "ferrous" derive from the Latin ferrum, iron.

Biologically, iron plays crucial roles in the transport and storage of oxygen and also in electron transport, and it is safe to say that, with only a few possible exceptions in the bacterial world, there would be no life without iron. Again, within the last three decades, the already rich organometallic chemistry of iron has been enormously expanded, and work in the whole field given an added impetus by the discovery and characterization of ferrocene.[2]

[1] V. G. CHILDE, What Happened in History, pp. 182–5, Penguin Books, London, 1942.
[2] J. S. THAYER, Organometallic chemistry: a historical perspective, Adv. Organometallic Chem. 13, 1–49 (1975).

Ruthenium and osmium, though interesting and useful, are in no way comparable with iron and are relative newcomers. They were discovered independently in the residues left after crude platinum had been dissolved in aqua regia; ruthenium in 1844 from ores from the Urals by K. Klaus who named it after *Ruthenia*, the Latin name for Russia; and osmium in 1803 by S. Tennant who named it from the Greek word for odour (ὀσμή, *osme*) because of the characteristic and pungent smell of the volatile oxide, OsO_4. (CAUTION: OsO_4 is very toxic.)

25.2 The Elements Iron, Ruthenium, and Osmium

25.2.1 *Terrestrial abundance and distribution*

Ruthenium and osmium are generally found in the metallic state along with the other "platinum" metals and the "coinage" metals. The major sources of the platinum metals are the nickel–copper sulfide ores found in South Africa and Sudbury (Canada), and in the river sands of the Urals, USSR. They are rare elements, ruthenium particularly so, their estimated abundances in the earth's crustal rocks being but 0.0001 (Ru) and 0.005 (Os) ppm. However, as in Group VIIA, there is a marked contrast between the abundances of the two heavier elements and that of the first.

The nuclei of iron are especially stable, giving it a comparatively high cosmic abundance (Chap. 1, p. 15), and it is thought to be the main constituent of the earth's core (which has a radius of approximately 3500 km, i.e. 2150 miles) as well as being the major component of "siderite" meteorites. About 0.5% of the lunar soil is now known to be metallic iron and, since on average this soil is 10 m deep, there must be $\sim 10^{12}$ tonnes of iron on the moon's surface. In the earth's crustal rocks (6.2%, i.e. 62 000 ppm) it is the fourth most abundant element (after oxygen, silicon, and aluminium) and the second most abundant metal. It is also widely distributed, as oxides and carbonates, of which the chief ones are: haematite (Fe_2O_3), magnetite (Fe_3O_4), limonite ($\sim 2Fe_2O_3.3H_2O$), and siderite ($FeCO_3$). Iron pyrite (FeS_2) is also common but is not used as a source of iron because of the difficulty in eliminating the sulfur. The distribution of iron has been considerably influenced by weathering. Leaching from sulfide and silicate deposits occurs readily as $FeSO_4$ and $Fe(HCO_3)_2$ respectively. In solution, these are quickly oxidized, and even mildly alkaline conditions cause the precipitation of iron(III) oxide. Because of their availability, production of iron ores can be confined to those of the highest grade in gigantic operations. The world's largest iron-ore mine is at Mt. Whaleback in Western Australia: this only opened in 1969 but is now producing 40 million tonnes pa of high-grade haematite (70% Fe).

25.2.2 *Preparation and uses of the elements*

Pure iron is produced on a small scale by the reduction of the pure oxide or hydroxide with hydrogen, or by the carbonyl process in which iron is heated with carbon monoxide under pressure and the $Fe(CO)_5$ so formed decomposed at 250°C to give the powdered metal. However, it is not in the pure state but in the form of an enormous variety of steels that iron finds its most widespread uses, the world's annual production being of the order of 700 million tonnes.

The first stage in the conversion of iron ore to steel is the *blast furnace* (see Panel), which accounts for the largest tonnage of any metal produced by man. In it the iron ore is reduced by coke,† while limestone removes any sand or clay as a slag. The molten iron is run off to be cast into moulds of the required shape or into ingots ("pigs") for further processing—hence the names "cast-iron" or "pig-iron". This is an impure form of iron, containing about 4% of carbon along with variable amounts of Si, Mn, P, and S. It is hard but notoriously

Iron and Steel[4]

About 1773, in order to overcome a shortage of timber for the production of charcoal, Abraham Darby developed a process for producing carbon (coke) from coal and used this instead of charcoal in his blast furnace at Coalbrookdale in Shropshire. The impact was dramatic. It so cheapened and increased the scale of ironmaking that in the succeeding decades Shropshire iron was used to produce for the first time: iron cylinders for steam-engines, iron rails, iron boats and ships, iron aqueducts, and iron-framed buildings. The iron bridge erected nearby over the River Severn in 1779, gave its name to the small town which grew around it and still stands, a monument to the process which "opened up" the iron industry to the Industrial Revolution.

The blast furnace (Fig. A) remains the basis of ironmaking[5] though the scale, if not the principle, has changed considerably since the eighteenth century: the largest modern blast furnaces have hearths 14 m in diameter and produce up to 10 000 tonnes of iron daily.

The furnace is charged with a mixture of the ore (usually haematite), coke, and limestone, then a blast of hot air, or air with fuel oil, is blown in at the bottom. The coke burns and such intense heat is generated that temperatures approaching 2000°C are reached near the base of the furnace and perhaps 200°C at the top. The net result is that the ore is reduced to iron, and silicaceous gangue forms a slag (mainly $CaSiO_3$) with the limestone:

$$2Fe_2O_3 + 3C \longrightarrow 4Fe + 3CO_2$$

$$SiO_2 + CaCO_3 \longrightarrow CaSiO_3 + CO_2$$

The molten iron, and the molten slag which floats on the iron, collect at the bottom of the furnace and are tapped off separately. As the charge moves down, the furnace is recharged at the top, making the process continuous. Of course the actual reactions taking place are far more numerous than this and only the more important ones are summarized in Fig. A. The details are exceedingly complex and still not fully understood. At least part of the reason for this complexity is the rapidity with which the blast passes through the furnace (~ 10 s) which does not allow the gas–solid reactions to reach equilibrium. The main reduction occurs near the top, as the hot rising gases meet the descending charge. Here too the limestone is converted to CaO. Reduction to the metal is completed at somewhat higher temperatures, after which fusion occurs and the iron takes up Si and P in addition to C. The deleterious uptake of S is considerably reduced if manganese is present, because of the formation of MnS which passes into the slag. For this the slag must be adequately fluid and to this end the ratio of base (CaO):acid (SiO_2, Al_2O_3) is maintained by the addition, if necessary, of gravel (SiO_2). The slag is subsequently used as a building material (breeze blocks, wall insulation) and in the manufacture of some types of cement.

† The actual reducing agent is, in the main, CO. Many attempts have been made to effect "direct reduction" of the ore using hot $CO + H_2$ gas (produced from natural gas or fossil fuels).[3] With a much lower operating temperature than that of the blast furnace, reduction is confined to the ore, producing a "sponge" iron and leaving the gangue relatively unchanged. This offers a potential economy in fuel providing that the quantity and composition of the gangue do not adversely affect the subsequent conversion to steel—which is most commonly by the electric arc furnace. The direct reduction/electric arc furnace route to steel may well become increasingly attractive as an alternative to the usual coke-based, blast furnace/basic oxygen route, especially for small-scale production and where coke is not readily available.

³ K. S. Pitt, Direct reduction of iron ore, *Iron and Steel Int.* 242–51 and 347–53 (1973).
⁴ *Kirk-Othmer Encyclopedia of Chemical Technology*, 3rd edn, Vol. 13, pp. 735–763, Interscience, New York, 1981, and M. Finniston. The technological future of the steel industry, *Chem. Ind.* 1976, 501–8.
⁵ C. B. Alcock, *Principles of Pyrometallurgy*, pp. 186–9, Academic Press, London, 1976.

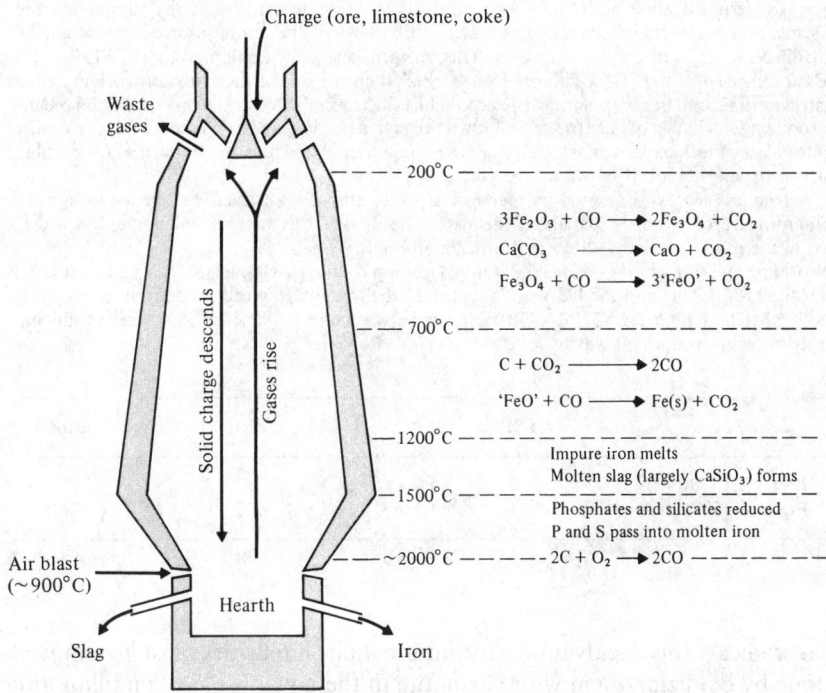

Charge (ore, limestone, coke)

Waste gases

Solid charge descends

Gases rise

$$3Fe_2O_3 + CO \longrightarrow 2Fe_3O_4 + CO_2$$

$$CaCO_3 \longrightarrow CaO + CO_2$$

$$Fe_3O_4 + CO \longrightarrow 3\text{'}FeO\text{'} + CO_2$$

$$C + CO_2 \longrightarrow 2CO$$

$$\text{'}FeO\text{'} + CO \longrightarrow Fe(s) + CO_2$$

Impure iron melts
Molten slag (largely $CaSiO_3$) forms

Phosphates and silicates reduced
P and S pass into molten iron

$$2C + O_2 \longrightarrow 2CO$$

Air blast (~900°C)

Hearth

Slag   Iron

Fig. A   Blast furnace (diagrammatic).

Bessemer process. Traditionally, pig-iron was converted to wrought-iron by the "puddling" process in which the molten iron was manually mixed with haematite and excess carbon and other impurities burnt out. Some wrought-iron was then converted to steel by essentially small-scale and expensive methods, such as the Cementation process (prolonged heating of wrought-iron bars with charcoal) and the crucible process (fusion of wrought-iron with the correct amount of charcoal). By the mid-nineteenth century demand had reached such a level that the puddling process had become a serious bottleneck. In 1856, however, H. Bessemer (born in Hertfordshire, 1813, and son of a French refugee) patented the process which bears his name: in this, molten pig-iron is poured into a "converter" and air blown through it until the carbon content has been sufficiently lowered. Basic impurities form a slag with the silaceous lining of the converter and the addition of manganese (in the form of spiegeleisen) controls the level of sulfur and occluded oxygen. The addition of manganese was suggested by R. Mushet and was absolutely vital to the success of Bessemer's process. Within 10 years the price of British steel plummeted from £70 to £15 per tonne. Even more spectacularly, Andrew Carnegie adopted the process in the USA using ores from the Lake Superior district and started making steel rails in 1872; by 1898 their price had fallen from $160 to $17 per tonne.[6] The process dominated world steelmaking until about 1910.

In 1879 the cousins S. G. Thomas and P. C. Gilchrist used limestone or dolomite as the lining of the converter (i.e. the "basic" Bessemer process), and this allowed the use of ores with a high phosphorus content by combining with the P and other acid impurities to form a basic slag (used as a fertilizer).

Open-hearth process. This was invented by the German-born Sir William Siemens and was introduced a few years after the Bessemer process. The oxidation is effected on a shallow hearth which can also be adapted for differing types of pig-iron feed by using acid or basic linings. It

[6] BERTRAND RUSSELL, *Freedom and Organization 1814–1914*, pp. 373–5, George Allen & Unwin, London, 1945.

requires external heating and, though not immediately so popular, eventually supplanted the Bessemer process, largely because of the greater ease with which the composition of the steel can be controlled and the greater fuel economy. This, in turn has been supplanted by the BOP.

Basic oxygen process (BOP). This process, of which there are several modifications, originated in Austria in 1952, and because of its greater speed has since become by far the most common means of producing steel. A jet of pure oxygen is blown through a retractable steel "lance" into, or over the surface of, the molten pig-iron which is contained in a basic-lined furnace. Impurities form a slag which is usually removed by tilting the converter.

Electron arc process. Patented by Siemens in 1878, this uses an electric current through the metal (direct-arc), or an arc just above the metal (indirect-arc), as a means of heating. It is widely used in the manufacture of alloy- and other high-quality steels.

World production of iron ore in 1980 was 904 million tonnes (Mt) (the USSR 27%, Brazil 13%, Australia 11%, China and the USA ~8% each). In the same year world production of raw steel was 706 Mt and pig-iron 507 Mt. Distribution between the major iron and steel producing countries (expressed in %) was as follows:

	USSR	Japan	USA	FRG	China	Others
Raw steel (706 Mt)	21.0	15.8	14.4	6.2	5.3	37.3
Pig-iron (507 Mt)	21.1	17.3	11.3	6.7	7.5	36.1

brittle. To eradicate this disadvantage the non-metallic impurities must be removed. This can be done by oxidizing them with haematite in the now obsolete "puddling process", producing the much purer "wrought-iron", which is tough and malleable and ideal for mechanical working. Nowadays, however, the bulk of pig-iron is converted into steel containing $\frac{1}{2}$–$1\frac{1}{2}$% C but very little S or P. The oxidation in this case is most commonly effected in one of a number of related processes by pure oxygen (basic oxygen process, or BOP), but open-hearth and electric arc furnaces are also used, while the Bessemer Converter (see Panel) was of great historical importance. This "mild steel" is cheaper than wrought-iron and stronger and more workable than cast-iron; it also has the advantage over both that it can be hardened by heating to redness and then cooling rapidly (quenching) in water or mineral oil, and "tempered" by re-heating to 200–300°C and cooling more slowly. The hardness, resilience, and ductility can be controlled by varying the temperature and the rate of cooling as well as the precise composition of the steel (see below). Alloy steels, with their enormous variety of physical properties, are prepared by the addition of the appropriate alloying metal or metals.

Ruthenium and osmium are produced from "platinum concentrates" which are commonly obtained as anode slimes in the electrolytic refinement of nickel. Treatment with aqua regia removes Pt, Pd, and Au (which is also present), and Ag is removed as its soluble nitrate by heating with lead carbonate and treating with nitric acid. The insoluble residue is then worked for Ru, Os, Rh, and Ir by methods such as the one outlined in Fig. 25.1. The metals are obtained in the form of powder or sponge and are usually consolidated by powder-metallurgical techniques. The main use for ruthenium is for hardening platinum and palladium; osmium is also used in the production of very hard alloys for such things as instrument pivots. Like all the platinum metals Ru and Os have considerable catalytic powers and where these are specific, as in the case of certain hydrogenations, they may be preferred. However, industrial applications are compara-

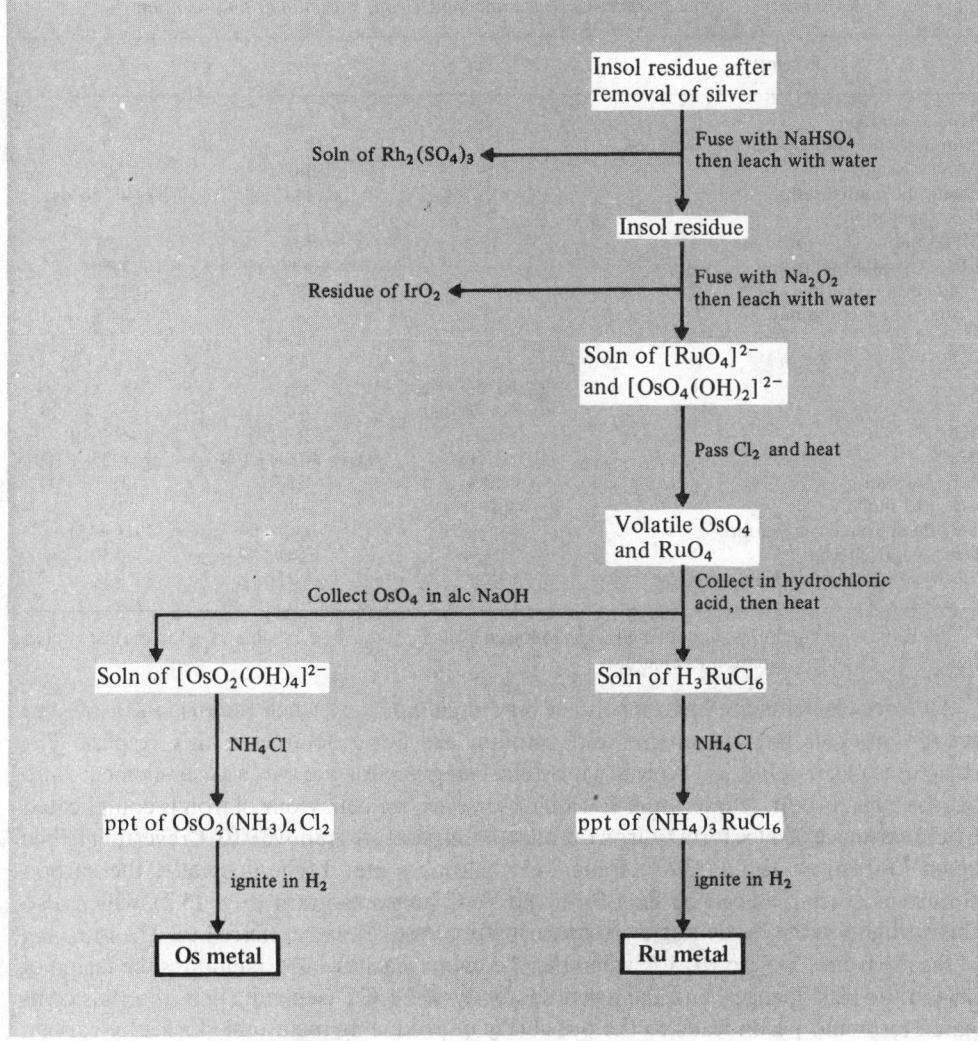

FIG. 25.1   Flow diagram for the extraction of ruthenium and osmium.

tively few, being limited by scarcity and consequent high cost (annual world production is only of the order of tonnes).†

25.2.3 *Properties of the elements*

Table 25.1 summarizes some of the important properties of Fe, Ru, and Os. The two heavier elements in particular have several naturally occurring isotopes, and difficulties in obtaining calibrated measurements of their relative abundances limit the precision with which their atomic weights can be determined.

† Weights of Ru and Os, as of most precious metals, are generally quoted in troy ounces: 1 troy ounce = 1.097 avoirdupois ounce = 31.103 g.

TABLE 25.1   *Some properties of the elements iron, ruthenium, and osmium*

Property		Fe	Ru	Os
Atomic number		26	44	76
Number of naturally occurring isotopes		4	7	7
Atomic weight[a]		55.847[a]	101.07[a]	190.2
Electronic configuration		[Ar]$3d^6 4s^2$	[Kr]$4d^7 5s^1$	[Xe]$4f^{14} 5d^6 6s^2$
Electronegativity		1.8	2.2	2.2
Metal radius (12-coordinate)/pm		126	134	135
Effective ionic radius/pm	VIII	—	36[b]	39[b]
(4-coordinate if marked[b],	VII	—	38[b]	52.5
otherwise 6-coordinate)	VI	25[b]	—	54.5
	V	—	56.5	57.5
	IV	58.5	62	63
	III	55 (ls), 64.5 (hs)	68	—
	II	61 (ls), 78 (hs)	—	—
MP/°C		1535	2282($\pm$20)	3045($\pm$30)
BP/°C		2750	extrap 4050($\pm$100)	extrap 5025($\pm$100)
ΔH_{fus}/kJ mol^{-1}		13.8	~25.5	31.7
ΔH_{vap}/kJ mol^{-1}		340($\pm$13)	—	738
ΔH_f (monatomic gas)/kJ mol^{-1}		398($\pm$17)	640	791($\pm$13)
Density (20°C)/g cm^{-3}		7.874	12.41	22.57
Electrical resistivity (20°C)/μohm cm		9.71	6.71	8.12

[a] Reliable to $\pm$3 in the last quoted digit when followed by an (a); otherwise, reliable to $\pm$1 in the last quoted digit. [b] Refers to coordination number 4.

All three elements are lustrous and silvery in colour. Iron when pure is fairly soft and readily worked, but ruthenium and osmium are less tractable in this respect. The structures of the solids are typically metallic, being hcp for the two heavier elements and bcc for iron at room temperature (α-iron). However, the behaviour of iron is complicated by the existence of a fcc form (γ-iron) at higher temperatures (above 910°), reverting to bcc again (δ-iron) at about 1390°, some 145° below its mp. Technologically, the carbon content is crucial, as can be seen from the Fe/C phase diagram (Fig. 25.2), which also throws light on the hardening and tempering processes already referred to. The lowering of the mp from 1535° to 1015°C when the C content reaches 4.3%, facilitates the fusion of iron in the blast furnace, and the lower solubility of Fe$_3$C ("cementite") in α-iron as compared to δ- and γ-iron leads to the possibility of producing metastable forms by varying the rate of cooling of hot steels. At elevated temperatures a solid solution of Fe$_3$C in γ-iron, known as "austenite", prevails. If 0.8% C is present, slow cooling below 723°C causes Fe$_3$C to separate forming alternate layers with the α-iron. Because of its appearance when polished, this is known as "pearlite" and is rather soft and malleable. If, however, the cooling is rapid (quenching) the separation is suppressed and the extremely hard and brittle "martensite" is produced. Reheating to an intermediate temperature tempers the steel by modifying the proportions of hard and malleable forms. If the C content of the steel is below 0.8% then slow cooling gives a mixture of pearlite and α-iron and, if higher than 0.8% a mixture of pearlite and Fe$_3$C. Varying the proportion of carbon in the steel thereby further extends the range of physical properties which can be attained by appropriate heat treatment.

The magnetic properties of iron are also dependent on purity and heat treatment. Up to 768°C (the Curie point) pure iron is ferromagnetic as a result of extensive magnetic interactions between unpaired electrons on adjacent atoms, which cause the electron spins

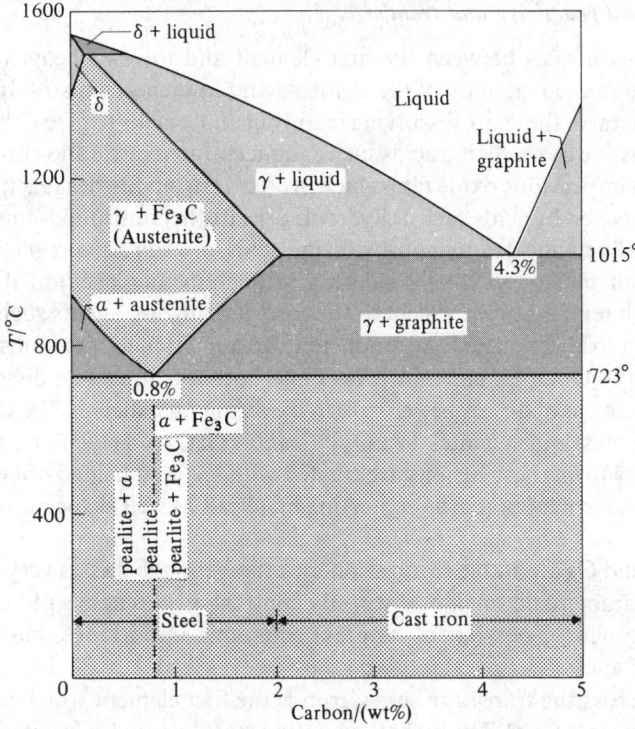

FIG. 25.2   The iron–carbon phase diagram for low concentrations of carbon.

to be aligned in the same direction, so producing exceedingly high magnetic susceptibilities and the characteristic ferromagnetic properties of "saturation" and "hysteresis". The existence of unpaired electrons on the individual atoms, as opposed to being delocalized in bands permeating the lattice, can be rationalized at least partly by supposing that in the bcc lattice the metal d_{z^2} and $d_{x^2-y^2}$ orbitals, which are not directed towards nearest neighbours, are therefore nonbonding and so can retain 2 unpaired electrons on the atom. On the other hand, electrons in the remaining three d orbitals participate in the formation of a conduction band of predominantly paired electrons. At temperatures above the Curie temperature, thermal energy overcomes the interaction between the electrons localized on individual atoms; their mutual alignment is broken, and normal paramagnetic behaviour ensues. This is sometimes referred to as β-iron (768–910°) though the crystal structure remains bcc as in ferromagnetic α-iron. For the construction of permanent magnets, cobalt steels are particularly useful, whereas for the "soft" irons used in electric motors and transformer cores (where the magnetization undergoes rapid reversal) silicon steels are preferred.

The mps and bps and enthalpies of atomization indicate that the $(n\text{-}1)\mathrm{d}$ electrons are contributing to metal bonding less than in earlier groups although, possibly due to an enhanced tendency for metals with a d^5 configuration to resist delocalization of their d electrons, Mn and to a lesser extent Tc occupy "anomalous" positions so that for Fe and Ru the values of these quantities are actually higher than for the elements immediately preceding them. In the third transition series Re appears to be "well-behaved" and the changes from W→Re→Os, are consequently smooth.

25.2.4 *Chemical reactivity and trends*

As expected, contrasts between the first element and the two heavier congeners are noticeable, both in the reactivity of the elements and in their chemistry. Iron is much the most reactive metal of the triad, dissolving readily in dilute acids to give Fe^{II} salts; however, it is rendered passive by oxidizing acids such as concentrated nitric and chromic, due to the formation of an impervious oxide film which protects it from further reaction but which is immediately removed by acids such as hydrochloric. Ruthenium and osmium, on the other hand, are virtually unaffected by non-oxidizing acids, or even aqua regia. Iron also reacts fairly easily with most non-metals whereas ruthenium and osmium do so only with difficulty at high temperatures, except in the case of oxidizing agents such as F_2 and Cl_2. Indeed, it is with oxidizing agents generally that Ru and Os metals are most reactive. Thus Os is converted to OsO_4 by conc nitric acid and both metals can be dissolved in molten alkali in the presence of air or, better still, in oxidizing flux such as Na_2O_2 or $KClO_3$ to give the ruthenates and osmates $[RuO_4]^{2-}$ and $[OsO_2(OH)_4]^{2-}$ respectively. If the aqueous extracts from these fusions are treated with Cl_2 and heated, the tetroxides distil off, providing convenient preparative starting materials as well as the means of recovering the elements.

Ruthenium and Os are stable to atmospheric attack though if Os is very finely divided it gives off the characteristic smell of OsO_4. By contrast, iron is subject to corrosion in the form of rusting which, because of its great economic importance, has received much attention (see Panel).

In moving across the transition series, iron is the first element which fails to attain its group oxidation state ($+8$). The highest oxidation state known (so far) is $+6$ in $[FeO_4]^{2-}$ and even this is extremely easily reduced. On the other hand, ruthenium and osmium do attain the group oxidation state of $+8$, though they are the last elements to do so in the

---

Rusting of Iron[7]

The economic importance of rusting can scarcely be overestimated. In Britain alone it was estimated[8] that the cost of corrosion in 1971 was £1367 million, or $3\frac{1}{2}\%$ of the gross national product. This was 3 times the level in the USA or Sweden and 7 times that in the Netherlands.

Rusting of iron consists of the formation of hydrated oxide, $Fe(OH)_3$ or $FeO(OH)$, and is evidently an electrochemical process which requires the presence of water, oxygen, and an electrolyte—in the absence of any one of these rusting does not occur to any significant extent. In air, a relative humidity of over 50% provides the necessary amount of water. The mechanism is complex and will depend in detail on the prevailing conditions, but may be summarized as:

the cathodic reduction: $3O_2 + 6H_2O + 12e^- \longrightarrow 12(OH)^-$

and the anodic oxidations: $4Fe \longrightarrow 4Fe^{2+} + 8e^-$

and $4Fe^{2+} \longrightarrow 4Fe^{3+} + 4e^-$

i.e. overall: $4Fe + 3O_2 + 6H_2O \longrightarrow 4Fe^{3+} + 12(OH)^-$

$\underbrace{\qquad\qquad}$

$4Fe(OH)_3$ or $4FeO(OH) + 4H_2O$

The presence of the electrolyte is required to provide a pathway for the current and, in urban areas, this is commonly iron(II) sulfate formed as a result of attack by atmospheric SO_2 but, in seaside

---

[7] U. R. Evans, *An Introduction to Metallic Corrosion*, Arnold, London. 3rd edn., 1981, 320 pp.
[8] D. M. Yorke, More corrosion on the curriculum, *Sch. Sci. Rev.* **58**, 235–50 (1976); see also Report of the Hoare Committee on *Corrosion and Protection*, Department of Trade and Industry, London, HMSO, 1971.

areas, airborne particles of salt are important. Because of its electrochemical nature, rusting may continue for long periods at a more or less constant rate, in contrast to the formation of an anhydrous oxide coating which under dry conditions slows down rapidly as the coating thickens.

The anodic oxidation of the iron is usually localized in surface pits and crevices which allow the formation of adherent rust over the remaining surface area. Eventually the lateral extension of the anodic area undermines the rust to produce loose flakes. Moreover, once an adherent film of rust has formed, simply painting over gives but poor protection. This is due to the presence of electrolytes such as iron(II) sulfate in the film so that painting merely seals in the ingredients for anodic oxidation. It then only requires the exposure of some other portion of the surface, where cathodic reduction can take place, for rusting beneath the paint to occur.

The protection of iron and steel against rusting takes many forms, including: simple covering with paint; coating with another metal such as zinc (galvanizing) or tin; treating with "inhibitors" such as chromate(VI) or (in the presence of air) phosphate or hydroxide, all of which produce a coherent protective film of Fe_2O_3. Another method uses sacrificial anodes, most usually Mg or Zn which, being higher than Fe in the electrochemical series, are attacked preferentially. In fact, the Zn coating on galvanized iron is actually a sacrificial anode.

second and third transition series, and this is consequently the highest oxidation state for any element (see also Xe^{VIII}, p. 1048). Ruthenium(VIII) is significantly less stable than Os^{VIII} and it is clear that the second- and third-row elements, though similar, are by no means as alike as for earlier element-pairs in the transition series. The same gradation within the triad is well illustrated by the reactions of the metals with oxygen. All react on heating, but their products are, respectively, Fe_2O_3 and Fe_3O_4, $Ru^{IV}O_2$ and $Os^{VIII}O_4$. In general terms it can be said that the most common oxidation states for the three elements are $+2$ and $+3$ for Fe, $+3$ for Ru, and $+4$ for Os. And, while Fe (and to a lesser extent Ru) has an extensive aqueous cationic chemistry in its lower oxidation states, Os has none. Table 25.2

TABLE 25.2 *Standard reduction potentials for iron, ruthenium, and osmium in acidic aqueous solution*[a]

Half reaction	$E°/V$	Volt-equivalent
$Fe^{2+} + 2e^- \rightleftharpoons Fe$	-0.447	-0.894
$Fe^{3+} + 3e^- \rightleftharpoons Fe$	-0.037	-0.111
$(FeO_4)^{2-} + 8H^+ + 3e^- \rightleftharpoons Fe^{3+} + 4H_2O$	2.20	6.49
$Ru^{2+} + 2e^- \rightleftharpoons Ru$	0.455	0.910
$Ru^{3+} + e^- \rightleftharpoons Ru^{2+}$	0.249	1.159
$RuO_2 + 4H^+ + 2e^- \rightleftharpoons Ru^{2+} + 2H_2O$	1.120	3.150
$(RuO_4)^{2-} + 8H^+ + 4e^- \rightleftharpoons Ru^{2+} + 4H_2O$	1.563	7.162
$(RuO_4)^- + 8H^+ + 5e^- \rightleftharpoons Ru^{2+} + 4H_2O$	1.368	7.750
$RuO_4 + 4H^+ + 4e^- \rightleftharpoons RuO_2 + 2H_2O$	1.387	8.698
$OsO_2 + 4H^+ + 4e^- \rightleftharpoons Os + 2H_2O$	0.687	2.748
$(OsO_4)^{2-} + 8H^+ + 6e^- \rightleftharpoons Os + 4H_2O$	0.994	5.964
$OsO_4 + 8H^+ + 8e^- \rightleftharpoons Os + 4H_2O$	0.85	6.80

(a) See also Table A (p. 1270) and Table 25.8.

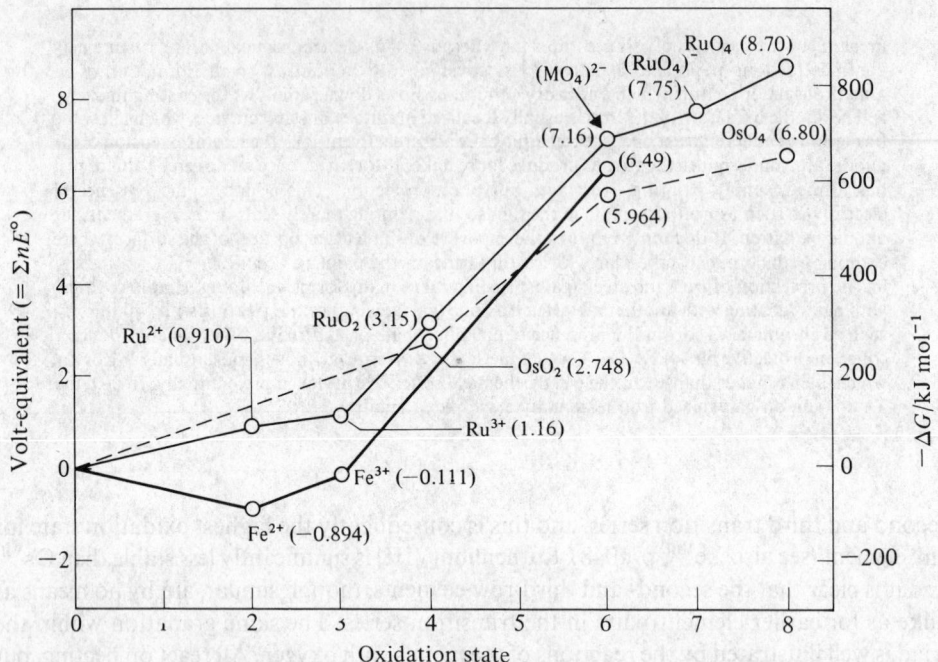

FIG. 25.3   Plot of volt-equivalent against oxidation state for Fe, Ru, and Os in acidic aqueous solution.

and Fig. 25.3 summarize the relative stabilities of the various oxidation states in acidic aqueous solution.

A selection of representative examples of compounds of the three elements is given in Table 25.3. As in the preceding group there is a remarkably wide range of oxidation states, particularly for Ru and Os, and, although it is now evident that as the size of the atoms decreases across each period the tendency to form compounds with high coordination numbers is diminishing, Os has a greater tendency than Ru to adopt a coordination number of 6 in the higher oxidation states. Thus OsO_4 expands its coordination sphere far more readily than RuO_4 to form complexes such as $[OsO_4(OH)_2]^{2-}$, and Os has no 4-coordinate analogues of $[RuO_4]^-$ or $[RuO_4]^{2-}$.

Iron is notable for the range of electronic spin states to which it gives rise. The values of S which are found include every integral and half-integral value from 0 to $\frac{5}{2}$ i.e. every value possible for a d-block element (Table 25.4).

25.3 Compounds of Iron, Ruthenium, and Osmium[9, 10]

The borides (p. 162), carbides (pp. 318, 1248), and nitrides (p. 479) have been discussed previously.

[9] D. NICHOLLS, Iron, Chap. 40 in *Comprehensive Inorganic Chemistry*, Vol. 3, pp. 979–1051, Pergamon Press, Oxford, 1973. S. E. LIVINGSTONE, The second- and third-row elements of Group VIIIA, B, and C, Chap. 43, ibid., pp. 1163–1370.
[10] W. P. GRIFFITH, *The Chemistry of the Rarer Platinum Metals (Os, Ru, Ir and Rh)*, Interscience, London, 1967, 491 pp.

TABLE 25.3   *Oxidation states and stereochemistries of some compounds of iron, ruthenium, and osmium*

Oxidation state	Coordination number	Stereochemistry	Fe	Ru, Os
-2 (d^{10})	4	Tetrahedral	$[Fe(CO)_4]^{2-}$	$[M(CO)_4]^{2-}$
-1 (d^9)	5	Trigonal bipyramidal	$[Fe_2(CO)_8]^{2-}$	
0 (d^8)	5	Trigonal bipyramidal	$[Fe(CO)_5]$	$[M(CO)_5](?)$
	6	Octahedral (D_3)	$[Fe(bipy)_3]$	
	7	Face-capped octahedral	$[Fe_2(CO)_9]$	
1 (d^7)	6	Octahedral	$[Fe(NO)(H_2O)_5]^{2+}$	$[Os(NH_3)_6]^+$
	9	(See Fig. 25.15(a))	$[\{Fe(\eta^5\text{-}C_5H_5)(CO)(\mu\text{-}CO)\}_2]$	
2 (d^6)	4	Tetrahedral	$[FeCl_4]^{2-}$	$[RuH\{N(SiMe_3)_2\}(PPh_3)_2]$
		Square planar	$BaFeSi_4O_{10}$	
	5	Trigonal bipyramidal	$[FeBr\{N(C_2H_4NMe_2)_3\}]^+$	
		Square pyramidal	$[Fe(OAsMePh_2)(ClO_4)]^+$	$[RuCl_2(PPh_3)_3]$
	6	Octahedral	$[Fe(H_2O)_6]^{2+}$	$[M(CN)_6]^{4-}$
	7	(p. 197)	$[Fe(\eta^4\text{-}B_4H_8)(CO)_3]$	
	8	(See Fig. 25.15c)	$[Fe(\eta^1\text{-}C_5H_5)(\eta^5\text{-}C_5H_5)(CO)_2]$	
	10	Sandwich	$[Fe(\eta^5\text{-}C_5H_5)_2]$	$[M(\eta^5\text{-}C_5H_5)_2]$
3 (d^5)	3	Planar	$[Fe\{N(SiMe_3)_2\}_3]$	
	4	Tetrahedral	$[FeCl_4]^-$	
	5	Square pyramidal	$[Fe(acac)_2Cl]$	
	6	Octahedral	$[Fe(CN)_6]^{3-}$	$[MCl_6]^{3-}$
	7	Pentagonal bipyramidal	$[Fe(EDTA)(H_2O)]^-$	
	8	Dodecahedral	$[Fe(NO_3)_4]^-$	
4 (d^4)	6	Octahedral	$[Fe(diars)_2Cl_2]^{2+}$	$[MCl_6]^{2-}$
5 (d^3)	6	Tetrahedral	$[FeO_4]^{3-}$	
		Octahedral		$[MCl_6]^{3-}$
6 (d^2)	4	Tetrahedral	$[FeO_4]^{2-}$	$[RuO_4]^{2-}$
	5	Square pyramidal		$[OsNCl_4]^-$
	6	Octahedral		$[OsO_2(OH)_4]^{2-}$
7 (d^1)	4	Tetrahedral		$[RuO_4]^-$
	6	Octahedral		$[OsOF_5]$
	7	Pentagonal bipyramidal		OsF_7
8 (d^0)	4	Tetrahedral		MO_4
	6	Octahedral		$[OsO_4F_2]^{2-}$

TABLE 25.4   *Electronic spin-states of iron*

Spin quantum number (S)	Ion	Electronic configuration	Typical compounds
0 (diamagnetic)	Low-spin Fe^{II}	t_{2g}^6	$K_4[Fe(CN)_6].3H_2O$ HbO_2 (oxygenated haemoglobin)
$\frac{1}{2}$ (1 unpaired e^-)	Low-spin Fe^{III}	t_{2g}^5	$K_3[Fe(CN)_6]$, HbCN
	Low-spin Fe^I	$t_{2g}^6 e_g^1$	$[Fe(diars)(CO)_2I]$
1 (2 unpaired e^-)	Low-spin Fe^{IV}	t_{2g}^4	$[Fe(diars)_2Cl_2](ClO_4)_2$
	Tetrahedral Fe^{VI}	e^2	$Ba[FeO_4]$
$\frac{3}{2}$ (3 unpaired e^-)	Distorted square pyramidal Fe^{III}	$d_{x^2-y^2}^2 d_{yz}^1 d_{xz}^1 d_{z^2}^1$	$[Fe(S_2CNR_2)_2Cl]^{(11)}$
2 (4 unpaired e^-)	High-spin Fe^{II}	$t_{2g}^4 e_g^2$	$[Fe(H_2O)_6]^{2+}$, deoxyhaemoglobin
$\frac{5}{2}$ (5 unpaired e^-)	High-spin Fe^{III}	$t_{2g}^3 e_g^2$	$[Fe(acac)_3]$, iron-transport proteins

[11] P. Ganguli, V. R. Marathi, and S. Mitra, Paramagnetic anisotropy and electronic structure of $S=\frac{3}{2}$ halobis(diethyldithiocarbamate)iron(III). I. Spin Hamiltonian formalism and ground-state zero-field splitting of ferric ion, *Inorg. Chem.* **14,** 970–3 (1975).

25.3.1 *Oxides and other chalcogenides*[11a]

The principal oxides of the elements of this group are given in Table 25.5.

TABLE 25.5   *The oxides of iron, ruthenium, and osmium*

Oxidation state	+8	+4	+3	+2
Fe			Fe_2O_3 Fe_3O_4 'FeO'	
Ru	RuO_4	RuO_2		
Os	OsO_4	OsO_2		

Three oxides of iron may be distinguished, but are all subject to nonstoichiometry.[12] The lowest is FeO which is obtained by heating iron in a low partial pressure of O_2 or as a fine, black pyrophoric powder by heating iron(II) oxalate *in vacuo*. Below about 575°C it is unstable towards disproportionation into Fe and Fe_3O_4 but can be obtained as a metastable phase if cooled rapidly. It has a rock-salt structure but is always deficient in iron, with a homogeneity range of $Fe_{0.84}O$ to $Fe_{0.95}O$. Treatment of any aqueous solution of Fe^{II} with alkali produces a flocculent precipitate. If air is rigorously excluded this is the virtually white $Fe(OH)_2$ which is almost entirely basic in character, dissolving readily in non-oxidizing acids to give Fe^{II} salts but showing only slight reactivity towards alkali. It gradually decomposes, however, to Fe_3O_4 with evolution of hydrogen and in the presence of oxygen darkens rapidly and eventually forms the reddish-brown hydrated iron(III) oxide. Fe_3O_4 is a mixed Fe^{II}/Fe^{III} oxide which can be obtained by partial oxidation of FeO or, more conveniently, by heating Fe_2O_3 above about 1400°C. It has the inverse spinel structure. Spinels are of the form $M^{II}M^{III}_2O_4$ and in the normal spinel (p. 279) the oxide ions form a ccp lattice with M^{II} ions occupying tetrahedral sites and M^{III} ions octahedral sites. In the inverse structure half the M^{III} ions occupy tetrahedral sites, with the M^{II} and the other half of the M^{III} occupying octahedral sites.† Fe_3O_4 occurs naturally as the mineral magnetite or lodestone. It is a black, strongly ferromagnetic substance (or, more strictly, "ferrimagnetic"—see p. 1256), insoluble in water and acids. Its electrical properties are not simple, but its rather high conductivity may be ascribed to electron transfer between Fe^{II} and Fe^{III}.

Fe_2O_3 is known in a variety of modifications of which the more important are the α- and γ-forms. When aqueous solutions of iron(III) are treated with alkali, a gelatinous reddish-brown precipitate of hydrated oxide is produced (this is amorphous to X-rays and is not simple $Fe(OH)_3$, but probably $FeO(OH)$); when heated to 200°C, this gives the red-brown α-Fe_2O_3. Like V_2O_3 and Cr_2O_3 this has the corundum structure (p. 274) in which the oxide ions are hcp and the metal ions occupy octahedral sites. It occurs naturally as the

---

† Although Fe_3O_4 is an inverse spinel it will be recalled that Mn_3O_4 (p. 1219) is normal. This contrast can be explained on the basis of crystal field stabilization. Manganese^{II} and Fe^{III} are both d^5 ions and, when high-spin, have zero CFSE whether octahedral or tetrahedral. On the other hand, Mn^{III} is a d^4 and Fe^{II} a d^6 ion, both of which have greater CFSEs in the octahedral rather than the tetrahedral case. The preference of Mn^{III} for the octahedral sites therefore favours the spinel structure, whereas the preference of Fe^{II} for these octahedral sites favours the inverse structure.

[11a] C. N. R. RAO and G. V. S. RAO, *Transition Metal Oxides*, National Standard Reference Data System NSRDS-NBS49, Washington, 1964, 130 pp.
[12] N. N. GREENWOOD, Chap. 6, in *Ionic Crystals, Lattice Defects and Nonstoichiometry*, pp. 111–47, Butterworths, London, 1968.

mineral haematite and, besides its overriding importance as a source of the metal (p. 1244), it is used (a) as a pigment, (b) in the preparation of rare earth/iron garnets and other ferrites (p. 1256), and (c) as a polishing agent—jewellers' rouge. The second variety γ-Fe_2O_3 is metastable and is obtained by careful oxidation of Fe_3O_4, like which it is cubic and ferrimagnetic. If heated *in vacuo* it reverts to Fe_3O_4 but heating in air converts it to α-Fe_2O_3. It is the most widely used magnetic material in the production of magnetic recording tapes.[13]

The interconvertibility of FeO, Fe_3O_4, and γ-Fe_2O_3 arises because of their structural similarity. Unlike α-Fe_2O_3, which is based on a hcp lattice of oxygen atoms, these three compounds are all based on ccp lattices of oxygen atoms. In FeO, Fe^{II} ions occupy the octahedral sites and nonstoichiometry arises by oxidation, when some Fe^{II} ions are replaced by two-thirds their number of Fe^{III} ions. Continued oxidation produces Fe_3O_4 in which the Fe^{II} ions are in octahedral sites, but the Fe^{III} ions are distributed between both octahedral and tetrahedral sites. Eventually, oxidation leads to γ-Fe_2O_3 in which all the cations are Fe^{III} which are randomly distributed between octahedral and tetrahedral sites. The oxygen lattice remains intact throughout but contracts somewhat as the number of iron atoms which it accommodates diminishes.

Ruthenium and osmium have no oxides comparable to those of iron and, indeed, the lowest oxidation state in which they form oxides is +4. RuO_2 is a blue to black solid, obtained by direct action of the elements at 1000°C, and has the rutile (p. 1118) structure. The intense colour has been suggested as arising from the presence of small amounts of Ru in another oxidation state, possibly +3. OsO_2 is a yellowish-brown solid, usually prepared by heating the metal at 650°C in NO. It, too, has the rutile structure.

The most interesting oxides of Ru and Os, however, are the volatile, yellow tetroxides, RuO_4 (mp 25°C, bp 40°C) and OsO_4 (mp 40°C, bp 130°C). The latter is perhaps the best-known compound of osmium. It is a tetrahedral molecule and is produced by aerial oxidation of the heated metal or by oxidizing other compounds of osmium with nitric acid. It dissolves in aqueous alkali to give $[Os^{VIII}O_4(OH)_2]^{2-}$ and oxidizes conc (but not dil) hydrochloric acid to Cl_2, being itself reduced to H_2OsCl_6. It is used in organic chemistry to oxidize C=C bonds to *cis*-diols and is also employed as a biological stain. Unfortunately, it is extremely toxic and its volatility renders it particularly dangerous. RuO_4 is presumed, by analogy, to be tetrahedral also. It is appreciably less stable and will oxidize dil as well as conc HCl, while in aqueous alkali it is reduced to $[Ru^{VI}O_4]^{2-}$. If heated above 100°C it decomposes explosively to RuO_2 and is liable to do the same at room temperature if brought into contact with oxidizable organic solvents such as ethanol. Its preparation obviously requires stronger oxidizing agents than that of OsO_4; nitric acid alone will not suffice and instead the action of $KMnO_4$, KIO_4, or Cl_2 on acidified solutions of a convenient Ru compound is used.

The sulfides are fewer in number than the oxides and favour lower metal oxidation states. Iron forms 3 sulfides (p. 804). FeS is a grey, nonstoichiometric material, obtained by direct action of the elements or by treating aqueous Fe^{II} with alkali metal sulfide. It has a NiAs structure (p. 803) in which each metal atom is octahedrally surrounded by anions but is also quite close to 2 other metal atoms. It oxidizes readily in air and dissolves in aqueous acids with evolution of H_2S. FeS_2 can be prepared by heating Fe_2O_3 in H_2S but is most commonly encountered as the yellow mineral pyrites. This does not contain Fe^{IV} but

[13] J. CRANGLE, *The Magnetic Properties of Solids*, Arnold, London, 1977, 194 pp.

is composed of Fe^{II} and S_2^{2-} ions in a distorted rock-salt arrangement, its diamagnetism indicating low-spin Fe^{II} (d^6). It is very unreactive unless heated, when it gives $Fe_2O_3 + SO_2$ in air, or $FeS + S$ in a vacuum. Fe_2S_3 is the unstable black precipitate resulting when aqueous Fe^{III} is treated with S^{2-}, and is rapidly oxidized in moist air to Fe_2O_3 and S.

Ruthenium and osmium form only disulfides. These have the pyrite structure and are diamagnetic semiconductors; this implies that they contain M^{II}. $RuSe_2$, $RuTe_2$, $OsSe_2$, and $OsTe_2$ are very similar. All 6 dichalcogenides are obtained directly from the elements.

25.3.2 *Mixed metal oxides and oxoanions*[13a]

The "ferrites" and "garnets" of iron are mixed metal oxides of considerable technological importance. They are obtained by heating Fe_2O_3 with the carbonate of the appropriate metal. The ferrites have the general form $M^{II}Fe_2^{III}O_4$. Some adopt the *normal* spinel structure and others the *inverse* spinel structure (p. 280) as just described for Fe_3O_4 (which can itself be regarded as the ferrite $Fe^{II}Fe_2^{III}O_4$). In inverse spinels the unpaired electrons of all the cations in octahedral sites (M^{II} and half the M^{III}) are magnetically coupled parallel to give a ferromagnetic sublattice, while the unpaired electrons of all the cations in tetrahedral sites (the remaining M^{III}) are similarly but independently coupled parallel to give a second ferromagnetic sublattice. The spins of one sublattice, however, are antiparallel to those of the other. If the cations in the octahedral sites have the same total number of unpaired electrons as those in the tetrahedral sites, then the effects of 2 ferromagnetic sublattices are mutually compensating and "antiferromagnetism" results; but where the sublattices are not balanced then a type of ferromagnetism known as ferrimagnetism results,[13] the explanation of which was first given by L. Néel in 1948 (Nobel Prize for Physics, 1970). Important applications of inverse spinel ferrites are as cores in high-frequency transformers (where they have the advantage over metals of being free from eddy-current losses), and in computer memory systems.

So-called "hexagonal ferrites" such as $BaFe_{12}O_{19}$ are ferrimagnetic and are used to construct permanent magnets. A third type of ferrimagnetic mixed oxides are the garnets, $M_3^{III}Fe_5O_{12}$, of which the best known is yttrium iron garnet (YIG) used as a microwave filter in radar.

Mixed oxides of Fe^{IV} such as $M_4^IFeO_4$ and $M_2^{II}FeO_4$ can be prepared by heating Fe_2O_3 with the appropriate oxide or hydroxide in oxygen. These do not contain discrete $[FeO_4]^{4-}$ anions and, as was seen above, mixed oxides of Fe^{III} are generally based on close-packed oxide lattices with no iron-containing anions. However, oxoanions of iron are known and are based on the FeO_4 tetrahedron. Thus for iron(III), Na_5FeO_4, $K_6[Fe_2O_6]$ (2 edge-sharing tetrahedral), and more recently $Na_{14}[Fe_6O_{16}]$ (rings of 6 corner-sharing tetrahedra), have been prepared.[14] But the best-known oxoanion of iron is the ferrate(VI) prepared by oxidizing a suspension of hydrous Fe_2O_3 in conc alkali with chlorine, or by the anodic oxidation of iron in conc alkali. The tetrahedral $[FeO_4]^{2-}$ ion is red-purple and is an extremely strong oxidizing agent. It oxidizes NH_3 to N_2 even at room temperature and, although it can be kept for a period of hours in alkaline solution, in acid

[13a] A. F. WELLS, *Structural Inorganic Chemistry*, 4th edn., Chap. 13, Complex oxides, pp. 476–515, Oxford University Press, Oxford, 1975.

[14] G. BRACHTEL and R. HOPPE, The first oxoferrate(III) having a ring structure: $Na_{14}[Fe_6O_{16}]$, *Angew. Chem.*, Int. Edn. (Engl.) **16**, 43 (1977).

or neutral solutions it rapidly oxidizes the water, so liberating oxygen:

$$2[FeO_4]^{2-} + 5H_2O \longrightarrow 2Fe^{3+} + 10(OH)^- + \tfrac{3}{2}O_2$$

The distinction between the first member of the group and the two heavier members, which was seen to be so sharp in the early groups of transition metals, is much less obvious here. The only unsubstituted, discrete oxoanions of the heavier pair of metals are the tetrahedral $[Ru^{VII}O_4]^-$ and $[Ru^{VI}O_4]^{2-}$. This behaviour is akin to that of iron or, even more, to that of manganese, whereas in the osmium analogues the metal always increases its coordination number by the attachment of extra OH^- ions. If RuO_4 is dissolved in cold dilute KOH, or aqueous K_2RuO_4 is oxidized by chlorine, virtually black crystals of $K[Ru^{VII}O_4]$ ("perruthenate") are deposited. These are unstable unless dried and are reduced by water, especially if alkaline, to the orange $[Ru^{VI}O_4]^{2-}$ ("ruthenate") by a mechanism which is thought to involve octahedral intermediates of the type $[RuO_4(OH)_2]^{3-}$ and $[RuO_4(OH)_2]^{2-}$. $K_2[RuO_4]$ is obtained by fusing Ru metal with KOH and KNO_3.

By contrast, dissolution of OsO_4 in cold aqueous KOH produces deep-red crystals of $K_2[Os^{VIII}O_4(OH)_2]$ ("perosmate"), which is easily reduced to the purple "osmate", $K_2[Os^{VI}O_2(OH)_4]$. The anions in both cases are octahedral with, respectively, *trans* OH and *trans* O groups.

By heating the metal with appropriate oxides or carbonates of alkali or alkaline earth metals, a number of mixed oxides of Ru and Os have been made. They include $Na_5Os^{VII}O_6$, $Li_6Os^{VI}O_6$, and the "ruthenites", $M^{II}Ru^{IV}O_3$, in all of which the metal is situated in octahedral sites of an oxide lattice. Ru^V (octahedral) has now also been established by ^{99}Ru Mössbauer spectroscopy as a common stable oxidation state in mixed oxides such as $Na_3Ru^VO_4$, $Na_4Ru_2^VO_7$, and the ordered perovskite-type phases $M_2^{II}Ln^{III}Ru^VO_6$.[14a]

25.3.3 *Halides and oxohalides*[10, 15]

The known halides of this group are listed in Table 25.6. As in the preceding group the highest halide is a heptafluoride, but OsF_7 (unlike ReF_7) is thermally unstable. It was for many years thought that OsF_8 existed but the yellow crystalline material to which the formula had been ascribed turned out to be OsF_6, the least unstable of the platinum metal hexafluorides. (In view of the propensity of higher fluorides to attack the vessels containing them, to disproportionate, and to hydrolyse,[15a] it is not surprising that early reports on them sometimes proved to be erroneous.) The highest chloride is $OsCl_5$[16] and, rather

[14a] T. C. GIBB, R. GREATREX, and N. N. GREENWOOD, A novel case of magnetic relaxation in the ^{99}Ru Mössbauer spectrum of Na_3RuO_4, *J. Solid State Chem.* **31**, 153–69 (1980). I. FERNANDEZ, R. GREATREX, and N. N. GREENWOOD, ^{99}Ru Mössbauer spectra of quaternary ruthenium(V) oxides with hexagonal barium titanate structure, *J Solid State Chem.* **34**, 121–8 (1980). R. GREATREX, N. N. GREENWOOD, M. LAL, and I. FERNANDEZ, A study of ruthenium(V) perovskites M_2LnRuO_6, etc., *J. Solid State Chem.* **30**, 137–48 (1979). I. FERNANDEZ, R. GREATREX, and N. N. GREENWOOD, A study of the new cubic, ordered perovskites $BaLaMRuO_6$, etc., *J. Solid State Chem.* **32**, 97–104 (1980).

[15] R. COLTON and J. H. CANTERFORD, *Halides of the First Row Transition Metals*, Chap. 6, pp. 271–326, Wiley, London, 1969, and *Halides of the Second and Third Row Transition Metals*, Chap. 8, pp. 322–46, Wiley, London, 1968.

[15a] J. H. CANTERFORD and T. A. O'DONNELL, Manipulations of volatile fluorides and other corrosive compounds, *Technique Inorg. Chem.* **7**, 273–306 (1968).

[16] R. C. BURNS and T. A. O'DONNELL, Preparation and characterization of osmium pentachloride, a new binary chloride of osmium, *Inorg. Chem.* **18**, 3081–3 (1979).

TABLE 25.6   *Halides of iron, ruthenium, and osmium (mp/°C)*

Oxidation state	Fluorides	Chlorides	Bromides	Iodides
+7	OsF_7 yellow			
+6	RuF_6 dark brown (54°) OsF_6 yellow (33°)			
+5	RuF_5 dark green (86.5°) OsF_5 blue (70°)	$OsCl_5$ black (d > 160°)		
+4	RuF_4 yellow OsF_4 yellow (230°)	$OsCl_4$ red (also black form)	$OsBr_4$ black (d 350°)	
+3	FeF_3 pale green (> 1000°) RuF_3 dark brown (d > 650°)	$FeCl_3$ brown-black (306°) $RuCl_3$ black (α) dark brown (β) $OsCl_3$ dark grey (d 450°)	$FeBr_3$ red-brown (d > 200°) $RuBr_3$ dark brown (d > 400°) $OsBr_3$ dark grey (d 340°)	RuI_3 black OsI_3 black
+2	FeF_2 white (> 1000°)	$FeCl_2$ pale yellow (674°) $RuCl_2$ brown	$FeBr_2$ yellow-green (d 684°) $RuBr_2$ black	FeI_2 grey RuI_2 blue OsI_2 black
+1				OsI metallic grey

unexpectedly perhaps, neither ruthenium nor iron form a chloride in an oxidation state higher than +3. Iron in fact does not form even a fluoride in an oxidation state higher than this and its halides are confined to the +3 and +2 states.

OsF_7 has been obtained as a yellow solid by direct action of the elements at 600°C and a pressure of 400 atm, but under less drastic conditions OsF_6 is produced, as is RuF_6. This latter pair are low-melting, yellow and brown solids, respectively, hydrolysing violently with water and with a strong tendency to disproportionate into F_2 and lower halides. The pentafluorides are both polymeric, easily hydrolysed solids obtained by specific oxidations or reduction of other fluorides, and their structures involve $[MF_5]_4$ units in which 4 corner-sharing MF_6 octahedra form a ring (Fig. 25.4).

The tetrafluorides are yellow solids, probably polymeric, and are obtained by reducing RuF_5 with I_2, and OsF_6 with $W(CO)_6$. The tetrachloride and tetrabromide of osmium require pressure as well as heat in their preparations from the elements and are black solids of uncertain structure.

In the +3 and +2 oxidation states those halides of osmium which have been reported are poorly characterized, grey or black solids. Apart from $RuCl_3$ the same could reasonably be said of ruthenium. $RuCl_3$, however, is well known and, as the anhydrous compound, exists in two forms: heating Ru metal at 330°C in CO and Cl_2 produces the

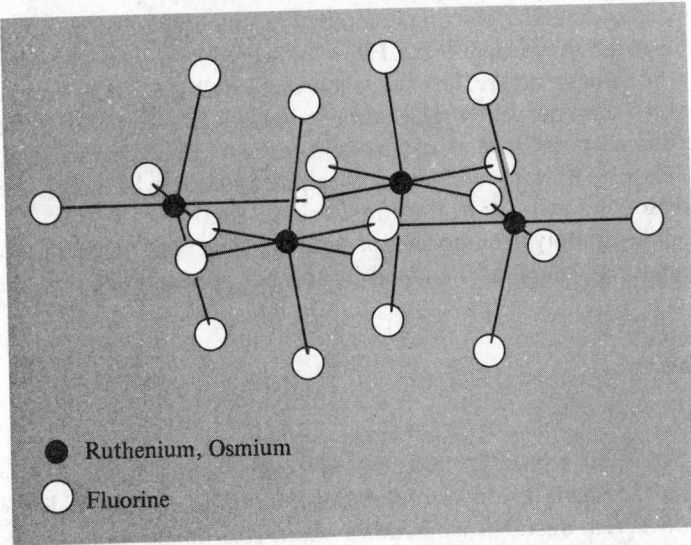

FIG. 25.4   Tetrameric pentafluorides of Ru and Os, $[M_4F_{20}]$. Their structures are similar to, but more distorted than, those of the pentafluorides of Nb and Ta (see Fig. 22.3).

dark-brown β-form which if heated above 450°C in Cl_2 is converted to the black α-form which is isomorphous with $CrCl_3$ (p. 1189). Evaporation of a solution of RuO_4 in hydrochloric acid in a stream of HCl gas, produces red $RuCl_3.3H_2O$; aqueous solutions contain both $[Ru(H_2O)_6]^{3+}$ and chloro-substituted species and are easily hydrolysed and oxidized to Ru^{IV}. Where impurities due to such reactions are suspected, conversion back to Ru^{III} chloride can be effected by repeated evaporations to dryness with conc HCl. This gives a uniform though rather poorly characterized product that is widely used as the starting material in ruthenium chemistry.

The halides of iron are comparatively straightforward. All the anhydrous +3 and +2 halides of iron are readily obtained, except for iron(III) iodide, (since the oxidizing properties of Fe^{III} and the reducing properties of I^- are incompatible: even in aqueous solution, $Fe^{III} + I^- \rightarrow Fe^{II} + \frac{1}{2}I_2$). The other anhydrous FeX_3 can be prepared by heating the elements (though in the case of $FeBr_3$ the temperature must not rise above 200°C otherwise $FeBr_2$ is formed). The fluoride, chloride, and bromide are respectively white, dark brown, and reddish-brown. The crystalline solids all contain Fe^{III} ions octahedrally surrounded by halide ions and decompose to $FeX_2 + \frac{1}{2}X_2$ if heated strongly under vacuum. $FeCl_3$ sublimes above 300°C and vapour pressure measurements show the vapour to contain dimeric Fe_2Cl_6, like Al_2Cl_6 consisting of 2 edge-sharing tetrahedra. The trifluoride is sparingly soluble, and the chloride and bromide very soluble in water and they crystallize as pale-pink $FeF_3.4.5H_2O$, yellow-brown $FeCl_3.6H_2O$, and dark-green $FeBr_3.6H_2O$. The chloride is probably the most widely used etching material, being particularly important for etching copper in the production of electrical printed circuits. It is also used in water treatment as a coagulant (by producing a "ferric hydroxide" floc which removes organic matter and suspended solids) in cases where the SO_4^{2-} of the more widely used iron(III) sulphate is undesirable.

Of the anhydrous dihalides of iron the iodide is easily prepared from the elements but the others are best obtained by passing HX over heated iron. The white (or pale-green) difluoride has the rutile structure the pale-yellow dichloride the $CdCl_2$ structure (based on ccp anions, p. 1407) and the yellow-green dibromide and grey diiodide the CdI_2 structure (based on hcp anions, p. 1407), in all of which the metal occupies octahedral sites. All these iron dihalides dissolve in water and form crystalline hydrates which may alternatively be obtained by dissolving metallic iron in the aqueous acid.

A few oxohalides, mainly oxo fluorides of osmium, have been obtained by the action of various halogenating agents on the metal but do not appear to be well characterized.

25.3.4 *Complexes*

Oxidation state VIII (d^0)

Iron forms barely any complexes in oxidation states above $+3$, and in the $+8$, $+7$, and $+6$ states those of ruthenium are less numerous than those of osmium. Ru^{VIII} complexes are confined to a few unstable (sometimes explosive) amine adducts of RuO_4. The "perosmates" (p. 1257) are, of course, adducts of OsO_4, but the most stable Os^{VIII} complexes are the "osmiamates", $[OsO_3N]^-$. Pale-yellow crystals of $K[OsO_3N]$ are obtained when solutions of OsO_4 in aqueous KOH (i.e. the perosmate) are treated with ammonia: the compound has been known since 1847 and A. Werner formulated it correctly in 1901. The anion is isoelectronic with OsO_4 and has a distorted tetrahedral structure (C_{3v}), while its infrared spectrum shows $\nu_{Os-N} = 1023$ cm^{-1}, consistent with an $Os\equiv N$ triple bond. Hydrochloric and hydrobromic acids reduce $K[OsO_3N]$ to red, $K_2[Os^{VI}NX_5]$.

Oxidation state VII (d^1)

Compounds of Ru^{VII} and Os^{VII} are largely confined to the fluorides and oxo compounds already referred to.

Oxidation state VI (d^2)

The most important members of this class are the osmium nitrido, and the "osmyl" complexes. The reddish-purple $K_2[OsNCl_5]$ mentioned above is the result of reducing the osmiamate. The anion has a distorted octahedral structure with a formal triple bond $Os\equiv N$ (161 pm) and a pronounced "*trans*-influence" (p. 1353), i.e. the Os–Cl distance *trans* to Os–N is much longer than the Os–Cl distances *cis* to Os–N (261 and 236 pm respectively). The anion $[OsNCl_5]^{2-}$ also shows a "*trans*-effect" in that the Cl opposite the N is more labile than the others, leading, for instance, to the formation of $[Os^{VI}NCl_4]^-$, which has a square-pyramidal structure with the N occupying the apical position.

The osmyl complexes, of which the osmate ion $[Os^{VI}O_2(OH)_4]^{2-}$ may be regarded as the precursor, are a series of diamagnetic complexes containing the linear $O=Os=O$ group together with 4 other, more remote, donor atoms which occupy the equatorial plane. That the Os–O bonds are double (i.e. 1σ and 1π) is evident from the bond lengths of 175 pm—very close to those of 172 pm in OsO_4. The diamagnetism can then be explained using the MO approach outlined in Chapter 19, but modified to allow for the tetragonal compression along the axis of the osmyl group (taken to define the z-axis). On this model, the effect of 6 σ interactions produces the molecular orbitals shown in Fig. 19.22. The tetragonal compression then splits the essentially metallic t_{2g} and $e_g{}^*$ sets, as shown to the left of Fig. 25.5b. Two 3-centre π bonds are then formed, one by overlap of the metal d_{xz} orbital with the p_x orbitals of the 2 oxygen atoms (Fig. 25.5a), the second similarly by d_{yz} and p_y overlap. Each 3-centre interaction produces 1 bonding, 1 virtually nonbonding, and 1 antibonding MO, as shown. The metal d_{xy} orbital remains unchanged and, in effect, the 2 d electrons of the Os are obliged to pair-up in it since other empty orbitals are inaccessible to them.

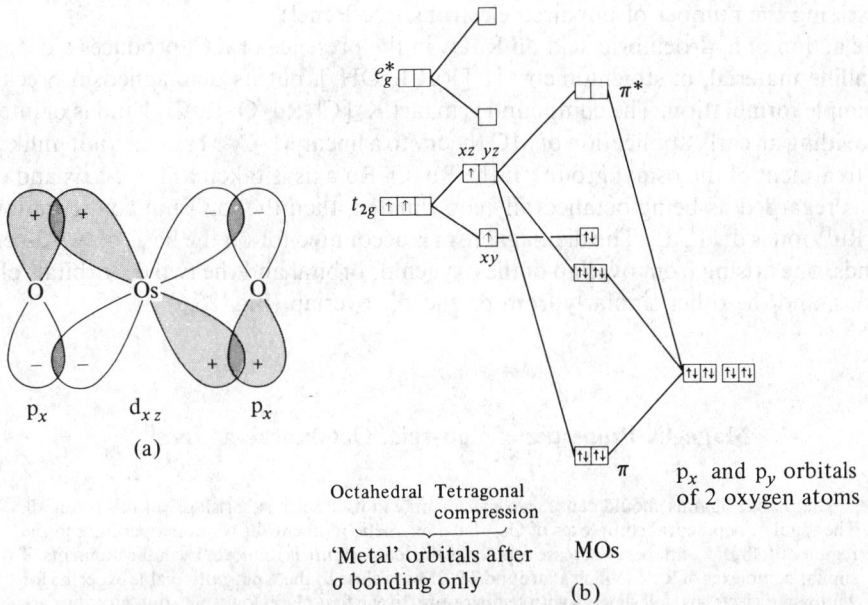

FIG. 25.5 Proposed π bonding in osmyl complexes: (a) 3-centre π bond formed by overlap of ligand p_x and metal d_{zx} orbitals (a similar bond is produced by p_y and d_{yz} overlap), and (b) MO diagram (see text).

The $\{Os^{VI}O_2\}^{2+}$ group has a formal similarity to the more familiar uranyl ion $[UO_2]^{2+}$ and is present in a variety of octahedral complexes in which the 4 equatorial sites are occupied by ligands such as OH^-, halides, CN^-, $(C_2O_4)^{2-}$, $NO_2{}^-$, NH_3, and phthalocyanine. These are obtained from OsO_4 or potassium osmate.

Corresponding "ruthenyl" complexes are known but are much less prevalent. An example is the deep reddish-purple, diamagnetic compound $Cs_2[RuCl_4O_2]$, prepared by adding CsCl to a solution of RuO_4 in conc HCl.

Oxidation state V (d^3)

The only established examples of coordination complexes with the metals in this oxidation state are $[MF_6]^-$ and it is clearly not a very stable state for this group of metals in solution. It is, however, well-characterized and stable in numerous solid-state oxide systems (p. 1257).

Oxidation state IV (d⁴)

Under normal circumstances this is the most stable oxidation state for osmium and the $[OsX_6]^{2-}$ complexes (X = F, Cl, Br I) are particularly well known. $[RuX_6]^{2-}$ (X = F, Cl, Br) are also familiar but can more readily be reduced to Ru^{III}. All these $[MX_6]^{2-}$ ions are octahedral and low-spin, with 2 unpaired electrons. Their magnetic properties are interesting and highlight the limitations of using "spin-only" values of magnetic moments in assessing the number of unpaired electrons (see Panel).

The action of hydrochloric acid on RuO_4 in the presence of KCl produces a deep-red crystalline material, of stoichiometry $K_2[RuCl_5(OH)]$, but its diamagnetism precludes this simple formulation. The compound is in fact $K_4[Cl_5Ru-O-RuCl_5]$ and is of interest as providing an early application of MO theory to a linear M–O–M system (not unlike the later treatment of the osmyl group). If the Ru–O–Ru axis is taken as the z-axis and each Ru^{IV} is regarded as being octahedrally coordinated, then the low-spin configuration of each Ru^{IV} ion is $d_{xy}^2 d_{xz}^1 d_{yz}^1$. The diamagnetism is accounted for on the basis of two 3-centre π bonds, one arising from overlap of the oxygen p_x orbital and the two d_{xz} orbitals of the Ru ions, and the other similarly from p_y and d_{yz} overlap (Fig. 25.6).

Magnetic Properties of Low-spin, Octahedral, d⁴ Ions[17]

That halide ligands should cause spin-pairing may in itself seem surprising, but this is not all. The regular, octahedral complexes of Os^{IV} have magnetic moments at room temperature in the region of 1.48 BM and these decrease rapidly as the temperature is reduced. Even the moments of similar complexes of Ru^{IV} (which at around 2.9 BM are close to the "spin-only" value expected for 2 unpaired electrons) fall sharply with temperature. In the first place, low-spin configurations are much more common for the second- and third-row than for first-row transition elements and this is due to (a) the higher nuclear charges of the heavier elements which exert stronger attractions on the ligands so that a given set of ligands produces a greater splitting of the metal d orbitals, and (b) the larger sizes of 4d and 5d orbitals compared to 3d orbitals, with the result that interelectronic repulsions, which tend to oppose spin-pairing, are lower in the former cases. These factors explain why the halide complexes of Os^{IV} and Ru^{IV} are low-spin but what of the temperature dependence and their magnetic behaviour? This arises from the effect of "spin–orbit coupling", briefly alluded to in Chapter 19 (p. 1098), and can be summarized in a plot of μ_e versus $kT/|\lambda|$ (Fig. A). λ is the spin-orbit coupling constant for a particular ion and is indicative of the strength of the coupling between the angular momentum vectors associated with S and L, and also of the magnitude of the splitting of the ground term of the ion (^3T, in the case of low-spin d^4). When $|\lambda|$ is of comparable magnitude to kT, $\mu_e \sim 3.6$ BM, which is the spin-only moment (2.83 BM) plus the orbital contribution which is expected (p. 1098) for a low-spin d^4 ion. Thus, Cr^{II} ($\lambda = -115 \text{ cm}^{-1}$) and Mn^{III} ($\lambda = -178 \text{ cm}^{-1}$) at room temperature ($kT \sim 200 \text{ cm}^{-1}$), lie on the flat portion of the curve and so have magnetic moments of about 3.6 BM which only begin to fall at appreciably lower temperatures. On the other hand, Ru^{IV} ($\lambda = -700 \text{ cm}^{-1}$) and Os^{IV} ($\lambda \sim -2000 \text{ cm}^{-1}$) have moments which at room temperature are already on the steep portion of the curve and so are

---

[17] A. EARNSHAW, *Introduction to Magnetochemistry*, pp. 60–72 Academic Press, London, 1968. B. N. FIGGIS, The magnetic properties of complex ions, Chap. 10 in *Introduction to Ligand Fields*, pp. 248–92, Interscience, New York, 1966.

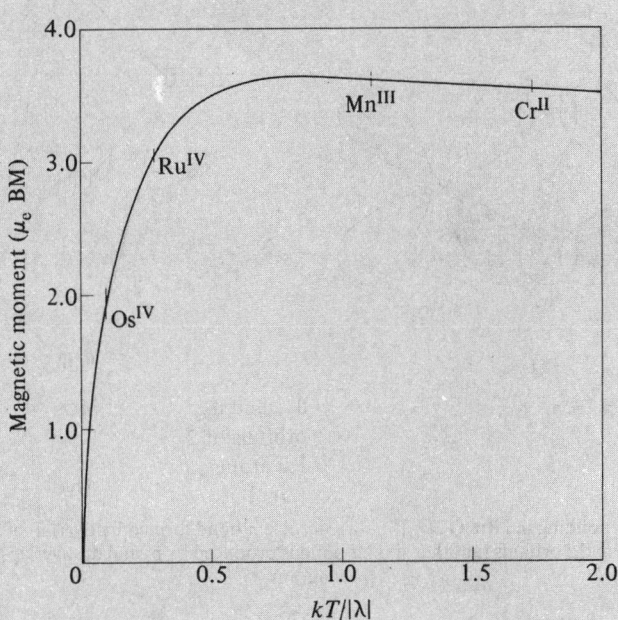

F<small>IG</small>. A   The variation with temperature and spin–orbit coupling constant, of the magnetic moments of octahedral, low-spin, d^4 ions. (The values of μ_e at 300 K are marked for individual ions.)

extremely dependent on temperature. In each case, as the temperature→0, so also μ_e→0, corresponding to a coupling of L and S vectors in opposition and their associated magnetic moments therefore cancelling each other.

All d^n configurations with T ground terms give rise to magnetic moments which are lower for second- and third-row than for first-row transition elements and are temperature dependent, but in no case so dramatically as for low-spin d^4.

Ruthenium(IV) produces few other complexes of interest but osmium(IV) yields several sulfito complexes (e.g. $[Os(SO_3)_6]^{8-}$ and substituted derivatives) as well as a number of complexes, such as $[Os(bipy)Cl_4]$ and $[Os(diars)_2X_2]^{2+}$ (X = Cl, Br, I), with mixed halide and Group V donor atoms. The iron analogues of the latter complexes (with X = Cl, Br), are obtained by oxidation of $[Fe(diars)_2X_2]^+$ with conc HNO_3 and provide rare examples of complexes containing iron in an oxidation state higher than $+3$. The halide ions are *trans* to each other and a reduction in the magnetic moment at 293 K from a value of ~3.6 BM (which might have been expected, since $\lambda = -260$ cm^{-1} for Fe^{IV}—see Panel) to 2.98 BM is explained by a large tetragonal distortion.

Oxidation state III (d⁵)

Ruthenium(III) and osmium(III) complexes are all octahedral and low-spin with 1 unpaired electron. Iron(III) complexes, on the other hand, may be high or low spin, and even though an octahedral stereochemistry is the most common, a number of other

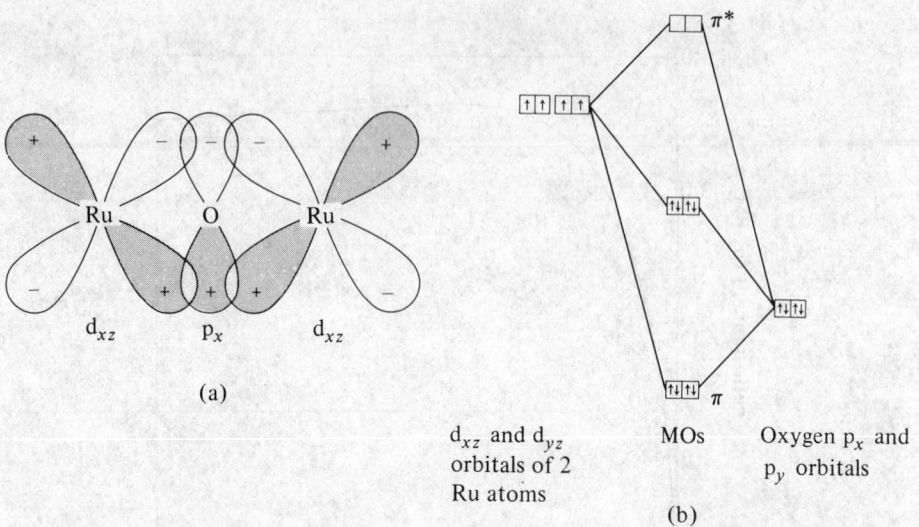

d_{xz} and d_{yz} MOs   Oxygen p_x and
orbitals of 2 p_y orbitals
Ru atoms

(b)

FIG. 25.6 π bonding in $[Ru_2OCl_{10}]^{4-}$: (a) 3-centre π bond formed by overlap of an oxygen p_x and ruthenium d_{xz} orbitals (another similar bond is produced by p_y and d_{yz} overlap), and (b) MO diagram.

geometries are also found. In other respects, however there is a gradation down the triad, with Ru^{III} occupying an intermediate position between Fe^{III} and Os^{III}. For iron the oxidation state $+3$ is one of its two most common and for it there is an extensive, simple, cationic chemistry (though the aquo ion, $[Fe(H_2O)_6]^{3+}$, is too readily hydrolysed to be really common). For ruthenium it is the best-known oxidation state and, although the $[Ru(H_2O)_6]^{3+}$ ion has not been isolated in solids, there is evidence that it can exist in solution. For osmium, however, Os^{III} is a distinctly less-common oxidation state, being readily oxidized to Os^{IV} or even, in the presence of π-acceptor ligands such as CN^-, reduced to Os^{II}. There is no evidence of a simple aquo ion of osmium in this, or, indeed, in any other oxidation state.

Except with anions such as iodide which have reducing tendencies, iron(III) forms salts with all the common anions, and these may be crystallized in pale-pink or pale-violet hydrated forms. These presumably contain the $[Fe(H_2O)_6]^{3+}$ cation, and the iron alums certainly do. These alums have the composition $Fe_2(SO_4)_3M_2^ISO_4.24H_2O$ and can be formulated $[M^I(H_2O)_6][Fe^{III}(H_2O)_6][SO_4]_2$. Like the analogous chrome alums they find use as mordants in dying processes. The sulfate is the cheapest salt of Fe^{III} and forms no less than 6 different hydrates (12, 10, 9, 7, 6, and 3 mols of H_2O of which $9H_2O$ is the most common); it is widely used as a coagulent in the treatment not only of potable water but also of sewage and industrial effluents.

In the crystallization of these hydrated salts from aqueous solutions it is essential that a low pH (high level of acidity) is maintained, otherwise hydrolysis occurs and yellow impurities contaminate the products. Similarly, if the salts are redissolved in water, the solutions turn yellow/brown. The hydrolytic processes are complicated, and, in the presence of anions with appreciable coordinating tendencies, are further confused by displacement of water from the coordination sphere of the iron. However, in aqueous solutions of salts such as the perchlorate the following equilibria are important:

$$[Fe(H_2O)_6]^{3+} \rightleftharpoons [Fe(H_2O)_5(OH)]^{2+} + H^+; \quad K = 10^{-3.05}$$
$$[Fe(H_2O)_5(OH)]^{2+} \rightleftharpoons [Fe(H_2O)_4(OH)_2]^+ + H^+; \quad K = 10^{-3.26}$$

and also $2[Fe(H_2O)_6]^{3+} \rightleftharpoons [Fe(H_2O)_4(OH)]_2^{4+} + 2H^+ + 2H_2O; \quad K = 10^{-2.91}$

(The dimer in the third equation is actually[18]) $[(H_2O)_4Fe\underset{OH}{\overset{OH}{\diagup\diagdown}}Fe(H_2O)_4]^{4+}$ and weakly coupled electron spins on the 2 metal ions reduce the magnetic moment per iron below the spin-only value for 5 unpaired electrons.)

It is evident therefore that Fe^{III} salts dissolved in water produce highly acidic solutions and the simple, pale-violet, hexaquo ion only predominates if further acid is added to give pH ~ 0. At somewhat higher values of pH the solution becomes yellow due to the appearance of the above hydrolysed species and if the pH is raised above 2–3, further condensation occurs, colloidal gels begin to form, and eventually a reddish-brown precipitate of hydrous iron(III) oxide is formed (see p. 1254).

The colours of these solutions are of interest. Iron(III) like manganese(II), has a d^5 configuration and its absorption spectrum might therefore be expected to consist similarly (p. 1232) of weak spin-forbidden bands. However, a crucial difference between the ions is that Fe^{III} carries an additional positive charge, and its correspondingly greater ability to polarize coordinated ligands produces intense, charge-transfer absorptions at much lower energies than those of Mn^{II} compounds. As a result, only the hexaquo ion has the pale colouring associated with spin-forbidden bands in the visible region of the spectrum, while the various hydrolysed species have charge transfer bands, the edges of which tail from the ultraviolet into the visible region producing the yellow colour and obscuring weak d–d bands. Even the hexaquo ion's spectrum is dominated in the near ultraviolet by charge transfer, and a full analysis of the d–d spectrum of this and of other Fe^{III} complexes is consequently not possible.

Iron(III) forms a variety of cationic, neutral, and anionic complexes, but an interesting feature of its coordination chemistry is a marked preference (not shown by Cr^{III} with which in many other respects it is similar) for O-donor as opposed to N-donor ligands. Ammines of Fe^{III} are unstable and dissociate in water; chelating ligands such as bipy and phen which induce spin-pairing produce more stable complexes, but even these are less stable than their Fe^{II} analogues. Thus, whereas deep-red aqueous solutions of $[Fe(phen)_3]^{2+}$ are indefinitely stable, the deep-blue solutions of $[Fe(phen)_3]^{3+}$ slowly turn khaki-coloured as polymeric hydroxo species form. By contrast, the intense colours produced when phenols or enols are treated with Fe^{III}, and which are used as characteristic tests for these organic materials, are due to the formation of Fe–O complexes. Again, the addition of phosphoric acid to yellow, aqueous solutions of $FeCl_3$, for instance, decolourizes them because of the formation of phosphato complexes such as $[Fe(PO_4)_3]^{6-}$ and $[Fe(HPO_4)_3]^{3-}$. The deep-red $[Fe(acac)_3]$ and the green $[Fe(C_2O_4)_3]^{3-}$ are other examples of complexes with oxygen-bonded ligands although the latter, whilst very stable towards dissociation, is photosensitive due to oxidation of the oxalate ion by Fe^{III} and so decomposes to $Fe(C_2O_4)$ and CO_2.

[18] T. I. MORRISON, A. H. REIS, G. S. KNAPP, F. Y. FRADIN, H. CHEN, and E. KLIPPERT, Extended X-ray absorption fine structure studies of the hydrolytic polymerization of iron(III). 1. Structural characterization of the μ-dihydroxo-octaaquodiiron(III) dimer, *J. Am. Chem. Soc.* **100**, 3262–4 (1978).

Complexes with mixed *O*- and *N*-donor ligands such as EDTA and Schiff bases are well known and $[Fe(EDTA)(H_2O)]^-$ and $[FesalenCl]$ are examples of 7-coordinate (pentagonal bipyramidal) and 5-coordinate (square-pyramidal) stereochemistries respectively.

As in the case of Cr^{III}, oxo- bridged species with magnetic moments reduced below the spin-only value (5.9 BM in the case of high-spin Fe^{III}) are known.[19] $[Fe salen]_2O$, for instance, has a moment of 1.9 BM at 298 K which falls to 0.6 BM at 80 K and the interaction between the electron spins on the 2 metal ions is transmitted across an Fe–O–Fe bridge, bent at an angle of 140°. Trinuclear, basic carboxylates, $[Fe_3O(O_2CR)_6L_3]^+$, are, however, entirely analogous to their Cr^{III} counterparts (p. 1200).

Halide complexes decrease markedly in stability from F^- to Br^- (there are no I^- complexes). Fluoro complexes are quite stable and in aqueous solutions the predominant species is $[FeF_5(H_2O)]^{2-}$ while isolation of the solid and fusion with KHF_2 yields $[FeF_6]^{3-}$. Chloro complexes are appreciably less stable, and tetrahedral rather than octahedral coordination is favoured. $[FeCl_4]^-$ can be isolated in yellow salts with large cations such as $[RN_4]^+$ from ethanolic solutions or conc HCl. $[FeBr_4]^-$ is also known but is thermally unstable due to internal oxidation-reduction producing Fe^{II} and Br_2.

The blood-red colour produced by mixing aqueous solutions of Fe^{III} and SCN^- (and which provides a well-known test for Fe^{III}) is largely due to $[Fe(SCN)(H_2O)_5]^{2+}$ but, in addition to this, the simple salt $Fe(SCN)_3$ and salts of complexes such as $[Fe(SCN)_4]^-$ and $[Fe(SCN)_6]^{3-}$ can also be isolated.

The high-spin d^5 configuration of Fe^{III}, like that of Mn^{II}, confers no advantage by virtue of CFSE on any particular stereochemistry. Some examples of its consequent ability to adopt stereochemistries other than octahedral have just been mentioned and further examples are given in Table 25.3. These cover the range of coordination numbers from 3 to 8.

Further similarity with Mn^{II} may be seen in the fact that the vast majority of the compounds of Fe^{III} are high-spin. Only ligands such as bipy and phen (already mentioned) and CN^-, which are high in the spectrochemical series, can induce spin-pairing. The low-spin $[Fe(CN)_6]^{3-}$, which is best known as its red, crystalline potassium salt, is usually prepared by oxidation of $[Fe(CN)_6]^{4-}$ with, for instance, Cl_2. It should be noted that in $[Fe(CN)_6]^{3-}$ the CN^- ligands are sufficiently labile to render it poisonous, in apparent contrast to $[Fe(CN)_6]^-$, which is kinetically more inert. Dilute acids produce $[Fe(CN)_5(H_2O)]^{2-}$, and other pentacyano complexes are known.

Fe^{III} complexes in general have magnetic moments at room temperature which are close to 5.92 BM if they are high-spin and somewhat in excess of 2 BM (due to orbital contribution) if they are low-spin. A number of complexes, however, were prepared in 1931 by L. Cambi and found to have moments intermediate between these extremes. They are the iron(III)-*N,N*-dialkyldithiocarbamates, $[Fe(S_2CNR_2)_3]$, in which the ligands are

so that the Fe^{III} is surrounded octahedrally by 6 sulfur atoms. They provide probably the best-documented examples of the high-spin/low-spin crossover (i.e. spin equilibria) referred to in Chapter 19 (p. 1094).

[19] K. S. MURRAY, Binuclear oxo-bridged iron(III) complexes, *Coord. Chem. Revs.* **12**, 1–35 (1974).

d^5 Spin Equilibria[20]

It has been suggested that Fe^{III} in the dithiocarbamato complexes might exist in a single intermediate spin state but it seems more likely that two distinct species, high-spin and low-spin, are present in thermal equilibrium. The simplified Tanabe–Sugano diagram (Fig. A) shows how, as the crystal field increases, it is designated as weak or strong (i.e. the configurations as high-spin or low-spin) depending on whether the ground term is a $^6A_1$ $(t_{2g}^3 e_g^2)$ or $^2T_2$ (t_{2g}^5) term. Where the CF is

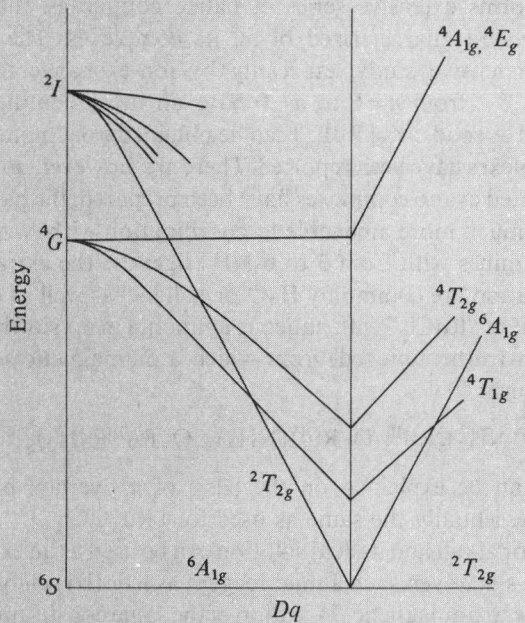

Fig. A   Simplified Tanabe–Sugano diagram for d^5 ions.

close to this critical boundary value both terms are thermally accessible and the observed magnetic moment will depend on the precise distribution between them. As this distribution will be temperature-dependent, so too will be the magnetic moment. For most of the dithiocarbamato complexes the CF is on the high-field side of the crossover, with the A term a few hundred wavenumbers above the T. (Note that at room temperature kT the thermal energy $\sim 200\ cm^{-1}$.) At very low temperatures therefore the moments are around 2.3 BM as expected for the low-spin configuration but, as the temperature rises, so does the population of the A term and with it the magnetic moment until μ_e is between 4 and 5 BM, consistent with equal populations of the two levels. Application of high pressure tends to reduce the moment (i.e. favour the low-spin configuration) as is expected since compression will shorten the metal–donor distances and so increase the CF.

Expressions for the μ_e can be derived by assuming a Boltzmann distribution between the two levels, weighted according to their respective degeneracies, and allowing for the fact that the moment of the low-spin configuration is itself dependent on temperature. Agreement with experiment is, however, not exact, but can be improved considerably, principally by allowing for the different metal–donor vibrational energies of the two configurations.

Continued

---

[20] C. A. Tsipis, C. C. Hadjikostas, and G. E. Manoussakis, Mixed-ligand iron(III) dithiocarbamates. Calculation of ligand field parameters in electron spin crossover d^5 systems by combination of spectral and magnetic data, *Inorg. Chimica Acta* **23,** 163–71 (1977); 40 references given.

These compounds have been studied by a variety of physical techniques besides magnetic, notably Mössbauer, ir, and nmr spectroscopy.

Similar behaviour is found in monothio-β-diketonates,[21] and other d^5 systems displaying this type of spin equilibria include a number of haem-type compounds.[22] Spin cross-over transitions have also been observed in the d^6 systems Fe^{II} and Co^{III} {low-spin $^1A_{1g}(t_{2g}^6)$—high-spin $^5T_{2g}$ $(t_{2g}^4e_g^2)$} (p. 1272) and in the d^7 Co^{II} system {$^2E_g(t_{2g}^6e_g^1)-{}^4T_{1g}(t_{2g}^5e_g^2)$}.

Ruthenium(III) forms extensive series of halide complexes, the aquo-chloro series being probably the best characterized of all its complexes. The $Ru^{III}/Cl^-/H_2O$ system has received extensive study, especially by ion-exchange techniques. The ions $[RuCl_n(H_2O)_{6-n}]^{(n-3)-}$ from $n=6$ to $n=0$ have all been identified in solution and a number also isolated as solids. $K_3[RuF_6]$ can be obtained from molten $RuCl_3/KHF_2$, and several bromo complexes have been reported. There are, however, no iodo complexes and, whilst some substituted cyano complexes have been prepared, the parent $[Ru(CN)_6]^{3-}$ is not known. Ru^{III} is much more amenable to coordination with N-donor ligands than is Fe^{III}, and forms ammines with from 3 to 6 NH_3 ligands (the extra ligands making up octahedral coordination are commonly H_2O or halides) as well as complexes with bipy and phen. Treatment of "$RuCl_3$" with aqueous ammonia in air slowly yields an extremely intense red solution (ruthenium red) from which a diamagnetic solid can be isolated, apparently of the form:

$$[(NH_3)_5Ru^{III}-O-Ru^{IV}(NH_3)_4-O-Ru^{III}(NH_3)_5]^{6+}$$

Its diamagnetism can be explained on the basis of π overlap, producing polycentre molecular orbitals essentially the same as used for $[Ru_2OCl_{10}]^{4-}$ (see Fig. 25.6). It is stable in either acid or alkali and its acid solution can be used as an extremely sensitive test for oxidizing agents since even such a mild reagent as iron(III) chloride oxidizes the red, $6+$ cation to a yellow, paramagnetic, $7+$ cation of the same constitution (a change which is detectable in solutions containing less than 1 ppm Ru).

Trinuclear basic acetates $[Ru_3O(O_2CMe)_6L_3]^+$ have also been prepared apparently with the same constitution as the analogous Fe^{III} and Cr^{III} compounds (p. 1200).

Oxidation state II (d^6)

This is the second of the common oxidation states for iron and is familiar for ruthenium, particularly with group V-donor ligands (Ru^{II} probably forms more nitrosyl complexes than any other metal). Osmium(II) also produces a considerable number of complexes which may seem surprising in view of the paucity of Os^{III} complexes.

Iron(II) forms salts with nearly all the common anions.† These are usually prepared in aqueous solution either from the metal or by reduction of the corresponding Fe^{III} salt. In

---

† An exception is NO_2^- which instantly oxidizes Fe^{II} to Fe^{III} and liberates NO. $Fe(BrO_3)_2$ and $Fe(IO_3)_2$ also are unstable.

[21] M. Cox, J. Darken, B. W. Fitzsimmons, A. W. Smith, L. F. Larkworthy, and K. A. Rodgers, Spin isomerism in tris(monothio-β-diketonato) iron(III) complexes, *JCS Dalton* 1972, 1192–5.

[22] See for instance, W. R. Scheidt and D. K. Geiger, Molecular stereochemistry and crystal structure of a "spin-equilibrium" ($S=\frac{1}{2}$, $S=\frac{5}{2}$) haemichrome salt: bis-(3-chloropyridine)octaethylporphinatoiron(III) perchlorate, *JCS Chem. Comm.* 1979, 1154–5.

the absence of other coordinating groups these solutions contain the pale-green $[Fe(H_2O)_6]^{2+}$ ion which is also present in solids such as $Fe(ClO_4)_2.6H_2O$, $FeSO_4.7H_2O$, and the well-known "Mohr's salt", $(NH_4)_2SO_4FeSO_4.6H_2O$ introduced into volumetric analysis by K. F. Mohr in the 1850s.† The hydrolysis (which in the case of Fe^{III} produces acidic solutions) is virtually absent, and in aqueous solution the addition of CO_3^{2-} does not result in the evolution of CO_2 but simply in the precipitation of white $FeCO_3$. The moist precipitate oxidizes rapidly on exposure to air but in the presence of excess CO_2 the slightly soluble $Fe(HCO_3)_2$ is formed. It is the presence of this in natural underground water systems, leading to the production of $FeCO_3$ on exposure to air, followed by oxidation to iron(III) oxide, which leads to the characteristic brown deposits found in many streams.

The possibility of oxidation to Fe^{III} is a crucial theme in the chemistry of Fe^{II} and most of its salts are unstable with respect to aerial oxidation, though double sulfates are much less so (e.g. Mohr's salt above). However, the susceptibility of Fe^{II} to oxidation is dependent on the nature of the ligands attached to it and, in aqueous solution, on the pH. Thus the solid hydroxide and alkaline solutions are very readily oxidized whereas acid solutions are much more stable (see Panel).

The Fe^{III}/Fe^{II} Couples

A selection of the standard reduction potentials for some iron couples is given in Table A, from which the importance of the participating ligand can be judged (see also Table 25.8 for biologically important iron proteins). Thus Fe^{III}, being more highly charged than Fe^{II} is stabilized (relatively) by negatively charged ligands such as the anions of EDTA and derivatives of 8-hydroxyquinoline, whereas Fe^{II} is favoured by neutral ligands which permit some charge delocalization in π-orbitals (e.g. bipy and phen).

The value of E° for the couple involving the simple aquated ions, shows that $Fe^{II}(aq)$ is thermodynamically stable with respect to hydrogen; which is to say that $Fe^{III}(aq)$ is spontaneously reduced by hydrogen gas (see p. 498). However, under normal circumstances, it is not hydrogen but atmospheric oxygen which is important and, for the process $\frac{1}{2}O_2 + 2H^+ + 2e^- \rightleftharpoons H_2O$, $E^\circ = 1.229$ V, i.e. oxygen gas is sufficiently strong an oxidizing agent to render $[Fe(H_2O)_6]^{2+}$ (and, indeed, all other Fe^{II} species in Table A) unstable wrt atmospheric oxidation. In practice the oxidation in acidic solutions is slow and, if the pH is increased, the potential for the Fe^{III}/Fe^{II} couple remains fairly constant until the solution becomes alkaline and hydrous Fe_2O_3 (considered here for convenience to be $Fe(OH)_3$) is precipitated. But here the change is dramatic.

The actual potential E of the couple is given by the Nernst equation,

$$E = E^\circ - \frac{RT}{nF} \ln \frac{[Fe^{II}]}{[Fe^{III}]}$$

where $E = E^\circ$ when all activities are unity. However, once precipitation occurs, the activities of the iron species are far from unity; they are determined by the solubility products of the 2 hydroxides. These are:

$$[Fe^{II}][OH^-]^2 \sim 10^{-14} \text{ (mol dm}^{-3})^3 \text{ and } [Fe^{III}][OH^-]^3 \sim 10^{-36} \text{ (mol dm}^{-3})^4$$

Continued

† K. F. Mohr (1806–79) was Professor of Pharmacy at the University of Bonn. Among his many inventions were the specific gravity balance, the burette, the pinch clamp, the cork borer, and the use of the so-called Liebig condenser for refluxing. In addition to his introduction of iron(II) ammonium sulfate as a standard reducing agent he devised Mohr's method for titrating halide solutions with silver ions in the presence of chromate as indicator, and was instrumental in establishing titrimeric methods generally for quantitative analysis.

$$\therefore \text{ when } [OH^-] = 1 \text{ mol dm}^{-3}, \frac{[Fe^{II}]}{[Fe^{III}]} \sim 10^{22}$$

$$\therefore E \sim 0.771 - 0.05916 \log_{10}(10^{22}) = 0.771 - 1.301 = -0.530 \text{ V}$$

Thus by making the solution alkaline the sign of E has been reversed and the susceptibility of Fe^{II} (aq) to oxidation (i.e. its reducing power) enormously increased. This is why white, precipitated $Fe(OH)_2$ and $FeCO_3$ are rapidly darkened by aerial oxidation and why Fe^{II} in alkaline solution will reduce nitrates to ammonia and copper(II) salts to metallic copper.

TABLE A　$E°$ at 25°C for some Fe^{III}/Fe^{II} couples in acid solution

Fe^{III}	Fe^{II}	$E°/V$
$[Fe(phen)_3]^{3+} + e^- \rightleftharpoons [Fe(phen)_3]^{2+}$		1.12
$[Fe(bipy)_3]^{3+} + e^- \rightleftharpoons [Fe(bipy)_3]^{2+}$		0.96
$[Fe(H_2O)_6]^{3+} + e^- \rightleftharpoons [Fe(H_2O)_6]^{2+}$		0.77
$[Fe(CN)_6]^{3-} + e^- \rightleftharpoons [Fe(CN)_6]^{4-}$		0.36
$[Fe(C_2O_4)_3]^{3-} + e^- \rightleftharpoons [Fe(C_2O_4)_2]^{2-} + (C_2O_4)^{2-}$		0.02
$[Fe(EDTA)]^- + e^- \rightleftharpoons [Fe(EDTA)]^{2-}$		-0.12
$[Fe(quin)_3] + e^- \rightleftharpoons [Fe(quin)_2] + quin^-$ [a]		-0.30

(a) quin⁻ = 5-methyl-8-hydroxyquinolinate,

Iron(II) forms complexes with a variety of ligands. As is to be expected, in view of its smaller cationic charge, these are usually less stable than those of Fe^{III} but the antipathy to N-donor ligands is less marked. Thus $[Fe(NH_3)_6]^{2+}$ is known whereas the Fe^{III} analogue is not; also there are fewer Fe^{II} complexes with O-donor ligands such as acac and oxalate, and they are less stable than those of Fe^{III}. High-spin octahedral complexes of Fe^{II} have a free-ion ^5D ground term, split by the crystal field into a ground $^5T_{2g}$ and an excited ^5E_g term. A magnetic moment of around 5.5 BM (i.e. 4.90 BM + orbital contribution) is expected for pure octahedral symmetry but, in practice, distortions produce values in the range 5.2–5.4 BM. Similarly, in the electronic spectrum, the expected single band due to the ^5E_g $(t_{2g}^3 e_g^3) \leftarrow {}^5T_{2g}$ $(t_{2g}^4 e_g^2)$ transition is broadened or split. Besides stereochemical distortions, spin–orbit coupling and a Jahn–Teller effect in the excited configuration (p. 1087) help to broaden the band, the main part of which is usually found between 10 000 and 11 000 cm^{-1}. The d–d spectra of low-spin Fe^{II} (which is isoelectronic with Co^{III}) are not so well documented, being generally obscured by charge-transfer absorption (p. 1309).

Most Fe^{II} complexes are octahedral but several other stereochemistries are known (Table 25.3). $[FeX_4]^{2-}$ (X = Cl, Br, I, NCS) are tetrahedral. A single absorption around 4000 cm^{-1} due to the $^5T_2 \leftarrow {}^5E$ transition is as expected, but magnetic moments of these and other apparently tetrahedral complexes are reported to lie in the range 5.0–5.4 BM, and the higher values are difficult to explain.

Low-spin octahedral complexes are formed by ligands such as bipy, phen, and CN⁻, and their stability is presumably enhanced by the symmetrical t_{2g}^6 configuration. $[Fe(bipy)_3]^{2+}$ and $[Fe(phen)_3]^{2+}$ are stable, intensely red complexes, the latter being employed as the redox indicator, "ferroin", due to the sharp colour change which occurs when strong oxidizing agents are added to it:

$$[Fe(phen)_3]^{2-} \rightleftharpoons [Fe(phen)_3]^{3+} + e^-$$
$$\text{red} \qquad\qquad\qquad \text{blue}$$

Mono- and bis-phenanthroline complexes can be prepared but these are both high-spin, and it is noteworthy that addition of phenanthroline to aqueous Fe^{II} leads almost entirely to the formation of the tris complex rather than mono or bis, even though the Fe^{II} is initially in great excess. The reason lies in the relative CFSEs of the high- and low-spin configurations ($\frac{2}{5}\Delta_o$ and $\frac{12}{5}\Delta_o$ respectively) giving an advantage of $2\Delta_o$ to the latter. Although this is partly counterbalanced by the extra energy required to pair electrons in the t_{2g} set, there is nevertheless a considerable gain in forming the tris complex. Pale-yellow $K_4[Fe(CN)_6].3H_2O$ can be crystallized from aqueous solutions of iron(II) sulfate and an excess of KCN: this is clearly more convenient than the traditional method of fusing nitrogeneous animal residues (hides, horn, etc.) with iron and K_2CO_3. The hexacyanoferrate(II) anion (ferrocyanide) is kinetically inert and is said to be non-toxic, but HCN is liberated by the addition of dilute acids.

Addition of $K_4[Fe^{II}(CN)_6]$ to aqueous Fe^{III} produces the intensely blue precipitate, Prussian blue. The X-ray powder pattern and Mössbauer spectrum of this are the same as those of Turnbull's blue which is produced by the converse addition of $K_3[Fe^{III}(CN)_6]$ to aqueous Fe^{II}. By varying the conditions and proportions of the reactants, a whole range of these blue materials can be produced of varying composition with some, which are actually colloidal, described as soluble Prussian blue. They have found application as pigments in the manufacture of inks and paints since their discovery in 1704 and, in 1840, their formation on sensitized paper was utilized in the production of blueprints. It appears that all these materials have the same basic structure. This consists of a cubic lattice of low-spin Fe^{II} and high-spin Fe^{III} ions with cyanide ions lying linearly along the cube edges, and water molecules situated inside the cubes. The intense colour is due to charge-transfer from Fe^{II} to Fe^{III}. Unfortunately, detailed characterizations have been bedevilled by difficulties in obtaining satisfactory single crystals and reproducible compositions. In a recent study[22a] good quality single crystals were produced and formulated as $Fe_4[Fe(CN)_6]_3.xH_2O$ ($x = 14–16$). This is compatible with the basic lattice arrangement if some of the Fe^{II} and CN^- sites are occupied by water. However these crystals were grown by the slow diffusion of H_2O vapour into a solution of Fe^{III} and $[Fe(CN)_6]^{4-}$ in conc HCl. Less delicate methods lead to the absorption of alkali metal ions (particularly K^+) and to formulations such as $M^I Fe^{II} Fe^{III}(CN)_6 \cdot xH_2O$. The same structure motif is found in $Fe^{III}Fe^{III}(CN)_6$ and in the virtually white, readily oxidizable $K_2 Fe^{II} Fe^{II}(CN)_6$, the former having no counter cations while the K^+ ions of the latter fill all the lattice cubes. Having all their iron atoms in a uniform oxidation state, however, these two compounds lack the intense colour of the Prussian blues.

It is possible to replace 1 CN^- in the hexacyanoferrate(II) ion with H_2O, CO, NO_2^-, and, most importantly, with NO^+. The "nitroprusside" ion $[Fe(CN)_5NO]^{2-}$ can be produced by the action of 30% nitric acid on either $[Fe(CN)_6]^{4-}$ or $[Fe(CN)_6]^{3-}$. That it formally

[22a] H. J. Buser, D. Schwarzenbach, W. Petter, and A. Ludi, The crystal structure of Prussian blue: $Fe_4[Fe(CN)_6]_3.xH_2O$ ($x = 14–16$), *Inorg. Chem.* **16**, 2704–10 (1973).

contains Fe^{II} and NO^+ (rather than Fe^{III} and NO) is evident from its diamagnetism, although Mössbauer studies indicate that there is appreciable π delocalization of charge from the t_{2g} orbitals of the Fe^{II} to the nitrosyl and cyanide groups. The red colour of $[Fe(CN)_5(NOS)]^{4-}$, formed by the addition of sulfide ion, is used in a common qualitative test for sulfur.

Another qualitative test involving an iron nitrosyl is the "brown ring" test for NO_3^-, using iron(II) sulfate and conc H_2SO_4 in which NO is produced. The brown colour, which appears to be due to charge-transfer, evidently arises from a cationic iron nitrosyl complex which has a magnetic moment ~3.9 BM; it is therefore formulated as $[Fe(NO)(H_2O)_5]^{2+}$ in which the iron can be considered formally to be in the $+1$ oxidation state.

Roussin's "red" and "black" salts, formulated respectively as $K_2[Fe_2(NO)_4S_2]$ and $K[Fe_4(NO)_7S_3]$, are obtained by the action of NO on Fe^{II} in the presence of S^{2-} and are structurally interesting. In both cases the iron atoms are pseudo-tetrahedrally coordinated (Fig. 25.7) and, though the assignment of formal oxidation states has only doubtful significance, their diamagnetism and the presence of rather short Fe–Fe distances are indicative of some direct metal–metal interaction.[23]

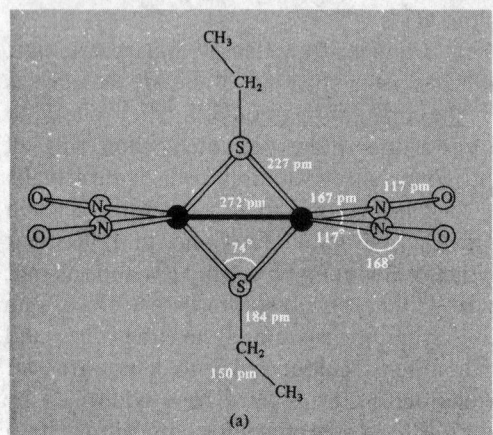
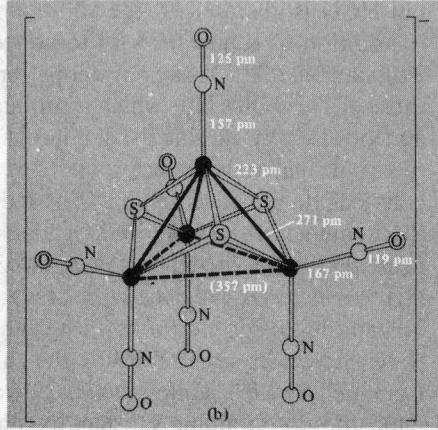

Fig. 25.7 The structure of Roussin's salts: (a) the ethyl ester $[Fe(NO)_2SEt]_2$ of the red salt showing pseudo-tetrahedral coordination of each iron (Fe–Fe = 272 pm), and (b) the anion of the black salt $Cs[Fe_4(NO)_7S_3].H_2O$ showing a pyramid of 4 Fe atoms with an S atom above each of its three non-horizontal faces ($Fe_{apex} - Fe_{base} = 271$ pm, $Fe_{base} \cdots Fe_{base}$ 357 pm). Note that even the short Fe–Fe distances are appreciably greater than the Fe–Fe "single-bond" distance of ~ 250 pm.

In addition to high-spin octahedral complexes with magnetic moments in excess of 5 BM, and diamagnetic, low-spin octahedral complexes, Fe^{II} affords further examples of high-spin/low-spin transitions within a given compound.[24] This can be understood in terms of the Tanabe–Sugano diagram for d^6 ions (Fig. 25.8) which shows that a strong crystal field changes the ground term from $^5T_2$ $(t_{2g}^4 e_g^2)$ to $^1A_1$ (t_{2g}^6). It has already been

[23] W. P. GRIFFITH, Carbonyls, cyanides, isocyanides, and nitrosyls, Chap. 46 in *Comprehensive Inorganic Chemistry*, Vol. 4, pp. 105–95, Pergamon Press, Oxford, 1973.
[24] H. A. GOODWIN, Spin transitions in six coordinate iron(II) complexes, *Coord. Chem. Revs.* **18**, 293–325 (1976). P. GÜTLICH, Spin crossover in iron(II)-complexes, *Structure and Bonding* **44**, 83–195 (1981).

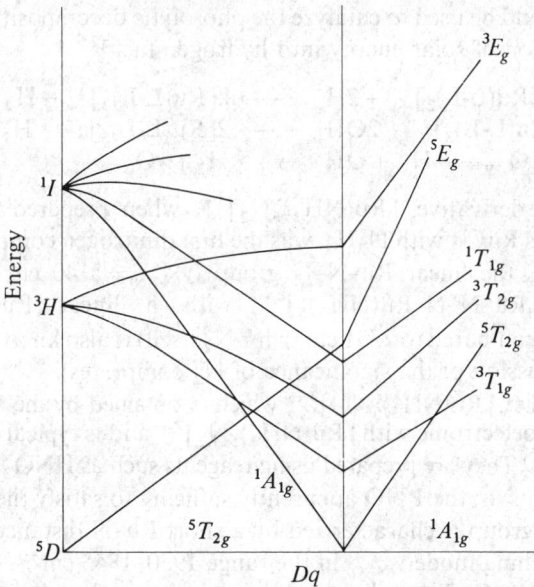

FIG. 25.8 Simplified Tanabe–Sugano diagram for d^6 ions such as Fe^{II}.

noted that a change from high-spin to low-spin accompanies the change,

$$[Fe(phen)_2(H_2O)_2]^{2+} \longrightarrow [Fe(phen)_3]^{2+}$$

so it is no great surprise that spin transitions have been found in $[Fe(phen)_2X_2]$ ($X = NCS$, NCSe) complexes and their bipy analogues. These evidently lie just to the high-field side of the crossover since at temperatures below $-125°C$ the compounds are almost diamagnetic (what paramagnetism there is is probably due to impurity), while at some temperature between $-125°C$ and $-75°C$ depending on the compound, the moment quite suddenly rises to over 5 BM. Confirmation of the transition in these and other Fe^{II} complexes has been provided by electronic and Mössbauer spectroscopy.

Apart from compounds such as $[RuCl_2(PPh_3)_3]$, which is square pyramidal because the sixth coordinating position is stereochemically blocked, Ru^{II} compounds (and also Os^{II} compounds) are octahedral and diamagnetic. The $[Ru(H_2O)_6]^{2+}$ ion can definitely be prepared in aqueous solution by electrolytic reduction of "$RuCl_3$" but is readily oxidized to Ru^{III}. The cyano complexes $[Ru(CN)_6]^{4-}$ and $[Ru(CN)_5NO]^{2-}$, analogous to their iron counterparts, are also known but the most notable compounds of Ru^{II} are undoubtedly its complexes with Group V donor ligands, such as the ammines and nitrosyls.

$[Ru(NH_3)_6]^{2+}$ and corresponding tris chelates with en, bipy, and phen, etc., are obtained from "$RuCl_3$" with Zn powder as a reducing agent. The hexammine is a strongly reducing substance and $[Ru(bipy)_3]^{2+}$, although thermally very stable, is capable of photochemical excitation involving the promotion of an electron from a molecular orbital of essentially metal character to one of an essentially ligand character, after which its oxidation is possible. A number of similar complexes with substituted bipyridyl ligands luminesce in visible light, and considerable effort is being devoted to preparing suitable derivatives which could be used to catalyze the photolytic decomposition of water, with a view to the conversion of solar energy into hydrogen fuel:[25]

$$2[Ru(L-L)_3]^{2+} + 2H^+ \longrightarrow 2[Ru(L-L)_3]^{3+} + H_2$$

then

$$2[Ru(L-L)_3]^{3+} + 2OH^- \longrightarrow 2[Ru(L-L)_3]^{2+} + H_2O + \tfrac{1}{2}O_2$$

i.e.

$$H_2O \rightleftharpoons H^+ + OH^- \longrightarrow H_2 + \tfrac{1}{2}O_2$$

The pentammine derivative, $[Ru(NH_3)_5N_2]^{2+}$, when prepared[26] in 1965 by the reduction of aqueous $RuCl_3$ with N_2H_4, was the first dinitrogen complex to be produced (p. 475). It contains the linear Ru–N–N group ($v_{(N-N)} = 2140$ cm^{-1}). The dinuclear derivative $[(NH_3)_5Ru-N-N-Ru(NH_3)_5]^{4+}$ with a linear Ru–N–N–Ru bridge ($v_{(N-N)} = 2100$ cm^{-1} compared to 2331 cm^{-1} for N_2 itself) is also known (see pp. 1207 and 475 for a fuller discussion of the significance of N_2 complexes).

The nitrosyl complex $[Ru(NH_3)_5NO]^{3+}$, which is obtained by the action of HNO_2 on $[Ru(NH_3)_6]^{2+}$, is isoelectronic with $[Ru(NH_3)_5N_2]^{2+}$ and is typical of a whole series of Ru^{II} nitrosyls.[10,23,27] They are prepared using reagents such as HNO_3 and NO_2^- and are invariably mononitrosyls, the 1 NO apparently sufficing to satisfy the π-donor potential of Ru^{II}. The RuNO group is characterized by a short Ru–N distance in the range 171–176 pm, and a stretching mode $v_{(N-O)}$ in the range 1930–1845 cm^{-1}, consistent with the formulation $Ru^{II}=\overset{+}{N}=O$. The other ligands making up the octahedral coordination include halides, O-donor anions, and neutral, mainly Group V donor ligands.

The stability of ruthenium nitrosyl complexes poses a practical problem in the processing of wastes from nuclear power stations. ^{106}Ru is a major fission product of uranium and plutonium and is a β^- and γ emitter with a half-life of 1 year (367d). The processing of nuclear wastes depends largely on the solvent extraction of nitric acid media, using tri-n-butyl phosphate (TBP) as the solvent (p. 1462). In the main, the uranium and plutonium enter the organic phase while fission products such as Cs, Sr, and lanthanides remain in the aqueous phase. Unfortunately, by this procedure Ru is less effectively removed from the U and Pu than any other contaminant. The reason for this problem is the coordination of TBP to stable ruthenium nitrosyl complexes which are formed under these conditions. This confers on the ruthenium an appreciable solubility in the organic phase, thereby necessitating several extraction cycles for its removal.

Osmium(II) forms no hexaquo complex and $[Os(NH_3)_6]^{2+}$, which may possibly be present in potassium/liquid NH_3 solutions, is also unstable. $[Os(NH_3)_5N_2]^{2+}$ and other dinitrogen complexes are known but only ligands with good π-acceptor properties, such as CN^-, bipy, phen, phosphines, and arsines, really stabilize Os^{II}, and these form complexes similar to their Ru^{II} analogues.

Lower oxidation states

With rare exceptions, such as $[Fe(bipy)_3]^0$, oxidation states lower than $+2$ are represented only by carbonyls, phosphines, and their derivatives. These will be considered together with other organometallic compounds in Section 25.3.6.

[25] D. M. WATKINS, Solar energy conversion using platinum group metal coordination complexes, *Platinum Metals Rev.* **22**, 118–25 (1978).

[26] A. D. ALLEN and C. SENOFF, Nitrogenopentammineruthenium (II) complexes, *Chem. Comm.* 1965, 621–2.

[27] F. BOTTOMLEY, Nitrosyl complexes of ruthenium, *Coord. Chem. Revs.* **26**, 7–32 (1978).

25.3.5 *The biochemistry of iron*[28]

Iron is the most important transition element involved in living systems, being vital to both plants and animals. The stunted growth of the former is well known on soils which are either themselves deficient in iron, or in which high alkalinity renders the iron too insoluble to be accessible to the plants. The problem is overcome by the application of simple iron salts or by the addition of a complex such as $[Fe(EDTA)]^{2-}$ which is sufficiently stable to prevent precipitation of the iron. In the case of man, iron was the first minor element to be recognized as being essential when, in 1681, the physician T. Sydenham used iron "steeped in cold Rhenish wine" to treat anaemia. The adult human body contains about 4 g of iron (i.e. $\sim 0.005\%$ of body weight), of which about 3 g are in the form of haemoglobin, and this level is maintained by absorbing a mere 1 mg of iron per day—a remarkably economical utilization.

Proteins involving iron have two major functions:

(a) oxygen transport and storage;
(b) electron transfer.

Ancillary to the proteins performing these functions are others which transport and store the iron itself. All these proteins are conveniently categorized according to whether or not they contain haem, and the more important classes found in nature are listed in Table 25.7.

TABLE 25.7 *Naturally occurring iron proteins*

Name	Donor atoms. Stereochemisty of Fe	Function	Source	Approximate Mol wt	No. of Fe atoms
Haem proteins					
Haemoglobin	$5 \times N$ Square pyramidal	O_2 transport	Animals	64 500	4
Myoglobin	$5 \times N$ Square pyramidal	O_2 storage	Animals	17 000	1
Cytochromes	$5 \times N + S$ Octahedral	Electron transfer	Bacteria, plants, animals	12 400	4
NHIP (non-haem iron proteins)					
Transferrin		Scavenging Fe	Animals	80 000	2
Ferritin		Storage of Fe	Animals	460 000	20% Fe
Ferredoxins	$4 \times S$ Distorted tetrahedral	Electron transfer	Bacteria, plants	6000–12 000	2–8
Rubridoxins	$4 \times S$ Distorted tetrahedral		Bacteria	6000	1
"MoFe protein"		Nitrogen fixation (see p. 1205)	In nitrogenase	220 000–230 000	24–36
"Fe protein"	$4 \times S$ Distorted tetrahedral			50 000–70 000	4

[28] *Iron in Model and Natural Compounds*, Vol. 7 of *Metal Ions in Biological Systems* (H. SIGEL, ed.), Marcel Dekker, New York, 1978, 417 pp.

Haemoglobin and myoglobin

Haemoglobin is the oxygen-carrying protein in red blood-cells (erythrocytes) and is responsible for their colour. Its biological function is to carry O_2 in arterial blood from the lungs to the muscles, where the oxygen is transferred to the immobile myoglobin, which stores it so that it is available as and when required for the generation of energy by the metabolic oxidation of glucose. At this point the haemoglobin picks up CO_2, which is a product of the oxidation of glucose, and transports it in venous blood back to the lungs.†

In haemoglobin which has no O_2 attached (and is therefore known as deoxyhaemo-globin or reduced haemoglobin), the iron is present as high-spin Fe^{II} and the reversible attachment of O_2 (giving oxyhaemoglobin) changes this to diamagnetic, low-spin Fe^{II} without affecting the metal's oxidation state. This is remarkable, the more so because, if the globin is removed by treatment with HCl/acetone, the isolated haem in water entirely loses its O_2-carrying ability, being instead oxidized by air to haematin in which the iron is high-spin Fe^{III}:

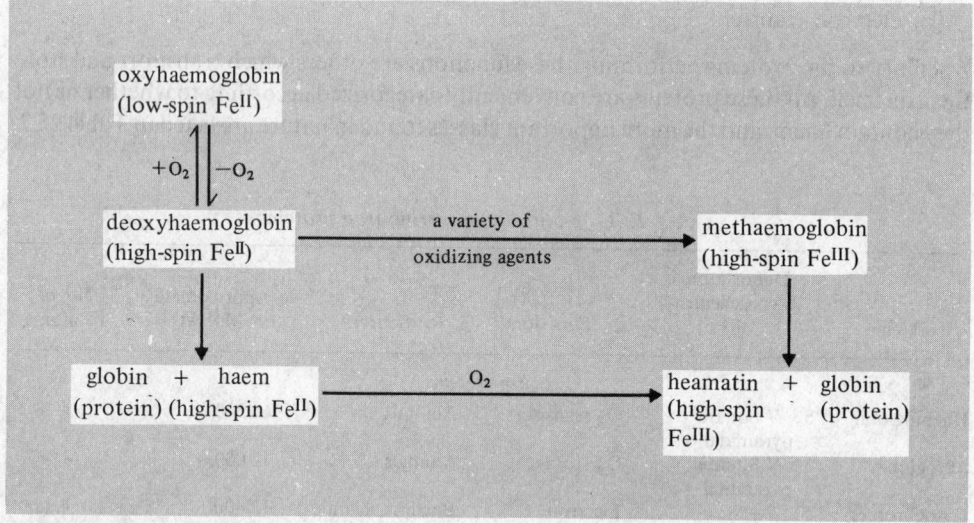

The key to the explanation lies in (a) the observation that in general the ionic radius of Fe^{II} (and also, for that matter, of Fe^{III}) decreases by roughly 20% when the configuration changes from high- to low-spin (see Table 25.1), and (b) the structure of haemoglobin.

As a result of intensive study, enough is now known about the structure of haemoglobin to allow the broad principles of its operation to be explained. It is made up of 4 subunits, each of which consists of a protein (globin), in the form of a folded helix or spiral, attached to 1 iron-containing group (haem). Within the haem group the iron is coordinated to 4 nitrogen atoms of the rigidly planar molecule known as protoporphyrin IX (PIX) and the haem is attached to the globin by coordination of the iron to an imidazole-nitrogen of one of the globin's histidine residues. Figure 25.9a shows the coordination sphere of the iron and Fig. 25.9b a diagramatic view of its relationship with the globin chain.[29] The polar

---

† Human arterial blood can absorb over 50 times more oxygen than can water, and venous blood can absorb 20 times more CO_2 than water can.

[29] M. F. PERUTZ, Stereochemistry of cooperative effects in haemoglobin, *Nature* **228**, 726–39 (1970).

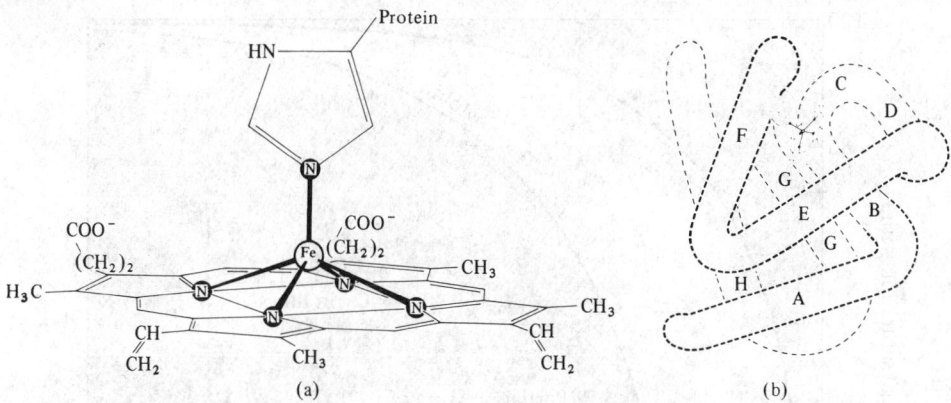

FIG. 25.9   Haemoglobin: (a) The haem group, composed of the planar PIX molecule and iron, and shown here attached to the globin via an imidazole-nitrogen which completes the square pyramidal coordination of the Fe^{II}, and (b) one of the 4 subunits of haemoglobin showing, diagrammatically, the haem group in a "pocket" formed by the folded protein. The globin chain is actually in the form of 8 helical sections, labelled A to H, and the haem is situated between the E and F sections.

groups of the protein are on the outside of the structure leaving a hydrophobic interior. The haem group, which is held in a protein "pocket", is therefore in a hydrophobic environment, and the high-spin Fe^{II} within the haem, being too large to fit inside the hole provided by the rigid porphyrin ring, is situated about 80 pm above the ring and is essentially 5-coordinate. When O_2 is bonded to the vacant coordination site, opposite the imidazole-nitrogen, the Fe^{II} becomes low-spin and the accompanying reduction in size allows it to slip down into the plane of the ring.

Evidently the globin prevents the $Fe^{II} \rightarrow Fe^{III}$ oxidation and, although the precise reasons are not yet certain, it appears likely that the hydrophobic environment of the iron, by effectively acting as a non-polar solvent of low dielectric constant, inhibits the electron transfer entailed in the oxidation. In addition the formation of a dimer involving an Fe–O–Fe or Fe–O_2–Fe bridge, which is believed to be an intermediate in $Fe^{II} \rightarrow Fe^{III}$ oxidations, would clearly be sterically hindered.

Now the 4 subunits of haemoglobin form an approximately tetrahedral arrangement, held together by $-NH_3^+ \cdots \bar{O}OC-$ "salt bridges", and these are disturbed by the attachment of O_2 in such a way that the "pockets" holding the haem groups are opened, and the sixth coordination site of their iron atoms exposed. It seems that this is a result of the movement of the low-spin Fe^{II} into the porphyrin plane, pulling the attached histidine residue with it and triggering off conformational changes throughout the globin chain which are then transmitted, because of the salt bridges, to the 3 adjoining subunits. A consequence of this is that the attachment of one O_2 molecule to haemoglobin actually increases its affinity for more. Conversely, as O_2 is removed from oxyhaemoglobin the reverse conformational changes occur and successively decrease its affinity for oxygen. This is the phenomenon of *cooperativity*, and its physiological importance lies in the fact that it allows the efficient transfer of oxygen from oxyhaemoglobin to myoglobin. This is because myoglobin contains only 1 haem group and can be regarded crudely as a single haemoglobin subunit. It therefore cannot display a cooperative effect and at lower partial pressures of oxygen it has a greater affinity than haemoglobin for oxygen. This can be seen

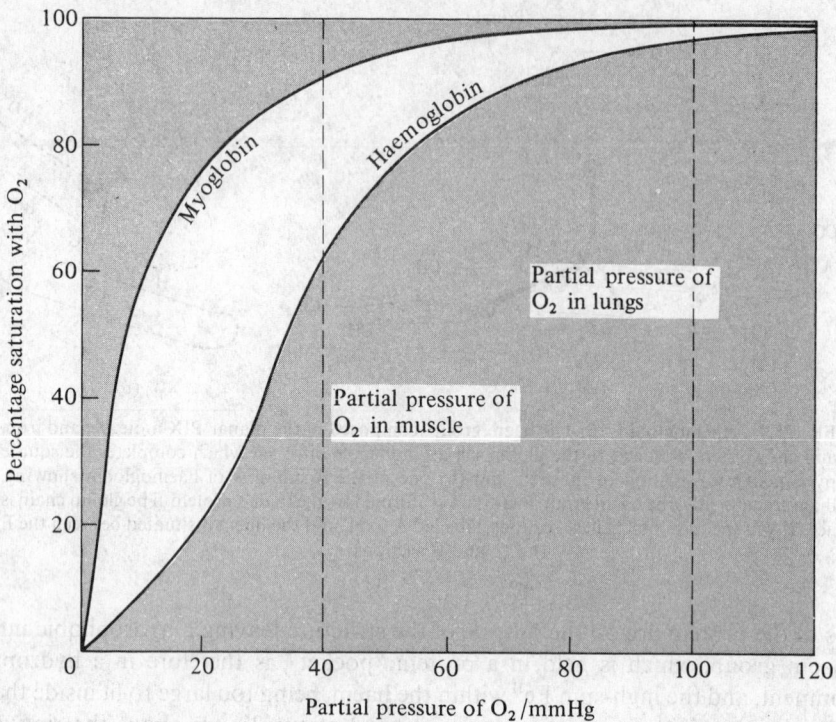

FIG. 25.10 Oxygen dissociation curves for haemoglobin and myoglobin, showing how haemoglobin is able to absorb O_2 efficiently in the lungs yet transfer it to myoglobin in muscle tissue.

in Fig. 25.10 which shows that, while haemoglobin is virtually saturated with O_2 in the lungs, when it experiences the lower partial pressures of oxygen in the muscle tissue its affinity for O_2 has fallen off so much more rapidly than that of myoglobin that oxygen transfer ensues. Indeed, the actual situation is even more effective than this, because the affinity of haemoglobin for oxygen decreases when the pH is lowered (this is called the Bohr effect and arises in a complicated manner from the effect of pH on the salt bridges holding the subunits together). Since the CO_2 released in the muscle lowers the pH, it thereby facilitates the transfer of oxygen from the oxyhaemoglobin, and the greater the muscular activity the more the release of CO_2 helps to meet the increased demand for oxygen. Excess CO_2 is then removed from the tissue, predominantly in the form of soluble HCO_3^- ions whose formation is facilitated by the protein chain of deoxyhaemoglobin which acts as a buffer by picking up the accompanying protons.

$$CO_2 + H_2O \rightleftharpoons HCO_3^- + H^+$$

The mode of bonding of the O_2 to Fe has understandably received much attention but, despite this, it is still not possible to say with absolute certainty whether the bond is end-on linear Fe-O-O, bent Fe-O$\diagup$O , or side-on Fe$\Big\langle\begin{smallmatrix}O\\|\\O\end{smallmatrix}$ (see p. 719). One approach has been to study model compounds (i.e. simpler compounds containing what is hoped are the main features of the original) not only of Fe^{2+} but of Co^{2+} as well. The similarity between oxyhaemoglobin and model systems which are known to have the bent structure in such characteristics as their Mössbauer quadrupole splitting, provides persuasive evidence for the bent structure. Models have been similarly used[30] in attempts to understand the role of the protein, globin, in preventing oxidation of haem to haematin, while still allowing the reversible uptake of O_2.

The poisoning effect of molecules such as CO and PF_3 arises simply from their ability to bond reversibly to haem in the same manner as O_2, but much more strongly, so that oxygen transport is prevented. The cyanide ion CN^- can also displace O_2 from oxyhaemoglobin but its very much greater toxicity at small concentrations stems not from this but from its interference with the action of cytochrome a.

Cytochromes

The haem unit was evidently a most effective evolutionary development since it is found not only in the oxygen-transporting substances but also in electron transporters such as the cytochromes which are scattered widely throughout nature. There are numerous varieties of cytochromes, many differing only in minor details, but they are characterized by iron held in a porphyrin ring with the fifth coordination site occupied by an imidazole N from the associated protein, just as in haemoglobin and myoglobin. However, the function of cytochromes being to transfer electrons not oxygen, the sixth site is in most cases occupied by a strongly bonded S from a methionine group in the protein, so rendering them inert not only to oxygen but also the poisons which effect the oxygen carriers. Their role is as intermediaries in the metabolic oxidation of glucose by molecular oxygen. This apparently is effected in a series of steps, in each of which the oxidation state of the iron which is normally in a low-spin configuration oscillates between $+2$ and $+3$. Since different cytochromes are involved in the order b, c, a, the reduction potential of each step is successively increased (Table 25.8), so forming a "redox gradient". This allows energy from the glucose oxidation to be released gradually and to be stored in the form of adenosine triphosphate (ATP) (see also p. 611). The details of the sequence are far from clear but the link with the final electron acceptor O_2 is cytochrome oxidase, a complex material made up of two a-cytochromes and also containing copper (p. 1392). Its reducing ability seems to depend on the interaction of the Fe and Cu and, as the end member of the redox gradient, it apparently differs from the other members in bonding O_2 directly and so being extremely susceptible to poisoning by CN^-.

TABLE 25.8　*Reduction potentials of some iron proteins*

Iron protein	Oxidation states of Fe	$E°/V$
Cytochrome a	Fe^{III}/Fe^{II}	0.29
Cytochrome b	Fe^{III}/Fe^{II}	0.04
Cytochrome c	Fe^{III}/Fe^{II}	0.26
Rubredoxin	Fe^{III}/Fe^{II}	-0.06
2-Fe plant ferredoxins	$Fe^{III}/$fractional	-0.40
4-Fe bacterial ferredoxins	Fractional/fractional	-0.37
8-Fe bacterial ferredoxins	Fractional/fractional	-0.42

[30] R. D. JONES, D. A. SUMMERVILLE, and F. BASOLO, Synthetic oxygen carriers related to biological species, *Chem. Revs.* **79**, 139–79 (1979).

Iron–sulfur proteins

In spite of the obvious importance and diversity of haem proteins, comparable functions, especially that of electron transfer, are performed by non-haem iron proteins (NHIP). These too are widely distributed, different types being involved in nitrogen fixation (p. 1205) and photosynthesis as well as in the metabolic oxidation of sugars prior to the involvement of the cytochromes mentioned above. The NHIP responsible for electron transfer are the iron–sulfur proteins which are of relatively low molecular weight (6000–12 000) and contain 1, 2, 4, or 8 Fe atoms which, in all the structures which have been definitely established, are each coordinated to 4 S atoms in an approximately tetrahedral manner. Nearly all of them are notable for reduction potentials in the unusually low range -0.05 to -0.49 V (Table 25.8), indicating their ability to act as reducing agents at the low-potential end of biochemical processes.

The simplest NHIP is rubredoxin, in which the single iron atom is coordinated (Fig. 25.11a) to 4 S atoms belonging to cysteine residues in the protein chain, and has high-spin configurations in both the oxidized (Fe^{III}) and reduced (Fe^{II}) forms. It differs from the other Fe–S proteins in having no labile sulfur (i.e. inorganic sulfur which can be liberated

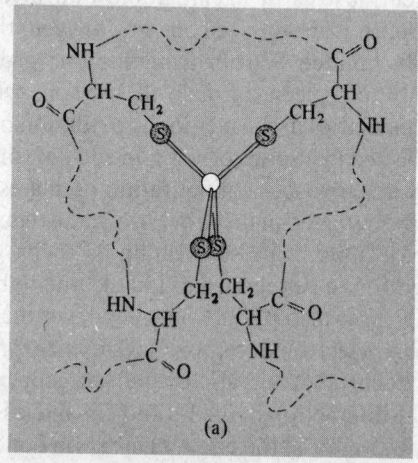

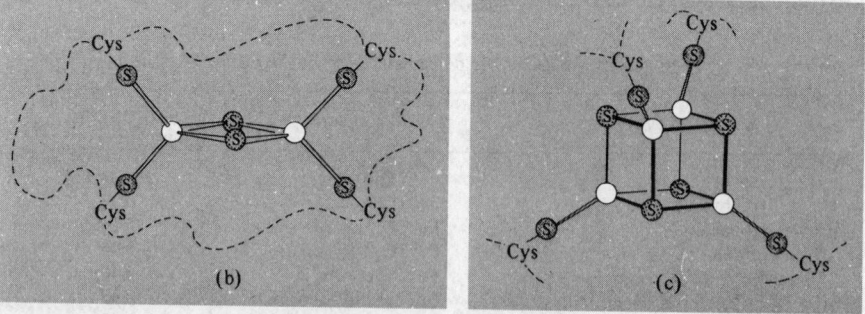

Fig. 25.11 Some non-haem iron proteins: (a) rubredoxin in which the single Fe is coordinated, almost tetrahedrally, to 4 cysteine-sulfurs, (b) plant ferredoxin, $[Fe_2S_2^*(S\text{-}Cys)_4]$, (c) $[Fe_4S_4^*(S\text{-}Cys)_4]$ cube of bacterial ferredoxins. (This is in fact distorted, the Fe_4 and S_4^* making up the two interpenetrating tetrahedra, of which the latter is larger than the former.)

as H_2S by treatment with mineral acid; such sulfur is conventionally designated S* in abbreviated formulae of these proteins).

The "labile sulfur"-NHIP, or ferredoxins, can be classified as $[Fe_2S_2^*(S-Cys)_4]$, $[Fe_4S_4^*(S-Cys)_4]$ and $[Fe_4S_4^*(S-Cys)_4]_2$ types. The structure of the first of these, the plant-type ferredoxins, has not been established with certainty but is believed to consist of 2 Fe atoms joined by S* bridges and with terminal cysteine groups (Fig. 25.11b), the 2 irons in the oxidized form being high-spin Fe(III). The observed very low magnetic moment, implying considerable spin–spin interaction via the bridging atoms, and the operation of these proteins as 1-electron transfer agents, implying a *formal* oxidation state of +2.5 for the Fe in the reduced form, indicate that the dimer must be considered as a single unit rather than as 2 independent atoms.

The structures of the 4-Fe and 8-Fe proteins, the bacterial ferredoxins, have been more definitely established. They contain, respectively, 1 and 2 $Fe_4S_4^*$ clusters (Fig. 25.11c), the centres of those in the 8-Fe proteins being about 1200 pm apart. In these clusters the $4 \times Fe$ and $4 \times S^*$ atoms form 2 interpenetrating tetrahedra which together make up a distorted cube in which each Fe atom is additionally coordinated to a cysteine sulfur to give it an approximately tetrahedral coordination sphere. The cluster, like the 2-Fe dimer, acts as a 1-electron transfer agent so that the 8-Fe protein can effect a 2-electron transfer. Why 4 Fe atoms are required to transfer 1 electron is not obvious. Synthetic analogues, prepared by reacting $FeCl_3$, NaHS, and an appropriate thiol (or still better[31] $FeCl_3$, elemental sulfur, and the Li salt of a thiol), have properties similar to those of the natural proteins and have been used extensively in attempts to solve this problem.[32] The esr, electronic, and Mössbauer spectra, as well as the magnetic properties of the synthetic $[Fe_4S_4(SR)_4]^{3-}$ anions, are similar to those of the reduced ferredoxins, whose redox reaction is therefore mirrored by:

$$[Fe_4S_4(SR)_4]^{3-} \rightleftharpoons [Fe_4S_4(SR)_4]^{2-}$$

$$\text{reduced ferredoxin} \underset{+e}{\overset{-e}{\rightleftharpoons}} \text{oxidized ferredoxin}$$

This indicates a change in the formal oxidation state of the iron from +2.25 to +2.5, and mixed Fe^{III}/Fe^{II} species have been postulated. However, it is evident from various spectroscopic measurements that, if these do actually occur, they must be exceedingly short-lived and, as with the 2-Fe ferredoxins, the clusters are best regarded as electronically delocalized systems in which all the Fe atoms are equivalent.

Other non-haem proteins, distinct from the above iron–sulfur proteins are involved in the roles of iron transport and storage, but knowledge about their structures and mode of operation is rather sketchy. Iron is absorbed as Fe^{II} in the human duodenum and passes into the blood as the Fe^{III} protein, transferrin. This has a stability constant sufficiently high for the uncombined protein to strip Fe^{III} from such stable complexes as those with phosphate and citrate ions, and so it very efficiently scavenges iron from the blood plasma. The iron is then transported to the bone marrow where it is released from the transferrin (presumably after the temporary reduction of Fe^{III} to Fe^{II} since the latter's is a much less-stable complex), to be stored as ferritin, prior to its incorporation into haemoglobin. Ferritin is a water-soluble material consisting of a layer of protein encapsulating iron(III) hydroxyphosphate to give an overall iron content of about 20%.

[31] G. CHRISTOU and C. D. GARNER, A convenient synthesis of tetrakis[thiolato-μ_3-sulphido-iron](2-) clusters, *JCS Dalton* 1979, 1093–4.

[32] R. H. HOLM, Iron-sulphur clusters in natural and synthetic systems, *Endeavour* **34**, 38–43 (1975).

25.3.6 *Organometallic compounds*

In this triad the simple alkyls and aryls are not known, and M–C σ bonds require the stabilizing presence of π-bonding ligands such as CO, as in $[MR(CO)_2(C_5H_5)]$ (M = Fe, Ru) and $[MRX(P-P)_2]$ (M = Ru, Os; X = halogen; P–P = chelating phosphine). The areas of main interest and importance are the carbonyls and the metallocenes; two fields in which iron has historically played a crucial central role.

Carbonyls[23] (see p. 349)

Having the d^6s^2 configuration, the elements of this triad are able to conform with the 18-electron rule by forming mononuclear carbonyls of the type $M(CO)_5$. These are volatile liquids which can be prepared by the direct action of CO on the powdered metal (Fe and Ru), or by the action of CO on the tetroxide (Os), in each case at elevated temperatures and pressures.

$Fe(CO)_5$ is a highly toxic substance discovered in 1891, the only previously known metal carbonyl being $Ni(CO)_4$. Whether its structure is trigonal bipyramidal or square pyramidal was the subject of prolonged debate, but it is now established to be the former (Fig. 25.12a) with Fe–CO_{ax} 180.6, Fe–CO_{eq} 183.3, C–O 114.5 pm. On the basis of their similar infrared spectra, the Ru and Os analogues are presumed to have the same structure. The ^{13}C nmr spectrum of $Fe(CO)_5$, however, indicates that all 5 carbon atoms are equivalent and this is explained by the molecules' fluxional behaviour (p. 1071).

Exposure of $Fe(CO)_5$ in organic solvents to ultraviolet light produces volatile orange crystals of the enneacarbonyl, $Fe_2(CO)_9$. Its structure consists of two face-sharing octahedra (Fig. 25.12b).[32a] An electron count shows that the dimer has a total of 34 valence electrons, i.e. 17 per iron atom. The observed diamagnetism is therefore explained by the presence of an Fe–Fe bond which is consistent with an interatomic separation virtually the same as in the metal itself. It is of interest that Ru and Os counterparts of $Fe_2(CO)_9$ have only recently been prepared and (the former especially) are only stable at reduced temperatures. The structure of $Os_2(CO)_9$ is suggested to involve an Os–Os bond, but with only 1 CO bridge. The carbonyls, which are produced along with the pentacarbonyls of Ru and Os and were initially thought to be enneacarbonyls, are in fact trimers, $M_3(CO)_{12}$, which also differ structurally from $Fe_3(CO)_{12}$ (Fig. 25.12c and d). This dark-green solid, which is best obtained by oxidation of $[FeH(CO)_4]^-$ (see below), has a triangular structure in which 2 of the iron atoms are bridged by a pair of carbonyl groups, and can be regarded as being derived from $Fe_2(CO)_9$ by replacing a bridging CO with $Fe(CO)_4$. The Ru and Os compounds (orange and yellow respectively), on the other hand, have a more symmetrical structure in which all the metal atoms are equivalent and are held together solely by M–M bonds. It has been suggested[33] that the reason for these structural differences is that cluster compounds of the larger Ru and Os entail M–M distances too great to support CO bridges.

The chemistry of these carbonyls, especially those of iron, has been extensively studied and gives rise to a wide variety of products, some of which are discussed below. The polymerization of $Os_3(CO)_{12}$ is also important. When this is heated at 190°C in a sealed

---

[32a] F. A. COTTON and J. M. TROUP, Accurate determination of a classic structure in the metal carbonyl field: nonacarbonyl di-iron, *JCS Dalton* 1974, 800–2.

[33] B. F. G. JOHNSON, The structures of simple binary carbonyls, *JCS Chem. Comm.* 211–13 (1976).

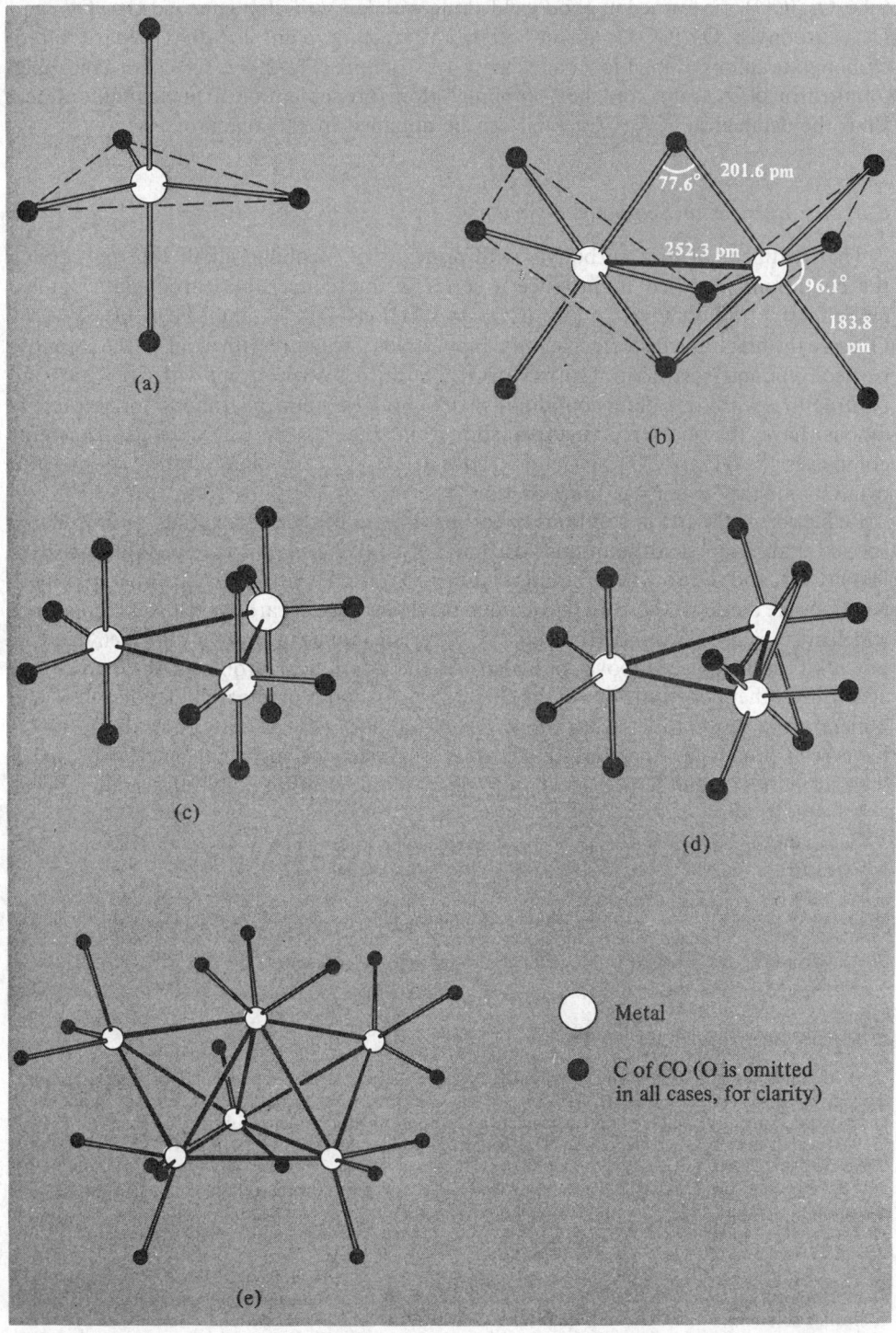

FIG. 25.12 Carbonyls of Fe, Ru, and Os: (a) M(CO)$_5$; M = Fe, Ru, Os. (b) Fe$_2$(CO)$_9$;
Fe–Fe = 252.3 pm. (c) M$_3$(CO)$_{12}$; M = Ru, Os. (d) Fe$_3$(CO)$_{12}$; 1 Fe–Fe = 256 pm, 2
Fe–Fe = 268 pm. (e) Os$_6$(CO)$_{18}$.

Metal

C of CO (O is omitted
in all cases, for clarity)

tube, $Os_6(CO)_{18}$ is formed in good yield along with less well-characterized Os_5, Os_7, and Os_8 carbonyls. $Os_6(CO)_{18}$ is an important starting point for the preparation of carbonylate anions of still higher nuclearity. Its structure (Fig. 25.12e) is that of a bicapped tetrahedron of Os atoms, and the 2 "capping" atoms are susceptible to nucleophilic attack. Thus the octahedral $[Os_6(CO)_{18}]^{2-}$ can be obtained by the action of I^-.

Carbonyl hydrides and carbonylate anions

The treatment of iron carbonyls with aqueous or alcoholic alkali can, by varying the conditions, be used to produce a series of interconvertible carbonylate anions: $[HFe(CO)_4]^-$, $[Fe(CO)_4]^{2-}$, $[Fe_2(CO)_8]^{2-}$, $[HFe(CO)_{11}]^-$, and $[Fe_4(CO)_{13}]^{2-}$.[33a] Of these the first has a distorted trigonal bipyramidal structure with axial H, the second is isoelectronic and isostructural with $Ni(CO)_4$, the third is isoelectronic with $Co_2(CO)_8$ and isostructural with the isomer containing no CO bridges, while the trimeric and tetrameric anions have the cluster structures shown in Fig. 25.13. The related ruthenium complexes[34] $[HRu_3(CO)_{11}]^-$ and $[H_3Ru_4(CO)_{12}]^-$, are of interest[35] as possible catalysts for the "water-gas shift reaction".†

Reduction of the pH of solutions of carbonylate anions yields a variety of protonated species and, from acid solutions, carbonyl hydrides such as the unstable, gaseous $H_2Fe(CO)_4$ and the polymeric liquids $H_2Fe_2(CO)_8$ and $H_2Fe_3(CO)_{11}$ are liberated. Such reactions are amongst those used to produce a whole range of carbonyl cluster compounds which are typical of Group VIII metals.[35b] They are interesting not only for their catalytic potential, as instanced above, but also for the theoretical problems they pose, and accordingly they constitute one of the currently active fields of research. Besides the 3- and 4-metal atom clusters referred to above, 5-, 6-, 8- and 10-metal atom clusters of Ru and Os have been produced. An interesting feature of clusters of such high nuclearity is that their structures cannot be predicted precisely by simple electron-counting arguments. The

---

† Water-gas is produced by the high-temperature reaction of water and C:

$$C + H_2O \longrightarrow CO + H_2$$

and is therefore a mixture of H_2O, CO, and H_2. By suitably adjusting the relative proportions of CO and H_2, "synthesis gas" is obtained which can be used for the synthesis of methanol and hydrocarbons (the Fischer–Tropsch process[35a]). It is this catalytically controlled adjustment:

$$H_2O + CO \rightleftharpoons H_2 + CO_2$$

which is the water-gas shift reaction (WGSR) (see p. 483).

[33a] R. GREATREX and N. N. GREENWOOD, Mössbauer spectra, structure and bonding in iron carbonyl derivatives, *Disc. Faraday Soc.* **47**, 126–35 (1969). K FARMERY, M. KILNER, R. GREATREX, and N. N. GREENWOOD, Structural studies of the carbonylate and carbonyl hydride anions of iron, *J. Chem. Soc.* (A) 1969, 2339–345, and references therein. See also M. B. SMITH and R. BAU, Structure of the $[HFe(CO)_4]^-$ anion, *J. Am. Chem. Soc.* **95**, 2385–9 (1973).

[34] B. F. G. JOHNSON, J. LEWIS, P. R. RAITHBY, and G. SÜSS, The triruthenium cluster anion $[Ru_3H(CO)_{11}]^-$: preparation, structure and fluxionality, *JCS Dalton* 1979, 1356–61. P. F. JACKSON, B. F. G. JOHNSON, J. LEWIS, M. MCPARTLIN, and W. J. H. NELSON, $[H_3Ru_4(CO)_{12}]^-$: the X-ray crystallographic determination of two structural isomers, *JCS Chem. Comm.* 1978, 920–1.

[35] C. UNGERMANN, V. LANDIS, S. A. MOYA, H. COHEN, H. WALKER, R. G. PEARSON, R. G. RINKER, and P. C. FORD, Homogeneous catalysis of the water gas shift reaction by ruthenium and other metal carbonyls. Studies in alkaline solutions, *J. Am. Chem. Soc.* **101**, 5922–9 (1979).

[35a] C. MASTERS, The Fischer–Tropsch reaction, *Adv. Organomet. Chem.* **17**, 61–103 (1979).

[35b] B. F. G. JOHNSON (ed.), *Transition Metal Clusters*, Wiley, Chichester, 1980, 681 pp.

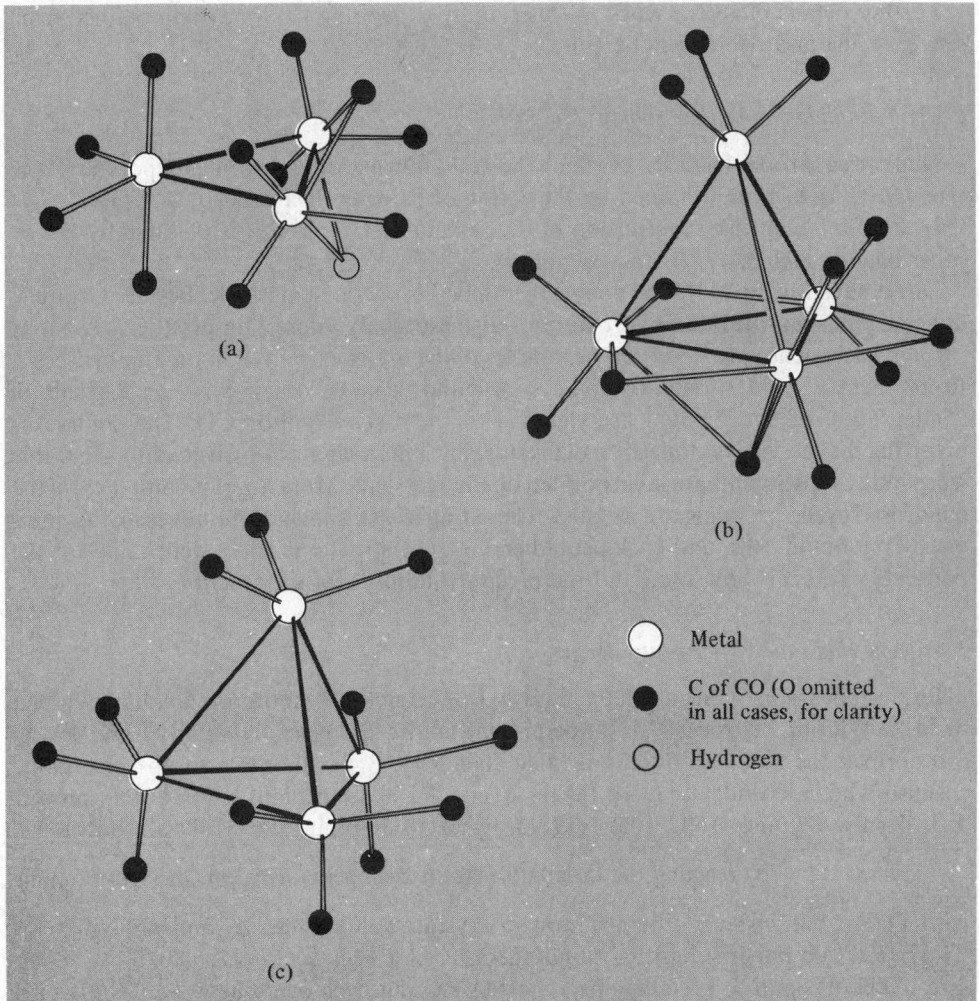

(a)

(b)

(c)

○ Metal

● C of CO (O omitted
  in all cases, for clarity)

○ Hydrogen

Fig. 25.13 Some carbonylate anion clusters of Fe, Ru, and Os: (a) $[HM_3(CO)_{11}]^-$; M = Fe, Ru.
(b) $[Fe_4(CO)_{13}]^{2-}$. (c) $[H_3Ru_4(CO)_{12}]^-$. The H atoms are not shown because this ion exists in
two isomeric forms: (i) the 3 H atoms bridge the edges of a single face of the tetrahedron, and (ii)
the 3 H atoms bridge three edges of the tetrahedron which do not form a face.

$[H_2M_6(CO)_{18}$, $[HM_6(CO)_{18}]^-$, and $[M_6(CO)_{18}]^{2-}$ clusters, for instance, while being
stoichiometrically the same for M = Ru and M = Os, and having the same essentially
octahedral skeletons, nevertheless differ appreciably in the disposition of the attached
carbonyl groups.[36] The incorporation of interstitial atoms such as C, H, S, and P into
these clusters is now a well-established, and frequently stabilizing, feature (pp. 354–356).

[36] C. R. Eady, P. F. Jackson, B. F. G. Johnson, J. Lewis, M. C. Malatesta, M. McPartlin, and W. J. H.
Nelson, Improved synthesis of the hexanuclear clusters $[Ru_6(CO)_{18}]^{2-}$, $[HRu_6(CO)_{18}]^-$, and $[H_2Ru(CO)_{18}]$.
The X-ray analysis of $[HRu_6(CO)_{18}]^-$, a polynuclear carbonyl containing an interstitial hydrogen ligand, *JCS
Dalton* 1980, 383–92.

Further details of recent work in this rapidly expanding field may be obtained from reference 36a and references therein.

Carbonyl halides and other substituted carbonyls

Numerous carbonyl halides, of which the best known are octahedral compounds of the type $[M(CO)_4X_2]$ are obtained by the action of halogen on $Fe(CO)_5$, or CO on MX_3 ($M = Ru$, Os). Stepwise substitution of the remaining CO groups is possible by X^- or other ligands such as N, P, and As donors.

Direct substitution of the carbonyls themselves is of course possible. Besides Group V donor ligands, unsaturated hydrocarbons give especially interesting products. The iron carbonyl acetylenes provided early examples of the use of carbonyls in organic synthesis. From them a wide variety of cyclic compounds can be obtained[37] as a result of condensation of coordinated acetylenes with themselves and/or CO. The complexes involving the acetylenes alone are usually unstable intermediates which are only separable when bulky substituents are incorporated on the acetylene. More usually, complexes of the condensed cyclic products are isolated. These ring systems include quinones, hydroquinones, cyclobutadienes, and cyclopentadienones, the specific product depending on the particular iron carbonyl used and the precise conditions of the reaction.

Ferrocene and other cyclopentadienyls

Bis(cyclopentadienyl)iron, $[Fe(\eta^5\text{-}C_5H_5)_2]$, or, to give it the more familiar name coined by M. C. Whiting, "ferrocene", is the compound whose discovery in the early 1950s utterly transformed the study of organometallic chemistry.[2] Yet the two groups of organic chemists who independently made the discovery, did so accidentally.[38] P. L. Pauson and T. J. Kealy (*Nature* **168**, 1039 (1951)) were attempting to synthesize fulvalene,

,by reacting the Grignard reagent cyclopentadienyl magnesium bromide

with $FeCl_3$, but instead obtained orange crystals containing Fe^{II} and analysing for $C_{10}H_{10}Fe$. In a paper submitted simultaneously (*J. Chem. Soc.* 1952, 632), S. A. Miller, T. A. Tebboth, and J. F. Tremaine reported passing cyclopentadiene and N_2 over a reduced iron catalyst as part of a programme to prepare amines and they too obtained $C_{10}H_{10}Fe$.†

The initial structural formulation was , but the correct formulation,

an unprecedented "sandwich" compound, was soon to follow.[39] For this and for

---

† In retrospect it seems likely that ferrocene was actually first prepared in the 1930s by chemists at Union Carbide who passed dicyclopentadiene through a heated iron tube, but the significance was not then realized.

36a J. N. NICHOLLS, D. H. FARRAR, P. F. JACKSON, B. F. G. JOHNSON, and J. LEWIS, A high-pressure infrared study of the stability of some ruthenium and osmium clusters to CO and H_2 under pressure, *JCS Dalton* 1982, 1395–1400. P. F. JACKSON, B. F. G. JOHNSON, J. LEWIS, M. MCPARTLIN, and W. J. H. NELSON, Synthesis and X-ray analysis of the hydrido-carbido-monoanion $[HOs_{10}C(CO)_{24}]^-$; a cluster compound with an interstitial hydrogen ligand in a tetrahedral site. *JCS Chem. Comm.* 1982, 49–51.

37 B. L. SHAW and N. I. TUCKER, Compounds formed by condensation of acetylenes with carbon monoxide, and related organometallic complexes, Chap. 53 in *Comprehensive Inorganic Chemistry*, Vol. 4, pp. 884–92, Pergamon Press, Oxford, 1973.

38 C. B. HUNT, Metallocenes—the first 25 years, *Educ. Chem.* **14**, 110–13 (1977).

39 G. WILKINSON, M. ROSENBLUM, M. C. WHITING, and R. B. WOODWARD, The structure of iron bis-cyclopentadienyl, *J. Am. Chem. Soc.* **74**, 2125–6 (1952).

subsequent independent work in this field, G. Wilkinson and E. O. Fischer shared the 1973 Nobel Prize for Chemistry.

The structure of ferrocene and a recent MO description of its bonding have already been given (p. 369) but it is perhaps worth mentioning an oversimplified VB description; in this each $C_5H_5^-$ group is assumed to donate 3 pairs of electrons (these are the electrons not involved in C–C or C–H σ bonding) to the Fe^{II}, which accommodates the 6 pairs in d^2sp^3 hybrid orbitals while its d^6 electrons pair-up in the remaining 3d orbitals, thus satisfying the 18-electron rule and explaining the observed diamagnetism. Although the energy barrier to rotation of the rings is small, it was thought that crystal packing forces favoured the staggered configuration for crystalline ferrocene. However it is now known (p. 368) that the rings are virtually eclipsed as they are in the analogous ruthenocene and osmocene.

Ferrocene is thermally stable and is also stable to water and to air, though it can be oxidized to the ferricinium ion, $[Fe^{III}(\eta^5\text{-}C_5H_5)_2]^+$. However, its most notable chemistry results from the aromaticity of the cyclopentadienyl rings. This is now far too extensively documented to be described in full but an outline of some of its manifestations is in Fig. 25.14. Ferrocene resists catalytic hydrogenation and does not undergo the typical reactions of conjugated dienes, such as the Diels–Alder reaction. Nor are direct nitration and halogenation possible because of oxidation to the ferricinium ion. However, Friedel–Crafts acylation as well as alkylation and metallation reactions, are readily effected. Indeed, electrophilic substitution of ferrocene occurs with such facility compared to, say, benzene that some explanation is called for. It has been suggested that, in general, electrophilic substituents (E^+) interact first with the metal atom and then transfer to the C_5H_5 ring with proton elimination. Similar reactions are possible for ruthenocene and osmocene but usually occur less readily, and it appears that reactivity decreases with increasing size of the metal.

A number of interesting cyclopentadienyl iron carbonyls have been prepared, the best known being the purple dimer, $[Fe(\eta^5\text{-}C_5H_5)(CO)_2]_2$ (Fig. 25.15a), prepared by reacting $Fe(CO)_5$ and dicyclopentadienyl at 135°C in an autoclave. Diamagnetism and an Fe–Fe distance of only 249 pm indicate the presence of an Fe–Fe bond. Prolonged reaction of the same reactants produces the very dark green, tetrameric cluster compound, $[Fe(\eta^5\text{-}C_5H_5)(CO)]_4$ (Fig. 25.15b), which involves CO groups which are triply bridging and so give rise to an exceedingly low $(1620\ cm^{-1})$ ν_{CO} absorption. $[Fe(\eta^1\text{-}C_5H_5)(\eta^5\text{-}C_5H_5)(CO)_2]$ (Fig. 25.15c) is also of note as an early example of a fluxional organometallic compound. The ^1H nmr spectrum consists of only two sharp lines, one for each ring. A single line is expected for the pentahapto ring since all its protons are equivalent, but it is clear that some averaging process must be occurring for the non-equivalent protons of the monohapto ring to produce just one line. It is concluded that the

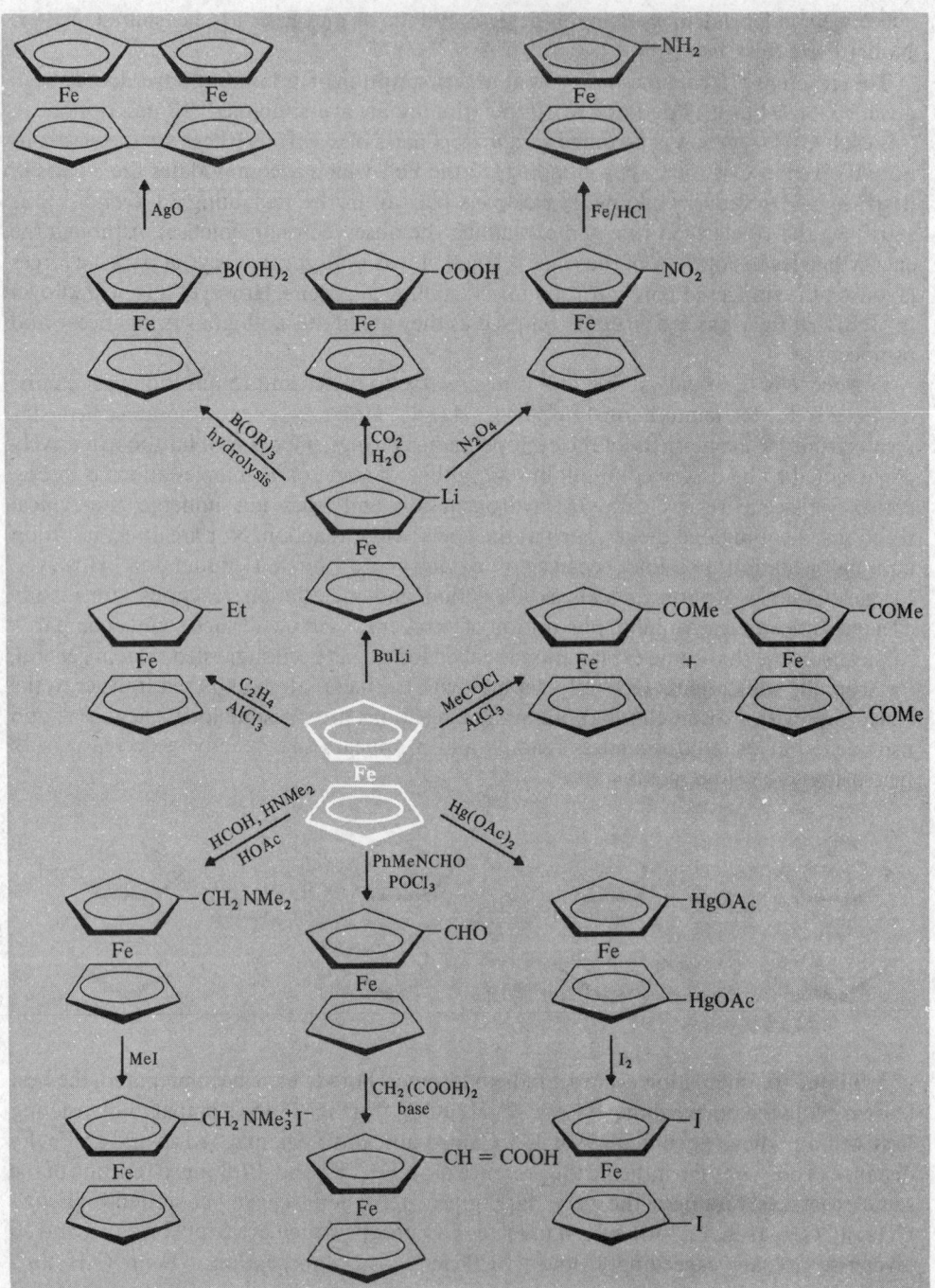

FIG. 25.14 Some reactions of ferrocene.

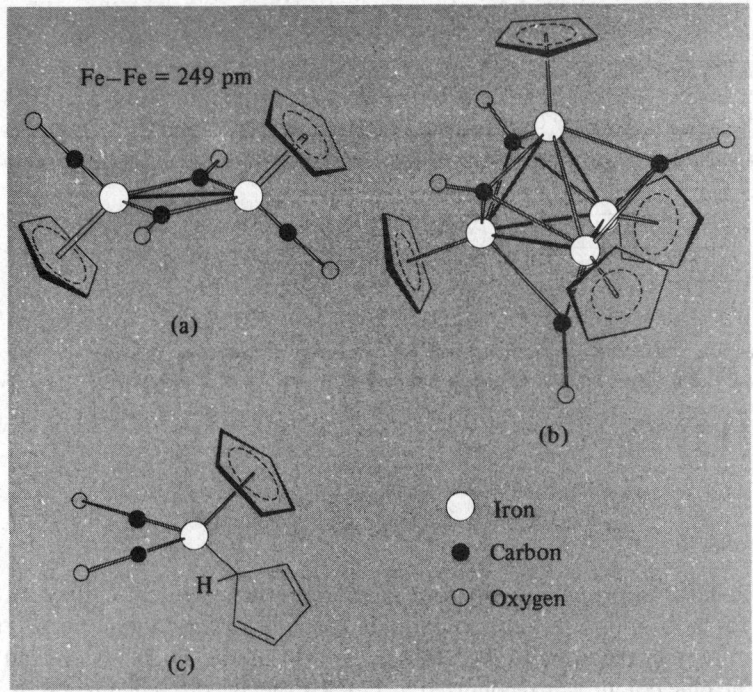

Fe–Fe = 249 pm

(a)

(b)

(c)

Iron

Carbon

Oxygen

FIG. 25.15   Some cyclopentadienyl iron carbonyls: (a) $[(\eta^5\text{-}C_5H_5)Fe(CO)_2]_2$,
(b) $[(\eta^5\text{-}C_5H_5)Fe(CO)]_4$, and (c) $(\eta^1\text{-}C_5H_5)(\eta^5\text{-}C_5H_5)Fe(CO)_2$.

point of attachment of the monohapto ring to the metal must change repeatedly and rapidly ("ring whizzing") thus averaging the protons.

Other aspects of the organometallic chemistry of iron have been referred to in Chapter 8 but for fuller details more extensive reviews should be consulted.[40]

[40] *The Organic Chemistry of Iron* (E. A. K. von GUSTORF, F.-W. GREVELS, and I. FISCHER, eds), Academic Press, New York, Vol. 1, 1978, 673 pp.; Vol. 2, 1981, 340 pp. *MTP International Review of Science, Inorganic Chemistry, Series 1, Vol. 6, Transition Metals, Part 2* (M. J. MAY, ed.), Butterworths, London, 1972, 442 pp.

The periodic table grid is shown with element symbols including:

3 Li	4 Be											5 B	6 C	7 N	8 O	9 F	10 Ne
11 Na	12 Mg											13 Al	14 Si	15 P	16 S	17 Cl	18 Ar
19 K	20 Ca	21 Sc	22 Ti	23 V	24 Cr	25 Mn	26 Fe	27 Co	28 Ni	29 Cu	30 Zn	31 Ga	32 Ge	33 As	34 Se	35 Br	36 Kr
37 Rb	38 Sr	39 Y	40 Zr	41 Nb	42 Mo	43 Tc	44 Ru	45 Rh	46 Pd	47 Ag	48 Cd	49 In	50 Sn	51 Sb	52 Te	53 I	54 Xe
55 Cs	56 Ba	57 La	72 Hf	73 Ta	74 W	75 Re	76 Os	77 Ir	78 Pt	79 Au	80 Hg	81 Tl	82 Pb	83 Bi	84 Po	85 At	86 Rn
87 Fr	88 Ra	89 Ac	104 Unq	105 Unp	106 Unh	107 Uns											

58 Ce	59 Pr	60 Nd	61 Pm	62 Sm	63 Eu	64 Gd	65 Tb	66 Dy	67 Ho	68 Er	69 Tm	70 Yb	71 Lu
90 Th	91 Pa	92 U	93 Np	94 Pu	95 Am	96 Cm	97 Bk	98 Cf	99 Es	100 Fm	101 Md	102 No	103 Lr

26

Cobalt, Rhodium, and Iridium

26.1 Introduction

Although hardly any metallic cobalt was used until the twentieth century, its ores have been used for thousands of years to impart a blue colour to glass and pottery. It is present in Egyptian pottery dated at around 2600 BC and Iranian glass beads of 2250 BC.† The source of the blue colour was recognized in 1735 by the Swedish chemist G. Brandt, who isolated a very impure metal, or "regulus", which he named "cobalt rex". In 1780 T. O. Bergman showed this to be a new element. Its name has some resemblance to the Greek word for "mine" but is almost certainly derived from the German word *Kobold* for "goblin" or "evil spirit". The miners of northern European countries thought that the spitefulness of such spirits was responsible for ores which, on smelting, not only failed unexpectedly to yield the anticipated metal but also produced highly toxic fumes (As_4O_6).

In 1803 both rhodium and iridium were discovered, like their preceding neighbours in the periodic table, ruthenium and osmium, in the black residue left after crude platinum had been dissolved in aqua regia. W. H. Wollaston discovered rhodium, naming it after the Greek word ῥόδον for "rose" because of the rose-colour commonly found in aqueous solutions of its salts. S. Tenant discovered iridium along with osmium, and named it after the Greek goddess Iris (ἶρις, ἰριδ-), whose sign was the rainbow, because of the variety of colours of its compounds.

26.2 The Elements

26.2.1 *Terrestrial abundance and distribution*

Rhodium and iridium are exceedingly rare elements, comprising only 0.0001 and 0.001 ppm of the earth's crust respectively, and even cobalt (29 ppm, i.e. 0.0029%), though widely distributed, stands only thirtieth in order of abundance and is less common than all other elements of the first transition series except scandium (25 ppm).

† "Smalt", produced by fusing potash, silica, and cobalt oxide, can be used for colouring glass or for glazing pottery. The secret of making this brilliant blue pigment was apparently lost, to be rediscovered in the fifteenth century. Leonardo da Vinci was one of the first to use powdered smalt as a "new" pigment when painting his famous "The Madonna of the Rocks".

More than 200 ores are known to contain cobalt but only a few are of commercial value. The more important are arsenides and sulfides such as smaltite, $CoAs_2$, cobaltite (or cobalt glance), CoAsS, and linnaeite, Co_3S_4. These are invariably associated with nickel, and often also with copper and lead, and it is usually obtained as a byproduct or coproduct in the recovery of these metals. The world's major sources of cobalt are the African continent and Canada with smaller reserves in Australia and the USSR. All the platinum metals are generally associated with each other and rhodium and iridium therefore occur wherever the other platinum metals are found. However, the relative proportions of the individual metals are by no means constant and the more important sources of rhodium are the nickel–copper–sulfide ores found in South Africa and in Sudbury, Canada, which contain about 0.1% Rh. Iridium is usually obtained from native osmiridium (Ir ~50%) or iridiosmium (Ir ~70%) found chiefly in Alaska as well as South Africa.

26.2.2 *Preparation and uses of the elements*

The production of cobalt[1] is usually subsidiary to that of copper, nickel, or lead and the details of its extraction depend on which of these it is associated with. In general the ore is subjected to appropriate roasting treatment so as to remove gangue material as a slag and produce a "speiss" of mixed metal and oxides. In the case of arsenical ores, As_2O_6 is condensed and provides a valuable byproduct. Leaching the speiss with sulfuric acid leaves behind metallic copper but dissolves out iron, cobalt, and nickel. Iron is then precipitated with lime, and cobalt with sodium hypochlorite:

$$2Co^{II} + OCl^- + 4OH^- + H_2O \longrightarrow 2Co(OH)_3\downarrow + Cl^-$$

Heating of the "hydroxide" converts it to oxide which can then be reduced to the metal by heating with charcoal.

World production of cobalt ores in 1980 was 32 700 tonnes of contained cobalt, of which 14 700 tonnes were from Zaire, approximately 3000 tonnes each from Zambia, Australia, and New Caledonia, 2000 tonnes from the USSR, and 1600 tonnes from Canada. About 30% of the output is devoted to the production of chemicals for the ceramic and paint industries. In ceramics the main use now is not to provide a blue colour, but rather white by counterbalancing the yellow tint arising from iron impurities. Blue pigments are, however, used in paints and inks, and cobalt compounds are used to hasten the oxidation and hence the drying of oil-based paints. Cobalt compounds are also employed as catalysts in a range of organic reactions of which the "OXO" (or hydroformylation) reaction and hydrogenation and dehydrogenation reactions are the most important (p. 1317).

A further 30% of output is devoted to the production of high-temperature alloys used primarily in the construction of gas turbines. Again, cobalt, like iron and nickel, is ferromagnetic, and accordingly over 20% of output is used for the manufacture of magnetic alloys. Of these the best known is "Alnico", a steel containing, as its name implies, aluminium and nickel, as well as cobalt. It is used for permanent magnets which are up to 25 times more powerful than ordinary steel magnets.

[1] F. PLANINSEK and J. B. NEWKIRK, Cobalt and cobalt alloys, *Kirk–Othmer Encyclopedia of Chemical Technology*, 3rd edn., Vol. 6, pp. 481–94; Cobalt compounds (by F. R. MORRAL), pp. 495–510, Interscience, New York, 1979.

An important, if smaller, use of cobalt is as a cement for tungsten carbide in cutting tools (p. 1169). The carbide is initially in the form of a powder, and to convert it to a more massive form suitable for a tool tip it is mixed with about 10% of powdered cobalt and sintered under hydrogen at 1400–1500°C.

Rhodium and iridium, like ruthenium and osmium (p. 1246), are obtained by procedures designed to separate silver, gold, and all the platinum metals.[2] After Pt, Pd, and Au have been dissolved out in aqua regia and Ag removed as its soluble nitrate, the residue is worked for Ru, Os, Rh, and Ir, the scheme relevant to the last two metals being outlined in Fig. 26.1. The metals, which are in the form of a powder or sponge, can be consolidated by the techniques of powder metallurgy. Because of their scarcity (only a few tonnes of each are produced annually) the uses of these elements are essentially specialist in nature. Iridium, like osmium, is used in the hard alloys employed for instrument pivots and suchlike. It is also incorporated in long-life Pt/Ir electrodes in sparking plugs and because of the use of these in US helicopters demand for Ir was inflated for the duration of the Vietnam war. It is occasionally used to fabricate crucibles and other apparatus in cases where its greater resistance to chemical attack makes it preferable to platinum.

Unquestionably the main uses of rhodium are now catalytic, e.g. for the control of exhaust emissions in the car (automobile) industry and, in the form of phosphine complexes, in hydrogenation and hydroformylation reactions where it is frequently more efficient than the more commonly used cobalt catalysts.

26.2.3 *Properties of the elements*

Some of the important properties of these three elements are summarized in Table 26.1. From an archival point of view it is interesting to note that iridium is now known to be the densest of all elements, surpassing osmium, which was for a long time believed to have this distinction, by the tiniest of margins.

The metals are lustrous and silvery with, in the case of cobalt, a bluish tinge. Rhodium and iridium are both hard, cobalt less so but still appreciably harder than iron. Rhodium and Ir have fcc structures, the first elements in the transition series to do so; this is in keeping with the view, based on band-theory calculations, that the fcc structure is more stable than either bcc or hcp when the outer d orbitals are nearly full. Cobalt, too, has an allotrope (the β-form) with this structure but this is only stable above 417°C; below this temperature an hcp, α-form is the more stable. However, the transformation between these allotropes is generally slow and the β-form, which can be stabilized by the addition of a few per cent of iron, is often present at room temperature. This, of course, has an effect on physical properties and is no doubt responsible for variations in reported values for some properties even in the case of very pure cobalt. By contrast the atomic weights of cobalt and rhodium at least are known with considerable precision, since these elements each have but one naturally occurring isotope. In the case of cobalt this is ^{59}Co, but bombardment by thermal neutrons converts this to the radioactive ^{60}Co. The latter has a half-life of 5.271 y and decays by means of β^- and γ emission to non-radioactive ^{60}Ni. It is used in many fields of research as a concentrated source of γ-radiation, and also

² W. P. GRIFFITH, *The Chemistry of the Rare Platinum Metals (Os, Ru, Ir and Rh)*, Interscience, London, 1967, 491 pp.

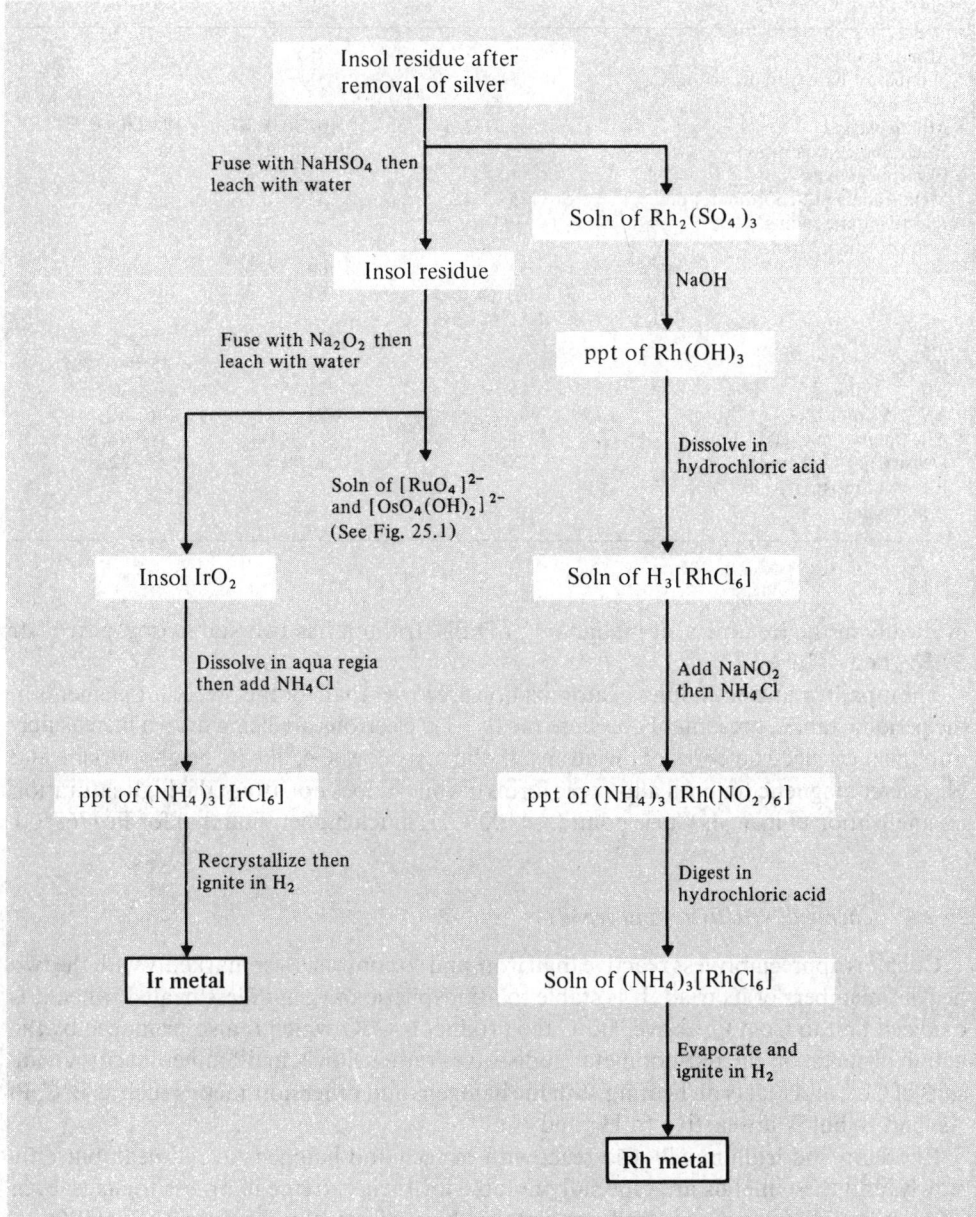

Fig. 26.1   Flow diagram for the extraction of rhodium and iridium.

TABLE 26.1   *Some properties of the elements cobalt, rhodium, and iridium*

Property		Co	Rh	Ir
Atomic number		27	45	77
Number of naturally occurring isotopes		1	1	2
Atomic weight		58.9332	102.9055	192.22(± 0.03)
Electronic configuration		$[Ar]3d^74s^2$	$[Kr]4d^85s^1$	$[Xe]4f^{14}5d^76s^2$
Electronegativity		1.8	2.2	2.2
Metal radius (12-coordinate)/pm		125	134	135.5
Effective ionic radius				
(6-coordinate)/pm	V	—	55	57
	IV	53	60	62.5
	III	54.5 (ls), 61 (hs)	66.5	68
	II	65 (ls), 74.5 (hs)	—	—
MP/°C		1495	1960	2443
BP/°C		3100	3760	4550(± 100)
ΔH_{fus}/kJ mol^{-1}		16.3	21.6	26.4
ΔH_{vap}/kJ mol^{-1}		382	494	612(± 13)
ΔH_f (monatomic gas)/kJ mol^{-1}		425(± 17)	556(± 11)	669(± 8)
Density (20°C)/g cm^{-3}		8.90	12.39	22.61
Electrical resistivity (20°C)/ μohm cm		6.24	4.33	4.71

medically in the treatment of malignant growths. Iridium has two stable isotopes: ^{191}Ir 37.3% and ^{193}Ir 62.7%.

The mps, bps, and enthalpies of atomization are lower than for the preceding elements in the periodic tables, presumably because the $(n-1)$d electrons are being drawn increasingly into the inert electron cores of the atoms. In the first series Co, like its neighbours Fe and Ni, is ferromagnetic (in both allotropic forms); while it does not attain the high saturation magnetization of iron, its Curie point (> 1100°C) is much higher than that for Fe (768°C).

26.2.4   *Chemical reactivity and trends*

Cobalt is appreciably less reactive than iron, and so contrasts less markedly with the two heavier members of its triad. It is stable to atmospheric oxygen unless heated, when it is oxidized first to Co_3O_4; above 900°C the product is CoO which is also produced by the action of steam on the red-hot metal. It dissolves rather slowly in dil mineral acids giving salts of Co^{II}, and reacts on heating with the halogens and other non-metals such as B, C, P, As, and S, but is unreactive to H_2 and N_2.

Rhodium and iridium will also react with oxygen and halogens at red-heat, but only slowly, and these metals are especially notable for their extreme inertness to acids, even aqua regia. Dissolution of rhodium metal is best effected by fusion with $NaHSO_4$, a process used in its commercial separation. In the case of iridium oxidizing, molten alkalis such as Na_2O_2 or $KOH + KNO_3$ will produce IrO_2 which can then be dissolved in aqua regia. Alternatively, a rather extreme measure which is efficacious with both metals, is to heat them with conc $HCl + NaClO_3$ in a sealed tube at 125–150°C.

Table 26.2 is a list of examples of compounds of these elements in various oxidation states. The most striking feature of this, as compared to the corresponding lists for

TABLE 26.2 *Oxidation states and stereochemistries of some compounds of cobalt, rhodium, and iridium*

Oxidation state	Coordination number	Stereochemistry	Co	Rh/Ir
-1 (d^{10})	4	Tetrahedral	$[Co(CO)_4]^-$	$[Rh(CO)_4]^-$, $[Ir(CO)_3(PPh_3)]^-$
0 (d^9)	4	Tetrahedral	$[Co(PMe_3)_4]$	
	6	Octahedral	$[Co_2(CO)_8]$	$[M_4(CO)_{12}]$
1 (d^8)	3	Planar (?)		$[RhCl(PCy_3)_2]$
		T-shaped		$[Rh(PPh_3)_3]^+$
	4	Square planar		$[RhCl(PPh_3)_3]$
				$[Ir(CO)Cl(PPh_3)_2]$
	5	Trigonal bipyramidal	$[Co(NCMe)_5]^+$	$[RhH(PF_3)_4]$, $[Ir(CO)H(PPh_3)_3]$
		Square pyramidal	$[Co(NCPh)_5]^+$	
	6	Octahedral	$[Co(bipy)_3]^+$	
2 (d^7)	2	Linear	$[Co\{N(SiMe_3)_2\}_2]$	
	3	Planar	$[Co\{N(SiMe_3)_2\}_2(PPh_3)]$	
	4	Tetrahedral	$[CoCl_4]^{2-}$	
		Square planar	$[Co(phthalocyanine)]$	$[RhCl_2\{P(o\text{-}MeC_6H_4)_3\}_2]$
	5	Trigonal bipyramidal	$[CoBrN(C_2H_4NMe_2)_3]^+$	
		Square pyramidal	$[Co(CN)_5]^{3-}$	$[Rh_2(O_2CMe)_4]$
	6	Octahedral	$[Co(H_2O)_6]^{2+}$	$[Rh_2(O_2CMe)_4(H_2O)_2]$
	8	Dodecahedral	$[Co(NO_3)_4]^{2-}$	
3 (d^6)	4	Tetrahedral	$[CoW_{12}O_{40}]^{5-}$	
	5	Trigonal bipyramidal		$[IrH_3(PR_3)_2]$
		Square planar	$[Co(corrole)(PPh_3)]^{(a)}$	$[RhI_2Me(PPh_3)_2]$
	6	Octahedral	$[Co(NH_3)_6]^{3+}$	$[MCl_6]^{3-}$
4 (d^5)	4	Tetrahedral	$[Co(1\text{-norbornyl})_4]^{(b)}$	
	6	Octahedral	$[CoF_6]^{2-}$	$[MCl_6]^{2-}$
5 (d^4)	6	Octahedral		$[MF_6]^-$
	7	?		$[IrH_5(PEt_2Ph)_2]$
6 (d^3)	6	Octahedral		$[MF_6]$

(a) Corrole is a tetrapyrrolic macrocycle (P. B. HITCHCOCK and G. M. M. McLAUGHLIN, *JCS (Dalton)* 1976; 1927.)

(b) 1-Norbornyl is a bicyclo[2.2.1]hept-1-yl (B. K. BOWER and H. G. TENNENT, *J. Am. Chem. Soc.* **94**, 2512 (1972).)

preceding triads, is that for the first time the range of oxidation states has diminished. This is a manifestation of the increasing stability of the $(n-1)d$ electrons, whose attraction to the atomic nucleus is now sufficient to prevent the elements attaining the highest oxidation states and so to render irrelevant the concept of a "group" oxidation state. No oxidation states are found above $+6$ for Rh and Ir, or above $+5$ for Co and, indeed, examples of cobalt in $+4$ and $+5$ and of rhodium or iridium in $+5$ and $+6$ oxidation states are rare and sometimes poorly characterized.

The most common oxidation states of cobalt are $+2$ and $+3$. $[Co(H_2O)_6]^{2+}$ and $[Co(H_2O)_6]^{3+}$ are both known but the latter is a strong oxidizing agent and in aqueous solution, unless it is acidic, it decomposes rapidly as the Co^{III} oxidizes the water with evolution of oxygen. Consequently, in contrast to Co^{II}, Co^{III} provides few simple salts, and those which do occur are unstable. However, Co^{III} is unsurpassed in the number of coordination complexes which it forms, especially with N-donor ligands. Virtually all of these complexes are low-spin, the t_{2g}^6 configuration producing a particularly high CFSE (p. 1097).

The effect of the CFSE is expected to be even more marked in the case of the heavier elements because for them the crystal field splittings are much greater (p. 1096). As a result the $+3$ state is the most important one for both Rh and Ir (in contrast to the $+2$ state which is relatively unimportant) and $[Rh(H_2O)_6]^{3+}$ is the only simple aquo ion formed by either of these elements. With π-acceptor ligands the $+1$ oxidation state is also well known for Rh and Ir. It is noticeable, however, that the similarity of these two heavier elements is less than is the case earlier in the transition series and, although rhodium is more like iridium than cobalt, nevertheless there are significant differences. One example is provided by the $+4$ oxidation state which occurs to an appreciable extent in iridium but not in rhodium. (The ease with which $Ir^{IV} \rightleftharpoons Ir^{III}$ sometimes occurs can be a source of annoyance to preparative chemists.)

Table 26.2 also reveals a diminished tendency on the part of these elements to form compounds of high coordination number when compared with the iron group and, apart from $[Co(NO_3)_4]^{2-}$, a coordination number of 6 is rarely exceeded. Also, with the exception of the unstable $[Co^VO_4]^{3-}$ and the recently made $[Co^{II}O_3]^{4-}$ (p. 1300), oxoanions are not found in this triad, Rh and Ir being the first heavy transition elements not to form them. This is presumably because their formation requires the donation of π electrons from the oxygen atoms to the metal and the metals become progressively less able to act as π acceptors as their d orbitals are filled.

26.3   Compounds of Cobalt, Rhodium, and Iridium[2-4]

Binary borides (p. 162) and carbides (p. 318) have been discussed already.

26.3.1   *Oxides*[5] *and sulfides*

As a result of the diminution in the range of oxidation states which has already been mentioned, the number of oxides formed by these elements is less than in the preceding groups, being confined to two each for cobalt (CoO, Co_3O_4) and rhodium (Rh_2O_3, RhO_2) and to just one for iridium (IrO_2) (though an impure sesquioxide Ir_2O_3 has been reported—see below). No trioxides are known.

The only oxide formed by any of these metals in the divalent state is CoO; this is prepared as an olive-green powder by strongly heating the metal in air or steam, or alternatively by heating the hydroxide, carbonate, or nitrate in the absence of air. It has the rock-salt structure and is antiferromagnetic below 289 K. By reacting it with silica and alumina, pigments are produced which are used in the ceramics industry. CoO is stable in

[3] D. NICHOLLS, Cobalt, Chap. 41 in *Comprehensive Inorganic Chemistry*, Vol. 3, pp. 1053–1107, Pergamon Press, Oxford, 1973.

[4] S. E. LIVINGSTONE, The second- and third-row elements of Group VIII A, B, and C, Chap. 43, ibid., pp. 1163–1370.

[5] C. N. R. RAO and G. V. S. RAO, Transition metal oxides, *National Standard Reference Data System*, NSRDS-NBS49, Washington, 1964, 130 pp.

air at ambient temperatures and above 900°C but if heated at, say, 600–700°C, it is converted into the black Co_3O_4. This is $Co^{II}Co_2^{III}O_4$ and has the normal spinel structure with Co^{II} ions in tetrahedral and Co^{III} in octahedral sites within the ccp lattice of oxide ions. This is to be expected (p. 1254) because of the dominating advantage of placing the d^6 ions in octahedral sites, where adoption of the low-spin configuration gives it a decisively favourable CFSE. The ability of Co_3O_4 to absorb oxygen, and possibly also the retention of water in preparations from the hydroxide, have led to claims for the existence of Co_2O_3, but it is doubtful if these claims are valid. Oxidation of $Co(OH)_2$, or addition of aqueous alkali to a cobalt(III) complex, produces a dark-brown material which on drying at 150°C in fact gives cobalt(III) oxide hydroxide, $CoO(OH)$.

Heating rhodium metal or the trichloride in oxygen at 600°C, or simply heating the trinitrate, produces dark-grey Rh_2O_3 which has the corundum structure (p. 274); it is the only stable oxide formed by this metal. The yellow precipitate formed by the addition of alkali to aqueous solutions of rhodium(III) is actually $Rh_2O_3.5H_2O$ rather than a genuine hydroxide. Electrolytic oxidation of Rh^{III} solutions and addition of alkali gives a yellow precipitate of $RhO_2.2H_2O$, but attempts to dehydrate this produce Rh_2O_3. Black anhydrous RhO_2 is best obtained by heating Rh_2O_3 in oxygen under pressure; it has the rutile structure, but it is not well characterized.

For iridium the position is reversed. This time it is the black dioxide, IrO_2, with the rutile structure (p. 1118), which is the only definitely established oxide. It is obtained by heating the metal in oxygen or by dehydrating the precipitate produced when alkali is added to an aqueous solution of $[IrCl_6]^{2-}$. Contamination either by unreacted metal or by alkali is, however, difficult to avoid. The other oxide, Ir_2O_3, is said to be obtained by igniting K_2IrCl_6 with $NaCO_3$ or, as its hydrate, by adding KOH to aqueous $K_3[IrCl_6]$ under CO_2. However, even if it is a true compound, it is always impure and is readily oxidized to IrO_2.

A larger number of sulfides have been reported but not all of them have been fully characterized. Cobalt gives rise to CoS_2 with the pyrites structure (p. 804), Co_3S_4 with the spinel structure (p. 279), and $Co_{1-x}S$ which has the NiAs structure (p. 648) and is cobalt-deficient. All are metallic, as is Co_9S_8 and the corresponding selenides and tellurides. The sulfides of rhodium and iridium are notable mainly for their inertness especially towards acids, and most of them are semiconductors. They are the disulfides MS_2, obtained from the elements; the "sesquisulfides" M_2S_3, obtained by passing H_2S through aqueous solutions of M^{III}; and Rh_2S_5 and IrS_3, obtained by heating $MCl_3 + S$ at 600°C. Numerous nonstoichiometric selenides and tellurides are also known.[5a]

26.3.2 *Halides*[6]

The known halides of this triad are listed in Table 26.3. It can be seen that, apart from CoF_3 and the doubtful iridium tetrahalides, they fall into three categories:

[5a] F. HULLIGER, Crystal chemistry of chalcogenides and pnictides of the transition elements, *Struct. Bonding* **4**, 83–229 (1968).

[6] R. COLTON and J. H. CANTERFORD, *Halides of the First Row Transition Metals*, Chap. 7, pp. 327–405, Wiley, London, 1969; and *Halides of the Second and Third Row Transition Metals*, Chap. 9, pp. 346–58, Wiley, London, 1968.

(a) higher fluorides of Ir and Rh;

(b) a full complement of trihalides of Ir and Rh;

(c) dihalides of cobalt

The octahedral hexafluorides are obtained directly from the elements and both are volatile, extremely reactive, and corrosive solids, RhF_6 being the least stable of the platinum metal hexafluorides and reacting with glass even when carefully dried. They are thermally unstable and must be frozen out from the hot gaseous reaction mixtures, otherwise they dissociate.

The pentafluorides of Rh and Ir may be prepared by the deliberate thermal dissociation of the hexafluorides. They also are highly reactive and are respectively dark-red and yellow solids, with the same tetrameric structure as $[RuF_5]_4$ and $[OsF_5]_4$ (p. 1259).

RhF_4 is a purple-red solid, usually prepared by the reaction of the strong fluorinating agent BrF_3 on $RhBr_3$. The corresponding compound IrF_4 has had an intriguing and instructive history.[6a] It was first claimed in 1929 and again in 1956 but this material was shown in 1965 to be, in reality, the previously unknown IrF_5. IrF_4 can now be made (1974) by reducing IrF_5 with the stoichiometric amount of iridium-black:

$$4IrF_5 + Ir \xrightarrow{400°} 5IrF_4$$

TABLE 26.3   *Halides of cobalt, rhodium, and iridium* $(mp/°C)$

Oxidation state	Fluorides	Chlorides	Bromides	Iodides
+6	RhF_6 black (70°) IrF_6 yellow (44°) bp 53°			
+5	$[RhF_5]_4$ dark red $[IrF_5]_4$ yellow (104°)			
+4	RhF_4 purple-red IrF_4 dark brown	$IrCl_4$?	$IrBr_4$?	IrI_4?
+3	CoF_3 light brown RhF_3 red IrF_3 black	$RhCl_3$ red $IrCl_3$ red	$RhBr_3$ red-brown $IrBr_3$ red-brown	RhI_3 black IrI_3 dark brown
+2	CoF_2 pink (1200°)	$CoCl_2$ blue (724°)	$CoBr_2$ green (678°)	CoI_2 blue-black (515°)

[6a] N. BARTLETT and A. TRESSAUD, A novel structure type for transition-metal tetrafluorides: synthesis crystallographic and magnetic study of IrF_4, *Comptes Rendus* **278C**, 1501–4 (1974).

The dark-brown product disproportionates above 400° into IrF_3 and the volatile IF_5. The structure features $\{IrF_6\}$ octahedra which share 4 F atoms, each with one other $\{IrF_6\}$ group, leaving a pair of *cis* vertices unshared: this is essentially a rutile type structure (p. 1118) from which alternate metal atoms have been removed from each edge-sharing chain and was the first 3D structure to have been found for a tetrafluoride.[6a] Claims have been made for the isolation of all the other iridium tetrahalides, but there is some doubt as to whether these can be substantiated.[2] This is an unexpected situation since +4 is one of iridium's common oxidation states and, indeed, the derived anions $[IrX_6]^{2-}$ (X = F, Cl, Br) are well known.

The most familiar and most stable of the halides of Rh and Ir, however, are the trihalides. Those of Rh range in colour from the red RhF_3 to black RhI_3 and, apart from the latter, which is obtained by the action of aqueous KI on the tribromide, they may be obtained in the anhydrous state directly from the elements. RhF_3 has a structure similar to that of ReO_3 (p. 1219), while $RhCl_3$ is isomorphous with $AlCl_3$ (p. 263). The anhydrous trihalides are generally unreactive and insoluble in water but, excepting the tri-iodide which is only known in this form, water-soluble hydrates can be produced by wet methods. $RhF_3.6H_2O$ and $RhF_3.9H_2O$ can be isolated from aqueous solutions of Rh^{III} acidified with HF. Their aqueous solutions are yellow, possibly due to the presence of $[Rh(H_2O)_6]^{3+}$. The dark-red deliquescent $RhCl_3.3H_2O$ is the most common compound of rhodium and the usual starting point for the preparation of other rhodium compounds, and is itself best prepared from the metal sponge. This is heated with KCl in a stream of Cl_2 and the product extracted with water. The solution contains $K_2[Rh(H_2O)Cl_5]$ and treatment with KOH precipitates the hydrous Rh_2O_3 which can be dissolved in hydrochloric acid and the solution evaporated to dryness. $RhBr_3.2H_2O$ also is formed from the metal by treating it with hydrochloric acid and bromine.

The iridium trihalides are rather similar to those of rhodium. Anhydrous IrF_3 is obtained by reducing IrF_6 with the metal, $IrCl_3$ and $IrBr_3$ by heating the elements, and IrI_3 by heating its hydrate *in vacuo*. Water-soluble hydrates of the tri-chloride, -bromide, and -iodide are produced by dissolving hydrous Ir_2O_3 in the appropriate acid and, like its rhodium analogue, $IrCl_3.3H_2O$ provides a convenient starting point in iridium chemistry.

Lower halides of Rh and Ir have been reported and, whilst their existence cannot be denied with certainty, further substantiation is needed. Unquestionably, the divalent state is the preserve of cobalt. Apart from the strongly oxidizing CoF_3 (a light-brown powder isomorphous with $FeCl_3$ and the product of the action of F_2 on $CoCl_2$ at 250°C), the only known halides of cobalt are the dihalides. In all of these the cobalt is octahedrally coordinated. The anhydrous compounds are prepared by dry methods: CoF_2 (pink) by heating $CoCl_2$ in HF, $CoCl_2$ (blue), and $CoBr_2$ (green) by the action of the halogens on the heated metal, and CoI_2 (blue-black) by the action of HI on the heated metal. The fluoride is only slightly soluble in water but the others dissolve readily to give solutions from which pink or red hexahydrates can be crystallized. These solutions can alternatively and more conveniently be made by dissolving the metal, oxide, or carbonate in the appropriate hydrohalic acid. The chloride is widely used as an indicator in the desiccant, silica gel, since its blue anhydrous form turns pink as it hydrates.

The disinclination of these metals to form oxoanions has already been remarked and the same is evidently true of oxohalides: none have been authenticated.

26.3.3 *Complexes*

The chemistry of oxidation states above IV is sparse. Apart from RhF_6 and IrF_6, such chemistry as there is, is mainly confined to salts of $[IrF_6]^-$; these are made by fluorinating a lower halide of iridium with BrF_3 in the presence of a halide of the counter cation. In the case of cobalt, heating mixtures of the appropriate oxides in oxygen, or under pressure, produces materials with the stoichiometry, $M_3^I CoO_4$, which, together with their oxidizing properties, suggests the presence of Co^V. When CoO is heated with 2.2 moles of Na_2O at 550° in a sealed tube under argon, bright-red crystals of $Na_4Co^{II}O_3$ are formed.[6b] The compound hydrolyses immediately on contact with atmospheric moisture and is notable in containing discrete planar $[CoO_3]^{4-}$ ions reminiscent of the carbonate ion (Co–O 186 ± 6 pm) and is similar to the recently prepared red oxoferrate(II), $Na_4[FeO_3]$. The lustrous red tetracobaltate(II) $Na_{10}[Co_4^{II}O_9]$, with an anion analogous to the *catena*-tetracarbonate $[C_4O_9]^{2-}$, is also known.[6b]

Oxidation state IV(d⁵)

Cobalt provides only a few examples of this oxidation state, namely some fluoro compounds and mixed metal oxides, but their purity is questionable. Rhodium(IV) complexes are somewhat better documented, but examples are confined to salts of the oxidizing and readily hydrolysed $[RhX_6]^{2-}$ (X = F, Cl). Only iridium(IV) shows appreciable stability.

The salts of $[IrX_6]^{2-}$ (X = F, Cl, Br) are comparatively stable and their colour deepens from red, through reddish-black to bluish-black with increasing atomic weight of the halogen. $[IrF_6]^{2-}$ is obtained by reduction of $[IrF_6]^-$, $[IrCl_6]^{2-}$ by oxidation of $[IrCl_6]^{3-}$ with chlorine, and $[IrBr_6]^{2-}$ by Br^- substitution of $[IrCl_6]^{2-}$ in aqueous solution. The hexachloroiridates in particular have been the subject of many magnetic investigations. They have magnetic moments at room temperature somewhat below the spin-only value for the t_{2g}^5 configuration (1.73 BM), and this falls with temperature. This has been interpreted as the result of antiferromagnetic interaction operating by a superexchange mechanism between adjacent Ir^{IV} ions via intervening chlorine atoms. More importantly, in 1953 in a short but classic paper,[7] J. Owen and K. W. H. Stevens reported the observation of hyperfine structure in the esr signal obtained from solid solutions of $(NH_4)_2[IrCl_6]$ in the isomorphous, but diamagnetic, $(NH_4)_2[PtCl_6]$. This arises from the influence of the chlorine nuclei and, from the magnitude of the splitting, it was inferred that the single unpaired electron, which is ostensibly one of the metal d⁵ electrons, in fact spends only 80% of its time on the metal, the rest of the time being divided equally between the 6 chlorine ligands. This was the first unambiguous evidence that metal d electrons are able to move in molecular orbitals over the whole complex, and implies the presence of π as well as σ bonding.

In aqueous solution, the halide ions of $[IrX_6]^{2-}$ may be replaced by solvent and a number of aquo substituted derivatives have been reported. Other Ir^{IV} complexes with

⁶ᵇ W. Burow and R. Hoppe, Co^{II} with a coordination number 3: $Na_4[CoO_3]$—the first oxocobaltate(II) with an island structure, *Angew. Chem.*, Int. Edn. (Engl.) **18**, 542–3 (1979). See also R. Hoppe and H. Rieck, *Z. anorg. allgem. Chem.* **437**, 95–104 (1977) for Na_4FeO_3; W. Burow and R. Hoppe, *Angew. Chem.*, Int. Edn. (Engl.) **18**, 61–62 (1979) for $Na_{10}Co_4O_9$.

⁷ J. Owen and K. W. H. Stevens, Paramagnetic resonance and covalent bonding, *Nature* **171**, 836 (1953).

O-donor ligands are $[IrCl_4(C_2O_4)]^{2-}$, obtained by oxidizing Ir^{III} oxalato complexes with chlorine, and Na_2IrO_3, obtained by fusing Ir and Na_2CO_3.

Two interesting trinuclear complexes must also be mentioned. They are $K_{10}[Ir_3O(SO_4)_9].3H_2O$, obtained by boiling Na_2IrCl_6 and K_2SO_4 in conc sulfuric acid, and $K_4[Ir_3N(SO_4)_6(H_2O)_3]$, obtained by boiling Na_3IrCl_6 and $(NH_4)_2SO_4$ in conc sulfuric acid. They are believed to have the structure shown in Fig. 26.2, analogous to that of the basic carboxylates, $[M_3^{III}O(O_2CR)_6L_3]^+$ (see Fig. 23.11). The oxo species formally contains 1 Ir^{IV} and 2 Ir^{III} ions and the nitride species 2 Ir^{IV} ions and 1 Ir^{III} ion, but in each case the charges are probably delocalized over the whole complex.

Oxidation state III(d⁶)

For all three elements this is the most prolific oxidation state, providing a wide variety of kinetically inert complexes. As has already been pointed out, these are virtually all low-spin and octahedral, a major stabilizing influence being the high CFSE associated with the t_{2g}^6 configuration ($\frac{12}{5}\Delta_o$, the maximum possible for any d^x configuration). Even $[Co(H_2O)_6]^{3+}$ is low-spin but it is such a powerful oxidizing agent that it is unstable in aqueous solutions and only a few simple salt hydrates, such as the blue $Co_2(SO_4)_3.18H_2O$ and $MCo(SO_4)_2.12H_2O$ (M = K, Rb, Cs, NH_4), which contain the hexaquo ion, and $CoF_3.3\frac{1}{2}H_2O$ can be isolated. This paucity of simple salts of cobalt(III) contrasts sharply with the great abundance of its complexes, expecially with *N*-donor ligands, and it is

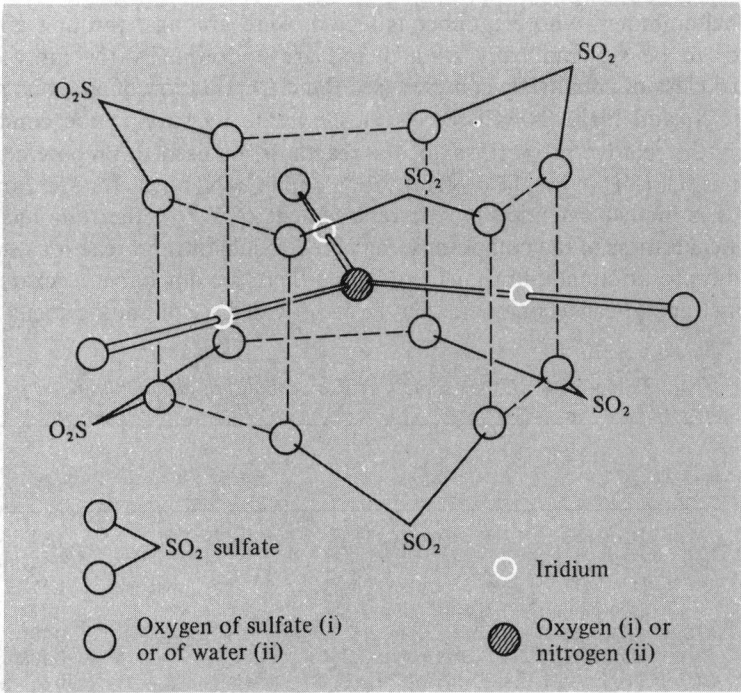

FIG. 26.2   Trinuclear structure of (i) $[Ir_3O(SO_4)_9]^{10-}$ and (ii) $[Ir_3N(SO_4)_6(H_2O)_3]^{4-}$.

evident that the high CFSE is not the only factor affecting the stability of this oxidation state.

Table 26.4 illustrates the remarkable sensitivity of the reduction potential of the Co^{III}/Co^{II} couple to different ligands whose presence renders Co^{II} unstable to aerial oxidation. The extreme effect of CN^- can be thought of as being due, on the one hand, to the ability of its empty π^* orbitals to accept "back-donated" charge from the metal's filled t_{2g} orbitals and, on the other, to its effectiveness as a σ donor (enhanced partly by its negative charge). The magnitudes of the changes in E° are even greater than those noted for the Fe^{III}/Fe^{II} couple (p. 1269), though if the two systems are compared it must be remembered that the oxidation state which can be stabilized by adoption of the low-spin t_{2g}^6 configuration is $+3$ for cobalt but only $+2$ for iron. Nevertheless, the effect of increasing pH is closely similar, the M^{III} "hydroxide" of both metals being far less soluble than the M^{II} "hydroxide". In the case of cobalt this reduces E° from 1.83 to 0.17 V:

$$Co^{III}O(OH) + H_2O + e^- \rightleftharpoons Co^{II}(OH)_2 + OH^-; \quad E^\circ = 0.17 \text{ V}$$

thereby facilitating oxidation to the $+3$ state.

Complexes of cobalt(III), like those of chromium(III) (p. 1196), are kinetically inert and so, again, indirect methods of preparation are to be preferred. Most commonly the ligand is added to an aqueous solution of an appropriate salt of cobalt(II), and the cobalt(II) complex thereby formed is oxidized by some convenient oxidant, frequently (if an N-donor ligand is involved) in the presence of a catalyst such as active charcoal. Molecular oxygen is often used as the oxidant simply by drawing a stream of air through the solution for a few hours, but the same result can, in many cases, be obtained more quickly by using aqueous solutions of H_2O_2.

The cobaltammines, whose number is legion, were amongst the first coordination compounds to be systematically studied and are undoubtedly the most extensively investigated class of cobalt(III) complex (see Panel). Oxidation of aqueous mixtures of CoX_2, NH_4X, and NH_3 ($X=Cl$, Br, NO_3, etc.) can, by varying the conditions and particularly the relative proportions of the reactants, be used to prepare complexes of types such as $[Co(NH_3)_6]^{3+}$, $[Co(NH_3)_5X]^{2+}$, and $[Co(NH_3)_4X_2]^+$. The range of these compounds is further extended by the replacement of X by other anionic or neutral ligands. The inertness of the compounds makes such substitution reactions slow (taking hours or days to attain equilibrium) and, being therefore amenable to examination by conventional analytical techniques, they have provided a continuing focus for kinetic

TABLE 26.4 E° *for some* Co^{III}/Co^{II} *couples in acid solution*

Couple	E°/V
$[Co(H_2O)_6]^{3+} + e^- \rightleftharpoons [Co(H_2O)_6]^{2+}$	1.83
$[Co(C_2O_4)_3]^{3-} + e^- \rightleftharpoons [Co(C_2O_4)_3]^{4-}$	0.57
$[Co(EDTA)]^- + e^- \rightleftharpoons [Co(EDTA)]^{2-}$	0.37
$[Co(bipy)_3]^{3+} + e^- \rightleftharpoons [Co(bipy)_3]^{2+}$	0.31
$[Co(en)_3]^{3+} + e^- \rightleftharpoons [Co(en)_3]^{2+}$	0.18
$[Co(NH_3)_6]^{3+} + e^- \rightleftharpoons [Co(NH_3)_6]^{2+}$	0.108
$[Co(CN)_6]^{3-} + H_2O + e^- \rightleftharpoons [Co(CN)_5(H_2O)]^{3-} + CN^-$	-0.8
$\frac{1}{2}O_2 + 2H^+ + 2e^- \rightleftharpoons H_2O$	1.229

Naming Complexes according to Colour

The observation by B. M. Tassaert in 1798 that solutions of cobalt(II) chloride in aqueous ammonia gradually turn brown in air, and then wine-red on being boiled, is generally accepted as the first preparation of a cobalt(III) complex. It was realized later that more than one complex was involved and that, by varying the relative concentrations of ammonia and chloride ion, the complexes $CoCl_3 \cdot xNH_3$ ($x = 6$, 5, and 4) could be separated. Prior to this it had been customary to name compounds after their discoverer or place of discovery, but the distinctive colours of these ammines and other derivatives provided, for a time, another convenient system of nomenclature.

To the name of the cobalt(III) salt, from which the ammine could be regarded as being derived, was added a prefix denoting its colour. Thus $CoCl_3 \cdot 6NH_3$ was "luteocobaltic chloride" because of its golden-brown colour. Further examples are given in Table A using the modern formulation of the complex ions. Since the corresponding chromium(III) complexes were frequently, if fortuitously, found to have similar colours, the prefixes were used to describe similarly formulated complexes first of chromium and then indeed of any metal regardless of colour. Thus $[Ir(NH_3)_6]Cl_3$ was luteoiridium chloride and $[Cr(NH_3)_5(H_2O)]Cl_3$ was roseochromic chloride despite their respective white and orange colours. For modern chemists this rather whimsical nomenclature has clearly outgrown its usefulness though a recognition of its origin is important for anyone reading the earlier literature.

TABLE A *Prefixes denoting colour*

Complex ion	Colour	Prefix to name of salt	Derivation
$[Co(NH_3)_6]^{3+}$	Golden-brown	Luteo-	L. *luteus* (yellow)
$[Co(NH_3)_5(H_2O)]^{3+}$	Bright red	Roseo-	L. *roseus* (rose-coloured)
$[Co(NH_3)_5Cl]^{2+}$	Purple	Purpureo-	L. *purpureus* (purple)
$[Co(NH_3)_5(NO_2)]^{2+}$	Orange	Xantho-	Gk. ξανθός, *xanthos* (yellow)
cis-$[Co(NH_3)_4Cl_2]^+$	Deep violet	Violeo-	Fr. *violet* (violet-coloured)
trans-$[Co(NH_3)_4Cl_2]^+$	Green	Praseo-	Gk. πράσιος, *praseos* (leek-coloured)
cis-$[Co(NH_3)_4(NO_2)_2]^+$	Golden-yellow	Flavo-	L. *flavus* (yellow)
trans-$[Co(NH_3)_4(NO_2)_2]^+$	Saffron	Croceo-	L. *croceus* (saffron-coloured)

studies. The forward (aquation) and backward (anation) reactions of the pentammines:

$$[Co(NH_3)_5X]^{2+} + H_2O \rightleftharpoons [Co(NH_3)_5(H_2O)]^{3+} + X^-$$

must be the most thoroughly studied substitution reactions, certainly of octahedral compounds. Furthermore, the isolation of *cis* and *trans* isomers of the tetrammines (p. 1072) was an important part of Werner's classical proof of the octahedral structure of 6-coordinate complexes.

Compounds analogous to the cobaltammines may be similarly obtained using chelating amines such as ethythenediamine or bipyridyl, and these too have played an important role in stereochemical studies. Thus cis-$[Co(en)_2(NH_3)Cl]^{2+}$ was resolved into $d(+)$ and $l(-)$ optical isomers by Werner in 1911 thereby demonstrating, to all but the most determined doubters, its octahedral stereochemistry.† More recently, the absolute configuration of one of the optical isomers of $[Co(en)_3]^{3+}$ was determined (see Panel).[8]

Another N-donor ligand, which forms extremely stable complexes, is the NO_2^- ion: its best-known complex is the orange "sodium cobaltinitrite", $Na_3[Co(NO_2)_6]$, aqueous solutions of which can be used for the quantitative precipitation of K^+ as $K_3[Co(NO_2)_6]$. Treatment of this with fluorine yields $K_3[CoF_6]$, whose anion is notable not only as the only hexahalo complex of cobalt(III) but also for being high-spin and hence paramagnetic with a magnetic moment at room temperature of nearly 5.8 BM.

Determination of Absolute Configuration

Because they rotate the plane of polarized light in opposite directions (p. 1078) it is a relatively simple matter to distinguish an optical isomer from its mirror image. But to establish their absolute configurations is a problem which for long defeated the ingenuity of chemists. Normal X-ray diffraction techniques do not distinguish between them, but J. M. Bijvoet developed the absorption edge, or anomalous, diffraction technique which does. In this method[9] the wavelength of the X-rays is chosen so as to correspond to an electronic transition of the central metal atom, and under these circumstances phase changes are introduced into the diffracted radiation which are different for the two isomers. An understanding of the phenomenon not only allows the isomers to be distinguished but also their configurations to be identified. Once the absolute configuration of one complex has been determined in this way, it can then be used as a standard to determine the absolute configuration of other, similar, complexes by the relatively simpler method of comparing their *optical rotary dispersion* (ORD) and *circular dichroism* (CD) curves.[10]

Normal measurements of optical activity are concerned with the ability of the optically active substance to rotate the plane of polarization of plane polarized light, its specific optical rotatory power (α_m) being given by

$$\alpha_m = \frac{\alpha V}{m l} \text{ rad m}^2 \text{ kg}^{-1}$$

where α is the observed angle of rotation, V is the volume, m is the mass, and l is path length.

The reason why this phenomenon occurs is that plane polarized light can be considered to be made up of left- and of right-circularly polarized components, and the nature of an optically active substance is such that, in passing through it, one component passes through greater electron density than does the other. As a result, that component is slowed down relative to the other and the two components emerge somewhat out-of-phase, i.e. the plane of polarization of the light has been rotated. If the wavelength of the polarized light is varied, and α_m then plotted against wavelength, the result is known as an *optical rotary dispersion* curve. For those wavelengths at which the substance is transparent, α_m is virtually constant, which is to say the ORD curve is flat.

---

† So deep-seated at that time was the conviction that optical activity could arise only from carbon atoms that it was argued that the ethylenediamine must be responsible, even though it is itself optically inactive. The opposition was only finally assuaged by Werner's subsequent resolution of an entirely inorganic material (p. 1073).

[8] Y. SAITO, K. NAKATSU, M. SHIRO, and H. KUROYA, Determination of the absolute configuration of optically active complex $[Co(en)_3]^{3+}$ by means of X-rays, *Acta Cryst.* **8**, 729–30 (1955).

[9] J. M. BIJVOET, Determination of the absolute configuration of optical antipodes, *Endeavour* **14**, 71–77 (1955).

[10] R. D. GILLARD, The Cotton effect in coordination compounds, *Prog. Inorg. Chem.* **7**, 215–76 (1966).

But what happens when the wavelength of the light is such that it is absorbed by the substance in question?

In absorbing light the molecules of a substance undergo electronic excitations which involve displacement of electron charge. Because of their differing routes through the molecules, the two circularly polarized components of the light produce these excitations to different extents and are consequently absorbed to different extents. The difference in extinction coefficients, $\varepsilon_{left}-\varepsilon_{right}$, can be measured and is known as the *circular dichroism*. If the CD is plotted against wavelength it is therefore zero at wavelengths where there is no absorption but passes through a maximum, or a minimum, where absorption occurs. Accompanying these changes in CD it is found that the ORD curve is like a first derivative, passing through zero at the absorption maximum (Fig. A). Such a change in sign of α_m highlights the importance of quoting the wavelength of the light used when classifying optical isomers as (+) or (−), since the classification could be reversed by simply using light of a different wavelength.[†]

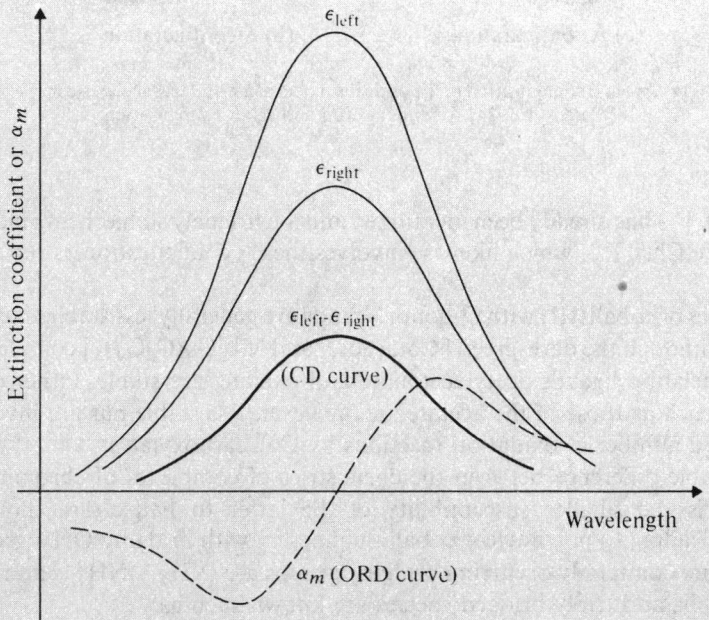

Fig. A Diagrammatic representation of the Cotton effect (actually "positive" Cotton effect. The "negative" effect occurs when the CD curve shows a minimum and the ORD curve is the reverse of the above).

The behaviours of CD and ORD curves in the vicinity of an absorption band are collectively known as the *Cotton effect* after the French physicist A. Cotton who discovered them in 1895. Their importance in the present context is that molecules with the same absolute configuration will exhibit the same Cotton effect for the same d–d absorption and, if the configuration of one compound is known, that of *closely similar* ones can be established by comparison.

The optical isomer of $[Co(en)_3]^{3+}$ referred to in the main text is the $(+)_{NaD}$ isomer, which has a left-handed (*laevo*) screw axis as shown in Fig. Ba, and according to the convention recommended

[†] The situation is perhaps not quite so bad as is implied here, since *single* measurements of α_m are usually made at the sodium D line, 589.6 nm. Nevertheless, it is clearly better to state the wavelength than to assume that this will be understood.

by IUPAC is given the symbol Λ. This is in contrast to its mirror image (Fig. Bb) which has a right-handed (*dextro*) screw axis and is given the symbol Δ.

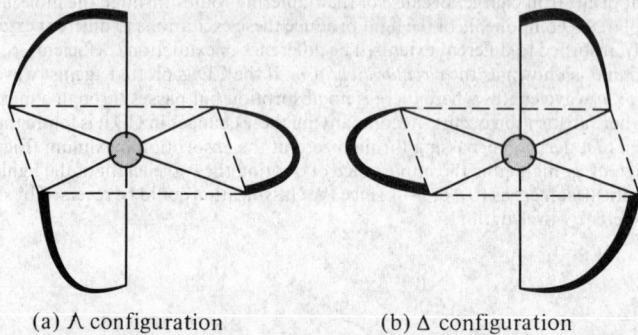

(a) Λ configuration   (b) Δ configuration

FIG. B   The absolute configuration of the optical isomers of a metal tris-chelates complex such as $[Co(en)_3]^{3+}$. (a) Λ configuration and (b) Δ configuration.

$[Co(CN)_6]^{3-}$ has already been mentioned and is extremely stable, being inert to alkalis and, like $[Fe(CN)_6]^{4-}$, which likewise involves the t_{2g}^6 configuration, is reportedly non-toxic.

Complexes of cobalt(III) with O-donor ligands are generally less stable than those with N-donors although the dark-green $[Co(acac)_3]$ and $M_3^I[Co(C_2O_4)_3]$ complexes, formed from the chelating ligands acetylacetonate and oxalate, are stable. Other carboxylato complexes such as those of the acetate are, however, less stable but are involved in the catalysis of a number of oxidation reactions by Co^{II} carboxylates.

A noticeable difference between the chemistries of complexes of chromium(III) and cobalt(III) is the smaller susceptibility of the latter to hydrolysis, though limited hydrolysis, leading to polynuclear cobaltammines[11] with bridging OH^- groups, is well known. Other commonly occurring bridging groups are NH_2^-, NH^{2-}, and NO_2^-, and singly, doubly, and triply bridged species are known such as

the bright-blue $[(NH_3)_5Co\text{–}NH_2\text{–}Co(NH_3)_5]^{5+}$,

$$\text{garnet-red } [(NH_3)_4Co\overset{\displaystyle /OH\diagdown}{\underset{\displaystyle \diagdown OH/}{}}Co(NH_3)_4]^{4+},$$

$$\text{and red } [(NH_3)_3Co\overset{\displaystyle /NH_2\diagdown}{\underset{\diagdown OH/}{-OH-}}Co(NH_3)_3]^{3+}$$

But probably the most interesting of the polynuclear complexes are those containing $-O\text{–}O-$ bridges (see also p. 719).

In the preparation of cobalt(III) hexammine salts by the aerial oxidation of cobalt(II) in aqueous ammonia it is possible, in the absence of a catalyst, to isolate a brown

[11] R. DAVIES, M. MORI, A. G. SYKES, and J. A. WEIL, Binuclear complexes of cobalt(III), *Inorg. Synth.* **12**, 197–214 (1970).

intermediate, $[(NH_3)_5Co-O_2-Co(NH_3)_5]^{4+}$. This is moderately stable in conc aqueous ammonia and in the solid, but decomposes readily in acid solutions to Co^{II} and O_2, while oxidizing agents such as $(S_2O_8)^{2-}$ convert it to the green, paramagnetic $[(NH_3)_5Co-O_2-Co(NH_3)_5]^{5+}$ ($\mu_{300} \sim 1.7$ BM). The formulation of the brown compound poses no problems. The 2 cobalt atoms are in the +3 oxidation state and are joined by a peroxo group, O_2^{2-}, all of which accords with the observed diamagnetism; moreover, the stereochemistry of the central Co–O–O–Co group (Fig. 26.3a) is akin to that of H_2O_2 (p. 744). By contrast, the green compound has been the subject of considerable debate.[12] Werner thought that it also involved a peroxo group but in this instance bridging Co^{III} and Co^{IV} atoms. This could account for the paramagnetism, but esr evidence shows that the 2 cobalt atoms are actually equivalent, and X-ray evidence shows the central Co–O–O–Co group to be planar with an O–O distance of 131 pm, which is very close to the 128 pm of the superoxide, O_2^-, ion. A more satisfactory formulation therefore is that of 2 Co^{III} atoms joined by a superoxide bridge.[13] Molecular orbital theory predicts that the unpaired electron is situated in a π orbital extending over all 4 atoms. If this is the case, then the π orbital is evidently concentrated very largely on the bridging oxygen atoms.

If $[(NH_3)_5Co-O_2-Co(NH_3)_5]^{4+}$ is treated with aqueous KOH another brown complex,

$$[(NH_3)_4Co\underset{O_2}{\overset{NH_2}{<}}Co(NH_3)_4]^{3+}$$

is obtained and, again, a 1-electron oxidation yields a green superoxo species,

$$[(NH_3)_4Co\underset{O_2}{\overset{NH_2}{<}}Co(NH_3)_4]^{4+}$$

The sulfate of this latter is actually one component of Vortmann's sulfate—the other is the red

$$[(NH_3)_4Co\underset{OH}{\overset{NH_2}{<}}Co(NH_3)_4](SO_4)_2$$

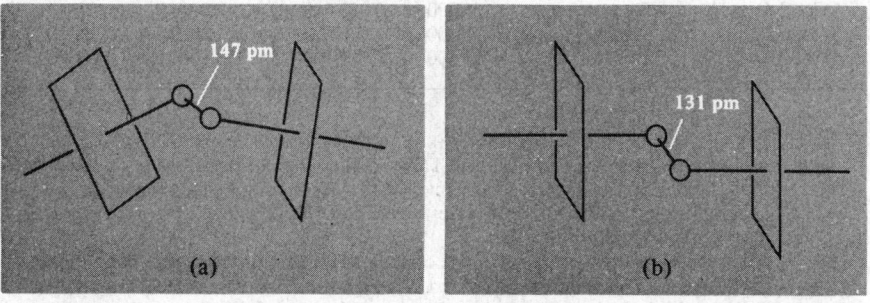

FIG 26.3 O_2 bridges in dinuclear cobalt complexes: (a) peroxo (O_2^{2-}) bridge, and (b) superoxo (O_2^-) bridge.

[12] A. G. SYKES and J. A. WEIL, The formation, structure and reactions of binuclear complexes of cobalt, *Prog. Inorg. Chem.* **13**, 1–106 (1970).
[13] V. M. MISKOWSKI, J. L. ROBBINS, I. M. TREITEL, and H. B. GRAY, Electronic structure and spectra of μ-superoxo-dicobalt(III) complexes, *Inorg. Chem.* **14**, 2318–21 (1975).

They are obtained by aerial oxidation of ammoniacal solutions of cobalt(II) nitrate followed by neutralization with H_2SO_4.

Apart from the above green superoxo-bridged complexes and the blue fluoro complexes, $[CoF_6]^{3-}$ and $[CoF_3(H_2O)_3]$, octahedral complexes of cobalt(III) are low-spin and hence diamagnetic. Their magnetic properties are therefore of little interest but, somewhat unusually for low-spin compounds (p. 1096), their electronic spectra have received a good deal of attention (see Panel).

Electronic Spectra of Octahedral Low-spin Complexes of Co(III)

It is possible to observe spin-allowed, d–d bands in the visible region of the spectra of low-spin cobalt(III) complexes because of the small value of $10Dq$, (Δ), which is required to induce spin-pairing in the cobalt(III) ion. This means that the low-spin configuration occurs in complexes with ligands which do not cause the low-energy charge transfer bands which so often dominate the spectra of low-spin complexes.

In practice two bands are generally observed and are assigned to the transitions:

$$v_1 = {}^1T_{1g} \leftarrow {}^1A_{1g}$$

and

$$v_2 = {}^1T_{2g} \leftarrow {}^1A_{1g} \quad \text{(see Fig. A)}$$

These transitions correspond to the electronic promotion $t_{2g}^6 \rightarrow t_{2g}^5 e_g^1$ with the promoted electron maintaining its spin unaltered. The orbital multiplicity of the $t_{2g}^5 e_g^1$ configuration is 6 and so corresponds to two orbital triplet terms ${}^1T_{1g}$ and ${}^1T_{2g}$. If, on the other hand, the promoted electron changes its spin, the orbital multiplicity is again 6 but the two T terms are now spin triplets, ${}^3T_{1g}$ and ${}^3T_{2g}$. A weak band attributable to the spin-forbidden ${}^3T_{1g} \leftarrow {}^1A_{1g}$ transition is indeed observed in some cases in the region of 11 000–14 000 cm^{-1}.

TABLE A *Spectra of octahedral low-spin complexes of cobalt(III)*

Complex	Colour	v_1/cm^{-1}	v_2/cm^{-1}	$10Dq$/cm^{-1}	B/cm^{-1}
$[Co(H_2O)_6]^{3+}$	Blue	16 600	24 800	18 200	670
$[Co(NH_3)_6]^{3+}$	Golden-brown	21 000	29 500	22 900	620
$[Co(C_2O_4)_3]^{3-}$	Dark green	16 600	23 800	18 000	540
$[Co(en)_3]^{3+}$	Yellow	21 400	29 500	23 200	590
$[Co(CN)_6]^{3-}$	Yellow	32 400	39 000	33 500	460

Data for some typical complexes are given in Table A. The assignments are made, as described on p. 1197, by fitting the ratio of the frequencies of the observed spin-allowed bands to the Tanabe–Sugano diagram, which produces values of the interelectronic repulsion parameter B as well as of the crystal-field splitting, $10Dq$.

As is evident from Table A on p. 1303, the colours of *cis* and *trans* isomers of complexes $[CoL_4X_2]$ or $[Co(L-L)_2X_2]$ frequently differ and, although simple observation of colour will not alone suffice to establish a *cis* or *trans* geometry, an examination of the electronic spectra does have diagnostic value. Calculations of the effect of low-symmetry components in the crystal field show that the *trans* isomer will split the excited terms appreciably more than the *cis*, and the effect is most marked for ${}^1T_{1g}$, the lowest of the excited terms. In practice, if L–L and X are sufficiently far apart in the spectromchemical series (e.g. L–L=en and X=F which has been thoroughly examined), the v_1 band splits completely, giving rise to three separate bands for the *trans* complex whereas the *cis* merely shows slight asymmetry in the lower energy band. Furthermore, because

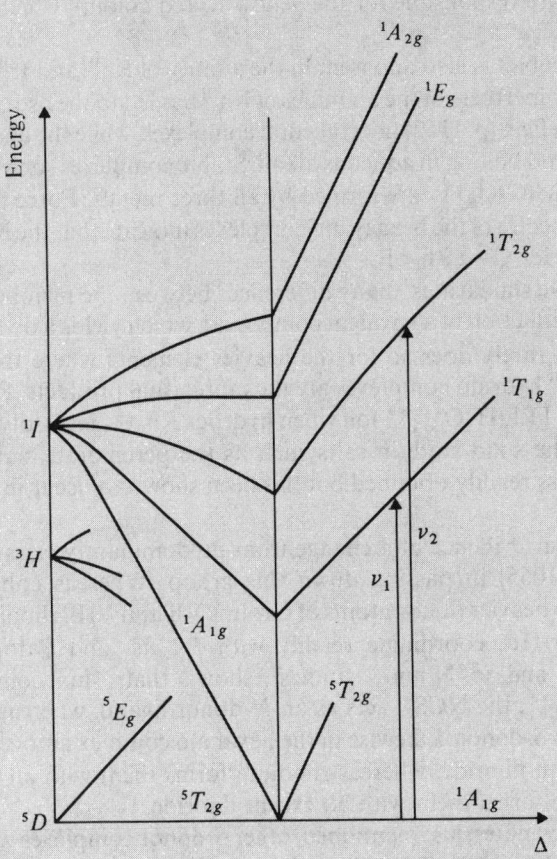

Fig. A Simplified Tanabe–Sugano diagram for d^6 ions showing possible spin-allowed transitions in complexes of low-spin cobalt(III).

(like tetrahedral complexes, p. 1094) a *cis* isomer lacks a centre of symmetry, its spectrum is more intense than that of the centrosymmetric *trans* isomer.

It is relevant to note at this point that, because the metal ions are isoelectronic, the spectra of low-spin Fe^{II} complexes might be expected to be similar to those of low-spin Co^{III}. However, Fe^{II} requires a much stronger crystal field to effect spin-pairing and the ligands which provide such a field also give rise to low-energy charge-transfer bands which almost always obscure the d–d bands. Nevertheless, the spectrum of the pale-yellow $[Fe(CN)_6]^{4-}$ shows a shoulder at $31\,000\ cm^{-1}$ on the side of a charge transfer absorption and this is attributed to the $^1T_{1g} \leftarrow {}^1A_{1g}$ transition.

Complexes of rhodium(III) are usually derived, directly or indirectly, from $RhCl_3.3H_2O$ and those of iridium(III) from $(NH_4)_3[IrCl_6]$. All the compounds of Rh^{III} and Ir^{III} are diamagnetic and low-spin, the vast majority of them being octahedral with the t_{2g}^6 configuration. Their electronic spectra can be interpreted in the same way as the spectra of Co^{III} complexes, though the second d–d band, especially in the case of Ir^{III}, is frequently obscured by charge-transfer absorption. The d–d absorptions at the blue end of the visible region are responsible for the yellow to red colours which characterize Rh^{III} complexes.

Similarity with cobalt is also apparent in the affinity of Rh^{III} and Ir^{III} for ammonia and amines. The kinetic inertness of the ammines of Rh^{III} has led to the use of several of them in studies of the *trans* effect (p. 1352) in octahedral complexes, while the ammines of Ir^{III} are so stable as to withstand boiling in aqueous alkali. Stable complexes such as $[M(C_2O_4)_3]^{3-}$, $[M(acac)_3]$, and $[M(CN)_6]^{3-}$ are formed by all three metals. Force constants obtained from the infrared spectra of the hexacyano complexes indicate that the M–C bond strength increases in the order $Co < Rh < Ir$.

Despite the above similarities, many differences between the members of this triad are also to be noted. Reduction of a trivalent compound, which yields a divalent compound in the case of cobalt, rarely does so for the heavier elements where the metal, univalent compounds, or M^{III} hydrido complexes are the more usual products. Rhodium forms the quite stable, yellow $[Rh(H_2O)_6]^{3+}$ ion when hydrous Rh_2O_3 is dissolved in mineral acid, and it occurs in the solid state in salts such as the perchlorate, sulphate, and alums. $[Ir(H_2O)_6]^{3+}$ is less readily obtained but has been shown to occur in solutions of Ir^{III} in conc $HClO_4$.[13a]

There is also clear evidence of a change from predominantly class-a to class-b metal characteristics (p. 1065) in passing down this group. Whereas cobalt(III) forms few complexes with the heavier donor atoms of Groups VB and VIB, rhodium(III), and more especially iridium (III), coordinate readily with *P*-, *As*- and *S*-donor ligands. Thus infrared, X-ray, and ^{14}N nmr studies show that, in complexes such as $[Co(NH_3)_4(NCS)_2]^+$, the NCS^- acts as an *N*-donor ligand, whereas in $[M(SCN)_6]^{3-}$ (M = Rh, Ir) it is an *S*-donor. Likewise in the hexahalo complex anions, $[MX_6]^{3-}$, cobalt forms only that with fluoride, whereas rhodium forms them with all the halides except iodide, and iridium forms them with all except fluoride.

Besides the thiocyanates, just mentioned, other *S*-donor complexes which are of interest are the dialkyl sulfides, $[MCl_3(SR_2)_3]$, produced by the action of SR_2 on ethanolic $RhCl_3$ or on $[IrCl_6]^{3-}$. Phosphorus and arsenic compounds are obtained in similar fashion, and the best known are the yellow to orange complexes, $[ML_3X_3]$, (M = Rh, Ir; X = Cl, Br, I; L = trialkyl or triaryl phosphine or arsine). For many of these complexes the measurement of dipole moments and, more recently of proton nmr spectra have proved convenient tools with which to distinguish *mer* and *fac* isomers. An especially interesting feature of their chemistry is the ease with which they afford hydride and carbonyl derivatives. For instance, the colourless, air-stable $[RhH(NH_3)_5]SO_4$ is produced by the action of Zn powder on ammoniacal $RhCl_3$ in the presence of $(NH_4)_2SO_4$:

$$[RhCl(NH_3)_5]Cl_2 \xrightarrow[SO_4^{2-}]{Zn} [RhH(NH_3)_5]SO_4$$

It is unusual for hydrides of metals in such a high *formal* oxidation state as $+3$ to be stable in the absence of π-acceptor ligands and, indeed, in the presence of π-acceptor ligands such as tertiary phosphines and arsines, the stability of rhodium(III) hydrides is enhanced. Thus H_3PO_2 reduces $[RhCl_3L_3]$ to either $[RhHCl_2L_3]$ or $[RhH_2ClL_3]$, depending on L;

[13a] P. BEUTLER and H. GAMSJÄGER, Preparation and ultraviolet-visible spectrum of hexa-aquairidium(III), *JCS Chem. Comm.* 1976, 554–5; The hydrolysis of iridium(III), *JCS Dalton* 1979, 1415–8.

and the action of H_2 on $[Rh^I(PPh_3)_3X]$ (X = Cl, Br, I) yields $[RhH_2(PPh_3)_3X]$ which is, formally at least, an oxidation by molecular hydrogen. However, it is iridium(III) that forms more hydrido-phosphine and hydrido-arsine complexes than any other platinum metal. Using $NaBH_4$, $LiAlH_4$, EtOH, or even $SnCl_2 + H^+$ to provide the hydride ligand, complexes of the type $[MH_nL_3X_{3-n}]$ can be formed for very many of the permutations which are possible from L = trialkyl or triaryl phosphine or arsine; X = Cl, Br, or I.

Oxidation state II (d^7)

There is a very marked contrast in this oxidation state between cobalt on the one hand, and the two heavier members of the group on the other. For cobalt it is one of the two most stable oxidation states, whereas for the others it is of only minor importance.

Many early reports of Rh^{II} and Ir^{II} complexes have not been verified and in some cases may have involved M^{III} hydrides. The very few substantiated examples of Ir^{II} require stabilization with phosphines, but Rh^{II} is somewhat more widely represented. Besides a few square planar complexes, $[RhCl_2L_2]$, which are paramagnetic and apparently owe their existence to the stabilizing effect of the bulky phosphines which block the octahedral fifth and sixth positions, Rh^{II} is found in a number of green dimeric, diamagnetic compounds. Hydrous Rh_2O_3 reacts with carboxylic acids to form $[Rh(O_2CR)_2]_2$ with the same bridged structure as the carboxylates of Cr^{II}, Mo^{II}, and Cu^{II}; in the case of the acetate this involves a Rh–Rh distance of 239 pm which is consistent with a Rh–Rh bond. If rhodium acetate is treated with a strong acid such as HBF_4, whose anion has little tendency to coordinate, green solutions apparently containing the diamagnetic Rh_2^{4+} ion are obtained but no solid salt of this has been isolated. Why no comparable Ir^{II} carboxylates, or other dimeric species stabilized by metal–metal bonding, have yet been prepared is not clear.

By contrast, Co^{II} carboxylates such as the red acetate, $Co(O_2CMe)_2.4H_2O$, are monomeric and in some cases the carboxylate ligands are unidentate. The acetate is employed in the production of catalysts used in certain organic oxidations, and also as a drying agent in oil-based paints and varnishes. Cobalt(II) gives rise to simple salts with all the common anions and they are readily obtained as hydrates from aqueous solutions. The parent hydroxide, $Co(OH)_2$, can be precipitated from the aqueous solutions by the addition of alkali and is somewhat amphoteric, not only dissolving in acid but also redissolving in excess of conc alkali, in which case it gives a deep-blue solution containing $[Co(OH)_4]^{2-}$ ions. It is obtainable in both blue and pink varieties: the former is precipitated by slow addition of alkali at 0°C, but it is unstable and, in the absence of air, becomes pink on warming.

Complexes of cobalt(II) are less numerous than those of cobalt(III) but, lacking any configuration comparable in stability with the t_{2g}^6 of Co^{III}, they show a greater diversity of types and are more labile. The redox properties have already been referred to and the possibility of oxidation must always be considered when preparing Co^{II} complexes. However, providing solutions are not alkaline and the ligands not too high in the spectrochemical series, a large number of complexes can be isolated without special precautions. The most common type is high-spin octahedral, though spin-pairing can be achieved by ligands such as CN^- which also favour the higher oxidation state.

Appropriate choice of ligands can however lead to high-spin–low-spin equilibria as shown by the recently investigated $[Co(terpy)_2]X_2.nH_2O$.[13b]

Many of the hydrated salts and their aqueous solutions contain the octahedral, pink $[Co(H_2O)_6]^{2+}$ ion, and bidentate N-donor ligands such as en, bipy, and phen form octahedral cationic complexes $[Co(L-L)_3]^{3+}$, which are much more stable to oxidation than is the hexammine $[Co(NH_3)_6]^{2+}$. Acac yields the orange $[Co(acac)_2(H_2O)_2]$ which has the *trans* octahedral structure and can be dehydrated to $[Co(acac)_2]$ which attains octahedral coordination by forming the tetrameric species shown in Fig. 26.4. This is comparable with the trimeric $[Ni(acac)_2]_3$ (p. 1343), like which it shows evidence of weak ferromagnetic interactions at very low temperatures. $[Co(EDTA)(H_2O)]^{2+}$ is ostensibly analogous to the 7-coordinate Mn^{II} and Fe^{II} complexes with the same stoichiometry, but in fact the cobalt is only 6-coordinate, 1 of the oxygen atoms of the EDTA being too far away from the cobalt (272 compared to 223 pm for the other EDTA donor atoms) to be considered as coordinated.

Tetrahedral complexes are also common, being formed more readily with cobalt(II) than with the cation of any other truly transitional element (i.e. excluding Zn^{II}). This is consistent with the CFSEs (p. 1097) of the two stereochemistries (Table 26.5). Quantitative comparisons between the values given for CFSE(oct) and CFSE(tet) are not possible because of course the crystal field splittings, Δ_o and Δ_t differ. Nor is the CFSE by any means the most important factor in determining the stability of a complex. Nevertheless, where other factors are comparable, it can have a decisive effect and it is apparent that no configuration is more favourable than d^7 to the adoption of a tetrahedral as opposed to an octahedral stereochemistry. Thus, in aqueous solutions containing $[Co((H_2O)_6]^{2+}$ there are also present in equilibrium, small amounts of tetrahedral $[Co(H_2O)_4]^{2+}$, and in acetic acid the tetrahedral $[Co(O_2CMe)_4]^{2-}$ occurs. The anionic

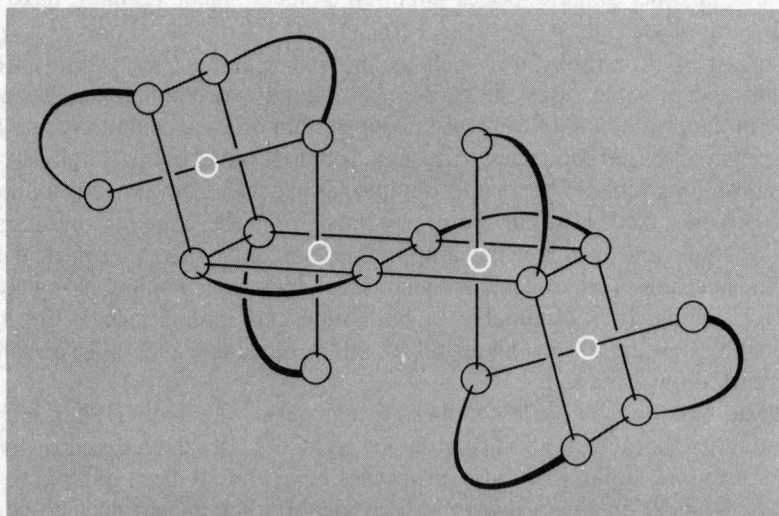

FIG. 26.4   The terameric structure of $[Co(acac)_2]_4$.

[13b] S. KREMER, W. HENKE, and D. REINEN, High-spin–low-spin equilibria of cobalt(2+) in the terpyridine complexes Co(terpy)$_2$X$_2$.nH$_2$O, *Inorg. Chem.* **21**, 3013–22 (1982).

TABLE 26.5 *CFSE values for high-spin complexes of d^0 to d^{10} ions*

No. of d electrons	0	1	2	3	4	5	6	7	8	9	10
CFSE(oct)/(Δ_o)	0	$\frac{2}{5}$	$\frac{4}{5}$	$\frac{6}{5}$	$\frac{3}{5}$	0	$\frac{2}{5}$	$\frac{4}{5}$	$\frac{6}{5}$	$\frac{3}{5}$	0
CFSE(tet)/(Δ_t)	0	$\frac{3}{5}$	$\frac{6}{5}$	$\frac{4}{5}$	$\frac{2}{5}$	0	$\frac{3}{5}$	$\frac{6}{5}$	$\frac{4}{5}$	$\frac{2}{5}$	0

complexes $[CoX_4]^{2-}$ are formed with the unidentate ligands, X=Cl, Br, I, SCN, and OH, and a whole series of complexes, $[CoL_2X_2]$ (L=ligand with group VB donor atom; X=halide, NCS), has been prepared in which both stereochemistries are found. $[CoCl_2py_2]$ exists in two isomeric forms: a blue metastable variety which is monomeric and tetrahedral, and a violet, stable form which is polymeric and achieves octahedral coordination by means of chloride bridges. Ligand polarizability is an important factor determining which stereochemistry is adopted, the more polarizable ligands favouring the tetrahedral form since fewer of them are required to neutralize the metal's cationic charge. Thus, if L=py, replacement of Cl⁻ by I⁻ makes the stable form tetrahedral and if L=phosphine or arsine the tetrahedral form is favoured irrespective of X.

The most obvious distinction between the octahedral and tetrahedral compounds is that *in general* the former are pink to violet in colour whereas the latter are blue, as exemplified by the well-known equilibrium:

$$[Co(H_2O)_6]^{2+} + 4Cl^- \rightleftharpoons [CoCl_4]^{2-} + 6H_2O$$

pink		blue
octahedral		tetrahedral

This is not an infallible distinction (as the blue but octahedral $CoCl_2$ demonstrates) but is a useful empirical guide whose reliability is improved by a more careful analysis of the electronic spectra (see Panel).

Electronic Spectra and Magnetic Properties of High-spin Octahedral and Tetrahedral Complexes of Cobalt(II)[14]

Cobalt(II) is the only common d^7 ion and because of its stereochemical diversity its spectra have been widely studied. In a cubic field, three spin-allowed transitions are anticipated (p. 1094) because of the splitting of the free-ion, ground ^4F term, and the accompanying ^4P term. In the octahedral case the splitting is the same as for the octahedral d^2 ion and the spectra can therefore be interpreted by using the same Tanabe–Sugano diagram as was used for V^{3+} (Fig. 22.8, p. 1162). In the present case the spectra usually consist of a band in the near infrared, which may be assigned as

$$\nu_1 = {}^4T_{2g}(F) \leftarrow {}^4T_{1g}(F),$$

and another in the visible, often with a shoulder on the low energy side. Since the transition $^4A_{2g}(F) \leftarrow {}^4T_{1g}(F)$ is essentially a 2-electron transition from $t_{2g}^5 e_g^2$ to $t_{2g}^3 e_g^4$ it is expected to be weak, and the usual assignment is

---

[14] R. L. CARLIN, Electronic structure and stereochemistry of cobalt(II), *Transition Metal Chemistry* **1**, 1–33 (1965).

$$v_2(\text{shoulder}) = {}^4A_{2g}(F) \leftarrow {}^4T_{1g}(F)$$
$$v_3 = {}^4T_{1g}(P) \leftarrow {}^4T_{1g}(F)$$

Indeed, in some cases it is probable that v_2 is not observed at all, but that the fine structure arises from term splitting due to spin–orbit coupling or to distortions from regular octahedral symmetry.

For d^7 ions in tetrahedral crystal fields, the splitting of the free-ion, ground F term is the reverse of that in octahedral fields so that ${}^4A_{2g}(F)$ lies lowest. Now it has already been shown (Table 19.3, p. 1093), on the basis of the positive hole formalism, that octahedral d^7 splitting is the exact reverse of the octahedral d^3. It therefore follows that the tetrahedral d^7 case is the same as the octahedral d^3 and so spectra of tetrahedral Co^{II} complexes are to be interpreted in the same way as for octahedral Cr^{III}, i.e. on the basis of Fig. 23.10 (p. 1198). In fact, the observed spectra usually consist of a broad, intense band in the visible region (responsible for the colour and often about 10 times as intense as in octahedral compounds) with a weaker one in the infrared. The only satisfactory interpretation is to assign these, respectively, as,

$$v_3 = {}^4T_1(P) \leftarrow {}^4A_2(F)$$
$$v_2 = {}^4T_1(F) \leftarrow {}^4A_2(F)$$

in which case $v_1 = {}^4T_2(F) \leftarrow {}^4A_2(F)$ should be in the region 3000–5000 cm^{-1}. Examination of this part of the infrared has sometimes indicated the presence of a band, though overlying vibrational bands make interpretation difficult.

Table A gives data for a number of octahedral and tetrahedral complexes, the values of $10Dq$ and B having been derived as described for V^{II} (p. 1163) and Cr^{III} (p. 1198). The ability to assign the absorption bands reasonably in this manner, bearing in mind the greater intensity expected in the tetrahedral case (p. 1094), is a more reliable guide to stereochemistry than simple colour. It is, for instance, clear from these data that the "anomolous" blue colour of octahedral $CoCl_2$ arises because 6 Cl$^-$ ions generate such a weak crystal field that the main band in its spectrum is at an unusually low energy, extending into the red region (hence giving a blue colour) rather than the green-blue region (which would give a red colour) more commonly observed for octahedral Co^{II} compounds.

Magnetic properties provide a complementary means of distinguishing stereochemistry. The T ground term of the octahedral ion is expected to give rise to a temperature-dependent orbital contribution to the magnetic moment whereas the A ground term of the tetrahedral ion is not (see

TABLE A   *Electronic spectra of complexes of cobalt(II)*

(a) Octahedral

Complex	v_1/cm^{-1}	v_2/cm^{-1} (weak)	v_3/cm^{-1} (main)	$10Dq$/cm^{-1}	B/cm^{-1}
$[Co(bipy)_3]^{2+}$	11 300		22 000	12 670	791
$[Co(NH_3)_6]^{2+}$	9000		21 100	10 200	885
$[Co(H_2O)_6]^{2+}$	8100	16 000	19 400	9200	825
$CoCl_2$	6600	13 300	17 250	6900	780

(b) Tetrahedral

Complex	v_2/cm^{-1}	v_3/cm^{-1} (main)	$10Dq$/cm^{-1}	B/cm^{-1}
$[Co(NCS)_4]^{2-}$	7780	16 250	4550	691
$[Co(N_3)_4]^{2-}$	6750	14 900	3920	658
$[CoCl_4]^{2-}$	5460	14 700	3120	710
$[CoI_4]^{2-}$	4600	13 250	2650	665

p. 1098). As a matter of fact, in a tetrahedral field the excited $^4T_2(F)$ term is "mixed into" the ground $^4A_2$ term because of spin–orbit coupling and tetrahedral complexes of Co^{II} are expected to have magnetic moments given by

$$\mu_e = \mu_{\text{spin-only}}\left(1 - \frac{4\lambda}{10Dq}\right)$$

where $\lambda = -170 \text{ cm}^{-1}$ and $\mu_{\text{spin-only}} = 3.87$ BM.

Thus the magnetic moments of tetrahedral complexes lie in the range 4.4–4.8 BM, whereas those of octahedral complexes are around 4.8–5.2 BM at room temperature, falling off appreciably as the temperature is reduced.

Square planar complexes are also well authenticated if not particularly numerous and include [Co(phthalocyanine)] as well as [Co(salen)] and complexes with other Schiff bases.[15] These are invariably low-spin with magnetic moments at room temperature in the range 2.1–2.9 BM, indicating 1 unpaired electron. They are primarily of interest because of their oxygen-carrying properties, discussed already in Chapter 14 where numerous reviews on the subject are cited. The uptake of dioxygen, which bonds in the bent configuration,

$$Co-O \overset{\displaystyle O}{\diagup}\;,$$

is accompanied by the attachment of a solvent molecule *trans* to the O_2 and the retention of the single unpaired electron. There is fairly general agreement, based on esr evidence, that electron transfer from metal to O_2 occurs just as in the bridged complexes referred to on p. 1307, producing a situation close to the extreme represented by low-spin Co^{III} attached to a superoxide ion, O_2^-. (The opposite extreme, represented by Co^{II}–O_2, implies that the unpaired electron resides on the metal with the dioxygen being rendered diamagnetic by the loss of the degeneracy of its π^* orbitals with consequent spin pairing.) However, the precise extent of the electron transfer is probably determined by the nature of the ligand *trans* to the O_2.

The difficulty of assigning a formal oxidation state is more acutely seen in the case of 5-coordinate NO adducts of the type [Co(NO)(salen)]. These are effectively diamagnetic and so have no unpaired electrons. They may therefore be formulated either as Co^{III}–NO^- or Co^I–NO^+. The infrared absorptions ascribed to the N–O stretch lie in the range 1624–1724 cm^{-1}, which is at the lower end of the range said to be characteristic of NO^+. But, as in all such cases which are really concerned with the differing polarities of covalent bonds, such formalism should not be taken literally.

Other 5-coordinate Co^{II} compounds which have been characterized include $[CoBr\{N(C_2H_4NMe_2)_3\}]^+$, which is high-spin with 3 unpaired electrons and is trigonal bipyramidal (imposed by the "tripod" ligand), and $[Co(CN)_5]^{3-}$, which is low-spin with 1 unpaired electron and is square pyramidal. The latter complex has only fairly recently been isolated[16] from solutions of $Co(CN)_2$ and KCN as the yellow $[NEt_2Pr^i_2]^+$ salt, an

[15] C. DAUL, C. W. SCHLÄPFER, and A. VON ZELEWSKY, The electronic structure of cobalt(II) complexes with Schiff bases and related ligands, *Struct. Bonding* **36**, 129–71 (1979).

[16] L. D. BROWN and K. N. RAYMOND, Structural characterization of the pentacyanocobaltate(II) anion in the salt [NEt$_2$Pr$^i_2$]$_3$[Co(CN)$_5$], *Inorg. Chem.* **14**, 2590–4 (1975).

extremely oxygen-sensitive and hygroscopic material. A further difficulty which hindered its isolation is its tendency to dimerize to the more familiar deep-violet, $[(CN)_5Co\text{–}Co(CN)_5]^{6-}$. The absence of a simple hexacyanocomplex is significant as it seems to be generally the case that ligands such as CN^-, which are expected to induce spin-pairing, favour a coordination number for Co^{II} of 4 or 5 rather than 6; the planar $[Co(diars)_2]ClO_4$ is a further illustration of this. Presumably the Jahn–Teller distortion, which is anticipated for the low-spin $t_{2g}^6 e_g^1$ configuration is largely responsible.

Oxidation state I (d^8)

Oxidation states lower than $+2$ require the stabilizing effect of π-acceptor ligands and some of these are appropriately considered along with organometallic compounds in Section 26.3.5. However, although $+1$ is not a common oxidation state for cobalt, it is one of the two most common states for both rhodium and iridium and as such merits separate consideration.

Simple ligand-field arguments, which will be elaborated when M^{II} ions of the Ni, Pd, Pt triad are discussed on p. 1346, indicate that the d^8 configuration favours a 4-coordinate, square-planar stereochemistry. In the present group, however, the configuration is associated with a lower oxidation state and the requirements of the 18-electron rule,† which favour 5-coordination, are also to be considered. The upshot is that most Co^I complexes are 5-coordinate, like $[Co(CNR)_5]^+$, and square-planar Co^I is apparently unknown. On the other hand, complexes of Rh^I and Ir^I are predominantly square planar, although 5-coordination does also occur.

These complexes are usually prepared by the reduction of compounds such as $RhCl_3.3H_2O$ and K_2IrCl_6 in the presence of the desired ligand. It is often unnecessary to use a specific reductant, the ligand itself or alcoholic solvent being adequate, and not infrequently leading to the presence of CO or H in the product. A considerable proportion of the complexes of Rh^I and Ir^I are phosphines and of these, two in particular demand attention. They are Wilkinson's catalyst, $[RhCl(PPh_3)_3]$, and Vaska's compound, *trans*-$[IrCl(CO)(PPh_3)_2]$, both essentially square planar.

Wilkinson's catalyst, $[RhCl(PPh_3)_3]$. This red-violet compound, which is readily obtained by refluxing ethanolic $RhCl_3.3H_2O$ with an excess of PPh_3, was discovered[17] in 1965. It undergoes a variety of reactions, most of which involve either replacement of a phosphine ligand (e.g. with CO, CS, C_2H_4, O_2 giving *trans* products) or oxidative addition (e.g. with H_2, MeI) to form Rh^{III}, but its importance arises from its effectiveness as a catalyst for highly selective hydrogenations of complicated organic molecules which are of

---

† The filling-up of the bonding MOs of the molecule may be regarded, more simply, as the filling of the outer 9 orbitals of the metal ion with its own d electrons plus a pair of σ electrons from each ligand. A 4-coordinate d^8 ion is thus a "16-electron" species and is "coordinatively unsaturated". Saturation in this sense requires the addition of 10 electrons, i.e. 5 ligands, to the metal ion. By contrast rhodium(III) is a d^6 ion and so can expand its coordination sphere to accommodate 6 ligands with important consequences in catalysis which will be seen below.

[17] J. F. YOUNG, J. A. OSBORNE, F. H. JARDINE, and G. WILKINSON, Hydride intermediates in homogenous hydrogenation reactions of olefins and acetylenes using rhodium catalysts, *Chem. Comm.* 1965, 131–2.

great importance in the pharmaceutical industry. Its use allowed, for the first time, rapid *homogeneous* hydrogenation at ambient temperatures and pressures:

$$
\begin{array}{c} \diagdown \\ C \end{array}\!\!=\!\!\begin{array}{c} \diagup \\ C \\ \diagdown \end{array} + H_2 \quad \xrightarrow{\text{catalyst}} \quad
\begin{array}{c} | \quad | \\ -C-C- \\ | \quad | \\ H \quad H \end{array}
$$

The precise mechanism is complicated and has been the subject of much speculation and controversy, but Fig. 26.5 shows a simplified but reasonable scheme. The essential steps in this are the oxidative addition of H_2 (if the hydrogen atoms are regarded as "hydridic", i.e. as H^-, the metal's oxidation state increases from $+1$ to $+3$); the formation of an alkene complex; alkene insertion, and, finally, the reductive elimination of the alkane (i.e. the metal's oxidation state reverting to $+1$). The rhodium catalyst is able to fulfil its role because the metal is capable of changing its coordination number (loss of phosphine from the dihydro complex being encouraged by the large size of the ligand) and it possesses oxidation states ($+1$ and $+3$) which differ by 2 and are of comparable stability.

The discovery of the catalytic properties of $[RhCl(PPh_3)_3]$ naturally brought about a widespread search for other rhodium phosphines with catalytic activity. One of those which was found, also in Wilkinson's laboratory,[18] was *trans*-$[Rh(CO)H(PPh_3)_3]$ which

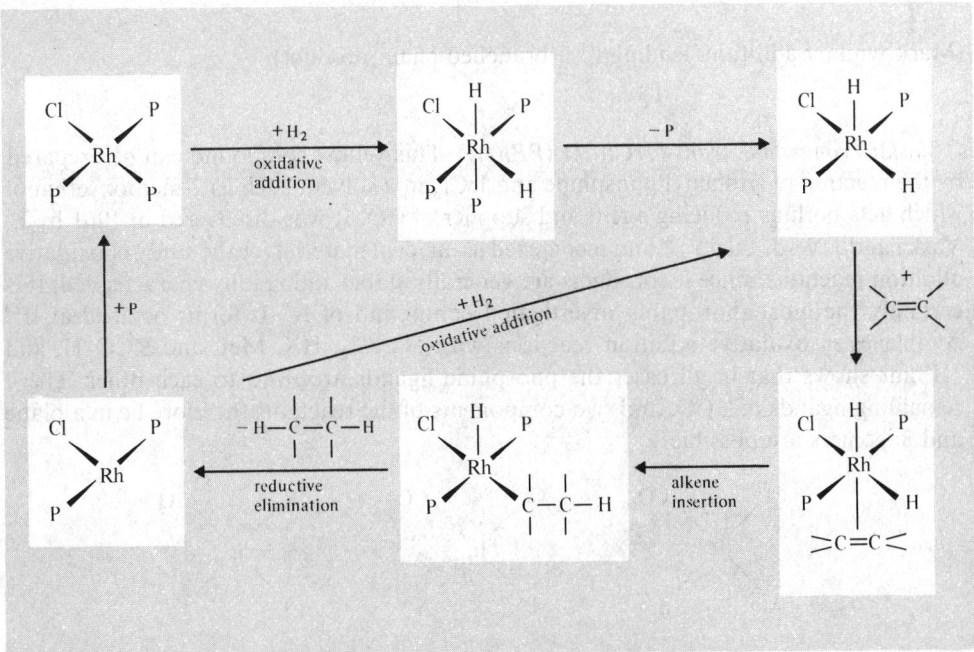

FIG. 26.5   The catalytic cycle for the hydrogenation of an alkene, catalysed by $[RhCl(PPh_3)_3]$ in benzene; possible coordination of solvent molecules has been ignored and the ligand PPh_3 has been represented as P throughout, for clarity.

18 C. O'CONNOR and G. WILKINSON, Selective homogeneous hydrogenation of alk-1-enes using hydrido-carbonyltris(triphenylphosphine)rhodium(I) as catalyst, *J. Chem. Soc.* (A) 1968, 2665–71.

can conveniently be dealt with here. It was found that, for steric reasons, it selectively catalyses the hydrogenation of alk-1-enes (i.e. terminal olefins) rather than alk-2-enes and it has recently been used in the *hydroformylation*[18a] of alkenes, (i.e. the addition of H and the formyl group, CHO) also known as the OXO process because it introduces oxygen into the hydrocarbon. This is a process of enormous industrial importance, being used to convert alk-1-enes into aldehydes which can then be converted to alcohols for the production of polyvinylchloride (PVC) and polyalkenes and, in the case of the long-chain alcohols, in the production of detergents:

$$RCH{=}CH_2 + H_2 + CO \xrightarrow{\text{catalyst}} RCH_2CH_3CHO$$

A simplified reaction scheme is shown in Fig. 26.6. Again, the ability of rhodium to change its coordination number and oxidation state is crucial, and this catalyst has the great advantage over the conventional cobalt carbonyl catalyst that it operates efficiently at much lower temperatures and pressures and produces straight-chain as opposed to branched-chain products. The reason for its selectivity lies in the insertion step of the cycle. In the presence of the two bulky PPh_3 groups, the attachment to the metal of $-CH_2CH_2R$ (anti-Markovnikov addition, leading to a straight chain product) is easier than the attachment of

$$-CH{\Big\langle}{\begin{matrix} CH_3 \\ R \end{matrix}}$$

(Markovnikov addition, leading to a branched-chain product).

Vaska's compound, trans-$[IrCl(CO)(PPh_3)_2]$. This yellow compound can be prepared by the reaction of triphenyl phosphine and $IrCl_3$ in a solvent such as 2-methoxyethanol which acts both as reducing agent and supplier of CO. It was discovered in 1961 by L. Vaska and J. W. di Luzio[19] and recognized as an ideal material for the study of oxidative addition reactions, since its products are generally stable and readily characterized. It is certainly the most thoroughly investigated compound of Ir^I. It forms octahedral Ir^{III} complexes in oxidative addition reactions with H_2, Cl_2, HX, MeI, and RCO_2H, and ^1H nmr shows that in all cases the phosphine ligands are *trans* to each other. The 4 remaining ligands (Cl, CO, and two components of the reactant) therefore lie in a plane and 3 isomers are possible:

There is apparently no simple way of predicting which of these will be formed and each case must be examined individually. Addition reactions with ligands such as CO and SO_2

[18a] R. L. PRUETT, Hydroformylation, *Adv. Organometallic Chem.* **17**, 1–60 (1979).

[19] L. VASKA and J. W. DI LUZIO, Carbonyl and hydrido-carbonyl complexes of iridium by reaction with alcohols. Hydrido complexes by reaction with acid, *J. Am. Chem. Soc.* **83**, 2784–5 (1961).

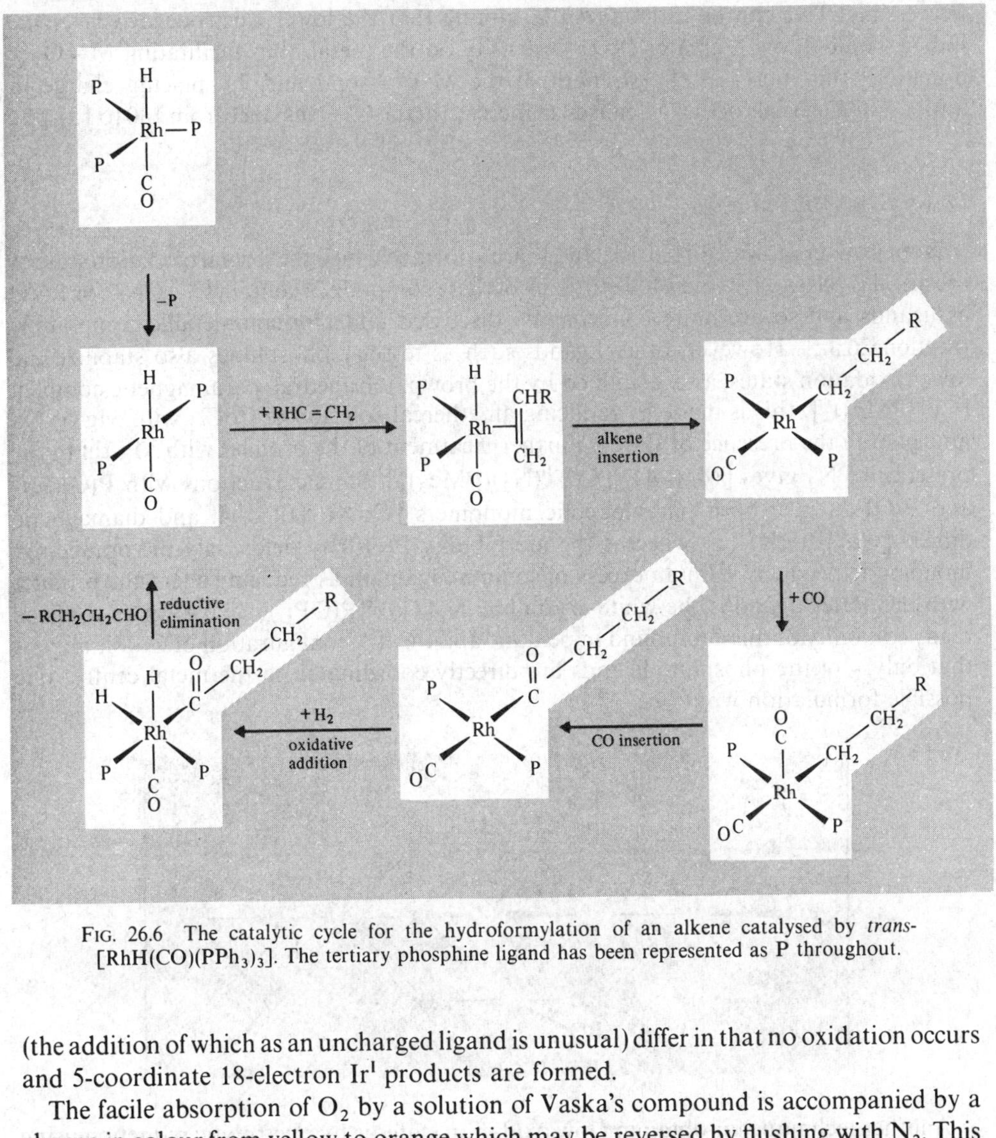

FIG. 26.6 The catalytic cycle for the hydroformylation of an alkene catalysed by *trans*-[RhH(CO)(PPh$_3$)$_3$]. The tertiary phosphine ligand has been represented as P throughout.

(the addition of which as an uncharged ligand is unusual) differ in that no oxidation occurs and 5-coordinate 18-electron Ir^{I} products are formed.

The facile absorption of O$_2$ by a solution of Vaska's compound is accompanied by a change in colour from yellow to orange which may be reversed by flushing with N$_2$. This is one of the most widely studied synthetic oxygen-carrying systems and has been discussed earlier (p. 718). The O–O distance of 130 pm in the oxygenated product (see Fig. 14.6b, p. 721) is rather close to the 128 pm of the superoxide ion, O$_2^-$, but this would imply Ir^{II} which is paramagnetic whereas the compound is actually diamagnetic. The oxygenation is instead normally treated as an oxidative addition with the O$_2$ acting as a bidentate peroxide ion, O$_2^{2-}$, to give a 6-coordinate Ir^{III} product. However, in view of the small "bite" of this ligand the alternative formulation in which the O$_2$ acts as a neutral unidentate ligand giving a 5-coordinate Ir^{I} product has also been proposed.

Oxygen-carrying properties are evidently critically dependent on the precise charge distribution and steric factors within the molecule. Replacement of the Cl in Vaska's compound with I causes loss of oxygen-carrying ability, the oxygenation being irreversible. This can be rationalized by noting that the lower electronegativity of the iodine would allow a greater electron density on the metal, thus facilitating $M \rightarrow O_2$ π donation: this increases the strength of the $M-O_2$ bond and, by placing charge in antibonding orbitals of the O_2, causes an increase in the O–O distance from 130 to 151 pm.

Lower oxidation states

Numerous complexes of Co, Rh, and Ir are known in which the formal oxidation state of the metal is zero, -1, or even lower. Many of these compounds contain CO, CN^-, or RNC as ligands and so are more conveniently discussed under organometallic compounds (Section 26.3.5). However, other ligands such as tertiary phosphines also stabilize the lower oxidation states, as exemplified by the brown, tetrahedral, paramagnetic complex $[Co^0(PMe_3)_4]$: this is made by reducing an ethereal solution of $CoCl_2$ with Mg or Na amalgam in the presence of PMe_3. Further treatment of the product with Mg/thf in the presence of N_2 gives $[Mg(thf)_4][Co^{-II}(N_2)(PMe_3)_4]$. Similar reactions with $P(OMe)_3$ and $P(OEt)_3$ give both paramagnetic monomers $[Co^0\{P(OR)_3\}_4]$, and diamagnetic dimers $[Co_2^0\{P(OR)_3\}_8]$, whereas the more bulky $P(OPr^i)_3$ yields only the orange-red monomeric product. With an excess of sodium amalgam as reducing agent the product with this latter ligand is the white-crystalline $Na[Co^{-I}\{P(OPr^i)_3\}_5]$.[19a] In view of the ready solubility of this compound in pentane and the d^{10} configuration of Co^{-I} it may be that only 4 of the phosphite ligands are directly coordinated to the metal centre: one possible formulation would be

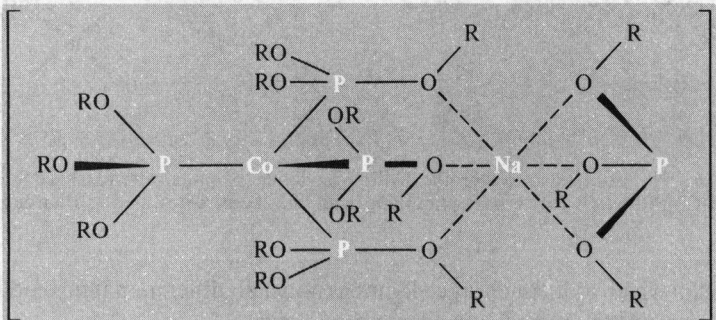

Another technique for obtaining low oxidation states is by electrolytic reduction using cyclic voltametry. Some spectacular series can be achieved of which, perhaps, the most notable is based on $[Ir^{III}(bipy)_3]^{3+}$: this, when dissolved in MeCN, can be oxidized to $[Ir^{IV}(bipy)_3]^{4+}$ and reduced in successive 1-electron steps to give every oxidation state down to $[Ir^{-III}(bipy)_3]^{3-}$, a total of 8 interconnected redox complexes. However, by no means all have been isolated as solid products from solution.[19b] Many other such redox series are known for these and other elements.

[19a] M. C. RAKOWSKI and E. L. MUETTERTIES, Low-valent cobalt triisopropyl phosphite complexes. Characterization of a catalyst for the hydrogenation of α,β-unsaturated ketones, *J. Am. Chem. Soc.* **99**, 739–43 (1977), and references therein.
[19b] J. L. KAHL, K. W. HANCK, and K. DEARMOND, Electrochemistry of iridium–bipyridine complexes, *J. Phys. Chem.* **82**, 540–5 (1978), and references therein.

26.3.4 *The biochemistry of cobalt*[20]

The wasting disease in sheep and cattle known variously as "pine" (Britain), "bush sickness" (New Zealand), "coast disease" (Australia), and "salt sick" (Florida) has been recognized since the late eighteenth century. When it was realized to be an anaemic condition it was thought to be due to iron deficiency and was therefore treated, with mixed success, by administering iron salts. Then, in the 1930s, it was found by workers in Australia and New Zealand that the efficacious principle in the iron treatment was actually an impurity (cobalt) but its role was not understood. This became more evident when vitamin B_{12} was extracted from raw liver and shown to be responsible for the latter's well-known effectiveness in treating pernicious anaemia. It is now known that vitamin B_{12} is a coenzyme† in a number of biochemical processes, the most important of which is the formation of erythrocytes (red blood-cells). It obviously functions extremely effectively, the human body for instance containing a mere 2–5 mg, concentrated in the liver.

The structure of the diamagnetic, yellow-orange vitamin B_{12} is shown in Fig. 26.7 and it can be seen that the coordination sphere of the cobalt has many similarities with that of iron in haem (see Fig. 25.9). In both cases the metal is coordinated to 4 coplanar nitrogen atoms (in this case part of a "corrin" ring which is less symmetrical and not so unsaturated as the porphyrin in haem) with an imidazole nitrogen in the fifth position. A major difference is apparent, however, in the sixth coordination position which, in haemoglobin, is either vacant or occupied by O_2. Here it is filled by a σ-bonded carbon,[21] making vitamin B_{12} the first, and so far the only, naturally occurring organometallic compound. The usual methods of isolation lead to a product known as *cyanocobalamin*, which is the same as vitamin B_{12} itself but with CN^- instead of deoxyadenosine in the sixth coordination position. This is a labile site, and other derivatives such as *aquocobalamin* can be prepared.

Incorporation of cobalt into the corrin ring system modifies the reduction potentials of cobalt giving it three accessible and consecutive oxidation states:

The reductions are effected in nature by ferredoxin (p. 1281). This behaviour can be reproduced surprisingly well by simpler, model compounds. Some of the best known of

---

† Enzymes are proteins which act as very specific catalysts in biological systems. Their activity may depend on the presence of substances, often metal complexes, of much lower molecular weight. These activators are known as "coenzymes".

[20] D. A. PHIPPS, *Metals and Metabolism*, pp. 112–20, Oxford University Press, Oxford, 1976. G. N. SCHRAUZER, New developments in the field of vitamin B_{12}: reactions of the cobalt atom in corrins and in vitamin B_{12} model compounds, *Angew. Chem.*, Int. Edn. (Engl.) **15**, 417–26 (1976). R. S. Young, *Cobalt in Biology and Biochemistry*, Academic Press, London, 1979, 144 pp.
[21] D. C. HODGKIN, The structure of the corrin nucleus from X-ray analysis, *Proc. Roy. Soc.* A **288**, 294–305 (1965).

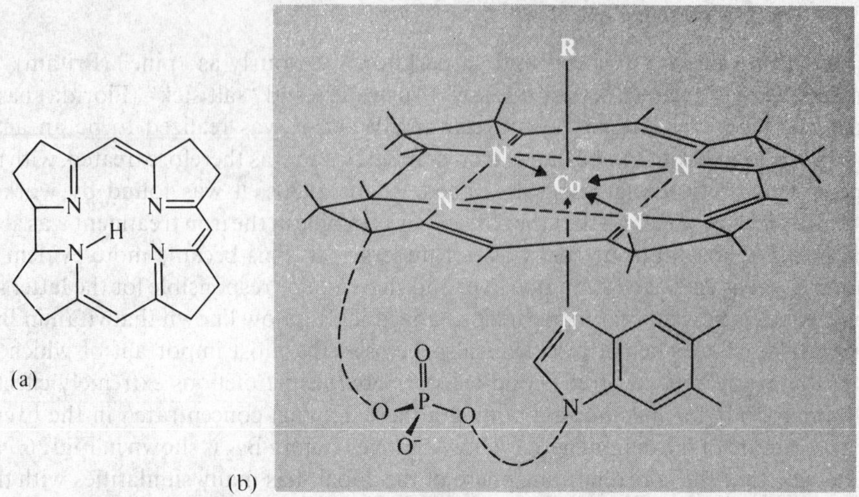

(a)

(b)

FIG. 26.7 Vitamin B_{12}: (a) a *corrin* ring showing a square-planar set of N atoms and a replaceable H, and (b) simplified structure of B_{12}. In view of the H displaced from the corrin ring, the Co–C bond, and the charge on the ribose phosphate, the cobalt is formally in the $+3$ oxidation state. This and related molecules are conveniently represented as:

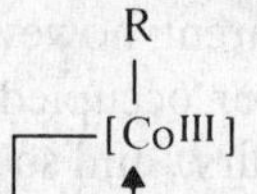

$$R$$
$$|$$
$$[Co^{III}]$$

these are obtained by the addition of axial groups to the square-planar complexes of Co^{II} with Schiff bases, or substituted glyoximes (giving cobaloximes) as illustrated in Fig. 26.8. The reduced Co^{I} species of these, along with vitamin B_{12s}, are amongst the most powerful nucleophiles known (hence, "supernucleophiles"), liberating H_2 from water.

Virtually all the biological processes, in which vitamin B_{12} is active, involve substituent exchange of the type:

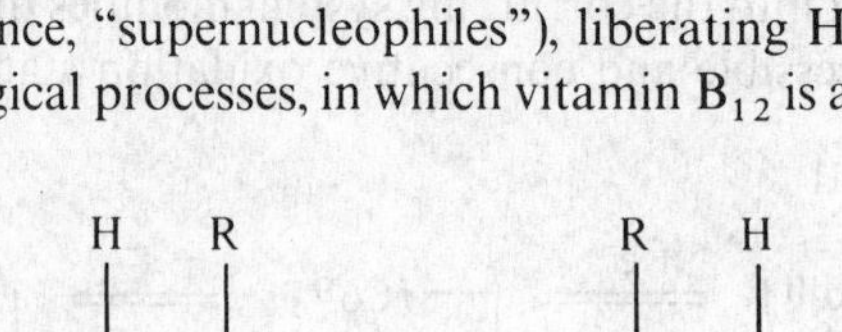

$$\begin{array}{cc} H & R \\ | & | \\ -C-C- \\ | & | \end{array} \rightleftharpoons \begin{array}{cc} R & H \\ | & | \\ -C-C- \\ | & | \end{array}$$

which, significantly, does not involve solvent protons. Though the precise mechanism of these reactions, and in particular the redox processes involved, are not yet settled, it is evident that vitamin B_{12} acts as a hydrogen carrier and the lability of the sixth coordination site is crucial to this. The investigation of model systems[21a] to distinguish between a variety of conceivable mechanisms is a currently very active area of research.

26.3.5 *Organometallic compounds*

Many of the organometallic compounds of the elements of this group show valuable catalytic activity and, as discussed above, much of the chemistry of vitamin B_{12} is the

[21a] R. D. JONES, D. A. SOMMERVILLE, and F. BASOLO, Synthetic oxygen carriers related to biological systems, *Chem. Revs.* **79**, 139–79 (1979).

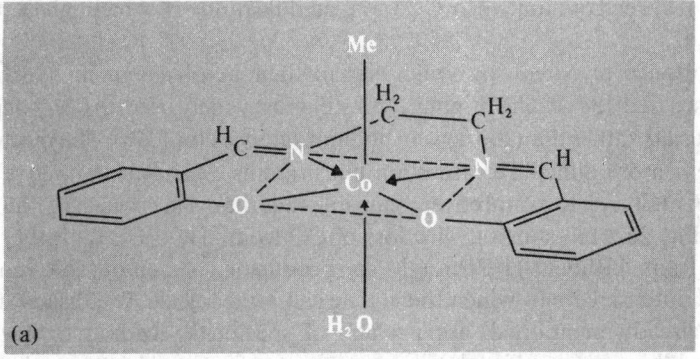

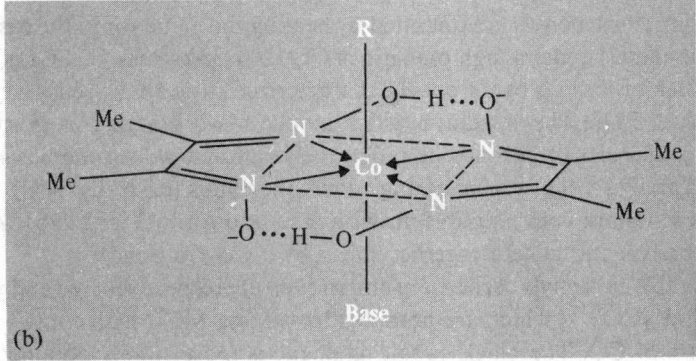

FIG. 26.8   Model vitamin B_{12} compounds: (a) a Schiff base derivative, and (b) a cobaloxime, in
this case derived from dimethylglyoxime.

chemistry of the Co–C σ bond. Simple homoleptic alkyls and aryls of cobalt, $[CoR_x]$, have
not in fact been prepared, but this is evidently not due to thermodynamic instability of the
Co–C bond since compounds containing such bonds can be prepared in abundance, not
only with π-bonding ligands such as phosphines and CO but also with non-π-bonding
ligands such as Schiff bases and glyoximes. These latter presumably owe their existence not
to electronic but rather to steric factors, the additional ligands blocking what might
otherwise be energetically favourable decomposition paths.

Carbonyls[22] (see p. 349)

   Because they possess an odd number of valence electrons the elements of this group can
only satisfy the 18-electron rule in their carbonyls if M–M bonds are present. In accord
with this, mononuclear carbonyls are not formed. Instead $[M_2(CO)_8]$, $[M_4(CO)_{12}]$, and
$[M_6(CO)_{16}]$ are the principal binary carbonyls of these elements. But reduction of
$[Co_2(CO)_8]$ with, for instance, sodium amalgam in benzene yields the monomeric and

22 W. P. GRIFFITH, Carbonyls, cyanides, isocyanides and nitrosyls, Chap. 46 in *Comprehensive Inorganic Chemistry*, Vol. 4, pp. 105–95, Pergamon Press, Oxford, 1973.

tetrahedral, 18-electron ion, $[Co(CO)_4]^-$, acidification of which gives the hydride, $[HCo(CO)_4]$.

The importance of cobalt carbonyls lies in their involvement in hydroformylation reactions discussed above. The original, and still most widely used, process depends on the use of cobalt salts rather than the newer rhodium catalysts (p. 1316). The mechanism of the cobalt cycle is more difficult to ascertain but it seems clear that the active agent is the hydride, $[HCo(CO)_4]$. It is, moreover, plausible that the cycle is basically the same as that outlined in Fig. 26.6 but starting with loss of CO from $[HCo(CO)_4]$ rather than loss of phosphine from $[Rh(CO)H(PPh_3)_3]$, so producing a comparable coordinatively unsaturated intermediate to which the alkene can attach itself. The disadvantages of the system, as already mentioned, are its lack of specificity, leading to branched-chain products, and the necessity of high temperatures ($> 150°C$) and pressure (~ 200 atm). In addition the volatility of $[HCo(CO)_4]$ poses recovery problems.

The dimeric octacarbonyls are obtained by heating the metal (or in the case of iridium, $IrCl_3$ + copper metal) under a high pressure of CO (200–300 atms). $Co_2(CO)_8$ is by far the best known, the other two being poorly characterized; it is an air-sensitive, orange-red solid melting at $51°C$. The structure, which involves two bridging carbonyl groups as shown in Fig. 26.9a), can perhaps be most easily rationalized on the basis of a "bent" Co–Co bond arising from overlap of angled metal orbitals (d^2sp^3 hybrids). However, in solution this structure is in equilibrium with a second form (Fig. 26.9b) which has no bridging carbonyls and is held together solely by a Co–Co bond.

The most stable carbonyls of rhodium and iridium are respectively red and yellow solids of the form $[M_4(CO)_{12}]$ which are obtained by heating MCl_3 with copper metal under about 200 atm of CO. The black cobalt analogue is more simply obtained by heating $[Co_2(CO)_8]$ in an inert atmosphere

$$2[Co_2(CO)_8] \xrightarrow{\ 50°C\ } [Co_4(CO)_{12}] + 4CO$$

The structures are shown in Fig. 26.9c and d and differ in that, whereas the Ir compound consists of a tetrahedron of metal atoms held together solely by M–M bonds, the Rh and Co compounds each incorporate 3 bridging carbonyls. A similar difference was noted in the case of the trinuclear carbonyls of Fe, Ru, and Os (p. 1283) and a similar explanation may be invoked,[23] namely that the Ir is too large for the Ir–Ir bond to support a CO bridge. Of the 3 $[M_6(CO)_{16}]$ carbonyls the very dark-brown Rh compound prepared simultaneously with and separated from $[Rh_4(CO)_{12}]$ is the best known. In the solid its structure consists of an octahedral array of $Rh(CO)_2$ units with the remaining 4 CO's bridging 4 faces of the octahedron (Fig. 26.9e). Theoretical perceptions of such clusters and their derivatives have been developed.[23a]

Prompted not only by their catalytic potential, but also by the intrinsic interest in their structural complexity, the cluster carbonyls of this group (especially of Rh) currently form an active area of research, and anionic species of very high nuclearity have been

[23] B. F. G. JOHNSON, The structures of simple binary carbonyls, *JCS Chem. Comm.* 1976, 211–13.

[23a] R. HOFFMANN, B. E. R. SCHILLING, R. BAU, H. D. KAESZ, and D. M. P. MINGOS, Electronic structures of $[M_4(CO)_{12}H_n]$ and $[M_4(\eta^5\text{-}C_5H_5)_4H_n]$ complexes, *J. Am. Chem. Soc.* **100**, 6088–93 (1978).

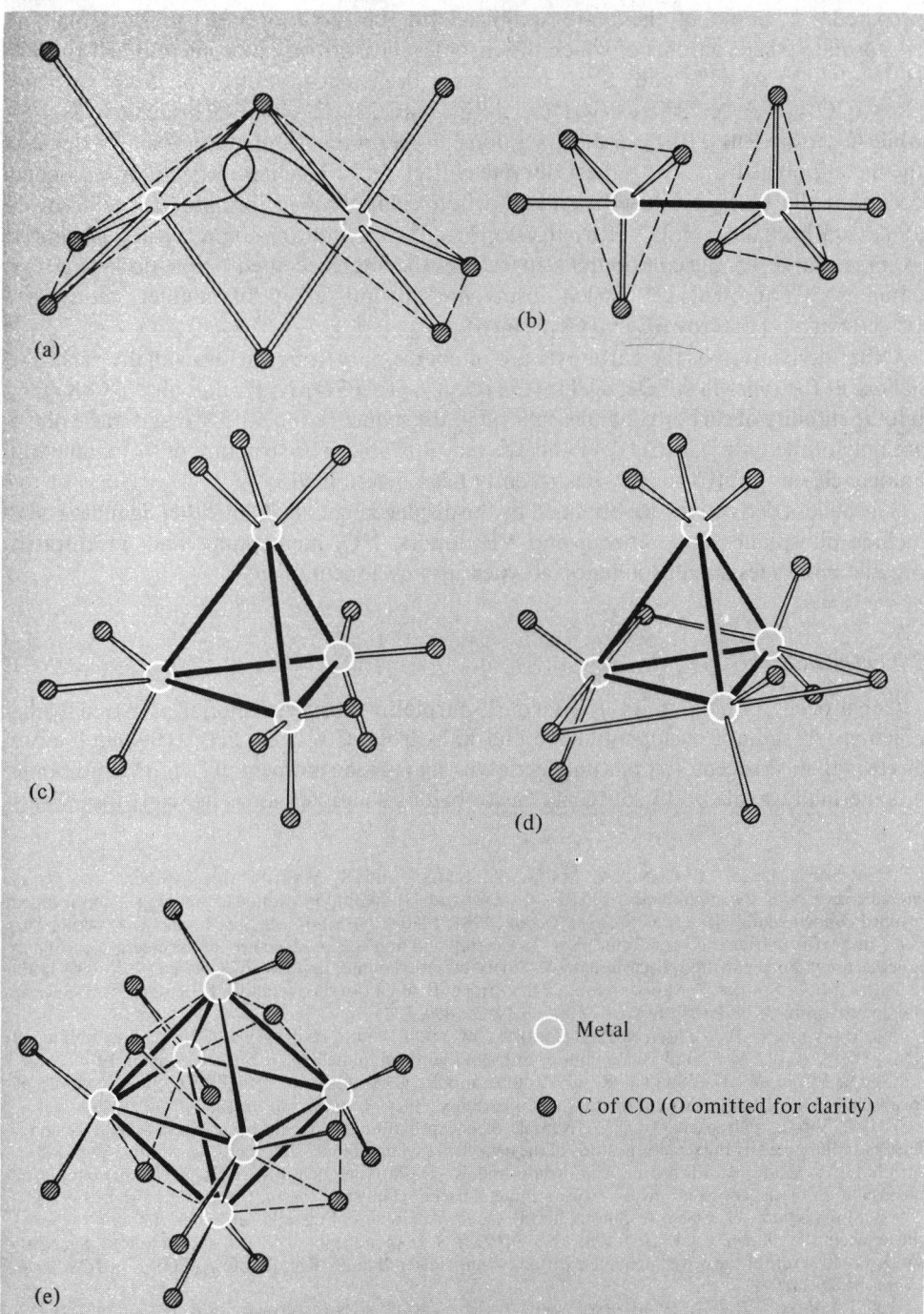

FIG. 26.9　Molecular structures of some binary carbonyls of Co, Rh, and Ir. (a) $Co_2(CO)_8$ in solid state, showing the formation of a "bent" Co–Co bond. (b) $Co_2(CO)_8$ in solution. (c) $Ir_4(CO)_{12}$. (d) $M_4(CO)_{12}$, M = Co, Rh. (e) $Rh_6(CO)_{16}$.

prepared.[23b] Some of these are stabilized by the incorporation of interstitial, or encapsulated, heteroatoms of which the carbides have already been mentioned (p. 354). H,[24] P,[24a] As,[24b] and S[25] have also been encapsulated in ions such as $[Rh_{13}(CO)_{24}H_3]^{2-}$, $[Rh_9(CO)_{21}P]^{2-}$, $[Rh_{10}As(CO)_{22}]^{3-}$, and $[Rh_{17}(CO)_{32}(S)_2]^{3-}$, while in certain Rh_{13}, Rh_{14}, and Rh_{15} anions[26] one of the metal atoms is situated inside the metal polyhedron. The largest discrete cluster yet to be characterized for any metal occurs in the black crown-ether alkali-cation compound of formula $[Cs_9(18\text{-crown-}6)_{14}]^{9+}$ $[Rh_{22}(CO)_{35}H_x]^{5-}$ $[Rh_{22}(CO)_{35}H_{x+1}]^{4-}$. In addition to the 2 Rh_{22} clusters of average charge 4.5, the compound also features the unprecedented "triple-decker" crown complex $[Cs_2(crown)_3]^{2+}$ which forms part of the array of counter cations:[27] $[Cs(crown)]^+{}_2[Cs(crown)_2]^+{}_3[Cs_2(crown)_3]^{2+}{}_2$.

Other derivatives of the carbonyls are of course numerous; Ir forms many carbonyl halides of the types $[Ir^I(CO)_3X]$, $[Ir^I(CO)_2X_2]^-$, $[Ir^{III}(CO)_2X_4]^-$, and $[Ir^{III}(CO)X_5]^{2-}$, but the stability of carbonyl halides falls off in the sequence Ir > Rh > Co and those of Co are only of the type $[Co(CO)_4X]$ and are very unstable. The structure of the octahedral anionic cluster $[Ir_6(CO)_{15}]^{2-}$ has recently been determined.[28]

The bulk of derivatives are obtained by the displacement of CO by other ligands. These include phosphines and other group VB donors, NO, mercaptans, and unsaturated organic molecules such as alkenes, alkynes, and cyclopentadienyls.

Cyclopentadienyls

Cobaltocene, $[Co^{II}(\eta^5\text{-}C_5H_5)_2]$, is a dark-purple air-sensitive material, prepared by the reactions of sodium cyclopentadiene and anhydrous $CoCl_2$ in THF. Having 1 more electron than ferrocene, it is paramagnetic with a magnetic moment of 1.76 BM and, while it is thermally stable up to 250°C, its most obvious characteristic is its ready loss of this

[23b] See for instance, G. CIANI, A. MAGNI, A. SIRONI, and S. MARTINENGO, Synthesis and X-ray characterization of the high-nuclearity $[Rh_{17}(\mu_3\text{-}CO)_3(\mu\text{-}CO)_{15}(CO)_{12}]^{3-}$ anion containing a tetracapped twinned cuboctahedral cluster, *JCS Chem. Comm.* 1981, 1280–2; G. CIANI, A. SIRONI, and S. MARTINENGO, High nuclearity clusters of rhodium. Part 3. Crystal and molecular structure of hexadeca-μ-carbonyl-enneacarbonyl-polyhedro-tetradecarhodate(4−) in its tetraethylammonium salt, *JCS Dalton* 1982, 1099–1102; B. T. HEATON, L. STRONA, S. MARTINENGO, D. STRUMOLO, R. J. GOODFELLOW, and J. H. SADLER, Hexanuclear rhodium hydrido-carbonyl clusters, *JCS Dalton* 1982, 1499–502.

[24] S. MARTINENGO, B. T. HEATON, R. J. GOODFELLOW, and P. CHINI, Hydrogen and carbonyl scrambling in $[Rh_{13}(CO)_{24}H_{5-n}]^{n-}$ ($n=2$ and 3); a unique example of hydrogen tunnelling, *JCS Chem. Comm.* 1977, 39–40.

[24a] J. L. VIDAL, W. E. WALKER, R. L. PRUETT, and R. C. SCHOENING, $[Rh_9P(CO)_{21}]^{2-}$. Example of encapsulation of phosphorus by transition-metal-carbonyl clusters, *Inorg. Chem.* **18**, 129–36 (1979).

[24b] J. L. VIDAL, $[Rh_{10}As(CO)_{22}]^{3-}$. Example of encapsulation of arsenic by transition-metal-carbonyl clusters as illustrated by the structural study of the benzyltriethylammonium salt, *Inorg. Chem.* **20**, 243–9 (1981).

[25] J. L. VIDAL, R. A. FIATO, L. A. CROSBY, and R. L. PRUETT, $[Rh_{17}(S)_2(CO)_{32}]^{3-}$. An example of encapsulation of chalcogen atoms by transition-metal-carbonyl clusters, *Inorg. Chem.* **17**, 2574–82 (1978).

[26] S. MARTINENGO, G. CIANI, A. SIRONI, and P. CHINI, Analogues of metallic lattices in rhodium-carbonyl cluster chemistry, *J. Am. Chem. Soc.* **100**, 7096–8 (1978). S. MARTINENGO, G. CIANI, and A. SIRONI, Synthesis and X-ray structure of the novel high nuclearity rhodium-cluster dianion $[Rh_{14}(\mu\text{-}CO)_{15}(CO)_{11}]^{2-}$, *JCS Chem. Comm.* 1980, 1140–1.

[27] J. L. VIDAL, R. C. SCHOENING, and J. M. TROUP, $[Cs_9(18\text{-crown-6})_{14}]^{9+}[Rh_{22}(CO)_{35}H_x]^{5-}$–$[Rh_{22}(CO)_{35}H_{x+1}]^{4-}$. Synthesis, structure, and reactivity of a rhodium-carbonyl cluster with a body-centred cubic arrangement of metal atoms, *Inorg. Chem.* **20**, 227–8 (1981).

[28] F. DEMARTIN, M. MANASSERO, M. SANSONI, L. GARLASCHELLI, S. MARTINENGO, and F. CANZIANI, A new synthesis and the X-ray crystal structure of bis(trimethylbenzylammonium) tri-μ-carbonyl-dodecacarbonyl-*octahedro*-hexairidate(2-), *JCS Chem. Comm.* 1980, 903–4.

electron to form the yellow-green cobalticenium ion, $[Co^{III}(\eta^5\text{-}C_5H_5)_2]^+$. This resists further oxidation, being stable even in conc HNO_3 but, like the isoelectronic ferrocene, is susceptible to nucleophilic attack on its rings.

Rhodocene, $[Rh(\eta^5\text{-}C_5H_5)_2]$, is also known but is unstable to oxidation and has a tendency to form dimeric species. Claims for the existence of iridocene probably refer to Ir^{III} complexes. However, the yellow rhodicenium and iridicenium cations are certainly known and are entirely analogous to the cobalticenium cation in their resistance to oxidation and susceptibility to nucleophilic attack.

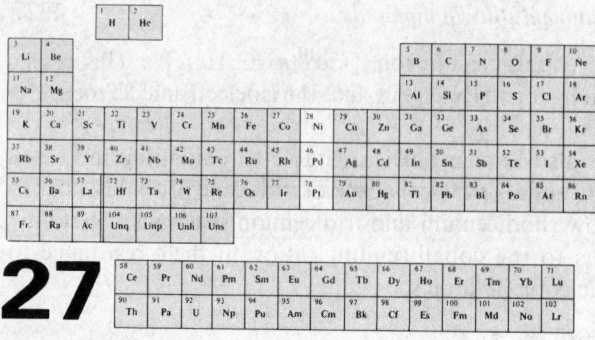

27

Nickel, Palladium, and Platinum

27.1 Introduction

An alloy of nickel was known in China over 2000 years ago, and Saxon miners were familiar with the reddish-coloured ore, NiAs, which superficially resembles Cu_2O. These miners attributed their inability to extract copper from this source to the work of the devil and named the ore "Kupfernickel" (Old Nick's copper). In 1751 A. F. Cronstedt isolated an impure metal from some Swedish ores and, identifying it with the metallic component of Kupfernickel, named the new metal "nickel". In 1804 J. B. Richter produced a much purer sample and so was able to determine its physical properties more accurately.

Impure, native platinum seems to have been used unwittingly by ancient Egyptian craftsmen in place of silver, and was certainly used to make small items of jewellery by the Indians of Ecuador before the Spanish conquest. The introduction of the metal to Europe is a complex and intriguing story.[1] In 1736 A. de Ulloa, a Spanish astronomer and naval officer, observed an unworkable metal, *platina* (Spanish, little silver), in the gold mines of what is now Colombia. Returning home in 1745 his ship was attacked by privateers and finally captured by the British navy. He was brought to London and his papers confiscated, but was fortunately befriended by members of the Royal Society and was indeed elected to that body in 1746 when his papers were returned. Meanwhile, in 1741, C. Wood brought to England the first samples of the metal and, following the eventual publication of de Ulloa's report in 1748, investigation of its properties began in England and Sweden. It became known as "white gold" (this term is now used to describe an Au/Pd alloy) and the "eighth metal" (the seven metals Au, Ag, Hg, Cu, Fe, Sn, and Pb having been known since ancient times), but great difficulty was experienced in working it because of its high mp and brittle nature (due to impurities of Fe and Cu). Powder metallurgical techniques of fabrication were developed† in great secrecy in Spain by the Frenchman P. F. Chabeneau, and subsequently in London by W. H. Wollaston,[2] who in the years 1800–21 produced well over 1 tonne of malleable platinum. These techniques were developed because the chemical methods used to isolate the metal produced an easily

---

† Precedence must in fact be given to the South American Indians to whom platinum was available only in the form of fine, hand-separated grains which must have been fabricated by ingenious, if crude, powder metallurgy.

[1] L. B. HUNT, Swedish contributions to the discovery of platinum, *Platinum Metals Rev.* **24**, 31–39 (1980).
[2] J. C. CHASTON, The powder metallurgy of platinum, *Platinum Metals Rev.* **24**, 70–79 (1980).

powdered spongy precipitate. Not until the availability, half a century later, of furnaces capable of sustaining sufficiently high temperatures was easily workable, fused platinum commercially available.

In 1803, in the course of his study of platinum, Wollaston isolated and identified palladium from the mother liquor remaining after platinum had been precipitated as $(NH_4)_2PtCl_6$ from its solution in aqua regia. He named it after the newly discovered asteroid, Pallas, itself named after the Greek goddess of wisdom (παλλάδιον, palladion, of Pallas).

27.2 The Elements

27.2.1 *Terrestrial abundance and distribution*

Nickel is the seventh most abundant transition metal and the twenty-second most abundant element in the earth's crust (99 ppm). Its commercially important ores are of two types:

(1) *Laterites*, which are oxide/silicate ores such as garnierite, $(Ni,Mg)_6Si_4O_{10}(OH)_8$, and nickeliferous limonite, $(Fe,Ni)O(OH).nH_2O$, which have been concentrated by weathering in tropical rainbelt areas such as New Caledonia, Cuba, and Queensland.
(2) *Sulfides* such as pentlandite, $(Ni,Fe)_9S_8$, associated with copper, cobalt, and precious metals so that the ores typically contain about $1\frac{1}{2}\%$ Ni. These are found in more temperate regions such as Canada, the USSR, and South Africa.

Arsenide ores such as niccolite (Kupfernickel (NiAs), smaltite ($(Ni,Co,Fe)As_2$), and nickel glance (NiAsS) are no longer of importance.

The most important single deposit of nickel is at Sudbury Basin, Canada. It was discovered in 1883 during the building of the Canadian Pacific Railway and consists of sulfide outcrops situated around the rim of a huge basin 17 miles wide and 37 miles long (possibly a meteoritic crater). Fifteen elements are currently extracted from this region (Ni, Cu, Co, Fe, S, Te, Se, Au, Ag, and the six platinum metals).

Although estimates of their abundances vary considerably, Pd and Pt (approximately 0.015 and 0.01 ppm respectively) are much rarer than Ni. They are generally associated with the other platinum metals and occur either native in placer (i.e. alluvial) deposits or as sulfides or arsenides in Ni, Cu, and Fe sulfide ores. Until the 1820s all platinum metals came from South America, but in 1819 the first of a series of rich placer deposits which were to make Russia the chief source of the metals for the next century, was discovered in the Urals. Since the 1920s however, the copper–nickel ores of Canada and South Africa, in which the platinum metals occur as sulfides and arsenides as well as in the metallic state, have become major sources, supplemented more recently by similar deposits in the USSR.

27.2.2 *Preparation and uses of the elements*[3]

Production methods for all three elements are complicated and dependent on the particular ore involved; they will therefore only be sketched in outline. In the case of nickel

---

[3] *Kirk–Othmer Encyclopedia of Chemical Technology*, 3rd edn., Vol. 15, pp. 787–801, Interscience, New York, 1981, and Vol. 18, pp. 228–253, loc. cit., 1982.

the oxide ores are not generally amenable to concentration by normal physical separations and so the whole ore has to be treated. By contrast the sulfide ores can be concentrated by flotation and magnetic separations, and for this reason provide the major part of the world's nickel, though the use of laterite ores is increasing.

Over a quarter of the world's nickel comes from Sudbury and there silica is added to the nickel/copper concentrates which are then subjected to a series of roasting and smelting operations which reduce the sulfide and iron contents by converting the iron sulfide first to the oxide and then to the silicate which is removed as a slag. The resulting "matte" of nickel and copper sulfides is allowed to cool over a period of days, when Ni_3S_2, Cu_2S, and Ni/Cu metal† form distinct phases which can be mechanically separated. (In the older, Orford, process the matte was heated with $NaHSO_4$ and coke, producing molten Na_2S which dissolved the copper sulfide and formed an upper layer, leaving the nickel sulfide below; on solidification the silvery upper layer was cut from the black lower layer—hence the process was commonly called the "tops and bottoms" process.) Roasting the nickel sulfide converts it to the oxide which is then either used directly for steelmaking or is converted to metallic nickel and refined. The latter can be achieved electrolytically or by the Mond process.

In the electrolytic process the oxide is first reduced with carbon and the impure metal cast into anodes. Using aqueous $NiSO_4$ and $NiCl_2$ as electrolyte with pure nickel sheet as cathode, electrolysis then dissolves nickel from the anode and deposits it, in a state of 99.9% purity, on the cathode.

Alternatively, in the process developed in 1899 by L. Mond and operated primarily at Clydach in Wales, the heated oxide is first reduced by the hydrogen in water gas ($H_2 + CO$). At atmospheric pressure and a temperature around $50°C$, the impure nickel is then reacted with the residual CO to give the volatile $Ni(CO)_4$. This is passed over nucleating pellets of pure nickel at a temperature of $230°C$ when it decomposes, depositing nickel of 99.95% purity and leaving CO to be recycled.

$$Ni + 4CO \underset{230\,C}{\overset{50\,C}{\rightleftharpoons}} Ni(CO)_4$$

Somewhat higher pressures and temperatures (e.g. 20 atm and $150°C$) are used to form the carbonyl in new Canadian plant, but the essential principle of the Mond process is retained.

Total world production of nickel is in the region of 750 000 tonnes pa of which (1980) 26% comes from Canada, 19% from the USSR, 11% from New Caledonia, and 8% from Australia. The bulk of this is used in the production of alloys both ferrous and non-ferrous. In 1889 J. Riley of Glasgow published a report on the effect of adding nickel to steel. This was noticed by the US Navy who initiated the use of nickel steels in armour plating. Stainless steels contain up to 8% Ni and the use of "Alnico" steel for permanent magnets has already been mentioned (p. 1291).

The non-ferrous alloys include the misleadingly named nickel silver (or German silver) which contains 10–30% Ni, 55–65% Cu, and the rest Zn; when electroplated with silver (*electroplated nickel silver*) it is familiar as EPNS tableware. *Monel* (68% Ni, 32% Cu, traces of Mn and Fe) is used in apparatus for handling corrosive materials such as F_2; cupro-nickels (up to 80% Cu) are used for "silver" coinage; *Nichrome* (60% Ni, 40% Cr),

---

† This metallic phase is worked for precious metals which are preferentially dissolved in it.

which has a very small temperature coefficient of electrical resistance, and *Invar*, which has a very small coefficient of expansion are other well-known Ni alloys. Electroplated nickel is an ideal undercoat for electroplated chromium, and smaller amounts of nickel are used as catalysts in the hydrogenation of unsaturated vegetable oils and in storage batteries such as the Ni/Fe batteries.

Ninety-eight per cent of the world's supply of platinum metals comes from three countries—the Republic of South Africa, (45%), the USSR (45%), and Canada (8%). Because of the high proportion of Pt itself in this deposit, no less than 65% of the world's Pt comes from the South African Merensky Reef alone, which is mined primarily for the platinum metals with copper and nickel as byproducts. In the other coutries it is the platinum metals which are byproducts. In all cases the concentration of platinum metals in the original ore is a mere fraction of a gramme per tonne, so that millions of tonnes of ore must be mined, milled, and smelted each year. Precious metal concentrates are obtained either from the metallic phase of the sulfide matte (see above) or as anode slimes in the electrolytic refinement of the baser metals. Platinum and palladium are then obtained by a composite process in which Ag, Au, and all six platinum metals are separated (Fig. 27.1; see also Figs. 25.1 and 26.1).

Current annual world production of all platinum metals is around 200 tonnes of which perhaps 90 tonnes is platinum and rather less is palladium. Both metals find their most extensive uses as catalysts: Pd for hydrogenation and dehydrogenation reactions, Pt in a wide variety of processes. These include the oxidation of ammonia for the production of nitric acid (p. 537), petroleum reforming, and, more recently, the oxidation of noxious organic vapours in the control of car-exhaust emissions. In addition, platinum is used in the chemical and glass industries as cladding to prevent attack by particularly corrosive materials such as hot hydrofluoric acid or molten glass; electrically for electrodes, contacts, and resistance wires; and in the manufacture of jewellery.†

27.2.3 *Properties of the elements*

Table 27.1 lists some of the important atomic and physical properties of these three elements. Being composed of several naturally occurring isotopes, their atomic weights cannot be quoted with great precision though the value for Pd was improved substantially in 1979. Difficulties in attaining high purities have also frequently led to disparate values for some physical properties, while mechanical history has considerable effect on such properties as hardness. The metals are silvery-white and lustrous, and are both malleable and ductile so that they are readily worked. They are also readily obtained in finely divided forms which are catalytically very active. *Platinum black*, for instance, is a velvety-black powder obtained by adding ethanol to a solution of $PtCl_2$ in aqueous KOH and warming. Another property of platinum which has led to numerous laboratory applications is its coefficient of expansion which is virtually the same as that of soda glass into which it can therefore be fused to give a permanent seal.

Like Rh and Ir, all three members of this triad have the fcc structure predicted by band

---

† It should not be overlooked that platinum has played a crucial role in the development of many branches of science even though the amounts of metal involved may have been small. Reliable Pt crucibles were vital in classical analysis on which the foundations of chemistry were laid. It was also widely used in the development of the electric telegraph, incandescent lamps, and thermionic valves.

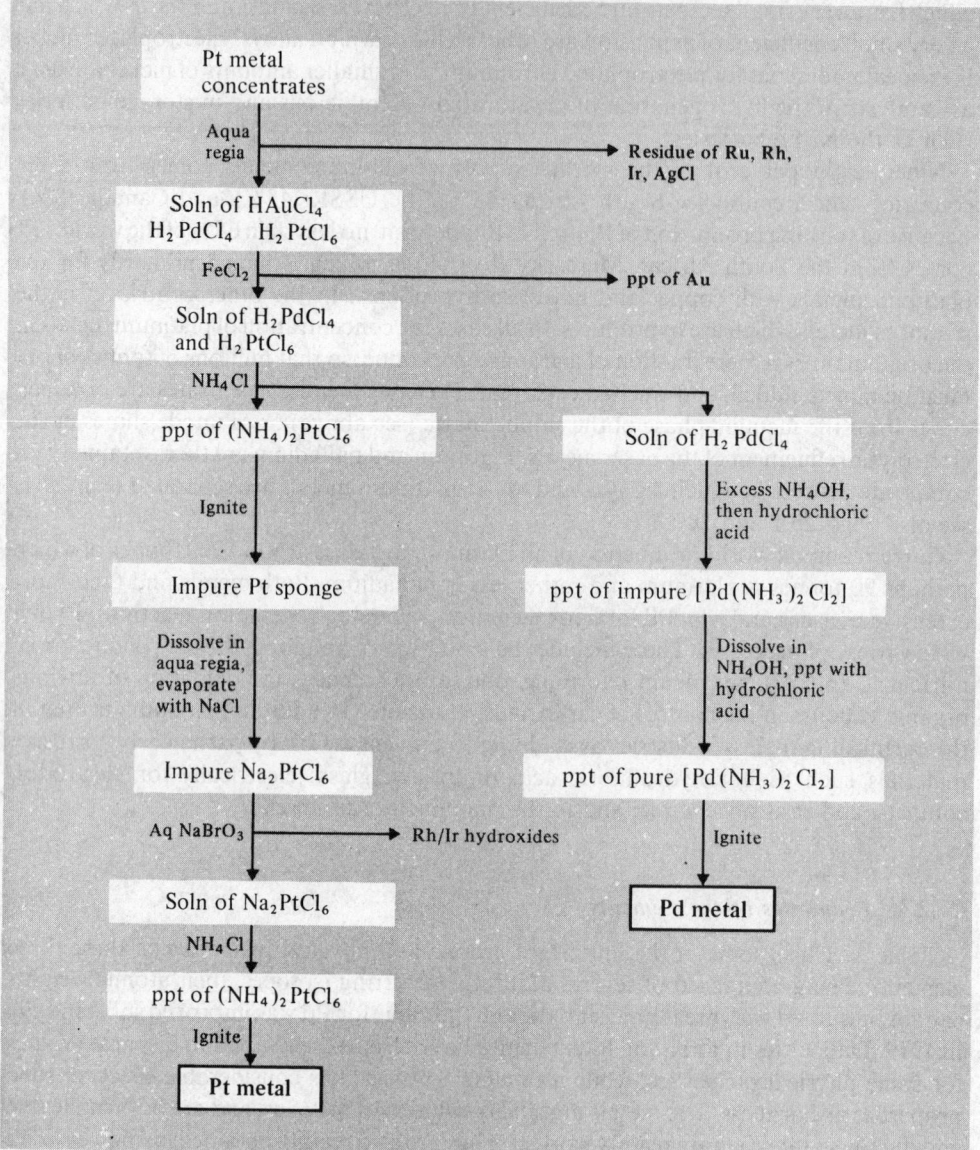

FIG. 27.1   Flow diagram for the extraction of palladium and platinum.

theory calculations for elements with nearly filled d shells. Also in this region of the periodic table, densities and mps are decreasing with increase in Z across the table: thus, although by comparison with the generality of members of the d block these elements are in each case to be considered as dense refractory metals, they are somewhat less so than their immediate predecessors, and palladium has the lowest density and melting point of any platinum metal.

Nickel is ferromagnetic, but less markedly so that either iron or cobalt and its Curie point (375°C) is also lower.

TABLE 27.1 *Some properties of the elements, nickel, palladium, and platinum*

Property		Ni	Pd	Pt
Atomic number		28	46	78
Number of naturally occurring isotopes		5	6	6[a]
Atomic weight		58.69	106.42	195.08($\pm$0.03)
Electronic configuration		[Ar]$3d^84s^2$	[Kr]$4d^{10}$	[Xe]$4f^{14}5d^96s^1$
Electronegativity		1.8	2.2	2.2
Metal radius (12-coordinate)/pm		124	137	138.5
Effective ionic radius (6-coordinate)/pm	V	—	—	57
	IV	48	61.5	62.5
	III	56 (ls), 60 (hs)	76	—
	II	69	86	80
MP/°C		1455	1552	1769
BP/°C		2920	2940	4170
ΔH_{fus}/kJ mol^{-1}		17.2($\pm$0.3)	17.6($\pm$2.1)	19.7($\pm$2.1)
ΔH_{vap}/kJ mol^{-1}		375($\pm$17)	362($\pm$11)	469($\pm$25)
ΔH_f (monatomic gas)/kJ mol^{-1}		429($\pm$13)	377($\pm$3)	545($\pm$21)
Density (20°C)/g cm^{-3}		8.908	11.99	21.41
Electrical resistivity (20°C)/μohm cm		6.84	9.93	9.85

[a] All have zero nuclear spin except ^{195}Pt (33.8% abundance) which has a nuclear spin quantum number $\frac{1}{2}$: this isotope finds much used in nmr spectroscopy both via direct observation of the ^{195}Pt resonance and even more by the observation of ^{195}Pt "satellites". Thus, a given nucleus coupled to ^{195}Pt will be split into a doublet symmetrically placed about the central unsplit resonance arising from those species containing any of the other 5 isotopes of Pt. The relative intensity of the three resonances will be ($\frac{1}{2} \times 33.8$):66.2:($\frac{1}{2} \times 33.8$), i.e. 1:4:1.

27.2.4 *Chemical reactivity and trends*

In the massive state none of these elements is particularly reactive and they are indeed very resistant to atmospheric corrosion at normal temperatures. However, nickel tarnishes when heated in air and is actually pyrophoric if very finely divided (finely divided Ni catalysts should therefore be handled with care). Palladium will also form a film of oxide if heated in air.

Nickel reacts on heating with B, Si, P, S, and the halogens, though more slowly with F_2 than most metals do. It is oxidized at red heat by steam, and will dissolve in dilute mineral acids: slowly in most but quite rapidly in dil HNO_3. Conc HNO_3, on the other hand, renders it passive and dry hydrogen halides have little effect. It has a notable resistance to attack by aqueous caustic alkalis and therefore finds used in apparatus for producing NaOH.

Palladium is oxidized by O_2, F_2, and Cl_2 at red heat and dissolves slowly in oxidizing acids. Platinum is generally more resistant to attack than Pd and is, for instance, barely affected by mineral acids except aqua regia. However, both metals dissolve in fused alkali metal oxides and peroxides. It is also wise to avoid heating compounds containing B, Si, Pb, P, As, Sb, or Bi in platinum crucibles under reducing conditions (e.g. the blue flame of a bunsen burner) since these elements form low-melting eutectics with Pt which cause the metal to collapse. All three elements absorb molecular hydrogen to an extent which depends on their physical state, but palladium does so to an extent which is unequalled by any other metal (section 27.3.1).

A list of typical compounds of these elements is given in Table 27.2 and it is noticeable that the reduction in the range of oxidation states compared to that in previous groups is

TABLE 27.2 *Oxidation states and stereochemistries of compounds of nickel, palladium, and platinum*

Oxidation state	Coordination number	Stereochemistry	Ni	Pd/Pt
−1	4	?	$[Ni_2(CO)_6]^{2-}$	
0 (d^{10})	3	Planar	$[Ni\{P(OC_6H_4\text{-}2\text{-}Me\}_3]$	$[M(PPh_3)_3]$
	4	Tetrahedral	$[Ni(CO)_4]$	$[M(PF_3)_4]$
1 (d^9)	4	Tetrahedral	$[NiBr(PPh_3)_3]$	
2 (d^8)	4	Tetrahedral	$[NiCl_4]^{2-}$	
		Square planar	$[Ni(CN)_4]^{2-}$	$[MCl_4]^{2-}$
	5	Trigonal bipyramidal	$[Ni(PPhMe_2)_3(CN)_2]$	$[M(QAS)I]^{+\,(b)}$
		Square pyramidal	$[Ni(CN)_5]^{3-}$	$[Pd(TPAS)Cl]^{+\,(c)}$
	6	Octahedral	$[Ni(H_2O)_6]^{2+}$	$[Pd(diars)_2I_2]$
		Trigonal prismatic	NiAs	
	7	Pentagonal bipyramidal	$[Ni(DAPBH)_2(H_2O)_2]^{2+\,(a)}$	
3 (d^7)	5	Trigonal bipyramidal	$[NiBr_3(PEt_3)_2]$	
	6	Octahedral	$[NiF_6]^{3-}$	
4 (d^6)	6	Octahedral	$[NiF_6]^{2-}$	$[MCl_6]^{2-}$
	8	—		$[Pt(\eta^5\text{-}C_5H_5)Me_3]$
5 (d^5)	6	Octahedral	—	$[PtF_6]^{-}$
6 (d^4)	6	Octahedral	—	PtF_6

(a) DAPBH, 2,6-diacetylpyridinebis(benzoic acid hydrazone) (see G. J. Palenik *et al.*, *Inorg. Chem.* **18**, 2445 (1979).

(b) QAS, tris-(2-diphenylarsinophenyl)arsine, $As(C_6H_4\text{-}2\text{-}AsPh_2)_3$.

(c) TPAS, 1,2-phenylenebis{(2-dimethylarsinophenyl)-methylarsine}.

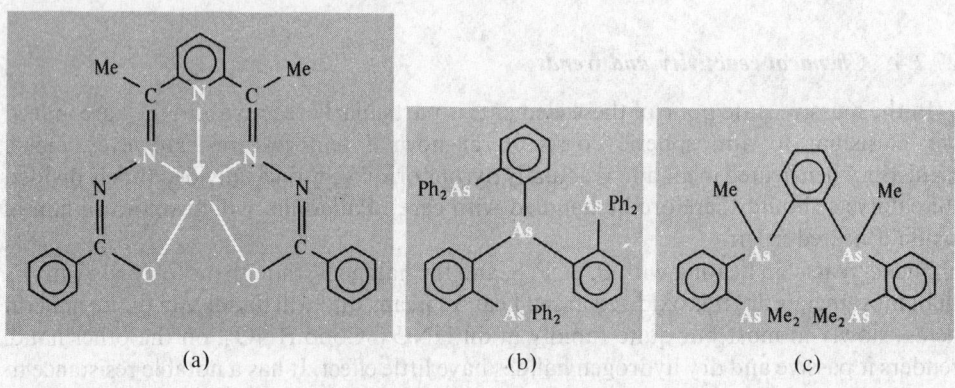

(a)   (b)   (c)

continuing and differences between the two heavier elements are becoming increasingly evident. The maximum oxidation state is +6 but this is attained only by the heaviest element, platinum, in PtF_6; nickel and palladium only reach +4. At the other extreme, palladium and platinum provide no oxidation state below zero. The changes down the triad implied by these facts are also evidenced by those oxidation states which are the most stable for each element. For nickel, +2 is undoubtedly the most common and provides that element's most extensive aqueous chemistry. For palladium, +2 is again the most common, and $[Pd(H_2O)_4]^{2+}$ occurs in aqueous solutions from which potential ligands are excluded. For platinum, however, both +2 and +4 are prolific and form a vital part of early as well as more recent coordination chemistry.

Table 27.2 also reveals the reluctance of these elements to form compounds with high coordination numbers, a coordination number of 6 being rarely exceeded. In the divalent state nickel exhibits a wide and interesting variety of coordination numbers and stereochemistries which often exist simultaneously in equilibrium with each other, whereas palladium and platinum have a strong preference for the square planar geometry. The kinetic inertness of Pt^{II} complexes has led to their extensive use in studies of geometrical isomerism and reaction mechanisms. As will be seen presently, these differences between the lightest and heaviest members of the triad can be largely rationalized by reference to their CFSEs.

Also in the divalent state, Pd and Pt show the class-b characteristic of preferring CN$^-$ and ligands with nitrogen or heavy donor atoms rather than oxygen or fluorine (Pt forms no aquo ion). Platinum(IV) by contrast is more nearly class-a in character and is frequently reduced to Pt^{II} by P- and As-donor ligands. The organometallic chemistry of these metals is rich and varied and that involving unsaturated hydrocarbons is the most familiar of its type.

27.3 Compounds of Nickel, Palladium and Platinum[4, 5]

Such binary borides (p. 162), carbides (p. 318), and nitrides (p. 479) as are formed have been referred to already. The ability of the metals to absorb molecular hydrogen has also been alluded to above. While the existence of definite hydrides of nickel and platinum is in doubt the existence of definite palladium hydride phases is not.

27.3.1 *The Pd/H₂ system*[6]

The absorption of molecular hydrogen by metallic palladium has been the subject of theoretical and practical interest ever since 1866 when T. Graham reported that, on being cooled from red heat, Pd can absorb (or "occlude" as he called it) up to 935 times its own volume of H$_2$.† The gas is given off again on heating and this provides a convenient means of weighing H$_2$—a fact utilized by E. W. Morley in his classic work on the composition of water (1895).

As hydrogen is absorbed, the metallic conductivity falls until the material becomes a semiconductor at a composition of about PdH$_{0.5}$. Palladium is unique in that it does not lose its ductility until large amounts of H$_2$ have been absorbed. The hydrogen is first chemisorbed at the surface of the metal but at increased pressures it enters the metal lattice and the so-called α- and β-phase hydrides are formed (Fig. 27.2). The basic lattice structure is not altered but, whereas the α-phase causes only a slight expansion, the β-phase causes an expansion of up to 10% by volume. The precise nature of the metal–hydrogen interaction is still unclear but the hydrogen has a high mobility within the lattice and diffuses rapidly through the metal. This process is highly specific to H$_2$ and D$_2$, palladium

† This approximates to a composition of Pd$_4$H$_3$ and represents a concentration of hydrogen approaching that in liquid hydrogen!

[4] D. NICHOLLS, Nickel, Chap. 42 in *Comprehensive Inorganic Chemistry*, Vol. 3, pp. 1109–61, Pergamon Press, Oxford, 1973.
[5] S. E. LIVINGSTONE, The second- and third-row transition elements of Group VIIIA, B, and C, in *Comprehensive Inorganic Chemistry*, Vol. 3, pp. 1163–1370, Pergamon Press, Oxford, 1973.
[6] F. A. LEWIS, *The Palladium Hydrogen System*, Academic Press, London, 1967, 178 pp.

being virtually impervious to all other gases, even He, a fact which is utilized in the separation of hydrogen from mixed gases. Industrial installations with outputs of up to 9 million ft³/day (255 million litres/day) are operated and it is of great importance in these that formation of the β-phase hydride is avoided, since the gross distortions and hardening which accompany it may result in splitting of the diffusion membrane. This can be done by maintaining the temperature above 300°C (Fig. 27.2), or alternatively by alloying the Pd with about 20% Ag which has the additional advantage of actually increasing the permeability of the Pd to hydrogen (p. 45).

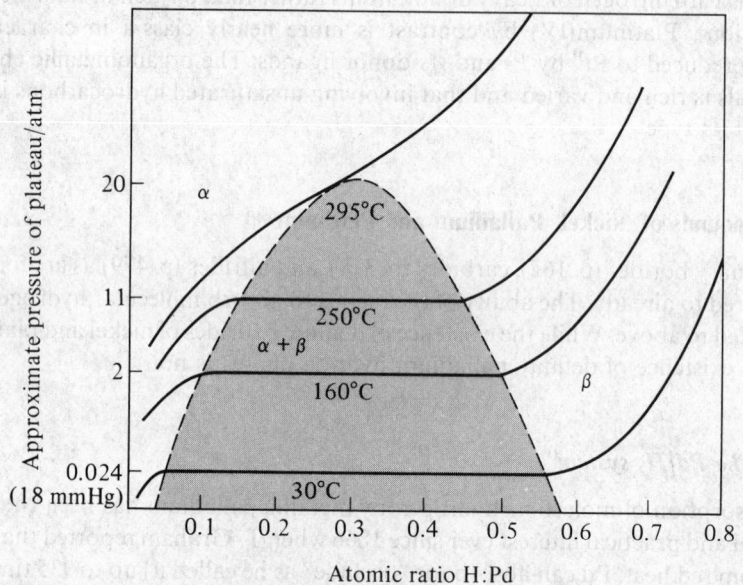

FIG. 27.2   Pressure–concentration isotherms for the Pd/H₂ system: the biphasic region (in which the α- and β-phases coexist) is shaded. (From A. G. Knapton, *Plat. Met. Revs.* **21**, 44 (1977).)

27.3.2   *Oxides*[7] *and chalcogenides*[7a]

The elements of this group form only one reasonably well-characterized oxide each, namely NiO, PdO, and PtO₂, although claims for the existence of many others have been made. Formation of NiO by heating the metal in oxygen is difficult to achieve and incomplete conversion may well account for some of the claims for other nickel oxides, while grey to black colours probably arise from slight nonstoichiometry. It is best prepared as a green powder with the rock-salt structure (p. 273) by heating the hydroxide, carbonate, or nitrate. Ni(OH)₂ is a green precipitate obtained by adding alkali to aqueous solutions of Ni^{II} salts and, like NiO, is entirely basic, dissolving easily in acids.

Black PdO can be produced by heating the metal in oxygen but dissociates above about

---

[7] C. N. R RAO and G. V. S. RAO, *Transition Metal Oxides*, National Standard Reference Data System NSRDS-NBS49, Washington, 1964, 130 pp.

[7a] F. HULLINGER, Crystal chemistry of the chalcogenides and pnictides of the transition elements, *Struct. Bonding* **4**, 83–229 (1968).

900°C. It is insoluble in acids. However, addition of alkali to aqueous solutions of $Pd(NO_3)_2$ produces a gelatinous dark-yellow precipitate of the hydrous oxide which is soluble in acids but cannot be fully dehydrated without loss of oxygen. No other palladium oxide has been characterized although the addition of alkali to aqueous solutions of Pd^{IV} produces a strongly oxidizing, dark-red, precipitate which slowly looses oxygen and, at 200°C, forms PdO.

Addition of alkali to aqueous solutions of Pt^{II} produces a black precipitate which is rapidly oxidized by air and has been formulated as $PtO_x.H_2O$, but it is too unstable to be properly characterized. The stable oxide of platinum is found, instead, in the higher oxidation state. Addition of alkali to aqueous solutions of $PtCl_4$ yields a yellow amphoteric precipitate of the hydrated dioxide which redissolves on being boiled with an excess of strong alkali to give solutions of $[Pt(OH)_6]^{2-}$; it also dissolves in acids. Dehydration by heating produces almost black PtO_2 but this decomposes to the elements above 650°C and cannot be completely dehydrated without some loss of oxygen.

Nickel sulfides are very similar to those of cobalt, consisting of NiS_2 (pyrites structure, p. 804), Ni_3S_4 (spinel structure, p. 279), and the black, nickel-deficient $Ni_{1-x}S$ (NiAs structure, p. 648), which is precipitated from aqueous solutions of Ni^{II} by passing H_2S. There are also numerous metallic phases having compositions between NiS and Ni_3S_2.

Palladium and platinum both form a mono- and a di-sulfide. Brown PdS and black PtS_2 are obtained when H_2S is passed through aqueous solutions of Pd^{II} and Pt^{IV} respectively. Grey PdS_2 and green PtS are best obtained by respectively heating PdS with excess S and by heating $PtCl_2$, Na_2CO_3. and S. The complex crystal chemistry and electrical (and magnetic) properties of these phases and the many selenides and tellurides of Ni, Pd, and Pt have been extensively reviewed.[7a]

27.3.3 *Halides*[8]

The known halides of this group are listed in Table 27.3. This list differs from that of the halides of Co, Rh, and Ir (Table 26.3) most obviously in that the +2 rather than the +3 oxidation state is now well represented for the heavier elements as well as for the lightest. The only hexa- and penta-halides are the dark-red PtF_6 and $(PtF_5)_4$ which are both obtained by controlled heating of Pt and F_2. The former is a volatile solid and, after RhF_6, is the least-stable platinum-metal hexafluoride and one of the strongest oxidizing agents known, oxidizing both O_2 (to $O_2^+[PtF_6]^-$) and Xe (to $XePtF_6$) (p. 1047). The pentafluoride is also very reactive and has the same tetrameric structure as the pentafluorides of Ru, Os, Rh, and Ir (Fig. 25.4). It readily disproportionates into the hexa- and tetra-fluorides.

Platinum alone forms all 4 tetrahalides and these vary in colour from the light-brown PtF_4 to the very dark-brown PtI_4. PtF_4 is obtained by the action of BrF_3 on $PtCl_2$ at 200°C and is violently hydrolysed by water. The others are obtained directly from the elements, the chloride being recrystallizable from water but the bromide and iodide being more soluble in alcohol and in ether. The only other tetrahalide is the red PdF_4 which is similar to its platinum analogue.

[8] R. COLTON and J. H. CANTERFORD, *Halides of the First Row Transition Metals*, Chap. 8, pp. 406–84, Wiley, London, 1969; and *Halides of the Second and Third Row Transition Metals*, Chap. 10, pp. 358–90, Wiley, London, 1968. See also *Transit. Met. Chem.*, **1**, 41–7 (1975).

TABLE 27.3   *Halides of nickel, palladium, and platinum (mp/°C)*

Oxidation State	Fluorides	Chlorides	Bromides	Iodides
+6	PtF_6 [(a)] dark red (61.3°)			
+5	$[PtF_5]_4$ deep red (80°)			
+4	PdF_4 brick-red			
	PtF_4 yellow-brown (600°)	$PtCl_4$ red-brown (d 370°)	$PtBr_4$ brown-black (d 180°)	PtI_4 brown-black (d 130°)
+3		$PtCl_3$ green-black (d 400°)	$PtBr_3$ green-black (d 200°)	PtI_3 black (d 310°)
+2	NiF_2 yellow (1450°)	$NiCl_2$ yellow (1001°)	$NiBr_2$ yellow (965°)	NiI_2 black (780°)
	PdF_2 pale violet	α-$PdCl_2$ [(b)] dark red (d 600°)	$PdBr_2$ red-black	PdI_2 black
		α-$PtCl_2$ [(b)] olive-green (d 581°)	$PtBr_2$ brown (d 250°)	PtI_2 black (d 360°)

[(a)] PtF_6 boils at 69.1°.
[(b)] β-$PdCl_2$ and β-$PtCl_2$ (reddish-black) contain M_6Cl_{12} clusters (Fig. 27.3b).

The most stable product of the action of fluorine on metallic palladium is actually $Pd^{II}[Pd^{IV}F_6]$, and true trihalides of Pd do not occur. The diamagnetic $PtCl_3$ and $PtBr_3$ contain Pt^{II} and Pt^{IV} and the triiodide probably does also. Trihalides of nickel are confined to impure specimens of NiF_3.

All the dihalides, except PtF_2, are known, fluorine perhaps being too strongly oxidizing to be readily compatible with the metal in the lower of its two major oxidation states. Except for NiF_2, the yellow to dark-brown dihalides of nickel can be obtained directly from the elements; they dissolve in water from which hexahydrates containing the $[Ni(H_2O)_6]^{2+}$ ion can be crystallized. These solutions may also be prepared more conveniently by dissolving $Ni(OH)_2$ in the appropriate hydrohalic acid. NiF_2 is best formed by the reaction of F_2 on $NiCl_2$ at 350°C and is only slightly soluble in water, from which the trihydrate crystallizes.

Violet, easily hydrolysed, PdF_2 is produced when $Pd^{II}[Pd^{IV}F_6]$ is refluxed with SeF_4 and is notable as one of the very few paramagnetic compounds of Pd^{II}. The paramagnetism arises from the $t_{2g}^6 e_g^2$ configuration of Pd^{II} which is consequent on its octahedral coordination in the rutile-type structure (p. 1120). The dichlorides of both Pd and Pt are obtained from the elements and exist in two isomeric forms: which form is produced depends on the exact experimental conditions used. The more usual α-form of $PdCl_2$ is a red material with a chain structure (Fig. 27.3a) in which each Pd has a square planar geometry. It is hygroscopic and its aqueous solution provides a useful starting point for studying the coordination chemistry of Pd^{II}. β-$PdCl_2$ is also known and its structure is based on Pd_6Cl_{12} units in which, nevertheless, the preferred square-planar coordination of the Pd^{II} is still retained (Fig. 27.3b). Brownish-green α-$PtCl_2$ is insoluble in water but dissolves in hydrochloric acid forming $[PtCl_4]^{2-}$ ions. The dark-red β-$PtCl_2$ is isomorphous with β-$PdCl_2$ and the Pt_6Cl_{12} unit is retained on dissolution in benzene. Red $PdBr_2$ and black PdI_2, obtained respectively by the action of Br_2 on Pd and the addition of I^- to aqueous solutions of $PdCl_2$, are both insoluble in water but form $[PdX_4]^{2-}$ ions on

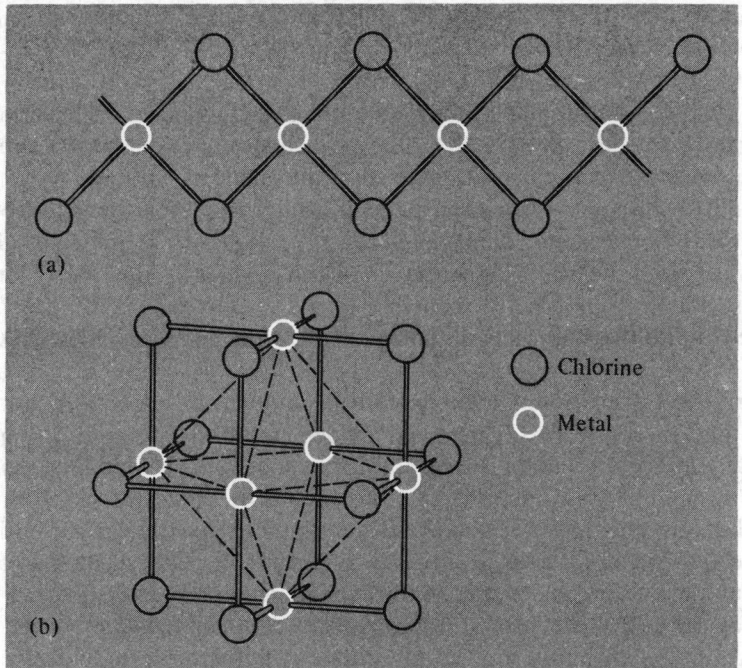

FIG. 27.3   (a) The chain structure of α-PdCl₂, and (b) the M_6Cl_{12} structural unit of β-PdCl₂ and
β-PtCl₂. (Note its broad similarity with the $[M_6X_{12}]^{n+}$ unit of the lower halides of Nb and Ta
shown in Fig. 22.5 and to the unit cell of the three-dimensional structure of NbO.)

addition of HX (X = Br, I). PtBr₂ and PtI₂ are obtained by thermal decomposition of
the tetrahalides but are rather unstable and their purity is often doubtful.

Oxohalides in this group are apparently confined to the strongly oxidizing PtOF₃. The
existence of PtOF₄ is doubtful.

27.3.4 *Complexes* (4, 5, 9, 10, 10a)

Apart from the few Pt^{VI} and Pt^V fluoro and oxofluoro compounds mentioned above,
there is no chemistry in oxidation states above IV.

Oxidation state IV (d⁶)

All complexes in this oxidation state which have been characterized are octahedral and
diamagnetic with the low-spin t_{2g}^6 configuration.

⁹ F R. HARTLEY, *The Chemistry of Platinum and Palladium*, Applied Science Publishers, London, 1973,
544 pp.
¹⁰ U. BELLUCO, *Organometallic and Coordination Chemistry of Platinum*, Academic Press, London, 1974,
701 pp.
¹⁰ᵃ K. NAG and A. CHAKRAVORTY, Monovalent, trivalent, and tetravalent nickel, *Coord. Chem. Rev.* **33**,
87–147 (1980).

Fluorination of $NiCl_2 + KCl$ produces red K_2NiF_6 which is strongly oxidizing and will liberate O_2 from water. Dark red complexes of the type $[Ni^{IV}(L)](ClO_4)_2$ (H_2L is a sexadentate oxime) have been obtained[10b] by the action of conc HNO_3 on $[Ni^{II}(H_2L)](ClO_4)_2$ and are stable indefinitely under vacuum but are reduced in moist air. Neutral complexes of the type $[Ni(S_2C_2R_2)_2]$ which formally contain Ni^{IV} are probably best described as Ni^{II} with an oxidized form of the ligand.

Palladium(IV) complexes are rather sparse and much less stable than those of Pt^{IV}. The best known are the hexahalo complexes $[PdX_6]^{2-}$ ($X = F$, Cl, Br) of which $[PdCl_6]^{2-}$, formed when the metal is dissolved in aqua regia, is the most familiar. In all of these the Pd^{IV} is readily reducible to Pd^{II}. In water, $[PdF_6]^{2-}$ hydrolyses immediately to $PdO \cdot xH_2O$ while the chloro and bromo complexes give $[PdX_4]^{2-}$ plus X_2.

By contrast Pt^{IV} complexes rival those of Pt^{II} in number, and are both thermodynamic-ally stable and kinetically inert. Those with halides, pseudo-halides, and N-donor ligands are especially numerous and, of the multitude of conceivable compounds ranging from $[PtX_6]^{2-}$ through $[PtX_4L_2]$ to $[PtL_6]^{4+}$, ($X = F$, Cl, Br, I, CN, SCN, SeCN; $L = NH_3$, amines) a large number have been prepared and characterized though, curiously, they do not include the $[Pt(CN)_6]^{2-}$ ion. K_2PtCl_6 is commercially the most common compound of platinum and the brownish-red, "chloroplatinic acid", $H_2[PtCl_6](aq)$, is the usual starting material in Pt^{IV} chemistry. It is prepared by dissolving platinum metal sponge in aqua regia, followed by one or more evaporations with hydrochloric acid. A route to Pt^{II} chemistry also is provided by precipitation of the sparingly soluble K_2PtCl_6 followed by its reduction with hydrazine to K_2PtCl_4. The chloroammines were extensively used by Werner and other early coordination chemists in their studies on the nature of the coordinate bond in general and on the octahedral geometry of Pt^{IV} in particular.

O-donor ligands such as OH^- and acac also coordinate to Pt^{IV}, but S- and Se-, and more especially P- and As-donor ligands, tend to reduce it to Pt^{II}.

Oxidation state III (d^7)

Perhaps surprisingly, this unimportant oxidation state is better represented by nickel than by either palladium or platinum. K_3NiF_6 has been prepared by fluorinating $KCl + NiCl_2$ at high temperatures and pressures. It is a violet crystalline material which is reduced by water with evolution of oxygen. The observed elongation of the $[NiF_6]^{3-}$ octahedron has been ascribed to the Jahn–Teller effect (p. 1087) to be expected for a $t_{2g}^6 e_g^1$ configuration although the reported magnetic moment of 2.5 BM at room temperature seems rather high for this configuration. Five-coordinate Ni^{III} complexes such as $[NiBr_3(PEt_3)_2]$ are known. This is a black solid with a trigonal bipyramidal structure in which the 3 bromine atoms form the equatorial plane.

No Pd^{III} or Pt^{III} complex has been unambiguously established, and a number of compounds which have in the past been claimed to contain the trivalent metals have later turned out to contain them in more than one oxidation state. One such is H. Wolffram's red salt, $Pt(EtNH_2)_4Cl_3 \cdot 2H_2O$, which is now known to have a structure (Fig. 27.4a) consisting of alternate octahedral Pt^{IV} and square-planar Pt^{II} linked by Cl bridges, i.e.

[10b] J. G. MOHANTY, R. P. SINGH, and A. CHAKRAVORTY, Chemistry of tetravalent nickel and related species.
1. MN$_6$ coordination octahedra generated by hexadentate oxime ligands, *Inorg. Chem.* **14**, 2178–83 (1975).

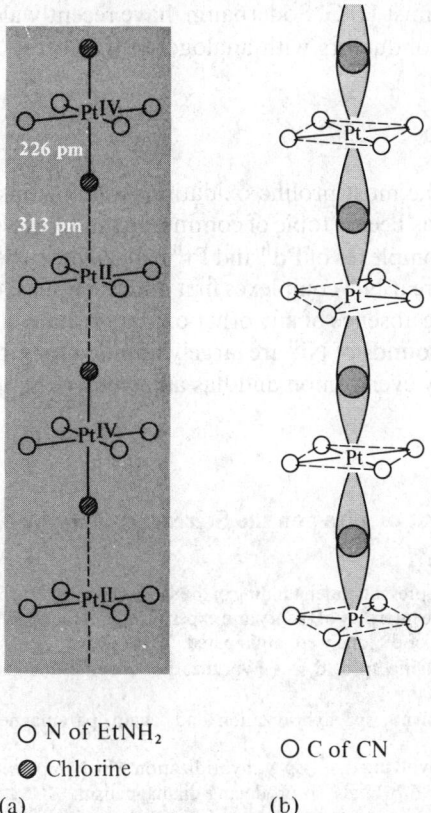

○ N of EtNH₂   ○ C of CN

◎ Chlorine

(a)   (b)

FIG. 27.4   Linear chain polymers of Pt. (a) The coordination of platinum in Wolffram's red salt, $Pt(EtNH_2)_4Cl_3.2H_2O$, showing alternating Pt^{II} and Pt^{IV} linked by Cl bridges. Four remaining Cl^- ions and 4 H_2O are situated within the lattice. (b) Stacking of square planar units in $[Pt(CN)_4]^{n-}$ showing the possible overlap of d_{z^2} orbitals. (Note the successive 45° rotations, or "staircase staggering", of these units.)

$[Pt^{II}(EtNH_2)_4]^{2+}[trans\text{-}(\mu\text{-}Cl)_2Pt^{IV}(EtNH_2)_4]^{2+}Cl^-_4.4H_2O$. Other examples are provided by the one-dimensional conductors of platinum,[11] of which the cyano complexes are the best known. Thus $K_2[Pt(CN)_4].3H_2O$ is a very stable colourless solid, but by appropriate partial oxidation it is possible to obtain bronze-coloured, "cation deficient", $K_{1.75}Pt(CN)_4.1.5H_2O$, and other partially oxidized compounds such as $K_2Pt(CN)_4Cl_{0.3}.3H_2O$. In these, square-planar $[Pt(CN)_4]^{n-}$ ions are stacked (Fig. 27.4b) to give a linear chain of Pt atoms in which the Pt–Pt distances of 280–300 pm (compared to 348 pm in the original $K_2[Pt(CN)_4].3H_2O$ and 278 pm in the metal itself) allow strong overlap of the d_{z^2} orbitals. This accounts for the metallic conductance of these materials along the crystal axis. Indeed, there is considerable current interest in such "one-dimensional" electrical conductors.

Oxalato complexes [e.g. $K_{1.6}Pt(C_2O_4)_2.1.2H_2O$] originally prepared as long ago as

[11] J. S. MILLER and A. J. EPSTEIN, One-dimensional inorganic complexes, *Prog. Inorg. Chem.* **20**, 1–151 (1976).

1888 by the German chemist H. G. Söderbaum, have recently also been recognized[12] as being one-dimensional conductors with analogous structures.

Oxidation state II (d^8)

This is undoubtedly the most prolific oxidation state for this group of elements. The stereochemistry of Ni[II] has been a topic of continuing interest (see Panel), and kinetic and mechanistic studies on complexes of Pd[II] and Pt[II] have likewise been of major importance. It will be convenient to treat Ni[II] complexes first and then those of Pd[II] and Pt[II] (p. 1349).

Complexes of Ni[II]. The absence of any other oxidation state of comparable stability for nickel implies that compounds of Ni[II] are largely immune to normal redox reactions. Ni[II] forms salts with virtually every anion and has an extensive aqueous chemistry based on

---

Development of Ideas on the Stereochemistry of Nickel(II)

The way in which our present understanding of the stereochemical intricacies of Ni[II] has evolved illustrates rather well the interplay of theory and experiment. On the basis of valence-bond theory, three types of complex of d^8 ions were anticipated. These were:

    (i) *octahedral*, involving $sp^3d_{z^2}d_{x^2-y^2}$ hybridization, and paramagnetism from 2 unpaired electrons;

    (ii) *tetrahedral*, involving sp^3 hybridization and, again, paramagnetism from 2 unpaired electrons;

    (iii) *square planar*, involving $d_{x^2-y^2}sp_xp_y$ hybridization which implies the confinement of all 8 electrons in four d orbitals, so producing diamagnetism.

Since X-ray determinations of structure were too time-consuming to be widely used in the 1930s and 1940s and, in addition, square-planar geometry was a comparative rarity, any paramagnetic compound, which on the basis of stoichiometry appeared to be 4-coordinate, was presumed to be tetrahedral.

However, with the application in the 1950s of crystal field theory to transition-metal chemistry it was realized that CFSEs were unfavourable to the formation of tetrahedral d^8 complexes, and previous assignments were re-examined. A typical case was [Ni(acac)$_2$], which had often been cited as an example of a tetrahedral nickel complex, but which was shown[13] in 1956 to be trimeric and octahedral. The over-zealous were then inclined to regard tetrahedral d^8 as non-existent until first L. M. Venanzi[14] and then N. S. Gill and R. S. Nyholm[15] demonstrated the existence of discrete tetrahedral species which in some cases were also rather easily prepared.

More comprehensive examination of spectroscopic and magnetic properties of d^8 ions followed which provided an explanation for the different types of Lifschitz salts (p. 1348) and led to studies[16] of systems exhibiting anomalous properties. Rational explanations of these properties were eventually forthcoming, though current understanding of the factors deciding the stereochemistry of nickel(II) is by no means yet complete (see Table 27.2).

---

[12] A. H. REIS, S. W. PETERSON, and S. C. LIN, A novel partially oxidized distorted one-dimensional platinum chain in K$_{1.6}$Pt(C$_2$O$_4$)$_2$1.2H$_2$O, *J Am. Chem. Soc.* **98,** 7839–40 (1976).

[13] G. J. BULLEN, Trinuclear molecules in the crystal structure of bisacetylacetonatenickel(II), *Nature* **177,** 537–8 (1956).

[14] L. M. VENANZI, Tetrahedral nickel(II) complexes and the factors determining their formation. Pt. I. Bistriphenylphosphine nickel(II) compounds, *J. Chem. Soc.* 1958, 719–24.

[15] N. S. GILL and R. S. NYHOLM, Complex halides of the transition metals. Pt. I. Tetrahedral nickel complexes, *J. Chem. Soc.* 1959, 3997–4007.

[16] L. SACCONI, Electronic structure and stereochemistry of nickel(II), *Transit. Metal Chem.* **4,** 199–298 (1968).

the green† $[Ni(H_2O)_6]^{2+}$ ion which is always present in the absence of strongly complexing ligands.

The coordination number of Ni^{II} rarely exceeds 6 and its principal stereochemistries are octahedral and square planar (4-coordinate) with rather fewer examples of trigonal bipyramidal (5), square pyramidal (5), and tetrahedral (4). Octahedral complexes of Ni^{II} are obtained (often from aqueous solution by replacement of coordinated water) especially with neutral N-donor ligands such as NH_3, en, bipy, and phen, but also with NCS^-, NO_2^-, and the O-donor dimethylsulfoxide, DMSO (Me_2SO).

An interesting example of the attainment of octahedral coordination by polymerization is provided by the green acac complex. This is obtained as the monomeric octahedral *trans*-dihydrate, $[Ni(acac)_2(H_2O)_2]$, from aqueous solution, but on dehydration it gives the trimer, $[Ni(acac)_2]_3$ (Fig. 27.5), which has been the subject of extensive magnetic studies by R. L. Martin and co-workers.[17] It behaves as a normal paramagnet down to about 80 K, but at lower temperatures the magnetic moment per nickel atom rises from about 3.2 BM (as expected for 2 unpaired electrons, i.e. $S = 1$) to 4.1 BM at 4.3 K. This corresponds to the ferromagnetic coupling of all 6 unpaired electrons in the trimer (i.e. $S' = 3$). Refined calculations actually suggest that, while adjacent Ni atoms experience weak (and hence only apparent at very low temperatures) ferromagnetic interactions, the 2 terminal nickels experience an additional and still weaker antiferromagnetic interaction. Replacement of the $-CH_3$ groups of acetylacetone by the bulkier $-C(CH_3)_3$ apparently prevents the above polymerization and leads instead to the red square-planar monomer (Fig. 27.6a).

Of the four-coordinate complexes of Ni^{II}, those with the square planar stereochemistry are the most numerous. They include the yellow $[Ni(CN)_4]^{2-}$, the red bis(N-methylsalicylaldiminato)nickel(II) and the well-known bis(dimethylglyoximato)nickel(II) (Fig. 27.6b and c) obtained as a flocculent red precipitate in gravimetric determinations

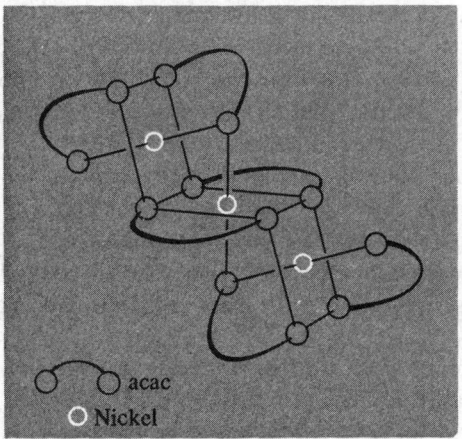

acac
Nickel

FIG. 27.5   Trimeric structure of $[Ni(acac)_2]_3$.

† It was work on the absorption of light by solutions of nickel(II) which led A. Beer in 1852 to formulate the law which bears his name.

[17] P. D. W. BOYD and R. L. MARTIN, Spin–spin interactions in polynuclear nickel(II) complexes. Susceptibility and low temperature magnetization studies of $[Ni_3(pd)_6]$, *JCS Dalton* 1979, 92–95.

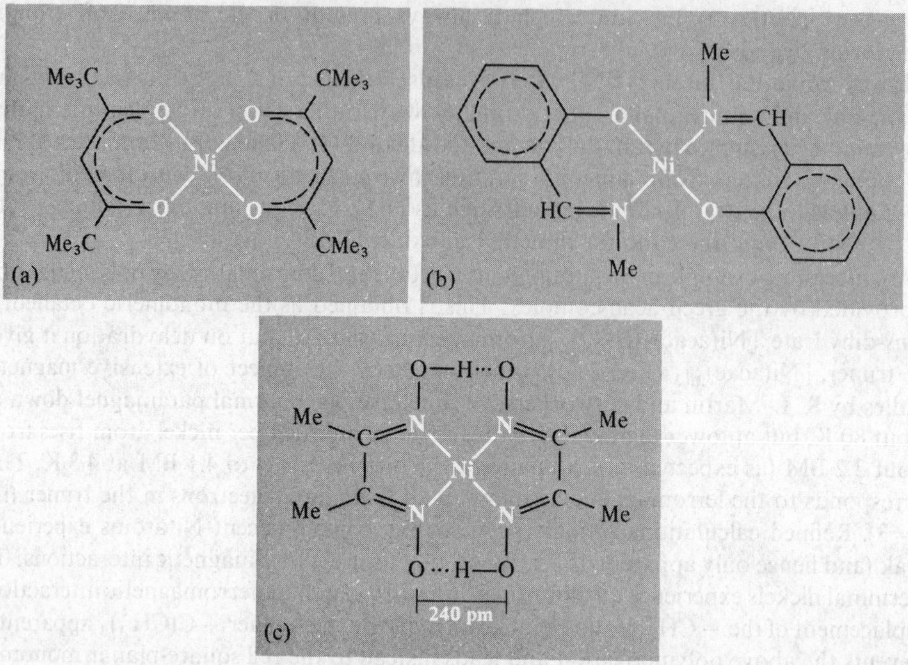

FIG. 27.6   Some typical planar complexes of nickel(II): (a) [Ni(Me$_6$-acac)$_2$], (b) [Ni(Me-sal)$_2$], and (c) [Ni(DMGH)$_2$]. (Note the short O–O distance which is due to strong hydrogen bonding.)

of nickel. Actually, in the solid state, this last compound consists of planar molecules stacked above each other so that Ni–Ni interactions occur (Ni–Ni 325 pm), and the nickel atoms should therefore be described as octahedrally coordinated. However, in non-coordinating solvents it dissociates into the square-planar monomer, while in bis-(ethylmethylglyoximato)nickel(II) a much longer Ni–Ni separation (475 pm) indicates that even in the solid it must be regarded as square planar.

Although less numerous than the square-planar complexes, tetrahedral complexes of nickel(II) also occur. The simplest of these are the blue [NiX$_4$]$^{2-}$ (X = Cl, Br, I) ions, precipitated[15] from ethanolic solutions by large cations such as [NR$_4$]$^+$, [PR$_4$]$^+$, and [AsR$_4$]$^+$. Other examples include a number of those of the type [NiL$_2$X$_2$] (L = PR$_3$, AsR$_3$, OPR$_3$, OAsR$_3$) amongst which were the first authenticated examples of tetrahedral nickel(II).[14]

The square-planar compounds are diamagnetic (see below) and commonly red to yellow, whereas the octahedral and tetrahedral are paramagnetic because of the 2 unpaired electrons in the $t_{2g}^6 e_g^2$ and $e^4 t_2^4$ configurations respectively, and are commonly green to blue. Colour is not a reliable criterion however, as witnessed, for instance, by the brown-red, but paramagnetic and octahedral, [Ni(NO$_2$)$_6$]$^{4-}$, and the green, but square-planar and diamagnetic, [NiI$_2$(quinoline)$_2$]. In the absence of an X-ray analysis, satisfactory assignment of the electronic spectrum and confirmation of the expected magnetic properties may be used as a means of distinguishing these stereochemistries, as elaborated in the Panel.

Electronic Spectra and Magnetic Properties of Complexes of Nickel(II)[4, 16]

Nickel(II) is the only common d^8 ion and its spectroscopic and magnetic properties have accordingly been extensively studied.

In a cubic field three spin-allowed transitions are expected (p. 1094) because of the splitting of the free-ion, ground ^3F term and the presence of the ^3P term. In an octahedral field the splitting is the same as for the octahedral d^3 ion and the same Tanabe–Sugano diagram (i.e. Fig. 23.10) can be used to interpret the spectra as was used for octahedral Cr^{III}. Spectra of octahedral Ni^{II} usually do consist of three bands which are accordingly assigned as:

$$\nu_1 = {}^3T_{2g}(F) \leftarrow {}^3A_{2g}(F) = 10\,Dq$$
$$\nu_2 = {}^3T_{1g}(F) \leftarrow {}^3A_{2g}(F)$$
$$\nu_3 = {}^3T_{1g}(P) \leftarrow {}^3A_{2g}(F)$$

with ν_1 giving the value of Δ, or $10Dq$, directly. Quite often there is also evidence of weak spin-forbidden (i.e. spin triplet→singlet) absorptions and, in $[Ni(H_2O)_6]^{2+}$ and $[Ni(DMSO)_6]^{2+}$, for instance, the ν_2 absorption has a strong shoulder on it. This has been ascribed to the influence of spin–orbit coupling in "mixing" a spin singlet (^1E_g) with the $^3T_{1g}(F)$, spin triplet, thereby allowing the spin-forbidden transition to gain intensity from the spin-allowed transition. This effect is greatly enhanced when the crystal field is about $Dq/B = 1$, which is the case in the aquo and DMSO complexes, for here the ^1E_g and $^3T_{1g}(F)$ terms are in close proximity.[18]

For d^8 ions in tetrahedral fields the splitting of the free-ion ground term is the inverse of its splitting in an octahedral field, so that $^3T_{1g}(F)$ lies lowest. On the basis of the positive-hole formalism, however, the splitting for the octahedral d^2 ion is also the inverse of that for the octahedral d^8 ion (see Table 19.3). Therefore tetrahedral d^8 ions and octahedral d^2 ions can be treated similarly, and the spectra of tetrahedral Ni^{II} interpreted in the same way as those of octahedral V^{III} (i.e. Fig. 22.8) so that three bands are to be expected, arising from the transitions:

$$\nu_1 = {}^3T_2(F) \leftarrow {}^3T_1(F)$$
$$\nu_2 = {}^3A_2(F) \leftarrow {}^3T_1(F)$$
$$\nu_3 = {}^3T_1(P) \leftarrow {}^3T_1(F)$$

In practice the spectra are a good deal more intense than those of octahedral complexes and bands are often split by spin–orbit coupling to an extent which can make unambiguous assignments difficult. The broad, or double, band around 15 000 cm^{-1} is normally assumed to represent one rather than two transitions, with the corollary that ν_1 must often be assumed to be beyond the lowest energies scanned in the spectrum. This, however, is reasonable in view of the much lower field anticipated in tetrahedral as compared with octahedral complexes. Table A gives data for a number of octahedral and tetrahedral complexes.

TABLE A   *Electronic spectra of some complexes of nickel(II)*

Complex	ν_1/cm^{-1}	ν_2/cm^{-1}	ν_3/cm^{-1}	$10Dq$/cm^{-1}
Octahedral				
$[Ni(DMSO)_6]^{2+}$	7 730	12 970	24 040	7 730
$[Ni(H_2O)_6]^{2+}$	8 500	13 800	25 300	8 500
$[Ni(NH_3)_6]^{2+}$	10 750	17 500	28 200	10 750
$[Ni(en)_3]^{2+}$	11 200	18 350	29 000	11 200
$[Ni(bipy)_3]^{2+}$	12 650	19 200	(a)	
Tetrahedral				
$[NiI_4]^{2-}$		7 040	14 030	3 820
$[NiBr_4]^{2-}$		7 000	13 230, 14 140	3 790
$[NiCl_4]^{2-}$		7 549	14 250, 15 240	4 090
$[NiBr_2(OPPh_3)_2]$		7 250	15 580	3 950

(a) Obscured by intense charge-transfer absorptions.

Continued

[18] S. F. A. KETTLE, *Coordination Compounds*, pp. 116–120. Nelson, London, 1969.

The T ground term of the tetrahedral ion is expected to lead to a temperature-dependent orbital contribution to the magnetic moment, whereas the A ground term of the octahedral ion is not, though "mixing" of the excited $^3T_{2g}(F)$ term into the $^3A_{2g}(F)$ ground term is expected to raise its moment to:

$$\mu_e = \mu_{\text{spin-only}}\left(1 - \frac{4\lambda}{10Dq}\right)$$

where $\lambda = -315$ cm^{-1} and $\mu_{\text{spin-only}} = 2.83$ BM. (This is the exact reverse of the situations found for Co^{II}; p. 1315.) The upshot is that the magnetic moments of tetrahedral compounds are found to lie in the range 3.2–4.1 BM (and are dependent on temperature, and are reduced towards $\mu_{\text{spin-only}}$ by electron delocalization on to the ligands and by distortions from ideal tetrahedral symmetry) whereas those of octahedral compounds lie in the range 2.9–3.3 BM.

Although, in principle, a square-planar d^8 complex could be high-spin with unpaired electrons in the $d_{x^2-y^2}$ and d_{xy} orbitals (see Fig. 27.7 below) this is not found, the separation of these 2 orbitals evidently always being sufficient to force spin-pairing and consequent diamagnetism. Their spectra are usually characterized by a fairly strong band in the yellow to blue (i.e. about 17 000–22 000 cm^{-1} or 600–450 nm) region which is responsible for the reddish colour, and another band near the ultraviolet. The likelihood of π-bonding and attendant charge transfer makes a simple crystal-field treatment inappropriate, and unambiguous assignments are difficult.

A partial explanation, at least, can be provided for the relative abundances and ease of formation of the above stereochemical varieties of Ni^{II} complexes. It can be seen from Table 26.5 that the CFSEs of the d^8 configuration, unlike those of the d^7 configuration (e.g. Co^{II}), favour an octahedral as opposed to a tetrahedral stereochemistry. It might also appear that the symmetrical $t_{2g}^6 e_g^2$ configuration, for which the Jahn–Teller effect is inoperative, offers no possibility of increasing its stability by a tetragonal distortion and so should favour an octahedral as opposed this time to a square-planar stereochemistry. This, however, is an over-simplified and misleading analysis, as can be seen by reference to Fig. 27.7 which shows that the square-planar geometry offers the possibility, not available in either the octahedral or tetrahedral cases, of accommodating all 8 d electrons in 4 lower orbitals, thus leaving the uppermost ($d_{x^2-y^2}$) orbital† empty. Providing therefore that the crystal field is of sufficiently low symmetry (or is sufficiently strong) to split the d orbitals enough to offset the energy required to pair-up 2 electrons, then the 4-coordinate, square-planar extreme can be energetically preferable not only to the tetrahedral but also to the 6-coordinate octahedral extreme. Thus, with the CN$^-$ ligand which produces an exceptionally strong field, the square-planar [Ni(CN)$_4$]$^{2-}$ is formed rather than the tetrahedral isomer or the octahedral [Ni(CN)$_6$]$^{4-}$. Also, many compounds of the type [NiL$_2$X$_2$], in which low-symmetry crystal fields are clearly possible, are planar. However, this selfsame formulation was mentioned above as including examples of tetrahedral complexes: evidently the factors which determine the geometry of a particular complex are finely balanced[19] so that predictions should be made with caution.

This balance, which apparently involves both steric and electronic factors, is well

---

† This is the d orbital used, according to VB theory, for the formation of dsp^2 hybrids to accommodate the 4 σ-pairs of donated electrons in square-planar complexes.

[19] R. H. HOLM and M. J. O'CONNOR, The stereochemistry of bischelate metal(II) complexes, *Prog. Inorg. Chem.* **14**, 241–401 (1971).

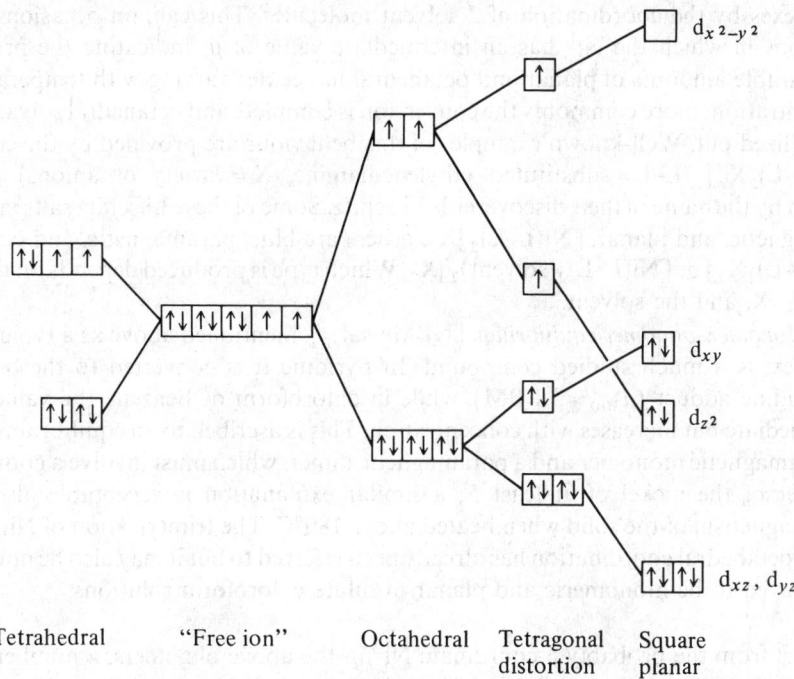

| Tetrahedral | "Free ion" | Octahedral | Tetragonal | Square |
| | | | distortion | planar |

FIG. 27.7   The splitting of d orbitals in fields of different symmetries, and the resulting electronic configurations of the d^8, Ni^{II}, ion.

illustrated by the series of complexes [Ni(PR$_3$)$_2$X$_2$] (X = Cl, Br, I). The diamagnetic, planar forms are favoured by R = alkyl, X = I, and the paramagnetic tetrahedral forms by R = aryl, X = Cl. When mixed alkylarylphosphines are involved, conformational isomerism may occur. For example, [NiBr$_2$(PEtPh$_2$)$_2$] has been isolated in both a green, paramagnetic (μ_{300} = 3.20 BM), tetrahedral form and a brown, diamagnetic, planar form.[19a] Indeed, in solution, this and analogous compounds can exist as equilibrium mixtures of the 2 isomers, so giving rise to intermediate values of μ_e(0–3.2 BM). This type of behaviour in solution is typical of an increasing number of nickel(II) complexes which were considered "anomalous" until reasonable explanations were forthcoming.

"Anomalous" behaviour of nickel(II) complexes. It is now recognized that the origin of this behaviour lies in one of three possible types of equilibria:

1. *Planar–tetrahedral equilibria.* Besides the bis(triphosphine) derivatives just mentioned, a number of *sec*-alkylsalicyladiminato derivatives (i.e. –CH$_3$ in Fig. 27.6b replaced by a *sec*-alkyl group) dissolve in non-coordinating solvents such as chloroform or toluene to give solutions whose spectra and magnetic properties are temperature-dependent and indicate the presence of an equilibrium mixture of planar and tetrahedral molecules.

2. *Planar–octahedral equilibria.* Dissolution of planar Ni^{II} compounds in coordinating solvents such as water or pyridine frequently leads to the formation of octahedral

[19a] R. G. HAYTER and F. S. HUMIEC, Square-planar–tetrahedral isomerism of nickel halide complexes of diphenylaklylphosphines, *Inorg. Chem.* **4**, 1701–6 (1965).

complexes by the coordination of 2 solvent molecules. This can, on occasions, lead to solutions in which the Ni^{II} has an intermediate value of μ_e indicating the presence of comparable amounts of planar and octahedral molecules varying with temperature and concentration; more commonly the conversion is complete and octahedral solvates can be crystallized out. Well-known examples of this behaviour are provided by the complexes $[Ni(L-L)_2X_2]$ (L–L = substituted ethylenediamine, X = variety of anions) generally known by the name of their discoverer I. Lifschitz. Some of these Lifschitz salts are yellow, diamagnetic, and planar, $[Ni(L-L)_2]X_2$, others are blue, paramagnetic, and octahedral, $[Ni(L-L)_2X_2]$ or $[Ni(L-L)_2(\text{solvent})_2]X_2$. Which type is produced depends on the nature of L–L, X, and the solvent.

3. *Monomer–oligomer equilibrium.* $[Ni(Me-sal)_2]$, mentioned above as a typical planar complex, is a much studied compound. In pyridine it is converted to the octahedral bispyridine adduct ($\mu_{300} = 3.1$ BM), while in chloroform or benzene the value of μ_e is intermediate but increases with concentration. This is ascribed to an equilibrium between the diamagnetic monomer and a paramagnetic dimer, which must involve a coordination number of the nickel of at least 5; a similar explanation is acceptable also for the paramagnetism of the solid when heated above 180°C. The trimerization of $Ni(acac)_2$ to attain octahedral coordination has already been referred to but it may also be noted that it is reported to be monomeric and planar in dilute chloroform solutions.

Apart from the probably 5-coordinate Ni^{II} in the above oligomers, a number of well-characterized 5-coordinate complexes have been produced in recent years. These may be classified as either trigonal bipyramidal or square pyramidal, though the two forms are energetically similar and the stereochemistry is often imposed by the ligands. Thus the trigonal bipyramidal complexes, which are the more common, often involve tripod ligands (p. 1061), while the quadridentate chain ligand,

$$Me_2As(CH_2)_3As\text{-}CH_2\text{-}As(CH_2)_3AsMe_2$$
$$\underset{Ph}{|} \qquad \underset{Ph}{|}$$

(tetars) produces square pyramidal complexes of the type $[Ni(tetars)X]^+$. Not surprisingly, since they may be considered intermediate between high-spin, paramagnetic, octahedral complexes and low-spin, diamagnetic, planar complexes, these 5-coordinate complexes can be of either high-spin or low-spin type. The former is found in $[NiBr\{N(C_2H_4NMe_2)_3\}]^+$ but with *P*- or *As*-donor ligands low-spin configurations are found.

The Ni^{II}/CN^- system illustrates nicely the ease of conversion of the two stereochemistries. Although, as already pointed out, there is no evidence of a hexacyano complex, a square pyramidal pentacyano complex is known:

$$Ni \xrightarrow{\text{CN}} \underset{\text{green ppt}}{[Ni(CN)_2aq]} \xrightarrow[\text{redissolves}]{\text{CN}} \underset{\text{yellow}}{[Ni(CN)_4]^{2-}} \xrightarrow[\text{excess}]{\text{CN}} \underset{\text{red}}{[Ni(CN)_5]^{3-}}$$

The fascinating crystalline compound $[Cr(en)_3][Ni(CN)_5].1\frac{1}{2}H_2O$ contains both square pyramidal and trigonal bipyramidal anions though each is distorted from true C_{4v} or D_{3h} symmetry as shown in Fig. 27.8.[19b]

[19b] K. N. RAYMOND, P. W. R. CORFIELD, and J. A. IBERS, The structure of tris(ethylenediamine)chromium(III) pentacyanonickelate(II) sesquihydrate, $[Cr(en)_3][Ni(CN)_5].1\frac{1}{2}H_2O$, *Inorg. Chem.* **7**, 1362–72 (1968).

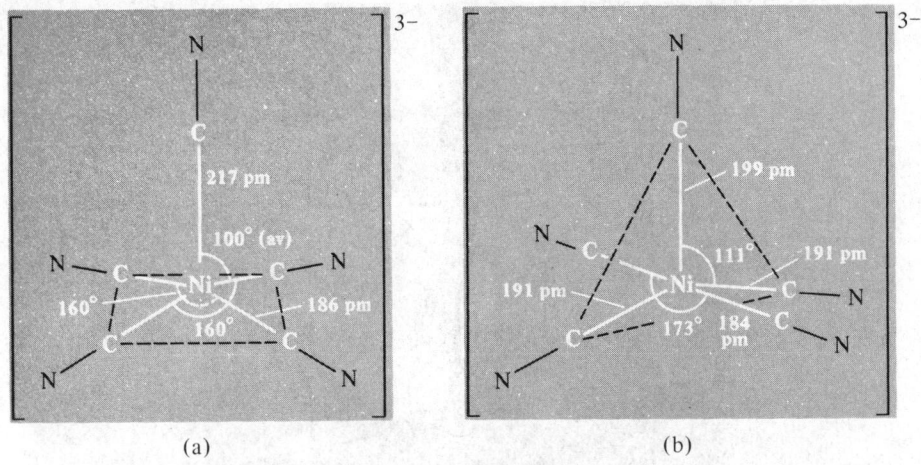

Fig. 27.8   The structure of the distorted (a) square-pyramidal and (b) trigonal bipyramidal $[Ni(CN)_5]^{3-}$ ions in $[Cr(en)_3][Ni(CN)_5].1\tfrac{1}{2}H_2O$.

Another interesting cyano derivative of nickel(II) which may conveniently be mentioned here is the *clathrate* compound, $[Ni(CN)_2(NH_3)].xC_6H_6$ ($x \leqslant 1$). If CN^- is added to the blue-violet solution obtained by mixing aqueous solutions of Ni^{II} and NH_3, and this is then shaken with benzene, a pale-violet precipitate is obtained. This precipitate is soluble in conc NH_3. The benzene and ammonia can be removed by heating it above 150°C but not by washing or by application of reduced pressure. The benzene molecule is, in fact, trapped inside a cage formed by the lattice in which the nickel ions are coordinated to 4 cyanides situated in a square plane, and half are additionally coordinated to 2 ammonias (Fig. 27.9). The observed magnetic moment per Ni atom of 2.2 BM is entirely consistent with this since this average moment arises solely from the octahedrally coordinated Ni atoms, the square-planar Ni being diamagnetic.

Complexes of Pd^{II} and Pt^{II}. The effect of complexation on the splitting of d orbitals is much greater in the case of second- and third- than for first-row transition elements, and the associated effects already noted for Ni^{II} are even more marked for Pd^{II} and Pt^{II}; as a result, their complexes are diamagnetic and the vast majority are planar also. Not many complexes are formed with *O*-donor ligands but, of the few that are, $[Pd(H_2O)_4]^{2+}$ present in the brown $Pd(ClO_4)_2.4H_2O$, and the polymeric anhydrous acetates $[Pd(O_2CMe)_2]_3$ and $[Pt(O_2CMe)_2]_4$ (Fig. 27.10), may be noted.[19c] Fluoro complexes are even less prevalent, the preference of these cations being for the other halides, cyanide, *N*- and heavy atom-donor ligands.

The complexes $[MX_4]^{2-}$ (M = Pd, Pt; X = Cl, Br, I, SCN, CN) are all easily obtained and may be crystallized as salts of $[NH_4]^+$ and the alkali metals. By using $[NR_4]^+$ cations it is possible to isolate binuclear halogen-bridged anions $[M_2X_6]^{2-}$ (X = Br, I) which retain the square-planar coordination of M. Aqueous solutions of yellowish-brown

[19c] A. C. Skapski and M. L. Smart, The crystal structure of trimeric palladium(II) acetate, *Chem. Comm.* 1970, 658–9. M. A. A. F. de C. T. Carrondo and A. C. Skapski, X-ray crystal structure of tetrameric platinum(II) acetate: a square cluster complex with short Pt–Pt bonds and octahedral co-ordination geometry, *JCS Chem. Comm.* 410–11 (1976).

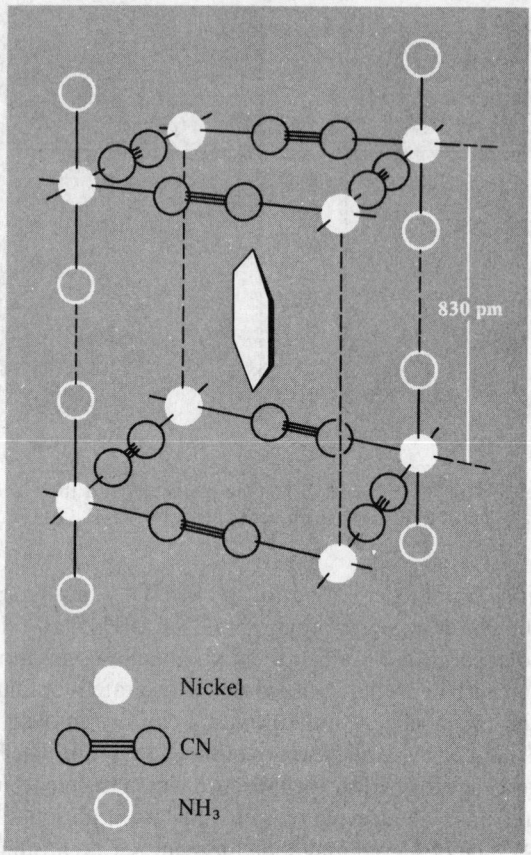

FIG. 27.9   A "cage" in the structure of $[Ni(CN)_2(NH_3)].x(C_6H_6)$, showing a trapped benzene molecule.

$[PdCl_4]^{2-}$ and red $[PtCl_4]^{2-}$ are common starting materials for the preparation of other Pd^{II} and Pt^{II} complexes by successive substitution of the chloride ligands. In both $[M(SCN)_4]^{2-}$ complexes the ligands bond through their π-acceptor (S) ends, though in the presence of stronger π-acceptor ligands such as PR_3 and AsR_3 they tend to bond through their N ends.† Not surprisingly, therefore, several instances of linkage isomerism (p. 1081) have been established in compounds of the type $trans$-$[M(PR_3)_2(SCN)_2]$.

Complexes with ammonia and amines, especially those of the types $[ML_4]^{2+}$ and $[ML_2X_2]$, are numerous for Pd^{II} and even more so for Pt^{II}. Many of them were amongst the first complexes of these metals to be prepared and interest in them has continued since. For example, the colourless $[Pt(NH_3)_4]Cl_2.H_2O$ can be obtained by adding NH_3 to an aqueous solution of $PtCl_2$ and, in 1828, was the first of the platinum ammines to be

† Steric, as well as electronic, effects are probably involved. When the ligand is N-bonded, $M\leftarrow N\equiv C\text{-}S$ is linear and so sterically undemanding. However, when the ligand is S-bonded, $M\text{-}S\text{-}C$ is nonlinear, the bonding and nonbonding electron pairs around the sulfur being more or less tetrahedrally disposed. On purely steric grounds, therefore, the latter type of bonding is expected to be less favoured than the former when bulky ligands such as PR_3 and AsR_3 are present.

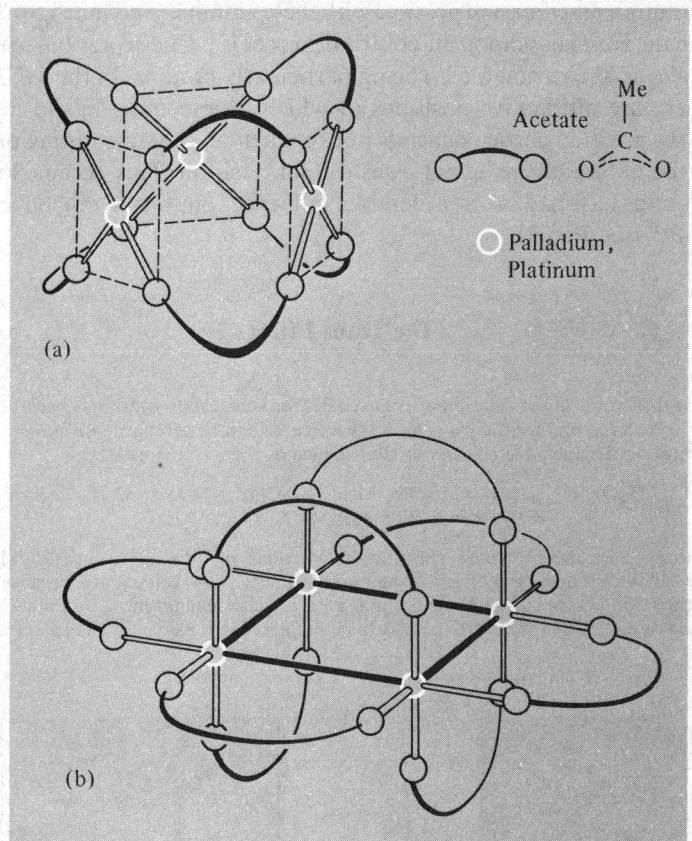

FIG. 27.10   Anhydrous acetates of Pd^{II} and Pt^{II}: (a) trimeric $[Pd(O_2CMe)_2]_3$ involving square-planar coordinated Pd but no metal metal bonding (average Pd Pd = 315 pm), and (b) tetrameric $[Pt(O_2CMe)_2]_4$ involving octahedrally coordinated Pt and metal metal bonds (average Pt Pt = 249.5 pm).

discovered (by G. Magnus). Other salts of the $[Pt(NH_3)_4]^{2+}$ ion are easily derived, the best known being Magnus's green salt $[Pt(NH_3)_4][PtCl_4]$. That a green salt should result from the union of a colourless cation and a red anion was unexpected and is a consequence of the crystal structure, which consists of the square-planar anions and cations stacked alternately to produce a linear chain of Pt atoms only 325 pm apart. Interaction between these metal atoms shifts the d–d absorption of the $[PtCl_4]^{2-}$ ion from the green region (whence the normal red colour) towards the red, so producing the green colour.

Magnus's salt is an electrolyte and non-ionized polymerization isomers (p. 1081) of the stoichiometry $PtCl_2(NH_3)_2$ are also known which can be prepared as monomeric *cis* and *trans* isomers:

$$[PtCl_4]^{2-} \xrightarrow{\text{NH}_3} \textit{cis-}[PtCl_2(NH_3)_2]$$

$$[Pt(NH_3)_4]^{2+} \xrightarrow{\text{HCl}} \textit{trans-}[PtCl_2(NH_3)_2]$$

Their existence led Werner to infer a square-planar geometry for Pt^{II}.

Many substitution reactions are possible with these ammines and much work in this field has come from the Russian school, the contributions of I. I. Chernyaev (also transliterated from the Cyrillic as Tscherniaev, etc.) being particularly notable. In the 1920s he noticed that when there are alternative positions at which an incoming ligand might effect a substitution, the position chosen depends not so much on the substituting or substituted ligand as on the nature of the ligand *trans* to that position. This became known as the "*trans*-effect" and has had a considerable influence on the synthetic coordination chemistry of Pt[II] (see Panel).

The Trans Effect

Because their rates of substitution are convenient for study, most work has been done with platinum complexes, and for these it is found that ligands can be arranged in a fairly consistent order indicating their relative abilities to labilize ligands *trans* to themselves:

$$F^-, \ OH^-, \ H_2O, \ NH_3, \ py < Cl^- < Br^- < I^-, \ -SCN^-, \ NO_2^-, \ C_6H_5^- < S{=}C(NH_2)_2,$$
$$CH_3^- < H^-, \ PR_3, \ AsR_3 < CN^-, \ CO, \ C_2H_4$$

The reason why the particular substitution reactions of $[PtCl_4]^{2-}$ and $[Pt(NH_3)_4]^{2+}$, mentioned above, produce respectively *cis* and *trans* isomers is now evident. It is because, in both cases, in the second of the stepwise substitutions there is a choice of positions for the substitution and in each case it is a ligand *trans* to a Cl⁻ which is replaced in preference to a ligand *trans* to an NH₃:

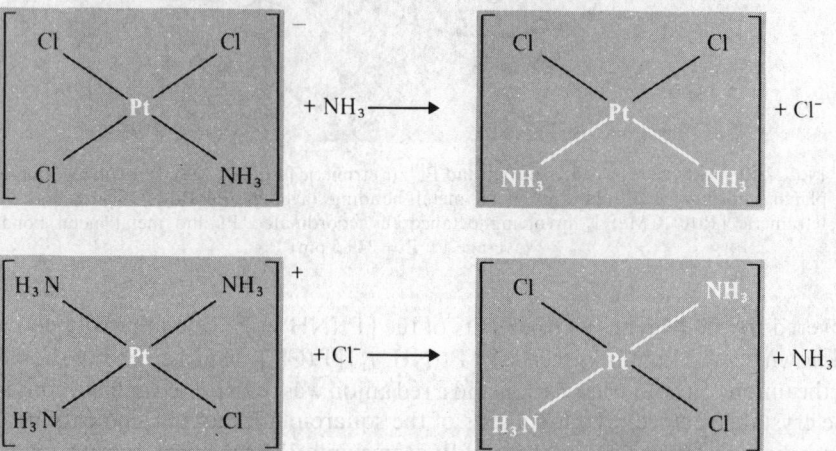

Similar considerations have been invaluable in devising synthetic routes to numerous other isomeric complexes of Pt[II] but, as can be seen in Fig. A, other considerations such as the relative stabilities of the different Pt-ligand bonds are also involved.

Explanations for the *trans*-effect abound (ref. 10, p. 44) and it seems that either π or σ effects, or both, are involved. The ligands exerting the strongest *trans*-effects are just those (e.g. C_2H_4, CO, PR_3, etc.) whose bonding to a metal is thought to have most π-acceptor character and which therefore remove most π-electron density from the metal. This reduces the electron density most at the coordination site directly opposite, i.e. *trans*, and it is there that nucleophilic attack is most likely. This interpretation is not directly concerned with labilizing a particular ligand but rather

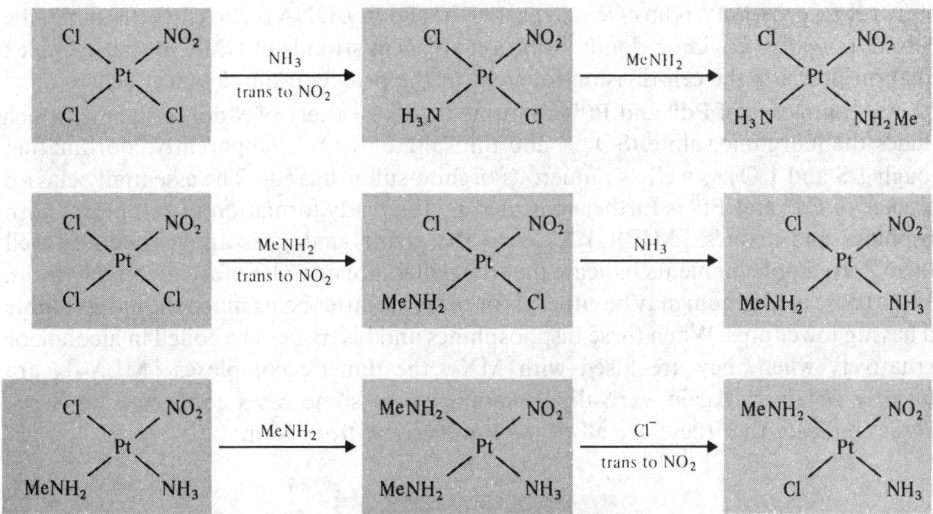

FIG. A   Preparation of the three isomers of [PtCl(NH₃)(NH₂Me)(NO₂)]. Where indicated these steps can be explained by the greater *trans*-effect of the NO_2^- ligand. Elsewhere the weakness of the Pt–Cl as compared to the Pt–N bond must be invoked.

with encouraging the attachment of a further ligand. It has consequently been applied most successfully to explaining kinetic phenomena such as reaction rates and the proportions of different isomers formed in a reaction (which depend on the rates of reaction), due to the stabilization of 5-coordinate reaction intermediates.

On the other hand, a ligand which is a strong σ-donor is expected to produce an axial polarization of the metal, its lone-pair inducing a positive charge on the near side of the metal and a concomitant negative charge on the far side. This will weaken the attachment to the metal of the *trans* ligand. This interpretation has been most successfully used to explain thermodynamic, "ground state" properties such as the bond lengths and vibrational frequencies of the *trans* ligands and their nmr coupling constants with the metal.

In order to distinguish between kinetic and thermodynamic phenomena it is convenient to refer to the former as the "*trans*-effect" and the latter as the "*trans*-influence" or "static *trans*-effect", though this nomenclature is by no means universally accepted. However, it appears that to account satisfactorily for the kinetic "*trans*-effect", both π (kinetic) and σ (thermodynamic) effects must be invoked to greater or lesser extents. Thus, for ligands which are low in the *trans* series (e.g. halides), the order can be explained on the basis of a σ effect whereas for ligands which are high in the series the order is best interpreted on the basis of a π effect. Even so, the relatively high position of H⁻, which can have no π-acceptor properties, seems to be a result of a σ mechanism or some other interaction.

A resurgence of interest in these seemingly simple complexes of platinum started in 1969 when B. Rosenberg and co-workers discovered[20] the anti-tumour activity of *cis*-[PtCl₂(NH₃)₂] ("cisplatin"). Along with [PtCl₂(en)] this compound provides one of the best current treatments for some types of cancer.[21] Unfortunately the mode of action is not properly understood, but the fact that only *cis* isomers of these and related compounds

[20] B. ROSENBERG, L. VAN CAMP, J. E. TRASKO, and V. H. MANSOUR, Platinum compounds: a new class of potent antitumour agents, *Nature* **222**, 385–6 (1969).
[21] E. WILTSHAW, Cisplatin in the treatment of cancer, *Platinum Metals Rev.* **23**, 90–98 (1979).

are active suggests that the formation of a chelate ring, or at least coordination to donor atoms in close proximity, is an essential part of the activity. DNA is currently thought to be involved, possibly providing donor atoms on adjacent strands of DNA so that complex formation prevents the cell division inherent in the proliferation of cancer cells.

Stable complexes of Pd^{II} and Pt^{II} are formed with a variety of S-donor ligands which includes the inorganic sulfite (SO_3^-) and thiosulfate ($S_2O_3^{2-}$ apparently coordinating through 1 S and 1 O) as well as numerous organo-sulfur ligands. The essentially class-b character of Pd^{II} and Pt^{II} is further indicated by the ready formation of complexes with phosphines and arsines. $[M(PR_3)_2X_2]$ and the arsine analogues are particularly well known. Zero dipole moments indicate that the palladium complexes are invariably *trans*, whereas those of platinum may be either *cis* or *trans* the latter being much the more soluble and having lower mps. When these bisphosphines and bisarsines are boiled in alcohol, or alternatively when they are fused with MX_2, the dimeric complexes $[MLX_2]_2$ are frequently obtained. Again, zero dipole moments (in some cases confirmed by X-ray analysis) indicate that these are all of the symmetrical *trans* form:

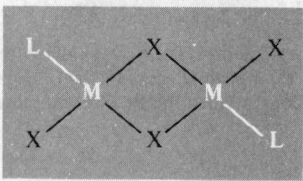

By involving SCN^- a novel 8-membered ring system has been produced:

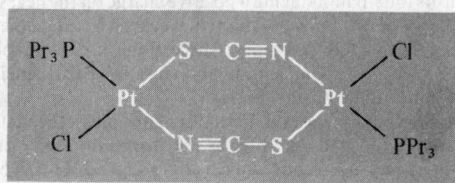

Nuclear magnetic resonance has proved to be a particularly useful tool in studying phosphines of platinum. The nuclear spins of ^{31}P and ^{195}Pt (both equal to $\frac{1}{2}$) couple, and the strength of this coupling (as measured by the "coupling constant" J) is affected much more by ligands *trans* to the phosphine than by those which are *cis*. This has helped in determinations of structure and also in studies of the "*trans* influence". Platinum(II) also forms a number of quite stable monohydrido (H^-) phosphines which have proved similarly interesting, the $^1H-^{195}Pt$ coupling constants being likewise sensitive to the *trans* ligand.

It has already been pointed out that the overwhelming majority of complexes of Pd^{II} and Pt^{II} are square planar. However, 5-coordinate intermediates are almost certainly involved in many of the substitution reactions of these 4-coordinate complexes, and 5-coordinate trigonal bipyramidal complexes with "tripod" ligands (p. 1061) are well established. These ligands include the tetraarsine, $As(C_6H_4\text{-}2\text{-}AsPh_2)_3$ (QAS, p. 1334) and its phosphine analogue[22] and more recently[23] $N(CH_2CH_2NMe_2)_3$ i.e.(Me_6tren). The somewhat less-rigid tetraarsine,

$$1,2\text{-}C_6H_4(\overset{\displaystyle Me}{\overset{\displaystyle |}{As}}\text{-}C_6H_4\text{-}AsMe_2)_2$$

(TPAS), however forms a red, square-pyramidal complex [Pd(TPAS)Cl]ClO$_4$ with palladium.

Oxidation state I (d^9)

The only complexes of any importance formed by these elements in this oxidation state are the yellow to orange nickel(I) phosphines of the type [Ni(PPh$_3$)$_3$X] (X = Cl, Br, I). These are relatively stable and are tetrahedral and paramagnetic, as anticipated for a d^9 configuration. Orange-red crystals of the tetrahedral Ni^I complex [Ni(PMe$_3$)$_4$][BPh$_4$] have also been structurally characterized.[23a]

Oxidation state 0 (d^{10})

Besides [Ni(CO)$_4$] and organometallic compounds discussed in the next section, nickel is found in the formally zero oxidation state with ligands such as CN$^-$ and phosphines. Reduction of K$_2$[Ni^{II}(CN)$_4$] with potassium in liquid ammonia precipitates yellow K$_4$[Ni^0(CN)$_4$], which is sensitive to aerial oxidation. Being isoelectronic with [Ni(CO)$_4$] it is presumed to be tetrahedral. Similarity with the carbonyl is still more marked in the case of the gaseous [Ni(PF$_3$)$_4$], which also can be prepared directly from the metal and ligand:

$$Ni + 4PF_3 \xrightarrow[350 \text{ atm}]{1000 \text{ C}} [Ni(PF_3)_4]$$

In sharp contrast to nickel, palladium and platinum form no simple carbonyls, and reports of K$_4$[M(CN)$_4$] (M = Pd, Pt) may well refer to hydrido complexes; in any event they are very unstable. The chemistry of these two metals in the zero oxidation state is in fact essentially that of their phosphine and arsine complexes and was initiated[24] by L. Malatesta and his school in the 1950s. Compounds of the type [M(PR$_3$)$_4$], of which [Pt(PPh$_3$)$_4$] has been most thoroughly studied, are in general yellow, air-stable solids or liquids obtained by reducing M^{II} complexes in H$_2$O or H$_2$O/EtOH solutions with hydrazine or sodium borohydride. They are tetrahedral molecules whose most important property is their readiness to dissociate in solution to form 3-coordinate,

[22] L. M. Venanzi, Complexes of tetradentate ligands containing phosphorus and arsenic, *Angew. Chem.*, Int. Edn. (Engl.) **3**, 453–60 (1964).

[23] C. Senoff, A five-coordinate palladium(II) complex containing tris{2-dimethylamino)ethyl}amine, N{CH$_2$CH$_2$N(CH$_3$)$_2$}$_3$, *Inorg. Chem.* **17**, 2320–2 (1978).

[23a] A. Gleizes, M. Dartiguenave, Y. Dartiguenave, J. Galy and H. F. Klein, Nickel phosphine complexes. Crystal and molecular structure of a new nickel(I) complex: [Ni(PMe$_3$)$_4$][BPh$_4$], *J. Am. Chem. Soc.* **99**, 5187–9 (1977).

[24] L. Malatesta and M. Angoletta, Palladium(0) compounds. Part II. Compounds with triarylphosphines, triarylphosphites and triarylarsines, *J. Chem. Soc.* 1957, 1186–8.

planar $[M(PR_3)_3]$ and, in traces, probably also $[M(PR_3)_2]$ species. The latter are intermediates in an extensive range of addition reactions (many of which may properly be regarded as *oxidative* additions) giving such products as $[Pt^{II}(PPh_3)_2L_2]$, $(L = O, CN, N_3)$ and $[Pt^{II}(PPh_3)_2LL']$, $(L,L' = H,Cl; R,I)$ as well as $[Pt^0(C_2H_4)(PPh_3)_2]$ and $[Pt^0(CO)_2(PPh_3)_2]$.

The mechanism by which this low oxidation state is stabilized for this triad has been the subject of some debate. That it is not straightforward is clear from the fact that nickel is alone in forming a simple binary carbonyl whereas palladium and platinum require the stabilizing presence of phosphines before CO can be attached. For most transition metals the π-acceptor properties of the ligand are thought to be of considerable importance and there is no reason to doubt that this is true for Ni^0. For Pd^0 and Pt^0, however, it appears that σ-bonding ability is also important, and the smaller importance of π backbonding which this implies is in accord with the higher ionization energies of Pd and Pt [804 and 865 kJ mol^{-1} respectively] compared with that for Ni [737 kJ mol^{-1}].

27.3.5 *Organometallic compounds*[9, 10, 25, 26]

All three of these metals have played major roles in the development of organometallic chemistry. The first compound containing an unsaturated hydrocarbon attached to a metal (and, indeed, the first organometallic compound if one excludes the cyanides) was $[Pt(C_2H_4)Cl_2]_2$, discovered by the Danish chemist W. C. Zeise as long ago as 1827 and followed 4 years later by the salt which bears his name, $K[Pt(C_2H_4)Cl_3].H_2O$. $[Ni(CO)_4]$ was the first metal carbonyl to be prepared when L. Mond and his co-workers discovered it in 1888. The platinum methyls, prepared in 1907 by W. J. Pope, were amongst the first-known transition metal alkyls, and the discovery by W. Reppe in 1940 that Ni^{II} complexes catalyse the cyclic oligomerization of acetylenes produced a surge of interest which was reinforced by the discovery in 1960 of the π-allylic complexes of which those of Pd^{II} are by far the most numerous.

σ-Bonded compounds

These are of two main types: compounds of Pt^{IV}, which have been known since the beginning of this century and commonly involve the stable $\{PtMe_3\}$ group; and compounds of the divalent metals, which were first studied by J. Chatt and co-workers in the late 1950's and are commonly of the type $[ML_2R_2](L = phosphine)$. In the Pt^{IV} compounds the metal is always octahedrally coordinated and this is frequently achieved in interesting ways. Thus the trimethyl halides, conveniently obtained by treating $PtCl_4$ with MeMgX in benzene, are tetramers, $[PtMe_3X]_4$, in which the 4 Pt atoms form a cube

[25] P. W. JOLLY and G. WILKE, *The Organic Chemistry of Nickel*, Academic Press, London, Vol. I, 1974, 517 pp.; Vol. II, 1975, 400 pp.
[26] P. M. MAITLIS, *The Organic Chemistry of Palladium*, Academic Press, London, 1971, Vol. I, 319 pp; Vol. II, 216 pp.

involving triply-bridging halogen atoms† (Fig. 27.11a). The dimeric [PtMe₃(acac)]₂ is also unusual in that the acac is both *O*- and *C*-bonded (Fig. 27.11b), while in [PtMe₃(acac)(bipy)] 7-coordination is avoided because the acac coordinates merely as a unidentate *C* donor.

The stabilities of the [ML₂R₂] phosphines increase from Ni^{II} to Pt^{II} and for Ni^{II} they are only isolable when R is an *o*-substituted aryl. Those of Pt^{II}, on the other hand, are amongst the most stable σ-bonded organo-transition metal compounds while those of Pd^{II} occupy an intermediate position.

Carbonyls (see p. 349)

On the basis of the 18-electron rule, the d⁸s² configuration is expected to lead to carbonyls of formula [M(CO)₄] and this is found for nickel. [Ni(CO)₄], the first metal carbonyl to be discovered (by C. Langer in L. Mond's laboratory, 1888), is an extremely toxic, colourless liquid (mp − 19.3°, bp 42.2°) which is tetrahedral in the vapour and in the solid (Ni–C 184 pm, C–O 115 pm). Its importance in the Mond process for manufacturing nickel metal has already been mentioned as has the absence of stable analogues of Pd and Pt. It may be germane to add that the introduction of halides (which are σ-bonded) reverses the situation: no carbonyl halides of Ni are known whereas the yellow [Pd^{II}(CO)Cl₂]ₙ, even though unstable, has been prepared and the colourless [Pt^{II}(CO)₂Cl₂] is quite stable.

---

† The chequered history of compounds of this type makes salutory reading. H. Gilman and M. Lichtenwalter (1938, 1953) reported the synthesis of PtMe₄ in 46% yield by reacting Me₃PtI with NaMe in hexane. R. E. Rundle and J. H. Sturdivant determined the X-ray crystal structure of this product in 1947 and described it as a tetramer [(PtMe₄)₄]: this required the concept of a multicentred, 2-electron bond, and was one of the first attempts to interpret the bonding in a presumed electron-deficient cluster compound. In fact, tetramethylplatinum cannot be prepared in this way and is unknown; Gilman's compound was actually a hydrolysis product [{PtMe₃(OH)}₄] and the mistaken identity of the crystal went undetected by the X-ray work because, at that time, the scattering curves for the 9-electron groups CH₃ and OH were indistinguishable in the presence of Pt. Interestingly, the compound PtMe₃(OH) had, in reality, already been synthesized by W. J. Pope and S. J. Peachey as long ago as 1909, and its tetrameric structure was confirmed by subsequent X-ray work.[26a] In a parallel study,[26b] the transparent tan-coloured tetramer [(PtMe₃I)₄] has now been shown to be the same compound as was previously erroneously reported in 1938 as hexamethyldiplatinum, [Me₃Pt-PtMe₃]. This was equally erroneously described in 1949 on the basis of an incomplete X-ray structural study as a methyl-bridged oligomer [(PtMe₃)₁₂] or an infinite chain of methyl-bridged 6-coordinate {PtMe₃}-groups. A qualitative test for iodine would have revealed the error 30 years earlier.

Although PtMe₄ remains unknown it has recently been shown that reaction of [PtMe₂(PPh₃)₂] with LiMe yields the square-planar Pt^{II} complex Li₂[PtMe₄], whereas reaction of [PtMe₃I)₄] with LiMe affords the octahedral Pt^{IV} complex Li₂[PtMe₆].[26c] The thermally stable, colourless, 8-coordinate complex [PtMe₃(η⁵-C₅H₅)] is also known.[26d]

²⁶ᵃ D. O. COWAN, N. G. KRIEGHOFF, and G. DONNAY, Reinvestigation of the reaction of trimethylplatinum(IV) iodide with methylsodium, *Acta Cryst.* **B24**, 287–8 (1968), and references therein.

²⁶ᵇ G. DONNAY, L. B. COLEMAN, N. G. KRIEGHOFF, and D. O. COWAN, Trimethylplatinum(IV) iodide and its misrepresentation as hexamethyldiplatinum, *Acta Cryst.* **B24**, 157–9 (1968), and references therein.

²⁶ᶜ G. W. RICE and R. S. TOBIAS, Synthesis of hexamethyl- and pentamethyl-platinate(IV) and tetramethyl- and trimethyl-platinate(II) complexes. Spectroscopic studies of organo-platinum complexes, *J. Am. Chem. Soc.* **99**, 2141–9 (1977).

²⁶ᵈ O. HACKELBERG and A. WOJCICKI, The photolysis of η⁵-cyclopentadienyltrimethylplatinum(IV), *Inorg. Chim. Acta* **44**, L63–L64 (1980).

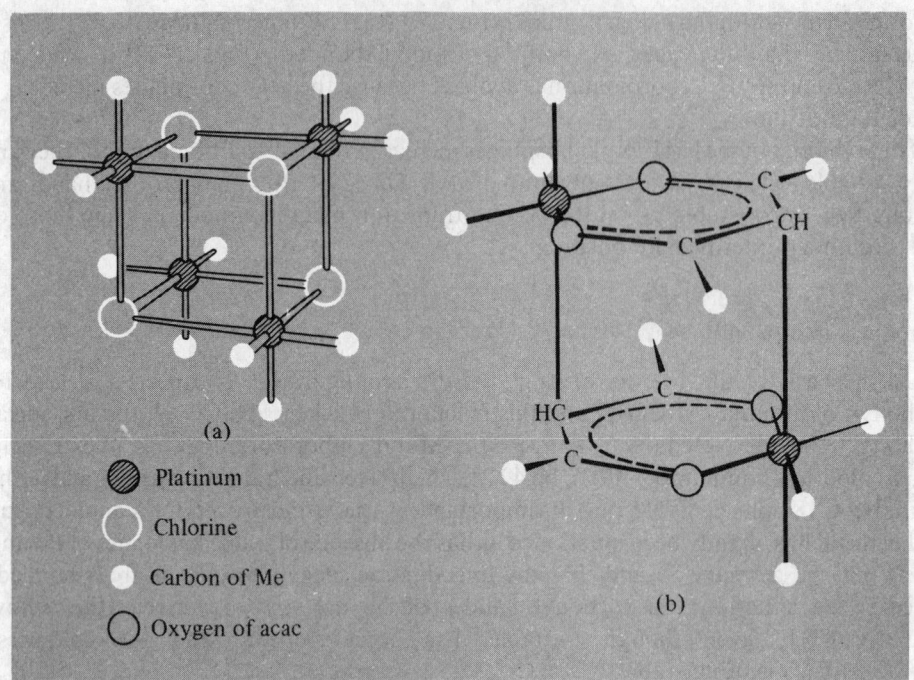

(a)

- ◑ Platinum
- ◯ Chlorine
- ◔ Carbon of Me
- ◯ Oxygen of acac

(b)

Fig. 27.11　Schematic representation of the structures of compounds containing octahedrally coordinated Pt^{IV}: (a) the tetramer, $[PtMe_3Cl]_4$, and (b) the dimer, $[PtMe_3(acac)]_2$.

$[Ni(CO)_4]$ is readily oxidized by air and can also be reduced by alkali metals in liquid ammonia to the red, dimeric carbonyl hydride, $[\{Ni\,H(CO)_3\}_2]\cdot 4NH_3$. Other reductions have yielded a number of polynuclear carbonylate anion clusters of which $[Ni_5(CO)_{12}]^{2-}$ and $[Ni_6(CO)_{12}]^{2-}$ (Fig. 27.12) may be mentioned.

Alkaline reduction of $[PtCl_6]^{2-}$ in an atmosphere of CO yields a series of platinum carbonylate anion clusters, $[Pt_3(CO)_6]_n^{2-}$, which adopt more nearly "prismatic", or "eclipsed" structures than the corresponding nickel compounds. By refluxing salts of the $n=3$ anion in acetonitrile, the spectacular, brown cluster anion $[Pt_{19}(CO)_{22}]^{4-}$ containing 2 totally encapsulated metal atoms has recently[27] been obtained (Fig. 27.12e). Apparently it has not yet been possible to prepare any analogous palladium compounds.

Cyclopentadienyls

Nickelocene, $[Ni^{II}(\eta^5\text{-}C_5H_5)_2]$, is a bright green, reactive solid, conveniently prepared by adding a solution of $NiCl_2$ in dimethylsulfoxide to a solution of KC_5H_5 in 1,2-dimethoxyethane. It has the sandwich structure of ferrocene, and is similarly susceptible to

27 D. M. WASHECHECK, E. J. WUCHERER, L. F. DAHL, A. CERIOTTI, G. LONGONI, M. MANASSERO, M. SANSONI, and P. CHINI, Synthesis, structure and stereochemical implications of the $[Pt_{19}(CO)_{12}(\mu_2\text{-}CO)_{10}]^{4-}$ tetranion, J. Am. Chem. Soc. 101, 6110–12 (1979).

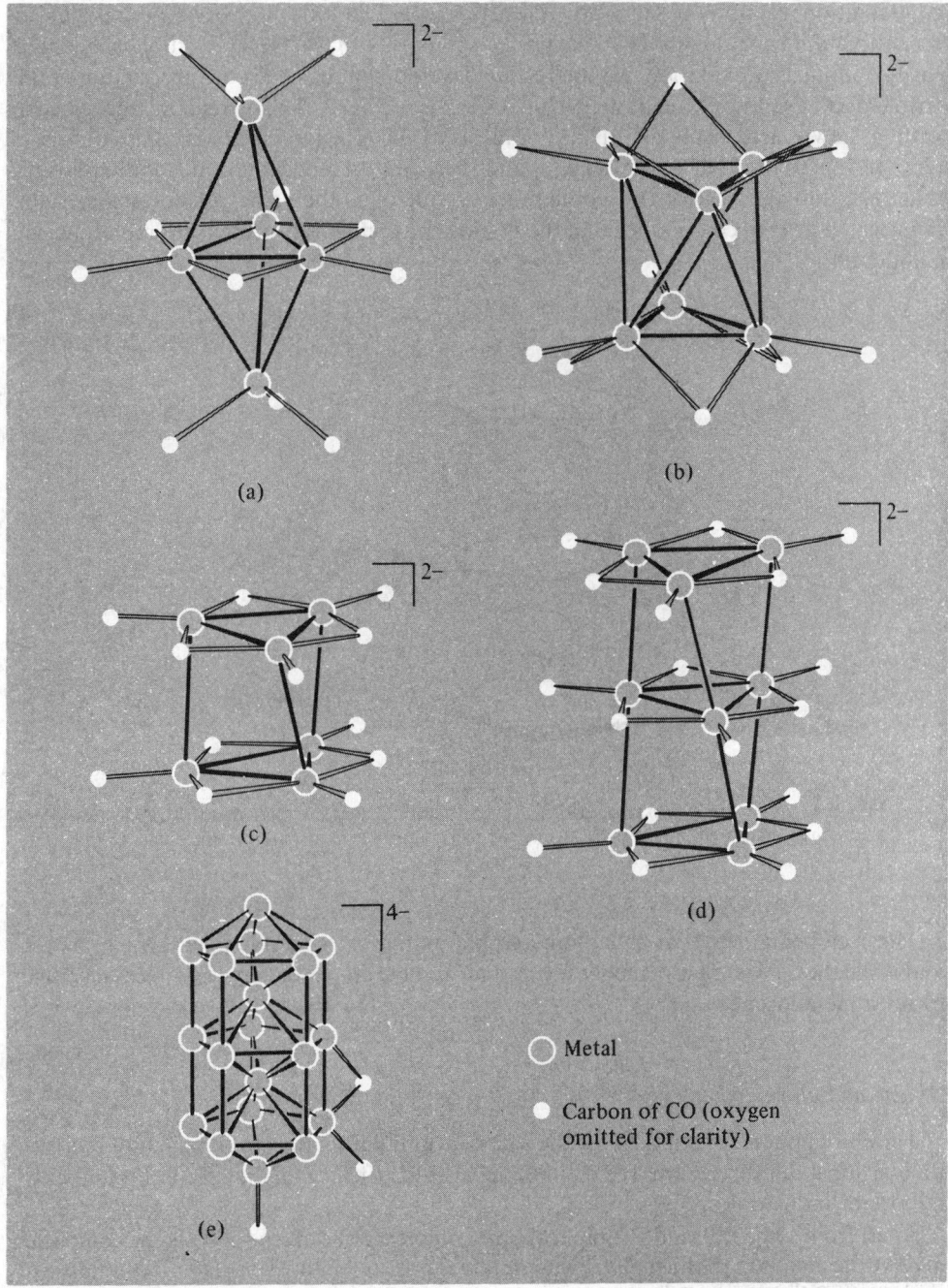

Fig. 27.12 Some carbonylate anion clusters of nickel and platinum: (a) $[Ni_5(CO)_{12}]^{2-}$, (b) $[Ni_6(CO)_{12}]^{2-}$, (c) $[Pt_6(CO)_{12}]^{2-}$, (d) $[Pt_9(CO)_{18}]^{2-}$, and (e) the Pt_{19} core of $[Pt_{19}(CO)_{22}]^{4-}$ showing one of the 10 bridging COs and 2 of the 12 terminal COs (which are attached to each of the 6 metal atoms at each end of the ion).

ring-addition reactions, but its 2 extra electrons ($\mu_e = 2.86$ BM) must be accommodated in an antibonding orbital (p. 371). The orange-yellow, $[Ni(\eta^5\text{-}C_5H_5)_2]^+$, cation is therefore easily obtained by oxidation. An interesting development has been the production of the "triple-decker sandwich" cation, $[Ni_2(\eta^5\text{-}C_5H_5)_3]^+$ (Fig. 27.13), by reacting nickelocene with a Lewis acid such as BF_3.[28] This is a 34 valence electron compound [i.e. $(2 \times 8) + (3 \times 6)$ for $2Ni^{II}$ and $3C_5H_5^-$] and there are theoretical grounds for supposing that this, and the 30-electron configuration, will offer the same sort of stability for binuclear sandwich compounds that the 18-electron configuration offers for mononuclear compounds.

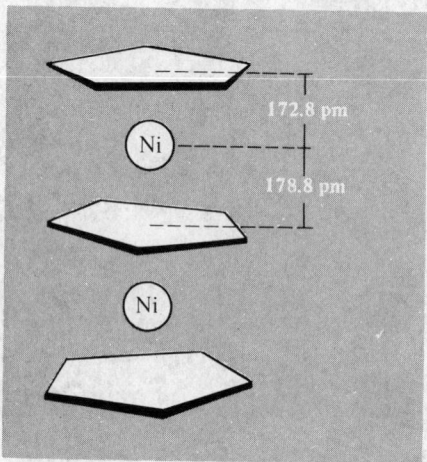

172.8 pm

178.8 pm

FIG. 27.13   The "triple-decker sandwich", $[Ni_2(\eta^5\text{-}C_5H_5)_3]^+$ cation. Note that the C_5H_5 rings are neither "staggered" nor "eclipsed", and the nickel atoms are closer to the outer than to the central ring.

The cyclopentadienyls of palladium and platinum are less stable than those of nickel, and while the heavier pair of metals form some monocyclopentadienyl complexes, neither forms a metallocene.

Alkene and alkyne complexes[29]

These are important not only for their part in stimulating the development of bonding theory (for a full discussion, see p. 360) but also for their catalytic role in a number of important industrial processes.

Apart from some Pd^0 and Pt^0 biphosphine complexes, the alkene and alkyne complexes involve the metals in the formally divalent state. Those of Ni^{II} are few in number compared to those of Pd^{II}, but it is Pt^{II} which provides the most numerous and stable compounds of this type. These are of the forms $[PtCl_3Alk]^-$, $[PtCl_2Alk]_2$, and $[PtCl_2Alk_2]$. They are

[28] H. WERNER, New varieties of sandwich complexes, *Angew. Chem.*, Int. Edn. (Engl.) **16**, 1–9 (1977).
[29] B. L. SHAW and N. I. TUCKER, Organo-transition metal compounds and related aspects of homogeneous catalysis, Chap. 53 in *Comprehensive Inorganic Chemistry*, Vol. 4, pp. 781–994, Pergamon Press, Oxford 1973.

generally prepared by treating an M^{II} salt with the hydrocarbon when a less strongly bonded anion is displaced. Thus, Zeise's salt (p. 357) may be obtained by prolonged shaking of a solution of K_2PtCl_4 in dil HCl with C_2H_4, though the reaction can be speeded-up by the addition of a small amount of $SnCl_2$ (see p. 462 of ref. 9). Treatment of an ethanolic solution of the product with conc HCl then affords the orange dimer, $[\{PtCl_2(C_2H_4)\}_2]$, and if this is then dissolved in acetone at $-70°C$ and further treated with C_2H_4, yellow, unstable crystals of the *trans*-bis(ethene) are formed:

cis-Substituted dichloro complexes are obtained if chelating dialkenes such as *cis-cis*-cyclooocta-1,5-diene (COD) are used (p. 361).

A common property of coordinated alkenes is their susceptibility to attack by nucleophiles such as OH^-, OMe^-, $MeCO_2^-$, and Cl^-, and it has long been known that Zeise's salt is slowly attacked by non-acidic water to give MeCHO and Pt metal, while corresponding Pd complexes are even more reactive. This forms the basis of the Wacker process (developed by J. Smidt and his colleagues at Wacker Chemie, 1959–60) for converting ethene (ethylene) into ethanal (acetaldehyde)—see Panel.

Catalytic Applications of Alkene and Alkyne Complexes

The Wacker process

Ethanal is produced by the aerial oxidation of ethene in the presence of $PdCl_2/CuCl_2$ in aqueous solution. The main reaction is the oxidative hydrolysis of ethene:

$$C_2H_4 + PdCl_2 + H_2O \longrightarrow MeCHO + Pd\downarrow + 2HCl$$

The mechanism of this reaction is not straightforward but the crucial step appears to be nucleophilic attack by water or OH^- on the coordinated ethene to give σ-bonded $-CH_2CH_2OH$ which then rearranges and is eventually eliminated as MeCHO with loss of a proton.

The commercial viability of the reaction depends on the formation of a catalytic cycle by reoxidizing the palladium metal *in situ*. This is achieved by the introduction of $CuCl_2$:

$$Pd + CuCl_2 \longrightarrow PdCl_2 + 2CuCl$$

Because the solution is slightly acidic, the $CuCl_2$ itself can be regenerated by passing in oxygen:

$$2CuCl + 2HCl + \tfrac{1}{2}O_2 \longrightarrow 2CuCl_2 + H_2O$$

The overall reaction is thus:

$$C_2H_4 + \tfrac{1}{2}O_2 \longrightarrow MeCHO$$

Continued

The Reppe Synthesis

Polymerization of alkynes by Ni^{II} complexes produces a variety of products which depend on conditions and especially on the particular nickel complex used. If, for instance, O-donor ligands such as acetylacetone or salicaldehyde are employed in a solvent such as tetrahydrofuran or dioxan, 4 coordination sites are available and cyclotetramerization occurs to give mainly cyclo-octatetraene (COT). If a less-labile ligand such as PPh_3 is incorporated, the coordination sites required for tetramerization are not available and cyclic trimerization to benzene predominates (Fig. A). These syntheses are amenable to extensive variation and adaptation. Substituted ring systems can be obtained from the appropriately substituted alkynes while linear polymers can also be produced.

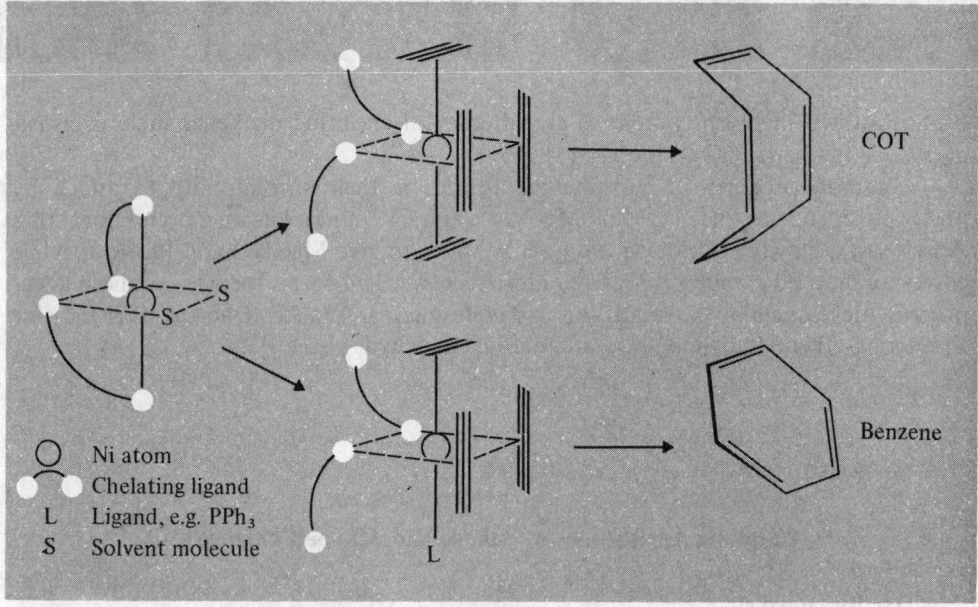

FIG. A   Cyclic oligomerizations of acetylene: tetramerization producing COT and trimerization producing benzene.

Alkyne complexes are essentially similar to the alkenes (p. 361) and those of Pt^{II}, particularly when the alkyne incorporates the t-butyl group, are the most stable. Ni^{II} alkyne complexes are less numerous and generally less stable but are of greater practical importance because of their role as intermediates in the cyclic oligomerization of alkynes, discovered by Reppe (see Panel).

π-Allylic complexes[29]

The preparation and bonding of complexes of the η^3-allyl radical, $CH_2{=}CH{-}CH_2{-}$, has already been discussed (p. 362). This group, and substituted derivatives of it, may act as σ-bonded ligands, but it is as 3-electron π-donor ligands that they are most important. Crudely:

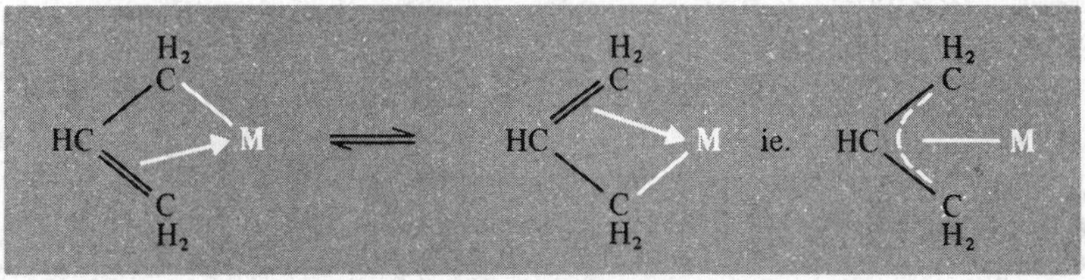

The π-allyl complexes of Pd^{II}, e.g. $[Pd(\eta^3\text{-}C_3H_5)X]_2$ (X = Cl, Br, I), are very stable and more numerous than for any other metal, and neither Ni nor Pt form as many of these complexes. Indeed, the contrast between Pd and Pt is such that in reactions with alkenes, where a particular compound of Pt is likely to form an alkene complex, the corresponding compound of Pd is more likely to form a π-allyl complex. The role of the Pd and Ni complexes as intermediates in the oligomerization of conjugated dienes (of which 1,3-butadiene, C_4H_6, is the most familiar) have been extensively studied, particularly by G. Wilke and his group. For instance in the presence of $[Ni(\eta^3\text{-}C_3H_5)_2]$ (or $[Ni(acac)_2]_3 + Al_2Et_6$), butadiene trimerizes, probably via the catalytic cycle:

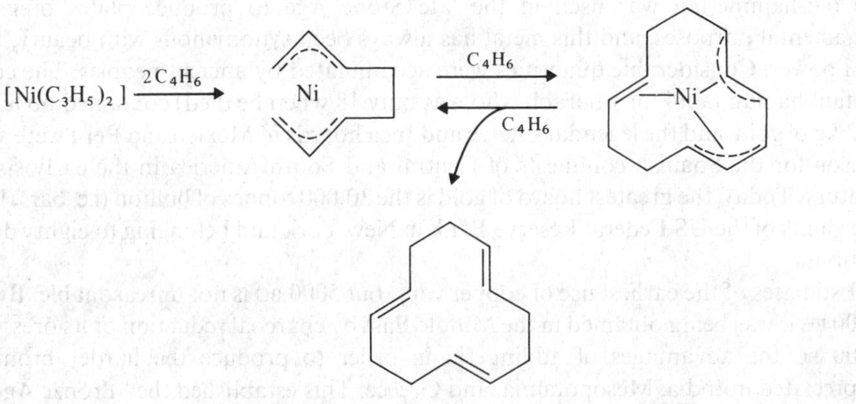

trans, trans, trans-cyclododeca-1,5,9-triene (CDT)

Other isomers of CDT are also obtained and, if a coordination site on the nickel is blocked by the addition of a ligand such as a tertiary phosphine, dimerization of the butadiene, rather than trimerization, occurs.

28

Copper, Silver, and Gold

28.1 Introduction[1]

Collectively known as the "coinage metals" because of their former usage, these elements were almost certainly the first three metals known to man. All of them occur in the elemental, or "native", form and must have been used as primitive money long before the introduction of gold coins in Egypt around 3400 BC.

Cold-hammering was used in the late Stone Age to produce plates of gold for ornamental purposes, and this metal has always been synonymous with beauty, wealth, and power. Considerable quantities were accumulated by ancient peoples. The coffin of Tutankhamun (a minor Pharaoh who was only 18 when he died) contained no less than 112 kg of gold, and the legendary Aztec and Inca hoards in Mexico and Peru were a major reason for the Spanish conquests of Central and South America in the early sixteenth century. Today, the greatest hoard of gold is the 30 000 tonnes of bullion (i.e. bars) lying in the vaults of the US Federal Reserve Bank in New York and belonging to eighty different nations.

Estimates of the earliest use of copper vary, but 5000 BC is not unreasonable. By about 3500 BC it was being obtained in the Middle East by charcoal reduction of its ores, and by 3000 BC the advantages of adding tin in order to produce the harder bronze was appreciated in India, Mesopotamia, and Greece. This established the "Bronze Age", and copper has continued to be one of man's most important metals.

The monetary use of silver may well be as old as that of gold but the abundance of the native metal was probably far less, so that comparable supplies were not available until a method of winning the metal from its ores had been discovered. It appears, however, that by perhaps 3000 BC a form of cupellation† was in operation in Asia Minor and its use gradually spread, so that silver coinage was of crucial economic importance to all subsequent classical Mediterranean civilizations.

The name *copper* and the symbol Cu are derived from *aes cyprium* (later Cuprum), since it was from Cyprus that the Romans first obtained their copper metal. The words *silver* and *gold* are Anglo-Saxon in origin but the chemical symbols for these elements (Ag and

---

† Cupellation processes vary but consist essentially of heating a mixture of precious and base (usually lead) metals in a stream of air in a shallow hearth, when the base metal is oxidized and removed either by blowing away or by absorption into the furnace lining. In the early production of silver, the sulfide ores must have been used to give first a silver/lead alloy from which the lead was then removed.

[1] R. F. TYLECOTE, *History of Metallurgy*, The Metals Society, London, 1976, 182 pp.

Au) are derived from the Latin *argentum* (itself derived from the Greek ἀργός, *argos*, shiny or white) and *aurum*, gold.

28.2   The Elements

28.2.1   *Terrestrial abundance and distribution*

The relative abundances of these three metals in the earth's crust (Cu 68 ppm, Ag 0.08 ppm, Au 0.004 pm) are comparable to those of the preceding triad—Ni, Pd, and Pt. Copper is found mainly as the sulfide, oxide, or carbonate, its major ores being copper pyrite (chalcopyrite), $CuFeS_2$, which is estimated to account for about 50% of all Cu deposits; copper glance (chalcocite), Cu_2S; cuprite, Cu_2O, and malachite, $Cu_2CO_3(OH)_2$. Large deposits are found in various parts of North and South America, and in Africa and the USSR. The native copper found near Lake Superior is extremely pure but the vast majority of current supplies of copper are obtained from low-grade ores containing only about 1% Cu.

Silver is widely distributed in sulfide ores of which silver glance (argentite), Ag_2S, is the most important. Native silver is sometimes associated with these ores as a result of their chemical reduction, while the action of salt water is probably responsible for their conversion into "horn silver", AgCl, which is found in Chile and New South Wales. The Spanish Americas provided most of the world's silver for the three centuries after about 1520, to be succeeded in the nineteenth century by Russia. Appreciable quantities are now obtained as a byproduct in the production of other metals such as copper, and the main "Western" producers are Mexico, Peru, Canada, the USA, and Australia.

Gold, too, is widely, if sparsely, distributed both native† and in tellurides, and is almost invariably associated with quartz or pyrite, both in veins and in alluvial or placer deposits laid down after the weathering of gold-bearing rocks. It is also present in sea water to the extent of around 1×10^{-3} ppm, depending on location, but no economical means of recovery has yet been devised. Prior to about 1830 a large proportion of the world's stock of gold was derived from ancient and South American civilizations (recycling is not a new idea), and the annual output of new gold was no more than 12 tonnes pa. This supply gradually increased with the discovery of gold in Siberia followed by "gold rushes" in 1849 (California: as a result of which the American West was settled), 1851 (New South Wales: within 7 y the population of Australia doubled to 1 million), 1884 (Transvaal), 1896 (Klondike, North-west Canada), and, finally, 1900 (Nome area of Alaska) as a result of which by 1890 world production had risen to 150 tonnes pa. It is now 8 times that amount, ~1200 tonnes pa.

28.2.2   *Preparation and uses of the elements*[2]

A few of the oxide ores of copper can be reduced directly to the metal by heating with coke, but the bulk of production is from sulfide ores containing iron, and they require

---

† The "Welcome Stranger" nugget found in Victoria, Australia, in 1869 weighed over 71 kg and yielded nearly 65 kg of refined gold but was, unfortunately, exceptional.

² *Kirk–Othmer Encyclopedia of Chemical Technology*, 3rd edn., Interscience, New York; for Cu see Vol. 6, 1979, pp. 819–69; for Ag see Vol. 21, 1983, pp. 1–32; for Au see Vol. 11, 1980, pp. 972–95. See also G. THORSEN, Extractive metallurgy of copper, pp. 1–41 of *Topics in Non-ferrous Extractive Metallurgy* (ed. A. R. Burkin), *Critical Reports in Applied Chemistry*, Vol. 1, Blackwell, Oxford, 1980; and *Gold*, Mineral Dossier No. 14, HMSO, 1975, 66 pp.

more complicated treatment. These ores are comparatively lean (often $\sim\frac{1}{2}\%$ Cu) and their exploitation requires economies of scale. They are therefore obtained in huge, open-pit operations employing shovels of up to 25 m^3 (900 ft^3) and trucks of up to 250 tonnes capacity, followed by crushing and concentration (up to 15–20% Cu) by froth-flotation. (The environmentally acceptable disposal of the many millions of tonnes of finely ground waste poses serious problems.) Silica is added to the concentrate which is then heated in a reverberatory furnace (blast furnaces are unsuitable for finely powdered ores) to about 1400°C when it melts. FeS is more readily converted to the oxide than is Cu$_2$S and so, with the silica, forms an upper layer of iron silicate slag leaving a lower layer of copper matte which is largely Cu$_2$S and FeS. The liquid matte is then placed in a converter (similar to the Bessemer converter) with more silica and a blast of air forced through it. This transforms the remaining FeS first to FeO and then to slag, while the Cu$_2$S is partially converted to Cu$_2$O and then to metallic copper:

$$2FeS + 3O_2 \longrightarrow 2FeO + 2SO_2$$

$$2Cu_2S + 3O_2 \longrightarrow 2Cu_2O + 2SO_2$$

$$2Cu_2O + Cu_2S \longrightarrow 6Cu + SO_2$$

The major part of this "blister" copper is further purified electrolytically by casting into anodes which are suspended in acidified CuSO$_4$ solution along with cathodes of purified copper sheet. As electrolysis proceeds the pure copper is deposited on the cathodes while impurities collect below the anodes as "anode slime" which is a valuable source of Ag, Au, and other precious metals.

About one-third of the copper used is secondary copper (i.e. scrap) but the annual production of new metal is nearly 8 million tonnes, the chief sources (1980) being the USA (15%), the USSR (15%), Chile (14%), Canada (9%), Zambia (8%) and Zaire (6%). The major use is as an electrical conductor but it is also widely employed in coinage alloys as well as the traditional bronze (Cu plus 7–10% Sn), brass (Cu–Zn), and special alloys such as *Monel* (Ni–Cu).

Most silver is nowadays produced as a byproduct in the manufacture of non-ferrous metals such as copper, lead, and zinc, when the silver follows the base metal through the concentration and smelting processes. In the case of copper production, for instance, the anode slimes mentioned above are treated with hot, aerated dilute H$_2$SO$_4$, which dissolves some of the base metal content, then heated with a flux of lime or silica to slag-off most of the remaining base metals, and, finally, electrolysed in nitrate solution to give silver of better than 99.9% purity. As with copper, much of the metal used is salvage but over 10 000 tonnes of new metal were produced in 1980, mainly from Mexico and the USSR (14% each), Peru (12%), Canada and the USA (10% each), Australia and Poland (7$\frac{1}{2}$% each). Photography accounts for the use of about one-third of this and it is also used in silverware and jewellery, electrically, for silvering mirrors, and in the high-capacity Ag–Zn, Ag–Cd batteries. A minor though important use from 1826 until recent times was as dental amalgam (Hg/γ-Ag$_3$Sn).

Traditionally, gold was recovered from river sands by methods such as "panning" which depend on the high density of gold (19.3 g cm^{-3}) compared with sand ($\sim$2.5 g cm^{-3})† but as such sources are largely worked out, modern production depends on

---

† In ancient times, gold-bearing river sands were washed over a sheep's fleece which trapped the gold. It seems likely that this was the origin of the Golden Fleece of Greek mythology.

the mining of the gold-containing rock (perhaps 25 ppm of Au). This is crushed to a fine powder to liberate the metallic grains and these are extracted either by amalgamation with mercury (after which the Hg is distilled off), or by the cyanide process (p. 1388). In the latter, the gold and any silver present is leached from the crushed rock with an aerated, dilute solution of cyanide.

$$4Au + 8NaCN + O_2 + 2H_2O \longrightarrow 4Na[Au(CN)_2] + 4NaOH$$

and then precipitated by adding Zn dust. Electrolytic refining may then be used to provide gold of 99.95% purity.†

Total annual production of new gold is now about 1200 tonnes of which (1980) 56% comes from South Africa, 25% from the USSR, 3.8% from Canada, and 2.5% from the USA. Over 80% of the gold from "Western" countries now passes through the London Bullion Market which was established in 1666. Prices were formerly stable at $35 per troy oz‡ but showed an astonishing increase from $225 per oz in January 1979 to $855 per oz one year later before settling back to ~$450 per oz in 1982.

The two main uses for gold are in settling international debts and in the manufacture of jewellery, but other important uses are in dentistry, the electronics industry (corrosion-free contacts), and the aerospace industry (brazing alloys and heat reflection), while in office buildings it has been found that a mere 20 pm film on the inside face of windows cuts down heat losses in winter and reflects unwanted infrared radiation in summer.

28.2.3 *Atomic and physical properties of the elements*

Some important properties are listed in Table 28.1. As gold has only one naturally occurring isotope, its atomic weight is known with considerable accuracy; Cu and Ag each have 2 stable isotopes, and a slight variability of their abundance in the case of Cu prevents its atomic weight being quoted with greater precision. This is the first triad since Ti, Zr, and Hf in which the ground-state electronic configuration of the free atoms is the same for the outer electrons of all three elements. Gold is the most electronegative of all metals: the value of 2.4 equals that for Se and approaches the value of 2.5 for S and I. Estimates of electron affinity vary considerably but typical values (kJ mol^{-1}) are Cu 87, Ag 97, and Au 193. These may be compared with values for H 87, O 143, and I 296 kJ mol^{-1}. Consistent with this the compound CsAu has many salt-like rather than alloy-like properties and, when fused, behaves much like other molten salts. Similarly when Au is dissolved in solutions of Cs, Rb, or K in liquid ammonia, the spectroscopic and other properties are best interpreted in terms of the solvated Au^- ion ($d^{10}s^2$) analogous to a halide ion (s^2p^6).[2a]

The elements are obtainable in a state of very high purity but some of their physical properties are nonetheless variable because of their dependence on mechanical history.

---

† Gold is commonly alloyed with other metals (an appropriate mixture of Ag and Cu will maintain the golden hue) in order to make it harder and cheaper. The proportion of gold is expressed in *carats*, a *carat* being a twenty-fourth part by weight of the metal so that pure gold is 24 *carats*. In the case of precious stones the *carat* expresses mass not purity and is then defined as 200 mg. The term is derived from the name of the small and very uniform seeds of the carob tree which in antiquity were used to weigh precious metals and stones (p. 300).

‡ 1 troy (or fine) oz = 31.1035 g as distinct from 1 oz avoidupois = 28.3495 g.

2a W. J. PEER and J. J. LAGOWSKI, Metal-ammonia solutions. 11. Au⁻, a solvated transition metal anion, *J. Am. Chem. Soc.* **100,** 6260–1 (1978), and references therein.

TABLE 28.1 *Some properties of the elements copper, silver, and gold*

Property		Cu	Ag	Au
Atomic number		29	47	79
Number of naturally occurring isotopes		2	2	1
Atomic weight		63.546 ($\pm$0.003)	107.8682($\pm$3)	196.9665
Electronic configuration		$[Ar]3d^{10}4s^1$	$[Kr]4d^{10}5s^1$	$[Xe]4f^{14}5d^{10}6s^1$
Electronegativity		1.9	1.9	2.4
Metal radius (12-coordinate)/pm		128	144	144
Effective ionic radius (6-coordinate)/pm	V	—	—	57
	III	54	75	85
	II	73	94	—
	I	77	115	137
Ionization energy/kJ mol^{-1}	1st	745.3	730.8	889.9
	2nd	1957.3	2072.6	1973.3
	3rd	3577.6	3359.4	(2895)
MP/°C		1083	961	1064
BP/°C		2570	2155	2808
ΔH_{fus}/kJ mol^{-1}		13.0	11.1	12.8
ΔH_{vap}/kJ mol^{-1}		307($\pm$6)	258($\pm$6)	343($\pm$11)
$\Delta H_{(monatomic\ gas)}$/kJ mol^{-1}		337($\pm$6)	284($\pm$4)	379($\pm$8)
Density$^{(a)}$ (20°C)/g cm^{-3}		8.95	10.49	19.32
Electrical resistivity (20°C)/μohm cm		1.673	1.59	2.35

$^{(a)}$ Depends on mechanical history of sample.

Their colours (Cu reddish, Ag white, and Au yellow) and sheen are so characteristic that the names of the metals are used to describe them.† Gold can also be obtained in red, blue, and violet colloidal forms by the addition of various reducing agents to very dilute aqueous solutions of gold(III) chloride. A remarkably stable example is the "Purple of Cassius", obtained by using $SnCl_2$ as reductant, which not only provides a sensitive test for Au^{III} but is also used to colour glass and ceramics. Colloidal silver and copper are also obtainable but are less stable.

The solid metals all have the fcc structure, like their predecessors in the periodic table, Ni, Pd, and Pt, and they continue the trend of diminishing mp and bp. They are soft, and extremely malleable and ductile, gold more so than any other metal. One gram of gold can be beaten out into a sheet of ~1.0 m² only 230 atoms thick (i.e. 1 cm³ to 18 m²); likewise 1 g Au can be drawn into 165 m of wire of diameter 20 μm. The electrical and thermal conductances of the three metals are also exceptional, pre-eminence in this case belonging to silver. All these properties can be directly related to the $d^{10}s^1$ electronic configuration.

28.2.4 *Chemical reactivity and trends*

Because of the traditional designation of Cu, Ag, and Au as a subdivision of the group containing the alkali metals (justified by their respective $d^{10}s^1$ and p^6s^1 electron

† The colours arise from the presence of filled d bands near the electron energy surface of the s–p conduction band of the metals (Fermi surface). X-ray data indicate that the top of the d-band is ~220 kJ mol^{-1} (2.3 eV/atom) below the Fermi surface for Cu so electrons can be excited from the d band to the s–p band by absorption of energy in the green and blue regions of the visible spectrum but not in the orange or red regions. For silver the excitation energy is rather larger (~385 kJ mol^{-1}) corresponding to absorption in the ultraviolet region of the spectrum. Gold is intermediate but much closer to Cu, the absorption in the near ultraviolet and blue region of the spectrum giving rise to the characteristic golden yellow colour of the metal.

configurations) some similarities in properties might be expected. Such similarities as do occur, however, are confined almost entirely to the stoichiometries (as distinct from the chemical properties) of the compounds of the +1 oxidation state. The reasons are not hard to find. A filled d shell is far less effective than a filled p shell in shielding an outer s electron from the attraction of the nucleus. As a result the first ionization energies of the coinage metals are much higher, and their ionic radii smaller than those of the corresponding alkali metals (Table 28.1 and p. 86). They consequently have higher mps, are harder, denser, less reactive, less soluble in liquid ammonia, and their compounds more covalent. Again, whereas the alkali metals stand at the top of the electrochemical series (with $E°$ between -3.045 and -2.714 V), the coinage metals are near the bottom: Cu^+/Cu $+0.521$, Ag^+/Ag $+0.799$, $Au^+/Au + 1.691$ V. On the other hand, a filled d shell is more easily disrupted than a filled p shell and the second and third ionization energies of the coinage metals are therefore *lower* than those of the alkali metals so that they are able to adopt oxidation states higher than $+1$. They also more readily form coordination complexes. In short, Cu, Ag, and Au are transition metals whereas the alkali metals are not. Indeed, the somewhat salt-like character of CsAu and the formation of the solvated Au^- ion in liquid ammonia, mentioned above, can be regarded as halogen-like behaviour arising because the $d^{10}s^1$ configuration is 1 electron short of the closed configuration $d^{10}s^2$ (cf hydrogen, p. 51).

Copper, silver, and gold are notable in forming an extensive series of alloys with many other metals and many of these have played an important part in the development of technology through the ages (p. 1364). In many cases the alloys can be thought of as nonstoichiometric intermetallic compounds of definite structural types and, despite the apparently bizarre formulae that emerge from the succession of phases, they can readily be classified by a set of rules first outlined by W. Hume-Rothery in 1926. The determining feature is the ratio of the number of electrons to the number of atoms ("electron concentration"), and because of this the phases are sometimes referred to as "electron compounds".

The fcc lattice of the coinage metals has 1 valency electron per atom ($d^{10}s^1$). Admixture with metals further to the right of the periodic table (e.g. Zn) increases the electron concentration in the primary alloy (α-phase) which can be described as an fcc solid solution of M in Cu, Ag, or Au. This continues until, as the electron concentration approaches 1.5, the fcc structure becomes less stable than a bcc arrangement which therefore crystallizes as the β-phase (e.g. β-brass, CuZn; see phase diagram). Further increase in electron concentration results in formation of the more complex γ-brass phase of nominal formula Cu_5Zn_8 and electron concentration of $\{(5 \times 1)+(8 \times 2)\}/13 = 21/13 = 1.615$. The phase is still cubic but has 52 atoms in the unit cell (i.e. $4Cu_5Zn_8$). This γ-phase can itself take up more Zn until a third critical concentration is reached near 1.75 (i.e. 7/4) when the hcp ε-phase of $CuZn_3$ is formed. Hume-Rothery showed that this succession of phases is quite general (and also held for the Group VIII and platinum metals to the left of the coinage metals if they were taken to contribute no electrons to the lattice). The results for the coinage metals are summarized in Table 28.2.

The reactivity of Cu, Ag, and Au decreases down the group, and in its inertness gold resembles the platinum metals. All three metals are stable in pure dry air at room temperature but copper forms Cu_2O at red heat.† Copper is also attacked by sulfur and

† It was because of their resistance to attack by air, even when heated, that gold and silver were referred to as *noble* metals by the alchemists.

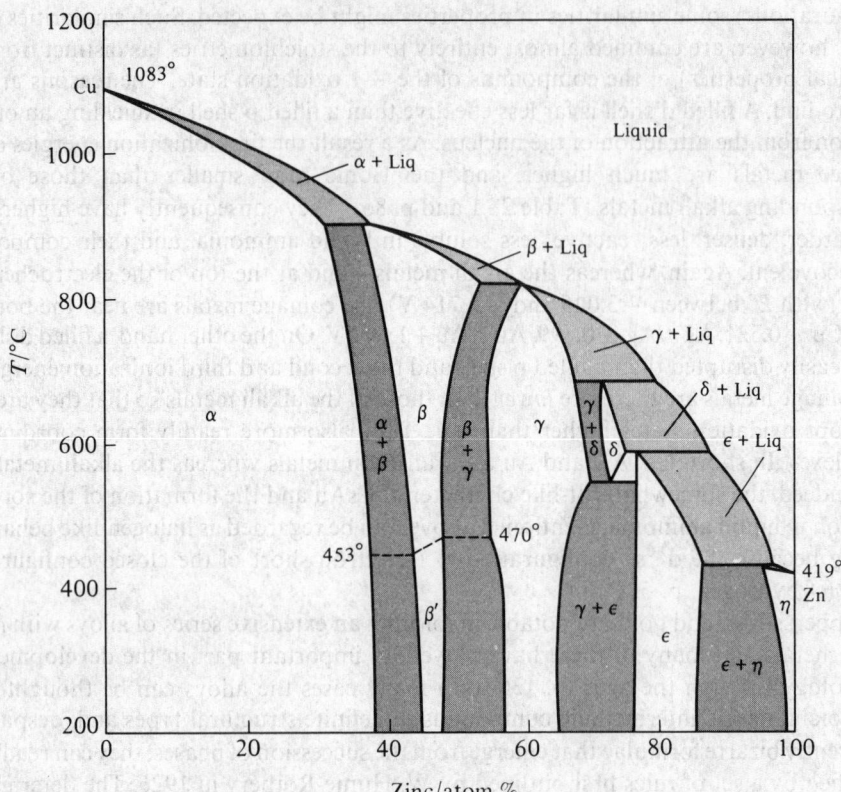

Phase diagram of the system Cu/Zn.

TABLE 28.2   *Hume-Rothery classification of alloy phases*

Electron concentration	Phase (structure)	Examples
1.0 to ~1.4	α-solid soln (fcc)	CuM_x, AgM_x, AuM_x
~1.5 (i.e. 3:2 or 21:14)	β-phase (bcc)	CuBe, CuZn; Cu_3Al, Cu_3Ga, Cu_3In; Cu_5Si, Cu_5Sn
		AgMg, AgZn, AgCd; Ag_3Al, Ag_3In
		AuMg, AuZn, AuCd; Au_3Al (complex cubic β-Mn structure)
	β-phase (hcp)	Ag_3Ga, Au_3In; Cu_5Ge, Ag_5Sn, Au_5Sn; Ag_7Sb
~1.62 (i.e. 21:13)	γ-phase (complex cubic with $4 \times 13 = 52$ atoms in unit cell)	Cu_5Zn_8, Cu_5Cd_8, Cu_5Hg_8; Cu_9Al_4, Cu_9Ga_4, Cu_9In_4; $Cu_{31}Si_8$, $Cu_{31}Sn_8$
		Ag_5Zn_8, Ag_5Cd_8, Ag_5Hg_8; Ag_9In_4
		Au_5Zn_8, Au_5Cd_8; Au_9In_4
~1.75 (i.e. 7:4 or 21:12)	ε-phase (hcp)	$CuBe_3$, $CuZn_3$, $CuCd_3$; Cu_3Si, Cu_3Ge, Cu_3Sn; $Cu_{13}Sb_3$
		$AgZn_3$, $AgCd_3$; Ag_5Al_3; Ag_3Sn; $Ag_{13}Sb_3$
		$AuZn_3$, $AuCd_3$; Au_5Al_3; Au_3Sn

halogens, and the sensitivity of silver to sulfur and its compounds is responsible for the familiar tarnishing of the metal (black AgS) when exposed to air containing such substances. Under similar circumstances copper forms a green coating of a basic sulfate. In sharp contrast, gold is the only metal which will not react directly with sulfur. In general the reactions of the metals are assisted by the presence of oxidizing agents. Thus, in the absence of air, non-oxidizing acids have little effect, but Cu and Ag dissolve in hot conc H_2SO_4 and in both dil and conc HNO_3, while Au dissolves in conc HCl if a strong oxidizing agent is present (e.g. in *aqua regia*, a 3:1 mixture of conc HCl and conc HNO_3, so named by alchemists because it dissolves gold, the king of metals). In addition, the metals dissolve readily in aqueous cyanide solutions in the presence of air or, better still, H_2O_2.

Table 28.3 is a list of typical compounds of the elements, which reveals a further reduction in the range of oxidation states consequent on the stabilization of d orbitals at the end of the transition series. Apart from a single Cu^{IV} fluoro-complex, neither Cu nor Ag is known to exceed the oxidation state $+3$ and even Au does so only in a few Au^V fluoro-compounds (see below): these may owe their existence at least in part to the stabilizing effect of the t_{2g}^6 configuration. It is also significant that, in a number of

TABLE 28.3   *Oxidation states and stereochemistries of compounds of copper, silver, and gold*

Oxidation state	Coordination number	Stereochemistry	Cu	Ag/Au
-1 ($d^{10}s^2$)	?	?		$[Au(NH_3)_n]^-$ (liq NH_3)
0 ($d^{10}s^1$)	3	Planar	$[Cu(CO)_3]$ (10 K)	$[Ag(CO)_3]$ (10 K)
	4	—	$[(CO)_3CuCu(CO)_3]$ (30 K)	$[(CO)_3AgAg(CO)_3]$ (30 K)
$< +1$	8	See Fig. 28.8(a)		$[(Ph_3P)Au\{Au(PPh_3)\}_7]^{2+}$
	10	See Fig. 28.8(c)		$[Au_{11}I_3\{P(C_6H_4-4-F)_3\}_7]$
	12	Icosahedral		$[Au_{13}Cl_2(PMe_2Ph)_{10}]^{3+}$
1 (d^{10})	2	Linear	$[CuCl_2]^-$, Cu_2O	$[M(CN)_2]^-$
	3	Trigonal planar	$[Cu(CN)_3]^{2-}$	$[AgI(PEt_2Ar)_2]$, $[AuCl(PPh_3)_2]$
	4	Tetrahedral	$[Cu(py)_4]^+$	$[M(diars)_2]^+$, $[Au(PMePh_2)_4]^{+}$ (d)
		Square planar		$[Au\{\eta^2-Os_3(CO)_{10}H\}_2]^-$ (e)
	6	Octahedral		AgX (X = F, Cl, Br)
2 (d^9)	4	Tetrahedral	$Cs_2[CuCl_4]$ (a)	
		Square planar	$[EtNH_3]_2[CuCl_4]$ (a)	$[Ag(py)_4]^{2+}$
				$[Au\{S_2C_2(CN)_2\}_2]^{2-}$
	5	Trigonal bipyramidal	$[Cu(bipy)_2I]^+$	
		Square pyramidal	$[\{Cu(DMGH)_2\}_2]$ (b)	
	6(c)	Octahedral	$K_2Pb[Cu(NO_2)_6]$	
3 (d^8)	4	Square planar	$[CuBr_2(S_2CNBu_2^i)]$	$[AgF_4]^-$, $[AuBr_4]^-$
	6	Octahedral	$[CuF_6]^{3-}$	$[AgF_6]^{3-}$, $[AuI_2(diars)_2]^+$
4 (d^7)	6	?	$[CuF_6]^{2-}$	
5 (d^6)	6	Octahedral (?)		$[AuF_6]^-$

(a) See text, p. 1385.

(b) $DMGH_2$ = dimethylglyoxime; see also Fig. 28.4.

(c) Higher coordination numbers for Cu^{II} are found in the 7-coordinate pentagonal bipyramidal complex $[Cu(H_2O)_2(dps)]^{2+}$ (dps = 2,6-diacetylpyridinebissemicarbazone) and the 8-coordinate, very distorted dodecahedral tetrakischelate $[Cu(O_2CMe)_4]^{2+}$.

(d) R. C. ELDER, E. H. K. ZEIHER, M. ONADY, and R. R. WHITTLE, *JCS Chem. Comm.* 1981, 900–1. See also R. V. PARISH, O. PARRY, and C. A. McAULIFFE, *JCS Dalton* 1981, 2098–2104.

(e) B. F. G. JOHNSON, D. A. KANER, J. LEWIS, and P. RAITHBY, *JCS Chem. Comm.* 1981, 753–5.

instances, the $+1$ oxidation state no longer requires the presence of presumed π-acceptor ligands even though the M^I metals are to be regarded mainly as class b in character. Stable, zero-valent compounds are not found, but a number of cluster compounds with the metal in a fractional (<1) oxidation state, especially of gold, have excited recent interest. The only aquo ions of this group are those of Cu^I (unstable), Cu^{II}, Ag^I, and Ag^{II} (unstable), and the best-known oxidation states, particularly in aqueous solution, are $+2$ for Cu, $+1$ for Ag, and $+3$ for Au. This accords with their ionization energies (Table 28.1) though, of course, few of the compounds are completely ionic. Silver has the lowest first ionization energy, while the sum of first and second is lowest for Cu and the sum of first, second, and third is lowest for Au. This is an erratic sequence, however, and illustrates what is perhaps the most notable feature of the triad from a chemical point of view, namely that the elements are not well related either as three elements showing a monotonic gradation in properties or as a triad comprising a single lighter element together with a pair of closely similar heavier elements. "Horizontal" similarities with their neighbours in the periodic table are in fact more noticable than "vertical" ones.

The reasons are by no means certain but no doubt involve several factors, of which size is probably a major one. Thus the Cu^{II} ion is smaller than Cu^I and, having twice the charge, interacts much more strongly with solvent water (heats of hydration are -2100 and -580 kJ mol^{-1} respectively). The difference is evidently sufficient to outweigh the second ionization energy of copper and to render Cu^{II} more stable in aqueous solution (and in ionic solids) than Cu^I, in spite of the stable d^{10} configuration of the latter. In the case of silver, however, the ionic radii are both much larger and so the difference in hydration energies will be much smaller; in addition the second ionization energy is even greater than for copper. The $+1$ ion with its d^{10} configuration is therefore the more stable. For gold, it is evident that the effect of the lanthanide contraction is now much less than in preceding groups, since the ionic radii of Au are appreciably larger than those of Ag.† The third electron is therefore more readily removed in gold and, coupled with the high CFSE associated with square planar d^8 ions (see p. 1346), this makes $+3$ the favoured oxidation state for gold.

Coordination numbers in this triad are again rarely higher than 6, but the univalent metals provide examples of the coordination number 2 which tends to be uncommon in transition metals proper (i.e. excluding Zn, Cd, and Hg).

Organometallic chemistry is not particularly extensive even though gold alkyls were amongst the first organo-transition metal compounds to be prepared. Those of Au^{III} are the most stable in this group, while Cu^I and Ag^I (but not Au^I) form complexes, of lower stability, with unsaturated hydrocarbons.

28.3 Compounds of Copper, Silver, and Gold [3-6]

Binary carbides, M_2C_2 (i.e. acetylides), are obtained by passing C_2H_2 through ammoniacal solutions of Cu^+ and Ag^+. Both are explosive when dry but regenerate

---

† Note, however, the similarity in 12-coordinate-metallic radii: Ag 144.2, Au 143.9 pm.

[3] A. G. MASSEY, Copper, Chap. 27 in *Comprehensive Inorganic Chemistry*, Vol. 3, pp. 1–78, Pergamon Press, Oxford, 1973.
[4] N. R. THOMPSON, Silver, Chap. 28 in *CIC*, Vol. 3, pp. 79–128, Pergamon Press, Oxford, 1973.
[5] B. F. G JOHNSON and R. DAVIS, Gold, Chap. 29 in *CIC*, Vol. 3, pp. 129–86, Pergamon Press, 1973.
[6] R. J. PUDDEPHAT, *The Chemistry of Gold*, Elsevier, Amsterdam, 1978, 274 pp.

acetylene if treated with a dilute acid. Copper and silver also form explosive azides while the even more dangerous "fulminating" silver and gold, which probably contain M_3N, are produced by the action of aqueous ammonia on the metal oxides. None of the metals reacts significantly with H_2 but the reddish-brown precipitate, obtained when aqueous $CuSO_4$ is reduced by hypophosphorous acid (H_3PO_2), is largely CuH.

28.3.1 *Oxides and sulfides*

Two oxides of copper, Cu_2O (yellow or red) and CuO (black), are known, both with narrow ranges of homogeneity and both formed when the metal is heated in air or O_2, Cu_2O being favoured by high temperatures. Cu_2O (mp 1230°) is conveniently prepared by the reduction in alkaline solution of a Cu^{II} salt using hydrazine or a sugar.† CuO is best obtained by igniting the nitrate or basic carbonate of Cu^{II}. Addition of alkali to aqueous solutions of Cu^{II} gives a pale-blue precipitate of $Cu(OH)_2$. This will redissolve in acids but is somewhat amphoteric since it is also soluble in conc alkali giving deep-blue solutions probably containing species of the type $[Cu(OH)_4]^{2-}$. The lower affinities of silver and gold for oxygen lead to oxides of lower thermal stabilities than those of copper. Ag_2O is a dark-brown precipitate produced by adding alkali to a soluble Ag^I salt; AgOH is probably present in solution but not in the solid. It is readily reduced to the metal, and decomposes to the elements if heated above 160°C. The action of the vigorous oxidizing agent, $S_2O_8{}^{2-}$, on Ag_2O or other Ag^I compounds, produces a black oxide of stoichiometry AgO. That this is not a compound of Ag^{II} is, however, evidenced by its diamagnetism and by diffraction studies which show it to contain two types of silver ion, one with 2 colinear oxygen neighbours (Ag^I–O 218 pm) and the other with square-planar coordination (Ag^{III}–O 205 pm). It is therefore formulated as $Ag^IAg^{III}O_2$. Hydrothermal treatment of AgO in a silver tube at 80°C and 4 kbar leads to crystals of a new oxide which was originally (1963) incorrectly designated as $Ag_2O(II)$. The compound has a metallic conductivity and a recent single-crystal X-ray study has shown that the stoichiometry is, in fact, Ag_3O; it can be described as an anti-BiI_3 structure (p. 652) in which oxide ions fill two-thirds of the octahedral sites in a hcp arrangement of Ag atoms (Ag-O 229 pm; Ag-Ag 276, 286, and 299 pm).[6a]

The action of alkali on aqueous Au^{III} solutions produces a precipitate, probably of $Au_2O_3.xH_2O$, which on dehydration yields brown Au_2O_3. This is the only confirmed oxide of gold. It decomposes if heated above about 160°C and, when hydrous, is weakly acidic, dissolving in conc alkali and probably forming salts of the $[Au(OH)_4]^-$ ion.

The sulfides are all black, or nearly so, and those with the metal in the +1 oxidation state are the more stable (p. 1365). Cu_2S (mp 1130°) is formed when copper is heated strongly in sulfur vapour or H_2S, and CuS is formed as a colloidal precipitate when H_2S is passed through aqueous solutions of Cu^{2+}. CuS, however, is not a simple copper(II) compound since it contains the S_2 unit and is better formulated as $Cu^I_2Cu^{II}(S_2)S$. Ag_2S is very readily formed from the elements or by the action of H_2S on the metal or on aqueous

---

† This is the basis of the very sensitive Fehling's test for sugars and other reducing agents. A solution of a copper(II) salt dissolved in alkaline tartrate solution is added to the substance in question. If this is a reducing agent then a characteristic red precipitate is produced.

6a W. BEESK, P. G. JONES, H. RUMPEL, E. SCHWARZMANN and G. M. SHELDRICK, X-ray crystal structure of Ag_6O_2, *JCS Chem. Comm.* 1981, 664–5.

Ag^l. The action of H_2S on aqueous Au^l precipitates Au_2S whereas passing H_2S through cold solutions of $AuCl_3$ in dry ether yield Au_2S_3, which is rapidly reduced to Au^l or the metal on addition of water. The relationships between the crystal structures of the oxides and sulfides of Cu, Ag, and Au and the binding energies of the metal's d and p valence orbitals have been reviewed.[6b]

The selenides and tellurides of the coinage metals are all metallic and some, such as $CuSe_2$, $CuTe_2$, $AgTe_{\sim 3}$, and Au_3Te_5 are superconductors at low temperature (as also are CuS and CuS_2). Other phases are CuSe, CuTe; Cu_3Se_2, Cu_3Te_2; AgSe, AgTe; Ag_2Se_3, $AgSe_2$; Ag_5Te_3; Au_2Te_3 and $AuTe_2$. Most of these are nonstoichiometric.

28.3.2 *Halides*[3-7]

Table 28.4 is a list of the known halides: only gold forms a pentahalide and trihalides and, with the exception of AgF_2, only copper forms dihalides (as yet).

AuF_5 is an unstable, polymeric, diamagnetic, dark-red powder, produced by heating $[O_2][AuF_6]$ under reduced pressure and condensing the product on to a "cold finger":[8]

$$Au + O_2 + 3F_2 \xrightarrow[8 \text{ atm}]{370} O_2AuF_6 \xrightarrow[\text{(hot/cold)}]{180/20} AuF_5 + O_2 + \tfrac{1}{2}F_2$$

TABLE 28.4   *Halides of copper, silver, and gold (mp/°C)*

Oxidation state	Fluorides	Chlorides	Bromides	Iodides
+5	AuF_5 red (d > 60°)			
+3	AuF_3 orange-yellow (subl 300°)	$AuCl_3$ red (d > 160°)	$AuBr_3$ red-brown	
+2	CuF_2 white (785°) AgF_2 brown (690°)	$CuCl_2$ yellow-brown (630°)	$CuBr_2$ black (498°)	
+1	 AgF yellow (435°)	CuCl white (422°) AgCl white (455°) AuCl yellow (d > 420°)	CuBr white (504°) AgBr pale yellow (430°)	CuI white (606°) AgI yellow (556°) AuI yellow
$+\tfrac{1}{2}(0,+1)$	Ag_2F yellow-green (d > 100°)			

[6b] J. A. TOSSELL and D. J. VAUGHAN, Relationships between valence orbital binding energies and crystal structure in compounds of copper, silver, gold, zinc, cadmium, and mercury, *Inorg. Chem.* **20**, 3333–40 (1981).

[7] R. COLTON and J. H. CANTERFORD, *Halides of the First Row Transition Metals*, Chap. 9, pp. 485–574, Wiley, London, 1969; and *Halides of the Second and Third Row Transition Metals*, Chap. 11, pp. 390–403, Wiley, London, 1968.

[8] M. J. VASILE, T. J. RICHARDSON, F. A. STEVIE, and W. E. FALCONER, Preparation and characterization of gold pentafluoride, *JCS Dalton* 1976, 351–3.

The compound tends to dissociate into AuF_3 and, when treated with XeF_2 in anhydrous HF solution below room temperature, yields yellow-orange crystals of the complex $[Xe_2F_3][AuF_6]$:

$$AuF_5 + 2XeF_2 \xrightarrow{\text{HF/0}} [Xe_2F_3][AuF_6] \xrightarrow{>60} AuF_3 + XeF_2 + XeF_4$$

Again, in the $+3$ oxidation state, only gold is known to form binary halides, though AuI_3 has not been isolated. The chloride and the bromide are red-brown solids prepared directly from the elements and have a planar dimeric structure in both the solid and vapour phases. Dimensions for the chloride are as shown in Structure (1). On being heated

(**1**) Au_2Cl_6

(**2**) Unique helical chain structure of AuF_3.

both compounds lose halogen to form first the monohalide and finally metallic gold. Au_2Cl_6 is one of the best-known compounds of gold and provides a convenient starting point for much coordination chemistry, dissolving in hydrochloric acid to give the stable $[AuCl_4]^-$ ion. Treatment of Au_2Cl_6 with F_2 or BrF_3 also affords a route to AuF_3, a powerful fluorinating agent. This orange solid consists of square-planar AuF_4 units which share *cis*-fluorine atoms with 2 adjacent AuF_4 units so as to form a helical chain (Structure (2)).

No halides are known for gold in the $+2$ oxidation state and silver only forms the difluoride; this is obtained by direct heating of silver in a stream of fluorine. AgF_2 is thermally stable but is a vigorous fluorinating agent used especially to fluorinate hydrocarbons. For copper, on the other hand, 3 dihalides are stable and the anhydrous difluoride, dichloride, and dibromide can all be obtained by heating the elements. The white ionic CuF_2 has a distorted rutile structure (p. 1120) with four shorter equatorial distances (Cu–F 193 pm) and two longer axial distances (Cu–F 227 pm). A similar distortion is found in the d^4 compound CrF_2 (p. 1190). When prepared from aqueous solution by dissolving copper(II) carbonate or oxide in 40% hydrofluoric acid, blue crystals of the dihydrate are obtained; these are composed of puckered sheets of planar *trans*-$[CuF_2(H_2O)_2]$ groups linked by strong H bonds to give distorted octahedral coordination about Cu with 2 Cu–O 194 pm, 2 Cu–F 190 pm, and 2 further Cu–F at 246.5 pm; the O–H$\cdots$F distance is 271.5 pm. With anhydrous $CuCl_2$ and $CuBr_2$ their increasing covalency is reflected in their polymeric chain structure, consisting of planar

CuX$_4$ units with opposite edges shared, and by the deepening colours of brown and black respectively. The chloride and bromide are both very soluble in water, and various hydrates and complexes can be recrystallized. The solutions are more conveniently obtained by dissolution of the metal or Cu(OH)$_2$ in the relevant hydrohalic acid.

Iodide ions reduce Cu^{II} to Cu^I, and attempts to prepare copper(II) iodide therefore result in the formation of CuI. (In a quite analogous way attempts to prepare copper(II) cyanide yield CuCN instead.) In fact it is the electronegative fluorine which fails to form a salt with copper(I), the other 3 halides being white insoluble compounds precipitated from aqueous solutions by the reduction of the Cu^{II} halide. By contrast, silver(I) provides (for the only time in this triad) 4 well-characterized halides. All except AgI have the rock-salt structure (p. 273).† Increasing covalency from chloride to iodide is reflected in the deepening colour white→yellow, as the energy of the charge transfer (X$^-$Ag$^+$→XAg) is lowered, and also in increasing insolubility. In the latter respect, however, AgF is quite anomalous in that it is one of the few silver(I) salts which form hydrates (2H$_2$O and 4H$_2$O). That it is soluble in water is understandable in view of its ionic character and the high solvation energy of the small fluoride ion, but the *extent* of its solubility (1800 g l^{-1} at 25°C) is surprising. All 4 AgX can be prepared directly from the elements but it is more convenient to prepare AgF by dissolving AgO in hydrofluoric acid and evaporating the solution until the solid crystallizes; the others can be made by adding X$^-$ to a solution of AgNO$_3$ or other soluble Ag^I compound, when AgX is precipitated. The most important property of these halides, particularly AgBr, is their sensitivity to light (AgF only to ultraviolet) which is the basis for their use in photography, discussed below.

Gold, like copper, forms three monohalides, the high electronegativity of fluorine being incompatible with the covalency required by gold(I). AuCl and AuBr are formed by heating the trihalides to no more than 150°C and AuI by heating the metal and iodine. At higher temperatures they all dissociate into the elements. AuI is a chain polymer which features linear 2-coordinate Au with Au–I 262 pm and the angle Au–I–Au 72°.

28.3.3 *Photography*[9]

Photography is a good example of a technology which evolved well in advance of a proper understanding of the principles involved (see Panel). Most of the basic processes

---

† At room temperature the stable form of silver iodide is γ-AgI which has the cubic zinc blende structure (p. 1405). β-AgI, which has the hexagonal ZnO (or wurtzite) structure (p. 1405), is the stable form between 136° and 146°. This structure is closely related to that of hexagonal ice (p. 731) and AgI has been found to be particularly effective in nucleating ice crystals in super-cooled clouds, thereby inducing the precipitation of rain. β-AgI has another remarkable property: at 146° it undergoes a phase change to cubic α-AgI in which the iodide sublattice is rigid but the silver sublattice "melts". This has a dramatic effect on the (ionic) electrical conductivity of the solid which leaps from 3.4×10^{-4} to 1.31 ohm^{-1} cm^2, a factor of nearly 4000. The iodide sublattice in α-AgI is bcc and this provides 42 possible sites for each 2Ag$^+$, distributed as follows:

6 sites having 2I$^-$ neighbours at 252 pm
12 sites having 3I$^-$ neighbours at 267 pm
24 sites having 4I$^-$ neighbours at 286 pm

The silver ions are almost randomly distributed on these sites, thus accounting for their high mobility. Many other fast ion conductors have subsequently been developed on this principle, e.g.

$$Ag_2HgI_4 \text{ yellow hexag} \xrightarrow{50.7°} \text{orange-red cubic.}$$

9 J. F. HAMILTON, The photographic process, *Prog. Solid State Chem.* **8**, 167–88 (1973). C. E. K. MEES and T. H. JAMES (eds.), *The Theory of the Photographic Process*, 3rd edn., Macmillan, New York, 1966, 591 pp.

History of Photography[10]

In 1727 J. H. Schulze, a German physician, found that a paste of chalk and $AgNO_3$ was blackened by sunlight and, using stencils, he produced black images. At the end of the eighteenth century Thomas Wedgwood (son of the potter Josiah) and Humphry Davy used a lens to form an image on paper and leather treated with $AgNO_3$, and produced pictures which unfortunately faded rather quickly.

The first permanent images were obtained by the French landowner J. N. Niépce using bitumen-coated pewter (bitumen hardens when exposed to light for *several hours* and the unexposed portions can then be dissolved away in oil of turpentine). He then helped the portrait painter, L. J. M. Daguerre, to perfect the "daguerreotype" process which utilized plates of copper coated with silver sensitized with iodine vapour. The announcement of this process in 1839 was greeted with enormous enthusiasm but it suffered from the critical drawback that each picture was unique and could not be duplicated.

Reproducibility was provided by the "calotype" process, patented in 1841 by the English landowner W. H. Fox Talbot, which used semi-transparent paper treated with AgI and a "developer", gallic acid. This produced a "negative" from which any number of "positive" prints could subsequently be obtained. Furthermore it embodied the important discovery of the "latent image" which could be fully developed later. Even with Talbot's very coarse papers, exposure times were reduced to a few minutes and portraits became feasible, even if uncomfortable for the subject.

Though Talbot's pictures were undoubtedly much inferior in quality to Daguerre's, the innovations of his process were the ones which facilitated further improvements and paved the way for photography as we now know it. Sir John Herschel, who first coined the terms "photography", "negative", and "positive", suggested the use of "hyposulphite" (sodium thiosulphate) for "fixing" the image, and later the use of glass instead of paper—hence, photographic "plates". F. S. Archer's "wet collodion" process (1851) reduced the exposure time to about 10 s and R. L. Maddox's "dry gelatin" plates reduced it to only $\frac{1}{2}$ s. In 1889 G. Eastman used a roll film of celluloid and founded the American Eastman Kodak Company.

Meanwhile the Scottish physicist, Clerk Maxwell (1861), recognizing that the sensitivities of the silver halides are not uniform across the spectrum, proposed a three-colour process in which separate negatives were exposed through red, green, and blue filters, and thereby provided the basis for the later development of colour photography. Actually, the sensitivity is greatest at the blue end of the spectrum; a fact which seriously affected all early photographs. This problem was overcome when the German, H. W. Vogel, discovered that sensitivity could be extended by incorporating certain dyes into the photographic emulsion. "Spectral sensitization" at the present time is able to extend the sensitivity not only across the whole visible region but far into the infrared as well.

were established almost a century and a half ago, but a coherent theoretical explanation was not available until the publication in 1938 of the classic paper by R. W. Gurney and N. F. Mott.[11] Since then the subject has stimulated a vast amount of fundamental research in wide areas of solid-state chemistry and physics.

A photograph is the permanent record of an image formed on a light-sensitive surface, and the essential steps in producing it are:

(a) the production of the light-sensitive surface;
(b) exposure to produce a "latent image";
(c) development of the image to produce a "negative";
(d) making the image permanent, i.e. "fixing" it;
(e) making "positive" prints from the negative.

[10] H. GERNSHEIM, W. H. Fox Talbot and the history of photography, *Endeavour* **1**, 18–22 (1977).

[11] R. W. GURNEY and N. F. MOTT, The theory of the photolysis of silver bromide and the photographic latent image, *Proc. Roy. Soc.* **A164**, 151–67 (1938).

(a) In modern processes the light-sensitive surface is an "emulsion" of silver halide in gelatine, coated on to a suitable transparent film, or support. The halide is carefully precipitated so as to produce small uniform crystals, ($<1\ \mu$m diameter, containing $\sim 10^{12}$ Ag atoms), or "grains" as they are normally called. The particular halide used depends on the sensitivity required, but AgBr is most commonly used on films; AgI is used where especially fast film is required and AgCl and certain organic dyes are also incorporated in the emulsion.

(b) When, on exposure of the emulsion to light, a photon of energy impinges on a grain of AgX, a halide ion is excited and loses its electron to the conduction band, through which it passes rapidly to the surface of the grain where it is able to liberate an atom of silver:

$$X^- + h\nu \longrightarrow X + e^-$$

$$Ag^+ + e^- \longrightarrow Ag$$

These steps are, in principle, reversible but in practice are not because the Ag is evidently liberated on a crystal dislocation, or defect, or at an impurity site such as may be provided by Ag_2S, all of which allow the electron to reduce its energy and so become "trapped". The function of the dye sensitizers is to extend the sensitivity of the emulsion across the whole visible spectrum, by absorbing light of characteristic frequency and providing a mechanism for transferring the energy to X^- in order to excite its electron. As more photons are incident on the grain, so more electrons migrate and discharge Ag atoms at the same point. A collection of just a few silver atoms on a grain (in especially sensitive cases a mere 4–6 atoms but, more usually, perhaps 10 times that number) constitutes a "speck", too small to be visible, but the concentration of grains possessing such specks varies across the film according to the varying intensity of the incident light thereby producing an invisible "latent image". The parallel formation of X atoms leads to the formation of X_2 which is absorbed by the gelatine.

(c) The "development" or intensification of the latent image is brought about by the action of a mild reducing agent whose function is to selectively reduce those grains which possess a speck of silver, while leaving unaffected all unexposed grains. To this end, such factors as temperature and concentration must be carefully controlled and the reduction

stopped before any unexposed grains are affected. Hydroquinone,   HO—⟨ ⟩—OH,

is a common "developer" and the reduction is a good example of a catalysed solid-state reaction. Its mechanism is imperfectly understood but the complete reduction to metal of a grain (say 10^{12} atoms of Ag), starting from a single speck (say 10 or 100 atoms of Ag), represents a remarkable intensification of the latent image of about 10^{11} or 10^{10} times, allowing vastly reduced exposure times; this is the real reason for the superiority of silver halides over all other photosensitive materials, though an intensive search for alternative systems still continues.

(d) After development, the image on the negative is "fixed" by dissolving away all remaining silver salts to prevent their further reduction. This requires an appropriate complexing agent and sodium thiosulphate is the usual one since the reaction

$$AgX(s) + 2Na_2S_2O_3 \longrightarrow Na_3[Ag(S_2O_3)_2] + NaX$$

goes essentially to completion and both products are water-soluble.

(e) A positive print is the reverse of the negative and is obtained by passing light through the negative and repeating the above steps using a printing paper instead of a transparent film.

28.3.4 *Complexes*[3-6, 12]

Oxidation states above $+3$ have been attained only relatively recently and are confined to AuF_5, mentioned above, together with salts of the octahedral anion $[AuF_6]^-$, and to $Cs_2[Cu^{IV}F_6]$, prepared by fluorinating $CsCuCl_3$ at high temperature and pressure.[13]

Oxidation state III (d^8)

Copper(III) is generally regarded as uncommon, being very easily reduced, but it has recently received attention because of its apparent involvement in some biological processes. The pale-green, paramagnetic (2 unpaired electrons), K_3CuF_6, is obtained by the reaction of F_2 on $3KCl + CuCl$ and is readily reduced. This is the only high-spin Cu^{III} complex, the rest being low-spin, diamagnetic, and usually square planar, as is to be expected for a cation which, like Ni^{II}, has a d^8 configuration and is more highly charged. Examples are violet $[CuBr_2(S_2CNBu_2^i)]$, obtained by reacting $[Cu(S_2CNBu_2^i)]$ with Br_2 in CS_2, and bluish $MCuO_2$ (M = alkali metal), obtained by heating CuO and MO_2 in oxygen. The oxidation of Cu^{II} by alkaline ClO^- in the presence of periodate or tellurate ions yields salts in which chelated ligands apparently produce square-planar coordinated copper:

(X = I, Te) in complex ions such as, $[Cu(IO_5OH)_2]^{5-}$ and $[Cu\{TeO_4(OH)_2\}_2]^{5-}$

Silver(III) is quite similar to copper(III) and analogous, though more stable, periodate and tellurate complexes can be produced by the oxidation of Ag^I with alkaline $S_2O_8^{2-}$. The diamagnetic, red ethylenedibiguanide complex (Fig. 28.1) is also obtained by peroxodisulfate oxidation and is again quite stable to reduction. However, yellow, diamagnetic, square-planar fluoro-complexes such as $K[AgF_4]$, obtained by fluorinating $AgNO_3 + KCl$ at 300°C, are much less stable; they attack glass and fume in moist air.

For gold, by contrast, $+3$ is the element's best-known oxidation state and Au^{III} is often compared with the isoelectronic Pt^{II} (p. 1349). The usual route to gold(III) chemistry is by dissolving the metal in aqua regia, or the compound Au_2Cl_6 in conc HCl, after which evaporation yields yellow chloroauric acid, $HAuCl_4 . 4H_2O$, from which numerous salts of the square-planar ion $[AuCl_4]^-$ can be obtained. Other square-planar ions of the type $[AuX_4]^-$ can then be derived in which X = F, Br, I, CN, SCN, and NO_3, the last of these

[12] H SCHMIDBAUER, Is gold chemistry a topical field of study?, *Angew. Chem.*, Int. Edn. (Engl.) **15**, 728–40 (1976).

[13] W. HARNISCHMACHER and R. HOPPE, Tetravalent copper: $Cs_2[CuF_6]$, *Angew. Chem.*, Int. Edn. (Engl.) **12**, 582–3 (1973).

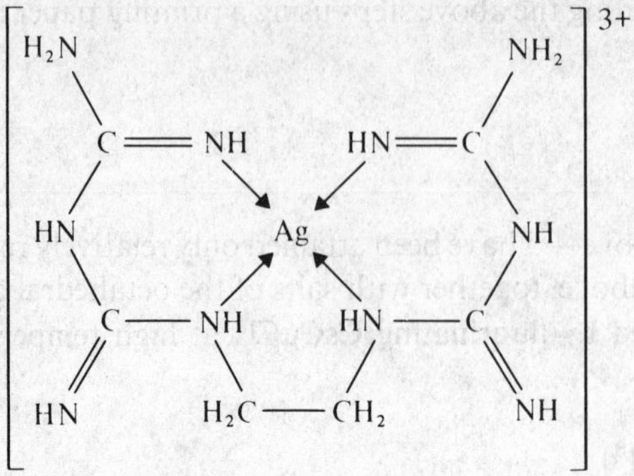

FIG. 28.1. Silver (III) ethylenedibiguanide complex ion; the counter anion can be HSO_4^-, ClO_4^-, NO_3^-, or OH^-.

being of interest as one of the few authenticated examples of the unidentate nitrate ion.[14] $[Au(SCN)_4]^-$ contains S-bonded SCN^- but, as with Pt^{II} (p. 1350), this ligand also gives rise to linkage isomers, this time[15] in the K^+ and $(NEt_4)^+$ salts of $[Au(CN)_2(SCN)_2]^-$ and $[Au(CN)_2(NCS)_2]^-$. Numerous cationic complexes have been prepared with amines, both unidentate (e.g. py, quinoline, as well as NH_3) and chelating (e.g. en, bipy, phen). Complexes with phosphines and arsines are also obtainable, the octahedral $[AuI_2(diars)_2]^+$, for instance, providing a rare example of Au^{III} with a coordination number in excess of 4, but are generally readily reduced to Au^I species. In forming the fluoro complex $[AuF_4]^-$ mentioned above, and indeed the simple fluoride also, Au^{III} differs from Pt^{II}. A further difference is evident in the formation of complexes with O-donor ligands, e.g. $[AuCl_3(OH)]^-$, which is a product of the hydrolysis of salts of $[AuCl_4]^-$ in aqueous solution.

Oxidation state II (d⁹)

Oxidation state II (d^9)

The importance of this oxidation state falls off with increase in atomic number in the group, and most of the compounds ostensibly of Au^{II} are actually mixed valency Au^I/Au^{III} compounds. Examples include the sulfate $Au^IAu^{III}(SO_4)_2$ and the chlorocomplex, $Cs_2[Au^ICl_2][Au^{III}Cl_4]$, the anions of the latter being arranged so as to give linearly coordinated Au^I and tetragonally coordinated Au^{III} (Fig. 28.2). The analogous mixed-metal complex, $Cs_2AgAuCl_6$, has the same structure with Ag^I instead of Au^I. One of the few authenticated examples of Au^{II} is the maleonitriledithiolato complex

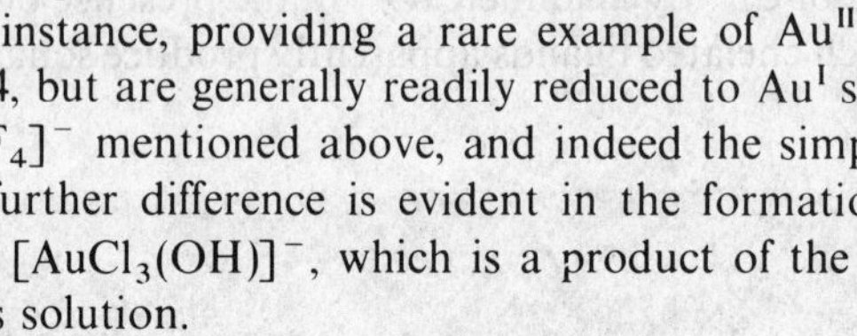

[14] C. D. GARNER and S. C. WALLWORK, The structure of anhydrous nitrates and their complexes. Part V. Potassium tetranitratoaurate(III), *J. Chem. Soc.* (A) 1970, 3092–5.

[15] D. NEGOIU and L. M. BALOIU, The synthesis of linkage isomers of the dicyanodithiocyanatoaurate(III) complex ion, *Z. anorg. Chem.* **382**, 92–6 (1971).

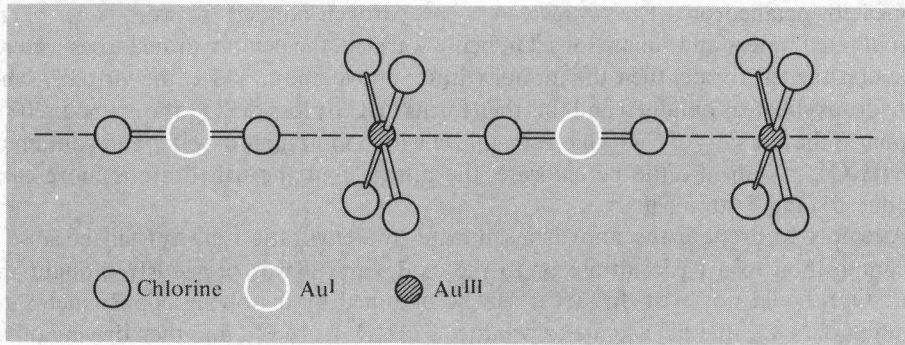

FIG. 28.2. The anions of the chlorocomplex of stoichiometry, $CsAuCl_3$, showing linearly coordinated Au^I and $(4+2)$ tetragonally coordinated Au^{III}, i.e. $Cs_2[Au^ICl_2][Au^{III}Cl_4]$.

which has a magnetic moment at room temperature of 1.85 BM. Even here, however, esr evidence indicates appreciable delocalization of the unpaired electron on to the ligands and, in solution, the complex is readily oxidized to Au^{III}.

Compounds of Ag^{II} are more familiar and are, in general, square planar and paramagnetic ($\mu_e \sim 1.7\text{--}2.2$ BM); this is as expected for an ion which is isoelectronic with Cu^{II} (see below), particularly in view of the greater crystal field splitting associated with 4d (as opposed to 3d) electrons. The $Ag^{II}(aq)$ ion has a transitory existence when an Ag^I salt is oxidized by ozone in a strongly acid solution, but it is an appreciably stronger oxidizing agent than MnO_4^- $[E°(Ag^{2+}/Ag^+) = +1.980$ V in 4N $HClO_4$; $E°(MnO_4^-/Mn^{2-})$ $= 1.507$ V] and oxidizes water even when strongly acidic. Of the acidic solutions the most stable is that in phosphoric acid, no doubt because of complex formation, and even NO_3^- and ClO_4^- ions appear to coordinate in solution since the colours of these solutions depend on their concentrations. A variety of complexes, particularly with heterocyclic amines, has been obtained by oxidation of Ag^I salts with $[S_2O_8]^{2-}$ in aqueous solution in the presence of the ligand. They include $[Ag(py)_4]^{2+}$ and $[Ag(bipy)_2]^{2+}$ and are comparatively stable providing the counter-anion is a non-reducing ion such as NO_3^-, ClO_4^-, or $S_2O_8^{2-}$. Other complexes include some with N-, O-donor ligands such as pyridine carboxylates, and also the violet, $Ba[AgF_4]$.

However, in this oxidation state it is copper which provides by far the most familiar and extensive chemistry. Simple salts are formed with most anions, except CN^- and I^-, which instead form covalent Cu^I compounds which are insoluble in water. The salts are predominantly water-soluble, the blue colour of their solutions arising from the $[Cu(H_2O)_6]^{2+}$ ion, and they frequently crystallize as hydrates. The aqueous solutions are prone to slight hydrolysis and, unless stabilized by a small amount of acid, are liable to deposit basic salts. Basic carbonates occur in nature (p. 1365), basic sulfates and chlorides are produced by atmospheric corrosion of copper, and basic acetates (verdigris) find use as pigments.

The best-known simple salt is the sulfate pentahydrate ("blue vitriol"), $CuSO_4 \cdot 5H_2O$, which is widely used in electroplating processes, as a fungicide (Bordeaux mixture) to protect crops such as potatoes, and as an algicide in water purification. It is also the starting material in the production of most other copper compounds. It is significant, as will be seen presently, that in the crystalline salt 4 of the water molecules form a square plane around the Cu^{II} and 2, more remote, oxygen atoms from SO_4^{2-} ions complete an elongated octahedron. The fifth water is hydrogen-bonded between one of the coordinated waters and sulfate ions. On being warmed, the pentahydrate looses water to give first the trihydrate, then the monohydrate; above about 200°C the virtually white anhydrous sulfate is obtained and this then forms CuO by loss of SO_3 above about 700°C. Amongst the few salts of Cu^{II} which crystallize with 6 molecules of water and contain the $[Cu(H_2O)_6]^{2+}$ ion are the perchlorate, the nitrate (but the trihydrate is more easily produced), and Tutton salts.†

Attempts to prepare the anhydrous nitrate by dehydration always fail because of decomposition to a basic nitrate or to the oxide, and it was previously thought that $Cu(NO_3)_2$ could not exist. In fact it can be obtained by dissolving copper metal in a solution of N_2O_4 in ethyl acetate to produce $Cu(NO_3)_2 . N_2O_4$, and then driving off the N_2O_4 by heating this at 85–100°C. The observation by C. C. Addison and B. J. Hathaway in 1958[16] that the blue $Cu(NO_3)_2$ could be sublimed (at 150–200°C under vacuum) and must therefore involve covalently bonded NO_3^-, was completely counter to current views on the bonding of nitrates and initiated a spate of work on the coordination chemistry of the ion (p. 541).[17] Solid $Cu(NO_3)_2$ actually exists in two forms, both of which involve chains of copper atoms bridged by NO_3 groups, but its vapour is monomeric (Fig. 28.3).

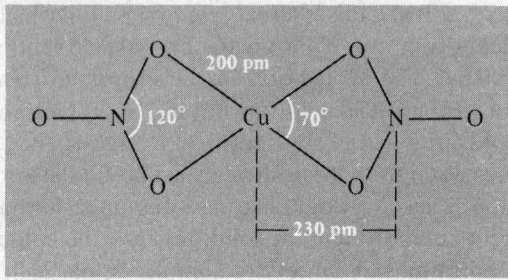

FIG. 28.3.   The $Cu(NO_3)_2$ molecule in the vapour phase (dimensions are approximate).

The most common coordination numbers of copper(II) are 4, 5, and 6, but regular geometries are rare and the distinction between square-planar and tetragonally distorted octahedral coordination is generally not easily made. The reason for this is ascribed to the Jahn–Teller effect (p. 1087) arising from the unequal occupation of the e_g pair of orbitals (d_{z^2} and $d_{x^2-y^2}$) when a d^9 ion is subjected to an octahedral crystal field. Occasionally, as in solid K_2CuF_4 for instance, this results in a compression of the octahedron, i.e. "2+4" coordination (2 short and 4 long bonds). The usual result, however, is an elongation of the octahedron, i.e. "4+2" coordination (4 short and 2 long bonds), as is expected if the metal's d_{z^2} orbital is filled and its $d_{x^2-y^2}$ half-filled, and in its most extreme form is equivalent to the complete loss of the axial ligands leaving a square-planar complex. The

---

† Tutton salts are the double sulfates $M_2^I Cu(SO_4)_2 . 6H_2O$ which all contain $[Cu(H_2O)_6]^{2+}$ and belong to the more general class of double sulfates of M^I and M^{II} cations which are known as schönites after the naturally occurring K^I/Mg^{II} compound.

[16] C. C. ADDISON and B. J. HATHAWAY, The vapour pressure of anhydrous copper nitrate and its molecular weight in the vapour state, *J. Chem. Soc.* 1958, 3099–3106.

[17] C. C. ADDISON, N. LOGAN, and S. C. WALLWORK, Structural aspects of co-ordinate nitrate groups, *Q. Revs.* **25**, 289–322 (1971).

effect of configurational mixing of higher-lying s orbitals into the ligand field d-orbital basis set is also likely to favour elongation rather than contraction.[17a] Elongation has the further consequence that the fifth and sixth stepwise stability constants (p. 1064) are invariably much smaller than the first 4 for Cu^{II} complexes. This is clearly illustrated by the preparation of the ammines. Tetraammines are easily isolated by adding ammonia to aqueous solutions of Cu^{II} until the initial precipitate of $Cu(OH)_2$ redissolves, and then adding ethanol to the deep blue solution,† when $Cu(NH_3)_4SO_4.xH_2O$ slowly precipitates. Recrystallization of tetrammines from 0.880 ammonia yields violet-blue pentammines, but the fifth NH_3 is easily lost; hexammines can only be obtained from liquid ammonia and must be stored in an atmosphere of ammonia. Pyridine and other monoamines are similar in behaviour to ammonia. Likewise, chelating N-donor ligands such as en, bipy, and phen show a reluctance to form tris complexes (though these can be obtained if a high concentration of ligand is used) and a number of 5-coordinate complexes such as $[Cu(bipy)_2I]^+$ with a trigonal bipyramidal structure are known. The macrocyclic N-donor, phthalocyanine, forms a square-planar complex and substituted derivatives are used to produce a range of blue to green pigments which are thermally stable to over 500°C, and are widely used in inks, paints, and plastics.

In alkaline solution biuret, $HN(CONH_2)_2$ reacts with copper(II) sulfate to give a characteristic violet colour due to the formation of the complexes:

and $[Cu(NHCONHCONH)_2]^{2-}$

This is the basis of the "biuret test" in which an excess of NaOH solution is added to the unknown material together with a little $CuSO_4$ soln: a violet colour indicates the presence of a protein or other compound containing a peptide linkage.

Copper(II) also forms stable complexes with O-donor ligands. In addition to the hexaquo ion, the square planar β-diketonates, such as $[Cu(acac)_2]$ which can be precipitated from aqueous solution and recrystallized from non-aqueous solvents, are well known, and tartrate complexes are used in Fehling's test (p. 1373).

Mixed O,N-donor ligands such as Schiff bases[18] are of interest in that they provide examples not only of square-planar coordination but also, in the solid state, examples of square-pyramidal coordination by dimerization (Fig. 28.4). A similar situation occurs in the bis-dimethylglyoximato complex, which dimerizes by sharing oxygen atoms, though

---

† This solution will dissolve cellulose which can be re-precipitated by acidification, a fact used in one of the processes for producing rayon.

[17a] M. GERLOCH, The sense of Jahn–Teller distortions in octahedral copper(II) and other transition metal complexes, *Inorg. Chem.* **20**, 638–40 (1981).

[18] H. S. MASLEN and T. N. WATERS, The conformation of Schiff-base complexes of copper(II): A stereo-electronic view, *Coord. Chem. Revs.* **17**, 137–76 (1975).

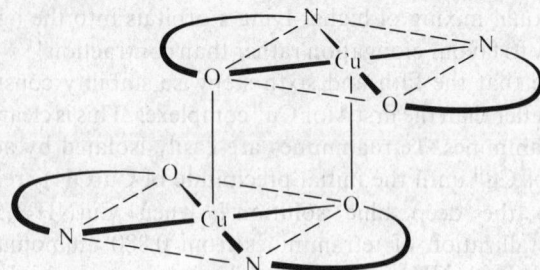

FIG. 28.4.　Schematic representation of square-pyramidal coordination of Cu^{II} in dimeric Schiff-base complexes.

the 4 planar donor atoms are all nitrogen atoms. Copper(II) carboxylates[19] are easily obtained by crystallization from aqueous solution or, in the case of the higher carboxylates, by precipitation with the appropriate acid from ethanolic solutions of the acetate. In the early 1950s it was found that the magnetic moment of green, copper(II) acetate monohydrate is lower than the spin-only value (1.4 BM at room temperature as opposed to 1.73 BM) and that, contrary to the Curie law, its susceptibility reaches a maximum around 270 K but falls rapidly at lower temperatures. Furthermore, the compound has a dimeric structure in which 2 copper atoms are held together by 4 acetate bridges (Fig. 28.5a). Clearly the single unpaired electrons on the copper atoms interact, or "couple", antiferromagnetically to produce a low-lying singlet (diamagnetic) and an excited but thermally accessible triplet (paramagnetic) level (Fig. 28.5b). The separation is therefore only a few kJ mol^{-1} (at room temperature, RT the thermal energy available to populate the higher level ∼2.5 kJ mol^{-1}) and as the temperature is reduced the population of the ground level increases and diamagnetism is eventually approached.

Similar behaviour is found in many other carboxylates of Cu^{II} as well as their adducts in

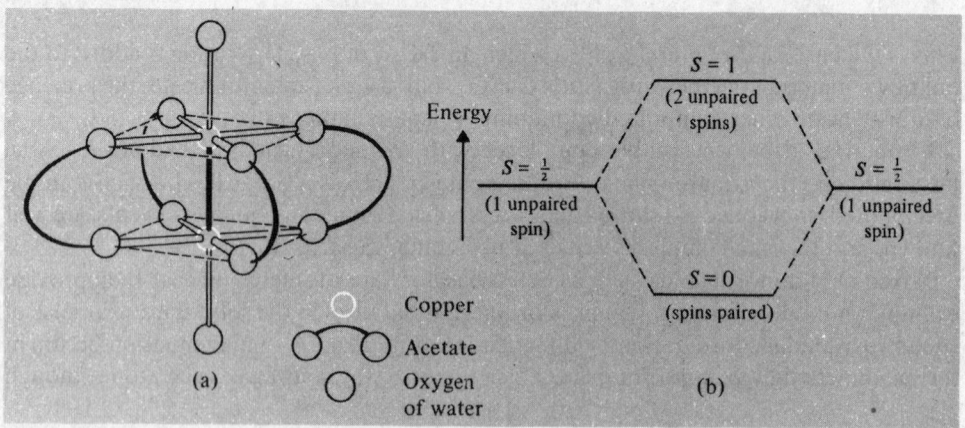

FIG. 28.5. (a) Dinuclear structure of copper(II) acetate, and (b) spin singlet ($2S+1=1$) and spin triplet ($2S+1=3$) energy levels in dinuclear Cu^{II} carboxylates.

[19] J. CATTERICK and P. THORNTON, Structures and physical properties of polynuclear carboxylates, *Adv. Inorg. Chem. Radiochem.* **20**, 291–362 (1977). R. J. DOEDENS, Structure and metal–metal interactions in copper(II) carboxylate complexes, *Prog. Inorg. Chem.* **21**, 209–31 (1976).

which axial water is replaced by other *O*- or *N*-donor ligands. In spite of a continuous flow of work on these compounds there is still no general agreement as to the actual mechanism of the interaction nor on possible correlations of its magnitude with relevant properties of the carboxylate and axial ligands. The simplest interpretation is to assume[20] that the singlet and triplet levels arise from a single interaction between the unpaired spins of the copper atoms and, with B. N. Figgis and R. L. Martin,[21] that this takes the form of "face-to-face" or δ overlap of the copper $d_{x^2-y^2}$ orbitals. However, σ overlap of d_{z^2} orbitals, or even a "superexchange" interaction transmitted via the π orbitals of the bridging carboxylates, are also feasible. A distinction should also be made[22] between interactions which leave the electrons essentially on the individual metal atoms and those which involve covalent pairing of the 2 electrons in a single molecular orbital. This introduces an extra adjustable parameter into the treatment and it has been argued that this extra parameter, rather than any inherent correctness in the model, is the reason for any better agreement with experiment which might be obtained.

It seems generally true that the magnetic interaction is greater for alkylcarboxylates than arylcarboxylates and for *N*-donor rather than *O*-donor axial ligands. More extensive correlations are unfortunately difficult to deduce from published results because of the existence of polymeric or other isomeric forms beside the dinuclear, and because of the possible presence of mononuclear impurities. Precise structural data are therefore desirable but are all too rarely available.

Other copper(II) complexes of stereochemical interests are the halocuprate(II) anions[23] which can be crystallized from mixed solutions of the appropriate halides. The structures of the solids are markedly dependent on the counter cation. The compounds $MCuCl_3$ (M = Li, K, NH$_4$) contain red, planar $[Cu_2Cl_6]^{2-}$ ions, and $CsCuCl_3$ has a polymeric structure in which chains of $CuCl_6$ octahedra (4 + 2 coordination) share opposite faces. The $[CuCl_5]^-$ salts present an even greater variety[24] which includes 5-coordinate trigonal bipyramidal and square-pyramidal coordination, as well as $[dienH_3][CuCl_4]Cl$ which contains a square-planar anion and exhibits a curious mixture of ferro- and antiferro-magnetic properties.[25] But it is the salts of $[CuX_4]^{2-}$ which have received most attention: e.g. depending on the cation, $[CuCl_4]^{2-}$ displays structures ranging from square planar to almost tetrahedral (p. 1071), the former being usually green and the latter orange in colour. $(NH_4)_2[CuCl_4]$ is an oft-quoted example of planar geometry, but 2 long Cu–Cl distances of 279 pm (compared to 4 Cu–Cl distances of 230 pm) make 4 + 2 coordination a more reasonable description. In the $[EtNH_3]^+$ salt the longer Cu–Cl distances increase still further to 298 pm, but the clearest example of

[20] B. BLEANEY and K. D. BOWERS, Anomalous paramagnetism of copper acetate, *Proc. Roy. Soc.* A, **214**, 451–65 (1952).

[21] B. N. FIGGIS and R. L. MARTIN, Magnetic studies with copper(II) salts. Part I. Anomalous paramagnetism and δ-bonding in anhydrous and hydrated copper(II) acetate, *J. Chem. Soc.* 1956, 3837–46.

[22] R. W. JOTHAM and S. F. A. KETTLE, Antiferromagnetism in transition-metal complexes. Part I. Metal–metal bonding and spin exchange in binuclear copper(II) complexes, *J Chem. Soc.* (A) 1969, 2816–20; Part II, ibid. 1969, 2821–5.

[23] D. W. SMITH, Chlorocuprates(II), *Coord. Chem. Revs.* **21**, 93–158 (1976).

[24] L. ANTOLINI, G. MARCOTRIGIANE, L. MENABUE, and G. C. PELLACINI, Spectroscopic and structural investigation on the pentachloro- and (mixed pentahalo) cuprates(II) of the N-(2-ammoniumethyl)-piperazinium cation. The first case of monomeric pentahalocuprate(II) having square pyramidal structure, *J. Am. Chem. Soc.* **102**, 130–9 (1980).

[25] D. B. LOSEE and W. E. HATFIELD, Field dependent magnetic susceptibilities of diethylenetriammonium chlorocuprate(II), $[(NH_3CH_2CH_2)_2NH_2][CuCl_4]Cl$, *J. Am. Chem. Soc.* **95**, 8169–70 (1973).

square-planar $[CuCl_4]^{2-}$ is the recently prepared methadone salt[26] in which the fifth and sixth Cl atoms are more than 600 pm from the Cu^{II}. At the other extreme, $Cs[CuX_4]$ $(X = Cl, Br)$ and $[NMe_4]_2[CuCl_4]$ approach a tetrahedral geometry and it appears that this geometry is retained in aqueous solution since the electronic spectra in the two phases are the same. For $[CuCl_4]^{2-}$ the Cu–Cl distance is close to 223 pm and the somewhat flattened (Jahn–Teller distorted) tetrahedron has four Cl–Cu–Cl angles in the range 100–103° and the other 2 enlarged to 124° or 130°. The angular distortions in $[CuBr_4]^{2-}$ are almost identical: 4 at 100–102° and the others at 126° and 130°.

Electronic spectra and magnetic properties of copper(II)[27]

Because the d^9 configuration can be thought of as an inversion of d^1 (see p. 1092), relatively simple spectra might be expected, and it is indeed true that the great majority of Cu^{II} compounds are blue or green because of a single broad absorption band in the region 11 000–16 000 cm^{-1}. However, as already noted, the d^9 ion is characterized by large distortions from octahedral symmetry and the band is unsymmetrical, being the result of a number of transitions which are by no means easy to assign unambiguously. The free-ion ground ^2D term is expected to split in a crystal field in the same way as the ^5D term of the d^4 ion (p. 1202) and a similar interpretation of the spectra is likewise expected. Unfortunately this is now more difficult because of the greater overlapping of bands which occurs in the case of Cu^{II}.

The T ground term of the tetrahedrally coordinated ion implies (p. 1098) an orbital contribution to the magnetic moment, and therefore a value in excess of $\mu_{\text{spin-only}}$ (1.73 BM). But the E ground term of the octahedrally coordinated ion is also expected to yield a moment ($\mu_e = \mu_{\text{spin-only}} (1 - 2\lambda/10Dq)$) in excess of 1.73 BM, because of "mixing" of the excited T term into the ground term, and the high value of λ (-850 cm^{-1}) makes the effect significant. In practice, moments of magnetically dilute compounds are in the range 1.9–2.2 BM, with compounds whose geometry approaches octahedral having moments at the lower end, and those with geometries approaching tetrahedral having moments at the higher end, but their measurements cannot be used diagnostically with safety unless supported by other evidence.

Oxidation state I (d^{10})

All 3 cations are diamagnetic and, unless coordinated to easily polarized ligands, colourless too. In aqueous solution the Cu^I ion is very unstable with respect to disproportionation ($2Cu^I \rightleftharpoons Cu^{II} + Cu(s)$) largely because of the high heat of hydration of the divalent ion as already mentioned. Recent studies[28] show that at 25°C, K ($= [Cu^{II}][Cu^I]^{-2}$) is large, $(5.38 \pm 0.37) \times 10^5$ l mol^{-1}, and standard reduction potentials

[26] H. C. NELSON, S. H. SIMONSEN and G. W. WATT, Novel methadone salt containing a discrete, square-planar $CuCl_4^{2-}$ anion; X-ray crystal and molecular structure of bis(1-methyl-4-oxo-3,3-diphenylhexyl-dimethylammonium)tetrachlorocuprate(II), *JCS Chem. Comm.* 1979, 632.

[27] B. J. HATHAWAY, Stereochemistry and electronic properties of the copper(II) ion, *Essays in Chemistry* **2**, 61–92 (1971), Academic Press, London.

[28] L. CRAVATTA, D. FERRI, and R. PALOMBARI, On the equilibrium $Cu^{2+} + Cu(s) \rightleftharpoons 2Cu^+$, *J. Inorg. Nucl. Chem.* **42**, 593–8 (1980).

were calculated to be:

$$E°(Cu^+/Cu) = +0.5072 \text{ V} \quad \text{and} \quad E°(Cu^{2+}/Cu^+) = 0.1682 \text{ V}$$

Nevertheless, Cu^I can be stabilized either in compounds of very low solubility or by complexing with ligands having π-acceptor character. The usual stereochemistry is tetrahedral as in complexes such as $[Cu(CN)_4]^{3-}$, $[Cu(py)_4]^+$, and $[Cu(L-L)_2]^+$ (e.g. $L-L = $ bipy, phen), but lower coordination numbers are possible such as 2, in linear $[CuCl_2]^-$ formed when CuCl is dissolved in hydrochloric acid and 3, as in $K[Cu(CN)_2]$, which in the solid contains trigonal, almost planar, $Cu(CN)_3$ units linked in a polymeric chain (Fig. 28.6). Recently,[28a] the first discrete planar anion $[Cu(CN)_3]^{2-}$ has been established, in $Na_2[Cu(CN)_3] \cdot 3H_2O$.

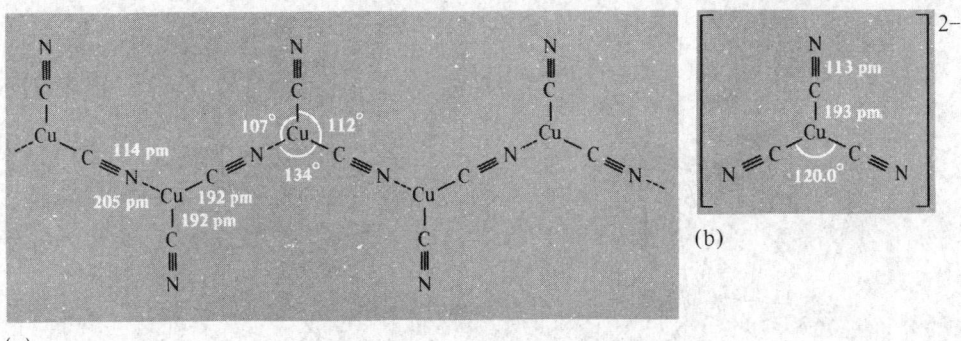

(a)   (b)

FIG. 28.6.   (a) Chain of Cu^I atoms linked by CN bridges to form the helical anion $[Cu(CN)_2^-]_x$ in $KCu(CN)_2$, and (b) one of the two types of $[Cu(CN)_3]^{2-}$ ions in $Na_2[Cu(CN)_3] \cdot 3H_2O$—the other set have Cu–C 195 pm and C–N 116 pm.

Polymeric and oligomeric complexes of Cu^I are becoming increasingly familiar, the most important being of the type $[CuXL]_4$ (X = halide L = phosphine, arsine)[29] which do not involve any M–M bonds. These are made up of

$$\begin{array}{ccc} Cu & — & X \\ | & & | \\ X & — & Cu \end{array}$$

units arranged so as to produce a "cubane" type of structure consisting of a tetrahedron of copper atoms with halide bridges over each face or (if L and X are sufficiently bulky) of an open "step" (or "chair") structure (Fig. 28.7a and b). Cu_4 tetrahedra are also found in complexes such as $[Cu_4(SPh)_6]^{2-}$ in which S-donor ligands bridge the edges of the tetrahedron,[30] and in $[Cu_4OCl_6(OPPh_3)_4]$ in which Cl^- ions bridge the edges and the O atom is situated at the centre of the tetrahedron[31] (Fig. 28.7c and d).

[28a] C. KAPPENSTEIN and R. P. HUGEL, Existence of the monomeric $[Cu(CN)_3]^{2-}$ anion in the solid state. Molecular structure and disorder of sodium tricyanocuprate(I) trihydrate, *Inorg. Chem.* **17**, 1945–9 (1978).

[29] M. R. CHURCHILL, B. G. DEBOER, and S. J. MENDAK, Molecules with an M_4X_4 core. V. Crystallographic characterization of the tetrameric "cubane-like" species triethylphosphinecopper(I) chloride and triethylphosphinecopper(I) bromide, *Inorg. Chem.* **14**, 2041–7 (1975).

[30] I. DANCE and J. C. CALABRESE, The crystal and molecular structure of the hexa(μ_2-benzenethiolato)tetracuprate(I) dianion, *Inorg. Chim. Acta* **19**, L41–L42 (1976).

[31] J. A. BERTRAND and J. A. KELLEY, Preparation, structure, and properties of the tetramethylammonium salt of μ_4-oxo-hexa-μ-chloro-tetra(chlorocuprate(II)), *Inorg. Chem.* **8**, 1982–5 (1969).

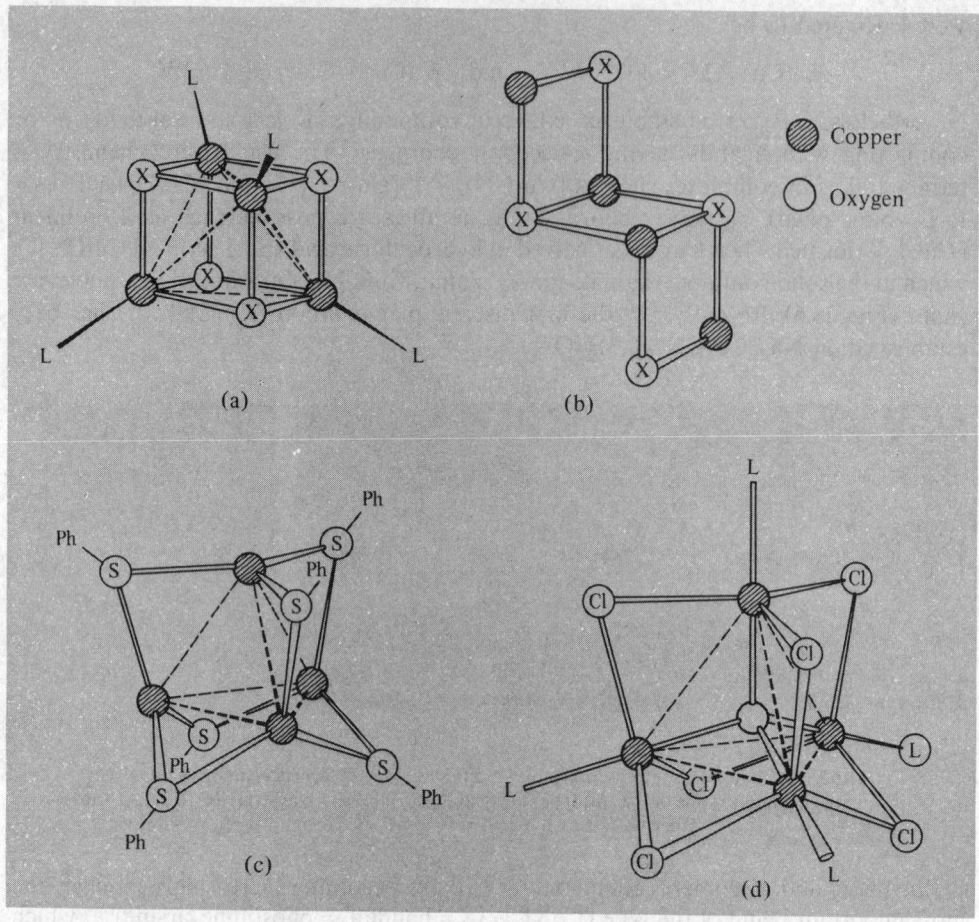

FIG. 28.7. Some Cu^I_4 cluster compounds. (a) "cubane" complexes $[CuXL]_4$; X = halide, L = phosphine or arsine. (b) "step" complexes, $[CuXL]_4$; X = halide, L = phosphine or arsine. (c) $[Cu_4(SPh)_6]^{2-}$. (d) $[Cu_4OCl_6L_4]$, L = $OPPh_3$.

The $+1$ state is by far the best-known oxidation state of silver and salts with most anions are formed. These reveal the reluctance of Ag^I to coordinate to oxygen for, with the exceptions of the nitrate, perchlorate, and fluoride, most are insoluble in water. The last two of these salts are also among the very few Ag^I salts which form hydrates and, paradoxically, their solubilities are actually noted for their astonishingly high values (respectively 5570 and 1800 g l^{-1} at 25°C). The hydrated ion is present in aqueous solution but apparently only as the dihydrate, $[Ag(H_2O)_2]^+$. A coordination number of 2 is typical of Ag^I, which, unlike Cu^I, forms 4-coordinate tetrahedral complexes less readily; a wide variety of linear complexes are formed with N-, P-, and S-donor ligands, some of them of great practical importance. The familiar dissolution of AgCl in aqueous ammonia is due to the formation of $[Ag(NH_3)_2]^+$; the formation of $[Ag(S_2O_3)_2]^{3-}$ in photographic "fixing" has already been discussed (p. 1378), and the cyanide extraction process depends upon the formation of $[M(CN)_2]^-$ (M = Ag, Au) (contrast polymeric $[Cu(CN)_2]^-$, Fig. 28.6). AgCN itself is a linear polymer, {Ag–C≡N→Ag–C≡N→} but AgSCN is non-linear

Complexes

mainly because the sp^3 hybridization of the sulfur forces a zigzag structure; there is also slight bending about the Ag^I atom.

Because of their inability to form linear complexes, chelating ligands tend instead to produce polymeric species, but compounds with coordination numbers higher than 2 can be produced as in the almost tetrahedral diphosphine and diarsine complexes $[Ag(L-L)_2]^+$. Four-coordination is also found in tetrameric phosphine and arsine halides $[AgXL]_4$ which occur in "cubane" and "step" (or "chair") forms like their copper analogues (Fig. 28.7). Indeed, $[AgI(PPh_3)]_4$ has been shown to exist in both forms.[32] Still higher nuclearity is possible for instance in the cyclohexanethiolato complex, $[Ag(SC_6H_{11})]_{12}$, which was recently reported[33] to consist of a 24-membered puckered ring of alternate Ag and S atoms.

Like Ag^I, Au^I also readily forms 2-coordinate complexes of which $[Au(CN)_2]^-$ is of great technological importance. But it is much more susceptible to oxidation and to disproportionation into Au^{III} and Au^0 which renders all its binary compounds, except AuCN, unstable to water. It is also more clearly a class b or "soft" metal with a preference for the heavier donor atoms P, As, and S. Stable, linear complexes are obtained when tertiary phosphines reduce Au^{III} in ethanol,

$$[AuCl_4]^- \xrightarrow{\text{PR}_3/\text{EtOH}} [AuCl(PR_3)].$$

The Cl ligand can be replaced by other halides and pseudo-halides by metathetical reactions. Trigonal planar coordination is found in phosphine complexes of the stoichiometry $[AuL_2X]$ but 4-coordination, though possible, is less prevalent. Diarsine gives the almost tetrahedral complex $[Au(diars)_2]^+$ but, for reasons which are not clear, the colourless complexes $[AuL_4]^+[BPh_4]^-$ with monodentate phosphines fail to achieve a regular tetrahedral geometry.[34]

Complexes with dithiocarbamates involve linear S–Au–S coordination but are dimeric and the Au–Au distance of 276 pm compared with 288 pm in the metal is indicative of

[32] B-K Teo and T. C. Calabrese, Stereochemical systematics of metal clusters. Crystallographic evidence for a new cubane⇌chair isomerism in tetrameric triphenylphosphine silver iodide, $(Ph_3P)_4Ag_4I_4$, *Inorg. Chem.* **15**, 2474–86 (1976).

[33] I. G. Dance, On the molecularity of crystalline cyclohexanethiolato-silver(I), *Inorg. Chim. Acta* **25**, L17–L18 (1977).

[34] P. G. Jones, Is regular tetrahedral geometry possible in gold(I)–phosphine complexes? X-ray crystal structures of three modifications of $(PPh_3)_4Au^+BPh_4^-$ *JCS Chem. Comm.* 1980, 1031–3. But see also, R. C. Elder, E. H. K. Zeiher, M. Onady and R. R. Whittle. Nearly regular tetrahedral geometry in a gold(I)-phosphine complex. X-ray crystal structure of tetrakis(methyldiphenylphosphine) gold(I) hexafluorophosphate, *JCS Chem. Comm.* 1981, 900–1.

metal–metal bonding,

$$S-Au-S$$
$$R_2N-C \quad \begin{vmatrix} 276 \\ pm \end{vmatrix} \quad C-NR_2$$
$$S-Au-S$$

In the thermal production of gold coatings on ceramics and glass, paints are used which comprise Au^{III} chloro-complexes and sulphur-containing resins dissolved in an organic solvent. It seems likely that polymeric species are responsible for rendering the gold soluble.

Gold cluster compounds[12, 35]

Polymeric complexes of the types formed by copper and silver are not found for gold but instead a range of variously coloured cluster compounds, with gold in an average oxidation state <1 and involving M–M bonds, can be obtained by the general process of reducing a gold phosphine halide, usually with sodium borohydride. Yellow, $[Au_6\{P(C_6H_4\text{-}4\text{-}Me)_3\}_6]^{2+}$ consists of an octahedron of 6 gold atoms with a phosphine attached to each. In red, $[Au_8(PPh_3)_8]^{2+}$ each gold has an attached phosphine, but 1 gold can be regarded as central to the other 7, 6 of which form a chair-like hexagon.[36] In the green, $[Au_9\{P(C_6H_4\text{-}4\text{-}Me)_3\}_8]^{3+}$ and in $[Au_{11}\{P(C_6H_4\text{-}4\text{-}F)_3\}_7I_3]$ 1 metal atom is completely surrounded by the others and has no ligand attached to it (Fig. 28.8). The nucleation is approaching that in metallic gold, which is not surprising since the preparative methods are of the type used to produce colloidal gold. Even more recently the dark-red, centred icosahedral cluster cation $[Au_{13}Cl_2(PMe_2Ph)_{10}]^{3+}$ has been isolated as its $[PF_6]^-$ salt following the reduction of $[AuCl(PMe_2Ph)$ with $[Ti(\eta\text{-}C_7H_8)_2]$. The most recent publications in this area[36a] report the preparation and structural characterization

[35] D. M. P. MINGOS, Molecular-orbital calculations on cluster compounds of gold, JCS Dalton 1976, 1163–9.

[36] M. MANASSERO, L. NALDINI, and M. SANSONI, A new class of gold cluster compounds. Synthesis and X-ray structure of the octakis(triphenylphosphinegold) dializarinsulphonate, $[Au_8(PPh_3)_8](aliz)_2$, JCS Chem. Comm. 1979, 385–6. F. A. VOLLENBROEK, W. P. BOSMAN, J. J. BOUR, J. H. NOORDIK, and P. T. BEURSKENS, Reactions of gold–phosphine cluster compounds. Preparation and X-ray structure determination of octakis(triphenylphosphine)octagold bis(hexafluorophosphate), JCS Chem. Comm. 1979, 387–8; see also Inorg. Chem. 19, 701–4 (1980); JCS Chem. Comm. 1979, 1162–3; C. E. BRIANT, B. R. C. THEOBALD, J. W. WHITE, L. K. BELL, D. M. P. MINGOS, and A. J. WELCH, Synthesis and X-ray structural characterization of the centred icosahedral gold cluster compound $[Au_{13}(PMe_2Ph)_{10}Cl_2]$ $[PF_6]_3$; the realization of a theoretical prediction, JCS Chem. Comm. 1981, 201–2.

[36a] J. W. A. VAN DER VELDEN, J. J. BOUR, W. P. BOSMAN and J. H. NOORDIK, Synthesis and X-ray crystal structure determination of the cationic gold cluster compound $[Au_8(PPh_3)_7](NO_3)_2$, JCS Chem. Comm. 1981, 1218–9. J. W. A. VAN DER VELDEN, J. J. BOUR, B. F. OTTERLOO, W. P. BOSMAN, and J. H. NORDIK, Synthesis of a new heteronuclear gold-cobalt cluster. Preparation and X-ray structure determination of tetrakis(triphenyl-phosphine)bis(tetracarbonyl-cobalt)hexagold, JCS Chem. Comm. 1981, 583–4. K. P. HALL, B. C. R. THEOBOLD, D. I. GILMOUR, D. M. P. MINGOS, and A. J. WELCH, Synthesis and structural characterization of $[Au_9\{P(p\text{-}C_6H_4OMe)_3\}_8](BF_4)_3$; a cluster with a centred crown of gold atoms, JCS Chem. Comm., 1982, 528–30. M. GREEN, A. G. ORPEN, I. D. SALTER, and F. G. A. STONE, Hydrido-bridged complexes of gold and silver: X-ray crystal structure of $[AuCr(\mu\text{-}H)(CO)_5(PPh_3)]$, JCS Chem. Comm. 1982, 813–4. See also L. W. BATEMAN, M. GREEN, J. A. K. HOWARD, K. A. MEAD, R. M. MILLS, I. D. SALTER, F. G. A. STONE, and P. WOODWARD, Replacement of hydrido-ligands in $[Ru_3(\mu\text{-}H)_3(\mu_3\text{-}COMe)(CO)_9]$ and $[Ru_4(\mu\text{-}H)_4(CO)_{12}]$ by triphenyl-phosphinegold groups: X-ray crystal structures of $[AuRu_3(\mu\text{-}H)_2(\mu\text{-}COMe)(CO)_9(PPh_3)]$, $[Au_3Ru_3(\mu_3\text{-}COMe)(CO)_9(PPh_3)_3]$, and $[Au_3Ru_4(\mu\text{-}H)(CO)_{12}(PPh_3)_3]$, JCS Chem. Comm. 1982, 773–5.

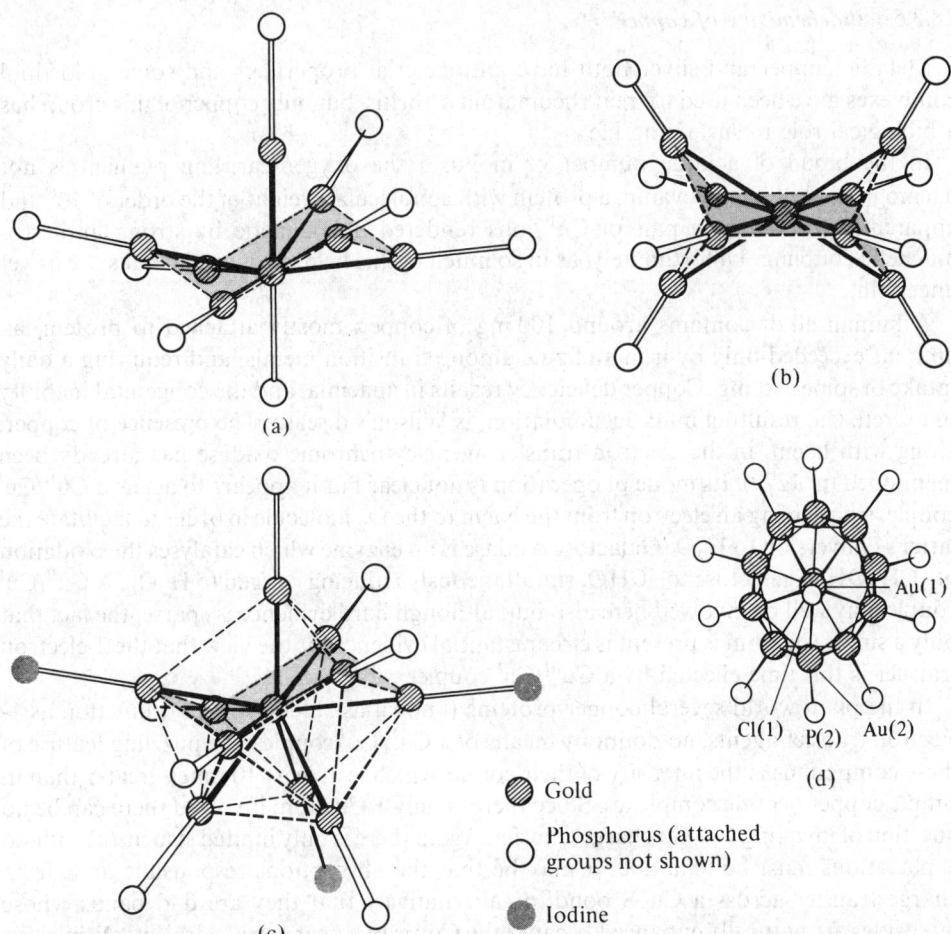

FIG. 28.8 Some gold cluster compounds. Note that a chair-like centred hexagon of gold atoms persists throughout these structures and is shaded in (a), (b), and (c): (a) $[Au_8(PPh_3)_8]^{2+}$, (b) $Au_9\{P(C_6H_4-4-Me)_3\}_8]^{3+}$, (c) $Au_{11}I_3\{P(C_6H_4-4-F)_3\}_7]$, and (d) $[Au_{13}Cl_2(PMe_2Ph)_{10}]^{3+}$. In (d) the 12th icosahedral gold atom and the 13th (central) gold atom are obscured by Au(1).

of the dark red-brown cluster $[Au_8(PPh_3)_7]^{2+}$, the dark red-brown, air-stable, mixed metal cluster $[Au_6(PPh_3)_4Co_2(CO)_8]$, the golden brown, centred-crown cluster $[Au_9\{P(C_6H_4-4-OMe)_3\}_8]^{3+}$, and the first example of an H-bridged gold complex: yellow $[AuCr(\mu-H)(CO)_5(PPh_3)]$.

Copper cluster compounds are also beginning to appear, for example the pale yellow salt $[Li(thf)_4]^+[Cu_5Ph_6]^-$, prepared by treating a suspension of CuBr in Et_2O with LiPh at $-20°$ and recrystallizing from thf/Et_2O; the anion is a "squashed trigonal bipyramid" of 5 Cu atoms with bridging Ph groups above the 6 slant edges: $Cu_{ax}-Cu_{eq}$ 245 pm, $Cu_{eq}-Cu_{eq}$ 315 pm.[36b]

[36b] P. G. EDWARDS, R. W. GELLERT, M. W. MARKS, and R. BAU, Preparation and structure of the $[Cu_5(C_6H_5)_6]^-$ anion, *J. Am. Chem. Soc.* **104,** 2072–3 (1982).

28.3.5 *Biochemistry of copper*[37]

Metallic copper and silver both have antibacterial properties† and some gold thiol complexes have been used to treat rheumatoid arthritis, but only copper of this group has a biological role in sustaining life.

In the blood of a large number of molluscs the oxygen-carrying pigment is not haemoglobin but haemocyanin, a protein with a molecular weight of the order of 10^6 and apparently containing a pair of Cu^{II} ions rendered diamagnetic by strong antiferro-magnetic coupling. Unfortunately, as in so much of this field, structural details are as yet uncertain.

A human adult contains around 100 mg of copper, mostly attached to protein, an amount exceeded only by iron and zinc among transition metals, and requiring a daily intake of some 3–5 mg. Copper deficiency results in anaemia, and the congenital inability to excrete Cu, resulting in its accumulation, is Wilson's disease. The presence of copper, along with haem, in the electron transfer agent cytochrome oxidase has already been mentioned (p. 1279). Its mode of operation is not clear but it appears to act as a Cu^{II}/Cu^{I} couple, transferring an electron from the haem to the O_2 molecule in order to facilitate the latter's conversion to H_2O. Galactose oxidase is an enzyme which catalyses the oxidation of $-CH_2OH$ in galactose to $-CHO$, simultaneously reducing oxygen to H_2O_2. A Cu^{II}/Cu^{I} couple may well be involved here also but, although hard evidence is sparse, the fact that only a single Cu atom is present is circumstantial evidence for the view that the 2-electron transfer is this time effected by a Cu^{III}/Cu^{I} couple.

In the plant world several copper proteins (known as "blue proteins") function as 1-electron transfer agents, no doubt by means of a Cu^{II}/Cu^{I} couple. The puzzling feature of these compounds is the intensity of their colour which is at least 10 times greater than in simple copper peptide complexes. Since there is only 1 Cu atom involved there can be no question of invoking Cu–Cu charge transfer. Again there is only limited structural data, so explanations must be tentative. It may be that the absorptions responsible arise from charge transfer across a Cu–S bond, or alternatively that they are d–d bands whose intensities are naturally enhanced because the Cu^{II} is in a near-tetrahedral site. While this is unusual for Cu^{II} it is the preferred geometry for Cu^{I}.

The field is one of intense current interest and activity.

28.3.6 *Organometallic compounds*[3, 6, 12]

Neutral binary carbonyls are not formed by these metals at normal temperatures‡ but copper and gold each form an unstable carbonyl halide, $[M(CO)Cl]$. These colourless compounds can be obtained by passing CO over MCl or, in the case of copper only (since

---

† This was unknowingly utilized in ancient Persia where, by law, drinking water had to be stored in bright copper vessels.

‡ Some have been synthesized by the condensation of Cu or Ag vapour and CO at temperatures of 6–15 K : e.g. $M(CO)_3$, $M_2(CO)_6$, $M(CO)_2$, and $M(CO)$. Thus $[Ag(CO)_3]$ is green, planar, and paramagnetic; above 25–30 K it dimerizes, perhaps by formation of an Ag–Ag bond.[38]

[37] H. BEINERT, Structure and function of copper proteins, *Coord. Chem. Rev.* **23**, 119–29 (1977). H. Sigel (ed.) *Metal Ions in Biological Systems. Vol 13 Copper Proteins.* Marcel Dekker, New York, 1981, 394 pp.

[38] D. McINTOSH and G. A. OZIN, Synthesis using metal vapours. Silver carbonyls. Matrix infrared, ultraviolet-visible, and electron spin resonance spectra, structures and bonding of $Ag(CO)_3$, $Ag(CO)_2$, $Ag(CO)$, and $Ag_2(CO)_6$, *J. Am. Chem. Soc.* **98**, 3167–75 (1976), and references therein.

the gold compound is very sensitive to moisture) by bubbling CO through a solution of CuCl in conc HCl or in aqueous NH_3. The latter reactions can in fact be used for the quantitative estimation of the CO content of gases. In a similar manner complexes of the type [MLX], which are often polymeric, can be obtained for Cu^I and Ag^I with many olefins (alkenes) and acetylenes (alkynes) either by anhydrous methods or in solution. They are generally rather labile, often decomposing when isolated. The silver complexes have received most attention and the silver–olefin bonds are found to be thermo-dynamically weaker than, for instance, corresponding platinum–olefin bonds. Since the former bonds are also found to be somewhat unsymmetrical it seems likely that π bonding is weaker for the group IB metals. Gold also forms olefin complexes, but not nearly so readily as silver and then only with high molecular weight olefins.

M–C σ bonds can be formed by each of the M^I metals. The simple alkyls and aryls of Ag^I are less stable than those of Cu^I, while those of Au^I have not been isolated, the Au^I–R bond evidently requiring the stabilizing presence of a ligand such as a phosphine. Copper alkyls and aryls are prepared by the action of LiR or a Grignard reagent on a Cu^I halide:

$$CuX + LiR \longrightarrow CuR + LiX$$

$$CuX + RMgX \longrightarrow CuR + MgX_2$$

CuMe is a yellow polymeric solid which explodes if allowed to dry in air, and CuPh, which is white and also polymeric, though more stable, is still sensitive to both air and water. Much greater stability is achieved by the σ-cyclopentadienyl complex $[Cu(\eta^1\text{-}C_5H_5)(PEt_3)]$ prepared by the reaction of C_5H_6, CuO, and PEt_3 in petroleum ether; a similar Au^I compound, $[Au(\eta^1\text{-}C_5H_4Me)(PPh_3)]$, is also known. The Au^I alkyls are obtained like those of copper but with an appropriate ligand present, e.g.:

$$[Au(PEt_3)X] + LiR \longrightarrow [Au(PEt_3)R] + LiX$$

The colourless solids are composed of linear monomers.

The alkyl derivatives of Au^{III}, discovered by W. J. Pope and C. S. Gibson in 1907 include some of the most familiar and stable organo compounds of the group, and are notable for not requiring the stabilizing presence of π-bonding ligands. They are of three types:

AuR$_3$ (stable, when they occur at all, only in ether below $-35°C$);
AuR$_2$X   (much the most stable); X = anionic ligand especially Br
AuRX$_2$ (unstable, only dibromides characterized)

Corresponding aryl derivatives are rare and unstable. Thus, while AuMe$_3$ decomposes above $-35°C$ but is stabilized in $[AuMe_3(PPh_3)]$, AuPh$_3$ is unknown.

The dialkylgold(III) halides are generally prepared from the tribromide and a Grignard reagent:

$$AuBr_3 + 2RMgBr \longrightarrow AuR_2Br + 2MgBr_2$$

and many other anions can be substituted by metathetical reactions with the appropriate silver salt:

$$AuR_2Br + AgX \longrightarrow AuR_2X + AgBr$$

In all cases where the structure has been determined, the Au^{III} attains planar four-fold coordination and polymerizes as appropriate to achieve this. The halides for instance are dimeric but with the cyanide, which forms linear rather than bent bridges, tetramers are produced:

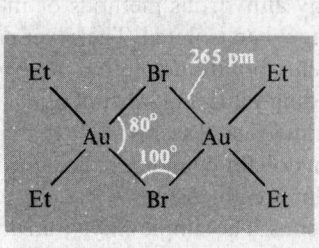

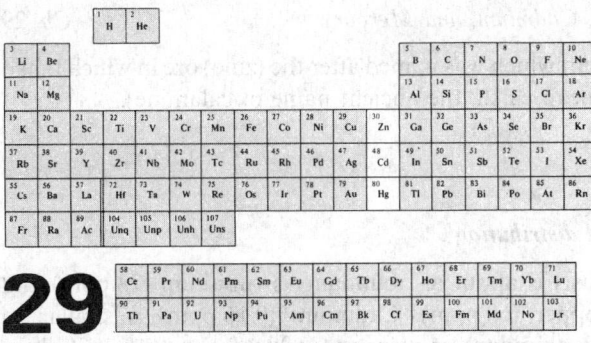

29

Zinc, Cadmium, and Mercury

29.1 Introduction

The reduction of ZnO by charcoal requires a temperature of 1000°C or more and, because the metal is a vapour at that temperature and is liable to reoxidation, its collection requires some form of condenser and the exclusion of air. This was apparently first achieved in India in the thirteenth century. The art then passed to China where zinc coins were used in the Ming Dynasty (1368–1644). The preparation of alloyed zinc by smelting mixed ores does not require the isolation of zinc itself and is much more easily achieved. The small amounts of zinc present in samples of early Egyptian copper no doubt simply reflect the composition of local ores, but Palestinian brass dated 1400–1000 BC and containing about 23% Zn must have been produced by the deliberate mixing of copper and zinc ores. Brass was similarly produced by the Romans in Cyprus and later in the Cologne region of Germany.

Zinc was not intentionally made in medieval Europe, though small amounts were obtained by accidental condensation in the production of lead, silver, and brass; it was imported from China by the East India Company after about 1605. The English zinc industry started in the Bristol area in the early eighteenth century and production quickly followed in Silesia and Belgium. The origin of the name is obscure but may plausibly be thought to be derived from *Zinke* (German for spike, or tooth) because of the appearance of the metal.

Mercury is more easily isolated from its ore, cinnabar, and was used in the Mediterranean world for extracting metals by amalgamation as early as 500 BC, possibly even earlier. Cinnabar, HgS, was widely used in the ancient world as a pigment (vermilion). For over a thousand years, up to AD 1500, alchemists regarded the metal as a key to the transmutation of base metals to gold and employed amalgams both for gilding and for producing imitation gold and silver. Because of its mobility, mercury is named after the messenger of the gods in Roman mythology, and the symbol, Hg, is derived from *hydrargyrum* (Latin, liquid silver).

Cadmium made its appearance much later. In 1817 F. Stromeyer of Göttingen noticed that a sample of "cadmia" (now known as "calamine"), used in a nearby smelting works, was yellow instead of white. The colour was not due to iron, which was shown to be absent, but arose instead from a new element which was named after the (zinc) ore in which it had been found (Greek καδμεία, cadmean earth, the ancient name of calamine).

29.2 The Elements

29.2.1 *Terrestrial abundance and distribution*

Zinc (76 ppm of the earth's crust) is about as abundant as rubidium (78 ppm) and slightly more abundant than copper (68 ppm). Cadmium (0.16 ppm) is similar to antimony (0.2 ppm); it is twice as abundant as mercury (0.08 ppm), which is itself as abundant as silver (0.08 ppm) and close to selenium (0.05 ppm). These elements are "chalcophiles" (p. 760) and so, in the reducing atmosphere prevailing when the earth's crust solidified, they separated out in the sulfide phase, and their most important ores are therefore sulfides.[1, 2] Subsequently, as rocks were weathered zinc was leached out to be precipitated as carbonate, silicate, or phosphate.

The major ores of zinc are ZnS (which is known as zinc blende in Europe and as sphalerite in the USA) and $ZnCO_3$ (calamine in Europe, smithsonite in the USA†). Large deposits are situated in Canada, the USA, and Australia. Less important ores are hemimorphite, $Zn_4Si_2O_7(OH)_2.H_2O$ and franklinite, $(Zn,Fe)O.Fe_2O_3$. Cadmium is found as greenockite, CdS, but its only commercially important source is the 0.2–0.4% found in most zinc ores. Cinnabar, HgS, is the only important ore and source of mercury and is found along lines of previous volcanic activity. The most famous and extensive deposits are at Almaden in Spain; these contain up to 6–7% Hg and have been worked since Roman times. Other deposits, usually containing $<1\%$ Hg, are situated in the USSR, Algeria, Mexico, Yugoslavia, and Italy.

29.2.2 *Preparation and uses of the elements*[3, 4]

The isolation of zinc, over 90% of which is from sulfide ores, depends on conventional physical concentration of the ore by sedimentation or flotation techniques, followed by roasting to produce the oxides; the SO_2 which is generated is used to produce sulfuric acid. The ZnO is then either treated electrolytically or smelted with coke. In the former case the zinc is leached from the crude ZnO with dil H_2SO_4, at which point cadmium is precipitated by the addition of zinc dust. The $ZnSO_4$ solution is then electrolysed and the metal deposited—in a state of 99.95% purity—on to aluminium cathodes.

A variety of smelting processes have been employed to effect the reduction of ZnO by coke:

$$ZnO + C \longrightarrow Zn + CO$$

---

† After James Smithson, founder of the Smithsonian Institution, Washington. The name calamine is applied in the USA to a basic carbonate.

[1] C. S. G. PHILLIPS and R. J. P. WILLIAMS, *Inorganic Chemistry*, Vol. II, pp. 605–15, Oxford University Press, Oxford, 1966.
[2] R. G. BURNS, *Mineralogical Applications of Crystal Field Theory*, Cambridge University Press, Cambridge, 1970, 224 pp.
[3] *Kirk–Othmer Encyclopedia of Chemical Technology*, Interscience, New York. For Zn, see Vol. 24, 3rd edn., 1984, pp. 807–51. For Cd, see Vol. 4, 3rd edn., 1978, pp. 387–410. For Hg, see Vol. 15, 3rd edn., 1981, pp. 143–71.
[4] A. W. RICHARDS, Zinc extraction metallurgy in the UK, *Chem. Br.* **5**, 203–6 (1969).

These formerly involved the use of banks of externally heated, horizontal retorts, operated on a batch basis. They were replaced by continuously operated vertical retorts, in some cases electrically heated. Unfortunately none of these processes has the thermal efficiency of a blast furnace process (p. 1244) in which the combustion of the fuel for heating takes place in the same chamber as the reduction of the oxide. The inescapable problem posed by zinc is that the reduction of ZnO by carbon is not spontaneous below the boiling point of Zn (a problem not encountered in the smelting of Fe, Cu, or Pb, for instance), and the subsequent cooling to condense the vapour is liable, in the presence of the combustion products, to result in the reoxidation of the metal:

$$Zn + CO_2 \rightleftharpoons ZnO + CO$$

A major breakthrough was accomplished in the 1950s by the Imperial Smelting Co. of Bristol who developed a blast furnace[3] which overcomes this difficulty. The zinc vapour leaving the top of the furnace is chilled and dissolved so rapidly by a spray of lead that reoxidation is minimal. The zinc then separates as a liquid of nearly 99% purity and is further refined by vacuum distillation to give a purity of 99.99%. Any cadmium present is recovered in the course of this distillation. The use of a blast furnace has the further advantage that the composition of the charge is not critical, and mixed Zn/Pb ores can be used (ZnS and PbS are commonly found together) to achieve the simultaneous production of both metals, the lead being tapped from the bottom of the furnace.

World production of zinc is fairly steady at around 6 million tonnes pa (i.e. slightly more than that of Pb). The largest supplier of ores is Canada but most of these are refined elsewhere. Cadmium is produced in much smaller quantities ($\sim$18 000 tonnes pa) and these are dependent on the supply of zinc.

Zinc finds a wide range of uses, the most important, accounting for 35–40% of output, is as an anti-corrosion coating. The application of the coating takes various forms: immersion in molten zinc (hot-dip galvanizing), electrolytic deposition, spraying with liquid metal, heating with powdered zinc ("Sherardizing"), and applying paint containing zinc powder. In addition to brasses (Cu + 20–50% Zn), a rapidly increasing number of special alloys, predominantly of zinc, are used for diecasting and, indeed, the vast majority of *pressure* diecastings are now made in these alloys. Zinc sheeting is used in roof cladding and large quantities are used in the manufacture of dry batteries of which the carbon–zinc type is the most common (see Panel).

Dry Batteries

A portable source of electricity, if not a necessity, is certainly a great convenience in modern life and is dependent on compact, sealed, dry batteries. The main types are listed below and they incorporate the metals Zn, Ni, Hg, and Cd as well as MnO_2

(a) Carbon–zinc cell

The first dry battery was that patented in 1866 by the young French engineer, G. Leclanché. The positive pole consisted of carbon surrounded by MnO_2 (p. 1219) contained in a porous pot, and the negative pole was simply a rod of zinc. These were situated inside a glass jar containing the electrolyte, ammonium chloride solution thickened with sand or sawdust. This is still the basis of the most common type of modern dry cell in which a carbon rod is the positive pole, surrounded by a paste of MnO_2, carbon black, and NH_4Cl, inside a zinc can which is both container and negative pole. The reactions are:

negative pole: $Zn \longrightarrow Zn^{2+} + 2e^-$
electrolyte: $Zn^{2+} + 2NH_4Cl + 2OH^- \longrightarrow [ZnCl_2(NH_3)_2] + 2H_2O$
positive pole: $2MnO_2 + 2H_2O + 2e^- \longrightarrow 2MnO(OH) + 2OH^-$

net reaction: $Zn + 2NH_4Cl + 2MnO_2 \longrightarrow [ZnCl_2(NH_3)_2] + 2MnO(OH)$

(b) Mercury cell

The negative pole of pressed amalgamated zinc powder and the positive pole of mercury(II) oxide and graphite are separated by an absorbent impregnated with the electrolyte, conc KOH:

negative pole: $Zn + 2OH^- \longrightarrow ZnO + H_2O + 2e^-$
positive pole: $HgO + H_2O + 2e^- \longrightarrow Hg + 2OH^-$

net reaction: $Zn + HgO \longrightarrow Hg + ZnO$

(c) Alkaline manganese cell

This is similar in principle to (a) but is constructed in a manner akin to (b). The negative pole of powdered zinc, formed into a paste with the electrolyte KOH, and the positive pole of compressed graphite and MnO_2 are separated by an absorbent impregnated with the electrolyte:

negative pole: $Zn + 2OH^- \longrightarrow ZnO + H_2O + 2e^-$
positive pole: $2MnO_2 + H_2O + 2e^- \longrightarrow Mn_2O_3 + 2OH^-$

net reaction: $Zn + 2MnO_2 \longrightarrow ZnO + Mn_2O_3$

(d) Nickel–cadmium cell

Unlike the cells above, which are all primary cells, this is a secondary, i.e. rechargeable, cell, and the two poles are composed in the uncharged condition of nickel and cadmium hydroxides respectively. These are each supported on microporous nickel, made by a sintering process, and separated by an absorbent impregnated with electrolyte. The charging reactions are:

negative pole: $Cd(OH)_2 + 2e^- \longrightarrow Cd + 2OH^-$
positive pole: $Ni(OH)_2 + 2OH^- \longrightarrow NiO(OH) + 2H_2O + 2e^-$

net reaction: $Ni(OH)_2 + Cd(OH)_2 \longrightarrow NiO(OH) + Cd + 2H_2O$

During discharge these reactions are reversed. A crucial feature of the construction of this cell is that oxygen produced at the positive pole during charging by the side-reaction:

$$4OH^- \longrightarrow 2H_2O + O_2 + 4e^-$$

can migrate readily to the negative pole to be recombined in the reaction:

$$O_2 + H_2O + 2Cd \longrightarrow 2Cd(OH)_2$$

But for this rapid migration and recombination, the cell could not be sealed.

As with zinc, the major use of cadmium, and one which would probably be even more extensive but for environmental concern (p. 1421) is as a protective coating. It is also used in small quantities in alloys and batteries, and some of its compounds are used as stabilizers, in PVC for instance, to prevent degradation by heat or ultraviolet radiation.

The isolation of mercury is comparatively straightforward. The most primitive method consisted simply of heating cinnabar in a fire of brushwood. The latter acted as fuel and condenser, and metallic mercury collected in the ashes. Modern techniques are of course less crude than this but the basic principle is much the same. After being crushed and concentrated by flotation, the ore is roasted in a current of air and the vapour condensed:

$$HgS + O_2 \xrightarrow{600°C} Hg + SO_2$$

Alternatively, in the case of especially rich ores, roasting with scrap iron or quicklime is used:

$$HgS + Fe \longrightarrow Hg + FeS$$

$$4HgS + CaO \longrightarrow 4Hg + 3CaS + CaSO_4$$

Blowing air through the hot, crude, liquid metal oxidizes traces of metals such as Fe, Cu, Zn, and Pb which form an easily removable scum. Further purification is by distillation under reduced pressure. About 6500 tonnes† of mercury are produced annually of which (1980) 33% comes from the USSR, 17% from Spain, and 16% each from Algeria and Mexico.

The use of mercury for extracting precious metals by amalgamation has a long history and was extensively used by Spain in the sixteenth century when her fleet carried mercury from Almaden to Mexico and returned with silver. Now, however, the greatest use is in the Castner–Kellner process for manufacturing chlorine and NaOH (p. 81), and increasing amounts are consumed in the electrical and electronic industries, as, for instance, in street lamps and AC rectifiers. Its small-scale use in thermometers, barometers, and gauges of different kinds, are familiar in many laboratories. In the form of its compounds, it has widespread germicidal and fungicidal applications.

29.2.3 *Properties of the elements*

A selection of some important properties of the elements is given in Table 29.1. Because the elements each have several naturally occurring isotopes their atomic weights cannot be quoted with a precision much greater than 1 part in 6000. Their most noticeable features compared with other metals are their low melting and boiling points, mercury being unique as a metal which is a liquid at room temperature. Zinc and cadmium are silvery solids with a bluish lustre when freshly formed. Mercury is also unusual in being the only element, apart from the noble gases, whose vapour is almost entirely monatomic, while its appreciable vapour pressure (1.9×10^{-3} mmHg at 25°C), coupled with its toxicity, makes it necessary to handle it with care. The electrical resistivity of liquid mercury is exceptionally high for a metal, and this facilitates its use as an electrical standard (the international ohm is defined as the resistance of 14.4521 g of Hg in a column 106.300 cm long and 1 mm^2 cross-sectional area at 0°C and a pressure of 760 mmHg.

The structures of the solids, although based on the typically metallic hexagonal close-packing, are significantly distorted. In the case of Zn and Cd the distortion is such that, instead of having 12 equidistant neighbours, each atom has 6 nearest neighbours in the close-packed plane with the 3 neighbours in each of the adjacent planes being about 10%

† Mercury is sold in iron *flasks* holding 76 lb of mercury and this is the unit in which output is normally measured.

TABLE 29.1   *Some properties of the elements zinc, cadmium, and mercury*

Property		Zn	Cd	Hg
Atomic number		30	48	80
Number of naturally occurring isotopes		5	8	7
Atomic weight		65.38	112.41	200.59($\pm$0.03)
Electronic configuration		$[Ar]3d^{10}4s^2$	$[Kr]4d^{10}5s^2$	$[Xe]4f^{14}5d^{10}6s^2$
Electronegativity		1.6	1.7	1.9
Metal radius (12 coordinate)/pm		134	151	151
Effective ionic radius/pm II		74	95	102
I		—	—	119
Ionization energies/kJ mol^{-1}	1st	906.1	876.5	1007
	2nd	1733	1631	1809
	3rd	3831	3644	3300
$E°(M^{2+}/M)$/V		−0.7619	−0.4030	+0.8545
MP/°C		419.5	320.8	−38.9
BP/°C		907	765	357
ΔH_{fus}/kJ mol^{-1}		7.28($\pm$0.01)	6.4($\pm$0.2)	2.30($\pm$0.02)
ΔH_{vap}/kJ mol^{-1}		114.2($\pm$1.7)	100.0($\pm$2.1)	59.1($\pm$0.4)
$\Delta H_{(monatomic\ gas)}$/kJ mol^{-1}		129.3($\pm$2.9)	111.9($\pm$2.1)	61.3
Density (25°C)/g cm^{-3}		7.14	8.65	13.534 (1)
Electrical resistivity (20°C)/μohm cm		5.8	7.5	95.8

more distant. In the case of (rhombohedral) Hg the distortion, again uniquely, is the reverse, with the coplanar atoms being the more widely separated (by some 16%). The consequence is that these elements are much less dense and have a lower tensile strength than their predecessors in Group IB. These facts have been ascribed to the stability of the d electrons which are now tightly bound to the nucleus: the metallic bonding therefore involves only the outer s electrons, and is correspondingly weakened.

29.2.4   *Chemical reactivity and trends*[5]

Zinc and cadmium tarnish quickly in moist air and combine with oxygen, sulfur, phosphorus, and the halogens on being heated. Mercury also reacts with these elements, except phosphorus, and its reaction with oxygen was of considerable practical importance in the early work of J. Priestley and A. L. Lavoisier on oxygen (p. 698). The reaction only becomes appreciable at temperatures of about 350°C, but above about 400°C HgO decomposes back into the elements. None of the three metals reacts with hydrogen, carbon, or nitrogen.

Non-oxidizing acids dissolve both Zn and Cd with the evolution of hydrogen. With oxidizing acids the reactions are more complicated, nitric acid for instance producing a variety of oxides of nitrogen dependent on the concentration and temperature. Mercury is unreactive to non-oxidizing acids but dissolves in conc HNO_3 and in hot conc H_2SO_4 forming the Hg^{II} salts along with oxides of nitrogen and sulfur. Dilute HNO_3 slowly produces $Hg_2(NO_3)_2$. Zinc is the only element in the group which dissolves in aqueous alkali to form ions such as aquated $[Zn(OH)_4]^{2-}$ (zincates).

All three elements form alloys with a variety of other metals. Those of zinc include the

---

[5] Chapters 30, 31, and 32, pp. 459–549, of Ref. 1.

brasses (p. 1369) and, as mentioned above, are of considerable commercial importance. Those of mercury are known as amalgams and some, such as sodium and zinc amalgams, are valuable reducing agents: in a number of cases, high heats of formation and stoichiometric compositions (e.g. Hg_2Na) suggest chemical combination. Amalgams are most readily formed by heavy metals, whereas the lighter metals of the first transition series (with the exception of manganese and copper) are insoluble in mercury. Hence iron flasks can be used for its storage.

Chemically, it is clear that Zn and Cd are rather similar and that Hg is somewhat distinct. The lighter pair are more electropositive, as indicated both by their electronegativity coefficients and electrode potentials (Table 29.1), while Hg has a positive electrode potential and is comparatively inert. With the exception of the metallic radii, all the evidence indicates that the effects of the lanthanide contraction have died out by the time this group is reached. Compounds are characterized by the d^{10} configuration and, with the exception of derivatives of the Hg_2^{2+} ion, which formally involved Hg^I, they almost exclusively involve M^{II}. The ease with which the s^2 electrons are removed compared with the more firmly held d electrons is shown by the ionization energies. The sum of the first and second is in each case smaller than for the preceding element in Group IB, whereas the third is appreciably higher. Even so the first two ionization energies are high for mercury (as they are for gold)—perhaps reflecting the poor nuclear shielding afforded by the filled 4f shell—and this, coupled with the small hydration energy associated with the large Hg^{II} cation, accounts for the positive value of its electrode potential.

In view of the stability of the filled d shell, these elements show few of the characteristic properties of transition metals (p. 1060) despite their position in the d block of the periodic table. Thus zinc shows similarities with the main-group metal magnesium, many of their compounds being isomorphous, and it displays the class-a characteristic of complexing readily with O-donor ligands. On the other hand, zinc has a much greater tendency than magnesium to form covalent compounds, and its resembles the transition elements in forming stable complexes not only with O-donor ligands but with N- and S-donor ligands and with halides and CN^- as well. As mentioned above, cadmium is rather similar to zinc and may be regarded as on the class-a/b borderline. However, mercury is undoubtedly class b: it has a much greater tendency to covalency and a preference for N-, P-, and S-donor ligands, with which Hg^{II} forms complexes whose stability is rarely exceeded by those of any other divalent cation. Compounds of the M^{II} ions of this group are characteristically diamagnetic and those of Zn^{II}, like those of Mg^{II}, are colourless. By contrast, many compounds of Hg^{II}, and to a lesser extent those of Cd^{II}, are highly coloured due to the greater ease of charge transfer from ligands to the more polarizing cations. The increasing polarizing power and covalency of their compounds in the sequence, $Mg^{II} < Zn^{II} < Cd^{II} < Hg^{II}$, is a reflection of the decreasing nuclear shielding and consequent increasing power of distortion in the sequence filled p shell < filled d shell < filled f shell.

A further manifestation of these trends is the increasing stability of σ-bonded alkyls and aryls in passing down the group (p. 1416). Those of Zn and Cd are rather reactive and unstable to both air and water, whereas those of Hg are stable to both. (The Hg–C bond is not in fact strong but the competing Hg–O bond is weaker.) However, the M^{II} ions do not form π complexes with CO, NO, or olefins (alkenes), no doubt because of the stability of their d^{10} configurations and their consequent inability to provide electrons for "back bonding". Likewise their cyanides presumably owe their stability primarily to σ rather

TABLE 29.2   *Stereochemistries of compounds of Zn^{II}, Cd^{II}, and Hg^{II}*

Coordination number	Stereochemistry	Zn	Cd	Hg
2	Linear	$ZnEt_2$	$CdEt_2$	$[Hg(NH_3)_2]^{2+}$
3	Planar	$[ZnMe(NPh_3)]_2$		$[HgI_3]^-$
4	Tetrahedral	$[M(H_2O)_4]^{2+}$, $[M(NH_3)_4]^{2+}$	$[MCl_4]^{2-}$	$[Hg(SCN)_4]^{2-}$
	Planar	$[Zn(glycinyl)_2]$		
5	Trigonal bipyramidal	$[Zn(terpy)Cl_2]$	$[CdCl_5]^{3-}$	$[Hg(terpy)Cl_2]$
	Square pyramidal	$[Zn(S_2CNEt_2)_2]_2$	$[Cd(S_2CNEt_2)_2]_2$	$[Hg\{N(C_2H_4NMe_2)_3\}I]^+$
6	Octahedral	$[Zn(en)_3]^{2+}$	$[Cd(NH_3)_6]^{2+}$	$[Hg(C_5H_5NO)_6]^{2+}$
7	Pentagonal bipyramidal	$[Zn(H_2dapp)-(H_2O)_2]^{2+(a)}$	$[Cd(quin)_2-(NO_3)_2H_2O]^{(b)}$	
8	Distorted dodecahedral	$[Zn(NO_3)_4]^{2-(c)}$		
	Distorted square antiprismatic			$[Hg(NO_2)_4]^{2-}$

[a] H_2dapp, 2,6-diacetylpyridinebis(2'-pyridylhydrazone). (See D. WESTER and G. T. PALENIK, *Inorg. Chem.* **15**, 755 (1976).)

[b] The 2 nitrate ions are not equivalent (both are bidentate but one is coordinated symmetrically, the other asymmetrically) and the structure of the complex is by no means regular (p. 1413).

[c] The distortion arises because the bidentate nitrate ions are coordinated asymmetrically to such an extent that the stereochemistry may alternatively be regarded as approaching tetrahedral (p. 1413).

than π bonding. The filled d shell also prevents π acceptance and complexes with cyclopentadienide ions (which are good π donors) are σ- rather than π-bonded.

The range of stereochemistries found in compounds of the M^{II} ions is illustrated in Table 29.2. Since the d^{10} configuration affords no crystal field stabilization, the stereochemistry of a particular compound depends on the size and polarizing power of the M^{II} cation and the steric requirements of the ligands. Thus both Zn^{II} and Cd^{II} favour 4-coordinate tetrahedral complexes but Cd^{II}, being the larger, forms 6-coordinate octahedral complexes more readily than does Zn^{II}. However, although the still larger Hg^{II} also commonly adopts a tetrahedral stereochemistry, octahedral 6-coordination is less prevalent than for either of its congeners.† When it does occur it is usually highly distorted with 2 short and 4 long bonds, a distortion which in its extreme form produces the 2-coordinate, linear stereochemistry which is characteristic of Hg^{II}. This is also found in organozinc and organocadmium compounds but only with Hg^{II} is it one of the predominant stereochemistries. Explanations of this fact have been given[6] in terms of the promotional energies involved in various hybridization schemes, but it may be regarded pictorially as a consequence of the greater deformability of the d^{10} configuration of the large Hg^{II} ion. Thus, if 2 ligands are considered to approach the cation from opposite ends of the z-axis, the resulting deformation increases the electron density in the xy-plane and

† A single example of trigonal prismatic coordination has been reported for Hg in the green, zero-valent mixed-metal cluster $[Hg\{Pt(2,6-Me_2C_6H_3NC)\}_6]$ (Y. YAMAMOTO, H. YAMAZAKI, and T. SAKURAI, *J. Am. Chem. Soc.* **104**, 2329–30 (1982).

[6] D. GRDENIC, The structural chemistry of mercury, *Q. Revs.* **19**, 303–28 (1965).

so discourages the close approach of other ligands. Coordination numbers greater than 6 are rare and generally involve bidentate, *O*-donor ligands with a small "bite", such as NO_3^- and NO_2^-.

29.3 Compounds of zinc, cadmium, and mercury[7-10]

Zinc hydride has recently been isolated from the reaction of LiH with $ZnBr_2$ or NaH with ZnI_2:

$$2MH + ZnX_2 \xrightarrow{\text{thf}} ZnH_2 + 2MX$$

The alkali metal halide remains in solution and ZnH_2 is precipitated as a white solid of moderate stability at or below room temperature.[10a] CdH_2 and HgH_2 are much less stable and decompose rapidly even below 0°. The complex metal hydrides $LiZnH_3$, Li_2ZnH_4, and Li_3ZnH_5 have each been prepared as off-white powders by the reaction of $LiAlH_4$ with the appropriate organometallic complex Li_nZnR_{n+2}.

The carbides of these metals (which are actually acetylides, MC_2, p. 319) and also the nitrides are unstable materials, those of mercury explosively so.

29.3.1 *Oxides and chalcogenides*

The principal compounds in this category are the monochalcogenides, which are formed by all three metals. It is a notable indication of the stability of tetrahedral coordination for the elements of Group IIB that, of the 12 compounds of this type, only CdO, HgO, and HgS adopt a structure other than wurzite or zinc blende, both of which involve tetrahedral coordination of the cation (see below): CdO adopts the 6-coordinate rock-salt structure; HgO features zigzag chains of almost linear O–Hg–O units; and HgS exists in both a zinc-blende form and in a rock-salt form.

The normal oxide, formed by each of the elements of this group, is MO, and peroxides MO_2 are known for Zn and Cd. Reported lower oxides, M_2O, are apparently mixtures of the metal and MO.

ZnO is by far the most important manufactured compound of zinc and, being an inevitable byproduct of primitive production of brass, has been known longer than the metal itself. It is manufactured by burning in air the zinc vapour obtained on smelting the ore or, for a purer and whiter product, the vapour obtained from previously refined zinc. It is normally a white, finely divided material with the wurtzite structure. On heating, the colour changes to yellow due to the evaporation of oxygen from the lattice to give a nonstoichiometric phase $Zn_{1+x}O$ ($x \leqslant 70$ ppm); the supernumerary Zn atoms produce

[7] M. FARNSWORTH and C. H. KLINE, *Zinc Chemicals*, International Lead Zinc Research Org. Inc., New York, 1973, 243 pp.

[8] M. FARNSWORTH, *Cadmium Chemicals*, International Lead Zinc Research Org. Inc., New York, 1980, 158 pp.

[9] C. A. MCAULIFFE (ed.), *The Chemistry of Mercury*, Macmillan, London, 1977, 288 pp.

[10] B. J. AYLETT, Group IIB, Chap. 30, pp. 187–328, in *Comprehensive Inorganic Chemistry*, Vol. 3, Pergamon Press, Oxford, 1973.

[10a] J. J. WATKINS and E. C. ASHBY, Reactions of alkali metal hydrides with zinc halides in tetrahydrofuran. A convenient and economical preparation of zinc hydride, *Inorg. Chem.* **13**, 2350–4 (1974). See also E. C. ASHBY and J. J. WATKINS, *Inorg. Chem.* **12**, 2493–2503 (1973).

lattice defects which trap electrons which can subsequently be excited by absorption of visible light.[10b] Indeed, by "doping" ZnO with an excess of 0.02–0.03% Zn metal, a whole range of colours—yellow, green, brown, red—can be obtained. The reddish hues of the naturally occurring form, zincite, arise, however, from the presence of Mn or Fe.

The major industrial use of ZnO is in the production of rubber where it shortens the time of vulcanization. As a pigment in the production of paints it has the advantage over the traditional "white lead" (basic lead carbonate) that it is non-toxic and is not discoloured by sulfur compounds, but it has the disadvantage compared to TiO_2 of a lower refractive index and so a reduced "hiding power" (p. 1118). It improves the chemical durability of glass and so is used in the production of special glasses, enamels, and glazes. In the chemical industry it is the usual starting material for other zinc chemicals of which the soaps (i.e. salts of fatty acids, such as Zn stearate, palmitate, etc.) are the most important, being used as paint driers, stabilizers in plastics, and as fungicides. An important small scale use is in the production of "zinc ferrites". These are spinels of the type $Zn_x^{II}M_{1-x}^{II}Fe_2^{III}O_4$ involving a second divalent cation (usually Mn^{II} or Ni^{II}), such that when $x = 0$ the structure is that of an inverse spinel (i.e. half the Fe^{III} ions occupy tetrahedral sites—see p. 1256) and, where $x = 1$, a normal spinel (i.e. all the Fe^{III} ions occupy octahedral sites), since Zn^{II} displaces Fe^{III} from the tetrahedral sites. Reducing the proportion of Fe^{III} ions in tetrahedral sites lowers the Curie temperature. The magnetic properties of the ferrite can therefore be controlled by adjustment of the zinc content.

ZnO is amphoteric (p. 752), dissolving in acids to form salts and in alkalis to form zincates, such as $[Zn(OH)_3]^-$ and $[Zn(OH)_4]^{2-}$. The gelatinous, white precipitate obtained by adding alkali to aqueous solutions of Zn^{II} salts is $Zn(OH)_2$ which, like ZnO, is amphoteric.

CdO is produced from the elements and, depending on its thermal history, may be greenish-yellow, brown, red, or nearly black. This is partly due to particle size but more importantly, as with ZnO, is a result of lattice defects—this time in an NaCl lattice. It is more basic than ZnO, dissolving readily in acids but hardly at all in alkalis. White, $Cd(OH)_2$ is precipitated from aqueous solutions of Cd^{II} salts by the addition of alkali and treatment with very concentrated alkali yields hydroxocadmiates such as $Na_2[Cd(OH)_4]$. Cadmium oxide and hydroxide find important applications in decorative glasses and enamels and in Ni–Cd storage cells. CdO also catalyses a number of hydrogenation and dehydrogenation reactions.

Treatment of the hydroxides of Zn and Cd with aqueous H_2O_2 produces hydrated peroxides of rather variable composition. That of Zn has antiseptic properties and is widely used in cosmetics.

HgO exists in a red and a yellow variety. The former is obtained by pyrolysis of $Hg(NO_3)_2$ or by heating the metal in O_2 at about 350°C; the latter by cold methods such as precipitation from aqueous solutions of Hg^{II} by addition of alkali ($Hg(OH)_2$ is not known). The difference in colour is entirely due to particle size, both forms having the same structure which consists of zigzag chains of virtually linear O–Hg–O units with Hg–O 205 pm and angle Hg–O–Hg 107°. The shortest Hg⋯O distance between chains is 282 pm.

Zinc blende, ZnS, is the most widespread ore of zinc and the main source of the metal,

[10b] N. N. GREENWOOD, *Ionic Crystals, Lattice Defects, and Nonstoichiometry*, Chaps. 6 and 7, pp. 111–81, Butterworths, London, 1968.

but ZnS is also known in a second naturally occurring, though much rarer, form, wurtzite, which is the more stable at high temperatures. The names of these minerals are now also used as the names of their crystal structures which are important structure types found in many other AB compounds. In both structures each Zn is tetrahedrally coordinated by 4 S and each S is tetrahedrally coordinated by 4 Zn; the structures differ significantly only in the type of close-packing involved, being cubic in zinc-blende and hexagonal in wurtzite (Fig. 29.1). Pure ZnS is white and, like ZnO, finds use as a pigment for which purpose it is often obtained (as "lithopone") along with $BaSO_4$ from aqueous solution of $ZnSO_4$ and BaS:

$$ZnSO_4 + BaS \longrightarrow ZnS\downarrow + BaSO_4\downarrow$$

Freshly precipitated ZnS dissolves readily in mineral acids with evolution of H_2S, but roasting renders it far less reactive and it is then an acceptable pigment in paints for children's toys since it is harmless if ingested. ZnS also has interesting optical properties. It turns grey on exposure to ultraviolet light, probably due to dissociation to the elements, but the process can be inhibited by trace additives such as cobalt salts. Cathode rays, X-rays, and radioactivity also produce fluorescence or luminescence in a variety of colours which can be extended by the addition of traces of various metals or the replacement of Zn by Cd and S by Se. It is widely used in the manufacture of cathode-ray tubes and radar screens.

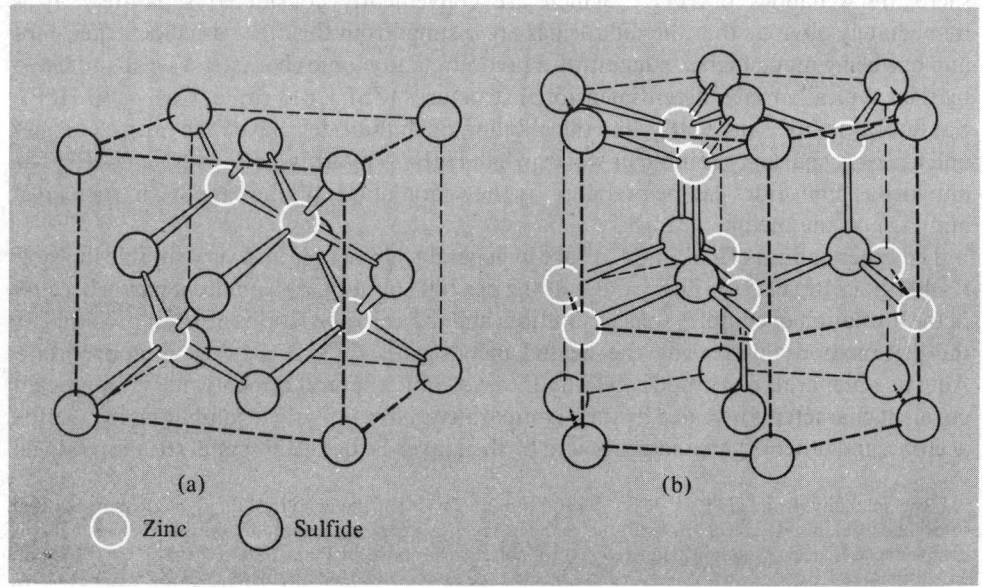

(a)   (b)

◯ Zinc   ⬤ Sulfide

Fig. 29.1 Crystal structures of ZnS. (a) Zinc blende, consisting of two, interpenetrating, ccp lattices of Zn and S atoms displaced with respect to each other so that the atoms of each achieve 4-coordination (Zn–S = 235 pm) by occupying tetrahedral sites of the other lattice. The face-centred cube, characteristic of the ccp lattice, can be seen—in this case composed of S atoms, but an extended diagram would reveal the same arrangement of Zn atoms. Note that if all the atoms of this structure were C instead of Zn and S, the structure would be that of diamond (p. 305). (b) Wurtzite. As with zinc blende, tetrahedral coordination of both Zn and S is achieved (Zn–S = 236 pm) but this time the interpenetrating lattices are hexagonal, rather than cubic, close-packed.

Yellow ZnSe and brown ZnTe are structurally akin to the sulfide and the former especially is used mainly in conjunction with ZnS as a phosphor.

Chalcogenides of Cd are similar to those of Zn and display the same duality in their structures. The sulfide and selenide are more stable in the hexagonal form whereas the telluride is more stable in the cubic form. CdS is the most important compound of cadmium and, by addition of CdSe, ZnS, HgS, etc., it yields thermally stable pigments of brilliant colours from pale yellow to deep red, while colloidal dispersions are used to colour transparent glasses. CdS and CdSe are also useful phosphors and CdTe is believed to have considerable potential as a semiconductor.[11]

HgS is polymorphic. The red α-form is the mineral cinnabar, or vermilion, which has a distorted rock-salt structure and can be prepared from the elements. β-HgS is the rare, black, mineral metacinnabar which has the zinc-blende structure and is converted by heat to the stable α-form. In the laboratory the most familiar form is the highly insoluble† black precipitate obtained by the action of H_2S on aqueous solutions of Hg^{II}. HgS is an unreactive substance, being attacked only by conc HBr, HI, or aqua regia. HgSe and HgTe are easily obtained from the elements and have the zinc-blende structure.

29.3.2 *Halides*

The known halides are listed in Table 29.3. All 12 dihalides are known and in addition there are 4 halides of Hg_2^{2+} which are conveniently considered separately. It is immediately obvious that the difluorides are distinct from the other dihalides, their mps and bps being much higher, suggesting a predominantly ionic character, as also indicated by their typically ionic three-dimensional structures (ZnF_2, 6:3 rutile; CdF_2 and HgF_2, 8:4 fluorite). ZnF_2 and CdF_2, like the alkaline earth fluorides, have high lattice energies and are only sparingly soluble in water, while HgF_2 is hydrolysed to HgO and HF. The anhydrous difluorides can be prepared by the action of HF (in the case of Zn) or F_2 (Cd and Hg) on the metal.

The other halides of Zn^{II} and Cd^{II} are in general hygroscopic and very soluble in water (~ 400 g per 100 cm^3 for ZnX_2 and ~ 100 g per 100 cm^3 for CdX_2), at least partly because of the formation of complex ions in solution, and the anhydous forms are best prepared by the dry methods of treating the heated metals with HCl, Br_2, or I_2 as appropriate. Aqueous preparative methods yield hydrates of which several are known.[12] Significant covalent character is revealed by their comparatively low mps, their solubilities in ethanol, acetone, and other organic solvents, and by their layer-lattice (2D) crystal structures. In all

---

† The solubility product, $[Hg^{2+}][S^{2-}] = 10^{-52}$ mol^2 dm^{-6} but the actual solubility is greater than that calculated from this extremely low figure, since the mercury in solution is present not only as Hg^{2+} but also as complex species. In acid solution $[Hg(SH)_2]$ is probably formed and in alkaline solution, $[HgS_2]^{2-}$: the relevant equilibria are:

$$HgS(s) \rightleftharpoons Hg^{2+}(aq) + S^{2-}(aq); \qquad pK_s \ 51.8$$

$$HgS(s) + 2H^+(aq) \rightleftharpoons Hg^{2+}(aq) + H_2S \ (1 \ atm); \quad pK' \ 30.8$$

$$HgS(s) + H_2S(g) \rightleftharpoons [Hg(SH)_2](aq); \qquad pK \ \ 6.2$$

$$HgS(s) + S^{2-}(aq) \rightleftharpoons [HgS_2]^{2-}(aq); \qquad pK \ \ 1.5$$

[11] K. ZANIO, *Semiconductors and Semimetals*, Vol. 13, Academic Press, New York, 1978, 235 pp.
[12] N. V. SIDGWICK, *The Chemical Elements and their Compounds*, p. 271, Oxford University Press, Oxford, 1950.

TABLE 29.3   *Halides of zinc, cadmium, and mercury (mp, bp, in parentheses)*

Fluorides	Chlorides	Bromides	Iodides
ZnF_2 white (872°, 1500°)	$ZnCl_2$ white (275°, 756°)	$ZnBr_2$ white (394°, 702°)	ZnI_2 white (446°, d > 700°)
CdF_2 white (1049°, 1748°)	$CdCl_2$ white (568°, 980°)	$CdBr_2$ pale yellow (566°, 863°)	CdI_2 white (388°, 787°)
HgF_2 white (d > 645°)	$HgCl_2$ white (280°, 303°)	$HgBr_2$ white (238°, 318°)	HgI_2 α red, β yellow (257°, 351°)
Hg_2F_2 yellow (d > 570°)	Hg_2Cl_2 white (subl 383°)	Hg_2Br_2 White (subl 345°)	Hg_2I_2 yellow (subl 140°)

cases these may be regarded as close-packed lattices of halides ions in which the Zn^{II} ions occupy tetrahedral, and the Cd^{II} ions octahedral, sites. The structures of $CdCl_2$ ($CdBr_2$ is similar) and CdI_2 are of importance (Fig. 29.2) since they are typical of MX_2 compounds in which marked polarization effects are expected.[13]

Concentrated, aqueous solutions of $ZnCl_2$ dissolve starch, cellulose (and therefore cannot be filtered through paper!), and silk. Commercially $ZnCl_2$ is one of the important compounds of zinc, with applications in textile processing and, because when fused it readily dissolves other oxides, is used in a number of metallurgical fluxes as well as in the

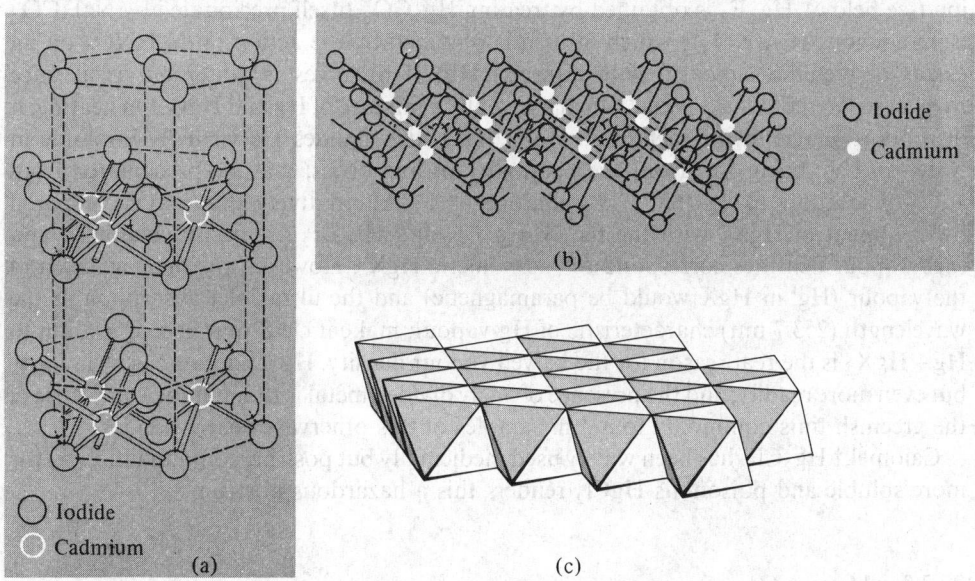

FIG. 29.2   The layer structure of crystalline CdI_2: (a) Shows the hexagonal close-packing of I atoms with Cd atoms in alternate layers of octahedral sites sandwiched between layers of I atoms. In $CdCl_2$ the individual composite layers are identical with those in CdI_2, but they are arranged so that the Cl atoms are ccp. (b) Shows a portion of an individual composite layer of CdI_6 (or $CdCl_6$) octahedra. (c) Shows the same portion of a composite layer as in (b) and viewed from the same angle, but with CdI_2 (or $CdCl_2$) units represented by solid, edge-sharing octahedra.

[13] N. N. GREENWOOD, Chap. 3 of ref. 10b, pp. 37–61.

manufacture of magnesia cements in dental fillings. Cadmium halides are used in the preparation of electroplating baths and in the production of pigments.

Covalency is still more pronounced in HgX_2 (X = Cl, Br, I) than in the corresponding halides of Zn and Cd. These compounds are readily prepared from the elements and are low melting volatile solids, soluble in many organic solvents. Their solubilities in water, where they exist almost entirely as HgX_2 molecules, decrease with increasing molecular weight, HgI_2 being only slightly soluble, and they may be precipitated anhydrous from aqueous solutions by metathetical reactions. Their crystalline structures[6] reveal an interesting gradation. $HgCl_2$ is composed of linear Cl–Hg–Cl molecules (Hg–Cl = 225 pm and the next shortest Hg$\cdots$Cl distance is 334 pm); $HgBr_2$ and HgI_2 have layer structures. However, in the bromide although the Hg^{II} may be regarded as 6-coordinated, two Hg–Br distances are much shorter than the other four (248 pm compared to 323 pm), while in the red variety of the iodide the Hg^{II} is unambiguously tetrahedrally 4-coordinated (Hg–I = 278 pm). At temperatures above 126°C HgI_2 exists as a less-dense, yellow form similar to $HgBr_2$. In the gaseous phase all 3 of these Hg^{II} halides exist as discrete linear HgX_2 molecules. Comparison of the Hg–X distances in these molecules (Hg–Cl = 228 pm, Hg–Br = 240 pm; Hg–I = 257 pm) with those given above, indicate an increasing departure from molecularity in passing from the solid chloride to the solid iodide.

$HgCl_2$ is the "corrosive sublimate" of antiquity, formerly obtained by sublimation from $HgSO_4$ and NaCl and used as an antiseptic. It is, however, a violent poison and was widely used as such in the Middle Ages.[9]

The halides are the most familiar compounds of mercury(I) and all contain the Hg_2^{2+} ion (see below). Hg_2F_2 is obtained by treating Hg_2CO_3 (itself precipitated by $NaHCO_3$ from aqueous $Hg_2(NO_3)_2$ which in turn is obtained by the action of dil HNO_3 on an excess of metallic mercury) with aqueous HF. It dissolves in water but is at once hydrolysed to the "black oxide" which is actually a mixture of Hg and HgO. On heating, it disproportionates to the metal and HgF_2. The other halides are virtually insoluble in water and so, being free from the possibility of hydrolysis, may be precipitated from aqueous solutions of $Hg_2(NO_3)_2$ by addition of X^-. Alternatively, they may be prepared by treatment of HgX_2 with the metal. Hg_2Cl_2 and Hg_2Br_2 are easily volatilized and their vapour densities correspond to "monomeric HgX". However, the diamagnetism of the vapour (Hg^I in HgX would be paramagnetic) and the ultraviolet absorption at the wavelength (253.7 nm) characteristic of Hg vapour, make it clear that decomposition to $Hg + HgX_2$ is the real reason for the halved vapour density. Hg_2I_2 decomposes similarly but even more readily, and the presence of finely divided metal is thought to be the cause of the greenish tints commonly found in samples of this otherwise yellow solid.

Calomel,† Hg_2Cl_2, has been widely used medicinally but possible contamination by the more soluble and poisonous $HgCl_2$ renders this a hazardous nostrum.

29.3.3 *Mercury(I)*

Although $Cd_2(AlCl_4)_2$, isolated from melts of Cd and $CdCl_2$ in $NaAlCl_4$, and the glass obtained from the melt of Zn in $ZnCl_2$ give Raman spectra which indicate the presence of

---

† Calomel, derived from the Greek words καλό-ς (beautiful) and μέλας (black), seems an odd name for a white solid. It might arise from the colour of the material obtained when Hg_2Cl_2 is treated with ammonia; this is a product of variable composition (see below) which owes its colour to the presence of metallic mercury. Other more fanciful derivations are listed in the *Oxford English Dictionary* **2,** 41 (1970).

$(Cd-Cd)^{2+}$ and $(Zn-Zn)^{2+}$ respectively, it is only for mercury that the formal oxidation state 1 is of importance.

Mercury(I) compounds in general may be prepared, like the halides just discussed, by the reduction of the corresponding Hg^{II} salt, often by the metal itself, or by precipitation from aqueous solutions of the nitrate. The nitrate is known as the dihydrate, $Hg_2(NO_3)_2 . 2H_2O$, and is stable in water if this is acidified, otherwise basic salts such as $Hg(OH)(NO_3)$ are precipitated. The perchlorate is the only other appreciably soluble salt, the rest being either insoluble or, like the sulfate, chlorate, and salts of organic acids, only sparingly soluble. In all cases the dinuclear Hg_2^{2+} ion is present rather than mononuclear Hg^+. The evidence for this is overwhelming and includes the following:

(1) In crystalline mercury(I) compounds, instead of the sequence of alternate M^+ and X^- expected for MX compounds, Hg–Hg pairs are found in which the separation, though not constant, lies in the range 250–270 pm[9] which is shorter than the Hg–Hg separation of 300 pm found in the metal itself.

(2) The Raman spectrum of aqueous mercury(I) nitrate has, in addition to lines characteristic of the NO_3^- ion, a strong absorption at 171.7 cm^{-1} which is not found in the spectra of other metal nitrates and is not active in the infrared; it is therefore diagnostic of the Hg–Hg stretching vibration since homonuclear diatomic vibrations are Raman active not infrared active.† Similar data have recently been produced for a number of other compounds in the solid state and solution.

(3) Mercury(I) compounds are diamagnetic, whereas the monatomic Hg^+ ion would have a $d^{10}s^1$ configuration and so be paramagnetic.

(4) The measured emfs of concentration cells of mercury(I) salts are only explicable on the assumption that a 2-electron transfer is involved. This would not be the case if Hg^+ were involved: $[E = (2.303 RT/nF) \log a_1/a_2$ where $n=2$ for Hg_2^{2+} and $n=1$ for $Hg^+]$.

(5) It is found that "equilibrium constants" are in fact only constant if the concentration $[Hg_2^{2+}]$ is employed rather than $[Hg^+]^2$, i.e. the equilibria must be of the type:

$$2Hg + 2Ag^+ \rightleftharpoons Hg_2^{2+} + 2Ag \quad NOT \quad Hg + Ag^+ \rightleftharpoons Hg^+ + Ag$$

$$\text{or } Hg + Hg^{2+} \rightleftharpoons Hg_2^{2+} \qquad NOT \quad Hg + Hg^{2+} \rightleftharpoons 2Hg^+$$

In order to understand the formation and stability of mercury(I) compound it is helpful to consider the relevant reduction potentials:

$$Hg_2^{2+} + 2e^- \rightleftharpoons 2Hg(l); \quad E^\circ +0.7889 \text{ V}$$

and

$$2Hg^{2+} + 2e^- \rightleftharpoons Hg_2^{2+}; \quad E^\circ +0.920 \text{ V}$$

From this it follows that

$$Hg^{2+} + 2e^- \rightleftharpoons Hg(l); \quad E^\circ +0.8545 \text{ V}$$

and

$$Hg_2^{2+} \rightleftharpoons Hg(l) + Hg^{2+}; \quad E^\circ -0.131 \text{ V}$$

Now, $E^\circ = (RT/nF) \ln K$, i.e. $E^\circ = (0.0591/n) \log_{10} K$.

---

† Indeed, this is perhaps the earliest example of a new structural species to be established by Raman spectroscopy. (L. A. WOODWARD, Raman effect and the complexity of mercurous and thallous ions, *Phil. Mag.* **18**, 823–7 (1934).)

Hence, $\log_{10} K = -(0.131/0.0591) = -2.217$,

i.e. $K = [Hg^{2+}]/[Hg_2^{2+}] = 0.0061$

Thus, at equilibrium, aqueous solutions of mercury(I) salts will contain around 0.6% of mercury(II) and the rather finely balanced equilibrium is easily displaced. The presence of any reagent which reduces the activity (in effect the concentration) of Hg^{2+} more than that of Hg_2^{2+}, either by forming a less-soluble salt or a more-stable complex of Hg^{2+} will displace the equilibrium to the right and cause the disproportionation of the Hg_2^{2+}. There are many such reagents, including S^{2-}, OH^-, CN^-, NH_3, and acetylacetone. This is why the most stable Hg_2^{2+} salts are the insoluble ones and why there are few stable complexes. Those which are known all involve either O- or N-donor ligands,† the linear O–Hg–Hg–O group being a common feature of the former.

Polycations of mercury

The Hg–Hg bond in Hg_2^{2+} may be ascribed to overlap of the 6s orbitals with little involvement of 6p orbitals or of the filled d^{10} shell of each atom. If this is regarded as the coordination of Hg to an Hg^{2+} cation, the coordination of a second Hg ligand is reasonable. Accordingly, $Hg_3(AlCl_4)_2$ can be obtained from a molten mixture of $HgCl_2$, Hg, and $AlCl_3$ and contains the discrete, virtually linear cation

$$[Hg \underset{255\ pm}{\rule{2cm}{0.4pt}} Hg \rule{2cm}{0.4pt} Hg]^{2+}$$

in which the formal oxidation state of Hg is $+\frac{2}{3}$.

Still more interesting is the oxidation of Hg by AsF_5 in liquid SO_2:[14] in this process the AsF_5 serves both as an oxidant (being reduced to AsF_3) and also as a fluoride-ion acceptor to give AsF_6^-. In a matter of minutes the colour of the solution becomes bright yellow then deepens to red as the Hg simultaneously turns to a shiny golden-yellow solid; the solid then begins to dissolve to give an orange and, finally, a colourless solution. By controlling the quantity of oxidant, AsF_5, and removing the solution at the appropriate stages, it is possible to crystallize a series of extremely moisture-sensitive materials:

deep red-black $Hg_4(AsF_6)_2$, the cation of which is the almost linear

$$[Hg \underset{259\ pm}{\rule{1.5cm}{0.4pt}} Hg \underset{262\ pm}{\rule{1.5cm}{0.4pt}} Hg \underset{259\ pm}{\rule{1.5cm}{0.4pt}} Hg]^{2+}$$

with Hg in the formal average oxidation state $+\frac{1}{2}$;
orange $Hg_3(AsF_6)_2$, containing the trimeric cation mentioned above; and
colourless $Hg_2(AsF_6)_2$, containing the dimeric Hg^I cation.

By working at lower temperatures ($-20°C$) to reduce the reaction rate, or by using

† For this reason, although Hg_2^{2+} must be regarded as a class-b cation (e.g. the solubilities of its halides decrease in the order F^- to I^-), it is evidently less so than Hg^{2+} which has a notable preference for S donors.

14 R. J. GILLESPIE, private communication (October 1981) but see B. D. CUTFORTH and R. J. GILLESPIE, Polymercury cations, *Inorg. Syntheses* **19**, 22–27 (1979). I. D. BROWN, B. D. CUTFORTH, C. G. DAVIES, P. R. IRELAND, and J. E. VEKRIS, Alchemists' gold, $Hg_{2.86}AsF_6$: an X-ray crystallographic study of a novel disordered mercury compound containing metallically bonded infinite chains, *Can. J. Chem.* **52**, 791–3 (1974).

specially designed apparatus which limits the access of AsF_5 to the Hg, it has been possible to isolate large single crystals of the intermediate golden-yellow solid having dimensions up to $35 \times 35 \times 2$ mm^3. X-ray analysis,[14] supported by neutron diffraction,[14a] shows that it consists of a tetragonal lattice ($a=b \neq c$) of octahedral AsF_6^- anions with two non-intersecting and mutually perpendicular chains of Hg atoms running through it in the a and b directions. Chemical analysis suggests the composition $Hg_3(AsF_6)$ and a formal oxidation state of $Hg = +\frac{1}{3}$, but the measured Hg–Hg separation of 264 pm along the chains is not commensurate with the parallel dimensions of the lattice unit cell, $a=b=754$ pm (cf. 3×264 pm $= 792$ pm) and implies instead the nonstoichiometric composition $Hg_{2.86}(AsF_6)$ or more generally $Hg_{3-\delta}(AsF_6)$ since the composition apparently varies with temperature. An alternative explanation[14b] is that the incommensurate dimensions are a result of anion vacancies and that the compound should be formulated more correctly as $Hg_{2.82}(AsF_6)_{0.94}$. Partially filled conduction bands formed by overlap of Hg orbitals produce a conductivity in the a–b plane which approaches that of liquid mercury and the material becomes superconducting at 4 K.

The analogous $(SbF_6)^-$ compound[14] has also been characterized but because the unit cell of the $(SbF_6)^-$ lattice is somewhat larger than that of $(AsF_6)^-$, it is formulated as $Hg_{2.91}(SbF_6)$.

29.3.4 *Zinc(II) and cadmium(II)*

The almost invariable oxidation state of these elements is $+2$ and, in addition to the oxides, chalcogenides, and halides already discussed, salts of most anions are known. Oxo-salts are often isomorphous with those of Mg^{II} but with lower thermal stabilities. The carbonates, nitrates, and sulfates all decompose to the oxides on heating. Several, such as the nitrates, perchlorates, and sulfates, are very soluble in water and form more than one hydrate. $[Zn(H_2O)_6]^{2+}$ is probably the predominant aquo species in solutions of Zn^{II} salts. Aqueous solutions are appreciably hydrolysed to species such as $[M(OH)(H_2O)_x]^+$ and $[M_2(OH)(H_2O)_x]^{3+}$ and a number of basic (i.e. hydroxo) salts such as $ZnCO_3.2Zn(OH)_2.H_2O$ and $CdCl_2.4Cd(OH)_2$ can be precipitated. Distillation of zinc acetate under reduced pressure yields a crystalline basic acetate, $[Zn_4O(OCOMe)_6]$. The molecular structure of this consists of an oxygen atom surrounded by a tetrahedron of Zn atoms bridged across each edge by acetates. It is isomorphous with the basic acetate of beryllium (p. 135) but, in contrast, the Zn^{II} compound hydrolyses rapidly in water, no doubt because of the ability of Zn^{II} to increase its coordination number above 4.

The coordination chemistry of Zn^{II} and Cd^{II}, although much less extensive than for preceding transition metals, is still appreciable. Neither element forms stable fluoro complexes but, with the other halides, they form the complex anions $[MX_3]^-$ and $[MX_4]^{2-}$, those of Cd^{II} being moderately stable in solution.[7, 8] By using the large cation $[Co(NH_3)_6]^{3+}$ it is also possible to isolate the trigonal bipyramidal $[CdCl_5]^{2-}$. Tetrahedral complexes are the most common type and are formed with a variety of

[14a] A. J. SCHULTZ, J. M. WILLIAMS, N. D. MIRO, A. G. MACDIARMID, and A. J. HEEGER, A neutron diffraction investigation of the crystal and molecular structure of the anisotropic superconductor Hg_3AsF_6, *Inorg. Chem.*, **17**, 646–9 (1978).
[14b] N. D. MIRO, A. G. MACDIARMID, A. J. HEEGER, A. F. GARITO, C. K. CHIANG, A. J. SCHULTZ, and J. M. WILLIAMS, Synthesis and constitution of the metallic mercury compound $Hg_{2.82}(AsF_6)_{0.94}$, *J. Inorg. Nucl. Chem.* **40**, 1351–5 (1978).

O-donor ligands (more readily with Zn[II] than Cd[II]), more stable ones with *N*-donor ligands such as NH_3 and amines, and also with CN^-. Some of the apparently 3-coordinate complexes have a higher coordination number because of aquation or association.

Complexes of higher coordination number are often in equilibrium with the tetrahedral form and may be isolated by increasing the ligand concentration or by adding large counter ions, e.g. $[M(NH_3)_6]^{2+}$, $[M(en)_3]^{2+}$, or $[M(bipy)_3]^{2+}$. With acetylacetone, zinc achieves both 5- and 6-coordination by trimerizing to $[Zn(acac)_2]_3$ (Fig. 29.3a). Five-coordination is also found in adducts such as the distorted trigonal bipyramidal $[Zn(acac)_2(H_2O)]$ and $[Zn(glycinate)_2(H_2O)]$ while the hydrazinium sulphate $[N_2H_5]_2Zn[SO_4]_2$ contains 6-coordinated zinc. This is isomorphous with the Cr[II] compound (p. 1201) and in the crystalline form consists of chains of Zn[II] bridged by $SO_4{}^{2-}$ ions, with each Zn[II] additionally coordinated to two *trans*-$N_2H_5{}^+$ ions.[15]

Complexes with SCN^- throw light on the relative affinities of the two metals for *N*- and *S*-donors. In $[Zn(NCS)_4]^{2-}$ the ligand is *N*-bonded whereas in $[Cd(SCN)_4]^{2-}$ it is *S*-bonded. SCN^- can also act as a bridging group, as in $[Cd\{S=C(NHCH_2)_2\}_2(SCN)_2]$ when linear chains of octahedrally coordinated Cd[II] are formed (Fig. 29.3b). A number of zinc–sulfur compounds are used as accelerators in the vulcanization of rubber. Among

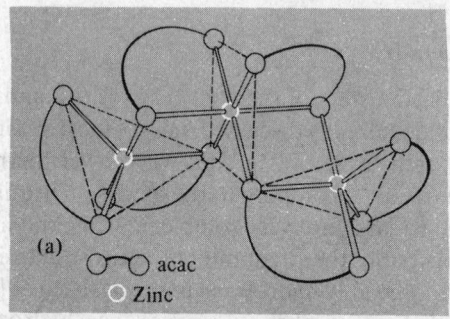

FIG. 29.3 Some polymeric complexes: (a) $[Zn(acac)_2]_3$, (b) $[Cd\{S=C(NHCH_2)_2\}_2(SCN)_2]$, and (c) $[M(S_2CNEt_2)_2]_2$, M = Zn, Cd, Hg. (Note that M is 5-coordinate but with one M–S distance appreciably greater than the other four.)

[15] D. W. HAND and C. K. PROUT, The unit-cell dimensions of the metal hydrazinium sulphates $(N_2H_5)_2M^{II}(SO_4)_2$, *J. Chem. Soc.* (A) 1966, 168–70.

these are the dithiocarbamates, of which $[Zn(S_2CNEt_2)_2]_2$, and the isostructural Cd^{II} and Hg^{II} compounds achieve 5-coordination by dimerizing (Fig. 29.3c).

Coordination numbers higher than 6 are rare and in some cases are known to involve chelating NO_3^- ions which not only have a small "bite" but, may also be coordinated asymmetrically so that the coordination number is not well defined.[16, 17]

29.3.5 *Mercury(II)*

The oxide (p. 1404), chalcogenides (p. 1406), and halides (p. 1406) have already been described. Of them, the only ionic compound is HgF_2 but other compounds in which there is appreciable charge separation are the hydrated salts of strong oxoacids, e.g. the nitrate, perchlorate, and sulfate. In aqueous solution such salts are extensively hydrolysed (HgO is only very weakly basic) and they require acidification to prevent the formation of polynuclear hydroxo-bridged species or the precipitation of basic salts such as $Hg(OH)(NO_3)$ which contains infinite zigzag chains:

Their ionic character is symptomatic of the marked reluctance of Hg^{II} to form covalent bonds to oxygen. In the presence of excess NO_3^- ions the aqueous nitrate forms the complex anion $[Hg(NO_3)_4]^{2-}$ in which 8 oxygen atoms from the bidentate nitrate groups are equidistant from the mercury at 240 pm, which is almost precisely the sum of the ionic radii $(140 + 102$ pm$)$. Also, the unusual regular octahedral coordination is found in complexes with *O*-donor ligands:[18] $[Hg(C_5H_5NO)_6]^{2+}$ (Hg–O $= 235$ pm), $[Hg(H_2O)_6]^{2+}$ (Hg–O $= 234$ pm), and $[Hg(Me_2SO)_6]^{2+}$ (Hg–O $= 234$ pm). In contrast, the more covalently bonding β-diketonates do not form complexes.

The most usual type of coordination in compounds of Hg^{II} with other donor atoms is a distorted octahedron with 2 bonds much shorter than the other 4.† In the extreme, this results in linear 2-coordination in which case the bonds are largely covalent. $Hg(CN)_2$ is actually composed of discrete linear molecules, whereas crystalline‡ $Hg(SCN)_2$ is built up of distorted octahedral units, all SCN groups being bridging.

---

† This has led to the distinction[6] by several workers between the *characteristic coordination number*, which is the number of shortest equal, Hg–L distances and the *effective coordination number*, which is the number of all Hg–L distances which are less than the sum of the van der Waals radii of Hg and L.

‡ Pellets of the dry powder, when ignited in air, form snake-like tubes of spongy ash of unknown composition—the so-called "Pharaoh's serpents".

¹⁶ A. F. CAMERON, D. W. TAYLOR, and R. H. NUTTALL, Structural investigations of metal-nitrate complexes. Part VII. Crystal and molecular structures of aquodinitratobis(quinoline)cadmium(II), *JCS (Dalton)* 1973, 2130–4.

¹⁷ C. BELLITO, L. GASTALDI, and A. A. G. TOMLINSON, Crystal structure of tetraphenylarsonium tetranitratozincate(2−) and optical spectrum of tetranitratocobaltate(2−), *JCS (Dalton)* 1976, 989–92.

¹⁸ D. L. KEPERT, D. TAYLOR, and A. H. WHITE, Crystal structure of hexakis(pyridine-1-oxide)mercury(II) bisperchlorate, *JCS (Dalton)* 1973, 670–3. G. JOHANSSON and M. SANDSTRÖM, The crystal structure of hexaaquamercury(II) perchlorate, $[Hg(H_2O)_6](ClO_4)_2$, *Acta Chem. Scand.* **A32**, 107–13 (1978). M. SANDSTRÖM and I. PERSSON, Crystal and molecular structure of hexakis(dimethylsulfoxide)mercury(II) perchlorate, $[Hg\{(CH_3)_2SO\}_6](ClO_4)_2$, ibid, 95–100 (1978).

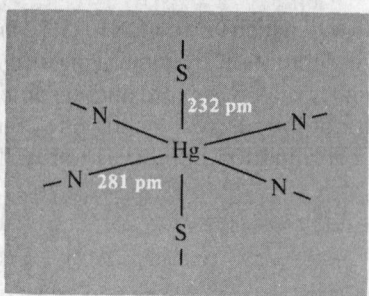

With both these pseudo halides, an excess produces complex anions $[HgX_3]^-$ and the tetrahedral $[HgX_4]^{2-}$.

Similar halo complexes are produced in solution, and several salts of $[HgX_3]^-$ have been isolated and characterized; they display a variety of stereochemistries. In $[HgCl_3]^-$ the environment of the Hg^{II} is either distorted octahedral (with small cations such as NH_4^+ or Na^+) or distorted trigonal bipyramidal (with larger cations such as $[NEt_4]^+$ or $[SMe_3]^+$), whereas in salts of $[HgBr_3]^-$ and $[HgI_3]^-$ the coordination is more commonly distorted tetrahedral. $[NBu_4^n][HgI_3]$ has recently been characterized[18a] and affords the first example of monomeric $[HgX_3]^-$. The anion is planar but, with one I–Hg–I angle 115°, its symmetry is C_{2v} rather than D_{3h}. In the presence of excess halide, $[HgX_4]^{2-}$ complex ions are produced and in comparison with those of Zn^{II} and Cd^{II} it can be seen that their stabilities increase with the sizes of the anion and the cation so that $[HgI_3]^{2-}$ is the most stable of all. Thus, the normally very insoluble HgI_2 will dissolve in aqueous solutions of I^- which can then be made strongly alkaline without precipitation occurring. Such solutions are known as Nessler's reagent, which is used as a sensitive test for ammonia since this produces a yellow or brown coloration due to the formation of the iodo salt, $Hg_2NI.H_2O$, of Millon's base (see below). Adducts of the halides HgX_2, with N-, S-, and P-donor ligands are known, those with N-donors being especially numerous.[19] Their stereochemistries are largely of the expected tetrahedral, or grossly distorted octahedral, types.

Hg^{II}–N compounds[9, 20]

Mercury has a characteristic ability to form not only conventional ammine and amine complexes but also, by the displacement of hydrogen, direct covalent bonds to nitrogen, e.g.:

$$Hg^{2+} + 2NH_3 \rightleftharpoons [Hg-NH_2]^+ + NH_4^+$$

Thus in the presence of an excess of NH_4^+, which suppresses this forward reaction, and counteranions such as NO_3^- and ClO_4^-, which have little tendency to coordinate, complexes such as $[Hg(NH_3)_4]^{2+}$, $[Hg(L–L)_2]^{2+}$, and even $[Hg(L–L)_3]^{2+}$ (L = en, bipy,

[18a] P. L. GOGGIN, P. KING, D. M. McEWAN, G. E. TAYLOR, P. WOODWARD and M. SANDSTRÖM, Vibrational spectroscopic studies of tetra-n-butylammonium trihalogenomercurates; crystal structures of $[NBu_4^n](HgCl_3)$ and $[NBu_4^n][HgI_3]$, *JCS (Dalton)* 1982, 875–82.

[19] P. A. W. DEAN, The coordination chemistry of the mercuric halides, *Prog. Inorg. Chem.* **24**, 109–78 (1978).

[20] D. BREITINGER and K. BRODERSEN, Development of, and problems in, the chemistry of mercury–nitrogen compounds, *Angew. Chem.*, Int. Edn. (Engl.) **9**, 357–67 (1970).

phen) can be prepared. But, in the absence of such precautions, amino, or imino-compounds are likely to be formed, often together. Because of this variety of simultaneous reactions and their dependence on the precise conditions, many reactions between Hg^{II} and amines, although first performed by alchemists in the Middle Ages, remained obscure until the application of X-ray crystallography and, still more recently, spectroscopic techniques such as nmr, infrared, and Raman.

The action of aqueous ammonia on $HgCl_2$, for instance, may be described by the three reactions:

$$HgCl_2 + 2NH_3 \rightleftharpoons [Hg(NH_3)_2Cl_2] \tag{1}$$

$$[Hg(NH_3)_2Cl_2] \rightleftharpoons [Hg(NH_2)Cl] + NH_4Cl \tag{2}$$

$$2[Hg(NH_2)Cl] + H_2O \rightleftharpoons [Hg_2NCl(H_2O)] + NH_4Cl \tag{3}$$

In general, all these products are obtained in proportions which depend on the concentrations of NH_3 and NH_4^+ and on the temperature, but more or less pure products can be prepared by suitably adjusting the conditions.

The diammine $[Hg(NH_3)_2Cl_2]$, descriptively known as "fusible white precipitate", can be isolated by maintaining a high concentration of NH_4^+, since reactions (2) and (3) are thereby inhibited, or better still by using non-polar solvents.[21] It is made up of a cubic lattice of Cl^- ions with linear $H_3N-Hg-NH_3$ groups inserted so as to give the common, distorted octahedral coordination about Hg^{II} (Hg–N = 203 pm, Hg–Cl = 287 pm) (Fig. 29.4).

By using a low concentration of NH_3 and with no NH_4^+ initially present, the amide $[Hg(NH_2)Cl]$, "infusible white precipitate" is obtained. This consists of parallel chains of

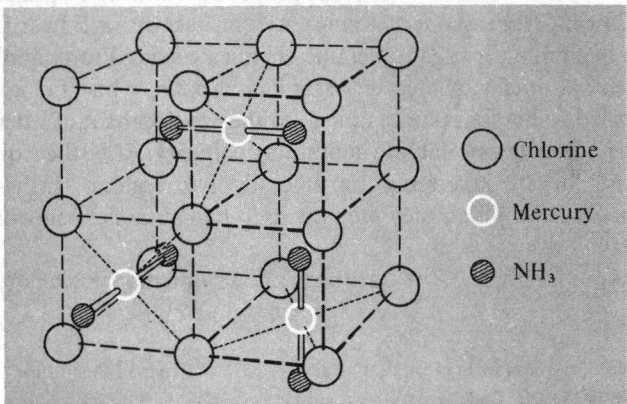

separated by Cl^- ions.

FIG. 29.4   Crystal structure of $Hg(NH_3)_2Cl_2$ showing linear NH_3–Hg–NH_3 groups inside a lattice of chloride ions.

[21] W. G. PALMER, *Experimental Inorganic Chemistry*, p. 184, Cambridge University Press, Cambridge, 1954.

$[Hg_2NCl(H_2O)]$ is the chloride of Millon's base, $[Hg_2N(OH).(H_2O)_2]$, and can be obtained either by heating the diammine, or amide with water or, better still, by the action of hydrochloric acid on Millon's base which is best prepared by the method, used in 1845 by its discoverer, of warming yellow HgO with aqueous NH_3. Replacement of the OH^- yields a series of salts, $[Hg_2NX(H_2O)]$, the structures of which (and that of the base itself) consist, with only minor variations, of a network of $\{Hg_2N\}^+$ units linked so that each N is tetrahedrally linked to 4 Hg and each Hg is linearly linked to 2 N (Hg–N = 204–209 pm depending on X).[21a] The X^- ions and water molecules are accommodated interstitially and these materials behave as anion exchangers.

When Hg_2Cl_2 is treated with aqueous NH_3 disproportionation occurs $(Hg_2Cl_2 \rightarrow HgCl_2 + Hg)$; the $HgCl_2$ then reacts as outlined above to give a precipitate of variable composition. The liberated mercury, however, renders the precipitate black, as previously mentioned, and so forms the basis of a distinctive qualitative test for Hg_2^{2+}.

As indicated by the insolubility and inertness of HgS, Hg^{II} has a great affinity for sulfur. HgO reacts vigorously with mercaptans (which is why RSH were given the name mercaptans†), displacing the H as with amines:

$$HgO + 2RSH \longrightarrow Hg(SR)_2 + H_2O$$

These mercaptides are low-melting solids, soluble in $CHCl_3$ and C_6H_6, and, though their structures depend on R, they all apparently contain linear S–Hg–S or, in $Hg(SBu^t)_2$, tetrahedral HgS_4 units. Most of the thioether (SR_2) complexes which have been prepared are adducts of the Hg^{II} halides and include both monomeric and polymeric (i.e. X-bridged) species. The dithiocarbamate $[Hg(S_2CNEt_2)_2]$ exists in two forms, one of which has the same structure as the corresponding Zn^{II} and Cd^{II} compounds (Fig. 29.3c), while the other is polymeric.

29.3.6 *Organometallic compounds*[21b]

Although they were not the first organometallic compounds to be prepared (Zeise's salt was discovered in 1827) the discovery of zinc alkyls in 1849 by Sir Edward Frankland may be taken as the beginning of organometallic chemistry since Frankland's studies led to their employment as intermediates in organic synthesis, while the measurements of vapour densities led to his suggestion, crucial to the development of valency theory, that each element has a limited but definite combining capacity. After their discovery in 1900 Grignard reagents largely superseded the zinc alkyls in organic synthesis, but by then many of the reactions for which they are now used had already been worked out on the zinc compounds.

Alkyls of the types RZnX and ZnR_2 are both known and may be prepared by essentially

---

† Mercaptans were discovered in 1834 by W. C. Zeise (p. 357) who named them from the Latin *mercurium captans*, catching mercury.

21a A. F. WELLS, *Structural Inorganic Chemistry*, 4th edn., Oxford University Press, Oxford, 1975: the structural chemistry of mercury is reviewed on pp. 916–26.
21b G. E. COATES and K. WADE, *Organometallic Compounds*, 3rd edn., Vol. 1, pp. 121–76 (Zinc, cadmium, and mercury), Methuen, London, 1967.

the original method of heating Zn with boiling RX in an inert atmosphere (CO_2 or N_2):

$$Zn + RX \longrightarrow RZnX$$

and then raising the temperature to distil the dialkyl:

$$2RZnX \longrightarrow ZnR_2 + ZnX_2$$

These reactions work best with $X = I$ but the less-expensive RBr can be used in conjunction with a Zn–Cu alloy instead of pure Zn. Diaryls are best obtained from appropriate organoboranes or organomercury compounds:

$$3ZnMe_2 + 2BR_3 \longrightarrow 3ZnR_2 + 2BMe_3$$

or

$$Zn + HgR_2 \longrightarrow ZnR_2 + Hg$$

ZnR_2 are covalent, non-polar liquids or low-boiling solids (Table 29.4). They are invariably monomeric in solution with linear C–Zn–C coordination at the Zn atom. They are very susceptible to attack by air and those of low molecular weight are spontaneously flammable, producing a smoke of ZnO. Their reactions with water, alcohols, and ammonia, etc., are generally similar to, but less vigorous than, those of Grignard reagents (p. 147) with the important difference that they are unaffected by CO_2; indeed, they are often prepared under an atmosphere of this gas.

TABLE 29.4 *Comparison of some typical organometallic compounds* MR_2

R	Zn		Cd		Hg	
	MP/°C	BP/°C	MP/°C	BP/°C	MP/°C	BP/°C
Me	−29.2	46	−4.5	105.5	—	92.5
Et	−28	117	−21	64 (19 mmHg)	—	159
Ph	107	d 280	173	—	121.8 (subl)	204 (10 mmHg)

Organocadmium compounds are normally prepared from the appropriate Grignard reagents:

$$CdX_2 + 2RMgX \longrightarrow CdR_2 + 2MgX_2$$

and then if desired:

$$CdR_2 + CdX_2 \longrightarrow 2RCdX$$

They are thermally less stable than their Zn counterparts but generally less reactive (not normally catching fire in air), and so their most important use (but see also ref. 22) is to prepare ketones from acid chlorides:

$$2R'COCl + CdR_2 \longrightarrow 2R'COR + CdCl_2$$

The use of Grignard reagents is impracticable here since they react further with the ketone.

[22] P. R. JONES and P. J. DESIO, The less familiar reactions of organocadmium reagents, *Chem. Revs.* **78**, 491–516 (1978).

An enormous number of organomercury compounds are known. They are predominantly of the same stoichiometries as those of Zn and Cd, viz. RHgX and HgR_2, and may be prepared by the action of sodium amalgam on RX:

$$2Hg + 2RX \longrightarrow HgR_2 + HgX_2$$

$$HgCl_2 + 2Na \longrightarrow Hg + 2NaCl$$

More usually they are made by the reaction of Grignard reagents on $HgCl_2$ in thf:

$$RMgX + HgCl_2 \longrightarrow RHgCl + MgXCl$$

$$RMgX + RHgCl \longrightarrow HgR_2 + MgXCl$$

or simply by the action of HgX_2 on a hydrocarbon:

$$HgX_2 + RH \longrightarrow RHgX + HX \text{ (mercuration)}.$$

RHgX are crystalline solids, and HgR_2 are extremely toxic liquids or low-melting solids (Table 29.4). They are essentially covalent materials except when $X^- = F^-$, NO_3^-, or SO_4^{2-}, in which cases the former are water-soluble and apparently ionic, $[RHg]^+X^-$. There are several reasons for the attention which these compounds have received. The search for pharmacologically valuable drugs has provided a continuing stimulus, and the existence of convenient preparative methods, coupled with their remarkable stability to air and water, has led to their extensive use in mechanistic studies.[23] Their stability sets them apart from the organic derivatives of other Group II metals but arises from the extreme weakness of the Hg–O bond rather than an inherently strong Hg–C bond. In fact the latter is weak, being commonly only ~ 60 kJ mol^{-1} and organomercury compounds are thermally and photochemically unstable, in some cases requiring to be stored at low temperatures in the dark. Indeed, because of the weakness of the bond, Hg can be replaced by many metals which give stronger M–C bonds and the preparation of organo derivatives of other metals (e.g. of Zn and Cd as referred to above) is the most important application of these compounds.

It appears that all RHgX and HgR_2 compounds are made up of linear R–Hg–X or R–Hg–R units, which could arise from sp hybridization of the metal. In some cases polymerization is necessary to achieve this linearity. Thus, for instance, o-phenylenemercury, which could conceivably be formulated as

is in fact a cyclic trimer (Fig. 29.5a). These organomercury compounds generally have little tendency to coordinate to further ligands but a few complexes involving irregular 3-coordination are known: these appear to be incompatible with sp² hybridization since this would produce symmetrical planar (local C_{2v}) geometry at Hg. Of these $[HgMe(bipy)]NO_3^{(24)}$ and $[HgR(HDz)]^{(25)}$ (Fig. 29.5b and c) may be mentioned.

[23] M. H. ABRAHAM, Chaps. 4–9, pp. 23–177, Electrophilic substitution at a saturated carbon atom, Vol. 12 of *Comprehensive Chemical Kinetics* (eds. C. H. BAMFORD and C. F. H. TIPPER), Elsevier, Amsterdam, 1973.

[24] A. J. CANTY and B. M. GATEHOUSE, Crystal and molecular structures of (2,2'-bipyridyl(methylmercury(II) nitrate, a complex with irregular three-coordination, *JCS (Dalton)* 1976, 2018–20.

[25] A. T. HUTTON and H. M. N. H. IRVING, Three-coordinate mercury complex: photochromism and molecular structure of phenylmercury(II)dithizonate, *JCS Chem. Comm.* 1979, 1113–14.

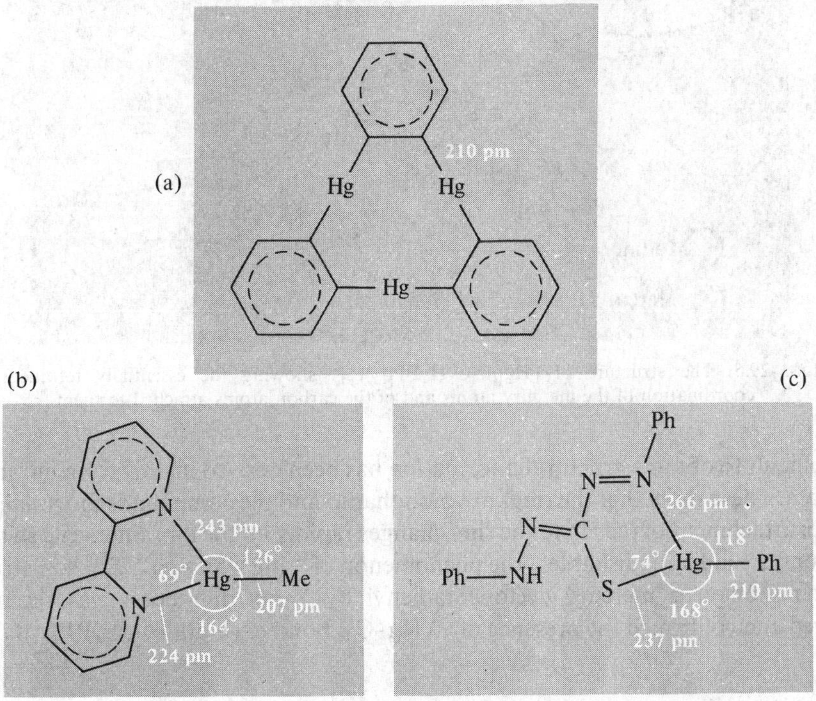

FIG 29.5 (a) *o*-Phenylenemercury trimer,[25a] (b) the planar cation in [HgMe(bipy)]NO$_3$,[24] and (c) phenylmercury(II)dithizonate.[25]

Among the versatile and synthetically useful reactions are those typified by the absorption of alkenes (olefins) by methanolic solutions of salts, particularly, the acetate of Hg^{II}. The products are not π complexes, but σ-bonded addition compounds, e.g.:

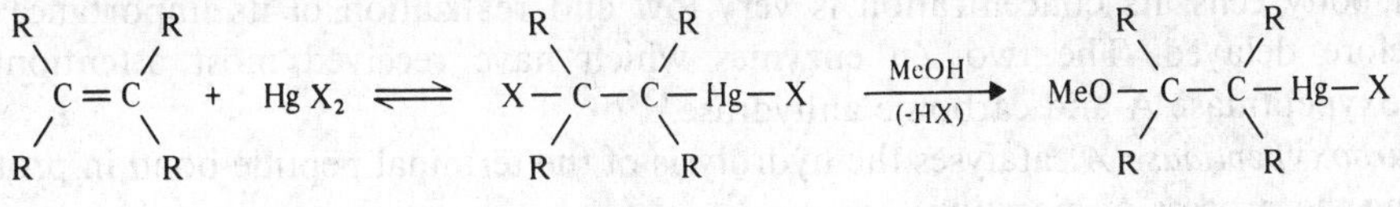

Oxomercuration

Regeneration of the alkene occurs on acidification, e.g. with HCl:

$$R_2C(OMe)C(HgX)R_2 + HCl \longrightarrow R_2C{=}CR_2 + MeOH + HgXCl$$

Methanolic solutions of Hg^{II} also absorb CO and the products, of the type

$$X{-}Hg{-}\underset{\underset{O}{\|}}{C}{-}OMe,$$

are again σ-bonded.

A similar reluctance to form π bonds is seen in the cyclopentadienyls of mercury such as [Hg(η^1-C$_5$H$_5$)$_2$] and [Hg(η^1-C$_5$H$_5$)X]. As they are photosensitive and single crystals are

[25a] D. S. Brown, A. G. Massey, and D. A. Wickens, Structure of tribenzo[*b,e,h*]-[1,4,7]trimercuronin, *Acta Cryst.*, **B34**, 1695–97 (1978).

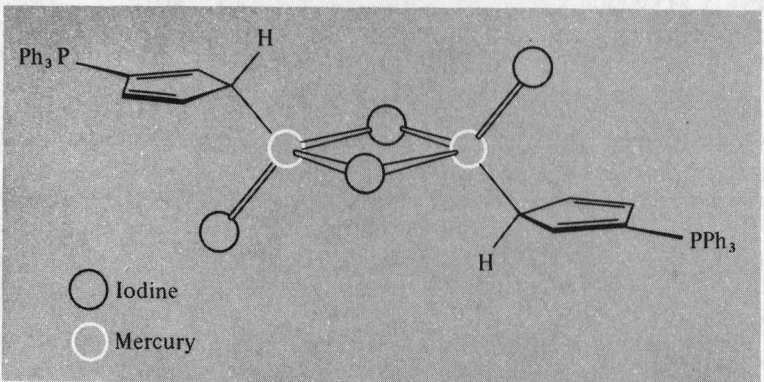

FIG. 29.6   The structure of $[Hg(\eta^1\text{-}C_5H_4PPh_3)I_2]_2$ showing the essentially tetrahedral coordination of the mercury atoms and of the carbon atoms attached to them.

very difficult to obtain, structural information has been derived mainly from infrared and nmr data. These show that the rings are monohapto and the compounds fluxional, i.e. the point of attachment of the Hg to the ring changes rapidly on the nmr time scale so that the 5 carbons are indistinguishable—the phenomenon of "ring whizzing". The first structural determination of a mercury cyclopentadienyl by X-ray diffraction was recently[25a] achieved and confirmed the presence of an Hg–C σ bond in $[Hg(\eta^1\text{-}C_5H_4PPh_3)I_2]_2$ (Fig. 29.6).

29.3.7 *Biological and environmental importance*

It is a remarkable contrast that, whereas Zn is biologically one of the most important metals and is apparently necessary to all forms of life, Cd and Hg have no known biological role and are amongst the most toxic of elements.

The body of an adult human contains about 2 g of Zn but, as Zn enzymes are present in most body cells, its concentration is very low and realization of its importance was therefore delayed. The two Zn enzymes which have received most attention are carboxypeptidase A and carbonic anhydrase.[26]

Carboxypeptidase A catalyses the hydrolysis of the terminal peptide bond in proteins during the process of digestion:

[25a] N. L. HOLY, N. C. BAENZIGER, R. M. FLYNN, and D. C. SWENSON, The synthesis and structure of mercuric halide complexes of triphenylphosphonium cyclopentadienylide. The first X-ray structure of a mercury cyclopentadienyl, *J. Am. Chem. Soc.* **98**, 7823–4 (1976).

[26] R. H. PRINCE, Some aspects of the bioinorganic chemistry of zinc, *Adv. Inorg. Chem. Radiochem.* **22**, 349–440 (1979).

It has a molecular weight of about 30 000–35 000 and contains one Zn tetrahedrally coordinated to 2 histidine N atoms, a carboxyl O of a glutamate residue, and a water molecule. The precise mechanism of its action is not finally settled, but it is agreed that the first step is the replacement of the water molecule attached to the Zn by the

of the terminal peptide which is thereby polarized, giving the C a positive charge and making it susceptible to nucleophilic attack. This attack is probably by the –OH of a water molecule, followed by proton-rearrangement and breaking of the C–N peptide bond, though alternative possibilities, such as attack by the carboxyl group of a second glutamate residue in the enzyme have also been considered. In any event it is evident that the conformation of the enzyme provides a hydrophobic pocket, close to the Zn, which accommodates the non-polar side-chain of the protein being hydrolysed, and that this protein is, throughout, held in the correct position by H bonding to appropriate groups in the enzyme.

Carbonic anhydrase was the first Zn metalloenzyme to be discovered (1940) and in its several forms is widely distributed in plants and animals. Three closely related forms (A, B, and C), are found in mammalian erythrocytes (red blood-cells) where they catalyse the equilibrium reaction:

$$CO_2 + H_2O \rightleftharpoons HCO_3^- + H^+$$

The forward (hydration) reaction occurs during the uptake of CO_2 by blood in tissue, while the backward (dehydration) reaction takes place when the CO_2 is subsequently released in the lungs. The enzyme increases the rates of these reactions by a factor of about one million.

The molecular weight of the enzyme is in the region of 28 000–30 000 and the roughly spherical molecule contains just 1 zinc atom situated in a deep protein pocket, which also contains a number of water molecules arranged in an ice-like order. This Zn is coordinated to 3 imidazole nitrogen atoms of histidine residues and to a water molecule. Once again the precise details of the enzyme's action are not settled, but it seems probable that the coordinated H_2O ionizes to give Zn–OH$^-$ and the nucleophilic OH$^-$ then interacts with the C of CO_2 (which may be held in the correct position by H bonds to its 2 oxygen atoms) to yield HCO_3^-. This is equivalent to replacing the slow hydration of CO_2 with H_2O, by the fast reaction:

$$CO_2 + OH^- \longrightarrow HCO_3^-$$

The latter would normally require a high pH and the contribution of the enzyme is therefore presumed to be the provision of a suitable environment, within the protein pocket, which allows the dissociation of the coordinated H_2O to occur in a medium of pH 7 which would otherwise be much too low.

Cadmium is extremely toxic[27] and accumulates in humans mainly in the kidneys and liver; prolonged intake, even of very small amounts, leads to dysfunction of the kidneys. It acts by binding to the –SH group of cysteine residues in proteins and so inhibits SH enzymes. It can also inhibit the action of zinc enzymes by displacing the zinc.

[27] J. H. MENNEAR, *Cadmium Toxicity*, Dekker, New York, 1979, pp. 224.

The history of the toxic effects of mercury[9] is long and the use of $HgCl_2$, for instance, as a poison has already been mentioned. The use of mercury salts in the production of felt† for hats and the dust generated in ill-ventilated workshops by the subsequent drying process, led to the nervous disorder known as "hatter's shakes" and possibly also to the expression "mad as a hatter".

The metal itself, having an appreciable vapour pressure, is also toxic, and produces headaches, tremors, inflammation of the bladder, and loss of memory. The best documented case is that of Alfred Stock (p. 171) whose constant use of mercury in the vacuum lines employed in his studies of boron and silicon hydrides, caused him to suffer for many years. The cause was eventually recognized and it is largely due to Stock's publication in 1926 of details of his experiences that the need for care and adequate ventilation is now fully appreciated.

Still more dangerous than metallic mercury or inorganic mercury compounds are organomercury compounds of which the methyl mercury ion $HgMe^+$ is probably the most ubiquitous.[28] This and other organomercurials are more readily absorbed in the gastrointestinal tract than Hg^{II} because of their greater permeability of biomembranes. They concentrate in the blood and have a more immediate and permanent effect on the brain and central nervous system, no doubt acting by binding to the –SH groups in proteins. Naturally occurring, anaerobic, bacteria in the sediments of sea or lake floors are able to methylate inorganic mercury (Co–Me groups in vitamin B_{12} are able to transfer the Me to Hg^{II}) which is then concentrated in plankton and so enters the fish food chain.

The Minamata disaster in Japan, when 52 people died in 1952, occurred because fish, which formed the staple diet of the small fishing community, contained abnormally high concentrations of mercury in the form of MeHgSMe. This was found to originate from a local chemical works where Hg^{II} salts were used (inefficiently) to catalyse the production of acetylene from acetaldehyde, and the effluent then discharged into the shallow sea. Evidence of a similar bacterial production of organomercury is available from Sweden where methylation of Hg^{II} in the effluent from paper mills has been shown to occur. The use of organomercurials as fungicidal seed dressings has also resulted in fatalities in many parts of the world when the seed was subsequently eaten.

Public concern about mercury poisoning has also led to more stringent regulations for the use of mercury cells in the chlor-alkali industry (pp. 82, 931). The health record of this industry has, in fact, been excellent, but the added costs of conforming to still higher standards have led manufacturers to move from mercury cells to diaphragm cells, and this change has been made a legal requirement in Japan.

---

† It was apparently helpful to add Hg^{II} to the dil HNO_3 used to roughen the surface of the animal hair employed in the making of felt which is a non-woven fabric of randomly oriented hairs.

28 L. T. FRIBERG and J. J. VOSTAL, *Mercury in the Environment*, CRC Press, Cleveland, 1972, 232 pp.

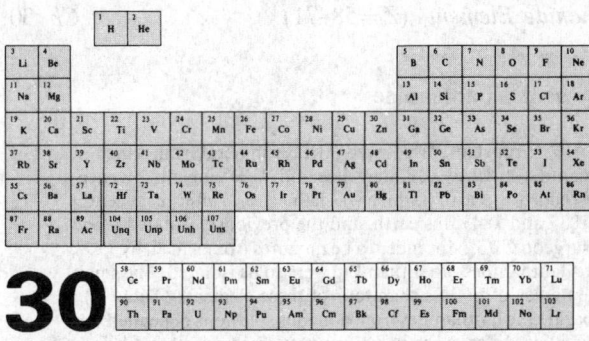

30

The Lanthanide Elements (Z=58-71)

30.1 Introduction

Not least of the confusions associated with this group of elements is that of terminology. The name "rare earth" was originally used to describe almost any naturally occurring but unfamiliar oxide and even until about 1920 generally included both ThO_2 and ZrO_2. About that time the name began to be applied to the elements themselves rather than their oxides, and also to be restricted to that group of elements which could only be separated from each other with great difficulty. On the basis of their separability it was convenient to divide these elements into the "cerium group" or "light earths" (La to about Eu) and the "yttrium group" or "heavy earths" (Gd to Lu plus Y which, though much lighter than the others, has a comparable ionic radius and is consequently found in the same ores, usually as the major component). It is now accepted that the "rare-earth elements" comprise the fourteen elements from $_{58}$Ce to $_{71}$Lu, but are commonly taken to include $_{57}$La and sometimes Sc and Y as well.

To avoid this confusion, and because many of the elements are actually far from rare, the terms "lanthanide", "lanthanon", and "lanthanoid" have been introduced. Even now, however, there is no general agreement about the position of La, i.e. whether the group is made up of the elements La to Lu or Ce to Lu. Throughout this chapter the term "lanthanide" and the general symbol, Ln, will be used to refer to the fourteen elements cerium to lutetium inclusive, the Group IIIA elements, scandium, yttrium, and lanthanum, having been dealt with in Chapter 20.

The lanthanides comprise the largest naturally-occurring group in the periodic table with properties so similar that from 1794, when J. Gadolin isolated "yttria" which he thought was the oxide of a single new element, until 1907, when lutetium was discovered, nearly a hundred claims were made for the discovery of elements belonging to this group. In view of the absence at that time of a conclusive test to determine whether or not a mixture was involved, this is not surprising. Indeed, there was a general lack of understanding of the large number of elements involved since the periodic table of the time could accommodate only one element, namely La. Not until 1913, as a result of H. G. J. Moseley's work on atomic numbers, was it realized that there were just fourteen elements between La and Hf, and in 1918 Niels Bohr interpreted this as an expansion of the fourth quantum group from 18 to 32 electrons. More information is in the Panel.

History of the Lanthanides[1, 2]

In 1751 the Swedish mineralogist, A. F. Cronstedt, discovered a heavy mineral from which in 1803 M. H. Klaproth in Germany and, independently, J. J. Berzelius and W. Hisinger in Sweden, isolated what was thought to be a new oxide (or "earth") which was named *ceria* after the recently discovered asteroid, Ceres. Between 1839 and 1843 this earth, and the previously isolated *yttria* (p. 1102), were shown by the Swedish surgeon C. G. Mosander to be mixtures from which, by 1907, the oxides of Sc, Y, La, and the thirteen lanthanides other than Pm were to be isolated. The small village of Ytterby is celebrated in the names of no less than four of these elements (Table A).

The classical methods[3] used to separate the lanthanides from aqueous solutions depended on: (i) differences in basicity, the less-basic hydroxides of the heavy lanthanides precipitating before those of the lighter ones on gradual addition of alkali (ii) differences in solubility of salts such as oxalates, double sulfates, and double nitrates, and (iii) conversion, if possible, to an oxidation state other than +3, e.g. Ce(IV), Eu(II). This latter process provided the cleanest method but was only occasionally applicable. Methods (i) and (ii) required much repetition to be effective, and fractional recrystallizations were sometimes repeated thousands of times. (In 1911 the American C. James performed 15 000 recrystallizations in order to obtain pure thulium bromate).

TABLE A   *The discovery of the oxides of Group IIIA and the lanthanide elements*

Element	Discoverer	Date	Origin of name
From ceria			
Cerium, Ce	C. G. Mosander	1839	The asteroid, Ceres
Lanthanum, La	C. G. Mosander	1839	Greek λανθάvein, lanthanein, to escape notice
Praseodymium, Pr	C. A. von Welsbach	1885	Greek πρασιος + διδυμος, praseos + didymos, leek green + twin
Neodymium, Nd	C. A. von Welsbach	1885	Greek νέος + δίδυμος, neos + didymos, new twin
Samarium, Sm	L. de Boisbaudran	1879	The mineral, samarskite
Europium, Eu	E. A. Demarcay	1901	Europe
From yttria			
Yttrium, Y	C. G. Mosander	1843	Ytterby
Terbium, Tb[a]	C. G. Mosander	1843	Ytterby
Erbium, Er[a]	C. G. Mosander	1843	Ytterby
Ytterbium, Yb	J. C. G. de Marignac	1878	Ytterby
Scandium, Sc	L. F. Nilson	1879	Scandinavia
Holmium, Ho	P. T. Cleve	1879	Latin *Holmia*: Stockholm
Thulium, Tm	P. T. Cleve	1879	Latin *Thule*, "most northerly land"
Gadolinium, Gd	J. C. G. de Marignac	1880	Finnish chemist, J. Gadolin
Dysprosium, Dy	L. de Boisbaudran	1886	Greek δυσπροσιτος, dysprositos, hard to get
Lutetium, Lu[b]	G. Urbain C. A. von Welsbach C. James	1907	Latin *Lutetia*: Paris

[a] Terbium and erbium were originally named in the reverse order.
[b] Originally spelled lutecium, but changed to lutetium in 1949.

[1] J. W. MELLOR, *A Comprehensive Treatise on Inorganic and Theoretical Chemistry* Vol. V, pp. 494–586, Longman Green, London, 1924.
[2] R. J. CALLOW, *The Rare Earth Industry*, Pergamon Press, Oxford, 1966, 84 pp.
[3] H. F. V. LITTLE, Chap. XI, pp. 317–75, of *A Text-book of Inorganic Chemistry* (J. N. FRIEND, ed.), Vol. IV, 2nd edn., Griffin, London, 1921.

The minerals on which the work was performed during the nineteenth century were indeed rare, and the materials isolated were of no interest outside the laboratory. By 1891, however, the Austrian chemist C. A. von Welsbach had perfected the thoria gas "mantle" to improve the low luminosity of the coal-gas flames then used for lighting. Woven cotton or artificial silk of the required shape was soaked in an aqueous solution of the nitrates of appropriate metals and the fibre then burned off and the nitrates converted to oxides. A mixture of 99% ThO_2 and 1% CeO_2 was used and has not since been bettered. CeO_2 catalyses the combustion of the gas and apparently, because of the poor thermal conductivity of the ThO_2, particles of CeO_2 become hotter and so brighter than would otherwise be possible. The commercial success of the gas mantle was immense and produced a worldwide search for thorium. Its major ore is monazite, which rarely contains more than 12% ThO_2 but about 45% Ln_2O_3. Not only did the search reveal that thorium, and hence the lanthanides, are more plentiful than had previously been thought, but the extraction of the thorium produced large amounts of lanthanides for which there was at first little use.

Applications were immediately sought and it was found that electrolysis of the fused chloride of the residue left after the removal of Th yielded the pyrophoric "mischmetall" (approximately 50% Ce, 25% La, 25% other light lanthanides) which, when alloyed with 30% Fe, is ideal as a lighter flint. Besides small amounts of lanthanides used in special glasses to control absorption at particular wavelengths, this was the pattern of usage until the 1940s. Before then there was little need for the pure metals and, because of the difficulty in obtaining them (high mps and very easily oxidized), such samples as were produced were usually impure. Attempts were also made to find element 61, which had not been found in the early studies, and in 1926 unconfirmed reports of its discovery from Illinois and Florence produced the temporary names *illinium* and *florentium*.

During the 1939–45 war, Mg-based alloys incorporating lanthanides were developed for aeronautical components and the addition of small amounts of mischmetall to cast-iron, by causing the separation of carbon in nodular rather than flake form, was found to improve the mechanical properties. But, more significantly from the chemical point of view, work on nuclear fission requiring the complete removal of the lanthanide elements from uranium and thorium ores, coupled with the fact that the lanthanides constitute a considerable proportion of the fission products, stimulated a great surge of interest. Solvent extraction and, more especially, ion-exchange techniques were developed, the work of F. H. Spedding and coworkers at Iowa State University being particularly notable.

As a result, in 1947, J. A. Marinsky, L. E. Glendenin, and C. D. Coryell at Oak Ridge, Tennessee, finally established the existence of element 61 in the fission products of ^{235}U and at the suggestion of Coryell's wife it was named *prometheum* (later *promethium*) after Prometheus who, according to Greek mythology, stole fire from heaven for the use of mankind. Since about 1955, individual lanthanides have been obtainable in increasing amounts in elemental as well as combined forms.

30.2　The Elements

30.2.1　*Terrestrial abundance and distribution*

Apart from the unstable ^{147}Pm (half-life 2.62 y) of which traces occur in uranium ores, the lanthanides are actually not rare (Table 30.1). Cerium is the twenty-sixth most abundant of all elements, being half as abundant as Cl and 5 times as abundant as Pb. Even Tm, the rarest after Pm, is rather more abundant than I.

There are over 100 minerals known to contain lanthanides but the only two of commercial importance are monazite, a mixed La, Th, Ln phosphate, and bastnaesite, an La, Ln fluorocarbonate ($M^{III}CO_3F$). Monazite is widely but sparsely distributed in many rocks but, because of its high density and inertness, it is concentrated by weathering into sands on beaches and river beds, often in the presence of other similarly concentrated minerals such as ilmenite ($FeTiO_3$) and cassiterite (SnO_2). Deposits occur in southern India, South Africa, Brazil, Australia, and Malaysia and, until the 1960s, these provided the bulk of the world's La, Ln, and Th. Then, however, a vast deposit of bastnaesite, which

TABLE 30.1   *Natural abundances (ppm) in earth's crust*

Element	Abundance	Element	Abundance
Ce	66	Tb	1.2
Pr	9.1	Dy	4.5
Nd	40	Ho	1.4
Pm	4.5×10^{-20}	Er	3.5
Sm	7.0	Tm	0.5
Eu	2.1	Yb	3.1
Gd	6.1	Lu	0.8

had been discovered in 1949 in the Sierra Nevada Mountains in the USA, came into production and has since become the most important single source of La and Ln.

The distribution of the metals in the two minerals is quite similar, the bulk being Ce, La, Nd, and Pr (in that order) but with the difference that monazite typically contains around 5–10% ThO_2 and 3% yttrium earths which are virtually absent in bastnaesite. Although thorium is only weakly radioactive it is contaminated with daughter elements such as ^{228}Ra which are more active and therefore require careful handling during the processing of monazite. This is a complication not encountered in the processing of bastnaesite.

30.2.2   *Preparation and uses of the elements*[2–6]

Conventional mineral dressing yields concentrates of the minerals of better than 90% purity. These are then broken down ("opened") by either acidic or alkaline attack. Details vary considerably, since they depend on the ore being used and on the extent to which the metals are to be separated from each other, but Fig. 30.1 outlines two schemes which are typical of the treatment of monazite. They make use of the different solubilities of $Ln_2(SO_4)_3.Na_2SO_4.xH_2O$ for the light and heavy lanthanides, and of the low solubility of the hydrous oxide of thorium. Because only small amounts of thorium and heavy lanthanides are present in bastnaesite, its chemical treatment is less involved. It is commonly treated with hot conc H_2SO_4, when CO_2, HF, and SiF_4 are evolved; the dried product is then leached with water to obtain a solution containing lanthanum and the light lanthanides.

The largest, and most easily separated component of the mixtures obtained from all these ore-opening processes, is cerium, and its removal is therefore invariably the next step. This is feasible because Ce^{IV}, being less basic than Ln^{III}, is more easily hydrolysed and precipitates as a basic salt or as the hydroxide when the aqueous solution is oxidized with, for instance, $KMnO_4$ or bleaching powder (p. 1006). At this point the classical methods already referred to were formerly employed to separate the individual elements, and the separation of lanthanum by the fractional crystallization of

[4] K. A. GSCHNEIDER and L. R. EYRING (eds.), *Handbook on the Physics and Chemistry of Rare Earths*: Vol. 1, *Metals*, 894 pp.; Vol. 2, *Alloys and Intermetallics*, 620 pp.; Vol. 3, *Non-metallic Compounds—I*, 664 pp.; Vol. 4, *Non-metallic Compounds—II*, 602 pp.; North-Holland, Amsterdam, 1979. E. C. SUBBARAO and W. E. WALLACE (eds.), *Science and Technology of Rare Earth Materials*, Academic Press, New York, 1980, 448 pp.

[5] F. H. SPEDDING and A. H. DAANE, *The Rare Earths*, Wiley, New York, 1961, 641 pp.

[6] N. E. TOPP, *The Chemistry of the Rare-earth Elements*, Elsevier, Amsterdam, 1965, 164 pp.

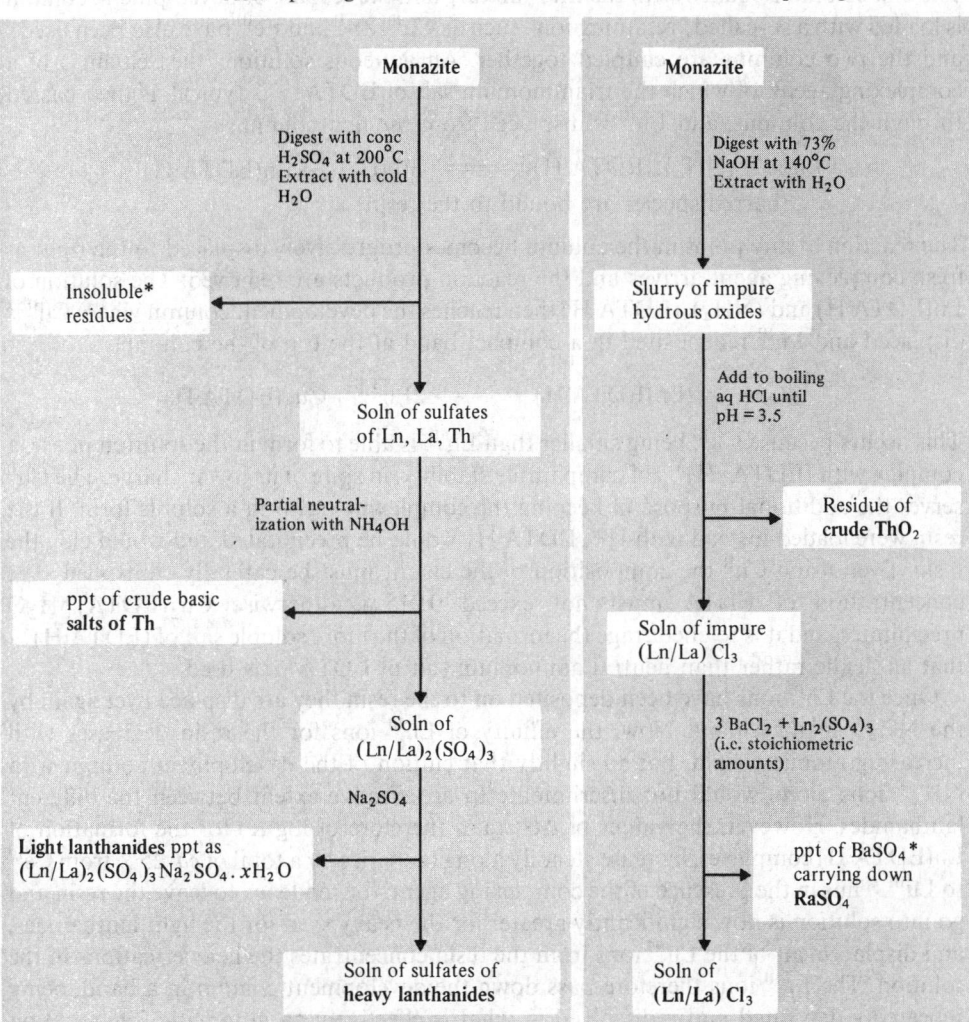

Acid opening   Alkali opening

Monazite   **Monazite**

Digest with conc   Digest with 73%
H_2SO_4 at 200°C   NaOH at 140°C
Extract with cold   Extract with H_2O
H_2O

Insoluble*   Slurry of impure
residues   hydrous oxides

                                                    Add to boiling
                                                    aq HCl until
                                                    pH = 3.5

Soln of sulfates
of Ln, La, Th   Residue of
                                                    crude ThO$_2$

Partial neutral-
ization with NH_4OH Soln of impure
                                                    (Ln/La) Cl$_3$

ppt of crude basic
salts of Th

                                                    $3\ BaCl_2 + Ln_2(SO_4)_3$
Soln of   (i.e. stoichiometric
(Ln/La)$_2$(SO$_4$)$_3$ amounts)

Na_2SO_4 ppt of BaSO$_4$*
                                                    carrying down
Light lanthanides ppt as   **RaSO$_4$**
(Ln/La)$_2$(SO$_4$)$_3$Na$_2$SO$_4$·xH$_2$O

Soln of sulfates of   Soln of
heavy lanthanides   (Ln/La) Cl$_3$

*These residues contain ^{228}Ra, a daughter element of Th and an active γ-emitter,
and must therefore be handled with care

FIG. 30.1   Flow diagram for the extraction of the lanthanide elements.

La(NO$_3$)$_3$·2NH$_4$NO$_3$·4H$_2$O is still used. However the separations can now be effected
on a large scale by solvent extraction using aqueous solutions of the nitrates and a solvent
such as tri-n-butylphosphate, (Bu^{n}O)$_3$PO (often with kerosene as an inert diluent), in
which the solubility of Ln^{III} increases with its atomic weight. This type of process has the
advantage of being continuous and is ideal where the product and feed are not to be
changed.

Alternatively, for high-purity or smaller-scale production the more easily adapted ion-
exchange techniques are ideal, the best of these being "displacement chromatography".
Two separate columns of cation exchange resin are generally employed for this purpose.

The first column is loaded with the Ln^{III} mixture and the second, or development, column is loaded with a so-called "retaining ion" such as Cu^{II} (Zn^{II} and Fe^{III} have also been used), and the two columns are coupled together. An aqueous solution (the "eluant") of a complexing agent, of which the triammonium salt of $EDTA^{4-}$ is typical, is then passed through the columns, and Ln^{III} is displaced from the first column,

$$\overline{Ln^{3+}} + (NH_4)_3(EDTA \cdot H) \rightleftharpoons 3(NH_4^+) + \overline{Ln(EDTA \cdot H)}$$

(barred species are bound to the resin)

The reaction at any point in the column becomes progressively displaced to the right as fresh complexing agent arrives and the reaction products are removed. The solution of $Ln(EDTA \cdot H)$ and $(NH_4)_3(EDTA \cdot H)$ then reaches the development column where Cu^{II} is displaced and Ln^{III} redeposited in a compact band at the top of the column,

$$\overline{3Cu^{2+}} + 2Ln(EDTA \cdot H) \longrightarrow 2\overline{Ln^{3+}} + Cu_3(EDTA \cdot H)_2$$

This occurs because Cu^{II}, being smaller than Ln^{III}, is able to form in the solution phase, a complex with $(EDTA \cdot H)^{3-}$ of comparable stability, in spite of its lower charge. The Cu^{II} serves the additional purpose of keeping the complexing agent in a soluble form. If the resin were loaded instead with H^+, $EDTA \cdot H_4$ would be precipitated and would clog the resin. Even using Cu^{II} the composition of the eluant must be carefully controlled. The concentration of EDTA must not exceed 0.015 M, otherwise $Cu_2(EDTA).5H_2O$ precipitates, and it is to encourage the formation of the more soluble salt of $(EDTA \cdot H)^{3-}$ that an acidic rather than neutral ammonium salt of $EDTA^{4-}$ is used.

Once the Ln^{III} ions have been deposited on to the resin they are displaced yet again by the NH_4^+ in the eluant. Now, the affinity of Ln^{III} ions for the resin decreases with increasing atomic weight, but so slightly that elution of the development column with NH_4^+ ions alone would not discriminate to an effective extent between the different lanthanides. However, the values of ΔG° (and therefore of $\log K$) for the formation of $Ln(EDTA \cdot H)$ complexes, increase steadily along the series by a total of *ca.* 25% from Ce^{III} to Lu^{III}. Thus in the presence of the complexing agent, the tendency to leave the resin and go into solution is now significantly greater for the heavy than for the light lanthanides, and displacement of the Ln^{III} ions from the resin concentrates the heavier cations in the solution. The Ln^{III} ions therefore pass down the development column in a band, being repeatedly deposited and redissolved in what is effectively an automatic fractionation process, concentrating the heavier members in the solution phase. The result is that when all the copper has come off the column the lanthanides emerge in succession, heaviest first. They may then be precipitated from the eluant as insoluble oxalates and ignited to the oxides.

The production of mischmetall by the electrolysis of fused $(Ln,La)Cl_3$, and the difficulties in obtaining pure metals because of their high mps and ease of oxidation, have already been mentioned (p. 1425). Two methods are in fact available for producing the metals.

(i) *Electrolysis of fused salts*. A mixture of $LnCl_3$ with either NaCl or $CaCl_2$ is fused and electrolysed in a graphite or refractory-lined steel cell, which serves as the cathode, with a graphite rod as anode. This is used primarily for mischmetall, the lighter, lower-melting Ce, and for Sm, Eu, and Yb for which method (ii) yields Ln^{II} ions.

(ii) *Metallothermic reduction*. This consists of the reduction of the anhydous halides with calcium metal. Fluorides are preferred, since they are non-hygroscopic and the CaF_2

produced is stable, unlike the other Ca halides which are liable to boil at the temperatures reached in the process. $LnF_3 + Ca$ are heated in a tantalum crucible to a temperature $50°$ above the mp of Ln under an atmosphere of argon. After completion of the reaction, the charge is cooled and the slag and metal (of 97–99% purity) broken apart. The main impurity is Ca which is removed by melting under vacuum. With the exceptions of Sm, Eu, and Yb this method has general applicability.

In 1979, 16 500 tonnes of bastnaesite concentrate were produced in the USA and roughly the same amount of monazite concentrate in the rest of the world. Half of the latter was a byproduct of Australian TiO_2 production and the rest came mainly from Brazil, India, and Malaysia. This corresponds to a total world production of about 13 000 tonnes of contained Ln and 5600 tonnes of contained La. The bulk of output is used without separation of individual lanthanides, the most important single use being in the production of low alloy steels for plate and pipe, where $<1\%$ Ln/La added in the form of mischmetall or silicides greatly improves strength and workability. A good deal of mischmetall is also used in Mg-based alloys and to produce lighter flints, while Ln/Co alloys are increasingly in demand for the construction of permanent magnets. Various mixed oxides are employed as catalysts in petroleum "cracking". The walls of domestic "self-cleaning" ovens are treated with CeO_2 which prevents the formation of tarry deposits, and CeO_2 of varying purity is used to polish glass. Other individual Ln oxides are used as phosphors in television screens and similar fluorescing surfaces.

30.2.3 *Properties of the elements*

The metals are silvery in appearance and rather soft, but become harder across the series. Most of them exist in more than one crystallographic form, of which hcp is the most common; all are based on typically metallic close-packed arrangements, but their conductivities are appreciably lower than those of other close-packed metals.

The more important physical properties of the elements are summarized in Table 30.2. The alternation between several and few stable isotopes for even and odd atomic number respectively, mirrors the even–odd variation in the natural abundances of the elements already seen in Table 30.1 (see also p. 4).

The electronic configuration of the free atoms are determined only with difficulty because of the complexity of their atomic spectra, but it is generally agreed that they are nearly all $[Xe]4f^n5d^06s^2$. The exceptions are:

(1) Cerium, for which the sudden contraction and reduction in energy of the 4f orbitals immediately after La is not yet sufficient to avoid occupancy of the 5d orbital.
(2) Gd, which reflects the stability of the half-filled 4f shell;
(3) Lu, at which point the shell has been filled.

Only in the case of cerium, however (see below), does this have any marked effect on the aqueous solution chemistry, which is otherwise dominated by the $+3$ oxidation state, for which the configuration varies regularly from $4f^1$ (Ce^{III}) to $4f^{14}$ (Lu^{III}). It is notable that a regular variation is found for any property for which this $4f^n$ configuration is maintained across the series, whereas the variation in those properties for which this configuration is not maintained can be highly irregular. This is illustrated dramatically in size variations (Fig. 30.2). On the one hand, the radii of Ln^{III} ions decrease regularly from La^{III} (included

TABLE 30.2   Some properties of the lanthanide elements

Property	Ce	Pr	Nd	Pm	Sm	Eu	Gd	Tb	Dy	Ho	Er	Tm	Yb	Lu
Atomic number	58	59	60	61	62	63	64	65	66	67	68	69	70	71
Number of naturally occurring isotopes	4	1	7	—	7	2	7	1	7	1	6	1	7	2
Outer electron configuration	$4f^15d^16s^2$	$4f^36s^2$	$4f^46s^2$	$4f^56s^2$	$4f^66s^2$	$4f^76s^2$	$4f^75d^16s^2$	$4f^96s^2$	$4f^{10}6s^2$	$4f^{11}6s^2$	$4f^{12}6s^2$	$4f^{13}6s^2$	$4f^{14}6s^2$	$4f^{14}5d^16s^2$
Atomic weight	140.12	140.9077	144.24(±3)	—	150.36(±3)	151.96	157.25(±3)	158.9254	162.50(±3)	164.9304	167.26(±3)	168.9342	173.04(±3)	174.967(±3)
Metal radius (CN 6)/pm	181.8	182.4	181.4	183.4	180.4	208.4	180.4	177.3	178.1	176.2	176.1	175.9	193.3	173.8
Ionic radius (CN 6)/pm IV	87	85						76						
III	102	99	98.3	97	95.8	94.7	93.8	92.3	91.2	90.1	89.0	88.0	86.8	86.1
II			129[a]		122[b]	117			107			103	102	
$E^\ominus(M^{4+}/M^{3+})$/V	1.61	2.860[c]						2.7						
$E^\ominus(M^{3+}/M^{2+})$/V					−1.000	−0.360							−1.205	
$E^\ominus(M^{3+}/M)$/V	−2.483	−2.462	−2.431	−2.423	−2.414	−2.407	−2.397	−2.391	−2.353	−2.319	−2.296	−2.278	−2.267	−2.255
MP/°C	804	935	1024	1042	1072	826	1312	1356	1407	1461	1497	1545	824	1652
BP/°C	(3470)	(3020)	(3027)		1800	(1439)	(3000)	(2800)	(2600)	(2600)	(2900)	1727	(1427)	(3327)
ΔH_{fus}/kJ mol⁻¹	5.2(±1.2)	11.3(±2.1)	7.13		8.9(±0.4)									
ΔH_{vap}/kJ mol⁻¹	398	331	289		165(±17)	176	301	293	280	280	280	247	159	
ΔH_f(monatomic gas)/kJ mol⁻¹	419	356	328	301	207	178	398	389	291	301	317	232	152	414
Ionization energies/kJ mol⁻¹ 1st	541	522	530	536	542	547	595	569	567	574	581	589	603	513
2nd	1047	1018	1034	1052	1068	1085	1172	1112	1126	1139	1151	1163	1175	1341
3rd	1940	2090	2128	2140	2285	2425	1999	2122	2230	2221	2207	2305	2408	2054
ΔH(hydration Ln³⁺)/kJ mol⁻¹	3370	3413	3442	3478	3515	3547	3571	3605	3637	3667	3691	3717	3739	3760
Density (25°C)/g cm⁻³	6.773	6.475	7.003	7.2	7.536	5.245	7.886	8.253	8.559	8.78	9.045	9.318	6.973	9.84
Electrical resistivity (25°C)/μ ohm cm	75	68	64		92	81	134	116	91	94	86	90	28	68

(a) CN = 8.

(b) CN = 7.

(c) Estimated values since Pr^{IV} and Tb^{IV} are not stable in aqueous solution.

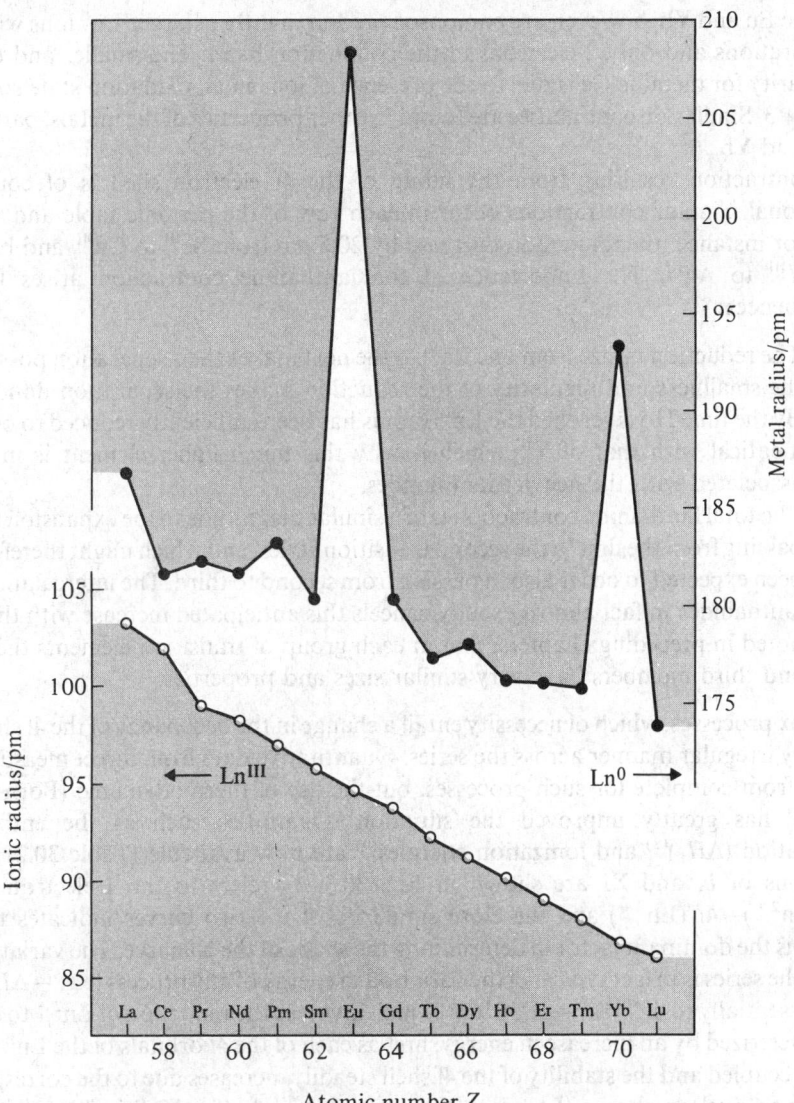

FIG. 30.2   Variation of metal radius and 3+ ionic radius for La and the lanthanides. Other data
for Ln^{II} and Ln^{IV} are in Table 30.2.

for completeness) to Lu^{III}. This "lanthanide contraction" occurs because, although each
increase in nuclear charge is exactly balanced by a simultaneous increase in electronic
charge, the directional characteristics of the 4f orbitals cause the 4f^n electrons to shield
themselves and other electrons from the nuclear charge only imperfectly. Thus, each unit
increase in nuclear charge produces a net increase in attraction for the whole extranuclear
electron charge cloud and each ion shrinks slightly in comparison with its predecessor. On
the other hand, although a similar overall reduction is seen in the metal radii, Eu and Yb
are spectacularly irregular. The reason is that most of the metals are composed of a lattice
of Ln^{III} ions with a 4f^n configuration and 3 electrons in the 5d/6s conduction band.

Metallic Eu and Yb, however, are composed predominantly of larger Ln^{II} ions with $4f^{n+1}$ configurations and only 2 electrons in the conduction band. The smaller and opposite irregularity for metallic Ce is due to the presence of ions in an oxidation state somewhat above $+3$. Similar discontinuities are found in other properties of the metals, particularly at Eu and Yb.

A contraction resulting from the filling of the 4f electron shell is of course not exceptional. Similar contractions occur in each row of the periodic table and, in the d block for instance, the ionic radii decrease by 20.5 pm from Sc^{III} to Cu^{III}, and by 15 pm from Y^{III} to Ag^{III}. The importance of the lanthanide contraction arises from its consequences:

(1) The reduction in size from one Ln^{III} to the next makes their separation possible, but the smallness and regularity of the reduction makes the separation difficult.
(2) By the time Ho is reached the Ln^{III} radius has been sufficiently reduced to be almost identical with that of Y^{III} which is why this much ligher element is invariably associated with the heavier lanthanides.
(3) The total lanthanide contraction is of a similar magnitude to the expansion found in passing from the first to the second transition series, and which might therefore have been expected to occur also in passing from second to third. The intercalation of the lanthanides in fact almost exactly cancels this anticipated increase with the result, noted in preceding chapters, that in each group of transition elements the second and third members have very similar sizes and properties.

Redox processes, which of necessity entail a change in the occupancy of the 4f shell, vary in a very irregular manner across the series. Quantitative data from direct measurements are far from complete for such processes, but the use of thermodynamic (Born–Haber) cycles[7] has greatly improved the situation. Quantities such as the enthalpy of atomization (ΔH_f)[8] and ionization energies[9] are now available (Table 30.2) and the variations of I_3 and ΣI are shown in Fig. 30.3. I_3 refers to the 1-electron change, $4f^{n+1}(Ln^{2+}) \rightarrow 4f^n(Ln^{3+})$ and the close similarity of the two curves indicates that this change is the dominant factor in determining the shape of the ΣI curve. The variation of I_3 across the series is in fact typical of the variation in energy of any process (e.g. $-\Delta H_f$ which refers essentially to $4f^{n+1}6s^2 \rightarrow 4f^n 5d^1 6s^2$) which involves the reduction of Ln^{3+} to Ln^{2+}. It is characterized by an increase in energy, first as each of the 4f orbitals of the Ln^{II} ions are singly occupied and the stability of the 4f shell steadily increases due to the corresponding increase in nuclear charge (La→Eu), then again as each 4f orbital is doubly occupied (Gd→Lu). The sudden falls at Gd and Lu reflect the ease with which it is possible to remove the single electrons in excess of the stable $4f^7$ and $4f^{14}$ configurations. Explanations have been given for the smaller irregularities at the quarter- and three-quarter-shell stages, but require a careful consideration of interelectronic repulsion, as well as exchange energy, terms.[8]

[7] D. A. JOHNSON, Principles of lanthanide chemistry, *J. Chem. Ed.* **57**, 475–7 (1980).
[8] D. A. JOHNSON, Recent advances in the chemistry of the less-common oxidation states of the lanthanide elements, *Adv. Inorg. Chem. Radiochem.* **20**, 1–132 (1977).
[9] L. R. MORSS, Thermochemistry of some chlorocomplex compounds of the rare-earths. Third ionization potentials and hydration enthalpies of the trivalent ions, *J. Phys. Chem.* **75**, 392–9 (1971).

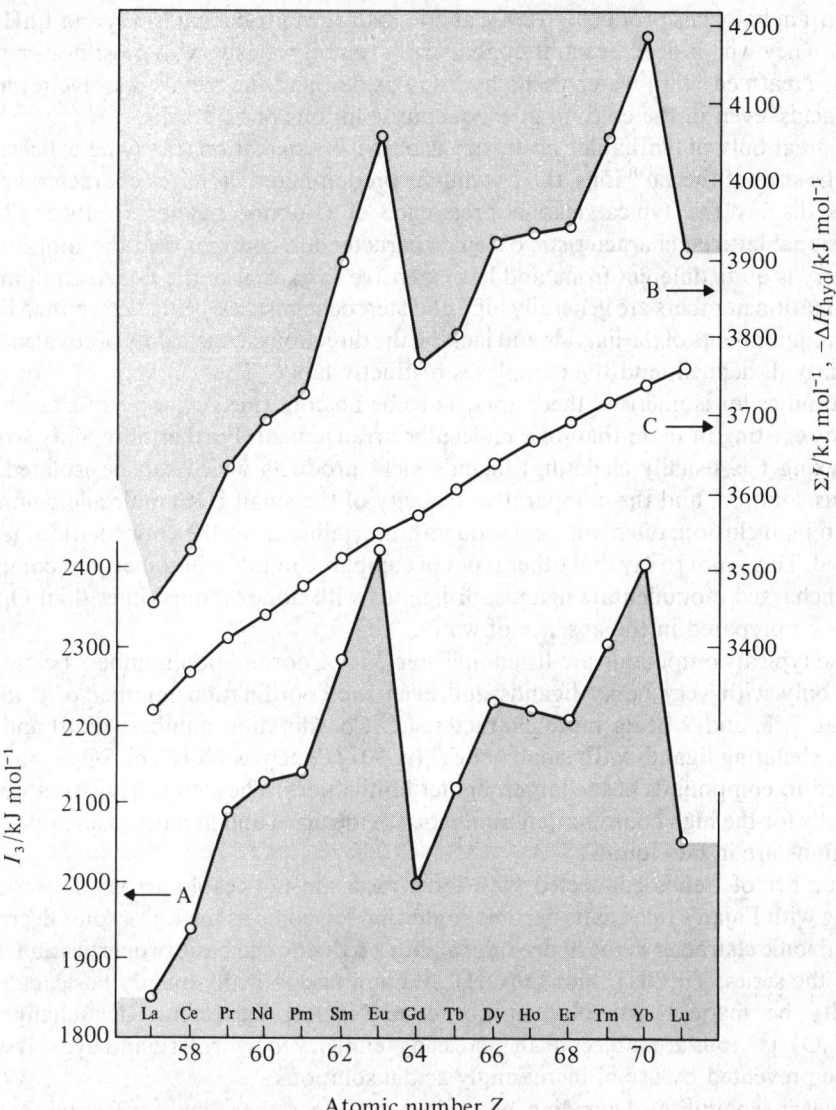

FIG. 30.3 Variation with atomic number of some properties of La and the lanthanides: A, the third ionization energy (I_3); B, the sum of the first three ionization energies (ΣI); C, the enthalpy of hydration of the gaseous trivalent ions ($-\Delta H_{hyd}$). The irregular variations in I_3 and ΣI, which refer to redox processes, should be contrasted with the smooth variation in ΔH_{hyd}, for which the $4f^n$ configuration of Ln^{III} is unaltered.

30.2.4 *Chemical reactivity and trends*

The lanthanides are very electropositive and reactive metals. With the exception of Yb their reactivity apparently depends on size so that Eu which has the largest metal radius is much the most reactive. They tarnish in air and, if ignited in air or O_2, burn readily to give Ln_2O_3, or, in the case of cerium, CeO_2 (praseodymium and terbium yield nonstoichiometric products approximating to Pr_6O_{11} and Tb_4O_7 respectively). When heated, they also burn in halogens producing LnX_3, and in hydrogen producing LnH_2 and LnH_3 (see below). They will, indeed, react, though usually less vigorously, with most non-metals if heated. Treatment with water yields hydrous oxides, and the metals dissolve rapidly in dilute acids, even in the cold, to give aqueous solutions of Ln^{III} salts.

The great bulk of lanthanide chemistry is of the $+3$ oxidation state where, because of the large sizes of the Ln^{III} ions, the bonding is predominantly ionic in character, and the cations display the typical class-a preference of *O*-donor ligands (p. 1065). Three-dimensional lattices, characteristic of ionic character, are common and the coordination chemistry is quite different from, and less extensive than, that of the d-transition metals. Coordination numbers are generally high and stereochemistries, being determined largely by the requirements of the ligands and lacking the directional constraints of covalency, are frequently ill-defined, and the complexes distinctly labile. Thus, in spite of widespread opportunities for isomerism, there appears to be no confirmed example of a lanthanide complex existing in more than one molecular arrangement. Furthermore, only strongly complexing (i.e. usually chelating) ligands yield products which can be isolated from aqueous solution, and the comparative tenacity of the small H_2O molecule commonly leads to its inclusion, often with consequent uncertainty as to the coordination number involved. This is not to say that other types of complex cannot be obtained, but complexes with uncharged monodentate ligands, or ligands with donor atoms other than O, must usually be prepared in the absence of water.

Some typical compounds are listed in Table 30.3. Coordination numbers below 6 are found only with very bulky ligands and even the coordination number of 6 itself is unusual, 7, 8, and 9 being more characteristic. Coordination numbers of 10 and over require chelating ligands with small "bites" (p. 1077), such as NO_3^- or SO_4^{2-}, and are confined to compounds of the larger, lighter lanthanides. The stereochemistries quoted, especially for the high coordination numbers, are idealized and in most cases appreciable distortions are in fact found.

A number of trends connected with ionic radii are noticeable across the series. In keeping with Fajan's rules, salts become somewhat less ionic as the Ln^{III} radius decreases; reduced ionic character in the hydroxide implies a reduction in basic properties and, at the end of the series, $Yb(OH)_3$ and $Lu(OH)_3$, though undoubtedly mainly basic, can with difficulty be made to dissolve in hot conc NaOH. Paralleling this change, the $[Ln(H_2O)_x]^{3+}$ ions are subject to an increasing tendency to hydrolyse, and hydrolysis can only be prevented by use of increasingly acidic solutions.

However, solubility, depending as it does on the rather small difference between solvation energy and lattice energy (both large quantities which themselves increase as cation size decreases) and on entropy effects, cannot be simply related to cation radius. No consistent trends are apparent in aqueous, or for that matter nonaqueous, solutions but an empirical distinction can often be made between the lighter "cerium" lanthanides and the heavier "yttrium" lanthanides. Thus oxalates, double sulfates, and double nitrates of the former are rather less soluble and basic nitrates more soluble than those of the latter. The differences are by no means sharp, but classical separation procedures depended on them.

Although lanthanide chemistry is dominated by the $+3$ oxidation state, and a number of binary compounds which ostensibly involve Ln^{II} are actually better formulated as involving Ln^{III} with an electron in a delocalized conduction band, genuine oxidation states of $+2$ and $+4$ can be obtained. Ce^{IV} and Eu^{II} are stable in water and, though they are respectively strongly oxidizing and strongly reducing, they have well-established aqueous

TABLE 30.3   Oxidation states and stereochemistries of compounds of the lanthanides[a]

Oxidation state	Coordination number	Stereochemistry	Examples
2	6	Octahedral	LnZ (Ln = Sm, Eu, Yb; Z = S, Se, Te)
	8	Cubic	LnF_2 (Ln = Sm, Eu, Yb)
3	3	Pyramidal	$[Ln\{N(SiMe_3)_2\}_3]$ (Ln = Nd, Eu, Yb)
	4	Tetrahedral	$[Lu(2,6\text{-dimethylphenyl})_4]^-$
		Distorted tetrahedral	$[Ln\{N(SiMe_3)_2\}_3(OPPh_3)]$ (Ln = Eu, Lu)
	6	Octahedral	$[LnX_6]^{3-}$ (X = Cl, Br); $LnCl_3$ (Ln = Dy–Lu)
	7	Capped trigonal prismatic	$[Dy(dpm)_3(H_2O)]$[b]
		Capped octahedral	$[Ho\{PhC(O)CH=C(O)Ph\}_3(H_2O)]$
	8	Dodecahedral	$[Ho(tropolonate)_4]^-$
		Square antiprismatic	$[Eu(acac)_3(phen)]$
		Bicapped trigonal prismatic	LnF_3 (Ln = Sm–Lu)
	9	Tricapped trigonal prismatic	$[Ln(H_2O)_9]^{3+}$, $[Eu(terpy)_3]^{3+}$
		Capped square antiprismatic	$[Pr(terpy)Cl_3(H_2O)_5] \cdot 3H_2O$
		Bicapped dodecahedral	$[Ln(NO_3)_5]^{2-}$ (Ln = Ce, Eu)
		Irregular	$[Gd(NO_3)_3(DPAE)]$[c]
	12	Icosahedral	$[Ce(NO_3)_6]^{3-}$[d]
4	6	Octahedral	$[CeCl_6]^{2-}$
	8	Cubic	LnO_2 (Ln = Ce, Pr, Tb)
		Square antiprismatic	$[Ce(acac)_4]$, LnF_4 (Ln = Ce, Pr, Tb)
	10	Complex	$[Ce(NO_3)_4(OPPh_3)_2]$[d]
	12	Icosahedral	$[Ce(NO_3)_6]^{2-}$[d]

[a] Except where otherwise stated, Ln is used rather loosely to mean most of the lanthanides; the Pm compound, for instance, is usually missing simply because of the scarcity and consequent expense of Pm.

[b] dpm = dipivaloylmethane, $Me_3C-\overset{\overset{O}{\|}}{C}-CH=\overset{\overset{O^-}{\|}}{C}-CMe_3$

[c] DPAE = 1,2-di(pyridine-2-aldimino)ethane,

$$\begin{array}{c} CH_2 - CH_2 \\ | \quad\quad | \\ CH \overset{N}{=} \quad N \overset{}{=} CH \end{array}$$

[d] The structure can be visualized as octahedral if each NO_3^- is considered to occupy a single coordination site (p. 1444).

chemistries. Ln^{IV} (Ln = Pr, Tb) and Ln^{II} (Ln = Nd, Sm, Eu, Dy, Tm, Yb) also are known in the solid state but are unstable in water. The rather restricted aqueous redox chemistry which this implies is summarized in the oxidation state diagram (Fig. 30.4).

The prevalence of the +3 oxidation state is a result of the stabilizing effects exerted on different orbitals by increasing ionic charge. As successive electrons are removed from a neutral lanthanide atom, the stabilizing effect on the orbitals is in the order 4f > 5d > 6s, this being the order in which the orbitals penetrate through the inert core of electrons towards the nucleus. By the time an ionic charge of +3 has been reached, the preferential stabilization of the 4f orbitals is such that in all cases the 6s and 5d orbitals have been emptied, and in most cases the electrons remaining in the 4f orbitals are themselves so far embedded in the inert core as to be immovable by chemical means. Exceptions are Ce and, to a lesser extent, Pr which are at the beginning of the series where, as already noted, the 4f orbitals are still at a comparatively high energy and can therefore lose a further electron. Tb^{IV} presumably owes its existence to the stability of the $4f^7$ configuration.

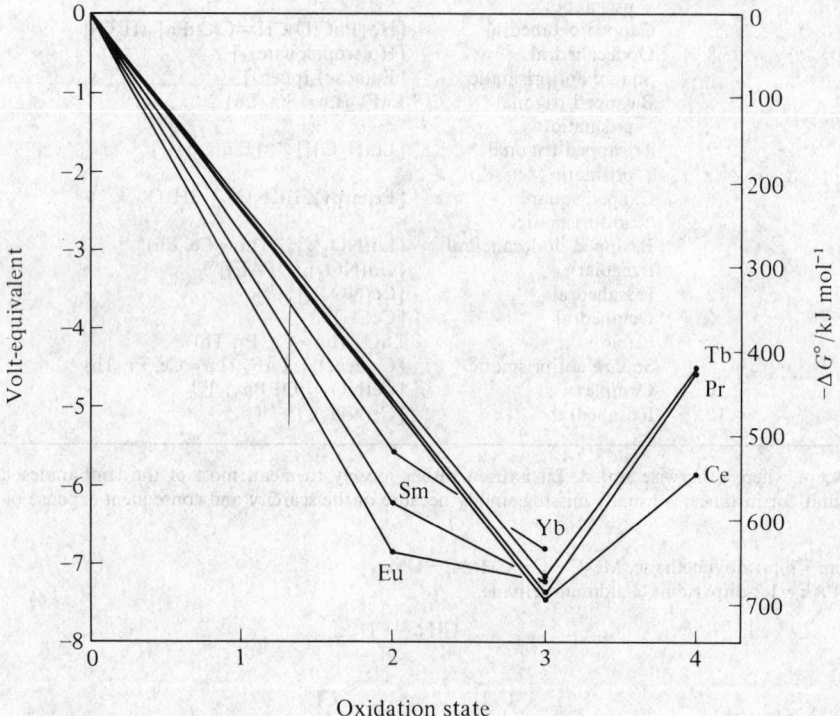

Fig. 30.4 Volt-equivalent versus oxidation state for lanthanides with more than one oxidation state.

The stabilizing effects of half, and completely, filled shells can be similarly invoked to explain the occurrence of the divalent state in $Eu^{II}(4f^7)$ and $Yb^{II}(4f^{14})$ while these, and the other known divalent ions are of just those elements which occupy elevated positions on the I_3 plot (Fig. 30.3).

The absence of 5d electrons and the inertness of the lanthanides' 4f shell makes π backbonding energetically unfavourable and simple carbonyls, for instance, have only been obtained in argon matrices at 8–12 K. On the other hand, essentially ionic cyclopentadienides are well known and an increasing number of σ- bonded Ln–C compounds have recently been produced (see section 30.3.5).

30.3 Compounds of the Lanthanides[6, 10, 11]

The reaction between H_2 and the gently heated (300–350°C) metals produces black, reactive and highly conducting solids, LnH_2. These hydrides have the fcc fluorite structure and are evidently composed of Ln^{III}, $2H^-$, e^-, the electron being delocalized in a metallic conduction band. Further hydrogen can be accommodated in the interstices of the lattice and, with the exceptions of Eu and Yb, which are the two lanthanides most favourably disposed to divalency, a limiting stoichiometry of LnH_3 can be achieved if high pressures are employed (p. 72). The composition of LnH_3 is Ln^{III}, $3H^-$ with conductivity correspondingly reduced as the additional H atom traps the previously delocalized electron (to form H^-).

Metallic conductivity, arising from the presence of Ln^{III} ions with the balance of electrons situated in a conduction band, is also found in some of the borides (p. 162) and carbides (p. 318).

30.3.1 *Oxides and chalcogenides*

Ln_2O_3 are all well characterized. With three exceptions they are the final products of combustion of the metals or ignition of the hydroxides, carbonate, nitrate, etc. The exceptions are Ce, Pr, and Tb, the most stable products of which are respectively, CeO_2, Pr_6O_{11}, and Tb_4O_7, from which the sesquioxides can be obtained by controlled reduction with H_2. Ln_2O_3 adopt three structure types conventionally classified as:[12]

A-type, consisting of LnO_7 units which approximate to capped octahedral geometry, and favoured by the lightest lanthanides.

B-type, also consisting of LnO_7 units but now of three types, two are capped trigonal prisms and one is a capped octahedron; favoured by the middle lanthanides.

C-type, related to the fluorite structure but with one-quarter of the anions removed in such a way as to reduce the metal coordination number from 8 to 6 (but not octahedral); favoured by the middle and heavy lanthanides.

Ln_2O_3 are strongly basic and the lighter, more basic, ones resemble the oxides of Group IIA in this respect. All are insoluble in water but absorb it to form hydroxides. They dissolve readily in aqueous acids to yield solutions which, providing they are kept on the acid side of pH5 to avoid hydrolysis, contain the $[Ln(H_2O)_x]^{3+}$ ions. Hydrous hydroxides can be precipitated from these solutions by addition of ammonia or aqueous alkali. Crystalline $Ln(OH)_3$ have a 9-coordinate, tricapped, trigonal prismatic structure, and may be obtained by prolonged treatment of Ln_2O_3 with conc NaOH at high temperature and pressure[13] (hydrothermal ageing).

The pale-yellow CeO_2 is a rather inert material when prepared by ignition, but in the hydrous, freshly precipitated form it redissolves quite easily in acids. The analogous dark-

[10] T. MOELLER, The lanthanides, Chap. 44, pp. 1–101, in *Comprehensive Inorganic Chemistry*, Vol. 4, Pergamon Press, Oxford, 1973; Lanthanides and actinides, Vol. 7, *MTP International Review of Science*, *Inorganic Chemistry* (Series Two) (K. W. BAGNALL, ed.), Butterworths, London, 1975, 329 pp.

[11] S. A. COTTON and F. A. HART, *The Heavy Transition Elements*, Chap. 10, pp. 188–219, Macmillan, London, 1975.

[12] A. F. WELLS, *Structural Inorganic Chemistry*, 4th edn., pp. 450–3, Oxford University Press, Oxford, 1975.

[13] D. F. MULLICA, W. O. MILLIGAN, and G. W. BEALL, Crystal structures of $Pr(OH)_3$, $Eu(OH)_3$, and $Tm(OH)_3$, *J. Inorg. Nucl. Chem.* **41**, 525–32 (1979).

coloured PrO_2 and TbO_2 can be obtained from their stable oxides Pr_6O_{11} and Tb_4O_7 but require more extreme conditions (O_2 at 282 bar and 400°C for PrO_2, and atomic oxygen at 450°C for TbO_2). All three dioxides have the fluorite structure. Since this is the structure on which C-type Ln_2O_3 is based (by removing a quarter of the anions) it is not surprising that these three oxide systems involve a whole series of nonstoichiometric phases[10, 14, 15] between the extremes represented by $LnO_{1.5}$ and LnO_2 (p. 755).

Claims for the existence of several lower oxides, LnO, have been made but most have been rejected, and it seems that only NdO, SmO (both lustrous golden yellow), EuO (dark red), and YbO (greyish-white) are genuine. They are obtained by reducing Ln_2O_3 with the metal at high temperatures and, except for EuO, at high pressures.[16] They have the NaCl structure but whereas EuO and YbO are composed of Ln^{II} and are insulators or semiconductors, NdO and SmO like the dihydrides consist essentially of Ln^{III} ions with the extra electrons forming a conduction band. EuO was unexpectedly found to be ferromagnetic at low temperatures. The absence of conduction electrons, and the presence of 4f orbitals which are probably too contracted to allow overlap between adjacent cations, makes it difficult to explain the mechanism of the ferromagnetic interaction. Because of this and potential applications in memory devices, considerable interest[17] has centred on EuO and the monochalcogenides since the 1960s.

Chalcogenides of similar stoichiometry to the oxides, but for a wider range of metals, are known, though their characterization is made more difficult by the prevalence of nonstoichiometry and the occurrence of phase changes in several instances. In general the chalcogenides are stable in dry air but are hydrolysed if moisture is present. If heated in air they oxidize (sulfides especially) to basic salts of the corresponding oxo-anion and they show varying susceptibility to attack by acids with evolution of H_2Z.

Monochalcogenides, LnZ (Z = S, Se, Te), have been prepared for all the lanthanides except Pm, mostly by direct combination.[6, 8] They are almost black and, like the monoxides, have the NaCl structure. However, with the exceptions of SmZ, EuZ, YbZ, TmSe, and TmTe, they have metallic conductivity and evidently consist of $Ln^{III} + Z^{2-}$ ions with 1 electron from each cation delocalized in a conduction band. EuZ and YbZ, by contrast, are semiconductors or insulators with genuinely divalent cations, but SmZ seem to be intermediate and may involve the equilibrium:

$$Sm^{II} + Z^{2-} \rightleftharpoons Sm^{III} + Z^{2-} + e^-$$

Trivalent chalcogenides, Ln_2Z_3, can be obtained by a variety of methods which include direct combination and, in the case of the sulfides, the action of H_2S on the chloride or oxide. As with Ln_2O_3, various crystalline modifications occur. When Ln_2Z_3 are heated with an excess of chalcogen in a sealed tube at 600°C, products with compositions up to or nearing LnZ_2 are obtained. They seem to be polychalcogenides, however, with the metal uniformly in the +3 state.

[14] D. J. M. BEVAN, Non-stoichiometric compounds: an introductory essay, Chap. 49, in *Comprehensive Inorganic Chemistry*, Vol. 4, pp. 453–540, Pergamon Press, Oxford, 1973.

[15] N. N. GREENWOOD, *Ionic Crystals, Lattice Defects, and Nonstoichiometry*, Chap. 6, pp. 111–47, Butterworths, London, 1968.

[16] J. M. LEGER, N. YACOUBI, and J. LORIERS, Synthesis of neodymium and samarium monoxides under high pressure, *Inorg. Chem.* **19**, 2252–4 (1980).

[17] Pages 23–41 of ref. 8.

30.3.2   *Halides*[18]

Halides of the lanthanides are listed in Table 30.4 and are of the types LnX_4, LnX_3, and LnX_2. Not surprisingly, LnX_4 occur only as the fluorides of Ce^{IV}, Pr^{IV}, and Tb^{IV}. CeF_4 is comparatively stable and can be prepared either directly from the elements or by the action of F^- on aqueous solutions of Ce^{IV} when it crystallizes as the monohydrate. The other tetrafluorides are thermally unstable and, as they oxidize water, can only be prepared dry; TbF_4 from $TbF_3 + F_2$ at 320°C, and PrF_4 by the rather complex procedure of fluorinating a mixture of NaF and PrF_3 with F_2 ($\rightarrow Na_2PrF_6$) and then extracting NaF from the reaction mixture with liquid HF.

Promethium apart, all possible trihalides (52) are known. The trifluorides, being very insoluble, can be precipitated as $LnF_3.\frac{1}{2}H_2O$ by the action of HF on aqueous $Ln(NO_3)_3$. Aqueous solutions of the other trihalides are obtained by simply dissolving the oxides or carbonates in aqueous HX, and hydrated ($6-8H_2O$) salts can be crystallized, though with difficulty because of their high solubilities. Preparation of the anhydrous trihalides by thermal dehydration of these hydrates is possible for fluorides and chlorides of the lighter lanthanides, and an atmosphere of HX extends the applicability of the method to heavier lanthanides. Unfortunately the bromides and iodides are more susceptible to hydrolysis, and dehydration inevitably leads to contamination with oxyhalides. As a result, direct combination, which is a completely general method, is preferred for the anhydrous tribromides and triiodides.[19]

The anhydrous trihalides are ionic, high melting, crystalline substances which, apart from the trifluorides are extremely deliquescent. As can be seen from Table 30.4, the coordination number of the Ln^{III} changes with the radii of the ions, from 9 for the trifluorides of the large lanthanides to 6 for the triiodides of the smaller lanthanides. Their chief importance has been as materials from which the pure metals can be prepared.

The dihalides are obtained from the corresponding trihalides, most generally by reduction with the lanthanide metal itself but also, in the case of the more stable of the diiodides (SmI_2, EuI_2, YbI_2), by thermal decomposition.† SmI_2 and YbI_2 can also be conveniently prepared in quantitative yield by reacting the metal with 1,2-diiodoethane in anhydrous tetrahydrofuran at room temperature:[19a] $Ln + ICH_2CH_2I \rightarrow LnI_2 + CH_2{=}CH_2$. With the exception of EuX_2, all the dihalides are very easily oxidized and will liberate hydrogen from water. The occurrence of these dihalides parallels the occurrence of high values of the third ionization energy amongst the metals[7, 8] (Fig. 30.3), with the reasonable qualification that, in this low oxidation state, iodides are more numerous than fluorides. The same types of structures are found as for the alkaline earth dihalides with the coordination number of the cation ranging from 9 to 6 and, like CaI_2, the diiodides of Dy, Tm and Yb form layer structures ($CdCl_2$, CdI_2; see Fig. 29.2) typical of compounds with large anions where marked polarization effects are expected.

---

† Dilute solid solutions of Ln^{III} ions in CaF_2 may be reduced by Ca vapour to produce Ln^{II} ions trapped in the crystal lattice. By their use it has been possible to obtain the electronic spectra of Ln^{II} ions.

[18] D. BROWN, *Halides of Lanthanides and Actinides*, Wiley, London, 1968, 280 pp.
[19] Pages 77–101 of ref. 5.
[19a] P. GIRARD, J. L. NAMY, and H. B. KAGAN, Divalent lanthanide derivatives in organic synthesis. 1. Mild preparation of SmI_2 and YbI_2 and their use as reducing or coupling agents, *J. Am. Chem. Soc.* **102**, 2693–8 (1980).

TABLE 30.4 *Properties of lanthanide halides: colour, mp/°C, and coordination*[a]

	Ce	Pr	Nd	Sm	Eu	Gd	Tb	Dy	Ho	Er	Tm	Yb	Lu
LnF₄	white dec 8 sa	white dec 8 sa	—	—	—	—	white dec 8 sa	—	—	—	—	—	—
LnF₃	white 1430 9 ttp	green 1395 9 ttp	violet 1374 9 ttp	white 1306 9 ttp	white 1276 9 ttp	white 1231 8 btp	white 1172 8 btp	green 1154 8 btp	pink 1143 8 btp	pink 1140 8 btp	white 1158 8 btp	white 1157 8 btp	white 1182 8 btp
LnCl₃	white 817 9 ttp	green 786 9 ttp	mauve 758 9 ttp	yellow 682 9 ttp	yellow dec 9 ttp	white 602 9 ttp	white 582 8 btp	white 647 6 o	yellow 720 6 o	violet 776 6 o	yellow 824 6 o	white 865 6 o	white 925 6 o
LnBr₃	white 733 9 ttp	green 691 9 ttp	violet 682 8 btp	yellow 640 8 btp	grey dec 8 btp	white 770 6 o	white 828 6 o	white 879 6 o	yellow 919 6 o	violet 923 6 o	white 954 6 o	white dec 6 o	white 1025 6 o
LnI₃	yellow 766 8 btp	737 8 btp	green 784 8 btp	orange 850 6 o	dec 6 o	yellow 925 6 o	957 6 o	green 978 8,7	yellow 994 6 o	violet 1015 6 o	yellow 1021 6 o	white dec 6 o	brown 1050 6 o
LnF₂				purple 8 c	greenish yellow 8 c							grey 8 c	
LnCl₂			green 841 9 ttp	brown 855 9 ttp	white 731 9 ttp			black 721 dec 8,7			green 718 7 co	green 702 7 co	
LnBr₂			9 ttp	brown 669 8,7	white 683 8,7			7 co			7 co	yellow 673 6 o	
LnI₂	bronze 808	bronze 758	violet 562 8,7	green 520 7 co	green 580 7 co	bronze 831		purple 659 6 ol			black 756 6 ol	black 772 6 ol	

[a] 9 ttp = 9-coordinate tricapped trigonal prismatic; 8 sa = 8-coordinate square antiprismatic; 8 btp = 8-coordinate bicapped trigonal prismatic; 8 c = 8-coordinate cubic (fluorite); 8,7 = mixed 8- and 7-coordinate (SrBr₂ structure); 7 co = 7-coordinate capped octahedral; 6 o = 6-coordinate octahedral; 6 ol = 6-coordinate octahedral layered.

The diiodides of Ce, Pr, and Gd stand apart from all the other, salt-like, dihalides. These three, like LaI_2, are notable for their metallic lustre and very high conductivities and are best formulated as $Ln^{III}, 2I^-, e^-$, the electron being in a delocalized conduction band.

Other reduced phases (LnX_x) have been identified in LnX_3/Ln systems as well as the dihalides. $SmBr_3/Sm$ for instance yields Sm_5Br_{11} $(x=2.20)$, $Sm_{11}Br_{24}$ $(x=2.182)$ and Sm_6Br_{13} $(x=2.167)$.[19b] They have fluorite-related structures (p. 129) in which the anionic sublattice is partially rearranged to accommodate additional anions, leading to irregular 7- and 8-coordination of the cations. Ln_5X_{11} phases, isostructural with Sm_5Br_{11}, have also been characterized[19c] in Gd_5Cl_{11} and Ho_5Cl_{11}. The $TmCl_3/Tm$ system is apparently much more complex. Reduction below the $+2$ oxidation state in $LnCl_3/Ln$ $(Ln=Gd, Tb)$ melts gives rise to Ln_2Cl_3 and finally "graphite-like" $LnCl$ phases.[20] In both types, extensive metal–metal bonding occurs, Ln_2Cl_3 containing close-packed double layers of Ln atoms as in ScCl (p. 1108) and ZrCl (p. 1125). Other stoichiometries that have been structurally characterized are Ln_7X_{12}, Ln_5X_8, and Ln_6X_7 (see Table 30.5).[20a]

30.3.3 *Magnetic and spectroscopic properties*[21]

In principle the electronic configurations of the lanthanides are described by using the Russell–Saunders coupling scheme, as was done in Chapter 19 for the d-transition elements. Values of S and L corresponding to the lowest energy are derived as on p. 1090 with the difference that there are seven f orbitals corresponding to the integral values $+3$ to -3 for m_l. These are then expressed for each ion in the form of a "ground term" with the symbolism extended so that $S, P, D, F, G, H, I, \ldots$, correspond to $L=0, 1, 2, 3, 4, 5, 6, \ldots$, in that order. At this point, however, a crucial difference emerges. Because the d electrons of transition-metal ions are exposed directly to the influence of neighbouring groups, the effect of the crystal field on the free-ion ground term has to be considered before that of spin–orbit coupling (i.e. the coupling of the angular momentum vectors associated with the quantum numbers S and L). But the 4f electrons of lanthanide ions are largely buried in the inner electron core and so are effectively shielded from their chemical environments. Spin–orbit coupling (mostly of the order of 2000 cm^{-1}), which gives rise to a resultant angular momentum associated with an overall quantum number J, is therefore much larger than the crystal field (~ 100 cm^{-1}) and its effect must be considered first.

J can take the values $J=L+S, L+S-1, \ldots, L-S$ (or $S-L$ if $S>L$), each corresponding to a different energy, so that a "term" (defined by a pair of S and L values) is said to split into a number of component "states" (each defined by the same S and L values plus a value of J). The "ground state" of the ion is that with $J=L-S$ (or $S-L$) if the f shell

[19b] H. BÄRNIGHAUSEN and J. M. HASCHKE, Compositions and crystal structures of the intermediate phases in the samarium–bromine system, *Inorg. Chem.* **17**, 18–21 (1978).

[19c] U. LÖCHNER, H. BÄRNIGHAUSEN, and J. D. CORBETT, Rare earth metal–metal halide systems. 19. Structural characterization of the reduced holmium chloride Ho_5Cl_{11}. *Inorg. Chem.* **16**, 2134–5 (1977).

[20] A. SIMON, H. MATTAUSCH, and N. HOLZER, Monochlorides of lanthanides: GdCl and TbCl, *Angew. Chem.*, Int. Edn. (Engl.) **15**, 624–5 (1976).

[20a] H. MATTAUSCH, J. B. HENDRICKS, R. EGER, J. D. CORBETT, and A. SIMON, Reduced halides of yttrium with strong metal–metal bonding: YCl, YBr, Y_2Cl_3, and Y_2Br_3, *Inorg. Chem.* **19**, 2128–32 (1980), and references cited therein.

[21] K. B. YATSIMIRSKII and N. K. DAVIDENKO, Absorption spectra and structure of lanthanide coordination compounds in solution, *Coord. Chem. Revs.* **27**, 223–73 (1979).

TABLE 30.5   *Stoichiometries and structures of reduced halides* ($X/M < 2$) *of scandium, yttrium, lanthanum, and the lanthanides*

Average oxidation state	Examples	Structural features
1.714	Sc_7Cl_{12}	Discrete M_6X_{12} clusters
	M_7I_{12} (La, Pr, Tb)	Discrete M_6X_{12} clusters
1.600	Sc_5Cl_8	Single chains of edge-sharing metal octahedra with
	M_5Br_8 (Gd, Tb)	M_6X_{12}-type environment (edge-capped by X) along with parallel chains of edge-sharing MX_6 octahedra
1.500	M_2Cl_3 (Y, Gd, Tb, Er, Lu)	Single chains of edge-sharing metal octahedra with
	M_2Br_3 (Y, Gd)	M_6X_8-type environment (face-capped by X)
1.429	Sc_7Cl_{10}	Double chains of edge-sharing metal octahedra with M_6X_8-type environment with parallel chains of edge-sharing MCl_6 octahedra
1.167	Er_6I_7, Tb_6Br_7	Double chains of edge-sharing metal octahedra with M_6X_8-type environment
1.000	MCl (Sc, Y, Gd, Tb)	Double metal layers of edge-sharing metal octahedra,
	MBr (Y, La, Pr, Gd, Tb, Ho, Er)	M_6X_8-type environment

is less than half-full, and that with $J = L + S$ if the f shell is more than half-full. It is indicated simply by adding this value of J as a subscript to the symbol for the "ground term".

The magnitude of the separation between the adjacent states of a term indicates the strength of the spin–orbit coupling, and in all but two cases (Sm^{III} and Eu^{III}) it is sufficient to render the first excited state of the Ln^{III} ions thermally inaccessible, and so the magnetic properties are determined solely by the ground state. It can be shown that the magnetic moment expected for such a situation is given by:

$$\mu_e = g\sqrt{J(J+1)} \text{ BM}$$

where
$$g = \frac{3}{2} + \frac{S(S+1) - L(L+1)}{2J(J+1)}$$

As can be seen in Table 30.6, this agrees very well with experimental values except for Sm^{III} and Eu^{III} and agreement is reasonable for these also if allowance is made for the temperature-dependent population of excited states.

Electronic absorption spectra are produced when electromagnetic radiation promotes the ions from their ground state to excited states. For the lanthanides the most common of such transitions involve excited states which are either components of the ground term† or else belong to excited terms which arise from the same $4f^n$ configuration as the ground term. In either case the transitions therefore involve only a redistribution of electrons within the 4f orbitals (i.e. f→f transitions) and so are orbitally forbidden just like d→d transitions (p. 1094). In the case of the latter the rule is partially relaxed by a mechanism which depends on the effect of the crystal field in distorting the symmetry of the metal ion. However, it has already been pointed out that crystal field effects are very much smaller in

---

† The separation of these component states being, as pointed out above, of the order of a few thousand wavenumbers, such transitions produce absorptions in the infrared region of the spectrum. Ions which have no terms other than the ground term will therefore be colourless, having no transitions of sufficiently high energy to absorb in the visible region. This accounts for the colourless ions listed in Table 30.6.

TABLE 30.6 *Magnetic and spectroscopic properties of* Ln^{III} *ions in hydrated salts*

Ln	Unpaired electrons	Ground state	Colour	μ_e/BM $g\sqrt{J(J+1)}$	Observed
Ce	1 (4f^1)	$^2F_{5/2}$	Colourless	2.54	2.3–2.5
Pr	2 (4f^2)	$^3H_4$	Green	3.58	3.4–3.6
Nd	3 (4f^3)	$^4I_{9/2}$	Lilac	3.62	3.5–3.6
Pm	4 (4f^4)	$^5I_4$	Pink	2.68	—
Sm	5 (4f^5)	$^6H_{5/2}$	Yellow	0.85	1.4– 1.7[a]
Eu	6 (4f^6)	$^7F_0$	Very pale pink	0	3.3– 3.5[a]
Gd	7 (4f^7)	$^8S_{7/2}$	Colourless	7.94	7.9– 8.0
Tb	6 (4f^8)	$^7F_6$	Very pale pink	9.72	9.5– 9.8
Dy	5 (4f^9)	$^6H_{15/2}$	Yellow	10.65	10.4–10.6
Ho	4 (4f^{10})	$^5I_8$	Yellow	10.60	10.4–10.7
Er	3 (4f^{11})	$^4I_{15/2}$	Rose-pink	9.58	9.4– 9.6
Tm	2 (4f^{12})	$^3H_6$	Pale green	7.56	7.1– 7.5
Yb	1 (4f^{13})	$^2F_{7/2}$	Colourless	4.54	4.3– 4.9
Lu	0 (4f^{14})	$^1S_0$	Colourless	0	0

[a] These are the values of μ_e at room temperature. The values fall as the temperature is reduced (see text).

the case of Ln^{III} ions and they cannot therefore produce the same relaxation of the selection rule. Consequently, the colours of Ln^{III} compounds are usually less intense. A further consequence of the relatively small effect of the crystal field is that the energies of the electronic slates are only slightly affected by the nature of the ligands or by thermal vibrations, and so the absorption bands are very much sharper than those for d→d transitions. Because of this they provide a useful means of characterizing, and quantitatively estimating, Ln^{III} ions.

Nevertheless, crystal fields cannot be completely ignored. The intensities of a number of bands ("hypersensitive" bands)[22] show a distinct dependence on the actual ligands which are coordinated. Also, in the same way that crystal fields lift some of the orbital degeneracy $(2L+1)$ of the terms of d^n ions, so they lift some of the $2J+1$ degeneracy of the sates of f^n ions, though in this case only by the order of 100 cm^{-1}. This produces fine structure in some bands of Ln^{III} spectra.

Ce^{III} and Tb^{III} are exceptional in providing (in the ultraviolet) bands of appreciably higher intensity than usual. The reason is that the particular transitions involved are of the type 4f^n→4f^{n-1} 5d^1, and so are not orbitally forbidden. These 2 ions have 1 electron more than an empty f shell and 1 electron more than a half-full f shell, respectively, and the promotion of this extra electron is thereby easier than for other ions.

It has been possible, as already noted (p. 1439), to study the spectra of Ln^{II} ions stabilized in CaF_2 crystals. It might be expected that these spectra would resemble those of the $+3$ ions of the next element in the series. However, because of the lower ionic charge of the Ln^{II} ions their 4f orbitals have not been stabilized relative to the 5d to the same extent as those of the Ln^{III} ions. Ln^{II} spectra therefore consist of rather broad, orbitally allowed, 4f→5d bands overlaid with weaker and much sharper f→f bands.

[22] D. E. HENRIE, R. L. FELLOWS, and G. R. CHOPPIN, Hypersensitivity in the electronic transitions of lanthanide and actinide complexes, *Coord. Chem. Revs.* **18**, 199–224 (1976).

30.3.4 *Complexes*[10, 11, 18, 23]

Oxidation state IV

Cerium is the only lanthanide with a significant aqueous or coordination chemistry in the +4 oxidation state. In the case of such other lanthanides as attain it, the +4 oxidation state is confined to oxides (p. 1438), fluorides (p. 1439), and a few fluoro complexes such as Na_2PrF_6, $Na_7Pr_6F_{31}$, Cs_3NdF_7, and Cs_3DyF_7.

Aqueous "ceric" solutions are widely used as oxidants in quantitative analysis; they can be prepared by the oxidation of Ce^{III} ("cerous") solutions with strong oxidizing agents such as peroxodisulfate, $S_2O_8{}^{2-}$, or bismuthate, $BiO_3{}^-$. Complexation and hydrolysis combine to render $E(Ce^{4+}/Ce^{3+})$ markedly dependent on anion and acid concentration. In relatively strong perchloric acid the aquo ion is present but in other acids coordination of the anion is likely. Also, if the pH is increased, hydrolysis to $Ce(OH)^{3+}$ occurs followed by polymerization and finally, as the solution becomes alkaline, by precipitation of the yellow, gelatinous $CeO_2 \cdot xH_2O$.

Of the various salts which can be isolated from aqueous solution, probably the most important is the water-soluble double nitrate, $(NH_4)_2[Ce(NO_3)_6]$, which is the compound generally used in Ce^{IV} oxidations. The anion involves 12-coordinated Ce (Fig. 30.5a). Two *trans*-nitrates of this complex can be replaced by Ph_3PO to give the orange 10-coordinate neutral complex $[Ce(NO_3)_4(OPPh_3)_2]$ (Fig. 30.5b). The sulphates $Ce(SO_4)_2 \cdot nH_2O$ ($n = 0, 4, 8, 12$) and $(NH_4)_2Ce(SO_4)_3$, and the iodate are also known. Also obtainable from aqueous

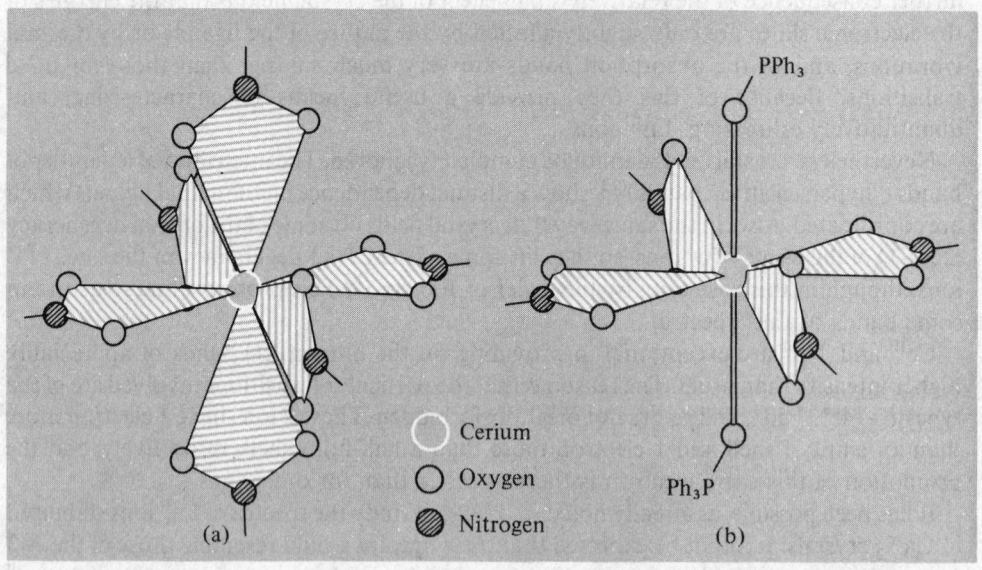

FIG. 30.5   Nitrato complexes of Ce^{IV}. (a) $[Ce(NO_3)_6]^{2-}$. The Ce^{IV} is surrounded by 12 oxygen atoms from 6 bidentate nitrate ions (in each case the third oxygen is omitted for clarity) in the form of an icosahedron. (Note that this implies an octahedral disposition of the 6 nitrogens.) (b) $[Ce(NO_3)_4(OPPh_3)_2]$.

[23] D. K. KOPPIKAR, P. V. SIVAPULLAIAH, L. RAMAKRISHNAN, and S. SOUNDARARAJAN, Complexes of the lanthanides with neutral oxygen donor ligands, *Struct. Bonding* (Berlin) **34**, 135–213 (1978).

solutions are complexes with other *O*-donor ligands such as β-diketonates, and fluoro complexes such as $[CeF_8]^{4-}$ and $[CeF_6]^{2-}$. This last ion is not in fact 6-coordinated but achieves an 8-coordinate, square-antiprismatic geometry with the aid of fluoride bridges. In the orange $[CeCl_6]^{2-}$ by contrast, the larger halide is able to stabilize a 6-coordinate, octahedral geometry. It is prepared by treatment of CeO_2 with HCl but, because Ce^{IV} in aqueous solution oxidizes HCl to Cl_2, the reaction must be performed in a nonaqueous solvent such as pyridine or dioxan.

Oxidation state III

The coordination chemistry of the large, electropositive Ln^{III} ions is complicated, especially in solution, by ill-defined stereochemistries and uncertain coordination numbers. This is well illustrated by the aquo ions themselves.[24] These are known for all the lanthanides, providing the solutions are moderately acidic to prevent hydrolysis, with hydration numbers probably about 8 or 9 but with reported values depending on the methods used to measure them. It is likely that the primary hydration number decreases as the cationic radius falls across the series. However, confusion arises because the polarization of the H_2O molecules attached directly to the cation facilitates hydrogen bonding to other H_2O molecules. As this tendency will be the greater, the smaller the cation, it is quite reasonable that the secondary hydration number increases across the series.

Hydrated salts with all the common anions can be crystallized from aqueous solutions and frequently, but by no means invariably, they contain the $[Ln(H_2O)_9]^{3+}$ ion. An enormous number of salts of organic acids such as oxalic, citric, and tartaric have been studied, often for use in separation methods. These anions are, in fact, chelating O ligands which as a class provide the most extensive series of Ln^{III} complexes. NO_3^- is an inorganic counterpart and is notable for the high coordination numbers it yields, as in the 10-coordinate bicapped dodecahedral $[Ce(NO_3)_5]^{2-}$, and in $[Ce(NO_3)_6]^{3-}$ which like its Ce^{IV} analogue, has the 12-coordinate icosahedral geometry (Fig. 30.5a) (see also p. 541).

β-diketonates (L–L) provide further important examples of this class of ligand, and yield complexes of the type $[Ln(L-L)_3L']$ (L'$=H_2O$, py, etc.) and $[Ln(L-L)_4]^-$, which are respectively 7- and 8-coordinate. Dehydration, under vacuum, of the hydrated tris-diketonates produces $[Ln(L-L)_3]$ complexes which probably increase their coordination by dimerizing or polymerizing. They may be sublimed, the most volatile and thermally stable being those with bulky alkyl groups R in $[RC(O)CHC(O)R]^-$; they are soluble in non-polar solvents, and have received much attention as "nmr shift" reagents. Thus, in the case of organic molecules which are able to coordinate to Ln^{III} (i.e. if they contain groups such as –OH or –COO$^-$), the addition of one of these coordinatively unsaturated reagents produces a labile adduct; because this adduct is anisotropic the paramagnetic Ln^{III} ion shifts the resonance line of each proton by an amount which is critically dependent on the spatial relationship of the Ln^{III} and the proton. Greatly improved resolution is thereby obtained along with the possibility of distinguishing between alternative structures of the organic molecule.[25]

Various crown ethers (p. 105) with differing cavity diameters provide a range of

[24] J. BURGESS, *Metal ions in Solution*, Ellis Horwood, Chichester, 1978, 481 pp.
[25] G. R. SULLIVAN, Chiral lanthanide shift reagents, *Topics in Stereochem.* **10**, 287–329 (1978).

coordination numbers and stoichiometries, although crystallographic data are sparse. An interesting series, illustrating the dependence of coordination number on cationic radius and ligand cavity diameter, is provided by the complexes formed by the lanthanide nitrates and the 18-crown-6 ether (i.e. 1,4,7,10,13,16-hexaoxacyclo-octadecane). For Ln = La–Gd the most thermally-stable product is that with a ratio of Ln: crown ether = 4:3, but the larger of these lanthanides (i.e. La, Ce, Pr, and Nd) also form a 1:1 complex. This is $[Ln(NO_3)_3L]$ in which the Ln^{III} is 12-coordinate[26] (Fig. 30.6a). The 4:3 complex, on the other hand, is probably $[Ln(NO_3)_2L]_3[Ln(NO_3)_6]$ in which, compared to the 1:1 complex, the Ln^{III} in the complex cation has lost one NO_3^-, so reducing its coordination number to 10. The remaining, still smaller lanthanides (Tb–Lu) find the cavity of this ligand too large and form $[Ln(NO_3)_3(H_2O)_3]L$, in which the ligand is uncoordinated.

Unidentate O donors such as pyridine-N-oxide and triphenylphosphine oxide also form many complexes, as do alkoxides. A particularly interesting example is $[Nd_6(OPr^i)_{17}Cl]$, made up of 6 Nd atoms held together around a central Cl atom by means of bridging $OCHMe_2$ groups[27] (Fig. 30.6b).

That complexes with O-donor ligands are more numerous than those with N donors is probably because the former ligands are more often negatively charged—a clear advantage when forming essentially ionic bonds. However, by using polar organic solvents such as ethanol, acetone, or acetonitrile in order to avoid competitive coordination by water, complexes with ligands such as en, dien, bipy, and terpy can be prepared.[28] Coordination numbers of 8, 9, and 10 as in $[Ln(en)_4]^{3+}$, $[Ln(terpy)_3]^{3+}$, and $[Ln(dien)_4(NO_3)]^{2+}$ are typical. Nor do complexes such as the well-known $[Ln(EDTA)(H_2O)_3]^-$ show any destabilization because of the N donor atoms (EDTA has 4 oxygen and 2 nitrogen donor atoms). More pertinently, whereas the complexes of 18-crown-6-ethers mentioned above dissociate instantly in water, complexes of the N-donor analogues are sufficiently stable to remain unchanged.

As with other transition elements, the lanthanides can be induced to form complexes with exceptionally low coordination numbers by use of the very bulky ligand, $N(SiMe_3)_2^-$:[29]

$$3LiN(SiMe_3)_2 + LnCl_3 \longrightarrow [Ln\{N(SiMe_3)_2\}_3] + 3LiCl$$

The volatile, but air-sensitive, and very easily hydrolysed products have a coordination number of 3, the lowest found for the lanthanides; they are apparently planar in solution (zero dipole moment) but pyramidal in the solid state. With Ph_3PO the 4-coordinate distorted tetrahedral adducts $[Ln\{N(SiMe_3)_2\}_3(OPPh_3)]$ are obtained. So difficult is it to expand the coordination sphere that attempts to prepare bis(Ph_3PO) adducts produce instead the dimeric peroxo bridged complex, $[(Ph_3PO)\{(Me_3Si)_2N\}_2LnO_2Ln\{N(SiMe_3)_2\}_2(OPPh_3)]$ (see p. 723).

[26] M. E. HARMAN, F. A. HART, M. B. HURSTHOUSE, G. P. MOSS, and P. R. RAITHBY, 12-coordinated crown ether complex of lanthanum; X-ray crystal structure, *JCS Chem. Comm.* 396–7 (1976). J.-C. G. BÜNZLI, B. KLEIN, and D. WESSNER, Crystal and molecular structure of 18-crown-6 ether with neodymium nitrate, *Inorg. Chim. Acta* **44**, L147–9 (1980).

[27] R. A. ANDERSEN, D. H. TEMPLETON, and A. ZALKIN, Synthesis and crystal structure of a neodymium isopropoxide chloride, $Nd_6[OCH(CH_3)_2]_{17}Cl$, *Inorg. Chem.* **17**, 1962–5 (1978).

[28] J. H. FORSBERG, Complexes of lanthanide(III) ions with nitrogen donor ligands, *Coord. Chem. Revs.* **10**, 195–226 (1973).

[29] D. C. BRADLEY, J. S. GHOTRA, F. A. HART, M. B. HURSTHOUSE, and P. R. RAITHBY, Low coordination numbers in lanthanoid and actinoid complexes, *JCS Dalton* 1977, 1166–72.

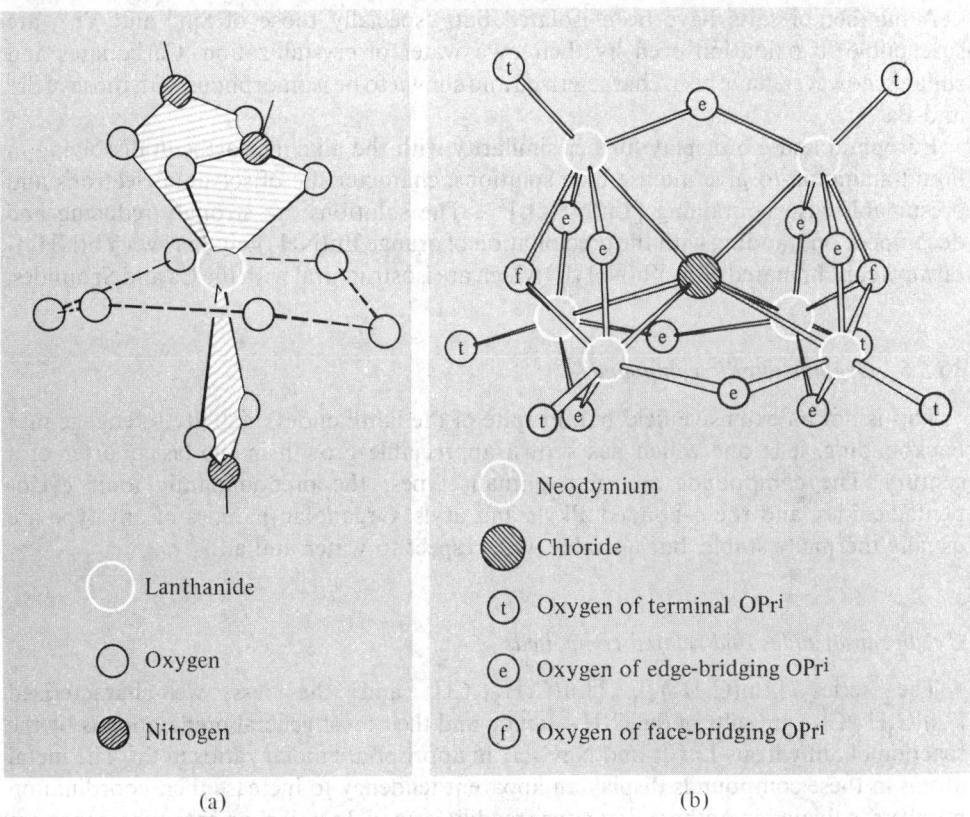

(a)   (b)

Lanthanide

Oxygen

Nitrogen

Neodymium

Chloride

(t) Oxygen of terminal OPr^i

(e) Oxygen of edge-bridging OPr^i

(f) Oxygen of face-bridging OPr^i

FIG. 30.6 (a) Ln(NO$_3$)$_3$ (18-crown-6). For clarity only 2 of the oxygen atoms of each nitrate ion are shown, and only the 6 oxygen atoms of the crown ether. (Note the boat conformation of the crown ether which allows access to two NO$_3^-$ on the open side and only one on the hindered side.) (b) [Nd$_6$(OPr^i)$_{17}$Cl]. Only the oxygens of the OPr^i groups are shown. Note that the 6 Nd atoms surrounding the Cl atom are situated at the corners of a trigonal prism, held together by 2 face-bridging and 9 edge-briding alkoxides.

Coordination by halide ions is rather weak, that of I$^-$ especially so, but from nonaqueous solutions it is possible to isolate anionic complexes of the type [LnX$_6$]$^{3-}$ which are apparently, and unusually for Ln^{III}, 6-coordinate and octahedral. The heavier donor atoms S, Se, P, and As have little tendency to coordinate, dithiocarbamates such as [Ln(S$_2$CNMe$_2$)$_3$] and [Ln(S$_2$CNMe$_2$)$_4$]$^-$ being amongst the few confirmed examples.

Oxidation state II[8]

The coordination chemistry in this oxidation state is essentially confined to the ions Sm^{II}, Eu^{II}, and Yb^{II}. These are the only ones with an aqueous chemistry and their solutions may be prepared by electrolytic reduction of the Ln^{III} solutions or, in the case of Eu^{II}, by reduction with amalgamated Zn. These solutions are blood-red for Sm^{II}, colourless or pale greenish-yellow for Eu^{II}, and yellow for Yb^{II}, and presumably contain the aquo ions. All are rapidly oxidized by air, and Sm^{II} and Yb^{II} are also oxidized by water itself although aqueous Eu^{II} is relatively stable, especially in the dark.

A number of salts have been isolated but, especially those of Sm^{II} and Yb^{II}, are susceptible to oxidation even by their own water of crystallization. Carbonates and sulfates, however, have been characterized and shown to be isomorphous with those of Sr^{II} and Ba^{II}.

Europium and Yb display further similarity with the alkaline earths in dissolving in liquid ammonia to give intense blue solutions, characteristic of solvated electrons and presumably also containing $[Ln(NH_3)_x]^{2+}$. The solutions are strongly reducing and decompose on standing with the precipitation of orange $Eu(NH_2)_2$ and brown $Yb(NH_2)_2$ (always contaminated with $Yb(NH_2)_3$) which are isostructural with the Ca and Sr amides.

30.3.5 *Organometallic compounds*[30, 31]

This is not an extensive field but, in spite of the lanthanides' inability to engage in π backbonding, it is one which has shown appreciable growth in the last quarter of a century. The compounds are of two main types: the predominantly ionic cyclopentadienides, and the σ-bonded alkyls and aryls. Organolanthanides of any type are usually thermally stable, but unstable with respect to water and air.

Cyclopentadienides and related compounds

The series $[Ln(C_5H_5)_3]$, $[Ln(C_5H_5)_2Cl]$, and the less well-characterized $[Ln(C_5H_5)Cl_2]$ are salts of the $C_5H_5{}^-$ anion and their most general preparation is by the reaction of anhydrous $LnCl_3$ and NaC_5H_5 in appropriate molar ratios in thf. The metal atoms in these compounds display an apparent tendency to increase their coordination numbers: solvates and other adducts are readily formed. In polar solvents, where they are no doubt solvated, they are monomeric but, in non-polar solvents the tris(C_5H_5) compounds are insoluble, while the bis(C_5H_5) compounds dimerize. Structural data for these solids is sparse but, where it exists, indicates rather complicated structures. Thus, in orange $[Sm(C_5H_5)_3]$, each $C_5H_5{}^-$ ion is pentahapto towards 1 metal atom but some also act as bridges by presenting a ring vertex (η^1) or edge (η^2) towards an adjacent metal atom, so producing a chain structure.† In a less complicated way the blue, $[Nd(C_5H_4Me)_3]$ is actually tetrameric, each Nd being attached to three rings in a pentahapto mode with one ring being further attached in a monohapto manner to an adjacent Nd. The red solid, $[Yb(C_5H_4Me)_2Cl]$ is a dimer with simple pentahapto rings and two chloride bridges.

Complexes with the two analogous ligands, indenide, $C_9H_7{}^-$,

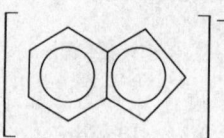

---

† Ring bridges of these types are found in alkaline earth cyclopentadienides such as $[Ca(C_5H_5)_2]$ and are characteristic of the electrostatic nature of the bonding.

[30] T. J. Marks, Chemistry and spectroscopy of f-element organometallics, part I: the lanthanides, *Prog. Inorg. Chem.* **24**, 51–107 (1978).

[31] T. J. Marks and R. D. Fischer (eds.), *Organometallics of the f-elements*, Reidel, Dordrecht, Holland, 1979, 517 pp. Proceedings of a NATO Adv. Study Inst. meeting, September 1978.

and cyclooctatetraenide, COT, $C_8H_8^{2-}$, ions can be prepared by similar means. In solid $[Sm(C_9H_7)_3]$ the 5-membered rings of the 3 ligands are bonded in a pentahapto manner and the compound shows little tendency to solvate, presumably because of the bulky nature of the $C_9H_7^{-}$ ions. The lighter (and therefore larger) Ln^{III} ions form $K[Ln(\eta^8\text{-}C_8H_8)_2]$. The Ce^{III} member of the series has a similar "sandwich" structure to the so-called "uranocene" (p. 1486). The other members of the series have the same infrared spectrum and so also are presumed to have this structure.

Alkyls and aryls

These are prepared by metathesis in thf or ether solutions:

$$LnCl_3 + 3LiR \longrightarrow LnR_3 + 3LiCl$$

$$LnCl_3 + 4LiR \longrightarrow Li[LnR_4] + 3LiCl_3$$

The triphenyls are probably polymeric and the first fully-characterized compound was $[Li(thf)_4][Lu(C_6H_3Me_2)_4]$ in which the Lu is tetrahedrally coordinated to four σ-aryl groups. More stable products, of the form $[LnR_3(thf)_2]$, are obtained by use of the bulky alkyl groups such as $-CH_2CMe_3$ and $-CH_2SiMe_3$.

Methyl derivatives have only recently been obtained but H. Schumann and co-workers have now isolated octahedral $[LnMe_6]^{3-}$ species for most of the lanthanides.[31]

Novel, mixed alkyl cyclopentadienides have also been prepared for the heavy lanthanides:[32, 33]

[32] N. M. ELY and M. TSUTSUI, Organolanthanides and organoactinides. XV. Synthesis and properties of new σ-bonded organolanthanide complexes, *Inorg. Chem.* **14**, 2680–7 (1975).

[33] J. HOLTON, M. F. LAPPERT, D. G. H. BALLARD, R. PEARCE, J. L. ATWOOD, and W. E. HUNTER, Alkyl-bridged complexes of the d- and f-block elements. Part I. Di-μ-alkyl-bis(η-cyclopentadienyl)metal(III) dialkylaluminium(III) complexes and the crystal and molecular structure of the ytterbium methyl species. Part II. Bis[bis(η-cyclopentadienyl)methylmetal(III)] complexes and the crystal and molecular structures of the yttrium and ytterbium species, *JCS Dalton* 1979, 45–61.

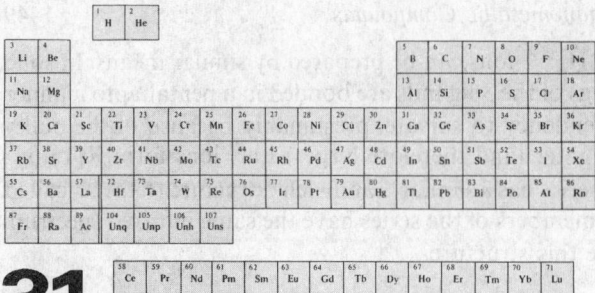

31

The Actinide Elements (*Z*=90-103)

31.1 Introduction

The "actinides" ("actinons" or "actinoids") are the fourteen elements from thorium to lawrencium inclusive, which follow actinium in the periodic table. They are analogous to the lanthanides and result from the filling of the 5f orbitals, as the lanthanides result from the filling of 4f. The position of actinium, like that of lanthanum, is somewhat equivocal and, although not itself an actinide, it is often included with them for comparative purposes.

Prior to 1940 only the naturally occurring actinides, thorium, protactinium, and uranium were known; the remainder have been produced artificially since then. Indeed, manmade transuranium elements now constitute some 15% of all the known chemical elements.

In 1789 M. H. Klaproth examined pitchblende, thought at the time to be a mixed oxide ore of zinc, iron, and tungsten, and showed that it contained a new element which he named *uranium* after the recently discovered planet, Uranus. Then in 1828 J. J. Berzelius obtained an oxide, from a Norwegian ore now known as "thorite"; he named this *thoria* after the Scandanavian god of war and, by reduction of its tetrachloride with potassium, isolated the metal *thorium*. The same method was subsequently used in 1841 by B. Peligot to effect the first preparation of metallic uranium.

The much rarer, protactinium, was not found until 1913 when K. Fajans and O. Göhring identified ^{234}Pa as an unstable member of the ^{238}U decay series:

$$^{238}_{92}U \xrightarrow{-\alpha} {}^{234}_{90}Th \xrightarrow{-\beta^-} {}^{234}_{91}Pa \xrightarrow{-\beta^-} {}^{234}_{92}U$$

They named it brevium because of its short half-life (6.66 h). The more stable isotope ^{231}Pa ($t_{\frac{1}{2}}$ 32 800 y) was identified 3 years later by O. Hahn and L. Meitner and independently by F. Soddy and J. A. Cranston as a product of ^{235}U decay:

$$^{235}_{92}U \xrightarrow{-\alpha} {}^{231}_{90}Th \xrightarrow{-\beta^-} {}^{231}_{91}Pa \xrightarrow{-\alpha} {}^{227}_{89}Ac$$

As the parent of actinium in this series it was named protoactinium, shortened in 1949 to *protactinium*. Because of its low natural abundance its chemistry was obscure until 1960 when A.G. Maddock and co-workers at the UK Atomic Energy Authority worked up about 130 g from 60 tons of sludge which had accumulated during the extraction of uranium from UO_2 ores. It is from this sample, distributed to numerous laboratories throughout the world, that the bulk of our knowledge of the element's chemistry has been gleaned.

In the early years of this century the periodic table ended with element 92 but, with J. Chadwick's discovery of the neutron in 1932 and the realization that neutron-capture by a heavy atom is frequently followed by β^- emission yielding the next higher element, the synthesis of new elements became an exciting possibility. E. Fermi and others were quick to attempt the synthesis of element 93 by neutron bombardment of ^{238}U, but it gradually become evident that the main result of the process was not the production of element 93 but nuclear fission, which produces lighter elements. However, in 1940, E. M. McMillan and P. H. Abelson in Berkeley, California, were able to identify, along with the fission products, a short-lived isotope of element 93 ($t_{\frac{1}{2}}$ 2.35 days):†

$$^{238}_{92}U + {}^1_0n \longrightarrow {}^{239}_{92}U \xrightarrow{-\beta} {}^{239}_{93}Np$$

Coming after uranium it was appropriate to name the new element *neptunium* after Neptune, which is the next planet beyond Uranus.

The remaining actinide elements were prepared[1, 2] by various "bombardment" techniques fairly regularly over the next 20 years (Table 31.1) though, for reasons of security, publication of the results was sometimes delayed. The dominant figure in this field has been G. T. Seaborg, of the University of California, Berkeley, in early recognition of which, he and E. M. McMillan were awarded the 1951 Nobel Prize for Chemistry.

The isolation and characterization of these elements, particularly the heavier ones, posed enormous problems. Individual elements are not produced cleanly in isolation, but must be separated from other actinides as well as from lanthanides produced simultaneously by fission. In addition, all the actinides are radioactive, their stability decreasing with increasing atomic number, and this has two serious consequences. Firstly, it is necessary to employ elaborate radiation shielding and so, in many cases, operations must be carried out by remote control. Secondly, the heavier elements are produced only in the minutest amounts. Thus mendelevium was prepared in almost unbelievably small yields of the order of 1 to 3 atoms per experiment! Paradoxically, however, the intense radioactivity also facilitated the detection of these minute amounts, firstly by the development and utilization of radioactive decay systematics, which enabled the detailed properties of the expected radiation to be predicted, and, secondly, by using the radioactive decay itself to detect and count the individual atoms synthesized. Accordingly, the separations were effected by ion-exchange techniques, and the elements identified by chemical tracer methods and by their characteristic nuclear decay properties. In view of the quantities involved, especially of californium and later elements, it is clear that this would not have been feasible without accurate predictions of the chemical properties also.

---

† $^{239}_{93}Np$ itself also decays by β^- emission to produce element 94 but this was not appreciated until after that element (plutonium) had been prepared from $^{238}_{93}Np$.

[1] G. T. Seaborg (ed.), *Transuranium Elements; Products of Modern Alchemy*, Dowden, Hutchinson & Ross, Stroudsburg, 1978. This reproduces, in their original form, 122 key papers in the story of man-made elements.
[2] J. J. Katz and G. T. Seaborg, *The Chemistry of the Actinide Elements*, Methuen, London, 1957, 508 pp. An early but authoritative summary up to element 102.

TABLE 31.1 *The discovery (synthesis) of the artificial actinides*

Element	Discoverers	Date	Synthesis	Origin of name
93 Neptunium, Np	E. M. McMillan and P. Abelson	1940	Bombardment of $^{238}_{92}U$ with $^{1}_{0}n$	The planet Neptune
94 Plutonium, Pu	G. T. Seaborg, E. M. McMillan, J. W. Kennedy, and A. Wahl	1940	Bombardment of $^{238}_{92}U$ with $^{2}_{1}H$	The planet Pluto (next planet beyond Neptune)
95 Americium, Am	G. T. Seaborg, R. A. James, L. O. Morgan, and A. Ghiorso	1944	Bombardment of $^{239}_{94}Pu$ with $^{1}_{0}n$	America (by analogy with Eu, named after Europe)
96 Curium, Cm	G. T. Seaborg, R. A. James, and A. Ghiorso	1944	Bombardment of $^{239}_{94}Pu$ with $^{4}_{2}He$	P. and M. Curie (by analogy with Gd, named after J. Gadolin)
97 Berkelium, Bk	S. G. Thompson, A. Ghiorso, and G. T. Seaborg	1949	Bombardment of $^{241}_{95}Am$ with $^{4}_{2}He$	Berkeley (by analogy with Tb, named after the village of Ytterby)
98 Californium, Cf	S. G. Thompson, K. Street, A. Ghiorso, and G. T. Seaborg	1950	Bombardment of $^{242}_{96}Cm$ with $^{4}_{2}He$	California (location of the laboratory)
99 Einsteinium, Es	Workers at Berkeley, Argonne, and Los Alamos	1952	Found in debris of first thermo-nuclear explosion as a result of irradiation of U with fast neutrons	Albert Einstein (relativistic relation between mass and energy)
100 Fermium, Fm	Workers at Berkeley, Argonne, and Los Alamos	1952	Found in debris of first thermo-nuclear explosion as a result of irradiation of U with fast neutrons	Enrico Fermi (construction of first self-sustaining nuclear reactor)
101 Mendelevium, Md	A. Ghiorso, B. H. Harvey, G. R. Choppin, S. G. Thompson, and G. T. Seaborg	1955	Bombardment of $^{253}_{99}Es$ with $^{4}_{2}He$	Dimitri Mendeleev (periodic table of the elements)
102 Nobelium, No	A. Ghiorso, T. Sikkeland, J. R. Walton, and G. T. Seaborg[b]	1958	Bombardment of $^{246}_{96}Cm$ with $^{12}_{6}C$	Alfred Nobel (benefactor of science)[b]
103 Lawrencium, Lr[a]	A. Ghiorso, T. Sikkeland, A. E. Larsch, and R. M. Latimer	1961	Bombardment of mixed isotopes of Cf with $^{10}_{5}B$ and $^{11}_{5}B$	Ernest Lawrence (developer of the cyclotron)

[a] Formerly Lw; present symbol adopted by IUPAC in 1965.

[b] The first report of element 102 was in 1957 by an international team of scientists working at the Nobel Institute for Physics in Stockholm. They bombarded $^{244}_{96}Cm$ with $^{13}_{6}C$ but their result could not be repeated elsewhere. Their suggested name for the element has however been accepted.

It was Seaborg's realization in 1944 that these elements should be regarded as a second f series akin to the lanthanides that made this possible. (Thorium, protactinium, and uranium had previously been regarded as transition elements belonging to groups IVA, VA, and VIA, respectively.)

Elements beyond 103 are expected to be 6d elements forming a fourth transition series, and attempts to synthesize them of course continued after lawrencium had been discovered. The work requires the commitment of extensive facilities and is at present being carried out by A. Ghiorso and others in California and by G. N. Flerov's group at Dubna, near Moscow.

There is apparently little doubt that element 104 can be made and that it forms a chloride and chloro-complex similar to hafnium's,[3] but priority in its discovery is disputed. The Berkeley workers propose the name *rutherfordium* (Rf) after Ernest Rutherford, while the Dubna workers proposed *kurchatovium* (Ku) after Igor Kurchatov. In the absence of agreement on priorities, and on the details of the experiments, the IUPAC recommended that a systematic name should be adopted: *un-nil-quadium*, Unq (p. 35). Later elements are less certain. The names *hahnium* (Ha) after Otto Hahn and *nielsbohrium* (Ns) after Niels Bohr have been suggested by the Berkeley and Dubna groups respectively, but neither has been approved internationally; the systematic name is *unnilpentium*, Unp. Both groups claim also to have prepared element 106 but neither have suggested a name; there is also growing evidence for the synthesis of element 107. The IUPAC recommended names are, respectively, *unnilhexium* (Unh) and *unnilseptium* (Uns).

Superheavy elements

Since the radioactive half-lives of the known transuranium elements and their resistance to spontaneous fission decrease with increase in atomic number, the outlook for the synthesis of further elements might appear increasingly bleak. However, theoretical calculations of nuclear stabilities, based on the concept of closed nucleon shells (p. 15) suggest the existence of an "island of stability" around $Z = 114$ and $N = 184$.[3a] Attention has therefore been directed towards the synthesis of element 114 (a congener of Pb in Group IVB) and adjacent, "superheavy" elements, by bombardment of heavy nuclides with a wide range of heavy ions, but so far without success.

Searches have been made for naturally occurring superheavies ($Z = 112-115$) in ores of Hg, Tl, Pb, and Bi, on the assumption that they would follow their homologues in their geochemical evolution and could be recognized by the radiation damage caused over geological time by their very energetic decay. Early claims to have detected such superheavies in natural ores have been convincingly discounted.[3b] More recent uncorroborated claims to success have been made but, even if confirmed, the

---

[3] A. K. Hulet, R. W. Lougheed, J. F. Wild, J. H. Landrum, J. M. Nitschke, and A. Ghiorso, Chloride complexation of element 104, *J. Inorg. Nucl. Chem.* **42**, 79–82 (1980).

[3a] B. Fricke, Superheavy elements, *Struct. Bonding, (Berlin)*, **21**, 89–144 (1975).

[3b] F. Bosh, A. ElGoresy, W. Krätschmer, B. Martin, B. Povh, R. Nobiling, K. Traxel, and D. Schwalm, Possible existence of superheavy elements in monazite, *Z. Physik* **A280**, 39–44 (1977); see also C. J. Sparks, S. Raman, H. L. Takel, R. V. Gentry, and M. O. Krause, *Phys. Rev. Letters* **38**, 205–8 (1977), for retraction of their earlier claim to have detected naturally occurring primordial superheavy elements.

concentrations found in the samples examined, are exceedingly small[4] (less than 1 in 10^{13}).

31.2 The Elements[5–10]

31.2.1 *Terrestrial abundance and distribution*

Every known isotope of the actinide elements is radioactive and the half-lives are such that only ^{232}Th, ^{235}U, ^{238}U, and possibly ^{244}Pu could have survived since the formation of the solar system. In addition, continuing processes produce equilibrium traces of some isotopes of which the most prominent is ^{234}U ($t_{\frac{1}{2}}$ 2.45 × 10^5 y, comprising 0.0054% of naturally occurring U isotopes). ^{231}Pa (and therefore ^{227}Ac) is formed as a product of the decay of ^{235}U, while ^{237}Np and ^{239}Pu are produced by the reactions of neutrons with, respectively, ^{235}U and ^{238}U. Traces of Pa, Np, and Pu are consequently found, but only Th and U occur naturally to any useful extent. Indeed, these two elements are far from rare: thorium comprises 8.1 ppm of the earth's crust, and is almost as abundant as boron, whilst uranium at 2.3 ppm is rather more abundant than tin. The radioactive decay schemes of the naturally occurring long-lived isotopes of ^{232}Th, ^{235}U, and ^{238}U, together with the artificially generated series based on ^{241}Pu, are summarized in the scheme on the page opposite.

Thorium is widely but rather sparsely distributed and its only commercial sources are monazite sands (see p. 1425) and the mineral conglomerates of Ontario. The former are found in India, South Africa, Brazil, Australia, and Malaysia, and in exceptional cases may contain up to 20% ThO$_2$ but more usually contain less than 10%. In the Canadian ores the thorium is present as uranothorite, a mixed Th,U silicate, which is accompanied by pitchblende. Even though present as only 0.4% ThO$_2$, the recovery of Th, as a co-product of the recovery of uranium, is viable.

Uranium, too, is widely distributed and, since it probably crystallized late in the formation of igneous rocks, tends to be scattered in the faults of older rocks. Some concentration by leaching and subsequent re-precipitation has produced a large number of oxide minerals of which the most important are pitchblende or uraninite, U_3O_8, and carnotite, $K_2(UO_2)_2(VO_4)_2.3H_2O$. However, even these are usually dispersed so that typical ores contain only about 0.1% U, and many of the more readily exploited deposits are nearing exhaustion. The principal sources are the USA, Canada, South Africa, and Australia.

The transuranium elements must all be prepared artificially.

[4] See, for instance, E. L. FIREMAN, B. H. KETELLE, and R. W. STOUGHTON, Superheavy element search in bismuthinite, Bi$_2$S$_3$, *J. Inorg. Nucl. Chem.* **41**, 613–15 (1979).

[5] *Kirk–Othmer Encyclopedia of Chemical Technology*, 3rd edn., Interscience, New York; for Actinides see Vol. 1, 1978, pp. 456–89; for Thorium see Vol. 22, 1983, pp. 989–1002; for Uranium, see Vol. 23, 1983, pp. 502–47.

[6] G. R. CHOPPIN and J. RYDBERG, *Nuclear Chemistry: Theory and Applications*, Pergamon Press, Oxford, 1980, 667 pp.

[7] *The Actinides*, Vol. 5 of *Comprehensive Inorganic Chemistry*, Pergamon Press, Oxford, 1973, 715 pp.

[8] *Lanthanides and Actinides*, Vol. 7 of *MTP International Review of Science, Inorganic Chemistry* (K. W. BAGNALL, Ed.), Butterworths, London, Series 1, 1972, pp. 367; series 2, 1975, pp. 329.

[9] J. F. SMITH, O. N. CARLSON, D. T. PETERSON, and T. E. SCOTT, *Thorium: Preparation and Properties*, Iowa State University Press, Iowa, 1975, 385 pp.

[10] M. TAUBE, *Plutonium: A General Survey*, Verlag Chemie, Weinheim, 1974, 242 pp.

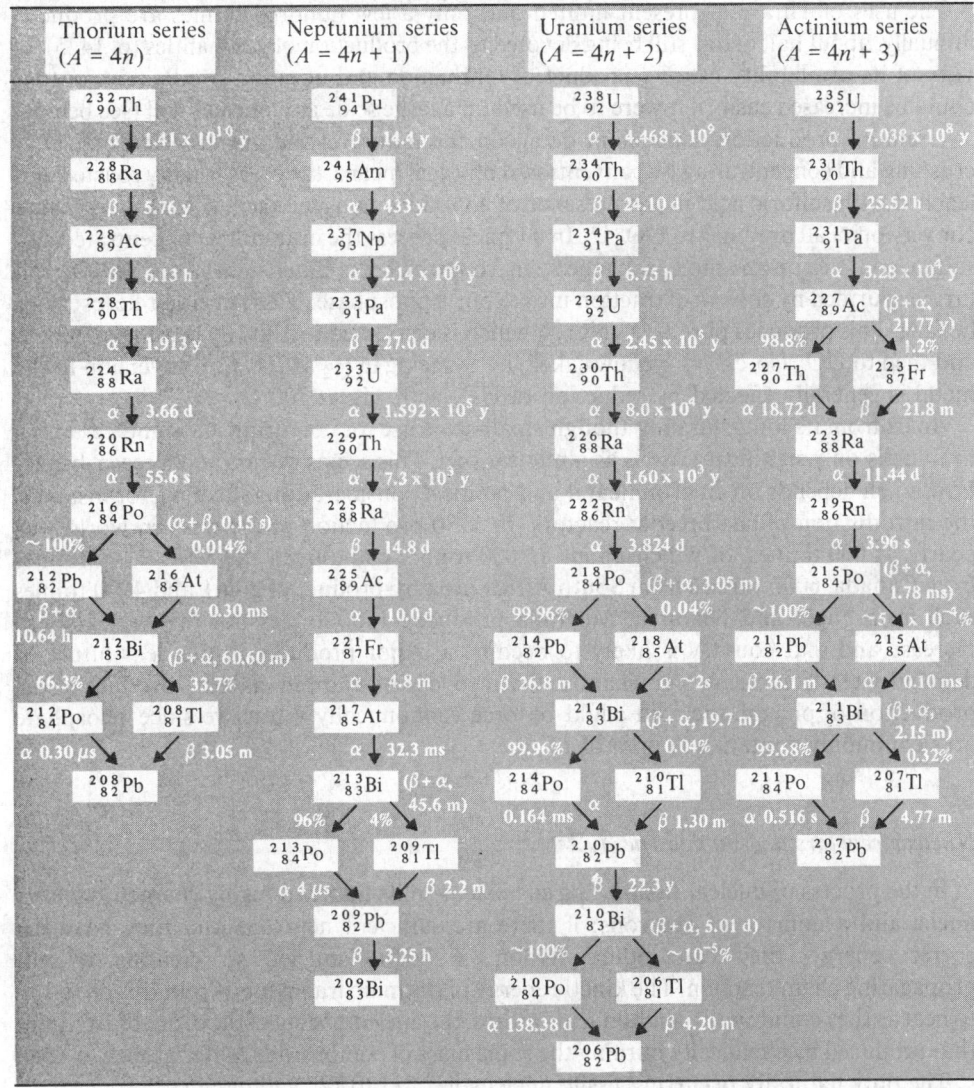

Thorium series ($A = 4n$)	Neptunium series ($A = 4n + 1$)	Uranium series ($A = 4n + 2$)	Actinium series ($A = 4n + 3$)

The radioactive decay series.

31.2.2 *Preparation and uses of the elements*

The separation of basic precipitates of hydrous ThO_2 from the lanthanides in monazite sands has been outlined in Fig. 30.1 (p. 1427). These precipitates may then be dissolved in nitric acid and the thorium extracted into tributyl phosphate, $(Bu^nO)_3PO$, diluted with kerosene. In the case of Canadian production, the uranium ores are leached with sulfuric acid and the anionic sulfato complex of U preferentially absorbed onto an anion exchange resin. The Th is separated from Fe, Al, and other metals in the liquor by solvent extraction.

Metallic thorium can be obtained by reduction of ThO_2 with Ca or reduction of $ThCl_4$ with Ca or Mg under an atmosphere of Ar (like Ti, finely divided Th is extremely reactive when hot).

The uses of Th are at present limited and only a few hundred tonnes are produced annually, about half of this still being devoted to the production of gas mantles (p. 1425). In view of its availability as a by-product of lanthanide and uranium production, output could be increased easily if it were to be used on a large scale as a nuclear fuel (see below).

Uranium production depends in detail on the nature of the ore involved but, after crushing and concentrating by conventional physical means, the ore is usually roasted and leached with sulfuric acid in the presence of an oxidizing agent such as MnO_2 to ensure conversion of all uranium to UO_2^{2+}. In a typical process the uranium is concentrated as a sulfato complex on an anion exchange resin from which it is eluted with strong HNO_3 and further purified by solvent extraction into tributyl phosphate (TBP) in either kerosene or hexane. This produces pure $UO_2(NO_3)_2$ which is converted to UO_3 by heating at 300°C, and then to UO_2 (so-called "yellow cake") by reducing in H_2 at 700°C. Conversion to the metal is generally effected by reduction of UF_4 with Mg at 700°C.

Apart from its long-standing though small-scale use for colouring glass and ceramics, uranium's only significant use is as a nuclear fuel. This seems certain to increase, but by how much depends on environmental and political considerations affecting, for instance, the introduction of fast breeder reactors. In 1980 production in the western world was nearly 44 000 tonnes, of which about 16 000 tonnes came from the USA, 7000 tonnes from Canada, 6000 tonnes from South Africa (as a by-product of gold), and 4000 tonnes each from Niger and Namibia. Australia produced 1500 tonnes but this was a sharp increase and that country is likely to become a major producer in the near future. If demand were to increase significantly, the recovery of uranium as a by-product of the production of phosphoric acid could become economically attractive since phosphate ores commonly contain 100 ppm of U.

Nuclear reactors and atomic energy [11, 12]

In the process of nuclear fission a large nucleus splits into two highly energetic smaller nuclei and a number of neutrons; if there are sufficient neutrons and they have the correct energy, they can induce fission of further nuclei, so creating a self-propagating chain reaction. The kinetic energy of the main fragments is rapidly converted to heat as they collide with neighbouring atoms, the amount being of the order of 10^6 times that produced by chemically burning the same mass of combustible material such as coal.

The only naturally occurring fissile nucleus is $^{235}_{92}U$ (0.72% abundant):

$$^{235}_{92}U + {}^{1}_{0}n \longrightarrow 2 \text{ fragments} + x{}^{1}_{0}n \quad (x = 2\text{–}3)$$

The so-called "fast" neutrons which this fission produces have energies of about 2 MeV, or 190×10^6 kJ mol^{-1}, and are not very effective in producing fission of further $^{235}_{92}U$ nuclei. Better in this respect are "slow" or "thermal" neutrons whose energies are of the order of 0.025 eV, or 2.4 kJ mol^{-1} (i.e. equivalent to the thermal energy available at ambient temperatures). In order to produce and sustain a chain reaction in uranium it is therefore necessary to counter the inefficiency of fast neutrons by either (a) increasing the

[11] D. CRABBE and R. McBRIDE, *The World Energy Book*, Kogan Page, London, 1978, 259 pp. See individual entries.

[12] J. R. FINDLAY, K. M. GLOVER, I. L. JENKINS, N. R. LARGE, J. A. C. MARPLES, P. E. POTTER, and P. W. SUTCLIFFE, The inorganic chemistry of nuclear fuel cycles, in R. THOMPSON (ed.), *The Modern Inorganic Chemicals Industry*, Special publication No. 31, The Chemical Society, London, 1977, pp. 419–66.

proportion of $^{235}_{92}$U (i.e. fuel enrichment) or (b) slowing down (i.e. moderating) the fast neutrons. In addition, there must be sufficient uranium to prevent excessive loss of neutrons from the surface (i.e. a "critical mass" must be exceeded). If the reaction is not to run out of control, an adjustable neutron absorber is also required to ensure that the rate of production of neutrons is balanced by the rate of their absorption.

The first manmade self-sustaining nuclear fission chain reaction was achieved on 2 December 1942 in a disused squash court at the University of Chicago by a team which included E. Fermi. This was before nuclear-fuel enrichment had been developed: alternate sections of natural-abundance UO_2 and graphite moderator were piled on top of each other (hence, nuclear reactors were originally known as "atomic piles") and the reaction was controlled by strips of cadmium which could be inserted or withdrawn as necessary. In this crude structure, 6 tonnes of uranium metal, 50 tonnes of uranium oxide, and nearly 400 tonnes of graphite were required to achieve criticality. The dramatic success of Fermi's team in achieving a self-sustaining nuclear reaction invited the speculation as to whether such a phenomenon could occur naturally.[12a] In one of the most spectacular pieces of scientific detection work ever conducted, it has now unambiguously been established that such natural chain reactions have indeed occurred in the geological past when conditions were far more favourable than at present (see Panel).

Natural Nuclear Reactors—The Oklo Phenomenon[13]

Natural uranium consists almost entirely of ^{238}U and the fissionable isotope ^{235}U. The proportions of these are determined by their relative rates of decay, which is by α emission. As ^{235}U decays more than six times faster than ^{238}U (half-lives 0.7038×10^9 and 4.468×10^9 y respectively), the proportion of ^{235}U is very slowly but inexorably decreasing with time. Prior to 1972, all analyses of naturally occurring uranium had shown this proportion to be notably constant at $0.7202 \pm 0.006\%$.† In that year, however, workers at the French Atomic Energy laboratories in Pierrelatte performing routine mass spectrometric analyses recorded a value of 0.7171%. The difference was small but significant.

Contamination with commercially depleted U was immediately assumed, but it was gradually realized that the depletion was characteristic of the ore, which came from a mine at Oklo in Gabon, near the west coast (1° 25'S, 13° 10'W). An intensive examination of the mine was quickly mounted and it was found that the depletion was not uniform but was greatest near those areas where the total U content was highest. The record depletion was an astonishingly low 0.296% ^{235}U from an area where the total U content of the ore rose to around 60%. Various explanations of what quickly became known as the "Oklo phenomenon" were considered. Incredible as it may appear in view of the diverse and exacting requirements for the construction of a manmade nuclear reactor, the only satisfactory explanation is that the Oklo mine is the site of at least six spent, natural, nuclear reactors!

The Oklo ore bed consists of sedimentary rocks believed to have been laid down about 1.8×10^9 y ago. U^{IV} minerals in the igneous rocks, formed in the early history of the earth when the

† If, as is believed (p. 16), the earth was formed about 4.6×10^9 y ago it follows that the proportion of ^{235}U at that time must have been about 25%.

12a P. K. KURODA, On the nuclear physical stability of the uranium minerals, *J. Chem. Phys.* **25**, 781–2 and 1295–6 (1956). P. K. KURODA, Nuclear fission in the early history of the earth, *Nature* **187**, 36–38 (1960).

13 *Le Phenomene d'Oklo*, Proceedings of a Symposium on the Oklo Phenomenon, International Atomic Energy Agency, Vienna, Proceedings Series, 1975. *Natural Fission Reactions*, IAEA, Vienna, Panel Proceedings Series STI/PUB/475, 1978, 754 pp. G. A. COWAN, A natural fission reactor, *Scientific American* **235**, 36–47 (1976). R. WEST, Natural nuclear reactors. The Oklo phenomenon, *J. Chem. Ed.* **53**, 336–40 (1976).

atmosphere was a reducing one, were converted to soluble U^{VI} salts by the atmosphere which had since become oxidizing. These were then re-precipitated as U^{IV} by bacterial reduction in the silt of a river delta and gradually buried under other sedimentary deposits. During this process the underlying granite rocks were tilted, the ores which contained about 0.5% U were fractured, and water percolating through the fissures created rich pockets of ore which in places consisted of almost pure UO_2. At that time the ^{235}U content of the uranium was about 3%, which is the value to which the fuel used in most modern water-moderated reactors is now artificially enriched.

Under these circumstances the critical mass could be attained and a nuclear chain reaction initiated, with water as the necessary moderator. The 15% water of hydration contained in the clays associated with the ore would be ideal for this purpose. As the reaction proceeded, the consequent rise in temperature would have driven off water, so producing "undermoderation" and slowing the reaction, thereby avoiding a "runway" reaction. As a result, a particular reactor may have operated in a steady manner or perhaps in a slowly pulsating manner, as water was alternately driven off (causing loss of criticality and cooling) and re-absorbed (recovering criticality and again heating).

Further control of the reactions must have been effected by neutron-absorbing "poisons", such as lithium and boron, which are nearly always present in clays. That these are present in the Oklo clays in comparatively low concentrations is one of the factors which allowed the reactions to take place. As the nuclear fuel in the original, rich pockets was being used up, the poisons in the surrounding ore would be simultaneously "burned out" by escaping neutrons. Thus ore of only slightly poorer quality, which was initially prevented only by the poisons from being critical, would gradually become so and the chain reaction would be propogated further through the ore bed. Just how long these reactions operated is difficult to judge, not least because their operation may not have been continuous, but it is thought to be about $(0.2–1) \times 10^6$ y with output in the region of 10–100 kW, consuming altogether 4–6 tonnes of ^{235}U from a total deposit of the order of 400 000 tonnes of uranium. Subsequent preservation of the fossil reactors is a result of continued burial which protected the uranium from redissolution.

Confirmation of this explanation is provided by the presence in the reactor zones of at least half of the more than 30 fission products of uranium. Although soluble salts, such as those of the alkali and alkaline earth metals, have been leached out, lanthanide and platinum metals remain along with traces of trapped krypton and xenon. Most decisively, the observed distribution of the various isotopes of these elements is that of fission products as opposed to the distribution normally found. The reasons for the retention of these elements on this particular site is clearly germane to the problem of the long-term storage of nuclear wastes, and is therefore the subject of continuing study.

The circumstances which led to the Oklo phenomenon may well have occurred in other, as yet unidentified, places, but in view of the intervening natural depletion of ^{235}U, the possibility of a natural chain reaction being initiated at the present time may be discounted. While a chain reaction in unenriched UO_2 is possible, it requires moderation not by ordinary water but by deuterium or graphite in a reactor precisely constructed to the most stringent specifications.

If a chain reaction is to provide useful energy, the heat it generates must be extracted by means of a suitable coolant and converted, usually by steam turbines, into electrical energy. The high temperatures and the intense radioactivity generated within a reactor pose severe, and initially totally new, constraints on the design. The choices of fuel and its immediate container (cladding), of the moderator, coolant, and controller involve problems in nuclear physics, chemistry, metallurgy, and engineering. Nevertheless, the first commercial power station (as opposed to experimental reactors or those whose function was to produce plutonium for bomb manufacture) was commissioned in 1956 at Calder Hall in Cumberland, UK. Since then a variety of different types has been developed in several countries, as summarized in Fig. 31.1. At the present time (1982) some 35 countries are operating nuclear power stations to supply energy.

Fuels. Although the concentration of ^{235}U in natural uranium is sufficient to sustain a chain reaction, its effective dilution by the fuel cladding and other materials used to

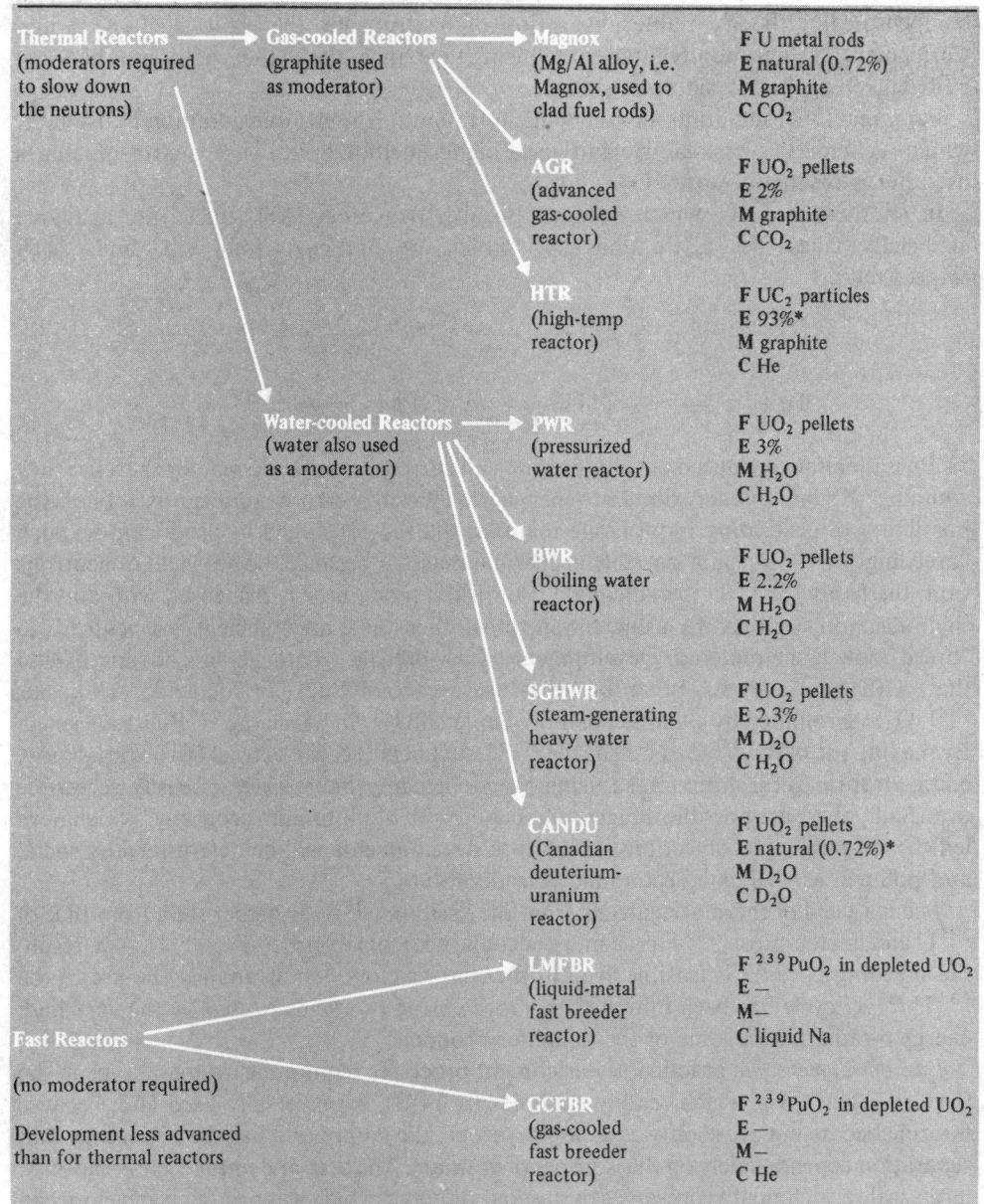

Thermal Reactors	Gas-cooled Reactors	Magnox	F U metal rods
(moderators required to slow down the neutrons)	(graphite used as moderator)	(Mg/Al alloy, i.e. Magnox, used to clad fuel rods)	E natural (0.72%) M graphite C CO_2
		AGR (advanced gas-cooled reactor)	F UO_2 pellets E 2% M graphite C CO_2
		HTR (high-temp reactor)	F UC_2 particles E 93%* M graphite C He
	Water-cooled Reactors (water also used as a moderator)	PWR (pressurized water reactor)	F UO_2 pellets E 3% M H_2O C H_2O
		BWR (boiling water reactor)	F UO_2 pellets E 2.2% M H_2O C H_2O
		SGHWR (steam-generating heavy water reactor)	F UO_2 pellets E 2.3% M D_2O C H_2O
		CANDU (Canadian deuterium-uranium reactor)	F UO_2 pellets E natural (0.72%)* M D_2O C D_2O
Fast Reactors (no moderator required) Development less advanced than for thermal reactors		LMFBR (liquid-metal fast breeder reactor)	F $^{239}PuO_2$ in depleted UO_2 E — M— C liquid Na
		GCFBR (gas-cooled fast breeder reactor)	F $^{239}PuO_2$ in depleted UO_2 E — M— C He

* HTR and CANDU could also possibly use ^{232}Th-^{233}U fuel

FIG. 31.1   Various types of nuclear reactor currently in use or being developed (F fuel; E enrichment, expressed as %^{235}U present; M moderator; C coolant).

construct the reactor make fuel enrichment advantageous. Indeed, if ordinary (light) water is used as moderator or coolant, a concentration of 2-3% ^{235}U is necessary to compensate for the inevitable absorption of neutrons by the protons of the water. Enrichment also has the advantage of reducing the critical size of the reactor but this must be balanced against its enormous cost.

Early reactors used uranium in metallic form but it is now more common to use UO_2 which is chemically less reactive and has a higher melting point. UC_2 is also sometimes used but is reactive towards O_2.

In addition to $^{235}_{92}$U, which occurs naturally, two other fissile nuclei are available artificially. These are $^{239}_{94}$Pu and $^{233}_{92}$U which are obtained from $^{238}_{92}$U and $^{232}_{90}$Th respectively:

$$^{238}_{92}U + ^1_0n \longrightarrow ^{239}_{92}U \xrightarrow[23.5 \text{ min}]{-\beta^-} ^{239}_{93}Np \xrightarrow[2.35 \text{ days}]{-\beta} ^{239}_{94}Pu \xrightarrow[2.41 \times 10^4 \text{ y}]{-\alpha}$$

$$^{232}_{90}Th + ^1_0n \longrightarrow ^{233}_{90}Th \xrightarrow[22.3 \text{ min}]{-\beta^-} ^{233}_{91}Pa \xrightarrow[27.0 \text{ days}]{-\beta} ^{233}_{92}U \xrightarrow[1.59 \times 10^5 \text{ y}]{-\alpha}$$

^{239}Pu is therefore produced to some extent in all currently operating reactors because they contain ^{238}U, and this contributes to the reactor efficiency. More significantly, it offers the possibility of generating more fissile material than is consumed in producing it. Such "breeding" of ^{239}Pu is not possible in thermal reactors because the net yield of neutrons from the fission of ^{235}U is inadequate. But, if the moderator is dispensed with and the chain reaction sustained by using enriched fuel, then there are sufficient fast neutrons to "breed" new fissile material. Development of fast-breeder reactors is less advanced than that of thermal reactors, but prototypes use a core of PuO_2 in "depleted" UO_2 (i.e. $^{238}UO_2$) surrounded by a blanket of more depleted UO_2 in which the ^{239}Pu is generated. By making use of the ^{238}U as well as the ^{235}U, such reactors can extract 50-60 times more energy from natural uranium, so using more efficiently the reserves of easily accessible ores. Sadly there are possible dangers associated with a "plutonium economy" which have led to well-publicized objections, and future developments will be determined by social and political as well as by economic considerations.

The net yield of thermal neutrons from the fission of ^{233}U is higher than from that of ^{235}U and, furthermore, ^{232}Th is a more effective neutron absorber than ^{238}U. As a result, the breeding of ^{233}U is feasible even in thermal reactors. Unfortunately the use of the ^{232}Th/^{233}U cycle has been inhibited by reprocessing problems caused by the very high energy γ-radiation of some of the daughter products.

Fuel enrichment. All practicable enrichment processes require the uranium to be in the form of a gas. UF_6, which readily sublimes (p. 1473), is universally used and, because fluorine occurs in nature only as a single isotope, the compound has the advantage that separation depends solely on the isotopes of uranium. The first, and until recently the only, large-scale enrichment process was by gaseous diffusion which was originally developed in the "Manhattan Project" to produce nearly pure ^{235}U for the first atomic bomb (exploded at Alamogordo, New Mexico, 5.30 a.m., 16 July 1945). UF_6 is forced to diffuse through a porous membrane and becomes very slightly enriched in the lighter isotope. This operation is repeated thousands of times by pumping, in a kind of cascade process in which at each stage the lighter fraction is passed forward and the heavier fraction backwards. Unfortunately gaseous diffusion plants are large, very demanding in terms of membrane technology, and extremely expensive in energy: alternatives have therefore actively been sought. The furthest developed of these is the gas-centrifuge process, prototypes of which have operated in the UK and The Netherlands since 1976. In cylindrical centrifuges rotating at about 1700 rev sec^{-1} $^{238}UF_6$ concentrates towards the walls and $^{235}UF_6$ towards the centres, where the fractions are scooped out.

Cladding. The Magnox reactors get their name from the magnesium-aluminium alloy used to clad the fuel elements, and stainless steels are used in other gas-cooled reactors. In water reactors zirconium alloys are the favoured cladding materials.

Moderators. Neutrons are most effectively slowed by collisions with nuclei of about the same mass. Thus the best moderators are those light atoms which do not capture neutrons. These are ^2H, ^4He, ^9Be, and ^{12}C. Of these He, being a gas, is insufficiently dense and Be is expensive and toxic, so the common moderators are highly purified graphite or the more expensive heavy water. In spite of its neutron-absorbing properties, which as mentioned above must be offset by using enriched fuel, ordinary water is also used because of its cheapness and excellent neutron-moderating ability.

Coolants. Because they must be mobile, coolants are either gases or liquids. CO_2 and He are appropriate gases and are used in conjunction with graphite moderators. The usual liquids are heavy and light water, with water also as moderator. In order to keep the water in the liquid phase it must be pressurized (PWR), otherwise it boils in the reactor core (BWR, etc.) in which case the coolant is actually steam. In the case of breeder reactors the higher temperatures of their more compact cores pose severe cooling problems and liquid Na (or Na/K alloy) is favoured, although highly compressed He is another possibility.

Control rods. These are usually made of boron steel or boron carbide (p. 167), but other good neutron absorbers which can be used are Cd and Hf.

Nuclear fuel reprocessing[10, 12]

Many of the fission products formed in a nuclear reactor are themselves strong neutron absorbers (i.e. "poisons") and so will stop the chain reaction before all the ^{235}U (and ^{239}Pu which has also been formed) has been consumed. To avoid this wastage the irradiated fuel elements must be removed periodically and the fission products separated from the remaining uranium and the plutonium. Such reprocessing is of course inherent in the operation of fast-breeder reactors, but is also necessary for prolonged operation even in thermal reactors.

Irradiated nuclear fuel is one of the most complicated high-temperature systems found in modern industry, and it has the further disadvantage of being intensely radioactive so that it must be handled exclusively by remote control. The composition of the irradiated nuclear fuel depends on the particular reactor in question, but in general it consists of uranium, plutonium, small amounts of other transuranium elements, and various isotopes of over 30 fission-product elements. The distribution of fission products is such as to produce high concentrations of elements with mass numbers in the regions 90–100 (second transition series) and 130–145 ($_{54}Xe$, $_{55}Cs$, $_{56}Ba$, and lanthanides). The more noble metals, such as $_{44}Ru$, $_{45}Rh$, and $_{46}Pd$, tend to form alloy pellets while class-a metals such as $_{38}Sr$, $_{56}Ba$, $_{40}Zr$, $_{41}Nb$ and the lanthanides are present in complex oxide phases.

The first step is to immerse the fuel elements in large "cooling ponds" of water for a hundred days or so, during which time the short-lived, intensely radioactive species such as $^{131}_{53}I$ ($t_{\frac{1}{2}} = 8.04$ days) lose most of their activity and the generation of heat subsides.

Then the fuel elements are dissolved in 7M HNO_3 to give a solution containing U^{VI} and Pu^{IV} which, in the widely used Purex process, are extracted into 20% tributyl phosphate (TBP) in kerosene leaving most of the fission products (FP) in the aqueous phase. Subsequent separation of U and Pu depends on their differing redox properties (Fig. 31.2). The separations are far from perfect (see p. 1274), and recycling or secondary purification by ion-exchange techniques is required to achieve the necessary overall separations.

This reprocessing requires the handling of kilogram quantities of Pu and must be adapted to avoid a chain reaction (i.e. a criticality accident). The critical mass for an isolated sphere of Pu is about 10 kg, but in saturated aqueous solutions may be little more than 500 g. (Because of the large amounts of "inert" ^{238}U present, U does not pose this problem.)

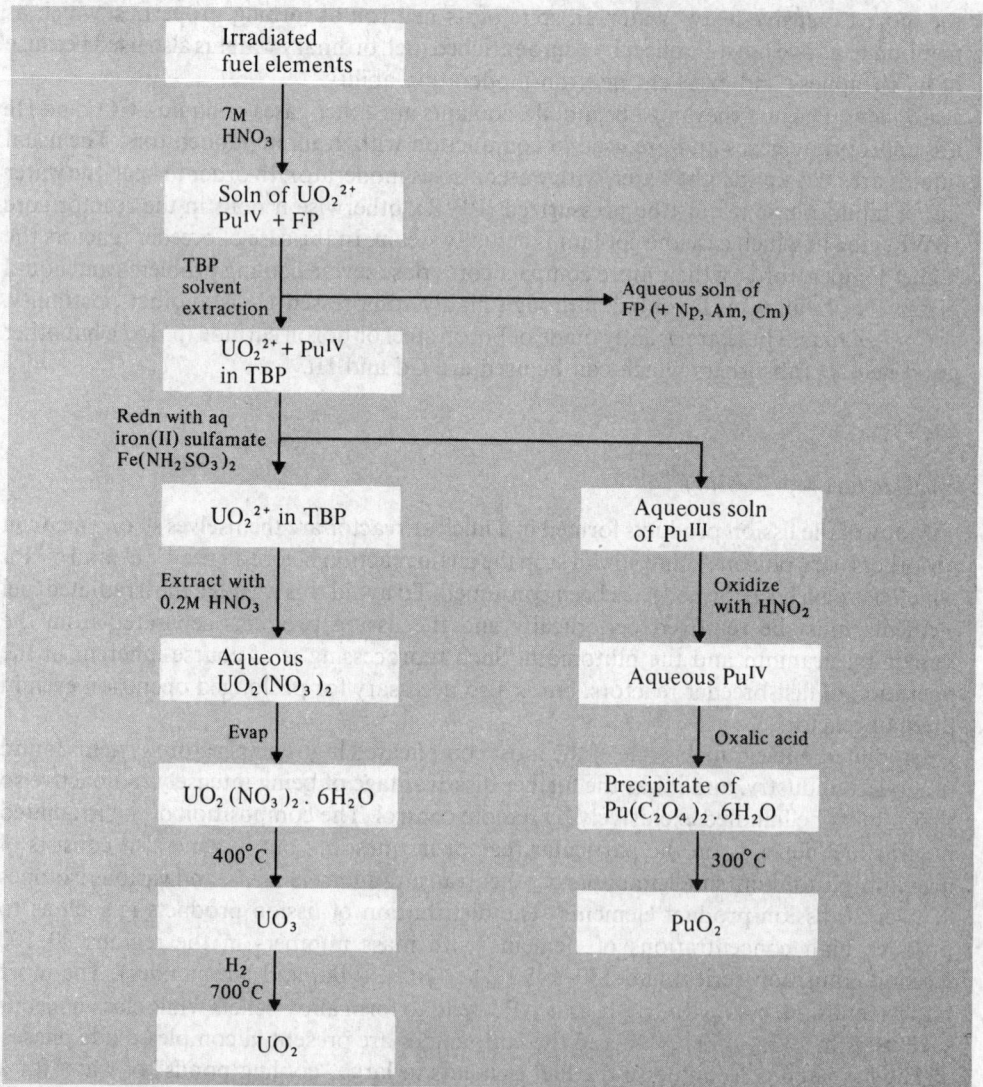

FIG. 31.2 Flow diagram for the reprocessing of nuclear fuel [FP = fission products; TBP = $(Bu^nO)_3PO$].

The solution of waste products is concentrated and stored in double-walled, stainless steel tanks shielded by a metre or more of concrete. Vitrification processes are being developed in several countries in which the waste is mixed with a slurry of silica in a solution of borax and alkali metal nitrates and slowly heated. A borosilicate glass is eventually produced and this will be stored or disposed of more permanently if there is agreement on suitable sites.

$^{237}_{93}Np$, $^{241}_{95}Am$, and $^{243}_{95}Am$ can be extracted from reactor wastes and these two elements are available in kg and 100 g quantities respectively. Prolonged neutron irradiation of $^{239}_{94}Pu$ is used at the Oak Ridge laboratories in Tennessee to produce: $^{244}_{96}Cm$ on a 100-g scale; $^{242}_{96}Cm$, $^{249}_{97}Bk$, $^{252}_{98}Cf$, $^{253}_{99}Es$, and $^{254}_{99}Es$ all on a mg scale; and $^{257}_{100}Fm$ on a μg scale. For trace amounts of these mixtures, dissolution in nitric acid, absorption of the $+3$ ions on to a cation exchange resin and elution with ammonium α-hydroxyisobutyrate provides an efficient separation of the elements from each other and from accompanying lanthanides, etc. The separation of macro amounts of these elements, however, is not feasible by this method because of radiolytic damage to the resin caused by their intense radioactivity. Much quicker solvent extraction processes similar to those used for reprocessing nuclear fuels are therefore used.[14]

Because the sequence of neutron captures inevitably leads to $^{258}_{100}Fm$ which has a fission half-life of only a few seconds, the remaining three actinides, $_{101}Md$, $_{102}No$, and $_{103}Lr$, can only be prepared by bombardment of heavy nuclei with the light atoms ^4_2He to $^{20}_{10}Ne$. This raises the mass number in multiple units and allows the $^{258}_{100}Fm$ barrier to be avoided; even so, yields are minute and are measured in terms of the individual number of atoms produced.

Apart from $^{239}_{94}Pu$, which is a nuclear fuel and explosive, the transuranium elements are produced mainly for research purposes, but a few specialized applications have been found where their high cost is not prohibitive. The most important of these is that of $^{238}_{94}Pu$ produced by the reaction

$$^{237}_{93}Np + ^1_0n \longrightarrow {}^{238}_{93}Np \xrightarrow{-\beta^-} {}^{238}_{94}Pu$$

This is used as a compact energy source by utilizing the heat generated during decay. In conjunction with PbTe thermoelectric elements, for instance, it provides a stable and totally reliable source of electricity with no moving parts, and has been used in the American Apollo space missions. Since its α-emission is harmless and is not accompanied by γ-radiation it has also been used in human heart pacers (~ 200 mg $^{238}_{94}Pu$) and for heating deep-sea diving suits (750 g ^{238}Pu without thermoelectric conversion).

31.2.3   *Properties of the elements*[15]

The dominant feature of the actinides is their nuclear instability, as manifest in their radioactivity (mostly α-decay) and tendency to spontaneous fission; both of these modes of decay become more pronounced (shorter half-lives) with the heavier elements. The

[14] J. D. NAVRATIL and W. W. SCHULZ (eds.), *Actinide Separation*, ACS Symposium Series. **117**, Am. Chem. Soc., Washington, 1979, 609 pp., and *Transplutonium Elements—Production and Recovery*, ibid. **161**, 1981. 302 pp. E. K. HULET and D. D. BODE, Separation chemistry of the lanthanides and transplutonium actinides, Chap. 1, pp. 1–45, in Vol. 7, Series 1, of ref. 8.

[15] J. A. LEE and M. B. WALDRON, The actinide metals, Chap. 7, pp. 221–56, in Vol. 7, Series 2, of ref. 8.

radioactivity of Th and U is probably responsible for much of the earth's internal heat, but is of a sufficiently low level to allow their compounds to be handled and transported without major problems. By contrast, the instability of the heavier elements not only imposes most severe handling problems[16] but drastically limits their availability. Thus, for instance, the crystal structures of Cf and Es were determined on only microgram quantities,[1] while the concept of "bulk" properties is not applicable at all to elements such as Md, No, and Lr which have never been seen and have only been produced in unweighably small amounts. Even where adequate amounts are available, the constant build-up of decay products and the associated generation of heat may seriously affect the measured properties (see also p. 888). An indication of the difficulties of working with these elements can be gained from the fact that two phases described in 1974 as two forms of Cf metal were subsequently shown, in fact, to be hexagonal Cf_2O_2S and fcc CfS.[16a]

Some of the more important known properties of the actinides are summarized in Table 31.2. The metals are silvery in appearance but display a variety of structures. All except Cf have more than one crystalline form (Pu has six) but most of these are based on typically metallic close-packed arrangements. Structural variability is mirrored by irregularities in metal radii (Fig. 31.3) which are far greater than are found in lanthanides and probably arise from a variability in the number of electrons in the metallic bands of the actinide elements. From Ac to U, since the most stable oxidation state increases from +3 to +6, it seems likely that the sharp fall in metal radius is due to an increasing number of electrons being involved in metallic bonding. Neptunium and Pu are much the same as U but thereafter increasing metal radius is presumably a result of fewer electrons being involved in metallic bonding since it roughly parallels the reversion to a lanthanide-like preference for tervalency in the heavier actinides.

By contrast, the ionic radius in a given oxidation state falls steadily and, though the available data are less extensive, it is clear than an "actinide contraction" exists, especially for the +3 state, which is closely similar to the "lanthanide contraction" (see p. 1431).

31.2.4 *Chemical reactivity and trends*[17, 18]

The actinide metals are electropositive and reactive, apparently becoming increasingly so with atomic number, although available data are restricted largely to Th, U, and Pu. They tarnish rapidly in air, forming an oxide coating which is protective in the case of Th but less so for the other elements. Because of the self-heating associated with its radioactivity (100 g ^{239}Pu generates ~ 0.2 watts of heat) Pu is best stored in circulating dried air. All are pyrophoric if finely divided.

The metals react with most non-metals especially if heated, but resist alkali attack and are less reactive towards acids than might be expected. Concentrated HCl probably reacts most rapidly, but even here insoluble residues remain in the cases of Th (black), Pa (white), and U (black). Those of Th and U have the approximate compositions HThO(OH) and

[16] R. A. BULMAN, Some aspects of the bioinorganic chemistry of the actinides, *Coord. Chem. Revs.* **31**, 221–50 (1980).

[16a] W. H. ZACHARIASEN, On californium metal, *J. Inorg. Nucl. Chem.* **37**, 1441–2 (1975).

[17] K. W. BAGNALL, *The Actinide Elements*, Elsevier, Amsterdam, 1972, 272 pp.

[18] S. A. COTTON and F. A. HART, *The Heavy Transition Elements*, Macmillan, London, 1975, Chap. 11, pp. 220–61.

TABLE 31.2 Some properties of the actinide elements

Property	Th	Pa	U	Np	Pu	Am	Cm	Bk	Cf	Es	Fm	Md	No	Lr
Atomic number	90	91	92	93	94	95	96	97	98	99	100	101	102	103
Number of naturally occurring isotopes	1	—	3											
Most common isotopes:														
Mass number	232	231	238	237	239	241	244	249	252	253	257	256	255	256
Half-life[a]	1.40×10^{10} y (α)	3.28×10^4 y (α)	4.47×10^9 y (α)	2.14×10^6 y (α)	2.41×10^4 y (α)	432 y (α)	18.11 y (α)	314 d (β⁻)	2.65 y (α)	20.5 d (α)	100.5 d (α)	75 min (β⁺/EC)	3.1 min (α, EC)	27 s (α)
Relative nuclidic mass	232.0381	231.0359	238.0289[b]	237.0482	239.052	241.057	244.063	249.075	252.082	253.085	257.095	(256.094)	(255.093)	(256.099)
Electronic configuration, [Rn] plus	$6d^27s^2$	$5f^26d^17s^2$ or $5f^16d^27s^2$	$5f^36d^17s^2$	$5f^46d^17s^2$ or $5f^57s^2$	$5f^67s^2$	$5f^77s^2$	$5f^76d^17s^2$	$5f^97s^2$ or $5f^86d^17s^2$	$5f^{10}7s^2$	$5f^{11}7s^2$	$5f^{12}7s^2$	$5f^{13}7s^2$	$5f^{14}7s^2$	$5f^{14}6d^17s^2$
Metal radius (CN 12)[c]/pm	179	163	156	155	159	173	174	170	186±2	186±2	—	—	—	—
Ionic radius (CN 6)/pm VII	—	—	—	71	71	—	—	—	—	—	—	—	—	—
VI	—	—	73	72	74	—	—	—	—	—	—	—	—	—
V	—	78	76	75	86	85	—	—	—	—	—	—	—	—
IV	94	90	89	87	100	85	85	83	82.1	—	—	—	—	—
III	—	104	102.5	101	—	97.5	97	96	95	—	—	—	—	—
II	—	—	—	110	—	126[d]	—	—	—	—	—	—	—	—
$E^\circ(\mathrm{MO_2^{2+}/MO_2^+})$/V	—	—	0.062	1.130	0.928	—	—	—	—	—	—	—	—	—
$E^\circ(\mathrm{MO_2^+/M^{4+}})$/V	—	—	0.612	0.749	1.157	—	—	—	—	—	—	—	—	—
$E^\circ(\mathrm{M^{4+}/M^{3+}})$/V	—	−1.0	−0.607	0.147	1.006	2.181	—	—	—	—	—	—	—	—
$E^\circ(\mathrm{M^{4+}/M})$/V	—	−1.7	−1.500	−1.355	−1.272	−1.24	—	—	—	—	—	—	—	—
$E^\circ(\mathrm{M^{3+}/M})$/V	−1.899	−1.95	−1.798	−1.856	−2.031	−2.38	—	—	—	—	—	—	—	—
MP/°C	1750	1552	1130	640	640	1170	1340	986	(900)	(860)	—	—	—	—
BP/°C	4850	4227	3930	5235	(3230)	2600	—	—	—	—	—	—	—	—
ΔH_{fus}/kJ mol⁻¹	16.11	16.7	12.6	(9.46)	2.80	(10.0)	—	—	—	—	—	—	—	—
ΔH_{vap}/kJ mol⁻¹	513.7	481	417	336	343.5	238.5	—	—	—	—	—	—	—	—
ΔH_f(monatomic gas)/kJ mol⁻¹[e]	575	—	482	—	352	—	—	—	—	—	—	—	—	—
Density (25°C)/g cm⁻³	11.78	15.37	19.05	20.45	19.86	13.67	13.51	14.78	—	—	—	—	—	—
μ ohm cm	15.4	19.1	30.8	122	150	—	—	—	—	—	—	—	—	—

[a] The rate of decay by spontaneous fission increases with atomic number and is an important additional cause of instability in the later actinides (*trans*-Np).

[b] This value refers to the natural mixture of uranium isotopes, i.e. it is the atomic weight. Variations are possible because (i) some geological samples have anomalous isotopic compositions, and (ii) commercially available samples may have been depleted in ²³⁵U.

[c] For Pa, CN=10 and for U, Np, and Pu the structures are rather irregular so that the coordination number is not a precise concept.

[d] For Amᴵᴵ, radius refers to CN=8.

[e] Polymorphism is common amongst the actinides and these data refer to the form most stable at room temperature.

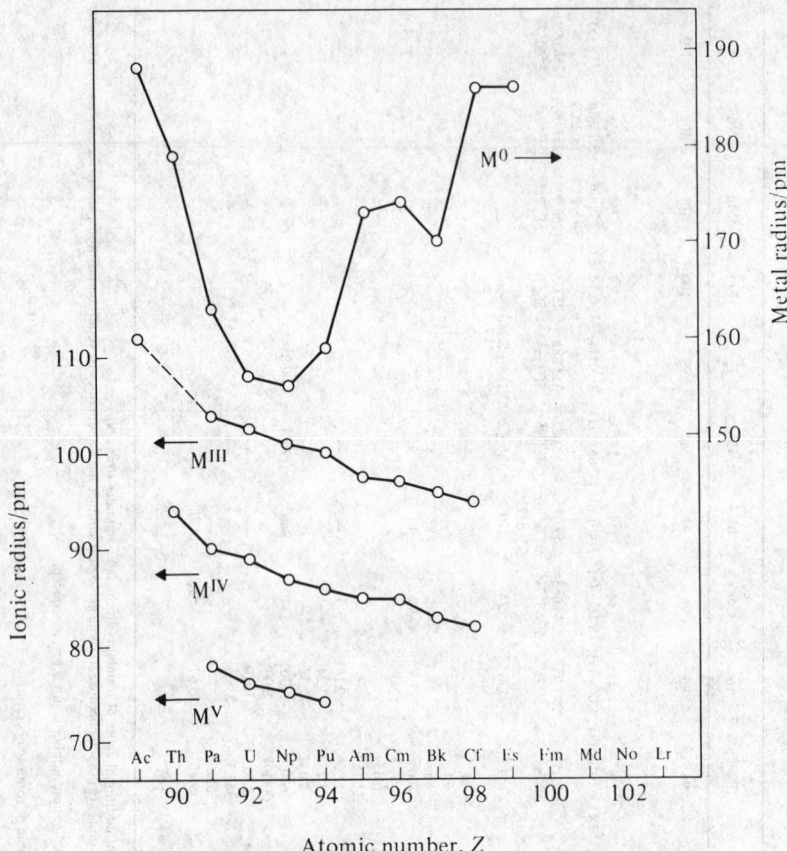

FIG. 31.3   Metal and ionic radii of Ac and the actinides.

UH(OH)$_2$. Concentrated HNO$_3$ passivates Th, U, and Pu, but the addition of F$^-$ ions avoids this and provides the best general method for dissolving these metals.

Reactions with water are complicated and are affected by the presence of oxygen. With boiling water or steam, oxide is formed on the surface of the metal and H$_2$ is liberated. Since the metals react readily with the latter, hydrides are produced which themselves react rapidly with water and so facilitate further attack on the metals.

Knowledge of the detailed chemistry of the actinides is concentrated mainly on U and, to a lesser extent, Th. Availability and safety are, of course, major problems for the remaining elements, but self-heating and radiolytic damage can be troublesome, the energy evolved in radioactive decay being far greater than that of chemical bonds. Thus in aqueous solutions of concentrations greater than 1 mg cm^{-3} (i.e. 1 g l^{-1}), isotopes with half-lives less than, say, 20 years, will produce sufficient H$_2$O$_2$ to produce appreciable oxidation or reduction where the redox behaviour of the element allows this. Fortunately, the nuclear instability which produces these problems also assists in overcoming them: by performing chemical reactions with appropriate, non-radioactive, carrier elements containing only trace amounts of the actinide in question, it is possible to detect the presence of the latter, and hence explore its chemistry because of the extreme sensitivity of radiation detectors. Such "tracer" techniques have provided remarkably extensive information particularly about the aqueous solution chemistry of the actinides.

Table 31.3 lists the known oxidation states. For the first three elements (Th, Pa, and U) the most stable oxidation state is that involving all the valence electrons, but after this the most stable becomes progressively lower until, in the second half of the series, the $+3$ state becomes dominant. Appropriate quantitative data for elements up to Am are summarized in Fig. 31.4. The highest oxidation state attainable by Th, and the only one occurring in solution, is $+4$. Data for Pa are difficult to obtain because of its propensity for hydrolysis which results in the formation of colloidal precipitates, except in concentrated acids or in the presence of complexing anions such as F^- or $C_2O_4^{2-}$. However, it is clear that $+5$ is its most stable oxidation state since its reduction to $+4$ requires rather strong reducing agents such as Zn/H^+, Cr^{II}, or Ti^{III} and the $+4$ state in solution is rapidly reoxidized to $+5$ by air. In the case of uranium the shape of the volt-equivalent versus oxidation state curve reflects the ready disproportionation of UO_2^+ into the more stable U^{IV} and UO_2^{2+}; it should also be possible for atmospheric oxygen ($\frac{1}{2}O_2 + 2H^+ + 2e^- \rightleftharpoons H_2O$, $E° = 1.229$ V)

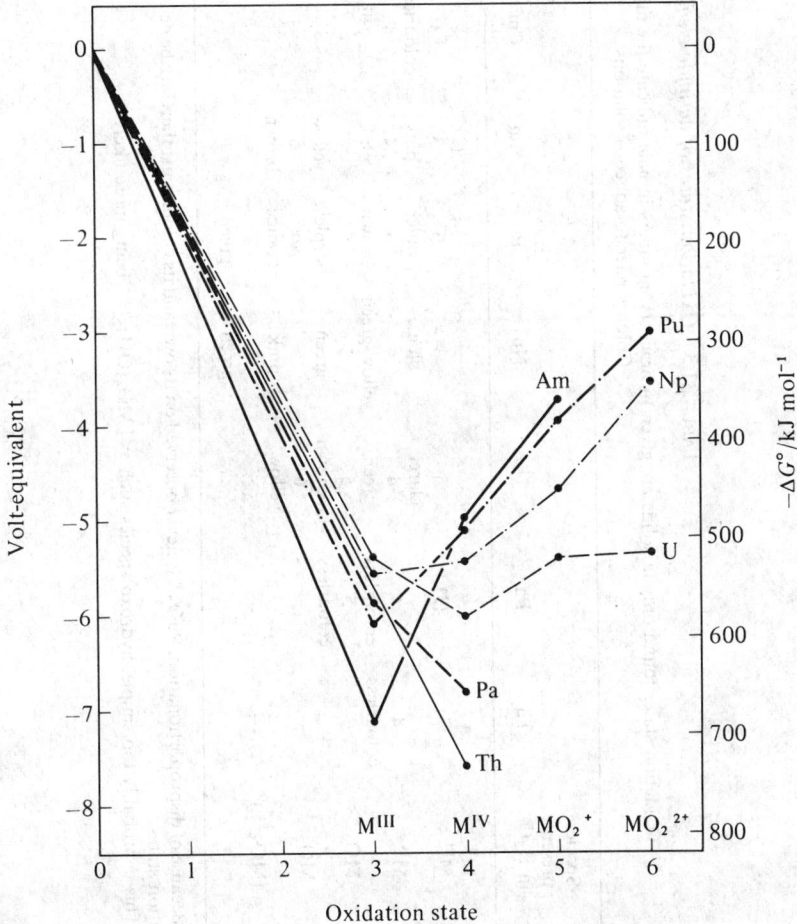

FIG. 31.4 Volt-equivalent versus oxidation state for actinide ions.

TABLE 31.3 *Oxidation states of actinide elements*

Oxidation states found only in solids are given in brackets; numbers in **bold** indicate the most stable oxidation states in aqueous solution. Colours refer to aqueous solutions

Species present in H₂O	Th	Pa	U	Np	Pu	Am	Cm	Bk	Cf	Es	Fm	Md	No	Lr
M^{II}	—					(2)	—		(2)	(2)	2	2	**2**	
M^{III}	(3)	(3)	3 claret	3 blue	3 violet	**3** pink	**3** colourless	**3**	**3**	**3**	**3**	**3**	3	**3**
M^{IV}	**4** colourless	4 colourless	4 green	4 yellow-green	**4** brown	4 pink	4 pale yellow	4	(4)					
MO_2^+		**5** colourless	5 unknown	**5** green	5 purple[a]	5 yellow								
MO_2^{2+}			**6** yellow	6 pink	6 orange	6 brown								
(MO_5^{3-})[b]				7 green	7 green									

[a] Because of disproportionation, PuO_2^+ is never observed on its own and its colour must therefore be deduced from the spectrum of a mixture involving Pu in several oxidation states.

[b] This is probably too simple, hydroxo species such as $[MO_4(OH)_2]^{3-}$ being more likely.

to oxidize U^{IV} to UO_2^{2+} though in practice this occurs only slowly. For the heavier elements the increasingly steep-sided trough indicates the increasing stability of the $+3$ state.

The redox behaviour of Th, Pa, and U is of the kind expected for d-transition elements which is why, prior to the 1940s, these elements were commonly placed respectively in groups IVA, VA, and VIA of the periodic table. Behaviour obviously like that of the lanthanides is not evident until the second half of the series. However, even the early actinides resemble the lanthanides in showing close similarities with each other and gradual variations in properties, providing comparisons are restricted to those properties which do not entail a change in oxidation state. The smooth variation with atomic number found for stability constants, for instance, is like that of the lanthanides rather than the d-transition elements, as is the smooth variation in ionic radii noted in Fig. 31.3. This last factor is responsible for the close similarity in the structures of many actinides and lanthanides, especially noticeable in the $+3$ oxidation state for which a given actinide ion is only about 4 pm larger than the corresponding Ln^{3+}.

It is evident from the above behaviour that the ionization energies of the early actinides, though not accurately known, must be lower than for the early lanthanides. This is quite reasonable since it is to be expected that, when the 5f orbitals of the actinides are beginning to be occupied, they will penetrate less into the inner core of electrons, and the 5f electrons will therefore be more effectively shielded from the nuclear charge than are the 4f electrons of the corresponding lanthanides (i.e. the relationship between 4f and 5f series may be compared to that between 3d and 4d). Because the outer electrons are less firmly held, they are all available for bonding in the actinide series as far as Np (4th member), but only for Ce (1st member) in the lanthanides, and the onset of the dominance of the $+3$ state is accordingly delayed in the actinides. That the 5f and 6d orbitals of the early actinides are energetically closer than the 4f and 5d orbitals of the early lanthanides is evidenced by the more extensive occupation of the 6d orbitals in the neutral atoms of the former (compare the outer electron configuration in Tables 31.2 and 30.2). These 5f orbitals also extend spatially further than the 4f and are able to make a covalent contribution to the bonding which is much greater than that in lanthanide compounds. This leads to a more extensive actinide coordination chemistry and to crystal-field effects, especially with ions in oxidation states above $+3$, much larger than those found for lanthanide complexes.

Table 31.4 is a list of typical compounds of the actinides and demonstrates the wider range of oxidation states compared to lanthanide compounds. High coordination numbers are still evident, and distortions from the idealized stereochemistries which are quoted are again general. However, no doubt at least partly because the early actinides have received most attention, the widest range of stereochemistries is now to be found in the $+4$ oxidation state rather than $+3$ as in the lanthanides.

31.3 Compounds of the Actinides[7, 8, 17]

Compounds with many non-metals are prepared, in principle simply, by heating the elements. Hydrides of the types AnH_2 (An = Th, Np, Pu, Am, Cm) and AnH_3 (Pa→Am), as well as Th_4H_{15} (i.e. $ThH_{3.75}$) have been so obtained but are not very stable thermally and are decidedly unstable with respect to air and moisture. Borides, carbides, silicides, and nitrides (q.v.) are mostly less sensitive chemically and, being refractory materials,

TABLE 31.4 *Oxidation states and stereochemistries of compounds of the actinides*

"An" is used as a general symbol for the actinides

Oxida-tion state	Coordina-tion number	Stereochemistry	Examples
3	6	Octahedral	$[AnCl_6]^{3-}$ (An = Np, Am, Bk)
	8	Bicapped trigonal prismatic	AnX_3 (X = Br, An = Pu→Bk; X = I, An = Pa→Pu)
	9	Tricapped trigonal prismatic	$AnCl_3$ (An = U→Cm)
4	4	Complex	$U(NPh_2)_4$
	5	Trigonal bipyramidal	$U_2(NEt_2)_8$
	6	Octahedral	$[AnX_6]^{2-}$ (An = U, Np, Pu; X = Cl, Br)
	7	Pentagonal bipyramidal	UBr_4
	8	Cubic	$[An(NCS)_8]^{4-}$ (An = Th→Pu)
		Dodecahedral	$[Th(C_2O_4)_4]^{4-}$, $[An(S_2CNEt_2)_4]$ (An = Th, U, Np, Pu)
		Square antiprismatic	$[An(acac)_4]$ (An = Th, U, Np, Pu)
	9	Tricapped trigonal prismatic	$(NH_4)_3[ThF_7]$
		Capped square antiprismatic	$[Th(tropolonate)_4(H_2O)]$
	10	Bicapped square antiprismatic	$K_4[Th(C_2O_4)_4].4H_2O$
		Complex	$[Th(NO_3)_4(OPPh_3)_2]^{(a)}$
	11	See Fig. 31.7a	$[Th(NO_3)_4(H_2O)_3].2H_2O$
	12	Icosahedral	$[Th(NO_3)_6]^{2-(a)}$
	14	Bicapped hexagonal antiprismatic	$[U(BH_4)_4]$
5	6	Octahedral	$Cs[AnF_6]$ (An = U, Np, Pu)
	7	Pentagonal bipyramidal	$PaCl_5$
	8	Cubic	$Na_3[AnF_8]$ (An = Pa, U, Np)
	9	Tricapped trigonal prismatic	$M_2[PaF_7]$ (M = NH_4, K, Rb, Cs)
6	6	Octahedral	AnF_6 (An = U, Np, Pu), UCl_6, $Cs_2[UO_2X_4]^{(b)}$ (X = Cl, Br)
	7	Pentagonal bipyramidal	$[UO_2(S_2CNEt_2)_2(ONMe_3)]^{(b)}$
	8	Hexagonal bipyramidal	$[UO_2(NO_3)_2(H_2O)_2]^{(b)}$
7	6	Octahedral	$Li_5[AnO_6]$ (An = Np, Pu)

(a) These compounds are isostructural with the corresponding compounds of Ce (see Fig. 30.5, p. 1444) and can be visualized as octahedral if each NO_3^- is considered to occupy a single coordination site.

(b) The polyhedra of these complexes are actually flattened because the two *trans* U–O bonds of the UO_2^{2+} group are shorter than the bonds to the remaining groups which form an equatorial plane.

those of Th, U, and Pu in particular have been studied extensively as possible nuclear fuels.[19] Their stoichiometries are very varied but the more important ones are the semi-metallic monocarbides, AnC, and mononitrides, AnN, all of which have the rock-salt structure: they are predominantly ionic but with supernumerary electrons in a delocalized conduction band.

31.3.1 *Oxides and chalcogenides*

Oxides of the actinides are refractory materials and, in fact, ThO_2 has the highest mp (3390°C) of any oxide. They have been extensively studied because of their importance as nuclear fuels.[19] However, they are exceedingly complicated because of the prevalence of

[19] K. NAITO and N. KAGEGASHIRA, High temperature chemistry of ceramic nuclear fuels with emphasis on nonstoichiometry, *Adv. Nucl. Sci. Tech.* **9**, 99–180 (1976).

polymorphism, nonstoichiometry, and intermediate phases. The simple stoichiometries quoted in Table 31.5 should therefore be regarded as idealized compositions.

The only anhydrous trioxide is UO_3, a common form of which (γ-UO_3) is obtained by heating $UO_2(NO_3) \cdot 6H_2O$ in air at $400°C$; six other forms are also known. Heating any of these, or indeed any other oxide of uranium, in air at $800-900°C$ yields U_3O_8 which contains pentagonal bipyramidal UO_7 units and can be used in gravimetric determinations of uranium. Reduction with H_2 or H_2S leads to a series of intermediate nonstoichiometric phases (of which U_2O_5 may be mentioned) and ending with UO_2. Pentoxides are known also for Pa and Np. Pa_2O_5 is prepared by igniting Pa^V hydroxide in air, and the nonstoichiometric Np_2O_5 by treating Np^{IV} hydroxide with ozone and heating the resulting $NpO_3 \cdot H_2O$ at $300°C$ under vacuum.

Dioxides are known for all the actinides as far as Cf. They have the fcc fluorite structure (p. 129) in which each metal atom has $CN = 8$; the most common preparative method is ignition of the appropriate oxalate or hydroxide in air. Exceptions are CmO_2 and CfO_2, which require O_2 rather than air, and PaO_2 and UO_2, which are obtained by reduction of higher oxides.

From Pu onwards, sesquioxides become increasingly stable with structures analogous to those of Ln_2O_3 (p. 1437); BkO_2 is out-of-sequence but this is presumably due to the stability of the f^7 configuration in Bk^{IV}. For each actinide the C-type M_2O_3 structure (metal $CN = 6$) is the most common but A and B types (metal $CN = 7$) are often also obtainable.

Monoxides have been prepared as surface layers on the metal for most actinides and all of them have the fcc rock-salt structure (p. 273). They are lustrous semi-metallic materials and it is clear that they feature cations in oxidation states above $+2$, with the supernumerary electrons in a delocalized conduction band, rather than being compounds of An^{II}.

The oxides are basic in character but their reactivity is usually strongly influenced by their thermal history, being much more inert if they have been ignited. Dioxides of Th, Np, and Pu are best dissolved in conc HNO_3 with added F^-, but all oxides of U dissolve readily in conc HNO_3 or conc $HClO_4$ to yield salts of UO_2^{2+}.

Hydroxides are not well-characterized but gelantinous precipitates, which redissolve in acid, are produced by the addition of alkali to aqueous solutions of the actinides. Those of Th^{IV}, Pa^V, Np^V, Pu^{IV}, Am^{III}, and Cm^{III} are stable to oxidation but lower oxidation states of these metals are rapidly oxidized. Aqueous solutions of hexavalent U, Np, and Pu yield hydrous precipitates of $AnO_2(OH)_2$, which contain AnO_2^{2+} units linked by OH bridges, but they are often formulated as hydrated trioxides $AnO_3 \cdot xH_2O$.

Actinide chalcogenides[20] can be obtained for instance by reaction of the elements, and thermal stability decreases $S > Se > Te$. Those of a given actinide differ from those of another in much the same way as do the oxides. Nonstoichiometry is again prevalent and, where the actinide appears to have an uncharacteristically low oxidation state, semi-metallic behaviour is usually observed.

[20] R. M. DELL and N. J. BRIDGER, Actinide chalcogenides and pnictides, Chap. 6, pp. 221–74, in Vol. 7, Series 1, of ref. 8. D. DAMIEN and R. BERGER, The crystal chemistry of some transuranium element chalcogenides, *Chemistry of Transuranium Elements*, pp. 109–16. (Proceedings of Moscow Symposium published as supplement to *J. Inorg. Nucl. Chem.*, 1976).

TABLE 31.5 *Oxides of the Actinides*

The most stable oxide of each element is printed in **bold**.

Formal oxidation state of metal	Th	Pa	U	Np	Pu	Am	Cm	Bk	Cf	Es
+6	—	—	UO_3 Orange-yellow	—	—	—				
			U_3O_8 Very dark green							
+5	—	Pa_2O_5 White	U_2O_5	Np_2O_5 Dark brown	—	—				
+4	**ThO_2** White	PaO_2 Black	UO_2 Dark brown	**NpO_2** Brown-green	**PuO_2** Yellow-brown	**AmO_2** Dark brown	CmO_2 Black	**BkO_2** Brown	CfO_2 Black	
+3					Pu_2O_3 Black	Am_2O_3 Red-brown	**Cm_2O_3** White	Bk_2O_3 Yellow-green	**Cf_2O_3** Pale green	**Es_2O_3**[a]
+2	—	PaO	UO	NpO	PuO	AmO	CmO		CfO	

Also printed bold: **Pa_2O_5** White.

[a] This is the only known oxide of Es. It is expected to be the most stable for this actinide but investigation of the Es/O system is hampered not only by low availability but also by the high α-activity ($t_{\frac{1}{2}}$ = 20.5 days) which causes crystals to disintegrate. Es_2O_3 was characterized by electron diffraction using microgram samples measuring only about 0.03 μm on edge.

31.3.2 *Mixed metal oxides*[21]

Alkali and alkaline earth metallates are obtained by heating the appropriate oxides, in the presence of oxygen where necessary. For instance, the reaction

$$5Li_2O + AnO_2 \xrightarrow[400-420\,°C]{O_2} Li_5AnO_6 \quad [An = Np, Pu]$$

provides a means of stabilizing Np^{VII} and Pu^{VII} in the form of isolated $[AnO_6]^{5-}$ octahedra.

By suitable adjustment of the proportions of the reactants, An^{VI} species (An = U→Am) are obtained of which the "uranates" are the best known. These are of the types $M_2^I U_2O_7$, $M_2^I UO_4$, $M_4^I UO_5$, and $M_3^{II} UO_6$ in each of which the U atoms are coordinated by 6 O atoms disposed octahedrally but distorted by the presence of 2 short *trans* U–O bonds, characteristic of the uranyl UO_2^{2+} group. In view of the earlier inclusion of U in Group VIA it is interesting to note that, unlike Mo^{VI} and W^{VI}, U^{VI} apparently does not show any tendency to form iso- or hetero-poly anions in aqueous solution.

$M^I An^V O_3$, $M_3^I An^V O_4$, and $M_7^I An^V O_6$ have been characterized for An = Pa→Am. Compounds of the first of these types have the perovskite structure (p. 1122), those of the second a defect-rock-salt structure (p. 273), and those of the third have structures based on hexagonally close-packed O atoms. In all cases, therefore, the actinide atom is octahedrally coordinated. It is also notable that magnetic and spectroscopic evidence shows that, for uranium, these compounds contain the usually unstable U^V and not, as might have been supposed, a mixture of U^{IV} and U^{VI}.

$BaAn^{IV}O_3$ (An = Th→Am) all have the perovskite structure and are obtained from the actinide dioxide. In accord with normal redox behaviour, the Pa and U compounds are only obtainable if O_2 is rigorously excluded, and the Am compound if O_2 is present. Actinide dioxides also yield an extensive series of nonstoichiometric, mixed oxide phases in which a second oxide is incorporated into the fluorite lattice of the AnO_2. The UO_2/PuO_2 system, for example, is of great importance in the fuel of fast-breeder reactors.

31.3.3 *Halides*[17, 22, 23]

The known actinide halides are listed in Table 31.6. They range from AnX_6 to AnX_2 and their distribution follows much the same trends as have been seen already in Tables 31.3 and 31.5. Thus the hexahalides are confined to the hexafluorides of U, Np, and Pu (which are volatile solids obtained by fluorinating AnF_4) and the hexachloride of U, which is obtained by the reaction of $AlCl_3$ and UF_6. All are powerful oxidizing agents and are extremely sensitive to moisture:

$$AnX_6 + 2H_2O \longrightarrow AnO_2X_2 + 4HX$$

[21] C. KELLER, Lanthanide and actinide mixed metal oxide systems with alkali and alkaline earth metals, Chap. 2, pp. 47–85, in Vol. 7, Series 1, of ref. 8; also Lanthanide and actinide mixed oxide systems of face-centred cubic symmetry, Chap. 1, pp. 1–39, in Vol. 7, Series 2, of ref. 8.

[22] D. BROWN, *Halides of the Lanthanides and Actinides*, Wiley, London, 1968, 280 pp.; also Halides, halates, perhalates, thiocyanates, selenocyanates, cyanates, and cyanides, pp. 151–217 of ref. 7.

[23] J. C. TAYLOR, Systematic features in the structural chemistry of the uranium halides, oxyhalides and related transition metal and lanthanide halides, *Coord. Chem. Rev.* **20**, 197–273 (1976).

TABLE 31.6 *Properties of actinide halides*[a]

	Th	Pa	U	Np	Pu	Am	Cm	Bk	Cf	Es
AnF_6			White 64° 6 o	Orange 54.7° 6 o	Brown 52° 6 o					
$AnCl_6$			Dark green 177° 6 o							
AnF_5		White na 7 pbp	Very pale Blue 348° 7 pbp	‡ na 7 pbp						
$AnCl_5$		Yellow 306° 7 pbp	Brown na 6 o							
$AnBr_5$		Dark red 6 o	Brown 6 o							
AnI_5		Black								
An_2F_9		Black (9)	Black (9)							
An_4F_{17}		‡			Red					
AnF_4	White 1110° 8 sa	Brown na 8 sa	Green 1036° 8 sa	Green na 8 sa	Brown 1027° 8 sa	Tan na 8 sa	Brown na 8 sa	‡ na 8 sa	Green na 8 sa	
$AnCl_4$	White 770° 8 d	Green-yellow na 8 d	Green 590° 8 d	Red-brown 517° 8 d						
$AnBr_4$	White 679° 8 d	Brown na 8 d	Brown 519° 7 pbp	Dark-red 464° 7 pbp						
AnI_4	White 570° 8 sa	Green na	Black 506° 6 ol							
AnF_3			Black decomp. 9 ttp	Purple 1425° 9 ttp	Violet 1396° 9 ttp	Pink 1395° 9 ttp	White 1406° 9 ttp	Yellow na 9 ttp	‡ na 8 btp	
$AnCl_3$			Green 837° 9 ttp	Green 800° 9 ttp	Green 767° 9 ttp	Pink 500° 9 ttp	White na 9 ttp	Green na 9 ttp	Green 575° 9 ttp	‡ na 9 ttp
$AnBr_3$			Red 727° 9 ttp	Green na 9 ttp	‡ 681° 8 btp	White na 8 btp	White 400° 8 bpt	Yellow-green na 8 btp	Pale green na 6 o	‡ na na
AnI_3	Black na na	Black na 8 btp	Black 766° 8 btp	Purple 760° 8 btp	Green (777°) 8 btp	Yellow na 8 btp	White na 6 o	Yellow na 6 o	Yellow na 6 o	
$AnCl_2$						Black 9 ttp				
$AnBr_2$						Black 8,7			Amber 8,7	
AnI_2	Gold complex					Black 7 co			Violet	

[a] Key: Colour (‡ indicates preparation but no report of colour); mp/°C (na indicates value not reported); coordination 9 ttp = tricapped trigonal prismatic; 8 d = dodecahedral; 8 sa = square antiprismatic; 8 btp = bicapped trigonal prismatic; 8,7 = mixed 8- and 7-coordination ($SrBr_2$ structure); 7 cc = capped octahedral; 7 pbp = pentagonal bipyramidal; 6 o = octahedral; 6 och = octahedral chain;[23a] 6 ol = octahedral layered.[23b]

UF_6 is important in the separation of uranium isotopes by gaseous diffusion (p. 1460).

   Pentahalides are, perhaps surprisingly, not found beyond Np (for which the pentafluoride alone is known) but all four are known for Pa. All the pentafluorides as well as $PaCl_5$ are polymeric and attain 7-coordination by means of double X-bridges between adjacent metal atoms (Fig. 31.5); by contrast UCl_5, $PaCl_5$, and $PaBr_5$ consist of halogen-bridged dimeric An_2X_{10} units, e.g. $Cl_4U(\mu\text{-}Cl)_2UCl_4$. All are very sensitive to water, the hydrolysis of the U^V halides being complicated by simultaneous disproportionation. Fluorides of intermediate compositions An_2F_9 (An = Pa, U) and An_4F_{17} (An = U, Pu) have also been reported. U_2F_9 is the best known; its black colour probably results from charge transfer between U^{IV} and U^V.[23a]

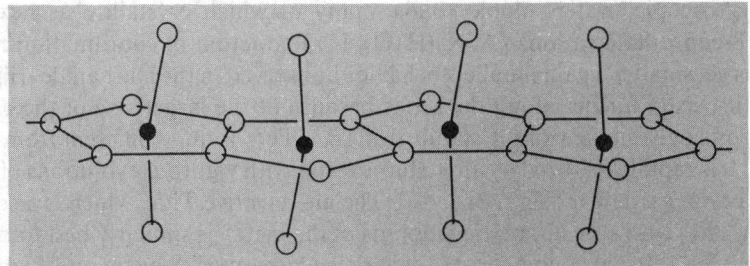

FIG. 31.5   The polymeric structure of AnF_5 (An = Pa, U, Np) and $PaCl_5$, showing the distorted pentagonal bipyramidal coordination of the metal.

   A much more extensive series is formed by the tetrahalides of which the tetrafluorides are known as far as Cf. The early tetrafluorides $ThF_4\rightarrow PuF_4$ are produced by heating the dioxides in HF in the presence of H_2 for PaF_4 in order to prevent oxidation, and in the presence of O_2 for NpF_4 and PuF_4 in order to prevent reduction. The later tetrafluorides $AmF_4\rightarrow CfF_4$ are obtained by heating the corresponding trifluoride with F_2. In all cases the metal is 8-coordinated, being surrounded by a slightly distorted square-antiprismatic array of F^- ions. The tetrachlorides (Th→Np) are prepared by heating the dioxides in CCl_4 or a similar chlorinated hydrocarbon, whereas the tetrabromides (Th→Np) and tetraiodides (Th→U) are obtained from the elements. Eight-coordination is again common (this time dodecahedral) but a reduction to 7-coordination occurs with UBr_4 and $NpBr_4$, and to octahedral coordination for UI_4.[23b] AnF_4 are insoluble in water and, for Th, U, and Pu at least, are precipitated as the hydrate $AnF_4\cdot2\frac{1}{2}H_2O$ when F^- is added to any aqueous solution of An^{IV}. $AnCl_4$, $AnBr_4$, and AnI_4 are rather hygroscopic, and dissolve readily in water and other polar solvents. An extensive coordination chemistry is based on the actinide tetrahalides and UCl_4 is one of the best-known compounds of uranium, providing the usual starting point for most studies of U^{IV} chemistry.

   The trihalides are the most nearly complete series, all members having been obtained for the actinides U→Cf, and the series could no doubt be extended. Preparative methods are varied and depend in particular on the actinide involved. For the heavier actinides (Am→Cf) heating the sesquioxide or dioxide in HX is generally applicable, but the lighter

[23a] P. G. ELLER, A. C. LARSON, J. R. PETERSEN, D. D. ENSER, and J. P. YOUNG, Crystal structures of α-UF_5 and U_2F_9 and spectral characterization of U_2F_9, *Inorg. Chim. Acta* **37**, 129–33 (1979).
   [23b] J. H. LEVY, J. C. TAYLOR, and A. B. WAUGH, Crystal structure of uranium(IV) tetraiodide by X-ray and neutron diffraction, *Inorg. Chem.* **19**, 672–4 (1980).

actinides require reducing conditions. For NpF_3 and PuF_3 the addition of H_2 to the reaction suffices, but UF_3 is best obtained by the reduction of UF_4 with metallic U or Al. Trichlorides and tribromides of these lighter actinides can be obtained by heating the metal with I_2 (U, Np) or HI (Pu). Triiodides of Th and Pa have also been prepared, the former by heating stoichiometric amounts of ThI_4 with Th metal and the latter, reportedly by heating PaI_5 in a vacuum.

With the exception of their redox properties, the actinide trihalides form a homogeneous group showing strong similarities with the lanthanide trihalides. The ionic, high-melting trifluorides are insoluble in water, from which they can be precipitated as monohydrates; at Cf^{III} (ionic radius = 95 pm) they show the same structural change from CN 9 to CN 8 as the lanthanides do at Gd^{III} (ionic radius = 93.8 pm). The other trihalides are all hygroscopic, water-soluble solids, many of which crystallize as hexahydrates featuring 8-coordinate cations $[AnX_2(H_2O)_6]^+$. Reduction in coordination number as the cations get smaller, again parallels behaviour observed in the lanthanide trihalides but of course it occurs further along the series because of the larger size of the actinides.

Not surprisingly, in view of the stability of Th^{IV}, ThI_3 is quite different from the above trihalides. It is rapidly oxidized by air, reduces water with vigorous evolutions of H_2, and is probably best regarded as $[Th^{IV}, 3I^-, e^-]$. The air-sensitive ThI_2, which is also obtained by heating ThI_4 with stoichiometric amounts of the metal, is similarly best formulated as $[Th^{IV}, 2I^-, 2e^-]$. It has a complicated layer structure and its lustre and high electrical conductivity indicate a close similarity with the diiodides of Ce, Pr, and Gd.

Halides of truly divalent americium, however, can be prepared:

$$Am + HgX_2 \xrightarrow{400-500°C} AmX_2 + Hg \ (X = Cl, Br, I)$$

Like those of Eu with which they are structurally similar, these dihalides presumably owe their existence to their f^7 configuration. $CfBr_2$ and CfI_2 are also known and it seems probable that actinide dihalides would be increasingly stable as far as No if the problems of availability, etc., were overcome.

Several oxohalides[17, 22, 23] are also known, mostly of the types $An^{VI}O_2X_2$, An^VO_2X, $An^{IV}OX_2$, and $An^{III}OX$, but they have been less thoroughly studied than the halides. They are commonly prepared by oxygenation of the halide with O_2 or Sb_2O_3, or in case of AnOX by hydrolysis (sometimes accidental) of AnX_3. As is to be expected, the higher oxidation states are formed more readily by the lighter actinides; thus AnO_2X_2, apart from the fluoro compounds, are confined to An = U. Conversely the lower oxidation states are favoured by the heavier actinides (from Am onwards).

31.3.4 *Magnetic and spectroscopic properties*[17, 24, 25]

As the actinides are a second f series it is natural to expect similarities with the lanthanides in their magnetic and spectroscopic properties. However, while previous treatments of the lanthanides (p. 1441) provide a useful starting point in discussing the actinides, important differences are to be noted. Spin–orbit coupling is again strong ($2000–4000$ cm^{-1}) but, because of the greater exposure of the 5f electrons, crystal-field

[24] T. J. MARKS, Chemistry and spectroscopy of f-element organometallics. Part II. The actinides, *Prog. Inorg. Chem.* **25**, 223–333 (1979).

[25] J. L. RYAN, Absorption spectra of actinide compounds, Chap. 9, pp. 323–67, Vol. 7, series 1, of ref. 8.

splittings are now of comparable magnitude and J is no longer such a good quantum number. Furthermore, as already mentioned (p. 1469), the energy levels of the 5f and 6d orbitals are sufficiently close for the lighter actinides at least, to render the 6d orbitals accessible. As a result, rigorous treatments of electronic properties must consider each actinide compound individually. They must allow for the mixing of "J levels" obtained from Russell–Saunders coupling and for the population of thermally accessible excited levels. Accordingly, the expression $\mu_e = g\sqrt{J(J+1)}$ is less applicable than for the lanthanides: the values of magnetic moment obtained at room temperature roughly parallel those obtained for compounds of corresponding lanthanides (see Table 30.6), but they are usually appreciably lower and are much more temperature-dependent.

The electronic spectra of actinide compounds arise from three types of electronic transition:

(i) *f→f transitions* (see p. 1442). These are orbitally forbidden, but the selection rule is partially relaxed by the action of the crystal field in distorting the symmetry of the metal ion. Because the field is stronger than for the lanthanides, the bands are more intense by about a factor of 10 and, though still narrow, are about twice as broad and are more complex than those of the lanthanides. They are observed in the visible and ultraviolet regions and produce the colours of aqueous solutions of simple actinide salts as given in Table 31.3.

(ii) *5f→6d transitions*. These are orbitally allowed and give rise to bands which are therefore much more intense than those of type (i) and are usually rather broader. They occur at lower energies than do the 4f→5d transitions of the lanthanides but are still normally confined to the ultraviolet region and do not affect the colour of the ion.

(iii) *Metal→ligand charge transfer*. These again are fully allowed transitions and produce broad, intense absorptions usually found in the ultraviolet but sometimes trailing into the visible region. They produce the intense colours which characterize many actinide complexes, especially those involving the actinide in a high oxidation state with readily oxidizable ligands.

In view of the magnitude of crystal-field effects it is not surprising that the spectra of actinide ions are sensitive to the latter's environment and, in contrast to the lanthanides, may change drastically from one compound to another. Unfortunately, because of the complexity of the spectra and the low symmetry of many of the complexes, spectra are not easily used as a means of deducing stereochemistry except when used as "fingerprints" for comparison with spectra of previously characterized compounds. However, the dependence on ligand concentration, of the positions and intensities, especially of the charge-transfer bands, can profitably be used to estimate stability constants.

31.3.5 *Complexes*[17, 18, 22, 23]

Because of the technical importance of solvent extraction, ion-exchange, and precipitation processes for the actinides, a major part of their coordination chemistry has been concerned with aqueous solutions,[26] particularly that involving uranium. It is,

[26] S. K. PATIL, V. V. RAMAKRISHNA, and M. V. RAMANIAH, Aqueous coordination complexes of neptunium, *Coord. Chem. Revs.* **25**, 133–71 (1978). J. M. CLEVELAND, Aqueous coordination complexes of plutonium, *Coord. Chem. Revs.* **5**, 101–37 (1970).

however, evident that the actinides as a whole have a much stronger tendency to form complexes than the lanthanides and, as a result of the wider range of available oxidation states, their coordination chemistry is more varied.

Oxidation state VII

This has been established only for Np and Pu, alkaline An^{VI} solutions of which can be electrolytically oxidized to give dark-green solutions probably containing species such as $[AnO_4(OH)_2]^{3-}$. Similar strongly oxidizing solutions (the more so if made acidic) are obtained when the mixed oxides Li_5AnO_6 are dissolved in water.

Oxidation state VI

Apart from UO_3 and the An^{VI} halides already discussed, this oxidation state is dominated by the dioxo, or "actinyl" AnO_2^{2+} ions which are found both in aqueous solutions and in solid compounds of U, Np, Pu, and Am. These dioxo ions retain their identity throughout a wide variety of reactions, and are present, for instance in the oxohalides AnO_2X_2. The An–O bond strength and the resistance of the group to reduction decreases in the order $U > Np > Pu > Am$. Thus yellow uranyl salts are the most common salts of uranium and are the final products when other compounds of the element are exposed to air and moisture. The nitrate is the most familiar and has the remarkable property, utilized in the extraction of U, of being soluble in nonaqueous solvents such as tributyl phosphate. On the other hand, the formation of AmO_2^{2+} requires the use of such strong oxidizing agents as peroxodisulfate, $S_2O_8^{2-}$. Similarly, whereas the oxofluorides AnO_2F_2 are known for U, Np, Pu, and Am, only U forms the corresponding oxochloride and oxobromide, Cl^- and Br^- reducing AmO_2^{2+} to Am^V species.

In aqueous solutions hydrolysis of the actinyl ions is important and such solutions are distinctly acidic. The reactions are complicated but, at least in the case of UO_2^{2+}, it appears that loss of H^+ from coordinated H_2O is followed by polymerization involving –OH– bridges and yielding species such as $[(UO_2)(OH)]^+$, $[(UO_2)_2(OH)_2]^{2+}$, and $[(UO_2)_3(OH)_5]^+$.

Actinyl ions seem to behave rather like divalent, class-a, metal ions of smaller size (or metal ions of the same size but higher charge) and, accordingly, they readily form complexes with F^- and O-donor ligands such as OH^-, SO_4^{2-}, NO_3^-, and carboxylates.[27] The $O=An=O$ groups are in all cases linear, and coordination of a further 4, 5, or 6 ligands is possible in the equatorial plane. Octahedral, pentagonal bipyramidal, and hexagonal bipyramidal geometries result;[28] some examples are shown in Fig. 31.6. These ligands lying in the plane may be neutral molecules such as H_2O, OPR_3, $OAsR_3$, py, or the anions mentioned above, many of which are bidentate.

The axial O–An bonds are clearly very strong. They cannot be protonated and are nearly always shorter than the equatorial bonds. In the case of UO_2^{2+}, for instance, it is likely that the U–O bond order is even greater than 2, since the U–O distance is only about 180

[27] U. CASELLATO, P. A. VIGATO, and M. VIDALI, Actinide complexes with carboxylic acids, *Coord. Chem. Revs.* **26**, 85–159 (1978).
[28] A. F. WELLS, *Structural Inorganic Chemistry*, 4th edn., Oxford University Press, Oxford, 1975, pp. 988–1007.

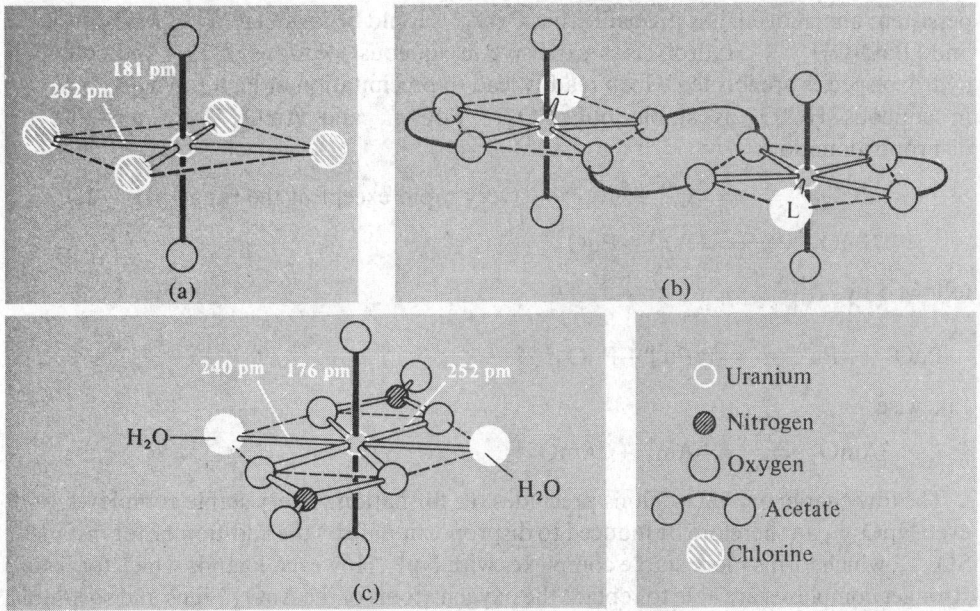

FIG. 31.6   (a) The octahedral anion in $Cs_2[UO_2Cl_4]$. (b) Pentagonal bipyramidal coordination
of U in dinuclear $[UO_2(O_2CMe)_2L]_2$ ($L=OPPh_3$, $OAsPh_3$). (c) Hexagonal bipyramidal
coordination of U in uranyl nitrate, $UO_2(NO_3)_2.6H_2O$.

pm which, in spite of the difference in the ionic radii of the metal ions ($U^{VI}=73$ pm,
$Os^{VI}=54.5$ pm), is close to that of the $Os=O$ double bond found in the isostructural,
osmyl group (175 pm, see p. 1261). It is usually assumed that combinations of appropriate
metal 6d and 5f orbitals overlap with the three p orbitals (or two p and one sp hybrid) of
each oxygen to produce one σ and two π bonds, i.e. $O\rightleftharpoons U\rightleftharpoons O$. This interpretation implies
that the change from bent to linear geometries in comparing MoO_2^{2+} (p. 1192) and
UO_2^{2+} is due to the involvement of empty f orbitals in the latter case. It has recently been
suggested[29] that the 4d orbitals of Mo and the 5f orbitals of U in fact behave quite
similarly, and that the important difference is actually the repulsive effect of the 6p
electrons of the U which were shown to favour the linear form, whereas the corresponding
4p electrons of Mo are too deeply embedded in the inert electron core to exert a similar
effect. This, however, has been disputed:[29a] matrix isolation studies have shown that the
unstable molecular species ThO_2, which is isoelectronic with UO_2^{2+}, is strongly bent
(angle O–Th–O 122°), and relativistic calculations ascribe this to back-bonding from 6d
orbitals rather than from 5f orbitals in Th, which favours bent rather than linear
geometry; the 6p orbitals are not implicated.

Oxidation state V

In aqueous solution the AnO_2^+ ions ($An=Pa\rightarrow Am$) may be formed, at least in the
absence of strongly coordinating ligands. They are linear cations like AnO_2^{2+} but are less

[29] K. TATSUMI and R. HOFFMANN, Bent cis d^0 MoO_2^{2+} vs. linear trans $d^0f^0UO_2^{2+}$: a significant role for
nonvalence 6p orbitals in uranyl, *Inorg. Chem.* **19**, 2656–8 (1980).
[29a] W. R. WADT, Why UO_2^{2+} is linear and isoelectronic ThO_2 is bent, *J. Am. Chem. Soc.* **103**, 6053–7 (1981).

persistent and, indeed, it is probable that $PaO_2{}^+$ should be formulated as $[PaO(OH)_2]^+$ and $[PaO(OH)]^{2+}$. Hydrolysis is extensive in aqueous solutions of Pa^V and colloidal hydroxo species are formed which readily lead to precipitation of $Pa_2O_5.nH_2O$. $NpO_2{}^+$ in aqueous $HClO_4$ is stable but $UO_2{}^+$, $PuO_2{}^+$, and $AmO_2{}^+$ are unstable to disproportionation:

$$2UO_2{}^+ \rightleftharpoons U^{IV} + UO_2{}^{2+} \quad \text{(very rapid except in the range pH 2–4)}$$

$$2PuO_2{}^+ \rightleftharpoons Pu^{IV} + PuO_2{}^{2+}$$

followed by

$$PuO_2{}^+ + Pu^{IV} \rightleftharpoons Pu^{III} + PuO_2{}^{2+}\dagger$$

Likewise

$$3AmO_2{}^+ \rightleftharpoons Am^{III} + 2AmO_2{}^{2+}$$

The low charge on $AnO_2{}^+$ ions precludes the formation of very stable complexes, and even $NpO_2{}^-$ can therefore be induced to disproportionate by the addition of, for instance, $SO_4{}^{2-}$, which forms more stable complexes with Np^{IV}. However, ligands which form still stronger complexes are able to replace the oxygen atoms of the $AnO_2{}^+$ ions and so inhibit disproportionation where this might otherwise occur. F^- is notable in this respect and complex ions, $AnF_6{}^-$ (An = Pa, U, Np, Pu), $PaF_7{}^{2-}$, and $PaF_8{}^{3-}$ can be precipitated from aqueous HF solutions, though nonaqueous preparations such as the oxidation of M^IF and AnF_2 by F_2 are more common for U, Np, and Pu and extend the range of complex ions to include $AnF_7{}^{2-}$ (An = U, Np, Pu) and $AnF_8{}^{3-}$ (An = U, Np). The stereochemistries of these anions[30] are dependent on the particular counter cation as well as on An, and involve 6-, 7-, 8-, and 9-coordination (see Tables 31.4 and pp. 107–8 of ref. 14). The most remarkable of these complexes are the compounds Na_3AnF_8 (An = Pa, U, Np) in which the actinide ion is surrounded by 8 F^- at the corners of a nearly perfect cube in spite of the large inter-ligand repulsions which this entails.

Finally, the alkoxides $U(OR)_5$ must be mentioned. Although easily hydrolysed, they are thermally stable and unusually resistant to disproportionation. They are usually dimeric

$$[(RO)_4U \overset{\displaystyle OR}{\underset{\displaystyle OR}{<>}} U(OR)_4]$$

and are best obtained by the reactions:

$$2[U(OEt)_4] \xrightarrow[(-2NaBr)]{Br_2 + 2NaOEt} [\{U(OEt)_5\}_2] \xrightarrow{10ROH} [\{U(OR)_5\}_2]$$

† In the diagram of volt-equivalent versus oxidation state of Pu (Fig. 31.4) the oxidation states III to VI inclusive lie on a virtually straight line. It follows that if either Pu^{IV} or Pu^V is dissolved in water, disproportionations are thermodynamically feasible, and within a matter of hours mixtures of Pu in all four oxidation states are obtained.

30 R. A. PENNEMAN, R. R. RYAN, and A. ROSENZWEIG, Structural systematics in actinide fluoride complexes, *Struct. Bonding* (Berlin) **13**, 1–52 (1973).

Oxidation state IV

This is the only important oxidation state for Th, and is one of the two for which U is stable in aqueous solution; it is moderately stable for Pa and Np also. In water Pu^{IV}, like Pu^V, disproportionates into a mixture of oxidation states III, IV, V, and VI, while Am^{IV} not only disproportionates into $Am^{III} + Am^V O_2^+$ but also (like the strongly oxidizing Cm^{IV}) undergoes rapid self-reduction due to its α-radioactivity. As a result, aqueous Am^{IV} and Cm^{IV} require stabilization with high concentrations of F^- ion. Berkelium(IV), though easily reduced, clearly has an enhanced stability, presumably due to its f^7 configuration, and the only other $+4$ ion is Cf^{IV}, found in the solids CfF_4 and CfO_2.

In aqueous solutions the hydrated cations are probably 8- or even 9-coordinated and, because they are the most highly charged ions in the actinide series, they have the greatest tendency to split-off protons and so function as quite strong acids (slightly stronger in most cases than H_2SO_3). This hydrolysis is followed by polymerization which has been most extensively studied in the case of Th. The aquated, dimeric ion, $[Th_2(OH)_2]^{6+}$, which probably involves two OH bridges, seems to predominate even in quite acidic solutions, but in solutions more alkaline than pH 3 polymerization increases considerably and eventually yields an amorphous precipitate of the hydroxide. Just before precipitation it is noticeable that the polymerization process slows down and equilibrium may take weeks to attain. The same effect is found also with Pu^{IV} where the persistence of polymers, even at acidities which would prevent their formation, can cause serious problems in the reprocessing of nuclear fuels.

The isolation of An^{IV} salts with oxoanions is limited by hydrolysis and redox compatibility. Thus, with the possible exception of $Pu(CO_3)_2$, carbonate ions furnish only basic carbonates or carbonato complexes such as $[An(CO_3)_5]^{6-}$ (An = Th, U, Pu). Stable tetranitrates are isolable only for Th, and Pu, but $Th(NO_3)_4.5H_2O$ is the most common salt of Th and is notable as the first confirmed example of 11-coordination (Fig. 31.7(a). $Pu(NO_3)_4.5H_2O$ is isomorphous, and stabilization of Pu^{IV} by strong nitric acid solutions is crucial in the recovery of Pu by solvent extraction. *O*-donor ligands such as DMSO and Ph_3PO form adducts, of which $[Th(NO_3)_4(OPPh_3)_2]$ is known to have a 10-coordinate structure like its Ce^{IV} analogue (Fig. 30.5b, p. 1444). Anionic complexes $[An(NO_3)_6]^{2-}$ (An = Th, U, Np, Pu) are also obtained, that of Th, and probably the others, having bidentate NO_3^- ions forming a slightly distorted icosahedron similar to that of the Ce^{IV} analogue (see Fig. 30.5a) $Th(ClO_4)_4.4H_2O$ is readily obtained from aqueous solutions but attempts to prepare the U^{IV} salt have produced a green explosive solid of uncertain composition. Hydrated sulfates are known for Th, U, Np, and Pu. That of Np is of uncertain hydration but the others can be prepared with both $4H_2O$ and $8H_2O$, $PuSO_4.4H_2O$ having possible use as an analytical standard.

The actinides provide a wider range of complexes in their $+4$ oxidation state than in any other, and these display the usual characteristics of actinide complexes, namely high coordination numbers and varied geometry. Complexes with halides and with *O*-donor chelating ligands are particularly numerous. The main fluoro-complexes are of the types $[AnF_5]^-$, $[AnF_6]^{2-}$, $[AnF_7]^{3-}$, $[AnF_8]^{4-}$, and $[An_6F_{31}]^{7-}$ which are nearly all known for An = Th→Bk. Their stoichiometries have not all been determined but, in some cases at least, are known to depend on the counter cation.[30] $[UF_6]^{2-}$, for instance, has a distorted cubic structure in its K^+ salt and a distorted dodecahedral structure in its Rb^+ salt.

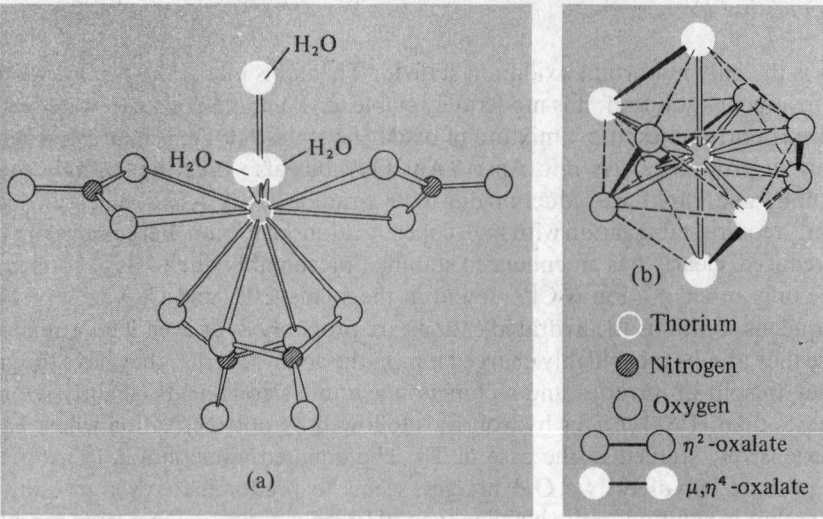

FIG. 31.7   (a) Eleven-coordinate Th in Th(NO$_3$)$_4$.5H$_2$O: average Th–O (of NO$_3$) = 257 pm;
average Th–O (of H$_2$O) = 246 pm. (b) The 10-coordinate bicapped square antiprismatic anion in
K$_4$[Th(C$_2$O$_4$)$_4$].4H$_2$O. Note that two pyramidal (267 pm) and three equatorial edges (276 pm) are
spanned by oxalate groups, but none of the longer edges of the squares (311 pm). The oxalate
groups on the pyramidal edges are actually quadridentate, being coordinated also to adjacent Th
atoms.

Several carboxylates, both simple salts and complex anions, have been prepared
often as a means of precipitating the An^{IV} ion from solution or, as in the case of simple
oxalates, in order to prepare the dioxides by thermal decomposition. In
K$_4$[Th(C$_2$O$_4$)$_4$].4H$_2$O the anion is known to have a 10-coordinate, bicapped square
antiprismatic structure (Fig. 31.7b). β-diketonates are precipitated from aqueous solutions
of An^{IV} and the ligand by addition of alkali, and nearly all are sublimable under vacuum.
[An(acac)$_4$], (An = Th, U, Np, Pu) are apparently dimorphic but both structures are
based on an 8-coordinate, distorted square antiprism.

Complexes with *S*-donor ligands are generally less stable and more liable to hydrolysis
than those with *O* donors but can be obtained if the ligand is anionic and chelating.[31]
Diethyldithiocarbamates [An(S$_2$CNEt$_2$)$_4$] (An = Th, U, Np, Pu) are the best known and
possess an almost ideal dodecahedral structure.

Finally, the borohydrides An(BH$_4$)$_4$ must be mentioned.[32] Those of Th and U were
originally prepared as part of the Manhattan project and those of Pa, Np, and Pu have
been prepared more recently, all by the general reaction

$$AnF_4 + 2Al(BH_4)_3 \longrightarrow An(BH_4)_4 + 2AlF_2BH_4$$

The compounds are isolated by sublimation from the reaction mixture. Perhaps
surprisingly the compounds fall into two quite distinct classes. Those of Np and Pu are

[31] U. CASELLATO, M. VIDALI, and P. A. VIGATO, Actinide complexes with chelating ligands containing sulfur
and amidic nitrogen donor atoms, *Coord. Chem. Revs.* **28**, 231–77 (1979).
   [32] R. H. BANKS and N. M. EDELSTEIN, Synthesis and characterization of protactinium(IV), neptunium(IV),
and plutonium(IV) borohydrides, *Lanthanide and Actinide Chemistry and Spectroscopy*, ACS Symposium,
Series 131, Am. Chem. Soc., Washington, 1980, pp. 331–48.

unstable, volatile, monomeric liquids which at low temperatures crystallize with the 12-coordinate structure of $Zr(BH_4)_4$ (Fig. 21.9, p. 1131). The borohydrides of Th, Pa, and U, on the other hand, are thermally more stable and less reactive solids. They possess a curious helical polymeric structure in which each An is surrounded by 6 BH_4^- ions, 4 being bridging groups attached by 2 H atoms and 2 being *cis* terminal groups attached by 3 H atoms. The coordination number of the actinide is therefore 14 and the stereochemistry may be described as bicapped hexagonal antiprismatic.

Oxidation state III

This is the only oxidation state which, with the possible exceptions of Th and Pa, is displayed by all actinides. From U onwards, its resistance to oxidation in aqueous solution increases progressively with increase in atomic number and it becomes the most stable oxidation state for Am and subsequent actinides (except No for which the f^{14} configuration confers greater stability on the $+2$ state).

U^{III} can be obtained by reduction of UO_2^{2+} or U^{IV}, either electrolytically or with Zn amalgam,[33] but is thermodynamically unstable to oxidation not only by O_2 and aqueous acids but by pure water also.† It is nevertheless possible to crystallize double sulfates or double chlorides from aqueous solution and these can then be used to prepare other U^{III} complexes in nonaqueous solvents. Crystallographic data are lacking but a number of cationic amide complexes, for instance, are known for which infrared evidence suggests[34] the low symmetry coordination of 8 oxygen atoms to each uranium atom.

Instability also limits the number of complexes of Np^{III} and Pu^{III} but for Am^{III} the number so far prepared is apparently limited mainly by unavailability of the element. As has already been pointed out, the problem becomes still more acute as the series is traversed. While it is clear that lanthanide-like dominance by the tervalent state occurs with the actinides after Pu, the experimental evidence though compelling, is understandably sparse, being largely restricted to solvent extraction and ion-exchange behaviour.[35]

Oxidation state II

This state is found for the six elements Am and Cf→No, though in aqueous solution only for Fm, Md, and No. However, for No, alone amongst all the f-series elements, it is the normal oxidation state in aqueous solution. The greater stabilization of the $+2$ at the end of the actinides as compared to that at the end of the lanthanides which this implies, has

---

† In pure water the activity of H^+ is only 10^{-7} mol dm^{-3}, and E for the $2H^+/H_2$ couple consequently falls to -0.414 V compared to $E° = 0$. However, $E°$ for U^{4+}/U^{3+} is even more negative (-0.607 V) and U^{III} will accordingly reduce water.

[33] R. BARNARD, J. I. BULLOCK, B. J. GELLATLY, and L. F. LARKWORTHY, The chemistry of the trivalent actinides. Part 2. Uranium(III) double chlorides and some complexes with oxygen donor ligands, *JCS Dalton* (1972). 1932–8. J. Drożdżynski, A method of preparation of uranium ($+3$) compounds from solutions, *Inorg. Chim. Acta* **32**, L83–L85 (1979).

[34] J. I. BULLOCK, A. E. STOREY, and P. THOMPSON, The chemistry of the trivalent actinoids. Part 5. Uranium(III) complexes with bidentate organic amides, *JCS Dalton*, 1979, 1040–4.

[35] See, for instance, E. K. HULET, Chemistry of the heaviest actinides: fermium, mendelevium, nobelium, and lawrencium, ref. 32, pp. 239–63.

been taken[35] to indicate a greater separation between the 5f and 6d than between the 4f and 5d orbitals at the ends of the two series. This is the reverse of the situation found at the beginnings of the series (p. 1469).

Reports of the observation of the $+1$ oxidation state in aqueous solutions of Md have not been substantiated despite attempts in several major laboratories, and it has been concluded that Md^I does not exist in either aqueous or ethanolic solutions.[36]

31.3.6 *Organometallic compounds*[24]

The growth of organoactinide chemistry, like that of organolanthanide chemistry, is comparatively recent. Attempts in the 1940s to prepare volatile carbonyls and alkyls of uranium for isotopic separations were unsuccessful† and subsequent work has mainly centred on cyclopentadienyls and, to a lesser extent, cyclooctatetraenyls; σ-bonded alkyl and aryl derivatives of the cyclopentadienyls have also been obtained. In general, these compounds are thermally stable, sublimable, but extremely air-sensitive solids which are sometimes water-sensitive also. Their bonding is evidently more covalent than that in organolanthanides, presumably because of involvement of 5f orbitals.

The cyclopentadienyls are of the three main types $[An^{III}(C_5H_5)_3]$, $[An^{IV}(\eta^5\text{-}C_5H_5)_4]$, and derivatives of the type $[An^{IV}(\eta^5\text{-}C_5H_5)_3X]$ where X is a halogen atom, an alkyl or alkoxy group, or BH_4.

(a) $[An^{III}(C_5H_5)_3]$ (An = U→Cf): the uranium compound is prepared from UCl_3 and $K(C_5H_5)$ but the others are best made by the reaction:

$$2AnCl_3 + 3Be(C_5H_5)_2 \xrightarrow{65^\circ C} 2An(C_5H_5)_3 + 3BeCl_2$$

All readily form adducts and, though complete structural data are lacking, X-ray powder diffraction patterns suggests that the tris cyclopentadienyls are isostructural with $[Sm(C_5H_5)_3]$, which incorporates both η^5 and bridging ring systems in a rather complicated manner (see p. 1448).

(b) $[An^{IV}(C_5H_5)_4]$ (An = Th→Np): the Pa compound is prepared by treating $PaCl_4$ with $Be(C_5H_5)_2$ but the general method of preparation is:

$$AnCl_4 + 4K(C_5H_5) \xrightarrow[\text{in } C_6H_6]{\text{reflux}} [An(C_5H_5)_4] + 4KCl$$

$[U(C_5H_5)_4]$ contains four identical η^5 rings arranged tetrahedrally around the metal atom (Fig. 31.8). The corresponding compounds of Th, Pa, and Np are probably the same since all four compounds have very similar nmr and ir spectra.

(c) Halide derivatives: the most plentiful are of the type $[An^{IV}(C_5H_5)_3X]$ (An = Th, U, Np); they can be prepared by the general reaction:

$$AnX_4 + 3M^I(C_5H_5) \longrightarrow [An(C_5H_5)_3X] + 3M^IX$$

† As with the lanthanides, simple carbonyls of uranium have since been obtained in argon matrices quenched to 4 K.

[36] E, K. HULET, R. W. LOUGHEED, P. A. BAISDEN, J. H. LANDRUM, J. F. WILD, and R. F. D. LUNDQVIST, Non-observation of monovalent Md, *J. Inorg. Nucl. Chem.* **41**, 1743–7 (1979). K. SAMHOUN, F. DAVID, R. L. HAHN, G. D. O'KELLEY, J. R. TARRANT, and D. E. HOBART, Electrochemical study of mendelevium in aqueous solution: no evidence of monovalent ions, *J. Inorg. Nucl. Chem.* **41**, 1749–54 (1979). F. DAVID, K. SAMHOUN, E. K. HULET, P. A. BAISEN, R. DOUGAN, J. H. LANDRUM, R. W. LOUGHEED, and J. F. WILD, Radiopolarography of mendelevium in aqueous solution, *J. Inorg. Nucl. Chem.* **43**, 2941–5 (1981).

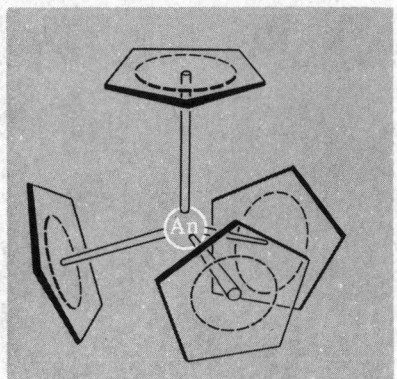

FIG. 31.8   Structure of $[An(\eta^5\text{-}C_5H_5)_4]$ showing the tetrahedral arrangement of the four rings around the metal atom.

Indeed, the first report of an organoactinide was that of the pale brown $[U(C_5H_5)_3Cl]$ by L. T. Reynolds and G. Wilkinson in 1956. They showed that, unlike $Ln(C_6H_5)_3$, this compound does not yield ferrocene on reaction with $FeCl_2$, suggesting greater covalency in the bonding of $C_5H_5^-$ to U^{IV} than to Ln^{III}.

Replacement of Cl in $[U(C_5H_5)_3Cl]$ and $[Th(C_5H_5)_3Cl]$, by other halogens or by alkoxy, alkyl, aryl, or BH_4 groups, provides the most extensive synthetic route in this field. An essentially tetrahedral disposition of three $(\eta^5\text{-}C_5H_5)$ rings and the fourth group around the metal appears to be general. The alkyl and aryl derivatives $[An(\eta^5\text{-}C_5H_5)_3R]$ $(An = Th, U)$, are of interest as they provide a means of investigating the An–C σ bond, and mechanistic studies of their thermal decomposition (thermolysis) have been prominent. The precise mechanism is not yet certain but it is clearly not β elimination† since the eliminated molecule is RH, the H of which originates from a cyclopentadienyl ring. The decomposition of the Th compounds are cleaner than those of U and a crystalline product has been isolated[37] from the thermolysis at 170°C of a solution of $[Th(\eta^5\text{-}C_5H_5)_3Bu^n]$. This product is a dimer with the 2 Th atoms bridged by a pair of $(\eta^5,\eta^1\text{-}C_5H_5)$ rings, i.e. $[Th(\eta^5\text{-}C_5H_5)_2\text{-}\mu\text{-}(\eta^5,\eta^1\text{-}C_5H_5)]_2$. This remarkable bridge system is like that in niobocene (Fig. 22.9, p. 1165) but each Th has two additional $(\eta^5\text{-}C_5H_5)$ rings instead of one $(\eta^5\text{-}C_5H_5)$ and an H atom.

The complexes $[An(\eta^8\text{-}C_8H_8)_2]$ of cyclooctatetraene (COT) have been prepared for $An = Th \rightarrow Pu$ by the reactions:

$$AnCl_4 + 2K_2C_8H_8 \xrightarrow{\text{THF}} [An(C_8H_8)_2] + 4KCl \quad (An = Th \rightarrow Np)$$

and

$$[NEt_4]_2[PuCl_6] + 2K_2C_8H_8 \xrightarrow{\text{THF}} [Pu(C_8H_8)_2] + 4KCl + 2[NEt_4]Cl$$

† In "β elimination" an H atom on a β-carbon atom is transferred to the metal and an olefin (alkene) is eliminated (see p. 348).

37 E. C. BAKER, K. N. RAYMOND, T. J. MARKS, and W. A. WACHTER, Isolation and structural characterization of a μ-di(η^5:η^1-cyclopentadienyl) dithorium(IV) complex *J. Am. Chem. Soc.* **96**, 7586–8 (1974).

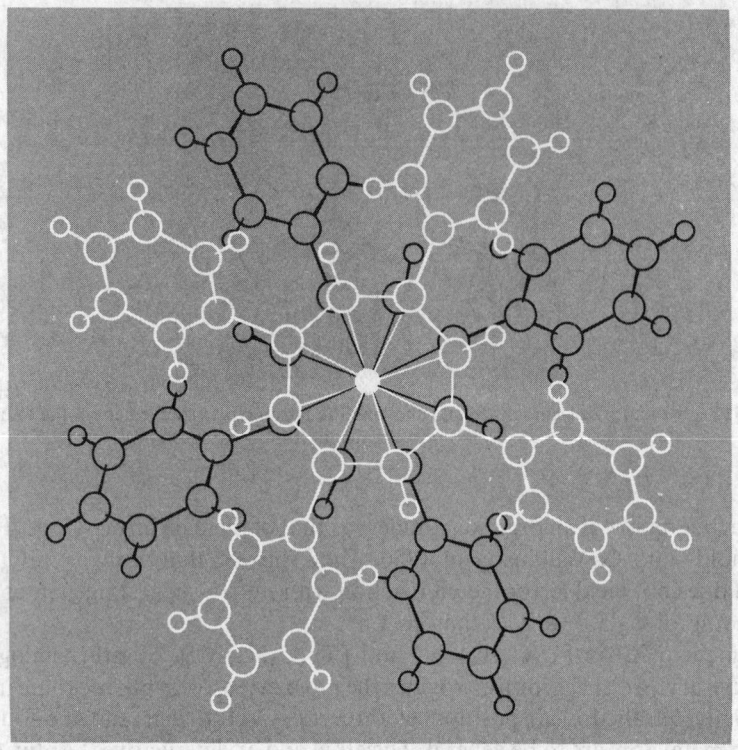

FIG. 31.9   The structure of $[U(\eta^8\text{-}C_8H_4Ph_4)_2]$.

followed by sublimation under vacuum. They are "sandwich" molecules with parallel and eclipsed rings (see Fig. 8.41, p. 376). This structure is strikingly similar to that of ferrocene (Fig. 8.37, p. 372), and extensive discussions on the nature of the metal-ring bonding suggests that this too is very similar.[38] In order to emphasise these resemblances with the d-series cyclopentadienyls, the names "uranocene", etc., have been coined. Although thermally stable, these compounds are extremely sensitive to air and, except for uranocene, are also decomposed by water. However, uranocene can be made more air-stable by use of sufficiently bulky substituents, and 1,3,5,7-tetraphenylcyclo-octatetraene yields the completely air-stable $[U(\eta^8\text{-}C_8H_4Ph_4)_2]$, in which the parallel ligands are virtually eclipsed but the phenyl substituents staggered and rotated on average 42° out of the C_8 ring plane[39] (Fig. 31.9).

[38] K. D. WARREN, Ligand field theory of f-orbital sandwich complexes, *Struct. Bonding* (*Berlin*) **33**, 97–138 (1977).

[39] L. K. TEMPLETON, D. H. TEMPLETON, and R. WALKER, Molecular structure and disorder in crystals of octaphenyluranocene, $U[C_8H_4(C_6H_5)_4]_2$, *Inorg. Chem.* **15**, 3000–3 (1976).

Appendix 1
Atomic Orbitals

THE spacial distribution of electron density in an atom is described by means of atomic orbitals $\psi(r, \theta, \phi)$ such that for a given orbital ψ the function $\psi^2 dv$ gives the probability of finding the electron in an element of volume dv at a point having the polar coordinates r, θ, ϕ. Each orbital can be expressed as a product of two functions, i.e. $\psi_{n,l,m}(r, \theta, \phi) = R_{n,l}(r)A_{l,m}(\theta, \phi)$, where

(a) $R_{n,l}(r)$ is a radial function which depends only on the distance r from the nucleus (independent of direction) and is defined by the two quantum numbers n, l;

(b) $A_{l,m}(\theta, \phi)$ is an angular function which is independent of distance but depends on the direction as given by the angles θ, ϕ; it is defined by the two quantum numbers l, m.

Normalized radial functions for a hydrogen-like atom are given in Table A1.1 and plotted graphically in Fig. A1.1 for the first ten combinations of n and l. It will be seen that the radial functions for 1s, 2p, 3d, and 4f orbitals have no nodes and are everywhere of the same sign (e.g. positive). In general $R_{n,l}(r)$ becomes zero $(n-l-1)$ times between r equals 0 and ∞. The probability of finding an electron at a distance r from the nucleus is given by $4\pi R_{n,l}^2(r)r^2 dr$, and this is also plotted in Fig. A1.1. However, the probability of finding an electron frequently depends also on the direction chosen. The probability of finding an electron in a given direction, independently of distance from the nucleus, is given by the square of the angular dependence function $A_{l,m}^2(\theta, \phi)$. The normalized functions $A_{l,m}(\theta, \phi)$ are listed in Table A1.2 and illustrated schematically by the models in Fig. A1.2. It will be seen that for s orbitals $(l=0)$ the angular dependence function A is constant, independent of θ, and ϕ, i.e. the function is spherically symmetrical. For p orbitals $(l=1)$ A comprises two spheres in contact, one being positive and one negative, i.e. there is one planar node. The d and f functions $(l=2, 3)$ have more complex angular dependence with 2 and 3 nodes respectively.

Appendix 1

Table A1.1 *Normalized radial functions $R_{n,l}(r)$ for hydrogen-like atoms*

$$R_{n,l}(r) = -\sqrt{\frac{4(n-l-1)!Z^3}{n^4[(n+l)!]^3 a_0^3}} \times \left(\frac{2Zr}{a_0 n}\right)^l L_{n+1}^{2l+1}\left(\frac{2Zr}{a_0 n}\right) \times e^{-Zr/a_0 n}$$

Orbital	n	l	$R_{n,l}$ =	Constant	×	Polynomial	×	Expon.
1s	1	0	$R_{1,0}$	$2(Z/a_0)^{3/2}$		1		e^{-Zr/a_0}
2s	2	0	$R_{2,0}$	$\dfrac{(Z/a_0)^{3/2}}{2\sqrt{2}}$		$\left(2-\dfrac{Zr}{a_0}\right)$		$e^{-Zr/2a_0}$
2p	2	1	$R_{2,1}$	$\dfrac{(Z/a_0)^{3/2}}{2\sqrt{6}}$		$\dfrac{Zr}{a_0}$		$e^{-Zr/2a_0}$
3s	3	0	$R_{3,0}$	$\dfrac{2(Z/a_0)^{3/2}}{81\sqrt{3}}$		$\left(27-18\dfrac{Zr}{a_0}+2\dfrac{Z^2r^2}{a_0^2}\right)$		$e^{-Zr/3a_0}$
3p	3	1	$R_{3,1}$	$\dfrac{4(Z/a_0)^{3/2}}{81\sqrt{6}}$		$\left(6\dfrac{Zr}{a_0}-\dfrac{Z^2r^2}{a_0^2}\right)$		$e^{-Zr/3a_0}$
3d	3	2	$R_{3,2}$	$\dfrac{4(Z/a_0)^{3/2}}{81\sqrt{30}}$		$\dfrac{Z^2r^2}{a_0^2}$		$e^{-Zr/3a_0}$
4s	4	0	$R_{4,0}$	$\dfrac{(Z/a_0)^{3/2}}{768}$		$\left(192-144\dfrac{Zr}{a_0}+\right.$ $\left.24\dfrac{Z^2r^2}{a_0^2}-\dfrac{Z^3r^3}{a_0^3}\right)$		$e^{-Zr/4a_0}$
4p	4	1	$R_{4,1}$	$\dfrac{(Z/a_0)^{3/2}}{265\sqrt{15}}$		$\left(80\dfrac{Zr}{a_0}-20\dfrac{Z^2r^2}{a_0^2}+\dfrac{Z^3r^3}{a_0^3}\right)$		$e^{-Zr/4a_0}$
4d	4	2	$R_{4,2}$	$\dfrac{(Z/a_0)^{3/2}}{768\sqrt{5}}$		$\left(12\dfrac{Z^2r^2}{a_0^2}-\dfrac{Z^3r^3}{a_0^3}\right)$		$e^{-Zr/4a_0}$
4f	4	3	$R_{4,3}$	$\dfrac{(Z/a_0)^{3/2}}{768\sqrt{35}}$		$\dfrac{Z^3r^3}{a_0^3}$		$e^{-Zr/4a_0}$

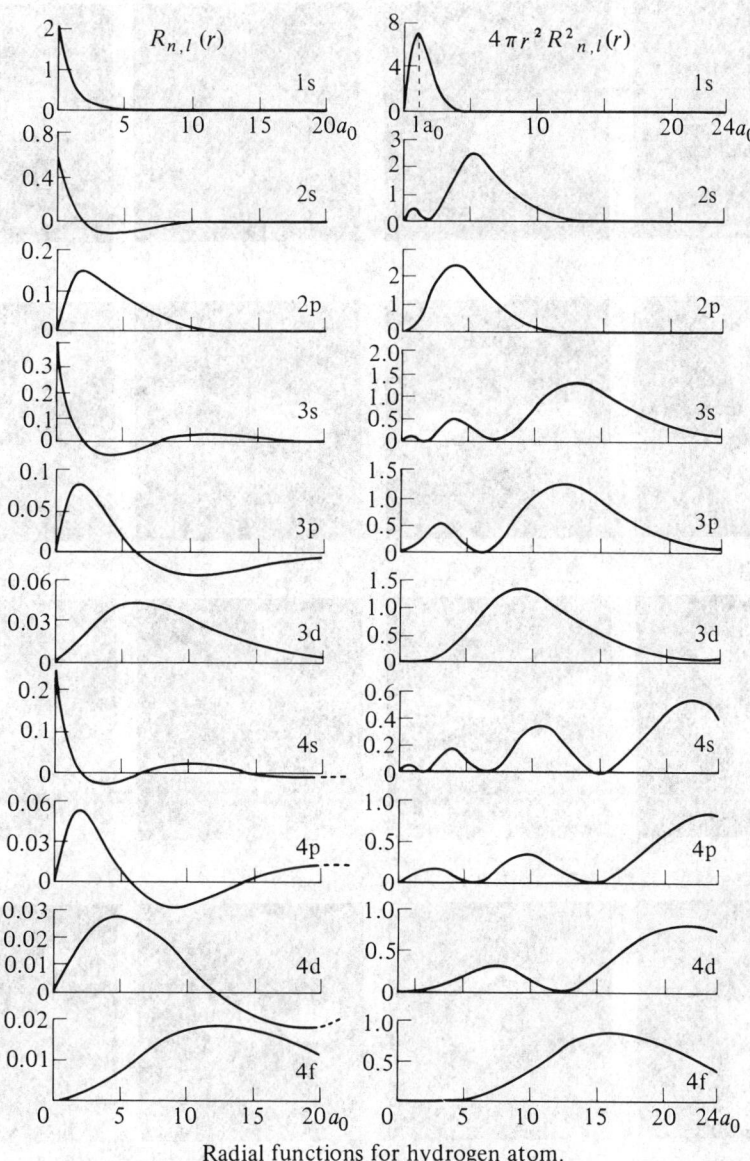

Radial functions for hydrogen atom.

FIG. A1.1   Radial functions for a hydrogen atom. (Note that the horizontal scale is the same in each graph but the vertical scale varies by as much as a factor of 100. The Bohr radius $a_0 = 52.9$ pm.)

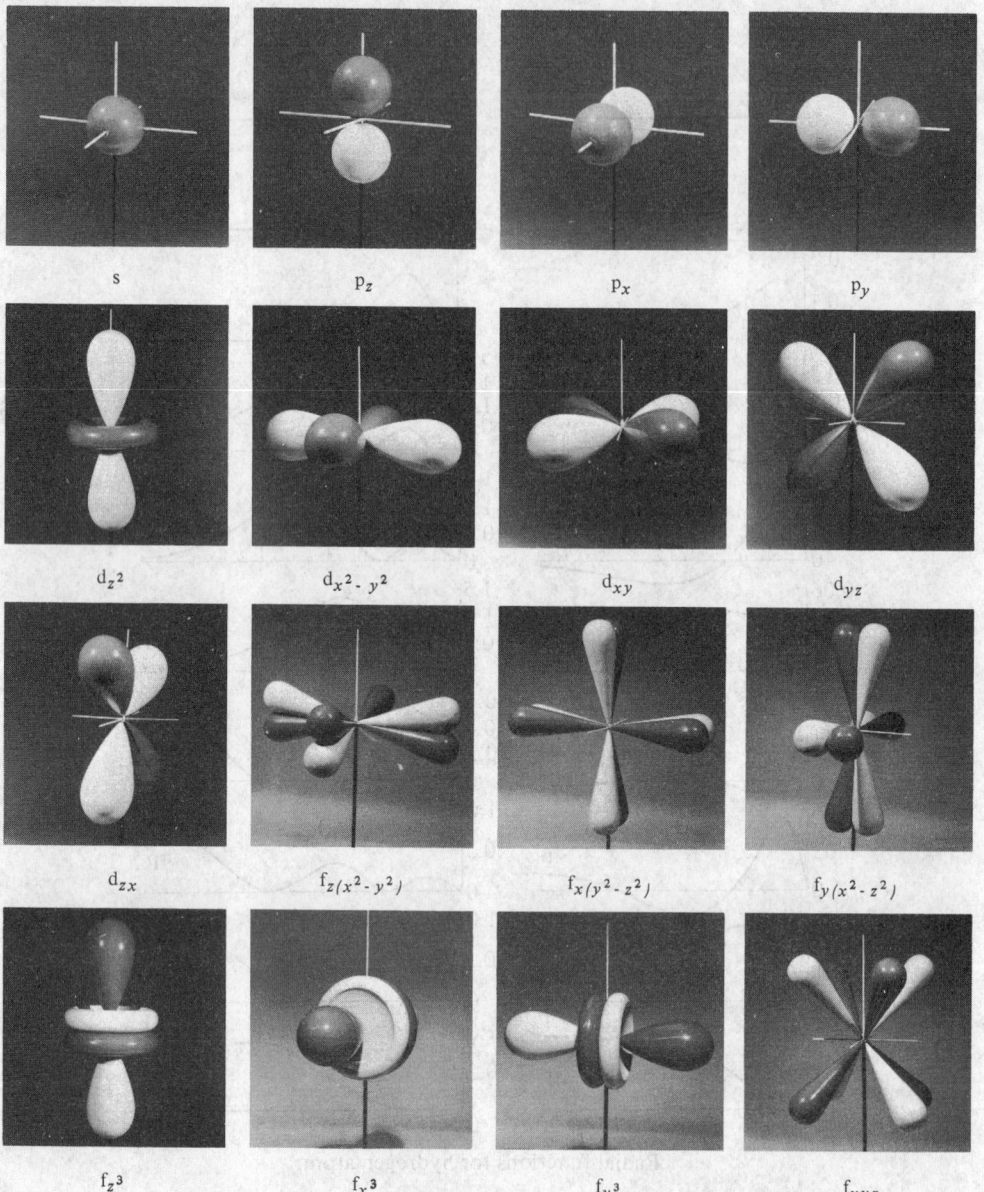

FIG. A1.2 Models schematically illustrating the angular dependence functions $A_{l,m}(\theta, \phi)$. There is no unique way of representing the angular dependence functions of all seven f orbitals. An alternative to the set shown is one f_{z^3}, three $f_{xz^2}, f_{yx^2}, f_{zy^2}$, and three $f_{x(x^2-3y^2)}, f_{y(y^2-3z^2)}$, and $f_{z(z^2-3x^2)}$.

TABLE A1.2   *Normalized angular dependence functions,* $A_{l,m}(\theta,\phi)=\Theta_{l,m}(\theta)\Phi_m(\phi)$

Orbital	Angular dependence function	Orbital	Angular dependence function
s	$\dfrac{1}{2\sqrt{\pi}}$		
p_z	$\dfrac{\sqrt{3}}{2\sqrt{\pi}}\cos\theta$		
p_x	$\dfrac{\sqrt{3}}{2\sqrt{\pi}}\sin\theta\cos\phi$	f_{z^3}	$\dfrac{\sqrt{7}}{4\sqrt{\pi}}(5\cos^3\theta-3\cos\theta)$
p_y	$\dfrac{\sqrt{3}}{2\sqrt{\pi}}\sin\theta\sin\phi$	f_{z^2x}	$\dfrac{\sqrt{42}}{8\sqrt{\pi}}(5\cos^2\theta-1)\sin\theta\cos\phi$
d_{z^2}	$\dfrac{\sqrt{5}}{4\sqrt{\pi}}(3\cos^2\theta-1)$	f_{z^2y}	$\dfrac{\sqrt{42}}{8\sqrt{\pi}}(5\cos^2\theta-1)\sin\theta\sin\phi$
$d_{x^2-y^2}$	$\dfrac{\sqrt{15}}{4\sqrt{\pi}}\sin^2\theta(2\cos^2\phi-1)$	$f_{z(x^2-y^2)}$	$\dfrac{\sqrt{105}}{4\sqrt{\pi}}\cos\theta\sin^2\theta(2\cos^2\phi-1)$
d_{zx}	$\dfrac{\sqrt{15}}{2\sqrt{\pi}}\cos\theta\sin\theta\cos\phi$	f_{zxy}	$\dfrac{\sqrt{105}}{2\sqrt{\pi}}\cos\theta\sin^2\theta\cos\phi\sin\phi$
d_{zy}	$\dfrac{\sqrt{15}}{2\sqrt{\pi}}\cos\theta\sin\theta\sin\phi$	f_{x^3}	$\dfrac{\sqrt{70}}{8\sqrt{\pi}}\sin^3\theta(4\cos^3\phi-3\cos\phi)$
d_{xy}	$\dfrac{\sqrt{15}}{2\sqrt{\pi}}\sin^2\theta\sin\phi\cos\phi$	f_{y^3}	$\dfrac{\sqrt{70}}{8\sqrt{\pi}}\sin^3\theta(3\sin\phi-4\sin^3\phi)$

Appendix 2
Symmetry Elements, Symmetry Operations, and Point Groups

An object has *symmetry* when certain parts of it can be interchanged with others without altering either the identity or the apparent orientation of the object. For a discrete object such as a molecule *5 elements of symmetry* can be envisaged:

> *axis* of symmetry, C;
> *plane* of symmetry, σ;
> *centre of inversion, i*;
> *improper axis* of symmetry, S; and
> *identity, E*.

These elements of symmetry are best recognized by performing various *symmetry operations*, which are geometrically defined ways of exchanging equivalent parts of a molecule. The 5 symmetry operations are:

> C_n, *rotation* of the molecule about a symmetry *axis* through an angle of $360°/n$; n is called the *order* of the rotation (twofold, threefold, etc.);
> σ reflection of all atoms through a *plane* of the molecule;
> i, inversion of all atoms through a *point* of the molecule;
> S_n, *Rotation* of the molecule through an angle $360°/n$ *followed by reflection* of all atoms through a plane perpendicular to the axis of rotation; the combined operation (which may equally follow the sequence *reflection then rotation*) is called *improper rotation*;
> E, the *identity* operation which leaves the molecule unchanged.

The rotation axis of highest order is called the *principal axis* of rotation; it is usually placed in the vertical direction and designated the z-axis of the molecule. Planes of reflection which are perpendicular to the principal axis are called *horizontal planes* (h). Planes of reflection which contain the principal axis are called *vertical planes* (v), or *dihedral planes* (d) if they bisect 2 twofold axes.

The complete set of symmetry operations that can be performed on a molecule is called the *symmetry group* or *point group* of the molecule and the *order* of the point group is the number of symmetry operations it contains. Table A2.1 lists the various point groups, together with their elements of symmetry and with examples of each. It is instructive to

TABLE A2.1   *Point groups*

Point group	Elements of symmetry	Examples
C_1	E	CHFClBr
C_s	E, σ	SO_2FBr, HOCl, BFClBr
C_i	E, i	CHClBr–CHClBr (staggered)
C_2	E, C_2	H_2O_2, *cis*-[Co(en)$_2$X$_2$]
C_3	E, C_3	PPh_3 (propeller)
C_{2v}	$E, C_2, 2\sigma_v$	H_2O (V-Shaped), H_2CO (Y-shaped), ClF_3 (T-shaped), SF_4 (see-saw), SiH_2Cl_2, *cis*-[Pt(NH$_3$)$_2$Cl$_2$], C_6H_5Cl
C_{3v}	$E, C_3, 3\sigma_v$	GeH_3Cl, PCl_3
C_{4v}	$E, C_4, 4\sigma_v$	SF_5Cl, IF_5, $XeOF_4$
C_{5v}	$E, C_5, 5\sigma_v$	[Ni(η^5-C$_5$H$_5$)(NO)]
C_{6v}	$E, C_6, 6\sigma_v$	[Cr(η^6-C$_6$H$_6$)(η^6-C$_6$Me$_6$)]
$C_{\infty v}$	$E, C_v, \infty\sigma_v$	NO, HCN, COS
C_{2h}	E, C_2, σ_h, i	*trans*-N$_2$F$_4$
C_{3h}	E, C_3, σ_h, i	$B(OH)_3$
C_{4h}	E, C_4, σ_h, i	[Re$_2$(μ, η^2-SO$_4$)$_4$]
D_3	$E, C_3, 3C_2'$	trischelates[M(chel)$_3$], C_2H_6 *(gauche)*
D_{2d}	$E, C_2, 2C_2', 2\sigma_d, S_4$	B_2Cl_4 (vapour, staggered), As_4S_4
D_{3d}	$E, C_3, 3C_2', 3\sigma_d, i, S_6$	$R_3W\equiv WR_3$ (staggered)
D_{4d}	$E, C_4, 4C_2', 4\sigma_d, S_8$	S_8 (crown), *closo*-B$_{10}$H$_{10}$$^{2-}$
D_{2h}	$E, C_2, 2C_2', 2\sigma_v, \sigma_h, i$	B_2Cl_4 (planar), B_2H_6, *trans*-[Pt(NH$_3$)$_2$Cl$_2$], *trans*-[Co(NH$_3$)$_2$Cl$_2$Br$_2$]$^-$, 1,4-C$_6$H$_4$Cl$_2$
D_{3h}	$E, C_3, 3C_2', 3\sigma_v, \sigma_h, S_3$	BCl_3, PF_5, $B_3N_3H_6$, [ReH$_9$]$^{2-}$
D_{4h}	$E, C_4, 4C_2', 4\sigma_v, \sigma_h, i, S_4$	XeF_4, PtCl$_4$$^-$, *trans*-[Co(NH$_3$)$_4Cl_2$]$^+$, [Re$_2Cl_8$]$^{2-}$, *closo*-1,6-C$_2B_4H_6$
D_{5h}	$E, C_5, 5C_2', 5\sigma_v, \sigma_h, S_5$	[Fe(η^5-C$_5$H$_5$)$_2$]eclipsed, B$_7$H$_7$$^{2-}$, IF$_7$
D_{6h}	$E, C_6, 6C_2', 6\sigma_v, \sigma_h, i, S_6$	C_6H_6
$D_{\infty h}$	$E, C_v, \infty C_2', \infty\sigma_v, i$	Cl_2, CO_2
S_4	E, S_4	*cyclo*-Cl$_4$B$_4$N$_4$R$_4$
T_d	$E, 4C_3, 6\sigma_d, 3S_4$	SiF_4, B_4Cl_4, [Ni(CO)$_4$], [Ir$_4$(CO)$_{12}$]
T_h	$E, 4C_3, 3C_2, 3\sigma_h, i, 4S_6$	[Co(NO$_2$)$_6$]$^{3-}$ *(trans* NO$_2$ groups eclipsed) [M(η^2-NO$_3$)$_6$]$^{n-}$, [W(NMe$_2$)$_6$]
O_h	$E, 3C_4, 4C_3, 6C_2', 3\sigma_h, 6\sigma_d, i, 3S_4, 4S_6$	SF_6, B$_6$H$_6$$^{2-}$ (octahedron)
I_h	$E, 6C_5, 10C_3, 15C_2, 15\sigma_v, i, 12S_{10}, 10S_6$	$B_{12}H_{12}$$^{2-}$ (icosahedron)

add to these examples from the numerous instances of point group symmetry mentioned throughout the text. In this way a facility will gradually be acquired in discerning the various elements of symmetry present in a molecule.

A convenient scheme for identifying the point group symmetry of any given species is set out in the flow chart.[1] Starting at the top of the chart each vertical line asks a question: if the answer is "yes" then move to the right, if "no" then move to the left until the correct point group is arrived at. Other similar schemes have been devised.[2–5]

[1] J. DONOHUE, A "tree" for identification of the point groups, *Sov. Phys. Crystallogr.* **26**, 516 (1981); *Kristallografiya* **26**, 908–9 (1981).

[2] R. L. CARTER, A flow chart approach to point group classification, *J. Chem. Educ.* **45**, 44 (1968).

[3] F. A. COTTON, *Chemical Applications of Group Theory*, 2nd edn., pp. 45–7, Wiley-Interscience, New York, 1971.

[4] J. D. DONALDSON and S. D. ROSS, *Symmetry and Stereochemistry*, pp. 35–49, Intertext Books, London, 1972.

[5] J. A. SALTHOUSE and M. J. WARE, *Point Group Character Tables and Related Data*, p. 29, Cambridge University Press, 1972.

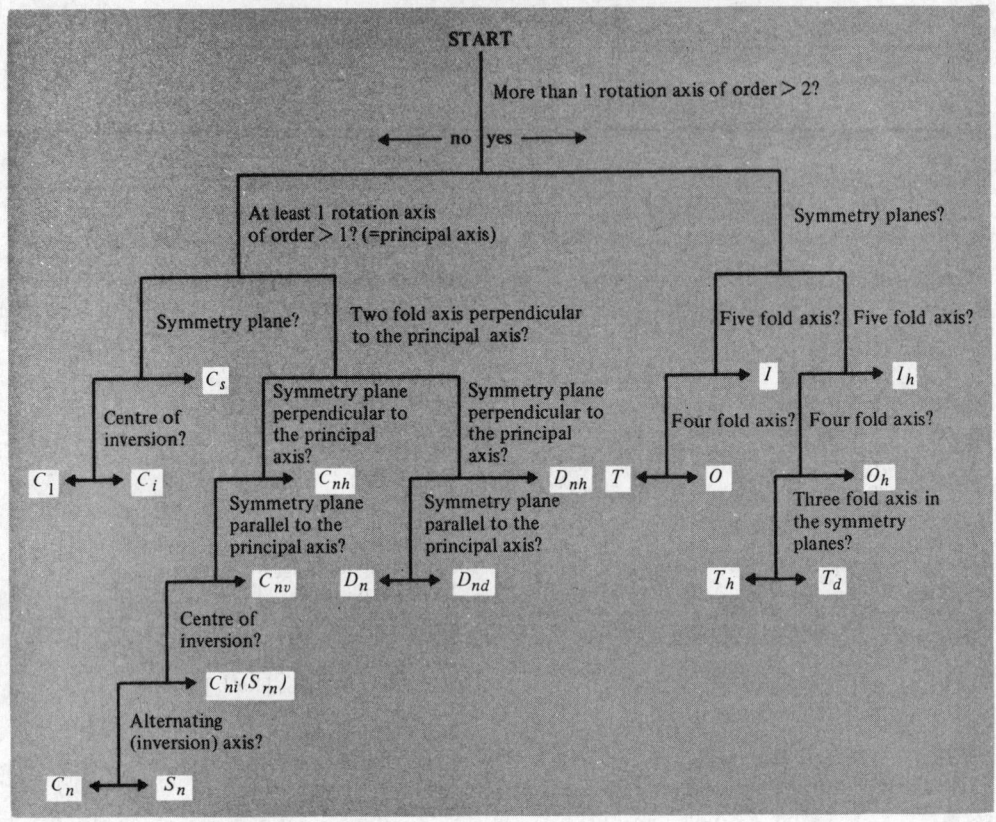

FIG. A2.1

Appendix 3
Some Non-SI Units†

Physical quantity	Name of unit	Symbol for unit	Definition of unit
Length	ångström	Å	10^{-8} m
Time	minute	min	60 s
	hour	h	3600 s
	day	d	86 400 s
Energy	erg	erg	10^{-7} J
	kilowatt hour	kWh	3.6×10^{6} J
	thermochemical calorie	cal_{th}	4.184 J
Force	dyne	dyn	10^{-5} N
Pressure	bar	bar	10^{5} Pa
	atmosphere	atm	101 325 Pa
	conventional millimetre of mercury	mmHg	$13.5951 \times 9.806\,65$ Pa i.e. 133.322 Pa
	torr	Torr	(101 325/760) Pa
Magnetic flux	maxwell	Mx	10^{-8} Wb
Magnetic flux density (magnetic induction)	gauss	G, Gs	10^{-4} T
Dynamic viscosity	poise	P	10^{-1} Pa s
Concentration	—	M	mol dm^{-3}
Radioactivity	curie	Ci	$3.7 \times 10^{10}\ s^{-1}$
Radioactive exposure	röntgen	R	2.58×10^{-4} C kg^{-1}
Absorbed dose	rad	rad	10^{-2} J kg^{-1}
Angle	degree	°	$1° = (\pi/180)$ radian

† The unit "degree Celsius" (°C) is identical with the kelvin (K). The Celsius temperature (t_C) is related to the thermodynamic temperature T by the definition: $t_C = T - 273.15$ K.

Some useful conversion factors:

1 m = 3.280 839 9 ft = 39.370 079 inches
1 inch = 25.4 mm (defined)
1 statute mile = 1.609 344 km
1 light year = $9.460\,55 \times 10^{12}$ km
1 acre = 4046.8564 m^2
1 gal (Imperial) = 1.200 949 gal (US) = 4.545 960 1
1 gal (US) = 0.832 674 7 gal (Imp.) = 3.785 411 8 1
1 lb (avoirdupois) = 0.453 592 37 kg
1 oz (avoirdupois) = 28.349 527 g
1 oz (troy, or apoth.) = 31.103 486 g
1 carat = 3.086 47 grains = 200 mg
1 tonne = 1000 kg = 2204.622 6 lb = 1.102 311 3 short tons
1 short ton = 907.184 74 kg = 2000 lb = 0.892 857 14 long tons
1 long ton = 1016.046 9 kg = 2240 lb = 1.120 short tons
1 atm = 101 325 Pa = 1.013 25 bar = 760 Torr = 14.695 95 lb/in^2
1 Pa = 10^{-5} bar $\sim 1.019\,716 \times 10^{-1}$ kg m^{-2} = $0.986\,923 \times 10^{-5}$ atm
1 mdyn $Å^{-1}$ = 100 N m^{-1}
1 calorie (thermochem) = 4.184 J (defined)
1 eV = $1.602\,19 \times 10^{-19}$ J
1 eV/molecule = 96.484 56 kJ mol^{-1} = 23.060 36 kcal mol^{-1}

Appendix 4
Abundance of Elements in Crustal Rocks/ppm (i.e. g/tonne)[†]

No.	Elt.	ppm	Σ%	No.	Elt.	ppm	No.	Elt.	ppm	No.	Elt.	ppm
1	O	455 000	45.50	20	Cl	126	39	Th	8.1	58	Tl	0.7
2	Si	272 000	72.70	21	Cr	122	40	Sm	7.0	59	Tm	0.5
3	Al	83 000	81.00	22	Ni	99	41	Gd	6.1	60	I	0.46
4	Fe	62 000	87.20	23	Rb	78	42	Er	3.5	61	In	0.24
5	Ca	46 600	91.86	24	Zn	76	43	Yb	3.1	62	Sb	0.2
6	Mg	27 640	94.62	25	Cu	68	44	Hf	2.8	63	Cd	0.16
7	Na	22 700	96.89	26	Ce	66	45	Cs	2.6	64	{ Ag	0.08
8	K	18 400	98.73	27	Nd	40	46	Br	2.5	64	{ Hg	0.08
9	Ti	6320	99.36	28	La	35	47	U	2.3	66	Se	0.05
10	H	1520	99.51	29	Y	31	48	{ Sn	2.1	67	Pd	0.015
11	P	1120	99.63	30	Co	29	48	{ Eu	2.1	68	Pt	0.01
12	Mn	1060	99.73	31	Sc	25	50	Be	2	69	Bi	0.008
13	F	544	99.79	32	Nb	20	51	As	1.8	70	Os	0.005
14	Ba	390	99.83	33	{ N	19	52	Ta	1.7	71	Au	0.004
15	Sr	384	99.86	33	{ Ga	19	53	Ge	1.5	72	{ Ir	0.001
16	S	340	99.90	35	Li	18	54	Ho	1.3	72	{ Te	0.001
17	C	180	99.92	36	Pb	13	55	{ Mo	1.2	74	Re	0.0007
18	Zr	162	99.93	37	Pr	9.1	55	{ W	1.2	75	{ Ru	0.0001
19	V	136	99.95	38	B	9	55	{ Tb	1.2	75	{ Rh	0.0001

† Taken from W. S. Fyfe, *Geochemistry*, Oxford University Press, 1974, with some modifications and additions to incorporate later data.

Appendix 5
Effective Ionic Radii in pm for Various Oxidation States (in parentheses)†

s Block

Li	(+1) 76
Be	(+2) 45
Na	(+1) 102
Mg	(+2) 72.0
K	(+1) 138
Ca	(+2) 100
Rb	(+1) 152
Sr	(+2) 118
Cs	(+1) 167
Ba	(+2) 135
Fr	(+1) 180
Ra	(+2)^{VIII} 148

p Block

B	(+3) 27
C	(+4) 16
N	(+3)^{IV} 146; (+3) 16; (+5) 13
O	(−2) 140
F	(−1) 133; (+7) 8
Al	(+3) 53.5
Si	(+4) 40
P	(+3) 44; (+5) 38
S	(−2) 184; (+4) 37; (+6) 29
Cl	(−1) 184; (+5)^{III} 12; (+7) 27
Ga	(+3) 62.0
Ge	(+2) 73; (+4) 53.0
As	(+3) 58; (+5) 46
Se	(−2) 198; (+4) 50; (+6) 42
Br	(−1) 196; (+3)^{IV} 59; (+5)^{III} 31; (+7) 39
In	(+3) 80.0
Sn	(+2) 118; (+4) 69.0
Sb	(+3) 76; (+5) 60
Te	(−2) 221; (+4) 97; (+6) 56
I	(−1) 220; (+5) 95; (+7) 53
Xe	(+8) 48
Tl	(+1) 150; (+3) 88.5
Pb	(+2) 119; (+4) 77.5
Bi	(+3) 103; (+5) 76
Po	(+4) 94; (+6) 67
At	(+7) 62

d Block

Sc	(+3) 74.5
Ti	(+2) 86; (+3) 67.0; (+4) 60.5
V	(+2) 79; (+3) 64.0; (+4) 58; (+5) 54
Cr	(+2) 73 ls, 80 hs; (+3) 61.5; (+4) 55; (+5) 49; (+6) 44
Mn	(+2) 67 ls, 83 hs; (+3) 58 ls, 64.5 hs; (+4) 53.0; (+5) 33^{IV}; (+6) 25.5^{IV}; (+7) 46
Fe	(+2) 61 ls, 78.0 hs; (+3) 55 ls, 64.5 hs; (+4) 58.5; (+6) 25^{IV}
Co	(+2) 65 ls, 74.5 hs; (+3) 54.5 ls, 61 hs; (+4) 53
Ni	(+2) 69; (+3) 56 ls, 60 hs; (+4) 48 ls
Cu	(+1) 77; (+2) 73; (+3) 54 ls
Zn	(+2) 74.0
Y	(+3) 90.0
Zr	(+4) 72
Nb	(+3) 72; (+4) 68; (+5) 64
Mo	(+3) 69; (+4) 65; (+5) 61; (+6) 59
Tc	(+4) 64.5; (+5) 60; (+7) 56
Ru	(+3) 68; (+4) 62.0; (+5) 56.5; (+7) 38^{IV}; (+8) 36^{IV}
Rh	(+3) 66.5; (+4) 60; (+5) 55
Pd	(+2) 59^{II}, 86; (+3) 76; (+4) 61.5
Ag	(+1) 115; (+2) 94; (+3) 75
Cd	(+2) 95
La	(+3) 103.2
Hf	(+4) 71
Ta	(+3) 72; (+4) 68; (+5) 64
W	(+4) 66; (+5) 62; (+6) 60
Re	(+4) 63; (+5) 58; (+6) 55; (+7) 53
Os	(+4) 63.0; (+5) 57.5; (+6) 54.5; (+7) 52.5; (+8) 39^{IV}
Ir	(+3) 68; (+4) 62.5; (+5) 57
Pt	(+2) 80; (+4) 62.5; (+5) 57
Au	(+1) 137; (+3) 85; (+5) 57
Hg	(+1) 119; (+2) 102
Ac	(+3) 112

f Block

Ce	(+3) 102; (+4) 87
Pr	(+3) 99; (+4) 85
Nd	(+2) 129^{VIII}; (+3) 98.3
Pm	(+3) 97
Sm	(+2) 122^{VII}; (+3) 95.8
Eu	(+2) 117; (+3) 94.7
Gd	(+3) 93.8
Tb	(+3) 92.3; (+4) 76
Dy	(+2) 107; (+3) 91.2
Ho	(+3) 90.1
Er	(+3) 89.0
Tm	(+2) 103; (+3) 88.0
Yb	(+2) 102; (+3) 86.8
Lu	(+3) 86.1
Th	(+4) 94
Pa	(+3) 104; (+4) 90; (+5) 78
U	(+3) 102.5; (+4) 89; (+5) 76; (+6) 73
Np	(+2) 110; (+3) 101; (+4) 87; (+5) 75; (+6) 72; (+7) 71
Pu	(+3) 100; (+4) 86; (+5) 74; (+6) 71
Am	(+2) 126^{VIII}; (+3) 97.5; (+4) 85
Cm	(+3) 97; (+4) 85
Bk	(+3) 96; (+4) 83
Cf	(+3) 95; (+4) 82.1

† Values refer to coordination number 6 unless otherwise indicated by superscript roman numerals ^{III}, ^{IV}, etc. (All data taken from R. D. Shannon, *Acta Cryst.* **A32**, 751–67 (1976).)

Appendix 6
Nobel Prize for Chemistry

1901 **J. H. van't Hoff** (Berlin): discovery of the laws of chemical dynamics and osmotic pressure in solutions.

1902 **E. Fischer** (Berlin): sugar and purine syntheses.

1903 **S. Arrhenius** (Stockholm): electrolytic theory of dissociation.

1904 **W. Ramsay** (University College, London): discovery of the inert gaseous elements in air and their place in the periodic system.

1905 **A. von Baeyer** (Munich): advancement of organic chemistry and the chemical industry through work on organic dyes and hydroaromatic compounds.

1906 **H. Moissan** (Paris): isolation of the element fluorine and development of the electric furnace.

1907 **E. Buchner** (Berlin): biochemical researches and the discovery of cell-free fermentation.

1908 **E. Rutherford** (Manchester): investigations into the disintegration of the elements and the chemistry of radioactive substances.

1909 **W. Ostwald** (Gross-Bothen): work on catalysis and investigations into the fundamental principles governing chemical equilibria and rates of reaction.

1910 **O. Wallach** (Göttingen): pioneer work in the field of alicyclic compounds.

1911 **Marie Curie** (Paris): discovery of the elements radium and polonium, the isolation of radium, and the study of the nature and compounds of this remarkable element.

1912 **V. Grignard** (Nancy): discovery of the Grignard reagent.
      P. Sabatier (Toulouse): method of hydrogenating organic compounds in the presence of finely disintegrated metals.

1913 **A. Werner** (Zürich): work on the linkage of atoms in molecules which has thrown new light on earlier investigations and opened up new fields of research especially in inorganic chemistry.

1914 **T. W. Richards** (Harvard): accurate determination of the atomic weight of a large number of chemical elements.

1915 **R. Willstätter** (Munich): plant pigments, especially chlorophyll.

1916 Not awarded

1917 Not awarded

1918 **F. Haber** (Berlin–Dahlem): the synthesis of ammonia from its elements.

1919 Not awarded

1920 **W. Nernst** (Berlin): work in thermochemistry.

1921   **F. Soddy** (Oxford): contributions to knowledge of the chemistry of radioactive substances and investigations into the origin and nature of isotopes.

1922   **F. W. Aston** (Cambridge): discovery, by means of the mass spectrograph, of isotopes in a large number of non-radioactive elements and for enunciation of the whole-number rule.

1923   **F. Pregl** (Graz): invention of the method of microanalysis of organic substances.

1924   Not awarded

1925   **R. Zsigmondy** (Göttingen): demonstration of the heterogeneous nature of colloid solutions by methods which have since become fundamental in modern colloid chemistry.

1926   **T. Svedberg** (Uppsala): work on disperse systems.

1927   **H. Wieland** (Munich): constitution of the bile acids and related substances.

1928   **A. Windaus** (Göttingen): constitution of the sterols and their connection with the vitamins.

1929   **A. Harden** (London) and **H. von Euler-Chelpin** (Stockholm): investigations on the fermentation of sugars and fermentative enzymes.

1930   **H. Fischer** (Munich): the constitution of haemin and chlorophyll and especially for the synthesis of haemin.

1931   **C. Bosch** and **F. Bergius** (Heidelberg): the invention and development of chemical high pressure methods.

1932   **I. Langmuir** (Schenectady, New York): discoveries and investigations in surface chemistry.

1933   Not awarded

1934   **H. C. Urey** (Columbia, New York): discovery of heavy hydrogen.

1935   **F. Joliot** and **Irène Joliot-Curie** (Paris): synthesis of new radioactive elements.

1936   **P. Debye** (Berlin–Dahlem): contributions to knowledge of molecular structure through investigations on dipole moments and on the diffraction of X-rays and electrons in gases.

1937   **W. N. Haworth** (Birmingham): investigations on carbohydrates and vitamin C.
P. Karrer (Zürich): investigations of carotenoids, flavins, and vitamins A and B_2.

1938   **R. Kuhn** (Heidelberg): work on carotenoids and vitamins.

1939   **A. F. J. Butenandt** (Berlin): work on sex hormones.
L. Ruzicka (Zürich): work on polymethylenes and higher terpenes.

1940   Not awarded.

1941   Not awarded.

1942   Not awarded.

1943   **G. Hevesy** (Stockholm): use of isotopes as tracers in the study of chemical processes.

1944   **O. Hahn** (Berlin–Dahlem): discovery of the fission of heavy nuclei.

1945   **A. J. Virtanen** (Helsingfors): research and inventions in agricultural and nutrition chemistry, especially fodder preservation.

1946   **J. B. Sumner** (Cornell): discovery that enzymes can be crystallized.
J. H. Northrop and **W. M. Stanley** (Princeton): preparation of enzymes and virus proteins in a pure form.

1947   **R. Robinson** (Oxford): investigations on plant products of biological importance, especially the alkaloids.

1948   **A. W. K. Tiselius** (Uppsala): electrophoresis and adsorption analysis, especially for discoveries concerning the complex nature of the serum proteins.

1949   **W. F. Giauque** (Berkeley): contributions in the field of chemical thermodynamics, particularly concerning the behaviour of substances at extremely low temperatures.

1950   **O. Diels** (Kiel) and **K. Alder** (Cologne): discovery and development of the diene synthesis.

1951   **E. M. McMillan** and **G. T. Seaborg** (Berkeley): discoveries in the chemistry of the transuranium elements.

1952   **A. J. P. Martin** (London) and **R. L. M. Synge** (Bucksburn): invention of partition chromatography.

1953   **H. Staudinger** (Freiburg): discoveries in the field of macromolecular chemistry.

1954   **L. Pauling** (California Institute of Technology, Pasadena): research into the nature of the chemical bond and its application to the elucidation of the structure of complex substances.

1955   **V. du Vigneaud** (New York): biochemically important sulfur compounds, especially the first synthesis of a polypeptide hormone.

1956   **C. N. Hinshelwood** (Oxford) and **N. N. Semenov** (Moscow): the mechanism of chemical reactions.

1957   **A. Todd** (Cambridge): nucleotides and nucleotide co-enzymes.

1958   **F. Sanger** (Cambridge): the structure of proteins, especially that of insulin.

1959   **J. Heyrovský** (Prague): discovery and development of the polarographic method of analysis.

1960   **W. F. Libby** (Los Angeles): use of carbon-14 for age determination in archeology, geology, geophysics, and other branches of science.

1961   **M. Calvin** (Berkeley): research on the carbon dioxide assimilation in plants.

1962   **J. C. Kendrew** and **M. F. Perutz** (Cambridge): the structures of globular proteins.

1963   **K. Ziegler** (Mülheim/Ruhr) and **G. Natta** (Milan): the chemistry and technology of high polymers.

1964   **Dorothy Crowfoot Hodgkin** (Oxford): determinations by X-ray techniques of the structures of important biochemical substances.

1965   **R. B. Woodward** (Harvard): outstanding achievements in the art of organic synthesis.

1966   **R. S. Mulliken** (Chicago): fundamental work concerning chemical bonds and the electronic structure of molecules by the molecular orbital method.

1967   **M. Eigen** (Göttingen), **R. G. W. Norrish** (Cambridge), and **G. Porter** (London): studies of extremely fast chemical reactions, effected by disturbing the equilibrium by means of very short pulses of energy.

1968   **L. Onsager** (Yale): discovery of the reciprocity relations bearing his name, which are fundamental for the thermodynamics of irreversible processes.

1969   **D. H. R. Barton** (Imperial College, London) and **O. Hassel** (Oslo): development of the concept of conformation and its application in chemistry.

1970   **L. F. Leloir** (Buenos Aires): discovery of sugar nucleotides and their role in the biosynthesis of carbohydrates.

1971   **G. Herzberg** (Ottawa): contributions to the knowledge of electronic structure and geometry of molecules, particularly free radicals.

1972   **C. B. Anfinsen** (Bethesda): work on ribonuclease, especially concerning the connection between the amino-acid sequence and the biologically active conformation.

S. Moore and **W. H. Stein** (Rockefeller, New York): contributions to the understanding of the connection between chemical structure and catalytic activity of the active centre of the ribonuclease molecule.

1973 **E. O. Fischer** (Münich) and **G. Wilkinson** (Imperial College, London): pioneering work, performed independently, on the chemistry of the organometallic so-called sandwich compounds.

1974 **P. J. Flory** (Stanford): fundamental achievements both theoretical and experimental in the physical chemistry of macromolecules.

1975 **J. W. Cornforth** (Sussex): stereochemistry of enzyme-catalysed reactions.
V. Prelog (Zürich): the stereochemistry of organic molecules and reactions.

1976 **W. N. Lipscomb** (Harvard): studies on the structure of boranes illuminating problems of chemical bonding.

1977 **I. Prigogine** (Brussels): non-equilibrium thermodynamics, particularly the theory of dissipative structures.

1978 **P. Mitchell** (Bodmin, Cornwall): contributions to the understanding of biological energy transfer through the formulation of the chemiosmotic theory.

1979 **H. C. Brown** (Purdue) and **G. Wittig** (Heidelberg): for their development of boron and phosphorus compounds, respectively, into important reagents in organic synthesis.

1980 **P. Berg** (Stanford): the biochemistry of nucleic acids, with particular regard to recombinant-DNA.
W. Gilbert (Harvard) and **F. Sanger** (Cambridge): the determination of base sequences in nucleic acids.

1981 **K. Fukui** (Kyoto) and **R. Hoffmann** (Cornell): quantum mechanical studies of chemical reactivity.

1982 **A. Klug** (Cambridge): development of crystallographic electron microscopy and the structural elucidation of biologically important nucleic acid-protein complexes.

1983 **H. Taube** (Stanford): mechanisms of electron transfer reactions of metal complexes.

1984 **R. B. Merrifield** (Rockefeller, New York): solid-phase synthesis of peptides.

Appendix 7
Nobel Prize for Physics

1901 **W. C. Röntgen** (Munich): discovery of the remarkable rays subsequently named after him.

1902 **H. A. Lorentz** (Leiden) and **P. Zeeman** (Amsterdam): influence of magnetism upon radiation phenomena.

1903 **H. A. Becquerel** (École Polytechnique, Paris): discovery of spontaneous radioactivity.
P. Curie and **Marie Curie** (Paris): researches on the radiation phenomena discovered by H. Becquerel.

1904 **Lord Rayleigh** (Royal Institution, London): investigations of the densities of the most important gases and for the discovery of argon in connection with these studies.

1905 **P. Lenard** (Kiel): work on cathode rays.

1906 **J. J. Thomson** (Cambridge): theoretical and experimental investigations on the conduction of electricity by gases.

1907 **A. A. Michelson** (Chicago): optical precision instruments and the spectroscopic and metrological investigations carried out with their aid.

1908 **G. Lippmann** (Paris): method of reproducing colours photographically based on the phenomenon of interference.

1909 **G. Marconi** (London) and **F. Braun** (Strasbourg): the development of wireless telegraphy.

1910 **J. D. van der Waals** (Amsterdam): the equation of state for gases and liquids.

1911 **W. Wien** (Würzburg): the laws governing the radiation of heat.

1912 **G. Dalén** (Stockholm): invention of automatic regulators for use in conjunction with gas accumulators for illuminating lighthouses and buoys.

1913 **H. Kamerlingh Onnes** (Leiden): properties of matter at low temperatures and production of liquid helium.

1914 **M. von Laue** (Frankfurt): discovery of the diffraction of X-rays by crystals.

1915 **W. H. Bragg** (University College, London) and **W. L. Bragg** (Manchester): analysis of crystal structure by means of X-rays.

1916 Not awarded.

1917 **C. G. Barkla** (Edinburgh): discovery of the characteristic Röntgen radiation of the elements.

1918 **M. Planck** (Berlin): services rendered to the advancement of physics by discovery of energy quanta.

1919 **J. Stark** (Greifswald): discovery of the Doppler effect on canal rays and of the splitting of spectral lines in electric fields.

1920 **C. E. Guillaume** (Sèvres): service rendered to precise measurements in physics by discovery of anomalies in nickel steel alloys.

1921 **A. Einstein** (Berlin): services to theoretical physics, especially discovery of the law of the photoelectric effect.

1922 **N. Bohr** (Copenhagen): investigations of the structure of atoms, and of the radiation emanating from them.

1923 **R. A. Millikan** (California Institute of Technology, Pasadena): work on the elementary charge of electricity and on the photo-electric effect.

1924 **M. Siegbahn** (Uppsala): discoveries and researches in the field of X-ray spectroscopy.

1925 **J. Franck** (Göttingen) and **G. Hertz** (Halle): discovery of the laws governing the impact of an electron upon an atom.

1926 **J. Perrin** (Paris): the discontinuous structure of matter, and especially for the discovery of sedimentation equilibrium.

1927 **A. H. Compton** (Chicago): discovery of the effect named after him.
C. T. R. Wilson (Cambridge): method of making the paths of electrically charged particles visible by condensation of vapour.

1928 **O. W. Richardson** (King's College, London): thermionic phenomenon and especially discovery of the law named after him.

1929 **L. V. de Broglie** (Paris): discovery of the wave nature of electrons.

1930 **V. Raman** (Calcutta): work on the scattering of light and discovery of the effect named after him.

1931 Not awarded.

1932 **W. Heisenberg** (Leipzig): the creation of quantum mechanics, the application of which has, inter alia, led to the discovery of the allotropic forms of hydrogen.

1933 **E. Schrödinger** (Berlin) and **P. A. M. Dirac** (Cambridge): discovery of new productive forms of atomic theory.

1934 Not awarded.

1935 **J. Chadwick** (Liverpool): discovery of the neutron.

1936 **V. F. Hess** (Innsbruck): discovery of cosmic radiation.
C. D. Anderson (California Institute of Technology, Pasadena): discovery of the positron.

1937 **C. J. Davisson** (New York) and **G. P. Thomson** (London): experimental discovery of the diffraction of electrons by crystals.

1938 **E. Fermi** (Rome): demonstration of the existence of new radioactive elements produced by neutron irradiation and for the related discovery of nuclear reactions brought about by slow neutrons.

1939 **E. O. Lawrence** (Berkeley): invention and development of the cyclotron and for results obtained with it, especially with regard to artificial radioactive elements.

1940 Not awarded.

1941 Not awarded.

1942 Not awarded.

1943 **O. Stern** (Pittsburgh): development of the molecular ray method and discovery of the magnetic moment of the proton.

1944 **I. I. Rabi** (Columbia, New York): resonance method for recording the magnetic properties of atomic nuclei.

1945 **W. Pauli** (Zürich): discovery of the Exclusion Principle, also called the Pauli Principle.

1946 **P. W. Bridgman** (Harvard): invention of an apparatus to produce extremely high pressures and discoveries in the field of high-pressure physics.

1947 **E. V. Appleton** (London): physics of the upper atmosphere, especially the discovery of the so-called Appleton layer.

1948 **P. M. S. Blackett** (Manchester): development of the Wilson cloud chamber method and discoveries therewith in the field of nuclear physics and cosmic radiation.

1949 **H. Yukawa** (Kyoto): prediction of the existence of mesons on the basis of theoretical work on nuclear forces.

1950 **C. F. Powell** (Bristol): development of the photographic method of studying nuclear processes and discoveries regarding mesons made with this method.

1951 **J. D. Cockroft** (Harwell) and **E. T. S. Walton** (Dublin): pioneer work on the transmutation of atomic nuclei by artificially accelerated atomic particles.

1952 **F. Bloch** (Stanford) and **E. M. Purcell** (Harvard): development of new methods for nuclear magnetic precision measurements and discoveries in connection therewith.

1953 **F. Zernike** (Groningen): demonstration of the phase contrast method and invention of the phase contrast microscope.

1954 **M. Born** (Edinburgh): fundamental research in quantum mechanics, especially for the statistical interpretation of the wave function.
W. Bothe (Heidelberg): the coincidence method and discoveries made therewith.

1955 **W. E. Lamb** (Stanford): the fine structure of the hydrogen spectrum.
P. Kusch (Columbia, New York): precision determination of the magnetic moment of the electron.

1956 **W. Shockley** (Pasadena), **J. Bardeen** (Urbana), and **W. H. Brattain** (Murray Hill): investigations on semiconductors and discovery of the transistor effect.

1957 **T. Lee** (Columbia) and **C. Yang** (Princeton): penetrating investigation of the so-called parity laws, which has led to important discoveries regarding the elementary particles.

1958 **P. A. Cherenkov, I. M. Frank,** and **I. E. Tamm** (Moscow): discovery and the interpretation of the Cherenkov effect.

1959 **E. Segrè** and **O. Chamberlain** (Berkeley): discovery of the antiproton.

1960 **D. A. Glaser** (Berkeley): invention of the bubble chamber.

1961 **R. Hofstadter** (Stanford): pioneering studies of electron scattering in atomic nuclei and discoveries concerning the structure of the nucleons.
R. L. Mössbauer (Munich): resonance absorption of gamma radiation and discovery of the effect which bears his name.

1962 **L. D. Landau** (Moscow): pioneering theories for condensed matter, especially liquid helium.

1963 **E. P. Wigner** (Princeton): the theory of the atomic nucleus and elementary particles, particularly through the discovery and application of fundamental symmetry principles.
Maria Goeppert-Mayer (La Jolla) and **J. H. D. Jensen** (Heidelberg): discoveries concerning nuclear shell structure.

1964 **C. H. Townes** (Massachusetts Institute of Technology) and **N. G. Basov** and **A. M.**

Prokhorov (Moscow): fundamental work in the field of quantum electronics, which led to the construction of oscillators and amplifiers based on the maser-laser-principle.

1965 **S. Tomonaga** (Tokyo), **J. Schwinger** (Cambridge, Mass.,), and **R. P. Feynman** (California Institute of Technology, Pasadena): fundamental work in quantum electrodynamics, with deep-ploughing consequences for the physics of elementary particles.

1966 **A. Kastler** (Paris): discovery and development of optical methods for studying hertzian resonances in atoms.

1967 **H. A. Bethe** (Cornell): contributions to the theory of nuclear reactions, especially discoveries concerning the energy production in stars.

1968 **L. W. Alvarez** (Berkeley): decisive contributions to elementary particle physics, in particular the discovery of a large number of resonance states, made possible by the hydrogen bubble chamber technique and data analysis.

1969 **M. Gell-Mann** (California Institute of Technology, Pasadena): contributions and discoveries concerning the classification of elementary particles and their interactions.

1970 **H. Alfvén** (Stockholm): discoveries in magneto-hydrodynamics with fruitful applications in different parts of plasma physics.
L. Néel (Grenoble): discoveries concerning antiferromagnetism and ferrimagnetism which have led to important applications in solid state physics.

1971 **D. Gabor** (Imperial College, London): invention and development of the holographic method.

1972 **J. Bardeen** (Urbana), **L. N. Cooper** (Providence), and **J. R. Schrieffer** (Philadelphia): theory of superconductivity, usually called the BCS theory.

1973 **L. Esaki** (Yorktown Heights) and **I. Giaever** (Schenectady): experimental discoveries regarding tunnelling phenomena in semiconductors and superconductors respectively.
B. D. Josephson (Cambridge): theoretical predictions of the properties of a supercurrent through a tunnel barrier, in particular those phenomena which are generally known as the Josephson effects.

1974 **M. Ryle** and **A. Hewish** (Cambridge): pioneering research in radioastrophysics: Ryle for his observations and inventions, in particular of the aperture-synthesis technique, and Hewish for his decisive role in the discovery of pulsars.

1975 **A. Bohr** (Copenhagen), **B. Mottelson** (Copenhagen), and **J. Rainwater** (New York): discovery of the connection between collective motion and particle motion in atomic nuclei and the development of the theory of the structure of the atomic nucleus based on this connection.

1976 **B. Richter** (Stanford) and **S. C. C. Ting** (Massachusetts Institute of Technology): discovery of a heavy elementary particle of a new kind.

1977 **P. W. Anderson** (Murray Hill), **N. F. Mott** (Cambridge), and **J. H. van Vleck** (Harvard): fundamental theoretical investigations of the electronic structure of magnetic and disordered systems.

1978 **P. L. Kapitsa** (Moscow): basic inventions and discoveries in the area of low-temperature physics.
A. A. Penzias and **R. W. Wilson** (Holmdel): discovery of cosmic microwave background radiation.

1979 **S. L. Glashow** (Harvard), **A. Salam** (Imperial College, London), and **S. Weinberg** (Harvard): contributions to the theory of the unified weak and electromagnetic interaction between elementary particles, including, inter alia, the prediction of the weak neutral current.

1980 **J. W. Cronin** (Chicago) and **V. L. Fitch** (Princeton): discovery of violations of fundamental symmetry principles in the decay of neutral K-mesons.

1981 **K. M. Siegbahn** (Uppsala): development of high-resolution electron spectroscopy. **N. Bloembergen** (Harvard) and **A. L. Schawlow** (Stanford): development of laser spectroscopy.

1982 **K. G. Wilson** (Cornell): theory for critical phenomena in connection with phase transitions.

1983 **S. Chandrasekar** (Chicago): theoretical studies of the physical processes of importance to the structure and evolution of the stars. **W. A. Fowler** (California Institute of Technology, Pasadena): theoretical and experimental studies of the nuclear reactions of importance in the formation of the chemical elements in the universe.

1984 **C. Rubbia** and **S. van der Meer** (CERN, Geneva): discovery of the intermediate vector bosons W and Z.

Index

α-Process in stars 13
Absolute configuration, determination of 1304 1306
Abundance of elements, Table of 1496
Acetylides *see* Carbides
Acidity function H_0 (Hammett) 56, 57
Acid strength of binary hydrides 52, 55
  of oxoacids (Pauling's rules) 54
  of HF 952
Actinide contraction 1464
Actinide elements
  abundance 1454
  alkyls and aryls 1484
  atomic and physical properties 1465
  carbonyls 1484
  chalcogenides 1471
  complexes
    +7 oxidation state 1478
    +6 oxidation state 1478
    +5 oxidation state 1479
    +4 oxidation state 1481
    +3 oxidation state 1483
    +2 oxidation state 1483
    +1 oxidation state 1484
  coordination numbers and stereochemistries 1470
  cyclopentadienyls 1484
  discovery 1452
  group trends 1464-1469
  halides 1473-1476
  hydrides 70. 1469
  lanthanide-like behaviour 1453, 1464, 1469
  magnetic properties 1476, 1477
  mixed metal oxides 1473
  organometallic compounds 1484-1486
  oxidation states 1468
  oxides 1470-1473
  preparation of artificial elements 1452-1463
  problems of isolation and characterization 1451, 1461, 1464, 1466
  redox behaviour 1467-1469
  separation from used nuclear fuels 1461-1463
  spectroscopic properties 1476-1477
Actinium
  abundance 1103
  discovery 1102
  radioactive decay series 1455
  see also Actinide elements, Group IIIA elements
Actinoid *see* Actinide elements
Actinon *see* Actinide elements

Actinyl ions 1478, 1479
Activated carbon 301
  see also Carbon
Adenine 611 614
Adenosine triphosphate (ATP)
  discovery in muscle fibre 547
  in life processes 610 612, 1279
  in nitrogen fixation 1206, 1207
  in phosphorus cycle 550
  in photosynthesis 139
Agate 393
Albite 414, 415
Alkali metals (Li, Na, K, Rb, Cs, Fr)
  abundance 76
  atomic properties 86
  biological systems containing 77, 79, 83, 108
  carbonates 102-104
  chelated complexes of 106, 110
  chemical reactivity of 87
  complexes of 105
  crown-ether complexes of 105
  cryptates of 105
  cyanates 342
  cyanides 339
  discovery of 75, 76
  flame colours of 86
  halides
    bonding in 92
    properties of 94-96
  hydrides
    reactions of 97
  hydroxides 100, 101
  hydrogen carbonates (bicarbonates) 102, 104
  intercalation compounds with graphite 313 316
  intermetallic compounds with As, Sb, Bi 646, 647
  isolation of 75, 76
  nitrates 103
  nitrites 105
  organometallic compounds 110 116
  oxoacid salts 101-105
  oxides 97, 98
  ozonides 99
  peroxides 97, 98
  physical properties 85, 86
  polysulfides 800-802, 803, 805
  reactions in liquid ammonia 89 91
  sesquioxides 99
  solubilities in liquid ammonia 89
  solutions in amines 91

Alkali metals—*contd*
  solutions in liquid ammonia 88–91, 455
  solutions in polyethers 91
  suboxides 97, 99, 100
  sulfides 800–803
  superoxides 97, 98
  see also individual metals; Li, Na, K, Rb, Cs, Fr
Alkaline earth metals (Be, Mg, Ca, Sr, Ba, Ra)
  abundance 118–120
  atomic properties 121, 122
  chemical reactivity of 123
  complexes of 135–141
  halides
    crystal structures of 128–130
    uses 130
  hydride halides, MHX 131
  hydrides 70, 125
  hydroxides 132, 133
  intermetallic compounds with As, Sb, Bi 646, 647
  organometallic compounds 141–153
  oxides 131, 132
  oxoacid salts 133
  ozonides 132
  peroxides 132
  physical properties 122
  sulfides 803
  superoxides 132
  thermal stability of oxoacid salts 124
  univalent compounds of 124
  see also individual metals: Be, Mg, Ca, Sr, Ba, Ra
Alkene insertion reactions (Ziegler) 291, 292
  polymerization (Ziegler-Natta) 292, 293, 1134
Alkene ligands 357, 359–362
Alkyne complexes 361, 362
Allotelluric acid 915
Allotropy *see* individual elements: B, C, P, S etc.
Allyl group as ligand 362–365
  η^1 to η^3 conversion 363
  bonding in complexes 363, 364
  fluxionality 365
Alnico steel 1291, 1330
Aluminium
  abundance 244
  alloys 248
  borohydride 258
  chalcogenides 285
  III–V compounds 288, 290
  halide complexes 262–266
  heterocyclic organo-AlN oligomers 294, 295
  history of 243
  hydrides 256
  hydrido complexes 257–261
  lower halides AlX, AlX$_2$ 261, 262
  organometallic compounds 289–295
  orthophosphate AlPO$_4$ 607
  production 243, 245–247
  ternary oxide phases 278–284
  trialkyls and triaryls 289, 290–292
    dimeric structure of 289
    preparation 291
    in Ziegler-Natta catalysis 291–293, 1134

Aluminium—*contd*
  trichloride
    Friedel-Crafts activity of 263, 266
    structure 263
  trichloride adducts, structure of 263, 264, 794
  trihalide complexes 262–266
  uses 248, 249
  see also Group III elements
Aluminium hydroxide
  bayerite and gibbsite 274, 277
Aluminium ion
  hydration number of 704
Aluminium oxide hydroxide
  diaspore and boehmite 274, 276
Aluminium oxides
  catalytic activity 277
  corundum 273, 274
  "Saffil" fibres 275, 276
  structural classification 273, 274
  see also Portland cement, High-alumina cement
Aluminium trimethyl, Al$_2$Me$_6$ 289, 291
  reactions with MgMe$_2$ 147
Alumino-silicates *see* Silicate minerals
Alumino-thermic process 1168
Amalgamation process 1367, 1395, 1399
Americium 1452, 1463
  see also Actinide elements
Amethyst 393
Amidopolyphosphates 580, 581
Aminoboranes 237
Ammonia
  adduct with NI$_3$ 506
  chemical reactions 485, 486
  fertilizer applications 484
  hydrazine, production from (Raschig synthesis) 490, 491
  industrial production 468, 481
  inversion frequency 484, 485
  inversion of, discovery 468, 484
  nitric acid production from 537, 538
  odour 481
  physical properties 484. 485
  production statistics 483
  synthesis of (Haber-Bosch) 468, 481
  uses of 484
Ammonia, liquid
  acid-base reactions in 487
  alkali metal solutions in 88–91, 455
  ammonolysis of PCl$_5$ 624
  amphoteric behaviour in 487
  discovery of coloured metal solutions 468
  H bonding in 57–59, 484
  metathesis reactions in 487
  redox reactions in 488
  solubility of compounds in 486, 487
  solvate formation in 487
  solvent properties of 88–91, 484, 486–489
  syntheses in 488, 489 and *passim*
  synthesis of metal cluster ions in 455, 456
Ammonium nitrate
  explosive decomposition of 538, 541
  thermolysis of 510, 541

Ammonium nitrite, decomposition to N$_2$ 471
Ammonium phosphates 604
Ammonium salts 484
Amosites 405
Amphiboles 405, 406
Amphoteric behaviour
  definition 253
  of Al and Ga 253, 254
  of SbCl$_5$ 823
  of SF$_4$ (Lewis acid-base) 811
  of V$_2$O$_5$ 1144
  of ZnO 1404
Amphoteric cations 56
Andrieux's phosphide synthesis 562
Angular functions 1487, 1490, 1491
Andrussow process for HCN 338
Angeli's salt 528
Anorthite 414, 415
Antiferroelectrics 62
Anti-knock additives 432, 932
Antimonates 674
Antimonides 646
Antimonious acid, H$_3$SbO$_3$ 671, 673
Antimonite esters 654
Antimony
  abundance and distribution 638-640
  alloys 639, 640, 646
  allotropes 643, 644
  amino derivatives 655
  atomic properties 641, 642
  catenation 678
  chalcogenides 677, 679
  chemical reactivity 645, 646, 673
  cluster anions 646, 686
  encapsulated 646
  extraction and production 639
  halide complexes 659-664
  halides 651-659
    see also trihalides, pentahalides, oxide-halides,
      mixed halides, halide complexes
  halogeno-organic compounds 695, 696
  history 637
  hydride 650, 651
  intermetallic compounds *see* Antimonides
  metal-metal bonded clusters 678, 685, 686
  mixed halides 656, 657
  organometallic compounds 690, 694-697
  organometallic halides 695, 696
  oxoacid salts of 689
  oxidation state diagram 673
  oxide, Sb$_2$O$_3$ 668-670
  oxide halides 665-668
  oxides and oxocompounds 668-673
  pentahalides 655-657, 662-665, 785
  pentaphenyl 636
  physical properties 643, 644
  selenium complex anions 678
  sulfide, Sb$_2$S$_3$ 637, 639, 677, 678, 761
  trichloride solvent system 654
  trifluoride as fluorinating agent 653
  trihalides 651-655, 659, 661, 663
  uses 639, 640

Antimony—*contd*
  volt-equivalent diagram 673
Anti-tumour activity of Pt^{II} complexes 1353
Apatites 119, 549, 554, 605, 606
  reduction to phosphides 562
Aqua regia 921, 923, 1371
Aragonite 119
Arenes as η^6 ligands 373
Argentite (silver glance) 1365
Argon
  atomic and physical properties 1045, 1046
  clathrates 1047, 1048
  discovery 1043
  production and uses 1044
  see also Noble gases
Arsenates 673
Arsenic
  abundance and distribution 638, 639
  allotropes 643
  alloys 639, 646
  amino derivatives 655
  atomic properties 641, 642
  catenation 678-687
  chalcogenide cluster cations 675
  chalcogenides 674-679
  chemical reactivity 644-646, 673
  cluster anions 646, 686-689
  coordination geometries 646
  diiodide 658
  encapsulated 646
  extraction and production 638, 639
  halide complexes 659-662
  halides 651-659
    see also trihalides, pentahalides, diiodide, oxide
      halides, mixed halides, halide complexes
  halogenoorganoarsenic compounds 691-693
  history 637
  hydrides 650, 651, 679
  intermetallic compounds *see* Arsenides
  metal-metal bonded species 678-687
  mixed halides 656
  organoarsenic(I) compounds 695
  organoarsenic(III) compounds 679-683, 690-
    692, 694
    arsabenzene 690, 691
    arsanaphthalene 690, 691
    physiological activity 694
    preparation of 691, 692
    reactions of 691, 692
  organoarsenic(V) compounds 693, 694
  organic compounds 646
  oxoacid salts of 689
  oxidation state diagram 673
  oxide, As$_2$O$_3$ 639, 668-671
    reactions 670
    structure and polymorphism 668
    uses 639, 670
  oxide halides 665-668
  oxides and oxocompounds 668-673
  pentahalides 655, 785
  physical properties 643, 644
  selenides 677

Arsenic—*contd*
  sulfide, As_2S_3 637, 638, 674, 675, 761
    chemical reactions 676, 677
    structure 674, 675
  sulfide, As_4S_4 638, 674–677, 761
  sulfides, As_4S_n 674–676, 679
  triangular species 679, 683–685
  trichloride solvent system 654
  trihalides 651–655, 659, 661, 663
  uses 639
  volt-equivalent diagram 673
Arsenicals, therapeutic uses 694
  see also Organoarsenic(III) and Organoarsenic(V)
    compounds
Arsenides 646–650
  stoichiometries 646
  structures 647–650
Arsenious acid $As(OH)_3$ 670
Arsenite 670, 671, 673
Arsenite esters 654
Arsenopyrite 761
Arsenidine complexes 695
Arsine 650
Arsinic acids $R_2AsO(OH)$ 694
Arsinous acids R_2AsOH 691
Arsonic acids $RAsO(OH)$ 693, 694
Arsonous acids $RAs(OH)_2$ 691
Asbestos 119
Asbestos minerals 405, 406
Astatate ion, $AtO_3{}^-$ 1040
Astatide ion, At^- 1040
Astatine
  abundance 928
  atomic properties 934
  chemistry 1040, 1041
  nuclear synthesis 922, 926
  radioactivity of isotopes 926, 1039
  redox systematics 1040
  trihalide ions 1041
  see also Halogens
Atactic polymer 1134
Atmophile elements 760
Atmosphere
  composition 471, 701
  industrial production of gases from 471, 702–704
  origin of O_2 in 700
Atomic energy 1456
Atomic orbitals 1487–1491
Atomic piles 1457
  see also Nuclear reactors
Atomic properties, periodic trends in 27–31
Atomic volume curve, periodicity of 27, 28
Atomic weight, definition 18
Atomic weights
  history of 18, 699
  precision of 18–22
  relative uncertainty of 20
  Table of *see* inside back cover
  variability of 20–22, 428
ATP *see* Adenosine triphosphate
Austenite 1248
Autoprotolysis constants of anhydrous acids 843

Azides 497
Azoferredoxin 1206, 1275

β-alumina *see* Sodium β-alumina
β-elimination reactions 348
Back (π) bonding 349, 351, 360, 1101, 1303
Baddeleyite 1112
  structure of 1121
Baking powders 604, 605
Barium
  history of 118
  organometallic compounds 153
  polysulfides 805, 806
  see also Alkaline earth metals
Basic oxygen steel process 134, 1246
Bastnaesite 1103, 1426, 1429
Batteries
  dry 1397, 1398
  lead 432, 640
  sodium-sulfur 801, 802
Bauxite 275
  production statistics 245
  in Al production 246
Bayerite 274, 277
Belousov-Zhabotinskii oscillating reactions 1012,
    1013
Bentonite *see* Montmorillonite
Benzene as η^6 ligand 373–375
Berkelium 1452, 1463
  see also Actinide elements
Berry pseudorotation 547, 572
  see also Stereochemical non-rigidity
Beryl 117, 119, 403
Beryllia *see* Beryllium oxide
Beryllium
  alkoxides etc 136, 144
  alkyls 141–145
  'anomalous' properties of 125, 135
  'basic acetate' 135, 136
  'basic nitrate' 135
  borohydride 126, 127
  chloride 128, 129
  complexes of 135–137
  cyclopentadienyl complexes 145, 146
  discovery 117, 118
  fluoride 128
  hydride 125, 126, 143
  oxide 117, 131, 132
  salts, hydrolysis of 132
  uses of metal and alloys 120
  see also Alkaline earth metals
Beryllium compounds, toxicity of 117
Beryllium ion, hydration number of 704
Bessemer process 1245
BHB bridge bond, comparison with H bond 70
  see also Three-centre bonds
"Big bang" 1, 5
Biotite *see* Micas
Bismuth
  abundance and distribution 638, 640, 641
  allotropes 643, 644

Bismuth—*contd*
  alloys 640, 646
  atomic properties 641, 642
  Bi^+ cation 659, 686
  catenation 678
  chalcogenides 677, 679
  chemical reactivity 645, 646, 673
  cluster cations 658, 659, 679, 686–688
  extraction and production 640, 641
  halide complexes 659–663
  halides 651–659
    see also trihalides, pentahalides, oxide halides,
      lower halides, mixed halides, halide complexes
  history 637
  hydride 650, 651
  hydroxide, $Bi(OH)_3$ 671
  hydroxo cluster cation, $[Bi_6(OH)_{12}]^{6+}$ 671, 689,
    690
  intermetallic compounds *see* Bismuthides
  lower halides 658
  metal–metal bonded clusters 678, 686–688
  mixed halides 659
  nitrate and related complexes 689, 690
  organometallic compounds 690, 694, 697
  oxidation state diagram 673
  oxide, Bi_2O_3 669, 670
  oxide halides 667, 668
  oxides and oxo compounds 669–672
  oxoacid salts of 689
  pentafluoride 655
  physical properties 643, 644
  trihalides 651–655, 659, 661
  uses 640
  volt-equivalent diagram 673
Bismuthates 674
Bismuthides 646
Bismuthine 650
Biuret 324, 1383
"Black oxide" of mercury 1408
Blast furnace 1245
Bleaching powder 921, 1006
"Blister" copper 1366
"Blue proteins" 1392
Blue vitriol 1381
Boehmite 274, 276
Boiling points, influence of H bonding 58
Borane adducts, LBH_3 188
  amine-boranes 235–237
Borane anions 171ff, 191, 201, 202, 686
  see also Boranes
Boranes
  arachno-structures 173, 174, 184
  bonding 179–185, 686
  classification 171
  closo-structures 171, 172, 184
  conjuncto-structures 173, 175–178
  hypho-structures 171, 173, 195
  nido-structures 173, 174, 184
  nomenclature 173
  optical resolution of i-$B_{18}H_{22}$ 791
  physical properties 185, 186

Boranes—*contd*
  topology 180–182, 199
  see also Diborane, Pentaborane, Decaborane,
    Metalloboranes, Carboranes, etc.
Borate minerals
  occurrence 156
  production and uses 156
  see also Borates
Borates 231–234
  B–O distances in 233
  industrial uses 233, 234
  structural principles 231, 232
Borax 155
  see also Borate minerals
Borazanes 238
Borazine 238, 239, 468
Bordeaux mixture (fungicide) 1381
Boric acids
  $B(OH)_3$ 229, 230
  HBO_2 230, 231
Borides
  bonding 170
  catenation in 166
  preparation 163, 164
  properties and uses 163
  stoichiometry 162, 165
  structure 165–170
Born–Haber cycle 91, 94–96
Borohydrides *see* Borane anions, Tetrahydro-
    borates, etc.
Boron
  abundance 155
  allotropes 157–160
  atomic properties 160, 161, 250
  chemical properties 161, 162
  crystal structures of allotropes 157
  hydrides *see* Boranes
  isolation 155–157, 160
  nitride 235, 236
  nuclear properties 161
  oxide 229
  physical properties 161, 250
  sulfides 240, 241
  variable atomic weight of 21
  see also Group III elements
Boron carbide
  B_4C 167, 168
  $B_{13}C_2$ 167, 168
  $B_{50}C_2$ 159, 160
Boron halides 220–228
  B_2X_4 225–227
  B_nX_n 227, 225
  lower halides 224–228
  see also Boron trihalides
Boron nitride, $B_{50}N_2$ 159, 160
Boron–nitrogen compounds 234–240
Boron–oxygen compounds 228–234
  organic derivatives 234
Boron trifluoride *see* Boron trihalides
Boron trihalides
  adducts 222–224
  bonding 220, 221

Boron trihalides—*contd*
  physical properties 220
  preparation 221
  scrambling reactions 222, 224
Brass 1366, 1369, 1370, 1395, 1397
Brimstone 757, 758
Brønsted's acid-base theory 38, 51ff
  in non-aqueous solutions 56
  in aqueous solutions 738
Bromates, BrO$_3^-$ 1010–1012
  reaction scheme 1014
  redox systematics 1000–1002
Bromic acid, HBrO$_3$ 1010
Bromides, synthesis of 961, 962
  see also individual elements
Bromine
  abundance and distribution 927
  atomic and physical properties 934–938
  cations Br$_n^+$ 986–988
  history 921, 922, 924, 925
  monochloride 964–966, 975
  monofluoride 965, 966, 975
  oxide fluorides 1032–1034
  oxides 996, 997
  oxoacid and oxoacid salts 1000ff
    nomenclature 1000
    redox properties 1000–1002 etc.
    see also individual compounds, bromic acid,
      bromates, perbromates, etc.
  pentafluoride 974–976
  production and uses 932, 933
  radioactive isotopes 935, 936
  reactivity 940
  standard reduction potentials 1001
  stereochemistry 941
  trifluoride 968–973, 975
  volt-equivalent diagram 1002
  see also Halogens
Bronze 1364, 1366
Bronzes
  molybdenum 1185
  titanium 1123
  tungsten 1185
  vanadium 1150
"Brown-ring" complex 514, 1272
Brucite 133, 407, 447
Buffer solutions 52, 53, 599, 603
Bulky tertiary phosphine ligands, special properties
  of 567
Butadiene as a η^4 ligand 366

Cacodylic acid 639
Cadmium
  abundance 1396
  chalcogenides 1403, 1406
  coordination chemistry 1411–1413
  discovery 1395
  halides 1406–1408
  organometallic compounds 1417
  oxides 1403, 1404
  production and uses 1396–1398

Cadmium—*contd*
  toxicity 1421
  see also Group IIB elements
Cadmium chloride structure 1407
Cadmium iodide structure type 649, 803, 804, 1407
  relation to NiAs 649, 803, 804
  nonstoichiometry in 804
Caesium
  abundance 79
  compounds with oxygen 97–100
  discovery 76
Calamine (smithsonite) 1396
Calcite 119
Calcium
  in biochemical processes 138
  carbide 319, 320
  carbonate *see* limestone
  cyclopentadienyl 153, 154
  history of 118
  organometallic compounds 153, 154
  phosphorus 604–606
  see also Alkaline earth metals, Lime, etc.
Caliche (Chilean nitre) 927
Californium 1452, 1463
  see also Actinide elements
Calomel 1408
Capped octahedral complexes 1074
Capped trigonal prismatic complexes 1074
Caprolactam, for nylon-6 484
Carat 300, 1367
Carbaboranes *see* Carboranes
Carbene ligands 352, 353
Carbides 318–322
  silicon (SiC) 386
Carbido complexes 352, 354–358
Carbohydrates, photosyntheses of 139
Carbon
  abundance 298
  allotropes 303–308
  atomic properties 306, 431–434
  bond lengths (interatomic distances) 310, 311
  chalcogenides 333–336
  chemical properties 308
  coordination numbers 310, 312, 313
  cycle (global) 302
  disulfide 333–335, 770
  halides 322, 323
  history of 296–298
  hydrides 322
  interatomic distances in compounds 310, 311
  occurrence and distribution 298–303
  oxides 325
    see also Carbon monoxide, Carbon dioxide
  "monofluoride" 309
  radioactive ^{14}C 306
  suboxides 325
Carbon dioxide 325–333
  aqueous solutions (acidity of) 329, 331
  atmospheric 301–303
  coordination chemistry 331, 332
  industrial importance 326, 327
  insertion into M-C bonds 150, 333

Carbon dioxide—*contd*
  in photosynthesis 139
  physical properties 325
  production and uses of 330, 331
  use in ^{14}C syntheses 329
Carbon monoxide 325–329
  bonding in 349, 350
  chemical reactions 328, 329
  industrial importance 326, 327
  as a ligand 349–352
  physical properties 325
  poisoning effect of 1275
  preparation of pure 326
  similarity to PF_3 as a ligand 568
Carbonates, terrestrial distribution 301, 302
Carbonic acid 329
Carbonic anhydrase 1421
Carbonyl fluoride, reaction with OF_2 751
Carbonyl halides *see* Carbon oxohalides
Carbonyls *see* Carbon monoxide as a ligand
  see also individual elements
Carboranes 185, 202ff
  bonding 203
  chemical reactions 208, 209
  isomerization 206–208
  preparation 204–206
  structures 203, 207
Carborundum *see* Silicon carbide
Carboxypeptidase A 1420
Carbyne ligands 352–354
Carnotite 1139
Caro's acid *see* Peroxomonosulfuric acid
Cast-iron 1244
Cassiterite *see* Tin dioxide
Castner-Kelner process (chlor-alkali) 922, 931, 1399
Catenation in Group IVB elements 435, 459, 461–463
Catena-polyarsanes 681–683
Catena-S_8 diradical 780, 783
Catena-S_x 774, 775, 779
  formation at λ point 780
Caustic soda *see* Sodium hydroxide
Cellophane 334
Cementite 1248
Cerium 1424
  diiodide 1441
  +4 oxidation state 1434, 1438, 1444
  production 1426
  see also Lanthanide elements
Chabazite *see* Zeolites
Chain polyphosphates 608–617
  diphosphates 608
  tripolyphosphates 609, 610
  see also Adenosine triphosphate, Sodium tri-polyphosphate, Graham's salt, Kurrol's salt, Maddrell's salt
Chain reactions, nuclear 1456, 1462
Chalcocite (copper glance) 1365
Chalcogens, group trends 890–896
  see also S, Se, Te, Po
Chalcophile elements 758, 760
Chalcopyrite (copper pyrite) $CuFeS_2$ 761, 1365

Chaoite 303, 306
Charge-transfer bands 1094
Chelate effect 1066–1068
Chelation 1061, 1066
Chemical periodicity 24ff
Chemical properties of the elements, periodic trends in 31
Chemical shear structures 755
  see also oxides of Ti, Mo, W, Re, etc.
Chemiluminescence of phosphorus 559
Chile saltpetre 469
  see also Sodium nitrate
Chlorates, ClO_3^- 1009–1012
  redox properties 1001, 1002
Chloric acid, $HClO_3$ 1010
Chlorides, synthesis of 961, 962
  see also individual elements
Chlorin 140
Chlorine
  abundance and distribution 927
  atomic and physical properties 934–938
  bleaching power 921, 924
  cations Cl_2^+, Cl_3^+ 986, 987
  dioxide, ClO_2 989, 990, 992–994
  history 921–924
  hydrate 921
  monofluoride 964–967, 975
  oxide fluorides 1026–1032
  oxides 988–996
  oxoacids and oxoacid salts 999ff
    nomenclature 1000
    redox properties 1000–1002 etc.
    see individual compounds, Hypochlorous acid, Hypochlorites, Chlorous acid, etc.
  pentafluoride 974–976
  production and uses 930, 931
  radioactive isotopes 935, 936
  reactivity 940
  standard reduction potentials 1001
  stereochemistry 941
  toxicity 924
  trifluoride 968–971, 975
  volt equivalent diagram
  see also Halogens
Chlorite 412, 413
Chlorites, ClO_2^- 1001, 1002, 1007–1009
Chlorophylls 119, 138–141, 614
Chloroplatinic acid 1340
Chlorosulfanes *see* Sulfur chlorides
Chlorous acid, $HClO_2$ 1001, 1002, 1007, 1008
Chromate ion 1175, 1193
Chrome alum 1197
Chrome ochre 1168
Chromic acid 1171
Chromite 1168
Chromium
  abundance 1167
  bis(cyclopentadienyl) 371, 1209
  borazine complex 239
  carbonyls 351, 353, 1208–1209
  carbyne complexes 353
  chalcogenides 804, 1186, 1187

Chromium—*contd*
  complexes
    +6 oxidation state 1191–1193
    +5 oxidation state 1193, 1194
    +4 oxidation state 1194–1196
    +3 oxidation state 1196–1200
    +2 oxidation state 1200–1205
    with S 788
  compounds with quadruple metal-metal bonds 1202–1205
  cyclooctatetraene complex 377
  cyclopentadienyls 371, 1209
  dibenzene "sandwich" compound 374, 1210
  discovery 1167
  dithiolene complexes, redox series 797
  halides and oxohalides 1187–1192
  hexacarbonyl 351, 353, 1208
  importance of Cr^{III} in early coordination chemistry 1072, 1196
  organometallic compounds 371, 373–375, 1207–1210
  oxides 1171–1175
  peroxo complexes 747, 748
  polynuclear complexes in dying and tanning 1200
  "polyphenyl" compounds 373
  production and uses 1168
  sulfides, nonstoichiometry in 804
  see also Group VIA elements
Chromocene 1209
Chrysotile 405, 408, 413
Cinnabar (vermillion), HgS 761, 1396, 1406
Circular dichroism (CD) 1304–1305
Cisplatin 1353
Class-a and class-b metal ions 1065
  see also Ligands
Clathrate compounds 1047, 1349
Claus process (S from H_2S) 768, 771
Clay minerals 406–413
  uses of 410, 411
Clock reaction (H. Landolt) 1012
Cluster compounds
  boron carbide 167, 168
  boron hydrides 171–202
  carbido metal carbonyls 355–357
  carboranes 202–209
  cobalt, rhodium, and iridium carbonyls 351ff, 1323–1326
  of germanium, tin, and lead 446, 455–458
  gold phosphines 1391
  iron, ruthenium, and osmium carbonyls 351ff, 1283–1286
  lithium alkyls 112–114
  metal borides 166, 168–170
  metalloboranes 195–198, 201
  metallocarboranes 209–220
  nickel and platinum carbonyls 1358, 1359
  niobium and tantalum halides 1155
  of phosphorus, arsenic, antimony and bismuth 658, 686–689
  rhenium alkyls 1240, 1241
  stabilization by encapsulated heteroatoms 355–357, 1285, 1326

Cluster compounds—*contd*
  Te_6^{4+} 897, 898
  tungsten and molybdenum halides 1189–1191
Cobalt
  abundance 1290
  allyl complexes 363
  arsenide 648, 649
  atomic and physical properties 1292–1294
  biochemistry of 1321, 1322
  carbidocarbonyls 355, 358
  carbonyls 351, 352, 355, 1323–1326
  complexes
    +5 oxidation state 1300
    +4 oxidation state 1300
    +3 oxidation state 1301–1310
    +2 oxidation state 1311–1316
    +1 oxidation state 1316
    lower oxidation states 1320
    with S 788, 789, 790, 792
    with SO_2 831
  coordination numbers and stereochemistries 1295
  cyclobutadiene complex 367
  cyclooctatetraene complexes 377
  cyclopentadienyls 1326–1327
  dithiolene complexes, redox series 798
  halides 1297–1299
  importance of Co^{III} in early coordination chemistry 1072, 1301, 1302
  nitrate (anhydrous) 541
  nitrato complexes 541, 543
  optical resolution of $[Co\{(\mu\text{-}OH)_2Co(NH_3)_4\}_3]^{6+}$ 791, 1073
  organometallic compounds 1322–1327
  oxidation states 1295
  oxides 1296–1297
  oxoanions 1299, 1300
  production and uses 1291, 1292
  reactivity of element 1294
  relationship with other transition elements 1294–1296
  standard reduction potentials 1302
  sulfides 1297
Cobaltite (cobalt glance) 1291
Cobaltocene 371, 1326
Coesite 394, 395
Coinage metals (Cu, Ag, Au) 1364
  see also Group IB elements and individual elements
Coke 301
  historical importance in steel making 1242, 1244
  see also Carbon
Colours of transition element compounds 1089–1096
  providing an early nomenclature 1303
Columbite 1139
Columbium 1138
  see also Niobium
Combustion 698–700, 713
Complexometric titration of Bi^{III} with EDTA 674
Contact process *see* Sulfuric acid manufacture
Cooperativity 1276, 1277
Coordinate bond 223, 1082
  see also Donor–acceptor complexes

Coordination number 1068
  two 1070
  three 1070
  four 1070
  five 1071
  six 1072
  seven 1074
  eight 1074
  nine 1076
  above nine 1077
Copper
  abundance 1365
  acetylide 1372
  alkenes and alkynes 1393
  alkyls and aryls 1393
  biochemistry of 1392
  carboxylates 1384–1386
  chalcogenides 1373–1374
  complexes
    +3 oxidation state 1379
    +2 oxidation state 1381–1386
    +1 oxidation state 1386, 1387
  $Cu-S-O$ system 299
  halides 1374–1376
  history 1364
  nitrate, structure of 544, 1382
  nitrato complexes 541, 542, 544
  organometallic compounds 348, 1392, 1393
  oxides 1373
  production and uses 1365, 1366
  see also Group IB elements
Copper oxide $Cu_{2-x}O$, nonstoichiometry in 754
"Corrosive sublimate" 1408
Corundum *see* Aluminium oxides
Cosmic black-body radiation 2
Cotton effect 1305
Cristobalite 394, 395
Critical mass 1457, 1458, 1462
Crocodilite 405
Crocoite 1168
Crown ethers 107, 108
  complexes with alkali metals 107, 108
  complexes with alkaline earth metals 137, 141
  "hole sizes" of 107
  "triple-decker" complex 1326
Cryolite 246, 247
Crypt, molecular structure of 109
  complexes with alkali metals 109, 455–457
  complexes with alkaline earth metals 137
Crystal field
  octahedral 1085
  strong 1086
  tetrahedral 1087
  weak 1085
Crystal field splittings 1084–1088
Crystal field stabilization energy 1097, 1313
Crystal field theory 1082, 1084–1089
Cubic, eight coordination 1074, 1480
Cupellation 1364
Cuprite 1365
Curium 1452, 1463
  see also Actinide elements

Cyanamide 337
  industrial production 341
Cyanates 337, 342
  as ligands 344
Cyanide ion as ligand 339, 349
Cyanide process 1367, 1388
Cyanogen 336–338
  halides 337, 339, 340
Cyanuric acid 324
Cyanuric compounds 337, 341
Cyclobutadiene as η^4 ligand 366–368
Cycloheptatrienyl as η^7 ligand 375
Cyclometaphosphimic acids 632
*Cyclo*metaphosphoric acids $(HPO_3)_n$ 588
  see also Cyclo-polyphosphoric acid, *Cyclo*-poly-
    phospates
Cycloocta-1,5-diene (COD) as ligand 361
Cyclooctatetraene as η^8 ligand 376
  as η^2, η^4, η^6, etc., ligand 377
Cyclopentadienyl
  as η^5 ligand 367–373
  as η^1 ligand 373
  see also Ferrocene, individual metals
Cyclo-polyarsanes 681–683
Cyclo polyphosphates 617–619
Cyclophosphazenes 621, 622
Cyclo polyphosphoric acids 617
  see also Cyclo-metaphosphoric acids
Cyclo-S_6 (ε-sulfur) 773–775
Cyclo-S_7 774–776
Cyclo-S_8 (α)
  crystal and molecular structure 772
  physical properties 772–775
  polymerization at λ point 780
  solubility 773
  transition α-$S_8 \rightleftharpoons \beta$-$S_8$ 773
  vapour pressure 781
Cyclo-S_9 775
Cyclo-S_{10} 774–776
Cyclo-S_{11} 775
Cyclo-S_{12} 774–777
Cyclo-S_{18} 774, 775, 777, 778
Cyclo-S_{20} 774, 775, 778, 779
Cyclo-silicates 400, 403
Cytochromes 1275, 1279, 1392
Cytosine 612–614

d-block contraction 32, 251, 655, 1432
d orbitals 1084, 1487
  splitting by crystal fields 1084–1088, 1349
Dalton's atomic theory 509
Dalton's law of multiple proportions 509
Decaborane, $B_{10}H_{14}$ 174, 182, 184, 186
  adduct formation 200
  Brønsted acidity 199
  chemical reactions 198–201
  preparation 187
  structure 199
Degussa process for HCN 338
Density of the elements
  periodic trends in 29

Deoxyribonucleic acids 550, 611
  genetic code 612–614
2′-Deoxyribose 612
Detergents
  polyphosphates in 547, 551
  sodium tripolyphosphate in 609, 610
Deuterium
  atomic properties 40
  discovery 38
  ortho- and *para-* 42
  physical properties 41
  preparation 47
Dewar-Chatt-Duncanson theory 360
Diagonal relationships 32
  B and Si 228, 400
  Be and Al 117
  Li and Mg 87, 110, 124
  N and S 855
Diamond
  chemical properties 307, 308
  occurrence and distribution 299, 300
  physical properties 307, 308
  production and uses 300
Diarsane, As_2H_4 679
  see also Arsenic hydrides
Diaspore 274, 276
Diatomaceous earth 394
Diatomic molecules (homonuclear), bond dissoci-
    ation energies of 680
Diazotization of aromatic amines 533
Dibenzenechromium 374, 1210
  see also Benzene as η^6 ligand
Diborane, B_2H_6 174, 181
  chemical reactions 186–193
  cleavage reactions 188
  hydroboration reactions 192, 193
  preparation 186
  pyrolysis 187
Dibromonium cation Br_2^+, compound with $Sb_3F_{16}^-$
    663
Dicacodyl, As_2Me_4 679–681
Dichlorine hexoxide, Cl_2O_6 989, 990, 995, 996
Dichlorine monoxide, Cl_2O 989–992
Dichromate ion, $Cr_2O_7^{2-}$ 1175
Dicyandiamide 337, 341
Dielectric constant, influence of H bonding 59
Dimethylaminophosphorus dihalides 620
Dimethyl sulfoxide as ionizing solvent 820
Dinitrogen
  complexes, synthesis of 475, 476
  coordination modes 477, 478
  discovery of donor properties 475, 1274
  isoelectronic with CO, C_2H_4 478
  as ligand 468, 475–478, 1274
Dinitrogen monoxide *see* Nitrous oxide
Dinitrogen pentoxide N_2O_5 509, 526, 527
Dinitrogen tetroxide 509, 521–526
    chemical reactions 524–526
  nonaqueous solvent properties 524–526
  physical properties 523, 525
  preparation 524

Dinitrogen tetroxide—*contd*
  structure 522
  see also Nitrogen dioxide
Dinitrogen trioxide, N_2O_3 509
Diopside 403
Dioxygen
  bonding in metal complexes 724–725
  chemical properties 713, 716–726
  coordination chemistry of 718–726
  difluoride 750
  molecular-orbital diagram 706
  paramagnetic behaviour 699, 706
  reactions of coordinated O_2 724–726
  reaction with haemoglobin 716, 1276–1279
  singlet state 699, 716
  singlet-triplet transitions 707
  superoxo and peroxo complexes 719, 720
  Vaska's discovery of reversible coordination 699,
    718, 1318
  see also Oxygen, Oxygen carriers, Singlet oxygen
Diphosphazenes 623, 624
Diphosphonic acid *see* Diphosphorous acid
Diphosphoric acid, $H_2P_2O_7$ 587, 595, 597, 601, 602
Diphosphorous acid, $H_4P_2O_5$ 589
Diphosphorus tetrahalides 570, 571
Disilenes 388
Disulfates, $S_2O_7^{2-}$ 835, 845
  imido derivative 878
  preparation in liquid SO_2 828
Disulfites, $S_2O_5^-$ 835, 853
Disulfuric acid, $H_2S_2O_7$ 835, 843
Disulfurous acid, $H_2S_2O_5$ 835, 853
Dithiocarbamates 334
  as ligands 786, 795, 796
Dithiolenes as ligands 786, 796–798
Dithionates, $S_2O_6^{2-}$ 835, 848, 849
Dithionic acid, $H_2S_2O_6$ 835, 848
Dithionites, $S_2O_4^{2-}$ 835, 853–855
Dithionous acid, $H_2S_2O_4$ 835, 853
DNA *see* Deoxyribonucleic acids
Döbereiner's triads 25
Dodecahedral complexes 1075
Dolomite 119, 301
Donor-acceptor complexes
  of AlX_3 263–266
  of AsX_3 645, 659
  of CN^- 339, 340
  of CO *see* Carbonyls
  of *cyclo*-polyarsanes 681–683
  of cyclopolyphosphazenes 631
  of dithiocarbamates and xanthates 795, 796, 1266
  of dithiolenes 796
  first (H_3NBF_3) 468
  of GaH_3 257, 255
  of Group III halides 267–269
  of H_2S 794, 795, 846
  of N_2 475–478, 1274
  of NO *see* Nitrosyls
  of O_2 718–726
  of PH_3 and tertiary phosphines 565–567
  of PX_3 568, 570
  of S_n 786–794

Donor-acceptor complexes—*contd*
    of SbF$_5$ 655, 662, 664, 665, 830
    of SCN⁻ 342–345
    of SF$_4$ 811
    of S$_4$N$_4$ 857
    of SO, S$_2$O$_2$, SO$_2$ 828–832
    of SO$_3$ 832, 833
    stability of 223, 224
    see also Class-a and class-b metal ions, Coordinate
        bond, Ligands, individual elements
Double-helix structure of nucleic acids 547
Dry batteries 1397, 1398
Dysprosium 1424
    +4 oxidation state 1444
    see also Lanthanide elements

e-Process in stars 14
Effective atomic number (EAN) rule 1082
    see also Eighteen electron rule
Effective ionic radii, Table of 1497
Eighteen electron rule 1082, 1208, 1287, 1316
Einsteinium 1452, 1463
    see also Actinide elements
Eka-silicon, Mendeleev's predictions 34
Electrical properties, influence of H bonding 57
Electric arc process of steelmaking 1246
Electrofluorination 960
Electron delocalization factor, k 1099
Electronegativity
    definition of 30
    periodic trends in 30
Electronic structure and chemical periodicity 25–27
Electronic structure of atoms 25–27
Elements
    abundance in crustal rocks 1496
    bond dissociation energies of gaseous diatomic
        680
    cosmic abundance 2ff, 14ff
    isotopic composition of 20
    table of atomic weights *see* inside back cover
    origin of 1, 5, 10ff, 15ff
    periodic table of *see* inside front cover
    periodicity in properties 24–36
Ellingham diagram 327, 328, 429
Emerald 117, 1168
Enstatite 403
Entropy and the chelate effect 1066
Equilibrium process in stars (e-process) 14
Erbium 1424
    see also Lanthanide elements
Ethene (ethylene) as a ligand 357, 359, 360
Eutrophication 552, 609
Europium 1424
    +2 oxidation state 1438, 1439, 1447
    magnetic properties of 1442
    see also Lanthanide elements
Exclusion principle (Pauli) 26
Extended X-ray absorption fine structure (EXAFS)
    1206

f-block contraction 655, 1431

Faraday's phosphide synthesis 562
Faujasite 415
Feldspars 414, 415
Fermium 1452, 1463
    see also Actinide elements
Fermi level, definition of 383
Ferredoxins 1206, 1207, 1275, 1279–1281
Ferricinium ion 1287
Ferrites 1256, 1404
Ferritin 1257, 1281
Ferrocene
    bonding 369–371
    historical importance of 345, 1242, 1286
    physical properties 367
    reactions 1287–1289
    structure 368, 372
    synthesis 372
Ferrochrome 1168
Ferroelectricity 60–62, 449, 665
Ferromanganese 1212
Ferromolybdenum 1168
Ferrophosphorus 554, 563, 605
Ferrosilicon alloys 381
"Ferrous oxide", Fe$_{1-x}$O, nonstoichiometry in
    754, 755
Ferrovanadium 1140
First short period, "anomalous" properties of 32
Fischer (Karl) reagent 736
Fischer-Tropsch process 1284
Fish population, relation to phosphate-rich waters
    553
Flint 379, 393
Fluorapatite *see* Apatites
Fluoridation and dental caries 551, 605, 922, 923
Fluorides 958–960
    solubility in HF 955
    synthesis 958–960
Fluorinated peroxo compounds 750, 751
Fluorinating agents 958–960
    AsF$_3$, SbF$_3$ 653, 654
    AsF$_5$, BiF$_5$, SbF$_5$ 656, 658
Fluorine
    abundance and distribution 927
    atomic and physical properties 934–938
    history 920–923
    isolation 920, 922
    oxides *see* Oxygen fluorides
    oxoacid, HOF 922, 999, 1003
    preparation of fluorides using 959
    production and uses 928–930
    radioactive isotopes 935, 936
    reactivity 938–940
    stereochemistry 941
    toxicity 923, 946
    see also Halogens
Fluorite, CaF$_2$ 119
    crystal structure of 129
Fluorspar 920, 921
    fluorescence 920, 921
    see also Fluorite
Fluorosulfuric acid 814
Fluxional behaviour *see* Stereochemical non-rigidity

Francium
  abundance 76
  discovery 76
  see also Alkali metals
Frasch process for sulfur 758
Freons (eg CCl$_2$F$_2$) 323, 922, 923
Friedel-Crafts catalysis
  AlX$_3$ complexes 263, 266
  BF$_3$ complexes 224
  SnCl$_4$ 448
Fuller's earth *see* Montmorillonite
"Fulminating" silver and gold 1373
Fulminate ion 337
Fundamental physical constants, Table of values
    back end paper
"Fusible white precipitate" 1415

g (gerade), definition 369
Gabbro rock 415
Gadolinium 1424
  diiodide 1441
  see also Lanthanide elements
Galena (Pb glance) 761
  roasting reactions 799
  see also Lead sulfide
Gallium
  abundance 245
  arsenide, semiconductor 247
  chalcogenides 285, 286
  III-V compounds 288, 290
  discovery 244
  as eka-aluminium 244
  hydride 257
  hydride halides 258
  lower halides 270, 271
  organometallic compounds 293-295
  oxides 277, 278
  production and uses 247
  sulfides 285, 286
  trihalides 266, 267
  see also Group III elements
Gallium ion, hydration number of 704
Garnets 401
  magnetic properties of 1104, 1256
Garnierite 1329
Germanes *see* Germanium hydrides
Germanium
  abundance 428
  atomic properties 431, 433
  chalcogenides 453, 454
  chemical reactivity and group trends 434, 435
  cluster anions 455
  dihalides 437-439
  dihydroxide 438, 445
  dioxide 445
  discovery 427
  halogeno complexes 439
  hydrides 436, 437
  hydrohalides 437
  isolation from flue dust 429

Germanium—*contd*
  monomeric Ge(OAr)$_2$ 454
  monoxide 438, 445
  organo compounds 439, 459, 464
  physical properties 433, 434
  silicate analogues 445
  sulfate 451
  tetraacetate 451
  tetrahalides 437, 439
  uses 429
German silver 1330
Gibbs' phase rule 798
Gibbsite 274, 277, 407
Girbotol process 330
Glassmaker's soap 1220
Gold
  abundance 1365
  alkyls 1372, 1395
  chalcogenides 1373, 1374
  cluster compounds 1390, 1391
  complexes
    +3 oxidation state 1379, 1380
    +2 oxidation state 1380, 1381
    +1 oxidation state 1389, 1390
    lower oxidation states 1390, 1391
    with S 788
  halides 1374-1376
  history 1364
  nitrato complexes 541, 543
  organometallic compounds 348, 1392-1394
  oxide 1373
  production and uses 1366, 1367
  see also Group IB elements
Goldschmidt's geochemical classification 760
Graham's salt 615-617, 619
Graphite
  alkali metal intercalates 313-316
  chemical properties 307-313
  halide intercalates 316, 317
  intercalation compounds 313-318
  monofluoride 309
  occurrence and distribution 298, 299
  oxide 308, 309
  oxide intercalates 315
  physical properties 307, 308
  production and uses 299
  structure 303, 304
  subfluoride 309, 310
Greek alphabet *see* back end paper
Greenhouse effect 301, 302
Grignard reagents 146-153
  allyl 362
  constitution of 147, 148
  crystalline adducts of 149
  preparation of 148
  Schlenk equilibrium 146, 148
  synthetic uses of 150, 151
Ground terms of transition element ions 1090, 1093
Group 0 elements *see* Noble gases
Group IA elements *see* Alkali metals
Group IB elements (Cu, Ag, Au)
  atomic and physical properties 1367, 1368

Group IB elements (Cu, Ag, Au)—*contd*
  coordination numbers and stereochemistries 1371,
    1372
  group trends 1368–1372
  oxidation states 1371
  see also Copper, Silver, Gold
Group IIA elements *see* Alkaline earth metals
Group IIB elements (Zn, Cd, Hg)
  atomic and physical properties 1399, 1400
  coordination numbers and stereochemistries 1402
  group trends 1400–1403
  see also Zinc, Cadmium, Mercury
Group III elements (B, Al, Ga, In, Tl)
  amphoteric behaviour of Al, Ga 253, 254
  atomic properties 250
  chemical reactivity 252–256
  group trends 251–256, 267
  +1 oxidation state 253, 256
  physical properties 251, 252
  trihalide complexes, stability of 267–269
  see also individual elements
Group IIIA elements (Sc, Y, La, Ac)
  atomic and physical properties 1104, 1105
  chemical reactivity 1105–1107
  group trends 1105–1107
  high coordination, numbers 1110
  oxidation states lower than +3 1107, 1108
  see also individual elements and Lanthanide
    elements
Group IV elements *see* Carbon, Silicon,
    Germanium, Tin, Lead
Group IVA elements (Ti, Zr, Hf)
  atomic and physical properties 1114–1116
  coordination numbers and stereochemistries 1117
  group trends 1114–1117
  oxidation states 1117
  see also Titanium, Zirconium, Hafnium
Group V elements
  see Nitrogen, Phosphorus, Arsenic, Antimony,
    Bismuth
Group VA elements (V, Nb, Ta)
  atomic and physical properties 1141
  coordination numbers and stereochemistries 1143
  group trends 1142, 1143
  oxidation states 1143
  see also Vanadium, Niobium, Tantalum
Group VI elements *see* Oxygen, Sulphur, Selenium,
    Tellurium, Polonium
Group VIA elements (Cr, Mo, W)
  atomic and physical properties 1169, 1170
  coordination numbers and stereochemistries 1172
  group trends 1170, 1171
  oxidation states 1172
  see also Chromium, Molybdenum, Tungsten
Group VII elements *see* Halogens
Group VIIA elements (Mn, Tc, Re)
  atomic and physical properties 1214
  coordination numbers and stereochemistries 1217
  group trends 1215, 1216
  oxidation states 1217
  oxoanions 1221, 1222

Group VIIA elements (Mn, Tc, Re)—*contd*
  redox properties 1215, 1216
  see also Manganese, Technetium, Rhenium
Group VIII elements (Fe, Ru, Os; Co, Rh, Ir;
    Ni, Pd, Pt) 1242
  see also individual elements
Guanidine 324
Guanine 612–614
Gunpowder 757, 758
Gypsum 119, 133
  diluent in superphosphate fertilizer 606
  occurrence in evaporites 759
  process for H_3PO_4 manufacture 599, 601
  S recovery from 768, 769, 771

H bridge-bond in boranes and carboranes 174ff
Haber-Bosch ammonia synthesis 468, 471, 481
  historical development 482
  production statistics 483
  technical details 482, 483
Haem 140, 1276
Haematite 1243
Haemocyanin 1392
Haemoglobin 1275–1279
Hafnates 1123
Hafnium
  abundance 1112
  alkyls and aryls 1135
  carbonyls 1135
  complexes
    +4 oxidation state 1126–1130
    +3 oxidation state 1130
    lower oxidation states 1133
  compounds with oxoanions 1125, 1126
  cyclopentadienyls 1135–1137
  dioxide 1120
  discovery 1111
  halides 1123–1125
  neutron absorber 1113, 1461
  organometallic compounds 1135–1137
  production and uses 1114
  sulfides 1121
  see also Group IVA elements
Hahnium 1453
Halates, XO_3^- 1009–1014
  astatate 1040
  disproportionation 1002
Halic acids, $HOXO_2$ 1009, 1010
Halides 957–964
  astatide 1040
  intercalation into graphite 316, 317
  synthesis of 958–962
  trends in properties 963
  see also individual elements: Al, As, Be, B, etc. *and*
    individual halides: Br^-, Cl^-, etc.
Halites, XO_2^- 1007–1009
Hall effect 290, 639, 644, 1185
Halogen cations, X_n^+ 986–988
  see also Polyhalonium cations
Halogen(I) fluorosulfates, $XOSO_2F$ 1037–1039
Halogen(I) nitrates, $XONO_2$ 1037, 1038

Halogen(I) perchlorates, XOClO₃ 1037, 1038
Halogens (F, Cl, Br, I, At) 920–1041
  abundance and distribution 927, 928
  atomic and physical properties 934–938
  charge-transfer complexes 940–944
  chemical reactivity and trends 938–943
  history (time charts) 921, 922
  origin of name 920, 921
  production and uses 928–934
  reactivity towards graphite 317
  sterochemistry 941
  see also individual elements, Interhalogen compounds
Halous acids, HOXO 1007, 1008
Hammett acidity function H_0 56, 57
Hapticity
  classification of organometallic compounds 346, 347
  definition 346
  distinction from connectivity 347, 351
Hausmannite 1212, 1219
Heat of vaporization, influence of H bonding 58
Heavy water 47
Helium
  abundance in universe 2
  atomic and physical properties 1045, 1046
  discovery 1042
  production and uses 1043, 1044
  thermonuclear reactions in stars 12
  see also Noble gases
Hemimorphite 401, 402
Hertzsprung-Russell diagrams 6ff
Heteropoly blues 1185
Heteropolymolybdates 1184, 1185
Heteropolyniobates 1150
Heteropolytungstates 1184, 1185
Heteropolyvanadates 1150
Hexamethylphosphoramide 619
Hexathionates, $S_6O_6^{2-}$, preparation and structure 850, 851
High-alumina cement 284
High-spin complexes 1086
Holmium 1424
  see also Lanthanide elements
"Horn silver" 1365
Hume-Rothery rules 1369
Hund's rules 1085, 1090
Hydrazido complexes with metals 493
  see also Dinitrogen as a ligand
Hydrazine 468, 484, 489–494
  acid-base properties 493
  hydrate 492, 493
  industrial production 491
  methyl derivatives as fuels 492
  molecular structure and conformation 490, 491
  oxidation of 497
  preparation 490
  properties of 489, 490
  reaction with nitrous acid 496
  reducing properties 493
  uses 491, 492
  water treatment using 492

Hydrides, binary 69–74
  acid strength of 52
  bonding in 70
  of boron *see* Boranes
  classification of 70
  complex 74
  covalent 70, 74
  interstitial 73
  ionic 70, 71
  nonstoichiometric 72, 73
  and periodic table 71
  of sulfur *see* Sulfanes
  see also individual elements
Hydroboration *see* Diborane
Hydrochloric acid, HCl(aq) 921, 923, 943, 948
  azeotrope 953
Hydrofluoric acid, HF(aq) 921–923
  acid strength of 952
  azeotrope 953
  preparation of fluorides using 958–960
Hydroformylation of alkenes 1317, 1324
Hydrogen
  abundance (terrestrial) 38
  abundance in universe 2
  atomic properties 40ff
  chemical properties 50ff
  cyanide, H bonding in 59
  as the essential element in acids 38
  history of 38, 39
  industrial production 44
  ionized forms of 42
  isotopes of 40
    see also Deuterium, Tritium
  ortho- and *para-* 38, 41
  physical properties of 40ff
  portable generator for 45
  preparation 43
  thermonuclear reactions in stars 11
  variable atomic weight of 21
Hydrogen azide, HN₃ 496, 497
Hydrogen bond 39, 57–69
  in ammonia 484
  in aquo complexes 733
  bond lengths (table) 65
  comparison with BHB bonds 70
  in DNA 614
  and ferroelectricity 60
  in HF 949, 950
  influence on properties 57
  influence on structure 63
  in proteins and nucleic acids 66, 67
  and proton nmr 62
  strength of 66, 68
  study by X-ray and neutron diffraction 62
  theory of 68, 69
  and vibrational spectroscopy 62
  in water 730
Hydrogen bromide, HBr
  azeotrope 953
  hydrates 952
  physical properties 949
  production and uses 948, 949

Hydrogen chloride, HCl
  hydrates 951, 952
  nonaqueous solvent properties 957
  physical properties 949
  production and uses 946–948
  see also Hydrochloric acid
Hydrogen dinitrate ion, $[H(NO_3)_2]^-$ 540
"Hydrogen economy" 44, 46
Hydrogen fluoride, HF 943–946
  H bonding in 57–59, 949, 950
  hydrates 951
  nonaqueous solvent properties 954–956
  physical properties 949, 950
  preparation of fluorides using 959
  production and uses 944–946
  skin burns, treatment of 946
  see also Hydrofluoric acid
Hydrogen halides, HX 943–957
  chemical reactivity 950–954
  nonaqueous solvent properties 954–957
  physical properties 949, 950
  preparation and uses 944–949
Hydrogen iodide, HI
  azeotrope 953
  hydrates 952
  physical properties 949
  production and uses 948, 949
Hydrogen-ion concentration *see* pH scale
Hydrogen peroxide 742
  acid-base properties 747, 748
  chemical properties 745–748
  physical properties 742, 744, 745
  preparation 742, 743
  production statistics 743
  redox properties 745–747
  structure 745
  uses of 743, 744
Hydrogen sulfate ion, HSO_4^- 835, 836, 843–845
Hydrogen sulfide, H_2S
  chemistry 807
  as ligand 786, 794, 795
  molecular properties 806, 900
  occurrence in nature 758–760, 769, 771
  physical properties 806, 900
  preparation (laboratory) 806
  protonated $[SH_3]^+$ 807
  see also Sulfanes, Wackenroder's solution
Hydrogen sulfite ion, HSO_3^- 835, 852, 853
Hydrometalation 348
Hydronitrous acid *see* Nitroxylic acid
Hydroperoxides 747
Hydroxonium ion, H_3O^+ 738–740, 951, 952
Hydroxyl ion, hydration of 738, 741
Hydroxylamine 484, 494–496
  configurational isomers 495, 496
  hydroxylamides of sulfuric acid 879–881
  preparation 494, 495
  properties 494, 495
Hypersensitive bands in spectra of lanthanides 1443
Hypobromous acid, HOBr 1000–1002, 1004, 1005
Hypochlorite, OCl^- 1001, 1002, 1004–1006
  molecular hypochlorites 1007

Hypochlorous acid, HOCl 1000–1007
  preparation 1004
  reactions 1005–1007
  uses 1006
Hypofluorites (covalent) 750, 814
Hypofluorous acid, HOF 749, 750
  preparation 922, 999, 1003
Hypohalates, OX^- 1000–1002, 1004–1006
Hypohalous acids, HOX 1000–1007
Hypoiodous acid, HOI 1000–1002, 1004, 1005
Hyponitric acid, $H_2N_2O_3$ 528, 530
Hyponitrites 527–531
Hyponitrous acid, $H_2N_2O_2$ 527–529
Hypophosphoric acid (diphosphoric(IV) acid),
    $H_4P_2O_6$ 588, 592–594
  isomerism to isohypophosphoric acid 593
Hypophosphorous acid, H_3PO_2 589, 591
Hypophosphites 591, 594
"Hyposulphite" used in photography 1377, 1388

Icosahedron, symmetry elements of 158
Ilmenite 1112, 1119, 1122
Indium
  abundance 245
  chalcogenides 285–287
  III–V compounds 288–290
  discovery 244
  lower halides 271
  organometallic compounds 293, 294
  oxide 278
  production and uses 247
  trihalides 267, 268
  see also Group III elements
Industrial chemicals, production statistics 467
  see also individual elements
Industrial Revolution 1242, 1244
Inert-pair effect 32
  in Al, Ga, In, Tl 255, 256
  in Ge, Sn, Pb 435, 443, 444
  in P, As, Sb, Bi 645, 646, 659, 661, 662
"Infusible white precipitate" 1415
Ino-silicates 400, 403–406
Interelectronic repulsion parameter 1099
  for hexaquochromium(III) 1198
  for hexaquovanadium(III) 1161
  for high-spin complexes of cobalt(III) 1308
  for low-spin complexes of cobalt(II) 1314
Interhalogen compounds 964–978
  diatomic, XY 964–968, 975
  first preparation 921, 922
  hexa-atomic, XF_5 974–978
  octa-atomic, IF_7 974–978
  tetra-atomic, XY_3 968–973, 975
  see also Polyhalide anions, Polyhalonium cations
Invar 1331
Iodates, IO_3^- 1010, 1011
  reaction scheme 1014
  redox systematics 1001, 1002
Iodic acid, HIO_3 1010, 1012, 1014
Iodides, synthesis of 962
    see also individual elements

Iodine
    abundance and distribution 927
    atomic and physical properties 934–938
    cations I_n^+ 986–988
    charge-transfer complexes 940–944
    colour of solutions 941–943
    crystal structure 937
    goitre treatment 921, 925, 926
    heptafluoride 974–978
    history 921, 922, 925
    Karl Fischer reagent for H_2O 736
    monohalides 964–968, 975
    oxoacids and oxoacid salts 999ff
        nomenclature 1000
        redox properties 1000–1002
        see also individual compounds, Iodic acid,
            Iodates, Periodic acids, Periodates, etc.
    oxide fluorides 1034–1036
    oxides 997–999
    pentafluoride 974–976
    "pentoxide", I_2O_5 997–999
    production and uses 933, 934
    radioactive isotopes 935, 936
    reactivity 940
    standard reduction potentials 1001
    stereochemistry 941
    trichloride 968–970, 973, 975
    trifluoride 968, 969, 975
    volt equivalent diagram 1002
    see also Halogens
Iodine trichloride, complex with $SbCl_5$ 662
Iodyl fluorosulfate, IO_2SO_2F 1035
Ionic-bond model 91–94, 963
    deviations from 94
Ionic radii 92, 93
    table of 1497
Ionization energy, periodicity of 27, 28
Iridium
    abundance 1290
    atomic and physical properties 1292–1294
    carbonyls 351, 1323–1326
    complexes
        +5 oxidation state 1300
        +4 oxidation state 1300, 1301
        +3 oxidation state 1301, 1311
        +2 oxidation state 1311
        +1 oxidation state 1316–1320
        lower oxidation states 1320
        with S, S_2O, S_2O_2 829
        with SO_2 831
    coordination numbers and stereochemistries 1295
    cyclopentadienyls 1326, 1327
    discovery 1290
    halides 1297–1299
    organometallic compounds 1322–1327
    oxidation states 1295
    oxides 1296, 1297
    production and uses 1292
    reactivity of element 1294
    relationship with other transition elements
        1294–1296
    sulfides 1297

Iron
    abundance 1243
    allyls 363
    alums 1264
    atomic and physical properties 1247–1249
    biochemistry of 1275–1281
    bis (cyclopentadienyl) *see* Ferrocene
    carbidocarbonyls 355, 356
    carbonyl halides 1286
    carbonyl hydrides and carbonylate anions 1284,
        1285
    carbonyls 351, 352, 354, 1243, 1282, 1283
    chalcogenides 1255, 1256
    complexes
        +3 oxidation state 1263–1268
        +2 oxidation state 1268–1273
        lower oxidation states 1274
        with S 788, 793
        with SO 829
        with SO_2 831
    coordination numbers and stereochemistries 1253
    cyclooctatetraene complexes 377
    cyclopentadienyls 1286–1289
        see also Ferrocene
    dithiocarbamate complexes 795, 1266
    electronic spin states 1253
    halides 1258–1260
    history 1242, 1244–1246
    mixed metal oxides (ferrites) 1256
    organometallic compounds 366, 372, 1282–1289
    oxidation states 1252, 1253
    oxides 1254, 1255
    oxoanions 1256, 1300
    production and uses 1243–1246
    proteins 1275–1281
    reactivity of element 1250
    relationship with other transition elements 1250–
        1252
    standard reduction potentials 1251, 1269, 1279
Iron age 1242
Iron pyrites
    reserves of 769
    source of S 759–761
    structure 648, 650, 804
Iron-sulfur proteins 1280–1281
Irving-Williams order 1065, 1097
Isocyanates 337, 341
    as polyurethane intermediates 324
Isohypophosphoric acid [diphosphoric (III, V) acid]
    $H_4P_2O_6$ 588, 593, 594
Isomerism 1077
    cis–trans 1078, 1308
    conformational 1078
    coordination 1081
    fac–mer 1078
    geometrical 1078
    ionization 1079
    ligand 1082
    linkage 1081
    optical 1078
    polymerization 1081
    polytopal 1078

Isopolymolybdates 1175–1183
Isopolytungstates 1175–1183
Isopolyvanadates 1146–1150
Isotopes, definition of 26

Jahn–Teller effect
   in Cu^{II} 1382
   in high-spin Cr^{II} 1190, 1201
   in high-spin Mn^{II} 1220, 1229
   in low-spin Co^{II} 1315
   in Ti^{III} 1132
Jasper 395

Kaolinite 406, 407–410, 413
Karl Fischer reagent 736
Keatite 394, 395
Keggin structure 1183
Kieselguhr 394
Kinetic inertness
   of chromium(III) complexes 1196
   of cobalt(III) complexes 1302
Kirsanov reactions 623
Kraft cheese process 604
Kraft paper process 103
Kreb's cycle 614
Kroll process 1112, 1113
Krypton
   atomic and physical properties 1045, 1046
   clathrates 1047, 1048
   discovery 1043
   see also Noble gases
Kupfernickel 1328, 1329
Kurchatovium 1453
Kurrol's salt 615–617

Landolt's chemical clock 1012
Lanthanide contraction 32, 1431, 1432
Lanthanide elements
   abundance 1425
   alkyls and aryls 1449
   aquo ions 1445
   arsenides 648
   atomic and physical properties 1429–1433
   carbonyls 1436
   chalcogenides 803, 1437, 1438
   complexes 1444–1448
   coordination numbers and stereochemistries
      1434, 1435
   cyclopentadienides 1436
   group trends 1431–1436
   halides 1439–1441
   history 1424, 1425
   magnetic properties 1441–1444
   organometallic compounds 1448, 1449
   +2 oxidation state 1435, 1439, 1447
   +4 oxidation state 1435, 1444
   oxides 1437, 1438
   production and uses 1426–1429
   as products of nuclear fission 1425, 1451, 1461
   separation of individual elements 1424, 1426–1428

Lanthanide elements—*contd*
   spectroscopic properties 1441–1444
   see also individual elements and Group IIIA
      elements
Lanthanoid *see* Lanthanide elements
Lanthanon *see* Lanthanide elements
Lanthanum
   abundance 1103
   complexes 1108–1110
   discovery 1102, 1424
   halides 1107–1108
   organometallic compounds 1110
   oxide 1107
   production and uses 1103–1104
   salts with oxoanions 1107
   see also Group IIIA elements, Lanthanide elements
Lattice defects, nonstoichiometry 754, 755
   see also Spinels (inverse) and individual elements
Lattice energy
   calculation of 95
   of hypothetical compounds 96
Lawrencium 1452, 1463
   see also Actinide elements
Lead
   abundance 428
   alloys 432
   in antiquity 427
   atomic properties 431, 433
   atomic weight variability 428
   benzene complex with Pb^{II} 465
   bis(cyclopentadienyl) 458, 464
   chalcogenides 453
   chemical reactivity and group trends 434, 435
   cluster anions 435, 455, 456
   cluster complexes 451, 458
   dihalides 437, 444, 445
   dinitrate 452
   halogeno complexes 444, 445
   hydride 437
   hydroxo cluster cation 458
   isolation and purification 429–431
   metal-metal bonded compounds 455–458
   mixed dihalides 444
   monomeric Pb(OAr)$_2$ 454
   monoxide 446, 448, 449, 458
   nitrate, thermolysis of 524, 541
   nonstoichiometric oxides 448–451
   organohydrides 437
   organometallic compounds 463–465
   oxides, nonstoichiometric 448–451
   oxides, uses of 449
   oxoacid salts 457
   Pb-S-O system 799
   physical properties 433, 434
   pigments 449, 453
   production statistics 429, 430, 432
   pseudohalogen derivatives 453
   radiogenic origin 428
   sulfate 452
   tetraacetate 452
   tetrafluoride 453
   tetrahalides 437, 444

Lead—*contd*
  toxicity 427, 428
  uses 432, 449
Lead chamber process for H_2SO_4 758
Leblanc process for NaOH 82, 921
Lewis acid (acceptor) 1060
  see also Donor-acceptor complexes, Class-a and
    class-b metals
Lewis base (donor) 224, 1060
  see also Donor-acceptor complexes, Ligands
Lifschitz salts 1342, 1348
Ligand field theory 1082, 1099
Ligands
  ambidentate 1063, 1081
  chelating 1061
  classification as "hard" and "soft" 343, 345, 1065
  classification by number of donor atoms 1061-
    1063
  macrocyclic 1061
  non-innocent 1227
  "octopus" 110
  tripod 1061
  see also Class-a and class-b metals, Hapticity,
    Linkage isomerism, Synergic bonding
Lime, production and uses of 131-135
Limestone, occurrence and uses 119, 133-135, 301,
  303
Limonite 1243, 1329
Linde synthetic zeolites 415, 416
Linkage isomerism
  of nitrite ion 534, 535, 1081
  of SO_2 830. 831
  of thiocyanate ion 343, 1081
Litharge *see* Lead monoxide
Lithium
  abundance 76
  acetylide, synthetic use of 112
  alkyls and aryls 110-115
  "anomalous" properties of 86-88
  compounds with oxygen 97, 98
  diagonal relationship with Mg 87, 110, 124
  discovery of 75
  methyl, bonding in 112, 114
  methyl, structure of tetrameric cluster 112, 113
  organometallic compounds 110-115
  production of metal 82, 84
  reduction potential of 86, 87
  terrestrial distribution of 76
  variable atomic weight of 21
  see also Alkali metals
Lithium compounds
  industrial uses of 77
  organometallics, synthesis of 110, 111
  organometallics, synthetic uses of 114-116
Lithium tetrahydroaluminate 258
  synthetic reactions of 259, 260
Lithophile elements 760
Lodestone 1254
Loellingite structure 650
Lone pair, stereochemical influence 439, 904, 907-
  909
Lonsdaleite 303, 305

Low-spin complexes 1086
  of cobalt(III), electronic spectra 1308
  of octahedral, d^4 ions 1262
Lutetium 1424
  see also Lanthanide elements

Macrocyclic polyethers *see* Crown ethers
Maddrell's salt 615-617
Madelung constant 95
"Magic numbers" in nuclear structure 4, 16
Magma, crystallization of silicates from 380
Magnéli-type phases
  of molybdenum and tungsten oxides 1174
  of titanium oxides 1119
  of vanadium oxides 1145
Magnesium
  alkyl alkoxides 148, 152
  in biochemical processes 138-141
  complexes of 137-140
  cyclopentadienyl 152, 153
  diagonal relationship with Li 87, 110, 124
  dialkyls and diaryls 146-148
  history of 118
  organometallic compounds 146-153
    see also Grignard reagents
  porphyrin complexes of *see* Chlorophyl
  production and uses of 120, 121
  see also Alkaline earth metals
Magnetic moment 1098
  of low-spin, octahedral, d^4 ions 1262
  orbital contribution to 1098, 1314, 1346
  spin-only 1098
  see also individual transition elements, Spin
    equilibria
Magnetic quantum number m 26
Magnetite, Fe_3O_4 1243
  inverse spinel structure 280, 1254
Magnus's salt 1351
Malachite 1365
Malathion 585
Manganates 1221, 1222
Manganese
  abundance 1212
  allyl complexes 363
  carbonyls 351, 1234-1237
  chalcogenides 1221
  complexes
    +4 oxidation state 1228
    +3 oxidation state 1228, 1229
    +2 oxidation state 1231-1234
    lower oxidation states 1234
    with S 789, 790, 792
    with SO 828
    with SO_2 831
  cyclooctatetraene complex 377
  cyclopentadienyls 1238, 1239
  dioxide 1218-1220
    uses of 1220
  halides and oxohalides 1222-1226
  nodules 1212
  organometallic compounds 363, 377, 1234-1239

Manganese—*contd*
  oxides 1218
  production and uses 1212–1213
  see also Group VIIA elements
Manganin 1213
Manganocene 1238, 1239
Marcasite structure 648–650, 804
  mineral FeS₂ 761
Marine acid 924
Marsh's test for As 650
Martensite 1248
Masurium 1211
  see also Technetium
Mass number of atom 26
Matches 547, 585
Melamine 341
Mellitic acid 308
Melting points, influence of H bonding 58
Mendeleev's periodic table 24ff
Mendeleev, prediction of new elements 32, 34, 244
Mendelevium 1452, 1463
  see also Actinide elements
Mercaptans, origin of name 1416
Mercuration 1418
Mercury
  abundance 1396
  alkyls and aryls 1418, 1419
  chalcogenides 1403, 1406
  cyclopentadienyls 1420
  "effective coordination number" 1413
  halides 1406–1408
  history 1395
  organometallic compounds 348, 1418–1420
  +1 oxidation state 1408–1410
  +2 oxidation state 1413–1416
  oxide 1403, 1404
  polycations 1410, 1411
  production and uses 1399
  sulfide, solubility of 802, 803
  toxicity 1422
Mesoperiodic acid *see* Periodic acids
Metal cations
  amphoteric 56
  hydrolysis of 55
Metalloboranes 195–198, 201
Metallocarboranes 209–220
  bonding 210, 215, 217
  chemical reactions 219
  structures 210ff
  synthesis 209–215
Metaperiodic acid *see* Periodic acids
Metaphosphates *see* Chain polyphosphates, Cyclo-
  polyphosphates
Metaphosphimic acid tautomers 632
  see also Tetrametaphosphimates
Metatelluric acid (H₂TeO₄)ₙ 915
Methane 322
  in Haber-Bosch NH₃ synthesis 482
Methanides *see* Carbides
Methyl bridges
  in BeMe₂ 142
  in MgMe₂ and Mg(AlMe₄)₂ 146, 147

Methyl methacrylate 338
Methylene complexes *see* Carbene ligands
Methylparathion 585
Meyer's periodic table 25, 27
Mica 119, 407, 411–413
Millon's base 1414, 1416
Mischmetall 1104, 1425, 1429
Mohorovicic discontinuity 415
Mohr's salt 1269
Molecular orbital theory of coordination compounds
  1082, 1099–1101
Molecular sieves *see* Zeolites
Molybdates 1175–1185
Molybdenite, MoS₂ 761, 1168
Molybdenum
  abundance 1167
  benzene tricarbonyl 374
  blues 1175
  bronzes 1185
  carbonyls 351, 1208, 1209
  carbyne complexes 353
  chalcogenides 1187
  complexes
    +6 oxidation state 1191–1193
    +5 oxidation state 1193, 1194
    +4 oxidation state 1194–1196
    +3 oxidation state 1196
    +2 oxidation state 1200
    with S 787–792, 794
    with SO₂ 831
  compounds with quadruple metal-metal bonds
    1201, 1203–1205
  cyclopentadienyl compounds 372, 375, 1209
  discovery 1167
  halides and oxohalides 1187
  heteropolyacids and salts 1184–1185
  isopolyacids and salts 1175–1183
  nitrogen fixation, role in 1205–1207
  nonstoichiometric oxides 1174
  organometallic compounds 1207–1210
  oxides 1171–1175
  production and uses 1168, 1169
  see also Group VIA elements
Molybdic acid 1176
Molybdoferredoxin 1206, 1275
Monactin 108
Monazite 1103, 1427, 1429, 1455
Mond process 1330
Monel 1330
γ-Monoclinic sulfur 773
Montmorillonite 407, 411, 413
Mössbauer spectroscopy
  with ⁵⁷Fe 1268, 1271, 1272, 1273, 1279, 1284
  with ¹²⁷I, ¹²⁹I 935, 983, 985
  of nonstoichiometric oxides 754
  with ⁹⁹Ru 1257
  with ¹²¹Sb 696
  with ¹¹⁹Sn 431
  with ¹²⁵Te 889, 909
  with ¹²⁹Xe 1054
Muriatic acid 924

Muscovite *see* Mica
Myoglobin 1275-1279

n-p-n junction *see* Transistor
n-type semiconductor *see* Semiconductor, Transistor
Nacreous sulfur 773
NADP 139
NbS_2Cl_2 structure 793
NbS_2X_2 787
Neodymium 1424
    +2 oxidation state 1435, 1438
    +4 oxidation state 1444
    see also Lanthanide elements
Neon
    atomic and physical properties 1045, 1046
    discovery 1043
    see also Noble gases
Nephelauxetic series 1099
Neptunium 1452, 1463
    bis(cyclooctatetraene) 376
    radioactive decay series 1455
    see also Actinide elements
Nernst equation (for electrode potentials) 498
Neso-silicates 400, 401
Nessler's reagent 1414
Neutrons
    fast 1456
    slow, thermal 1456
Newnham process for roasting PbS 799
Niccolite (Kupfernickel) 1329
Nichrome 1330
Nickel
    abundance 1329
    alkene and alkyne complexes 1360-1362
    π-allylic complexes 362, 363, 1362, 1363
    "anomalous" behaviour of Ni^{II} 1347, 1348
    aryls 1357
    atomic and physical properties 1331-1333
    carbonyls 351, 352, 1357-1359
    chalcogenides 1337
    complexes
        +4 oxidation state 1339-1340
        +3 oxidation state 1340
        +2 oxidation state 1342-1349
        +1 oxidation state 1355
        zero oxidation state 1355, 1356
        with SO_2 831
    coordination numbers and stereochemistries 1334
    cyclobutadiene complexes 367
    cyclopentadienyls 1358-1360
    dithiolene complexes, redox series 797
    halides 1337, 1338
    organometallic compounds 1356-1363
    oxidation states 1334
    oxides 1336, 1337
    phosphides 562
    production and uses 1329-1331
    reactivity of elements 1333
    tetracarbonyl 351, 352, 1357
Nickel arsenide 648, 649

Nickel arsenide—*contd*
    relation to CdI_2 649, 803
    structure type 648, 649, 803
Nickelocene 371, 1358, 1360
Nickel silver 1330
Nicotinamide-adenine dinucleotide, NAD^+ 614, 615
Nicotinamide-adenine dinucleotide-2-phosphate, NADPT 614, 615
Nielsbohrium 1453
Niobates 1150
Niobium
    abundance 1139
    alkyls and aryls 1163
    bronzes 1150
    carbonylate anions 1164
    chalcogenides 1151
    complexes
        +5 oxidation state 1157, 1158
        +4 oxidation state 1158, 1159
    compounds with oxoanions 1156, 1157
    cyclopentadienyls 373, 1165
    discovery 1138
    halides and oxohalides 1152-1157
    nonstoichiometric oxides 1145, 1146
    organometallic compounds 1163-1166
    oxides 1145, 1146
    production and uses 1140
    see also Group VA elements
Nitramide, H_2NNO_2 528
Nitrates 536, 539-545
    coordination modes 541-544
    thermal stability 540, 541
    see also individual elements
Nitric acid 484, 524, 526-528, 536-539
    anhydrous 536, 538, 539
    hydrates 538, 539
    industrial production 537, 538
    industrial uses 538
    ionization in H_2SO_4 844
    self-ionic dissociation 538, 539
Nitric oxide, NO 484, 508
    bonding in paramagnetic molecule 512
    catalytic production from NH_3 537
    chemical reactions 513
    colourless, not blue 512
    complexes with transition metals *see* Nitrosyl complexes
    crystal structure 512, 513
    dimeric 513
    physical properties 512
    preparation 511
    reaction with atomic N 474
Nitride ion, N^{3-} 479
    as ligand 480, 481
Nitrides 479-481
Nitrido complexes *see* Nitride ion as ligand, also individual elements
Nitriles *see* Cyanides
Nitrite ion, NO_2^-
    coordination modes 534
    nitro-nitrito isomerism 534, 535, 1081

Nitrites 484, 531–536
  see also Nitro-nitrito isomerism
Nitrogen
  abundance in atmosphere 466, 469–471
  abundance in crustal rocks 469
  active *see* atomic
  atomic, production and reactivity of 474, 475
  atomic properties 472, 473, 642
  atypical group properties 478, 641, 642
  chemical reactivity 473–478
  comparison with C and O 478
  comparison with heavier Group VB elements 478, 644, 673
  cycle in nature 466, 469, 470
  dinitrogen tetrafluoride 504, 505
  dioxide 509, 522, 713
    see also dinitrogen tetroxide
  discovery 466
  fixation, industrial 537
    see also Haber-Bosch ammonia synthesis
  fixation, natural 1205–1207, 1275, 1280
  halides 503–506
  history 467, 468
  hydrides of 489–497
    see also Ammonia, Hydrazine, Hydroxylamine
  industrial uses 471
    isotopes, discovery of 468
  isotopes, separation of 473
  ligand 468
  monoxide *see* nitric oxide
  multiple bond formation 478, 480
  oxidation states 497, 501
  oxides 508–527
    see also individual oxides
  oxoacids 527–537
    seel also individual oxoacids
  oxoanions 527–535, 539–545
    see also individual oxoanions
  oxoanion salts *see* oxoanions, Hyponitrites, Nitrates, Nitrites
  oxohalides *see* Nitrosyl halides, Nitryl halides
  physical properties 473
  production 472
  standard reduction potential for N species 500
  stereochemistry 475
  synthesis of pure 471
  tribromide 506
  trichloride 505
  trifluoride 503
  triiodide, ammonia adduct 506
  trioxide 509, 526, 527
  see also Dinitrogen
Nitrogenase 1206, 1275
Nitro-nitrito isomerism 534, 535, 1081
Nitronium ion 527, 844
Nitroprusside ion 1271, 1272
Nitrosyl trifluoride, ONF_3 503, 504
Nitrosyl halides 507, 508
Nitrosyl complexes 514–520
  coordination modes 514, 517–520
  electronic structure of 517, 519

Nitrosyl complexes—*contd*
  preparation 515, 516
  see also individual elements
Nitrous acid 528, 531–533
  reaction with hydrazine 496
Nitrous oxide, N_2O 508–511
  chemical reactions 510, 511
  isotopically labelled 510
  physical properties 510, 511
  preparation 510
  use in "whipped" ice cream 511
Nitroxyl 528, 530
Nitroxylic acid 528
Nitryl halides, XNO_2 507, 508
Nmr spectroscopy with:
  ^{10}B 161
  ^{11}B 161, 222
  $^{79,81}Br$ 936, 937
  ^{13}C 306, 343, 1071, 1159, 1282
  $^{35,37}Cl$ 922, 936, 937
  ^{19}F 222, 572, 655, 656, 809, 874, 922, 935–937, 955, 985
  ^{1}H 40, 62, 258, 260, 362, 365, 373, 620, 1135, 1287, 1310, 1318, 1354
  $^{2,3}H$ 40
  ^{127}I 936, 937
  ^{14}N 343, 468, 472, 473
  ^{15}N 468, 472
  ^{17}O 699, 704, 705, 738, 1147
  ^{31}P 547, 558, 569, 594, 1354
  ^{195}Pt 1354
  ^{33}S 782
  ^{29}Si 382
  ^{119}Sn 431
NO *see* Nitric oxide
Nobel prize for Chemistry, list of laureates 1498–1501
Nobel prize for Physics, list of laureates 1502–1506
Nobelium 1452, 1463
  see also Actinide elements
Noble gases (He, Ne, Ar, Kr, Xe, Rn) 1042–1059
  atomic and physical properties 1045, 1046
  chemical properties 1047–1059
  clathrates 1047, 1048
  discovery 1042, 1043
  production and uses 1043, 1044
  see also individual elements
Nomenclature of elements having Z > 100 34, 35
Nonactin 108
Nonaqueous solvent systems
  $AsCl_3$ 654, 655
  BrF_3 972, 973
  ClF_3 971
  HCl 957
  HF 664, 665, 954–956
  H_2SO_4 785, 843, 844, 896
  ICl 968
  IF_5 977
  NH_3 88–91, 486–489
  $SbCl_3$ 654, 655
  SO_2 785, 828, 896
  superacids 664, 665

Non-haem iron proteins (NHIP) 1280, 1281
Nonstoichiometry
    in chalcogenides 898, 899
    in oxides 753–756
    in sulfides 804
    see also individual compounds
Nuclear fission 1456, 1457
    products 1458, 1461
    spontaneous 1453, 1463
Nuclear fuels 1458–1463
    breeding 1460
    enrichment 1460
    reprocessing 1274, 1461–1463
Nuclear reactions in stars 8–17
Nuclear reactors 1456–1461
    different types 1459
    natural 1457
Nuclear structure of atoms 26
Nuclear waste, storage 1458, 1463
Nucleic acid
    definition 612
    double helix structure of 547, 613
    and H bonding 66, 67
Nucleogenesis 2ff
Nucleoside, definition 612
Nucleotide, definition 612
Nylon-6 and -66 484

Obsidian 394
Octahedral complexes 1073, 1085, 1100
    distortions in 1073, 1087
"Octopus" ligands 110
Oddo's rule 4
Oklo phenomenon 1457, 1458
Oligomerization of acetylene 1362
Olivine 119, 400
One-dimensional conductors 1341
Onyx 393
Opal 394
Open-hearth process 1245
Optical activity 1078, 1304
Optical isomers 1073, 1078, 1304
Optically active metal cluster compound 787
Optical rotatory dispersion (ORD) 1304, 1305
Orbital contribution to magnetic moment 1098
    of high-spin complexes of Co^{II} 1314
    of octahedral d^4 ions 1262
    of tetrahedral complexes of Ni^{II} 1346
Orbital multiplicity 1091
Orbital quantum number, l 26, 1090
Orbital reduction factor, k 1099
Orbital selection rule 1094
Orford process 1330
Organometallic compounds 345–378
    classification 346, 347
    definition 345
    dihapto ligands 357–362
    heptahapto ligands 375
    hexahapto ligands 373–375
    monohapto ligands 347–357

Organometallic compounds—*contd*
    octahapto ligands 375–378
    pentahapto ligands 367–373
    tetrahapto ligands 366–367
    trihapto ligands 362–365
    see also individual elements and ligands
Orpiment *see* Arsenic sulfide, As$_2$S$_3$
Orthonitrate ion, NO$_4^{3-}$ 545
Orthoperiodic acid *see* Periodic acids
Orthophosphates 603–608
    AlPO$_4$, structural analogy with SiO$_2$ 607, 608
    uses of 603–606
    see also individual metals
Orthophosphoric acid *see* Phosphoric acid
Osmates 1257
Osmiamates 1260
Osmium
    abundance 1243
    anomalous atomic weight of in Re ores 22
    atomic and physical properties 1247-1249
    carbidocarbonyls 355
    carbonyl halides 1286
    carbonyl hydrides and carbonylate anions 1284-1286
    carbonyls 351, 352, 1282–1284
    chalcogenides 1256
    complexes
        +8 oxidation state 1260
        +7 oxidation state 1260
        +6 oxidation state 1260, 1261
        +5 oxidation state 1261
        +4 oxidation state 1262, 1263
        +3 oxidation state 1263, 1264
        +2 oxidation state 1268–1274
        with SO$_2$ 831
    coordination numbers and stereochemistries 1253
    cyclooctatetraene complex 377
    discovery 1243
    halides and oxohalides 1257–1260
    organometallic compounds 1282–1287
    oxidation states 1252, 1253
    oxides 1254, 1255
    oxoanions 1257
    production and uses 1246, 1247
    reactivity of element 1250
    relationship with other transition elements 1250–1252
    standard reduction potentials 1251
Osmocene 368, 1287
Osmyl complexes 1260, 1261
Oxidation state
    periodic trends in 31–33
    variability of 32, 1060
Oxides 751–756
    acid-base properties 738, 751, 752
    classification 751–753
    nonstoichiometry in 753–756
    structure types 752, 753
    see also individual elements
Oxonium ion 51
OXO process 1317, 1324
Oxovanadium (vanadyl) ion 1144, 1159

Oxygen
  abundance 698, 700
  allotropes 707
  atomic 712, 713
  atomic properties 704, 705
  chemical properties 713–718
  coordination geometries 714–716
  crown ether compounds 105, 107, 137, 138, 141, 699
  difluoride 748, 749
  fluorides 748–751
  history 698, 699, 702
  industrial production 702–704
  industrial uses 702, 704
  isotopes, separation of 704
  liquefaction 699, 702
    see also industrial production
  liquid 699, 702–704, 706
  occurrence in atmosphere, hydrosphere and lithosphere 698, 700, 701
  origin of, in atmosphere 700
  origin of blue colour in liquid 707
  oxidation states 714
  physical properties 705
  preparation 701, 702
  radioactive isotopes 704
  reduction potential, pH dependence 757
  roasting of metal sulfides 798–800
  standard reduction potentials 736, 737
    see also Dioxygen, Ozone
Oxygen carriers
  complexes of cobalt 1315
  haemocyanin 1392
  haemoglobin and synthetic models 1275–1279
  Vaska's compound 718, 721, 1319, 1320
    see also Dioxygen
Oxyhyponitrous acid *see* Hyponitric acid
Ozone, O_3
  bonding 707, 708
  chemical reactions 709–712, 994, 995
  discovery 707
  environmental implications 708, 994
  molecular structure 707, 708
  physical properties 707, 708
  preparation 709, 712
Ozonide ion, O_3^- 710
Ozonides (organic) 711, 712
Ozonolysis 711, 712, 995

p-Process in stars 16
p-type semiconductors *see* Semiconductor, Transistor
Palladium
  absorption of hydrogen 1335, 1336
  abundance 1329
  alkene and alkyne complexes 1360–1362
  alkyls and aryls 1356, 1357
  π-allylic complexes 363, 1362, 1363
  atomic and physical properties 1331–1333
  carbonyl chloride 1357
  chalcogenides 1337

Palladium—*contd*
  complexes
    +4 oxidation state 1339, 1340
    +3 oxidation state 1340, 1341
    +2 oxidation state 1349–1355
    zero oxidation state 1355–1356
  coordination numbers and stereochemistries 1334
  discovery 1329
  halides 1337–1339
  organometallic compounds 1356–1363
  oxidation states 1334
  oxides 1336, 1337
  production and uses 1331, 1332
  reactivity of element 1333
Paraperiodic acid *see* Periodic acids
Parathion 585
Patronite 1139
Pauli exclusion principle 26
Pearlite 1248
Pentaborane, B_5H_9 174, 181, 186
  Brønsted acidity 194, 195
  chemical reactivity 194
  metalloborane derivatives 195–198
  preparation 187
  properties 193
  structure 194
Pentagonal bipyramidal complexes 1074
Pentathionates, $S_5O_6^{2-}$, preparation and structure 850, 851
Pentlandite 1329
Perbromates, BrO_4^- 1020–1022
  discovery 922, 1020
  radiochemical synthesis 1020
  redox systematics 1001, 1002, 1021
  structure 1021
Perbromic acid, $HBrO_4$ 1020
Perbromyl fluoride, $FBrO_3$ 1033, 1034
Perchlorates, ClO_4^- 1013–1020
  bridging ligand 922, 1017–1020
  chelating ligand 1017–1020
  coordinating ability 922, 1017–1020
  monodentate ligand 922, 1017–1020
  production and uses 1013, 1015
  redox systematics 1001, 1002
  structure 1016
Perchloric acid, $HClO_4$ 1013–1017
  chemical reactions 1016, 1017
  hydrates 1016
  physical properties 1015, 1016
  preparation 1013, 1015
  structure 1016
Perchloryl fluoride, $FClO_3$ 1027, 1031, 1032
Perhalates, XO_4^- 1001, 1002, 1013–1026
Perhalic acids, $HOXO_3$ 1013–1025
Periodates 1022–1026
  redox systematics 1001, 1002
  reaction schemes 1024, 1025
  structural relations 1023
  synthesis 1022, 1024
  transition metal complexes 1025, 1026
Periodic acids 1022–1025
  acid-base systematics 1024, 1025

Periodic acids—*contd*
  nomenclature 1022
  preparation 1024
  redox systematics 1001, 1002
  structural relations 1023
Periodic reactions
  Belousov-Zhabotinskii reactions 1012, 1013
  Bray's reaction 1012
Periodic table
  and atomic structure 24–27
  history of 24, 25
  and predictions of new elements 32–36
Permanganates 1221, 1222
Perosmate 1257, 1260
Perovskite structure 1122
  in ternary sulfides 805
Peroxo anions 748
Peroxodisulfates, $S_2O_8^{2-}$ 846
Peroxodisulfuric acid, $H_2S_2O_8$ 846
Peroxo complexes of O_2 719, 720
Peroxo compounds, fluorinated 750, 751
Peroxochromium complexes 748, 1193
Peroxodiphosphoric acid, $H_4P_2O_8$ 588
Peroxomonophosphoric acid, H_3PO_5 588
Peroxomonosulfuric acid, H_2SO_5 835, 846
Peroxonitric acid, $HOONO_2$ 527
Peroxonitrous acid, $HOONO$ 528
Peroxoselenous acid 916
Peroxotellurates 916
Perrhenates 1222
Perruthenates 1257
Pertechnetate ion 1222
Perxenates, XeO_6^{4-} 1057
pH scale 38, 53
"Pharaoh's serpents" 1413
Phase rule 798
Phase-transfer catalysis 109
Phenacite 400
Phlogiston theory 38, 698, 699, 924
Phlogopite *see* Micas
Phosgene 324
Phospha-alkenes 635
Phospha-alkynes 633
Phosphate cycles in nature 549–553
Phosphate rock
  occurence and reserves 549
  phosphorus production from 554, 600, 605
  statistics of uses 605
Phosphates *see* Chain phosphates, Cyclophos-
    phates, Orthophosphates, Superphosphates,
    Tripolyphosphates
Phosphatic fertilizers 547, 551–553, 600, 604–606,
    771
Phosphazenes 622–624
Phosphides 562–564
Phosphine, PH_3
  chemical reactions 564, 565
  comparison with NH_3, AsH_3, SbH_3, BiH_3 650
  inversion frequency 565
  Lewis base activity 565–567
  molecular structure 564, 565
  preparation 564

Phosphine, PH_3—*contd*
  tertiary phosphine ligands '566, 567
Phosphinic acid *see* Hypophosphorous acid
Phosphinoboranes 240
Phosphites 591, 592
Phosphocreatine 614, 615
Phosphonic acid *see* Phosphorous acid
Phosphonitrilic chloride $(NPCl_2)_x$ 468
  see also Polyphosphazenes
Phosphoramidic acid 619
Phosphorescence
  of arsenic 645
  of phosphorus 546, 559
Phosphoric acid, H_3PO_4 587, 595, 597–602
  autoprotolysis 597
  in colas and soft drinks 601
  hemihydrate 597–599
  industrial production 599–601
  industrial uses 600
  polyphosphoric acids in 601, 602
  proton-switch conduction in 597, 598
  self-dehydration to diphosphoric acid 595
  self-ionization 597
  structure 595
  successive replacement of H in 598, 599, 603
  "thermal" process 599, 600
  trideutero 597
  "wet" process 599, 600, 605
Phosphoric triamide 619
Phosphorous acid, H_3PO_3 588, 591
Phosphorus
  abundance and distribution 548, 549
  allotropes 546, 547, 553–558
  alloys 564
  atomic properties 557–559, 642
  black allotropes 555–558, 643
  bond energies 559
  catenation 546, 559
  chemical reactivity 559, 673
  cluster anions 686
  coordination geometries 559–561
  disproportionation in aqueous solutions 589–591
  encapsulated 646
  fertilizers 547, 551–553, 600, 604
  halides 567–575
    see also individual trihalides and pentahalides
  history 546, 547
  Hittorf's violet allotrope 555, 557
  hydrides 564–567
    see also Phosphine
  mixed halides 567
  multiple bond formation 546
  organic compounds 633–636
  oxides 577–581
    see also "Phosphorus trioxide", "Phosphorus
      pentoxide"
  oxohalides 575, 576
  oxosulfides 582, 586
  pentaphenyl 636
  peroxide, P_2O_6 581
  production and uses 553, 554, 600, 605
  pseudohalides 567, 574

Phosphorus—*contd*
  radioactive ^{32}P 558
  red, amorphous 555, 556, 559
  stereochemistry 559–561
  sulfides 581–585
    industrial uses of 585
    organic derivatives 585
    physical properties 583
    stoichiometry 581
    structures 582
    synthesis 581, 583, 584
  thiohalides 571, 575–577, 583
  triangulo-μ_3-P_3 species 684, 685
  white, α-P_4 554–556, 559, 644
  ylides 635, 636
Phosphorus oxoacids 586–619
  lower oxoacids 594–596
  nomenclature 587–589, 596
  standard reduction potentials 589, 590
  structural principles 586–589, 596
  volt-equivalent diagram 590
  see also individual acids and their salts, e.g.
    Phosphoric acid, Phosphates, etc.
Phosphorus pentahalides 568, 571–574
  ammonolysis of PCl_5 with liquid NH_3 624
  fluxionality of PF_5 572
  industrial production of PCl_5 574
  ionic and covalent forms 571–574, 626
  mixed pentahalides 572, 573
  organo derivatives 573–575
  reactions of PCl_5 with NH_4Cl-diphosphazenes
    623
  structural isomerism 572, 573
"Phosphorus pentoxide", P_4O_{10}
  chemical reactions 580
  cyclo-phosphates, relation to 618
  hydrolysis 580, 600
  polymorphism 578, 580
  preparation 578
  structure 579
Phosphorus tribromide 568, 570, 574
Phosphorus trichloride
  chemical reactions 570
  hydrolysis to phosphates 591, 592
  industrial production 569
  organophosphorus derivatives 569, 570, 592
Phosphorus trifluoride 568, 569
  as a poison 1279
  similarity to CO as ligand 568
Phosphorus triiodide 568, 570
"Phosphorus trioxide", P_4O_6
  disproportionation to P_4O_n 578
  hydrolysis 578
  preparation 577, 578
  structure 579
Phosphoryl
  halides, POX_3 575, 576
  pseudohalides 575
Photographic image intensification with ^{35}S 782
Photographic process 921, 925, 926, 1376–1379
Photosynthesis
  $NADP^+$ in 614

Photosynthesis—*contd*
  ^{18}O tracer experiments 699, 700
  as origin of atmospheric O_2 700
  see also Chlorophylls
Phyllo-silicates 400, 406–413
Physical constants, table of consistent values: back
  end paper
Physical properties, periodic trends in 27–31
Piezoelectricity 62, 396
Pig-iron 1245
Pitchblende 1450, 1454
Plagioclase *see* Feldspars
Plaster of Paris 133
Platinum
  abundance 1329
  alkene and alkyne complexes 1360–1362
  alkyls and aryls 1356–1357
  π-allylic complexes 364, 365, 1362–1363
  anti-tumour compounds 1353
  atomic and physical properties 1331–1333
  β-elimination in alkyls and aryls 348
  black 1331
  carbonylate anions 1358, 1359
  catalytic uses 537, 538, 1331
  chalcogenides 1337
  complexes
    +6 and +5 oxidation states 1339
    +4 oxidation state 1339, 1340
    +3 oxidation state 1340–1342
    +2 oxidation state 1349–1355
    zero oxidation state 1355, 1356
    with S 788, 791, 794
    with SO_2 830, 831
  coordination numbers and stereochemistries 1334
  halides 1337–1339
  optical resolution of $[Pt(S_5)_3]^{2-}$ 791, 794
  organometallic compounds 361, 362, 1356–1363
  see also Zeise's salt
  oxidation states 1334
  oxides 1336, 1337
  production and uses 1331, 1332
  reactivity of element 1333
  sulfide, structure 803
Platinum metals, definition 1242
Plutonium
  "breeding" 1460
  bis(cyclooctatetraene) 378
  critical mass 1462
  discovery 1452
  extraction from irradiated nuclear fuel 1461, 1462
  natural abundance 1454
  redox behaviour 1467–1469
  self-heating 1464
  see also Actinide elements
Plutonium economy 1460
P–N compounds 619–633
Point groups 1492–1494
Pollution, atmospheric by SO_2 758, 824–827, 842
  see also Eutrophication, Water
Polonium
  abundance 883
  allotropy 888

Polonium—contd
  atomic and physical properties   889, 890
  chemical reactivity   890–896
  coordination geometries   891–893
  dioxide   912
  discovery   882
  halides   901, 902, 904, 907
  hydride, H_2Po 899, 900
  hydroxide   913, 914
  nitrate   917
  oxides   911–913
  polonides   898, 899
  production and uses   886
  radioactivity   883, 886
  redox properties   891, 893
  selenate   917
  sulfate   917
  toxicity   894–896
Polyethene (polythene)   292, 293
Polyhalide anions   967, 968, 971–973, 977–983
  bonding   982, 983
  containing astatine   1041
  structural data   979–982
Polyhalonium cations   967, 968, 971–973, 976–978,
    983–988
  stoichiometries   983
  structures   984
Polyiodides   941, 978–983
Polypeptide chains   66
Polypeptide links   612
Polymetaphosphoric acid   588
Polyphosphates, factors affecting rate of degradation
    602
  see also Chain polyphosphates, Cyclo-polyphos-
    phates
Polyphosphazenes   624
  analogy with silicones   624
  applications   632–634
  basicities   630
  bonding in   627–630
  hydrolysis   632
  melting points of $(NPX_2)_n$ 628
  pentameric $(NPCl_2)_5$ 627
  preparation   625, 626
  reactions   630–632
  structure   625–627
  tetrameric $(NPCl_2)_4$ 626, 627
  trimeric $(NPX_2)_3$ 626
Polyphosphoric acid   587
  catalyst for petrochemical processes   601
Polysulfanes   see Sulfanes
Polysulfates, $S_nO_{3n+1}{}^{2-}$ 845
Polysulfides
  of chlorine   see Sulfur chlorides
  of hydrogen   see Sulfanes
  use in Na/S batteries   802
Polythiazyl   see $(SN)_x$
Polythionates, $S_nO_6{}^{2-}$ 835, 847, 849–851
  seleno- and telluro- derivatives   850, 916
Polythionic acids, $H_2S_nO_6$ 835, 849
Polyurethane   324, 484
Polywater   741, 742

Porphin   140
Porphyrin complexes of Mg   see Chlorophylls
Portland cement
  constitution of   282, 283
  manufacture of   283, 284
  see also High-alumina cement
Potassium
  abundance   76
  compounds with oxygen   97–99
  discovery   75
  graphite intercalates   313–316
  nitrate, thermolysis of   540, 541
  orthonitrate   545
  phosphates   604
  polysulfides   805, 806
  polythionates, preparation and structure   850, 851
  production of metal   84, 85
  silyl   390, 391
  terrestrial distribution of   79
  see also Alkali metals
Potassium chlorate, thermal decomposition to give
    O_2 701
Potassium compounds
  as fertilizers   83
  production and uses of   83, 84
Potassium permanganate, thermal decomposition to
    give O_2 702
Powder metallurgy   1169, 1328
Praseodymium   1424
  diiodide   1441
  +4 oxidation state   1435, 1438, 1444
  see also Lanthanide elements
Praseodymium-oxygen system
  ordered defects and nonstoichiometry in   755, 756
Principal quantum number n 26
Promethium   1425
  see also Lanthanide elements
Protactinium
  abundance   1454
  bis(cyclooctatetraene)   376
  discovery   1450
  redox behaviour   1467–1469
  see also Actinide elements
Proton, hydration of   738–740, 951, 952
  see also Hydrogen, ionized forms, and pH
Proton-switch conduction
  in H_2O 730
  in H_3PO_4 895
  in H_2SO_4 843
Protoporphyrin IX (PIX)   1276
Prout's hypothesis   1042
Prussian blue   1271
Pseudohalogen concept   336, 342
  see also individual elements for pseudohalogen
    derivatives
PTFE   see Teflon
Purex process   1462
Purple of Cassius   1368
PVC plastics, organotin stabilizers for   460
Pyrite structure   648, 650
  see also Iron pyrites
Pyrophosphoric acid   see Diphosphoric acid

Pyrophosphoryl halides 576, 577, 581
Pyrophyllite 408-410, 413
Pyrrhotite $Fe_{1-x}S$ 761

Quadruple metal-metal bonds 1201-1205
Quantum numbers 26
Quartz 393-395
   enantiomorphism 394
   uses 396
Quaternary arsonium compounds 693
Quaternary bismuth cations, BiR_4^+ 697
Quaternary phosphonium cations 560, 565, 571-574, 635, 636

r-Process in stars 15
Racah parameter *see* Interelectronic repulsion parameter
Radial functions 1487-1489
Radioactive decay series 1455
Radioactive elements
   discovery 32
   varying atomic weights of 21
Radiocarbon dating 306
Radium, history of 118
   see also Alkaline earth elements
Radius ratio rules 92
Radon
   atomic and physical properties 1045, 1046
   difluoride 1059
   discovery 1043
   fluoro complexes 1059
   see also Noble gases
Rare earths *see* Lanthanide elements
Raschig synthesis of hydrazine 490, 491
Rayon 334, 484, 770
Realgar *see* Arsenic sulfide, As_4S_4
Red cake 1140
Red lead, Pb_3O_4 450, 452
Reduction potentials *see* Standard reduction potentials
Reinecke's salt 1197
Reppe synthesis 1362
Rhenates 1222
Rhenium
   abundance 1212
   alkyls 1234, 1239-1241
   carbidocarbonyls 355
   carbonyls 351, 1234-1238
   chalcogenides 1221
   complexes
     +7 oxidation state 1226
     +6 oxidation state 1226, 1227
     +5 oxidation state 1227, 1228
     +4 oxidation state 1228
     +3 oxidation state 1229-1231
     +2 oxidation state 1231
     lower oxidation states 1234
   compounds with metal-metal multiple bonds 1229, 1230

Rhenium—*contd*
   cyclopentadienyls 372, 1239
   discovery 1211
   halides and oxohalides 1222-1226
   nine-coordinate hydrido complex 1217, 1226, 1227
   organometallic compounds 1234-1241
   oxides 1218-1219
   production and uses 1213
   trioxide 1218
   structure of 1219
Rhodium
   abundance 1290
   atomic and physical properties 1292-1294
   carbidocarbonyls 354, 356-359
   carbonyls 351, 354, 1323-1326
   complexes
     +4 oxidation state 1300
     +3 oxidation state 1301-1311
     +2 oxidation state 1311
     +1 oxidation state 1316-1319
     lower oxidation states 1320
     with SO_2 831
   coordination numbers and stereochemistries 1295
   cyclopentadienyls 1326, 1327
   discovery 1290
   halides 1297-1299
   optical resolution of
     cis-$[Rh\{\eta^2\text{-}(NH)_2SO_2\}_2(OH_2)_2]^-$ 791
   organometallic compounds 1322-1327
   oxidation states 1295
   oxides 1296, 1297
   production and uses 1292
   reactivity of element 1294
   relationship with other transition elements 1294-1296
   sulfides 1297
Rhodocene 1326
Ribonucleic acids 550
   messenger RNA 614
   ribosomal RNA 614
   transfer RNA 614
β-D-Ribose 611
"Ring whizzing" 1135, 1289, 1420
RNA *see* Ribonucleic acids
Rochelle salt, discovery of ferroelectricity in 60, 1122
Roussin's salts 514, 1272
Rubidium
   abundance 79
   compounds with oxygen 77-100
   discovery 76
   see also Alkali metals
Rubridoxins 1275, 1279, 1280
Russell-Saunders coupling 1090
Rusting of iron 1250, 1251
Ruthenates 1257
Ruthenium
   abundance 1243
   atomic and physical properties 1247-1249
   bipyridyl complexes and solar energy conversion 1274
   carbidocarbonyls 354, 355, 1285

Ruthenium—*contd*
  carbonyl halides 1286
  carbonyl hydrides and carbonylate anions 1284–1286
  carbonyls 351, 352, 354, 1282–1284
  chalcogenides 1256
  complexes
    +8 oxidation state 1260
    +7 oxidation state 1260
    +6 oxidation state 1261
    +5 oxidation state 1261
    +4 oxidation state 1262, 1263
    +3 oxidation state 1263, 1264, 1268
    +2 oxidation state 1268–1274
    with S 789, 790, 792
    with SO_2 831
  coordination numbers and stereochemistries 1253
  discovery 1243
  halides 1257–1259
  nitrosyl complexes 1274
  organometallic compounds 1282–1287
  oxidation states 1252, 1253
  oxides 1254, 1255
  oxoanions 1257
  production and uses 1246, 1247
  reactivity of element 1250
  relationship to other transition elements 1250–1252
  standard reduction potentials 1251
Ruthenium red 1268
Ruthenocene 368, 1287
Rutherfordium 1453
Rutile 1112, 1119
  structure type 1120

s-Process in stars 15
"Saffil" fibres 275
Salt (NaCl)
  history 921, 923
  location of deposits 78, 924, 927
  uses in chemical industry 81, 82
  world production statistics 80
Salt cake *see* Sodium sulfate
Saltpetre 469
  see also Potassium nitrate
Samarium 1424
  magnetic properties 1442
  +2 oxidation state 1438, 1439, 1447
  see also Lanthanide elements
"Sandwich" molecules 345, 1286
  see also Ferrocene, cyclopentadienyls of individual elements, Dibenzenechromium, Uranocene
Scandium
  abundance 1103
  complexes 1108–1110
  discovery 1102, 1424
  as eka-boron 1102
  halides 1107, 1108
  organometallic compounds 1110
  oxide 1107
  production 1103

Scandium—*contd*
  salts with oxoanions 1107
  see also Group IIIA elements
Scheelite 1168, 1169
Schönites 1382
Scotch hearth process for roasting PbS 799
Se_2 as ligand 894, 895
Secondary valency 1068
Selenates 914
Selenic acid, H_2SeO_4 914
Selenides 898, 899
Selenites and diselenites 914
Selenium
  abundance 883
  allotropy 886–888
  atomic and physical properties 889, 890
  chemical reactivity 890–895
  coordination geometries 891–894
  dioxide 911, 912
  discovery 882
  halide complexes 908
  halides 901, 902, 904–908
  hydride, H_2Se 894, 899, 900
  nitride, Se_4N_4 916
  organocompounds 895, 917, 918
  oxides 911–913
  oxoacids 913–916
  oxohalides 909, 910
  polyatomic cations, Se_n^{2+}, $[Te_nSe_{4-n}]^2$ 896–898
  production and uses 883, 884
  pseudohalides 910, 911
  redox properties 891, 893
  sulfate 917
  sulfides 916
  toxicity 894, 895
  trioxide 912, 913
Selenocyanate ion 337, 342–344
  ambidentate properties 894
Selenopolythionates 916
Selenosulfates 916
Selenous acid, H_2SeO_3 913, 914
Semiconductors
  II–VI 288
  III–V 247, 288, 290, 639, 640
  As, Sb and Bi chalcogenides 677, 679
  nonstoichiometric oxides 756
Shear plane 1174
SI prefixes, origin of front end paper
SI units front end paper
  conversion to non SI units 1495
Siderite 1243
Siderophite elements 760
Silanes
  chemical reactions 389–391
  homocyclic polysilanes 421
  physical properties 388
  silyl halides 390, 391
  silyl potassium 390, 391
  synthesis 389
Silica
  fumed 397
  gel 397

Silica—*contd*
  historical importance 379
  hydrated 398
  phase diagram 395
  polymorphism 393–396
  in transistor technology 383
  uses 396, 397
  vitreous 396
  see also Quartz, Tridymite, Cristobalite, Coesite,
    Stishovite
Silaethenes 419
Silicate minerals 399–416
  comparison with silicones 423
Silicates 379–381, 399–416
  with chain structures (metasilicates) 403, 404
  with discrete units 400–402
  disilicates 401, 402
  with framework structures 414–416
  with layer structures 406–413
  metasilicates 403, 404
  orthosilicates 400, 401
  soluble (Na, K) 396, 398
Silicides 386–388
  preparation 387
  structural units in 387
Silicomanganese 1213
Silicon
  abundance and distribution 380
  atomic properties 382, 431
  carbide 386
  chemical properties of 379, 385, 434
  coordination numbers 385
  dioxide *see* Silica
  double bonds to 388, 419, 420
  halides 391–393
  history 379
  hydrides *see* Silanes
  isolation 379, 381
  nitride 417
  organic compounds 419–426
  physical properties 382, 433, 434
  purification 381, 382
  sulfide 417
Silicones 419, 422–426
  comparison with mineral silicates 423
  elastomers 424, 425
  oils 424
  organotin, curing agents for 460
  resins 425
  synthesis 422, 424
  uses 424, 425
Siloxanes 422, 423
Silylamides 417, 418
Silver
  abundance 1365
  acetylide 1372
  alkenes and alkynes 1393
  chalcogenides 1373, 1374
  complexes
    +3 oxidation state 1379, 1380
    +2 oxidation state 1381
    +1 oxidation state 1388, 1389

Silver—*contd*
  halides 1374–1376
  history 1364
  nitrate, thermolysis of 541
  organometallic compounds 1392–1394
  oxides 1373
  production and uses 1366
Silver halides in photographic emulsions 1378
Singlet oxygen 699
  generation of 716
  reactions of 717, 718
Skutterudite *see* Cobalt arsenide
Smaltite 1291, 1329
S–N heterocycles incorporating a third element 869,
    871
$(SN)_x$ polymer
  partially halogenated derivatives 861
  structure 860, 861
  superconducting properties of 468, 758, 854, 860,
    861
  synthesis 860, 861
S_2N_2 859, 860
  polymerization to $(SN)_x$ *qv*
  preparation 859
  structure and bonding 860
S_4N_2 861, 862
S_4N_4 468, 758, 855–859
  preparation 856
  reactions 858, 859, 864, 868, 871, 872
  structure and bonding 856, 857
S_5N_6 863
$S_{11}N_2$ 861–863
$S_{14+x}N_2$ 861–863
Soapstone *see* Talc
Sodalite *see* Ultramarines
Sodanitre *see* Chile saltpetre
Sodide anion, Na^- 110
Sodium
  abundance 76
  β-alumina
    structure and properties 281, 282
    use in Na/S batteries 802
  arsenide 646, 647
  azide 471, 497, 505
  bismuthate 647
  carbonate
    hydrates of 102, 104
    production and uses of 102
  chlorate 1009
  compounds with oxygen 97–99
  diphosphates 608, 609
  discovery 75
  distribution 76, 77
  dithionite 854, 855
  hydroxide, production and uses of 81, 103
  hypophosphite 591
  nitrate, thermolysis of 540, 541
  nitroprusside 514
  nitroxylate 528
  orthonitrate 545
  phosphates 589, 599, 603–605, 607
  polysulfides 800–802, 805, 806

Sodium—*contd*
  polythionates, preparation and structure   850, 851
  production of metal   84
  silicates, soluble   396, 398
  solutions in liquid ammonia   88–91, 455, 456
  sulfide batteries   801, 802
  thiosulfate, in photography   847, 1377, 1378
  tripolyphosphate   609, 610
  see also Alkali metals
Sodium chloride structure   92, 273, 1146
  see also Salt
Sodium hypochlorite
  industrial uses   1006
  Raschig synthesis of hydrazine using   490, 491
Solar energy conversion   1274
Solvo-acids and bases
  in anhydrous H_2SO_4 844
  in liquid $AsCl_3$, $SbCl_3$ 654
  in liquid BrF_3 973
  in liquid NH_3 487
  in liquid N_2O_4 525
  in water   738
Soro-silicates   400–402
Solvay process   82
  byproduct Ca from   121
Spallation   17
Spectral sensitization of photographic emulsions
  1377
Spectrochemical series   1096
Spectroscopic terms   1090
Sphalerite (Zinc blende)   761, 1396
  structure of   803, 1405
Spiegeleisen   1213, 1245
Spin crossover   *see* Spin equilibria
Spin equilibria   1094
  in Fe^{II} compounds   1272
  in Fe^{III} compounds   1267
  in Mn^{II} compounds   1239
Spinel   119
Spinel structure
  in Co_3O_4 1297
  defect structure of γ-Al_2O_3 274
  in Fe_3O_4 (inverse)   1254
  in ferrites and garnets   1256
  in Mn_3O_4 1219
  normal and inverse   279–281, 1254
  in ternary sulfides   805
  valence disordered types   280
Spin-forbidden bands
  in compounds of Fe^{III} 1265
  in compounds of Mn^{II} 1232
  in compounds of Ni^{II} 1345
Spin multiplicity   1091
Spin-orbit coupling
  in actinide ions   1476
  in d^4 ions   1262
  in lanthanide ions   1441
  in transition element ions   1094
Spin quantum number m_s 26, 1090
Spin selection rule   1092
  see also Spin-forbidden bands
Spodumene   76, 403

Square antiprismatic complexes   1075
Square planar complexes   1070, 1088
Square pyramidal complexes   1071
Stability constants of coordination compounds
  factors affecting   1064–1068
  overall   1064
  stepwise   1064
Staging   *see* Graphite intercalation compounds
Standard reduction potentials   497–502
  IUPAC sign convention   499
  see also individual elements
Stannates   447
Starch/iodine reaction   921, 1012
Stars
  spectral classification of   4ff
  temperatures of   5ff
Steel   1244–1249
Stellar evolution   4ff
Stereochemical non-rigidity (fluxional behaviour)
  $Al(BH_4)_3$, $\{Al(BH_4)_2H_\mu\}_2$ 258, 260
  allyl complexes   365
  Berry pseudorotation mechanism   547, 572
  5-coordinate compounds   1071
  8-coordinate compounds   1158
  $Fe(CO)_5$ 1071, 1282
  iron cyclopentadienyl complexes   1287
  PF_5 572
  SF_4 809
  titanium cyclopentadienyl   1135
Stibine   650
Stibinidene complexes   694, 695
Stibnite   *see* Antimony sulfide, Sb_2S_3
Stishovite   394, 395
Strength of oxoacids, Pauling's rules   54
Strontium
  history   118
  organometallic compounds   153
  polysulfides   806
  see also Alkaline earth metals
styx numbers   *see* Boranes, topology
Sulfamic acid, $H[H_2NSO_3]$ 468, 876, 877
Sulfamide $(H_2N)_2SO_2$ 877, 878
Sulfanes   806–808
  nonexistence of SH_4 and SH_6 810
  physical properties   808
  preparation   807
  synthesis of polythionic acids from   849
  see also Hydrogen sulfide
Sulfate ion as ligand (η^1, η^2, μ) 845
Sulfates   844, 845
  see also individual elements
Sulfide minerals
  geochemical classification   760
  names and formulae   761
Sulfides
  anionic polysulfides   802, 805, 806
  applications and uses   800–802
  electrical properties   805
  hydrolysis   800, 803, 806
  industrial production   800
  magnetic properties   805
  preparation (laboratory)   800

Sulfides—*contd*
  roasting in air 798–800
  solubility in water 800, 802, 803
  S_n^{2-} structures 805, 806
  structural chemistry 803–805
  see also individual elements
Sulfinates 832
Sulfites, SO_3^{2-} 835, 852, 853
  in paper manufacture 771
  protonation to HSO_3^- 852
Sulfoxylates, MS(O)OR 832
Sulfur
  abundance 759
  allotropes 758, 769–781
    see also individual allotropes, e.g. *Cyclo*-S_n,
      Catena-S_n, etc.
  atomic properties 781
  atomic S 783
  atomic weight, variability of 21, 781, 782
  in biological complexes 787
  bromides S_nBr_2 816, 817
  catenation 769, 774ff, 805–808, 815, 816, 849–851
  chemical reactivity 782–785
  chiral helices 780
  chlorides 815–817, 849
    industrial applications of SCl_2 and S_2Cl_2 815
    preparation 815, 816
    properties of S_nCl_2 815–816
    $[SCl_3]^+$ 816, 819
    SCl_7I 819
  chlorofluorides 751, 811–814
  conformations (c, dt, lt) 774, 778, 779, 786
  conversion to SO_2/SO_3 for H_2SO_4 *see* Sulfuric
    acid
  coordination geometries 784
  Crystex 779
  dihedral angles in S_n 771, 772, 774
  fibrons (ψ, ϕ) 779, 780
  fluorides 808–814
    chemical reactions 810–814
    fluxionality of SF_4 809
    isomeric S_2F_2 809
    physical properties 810, 813
    stoichiometries 809
    structures 808–810
    synthesis 810–813
  gaseous species 781
  halides 808–819
    see also individual halides
  hexafluoride 810, 813
    applications as a dielectric gas 813
    reaction with SO_3 821
  history 757, 758
  iodides 817–819
    bond energy relations in 817
    SCl_7I 819
    $[S_2I_4]^{2+}$ 818, 819
    $[S_7I]^+$ 817
    $[S_{14}I_3]^{3+}$ 818
  λ point 780
  as ligands 828–832
  ligand properties of chelating $-S_n-$ 786, 791, 794

Sulfur—*contd*
  ligand properties of S atom 786–788
  ligand properties of S_2^{2-} 786–790, 792, 793
  liquid 772, 780
  monoxide 824
  nitrides 855–863
  organic thio ligands 795
  origin in caprock of salt domes 759
  oxidation states 785
  oxidation state diagram of species 836
  oxides 821–834
    higher, SO_{3+x}, SO_4 833, 834
    lower dioxides S_nO_2 821–824
    lower oxides S_nO 821–824
    see also Sulfur dioxide, Sulfur trioxide
  oxoacids 834–854
    schematic classification 837
    table of 835
    thermodynamic interrelations 836
    see also individual oxoacids and oxoacid anions
  oxofluorides 814
    see also Thionyl fluorides, Sulfuryl fluorides
  peroxofluorides 814
  plastic (χ) 779
  polyatomic cations 785, 786
  polymeric (μ) 779
  production 761–769
    Frasch process 761–763
    from pyrite 768
    from sour gas and crude oil 762, 768
    statistics 762, 768
  radioactive isotopes 782
  reserves 769
  rhombohedral *see Cyclo-α-*S_8
  rubbery S 779
  S–S bonds 769, 771, 772, 782, 787, 805–808,
    849–851
  S_2 774, 775, 781
  S_3 775, 781
  S_4 781
  S_4^{2+} *see* polyatomic cations
  S_8 *see Cyclo-*S_8
  S_8^{2+} *see* polyatomic cations
  S_n *see Cyclo-*S_n
  S_{19}^{2+} *see* polyatomic cations
  SN compounds 811, 854–881
    see also S_4N_4, S_2N_2, $(SN)_x$, S–N–X compounds,
      Sulfur imides, S–N–O compounds, Sulfur-
      nitrogen anions, Sulfur-nitrogen cations
  singlet state S_2 781
  standard reduction potentials of S species 834,
    836
  terrestrial distribution 759
  triplet state S_2 781
  uses 769, 770
  uses of radioactive ^{35}S 782, 847
  volt-equivalent diagram of S species 836
  see also Chalcogenides
Sulfur chloride pentafluoride SF_5Cl
  photolytic reduction to S_2F_{10} 812
  reaction with O_2 751
  synthetically useful reactions of 813, 814

Sulfur dioxide 824–828
  atmospheric pollution by 758, 824–827, 842
  chemical reactions 824
  clathrate hydrate 827
  industrial production 824, 838
  insertion into M–C bonds 831, 832
  as ligand 829–832
  molecular and physical properties 824, 827, 912
  in M–S–O phase diagrams 799
  solvent for chemical reactions 783, 828
  toxicity 824
  uses 824
  see also Wackenroder's solution, Sulfuric acid
    production
Sulfur imides, $S_{8-n}(NH)_n$ 868–870
Sulfur-nitrogen anions, $S_xN_y^-$ 866–868
Sulfur-nitrogen cations, $S_xN_y^+$ 864–866
Sulfur-nitrogen-halogen compounds 869–875
  cyclo-$(NSF)_n$ 872, 873
  $N_3S_3Cl_3$ 873
  $N_3S_3X_3O_3$ 873
  S_4N_3Cl and $S_4N_4Cl_2$ 874
  thiazyl halides NSX 870–872
Sulfur-nitrogen-oxygen compounds 875
  amides of H_2SO_4 876
    see also Sulfamic acid, Sulfamide
  hydrazine derivatives of H_2SO_4 879
  hydroxylamine derivatives of H_2SO_4 879–881
  imido and nitrido derivatives of H_2SO_4 878, 879
  sulfur-nitrogen oxides 875, 876
Sulfur trioxide
  chemical reactions 832, 833
  molecular and physical properties 832, 833
  monomeric 832, 833
  polymeric 832, 833
  polymorphism 832, 833
  preparation by catalytic oxidation of SO_2 824,
    838
  reaction with F_2 751
  reaction with SF_6 821
  trimeric 832, 833
  see also Sulfuric acid production
Sulfuric acid 834–845
  amides of 876–878
  autoprotolysis in anhydrous 843
  contact process 758, 838–842
  D_2SO_4 837, 843
  history 758, 838
  hydrates 842
  hydrazine derivatives of 879
  hydroxylamine derivatives of 879–881
  imido derivatives of 878, 879
  ionic dissociation equilibria in anhydrous 843,
    844
  lead chamber process 758, 838
  nitrido derivatives of 878, 879
  physical properties 837
  physical properties of D_2SO_4 837
  production from sulfide ores 838
  production from sulfur 769–771, 838
  production statistics 467, 838, 841, 842
  solvent system 844

Sulfuric acid—*contd*
  uses 842
Sulfurous acid, H_2SO_3 771, 827, 835, 850, 851
Sulfuryl chloride 820, 821
Sulfuryl fluoride 814, 820, 821
  mixed fluoride halides 821
Super acids
  $HF/SO_3/SbF_5$ 664, 665
  HSO_3F/SbF_5 664
Superconductivity of metal sulfides 805
Superheavy elements 34, 1453
Supernucleophiles 1322
Superoxo complexes of O_2 719, 720, 1307
Superphosphate fertilizer 547, 605, 606
Symmetry elements 1492–1494
Symmetry operations 1492–1494
Synergic bonding
  in alkene complexes 360
  in CO and CN^- complexes 349, 351
  in cobalt cyanides 1302
Synthesis gas 1284

Talc 119, 408, 411, 413
Tanabe-Sugano diagrams
  for d^2 ions 1162
  for d^3 ions 1198
  for d^6 ions 1273, 1309
  for d^8 ions 1345
  for Mn^{II} 1233
Tantalates 1150
Tantalite 1139
Tantalum
  abundance 1139
  alkyls and aryls 1163
  carbene complex 348
  carbonylate anions 1164
  chalcogenides 1151
  complexes
    +5 oxidation state 1157–1158
    +4 oxidation state 1158–1159
  compounds with oxoanions 1156, 1157
  cyclopentadienyls 1165–1166
  discovery 1138
  halides and oxohalides 1152–1157
  organometallic compounds 353, 1163–1166
  oxides 1145, 1146
  production and uses 1140
  see also Group VA elements
Technetates 1222
Technetium
  abundance 1212
  carbonyls 351, 1234–1237
  chalcogenides 1221
  complexes
    +7 oxidation state 1226
    +6 oxidation state 1226, 1227
    +5 oxidation state 1228
    +4 oxidation state 1228
    +3 oxidation state 1229–1231
    +2 oxidation state 1231
  lower oxidation states 1234

Technetium—*contd*
  cyclopentadienyls 1239
  discovery 1211
  halides and oxohalides 1222–1226
  organometallic compounds 1234–1239
  oxides 1218
  production and uses 1213
  see also Group VIIA elements
Tectites 394
Tecto-silicates 400, 414–416
Teflon (PTFE) 323, 923
Tellurates 915, 916
Telluric acid, Te(OH)$_6$ 915
Tellurides 898, 899
Tellurites 914
Tellurium
  abundance 883
  allotropy 887, 888
  atomic and physical properties 889, 890
  chemical reactivity 890—895
  coordination geometries 891–894
  dioxide 911, 912
  discovery 882
  halide complexes 908
  halides 901–908
  hydride, H$_2$Te 894, 899, 900
  nitrate 917
  nitride, Te$_3$N$_4$ 916
  organo compounds 917–919
  oxides 911–913
  oxoacids 913–916
  oxohalides 910
  polyatomic cations
    Te$_n^{m+}$ 896–898
    [Te$_n$Se$_{4-n}$]$^{2+}$ 898
  production and uses 883, 884
  redox properties 891–893
  sulfide, TeS$_7$ 916
  toxicity 894, 895
  trioxide 912, 913
Tellurocyanate ion, TeCN$^-$ 911
Telluropolythionates 916
Tellurous acid 913, 914
Terbium 1424
  +4 oxidation state 1435, 1438
  see also Lanthanide elements
Tetracyanoethylene complexes, bonding in 361
Tetrafluoroethylene complexes, bonding in 361
Tetrafluoronitronium cation NF$_4^+$ 503
Tetrahalogenophosphonium cations PX$_4^+$ 572–574
Tetrahedral complexes 1070, 1087
Tetrahydroborates 188, 189
  of Al 258, 260
  of Ga 261
  use in synthesis 189–191
Tetrametaphosphimate conformers 632
Tetrathionates, S$_4$O$_6^{2-}$, preparation and structure 850, 851
Thallium
  abundance 245
  chalcogenides 285–287
  III–V compounds 288–290

Thallium—*contd*
  discovery 244
  halide complexes 270
  lower halides 272
  monohalides 272, 273
  organometallic compounds 293, 294
  oxides 278
  production 250
  similarity of Tl^I to alkali metals 255
  trihalides 269
  triiodide Tl^I[I$_3$]$^-$ 269, 270
  see also Group III elements
Thiazyl halides NSX 870–872
Thioarsenites 677
Thiocarbonyl (CS) complexes 335, 336
Thiocyanates 337, 342
  as ambidentate ligands 342–345, 1063, 1081
Thioethers as ligands 795
Thionitrosyl (NS) complexes 520, 521
Thionyl bromide 820
Thionyl chloride 819, 820
  relation to SO$_2$ and Me$_2$SO as ionizing solvents 820
Thionyl fluoride 814, 819, 820
  mixed fluoride chloride 820
Thiophosphoryl
  halides, PSX$_3$ 573, 576
  pseudohalides 575
Thioselenates 916
Thiosulfates, S$_2$O$_3^{2-}$ 835, 847, 848
  as ligands 847, 848
  redox reactions in analysis 847, 848
  structure 847, 848
  use in photography 847, 1377, 1378
Thiosulfuric acid, H$_2$S$_2$O$_3$ 835, 846
  isomeric H$_2$S.SO$_3$ 846
  redox interconversions in water 846
Thio-urea 334
Thiovanadyl ion 1151
Thixotropy 411, 1128
Thorium
  abundance 1454
  bis(cyclooctatetraene) 376
  production and uses 1455, 1456
  radioactive decay series 1455
  redox behaviour 1467–1469
  use as a nuclear fuel 1459, 1460
  see also Actinide elements
Thortveitite 401
Three-centre bonds
  in Al trialkyls and triaryls 289
  in beryllium alkyls 142
  BHB bond 70, 171ff
  BBB bond 180ff
  in magnesium alkyls 142
Thulium 1424
  see also Lanthanide elements
Thymine 612–614
Thyroxine 925, 926
Tin
  abundance 428
  allotropes 434

Tin—*contd*
  alloys 430
  in antiquity 427, 428
  atomic properties 431, 433
  bis(cyclopentadienyl) 462
  chalcogenides 453
  chemical reactivity and group trends 434, 435
  cluster anions 435, 455, 456
  cluster complexes 446, 458
  compounds, use of 448, 460
  dibromide 442
  dichloride 441, 442
  difluoride 440
  dihalides 437, 439–443
  diiodide 443
  dioxide 447, 448, 451, 452, 460
  halogeno complexes 440–444, 461
  hydrides 437
  hydroxo species 446, 458
  isolation and purification 429
  metal-metal bonded compounds 454–458, 461–464
  monomeric Sn(OAr)$_2$ 454
  monoxide 440, 445, 446, 451, 452
  nitrates 451
  organometallic compounds 459–463
    oligomerization 461
    production statistics 460
    toxic action 460
    uses of 460, 461
  oxoacid salts 451, 452
  physical properties 433, 434
  production statistics 428, 429
  pseudohalogen derivatives 453
  sulfide 428
  tetrahalides 437, 443, 444, 448
  uses 430, 448
Titanates 1121–1123
  ferroelectric properties 1122
Titanium
  abundance 1112
  alkoxides 1127
  alkyls and aryls 1135
  alum 1131
  bronzes 1123
  carbonyls 1135
  complexes
    +4 oxidation state 1126–1130
    +3 oxidation state 1130–1133
    lower oxidation states 1133–1137
    with S 791, 794
  compounds with oxoanions 1125
  cyclooctatetraene complex 377
  cyclopentadienyls 1135–1137
  dioxide 1118
  discovery 1111
  estimation using H$_2$O$_2$ 1128
  halides 1123–1125
  mixed metal oxides (titanates) 1121–1123
  nonstoichiometric oxide phases 754, 1119
  organometallic compounds 1133–1137
  production and uses 1112–1113

Titanium—*contd*
  "sponge" 1113
  sulfides 1121
  see also Group IVA elements
"Titanocene" 1135–1137
Tolman's cone angle 566, 567
Tooth enamel 551
Toothpastes, calcium compounds in 605
"Tops and bottoms" process 1330
Trans-effect 1352, 1353
  in [OsNCl$_5$]$^{2-}$ 1260
  in Pt^{II} complexes 1352, 1353
  in Rh^{III} ammines 1310
Trans-influence 1353, 1354
  in [OsNCl$_5$]$^{2-}$ 1260
  in Pt^{II} complexes 1353
Transferrin 1281
Transistor action 383, 384
  chemistry of manufacture 383
  discovery 383
Transition element ions
  colours 1089–1096
  coordination chemistry 1060–1101
  electronic absorption spectra 1089–1096
  hydration energies of bivalent ions 1097
  magnetic properties 1098, 1099
  spectroscopic ground terms 1090, 1093
  see also individual elements
Transition elements
  definition of 1060
  characteristic properties 1060
  see also Transition element ions and individual elements
Transuranium elements
  discovery of 25, 34, 1452
  extraction from reactor wastes 1463
  see also Actinide elements
Tricalcium aluminate, Ca$_3$Al$_2$O$_6$, structure 282, 283
  see also Portland cement
Tricalcium phosphate [Ca$_5$(PO$_4$)$_3$OH] 604
Tri-capped trigonal prismatic complexes 1076
Tridymite 394, 395
Trigonal bipyramidal complexes 1071
Trigonal prismatic complexes 1073
Triperiodic acid *see* Periodic acids
Triphosphoric acid H$_5$P$_3$O$_{10}$ 587
"Triple-decker" complexes 1326, 1360
Tris(dimethylamino)phosphine 620
Trithiocarbonates 334
Trithionates, S$_3$O$_6$$^{2-}$, preparation and structure 850, 851
Tritium
  atomic properties 40
  discovery 38
  physical properties 41
  preparation of tritiated compounds 49, 50
  radioactivity of 49
  synthesis 48
  uses as a tracer 49
Tropylium (cycloheptatrienyl) 375
Tungstates 1175–1185

Tungsten
  abundance 1167
  benzene tricarbonyl 374
  bronzes 1185
  carbonyls 351, 352, 1208–1209
  carbyne complexes 353
  chalcogenides 1187
  complexes
    +6 oxidation state 1191–1193
    +5 oxidation state 1193, 1194
    +4 oxidation state 1194–1196
    +3 oxidation state 1196
    +2 oxidation state 1200
    with S 791
  cyclopentadienyl derivatives 1209
  discovery 1167
  halides and oxohalides 1187–1191
  hexacarbonyl 351, 352, 1208
  heteropolyacids and salts 1184, 1185
  isopolyacids and salts 1175–1183
  nonstoichiometric oxides 1174
  organometallic compounds 353, 374, 375, 1207–1210
  oxides 1171–1175
  production and uses 1169
  see also Group VIA elements
Tungstic acid 1177
Turnbull's blue 1271
Tutton salts
  of copper 1382
  of vanadium 1156
Type metal 637, 639
Tyrian purple 921, 922, 924, 925

u (ungerade), definition of 369
Ultramarines 414, 416
Units
  conversion factors 1495
  non-SI 1495
  SI, definitions front end paper
  SI, derived front end paper
Universe
  expansion of 2, 5
  origin of 1, 2
Uranium
  abundance 1454
  bis (cyclooctatetraene) 376
  isotopic enrichment 1460
  production 1456
  radioactive decay series 1455
  redox behaviour 1467–1469
  variable atomic weight of 21
  see also Actinide elements
Uranium hexafluoride 1460, 1473–1475
Uranium oxides, nonstoichiometry in 754
"Uranocene" 1486
Uranyl ion 1467, 1471, 1478, 1479
Urea 324, 331, 342, 484
  hydrazine production from 491
  phosphate 604
  Wöhler's synthesis 468

Uridine triphosphate, UTP 614, 615

Valence, periodic trends in 31
Valence bond theory of transition metal complexes 1082–1084
Valinomycin 108
Vanadates 1144, 1146–1150
Vanadium
  abundance 1139
  accumulation in blood of invertebrates 1139
  alkyls and aryls 1163
  bronzes 1150
  carbonyl 351, 1164
  chalcogenides 1151
  complexes
    +5 oxidation state 1157, 1158
    +4 oxidation state 1158–1160
    +3 oxidation state 1160–1163
    +2 oxidation state 1163
  compounds with oxoanions 1156, 1157
  cyclopentadienyls 371, 1165
  discovery 1138
  dithiolene complexes 796, 797
  halides and oxohalides 1152–1157
  hexacarbonyl 351, 1164
  isopolyacids and salts 1146–1150
  nonstoichiometric oxides 1145
  organometallic compounds 351, 371, 375, 376, 1163–1165
  oxides 1144–1146
  production and uses 1140
  see also Group VA elements
Vanadocene 1165
Vanadyl compounds 1144, 1159, 1160
Van Arkel-de Boer process 1113
Vaska's compound 718, 720, 1318–1320
Venus, atmosphere of 757, 758
Vermiculite 407, 413
Viscose rayon 334
Vitamin B_{12} 1321, 1322, 1422
Volt equivalent
  definition 499
  diagrams 449–502
  see also individual elements for volt equivalent diagrams
Vortmann's sulfate 1307
Vulcanization of rubber 758

Wackenroder's solution 849, 850, 853
Wacker process 1361
Wade's rules 184, 185, 204, 646, 688, 689
Water
  acid-base behaviour 51, 738
  aquo complexes 733
  autoprotolysis constant 51
  chemical properties 735
  clathrate hydrates 734, 735
  distribution and availability 726–729
  H bonding in 57–59
  heavy (D_2O) 730
  history 726

Water—*contd*
  hydrates 733–735
  hydrolysis reactions 735
  ice, polymorphism of 731, 732
  ionic product of 52
  Karl Fischer reagent for 736
  lattice water 734
  physical properties 729–732, 900
  pollution of 728
  polywater 741, 742
  purification and recycling 728, 729
  self ionic dissociation of 51
  tritiated (T_2O) 730
  zeolitic water 734
Water-gas shift reaction 330, 1284
  in Haber-Bosch NH_3 synthesis 483
Water supplies, treatment of 134
White arsenic *see* Arsenic oxide, As_2O_3
"White gold" 1328
"White lead" 453, 1404
Wilkinson's catalyst 1316–1318
Wilson's disease 1392
Wittig reaction 547, 548, 635
  with arsenic ylides 693
Wolffram's red salt 1340
Wolfram 1167
  see also Tungsten
Wolframite 1168, 1169
Wrought-iron 1246
Wurtzite structure (ZnS) 803, 1405

x-Process in stars 16
X-ray absorption spectroscopy 1206
Xanthates 334, 758
  γ-S_8 from Cu^I ethyl xanthate 773
  as ligands 795
XeF_2 1048–1050
  bonding 1053, 1054
  bonding compared with H bond 70
XeF_4 1048–1051
XeF_6 1048, 1049, 1051, 1052, 1056, 1057
Xenate ion, $HXeO_4^-$ 1057
Xenon
  atomic and physical properties 1045, 1046
  carbon bonds 1058, 1059
  chemical reactivity discovered 1047
  chloride 1054
  clathrates 1047, 1048
  discovery 1043
  fluorides 1048–1059
  fluorocomplexes 1054, 1055, 1058
  fluorosulfate 1055, 1056
  nitrogen bonds 1058
  oxidation states 1048
  oxides 1048–1051
  oxoanions (xenate, perxenate) 1057
  oxofluorides 1056
  perchlorate 1055
  stereochemistry 1048, 1049
  trifluoromethyl compounds 1059
Xerography 885
Xerox process (xerography) 885

"Yellow cake" 1456
Ylides 635
  arsonium 693
Ytterbium 1424
  +2 oxidation state 1438, 1439, 1447
  see also Lanthanide elements
Yttrium
  abundance 1103
  complexes 1108–1110
  discovery 1102, 1424
  halides 1107–1108
  organometallic compounds 1110
  oxide 1107
  oxo salts 1107
  production and uses 1103, 1104
  see also Group IIIA elements, Lanthanide elements

Zeise's salt 357, 359, 1356, 1361
Zeolite 414–416
Ziegler-Natta catalysis 292, 293, 1134
Zinc
  abundance 1396
  alkyls and aryls 1416, 1417
  biochemistry 1420, 1421
  chalcogenides 1403, 1405, 1406
  coordination chemistry 1411–1413
  ferrites 1404
  halides 1406–1408
  history 1395
  organometallic compounds 1416, 1417
  +2 oxidation state 1411–1413
  oxides 1397, 1403, 1404
    nonstoichiometry in 754, 1403
  production and uses 1396–1398
  see also Group IIB elements
Zinc blende (sphalerite) 1396
  structure 803, 1405
Zircon 400, 1112
Zirconates 1123
Zirconium dioxide 275, 1112, 1121
  "Saffil" fibres 275, 276
Zirconium
  abundance 1112
  alkyls and aryls 1135
  borohydride 189, 1130
  carbonyls 1135
  complexes
    +4 oxidation state 1126–1130
    +3 oxidation state 1130
    lower oxidation states 1130
  compounds with oxoanions 1125, 1126
  cyclopentadienyls 1135–1137
  dioxide (baddeleyite) 275, 1112, 1121
  discovery 1111
  disulfide 1121
  halides 1123 1125
  in nuclear reactors 1113, 1461
  organometallic compounds 1135–1137
  production and uses 1113
  tetrahydroborate 189, 1130

Recommended Consistent Values of Some Fundamental Physical Constants

(The numbers in parentheses are the standard deviation in the last digits of the quoted value.)

Quantity	Symbol	Value	Uncertainty (ppm)
Permeability of vacuum	μ_0	$4\pi \times 10^{-7}$ H m^{-1} $= 12.566\ 370\ 614\ 4 \times 10^{-7}$ H m^{-1}	
Speed of light in vacuum	c	$2.997\ 924\ 58(1.2) \times 10^8$ m s^{-1}	0.004
Permittivity of vacuum	$\varepsilon_0 = (\mu_0 c^2)^{-1}$	$8.854\ 187\ 82(7) \times 10^{-12}$ F m^{-1}	0.008
Elementary charge	e	$1.602\ 189\ 2(46) \times 10^{-19}$ C	2.9
Planck constant	h	$6.626\ 176(36) \times 10^{-34}$ J Hz^{-1}	5.4
	$\hbar = h/2\pi$	$1.054\ 588\ 7(57) \times 10^{-34}$ J s	5.4
Avogadro constant	N_A	$6.022\ 045(31) \times 10^{23}$ mol^{-1}	5.1
Atomic mass unit	$1u = (10^{-3}$ kg mol$^{-1})/N_A$	$1.660\ 565\ 5(86) \times 10^{-27}$ kg	5.1
Electron rest mass	m_e	$0.910\ 953\ 4(47) \times 10^{-30}$ kg	5.1
		$0.511\ 003\ 4(14)$ MeV	2.8
Proton rest mass	m_p	$1.672\ 648\ 5(86) \times 10^{-27}$ kg	5.1
		$935.279\ 6(26)$ MeV	2.8
Neutron rest mass	m_n	$1.674\ 954\ 3(86) \times 10^{-27}$ kg	5.1
		$939.573\ 1(26)$ MeV	2.8
Ratio, proton mass to electron mass	m_p/m_e	$1836.151\ 52(70)$	0.38
Faraday constant	$\mathscr{F} = N_A e$	$9.648\ 456(27) \times 10^4$ C mol^{-1}	2.8
Rydberg constant	R_∞	$1.097\ 373\ 177(83) \times 10^7$ m^{-1}	0.075
Bohr radius	$a_0 = \alpha/4\pi R_\infty$	$0.529\ 177\ 06(44) \times 10^{-10}$ m	0.82
Electron g factor	$\tfrac{1}{2}g_e = \mu_e/\mu_B$	$1.001\ 159\ 656\ 7(35)$	0.0035
Bohr magneton	$\mu_B = eh/2m_e$	$9.274\ 078(36) \times 10^{-24}$ J T^{-1}	3.9
Nuclear magneton	$\mu_N = eh/2m_p$	$5.050\ 824(20) \times 10^{-27}$ J T^{-1}	3.9
Electron magnetic moment	μ_e	$9.284\ 832(36) \times 10^{-24}$ J T^{-1}	3.9
Proton magnetic moment	μ_p	$1.410\ 617\ 1(55) \times 10^{-26}$ J T^{-1}	3.9
Proton magnetic moment in Bohr magnetons	μ_p/μ_B	$1.521\ 032\ 209(16) \times 10^{-3}$	0.011
Ratio, electron to proton magnetic moments	μ_e/μ_p	$658.210\ 688\ 0(66)$	0.010
Proton gyromagnetic ratio	γ_p	$2.675\ 198\ 7(75) \times 10^8$ s^{-1} T^{-1}	2.8
Molar gas constant	R	$8.314\ 41(26)$ J mol^{-1} K^{-1}	31
Molar volume, ideal gas ($T_0 = 273.15$ K, $p_0 = 1$ atm)	$V_m = RT_0/p_0$	$0.022\ 413\ 83(70)$ m^3 mol^{-1}	31
Boltzmann constant	$k = R/N_A$	$1.380\ 662(44) \times 10^{-23}$ J K^{-1}	32
Gravitational constant	G	$6.6720(41) \times 10^{-11}$ N m^2 kg^{-2}	615

Greek Alphabet

α	A	Alpha	η	H	Eta	ν	N	Nu	τ	T	Tau
β	B	Beta	θ	Θ	Theta	ξ	Ξ	Xi	υ	Y	Upsilon
γ	Γ	Gamma	ι	I	Iota	o	O	Omicron	ϕ	Φ	Phi
δ	Δ	Delta	κ	K	Kappa	π	Π	Pi	χ	X	Chi
ε	E	Epsilon	λ	Λ	Lambda	ρ	P	Rho	ψ	Ψ	Psi
ζ	Z	Zeta	μ	M	Mu	$\sigma\ \varsigma$	Σ	Sigma	ω	Ω	Omega